Cowan

Microbiology: A Systems Approach

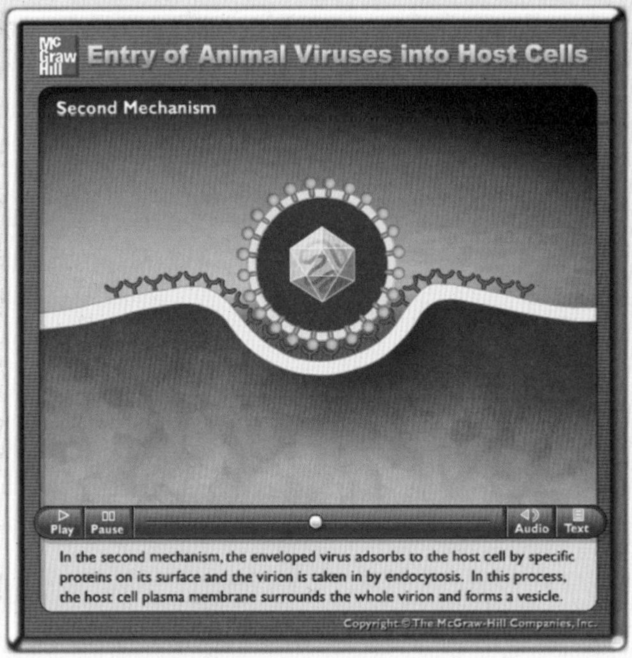
Mc Graw Hill
Entry of Animal Viruses into Host Cells
Second Mechanism
Play
Pause
Audio
Text
In the second mechanism, the enveloped virus adsorbs to the host cell by specific proteins on its surface and the virion is taken in by endocytosis. In this process, the host cell plasma membrane surrounds the whole virion and forms a vesicle.
Copyright © The McGraw-Hill Companies, Inc.

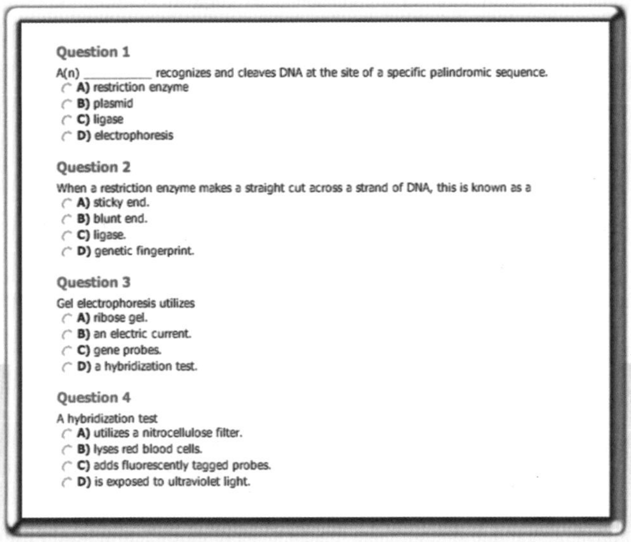
Question 1
A(n) __________ recognizes and cleaves DNA at the site of a specific palindromic sequence.
A) restriction enzyme
B) plasmid
C) ligase
D) electrophoresis
Question 2
When a restriction enzyme makes a straight cut across a strand of DNA, this is known as a
A) sticky end.
B) blunt end.
C) ligase.
D) genetic fingerprint.
Question 3
Gel electrophoresis utilizes
A) ribose gel.
B) an electric current.
C) gene probes.
D) a hybridization test.
Question 4
A hybridization test
A) utilizes a nitrocellulose filter.
B) lyses red blood cells.
C) adds fluorescently tagged probes.
D) is exposed to ultraviolet light.

ON TRACK . . .

These interactive activities and quizzing keep you motivated and on track in mastering key concepts:

- **Animations** Access to over 100 animations of key microbial processes will help you visualize and comprehend important concepts depicted in the text. The animations even include quiz questions to help ensure that you are retaining the information.
- **Test Yourself** Take a chapter quiz at the ARIS website to gauge your mastery of chapter content. Each quiz is specially constructed to test your comprehension of key concepts. Immediate feedback explains incorrect responses. You can even e-mail your quiz results to your professor!
- **Learning Activities** Helpful and engaging learning experiences await you at the *Microbiology, A Systems Approach* ARIS site. In addition to interactive online quizzing and animations, each chapter offers relevant case study presentations, digital images for creating PowerPoints®, vocabulary flash cards, and other activities designed to reinforce learning.

McGraw-Hill Higher Education

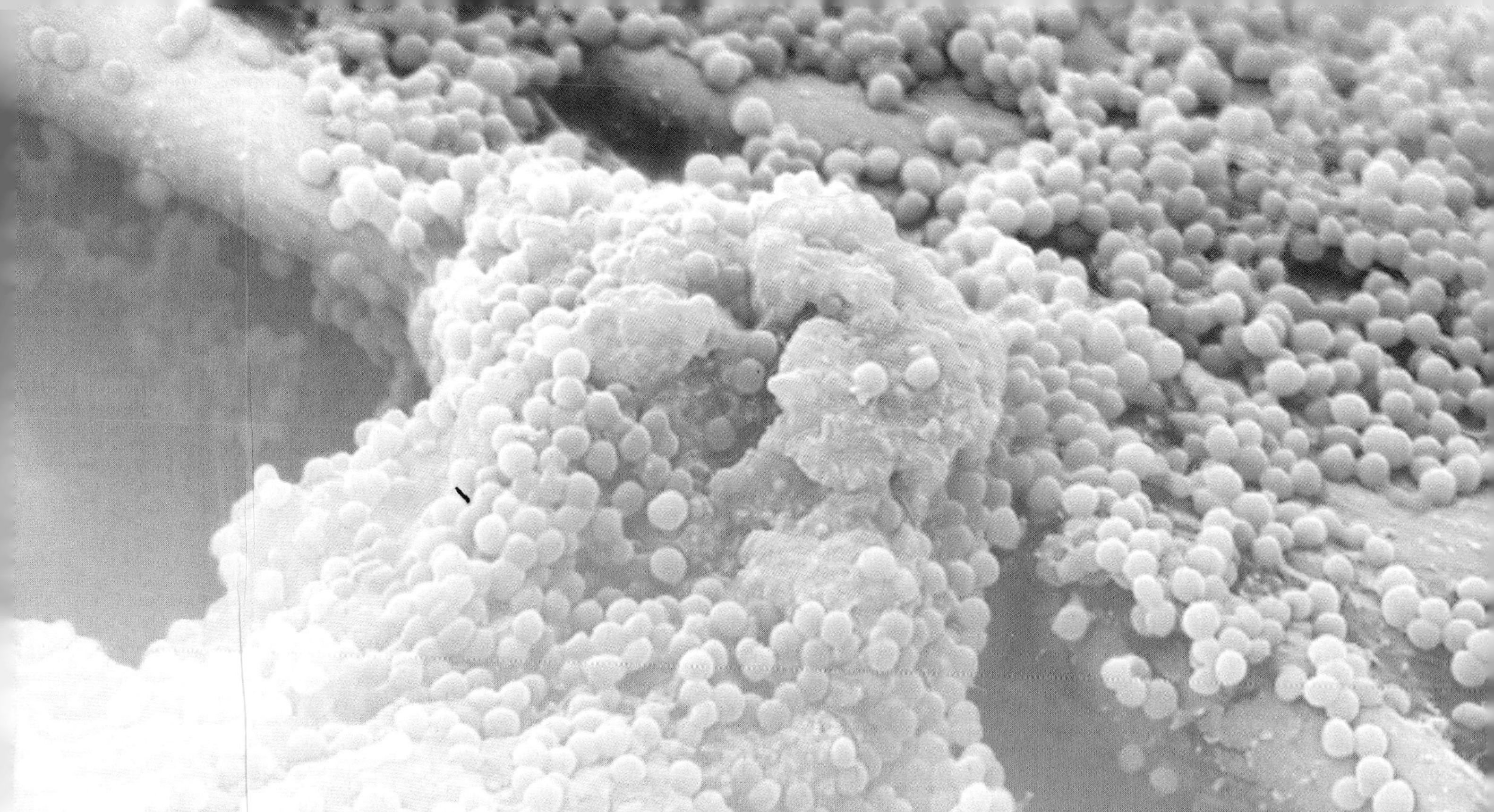

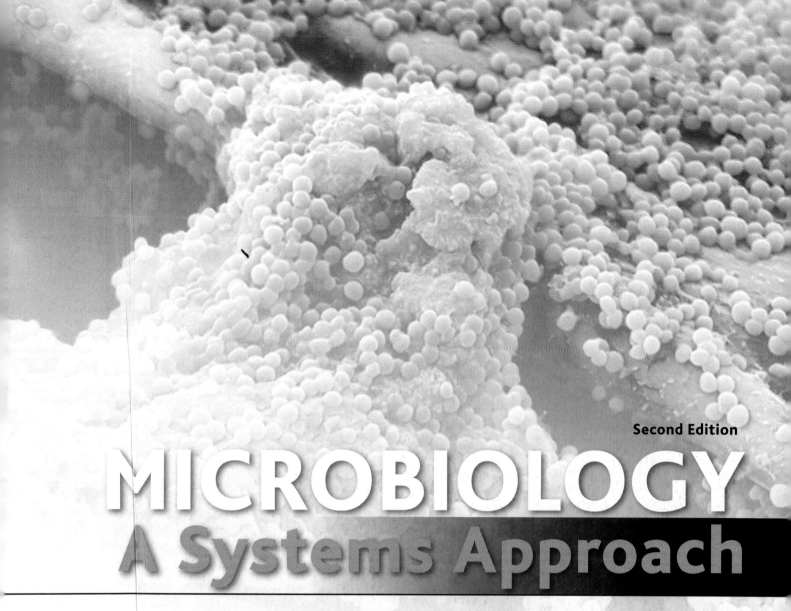

Second Edition

MICROBIOLOGY
A Systems Approach

Marjorie Kelly Cowan
Miami University

Kathleen Park Talaro
Pasadena City College

Mc Graw Hill **Higher Education**

Boston Burr Ridge, IL Dubuque, IA New York San Francisco St. Louis
Bangkok Bogotá Caracas Kuala Lumpur Lisbon London Madrid Mexico City
Milan Montreal New Delhi Santiago Seoul Singapore Sydney Taipei Toronto

The McGraw·Hill Companies

Higher Education

MICROBIOLOGY: A SYSTEMS APPROACH, SECOND EDITION

Published by McGraw-Hill, a business unit of The McGraw-Hill Companies, Inc., 1221 Avenue of the Americas, New York, NY 10020.

This book is printed on acid-free paper.

1 2 3 4 5 6 7 8 9 0 QPV/QPV 0 9 8

ISBN 978-0-07-726603-5
MHID 0-07-726603-X

Publisher: *Michelle Watnick*
Senior Sponsoring Editor: *James F. Connely*
Senior Developmental Editor: *Kathleen R. Loewenberg*
Senior Marketing Manager: *Tami Petsche*
Lead Project Manager: *Peggy J. Selle*
Senior Production Supervisor: *Laura Fuller*
Senior Media Project Manager: *Jodi K. Banowetz*
Cover/Interior Designer: *Laurie B. Janssen*
Senior Photo Research Coordinator: *John C. Leland*
Photo Research: *Editorial Image, LLC*
Supplement Producer: *Melissa M. Leick*
Compositor: *Aptara*
Typeface: 10/12 *Palatino*
Printer: *Quebecor World Versailles, KY*
(USE) Cover Image: *Ellen Swogger and Garth James,*
Center for Biofilm Engineering, Montana State University.

The credits section for this book begins on page 841 and is considered an extension of the copyright page.

Library of Congress Cataloging-in-Publication Data

Cowan, M. Kelly.
Microbiology : a systems approach / Marjorie Kelly Cowan, Kathleen Park Talaro.—2nd ed.
p. cm.
Includes index.
ISBN 978-0-07-299528-2 ISBN 0-07-299528-9 (acid-free paper)
1. Microbiology. I. Talaro, Kathleen P. II. Title.

QR41.2.C69 2009
616.9'041—dc22

2007026927
CIP

www.mhhe.com

About the Authors

Kelly Cowan has been a microbiologist at Miami University since 1993. She received her Ph.D. at the University of Louisville, and later worked at the University of Maryland Center of Marine Biotechnology and the University of Groningen in The Netherlands.

Her first love is teaching—both doing it and studying how to do it better. She is chair of the Undergraduate Education Committee of the American Society for Microbiology (ASM) and also Chair of the Education Division of ASM. In 1997, she won a Celebration of Teaching Award from the Greater Cincinnati Consortium of Colleges and Universities. She is currently the Campus Dean at Miami University's Middletown campus.

Kelly has published (with her students) twenty-four research articles stemming from her work on bacterial adhesion mechanisms and plant-derived antimicrobial compounds. She holds two patents for strategies to block microbial attachment. Kelly also travels extensively to present her research and to talk to other professors about teaching.

Kathleen Park Talaro is a microbiologist, author, illustrator, photographer, and educator at Pasadena City College. She began her college education at Idaho State University in Pocatello. There, she found a niche that fit her particular abilities and interests, spending part of her time as a scientific illustrator and part as a biology lab assistant. After graduation with a B.S. in biology, she entered graduate school at Arizona State University, majoring in physiological ecology. During her graduate studies she participated in two research expeditions to British Columbia with the Scripps Institution of Oceanography. Kathy continued to expand her background, first finishing a Master's degree at Occidental College and later taking additional specialized coursework in microbiology at California Institute of Technology and California State University.

If there is one continuing theme reverberating through Kathy's experiences, it is the love of education and teaching. She has been teaching allied health microbiology and majors biology courses for nearly 30 years. Kathy finds great joy in watching her students develop their early awareness of microorganisms—when they first come face-to-face with the reality of them on their hands, in the air, in their food, and, of course, nearly everywhere.

Kathy is a member of the American Society for Microbiology and the American Association for the Advancement of Science. She keeps active in self-study and research, and continues to attend workshops and conferences to remain current in her field. Kathy has also been active in science outreach programs by teaching Saturday workshops in microbiology and DNA technology to high school and junior high students.

We dedicate this book to all public health workers who devote their lives to bringing the advances and medicines enjoyed by the industrialized world to all humans.

Brief Contents

CHAPTER 1
The Main Themes of Microbiology 1

CHAPTER 2
The Chemistry of Biology 27

CHAPTER 3
Tools of the Laboratory: The Methods for Studying Microorganisms 57

CHAPTER 4
Prokaryotic Profiles: The Bacteria and Archaea 86

CHAPTER 5
Eukaryotic Cells and Microorganisms 117

CHAPTER 6
An Introduction to the Viruses 149

CHAPTER 7
Microbial Nutrition, Ecology, and Growth 179

CHAPTER 8
Microbial Metabolism: The Chemical Crossroads of Life 211

CHAPTER 9
Microbial Genetics 246

CHAPTER 10
Genetic Engineering: A Revolution in Molecular Biology 282

CHAPTER 11
Physical and Chemical Control of Microbes 312

CHAPTER 12
Drugs, Microbes, Host—The Elements of Chemotherapy 343

CHAPTER 13
Microbe-Human Interactions: Infection and Disease 379

CHAPTER 14
Host Defenses I: Overview and Nonspecific Defenses 414

CHAPTER 15
Host Defenses II: Specific Immunity and Immunization 442

CHAPTER 16
Disorders in Immunity 478

CHAPTER 17
Diagnosing Infections 511

CHAPTER 18
Infectious Diseases Affecting the Skin and Eyes 535

CHAPTER 19
Infectious Diseases Affecting the Nervous System 572

CHAPTER 20
Infectious Diseases Affecting the Cardiovascular and Lymphatic Systems 608

CHAPTER 21
Infectious Diseases Affecting the Respiratory System 649

CHAPTER 22
Infectious Diseases Affecting the Gastrointestinal Tract 686

CHAPTER 23
Infectious Diseases Affecting the Genitourinary System 733

CHAPTER 24
Environmental Microbiology 766

CHAPTER 25
Applied and Industrial Microbiology 789

Contents

Preface xvi

CHAPTER 1

The Main Themes of Microbiology 1

1.1 The Scope of Microbiology 2

1.2 The Impact of Microbes on Earth: Small Organisms with a Giant Effect 3

Microbial Involvement in Energy and Nutrient Flow 4

1.3 Human Use of Microorganisms 5

1.4 Infectious Diseases and the Human Condition 6

1.5 The General Characteristics of Microorganisms 7

Cellular Organization 7

A Note on Viruses 9

Microbial Dimensions: How Small Is Small? 10

Lifestyles of Microorganisms 11

1.6 The Historical Foundations of Microbiology 11

The Development of the Microscope: "Seeing Is Believing" 11

The Establishment of the Scientific Method 13

The Development of Medical Microbiology 16

1.7 Taxonomy: Naming, Classifying, and Identifying Microorganisms 17

The Levels of Classification 17

Assigning Specific Names 20

The Origin and Evolution of Microorganisms 20

Systems of Presenting a Universal Tree of Life 21

INSIGHT 1.1 The More Things Change . . . 8

INSIGHT 1.2 The Fall of Superstition and the Rise of Microbiology 12

INSIGHT 1.3 Martian Microbes and Astrobiology 18

Chapter Summary with Key Terms 23

Multiple-Choice and True-False Questions 24

Writing to Learn 25

Concept Mapping 25

Critical Thinking Questions 26

Visual Understanding 26

Internet Search Topics 26

CHAPTER 2

The Chemistry of Biology 27

2.1 Atoms, Bonds, and Molecules: Fundamental Building Blocks 28

Different Types of Atoms: Elements and Their Properties 28

The Major Elements of Life and Their Primary Characteristics 30

Bonds and Molecules 30

2.2 Macromolecules: Superstructures of Life 40

Carbohydrates: Sugars and Polysaccharides 40

Lipids: Fats, Phospholipids, and Waxes 44

Proteins: Shapers of Life 46

The Nucleic Acids: A Cell Computer and Its Programs 49

The Double Helix of DNA 50

2.3 Cells: Where Chemicals Come to Life 51

Fundamental Characteristics of Cells 52

INSIGHT 2.1 Redox: Electron Transfer and Oxidation-Reduction Reactions 34

INSIGHT 2.2 Membranes: Cellular Skins 46

Chapter Summary with Key Terms 52

Multiple-Choice and True-False Questions 53

Writing to Learn 54

Concept Mapping 54

Critical Thinking Questions 55

Visual Understanding 55

Internet Search Topics 56

CHAPTER 3

Tools of the Laboratory: The Methods for Studying Microorganisms 57

3.1 Methods of Culturing Microorganisms—The Five I's 58

Inoculation: Producing a Culture 58

Isolation: Separating One Species from Another 58

Media: Providing Nutrients in the Laboratory 60

Back to the Five I's: Incubation, Inspection, and Identification 68

3.2 The Microscope: Window on an Invisible Realm 69

Magnification and Microscope Design 69

Variations on the Optical Microscope 72

Electron Microscopy 76

Preparing Specimens for Optical Microscopes 78

INSIGHT 3.1 Animal Inoculation: "Living Media" 62

INSIGHT 3.2 The Evolution in Resolution: Probing Microscopes 77

Chapter Summary with Key Terms 81

Multiple-Choice and True-False Questions 82

Writing to Learn 83

Concept Mapping 84

Critical Thinking Questions 84

Visual Understanding 85

Internet Search Topics 85

CHAPTER 4

Prokaryotic Profiles: The Bacteria and Archaea 86

4.1 Prokaryotic Form and Function 87
The Structure of a Generalized Prokaryotic Cell 87

4.2 External Structures 88
Appendages: Cell Extensions 88

4.3 The Cell Envelope: The Boundary Layer of Bacteria 94
Differences in Cell Envelope Structure 94
Structure of the Cell Wall 94
Mycoplasmas and Other Cell-Wall-Deficient Bacteria 97
The Gram-Negative Outer Membrane 98
Cell Membrane Structure 98

4.4 Bacterial Internal Structure 99
Contents of the Cell Cytoplasm 99
Bacterial Endospores: An Extremely Resistant Stage 101

4.5 Bacterial Shapes, Arrangements, and Sizes 103

4.6 Classification Systems in the Prokaryotae 106
Taxonomic Scheme 106
Diagnostic Scheme 106
Species and Subspecies in Bacteria 108

4.7 Survey of Prokaryotic Groups with Unusual Characteristics 108
Unusual Forms of Medically Significant Bacteria 108
Free-Living Nonpathogenic Bacteria 110
Archaea: The Other Procaryotes 111

INSIGHT 4.1 Biofilms—The Glue of Life 92

INSIGHT 4.2 The Gram Stain: A Grand Stain 95

INSIGHT 4.3 Redefining Bacterial Size 109

Chapter Summary with Key Terms 113

Multiple-Choice and True-False Questions 113

Writing to Learn 114

Concept Mapping 115

Critical Thinking Questions 115

Visual Understanding 116

Internet Search Topics 116

CHAPTER 5

Eukaryotic Cells and Microorganisms 117

5.1 The History of Eukaryotes 118

5.2 Form and Function of the Eukaryotic Cell: External Structures 120
Locomotor Appendages: Cilia and Flagella 121
The Glycocalyx 121
Form and Function of the Eukaryotic Cell: Boundary Structures 122

5.3 Form and Function of the Eukaryotic Cell: Internal Structures 123
The Nucleus: The Control Center 123
Endoplasmic Reticulum: A Passageway in the Cell 123
Golgi Apparatus: A Packaging Machine 125
Mitochondria: Energy Generators of the Cell 126
Chloroplasts: Photosynthesis Machines 126
Ribosomes: Protein Synthesizers 128
The Cytoskeleton: A Support Network 128
Survey of Eukaryotic Microorganisms 129

5.4 The Kingdom of the Fungi 130
Fungal Nutrition 130
Organization of Microscopic Fungi 131
Reproductive Strategies and Spore Formation 133
Fungal Identification and Cultivation 134
The Roles of Fungi in Nature and Industry 134

5.5 The Protists 136
The Algae: Photosynthetic Protists 136
Biology of the Protozoa 137

5.6 The Parasitic Helminths 143
General Worm Morphology 143
Life Cycles and Reproduction 144
A Helminth Cycle: The Pinworm 145
Helminth Classification and Identification 145
Distribution and Importance of Parasitic Worms 145

INSIGHT 5.1 The Extraordinary Emergence of Eukaryotic Cells 119

INSIGHT 5.2 The Many Faces of Fungi 132

Chapter Summary with Key Terms 145

Multiple-Choice and True-False Questions 146

Writing to Learn 147

Concept Mapping 147

Critical Thinking Questions 147

Visual Understanding 148

Internet Search Topics 148

CHAPTER 6

An Introduction to the Viruses 149

6.1 The Search for the Elusive Viruses 150

6.2 The Position of Viruses in the Biological Spectrum 150

6.3 The General Structure of Viruses 151
Size Range 151
Viral Components: Capsids, Nucleic Acids, and Envelopes 152

6.4 **How Viruses Are Classified and Named 157**
6.5 **Modes of Viral Multiplication 160**
Multiplication Cycles in Animal Viruses 160
Viruses That Infect Bacteria 167
6.6 **Techniques in Cultivating and Identifying Animal Viruses 170**
Using Live Animal Inoculation 170
Using Bird Embryos 170
Using Cell (Tissue) Culture Techniques 171
6.7 **Medical Importance of Viruses 173**
6.8 **Other Noncellular Infectious Agents 173**
6.9 **Treatment of Animal Viral Infections 174**

INSIGHT 6.1 A Positive View of Viruses 153
INSIGHT 6.2 Replication Strategies in Animal Viruses 164
INSIGHT 6.3 Artificial Viruses Created! 172
INSIGHT 6.4 A Vaccine for Obesity? 173

Chapter Summary with Key Terms 174
Multiple-Choice and True-False Questions 175
Writing to Learn 176
Concept Mapping 177
Critical Thinking Questions 177
Visual Understanding 178
Internet Search Topics 178

CHAPTER 7

Microbial Nutrition, Ecology, and Growth 179

7.1 **Microbial Nutrition 180**
Chemical Analysis of Microbial Cytoplasm 181
Sources of Essential Nutrients 182
Transport Mechanisms for Nutrient Absorption 187
The Movement of Water: Osmosis 187
The Movement of Molecules: Diffusion and Transport 189
Endocytosis: Eating and Drinking by Cells 192
7.2 **Environmental Factors That Influence Microbes 192**
Temperature Adaptations 193
Gas Requirements 194
Effects of pH 196
Osmotic Pressure 196
Miscellaneous Environmental Factors 196
Ecological Associations Among Microorganisms 197
Interrelationships Between Microbes and Humans 199
7.3 **The Study of Microbial Growth 200**
The Basis of Population Growth: Binary Fission 200
The Rate of Population Growth 200
The Population Growth Curve 202
Stages in the Normal Growth Curve 202
Other Methods of Analyzing Population Growth 204

INSIGHT 7.1 Dining with an Amoeba 181
INSIGHT 7.2 Light-Driven Organic Synthesis 184
INSIGHT 7.3 Life in the Extremes 186
INSIGHT 7.4 Cashing In on "Hot" Microbes 194
INSIGHT 7.5 Life Together: Mutualism 198
INSIGHT 7.6 Steps in a Viable Plate Count—Batch Culture Method 203

Chapter Summary with Key Terms 205
Multiple-Choice and True-False Questions 207
Writing to Learn 207
Concept Mapping 208
Critical Thinking Questions 209
Visual Understanding 210
Internet Search Topics 210

CHAPTER 8

Microbial Metabolism: The Chemical Crossroads of Life 211

8.1 **The Metabolism of Microbes 212**
Enzymes: Catalyzing the Chemical Reactions of Life 212
Regulation of Enzymatic Activity and Metabolic Pathways 218
8.2 **The Pursuit and Utilization of Energy 221**
Energy in Cells 221
A Closer Look at Biological Oxidation and Reduction 221
Adenosine Triphosphate: Metabolic Money 222
8.3 **The Pathways 224**
Catabolism: Getting Materials and Energy 224
Aerobic Respiration 225
Pyruvic Acid—A Central Metabolite 227
The Krebs Cycle—A Carbon and Energy Wheel 227
The Respiratory Chain: Electron Transport and Oxidative Phosphorylation 229
Summary of Aerobic Respiration 232
Anaerobic Respiration 233
Fermentation 234
Catabolism of Noncarbohydrate Compounds 236
8.4 **Biosynthesis and the Crossing Pathways of Metabolism 236**
The Frugality of the Cell—Waste Not, Want Not 236
Anabolism: Formation of Macromolecules 238
Assembly of the Cell 238
8.5 **It All Starts with the Sun 239**

INSIGHT 8.1 Unconventional Enzymes 214
INSIGHT 8.2 The Enzyme Name Game 216
INSIGHT 8.3 Steps in the Krebs Cycle 228
INSIGHT 8.4 Pasteur and the Wine-to-Vinegar Connection 235

Chapter Summary with Key Terms 242
Multiple-Choice and True-False Questions 243
Writing to Learn 243
Concept Mapping 244
Critical Thinking Questions 245
Visual Understanding 245
Internet Search Topics 245

CHAPTER 9

Microbial Genetics 246

9.1 Introduction to Genetics and Genes: Unlocking the Secrets of Heredity 247
The Nature of the Genetic Material 247
The DNA Code: A Simple Yet Profound Message 249
The Significance of DNA Structure 249
DNA Replication: Preserving the Code and Passing It On 252

9.2 Applications of the DNA Code: Transcription and Translation 255
The Gene-Protein Connection 255
The Major Participants in Transcription and Translation 256
Transcription: The First Stage of Gene Expression 259
Translation: The Second Stage of Gene Expression 260
Eukaryotic Transcription and Translation: Similar Yet Different 264
The Genetics of Animal Viruses 265

9.3 Genetic Regulation of Protein Synthesis and Metabolism 265
The Lactose Operon: A Model for Inducible Gene Regulation in Bacteria 265
A Repressible Operon 267
Antibiotics That Affect Transcription and Translation 268

9.4 Mutations: Changes in the Genetic Code 268
Causes of Mutations 269
Categories of Mutations 269
Repair of Mutations 270
The Ames Test 271
Positive and Negative Effects of Mutations 271

9.5 DNA Recombination Events 272
Transmission of Genetic Material in Bacteria 273

INSIGHT 9.1 Deciphering the Structure of DNA 250

INSIGHT 9.2 Small RNAs: An Old Dog Shows Off Some New (?) Tricks 257

Chapter Summary with Key Terms 278
Multiple-Choice and True-False Questions 279
Writing to Learn 279
Concept Mapping 280
Critical Thinking Questions 280
Visual Understanding 281
Internet Search Topics 281

CHAPTER 10

Genetic Engineering: A Revolution in Molecular Biology 282

10.1 Basic Elements and Applications of Genetic Engineering 283

10.2 Tools and Techniques of Genetic Engineering 284
DNA: The Raw Material 284

10.3 Methods in Recombinant DNA Technology: How to Imitate Nature—or to "One-Up" It 292
Technical Aspects of Recombinant DNA and Gene Cloning 293
Construction of a Recombinant, Insertion into a Cloning Host, and Genetic Expression 294

10.4 Biochemical Products of Recombinant DNA Technology 296

10.5 Genetically Modified Organisms 297
Recombinant Microbes: Modified Bacteria and Viruses 297
Transgenic Plants: Improving Crops and Foods 299
Transgenic Animals: Engineering Embryos 300

10.6 Genetic Treatments: Introducing DNA into the Body 302
Gene Therapy 302
DNA Technology as Genetic Medicine 303

10.7 Genome Analysis: Maps, Fingerprints, and Family Trees 304
Genome Mapping and Screening: An Atlas of the Genome 304
DNA Fingerprinting: A Unique Picture of a Genome 305

INSIGHT 10.1 OK, the Genome's Sequenced—What's Next? 289

INSIGHT 10.2 A Moment to Think 298

INSIGHT 10.3 Better Bioterrorism Through Biotechnology? 299

Chapter Summary with Key Terms 307
Multiple-Choice and True-False Questions 308
Writing to Learn 309
Concept Mapping 310
Critical Thinking Questions 310
Visual Understanding 311
Internet Search Topics 311

CHAPTER 11

Physical and Chemical Control of Microbes 312

11.1 Controlling Microorganisms 313
General Considerations in Microbial Control 313
Relative Resistance of Microbial Forms 313
Terminology and Methods of Microbial Control 315
What Is Microbial Death? 316
How Antimicrobial Agents Work: Their Modes of Action 318

11.2 Methods of Physical Control 320
Heat as an Agent of Microbial Control 320
The Effects of Cold and Desiccation 323
Radiation as a Microbial Control Agent 324
Decontamination by Filtration: Techniques for Removing Microbes 326
11.3 Chemical Agents in Microbial Control 327
Choosing a Microbicidal Chemical 328
Factors That Affect the Germicidal Activity of Chemicals 329
Germicidal Categories According to Chemical Group 329

INSIGHT 11.1 Microbial Control in Ancient Times 314

INSIGHT 11.2 Pathogen Paranoia: "The Only Good Microbe Is a Dead Microbe" 328

INSIGHT 11.3 The Quest for Sterile Skin 334

INSIGHT 11.4 Decontaminating Congress 337

Chapter Summary with Key Terms 338
Multiple-Choice and True-False Questions 339
Writing to Learn 340
Concept Mapping 340
Critical Thinking Questions 341
Visual Understanding 341
Internet Search Topics 342

CHAPTER 12

Drugs, Microbes, Host—The Elements of Chemotherapy 343

12.1 Principles of Antimicrobial Therapy 344
The Origins of Antimicrobial Drugs 344
12.2 Interactions Between Drug and Microbe 346
Mechanisms of Drug Action 346
12.3 Survey of Major Antimicrobial Drug Groups 352
Antibacterial Drugs Targeting the Cell Wall 352
Antibacterial Drugs Targeting Protein Synthesis 356
Antibacterial Drugs Targeting Folic Acid Synthesis 357
Antibacterial Drugs Targeting DNA or RNA 358
Antibacterial Drugs Targeting Cell Membranes 358
Agents to Treat Fungal Infections 358
Antiparasitic Chemotherapy 359
Antiviral Chemotherapeutic Agents 360
Interactions Between Microbes and Drugs: The Acquisition of Drug Resistance 362
New Approaches to Antimicrobial Therapy 364
12.4 Interaction Between Drug and Host 368
Toxicity to Organs 368
Allergic Responses to Drugs 369
Suppression and Alteration of the Microbiota by Antimicrobials 369
12.5 Considerations in Selecting an Antimicrobial Drug 370
Identifying the Agent 370
Testing for the Drug Susceptibility of Microorganisms 371
The MIC and Therapeutic Index 372
An Antimicrobial Drug Dilemma 374

INSIGHT 12.1 From Witchcraft to Wonder Drugs 345

INSIGHT 12.2 A Quest for Designer Drugs 350

INSIGHT 12.3 The Rise of Drug Resistance 366

Chapter Summary with Key Terms 375
Multiple-Choice and True-False Questions 376
Writing to Learn 377
Concept Mapping 377
Critical Thinking Questions 377
Visual Understanding 378
Internet Search Topics 378

CHAPTER 13

Microbe-Human Interactions: Infection and Disease 379

13.1 The Human Host 380
Contact, Infection, Disease—A Continuum 380
Resident Biota: The Human as a Habitat 380
Indigenous Biota of Specific Regions 383
13.2 The Progress of an Infection 383
Becoming Established: Step One—Portals of Entry 385
The Size of the Inoculum 388
Becoming Established: Step Two—Attaching to the Host 388
Becoming Established: Step Three—Surviving Host Defenses 389
Causing Disease 390
The Process of Infection and Disease 393
Signs and Symptoms: Warning Signals of Disease 394
The Portal of Exit: Vacating the Host 395
The Persistence of Microbes and Pathologic Conditions 396
Reservoirs: Where Pathogens Persist 397
The Acquisition and Transmission of Infectious Agents 399
Nosocomial Infections: The Hospital as a Source of Disease 401
Universal Blood and Body Fluid Precautions 403
Which Agent Is the Cause? Using Koch's Postulates to Determine Etiology 404
13.3 Epidemiology: The Study of Disease in Populations 405
Who, When, and Where? Tracking Disease in the Population 405

INSIGHT 13.1 Life Without Microbiota 384

INSIGHT 13.2 Laboratory Biosafety Levels and Classes of Pathogens 386

INSIGHT 13.3 The Classic Stages of Clinical Infections 393

INSIGHT 13.4 The History of Human Guinea Pigs 405

INSIGHT 13.5 Koch's Postulates Still Critical in SARS Era 406

Chapter Summary with Key Terms 409
Multiple-Choice and True-False Questions 410
Writing to Learn 411
Concept Mapping 412
Critical Thinking Questions 412
Visual Understanding 413
Internet Search Topics 413

CHAPTER 14

Host Defenses I: Overview and Nonspecific Defenses 414

14.1 Defense Mechanisms of the Host in Perspective 415
Barriers at the Portal of Entry: A First Line of Defense 415

14.2 The Second and Third Lines of Defense: An Overview 418

14.3 Systems Involved in Immune Defenses 418
The Communicating Body Compartments 419

14.4 The Second Line of Defense 426
The Inflammatory Response: A Complex Concert of Reactions to Injury 427
The Stages of Inflammation 427
Phagocytosis: Cornerstone of Inflammation and Specific Immunity 432
Interferon: Antiviral Cytokines and Immune Stimulants 434
Complement: A Versatile Backup System 435
Overall Stages in the Complement Cascade 436

INSIGHT 14.1 When Inflammation Gets Out of Hand 427

INSIGHT 14.2 The Dynamics of Inflammatory Mediators 429

INSIGHT 14.3 Some Facts About Fever 431

Chapter Summary with Key Terms 438
Multiple-Choice and True-False Questions 439
Writing to Learn 439
Concept Mapping 440
Critical Thinking Questions 440
Visual Understanding 441
Internet Search Topics 441

CHAPTER 15

Host Defenses I: Specific Immunity and Immunization 442

15.1 Specific Immunity: The Third and Final Line of Defense 443

15.2 An Overview of Specific Immune Responses 445
Development of the Dual Lymphocyte System 445
Entrance and Presentation of Antigens and Clonal Selection 445
Activation of Lymphocytes and Clonal Expansion 445
Products of B Lymphocytes: Antibody Structure and Functions 445
How T Cells Respond to Antigen: Cell-Mediated Immunity (CMI) 445
Essential Preliminary Concepts for Understanding Immune Reactions 446
Receptors on Cell Surfaces Involved in Recognition of Self and Nonself 446
The Origin of Diversity and Specificity in the Immune Response 447

15.3 The Lymphocyte Response System in Depth 449
Specific Events in B-Cell Maturation 449
Specific Events in T-Cell Maturation 449
Entrance and Processing of Antigens and Clonal Selection 450

15.4 Cooperation in Immune Reactions to Antigens 451
The Role of Antigen Processing and Presentation 451
Presentation of Antigen to the Lymphocytes and Its Early Consequences 452

15.5 B-Cell Response 453
Activation of B Lymphocytes: Clonal Expansion and Antibody Production 453
Products of B Lymphocytes: Antibody Structure and Functions 453

15.6 T-Cell Response 458
Cell-Mediated Immunity 458

15.7 A Practical Scheme for Classifying Specific Immunities 462
Natural Active Immunity: Getting the Infection 462
Natural Passive Immunity: Mother to Child 462
Artificial Immunity: Immunization 463
Vaccination: Artificial Active Immunization 463
Immunotherapy: Artificial Passive Immunization 463

15.8 Immunization: Methods of Manipulating Immunity for Therapeutic Purposes 464
Passive Immunization 464
Artificial Active Immunity: Vaccination 466
Development of New Vaccines 467
Route of Administration and Side Effects of Vaccines 470
To Vaccinate: Why, Whom, and When? 471

INSIGHT 15.1 Monoclonal Antibodies: Variety Without Limit 460

INSIGHT 15.2 Breast Feeding: The Gift of Antibodies 464

INSIGHT 15.3 The Lively History of Active Immunization 465

INSIGHT 15.4 They Said It Couldn't Be Done 469

Chapter Summary with Key Terms 473
Multiple-Choice and True-False Questions 474
Writing to Learn 475
Concept Mapping 476
Critical Thinking Questions 476
Visual Understanding 477
Internet Search Topics 477

CHAPTER 16

Disorders in Immunity 478

16.1 The Immune Response: A Two-Sided Coin 479
Overreactions to Antigens: Allergy/Hypersensitivity 480

16.2 Type I Allergic Reactions: Atopy and Anaphylaxis 480
Epidemiology and Modes of Contact with Allergens 481
The Nature of Allergens and Their Portals of Entry 481
Mechanisms of Type I Allergy: Sensitization and Provocation 481
Cytokines, Target Organs, and Allergic Symptoms 483
Specific Diseases Associated with IgE- and Mast-Cell-Mediated Allergy 485
Anaphylaxis: An Overpowering Systemic Reaction 486
Diagnosis of Allergy 486
Treatment and Prevention of Allergy 487

16.3 Type II Hypersensitivities: Reactions That Lyse Foreign Cells 489
The Basis of Human ABO Antigens and Blood Types 489
Antibodies Against A and B Antigens 489
The Rh Factor and Its Clinical Importance 491
Other RBC Antigens 493

16.4 Type III Hypersensitivities: Immune Complex Reactions 494
Mechanisms of Immune Complex Disease 494
Types of Immune Complex Disease 495

16.5 Type IV Hypersensitivities: Cell-Mediated (Delayed) Reactions 495
Delayed-Type Hypersensitivity 495
Contact Dermatitis 496
T Cells and Their Role in Organ Transplantation 497

16.6 An Inappropriate Response Against Self, or Autoimmunity 499
Genetic and Gender Correlation in Autoimmune Disease 499
The Origins of Autoimmune Disease 500
Examples of Autoimmune Disease 501

16.7 Immunodeficiency Diseases: Hyposensitivity of the Immune System 503
Primary Immunodeficiency Diseases 503
Secondary Immunodeficiency Diseases 506

INSIGHT 16.1 Of What Value Is Allergy? 485

INSIGHT 16.2 Why Doesn't a Mother Reject Her Fetus? 493

INSIGHT 16.3 Pretty, Pesky, Poisonous Plants 496

INSIGHT 16.4 The Mechanics of Bone Marrow Transplantation 499

INSIGHT 16.5 An Answer to the Bubble Boy Mystery 505

Chapter Summary with Key Terms 506
Multiple-Choice and True-False Questions 507
Writing to Learn 508
Concept Mapping 508
Critical Thinking Questions 509
Visual Understanding 509
Internet Search Topics 510

CHAPTER 17

Diagnosing Infections 511

17.1 Preparation for the Survey of Microbial Diseases 512
Phenotypic Methods 512
Genotypic Methods 512
Immunologic Methods 513

17.2 On the Track of the Infectious Agent: Specimen Collection 513
Overview of Laboratory Techniques 514

17.3 Phenotypic Methods 516
Immediate Direct Examination of Specimen 516
Cultivation of Specimen 516

17.4 Genotypic Methods 518
DNA Analysis Using Genetic Probes 518
Nucleic Acid Sequencing and rRNA Analysis 518
Polymerase Chain Reaction 519

17.5 Immunologic Methods 519
General Features of Immune Testing 519
Agglutination and Precipitation Reactions 521
The Western Blot for Detecting Proteins 524
Complement Fixation 525
Miscellaneous Serological Tests 526
Fluorescent Antibodies and Immunofluorescence Testing 526
Immunoassays 527
Tests That Differentiate T Cells and B Cells 529
In Vivo Testing 529
A Viral Example 529

INSIGHT 17.1 The Uncultured 513

INSIGHT 17.2 When Positive Is Negative: How to Interpret Serological Test Results 522

Chapter Summary with Key Terms 531
Multiple-Choice and True-False Questions 532
Writing to Learn 533
Concept Mapping 533
Critical Thinking Questions 534
Visual Understanding 534
Internet Search Topics 534

CHAPTER 18

Infectious Diseases Affecting the Skin and Eyes 535

18.1 The Skin and Eyes 536

18.2 The Skin and Its Defenses 536

18.3 Normal Biota of the Skin 537

18.4 Skin Diseases Caused by Microorganisms 538
Acne 538
Impetigo 540
Cellulitis 544
Staphylococcal Scalded Skin Syndrome (SSSS) 544
Gas Gangrene 545
Vesicular or Pustular Rash Diseases 546
Maculopapular Rash Diseases 551
Wartlike Eruptions 555
Large Pustular Skin Lesions 557
Ringworm (Cutaneous Mycoses) 559
Superficial Mycoses 561

18.5 The Surface of the Eye and Its Defenses 562

18.6 Normal Biota of the Eye 562

18.7 Eye Diseases Caused by Microorganisms 562
Conjunctivitis 563
Trachoma 563
Keratitis 564
River Blindness 565

INSIGHT 18.1 The Skin Predators: *Staphylococcus* and *Streptococcus* 538

INSIGHT 18.2 Smallpox: An Ancient Scourge Revisited 549

INSIGHT 18.3 Naming Skin Lesions 552

Chapter Summary with Key Terms 568
Multiple-Choice and True-False Questions 569
Writing to Learn 570
Concept Mapping 570
Critical Thinking Questions 571
Visual Understanding 571
Internet Search Topics 571

CHAPTER 19

Infectious Diseases Affecting the Nervous System 572

19.1 The Nervous System and Its Defenses 573

19.2 Normal Biota of the Nervous System 573

19.3 Nervous System Diseases Caused by Microorganisms 574
Meningitis 574
Neonatal Meningitis 581
Meningoencephalitis 583
Acute Encephalitis 584
Subacute Encephalitis 586
Rabies 590
Poliomyelitis 592
Tetanus 596
Botulism 597
African Sleeping Sickness 599

INSIGHT 19.1 Baby Food and Meningitis 583

INSIGHT 19.2 A Long Way from Egypt: West Nile Virus in the United States 585

INSIGHT 19.3 Cheating Death 591

INSIGHT 19.4 Polio 593

INSIGHT 19.5 Botox: No Wrinkles. No Headaches. No Worries? 600

Chapter Summary with Key Terms 604
Multiple-Choice and True-False Questions 605
Writing to Learn 606
Concept Mapping 606
Critical Thinking Questions 606
Visual Understanding 607
Internet Search Topics 607

CHAPTER 20

Infectious Diseases Affecting the Cardiovascular and Lymphatic Systems 608

20.1 The Cardiovascular and Lymphatic Systems and Their Defenses 609
The Cardiovascular System 609
The Lymphatic System 609
Defenses of the Cardiovascular and Lymphatic Systems 610

20.2 Normal Biota of the Cardiovascular and Lymphatic Systems 610

20.3 Cardiovascular and Lymphatic System Diseases Caused by Microorganisms 612
Endocarditis 612
Septicemias 613
Plague 615
Tularemia 617
Lyme Disease 618
Infectious Mononucleosis 621
Hemorrhagic Fever Diseases 622
Nonhemorrhagic Fever Diseases 624
Malaria 627
Anthrax 632

HIV Infection and AIDS 634
Adult T-Cell Leukemia and Hairy-Cell Leukemia 642

INSIGHT 20.1 Atherosclerosis 612

INSIGHT 20.2 The Arthropod Vectors of Infectious Disease 619

INSIGHT 20.3 Computer Geek + Investment Guru = World Health Revolution 631

INSIGHT 20.4 AIDS-Defining Illnesses (ADIs) 636

Chapter Summary with Key Terms 645
Multiple-Choice and True-False Questions 646
Writing to Learn 647
Concept Mapping 647
Critical Thinking Questions 648
Visual Understanding 648
Internet Search Topics 648

CHAPTER 21

Infectious Diseases Affecting the Respiratory System 649

21.1 The Respiratory Tract and Its Defenses 650

21.2 Normal Biota of the Respiratory Tract 651

21.3 Upper Respiratory Tract Diseases Caused by Microorganisms 651
Rhinitis, or the Common Cold 651
Sinusitis 652
Acute Otitis Media (Ear Infection) 653
Pharyngitis 654
Diphtheria 657

21.4 Diseases Caused by Microorganisms Affecting the Upper and Lower Respiratory Tract 659
Whooping Cough 660
Respiratory Syncytial Virus Infection 661
Influenza 662

21.5 Lower Respiratory Tract Diseases Caused by Microorganisms 666
Tuberculosis 666
Pneumonia 671

INSIGHT 21.1 The World Health Organization Responds to the Threat of Bird Flu 664

INSIGHT 21.2 Fungal Lung Diseases 669

INSIGHT 21.3 Bioterror in the Lungs 676

Chapter Summary with Key Terms 682
Multiple-Choice and True-False Questions 683
Writing to Learn 684
Concept Mapping 684
Critical Thinking Questions 685
Visual Understanding 685
Internet Search Topics 685

CHAPTER 22

Infectious Diseases Affecting the Gastrointestinal Tract 686

22.1 The Gastrointestinal Tract and Its Defenses 687

22.2 Normal Biota of the Gastrointestinal Tract 688

22.3 Gastrointestinal Tract Diseases Caused by Microorganisms 689
Tooth and Gum Infections 689
Dental Caries (Tooth Decay) 689
Periodontal Diseases 691
Periodontitis 691
Necrotizing Ulcerative Gingivitis and Periodontitis 693
Mumps 693
Gastritis and Gastric Ulcers 695
Acute Diarrhea 696
Acute Diarrhea with Vomiting (Food Poisoning) 706
Chronic Diarrhea 710
Hepatitis 714
Helminthic Intestinal Infections 717

INSIGHT 22.1 Stools: To Culture or Not to Culture? 699

INSIGHT 22.2 A Little Water, Some Sugar, and Salt Save Millions of Lives 704

INSIGHT 22.3 Microbes Have Fingerprints, Too 707

INSIGHT 22.4 Treating Inflammatory Bowel Disease with Worms? 719

Chapter Summary with Key Terms 729
Multiple-Choice and True-False Questions 731
Writing to Learn 731
Concept Mapping 732
Critical Thinking Questions 732
Visual Understanding 732
Internet Search Topics 732

CHAPTER 23

Infectious Diseases Affecting the Genitourinary System 733

23.1 The Genitourinary Tract and Its Defenses 734

23.2 Normal Biota of the Urinary Tract 735
Normal Biota of the Male Genital Tract 735
Normal Biota of the Female Genital Tract 736

23.3 Urinary Tract Diseases Caused by Microorganisms 737
Urinary Tract Infections (UTIs) 737
Leptospirosis 738
Urinary Schistosomiasis 739

23.4 Reproductive Tract Diseases Caused by Microorganisms 740
Vaginitis and Vaginosis 740
Prostatitis 744

Discharge Diseases with Major Manifestation in the Genitourinary Tract 744
Genital Ulcer Diseases 748
Wart Diseases 755
Group B *Streptococcus* "Colonization"—Neonatal Disease 759

INSIGHT 23.1 Pelvic Inflammatory Disease and Infertility 743

INSIGHT 23.2 The Pap Smear 758

Chapter Summary with Key Terms 762
Multiple-Choice and True-False Questions 763
Writing to Learn 764
Concept Mapping 764
Critical Thinking Questions 764
Visual Understanding 765
Internet Search Topics 765

CHAPTER 24

Environmental Microbiology 766

24.1 Ecology: The Interconnecting Web of Life 767
The Organization of Ecosystems 768
Energy and Nutritional Flow in Ecosystems 769
Ecological Interactions Between Organisms in a Community 771

24.2 The Natural Recycling of Bioelements 772
Atmospheric Cycles 772
Sedimentary Cycles 776

24.3 Microbes on Land and in Water 778
Soil Microbiology: The Composition of the Lithosphere 778
Aquatic Microbiology 780

INSIGHT 24.1 Greenhouse Gases, Fossil Fuels, Cows, Termites, and Global Warming 774

INSIGHT 24.2 Cute Killer Whale—Or Swimming Waste Dump? 778

INSIGHT 24.3 The Waning Days of a Classic Test? 783

Chapter Summary with Key Terms 785
Multiple-Choice and True-False Questions 786
Writing to Learn 786
Concept Mapping 787
Critical Thinking Questions 787
Visual Understanding 788
Internet Search Topics 788

CHAPTER 25

Applied and Industrial Microbiology 789

25.1 Applied Microbiology and Biotechnology 790
Microorganisms in Water and Wastewater Treatment 790

25.2 Microorganisms and Food 793
Microbial Fermentations in Food Products from Plants 793
Microbes in Milk and Dairy Products 797
Microorganisms as Food 798
Microbial Involvement in Food-Borne Diseases 798
Prevention Measures for Food Poisoning and Spoilage 799

25.3 General Concepts in Industrial Microbiology 803
From Microbial Factories to Industrial Factories 805
Substance Production 806

INSIGHT 25.1 Bioremediation: The Pollution Solution? 791

INSIGHT 25.2 Wood or Plastic: On the Cutting Edge of Cutting Boards 801

INSIGHT 25.3 Microbes Degrade—and Repair—Ancient Works of Art 808

Chapter Summary with Key Terms 809
Multiple-Choice and True-False Questions 809
Writing to Learn 810
Concept Mapping 810
Critical Thinking Questions 810
Visual Understanding 811
Internet Search Topics 811

APPENDIX A: Exponents 812
APPENDIX B: Significant Events in Microbiology 814
APPENDIX C: Answers to Multiple-Choice and Selected True-False Matching Questions 815
APPENDIX D: An Introduction to Concept Mapping 817
Glossary 820
Credits 841
Index 844

Preface

Students: You just opened this book! Thank you! As the authors, we have poured our hearts and souls into helping microbiology come to life in these pages. (We can't help but point out that there are dozens of microbes on these pages that are, in fact, alive.) The interesting thing is that each of you has already had a lot of experience with microbiology. You are populated with microbes right now, and have probably had some bad experiences with quite a few. You have certainly been greatly benefited by many as well.

Many of you are interested in entering the health care profession in some way. It is absolutely indispensable for you to have a good background in the biology of microorganisms. But a grasp of this topic is important for everyone, not just health care workers. This is, after all, the Age of Biology. The 20th century was often thought of as the Age of Physics, with the development of quantum theories and the theory of relativity. The Human Genome Project is just the most visible sign of the Biology Age; in the 21st century we have an unprecedented understanding of genes and DNA, and a new respect for the beauty and power of microorganisms. But there is much more to learn.

What Sets This Book Apart?

Distinctive Organization of Infectious Disease Chapters

Following the tradition of microbiology textbooks, the first 16 chapters of ***Microbiology: A Systems Approach*** provide the basics about microorganisms: what they are, the methods used to study them, human attempts to control them, and our bodies' defenses against them. For chapters 17-23, we have developed an unequaled level of organization in our presentation of the infectious disease material.

Exclusive Chapter Chapter 17, "Diagnosing Infections," is unique among microbiology textbooks: it brings together in one place the methods used to diagnose infectious diseases. It starts with collecting samples from the patient, and details the biochemical, serological, and molecular methods used to identify causative microbes.

Highly Organized Disease Chapters Like other books, chapters 18-23 present the diseases according to the human organ systems. However, the organization of the material within each of these chapters has been taken to a new level.

The traditional organ system approach makes sense (to anyone who has experienced an infection!), but still leaves organizational threads hanging. Within a given organ system chapter, diseases are discussed in random order, and there is often no consistent pattern to what is said about each disease.

This book improves upon the approach by organizing the infectious agents according to the symptoms or condition they cause, instead of in a random order. For example, in the respiratory disease chapter, there is a major heading called "Causative Agents of Community-Acquired Pneumonia"—a condition that can be caused by several different microbes. Each of those microbes is discussed under that heading, in a systematic manner. At the end of the section, the microbes are summarized in a **Checkpoint table** called "Pneumonia by the Causative Organism." Conditions with only one possible cause, such as pertussis, also end with a Checkpoint table that includes the single causative agent.

CHECKPOINT 21.10	Pneumonia		
Causative Organism(s)	*Streptococcus pneumoniae*	*Legionella* species	*Mycoplasma pneumoniae*
Most Common Modes of Transmission	Droplet contact or endogenous transfer	Vehicle (water droplets)	Droplet contact
Virulence Factors	Capsule	–	Adhesins
Culture/Diagnosis	Gram stain often diagnostic, alpha-hemolytic on blood agar	Requires selective charcoal yeast extract agar; serology unreliable	Rule out other etiologic agents
Prevention	Pneumococcal polysaccharide vaccine (23-valent)	–	No vaccine, no permanent immunity
Treatment	Cefotaxime, ceftriaxone, ketek; much resistance	Fluoroquinolone, azithromycin, clarithromycin	Recommended not to treat in most cases, doxycycline or macrolides may be used if necessary
Distinctive Features	Patient usually severely ill	Mild pneumonias in healthy people; can be severe in elderly or immunocompromised	Usually mild; "walking pneumonia"

This approach is refreshingly logical, systematic, and intuitive, as it encourages clinical and critical modes of thinking in students—the type of thinking they will be using if their eventual careers are in health care. Students learn to examine multiple possibilities for a given condition and grow accustomed to looking for commonalities and differences among the various organisms that cause a given condition. In addition, they learn to consider the kinds of conditions that are caused by only one microbe.

Along with the higher level of organization offered in this book, students are provided with key pedagogical tools at the end of each disease chapter to reinforce and tie together the information they've just learned. Each disease chapter ends with a **system summary figure**—a "glass body" that highlights the affected organisms discussed in the chapter—and a **taxonomic list of organisms**. The distinctive summary figure lists the diseases that were presented in the chapter, with the microbes that could cause them *color-coded by type of microorganism*. The taxonomic list of organisms is presented in tabular form so students can see the diversity of microbes causing diseases in that system, and also appreciate their taxonomic positions

In summary, the disease presentation in this book makes the world of infectious diseases come together for the student. It presents the information within a consistent organizational structure (known to facilitate learning) and embeds it within a structure that teaches clinical and critical modes of thinking.

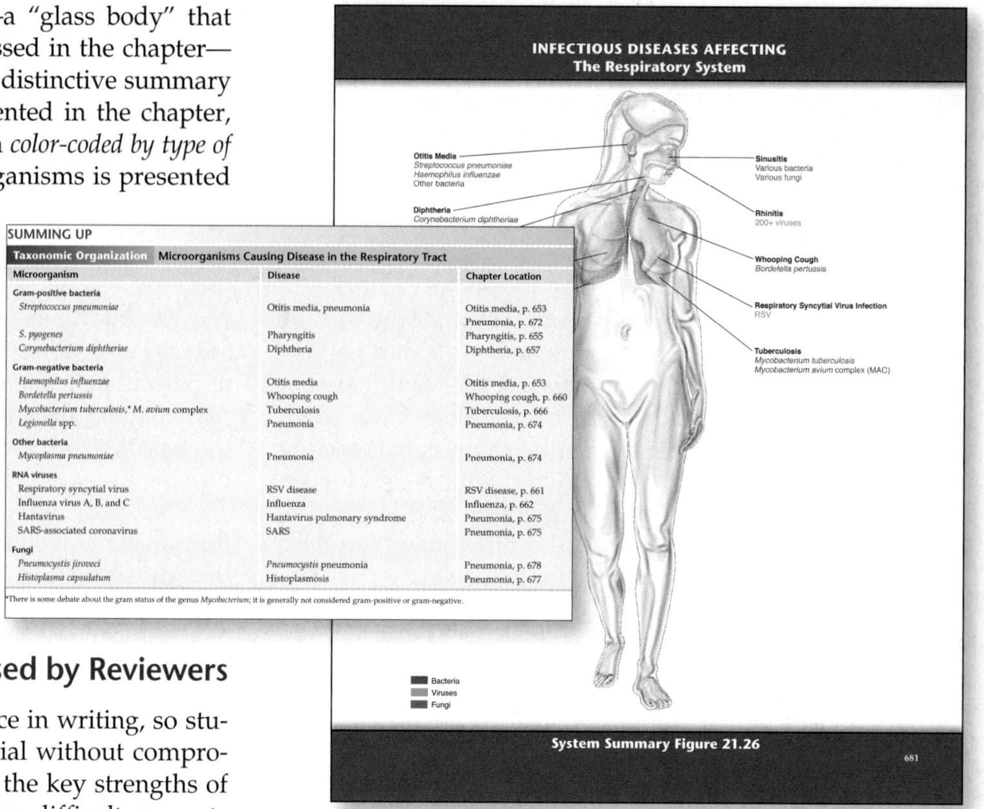

Microorganism	Disease	Chapter Location
Gram-positive bacteria		
Streptococcus pneumoniae	Otitis media, pneumonia	Otitis media, p. 653 Pneumonia, p. 672
S. pyogenes	Pharyngitis	Pharyngitis, p. 655
Corynebacterium diphtheriae	Diphtheria	Diphtheria, p. 657
Gram-negative bacteria		
Haemophilus influenzae	Otitis media	Otitis media, p. 653
Bordetella pertussis	Whooping cough	Whooping cough, p. 660
Mycobacterium tuberculosis,* *M. avium* complex	Tuberculosis	Tuberculosis, p. 666
Legionella spp.	Pneumonia	Pneumonia, p. 674
Other bacteria		
Mycoplasma pneumoniae	Pneumonia	Pneumonia, p. 674
RNA viruses		
Respiratory syncytial virus	RSV disease	RSV disease, p. 661
Influenza virus A, B, and C	Influenza	Influenza, p. 662
Hantavirus	Hantavirus pulmonary syndrome	Pneumonia, p. 675
SARS-associated coronavirus	SARS	Pneumonia, p. 675
Fungi		
Pneumocystis jiroveci	*Pneumocystis* pneumonia	Pneumonia, p. 678
Histoplasma capsulatum	Histoplasmosis	Pneumonia, p. 677

*There is some debate about the gram status of the genus *Mycobacterium*; it is generally not considered gram-positive or gram-negative.

An Engaging Writing Style, Praised by Reviewers

Our goal was to achieve a precise balance in writing, so students will easily comprehend the material without compromising the level of presentation. One of the key strengths of this text comes from our efforts in making difficult concepts understandable, as well as intriguing and exciting, for students. We use this consistent, direct approach throughout the text—in the narrative, the illustrations, and throughout the pedagogical aids. Analogies, case studies, and real-world examples also help students relate microbiology to their world.

"*The writing style of Cowan and Talaro is excellent. I feel the information is described in a very easily understood fashion. I really like the way concepts are introduced and then immediately fully explained.*" —Carl David Gilbert, University of Louisiana at Monroe

A Vivid Art Program That Explains Itself

Kathy Talaro brings her experience as a teacher, microbiologist, *and* illustrator to this text. Her insight and expertise provide an inimitable blend of scientific accuracy and aesthetics. Vivid, multi-dimensional illustrations complement self-contained, concept-specific narrative; it is not necessary to read page content surrounding the artwork to grasp concepts being illustrated. Development of the art in this manner further enhances learning and helps to build a solid foundation of understanding.

"*This book is so well organized, written, and illustrated that it is hard to identify weaknesses.*" —Larry Weiskirch, Onondaga Community College

This second edition has given us the opportunity to hone and improve the art even more. In addition to many new and revised figures, the Process Figures are now clearly defined as such and include colored steps that correlate the art to step-by-step explanations. Art has also been pulled into special Visual Understanding study tools to help students make connections between concepts presented in different chapters.

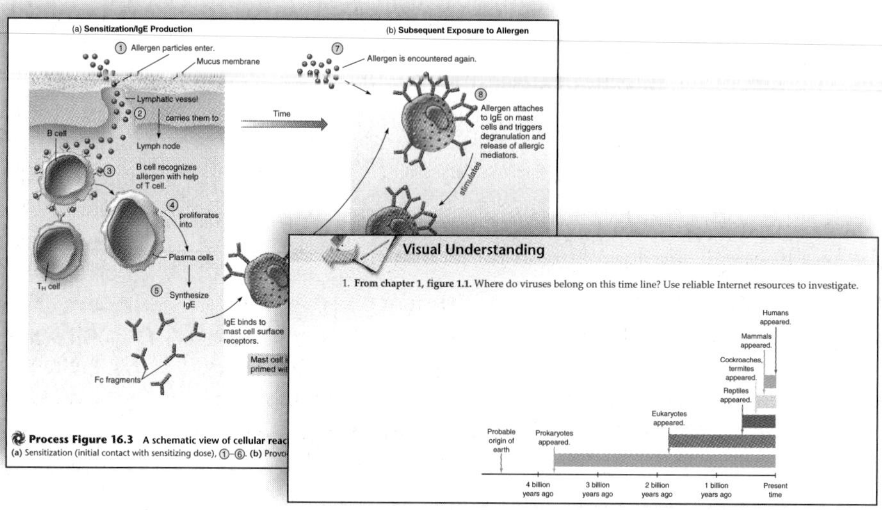

Pedagogy Designed for the Way Students Learn

Microbiology: A Systems Approach makes learning easier through its carefully crafted pedagogical system. Following is a closer look at some of the key features that our students have taught us are useful.

- All chapters open with **Case File** mysteries to solve. These real-world case studies help students appreciate and understand how microbiology impacts our lives on a daily basis. The solutions appear later in the chapter, after the necessary elements have been presented.

CHAPTER 13

Microbe-Human Interactions

Infection and Disease

CASE FILE 13

Acinetobacter baumannii is a gram-negative bacterium commonly found in soil and water. It has occasionally been linked with health-care-associated infections. Among hospitalized patients, infections caused by *A. baumannii* have become increasingly difficult to treat due to an increasing number of isolates showing multiple drug resistance. Military medical facilities have seen an increasing number of patients with bloodstream infections caused by *A. baumannii*. Many of these patients had received traumatic injuries in Iraq, Kuwait, or Afghanistan. Antimicrobial testing of isolates from patients being treated at Landstuhl Regional Medical Center in Germany and Walter Reed Army Medical Center in the District of Columbia showed widespread resistance to antimicrobials commonly used to treat this organism.

- *What could be the source of the infections seen in the military facilities?*
- *What do we call infections that are acquired in a medical setting?*
- *What is the significance in these isolates being resistant to multiple antibiotics?*

Case File 13 Wrap-Up appears on page 402.

CHAPTER OVERVIEW

- The normal biota of humans includes bacteria, fungi, and protozoa that live on the body without causing disease. These microbes can be found in areas exposed to the outside environment, such as the gastrointestinal tract, skin, and respiratory tract, and are generally beneficial to humans.
- Pathogens are those microbes that infect the body and cause disease. Disease results when an adequate number of pathogenic cells enter the body through a specific route, grow, and disrupt tissues.
- Pathogens produce virulence factors such as toxins and enzymes that help them invade and damage the cells of their host. The effects of infection and disease are seen in the host as signs and symptoms, which may include both short- and long-term damage.
- Pathogens may be spread by direct or indirect means involving overtly infected people, carriers, vectors, and vehicles. A significant source of infection is exposure to the hospital environment.
- Pathogens may be found residing in humans, animals, food, soil, and water.
- The field of epidemiology is concerned with the patterns of disease occurrence in a population.

379

- A **Chapter Overview** at the beginning of each chapter provides students with a framework from which to begin their study of a chapter.
- In chapters 1–16 and 24–25, major sections of the chapter are followed by **Checkpoints** that repeat and summarize the concepts of that section. In the disease chapters (18–23), the Checkpoints are in the form of the **disease tables** described earlier.
- **Insight** readings allow students to delve into material that goes beyond the chapter concepts and consider the application of those concepts. The Insight readings are divided into four categories: Discovery, Historical, Medical, and Microbiology.

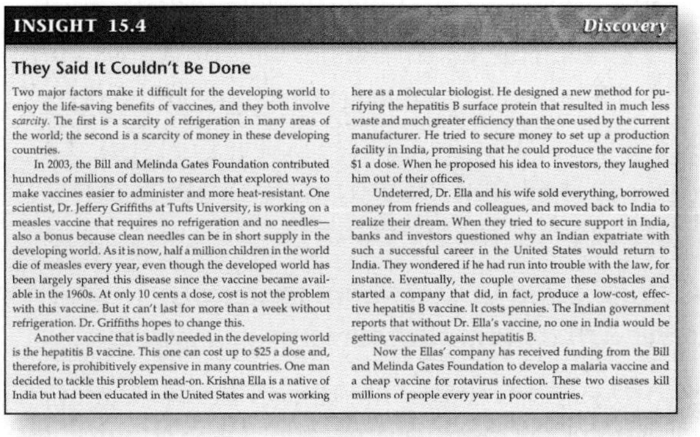

INSIGHT 15.4 *Discovery*

They Said It Couldn't Be Done

Two major factors make it difficult for the developing world to enjoy the life-saving benefits of vaccines, and they both involve *scarcity*. The first is a scarcity of refrigeration in many areas of the world; the second is a scarcity of money in these developing countries.

In 2003, the Bill and Melinda Gates Foundation contributed hundreds of millions of dollars to research that explored ways to make vaccines easier to administer and more heat-resistant. One scientist, Dr. Jeffery Griffiths at Tufts University, is working on a measles vaccine that requires no refrigeration and no needles—also a bonus because clean needles can be in short supply in the developing world. As it is now, half a million children in the world die of measles every year, even though the developed world has been largely spared this disease since the vaccine became available in the 1960s. At only 10 cents a dose, cost is not the problem with this vaccine. But it can't last for more than a week without refrigeration. Dr. Griffiths hopes to change this.

Another vaccine that is badly needed in the developing world is the hepatitis B vaccine. This one can cost up to $25 a dose and, therefore, is prohibitively expensive in many countries. One man decided to tackle this problem head-on. Krishna Ella is a native of India but had been educated in the United States and was working here as a molecular biologist. He designed a new method for purifying the hepatitis B surface protein that resulted in much less waste and much greater efficiency than the one used by the current manufacturer. He tried to secure money to set up a production facility in India, promising that he could produce the vaccine for $1 a dose. When he proposed his idea to investors, they laughed him out of their offices.

Undeterred, Dr. Ella and his wife sold everything, borrowed money from friends and colleagues, and moved back to India to realize their dream. When they tried to secure support in India, banks and investors questioned why an Indian expatriate with such a successful career in the United States would return to India. They wondered if he had run into trouble with the law, for instance. Eventually, the couple overcame these obstacles and started a company that did, in fact, produce a low-cost, effective hepatitis B vaccine. It costs pennies. The Indian government reports that without Dr. Ella's vaccine, no one in India would be getting vaccinated against hepatitis B.

Now the Ellas' company has received funding from the Bill and Melinda Gates Foundation to develop a malaria vaccine and a cheap vaccine for rotavirus infection. These two diseases kill millions of people every year in poor countries.

- All chapters end with a **summary**, and a comprehensive array of **end-of-chapter questions** that are not just multiple-choice, but also critical thinking questions, often with no correct answer. Considering and answering these questions, and even better, discussing them with fellow students, can make the difference between temporary (or limited) learning and true knowledge of the concepts. **Visual Understanding** questions incorporate art to help students connect important concepts from chapter to chapter, and **Concept Mapping** assists in retention as well as contextual organization.

> *"Cowan and Talaro have created a perfect tool for the instruction of microbiology to the non-major, allied health student! The text is well-designed in its layout of chapters, beginning with basic information about the discipline and building to the application of those principles in the infectious disease chapters. The authors do a great job of reminding the reader of material they have previously encountered by referencing specific pages. The writing style is very approachable, yet provides enough detail to please many instructors. This is certainly a text that would work for my teaching style."* —Angela Spence, Missouri State University

What's New?

We are committed to two goals for this book: making it the most current and scientifically accurate book in the field, and turning what could be a passive educational experience into an active learning opportunity.

Up-to-Date Content

- This edition, like microbiology itself, is **full of changes in content**. Probably the most important update is the new understanding of the "central dogma" of biology: that DNA is made into RNA, which is made into proteins. With the advances in genomics of the last decade we now know that the characteristics of all organisms are influenced just as strongly by the pieces of RNA that aren't made into protein, but that are used to regulate the DNA and proteins. We address this in chapter 9.
- We've also updated content on the new "-omics": genomics, proteomics and even metabolomics.
- Throughout the book there is much more emphasis on polymicrobial infections and biofilms.
- Also, in multiple chapters we discuss a new initiative to identify the sequences present in "normal biota" body sites, a project that is likely to revolutionize the way we think of normal biota.
- We even tackle the old laboratory warhorse, the coliform test, since nearly all experts believe the test is terribly outdated, even though we continue to teach it in introductory microbiology labs.

- Finally, we have split the environmental microbiology chapter into two new chapters. One of these focuses on microbes in the environment and the other examines ways we use microbes to get the things we need and want, such as food and medicines.

For a complete listing of chapter-by-chapter changes, please visit the text's ARIS website.

Active Learning Experience

Capitalizing on sound research in how students learn, we have added several new features to this edition:

- "**Visual Understanding**" is an exercise that does two things. First, it supplies a photo or a graphic that students have already seen, along with a thought-provoking question. Second, many of the Visual Understanding questions use images from previous chapters and pose queries that require students to combine knowledge from the new chapter with the knowledge they already have from the previous chapter. This encourages the making of connections and the weaving of a whole cloth of understanding, a task indispensable to real learning but very often neglected in courses and books.
- To offer different perspectives on similar topics, many **figures** in the text are now **correlated to digital animations**. Students may examine the figure's details in the book and then watch the concept in motion on their computer or download it to their portable player to study on the fly!
- **Process figures** now have matching numbered steps for easy to see explanations of complex processes.
- Perhaps most exciting of all of the changes: this is the first microbiology textbook that uses **concept maps**! Concept maps present ways for students to organize information in more meaningful forms than just simple lists. They appeal to a wide variety of learning styles and help readers get in the habit of putting facts in contextual form. Our concept maps build in varying degrees of complexity and are accompanied by an appendix on how to get started.

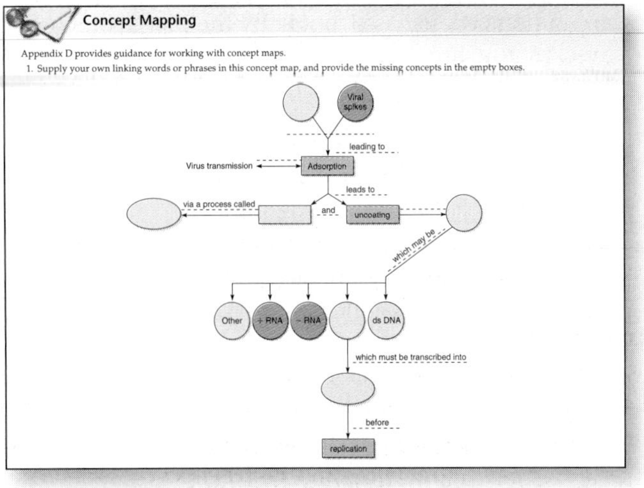

Teaching and Learning Supplements

McGraw-Hill offers various tools and technology products to support *Microbiology: A Systems Approach,* **2/e.** Instructors can obtain teaching aids by calling the McGraw-Hill Customer Service Department at 1-800-338-3987, visiting our Microbiology catalog at **www.mhhe.com/microbiology**, or contacting their local McGraw-Hill sales representative.

ARIS Text Website

The ARIS website that accompanies this textbook includes tutorials, animations, practice quizzing, helpful Internet links and more—a whole semester's worth of study help for students. Instructors will find a complete electronic homework and course management system where they can create and share course materials and assignments with colleagues in just a few clicks of the mouse. Instructors can also edit questions, import their own content, and create announcements and/or due dates for assignments. ARIS offers automatic grading and reporting of easy-to-assign homework, quizzing, and testing.

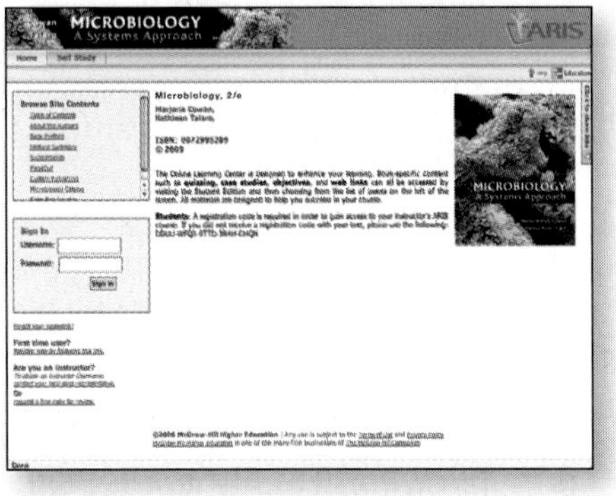

Check out **www.aris.mhhe.com**, select your subject and textbook, and start benefiting today!

NEW! Downloadable content for portable players!
Now students can study anywhere, anytime.

- Audio chapter summaries with quiz questions
- Animations (correlated to figures in the text)

Complete Set of Electronic Images and Assets for Instructors

Instructors, build instructional materials wherever, whenever, and however you want!

Part of the ARIS website, the digital library contains assets such as photos, artwork, animations, PowerPoints, and other media resources that can be used to create customized lectures, visually enhance tests and quizzes, and design compelling course websites or attractive printed support materials. All assets are copyrighted by McGraw-Hill Higher Education but can be used by instructors for classroom purposes. The visual resources in this collection include:

- **Art** Full-color digital files of all illustrations in the book can be readily incorporated into lecture presentations, exams, or custom-made classroom materials. In addition, all files are pre-inserted into blank PowerPoint slides for ease of lecture preparation.

- **Photos** The photos collection contains digital files of photographs from the text, which can be reproduced for multiple classroom uses.
- **Tables** Every table that appears in the text has been saved in electronic form for use in classroom presentations and/or quizzes.
- **Animations** Numerous full-color animations illustrating important microbial or physiological processes are also provided. Harness the visual impact of concepts in motion by importing these files into classroom presentations or online course materials.
- **Lecture Outlines** Specially prepared custom outlines for each chapter offered in easy-to-use PowerPoint slides.

Computerized Test Bank Online

A comprehensive bank of test questions is provided within a computerized test bank powered by McGraw-Hill's flexible electronic testing program EZ Test Online. EZ Test Online allows instructors to create and access paper or online tests or quizzes in an easy to use program anywhere, at any time without installing the testing software. Now, with EZ Test Online, instructors can select questions from multiple McGraw-Hill test banks or author their own, and then either print the test for paper distribution or give it online. Visit: **www.eztestonline.com** to learn more about creating and managing tests, online scoring and reporting, and support resources.

Electronic Books

If you, or your students, are ready for an alternative version of the traditional textbook, McGraw-Hill and VitalSource have partnered to introduce innovative and inexpensive electronic textbooks. By purchasing E-books, students can save as much as **50%** on selected titles delivered on the **most advanced** E-book platform available, VitalSource Bookshelf.

E-books from McGraw-Hill and VitalSource are smart, interactive, searchable and portable. VitalSource Bookshelf comes with a powerful suite of built-in tools that allow detailed searching, highlighting, note taking, and student-to-student or instructor-to-student note sharing. In addition, the media-rich E-book for *Microbiology, A Systems Approach* integrates relevant animations and videos into the textbook content for a true multimedia learning experience.

E-books from McGraw-Hill and VitalSource will help students study smarter and quickly find the information they need while saving money. Instructors: contact your McGraw-Hill sales representative to discuss E-book packaging options.

e-Instruction with CPS

The Classroom Performance System (CPS) is an interactive system that allows the instructor to administer in-class questions electronically. Students answer questions via hand-held remote control keypads (clickers), and their individual responses are logged into a grade book. Aggregated responses can be displayed in graphical form. Using this immediate feedback, the instructor can quickly determine if students understand the lecture topic, or if more clarification is needed. CPS promotes student participation, class productivity, and individual student confidence and accountability. Specially designed questions for e-Instruction to accompany *Microbiology, A Systems Approach* are provided through the book's ARIS website.

Course Delivery Systems

In addition to McGraw-Hill's ARIS course management options, instructors can also design and control their course content with help from our partners WebCT, Blackboard, Top-Class, and eCollege. Course cartridges containing website content, online testing, and powerful student tracking features are readily available for use within these or any other HTML-based course management platforms.

Acknowledgments

Textbooks are never written by just one person, or two people, in this case. Textbooks are the accumulation of good suggestions, corrections, and brainstorming from a large team of faculty from all over the country who teach in all kinds of institutions. The faculty below were involved in multiple ways to improve this book and we are deeply grateful to them. All of their names should be on the cover!

Board of Advisors

Gail Stewart, *Camden County College*
Kelley Black, *Jefferson State Community College*
Terri Lindsey, *Tarrant County College—South Campus*
Kathleen (Kate) Richardson, *Portland Community College*
Sheila Wise, *Ivy Tech Community College*
Jennifer Freed, *Rio Salado College*
Tracey Mills, *Ivy Tech Community College*
Cathy Murphy, *Ocean Community College*
Judy Haber, *California State University – Fresno*
Alison Davis, *East Los Angeles College*

Symposium Attendees

Hazel Barton, *Northern Kentucky University*
Karen Bentz, *Albuquerque Technical Vocational Institute*
Kelley Black, *Jefferson State Community College*
Elaina Bleifield, *North Hennepin Community College*
Rita Connolly, *Camden City College*
Janet Decker, *University of Arizona*
Carl D. Gilbert, *University of Louisiana—Monroe*
Andrew Henderson, *Horry-Georgetown Technical College*
Jeff G. Leid, *Northern Arizona University*
Timothy E. Secott, *Minnesota State University—Mankato*
Jeanne Weidner, *San Diego State University*

Reviewers

Joel Adams-Stryker, *Evergreen Valley College*
Shelley Aguilar, *Mt. San Jacinto College*
Cindy B. Anderson, *Mt. San Antonio College*
Karen L. Anderson, *Madison Area Technical College*

Arden Aspedon, *Southwestern Oklahoma State University*
Dave Bachoon, *Georgia College & State University*
S. E. Barbaro, *Rivier College*
Dale L. Barnard, *Utah State University*
Charles Lee Biles, *East Central University*
Susan Bjerke, *Washburn University*
Elaina M. Bleifield, *North Hennepin Community College*
Chad Brooks, *Austin Peay State University*
Barbara Y. Bugg, *Northwest Mississippi Community College*
D. Kim Burnham, *Oklahoma State University*
Suzanne Butler, *Miami-Dade College*
Misty Gregg Carriger, *Northeast State Community College*
Carol L. Castaneda, *Indiana University Northwest*
Erin Christensen, *Middlesex County College*
Kathy Ann Clark, *College of Southern Idaho*
James K. Collins, *University of Arizona*
James Constantine, *Bristol Community College*
Don C. Dailey, *Austin Peay State University*
Kristina Dameron, *College of Lake County*
RoxAnn Davenport, *Tulsa Community College*
Charles J. Dick, *Pasco-Hernando Community College*
Deborah A. Dixon, *Laredo Community College*
Nancy B. Dunning, *San Juan College*
Mohamed Elasri, *University of Southern Mississippi*
Debra Ellis, *Frederick Community College*
S. Marvin Friedman, *Hunter College of the City University of New York*
Carl D. Gilbert, *University of Louisiana at Monroe*
Brinda Govindan, *San Francisco State University*
W. Michael Gray, *Bob Jones University*
Judy Haber, *California State University—Fresno*
Robert C. Hairston, *Harrisburg Area Community College*
Julie Harless, *Montgomery College*
Randall K. Harris, *William Carey College*
Diane Hartman, *Baylor University*
Keith R. Hench, *Kirkwood Community College*
Joan M. Henson, *Montana State University*
Marian Hill, *St. Petersburg College*
Carolyn Holcroft-Burns, *Foothill College*
Jacob M. Hornby, *Lewis-Clark State College*
Janice Ito, *Leeward Community College*
Gilbert H. John, *Oklahoma State University*
Judy Kaufman, *Monroe Community College*
Robert A. Keeton, *University of Arkansas Community College—Morrilton*
Karen Kendall-Fite, *Columbia State Community College*
Kevin Kiser, *Cape Fear Community College*
Dennis J. Kitz, *Southern Illinois University—Edwardsville*
Carly L. Langlais, *Portland Community College*
Michael A. Lawson, *Missouri Southern State University*
Jeff G. Leid, *Northern Arizona University*
Kimberly G. Lyle-Ippolito, *Anderson University*
Rene Massengale, *Baylor University*
Ethel M. Matthews, *Midland College*
Mary Colleen McNamara, *Albuquerque TVI Community College*
Stephen Miller, *Golden West College*
Fernando P. Monroy, *Northern Arizona University*
Jonathan Morris, *Manchester Community College*
Richard L. Myers, *Missouri State University*
Russell Nordeen, *University of Arkansas—Monticello*
Lourdes P. Norman, *Florida Community College—Jacksonville*
Natalie Osterhoudt, *Broward Community College*
Clark L. Ovrebo, *University of Central Oklahoma*
Vanessa Passler, *Wallace State Community College*
R. Kevin Pegg, *Florida Community College—Jacksonville*
Inga B. Pinnix, *Florida Community College—Jacksonville*
Edith Porter, *California State University—Los Angeles*
Shelby C. Powell, *College of Eastern Utah*
Nirmala V. Prabhu, *Edison College*
Rolf Prade, *Oklahoma State University*
Davis W. Prichett, *University of Louisiana—Monroe*
Gregory S. Pryor, *Francis Marion University*
Sabine A. Rech, *San Jose State University*
Harold W. Reed , *Georgia College & State University*
Thomas F. Reed, *Brevard Community College*
Amy J. Reese, *Cedar Crest College*
Jackie S. Reynolds, *Richland College*
Kay Rezanka, *Central Lakes College*
Kenda L. Rigdon, *Jefferson State Community College*
Paulette W. Royt, *George Mason University*
Andrew M. Scala, *Dutchess Community College*
Gene M. Scalarone, *Idaho State University*
Timothy Secott, *Minnesota State University—Mankato*
Jerred Seveyka, *Yakima Valley Community College*
Michele Shuster, *New Mexico State University*
Edward Simon, *Purdue University*
Robert A. Smith, *University of the Sciences in Philadelphia*
Angela L. Spence, *Missouri State University*
Juliet V. Spencer, *University of San Francisco*
Timothy Steele, *Des Moines University*
Gail A. Stewart, *Camden County College*
Kathryn Sutton, *Clarke College*
Teresa Thomas, *Southwestern College—Chula Vista*
Andrew A. Thompson, *Central Florida Community College*
Juliette K. Tinker, *Boise State University*
Coe A. Vander Zee, *Austin Community College*
Manuel Varela, *Eastern New Mexico University*
Stephen C. Wagner, *Stephen F. Austin State University*
Valerie A. Watson, *West Virginia University*
Larry Weiskirch, *Onondaga Community College*
Carola Z. Wright, *Mt. San Antonio College*
Shawn B. Wright, *Albuquerque TVI Community College*
Karen R. Zagula, *Wake Technical Community College*

A Note of Thanks from Kelly Cowan

I am grateful to my many students who have tried to teach me how to most effectively communicate a subject I love to them. My partner, Kathy, has been a constant inspiration to me. I had significant content help with chapters 1, 8, 24, and 25 from Martin Klotz from the University of Louisville and Bob Findlay from the University of Alabama. I am grateful to Kathy Loewenberg at McGraw-Hill for being polite enough not to point out how often she had to fix things for me and for putting her heart into this project. Peggy Selle, Jim Connely, Jeanne Patterson, Tami Petsche, Laurie Janssen, and Alison Hammond were indispensable members of the team that helped this edition come together. A special thank you to Trina Zimmerman and Greg Duncan for believing in me and, when all else failed, buying me a burrito. Donna Hensley, Danielle Blevins, Tara Eagle, and Brittany Brewer provided vital logistical help. The real heroes in all of this are my sons Taylor and Sam who, over the course of two editions, have grown used to looking for their mom behind a stack of papers in the study. Their patience and understanding—and their awesomeness—know no bounds. Finally, Ted, all I can say is: thank you.

A Message from Kathy Talaro

In the second edition of this text, Kelly Cowan has continued to distinguish herself by bringing a fresh perspective and novel ideas for organizing and presenting the subject of microbiology. She has been a successful advocate and developer of case studies as an integral component of microbiology textbooks, and in this new edition, she has introduced concept maps and visual understanding exercises as well. These alternative learning aids will add immeasurably to the value of the textbook for teachers and students seeking different ways to explore its concepts.

Many thanks to the team of editors, researchers, designers, artists, and reviewers who helped bring this book to fruition. Without your expert guidance and insights, our endeavors would be a lot more work and a lot less fun. We salute you.

A special thanks to my family and friends for your unflagging support and understanding when taking all those little side trips to get pond samples, take photographs of relevant scenes or intriguing microbes, and just generally playing second fiddle to the demands of book creation. You're such good sports to endure my experiments incubating in the laundry room and admiring "beautiful" molds growing on food left in the refrigerator too long. I'm sure that often you know more about the microbial world than you wanted to.

Unique Systems-Based Approach Enhances Comprehension

An Unequaled Level of Organization in the Infectious Disease Material

Microbiology, A Systems Approach takes a unique approach to diseases by consistently covering multiple causative agents of a particular disease in the same section and summarizing this information in Checkpoint tables. The causative agents are categorized in a logical manner based on the presenting symptoms in the patient. Through this approach, students study how diseases affect patients—the way future healthcare professionals will encounter the material on the job.

"The systems approach to the study of microbiology and the organization of the important concepts and sub-concepts by Cowan and Talaro's textbook are the strongest features of the book and should definitely be retained in future editions. Furthermore, the chapter dealing with 'Diagnosing Infections' represents a novel approach. Often I tell my pre-med, pre-dental, and pre-vet students that they'll eventually have to think critically when diagnosing their patients."
—Manuel Varela, Eastern New Mexico University

Consistent, Clinical Presentation of Diseases

For each disease, the discussion begins with an introduction to the disease and its signs and symptoms. Next, the causative agent or agents of that disease are presented and the following areas are discussed:

- pathogenesis and virulence factors
- transmission and epidemiology
- culture and diagnosis
- prevention and treatment

Summary Checkpoint Tables

Following the textual discussion of each disease, a table summarizes the characteristics of agents that can cause that disease.

CHECKPOINT 19.5 Subacute Encephalitis

Causative Organism(s)	*Toxoplasma gondii*	Subacute sclerosing panencephalitis	Prions
Most Common Modes of Transmission	Vehicle (meat) or fecal-oral	Persistence of measles virus	CJD = direct/parenteral contact with infected tissue; or inherited vCJD = vehicle (meat, parenteral)
Virulence Factors	Intracellular growth	Cell fusion, evasion of immune system	Avoidance of host immune response
Culture/Diagnosis	Serological detection of IgM, culture, histology	EEGs, MRI, serology (Ab versus measles virus)	Biopsy, image of brain
Prevention	Personal hygiene, food hygiene	None	Avoiding tissue
Treatment	Pyrimethamine and/or sulfadiazine	None	None
Distinctive Features	Subacute, slower development of disease	History of measles	Long incubation period; fast progression once it begins

"The strong points of this book are the writing style and the attention to detail. The coverage is exceptional. There is nothing missing!! I found it easy to read and it makes difficult concepts understandable. I especially like the way the disease chapters are handled with the Checkpoints summarizing the diseases. It is a great feature." —Judy Kaufman, Monroe Community College

System Summary Figures

After the diseases of a particular body system have been discussed, students are invited to study the system summary figure at the end of the chapter—a "glass body" that highlights the affected organs and lists the diseases that were presented in the chapter. In addition, the microbes that could cause the diseases are color-coded by type of microorganism. The System Summary figures, along with the Checkpoint tables, provide an excellent set of study tools.

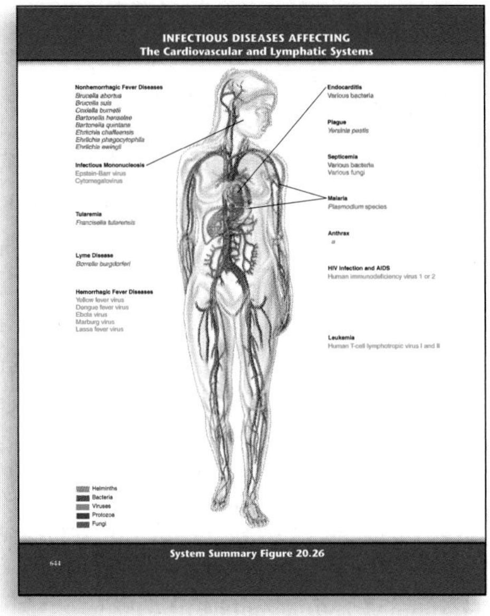

Taxonomic List of Organisms

A taxonomic list of organisms is also presented at the end of each disease chapter so students can see the diversity of microbes causing diseases in that system.

SUMMING UP

Taxonomic Organization Microorganisms Causing Disease in the Genitourinary Tract

Microorganism	Disease	Chapter Location
Gram-positive bacteria		
Staphylococcus saprophyticus	Urinary tract infection	UTI, p. 737
Gardnerella (note: stains gram-negative)	Vaginosis	Vaginitis or vaginosis, p. 741
Group B *Streptococcus*	Neonatal disease	Group B strep neonatal disease, p. 759
Gram-negative bacteria		
Escherichia coli	Urinary tract infection	UTI, p. 737
Leptospira interrogans (spirochete)	Leptospirosis	Leptospirosis, p. 738
Proteus mirabilis	Urinary tract infection plus kidney stones	UTI, p. 737
Neisseria gonorrhoeae	Gonorrhea	Discharge diseases, p. 744
Chlamydia trachomatis	"Chlamydia"	Discharge diseases, p. 747
Treponema pallidum (spirochete)	Syphilis	Genital ulcer diseases, p. 748
Haemophilus ducreyi	Chancroid	Genital ulcer diseases, p. 752
DNA viruses		
Herpes simplex viruses 1 and 2	Genital herpes	Genital ulcer diseases, p. 752
Human papillomaviruses	Genital warts, cervical carcinoma	Wart diseases, p. 755
Poxviruses	Molluscum contagiosum	Wart diseases, p. 758
Fungi		
Candida albicans	Vaginitis	Vaginitis or vaginosis, p. 740
Protozoa		
Trichomonas vaginalis	Trichomoniasis (vaginitis)	Vaginitis or vaginosis, p. 742
Helminth—trematode		
Schistosoma haematobium	Urinary schistosomiasis	Urinary schistosomiasis, p. 739

Unparalleled Art Program

Instructional Art Program Clarifies Concepts

Microbiology, A Systems Approach provides visually powerful artwork that paints conceptual pictures for students. The art combines vivid colors, multi-dimensionality, and self-contained narrative to help students study the challenging concepts of microbiology from a visual perspective—a proven study technique. Art is often coupled with photographs to enhance visualization and comprehension.

> “*Illustrations are strong and appropriately used to explain more difficult topics. I think this is one of the great strengths of the text.*”
> —Kay Rezanka, Central Lakes College

> “*The figures and illustrations are some of the best that I have seen.*”
> —Juliette Tinker, Boise State University

Process Figures

Microbiology, A Systems Approach illustrates many difficult microbiological concepts in steps that students find easy to follow. Each step is clearly marked with a yellow, numbered circle and correlated to accompanying narrative, also marked with yellow, numbered circles, to benefit all types of learners. Process Figures are now identified next to the figure number. The accompanying legend provides additional explanation.

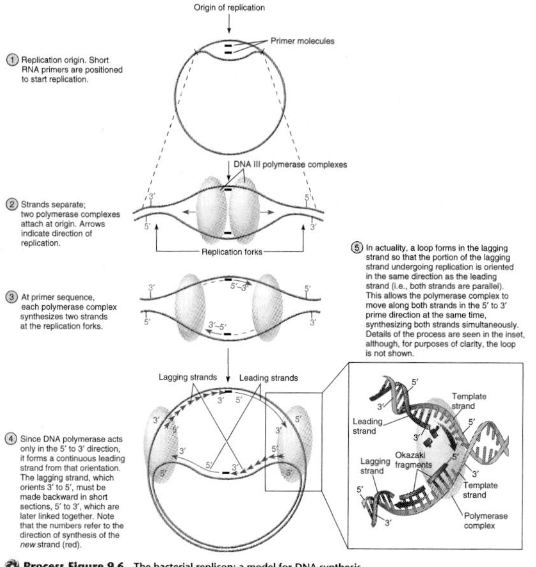

Origin of replication
Primer molecules
① Replication origin. Short RNA primers are positioned to start replication.
DNA III polymerase complexes
② Strands separate; two polymerase complexes attach at origin. Arrows indicate direction of replication.
Replication forks
③ At primer sequence, each polymerase complex synthesizes two strands at the replication forks.
⑤ In actuality, a loop forms in the lagging strand so that the portion of the lagging strand undergoing replication is oriented in the same direction as the leading strand (i.e., both strands are parallel). This allows the polymerase complex to move along both strands in the 5′ to 3′ prime direction at the same time, synthesizing both strands simultaneously. Details of the process are seen in the inset, although, for purposes of clarity, the loop is not shown.
Lagging strands
Leading strands
④ Since DNA polymerase acts only in the 5′ to 3′ direction, it forms a continuous leading strand from that orientation. The lagging strand, which orients 3′ to 5′, must be made backward in short sections, 5′ to 3′, which are later linked together. Note that the numbers refer to the direction of synthesis of the *new* strand (red).
Template strand
Leading strand
Okazaki fragments
Lagging strand
Template strand
Polymerase complex

Process Figure 9.6 **The bacterial replicon: a model for DNA synthesis.**
① Circular DNA has a special origin site where replication originates. ② When strands are separated, two replication forks form, and a DNA polymerase III complex enters at each fork. ③ Starting at the primer sequence, each polymerase moves along the template strands (blue), synthesizing the new strands (red) at each fork. ④ DNA polymerase works only in the 5′ to 3′ direction, necessitating a different pattern of replication at each fork. Because the leading strand orients in the 5′ to 3′ direction, it will be synthesized continuously. The lagging strand, which orients in the opposite direction, can only be synthesized in short sections, 5′ to 3′, which are later linked together. ⑤ Inset presents details of process at one replication fork and shows the Okazaki fragments and the relationship of the template, leading, and lagging strands.

Combination Figures

Line drawings combined with photos give students two perspectives: the realism of photos and the explanatory clarity of line drawings. The authors choose this method of presentation often in every chapter.

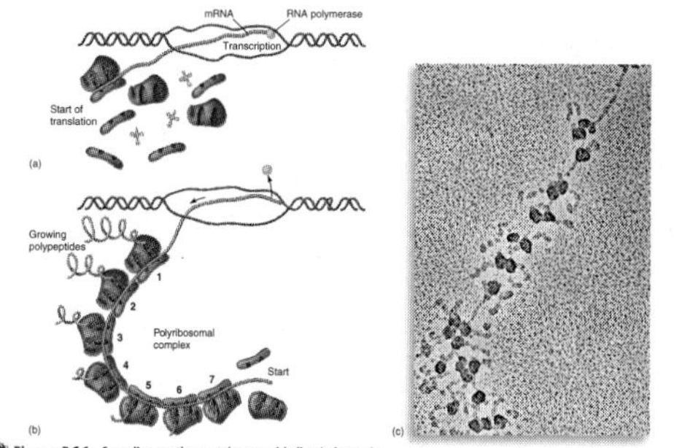

mRNA
RNA polymerase
Transcription
Start of translation
(a)
Growing polypeptides
Polyribosomal complex
Start
(b)
(c)

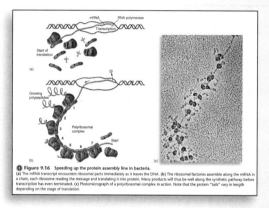

Figure 9.16 **Speeding up the protein assembly line in bacteria.**
(a) The mRNA transcript encounters ribosomal parts immediately as it leaves the DNA. **(b)** The ribosomal factories assemble along the mRNA in a chain, each ribosome reading the message and translating it into protein. Many products will thus be well along the synthetic pathway before transcription has even terminated. **(c)** Photomicrograph of a polyribosomal complex in action. Note that the protein "tails" vary in length depending on the stage of translation.

Clinical Photos

Color photos of individuals affected by disease provide students with a real-life, clinical view of how microorganisms manifest themselves in the human body.

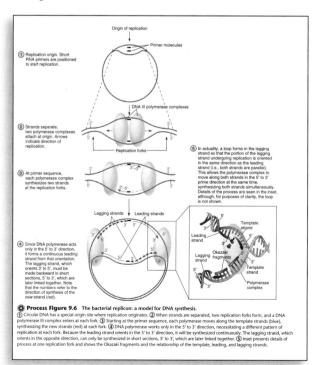

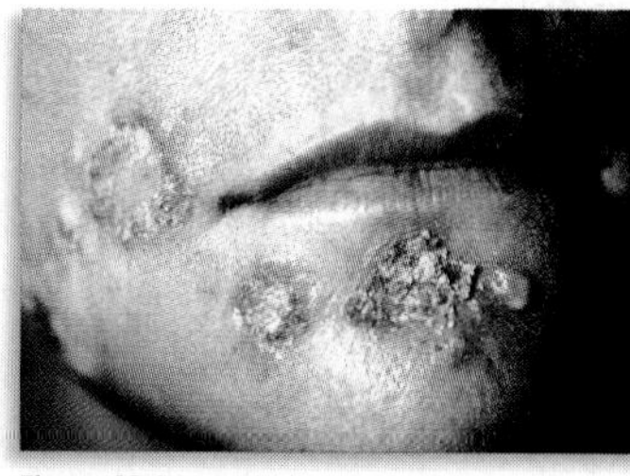

Figure 18.2 Impetigo lesions on the face.

Overview Figures

Many challenging concepts of microbiology consist of numerous interrelated activities. *Microbiology, A Systems Approach* visually summarizes these concepts to help students piece the activities together for a complete, conceptual picture.

> “*I really like the illustrations in this chapter. They are clearly tied to the text and they are effective in presenting the information. I found them easy to understand. The photographs are great too.*”
> —Carola Wright, Mt. San Antonio College

Pedagogy Designed for the Way Students Learn

Pedagogical Aids Promote Systematic Learning

Microbiology, A Systems Approach organizes each chapter with consistent pedagogical tools. Such tools enable students to develop a consistent learning strategy and enhance their understanding and retention of the concepts.

Case Files

All chapters open with a real-world case file to help students appreciate and understand how microbiology impacts lives on a daily basis. The solution to the case file appears later in the chapter, near where relevant material is being discussed.

CASE FILE 22 *WRAP-UP*

Stool samples or rectal swabs from 44 of the symptomatic patients were tested, and norovirus was identified in 22 of these samples. No other enteric pathogen was isolated. Noroviruses are often identified as the causative agent in outbreaks of gastroenteritis in the United States. Such outbreaks are often associated with contaminated food or water. In addition, outbreaks can be associated with persons living in crowded living conditions such as those present at the Reliant Park Complex.

Norovirus is highly contagious (I.D. < 100 organisms) and is easily spread person to person and by contact with contaminated materials. The typical incubation period is 24 to 48 hours, and the resulting symptoms persist for 12 to 60 hours. In this situation, it is likely that one or more individuals were infected with the virus when they arrived at the shelter. Although the source of the initial infection was unknown, contact with contaminated floodwaters was certainly a possibility. The infection spread quickly due to the crowded living conditions and shared facilities. Implementation of infection-control measures including isolation of symptomatic individuals, distribution of gel hand sanitizer, and education of staff and evacuees quickly brought the outbreak under control.

See: CDC. 2005. Norovirus outbreak among evacuees from Hurricane Katrina—Houston, Texas, September 2005. MMWR *53(40):1016–1018.*

Insight Readings

Current, real-world readings allow students to consider applications of the concepts they are studying. The Insight readings are divided into four interesting categories: Discovery, Historical, Medical, and Microbiology.

INSIGHT 9.2 *Discovery*

Small RNAs: An Old Dog Shows Off Some New(?) Tricks

Since the earliest days of molecular biology, RNA has been an overlooked worker of the cell, quietly ferrying the information in DNA to ribosomes to direct the formation of proteins. Current research however is showing a new, dynamic role for RNA in the cell that may forever change the reputation of this humble molecule.

Short lengths of RNA seem to have the ability to control the expression of certain genes. Some of these are called micro RNAs and some are called small interfering RNAs. They do this by folding back on themselves after being transcribed, and by doing so they activate a system inside cells that degrades dsRNA. Cells do this in order to rid themselves of invading viruses (which are organisms that might have dsRNA). The micro RNAs also bind with mRNA of certain genes, thereby causing them to be degraded as well. The repressing nature of dsRNA was discovered quite accidentally when researchers were trying to induce expression of genes by providing them in dsRNA form; instead, genes matching those RNA sequences were shut down entirely through this clever regulatory system. In 2006, the Nobel Prize for Medicine or Physiology was awarded to the two American scientists, Andrew Fire and Craig Mello, who discovered this phenomenon.

A second type of regulation seems to occur when small RNAs alter the structure of chromosomes. As DNA and proteins coil together to form chromatin, small RNAs direct how tightly or loosely the chromatin is constructed. Just as a closed book cannot be read, DNA sequences contained within tightly coiled chromatin are generally inaccessible to the cell, silencing the expression of those genes. Antisense RNA is produced from the opposite strand of the DNA that produces mRNA. This antisense molecule has the ability to pair with the "sense," or messenger, RNA and thus keep it from being transcribed. Riboswitches, RNAs that attach to a chemical with one end and only then become available for translation on the other end, were isolated for the first time in 2002. One riboswitch has been found to regulate the expression of 26 important genes in the bacterium *Bacillus subtilis*. Riboswitches have probably been around since the early days of life on the planet. So, although they are new to us, they have been used to regulate gene expression for billions of years.

These newly discovered RNA molecules have answered some vexing questions that came out of the genome sequencing studies (led by the Human Genome Project). Most of the DNA in organisms was found *not* to code for functional proteins. In humans, the "junk" percentage was 98%! Yet, in bacteria as well as humans, the junk DNA was preserved in the same form for the last millions of years of evolution, suggesting it had a very important function. We now know that much of this "junk" DNA codes for these important RNA regulatory molecules.

The RNA regulatory molecules are being heavily exploited to accomplish research tasks that were never before possible. More important, molecules such as antisense RNA are being explored for their therapeutic uses in cases where defective genes need to be shut down in order to restore a patient to health.

Our knowledge of the full role of small RNAs in the cell is just beginning. In the meantime, scientists will keep studying small RNAs while being mindful of the old adage, "Good things come in small packages."

New! Text Art Correlated to Animations

This symbol indicates to readers that the material presented in the text is also accompanied by an animation on the book's website. Students may view the animation on their computers or download it to their portable player and watch it on the fly!

Notes

"Heads-up" type material appears when appropriate letting students know about various terminology, exceptions to the rule, or, in the case of chapter 18, differences in chapter organization and pedagogy.

7.1 Microbial Nutrition 185

A NOTE ON TERMINOLOGY

Much of the vocabulary for describing microbial adaptations is based on some common root words. These are combined in various ways that assist in discussing the types of nutritional or ecological adaptations, as shown in this partial list:

Root	Meaning	Example of Use
troph-	Food, nourishment	Trophozoite—the feeding stage of protozoa
-phile	To love	Extremophile—an organism that has adapted to ("loves") extreme environments
-obe	To live	Microbe—to live "small"
hetero-	Other	Heterotroph—an organism that requires nutrients from other organisms
auto-	Self	Autotroph—an organism that does not need other organisms for food (obtains nutrients from a nonliving source)
photo-	Light	Phototroph—an organism that uses light as an energy source
chemo-	Chemical	Chemotroph—an organism that uses chemicals for energy, rather than light
sapro-	Rotten	Saprobe—an organism that lives on dead organic matter
halo-	Salt	Halophile—an organism that can grow in high-salt environments
thermo-	Heat	Thermophile—an organism that grows best at high temperatures
psychro-	Cold	Psychrophile—an organism that grows best at cold temperatures
aero-	Air (O_2)	Aerobe—an organism that uses oxygen in metabolism

Modifier terms are also used to specify the nature of an organism's adaptations. **Obligate** or **strict** refers to being restricted to a narrow niche or habitat, such as an obligate thermophile that requires high temperatures to grow. By contrast, **facultative** means not being so restricted but being able to adapt to a wider range of metabolic conditions and habitats. A facultative halophile can grow with or without high salt concentration.

Methane, sometimes called "swamp gas" or "natural gas" is formed in anaerobic, hydrogen-containing microenvironments of soil, swamps, mud, and even in the intestines of some animals. Methanogens are archaea, some of which live in extreme habitats such as ocean vents and hot springs, where temperatures reach up to 125°C **(Insight 7.3)**. Methane, which is used as a fuel in a large percentage of homes, can also be produced in limited quantities using a type of generator primed with a mixed population of microbes (including methanogens) and fueled with various waste materials that can supply enough methane to drive a steam generator. Methane also plays a role as one of the greenhouse gases that is currently an environmental concern (see chapter 24).

Heterotrophs and Their Energy Sources

The majority of heterotrophic microorganisms are **chemoheterotrophs** that derive both carbon and energy from organic compounds. Processing these organic molecules by respiration or fermentation releases energy in the form of ATP. An example of chemoheterotrophy is *aerobic respiration*, the principal energy-yielding pathway in animals, most protozoa and fungi, and aerobic bacteria. It can be simply represented by the equation:

$$\text{Glucose } [(CH_2O)_n] + O_2 \rightarrow CO_2 + H_2O + \text{Energy (ATP)}$$

This reaction is complementary to photosynthesis. Here, glucose and oxygen are reactants, and carbon dioxide is given off. Indeed, the earth's balance of both energy and metabolic gases is greatly dependent on this relationship. Chemoheterotrophic microorganisms belong to one of two main categories that differ in how they obtain their organic nutrients: **Saprobes** are free-living microorganisms that feed primarily on organic detritus from dead organisms, and **parasites** ordinarily derive nutrients from the cells or tissues of a host.

Saprobic Microorganisms Saprobes occupy a niche as decomposers of plant litter, animal matter, and dead microbes. If not for the work of decomposers, the earth would gradually fill up with organic material, and the nutrients it contains would not be recycled. Most saprobes, notably bacteria and fungi, have a rigid cell wall and cannot engulf large particles of food. To compensate, they release enzymes to the extracellular environment and digest the food particles into smaller molecules that can be transported into the cell **(figure 7.2)**. *Obligate saprobes* exist strictly on dead organic matter in soil and water and are unable to adapt to the body of a live host. This group includes many free-living protozoa, fungi, and bacteria. Apparently, there are fewer of these strict species than was once thought, and many supposedly nonpathogenic saprobes can infect a susceptible host. When a saprobe does infect a host, it is considered a *facultative parasite*. Such an infection usually occurs when the host is compromised, and the microbe is considered an *opportunistic*

Checkpoints

Major sections within the chapters end with a summary of the significant concepts covered. In the disease chapters, the Checkpoints take the form of tables that summarize the characteristics of the infectious agent(s) discussed. Students can use these as self-testing tools as well.

CHECKPOINT

- The fungi are nonphotosynthetic haploid species with cell walls. They are either saprobes or parasites and may be unicellular, colonial, or multicellular.
- All fungi are heterotrophic.
- Fungi have many reproductive strategies, including both asexual and sexual.
- Fungi have asexual spores called sporangiospores and conidiospores.
- Fungal sexual spores enable the organisms to incorporate variations in form and function.
- Fungi are often identified on the basis of their microscopic appearance.
- There are two categories of fungi that cause human disease: the primary pathogens, which infect healthy persons, and the opportunistic pathogens, which cause disease only in compromised hosts.

"*I think the checkpoints in the chapters are effective tools in that they reinforce what students have just finished reading. [They] clearly state the concepts the students should take from each section of the text.*"

—Suzanne Butler, Miami-Dade College

Chapter Summary with Key Terms

A brief outline of the main chapter concepts is provided with important terms highlighted. Key terms are also included in the glossary at the end of the book.

Multiple-Choice and New True-False Questions

Students can assess their knowledge of basic concepts by answering these two sets of questions. Other types of questions and activities build on this foundational knowledge.

Writing-to-Learn Questions

Using the facts and concepts they just studied, students must reason and problem-solve to answer these critical-thinking questions. Such questions do not have a single correct answer, and thus, open doors to discussion and serious thought.

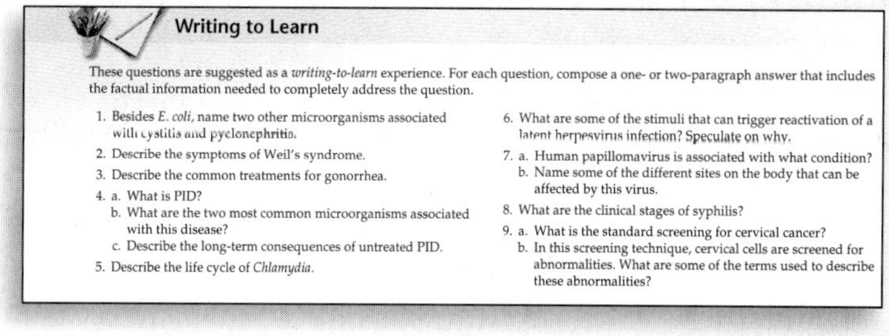
Writing to Learn

These questions are suggested as a *writing-to-learn* experience. For each question, compose a one- or two-paragraph answer that includes the factual information needed to completely address the question.

1. Besides *E. coli*, name two other microorganisms associated with cystitis and pyelonephritis.
2. Describe the symptoms of Weil's syndrome.
3. Describe the common treatments for gonorrhea.
4. a. What is PID?
   b. What are the two most common microorganisms associated with this disease?
   c. Describe the long-term consequences of untreated PID.
5. Describe the life cycle of *Chlamydia*.
6. What are some of the stimuli that can trigger reactivation of a latent herpesvirus infection? Speculate on why.
7. a. Human papillomavirus is associated with what condition?
   b. Name some of the different sites on the body that can be affected by this virus.
8. What are the clinical stages of syphilis?
9. a. What is the standard screening for cervical cancer?
   b. In this screening technique, cervical cells are screened for abnormalities. What are some of the terms used to describe these abnormalities?

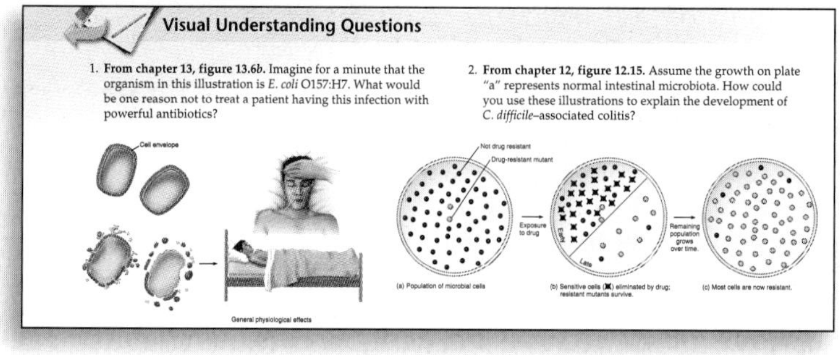
Visual Understanding Questions

1. **From chapter 13, figure 13.6*b*.** Imagine for a minute that the organism in this illustration is *E. coli* O157:H7. What would be one reason not to treat a patient having this infection with powerful antibiotics?
2. **From chapter 12, figure 12.15.** Assume the growth on plate "a" represents normal intestinal microbiota. How could you use these illustrations to explain the development of *C. difficile*–associated colitis?

New! Visual Understanding

Images from previous chapters are combined with integrated learning questions based on material in the current chapter to encourage an understanding of how important explanations and concepts are linked.

New! Concept Mapping

Three different types of concept mapping activities are used throughout the text in the end-of-chapter material to help students learn and retain what they've read.

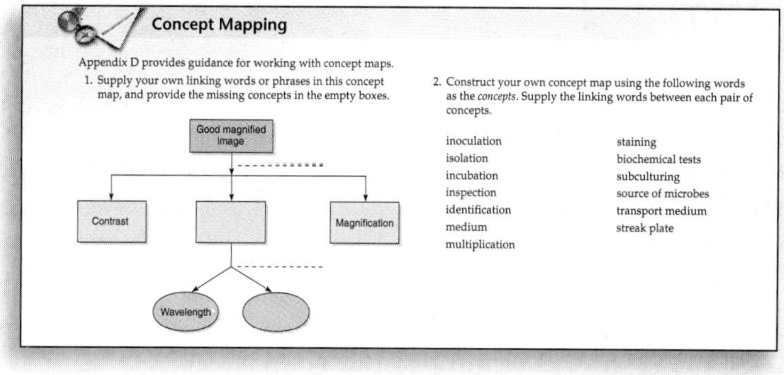
Concept Mapping

Appendix D provides guidance for working with concept maps.

1. Supply your own linking words or phrases in this concept map, and provide the missing concepts in the empty boxes.

2. Construct your own concept map using the following words as the *concepts*. Supply the linking words between each pair of concepts.

inoculation, isolation, incubation, inspection, identification, medium, multiplication, staining, biochemical tests, subculturing, source of microbes, transport medium, streak plate

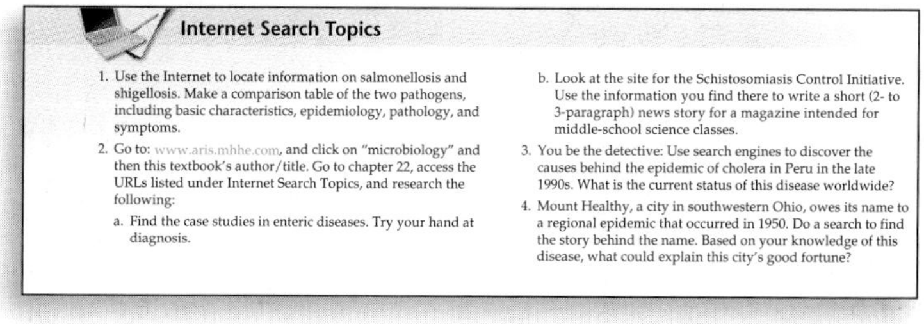
Internet Search Topics

1. Use the Internet to locate information on salmonellosis and shigellosis. Make a comparison table of the two pathogens, including basic characteristics, epidemiology, pathology, and symptoms.
2. Go to: www.aris.mhhe.com, and click on "microbiology" and then this textbook's author/title. Go to chapter 22, access the URLs listed under Internet Search Topics, and research the following:
   a. Find the case studies in enteric diseases. Try your hand at diagnosis.
   b. Look at the site for the Schistosomiasis Control Initiative. Use the information you find there to write a short (2- to 3-paragraph) news story for a magazine intended for middle-school science classes.
3. You be the detective: Use search engines to discover the causes behind the epidemic of cholera in Peru in the late 1990s. What is the current status of this disease worldwide?
4. Mount Healthy, a city in southwestern Ohio, owes its name to a regional epidemic that occurred in 1950. Do a search to find the story behind the name. Based on your knowledge of this disease, what could explain this city's good fortune?

Internet Search Topics

Opportunities for further research into the material just covered are outlined at the end of each chapter, in addition to the numerous resources available on the ARIS website accompanying the textbook.

CHAPTER 1

The Main Themes of Microbiology

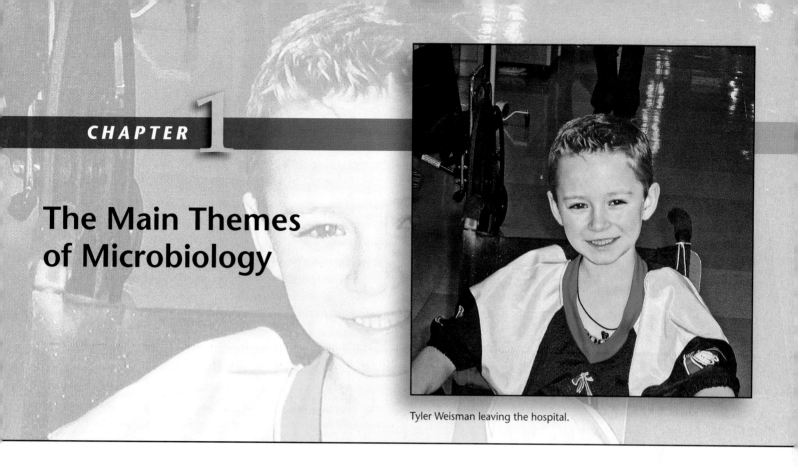

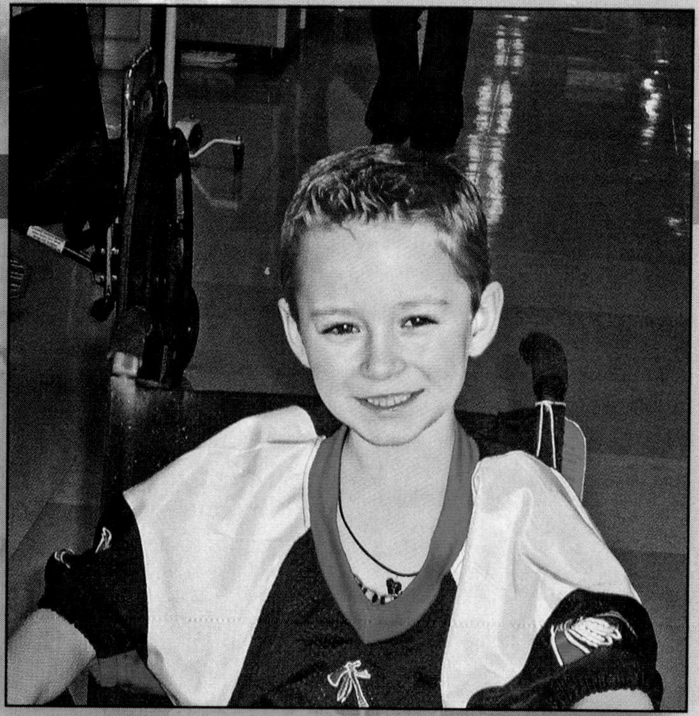

Tyler Weisman leaving the hospital.

CASE FILE 1

When 8-year-old Tyler from West Chester, Ohio, returned from a basketball tournament in early March 2006, he had no idea what the next few days, weeks, and months had in store for him. When he awoke the next morning, he had a pain in his hip and upper leg. Suspecting the pain was associated with an unseen bump from the previous day, his father treated the symptoms. As the day progressed, the pain continued and Tyler developed a fever. As a precaution, Tyler was taken to a local medical facility for evaluation. Other than pain and continuing fever, nothing unusual was found. A test for bacterial infection was negative and he was sent home. His condition did not improve, and by midnight he had become delirious. A call to the doctor, followed by a 911 call, landed Tyler in Children's Hospital Medical Center in Cincinnati.

By the next morning, his condition had improved but the worst was not over. What followed was a seemingly unending array of pokes, prods, and procedures to test for "every disease imaginable" including cancer. The most significant finding was a positive test for "Strep." A decision was made to transfer him to the ICU where he spent the next 6 weeks.

- *What is the organism commonly referred to as "Strep"?*
- *What conditions are usually associated with this organism?*

Case File 1 Wrap-Up appears on page 14

CHAPTER OVERVIEW

- Microorganisms, also called microbes, are organisms that—when single-celled—require a microscope to be readily observed. On the other hand, the largest living organism on earth is a mold fungus, a multicellular eukaryotic microbe that extends in size throughout the states of Michigan and Wisconsin.
- Major groups of microorganisms include bacteria, algae, protozoa, fungi, parasitic worms, and viruses.
- In terms of numbers and range of distribution, microbes are the dominant organisms on earth.

- Microbiology involves the study of numerous aspects of microbes involving their cell structure and function, growth and physiology, genetics, taxonomy and evolutionary history, and ecology.
- Microorganisms are essential to the operation of the earth's ecosystems.
- Microorganisms are the oldest organisms, having evolved over more than 3.5 billion years of earth's history to the modern varieties we now observe. Plants and animals as we know them are the successful product of a coevolution with all the microbes with which they share a habitat.
- Compared to the vast number of microorganisms that are benign or beneficial, only a small number of microbes are parasitic on plants and animals. An even lesser number of microbes are pathogens that cause infectious diseases.
- Humans use the versatility of microbes to improve the quality of their lives through industrial production, agriculture, medicine, and environmental reclamation and protection.
- Microbiologists use the scientific method to develop theories and explanations for microbial phenomena.
- The history of microbiology is marked by numerous significant discoveries and events in microscopy, culture techniques, and other methods of handling or controlling microbes.
- Microbes are classified into groups according to evolutionary relationships, provided with standard scientific names, and identified by specific characteristics.

1.1 The Scope of Microbiology

Microbiology is a specialized area of biology that deals with living things ordinarily too small to be seen without magnification. Such **microscopic** organisms are collectively referred to as **microorganisms** (my″-kroh-or′-gun-izms), **microbes,** or several other terms depending upon the kind of microbe or the purpose. In the context of infection and disease, some people call them germs, viruses, or agents; others even call them "bugs"; but none of these terms are clear. In addition, some of these terms place undue emphasis on the disagreeable reputation of microorganisms. But, as we will learn throughout the course of this book, only a small minority of microorganisms are implicated in causing harm to other living beings. There are several major groups of microorganisms that we'll be studying. They are **bacteria, algae, protozoa, helminths** (parasitic invertebrate animals such as worms), and **fungi**. All of these microbes—just like plants and animals—can be infected by **viruses,** which are noncellular, **parasitic,** protein-coated genetic elements, dependent on their infected host. They can cause harm to the host they infect. Although viruses are not strictly speaking microorganisms—namely, cellular beings—their evolutionary history and impact are intimately connected with the evolution of microbes and their study is thus integrated in the science of microbiology. As we will see in subsequent chapters, each group of microbes exhibits a distinct collection of biological characteristics.

The nature of microorganisms makes them both very easy and very difficult to study—easy because they reproduce so rapidly and we can quickly grow large populations in the laboratory and difficult because we can't see them directly. We rely on a variety of indirect means of analyzing them in addition to using microscopes.

Microbiology is one of the largest and most complex of the biological sciences because it includes many diverse biological disciplines. Microbiologists study every aspect of microbes—their cell structure and function, their growth and physiology, their genetics, their taxonomy and evolutionary history, and their interactions with the living and nonliving environment. The latter includes their uses in industry and agriculture and the way they interact with mammalian hosts, in particular, their properties that may cause disease or lead to benefits.

Some descriptions of different branches of study within microbiology follow.

Agricultural microbiology is concerned with the relationships between microbes and crops, with an emphasis on improving yields and combating plant diseases.

Biotechnology includes any processes in which humans use the metabolism of living things to arrive at a desired product, ranging from bread making to gene therapy. It is a tool used in industrial microbiology, which is concerned with the uses of microbes to produce or harvest large quantities of substances such as amino acids, beer, drugs, enzymes, and vitamins (see chapters 10 and 25).

Food microbiology, dairy microbiology, and *aquatic microbiology* examine the ecological and practical roles of microbes in food and water (see chapter 24).

Genetic engineering and *recombinant DNA technology* involve techniques that deliberately alter the genetic makeup of organisms to mass-produce human hormones and pharmaceuticals, create totally new substances, and develop organisms with unique methods of synthesis and adaptation. These comprise the most powerful and rapidly growing area in modern microbiology (see chapter 10).

Public health microbiology and *epidemiology* (aka *medical ecology*) aim to monitor and control the spread of diseases in communities. The principal U.S. and global institutions involved in this concern are the U.S. Public Health Service (USPHS) with its main agency, the Centers for Disease Control and Prevention (CDC), located in Atlanta, Georgia, and the World Health Organization (WHO), the medical limb of the United Nations. The CDC collects information on disease from around the United States and publishes the results in a weekly newsletter called the *Morbidity and Mortality Weekly Report*. The CDC also develops general guidelines and designs emergency strategies to contain identified outbreaks of infectious diseases. The CDC website, http://www.cdc.gov, is also one of the most reliable sources of information about diseases you will study in this book.

Immunology includes the study of the complex web of immune responses to infection by microorganisms. It also concerns itself with the study of *autoimmunity* and *hypersensitivity*, inappropriate immune responses that can be harmful to the human host. *Allergy* is one example of hypersensitivity (see chapter 16).

Each major discipline in microbiology contains numerous subdivisions or specialties that deal with a specific subject area or field. In fact, many areas of this science have become so specialized that it is not uncommon for a microbiologist to spend his or her whole life concentrating on a single group or type of microbe, biochemical process, or disease.

Among the specialty professions of microbiology are newer ones such as:

- *geomicrobiology*, which focuses on the roles microbes play in the earth's crust;
- *marine microbiology*, a study of the oceans and its smallest inhabitants; and
- *astromicrobiology*, which studies the potential for microbial life in space (see Insight 1.4).

Studies in microbiology have led to greater understanding of many general biological principles. For example, the study of microorganisms established universal concepts concerning the chemistry of life (see chapters 2 and 8); systems of inheritance (see chapter 9); and the global cycles of nutrients, minerals, and gases (see chapter 24).

1.2 The Impact of Microbes on Earth: Small Organisms with a Giant Effect

The most important knowledge that should emerge from a microbiology course is the profound influence microorganisms have on all aspects of the earth and its residents. For billions of years, microbes have extensively shaped the development of the earth's habitats and the evolution of other life forms. It is understandable that scientists searching for life on other planets first look for signs of microorganisms.

Bacterial-type organisms have been on this planet for about 3.5 billion years, according to the fossil record. It appears that they were the only living inhabitants on earth for almost 2 billion years. At that time (about 1.8 billion years ago), a more complex type of single-celled organism arose, of a **eukaryotic** (yoo″-kar-ee-ah′-tik) cell type. Eu-kary means *true nucleus*, which gives you a hint that those first inhabitants, the bacteria, had no true nucleus. For that reason they are called **prokaryotes** (proh″-kar-ee-otes) (prenucleus).

A NOTE ABOUT "-KARYOTE" VERSUS "-CARYOTE"

You will see the terms *prokaryote* and *eukaryote* spelled with *c* as well as (*procaryote* and *eucaryote*). Both spellings are accurate. This book uses the *k* spelling.

The early eukaryotes were the precursors of the cell type that eventually formed multicellular animals, including humans. But you can see from **figure 1.1** how long that took! On the scale pictured in the figure, humans seem to have just appeared. The bacteria preceded even the earliest animals by about 3 billion years. This is a good indication that humans are not likely to—nor should we try to—

Figure 1.1 Evolutionary timeline.
The first bacteria appeared approximately 3.5 billion years ago. They were the only form of life for half of the earth's history.

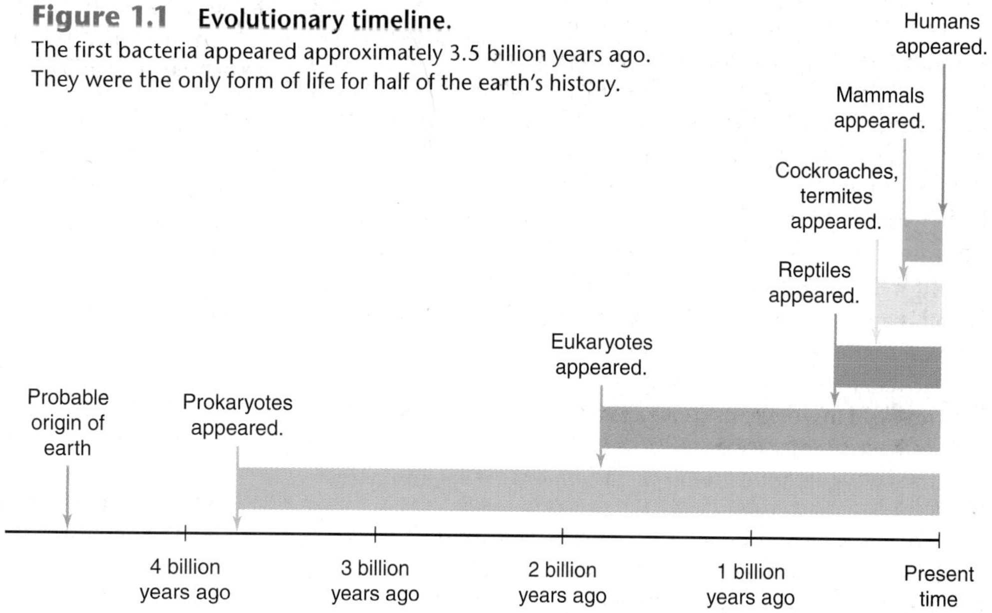

Figure 1.2 Examples of microbial habitats.
(a) Summer pond with a thick mat of algae—a rich photosynthetic community. **(b)** An orange being decomposed by a common soil fungus.

eliminate bacteria from our environment. They've survived and adapted to many catastrophic changes over the course of their geologic history.

Another indication of the huge influence bacteria exert is how **ubiquitous** they are. Microbes can be found nearly everywhere, from deep in the earth's crust, to the polar ice caps and oceans, to the bodies of plants and animals. Being mostly invisible, the actions of microorganisms are usually not as obvious or familiar as those of larger plants and animals. They make up for their small size by occurring in large numbers and living in places that many other organisms cannot survive. Above all, they play central roles in the earth's landscape that are essential to life.

Microbial Involvement in Energy and Nutrient Flow

Microbes are deeply involved in the flow of energy and food through the earth's ecosystems.[1] Most people are aware that plants carry out **photosynthesis,** which is the light-fueled conversion of carbon dioxide to organic material, accompanied by the formation of oxygen (called oxygenic photosynthesis). However, bacteria invented photosynthesis long before first plants appeared, first as a process that did not produce oxygen (*anoxygenic photosynthesis*). This anoxygenic photosynthesis later evolved into oxygenic photosynthesis, which not only produced oxygen but also was much more efficient in extracting energy from sunlight. Hence, bacteria were responsible for changing the atmosphere of the earth from one without oxygen to one with oxygen. The production of oxygen also allowed the use of oxygen for aerobic respiration and the formation of ozone, both of which set off an explosion in species diversification. Today, photosynthetic microorganisms (bacteria and algae) account for more than 50% of the earth's photosynthesis, contributing the majority of the oxygen to the atmosphere **(figure 1.2*a*).**

Another process that helps keep the earth in balance is the process of biological **decomposition** and nutrient recycling. Decomposition involves the breakdown of dead matter and wastes into simple compounds that can be directed back into the natural cycles of living things **(figure 1.2*b*).** If it were not for multitudes of bacteria and fungi, many chemical elements would become locked up and unavailable to organisms; we humans would drown in our own industrial and personal wastes! In the long-term scheme of things, microorganisms are the main forces that drive the structure and content of the soil, water, and atmosphere. For example:

- The very temperature of the earth is regulated by "greenhouse gases," such as carbon dioxide, nitrous oxide, and methane, which create an insulation layer in the atmosphere and help retain heat. Many of these gases are produced by microbes living in the environment and the digestive tracts of animals.

1. Ecosystems are communities of living organisms and their surrounding environment.

- Recent estimates propose that, based on weight and numbers, up to 50% of all organisms exist within and beneath the earth's crust in sediments, rocks, and even volcanoes. It is increasingly evident that this enormous underground community of microbes is a significant influence on weathering, mineral extraction, and soil formation.
- Bacteria and fungi live in complex associations with plants that assist the plants in obtaining nutrients and water and may protect them against disease. Microbes form similar interrelationships with animals, notably, in the stomach of cattle, where a rich assortment of bacteria digest the complex carbohydrates of the animals' diets.

1.3 Human Use of Microorganisms

Microorganisms clearly have monumental importance to the earth's operation. It is this very same diversity and versatility that also makes them excellent candidates for solving human problems. By accident or choice, humans have been using microorganisms for thousands of years to improve life and even to shape civilizations. Baker's and brewer's yeast, types of single-celled fungi, cause bread to rise and ferment sugar into alcohol to make wine and beers. Other fungi are used to make special cheeses such as Roquefort or Camembert. These and other "home" uses of microbes have been in use for thousands of years. For example, historical records show that households in ancient Egypt kept moldy loaves of bread to apply directly to wounds and lesions. When humans manipulate microorganisms to make products in an industrial setting, it is called biotechnology. For example, some specialized bacteria have unique capacities to mine precious metals **(figure 1.3*a*)** or to produce enzymes that are used in laundry detergents.

Genetic engineering is a newer area of biotechnology that manipulates the genetics of microbes, plants, and animals for the purpose of creating new products and genetically modified organisms (GMOs). One powerful technique for designing GMOs is termed **recombinant DNA** technology. This technology makes it possible to transfer genetic material from one organism to another and to deliberately alter DNA.[2] Bacteria and fungi were some of the first organisms to be genetically engineered. This was possible because they are single-celled organisms and they are so adaptable to changes in their genetic makeup. Recombinant DNA technology has unlimited potential in terms of medical, industrial, and agricultural uses. Microbes can be engineered to synthesize desirable proteins such as drugs, hormones, and enzymes **(figure 1.3*b*)**.

Among the genetically unique organisms that have been designed by bioengineers are bacteria that mass produce antibiotic-like substances, yeasts that produce human insulin, pigs that produce human hemoglobin, and plants that

Figure 1.3 Microbes at work.
(a) An aerial view of a copper mine looks like a giant quilt pattern. The colored patches are bacteria in various stages of extracting metals from the ore. **(b)** Microbes as synthesizers. A large complex fermentor manufactures drugs and enzymes using microbial metabolism. **(c)** Members of a biohazard team from the National Oceanic and Atmospheric Agency (NOAA) participate in the removal and detoxification of 63,000 tons of crude oil released by a wrecked oil tanker on the coast of Spain. The bioremediation of this massive spill made use of naturally occurring soil and water microbes as well as commercially prepared oil-eating species of bacteria and fungi. Complete restoration will take several years to complete.

2. DNA, or deoxyribonucleic acid, the chemical substance that comprises the genetic material of organisms.

contain natural pesticides or fruits that do not ripen too rapidly. The techniques also pave the way for characterizing human genetic material and diseases.

Another way of tapping into the unlimited potential of microorganisms is the relatively new science of **bioremediation** (by'-oh-ree-mee-dee-ay"-shun). This process involves the introduction of microbes into the environment to restore stability or to clean up toxic pollutants. Bioremediation is required to control the massive levels of pollution from industry and modern living. Microbes have a surprising capacity to break down chemicals that would be harmful to other organisms. This includes even man-made chemicals that scientists have developed and for which there are no natural counterparts. Most pollutants cannot be degraded by a single kind of microbe; instead, bioremediation relies on many different kinds of microbes working together.

Agencies and companies have developed microbes to handle oil spills and detoxify sites contaminated with heavy metals, pesticides, and other chemical wastes **(figure 1.3*c*).** The solid waste disposal industry is interested in developing methods for degrading the tons of garbage in landfills, especially human-made plastics and paper products. One form of bioremediation that has been in use for some time is the treatment of water and sewage. Because clean freshwater supplies are dwindling worldwide, it will become even more important to find ways to reclaim polluted water.

1.4 Infectious Diseases and the Human Condition

One of the most fascinating aspects of the microorganisms with which we share the earth is that, despite all of the benefits they provide, they also contribute significantly to human misery as **pathogens** (path'-oh-jenz). The vast majority of microorganisms that associate with humans cause no harm. In fact, they provide many benefits to their human hosts. There is little doubt that a diverse microbial biota living in and on humans is an important part of human well-being. However, humankind is also plagued by nearly 2,000 different microbes that can cause various types of disease. Infectious diseases still devastate human populations worldwide, despite significant strides in understanding and treating them. The most recent estimates from the World Health Organization (WHO) point to a total of 10 billion new infections across the world every year; that is 2.5 per capita of the world's present population! (There are more infections than people because many people acquire more than one infection.) Infectious diseases are also among the most common causes of death in much of humankind, and they still kill a significant percentage of the U.S. population. **Table 1.1** depicts the 10 top causes of death per year (by all causes, infectious and noninfectious) in the United States and worldwide. The worldwide death toll from infections is about 13 million people per year. For example, the CDC reports that every 50 minutes a human life is ended in the world because of tuberculosis. In **figure 1.4,** you can see the top infectious causes of death displayed in a different way. Note that many of these infections are treatable with drugs or preventable with vaccines. Those hardest hit are residents in countries where access to adequate medical care is lacking. One-third of the earth's inhabitants live on less than $1 per day, are malnourished, and are not fully immunized.

Malaria, which kills more than a million people every year worldwide, is caused by a microorganism transmitted by mosquitoes (see chapter 20). Currently, the most effective way for citizens of developing countries to avoid infection with the causal agent of malaria is to sleep under a bed net, because the mosquitoes are most active in the evening. Yet even this inexpensive solution is beyond the reach of many. Mothers in Southeast Asia and elsewhere have to make nightly decisions about which of their children will sleep under the single family bed net, because a second one, priced at about $3 to $5, is too expensive for them.

TABLE 1.1 Top Causes of Death—All Diseases

United States	No. of Deaths	Worldwide	No. of Deaths
1. Heart disease	725,000	1. Heart disease	11.1 million
2. Cancer	550,000	2. Cancer	7.1 million
3. Stroke	167,000	3. Stroke	5.5 million
4. Chronic lower-respiratory disease	124,000	4. Respiratory infections*	3.9 million
5. Unintentional injury (accidents)	97,000	5. Chronic lower-respiratory disease	3.6 million
6. Diabetes	68,000	6. Accidents	3.5 million
7. Influenza and pneumonia	63,000	7. HIV/AIDS	2.9 million
8. Alzheimer's disease	45,000	8. Perinatal conditions	2.5 million
9. Kidney problems	35,000	9. Diarrheal diseases	2.0 million
10. Septicemia (bloodstream infection)	30,000	10. Tuberculosis	1.6 million

*Diseases in red are those most clearly caused by microorganisms.

Source: Data adapted from The World Health Report 2002 (World Health Organization).

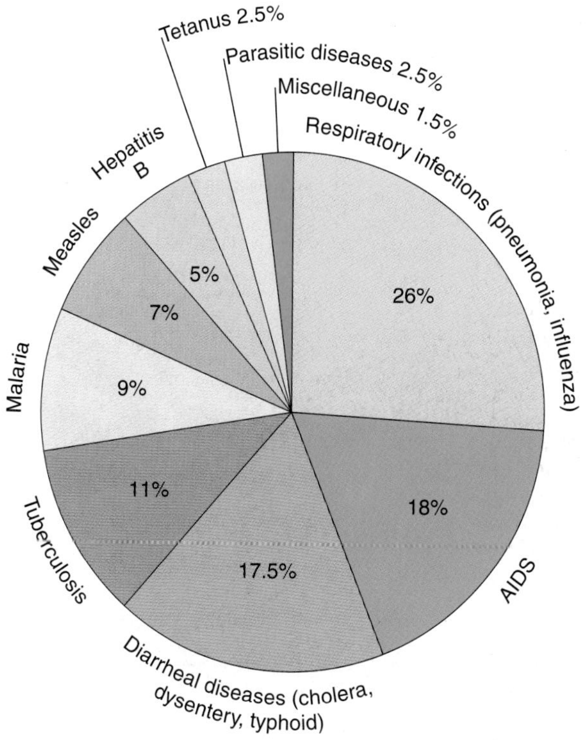

Figure 1.4 Worldwide infectious disease statistics. This figure depicts the 10 most common infectious causes of death.

Adding to the overload of infectious diseases, we are also witnessing an increase in the number of new (emerging) and older (reemerging) diseases. SARS, AIDS, hepatitis C, and viral encephalitis are examples of diseases that cause severe mortality and morbidity and are currently on the rise. To somewhat balance this trend, there have also been some advances in eradication of diseases such as polio, measles, leprosy, and diseases caused by certain parasitic worms. The WHO is currently on a global push to vaccinate children against the most common childhood diseases, which will reduce the reservoir for the causal agents of disease dramatically and could, eventually, lead to their eradication.

One of the most eye-opening discoveries in recent years is that many diseases that used to be considered noninfectious probably do involve microbial infection. The most famous of these is gastric ulcers, now known to be caused by a bacterium called *Helicobacter*. But there are more. An association has been established between certain cancers and viruses, between diabetes and the Coxsackie virus, and between schizophrenia and a virus called the borna agent. Diseases as disparate as multiple sclerosis, obsessive compulsive disorder, and coronary artery disease have been linked to chronic infections with microbes or viruses. It seems that the golden age of microbiological discovery, during which all of the "obvious" diseases were characterized and cures or preventions were devised for them, should more accurately be referred to as the *first* golden age. We're now discovering the subtler side of microorganisms. Their roles in quiet but slowly destructive diseases are now well known. These include female infertility caused by *Chlamydia* infection, and malignancies such as liver cancer (hepatitis viruses) and cervical cancer (human papillomavirus). Most scientists expect that, in time, many chronic conditions will be found to have some association with microbial agents.

As mentioned earlier, another important development in infectious disease trends is the increasing number of patients with weakened defenses that are kept alive for extended periods. They are subject to infections by common microbes that are not pathogenic to healthy people. There is also an increase in microbes that are resistant to drugs. It appears that even with the most modern technology available to us, microbes still have the "last word," as the great French scientist Louis Pasteur observed **(Insight 1.1).**

CHECKPOINT

- Microorganisms are defined as "living organisms too small to be seen with the naked eye." Among the members of this huge group of organisms are bacteria, algae, protozoa, fungi, parasitic worms (helminthes), and viruses.
- Microorganisms live nearly everywhere and influence many biological and physical activities on earth.
- There are many kinds of relationships between microorganisms and humans; most are beneficial, but some are harmful.
- The scope of microbiology is incredibly diverse. It includes basic microbial research, research on infectious diseases, study of prevention and treatment of disease, environmental functions of microorganisms, and industrial use of microorganisms for commercial, agricultural, and medical purposes.
- In the last 120 years, microbiologists have identified the causative agents for many infectious diseases. In addition, they have discovered distinct connections between microorganisms and diseases whose causes were previously unknown.
- Microorganisms: We have to learn to live with them because we cannot live without them.

1.5 The General Characteristics of Microorganisms

Cellular Organization Test 1

As discussed earlier, two basic cell lines appeared during evolutionary history. These lines, termed **prokaryotic cells** and **eukaryotic cells,** differ not only in the complexity of their cell structure **(figure 1.5*a*)** but also in contents and function.

In general, prokaryotic cells are about 10 times smaller than eukaryotic cells, and they lack many of the eukaryotic cell structures such as **organelles.** Organelles are small, double-membrane-bound structures in the eukaryotic cell that perform specific functions and include the nucleus,

INSIGHT 1.1 Historical

The More Things Change . . .

In 1967, the surgeon general of the United States delivered a speech to Congress: "It is time to close the book on infectious diseases," he said. "The war against pestilence is over."

In 1998, Surgeon General David Satcher had a different message. The *Miami Herald* reported his speech with this headline: "Infectious Diseases a Rising Peril; Death Rates in U.S. Up 58% Since 1980."

The middle of the last century was a time of great confidence in science and medicine. With the introduction of antibiotics in the 1940s, and a lengthening list of vaccines that prevented the most frightening diseases, Americans felt that it was only a matter of time before diseases caused by microorganisms (i.e., infectious diseases) would be completely manageable. The nation's attention turned to the so-called chronic diseases, such as heart disease, cancer, and stroke.

So what happened to change the optimism of the 1960s to the warning expressed in the speech from 1998? Dr. Satcher explained it this way: "Organisms changed and people changed." First, we are becoming more susceptible to infectious disease precisely because of advances in medicine. People are living longer. Sicker people are staying alive much longer than in the past. Older and sicker people have heightened susceptibility to what we might call garden-variety microbes. Second, the population has become more mobile. Travelers can crisscross the globe in a matter of hours, taking their microbes with them and introducing them into new "naive" populations. Third, there are growing numbers of microbes that truly are new (or at least, new to us). The conditions they cause are called **emerging diseases.** Changes in agricultural practices and encroachment of humans on wild habitats are just two probable causes of emerging diseases. The mass production and packing of food increases the opportunity for large outbreaks, especially if foods are grown in fecally contaminated soils or are eaten raw or poorly cooked. In the past several years, dozens of food-borne outbreaks have been associated with the bacterium *Escherichia coli* 0157:H7 in fresh vegetables, fruits, and meats. Fourth, microorganisms have demonstrated their formidable capacity to respond and adapt to our attempts to control them, most spectacularly by becoming resistant to the effects of our miracle drugs.

United States Surgeon General David Satcher in 1998.

United States Surgeon General Luther Terry addressing press conference in 1964.

And there's one more thing: Evidence is mounting that many conditions formerly thought to be caused by genetics or lifestyle, such as heart disease and cancer, can often be at least partially caused by microorganisms.

Microbes never stop surprising us—in their ability not only to harm but also to help us. The best way to keep up is to learn as much as you can about them. This book is a good place to start.

mitochondria, and chloroplasts. This does not mean that prokaryotic cells are totally devoid of internal structures; many of them contain inclusion bodies, which can be protein-lined compartments that can contain air or storage polymers or house selected metabolic pathways. The microorganisms that consist of these two different cell types (called prokaryotes and eukaryotes) are covered in more detail in chapters 4 and 5.

All prokaryotes are microorganisms, but only some eukaryotes are microorganisms. The majority of microorganisms are single-celled (all prokaryotes and some eukaryotes), but some consist of a few cells **(figure 1.6).** Because of their role in disease, certain invertebrate animals such as helminth worms and insects, many of which can be seen with the naked eye, are also included in the study of microorganisms. Even in its seeming simplicity, the microscopic world is every bit as complex and diverse as the macroscopic one. There is no doubt that microorganisms also outnumber macroscopic organisms in abundance and diversity by a factor of several million.

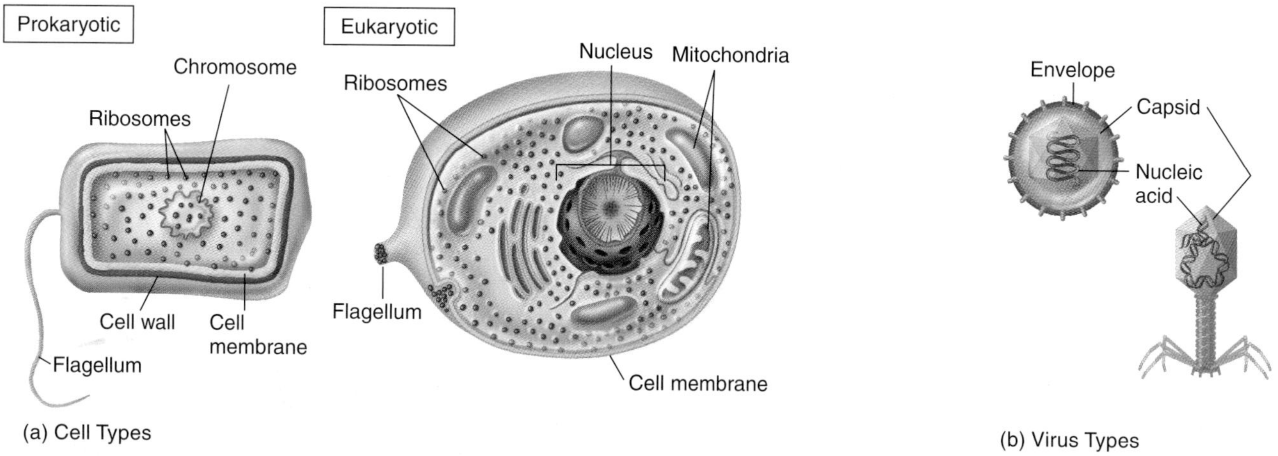

Figure 1.5 Cell structure.
(a) Cell Types Microbial cells are of the small, relatively simple prokaryotic variety (left) or the larger, more complex eukaryotic type (right). (Not to scale) **(b) Virus Types** Viruses are tiny particles, not cells, that consist of genetic material surrounded by a protective covering. Shown here are a human virus (top) and a bacterial virus (bottom). (Not to scale)

A Note on Viruses

Viruses are subject to intense study by microbiologists. As mentioned before, they are not independently living cellular organisms. Instead, they are small particles that exist at the level of complexity somewhere between large molecules and cells **(figure 1.5*b*).** Viruses are much simpler than cells; outside their host, they are composed essentially of a small amount of hereditary material (either DNA or RNA but never both) wrapped up in a protein covering that is sometimes enveloped by a protein-containing lipid membrane. In this extracellular state, they are individually referred to as

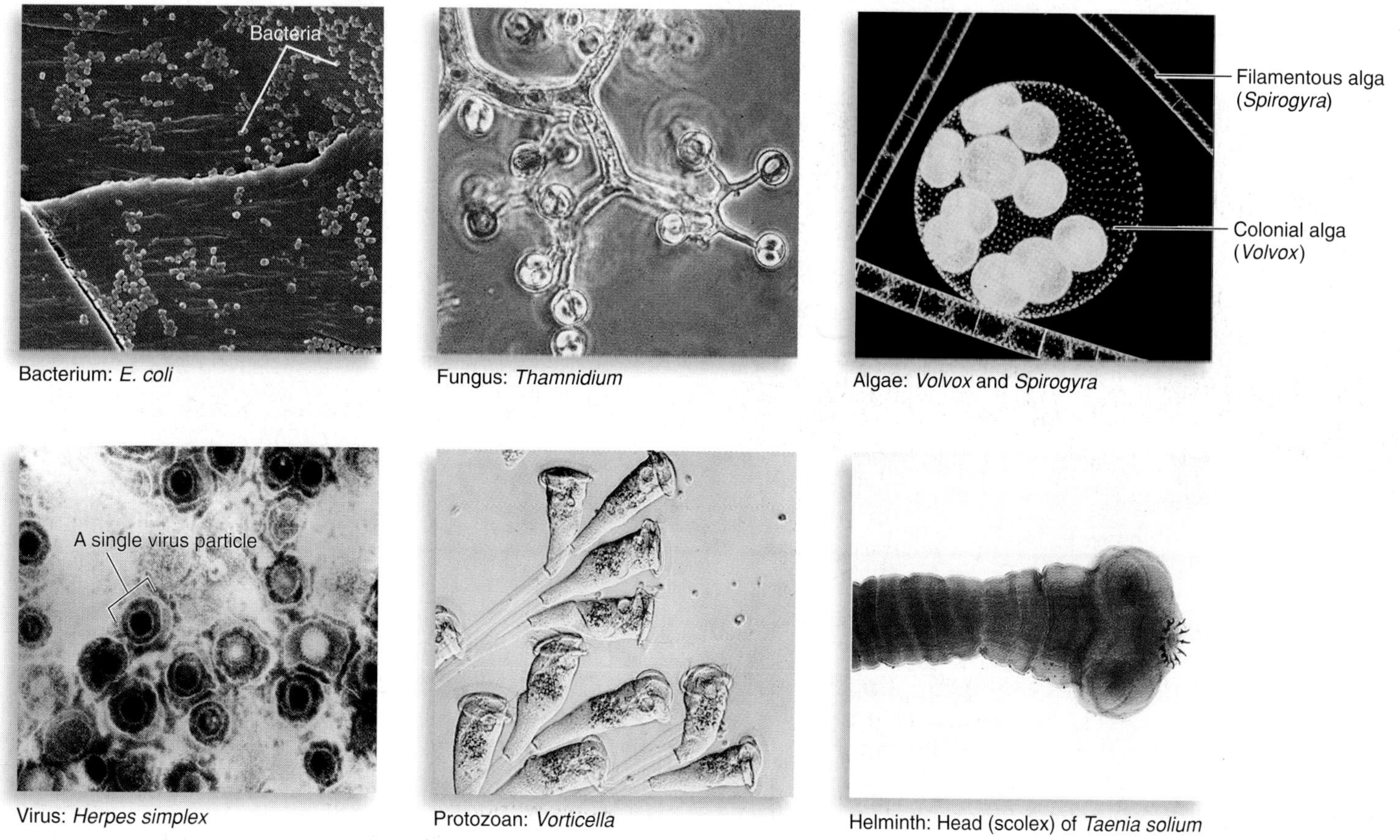

Figure 1.6 Six types of microorganisms.
(Organisms are not shown at the same magnifications.)

a **virus particle** or **virion.** When inside their host organism, in the intracellular state, viruses usually exist only in the form of genetic material that confers a partial genetic program on the host organisms. That is why many microbiologists refer to viruses as parasitic particles; however, a few consider them to be very primitive organisms. Nevertheless, all biologists agree that viruses are completely dependent on an infected host cell's machinery for their multiplication and dispersal.

Microbial Dimensions: How Small Is Small?

When we say that microbes are too small to be seen with the unaided eye, what sorts of dimensions are we talking about? The concept of thinking small is best visualized by comparing microbes with the larger organisms of the macroscopic world and also with the atoms and molecules of the molecular world **(figure 1.7).** Whereas the dimensions of macroscopic organisms are usually given in centimeters (cm) and meters (m), those of

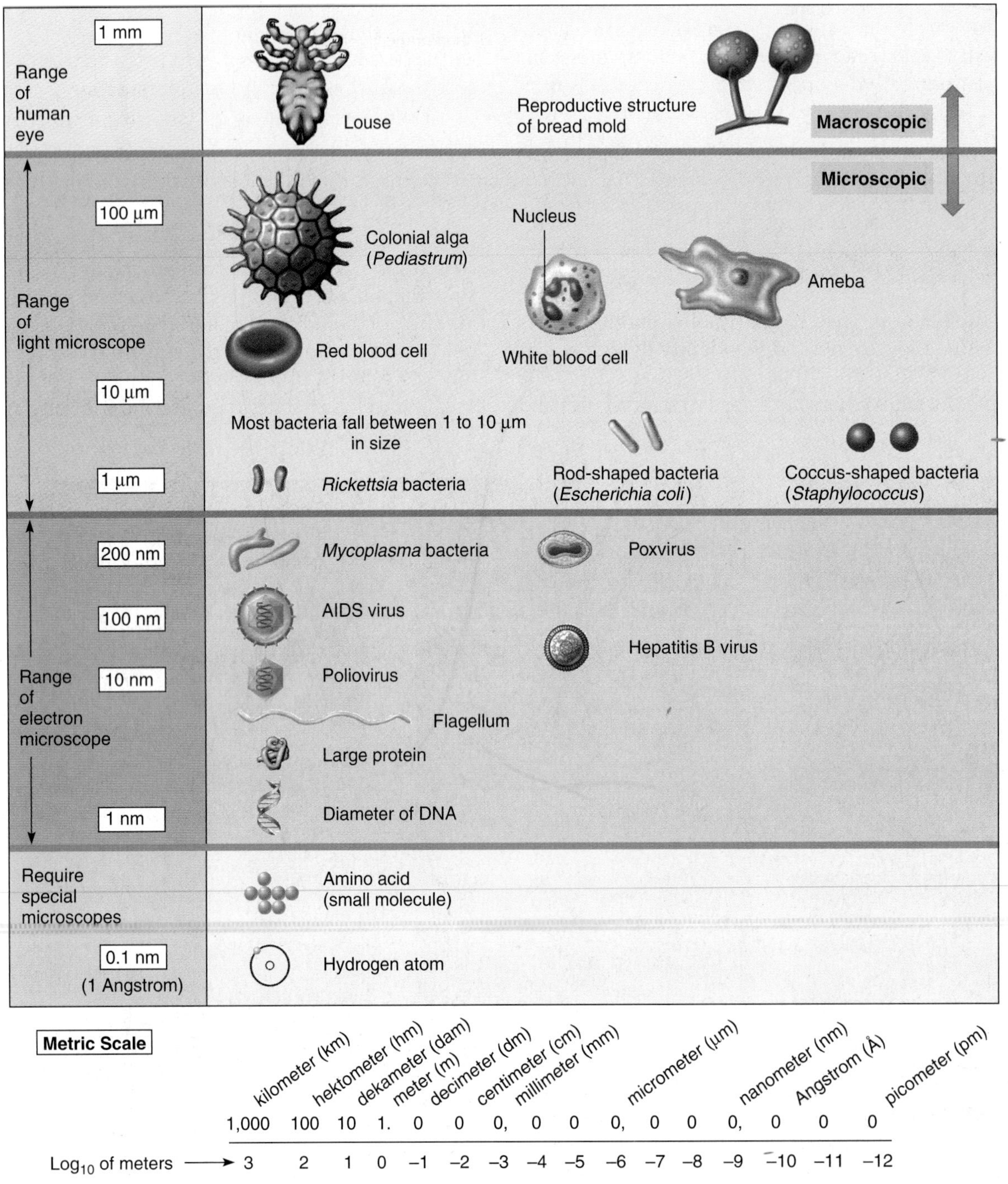

Figure 1.7 The size of things.
Common measurements encountered in microbiology and a scale of comparison from the macroscopic to the microscopic, molecular, and atomic. Most microbes encountered in our studies will fall between 100 μm and 10 nm in overall dimensions. The microbes shown are more or less to scale within size zone but not between size zones.

microorganisms fall within the range of millimeters (mm) to micrometers (µm) to nanometers (nm). The size range of most microbes extends from the smallest bacteria, measuring around 200 nm, to protozoa and algae that measure 3 to 4 mm and are visible with the naked eye. Viruses, which can infect all organisms including microbes, measure between 20 nm and 800 nm, and some of them are thus not much bigger than large molecules whereas others are just a tad larger than the smallest bacteria.

Lifestyles of Microorganisms

The majority of microorganisms live a free existence in habitats such as soil and water, where they are relatively harmless and often beneficial. A free-living organism can derive all required foods and other factors directly from the nonliving environment. Some microorganisms require interactions with other organisms. Sometimes these microbes are termed **parasites.** They are harbored and nourished by other living organisms called **hosts.** A parasite's actions cause damage to its host through infection and disease. Although parasites cause important diseases, they make up only a small proportion of microbes.

CHECKPOINT

- Excluding the viruses, there are two types of microorganisms: prokaryotes, which are small and lack a nucleus and organelles, and eukaryotes, which are larger and have both a nucleus and organelles.
- Viruses are not cellular and are therefore sometimes called particles rather than organisms. They are included in microbiology because of their small size and close relationship with cells.
- Most microorganisms are measured in micrometers, with two exceptions. The helminths are measured in millimeters, and the viruses are measured in nanometers.
- Contrary to popular belief, most microorganisms are harmless, free-living species that perform vital functions in both the enviroment and larger organisms. Comparatively few spacies are agents of disease.

1.6 The Historical Foundations of Microbiology

If not for the extensive interest, curiosity, and devotion of thousands of microbiologists over the last 300 years, we would know little about the microscopic realm that surrounds us. Many of the discoveries in this science have resulted from the prior work of men and women who toiled long hours in dimly lit laboratories with the crudest of tools. Each additional insight, whether large or small, has added to our current knowledge of living things and processes. This section summarizes the prominent discoveries made in the past 300 years: microscopy; the rise of the scientific method; and the development of medical microbiology, including the germ theory and the origins of modern microbiological techniques. Table B.1 in appendix B summarizes some of the pivotal events in microbiology, from its earliest beginnings to the present.

The Development of the Microscope: "Seeing Is Believing"

It is likely that from very earliest history, humans noticed that when certain foods spoiled they became inedible or caused illness, and yet other "spoiled" foods did no harm and even had enhanced flavor. Indeed, several centuries ago, there was already a sense that diseases such as the black plague and smallpox were caused by some sort of transmissible matter. But the causes of such phenomena were vague and obscure because the technology to study them was lacking. Consequently, they remained cloaked in mystery and regarded with superstition—a trend that led even well-educated scientists to believe in spontaneous generation **(Insight 1.2).**

True awareness of the widespread distribution of microorganisms and some of their characteristics was finally made possible by the development of the first microscopes. These devices revealed microbes as discrete entities sharing many of the characteristics of larger, visible plants and animals. Several early scientists fashioned magnifying lenses, but their microscopes lacked the optical clarity needed for examining bacteria and other small, single-celled organisms. The likely earliest record of microbes is in the works of Englishman Robert Hooke. In the 1660s, Hooke studied a great diversity of material from plants and trees, described for the first time cellular structures in tree bark, and drew sketches of and described "little structures" that seemed to be alive. The most careful and exacting observations of microbes, however, awaited the clever single-lens microscope hand-fashioned by Antonie van Leeuwenhoek, a Dutch linen merchant and self-made microbiologist **(figure 1.8).**

Figure 1.8 An oil painting of Antonie van Leeuwenhoek (1632–1723) sitting in his laboratory.

J. R. Porter and C. Dobell have commented on the unique qualities Leeuwenhoek brought to his craft: "He was one of the most original and curious men who ever lived. It is difficult to compare him with anybody because he belonged to a genus of which he was the type and only species, and when he died his line became extinct."

INSIGHT 1.2 Historical

The Fall of Superstition and the Rise of Microbiology

For thousands of years, people believed that certain living things arose from vital forces present in nonliving or decomposing matter. This ancient belief, known as **spontaneous generation,** was continually reinforced as people observed that meat left out in the open soon "produced" maggots, that mushrooms appeared on rotting wood, that rats and mice emerged from piles of litter, and other similar phenomena. Though some of these early ideas seem quaint and ridiculous in light of modern knowledge, we must remember that, at the time, mysteries in life were accepted, and the scientific method was not widely practiced.

Even after single-celled organisms were discovered during the mid-1600s, the idea of spontaneous generation continued to exist. Some scientists assumed that microscopic beings were an early stage in the development of more complex ones.

Over the subsequent 200 years, scientists waged an experimental battle over the two hypotheses that could explain the origin of simple life forms. Some tenaciously clung to the idea of **abiogenesis** (ah-bee"-oh-jen-uh-sis), which embraced spontaneous generation. On the other side were advocates of **biogenesis** saying that living things arise only from others of their same kind. There were serious proponents on both sides, and each side put forth what appeared on the surface to be plausible explanations of why their evidence was more correct. Gradually, the abiogenesis hypothesis was abandoned, as convincing evidence for biogenesis continued to mount. The following series of experiments were among the most important in finally tipping the balance.

Among the important variables to be considered in challenging the hypotheses were the effects of nutrients, air, and heat and the presence of preexisting life forms in the environment. One of the first people to test the spontaneous generation theory was Francesco Redi of Italy. He conducted a simple experiment in which he placed meat in a jar and covered it with fine gauze. Flies gathering at the jar were blocked from entering and thus laid their eggs on the outside of the gauze. The maggots subsequently developed without access to the meat, indicating that maggots were the offspring of flies and did not arise from some "vital force" in the meat. This and related experiments laid to rest the idea that more complex animals such as insects and mice developed through abiogenesis, but it did not convince many scientists of the day that simpler organisms could not arise in that way.

The Frenchman Louis Jablot reasoned that even microscopic organisms must have parents, and his experiments with infusions (dried hay steeped in water) supported that hypothesis. He divided into two containers an infusion that had been boiled to destroy any living things: a heated container that was closed to

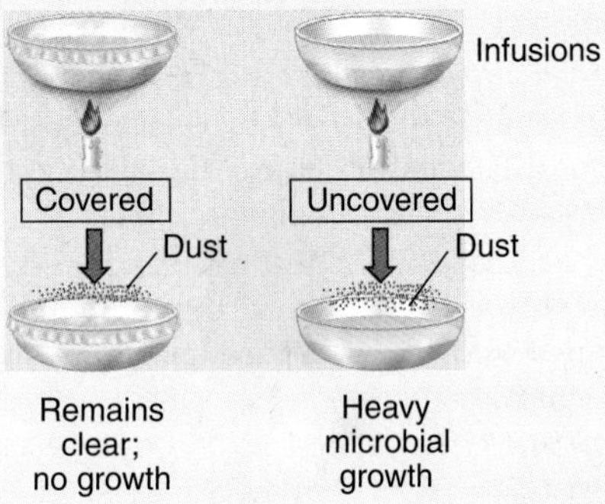

Paintings of historical figures like the one of Leeuwenhoek in figure 1.8 don't always convey a meaningful feeling for the event or person depicted. Imagine a dusty shop in Holland in the late 1600s. Ladies in traditional Dutch garb came in and out, choosing among the bolts of linens for their draperies and upholstery. Between customers, Leeuwenhoek retired to the workbench in the back of his shop, grinding glass lenses to ever-finer specifications. He could see with increasing clarity the threads in his fabrics. Eventually, he became interested in things other than thread counts. He took rainwater from a clay pot, smeared it on his specimen holder, and peered at it through his finest lens. He found "animals appearing to me ten thousand times less than those which may be perceived in the water with the naked eye."

He didn't stop there. He scraped the plaque from his teeth, and from the teeth of some volunteers who had never cleaned their teeth in their lives, and took a good close look at that. He recorded: "In the said matter there were many very little living animalcules, very prettily a-moving. . . . Moreover, the other animalcules were in such enormous numbers, that all the water . . . seemed to be alive." Leeuwenhoek started sending his observations to the Royal Society of London, and eventually he was recognized as a scientist of great merit.

Leeuwenhoek constructed more than 250 small, powerful microscopes that could magnify up to 300 times **(figure 1.9).** Considering that he had no formal training in science and that he was the first person ever to faithfully record this strange new world, his descriptions of bacteria and protozoa the air and a heated container that was freely open to the air. Only the open vessel developed microorganisms, which he presumed had entered in air laden with dust.

Additional experiments further defended biogenesis. Franz Shultze and Theodor Schwann of Germany felt sure that air was the source of microbes and sought to prove this by passing air through strong chemicals or hot glass tubes into heat-treated infusions in flasks. When the infusions again remained devoid of living things, the supporters of abiogenesis claimed that the treatment of the air had made it harmful to the spontaneous development of life.

Shultze and Schwann's Test

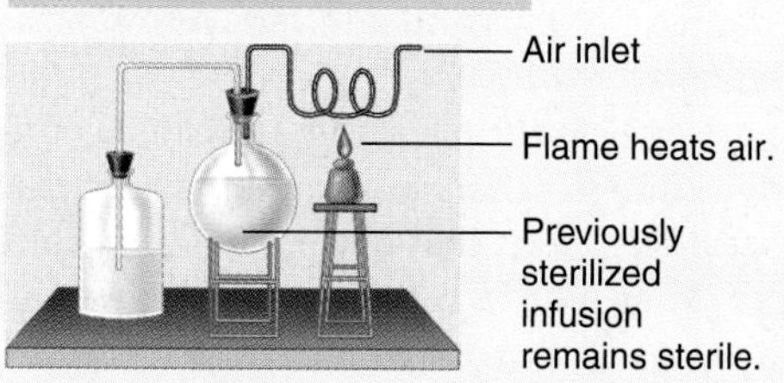

Air inlet
Flame heats air.
Previously sterilized infusion remains sterile.

Then, in the mid-1800s, the acclaimed chemist and microbiologist Louis Pasteur entered the arena. He had recently been studying the roles of microorganisms in the fermentation of beer and wine, and it was clear to him that these processes were brought about by the activities of microbes introduced into the beverage from air, fruits, and grains. The methods he used to discount abiogenesis were simple yet brilliant.

To further clarify that air and dust were the source of microbes, Pasteur filled flasks with broth and fashioned their openings into elongate, swan-neck-shaped tubes. The flasks' openings were freely open to the air but were curved so that gravity would cause any airborne dust particles to deposit in the lower part of the necks. He heated the flasks to sterilize the broth and then incubated them. As long as the flask remained intact, the broth remained sterile; but if the neck was broken off so that dust fell directly down into the container, microbial growth immediately commenced.

Pasteur summed up his findings, "For I have kept from them, and am still keeping from them, that one thing which is above the power of man to make; I have kept from them the germs that float in the air, I have kept from them life." Pasteur's contemporary, expert microscopist and pathologist Rudolph Virchow summarized the emerging theory of biogenesis with his famous statement: "omnis cellula e cellula"—a cell comes from a cell.

Pasteur's Experiment

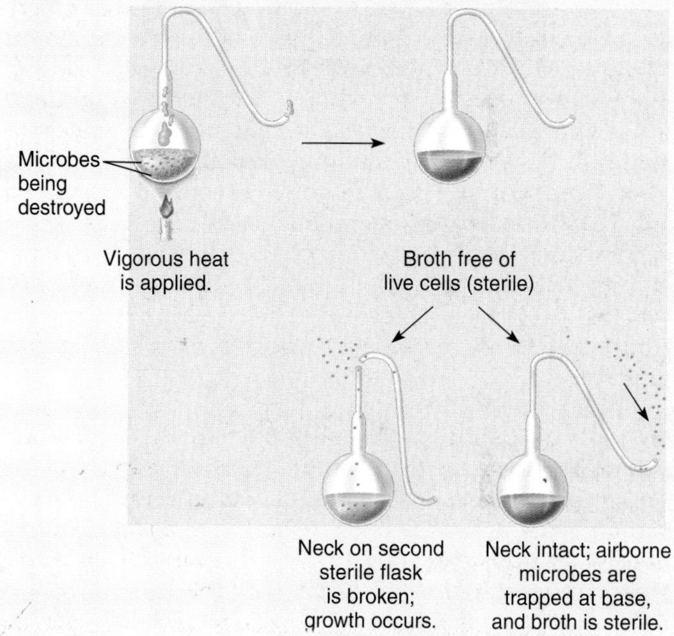

Microbes being destroyed
Vigorous heat is applied.
Broth free of live cells (sterile)
Neck on second sterile flask is broken; growth occurs.
Neck intact; airborne microbes are trapped at base, and broth is sterile.

(which he called "animalcules") were astute and precise. Because of Leeuwenhoek's extraordinary contributions to microbiology, he is known as the father of bacteriology and protozoology.

From the time of Leeuwenhoek, microscopes became more complex and improved with the addition of refined lenses, a condenser, finer focusing devices, and built-in light sources. The prototype of the modern compound microscope, in use from about the mid-1800s, was capable of magnifications of 1,000 times or more. Even our modern laboratory microscopes are not greatly different in basic structure and function from those early microscopes. The technical characteristics of microscopes and microscopy are a major focus of chapter 3.

The Establishment of the Scientific Method

A serious impediment to the development of true scientific reasoning and testing was the tendency of early scientists to explain natural phenomena by a mixture of belief, superstition, and argument. The development of an experimental system that answered questions objectively and was not based on prejudice marked the beginning of true scientific thinking. These ideas gradually crept into the consciousness of the scientific community during the 1600s. The general approach taken by scientists to explain a certain natural phenomenon is called the **scientific method.** A primary aim of this method is to formulate a **hypothesis,** a tentative explanation to account for what has been observed or

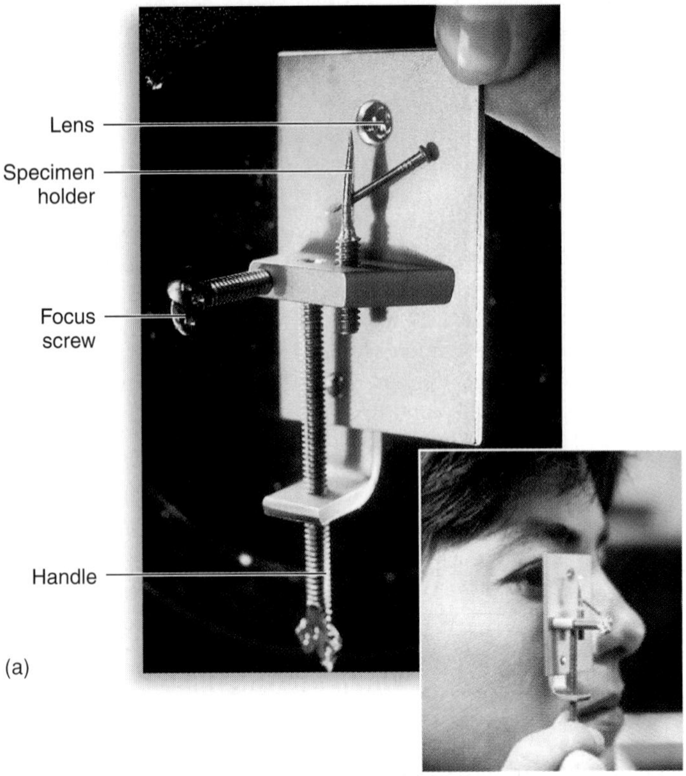

(a)

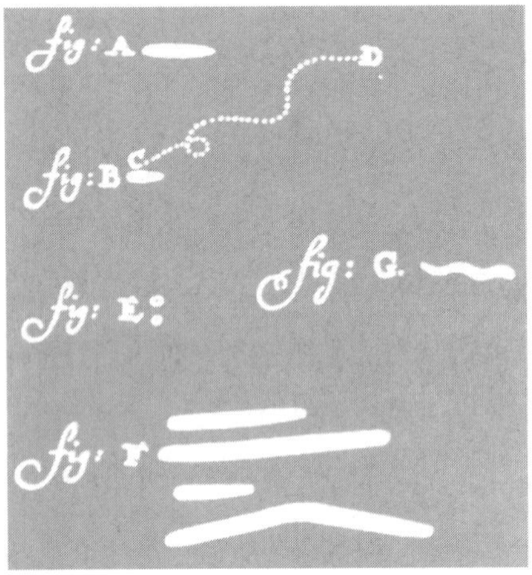

(b)

Figure 1.9 Leeuwenhoek's microscope.
(a) A brass replica of a Leeuwenhoek microscope and how it is held.
(b) Examples of bacteria drawn by Leeuwenhoek.

CASE FILE 1 *WRAP-UP*

The organism causing Tyler's condition was *Streptococcus* pyogenes, also known as Group A Strep or simply "Strep." Streptococci (*S. pyogenes, S. pneumoniae, S. agalactiae*) are organisms that can frequently colonize humans and sometimes result in serious, life-threatening illness. *S. pyogenes* is frequently associated with a sore throat but also causes scarlet fever and skin infections such as impetigo and the more serious "flesh-eating disease," necrotizing fasciitis. It sometimes travels through the bloodstream and establishes an infection in an unusual location. Doctors think that's what happened in Tyler's case with the organism invading his hip. His parents were informed about the seriousness of Tyler's condition, and doctors advised them that he might not survive. While in ICU, he had to be resuscitated. He developed septic shock, a serious condition that can damage organs and interfere with blood flow to the extremities. Exactly why Tyler's encounter with "Strep" turned out this way will likely never be known. According to the CDC, serious illness occurs in less than 1% of the millions of streptococcal infections that occur each year but the mortality rate can be as high as 50%. Tyler survived his battle with Strep and seems to have taken it all in stride, even though he lost 10 toes and the fingers from his right hand. He receives physical therapy to help him with his walking and to learn how to do things with his left hand. Tyler doesn't dwell on the fact that he's lucky to be alive. Cincinnati Bengals Quarterback Carson Palmer visited him in the hospital and brought him a game jersey. He was invited to throw out the first pitch for the Cincinnati Reds at the last home game of the 2006 season, and he did it with his left hand. When asked about his 142 days in the hospital, Tyler simply complains that the blood pressure cuff was too tight.

measured. A good hypothesis should be in the form of a statement. It must be capable of being either supported or discredited by careful, systematic observation or experimentation. For example, the statement that "microorganisms cause diseases" can be experimentally determined by the tools of science, but the statement that "diseases are caused by evil spirits" cannot.

There are various ways to apply the scientific method, but probably the most common is called the **deductive approach.** In the deductive approach, a scientist constructs a hypothesis, tests its validity by outlining particular events that are predicted by the hypothesis, and then performs experiments to test for those events **(figure 1.10).** The deductive process states: "*If* the hypothesis is valid, *then* certain specific events can be expected to occur."

A lengthy process of experimentation, analysis, and testing eventually leads to conclusions that either support or refute the hypothesis. If experiments do not uphold the hypothesis—that is, if it is found to be flawed—the hypothesis or some part of it is rejected; it is either discarded or modified to fit the results of the experiment **(figure 1.10*b*).** If the hypothesis is supported by the results from the experiment, it is not (or should not be) immediately accepted as fact. It then must be tested and retested. Indeed, this is an important guideline in the acceptance of a hypothesis. The results of the experiment must be published and then repeated by other investigators.

In time, as each hypothesis is supported by a growing body of data and survives rigorous scrutiny, it moves to the next level of acceptance—the **theory.** A theory is a collection of statements, propositions, or concepts that explains or

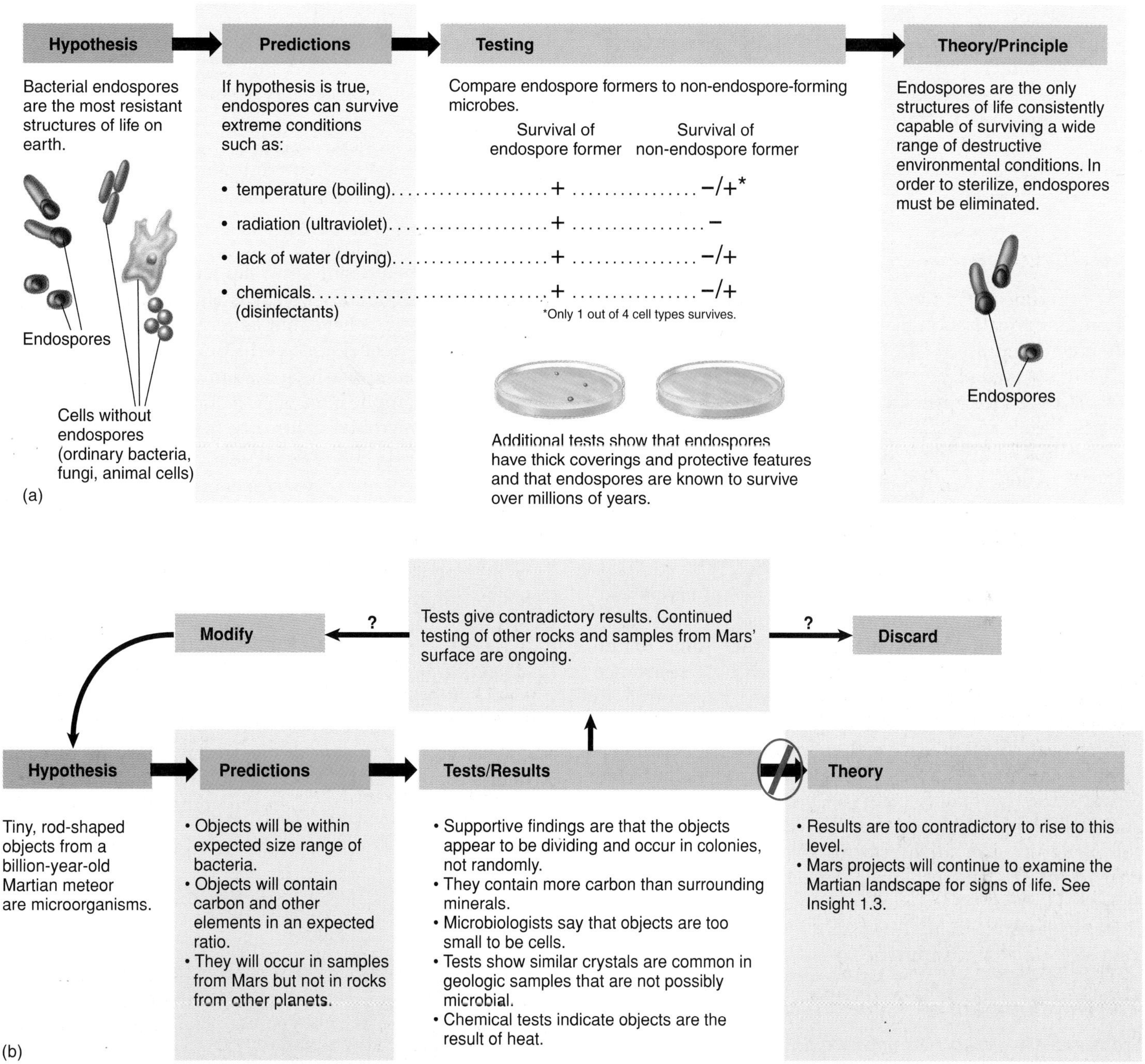

Figure 1.10 The pattern of deductive reasoning using two examples.
The deductive process starts with a general hypothesis that predicts specific expectations that may or may not be borne out by testing. **(a)** This example shows the reasoning behind a well-established principle that has been thoroughly tested over the past 150 years. **(b)** This example is based on a new hypothesis that is still in the early stages of scientific scrutiny.

accounts for a natural event. A theory is not the result of a single experiment repeated over and over again but is an entire body of ideas that expresses or explains many aspects of a phenomenon. It is not a fuzzy or weak speculation, as is sometimes the popular notion, but a viable declaration that has stood the test of time and has yet to be disproved by serious scientific endeavors. Often, theories develop and progress through decades of research and are added to and modified by new findings. At some point, evidence of the accuracy and predictability of a theory is so compelling that the next level of confidence is reached and the theory becomes a law, or principle. For example, although we still refer to the germ *theory* of disease, so little question remains that microbes can cause disease that it has clearly passed into the realm of law.

Science and its hypotheses and theories must progress along with technology. As advances in instrumentation allow new, more detailed views of living phenomena, old theories may be reexamined and altered and new ones proposed. But scientists do not take the stance that theories or even "laws" are ever absolutely proved.

The characteristics that make scientists most effective in their work are curiosity, open-mindedness, skepticism, creativity, cooperation, and readiness to revise their views of natural processes as new discoveries are made. The events described in Insight 1.2 provide important examples.

The Development of Medical Microbiology

Early experiments on the sources of microorganisms led to the profound realization that microbes are everywhere: Not only are air and dust full of them, but the entire surface of the earth, its waters, and all objects are inhabited by them. This discovery led to immediate applications in medicine. Thus the seeds of medical microbiology were sown in the mid to latter half of the 19th century with the introduction of the germ theory of disease and the resulting use of sterile, aseptic, and pure culture techniques.

The Discovery of Spores and Sterilization

Following Pasteur's inventive work with infusions (see Insight 1.2), it was not long before English physicist John Tyndall provided the initial evidence that some of the microbes in dust and air have very high heat resistance and that particularly vigorous treatment is required to destroy them. Later, the discovery and detailed description of heat-resistant bacterial endospores by Ferdinand Cohn, a German botanist, clarified the reason that heat would sometimes fail to completely eliminate all microorganisms. The modern sense of the word **sterile,** meaning completely free of all life forms (including spores) and virus particles, was established from that point on (see chapter 11). The capacity to sterilize objects and materials is an absolutely essential part of microbiology, medicine, dentistry, and some industries.

The Development of Aseptic Techniques

From earliest history, humans experienced a vague sense that "unseen forces" or "poisonous vapors" emanating from decomposing matter could cause disease. As the study of microbiology became more scientific and the invisible was made visible, the fear of such mysterious vapors was replaced by the knowledge and sometimes even the fear of "germs." About 125 years ago, the first studies by Robert Koch clearly linked a microscopic organism with a specific disease. Since that time, microbiologists have conducted a continuous search for disease-causing agents.

At the same time that abiogenesis was being hotly debated, a few physicians began to suspect that microorganisms could cause not only spoilage and decay but also infectious diseases. It occurred to these rugged individualists that even the human body itself was a source of infection. Dr. Oliver Wendell Holmes, an American physician, observed that mothers who gave birth at home experienced fewer infections than did mothers who gave birth in the hospital; and the Hungarian Dr. Ignaz Semmelweis showed quite clearly that women became infected in the maternity ward after examinations by physicians coming directly from the autopsy room.

The English surgeon Joseph Lister took notice of these observations and was the first to introduce **aseptic** (ay-sep'-tik) **techniques** aimed at reducing microbes in a medical setting and preventing wound infections. Lister's concept of asepsis was much more limited than our modern precautions. It mainly involved disinfecting the hands and the air with strong antiseptic chemicals, such as phenol, prior to surgery. It is hard for us to believe, but as recently as the late 1800s surgeons wore street clothes in the operating room and had little idea that hand washing was important. Lister's techniques and the application of heat for sterilization became the bases for microbial control by physical and chemical methods, which are still in use today.

The Discovery of Pathogens and the Germ Theory of Disease

Two ingenious founders of microbiology, Louis Pasteur of France **(figure 1.11)** introduced techniques that are still used today. Pasteur made enormous contributions to our understanding of the microbial role in wine and beer formation. He invented pasteurization and completed some of the first studies showing that human diseases could arise from infection. These studies, supported by the work of other scientists, became known as the **germ theory of disease.** Pasteur's contemporary, Koch, established *Koch's postulates,* a series of proofs that verified the germ theory and could

Figure 1.11 Louis Pasteur (1822–1895), one of the founders of microbiology.
Few microbiologists can match the scope and impact of his contributions to the science of microbiology.

establish whether an organism was pathogenic and which disease it caused (see chapter 13). About 1875, Koch used this experimental system to show that anthrax was caused by a bacterium called *Bacillus anthracis*. So useful were his postulates that the causative agents of 20 other diseases were discovered between 1875 and 1900, and even today, they are the standard for identifying pathogens of plants and animals.

Numerous exciting technologies emerged from Koch's prolific and probing laboratory work. During this golden age of the 1880s, he realized that study of the microbial world would require separating microbes from each other and growing them in culture. It is not an overstatement to say that he and his colleagues invented most of the techniques that are described in chapter 3: inoculation, isolation, media, maintenance of pure cultures, and preparation of specimens for microscopic examination. Other highlights in this era of discovery are presented in later chapters on microbial control (see chapter 11) and vaccination (see chapter 15).

CHECKPOINT

- Our current understanding of microbiology is the cumulative work of thousands of microbiologists, many of whom literally gave their lives to advance knowledge in this field.
- The microscope made it possible to see microorganisms and thus to identify their widespread presence, particularly as agents of disease.
- Antonie van Leeuwenhoek is considered the father of bacteriology and protozoology because he was the first person to produce precise, correct descriptions of these organisms.
- The theory of spontaneous generation of living organisms from "vital forces" in the air was disproved once and for all by Louis Pasteur.
- The scientific method is a process by which scientists seek to explain natural phenomena. It is characterized by specific procedures that either support or discredit an initial hypothesis.
- Knowledge acquired through the scientific method is rigorously tested by repeated experiments by many scientists to verify its validity. A collection of valid hypotheses is called a theory. A theory supported by much data collected over time is called a law.
- Scientific dogma or theory changes through time as new research brings new information. Scientists must be able and willing to change theory in response to new data.
- Medical microbiologists developed the germ theory of disease and introduced the critically important concept of aseptic technique to control the spread of disease agents.
- Koch's postulates are the cornerstone of the germ theory of disease. They are still used today to pinpoint the causative agent of a specific disease.
- Louis Pasteur and Robert Koch were the leading microbiologists during the golden age of microbiology (1875–1900); each had his own research institute.

1.7 Taxonomy: Naming, Classifying, and Identifying Microorganisms

Students just beginning their microbiology studies are often dismayed by the seemingly endless array of new, unusual, and sometimes confusing names for microorganisms. Learning microbial **nomenclature** is very much like learning a new language, and occasionally its demands may be a bit overwhelming. But paying attention to proper microbial names is just like following a baseball game or a theater production: You cannot tell the players apart without a program! Your understanding and appreciation of microorganisms will be greatly improved by learning a few general rules about how they are named.

The science of classifying living beings is **taxonomy.** It originated more than 250 years ago when Carl von Linné (also known as Linnaeus; 1701–1778), a Swedish botanist, laid down the basic rules for *classification* and established taxonomic categories, or **taxa** (singular: taxon).

Von Linné realized early on that a system for recognizing and defining the properties of living beings would prevent chaos in scientific studies by providing each organism with a unique name and an exact "slot" in which to catalog it. This classification would then serve as a means for future identification of that same organism and permit workers in many biological fields to know if they were indeed discussing the same organism. The von Linné system has served well in categorizing the 2 million or more different kinds of organisms that have been discovered since that time, including organisms that have gone extinct.

The primary concerns of modern taxonomy are still naming, classifying, and identifying. These three areas are interrelated and play a vital role in keeping a dynamic inventory of the extensive array of living and extinct beings. In general, nomenclature is the assignment of scientific names to the various taxonomic categories and individual organisms whereas *classification* attempts the orderly arrangement of organisms into a hierarchy of taxa. *Identification* is the process of discovering and recording the traits or organisms so that they may be recognized or named and placed in an overall taxonomic scheme. With the rapid increase in knowledge largely due to the mind-boggling pace of improvement in scientific instrumentation and analysis, taxonomy has never stood still. Instead, it has evolved from a science that artificially classified organisms from a viewpoint of the organism's usefulness, danger, or esthetic appeal to humans to a science that devised a system of natural relationships between organisms. A survey of some general methods of identification appears in chapter 3. Discovery of present or extinct life forms in space would certainly provide an ultimate test for our existing taxonomy and shed light on the origins of life on our planet earth **(Insight 1.3).**

The Levels of Classification

The main units, or taxa, of a classification scheme are organized into several descending ranks, beginning with a most general all-inclusive taxonomic category as a common denominator

INSIGHT 1.3 *Discovery*

Martian Microbes and Astrobiology

Professional and amateur scientists have long been intrigued by the possible existence of life on other planets and in the surrounding universe. This curiosity has given rise to a new discipline—astrobiology—that applies principles from biology, chemistry, geology, and physics to investigate extraterrestrial life. One of the few accessible places to begin this search is the planet Mars. It is relatively close to the earth and the only planet in the solar system besides earth that is not extremely hot, cold, or bathed in toxic gases.

The possibility that it could support at least simple life forms has been an important focus of NASA space projects stretching over 30 years. Several Mars explorations have included experiments and collection devices to gather evidence for certain life signatures or characteristics. One of the first experiments launched with the *Viking Explorer* was an attempt to culture microbes from Martian soil. Another used a gas chromatograph to check for complex carbon-containing (organic) compounds in the soil samples. No signs of life or organic matter were detected. But in scientific research, a single experiment is not sufficient to completely rule out a hypothesis, especially one as attractive as this one. Many astrobiologists reason that the nature of the "life forms" may be so different that they require a different experimental design.

In 1996, another finding brought considerable excitement and controversy to the astrobiology community. Scientists doing electron microscopic analyses of an ancient Martian meteorite from the Antarctic discovered tiny rodlike structures that resembled earth bacteria. Though the idea was appealing, many scientists argued that the rods did not contain the correct form of carbon and that geologic substances often contain crystals that mimic other objects. Another team of NASA researchers later discovered chains of magnetite crystals (tiny iron oxide magnets) in another Martian meteorite. These crystals bear a distinct resemblance to forms found in certain modern bacteria on earth and are generally thought to be formed only by living processes.

Obviously, these findings have added much fodder for speculation and further research. Current NASA projects in astrobiology are designed to bring larger samples of Martian rocks back to earth. A wider array of rocks could make isolation of microbes or microbial signatures more likely. Rocks will also make it possible to search for fossilized organisms that could provide a history of the planet. Perhaps it will be determined that living organisms were once present on Mars but have become extinct. After all, the Martian meteorites are billions of years old. Researchers will scrutinize the samples for phosphates,

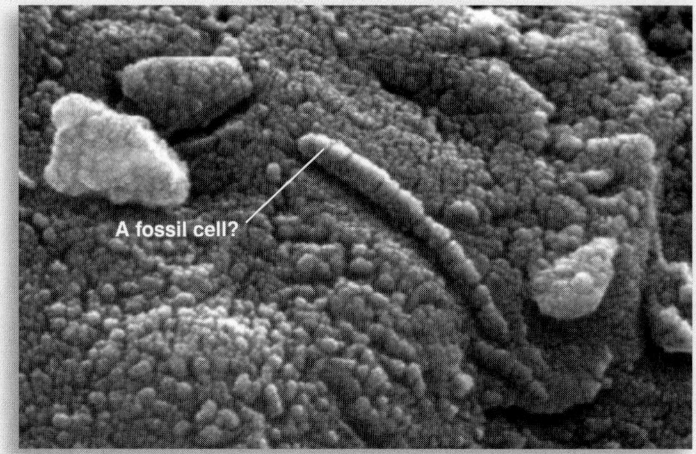

Martian microbes or mere molecules? Internal view of a section of a 4.5-billion-year-old Martian meteor shows an intriguing tiny cylinder (50,000×).

carbonates, and other molecules associated with life on earth. They are also testing for the presence of water, which is known to foster life. So far, tests have confirmed that there is a large amount of frozen water in the planet's crust.

Astrobiologists long ago put aside the quaint idea of meeting "little green men" when they got to the red planet, but they have not yet given up the possibility of finding "little green microbes." One hypothesis proposes that microbes hitchhiking on meteors and asteroids have seeded the solar system and perhaps universe with simple life forms. Certainly, of all organisms on earth, hardy prokaryotes are the ones most likely to survive the rigors of such travel. That is why several groups of astrobiologists are studying microbes called archaea that can survive the most intense conditions on earth to determine what the limits of life appear to be. American geologists working with very salty, acidic Australian lakes have found conditions and unusual rock sediments that are very similar to those found on Mars. During the next phase of their research, they will analyze the samples for the presence of microbes. Aside from our fascination with Mars, there is still much left on earth to discover!

* For more information on this subject, use a search engine to access the NASA Astrobiology Institute, NASA Mission to Mars, or NASA Exploration Program websites.

for organisms to exclude all others, and ending with the smallest and most specific taxon. This means that all members of the highest category share only one or a few general characteristics, whereas members of the lowest category are essentially the same kind of organism—that is, they share the majority of their characteristics. The taxonomic categories from top to bottom are: **Domain, kingdom, phylum** or **division,**[3] **class, order, family, genus,** and **species.** Thus, each kingdom can be subdivided into a series of phyla or divisions, each phylum is

3. The term *phylum* is used for bacteria, protozoa, and animals; the term *division* is used for algae, plants, and fungi.

made up of several classes, each class contains several orders, and so on. Because taxonomic schemes are to some extent artificial, certain groups of organisms may not exactly fit into the main taxa. In such a case, additional taxonomic levels can be imposed above (super) or below (sub) a taxon, giving us such categories as "superphylum" and "subclass."

Let's compare the taxonomic breakdowns of a human and a protozoan (proh-t*uh*-zoh-on′-*uh*n) to illustrate the fine points of this system **(figure 1.12).** Humans and protozoa are both organisms with nucleated cells (eukaryotes) but placed in different kingdoms. Humans are multicellular animals (Kingdom Animalia) whereas protozoa are single-cellular organisms that, together with algae, belong to the Kingdom Protista. To emphasize just how broad the category "kingdom" is, ponder the fact that we humans belong to the same kingdom as jellyfish. Of the several phyla within this kingdom, humans belong to the Phylum Chordata, but even a phylum is rather all-inclusive, considering that humans share it with other vertebrates as well as with creatures called sea squirts. The next level, Class Mammalia, narrows the field considerably by grouping only those vertebrates that have hair and suckle their young. Humans belong to the Order Primates, a group that also includes apes, monkeys, and lemurs. Next comes the

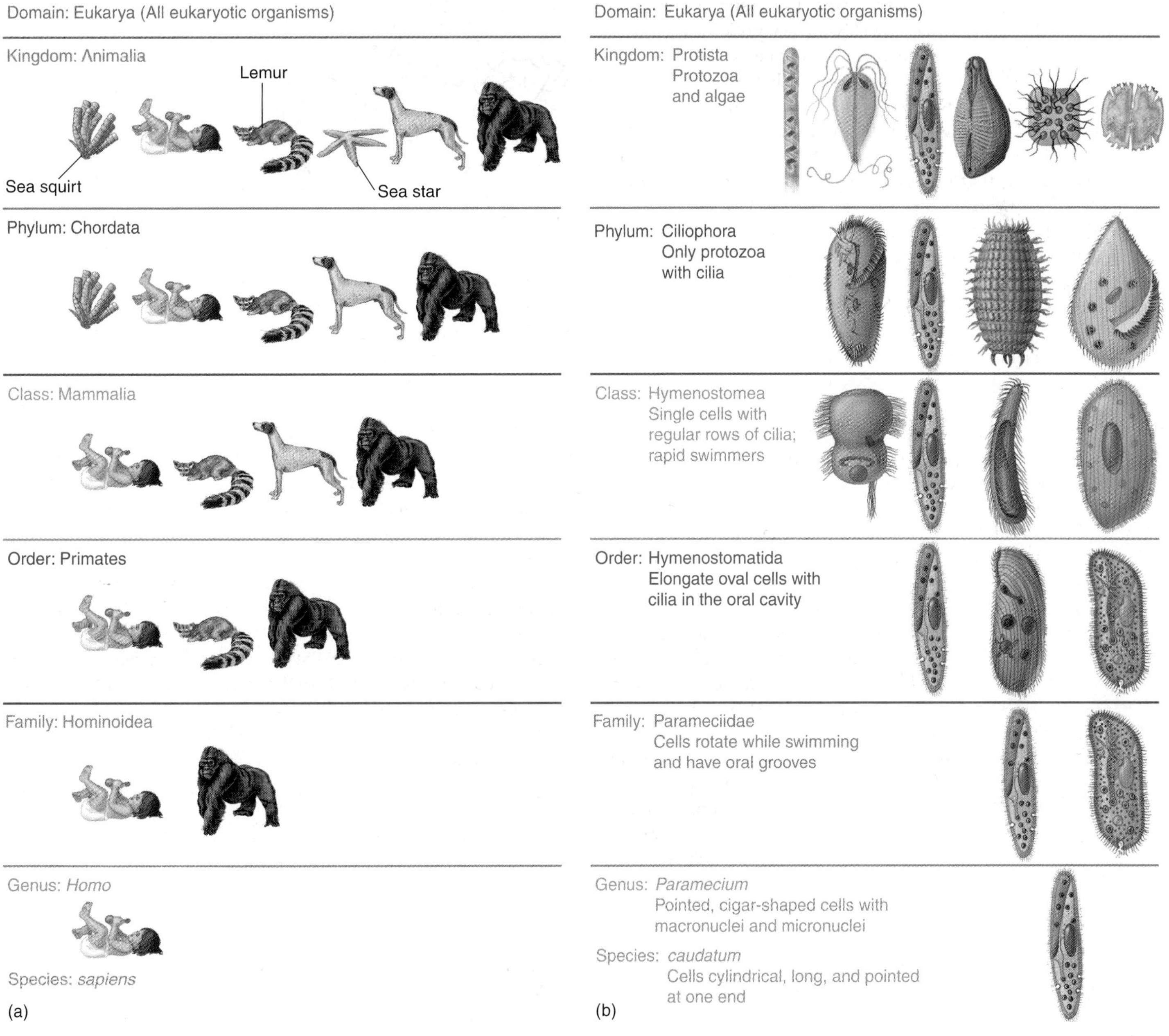

Figure 1.12 Sample taxonomy.
Two organisms belonging to the Eukarya domain, traced through their taxonomic series. **(a)** Modern humans, *Homo sapiens.* **(b)** A common protozoan, *Paramecium caudatum.*

Family Hominoidea, containing only humans and apes. The final levels are our genus, *Homo* (all races of modern and ancient humans), and our species, *sapiens* (meaning wise). Notice that for the human as well as the protozoan, the taxonomic categories in descending order become less inclusive and the individual members more closely related. Other examples of classification schemes are provided in sections of chapters 4 and 5 and in several later chapters.

We need to remember that all taxonomic **hierarchies** are based on the judgment of scientists with certain expertise in a particular group of organisms and that not all other experts may agree with the system being used. Consequently, no taxa are permanent to any degree; they are constantly being revised and refined as new information becomes available or new viewpoints become prevalent. In this text, we are usually concerned with only the most general (kingdom, phylum) and specific (genus, species) taxonomic levels.

Assigning Specific Names

Many macroorganisms are known by a common name suggested by certain dominant features. For example, a bird species might be called a red-headed blackbird or a flowering plant species a black-eyed Susan. Some species of microorganisms (especially those that directly or indirectly affect our well-being) are also called by informal names, including human pathogens such as gonococcus *(Neisseria gonorrhoeae)* and the tubercle bacillus *(Mycobacterium tuberculosis)*, or fermenters such as brewer's yeast *(Saccharomyces cerevisiae)*, but this is not the usual practice. If we were to adopt common names such as the "little yellow coccus" (*Micrococcus luteus* [my″-kroh-kok′-us loo′-tee-us] Gr. *micros*, small, and *kokkus*, berry; L. *luteus*, yellow) or the "club-shaped diphtheria bacterium," (*Corynebacterium diphtheriae* [kor-eye″-nee-bak-ter′-ee-yum dif′-theer-ee-eye] Gr. *coryne*, club, *bacterion*, little rod, and *diphtheriae*, the causative agent of the disease diphtheria) the terminology would become even more cumbersome and challenging than scientific names. Even worse, common names are notorious for varying from region to region, even within the same country. A decided advantage of standardized nomenclature is that it provides a universal language, thereby enabling scientists from all countries to freely exchange information.

The method of assigning scientific or specific name is called the **binomial** (two-name) **system** of **nomenclature.** The scientific name is always a combination of the generic (genus) name followed by the species name. The generic part of the scientific name is capitalized, and the species part begins with a lowercase letter. Both should be italicized (or underlined if italics are not available), as follows:

Staphylococcus aureus

Because other taxonomic levels are not italicized and consist of only one word, one can always recognize a scientific name. The binomial of an organism is sometimes abbreviated to save space, as in ***S. aureus,*** but only if the genus name has already been stated. The source for nomenclature is usually Latin or Greek. If other languages such as English or French are used, the endings of these words are revised to have Latin endings. In general, the name first applied to a species will be the one that takes precedence over all others assigned later.

An international group oversees the naming of every new organism discovered, making sure that standard procedures have been followed and that there is not already an earlier name for the organism or another organism with that same name. The inspiration for names is extremely varied and often rather imaginative. Some species have been named in honor of a microbiologist who originally discovered the microbe or who has made outstanding contributions to the field. Other names may designate a characteristic of the microbe (shape, color), a location where it was found, or a disease it causes. Some examples of specific names, their pronunciations, and their origins are:

- *Staphylococcus aureus* (staf′-i-lo-kok′-us ah′-ree-us) Gr. *staphule*, bunch of grapes, *kokkus*, berry, and Gr. *aureus*, golden. A common bacterial pathogen of humans.
- *Campylobacter jejuni* (cam′-peh-loh-bak-ter jee-joo′-neye) Gr. *kampylos*, curved, *bakterion*, little rod, and *jejunum*, a section of intestine. One of the most important causes of intestinal infection worldwide.
- *Lactobacillus sanfrancisco* (lak′-toh-bass-ill′-us san-fran-siss′-koh) L. *lacto*, milk, and *bacillus*, little rod. A bacterial species used to make sourdough bread.
- *Vampirovibrio chlorellavorus* (vam-py′-roh-vib-ree-oh klor-ell-ah′-vor-us) F. *vampire;* L. *vibrio*, curved cell; *Chlorella*, a genus of green algae; and *vorus*, to devour. A small, curved bacterium that sucks out the cell juices of *Chlorella.*
- *Giardia lamblia* (jee-ar′-dee-uh lam′-blee-uh) for Alfred Giard, a French microbiologist, and Vilem Lambl, a Bohemian physician, both of whom worked on the organism, a protozoan that causes a severe intestinal infection.

Here's a helpful hint: These names may seem difficult to pronounce and the temptation is to simply "slur over them." But when you encounter the name of a microorganism in the chapters ahead, it will be extremely useful to take the time to sound them out and repeat them until they seem familiar. You are much more likely to remember them that way—and they are less likely to end up in a tangled heap with all of the new language you will be learning.

The Origin and Evolution of Microorganisms

As we indicated earlier, *taxonomy*, the science of classification of biological species, is used to organize all of the forms of modern and extinct life. In biology today, there are different methods for deciding on taxonomic categories, but they all rely on the degree of relatedness among organisms. The scheme that represents the natural relatedness (relation by descent) between groups of living beings is called their *phylogeny* (Gr. *phylon*, race or class; L. *genesis*, origin or beginning), and—when unraveled—biologists use phylogenetic relationships to refine the system of taxonomy.

To understand the natural history of and the relatedness among organisms, we must understand some fundamentals of the process of evolution. **Evolution** is an important theme that underlies all of biology, including the biology of microorganisms. Put simply, evolution states that the hereditary information in living beings changes gradually through time (usually hundreds of millions of years) and that these changes result in various structural and functional changes through many generations. The process of evolution is selective in that those changes that most favor the survival of a particular organism or group of organisms tend to be retained whereas those that are less beneficial to survival tend to be lost. Charles Darwin called this process *natural selection*.

Evolution is founded on the two preconceptions that (1) all new species originate from preexisting species and (2) closely related organisms have similar features because they evolved from a common ancestor; hence, difference emerged by divergence. Usually, evolution progresses toward greater complexity but there are many examples of evolution toward lesser complexity (reductive evolution). This is because individual organisms never evolve in isolation but as populations of organisms in their specific environments, which exert the functional pressures of selection. Because such pressures may favor evolution toward lower complexity (for instance, loss of the ability to utilize a broad variety of energy and carbon sources), some modern organisms appear to be primitive and rather "ancient." However, it is very important to realize that all species presently residing on earth are modern, no matter how much they may differ from the original set of ancestral traits.

Because of the divergent nature of the evolutionary process, the phylogeny, or relatedness by descent, of organisms is often represented by a diagram of a tree. The trunk of the tree represents the origin of ancestral lines, and the branches show offshoots into specialized groups (elades) of organisms. This sort of arrangement places taxonomic groups with less divergence (less change in the heritable information) from the common ancestor closer to the root of the tree and taxa with lots of divergence closer to the top. The length of branches in such phylogentic trees may also indicate an approximate timescale for the evolutionary history in addition to how closely related various organisms are **(figures 1.13** and **1.14)**.

The volume of material to be covered in this book does not permit further introduction to the science of evolution, but the occurrence of this process is supported by a tremendous amount of evidence from the fossil record as well as the study of the **morphology** (structure), **physiology** (function), and **genetics** (inheritance) of organisms. Evolution, indeed, accounts for the millions of different species on planet earth and their adaptation to its many and diverse habitats.

Systems of Presenting a Universal Tree of Life

The first trees of life were constructed a long time ago on the basis of just two kingdoms, plants and animals, by Charles Darwin and Ernst Haeckel. These trees were chiefly based on visible morphological characteristics including the different stages of organism development. It became clear that certain (micro)organisms such as algae and protozoa, which only existed as single cells, did not truly fit either of those categories, so a third kingdom was recognized by Haeckel for these simpler organisms and named Protista. Eventually, when significant differences became evident among even the unicellular organisms, a fourth kingdom was established in the 1870s by Haeckel and named Monera. Almost a century passed before Robert Whittaker extended this work and added a fifth kingdom for fungi during the period of 1959 to 1969. The relationships that were used in Whittaker's tree were those based on structural similarities and differences, such as prokaryotic and eukaryotic cellular organization, and the way these organisms obtained their nutrition. These criteria indicated that there were five major taxonomic units, or kingdoms: the monera, protists, plants, fungi, and animals, all of which represented two major cell types, the prokaryotic and eukaryotic. Whittaker's five-kingdom system quickly became the standard and is in use for general taxonomic arrangements to date (see figure 1.13).

With the rise of genetics (defined as the study of genes) as a molecular science, newer methods for determining phylogeny have led to the development of a differently shaped tree—with important implications for our understanding of evolutionary relatedness. Molecular biology, in general, and molecular genetics, in particular, allowed an in-depth study of the structure and function of the genetic material at the molecular level. These studies have revealed that two of the four macromolecules that contribute to cellular structure and function, the proteins and nucleic acids, are very well suited to study how organisms differ from one another because their sequences can be aligned and compared. In 1975, Carl Woese discovered that one particular macromolecule, the ribonucleic acid in the small subunit of the ribosome (ssu rRNA), was highly conserved and nearly identical in organisms within the smallest taxonomic category, the species. Based on a vast amount of experimental data and the knowledge that protein synthesis proceeds in all organisms facilitated by the ribosome, Woese hypothesized that ssu rRNA provides a "biological chronometer" or a "living record" of the evolutionary history of a given organism. Extended analysis of this molecule in prokaryotic and eukaryotic cells indicated that all members in a group of bacteria, then known as archaeobacteria, had ssu rRNA with a sequence that was significantly different from the ssu rRNA found in other bacteria and in eukaryotes. This discovery lead Carl Woese and collaborator George Fox to propose a separate taxonomic unit for the archaeobacteria, which they named **Archaea.** At the time, these archaea were characterized by their ability to

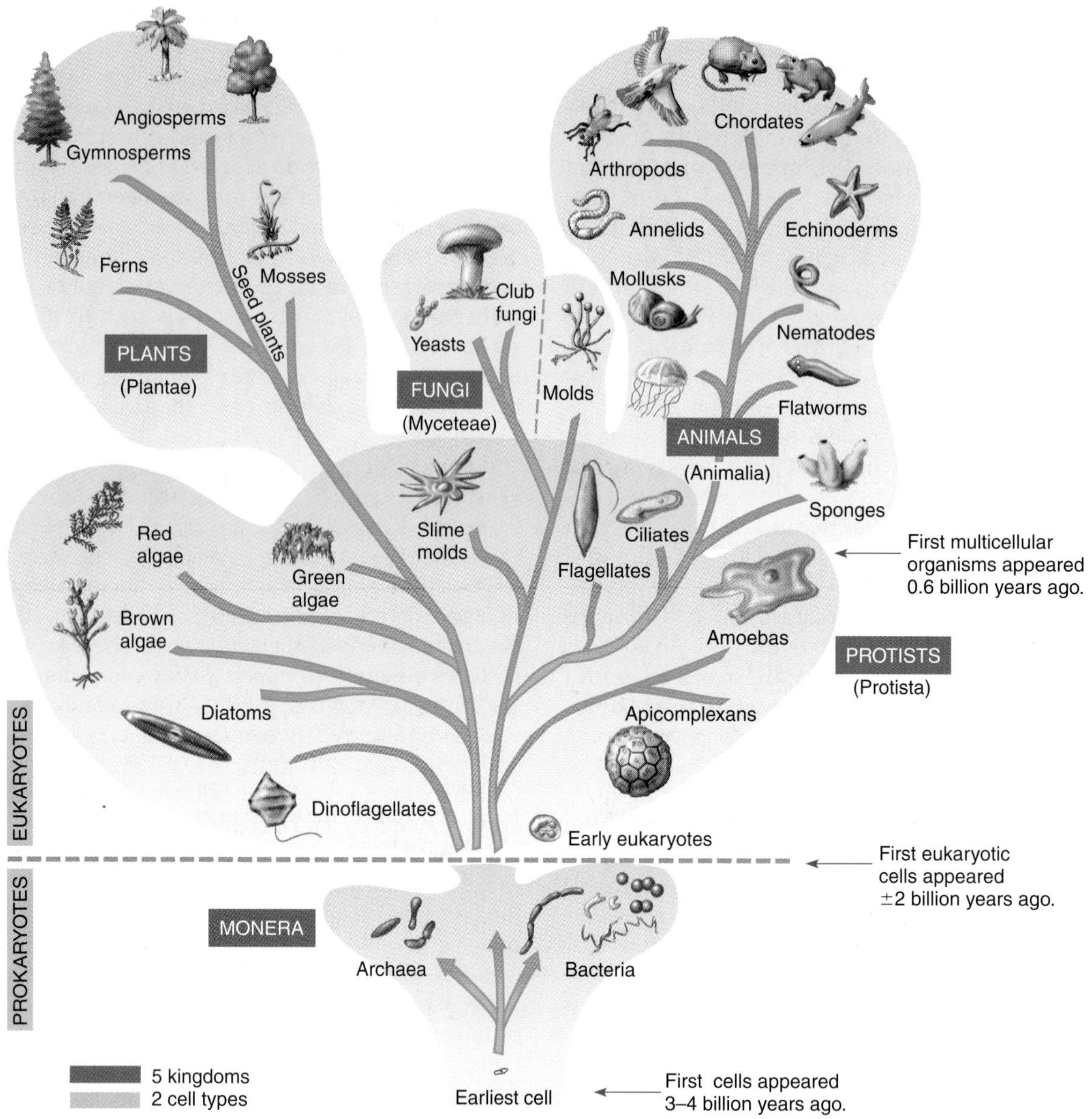

Figure 1.13 Traditional Whittaker system of classification.
In this system, kingdoms are based on cell structure and type, the nature of body organization, and nutritional type. Bacteria and Archaea (monerans) are made of prokaryotic cells and are unicellular. Protists are made of eukaryotic cells and are mostly unicellular. They can be photosynthetic (algae), or they can feed on other organisms (protozoa). Fungi are eukaryotic cells and are unicellular or multicellular; they have cell walls and are not photosynthetic. Plants have eukaryotic cells, are multicellular, have cell walls, and are photosynthetic. Animals have eukaryotic cells, are multicellular, do not have cell walls, and derive nutrients from other organisms.
After Dolphin, *Biology Lab Manual*, 4th ed., Fig. 14.1, p. 177, McGraw-Hill.

live in extreme environments, such as hot springs or highly salty environments. Under the microscope, they resembled the prokaryotic structure of bacteria, but molecular biology has revealed that the archaea, though prokaryotic in nature, were actually more closely related to eukaryotic cells than to bacterial cells (see table 4.6). To reflect these relationships, Carl Woese and George Fox have proposed an entirely new system that assigned all known organisms to one of the three major taxonomic units, the **domains,** each described by a different type of cell (see figure 1.14).

The domains are the highest level in hierarchy and can contain many kingdoms and superkingdoms. The prokaryotic cell types are represented by the domains Archaea and **Bacteria,** whereas eukaryotes are all placed in the domain **Eukarya.** Analysis of the ssu rRNAs from all organisms in these three domains suggests that all modern and extinct organisms on earth arose from a common ancestor.

The new three-domain system is still undergoing analysis. It somewhat complicates the presentation of organisms

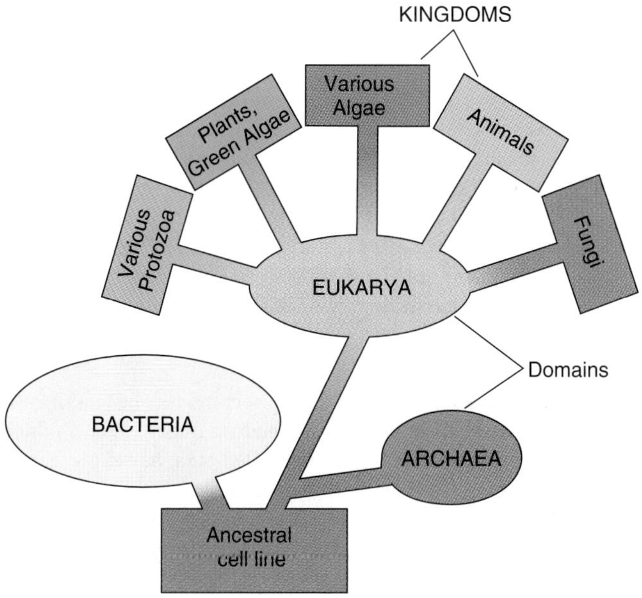

3 cell types, showing relationship with domains and kingdoms

Figure 1.14 **Woese-Fox system.**
A system for representing the origins of cell lines and major taxonomic groups as proposed by Carl Woese and colleagues. They propose three distinct cell lines placed in superkingdoms called domains. The first primitive cells, called progenotes, were ancestors of both lines of prokaryotes (Domains Bacteria and Archaea), and the Archaea emerged from the same cell line as eukaryotes (Domain Eukarya). Some of the traditional kingdoms are still present with this system (see figure 1.14).

in the original Kingdom Protista, which is now a collection of protozoa and algae that exist in several separate kingdoms (discussed in chapter 5). Nevertheless, this new scheme does not greatly affect our presentation of most microbes, because we will discuss them at the genus or species level. But be aware that biological taxonomy and, more important, our view of how organisms evolved on earth are in a period of transition. Keep in mind that our methods of classification or evolutionary schemes reflect our current understanding and will change as new information is uncovered.

Please note that viruses are not included in any of the classification or evolutionary schemes, because they are not cells or organisms and their position in a "tree of life" cannot be determined. The special taxonomy of viruses is discussed in chapter 6.

CHECKPOINT

- Taxonomy is the science used to classify living organisms. It assigns every organism a place and makes a place for every living organism.
- The taxonomic system has three primary functions: naming, classifying and identifying species.
- The major taxa, or groups, in the most advanced taxonomic system are (in descending order): domain, kingdom, phylum or division, class, order, family, genus, and species.
- Taxonomy groups organisms by their evolutionary histories, which in turn are based on evolutionary similarities in morphology, physiology, and genetics.
- Evolutionary patterns show a treelike branching thereby describing the diverging evolution of all life forms from the gene pool of a common ancestor.
- The Whittaker five-kingdom classification system places all bacteria in the kingdom Monera/Prokaryotae and subdivides the eukaryotes into kingdoms Protista, Myceteae, (Fungi) Animalia, and Plantae.
- The Woese-Fox classification system places all eukaryotes in the domain Eukarya and subdivides the prokaryotes into the two domains Archaea and Bacteria.

Chapter Summary with Key Terms

1.1 The Scope of Microbiology

A. **Microbiology** is the study of **bacteria, viruses, fungi, protozoa,** and **algae,** which are collectively called **microorganisms,** or **microbes.** In general, microorganisms are **microscopic** and, unlike **macroscopic** organisms, which are readily visible, they require magnification to be adequately observed or studied.

B. The simplicity, growth rate, and adaptability of microbes are some of the reasons that microbiology is so diverse and has branched out into many subsciences and applications. Important subsciences include **immunology, epidemiology,** public health, food, dairy, aquatic, and industrial microbiology.

1.2 The Impact of Microbes on Earth: Small Organisms with a Giant Effect

Microbes live in most of the world's habitats and are indispensable for normal, balanced life on earth. They play many roles in the functioning of the earth's ecosystems.

A. Microbes are **ubiquitous.**

B. Eukaryotes, which contain nuclei, arose from prokaryotes, which do not contain nuclei.

C. Microbes are involved in nutrient production and energy flow. Algae and certain bacteria trap the sun's energy to produce food through **photosynthesis.**

D. Other microbes are responsible for the breakdown and recycling of nutrients through **decomposition.** Microbes are essential to the maintenance of the air, soil, and water.

1.3 Human Use of Microorganisms

Microbes have been called upon to solve environmental, agricultural, and medical problems.

A. **Biotechnology** applies the power of microbes toward the manufacture of industrial products, foods, and drugs.

B. Microbes form the basis of **genetic engineering** and **recombinant DNA** technology, which alter genetic material to produce new products and modified life forms, genetically modified organisms (GMOs).
C. In **bioremediation,** microbes are used to clean up pollutants and wastes in natural environments.

1.4 Infectious Diseases and the Human Condition

A. Nearly 2,000 microbes are **pathogens** that cause infectious diseases. Infectious diseases result in high levels of mortality and morbidity (illness). Many infections are emerging, meaning that they are newly identified pathogens gaining greater prominence. Many older diseases are also increasing.
B. Some diseases previously thought to be noninfectious may involve microbial infections (e.g., *Helicobacter*, causing gastric ulcers, and Coxsackie viruses, causing diabetes).
C. An increasing number of individuals have weak immune systems, which makes them more susceptible to infectious diseases.

1.5 The General Characteristics of Microorganisms

A. Microbial cells are either the small, relatively simple, nonnucleated prokaryotic variety or the larger, more complex eukaryotic type that contain a nucleus and **organelles.**
B. **Viruses** are microorganisms but are not cells. They are smaller in size and infect their prokaryotic or eukaryotic hosts in order to reproduce themselves.
C. **Parasites** are free-living microorganisms that cause damage to their **hosts** through infection and disease.

1.6 The Historical Foundations of Microbiology

A. Microbiology as a science is about 200 years old. Hundreds of contributors have provided discoveries and knowledge to enrich our understanding.
B. With his simple microscope, Leeuwenhoek discovered organisms he called animalcules. As a consequence of his findings and the rise of the **scientific method,** the notion of **spontaneous generation,** or **abiogenesis,** was eventually abandoned for **biogenesis.** The scientific method develops rational **hypotheses** and **theories** that can be tested. Theories that withstand repeated scrutiny become law in time.
C. Early microbiology blossomed with the conceptual developments of **sterilization, aseptic techniques,** and the **germ theory of disease.**

1.7 Taxonomy: Naming, Classifying, and Identifying Microorganisms

A. **Taxonomy** is a hierarchical scheme for the classification, identification, and **nomenclature** of organisms, which are grouped in categories called **taxa,** based on features ranging from general to specific.
B. Starting with the broadest category, the taxa are **domain, kingdom, phylum** (or **division**), **class, order, family, genus,** and **species.** Organisms are assigned **binomial scientific names** consisting of their genus and species names.
C. The latest classification scheme for living things is based on the genetic structure of their ribosomes. The Woese-Fox system recognizes three domains: **Archaea,** simple prokaryotes that often live in extreme environments; **Bacteria,** typical prokaryotes; and **Eukarya,** all types of eukaryotic organisms.
D. An alternative classification scheme uses a five-kingdom organization: Kingdom Procaryotae (Monera), containing the **eubacteria** and the archaea; Kingdom Protista, containing primitive unicellular microbes such as algae and protozoa; Kingdom Myceteae, containing the fungi; Kingdom Animalia, containing animals; and Kingdom Plantae, containing plants.

Multiple-Choice and True-False Questions

Multiple-Choice Questions. Select the correct answer from the answers provided.

1. Which of the following is not considered a microorganism?
   a. alga c. protozoan
   b. bacterium d. mushroom
2. An area of microbiology that is concerned with the occurrence of disease in human populations is
   a. immunology c. epidemiology
   b. parasitology d. bioremediation
3. Which process involves the deliberate alteration of an organism's genetic material?
   a. bioremediation c. decomposition
   b. biotechnology d. recombinant DNA technology
4. Which of the following parts was absent from Leeuwenhoek's microscopes?
   a. focusing screw c. specimen holder
   b. lens d. condenser
5. Abiogenesis refers to the
   a. spontaneous generation of organisms from nonliving matter
   b. development of life forms from preexisting life forms
   c. development of aseptic technique
   d. germ theory of disease
6. A hypothesis can be defined as
   a. a belief based on knowledge
   b. knowledge based on belief
   c. a scientific explanation that is subject to testing
   d. a theory that has been thoroughly tested
7. When a hypothesis has been thoroughly supported by long-term study and data, it is considered
   a. a law c. a theory
   b. a speculation d. proved

8. Which is the correct order of the taxonomic categories, going from most specific to most general?
   a. domain, kingdom, phylum, class, order, family, genus, species
   b. division, domain, kingdom, class, family, genus, species
   c. species, genus, family, order, class, phylum, kingdom, domain
   d. species, family, class, order, phylum, kingdom
9. Which of the following are prokaryotic?
   a. bacteria
   b. archaea
   c. protists
   d. both a and b
10. Order the following items by size, using numbers: 1 = smallest and 8 = largest.
   ___ AIDS virus
   ___ amoeba
   ___ rickettsia
   ___ protein
   ___ worm
   ___ coccus-shaped bacterium
   ___ white blood cell
   ___ atom
11. How would you classify a virus?
   a. prokaryotic
   b. eukaryotic
   c. neither a nor b

True-False Questions. If the statement is true, leave as is. If it is false, correct it by rewriting the sentence.

12. Organisms in the same order are more closely related than those in the same family.
13. SARS is also known as "bird flu."
14. Prokaryotes have no nucleus.
15. In order to be called a theory, a scientific idea has to undergo a great deal of testing.
16. Microbes are ubiquitous.

Writing to Learn

These questions are suggested as a *writing-to-learn* experience. For each question, compose a one- or two-paragraph answer that includes the factual information needed to completely address the question.

1. Explain the important contributions microorganisms make in the earth's ecosystems.
2. Describe five different ways in which humans exploit microorganisms for our benefit.
3. Identify the groups of microorganisms included in the scope of microbiology, and explain the criteria for including these groups in the field.
4. Why was the abandonment of the spontaneous generation theory so significant? Using the scientific method, describe the steps you would take to test the theory of spontaneous generation.
5. a. Differentiate between a hypothesis and a theory.
   b. Is the germ theory of disease really a law? Why or why not?
6. a. Differentiate between taxonomy, classification, and nomenclature.
   b. What is the basis for a phylogenetic system of classification?
   c. What is a binomial system of nomenclature, and why is it used?
   d. Give the correct order of taxa, going from most general to most specific. Create a mnemonic (memory) device for recalling the order.
7. Compare the new domain system with the five-kingdom system. Does the newer system change the basic idea of prokaryotes and eukaryotes? What is the third cell type?
8. Evolution accounts for the millions of different species on the earth and their adaptation to its many and diverse habitats. Explain this. Cite examples in your answer.

Concept Mapping

Appendix D provides guidance for working with concept maps.

1. Supply your own linking words or phrases in this concept map, and provide the missing concepts in the empty boxes.

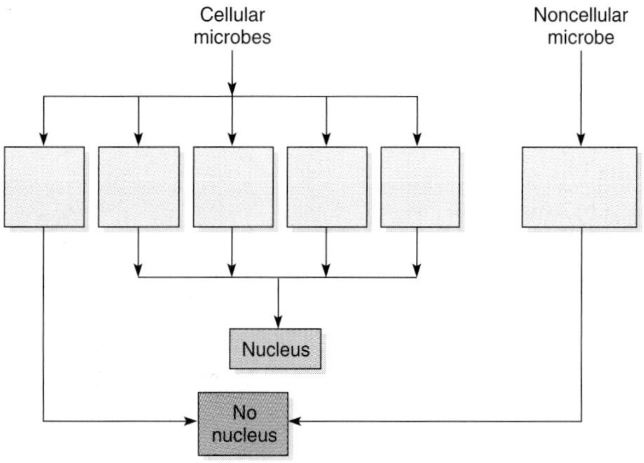

Critical Thinking Questions

Critical thinking is the ability to reason and solve problems using facts and concepts. These questions can be approached from a number of angles, and in most cases, they do not have a single correct answer.

1. a. Where do you suppose the "new" infectious diseases come from?
   b. Name some factors that could cause older diseases to show an increase in the number of cases.
   c. Comment on the sensational ways that some tabloid media portray infectious diseases to the public.
2. Using the index, look up each disease shown on figure 1.4 and see which ones could be prevented by vaccines or cured with drugs. Are there other ways (besides vaccines) to prevent any of these?
3. What events, discoveries, or inventions were probably the most significant in the development of microbiology and why?
4. Can you develop a scientific hypothesis and means of testing the cause of stomach ulcers? (Is it caused by an infection? By too much acid? By a genetic disorder?)
5. Where do you suppose viruses came from? Why do they require the host's cellular machinery?
6. Construct the scientific name of a newly discovered species of bacterium, using your name, a pet's name, a place, or aunique characteristic. Be sure to use proper notation and endings.
7. Archaea are often found in hot, sulfuric, acidic, salty habitats, much like the early earth's conditions. Speculate on the origin of life, especially as it relates to the archaea.

Visual Understanding

1. **Figure 1.1.** Look at the blue bar (the time that prokaryotes have been on earth) and at the pink arrow (the time that humans appeared). Speculate on the probability that we will be able to completely disinfect our planet or prevent all microbial diseases.

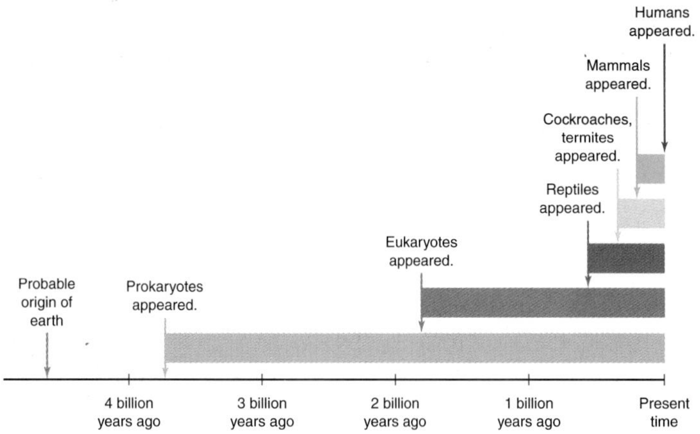

Internet Search Topics

The ARIS website that accompanies this textbook includes tutorials, animations, practice quizzes, downloadable study help for your portable media player and more. Just visit www.aris.mhhe.com and select your subject.

1. Using a search engine on the World Wide Web, search for the phrase *emerging diseases.* Adding terms like WHO and CDC will refine your search and take you to several appropriate websites. List the top 10 emerging diseases in the United States and worldwide.
2. Go to: www.aris.mhhe.com, and click on "microbiology" and then this textbook's author/title. Go to Chapter 1, access the URLs listed under Internet Search Topics, and research the following:
   a. Explore the "trees of life." Compare the main relationships among the three major domains.
   b. Observe the comparative sizes of microbes arrayed on the head of a pin.
   c. Look at the discussion of biology prefixes and suffixes. A little time spent here could make the rest of your microbiology studies much smoother.

CHAPTER 2

The Chemistry of Biology

CASE FILE 2

Berkeley Pit Lake in Butte, Montana, sits on the site of an abandoned open-pit copper mine. It would seem polite to describe the site as an environmental disaster. At a pH of 2.5, the more than 30 billion gallons of water in the lake are highly acidic. High levels of dissolved iron, copper, zinc, and aluminum compounds make the situation even worse. In a recent study involving the site, Dr. Andrea Stierle from the Department of Chemistry of the University of Montana and her colleagues found a compound called berkelic acid. This compound, known as a spiroketal, is produced by a member of the genus *Penicillium* that lives in the lake. Studies of this unique compound show that it has a selective inhibitory activity against OVCAR-3 cells. These cells are an ovarian cancer cell line belonging to the National Cancer Institute's cell library.

- *What name is given to organisms that live in such a seemingly inhospitable environment?*

Case File 2 Wrap-Up appears on page 33.

CHAPTER OVERVIEW

- The understanding of living cells and processes is enhanced by a knowledge of chemistry.
- The structure and function of all matter in the universe are based on atoms.
- Atoms have unique structures and properties that allow chemical reactions to occur.
- Atoms contain protons, neutrons, and electrons in combinations to form elements.
- Living things are composed of approximately 25 different elements.
- Elements interact to form bonds that result in molecules and compounds with different characteristics than the elements that form them.
- Atoms and molecules undergo chemical reactions such as oxidation/reduction, ionization, and dissolution.
- The properties of carbon have been critical in forming macromolecules of life such as proteins, fats, carbohydrates, and nucleic acids.
- The structure and shape of a macromolecule dictate its functions.
- Cells carry out fundamental activities of life, such as growth, metabolism, reproduction, synthesis, and transport, that are all essentially chemical reactions on a grand scale.

2.1 Atoms, Bonds, and Molecules: Fundamental Building Blocks

The universe is composed of an infinite variety of substances existing in the gaseous, liquid, and solid states. All such tangible materials that occupy space and have mass are called **matter.** The organization of matter—whether air, rocks, or bacteria—begins with individual building blocks called atoms. An **atom** is defined as a tiny particle that cannot be subdivided into smaller substances without losing its properties. Even in a science dealing with very small things, an atom's minute size is striking; for example, an oxygen atom is only 0.0000000013 mm (0.0013 nm) in diameter, and 1 million of them in a cluster would barely be visible to the naked eye.

Although scientists have not directly observed the detailed structure of an atom, the exact composition of atoms has been well established by extensive physical analysis using sophisticated instruments. In general, an atom derives its properties from a combination of subatomic particles called **protons** (p^+), which are positively charged; **neutrons** (n^0), which have no charge (are neutral); and **electrons** (e^-), which are negatively charged. The relatively larger protons and neutrons make up a central core, or *nucleus,*[1] that is surrounded by one or more electrons **(figure 2.1).** The nucleus makes up the larger mass (weight) of the atom, whereas the electron region accounts for the greater volume. To get a perspective on proportions, consider this: If an atom were the size of a football stadium, the nucleus would be about the size of a marble! The stability of atomic structure is largely maintained by (1) the mutual attraction of the protons and electrons (opposite charges attract each other) and (2) the exact balance of proton number and electron number, which causes the opposing charges to cancel each other out. At least in theory, then, isolated intact atoms do not carry a charge.

1. Be careful not to confuse the nucleus of an atom with the nucleus of a cell (discussed later).

Different Types of Atoms: Elements and Their Properties

All atoms share the same fundamental structure. All protons are identical, all neutrons are identical, and all electrons are identical. But when these subatomic particles come together in specific, varied combinations, unique types of atoms called **elements** result. Each element has a characteristic atomic structure and predictable chemical behavior. To date, about 115 elements, both naturally occurring and artificially produced by physicists, have been described. By convention, an element is assigned a distinctive name with an abbreviated shorthand symbol. The elements are often depicted in a periodic table. **Table 2.1** lists some of

Figure 2.1 Models of atomic structure.
(a) Three-dimensional models of hydrogen and carbon that approximate their actual structure. The nucleus is surrounded by electrons in orbitals that occur in levels called shells. Hydrogen has just one shell and one orbital. Carbon has two shells and four orbitals; the shape of the outermost orbitals is paired lobes rather than circles or spheres. **(b)** Simple models of the same atoms make it easier to show the numbers and arrangements of shells and electrons and the numbers of protons and neutrons in the nucleus. (Not to accurate scale.)

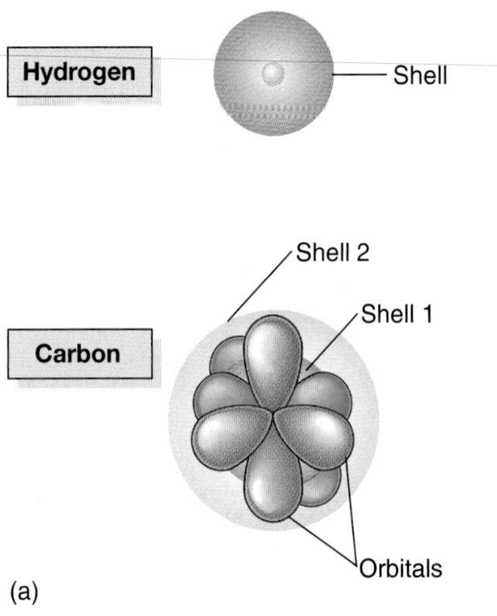

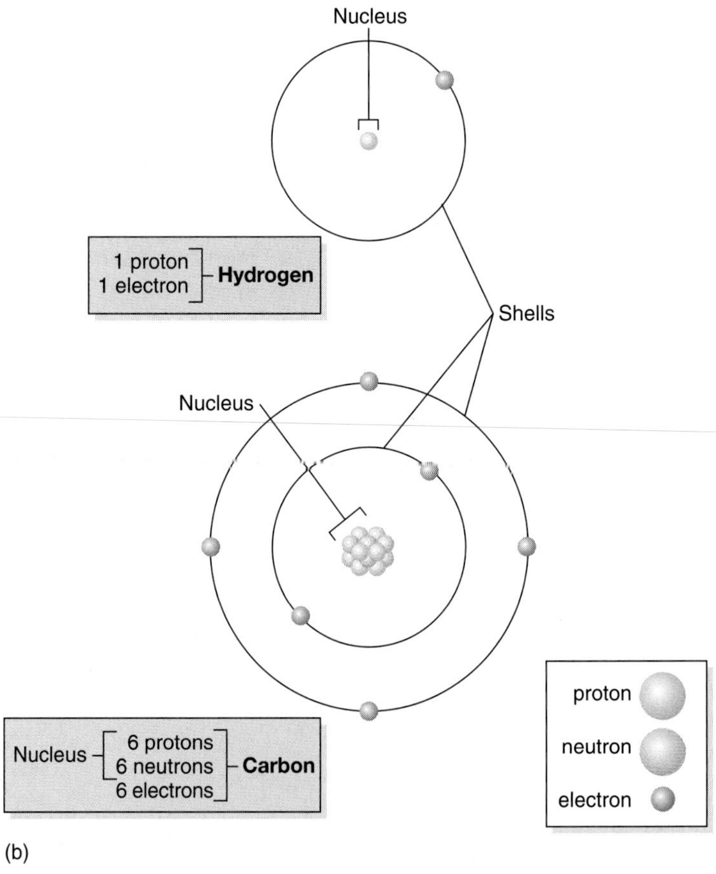

TABLE 2.1 The Major Elements of Life and Their Primary Characteristics

Element	Atomic Symbol*	Atomic Mass**	Examples of Ionized Forms	*Significance in Microbiology
Calcium	Ca	40.1	Ca^{2+}	Part of outer covering of certain shelled amoebas; stored within bacterial spores
Carbon	C	12.0	CO_3^{-2}	Principal structural component of biological molecules
Carbon•	C-14	14.0		Radioactive isotope used in dating fossils
Chlorine	Cl	35.5	Cl^-	Component of disinfectants, used in water purification
Cobalt	Co	58.9	Co^{2+}, Co^{3+}	Trace element needed by some bacteria to synthesize vitamins
Cobalt•	Co-60	60		An emitter of gamma rays; used in food sterilization; used to treat cancer
Copper	Cu	63.5	Cu^+, Cu^{2+}	Necessary to the function of some enzymes; Cu salts are used to treat fungal and worm infections
Hydrogen	H	1	H^+	Necessary component of water and many organic molecules; H_2 gas released by bacterial metabolism
Hydrogen•	H3	3		Has 2 neutrons; radioactive; used in clinical laboratory procedures
Iodine	I	126.9	I^-	A component of antiseptics and disinfectants; used in the Gram stain
Iodine•	I-131, I-125	131, 125		Radioactive isotopes for diagnosis and treatment of cancers
Iron	Fe	55.8	Fe^{2+}, Fe^{3+}	Necessary component of respiratory enzymes; required by some microbes to produce toxin
Magnesium	Mg	24.3	Mg^{2+}	A trace element needed for some enzymes; component of chlorophyll pigment
Manganese	Mn	54.9	Mn^{2+}, Mn^{3+}	Trace element for certain respiratory enzymes
Nitrogen	N	14.0	NO_3^-	Component of all proteins and nucleic acids; the major atmospheric gas
Oxygen	O	16.0		An essential component of many organic molecules; molecule used in metabolism by many organisms
Phosphorus	P	31	PO_4^{3-}	A component of ATP, nucleic acids, cell membranes; stored in granules in cells
Phosphorus•	P-32	32		Radioactive isotope used as a diagnostic and therapeutic agent
Potassium	K	39.1	K^+	Required for normal ribosome function and protein synthesis; essential for cell membrane permeability
Sodium	Na	23.0	Na^+	Necessary for transport; maintains osmotic pressure; used in food preservation
Sulfur	S	32.1	SO_4^{-2}	Important component of proteins; makes disulfide bonds; storage element in many bacteria
Zinc	Zn	65.4	Zn^{++}	An enzyme cofactor; required for protein synthesis and cell division; important in regulating DNA

*Based on the Latin name of the element. The first letter is always capitalized; if there is a second letter, it is always lowercased.

**The atomic mass or weight is equal to the average mass number for the isotopes of that element.

the elements common to biological systems, their atomic characteristics, and some of the natural and applied roles they play.

The Major Elements of Life and Their Primary Characteristics

The unique properties of each element result from the numbers of protons, neutrons, and electrons it contains, and each element can be identified by certain physical measurements.

Isotopes are variant forms of the same element that differ in the number of neutrons. These multiple forms occur naturally in certain proportions. Carbon, for example, exists primarily as carbon 12 with 6 neutrons; but a small amount (about 1%) is carbon 13 with 7 neutrons and carbon 14 with 8 neutrons. Although isotopes have virtually the same chemical properties, some of them have unstable nuclei that spontaneously release energy in the form of radiation. Such *radioactive isotopes* play a role in a number of research and medical applications. Because they emit detectable signs, they can be used to trace the position of key atoms or molecules in chemical reactions, they are tools in diagnosis and treatment, and they are even applied in sterilization procedures (see ionizing radiation in chapter 11). Another application of isotopes is in dating fossils and other ancient materials.

Electron Orbitals and Shells

The structure of an atom can be envisioned as a central nucleus surrounded by a "cloud" of electrons that constantly rotate about the nucleus in pathways (see figure 2.1). The pathways, called **orbitals,** are not actual objects or exact locations but represent volumes of space in which an electron is likely to be found. Electrons occupy energy shells, proceeding from the lower-level energy electrons nearest the nucleus to the higher-level energy electrons in the farthest orbitals.

Electrons fill the orbitals and shells in *pairs,* starting with the shell nearest the nucleus. The first shell contains one orbital and a maximum of 2 electrons; the second shell has four orbitals and up to 8 electrons; the third shell with nine orbitals can hold up to 18 electrons; and the fourth shell with 16 orbitals contains up to 32 electrons. The number of orbitals and shells and how completely they are filled depend on the numbers of electrons, so that each element will have a unique pattern. For example, helium has only a filled first shell of 2 electrons; oxygen has a filled first shell and a partially filled second shell of 6 electrons; and magnesium has a filled first shell, a filled second one, and a third shell that fills only one orbital, so is nearly empty. As we will see, the chemical properties of an element are controlled mainly by the distribution of electrons in the outermost shell. Figure 2.1 and **figure 2.2** present various simplified models of atomic structure and electron maps.

CHECKPOINT

- Protons (p^+) and neutrons (n^0) make up the nucleus of an atom. Electrons (e^-) orbit the nucleus.
- All elements are composed of atoms but differ in the numbers of protons, neutrons, and electrons they possess.
- Isotopes are varieties of one element that contain the same number of protons but different numbers of neutrons.
- The number of electrons in an element's outermost orbital (compared with the total number possible) determines the element's chemical properties and reactivity.

Bonds and Molecules

Most elements do not exist naturally in pure, uncombined form but are bound together as molecules and compounds. A **molecule** is a distinct chemical substance that results from the combination of two or more atoms. Some molecules such as oxygen (O_2) and nitrogen gas (N_2) consist of atoms of the same element. Molecules that are combinations of two or more *different* elements are termed **compounds.** Compounds such as water (H_2O) and biological molecules (proteins, sugars, fats) are the predominant substances in living systems. When atoms bind together in molecules, they lose the properties of the atom and take on the properties of the combined substance.

The **chemical bonds** of molecules and compounds result when two or more atoms share, donate (lose), or accept (gain) electrons **(figure 2.3).** The number of electrons in the outermost shell of an element is known as its **valence.** The valence determines the degree of reactivity and the types of bonds an element can make. Elements with a filled outer orbital are relatively stable because they have no extra electrons to share with or donate to other atoms. For example, helium has one filled shell, with no tendency either to give up electrons or to take them from other elements, making it a stable, inert (nonreactive) gas. Elements with partially filled outer orbitals are less stable and are more apt to form some sort of bond. Many chemical reactions are based on the tendency of atoms with unfilled outer shells to gain greater stability by achieving, or at least approximating, a filled outer shell. For example, an atom such as oxygen that can accept 2 additional electrons will bond readily with atoms (such as hydrogen) that can share or donate electrons. We explore some additional examples of the basic types of bonding in the following section.

In addition to reactivity, the number of electrons in the outer shell also dictates the number of chemical bonds an atom can make. For instance, hydrogen can bind with one other atom, oxygen can bind with up to two other atoms, and carbon can bind with four.

Covalent Bonds and Polarity: Molecules with Shared Electrons

Covalent (cooperative valence) **bonds** form between atoms that share electrons rather than donating or receiving them.

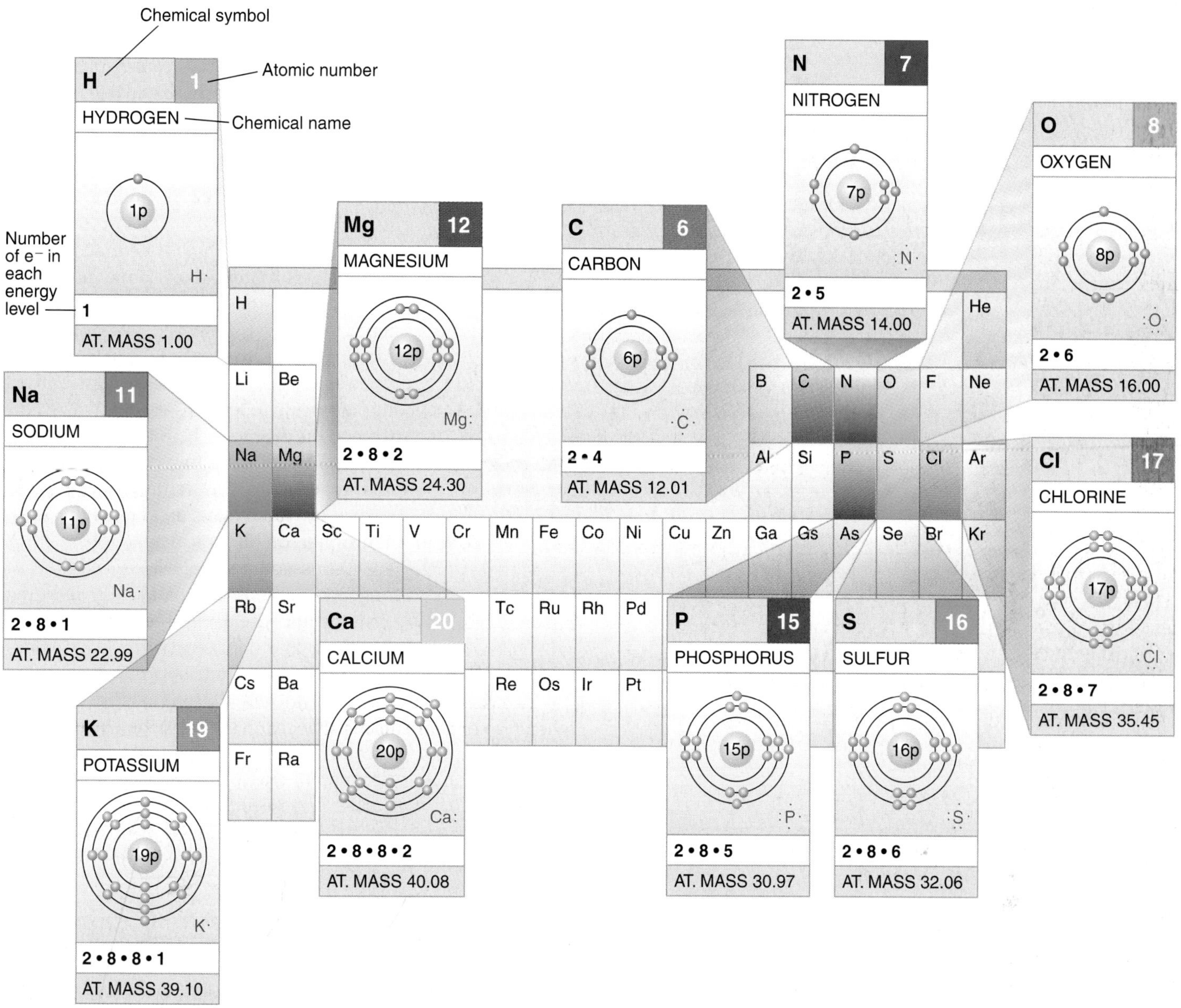

Figure 2.2 Examples of biologically important atoms.
Simple models show how the shells are filled by electrons as the atomic numbers increase. Notice that these elements have incompletely filled outer shells since they have less than 8 electrons. Chemists depict elements in shorthand form (red Lewis structures) that indicate only the valence electrons, since these are the electrons involved in chemical bonds. In the background is a partial display of the periodic table of elements, showing the position of these elements.

A simple example is hydrogen gas (H_2), which consists of two hydrogen atoms. A hydrogen atom has only a single electron, but when two of them combine, each will bring its electron to orbit about both nuclei, thereby approaching a filled orbital (2 electrons) for both atoms and thus creating a single **covalent** bond **(figure 2.4*a*).** Covalent bonding also occurs in oxygen gas (O_2) but with a difference. Because each atom has 2 electrons to share in this molecule, the combination creates two pairs of shared electrons, also known as a double covalent bond **(figure 2.4*b*).** The majority of the molecules associated with living things are composed of single and double covalent bonds between the most common biological elements (carbon, hydrogen, oxygen, nitrogen, sulfur, and phosphorus), which are discussed in more depth in chapter 7. Double bonds in molecules and compounds introduce more rigidity than single bonds. A slightly more complex pattern of covalent bonding is shown for methane gas (CH_4) in **figure 2.4*c*.**

Other effects of bonding result in differences in polarity. When atoms of different electronegativity[2] form covalent

2. Electronegativity—the ability to attract electrons.

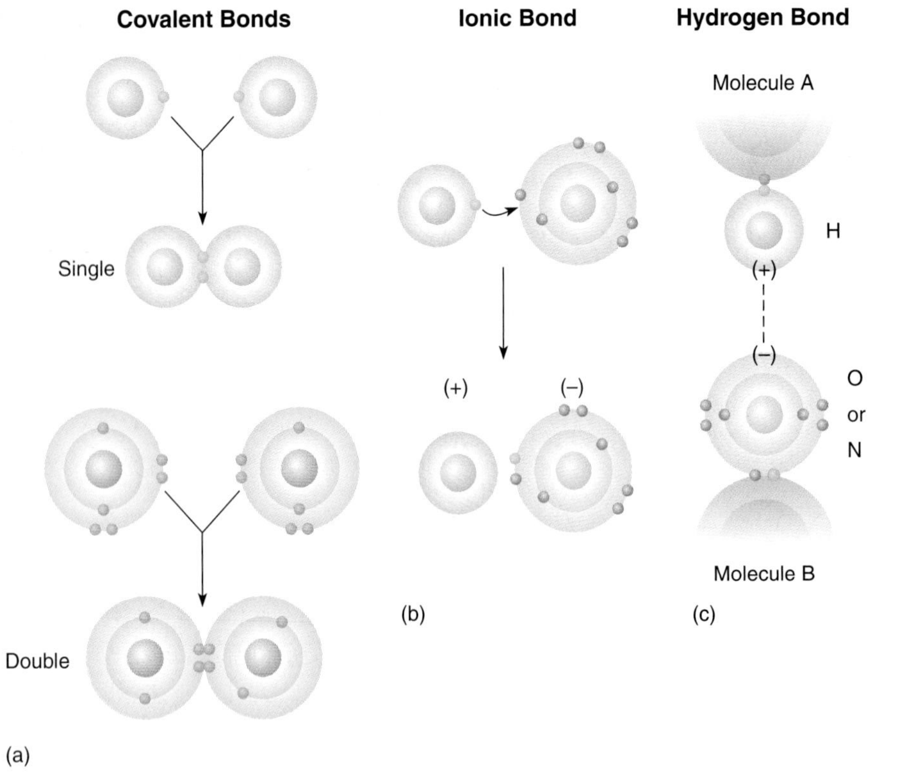

Figure 2.3 General representation of three types of bonding.
(a) Covalent bonds, both single and double. **(b)** Ionic bond. **(c)** Hydrogen bond. Note that hydrogen bonds are represented in models and formulas by dotted lines, as shown in **(c)**.

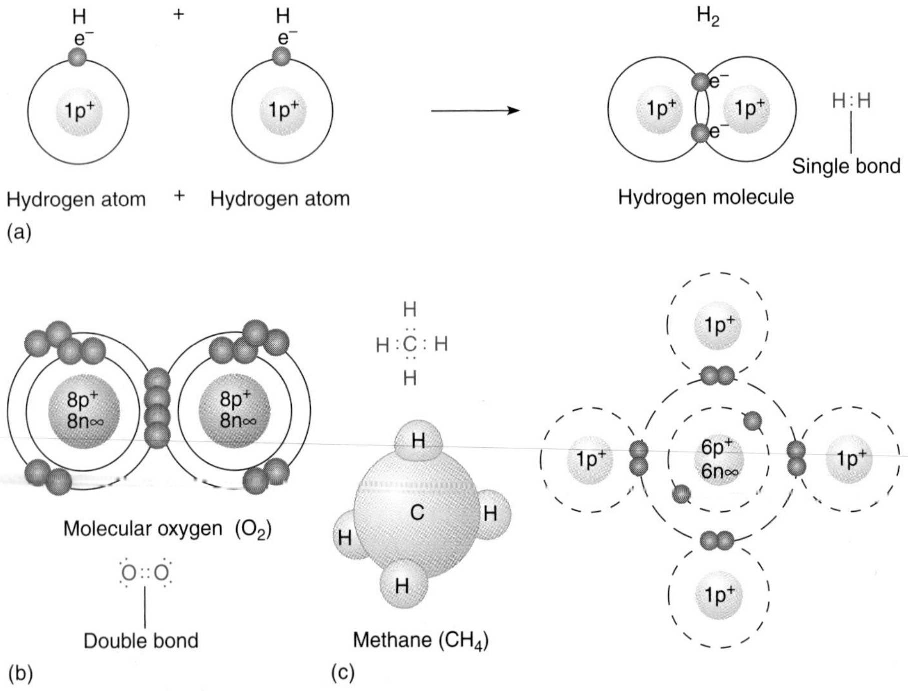

Figure 2.4 Examples of molecules with covalent bonding.
(a) A hydrogen molecule is formed when two hydrogen atoms share their electrons and form a single bond. **(b)** In a double bond, the outer orbitals of two oxygen atoms overlap and permit the sharing of 4 electrons (one pair from each) and the saturation of the outer orbital for both. **(c)** Simple, three-dimensional, and working models of methane. Note that carbon has 4 electrons to share and hydrogens each have one, thereby completing the shells for all atoms in the compound, and creating 4 single bonds.

bonds, the electrons are not shared equally and may be pulled more toward one atom than another. This pull causes one end of a molecule to assume a partial negative charge and the other end to assume a partial positive charge. A molecule with such an asymmetrical distribution of charges is termed **polar** and has positive and negative poles. Observe the water molecule shown in **figure 2.5** and note that, because the oxygen atom is larger and has more protons than the hydrogen atoms, it will tend to draw the shared electrons with greater force toward its nucleus. This unequal force causes the oxygen part of the molecule to express a negative charge (due to the electrons' being attracted there) and the hydrogens to express a positive charge (due to the protons). The polar nature of water plays an extensive role in a number of biological reactions, which are discussed later. Polarity is a significant property of many large molecules in living systems and greatly influences both their reactivity and their structure.

When covalent bonds are formed between atoms that have the same or similar electronegativity, the electrons are shared equally between the two atoms. Because of this balanced distribution, no part of the molecule has a greater attraction for the electrons. This sort of electrically neutral molecule is termed **nonpolar.**

Ionic Bonds: Electron Transfer Among Atoms

In reactions that form **ionic bonds,** electrons are transferred completely from one atom to another and are not shared. These reactions invariably occur between atoms with valences that complement each other, meaning that one atom has an unfilled shell that will readily accept electrons and the other atom has an unfilled shell that will readily lose electrons. A striking example is the reaction that occurs between sodium (Na) and chlorine (Cl). Elemental sodium is a soft, lustrous metal so reactive that it can burn flesh, and molecular chlorine is a very poisonous yellow gas. But when the two are combined, they form sodium chloride[3] (NaCl)—the familiar

3. In general, when a salt is formed, the ending of the name of the negatively charged ion is changed to *-ide*.

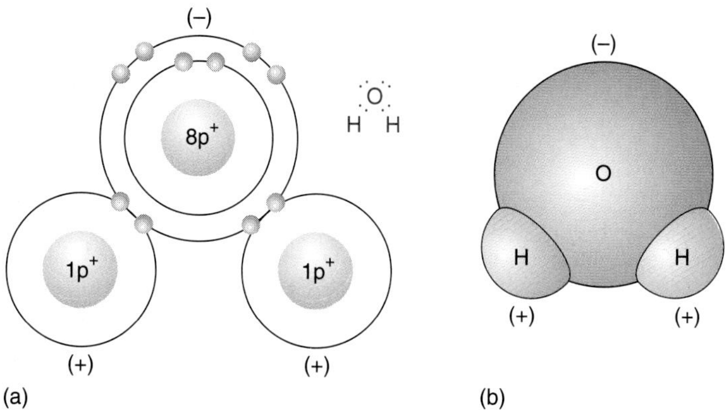

Figure 2.5 Polar molecule.
(a) A simple model and **(b)** a three-dimensional model of a water molecule indicate the polarity, or unequal distribution, of electrical charge, which is caused by the pull of the shared electrons toward the oxygen side of the molecule.

nontoxic table salt—a compound with properties quite different from either parent element **(figure 2.6)**.

How does this transformation occur? Sodium has 11 electrons (2 in shell one, 8 in shell two, and only 1 in shell three), so it is 7 short of having a complete outer shell. Chlorine has 17 electrons (2 in shell one, 8 in shell two, and 7 in shell three), making it 1 short of a complete outer shell. These two atoms are very reactive with one another, because a sodium atom will readily donate its single electron and a chlorine atom will avidly receive it. (The reaction is slightly more involved than a single sodium atom's combining with a single chloride atom **(Insight 2.1)**, but this complexity does not detract from the fundamental reaction as described here.) The outcome of this reaction is not many single, isolated molecules of NaCl but rather a solid crystal complex that interlinks millions of sodium and chloride ions (**figure 2.6*c*** and ***d***).

Ionization: Formation of Charged Particles Molecules with intact ionic bonds are electrically neutral, but they can produce charged particles when dissolved in a liquid called

CASE FILE 2 *WRAP-UP*

Dr. Stierle and her colleagues have been studying this EPA Superfund site for more than 10 years. In 1995, they discovered a diverse population of microbes living in the heavily contaminated water. These bacteria, fungi, and protozoans are described as extremophiles due to their ability to live in places seemingly unfit for habitation. Stierle's research has identified a number of unusual chemical wastes produced by these microbes. One variety of *Pithomyces* that they found produces a compound that may block migraine headaches. A strain of *Penicillium* produces a chemical that interferes with the growth of lung cancer cells. And this latest compound, berkelic acid, may prove useful in the treatment of ovarian cancer. More studies are needed to determine if these unusual compounds are the result of the extreme environment these organisms have chosen. Another question is whether microorganisms could actually play a role in healing this wasteland, a process called **bioremediation.**

See: Stierle, A. A., Stierle, D. B., and Kelly, K. 2006. Berkelic acid, a novel spiroketal with selective anticancer activity from an acid mine waste fungal extremophile. J. Organic Chemistry *71:5357–5360.*

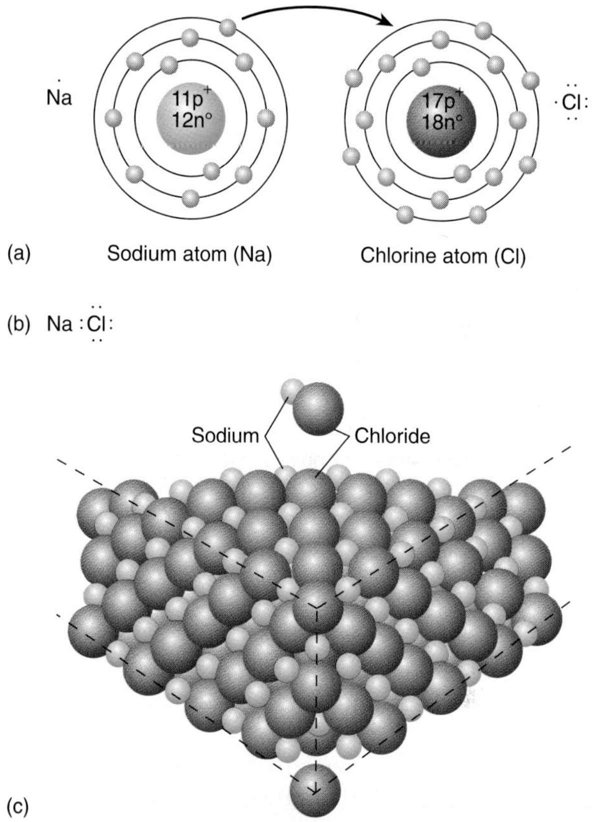

Figure 2.6 Ionic bonding between sodium and chlorine.
(a) When the two elements are placed together, sodium loses its single outer orbital electron to chlorine, thereby filling chlorine's outer shell. **(b)** Simple model of ionic bonding. **(c)** Sodium and chloride ions form large molecules, or crystals, in which the two atoms alternate in a definite, regular, geometric pattern. **(d)** Note the cubic nature of NaCl crystals at the macroscopic level.

INSIGHT 2.1 *Discovery*

Redox: Electron Transfer and Oxidation-Reduction Reactions

The metabolic work of cells, such as synthesis, movement, and digestion, revolves around energy exchanges and transfers. The management of energy in cells is almost exclusively dependent on chemical rather than physical reactions because most cells are far too delicate to operate with heat, radiation, and other more potent forms of energy. The outer-shell electrons are readily portable and easily manipulated sources of energy. It is in fact the movement of electrons from molecule to molecule that accounts for most energy exchanges in cells. Fundamentally, then, a cell must have a supply of atoms that can gain or lose electrons if they are to carry out life processes.

The phenomenon in which electrons are transferred from one atom or molecule to another is termed an **oxidation** and **reduction** (shortened to **redox**) reaction. The term *oxidation* was originally adopted for reactions involving the addition of oxygen. In current usage, the term oxidation can include any reaction causing electron loss, regardless of the involvement of oxygen. By comparison *reduction* is any reaction that causes an atom to receive electrons, because all redox reactions occur in pairs.

To analyze the phenomenon, let us again review the production of NaCl but from a different standpoint.

When these two atoms react to form sodium chloride, a sodium atom gives up an electron to a chlorine atom. During this reaction, sodium is oxidized because it loses an electron, and chlorine is reduced because it gains an electron (figure 2.6). With this system, an atom such as sodium that can donate electrons and thereby reduce another atom is a **reducing agent.** An atom that can receive extra electrons and thereby oxidize another molecule is an **oxidizing agent.** You may find this concept easier to keep straight if you think of redox agents as partners: The reducing partner gives its electrons away and is oxidized; the oxidizing partner receives the electrons and is reduced. (A mnemonic device to keep track of this is *LEO* says *GER: Lose Electrons Oxidized; Gain Electrons Reduced.*)

Redox reactions are essential to many of the biochemical processes discussed in chapter 8. In cellular metabolism, electrons are frequently transferred from one molecule to another as described here. In other reactions, oxidation and reduction occur with the transfer of a hydrogen atom (a proton and an electron) from one compound to another.

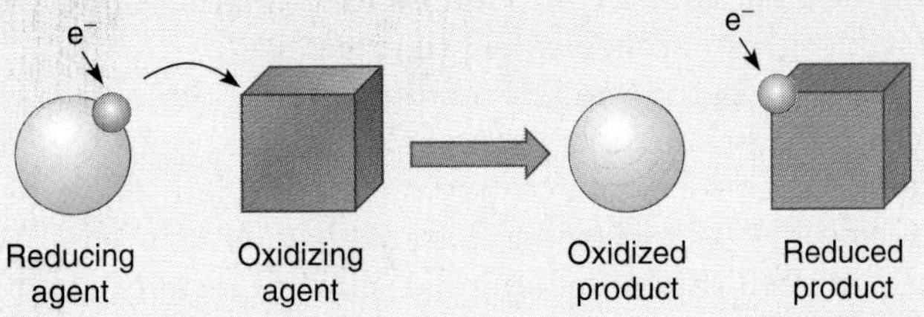

Simplified diagram of the exchange of electrons during an oxidation-reduction reaction.

a solvent. This phenomenon, called **ionization,** occurs when the ionic bond is broken and the atoms dissociate (separate) into unattached, charged particles called **ions (figure 2.7).** To illustrate what gives a charge to ions, let us look again at the reaction between sodium and chlorine. When a sodium atom reacts with chlorine and loses 1 electron, the sodium is left with one more proton than electrons. This imbalance produces a positively charged sodium ion (Na^+). Chlorine, on the other hand, has gained 1 electron and now has 1 more electron than protons, producing a negatively charged ion (Cl^-). Positively charged ions are termed **cations,** and negatively charged ions are termed **anions.** (A good mnemonic device is to think of the "t" in cation as a plus (+) sign and the first "n" in anion as a negative (−) sign.) Substances such as salts, acids, and bases that release ions when dissolved in water are termed **electrolytes** because their charges enable them to conduct an electrical current. Owing to the general rule that particles of like charge repel each other and those of opposite charge attract each other, we can expect ions to interact electrostatically with other ions and polar molecules. Such interactions are important in many cellular chemical reactions, in the formation of solutions, and in the reactions microorganisms have with dyes. The transfer of electrons from one molecule to another constitutes a significant mechanism by which biological systems store and release energy.

Hydrogen Bonding Some types of bonding do not involve sharing, losing, or gaining electrons but instead are due to attractive forces between nearby molecules or atoms. One such bond is a **hydrogen bond,** a weak type of bond that forms between a hydrogen covalently bonded to one molecule and an oxygen or nitrogen atom on the same molecule or on a different molecule. Because hydrogen in a covalent bond tends to be positively charged, it will attract a nearby negatively charged atom and form an easily disrupted bridge with it. This type of bonding is usually represented in molecular models with a dotted line. A simple example of hydrogen bonding occurs between water molecules **(figure 2.8).** More extensive hydrogen bonding is partly responsible for the structure and stability of proteins and nucleic acids, as you will see later on.

Other similar noncovalent associations between molecules are the **van der Waals forces.** These weak attractions occur between molecules that demonstrate low levels of polarity. Neighboring groups with slight attractions will interact and remain associated. These forces are an essential factor in maintaining the cohesiveness of large molecules with many packed atoms.

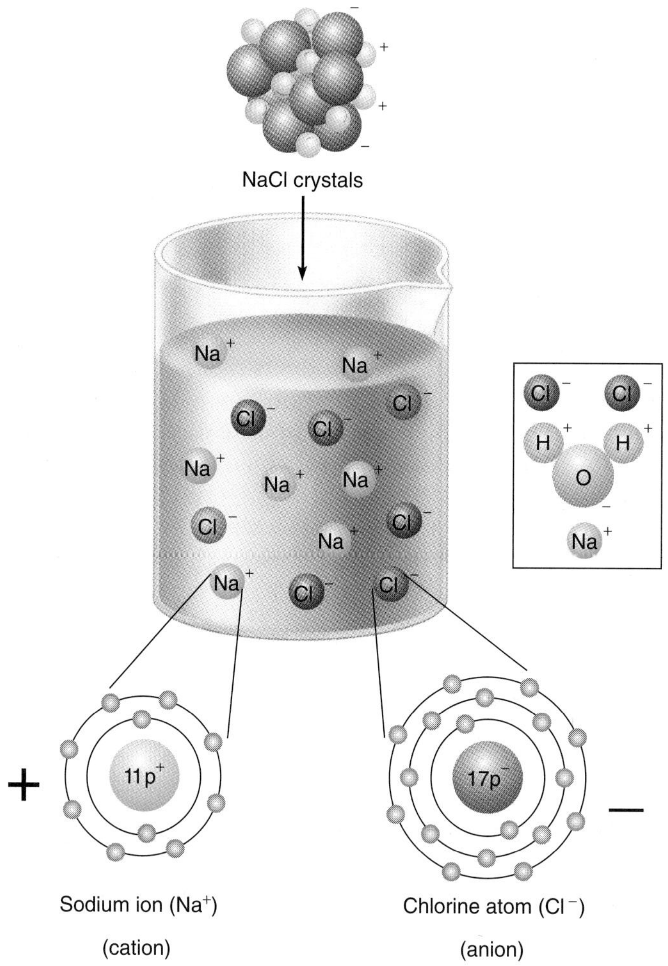

Figure 2.7 Ionization.
When NaCl in the crystalline form is added to water, the ions are released from the crystal as separate charged particles (cations and anions) into solution. (See also figure 2.11.) In this solution, Cl^- ions are attracted to the hydrogen component of water, and Na^+ ions are attracted to the oxygen (box).

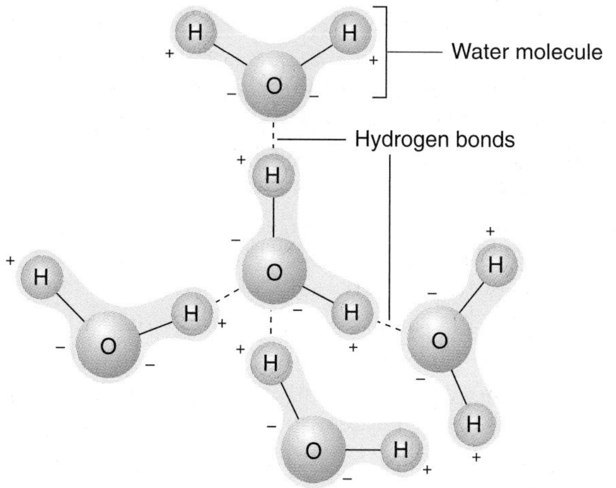

Figure 2.8 Hydrogen bonding in water.
Because of the polarity of water molecules, the negatively charged oxygen end of one water molecule is weakly attracted to the positively charged hydrogen end of an adjacent water molecule.

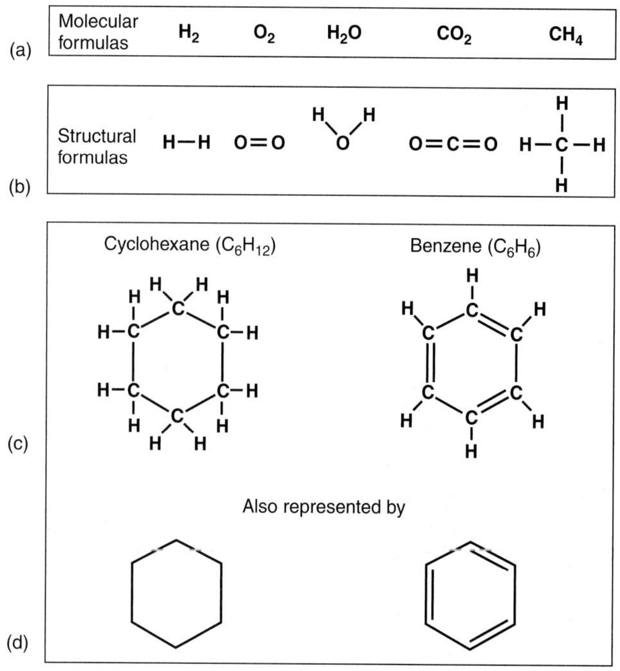

Figure 2.9 Comparison of molecular and structural formulas.
(a) Molecular formulas provide a brief summary of the elements in a compound. **(b)** Structural formulas clarify the exact relationships of the atoms in the molecule, depicting single bonds by a single line and double bonds by two lines. **(c)** In structural formulas of organic compounds, cyclic or ringed compounds may be completely labeled, or **(d)** they may be presented in a shorthand form in which carbons are assumed to be at the angles and attached to hydrogens. See figure 2.14 for structural formulas of three sugars with the same molecular formula, $C_6H_{12}O_6$.

Chemical Shorthand: Formulas, Models, and Equations

The atomic content of molecules can be represented by a few convenient formulas. We have already been using the molecular formula, which concisely gives the atomic symbols and the number of the elements involved in subscript (CO_2, H_2O). More complex molecules such as glucose ($C_6H_{12}O_6$) can also be symbolized this way, but this formula is not unique, because fructose and galactose also share it. Molecular formulas are useful, but they only summarize the atoms in a compound; they do not show the position of bonds between atoms. For this purpose, chemists use structural formulas illustrating the relationships of the atoms and the number and types of bonds **(figure 2.9).** Other structural models present the three-dimensional appearance of a molecule, illustrating the orientation of atoms (differentiated by color codes) and the molecule's overall shape **(figure 2.10).**

The printed page tends to make molecules appear static, but this picture is far from correct, because molecules are capable of changing through chemical reactions. For ease in tracing chemical exchanges between atoms or molecules, and to derive some sense of the dynamic character of reactions, chemists use shorthand equations containing symbols, numbers, and arrows to simplify or summarize

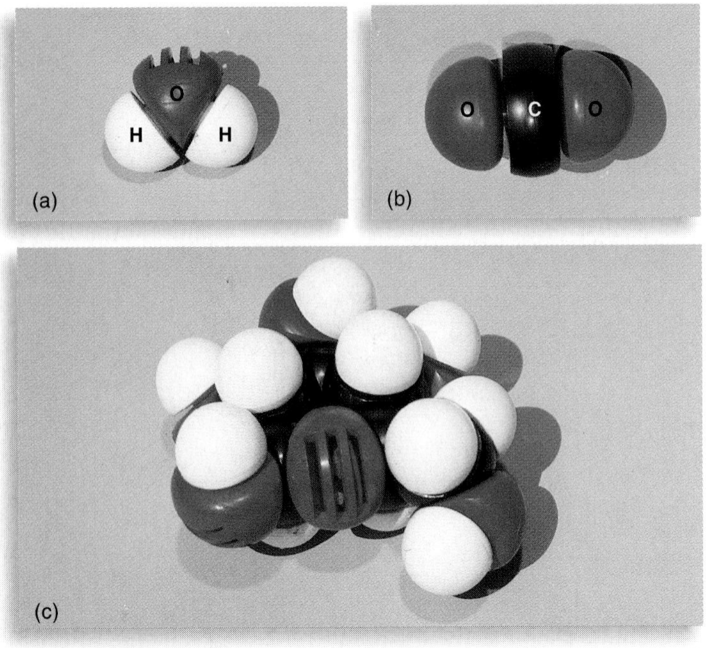

Figure 2.10 Three-dimensional, or space-filling, models of (a) water, (b) carbon dioxide, and (c) glucose.
The red atoms are oxygen, the white ones hydrogen, and the black ones carbon.

the major characteristics of a reaction. Molecules entering or starting a reaction are called **reactants,** and substances left by a reaction are called **products.** In most instances, summary chemical reactions do not give the details of the exchange, in order to keep the expression simple and to save space.

In a *synthesis reaction,* the reactants bond together in a manner that produces an entirely new molecule (reactant A plus reactant B yields product AB). An example is the production of sulfur dioxide, a by-product of burning sulfur fuels and an important component of smog:

$$S + O_2 \rightarrow SO_2$$

Some synthesis reactions are not such simple combinations. When water is synthesized, for example, the reaction does not really involve one oxygen atom combining with two hydrogen atoms, because elemental oxygen exists as O_2 and elemental hydrogen exists as H_2. A more accurate equation for this reaction is:

$$2H_2 + O_2 \rightarrow H_2O$$

The equation for reactions must be balanced—that is, the number of atoms on one side of the arrow must equal the number on the other side to reflect all of the participants in the reaction. To arrive at the total number of atoms in the reaction, multiply the prefix number by the subscript number; if no number is given, it is assumed to be 1.

In *decomposition reactions,* the bonds on a single reactant molecule are permanently broken to release two or more product molecules. One example is the resulting molecules when large nutrient molecules are digested into smaller units; a simpler example can be shown for the common chemical hydrogen peroxide:

$$2H_2O_2 \rightarrow 2H_2O + O_2$$

During *exchange reactions,* the reactants trade portions between each other and release products that are combinations of the two. This type of reaction occurs between acids and bases when they form water and a salt:

$$AB + XY \rightleftharpoons AX + BY$$

The reactions in biological systems can be reversible, meaning that reactants and products can be converted back and forth. These reversible reactions are symbolized with a double arrow, each pointing in opposite directions, as in the preceding exchange reaction. Whether a reaction is reversible depends on the proportions of these compounds, the difference in energy state of the reactants and products, and the presence of **catalysts** (substances that increase the rate of a reaction). Additional reactants coming from another reaction can also be indicated by arrows that enter or leave at the main arrow:

$$X + Y \xrightarrow{CD \curvearrowright C} XYD$$

Solutions: Homogeneous Mixtures of Molecules

A **solution** is a mixture of one or more substances called **solutes** uniformly dispersed in a dissolving medium called a **solvent.** An important characteristic of a solution is that the solute cannot be separated by filtration or ordinary settling. The solute can be gaseous, liquid, or solid, and the solvent is usually a liquid. Examples of solutions are salt or sugar dissolved in water and iodine dissolved in alcohol. In general, a solvent will dissolve a solute only if it has similar electrical characteristics as indicated by the rule of solubility, expressed simply as "like dissolves like." For example, water is a polar molecule and will readily dissolve an ionic solute such as NaCl, yet a nonpolar solvent such as benzene will not dissolve NaCl.

Water is the most common solvent in natural systems, having several characteristics that suit it to this role. The polarity of the water molecule causes it to form hydrogen bonds with other water molecules, but it can also interact readily with charged or polar molecules. When an ionic solute such as NaCl crystals is added to water, it is dissolved, thereby releasing Na^+ and Cl^- into solution. Dissolution occurs because Na^+ is attracted to the negative pole of the water molecule and Cl^- is attracted to the positive pole; in this way, they are drawn away from the crystal separately into solution. As it leaves, each ion becomes **hydrated,** which means that it is surrounded by a sphere of water molecules **(figure 2.11).** Molecules such as salt or sugar that attract water to their surface are termed **hydrophilic.** Nonpolar molecules, such as benzene, that repel water are considered **hydrophobic.** A third class of molecules, such as the phospholipids in cell membranes, are considered

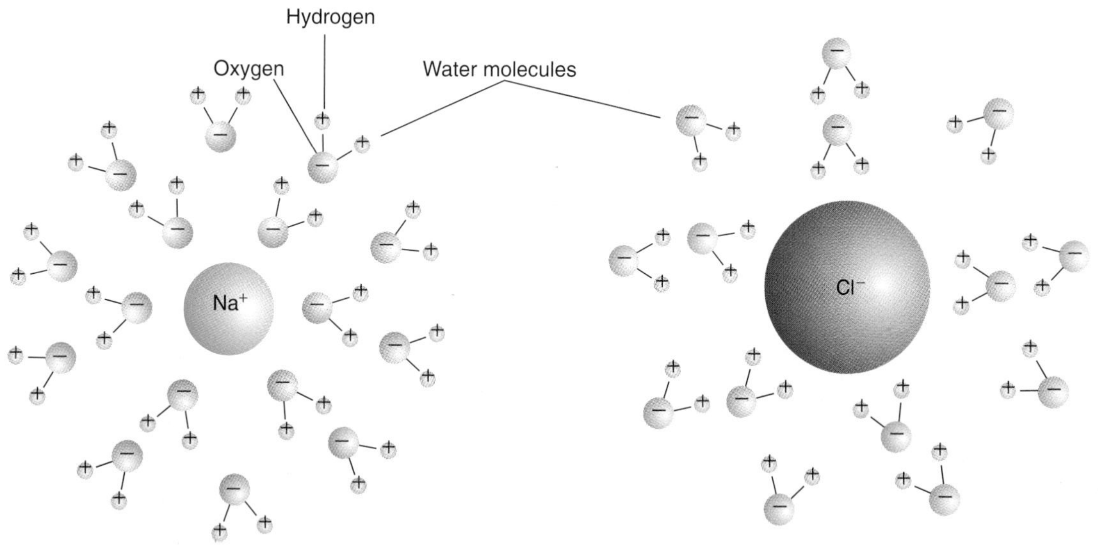

Figure 2.11 Hydration spheres formed around ions in solution.
In this example, a sodium cation attracts the negatively charged region of water molecules, and a chloride anion attracts the positively charged region of water molecules. In both cases, the ions become covered with spherical layers of specific numbers and arrangements of water molecules.

amphipathic because they have both hydrophilic and hydrophobic properties.

Because most biological activities take place in aqueous (water-based) solutions, the concentration of these solutions can be very important (see chapter 7). The **concentration** of a solution expresses the amount of solute dissolved in a certain amount of solvent. It can be calculated by weight, volume, or percentage. A common way to calculate percentage of concentration is to use the weight of the solute, measured in grams (g), dissolved in a specified volume of solvent, measured in milliliters (ml). For example, dissolving 3 g of NaCl in 100 ml of water produces a 3% solution; dissolving 30 g in 100 ml produces a 30% solution; and dissolving 3 g in 1,000 ml (1 liter) produces a 0.3% solution.

A common way to express concentration of biological solutions is by its molar concentration, or *molarity* (M). A standard molar solution is obtained by dissolving one *mole,* defined as the molecular weight of the compound in grams, in 1 liter (1,000 ml) of solution. To make a 1 mole solution of sodium chloride, we would dissolve 58 grams of NaCl to give 1 liter of solution; a 0.1 mole solution would require 5.8 grams of NaCl in 1 liter of solution.

Acidity, Alkalinity, and the pH Scale

Another factor with far-reaching impact on living things is the concentration of acidic or basic solutions in their environment. To understand how solutions develop acidity or basicity, we must look again at the behavior of water molecules. Hydrogens and oxygen tend to remain bonded by covalent bonds, but in certain instances, a single hydrogen can break away as the ionic form (H^+), leaving the remainder of the molecule in the form of an OH^- ion. The H^+ ion is positively charged because it is essentially a hydrogen ion that has lost its electron; the OH^- is negatively charged because it remains in possession of that electron. Ionization of water is constantly occurring, but in pure water containing no other ions, $H^+$and OH^- are produced in equal amounts, and the solution remains neutral. By one definition, a solution is considered **acidic** when a component dissolved in water (acid) releases excess hydrogen ions[4] (H^+); a solution is **basic** when a component releases excess hydroxyl ions (OH^-), so that there is no longer a balance between the two ions.

To measure the acid and base concentrations of solutions, scientists use the **pH** scale, a graduated numerical scale that ranges from 0 (the most acidic) to 14 (the most basic). This scale is a useful standard for rating relative acidity and basicity; use **figure 2.12** to familiarize yourself with the pH readings of some common substances. It is not an arbitrary scale but actually a mathematical derivation based on the negative logarithm (reviewed in appendix A) of the concentration of H^+ ions in moles per liter (symbolized as $[H^+]$) in a solution, represented as:

$$pH = -\log[H^+]$$

Acidic solutions have a greater concentration of H^+ than OH^-, starting with pH 0, which contains 1.0 mole H^+/liter. Each of the subsequent whole-number readings in the scale changes in $[H^+]$ by a tenfold reduction, so that pH 1 contains [0.1 mole H^+/liter], pH 2 contains [0.01 mole H^+/liter], and so on, continuing in the same manner up to pH 14, which contains [0.00000000000001 mole H^+/liter]. These same concentrations can be represented more manageably

4. Actually, it forms a hydronium ion (H_3O^+), but for simplicity's sake, we will use the notation of H^+.

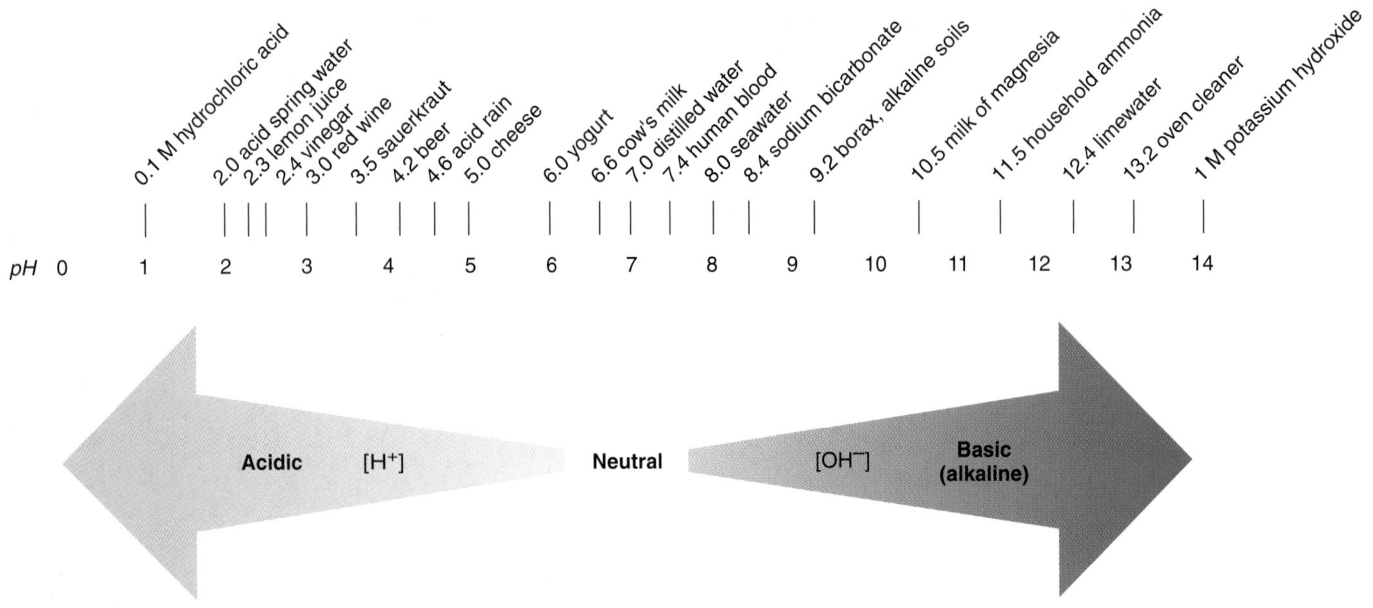

Figure 2.12 The pH scale.
Shown are the relative degrees of acidity and basicity and the approximate pH readings for various substances.

by exponents: pH 2 has an $[H^+]$ of 10^{-2} mole, and pH 14 has an $[H^+]$ of 10^{-14} mole **(table 2.2).** It is evident that the pH units are derived from the exponent itself. Even though the basis for the pH scale is $[H^+]$, it is important to note that, as the $[H^+]$ in a solution decreases, the $[OH^-]$ increases in direct proportion. At midpoint—pH 7, or neutrality—the concentrations are exactly equal and neither predominates, this being the pH of pure water previously mentioned.

TABLE 2.2 Hydrogen Ion and Hydroxide Ion Concentrations at a Given pH

Moles/Liter of Hydrogen Ions	Logarithm	pH	Moles/Liter of OH^-
1.0	10^{-0}	0	10^{-14}
0.1	10^{-1}	1	10^{-13}
0.01	10^{-2}	2	10^{-12}
0.001	10^{-3}	3	10^{-11}
0.0001	10^{-4}	4	10^{-10}
0.00001	10^{-5}	5	10^{-9}
0.000001	10^{-6}	6	10^{-8}
0.0000001	10^{-7}	7	10^{-7}
0.00000001	10^{-8}	8	10^{-6}
0.000000001	10^{-9}	9	10^{-5}
0.0000000001	10^{-10}	10	10^{-4}
0.00000000001	10^{-11}	11	10^{-3}
0.000000000001	10^{-12}	12	10^{-2}
0.0000000000001	10^{-13}	13	10^{-1}
0.00000000000001	10^{-14}	14	10^{-0}

In summary, the pH scale can be used to rate or determine the degree of acidity or basicity (also called alkalinity) of a solution. On this scale, a pH below 7 is acidic, and the lower the pH, the greater the acidity; a pH above 7 is basic, and the higher the pH, the greater the basicity. Incidentally, although pHs are given here in even whole numbers, more often, a pH reading exists in decimal form, for example, pH 4.5 or 6.8 (acidic) and pH 7.4 or 10.2 (basic). Because of the damaging effects of very concentrated acids or bases, most cells operate best under neutral, weakly acidic, or weakly basic conditions (see chapter 7).

Aqueous solutions containing both acids and bases may be involved in **neutralization** reactions, which give rise to water and other neutral by-products. For example, when equal molar solutions of hydrochloric acid (HCl) and sodium hydroxide (NaOH, a base) are mixed, the reaction proceeds as follows:

$$HCl + NaOH \rightarrow H_2O + NaCl$$

Here the acid and base ionize to H^+ and OH^- ions, which form water, and other ions, Na^+ and Cl^-, which form sodium chloride. Any product other than water that arises when acids and bases react is called a salt. Many of the organic acids (such as lactic and succinic acids) that function in **metabolism** are available as the acid and the salt form (such as lactate, succinate), depending on the conditions in the cell (see chapter 8).

The Chemistry of Carbon and Organic Compounds

So far, our main focus has been on the characteristics of atoms, ions, and small, simple substances that play diverse roles in the structure and function of living things. These substances are often lumped together in a category called **inorganic**

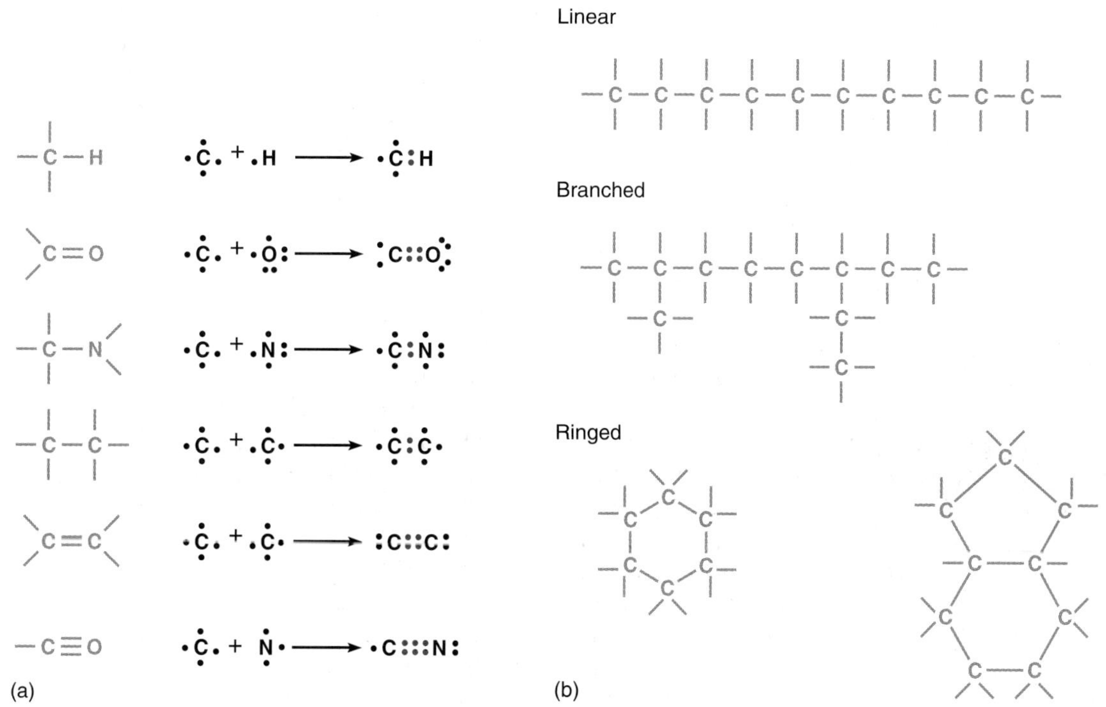

Figure 2.13 The versatility of bonding in carbon.
In most compounds, each carbon makes a total of four bonds. **(a)** Both single and double bonds can be made with other carbons, oxygen, and nitrogen; single bonds are made with hydrogen. Simple electron models show how the electrons are shared in these bonds. **(b)** Multiple bonding of carbons can give rise to long chains, branched compounds, and ringed compounds, many of which are extraordinarily large and complex.

chemicals. A chemical is usually inorganic if it does not contain both carbon and hydrogen. Examples of inorganic chemicals include NaCl (sodium chloride), $Mg_3(PO_4)_2$ (magnesium phosphate), $CaCO_3$ (calcium carbonate), and CO_2 (carbon dioxide). In reality, however, most of the chemical reactions and structures of living things occur at the level of more complex molecules, termed **organic chemicals.** These are carbon compounds with a basic framework of the element carbon bonded to other atoms. Organic molecules vary in complexity from the simplest, methane (CH_4; see figure 2.4*c*), which has a molecular weight of 16, to certain antibody molecules (produced by an immune reaction) that have a molecular weight of nearly 1,000,000 and are among the most complex molecules on earth.

The role of carbon as the fundamental element of life can best be understood if we look at its chemistry and bonding patterns. The valence of carbon makes it an ideal atomic building block to form the backbone of organic molecules; it has 4 electrons in its outer orbital to be shared with other atoms (including other carbons) through covalent bonding. As a result, it can form stable chains containing thousands of carbon atoms and still has bonding sites available for forming covalent bonds with numerous other atoms. The bonds that carbon forms are linear, branched, or ringed, and it can form four single bonds, two double bonds, or one triple bond **(figure 2.13).** The atoms with which carbon is most often associated in organic compounds are hydrogen, oxygen, nitrogen, sulfur, and phosphorus.

Functional Groups of Organic Compounds

One important advantage of carbon's serving as the molecular skeleton for living things is that it is free to bind with an unending array of other molecules. These special molecular groups or accessory molecules that bind to organic compounds are called **functional groups.** Functional groups help define the chemical class of certain groups of organic compounds and confer unique reactive properties on the whole molecule **(table 2.3).** Because each type of functional group behaves in a distinctive manner, reactions of an organic compound can be predicted by knowing the kind of functional group or groups it carries. Many synthesis, decomposition, and transfer reactions rely upon functional groups such as R—OH or R—NH_2. The —R designation on a molecule is shorthand for residue, and its placement in a formula indicates that the residue (functional group) varies from one compound to another.

CHECKPOINT

- Covalent bonds are chemical bonds in which electrons are shared between atoms. Equally distributed electrons form nonpolar covalent bonds, whereas unequally distributed electrons form polar covalent bonds.
- Ionic bonds are chemical bonds resulting from opposite charges. The outer electron shell either donates or receives electrons from another atom so that the outer shell of each atom is completely filled.

- Hydrogen bonds are weak chemical attractions that form between covalently bonded hydrogens and either oxygens or nitrogens on different molecules.
- Chemical equations express the chemical exchanges between atoms or molecules.
- Solutions are mixtures of solutes and solvents that cannot be separated by filtration or settling.
- The pH, ranging from a highly *acidic* solution to a highly *basic* solution, refers to the concentration of hydrogen ions. It is expressed as a number from 0 to 14.
- Biologists define organic molecules as those containing both carbon and hydrogen.
- Carbon is the backbone of biological compounds because of its ability to form single, double, or triple covalent bonds with itself and many different elements.
- Functional (R) groups are specific arrangements of organic molecules that confer distinct properties, including chemical reactivity, to organic compounds.

2.2 Macromolecules: Superstructures of Life

The compounds of life fall into the realm of **biochemistry**. Biochemicals are organic compounds produced by (or components of) living things, and they include four main families: carbohydrates, lipids, proteins, and nucleic acids **(table 2.4)**. The compounds in these groups are assembled from smaller molecular subunits, or building blocks, and because they are often very large compounds, they are termed **macromolecules**. All macromolecules except lipids are formed by polymerization, a process in which repeating subunits termed **monomers** are bound into chains of various lengths termed **polymers**. For example, proteins (polymers) are composed of a chain of amino acids (monomers). The large size and complex, three-dimensional shape of macromolecules enables them to function as structural components, molecular messengers, energy sources, enzymes (biochemical catalysts), nutrient stores, and sources of genetic information. In the following section and in later chapters, we consider numerous concepts relating to the roles of macromolecules in cells. Table 2.4 will also be a useful reference when you study metabolism in chapter 8.

Carbohydrates: Sugars and Polysaccharides

The term **carbohydrate** originates from the way that most members of this chemical class resemble combinations of carbon (carbo-) and water (-hydrate). Although carbohydrates can be generally represented by the formula $(CH_2O)_n$, in which n indicates the number of units of this combination of atoms, some carbohydrates contain additional atoms of sulfur or nitrogen. In molecular configuration, the

TABLE 2.3 Representative Functional Groups and Classes of Organic Compounds

Formula of Functional Group	Name	Class of Compounds
R* — O — H	Hydroxyl	Alcohols, carbohydrates
R — C(=O) — OH	Carboxyl	Fatty acids, proteins, organic acids
R — C(H)(H) — NH₂	Amino	Proteins, nucleic acids
R — C(=O) — O — R	Ester	Lipids
R — C(H)(H) — SH	Sulfhydryl	Cysteine (amino acid), proteins
R — C(=O) — H	Carbonyl, terminal end	Aldehydes, polysaccharides
R — C(=O) — C —	Carbonyl, internal	Ketones, polysaccharides
R — O — P(=O)(OH) — OH	Phosphate	DNA, RNA, ATP

*The R designation on a molecule is shorthand for residue, and it indicates that what is attached at that site varies from one compound to another.

carbons form chains or rings with two or more hydroxyl groups and either an aldehyde or a ketone group, giving them the technical designation of *polyhydroxy aldehydes* or *ketones* **(figure 2.14)**.

Carbohydrates exist in a great variety of configurations. The common term sugar **(saccharide)** refers to a simple carbo-hydrate such as a monosaccharide or a disaccharide that has a sweet taste. A **monosaccharide** is a simple polyhydroxy aldehyde or ketone molecule containing from 3 to 7 carbons; a **disaccharide** is a combination of two monosaccharides; and a **polysaccharide** is a polymer of five or more monosaccharides bound in linear or branched chain patterns

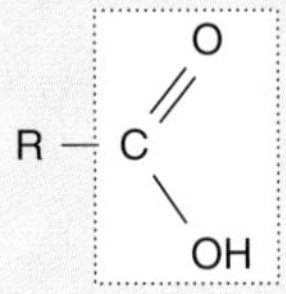

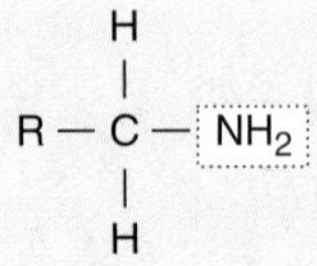

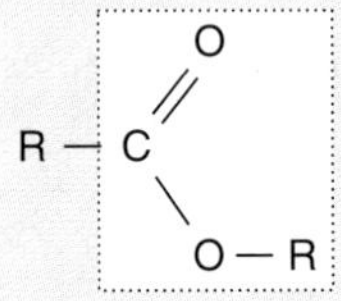

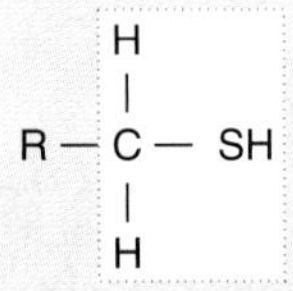

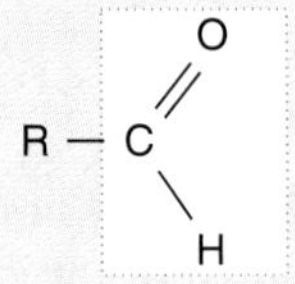

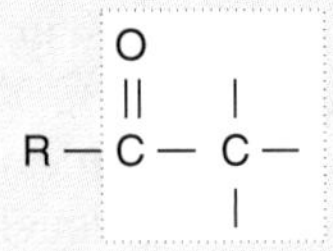

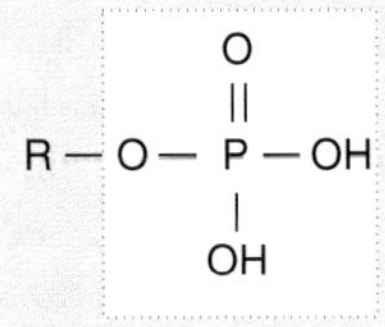

TABLE 2.4 Macromolecules and Their Functions

Macromolecule	Description/Basic Structure	Examples	Notes
Carbohydrates			
Monosaccharides	3- to 7-carbon sugars	Glucose, fructose	Sugars involved in metabolic reactions; building block of disaccharides and polysaccharides
Disaccharides	Two monosaccharides	Maltose (malt sugar)	Composed of two glucoses; an important breakdown product of starch
		Lactose (milk sugar)	Composed of glucose and galactose
		Sucrose (table sugar)	Composed of glucose and fructose
Polysaccharides	Chains of monosaccharides	Starch, cellulose, glycogen	Cell wall, food storage
Lipids			
Triglycerides	Fatty acids + glycerol	Fats, oils	Major component of cell membranes; storage
Phospholipids	Fatty acids + glycerol + phosphate	Membranes	
Waxes	Fatty acids, alcohols	Mycolic acid	Cell wall of mycobacteria
Steroids	Ringed structure	Cholesterol, ergosterol	Membranes of eukaryotes and some bacteria
Proteins			
	Amino acids	Enzymes; part of cell membrane, cell wall, ribosomes, antibodies	Metabolic reactions; structural components
Nucleic acids			
	Pentose sugar + phosphate + nitrogenous base Purines: adenine, guanine Pyrimidines: cytosine, thymine, uracil		
Deoxyribonucleic acid (DNA)	Contains deoxyribose sugar and thymine, not uracil	Chromosomes; genetic material of viruses	Inheritance
Ribonucleic acid (RNA)	Contains ribose sugar and uracil, not thymine	Ribosomes; mRNA, tRNA	Expression of genetic traits

(see figure 2.14). Monosaccharides and disaccharides are specified by combining a prefix that describes some characteristic of the sugar with the suffix *-ose*. For example, **hexoses** are composed of 6 carbons, and **pentoses** contain 5 carbons. **Glucose** (Gr. sweet) is the most common and universally important hexose; **fructose** is named for fruit (one of its sources); and xylose, a pentose, derives its name from the Greek word for wood. Disaccharides are named similarly: **lactose** (L. milk) is an important component of milk; **maltose** means malt sugar; and **sucrose** (Fr. sugar) is common table sugar or cane sugar.

The Nature of Carbohydrate Bonds

The subunits of disaccharides and polysaccharides are linked by means of **glycosidic bonds,** in which carbons (each is assigned a number) on adjacent sugar units are bonded to the same oxygen atom like links in a chain **(figure 2.15).** For example, maltose is formed when the number 1 carbon on a glucose bonds to the oxygen on the number 4 carbon on a second glucose; sucrose is formed when glucose and fructose bind oxygen between their number 1 and number 2 carbons; and lactose is formed when glucose and galactose connect by their number 1 and number 4 carbons. In order to form this bond, 1 carbon gives up its OH group and the other (the one contributing the oxygen to the bond) loses the H from its OH group. Because a water molecule is produced, this reaction is known as **dehydration synthesis,** a process common to most polymerization reactions (see proteins, page 48). Three polysaccharides (starch, cellulose, and glycogen) are structurally and biochemically distinct, even though all are polymers of the same monosaccharide—glucose. The basis for their differences lies primarily in the exact way the glucoses are bound together, which greatly

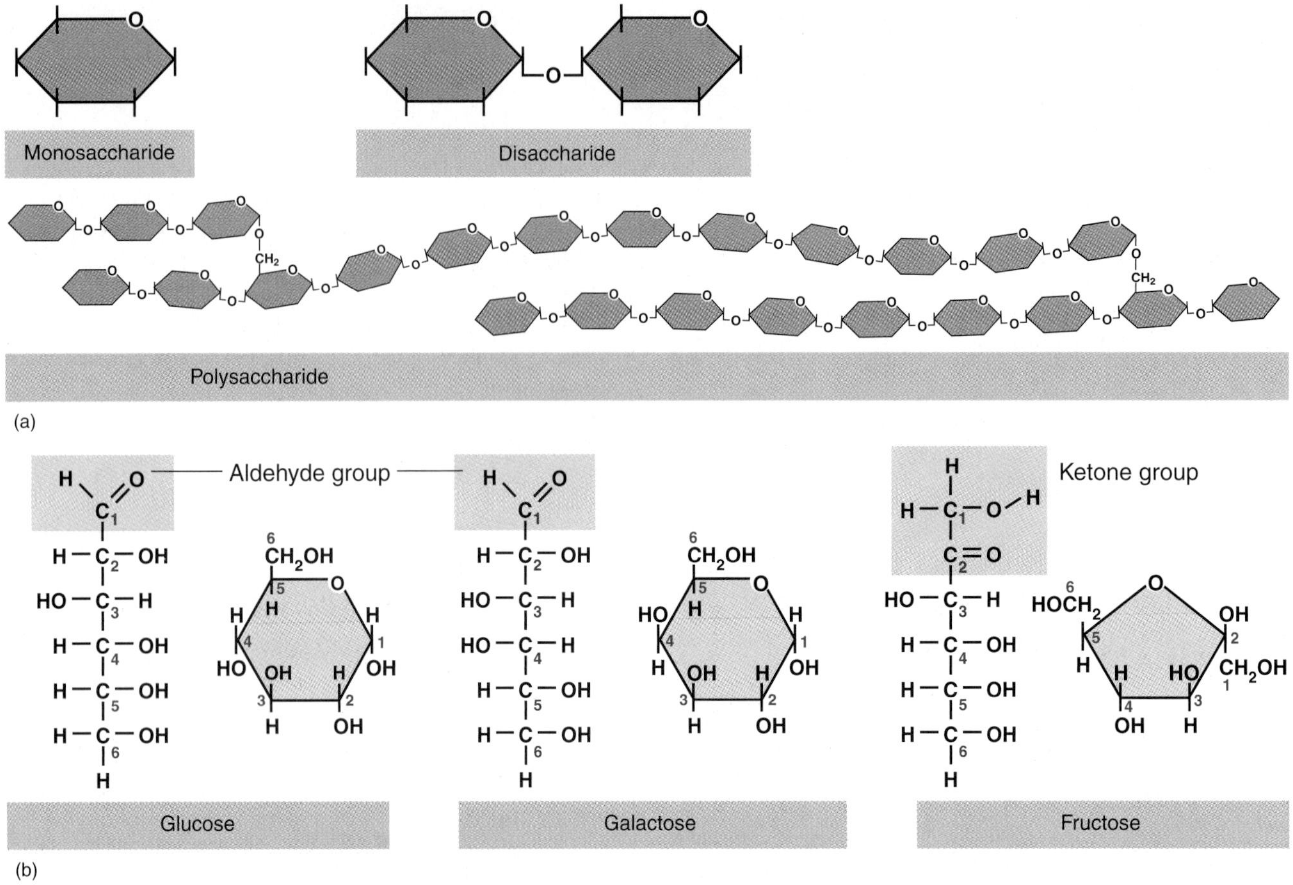

Figure 2.14 Common classes of carbohydrates.
(a) Major saccharide groups, named for the number of sugar units each contains. **(b)** Three hexoses with the same molecular formula and different structural formulas. Both linear and ring models are given. The linear form emphasizes aldehyde and ketone groups, although in solution the sugars exist in the ring form. Note that the carbons are numbered so as to keep track of reactions within and between monosaccharides.

affects the characteristics of the end product **(figure 2.16).** The synthesis and breakage of each type of bond requires a specialized catalyst called an enzyme (see chapter 8).

The Functions of Polysaccharides

Polysaccharides typically contribute to structural support and protection and serve as nutrient and energy stores. The cell walls in plants and many microscopic algae derive their strength and rigidity from **cellulose,** a long, fibrous polymer **(figure 2.16*a*).** Because of this role, cellulose is probably one of the most common organic substances on the earth, yet it is digestible only by certain bacteria, fungi, and protozoa. These microbes, called decomposers, play an essential role in breaking down and recycling plant materials (see figure 7.2). Some bacteria secrete slime layers of a glucose polymer called *dextran.* This substance causes a sticky layer to develop on teeth that leads to plaque, described later in chapter 22.

Other structural polysaccharides can be conjugated (chemically bonded) to amino acids, nitrogen bases, lipids, or proteins. **Agar,** an indispensable polysaccharide in preparing solid culture media, is a natural component of certain seaweeds. It is a complex polymer of galactose and sulfur-containing carbohydrates. The exoskeletons of certain fungi contain **chitin** (ky-tun), a polymer of glucosamine (a sugar with an amino functional group). **Peptidoglycan** (pep-tih-doh-gly′-kan) is one special class of compounds in which polysaccharides (glycans) are linked to peptide fragments (a short chain of amino acids). This molecule provides the main source of structural support to the bacterial cell wall. The cell wall of gram-negative bacteria also contains **lipopolysaccharide,** a complex of lipid and polysaccharide responsible for symptoms such as fever and shock (see chapters 4 and 13).

The outer surface of many cells has a "sugar coating" composed of polysaccharides bound in various ways to proteins (the combination is called mucoprotein or glycoprotein). This structure, called the **glycocalyx,** functions in attachment to other cells or as a site for *receptors*—surface molecules that receive external stimuli or act as binding sites. Small sugar molecules account for the differences in human blood types, and carbohydrates are a component of large protein molecules called antibodies. Viruses also have glycoproteins on their surface with which they bind to and invade their host cells.

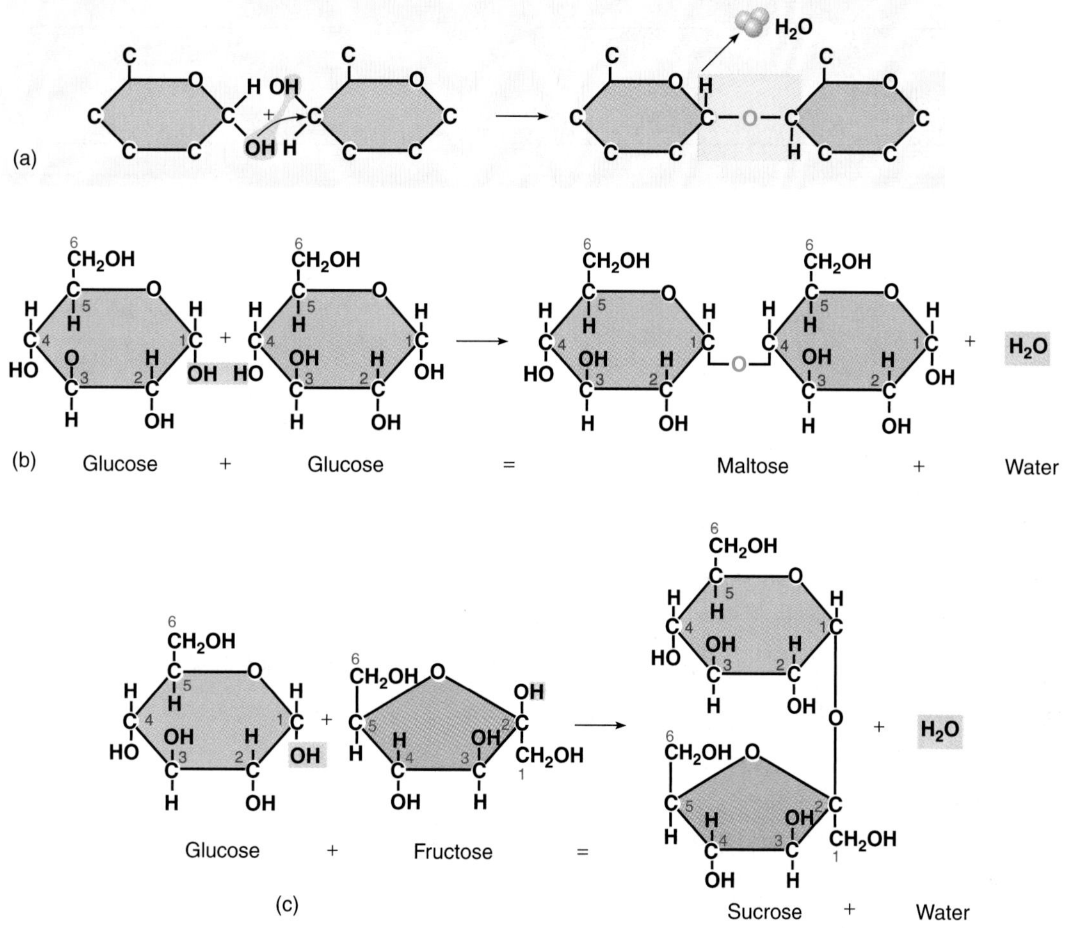

Figure 2.15 **Glycosidic bond.**
(a) General scheme in the formation of a glycosidic bond by dehydration synthesis. **(b)** Formation of the 1,4 bond between two α glucoses to produce maltose and water. **(c)** Formation of the 1,2 bond between glucose and fructose to produce sucrose and water.

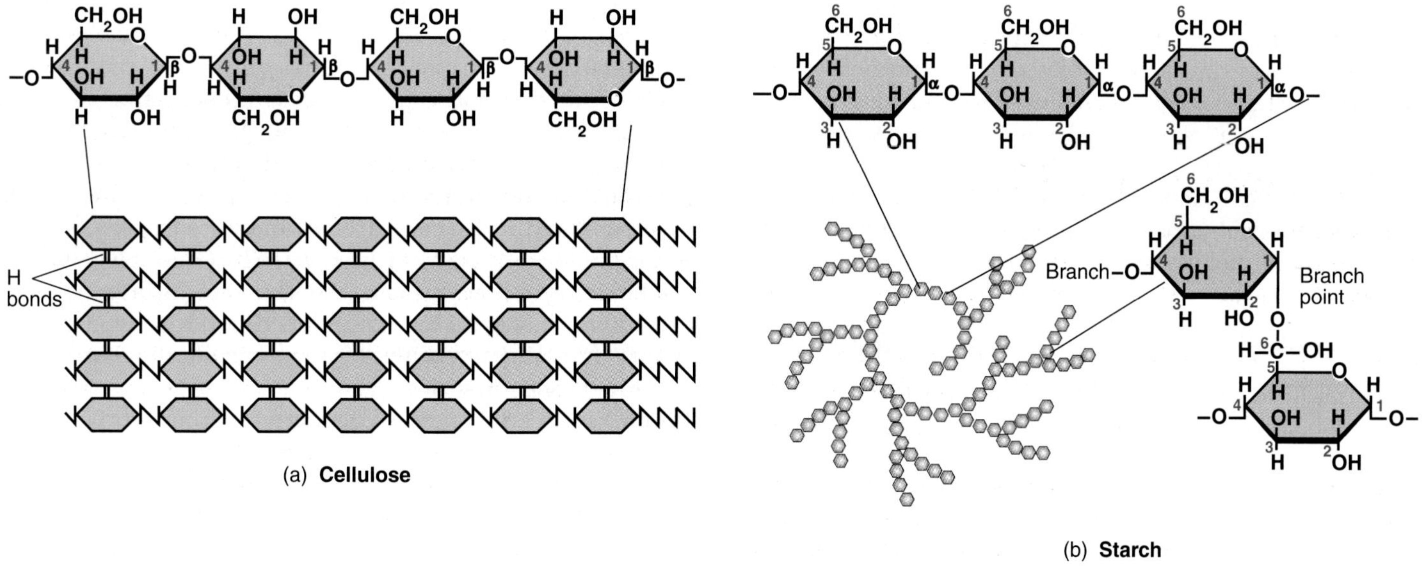

Figure 2.16 **Polysaccharides.**
(a) Cellulose is composed of β glucose bonded in 1,4 bonds that produce linear, lengthy chains of polysaccharides that are H-bonded along their length. This is the typical structure of wood and cotton fibers. **(b)** Starch is also composed of glucose polymers, in this case α glucose. The main structure is amylose bonded in a 1,4 pattern, with side branches of amylopectin bonded by 1,6 bonds. The entire molecule is compact and granular.

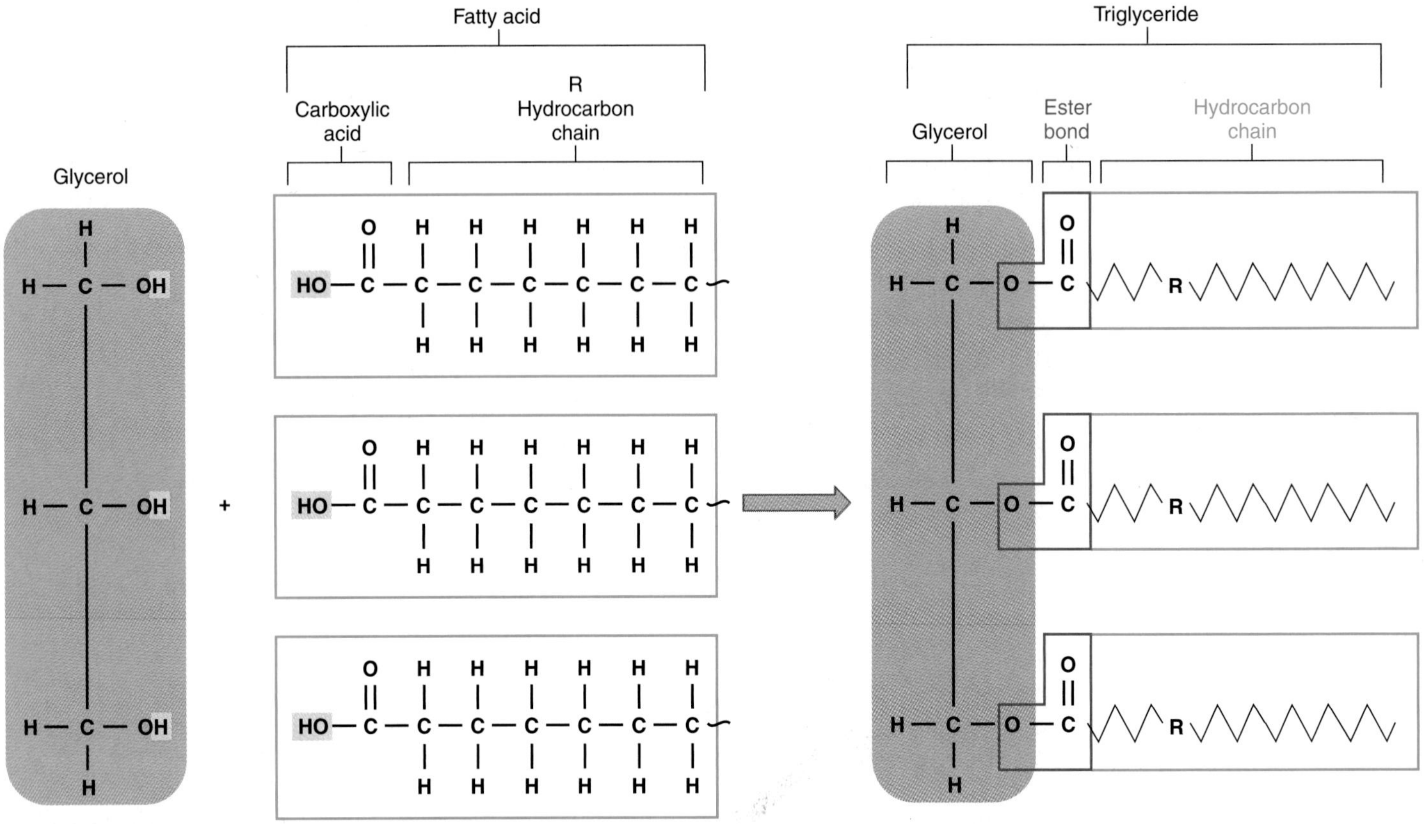

Figure 2.17 Synthesis and structure of a triglyceride.
Because a water molecule is released at each ester bond, this is another form of dehydration synthesis. The jagged lines and R symbol represent the hydrocarbon chains of the fatty acids, which are commonly very long.

Polysaccharides are usually stored by cells in the form of glucose polymers such as starch **(figure 2.16*b*)** or **glycogen,** but only organisms with the appropriate digestive enzymes can break them down and use them as a nutrient source. Because a water molecule is required for breaking the bond between two glucose molecules, digestion is also termed **hydrolysis.** Starch is the primary storage food of green plants, microscopic algae, and some fungi; glycogen (animal starch) is a stored carbohydrate for animals and certain groups of bacteria and protozoa.

Lipids: Fats, Phospholipids, and Waxes

The term **lipid,** derived from the Greek word *lipos,* meaning fat, is not a chemical designation but an operational term for a variety of substances that are not soluble in polar solvents such as water (recall that oil and water do not mix) but will dissolve in nonpolar solvents such as benzene and chloroform. This property occurs because the substances we call lipids contain relatively long or complex C—H (hydrocarbon) chains that are nonpolar and thus hydrophobic. The main groups of compounds classified as lipids are triglycerides, phospholipids, steroids, and waxes.

Important storage lipids are the **triglycerides,** a category that includes fats and oils. Triglycerides are composed of a single molecule of glycerol bound to three fatty acids **(figure 2.17). Glycerol** is a 3-carbon alcohol[5] with three OH groups that serve as binding sites, and fatty acids are long-chain hydrocarbon molecules with a carboxyl group (COOH) at one end that is free to bind to the glycerol. The hydrocarbon portion of a fatty acid can vary in length from 4 to 24 carbons; and, depending on the fat, it may be saturated or unsaturated. If all carbons in the chain are single-bonded to 2 other carbons and 2 hydrogens, the fat is saturated; if there is at least one C=C double bond in the chain, it is unsaturated. The structure of fatty acids is what gives fats and oils (liquid fats) their greasy, insoluble nature. In general, solid fats (such as beef tallow) are more saturated, and oils (or liquid fats) are more unsaturated. In most cells, triglycerides are stored in long-term concentrated form as droplets or globules. When they are acted on by digestive enzymes called lipases, the fatty acids and glycerol are freed to be used in metabolism. Fatty acids are a superior source of energy, yielding twice as much per gram as other storage molecules (starch). Soaps are K^+ or Na^+ salts of fatty acids whose qualities make them excellent grease removers and cleaners (see chapter 11).

5. Alcohols are carbon compounds containing OH groups.

Membrane Lipids

A class of lipids that serves as a major structural component of cell membranes is the **phospholipids.** Although phospholipids also contain glycerol and fatty acids, they have some significant differences from triglycerides. Phospholipids contain only two fatty acids attached to the glycerol, and the third glycerol binding site holds a phosphate group. The phosphate is in turn bonded to an alcohol, which varies from one phospholipid to another **(figure 2.18*a*).** These lipids have a hydrophilic region from the charge on the phosphoric acid–alcohol "head" of the molecule and a hydrophobic region that corresponds to the long, uncharged "tail" (formed by the fatty acids). When exposed to an aqueous solution, the charged heads are attracted to the water phase, and the nonpolar tails are repelled from the water phase **(figure 2.18*b*).** This property causes lipids to naturally assume single and double layers (bilayers), which contribute to their biological significance in membranes. When two single layers of polar lipids come together to form a double layer, the outer hydrophilic face of each single layer will orient itself toward the solution, and the hydrophobic portions will become immersed in the core of the bilayer. The structure of lipid bilayers confers characteristics on membranes such as selective permeability and fluid nature **(Insight 2.2).**

Miscellaneous Lipids

Steroids are complex ringed compounds commonly found in cell membranes and animal hormones. The best known of these is the sterol (meaning a steroid with an OH group) called **cholesterol (figure 2.19).** Cholesterol reinforces the structure of the cell membrane in animal cells and in an unusual group of cell-wall-deficient bacteria called the mycoplasmas (see chapter 4). The cell membranes of fungi also contain a sterol, called ergosterol. *Prostaglandins* are fatty acid derivatives found in trace amounts that function in inflammatory and allergic

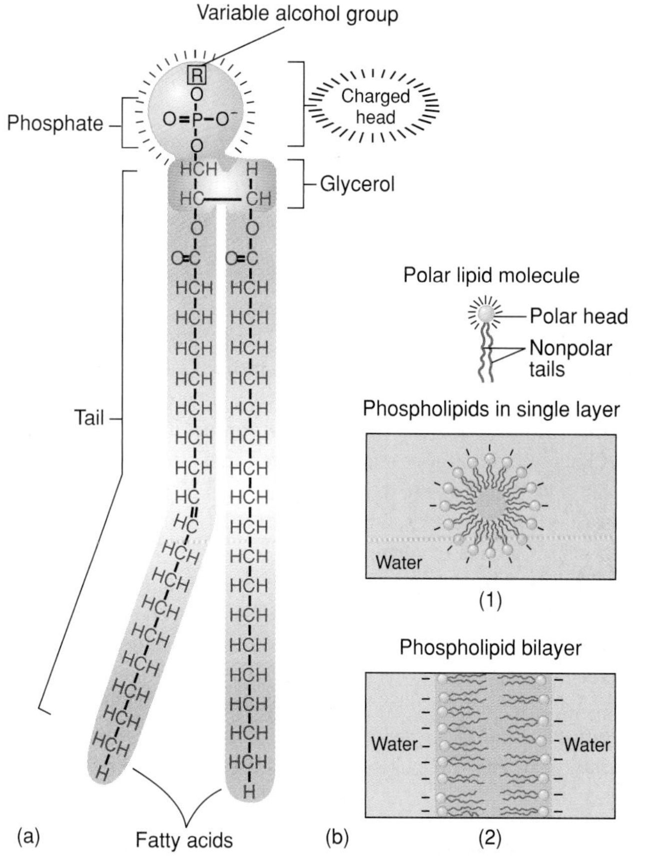

Figure 2.18 Phospholipids—membrane molecules. **(a)** A model of a single molecule of a phospholipid. The phosphate-alcohol head lends a charge to one end of the molecule; its long, trailing hydrocarbon chain is uncharged. **(b)** The behavior of phospholipids in water-based solutions causes them to become arranged **(1)** in single layers called micelles, with the charged head oriented toward the water phase and the hydrophobic nonpolar tail buried away from the water phase, or **(2)** in double-layered phospholipid systems with the hydrophobic tails sandwiched between two hydrophilic layers.

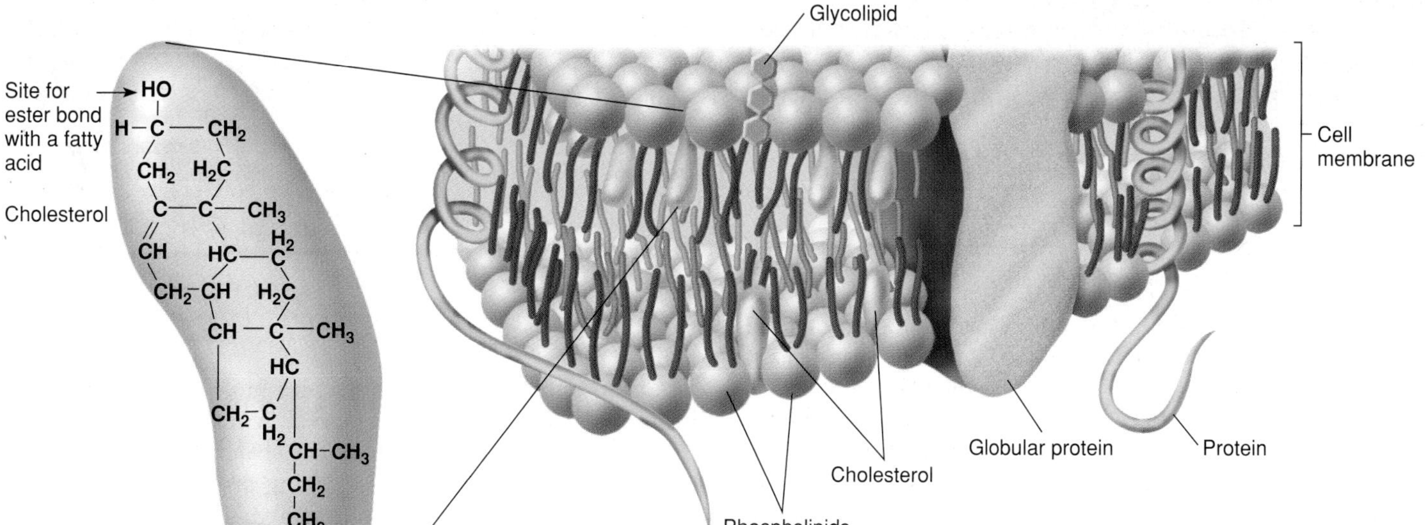

Figure 2.19 Formula for cholesterol, an alcoholic steroid that is inserted in some membranes. Cholesterol can associate with fatty acids at its OH group.

INSIGHT 2.2 — Discovery

Membranes: Cellular Skins

The word **membrane** appears frequently in descriptions of cells in this chapter and in chapters 4 and 5. The word itself describes any lining or covering, including such multicellular structures as the mucous membranes of the body. From the perspective of a single cell, however, a membrane is a thin, double-layered sheet composed of lipids such as phospholipids and sterols (averaging about 40% of membrane content) and protein molecules (averaging about 60%). The primary role of membranes is as a cell membrane that completely encases the cytoplasm. Membranes are also components of eukaryotic organelles such as nuclei, mitochondria, and chloroplasts, and they appear in internal pockets of certain prokaryotic cells. Even some viruses, which are not cells at all, can have a membranous protective covering.

Cell membranes are so thin—on the average, just 0.0070 µm (7 nm) thick—that they cannot actually be seen with an optical microscope. Even at magnifications made possible by electron microscopy (500,000×), very little of the precise architecture can be visualized, and a cross-sectional view has the appearance of railroad tracks. Following detailed microscopic and chemical analysis, S. J. Singer and C. K. Nicholson proposed a simple and elegant theory for membrane structure called the **fluid mosaic model.** According to this theory, a membrane is a continuous bilayer formed by lipids that are oriented with the polar lipid heads toward the outside and the nonpolar tails toward the center of the membrane. Embedded at numerous sites in this bilayer are various-size globular proteins. Some proteins are situated only at the surface; others extend fully through the entire membrane. The configuration of the inner and outer sides of the membrane can be quite different because of the variations in protein shape and position.

Membranes are dynamic and constantly changing because the lipid phase is in motion and many proteins can migrate freely about, somewhat as icebergs do in the ocean. This fluidity is essential to such activities as engulfment of food and discharge or secretion by cells. The structure of the lipid phase provides an impenetrable barrier to many substances. This property accounts for the selective permeability and capacity to regulate transport of molecules. It also serves to segregate activities within the cell's cytoplasm. Membrane proteins function in receiving molecular signals (receptors), in binding and transporting nutrients, and in acting as enzymes, topics to be discussed in chapters 7 and 8.

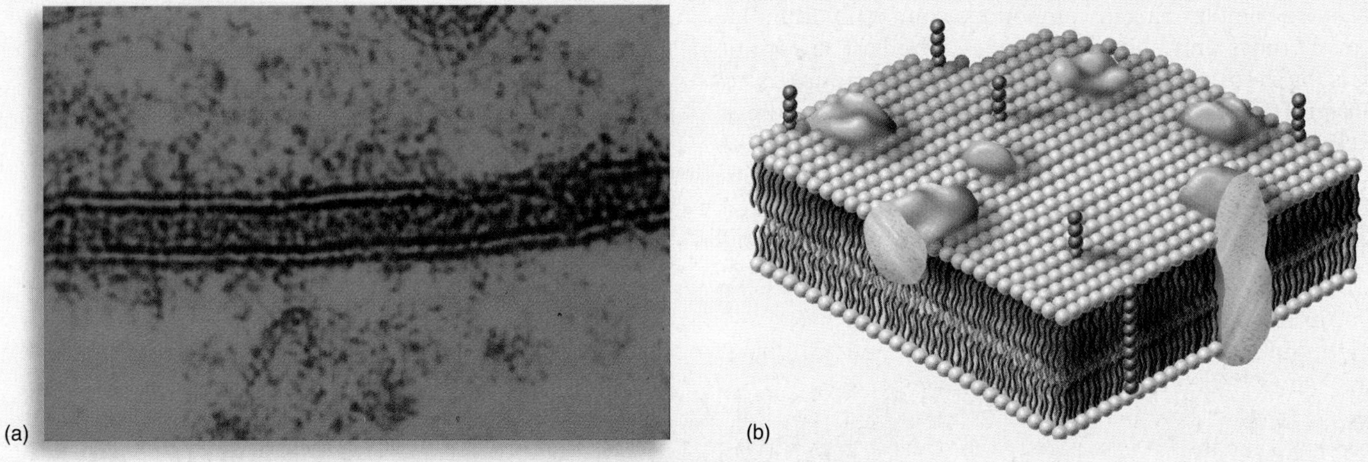

(a) Extreme magnification of a cross section of a cell membrane, which appears as double tracks. **(b)** A generalized version of the fluid mosaic model of a cell membrane indicates a bilayer of lipids with globular proteins embedded to some degree in the lipid matrix. This structure explains many characteristics of membranes, including flexibility, solubility, permeability, and transport.

reactions, blood clotting, and smooth muscle contraction. Chemically, a *wax* is an ester formed between a long-chain alcohol and a saturated fatty acid. The resulting material is typically pliable and soft when warmed but hard and water resistant when cold (paraffin, for example). Among living things, fur, feathers, fruits, leaves, human skin, and insect exoskeletons are naturally waterproofed with a coating of wax. Bacteria that cause tuberculosis and leprosy produce a wax that repels ordinary laboratory stains and contributes to their pathogenicity.

Proteins: Shapers of Life

The predominant organic molecules in cells are **proteins,** a fitting term adopted from the Greek word *proteios,* meaning first or prime. To a large extent, the structure, behavior, and unique qualities of each living thing are a consequence of the proteins they contain. To best explain the origin of the special properties and versatility of proteins, we must examine their general structure. The building blocks of proteins are **amino acids,** which exist in 20 different naturally occurring

TABLE 2.5 Twenty Amino Acids and Their Abbreviations

Acid	Abbreviation	Characteristic of R Groups*
Alanine	Ala	NP
Arginine	Arg	+
Asparagine	Asn	P
Aspartic acid	Asp	−
Cysteine	Cys	P
Glutamic acid	Glu	−
Glutamine	Gln	P
Glycine	Gly	P
Histidine	His	+
Isoleucine	Ile	NP
Leucine	Leu	NP
Lysine	Lys	+
Methionine	Met	NP
Phenylalanine	Phe	NP
Proline	Pro	NP
Serine	Ser	P
Threonine	Thr	P
Tryptophan	Trp	NP
Tyrosine	Tyr	P
Valine	Val	NP

*NP = nonpolar; P = polar; + = positively charged; − = negatively charged.

forms **(table 2.5).** Various combinations of these amino acids account for the nearly infinite variety of proteins. Amino acids have a basic skeleton consisting of a carbon (called the α carbon) linked to an amino group (NH_2), a carboxyl group (COOH), a hydrogen atom (H), and a variable R group. The variations among the amino acids occur at the R group, which is different in each amino acid and imparts the unique characteristics to the molecule and to the proteins that contain it **(figure 2.20).** A covalent bond called a **peptide bond** forms between the amino group on one amino acid and the carboxyl group on another amino acid. As a result of peptide bond formation, it is possible to produce molecules varying in length from two amino acids to chains containing thousands of them.

Various terms are used to denote the nature of compounds containing peptide bonds. **Peptide** usually refers to a molecule composed of short chains of amino acids, such as a dipeptide (two amino acids), a tripeptide (three), and a tetrapeptide (four). A **polypeptide** contains an unspecified number of amino acids but usually has more than 20 and is often a smaller subunit of a protein. A protein is the largest of this class of compounds and usually contains a minimum of 50 amino acids. It is common for the terms *polypeptide* and *protein* to be used interchangeably, though not all polypeptides are large enough to be considered proteins. In chapter 9, we see that protein synthesis is not just a random connection of amino acids; it is directed by information provided in DNA.

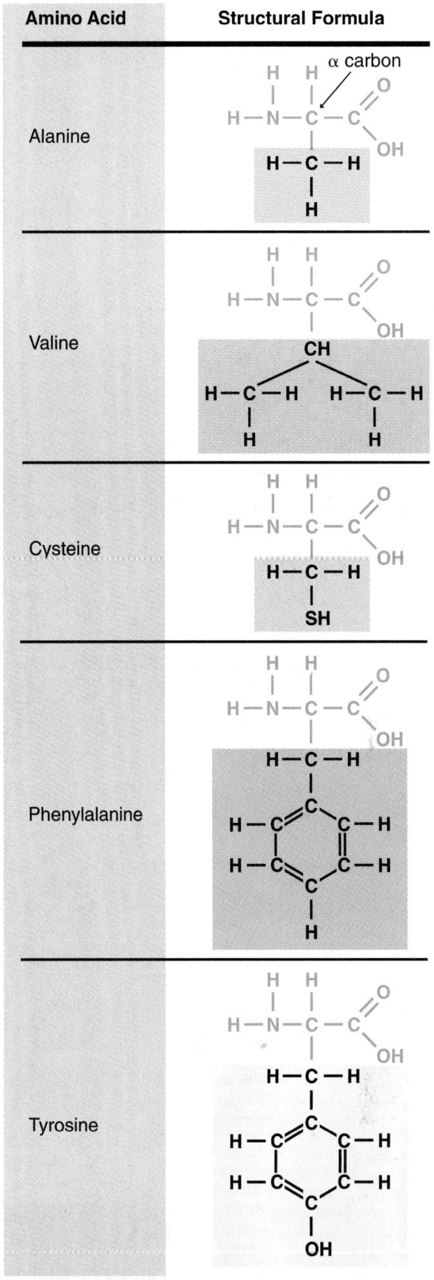

Figure 2.20 Structural formulas of selected amino acids. The basic structure common to all amino acids is shown in blue type; and the variable group, or R group, is placed in a colored box. Note the variations in structure of this reactive component.

Protein Structure and Diversity

The reason that proteins are so varied and specific is that they do not function in the form of a simple straight chain of amino acids (called the primary structure). A protein has a natural tendency to assume more complex levels of organization, called the secondary, tertiary, and quaternary structures **(figure 2.21).** The **primary (1°) structure** is the type, number, and order of amino acids in the chain, which varies extensively from protein to protein. The **secondary (2°) structure** arises when various functional groups exposed on the outer surface of the molecule

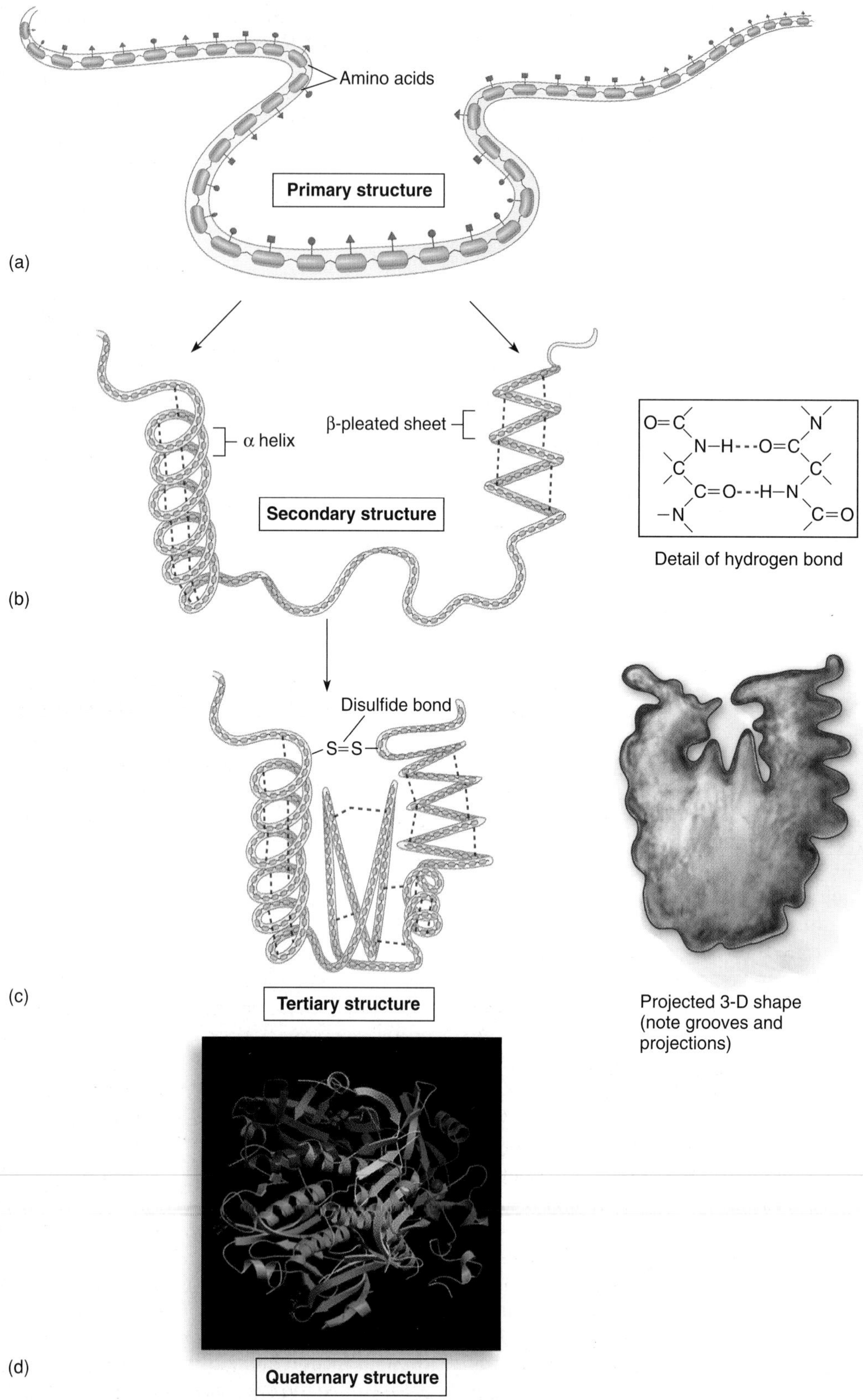

Figure 2.21 Stages in the formation of a functioning protein.
(a) Its primary structure is a series of amino acids bound in a chain. **(b)** Its secondary structure develops when the chain forms hydrogen bonds that fold it into one of several configurations such as an α helix or β-pleated sheet. Some proteins have several configurations in the same molecule. **(c)** A protein's tertiary structure is due to further folding of the molecule into a three-dimensional mass that is stabilized by hydrogen, ionic, and disulfide bonds between functional groups. **(d)** The quaternary structure exists only in proteins that consist of more than one polypeptide chain. The chains in this protein each have a different color.

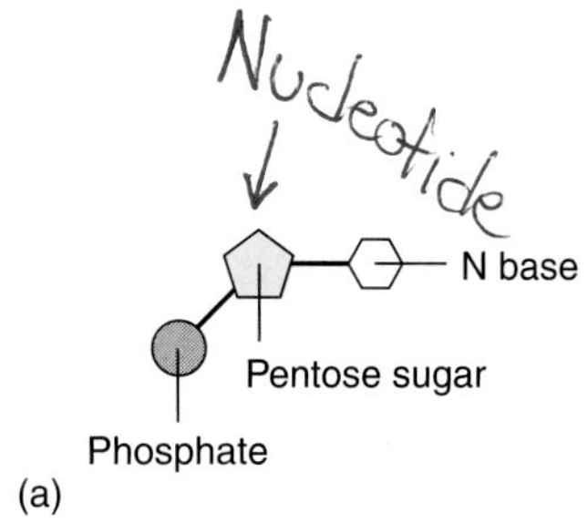

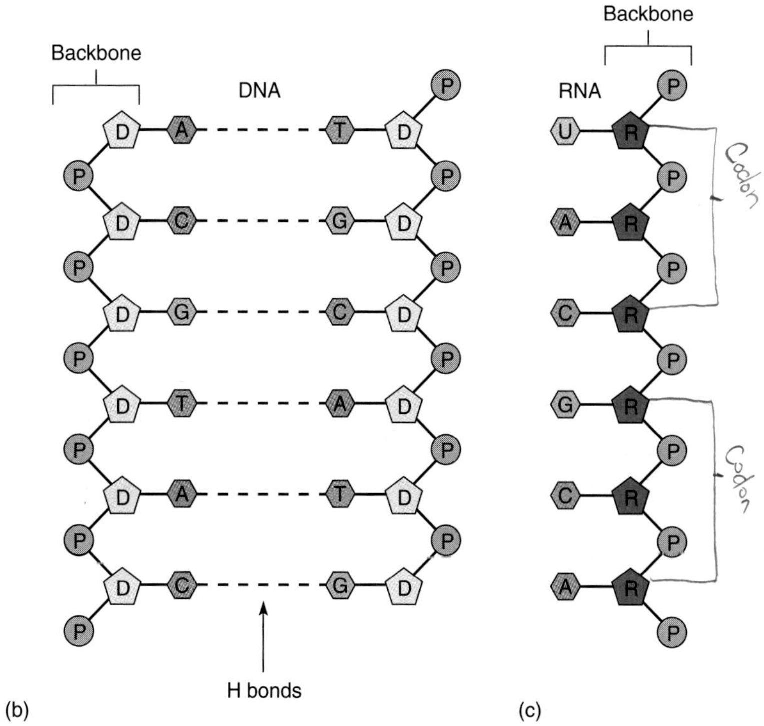

Figure 2.22 The general structure of nucleic acids. **(a)** A nucleotide, composed of a phosphate, a pentose sugar, and a nitrogen base (either A, T, U, C, or G), is the monomer of both DNA and RNA. **(b)** In DNA, the polymer is composed of alternating deoxyribose (D) and phosphate (P) with nitrogen bases (A, T, C, G) attached to the deoxyribose. DNA almost always exists in pairs of strands, oriented so that the bases are paired across the central axis of the molecule. **(c)** In RNA, the polymer is composed of alternating ribose (R) and phosphate (P) attached to nitrogen bases (A, U, C, G), but it is only a single strand.

interact by forming hydrogen bonds. This interaction causes the amino acid chain to twist into a coiled configuration called the *α helix* or to fold into an accordion pattern called a *β-pleated sheet*. Many proteins contain both types of secondary configurations. Proteins at the secondary level undergo a third degree of torsion called the **tertiary (3°) structure** created by additional bonds between functional groups **(figure 2.21*c*).** In proteins with the sulfur-containing amino acid **cysteine,** considerable tertiary stability is achieved through covalent disulfide bonds between sulfur atoms on two different parts of the molecule. Some complex proteins assume a **quaternary (4°) structure,** in which more than one polypeptide forms a large, multiunit protein. This is typical of antibodies (see chapter 15) and some enzymes that act in cell synthesis.

The most important outcome of intrachain[6] bonding and folding is that each different type of protein develops a unique shape, and its surface displays a distinctive pattern of pockets and bulges. As a result, a protein can react only with molecules that complement or fit its particular surface features like a lock and key. Such a degree of specificity can provide the functional diversity required for many thousands of different cellular activities. **Enzymes** serve as the catalysts for all chemical reactions in cells, and nearly every reaction requires a different enzyme (see chapter 8). **Antibodies** are complex glycoproteins with specific regions of attachment for bacteria, viruses, and other microorganisms; certain bacterial toxins (poisonous products) react with only one specific organ or tissue; and proteins embedded in the cell membrane have reactive sites restricted to a certain nutrient. The functional three-dimensional form of a protein is termed the *native state,* and if it is disrupted by some means, the protein is said to be *denatured*. Such agents as heat, acid, alcohol, and some disinfectants disrupt (and thus denature) the stabilizing intrachain bonds and cause the molecule to become nonfunctional, as described in chapter 11.

The Nucleic Acids: A Cell Computer and Its Programs

The nucleic acids, **deoxyribonucleic acid (DNA)** and **ribonucleic acid (RNA),** were originally isolated from the cell nucleus. Shortly thereafter, they were also found in other parts of nucleated cells, in cells with no nuclei (bacteria), and in viruses. The universal occurrence of nucleic acids in all known cells and viruses emphasizes their important roles as informational molecules. DNA, the master computer of cells, contains a special coded genetic program with detailed and specific instructions for each organism's heredity. It transfers the details of its program to RNA, "helper" molecules responsible for carrying out DNA's instructions and translating the DNA program into proteins that can perform life functions. For now, let us briefly consider the structure and some functions of DNA, RNA, and a close relative, adenosine triphosphate (ATP).

Both nucleic acids are polymers of repeating units called **nucleotides,** each of which is composed of three smaller units: a **nitrogen base,** a **pentose** (5-carbon) sugar, and a *phosphate* **(figure 2.22*a*).**[7] The nitrogen base is a cyclic compound that comes in two forms: *purines* (two rings) and *pyrimidines* (one ring). There are two types of purines—**adenine (A)** and **guanine (G)**—and three types of pyrimidines—**thymine (T), cytosine (C),** and **uracil (U) (figure 2.23).** A characteristic that

6. *Intra*chain means within the chain; *inter*chain would be between two chains.

7. The nitrogen base plus the pentose is called a *nucleoside*.

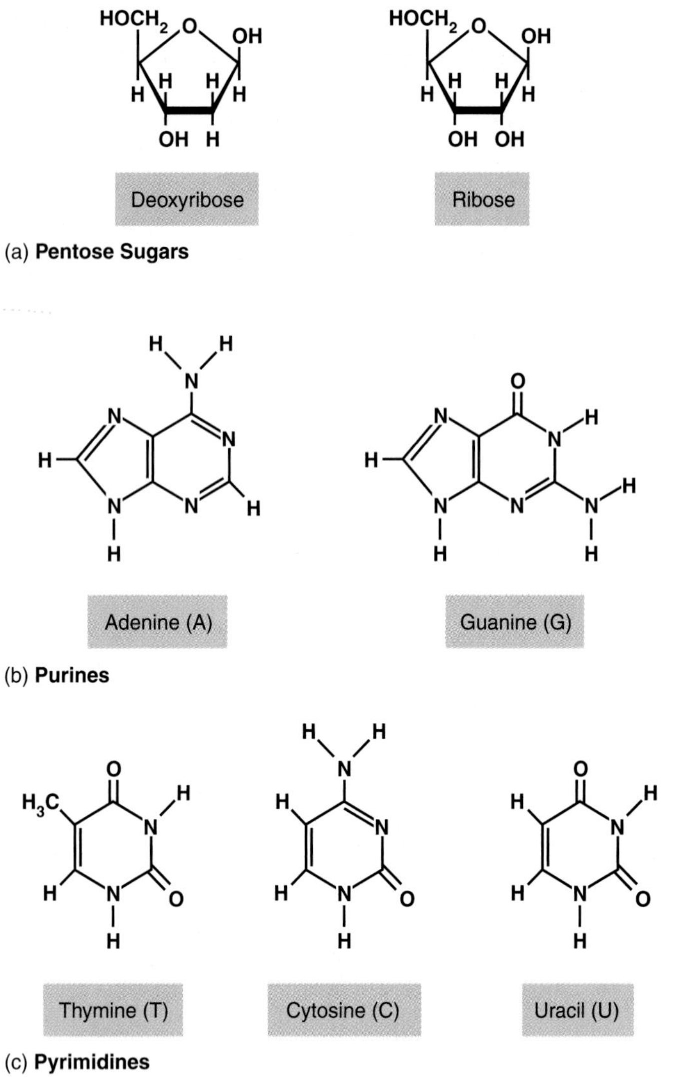

Figure 2.23 The sugars and nitrogen bases that make up DNA and RNA.
(a) DNA contains deoxyribose, and RNA contains ribose. **(b)** A and G purines are found in both DNA and RNA. **(c)** C pyrimidine is found in both DNA and RNA, but T is found only in DNA, and U is found only in RNA.

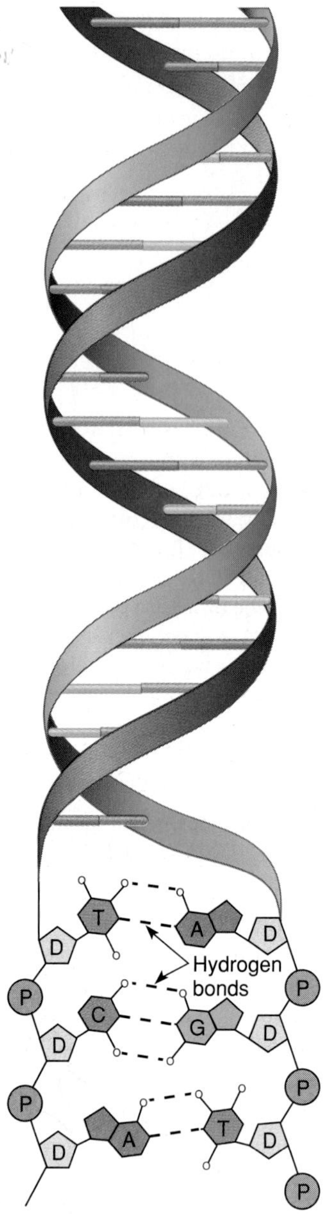

Figure 2.24 A structural representation of the double helix of DNA.
Shown are the details of hydrogen bonds between the nitrogen bases of the two strands.

differentiates DNA from RNA is that DNA contains all of the nitrogen bases except uracil, and RNA contains all of the nitrogen bases except thymine. The nitrogen base is covalently bonded to the sugar *ribose* in RNA and *deoxyribose* (because it has one less oxygen than ribose) in DNA. Phosphate (PO_4^{3-}), a derivative of phosphoric acid (H_3PO_4), provides the final covalent bridge that connects sugars in series. Thus, the backbone of a nucleic acid strand is a chain of alternating phosphate-sugar-phosphate-sugar molecules, and the nitrogen bases branch off the side of this backbone **(figure 2.22*b, c*).**

The Double Helix of DNA

DNA is a huge molecule formed by two very long polynucleotide strands linked along their length by hydrogen bonds between complementary pairs of nitrogen bases. The pairing of the nitrogen bases occurs according to a predictable pattern: Adenine ordinarily pairs with thymine, and cytosine with guanine. The bases are attracted in this way because each pair shares oxygen, nitrogen, and hydrogen atoms exactly positioned to align perfectly for hydrogen bonds **(figure 2.24).**

For ease in understanding the structure of DNA, it is sometimes compared to a ladder, with the sugar-phosphate backbone representing the rails and the paired nitrogen bases representing the steps. Owing to the manner of nucleotide pairing and stacking of the bases, the actual configuration of DNA is a *double helix* that looks somewhat like a spiral staircase. As is true of protein, the structure of DNA is intimately related to its function. DNA molecules are usually extremely long, a feature that satisfies a requirement for storing genetic information in the sequence of base pairs the molecule contains. The hydrogen bonds between pairs can be disrupted when DNA is being copied, and the fixed complementary base pairing is essential to maintain the genetic code.

RNA: Organizers of Protein Synthesis

Like DNA, RNA consists of a long chain of nucleotides. However, RNA is a single strand containing ribose sugar instead of deoxyribose and uracil instead of thymine (see figure 2.22). Several functional types of RNA are formed using the DNA template through a replicationlike process. Three major types of RNA are important for protein synthesis. Messenger RNA (mRNA) is a copy of a gene (a single functional part of the DNA) that provides the order and type of amino acids in a protein; transfer RNA (tRNA) is a carrier that delivers the correct amino acids for protein assembly; and ribosomal RNA (rRNA) is a major component of ribosomes (described in chapter 4). More information on these important processes is presented in chapter 9.

ATP: The Energy Molecule of Cells

A relative of RNA involved in an entirely different cell activity is **adenosine triphosphate (ATP).** ATP is a nucleotide containing adenine, ribose, and three phosphates rather than just one **(figure 2.25).** It belongs to a category of high-energy compounds (also including guanosine triphosphate [GTP]) that give off energy when the bond is broken between the second and third (outermost) phosphate. The presence of these high-energy bonds makes it possible for ATP to release and store energy for cellular chemical reactions. Breakage of the bond of the terminal phosphate releases energy to do cellular work and also generates adenosine diphosphate (ADP). ADP can be converted back to ATP when the third phosphate is restored, thereby serving as an energy depot. Carriers for oxidation-reduction activities (nicotinamide adenine dinucleotide [NAD], for instance) are also derivatives of nucleotides (see chapter 8).

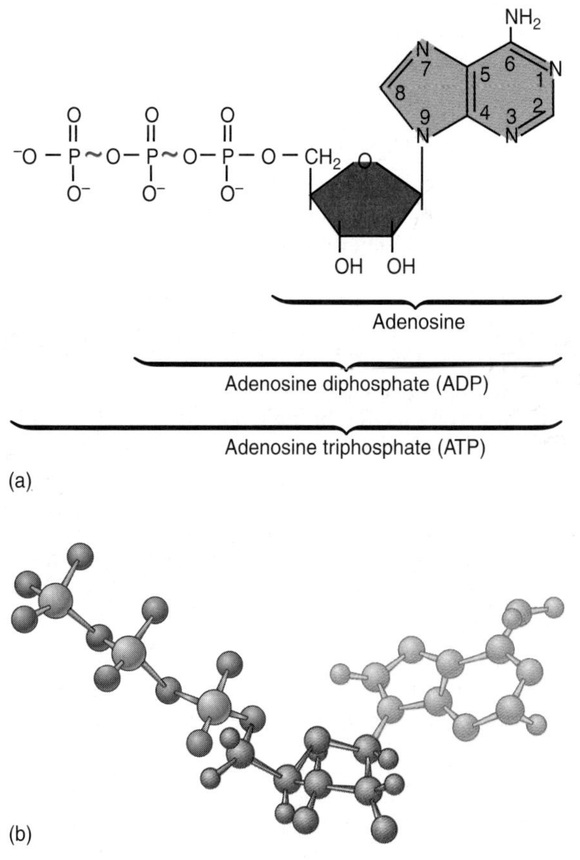

Figure 2.25 An ATP molecule.
(a) The structural formula. Wavy lines connecting the phosphates represent bonds that release large amounts of energy. **(b)** A model.

CHECKPOINT

- Macromolecules are very large organic molecules (polymers) built up by polymerization of smaller molecular subunits (monomers).
- Carbohydrates are biological molecules whose polymers are monomers linked together by glycosidic bonds. Their main functions are protection and support (in organisms with cell walls) and also nutrient and energy stores.
- Lipids are biological molecules such as fats that are insoluble in water. Their main functions are as cell components, cell secretions, and nutrient and energy stores.
- Proteins are biological molecules whose polymers are chains of amino acid monomers linked together by peptide bonds.
- Proteins are called the "shapers of life" because of the many biological roles they play in cell structure and cell metabolism.
- Protein structure determines protein function. Structure and shape are dictated by amino acid composition and by the pH and temperature of the protein's immediate environment.
- Nucleic acids are biological molecules whose polymers are chains of nucleotide monomers linked together by phosphate–pentose sugar covalent bonds. Double-stranded nucleic acids are linked together by hydrogen bonds. Nucleic acids are information molecules that direct cell metabolism and reproduction. Nucleotides such as ATP also serve as energy transfer molecules in cells.

2.3 Cells: Where Chemicals Come to Life

As we proceed in this chemical survey from the level of simple molecules to increasingly complex levels of macromolecules, at some point we cross a line from the realm of lifeless molecules and arrive at the fundamental unit of life called a **cell.**[8] A cell is indeed a huge aggregate of carbon, hydrogen, oxygen, nitrogen, and many other atoms, and it follows the basic laws of chemistry and physics, but it is much more. The combination of these atoms produces characteristics, reactions, and products that can only be described as *living*.

8. The word *cell* was originally coined from an Old English term meaning "small room" because of the way plant cells looked to early microscopists.

Fundamental Characteristics of Cells

The bodies of living things such as bacteria and protozoa consist of only a single cell, whereas those of animals and plants contain trillions of cells. Regardless of the organism, all cells have a few common characteristics. They tend to be spherical, polygonal, cubical, or cylindrical, and their protoplasm (internal cell contents) is encased in a cell or cytoplasmic membrane (see Insight 2.3). They have chromosomes containing DNA and ribosomes for protein synthesis, and they are exceedingly complex in function. Aside from these few similarities, most cell types fall into one of two fundamentally different lines (discussed in chapter 1): the small, seemingly simple prokaryotic cells and the larger, structurally more complicated eukaryotic cells.

Eukaryotic cells are found in animals, plants, fungi, and protists. They contain a number of complex internal parts called organelles that perform useful functions for the cell involving growth, nutrition, or metabolism. By convention, organelles are defined as cell components that perform specific functions and are enclosed by membranes. Organelles also partition the eukaryotic cell into smaller compartments. The most visible organelle is the nucleus, a roughly ball-shaped mass surrounded by a double membrane that contains the DNA of the cell. Other organelles include the Golgi apparatus, endoplasmic reticulum, vacuoles, and mitochondria.

Prokaryotic cells are possessed only by the bacteria and archaea. Sometimes it may seem that prokaryotes are the microbial "have nots" because, for the sake of comparison, they are described by what they lack. They have no nucleus or other organelles. This apparent simplicity is misleading, because the fine structure of prokaryotes is complex. Overall, prokaryotic cells can engage in nearly every activity that eukaryotic cells can, and many can function in ways that eukaryotes cannot. Chapters 4 and 5 delve deeply into the properties of prokaryotic and eukaryotic cells.

CHECKPOINT

- As the atom is the fundamental unit of matter, so is the cell the fundamental unit of life.

Chapter Summary with Key Terms

2.1 Atoms, Bonds, and Molecules: Fundamental Building Blocks

A. Atomic Structure and Elements
   1. All **matter** in the universe is composed of minute particles called **atoms.** Atoms are composed of smaller particles called **protons, neutrons,** and **electrons.**
   2. Atoms that differ in numbers of the protons, neutrons, and electrons are elements. Each **element** is known by a distinct name and symbol. Elements may exist in variant forms called **isotopes.**

B. Bonds and Molecules
   1. Atoms interact to form **chemical bonds** and **molecules.** If the atoms combining to make a molecule are different elements, then the substance is termed a **compound.**
   2. The type of bond is dictated by the electron makeup of the outer orbitals **(valence)** of the atoms. Bond types include:
      a. **Covalent bonds,** with shared electrons. The molecule shares the electrons; the balance of charge will be **polar** if unequal or **nonpolar** if equally shared/electrically neutral.
      b. **Ionic bonds,** where electrons are transferred to an atom that can come closer to filling up the outer orbital. Dissociation of these compounds leads to the formation of charged **cations** and **anions.**
      c. **Hydrogen bonds** involve weak covalent bonds between hydrogen and nearby electronegative oxygens and nitrogens.

C. Solutions, Acids, Bases, and pH
   1. A **solution** is a combination of a solid, liquid, or gaseous chemical (the **solute**) dissolved in a liquid medium (the **solvent**). Water is the most common solvent in natural systems.
   2. Ionization of water leads to the release of hydrogen ions (H^+) and hydroxyl (OH^-) ions. The **pH** scale expresses the concentration of H^+ such that a pH of less than 7.0 is considered **acidic,** and a pH of more than that, indicating fewer H^+, is considered **basic.**

2.2 Macromolecules: Superstructures of Life

A. Biochemistry studies those molecules that are found in living things. These are based on **organic** compounds, which usually consist of carbon and hydrogen covalently bonded in various combinations.

B. **Macromolecules** are very large compounds and are generally assembled from single units called **monomers** by polymerization.

C. Macromolecules of life fall into basic categories of **carbohydrates, lipids, proteins,** and nucleic acids.
   1. **Carbohydrates** are composed of carbon, hydrogen, and oxygen and contain aldehyde or ketone groups.
      a. **Monosaccharides** such as glucose are the simplest carbohydrates with 3 to 7 carbons; these are the monomers of carbohydrates.
      b. **Disaccharides** such as lactose consist of two monosaccharides joined by **glycosidic bonds. Polysaccharides** such as starch and **peptidoglycan** are chains of five or more monosaccharides.
   2. **Lipids** contain long hydrocarbon chains and are not soluble in polar solvents such as water due to their nonpolar, hydrophobic character. Examples are **triglycerides, phospholipids,** sterols, and waxes.
   3. **Proteins** are highly complex macromolecules that are crucial in most, if not all, life processes.

a. **Amino acids** are the basic building blocks of proteins. They all share a basic structure of an amino group, a carboxyl group, an R group, and hydrogen bonded to a carbon atom. There are 20 different R groups, which define the basic set of 20 amino acids found in all of life.

b. The structure of a protein is very important to the function it has. This is described by the **primary structure** (the chain of amino acids), the **secondary structure** (formation of helices and sheets due to hydrogen bonding within the chain), **tertiary structure** (cross-links, especially disulfide bonds, between secondary structures), and **quaternary structure** (formation of multisubunit proteins). The incredible variation in shapes is the basis for the diverse roles proteins play as **enzymes, antibodies,** receptors, and structural components.

4. Nucleic acids

a. **Nucleotides** are the building blocks of nucleic acids. They are composed of a **nitrogen base,** a **pentose** sugar, and phosphate. Nitrogen bases are ringed compounds: **adenine (A), guanine (G), cytosine (C), thymine (T),** and **uracil (U).** Pentose sugars may be deoxyribose or ribose.

b. **Deoxyribonucleic acid (DNA)** is a polymer of nucleotides that occurs as a double-stranded helix with hydrogen bonding in pairs between the helices. It has all of the bases except uracil, and the pentose sugar is deoxyribose. DNA is the master code for a cell's life processes.

c. **Ribonucleic acid (RNA)** is a polymer of nucleotides where the sugar is **ribose** and **uracil** is used instead of thymine. It is almost always found single stranded and is used to express the DNA code into proteins.

d. **Adenosine triphosphate (ATP)** is a nucleotide involved in the transfer and storage of energy in cells.

2.3 Cells: Where Chemicals Come to Life

A. All living things are composed of **cells,** which are aggregates of macromolecules that carry out living processes.

B. Cells can be divided into two basic types: prokaryotes and eukaryotes.

Multiple-Choice and True-False Questions

Multiple-Choice Questions. Select the correct answer from the answers provided.

1. The smallest unit of matter with unique characteristics is
   a. an electron c. an atom
   b. a molecule d. a proton

2. The ____ charge of a proton is exactly balanced by the ____ charge of a (an) ____.
   a. negative, positive, electron
   b. positive, neutral, neutron
   c. positive, negative, electron
   d. neutral, negative, electron

3. Electrons move around the nucleus of an atom in pathways called
   a. shells c. circles
   b. orbital d. rings

4. Bonds in which atoms share electrons are defined as ____ bonds.
   a. hydrogen c. double
   b. ionic d. covalent

5. Hydrogen bonds can form between ____ adjacent to each other.
   a. two hydrogen atoms
   b. two oxygen atoms
   c. a hydrogen atom and an oxygen atom
   d. negative charges

6. An atom that can donate electrons during a reaction is called
   a. an oxidizing agent c. an ionic agent
   b. a reducing agent d. an electrolyte

7. In a solution of NaCl and water, NaCl is the ____ and water is the ____.
   a. acid, base c. solute, solvent
   b. base, acid d. solvent, solute

8. A solution with a pH of 2 ____ than a solution with a pH of 8.
   a. has less H^+ c. has more OH^-
   b. has more H^+ d. is less concentrated

9. Fructose is a type of
   a. disaccharide c. polysaccharide
   b. monosaccharide d. amino acid

10. Bond formation in polysaccharides and polypeptides is accompanied by the removal of a
    a. hydrogen atom c. carbon atom
    b. hydroxyl ion d. water molecule

11. The monomer unit of polysaccharides such as starch and cellulose is
    a. fructose c. ribose
    b. glucose d. lactose

12. Proteins are synthesized by linking amino acids with ____ bonds.
    a. disulfide c. peptide
    b. glycosidic d. ester

13. DNA is a hereditary molecule that is composed of
    a. deoxyribose, phosphate, and nitrogen bases
    b. deoxyribose, a pentose, and nucleic acids
    c. sugar, proteins, and thymine
    d. adenine, phosphate, and ribose

14. Proteins can function as
    a. enzymes c. antibodies
    b. receptors d. a, b, and c

15. RNA plays an important role in what biological process?
    a. replication c. lipid metabolism
    b. protein synthesis d. water transport

True-False Questions. If the statement is true, leave as is. If it is false, correct it by rewriting the sentence.

16. Elements have varying numbers of protons, neutrons, and electrons.
17. Covalent bonds are those that are made between two different elements.
18. A compound is called "organic" if it is made of all-natural elements.
19. Cysteine is the amino acid that participates in disulfide bonds in proteins.
20. Membranes are mainly composed of macromolecules called carbohydrates.

Writing to Learn

These questions are suggested as a *writing-to-learn* experience. For each question, compose a one- or two-paragraph answer that includes the factual information needed to completely address the question.

1. How are the concepts of an atom and an element related? What causes elements to differ?
2. Distinguish between the general reactions in covalent, ionic, and hydrogen bonds.
3. a. Which kinds of elements tend to make covalent bonds?
   b. Distinguish between a single and a double bond.
   c. What is polarity?
   d. Why are some covalent molecules polar and others nonpolar?
   e. What is an important consequence of the polarity of water?
4. a. Which kinds of elements tend to make ionic bonds?
   b. Exactly what causes the charges to form on atoms in ionic bonds?
   c. Verify the proton and electron numbers for Na^+ and Cl^-.
   d. Differentiate between an anion and a cation.
   e. What kind of ion would you expect magnesium to make, on the basis of its valence?
5. Differentiate between an oxidizing agent and a reducing agent.
6. Why are hydrogen bonds relatively weak?
7. a. Compare the three basic types of chemical formulas.
   b. Review the types of chemical reactions and the general ways they can be expressed in equations.
8. a. What determines whether a substance is an acid or a base?
   b. Briefly outline the pH scale.
   c. How can a neutral salt be formed from acids and bases?
9. a. What atoms must be present in a molecule for it to be considered organic?
   b. What characteristics of carbon make it ideal for the formation of organic compounds?
   c. What are functional groups?
   d. Differentiate between a monomer and a polymer.
   e. How are polymers formed?
   f. Name several inorganic compounds.
10. a. Describe a nucleotide and a polynucleotide, and compare and contrast the general structure of DNA and RNA.
    b. Name the two purines and the three pyrimidines.
    c. Why is DNA called a double helix?
    d. What is the function of RNA?
    e. What is ATP, and what is its function in cells?

Concept Mapping

Appendix D provides guidance for working with concept maps.

1. Supply your own linking words or phrases in this concept map, and provide the missing concepts in the empty boxes.

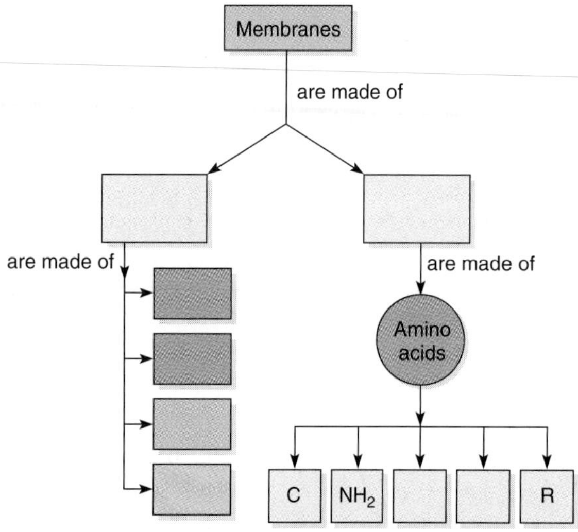

Critical Thinking Questions

Critical thinking is the ability to reason and solve problems using facts and concepts. These questions can be approached from a number of angles, and in most cases, they do not have a single correct answer.

1. The "octet rule" in chemistry helps predict the tendency of atoms to acquire or donate electrons from the outer shell. It says that those with fewer than 4 tend to donate electrons and those with more than 4 tend to accept additional electrons; those with exactly 4 can do both. Using this rule, determine what category each of the following elements falls into: N, S, C, P, O, H, Ca, Fe, and Mg. (You will need to work out the valence of the atoms.)
2. Predict the kinds of bonds that occur in ammonium (NH_3), phosphate (PO_4), disulfide (S—S), and magnesium chloride ($MgCl_2$). (Use simple models such as those in figure 2.4.)
3. Work out thc following problems:
   a. What is the number of protons in helium?
   b. Will an H bond form between $H_3C—CH=O$ and H_2O? Why or why not?
   c. Draw the following molecules and determine which are polar: Cl_2, NH_3, CH_4.
   d. What is the pH of a solution with a concentration of 0.00001 moles (M)/ml of H^+?
   e. What is the pH of a solution with a concentration of 0.00001 moles (M)/ml of OH^-?
4. a. Describe how hydration spheres are formed around cations and anions.
   b. What kinds of substances will be expected to be hydrophilic and hydrophobic, and what makes them so?
   c. Distinguish between polar and ionic compounds, using your own words.
5. In what way are carbon-based compounds like children's Tinker Toys or Lego blocks?
6. How many peptide bonds are in a tetrapeptide?
7. Looking at figure 2.24, can you see why adenine forms hydrogen bonds with thymine and why cytosine forms them with guanine?
8. Saturated fats are solid at room temperature and unsaturated fats are not. Is butter an example of a saturated or unsaturated fat? Is olive oil an example of a saturated or unsaturated fat? What are trans-fatty acids? Why is there currently a dietary trans-fatty acid debate?

Visual Understanding

1. **Figure 2.18*a*** and **Figure 2.19.** Speculate on why sterols like cholesterol can add "stiffness" to membranes that contain them.

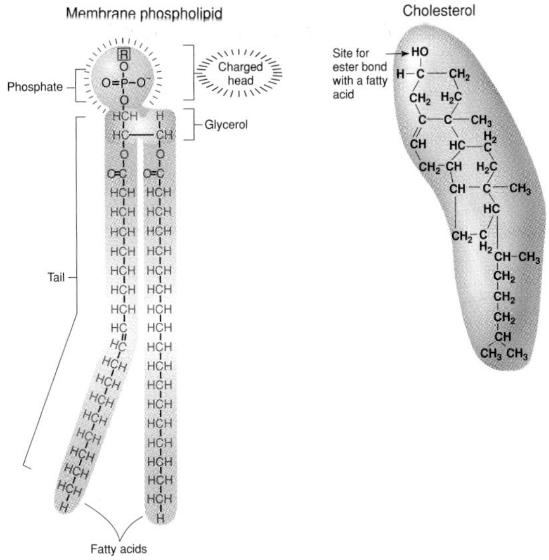

Internet Search Topics

1. Use a search engine to explore the topic of isotopes and dating ancient rocks. How can isotopes be used to determine if rocks contain evidence of life?
2. Use a search engine to try to determine exactly how many elements are currently recognized. Find at least two sites and compare what they have to say about the status of the elements numbered higher than 112.
3. Go to: www.aris.mhhe.com, and click on "microbiology" and then this textbook's author/title. Go to Chapter 2, access the URLs listed under Internet Search Topics, and research the following: Make a search for basic information on elements, using one or both of the websites listed in the Science Zone. Click on the icons for C, H, N, O, P, and S and list the source, biological importance, and other useful information about these elements.

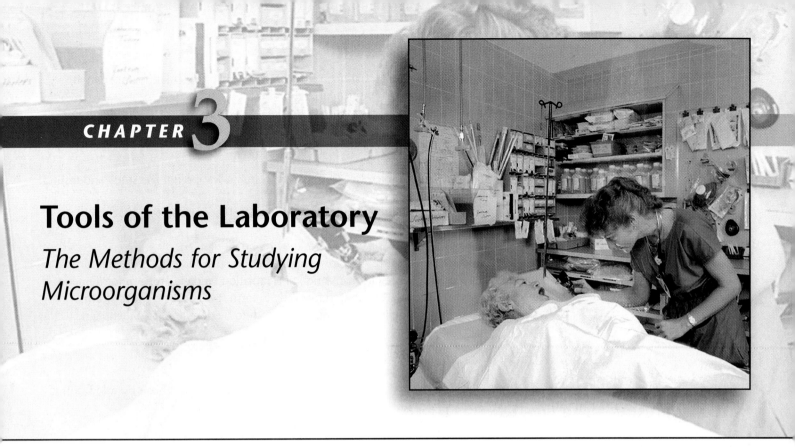

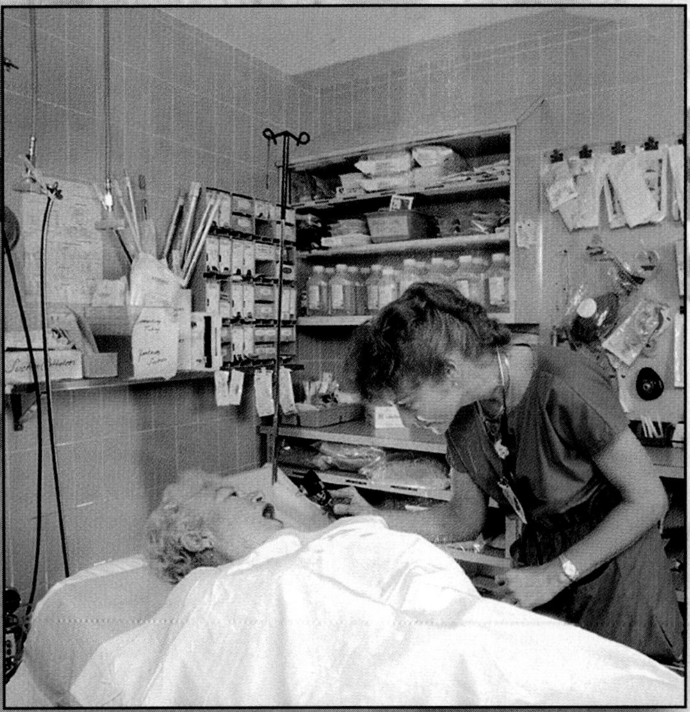

CHAPTER 3

Tools of the Laboratory

The Methods for Studying Microorganisms

CASE FILE 3

A 94-year-old woman went to her local hospital emergency department in mid-November 2001 complaining of a 5-day history of weakness, fever, nonproductive cough, and generalized myalgia (muscle aches). Otherwise, for a person her age she was fairly healthy, although she did suffer from chronic obstructive pulmonary disease, hypertension, and chronic kidney failure.

On physical examination, her heart rate was above normal and she had a fever of 102.3°F (39.1°C). The rest of her physical examination was normal. Initial laboratory studies of blood cell count, blood chemistries, and chest X ray were also normal except for the chemical urine testing. This finding along with the fever suggested an infection, so the patient was admitted to the hospital. Samples of blood and urine were sent to the microbiology laboratory and set up appropriately.

The next day, microscopic evaluation of the urine culture revealed rod-shaped bacteria that stained red, and the blood culture revealed rods that stained purple. The liquid blood culture was then transferred to appropriate solid media. This finding in the blood was unusual, so a sample culture was sent to the state health department laboratory. Antibiotic therapy was adjusted, yet the patient's condition deteriorated. Her most serious symptoms localized to her chest, and she was transferred to the intensive care unit. Four days after admission, the health department announced that the bacteria found in the patient's blood were *Bacillus anthracis*. She was suffering from inhalation anthrax. Further testing showed these bacteria to be of the same strain that had been involved in the recent bioterrorist attack. Despite treatment, the patient died on the fifth day after admission.

- *What techniques and equipment are used when the bacteria are observed as being purple and red? How are these findings reported?*
- *What are the stages of processing a blood sample?*

Case File 3 Wrap-Up appears on page 81.

CHAPTER OVERVIEW

- Microbes are managed and characterized with the Five I's—inoculation, incubation, isolation, inspection, and identification.
- Cultures are made by removing samples from a desired source and placing them in containers of media.
- Media can be varied in chemical and physical form and functional purposes, depending on the intention.
- Growth and isolation of microbes lead to pure cultures that permit the study and testing of single species.
- Cultures can be used to provide information on microbial morphology, biochemistry, and genetic characteristics.
- Unknown, invisible samples can become known and visible.
- The microscope is a powerful tool for magnifying and resolving cells and their parts.
- Microscopes exist in several forms, using light, radiation, and electrons to form images.
- Specimens and cultures are prepared for study in fresh (live) or fixed (dead) form.
- Staining procedures highlight cells and allow them to be described and identified.

3.1 Methods of Culturing Microorganisms—The Five I's

Biologists studying large organisms such as animals and plants can, for the most part, immediately see and differentiate their experimental subjects from the surrounding environment and from one another. In fact, they can use their senses of sight, smell, hearing, and even touch to detect and evaluate identifying characteristics and to keep track of growth and developmental changes. Because microbiologists cannot rely as much as other scientists on senses other than sight, they are confronted by some unique problems. First, most habitats (such as the soil and the human mouth) harbor microbes in complex associations, so it is often necessary to separate the species from one another. Second, to maintain and keep track of such small research subjects, microbiologists usually have to grow them under artificial (and thus distorting) conditions. A third difficulty in working with microbes is that they are invisible and widely distributed, and undesirable ones can be introduced into an experiment and cause misleading results. These impediments motivated the development of techniques to control microbes and their growth, primarily sterile, aseptic, and pure culture techniques.[1]

Microbiologists use five basic techniques to manipulate, grow, examine, and characterize microorganisms in the laboratory: inoculation, incubation, isolation, inspection, and identification (the Five I's; **figure 3.1**). Some or all of these procedures are performed by microbiologists, whether beginning laboratory students, researchers attempting to isolate drug-producing bacteria from soil, or clinical microbiologists working with a specimen from a patient's infection. These procedures make it possible to handle and maintain microorganisms as discrete entities whose detailed biology can be studied and recorded.

1. *Sterile* means the complete absence of viable microbes; *aseptic* refers to prevention of infection; *pure culture* refers to growth of a single species of microbe; *aseptic technique* is the process used to manipulate cultures without introducing contaminating microbes.

Inoculation: Producing a Culture

To cultivate, or **culture,** microorganisms, one introduces a tiny sample (the inoculum) into a container of nutrient **medium** (pl. media), which provides an environment in which they multiply. This process is called **inoculation.** Any instrument used for sampling and inoculation must initially be *sterile* (see footnote 1). The observable growth that appears in or on the medium after **incubation** is known as a culture. The nature of the sample being cultured depends on the objectives of the analysis. Clinical specimens for determining the cause of an infectious disease are obtained from body fluids (blood, cerebrospinal fluid), discharges (sputum, urine, feces), or diseased tissue. Other samples subject to microbiological analysis are soil, water, sewage, foods, air, and inanimate objects. Procedures for proper specimen collection are discussed in chapter 17.

Isolation: Separating One Species from Another

Certain **isolation** techniques are based on the concept that if an individual bacterial cell is separated from other cells and provided adequate space on a nutrient surface, it will grow into a discrete mound of cells called a **colony (figure 3.2).** If it was formed from a single cell, a colony consists of just that one species and no other. Proper isolation requires that a small number of cells be inoculated into a relatively large volume or over an expansive area of medium. It generally requires the following materials: a medium that has a relatively firm surface (see agar in "Physical States of Media," page 62), a Petri dish (a clear, flat dish with a cover), and inoculating tools. In the streak plate method, a small droplet of culture or sample is spread over the surface of the medium with an *inoculating loop* according to a pattern that gradually thins out the sample and separates the cells spatially over several sections of the

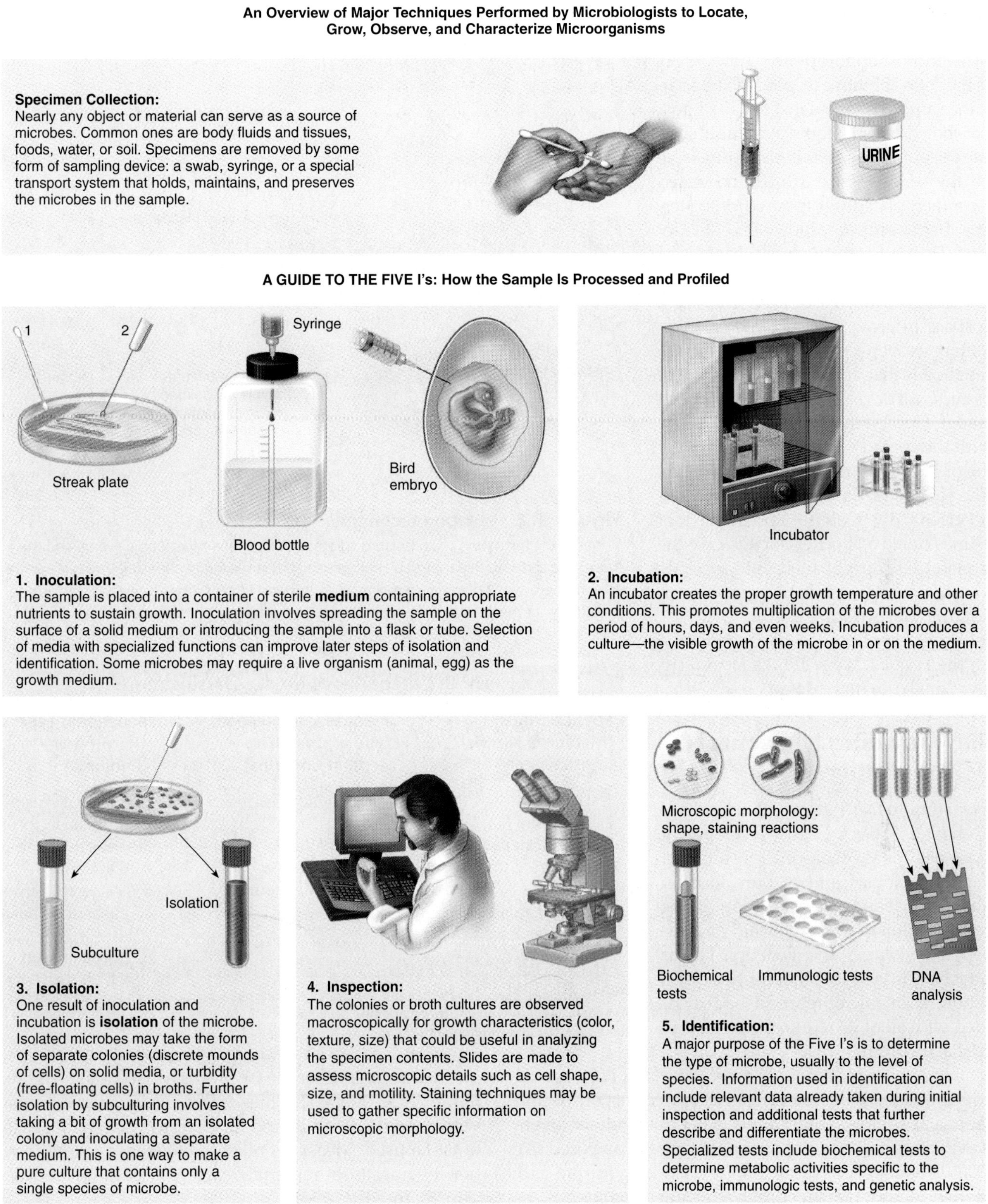

An Overview of Major Techniques Performed by Microbiologists to Locate, Grow, Observe, and Characterize Microorganisms
Specimen Collection:
Nearly any object or material can serve as a source of microbes. Common ones are body fluids and tissues, foods, water, or soil. Specimens are removed by some form of sampling device: a swab, syringe, or a special transport system that holds, maintains, and preserves the microbes in the sample.
URINE
A GUIDE TO THE FIVE I's: How the Sample Is Processed and Profiled
1
2
Syringe
Streak plate
Bird embryo
Blood bottle
1. Inoculation:
The sample is placed into a container of sterile medium containing appropriate nutrients to sustain growth. Inoculation involves spreading the sample on the surface of a solid medium or introducing the sample into a flask or tube. Selection of media with specialized functions can improve later steps of isolation and identification. Some microbes may require a live organism (animal, egg) as the growth medium.
Incubator
2. Incubation:
An incubator creates the proper growth temperature and other conditions. This promotes multiplication of the microbes over a period of hours, days, and even weeks. Incubation produces a culture—the visible growth of the microbe in or on the medium.
Isolation
Subculture
3. Isolation:
One result of inoculation and incubation is isolation of the microbe. Isolated microbes may take the form of separate colonies (discrete mounds of cells) on solid media, or turbidity (free-floating cells) in broths. Further isolation by subculturing involves taking a bit of growth from an isolated colony and inoculating a separate medium. This is one way to make a pure culture that contains only a single species of microbe.
4. Inspection:
The colonies or broth cultures are observed macroscopically for growth characteristics (color, texture, size) that could be useful in analyzing the specimen contents. Slides are made to assess microscopic details such as cell shape, size, and motility. Staining techniques may be used to gather specific information on microscopic morphology.
Microscopic morphology: shape, staining reactions
Biochemical tests
Immunologic tests
DNA analysis
5. Identification:
A major purpose of the Five I's is to determine the type of microbe, usually to the level of species. Information used in identification can include relevant data already taken during initial inspection and additional tests that further describe and differentiate the microbes. Specialized tests include biochemical tests to determine metabolic activities specific to the microbe, immunologic tests, and genetic analysis.

Figure 3.1 A summary of the general laboratory techniques carried out by microbiologists.
It is not necessary to perform all the steps shown or to perform them exactly in this order, but all microbiologists participate in at least some of these activities. In some cases, one may proceed right from the sample to inspection, and in others, only inoculation and incubation on special media are required.

plate **(figure 3.3a,b).** Because of its ease and effectiveness, the streak plate is the method of choice for most applications.

In the loop dilution, or pour plate, technique, the sample is inoculated serially into a series of cooled but still liquid agar tubes so as to dilute the number of cells in each successive tube in the series **(figure 3.3c,d).** Inoculated tubes are then plated out (poured) into sterile Petri dishes and are allowed to solidify (harden). The end result (usually in the second or third plate) is that the number of cells per volume is so decreased that cells have ample space to grow into separate colonies. One difference between this and the streak plate method is that in this technique some of the colonies will develop deep in the medium itself and not just on the surface.

With the spread plate technique, a small volume of liquid, diluted sample is pipetted onto the surface of the medium and spread around evenly by a sterile spreading tool (sometimes called a "hockey stick"). Like the streak plate, cells are pushed onto separate areas on the surface so that they can form individual colonies **(figure 3.3e,f).**

Before we continue to cover information on the Five I's, we will take a side trip to look at media in more detail.

Media: Providing Nutrients in the Laboratory

A major stimulus to the rise of microbiology in the late 1800s was the development of techniques for growing microbes out of their natural habitats and in pure form in the laboratory. This milestone enabled the close examination of a microbe and its morphology, physiology, and genetics. It was evident from the very first that for successful cultivation, each microorganism had to be provided with all of its required nutrients in an artificial medium.

Some microbes require only a very few simple inorganic compounds for growth; others need a complex list of specific inorganic and organic compounds. This tremendous diversity is evident in the types of media that can be prepared. At least 500 different types of media are used in culturing and identifying microorganisms. Culture media are contained in test tubes, flasks, or Petri dishes, and they are inoculated by such tools as loops, needles, pipettes, and swabs. Media are extremely varied in nutrient content and consistency and can be specially formulated for a particular purpose. Culturing microbes that cannot grow on artificial media (all viruses and certain bacteria) requires cell cultures or host animals **(Insight 3.1).**

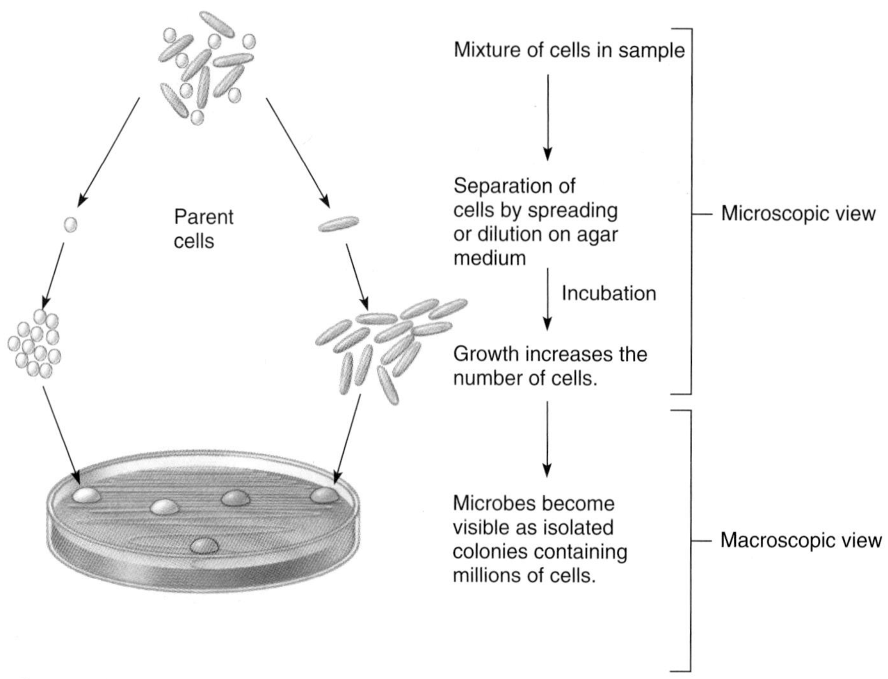

Figure 3.2 Isolation technique.
Stages in the formation of an isolated colony, showing the microscopic events and the macroscopic result. Separation techniques such as streaking can be used to isolate single cells. After numerous cell divisions, a macroscopic mound of cells, or a colony, will be formed. This is a relatively simple yet successful way to separate different types of bacteria in a mixed sample.

TABLE 3.1	Three Categories of Media Classification	
Physical State (Medium's Normal Consistency)	**Chemical Composition (Type of Chemicals Medium Contains)**	**Functional Type (Purpose of Medium)***
1. Liquid 2. Semisolid 3. Solid (can be converted to liquid) 4. Solid (cannot be liquefied)	1. Synthetic (chemically defined) 2. Nonsynthetic (complex; not chemically defined)	1. General purpose 2. Enriched 3. Selective 4. Differential 5. Anaerobic growth 6. Specimen transport 7. Assay 8. Enumeration

*Some media can serve more than one function. For example, a medium such as brain-heart infusion is general purpose and enriched; mannitol salt agar is both selective and differential; and blood agar is both enriched and differential.

For an experiment to be properly controlled, sterile technique is necessary. This means that the inoculation must start with a sterile medium and inoculating tools with sterile tips must be used. Measures must be taken to prevent introduction of nonsterile materials, such as room air and fingers, directly into the media.

Types of Media

Media can be classified according to three properties **(table 3.1):**

1. physical state,
2. chemical composition, and
3. functional type.

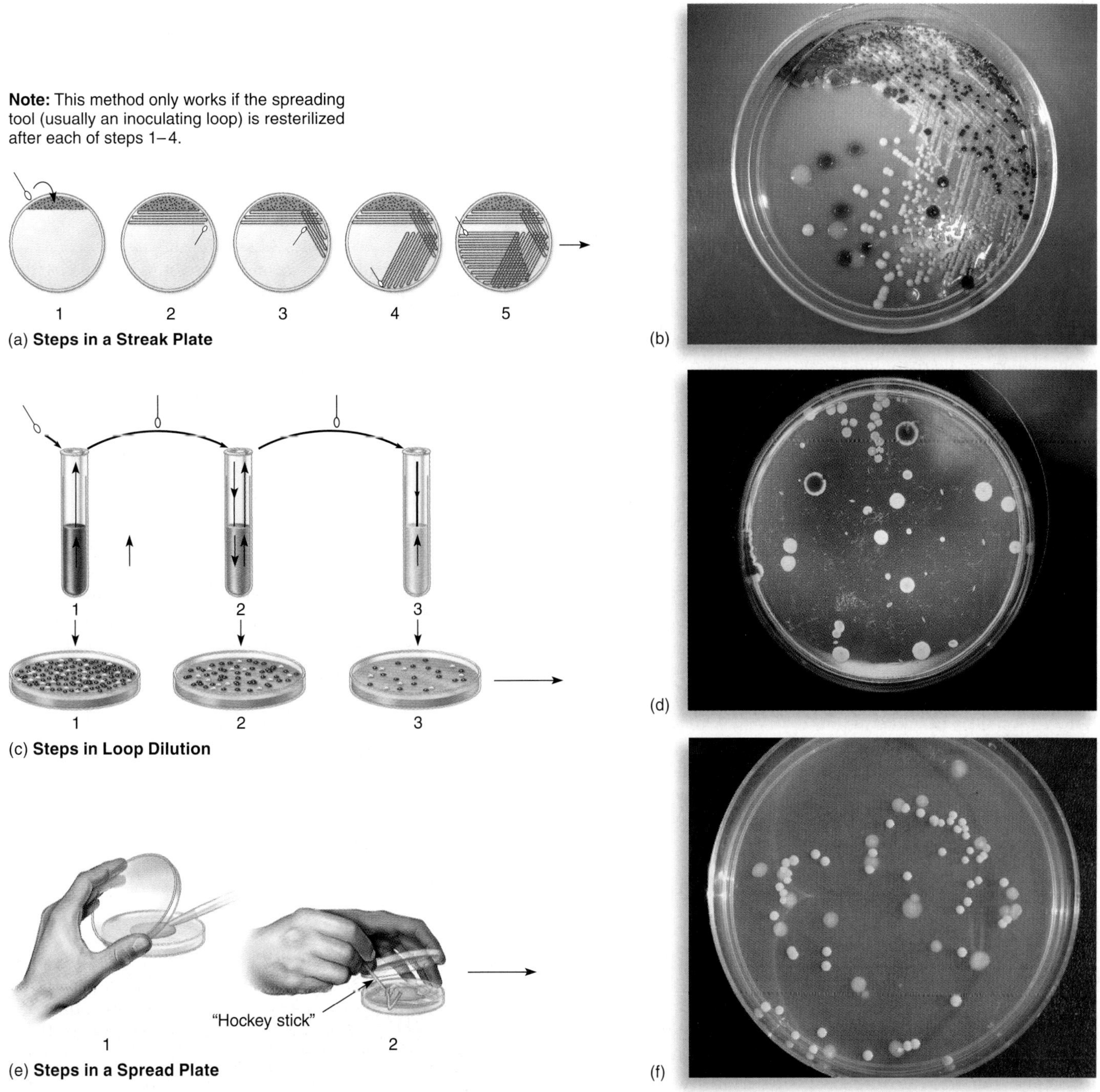

Figure 3.3 Methods for isolating bacteria.
(a) Steps in a quadrant streak plate and **(b)** resulting isolated colonies of bacteria. **(c)** Steps in the loop dilution method and **(d)** the appearance of plate 3. **(e)** Spread plate and **(f)** its result.

Most media discussed here are designed for bacteria and fungi, though algae and some protozoa can be propagated in media.

Physical States of Media

Liquid media are defined as water-based solutions that do not solidify at temperatures above freezing and that tend to flow freely when the container is tilted. These media, termed broths, milks, or infusions, are made by dissolving various solutes in distilled water. Growth occurs throughout the container and can then present a dispersed, cloudy, or particulate appearance. A common laboratory medium, *nutrient broth,* contains beef extract and peptone dissolved in water. Methylene blue milk and litmus milk are opaque liquids containing whole milk and dyes. Fluid thioglycollate is a slightly viscous broth used for determining patterns of growth in oxygen.

INSIGHT 3.1 *Medical*

Animal Inoculation: "Living Media"

A great deal of attention has been focused on the uses of animals in biology and medicine. Animal rights activists are vocal about practically any experimentation with animals and have expressed their outrage quite forcefully. Certain kinds of animal testing may seem trivial and unnecessary, but many times it is absolutely necessary to use animals bred for experimental purposes, such as guinea pigs, mice, chickens, and even armadillos. Such animals can be an indispensable aid for studying, growing, and identifying microorganisms. One special use of animals involves inoculation of the early life stages (embryos) of birds. Vaccines for influenza are currently produced in chicken embryos. The major rationales for live animal inoculation can be summarized as follows:

1. Animal inoculation is an essential step in testing the effects of drugs and the effectiveness of vaccines before they are administered to humans. It makes progress toward prevention, treatment, and cure possible without risking the lives of humans.
2. Researchers develop animal models for evaluating new diseases or for studying the cause or process of a disease. Koch's postulates are a series of proofs to determine the causative agent of a disease and require a controlled experiment with an animal that can develop a typical case of the disease. Researchers have also created hundreds of engineered animals to monitor the effects of genetic diseases and to study the actions of the immune system.
3. Animals are an important source of antibodies, antisera, antitoxins, and other immune products that can be used in therapy or testing.
4. Animals are sometimes required to determine the pathogenicity or toxicity of certain bacteria. One such test is the mouse neutralization test for the presence of botulism toxin in food. This test can help identify even very tiny amounts of toxin and thereby can avert outbreaks of this disease. Occasionally, it is necessary to inoculate an animal to distinguish between pathogenic or nonpathogenic strains of *Listeria* or *Candida* (a yeast).
5. Some microbes will not grow on artificial media but will grow in a suitable animal and can be recovered in a more or less pure form. These include animal viruses, the spirochete of syphilis, and the leprosy bacillus (grown in armadillos).

The nude or athymic mouse has genetic defects in hair formation and thymus development. It is widely used to study cancer, immune function, and infectious diseases.

At ordinary room temperature, **semisolid media** exhibit a clotlike consistency **(figure 3.4)** because they contain an amount of solidifying agent (agar or gelatin) that thickens them but does not produce a firm substrate. Semisolid media are used to determine the motility of bacteria and to localize a reaction at a specific site. Motility test medium and sulfur indole motility medium (SIM) both contain a small amount (0.3%–0.5%) of agar. In both cases the medium is stabbed carefully in the center and later observed for the pattern of growth around the stab line. In addition to motility, SIM can test for physiological characteristics used in identification (hydrogen sulfide production and indole reaction).

Solid media provide a firm surface on which cells can form discrete colonies (see figure 3.3) and are advantageous for isolating and culturing bacteria and fungi. They come in two forms: liquefiable and nonliquefiable. Liquefiable solid media, sometimes called reversible solid media, contain a solidifying agent that changes their physical properties in response to temperature. By far the most widely used and effective of these agents is **agar,** a complex polysaccharide isolated from the red alga *Gelidium*. The benefits of agar are numerous. It is solid at room temperature, and it melts (liquefies) at the boiling temperature of water (100°C). Once liquefied, agar does not resolidify until it cools to 42°C, so it can be inoculated and poured in liquid form at temperatures (45° to 50°C) that will not harm the microbes or the handler. Agar is flexible and moldable, and it provides a basic framework to hold moisture and nutrients, though it is not itself a digestible nutrient for most microorganisms.

Any medium containing 1% to 5% agar usually has the word *agar* in its name. *Nutrient agar* is a common one. Like nutrient broth, it contains beef extract and peptone, as well as 1.5% agar by weight. Many of the examples covered in the section on functional categories of media contain agar. Although gelatin is not nearly as satisfactory as agar, it will create a reasonably solid surface in concentrations of 10% to 15%. Agar and gelatin media are illustrated in **figure 3.5.**

Figure 3.4 Sample semisolid media.
(a) Semisolid media have more body than liquid media but less body than solid media. They do not flow freely and have a soft, clotlike consistency. **(b)** Sulfur indole motility medium (SIM). The **(1)** medium is stabbed with an inoculum and incubated. Location of growth indicates nonmotility **(2)** or motility **(3)**. If H_2S gas is released, a black precipitate forms **(4)**.

Nonliquefiable solid media have less versatile applications than agar media because they do not melt. They include materials such as rice grains (used to grow fungi), cooked meat media (good for anaerobes), and potato slices; all of these media start out solid and remain solid after heat sterilization. Other solid media containing egg and serum start out liquid and are permanently coagulated or hardened by moist heat.

Chemical Content of Media

Media whose compositions are precisely chemically defined are termed *synthetic*. Such media contain pure organic and inorganic compounds that vary little from one source to another and have a molecular content specified by means of an exact formula. Synthetic media come in many forms. Some media, such as minimal media for fungi, contain nothing more than a few essential compounds such as salts and amino acids dissolved in water. Others contain a variety of defined organic and inorganic chemicals **(table 3.2)**. Such standardized and reproducible media are most useful in research and cell culture when the exact nutritional needs of the test organisms are known. If even one component of a given medium is not chemically definable, the medium belongs in the "complex" category.

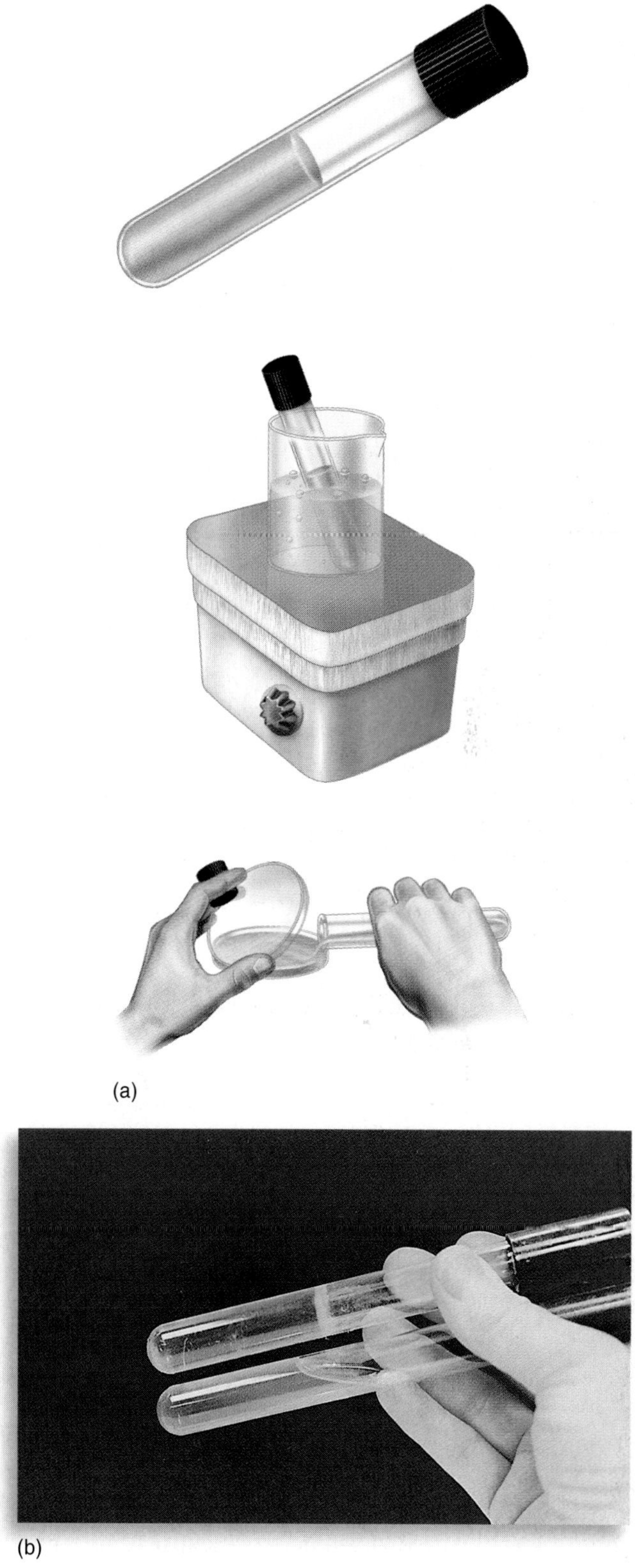

Figure 3.5 Solid media that are reversible to liquids.
(a) Media containing 1%–5% agar are solid enough to remain in place when containers are tilted or inverted. They are reversibly solid and can be liquefied with heat, poured into a different container, and resolidified. **(b)** Nutrient gelatin contains enough gelatin (12%) to take on a solid consistency. The top tube shows it as a solid. The bottom tube indicates what happens when it is warmed or when microbial enzymes digest the gelatin and liquefy it.

TABLE 3.2A Chemically Defined Synthetic Medium for Growth and Maintenance of Pathogenic *Staphylococcus aureus*

0.25 Grams Each of These Amino Acids	0.5 Grams Each of These Amino Acids	0.12 Grams Each of These Amino Acids
Cystine	Arginine	Aspartic acid
Histidine	Glycine	Glutamic acid
Leucine	Isoleucine	
Phenylalanine	Lysine	
Proline	Methionine	
Tryptophan	Serine	
Tyrosine	Threonine	
	Valine	

Additional ingredients

Ingredient	Category
0.005 mole nicotinamide	Vitamins
0.005 mole thiamine	Vitamins
0.005 mole pyridoxine	Vitamins
0.5 micrograms biotin	Vitamins
1.25 grams magnesium sulfate	Salts
1.25 grams dipotassium hydrogen phosphate	Salts
1.25 grams sodium chloride	Salts
0.125 grams iron chloride	Salts

Ingredients dissolved in 1,000 milliliters of distilled water and buffered to a final pH of 7.0.

TABLE 3.2B Brain Heart Infusion Broth: A Complex, Nonsynthetic Medium for Growth and Maintenance of Pathogenic *Staphylococcus aureus*

27.5 grams brain, heart extract, peptone extract
2 grams glucose
5 grams sodium chloride
2.5 grams di-sodium hydrogen phosphate
Ingredients dissolved in 1,000 milliliters of distilled water and buffered to a final pH of 7.0.

Complex, or nonsynthetic, media contain at least one ingredient that is *not* chemically definable—not a simple, pure compound and not representable by an exact chemical formula. Most of these substances are extracts of animals, plants, or yeasts, including such materials as ground-up cells, tissues, and secretions. Examples are blood, serum, and meat extracts or infusions. Other nonsynthetic ingredients are milk, yeast extract, soybean digests, and peptone. Peptone is a partially degraded protein, rich in amino acids, that is often used as a carbon and nitrogen source. Nutrient broth, blood agar, and MacConkey agar, though different in function and appearance, are all complex nonsynthetic media. They present a rich mixture of nutrients for microbes that have complex nutritional needs.

Table 3.2 provides a practical comparison of the two categories, using a *Staphylococcus* medium. Every substance in medium A is known to a very precise degree. The substances in medium B are mostly macromolecules that contain dozens of unknown (but required) nutrients. Both A and B will satisfactorily grow the bacterium.

Media to Suit Every Function

Microbiologists have many types of media at their disposal, with new ones being devised all the time. Depending upon what is added, a microbiologist can fine-tune a medium for nearly any purpose. Until recently, microbiologists knew of only a few species of bacteria or fungi that could not be cultivated artificially. Newer DNA detection technologies have shown us just how wrong we were; it is now thought that there are many times more microbes that we don't know how to cultivate in the lab than those that we do. Previous discovery and identification of microorganisms relied on our ability to grow them. Now we can detect a single bacterium in its natural habitat.

General-purpose media are designed to grow as broad a spectrum of microbes as possible. As a rule, they are nonsynthetic and contain a mixture of nutrients that could support the growth of a variety of microbial life. Examples include nutrient agar and broth, brain-heart infusion, and trypticase soy agar (TSA). TSA contains partially degraded milk protein (casein), soybean digest, NaCl, and agar.

An **enriched medium** contains complex organic substances such as blood, serum, hemoglobin, or special **growth factors** (specific vitamins, amino acids) that certain species must have in order to grow. Bacteria that require growth factors and complex nutrients are termed **fastidious.** Blood agar, which is made by adding sterile sheep, horse, or rabbit blood to a sterile agar base **(figure 3.6*a*)** is widely employed to grow fastidious streptococci and other pathogens. Pathogenic *Neisseria* (one species causes gonorrhea) are grown on Thayer-Martin medium or chocolate agar, which is made by heating blood agar **(figure 3.6*b*).**

Selective and Differential Media Some of the cleverest and most inventive media recipes belong to the categories of selective and differential media **(figure 3.7).** These media are designed for special microbial groups, and they have extensive applications in isolation and identification. They can permit, in a single step, the preliminary identification of a genus or even a species.

A **selective medium (table 3.3)** contains one or more agents that inhibit the growth of a certain microbe or microbes (call them A, B, and C) but not others (D) and thereby encourages, or *selects,* microbe D and allows it to grow. Selective media are very important in primary isolation of a specific type of microorganism from samples containing dozens of different species—for example, feces, saliva, skin, water, and soil. They hasten isolation

(a)

(b)

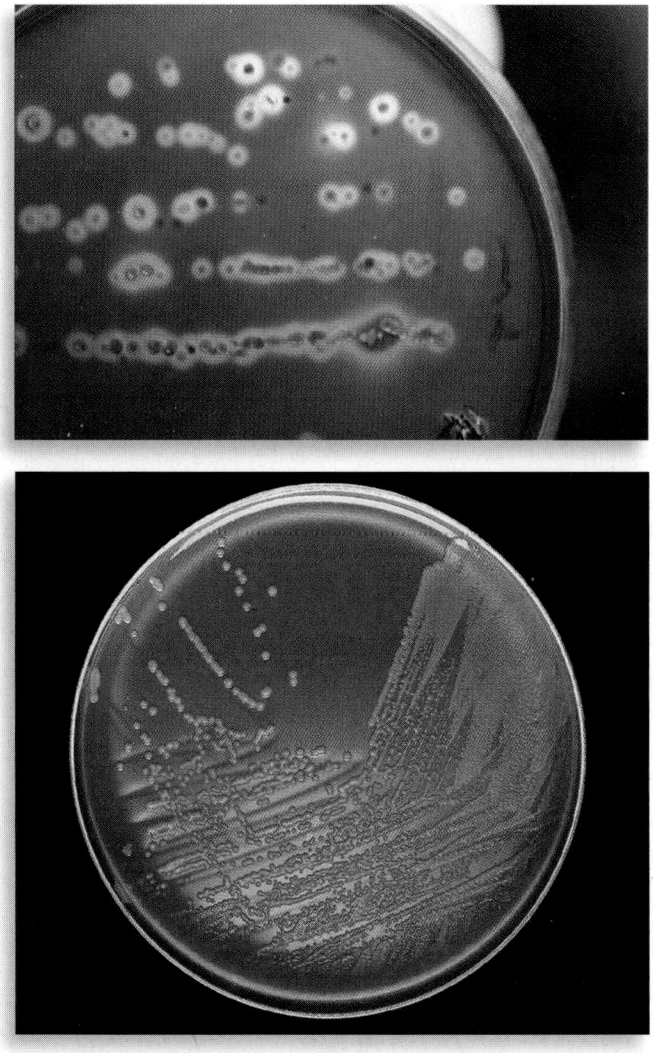

Figure 3.6 Examples of enriched media.
(a) Blood agar plate growing bacteria from the human throat. Note that this medium also differentiates among colonies by the zones of hemolysis (clear areas) they may show. **(b)** Chocolate agar, a medium that gets its brown color from heated blood, not from chocolate. It is commonly used to culture the fastidious *Haemophilus influenzae*, the causative agent of one type of meningitis.

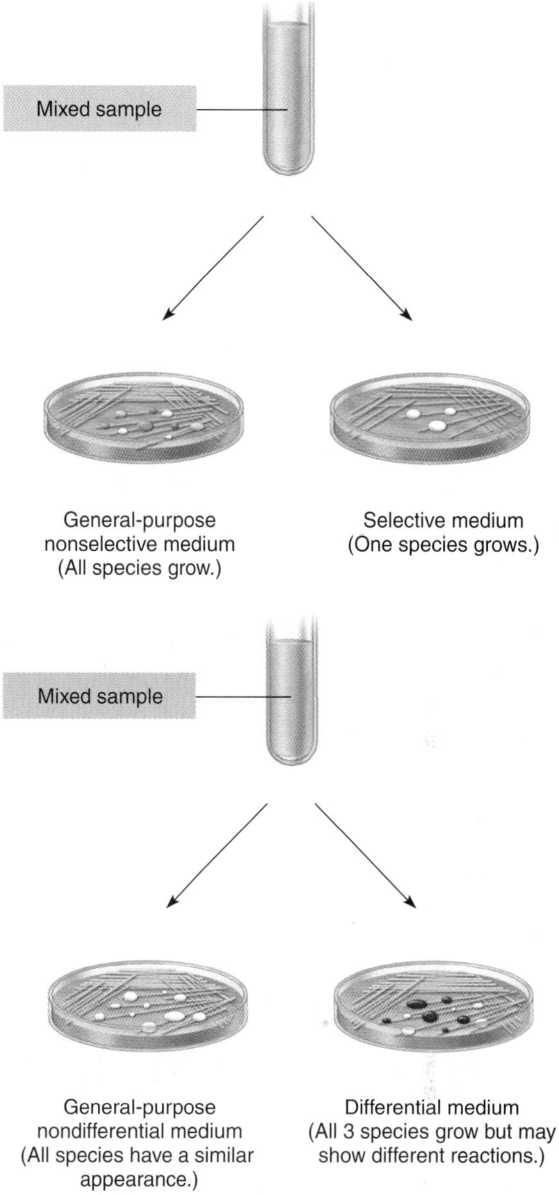

Figure 3.7 Comparison of selective and differential media with general-purpose media.
(a) A mixed sample containing three different species is streaked onto plates of general-purpose nonselective medium and selective medium. Note the results. **(b)** Another mixed sample containing three different species is streaked onto plates of general-purpose nondifferential medium and differential medium. Note the results.

by suppressing the unwanted background organisms and favoring growth of the desired ones.

Mannitol salt agar (MSA) **(figure 3.8*a*)** contains a high concentration of NaCl (7.5%) that is quite inhibitory to most human pathogens. One exception is the genus *Staphylococcus*, which grows well in this medium and consequently can be amplified in mixed samples. Bile salts, a component of feces, inhibit most gram-positive bacteria while permitting many gram-negative rods to grow. Media for isolating intestinal pathogens (MacConkey agar, Hektoen enteric [HE] agar) contain bile salts as a selective agent **(figure 3.8*b*)**. Dyes such as methylene blue and crystal violet also inhibit certain gram-positive bacteria. Other agents that have selective properties are antimicrobial drugs and acid. Some selective media contain strongly inhibitory agents to favor the growth of a pathogen that would otherwise be overlooked because of its low numbers in a specimen. Selenite and brilliant green dye are used in media to isolate *Salmonella* from feces, and sodium azide is used to isolate enterococci from water and food.

Differential media allow multiple types of microorganisms to grow but are designed to display visible differences among those microorganisms. Differentiation shows up as variations in colony size or color, in media color changes, or

TABLE 3.3 Selective Media, Agents, and Functions

Medium	Selective Agent	Used For
Mueller tellurite	Potassium tellurite	Isolation of *Corynebacterium diphtheriae*
Enterococcus faecalis broth	Sodium azide, tetrazolium	Isolation of fecal enterococci
Phenylethanol agar	Phenylethanol chloride	Isolation of staphylococci and streptococci
Tomato juice agar	Tomato juice, acid	Isolation of lactobacilli from saliva
MacConkey agar	Bile, crystal violet	Isolation of gram-negative enterics
Salmonella/Shigella (SS) agar	Bile, citrate, brilliant green	Isolation of *Salmonella* and *Shigella*
Lowenstein-Jensen	Malachite green dye	Isolation and maintenance of *Mycobacterium*
Sabouraud's agar	pH of 5.6 (acid)	Isolation of fungi—inhibits bacteria

(a)

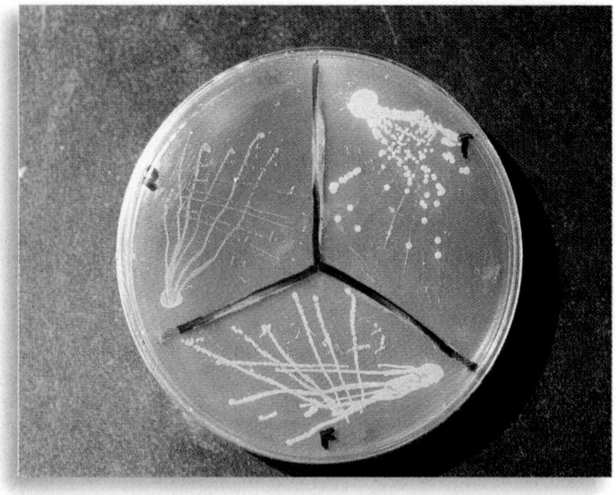

(b)

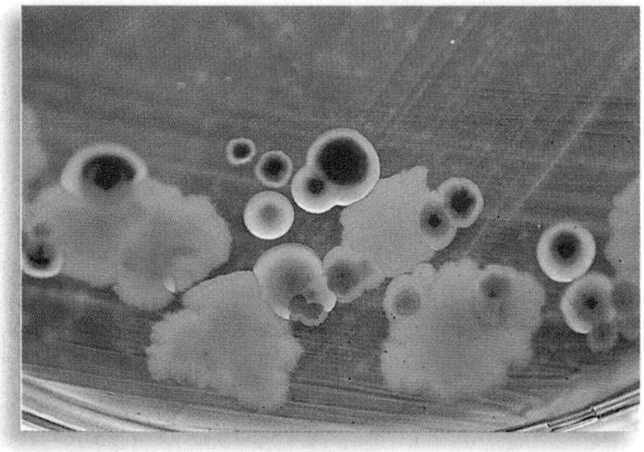

Figure 3.8 Examples of media that are both selective and differential.
(a) Mannitol salt agar is used to isolate members of the genus *Staphylococcus*. It is selective because *Staphylococcus* can grow in the presence of 7.5% sodium chloride, whereas many other species are inhibited by this high concentration. It contains a dye that also differentiates those species of *Staphylococcus* that produce acid from mannitol and turn the phenol red dye to a bright yellow.
(b) MacConkey agar differentiates between lactose-fermenting bacteria (indicated by a pink-red reaction in the center of the colony) and lactose-negative bacteria (indicated by an off-white colony with no dye reaction). It selects against gram-positive bacteria.

in the formation of gas bubbles and precipitates **(table 3.4).** These variations come from the type of chemicals these media contain and the ways that microbes react to them. For example, when microbe X metabolizes a certain substance not used by organism Y, then X will cause a visible change in the medium and Y will not. The simplest differential media show two reaction types such as the use or nonuse of a particular nutrient or a color change in some colonies but not in others. Some media are sufficiently complex to show three or four different reactions **(figure 3.9).** A single medium can be both selective and differential, owing to different ingredients in its composition. MacConkey agar, for example, appears in table 3.3 (selective media) and table 3.4 (differential media).

Dyes can be used as differential agents because many of them are pH indicators that change color in response to the production of an acid or a base. For example, MacConkey agar contains neutral red, a dye that is yellow when neutral and pink or red when acidic. A common intestinal bacterium such as *Escherichia coli* that gives off acid when it metabolizes the lactose in the medium develops red to pink colonies, and one like *Salmonella* that does not give off acid remains its natural color (off-white).

Miscellaneous Media A reducing medium contains a substance (thioglycollic acid or cystine) that absorbs oxygen or slows the penetration of oxygen in a medium, thus reducing its availability. Reducing media are important for growing anaerobic bacteria or for determining oxygen requirements of isolates (described in chapter 7). Carbohydrate fermentation media contain sugars that can be fermented (converted to acids) and a pH indicator to show this reaction (figure 3.8*a* and **figure 3.10**). Media for other biochemical reactions that provide the basis for identifying bacteria and fungi are presented in chapter 17.

Transport media are used to maintain and preserve specimens that have to be held for a period of time before clinical analysis or to sustain delicate species that die rapidly if not held under stable conditions. Transport media contain salts, buffers, and absorbants to prevent cell destruction by enzymes, pH changes, and toxic substances but will not support growth. Assay media are used by technologists to test the effectiveness of antimicrobial

TABLE 3.4 Differential Media

Medium	Substances That Facilitate Differentiation	Differentiates Between
Blood agar	Intact red blood cells	Types of hemolysis displayed by different species of *Streptococcus*
Mannitol salt agar	Mannitol, phenol red	Species of *Staphylococcus*
Hektoen enteric (HE) agar	Brom thymol blue, acid fuchsin, sucrose, salicin, thiosulfate, ferric ammonium citrate	*Salmonella*, *Shigella*, other lactose fermenters from nonfermenters
MacConkey agar	Lactose, neutral red	Bacteria that ferment lactose (lowering the pH) from those that do not
Urea broth	Urea, phenol red	Bacteria that hydrolyze urea to ammonia
Sulfur indole motility (STM)	Thiosulfate, iron	H_2S gas producers from nonproducers
Triple-sugar iron agar (TSIA)	Triple sugars, iron, and phenol red dye	Fermentation of sugars, H_2S production
XLD agar	Lysine, xylose, iron, thiosulfate, phenol red	*Enterobacter*, *Escherichia*, *Proteus*, *Providencia*, *Salmonella*, and *Shigella*
Birdseed agar	Seeds from thistle plant	*Cryptococcus neoformans* and other fungi

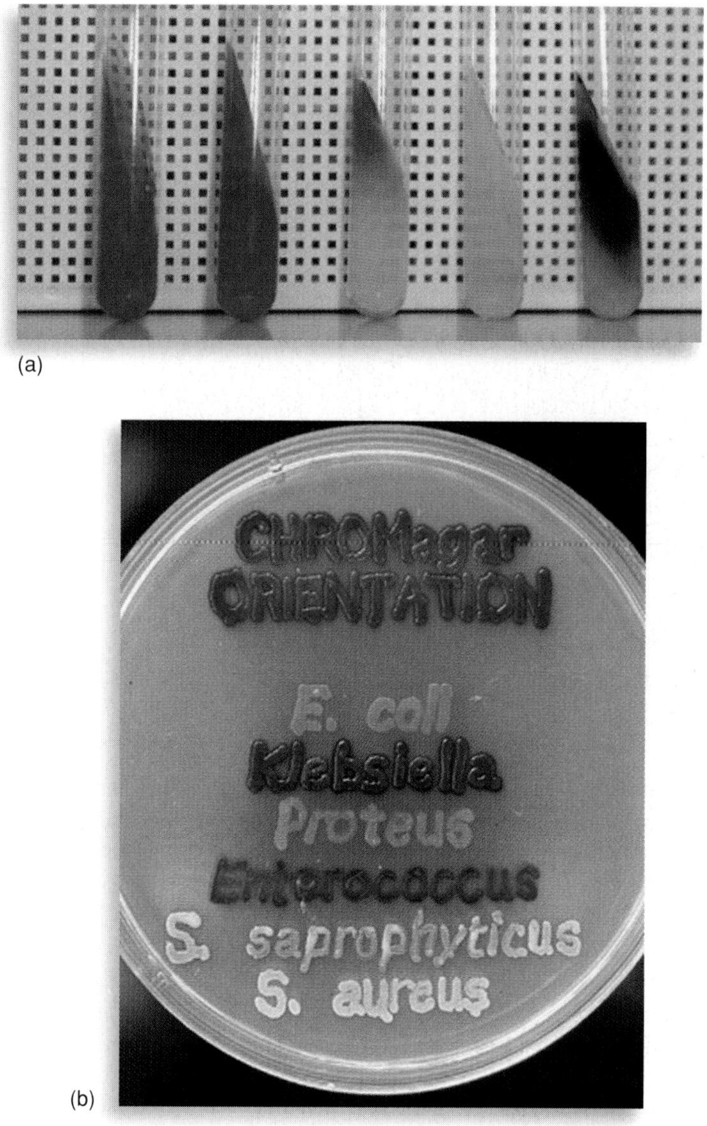

(a)

(b)

Figure 3.9 Media that differentiate characteristics. **(a)** Triple-sugar iron agar (TSIA) in a slant tube. This medium contains three fermentable carbohydrates, phenol red to indicate pH changes, and a chemical (iron) that indicates H_2S gas production. Reactions (from left to right) are: no growth; growth with no acid production; acid production in the bottom (butt) only; acid production all through the medium; and acid production in the butt with H_2S gas formation (black). **(b)** A state-of-the-art medium developed for culturing and identifying the most common urinary pathogens. CHROMagar Orientation™ uses color-forming reactions to distinguish at least seven species and permits rapid identification and treatment. In the example, the bacteria were streaked so as to spell their own names.

drugs (see chapter 12) and by drug manufacturers to assess the effect of disinfectants, antiseptics, cosmetics, and preservatives on the growth of microorganisms. Enumeration media are used by industrial and environmental microbiologists to count the numbers of organisms in milk, water, food, soil, and other samples.

CHECKPOINT

- Most microorganisms can be cultured on artificial media, but some can be cultured only in living tissue or in cells.
- Artificial media are classified by their *physical state* as either liquid, semisolid, liquefiable solid, or nonliquefiable solid.
- Artificial media are classified by their *chemical composition* as either *synthetic* or *nonsynthetic*, depending on whether the exact chemical composition is known.
- Artificial media are classified by their *function* as either general-purpose media or media with one or more specific purposes. Enriched, selective, differential, transport, assay, and enumerating media are all examples of media designed for specific purposes.

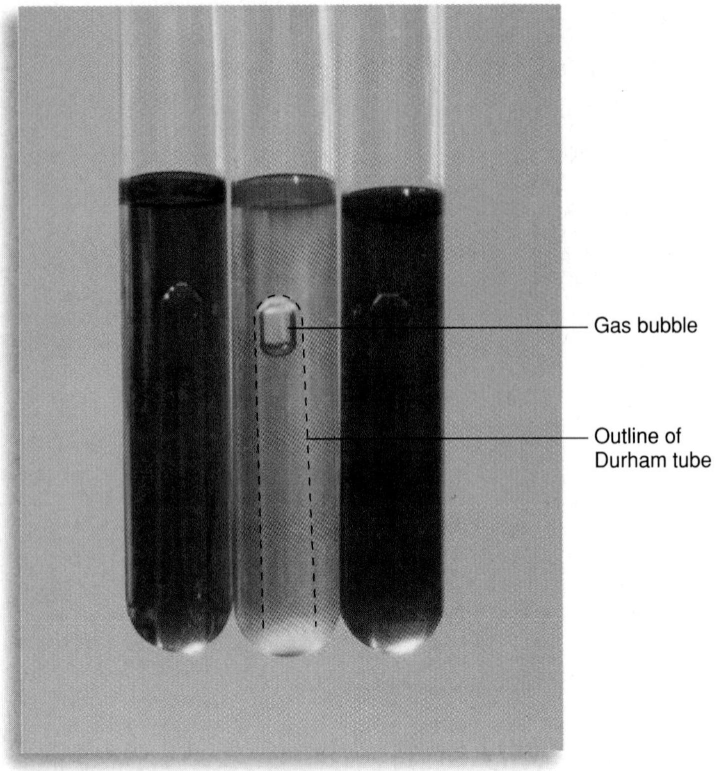

Figure 3.10 Carbohydrate fermentation in broths.
This medium is designed to show fermentation (acid production) and gas formation by means of a small, inverted Durham tube for collecting gas bubbles. The tube on the left is an uninoculated negative control; the center tube is positive for acid (yellow) and gas (open space); the tube on the right shows growth but neither acid nor gas.

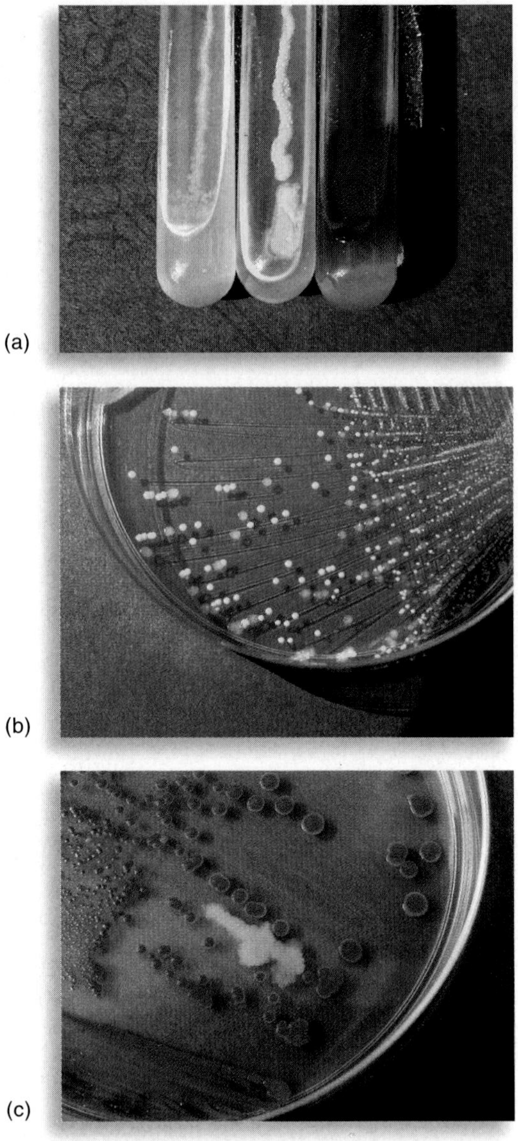

Figure 3.11 Various conditions of cultures.
(a) Three tubes containing pure cultures of *Escherichia coli* (white), *Micrococcus luteus* (yellow), and *Serratia marcescens* (red). **(b)** A mixed culture of *M. luteus* (bright yellow colonies) and *E. coli* (faint white colonies). **(c)** This plate of *S. marcescens* was overexposed to room air, and it has developed a large, white colony. Because this intruder is not desirable and not identified, the culture is now contaminated.

Back to the Five I's: Incubation, Inspection, and Identification

Once a container of medium has been inoculated, it is **incubated,** which means it is placed in a temperature-controlled chamber (incubator) to encourage multiplication. Although microbes have adapted to growth at temperatures ranging from freezing to boiling, the usual temperatures used in laboratory propagation fall between 20° and 40°C. Incubators can also control the content of atmospheric gases such as oxygen and carbon dioxide that may be required for the growth of certain microbes. During the incubation period (ranging from a day to several weeks), the microbe multiplies and produces growth that is observable macroscopically. Microbial growth in a liquid medium materializes as cloudiness, sediment, scum, or color. A common manifestation of growth on solid media is the appearance of colonies, especially in bacteria and fungi. Colonies are actually large masses of piled-up cells (see chapter 4).

In some ways, culturing microbes is analogous to gardening. Cultures are formed by "seeding" tiny plots (media) with microbial cells. Extreme care is taken to exclude weeds (contaminants). Once microbes have grown after incubation, the clinician must *inspect* the container (Petri dish, test tube, etc.). A **pure culture** is a container of medium that grows only a single known species or type of microorganism **(figure 3.11*a*).** This type of culture is most frequently used for laboratory study, because it allows the systematic examination and control of one microorganism by itself. Instead of the term *pure culture,* some microbiologists prefer the term **axenic,** meaning that the culture is free of other living things except for the one being studied. A standard method for preparing a pure culture is to **subculture,** or make a second-level culture from a well-isolated colony. A tiny bit of cells is transferred into a separate container of media and incubated (see figure 3.1, step 3).

A **mixed culture (figure 3.11*b*)** is a container that holds two or more *identified,* easily differentiated species of microorganisms, not unlike a garden plot containing both carrots and onions. A **contaminated culture (figure 3.11*c*)** was once pure or mixed (and thus a known entity) but has since had **contaminants** (unwanted microbes of uncertain identity) introduced into it, like weeds into a garden. Because contaminants have the potential for causing disruption, constant vigilance is required to exclude them from microbiology laboratories, as you will no doubt witness from your own experience. Contaminants get into cultures when the lids of tubes or Petri dishes are left off for too long, allowing airborne microbes to settle into the medium. They can also enter on an incompletely sterilized inoculating loop or on an instrument that you have inadvertently reused or touched to the table or your skin.

How does one determine (i.e., identify) what sorts of microorganisms have been isolated in cultures? Certainly, microscopic appearance can be valuable in differentiating the smaller, simpler prokaryotic cells from the larger, more complex eukaryotic cells. Appearance can be especially useful in identifying eukaryotic microorganisms to the level of genus or species because of their distinctive morphological features; however, bacteria are generally not identifiable by these methods because very different species may appear quite similar. For them, we must include other techniques, some of which characterize their cellular metabolism. These methods, called biochemical tests, can determine fundamental chemical characteristics such as nutrient requirements, products given off during growth, presence of enzymes, and mechanisms for deriving energy.

Several modern analytical and diagnostic tools that focus on genetic characteristics can detect microbes based on their DNA. Identification can also be accomplished by testing the isolate against known antibodies (immunologic testing). In the case of certain pathogens, further information on a microbe is obtained by inoculating a suitable laboratory animal. A profile is prepared by compiling physiological testing results with both macroscopic and microscopic traits. The profile then becomes the raw material used in final identification. In chapter 17, we present more detailed examples of identification methods.

Maintenance and Disposal of Cultures

In most medical laboratories, the cultures and specimens constitute a potential hazard and require prompt and proper disposal. Both steam sterilizing (see autoclave, chapter 11) and incineration (burning) are used to destroy microorganisms. On the other hand, many teaching and research laboratories maintain a line of *stock cultures* that represent "living catalogs" for study and experimentation. The largest culture collection can be found at the American Type Culture Collection in Manassas, Virginia, which maintains a voluminous array of frozen and freeze-dried fungal, bacterial, viral, and algal cultures.

CHECKPOINT

- The Five I's—inoculation, incubation, isolation, inspection, and identification—summarize the kinds of laboratory procedures used in microbiology.
- Following *inoculation,* cultures are *incubated* at a specified temperature to encourage growth.
- *Isolated colonies* that originate from single cells are composed of large numbers of cells piled up together.
- A culture may exist in one of the following forms: A *pure culture* contains only one species or type of microorganism. A *mixed culture* contains two or more known species. A *contaminated culture* contains both known and unknown (unwanted) microorganisms.
- During *inspection,* the cultures are examined and evaluated macroscopically and microscopically.
- Microorganisms are *identified* in terms of their macroscopic or immunologic morphology; their microscopic morphology; their biochemical reactions; and their genetic characteristics.
- Microbial cultures are usually disposed of in two ways: steam sterilization or incineration.

3.2 The Microscope: Window on an Invisible Realm

Imagine Leeuwenhoek's excitement and wonder when he first viewed a drop of rainwater and glimpsed an amazing microscopic world teeming with unearthly creatures. Beginning microbiology students still experience this sensation, and even experienced microbiologists remember their first view. The microbial existence is indeed another world, but it would remain largely uncharted without an essential tool: the microscope. Your efforts in exploring microbes will be more meaningful if you understand some essentials of **microscopy** and specimen preparation.

Magnification and Microscope Design

The two key characteristics of a reliable microscope are magnification, or the ability to make on object appear larger; and resolving power, or the ability to show detail.

A discovery by early microscopists that spurred the advancement of microbiology was that a clear, glass sphere could act as a lens to magnify small objects. Magnification in most microscopes results from a complex interaction between visible light waves and the curvature of the lens. When a beam or ray of light transmitted through air strikes and passes through the convex surface of glass, it experiences some degree of **refraction,** defined as the bending or change in the angle of the light ray as it passes through a medium such as a lens. The greater the difference in the composition of the two substances the light passes between, the more pronounced is the refraction. When an object is placed a certain distance from the spherical lens and illuminated with light, an optical replica, or image, of it is formed by the refracted light. Depending upon the size and curvature of the lens, the image appears enlarged to a particular degree, which is called its power of magnification and is usually identified with a number combined with × (read "times"). This behavior of light is evident if one looks through an everyday object such as a glass ball or a magnifying glass

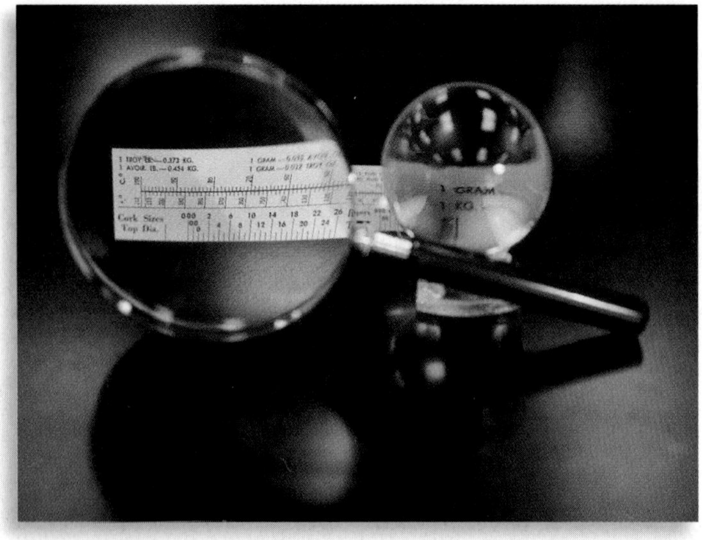

Figure 3.12 Effects of magnification.
Demonstration of the magnification and image-forming capacity of clear glass "lenses." Given a proper source of illumination, this magnifying glass and crystal ball magnify a ruler two to three times.

(figure 3.12). It is basic to the function of all optical, or light, microscopes, though many of them have additional features that define, refine, and increase the size of the image.

The first microscopes were simple, meaning they contained just a single magnifying lens and a few working parts. Examples of this type of microscope are a magnifying glass, a hand lens, and Leeuwenhoek's basic little tool shown earlier in figure 1.9*a*. Among the refinements that led to the development of today's compound microscope were the addition of a second magnifying lens system, a lamp in the base to give off visible light and illuminate the specimen, and a special lens called the condenser that converges or focuses the rays of light to a single point on the object. The fundamental parts of a modern compound light microscope are illustrated in **figure 3.13.**

Principles of Light Microscopy

To be most effective, a microscope should provide adequate magnification, resolution, and clarity of image. Magnification of the object or specimen by a compound microscope occurs in two phases. The first lens in this system (the one closest to the specimen) is the objective lens, and the second (the one closest to the eye) is the ocular lens, or eyepiece **(figure 3.14).** The objective forms the initial image of the specimen, called the **real image.** When this image is projected up through the microscope body to the plane of the eyepiece, the ocular lens forms a second image, the **virtual image.** The virtual image is the one that will be received by the eye and converted to a retinal and visual image. The magnifying power of the objective alone usually ranges from 4× to 100×, and the power of the ocular alone ranges from 10× to 20×. The total power of magnification of the final image formed by the combined lenses is a product of the separate powers of the two lenses:

Power of objective	×	*Usual power of ocular*	=	*Total magnification*
10× low power objective	×	10×	=	100×
40× high dry objective	×	10×	=	400×
100× oil immersion objective	×	10×	=	1,000×

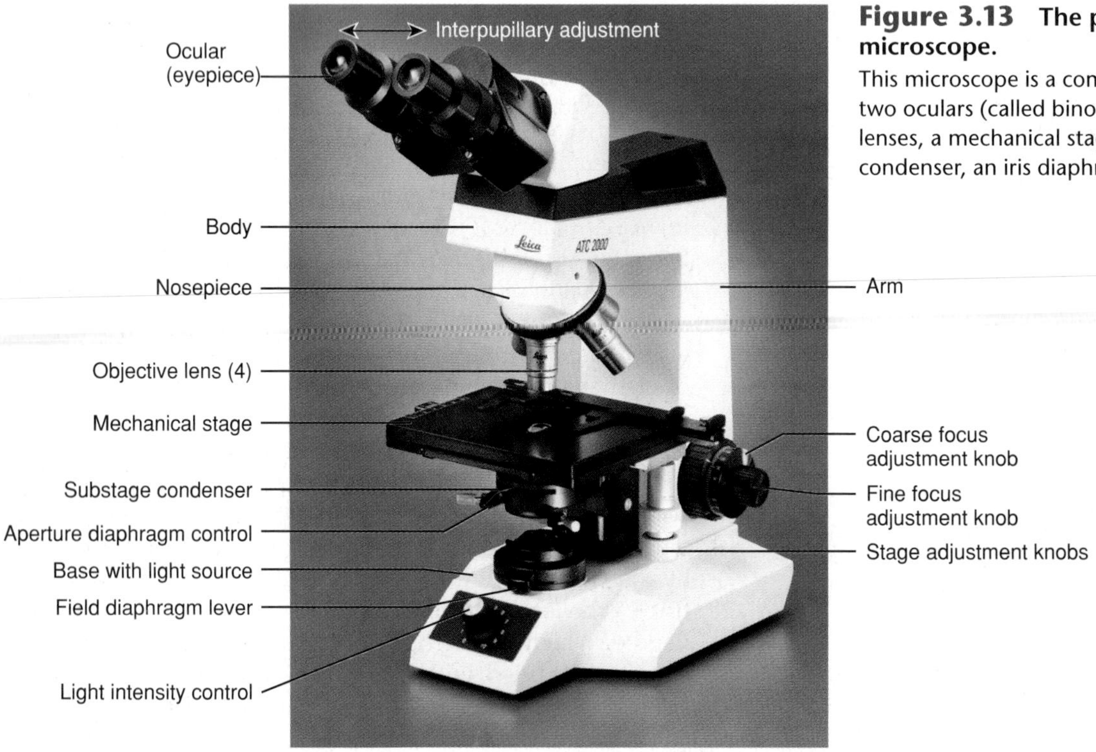

Figure 3.13 The parts of a student laboratory microscope.
This microscope is a compound light microscope with two oculars (called binocular). It has four objective lenses, a mechanical stage to move the specimen, a condenser, an iris diaphragm, and a built-in lamp.

Microscopes are equipped with a nosepiece holding three or more objectives that can be rotated into position as needed. The power of the ocular usually remains constant for a given microscope. Depending on the power of the ocular, the total magnification of standard light microscopes can vary from 40× with the lowest power objective (called the scanning objective) to 2,000× with the highest power objective (the oil immersion objective).

Resolution: Distinguishing Magnified Objects Clearly As important as magnification is for visualizing tiny objects or cells, an additional optical property is essential for seeing clearly. That property is resolution, or **resolving power.** Resolution is the capacity of an optical system to distinguish or separate two adjacent objects or points from one another. For example, at a certain fixed distance, the lens in the human eye can resolve two small objects as separate points just as long as the two objects are no closer than 0.2 millimeters apart. The eye examination given by optometrists is in fact a test of the resolving power of the human eye for various-size letters read at a distance of 20 feet. Because microorganisms are extremely small and usually very close together, they will not be seen with clarity or any degree of detail unless the microscope's lenses can resolve them.

A simple equation in the form of a fraction expresses the main determining factors in resolution:

$$\text{Resolving power (RP)} = \frac{\text{Wavelength of light in nm}}{2 \times \text{Numerical aperture of objective lens}}$$

This equation demonstrates that the resolving power is a function of the wavelength of light that forms the image, along with certain characteristics of the objective. The light source for optical microscopes consists of a band of colored wavelengths in the visible spectrum. The shortest visible wavelengths are in the violet-blue portion of the spectrum (400 nanometers), and the longest are in the red portion (750 nanometers). Because the wavelength must pass between the objects that are being resolved, shorter wavelengths (in the 400–500 nanometer range) will provide better resolution **(figure 3.15).** Some microscopes have a special blue filter placed over the lamp to limit the longer wavelengths of light from entering the specimen.

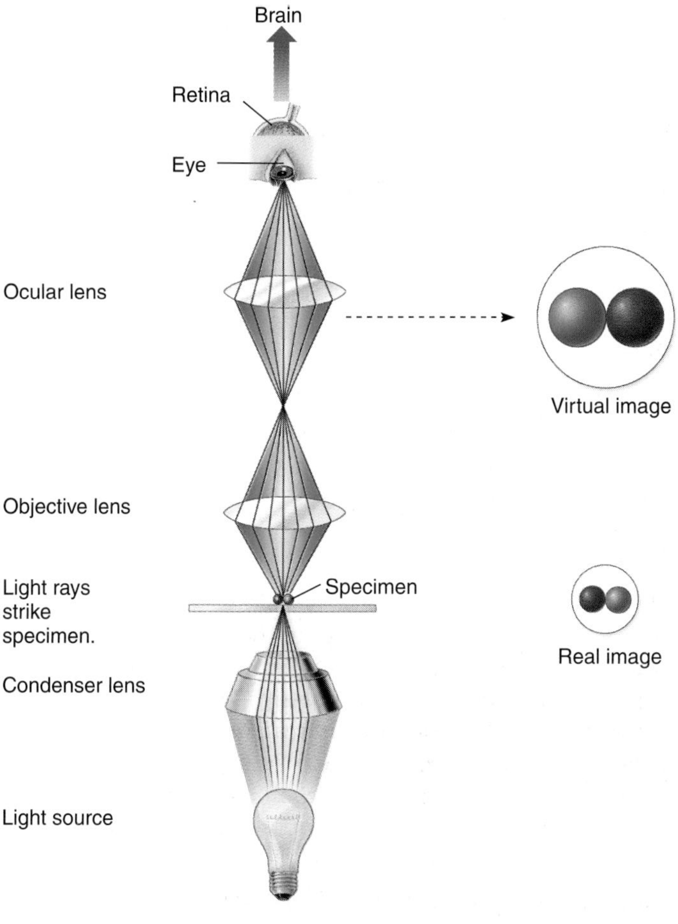

Figure 3.14 The pathway of light and the two stages in magnification of a compound microscope.
As light passes through the condenser, it forms a solid beam that is focused on the specimen. Light leaving the specimen that enters the objective lens is refracted so that an enlarged primary image, the real image, is formed. One does not see this image, but its degree of magnification is represented by the lower circle. The real image is projected through the ocular, and a second image, the virtual image, is formed by a similar process. The virtual image is the final magnified image that is received by the retina and perceived by the brain. Notice that the lens systems cause the image to be reversed.

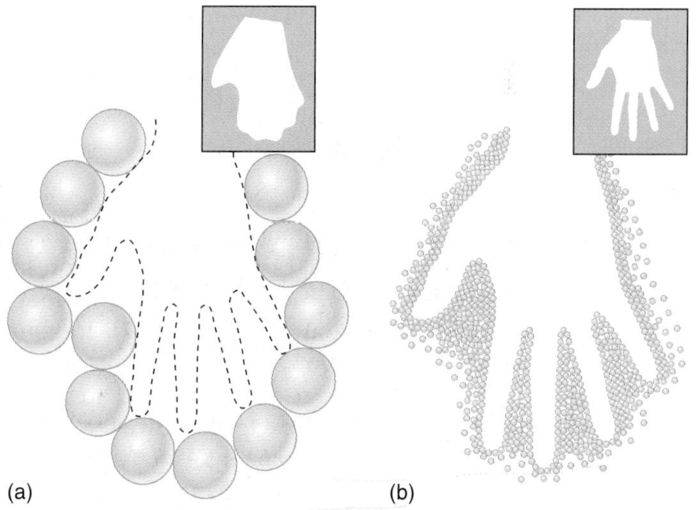

Figure 3.15 Effect of wavelength on resolution.
A simple model demonstrates how the wavelength influences the resolving power of a microscope. Here an outline of a hand represents the object being illuminated, and two different-size circles represent the wavelengths of light. In **(a)**, the longer waves are too large to penetrate between the finer spaces and produce a fuzzy, undetailed image. In **(b)**, shorter waves are small enough to enter small spaces and produce a much more detailed image that is recognizable as a hand.

The other factor influencing resolution is the **numerical aperture,** a mathematical constant that describes the relative efficiency of a lens in bending light rays. Without going into the mathematical derivation of this constant, it is sufficient to say that each objective has a fixed numerical aperture reading that is determined by the microscope design and ranges from 0.1 in the lowest power lens to approximately 1.25 in the highest power (oil immersion) lens.

A NOTE ABOUT OIL IMMERSION LENSES

The most important thing to remember is that a higher numerical aperture number will provide better resolution. In order for the oil immersion lens to arrive at its maximum resolving capacity, a drop of oil must be inserted between the tip of the lens and the specimen on the glass slide. Because oil has the same optical qualities as glass, it prevents refractive loss that normally occurs as peripheral light passes from the slide into the air; this property effectively increases the numerical aperture **(figure 3.16).**

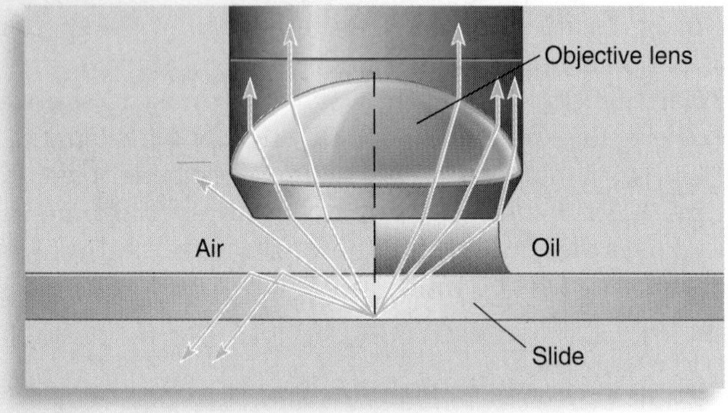

Figure 3.16 Workings of an oil immersion lens. Without oil, some of the peripheral light that passes through the specimen is scattered into the air or onto the glass slide; this scattering decreases resolution.

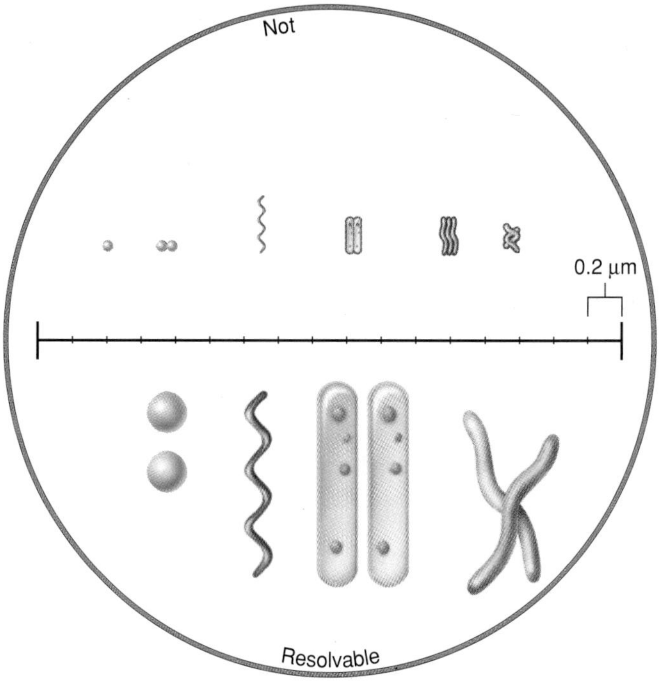

Figure 3.17 Effect of magnification. Comparison of cells that would not be resolvable versus those that would be resolvable under oil immersion at 1,000× magnification. Note that in addition to differentiating two adjacent things, good resolution also means being able to observe an object clearly.

In practical terms, the oil immersion lens can resolve any cell or cell part as long as it is at least 0.2 micron in diameter, and it can resolve two adjacent objects as long as they are at least 0.2 micron apart **(figure 3.17).** In general, organisms that are 0.5 micron or more in diameter are readily seen. This includes fungi and protozoa, some of their internal structures, and most bacteria. However, a few bacteria and most viruses are far too small to be resolved by the optical microscope and require electron microscopy (discussed later in this chapter). In summary then, the factor that most limits the clarity of a microscope's image is its resolving power. Even if a light microscope were designed to magnify several thousand times, its resolving power could not be increased, and the image it produced would simply be enlarged and fuzzy.

Because too much light can reduce contrast and burn out the image, an adjustable iris diaphragm on most microscopes controls the amount of light entering the condenser. The lack of contrast in cell components is compensated for by using special lenses (the phase-contrast microscope) and by adding dyes.

Variations on the Optical Microscope

Optical microscopes that use visible light can be described by the nature of their *field,* meaning the circular area viewed through the ocular lens. There are four types of visible-light microscopes: bright-field, dark-field, phase-contrast, and interference. A fifth type of optical microscope, the fluorescence microscope, uses ultraviolet radiation as the illuminating source; and another, the confocal microscope, uses a laser beam. Each of these microscopes is adapted for viewing specimens in a particular way, as described in the next sections and summarized in **table 3.5.**

Bright-Field Microscopy

The bright-field microscope is the most widely used type of light microscope. Although we ordinarily view objects like the words on this page with light reflected off the surface, a bright-field microscope forms its image when light is transmitted through the specimen. The specimen, being denser and more opaque than its surroundings, absorbs some of this light, and the rest of the light is transmitted directly up through the ocular into the field. As a result, the specimen will produce an image that is darker than the surrounding brightly

TABLE 3.5 Comparisons of Types of Microscopy

Microscope	Maximum Practical Magnification	Resolution	Important Features
Visible light as source of illumination			
Bright-field	2,000×	0.2 μm (200 nm)	Common multipurpose microscope for live and preserved stained specimens; specimen is dark, field is white; provides fair cellular detail
Dark-field	2,000×	0.2 μm	Best for observing live, unstained specimens; specimen is bright, field is black; provides outline of specimen with reduced internal cellular detail
Phase-contrast	2,000×	0.2 μm	Used for live specimens; specimen is contrasted against gray background; excellent for internal cellular detail
Differential interference	2,000×	0.2 μm	Provides brightly colored, highly contrasting, three-dimensional images of live specimens
Ultraviolet rays as source of illumination			
Fluorescent	2,000×	0.2 μm	Specimens stained with fluorescent dyes or combined with fluorescent antibodies emit visible light; specificity makes this microscope an excellent diagnostic tool
Confocal	2,000×	0.2 μm	Specimens stained with fluorescent dyes are scanned by laser beam; multiple images (optical sections) are combined into three-dimensional image by a computer; unstained specimens can be viewed using light reflected from specimen
Electron beam forms image of specimen			
Transmission electron microscope (TEM)	100,000×	0.5 nm	Sections of specimen are viewed under very high magnification; finest detailed structure of cells and viruses is shown
Scanning electron microscope (SEM)	650,000×	10 nm	Scans and magnifies external surface of specimen; produces striking three-dimensional image
Atomically sharp tip probes surface of specimen			
Atomic force microscope (AFM)	100,000,000×	0.01 Angstroms	Tip scans specimen and moves up and down with contour of surface; movement of tip is measured with laser and translated to image
Scanning tunneling microscope (STM)	100,000,000×	0.01 Angstroms	Tip moves over specimen while voltage is applied, generating current that is dependent on distance between tip and surface; atoms can be moved with tip

illuminated field. The bright-field microscope is a multipurpose instrument that can be used for both live, unstained material and preserved, stained material. The bright-field image is compared with that of other microscopes in **figure 3.18.**

Dark-Field Microscopy

A bright-field microscope can be adapted as a dark-field microscope by adding a special disc called a *stop* to the condenser. The stop blocks all light from entering the objective lens except peripheral light that is reflected off the sides of the specimen itself. The resulting image is a particularly striking one: brightly illuminated specimens surrounded by a dark (black) field **(figure 3.18*b*).** Some of Leeuwenhoek's more successful microscopes probably operated with dark-field illumination. The most effective use of dark-field microscopy is to visualize living cells that would be distorted by drying or heat or that cannot be stained with the usual methods. Dark-field microscopy can outline the organism's shape and permit rapid recognition of swimming cells that might appear in dental and other infections, but it does not reveal fine internal details.

Phase-Contrast and Interference Microscopy

If similar objects made of clear glass, ice, cellophane, or plastic are immersed in the same container of water, an observer would have difficulty telling them apart because they have similar optical properties. Internal components of a live, unstained cell also lack contrast and can be difficult to distinguish. But cell structures do differ slightly in density, enough that they can alter the light that passes through them in subtle ways. The phase-contrast microscope has been constructed to take advantage of this characteristic. This microscope contains devices that transform the subtle changes in light waves passing through the specimen into differences in light intensity. For example, denser cell parts such as organelles alter the pathway of light more than less dense regions (the cytoplasm). Light patterns coming from these regions will vary in contrast. The amount of internal detail visible by this method is greater than by either bright-field or dark-field methods. The phase-contrast microscope is most useful for observing intracellular structures such as bacterial spores, granules, and organelles, as well as the locomotor structures of eukaryotic cells such as cilia **(figure 3.18*c*).**

Like the phase-contrast microscope, the differential interference contrast (DIC) microscope provides a detailed view of unstained, live specimens by manipulating the light. But this microscope has additional refinements, including two prisms that add contrasting colors to the image and two beams of light rather than a single one. DIC microscopes produce extremely well-defined images that are vividly colored and appear three-dimensional **(figure 3.19).**

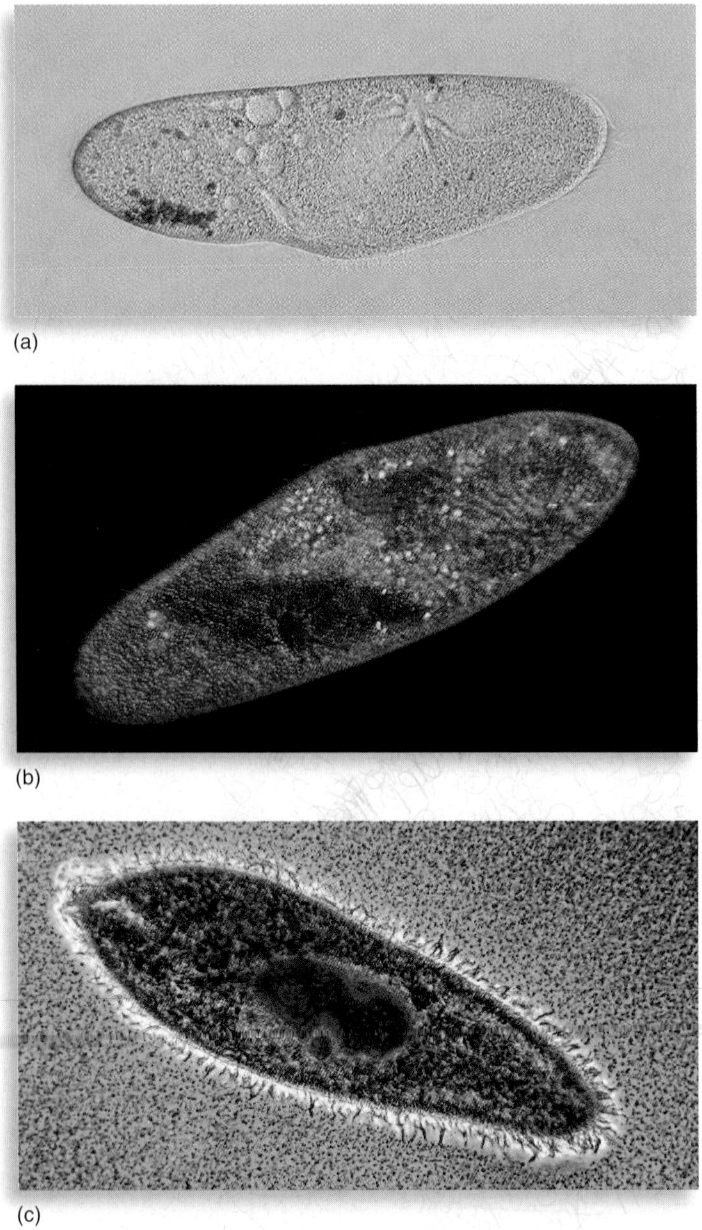

Figure 3.18 Three views of a basic cell.
A live cell of *Paramecium* viewed with **(a)** bright-field (400×), **(b)** dark-field (400×), and **(c)** phase-contrast (400×) microscopy. Note the difference in the appearance of the field and the degree of detail shown by each method of microscopy. Only in phase-contrast microscopy are the cilia on the cells noticeable. Can you see the nucleus? The oral groove?

Fluorescence Microscopy

The fluorescence microscope is a specially modified compound microscope furnished with an ultraviolet (UV) radiation source and a filter that protects the viewer's eye from injury by these dangerous rays. The name of this type of microscopy originates from the use of certain dyes (acridine, fluorescein) and minerals that show **fluorescence.** The dyes emit visible light when bombarded by short ultraviolet rays. For an image to be formed, the specimen must

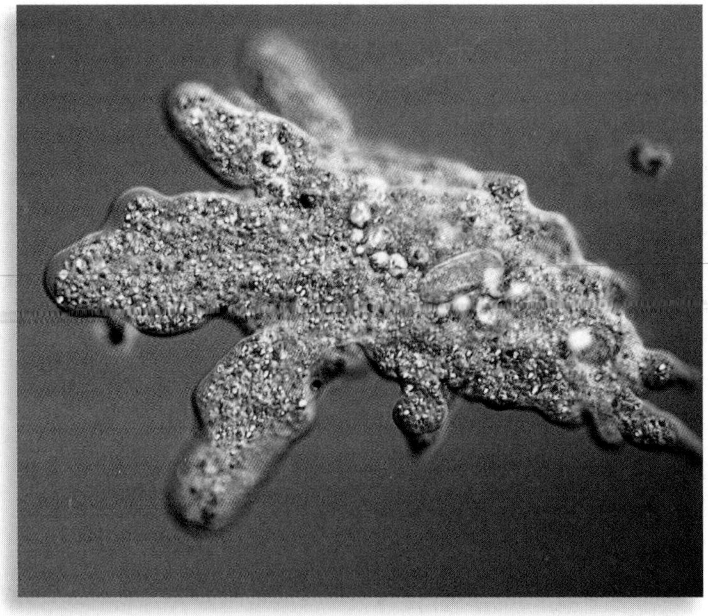

Figure 3.19 Differential interference contrast.
Differential interference micrograph of *Amoeba proteus,* a common protozoan. Note the outstanding internal detail, the depth of field, and the bright colors, which are not natural (160×).

first be coated or placed in contact with a source of fluorescence. Subsequent illumination by ultraviolet radiation causes the specimen to give off light that will form its own image, usually an intense yellow, orange, or red against a black field.

Fluorescence microscopy has its most useful applications in diagnosing infections caused by specific bacteria, protozoans, and viruses. A staining technique with fluorescent dyes is commonly used to detect *Mycobacterium tuberculosis* (the agent of tuberculosis) in patients' specimens (see figure 21.20). In a number of diagnostic procedures, fluorescent dyes are affixed to specific antibodies. These *fluorescent antibodies* can be used to detect the causative agents in such diseases as syphilis, chlamydiosis, trichomoniasis, herpes, and influenza. A technology using fluorescent nucleic acid stains can differentiate between live and dead cells in mixtures **(figure 3.20).** A fluorescence microscope can be handy for locating microbes in complex mixtures because only those cells targeted by the technique will fluoresce.

Most optical microscopes have difficulty forming a clear image of cells at higher magnifications, because cells are often too thick for conventional lenses to focus all levels of the cell simultaneously. This is especially true of larger cells with complex internal structures. A newer type of microscope that overcomes this impediment is called the *scanning confocal microscope*. This microscope uses a laser beam of light to scan various depths in the specimen and deliver a sharp image focusing on just a single plane. It is thus able to capture a highly focused view at any level, ranging from the surface to the middle of the cell. It is most often used on fluorescently stained specimens, but it can also be used to visualize live unstained cells and tissues **(figure 3.21).**

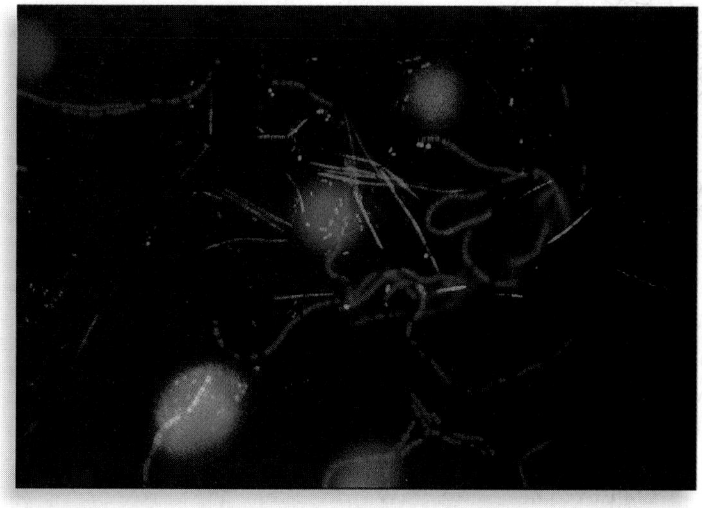

Figure 3.20 Fluorescent staining on a fresh sample of cheek scrapings from the oral cavity.

Cheek epithelial cells are the larger unfocused red or green cells. Bacteria appearing here are streptococci (tiny spheres in long chains) and filamentous rods. This particular staining technique also indicates whether cells are alive or dead; live cells fluoresce green, and dead cells fluoresce red.

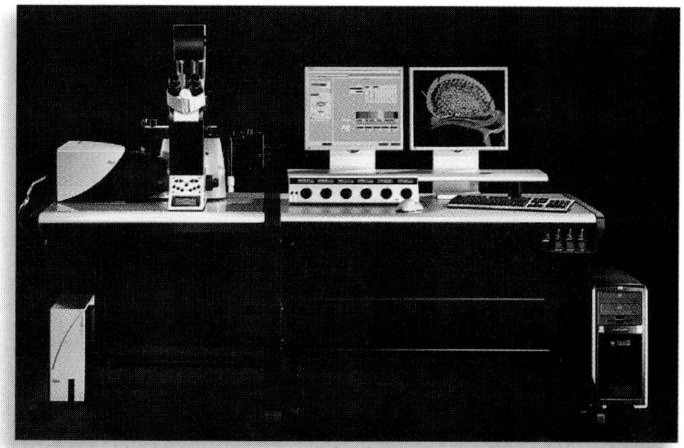

(a)

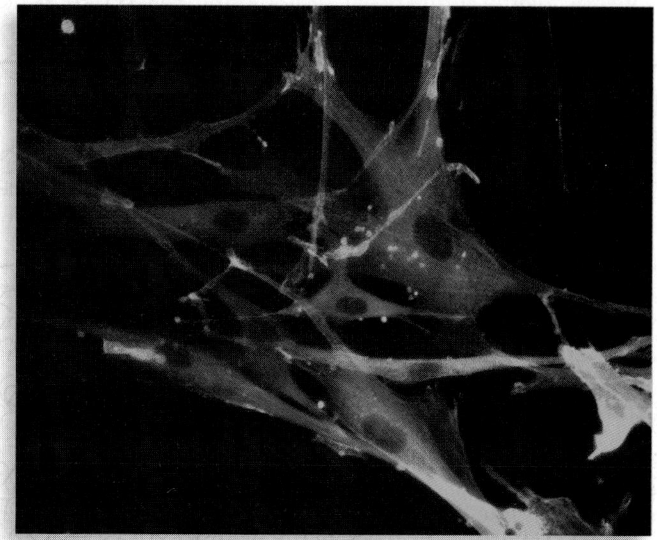

(b)

(c)

Figure 3.21 Confocal microscopy.

(a) A confocal microscope combines a specialized fluorescence microscope with a computer that converts light into electrical signals and displays the image. **(b)** Confocal image of myofibroblasts, cells involved in tissue repair. **(c)** *Paramecium* visualized by confocal microscope.

Electron Microscopy

If conventional light microscopes are our windows on the microscopic world, then the electron microscope (EM) is our window on the tiniest details of that world. Although this microscope was originally conceived and developed for studying nonbiological materials such as metals and small electronics parts, biologists immediately recognized the importance of the tool and began to use it in the early 1930s. One of the most impressive features of the electron microscope is the resolution it provides.

Unlike light microscopes, the electron microscope forms an image with a beam of electrons that can be made to travel in wavelike patterns when accelerated to high speeds. These waves are 100,000 times shorter than the waves of visible light. Because resolving power is a function of wavelength, electrons have tremendous power to resolve minute structures. Indeed, it is possible to resolve atoms with an electron microscope, though the practical resolution for biological applications is approximately 0.5 nanometers. Because the resolution is so substantial, it follows that magnification can also be extremely high—usually between 5,000× and 1,000,000× for biological specimens and up to 5,000,000× in some applications. Its capacity for magnification and resolution makes the EM an invaluable tool for seeing the finest structure of cells and viruses. If not for electron microscopes, our understanding of biological structure and function would still be in its early theoretical stages.

Two general forms of EM are the transmission electron microscope (TEM) and the scanning electron microscope (SEM) (see table 3.5). Transmission electron microscopes are the method of choice for viewing the detailed structure of cells and viruses. This microscope produces its image by transmitting electrons through the specimen. Because electrons cannot readily penetrate thick preparations, the specimen must be sectioned into extremely thin slices (20–100 nm thick) and stained or coated with metals that will increase image contrast. The darkest areas of TEM micrographs represent the thicker (denser) parts, and the lighter areas indicate the more transparent and less dense parts **(figure 3.22).**

The scanning electron microscope provides some of the most dramatic and realistic images in existence. This instrument is designed to create an extremely detailed three-dimensional view of all kinds of objects—from plaque on teeth to tapeworm heads. To produce its images, the SEM does not transmit electrons; it bombards the surface of a whole, metal-coated specimen with electrons while scanning back and forth over it. A shower of electrons deflected from the surface is picked up with great fidelity by a sophisticated detector, and the electron pattern is displayed as an image on a television screen. The contours of the specimens resolved with scanning electron micrography are very revealing and often surprising. Areas that look smooth and flat with the light microscope display intriguing surface features with the SEM **(figure 3.23).** Improved technology has continued to refine electron microscopes and to develop variations on the basic plan. One of the most inventive relatives of the EM is the scanning probe microscope **(Insight 3.2).**

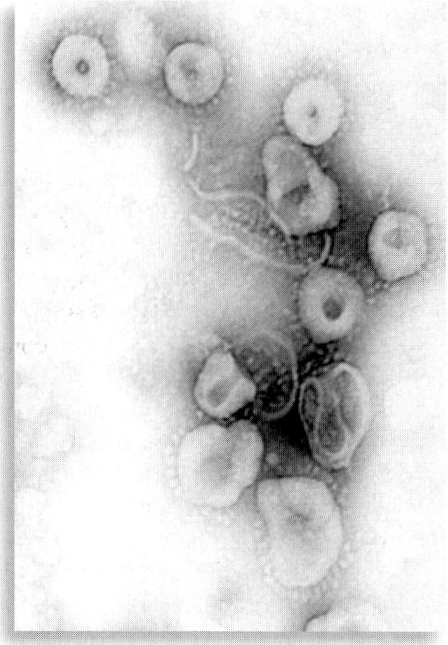

(a)

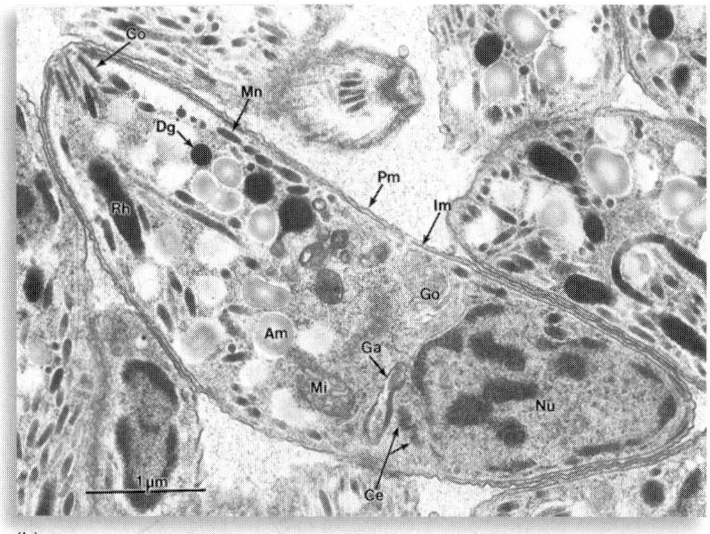

(b)

Figure 3.22 Transmission electron micrographs.
(a) A sample from the respiratory tract reveals coronaviruses (corona for the crownlike envelope) that cause infectious bronchitis (100,000×). A new form of this virus is responsible for severe acute respiratory syndrome (SARS) in humans. **(b)** A section through an infectious stage of *Toxoplasma gondii*, the cause of toxoplasmosis. Labels indicate fine structures such as cell membrane (Pm), Golgi complex (Go), nucleus (Nu), mitochondrion (Mi), centrioles (Ce), and granules (Am, Dg).

INSIGHT 3.2 — Discovery

The Evolution in Resolution: Probing Microscopes

In the past, chemists, physicists, and biologists had to rely on indirect methods to provide information on the structures of the smallest molecules. But technological advances have created a new generation of microscopes that "see" atomic structure by actually feeling it. *Scanning probe microscopes* operate with a minute needle tapered to a tip that can be as narrow as a single atom! This probe scans over the exposed surface of a material on the end of an arm and records an image of its outer texture. (Think of an old-fashioned record player. . . .) These revolutionary microscopes have such profound resolution that they have the potential to image single atoms (but not subatomic structure yet) and to magnify 100 million times. There are two types of scanning probe microscopes, the atomic force microscope (AFM) and the scanning tunneling microscope (STM). The STM uses a tungsten probe that hovers near the surface of an object and follows its topography while simultaneously giving off an electrical signal of its pathway, which is then imaged on a screen. The STM was used initially for detecting defects on the surfaces of electrical conductors and computer chips composed of silicon, but it has also provided the first incredible close-up views of DNA (see Insight 9.2).

The atomic force microscope (AFM) gently forces a diamond and metal probe down onto the surface of a specimen like a needle on a record. As it moves along the surface, any deflection of the metal probe is detected by a sensitive device that relays the information to an imager. The AFM is very useful in viewing the detailed structures of biological molecules such as antibodies and enzymes.

These powerful new microscopes can also move and position atoms, spawning a field called *nanotechnology*—the science of the "small." When this ability to move atoms was first discovered, scientists had some fun (see illustration *a*). But it has opened up an entirely new way to manipulate atoms in chemical reactions (illustration *b*) and to create nanoscale devices for computers and other electronics. In the future, it may be possible to use microstructures to deliver drugs and treat disease.

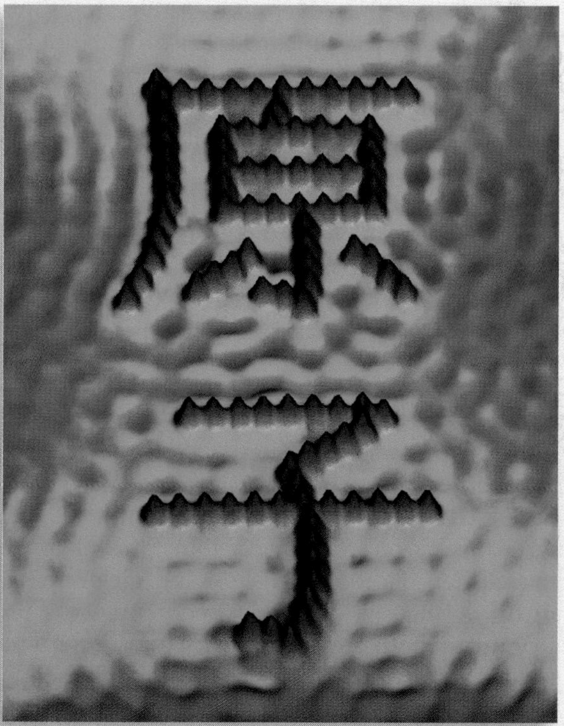

(a)

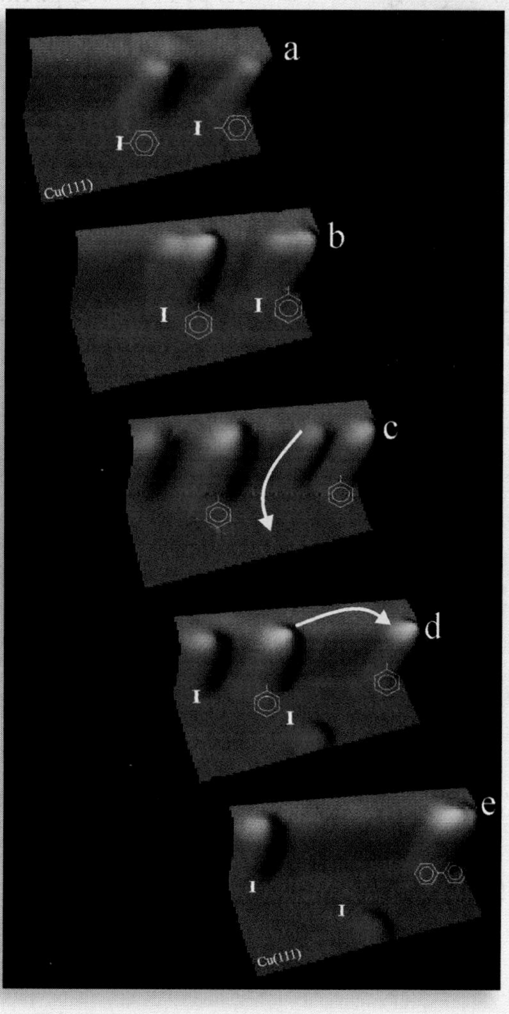

(b)

Scanning tunneling microscopy. (a) Scientists have dragged iron atoms over a copper matrix to spell (in kanji, a Japanese written alphabet) "atom" (literally: "original child"). **(b)** A chemical reaction performed by an STM microscope. At the top *(a)*, two iodobenzene molecules appear as two bumps on a copper surface. The STM tip emits a burst of electrons and causes the iodine groups to dissociate from each of the benzene groups *(b)*, The tip then drags away the iodine groups *(c)*, and the two carbon groups bind to one another (*d* and *e*).

Source: Insight 3.2a: http://www.almaden.ibm.com/vis/stm/atomo.html, page 80.

(a)

(b)

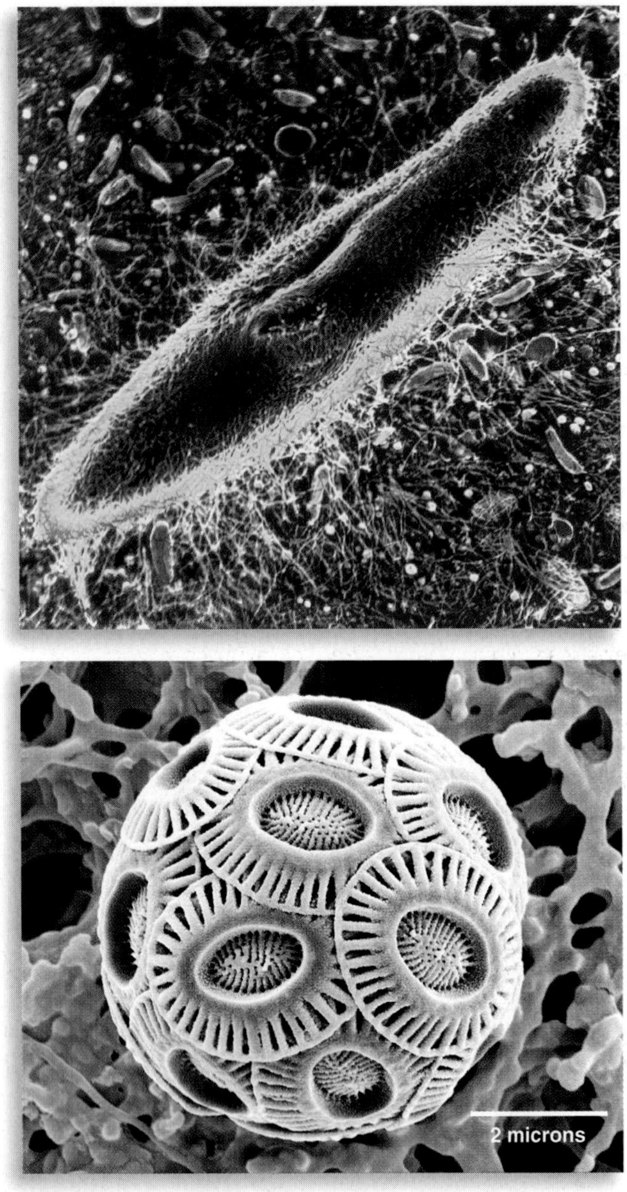

Figure 3.23 Scanning electron micrographs.
(a) A false-color scanning electron micrograph (SEM) of *Paramecium,* covered in masses of fine hairs (100×). These are actually its locomotor and feeding structures—the cilia. Cells in the surrounding medium are bacteria that serve as the protozoan's "movable feast." Compare this with figure 3.18 to appreciate the outstanding three-dimensional detail shown by an SEM. **(b)** A fantastic ornamental alga called a coccolithophore displays a complex cell wall formed by calcium discs. This alga often blooms in the world's oceans (see chapter 1).

Preparing Specimens for Optical Microscopes

A specimen for optical microscopy is generally prepared by mounting a sample on a suitable glass slide that sits on the stage between the condenser and the objective lens. The manner in which a slide specimen, or mount, is prepared depends upon: (1) the condition of the specimen, either in a living or preserved state; (2) the aims of the examiner, whether to observe overall structure, identify the microorganisms, or see movement; and (3) the type of microscopy available, whether it is bright-field, dark-field, phase-contrast, or fluorescence.

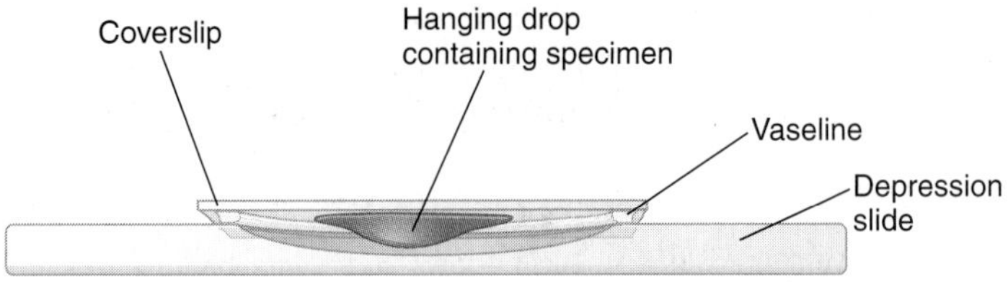

Figure 3.24 Hanging drop technique.
Cross-section view of slide and coverslip. (Vaseline actually surrounds entire well of slide.)

Fresh, Living Preparations

Live samples of microorganisms are placed in wet mounts or in hanging drop mounts so that they can be observed as near to their natural state as possible. The cells are suspended in a suitable fluid (water, broth, saline) that temporarily maintains viability and provides space and a medium for locomotion. A wet mount consists of a drop or two of the culture placed on a slide and overlaid with a cover glass. Although this type of mount is quick and easy to prepare, it has certain disadvantages. The cover glass can damage larger cells, and the slide is very susceptible to drying and can contaminate the handler's fingers. A more satisfactory alternative is the hanging drop preparation made with a special concave (depression) slide, a Vaseline adhesive or sealant, and a coverslip from which a tiny drop of sample is suspended **(figure 3.24).** These types of short-term mounts provide a true assessment of the size, shape, arrangement, color, and motility of cells. Greater cellular detail can be observed with phase-contrast or interference microscopy.

Fixed, Stained Smears

A more permanent mount for long-term study can be obtained by preparing fixed, stained specimens. The smear technique, developed by Robert Koch more than 100 years ago, consists of spreading a thin film made from a liquid suspension of cells on a slide and air-drying it. Next, the air-dried smear is usually heated gently by a process called heat fixation that simultaneously kills the specimen and secures it to the slide. Another important action of fixation is to preserve various cellular components in a natural state with minimal distortion. Fixation of some microbial cells is performed with chemicals such as alcohol and formalin.

Like images on undeveloped photographic film, the unstained cells of a fixed smear are quite indistinct, no matter how great the magnification or how fine the resolving power of the microscope. The process of "developing" a smear to create contrast and make inconspicuous features stand out requires staining techniques. Staining is any procedure that applies colored chemicals called dyes to specimens. Dyes impart a color to cells or cell parts by becoming affixed to them through a chemical reaction. In general, they are classified as basic (cationic) dyes, which have a positive charge, or acidic (anionic) dyes, which have a negative charge. Because chemicals of opposite charge are attracted to each other, cell parts that are negatively charged will attract basic dyes and those that are positively charged will attract acidic dyes **(table 3.7).** Many cells, especially those of bacteria, have numerous negatively charged acidic substances and thus stain more readily with basic dyes. Acidic dyes, on the other hand, tend to be repelled by cells, so they are good for negative staining (discussed in the next section).

Negative Versus Positive Staining Two basic types of staining technique are used, depending upon how a dye reacts with the specimen (summarized in table 3.7). Most procedures involve a **positive stain,** in which the dye actually sticks to the specimen and gives it color. A **negative stain,** on the other hand, is just the reverse (like a photographic negative). The dye does not stick to the specimen but settles around its outer boundary, forming a silhouette. In a sense, negative staining "stains" the glass slide to produce a dark background around the cells. Nigrosin (blue-black) and India ink (a black suspension of carbon particles) are the dyes most commonly used for negative staining. The cells themselves do not stain because these dyes are negatively charged and are repelled by the negatively charged surface of the cells. The value of negative staining is its relative simplicity and the reduced shrinkage or distortion of cells, as the smear is not heat fixed. A quick assessment can thus be made regarding cellular size, shape, and arrangement. Negative staining is also used to accentuate the capsule that surrounds certain bacteria and yeasts **(figure 3.25).**

TABLE 3.7 Comparison of Positive and Negative Stains

	Positive Staining	Negative Staining
Appearance of cell	Colored by dye	Clear and colorless
	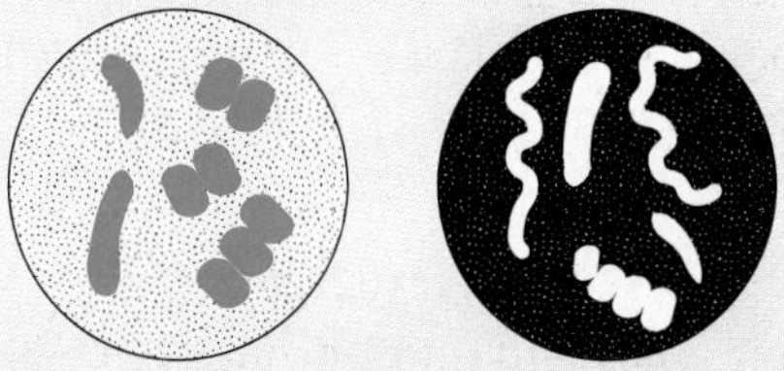	
Background	Not stained (generally white)	Stained (dark gray or black)
Dyes employed	Basic dyes: Crystal violet, Methylene blue, Safranin, Malachite green	Acidic dyes: Nigrosin, India ink
Subtypes of stains	Several types: Simple stain; Differential stains: Gram stain, Acid-fast stain, Spore stain; Special stains: Capsule, Flagella, Spore, Granules, Nucleic acid	Few types: Capsule, Spore

Simple Versus Differential Staining Positive staining methods are classified as simple, differential, or special (figure 3.25). Whereas **simple stains** require only a single dye and an uncomplicated procedure, **differential stains** use two differently colored dyes, called the *primary dye* and the *counterstain,* to distinguish between cell types or parts. These staining techniques tend to be more complex and sometimes require additional chemical reagents to produce the desired reaction.

Most simple staining techniques take advantage of the ready binding of bacterial cells to dyes like malachite green, crystal violet, basic fuchsin, and safranin. Simple stains cause all cells in a smear to appear more or less the same color, regardless of type, but they can still reveal bacterial characteristics such as shape, size, and arrangement.

Types of Differential Stains A satisfactory differential stain uses differently colored dyes to clearly contrast two cell types or cell parts. Common co mbinations are red and purple, red and green, or pink and blue. Differential stains can also pinpoint other characteristics, such as the size, shape, and arrangement of cells. Typical examples include Gram, acid-fast, and endospore stains. Some staining techniques (spore, capsule) fall into more than one category.

Gram staining, a century-old method named for its developer, Hans Christian Gram, remains the most universal diagnostic staining technique for bacteria. It permits ready differentiation of major categories based upon the color reaction of the cells: *gram-positive,* which stain purple, and *gram-negative,* which stain pink (red). The Gram stain is the basis of several important bacteriological topics, including bacterial taxonomy, cell wall structure, and identification and diagnosis of infection; in some cases, it even guides the selection of the correct drug for an infection. Gram staining is discussed in greater detail in Insight 4.2.

The **acid-fast** stain, like the Gram stain, is an important diagnostic stain that differentiates acid-fast bacteria (pink) from non-acid-fast bacteria (blue). This stain originated as a specific method to detect *Mycobacterium tuberculosis* in specimens. It was determined that these bacterial cells have a particularly impervious outer wall that holds fast (tightly or tenaciously) to the dye (carbol fuchsin) even when washed with a solution containing acid or acid alcohol. This stain is used for other medically important mycobacteria such as the Hansen's disease (leprosy) bacillus and for *Nocardia,* an agent of lung or skin infections.

The endospore stain (spore stain) is similar to the acid-fast method in that a dye is forced by heat into resistant bodies called spores or endospores (their formation and significance are discussed in chapter 4). This stain is designed to distinguish between spores and the cells that they come from (so-called

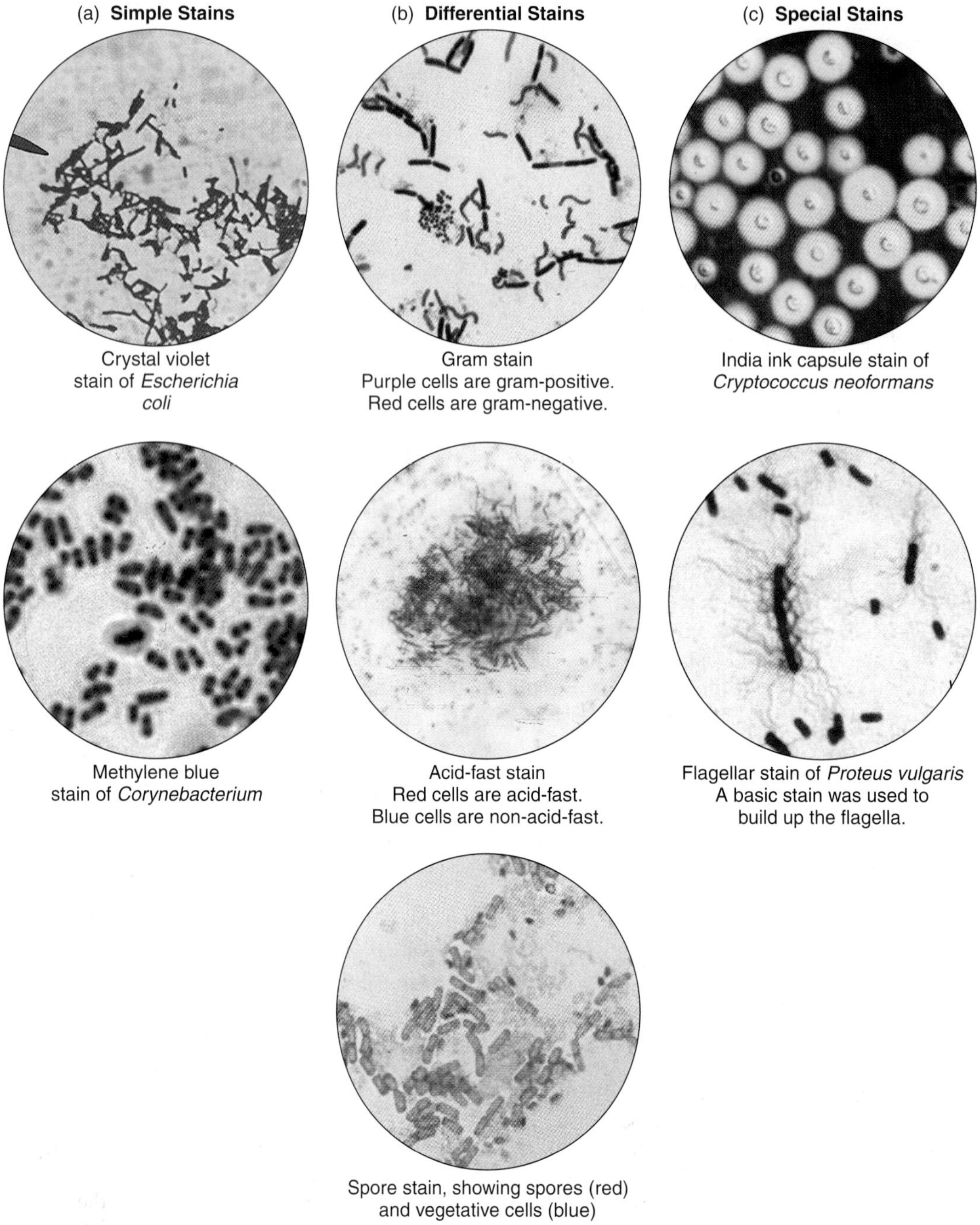

Figure 3.25 Types of microbiological stains.
(a) Simple stains. **(b)** Differential stains: Gram, acid-fast, and spore. **(c)** Special stains: capsule and flagellar.

vegetative cells). Of significance in medical microbiology are the gram-positive, spore-forming members of the genus *Bacillus* (the cause of anthrax) and *Clostridium* (the cause of botulism and tetanus)—dramatic diseases of universal fascination that we consider in later chapters.

Special stains are used to emphasize certain cell parts that are not revealed by conventional staining methods. Capsule staining is a method of observing the microbial capsule, an unstructured protective layer surrounding the cells of some bacteria and fungi. Because the capsule does not react with most stains, it is often negatively stained with India ink, or it may be demonstrated by special positive stains. The fact that not all microbes exhibit capsules is a useful feature for identifying pathogens. One example is *Cryptococcus,* which causes a serious fungal meningitis in AIDS patients (see chapter 19).

Flagellar staining is a method of revealing flagella, the tiny, slender filaments used by bacteria for locomotion. Because the width of bacterial flagella lies beyond the resolving power of the light microscope, in order to be seen, they must be enlarged by depositing a coating on the outside of the filament and then staining it. Their presence, number, and arrangement on a cell are taxonomically useful.

CASE FILE 3 *WRAP-UP*

The Gram stain is used to visualize and differentiate bacteria into broad categories. Purple-stained bacteria are called gram-positive, and red are called gram-negative. These classifications relate information about the cell wall structure of each. Because bacteria are so small, the highest magnification lens (100×) on a bright-field compound microscope is used to distinguish cells and determine their color and shape. The optic properties of immersion oil placed between the glass slide and this lens combine to resolve objects as small as 0.2 μm into view.

The blood sample is processed as follows: (1) inoculum—the blood is aseptically obtained and placed into liquid medium; (2) incubation—the specimen is left overnight in appropriate conditions (this is an ongoing process); (3) inspection—a Gram stain is prepared and viewed; (4) isolation—the culture growing in the liquid is transferred with proper technique to differential and selective solid media; (5) identification—the announcement of *Bacillus anthracis*. After isolation, further biotesting is performed to identify this bacterium. To further identify this particular strain as identical to those in other anthrax cases, a DNA study is also performed.

See: CDC. 2001. Update: Investigation of bioterrorism-related inhalational anthrax—Connecticut, 2001. MMWR *50:1049–1051.*

CHECKPOINT

- Magnification, resolving power, lens quality, and illumination source all influence the clarity of specimens viewed through the optical microscope.
- The maximum resolving power of the optical microscope is 200 nm, or 0.2 μm. This is sufficient to see the internal structures of eukaryotes and the morphology of most bacteria.
- There are six types of optical microscopes. Four types use visible light for illumination: bright-field, dark-field, phase-contrast, and interference microscopes. The fluorescence microscope uses UV light for illumination, but it has the same resolving power as the other optical microscopes. The confocal microscope can use UV light or visible light reflected from specimens.
- Electron microscopes (EM) use electrons, not light waves, as an illumination source to provide high magnification (5,000× to 1,000,000×) and high resolution (0.5 nm). Electron microscopes can visualize cell ultrastructure (transmission EM) and three-dimensional images of cell and virus surface features (scanning EM).
- The newest generation of microscope is called the scanning probe microscope and uses precision tips to image structures at the atomic level.
- Specimens viewed through optical microscopes can be either alive or dead, depending on the type of specimen preparation, but all EM specimens are dead because they must be viewed in a vacuum.
- Stains are important diagnostic tools in microbiology because they can be designed to differentiate cell shape, structure, and biochemical composition of the specimens being viewed.

Chapter Summary with Key Terms

3.1 Methods of Culturing Microorganisms—The Five I's

A. Microbiology as a science is very dependent on a number of specialized laboratory techniques. Laboratory steps routinely employed in microbiology are inoculation, incubation, isolation, inspection, and identification.
   1. Initially, a specimen must be collected from a source, whether environmental or a patient.
   2. **Inoculation** of a **medium** is the first step in obtaining a **culture** of the microorganisms present.
   3. **Isolation** of the microorganisms, so that each microbial cell present is separated from the others and forms discrete **colonies,** is aided by inoculation techniques such as streak plates, pour plates, and spread plates.
   4. **Incubation** of the medium with the microbes under the right conditions allows growth to visible colonies. Generally, isolated colonies would be **subcultured** for further testing at this point. The goal is a **pure culture** in most cases, or a **mixed culture. Contaminated cultures** can ruin correct analysis and study.
   5. **Inspection** begins with macroscopic characteristics of the colonies, and continues with microscopic analysis.
   6. *Identification* correlates the various morphological, physiological, genetic, and serological traits as needed to be able to pinpoint the actual species or even strain of microbe.

B. Growing microbes in the laboratory requires providing them with all their essential nutrients.
   1. Artificial media allow the growth and isolation of microorganisms in the laboratory, and can be classified by their physical state, chemical composition, and functional types. The nutritional requirements of microorganisms in the laboratory may be simple or complex.
   2. Physical types of media include those that are **liquid,** such as broths and milk, those that are **semisolid,** and those that are solid. Solid media may be liquefiable, containing a solidifying agent such as **agar** or gelatin.
   3. Chemical composition of a medium may be completely chemically defined, thus synthetic. Nonsynthetic, or complex, media contain ingredients that are not completely definable.
   4. Functional types of media serve different purposes, often allowing biochemical tests to be performed at the same time. Types include general-purpose,

enriched, selective, differential, anaerobic (reducing), assay, and enumeration **media.** Transport media are important for conveying certain clinical specimens to the laboratory.

5. In certain instances, microorganisms have to be grown in cell cultures or host animals.
6. Cultures are maintained by large collection facilities such as the American Type Culture Collection located in Manassas, Virginia.

3.2 The Microscope: Window on an Invisible Realm

A. Optical, or light, microscopy depends on lenses that **refract** light rays, drawing the rays to a focus to produce a magnified image.
   1. A simple microscope consists of a single magnifying lens, whereas a compound microscope relies on two lenses: the ocular lens and the objective lens.
   2. The total power of magnification is calculated from the product of the ocular and objective magnifying powers.
   3. Resolution, or the **resolving power,** is a measure of a microscope's capacity to make clear images of very small objects. Resolution is improved with shorter wavelengths of illumination and with a higher **numerical aperture** of the lens. Light microscopes are limited to magnifications around 2,000× by the resolution.
   4. Modifications in the lighting or the lens system give rise to the bright-field, dark-field, phase-contrast, interference, and fluorescence microscopes.

B. A transmission electron microscope (TEM) projects electrons through prepared sections of the specimen, providing detailed structural images of cells, cell parts, and viruses. Scanning electron microscopy (SEM) is more like dark-field microscopy, bouncing the electrons off the surface of the specimen to detectors.

C. Scanning probe microscopes use atomically sharp tips to achieve a resolution of up to 0.01 Angstroms.

D. Specimen preparation in optical microscopy is governed by the condition of the specimen, the purpose of the inspection, and the type of microscope being used.
   1. Wet mounts and hanging drop mounts permit examination of the characteristics of live cells, such as motility, shape, and arrangement.
   2. Fixed mounts are made by drying and heating a film of the specimen called a smear. This is then stained using dyes to permit visualization of cells or cell parts.

E. Staining uses either basic (cationic) dyes with positive charges or acidic (anionic) dyes with negative charges. The surfaces of microbes are negatively charged and attract basic dyes. This is the basis of positive staining. In **negative staining,** the microbe repels the dye and it stains the background. Dyes may be used alone and in combination.
   1. Simple stains use just one dye and highlight cell morphology.
   2. Differential stains require a primary dye and a contrasting counterstain in order to distinguish cell types or parts. Important differential stains include the **Gram stain, acid-fast stain,** and the endospore stain.
   3. Special stains are designed to bring out distinctive characteristics. Examples include capsule stains and flagellar stains.

Multiple-Choice and True-False Questions

Multiple-Choice Questions. Select the correct answer from the answers provided.

1. The term *culture* refers to the ____ growth of microorganisms in ____.
   a. rapid, an incubator
   b. macroscopic, media
   c. microscopic, the body
   d. artificial, colonies

2. A mixed culture is
   a. the same as a contaminated culture
   b. one that has been adequately stirred
   c. one that contains two or more known species
   d. a pond sample containing algae and protozoa

3. Resolution is ____ with a longer wavelength of light.
   a. improved
   b. worsened
   c. not changed
   d. not possible

4. A real image is produced by the
   a. ocular
   b. objective
   c. condenser
   d. eye

5. A microscope that has a total magnification of 1,500× when using the oil immersion objective has an ocular of what power?
   a. 150×
   b. 1.5×
   c. 15×
   d. 30×

6. The specimen for an electron microscope is always
   a. stained with dyes
   b. sliced into thin sections
   c. killed
   d. viewed directly

7. Motility is best observed with a
   a. hanging drop preparation
   b. negative stain
   c. streak plate
   d. flagellar stain

8. Bacteria tend to stain more readily with cationic (positively charged) dyes because bacteria
   a. contain large amounts of alkaline substances
   b. contain large amounts of acidic substances
   c. are neutral
   d. have thick cell walls

9. **Multiple Matching.** For each type of medium, select all descriptions that fit. For media that fit more than one description, briefly explain why this is the case.

   ____ mannitol salt agar
   ____ chocolate agar
   ____ MacConkey agar
   ____ nutrient broth

   a. selective medium
   b. differential medium
   c. chemically defined (synthetic) medium

____ Sabouraud's agar
____ triple-sugar iron agar
____ *Euglena* agar
____ SIM medium

d. enriched medium
e. general-purpose medium
f. complex medium
g. transport medium

10. A fastidious organism must be grown on what type of medium?
    a. general-purpose medium
    b. differential medium
    c. synthetic medium
    d. enriched medium
11. What type of medium is used to maintain and preserve specimens before clinical analysis?
    a. selective medium
    b. transport medium
    c. enriched medium
    d. differential medium

True-False Questions. If the statement is true, leave as is. If it is false, correct it by rewriting the sentence.

12. Agar has the disadvantage of being easily decomposed by microorganisms.
13. A subculture is a culture made from an isolated colony.
14. The factor that most limits the clarity of an image in a microscope is the *magnification*.
15. Living specimens can be examined either by light microscopy or electron microscopy.
16. The best stain to use to visualize a microorganism with a large capsule is a simple stain.

Writing to Learn

These questions are suggested as a *writing-to-learn* experience. For each question, compose a one- or two-paragraph answer that includes the factual information needed to completely address the question.

1. a. Describe briefly what is involved in the Five I's.
   b. Name three basic differences between inoculation and contamination.
2. a. Name two ways that pure, mixed, and contaminated cultures are similar and two ways that they differ from each other.
   b. What must be done to avoid contamination?
3. a. Explain what is involved in isolating microorganisms and why it is necessary to do this.
   b. Compare and contrast three common laboratory techniques for separating bacteria in a mixed sample.
   c. Describe how an isolated colony forms.
   d. Explain why an isolated colony and a pure culture are not the same thing.
4. a. Explain the two principal functions of dyes in media.
   b. Differentiate among the ingredients and functions of enriched, selective, and differential media.
5. Differentiate between microscopic and macroscopic methods of observing microorganisms, citing a specific example of each method.
6. a. Contrast the concepts of magnification, refraction, and resolution.
   b. Briefly explain how an image is made and magnified.
   c. Trace the pathway of light from its source to the eye, explaining what happens as it passes through the major parts of the microscope.
7. a. What does it mean in practical terms if the resolving power is 1.0 μm?
   b. How does a value greater than 1.0 μm compare? (Is it better or worse?)
   c. How does a value less than 1.0 μm compare?
   d. What can be done to a microscope to improve resolution?
8. Compare bright-field, dark-field, phase-contrast, and fluorescence microscopy as to field appearance, specimen appearance, light source, and uses.
9. a. Compare and contrast the optical compound microscope with the electron microscope.
   b. Why is the resolution so superior in the electron microscope?
   c. What will you never see in an unretouched electron micrograph?
10. a. Why are some bacteria difficult to grow in the laboratory? Relate this to what you know so far about metabolism.
    b. Why are viruses hard to cultivate in the laboratory?

Concept Mapping

Appendix D provides guidance for working with concept maps.

1. Supply your own linking words or phrases in this concept map, and provide the missing concepts in the empty boxes.

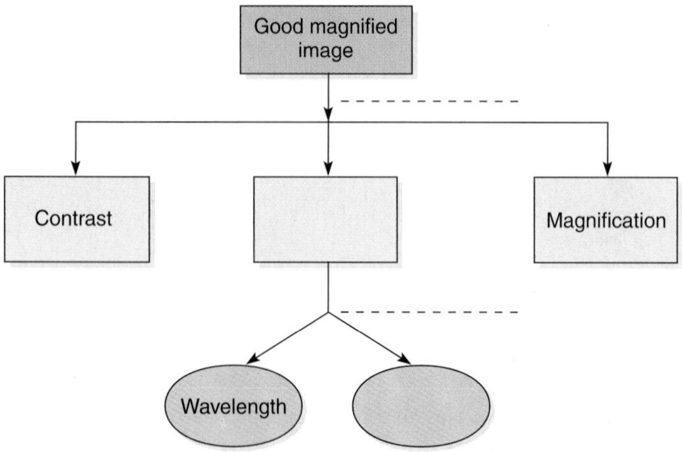
Good magnified image
Contrast
Magnification
Wavelength

2. Construct your own concept map using the following words as the *concepts*. Supply the linking words between each pair of concepts.

inoculation	staining
isolation	biochemical tests
incubation	subculturing
inspection	source of microbes
identification	transport medium
medium	streak plate
multiplication	

Critical Thinking Questions

Critical thinking is the ability to reason and solve problems using facts and concepts. These questions can be approached from a number of angles, and in most cases, they do not have a single correct answer.

1. Describe the steps you would take to isolate, cultivate, and identify a microbial pathogen from a urine sample. (Hint: Look at the Five I's.)
2. A certain medium has the following composition:

Glucose	15 g
Yeast extract	5 g
Peptone	5 g
KH_2PO_4	2 g
Distilled water	1,000 ml

   To what chemical category does this medium belong?
3. a. Name four categories that blood agar fits into.
   b. Name four differential reactions that TSIA shows.
   c. Can you tell what functional kind of medium *Enterococcus faecalis* medium is?
4. a. In what ways are dark-field microscopy and negative staining alike?
   b. How is the dark-field microscope like the scanning electron microscope?
5. Biotechnology companies have engineered hundreds of different types of mice, rats, pigs, goats, cattle, and rabbits to have genetic diseases similar to diseases of humans or to synthesize drugs and other biochemical products. They have patented these animals, and they sell them to researchers for study and experimentation.
   a. What do you think of creating new life forms just for experimentation?
   b. Comment on the benefits, safety, and ethics of this trend.
6. Some human pathogenic bacteria are resistant to most antibiotics. How would you prove a bacterium is resistant to antibiotics using laboratory culture techniques?
7. This is a test of *your* optical system's resolving power. Prop your book against a wall about 20 inches away and determine

So, Naturalists observe,
a flea has smaller
fleas that on him prey;
and these have smaller still
to bite 'em; and so proceed,
ad infinitum

Poem by Jonathan Swift.

the line in the following illustration that is no longer resolvable by your eye. See if you can determine your actual resolving power, using a millimeter ruler.

Visual Understanding

1. **Figure 3.3*a* and *b*.** If you were using the quadrant streak plate method to plate a very dilute broth culture (with many fewer bacteria than the broth used for 3*b*) would you expect to see single, isolated colonies in quadrant 4 or quadrant 3? Explain your answer.

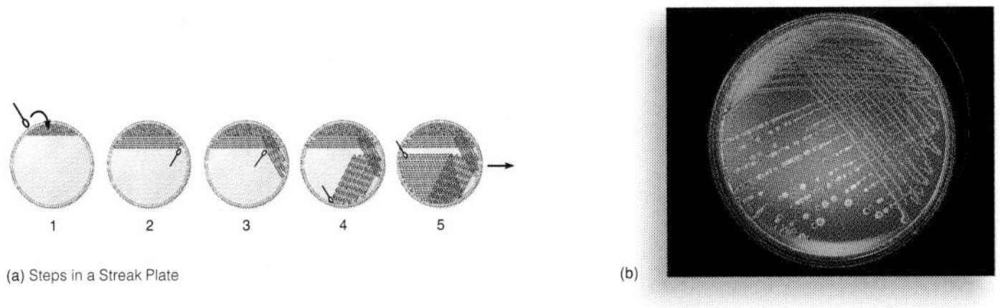

(a) Steps in a Streak Plate (b)

2. **From chapter 1, figure 1.6.** Which of these photos from chapter 1 is an SEM image? Which is a TEM image?

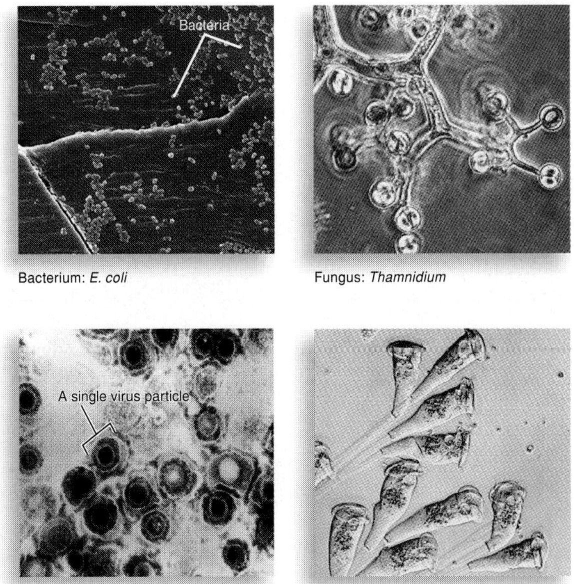

Bacterium: *E. coli* Fungus: *Thamnidium*
Virus: *Herpes simplex* Protozoan: *Vorticella*

Internet Search Topics

1. Search through several websites using the keywords "electron micrograph." Find examples of TEM and SEM micrographs and their applications in science and technology.
2. Search using the words "laboratory identification of anthrax" to make an outline of the basic techniques used in analysis of the microbe, under the headings of the Five I's.
3. Search for "carbon monoxide man." If you find a site about scanning probe microscopy, describe what the carbon monoxide man is.
4. Go to: www.aris.mhhe.com, and click on "microbiology" and then this textbook's author/title. Go to Chapter 3, access the URLs listed under Internet Search Topics, and research the following:
   a. Explore the website listed, which contains a broad base of information and images on microscopes and microscopy. Visit the photo gallery to compare different types of microscope images.
   b. Use the interactive website listed to see clearly how the numerical aperture changes with magnification.

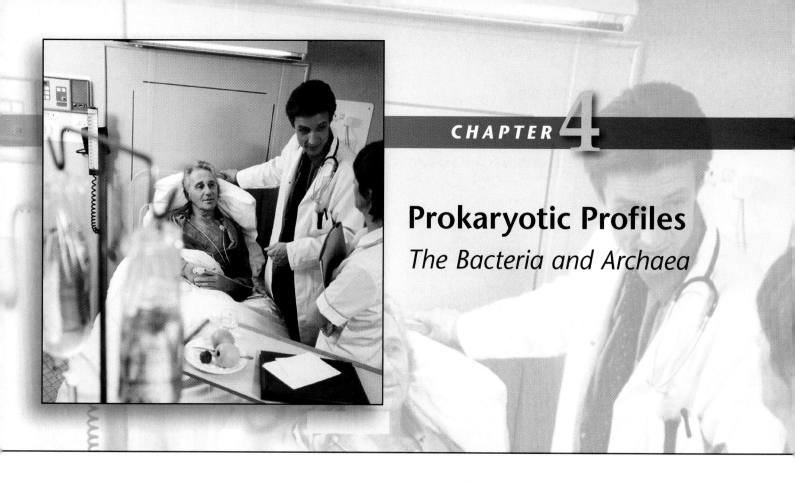

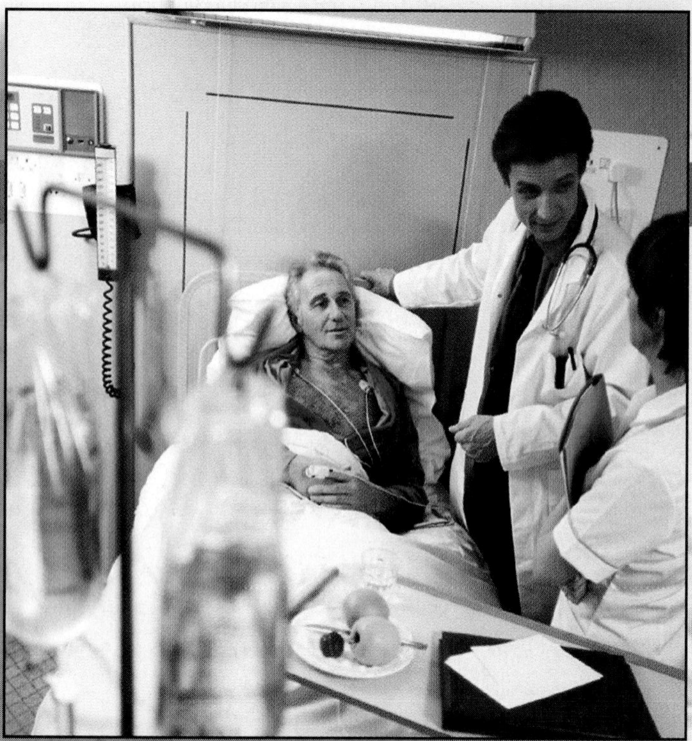

CHAPTER 4

Prokaryotic Profiles

The Bacteria and Archaea

CASE FILE 4

From April 3 to April 24, 2001, nine cases of pneumonia occurred in elderly residents (median age of 86 years) living at a long-term care facility in New Jersey. Seven of the nine patients had *Streptococcus pneumoniae* isolated from blood cultures, with capsular serotyping revealing that all isolates were serotype 14 and of the same clonal group. Seven of the nine patients also lived in the same wing of the nursing home. The two patients that were culture negative did contain gram-positive diplococci in their sputum and had chest X rays consistent with pneumonia. Epidemiological studies of the patients and controls revealed that all who developed pneumonia had no documented record of vaccination with the pneumococcal polysaccharide vaccine (PPV). In contrast, about 50% of the controls were vaccinated with PPV. Even though other risk factors were assessed, the lack of vaccination with PPV was the only one strongly associated with illness. Unfortunately, despite treatment, four of the nine patients with pneumonia died.

Once the outbreak was recognized, PPV was offered to those 55 residents who had not yet been vaccinated: 37 of these were vaccinated, whereas the other 18 were either ineligible or refused the vaccine. Other control measures included refusal to admit patients without a history of PPV vaccine.

- *What special advantage does the capsule confer on the pathogen* Streptococcus pneumoniae?
- *Why are those who have been vaccinated against* Streptococcus pneumoniae *more resistant to infection by this agent?*

Case File 4 Wrap-Up appears on page 98.

CHAPTER OVERVIEW

- Prokaryotic cells are the smallest, simplest, and most abundant cells on earth.
- Representative prokaryotes include bacteria and archaea, both of which lack a nucleus and organelles but are functionally complex.
- The structure of bacterial cells is compact and capable of adaptations to a multitude of habitats.
- The cell is encased in an envelope that protects, supports, and regulates transport.
- Bacteria have special structures for motility and adhesion in the environment.

- Bacterial cells contain genetic material in one or a few chromosomes, and ribosomes for synthesizing proteins.
- Bacteria have the capacity for reproduction, nutrient storage, dormancy, and resistance to adverse conditions.
- Shape, size, and arrangement of bacterial cells are extremely varied.
- Bacterial taxonomy and classification are based on their structure, metabolism, and genetics.
- Archaea are prokaryotes related to eukaryotic cells that possess unique biochemistry and genetics.

In chapter 1, we described prokaryotes as being cells with no true nucleus. (Eukaryotes have a membrane around their DNA, and this structure is called the nucleus.) Some microbiologists have recently been suggesting that we are not defining what a prokaryote is, only what it is *not*—and therefore we are not really defining it at all. This is one way scientists work. A previously accepted notion (i.e., what a prokaryote is) is questioned publicly, causing a variety of reactions ranging from surprise to dismissal. But usually other scientists begin discussing the question and the truth that might be behind the assertion is examined in a new way. But this whole chapter is about the type of cell we call a prokaryote. So how do we know whether a cell is prokaryotic or eukaryotic? A prokaryote can be distinguished from the other type of cell (a eukaryote) because of certain characteristics it possesses:

- *The way its DNA is packaged:* Prokaryotes have nuclear material that is not encased in a membrane (i.e., they do not have a nucleus). Eukaryotes have a membrane around their DNA (making up a nucleus). Prokaryotes don't wind their DNA around proteins called **histones;** eukaryotes do.
- *The makeup of its cell wall:* Prokaryotes (bacteria and archaea) generally have a wall structure that is unique compared to eukaryotes. Bacteria have sturdy walls made of a chemical called peptidoglycan. Archaeal walls are also tough and made of other chemicals, distinct from bacteria and distinct from eukaryotic cells.
- *Its internal structures:* Prokaryotes don't have complex, membrane-bounded organelles in their cytoplasm (eukaryotes do).

Both prokaryotic and eukaryotic microbes are found throughout nature. Both can cause infectious diseases. Examples of bacterial diseases include "strep" throat, Lyme disease, and ear infections. The medical response to them is informed by their "prokaryoteness." Eukaryotic infections (examples: histoplasmosis, malaria) often require a different approach. In this chapter and coming chapters, you'll discover why that is.

4.1 Prokaryotic Form and Function

The evolutionary history of prokaryotic cells extends back at least 3.8 billion years. It is now generally thought that the very first cells to appear on the earth were a type of archaea possibly related to modern forms that live on sulfur compounds in geothermal ocean vents. The fact that these organisms have endured for so long in such a variety of habitats indicates a cellular structure and function that are amazingly versatile and adaptable. The general cellular organization of a prokaryotic cell can be represented with this flowchart:

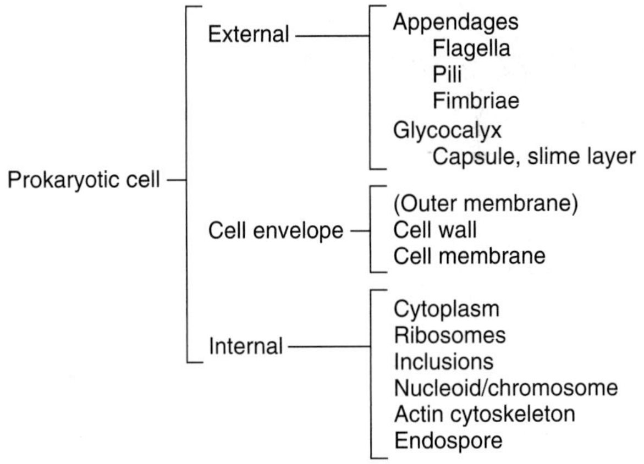

All bacterial cells invariably have a cell membrane, cytoplasm, ribosomes, and one (or a few) chromosome(s); the majority have a cell wall and some form of surface coating or glycocalyx. Specific structures that are found in some but not all bacteria are flagella, pili, fimbriae, capsules, slime layers, inclusions, an actin cytoskeleton, and endospores.

The Structure of a Generalized Prokaryotic Cell

Bacterial cells appear featureless and two-dimensional when viewed with an ordinary microscope. Not until they are subjected to the scrutiny of the electron microscope and biochemical studies does their intricate and functionally complex nature become evident. The descriptions of prokaryotic structure, except where otherwise noted, refer to the **bacteria,** a category of prokaryotes with peptidoglycan in their cell walls. **Figure 4.1** presents a three-dimensional anatomical view of a generalized, rod-shaped, bacterial cell. As we survey the principal anatomical features of this cell, we will perform a microscopic dissection of sorts, following a course that begins with the outer cell structures and proceeds to the internal contents.

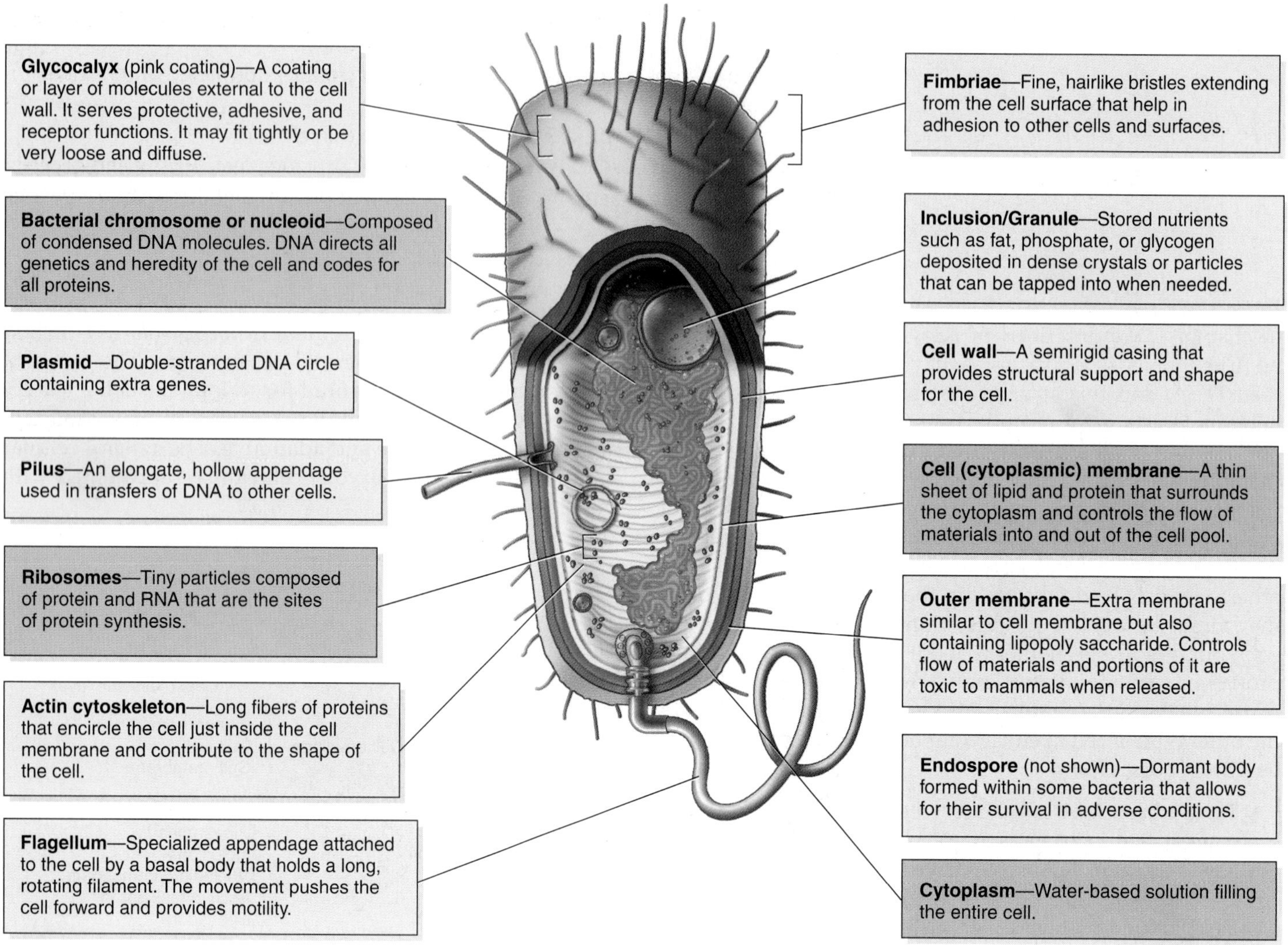
Glycocalyx (pink coating)—A coating or layer of molecules external to the cell wall. It serves protective, adhesive, and receptor functions. It may fit tightly or be very loose and diffuse.
Bacterial chromosome or nucleoid—Composed of condensed DNA molecules. DNA directs all genetics and heredity of the cell and codes for all proteins.
Plasmid—Double-stranded DNA circle containing extra genes.
Pilus—An elongate, hollow appendage used in transfers of DNA to other cells.
Ribosomes—Tiny particles composed of protein and RNA that are the sites of protein synthesis.
Actin cytoskeleton—Long fibers of proteins that encircle the cell just inside the cell membrane and contribute to the shape of the cell.
Flagellum—Specialized appendage attached to the cell by a basal body that holds a long, rotating filament. The movement pushes the cell forward and provides motility.
Fimbriae—Fine, hairlike bristles extending from the cell surface that help in adhesion to other cells and surfaces.
Inclusion/Granule—Stored nutrients such as fat, phosphate, or glycogen deposited in dense crystals or particles that can be tapped into when needed.
Cell wall—A semirigid casing that provides structural support and shape for the cell.
Cell (cytoplasmic) membrane—A thin sheet of lipid and protein that surrounds the cytoplasm and controls the flow of materials into and out of the cell pool.
Outer membrane—Extra membrane similar to cell membrane but also containing lipopoly saccharide. Controls flow of materials and portions of it are toxic to mammals when released.
Endospore (not shown)—Dormant body formed within some bacteria that allows for their survival in adverse conditions.
Cytoplasm—Water-based solution filling the entire cell.

Figure 4.1 Structure of a prokaryotic cell.
Cutaway view of a typical rod-shaped bacterium, showing major structural features. Note that not all components are found in all cells; dark-blue boxes indicate structures that all bacteria possess.

4.2 External Structures

Appendages: Cell Extensions

Several discrete types of accessory structures sprout from the surface of bacteria. These long **appendages** are common but are not present on all species. Appendages can be divided into two major groups: those that provide motility (flagella and axial filaments) and those that provide attachments or channels (fimbriae and pili).

Flagella—Bacterial Propellers

The prokaryotic **flagellum** (flah-jel′-em), an appendage of truly amazing construction, is certainly unique in the biological world. The primary function of flagella is to confer **motility,** or self-propulsion—that is, the capacity of a cell to swim freely through an aqueous habitat. The extreme thinness of a bacterial flagellum necessitates high magnification to reveal its special architecture, which has three distinct parts: the filament, the hook (sheath), and the basal body **(figure 4.2).** The **filament,** a helical structure composed of proteins, is approximately 20 nanometers in diameter and varies from 1 to 70 microns in length. It is inserted into a curved, tubular hook. The hook is anchored to the cell by the basal body, a stack of rings firmly anchored through the cell wall, to the cell membrane and the outer membrane. This arrangement permits the hook with its filament to rotate 360°, rather than undulating back and forth like a whip as was once thought.

One can generalize that all spirilla, about half of the bacilli, and a small number of cocci are flagellated (these bacterial shapes are shown in figure 4.22). Flagella vary both in number and arrangement according to two general patterns: (1) In a *polar* arrangement, the flagella are attached at one or both ends of the cell. Three subtypes of this pattern are: **monotrichous** (mah″-noh-trik′-us), with a single flagellum; **lophotrichous** (lo″-foh-), with small

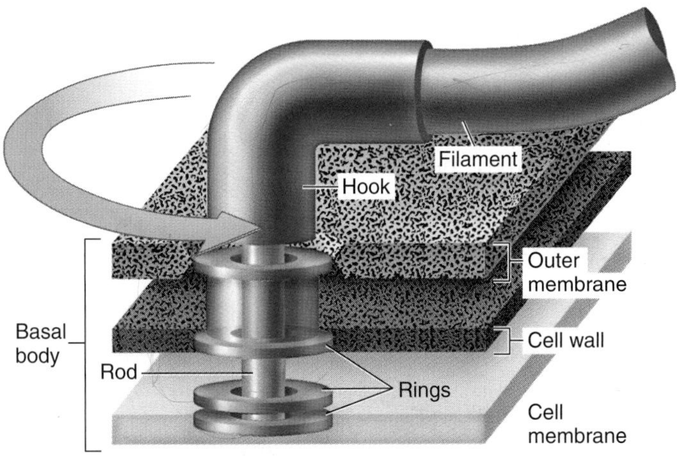

Figure 4.2 Details of the basal body of a flagellum in a gram-negative cell.
The hook, rings, and rod function together as a tiny device that rotates the filament 360°.

bunches or tufts of flagella emerging from the same site; and **amphitrichous** (am″-fee-), with flagella at both poles of the cell. (2) In a **peritrichous** (per″-ee-) arrangement, flagella are dispersed randomly over the surface of the cell **(figure 4.3).**

The presence of motility is one piece of information used in the laboratory identification or diagnosis of pathogens. Flagella are hard to visualize in the laboratory, but often it is sufficient to know simply whether a bacterial species is motile. One way to detect motility is to stab a tiny mass of cells into a soft (semisolid) medium in a test tube. Growth spreading rapidly through the entire medium is indicative of motility. Alternatively, cells can be observed microscopically with a hanging drop slide. A truly motile cell will flit, dart, or wobble around the field, making some progress, whereas one that is nonmotile jiggles about in one place but makes no progress.

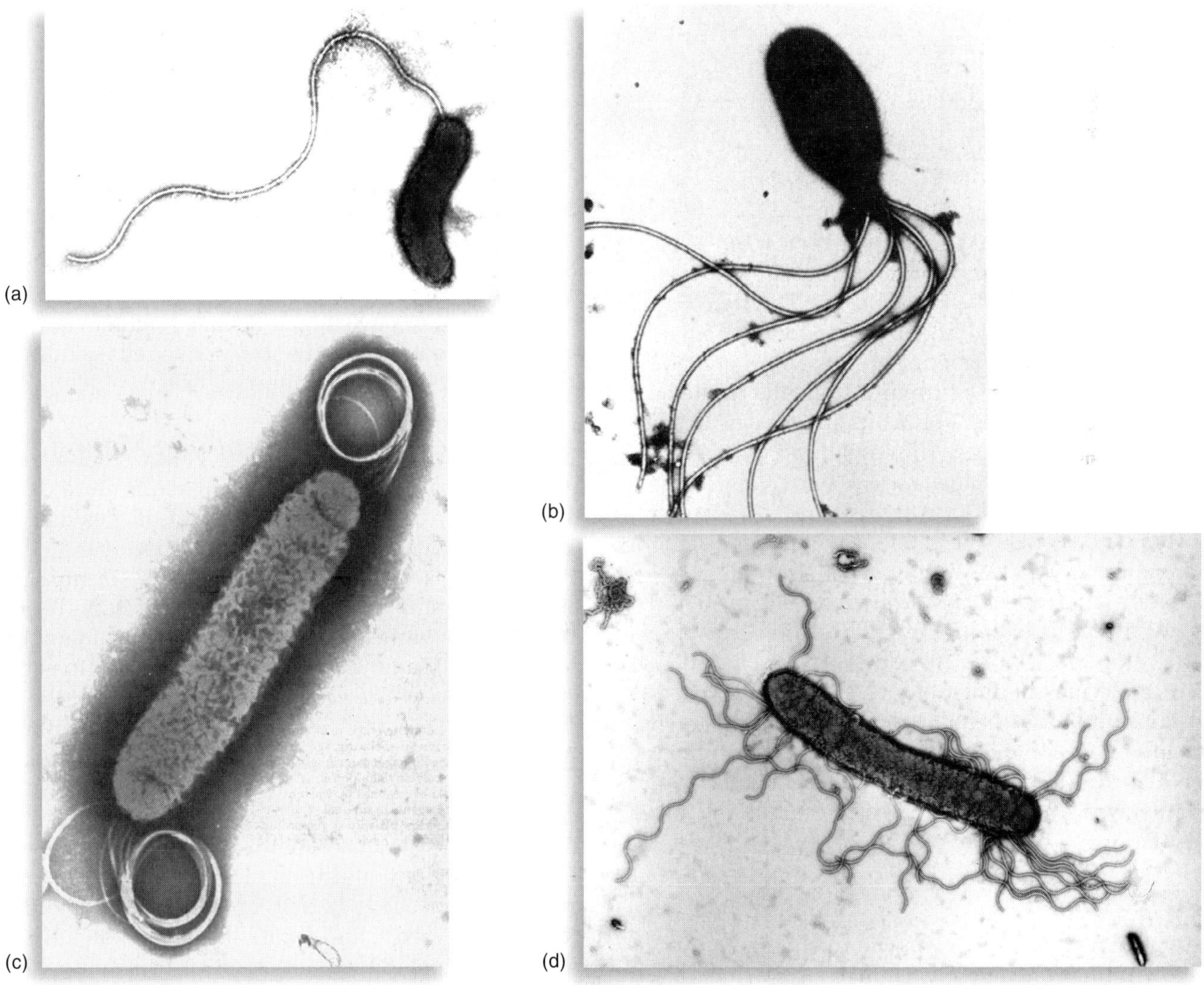

Figure 4.3 Electron micrographs depicting types of flagellar arrangements.
(a) Monotrichous flagellum on the predatory bacterium *Bdellovibrio.* **(b)** Lophotrichous flagella on *Vibrio fischeri,* a common marine bacterium (23,000×). **(c)** Unusual flagella on *Aquaspirillum* are amphitrichous (and lophotrichous) in arrangement and coil up into tight loops. **(d)** An unidentified bacterium discovered inside *Paramecium* cells exhibits peritrichous flagella.
(b) From Reichelt and Baumann, *Arch. Microbiol.* 94:283–330. © Springer-Verlag, 1973.

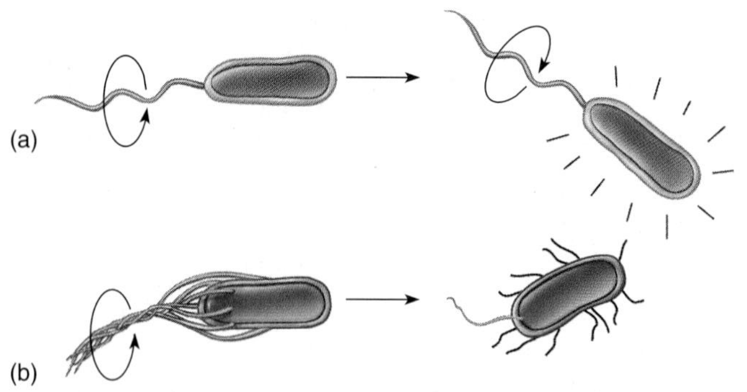

Figure 4.4 The operation of flagella and the mode of locomotion in bacteria with polar and peritrichous flagella.
(a) In general, when a polar flagellum rotates in a counterclockwise direction, the cell swims forward. When the flagellum reverses direction and rotates clockwise, the cell stops and tumbles.
(b) In peritrichous forms, all flagella sweep toward one end of the cell and rotate as a single group. During tumbles, the flagella lose coordination.

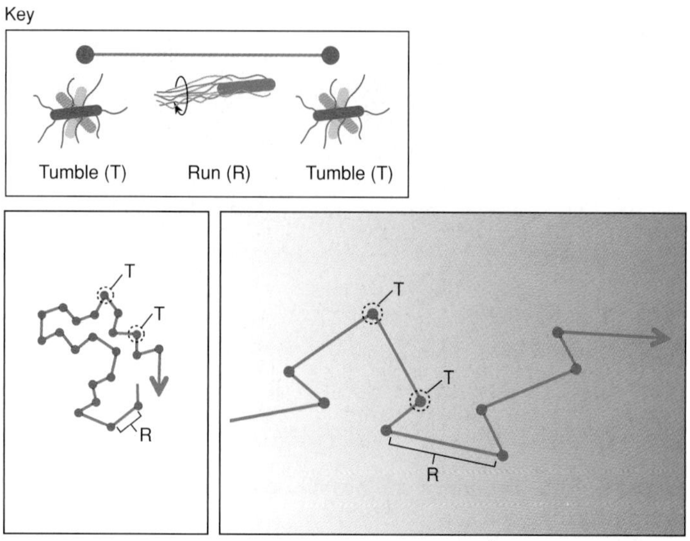

(a) No attractant or repellent **(b)** Gradient of attractant concentration

Figure 4.5 Chemotaxis in bacteria.
(a) A cell moves via a random series of short runs and tumbles when there is no attractant or repellent. **(b)** The cell spends more time on runs as it gets closer to the attractant.

Fine Points of Flagellar Function Flagellated bacteria can perform some rather sophisticated feats. They can detect and move in response to chemical signals—a type of behavior called **chemotaxis** (ke″-moh-tak′-sis). Positive chemotaxis is movement of a cell in the direction of a favorable chemical stimulus (usually a nutrient); negative chemotaxis is movement away from a repellent (potentially harmful) compound.

The flagellum is effective in guiding bacteria through the environment primarily because the system for detecting chemicals is linked to the mechanisms that drive the flagellum. Located in the cell membrane are clusters of receptors[1] that bind specific molecules coming from the immediate environment. The attachment of sufficient numbers of these molecules transmits signals to the flagellum and sets it into rotary motion. If several flagella are present, they become aligned and rotate as a group **(figure 4.4).** As a flagellum rotates counterclockwise, the cell itself swims in a smooth linear direction toward the stimulus; this action is called a *run.* Runs are interrupted at various intervals by *tumbles,* during which the flagellum reverses direction and causes the cell to stop and change its course. It is believed that attractant molecules inhibit tumbles and permit progress toward the stimulus. Repellents cause numerous tumbles, allowing the bacterium to redirect itself away from the stimulus **(figure 4.5).** Some photosynthetic bacteria exhibit *phototaxis,* a type of movement in response to light rather than chemicals.

Periplasmic Flagella

Corkscrew-shaped bacteria called **spirochetes** (spy′-roh-keets) show an unusual, wriggly mode of locomotion caused by two or more long, coiled threads, the periplasmic flagella or *axial filaments.* A periplasmic flagellum is a type of internal flagellum that is enclosed in the space between the cell wall and the cell membrane **(figure 4.6).** The filaments curl closely around the spirochete coils yet are free to contract and impart a twisting or flexing motion to the cell. This form of locomotion must be seen in live cells such as the spirochete of syphilis to be truly appreciated.

Appendages for Attachment and Mating

The structures termed **pilus** (pil-us) and **fimbria** (fim′-bree-ah) both refer to bacterial surface appendages that provide some type of adhesion, but not locomotion.

Fimbriae are small, bristlelike fibers sprouting off the surface of many bacterial cells **(figure 4.7).** Their exact composition varies, but most of them contain protein. Fimbriae have an inherent tendency to stick to each other and to surfaces. They may be responsible for the mutual clinging of cells that leads to biofilms and other thick aggregates of cells on the surface of liquids and for the microbial colonization of inanimate solids such as rocks and glass **(Insight 4.1).** Some pathogens can colonize and infect host tissues because of a tight adhesion between their fimbriae and epithelial cells **(figure 4.7*b*).** For example, the gonococcus (agent of gonorrhea) colonizes the genitourinary tract, and *Escherichia coli* colonizes the intestine by this means. Mutant forms of these pathogens that lack fimbriae are unable to cause infections.

A pilus (also called a *sex pilus*) is an elongate, rigid tubular structure made of a special protein, *pilin.* So far, true pili have been found only on gram-negative bacteria, where they are utilized in a "mating" process between cells

1. Cell surface molecules that bind specifically with other molecules.

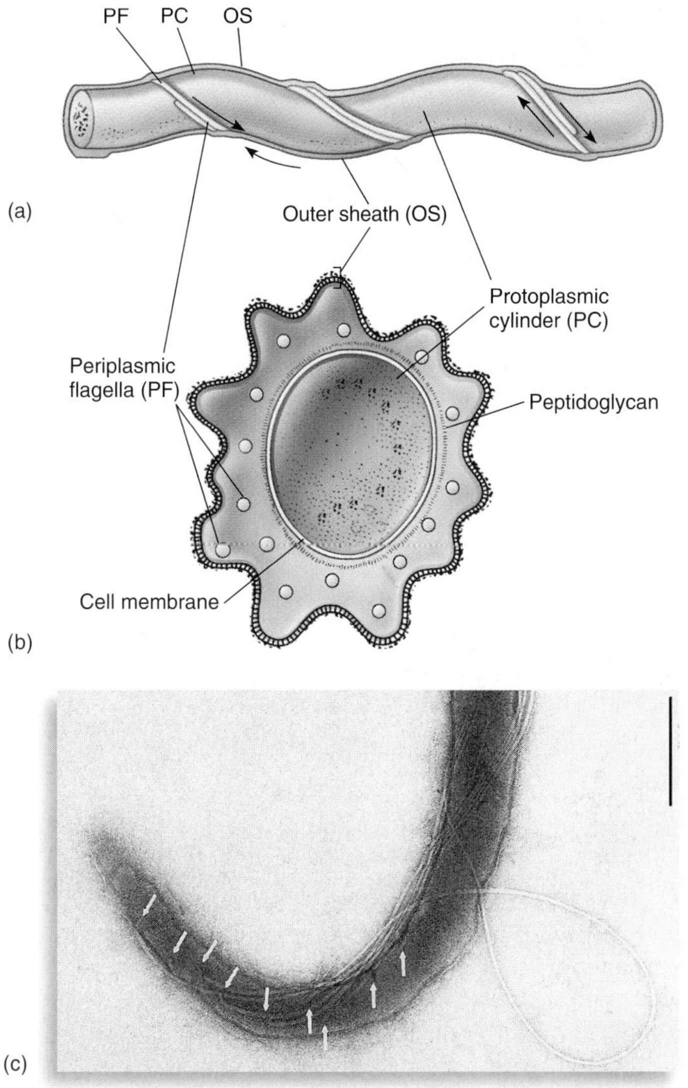

Figure 4.6 The orientation of periplasmic flagella on the spirochete cell.
(a) Longitudinal section. **(b)** Cross section. Contraction of the filaments imparts a spinning and undulating pattern of locomotion. **(c)** Electron micrograph captures the details of periplasmic flagella and their insertion points (arrows) in *Borrelia burgdorferi.* One flagellum has escaped the outer sheath, probably during preparation for EM. (Bar = 0.2 microns)

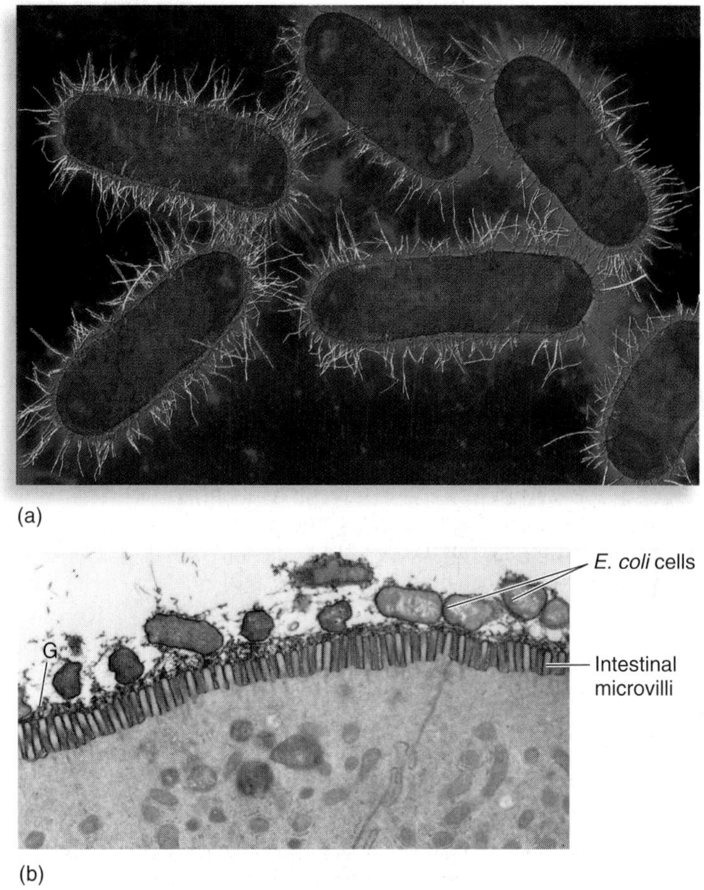

Figure 4.7 Form and function of bacterial fimbriae.
(a) Several cells of pathogenic *Escherichia coli* covered with numerous stiff fibers called fimbriae (30,000×). Note also the dark-blue granules, which are the chromosomes. **(b)** A row of *E. coli* cells tightly adheres by their fimbriae to the surface of intestinal cells (12,000×). This is how the bacterium clings to the body during an infection. (G = glycocalyx)

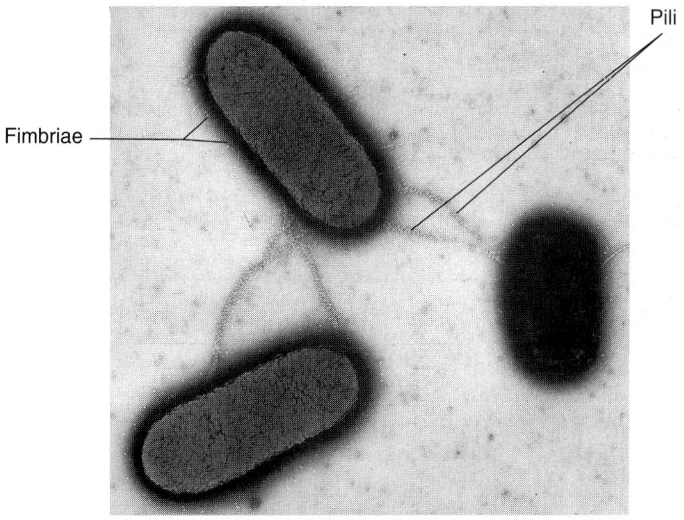

Figure 4.8 Three bacteria in the process of conjugating.
Clearly evident are the sex pili forming mutual conjugation bridges between a donor (upper cell) and two recipients (two lower cells). (Fimbriae can also be seen on the donor cell.)

called **conjugation,**[2] which involves partial transfer of DNA from one cell to another **(figure 4.8).** A pilus from the donor cell unites with a recipient cell thereby providing a cytoplasmic connection for making the transfer. Production of pili is controlled genetically, and conjugation takes place only between compatible gram-negative cells. Conjugation in gram-positive bacteria does occur but involves aggregation proteins rather than sex pili. The roles of pili and conjugation are further explored in chapter 9.

2. Although the term *mating* is sometimes used for this process, it is not a form of sexual reproduction.

INSIGHT 4.1 Discovery

Biofilms—The Glue of Life

Being aware of the widespread existence of microorganisms on earth, we should not be surprised that, when left undisturbed, they gather in masses, cling to various surfaces, and capture available moisture and nutrients. The formation of these living layers, called **biofilms,** is actually a universal phenomenon that all of us have observed. Consider the scum that builds up in toilet bowls and shower stalls in a short time if they are not cleaned; or the algae that collect on the walls of swimming pools; and, more intimately, the constant deposition of plaque on teeth. Microbes making biofilms is a primeval tendency that has been occurring for billions of years as a way to create stable habitats with adequate access to food, water, atmosphere, and other essential factors. Biofilms are often cooperative associations among several microbial groups (bacteria, fungi, algae, and protozoa) as well as plants and animals.

Substrates are most likely to accept a biofilm if they are moist and have developed a thin layer of organic material such as polysaccharides or glycoproteins on their exposed surface (see figure at right). This depositing process occurs within a few minutes to hours, making a slightly sticky texture that attracts primary colonists, usually bacteria. These early cells attach and begin to multiply on the surface. As they grow, various secreted substances in their glycocalyx (receptors, fimbriae, slime layers, capsules) increase the binding of cells to the surface and thicken the biofilm. (The image on the cover of this book is a biofilm growing on a gauze bandage. The extracellular matrix (the pink stuff in our drawing) is clearly visible.) As the biofilm evolves, it undergoes specific adaptations to the habitat in which it forms. In many cases, the earliest colonists contribute nutrients and create microhabitats that serve as a matrix for other microbes to attach and grow into the film, forming complete communities. The biofilm varies in thickness and complexity, depending upon where it occurs and how long it keeps developing. Complexity ranges from single cell layers to thick microbial mats with dozens of dynamic interactive layers.

Biofilms are a profoundly important force in the development of terrestrial and aquatic environments. They dwell permanently in bedrock and the earth's sediments, where they play an essential role in recycling elements, leaching minerals, and forming soil. Biofilms associated with plant roots promote the mutual exchange of nutrients between the microbes and roots. Invasive biofilms can wreak havoc with human-made structures such as cooling towers, storage tanks, air conditioners, and even stone buildings.

Biofilms also have serious medical implications. Most healthy human tissues do not accrue these thick layers of microbial life. Normal microbial inhabitants are generally limited to single-cell associations with skin and mucous membranes. But biofilms accumulate on damaged tissues (such as rheumatic heart valves), hard tissues (teeth), and foreign materials (catheters, IUDs, artificial hip joints). Microbes in a biofilm are extremely difficult to eradicate with antimicrobials. Previously it was assumed that the drugs had difficulty penetrating the viscous biofilm matrix. Now scientists have discovered that bacteria in biofilms turn on different genes when they are in a biofilm than when they are "free-floating." This altered gene expression gives the bacteria a different set of characteristics, often making them impervious to antibiotics. It is estimated that treating biofilm-related infections costs more than 1 billion dollars in the United States alone.

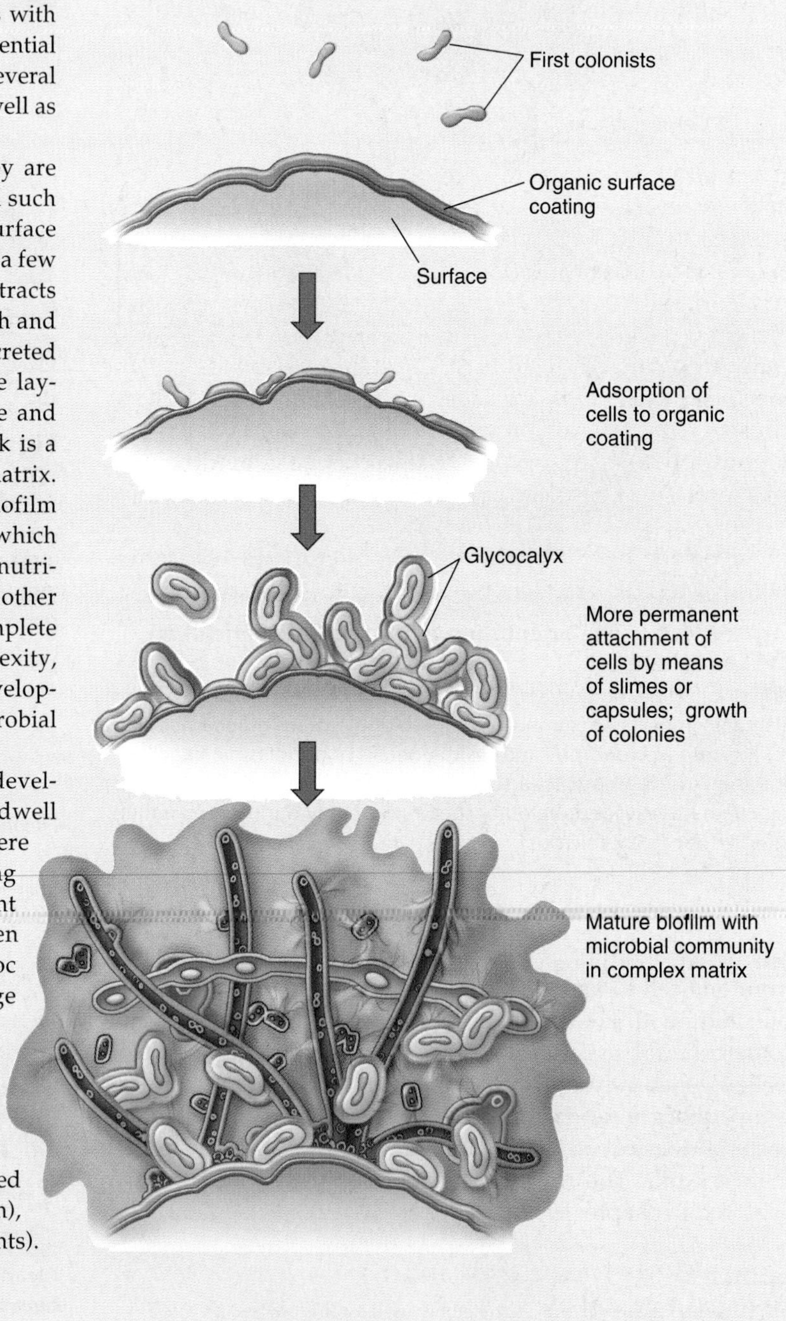

The Bacterial Surface Coating, or Glycocalyx

The bacterial cell surface is frequently exposed to severe environmental conditions. The **glycocalyx** develops as a coating of repeating polysaccharide units, protein, or both. This protects the cell and, in some cases, helps it adhere to its environment. Glycocalyces differ among bacteria in thickness, organization, and chemical composition. Some bacteria are covered with a loose shield called a slime layer that evidently protects them from loss of water and nutrients **(figure 4.9*a*).** A glycocalyx is called a **capsule** when it is bound more tightly to the cell than a slime layer is and it is denser and thicker **(figure 4.9*b*).** Capsules are often visible in negatively stained preparations **(figure 4.10*a*)** and produce a prominently sticky (mucoid) character to colonies on agar **(figure 4.10*b*).**

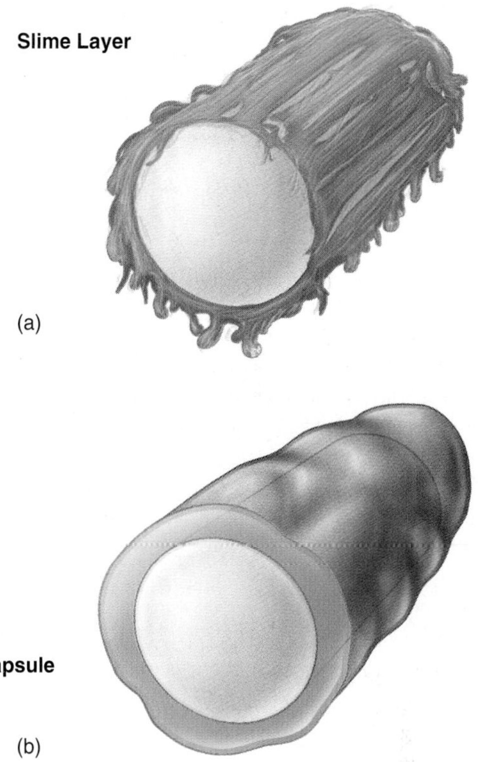

Figure 4.9 Drawing of sectioned bacterial cells to show the types of glycocalyces.
(a) The slime layer is a loose structure that is easily washed off. **(b)** The capsule is a thick, structured layer that is not readily removed.

Specialized Functions of the Glycocalyx Capsules are formed by many pathogenic bacteria, such as *Streptococcus pneumoniae* (a cause of pneumonia, an infection of the lung), *Haemophilus influenzae* (one cause of meningitis), and *Bacillus anthracis* (the cause of anthrax). Encapsulated bacterial cells generally have greater pathogenicity because capsules protect the bacteria against white blood cells called phagocytes. Phagocytes are a natural body defense that can engulf and destroy foreign cells through phagocytosis, thus preventing infection. A capsular coating blocks the mechanisms that phagocytes use to attach to and engulf bacteria. By escaping phagocytosis, the bacteria are free to multiply and infect body tissues. Encapsulated bacteria that mutate to nonencapsulated forms usually lose their pathogenicity.

Other types of glycocalyces can be important in formation of biofilms. The thick, white plaque that forms on teeth comes in part from the surface slimes produced by certain streptococci in the oral cavity. This slime protects them from being dislodged from the teeth and provides a niche for other oral bacteria that, in time, can lead to dental disease. The glycocalyx of some bacteria is so highly adherent that it is responsible for persistent colonization of nonliving materials such as plastic catheters, intrauterine devices, and metal pacemakers that are in common medical use **(figure 4.11).**

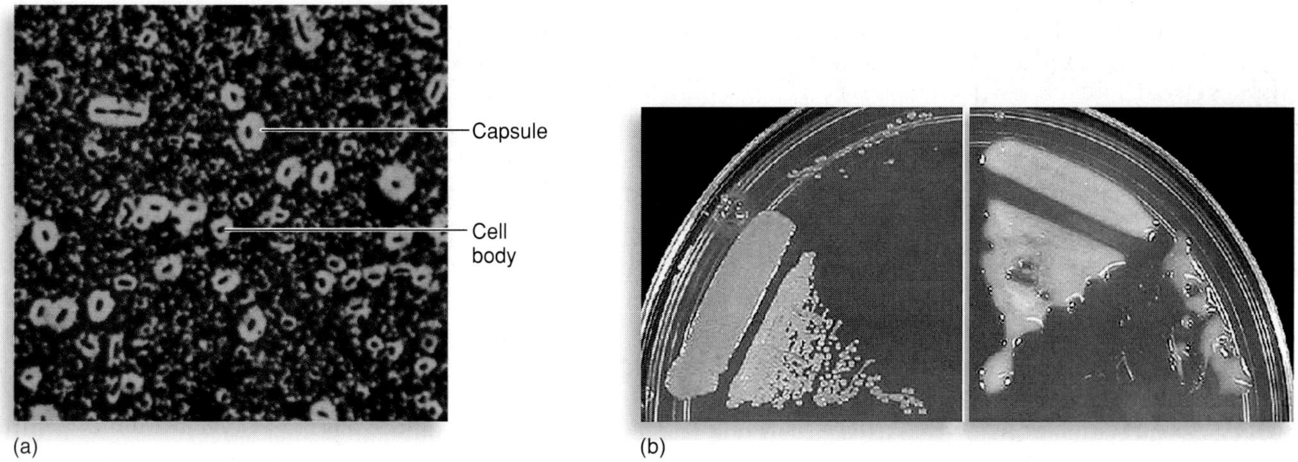

Figure 4.10 Encapsulated bacteria.
(a) Negative staining reveals the microscopic appearance of a large, well-developed capsule. **(b)** Colony appearance of a nonencapsulated (left) and encapsulated (right) version of a soil bacterium called *Sinorhizobium*.

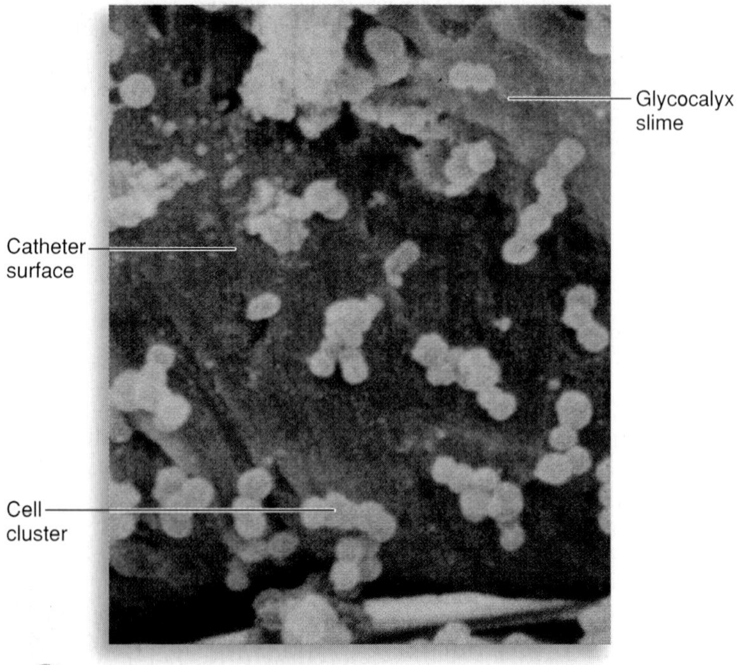

Figure 4.11 Biofilm.
Scanning electron micrograph of *Staphylococcus aureus* cells attached to a catheter by a slime secretion.

4.3 The Cell Envelope: The Boundary Layer of Bacteria

The majority of bacteria have a chemically complex external covering, termed the cell envelope, that lies outside of the cytoplasm. It is composed of two or three basic layers: the cell wall, the cell membrane, and, in some bacteria, the outer membrane. The layers of the envelope are stacked one upon another and are often tightly bonded together like the outer husk and casings of a coconut. Although each envelope layer performs a distinct function, together they act as a single protective unit.

Differences in Cell Envelope Structure

More than a hundred years ago, long before the detailed anatomy of bacteria was even remotely known, a Danish physician named Hans Christian Gram developed a staining technique, the **Gram stain,** that delineates two generally different groups of bacteria **(Insight 4.2).** The two major groups shown by this technique are the **gram-positive** bacteria and the **gram-negative** bacteria.

The structural difference denoted by the designations gram-positive and gram-negative lie in the cell envelope **(figure 4.12).** In gram-positive cells, a microscopic section resembles an open-faced sandwich with two layers: the thick cell wall, composed primarily of peptidoglycan (defined in the next section), and the cytoplasmic membrane. A similar section of a gram-negative cell envelope shows a complete sandwich with three layers: an outer membrane, a thin cell wall, and the cytoplasmic membrane.

Moving from outside to in, the outer membrane (if present) lies just under the glycocalyx. Next comes the cell wall. Finally, the innermost layer is always the cytoplasmic membrane. Because only some bacteria have an outer membrane, we discuss the cell wall first.

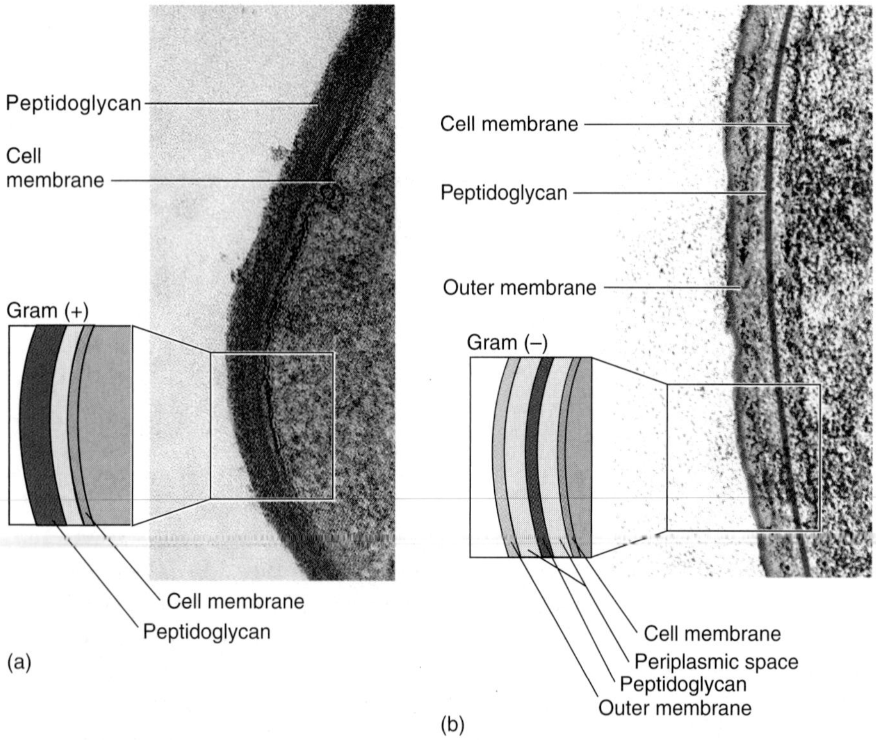

Figure 4.12 A comparison of the envelopes of gram-positive and gram-negative cells.
(a) A photomicrograph of a gram-positive cell wall/membrane and an artist's interpretation of its open-faced-sandwich-style layering with two layers.
(b) A photomicrograph of a gram-negative cell wall/membrane and an artist's interpretation of its complete-sandwich-style layering with three distinct layers.

Structure of the Cell Wall

The **cell wall** accounts for a number of important bacterial characteristics. In general, it helps determine the shape of a bacterium, and it also provides the kind of strong structural support necessary to keep a bacterium from bursting or collapsing because of changes in osmotic pressure. In this way, the cell wall functions like a bicycle tire that maintains the necessary shape and prevents the more delicate inner tube from bursting when it is expanded.

INSIGHT 4.2 *Discovery*

The Gram Stain: A Grand Stain

In 1884, Hans Christian Gram discovered a staining technique that could be used to make bacteria in infectious specimens more visible. His technique consisted of timed, sequential applications of crystal violet (the primary dye), Gram's iodine (IKI, the mordant), an alcohol rinse (decolorizer), and a contrasting counterstain. The initial counterstain used was yellow or brown and was later replaced by the red dye, safranin. Bacteria that stain purple are called gram-positive, and those that stain red are called gram-negative.

Although these staining reactions involve an attraction of the cell to a charged dye (see chapter 3), it is important to note that the terms *gram-positive* and *gram-negative* are not used to indicate the electrical charge of cells or dyes but whether or not a cell retains the primary dye-iodine complex after decolorization. There is nothing specific in the reaction of gram-positive cells to the primary dye or in the reaction of gram-negative cells to the counterstain. The different results in the Gram stain are due to differences in the structure of the cell wall and how it reacts to the series of reagents applied to the cells.

In the first step, crystal violet is added to the cells in a smear and stains them all the same purple color. The second and key differentiating step is the addition of the mordant—Gram's iodine. The mordant is a stabilizer that causes the dye to form large complexes in the peptidoglycan meshwork of the cell wall. Because the peptidoglycan layer in gram-positive cells is thicker, the entrapment of the dye is far more extensive in them than in gram-negative cells. Application of alcohol in the third step dissolves lipids in the outer membrane and removes the dye from the peptidoglycan layer and the gram-negative cells. By contrast, the crystals of dye tightly embedded in the peptidoglycan of gram-positive bacteria are relatively inaccessible and resistant to removal. Because gram-negative bacteria are colorless after decolorization, their presence is demonstrated by applying the counterstain safranin in the final step.

This century-old staining method remains the universal basis for bacterial classification and identification. It permits differentiation of four major categories based upon color reaction and shape: gram-positive rods, gram-positive cocci, gram-negative rods, and gram-negative cocci (see table 4.2). The Gram stain can also be a practical aid in diagnosing infection and in guiding drug treatment. For example, Gram staining a fresh urine or throat specimen can help pinpoint the possible cause of infection, and in some cases it is possible to begin drug therapy on the basis of this stain. Even in this day of elaborate and expensive medical technology, the Gram stain remains an important and unbeatable first tool in diagnosis.

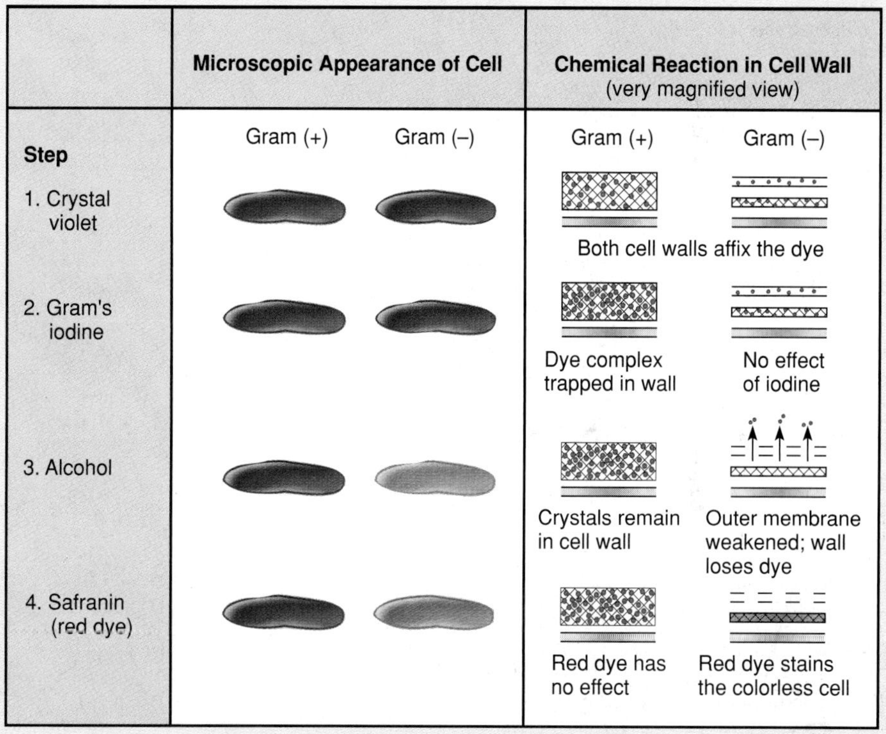
Microscopic Appearance of Cell
Chemical Reaction in Cell Wall (very magnified view)
Step
Gram (+)
Gram (–)
Gram (+)
Gram (–)
1. Crystal violet
Both cell walls affix the dye
2. Gram's iodine
Dye complex trapped in wall
No effect of iodine
3. Alcohol
Crystals remain in cell wall
Outer membrane weakened; wall loses dye
4. Safranin (red dye)
Red dye has no effect
Red dye stains the colorless cell

The cell walls of most bacteria gain their relatively rigid-quality from a unique macromolecule called **peptidoglycan (PG).** This compound is composed of a repeating framework of long *glycan* chains cross-linked by short peptide fragments to provide a strong but flexible support framework **(figure 4.13).** The amount and exact composition of peptidoglycan vary among the major bacterial groups.

Because many bacteria live in aqueous habitats with a low concentration of dissolved substances, they are constantly absorbing excess water by osmosis. Were it not for the strength and relative rigidity of the peptidoglycan in the cell wall, they would rupture from internal pressure. Understanding this function of the cell wall has been a tremendous boon to the drug industry. Several types of drugs used to treat infection (penicillin, cephalosporins) are effective because they target the peptide cross-links in the peptidoglycan, thereby disrupting its integrity. With their cell walls incomplete or missing, such cells have very little protection from **lysis** (ly'-sis). Lysozyme, an enzyme contained in tears and saliva, provides a natural defense against certain bacteria by hydrolyzing the bonds in the glycan chains and causing the wall to break down. (Chapter 11 discusses the actions of antimicrobial chemical agents.)

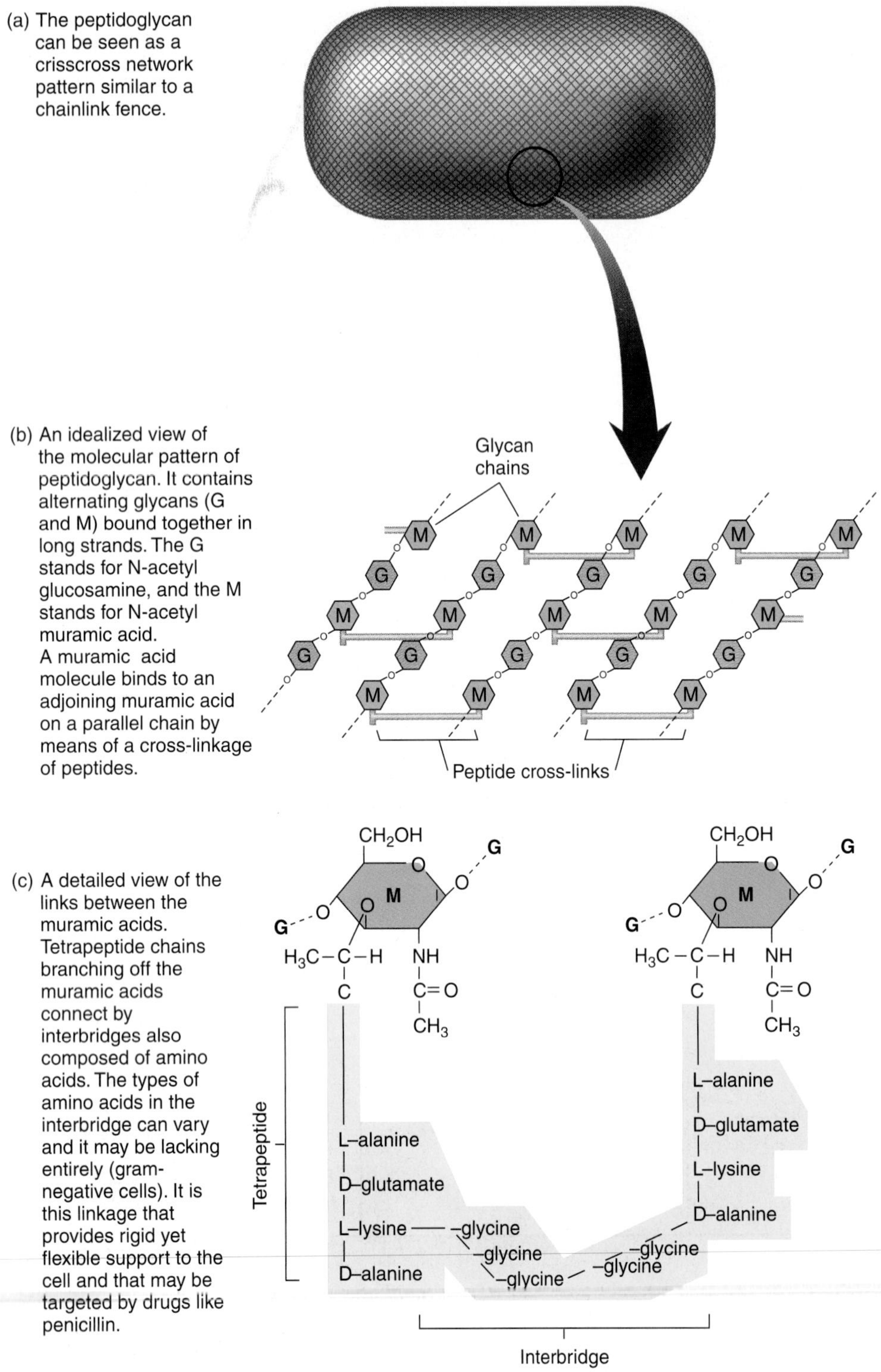
(a) The peptidoglycan can be seen as a crisscross network pattern similar to a chainlink fence.
(b) An idealized view of the molecular pattern of peptidoglycan. It contains alternating glycans (G and M) bound together in long strands. The G stands for N-acetyl glucosamine, and the M stands for N-acetyl muramic acid. A muramic acid molecule binds to an adjoining muramic acid on a parallel chain by means of a cross-linkage of peptides.
Glycan chains
Peptide cross-links
(c) A detailed view of the links between the muramic acids. Tetrapeptide chains branching off the muramic acids connect by interbridges also composed of amino acids. The types of amino acids in the interbridge can vary and it may be lacking entirely (gram-negative cells). It is this linkage that provides rigid yet flexible support to the cell and that may be targeted by drugs like penicillin.
CH2OH
G
M
H3C–C–H
NH
C=O
CH3
Tetrapeptide
L–alanine
D–glutamate
L–lysine
D–alanine
glycine
Interbridge

Figure 4.13 Structure of peptidoglycan in the cell wall.

The Gram-Positive Cell Wall

The bulk of the gram-positive cell wall is a thick, homogeneous sheath of peptidoglycan ranging from 20 to 80 nm in thickness. It also contains tightly bound acidic polysaccharides, including teichoic acid and lipoteichoic acid **(figure 4.14).** Teichoic acid is a polymer of ribitol or glycerol and phosphate embedded in the peptidoglycan sheath. Lipoteichoic acid is similar in structure but is attached to the lipids in the plasma membrane. These molecules appear to function in cell wall maintenance and enlargement during cell division, and they also contribute to the acidic charge on the cell surface.

The Gram-Negative Cell Wall

The gram-negative wall is a single, thin (1–3 nm) sheet of peptidoglycan. Although it acts as a somewhat rigid protective structure as previously described, its thinness gives gram-negative bacteria a relatively greater flexibility and sensitivity to lysis. A well-developed *periplasmic space* surrounds the peptidoglycan (see figure 4.14). This space is an important reaction site for a large and varied pool of substances that enter and leave the cell.

Nontypical Cell Walls

Several bacterial groups lack the cell wall structure of gram-positive or gram-negative bacteria, and some bacteria have no cell wall at all. Although these exceptional forms can stain positive or negative in the Gram stain, examination of their fine structure and chemistry shows that they do not really fit the descriptions for typical gram-negative or -positive cells. For example, the cells of *Mycobacterium* and *Nocardia* contain peptidoglycan and stain gram-positive, but the bulk of their cell wall is composed of unique types of lipids. One of these is a very-long-chain fatty acid called *mycolic acid,* or cord factor, that contributes to the pathogenicity of this group (see chapter 21). The thick, waxy nature imparted to the cell wall by these lipids is also responsible for a high degree of resistance to certain chemicals and dyes. Such resistance is the basis for the **acid-fast stain** used to diagnose tuberculosis and leprosy. In this stain, hot carbol fuchsin dye becomes tenaciously attached (is held fast) to these cells so that an acid-alcohol solution will not remove the dye (see chapter 3).

Because they are from a more ancient and primitive line of prokaryotes, the archaea exhibit unusual and chemically distinct cell walls. In some, the walls are composed almost

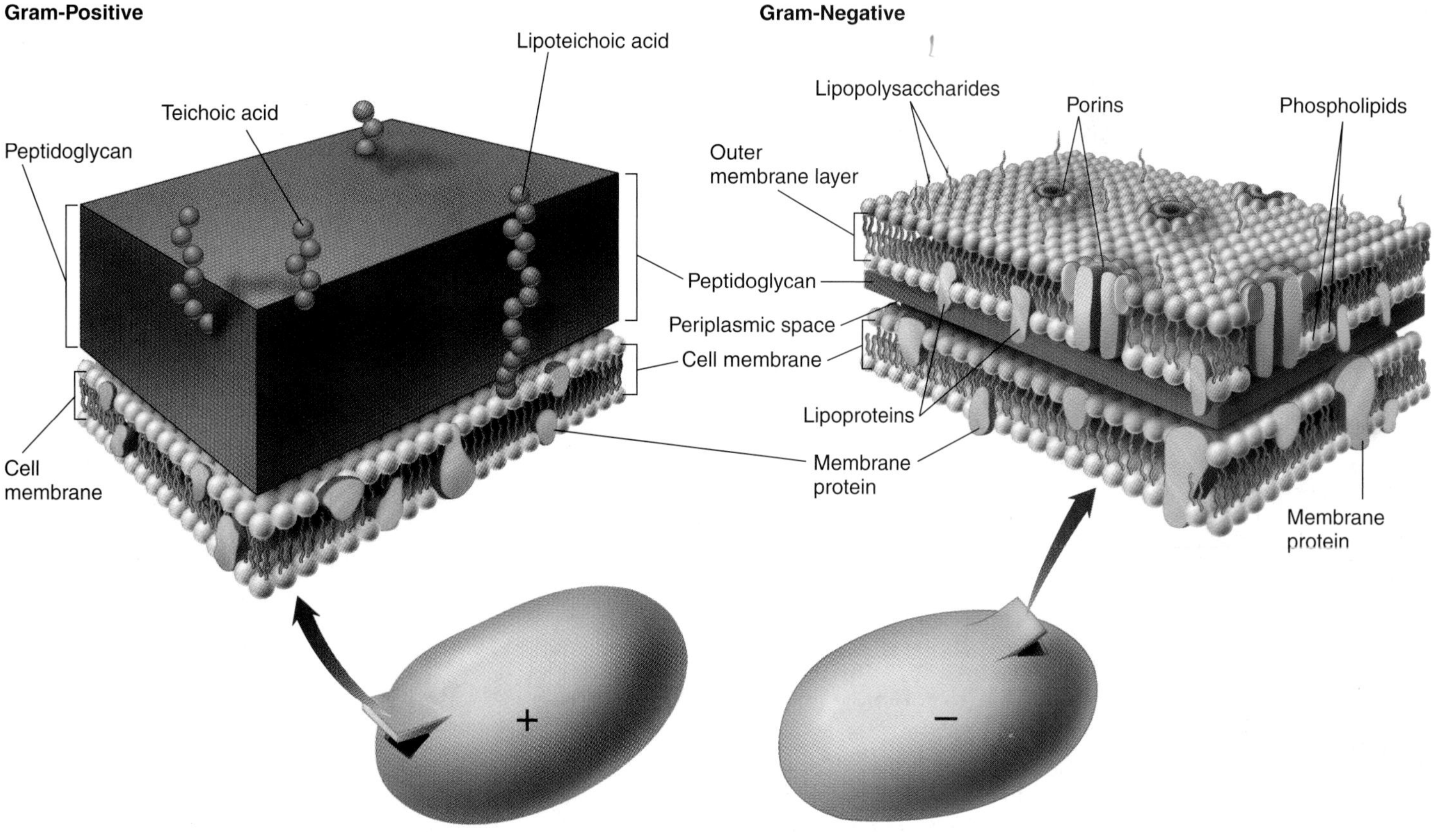

Figure 4.14 **A comparison of the detailed structure of gram-positive and gram-negative cell walls.**

entirely of polysaccharides, and in others, the walls are pure protein; but as a group, they all lack the true peptidoglycan structure described previously. Because a few archaea and all mycoplasmas (next section) lack a cell wall entirely, their cell membrane must serve the dual functions of support and transport.

Mycoplasmas and Other Cell-Wall-Deficient Bacteria

Mycoplasmas are bacteria that naturally lack a cell wall. Although other bacteria require an intact cell wall to prevent the bursting of the cell, the mycoplasma cell membrane is stabilized by sterols and is resistant to lysis. These extremely tiny, pleomorphic cells are very small bacteria, ranging from 0.1 to 0.5 μm in size. They range in shape from filamentous to coccus or doughnut-shaped. They are *not* obligate parasites and can be grown on artificial media, although added sterols are required for the cell membranes of some species. Mycoplasmas are found in many habitats, including plants, soil, and animals. The most important medical species is *Mycoplasma pneumoniae* **(figure 4.15)**, which adheres to the epithelial cells in the lung and causes an atypical form of pneumonia in humans (described in chapter 21).

Some bacteria that ordinarily have a cell wall can lose it during part of their life cycle. These wall-deficient forms are referred to as **L forms** or L-phase variants (for the Lister Institute, where they were discovered). L forms arise naturally from a mutation in the wall-forming genes, or they can be induced artificially by treatment with a chemical such as lysozyme or penicillin that disrupts the cell wall. When a gram-positive cell is exposed to either of these two chemicals, it will lose the cell wall completely and become

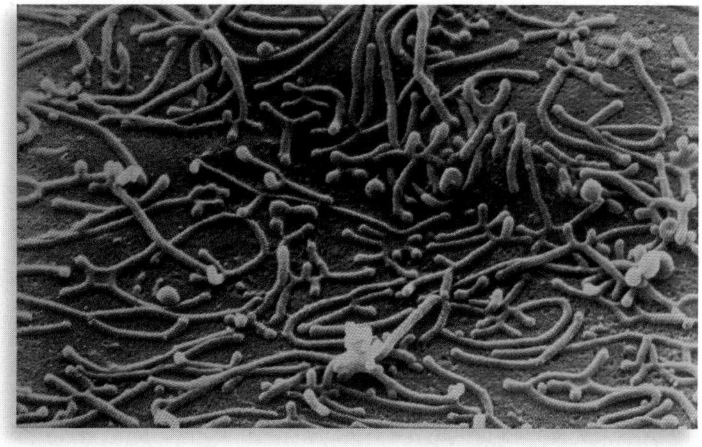

Figure 4.15 **Scanning electron micrograph of *Mycoplasma pneumoniae* (magnified 62,000×).** Cells like these that naturally lack a cell wall exhibit extreme variation in shape.

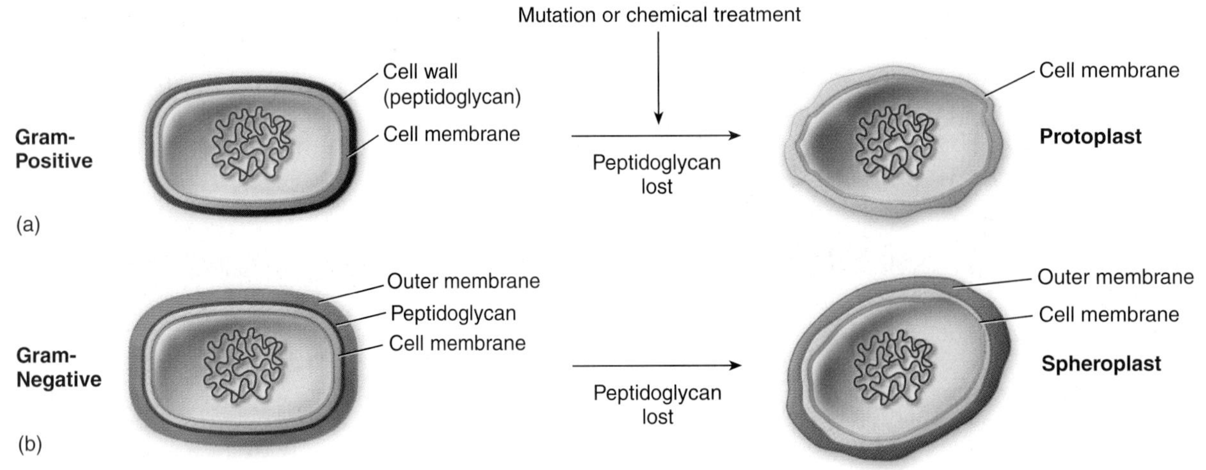

Figure 4.16 The conversion of walled bacterial cells to L forms.
(a) Gram-positive bacteria. **(b)** Gram-negative bacteria.

a **protoplast,** a fragile cell bounded only by a membrane that is highly susceptible to lysis **(figure 4.16*a*).** A gram-negative cell exposed to these same substances loses it peptidoglycan but retains at least part of its outer membrane, leaving a less fragile but nevertheless weakened **spheroplast (figure 4.16*b*).** Evidence points to a role for L forms in certain infections.

The Gram-Negative Outer Membrane

The **outer membrane** is somewhat similar in construction to the cell membrane, except that it contains specialized types of polysaccharides and proteins. The uppermost layer of the OM contains *lipopolysaccharide* (LPS). The polysaccharide chains extending off the surface function as antigens and receptors. The lipid portion of LPS has been referred to as *endotoxin* because it stimulates fever and shock reactions in gram-negative infections such as meningitis and typhoid fever. The innermost layer of the OM is a phospholipid layer anchored by means of lipoproteins to the peptidoglycan layer below. The outer membrane serves as a partial chemical sieve by allowing only relatively small molecules to penetrate. Access is provided by special membrane channels formed by *porin proteins* that completely span the outer membrane. The size of these porins can be altered so as to block the entrance of harmful chemicals, making them one defense of gram-negative bacteria against certain antibiotics (see figure 4.14).

Cell Membrane Structure

Appearing just beneath the cell wall is the cell, or **cytoplasmic, membrane,** a very thin (5–10 nm), flexible sheet molded completely around the cytoplasm. Its general composition was described in chapter 2 as a lipid bilayer with proteins embedded to varying degrees (see Insight 2.3). Bacterial cell

CASE FILE 4 *WRAP-UP*

The outbreak of pneumococcal pneumonia described at the beginning of the chapter points out that the presence of certain bacterial structures, such as a capsule, enhances virulence. Studies have shown that because the capsule allows the bacterium to resist host phagocytosis, encapsulated strains of *Streptococcus pneumoniae* are virulent, whereas those with no capsule are not. In fact, those individuals who have antibodies specific for the polysaccharide capsule of the *Streptococcus pneumoniae* strain will be resistant to attack by that strain. This knowledge has been used to make a vaccine for adults, using 23 types of polysaccharide capsular antigens, which, when injected, will elicit specific antibodies that protect from the most common strains causing pneumococcal pneumonia. The serum antibodies that arise after vaccination specifically coat the bacterial capsule and allow for uptake of the bacteria by the host phagocytes. Efficacy of the vaccine is shown in studies in which incidence of pneumococcal disease in the elderly is reduced in those vaccinated. This disease is significant, as the Centers for Disease Control and Prevention (CDC) estimates that about a half-million cases occur each year, resulting in about 40,000 deaths in the United States.

As was the case with this outbreak, the highest mortality rate (30%–40%) occurs in the elderly or in those with underlying medical conditions. CDC estimates that about half of these deaths could be prevented through use of the pneumococcal vaccine.

See: CDC. 2001. Outbreak of pneumococcal pneumonia among unvaccinated residents of a nursing home—New Jersey, April 2001. MMWR *50:707–710.*
CDC. 1997. Prevention of pneumococcal disease: Recommendations of the Advisory Committee on Immunization Practices (ACIP). MMWR *46, No. RR-09.*

membranes have this typical structure, containing primarily phospholipids (making up about 30%–40% of the membrane mass) and proteins (contributing 60%–70%). Major exceptions to this description are the membranes of mycoplasmas, which contain high amounts of sterols—rigid lipids that stabilize and reinforce the membrane—and the membranes of archaea, which contain unique branched hydrocarbons rather than fatty acids.

Photosynthetic prokaryotes such as cyanobacteria contain dense stacks of internal membranes that carry the photosynthetic pigments, which we describe later on.

Functions of the Cell Membrane

Because bacteria have none of the eukaryotic organelles, the cell membrane provides a site for functions such as energy reactions, nutrient processing, and synthesis. A major action of the cell membrane is to regulate *transport,* that is, the passage of nutrients into the cell and the discharge of wastes. Although water and small uncharged molecules can diffuse across the membrane unaided, the membrane is a *selectively permeable* structure with special carrier mechanisms for passage of most molecules (see chapter 7). The glycocalyx and cell wall can bar the passage of large molecules, but they are not the primary transport apparatus. The cell membrane is also involved in *secretion,* or the discharge of a metabolic product into the extracellular environment.

The membranes of prokaryotes are an important site for a number of metabolic activities. Most enzymes of respiration and ATP synthesis reside in the cell membrane since prokaryotes lack mitochondria (see chapter 8). Enzyme structures located in the cell membrane also help synthesize structural macromolecules to be incorporated into the cell envelope and appendages. Other products (enzymes and toxins) are secreted by the membrane into the extracellular environment.

Practical Considerations of Differences in Cell Envelope Structure

Variations in cell envelope anatomy contribute to several other differences between the two cell types. The outer membrane contributes an extra barrier in gram-negative bacteria that makes them impervious to some antimicrobial chemicals such as dyes and disinfectants, so they are generally more difficult to inhibit or kill than are gram-positive bacteria. One exception is for alcohol-based compounds, which can dissolve the lipids in the outer membrane and disturb its integrity. Treating infections caused by gram-negative bacteria often requires different drugs from gram-positive infections, especially drugs that can cross the outer membrane.

The cell envelope or its parts can interact with human tissues and contribute to disease. Proteins attached to the outer portion of the cell wall of several gram-positive species, including *Corynebacterium diphtheriae* (the agent of diphtheria) and *Streptococcus pyogenes* (the cause of strep throat), also have toxic properties. The lipids in the cell walls of certain *Mycobacterium* species are harmful to human cells as well. Because most macromolecules in the cell walls are foreign to humans, they stimulate antibody production by the immune system (see chapter 15).

CHECKPOINT

- Bacteria are the oldest form of cellular life. They are also the most widely dispersed, occupying every conceivable microclimate on the planet.
- The external structures of bacteria include appendages (flagella, fimbriae, and pili) and the glycocalyx.
- Flagella vary in number and arrangement as well as in the type and rate of motion they produce.
- The cell envelope is the complex boundary structure surrounding a bacterial cell. In gram-negative bacteria, the envelope consists of an outer membrane, the cell wall, and the cell membrane. Gram-positive bacteria have only the cell wall and cell membrane.
- In a Gram stain, gram-positive bacteria retain the crystal violet and stain purple. Gram-negative bacteria lose the crystal violet and stain red from the safranin counterstain.
- Gram-positive bacteria have thick cell walls of peptidoglycan and acidic polysaccharides such as teichoic acid. The cell walls of gram-negative bacteria are thinner and have a wide periplasmic space.
- The outer membrane of gram-negative cells contains lipopolysaccharide (LPS). LPS is toxic to mammalian hosts.
- The bacterial cell membrane is typically composed of phospholipids and proteins, and it performs many metabolic functions as well as transport activities.

4.4 Bacterial Internal Structure

Contents of the Cell Cytoplasm

Encased by the cell membrane is a gelatinous solution referred to as **cytoplasm,** which is another prominent site for many of the cell's biochemical and synthetic activities. Its major component is water (70%–80%), which serves as a solvent for the cell pool, a complex mixture of nutrients including sugars, amino acids, and salts. The components of this pool serve as building blocks for cell synthesis or as sources of energy. The cytoplasm also contains larger, discrete cell masses such as the chromatin body, ribosomes, granules, and actin strands that act as a cytoskeleton in bacteria that have them.

Bacterial Chromosomes and Plasmids: The Sources of Genetic Information

The hereditary material of most bacteria exists in the form of a single circular strand of DNA designated as the **bacterial chromosome.** (Some bacteria have multiple chromosomes.) By definition, bacteria do not have a nucleus; that is, their DNA is not

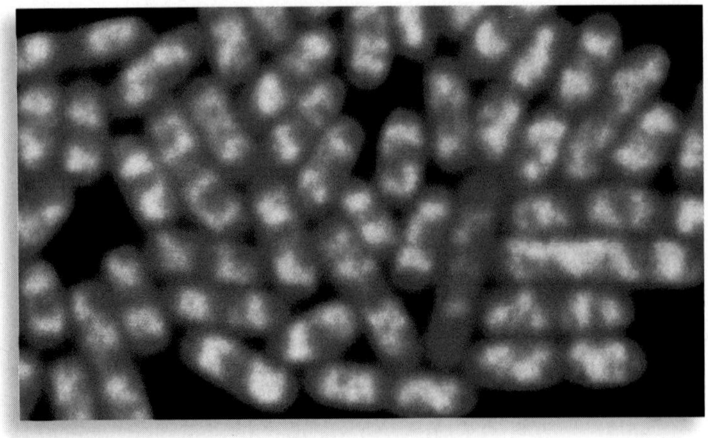

Figure 4.17 Chromosome structure.
Fluorescent staining highlights the chromosomes of the bacterial pathogen *Salmonella enteriditis*. The cytoplasm is orange, and the chromosome fluoresces bright yellow.

enclosed by a nuclear membrane but instead is aggregated in a dense area of the cell called the **nucleoid.** The chromosome is actually an extremely long molecule of double-stranded DNA that is tightly coiled around special basic protein molecules so as to fit inside the cell compartment. Arranged along its length are genetic units (genes) that carry information required for bacterial maintenance and growth. When exposed to special stains or observed with an electron microscope, chromosomes have a granular or fibrous appearance **(figure 4.17).**

Although the chromosome is the minimal genetic requirement for bacterial survival, many bacteria contain other, nonessential pieces of DNA called **plasmids.** These tiny strands exist as separate double-stranded circles of DNA, although at times they can become integrated into the chromosome. During conjugation, they may be duplicated and passed on to related nearby bacteria. During bacterial reproduction, they are duplicated and passed on to offspring. They are not essential to bacterial growth and metabolism, but they often confer protective traits such as resisting drugs and producing toxins and enzymes (see chapter 9). Because they can be readily manipulated in the laboratory and transferred from one bacterial cell to another, plasmids are an important agent in modern genetic engineering techniques.

Ribosomes: Sites of Protein Synthesis

A bacterial cell contains thousands of tiny **ribosomes,** which are made of RNA and protein. When viewed even by very high magnification, ribosomes show up as fine, spherical specks dispersed throughout the cytoplasm that often occur in chains (polysomes). Many are also attached to the cell membrane. Chemically, a ribosome is a combination of a special type of RNA called ribosomal RNA, or rRNA (about 60%), and protein (40%). One method of characterizing ribosomes is by S, or Svedberg,[3] units,

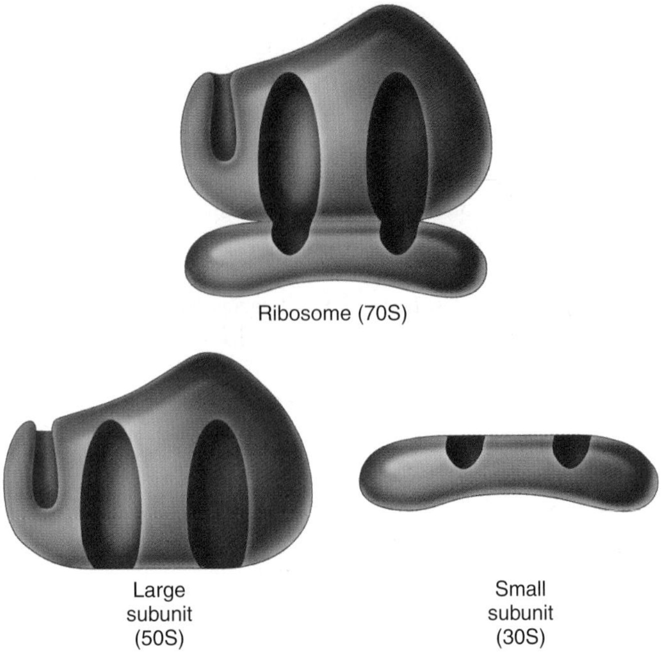

Figure 4.18 A model of a prokaryotic ribosome, showing the small (30S) and large (50S) subunits, both separate and joined.

which rate the molecular sizes of various cell parts that have been spun down and separated by molecular weight and shape in a centrifuge. Heavier, more compact structures sediment faster and are assigned a higher S rating. Combining this method of analysis with high-resolution electron micrography has revealed that the prokaryotic ribosome, which has an overall rating of 70S, is actually composed of two smaller subunits **(figure 4.18).** They fit together to form a miniature platform upon which protein synthesis is performed. We examine the more detailed functions of ribosomes in chapter 9.

Inclusions, or Granules: Storage Bodies

Most bacteria are exposed to severe shifts in the availability of food. During periods of nutrient abundance, some can compensate by laying down nutrients intracellularly in **inclusion bodies,** or **inclusions,** of varying size, number, and content. As the environmental source of these nutrients becomes depleted, the bacterial cell can mobilize its own storehouse as required. Some inclusion bodies enclose condensed, energy-rich organic substances, such as glycogen and poly β-hydroxybutyrate (PHB), within special single-layered membranes **(figure 4.19).** A unique type of inclusion found in some aquatic bacteria is gas vesicles that provide buoyancy and flotation. Other inclusions, also called granules, contain crystals of inorganic compounds and are not enclosed by membranes. Sulfur granules of photosynthetic bacteria and polyphosphate granules of *Corynebacterium* and *Mycobacterium,* described later, are of this type. The latter represent an important source of building blocks for nucleic acid and ATP

3. Named in honor of T. Svedberg, the Swedish chemist who developed the ultracentrifuge in 1926.

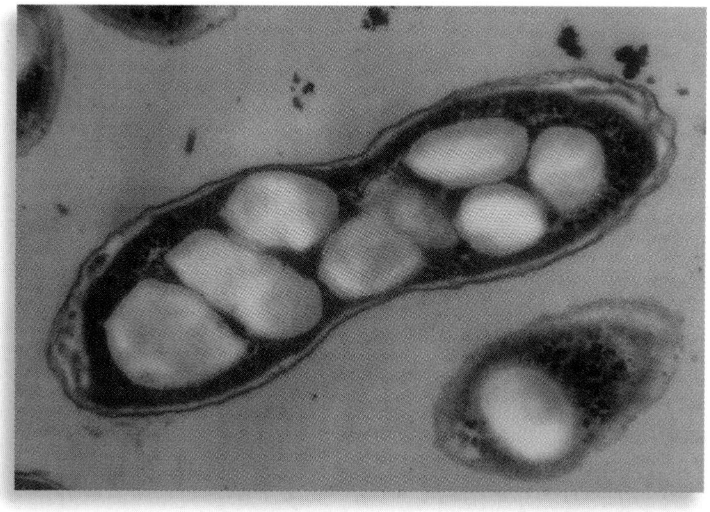

Figure 4.19 An example of a storage inclusion in a bacterial cell (magnified 32,500×).
Substances such as polyhydroxybutyrate can be stored in an insoluble, concentrated form that provides an ample, long-term supply of that nutrient.

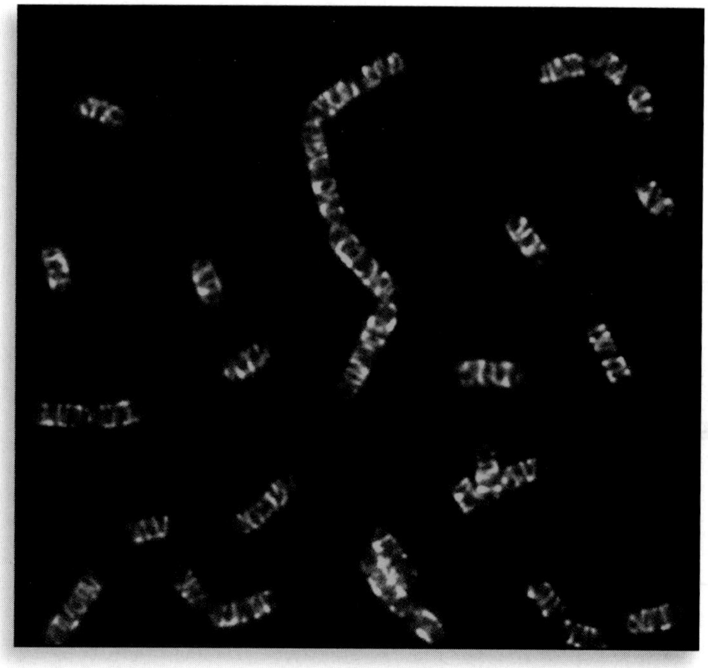

Figure 4.20 Bacterial cytoskeleton.
The actin fibers in these rod-shaped bacteria are fluorescently stained.

synthesis. They have been termed **metachromatic granules** because they stain a contrasting color (red, purple) in the presence of methylene blue dye.

Perhaps the most unique cell granule is not involved in cell nutrition but rather in cell orientation. Magnetotactic bacteria contain crystalline particles of iron oxide (magnetosomes) that have magnetic properties. Evidently the bacteria use these granules to be pulled by the polar and gravitational fields into deeper habitats with a lower oxygen content.

The Actin Cytoskeleton

Until very recently, scientists thought that the shape of all bacteria was completely determined by the peptidoglycan layer (cell wall). Although this is true of many bacteria, particularly the cocci, other bacteria produce long polymers of a protein called **actin,** arranged in helical ribbons around the cell just under the cell membrane **(figure 4.20).** These fibers contribute cell shape, perhaps by influencing the way peptidoglycan is manufactured. The fibers have been found in rod-shaped and spiral bacteria.

Bacterial Endospores: An Extremely Resistant Stage

Ample evidence indicates that the anatomy of bacteria helps them adjust rather well to adverse habitats. But of all microbial structures, nothing can compare to the bacterial **endospore** (or simply spore) for withstanding hostile conditions and facilitating survival.

Endospores are dormant bodies produced by the bacteria *Bacillus, Clostridium,* and *Sporosarcina.* These bacteria have a two-phase life cycle—a vegetative cell and an endospore **(figure 4.21).** The vegetative cell is a metabolically active and

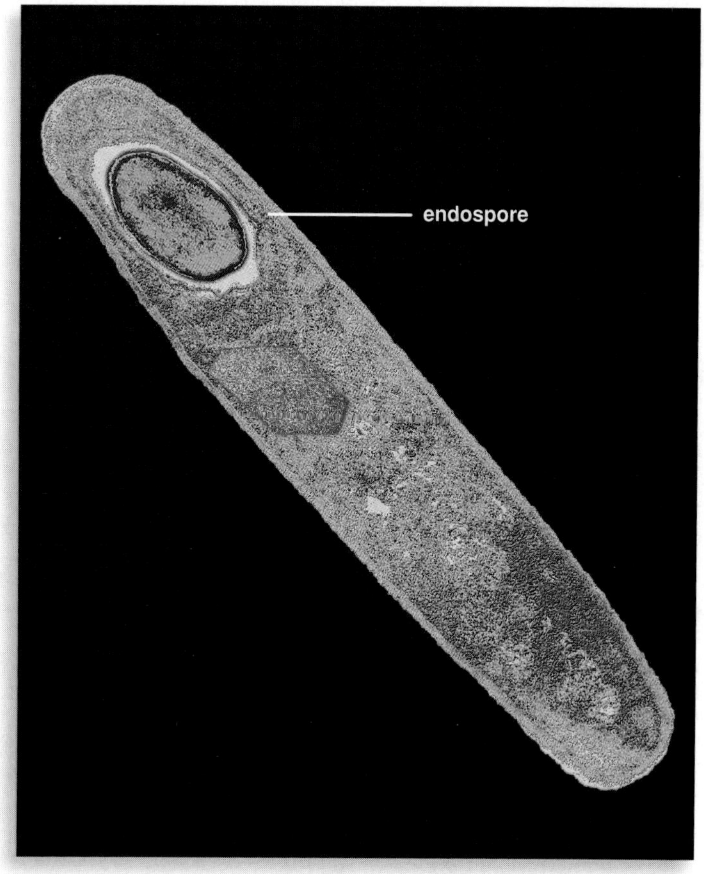

Figure 4.21 Endospore inside *Bacillus thuringiensis.*
The genus *Bacillus* forms endospores. *B. thuringiensis* additionally forms crystalline bodies (pink) that are used as insecticides.

TABLE 4.1 General Stages in Endospore Formation

Stage	State of Cell	Process/Event
1 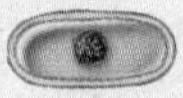	Vegetative cell	Cell in early stage of binary fission doubles chromosome.
2 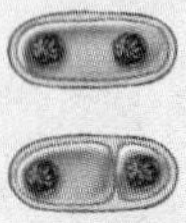	Vegetative cell becomes **sporangium** in preparation for sporulation.	One chromosome and a small bit of cytoplasm are walled off as a protoplast at one end of the cell. This core contains the minimum structures and chemicals necessary for guiding life processes. During this time, the sporangium remains active in synthesizing compounds required for spore formation.
3 	Sporangium	The protoplast is engulfed by the sporangium to continue the formation of various protective layers around it.
4 	Sporangium with prospore	Special peptidoglycan is laid down to form a cortex around the spore protoplast, now called the prospore; calcium and dipicolinic acid are deposited; core becomes dehydrated and metabolically inactive.
5 	Sporangium with prospore	Three heavy and impervious protein spore coats are added.
6 	Mature endospore	Endospore becomes thicker, and heat resistance is complete; sporangium is no longer functional and begins to deteriorate.
7 	Free spore	Complete lysis of sporangium frees spore; it can remain dormant yet viable for thousands of years.
8 	Germination	Addition of nutrients and water reverses the dormancy. The spore then swells and liberates a young vegetative cell.
9 	Vegetative cell	Restored vegetative cell

Fluorescent stain of *Bacillus subtilis*

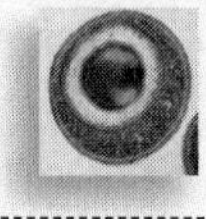
TEM of cross section of free endospore

growing entity that can be induced by environmental conditions to undergo spore formation, or **sporulation.** Once formed, the spore exists in an inert, resting condition that shows up prominently in a spore or Gram stain **(table 4.1).** Features of spores, including size, shape, and position in the vegetative cell, are somewhat useful in identifying some species. Both gram-positive and gram-negative bacteria can form endospores, but the medically relevant ones are all gram-positive.

Endospore Formation and Resistance

The depletion of nutrients, especially an adequate carbon or nitrogen source, is the stimulus for a vegetative cell to begin endospore formation. Once this stimulus has been received by the vegetative cell, it undergoes a conversion to a committed sporulating cell called a **sporangium.** Complete transformation of a vegetative cell into a sporangium and then into an endospore requires 6 to 8 hours in most spore-forming species. Table 4.1 illustrates some major physical and chemical events in this process. Bacterial endospores are the hardiest of all life forms, capable of withstanding extremes in heat, drying, freezing, radiation, and chemicals that would readily kill vegetative cells. Their survival under such harsh conditions is due to several factors. The heat resistance of spores has been linked to their high content of calcium and *dipicolinic acid,* although the exact role of these chemicals is not yet clear. We know, for instance, that heat destroys cells by inactivating proteins and DNA and that this process requires a certain amount of water in the protoplasm. Because the deposition of calcium dipicolinate in the endospore removes water and leaves the endospore very dehydrated, it is less vulnerable to the effects of heat. It is also metabolically inactive and highly resistant to damage from further drying. The thick, impervious cortex and spore coats also protect against radiation and chemicals (table 4.1). The longevity of bacterial spores verges on immortality. One record describes the isolation of viable endospores from a fossilized bee that was 25 million years old. More recently, microbiologists unearthed a viable endospore from a 250-million-year-old salt crystal. Initial analysis of this ancient microbe indicates it is a species of *Bacillus* that is genetically different from known species.

A NOTE ON TERMINOLOGY

The word *spore* can have more than one usage in microbiology. It is a generic term that refers to any tiny compact cells that are produced by vegetative or reproductive structures of microorganisms. Spores can be quite variable in origin, form, and function. The bacterial type discussed here is called an endospore, because it is produced inside a cell. It functions in *survival,* not in reproduction, because no increase in cell numbers is involved in its formation. In contrast, the fungi produce many different types of spores for both survival and reproduction (see chapter 5).

The Germination of Endospores

After lying in a state of inactivity for an indefinite time, endospores can be revitalized when favorable conditions arise. The breaking of dormancy, or germination, happens in the presence of water and a specific chemical or environmental stimulus (germination agent). Once initiated, it proceeds to completion quite rapidly ($1\frac{1}{2}$ hours). Although the specific germination agent varies among species, it is generally a small organic molecule such as an amino acid or an inorganic salt. This agent stimulates the formation of hydrolytic (digestive) enzymes by the endospore membranes. These enzymes digest the cortex and expose the core to water. As the core rehydrates and takes up nutrients, it begins to grow out of the endospore coats. In time, it reverts to a fully active vegetative cell, resuming the vegetative cycle.

Medical Significance of Bacterial Spores

Although the majority of spore-forming bacteria are relatively harmless, several bacterial pathogens are sporeformers. In fact, some aspects of the diseases they cause are related to the persistence and resistance of their spores. *Bacillus anthracis* is the agent of anthrax; its persistence in endospore form makes it an ideal candidate for bioterrorism. The genus *Clostridium* includes even more pathogens, such as *C. tetani,* the cause of tetanus (lockjaw), and *C. perfringens,* the cause of gas gangrene. When the spores of these species are embedded in a wound that contains dead tissue, they can germinate, grow, and release potent toxins. Another toxin-forming species, *C. botulinum,* is the agent of botulism, a deadly form of food poisoning. (Each of these disease conditions is discussed in the infectious disease chapters, according to the organ systems it affects.)

Because they inhabit the soil and dust, endospores are constant intruders where sterility and cleanliness are important. They resist ordinary cleaning methods that use boiling water, soaps, and disinfectants, and they frequently contaminate cultures and media. Hospitals and clinics must take precautions to guard against the potential harmful effects of endospores in wounds. Endospore destruction is a particular concern of the food-canning industry. Several endospore-forming species cause food spoilage or poisoning. Ordinary boiling (100°C) will usually not destroy such spores, so canning is carried out in pressurized steam at 120°C for 20 to 30 minutes. Such rigorous conditions ensure that the food is sterile and free from viable bacteria.

CHECKPOINT

- The cytoplasm of bacterial cells serves as a solvent for materials used in all cell functions.
- The genetic material of bacteria is DNA. Genes are arranged on large, circular chromosomes. Additional genes are carried on plasmids.
- Bacterial ribosomes are dispersed in the cytoplasm in chains (polysomes) and are also embedded in the cell membrane.
- Bacteria may store nutrients in their cytoplasm in structures called inclusions. Inclusions vary in structure and the materials that are stored.
- Some bacteria manufacture long actin filaments that help determine their cellular shape.
- A few families of bacteria produce dormant bodies called endospores, which are the hardiest of all life forms, surviving for hundreds or thousands of years.
- The genera *Bacillus* and *Clostridium* are sporeformers, and both contain deadly pathogens.

4.5 Bacterial Shapes, Arrangements, and Sizes

For the most part, bacteria function as independent single-celled, or unicellular, organisms. Each individual bacterial cell is fully capable of carrying out all necessary life activities, such as reproduction, metabolism, and nutrient processing, unlike the more specialized cells of a multicellular organism.

Bacteria exhibit considerable variety in shape, size, and colonial arrangement. It is convenient to describe most bac- teria by one of three general shapes as dictated by the configuration of the cell wall **(figure 4.22).** If the cell is spherical or ball-shaped, the bacterium is described as a **coccus** (kok′-us). Cocci can be perfect spheres, but they also can exist as oval, bean-shaped, or even pointed variants. A cell that is cylindrical (longer than wide) is termed a rod, or **bacillus** (bah-sil′-lus). There is also a genus named *Bacillus.* As might be expected, rods are also quite varied in their actual form. Depending on the bacterial species, they can be blocky, spindle-shaped, round-ended, long and thread-like (filamentous), or even club-shaped or drumstick-shaped. When a rod is short and plump, it is called a **coccobacillus;** if it is gently curved, it is a **vibrio** (vib′-ree-oh). A bacterium having the shape of a curviform or spiral-shaped cylinder is called a **spirillum** (spy-ril′-em), a rigid helix, twisted twice or more along its axis (like a corkscrew). Another spiral cell mentioned earlier in conjunction with periplasmic flagella is the spirochete, a more flexible form that resembles a spring. Because bacterial cells look two-dimensional and flat with traditional staining and microscope techniques, they are seen to best advantage with a scanning electron microscope, which emphasizes their striking three-dimensional forms **(figure 4.23).**

It is common for cells of a single species to vary to some extent in shape and size. This phenomenon, called **pleomorphism (figure 4.24),** is due to individual variations in cell wall structure caused by nutritional or slight hereditary differences. For example, although the cells of *Corynebacterium diphtheriae* are generally considered rod-shaped, in culture they display variations such as club-shaped, swollen, curved, filamentous, and coccoid. Pleomorphism reaches an extreme in the mycoplasmas, which entirely lack cell walls and thus display extreme variations in shape (see figure 4.15).

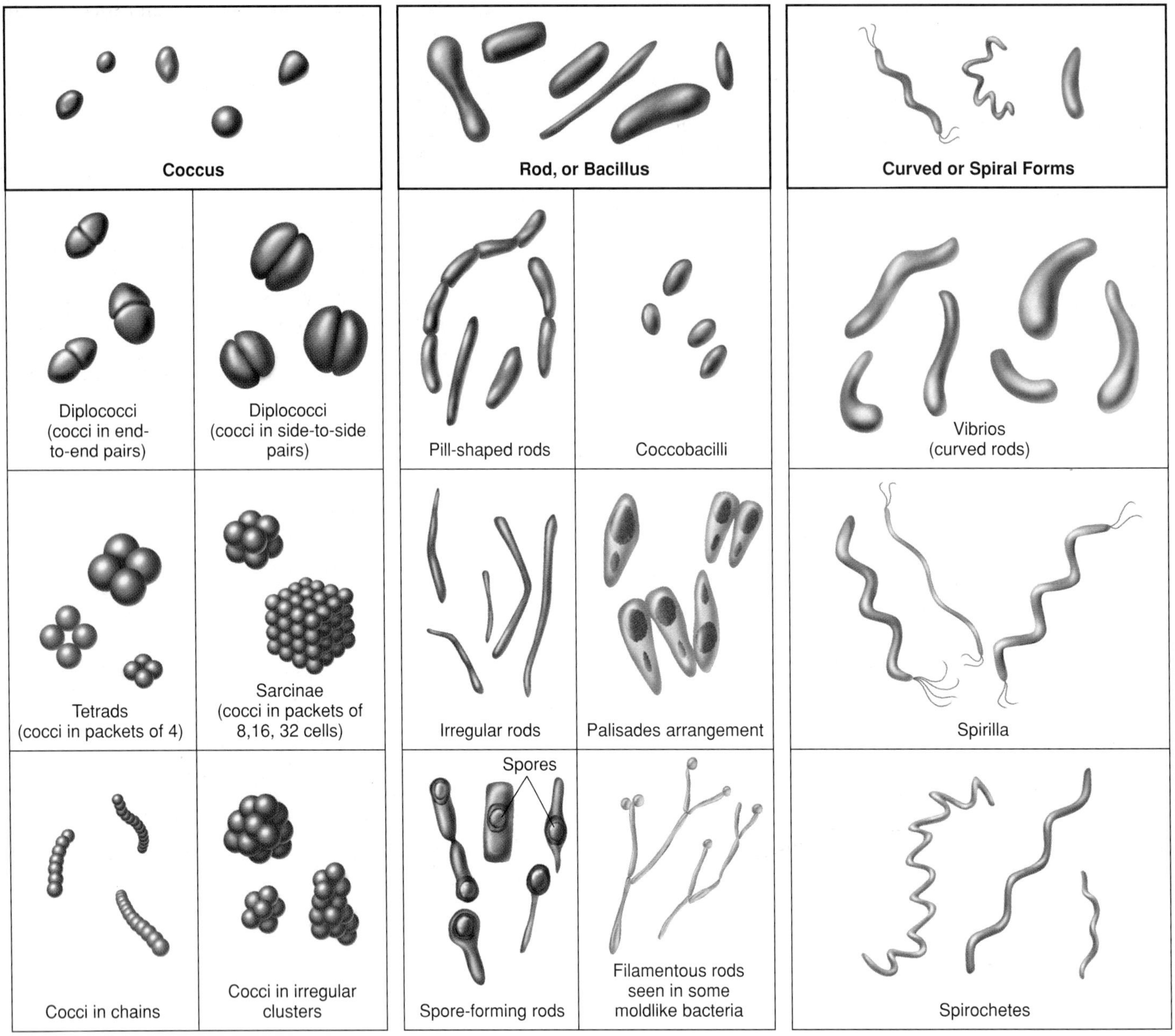

Figure 4.22 Bacterial shapes and arrangements.
May not be shown to exact scale.

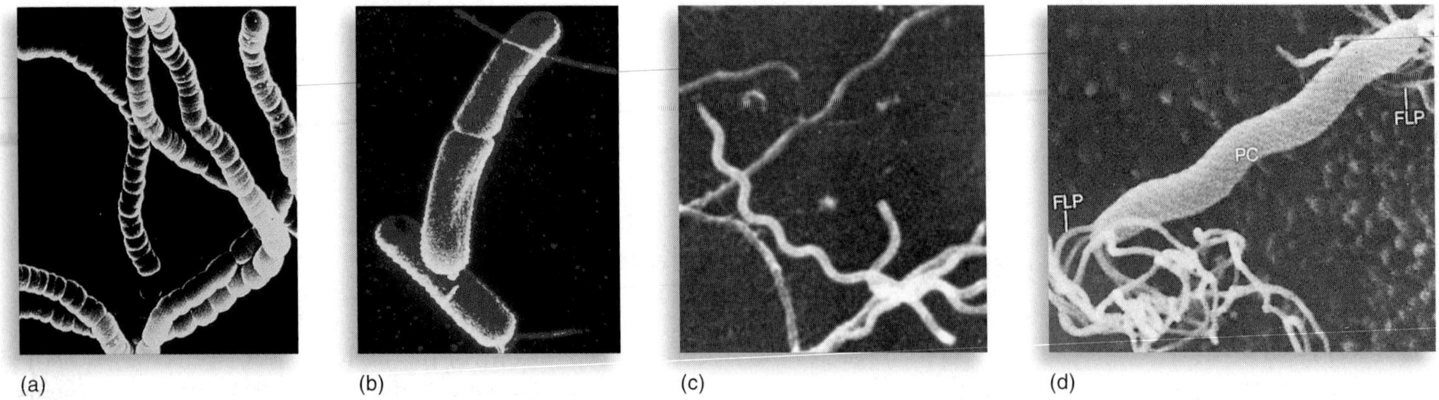

Figure 4.23 SEM photographs of basic bacterial shapes reveal their three dimensions and surface features.
(a) Cocci in chains. **(b)** A rod-shaped bacterium (*Escherichia coli*) in a diplobacillus arrangement. **(c)** A spirochete (*Borrelia burgdorferi*, the cause of Lyme disease) is a long, thin cell with irregular coils and no external flagella. **(d)** A spirillum is thicker with a few even coils (PC) and external flagella (FLP). Can you tell what the flagellar arrangement is?

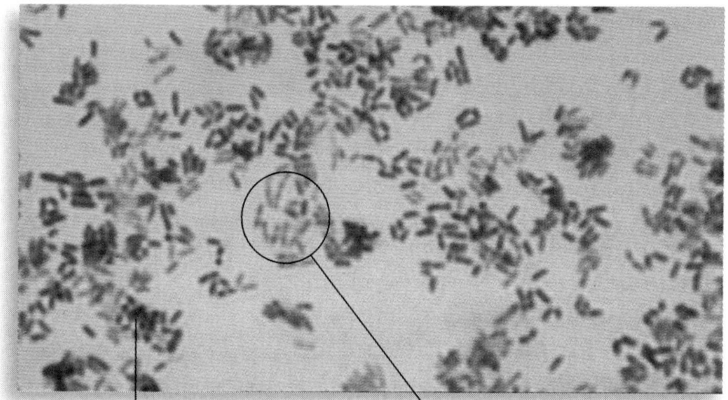

Figure 4.24 Pleomorphism in *Corynebacterium*.
Cells occur in a great variety of shapes and sizes (800×). This genus typically exhibits an unusual formation called a palisades arrangement, but some cells have other appearances. Close examination will also reveal darkly stained granules inside the cells.

The cells of bacteria can also be categorized according to arrangement, or style of grouping (see figure 4.22). The main factors influencing the arrangement of a particular cell type are its pattern of division and how the cells remain attached afterward. The greatest variety in arrangement occurs in cocci, which can be single, in pairs (diplococci), in **tetrads** (groups of four), in irregular clusters (both staphylococci and micrococci), or in chains of a few to hundreds of cells (streptococci). An even more complex grouping is a cubical packet of eight, sixteen, or more cells called a **sarcina** (sar′-sih-nah). These different coccal groupings are the result of the division of a coccus in a single plane, in two perpendicular planes, or in several intersecting planes; after division, the resultant daughter cells remain attached.

Bacilli are less varied in arrangement because they divide only in the transverse plane (perpendicular to the axis). They occur either as single cells, as a pair of cells with their ends attached (diplobacilli), or as a chain of several cells (streptobacilli). A **palisades** (pal′-ih-saydz) arrangement, typical of the corynebacteria, is formed when the cells of a chain remain partially attached by a small hinge region at the ends. The cells tend to fold (snap) back upon each other, forming a row of cells oriented side by side (see figures 4.22 and 4.24). The reaction can be compared to the behavior of boxcars on a jackknifed train, and the result looks superficially like an irregular picket fence. Spirilla are occasionally found in short chains, but spirochetes rarely remain attached after division. Comparative sizes of typical cells are presented in **figure 4.25.**

CHECKPOINT

- Most bacteria have one of three general shapes: coccus (round), bacillus (rod), or spiral, based on the configuration of the cell wall. Two types of spiral cells are spirochetes and spirilla.
- Shape and arrangement of cells are key means of describing bacteria. Arrangements of cells are based on the number of planes in which a given species divides.
- Cocci can divide in many planes to form pairs, chains, packets, or clusters. Bacilli divide only in the transverse plane. If they remain attached, they form chains or palisades.

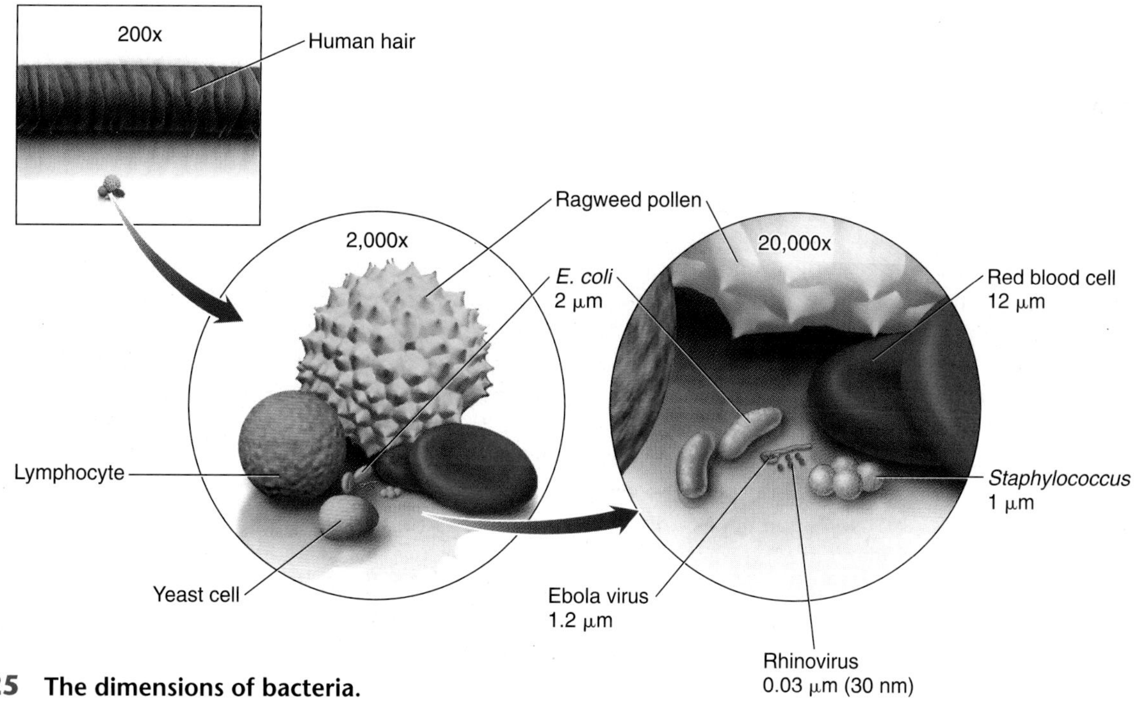

Figure 4.25 The dimensions of bacteria.
The sizes of bacteria range from those just barely visible with light microscopy (0.2 μm) to those measuring over a thousand times that size. Generally, cocci measure anywhere from 0.5 to 3.0 μm in diameter; bacilli range from 0.2 to 2.0 μm in diameter and from 0.5 to 20 μm in length; vibrios and spirilla vary from 0.2 to 2.0 μm in diameter and from 0.5 to 100 μm in length. Spirochetes range from 0.1 to 3.0 μm in diameter and from 0.5 to 250 μm in length. Note the range of sizes as compared with eukaryotic cells and viruses. Comparisons are given as average sizes.

4.6 Classification Systems in the Prokaryotae

Classification systems serve both practical and academic purposes. They aid in differentiating and identifying unknown species in medical and applied microbiology. They are also useful in organizing bacteria and as a means of studying their relationships and origins. Since classification was started around 200 years ago, several thousand species of bacteria and archaea have been identified, named, and cataloged.

For years scientists have had intense interest in tracing the origins of and evolutionary relationships among bacteria, but doing so has not been an easy task. One of the questions that has plagued taxonomists is, "What characteristics are the most indicative of closeness in ancestry"? Early bacteriologists found it convenient to classify bacteria according to shape, variations in arrangement, growth characteristics, and habitat. However, as more species were discovered and as techniques for studying their biochemistry were developed, it soon became clear that similarities in cell shape, arrangement, and staining reactions do not automatically indicate relatedness. Even though the gram-negative rods look alike, there are hundreds of different species, with highly significant differences in biochemistry and genetics. If we attempted to classify them on the basis of Gram stain and shape alone, we could not assign them to a more specific level than class. Increasingly, classification schemes are turning to genetic and molecular traits that cannot be visualized under a microscope or in culture.

One of the most viable indicators of evolutionary relatedness and affiliation is comparison of the sequence of nitrogen bases in ribosomal RNA, a major component of ribosomes. Ribosomes have the same function (protein synthesis) in all cells, and they tend to remain more or less stable in their nucleic acid content over long periods. Thus, any major differences in the sequence, or "signature," of the rRNA is likely to indicate some distance in ancestry. This technique is powerful at two levels: It is effective for differentiating general group differences (it was used to separate the three superkingdoms of life discussed in chapter 1), and it can be fine-tuned to identify at the species level (for example in *Mycobacterium* and *Legionella*). Elements of these and other identification methods are presented in more detail in chapter 17.

The definitive published source for bacterial classification, called *Bergey's Manual*, has been in print continuously since 1923. The basis for the early classification in *Bergey's* was the **phenotypic** traits of bacteria, such as their shape, cultural behavior, and biochemical reactions. These traits are still used extensively by clinical microbiologists or researchers who need to quickly identify unknown bacteria. As methods for RNA and DNA analysis became available, this information was used to supplement the phenotypic information. The current version of the publication, called *Bergey's Manual of Systematic Bacteriology*, presents a comprehensive view of bacterial relatedness, combining phenotypic information with rRNA sequencing information to classify bacteria; it is a huge, five-volume set. (We need to remember that all bacterial classification systems are in a state of constant flux; no system is ever finished.)

With the explosion of information about evolutionary relatedness among bacteria, the need for a *Bergey's Manual* that contained easily accessible information for identifying unknown bacteria became apparent. Now there is a separate book, called *Bergey's Manual of Determinative Bacteriology*, based entirely on phenotypic characteristics. It is utilitarian in focus, categorizing bacteria by traits commonly assayed in clinical, teaching, and research labs. It is widely used by microbiologists who need to identify bacteria but need not know their evolutionary backgrounds. This phenotypic classification is more useful for students of medical microbiology, as well.

Taxonomic Scheme

Bergey's Manual of Determinative Bacteriology organizes the Kingdom Prokaryotae into four major divisions. These somewhat natural divisions are based upon the nature of the cell wall. The **Gracilicutes** (gras″-ih-lik′-yoo-teez) have gram-negative cell walls and thus are thin-skinned; the **Firmicutes** have gram-positive cell walls that are thick and strong; the **Tenericutes** (ten″-er-ik′-yoo-teez) lack a cell wall and thus are soft; and the **Mendosicutes** (men-doh-sik′-yoo-teez) are the archaea (also called archaebacteria), primitive prokaryotes with unusual cell walls and nutritional habits. The first two divisions contain the greatest number of species. The 200 or so species that cause human and animal diseases can be found in four classes: the Scotobacteria, Firmibacteria, Thallobacteria, and Mollicutes. The system used in *Bergey's Manual* further organizes bacteria into subcategories such as classes, orders, and families, but these are not available for all groups.

Diagnostic Scheme

As mentioned earlier, many medical microbiologists prefer an informal working system that outlines the major families and genera. **Table 4.2** is an example of an adaptation of the phenotypic method of classification that might be used in clinical microbiology. This system is more applicable for diagnosis because it is restricted to bacterial disease agents, depends less on nomenclature, and is based on readily accessible morphological and physiological tests rather than on phylogenetic relationships. It also divides the bacteria into gram-positive, gram-negative, and those without cell walls and then subgroups them according to cell shape, arrangement, and certain physiological traits such as oxygen usage: *Aerobic* bacteria use oxygen in metabolism; *anaerobic* bacteria do not use oxygen in metabolism; and facultative bacteria may or may not use oxygen. Further tests not listed on the table would be required to separate closely related genera and species. Many of these are included in later chapters on specific bacterial groups.

TABLE 4.2 Medically Important Families and Genera of Bacteria, with Notes on Some Diseases*

I. Bacteria with gram-positive cell wall structure

Cocci in clusters or packets

- Family Micrococcaceae: *Staphylococcus* (members cause boils, skin infections)

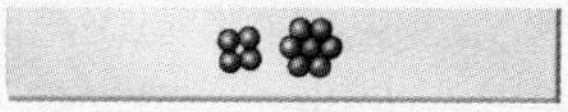

Cocci in pairs and chains

- Family Streptococcaceae: *Streptococcus* (species cause strep throat, dental caries)

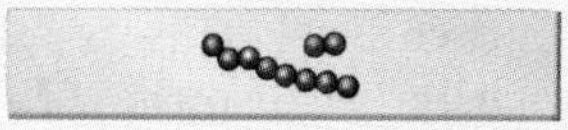

Anaerobic cocci in pairs, tetrads, irregular clusters

- Family Peptococcaceae: *Peptococcus, Peptostreptococcus* (involved in wound infections)

Spore-forming rods

- Family Bacillaceae: *Bacillus* (anthrax), *Clostridium* (tetanus, gas gangrene, botulism)

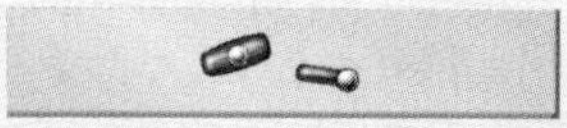

Non-spore-forming rods

- Family Lactobacillaceae: *Lactobacillus, Listeria, Erysipelothrix* (erysipeloid)
- Family Propionibacteriaceae: *Propionibacterium* (involved in acne)

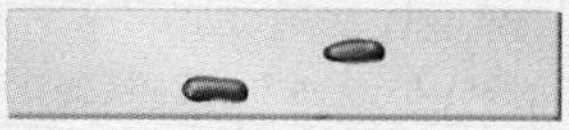

- Family Corynebacteriaceae: *Corynebacterium* (diphtheria)

- Family Mycobacteriaceae: *Mycobacterium* (tuberculosis, leprosy)

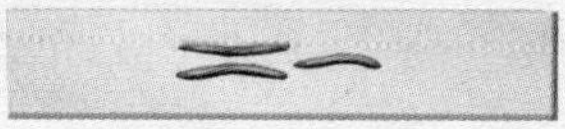

- Family Nocardiaceae: *Nocardia* (lung abscesses)

- Family Actinomycetaceae: *Actinomyces* (lumpy jaw), *Bifidobacterium*

- Family Streptomycetaceae: *Streptomyces* (important source of antibiotics)

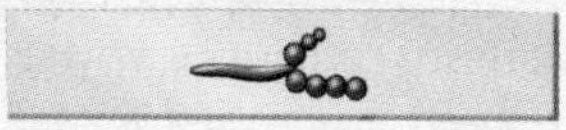

II. Bacteria with gram-negative cell wall structure

Aerobic cocci

- *Neisseria* (gonorrhea, meningitis), *Branhamella*

Aerobic coccobacilli

- *Moraxella, Acinetobacter*

Anaerobic cocci

- Family Veillonellaceae
- *Veillonella* (dental disease)

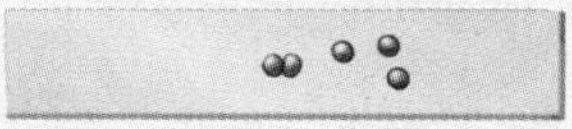

Miscellaneous rods

- *Brucella* (undulant fever), *Bordetella* (whooping cough), *Francisella* (tularemia)

Aerobic rods

- Family Pseudomonadaceae: *Pseudomonas* (pneumonia, burn infections)
- Miscellaneous: *Legionella* (Legionnaires' disease)

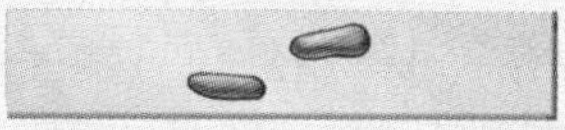

Facultative or anaerobic rods and vibrios

- Family Enterobacteriaceae: *Escherichia, Edwardsiella, Citrobacter, Salmonella* (typhoid fever), *Shigella* (dysentery), *Klebsiella, Enterobacter, Serratia, Proteus, Yersinia* (one species causes plague)

- Family Vibronaceae: *Vibrio* (cholera, food infection), *Campylobacter, Aeromonas*

- Miscellaneous genera: *Chromobacterium, Flavobacterium, Haemophilus* (meningitis), *Pasteurella, Cardiobacterium, Streptobacillus*

Anaerobic rods

- Family Bacteroidaceae: *Bacteroides, Fusobacterium* (anaerobic wound and dental infections)

Helical and curviform bacteria

- Family Spirochaetaceae: *Treponema* (syphilis), *Borrelia* (Lyme disease), *Leptospira* (kidney infection)

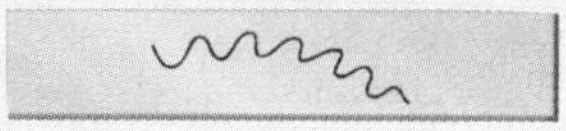

Obligate intracellular bacteria

- Family Rickettsiaceae: *Rickettsia* (Rocky Mountain spotted fever), *Coxiella* (Q fever)

- Family Bartonellaceae: *Bartonella* (trench fever, cat scratch disease)
- Family Chlamydiaceae: *Chlamydia* (sexually transmitted infection)

III. Bacteria with no cell walls

- Family Mycoplasmataceae: *Mycoplasma* (pneumonia), *Ureaplasma* (urinary infection)

*Details of pathogens and diseases in chapters 18 through 23.

Species and Subspecies in Bacteria

Among most organisms, the species level is a distinct, readily defined, and natural taxonomic category. In animals, for instance, a species is a distinct type of organism that can produce viable offspring only when it mates with others of its own kind. This definition does not work for bacteria primarily because they do not exhibit a typical mode of sexual reproduction. They can accept genetic information from unrelated forms, and they can also alter their genetic makeup by a variety of mechanisms. Thus, it is necessary to hedge a bit when we define a bacterial species. Theoretically, it is a collection of bacterial cells, all of which share an overall similar pattern of traits, in contrast to other groups whose patterns differ significantly. Although the boundaries that separate two closely related species in a genus are in some cases arbitrary, this definition still serves as a method to separate the bacteria into various kinds that can be cultured and studied. As additional information on bacterial genomes is discovered, it may be possible to define species according to specific combinations of genetic codes found only in a particular isolated culture.

Individual members of given species can show variations, as well. Therefore more categories within species exist, but they are not well defined. Microbiologists use terms like *subspecies*, *strain*, or *type* to designate bacteria of the same species that have differing characteristics. *Serotype* refers to representatives of a species that stimulate a distinct pattern of antibody (serum) responses in their hosts, because of distinct surface molecules.

CHECKPOINT

- Bacteria are formally classified by phylogenetic relationships and phenotypic characteristics.
- Medical identification of pathogens uses an informal system of classification based on Gram stain, morphology, biochemical reactions, and metabolic requirements.
- A bacterial species is loosely defined as a collection of bacterial cells that shares an overall similar pattern of traits different from other groups of bacteria.
- Variant forms within a species (subspecies) include strains and types.

4.7 Survey of Prokaryotic Groups with Unusual Characteristics

The bacterial world is so diverse that we cannot do complete justice to it in this introductory chapter. This variety extends into all areas of bacterial biology, including nutrition, mode of life, and behavior. Certain types of bacteria exhibit such unusual qualities that they deserve special mention. In this minisurvey, we consider some medically important groups and some more remarkable representatives of bacteria living free in the environment that are ecologically important. Many of the bacteria mentioned here do not have the morphology typical of bacteria discussed previously, and in a few cases, they are vividly different **(Insight 4.3).**

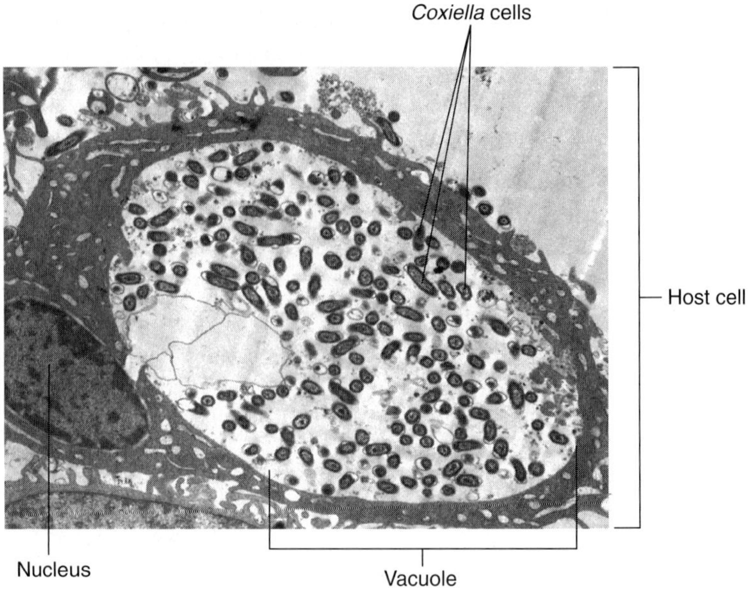

Figure 4.26 Transmission electron micrograph of the rickettsia *Coxiella burnetii.*
Its mass growth inside a host cell has filled a vacuole and displaced the nucleus to one side.

Unusual Forms of Medically Significant Bacteria

Most bacteria are free-living or parasitic forms that can metabolize and reproduce by independent means. Two groups of bacteria—the rickettsias and chlamydias—have adapted to life inside their host cells, where they are considered **obligate intracellular parasites.**

Rickettsias

Rickettsias[4] are distinctive, very tiny, gram-negative bacteria **(figure 4.26).** Although they have a somewhat typical bacterial morphology, they are atypical in their life cycle and other adaptations. Most are pathogens that alternate between a mammalian host and blood-sucking arthropods,[5] such as fleas, lice, or ticks. Rickettsias cannot survive or multiply outside a host cell and cannot carry out

4. Named for Howard Ricketts, a physician who first worked with these organisms and later lost his life to typhus.
5. An arthropod is an invertebrate with jointed legs, such as an insect, tick, or spider.

INSIGHT 4.3 — *Discovery*

Redefining Bacterial Size

Most microbiologists believe we are still far from having a complete assessment of the bacterial world, mostly because the world is so large and bacteria are so small. This fact becomes evident in the periodic discoveries of exceptional bacteria that are reported in newspaper headlines. Among the most remarkable are giant and dwarf bacteria.

Big Bacteria Break Records

In 1985, biologists discovered a new bacterium living in the intestine of surgeonfish that at the time was a candidate for the *Guinness Book of World Records*. The large cells, named *Epulopiscium fishelsoni* ("guest at a banquet of fish"), measure around 100 µm in length, although some specimens were as large as 300 µm. This record was recently broken when marine microbiologist Heide Schultz discovered an even larger species of bacteria living in ocean sediments near the African country of Namibia. These gigantic cocci are arranged in strands that look like pearls and contain hundreds of golden sulfur granules, inspiring their name, *Thiomargarita namibia* ("sulfur pearl of Namibia") (see photo). The size of the individual cells ranges from 100 up to 750 µm ($\frac{3}{4}$ mm), and many are large enough to see with the naked eye. By way of comparison, if the average bacterium were the size of a mouse, *Thiomargarita* would be as large as a blue whale!

Closer study revealed that they are indeed prokaryotic and have bacterial ribosomes and DNA, but that they also have some unusual adaptations to their life cycle. They live an attached existence embedded in sulfide sediments (H_2S) that are free of gaseous oxygen. They obtain energy through oxidizing these sulfides using dissolved nitrates (NO_3). Because the quantities of these substances can vary with the seasons, they must be stored in cellular depots. The sulfides are carried as granules in the cytoplasm, and the nitrates occupy a giant, liquid-filled vesicle that takes up a major proportion of cell volume. Due to their morphology and physiology, the cells can survive for up to 3 months without an external source of nutrients by tapping into their "storage tanks." These bacteria are found in such large numbers in the sediments that it is thought that they are essential to the ecological cycling of H_2S gas in this region, converting it to less toxic substances.

Miniature Microbes—The Smallest of the Small

At the other extreme, microbiologists are being asked to reevaluate the lower limits of bacterial size. Up until now it has been generally accepted that the smallest cells on the planet are some form of mycoplasma with dimensions of 0.2 to 0.3 µm, which is right at the limit of resolution with light microscopes. A new controversy is brewing over the discovery of tiny cells that look like dwarf bacteria but are 10 times smaller than mycoplasmas and a hundred times smaller than the average bacterial cell. These minute cells have been given the name **nanobacteria** or **nanobes** (Gr. *nanos*, one-billionth).

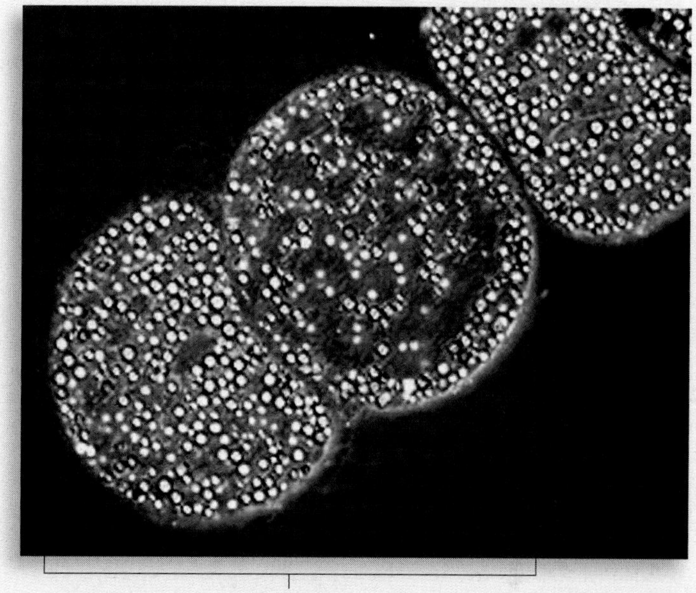

Nanobacterialike forms were first isolated from blood and serum samples. The tiny cells appear to grow in culture, have cell walls, and contain protein and nucleic acids, but their size range is only from 0.05 to 0.2 µm. Similar nanobes have been extracted by minerologists studying sandstone rock deposits in the ocean at temperatures of 100°C to 170°C and deeply embedded in billion-year-old minerals. The minute filaments were able to grow and are capable of depositing minerals in a test tube. Many geologists are convinced that these nanobes are real, that they are probably similar to the first microbes on earth, and that they play a strategic role in the evolution of the earth's crust. Microbiologists tend to be more skeptical. It has been postulated that the minimum cell size to contain a functioning genome and reproductive and synthetic machinery is approximately 0.14 µm. They believe that the nanobes are really just artifacts or bits of larger cells that have broken free.

Nanobe "believers" have recently been bolstered by a series of findings indicating that nanobes can infect humans and have been linked to diseases such as kidney stones and ovarian cancer. These diseases are influenced in some way by calcification that is catalyzed by nanobes.

It seems the real question is not whether nanobes exist but whether we should classify them as bacteria. One of the early nanobe discoverers, Olavi Kajander, blames himself for getting scientists distracted by that question by first coining the name "nanobacteria." "Calcifying self-propagating nanoparticles would have been much better," he says now.* Additional studies are needed to test this curious question of nanobes, and possibly to answer some questions about the origins of life on earth and even other planets.

*Wired.com news story, March 14, 2005.

metabolism completely on their own, so they are closely attached to their hosts. Several important human diseases are caused by rickettsias. Among these are Rocky Mountain spotted fever, caused by *Rickettsia rickettsii* (transmitted by ticks), and endemic typhus, caused by *Rickettsia typhi* (transmitted by lice).

Chlamydias

Bacteria of the genera *Chlamydia* and *Chlamydophila* are similar to the rickettsias in that they require host cells for growth and metabolism, but they are not closely related and are not transmitted by arthropods. Because of their tiny size and obligately parasitic lifestyle, they were at one time considered a type of virus. Species that carry the greatest medical impact are *Chlamydia trachomatis*, the cause of both a severe eye infection (trachoma) that can lead to blindness and one of the most common sexually transmitted diseases; and *Chlamydophila pneumoniae*, an agent in lung infections.

Diseases caused by rickettsias and by *Chlamydia* species are described in more detail in the infectious disease chapters according to the organ systems they affect.

Free-Living Nonpathogenic Bacteria

Photosynthetic Bacteria

The nutrition of most bacteria is heterotrophic, meaning that they derive their nutrients from other organisms. Photosynthetic bacteria, however, are independent cells that contain special light-trapping pigments and can use the energy of sunlight to synthesize all required nutrients from simple inorganic compounds. The two general types of photosynthetic bacteria are those that produce oxygen during photosynthesis and those that produce some other substance, such as sulfur granules or sulfates.

Cyanobacteria: Blue-Green Bacteria

The cyanobacteria were called blue-green algae for many years and were grouped with the eukaryotic algae. However, further study verified that they are indeed bacteria with a gram-negative cell wall and general prokaryotic structure. These bacteria range in size from 1 μm to 10 μm, and they can be unicellular or can occur in colonial or filamentous groupings **(figure 4.27*a, b*).** Some species occur in packets surrounded by a gelatinous sheath (figure 4.27*b*). A specialized adaptation of cyanobacteria is extensive internal membranes called **thylakoids,** which contain granules of chlorophyll *a* and other photosynthetic pigments **(figure 4.27*c*).** They also have gas inclusions, which permit them to float on the water surface and increase their light exposure, and cysts that convert gaseous nitrogen (N_2) into a form usable by plants. This group is sometimes called the blue-green bacteria in reference to their content of phycocyanin pigment that tints some members a shade of blue,

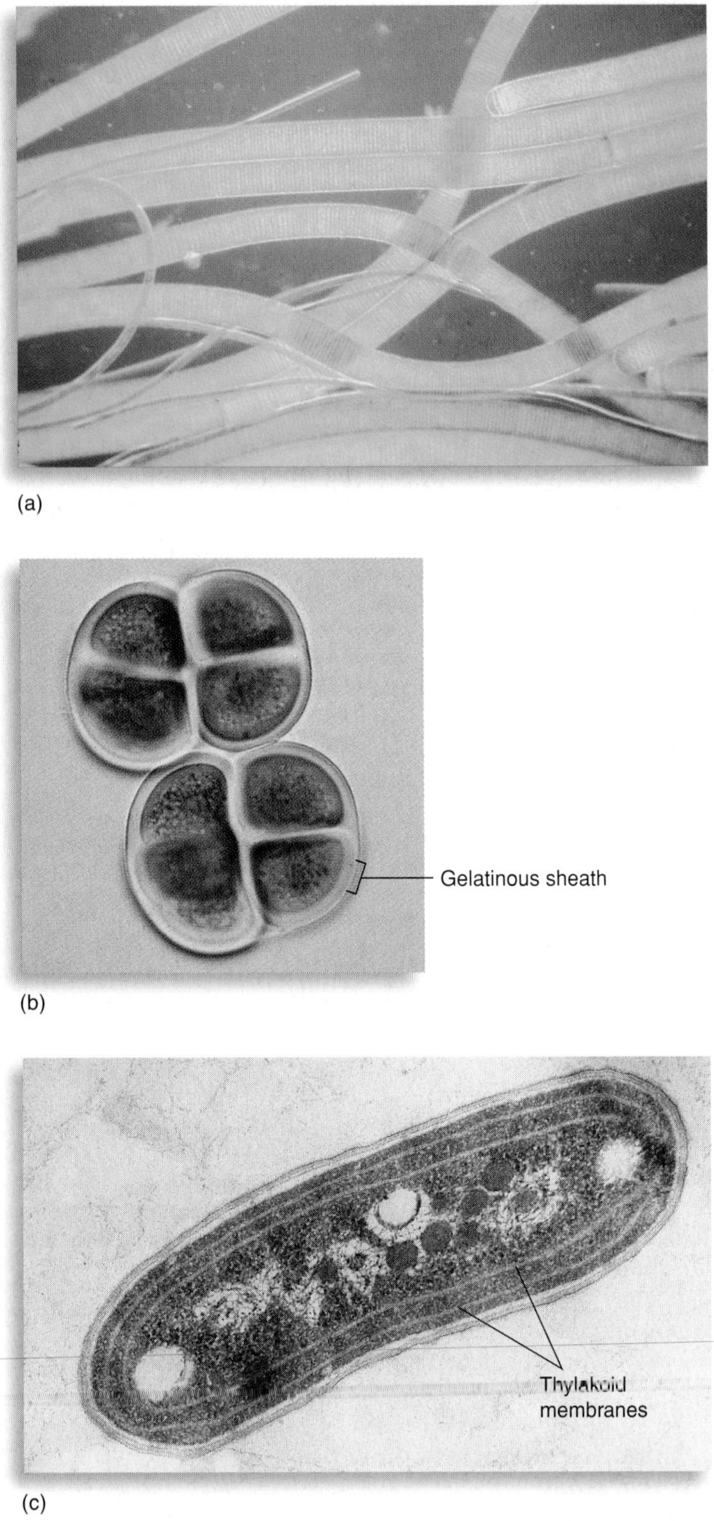

Figure 4.27 Structure and examples of cyanobacteria. **(a)** Two species of *Oscillatoria,* a gliding, filamentous form (100×). **(b)** *Chroococcus,* a colonial form surrounded by a gelatinous sheath (600×). **(c)** Electron micrograph of a cyanobacterial cell (80,000×) reveals folded stacks of membranes that contain the photosynthetic pigments and increased surface area for photosynthesis.

although other members are colored yellow and orange. Some representatives glide or sway gently in the water from the action of filaments in the cell envelope that cause wavelike contractions.

Cyanobacteria are very widely distributed in nature. They grow profusely in fresh water and seawater and are thought to be responsible for periodic blooms that kill off fish. Some members are so pollution-resistant that they serve as biological indicators of polluted water. Cyanobacteria inhabit and flourish in hot springs (see Insight 7.1) and have even exploited a niche in dry desert soils and rock surfaces.

Green and Purple Sulfur Bacteria

The green and purple bacteria are also photosynthetic and contain pigments. They differ from the cyanobacteria in having a different type of chlorophyll called *bacteriochlorophyll* and by not giving off oxygen as a product of photosynthesis. They live in sulfur springs, freshwater lakes, and swamps that are deep enough for the anaerobic conditions they require yet where their pigment can still absorb wavelengths of light **(figure 4.28).** These bacteria are named for their predominant colors, but they can also develop brown, pink, purple, blue, and orange coloration. Both groups utilize sulfur compounds (H_2S, S) in their metabolism.

Archaea: The Other Prokaryotes

The discovery and characterization of novel prokaryotic cells that have unusual anatomy, physiology, and genetics changed our views of microbial taxonomy and classification (see chapter 1). These single-celled, simple organisms, called **archaea,** are now considered a third cell type in a separate superkingdom (the Domain Archaea). We include them in this chapter because they are prokaryotic in general structure and they do share many bacterial characteristics.

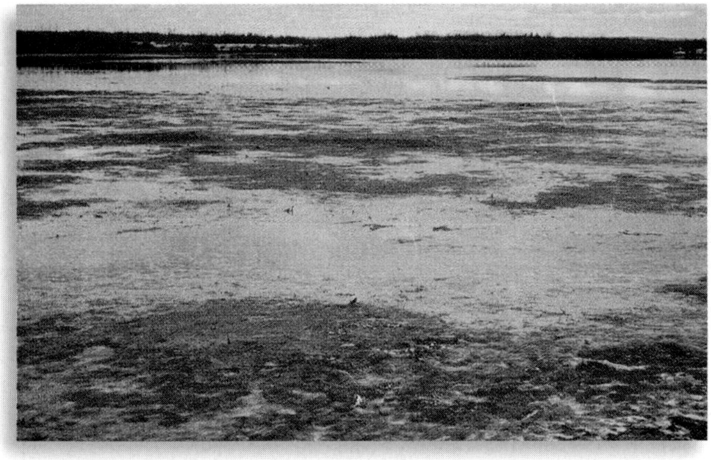

Figure 4.28 Behavior of purple sulfur bacteria. Floating purple mats are huge masses of purple sulfur bacteria blooming in the Baltic Sea. Photosynthetic bacteria can have significant effects on the ecology of certain habitats.

But evidence is accumulating that they are actually more closely related to Domain Eukarya than to bacteria. For example, archaea and eukaryotes share a number of ribosomal RNA sequences that are not found in bacteria, and their protein synthesis and ribosomal subunit structures are similar. **Table 4.3** outlines selected points of comparison of the three domains.

Among the ways that the archaea differ significantly from other cell types are that certain genetic sequences are found only in their rRNA, and that they have unique membrane lipids and cell wall construction. It is clear that the archaea are the most primitive of all life forms and are most closely related to the first cells that originated on the earth 4 billion years ago. The early earth is thought to have contained a hot, anaerobic "soup" with sulfuric gases and salts in abundance. The modern archaea still live in the remaining habitats on the earth that have these same ancient conditions—the most extreme habitats in nature. It is for this reason that they are often called extremophiles, meaning that they "love" extreme conditions in the environment.

Metabolically, the archaea exhibit incredible adaptations to what would be deadly conditions for other organisms. These hardy microbes have adapted to multiple combinations

TABLE 4.3 Comparison of Three Cellular Domains

Characteristic	Bacteria	Archaea	Eukarya
Cell type	Prokaryotic	Prokaryotic	Eukaryotic
Chromosomes	Single, or few, circular	Single, circular	Several, linear
Types of ribosomes	70S	70S but structure is similar to 80S	80S
Contains unique ribosomal RNA signature sequences	+	+	+
Number of sequences shared with Eukarya	1	3	(all)
Protein synthesis similar to Eukarya	−	+	
Presence of peptidoglycan in cell wall	+	−	−
Cell membrane lipids	Fatty acids with ester linkages	Long-chain, branched hydrocarbons with ether linkages	Fatty acids with ester linkages
Sterols in membrane	− (some exceptions)	−	+

of heat, salt, acid, pH, pressure, and atmosphere. Included in this group are methane producers, hyperthermophiles, extreme halophiles, and sulfur reducers.

Members of the group called *methanogens* can convert CO_2 and H_2 into methane gas (CH_4) through unusual and complex pathways. These archaea are common inhabitants of anaerobic swamp mud, the bottom sediments of lakes and oceans, and even the digestive systems of animals. The gas they produce collects in swamps and may become a source of fuel. Methane may also contribute to the "greenhouse effect," which maintains the earth's temperature and can contribute to global warming (see chapter 24).

Other types of archaea—the extreme halophiles—require salt to grow and may have such a high salt tolerance that they can multiply in sodium chloride solutions (36% NaCl) that would destroy most cells. They exist in the saltiest places on the earth—inland seas, salt lakes, salt mines, and salted fish. They are not particularly common in the ocean because the salt content is not high enough. Many of the "halobacteria" use a red pigment to synthesize ATP in the presence of light. These pigments are responsible for "red herrings," the color of the Red Sea, and the red color of salt ponds **(figure 4.29).**

Archaea adapted to growth at very low temperatures are called *psychrophilic* (loving cold temperatures); those growing at very high temperatures are *hyperthermophilic* (loving high temperatures). Hyperthermophiles flourish at temperatures between 80°C and 113°C and cannot grow at 50°C. They live in volcanic waters and soils and submarine vents and are also often salt- and acid-tolerant as well. One member, *Thermoplasma*, lives in hot, acidic habitats in the waste piles around coal mines that regularly sustain a pH of 1 and a temperature of nearly 60°C.

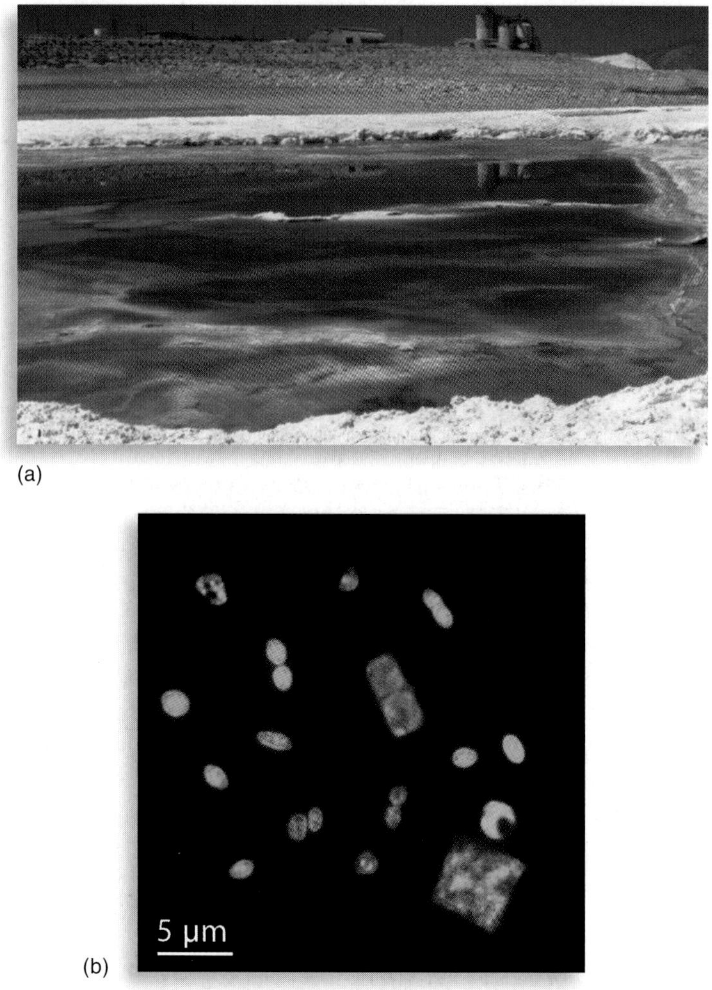

Figure 4.29 Halophiles around the world.
(a) A solar evaporation pond in Owens Lake, California, is extremely high in salt and mineral content. The archaea that dominate in this hot, saline habitat produce brilliant red pigments with which they absorb light to drive cell synthesis. **(b)** A sample taken from a saltern in Australia viewed by fluorescent microscopy (1,000×). Note the range of cell shapes (cocci, rods, and squares) found in this community.

CHECKPOINT

- The rickettsias are a group of bacteria that are intracellular parasites, dependent on their eukaryote host for energy and nutrients. Most are pathogens that alternate between arthropods and mammalian hosts.
- The chlamydias are also small, intracellular parasites that infect humans, mammals, and birds. They do not require arthropod vectors.
- Many bacteria are free-living, rather than parasitic. The photosynthetic bacteria encompass many subgroups that colonize specialized habitats, not other living organisms.
- Archaea are another type of prokaryotic cell that constitute the third domain of life. They exhibit unusual biochemistry and genetics that make them different from bacteria. Many members are adapted to extreme habitats with low or high temperature, salt, pressure, or acid.

Chapter Summary with Key Terms

4.1 Prokaryotic Form and Function
General Features of Prokaryotes
A. Prokaryotes consist of two major groups, the bacteria and the archaea. Life on earth would not be possible without them.
B. Prokaryotic cells lack the membrane-surrounded organelles and nuclear compartment of eukaryotic cells but are still complex in their structure and function. All prokaryotes have a cell membrane, cytoplasm, ribosomes, and a chromosome.

4.2 External Structures
Appendages: Cell Extensions
Some bacteria have projections that extend from the cell. **Flagella** (and internal **axial filaments** found in spirochetes) are used for motility. **Fimbriae** function in adhering to the environment; **pili** provide a means for genetic exchange. The **glycocalyx** may be a slime layer or a capsule.

4.3 The Cell Envelope: The Boundary Layer of Bacteria
A. Most prokaryotes are surrounded by a protective envelope that consists of either two or three parts: the **cytoplasmic membrane** and the **cell wall (peptidoglycan)** are present in almost all bacteria; the **outer membrane** is an additional layer present only in gram-negative bacteria.
B. The Gram stain differentiates two types of cells on the basis of their cell envelopes; gram-positive bacteria have a cytoplasmic membrane and a thick cell wall, whereas gram-negative bacteria have a cytoplasmic membrane, a thin cell wall, and an additional outer membrane.

4.4 Bacterial Internal Structure
The cell cytoplasm is a watery substance that holds some or all of the following internal structures in bacteria: the **chromosome**(s) condensed in the **nucleoid; ribosomes,** which serve as the sites of protein synthesis and are 70S in size; extra genetic information in the form of **plasmids;** storage structures known as **inclusions;** an **actin cytoskeleton,** which helps give the bacterium its shape; and in some bacteria an **endospore,** which is a highly resistant structure for survival. Bacterial endospores are not involved in reproduction.

4.5 Bacterial Shapes, Arrangements, and Sizes
A. Most bacteria are unicellular and are found in a great variety of shapes, arrangements, and sizes. General shapes include **cocci, bacilli,** and helical forms such as **spirilla** and **spirochetes.** Some show great variation within the species in shape and size and are **pleomorphic.** Other variations include **coccobacilli, vibrios,** and filamentous forms.
B. Prokaryotes divide by binary fission and do not utilize mitosis. Various arrangements result from cell division and are termed diplococci, streptococci, staphylococci, **tetrads,** and **sarcina** for cocci; bacilli may form pairs, chains, or **palisades.**

4.6 Classification Systems in the Prokaryotae
A. An important taxonomic system is standardized by *Bergey's Manual of Determinative Bacteriology*, which divides prokaryotes into four major groups:
  1. **Gracilicutes:** Bacteria with gram-negative cell walls.
  2. **Firmicutes:** Bacteria with gram-positive cell walls.
  3. **Tenericutes:** Bacteria without cell walls.
  4. **Mendosicutes:** Archaebacteria (archaea).
B. Bacterial species may also be classified on their observable characteristics which is more useful in clinical microbiology.

4.7 Survey of Prokaryotic Groups with Unusual Characteristics
Several groups of bacteria are so different that they have not always fit well in classification schemes.
A. Medically important bacteria: **Rickettsias** and chlamydias are within the gram-negative group but are small **obligate intracellular parasites** that replicate within cells of the hosts they invade.
B. Nonpathogenic bacterial groups: The majority of bacterial species are free-living and not involved in disease. Unusual groups include photosynthetic bacteria such as cyanobacteria, which provide oxygen to the environment, and the green and purple bacteria.
C. Archaea, the other major prokaryote group: Archaea share many characteristics of prokaryotes but do have some differences with bacteria in certain genetic aspects and some cell components. Many are adapted to extreme environments, as may have been found originally on earth. They are not considered medically important but are of ecological and potential economic importance.

Multiple-Choice and True-False Questions

Multiple-Choice Questions. Select the correct answer from the answers provided.

1. Which of the following is not found in all bacterial cells?
   a. cell membrane
   b. a nucleoid
   c. ribosomes
   d. actin cytoskeleton
2. The major locomotor structures in bacteria are
   a. flagella
   b. pili
   c. fimbriae
   d. cilia
3. Pili are tubular shafts in ______ bacteria that serve as a means of ______
   a. gram-positive, genetic exchange
   b. gram-positive, attachment
   c. gram-negative, genetic exchange
   d. gram-negative, protection

4. An example of a glycocalyx is
   a. a capsule   c. outer membrane
   b. pili   d. a cell wall
5. Which of the following is a primary bacterial cell wall function?
   a. transport   c. support
   b. motility   d. adhesion
6. Which of the following is present in both gram-positive and gram-negative cell walls?
   a. an outer membrane   c. teichoic acid
   b. peptidoglycan   d. lipopolysaccharides
7. Darkly-stained granules are concentrated crystals of ______ that are found in ______.
   a. fat, *Mycobacterium*   c. sulfur, *Thiobacillus*
   b. dipicolinic acid, *Bacillus*   d. PO_4, *Corynebacterium*
8. Bacterial endospores function in
   a. reproduction   c. protein synthesis
   b. survival   d. storage
9. A bacterial arrangement in packets of eight cells is described as a ______
   a. micrococcus   c. tetrad
   b. diplococcus   d. sarcina
10. To which division of bacteria do cyanobacteria belong?
    a. Tenericutes   c. Firmicutes
    b. Gracilicutes   d. Mendosicutes
11. Which stain is used to distinguish differences between the cell walls of medically important bacteria?
    a. simple stain   c. Gram stain
    b. acridine orange stain   d. negative stain

True-False Questions. If the statement is true, leave as is. If it is false, correct it by rewriting the sentence.

12. One major difference in the envelope structure between gram-positive bacteria and gram-negative bacteria is the presence or absence of a cytoplasmic membrane.
13. A research microbiologist looking at evolutionary relatedness between two bacterial species is more likely to use *Bergey's Manual for Determinative Bacteriology* than *Bergey's Manual of Systematic Bacteriology*.
14. Nanobes may or may not actually be bacteria.
15. Both bacteria and archaea are prokaryotes.
16. A collection of bacteria that share an overall similar pattern of traits is called a *species*.

Writing to Learn

These questions are suggested as a *writing-to-learn* experience. For each question, compose a one- or two-paragraph answer that includes the factual information needed to completely address the question.

1. a. Name several general characteristics that could be used to define the prokaryotes.
   b. Do any other microbial groups besides bacteria have prokaryotic cells?
   c. What does it mean to say that bacteria are ubiquitous? In what habitats are they found? Give some general means by which bacteria derive nutrients.
2. a. Describe the structure of a flagellum and how it operates. What are the four main types of flagellar arrangement?
   b. How does the flagellum dictate the behavior of a motile bacterium? Differentiate between flagella and periplasmic flagella.
   c. List some direct and indirect ways that one can determine bacterial motility.
3. Differentiate between pili and fimbriae.
4. a. Compare the cell envelopes of gram-positive and gram-negative bacteria.
   b. What function does peptidoglycan serve?
   c. To which part of the cell envelope does it belong?
   d. Give a simple description of its structure.
   e. What happens to a cell that has its peptidoglycan disrupted or removed?
   f. What functions does the LPS layer serve?
5. a. What is the Gram stain?
   b. What is there in the structure of bacteria that causes some to stain purple and others to stain red?
   c. How does the precise structure of the cell walls differ in gram-positive and gram-negative bacteria?
   d. What other properties besides staining are different in gram-positive and gram-negative bacteria?
   e. What is the periplasmic space, and how does it function?
   f. What characteristics does the outer membrane confer on gram-negative bacteria?
6. List five functions that the cell membrane performs in bacteria.
7. a. Compare the composition of the bacterial chromosome (nucleoid) and plasmids.
   b. What are the functions of each?
8. a. What is unique about the structure of bacterial ribosomes?
   b. How do they function?
   c. Where are they located?
9. a. Describe the vegetative stage of a bacterial cell.
   b. Describe the structure of an endospore, and explain its function.
   c. Describe the endospore-forming cycle.
   d. Explain why an endospore is not considered a reproductive body.
   e. Why are endospores so difficult to destroy?

10. a. Draw the three bacterial shapes.
    b. How are spirochetes and spirilla different?
    c. What is a vibrio? A coccobacillus?
    d. What is pleomorphism?
    e. What is the difference between the use of the term bacillus and the name *Bacillus*?
11. a. How is the species level in bacteria defined?
    b. Name at least three ways bacteria are grouped below the species level.
12. a. Explain the characteristics of archaea that indicate that they constitute a unique domain of living things that is neither bacterial nor eukaryotic.
    b. What leads microbiologists to believe the archaea are more closely related to eukaryotes than to bacteria?
    c. What is meant by the term *extremophile*? Describe some archaeal adaptations to extreme habitats.

Concept Mapping

Appendix D provides guidance for working with concept maps.

1. Construct your own concept map using the following words as the *concepts*. Supply the linking words between each pair of concepts.

genus	species
serotype	domain
Borrelia	*burgdorferi*
spirochete	

Critical Thinking Questions

Critical thinking is the ability to reason and solve problems using facts and concepts. These questions can be approached from a number of angles, and in most cases, they do not have a single correct answer.

1. What would happen if one stained a gram-positive cell only with safranin? A gram-negative cell only with crystal violet? What would happen to the two types if the mordant were omitted?
2. What is required to kill endospores? How do you suppose archaeologists were able to date some spores as being thousands (or millions) of years old?
3. Using clay, demonstrate how cocci can divide in several planes and show the outcome of this division. Show how the arrangements of bacilli occur, including palisades.
4. Under the microscope, you see a rod-shaped cell that is swimming rapidly forward.
   a. What do you automatically know about that bacterium's structure?
   b. How would a bacterium use its flagellum for phototaxis?
   c. Can you think of another function of flagella besides locomotion?
5. a. Name a bacterium that has no cell walls.
   b. How is it protected from osmotic destruction?
6. a. Name a bacterium that is aerobic, gram-positive, and spore-forming.
   b. What habitat would you expect this species to occupy?
7. a. Name an acid-fast bacterium.
   b. What characteristics make this bacterium different from other gram-positive bacteria?
8. a. Name two main groups of obligate intracellular parasitic bacteria.
   b. Why can't these groups live independently?
9. a. Name a bacterium that contains sulfur granules.
   b. What is the advantage in storing these granules?
10. a. Name a bacterium that uses chlorophyll to photosynthesize.
    b. Describe the two major groups of photosynthetic bacteria.
    c. How are they similar?
    d. How are they different?
11. a. What are some possible adaptations that the giant bacterium *Thiomargarita* has had to make because of its large size?
    b. If a regular bacterium were the size of an elephant, estimate the size of a nanobe at that scale.
12. Propose a hypothesis to explain how bacteria and archaea could have, together, given rise to eukaryotes.

Visual Understanding

1. **From chapter 3, Figure 3.10.** Do you believe that the bacteria spelling *"Klebsiella"* or the bacteria spelling *"S. aureus"* possess the larger capsule? Defend your answer.

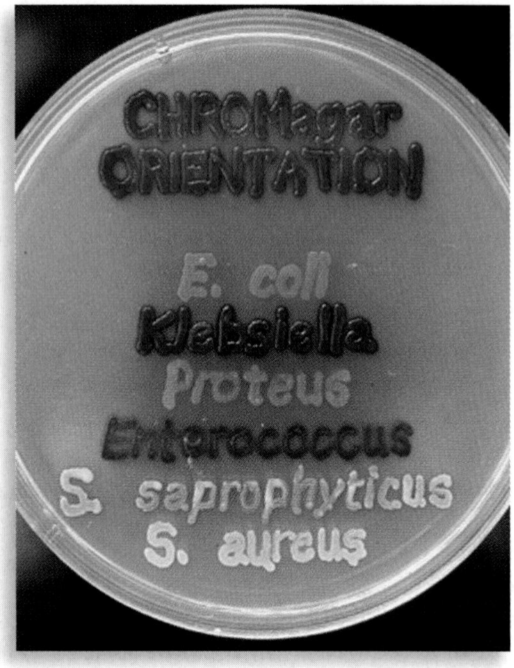

2. **From chapter 1, figure 1.15.** Study this figure. How would it be drawn differently if the archaea were more closely related to bacteria than to eukaryotes?

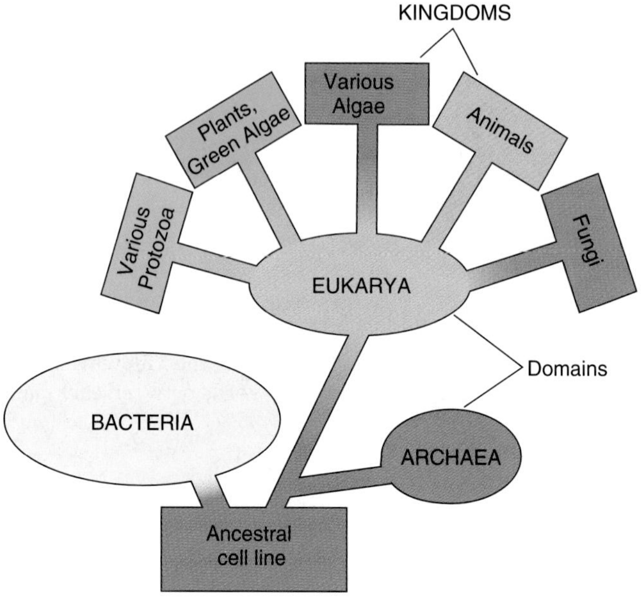

3 cell types, showing relationship with domains and kingdoms

Internet Search Topics

1. Go to a search engine and type in "Martian Microbes." Look for papers and information that support or reject the idea that fossil structures discovered in an ancient meteor from Mars could be bacteria. What are some of the reasons that microbiologists are skeptical of this possibility?
2. Search the Internet for information on nanobacteria. Give convincing reasons why these are or are not real organisms.
3. One of the premier institutes studying biofilms and their effects is at Montana State University. Search the Internet using two terms, "biofilm" and "Montana," and try to answer the question, What's the big deal about biofilms? Consider the question from either an industrial or a medical perspective.
4. Go to: www.aris.mhhe.com, and click on "microbiology" and then this textbook's author/title. Go to Chapter 4, access the URLs listed under Internet Search Topics, and research the following:

   Go to the Cells Alive website as listed. Click on "Microbiology" and go to the "Dividing Bacteria" and "Bacterial Motility" options to observe short clips on these topics.

CHAPTER 5

Eukaryotic Cells and Microorganisms

CASE FILE 5

During June of 2000, several children in Delaware, Ohio, were hospitalized at Grady Memorial General Hospital (GMH) after experiencing watery diarrhea, abdominal cramps, vomiting, and loss of appetite. Dr. McDermott, a new gastroenterologist at GMH, who also had a strong interest in infectious diseases, was asked to examine the children. Their illness lasted from 1 to 44 days, and nearly half of them complained of intermittent bouts of diarrhea. By July 20, over 150 individuals—mainly children and young adults between the ages of 20 and 40—experienced similar signs or symptoms. Dr. McDermott suspected that their illness was due to a microbial infection and queried the Delaware City County Health Department (DCCHD) to investigate this mysterious outbreak further.

Dr. McDermott helped the DCCHD team in surveying individuals hospitalized for intermittent diarrhea. They questioned individuals about recent travel, their sources of drinking water, visits to pools and lakes, swimming behaviors, contact with sick persons or young animals, and day-care attendance. The DCCHD's investigation reported that the outbreaks were linked to a swimming pool located at a private club in central Ohio. The swimming pool was closed on July 28. A total of 700 clinical cases among residents of Delaware County and three neighboring counties were identified during the entire span of the outbreak that began late June and continued through September. At least five fecal accidents were observed during that time period at the pool. Only one of these accidents was of diarrheal origin. Outbreaks of gastrointestinal distress associated with recreational water activities have increased in recent years, with most being caused by the organism in this case.

- *Do you know what microorganism might be the cause of the outbreak?*
- *How can a single fecal accident contaminate an entire pool and cause so many clinical cases of gastrointestinal distress?*

Case File 5 Wrap-Up appears on page 137.

CHAPTER OVERVIEW

- Eukaryotic cells are large complex cells divided into separate compartments by membrane-bound components called organelles.
- Major organelles—the nucleus, mitochondria, chloroplasts, endoplasmic reticulum, Golgi apparatus, and locomotor appendages—each serve an essential function to the cell, such as heredity, production of energy, synthesis, transport, and movement.
- Fungi, protozoa, algae, plants, and animals are made of eukaryotic cells, and exhibit single-celled and multicellular body plans.
- Fungi are eukaryotes that feed on organic substrates, have cell walls, reproduce asexually and sexually by spores, and exist in macroscopic or microscopic forms.

- Most fungi are free-living decomposers that are beneficial to biological communities; some may cause infections in animals and plants.
- Microscopic fungi include yeasts with spherical budding cells and molds with elongated filamentous hyphae in mycelia.
- Algae are aquatic photosynthetic protists with rigid cell walls and chloroplasts containing chlorophyll and other pigments.
- Protozoa are protists that feed by engulfing other cells, lack a cell wall, usually have some type of locomotor organelle, and may form dormant cysts.
- Subgroups of protozoa differ in their organelles of motility (flagella, cilia, pseudopods, nonmotile).
- Most protozoa are free-living aquatic cells that feed on bacteria and algae, and a few are animal parasites.
- The infective helminths are flatworms and roundworms that have greatly modified body organs so as to favor their parasitic lifestyle.

5.1 The History of Eukaryotes

Evidence from paleontology indicates that the first eukaryotic cells appeared on the earth approximately 2 billion years ago. Some fossilized cells that look remarkably like modern-day algae or protozoa appear in shale sediments from China, Russia, and Australia that date from 850 million to 950 million years ago **(figure 5.1).** Biologists have discovered convincing evidence to suggest that the eukaryotic cell evolved from prokaryotic organisms by a process of intracellular **symbiosis** (sim-beye-oh′-sis) **(Insight 5.1).** It now seems clear that some of the **organelles** that distinguish eukaryotic cells originated from prokaryotic cells that became trapped inside them. The structure of these first eukaryotic cells was so versatile that eukaryotic microorganisms soon spread out into available habitats and adopted greatly diverse styles of living.

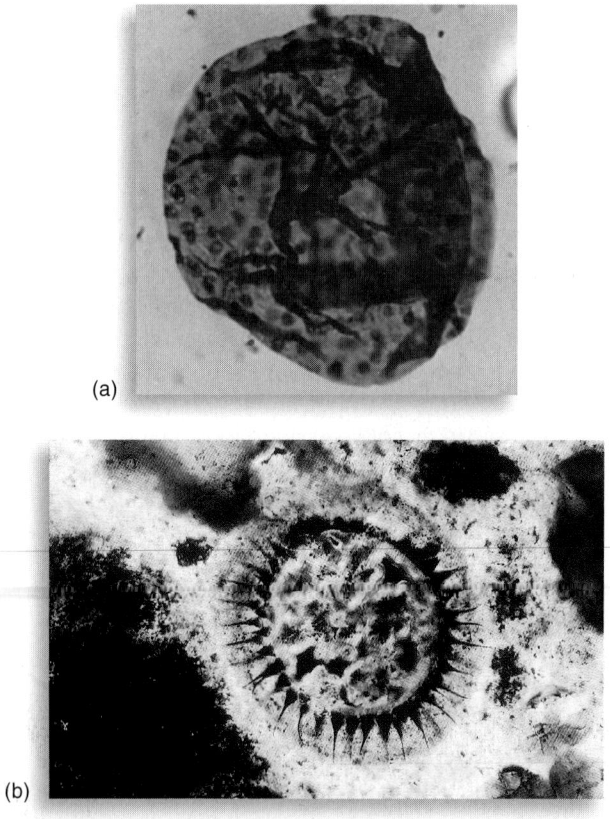

Figure 5.1 Ancient eukaryotic protists caught up in fossilized rocks.

(a) An alga-like cell found in Siberian shale deposits and dated from 850 million to 950 million years ago. **(b)** A large, disclike cell bearing a crown of spines is from Chinese rock dated 590 million to 610 million years ago.

The first primitive eukaryotes were probably single-celled and independent, but, over time, some forms began to aggregate, forming colonies. With further evolution, some of the cells within colonies became *specialized*, or adapted to perform a particular function advantageous to the whole colony, such as locomotion, feeding, or reproduction. Complex multicellular organisms evolved as individual cells in the organism lost the ability to survive apart from the intact colony. Although a multicellular organism is composed of many cells, it is more than just a disorganized assemblage of cells like a colony. Rather, it is composed of distinct groups of cells that cannot exist independently of the rest of the body. The cell groupings of multicellular organisms that have a specific function are termed *tissues*, and groups of tissues make up *organs*.

Looking at modern eukaryotic organisms, we find examples of many levels of cellular complexity **(table 5.1).** All protozoa, as well as numerous algae and fungi, are unicellular. Truly multicellular organisms are found only among plants and animals and some of the fungi (mushrooms) and algae (seaweeds). Only certain eukaryotes are

TABLE 5.1 Eukaryotic Organisms Studied in Microbiology

Always Unicellular	May Be Unicellular or Multicellular	Always Multicellular
Protozoa	Fungi Algae	Helminths (have unicellular egg or larval forms)

INSIGHT 5.1 — Historical

The Extraordinary Emergence of Eukaryotic Cells

For years, biologists have grappled with the problem of how a cell as complex as the eukaryotic cell originated. The explanation seems to be **endosymbiosis,** which suggests that eukaryotic cells arose when a much larger prokaryotic cell engulfed smaller bacterial cells that began to live and reproduce inside the prokaryotic cell rather than being destroyed. As the smaller cells took up permanent residence, they came to perform specialized functions for the larger cell, such as food synthesis and oxygen utilization, that enhanced the cell's versatility and survival. Over time, the engulfed bacteria gave up their ability to live independently and transferred some of their genes to the host cell.

The biologist responsible for early consideration of the theory of endosymbiosis is Dr. Lynn Margulis. Using molecular techniques, she accumulated convincing evidence of the relationships between the organelles of modern eukaryotic cells and the structure of bacteria. In many ways, the mitochondrion of eukaryotic cells is something like a tiny cell within a cell. It is capable of independent division, contains a circular chromosome that has bacterial DNA sequences, and has ribosomes that are clearly prokaryotic. Mitochondria also have bacterial membranes and can be inhibited by drugs that affect only bacteria.

Chloroplasts likely arose when endosymbiotic cyanobacteria provided their host cells with a built-in feeding mechanism. Margulis also found convincing evidence that eukaryotic cilia and flagella are the consequence of endosymbiosis between spiral bacteria and the cell membrane of early eukaryotic cells.

As molecular techniques improve, more evidence accumulates for the endosymbiont "theory," which is now widely accepted among evolutionary scientists.

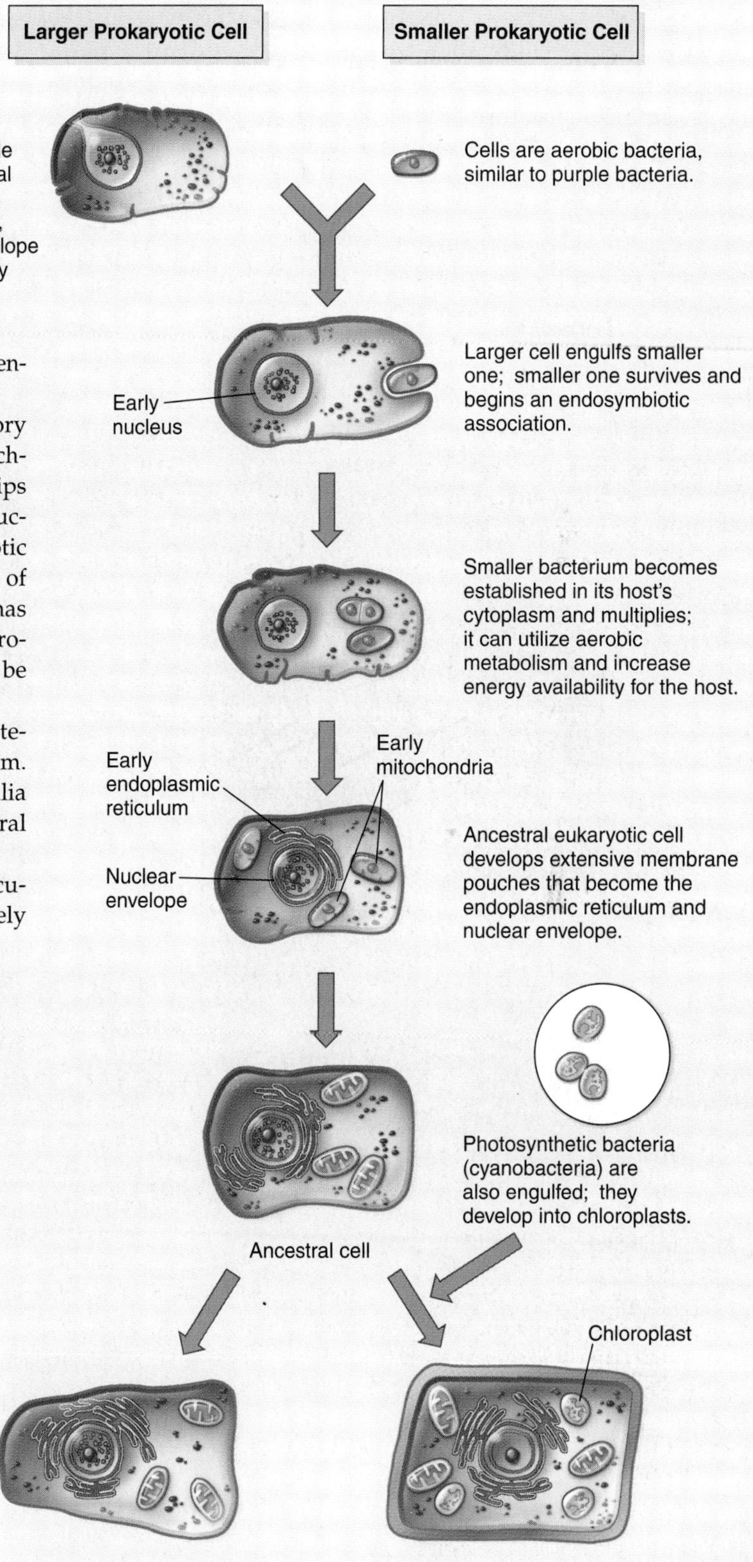
Larger Prokaryotic Cell
Smaller Prokaryotic Cell
Cell would have flexible membrane and internal extensions that could surround the nucleoid, forming a simple envelope that becomes the early nucleus.
Cells are aerobic bacteria, similar to purple bacteria.
Early nucleus
Larger cell engulfs smaller one; smaller one survives and begins an endosymbiotic association.
Smaller bacterium becomes established in its host's cytoplasm and multiplies; it can utilize aerobic metabolism and increase energy availability for the host.
Early endoplasmic reticulum
Early mitochondria
Nuclear envelope
Ancestral eukaryotic cell develops extensive membrane pouches that become the endoplasmic reticulum and nuclear envelope.
Photosynthetic bacteria (cyanobacteria) are also engulfed; they develop into chloroplasts.
Ancestral cell
Chloroplast
Protozoa, fungi, animals
Algae, higher plants

Dr. Lynn Margulis

traditionally studied by microbiologists—primarily the protozoa, the microscopic algae and fungi, and animal parasites, or helminths.

5.2 Form and Function of the Eukaryotic Cell: External Structures

The cells of eukaryotic organisms are so varied that no one member can serve as a typical example. **Figure 5.2** presents the generalized structure of typical algal, fungal, and protozoan cells. The following outline shows the organization of a eukaryotic cell. Compare this outline to the one found on page 87 in chapter 4.

In general, eukaryotic microbial cells have a cytoplasmic membrane, nucleus, mitochondria, endoplasmic reticulum, Golgi apparatus, vacuoles, cytoskeleton, and glycocalyx. A cell wall, locomotor appendages, and chloroplasts are found only in some groups. In the following sections, we cover the microscopic structure and functions of the eukaryotic cell. As with the prokaryotes, we begin on the outside and proceed inward through the cell.

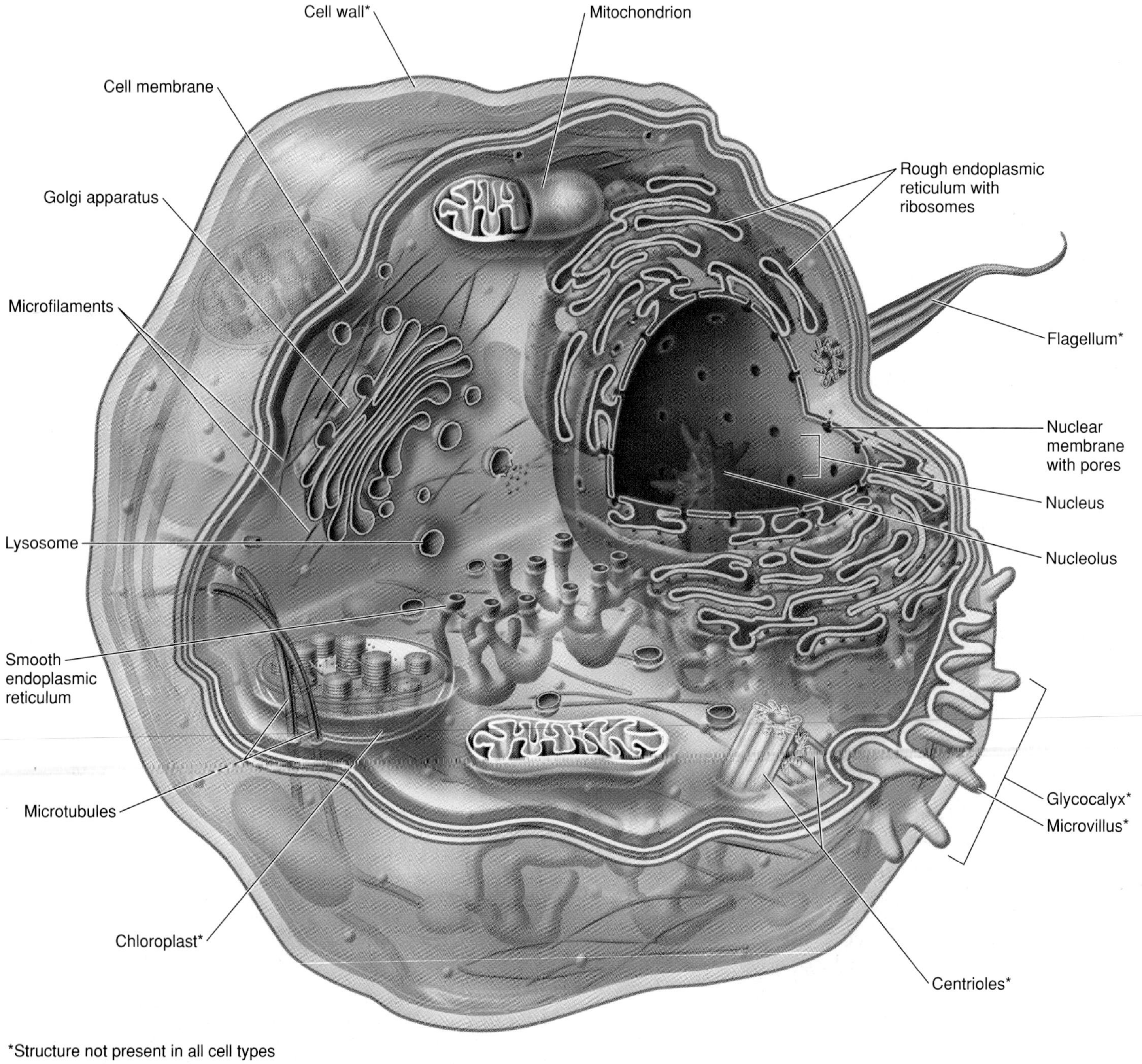

Figure 5.2 Structure of a eukaryotic cell.

Structure Flowchart

- Eukaryotic cell
  - External
    - Appendages
      - Flagella
      - Cilia
    - Glycocalyx
      - Capsules
      - Slimes
  - Boundary of cell
    - Cell wall
    - Cytoplasmic membrane
  - Internal
    - Cytoplasm
    - Nucleus
      - Nuclear envelope
      - Nucleolus
      - Chromosomes
    - Organelles
      - Endoplasmic reticulum — Ribosomes
      - Golgi apparatus — Lysosomes
      - Mitochondria
      - Chloroplasts
    - Ribosomes
    - Cytoskeleton
      - Microtubules
      - Microfilaments

Locomotor Appendages: Cilia and Flagella

Motility allows a microorganism to locate life-sustaining nutrients and to migrate toward positive stimuli such as sunlight; it also permits avoidance of harmful substances and stimuli. Locomotion by means of flagella or cilia is common in protozoa, many algae, and a few fungal and animal cells.

Although they share the same name, eukaryotic flagella are much different from those of prokaryotes. The eukaryotic flagellum is thicker (by a factor of 10), structurally more complex, and covered by an extension of the cell membrane. A single flagellum is a long, sheathed cylinder containing regularly spaced hollow tubules—microtubules—that extend along its entire length **(figure 5.3*a*).** A cross section reveals nine pairs of closely attached microtubules surrounding a single central pair. This scheme, called the 9 + 2 arrangement, is a universal pattern of flagella and cilia **(figure 5.3*b*).** During locomotion, the adjacent microtubules slide past each other, whipping the flagellum back and forth. Although details of this process are too complex to discuss here, it involves expenditure of energy and a coordinating mechanism in the cell membrane. The placement and number of flagella can be useful in identifying flagellated protozoa and certain algae.

Cilia are very similar in overall architecture to flagella, but they are shorter and more numerous (some cells have several thousand). They are found only on a single group of protozoa and certain animal cells. In the ciliated protozoa, the cilia occur in rows over the cell surface, where they beat back and forth in regular oarlike strokes (see **figure 5.4**). Such protozoa are among the fastest of all motile cells. The fastest ciliated protozoan can swim up to 2,500 microns per second—a meter and a half per minute! On some cells, cilia also function as feeding and filtering structures.

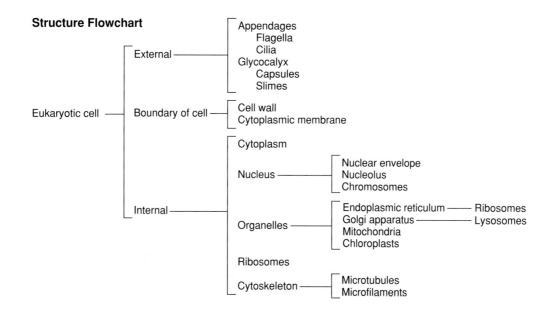

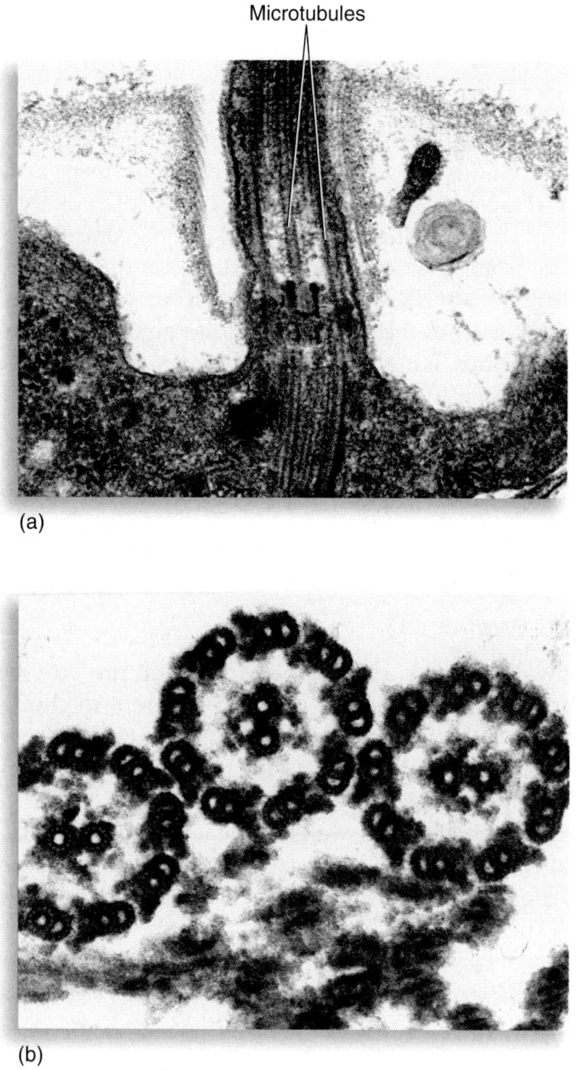

Figure 5.3 Microtubules in flagella.
(a) Longitudinal section through a flagellum, showing microtubules. **(b)** A cross section that reveals the typical 9 + 2 arrangement found in both flagella and cilia.

The Glycocalyx

Most eukaryotic cells have a **glycocalyx,** an outermost boundary that comes into direct contact with the

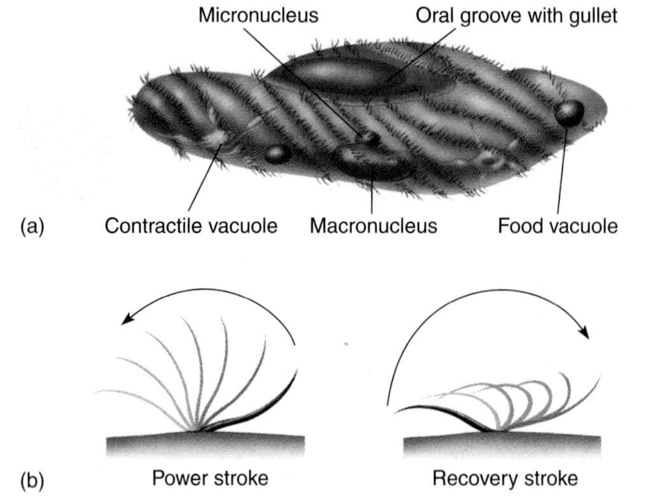

Figure 5.4 Structure and locomotion in ciliates. **(a)** The structure of a typical representative, *Paramecium.* **(b)** Cilia beat in coordinated waves, driving the cell forward and backward. View of a single cilium shows that it has a pattern of movement like a swimmer, with a power forward stroke and a repositioning stroke.

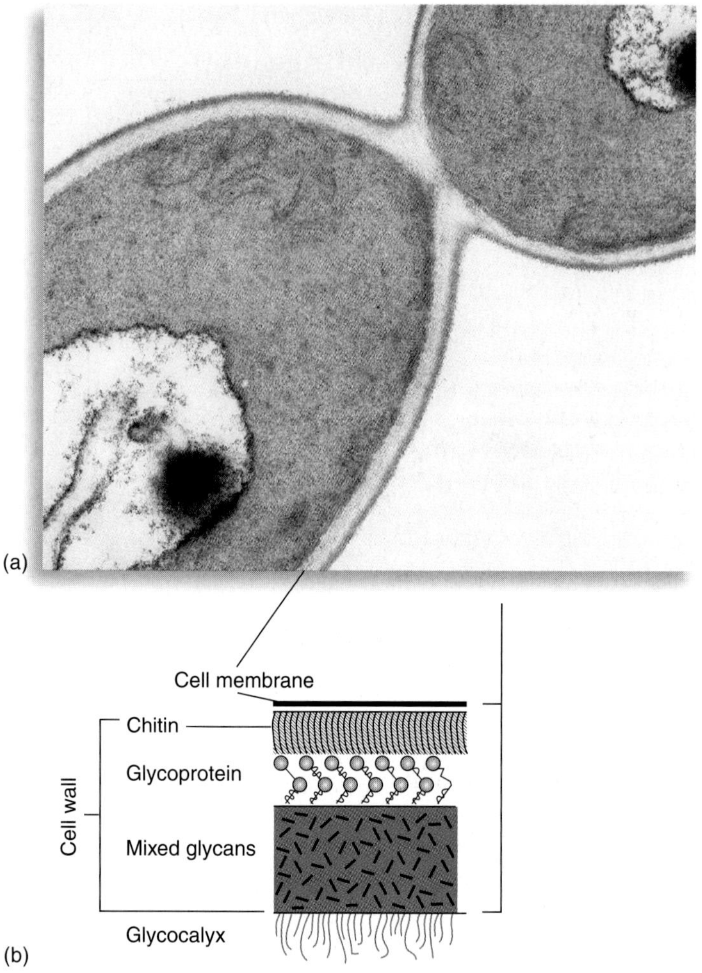

Figure 5.5 Cross-sectional views of fungal cell walls.

environment (see figure 5.2). This structure is usually composed of polysaccharides and appears as a network of fibers, a slime layer, or a capsule much like the glycocalyx of prokaryotes. Because of its positioning, the glycocalyx contributes to protection, adherence of cells to surfaces, and reception of signals from other cells and from the environment. The nature of the layer beneath the glycocalyx varies among the several eukaryotic groups. Fungi and most algae have a thick, rigid cell wall surrounding a cell membrane, whereas protozoa, a few algae, and all animal cells lack a cell wall and have only a cell membrane.

Form and Function of the Eukaryotic Cell: Boundary Structures

The Cell Wall

The cell walls of fungi and algae are rigid and provide structural support and shape, but they are different in chemical composition from prokaryotic cell walls. Fungal cell walls have a thick, inner layer of polysaccharide fibers composed of chitin or cellulose and a thin outer layer of mixed glycans **(figure 5.5).** The cell walls of algae are quite varied in chemical composition. Substances commonly found among various algal groups are cellulose, pectin,[1] mannans,[2] and minerals such as silicon dioxide and calcium carbonate.

The Cytoplasmic Membrane

The cytoplasmic (cell) membrane of eukaryotic cells is a typical bilayer of phospholipids in which protein molecules are embedded. In addition to phospholipids, eukaryotic membranes also contain *sterols* of various kinds. Sterols are different from phospholipids in both structure and behavior, as you may recall from chapter 2. Their relative rigidity confers stability on eukaryotic membranes. This strengthening feature is extremely important in cells that lack a cell wall. Cytoplasmic membranes of eukaryotes are functionally similar to those of prokaryotes, serving as selectively permeable barriers. Membranes have extremely sophisticated mechanisms for transporting nutrients *in* and waste and other products *out*. You'll read about these transport systems in prokaryotic membranes in chapter 7, but the systems in prokaryotes and eukaryotes are very similar.

CHECKPOINT

- Eukaryotes are cells with a nucleus and organelles compartmentalized by membranes. They might have originated from prokaryote ancestors about 2 billion years ago. Eukaryotic cell structure enabled eukaryotes to diversify from single cells into a huge variety of complex multicellular forms.

1. A polysaccharide composed of galacturonic acid subunits.
2. A polymer of the sugar known as mannose.

- The cell structures common to most eukaryotes are the cell membrane, nucleus, vacuoles, mitochondria, endoplasmic reticulum, Golgi apparatus, and a cytoskeleton. Cell walls, chloroplasts, and locomotor organs are present in some eukaryote groups.
- Microscopic eukaryotes use locomotor organs such as flagella or cilia for moving themselves or their food.
- The glycocalyx is the outermost boundary of most eukaryotic cells. Its functions are protection, adherence, and reception of chemical signals from the environment or from other organisms. The glycocalyx is supported by either a cell wall or a cell membrane.
- The cytoplasmic (cell) membrane of eukaryotes is similar in function to that of prokaryotes, but it differs in composition, possessing sterols as additional stabilizing agents.

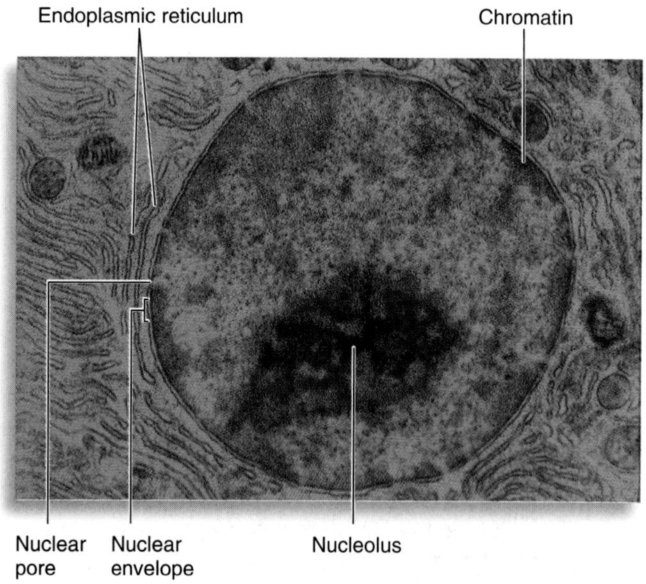

Figure 5.6 The nucleus.
Electron micrograph section of an interphase nucleus, showing its most prominent features.

5.3 Form and Function of the Eukaryotic Cell: Internal Structures

Unlike prokaryotes, eukaryotic cells contain a number of individual membrane-bound organelles that are extensive enough to account for 60% to 80% of their volume.

The Nucleus: The Control Center

The nucleus is a compact sphere that is the most prominent organelle of eukaryotic cells. It is separated from the cell cytoplasm by an external boundary called a nuclear envelope. The envelope has a unique architecture. It is composed of two parallel membranes separated by a narrow space, and it is perforated with small, regularly spaced openings, or pores, formed at sites where the two membranes unite **(figure 5.6).** The nuclear pores are passageways through which macromolecules migrate from the nucleus to the cytoplasm and vice versa. The nucleus contains a matrix called the nucleoplasm and a granular mass, the **nucleolus,** that can stain more intensely than the immediate surroundings because of its RNA content. The nucleolus is the site for ribosomal RNA synthesis and a collection area for ribosomal subunits. The subunits are transported through the nuclear pores into the cytoplasm for final assembly into ribosomes.

A prominent feature of the nucleoplasm in stained preparations is a network of dark fibers known as **chromatin.** Analysis has shown that chromatin actually comprises the eukaryotic **chromosomes,** large units of genetic information in the cell. The chromosomes in the nucleus of most cells are not readily visible because they are long, linear DNA molecules bound in varying degrees to **histone** proteins, and they are far too fine to be resolved as distinct structures without extremely high magnification. During **mitosis,** however, when the duplicated chromosomes are separated equally into daughter cells, the chromosomes themselves become readily visible as discrete bodies **(figure 5.7).** This happens when the DNA becomes highly condensed by forming coils and supercoils around the histones to prevent the chromosomes from tangling as they are separated into new cells. This process is described in more detail in chapter 9.

The nucleus, as you've just seen, contains instructions in the form of DNA. Elaborate processes have evolved for transcription and duplication of this genetic material. In addition to mitosis, some cells also undergo **meiosis,** the process by which sex cells are created. Much of the protein synthesis and other work of the cell takes place outside the nucleus in the cell's other organelles.

Endoplasmic Reticulum: A Passageway in the Cell

The **endoplasmic reticulum (ER)** is a microscopic series of tunnels used in transport and storage. Two kinds of endoplasmic reticulum are the **rough endoplasmic reticulum (RER) (figure 5.8)** and the **smooth endoplasmic reticulum (SER).** Electron micrographs show that the RER originates from the outer membrane of the nuclear envelope and extends in a continuous network through the cytoplasm, even out to the cell membrane. This architecture permits the spaces in the RER, or cisternae, to transport materials from the nucleus to the cytoplasm and ultimately to the cell's exterior. The RER appears rough because of large numbers of ribosomes partly attached to its membrane surface. Proteins synthesized on the ribosomes are shunted into the cavity of the reticulum and held there for later packaging and transport. In contrast to the RER, the SER is a closed tubular network

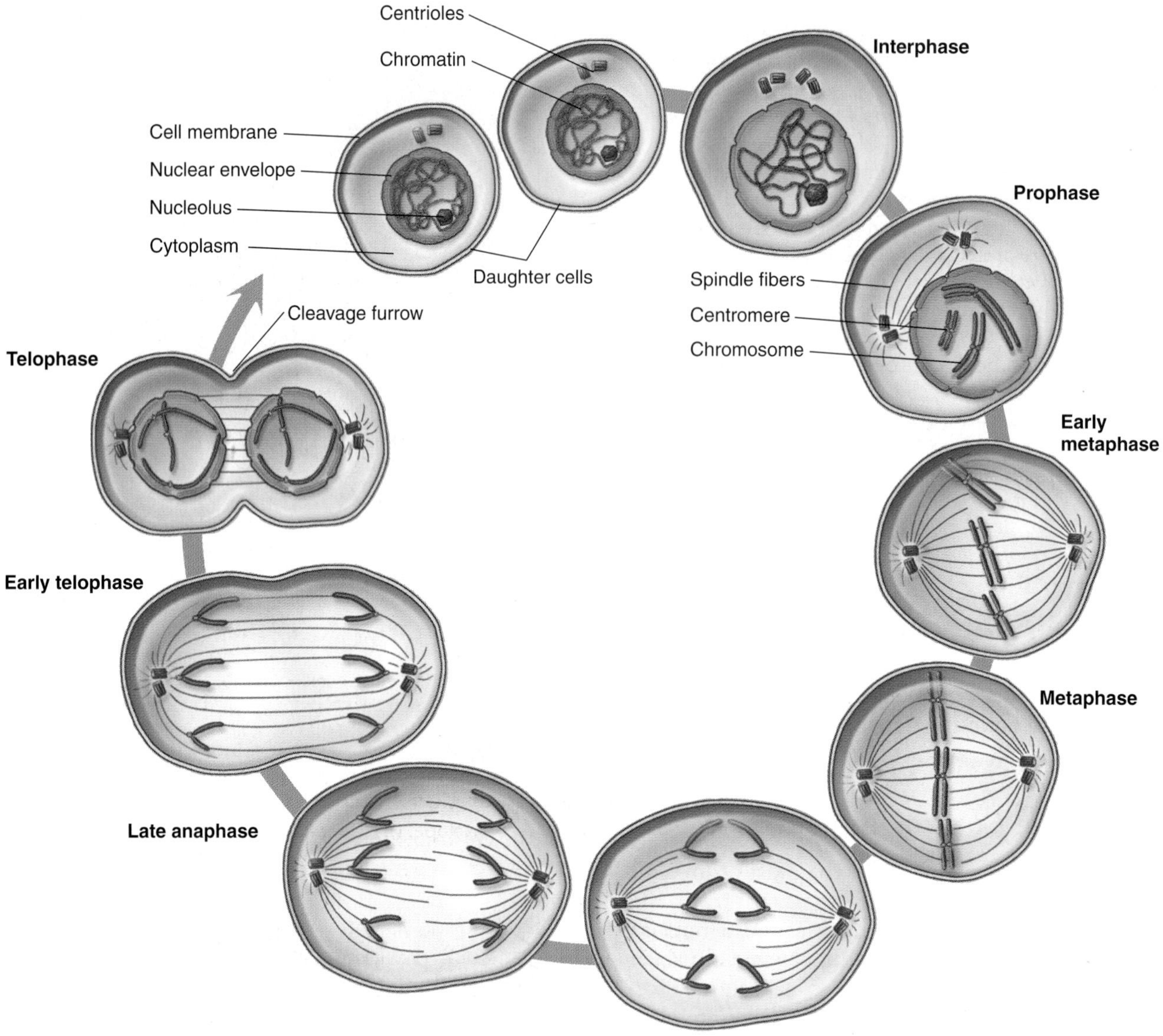

(a)

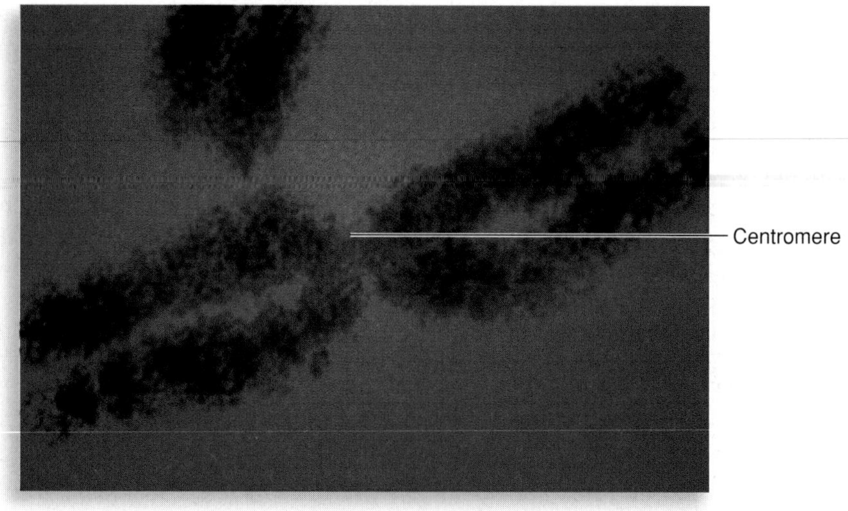

(b)

Figure 5.7 Changes in the cell and nucleus that accompany mitosis in a eukaryotic cell such as a yeast.
(a) Before mitosis (at interphase), chromosomes are visible only as chromatin. As mitosis proceeds (early prophase), chromosomes take on a fine, threadlike appearance as they condense, and the nuclear membrane and nucleolus are temporarily disrupted. **(b)** By metaphase, the chromosomes are fully visible as X-shaped structures. The shape is due to duplicated chromosomes attached at a central point, the centromere. Spindle fibers attach to these and facilitate the separation of individual chromosomes during metaphase. Later phases serve in the completion of chromosomal separation and division of the cell proper into daughter cells.

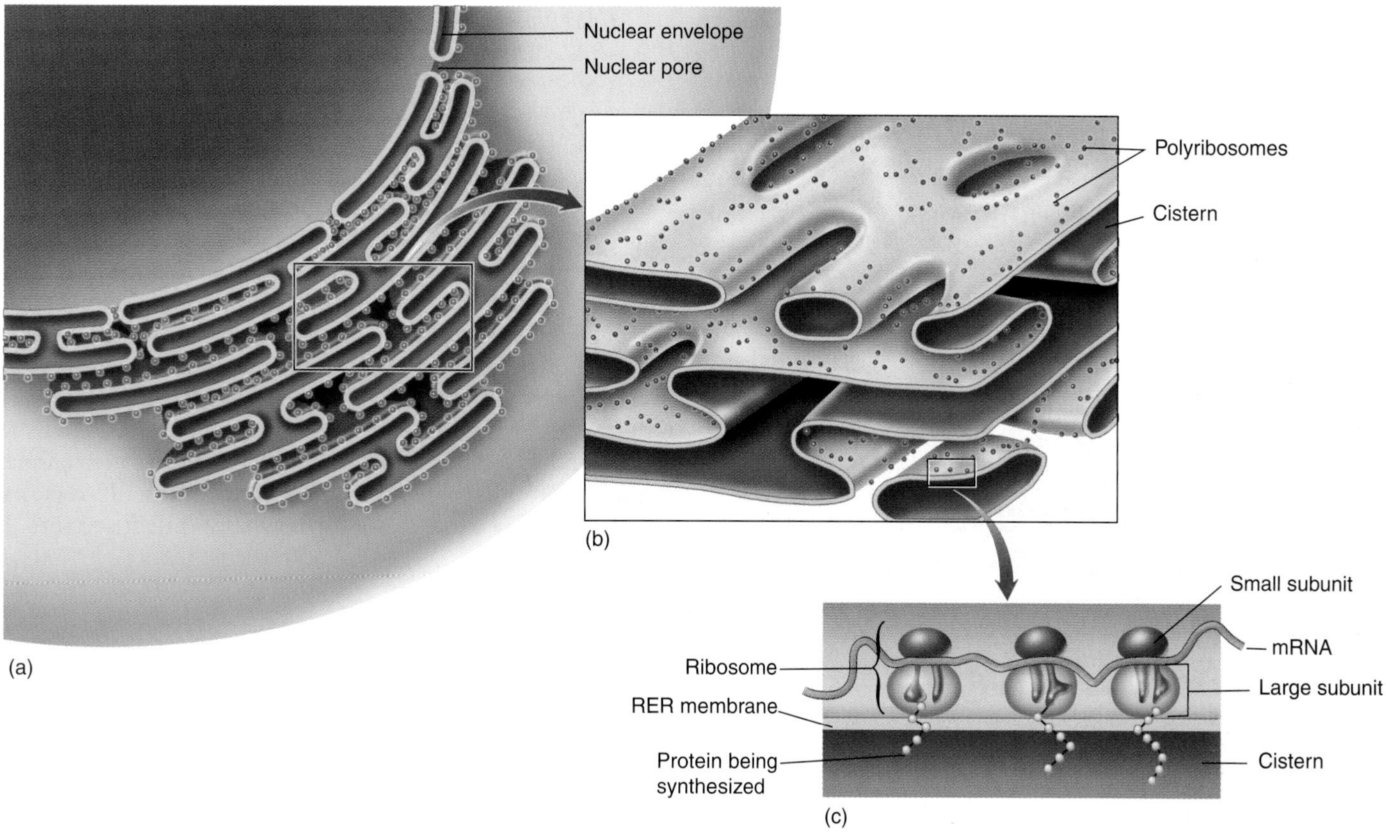

Figure 5.8 The origin and detailed structure of the rough endoplasmic reticulum (RER).
(a) Schematic view of the origin of the RER from the outer membrane of the nuclear envelope. **(b)** Three-dimensional projection of the RER. **(c)** Detail of the orientation of a ribosome on the RER membrane.

without ribosomes that functions in nutrient processing and in synthesis and storage of nonprotein macromolecules such as lipids.

Golgi Apparatus: A Packaging Machine

The **Golgi[3] apparatus,** also called the Golgi complex or body, is the site in the cell in which proteins are modified and then sent to their final destinations. It is a discrete organelle consisting of a stack of several flattened, disc-shaped sacs called cisternae. These sacs have outer limiting membranes and cavities like those of the endoplasmic reticulum, but they do not form a continuous network **(figure 5.9).** This organelle is always closely associated with the endoplasmic reticulum both in its location and function. At a site where it meets the Golgi apparatus, the endoplasmic reticulum buds off tiny membrane-bound packets of protein called *transitional vesicles* that are picked up by the forming face of the Golgi apparatus. Once in the complex itself, the proteins are often modified by the addition of polysaccharides and lipids. The final action of this apparatus is to pinch off finished *condensing vesicles* that will be conveyed to organelles

3. Named for C. Golgi, an Italian histologist who first described the apparatus in 1898.

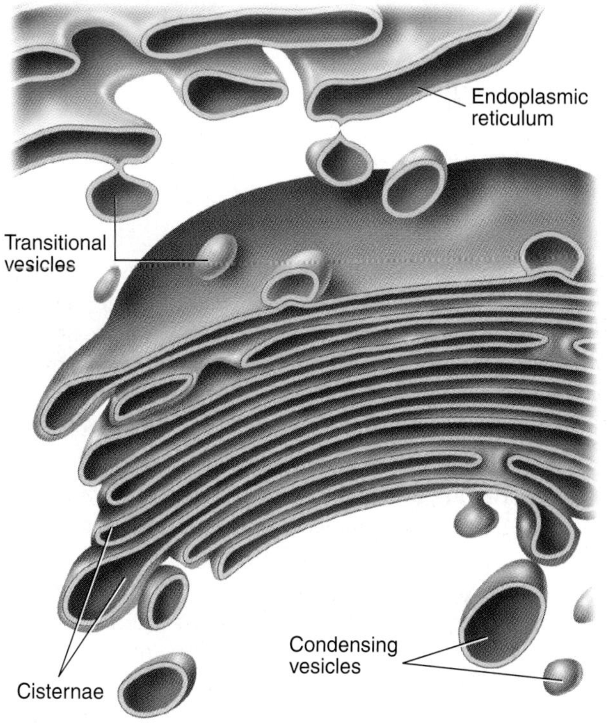

Figure 5.9 Detail of the Golgi apparatus.
The flattened layers are cisternae. Vesicles enter the upper surface and leave the lower surface.

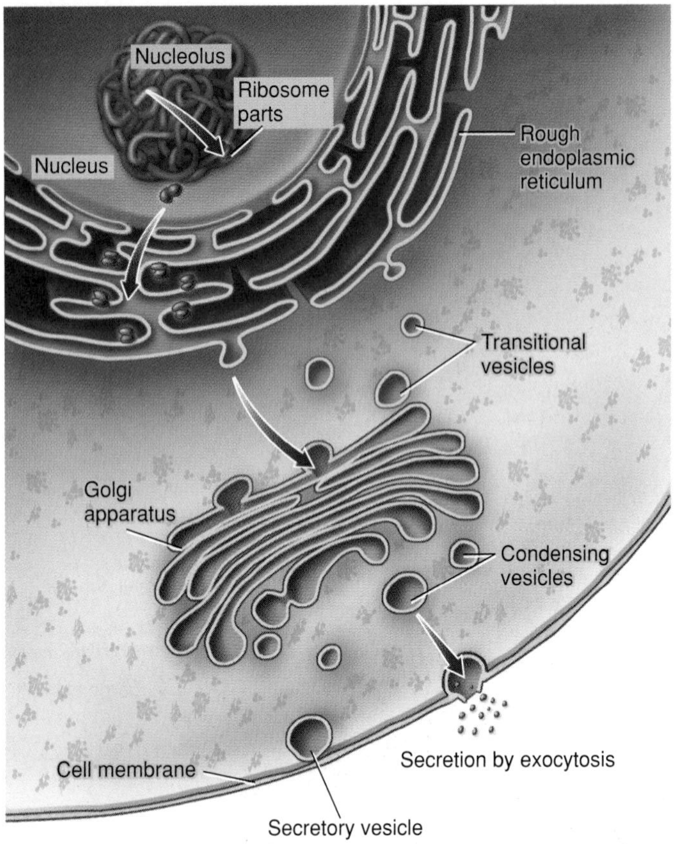

Figure 5.10 The transport process.
The cooperation of organelles in protein synthesis and transport: nucleus → RER → Golgi apparatus → vesicles → secretion.

such as lysosomes or transported outside the cell as secretory vesicles **(figure 5.10).**

Nucleus, Endoplasmic Reticulum, and Golgi Apparatus: Nature's Assembly Line

As the keeper of the eukaryotic genetic code, the nucleus ultimately governs and regulates all cell activities. But, because the nucleus remains fixed in a specific cellular site, it must direct these activities through a structural and chemical network (figure 5.10). This network includes ribosomes, which originate in the nucleus, and the rough endoplasmic reticulum, which is continuously connected with the nuclear envelope. Initially, a segment of the genetic code of DNA containing the instructions for producing a protein is copied into RNA and passed out through the nuclear pores directly to the ribosomes on the endoplasmic reticulum. Here, specific proteins are synthesized from the RNA code and deposited in the lumen (space) of the endoplasmic reticulum. After being transported to the Golgi apparatus, the protein products are chemically modified and packaged into vesicles that can be used by the cell in a variety of ways. Some of the vesicles contain enzymes to digest food inside the cell; other vesicles are secreted to digest materials outside the cell, and yet others are important in the enlargement and repair of the cell wall and membrane.

A **lysosome** is one type of vesicle originating from the Golgi apparatus that contains a variety of enzymes. Lysosomes are involved in intracellular digestion of food particles and in protection against invading microorganisms. They also participate in digestion and removal of cell debris in damaged tissue. Other types of vesicles include **vacuoles** (vak′-yoo-ohl), which are membrane-bound sacs containing fluids or solid particles to be digested, excreted, or stored. They are formed in phagocytic cells (certain white blood cells and protozoa) in response to food and other substances that have been engulfed. The contents of a food vacuole are digested through the merger of the vacuole with a lysosome. This merged structure is called a phagosome **(figure 5.11).** Other types of vacuoles are used in storing reserve food such as fats and glycogen. Protozoa living in freshwater habitats regulate osmotic pressure by means of contractile vacuoles, which regularly expel excess water that has diffused into the cell (described later).

Mitochondria: Energy Generators of the Cell

Although the nucleus is the cell's control center, none of the cellular activities it commands could proceed without a constant supply of energy, the bulk of which is generated in most eukaryotes by **mitochondria** (my″-toh-kon′-dree-uh). When viewed with light microscopy, mitochondria appear as round or elongated particles scattered throughout the cytoplasm. The internal ultrastructure reveals that a single mitochondrion consists of a smooth, continuous outer membrane that forms the external contour, and an inner, folded membrane nestled neatly within the outer membrane **(figure 5.12*a*).** The folds on the inner membrane, called **cristae** (kris′-te), may be tubular, like fingers, or folded into shelflike bands.

The cristae membranes hold the enzymes and electron carriers of aerobic respiration. This is an oxygen-using process that extracts chemical energy contained in nutrient molecules and stores it in the form of high-energy molecules, or ATP. More detailed functions of mitochondria are covered in chapter 8. The spaces around the cristae are filled with a chemically complex fluid called the **matrix,** which holds ribosomes, DNA, and the pool of enzymes and other compounds involved in the metabolic cycle. Mitochondria (along with chloroplasts) are unique among organelles in that they divide independently of the cell, contain circular strands of DNA, and have prokaryotic-sized 70S ribosomes. These findings have prompted some intriguing speculations on their evolutionary origins (see Insight 5.1).

Chloroplasts: Photosynthesis Machines

Chloroplasts are remarkable organelles found in algae and plant cells that are capable of converting the energy of sunlight into chemical energy through photosynthesis. The photosynthetic role of chloroplasts makes them the primary producers of organic nutrients upon which all

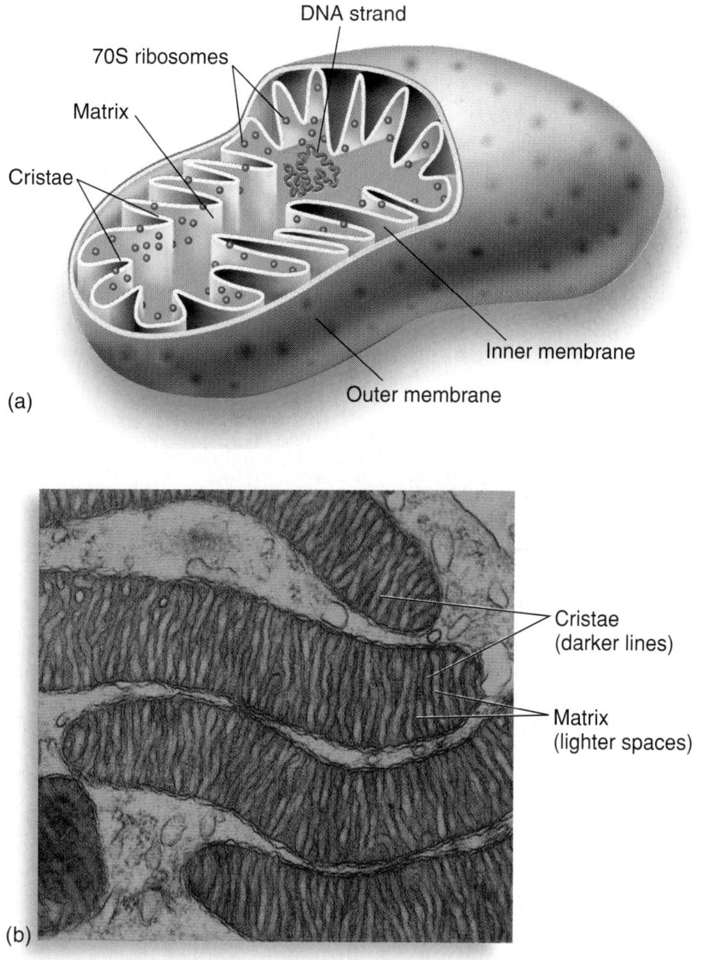

Figure 5.12 General structure of a mitochondrion. **(a)** A three-dimensional projection. **(b)** An electron micrograph. In most cells, mitochondria are elliptical or spherical, although in certain fungi, algae, and protozoa, they are long and filament-like.

other organisms (except certain bacteria) ultimately depend. Another important photosynthetic product of chloroplasts is oxygen gas. Although chloroplasts resemble mitochondria, chloroplasts are larger, contain special pigments, and are much more varied in shape.

There are differences among various algal chloroplasts, but most are generally composed of two membranes, one enclosing the other. The smooth, outer membrane completely covers an inner membrane folded into small, disclike sacs called **thylakoids** that are stacked upon one another into **grana.** These structures carry the green pigment chlorophyll and sometimes additional pigments as well. Surrounding the thylakoids is a ground substance called the **stroma (figure 5.13).** The role of the photosynthetic pigments is to absorb and transform solar energy into chemical energy, which is then used during reactions in the stroma to synthesize

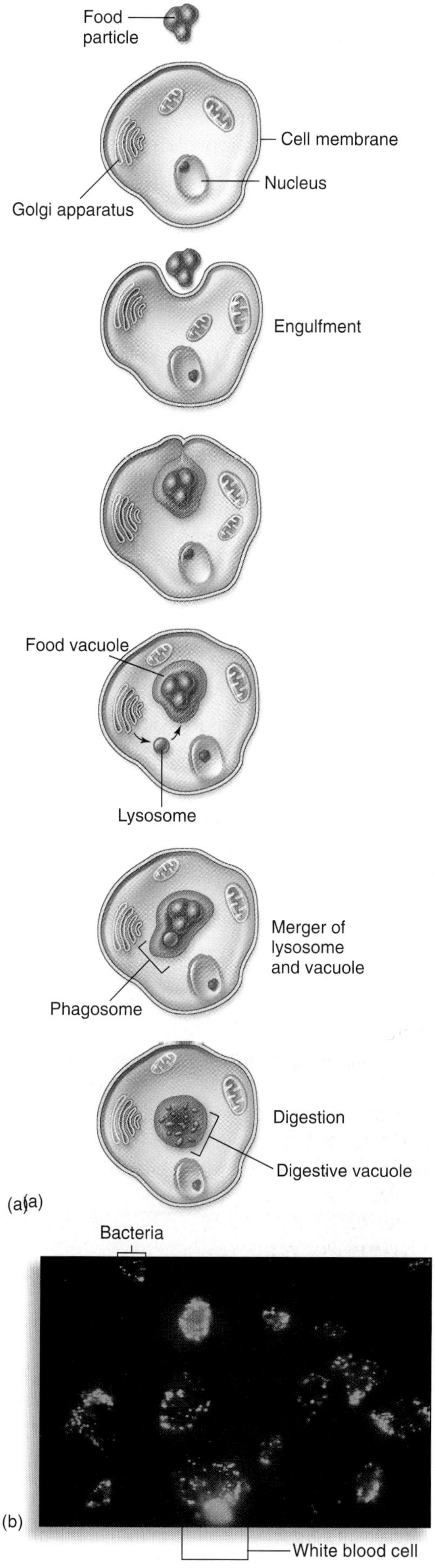

Figure 5.11 The origin and action of lysosomes in phagocytosis. **(a)** Schematic illustration. **(b)** Fluorescence micrograph of bacteria inside phagosomes of white blood cells.

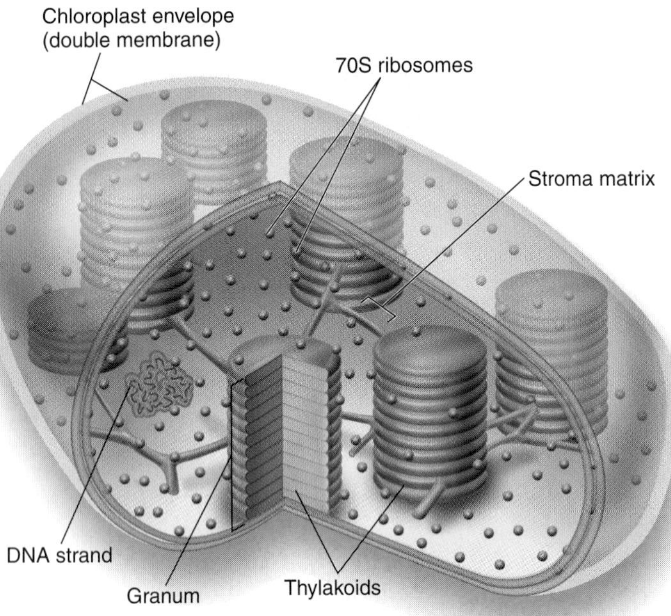

Figure 5.13 Detail of an algal chloroplast.

carbohydrates. We further explore some important aspects of photosynthesis in chapters 7 and 24.

Ribosomes: Protein Synthesizers

In an electron micrograph of a eukaryotic cell, ribosomes are numerous, tiny particles that give a "dotted" appearance to the cytoplasm. Ribosomes are distributed in two ways: Some are scattered freely in the cytoplasm and cytoskeleton; others are intimately associated with the rough endoplasmic reticulum as previously described. Multiple ribosomes are often found arranged in short chains called polyribosomes (polysomes). The basic structure of eukaryotic ribosomes is similar to that of prokaryotic ribosomes, described in chapter 4. Both are composed of large and small subunits of ribonucleoprotein (see figure 5.8). By contrast, however, the eukaryotic ribosome (except in the mitochondrion) is the larger 80S variety that is a combination of 60S and 40S subunits. As in the prokaryotes, eukaryotic ribosomes are the staging areas for protein synthesis.

The Cytoskeleton: A Support Network

The cytoplasm of a eukaryotic cell is criss-crossed by a flexible framework of molecules called the cytoskeleton **(figure 5.14).** This framework appears to have several functions, such as anchoring organelles, moving RNA and vesicles, and permitting shape changes and movement in some cells. The two main types of cytoskeletal elements are *microfilaments* and *microtubules.* **Microfilaments** are thin protein strands that attach to the cell membrane and form a network through the cytoplasm. Some microfilaments are responsible for movements of the cytoplasm, often made evident by the streaming of organelles around the cell in a cyclic pattern. Other microfilaments are active in *amoeboid motion,* a type of movement typical of cells such as amoebas and phagocytes that produces extensions of the cell membrane (pseudopods) into which the cytoplasm flows. **Microtubules** are long, hollow tubes that maintain the shape of eukaryotic cells without walls and transport substances from one part of a cell to another. The spindle fibers that play an essential role in mitosis are actually microtubules that attach to chromosomes and separate them into daughter cells. As indicated earlier,

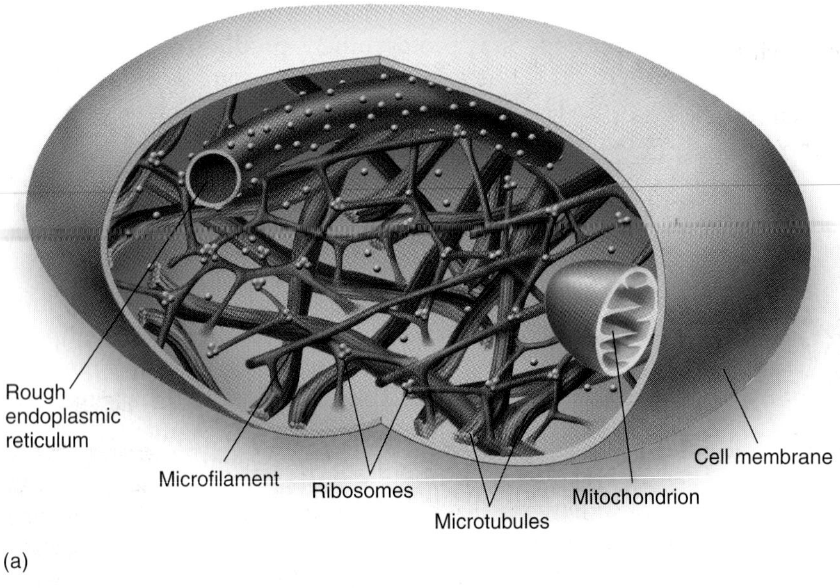

(a)

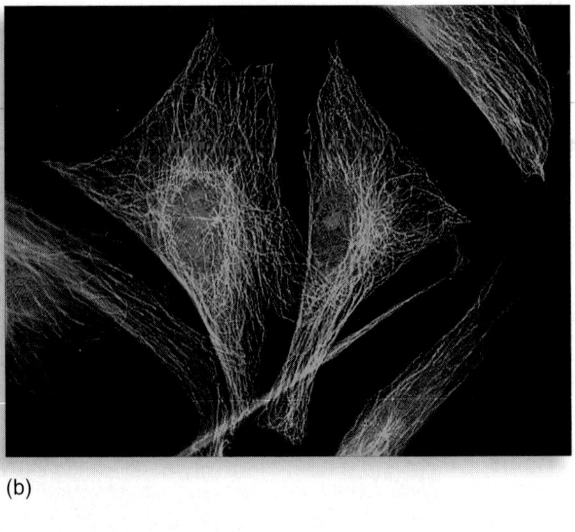

(b)

Figure 5.14 The cytoskeleton.
(a) Drawing of microtubules, microfilaments, and organelles. **(b)** Microtubules are dyed fluorescent green in this micrograph.

TABLE 5.2 A General Comparison of Prokaryotic and Eukaryotic Cells and Viruses*

Function or Structure	Characteristic	Procaryotic Cells	Eucaryotic Cells	Viruses**
Genetics	Nucleic acids	+	+	+
	Chromosomes	+	+	−
	True nucleus	−	+	−
	Nuclear envelope	−	+	−
Reproduction	Mitosis	−	+	−
	Production of sex cells	+/−	+	−
	Binary fission	+	+	−
Biosynthesis	Independent	+	+	−
	Golgi apparatus	−	+	−
	Endoplasmic reticulum	−	+	−
	Ribosomes	+***	+	−
Respiration	Enzymes	+	+	−
	Mitochondria	−	+	−
Photosynthesis	Pigments	+/−	+/−	−
	Chloroplasts	−	+/−	−
Motility/locomotor structures	Flagella	+/−***	+/−	−
	Cilia	−	+/−	−
Shape/protection	Membrane	+	+	+/−
	Cell wall	+***	+/−	− (have capsids instead)
	Capsule	+/−	+/−	−
Complexity of function		+	+	+/−
Size (in general)		0.5–3 μm****	2–100 μm	< 0.2 μm

*+ means most members of the group exhibit this characteristic; − means most lack it; +/− means some members have it and some do not.
**Viruses cannot participate in metabolic or genetic activity outside their host cells.
***The prokaryotic type is functionally similar to the eukaryotic, but structurally unique.
****Much smaller and much larger bacteria exist; see Insight 4.3.

microtubules are also responsible for the movement of cilia and flagella.

Table 5.2 summarizes the differences between eukaryotic and prokaryotic cells. Viruses (discussed in chapter 6) are included as well.

CHECKPOINT

- The genome of eukaryotes is located in the nucleus, a spherical structure surrounded by a double membrane. The nucleus contains the nucleolus, the site of ribosome synthesis. DNA is organized into chromosomes in the nucleus.
- The endoplasmic reticulum (ER) is an internal network of membranous passageways extending throughout the cell.
- The Golgi apparatus is a packaging center that receives materials from the ER and then forms vesicles around them for storage or for transport to the cell membrane for secretion.
- The mitochondria generate energy in the form of ATP to be used in numerous cellular activities.
- Chloroplasts, membranous packets found in plants and algae, are used in photosynthesis.
- Ribosomes are the sites for protein synthesis present in both eukaryotes and prokaryotes.
- The cytoskeleton maintains the shape of cells and produces movement of cytoplasm within the cell, movement of chromosomes at cell division, and, in some groups, movement of the cell as a unit.

Survey of Eukaryotic Microorganisms

With the general structure of the eukaryotic cell in mind, let us next examine the amazingly wide range of adaptations that this cell type has undergone. The following sections contain a general survey of the principal eukaryotic microorganisms—fungi, algae, protozoa, and parasitic worms—while also introducing elements of their structure, life history, classification, identification, and importance.

5.4 The Kingdom of the Fungi

The position of the **fungi** in the biological world has been debated for many years. Although they were originally classified with the green plants (along with algae and bacteria), they were later separated from plants and placed in a group with algae and protozoa (the Protista). Even at that time, however, many microbiologists were struck by several unique qualities of fungi that warranted their being placed into their own separate kingdom, and eventually they were.

The Kingdom Fungi, or Myceteae, is large and filled with forms of great variety and complexity. For practical purposes, the approximately 100,000 species of fungi can be divided into two groups: the *macroscopic fungi* (mushrooms, puffballs, gill fungi) and the *microscopic fungi* (molds, yeasts). Although the majority of fungi are either unicellular or colonial, a few complex forms such as mushrooms and puffballs are considered multicellular. Cells of the microscopic fungi exist in two basic morphological types: yeasts and hyphae. A yeast cell is distinguished by its round to oval shape and by its mode of asexual reproduction. It grows swellings on its surface called buds, which then become separate cells. **Hyphae** (hy′-fee) are long, threadlike cells found in the bodies of filamentous fungi, or molds **(figure 5.15).** Some species form a **pseudohypha,** a chain of yeasts formed when buds remain attached in a row **(figure 5.16).** Because of its manner of formation, it is not a true hypha like that of molds. While some fungal cells exist only in a yeast form and others occur primarily as hyphae, a few, called **dimorphic,** can take either form, depending upon growth conditions, such as changing temperature. This variability in growth form is particularly characteristic of some pathogenic molds.

Fungal Nutrition

All fungi are **heterotrophic**. They acquire nutrients from a wide variety of organic materials called **substrates (figure 5.17).** Most fungi are **saprobes,** meaning that they obtain these substrates from the remnants of dead plants and animals in soil or aquatic habitats. Fungi can also be **parasites** on the bodies of living animals or plants, although very few fungi absolutely require a living host. In general, the fungus penetrates the substrate and secretes enzymes that reduce it to small molecules that can be absorbed by the cells. Fungi have enzymes for digesting an incredible array of substances, including feathers, hair, cellulose, petroleum products, wood, and rubber. It has been said that every naturally occurring organic material on the earth can be attacked by some type of fungus. Fungi are often found in nutritionally poor or adverse environments. Various fungi thrive in substrates with high salt or sugar content, at relatively high temperatures, and even in snow and glaciers. Their medical and agricultural impact is extensive. A number of species cause **mycoses** (fungal infections) in animals, and

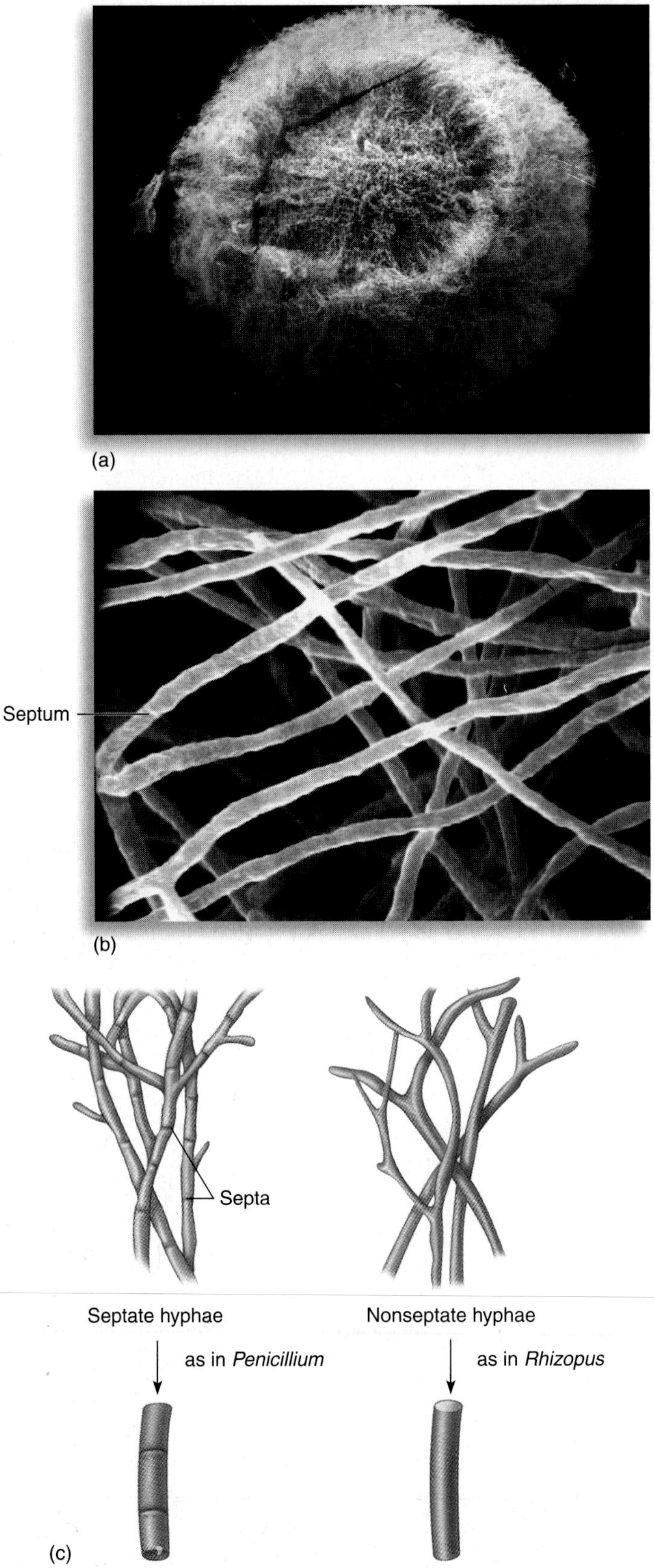

Figure 5.15 ***Diplodia maydis,*** **a pathogenic fungus of corn plants.**
(a) Scanning electron micrograph of a single colony showing its filamentous texture (24×). **(b)** Close-up of hyphal structure (1,200×). **(c)** Basic structural types of hyphae.

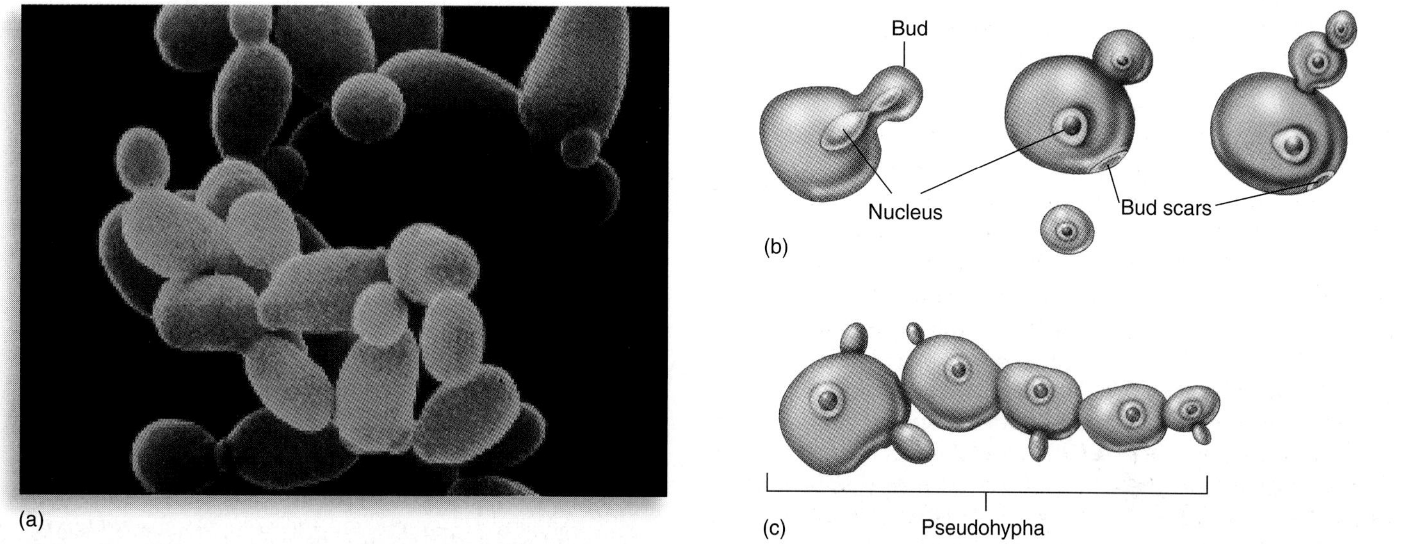

Figure 5.16 Microscopic morphology of yeasts.
(a) Scanning electron micrograph of the brewer's, or baker's, yeast *Saccharomyces cerevisiae* (21,000×). **(b)** Formation and release of yeast buds. **(c)** Formation of pseudohypha (a chain of budding yeast cells).

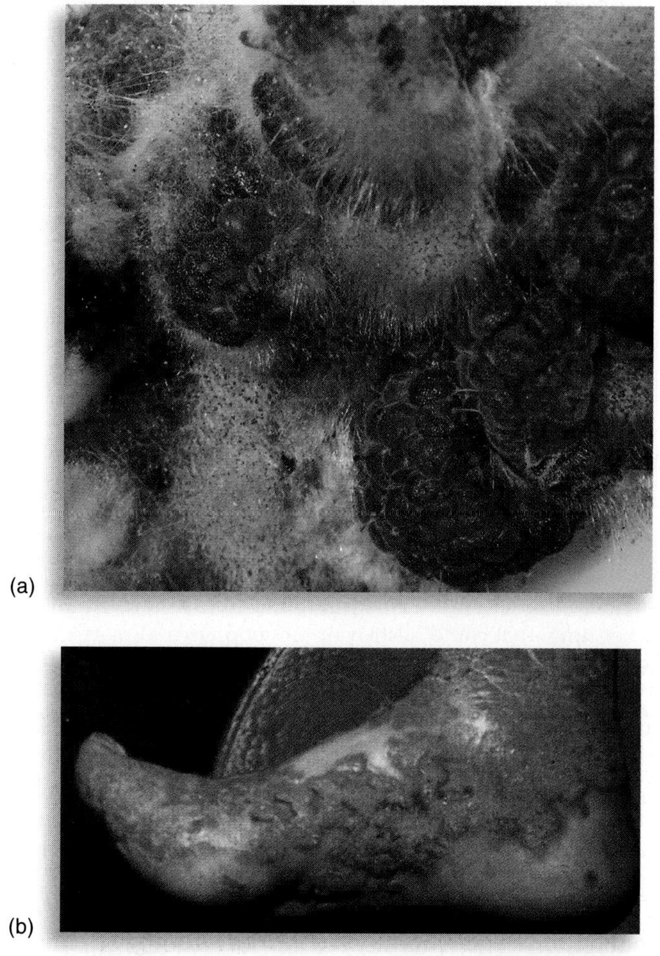

Figure 5.17 Nutritional sources (substrates) for fungi.
(a) A fungal mycelium growing on raspberries. The fine hyphal fillaments and black sporangia are typical of *Rhizopus*. **(b)** The skin of the foot infected by a soil fungus, *Fonsecaea pedrosoi*.

thousands of species are important plant pathogens. Fungal toxins may cause disease in humans, and airborne fungi are a frequent cause of allergies and other medical conditions **(Insight 5.2).**

Organization of Microscopic Fungi

The cells of most microscopic fungi grow in loose associations or colonies. The colonies of yeasts are much like those of bacteria in that they have a soft, uniform texture and appearance. The colonies of filamentous fungi are noted for the striking cottony, hairy, or velvety textures that arise from their microscopic organization and morphology. The woven, intertwining mass of hyphae that makes up the body or colony of a mold is called a **mycelium.**

Although hyphae contain the usual eukaryotic organelles, they also have some unique organizational features. In most fungi, the hyphae are divided into segments by cross walls, or **septa,** a condition called septate (figure 5.15*c*). The nature of the septa varies from solid partitions with no communication between the compartments to partial walls with small pores that allow the flow of organelles and nutrients between adjacent compartments. Nonseptate hyphae consist of one long, continuous cell *not* divided into individual compartments by cross walls. With this construction, the cytoplasm and organelles move freely from one region to another, and each hyphal element can have several nuclei.

Hyphae can also be classified according to their particular function. Vegetative hyphae (mycelia) are responsible for the visible mass of growth that appears on the surface of a substrate and penetrates it to digest and absorb nutrients. During the development of a fungal colony, the vegetative hyphae give rise to structures called reproductive, or fertile,

INSIGHT 5.2 Discovery

The Many Faces of Fungi

Fungi, Fungi, Everywhere

The importance of fungi in the ecological structure of the earth is well founded. They are essential contributors to complex environments such as soil, and they play numerous beneficial roles as decomposers of organic debris and as partners to plants. Fungi also have great practical importance due to their metabolic versatility. They are productive sources of drugs (penicillin) to treat human infections and other diseases, and they are used in industry to ferment foods and synthesize organic chemicals.

The fact that they are so widespread also means that they frequently share human living quarters, especially in locations that provide ample moisture and nutrients. Often their presence is harmless and limited to a film of mildew on shower stalls or other moist environments. In some cases, depending on the amount of contamination and the type of mold, these indoor fungi can also give rise to various medical problems. Such common air contaminants as *Penicillium*, *Aspergillus*, *Cladosporium*, and *Stachybotrys* all have the capacity to give off airborne spores and toxins that, when inhaled, cause a whole spectrum of symptoms sometimes referred to as "sick building syndrome." The usual source of harmful fungi is the presence of chronically water-damaged walls, ceilings, and other building materials that have come to harbor these fungi. People exposed to these houses or buildings report symptoms that range from skin rash, flulike reactions, sore throat, and headaches to fatigue, diarrhea, allergies, and immune suppression. Recent reports of sick buildings have been on the rise, affecting thousands of people, and some deaths have been reported in small children. The control of indoor fungi requires correcting the moisture problem, removing the contaminated materials, and decontaminating the living spaces. Mycologists are currently studying the mechanisms of toxic effects with an aim to develop better diagnosis and treatment.

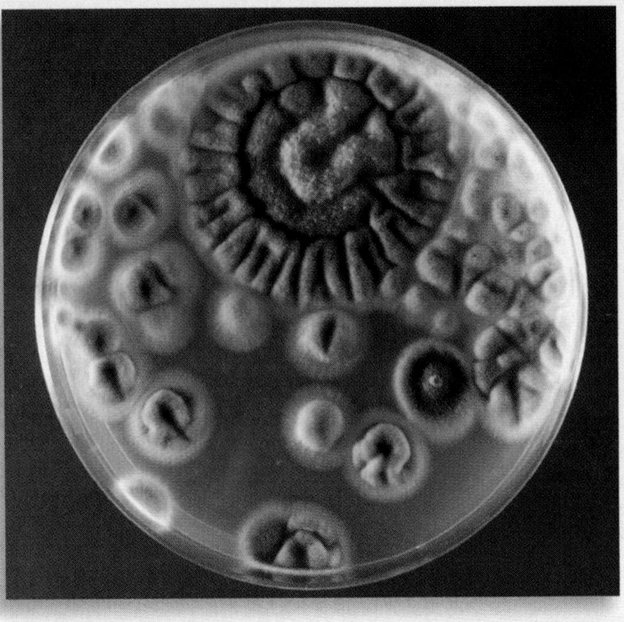

The penicillin-producing fungus *Penicillium*. Macroscopic view of a typical blue-green colony.

A microscopic view of *Penicillium* shows the brush arrangement of conidia (220×).

A Fungus in Your Future

Biologists are developing some rather imaginative uses for fungi as a way of controlling both the life and death of plants. Dutch and Canadian researchers studying ways to control a devastating fungus infection of elm trees (Dutch elm disease) have come up with a brand new use of an old method—they actually vaccinate the trees. Ordinarily, the disease fungus invades the plant vessels and chokes off the flow of water. The natural tendency of the elm to defend itself by surrounding and inactivating the fungus is too slow to save it from death. But treating the elm trees before they get infected helps them develop an immunity to the disease. Plants are vaccinated somewhat like humans and animals: nonpathogenic spores or proteins from fungi are injected into the tree over a period of time. So far it appears that the symptoms of disease and the degree of damage can be significantly reduced. This may be a whole new way to control fungal pests.

At the other extreme, government biologists working for narcotic control agencies have unveiled a recent plan to use fungi to kill unwanted plants. The main targets would be plants grown to produce illegal drugs like cocaine and heroin in the hopes of cutting down on these drugs right at the source. A fungus infection (*Fusarium*) that wiped out 30% of the coca crop in Peru dramatically demonstrated how effective this might be. Since then, at least two other fungi that could destroy opium poppies and marijuana plants have been isolated.

Purposefully releasing plant pathogens such as *Fusarium* into the environment has stirred a great deal of controversy. Critics in South America emphasize that even if the fungus appears specific to a particular plant, there is too much potential for it to switch hosts to food and ornamental plants and wreak havoc with the ecosystem. United States biologists who support the plan of using fungal control agents say that it is not as dangerous as massive spraying with pesticides, and that extensive laboratory tests have proved that the species of fungi being used will be very specific to the illegal drug plants and will not affect close relatives. Limited field tests will be started in the near future, paid for by a billion-dollar fund created by the U.S. government as part of its war on drugs.

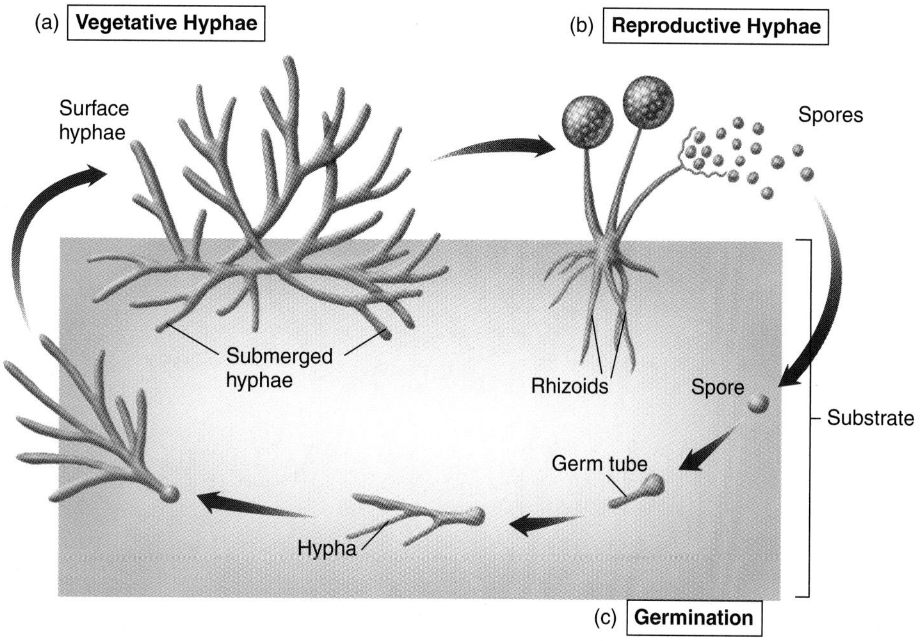

Figure 5.18 Functional types of hyphae using the mold *Rhizopus* as an example.
(a) Vegetative hyphae are those surface and submerged filaments that digest, absorb, and distribute nutrients from the substrate. This species also has special anchoring structures called rhizoids. **(b)** As the mold matures, it sprouts reproductive hyphae that produce asexual spores. **(c)** During the asexual life cycle, the free mold spores settle on a substrate and send out germ tubes that elongate into hyphae. Through continued growth and branching, an extensive mycelium is produced. So prolific are the fungi that a single colony of mold can easily contain 5,000 spore-bearing structures. If each of these released 2,000 single spores and if every spore were able to germinate, we would soon find ourselves in a sea of mycelia. Most spores do not germinate, but enough are successful to keep the numbers of fungi and their spores very high in most habitats.

hyphae, which branch off vegetative mycelium. These hyphae are responsible for the production of fungal reproductive bodies called **spores.** Other specializations of hyphae are illustrated in **figure 5.18.**

Reproductive Strategies and Spore Formation

Fungi have many complex and successful reproductive strategies. Most can propagate by the simple outward growth of existing hyphae or by fragmentation, in which a separated piece of mycelium can generate a whole new colony. But the primary reproductive mode of fungi involves the production of various types of spores. Do not confuse fungal spores with the more resistant, nonreproductive bacterial spores. Fungal spores are responsible not only for multiplication but also for survival, producing genetic variation, and dissemination. Because of their compactness and relatively light weight, spores are dispersed widely through the environment by air, water, and living things. Upon encountering a favorable substrate, a spore will germinate and produce a new fungus colony in a very short time (figure 5.18).

The fungi exhibit such a marked diversity in spores that they are largely classified and identified by their spores and spore-forming structures. There are elaborate systems for naming and classifying spores, but we won't cover them. The most general subdivision is based on the way the spores arise. Asexual spores are the products of mitotic division of a single parent cell, and sexual spores are formed through a process involving the fusing of two parental nuclei followed by meiosis.

Asexual Spore Formation

There are two subtypes of asexual spore, **sporangiospores** and **conidiospores,** also called conidia **(figure 5.19):**

1. Sporangiospores **(figure 5.19*a*)** are formed by successive cleavages within a saclike head called a **sporangium,** which is attached to a stalk, the sporangiophore. These spores are initially enclosed but are released when the sporangium ruptures.
2. Conidiospores or **conidia** are free spores not enclosed by a spore-bearing sac. They develop either by the pinching off of the tip of a special fertile hypha or by the segmentation of a preexisting vegetative hypha. There are many different forms of conidia, illustrated in **figure 5.19*b*.**

Sexual Spore Formation

Fungi can propagate themselves successfully with their millions of asexual spores. What is the function of their sexual

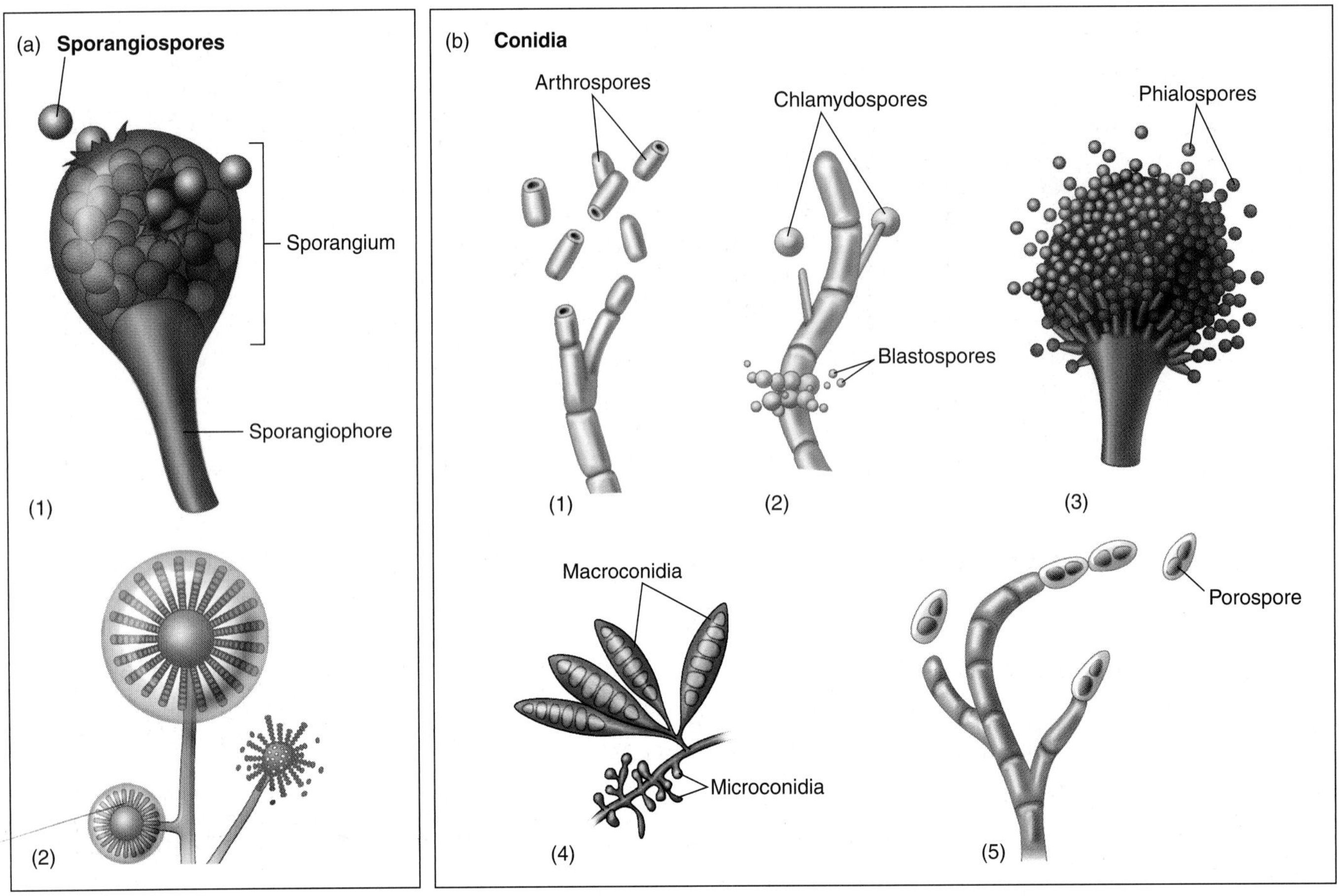

Figure 5.19 Types of asexual mold spores.
(a) Sporangiospores: (1) *Absidia,* (2) *Syncephalastrum.* **(b)** Conidial variations: (1) arthrospores (e.g., *Coccidioides*), (2) chlamydospores and blastospores (e.g., *Candida albicans*), (3) phialospores (e.g., *Aspergillus*), (4) macroconidia and microconidia (e.g., *Microsporum*), and (5) porospores (e.g., *Alternaria*).

spores? The answer lies in important variations that occur when fungi of different genetic makeup combine their genetic material. Just as in plants and animals, this linking of genes from two parents creates offspring with combinations of genes different from that of either parent. The offspring from such a union can have slight variations in form and function that are potentially advantageous in the adaptation and survival of their species.

The majority of fungi produce sexual spores at some point. The nature of this process varies from the simple fusion of fertile hyphae of two different strains to a complex union of differentiated male and female structures and the development of special fruiting structures. It may be a surprise to discover that the fleshy part of a mushroom is actually a fruiting body designed to protect and help disseminate its sexual spores.

Fungal Identification and Cultivation

Fungi are identified in medical specimens by first being isolated on special types of media and then being observed macroscopically and microscopically. Because the fungi are classified into general groups by the presence and type of sexual spores, it would seem logical to identify them in the same way, but sexual spores are rarely if ever detected in the laboratory setting. As a result, the asexual spore-forming structures and spores are usually used to identify organisms to the level of genus and species. Other characteristics that contribute to identification are hyphal type, colony texture and pigmentation, physiological characteristics, and genetic makeup. Even as bacterial and viral identification relies increasingly on molecular techniques, fungi are some of the most strikingly beautiful life forms, and their appearance under the microscope is still heavily relied upon to identify them **(figure 5.20*a,b*).**

The Roles of Fungi in Nature and Industry

Nearly all fungi are free-living and do not require a host to complete their life cycles. Even among those fungi that are pathogenic, most human infection occurs through accidental contact with an environmental source such as soil, water, or dust. Humans are generally quite resistant to fungal infection, except for two main types of fungal pathogens: the

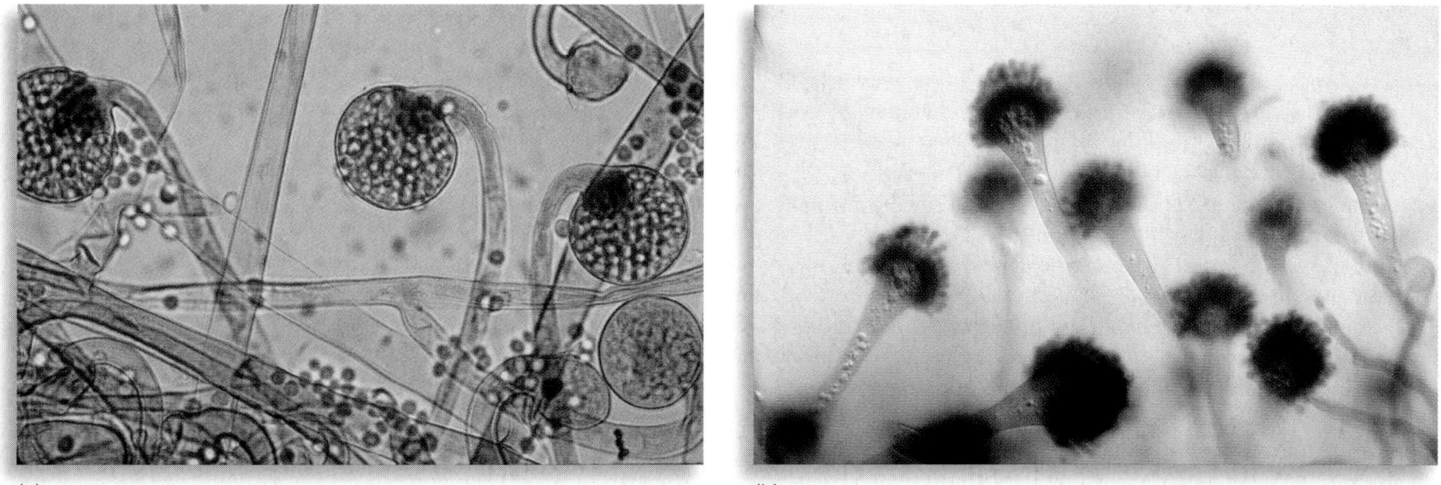

(a) (b)

Figure 5.20 Representative fungi.
(a) *Circinella,* a fungus associated with soil and decaying nuts. **(b)** *Aspergillus,* a ubiquitous environmental fungus that can be associated with human disease.

primary pathogens, which can sicken even healthy persons, and the opportunistic pathogens, which attack persons who are already weakened in some way. So far, about 270 species of fungi have been found to be able to cause human disease.

Mycoses (fungal infections) vary in the way the agent enters the body and the degree of tissue involvement **(table 5.3).** The list of opportunistic fungal pathogens has been increasing in the past few years because of newer medical techniques that keep immunocompromised patients alive. Even so-called harmless species found in the air and dust around us may be able to cause opportunistic infections in patients who already have AIDS, cancer, or diabetes (see Insight 21.1 in chapter 21).

Fungi are involved in other medical conditions besides infections (see Insight 5.2). Fungal cell walls give off chemical substances that can cause allergies. The toxins produced by poisonous mushrooms can induce neurological disturbances and even death. The mold *Aspergillus flavus* synthesizes a potentially lethal poison called aflatoxin, which is the cause of a disease in domestic animals that have eaten grain contaminated with the mold and is also a cause of liver cancer in humans.

Fungi pose an ever-present economic hindrance to the agricultural industry. A number of species are pathogenic to field plants such as corn and grain, and fungi also rot fresh produce during shipping and storage. It has been estimated that as much as 40% of the yearly fruit crop is consumed not by humans but by fungi. On the beneficial side, however, fungi play an essential role in decomposing organic matter and returning essential minerals to the soil. They form stable associations with plant roots (mycorrhizae) that increase the ability of the roots to absorb water and nutrients. Industry has tapped the biochemical potential of fungi to produce large quantities of antibiotics, alcohol, organic acids, and vitamins. Some fungi are eaten or used to impart flavorings to food. The yeast *Saccharomyces* produces the alcohol in beer and wine and the gas that causes bread to rise. Blue cheese, soy sauce, and cured meats derive their unique flavors from the actions of fungi (see chapter 24).

TABLE 5.3 Major Fungal Infections of Humans

Degree of Tissue Involvement and Area Affected	Name of Infection	Name of Causative Fungus
Superficial (not deeply invasive)		
Outer epidermis	Tinea versicolor	*Malassezia furfur*
Epidermis, hair, and dermis can be attacked.	Dermatophytosis, also called tinea or ringworm of the scalp, body, feet (athlete's foot), toenails	*Microsporum, Trichophyton,* and *Epidermophyton*
Mucous membranes, skin, nails	Candidiasis, or yeast infection	*Candida albicans*
Systemic (deep; organism enters lungs; can invade other organs)		
Lung	Coccidioidomycosis (San Joaquin Valley fever)	*Coccidioides immitis*
	North American blastomycosis (Chicago disease)	*Blastomyces dermatitidis*
	Histoplasmosis (Ohio Valley fever)	*Histoplasma capsulatum*
	Cryptococcosis (torulosis)	*Cryptococcus neoformans*
Lung, skin	Paracoccidioidomycosis (South American blastomycosis)	*Paracoccidioides brasiliensis*

CHECKPOINT

- The fungi are nonphotosynthetic haploid species with cell walls. They are either saprobes or parasites and may be unicellular, colonial, or multicellular.
- All fungi are heterotrophic.
- Fungi have many reproductive strategies, including both asexual and sexual.
- Fungi have asexual spores called sporangiospores and conidiospores.
- Fungal sexual spores enable the organisms to incorporate variations in form and function.
- Fungi are often identified on the basis of their microscopic appearance.
- There are two categories of fungi that cause human disease: the primary pathogens, which infect healthy persons, and the opportunistic pathogens, which cause disease only in compromised hosts.

5.5 The Protists

The algae and protozoa have been traditionally combined into the Kingdom Protista. The two major taxonomic categories of this kingdom are Subkingdom Algae and Subkingdom Protozoa. Although these general types of microbes are now known to occupy several kingdoms, it is still useful to retain the concept of a protist as any unicellular or colonial organism that lacks true tissues. We will only briefly mention algae, as they do not cause human infections for the most part.

The Algae: Photosynthetic Protists

The **algae** are a group of photosynthetic organisms usually recognized by their larger members, such as seaweeds and kelps. In addition to being beautifully colored and diverse in appearance, they vary in length from a few micrometers to 100 meters. Algae occur in unicellular, colonial, and filamentous forms, and the larger forms can possess tissues and simple organs. **Figure 5.21** depicts various types of algae. Algal cells as a group exhibit all of the eukaryotic organelles. The most noticeable of these are the chloroplasts, which contain, in addition to the green pigment chlorophyll, a number of other pigments that create the yellow, red, and brown coloration of some groups.

Algae are widespread inhabitants of fresh and marine waters. They are one of the main components of the large floating community of microscopic organisms called **plankton.** In this capacity, they play an essential role in the aquatic food web and produce most of the earth's oxygen. Other algal habitats include the surface of soil, rocks, and plants, and several species are even hardy enough to live in hot springs or snowbanks.

Animal tissues would be rather inhospitable to algae, so algae are rarely infectious. One exception is *Prototheca*, an unusual nonphotosynthetic alga, which has been associated with skin and subcutaneous infections in humans and animals.

The primary medical threat from algae is due to a type of food poisoning caused by the toxins of certain marine algae. During particular seasons of the year, the overgrowth of these motile algae imparts a brilliant red color to the water, which is referred to as a "red tide." When intertidal animals feed, their bodies accumulate toxins given off by the algae that can persist for several months. Paralytic shellfish poisoning is caused by eating exposed clams or other invertebrates. It is marked by severe neurological symptoms and can be fatal. Ciguatera is a serious intoxication caused by algal toxins that have accumulated in fish such as bass and mackerel. Cooking does not destroy the toxin, and there is no antidote.

(a) (b) (c)

Figure 5.21 Representative microscopic algae.
(a) *Spirogyra,* a colonial filamentous form with spiral chloroplasts. **(b)** A strew of beautiful algae called diatoms shows the intricate and varied structure of their silica cell wall. **(c)** *Pfiesteria piscicida.* Although it is free-living, it is known to parasitize fish and release potent toxins that kill fish and sicken humans.

Several episodes of a severe infection caused by *Pfiesteria piscicida,* a toxic algal form, have been reported over the past several years in the United States. The disease was first reported in fish and was later transmitted to humans. This newly identified species occurs in at least 20 forms, including spores, cysts, and amoebas (see **figure 5.21*c***), that can release potent toxins. Both fish and humans develop neurological symptoms and bloody skin lesions. The cause of the epidemic has been traced to nutrient-rich agricultural runoff water that promoted the sudden "bloom" of *Pfiesteria.* These microbes first attacked and killed millions of fish and later people whose occupations exposed them to fish and contaminated water.

Biology of the Protozoa

If a poll were taken to choose the most engrossing and vivid group of microorganisms, many biologists would choose the protozoa. Although their name comes from the Greek for "first animals," they are far from being simple, primitive organisms. The protozoa constitute a very large group (about 65,000 species) of creatures that although single-celled, have startling properties when it comes to movement, feeding, and behavior. Although most members of this group are harmless, free-living inhabitants of water and soil, a few species are parasites collectively responsible for hundreds of millions of infections of humans each year. Before we consider a few examples of important pathogens, let us examine some general aspects of protozoan biology.

Protozoan Form and Function

Most protozoan cells are single cells containing the major eukaryotic organelles except chloroplasts. Their organelles can be highly specialized for feeding, reproduction, and locomotion. The cytoplasm is usually divided into a clear outer layer called the **ectoplasm** and a granular inner region called the *endoplasm.* Ectoplasm is involved in locomotion, feeding, and protection. Endoplasm houses the nucleus, mitochondria, and food and contractile vacuoles. Some ciliates and flagellates[4] even have organelles that work somewhat like a primitive nervous system to coordinate movement. Because protozoa lack a cell wall, they have a certain amount of flexibility. Their outer boundary is a cell membrane that regulates the movement of food, wastes, and secretions. Cell shape can remain constant (as in most ciliates) or can change constantly (as in amoebas). Certain amoebas (foraminiferans) encase themselves in hard shells made of calcium carbonate. The size of most protozoan cells falls within the range of 3 to 300 μm. Some notable exceptions are giant amoebas and ciliates that are large enough (3 to 4 mm in length) to be seen swimming in pond water.

4. The terms *ciliate* and *flagellate* are common names of protozoan groups that move by means of cilia and flagella.

CASE FILE 5 *WRAP-UP*

The disease in the opening case study was ***cryptosporidiosis.*** It is caused by a single-celled protozoan parasite named *Cryptosporidium parvum. C. parvum* undergoes a complex life cycle. During one stage in the life cycle, thick-walled oocysts that are 3 to 5 μm in size are formed. It was these oocysts that were detected in water samples from the pool. When Dr. McDermott educated the patients about this disease, she emphasized that all it takes is the ingestion of one to ten oocysts to cause disease. This is referred to as a very "low infectious dose." It is why one fecal accident can sufficiently contaminate an entire swimming pool in which individuals may accidentally swallow only one or two mouthfuls of contaminated water. The oocysts are extremely resistant to disinfectants and the recommended concentrations of chlorine used in swimming pools. Because of their small size, they may not be removed efficiently by pool filters.

Cryptosporidium infected over 370,000 individuals in Milwaukee, Wisconsin, in 1993. This was the largest waterborne outbreak in U.S. history. During that epidemic, the public water supply was contaminated with human sewage containing *C. parvum* oocysts. Besides pool-associated and contaminated drinking water illness, there have been reports of *C. parvum* oocysts in oysters intended for human consumption that were harvested from sites where there were high levels of fecal contamination due to wastewater outfalls and cattle farms. Oysters remove oocysts from contaminated waters and retain them on their gills and within their body.

See: CDC. 2001. Protracted outbreaks of cryptosporidiosis associated with swimming pool use—Ohio and Nebraska, 2000. MMWR *50:406–410. Fayer, R., et al. 1999.* Cryptosporidium parvum *in oysters from commercial harvesting sites in the Chesapeake Bay.* Emerg. Infect. Dis. *5:706–710.*

Nutritional and Habitat Range Protozoa are heterotrophic and usually require their food in a complex organic form. Free-living species scavenge dead plant or animal debris and even graze on live cells of bacteria and algae. Some species have special feeding structures such as oral grooves, which carry food particles into a passageway or gullet that packages the captured food into vacuoles for digestion. Some protozoa absorb food directly through the cell membrane. Parasitic species live on the fluids of their host, such as plasma and digestive juices, or they can actively feed on tissues.

Although protozoa have adapted to a wide range of habitats, their main limiting factor is the availability of moisture. Their predominant habitats are fresh and marine water, soil, plants, and animals. Even extremes in temperature and pH are not a barrier to their existence; hardy species are found in hot springs, ice, and habitats with low or high pH. Many protozoa can convert to a resistant, dormant stage called a cyst.

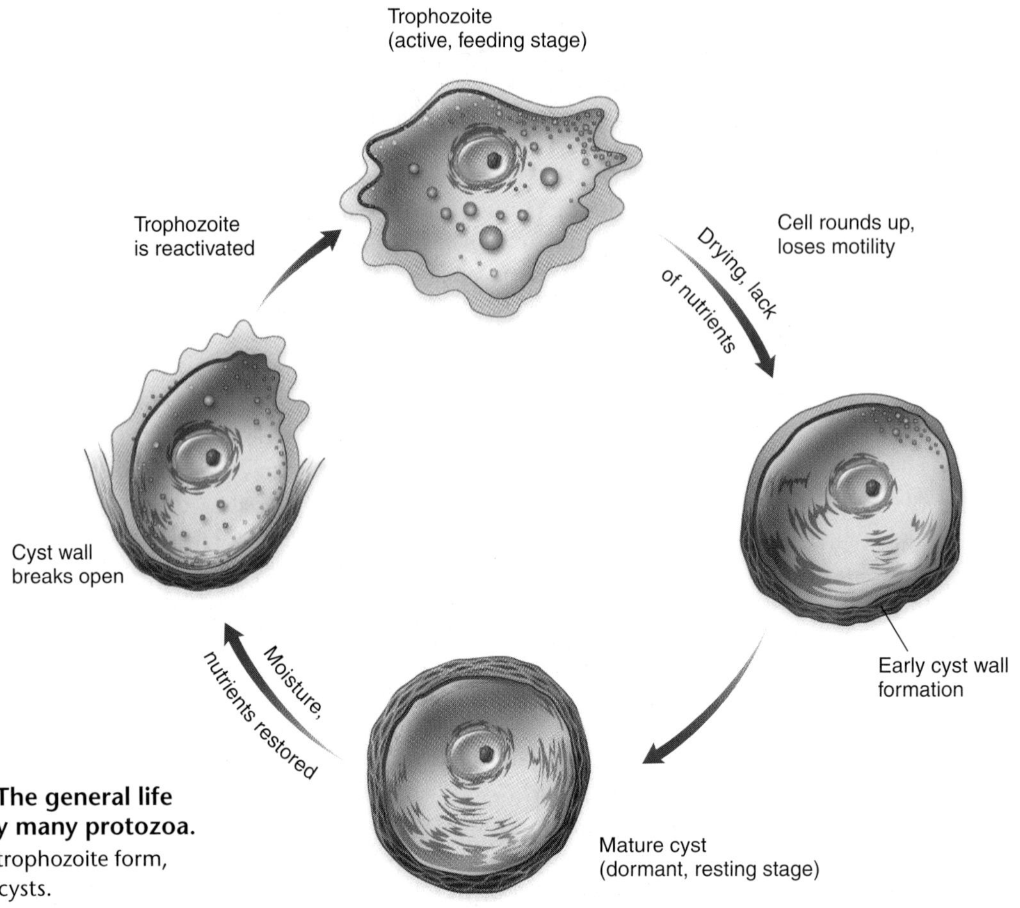

Figure 5.22 The general life cycle exhibited by many protozoa. All protozoa have a trophozoite form, but not all produce cysts.

Styles of Locomotion Except for one group (the Apicomplexa), protozoa are motile by means of **pseudopods** ("false feet"), **flagella,** or **cilia.** A few species have both pseudopods (also called pseudopodia) and flagella. Some unusual protozoa move by a gliding or twisting movement that does not appear to involve any of these locomotor structures. Pseudopods are blunt, branched, or long and pointed, depending on the particular species. The flowing action of the pseudopods results in amoeboid motion, and pseudopods also serve as feeding structures in many amoebas. (The structure and behavior of flagella and cilia were discussed in the first section of this chapter.) Flagella vary in number from one to several, and in certain species they are attached along the length of the cell by an extension of the cytoplasmic membrane called the *undulating membrane* (see figure 5.23). In most ciliates, the cilia are distributed over the entire surface of the cell in characteristic patterns. Because of the tremendous variety in ciliary arrangements and functions, ciliates are among the most diverse and awesome cells in the biological world. In certain protozoa, cilia line the oral groove and function in feeding; in others, they fuse together to form stiff props that serve as primitive rows of walking legs.

Life Cycles and Reproduction Most protozoa are recognized by a motile feeding stage called the **trophozoite** that requires ample food and moisture to remain active. A large number of species are also capable of entering into a dormant, resting stage called a **cyst** when conditions in the environment become unfavorable for growth and feeding. During *encystment,* the trophozoite cell rounds up into a sphere, and its ectoplasm secretes a tough, thick cuticle around the cell membrane **(figure 5.22).** Because cysts are more resistant than ordinary cells to heat, drying, and chemicals, they can survive adverse periods. They can be dispersed by air currents and may even be an important factor in the spread of diseases such as amoebic dysentery. If provided with moisture and nutrients, a cyst breaks open and releases the active trophozoite.

The life cycles of protozoans vary from simple to complex. Several protozoan groups exist only in the trophozoite state. Many alternate between a trophozoite and a cyst stage, depending on the conditions of the habitat. The life cycle of a parasitic protozoan dictates its mode of transmission to other hosts. For example, the flagellate *Trichomonas vaginalis* causes a common sexually transmitted disease. Because it does not form cysts, it is more delicate and must be transmitted by intimate contact between sexual partners. In contrast, intestinal pathogens such as *Entamoeba histolytica* and *Giardia lamblia* form cysts and are readily transmitted in contaminated water and foods.

All protozoa reproduce by relatively simple, asexual methods, usually mitotic cell division. Several parasitic species, including the agents of malaria and toxoplasmosis, reproduce asexually inside a host cell by multiple fission. Sexual reproduction also occurs during the life cycle of most protozoa. Ciliates participate in **conjugation,** a form of genetic exchange in which members of two different mating types fuse temporarily and exchange micronuclei. This process of sexual recombination yields new and different genetic combinations that can be advantageous in evolution.

Classification of Selected Medically Important Protozoa

Taxonomists have problems classifying protozoa. They, too, are very diverse and frequently frustrate attempts to generalize or place them in neat groupings. We use a simple system of four groups, based on method of motility, mode of reproduction, and stages in the life cycle, summarized here.

The Mastigophora (Flagellated) Motility is primarily by flagella alone or by both flagellar and amoeboid motion. Single nucleus. Sexual reproduction, when present, by syngamy; division by longitudinal fission. Several parasitic forms lack mitochondria and Golgi apparatus. Most species form cysts and are free-living; the group also includes several parasites. Some species are found in loose aggregates or colonies, but most are solitary. Members include: *Trypanosoma* and *Leishmania*, important blood pathogens spread by insect vectors; *Giardia*, an intestinal parasite spread in water contaminated with feces; *Trichomonas*, a parasite of the reproductive tract of humans spread by sexual contact **(figure 5.23).**

The Sarcodina (Amoebas) Cell form is primarily an amoeba **(figure 5.24).** Major locomotor organelles are pseudopods, although some species have flagellated reproductive states. Asexual reproduction by fission. Two groups have an external shell; mostly uninucleate; usually encyst. Most amoebas are free-living and not infectious; *Entamoeba* is a pathogen or parasite of humans; shelled amoebas called foraminifera and radiolarians are responsible for chalk deposits in the ocean.

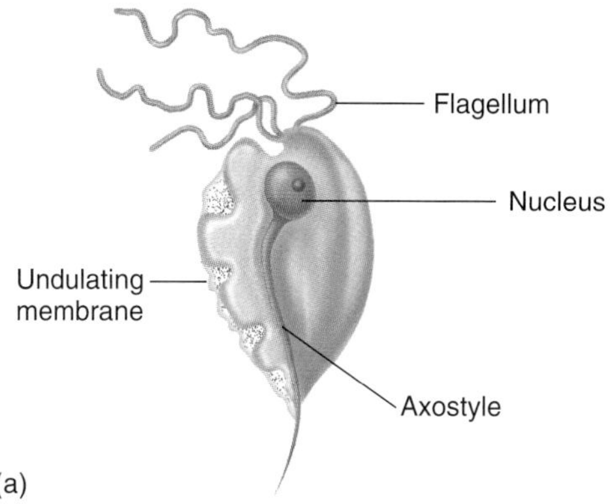

(a)

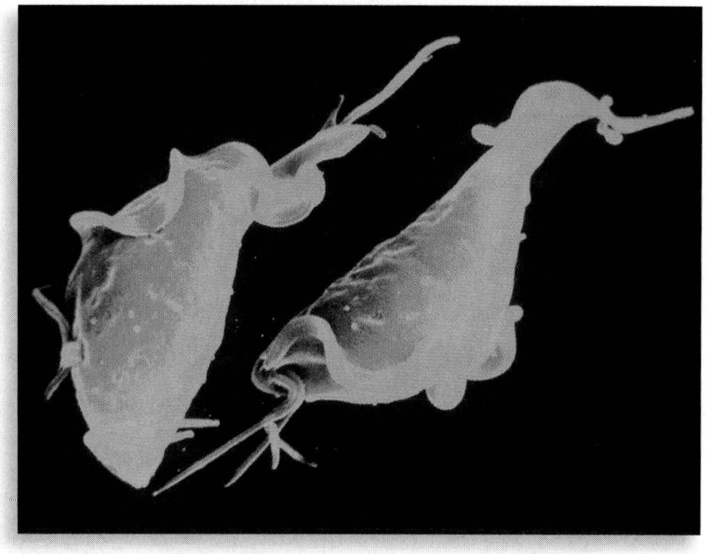

(b)

Figure 5.23 The structure of a typical mastigophoran, *Trichomonas vaginalis*.
This genital tract pathogen is shown in **(a)** a drawing and **(b)** a scanning electron micrograph.

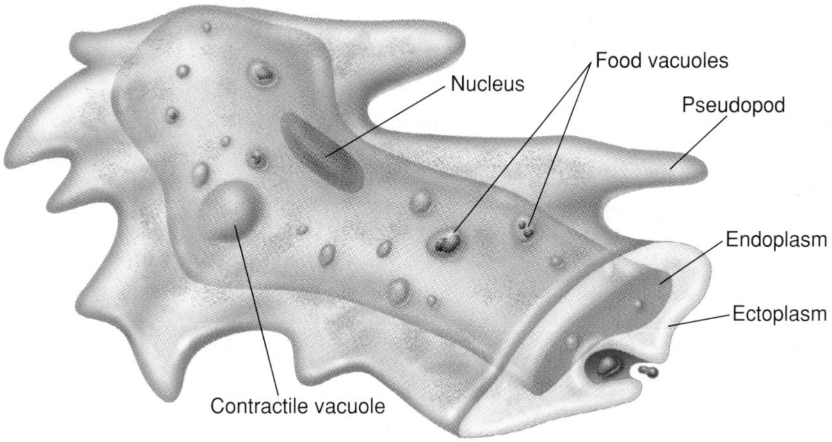

(a)

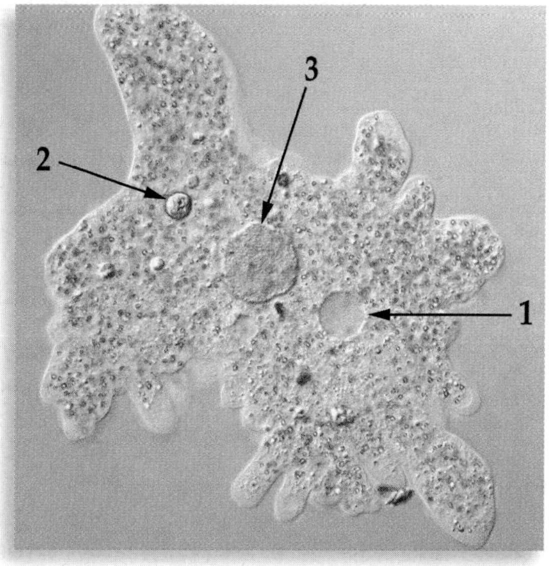

(b)

Figure 5.24 Amoebas.
(a) Artist's drawing. **(b)** Photomicrograph of the structure of an amoeba. (1) A water-expelling vacuole; (2) food vacuole; (3) nucleus.

The Ciliophora (Ciliated) Trophozoites are motile by cilia; some have cilia in tufts for feeding and attachment; most develop cysts; have both macronuclei and micronuclei; division by transverse fission; most have a definite mouth and feeding organelle; show relatively advanced behavior **(figure 5.25).** The majority of ciliates are free-living and harmless.

The Apicomplexa (Sporozoa) Motility is absent in most cells except male gametes. Life cycles are complex, with well-developed asexual and sexual stages. Sporozoa produce special sporelike cells called **sporozoites (figure 5.26)** following sexual reproduction, which are important in transmission of infections; most form thick-walled zygotes called oocysts; entire group is parasitic. *Plasmodium,* the most prevalent protozoan parasite, causes 100 million to 300 million cases of malaria each year worldwide. It is an intracellular parasite with a complex cycle alternating between humans and mosquitoes. *Toxoplasma gondii* causes an acute infection (toxoplasmosis) in humans, which is acquired from cats and other animals.

Just as with the prokaryotes and other eukaryotes, protozoans that cause disease produce symptoms in different

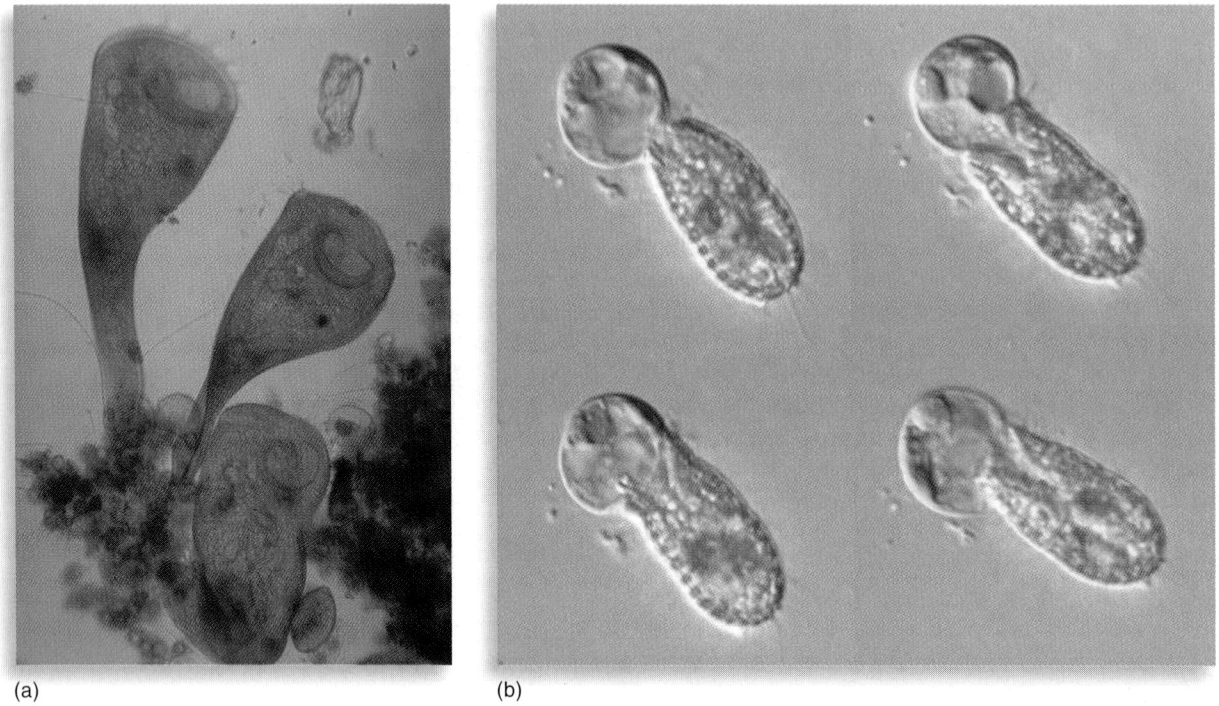

Figure 5.25 Selected ciliate representatives.
(a) Large, funnel-shaped *Stentor* with a rotating row of cilia around its oral cavity. Currents produced by the cilia sweep food particles into the gullet. **(b)** Stages in the process of *Coleps* feeding on an alga (round cell). The predaceous ciliate gradually pulls its prey into a large oral groove.

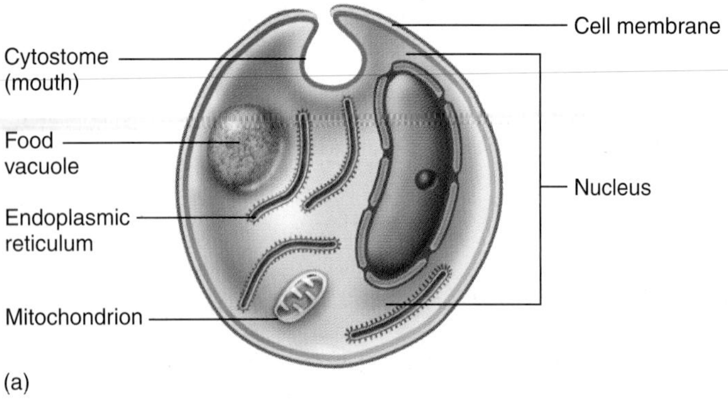

Figure 5.26 Sporozoan protozoan.
(a) General cell structure. Note the lack of specialized locomotor organelles. **(b)** Scanning electron micrograph of the sporozoite of *Cryptosporidium*, an intestinal parasite of humans and other mammals.

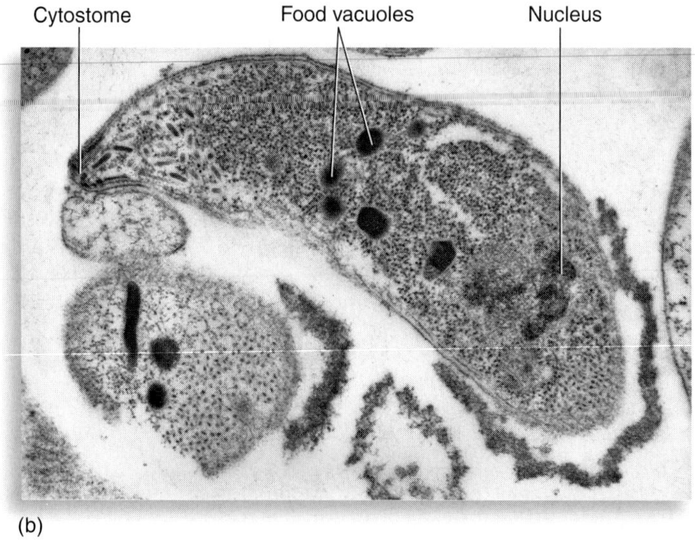

organ systems. These diseases are covered in chapters 18 through 23.

Protozoan Identification and Cultivation

The unique appearance of most protozoa makes it possible for a knowledgeable person to identify them to the level of genus and often species by microscopic morphology alone. Characteristics to consider in identification include the shape and size of the cell; the type, number, and distribution of locomotor structures; the presence of special organelles or cysts; and the number of nuclei. Medical specimens taken from blood, sputum, cerebrospinal fluid, feces, or the vagina are smeared directly onto a slide and observed with or without special stains. Occasionally, protozoa are cultivated on artificial media or in laboratory animals for further identification or study.

Important Protozoan Pathogens

Although protozoan infections are very common, they are actually caused by only a small number of species often restricted geographically to the tropics and subtropics **(table 5.4).** In this survey, we look at examples from two protozoan groups that illustrate some of the main features of protozoan diseases.

TABLE 5.4 Major Pathogenic Protozoa, Infections, and Primary Sources

Protozoan/Disease	Reservoir/Source
Amoeboid Protozoa	
Amoebiasis: *Entamoeba histolytica*	Human/water and food
Brain infection: *Naegleria, Acanthamoeba*	Free-living in water
Ciliated Protozoa	
Balantidiosis: *Balantidium coli*	Zoonotic in pigs
Flagellated Protozoa	
Giardiasis: *Giardia lamblia*	Zoonotic/water and food
Trichomoniasis: *T. hominis, T. vaginalis*	Human
Hemoflagellates	
Trypanosomiasis: *Trypanosoma brucei, T. cruzi*	Zoonotic/vector-borne
Leishmaniasis: *Leishmania donovani, L. tropica, L. brasiliensis*	Zoonotic/vector-borne
Apicomplexan Protozoa	
Malaria: *Plasmodium vivax, P. falciparum, P. malariae*	Human/vector-borne
Toxoplasmosis: *Toxoplasma gondii*	Zoonotic/vector-borne
Cryptosporidiosis: *Cryptosporidium*	Free-living/water, food
Cyclosporiasis: *Cyclospora cayetanensis*	Water/fresh produce

The study of protozoa and helminths is sometimes called *parasitology*. Although a parasite is more accurately defined as an organism that obtains food and other requirements at the expense of a host, the term *parasite* is often used to denote protozoan and helminth pathogens.

Pathogenic Flagellates: Trypanosomes Trypanosomes are protozoa belonging to the genus *Trypanosoma* (try″-pan-oh-soh′-mah). The two most important representatives are *T. brucei* and *T. cruzi,* species that are closely related but geographically restricted. *Trypanosoma brucei* occurs in Africa, where it causes approximately 35,000 new cases of sleeping sickness each year (see chapter 19). *Trypanosoma cruzi,* the cause of Chagas disease,[5] is endemic to South and Central America, where it infects several million people a year. Both species have long, crescent-shaped cells with a single flagellum that is sometimes attached to the cell body by an undulating membrane. Both occur in the blood during infection and are transmitted by blood-sucking vectors. We use *T. cruzi* to illustrate the phases of a trypanosomal life cycle and to demonstrate the complexity of parasitic relationships.

The trypanosome of Chagas disease relies on the close relationship of a warm-blooded mammal and an insect that feeds on mammalian blood. The mammalian hosts are numerous, including dogs, ats, opossums, armadillos, and foxes. The vector is the *reduviid* (ree-doo′-vee-id) *bug,* an insect that is sometimes called the "kissing bug" because of its habit of biting its host at the corner of the mouth. Transmission occurs from bug to mammal and from mammal to bug, but usually not from mammal to mammal, except across the placenta during pregnancy. The general phases of this cycle are presented in **figure 5.27.**

The trypanosome trophozoite multiplies in the intestinal tract of the reduviid bug and is harbored in the feces. The bug seeks a host and bites the mucous membranes, usually of the eye, nose, or lips. As it fills with blood, the bug soils the bite with feces containing the trypanosome. Ironically, the victims themselves inadvertently contribute to the entry of the microbe by scratching the bite wound. The trypanosomes ultimately become established and multiply in muscle and white blood cells. Periodically, these parasitized cells rupture, releasing large numbers of new trophozoites into the blood. Eventually, the trypanosome can spread to many systems, including the lymphoid organs, heart, liver, and brain. Manifestations of the resultant disease range from mild to very severe and include fever, inflammation, and heart and brain damage. In many cases, the disease has an extended course and can cause death.

Infective Amoebas: *Entamoeba* Several species of amoebas cause disease in humans, but probably the most common disease is amoebiasis, or amoebic dysentery, caused by *Entamoeba histolytica* (see chapter 22). This microbe is widely distributed in the world, from northern zones to the tropics,

5. Named for Carlos Chagas, the discoverer of *T. cruzi.*

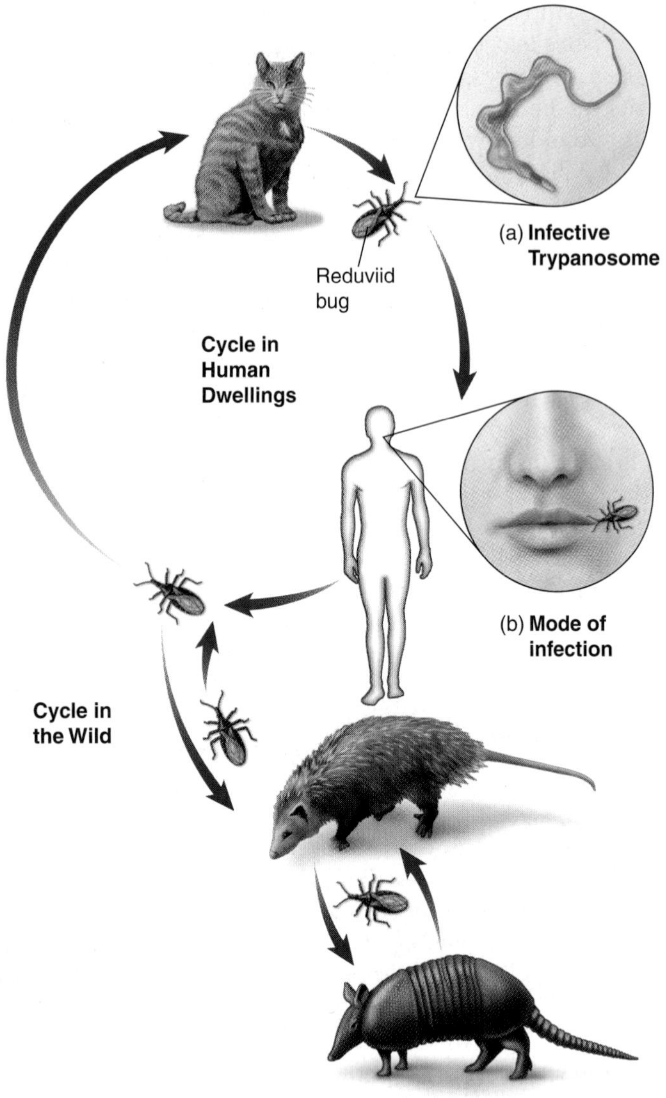

Figure 5.27 Cycle of transmission in Chagas disease. Trypanosomes (inset *a*) are transmitted among mammalian hosts and human hosts by means of a bite from the kissing bug (inset *b*).

and is nearly always associated with humans. Amoebic dysentery is the fourth most common protozoan infection in the world. This microbe has a life cycle quite different from the trypanosomes in that it does not involve multiple hosts and a blood-sucking vector. It lives part of its cycle as a trophozoite and part as a cyst. Because the cyst is the more resistant form and can survive in water and soil for several weeks, it is the more important stage for transmission. The primary way that people become infected is by ingesting food or water contaminated with human feces.

Figure 5.28 shows the major features of the amoebic dysentery cycle, starting with the ingestion of cysts. The viable, heavy-walled cyst passes through the stomach unharmed. Once inside the small intestine, the cyst germinates into a large multinucleate amoeba that subsequently divides to form small amoebas (the trophozoite stage). These trophozoites migrate to the large intestine and begin to feed and grow. From this site, they can penetrate the lining of the intestine and invade the liver, lungs, and skin. Common symptoms include gastrointestinal disturbances such as nausea, vomiting, and diarrhea, leading to weight loss and dehydration. Untreated cases with extensive damage to the organs experience a high death rate. The cycle is completed in the infected human when certain trophozoites in the feces begin to form cysts, which then pass out of the body with fecal matter. Knowledge of the amoebic cycle and role of cysts has been helpful in controlling the disease. Important preventive measures include sewage treatment, curtailing the use of human feces as fertilizers, and adequate sanitation of food and water.

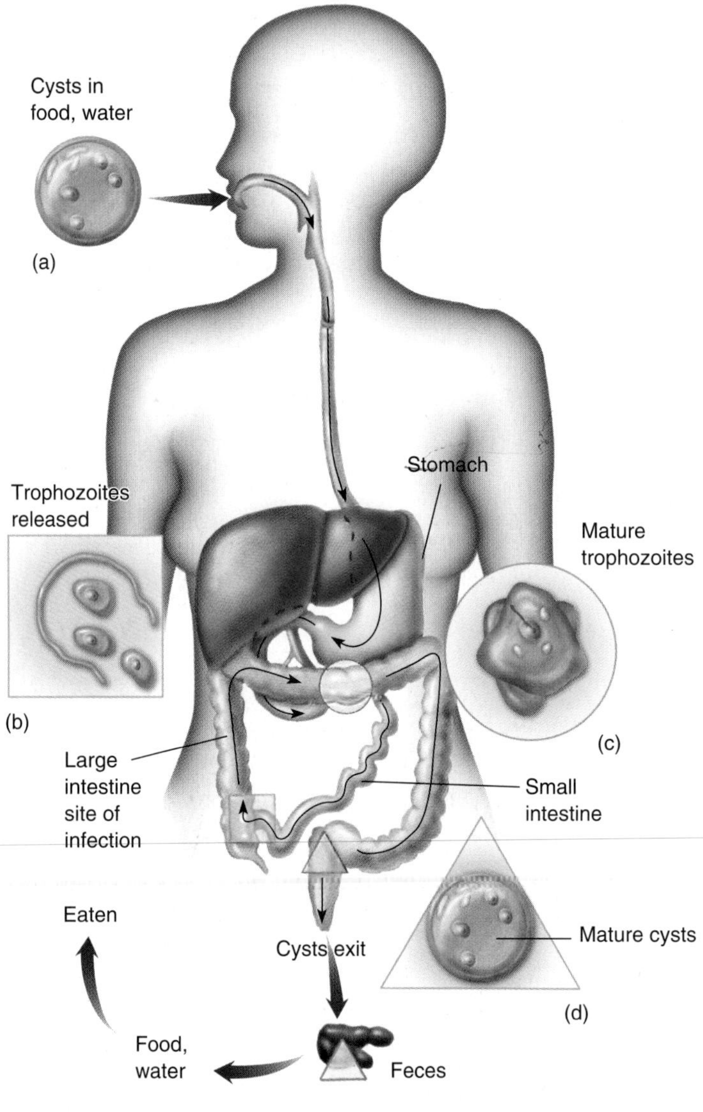

Figure 5.28 Stages in the infection and transmission of amoebic dysentery.
Arrows show the route of infection; insets show the appearance of *Entamoeba histolytica*. **(a)** Cysts are eaten. **(b)** Trophozoites (amoebas) emerge from cysts. **(c)** Trophozoites invade the large intestinal wall. **(d)** Mature cysts are released in the feces and may be spread through contaminated food and water.

5.6 The Parasitic Helminths

Tapeworms, flukes, and roundworms are collectively called helminths, from the Greek word meaning worm. Adult animals are usually large enough to be seen with the naked eye, and they range from the longest tapeworms, measuring up to about 25 m in length, to roundworms less than 1 mm in length. Nevertheless, they are included among microorganisms because of their infective abilities and because the microscope is necessary to identify their eggs and larvae.

On the basis of morphological form, the two major groups of parasitic helminthes are the flatworms (Phylum Platyhelminthes) and the roundworms (Phylum Aschelminthes, also called **nematodes**). Flatworms have a very thin, often segmented body plan **(figure 5.29),** and roundworms have an elongate, cylindrical, unsegmented body **(figure 5.30).** The flatworm group is subdivided into the **cestodes,** or tapeworms, named for their long, ribbonlike arrangement, and the **trematodes,** or flukes, characterized by flat, ovoid bodies. Not all flatworms and roundworms are parasites by nature; many live free in soil and water. Because most disease-causing helminths spend part of their lives in the gastrointestinal tract, they are discussed in chapter 22.

General Worm Morphology

All helminths are multicellular animals equipped to some degree with organs and organ systems. In parasitic helminths, the most developed organs are those of the reproductive tract, with some degree of reduction in the digestive, excretory, nervous, and muscular systems. In particular groups, such as the cestodes, reproduction is so dominant that the worms are reduced to little more than a series of flattened sacs filled with ovaries, testes, and eggs (see figure 5.29*a, b*). Not all worms have such extreme adaptations as cestodes, but most have a highly developed reproductive potential, thick cuticles for protection, and mouth glands for breaking down the host's tissue (figure 5.29*c*).

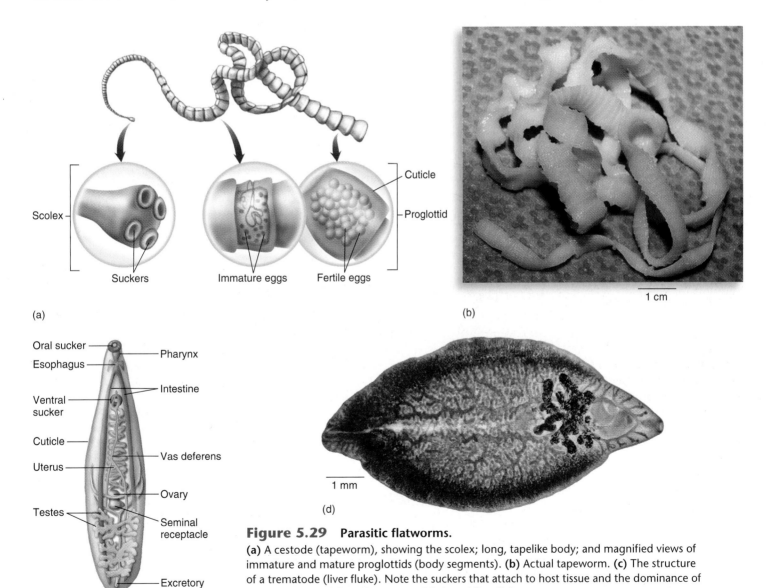

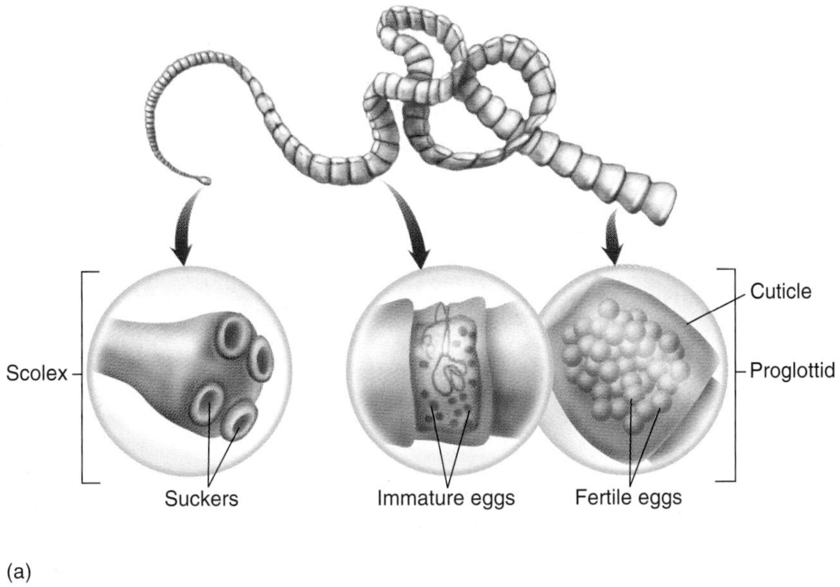

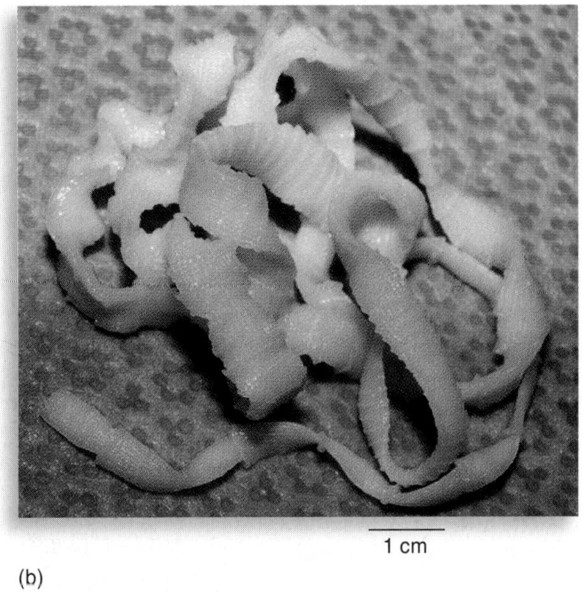

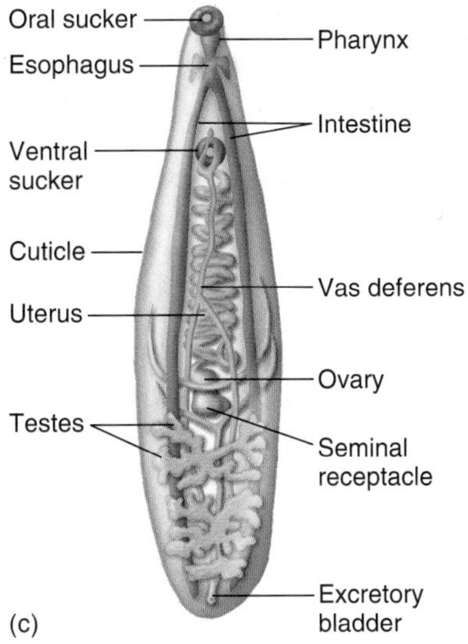

Figure 5.29 Parasitic flatworms.
(a) A cestode (tapeworm), showing the scolex; long, tapelike body; and magnified views of immature and mature proglottids (body segments). **(b)** Actual tapeworm. **(c)** The structure of a trematode (liver fluke). Note the suckers that attach to host tissue and the dominance of reproductive and digestive organs. **(d)** Actual liver fluke.

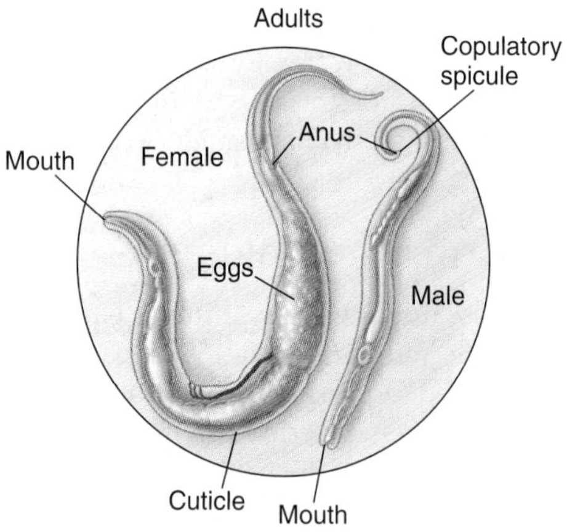

Figure 5.30 Adult pinworms.
These worms are approximately the size of a staple.

Life Cycles and Reproduction

The complete life cycle of helminths includes the fertilized egg (embryo), larval, and adult stages. In the majority of helminths, adults derive nutrients and reproduce sexually in a host's body. In nematodes, the sexes are separate and usually different in appearance; in trematodes, the sexes can be either separate or **hermaphroditic,** meaning that male and female sex organs are in the same worm; cestodes are generally hermaphroditic. For a parasite's continued survival as a species, it must complete the life cycle by transmitting an infective form, usually an egg or larva, to the body of another host, either of the same or a different species. The host in which larval development occurs is the intermediate (secondary) host, and adulthood and mating occur in the **definitive (final) host.** A transport host is an intermediate host that experiences no parasitic development but is an essential link in the completion of the cycle.

In general, sources for human infection are contaminated food, soil, and water or infected animals, and routes of infection are by oral intake or penetration of unbroken skin. Humans are the definitive hosts for many of the parasites listed in **table 5.5,** and in about half the diseases, they are also the sole biological reservoir. In other cases, animals or insect vectors serve as reservoirs or are required to complete worm development. In the majority of helminth infections, the worms must leave their host to complete the entire life cycle.

Fertilized eggs are usually released to the environment and are provided with a protective shell and extra food to aid their development into larvae. Even so, most eggs and larvae are vulnerable to heat, cold, drying, and predators and are destroyed or unable to reach a new host. To counteract this formidable mortality rate, certain wor ms have adapted a reproductive capacity that borders on the incredible: A single female *Ascaris*[6] can lay 200,000 eggs a day, and a large female can contain over 25 million eggs at varying stages of development! If only a tiny number of these eggs makes it to another host, the parasite will have been successful in completing its life cycle.

6. *Ascaris* is a genus of parasitic intestinal roundworms.

TABLE 5.5 Examples of Helminths and Their Modes of Transmission

Classification	Common Name of Disease or Worm	Life Cycle Requirement	Spread to Humans By
Roundworms			
Nematodes			
Intestinal Nematodes			Ingestion
Infective in egg (embryo) stage			
Ascaris lumbricoides	Ascariasis	Humans	Fecal pollution of soil with eggs
Enterobius vermicularis	Pinworm	Humans	Close contact
Infective in larval stage			
Trichinella spiralis	Trichina worm	Pigs, wild mammals	Consumption of meat containing larvae
Tissue Nematodes			Burrowing of larva into tissue
Onchocerca volvulus	River blindness	Humans, black flies	Fly bite
Dracunculus medinensis	Guinea worm	Humans and *Cyclops* (an aquatic invertebrate)	Ingestion of water containing *Cyclops*
Flatworms			
Trematodes			
Schistosoma japonicum	Blood fluke	Humans and snails	Ingestion of fresh water containing larval stage
Cestodes			
T. solium	Pork tapeworm	Humans, swine	Consumption of undercooked or raw pork
Diphyllobothrium latum	Fish tapeworm	Humans, fish	Consumption of undercooked or raw fish

A Helminth Cycle: The Pinworm

To illustrate a helminth cycle in humans, we use the example of a roundworm, *Enterobius vermicularis,* the pinworm or seatworm. This worm causes a very common infestation of the large intestine (see figure 5.30). Worms range from 2 to 12 mm long and have a tapered, curved cylinder shape. The condition they cause, enterobiasis, is usually a simple, uncomplicated infection that does not spread beyond the intestine.

A cycle starts when a person swallows microscopic eggs picked up from another infected person by direct contact or by touching articles that person has touched. The eggs hatch in the intestine and then release larvae that mature into adult worms within about 1 month. Male and female worms mate, and the female migrates out to the anus to deposit eggs, which cause intense itchiness that is relieved by scratching. Herein lies a significant means of dispersal: Scratching contaminates the fingers, which, in turn, transfer eggs to bedclothes and other inanimate objects. This person becomes a host and a source of eggs and can spread them to others in addition to reinfesting himself. Enterobiasis occurs most often among families and in other close living situations. Its distribution is worldwide among all socioeconomic groups, but it seems to attack younger people more frequently than older ones.

Helminth Classification and Identification

The helminths are classified according to their shape; their size; the degree of development of various organs; the presence of hooks, suckers, or other special structures; the mode of reproduction; the kinds of hosts; and the appearance of eggs and larvae. They are identified in the laboratory by microscopic detection of the adult worm or its larvae and eggs, which often have distinctive shapes or external and internal structures. Occasionally, they are cultured in order to verify all of the life stages.

Distribution and Importance of Parasitic Worms

About 50 species of helminths parasitize humans. They are distributed in all areas of the world that support human life. Some worms are restricted to a given geographic region, and many have a higher incidence in tropical areas. This knowledge must be tempered with the realization that jet-age travel, along with human migration, is gradually changing the patterns of worm infections, especially of those species that do not require alternate hosts or special climatic conditions for development. The yearly estimate of worldwide cases numbers in the billions, and these are not confined to developing countries. A conservative estimate places 50 million helminth infections in North America alone. The primary targets are malnourished children.

You have now learned about the variety of organisms that microbiologists study and classify. And as you've seen, many such organisms are capable of causing disease. In chapter 6, you'll learn about the "not-quite-organisms" that can cause disease, namely, viruses.

CHECKPOINT

- The protists are mostly unicellular or colonial eukaryotes that lack specialized tissues. There are two major organism types: the Algae and the Protozoa.
- Algae are photosynthetic organisms that contain chloroplasts with chlorophyll and other pigments.
- Protozoa are heterotrophs that usually display some form of locomotion. Most are single-celled trophozoites, and many produce a resistant stage, or cyst.
- The Kingdom Animalia has only one group that contains members that are (sometimes) microscopic. These are the helminths or worms. Parasitic members include flatworms and roundworms that are able to invade and reproduce in human tissues.

Chapter Summary with Key Terms

5.1 The History of Eukaryotes

5.2 and 5.3 Form and Function of the Eukaryotic Cell: External and Internal Structures

A. Eukaryotic cells are complex and compartmentalized into individual organelles.

B. Major organelles and other structural features include: Appendages (cilia, flagella), **glycocalyx,** cell wall, cytoplasmic (or cell) membrane, **organelles** (nucleus, **nucleolus, endoplasmic reticulum, Golgi apparatus, mitochondria,** chloroplasts), ribosomes, cytoskeleton (**microfilaments,** microtubules). A review comparing the major differences between eukaryotic and prokaryotic cells is provided in table 5.2, page 129.

5.4 The Kingdom of the Fungi

Common names of the macroscopic fungi are mushrooms, bracket fungi, and puffballs. Microscopic fungi are known as yeasts and molds.

A. *Overall Morphology:* At the cellular (microscopic) level, fungi are typical eukaryotic cells, with thick cell walls. Yeasts are single cells that form buds and **pseudohyphae. Hyphae** are long, tubular filaments that can be septate or nonseptate and grow in a network called a **mycelium;** hyphae are characteristic of the filamentous fungi called molds.

B. *Nutritional Mode/Distribution:* All are **heterotrophic.** The majority are harmless **saprobes** living off organic **substrates** such as dead animal and plant tissues. A few are **parasites,** living on the tissues of other organisms, but none are obligate. Distribution is extremely widespread in many habitats.

C. *Reproduction:* Primarily through **spores** formed on special reproductive hyphae. In asexual reproduction, spores are formed through budding, partitioning of one hypha, or in special sporogenous structures;

examples are **conidia** and **sporangiospores.** In sexual reproduction, spores are formed following fusion of male and female strains and the formation of a sexual structure.

D. *Importance:* Fungi are essential decomposers of plant and animal detritis in the environment; economically beneficial as sources of antibiotics; used in making foods and in genetic studies. Adverse impacts include: decomposition of fruits and vegetables; human infections, or **mycoses;** some produce substances that are toxic if eaten.

5.5 The Protists

A. The Algae

Include photosynthetic kelps and seaweeds.

1. *Overall Morphology:* Are unicellular, colonial, filamentous or larger forms.
2. *Nutritional Mode/Distribution:* Photosynthetic; fresh and marine water habitats; main component of **plankton.**
3. *Importance:* Provide the basis of the food web in most aquatic habitats. Certain algae produce neurotoxins that are harmful to humans and animals.

B. The Protozoa

Include large single-celled organisms; a few are pathogens.

1. *Overall Morphology:* Most are unicellular; lack a cell wall. The cytoplasm is divided into ectoplasm and endoplasm. Many convert to a resistant, dormant stage called a **cyst.**
2. *Nutritional Mode/Distribution:* All are heterotrophic. Most are free-living in a moist habitat (water, soil); feed by engulfing other microorganisms and organic matter.
3. *Reproduction:* Asexual by binary fission and **mitosis,** budding; sexual by fusion of free-swimming gametes, conjugation.
4. *Major Groups:* Protozoa are subdivided into four groups based upon mode of locomotion and type of reproduction: Mastigophora, the flagellates, motile by flagella; Sarcodina, the amoebas, motile by pseudopods; Ciliophora, the ciliates, motile by cilia; Apicomplexa, motility not well developed; produce unique reproductive structures.
5. *Importance:* Ecologically important in food webs and decomposing organic matter. Medical significance: hundreds of millions of people are afflicted with one of the many protozoan infections (malaria, trypanosomiasis, amoebiasis). Can be spread from host to host by insect vectors.

5.6 The Parasitic Helminths

Includes three categories: roundworms, tapeworms, and flukes.

A. *Overall Morphology:* Animal cells; multicellular; individual organs specialized for reproduction, digestion, movement, protection, though some of these are reduced.

B. *Reproductive Mode:* Includes embryo, larval, and adult stages. Majority reproduce sexually. Sexes may be hermaphroditic.

C. *Epidemiology:* Developing countries in the tropics hardest hit by helminth infections; transmitted via ingestion of larvae or eggs in food; from soil or water. They afflict billions of humans.

Multiple-Choice and True-False Questions

Multiple-Choice Questions. Select the correct answer from the answers provided.

1. Both flagella and cilia are found primarily in
   a. algae c. fungi
   b. protozoa d. both b and c
2. Features of the nuclear envelope include
   a. ribosomes
   b. a double membrane structure
   c. pores that allow communication with the cytoplasm
   d. b and c
   e. all of these
3. The cell wall is found in which eukaryotes?
   a. fungi c. protozoa
   b. algae d. a and b
4. Yeasts are _____ fungi, and molds are _____ fungi.
   a. macroscopic, microscopic
   b. unicellular, filamentous
   c. motile, nonmotile
   d. water, terrestrial
5. Algae generally contain some type of
   a. spore c. locomotor organelle
   b. chlorophyll d. toxin
6. Almost all protozoa have a
   a. locomotor organelle c. pellicle
   b. cyst stage d. trophozoite stage
7. All mature sporozoa are
   a. parasitic c. carried by vectors
   b. nonmotile d. both a and b
8. Parasitic helminths reproduce with
   a. spores c. mitosis
   b. eggs and sperm d. cysts
   e. all of these
9. Mitochondria likely originated from
   a. archaea
   b. invaginations of the cell membrane
   c. purple bacteria
   d. cyanobacteria
10. **Single Matching.** Select the description that best fits the word in the left column.

____ diatom	a. the cause of malaria
____ *Rhizopus*	b. single-celled alga with silica in its cell wall
____ *Histoplasma*	c. fungal cause of Ohio Valley fever
____ *Cryptococcus*	d. the cause of amoebic dysentery
____ euglenid	e. genus of black bread mold
____ dinoflagellate	f. helminth worm involved in pinworm infection

____ *Trichomonas*

____ *Entamoeba*

____ *Plasmodium*

____ *Enterobius*

g. motile flagellated alga with eyespots
h. a yeast that infects the lungs
i. flagellated protozoan genus that causes an STD
j. alga that causes red tides

11. Most helminth infections
    a. are localized to one site in the body
    b. spread through major systems of the body
    c. develop within the spleen
    d. develop within the liver

True-False Questions. If the statement is true, leave as is. If it is false, correct it by rewriting the sentence.

12. Vesicles are attached to the rough endoplasmic reticulum.
13. Hypha that are divided into compartments by cross walls are called septate hypha.
14. The infective stage of a protozoan is the trophozoite.
15. In humans, fungi can only infect the skin.
16. Fungi generally derive nutrients through photosynthesis.

Writing to Learn

These questions are suggested as a *writing-to-learn* experience. For each question, compose a one- or two-paragraph answer that includes the factual information needed to completely address the question.

1. Construct a chart that reviews the major similarities and differences between prokaryotic and eukaryotic cells.
2. a. Which kingdoms of the five-kingdom system contain eukaryotic microorganisms? How do unicellular, colonial, and multicellular organisms differ from each other?
   b. Give examples of each type.
3. a. Describe the anatomy and functions of each of the major eukaryotic organelles.
   b. How are flagella and cilia similar? How are they different?
   c. Compare and contrast the smooth ER, the rough ER, and the Golgi apparatus in structure and function.
4. Describe some of the ways that organisms use lysosomes.
5. For what reasons would a cell need a "skeleton"?
6. a. Differentiate between the yeast and hypha types of fungal cell.
   b What is a mold?
   c. What does it mean if a fungus is dimorphic?
7. a. How does a fungus feed?
   b. Where would one expect to find fungi?
8. What is a working definition of a "protist"?
9. a. Explain the general characteristics of the protozoan life cycle.
   b. Describe the protozoan adaptations for feeding.
   c. Describe protozoan reproductive processes.
10. a. Briefly outline the characteristics of the four protozoan groups.
    b. What is an important pathogen in each group?
11. a. Construct a chart that compares the four groups of eukaryotic microorganisms (fungi, algae, protozoa, helminths) in cellular structure.
    b. Indicate whether each group has a cell wall, chloroplasts, motility, or some other distinguishing feature.
    c. Include also the manner of nutrition and body plan (unicellular, colonial, filamentous, or multicellular) for each group.

Concept Mapping

Appendix D provides guidance for working with concept maps.

1. Construct your own concept map using the following words as the *concepts*. Supply the linking words between each pair of concepts.

Golgi apparatus
chloroplasts
cytoplasm
endospore
ribosomes
flagella
nucleolus

Critical Thinking Questions

Critical thinking is the ability to reason and solve problems using facts and concepts. These questions can be approached from a number of angles, and in most cases, they do not have a single correct answer.

1. Suggest some ways that one would go about determining if mitochondria and chloroplasts are a modified prokaryotic cell.
2. Give the common name of a eukaryotic microbe that is unicellular, walled, nonphotosynthetic, nonmotile, and bud-forming.
3. Give the common name of a microbe that is unicellular, nonwalled, motile with flagella, and has chloroplasts.
4. Which group of microbes has long, thin pseudopods and is encased in a hard shell?

5. What general type of multicellular parasite is composed primarily of thin sacs of reproductive organs?
6. a. Name two parasites that are transmitted in the cyst form.
   b. How must a non-cyst-forming pathogenic protozoan be transmitted? Why?
7. You just found an old container of food in the back in your refrigerator. You open it and see a mass of multicolored fuzz. As a budding microbiologist, describe how you would determine what types of organisms are growing on the food.
8. Explain what factors could cause opportunistic mycoses to be a growing medical problem.
9. a. How are bacterial endospores and cysts of protozoa alike?
   b. How do they differ?
10. Can you think of a way to determine if a child is suffering from pinworms? Hint: Scotch tape is involved.

Visual Understanding

1. **From chapter 4, figures 4.28 and 4.29*a*.** Although you may never have visited either of these two locations, you may have seen similar sites. List two locations that you have encountered that have shown colorful evidence of microbial growth.

2. **From chapter 1, figure 1.15.** Which of the groups of organisms from this figure will contain a nucleus? Why?

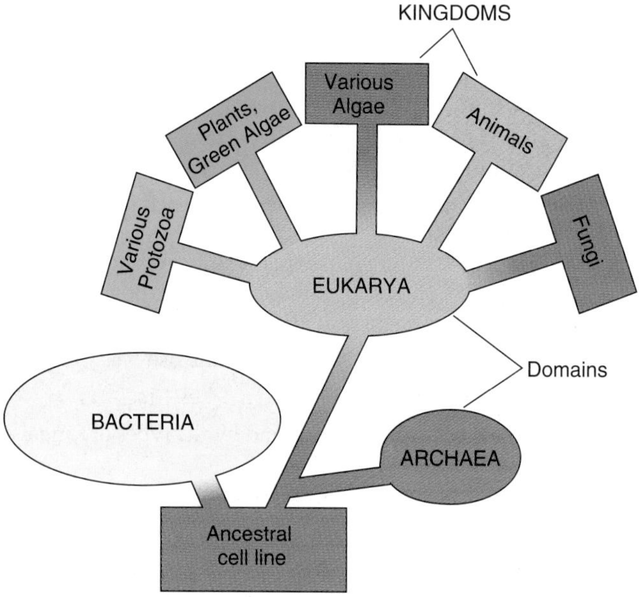

3 cell types, showing relationship with domains and kingdoms

Internet Search Topics

Go to http://www.aris.mhhe.com, and click on "microbiology" and then this textbook's author/title. Go to chapter 5, access the URLs listed under Internet Search Topics, and research the following:

1. The endosymbiotic theory of eukaryotic cell evolution. List data from studies that support this idea.
2. The medical problems caused by mycotoxins and "sick building" syndrome.
3. Several excellent websites provide information and animations on eukaryotic cells and protists. Access the websites listed, and survey these sources to observe the varied styles of feeding, reproduction, and locomotion seen among these microbes.
4. Fungi are sometimes associated with illness in animals. Search the Internet for information regarding a recent recall of dog food due to mold contamination. How serious was the health risk associated with this product?

CHAPTER 6

An Introduction to the Viruses

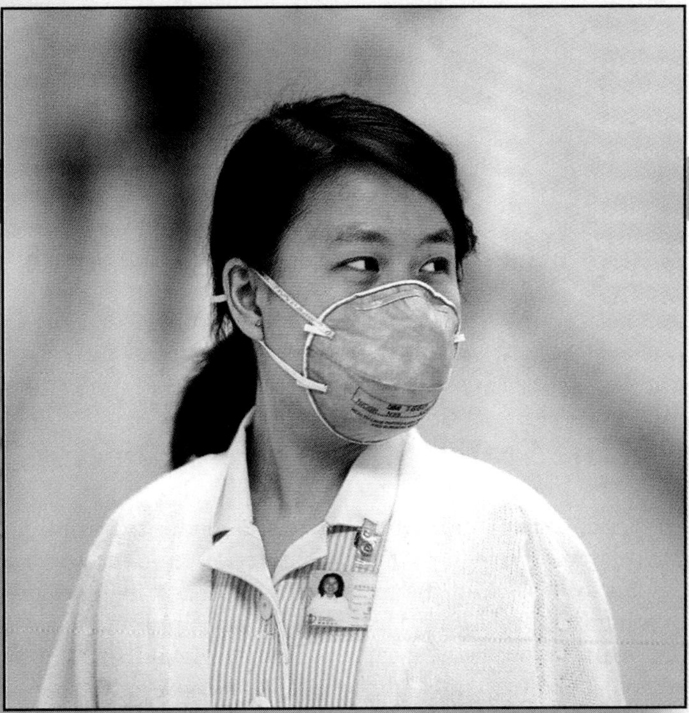

CASE FILE 6

Severe acute respiratory syndrome (SARS) is a newly identified respiratory infection caused by a novel coronavirus. The SARS pandemic is believed to have originated in the Guangdong Province of China during the fall of 2002. A SARS patient from this region traveled to Hong Kong on February 15, 2003, and may have infected several guests at a hotel where he resided. One of the hotel guests was a resident of Hong Kong. By February 24, the hotel resident came down with a fever, chills, dry cough, runny nose, and malaise. Over the next several days, his symptoms worsened to pneumonia, leading to his hospitalization at the Prince of Wales Hospital in Hong Kong.

The Prince of Wales Hospital is a large medical teaching hospital of the Chinese University of Hong Kong. By March 12, a large-scale outbreak of SARS occurred inside of the hospital. During the initial outbreak, March 15 through 25, 2003, 44% of the SARS cases (68 of 156) admitted to the Prince of Wales Hospital were hospital workers. SARS is a contagious disease that spreads from person to person primarily through contact with respiratory droplets containing the SARS virus. Chinese University researchers and the Hong Kong Hospital Authority conducted studies to determine why hospital workers were so vulnerable to SARS at this hospital.

- *Can you think of what factors contributed to the high rates of SARS transmission seen among hospital workers?*
- *What precautions would you take in caring for SARS patients?*

Case File 6 Wrap-Up appears on page 166.

CHAPTER OVERVIEW

Viruses:

- Are a unique group of tiny infectious particles that are obligate parasites of cells.
- Do not exhibit the characteristics of life but can regulate the functions of host cells.
- Infect all groups of living things and produce a variety of diseases.
- Are not cells but resemble complex molecules composed of protein and nucleic acid.
- Are encased in an outer shell or envelope and contain either DNA or RNA as their genetic material.

- Are genetic parasites that take over the host cell's metabolism and synthetic machinery.
- Can instruct the cell to manufacture new virus parts and assemble them.
- Are released in a mature, infectious form, followed by destruction of the host cell.
- May persist in cells, leading to slow progressive diseases and cancer.
- Are identified by structure, host cell, type of nucleic acid, outer coating, and type of disease.
- Are among the most common infectious agents, causing serious medical and agricultural impact.

6.1 The Search for the Elusive Viruses

The discovery of the light microscope made it possible to see firsthand the agents of many bacterial, fungal, and protozoan diseases. But the techniques for observing and cultivating these relatively large microorganisms were useless for viruses. For many years, the cause of viral infections such as smallpox and polio was unknown, even though it was clear that the diseases were transmitted from person to person. The French scientist Louis Pasteur was certainly on the right track when he postulated that rabies was caused by a "living thing" smaller than bacteria, and in 1884 he was able to develop the first vaccine for rabies. Pasteur also proposed the term **virus** (L. poison) to denote this special group of infectious agents.

The first substantial revelations about the unique characteristics of viruses occurred in the 1890s. First, D. Ivanovski and M. Beijerinck showed that a disease in tobacco was caused by a virus (tobacco mosaic virus). Then, Friedrich Loeffler and Paul Frosch discovered an animal virus that causes foot-and-mouth disease in cattle. These early researchers found that when infectious fluids from host organisms were passed through porcelain filters designed to trap bacteria, the filtrate remained infectious. This result proved that an infection could be caused by a cell-free fluid containing agents smaller than bacteria and thus first introduced the concept of a *filterable virus.*

Over the succeeding decades, a remarkable picture of the physical, chemical, and biological nature of viruses began to take form. Years of experimentation were required to show that viruses were noncellular particles with a definite size, shape, and chemical composition. Using special techniques, they could be cultured in the laboratory. By the 1950s, virology had grown into a multifaceted discipline that promised to provide much information on disease, genetics, and even life itself **(Insight 6.1)**.

6.2 The Position of Viruses in the Biological Spectrum

Viruses are a unique group of biological entities known to infect every type of cell, including bacteria, algae, fungi, protozoa, plants, and animals. Although the emphasis in this chapter is on animal viruses, much credit for our knowledge must be given to experiments with bacterial and plant viruses. The exceptional and curious nature of viruses prompts numerous questions, including:

1. Are they organisms; that is, are they alive?
2. What are their distinctive biological characteristics?
3. How can particles so small, simple, and seemingly insignificant be capable of causing disease and death?
4. What is the connection between viruses and cancer?

In this chapter, we address these questions and many others.

The unusual structure and behavior of viruses have led to debates about their connection to the rest of the microbial world. One viewpoint holds that viruses are unable to exist independently from the host cell, so they are not living things but are more akin to large, infectious molecules. Another viewpoint proposes that even though viruses do not exhibit most of the life processes of cells, they can direct them and thus are certainly more than inert and lifeless molecules. This view is the predominant one among scientists today. This debate has greater philosophical than practical importance because viruses are agents of disease and must be dealt with through control, therapy, and prevention, whether we regard them as living or not. In keeping with their special position in the biological spectrum, it is best to describe viruses as *infectious particles* (rather than organisms) and as either *active* or *inactive* (rather than alive or dead).

Viruses are different from their host cells in size, structure, behavior, and physiology. They are a type of *obligate intracellular parasites* that cannot multiply unless they invade a specific host cell and instruct its genetic and metabolic machinery to make and release quantities of new viruses. Because of this characteristic, viruses are capable of causing serious damage and disease. Other unique properties of viruses are summarized in **table 6.1.**

CHECKPOINT

- Viruses are noncellular entities whose properties have been identified through technological advances in microscopy and tissue culture.
- Viruses are infectious particles that invade every known type of cell. They are not alive, yet they are able to redirect the metabolism of living cells to reproduce virus particles.
- Viral replication inside a cell usually causes death or loss of function of that cell.

TABLE 6.1	Properties of Viruses

- Are obligate intracellular parasites of bacteria, protozoa, fungi, algae, plants, and animals.
- Ultramicroscopic size, ranging from 20 nm up to 450 nm (diameter).
- Are not cells; structure is very compact and economical.
- Do not independently fulfill the characteristics of life.
- Are inactive macromolecules outside the host cell and active only inside host cells.
- Basic structure consists of protein shell (capsid) surrounding nucleic acid core.
- Nucleic acid can be either DNA or RNA but not both.
- Nucleic acid can be double-stranded DNA, single-stranded DNA, single-stranded RNA, or double-stranded RNA.
- Molecules on virus surface impart high specificity for attachment to host cell.
- Multiply by taking control of host cell's genetic material and regulating the synthesis and assembly of new viruses.
- Lack enzymes for most metabolic processes.
- Lack machinery for synthesizing proteins.

6.3 The General Structure of Viruses

Size Range

As a group, viruses represent the smallest infectious agents (with some unusual exceptions to be discussed later in this chapter). Their size relegates them to the realm of the ultramicroscopic. This term means that most of them are so minute (<0.2 μm) that an electron microscope is necessary to detect them or to examine their fine structure. They are dwarfed by their host cells: More than 2,000 bacterial viruses could fit into an average bacterial cell, and more than 50 million polioviruses could be accommodated by an average human cell. Animal viruses range in size from the small parvoviruses[1] (around 20 nm [0.02 μm] in diameter) to mimiviruses[2] that are larger than small bacteria (up to 450 nm [0.4 μm] in length) **(figure 6.1).** Some cylindrical viruses are relatively long (800 nm [0.8 μm] in length) but so narrow in diameter (15 nm [0.015 μm]) that their visibility is still limited without

1. DNA viruses that cause respiratory infections in humans.
2. Mimivirus was just identified in 2003. Its name stands for "mimicking microbe."

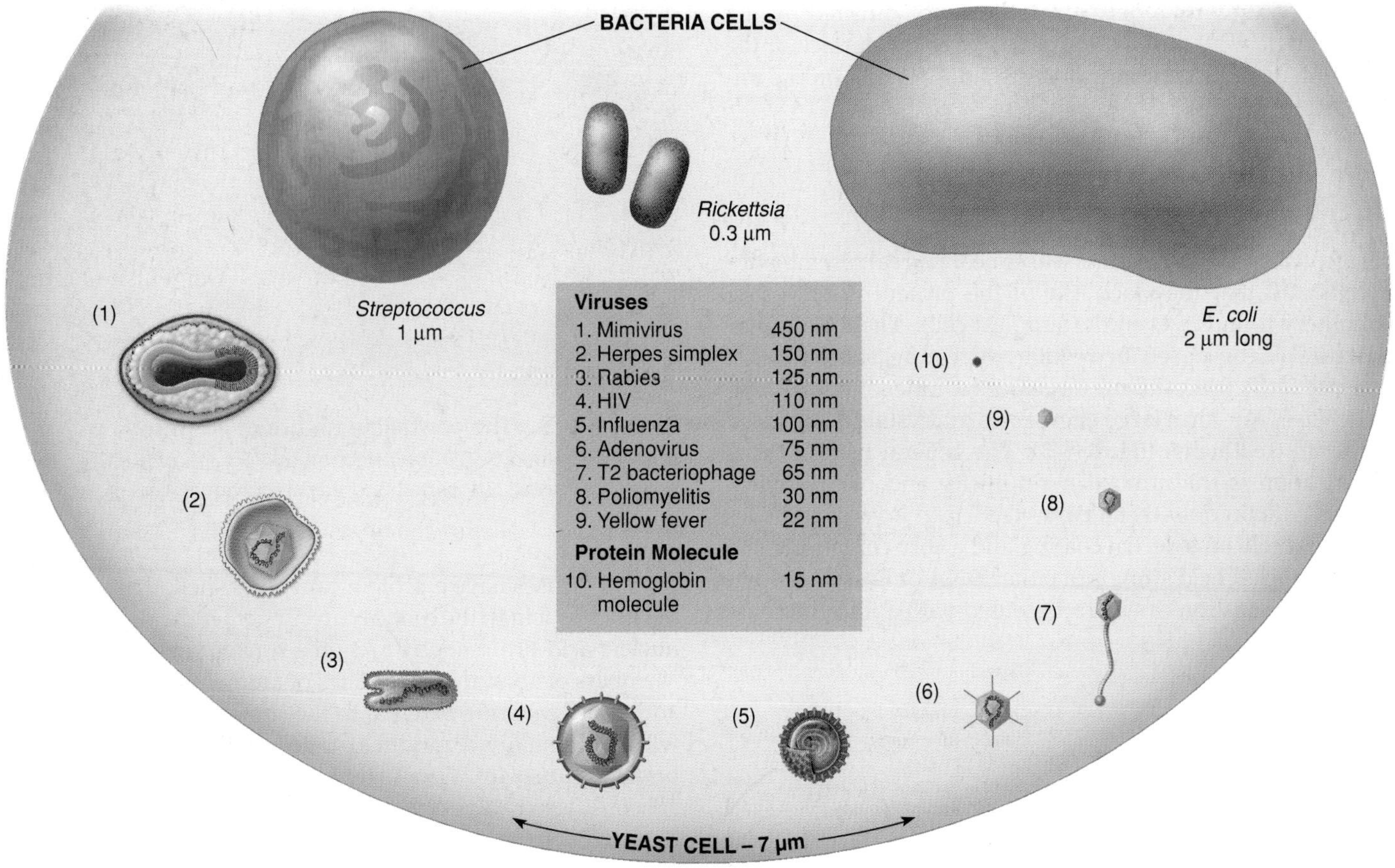

Figure 6.1 Size comparison of viruses with a eukaryotic cell (yeast) and bacteria.
Viruses range from largest (1) to smallest (9). A molecule of protein (10) is included to indicate proportion of macromolecules.

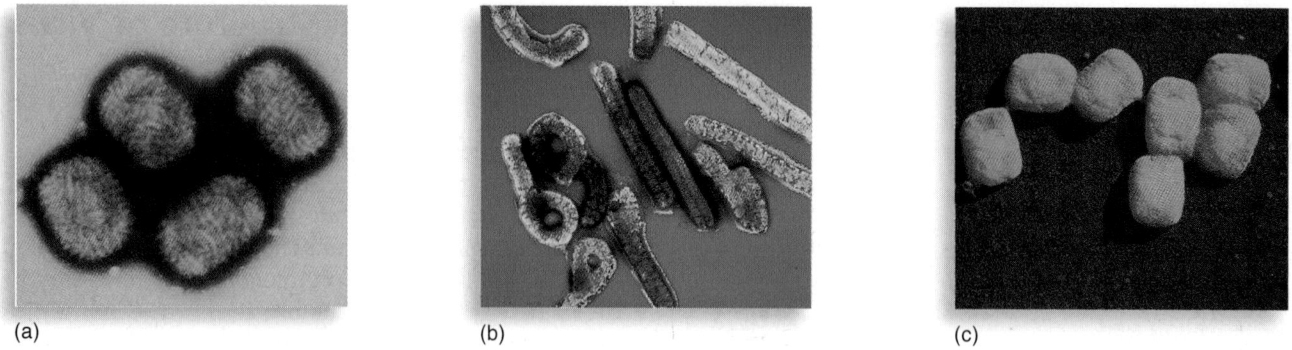

Figure 6.2 Methods of viewing viruses.
(a) Negative staining of an orfvirus (a type of poxvirus), revealing details of its outer coat. **(b)** Positive stain of the Ebola virus, a type of filovirus, so named because of its tendency to form long strands. Note the textured capsid. **(c)** Shadowcasting image of a vaccinia virus.

the high magnification and resolution of an electron microscope. Figure 6.1 compares the sizes of several viruses with prokaryotic and eukaryotic cells and molecules.

Viral architecture is most readily observed through special stains in combination with electron microscopy **(figure 6.2).** Negative staining uses very thin layers of an opaque salt to outline the shape of the virus against a dark background and to enhance textural features on the viral surface. Internal details are revealed by positive staining of specific parts of the virus such as protein or nucleic acid. The *shadowcasting* technique attaches a virus preparation to a surface and showers it with a dense metallic vapor directed from a certain angle. The thin metal coating over the surface of the virus enhances its contours, and a shadow is cast on the unexposed side.

Viral Components: Capsids, Nucleic Acids, and Envelopes

It is important to realize that viruses bear no real resemblance to cells and that they lack any of the protein-synthesizing machinery found in even the simplest cells. Their molecular structure is composed of regular, repeating subunits that give rise to their crystalline appearance. Indeed, many purified viruses can form large aggregates or crystals if subjected to special treatments **(figure 6.3).** The general plan of virus organization is the utmost in simplicity and compactness. Viruses contain only those parts needed to invade and control a host cell: an external coating and a core containing one or more nucleic acid strands of either DNA or RNA. This pattern of organization can be represented with a flowchart:

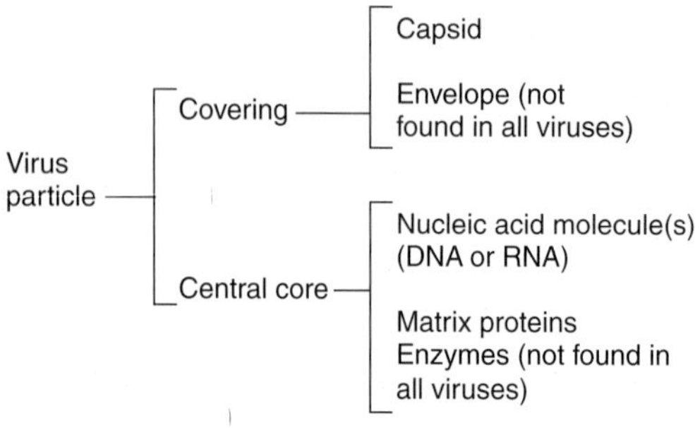

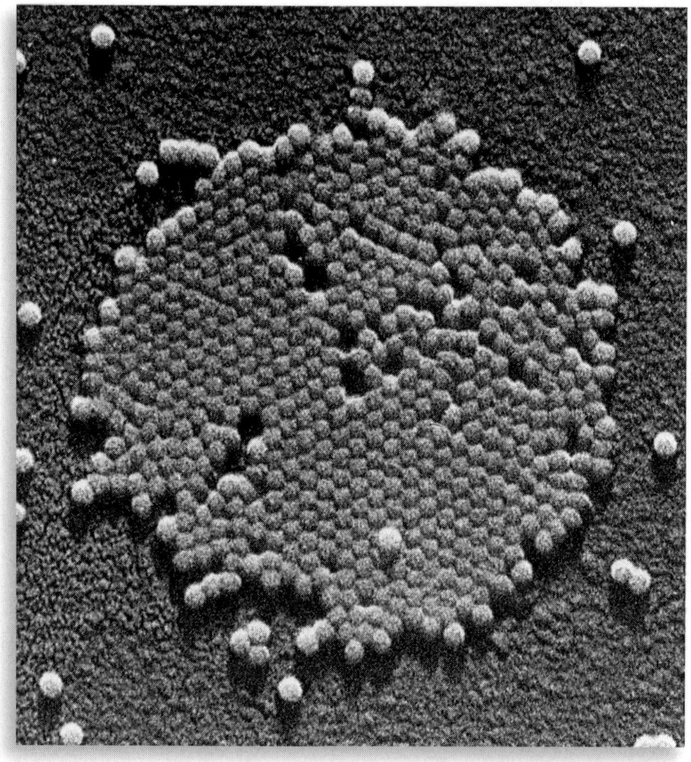

Figure 6.3 The crystalline nature of viruses.
Highly magnified (150,000×) electron micrograph of purified poliovirus crystals, showing hundreds of individual viruses.

All viruses have a protein **capsid,** or shell, that surrounds the nucleic acid in the central core. Together the capsid and the nucleic acid are referred to as the **nucleocapsid (figure 6.4).** Members of 13 of the 20 families of animal viruses possess an additional covering external to the capsid called an envelope, which is usually a modified piece of the host's cell membrane **(figure 6.4*b*).** Viruses that consist of only a nucleocapsid are considered *naked viruses* **(figure 6.4*a*).** As we shall see later, the enveloped viruses also differ from the naked viruses in the way that they enter and leave a host cell. A fully formed virus that is able to establish an infection in a host cell is often called a **virion.**

INSIGHT 6.1 — Discovery

A Positive View of Viruses

Looking at this beautiful tulip, one would never guess that it derives its pleasing appearance from a viral infection. It contains tulip mosaic virus, which alters the development of the plant cells and causes complex patterns of colors in the petals. Aside from this, the virus does not cause severe harm to the plants. Despite the reputation of viruses as cell killers, there is another side of viruses—that of being harmless, and in some cases, even beneficial.

Although there is no agreement on the origins of viruses, it is very likely that they have been in existence for billions of years. Virologists are convinced that viruses have been an important force in the evolution of living things. This is based on the fact that they interact with the genetic material of their host cells and that they carry genes from one host to another (transduction). It is convincing to imagine that viruses arose early in the history of cells as loose pieces of genetic material that became dependent nomads, moving from cell to cell. Viruses are also a significant factor in the functioning of many ecosystems. For example, it is documented that seawater can contain 10 million viruses per milliliter. Because viruses are made of the same elements as living cells, it is estimated that the sum of viruses in the ocean represents 270 million metric tons of organic matter.

Over the past several years, biomedical experts have been looking at viruses as vehicles to treat infections and disease. Viruses are already essential for production of vaccines to treat viral infections such as influenza, polio, and measles. Vaccine experts have also engineered new types of viruses by combining a less harmful virus such as vaccinia or adenovirus with some genetic material from a pathogen such as HIV or herpes simplex. This technique creates a vaccine that provides immunity but does not expose the person to the intact pathogen. Several of these types of vaccines are currently in development.

The "harmless virus" approach is also being used to treat genetic diseases such as cystic fibrosis and sickle-cell anemia. With gene therapy, the normal gene is inserted into a retrovirus, such as the mouse leukemia virus, and the patient is infected with this altered virus. It is hoped that the virus will introduce the needed gene into the cells and correct the defect. Dozens of experimental trials are currently underway to develop potential cures for diseases, with some successes (see chapter 10). One problem has been that infection with these mouse viruses has led to the development of cancer in some patients.

Virologists have also created mutant adenoviruses (ONYX) that target cancer cells. These viruses cannot spread among normal cells, but when they enter cancer cells, they immediately cause the cells to self-destruct.

An older therapy getting a second chance involves use of bacteriophages to treat bacterial infections. This technique was tried in the past with mixed success but was abandoned for more efficient antimicrobial drugs. The basis behind the therapy is that bacterial viruses would seek out only their specific host bacteria and would cause complete destruction of the bacterial cell. Newer experiments with animals have demonstrated that this method can control infections as well as traditional drugs can. Some potential applications being considered are adding phage suspension to grafts to control skin infections and to intravenous fluids for blood infections.

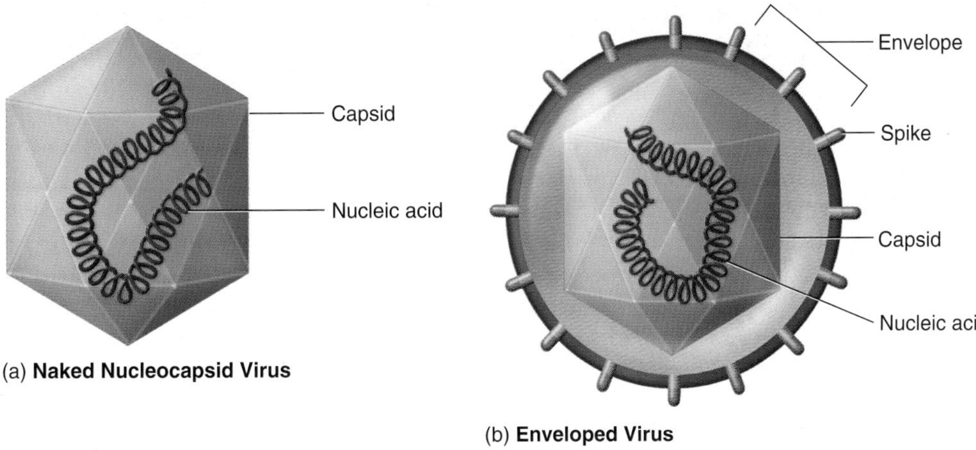

Figure 6.4 Generalized structure of viruses.
(a) The simplest virus is a naked virus (nucleocapsid) consisting of a geometric capsid assembled around a nucleic acid strand or strands.
(b) An enveloped virus is composed of a nucleocapsid surrounded by a flexible membrane called an envelope. The envelope usually has special receptor spikes inserted into it.

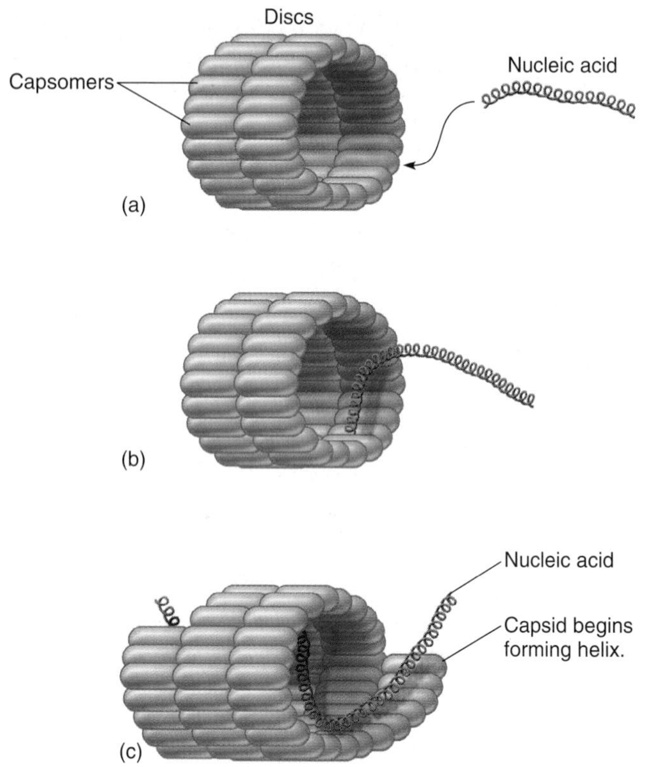

Figure 6.5 Assembly of helical nucleocapsids.
(a) Capsomers assemble into hollow discs. **(b)** The nucleic acid is inserted into the center of the disc. **(c)** Elongation of the nucleocapsid progresses from one or both ends, as the nucleic acid is wound "within" the lengthening helix.

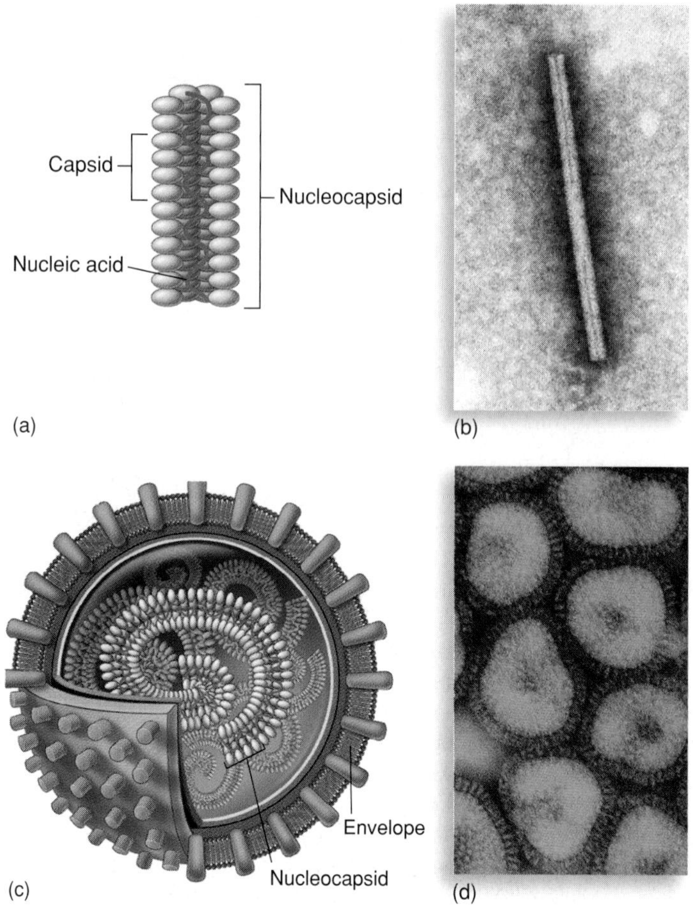

Figure 6.6 Typical variations of viruses with helical nucleocapsids.
Naked helical virus (tobacco mosaic virus): (a) a schematic view and **(b)** a greatly magnified micrograph. Note the overall cylindrical morphology.
Enveloped helical virus (influenza virus): (c) a schematic view and **(d)** an electron micrograph of the same virus (350,000×).

The Viral Capsid: The Protective Outer Shell

When a virus particle is magnified several hundred thousand times, the capsid appears as the most prominent geometric feature. In general, each capsid is constructed from identical subunits called **capsomers** that are constructed from protein molecules. The capsomers spontaneously self-assemble into the finished capsid. Depending on how the capsomers are shaped and arranged, this assembly results in two different types: helical and icosahedral.

The simpler **helical capsids** have rod-shaped capsomers that bond together to form a series of hollow discs resembling a bracelet. During the formation of the nucleocapsid, these discs link with other discs to form a continuous helix into which the nucleic acid strand is coiled **(figure 6.5).** In electron micrographs, the appearance of a helical capsid varies with the type of virus. The nucleocapsids of naked helical viruses are very rigid and tightly wound into a cylinder-shaped package **(figure 6.6*a,b*).** An example is the *tobacco mosaic virus,* which attacks tobacco leaves. Enveloped helical nucleocapsids are more flexible and tend to be arranged as a looser helix within the envelope **(figure 6.6*c,d*).** This type of morphology is found in several enveloped human viruses, including those of influenza, measles, and rabies.

The capsids of a number of major virus families are arranged in an **icosahedron** (eye″-koh-suh-hee′-drun)—a three-dimensional, 20-sided figure with 12 evenly spaced corners. The arrangements of the capsomers vary from one virus to another. Some viruses construct the capsid from a single type of capsomer while others may contain several types of capsomers **(figure 6.7).** Although the capsids of all icosahedral viruses have this sort of symmetry, they can have major variations in the number of capsomers; for example, a poliovirus has 32, and an adenovirus has 252 capsomers. Individual capsomers can look either ring- or dome-shaped, and the capsid itself can appear spherical or cubical **(figure 6.8).** During assembly of the virus, the nucleic acid is packed into the center of this icosahedron, forming a nucleocapsid. While most viruses have capsids that are either icosahedral or helical, there is another category of capsid that is simply called "complex." Complex capsids may have multiple types of proteins and take shapes that are not symmetrical. Two examples of complex

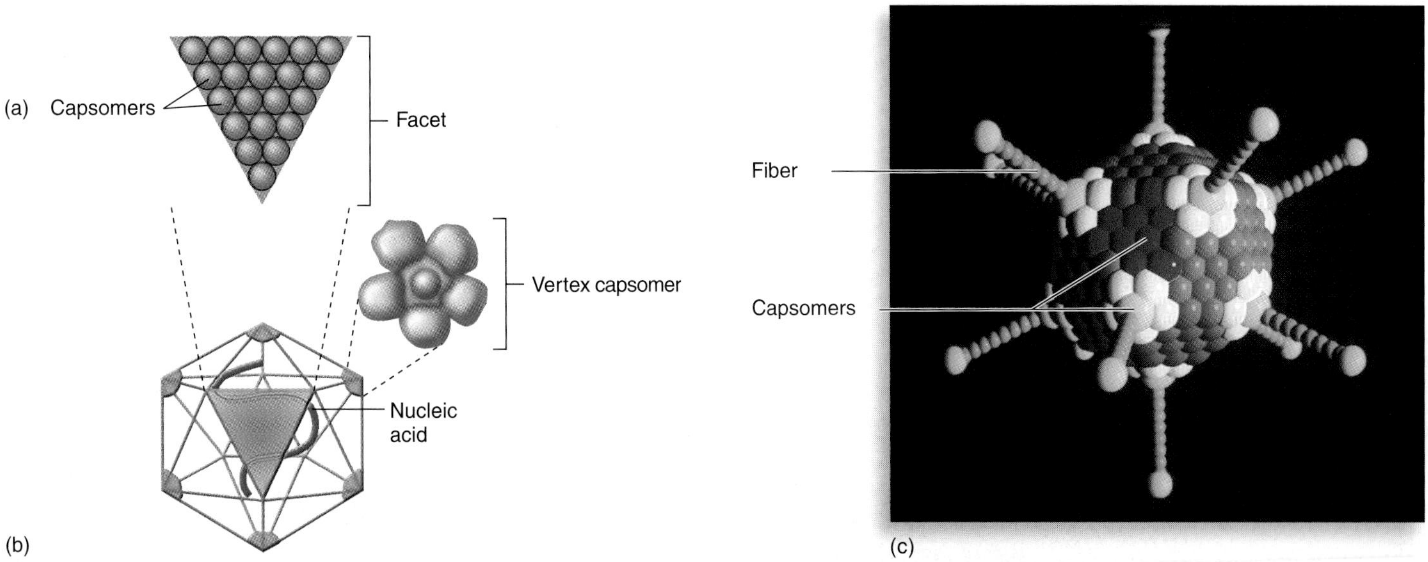

Figure 6.7 The structure and formation of an icosahedral virus (adenovirus is the model).
(a) A facet or "face" of the capsid is composed of 21 identical capsomers arranged in a triangular shape. Each vertex or "point" consists of a different type of capsomer with a single penton in the center. Other viruses can vary in the number, types, and arrangement of capsomers. **(b)** An assembled virus shows how the facets and vertices come together to form a shell around the nucleic acid. **(c)** A three-dimensional model of this virus shows fibers attached to the pentons.

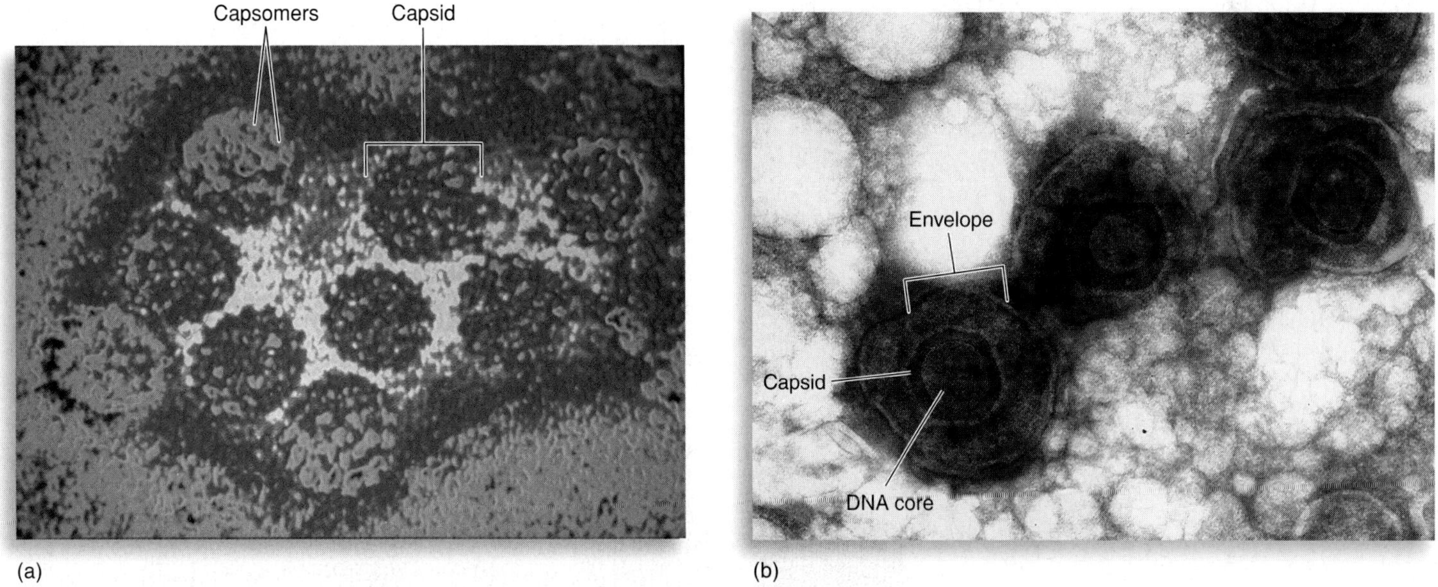

Figure 6.8 Two types of icosahedral viruses, highly magnified.
(a) Micrograph of papillomaviruses with unusual, ring-shaped capsomers. **(b)** Herpesvirus, an enveloped icosahedron (300,000×). Both micrographs have been colorized.

viruses are shown in **figure 6.9.** Another factor that alters the appearance of icosahedral viruses is whether or not they have an outer envelope; contrast a papillomavirus (warts) and its naked nucleocapsid with herpes simplex (cold sores) and its enveloped nucleocapsid **(figure 6.10).**

The Viral Envelope

When **enveloped viruses** (mostly animal) are released from the host cell, they take with them a bit of its membrane system in the form of an envelope, as described later on. Some viruses bud off the cell membrane; others leave via the nuclear envelope or the endoplasmic reticulum. Whichever avenue of escape, the viral envelope differs significantly from the host's membranes. In the envelope, some or all of the regular membrane proteins are replaced with special viral proteins. Some proteins form a binding layer between the envelope and capsid of the virus, and glycoproteins (proteins bound to a carbohydrate) remain exposed on the outside of the envelope. These protruding molecules, called **spikes** or

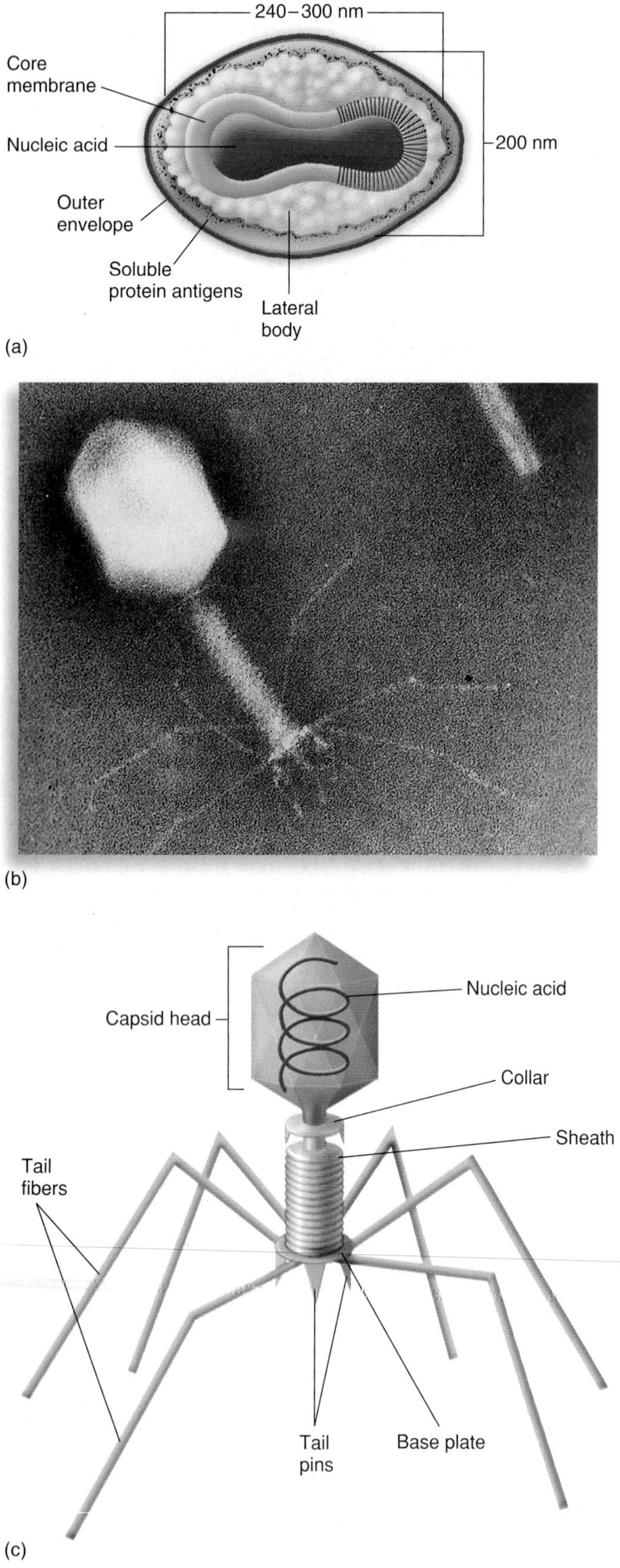

Figure 6.9 Structure of complex viruses.
(a) The vaccinia virus, a poxvirus. **(b)** Photomicrograph and **(c)** diagram of a T4 bacteriophage, a virus that infects bacteria.

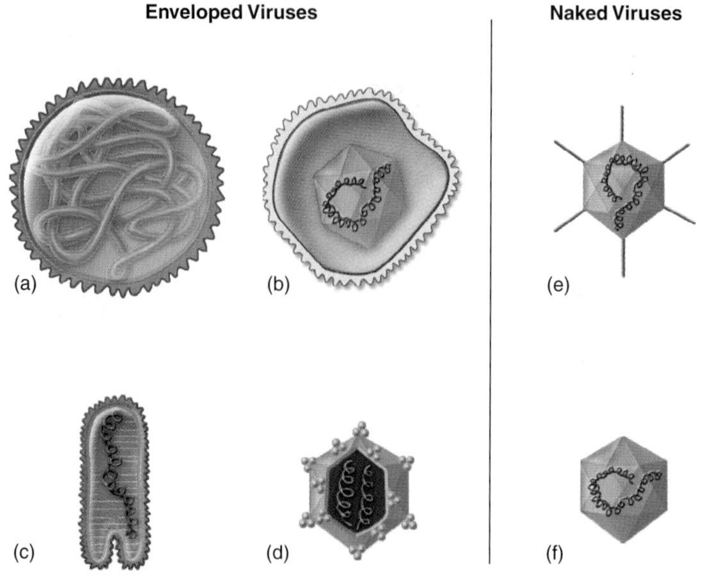

Figure 6.10 Morphology of viruses.
Enveloped viruses: (a) mumps virus, an enveloped RNA virus with a helical nucleocapsid; **(b)** herpesvirus, an enveloped DNA virus with an icosahedral nucleocapsid; **(c)** rhabdovirus, a helical RNA virus with a bullet-shaped envelope; **(d)** HIV, an RNA retrovirus with an icosahedral capsid.
Naked viruses: (e) adenovirus, a DNA virus with fibers on the capsid; **(f)** papillomavirus, a DNA virus that causes warts.

peplomers, are essential for the attachment of viruses to the next host cell. Because the envelope is more supple than the capsid, enveloped viruses are pleomorphic and range from spherical to filamentous in shape.

Functions of the Viral Capsid/Envelope

The outermost covering of a virus is indispensable to viral function because it protects the nucleic acid from the effects of various enzymes and chemicals when the virus is outside the host cell. For example, the capsids of enteric (intestinal) viruses such as polio and hepatitis A are resistant to the acid- and protein-digesting enzymes of the gastrointestinal tract. Capsids and envelopes are also responsible for helping to introduce the viral DNA or RNA into a suitable host cell, first by binding to the cell surface and then by assisting in penetration of the viral nucleic acid (to be discussed in more detail later in the chapter). In addition, parts of viral capsids and envelopes stimulate the immune system to produce antibodies that can neutralize viruses and protect the host's cells against future infections (see chapter 15).

Nucleic Acids: At the Core of a Virus

The sum total of the genetic information carried by an organism is known as its **genome.** So far, one biological constant is that the genetic information of living cells is carried by nucleic acids (DNA, RNA). Viruses, although neither alive nor cells, are no exception to this rule, but there is a significant difference. Unlike cells, which contain both

DNA and RNA, viruses contain either DNA or RNA *but not both*. Because viruses must pack into a tiny space all of the genes necessary to instruct the host cell to make new viruses, the number of viral genes is quite small compared with that of a cell. It varies from four genes in hepatitis B virus to hundreds of genes in some herpesviruses. Viruses possess only the genes needed to invade host cells and redirect their activity. By comparison, the bacterium *Escherichia coli* has approximately 4,000 genes, and a human cell has approximately 30,000–40,000 genes. These additional genes allow cells to carry out the complex metabolic activity necessary for independent life.

TABLE 6.2 Medically Relevant DNA Virus Groups

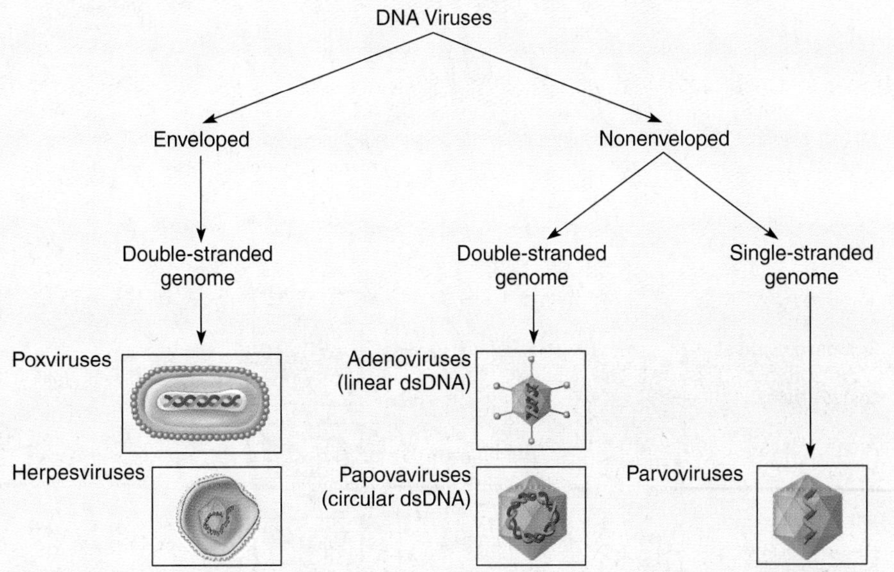

Source: Adapted from: *Poxviridae* from Buller et al., National Institute of Allergy & Infectious Disease, Department of Health & Human Services.

In chapter 2, you learned that DNA usually exists as a double-stranded molecule and that RNA is single-stranded. Although most viruses follow this same pattern, a few exhibit distinctive and exceptional forms. Notable examples are the parvoviruses, which contain single-stranded DNA, and reoviruses (a cause of respiratory and intestinal tract infections), which contain double-stranded RNA. In fact, viruses exhibit wide variety in how their RNA or DNA is configured. DNA viruses can have single-stranded (ss) or double-stranded (ds) DNA; the dsDNA can be arranged linearly or in ds circles. RNA viruses can be double-stranded but are more often single-stranded. You will learn in chapter 9 that all proteins are made by "translating" the nucleic acid code on a single strand of RNA into an amino acid sequence. Single-stranded RNA genomes that are ready for immediate translation into proteins are called *positive-sense* RNA. Other RNA genomes have to be converted into the proper form to be made into proteins, and these are called *negative-sense* RNA. RNA genomes may also be *segmented*, meaning that the individual genes exist on separate pieces of RNA. The influenza virus (an orthomyxovirus) is an example of this. A special type of RNA virus is called a *retrovirus*. We'll discuss it later. **Tables 6.2** and **6.3** summarize the structures of some medically relevant DNA and RNA viruses.

In all cases, these tiny strands of genetic material carry the blueprint for viral structure and functions. In a very real sense, viruses are genetic parasites because they cannot multiply until their nucleic acid has reached the internal habitat of the host cell. At the minimum, they must carry genes for synthesizing the viral capsid and genetic material, for regulating the actions of the host, and for packaging the mature virus.

Other Substances in the Virus Particle

In addition to the protein of the capsid, the proteins and lipids of envelopes, and the nucleic acid of the core, viruses can contain enzymes for specific operations within their host cell. They may come with preformed enzymes that are required for viral replication. Examples include *polymerases* (pol-im'-ur-ace-uz) that synthesize DNA and RNA, and replicases that copy RNA. The AIDS virus comes equipped with *reverse transcriptase* for synthesizing DNA from RNA. However, viruses completely lack the genes for synthesis of metabolic enzymes. As we shall see, this deficiency has little consequence, because viruses have adapted to completely take over their hosts' metabolic resources. Some viruses can actually carry away substances from their host cell. For instance, arenaviruses pack along host ribosomes, and retroviruses "borrow" the host's tRNA molecules.

6.4 How Viruses Are Classified and Named

Although viruses are not classified as members of the kingdoms discussed in chapter 1, they are diverse enough to require their own classification scheme to aid in their study and identification. In an informal and general way, we have already begun classifying viruses—as animal, plant, or bacterial viruses; enveloped or naked viruses; DNA or RNA viruses; and helical or icosahedral viruses. These introductory categories are certainly useful in organization and description, but the study of specific viruses requires a more standardized method of nomenclature. For many years, the animal viruses were classified mainly on the basis of their hosts and the kind of diseases they caused. Newer systems for naming viruses also take into account the actual nature of the virus particles themselves, with only partial emphasis on host and disease. The main criteria presently used to group viruses are structure, chemical composition, and similarities in genetic makeup.

In 2000, the International Committee on the Taxonomy of Viruses issued its latest report on the classification of viruses.

TABLE 6.3 Medically Relevant RNA Viruses

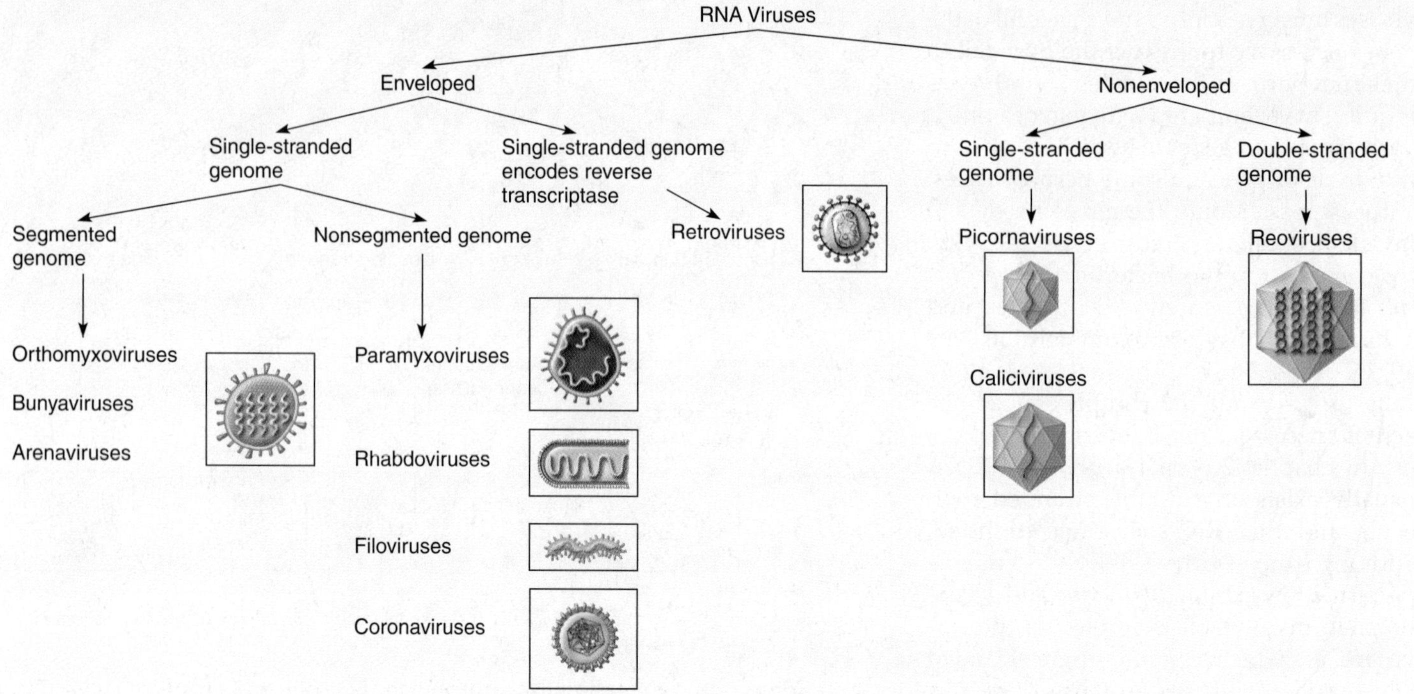

TABLE 6.4 Examples from the Three Orders of Viruses

Order	Family	Genus	Species	Host
Caudovirales	Poxviridae	Orthopoxvirus	Vaccinia virus	Animal
	Herpesviridae	Cytomegalovirus	Human herpesvirus 5	Animal
	Myoviridae	SPO1-like virus	Bacillus phage	Bacterium
Mononegavirales	Paramyxoviridae	Morbillivirus	Measles virus	Animal
	Filoviridae	Ebolavirus	Ebola virus	Animal
	Sequiviridae	Sequivirus	Parsnip yellow fleck virus	Plant
Nidovirales	Togaviridae	Rubivirus	Rubella virus	Animal
	Luteoviridae	Tobamovirus	Tobacco mosaic virus	Plant

Source: Adapted from Fauquet, C. M., et al. 2004. *Virus Taxonomy: Eighth Report of the International Committee on Taxonomy of Viruses.* New York: Academic Press.

The committee listed 3 orders, 63 families, and 263 genera of viruses. Previous to 2000, there had been only a single recognized order of viruses. Examples of each of the three orders of viruses are presented in **table 6.4.** Note the naming conventions—that is, virus families are written with "-viridae" on the end of the name, and genera end with "-virus."

Historically, some virologists had created an informal *species* naming system that mirrors the species names in higher organisms, using genus and species epithets such as *Measles morbillivirus.* This has not been an official designation, however. The species category has created a lot of controversy within the virology community, with many scientists arguing that nonorganisms such as viruses can never be speciated. Others argue that viruses are too changeable, and thus fine distinctions used for deciding on species classifications will quickly disappear. Over the past decade, virologists have largely accepted the concept of viral species, defining them as consisting of members that have a number of properties in common but have some variation in their properties. In other words, a virus is placed in a species on the basis of a collection of properties. For viruses that infect humans, species may be defined based on relatively minor differences in host range and how they affect their hosts. The important thing to remember is that viral species designations, in the words of one preeminent viral taxonomist, are "fuzzy sets with hazy boundaries."[3]

Because the use of standardized species names has not been widely accepted, the genus or common English vernacular names (for example, poliovirus and rabies virus) predominate in discussions of specific viruses in this text. **Table 6.5** illustrates the naming system for important viruses and the diseases they cause.

3. van Regenmortel, M. H. V., and Mahy, B. W. J. Emerging issues in virus taxonomy. *Emerg. Infect. Dis.* [serial online] 2004 Jan [*date cited*]. Available from http://www.cdc.gov/ncidod/EID/vol10no1/03-0279.htm.

TABLE 6.5 Important Human Virus Families, Genera, Common Names, and Types of Diseases

	Family	Genus of Virus	Common Name of Genus Members	Name of Disease
DNA Viruses				
	Poxviridae	*Orthopoxvirus*	Variola and vaccinia	Smallpox, cowpox
	Herpesviridae	*Simplexvirus*	Herpes simplex (HSV) 1 virus	Fever blister, cold sores
			Herpes simplex (HSV) 2 virus	Genital herpes
		Varicellovirus	Varicella zoster virus (VZV)	Chickenpox, shingles
		Cytomegalovirus	Human cytomegalovirus (CMV)	CMV infections
	Adenoviridae	*Mastadenovirus*	Human adenoviruses	Adenovirus infection
	Papovaviridae	*Papillomavirus*	Human papillomavirus (HPV)	Several types of warts
		Polyomavirus	JC virus (JCV)	Progressive multifocal leukoencephalopathy (PML)
	Hepadnaviridae	*Hepadnavirus*	Hepatitis B virus (HBV or Dane particle)	Serum hepatitis
	Parvoviridae	*Erythrovirus*	Parvovirus B19	Erythema infectiosum
RNA Viruses				
	Picornaviridae	*Enterovirus*	Poliovirus	Poliomyelitis
			Coxsackievirus	Hand-foot-mouth disease
		Hepatovirus	Hepatitis A virus (HAV)	Short-term hepatitis
		Rhinovirus	Human rhinovirus	Common cold, bronchitis
	Caliciviridae	*Calicivirus*	Norwalk virus	Viral diarrhea, Norwalk virus syndrome
	Togaviridae	*Alphavirus*	Eastern equine encephalitis virus	Eastern equine encephalitis (EEE)
			Western equine encephalitis virus	Western equine encephalitis (WEE)
			Yellow fever virus	Yellow fever
			St. Louis encephalitis virus	St. Louis encephalitis
		Rubivirus	Rubella virus	Rubella (German measles)
	Flaviviridae	*Flavivirus*	Dengue fever virus	Dengue fever
			West Nile fever virus	West Nile fever
	Bunyaviridae	*Bunyavirus*	Bunyamwera viruses	California encephalitis
		Hantavirus	Sin Nombre virus	Respiratory distress syndrome
		Phlebovirus	Rift Valley fever virus	Rift Valley fever
		Nairovirus	Crimean–Congo hemorrhagic fever virus (CCHF)	Crimean–Congo hemorrhagic fever
	Filoviridae	*Filovirus*	Ebola, Marburg virus	Ebola fever
	Reoviridae	*Coltivirus*	Colorado tick fever virus	Colorado tick fever
		Rotavirus	Human rotavirus	Rotavirus gastroenteritis
	Orthomyxoviridae	*Influenza virus*	Influenza virus, type A (Asian, Hong Kong, and swine influenza viruses)	Influenza or "flu"
	Paramyxoviridae	*Paramyxovirus*	Parainfluenza virus, types 1–5	Parainfluenza
			Mumps virus	Mumps
		Morbillivirus	Measles virus	Measles (red)
		Pneumovirus	Respiratory syncytial virus (RSV)	Common cold syndrome
	Rhabdoviridae	*Lyssavirus*	Rabies virus	Rabies (hydrophobia)
	Retroviridae	*Oncornavirus*	Human T-cell leukemia virus (HTLV)	T-cell leukemia
		Lentivirus	HIV (human immunodeficiency viruses 1 and 2)	Acquired immunodeficiency syndrome (AIDS)
	Arenaviridae	*Arenavirus*	Lassa virus	Lassa fever
	Coronaviridae	*Coronavirus*	Infectious bronchitis virus (IBV)	Bronchitis
			Enteric corona virus	Coronavirus enteritis
			SARS virus	Severe acute respiratory syndrome

6.5 Modes of Viral Multiplication

Viruses are closely associated with their hosts. In addition to providing the viral habitat, the host cell is absolutely necessary for viral multiplication. The process of viral multiplication is an extraordinary biological phenomenon. Viruses have often been aptly described as minute parasites that seize control of the synthetic and genetic machinery of cells. The nature of this cycle dictates the way the virus is transmitted and what it does to its host, the responses of the immune defenses, and human measures to control viral infections. From these perspectives, we cannot overemphasize the importance of a working knowledge of the relationship between viruses and their host cells.

Multiplication Cycles in Animal Viruses

The general phases in the life cycle of animal viruses are **adsorption, penetration, uncoating, synthesis, assembly,** and **release** from the host cell. The length of the entire multiplication cycle varies from 8 hours in polioviruses to 36 hours in some herpesviruses. See **figure 6.11** for the major phases of one type of animal virus.

Adsorption and Host Range

Invasion begins when the virus encounters a susceptible host cell and adsorbs specifically to receptor sites on the cell membrane. The membrane receptors that viruses attach to are usually glycoproteins the cell requires for its normal function. For example, the rabies virus affixes to the acetylcholine receptor of nerve cells, and the human immunodeficiency virus (HIV) attaches to the CD4 protein on certain white blood cells. The mode of attachment varies between the two general types of viruses. In enveloped forms such as influenza virus and HIV, glycoprotein spikes bind to the cell membrane receptors. Viruses with naked nucleocapsids (adenovirus, for example) use molecules on their capsids that adhere to cell membrane receptors **(figure 6.12).**

Because a virus can invade its host cell only through making an exact fit with a specific host molecule, the range of hosts it can infect in a natural setting is limited. This limitation, known as the **host range,** may be as restricted as hepatitis B, which infects only liver cells of humans; intermediate like the poliovirus, which infects intestinal and nerve cells of primates (humans, apes, and monkeys); or as broad as the rabies virus, which can infect various cells of all mammals. Cells that lack compatible virus receptors are resistant to adsorption and invasion by that virus. This explains why, for example, human liver cells are not infected by the canine hepatitis virus and dog liver cells cannot host the human hepatitis A virus. It also explains why viruses usually have tissue specificities called *tropisms* (troh-pizmz) for certain cells in the body. The hepatitis B virus targets the liver, and the mumps virus targets salivary glands. However, the fact that many viruses can be manipulated to infect cells that they would not infect naturally makes it possible to cultivate them in the laboratory.

Penetration/Uncoating of Animal Viruses

Animal viruses exhibit some impressive mechanisms for entering a host cell. The flexible cell membrane of the host is penetrated by the whole virus or its nucleic acid **(figure 6.13).** In penetration by **endocytosis (figure 6.13*a*),** the entire virus is engulfed by the cell and enclosed in a vacuole or vesicle. When enzymes in the vacuole dissolve the envelope and capsid, the virus is said to be **uncoated,** a process that releases the viral

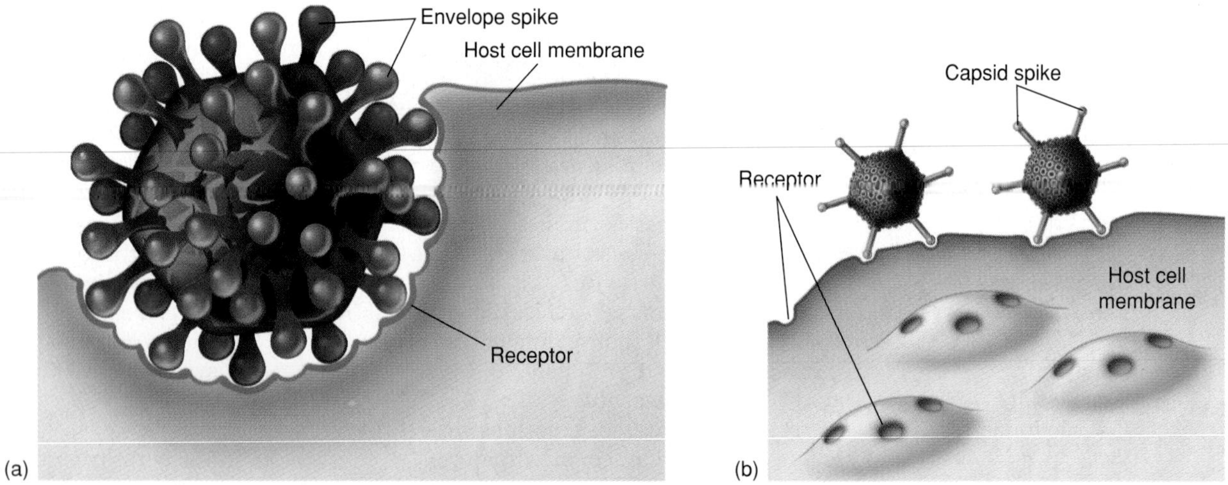

Figure 6.12 The mode by which animal viruses adsorb to the host cell membrane.
(a) An enveloped coronavirus with prominent spikes. The configuration of the spike has a complementary fit for cell receptors. The process in which the virus lands on the cell and plugs into receptors is termed docking. **(b)** An adenovirus has a naked capsid that adheres to its host cell by nestling surface molecules on its capsid into the receptors on the host cell's membrane.

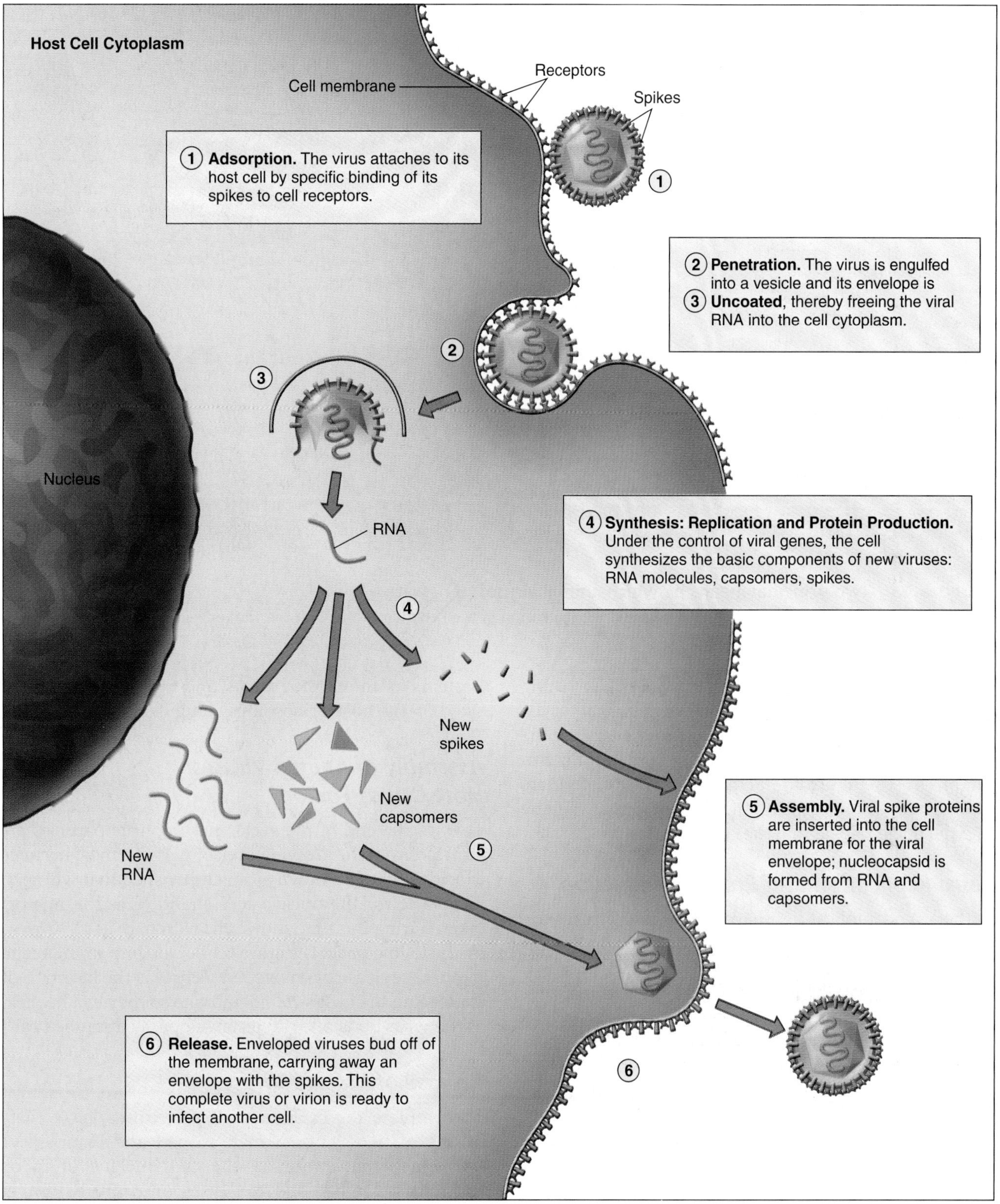
Host Cell Cytoplasm
Cell membrane
Receptors
Spikes
(1) Adsorption. The virus attaches to its host cell by specific binding of its spikes to cell receptors.
(1)
(2) Penetration. The virus is engulfed into a vesicle and its envelope is
(3) Uncoated, thereby freeing the viral RNA into the cell cytoplasm.
(2)
(3)
Nucleus
RNA
(4) Synthesis: Replication and Protein Production. Under the control of viral genes, the cell synthesizes the basic components of new viruses: RNA molecules, capsomers, spikes.
(4)
New spikes
New capsomers
New RNA
(5)
(5) Assembly. Viral spike proteins are inserted into the cell membrane for the viral envelope; nucleocapsid is formed from RNA and capsomers.
(6) Release. Enveloped viruses bud off of the membrane, carrying away an envelope with the spikes. This complete virus or virion is ready to infect another cell.
(6)

Process Figure 6.11 General features in the multiplication cycle of an enveloped RNA animal virus. Using an RNA virus (rubella virus), the major events are outlined, although other viruses will vary in exact details of the cycle.

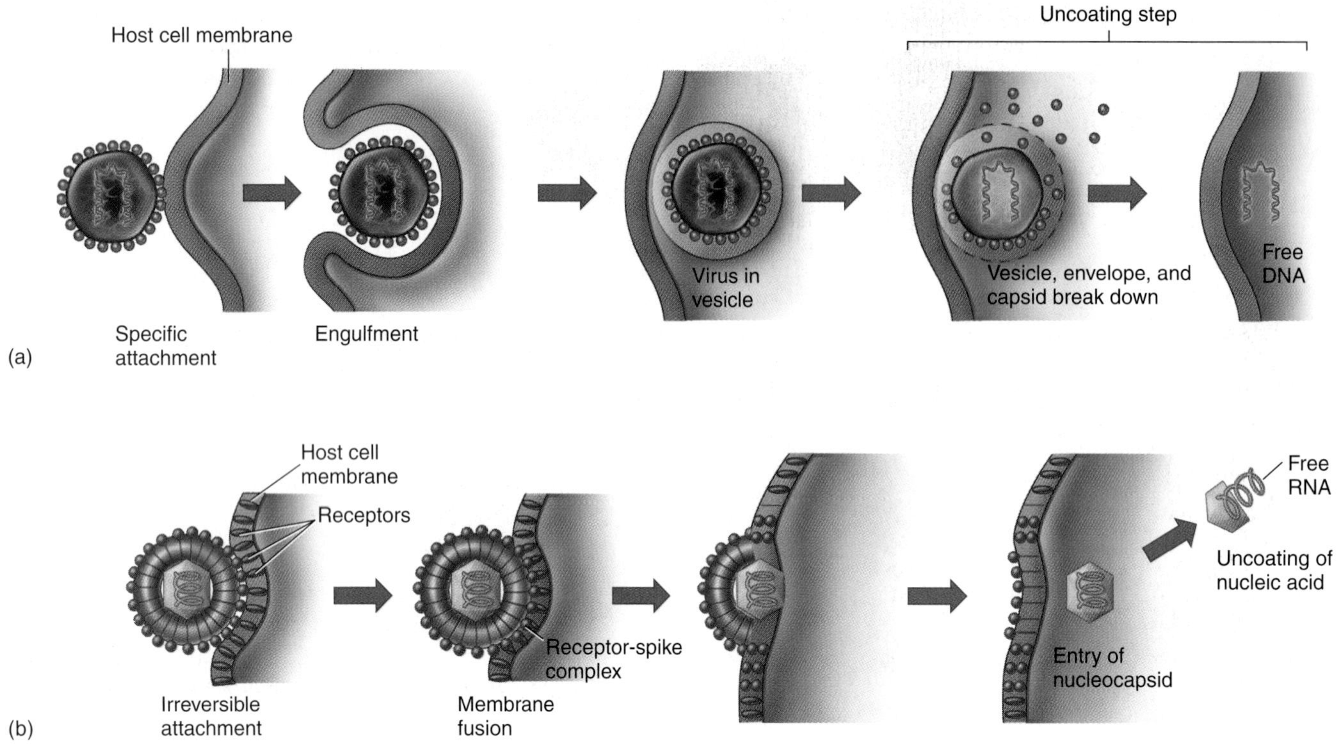

Figure 6.13 Two principal means by which animal viruses penetrate.
(a) Endocytosis (engulfment) and uncoating of a herpesvirus. **(b)** Fusion of the cell membrane with the viral envelope (mumps virus).

nucleic acid into the cytoplasm. The exact manner of uncoating varies, but in most cases, the virus fuses with the wall of the vesicle. Another means of entry involves direct fusion of the viral envelope with the host cell membrane (as in influenza and mumps viruses) **(figure 6.13*b*).** In this form of penetration, the envelope merges directly with the cell membrane, thereby liberating the nucleocapsid into the cell's interior.

Synthesis: Replication and Protein Production

The synthetic and replicative phases of animal viruses are highly regulated and extremely complex at the molecular level. Free viral nucleic acid exerts control over the host's synthetic and metabolic machinery. How this control proceeds will vary, depending on whether the virus is a DNA or an RNA virus. In general, the DNA viruses (except poxviruses) enter the host cell's nucleus and are replicated and assembled there. With few exceptions (such as retroviruses), RNA viruses are replicated and assembled in the cytoplasm.

The details of animal virus replication are discussed in **Insight 6.2.** Here we provide a brief overview of the process, using DNA viruses as a model. Almost immediately upon entry, the viral nucleic acid alters the genetic expression of the host and instructs it to synthesize the building blocks for new viruses. First, the DNA enters the nucleus and is transcribed by host machinery into RNA. This RNA becomes a message for synthesizing viral proteins (translation). Some viruses come equipped with the necessary enzymes for synthesis of viral components; others utilize those of the host. In the next phase, new DNA is synthesized using host nucleotides. Proteins for the capsid, spikes, and viral enzymes are synthesized on the host's ribosomes using its amino acids.

Assembly of Animal Viruses: Host Cell as Factory

Toward the end of the cycle, mature virus particles are constructed from the growing pool of parts. In most instances, the capsid is first laid down as an empty shell that will serve as a receptacle for the nucleic acid strand. Electron micrographs taken during this time show cells with masses of viruses, often in crystalline packets **(figure 6.14).** One important event leading to the release of enveloped viruses is the insertion of viral spikes into the host's cell membrane so they can be picked up as the virus buds off with its envelope, as discussed earlier.

Release of Mature Viruses

To complete the cycle, assembled viruses leave their host in one of two ways. Nonenveloped and complex viruses that reach maturation in the cell nucleus or cytoplasm are released when the cell lyses or ruptures. Enveloped viruses are liberated by **budding** or **exocytosis**[4] from the membranes of the cytoplasm, nucleus, endoplasmic reticulum, or vesicles. During this process, the nucleocapsid binds to the membrane, which curves completely around it

4. For enveloped viruses, these terms are interchangeable. They mean the release of a virus from an animal cell by enclosing it in a portion of membrane derived from the cell.

and forms a small pouch. Pinching off the pouch releases the virus with its envelope **(figure 6.15).** Budding of enveloped viruses causes them to be shed gradually, without the sudden destruction of the cell. Regardless of how the virus leaves, most active viral infections are ultimately lethal to the cell because of accumulated damage. Lethal damages include a permanent shutdown of metabolism and genetic expression, destruction of cell membrane and organelles, toxicity of virus components, and release of lysosomes.

The number of viruses released by infected cells is variable, controlled by factors such as the size of the virus and the health of the host cell. About 3,000 to 4,000 virions are released from a single cell infected with poxviruses, whereas a poliovirus-infected cell can release over 100,000 virions. If even a small number of these virions happen to meet another susceptible cell and infect it, the potential for rapid viral proliferation is immense.

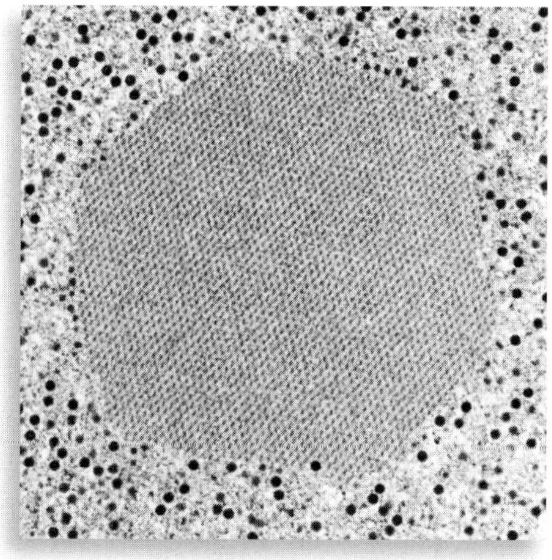

Figure 6.14 Nucleus of a cell, containing a crystalline mass of adenovirus (35,000×).

Damage to the Host Cell and Persistent Infections

The short- and long-term effects of viral infections on animal cells are well documented. **Cytopathic** (sy″-toh-path′-ik) **effects** (CPEs) are defined as virus-induced damage to the cell that alters its microscopic appearance. Individual cells can become disoriented, undergo gross changes in shape

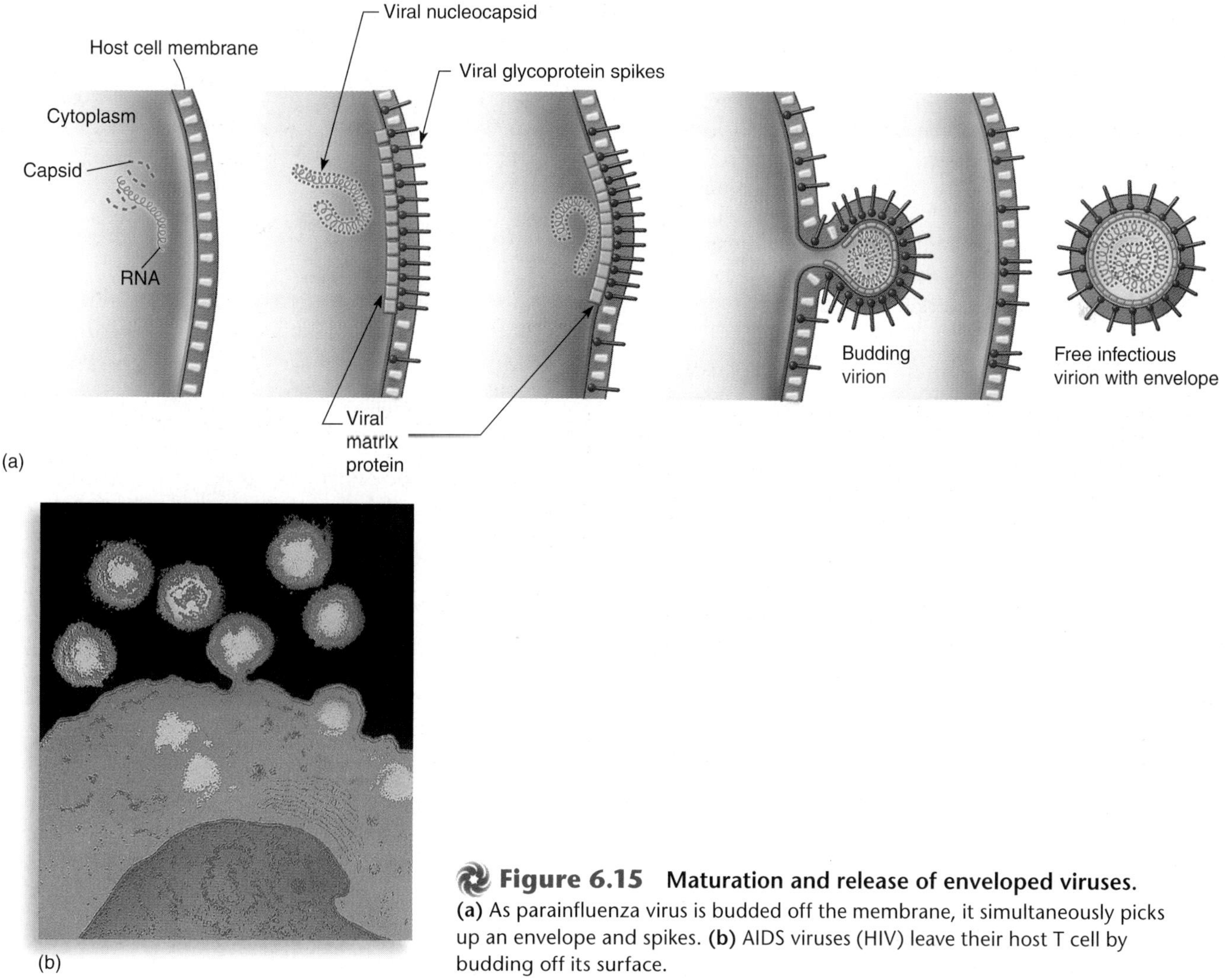

Figure 6.15 Maturation and release of enveloped viruses.
(a) As parainfluenza virus is budded off the membrane, it simultaneously picks up an envelope and spikes. **(b)** AIDS viruses (HIV) leave their host T cell by budding off its surface.

INSIGHT 6.2 Medical

Replication Strategies in Animal Viruses

Replication, Transcription, and Translation of dsDNA Viruses

Replication of dsDNA viruses is divided into phases (see illustration). During the early phase, viral DNA enters the nucleus, where several genes are transcribed into a messenger RNA. The newly synthesized RNA transcript then moves into the cytoplasm to be translated into viral proteins (enzymes) needed to replicate the viral DNA; this replication occurs in the nucleus. The host cell's own DNA polymerase is often involved, though some viruses (herpes, for example) have their own. During the late phase, other parts of the viral genome are transcribed and translated into proteins required to form the capsid and other structures. The new viral genomes and capsids are assembled, and the mature viruses are released by budding or cell disintegration.

Double-stranded DNA viruses interact directly with the DNA of their host cell. In some viruses, the viral DNA becomes silently *integrated* into the host's genome by insertion at a particular site on the host genome. This integration may later lead to the transformation of the host cell into a cancer cell and the production of a tumor. Several DNA viruses, including hepatitis B (HBV), the herpesviruses, and papillomaviruses (warts), are known to be initiators of cancers and are thus termed **oncogenic.*** The mechanisms of **transformation** and oncogenesis involve special genes called oncogenes that can regulate cellular genomes (see p. 167).

Replication, Transcription, and Translation of RNA Viruses

RNA viruses exhibit several differences from DNA viruses. Their genomes are smaller and less stable; they enter the host cell already in an RNA form; and the virus cycle occurs entirely in the cytoplasm for most viruses. RNA viruses can have one of the following genetic messages:

1. a positive-sense genome (+) that comes ready to be translated into proteins,
2. a negative-sense genome (−) that must be converted to positive sense before translation, and
3. a positive-sense genome (+) that can be converted to DNA, or a dsRNA genome.

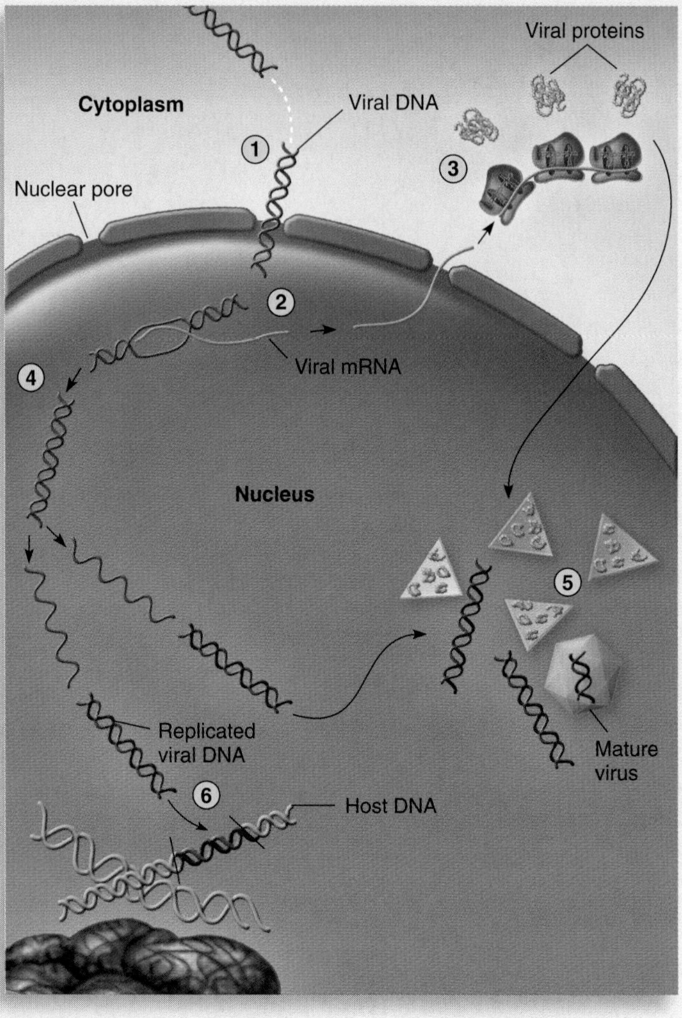

Genetic stages in the multiplication of double-stranded DNA viruses. The virus penetrates the host cell and releases DNA, which

(1) enters the nucleus and

(2) is transcribed.

Other events are:

(3) Viral mRNA is translated into structural proteins; proteins enter the nucleus.

(4) Viral DNA is replicated repeatedly in the nucleus.

(5) Viral DNA and proteins are assembled into a mature virus in the nucleus.

(6) Because it is double-stranded, the viral DNA can insert itself into host DNA (latency).

*__oncogenic__ (ahn-″koh′-jen′-ik): Gr. *onkos*, mass, and *gennan*, to produce. Refers to any cancer-causing process. Viruses that do this are termed oncoviruses.

Positive-Sense Single-Stranded RNA Viruses

Positive-sense RNA viruses such as polio and hepatitis A virus must first replicate a negative strand as a master template to produce more positive strands. Shortly after the virus uncoats in the cell, its positive strand is translated into a large protein that is soon cleaved into individual functional units, one of which is an RNA polymerase that initiates the replication of the viral strand (see illustration). Replication of a single-stranded positive-sense strand is done in two steps. First, a negative strand is synthesized using the parental positive strand as a template by the usual base-pairing mechanism. The resultant negative strand becomes a master template against which numerous positive daughter strands are made. Further translation of the viral genome produces large numbers of structural proteins for final assembly and maturation of the virus.

RNA Viruses with Reverse Transcriptase: Retroviruses

A most unusual class of viruses has a unique capability to reverse the order of the flow of genetic information. Thus far in our discussion, all genetic entities have shown the patterns DNA → DNA, DNA → RNA, or RNA → RNA. Retroviruses, including HIV, the cause of AIDS, and HTLV I, a cause of one type of human leukemia, synthesize DNA using their RNA genome as a template. They accomplish this by means of an enzyme, **reverse transcriptase,** that comes packaged with each virus particle. This enzyme synthesizes a single-stranded DNA against the viral RNA template and then directs the formation of a complementary strand of this ssDNA, resulting in a double strand of viral DNA. The dsDNA strand enters the nucleus, where it can be integrated into the host genome and transcribed by the usual mechanisms into new viral ssRNA. Translation of the viral RNA yields viral proteins for final virus assembly. The capacity of a retrovirus to become inserted into the host's DNA as a provirus has several possible consequences. In some cases, these viruses are oncogenic and are known to transform cells and produce tumors. It also allows the AIDS virus to remain latent in an infected cell until a stimulus activates it to continue a productive cycle.

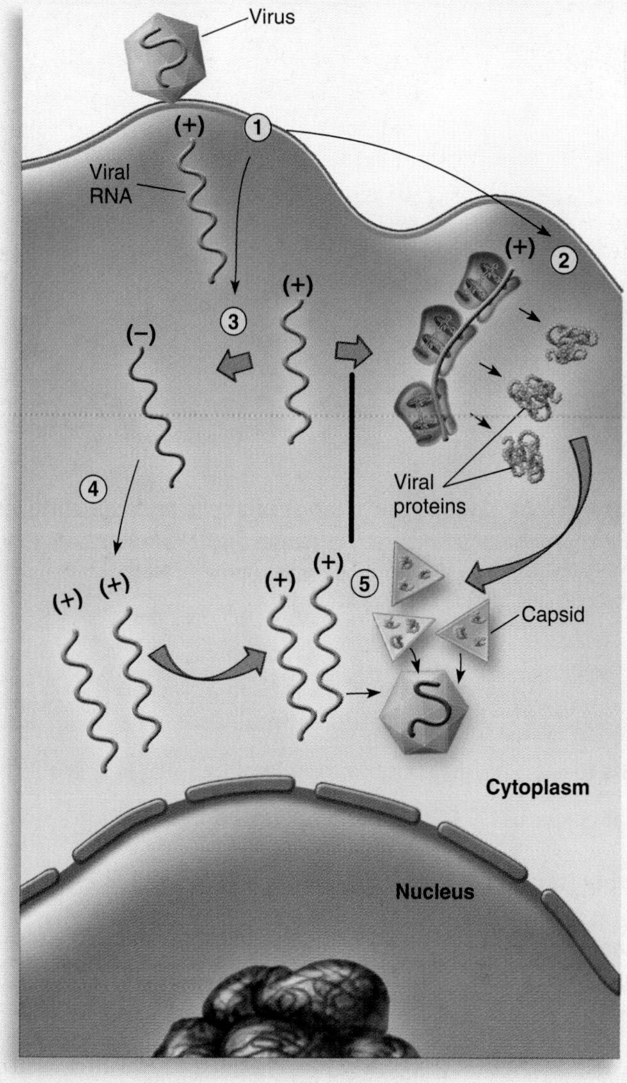

Replication of positive-sense, single-stranded RNA viruses.
In general, these viruses do not enter the nucleus.

(1) Virus penetrates host cell and its RNA is uncoated.

(2) Because it is positive in sense and single-stranded, the RNA can be directly translated on host cell ribosomes into various necessary viral proteins.

(3) A negative genome is synthesized against the positive template to produce large numbers of positive genomes for final assembly.

(4) The negative template is then used to synthesize a series of positive replicates.

(5) RNA strands and proteins assemble into mature viruses.

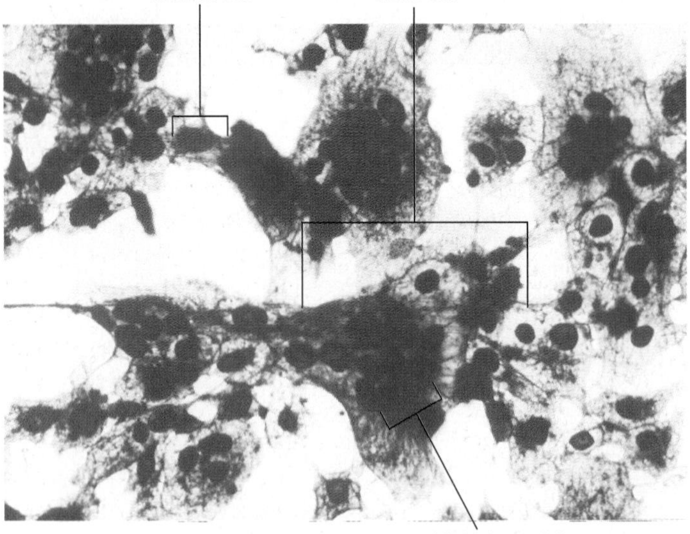

(a)

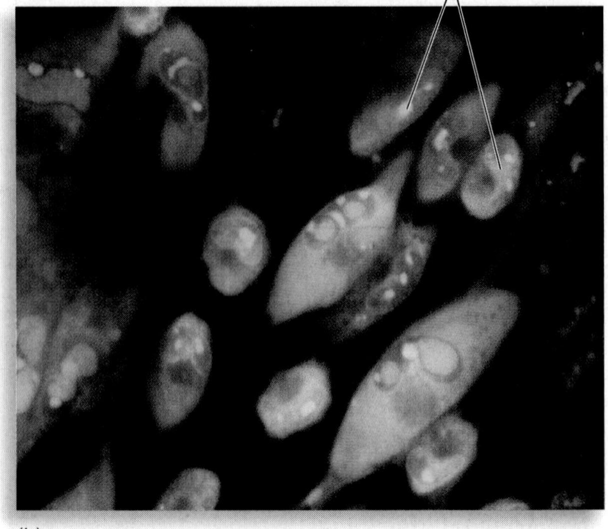

(b)

Figure 6.16 Cytopathic changes in cells and cell cultures infected by viruses.
(a) Human epithelial cells infected by herpes simplex virus demonstrate multinucleate giant cells. **(b)** Fluorescent-stained human cells infected with cytomegalovirus. Note the inclusion bodies (arrows). Note also that both viruses disrupt the cohesive junctions between cells, which would ordinarily be arranged side by side in neat patterns.

TABLE 6.6 Cytopathic Changes in Selected Virus-Infected Animal Cells

Virus	Response in Animal Cell
Smallpox virus	Cells round up; inclusions appear in cytoplasm
Herpes simplex	Cells fuse to form multinucleated syncytia; nuclear inclusions (see figure 6.16)
Adenovirus	Clumping of cells; nuclear inclusions
Poliovirus	Cell lysis; no inclusions
Reovirus	Cell enlargement; vacuoles and inclusions in cytoplasm
Influenza virus	Cells round up; no inclusions
Rabies virus	No change in cell shape; cytoplasmic inclusions (Negri bodies)
Measles virus	Syncytia form (multinucleate)

or size, or develop intracellular changes **(figure 6.16*a*)**. It is common to note *inclusion bodies*, or compacted masses of viruses or damaged cell organelles, in the nucleus and cytoplasm **(figure 6.16*b*)**. Examination of cells and tissues for cytopathic effects is an important part of the diagnosis of viral infections. **Table 6.6** summarizes some prominent cytopathic effects associated with specific viruses. One very common CPE is the fusion of multiple host cells into single large cells containing multiple nuclei. These **syncytia** are a result of some viruses' ability to fuse membranes. One virus (respiratory syncytial virus) is even named for this effect.

Although accumulated damage from a virus infection kills most host cells, some cells maintain a carrier relationship, in which the cell harbors the virus and is not immediately lysed. These so-called *persistent infections* can last from a few weeks to the remainder of the host's life. One of the more serious complications occurs with the measles virus. It may remain hidden in brain cells for many years, causing progressive damage and loss of function. Several viruses remain in a *chronic*

CASE FILE 6 *WRAP-UP*

During the initial outbreak of SARS at the Prince of Wales Hospital in Hong Kong, hospital workers were confronted with a new infectious disease caused by a virus. SARS was transmitted quickly among hospital workers. There were concerns that the new coronavirus was spreading through small aerosols or contact with contaminated surfaces in the hospital environment. The epidemiological investigation led by the Chinese University and Hong Kong Hospital Authority determined that hospital workers did not take special protective measures when in contact with SARS patients during the initial outbreak of the disease. Personal protection such as wearing masks, goggles, caps, and gowns was inadequate and workers had less than 2 hours of infection control training. Many did not understand infection control procedures and used personal protection equipment inconsistently.

The study revealed that 40% to 50% of hospital workers experienced difficulties with their masks fitting properly, fogging of protective goggles, and general compliance problems. This case provides an example of the consequences of inadequate infection control measures. Proper training and implementation of infection control measures reduce the risks of breakthrough transmission of the SARS virus.

See: Lau, J. T. F., et al. 2004. SARS transmission among hospital workers in Hong Kong. Emerg. Infect. Dis.

latent state,[5] periodically becoming reactivated. Examples of this are herpes simplex viruses (cold sores and genital herpes) and herpes zoster virus (chickenpox and shingles). Both viruses can go into latency in nerve cells and later emerge under the influence of various stimuli to cause recurrent symptoms. Specific damage that occurs in viral diseases is covered more completely in chapters 18 through 23.

Some animal viruses enter their host cell and permanently alter its genetic material, leading to cancer. These viruses are termed *oncogenic,* and their effect on the cell is called **transformation.** A startling feature of these viruses is that their nucleic acid is integrated into the host DNA. Transformed cells have an increased rate of growth; alterations in chromosomes; changes in the cell's surface molecules; and the capacity to divide for an indefinite period, unlike normal animal cells. Mammalian viruses capable of initiating tumors are called **oncoviruses.** Some of these are DNA viruses such as papillomavirus (genital warts are associated with cervical cancer), herpesviruses (Epstein-Barr virus causes Burkitt's lymphoma), and hepatitis B virus. Two viruses related to HIV—HTLV I and II[6]—are involved in human cancers. These findings have spurred a great deal of speculation on the possible involvement of viruses in cancers whose cause is still unknown. Additional information on the connection between viruses and cancer is found in chapters 9 and 20.

CHECKPOINT

- Virus size range is from 20 nm to 450 nm (diameter). Viruses are composed of an outer protein capsid enclosing either DNA or RNA plus a variety of enzymes. Some viruses also exhibit an envelope around the capsid.
- Viruses go through a multiplication cycle that generally involves adsorption, penetration (sometimes followed by uncoating), viral synthesis and assembly, and viral release by lysis or budding.
- These events turn the host cell into a factory solely for making and shedding new viruses. This results in the ultimate destruction of the cell.
- Animal viruses can cause acute infections or can persist in host tissues as chronic latent infections that can reactivate periodically throughout the host's life. Some persistent animal viruses are oncogenic.

Viruses That Infect Bacteria

We now turn to the life cycle of another type of virus called bacteriophage. When Frederick Twort and Felix d'Herelle discovered bacterial viruses in 1915, it first appeared that the bacterial host cells were being eaten by some unseen parasite, hence the name bacteriophage was used. Most bacteriophages (often shortened to *phage*) contain double-stranded DNA, though single-stranded DNA and RNA types exist as well. So far as is known, every bacterial species is parasitized by various specific bacteriophages. Bacteriophages are of great interest to medical microbiologists because they often make the bacteria they infect more pathogenic for humans. Probably the most widely studied bacteriophages are those of the intestinal bacterium *Escherichia coli*—especially the ones known as the T-even phages such as T_2 and T_4. They have an icosahedral capsid head containing DNA, a central tube (surrounded by a sheath), collar, base plate, tail pins, and fibers, which in combination make an efficient package for infecting a bacterial cell (see figure 6.9). Momentarily setting aside a strictly scientific and objective tone, it is tempting to think of these extraordinary viruses as minute spacecrafts docking on an alien planet, ready to unload their genetic cargo.

T-even bacteriophages go through similar stages as the animal viruses described earlier **(figure 6.17).** They *adsorb* to host bacteria using specific receptors on the bacterial surface. Although the entire phage does not enter the host cell, the nucleic acid *penetrates* the host after being injected through a rigid tube the phage inserts through the bacterial membrane and wall **(figure 6.18).** This eliminates the need for *uncoating*. Entry of the nucleic acid causes the cessation of host cell DNA replication and protein synthesis. Soon the host cell machinery is used for viral *replication* and synthesis of viral proteins. As the host cell produces new phage parts, the parts spontaneously *assemble* into bacteriophages.

An average-size *Escherichia coli* cell can contain up to 200 new phage units at the end of this period. Eventually, the host cell becomes so packed with viruses that it **lyses**—splits open—thereby releasing the mature virions **(figure 6.19).** This process is hastened by viral enzymes produced late in the infection cycle that digest the cell envelope, thereby weakening it. Upon release, the virulent phages can spread to other susceptible bacterial cells and begin a new cycle of infection.

Lysogeny: The Silent Virus Infection

The lethal effects of a virulent phage on the host cell present a dramatic view of virus-host interaction. Not all bacteriophages complete the lytic cycle just described, however. Special DNA phages, called **temperate phages,** undergo adsorption and penetration into the bacterial host but are not replicated or released immediately. Instead, the viral DNA enters an inactive **prophage** state, during which it is inserted into the bacterial chromosome. This viral DNA will be retained by the bacterial cell and copied during its normal cell division so that the "cell's progeny will also have the temperate phage DNA **(figure 6.20).** This condition, in which the host chromosome carries bacteriophage DNA, is termed **lysogeny** (ly-soj'-uhn-ee). Because viral particles are not produced, the bacterial cells carrying temperate phages do not lyse, and they appear entirely normal. On occasion, in a process called **induction,** the prophage in a lysogenic cell will be activated and progress directly into viral replication and the lytic cycle. Lysogeny is a less deadly form of parasitism than the full lytic cycle and is thought to be an advancement that allows the virus to spread without killing the host. Many bacteria that infect humans are lysogenized by phages. And sometimes that

5. Meaning that they exist in an inactive state over long periods.
6. Human T-cell lymphotropic viruses: cause types of leukemia.

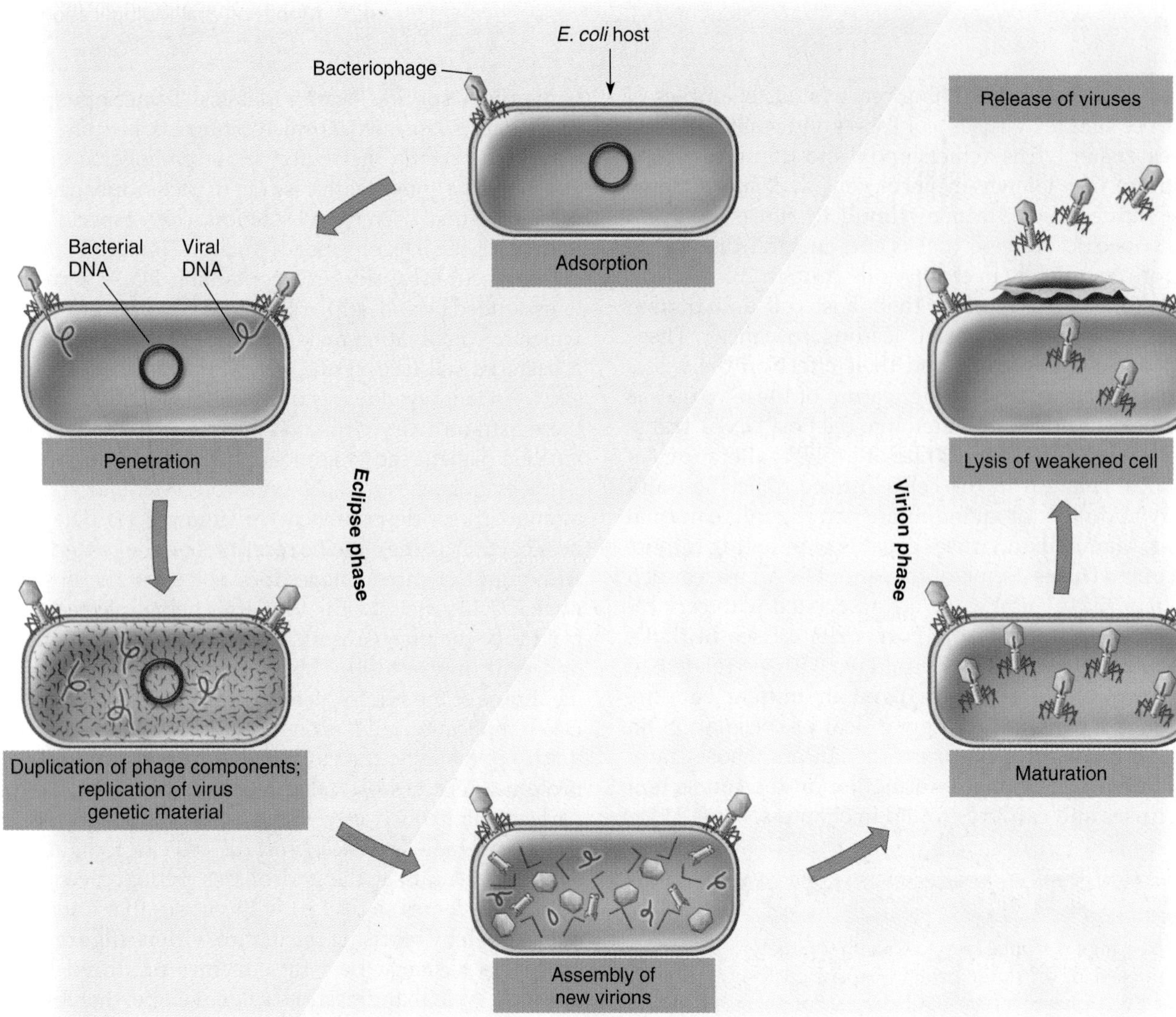

Figure 6.17 Events in the lytic cycle of T-even bacteriophages.
The cycle is divided into the eclipse phase (during which the phage is developing but is not yet infectious) and the virion phase (when the virus matures and is capable of infecting a host).

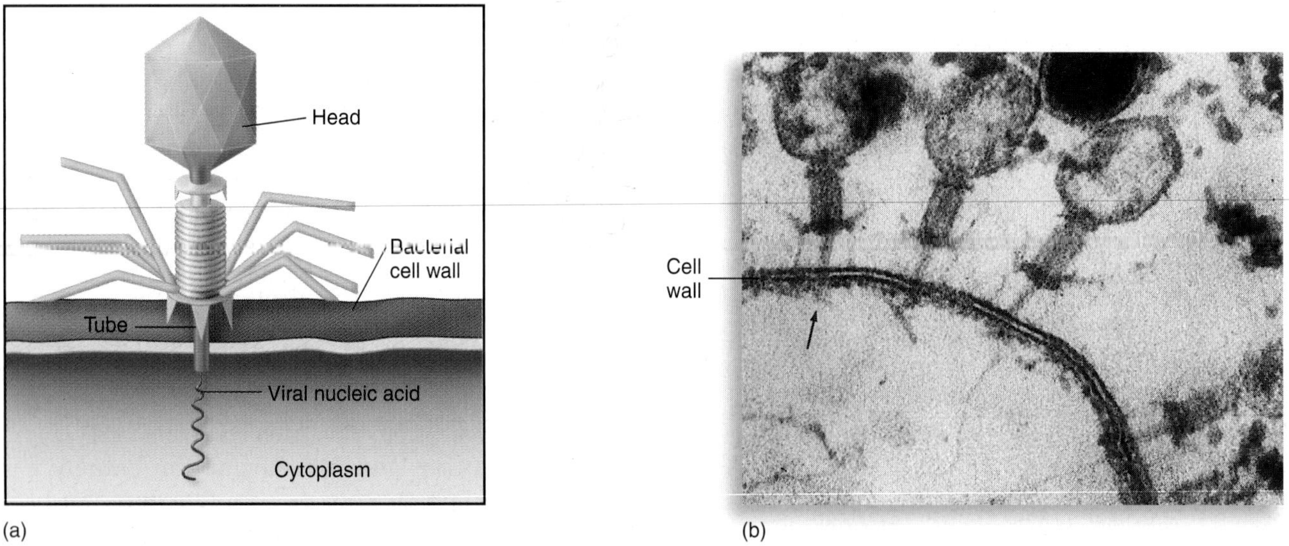

Figure 6.18 Penetration of a bacterial cell by a T-even bacteriophage.
(a) After adsorption, the phage plate becomes embedded in the cell wall and the sheath contracts, pushing the tube through the cell wall and releasing the nucleic acid into the interior of the cell. **(b)** Section through *Escherichia coli* with attached phages. Note that these phages have injected their nucleic acid through the cell wall and now have empty heads.

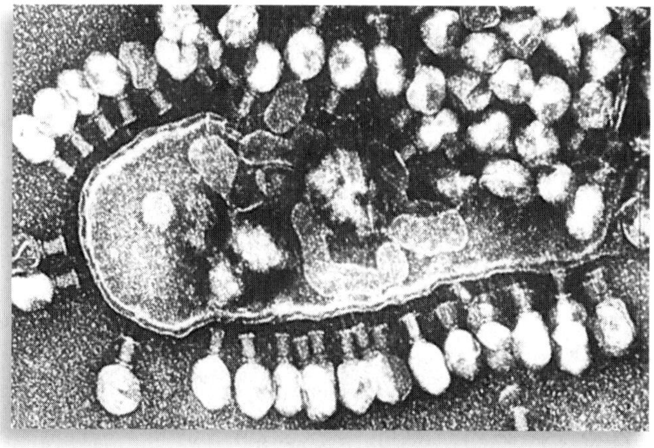

Figure 6.19 A weakened bacterial cell, crowded with viruses.
The cell has ruptured and released numerous virions that can then attack nearby susceptible host cells. Note the empty heads of "spent" phages lined up around the ruptured wall.

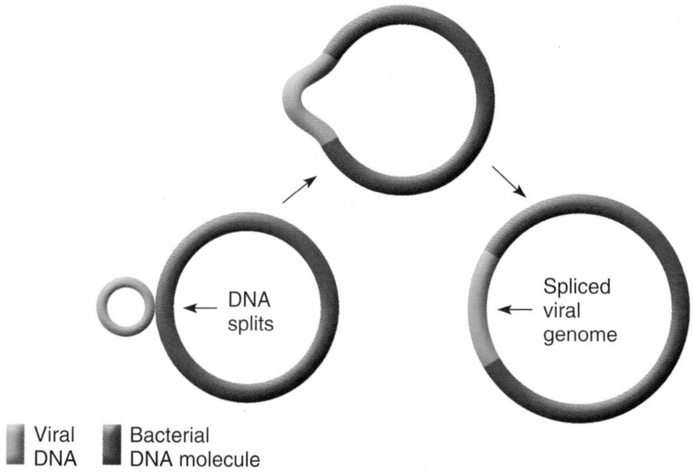

Figure 6.20 The lysogenic state in bacteria.
A bacterial DNA molecule can accept and insert viral DNA molecules at specific sites on its genome. This additional viral DNA is duplicated along with the regular genome and can provide adaptive characteristics for the host bacterium.

is very bad news for the human: Occasionally phage genes in the bacterial chromosome cause the production of toxins or enzymes that cause pathology in the human. When a bacterium acquires a new trait from its temperate phage, it is called **lysogenic conversion.** The phenomenon was first discovered in the 1950s in the bacterium that causes diphtheria, *Corynebacterium diphtheriae.* The diphtheria toxin responsible for the deadly nature of the disease is a bacteriophage product. *C. diphtheriae* without the phage are harmless. Other bacteria that are made virulent by their prophages are *Vibrio cholerae,* the agent of cholera, and *Clostridium botulinum,* the cause of botulism.

The cycle of animal and bacterial viruses (see figure 6.11 and figure 6.17) illustrates general features of viral multiplication in a very concrete and memorable way. The two cycles are compared in **table 6.7.** It is fascinating to realize that viruses are capable of lying "dormant" in their host cells, possibly becoming active at some later time. Because of the intimate association between the genetic material of the virus and host, phages occasionally serve as transporters of bacterial genes from one bacterium to another and consequently can play a profound role in bacterial genetics. This phenomenon, called transduction, is one way that genes for toxin production and drug resistance are transferred between bacteria (see chapters 9 and 12).

TABLE 6.7 Comparison of Bacteriophage and Animal Virus Multiplication

	Bacteriophage	Animal Virus
Adsorption	Precise attachment of special tail fibers to cell wall	Attachment of capsid or envelope to cell surface receptors
Penetration	Injection of nucleic acid through cell wall; no uncoating of nucleic acid	Whole virus is engulfed and uncoated, or virus surface fuses with cell membrane, nucleic acid is released
Synthesis and Assembly	Occurs in cytoplasm Cessation of host synthesis Viral DNA or RNA is replicated and begins to function Viral components synthesized	Occurs in cytoplasm and nucleus Cessation of host synthesis Viral DNA or RNA is replicated and begins to function Viral components synthesized
Viral Persistence	Lysogeny	Latency, chronic infection, cancer
Release from Host Cell	Cell lyses when viral enzymes weaken it	Some cells lyse; enveloped viruses bud off host cell membrane
Cell Destruction	Immediate	Immediate or delayed

CHECKPOINT

- Bacteriophages vary significantly from animal viruses in their methods of adsorption, penetration, site of replication, and method of exit from host cells.
- Lysogeny is a condition in which viral DNA is inserted into the bacterial chromosome and remains inactive for an extended period. It is replicated right along with the chromosome every time the bacterium divides.
- Some bacteria express virulence traits that are coded for by the bacteriophage DNA in their chromosomes. This phenomenon is called lysogenic conversion.

6.6 Techniques in Cultivating and Identifying Animal Viruses

One problem hampering earlier animal virologists was their inability to propagate specific viruses routinely in pure culture and in sufficient quantities for their studies. Virtually all of the pioneering attempts at cultivation had to be performed in an organism that was the usual host for the virus. But this method had its limitations. How could researchers have ever traced the stages of viral multiplication if they had been restricted to the natural host, especially in the case of human viruses? Fortunately, systems of cultivation with broader applications were developed, including *in vivo* (in vee′-voh) inoculation of laboratory-bred animals and embryonic bird tissues and *in vitro* (in vee′-troh) cell (or tissue) culture methods. Such use of substitute host systems permits greater control, uniformity, and wide-scale harvesting of viruses.

The primary purposes of viral cultivation are:

1. to isolate and identify viruses in clinical specimens;
2. to prepare viruses for vaccines; and
3. to do detailed research on viral structure, multiplication cycles, genetics, and effects on host cells.

Using Live Animal Inoculation

Specially bred strains of white mice, rats, hamsters, guinea pigs, and rabbits are the usual choices for animal cultivation of viruses. Invertebrates (insects) or nonhuman primates are occasionally used as well. Because viruses can exhibit some host specificity, certain animals can propagate a given virus more readily than others. The animal is exposed to the virus by injection of a viral preparation or specimen into the brain, blood, muscle, body cavity, skin, or footpads.

Using Bird Embryos

An embryo is an early developmental stage of animals marked by rapid differentiation of cells. Birds undergo their embryonic period within the closed protective case of an egg, which makes an incubating bird egg a nearly perfect system for viral propagation. It is an intact and self-supporting unit, complete with its own sterile environment and nourishment. Furthermore, it furnishes several embryonic tissues that readily support viral multiplication.

Chicken, duck, and turkey eggs are the most common choices for inoculation. The egg must be injected through the shell, usually by drilling a hole or making a small window. Rigorous sterile techniques must be used to prevent contamination by bacteria and fungi from the air and the outer surface of the shell. The exact tissue that is inoculated is guided by the type of virus being cultivated and the goals of the experiment **(figure 6.21)**.

Viruses multiplying in embryos may or may not cause effects visible to the naked eye. The signs of viral growth include death of the embryo, defects in embryonic development, and localized areas of damage in the

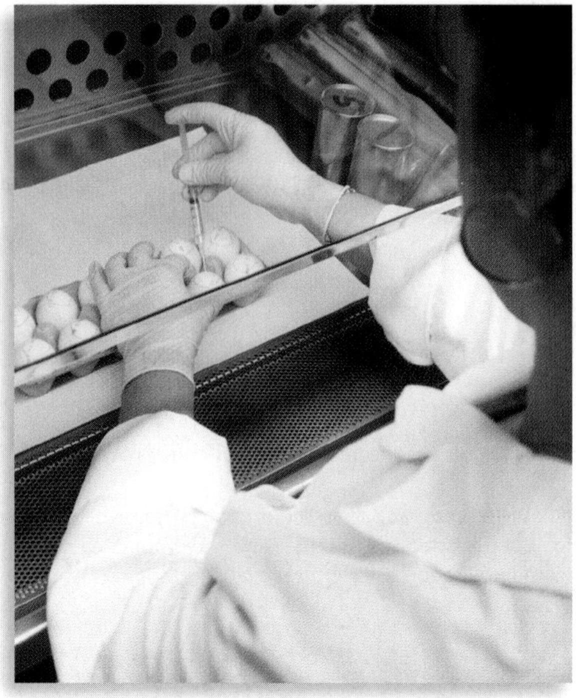

(a)

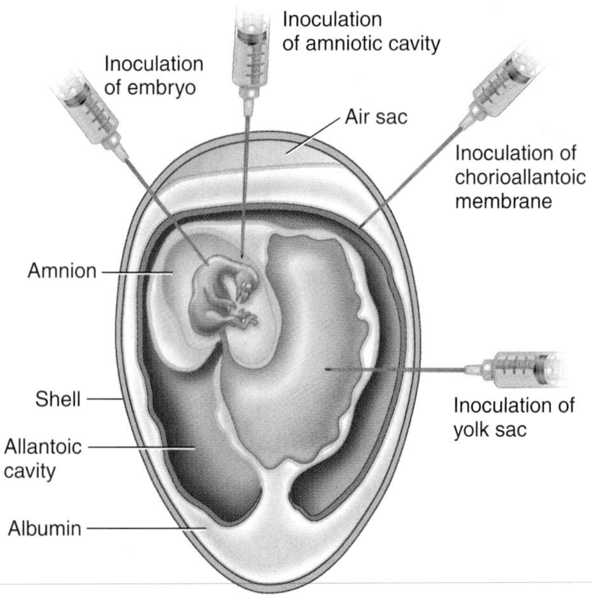

(b)

Figure 6.21 Cultivating animal viruses in a developing bird embryo.
(a) A technician inoculates fertilized chicken eggs with viruses in the first stage of preparing vaccines. This process requires the highest levels of sterile and aseptic precautions. Influenza vaccine is prepared this way. **(b)** The shell is perforated using sterile techniques, and a virus preparation is injected into a site selected to grow the viruses. Targets include the allantoic cavity, a fluid-filled sac that functions in embryonic waste removal; the amniotic cavity, a sac that cushions and protects the embryo itself; the chorioallantoic membrane, which functions in embryonic gas exchange; the yolk sac, a membrane that mobilizes yolk for the nourishment of the embryo; and the embryo itself.

membranes, resulting in discrete, opaque spots called pocks (a variant of *pox*). If a virus does not produce overt changes in the developing embryonic tissue, virologists have other methods of detection. Embryonic fluids and tissues can be prepared for direct examination with an electron microscope. Certain viruses can also be detected by their ability to agglutinate red blood cells (form big clumps) or by their reaction with an antibody of known specificity that will affix to its corresponding virus, if it is present.

Using Cell (Tissue) Culture Techniques

The most important early discovery that led to easier cultivation of viruses in the laboratory was the development of a simple and effective way to grow populations of isolated animal cells in culture. These types of in vitro cultivation systems are termed cell culture or tissue culture. (Although these terms are used interchangeably, cell culture is probably a more accurate description.) So prominent is this method that most viruses are propagated in some sort of cell culture, and much of the virologist's work involves developing and maintaining these cultures. Animal cell cultures are grown in sterile chambers with special media that contain the correct nutrients required by animal cells to survive. The cultured cells grow in the form of a *monolayer*, a single, confluent sheet of cells that supports viral multiplication and permits close inspection of the culture for signs of infection **(figure 6.22)**.

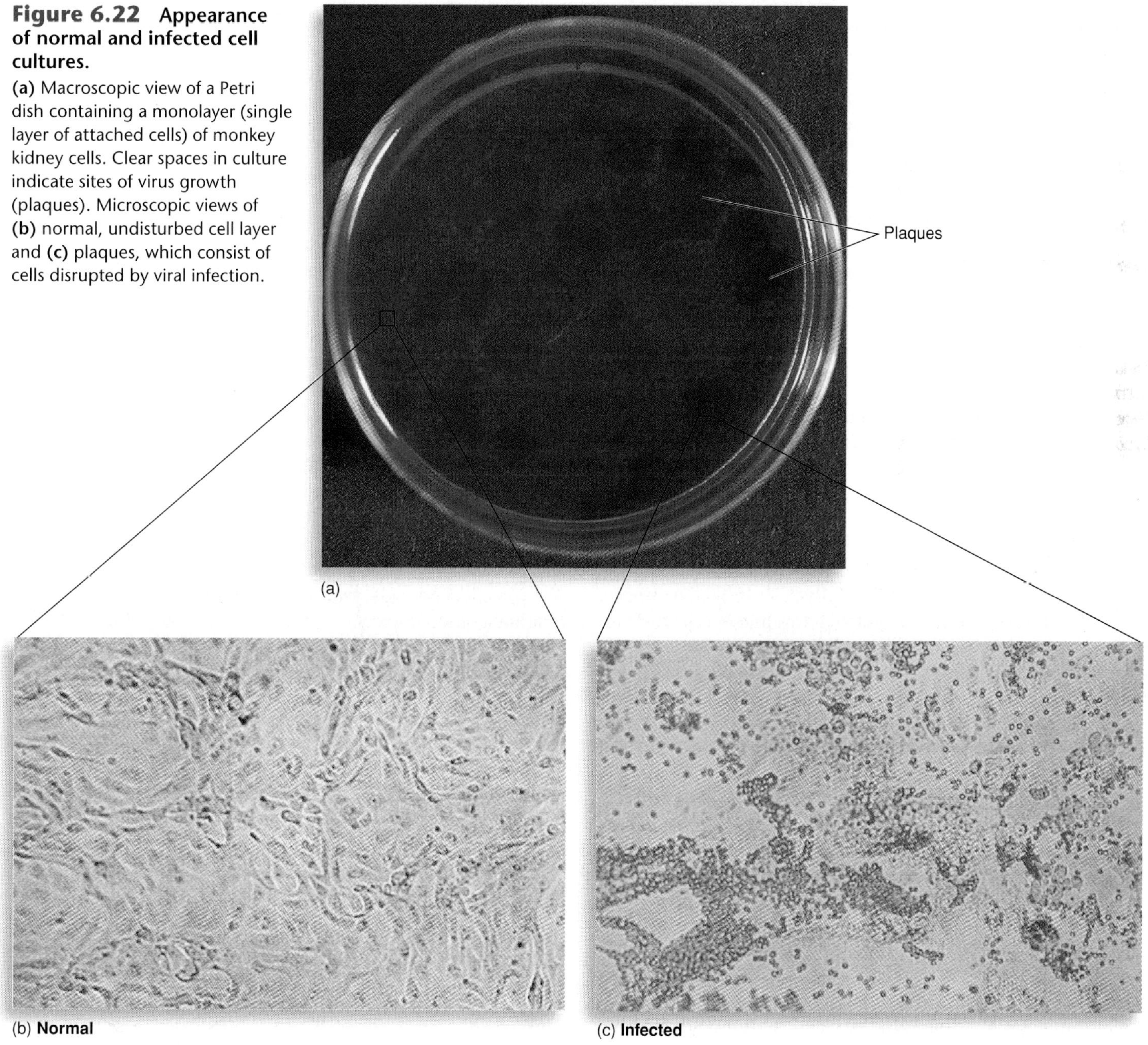

Figure 6.22 Appearance of normal and infected cell cultures.
(a) Macroscopic view of a Petri dish containing a monolayer (single layer of attached cells) of monkey kidney cells. Clear spaces in culture indicate sites of virus growth (plaques). Microscopic views of **(b)** normal, undisturbed cell layer and **(c)** plaques, which consist of cells disrupted by viral infection.

INSIGHT 6.3 Discovery

Artificial Viruses Created!

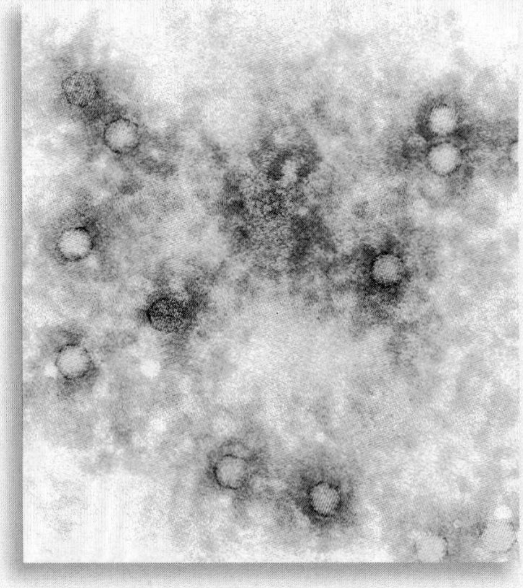

Newspapers are filled with stories of the debate over the ethics of creating life through cloning techniques. Dolly the cloned sheep and the cattle, swine, and goats that have followed in her footsteps have raised ethical questions about scientists "playing God" when they harvest genetic material from an animal and create an identical organism from it, as is the case with cloning.

Meanwhile, in a much less publicized event in 2002, scientists at the State University of New York at Stony Brook succeeded in artificially creating a virus that is virtually identical to natural poliovirus. They used DNA nucleotides they bought "off the shelf" and put them together according to the published poliovirus sequence. They then added an enzyme that would transcribe the DNA sequence into the RNA genome used by poliovirus. They ended up with a virus that was nearly identical to poliovirus (see illustration), with a similar capsid as well as a similar ability to infect host cells and reproduce itself.

The creation of the virus was greeted with controversy, particularly because poliovirus is potentially devastating to human health. The scientists, who were working on a biowarfare defense project funded by the Department of Defense, argued that they were demonstrating what could be accomplished if information and chemicals fell into the wrong hands.

In the fall of 2005, scientists at the Centers for Disease Control and Prevention and Mount Sinai School of Medicine reconstructed the strain of influenza that caused the worldwide flu pandemic of 1918. That pandemic killed 20–50 million people in the world and was noteworthy because of how deadly it was to otherwise healthy young adults. Scientists decided to recreate the virus so that they could determine the genetic basis of its extreme danger to human health, knowledge that could prove valuable as new influenza strains emerge. It was handled and stored in a very high security environment, and multiple safeguards were employed to make sure there was no possibility of an accidental release of the virus. But the prospect of harmful misuse of the new technology has prompted scientific experts to team with national security and bioethics experts to discuss the pros and cons of the new technology and ways to ensure its acceptable uses.

Cultures of animal cells usually exist in the primary or continuous form. *Primary cell cultures* are prepared by placing freshly isolated animal tissue in a growth medium. The cells undergo a series of mitotic divisions to produce a monolayer. Embryonic, fetal, adult, and even cancerous tissues have served as sources of primary cultures. A primary culture retains several characteristics of the original tissue from which it was derived, but this original line generally has a limited existence. Eventually, it will die out or mutate into a line of cells that can grow continuously. Continuous cell lines tend to have altered chromosome numbers, grow rapidly, and show changes in morphology; and they can be continuously subcultured, provided they are routinely transferred to fresh nutrient medium. One very clear advantage of cell culture is that a specific cell line can be available for viruses with a very narrow host range. The recent avian flu worries have prompted scientists to look for faster and more efficient ways to grow the vaccine strains of influenza virus, which has been grown in chicken eggs since the 1950s. Scientists have succeeded in propagating the viruses in a continuous cell line derived from dog kidney cells.

One way to detect the growth of a virus in culture is to observe degeneration and lysis of infected cells in the monolayer of cells. The areas where virus-infected cells have been destroyed show up as clear, well-defined patches in the cell sheet called **plaques (figure 6.22).** Plaques are essentially the macroscopic manifestation of cytopathic effects (CPEs), discussed earlier. This same technique is used to detect and count bacteriophages, because they also produce plaques when grown in soft agar cultures of their host cells (bacteria). A plaque develops when the viruses released by an infected host cell radiate out to adjacent host cells. As new cells become infected, they die and release more viruses, and so on. As this process continues, the infection spreads gradually and symmetrically from the original point of infection, causing the macroscopic appearance of round, clear spaces that correspond to areas of dead cells.

Even though growing viruses remains a challenge, scientists have recently succeeded in artificially creating viruses (**Insight 6.3**).

✓ CHECKPOINT

- Animal viruses must be studied in some type of host cell environment such as laboratory animals, bird embryos, or tissue cultures.
- Cell and tissue cultures are cultures of host cells grown in special sterile chambers containing correct types and proportions of growth factors using aseptic techniques to exclude unwanted microorganisms.
- Virus growth in cell culture is detected by the appearance of plaques.

INSIGHT 6.4 — Medical

A Vaccine for Obesity?

Could it be true? That it was not really the late-night brownies and lack of exercise that made you put on the 20 pounds? Researchers from several different labs are producing evidence that at least some types of obesity may be caused by viruses.

The evidence that viruses cause obesity in humans is somewhat indirect at this point, although animal models provide supporting evidence. So far, at least nine different viruses have been proven to cause obesity in animals, including dogs, rats, and birds. The viruses range from canine distemper virus, the Borna virus (in rats), to several adenoviruses that cause obesity in multiple species.

Of course researchers cannot inject humans with these viruses just to see if they cause them to get fat. So they use more indirect methods. One group, led by Nikhil Dhurnadha at Louisiana State University, tested stored blood from 500 people and found one particular adenovirus in 30% of obese people and in only 11% of nonobese people.

This group also studied 26 sets of twins and found that when one twin had evidence of the viral infection and the other did not, the infected twin always had a higher weight.

The researchers emphasize that obesity has many causes. Other factors considered to be important include a genetic predisposition and, yes, poor diet and exercise. But in the future, people may be offered a vaccine against these viruses to prevent at least some causes of obesity.

6.7 Medical Importance of Viruses

The number of viral infections that occur on a worldwide basis is nearly impossible to measure accurately. Certainly, viruses are the most common cause of acute infections that do not result in hospitalization, especially when one considers widespread diseases such as colds, hepatitis, chickenpox, influenza, herpes, and warts. If one also takes into account prominent viral infections found only in certain regions of the world, such as Dengue fever, Rift Valley fever, and yellow fever, the total could easily exceed several billion cases each year. Although most viral infections do not result in death, some, such as rabies, AIDS, and Ebola, have very high mortality rates, and others can lead to long-term debility (polio, neonatal rubella). Current research is focused on the possible connection of viruses to chronic afflictions of unknown cause, such as type I diabetes, multiple sclerosis, various cancers, and even obesity **(Insight 6.4)**. Additionally, several cancers have their origins in viral infection.

Don't forget that despite the reputation viruses have for being highly detrimental, in some cases, they may actually show a beneficial side (see Insight 6.1).

6.8 Other Noncellular Infectious Agents

Not all noncellular infectious agents have typical viral morphology. One group of unusual forms, even smaller and simpler than viruses, is implicated in chronic, persistent diseases in humans and animals. These diseases are called spongiform encephalopathies because the brain tissue removed from affected animals resembles a sponge. The infection has a long period of latency (usually several years) before the first clinical signs appear. Signs range from mental derangement to loss of muscle control. The diseases are progressive and universally fatal.

A common feature of these conditions is the deposition of distinct protein fibrils in the brain tissue. Researchers have hypothesized that these fibrils are the agents of the disease and have named them **prions** (pree'-onz).

Creutzfeldt-Jakob disease afflicts the central nervous system of humans and causes gradual degeneration and death. Cases in which medical workers developed the disease after handling autopsy specimens seem to indicate that it is transmissible, but by an unknown mechanism. Several animals (sheep, mink, elk) are victims of similar transmissible diseases. Bovine spongiform encephalopathy (BSE), or "mad cow disease," was recently the subject of fears and a crisis in Europe when researchers found evidence that the disease could be acquired by humans who consumed contaminated beef. This was the first incidence of prion disease transmission from animals to humans. Several hundred Europeans developed symptoms of a variant form of Creutzfeldt-Jakob disease, leading to strict governmental controls on exporting cattle and beef products. In 2003, isolated cows with BSE were found in Canada and in the United States. Extreme precautionary measures have been taken to protect North American consumers. (This disease is described in more detail in chapter 19.)

The exact mode of prion infection is currently being analyzed. The fact that prions are composed primarily of protein (no nucleic acid) has certainly revolutionized our ideas of what can constitute an infectious agent. One of the most compelling questions is just how a prion could be replicated, because all other infectious agents require some nucleic acid.

Other fascinating viruslike agents in human disease are defective forms called satellite viruses that are actually dependent on other viruses for replication. Two remarkable examples are the adeno-associated virus (AAV), which can replicate only in cells infected with adenovirus, and the delta agent, a naked strand of RNA that is expressed only in the presence of the hepatitis B virus and can worsen the severity of liver damage.

Plants are also parasitized by viruslike agents called **viroids** that differ from ordinary viruses by being very small (about one-tenth the size of an average virus) and being composed of only naked strands of RNA, lacking a capsid or any other type of coating. Viroids are significant pathogens in several economically important plants, including tomatoes, potatoes, cucumbers, citrus trees, and chrysanthemums.

6.9 Treatment of Animal Viral Infections

The nature of viruses has at times been a major impediment to effective therapy. Because viruses are not bacteria, antibiotics aimed at disrupting prokaryotic cells do not work on them. On the other hand, many antiviral drugs block virus replication by targeting the function of host cells and can cause severe side effects. Antiviral drugs are designed to target one of the steps in the viral life cycle you learned about earlier in this chapter. Azidothymide (AZT), a drug used to treat AIDS, targets the synthesis stage. A newer class of HIV drugs, the protease inhibitors, interrupts the assembly phase of the viral life cycle. Another compound that shows some potential for treating and preventing viral infections is a naturally occurring human cell product called *interferon* (see chapters 12 and 14). Vaccines that stimulate immunity are an extremely valuable tool but are available for only a limited number of viral diseases (see chapter 16).

We have completed our survey of prokaryotes, eukaryotes, and viruses and have described characteristics of different representatives of these three groups. Chapters 7 and 8 explore how microorganisms maintain themselves, beginning with nutrition (chapter 7) and then looking into microbial metabolism (chapter 8).

CHECKPOINT

- Viruses are easily responsible for several billion infections each year. It is conceivable that many chronic diseases of unknown cause will eventually be connected to viral agents.
- Other noncellular agents of disease are the prions, which are not viruses at all but protein fibers; viroids, extremely small lengths of protein-coated nucleic acid; and satellite viruses, which require larger viruses to cause disease.
- Viral infections are difficult to treat because the drugs that attack the viral replication cycle also cause serious side effects in the host.

Chapter Summary with Key Terms

6.1 The Search for the Elusive Viruses
Viruses, being much smaller than bacteria, fungi, and protozoa, had to be indirectly studied until the 20th century when they were finally seen with an electron microscope.

6.2 The Position of Viruses in the Biological Spectrum
Scientists don't agree about whether viruses are living or not. They are obligate intracellular parasites.

6.3 The General Structure of Viruses

A. Viruses are infectious particles and not cells; they lack organelles and locomotion of any kind; they are large, complex molecules; they can be crystalline in form. A virus particle is composed of a nucleic acid core (DNA or RNA, not both) surrounded by a geometric protein shell, or **capsid;** the combination is called a **nucleocapsid;** a capsid is **helical** or **icosahedral** in configuration; many are covered by a membranous envelope containing viral protein **spikes;** complex viruses have additional external and internal structures.

B. *Shapes/Sizes:* Icosahedral, helical, spherical, and cylindrical shaped. Smallest infectious forms range from the largest mimivirus (0.45 mm or 450 nm) to the smallest viruses (0.02 mm or 20 nm).

C. *Nutritional and Other Requirements:* Lack enzymes for processing food or generating energy; are tied entirely to the host cell for all needs (*obligate intracellular parasites*).

D. Viruses are known to parasitize all types of cells, including bacteria, algae, fungi, protozoa, animals, and plants.

6.4 How Viruses Are Classified and Named

A. The two major types of viruses are *DNA* and *RNA viruses*. These are further subdivided into families, depending on shape and size of capsid, presence or absence of an envelope, whether double- or single-stranded nucleic acid, and antigenic similarities.

B. The International Committee on the Taxonomy of Viruses oversees naming and classification of viruses. Viruses are classified into orders, families, and genera. These groupings are based on virus structure, chemical composition, and genetic makeup.

6.5 Modes of Viral Multiplication

A. *Multiplication Cycle: Animal Cells*
1. The life cycle steps of an animal virus are adsorption, penetration/uncoating, synthesis and assembly, and release from the host cell.
2. Each viral type is limited in its **host range** to a single species or group, mostly due to specificity of adsorption of virus to specific host receptors.

3. Some animal viruses cause chronic and persistent infections.
4. Viruses that alter host genetic material may cause *oncogenic* effects.

B. *Viruses That Infect Bacteria*
1. Bacteriophages are viruses that attack bacteria. They penetrate by injecting their nucleic acid and are released as virulent phage upon **lysis** of the cell.
2. Some viruses go into a latent, or **lysogenic,** phase in which they integrate into the DNA of the host cell and later may be active and produce a lytic infection.

6.6 Techniques in Cultivating and Identifying Animal Viruses

A. The need for an intracellular habitat makes it necessary to grow viruses in living cells, either in the intact host animal, in bird embryos, or in isolated cultures of host cells (cell culture).

B. *Identification:* Viruses are identified by means of **cytopathic effects** (CPEs) in host cells, direct examination of viruses or their components in samples, analyzing blood for antibodies against viruses, performing genetic analysis of samples to detect virus nucleic acid, growing viruses in culture, and symptoms.

6.7 Medical Importance of Viruses

A. *Medical:* Viruses attach to specific target hosts or cells. They cause a variety of infectious diseases, ranging from mild respiratory illness (common cold) to destructive and potentially fatal conditions (rabies, AIDS). Some viruses can cause birth defects and cancer in humans and other animals.

B. *Research:* Because of their simplicity, viruses have become an invaluable tool for studying basic genetic principles. Current research is also focused on the possible connection of viruses to chronic afflictions of unknown causes, such as type I diabetes and multiple sclerosis.

6.8 Other Noncellular Infectious Agents

A. Spongiform encephalopathies are chronic persistent neurological diseases caused by **prions.**

B. Examples of neurological diseases include "mad cow disease" and Creutzfeldt-Jakob disease.

C. Other noncellular infectious agents include satellite viruses and viroids.

6.9 Treatment of Animal Viral Infections

Viral infections are difficult to treat because the drugs that attack viral replication also cause serious side effects in the host.

Multiple-Choice and True-False Questions

Multiple-Choice Questions. Select the correct answer from the answers provided.

1. A virus is a tiny infectious
   a. cell c. particle
   b. living thing d. nucleic acid
2. Viruses are known to infect
   a. plants c. fungi
   b. bacteria d. all organisms
3. The nucleic acid of a virus is
   a. DNA only c. both DNA and RNA
   b. RNA only d. either DNA or RNA
4. The general steps in a viral multiplication cycle are
   a. adsorption, penetration, synthesis, assembly, and release
   b. endocytosis, uncoating, replication, assembly, and budding
   c. adsorption, uncoating, duplication, assembly, and lysis
   d. endocytosis, penetration, replication, maturation, and exocytosis
5. A prophage is an early stage in the development of a/an
   a. bacterial virus c. lytic virus
   b. poxvirus d. enveloped virus
6. In general, RNA viruses multiply in the cell ____, and DNA viruses multiply in the cell ____.
   a. nucleus, cytoplasm c. vesicles, ribosomes
   b. cytoplasm, nucleus d. endoplasmic reticulum, nucleolus
7. Enveloped viruses carry surface receptors called
   a. buds c. fibers
   b. spikes d. sheaths
8. Viruses cannot be cultivated in
   a. tissue culture c. live mammals
   b. bird embryos d. blood agar
9. Clear patches in cell cultures that indicate sites of virus infection are called
   a. plaques c. colonies
   b. pocks d. prions
10. Label the parts of this virus. Identify the capsid, nucleic acid, and other features of this virus. Can you identify it?

11. Circle the viral infections from this list: cholera, rabies, plague, cold sores, whooping cough, tetanus, genital warts, gonorrhea, mumps, Rocky Mountain spotted fever, syphilis, rubella, rat bite fever.

True-False Questions. If statement is true, leave as is. If it is false, correct it by rewriting the sentence.

12. In lysogeny, viral DNA is inserted into the host chromosome.
13. A viral capsid is composed of subunits called virions.
14. The envelope of an animal virus is derived from the cell wall of its lost cell.
15. The nucleic acid of animal viruses enters the cell through a process called translocation.
16. Viruses that persist in the (host) cell and cause recurrent disease are called latent.

Writing to Learn

These questions are suggested as a *writing-to-learn* experience. For each question, compose a one- or two-paragraph answer that includes the factual information needed to completely address the question.

1. a. Describe 10 *unique* characteristics of viruses (can include structure, behavior, multiplication).
   b. After consulting table 6.1, what additional statements can you make about viruses, especially as compared with cells?
2. a. What dictates the host range of animal viruses?
   b. What are two ways that animal viruses penetrate the host cell?
   c. What is uncoating?
   d. Describe the two ways that animal viruses leave their host cell.
3. a. What does it mean for a virus to be persistent or latent, and how are these events important?
   b. Briefly describe the action of an oncogenic virus.
4. a. What are bacteriophages and what is their structure?
   b. What is a tobacco mosaic virus?
   c. How are the poxviruses different from other animal viruses?
5. a. Since viruses lack metabolic enzymes, how can they synthesize necessary components?
   b. What are some enzymes with which the virus is equipped?
6. a. Compare and contrast the main phases in the lytic multiplication cycle in bacteriophages and animal viruses.
   b. When is a virus a virion?
   c. What is necessary for adsorption?
   d. Why is penetration so different in the two groups?
   e. What is eclipse?
   f. In simple terms, what does the virus nucleic acid do once it gets into the cell?
   g. What processes are involved in assembly?
7. a. What is a prophage or temperate phage?
   b. What is lysogeny?
8. a. Describe the three main techniques for cultivating viruses.
   b. What are the advantages of using cell culture?
   c. The disadvantages of using cell culture?
   d. What is a disadvantage of using live intact animals or embryos?
   e. What is a cell line? A monolayer?
   f. How are plaques formed?
9. a. What is the principal effect of the agent of Creutzfeldt-Jakob disease?
   b. How is the proposed agent different from viruses?
   c. What are viroids?
10. Why are virus diseases more difficult to treat than bacterial diseases?

Concept Mapping

Appendix D provides guidance for working with concept maps.

1. Supply your own linking words or phrases in this concept map, and provide the missing concepts in the empty boxes.

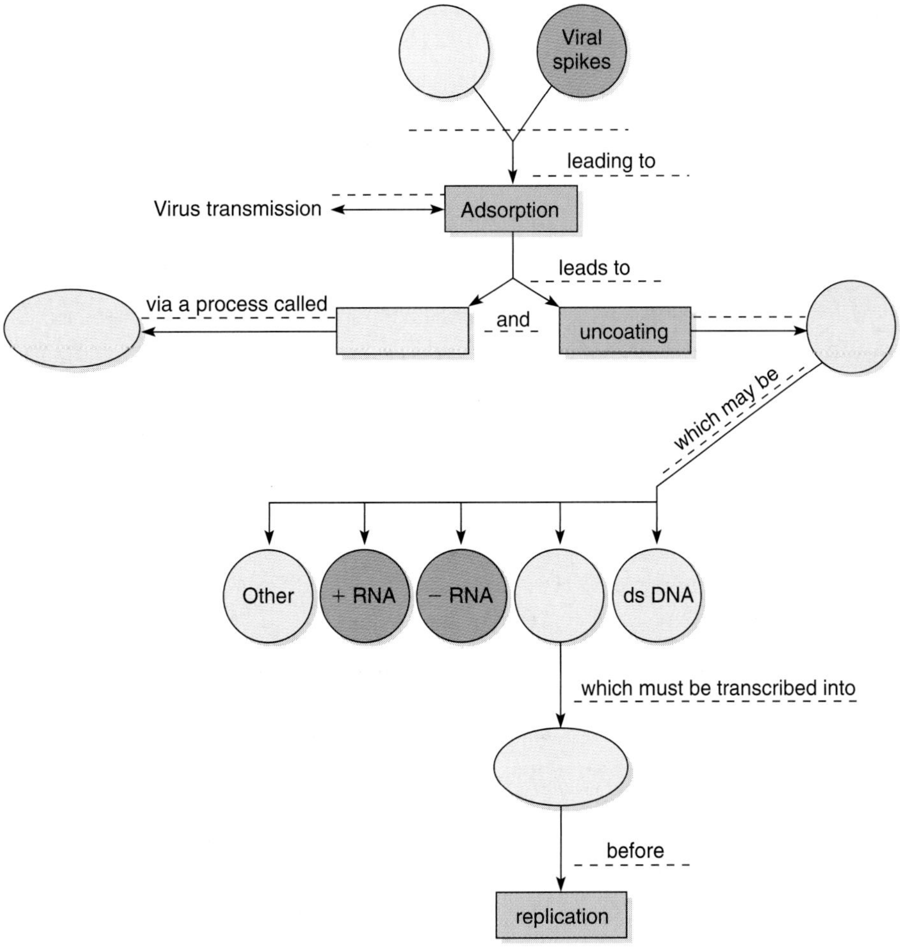

Critical Thinking Questions

Critical thinking is the ability to reason and solve problems using facts and concepts. These questions can be approached from a number of angles, and in most cases, they do not have a single correct answer.

1. a. What characteristics of viruses could be used to characterize them as life forms?
   b. What makes them more similar to lifeless molecules?
2. a. Comment on the possible origin of viruses. Is it not curious that the human cell welcomes a virus in and hospitably removes its coat as if it were an old acquaintance?
   b. How do spikes play a part in the action of the host cell?
3. a. If viruses that normally form envelopes were prevented from budding, would they still be infectious?
   b. If the RNA of an influenza virus were injected into a cell by itself, could it cause a lytic infection?
4. The end result of most viral infections is death of the host cell.
   a. If this is the case, how can we account for such differences in the damage that viruses do (compare the effects of the cold virus with those of the rabies virus)?
   b. Describe the adaptation of viruses that does not immediately kill the host cell and explain what its function might be.
5. a. Given that DNA viruses can actually be carried in the DNA of the host cell's chromosomes, comment on what this phenomenon means in terms of inheritance in the offspring.
   b. Discuss the connection between viruses and cancers, giving possible mechanisms for viruses that cause cancer.

6. HIV attacks only specific types of human cells, such as certain white blood cells and nerve cells. Can you explain why a virus can enter some types of human cells but not others?
7. a. Consult table 6.5 to determine which viral diseases you have had and which ones you have been vaccinated against.
   b. Which viruses would you investigate as possible oncoviruses?
8. One early problem in cultivating HIV was the lack of a cell line that would sustain indefinitely *in vitro*, but eventually one was developed. What do you think were the stages in developing this cell line?
9. a. If you were involved in developing an antiviral drug, what would be some important considerations? (Can a drug "kill" a virus?)
   b. How could multiplication be blocked?
10. a. Is there such a thing as a "good virus"? Explain why or why not. Consider both bacteriophages and viruses of eukaryotic organisms.
11. Why is an embryonic or fetal viral infection so harmful?
12. How are computer viruses analogous to real viruses?
13. Discuss some advantages and disadvantages of bacteriophage therapy in treating bacterial infections.

Visual Understanding

1. **From chapter 1, figure 1.1.** Where do viruses belong on this time line? Use reliable Internet resources to investigate.

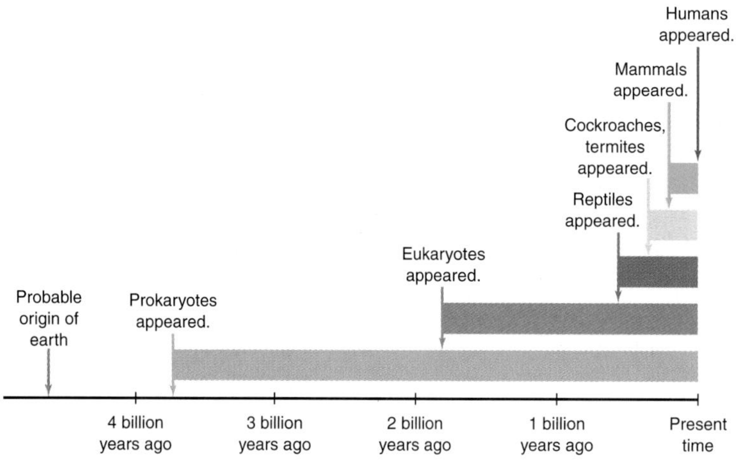

Internet Search Topics

Go to: www.aris.mhhe.com, and click on "microbiology" and then this textbook's author/title. Go to chapter 6, access the URLs listed under Internet Search Topics, and research the following:

1. Explore the excellent websites listed for viruses. Click on Principles of Virus Architecture and Virus Images and Tutorials.
2. Look up emerging viral diseases and make note of the newest viruses that have arisen since 1999. What kinds of diseases do they cause, and where did they possibly originate from?
3. Find websites that discuss prions and prion-based diseases. What possible way do the prions replicate?

Use your favorite search engine to find:

4. What wild game animals in the United States have shown evidence of chronic wasting disease? Is there any direct evidence of human disease caused by contact with these animals?
5. Look for information regarding virus-associated illness on cruise ships.

180-186
192-204

CHAPTER 7

Microbial Nutrition, Ecology, and Growth

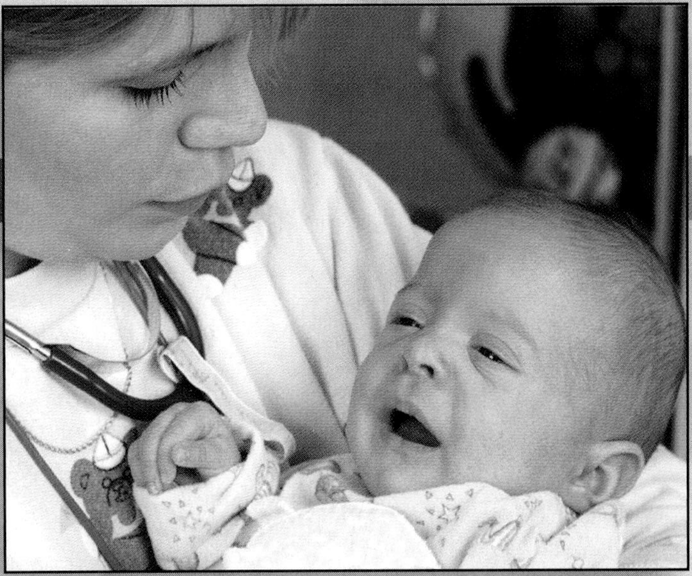

CASE FILE 7

In June 2003, frantic parents rushed a 3-month-old female infant to the emergency room of a regional medical center in rural Tennessee. On initial examination by a triage nurse, "Baby Caroline" appeared listless with unfocused eyes and labored breathing. Her parents reported that, over the past 72 hours, the infant had grown increasingly irritable and had cried weakly and seemed unable to nurse properly. Further questioning revealed that Baby Caroline had had no bowel movements for 3 days. Within 48 hours of admission, she developed flaccid paralysis and experienced respiratory failure. The child received supportive therapy, including the use of a ventilator and administration of antitoxin. Full recovery occurred in about 4 weeks.

Epidemiologists called in to determine the source of the disease examined the child's home. Baby Caroline's parents stated that they were feeding her a leading brand of powdered infant formula prepared with tap water. A week or so previously, Baby Caroline started to refuse the formula, so her mother sweetened it with fresh honey from the family apiary. Additional questioning revealed that a 2-year-old sibling often "borrowed" the baby's pacifier and played with it in the soil of the backyard. Baby Caroline's mother admitted that she had, on a few occasions, simply retrieved the pacifier and wiped it with tissue before returning it to the infant.

- *Based on the information given here, what is the diagnosis of Baby Caroline's illness?*
- *What culture methods could an epidemiologist use to determine the source of the causative agents of the disease?*

Case File 7 Wrap-Up appears on page 197.

CHAPTER OVERVIEW

- Microbes exist in every known natural habitat on earth.
- Microbes show enormous capacity to adapt to environmental factors.
- Factors that have the greatest impact on microbes are nutrients, temperature, pH, amount of available water, atmospheric gases, light, pressure, and other organisms.
- Nutrition involves absorbing required chemicals from the environment for use in metabolism.
- Autotrophs can exist solely on inorganic nutrients, while heterotrophs require both inorganic and organic nutrients.

- Energy sources for microbes may come from light or chemicals.
- Microbes can thrive at cold, moderate, or hot temperatures.
- Oxygen and carbon dioxide are primary gases used in metabolism.
- The water content of the cell versus its environment dictates the osmotic adaptations of cells.
- Transport of materials by cells across cell membranes involves movement by passive and active mechanisms.
- Microbes interact in a variety of ways with one another and with other organisms that share their habitats.
- The pattern of population growth in simple microbes is to double the number of cells in each generation.
- Growth rate is limited by availability of nutrients and buildup of waste products.

7.1 Microbial Nutrition

Nutrition is a process by which chemical substances called **nutrients** are acquired from the environment and used in cellular activities such as metabolism and growth. With respect to nutrition, microbes are not really so different from humans **(Insight 7.1).** Bacteria living in mud on a diet of inorganic sulfur or protozoa digesting wood in a termite's intestine seem to show radical adaptations, but even these organisms require a constant influx of certain substances from their habitat. In general, all living things require a source of elements such as carbon, hydrogen, oxygen, phosphorus, potassium, nitrogen, sulfur, calcium, iron, sodium, chlorine, magnesium, and certain other elements. But the ultimate source of a particular element, its chemical form, and how much of it the microbe needs are all points of variation between different types of organisms. Any substance, whether in elemental or molecular form, that must be provided to an organism is called an **essential nutrient.** Once absorbed, nutrients are processed and transformed into the chemicals of the cell.

Two categories of essential nutrients are **macronutrients** and **micronutrients.** Macronutrients are required in relatively large quantities and play principal roles in cell structure and metabolism. Examples of macronutrients are carbon, hydrogen, and oxygen. Micronutrients, or **trace elements,** such as manganese, zinc, and nickel are present in much smaller amounts and are involved in enzyme function and maintenance of protein structure. What constitutes a micronutrient can vary from one microbe to another.

Another way to categorize nutrients is according to their carbon content. An inorganic nutrient is an atom or simple molecule that contains a combination of atoms other than carbon and hydrogen. The natural reservoirs of inorganic compounds are mineral deposits in the crust of the earth, bodies of water, and the atmosphere. Examples include metals and their salts (magnesium sulfate, ferric nitrate, sodium phosphate), gases (oxygen, carbon dioxide), and water **(table 7.1).** In contrast, the molecules of organic nutrients contain carbon and hydrogen atoms and are usually the products of living things. They range from the simplest organic molecule, methane (CH_4), to large polymers (carbohydrates, lipids, proteins, and nucleic acids). The source of nutrients is extremely varied: Some microbes obtain their nutrients entirely from inorganic sources, and others require a combination of organic and inorganic sources. Parasites capable of invading and living on the human body derive all essential nutrients from host tissues, tissue fluids, secretions, and wastes.

TABLE 7.1 Principal Inorganic Reservoirs of Elements

Element	Inorganic Environmental Reservoir
Carbon	CO_2 in air; CO_3^{2-} in rocks and sediments
Oxygen	O_2 in air, certain oxides, water
Nitrogen	N_2 in air; NO_3^-, NO_2^-, NH_4^+ in soil and water
Hydrogen	Water, H_2 gas, mineral deposits
Phosphorus	Mineral deposits (PO_4^{3-}, H_3PO_4)
Sulfur	Mineral deposits, volcanic sediments (SO_4^{2-}, H_2S, S^0)
Potassium	Mineral deposits, the ocean (KCl, K_3PO_4)
Sodium	Mineral deposits, the ocean ($NaCl$, $NaSi$)
Calcium	Mineral deposits, the ocean ($CaCO_3$, $CaCl_2$)
Magnesium	Mineral deposits, geologic sediments ($MgSO_4$)
Chloride	The ocean ($NaCl$, NH_4Cl)
Iron	Mineral deposits, geologic sediments ($FeSO_4$)
Manganese, molybdenum, cobalt, nickel, zinc, copper, other micronutrients	Various geologic sediments

INSIGHT 7.1 **Discovery**

Dining with an Amoeba

An amoeba gorging itself on bacteria could be compared to a person eating a bowl of vegetable soup, because its nutrient needs are fundamentally similar to that of a human. Most food is a complex substance that contains many different types of nutrients. Some smaller molecules such as sugars can be absorbed directly by the cell; larger food debris and molecules must first be ingested and broken down into a size that can be absorbed. As nutrients are taken in, they add to a dynamic pool of inorganic and organic compounds dissolved in the cytoplasm. This pool will provide raw materials to be assimilated into the organism's own specialized proteins, carbohydrates, lipids, and other macromolecules used in growth and metabolism.

Food particles are phagocytosed into a vacuole that fuses with a lysosome containing digestive enzymes (E). Smaller subunits of digested macromolecules are transported out of the vacuole into the cell pool and are used in the anabolic and catabolic activities of the cell.

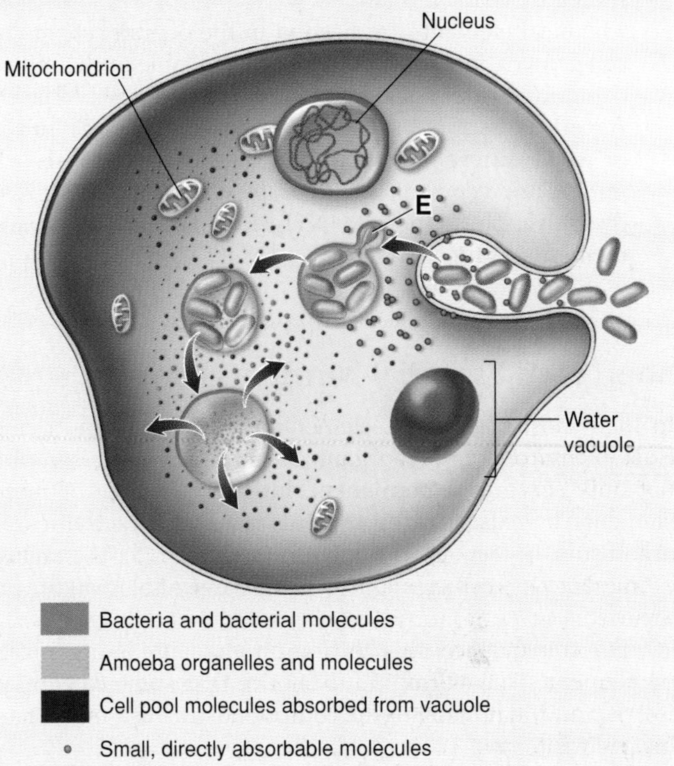

TABLE 7.2 Analysis of the Chemical Composition of an *Escherichia coli* Cell

	% Total Weight	% Dry Weight
Organic Compounds		
Proteins	15	50
Nucleic acids		
RNA	6	20
DNA	1	3
Carbohydrates	3	10
Lipids	2	Not determined
Miscellaneous	2	Not determined
Inorganic Compounds		
Water	70	
All others	1	3

	% Dry Weight
Elements	
Carbon (C)	50
Oxygen (O)	20
Nitrogen (N)	14
Hydrogen (H)	8
Phosphorus (P)	3
Sulfur (S)	1
Potassium (K)	1
Sodium (Na)	1
Calcium (Ca)	0.5
Magnesium (Mg)	0.5
Chlorine (Cl)	0.5
Iron (Fe)	0.2
Manganese (Mn), zinc (Zn), molybdenum (Mo), copper (Cu), cobalt (Co), zinc (Zn)	0.3

Chemical Analysis of Microbial Cytoplasm

Examining the chemical composition of a bacterial cell can indicate its nutritional requirements. **Table 7.2** lists the major contents of the intestinal bacterium *Escherichia coli*. Some of these components are absorbed in a ready-to-use form, and others must be synthesized by the cell from simple nutrients. Several important features of cell composition can be summarized as follows:

- Water content is the highest of all the components (70%).
- Proteins are the next most prevalent chemical.

- About 97% of the dry cell weight is composed of organic compounds.
- About 96% of the cell is composed of six elements (represented by CHONPS).
- Chemical elements are needed in the overall scheme of cell growth, but most of them are available to the cell as compounds and not as pure elements (see table 7.2).
- A cell as "simple" as *E. coli* contains on the order of 5,000 different compounds, yet it needs to absorb only a few types of nutrients to synthesize this great diversity. These include $(NH_4)_2SO_4$, $FeCl_2$, NaCl, trace elements, glucose, KH_2PO_4, $MgSO_4$, $CaHPO_4$, and water.

Sources of Essential Nutrients

In their most basic form, elements that make up nutrients exist in environmental inorganic reservoirs. These reservoirs not only serve as a permanent, long-term source of these elements but also can be replenished by the activities of organisms. In fact, as we shall see in chapter 24, the ability of microbes to keep elements cycling is essential to all life on the earth.

For convenience, this section on nutrients is organized by element. You will no doubt notice that some categories overlap and that many of the compounds furnish more than one element.

Carbon Sources

It seems worthwhile to emphasize a point about the *extracellular source* of carbon as opposed to the *intracellular function* of carbon compounds. Although a distinction is made between the type of carbon compound cells absorb as nutrients (inorganic or organic), the majority of carbon compounds involved in the normal structure and metabolism of all cells are organic.

A **heterotroph** is an organism that must obtain its carbon in an organic form. Because organic carbon originates from the bodies of other organisms, heterotrophs are dependent on other life forms (*hetero-* is a Greek prefix meaning "other"). Among the common organic molecules that can satisfy this requirement are proteins, carbohydrates, lipids, and nucleic acids. In most cases, these nutrients provide several other elements as well. Some organic nutrients available to heterotrophs already exist in a form that is simple enough for absorption (for example, monosaccharides and amino acids), but many larger molecules must be digested by the cell before absorption. Moreover, heterotrophs vary in their capacities to use different organic carbon sources. Some are restricted to a few substrates, whereas others (certain *Pseudomonas* bacteria, for example) are so versatile that they can metabolize more than 100 different substrates.

An **autotroph** ("self-feeder") is an organism that uses inorganic CO_2 as its carbon source. Because autotrophs have the special capacity to convert CO_2 into organic compounds, they are not nutritionally dependent on other living things.

Nitrogen Sources

The main reservoir of nitrogen is nitrogen gas (N_2), which makes up 79% of the earth's atmosphere. This element is indispensable to the structure of proteins, DNA, RNA, and ATP. Such nitrogenous compounds are the primary nitrogen source for heterotrophs, but to be useful, they must first be degraded into their basic building blocks (proteins into amino acids; nucleic acids into nucleotides). Some bacteria and algae utilize inorganic nitrogenous nutrients (NO_3^-, NO_2^-, or NH_3). A small number of prokaryotes can transform N_2 into compounds usable by other organisms through the process of nitrogen fixation (see chapter 24). Regardless of the initial form in which the inorganic nitrogen enters the cell, it must first be converted to NH_3, the only form that can be directly combined with carbon to synthesize amino acids and other compounds.

Oxygen Sources

Because oxygen is a major component of organic compounds such as carbohydrates, lipids, nucleic acids, and proteins, it plays an important role in the structural and enzymatic functions of the cell. Oxygen is likewise a common component of inorganic salts such as sulfates, phosphates, nitrates, and water. Free gaseous oxygen (O_2) makes up 20% of the atmosphere. It is absolutely essential to the metabolism of many organisms, as we shall see later in this chapter and in chapter 8.

Hydrogen Sources

Hydrogen is a major element in all organic and several inorganic compounds, including water (H_2O), salts ($Ca[OH]_2$), and certain naturally occurring gases (H_2S, CH_4, and H_2). These gases are both used and produced by microbes. Hydrogen performs these overlapping roles in the biochemistry of cells:

1. maintaining **pH,**
2. forming **hydrogen bonds** between molecules, and
3. serving as the source of **free energy** in oxidation-reduction reactions of respiration (see chapter 8).

Phosphorus (Phosphate) Sources

The main inorganic source of phosphorus is phosphate (PO_4^{3-}), derived from phosphoric acid (H_3PO_4) and found in rocks and oceanic mineral deposits. Phosphate is a key component of nucleic acids and is thereby essential to the genetics of cells and viruses. Because it is also found in ATP, it also serves in cellular energy transfers. Other phosphate-containing compounds are phospholipids in cell membranes and coenzymes such as NAD^+ (see chapter 8). Certain environments have very little available phosphate for use by organisms and therefore limit the ability of these organisms to grow. However, *Corynebacterium* is able to concentrate and store phosphate in metachromatic granules.

Sulfur Sources

Sulfur is widely distributed throughout the environment in mineral form. Rocks and sediments (such as gypsum) can contain sulfate (SO_4^{2-}), sulfides (FeS), hydrogen sulfide gas (H_2S), and elemental sulfur (S). Sulfur is an essential component of some vitamins (vitamin B_1) and the amino acids methionine and cysteine; the latter help determine shape and structural stability of proteins by forming unique linkages called disulfide bonds (described in chapter 2).

Other Nutrients Important in Microbial Metabolism

Other important elements in microbial metabolism include mineral ions. Potassium is essential to protein synthesis and membrane function. Sodium is important for certain types of cell transport. Calcium is a stabilizer of the cell wall and endospores of bacteria. Magnesium is a component of chlorophyll and a stabilizer of membranes and ribosomes. Iron is an important component of the cytochrome proteins of cell respiration. Zinc is an essential regulatory element for eukaryotic genetics. It is a major component of "zinc fingers"—binding factors that help enzymes adhere to specific sites on DNA. Copper, cobalt, nickel, molybdenum, manganese, silicon, iodine, and boron are needed in small amounts by some microbes but not others. On the other hand, in chapter 11 you will see that metals can also be very toxic to microbes. The concentration of metal ions can even influence the diseases microbes cause. For example, the bacteria that cause gonorrhea and meningitis grow more rapidly in the presence of iron ions.

Growth Factors: Essential Organic Nutrients

Few microbes are as versatile as *Escherichia coli* in assembling molecules from scratch. Many fastidious bacteria lack the genetic and metabolic mechanisms to synthesize every organic compound they need for survival. An organic compound such as an amino acid, nitrogenous base, or vitamin that cannot be synthesized by an organism and must be provided as a nutrient is a **growth factor.** For example, although all cells require 20 different amino acids for proper assembly of proteins, many cells cannot synthesize all of them. Those that must be obtained from food are called essential amino acids. A notable example of the need for growth factors occurs in *Haemophilus influenzae,* a bacterium that causes meningitis and respiratory infections in humans. It can grow only when hemin (factor X), NAD (factor V), thiamine and pantothenic acid (vitamins), uracil, and cysteine are provided by another organism or a growth medium.

How Microbes Feed: Nutritional Types

The earth's limitless habitats and microbial adaptations are matched by an elaborate menu of microbial nutritional schemes. Fortunately, most organisms show consistent trends and can be described by a few general categories **(table 7.3)** and a few selected terms (see "A Note on Terminology" on page 185). The main determinants of a microbe's nutritional type are its sources of carbon and energy. In a previous section, microbes were defined according to their carbon sources as autotrophs or heterotrophs. Now we will subdivide all bacteria according to their energy source as **phototrophs** or **chemotrophs.** Microbes that photosynthesize are phototrophs and those that gain energy from chemical compounds are chemotrophs. The terms for carbon and energy source are often merged into a single word for convenience (see table 7.3). The categories described here are meant to

TABLE 7.3 Nutritional Categories of Microbes by Energy and Carbon Source

Category	Energy Source	Carbon Source	Example
Autotroph	**Nonliving environment**	**CO_2**	
Photoautotroph	Sunlight	CO_2	Photosynthetic organisms, such as algae, plants, cyanobacteria
Chemoautotroph	Simple inorganic chemicals	CO_2	Only certain bacteria, such as methanogens, deep sea vent bacteria
Heterotroph	**Other organisms or sunlight**	**Organic**	
Photoheterotroph	Sunlight	Organic	Purple and green photosynthetic bacteria
Chemoheterotroph	Metabolic conversion of the nutrients from other organisms	Organic	Protozoa, fungi, many bacteria, animals
Saprobe	Metabolizing the organic matter of dead organisms	Organic	Fungi, bacteria (decomposers)
Parasite	Utilizing the tissues, fluids of a live host	Organic	Various parasites and pathogens; can be bacteria, fungi, protozoa, animals

INSIGHT 7.2 — *Discovery*

Light-Driven Organic Synthesis

Two equations sum up the reactions of photosynthesis in a simple way. The first equation shows a reaction that results in the production of oxygen:

$$CO_2 + H_2O \xrightarrow[\text{by chlorophyll}]{\text{Sunlight absorbed}} (CH_2O)_n{}^{*} + O_2$$

This oxygenic (oxygen-producing) type of photosynthesis occurs in plants, algae, and cyanobacteria. The function of chlorophyll is to capture light energy. Carbohydrates produced by the reaction can be used by the cell to synthesize other cell components, and they also become a significant nutrient for heterotrophs that feed on them. The production of oxygen is vital to maintaining this gas in the atmosphere.

A second equation shows a photosynthetic reaction that does not result in the production of oxygen:

$$CO_2 + H_2S \xrightarrow[\text{by bacteriochlorophyll}]{\text{Sunlight absorbed}} (CH_2O)_n + S^0 + H_2O$$

This anoxygenic (no oxygen produced) type of photosynthesis is found in bacteria such as purple and green sulfur bacteria. Note that the type of chlorophyll (bacteriochlorophyll, a substance unique to these microbes), one of the reactants (hydrogen sulfide gas), and one product (elemental sulfur) are different from those in the first equation. These bacteria live in anaerobic regions of aquatic habitats.

*$(CH_2O)_n$ is shorthand for a carbohydrate.

describe only the major nutritional groups and do not include unusual exceptions.

Autotrophs and Their Energy Sources

Autotrophs derive energy from one of two possible nonliving sources: sunlight (photoautotrophs) and chemical reactions involving simple chemicals (chemoautotrophs). **Photoautotrophs** are photosynthetic—that is, they capture the energy of light rays and transform it into chemical energy that can be used in cell metabolism **(Insight 7.2).** Because photosynthetic organisms (algae, plants, some bacteria) produce organic molecules that can be used by themselves and heterotrophs, they form the basis for most food webs. Their role as primary producers of organic matter is discussed in chapter 24.

Chemoautotrophs are of two types: one of these is the group called chemoorganic autotrophs. These use organic compounds for energy and inorganic compounds as a carbon source. The second type of chemoautotroph is a group called **lithoautotrophs,** which requires neither sunlight nor organic nutrients, relying totally on inorganic minerals. These bacteria derive energy in diverse and rather amazing ways. In very simple terms, they remove electrons from inorganic substrates such as hydrogen gas, hydrogen sulfide, sulfur, or iron and combine them with carbon dioxide and hydrogen. This reaction provides simple organic molecules and a modest amount of energy to drive the synthetic processes of the cell. Lithoautotrophic bacteria play an important part in recycling inorganic nutrients. For an example of lithoautotrophy and its importance to deep-sea communities, see Insight 7.5.

An interesting group of chemoautotrophs is **methanogens** (meth-an-oh-gen), which produce methane (CH_4) from hydrogen gas and carbon dioxide **(figure 7.1).**

$$4H_2 + CO_2 \rightarrow CH_4 + 2H_2O$$

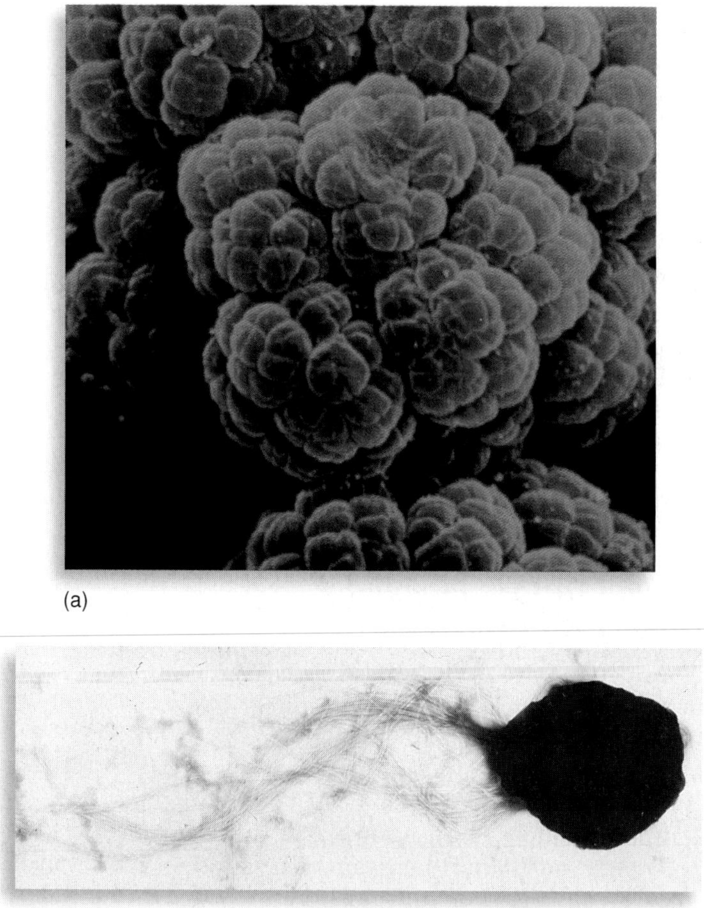

(a)

(b)

Figure 7.1 Methane-producing archaea.
Members of this group are primitive prokaryotes with unusual cell walls and membranes. **(a)** SEM of a small colony of Methanosarcina. **(b)** *Methanococcus jannaschii,* a motile archaea that inhabits hot vents in the seafloor and uses hydrogen gas as a source of energy.

A NOTE ON TERMINOLOGY

Much of the vocabulary for describing microbial adaptations is based on some common root words. These are combined in various ways that assist in discussing the types of nutritional or ecological adaptations, as shown in this partial list:

Root	Meaning	Example of Use
troph-	Food, nourishment	Trophozoite—the feeding stage of protozoa
-phile	To love	Extremophile—an organism that has adapted to ("loves") extreme environments
-obe	To live	Microbe—to live "small"
hetero-	Other	Heterotroph—an organism that requires nutrients from other organisms
auto-	Self	Autotroph—an organism that does not need other organisms for food (obtains nutrients from a nonliving source)
photo-	Light	Phototroph—an organism that uses light as an energy source
chemo-	Chemical	Chemotroph—an organism that uses chemicals for energy, rather than light
sapro-	Rotten	Saprobe—an organism that lives on dead organic matter
halo-	Salt	Halophile—an organism that can grow in high-salt environments
thermo-	Heat	Thermophile—an organism that grows best at high temperatures
psychro-	Cold	Psychrophile—an organism that grows best at cold temperatures
aero-	Air (O_2)	Aerobe—an organism that uses oxygen in metabolism

Modifier terms are also used to specify the nature of an organism's adaptations. **Obligate** or **strict** refers to being restricted to a narrow niche or habitat, such as an obligate thermophile that requires high temperatures to grow. By contrast, **facultative** means not being so restricted but being able to adapt to a wider range of metabolic conditions and habitats. A facultative halophile can grow with or without high salt concentration.

Methane, sometimes called "swamp gas" or "natural gas" is formed in anaerobic, hydrogen-containing microenvironments of soil, swamps, mud, and even in the intestines of some animals. Methanogens are archaea, some of which live in extreme habitats such as ocean vents and hot springs, where temperatures reach up to 125°C **(Insight 7.3).** Methane, which is used as a fuel in a large percentage of homes, can also be produced in limited quantities using a type of generator primed with a mixed population of microbes (including methanogens) and fueled with various waste materials that can supply enough methane to drive a steam generator. Methane also plays a role as one of the greenhouse gases that is currently an environmental concern (see chapter 24).

Heterotrophs and Their Energy Sources

The majority of heterotrophic microorganisms are **chemoheterotrophs** that derive both carbon and energy from organic compounds. Processing these organic molecules by respiration or fermentation releases energy in the form of ATP. An example of chemoheterotrophy is *aerobic respiration,* the principal energy-yielding pathway in animals, most protozoa and fungi, and aerobic bacteria. It can be simply represented by the equation:

$$\text{Glucose } [(CH_2O)_n] + O_2 \rightarrow CO_2 + H_2O + \text{Energy (ATP)}$$

This reaction is complementary to photosynthesis. Here, glucose and oxygen are reactants, and carbon dioxide is given off. Indeed, the earth's balance of both energy and metabolic gases is greatly dependent on this relationship. Chemoheterotrophic microorganisms belong to one of two main categories that differ in how they obtain their organic nutrients: **Saprobes** are free-living microorganisms that feed primarily on organic detritus from dead organisms, and **parasites** ordinarily derive nutrients from the cells or tissues of a host.

Saprobic Microorganisms Saprobes occupy a niche as decomposers of plant litter, animal matter, and dead microbes. If not for the work of decomposers, the earth would gradually fill up with organic material, and the nutrients it contains would not be recycled. Most saprobes, notably bacteria and fungi, have a rigid cell wall and cannot engulf large particles of food. To compensate, they release enzymes to the extracellular environment and digest the food particles into smaller molecules that can be transported into the cell **(figure 7.2).** *Obligate saprobes* exist strictly on dead organic matter in soil and water and are unable to adapt to the body of a live host. This group includes many free-living protozoa, fungi, and bacteria. Apparently, there are fewer of these strict species than was once thought, and many supposedly nonpathogenic saprobes can infect a susceptible host. When a saprobe does infect a host, it is considered a *facultative parasite.* Such an infection usually occurs when the host is compromised, and the microbe is considered an *opportunistic*

INSIGHT 7.3 Discovery

Life in the Extremes

Any extreme habitat—whether hot, cold, salty, acidic, alkaline, high pressure, arid, oxygen-free, or toxic—is likely to harbor microorganisms that have made special adaptations to their conditions. Although in most instances the inhabitants are archaea and bacteria, certain fungi, protozoans, and algae are also capable of living in harsh habitats. Microbiologists have termed such remarkable organisms **extremophiles.**

Hot and Cold

Some of the most extreme habitats are hot springs, geysers, volcanoes, and ocean vents, all of which support flourishing microbial populations. Temperatures in these regions range from 50°C to well above the boiling point of water, with some ocean vents even approaching 350°C. Many heat-adapted microbes are archaea whose genetics and metabolism are extremely modified for this mode of existence. A unique ecosystem based on hydrogen sulfide–oxidizing bacteria exists in the hydrothermal vents lying along deep oceanic ridges (see Insight 7.5). Heat-adapted bacteria even plague home water heaters and the heating towers of power and industrial plants.

A large part of the earth exists at cold temperatures. Microbes settle and grow throughout the Arctic and Antarctic, and in the deepest parts of the ocean, in temperatures that hover near the freezing point of water. Several species of algae and fungi thrive on the surfaces of snow and glacier ice (see figure 7.9). More surprising still is that some bacteria and algae are adapted to the sea ice of Antarctica. Although the ice appears to be completely solid, it is honeycombed by various-size pores and tunnels filled with liquid water. These frigid microhabitats harbor a microcosm of planktonic life, including predators (fish and shrimp) that live on these algae and bacteria.

Salt, Acidity, Alkalinity

The growth of most microbial cells is inhibited by high amounts of salt; for this reason, salt is a common food preservative. Yet whole communities of salt-dependent bacteria and algae occupy habitats in oceans, salt lakes, and inland seas, some of which are saturated with salt (30%). Most of these microbes have demonstrable metabolic requirements for high levels of minerals such as sodium, potassium, magnesium, chlorides, or iodides. Because of their salt-loving nature, some species are pesky contaminants in salt-processing plants, pickling brine, and salted fish.

Highly acidic or alkaline habitats are not common, but acidic bogs, lakes, and alkaline soils contain a specialized microbiota. A few species of algae and bacteria can actually survive at a pH near that of concentrated hydrochloric acid. They not only require such a low pH for growth, but particular bacteria (for example, *Thiobacillus*) actually help maintain the low pH by releasing strong acid.

Other Frontiers to Conquer

It was once thought that the region far beneath the soil and upper crust of the earth's surface was sterile. However, work with deep core samples (from 330 m down) indicates a vast microbial population in these zones. Myriad bacteria, protozoa, and fungi exist in this moist clay, which is high in minerals and complex organic substrates. Even deep mining deposits 2 miles into the earth's crust harbor a rich assortment of bacteria. They thrive in mineral deposits that are hot (90°C) and radioactive. Many biologists believe these are very similar to the first ancient microbes to have existed on earth.

Numerous species have carved a niche for themselves in the depths of mud, swamps, and oceans, where oxygen gas and sunlight cannot penetrate. The predominant living things in the deepest part of the oceans (10,000 m or below) are pressure- and cold-loving microorganisms. Even parched zones in sand dunes and deserts harbor a hardy brand of microbes, and thriving bacterial populations can be found in petroleum, coal, and mineral deposits containing copper, zinc, gold, and uranium.

As a rule, a microbe that has adapted to an extreme habitat will die if placed in a moderate one. And, except for rare cases, none of the organisms living in these extremes are pathogens, because the human body is a hostile habitat for them.

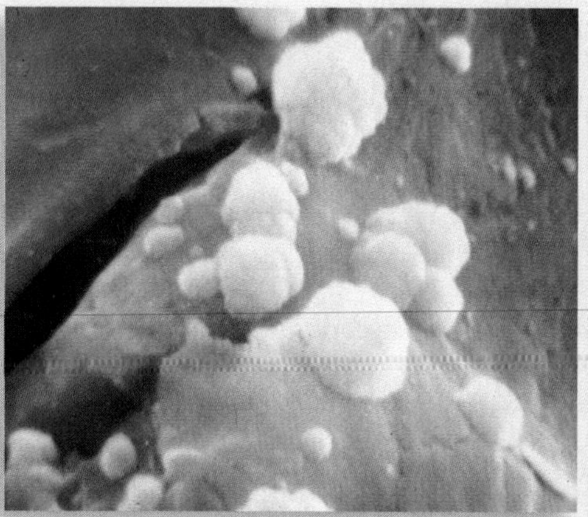
(a)

(b)

(a) Cells of *Sulfolobus*, an archaean that lives in mineral deposits of hot springs and volcanoes. It can survive temperatures of about 90°C and acidity of pH 1.5. **(b)** Clumps of bacteria (dark matter) growing on crystals of ice deep in the Antarctic sediments.

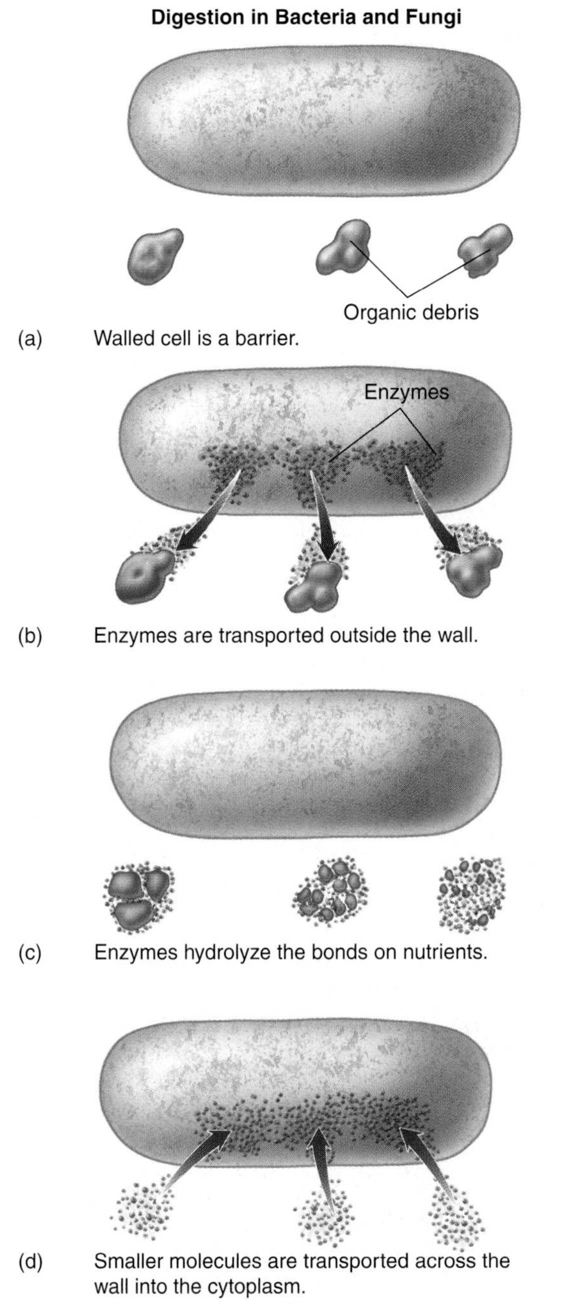
Digestion in Bacteria and Fungi
Organic debris
(a) Walled cell is a barrier.
Enzymes
(b) Enzymes are transported outside the wall.
(c) Enzymes hydrolyze the bonds on nutrients.
(d) Smaller molecules are transported across the wall into the cytoplasm.

Figure 7.2 Extracellular digestion in a saprobe with a cell wall (bacterium or fungus).
(a) A walled cell is inflexible and cannot engulf large pieces of organic debris. **(b)** In response to a usable substrate, the cell synthesizes enzymes that are transported across the wall into the extracellular environment. **(c)** The enzymes hydrolyze the bonds in the debris molecules. **(d)** Digestion produces molecules small enough to be transported into the cytoplasm.

pathogen. For example, although its natural habitat is soil and water, *Pseudomonas aeruginosa* frequently causes infections in hospitalized patients. The yeast *Cryptococcus neoformans* causes a severe lung and brain infection in AIDS patients (see chapter 19), yet its natural habitat is the soil.

Parasitic Microorganisms Parasites live in or on the body of a host, which they harm to some degree. Because parasites cause damage to tissues (disease) or even death, they are also called **pathogens.** Parasites range from viruses to helminths (worms) and they can live on the body (ectoparasites), in the organs and tissues (endoparasites), or even within cells (intracellular parasites, the most extreme type). *Obligate parasites* (for example, the leprosy bacillus and the syphilis spirochete) are unable to grow outside of a living host. Parasites that are less strict can be cultured artificially if provided with the correct nutrients and environmental conditions. Bacteria such as *Streptococcus pyogenes* (the cause of strep throat) and *Staphylococcus aureus* can grow on artificial media.

Obligate intracellular parasitism is an extreme but relatively common mode of life. Microorganisms that spend all or part of their life cycle inside a host cell include the viruses, a few bacteria (rickettsias, chlamydias), and certain protozoa (apicomplexa). Contrary to what one might think, the cell interior is not completely without hazards, and microbes must overcome some difficult challenges. They must find a way into the cell, keep from being destroyed, not destroy the host cell too soon, multiply, and find a way to infect other cells. Intracellular parasites obtain different substances from the host cell, depending on the group. Viruses are extreme, parasitizing the host's genetic and metabolic machinery. Rickettsias are primarily energy parasites, and the malaria protozoan is a hemoglobin parasite.

Transport Mechanisms for Nutrient Absorption

A microorganism's habitat provides necessary nutrients—some abundant, others scarce—that must still be taken into the cell. Survival also requires that cells transport waste materials out of the cell (and into the environment). Whatever the direction, transport occurs across the cell membrane, the structure specialized for this role. This is true even in organisms with cell walls (bacteria, algae, and fungi), because the cell wall is usually too nonselective to screen the entrance or exit of molecules. Before we talk about movement of nutrients (molecules, solutes) in and out of cells, we'll address the movement of water, or osmosis. You might want to take a moment to review solutes and solvents on page 36 in chapter 2.

The Movement of Water: Osmosis

Diffusion of water through a selectively permeable membrane, a process called **osmosis,** is also a physical phenomenon that is easily demonstrated in the laboratory with nonliving materials. It provides a model of how cells deal with various solute concentrations in aqueous solutions **(figure 7.3).** In an osmotic system, the membrane is *selectively,* or *differentially, permeable,* having passageways that allow free diffusion of water but can block certain other dissolved molecules. When this membrane is placed between solutions of differing concentrations and the solute is not diffusible (protein, for example), then under the laws of diffusion, water will diffuse at a faster rate from the side that has more

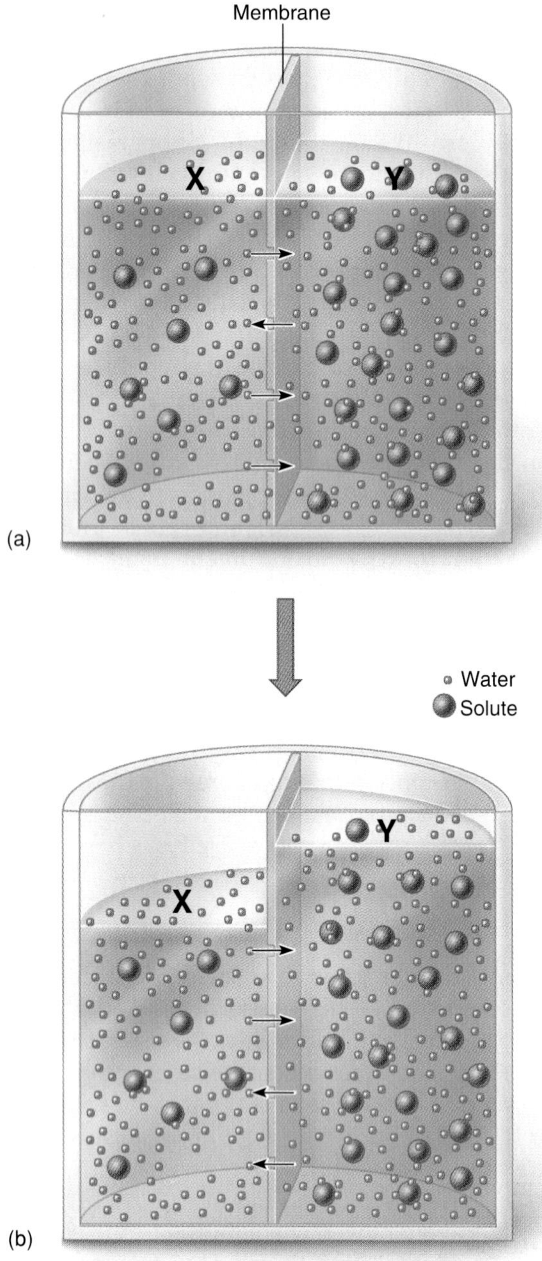

Figure 7.3 Osmosis, the diffusion of water through a selectively permeable membrane.
(a) A membrane has pores that allow the ready passage of water but not large solute molecules from one side to another. Placement of this membrane between solutions of different solute concentrations (X = less concentrated, and Y = more concentrated) results in a diffusion gradient for water. Water molecules undergo diffusion and move across the membrane pores in both directions. The result will be a net movement of water from X to Y. **(b)** The level of solution on the Y side rises as water continues to diffuse in. This process will continue until equilibration occurs and the rate of diffusion of water is equal on both sides.

water to the side that has less water. As long as the concentrations of the solutions differ, one side will experience a net loss of water and the other a net gain of water, until equilibrium is reached and the rate of diffusion is equalized.

Osmosis in living systems is similar to the model shown in figure 7.3. Living membranes generally block the entrance and exit of larger molecules and permit free diffusion of water. Because most cells are surrounded by some free water, the amount of water entering or leaving has a far-reaching impact on cellular activities and survival. This osmotic relationship between cells and their environment is determined by the relative concentrations of the solutions on either side of the cell membrane **(figure 7.4).** Such systems can be compared using the terms *isotonic, hypotonic,* and *hypertonic.* (The root *-tonic* means "tension." *Iso-* means "the same," *hypo-* means "less," and *hyper-* means "over" or "more.")

Under **isotonic** conditions, the environment is equal in solute concentration to the cell's internal environment, and because diffusion of water proceeds at the same rate in both directions, there is no net change in cell volume. Isotonic solutions are generally the most stable environments for cells, because they are already in an osmotic steady state with the cell. Parasites living in host tissues are most likely to be living in isotonic habitats.

Under **hypotonic** conditions, the solute concentration of the external environment is lower than that of the cell's internal environment. Pure water provides the most hypotonic environment for cells because it has no solute. The net direction of osmosis is from the hypotonic solution into the cell, and cells without walls swell and can burst.

A slightly hypotonic environment can be quite favorable for bacterial cells. The constant slight tendency for water to flow into the cell keeps the cell membrane fully extended and the cytoplasm full. This is the optimum condition for the many processes occurring in and on the membrane. Slight hypotonicity is tolerated quite well by bacteria because of their rigid cell walls.

Hypertonic[1] conditions are also out of balance with the tonicity of the cell's cytoplasm, but in this case, the environment has a higher solute concentration than the cytoplasm. Because a hypertonic environment will force water to diffuse out of a cell, it is said to have high *osmotic pressure* or potential. The growth-limiting effect of hypertonic solutions on microbes is the principle behind using concentrated salt and sugar solutions as preservatives for food, such as in salted hams.

Adaptations to Osmotic Variations in the Environment

Let us now see how specific microbes have adapted osmotically to their environments. In general, isotonic conditions pose little stress on cells, so survival depends on counteracting the adverse effects of hypertonic and hypotonic environments.

A bacterium and an amoeba living in fresh pond water are examples of cells that live in constantly hypotonic conditions. The rate of water diffusing across the cell membrane into the cytoplasm is rapid and constant, and the cells would die without a way to adapt. As just mentioned, the majority of bacterial cells compensate by having a cell wall that protects them from bursting even as the cytoplasmic membrane becomes *turgid* (ter'-jid) from pressure. The amoeba's adaptation is an anatomical and physiological one that requires the constant expenditure of energy. It has a water, or contractile, vacuole that moves excess water back out into the habitat like a tiny pump.

1. It will help you to recall these osmotic conditions if you remember that the prefixes iso-, hypo-, and hyper- refer to the environment *outside* of the cell.

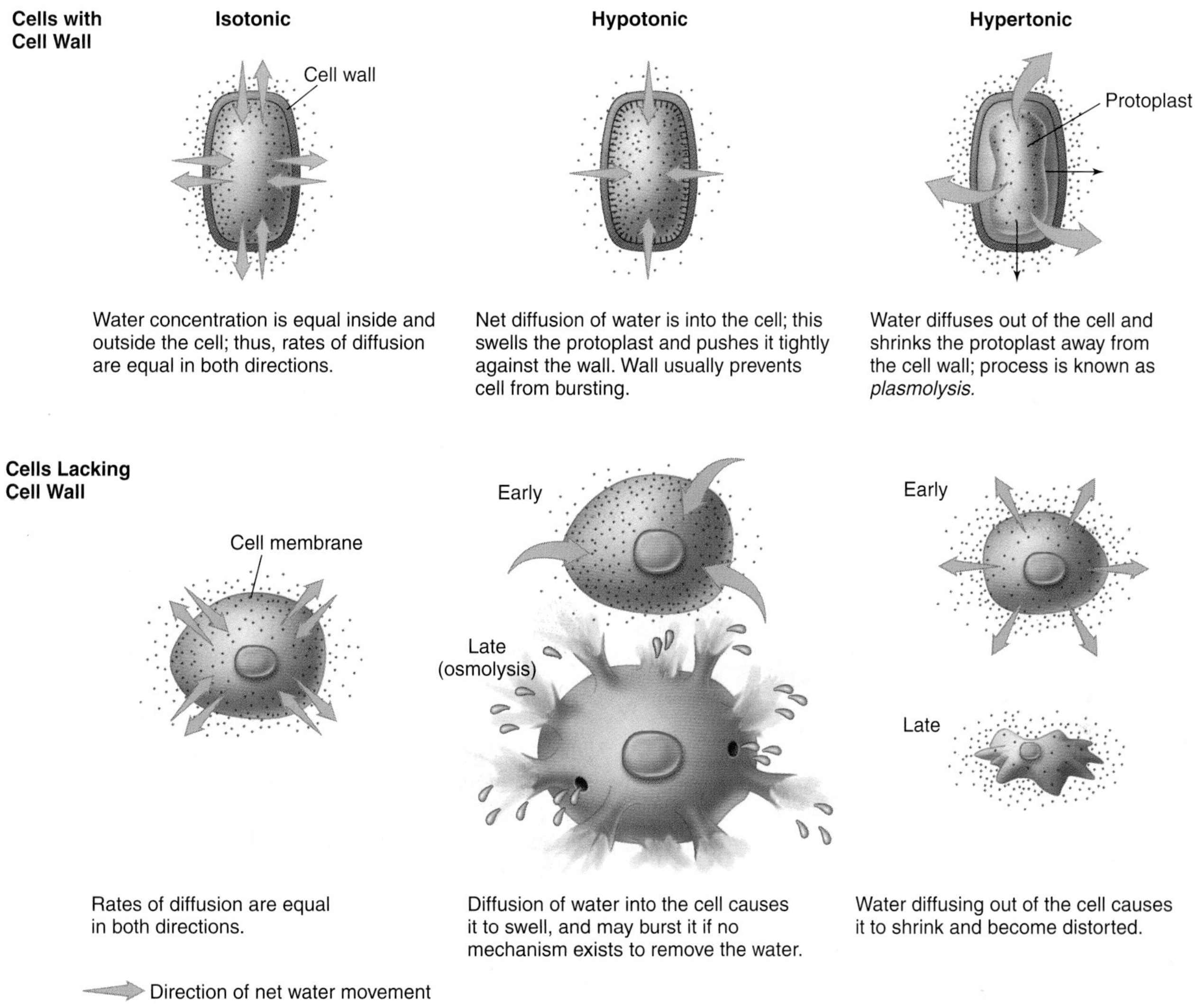

Cells with Cell Wall
Isotonic
Hypotonic
Hypertonic
Cell wall
Protoplast
Water concentration is equal inside and outside the cell; thus, rates of diffusion are equal in both directions.
Net diffusion of water is into the cell; this swells the protoplast and pushes it tightly against the wall. Wall usually prevents cell from bursting.
Water diffuses out of the cell and shrinks the protoplast away from the cell wall; process is known as plasmolysis.
Cells Lacking Cell Wall
Cell membrane
Early
Late (osmolysis)
Early
Late
Rates of diffusion are equal in both directions.
Diffusion of water into the cell causes it to swell, and may burst it if no mechanism exists to remove the water.
Water diffusing out of the cell causes it to shrink and become distorted.
Direction of net water movement

Figure 7.4 Cell responses to solutions of differing osmotic content.

A microbe living in a high-salt environment (hypertonic) has the opposite problem and must either restrict its loss of water to the environment or increase the salinity of its internal environment. Halobacteria living in the Great Salt Lake and the Dead Sea actually absorb salt to make their cells isotonic with the environment; thus, they have a physiological need for a high-salt concentration in their habitats (see halophiles on page 196).

The Movement of Molecules: Diffusion and Transport

The driving force of transport is atomic and molecular movement—the natural tendency of atoms and molecules to be in constant random motion. The existence of this motion is evident in Brownian movement of particles suspended in liquid. It can also be demonstrated by a variety of simple observations. A drop of perfume released into one part of a room is soon smelled in another part, or a lump of sugar in a cup of tea spreads through the whole cup without stirring. This phenomenon of molecular movement, in which atoms or molecules move in a gradient from an area of higher density or concentration to an area of lower density or concentration, is **diffusion (figure 7.5).**

Diffusion

All molecules, regardless of being in a solid, liquid, or gas, are in continuous movement, and as the temperature increases, the molecular movement becomes faster. This is called "thermal" movement. In any solution, including cytoplasm, these moving molecules cannot travel very far without having collisions with other molecules and, therefore, will bounce off each other like millions of pool balls every second. As a result of each collision, the directions of the colliding molecules are altered and the direction of any one molecule is unpredictable and is therefore "random." If we start with a solution in which the solute, or dissolved substance, is more concentrated in one area than another, then the random thermal movement of molecules in this solution will eventually distribute the molecules from the area of higher concentration to the area of lower concentration, thus evenly distributing the molecules. This net movement of molecules down their concentration gradient by

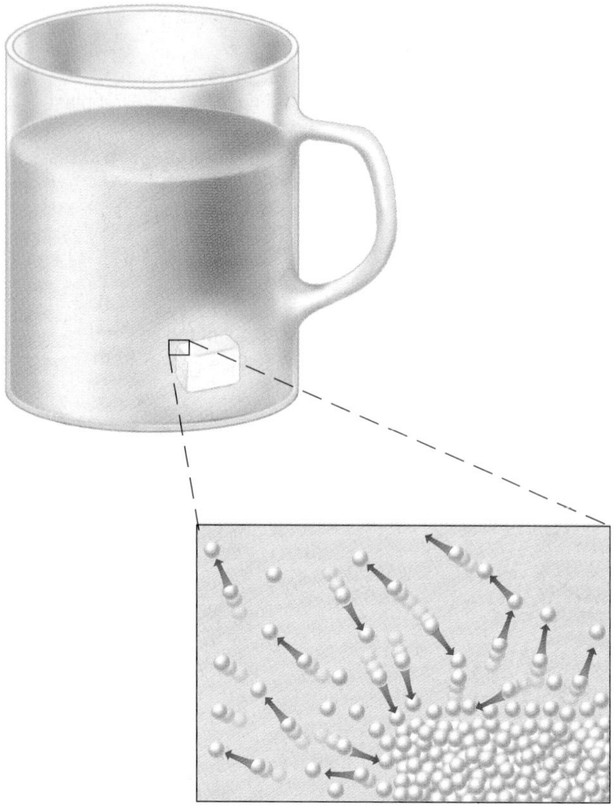

Figure 7.5 Diffusion of molecules in aqueous solutions. A high concentration of sugar exists in the cube at the bottom of the liquid. An imaginary molecular view of this area shows that sugar molecules are in a constant state of motion. Those at the edge of the cube diffuse from the concentrated area into more dilute regions. As diffusion continues, the sugar will spread evenly throughout the aqueous phase, and eventually there will be no gradient. At that point, the system is said to be in equilibrium.

random thermal motion is known as diffusion. Diffusion of molecules across the cell membrane is largely determined by the concentration gradient and permeability of the substance.

So far, the discussion of passive or simple diffusion has not included the added complexity of membranes or cell walls, which hinder simple diffusion by adding a physical barrier. Therefore, simple diffusion is limited to small nonpolar molecules like oxygen or lipid soluble molecules that may pass through the membranes. It is imperative that a cell be able to move polar molecules and ions across the plasma membrane, and given the greatly decreased permeability of these chemicals simple diffusion will not allow this movement. Therefore the concept of **facilitated diffusion** must be introduced **(figure 7.6).** This type of mediated transport mechanism utilizes a carrier protein that will bind a specific substance. This binding changes the conformation of the carrier proteins so that the substance is moved across the membrane. Once the substance is transported, the carrier protein resumes its original shape and is ready to transport again. These carrier proteins exhibit **specificity,** which means that they bind and transport only a single type of molecule. For example, a carrier protein that transports sodium will not bind glucose. A second characteristic exhibited by facilitated

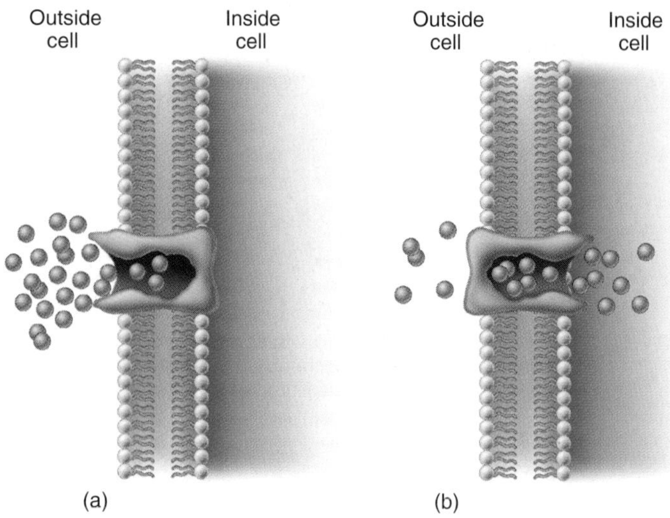

Figure 7.6 Facilitated diffusion. Facilitated diffusion involves the attachment of a molecule to a specific protein carrier. **(a)** Bonding of the molecule causes a conformational change in the protein that facilitates the molecule's passage across the membrane. **(b)** The membrane receptor opens into the cell and releases the molecule.

diffusion is **saturation.** The rate of transport of a substance is limited by the number of binding sites on the transport proteins. As the substance's concentration increases so does the rate of transport until the concentration of the transported substance is such that all of the transporters' binding sites are occupied. Then the rate of transport reaches a steady state and cannot move faster despite further increases in the substance's concentration. A third characteristic of these carrier proteins is that they exhibit competition. This is when two molecules of similar shape can bind to the same binding site on a carrier protein. The chemical with the higher binding affinity, or the chemical in the higher concentration, will be transported at a greater rate.

Neither simple diffusion nor facilitated diffusion requires energy, because molecules are moving down a concentration gradient.

Active Transport: Bringing in Molecules Against a Gradient

Free-living microbes exist under relatively nutrient-starved conditions and cannot rely completely on slow and rather inefficient passive transport mechanisms. To ensure a constant supply of nutrients and other required substances, microbes must capture those that are in extremely short supply and actively transport them into the cell. Features inherent in **active transport** systems are:

1. the transport of nutrients against the diffusion gradient or in the same direction as the natural gradient but at a rate faster than by diffusion alone,
2. the presence of specific membrane proteins (permeases and pumps; **figure 7.7*a***), and
3. the expenditure of energy. Examples of substances transported actively are monosaccharides, amino acids, organic acids, phosphates, and metal ions.

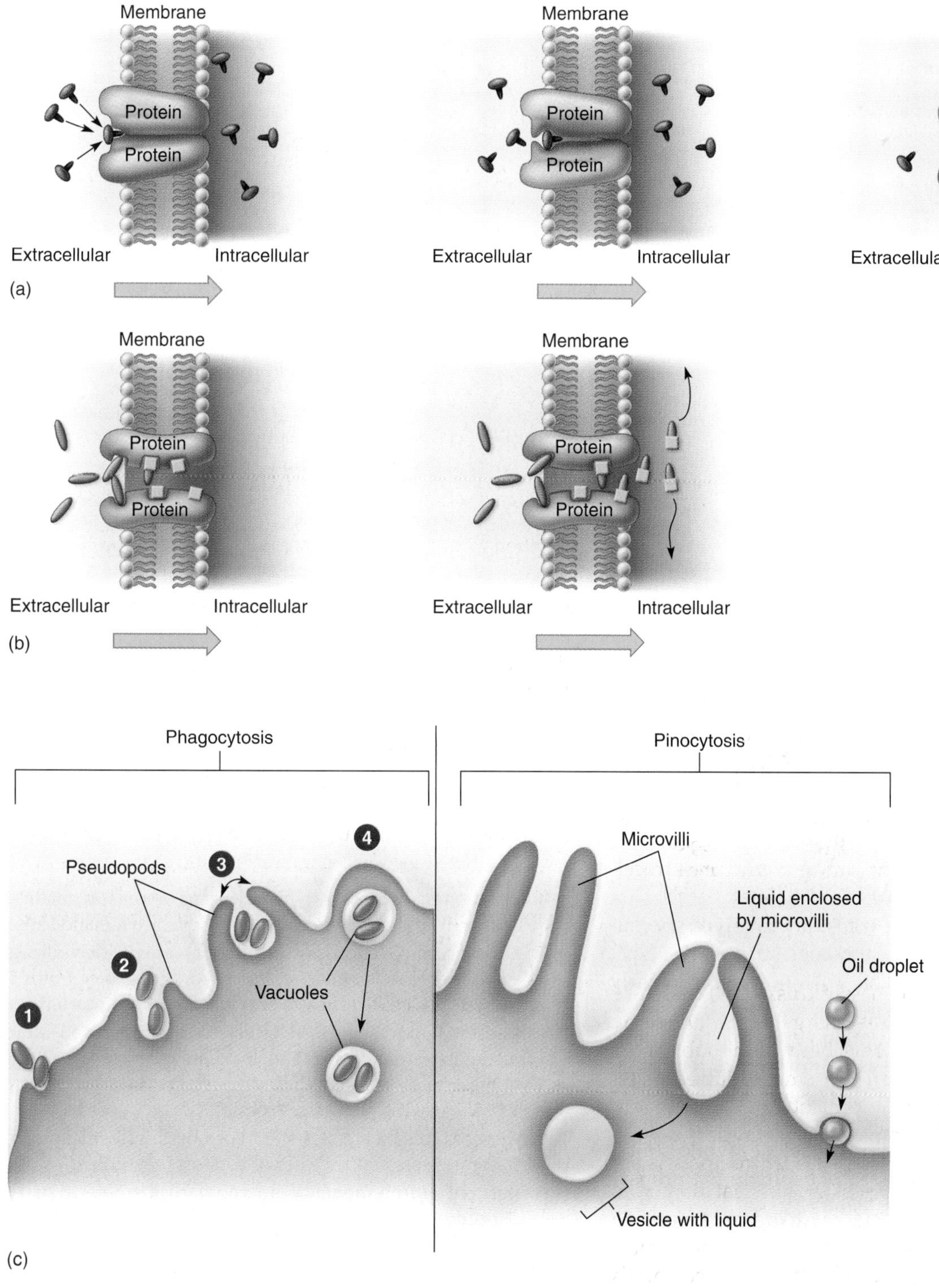

Figure 7.7 Active transport.

In active transport mechanisms, energy is expended to transport the molecule across the cell membrane. **(a)** Carrier-mediated active transport. The membrane proteins (permeases) have attachment sites for essential nutrient molecules. As these molecules bind to the permease, they are pumped into the cell's interior through special membrane protein channels. Microbes have these systems for transporting various ions (sodium, iron) and small organic molecules. **(b)** In group translocation, the molecule is actively captured, but along the route of transport, it is chemically altered. By coupling transport with synthesis, the cell conserves energy. **(c)** Endocytosis (phagocytosis and pinocytosis). Solid particles are phagocytosed by large cell extensions called pseudopods, and fluids and/or dissolved substances are pinocytosed into vesicles by very fine cell protrusions called microvilli. Oil droplets fuse with the membrane and are released directly into the cell.

TABLE 7.4 Summary of Transport Processes in Cells

General Process	Nature of Transport	Examples	Description	Qualities
Passive	Energy expenditure is not required. Substances exist in a gradient and move from areas of higher concentration toward areas of lower concentration in the gradient.	Diffusion	A fundamental property of atoms and molecules that exist in a state of random motion	Nonspecific Brownian movement
		Facilitated diffusion	Molecule binds to a receptor in membrane and is carried across to other side	Molecule specific; transports both ways
Active	Energy expenditure is required. Molecules need not exist in a gradient. Rate of transport is increased. Transport may occur against a concentration gradient.	Carrier-mediated active transport	Atoms or molecules are pumped into or out of the cell by specialized receptors. Driven by ATP or the proton motive force	Transports simple sugars, amino acids, inorganic ions (Na^+, K^+)
		Group translocation	Molecule is moved across membrane and simultaneously converted to a metabolically useful substance	Alternate system for transporting nutrients (sugars, amino acids)
		Bulk transport	Mass transport of large particles, cells, and liquids by engulfment and vesicle formation	Includes endocytosis, phagocytosis, pinocytosis

Some freshwater algae have such efficient active transport systems that an essential nutrient can be found in intracellular concentrations 200 times that of the habitat.

An important type of active transport involves specialized pumps, which can rapidly carry ions such as K^+, Na^+, and H^+ across the membrane. This behavior is particularly important in membrane ATP formation and protein synthesis, as described in chapter 8. Another type of active transport, **group translocation,** couples the transport of a nutrient with its conversion to a substance that is immediately useful inside the cell **(figure 7.7*b*).** This method is used by certain bacteria to transport sugars (glucose, fructose) while simultaneously adding molecules such as phosphate that prepare them for the next stage in metabolism.

Endocytosis: Eating and Drinking by Cells

Some eukaryotic cells transport large molecules, particles, liquids, or even other cells across the cell membrane. Because the cell usually expends energy to carry out this transport, it is also a form of active transport. The substances transported do not pass physically through the membrane but are carried into the cell by **endocytosis.** First the cell encloses the substance in its membrane, simultaneously forming a vacuole and engulfing it **(figure 7.7*c*).** Amoebas and certain white blood cells ingest whole cells or large solid matter by a type of endocytosis called **phagocytosis.** Liquids, such as oils or molecules in solution, enter the cell through **pinocytosis.** The mechanisms for transport of molecules into cells are summarized in **table 7.4.**

CHECKPOINT

- Nutrition is a process by which all living organisms obtain substances from their environment to convert to metabolic uses.
- Although the chemical form of nutrients varies widely, all organisms require six elements—carbon, hydrogen, oxygen, nitrogen, phosphorus, and sulfur—to survive, grow, and reproduce.
- Nutrients are categorized by the amount required (macronutrients or micronutrients), by chemical structure (organic or inorganic), and by their importance to the organism's survival (essential or nonessential).
- Microorganisms are classified both by the chemical form of their nutrients and the energy sources they utilize.
- Nutrient requirements of microorganisms determine their respective niches in the food webs of major ecosystems.
- Nutrients are transported into microorganisms by two kinds of processes: active transport that expends energy and passive transport that occurs independently of energy input.
- The molecular size and concentration of a nutrient determine the method of transport

7.2 Environmental Factors That Influence Microbes

Microbes are exposed to a wide variety of environmental factors in addition to nutrients. Microbial ecology focuses on ways that microorganisms deal with or adapt to such factors as heat, cold, gases, acid, radiation, osmotic and hydrostatic pressures, and even other microbes. Adaptation is a complex adjustment in biochemistry or genetics that enables long-term survival and growth. For most microbes, environmental factors fundamentally affect the function of metabolic enzymes. Thus, survival in a changing environment is largely a matter of whether the enzyme systems of microorganisms can adapt to alterations in their habitat. Incidentally, one must be careful to differentiate between growth in a given condition and tolerance, which implies survival without growth.

Temperature Adaptations

Microbial cells are unable to control their temperature and therefore assume the ambient temperature of their natural habitats. Their survival is dependent on adapting to whatever temperature variations are encountered in that habitat. The range of temperatures for the growth of a given microbial species can be expressed as three *cardinal temperatures.* The **minimum temperature** is the lowest temperature that permits a microbe's continued growth and metabolism; below this temperature, its activities are inhibited. The **maximum temperature** is the highest temperature at which growth and metabolism can proceed. If the temperature rises slightly above maximum, growth will stop, but if it continues to rise beyond that point, the enzymes and nucleic acids will eventually become permanently inactivated (otherwise known as denaturation) and the cell will die. This is why heat works so well as an agent in microbial control. The **optimum temperature** covers a small range, intermediate between the minimum and maximum, which promotes the fastest rate of growth and metabolism (rarely is the optimum a single point).

Depending on their natural habitats, some microbes have a narrow cardinal range, others a broad one. Some strict parasites will not grow if the temperature varies more than a few degrees below or above the host's body temperature. For instance, the typhus rickettsia multiplies only in the range of 32°C to 38°C, and rhinoviruses (one cause of the common cold) multiply successfully only in tissues that are slightly below normal body temperature (33°C to 35°C). Other organisms are not so limited. Strains of *Staphylococcus aureus* grow within the range of 6°C to 46°C, and the intestinal bacterium *Enterococcus faecalis* grows within the range of 0°C to 44°C.

Another way to express temperature adaptation is to describe whether an organism grows optimally in a cold, moderate, or hot temperature range. The terms used for these ecological groups are *psychrophile, mesophile,* and *thermophile* **(figure 7.8),** respectively.

A **psychrophile** (sy'-kroh-fyl) is a microorganism that has an optimum temperature below 15°C and is capable of growth at 0°C. It is obligate with respect to cold and generally cannot grow above 20°C. Laboratory work with true psychrophiles can be a real challenge. Inoculations have to be done in a cold room because room temperature can be lethal to the organisms. Unlike most laboratory cultures, storage in the refrigerator incubates, rather than inhibits, them. As one might predict, the habitats of psychrophilic bacteria, fungi, and algae are lakes and rivers, snowfields **(figure 7.9),** polar ice, and the deep ocean. Rarely, if ever, are they pathogenic. True psychrophiles must be distinguished from *psychrotrophs* or *facultative psychrophiles* that grow slowly in cold but have an optimum temperature above 20°C. Bacteria such as *Staphylococcus aureus* and *Listeria monocytogenes* are a concern because they can grow in refrigerated food and cause foodborne illness.

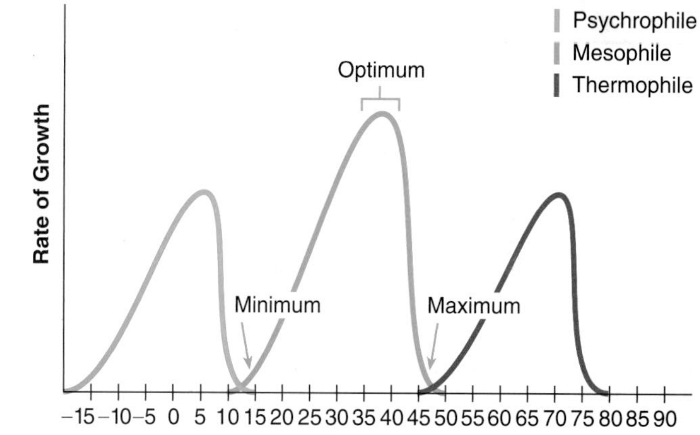

Figure 7.8 Ecological groups by temperature of adaptation.
Psychrophiles can grow at or near 0°C and have an optimum below 15°C. As a group, mesophiles can grow between 10°C and 50°C, but their optima usually fall between 20°C and 40°C. Generally speaking, thermophiles require temperatures above 45°C and grow optimally between this temperature and 80°C. Note that the extremes of the ranges can overlap to an extent.

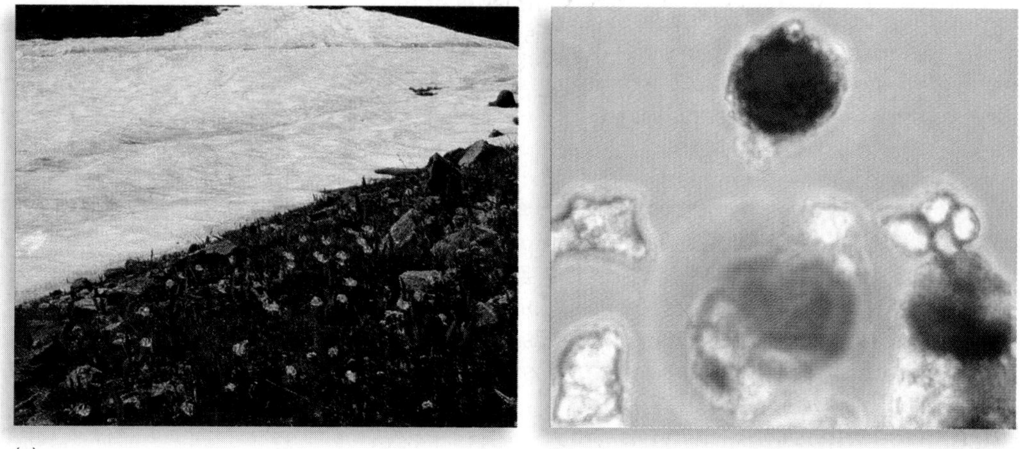
(a) (b)

Figure 7.9 Red snow.
(a) An early summer snowbank provides a perfect habitat for psychrophilic photosynthetic organisms like *Chlamydomonas nivalis*. **(b)** Microscopic view of this snow alga (actually classified as a "green" alga although a red pigment dominates at this stage of its life cycle).

INSIGHT 7.4 — Discovery

Cashing In on "Hot" Microbes

The smoldering thermal springs in Yellowstone National Park are more than just one of the geologic wonders of the world. They are also a hotbed of some of the most unusual microorganisms in the world. The thermophiles thriving at temperatures near the boiling point are the focus of serious interest from the scientific community. For many years, biologists have been intrigued that any living organism could function at such high temperatures. Such questions as these come to mind: Why don't they melt and disintegrate, why don't their proteins coagulate, and how can their DNA possibly remain intact?

One of the earliest thermophiles to be isolated was *Thermus aquaticus*. It was discovered by Thomas Brock in Yellowstone's Mushroom Pool in 1965 and was registered with the American Type Culture Collection. Interested researchers studied this species and discovered that it has extremely heat-stable proteins and nucleic acids, and its cell membrane does not break down readily at high temperatures. Later, an extremely heat-stable DNA-replicating enzyme was isolated from the species.

Biotechnology researchers harvesting samples in Yellowstone National Park.

What followed is a riveting example of how pure research for the sake of understanding and discovery also offered up a key ingredient in a multimillion-dollar process. Once an enzyme was discovered that was capable of copying DNA at very high temperatures (65°C to 72°C), researchers were able to develop a technique called the polymerase chain reaction (PCR), which could amplify a single piece of DNA into hundreds of thousands of identical copies. The enzyme, called **Taq polymerase** (from *Thermus aquaticus*), revolutionized forensic science, microbial ecology, and medical diagnosis. (Kary Mullis, who recognized the utility of Taq and developed the PCR technique in 1983, won the Nobel Prize in Chemistry for it in 1993.)

Spurred by this remarkable success story, biotechnology companies have descended on Yellowstone, which contains over 10,000 hot springs, geysers, and hot habitats. These industries are looking to unusual bacteria and archaea as a source of "extremozymes," enzymes that operate under high temperatures and acidity. Many other organisms with useful enzymes have been discovered. Some provide applications in the dairy, brewing, and baking industries for high-temperature processing and fermentations. Others are being considered for waste treatment and bioremediation.

This quest has also brought attention to questions such as: Who owns these microbes, and can their enzymes be patented? In the year 2000, the Park Service secured a legal ruling that allows it to share in the profits from companies and to add that money to its operating budget. The U.S. Supreme Court has also ruled that a microbe isolated from natural habitats cannot be patented. Only the technology that uses the microbe can be patented.

The majority of medically significant microorganisms are **mesophiles** (mez'-oh-fylz), organisms that grow at intermediate temperatures. Although an individual species can grow at the extremes of 10°C or 50°C, the optimum growth temperatures (optima) of most mesophiles fall into the range of 20°C to 40°C. Organisms in this group inhabit animals and plants as well as soil and water in temperate, subtropical, and tropical regions. Most human pathogens have optima somewhere between 30°C and 40°C (human body temperature is 37°C). *Thermoduric* microbes, which can survive short exposure to high temperatures but are normally mesophiles, are common contaminants of heated or pasteurized foods (see chapter 11). Examples include heat-resistant cysts such as *Giardia* or spore-formers such as *Bacillus* and *Clostridium*.

A **thermophile** (thur'-moh-fyl) is a microbe that grows optimally at temperatures greater than 45°C. Such heat- loving microbes live in soil and water associated with volcanic activity, in compost piles, and in habitats directly exposed to the sun. Thermophiles vary in heat requirements, with a general range of growth of 45°C to 80°C. Most eukaryotic forms cannot survive above 60°C, but a few thermophilic bacteria, called hyperthermophiles, grow between 80°C and 120°C (currently thought to be the temperature limit established by enzymes and cell structures). Strict thermophiles are so heat tolerant that researchers may use an autoclave to isolate them in culture. Currently, there is intense interest in thermal microorganisms on the part of biotechnology companies **(Insight 7.4).**

Gas Requirements

The atmospheric gases that most influence microbial growth are O_2 and CO_2. Of these, oxygen gas has the greatest impact on microbial growth. Not only is it an important respiratory gas, but it is also a powerful oxidizing agent that exists in many toxic forms. In general, microbes fall into one of three categories: those that use oxygen and can detoxify it; those that can neither use oxygen nor detoxify it; and those that do not use oxygen but can detoxify it.

How Microbes Process Oxygen

As oxygen enters into cellular reactions, it is transformed into several toxic products. Singlet oxygen (O) is an extremely reactive molecule produced by both living and nonliving processes. Notably, it is one of the substances produced by phagocytes to kill invading bacteria (see chapter 14). The buildup of singlet oxygen and the oxidation of membrane lipids and other molecules can damage and destroy a cell. The highly reactive superoxide ion (O_2^-), hydrogen peroxide (H_2O_2), and hydroxyl radicals (OH^-) are other destructive metabolic by-products of oxygen. To protect themselves against damage, most cells have developed enzymes that go about the business of scavenging and neutralizing these chemicals. The complete conversion of superoxide ion into harmless oxygen requires a two-step process and at least two enzymes:

Step 1. $2O_2^- + 2H^+ \xrightarrow[\text{dismutase}]{\text{Superoxide}} H_2O_2 \text{ (hydrogen peroxide)} + O_2$

Step 2. $2H_2O_2 \xrightarrow{\text{Catalase}} 2H_2O + O_2$

In this series of reactions (essential for aerobic organisms), the superoxide ion is first converted to hydrogen peroxide and normal oxygen by the action of an enzyme called superoxide dismutase. Because hydrogen peroxide is also toxic to cells (it is used as a disinfectant and antiseptic), it must be degraded by the enzyme catalase into water and oxygen. If a microbe is not capable of dealing with toxic oxygen by these or similar mechanisms, it is forced to live in habitats free of oxygen.

With respect to oxygen requirements, several general categories are recognized. An **aerobe** (air′-ohb) (aerobic organism) can use gaseous oxygen in its metabolism and possesses the enzymes needed to process toxic oxygen products. An organism that cannot grow without oxygen is an **obligate aerobe.** Most fungi and protozoa, as well as many bacteria (genera *Micrococcus* and *Bacillus*), have to have oxygen in their metabolism.

A **facultative anaerobe** is an aerobe that does not require oxygen for its metabolism and is capable of growth in the absence of it. This type of organism metabolizes by aerobic respiration when oxygen is present, but in its absence, it adopts an anaerobic mode of metabolism such as fermentation. Facultative anaerobes usually possess catalase and superoxide dismutase. A large number of bacterial pathogens fall into this group (for example, gram-negative intestinal bacteria and staphylococci). A **microaerophile** (myk″-roh-air′-oh-fyl) does not grow at normal atmospheric concentrations of oxygen but requires a small amount of it in metabolism. Most organisms in this category live in a habitat (soil, water, or the human body) that provides small amounts of oxygen but is not directly exposed to the atmosphere.

An **anaerobe** (anaerobic microorganism) lacks the metabolic enzyme systems for using oxygen in respiration. Because **strict,** or **obligate, anaerobes** also lack the enzymes for processing toxic oxygen, they cannot tolerate any free oxygen in the immediate environment and will die if exposed to it. Strict anaerobes live in highly reduced habitats, such as deep muds, lakes, oceans, and soil. Even though human cells use oxygen and oxygen is found in the blood and tissues, some body sites present anaerobic pockets or microhabitats where colonization or infection can occur. One region that is an important site for anaerobic infections is the oral cavity. Dental caries are partly due to the complex actions of aerobic and anaerobic bacteria, and most gingival infections consist of similar mixtures of oral bacteria that have invaded damaged gum tissues (see chapter 22). Another common site for anaerobic infections is the large intestine, a relatively oxygen-free habitat that harbors a rich assortment of strictly anaerobic bacteria. Anaerobic infections can occur following abdominal surgery and traumatic injuries (gas gangrene and tetanus). Growing anaerobic bacteria usually requires special media, methods of incubation, and handling chambers that exclude oxygen **(figure 7.10*a*).**

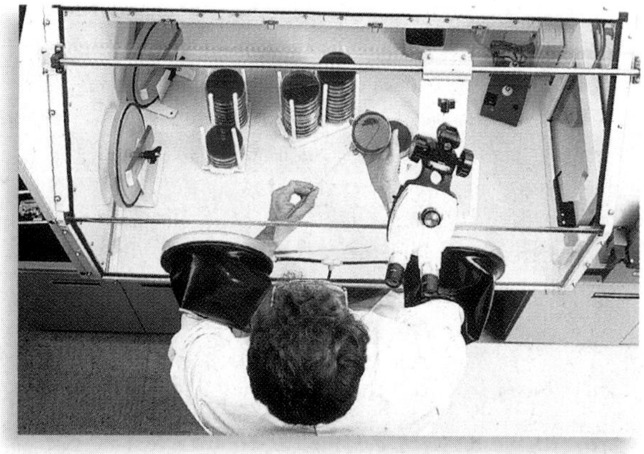

(a)

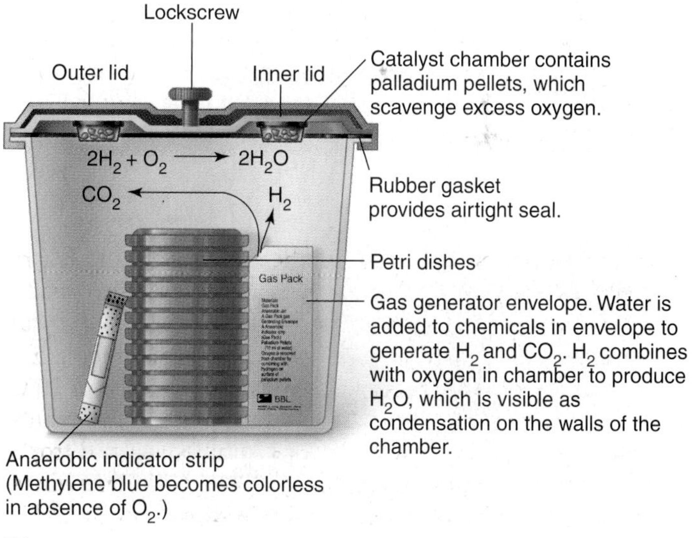

(b)

Figure 7.10 Culturing techniques for anaerobes.
(a) A special anaerobic environmental chamber makes it possible to handle strict anaerobes without exposing them to air. It also has provisions for incubation and inspection in a completely O_2-free system. **(b)** A simpler anaerobic, or CO_2, incubator system. To create an anaerobic environment, a packet is activated to produce hydrogen gas and the chamber is sealed tightly. The gas reacts with available oxygen to produce water. Carbon dioxide can also be added to the system for growth of organisms needing high concentrations of it.

Aerotolerant anaerobes do not utilize oxygen but can survive and grow to a limited extent in its presence. These anaerobes are not harmed by oxygen, mainly because they possess alternate mechanisms for breaking down peroxides and superoxide. Certain lactobacilli and streptococci use manganese ions or peroxidases to perform this task.

Determining the oxygen requirements of a microbe from a biochemical standpoint can be a very time-consuming process. Often it is illuminating to perform culture tests with reducing media (those that contain an oxygen-absorbing chemical). One such technique demonstrates oxygen requirements by the location of growth in a tube of fluid thioglycollate **(figure 7.11).**

Although all microbes require some carbon dioxide in their metabolism, *capnophiles* grow best at a higher CO_2 tension than is normally present in the atmosphere. This becomes important in the initial isolation of some pathogens from clinical specimens, notably *Neisseria* (gonorrhea, meningitis), *Brucella* (undulant fever), and *Streptococcus pneumoniae.* Incubation is carried out in a CO_2 incubator that provides 3% to 10% CO_2 (**figure 7.10*b***).

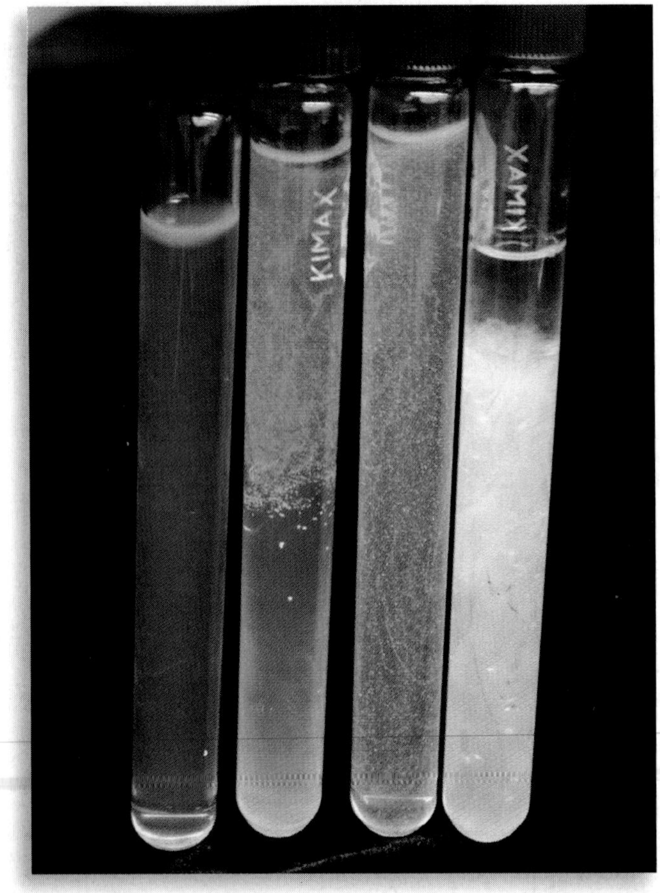

Figure 7.11 Use of thioglycollate broth to demonstrate oxygen requirements.
Thioglycollate is a reducing agent that allows anaerobic bacteria to grow in tubes exposed to air. Oxygen concentration is highest at the top of the tube. When a series of tubes is inoculated with bacteria that differ in O_2 requirements, the relative position of growth provides some indication of their adaptations to oxygen use. Tube 1 (on the left): aerobic (*Pseudomonas aeruginosa*); Tube 2: facultative (*Staphylococcus aureus*); Tube 3: facultative (*Escherichia coli*); Tube 4: obligate anaerobe (*Clostridium butyricum*).

Effects of pH

Microbial growth and survival are also influenced by the pH of the habitat. The pH was defined in chapter 2 as the degree of acidity or alkalinity (basicity) of a solution. It is expressed by the pH scale, a series of numbers ranging from 0 to 14. The pH of pure water (7.0) is neutral, neither acidic nor basic. As the pH value decreases toward 0, the acidity increases, and as the pH increases toward 14, the alkalinity increases. The majority of organisms live or grow in habitats between pH 6 and 8 because strong acids and bases can be highly damaging to enzymes and other cellular substances.

A few microorganisms live at pH extremes. Obligate *acidophiles* include *Euglena mutabilis,* an alga that grows in acid pools between 0 and 1.0 pH, and *Thermoplasma,* an archaea that lacks a cell wall, lives in hot coal piles at a pH of 1 to 2, and will lyse if exposed to pH 7. Because many molds and yeasts tolerate moderate acid, they are the most common spoilage agents of pickled foods. Alkalinophiles live in hot pools and soils that contain high levels of basic minerals (up to pH 10.0). Bacteria that decompose urine create alkaline conditions, because ammonium (NH_4^+) can be produced when urea (a component of urine) is digested. Metabolism of urea is one way that *Proteus* spp. can neutralize the acidity of the urine to colonize and infect the urinary system.

Osmotic Pressure

Although most microbes exist under hypotonic or isotonic conditions, a few, called osmophiles, live in habitats with a high solute concentration. One common type of osmophile prefers high concentrations of salt; these organisms are called **halophiles** (hay′-loh-fylz). Obligate *halophiles* such as *Halobacterium* and *Halococcus* inhabit salt lakes, ponds, and other hypersaline habitats. They grow optimally in solutions of 25% NaCl but require at least 9% NaCl (combined with other salts) for growth. These archaea have significant modifications in their cell walls and membranes and will lyse in hypotonic habitats. *Facultative halophiles* are remarkably resistant to salt, even though they do not normally reside in high-salt environments. For example, *Staphylococcus aureus* can grow on NaCl media ranging from 0.1% up to 20%. Although it is common to use high concentrations of salt and sugar to preserve food (jellies, syrups, and brines), many bacteria and fungi actually thrive under these conditions and are common spoilage agents.

Miscellaneous Environmental Factors

Various forms of electromagnetic radiation (ultraviolet, infrared, visible light) stream constantly onto the earth from the sun. Some microbes (phototrophs) can use visible light rays as an energy source, but nonphotosynthetic microbes tend to be damaged by the toxic oxygen products produced by contact with light. Some microbial species produce yellow carotenoid pigments to protect against the damaging effects of light by absorbing and dismantling toxic oxygen. Other types of radiation that can damage microbes are ultraviolet and ionizing rays (X rays and cosmic rays). In chapter 11, you will see just how these types of energy are applied in microbial control.

Descent into the ocean depths subjects organisms to increasing hydrostatic pressure. Deep-sea microbes called **barophiles** exist under pressures that range from a few times to over 1,000 times the pressure of the atmosphere. These bacteria are so strictly adapted to high pressures that they will rupture when exposed to normal atmospheric pressure.

Because of the high water content of cytoplasm, all cells require water from their environment to sustain growth and metabolism. Water is the solvent for cell chemicals, and it is needed for enzyme function and digestion of macromolecules. A certain amount of water on the external surface of the cell is required for the diffusion of nutrients and wastes. Even in apparently dry habitats, such as sand or dry soil, the particles retain a thin layer of water usable by microorganisms. Only dormant, dehydrated cell stages (for example, spores and cysts) tolerate extreme drying because of the inactivity of their enzymes.

Ecological Associations Among Microorganisms

Up to now, we have considered the importance of nonliving environmental influences on the growth of microorganisms. Another profound influence comes from other organisms that share (or sometimes are) their habitats. In all but the rarest instances, microbes live in shared habitats, which give rise to complex and fascinating associations. Some associations are between similar or dissimilar types of microbes; others involve multicellular organisms such as animals or plants. Interactions can have beneficial, harmful, or no particular effects on the organisms involved; they can be obligatory or nonobligatory to the members; and they often involve nutritional interactions. This outline provides an overview of the major types of microbial associations:

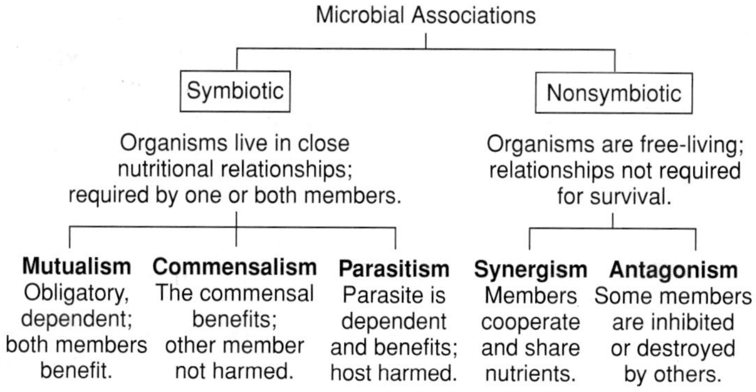
Microbial Associations
Symbiotic
Nonsymbiotic
Organisms live in close nutritional relationships; required by one or both members.
Organisms are free-living; relationships not required for survival.
Mutualism
Obligatory, dependent; both members benefit.
Commensalism
The commensal benefits; other member not harmed.
Parasitism
Parasite is dependent and benefits; host harmed.
Synergism
Members cooperate and share nutrients.
Antagonism
Some members are inhibited or destroyed by others.

CASE FILE 7 *WRAP-UP*

Baby Caroline suffered from infant botulism. Botulism is a neuroparalytic disease that in adults results from ingestion of or contamination of a wound by the preformed toxin of *Clostridium botulinum* or related species. In babies, the disease occurs differently. Infant botulism, often called "floppy baby syndrome," may occur when a toxin is produced in and absorbed from the gastrointestinal tract. In infants, this intoxication begins with ingestion of *Clostridium* spores. In the intestine, spores germinate and the organisms produce a toxin that the child absorbs. Substances such as honey, syrup, and soil may contain the spores.

Binding of the neurotoxin results in loss of muscle tone. The result is a flaccid paralysis ("floppy baby"), which may lead to respiratory and cardiac failure. Because an infant's intestinal biota are not well established and the immune system is immature, defense mechanisms that would stop the growth of the *Clostridium* may fail. Therefore, infants may develop this form of the disease, whereas symptoms in adults usually require consumption of preformed toxins in items such as improperly home-canned food.

To determine the source of the *Clostridium* spores, one might culture honey, prepared formula, and garden soil. Because organisms of the genus *Clostridium* are anaerobic, samples would be cultured in the absence of oxygen using culture media that select for and allow presumptive identification of these organisms. One suitable medium would be sulfite polymyxin sulfadiazine (SPS) medium on which clostridia produce blackened colonies. All cultures should be performed by state or national labs due to the danger associated with the bacterium.

See: CDC. 2003. Infant botulism—New York City, 2001–2002. MMWR *52:21–24.*

A general term used to denote a situation in which two organisms live together in a close partnership is **symbiosis,**[2] and the members are termed *symbionts*. Three main types of symbiosis occur. **Mutualism** exists when organisms live in an obligatory but mutually beneficial relationship. This association is rather common in nature because of the survival value it has for the members involved. **Insight 7.5** gives several examples to illustrate this concept. In other symbiotic relationships the relationship tends to be unequal, meaning it benefits one member and not the other, and it can be obligatory.

In a relationship known as **commensalism,** the member called the commensal receives benefits, while its coinhabitant is neither harmed nor benefited. A classic commensal

2. Note that *symbiosis* is a neutral term and does not by itself imply benefit or detriment.

INSIGHT 7.5 — Discovery

Life Together: Mutualism

A tremendous variety of mutualistic partnerships occurs in nature. These associations gradually evolve over millions of years as the participating members come to rely on some critical substance or habitat that they share. One of the earliest such associations is thought to have resulted in eukaryotic cells (see Insight 5.1).

Protozoan cells often receive growth factors from symbiotic bacteria and algae that, in turn, are nurtured by the protozoan cell. One peculiar ciliate propels itself by affixing symbiotic bacteria to its cell membrane to act as "oars." These relationships become so obligatory that some amoebas and ciliates require mutualistic bacteria for survival. This kind of relationship is especially striking in the complex mutualism of termites, which harbor protozoans specialized to live only inside them. The protozoans, in turn, contain endosymbiotic bacteria. Wood eaten by the termite gets processed by the protozoan and bacterial enzymes, and all three organisms thrive.

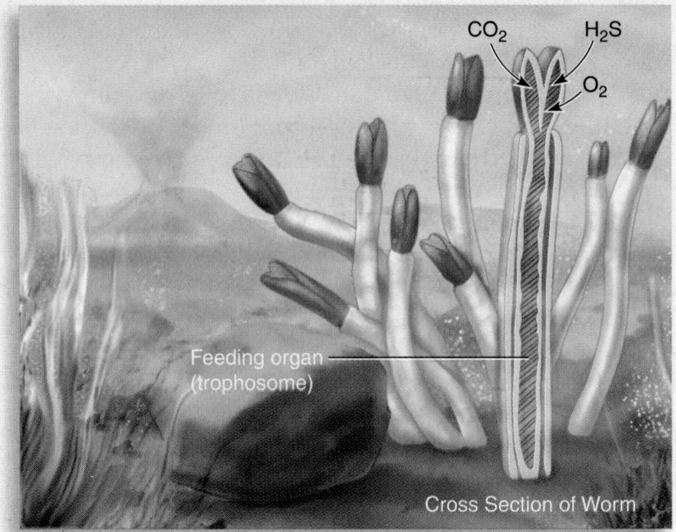

A view of a vent community based on mutualism and chemoautotrophy. The giant tube worm *Riftia* houses bacteria in its specialized feeding organ, the trophosome. The trophosome (gray) is filled with bacteria.

Symbiosis Between Microbes and Animals

Microorganisms carry on symbiotic relationships with animals as diverse as sponges, worms, and mammals. Bacteria and protozoa are essential in the operation of the rumen (a complex, four-chambered stomach) of cud-chewing mammals. These mammals produce no enzymes of their own to break down the cellulose that is a major part of their diet, but the microbial population harbored in their rumens does. The complex food materials are digested through several stages, during which time the animal regurgitates and chews the partially digested plant matter (the cud) and occasionally burps methane produced by the microbial symbionts.

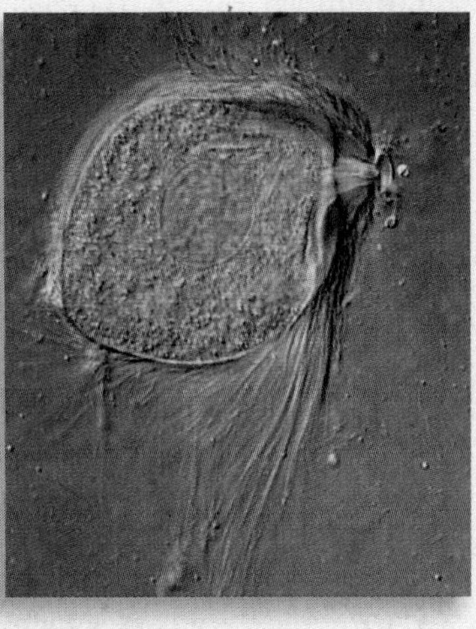

Termites are insects responsible for wood damage; however, it is the termite's endosymbiont (the protozoan pictured here) that provides the enzymes for digesting wood.

Thermal Vent Symbionts

Another fascinating symbiotic relationship has been found in the deep hydrothermal vents in the seafloor, where geologic forces spread the crustal plates and release heat and gas. These vents are a focus of tremendous biological and geologic activity. Discoveries first made in the late 1970s demonstrated that the source of energy in this community is not the sun, because the vents are too deep for light to penetrate (2,600 m). Instead, this ecosystem is based on a massive lithoautotrophic bacterial population that oxidizes the abundant hydrogen sulfide (H_2S) gas given off by the volcanic activity there. As the bottom of the food web, these bacteria serve as the primary producers of nutrients that service a broad spectrum of specialized animals.

interaction between microorganisms called **satellitism** arises when one member provides nutritional or protective factors needed by the other **(figure 7.12).** Some microbes can break down a substance that would be toxic or inhibitory to another microbe. Relationships between humans and resident commensals that derive nutrients from the body are discussed in a later section.

In an earlier section, we introduced the concept of **parasitism** as a relationship in which the host organism provides the parasitic microbe with nutrients and a habitat. Multiplication of the parasite usually harms the host to some extent. As this relationship evolves, the host may even develop tolerance for or dependence on a parasite, at which point we call the relationship commensalism or mutualism.

Synergism is an interrelationship between two or more free-living organisms that benefits them but is not necessary for their survival. Together, the participants cooperate to produce a result that none of them could do alone. Biofilms, which you read about in chapter 4, are the best examples of synergism.

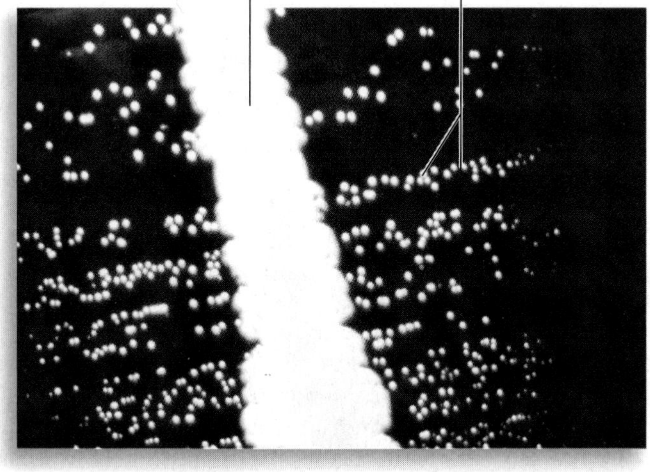

Figure 7.12 Satellitism, a type of commensalism between two microbes.
In this example, *Staphylococcus aureus* provides growth factors to *Haemophilus influenzae,* which grows as tiny satellite colonies near the streak of *Staphylococcus.* By itself, *Haemophilus* could not grow on blood agar. The *Staphylococcus* gives off several nutrients such as vitamins and amino acids that diffuse out to the *Haemophilus,* thereby promoting its growth.

A NOTE ON COEVOLUTION

Organisms that have close, ongoing relationships with each other participate in **coevolution,** the process whereby a change in one of the partners leads to a change in the other partner, which may in turn lead to another change in the first partner, and so on. This is another example of the interconnectedness of biological entities on this planet. There are many well-documented examples of the relationships between plants and insects. One of the earliest is the discovery by Charles Darwin of a plant that had a nectar tube that was 10 inches long. Knowing that the plant depended on insects for pollination, Darwin predicted the existence of an insect with a 10-inch tongue—and 41 years later one was discovered. The plant and the insect had influenced each other's evolution over time. Commensal gut bacteria are considered to have coevolved with their mammalian hosts, with the hosts evolving mechanisms to prevent the disease effects of their bacterial passengers, and the bacteria evolving mechanisms to be less pathogenic to their hosts.

Another example of synergism is observed in the exchange between soil bacteria and plant roots (see chapter 24). The plant provides various growth factors, and the bacteria help fertilize the plant by supplying it with minerals. In synergistic infections, a combination of organisms can produce tissue damage that a single organism would not cause alone. Gum disease, dental caries, and gas gangrene involve mixed infections by bacteria interacting synergistically.

Antagonism is an association between free-living species that arises when members of a community compete. In this interaction, one microbe secretes chemical substances into the surrounding environment that inhibit or destroy another microbe in the same habitat. The first microbe may gain a competitive advantage by increasing the space and nutrients available to it. Interactions of this type are common in the soil, where mixed communities often compete for space and food. *Antibiosis*—the production of inhibitory compounds such as antibiotics—is actually a form of antagonism. Hundreds of naturally occurring antibiotics have been isolated from bacteria and fungi and used as drugs to control diseases (see chapter 12).

Interrelationships Between Microbes and Humans

The human body is a rich habitat for symbiotic bacteria, fungi, and a few protozoa. Microbes that normally live on the skin, in the alimentary tract, and in other sites are called the *normal microbiota* (see chapter 13). These residents participate in commensal, parasitic, and synergistic relationships with their human hosts. For example, *Escherichia coli* living symbiotically in the intestine produce vitamin K, and species of symbiotic *Lactobacillus* residing in the vagina help maintain an acidic environment that protects against infection by other microorganisms. Hundreds of commensal species "make a living" on the body without either harming or benefiting it. For example, many bacteria and yeasts reside in the outer dead regions of the skin; oral microbes feed on the constant flow of nutrients in the mouth; and billions of bacteria live on the wastes in the large intestine. Because the normal microbiota and the body are in a constant state of change, these relationships are not absolute, and a commensal can convert to a parasite by invading body tissues and causing disease.

CHECKPOINT

- The environmental factors that control microbial growth are temperature, pH, moisture, radiation, gases, and other microorganisms.
- Environmental factors control microbial growth by their influence on microbial enzymes.
- Three cardinal temperatures for a microorganism describe its temperature range and the temperature at which it grows best. These are the minimum temperature, the maximum temperature, and the optimum temperature.
- Microorganisms are classified by their temperature requirements as psychrophiles, mesophiles, or thermophiles.
- Most eukaryotic microorganisms are aerobic, while bacteria vary widely in their oxygen requirements from obligately aerobic to anaerobic.
- Microorganisms live in association with other species that range from mutually beneficial symbiosis to parasitism and antagonism.

7.3 The Study of Microbial Growth

When microbes are provided with nutrients and the required environmental factors, they become metabolically active and grow. Growth takes place on two levels. On one level, a cell synthesizes new cell components and increases its size; on the other level, the number of cells in the population increases. This capacity for multiplication, increasing the size of the population by cell division, has tremendous importance in microbial control, infectious disease, and biotechnology. In the following sections, we will focus primarily on the characteristics of bacterial growth that are generally representative of single-celled microorganisms.

The Basis of Population Growth: Binary Fission

The division of a bacterial cell occurs mainly through **binary,** or **transverse, fission;** *binary* means that one cell becomes two, and *transverse* refers to the division plane forming across the width of the cell. During binary fission, the parent cell enlarges, duplicates its chromosome, and forms a central transverse septum that divides the cell into two daughter cells. This process is repeated at intervals by each newdaughter cell in turn, and with each successive round of division, the population increases. The stages in this continuous process are shown in greater detail in **figure 7.13** and **figure 7.14**.

The Rate of Population Growth

The time required for a complete fission cycle—from parent cell to two new daughter cells—is called the **generation,** or **doubling, time.** The term *generation* has a similar meaning as it does in humans. It is the period between an individual's birth and the time of producing offspring. In bacteria, each new fission cycle or generation increases the population by a factor of 2, or doubles it. Thus, the initial parent stage consists of 1 cell, the first generation consists of 2 cells, the second 4, the third 8, then 16, 32, 64, and so on. As long as the environment remains favorable, this doubling effect can continue at a constant rate. With the passing of each generation, the population will double, over and over again.

The length of the generation time is a measure of the growth rate of an organism. Compared with the growth rates of most other living things, bacteria are notoriously rapid. The average generation time is 30 to 60 minutes under optimum conditions. The shortest generation times can be 10 to 12 minutes, and longer generation times require days. For example, *Mycobacterium leprae,* the cause of Hansen's disease, has a generation time of 10 to 30 days—as long as

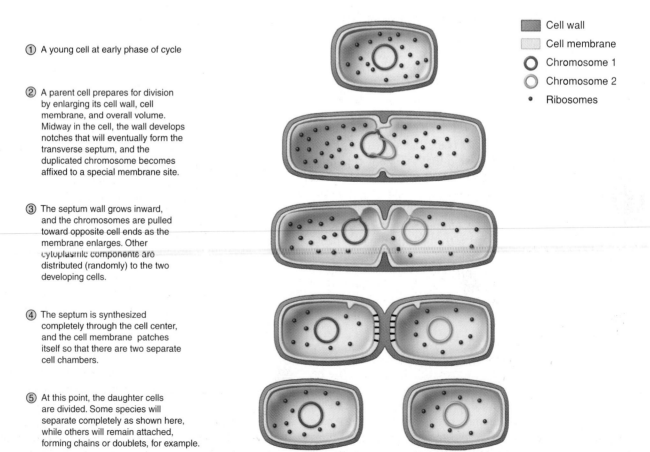

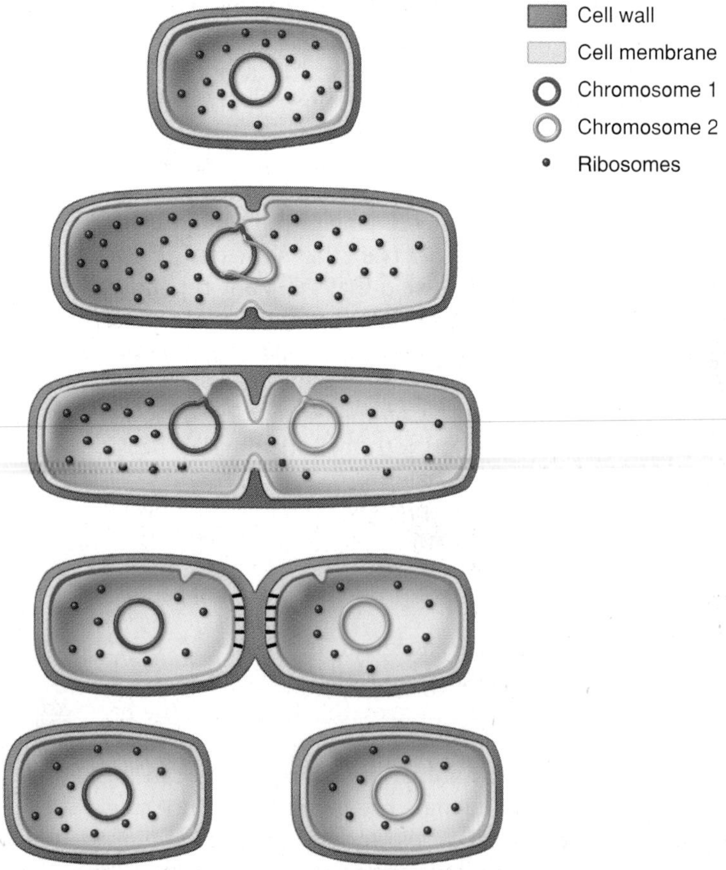

① A young cell at early phase of cycle

② A parent cell prepares for division by enlarging its cell wall, cell membrane, and overall volume. Midway in the cell, the wall develops notches that will eventually form the transverse septum, and the duplicated chromosome becomes affixed to a special membrane site.

③ The septum wall grows inward, and the chromosomes are pulled toward opposite cell ends as the membrane enlarges. Other cytoplasmic components are distributed (randomly) to the two developing cells.

④ The septum is synthesized completely through the cell center, and the cell membrane patches itself so that there are two separate cell chambers.

⑤ At this point, the daughter cells are divided. Some species will separate completely as shown here, while others will remain attached, forming chains or doublets, for example.

Process Figure 7.13 Steps in binary fission of a rod-shaped bacterium.

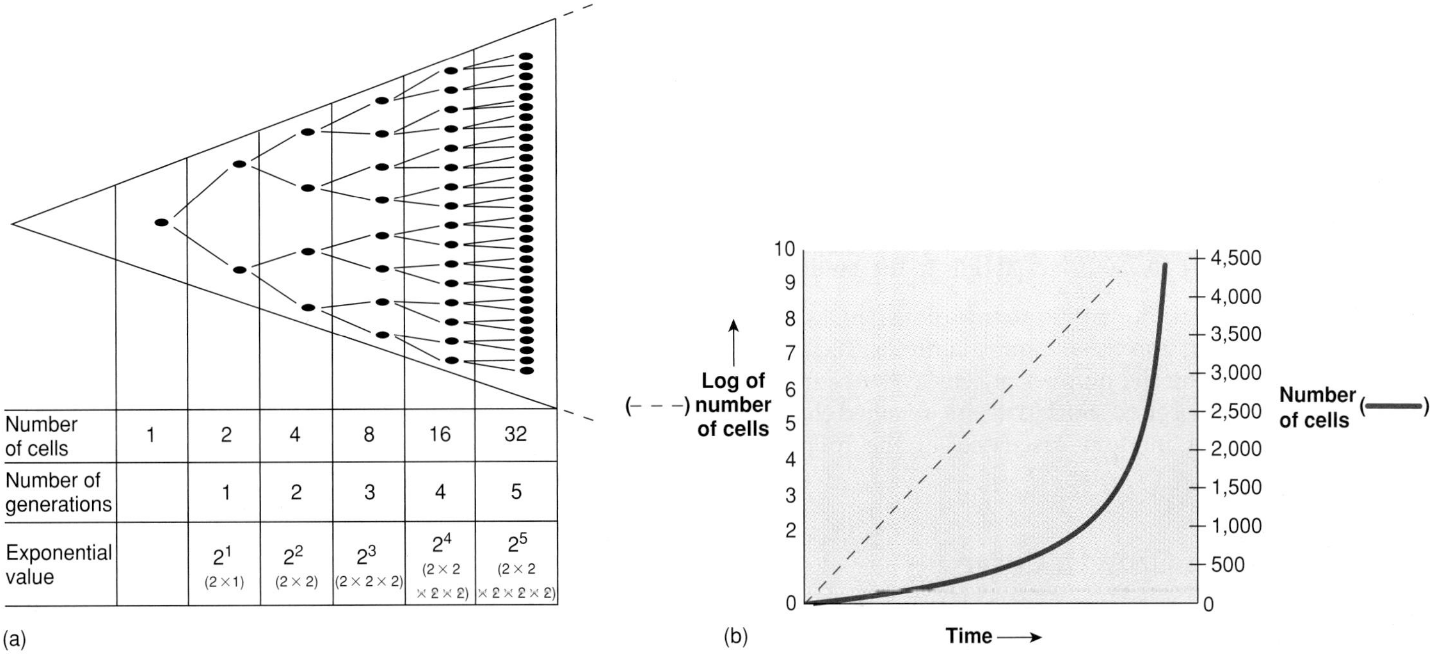

Figure 7.14 The mathematics of population growth.
(a) Starting with a single cell, if each product of reproduction goes on to divide by binary fission, the population doubles with each new cell division or generation. This process can be represented by logarithms (2 raised to an exponent) or by simple numbers. **(b)** Plotting the logarithm of the cells produces a straight line indicative of exponential growth, whereas plotting the cell numbers arithmetically gives a curved slope.

that of some animals. Environmental bacteria commonly have generation times measured in months. Most pathogens have relatively short doubling times. *Salmonella enteritidis* and *Staphylococcus aureus,* bacteria that cause food-borne illness, double in 20 to 30 minutes, which is why leaving food at room temperature even for a short period has caused many cases of food-borne disease. In a few hours, a population of these bacteria can easily grow from a small number of cells to several million.

A NOTE ON BACTERIAL REPRODUCTION—AND THE "CULTURE BIAS"

By far most of the bacteria that have ever been studied reproduce via binary fission, as described in this chapter. But there are important exceptions. In recent years, researchers have discovered bacteria that produce multiple offspring within their cytoplasm and then split open to release multiple new bacteria (killing the mother cell). Most of these bacteria have never been cultured but have been studied by dissecting the animals they colonize. The long-standing belief that bacteria always multiply by binary fission is another by-product of the "culture bias"—meaning that we understand most about the bacteria that we were able to cultivate in the lab, even though there are many more bacteria that exist in the biosphere that have not yet been cultivated.

Figure 7.14 shows several quantitative characteristics of growth: (A) The cell population size can be represented by the number 2 with an exponent (2^1, 2^2, 2^3, 2^4); (B) the exponent increases by one in each generation; and (C) the number of the exponent is also the number of the generation. This growth pattern is termed **exponential.** Because these populations often contain very large numbers of cells, it is useful to express them by means of exponents or logarithms (see appendix A). The data from a growing bacterial population are graphed by plotting the number of cells as a function of time. The cell number can be represented logarithmically or arithmetically. Plotting the logarithm number over time provides a straight line indicative of exponential growth. Plotting the data arithmetically gives a constantly curved slope. In general, logarithmic graphs are preferred because an accurate cell number is easier to read, especially during early growth phases.

Predicting the number of cells that will arise during a long growth period (yielding millions of cells) is based on a relatively simple concept. One could use the method of addition $2 + 2 = 4; 4 + 4 = 8; 8 + 8 = 16; 16 + 16 = 32$, and so on, or a method of multiplication (for example, $2^5 = 2 \times 2 \times 2 \times 2 \times 2$), but it is easy to see that for 20 or 30 generations, this calculation could be very tedious. An easier way to calculate the size of a population over time is to use an equation such as:

$$N_f = (N_i)2^n$$

In this equation, N_f is the total number of cells in the population at some point in the growth phase, N_i is the starting number, the exponent n denotes the generation number, and 2^n represents the number of cells in that generation. If we know any two of the values, the other values can be calculated. Let us use the example of *Staphylococcus aureus* to calculate how many cells (N_f) will be present in an egg salad sandwich after it sits in a warm car for 4 hours. We will assume that N_i is 10 (number of cells deposited in the sandwich while it was being prepared). To derive n, we need to divide 4 hours (240 minutes) by the generation time (we will use 20 minutes). This calculation comes out to 12, so 2^n is equal to 2^{12}. Using a calculator, we find that 2^{12} is 4,096.

$$\text{Final number } (N_f) = 10 \times 4{,}096$$
$$= 40{,}960 \text{ bacterial cells in the sandwich}$$

This same equation, with modifications, is used to determine the generation time, a more complex calculation that requires knowing the number of cells at the beginning and end of a growth period. Such data are obtained through actual testing by a method discussed in the following section.

The Population Growth Curve

In reality, a population of bacteria does not maintain its potential growth rate and does not double endlessly, because in most systems numerous factors prevent the cells from continuously dividing at their maximum rate. Laboratory studies indicate that a population typically displays a predictable pattern, or **growth curve,** over time. The method traditionally used to observe the population growth pattern is a viable count technique, in which the total number of live cells is counted over a given time period. In brief, this method entails

1. placing a tiny number of cells into a sterile liquid medium;
2. incubating this culture over a period of several hours;
3. sampling the broth at regular intervals during incubation;
4. plating each sample onto solid media; and
5. counting the number of colonies present after incubation.

Insight 7.6 gives the details of this process.

Stages in the Normal Growth Curve

The system of batch culturing described in Insight 7.6 is *closed,* meaning that nutrients and space are finite and there is no mechanism for the removal of waste products. Data from an entire growth period of 3 to 4 days typically produce a curve with a series of phases termed the lag phase, the exponential growth (log) phase, the stationary phase, and the death phase **(figure 7.15).**

The **lag phase** is a relatively "flat" period on the graph when the population appears not to be growing or is growing at less than the exponential rate. Growth lags primarily because

1. the newly inoculated cells require a period of adjustment, enlargement, and synthesis;
2. the cells are not yet multiplying at their maximum rate; and
3. the population of cells is so sparse or dilute that the sampling misses them.

The length of the lag period varies somewhat from one population to another. It is important to note that even though the population of cells is not increasing (growing), individual cells are metabolically active as they increase their contents and prepare to divide.

The cells reach the maximum rate of cell division during the **exponential growth (logarithmic or log) phase,** a period during which the curve increases geometrically. This phase will continue as long as cells have adequate nutrients and the environment is favorable.

At the **stationary growth phase,** the population enters a survival mode in which cells stop growing or grow slowly. The curve levels off because the rate of cell inhibition or death balances out the rate of multiplication. The decline in the growth rate is caused by depleted nutrients and oxygen

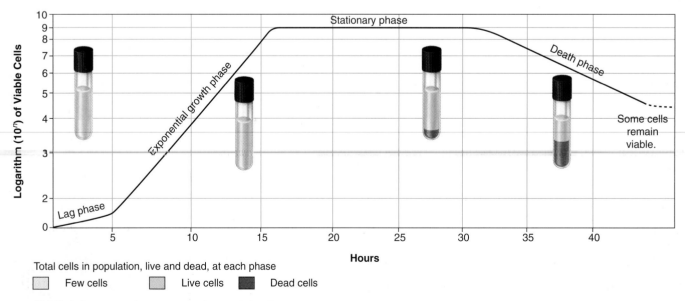

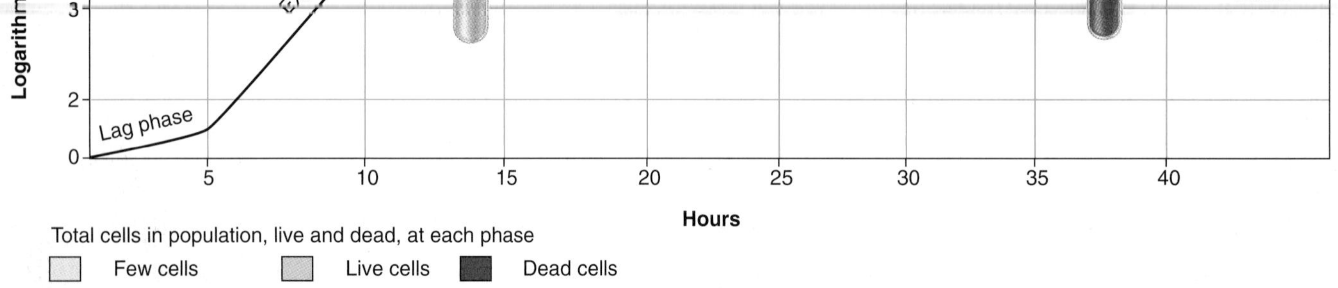

Figure 7.15 The growth curve in a bacterial culture.
On this graph, the number of viable cells expressed as a logarithm (log) is plotted against time. See text for discussion of the various phases. Note that with a generation time of 30 minutes, the population has risen from 10 (10^1) cells to 1,000,000,000 (10^9) cells in only 16 hours.

INSIGHT 7.6 **Microbiology**

Steps in a Viable Plate Count—Batch Culture Method

A growing population is established by inoculating a flask containing a known quantity of sterile liquid medium with a few cells of a pure culture. The flask is incubated at that bacterium's optimum temperature and timed. The population size at any point in the growth cycle is quantified by removing a tiny measured sample of the culture from the growth chamber and plating it out on a solid medium to develop isolated colonies. This procedure is repeated at evenly spaced intervals (i.e., every hour for 24 hours).

Evaluating the samples involves a common and important principle in microbiology: One colony on the plate represents one cell or colony-forming unit (CFU) from the original sample. Because the CFU of some bacteria is actually composed of several cells (consider the clustered arrangement of *Staphylococcus*, for instance), using a colony count can underestimate the exact population size to an extent. This is not a serious problem because, in such bacteria, the CFU is the smallest unit of colony formation and dispersal. Multiplication of the number of colonies in a single sample by the container's volume gives a fair estimate of the total population size (number of cells) at any given point. The growth curve is determined by graphing the number for each sample in sequence for the whole incubation period (see figure 7.15).

Because of the scarcity of cells in the early stages of growth, some samples can give a zero reading even if there are viable cells in the culture. Also, the sampling itself can remove enough viable cells to alter the tabulations, but since the purpose is to compare relative trends in growth, these factors do not significantly change the overall pattern.

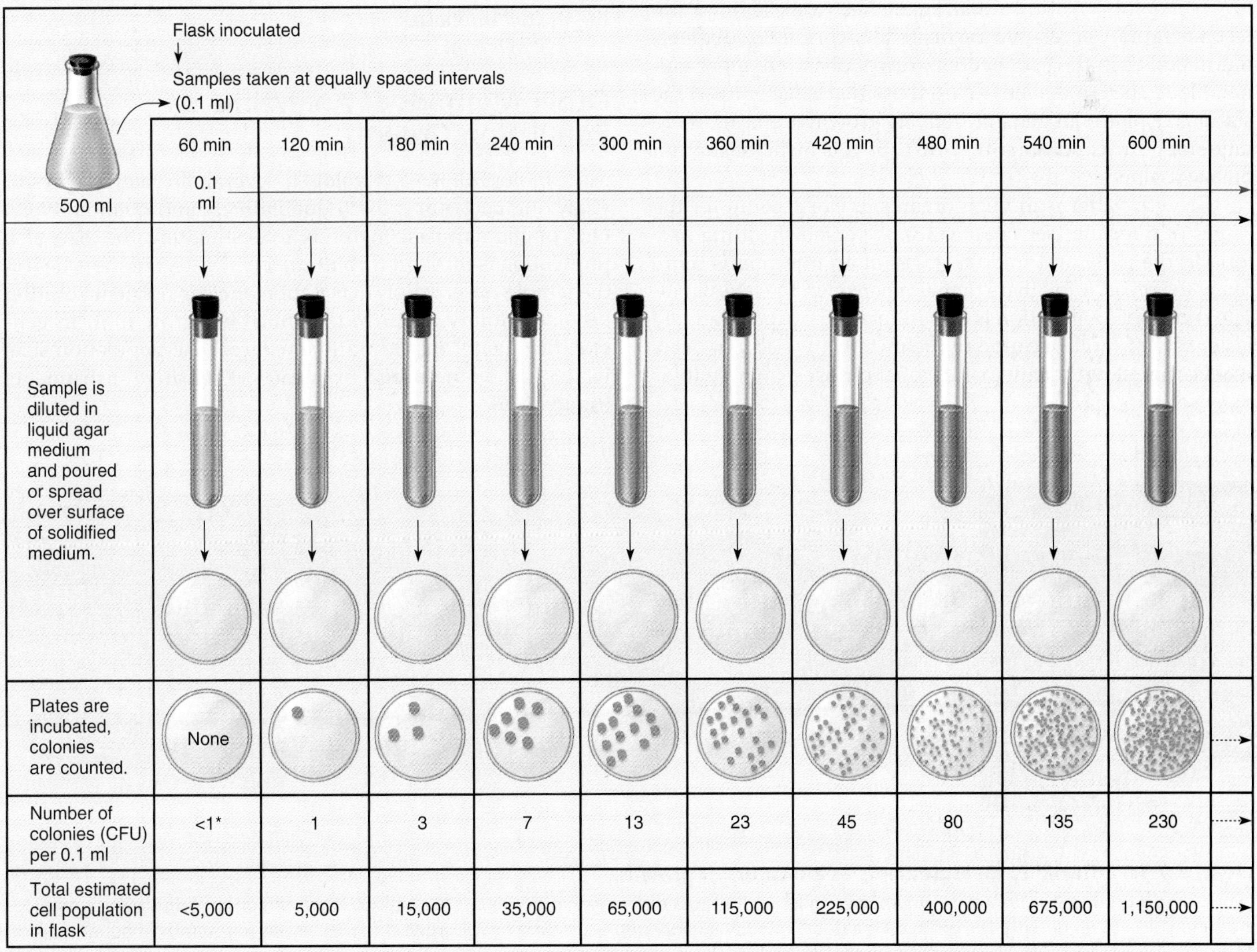
Flask inoculated
Samples taken at equally spaced intervals
(0.1 ml)
500 ml
60 min
120 min
180 min
240 min
300 min
360 min
420 min
480 min
540 min
600 min
0.1 ml
Sample is diluted in liquid agar medium and poured or spread over surface of solidified medium.
Plates are incubated, colonies are counted.
None
Number of colonies (CFU) per 0.1 ml
<1*
1
3
7
13
23
45
80
135
230
Total estimated cell population in flask
<5,000
5,000
15,000
35,000
65,000
115,000
225,000
400,000
675,000
1,150,000

* Only means that too few cells are present to be assayed.

plus excretion of organic acids and other biochemical pollutants into the growth medium, due to the increased density of cells.

As the limiting factors intensify, cells begin to die at an exponential rate (literally perishing in their own wastes), and they are unable to multiply. The curve now dips downward as the **death phase** begins. The speed with which death occurs depends on the relative resistance of the species and how toxic the conditions are, but it is usually slower than the exponential growth phase. Viable cells often remain many weeks and months after this phase has begun. In the laboratory, refrigeration is used to slow the progression of the death phase so that cultures will remain viable as long as possible.

Practical Importance of the Growth Curve

The tendency for populations to exhibit phases of rapid growth, slow growth, and death has important implications in microbial control, infection, food microbiology, and culture technology. Antimicrobial agents such as heat and disinfectants rapidly accelerate the death phase in all populations, but microbes in the exponential growth phase are more vulnerable to these agents than are those that have entered the stationary phase. In general, actively growing cells are more vulnerable to conditions that disrupt cell metabolism and binary fission.

Growth patterns in microorganisms can account for the stages of infection (see chapter 13). A person shedding bacteria in the early and middle stages of an infection is more likely to spread it to others than is a person in the late stages. The course of an infection is also influenced by the relatively faster rate of multiplication of the microbe, which can overwhelm the slower growth rate of the host's own cellular defenses.

Understanding the stages of cell growth is crucial for work with cultures. Sometimes a culture that has reached the stationary phase is incubated under the mistaken impression that enough nutrients are present for the culture to multiply. In most cases, it is unwise to continue incubating a culture beyond the stationary phase, because doing so will reduce the number of viable cells and the culture could die out completely. It is also preferable to use young cultures to do stains (an exception is the spore stain) and motility tests, because the cells will show their natural size and correct reaction and motile cells will have functioning flagella.

For certain research or industrial applications, closed batch culturing with its four phases is inefficient. The alternative is an automatic growth chamber called the **chemostat,** or continuous culture system. This device can admit a steady stream of new nutrients and siphon off used media and old bacterial cells, thereby stabilizing the growth rate and cell number. The chemostat is very similar to the industrial fermenters used to produce vitamins and antibiotics (see chapter 24). It has the advantage of maintaining the culture in a biochemically active state and preventing it from entering the death phase.

Other Methods of Analyzing Population Growth

Microbiologists have developed several alternative ways of analyzing bacterial growth qualitatively and quantitatively. One of the simplest methods for estimating the size of a population is through turbidometry. This technique relies on the simple observation that a tube of clear nutrient solution becomes cloudy, or **turbid,** as microbes grow in it. In general, the greater the turbidity, the larger the population size, which can be measured by means of sensitive instruments **(figure 7.16).**

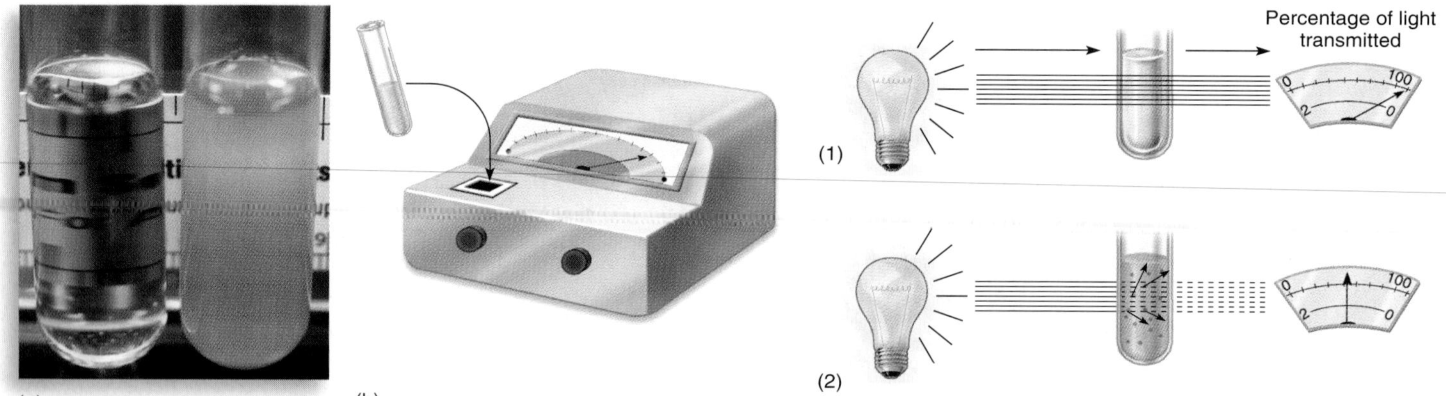

Figure 7.16 Turbidity measurements as indicators of growth.
(a) Holding a broth to the light is one method of checking for gross differences in cloudiness (turbidity). The broth on the left is transparent, indicating little or no growth; the broth on the right is cloudy and opaque, indicating heavy growth. **(b)** The eye is not sensitive enough to pick up fine degrees in turbidity; more sensitive measurements can be made with a spectrophotometer. (*1*) A tube with no growth will allow light to easily pass. Therefore more light will reach the photodetector and give a higher transmittance value. (*2*) In a tube with growth, the cells scatter the light, resulting in less light reaching the photodetector, and, therefore, gives a lower transmittance valve.

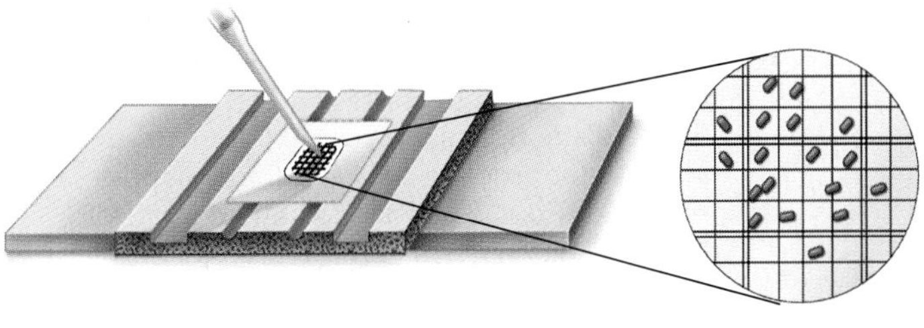

Figure 7.17 Direct microscopic count of bacteria. A small sample is placed on the grid under a cover glass. Individual cells, both living and dead, are counted. This number can be used to calculate the total count of a sample.

Enumeration of Bacteria

Turbidity readings are useful for evaluating relative amounts of growth, but if a more quantitative evaluation is required, the viable colony count described in Insight 7.6 or some other enumeration (counting) procedure is necessary. The **direct,** or **total, cell count** involves counting the number of cells in a sample microscopically **(figure 7.17).** This technique, very similar to that used in blood cell counts, employs a special microscope slide (cytometer) calibrated to accept a tiny sample that is spread over a premeasured grid. The cell count from a cytometer can be used to estimate the total number of cells in a larger sample (for instance, of milk or water). One inherent inaccuracy in this method as well as in spectrophotometry is that no distinction can be made between dead and live cells, both of which are included in the count.

Counting can be automated by sensitive devices such as the *Coulter counter,* which electronically scans a culture as it passes through a tiny pipette. As each cell flows by, it is detected and registered on an electronic sensor **(figure 7.18).** A *flow cytometer* works on a similar principle, but in addition to counting, it can measure cell size and even differentiate between live and dead cells. When used in conjunction with fluorescent dyes and antibodies to tag cells, it has been used to differentiate between gram-positive and gram-negative bacteria. It has been adapted for use as a rapid method to identify pathogens in patient specimens and to differentiate blood cells. More sophisticated forms of the flow cytometer can actually sort cells of different types into separate compartments of a collecting device.

Although flow cytometry can be used to count bacteria in natural samples without the need for culturing them, it requires fluorescent labeling of the cells you are interested in detecting, which is not always possible.

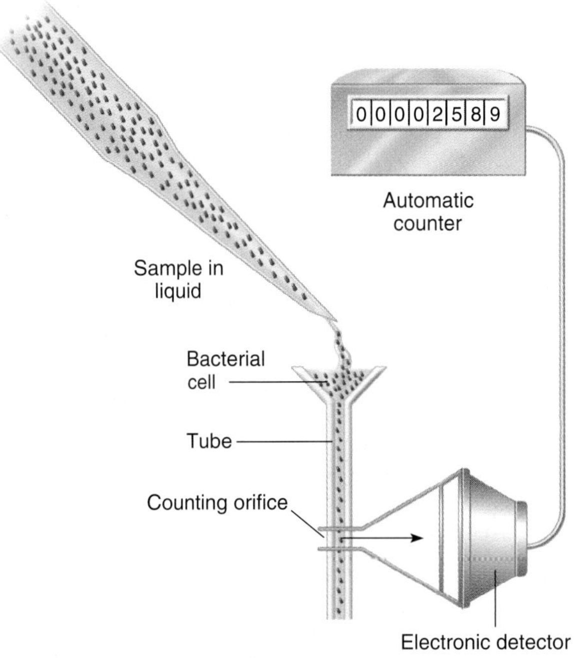

Figure 7.18 Coulter counter. As cells pass through this device, they trigger an electronic sensor that tallies their numbers.

A variation of the polymerase chain reaction (PCR) (see Insight 7.4), called real-time PCR, allows scientists to quantify bacteria and other microorganisms that are present in environmental or tissue samples without isolating them and without culturing them.

CHECKPOINT

- Microbial growth refers both to increase in cell size and increase in number of cells in a population.
- The generation time is a measure of the growth rate of a microbial population. It varies in length according to environmental conditions.
- Microbial cultures in a nutrient-limited batch environment exhibit four distinct stages of growth: the lag phase, the exponential growth (log) phase, the stationary phase, and the death phase.
- Microbial cell populations show distinct phases of growth in response to changing nutrient and waste conditions.
- Population growth can be quantified by measuring turbidity, colony counts, and direct cell counts. Other techniques can be used to count bacteria without growing them.

Chapter Summary with Key Terms

7.1 Microbial Nutrition

Nutrition consists of taking in chemical substances (nutrients) and assimilating and extracting energy from them.

A. Substances required for survival are **essential nutrients**—usually containing the elements (C, H, N, O, P, S, Na, Cl, K, Ca, Fe, Mg). Essential nutrients are considered **macronutrients** (required in larger amounts) or **micronutrients** (trace elements required in smaller amounts—Zn, Mn, Cu). Nutrients are classed as either inorganic or organic.

B. A **growth factor** is an organic nutrient that cannot be synthesized and must be provided.

C. *Nutritional Types*

1. An **autotroph** depends on carbon dioxide for its carbon needs. An autotroph that derives energy from light is a **photoautotroph;** one that extracts energy from inorganic substances is a **chemoautotroph. Methanogens** are chemoautotrophs that produce methane.
2. A **heterotroph** acquires carbon from organic molecules. A **saprobe** is a decomposer that feeds upon dead organic matter, and a **parasite** feeds from a live host and usually causes harm. Parasites that cause damage or death are called **pathogens.**

D. *Transport Mechanisms*

1. A microbial cell must take in nutrients from its surroundings by transporting them across the cell membrane.
2. **Osmosis** is diffusion of water through a selectively permeable membrane.
3. Osmotic changes that affect cells are **hypotonic** solutions, which contain a lower solute concentration, and **hypertonic** solutions, which contain a higher solute concentration. **Isotonic** solutions have the same solute concentration as the inside of the cell.
4. Passive transport involves the natural movement of substances down a concentration gradient and requires no additional energy **(diffusion).** A form of passive transport that can move specific substances is **facilitated diffusion.**
5. In **active transport,** substances are taken into the cell by a process that consumes energy. In **group translocation,** molecules are altered during transport.
6. **Phagocytosis** and **pinocytosis** are forms of active transport in which bulk quantities of solid and fluid material are taken into the cell.

7.2 Environmental Factors That Influence Microbes

Every organism adapts to a particular habitat and niche.

A. *Temperature:* An organism exhibits **optimum, minimum,** and **maximum temperatures.** Organisms that cannot grow above 20°C but thrive below 15°C and continue to grow even at 0°C are known as **psychrophiles. Mesophiles** grow from 10°C to 50°C, having temperature optima from 20°C to 40°C. The growth range of **thermophiles** is 45°C to 80°C. Extreme thermophiles may grow to temperatures of 120°C.

B. *Oxygen Requirements:* The ecological need for free oxygen (O_2) is based on whether a cell can handle toxic by-products such as superoxide and peroxide.

1. **Aerobes** grow in normal atmospheric oxygen and have enzymes to handle toxic oxygen by-products. An aerobic organism capable of living without oxygen if necessary is a **facultative anaerobe.** An aerobe that requires a small amount of oxygen but does not grow under anaerobic conditions is a **microaerophile.**
2. **Strict (obligate) anaerobes** do not use free oxygen and cannot produce enzymes to dismantle reactive oxides. They are actually damaged or killed by oxygen. An **aerotolerant anaerobe** cannot use oxygen for respiration, yet is not injured by it.

C. *Effects of pH:* Acidity and alkalinity affect the activity and integrity of enzymes and the structural components of a cell. Optimum pH for most microbes ranges approximately from 6 to 8. Acidophiles prefer lower pH, and alkalinophiles prefer higher pH.

D. *Other Environmental Factors:* Radiation and barometric pressure affect microbial growth.

E. *Microbial Interactions:* Microbes coexist in varied relationships in nature.

1. Types of **symbiosis** are **mutualism,** a reciprocal, obligatory, and beneficial relationship between two organisms, and **commensalism,** an organism receiving benefits from another without harming the other organism in the relationship. **Parasitism** occurs between a host and an infectious agent; the host is harmed in the interaction.
2. **Synergism** is a mutually beneficial but not obligatory coexistence. **Antagonism** entails competition, inhibition, and injury directed against the opposing organism. A special case of antagonism is antibiotic production.

7.3 The Study of Microbial Growth

A. The splitting of a parent bacterial cell to form a pair of similar-size daughter cells is known as **binary,** or **transverse, fission.**

B. The duration of each division is called the **generation,** or **doubling, time.** A population theoretically doubles with each generation, so the growth rate is **exponential,** and each cycle increases in geometric progression.

C. A **growth curve** is a graphic representation of a closed population over time. Plotting a curve requires an estimate of live cells, called a viable count. The initial flat period of the curve is called the **lag phase,** followed by the **exponential growth phase,** in which viable cells increase in logarithmic progression. Adverse environmental conditions combine to inhibit the growth rate, causing a plateau, or **stationary growth phase.** In the **death phase,** nutrient depletion and waste buildup cause increased cell death.

D. Cell numbers can be counted directly by a microscope counting chamber, Coulter counter, flow cytometer, or newer methods that detect organisms in their natural settings. Cell growth can also be determined by turbidometry and a total cell count.

Multiple-Choice and True-False Questions

Multiple-Choice Questions. Select the correct answer from the answers provided.

1. The source of the necessary elements of life is
   a. an inorganic environmental reservoir
   b. the sun
   c. rocks
   d. the air
2. An organism that can synthesize all its required organic components from CO_2 using energy from the sun is a
   a. photoautotroph
   b. photoheterotroph
   c. chemoautotroph
   d. chemoheterotroph
3. Chemoautotrophs can survive on ____ alone.
   a. minerals
   b. CO_2
   c. minerals and CO_2
   d. methane
4. Which of the following statements is true for *all* organisms?
   a. they require organic nutrients
   b. they require inorganic nutrients
   c. they require growth factors
   d. they require oxygen gas
5. A pathogen would most accurately be described as a
   a. parasite
   b. commensal
   c. saprobe
   d. symbiont
6. Which of the following is true of passive transport?
   a. it requires a gradient
   b. it uses the cell wall
   c. it includes endocytosis
   d. it only moves water
7. A cell exposed to a hypertonic environment will ____ by osmosis.
   a. gain water
   b. lose water
   c. neither gain nor lose water
   d. burst
8. Psychrophiles would be expected to grow
   a. in hot springs
   b. on the human body
   c. at refrigeration temperatures
   d. at low pH
9. Superoxide ion is toxic to strict anaerobes because they lack
   a. catalase
   b. peroxidase
   c. dismutase
   d. oxidase
10. In a viable plate count, each ____represents a ____ from the sample population.
    a. cell, colony
    b. colony, cell
    c. hour, generation
    d. cell, generation

True-False Questions. If the statement is true, leave as is. If it is false, correct it by rewriting the sentence.

11. Active transport of a substance across a membrane requires a concentration gradient.
12. An organic nutrient essential to an organism's metabolism that it cannot synthesize is called a growth factor.
13. The time required for a cell to undergo binary fission is called the growth curve.
14. An obligate halophile is an organism that requires high osmotic pressure.
15. A facultative anaerobe can grow with or without oxygen.

Writing to Learn

These questions are suggested as a *writing-to-learn* experience. For each question (except 3), compose a one- or two-paragraph answer that includes the factual information needed to completely address the question.

1. Differentiate between micronutrients and macronutrients.
   a. What elements do the letters CHONPS stand for?
   b. Briefly describe the general function of these elements in the cell.
   c. Define growth factors, and give examples of them.
2. Name some functions of metallic ions in cells.
3. Fill in the following table:

	Source of Carbon	Usual Source of Energy	Example
Photoautotroph			
Photoheterotroph			
Chemoautotroph			
Chemoheterotroph			
Saprobe			
Parasite			

4. a. Compare and contrast passive and active forms of transport, using examples of what is being transported and the requirements for each.
   b. How are phagocytosis and pinocytosis similar? How are they different?
5. Compare the effects of isotonic, hypotonic, and hypertonic solutions on an amoeba and on a bacterial cell. If a cell lives in a hypotonic environment, what will occur if it is placed in a hypertonic one? Answer for the opposite case as well.
6. Look at the following diagrams and predict in which direction osmosis will take place. Use arrows to show the net direction of osmosis. Is one of these microbes a halophile? Which one?

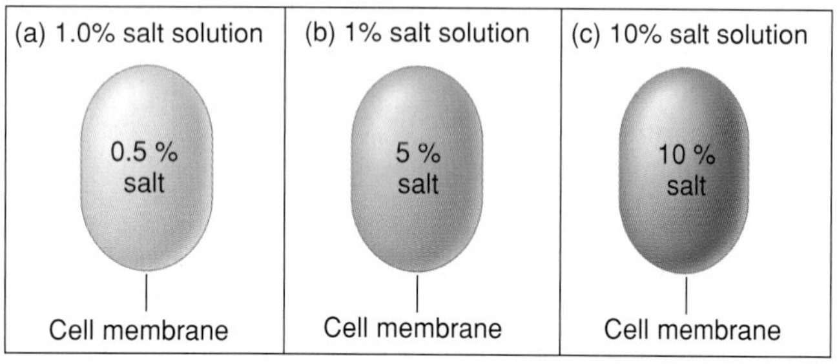

7. a. Explain what it means to be an obligate intracellular parasite.
   b. Name three groups of obligate intracellular parasites.
8. a. Classify a human with respect to oxygen requirements.
   b. What might be the habitat of an aerotolerant anaerobe?
   c. Where in the body are anaerobic habitats apt to be found?
9. Where do superoxide ions and hydrogen peroxide originate? What are their toxic effects?
10. Why is growth called exponential? What is the size of a population in 20 generations? Explain what is happening to the population at points A, B, C, and D in the following diagram.

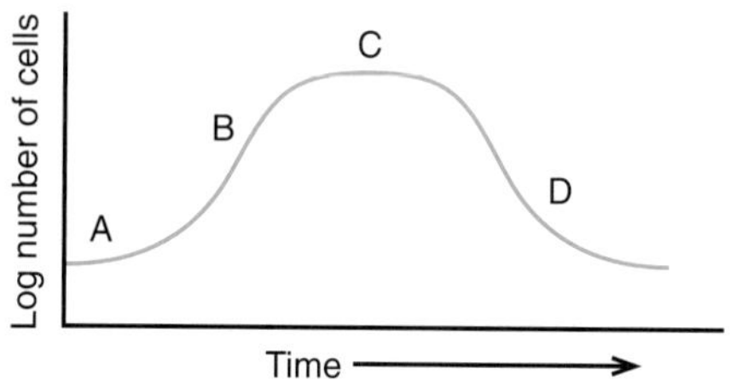

Concept Mapping

Appendix D provides guidance for working with concept maps.

1. Supply your own lines (linkers) and linking words or phrases in this concept map, and provide the missing concepts in the empty box.

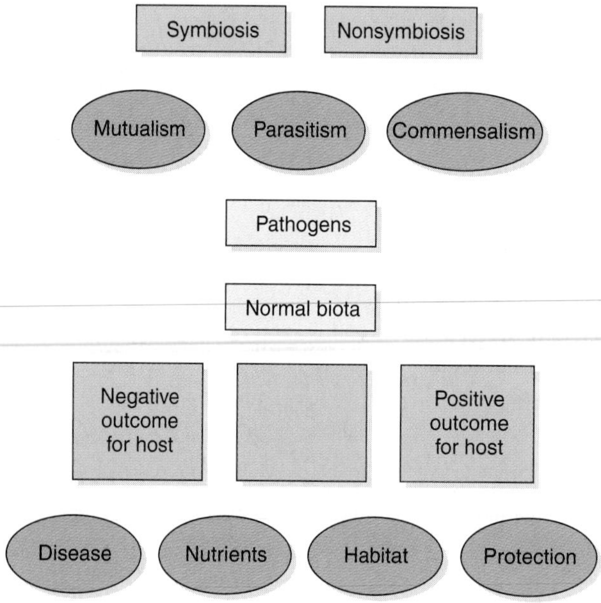

Critical Thinking Questions

Critical thinking is the ability to reason and solve problems using facts and concepts. These questions can be approached from a number of angles, and in most cases, they do not have a single correct answer.

1. a. Is there a microbe that could grow on a medium that contains only the following compounds dissolved in water: $CaCO_3$, $MgNO_3$, $FeCl_2$, $ZnSO_4$, and glucose? Defend your answer.
   b. Check the last entry in the chemical analysis summary on page 182. Are all needed elements present here? Where is the carbon?
2. Describe how one might determine the nutrient requirements of a microbe from Mars. If, after exhausting various nutrient schemes, it still does not grow, what other factors might one take into account?
3. a. The noted microbial ecologist Martin W. Beijerinck has stated: "Everything is everywhere, the environment selects." Use this concept to explain what ultimately determines whether a microorganism can live in a certain habitat.
   b. Give two examples of ways that microbes become modified to survive.
4. How can you explain the observation that unopened milk will spoil even while refrigerated?
5. What would be the effect of a fever on a thermophilic pathogen?
6. Patients with ketoacidosis associated with diabetes are especially susceptible to fungal infections. Can you explain why?
7. Using the concept of synergism, can you describe a way to grow a fastidious microbe?
8. a. If an egg salad sandwich sitting in a warm car for 4 hours develops 40,960 bacterial cells, how many more cells would result with just one more hour of incubation? (Use the same criteria that were stated in the sample problem.)
   b. With 10 additional hours of incubation?
   c. What would the cell count be after 4 hours if the initial bacterial dose were 100?
   d. What do your answers tell you about using clean techniques in food preparation and storage (other than aesthetic considerations)?
9. Discuss the idea of biotechnology companies being allowed to isolate and own microorganisms taken from the earth's habitats and derive profits from them. Can you come up with a solution that encourages exploration yet also benefits the public?
10. Place appropriate points on the axes and draw a graph that fits an obligate thermophile.

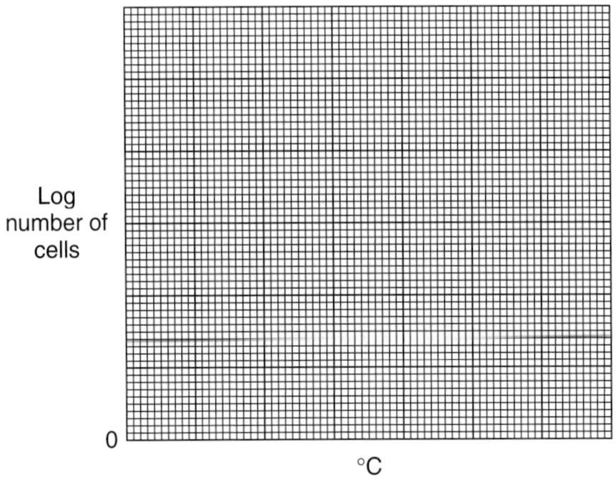

Enter points on the axes and show the growth results of an acidophilic microbe.

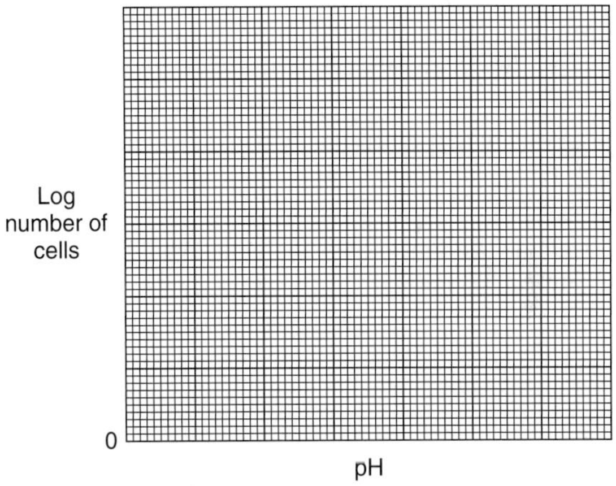

Visual Understanding

1. **Figure 7.8.** Draw a growth curve for a psychrotroph on the following diagram.

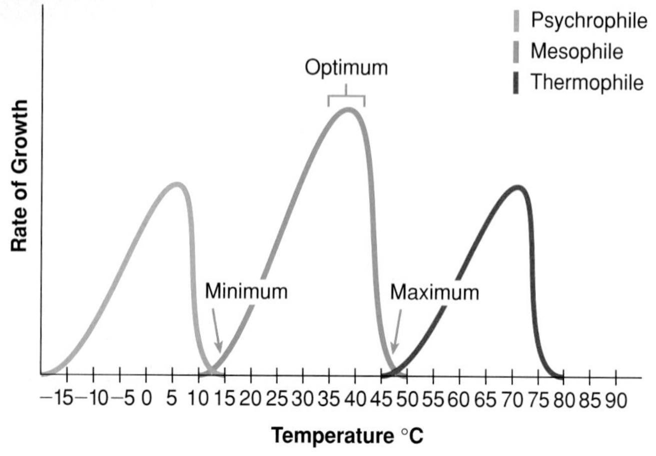

2. **Figure 7.18.** What effect will noncellular particulate material have on the automated cell counter shown below? Why?

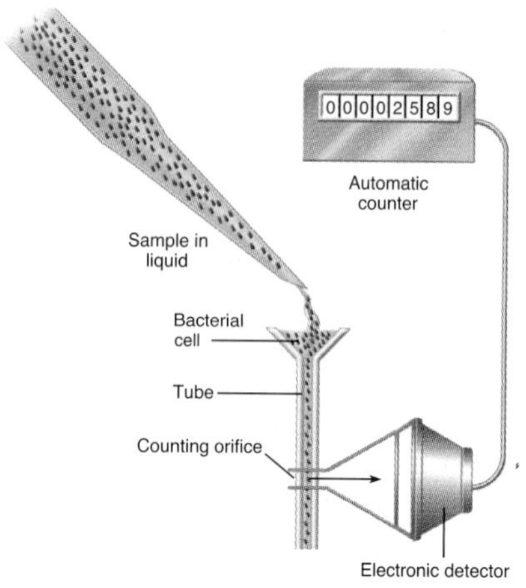

Internet Search Topics

1. Look up the term **extremophile** on a search engine. Find examples of microbes that have adapted to various extremes. Report any species that exist in several extremes simultaneously.
2. Find a website that discusses the use of microbes living in high temperatures to bioremediate (clean up) the environment.
3. Research the Rio Tinto River in Spain. What kinds of microbes dominate in this habitat?
4. Go to: www.aris.mhhe.com, and click on "microbiology" and then this textbook's author/title. Go to chapter 7, access the URLs listed under Internet Search Topics, and research the following:
   a. Log on to the website listed. Research the methods used by the laboratory to quantify microbes in water samples.
   b. Search in one or both websites listed to discover some very interesting symbiotic relationships.

CHAPTER 8

Microbial Metabolism

The Chemical Crossroads of Life

CASE FILE 8

Polystyrene foam, commonly known as Styrofoam®, is an amazing product. It's strong, lightweight, and durable. Its common uses range from building insulation to food and beverage containers. What happens to polystyrene when it is no longer needed? According to the EPA, most of it ends up in landfills and that's where it will remain basically unchanged for the next 500 years. That Styrofoam beverage cup you discard today will join some 25 billion annually that become part of more than 2.3 million tons of polystyrene waste generated each year in the United States.

Kevin O'Connor, Ph.D., at the University College Dublin, along with colleagues from Ireland and Germany, developed a method that uses a bacterium to convert polystyrene into a useful and biodegradable plastic known as polyhydroxyalkanoate (PHA). The organism, *Pseudomonas putida,* is able to utilize styrene oil as both an energy and carbon source.

- *What is the name for the process that utilizes microorganisms to reduce or degrade waste or pollutants?*
- *What other problems might pseudomonads be able to help us deal with?*

Case File 8 Wrap-Up appears on page 241.

CHAPTER OVERVIEW

- Cells are constantly involved in an orderly activity called metabolism that encompasses all of their chemical and energy transactions.
- Enzymes are protein catalysts that speed up chemical processes by lowering the required energy.
- Enzymes have a specific shape tailored to perform their actions on a single type of molecule called a substrate.
- Enzymes derive some of their special characteristics from cofactors such as vitamins, and they show sensitivity to environmental factors.
- Enzymes are regulated by several mechanisms that alter the structure or synthesis of the enzyme.
- The energy of living systems resides in the atomic structure of chemicals that can be acted upon and changed.
- Cell energetics involve the release of energy that powers the formation of bonds.
- The energy of electrons is transferred from one molecule to another in coupled redox reactions.
- Electrons are transferred from substrates such as glucose to coenzyme carriers and ultimately captured in high-energy adenosine triphosphate (ATP).

- Cell pathways involved in extracting energy from fuels are glycolysis, the tricarboxylic acid cycle, and electron transport.
- The molecules used in aerobic respiration are glucose and oxygen, and the products are CO_2, H_2O, and ATP.
- Microbes have evolved alternate pathways such as fermentation and anaerobic respiration.
- Photosynthesis makes energy from the sun available to chemical processes and also produces carbon for metabolism.
- Cells manage their metabolites through linked pathways that have numerous functions and can proceed in more than one direction.
- Pathways of anabolism create the molecules and macromolecules needed for cell reproduction and almost always require an input of energy.

8.1 The Metabolism of Microbes

Metabolism, from the Greek term *metaballein,* meaning change, pertains to all chemical reactions and physical workings of the cell. Although metabolism entails thousands of different reactions, most of them fall into one of two general categories. **Anabolism,** sometimes also called *biosynthesis,* is any process that results in synthesis of cell molecules and structures. It is a building and bond-making process that forms larger macromolecules from smaller ones, and it usually requires the input of energy. **Catabolism** is the opposite of anabolism. Catabolic reactions are degradative; they break the bonds of larger molecules into smaller molecules and often release energy. The linking of anabolism to catabolism ensures the efficient completion of many thousands of cellular processes.

In summary, metabolism performs these functions:

1. assembles smaller molecules into larger macromolecules needed for the cell; in this process, ATP is utilized to form bonds (anabolism) **(figure 8.1),**
2. degrades macromolecules into smaller molecules and yields energy (catabolism),
3. energy is conserved in the form of ATP (adenosine triphosphate) or heat.

It has built-in controls for reducing or stopping a process that is not in demand and other controls for storing excess nutrients. The metabolic workings of the cell are indeed intricate and complex, but they are also elegant and efficient. It is this very organization that sustains life.

Enzymes: Catalyzing the Chemical Reactions of Life

A microbial cell could be viewed as a microscopic factory, complete with basic building materials, a source of energy, and a "blueprint" for running its extensive network of metabolic reactions. But the chemical reactions of life, even when highly organized and complex, cannot proceed without a special class of macromolecules called **enzymes.** Enzymes are a remarkable example of **catalysts,** chemicals that increase the rate of a chemical reaction without becoming part of the products or being consumed in the reaction. Do not make the mistake of thinking that an enzyme creates a reaction. Because of the great energy of some molecules, a reaction could occur spontaneously at some point even without an enzyme, but at a very slow rate. A study of the enzyme urease shows that it increases the rate of the breakdown of urea by a factor of 100 trillion as compared to an uncatalyzed reaction. Because most uncatalyzed metabolic reactions do not occur fast enough to sustain cell processes, enzymes, which speed up the rate of reactions, are indispensable to life. Other major characteristics of enzymes are summarized in **table 8.1.**

How Do Enzymes Work?

We have said that an enzyme speeds up the rate of a metabolic reaction, but just how does it do this? During a chemical reaction, reactants are converted to products by bond formation or breakage. A certain amount of energy is required to initiate every such reaction, which limits its rate. This resistance to a reaction, which must be overcome for a reaction to proceed, is

TABLE 8.1 Checklist of Enzyme Characteristics

- Most composed of protein; may require cofactors
- Act as organic catalysts to speed up the rate of cellular reactions
- Lower the activation energy required for a chemical reaction to proceed (see Insight 8.1)
- Have unique characteristics such as shape, specificity, and function
- Enable metabolic reactions to proceed at a speed compatible with life
- Have an active site for target molecules called substrates
- Are much larger in size than their substrates
- Associate closely with substrates but do not become integrated into the reaction products
- Are not used up or permanently changed by the reaction
- Can be recycled, thus function in extremely low concentrations
- Are greatly affected by temperature and pH
- Can be regulated by feedback and genetic mechanisms

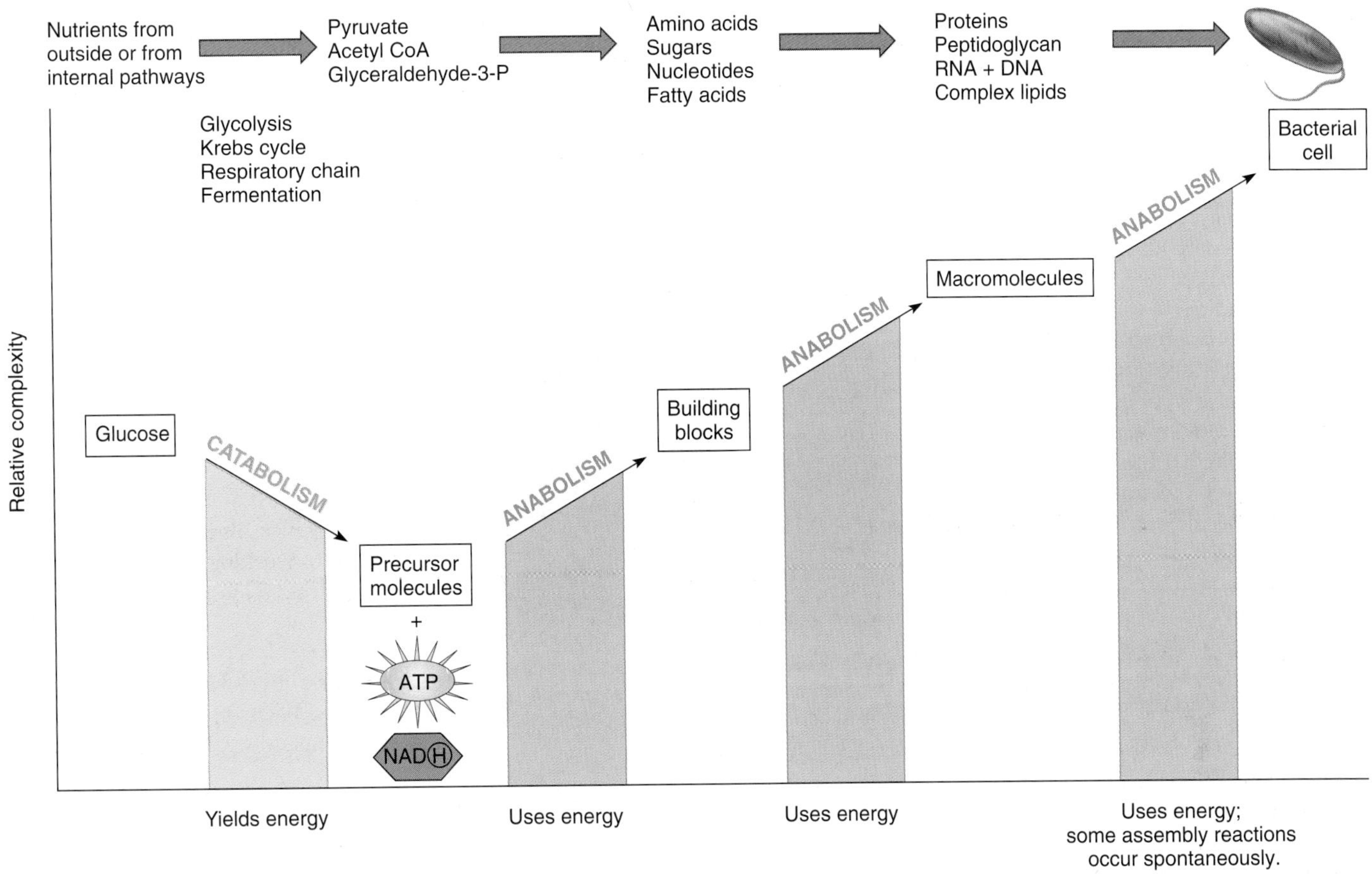

Figure 8.1 Simplified model of metabolism.
Cellular reactions fall into two major categories. Catabolism involves the breakdown of complex organic molecules to extract energy and form simpler end products. Anabolism uses the energy to synthesize necessary macromolecules and cell structures from precursors.

measurable and is called the **energy of activation** or activation energy. In the laboratory, overcoming this initial resistance can be achieved by:

1. increasing thermal energy (heating) to increase molecular velocity,
2. increasing the concentration of reactants to increase the rate of molecular collisions, or
3. adding a catalyst.

In most living systems, the first two alternatives are not feasible, because elevating the temperature is potentially harmful and higher concentrations of reactants are not practical. This leaves only the action of catalysts, and enzymes fill this need efficiently and potently.

At the molecular level, an enzyme promotes a reaction by serving as a physical site upon which the reactant molecules, called **substrates,** can be positioned for various interactions. The enzyme is much larger in size than its substrate, and it presents a unique active site that fits only that particular substrate. Although an enzyme binds to the substrate and participates directly in changes to the substrate, it does not become a part of the products, is not used up by the reaction, and can function over and over again. Enzyme speed, defined as the number of substrate molecules converted per enzyme per second, is well documented. Speeds range from several million for catalase to a thousand for lactate dehydrogenase. To further visualize the roles of enzymes in metabolism, we must next look at their structure.

Enzyme Structure

The primary structure of most enzymes is protein—with some exceptions **(Insight 8.1),** and they can be classified as simple or conjugated. Simple enzymes consist of protein alone, whereas conjugated enzymes **(figure 8.2)** contain

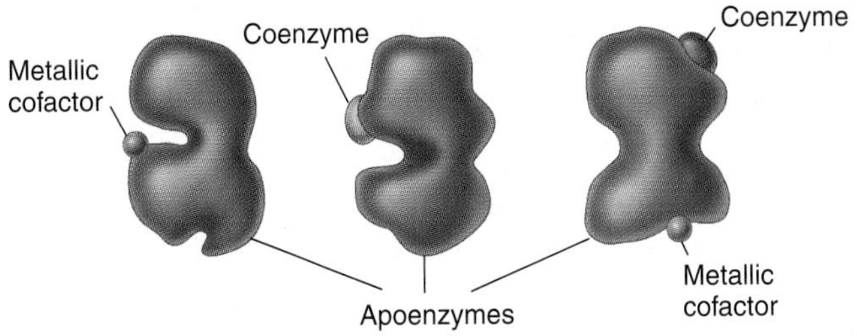

Figure 8.2 Conjugated enzyme structure.
All have an apoenzyme (polypeptide or protein) component and one or more cofactors.

INSIGHT 8.1 Discovery

Unconventional Enzymes

The molecular reactions of cells are still an active and rich source of discovery, and new findings come along nearly every few months that break older "rules" and change our understanding. It was once an accepted fact that proteins were the only biological molecules that act as catalysts, until biologists found a novel type of RNA termed **ribozymes.** Ribozymes are associated with many types of cells and a few viruses, and they display some of the properties of protein catalysts, such as having a specific active site and interacting with a substrate. But these molecules are remarkable because their substrate is other RNA. Ribozymes are thought to be remnants of the earliest molecules on earth that could have served as both catalysts and genetic material. In natural systems, ribozymes are involved in self-splicing or cutting of RNA molecules during final processing of the genetic code (see chapter 9). Further research has shown that ribozymes can be designed to handle other kinds of activities, such as inhibiting gene expression. Several companies have developed and are testing ribozyme-based therapies for treating cancer and AIDS.

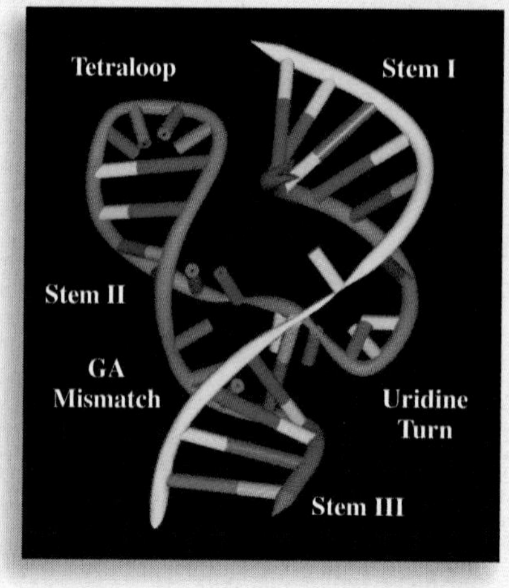

A "hammerhead" ribozyme consisting of a single RNA strand curved around to form an active site (indentation between tetraloop and stem I).

Another focus of study that could lead to new applications is the search for *extremozymes,* enzymes from microbes that live under rigorous conditions of temperature, pH, salt, and pressure. Because these enzymes are adapted to working under such extreme conditions, they may well be useful in industrial processes that require work in heat or cold or very low pH. Several enzymes that can work under high temperature and low pH have been isolated and are either currently in use (PCR technique) or in development.

protein and nonprotein molecules. A conjugated enzyme, sometimes referred to as a **holoenzyme,** is a combination of a protein, now called the **apoenzyme,** and one or more **cofactors (table 8.2).** Cofactors are either organic molecules, called **coenzymes,** or inorganic elements (metal ions).

Apoenzymes: Specificity and the Active Site

Apoenzymes range in size from small polypeptides with about 100 amino acids and a molecular weight of 12,000 to large polypeptide conglomerates with thousands of amino acids and a molecular weight of over 1 million. Like all proteins, an apoenzyme exhibits levels of molecular complexity called the primary, secondary, tertiary, and, in larger enzymes, quaternary organization **(figure 8.3).** As we saw in chapter 2, the first three levels of structure arise when a single polypeptide chain undergoes an automatic folding process and achieves stability by forming disulfide and other types of bonds. Folding causes the surface of the apoenzyme to acquire three-dimensional features that result in the enzyme's specificity for substrates. The actual site where the substrate binds is a crevice or groove called the **active site,** or **catalytic site,** and there can be from one to several such sites (see figure 8.3). Each type of enzyme has a different primary structure (type and sequence of amino acids), variations in folding, and unique active sites.

TABLE 8.2 Selected Enzymes, Catalytic Actions, and Cofactors

Enzyme	Action	Metallic Cofactor Required
Catalase	Breaks down hydrogen peroxide	Iron (Fe)
Oxidase	Adds electrons to oxygen	Iron (Fe), copper (Cu)
Hexokinase	Transfers phosphate to glucose	Magnesium (Mg)
Urease	Splits urea into an ammonium ion	Nickel (Ni)
Nitrate reductase	Reduces nitrate to nitrite	Molybdenum (Mo)
DNA polymerase complex	Synthesis of DNA	Zinc (Zn) and magnesium (Mg)

Enzyme-Substrate Interactions

For a reaction to take place, a temporary enzyme-substrate union must occur at the active site. There are several explanations for the process of attachment **(figure 8.4).** The specificity is often described as a "lock-and-key" fit in which the substrate is inserted into the active site's pocket.

The bonds formed between the substrate and enzyme are weak and, of necessity, easily reversible. Once the enzyme-substrate complex has formed, appropriate reactions occur on the substrate, often with the aid of a cofactor, and a product is formed and released. The enzyme can then attach to another substrate molecule and repeat this action. Although enzymes can potentially catalyze reactions in

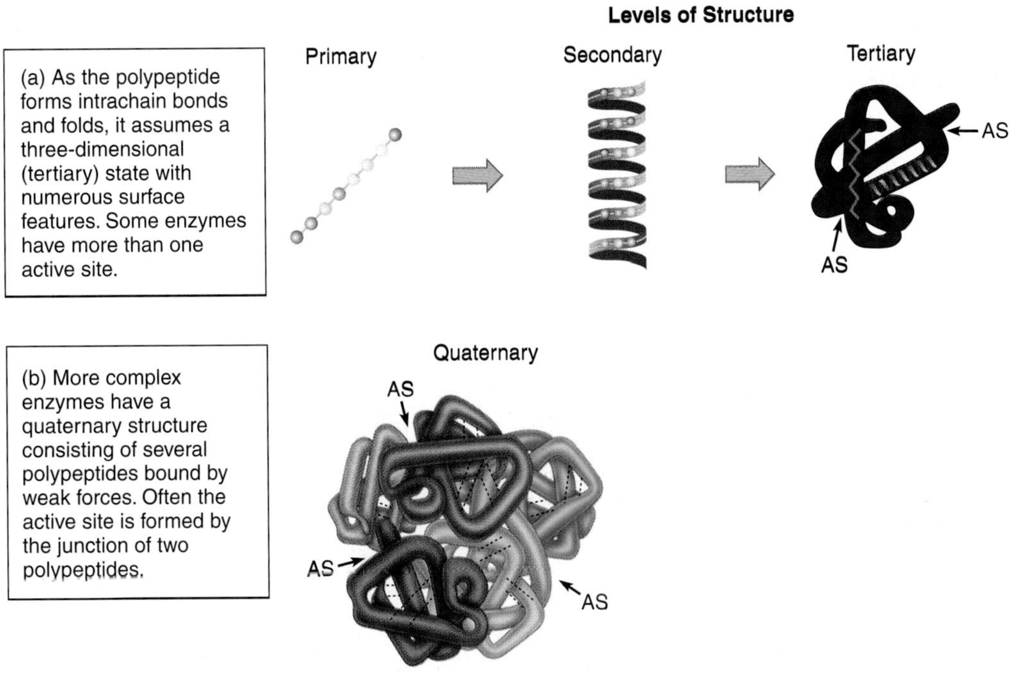

Levels of Structure
Primary
Secondary
Tertiary
AS
AS
(a) As the polypeptide forms intrachain bonds and folds, it assumes a three-dimensional (tertiary) state with numerous surface features. Some enzymes have more than one active site.
Quaternary
AS
AS
AS
(b) More complex enzymes have a quaternary structure consisting of several polypeptides bound by weak forces. Often the active site is formed by the junction of two polypeptides.

Figure 8.3 How the active site and specificity of the apoenzyme arise.

both directions, most examples in this chapter depict them working in one direction only.

Cofactors: Supporting the Work of Enzymes

In chapter 7, you learned that microorganisms require specific metal ions called trace elements and certain organic growth factors. In many cases, the need for these substances arises from their roles as cofactors. The metallic cofactors, including iron, copper, magnesium, manganese, zinc, cobalt, selenium, and many others, participate in precise functions between the enzyme and its substrate. In general, metals activate enzymes, help bring the active site and substrate close together, and participate directly in chemical reactions with the enzyme-substrate complex.

Coenzymes are a type of cofactor. They are organic compounds that work in conjunction with an apoenzyme to perform a necessary alteration of a substrate. The general function of a coenzyme is to remove a chemical group from one substrate molecule and add it to another substrate, thereby serving as a transient carrier of this group. The specific activities of coenzymes are many and varied. In a later section of this chapter, we shall see that coenzymes carry and transfer hydrogen atoms, electrons, carbon dioxide, and amino groups. One of the most important components of coenzymes is **vitamins,** which explains why vitamins are important to nutrition and may be required as growth factors for living things. Vitamin deficiencies prevent the complete holoenzyme from forming. Consequently, both the chemical reaction and the structure or function dependent upon that reaction are compromised.

Classification of Enzyme Functions

Enzymes are classified and named according to characteristics such as site of action, type of action, and substrate **(Insight 8.2).**

Location and Regularity of Enzyme Action

Enzymes perform their tasks either inside or outside of the cell in which they were produced. After initial synthesis

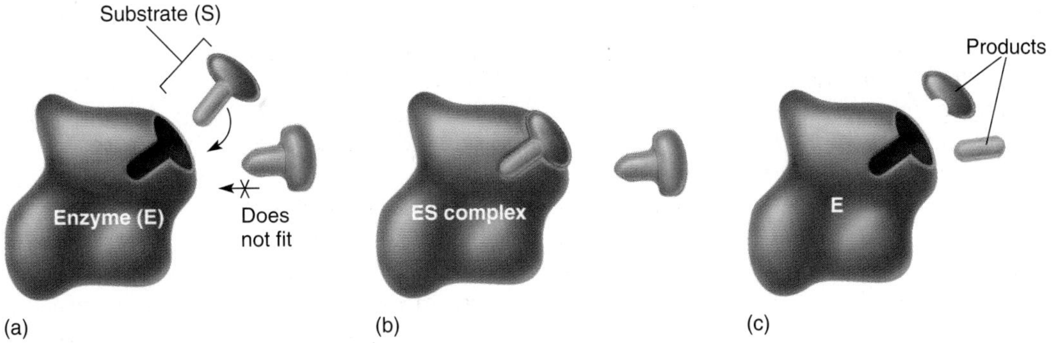

Figure 8.4 Enzyme-substrate reactions.
(a) When the enzyme and substrate come together, the substrate (S) must show the correct fit and position with respect to the enzyme (E). **(b)** When the ES complex is formed, it enters a transition state. During this temporary but tight interlocking union, the enzyme participates directly in breaking or making bonds. **(c)** Once the reaction is complete, the enzyme releases the products.

INSIGHT 8.2 — Discovery

The Enzyme Name Game

Most metabolic reactions require separate and unique enzymes. A standardized system of nomenclature and classification was developed to prevent discrepancies.

In general, an enzyme name is composed of two parts: a prefix or stem word derived from a certain characteristic—usually the substrate acted upon or the type of reaction catalyzed, or both—followed by the ending *-ase*.

The system classifies the enzyme in one of these six classes, on the basis of its general biochemical action:

1. *Oxidoreductases* transfer electrons from one substrate to another, and *dehydrogenases* transfer a hydrogen from one compound to another.
2. *Transferases* transfer functional groups from one substrate to another.
3. *Hydrolases* cleave bonds on molecules with the addition of water.
4. *Lyases* add groups to or remove groups from double-bonded substrates.
5. *Isomerases* change a substrate into its isomeric* form.
6. *Ligases* catalyze the formation of bonds with the input of ATP and the removal of water.

Each enzyme is also assigned a common name that indicates the specific reaction it catalyses. With this system, an enzyme that digests a carbohydrate substrate is a *carbohydrase;* a specific carbohydrase, *amylase,* acts on starch (amylose is a major component of starch). The enzyme *maltase* digests the sugar maltose. An enzyme that hydrolyzes peptide bonds of a protein is a *proteinase, protease,* or *peptidase,* depending on the size of the protein substrate. Some fats and other lipids are digested by *lipases.* DNA is hydrolyzed by *deoxyribonuclease,* generally shortened to *DNase.* A *synthetase* or *polymerase* bonds together many small molecules into large molecules. Other examples of enzymes are presented in table 8.A. (See also table 8.2.)

*An isomer is a compound that has the same molecular formula as another compound but differs in arrangement of the atoms.

TABLE 8.A A Sampling of Enzymes, Their Substrates, and Their Reactions

Common Name	Systematic Name	Enzyme Class	Substrates	Action
Lactase	β-D-galactosidase	Hydrolase	Lactose	Breaks lactose down into glucose and galactose
Penicillinase	Beta-lactamase	Hydrolase	Penicillin	Hydrolyzes beta-lactam ring
DNA polymerase	DNA nucleotidyl-transferase	Transferase	DNA nucleosides	Synthesizes a strand of DNA using the complementary strand as a model
Lactate dehydrogenase	Same as common name	Oxidoreductase	Pyruvic acid	Catalyzes the conversion of pyruvic acid to lactic acid
Oxidase	Cytochrome oxidase	Oxidoreductase	Molecular oxygen	Catalyzes the reduction (addition of electrons and hydrogen) to O_2

in the cell, **exoenzymes** are transported extracellularly, where they break down (hydrolyze) large food molecules or harmful chemicals. Examples of exoenzymes are cellulase, amylase, and penicillinase. By contrast, **endoenzymes** are retained intracellularly and function there. Most enzymes of the metabolic pathways are of this variety **(figure 8.5).**

In terms of their presence in the cell, enzymes are not all produced in equal amounts or at equal rates. Some, called **constitutive enzymes (figure 8.6*a*),** are always present and in relatively constant amounts, regardless of the amount of substrate. The enzymes involved in utilizing glucose, for example, are very important in metabolism and thus are constitutive. Other enzymes are **regulated enzymes (figure 8.6*b*),** the production of which is either turned on (induced) or turned off (repressed) in response to changes in concentration of the substrate. The level of inducible and repressible enzymes is controlled by the degree to which the genes for these proteins are transcribed into proteins, discussed in chapter 9.

Synthesis and Hydrolysis Reactions A growing cell is in a frenzy of activity, constantly synthesizing proteins, DNA, and RNA; forming storage polymers such as starch and glycogen; and assembling new cell parts. Such anabolic reactions require enzymes (ligases) to form covalent bonds between smaller substrate molecules. Also known as *dehydration reactions,* synthesis reactions typically require ATP and release one water molecule for each bond

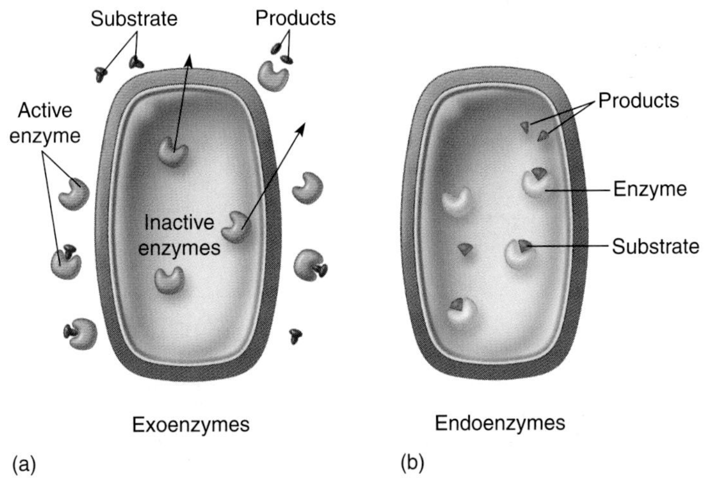

Figure 8.5 Types of enzymes, as described by their location of action.
(a) Exoenzymes are released outside the cell to function.
(b) Endoenzymes remain in the cell and function there.

made **(figure 8.7*a*).** Catabolic reactions involving energy transactions, remodeling of cell structure, and digestion of macromolecules are also very active during cell growth. For example, digestion requires enzymes to break down substrates into smaller molecules so they can be used by the cell. Because the breaking of bonds requires the input of water, digestion is often termed a *hydrolysis* (hy-drol′-uh-sis) *reaction* **(figure 8.7*b*).**

Transfer Reactions by Enzymes Other enzyme-driven processes that involve the simple addition or removal of a functional group are important to the overall economy of the cell. Oxidation-reduction and other transfer activities are examples of these types of reactions.

Some atoms and compounds readily give or receive electrons and participate in oxidation (the loss of electrons) or reduction (the gain of electrons). The compound that loses the electrons is **oxidized,** and the compound that receives the electrons is **reduced.** Such oxidation-reduction (redox) reactions are common in the cell and indispensable to the energy transformations discussed later in this chapter (see also Insight 2.2). Important components of cellular redox reactions are oxidoreductases, which remove electrons from one substrate and add them to another. Their coenzyme carriers are nicotinamide adenine dinucleotide (NAD) and flavin adenine dinucleotide (FAD). Oxidation and reduction are covered in more detail later in this chapter.

Other enzymes play a role in the molecular conversions necessary for the economical use of nutrients by directing the transfer of functional groups from one molecule to another. For example, *aminotransferases* convert one type of amino acid to another by transferring an amino group; *phosphotransferases* participate in the transfer of phosphate groups and are involved in energy transfer; *methyltransferases* move a methyl (CH_3) group from substrate to substrate; and *decarboxylases* (also called carboxylases) catalyze the removal of carbon dioxide from organic acids in several metabolic pathways.

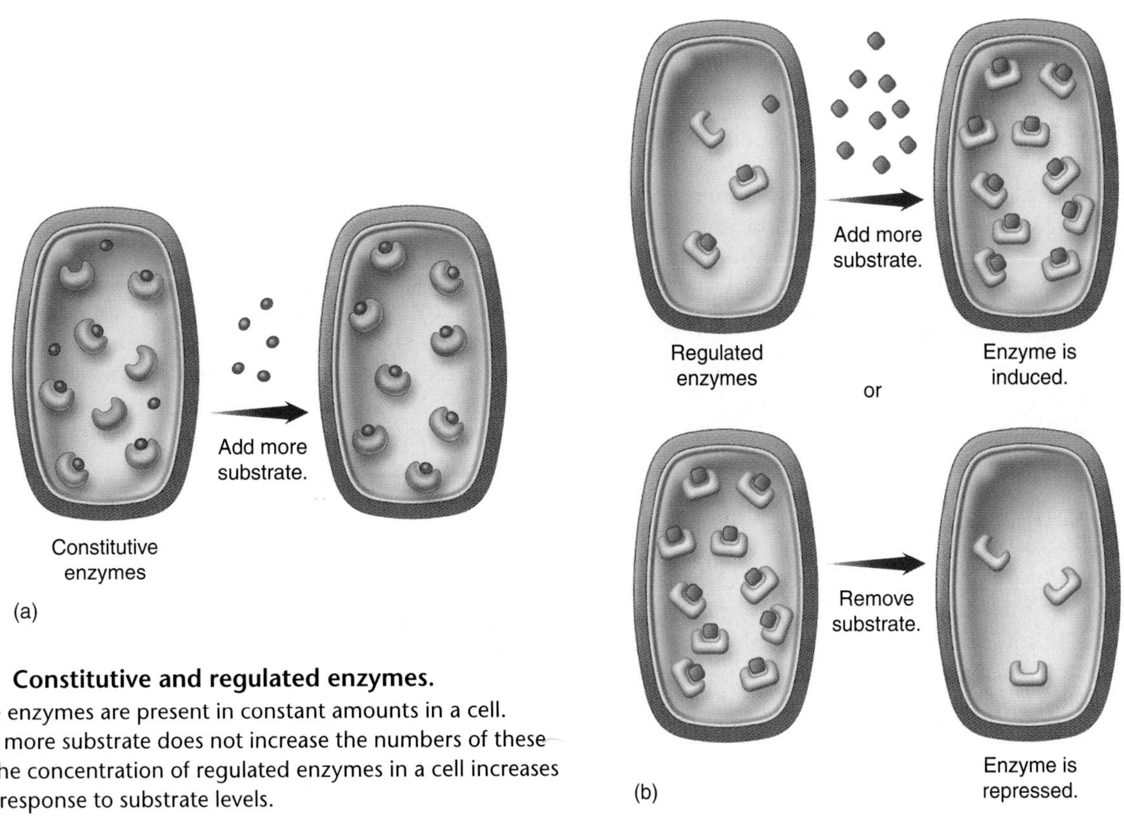

Figure 8.6 Constitutive and regulated enzymes.
(a) Constitutive enzymes are present in constant amounts in a cell. The addition of more substrate does not increase the numbers of these enzymes. **(b)** The concentration of regulated enzymes in a cell increases or decreases in response to substrate levels.

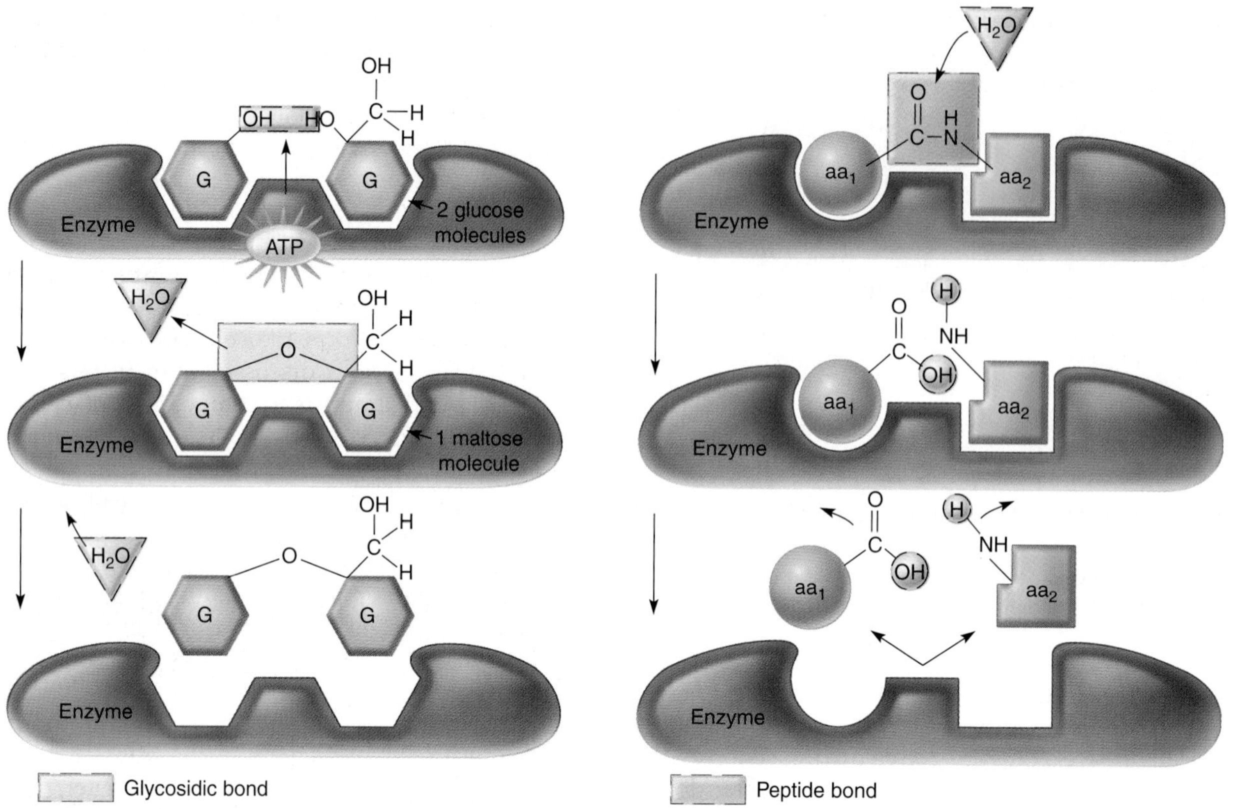

(a) **Dehydration Reaction.** Forming a glycosidic bond between two glucose molecules to generate maltose requires the removal of a water molecule and energy from ATP.

(b) **Hydrolysis Reaction.** Breaking a peptide bond between two amino acids requires a water molecule that adds an H and an OH to the amino acids.

Figure 8.7 Examples of enzyme-catalyzed synthesis and hydrolysis reactions.

The Role of Microbial Enzymes in Disease Many pathogens secrete unique exoenzymes that help them avoid host defenses or promote their multiplication in tissues. Because these enzymes contribute to pathogenicity, they are referred to as virulence factors, or toxins in some cases. *Streptococcus pyogenes* (a cause of throat and skin infections) produces a streptokinase that digests blood clots and apparently assists in invasion of wounds. Another exoenzyme from this bacterium is called streptolysin.[1] In mammalian hosts, streptolysin damages blood cells and tissues. It is also responsible for lysing red blood cells used in blood agar dishes, and this trait is used for identifying the bacteria growing in culture (see chapter 21). *Pseudomonas aeruginosa*, a respiratory and skin pathogen, produces elastase and collagenase, which digest elastin and collagen, two proteins found in connective tissue. These increase the severity of certain lung diseases and burn infections. *Clostridium perfringens*, an agent of gas gangrene, synthesizes lecithinase C, a lipase that profoundly damages cell membranes and accounts for the tissue death associated with this disease. Not all enzymes digest tissues; some, such as penicillinase, inactivate penicillin and thereby protect a microbe from its effects.

1. Even though most enzyme names end in *-ase* (see Insight 8.3), not all do.

The Sensitivity of Enzymes to Their Environment

The activity of an enzyme is highly influenced by the cell's environment. In general, enzymes operate only under the natural temperature, pH, and osmotic pressure of an organism's habitat. When enzymes are subjected to changes in these normal conditions, they tend to be chemically unstable, or **labile.** Low temperatures inhibit catalysis, and high temperatures denature the apoenzyme. **Denaturation** is a process by which the weak bonds that collectively maintain the native shape of the apoenzyme are broken. This disruption causes extreme distortion of the enzyme's shape and prevents the substrate from attaching to the active site (described in chapter 11). Such nonfunctional enzymes block metabolic reactions and thereby can lead to cell death. Low or high pH or certain chemicals (heavy metals, alcohol) are also denaturing agents.

Regulation of Enzymatic Activity and Metabolic Pathways

Metabolic reactions proceed in a systematic, highly regulated manner that maximizes the use of available nutrients and energy. The cell responds to environmental conditions by using those metabolic reactions that most favor growth and survival. Because enzymes are critical to these reactions, the regulation of metabolism is largely the regulation of enzymes by an elaborate system of checks and balances. Let us take a look at some general features of metabolic pathways.

Metabolic Pathways

Metabolic reactions rarely consist of a single action or step. More often, they occur in a multistep series or pathway, with each step catalyzed by an enzyme. An individual reaction is shown in various ways, depending on the purpose at hand **(figure 8.8).** The product of one reaction is often the reactant (substrate) for the next, forming a linear chain of reactions. Many pathways have branches that provide alternate methods for nutrient processing. Others take a cyclic form, in which the starting molecule is regenerated to initiate another turn of the cycle. Pathways generally do not stand alone; they are interconnected and merge at many sites.

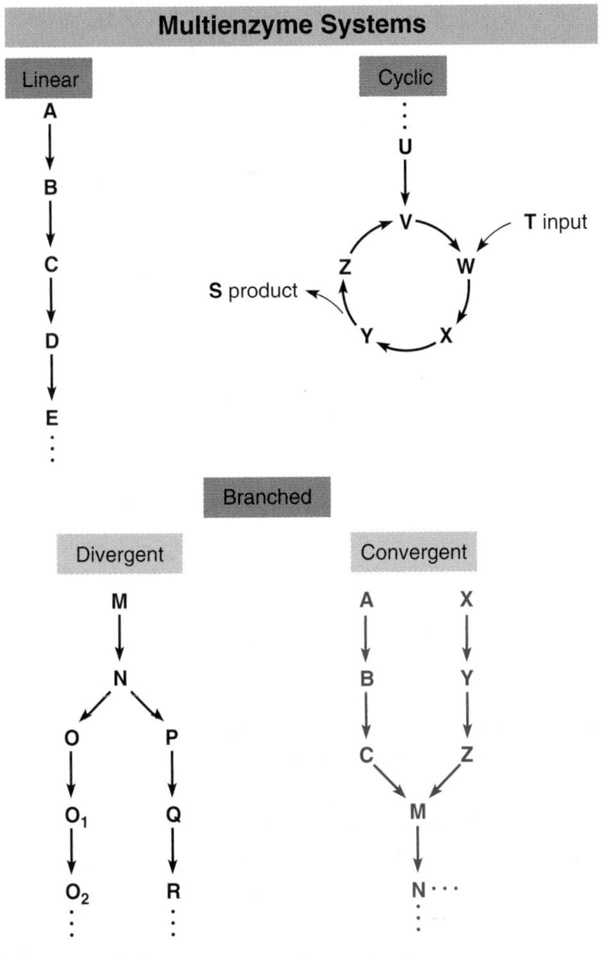

Figure 8.8 Patterns of metabolism.
In general, metabolic pathways consist of a linked series of individual chemical reactions that produce intermediary metabolites and lead to a final product. These pathways occur in several patterns, including linear, cyclic, and branched. Anabolic pathways involved in biosynthesis result in a more complex molecule, each step adding on a functional group, whereas catabolic pathways involve the dismantling of molecules and can generate energy. Virtually every reaction in a series—represented by an arrow—involves a specific enzyme.

Every pathway has one or more enzyme pacemakers (usually the first enzyme in the series) that set the rate of a pathway's progression. These enzymes respond to various control signals and, in so doing, determine whether a pathway proceeds. Regulation of pacemaker enzymes proceeds on two fundamental levels. Either the enzyme itself is directly inhibited or activated, or the amount of the enzyme in the system is altered (decreased or increased). Factors that affect the enzyme directly provide a means for the system to be finely controlled or tuned, whereas regulation at the genetic level (enzyme synthesis) provides a slower, less sensitive control.

Direct Controls on the Action of Enzymes

The bacterial cell has many ways of directly influencing the activity of its enzymes. It can inhibit enzyme activity by supplying a molecule that resembles the enzyme's normal substrate. The "mimic" can then occupy the enzyme's active site, preventing the actual substrate from binding there. Because the mimic cannot actually be acted on by the enzyme or function in the way the product would have, the enzyme is effectively shut down. This form of inhibition is called **competitive inhibition,** because the mimic is competing with the substrate for the binding site **(figure 8.9).** (In chapter 12, you will see that some antibiotics use the same strategy of competing with enzymatic active sites to shut down metabolic processes.)

Another form of inhibition can occur with special types of enzymes that have two binding sites—the active site and another area called the regulatory site (see figure 8.9). These enzymes are regulated by the binding of molecules other than the substrate in their regulatory sites. Often the regulatory molecule is the product of the enzymatic reaction itself. This provides a negative feedback mechanism that can slow down enzymatic activity once a certain concentration of product is produced. This is **noncompetitive inhibition,** because the regulator molecule does not bind in the same site as the substrate.

Controls on Enzyme Synthesis

Controlling enzymes by controlling their synthesis is another effective mechanism, because enzymes do not last indefinitely. Some wear out, some are deliberately degraded, and others are diluted with each cell division. For catalysis to continue, enzymes eventually must be replaced. This cycle works into the scheme of the cell, where replacement of enzymes can be regulated according to cell demand. The mechanisms of this system are genetic in nature; that is, they require regulation of DNA and the protein synthesis machinery, topics we shall encounter once again in chapter 9.

Enzyme repression is a means to stop further synthesis of an enzyme somewhere along its pathway. As the level of the end product from a given enzymatic reaction has built to excess, the genetic apparatus responsible for replacing these enzymes is automatically suppressed **(figure 8.10).** The

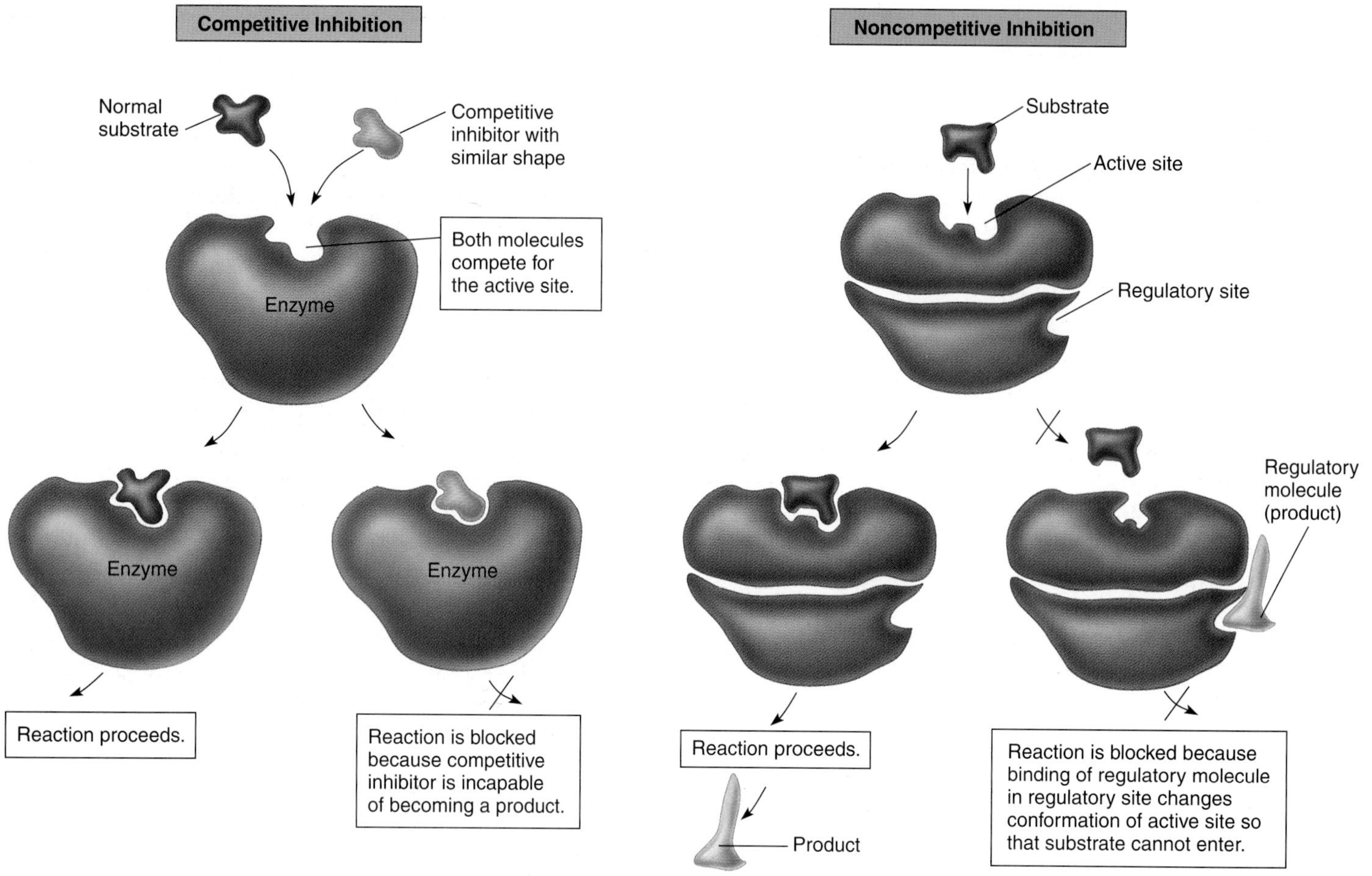

Competitive Inhibition
Normal substrate
Competitive inhibitor with similar shape
Both molecules compete for the active site.
Enzyme
Enzyme
Enzyme
Reaction proceeds.
Reaction is blocked because competitive inhibitor is incapable of becoming a product.
Noncompetitive Inhibition
Substrate
Active site
Regulatory site
Regulatory molecule (product)
Reaction proceeds.
Product
Reaction is blocked because binding of regulatory molecule in regulatory site changes conformation of active site so that substrate cannot enter.

Figure 8.9 **Examples of two common control mechanisms for enzymes.**

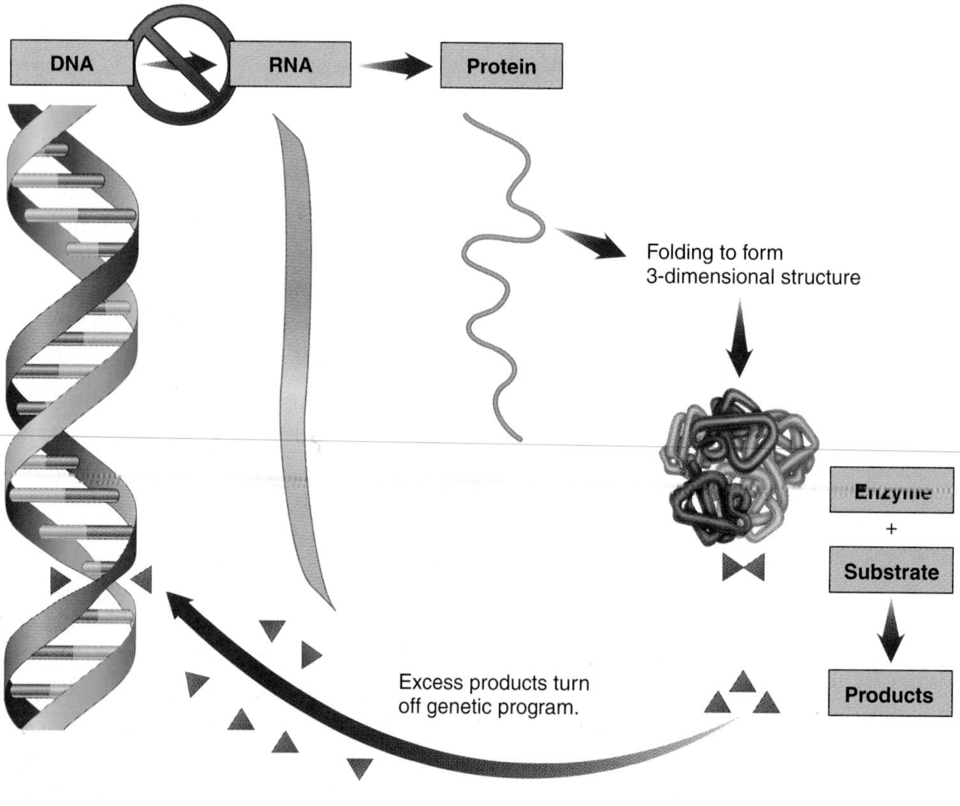

DNA
RNA
Protein
Folding to form 3-dimensional structure
Enzyme
+
Substrate
Products
Excess products turn off genetic program.

Figure 8.10 **One type of genetic control of enzyme synthesis: enzyme repression.**
The enzyme is synthesized continuously until enough product has been made, at which time the excess product reacts with a site on DNA that regulates the enzyme's synthesis, thereby inhibiting further enzyme production.

response time is longer than for feedback inhibition, but its effects are more enduring.

The inverse of enzyme repression is **enzyme induction.** In this process, enzymes appear (are induced) only when suitable substrates are present—that is, the synthesis of an enzyme is induced by its substrate. Both mechanisms are important genetic control systems in bacteria.

A classic model of enzyme induction occurs in the response of *Escherichia coli* to certain sugars. For example, if a particular strain of *E. coli* is inoculated into a medium whose principal carbon source is lactose, it will produce the enzyme lactase to hydrolyze it into glucose and galactose. If the bacterium is subsequently inoculated into a medium containing only sucrose as a carbon source, it will cease synthesizing lactase and begin synthesizing sucrase. This response enables the organism to adapt to a variety of nutrients, and it also prevents a microbe from wasting energy, making enzymes for which no substrates are present.

CHECKPOINT

- Metabolism includes all the biochemical reactions that occur in the cell. It is a self-regulating complex of interdependent processes that encompass many thousands of chemical reactions.
- Anabolism is the energy-requiring subset of metabolic reactions, which synthesize large molecules from smaller ones.
- Catabolism is the energy-releasing subset of metabolic reactions, which degrade or break down large molecules into smaller ones.
- Enzymes are proteins or RNA that catalyze all biochemical reactions by forming enzyme-substrate complexes. The binding of the substrate by an enzyme makes possible both bond-forming and bond-breaking reactions, depending on the pathway involved. Enzymes may utilize cofactors as carriers and activators.
- Enzymes are classified and named according to the kinds of reactions they catalyze.
- To function effectively, enzymes require specific conditions of temperature, pH, and osmotic pressure.
- Enzyme activity is regulated by processes that affect the enzymes directly, or by altering the amount of enzyme produced during protein synthesis.

8.2 The Pursuit and Utilization of Energy

In order to carry out the work of an array of metabolic processes, cells require constant input and expenditure of some form of usable energy. The energy comes directly from light or is contained in chemical bonds and released when substances are catabolized, or broken down. The energy is stored in ATP. For the most part, only chemical energy can routinely drive cell transactions, and chemical reactions are the universal basis of cellular energetics.

Energy in Cells

Cells manage energy in the form of chemical reactions that change molecules. This often involves activities such as the making or breaking of bonds and the transfer of electrons. Not all cellular reactions are equal with respect to energy. Some release energy, and others require it to proceed. For example, a reaction that proceeds as follows:

$$\text{X} + \text{Y} \xrightarrow{\text{Enzyme}} \text{Z} + \text{Energy}$$

releases energy as it goes forward. This type of reaction is termed **exergonic** (ex-er-gon′-ik). Energy of this type is considered free—it is available for doing cellular work. Energy transactions such as the following:

$$\text{Energy} + A + B \xrightarrow{\text{Enzyme}} C$$

are called **endergonic** (en-der-gon′-ik), because they are driven forward with the addition of energy. In cells, exergonic and endergonic reactions are often coupled, so that released energy is immediately put to use.

Summaries of metabolism might make it seem that cells "create" energy from nutrients, but they do not. What they actually do is extract chemical energy already present in nutrient fuels and apply that energy toward useful work in the cell, much like a gasoline engine releases energy as it burns fuel. The engine does not actually produce energy, but it converts some of the potential energy to do work.

At the simplest level, cells possess specialized enzyme systems that trap the energy present in the bonds of nutrients as they are progressively broken **(figure 8.11).** During exergonic reactions, energy released by bonds is stored in certain high-energy phosphate bonds such as in ATP. As we shall see, the ability of ATP to temporarily store and release the energy of chemical bonds fuels endergonic cell reactions. Before discussing ATP, we examine the process behind electron transfer: redox reactions.

A Closer Look at Biological Oxidation and Reduction

We stated earlier that biological systems often extract energy through **redox reactions.** Such reactions always occur in pairs, with an electron donor and an electron acceptor, which constitute a *redox pair*. The reaction can be represented as follows:

Electron donor	+ Electron acceptor	⟶ Electron donor	+ Electron acceptor
•e^-			•e^-
(Reduced)	(Oxidized)	(Oxidized)	(Reduced)

(Review Insight 2.2 for detailed figure.)

This process salvages electrons along with their inherent energy, and it changes the energy balance, leaving the previously reduced compound with less energy than the now oxidized one. The energy now present in the electron acceptor can be captured to **phosphorylate** (add an inorganic phosphate) to ADP or to some other compound. This

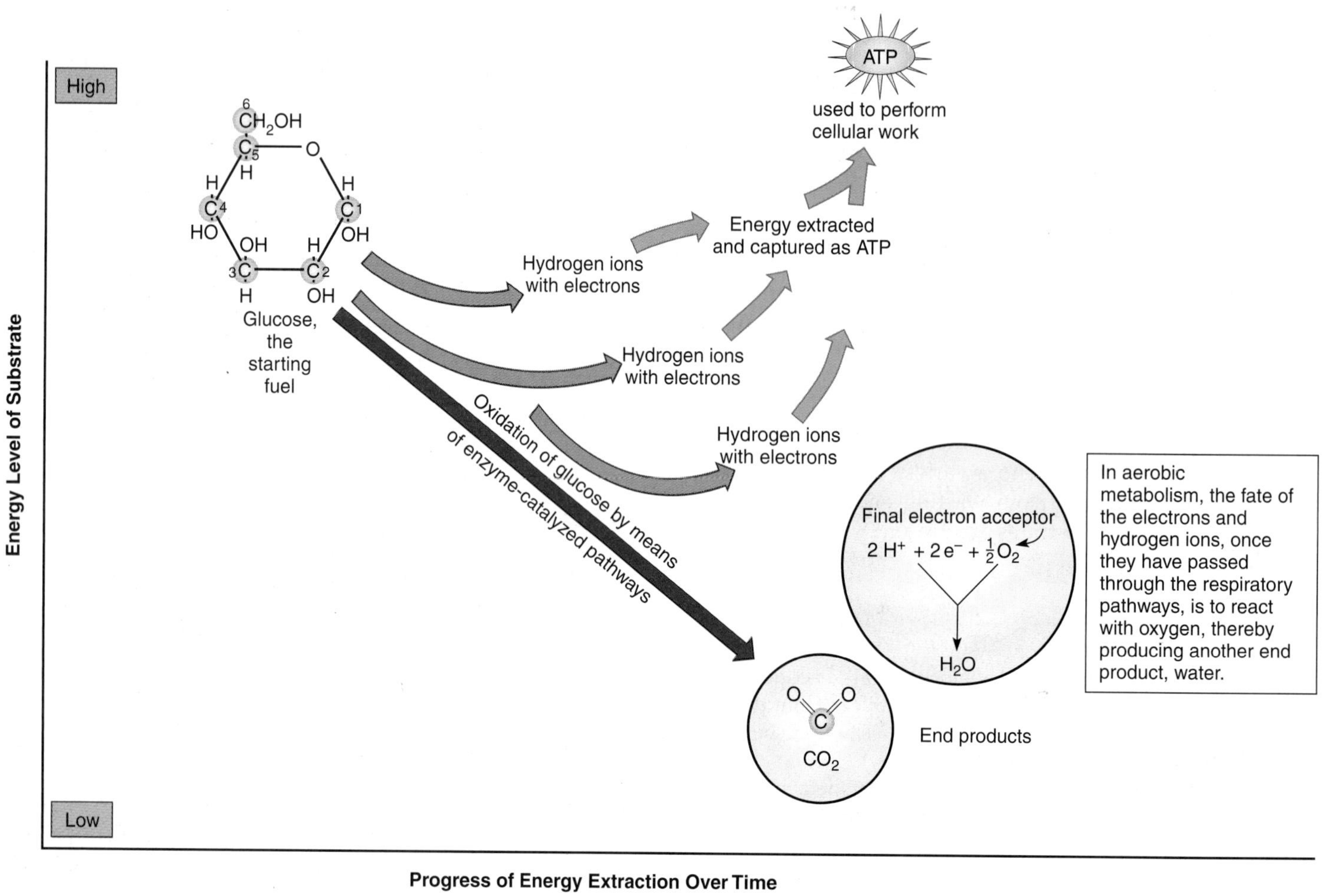

Figure 8.11 A simplified model of energy production.
The central events of cell energetics include the release of energy during the systematic dismantling of a fuel such as glucose. This is achieved by the shuttling of hydrogens and electrons to sites in the cell where their energy can be transferred to ATP. In aerobic metabolism, the final products are CO_2 and H_2O molecules.

process stores the energy in a high-energy molecule (ATP, for example). In many cases, the cell does not handle electrons as discrete entities but rather as parts of an atom such as hydrogen. For simplicity's sake, we will continue to use the term *electron transfer*, but keep in mind that hydrogens are often involved in the transfer process. The removal of hydrogens (a hydrogen atom consists of a single proton and a single electron) from a compound during a redox reaction is called dehydrogenation. The job of handling these protons and electrons falls to one or more carriers, which function as short-term repositories for the electrons until they can be transferred. As we shall see, dehydrogenations are an essential supplier of electrons for the respiratory electron transport system.

Electron Carriers: Molecular Shuttles

Electron carriers resemble shuttles that are alternately loaded and unloaded, repeatedly accepting and releasing electrons and hydrogens to facilitate the transfer of redox energy. Most carriers are coenzymes that transfer both electrons and hydrogens, but some transfer electrons only. The most common carrier is NAD (nicotinamide adenine dinucleotide), which carries hydrogens (and a pair of electrons) from dehydrogenation reactions **(figure 8.12).** Reduced NAD can be represented in various ways. Because 2 hydrogens are removed, the actual carrier state is NADH + H^+, but this is somewhat cumbersome, so we will represent it as "NADH." In catabolic pathways, electrons are extracted and carried through a series of redox reactions until the final electron acceptor at the end of a particular pathway is reached (see figure 8.11). In aerobic metabolism, this acceptor is molecular oxygen; in anaerobic metabolism, it is some other inorganic or organic compound. Other common redox carriers are FAD, NADP (NAD phosphate), and the compounds of the respiratory chain, which are fixed into membranes.

Adenosine Triphosphate: Metabolic Money

In what ways do cells extract chemical energy from electrons, store it, and then tap the storage sources? To answer these

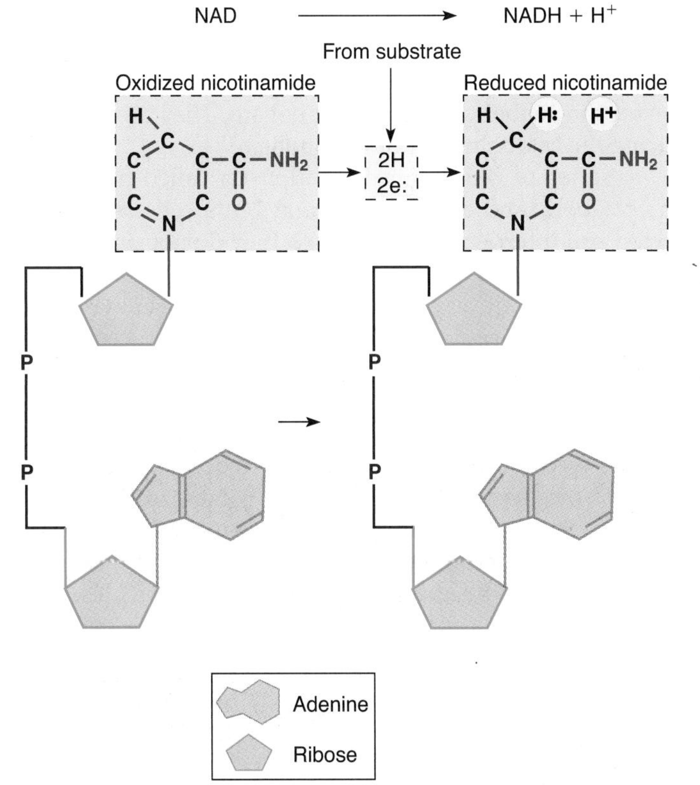

Figure 8.12 Details of NAD reduction.
The coenzyme NAD contains the vitamin nicotinamide (niacin) and the purine adenine attached to double ribose phosphate molecules (a dinucleotide). The principal site of action is on the nicotinamide (boxed areas). Hydrogens and electrons donated by a substrate interact with a carbon on the top of the ring. One hydrogen bonds there, carrying two electrons (H:), and the other hydrogen is carried in solution as H^+ (a proton).

questions, we must look more closely at the powerhouse molecule, adenosine triphosphate. ATP has also been described as metabolic money because it can be earned, banked, saved, spent, and exchanged. As a temporary energy repository, ATP provides a connection between energy-yielding catabolism and the other cellular activities that require energy. Some clues to its energy-storing properties lie in its unique molecular structure.

The Molecular Structure of ATP

ATP is a three-part molecule consisting of a nitrogen base (adenine) linked to a 5-carbon sugar (ribose), with a chain of three phosphate groups bonded to the ribose **(figure 8.13).** The type, arrangement, and especially the proximity of atoms in ATP combine to form a powerful high-energy molecule. The high energy of ATP originates in the orientation of the phosphate groups, which are relatively bulky and carry negative charges. The proximity of these repelling electrostatic charges imposes a strain that is most acute on the bonds between the last two phosphate groups. The strain on the phosphate bonds accounts for the energetic quality

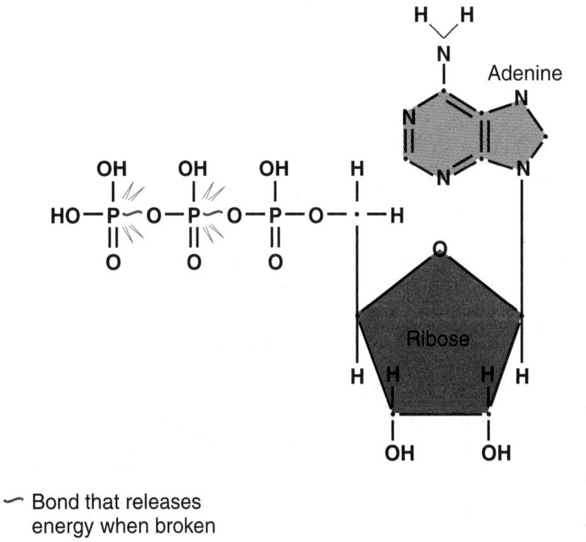

Figure 8.13 The structure of adenosine triphosphate (ATP).
Removing the left-most phosphate group yields ADP; removing the next one yields AMP.

of ATP because removal of the terminal phosphates releases free energy.

Breaking the bonds between two successive phosphates of ATP yields adenosine diphosphate (ADP), which is then converted to adenosine monophosphate (AMP). AMP derivatives help form the backbone of RNA and are also a major component of certain coenzymes (NAD, FAD, and coenzyme A).

The Metabolic Role of ATP

ATP is the primary energy currency of the cell; and when it is used in a chemical reaction, it must then be replaced. Therefore, ATP utilization and replenishment is an ongoing cycle. In many instances, the energy released during ATP hydrolysis powers biosynthesis by activating individual subunits before they are enzymatically linked together. ATP is also used to prepare molecules for catabolism such as the phosphorylation of a 6-carbon sugar during the early stages of glycolysis.

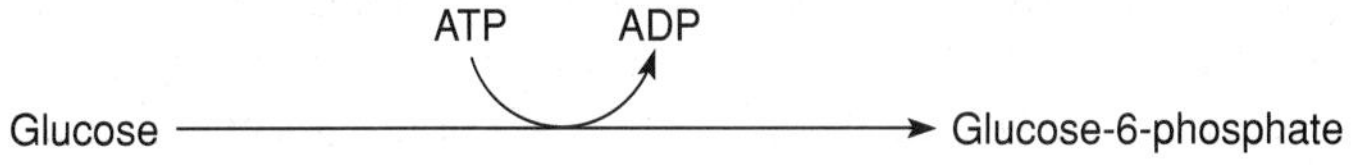

When ATP is utilized by the removal of the terminal phosphate to release energy plus ADP, ATP then needs to be re-created. The reversal of this process, that is, adding the terminal phosphate to ADP, will replenish ATP, but it requires an input of energy.

$$\text{ATP} \leftrightarrows \text{ADP} + P_i + \text{Energy}$$

In heterotrophs, the energy infusion that regenerates a high-energy phosphate comes from certain steps of catabolic pathways, in which nutrients such as carbohydrates are degraded

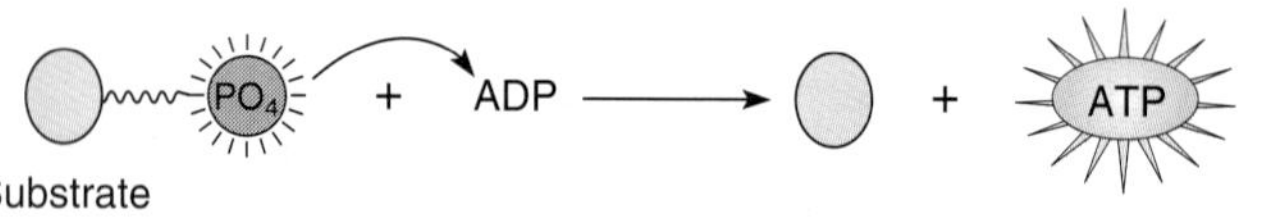

Figure 8.14 ATP formation by substrate-level phosphorylation.
The inorganic phosphate and the substrates form a bond with high potential energy. In a reaction catalyzed enzymatically, the phosphate is transferred to ADP, thereby producing ATP.

and yield energy. ATP is formed when substrates or electron carriers provide a high-energy phosphate that becomes bonded to ADP. Some ATP molecules are formed through *substrate-level phosphorylation.* In substrate-level phosphorylation, ATP is formed by transfer of a phosphate group from a phosphorylated compound (substrate) directly to ADP to yield ATP **(figure 8.14).**

Other ATPs are formed through *oxidative phosphorylation,* a series of redox reactions occurring during the final phase of the respiratory pathway. Phototrophic organisms have a system of *photophosphorylation,* in which the ATP is formed through a series of sunlight-driven reactions (discussed in more detail in chapter 24).

CHECKPOINT

- All metabolic processes require the constant input and expenditure of some form of usable energy. Chemical energy is the currency that runs the metabolic processes of the cell.
- Chemical energy is obtained from the electrons of nutrient molecules through catabolism. It is used to perform the cellular "work" of biosynthesis, movement, membrane transport, and growth.
- Energy is extracted from nutrient molecules by redox reactions. A redox pair of substances passes electrons and hydrogens between them. The donor substance loses electrons, becoming oxidized. The acceptor substance gains electrons, becoming reduced.
- ATP is an important energy molecule of the cell. It donates free energy to anabolic reactions and is continuously regenerated by three phosphorylation processes: substrate-level phosphorylation, oxidative phosphorylation, and (in certain organisms) photophosphorylation.

8.3 The Pathways

Now you have an understanding of all the tools a cell needs to *metabolize.* Metabolism uses *enzymes* to catalyze reactions that break down (*catabolize*) organic molecules to materials (*precursor molecules*) that cells can then use to build (*anabolize*) larger, more complex molecules that are particularly suited to them. Figure 8.1 represents this process symbolically. Another very important point about metabolism is that *reducing power* (the electrons available in NADH and $FADH_2$) and *energy* (stored in the bonds of ATP) are needed in large quantities for the anabolic parts of metabolism. They are produced during the catabolic part of metabolism.

A series of biochemical reactions is called a pathway. The catabolic and anabolic pathways in a cell are interconnected and interdependent, though they do not simply work "backward and forward." It might seem more economical to use identical pathways, but having different enzymes and reactants allows anabolism and catabolism to proceed simultaneously without interference.

Metabolism starts with "nutrients" from the environment, usually discarded molecules from other organisms. Cells have to get the nutrients inside; to do this, they use the mechanisms discussed in chapter 7. Some of these require energy, which is available from catabolism already occurring in the cell. In the next step, intracellular nutrients have to be broken down to the appropriate precursor molecules. These catabolic pathways are discussed next.

Catabolism: Getting Materials and Energy

Nutrient processing is extremely varied, especially in bacteria, yet in most cases it is based on three basic catabolic pathways. Frequently, the nutrient is glucose. There are several pathways that can be used to break down glucose, but the most common one is **glycolysis** (gly-kol'-ih-sis). In previous discussions, microorganisms were categorized according to their requirement for oxygen gas, and this requirement is related directly to their mechanisms of energy release. **Figure 8.15** provides an overview of the three major pathways for producing the needed precursors and energy (i.e., catabolism).

As we shall see, **aerobic respiration** is a series of reactions (glycolysis, the Krebs[2] cycle, and the respiratory chain) that convert glucose to CO_2 and allows the cell to recover significant amounts of energy (review figure 8.11). Aerobic respiration relies on free oxygen as the final acceptor for electrons and hydrogens and produces a relatively large amount of ATP. Aerobic respiration is characteristic of many bacteria, fungi, protozoa, and animals, and it is the system we emphasize here. Facultative and aerotolerant anaerobes may use only the glycolysis scheme to incompletely oxidize (ferment) glucose. In this case, oxygen is not required, organic compounds are the final electron acceptors, and a relatively small amount of ATP is produced. While the growth of aerobic bacteria is usually limited by the availability of substrates, the growth of anaerobes is likely to be stopped when final electron acceptors

2. Krebs is in honor of Sir Hans Krebs who, with F. A. Lipmann, delineated this pathway, an achievement for which they won the Nobel Prize in Physiology or Medicine in 1953.

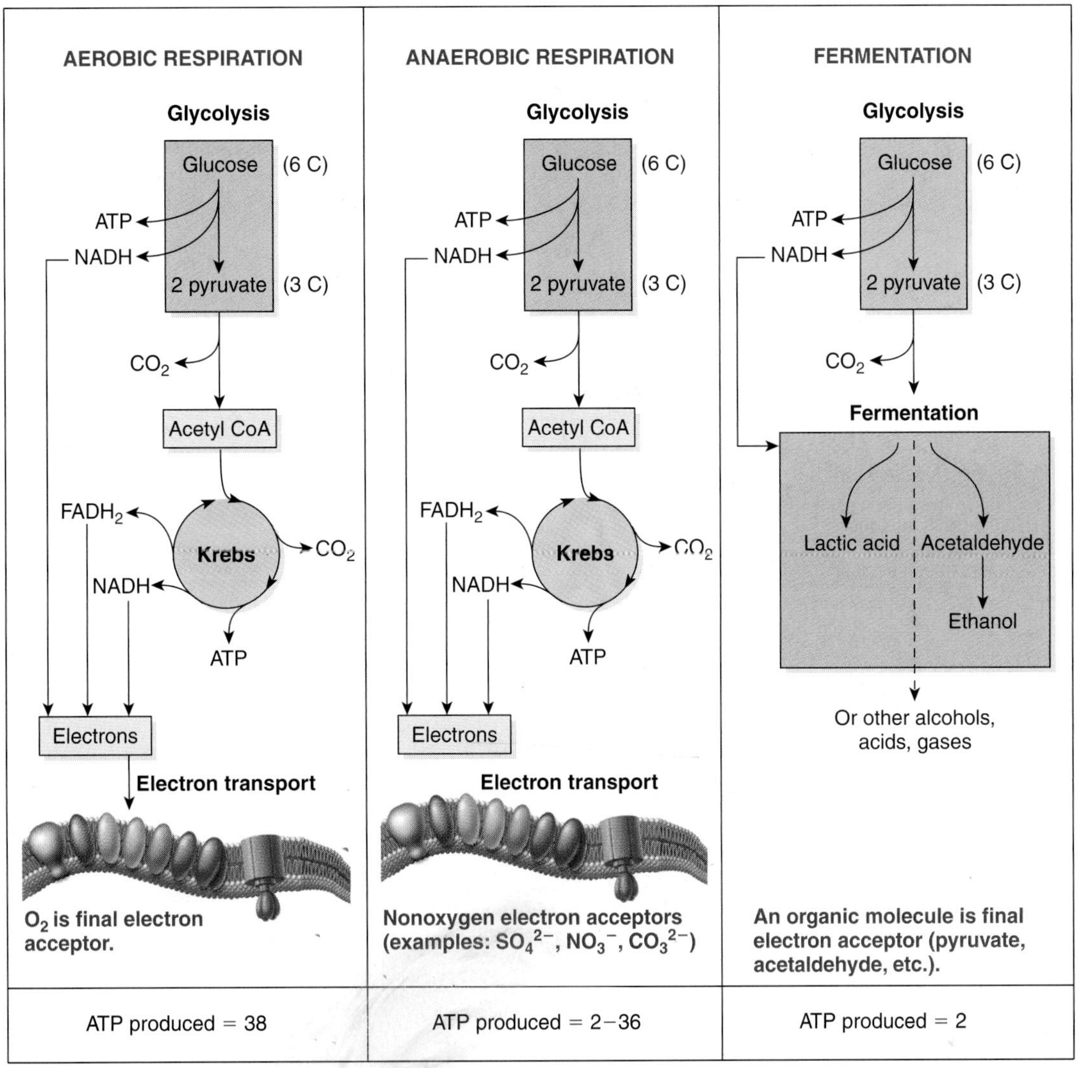

Figure 8.15 Summary of the most common pathways of glucose metabolism.
Glycolysis is the most common first step of metabolism. It yields two molecules of pyruvate as well as two ATPs and two NADHs. These products feed into either the Krebs cycle or fermentation.

run out. Some strictly anaerobic microorganisms metabolize by means of **anaerobic respiration.** This system involves the same three pathways as aerobic respiration, but it does not use molecular oxygen as the final electron acceptor; instead, NO_3^-, SO_4^{2-}, CO_3^{3-}, and other oxidized compounds are utilized. Aspects of fermentation and anaerobic respiration are covered in subsequent sections of this chapter.

Aerobic Respiration

Aerobic respiration is a series of enzyme-catalyzed reactions in which electrons are transferred from fuel molecules such as glucose to oxygen as a final electron acceptor. This pathway is the principal energy-yielding scheme for aerobic heterotrophs, and it provides both ATP and metabolic intermediates for many other pathways in the cell, including those of protein, lipid, and carbohydrate synthesis (see figure 8.15).

Glucose: The Starting Compound

Carbohydrates such as glucose are good fuels because these compounds are readily oxidized; that is, they are superior hydrogen and electron donors. The enzymatic withdrawal of hydrogen from them also removes electrons that can be used in energy transfers. The end products of the conversion of these carbon compounds are energy-rich ATP and energy-poor carbon dioxide and water. Polysaccharides (starch, glycogen) and disaccharides (maltose, lactose) are stored sources of glucose for the respiratory pathways. Although we use glucose as the main starting compound, other hexoses (fructose, galactose) and fatty acid subunits can enter the pathways of aerobic respiration as well.

Glycolysis: The Starting Lineup

Glycolysis enzymatically converts glucose through several steps into pyruvic acid. Depending on the organism and

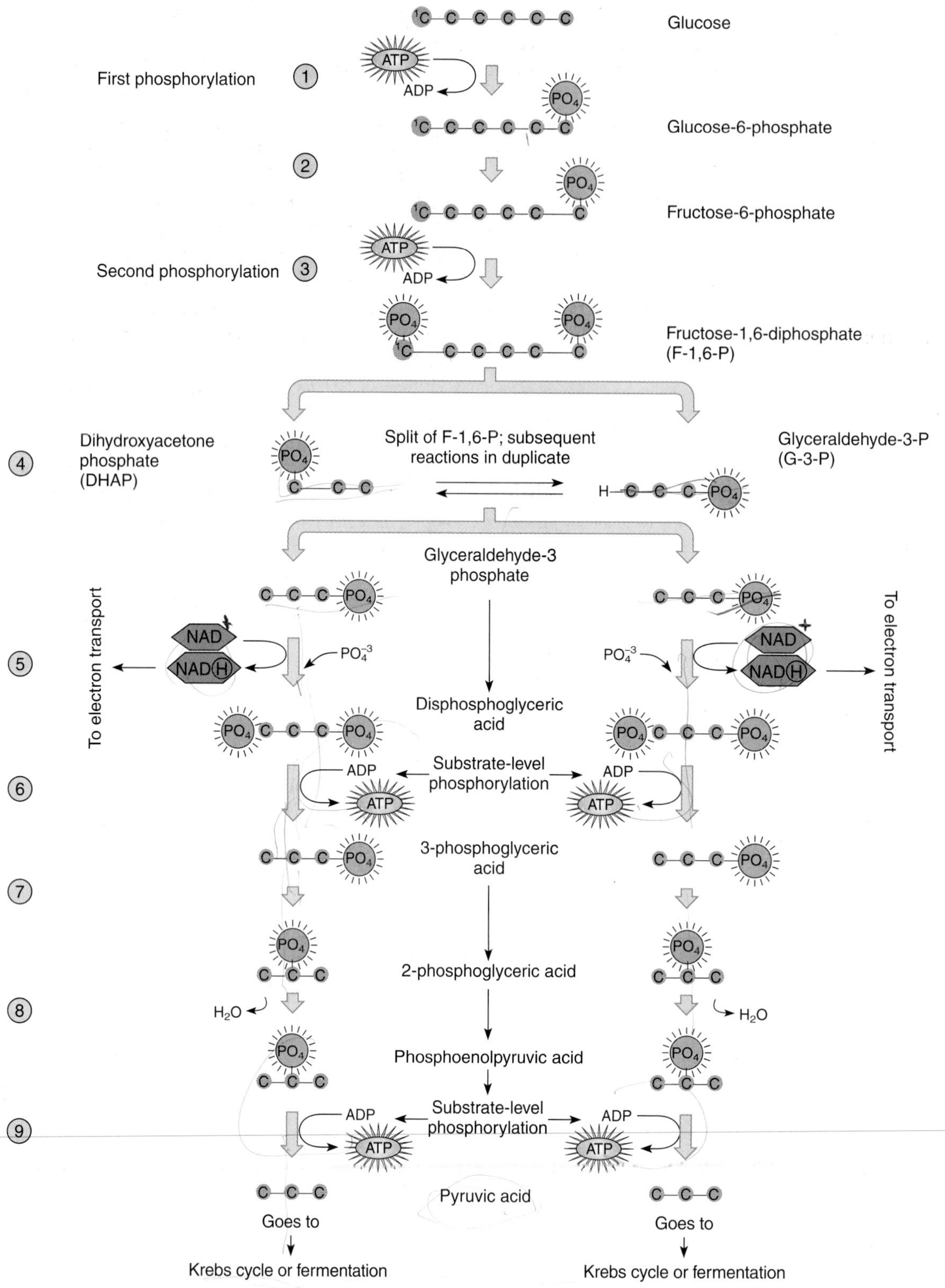

Process Figure 8.16 Summary of glycolysis.

the conditions, it may be only the first phase of aerobic respiration, or it may serve as the primary metabolic pathway (fermentation). Glycolysis provides a significant means to synthesize a small amount of ATP anaerobically and also to generate pyruvic acid, an essential intermediary metabolite.

Glycolysis proceeds along nine steps, starting with glucose and ending with pyruvic acid (pyruvate[3]). An overview of glycolysis will be presented here; **figure 8.16** contains the

3. In biochemistry, the terms used for organic acids appear as either the acid form (pyruvic acid) or its salt (pyruvate).

chemical structures and a visual representation of the reactions. Each of the nine reactions is catalyzed by a specific enzyme with a specific name, but we will not mention them here.

First, glucose is activated by adding a phosphate to it, resulting in glucose-6-phosphate. It is then converted (another reaction, another enzyme) to fructose-6-phosphate, and another phosphate is added. The resulting molecule—fructose diphosphate—is more symmetrical and can be split into two 3-carbon molecules **(figure 8.16, step 4).** At this point, no oxidation-reduction has occurred and, in fact, 2 ATPs have been used. The two 3-carbon molecules are isomers of each other, and the next step involves converting the one (DHAP) to glyceraldehyde-3-P (G-3-P), resulting in two G-3-Ps.

From here to the end, everything that happens in glycolysis happens twice—once to each of the 3-C molecules. First, the G-3-Ps each receive another phosphate. At the same time, 2 NADs in the vicinity are reduced to NADHs. These NADHs will be used in the last step of catabolism (the electronic transport system) to produce ATP.

In the last four steps of glycolysis **(figure 8.16, steps 6–9),** the 3-carbon molecule is manipulated enzymatically to donate both of its phosphates to ADPs via substrate-level phosphorylation. This results in four new ATPs and two 3-carbon molecules with no phosphates, called pyruvic acid. But because 2 ATPs were expended in the early steps of glycolysis, the net yield of ATP from glycolysis of one glucose molecule is 2.

Pyruvic Acid—A Central Metabolite

Pyruvic acid occupies an important position in several pathways, and different organisms handle it in different ways **(figure 8.17).** In strictly aerobic organisms and some anaerobes, pyruvic acid enters the Krebs cycle for further processing and energy release. Facultative anaerobes can adopt a fermentative metabolism, in which pyruvic acid is re-reduced into acids or other products.

The Krebs Cycle—A Carbon and Energy Wheel

As you have seen, the oxidation of glucose yields a comparatively small amount of energy and gives off pyruvic acid. Pyruvic acid is still energy-rich, containing a number of extractable hydrogens and electrons to power ATP synthesis, but this can be achieved only through the work of the second and third phases of respiration, in which pyruvic acid's hydrogens are transferred to oxygen, producing CO_2 and H_2O. In the following section, we examine the next phase of this process, which takes place in the cytoplasm of bacteria and in the mitochondrial matrix in eukaryotes.

To connect the glycolysis pathway to the Krebs cycle, for either aerobic or anaerobic respiration, the pyruvic acid is first converted to a starting compound for that cycle **(Insight 8.3).** This step involves the first oxidation-reduction reaction of this phase of respiration, and it also releases the first carbon dioxide molecule. It involves a cluster of enzymes and coenzyme A that participate in the dehydrogenation (oxidation) of pyruvic acid, the reduction of NAD to NADH, and the decarboxylation of pyruvic acid to a 2-carbon acetyl group. The acetyl group remains attached to coenzyme A, forming acetyl coenzyme A (acetyl CoA) that feeds into the Krebs cycle.

The NADH formed during this reaction will be shuttled into electron transport and used to generate ATP via oxidative phosphorylation. **Keep in mind that all reactions described actually happen twice for each glucose because of the two pyruvates that are released during glycolysis.**

The Krebs cycle as depicted in the box figure always looks intimidating. Think of it as a series of eight reactions catalyzed by eight different enzymes. The detailed narrative of how it works is in Insight 8.3 and we will summarize it here. The Krebs cycle serves to transfer the energy stored in acetyl CoA to NAD^+ and FAD by reducing them (transferring hydrogen ions to them). Thus, the main products of the

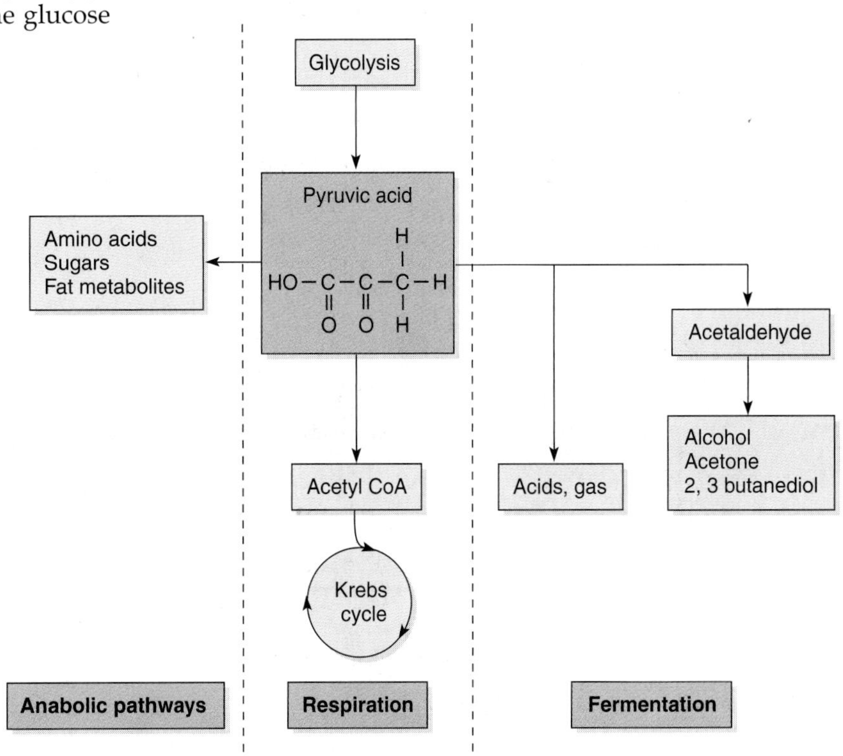

Figure 8.17 The fates of pyruvic acid (pyruvate).
This metabolite is an important hub in the processing of nutrients by microbes. It may be fermented anaerobically to several end products or oxidized completely to CO_2 and H_2O through the Krebs cycle and the electron transport system. It can also serve as a source of raw material for synthesizing amino acids and carbohydrates.

INSIGHT 8.3 Discovery

Steps in the Krebs Cycle

The reactions of a single turn of the Krebs cycle. Each glucose will produce two spins. Note that this is an enlarged, more detailed view of the middle phase depicted in figure 8.16.

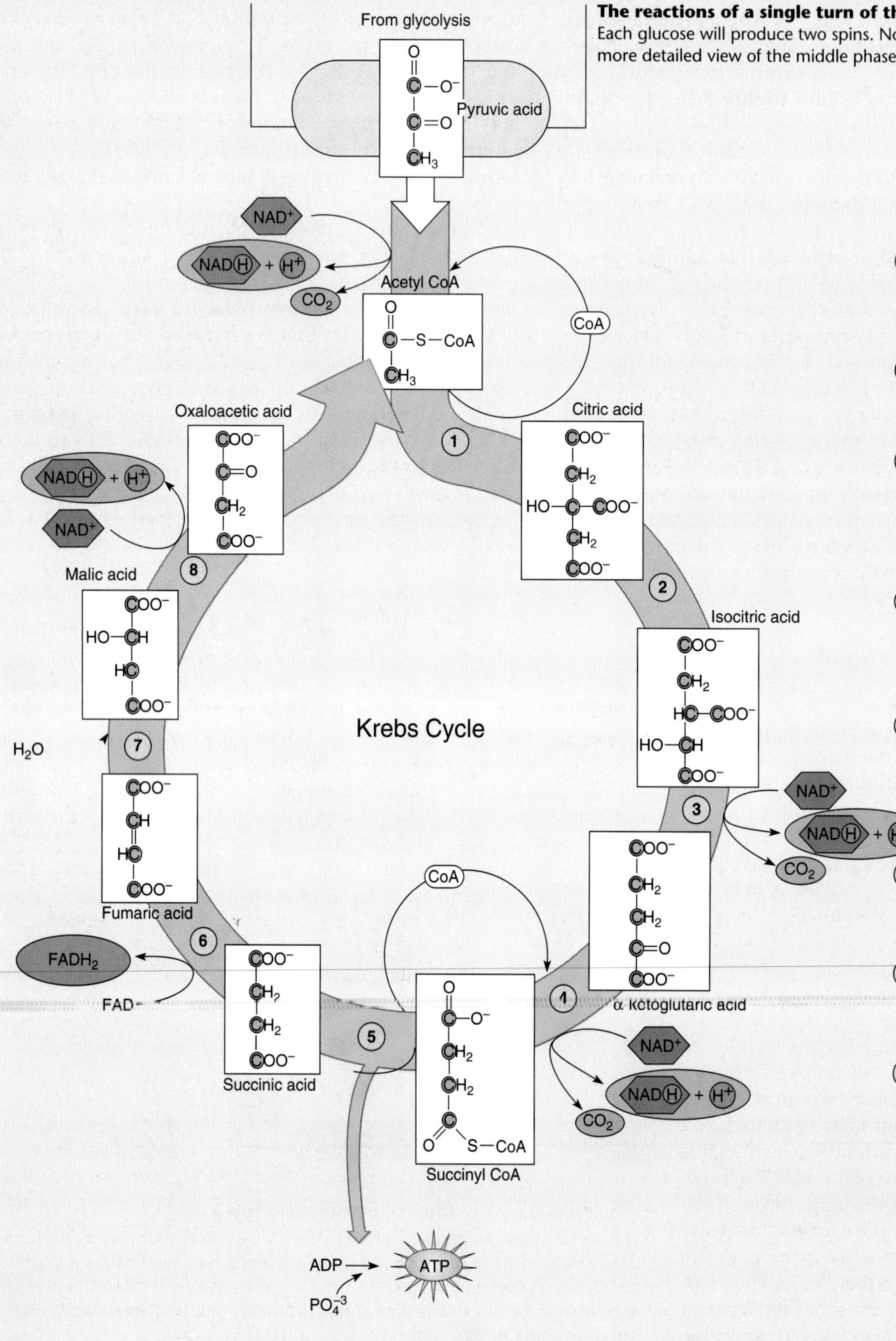

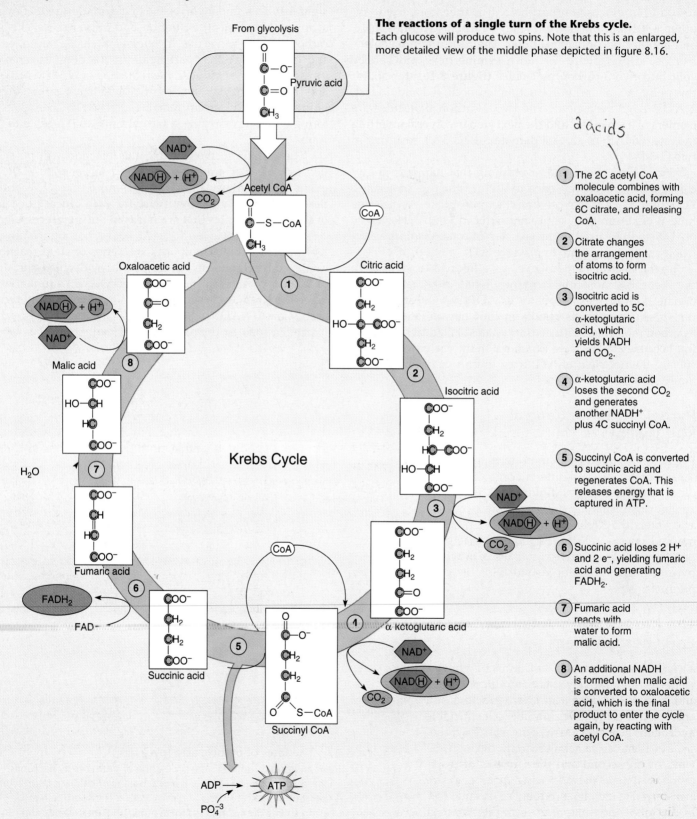

1. The 2C acetyl CoA molecule combines with oxaloacetic acid, forming 6C citrate, and releasing CoA.
2. Citrate changes the arrangement of atoms to form isocitric acid.
3. Isocitric acid is converted to 5C α-ketoglutaric acid, which yields NADH and CO_2.
4. α-ketoglutaric acid loses the second CO_2 and generates another $NADH^+$ plus 4C succinyl CoA.
5. Succinyl CoA is converted to succinic acid and regenerates CoA. This releases energy that is captured in ATP.
6. Succinic acid loses 2 H^+ and 2 e^-, yielding fumaric acid and generating $FADH_2$.
7. Fumaric acid reacts with water to form malic acid.
8. An additional NADH is formed when malic acid is converted to oxaloacetic acid, which is the final product to enter the cycle again, by reacting with acetyl CoA.

As you learned earlier, a cyclic pathway is one in which the starting compound is regenerated at the end. The Krebs cycle has eight steps, beginning with citric acid formation and ending with oxaloacetic acid. As we take a single spin around the Krebs cycle, it will be helpful to keep track of

1. the numbers of carbons (#C) of each substrate and product,
2. reactions where CO_2 is generated,
3. the involvement of the electron carriers NAD and FAD, and
4. the site of ATP synthesis.

The reactions in the Krebs cycle follow:

1. Oxaloacetic acid (oxaloacetate; 4C) reacts with the acetyl group (2C) on acetyl CoA, thereby forming citric acid (citrate; 6C) and releasing coenzyme A so it can join with another acetyl group.
2. Citric acid is converted to its isomer, isocitric acid (isocitrate; 6C), to prepare this substrate for the decarboxylation and dehydrogenation of the next step.
3. Isocitric acid is acted upon by an enzyme complex including NAD or NADP (depending on the organism) in a reaction that generates NADH or NADPH, splits off a carbon dioxide, and leaves α-ketoglutaric acid (α-ketoglutarate; 5C).
4. Alpha-ketoglutaric acid serves as a substrate for the last decarboxylation reaction and yet another redox reaction involving coenzyme A and yielding NADH. The product is the high-energy compound succinyl CoA (4C).

At this point, the cycle has completed the formation of 3 CO_2 molecules that balance out the original 3-carbon pyruvic acid that began the Krebs. The remaining steps are needed not only to regenerate the oxaloacetic acid to start the cycle again but also to extract more energy from the intermediate compounds leading to oxaloacetic acid.

5. Succinyl CoA is the source of the one substrate level phosphorylation in the Krebs cycle. In most microbes, it proceeds with the formation of ATP. The product of this reaction is succinic acid (succinate; 4C).
6. Succinic acid next becomes dehydrogenated, but in this case, the electron and H^+ acceptor is flavin adenine dinucleotide (FAD). The enzyme that catalyzes this reaction, succinyl dehydrogenase, is found in the bacterial cell membrane and mitochondrial crista of eukaryotic cells. $FADH_2$ then directly enters the electron transport system. Fumaric acid (fumarate; 4C) is the product of this reaction.
7. The addition of water to fumaric acid (called hydration) results in malic acid (malate; 4C). This is one of the few reactions in respiration that directly incorporates water.
8. Malic acid is dehydrogenated (with formation of a final NADH), and oxaloacetic acid is formed. This step brings the cycle back to its original starting position, where oxaloacetic acid can react with acetyl coenzyme A.

Krebs cycle are these reduced molecules (as well as 2 ATPs for each glucose molecule). The reduced coenzymes NADH and $FADH_2$ are vital to the energy production that will occur in electron transport. Along the way the 2-carbon acetyl CoA joins with a 4-carbon compound, oxaloacetic acid, and then participates in seven additional chemical transformations while "spinning off" the NADH and $FADH_2$. That's why we called the Krebs cycle the "carbon and energy wheel" in the preceding heading.

The Respiratory Chain: Electron Transport and Oxidative Phosphorylation

We now come to the energy chain, which is the final "processing mill" for electrons and hydrogen ions and the major generator of ATP. Overall, the electron transport system (ETS) consists of a chain of special redox carriers that receives electrons from reduced carriers (NADH, $FADH_2$) generated by glycolysis and the Krebs cycle and passes them in a sequential and orderly fashion from one redox molecule to the next (see figure 8.15). The flow of electrons down this chain is highly energetic and allows the active transport of hydrogen ions to the outside of the membrane where the respiratory chain is located. The step that finalizes the transport process is the acceptance of electrons and hydrogen by oxygen, producing water. Some variability exists from one organism to another, but the principal compounds that carry out these complex reactions are NADH dehydrogenase, flavoproteins, coenzyme Q (ubiquinone), and **cytochromes** (sy'-toh-krohm). The cytochromes contain a tightly bound metal atom at their center that is actively involved in accepting electrons and donating them to the next carrier in the series. The highly compartmentalized structure of the respiratory chain is an important factor in its function. Note in **figure 8.18** that the electron transport carriers and enzymes are embedded in the inner mitochondrial membranes in eukaryotes. The equivalent structure for housing them in bacteria is the cell membrane pictured in **figure 8.19.** We will describe the electron transport system in both eukaryotes and prokaryotes.

Elements of Electron Transport: The Energy Cascade

The principal questions about the electron transport system are: How are the electrons passed from one carrier to another in the series? How is this progression coupled to ATP synthesis? and Where and how is oxygen utilized? Although the biochemical details of this process are rather complicated, the basic reactions consist of a number of redox reactions now familiar to us. In general, the carrier compounds and their enzymes are arranged in linear sequence and are reduced and oxidized in turn.

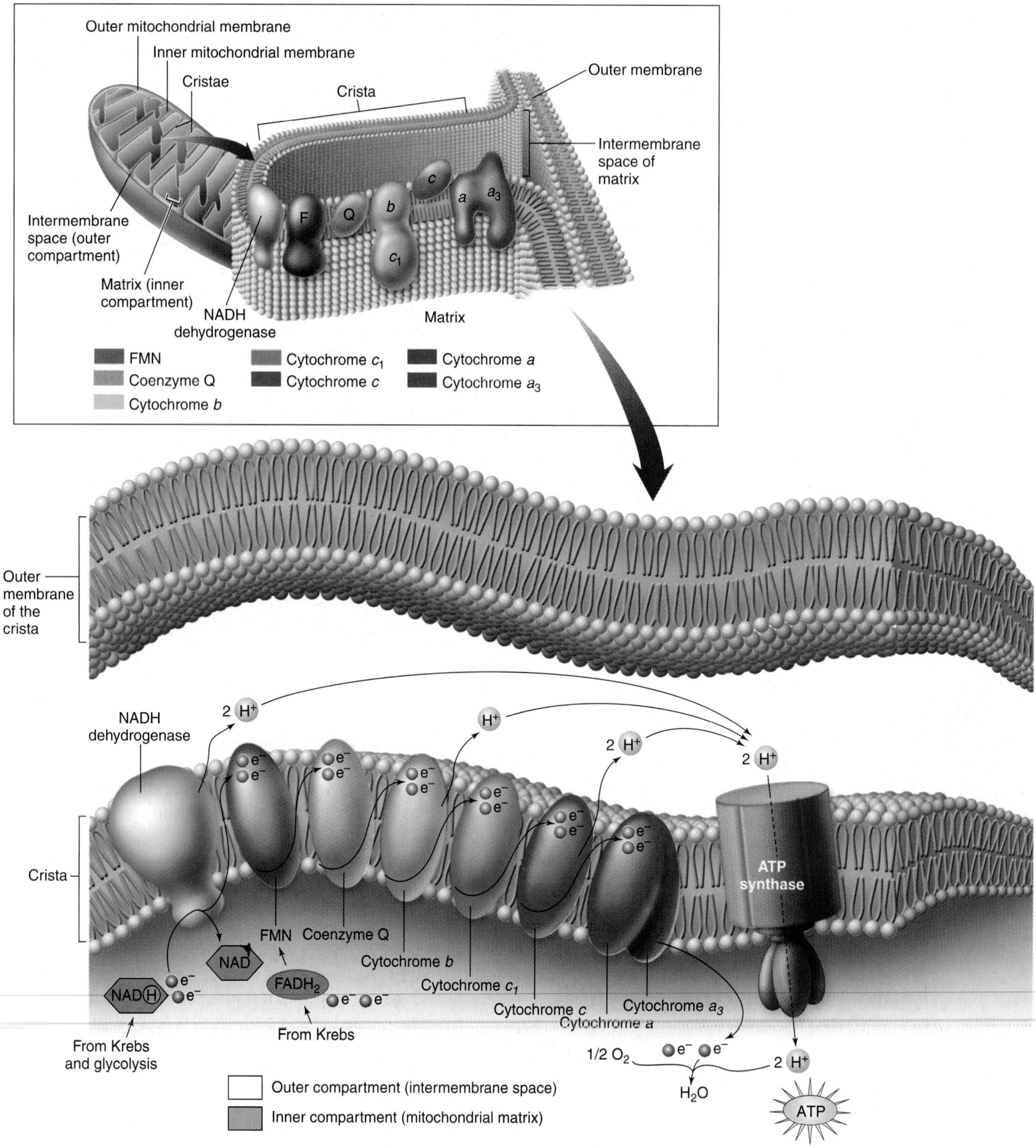

Figure 8.18 The electron transport system and oxidative phosphorylation on the mitochondrial crista.
Starting at NADH dehydrogenase, electrons brought in from the Krebs cycle by NADH are passed along the chain of electron transport carriers. Each adjacent pair of transport molecules undergoes a redox reaction. Coupled to the transport of electrons is the simultaneous active transport of H^+ into the outer compartment by specific carriers. These processes set the scene for ATP synthesis and final H^+ and e^- acceptance by oxygen (see figure 8.19 for details).

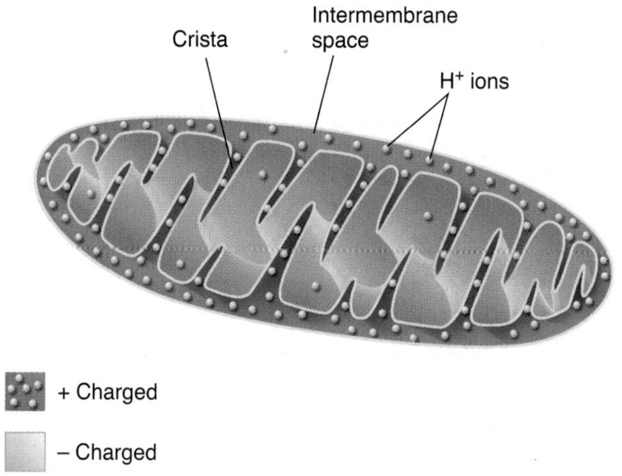

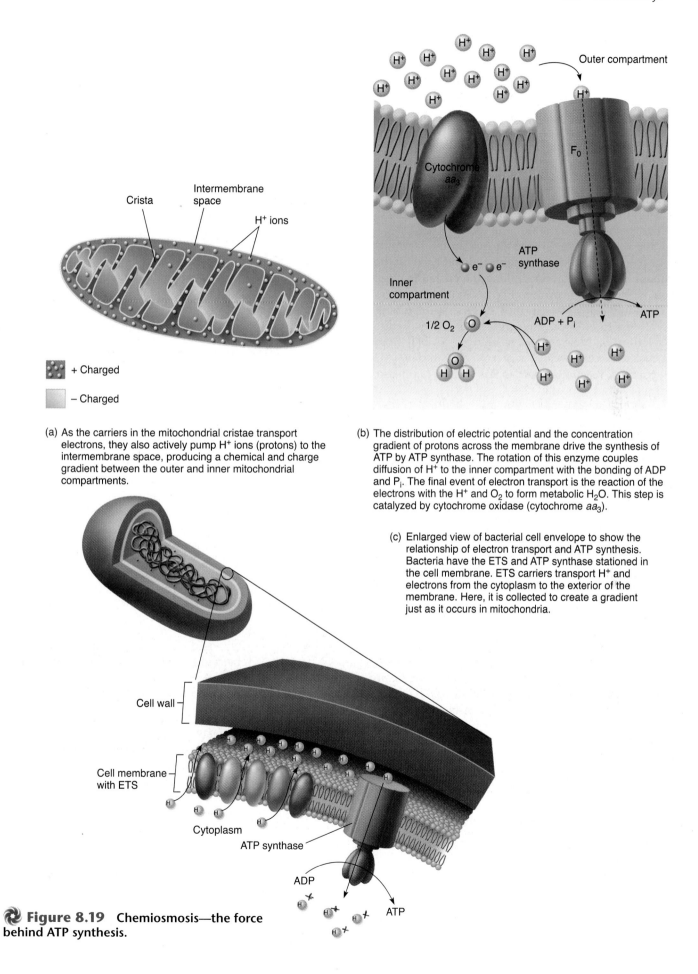

(a) As the carriers in the mitochondrial cristae transport electrons, they also actively pump H^+ ions (protons) to the intermembrane space, producing a chemical and charge gradient between the outer and inner mitochondrial compartments.

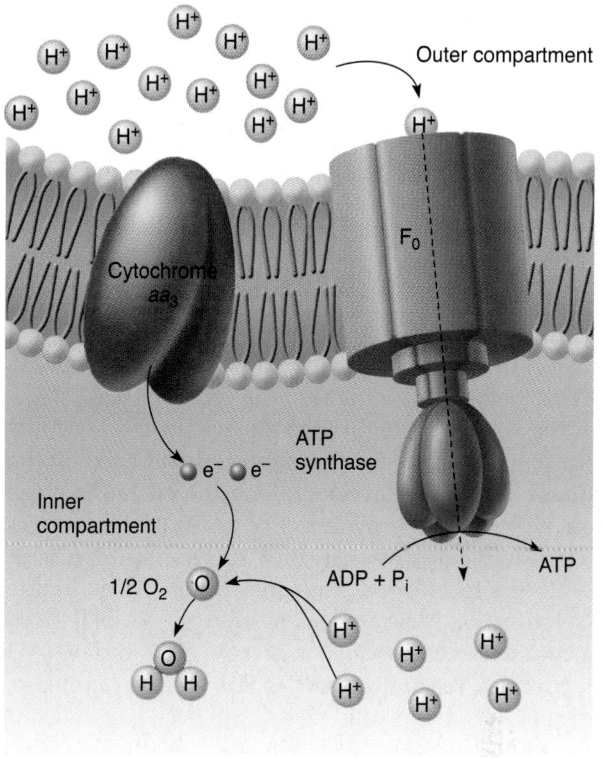

(b) The distribution of electric potential and the concentration gradient of protons across the membrane drive the synthesis of ATP by ATP synthase. The rotation of this enzyme couples diffusion of H^+ to the inner compartment with the bonding of ADP and P_i. The final event of electron transport is the reaction of the electrons with the H^+ and O_2 to form metabolic H_2O. This step is catalyzed by cytochrome oxidase (cytochrome aa_3).

(c) Enlarged view of bacterial cell envelope to show the relationship of electron transport and ATP synthesis. Bacteria have the ETS and ATP synthase stationed in the cell membrane. ETS carriers transport H^+ and electrons from the cytoplasm to the exterior of the membrane. Here, it is collected to create a gradient just as it occurs in mitochondria.

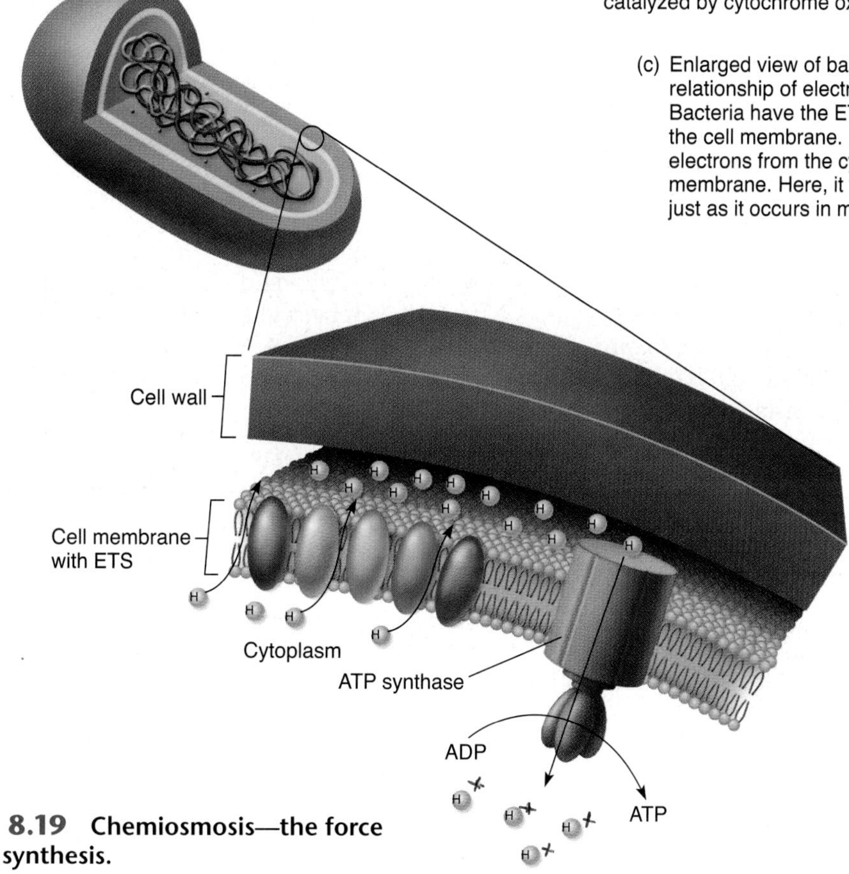

Figure 8.19 Chemiosmosis—the force behind ATP synthesis.

The sequence of electron carriers in the respiratory chain of most aerobic organisms is

1. NADH dehydrogenase, which is closely associated in a complex with the adjacent carrier, which is
2. flavin mononucleotide (FMN);
3. coenzyme Q;
4. cytochrome *b*;
5. cytochrome c_1;
6. cytochrome *c*; and
7. cytochromes *a* and a_3, which are complexed together.

Conveyance of the NADHs from glycolysis and the Krebs cycle to the first carrier sets in motion the remaining six steps. With each redox exchange, the energy level of the reactants is lessened. The released energy is captured and used by the **ATP synthase** complex, stationed along the cristae in close association with the ETS carriers. Each NADH that enters electron transport can give rise to 3 ATPs. This coupling of ATP synthesis to electron transport is termed **oxidative phosphorylation.** Because the electrons from $FADH_2$ from the Krebs cycle enter the cycle at a later point than the NAD and FMN complex reactions, there is less energy to release, and only 2 ATPs are the result.

The Formation of ATP and Chemiosmosis

What biochemical processes are involved in coupling electron transport to the production of ATP? We will first look at the system in eukaryotes, which have the components of electron transport embedded in a precise sequence on mitochondrial membranes. They are stationed between the inner mitochondrial matrix and the outer intermembrane space (see figure 8.18). According to a widely accepted concept called **chemiosmosis,** as the electron transport carriers shuttle electrons, they actively pump hydrogen ions (protons) into the outer compartment of the mitochondrion. This process sets up a concentration gradient of hydrogen ions called the *proton motive force (PMF)*. The PMF consists of a difference in charge between the outer membrane compartment (+) and the inner membrane compartment (−) **(figure 8.19*a*).**

Separating the charge has the effect of a battery, which can temporarily store potential energy. This charge will be maintained by the impermeability of the inner cristae membranes to H^+. The only site where H^+ can diffuse into the inner compartment is at the ATP synthase complex, which sets the stage for the final processing of H^+ leading to ATP synthesis.

ATP synthase is a complex enzyme composed of two large units, F_0 and F_1 **(figure 8.19*b*).** It is embedded in the membrane, but part of it rotates like a motor and traps chemical energy. As the H^+ ions flow through the F_0 center of the enzyme by diffusion, the F_1 compartments pull in ADP and P_i. Rotation causes a three-dimensional change in the enzyme that bonds these two molecules, thereby releasing ATP into the inner compartment (see figure 8.19*b*). The enzyme is then rotated back to the start position and will continue the process.

Bacterial ATP synthesis occurs by means of this same overall process. However, bacteria have the ETS stationed in the cell membrane, and the direction of the proton movement is from the cytoplasm to the periplasmic space in gram-negative bacteria, and to the area occupied by the cell wall in gram-positives **(figure 8.19*c*).** This difference will affect the amount of ATP produced (discussed in the next section). In both cell types, the chemiosmotic theory has been supported by tests showing that oxidative phosphorylation is blocked if the mitochondrial or bacterial cell membranes are disrupted.

Potential Yield of ATPs from Oxidative Phosphorylation

The total of five NADHs (four from the Krebs cycle and one from glycolysis) can be used to synthesize:

15 ATPs for ETS (5 × 3 per electron pair)

and

15 × 2 = 30 ATPs per glucose

The single FADH produced during the Krebs cycle results in:

2 ATPs per electron pair

and

2 × 2 = 4 ATPs per glucose

Table 8.3 summarizes the total of ATP and other products for the entire aerobic pathway. These totals are the potential yields possible but may not be fulfilled by many organisms.

Summary of Aerobic Respiration

Originally, we presented a summary equation for respiration. We are now in a position to tabulate the input and output of this equation at various points in the pathways and sum up the final ATP. Close examination of table 8.3 will review several important facets of aerobic respiration:

1. The total possible yield of ATP is 40: 4 from glycolysis, 2 from the Krebs cycle, and 34 from electron transport. However, because 2 ATPs were expended in early glycolysis, this leaves a maximum of **38 ATPs.**

The actual totals may be lower in certain eukaryotic cells because energy is expended in transporting the NADH produced during glycolysis across the mitochondrial membrane. Certain aerobic bacteria come closest to achieving the full total of 38 because they lack mitochondria and thus do not have to use ATP in transport of NADH across the outer mitochondrial membrane.

2. Six carbon dioxide molecules are generated during the Krebs cycle.
3. Six oxygen molecules are consumed during electron transport.
4. Six water molecules are produced in electron transport and 2 in glycolysis, but because 2 are used in the Krebs cycle, this leaves a net number of 6.

TABLE 8.3 Summary of Aerobic Respiration for One Glucose Molecule

Output	Glycolysis*	Net Output	Krebs Cycle*	Net Output	Respiratory Chain	Net Output	Total Net Output per Glucose
ATP produced	2 × 2 =	2	1 × 2 =	2	17 × 2 =	34	2 + 2 + 34 = 38**
ATP used	2		0		0		
NADH produced	1 × 2 =	2	4 × 2 =	8	0		10
FADH produced	0		1 × 2 =	2	0		2
CO_2 produced	0		3 × 2 =	6	0		6
O_2 used	0		0		3 × 2 =	6	
H_2O produced	2	2	0		3 × 2 =	6	2 − 2 + 6 = 6
H_2O used	0		2	−2	0		

*Products are multiplied by 2 because the first figure represents the amount for only one trip through the pathway, and two molecules make this trip for each glucose.
**This amount can vary among microbes.

The Terminal Step

The terminal step, during which oxygen accepts the electrons, is catalyzed by cytochrome aa_3, also called cytochrome oxidase. This large enzyme complex is specifically adapted to receive electrons from cytochrome *c*, pick up hydrogens from the solution, and react with oxygen to form a molecule of water (see figure 8.19*b*). This reaction, though in actuality more complex, is summarized as follows:

$$2\,H^+ + 2\,e^- + \tfrac{1}{2}O_2 \rightarrow H_2O$$

Most eukaryotic aerobes have a fully functioning cytochrome system, but bacteria exhibit wide-ranging variations in this part of the system. Some species lack one or more of the redox steps; others have several alternative electron transport schemes. Because many bacteria lack cytochrome *c* oxidase, this variation can be used to differentiate among certain genera of bacteria. An oxidase detection test can be used to help identify members of the genera *Neisseria* and *Pseudomonas* and some species of *Bacillus*. Another variation in the cytochrome system is evident in certain bacteria (*Klebsiella, Enterobacter*) that can grow even in the presence of cyanide because they lack cytochrome oxidase. Cyanide will cause rapid death in humans and other eukaryotes because it blocks cytochrome oxidase, thereby completely terminating aerobic respiration, but it is harmless to these bacteria.

A potential side reaction of the respiratory chain in aerobic organisms is the incomplete reduction of oxygen to superoxide ion (O_2^-) and hydrogen peroxide (H_2O_2). As mentioned in chapter 7, these toxic oxygen products can be very damaging to cells. Aerobes have neutralizing enzymes to deal with these products, including *superoxide dismutase* and *catalase*. One exception is the genus *Streptococcus*, which can grow well in oxygen yet lacks both cytochromes and catalase. The tolerance of these organisms to oxygen can be explained by the neutralizing effects of a special peroxidase. The lack of cytochromes, catalase, and peroxidases in anaerobes as a rule limits their ability to process free oxygen and contributes to its toxic effects on them.

Anaerobic Respiration

Some bacteria have evolved an anaerobic respiratory system that functions like the aerobic cytochrome system except that it utilizes oxygen-containing ions, rather than free oxygen, as the final electron acceptor in electron transport (see figure 8.16). Of these, the nitrate (NO_3^-) and nitrite (NO_2^-) reduction systems are best known. The reaction in species such as *Escherichia coli* is represented as:

$$NO_3^- + NADH \xrightarrow{\text{Nitrate reductase}} NO_2^- + H_2O + NAD^+$$

The enzyme nitrate reductase catalyzes the removal of oxygen from nitrate, leaving nitrite and water as products. A test for this reaction is one of the physiological tests used in identifying bacteria.

Some species of *Pseudomonas* and *Bacillus* possess enzymes that can further reduce nitrite to nitric oxide (NO), nitrous oxide (N_2O), and even nitrogen gas (N_2). This process, called **denitrification,** is a very important step in recycling nitrogen in the biosphere. Other oxygen-containing nutrients reduced anaerobically by various bacteria are carbonates and sulfates. None of the anaerobic pathways produce as much ATP as aerobic respiration.

CHECKPOINT

- Catabolic pathways release energy through three pathways: glycolysis, the Krebs cycle, and the respiratory electron transport system.
- Cellular respiration is described by the nature of the final electron acceptor. Aerobic respiration implies that O_2 is the final electron acceptor. Anaerobic respiration implies that some other molecule is the final electron acceptor. If the final electron acceptor is an organic molecule, the anaerobic process is considered fermentation.
- Carbohydrates are preferred cell energy sources because they are superior hydrogen (electron) donors.

- Glycolysis is the catabolic process by which glucose is oxidized and converted into two molecules of pyruvic acid, with a net gain of 2 ATPs. The formation of ATP is via substrate-level phosphorylation.
- The Krebs cycle processes the 3-carbon pyruvic acid and generates three CO_2 molecules. The electrons it releases are transferred to redox carriers for energy harvesting. It also generates 2 ATPs.
- The electron transport chain generates free energy through sequential redox reactions collectively called oxidative phosphorylation. This energy is used to generate up to 38 ATPs for each glucose molecule catabolized.

Fermentation

Of all the results of pyruvate metabolism, probably the most varied is fermentation. Technically speaking, **fermentation** is the incomplete oxidation of glucose or other carbohydrates in the absence of oxygen. This process uses organic compounds as the terminal electron acceptors and yields a small amount of ATP (see figure 8.15).

Over time, the term *fermentation* has acquired several looser connotations. Originally, Pasteur called the microbial action of yeast during wine production *ferments,* and to this day, biochemists use the term in reference to the production of ethyl alcohol by yeasts acting on glucose and other carbohydrates. Fermentation is also what bacteriologists call the formation of acid, gas, and other products by the action of various bacteria on pyruvic acid. The process is a common metabolic strategy among bacteria. Industrial processes that produce chemicals on a massive scale through the actions of microbes are also called fermentations (see chapter 25). Each of these usages is acceptable for one application or another.

It may seem that fermentation would yield only meager amounts of energy (2 ATPs maximum per glucose) and that would slow down growth. What actually happens, however, is that many bacteria can grow as fast as they would in the presence of oxygen. This rapid growth is made possible by an increase in the rate of glycolysis. From another standpoint, fermentation permits independence from molecular oxygen and allows colonization of anaerobic environments. It also enables microorganisms with a versatile metabolism to adapt to variations in the availability of oxygen. For them, fermentation provides a means to grow even when oxygen levels are too low for aerobic respiration.

Bacteria that digest cellulose in the rumens of cattle are largely fermentative. After initially hydrolyzing cellulose to glucose, they ferment the glucose to organic acids, which are then absorbed as the bovine's principal energy source. Even human muscle cells can undergo a form of fermentation that permits short periods of activity after the oxygen supply in the muscle has been exhausted. Muscle cells convert pyruvic acid into lactic acid, which allows anaerobic production of ATP to proceed for a time. But this cannot go on indefinitely, and after a few minutes, the accumulated lactic acid causes muscle fatigue.

Products of Fermentation in Microorganisms

Alcoholic beverages (wine, beer, whiskey) are perhaps the most prominent among fermentation products; others are solvents (acetone, butanol), organic acids (lactic, acetic), dairy products, and many other foods. Derivatives of proteins, nucleic acids, and other organic compounds are fermented to produce vitamins, antibiotics, and even hormones such as hydrocortisone.

Fermentation products can be grouped into two general categories: alcoholic fermentation products and acidic fermentation products **(figure 8.20). Alcoholic fermentation** occurs in yeast or bacterial species that have metabolic pathways for converting pyruvic acid to ethanol. This process involves a decarboxylation of pyruvic acid to acetaldehyde, followed by a reduction of the acetaldehyde to ethanol. In oxidizing the NADH formed during glycolysis, NAD is regenerated, thereby allowing the glycolytic pathway to continue. These processes are crucial in the production of beer and wine, though the actual techniques for arriving at the desired amount of ethanol and the prevention of unwanted side reactions are important

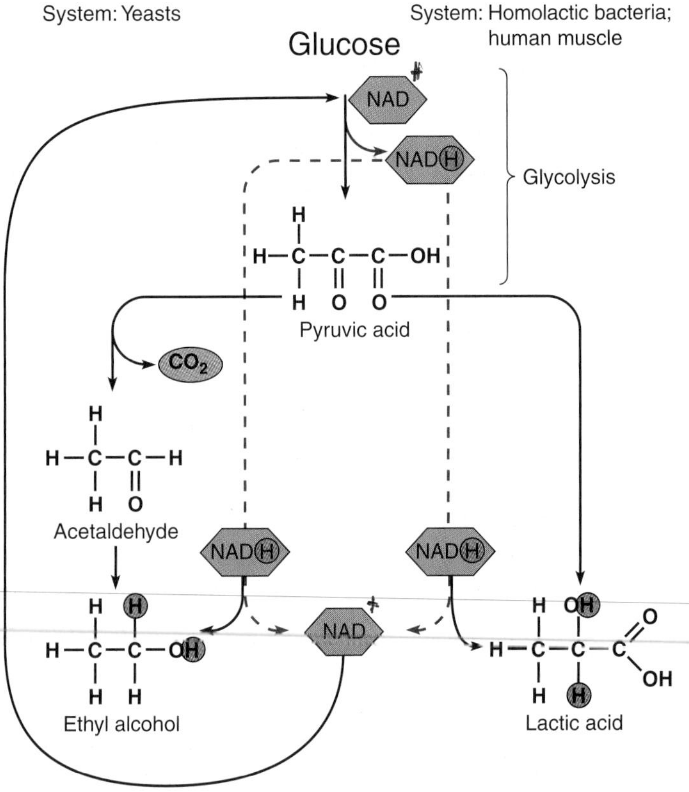

Figure 8.20 The chemistry of fermentation systems that produce acid and alcohol.
In both cases, the final electron acceptor is an organic compound. In yeasts, pyruvic acid is decarboxylated to acetaldehyde, and the NADH given off in the glycolytic pathway reduces acetaldehyde to ethyl alcohol. In homolactic fermentative bacteria, pyruvic acid is reduced by NADH to lactic acid. Both systems regenerate NAD to feed back into glycolysis or other cycles.

INSIGHT 8.4 *Historical*

Pasteur and the Wine-to-Vinegar Connection

The microbiology of alcoholic fermentation was greatly clarified by Louis Pasteur after French winemakers hired him to uncover the causes of periodic spoilage in wines. Especially troublesome was the conversion of wine to vinegar and the resultant sour flavor. Up to that time, wine formation had been considered strictly a chemical process. After extensively studying beer making and wine grapes, Pasteur concluded that wine, both fine and not so fine, was the result of microbial action on the juices of the grape and that wine "disease" was caused by contaminating organisms that produced undesirable products such as acid. Although he did not know it at the time, the bacterial contaminants responsible for the acidity of the spoiled wines were likely to be *Acetobacter* or *Gluconobacter* introduced by the grapes, air, or wine-making apparatus. These common gram-negative genera further oxidized ethanol to acetic acid and are presently used in commercial vinegar production. The following formula shows how this is accomplished:

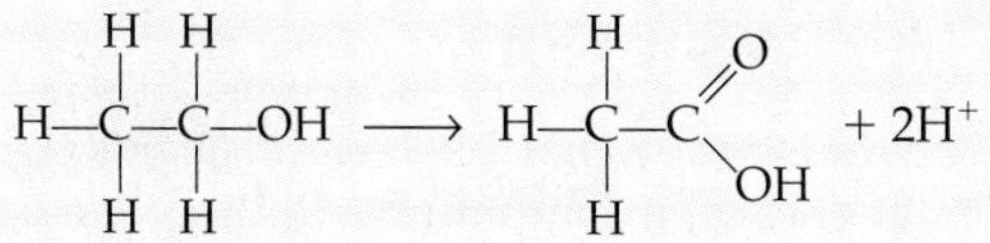

Ethanal Acetic acid

Pasteur's far-reaching solution to the problem is still with us today—mild heating, or *pasteurization*, of the grape juice to destroy the contaminants, followed by inoculation of the juice with a pure yeast culture. The topic of wine making is explored further in chapter 25.

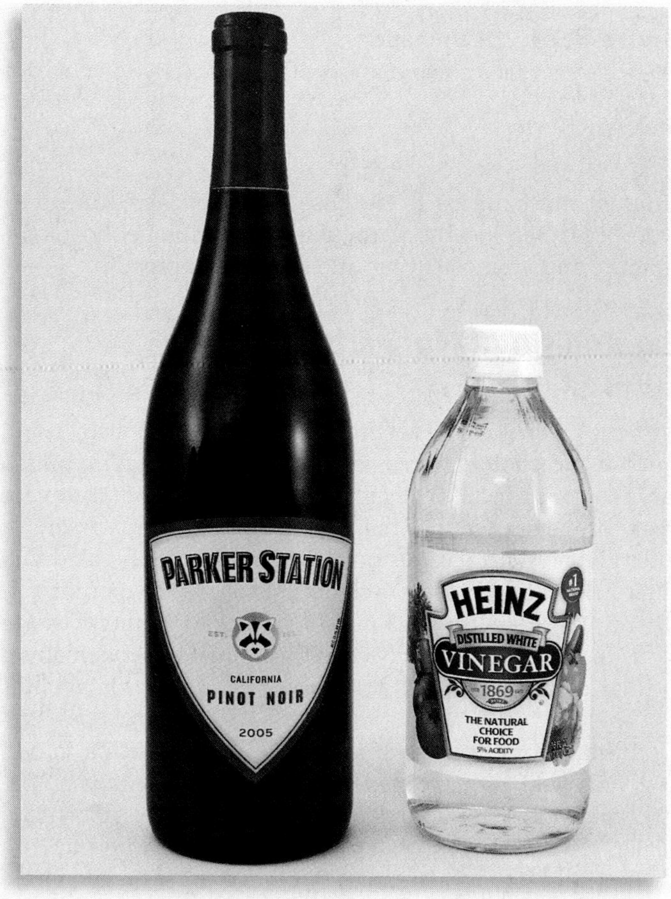

tricks of the brewer's trade **(Insight 8.4).** Note that the products of alcoholic fermentation are not only ethanol but also CO_2, a gas that accounts for the bubbles in champagne and beer (and the rising of bread dough).

Alcohols other than ethanol can be produced during bacterial fermentation pathways. Certain clostridia produce butanol and isopropanol through a complex series of reactions. Although this process was once an important source of alcohols for industrial use, it has been largely replaced by a nonmicrobial petroleum process.

The pathways of **acidic fermentation** are extremely varied. Lactic acid bacteria ferment pyruvate in the same way that humans do—by reducing it to lactic acid. If the product of this fermentation is mainly lactic acid, as in certain species of *Streptococcus* and *Lactobacillus*, it is termed *homolactic*. The souring of milk is due largely to the production of this acid by bacteria. When glucose is fermented to a mixture of lactic acid, acetic acid, and carbon dioxide, as is the case with *Leuconostoc* and other species of *Lactobacillus*, the process is termed *heterolactic fermentation*.

Many members of the family Enterobacteriaceae (*Escherichia*, *Shigella*, and *Salmonella*) possess enzyme systems for converting pyruvic acid to several acids simultaneously. **Mixed acid fermentation** produces a combination of acetic, lactic, succinic, and formic acids, and it lowers the pH of a medium to about 4.0. *Propionibacterium* produces primarily propionic acid, which gives the characteristic flavor to Swiss cheese while fermentation gas (CO_2) produces the holes. Some members also further decompose formic acid completely to carbon dioxide and hydrogen gases. Because enteric bacteria commonly occupy the intestine, this fermentative activity accounts for the accumulation of some types of gas—primarily CO_2 and H_2—in the intestine. Some bacteria reduce the organic acids and produce the neutral end product 2,3-butanediol.

We have provided only a brief survey of fermentation products, but it is worth noting that microbes can be harnessed to synthesize a variety of other substances by varying the raw materials provided them. In fact, so broad is the meaning of the word *fermentation* that the large-scale industrial syntheses by microorganisms often utilize entirely

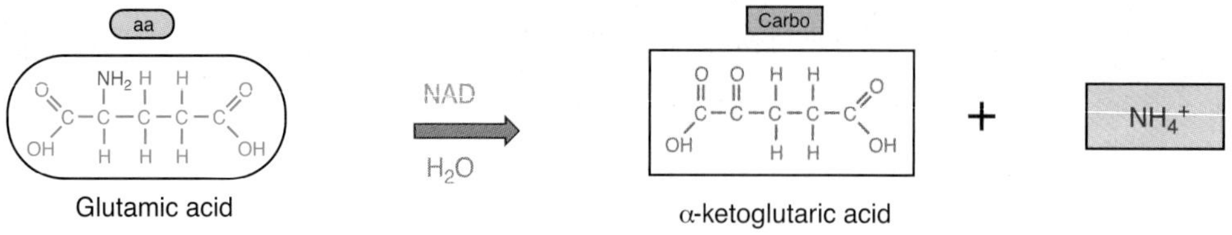

Figure 8.21 Deamination.
Removal of an amino group converts an amino acid to an intermediate of carbohydrate metabolism. Ammonium is a waste product.

different mechanisms from those described here, and they even occur aerobically, particularly in antibiotic, hormone, vitamin, and amino acid production (see chapter 25).

Catabolism of Noncarbohydrate Compounds

We have given you one version of events for catabolism, using glucose, a carbohydrate, as our example. Other compounds serve as fuel, as well. The more complex polysaccharides are easily broken down into their component sugars, which can enter glycolysis at various points. Microbes also break down other molecules for their own use, of course. Two other major sources of energy and building blocks for microbes are lipids (fats) and proteins. Both of these must be broken down to their component parts to produce precursor metabolites and energy.

Recall from chapter 2 that fats are fatty acids joined to glycerol. Enzymes called **lipases** break these apart. The glycerol is then converted to dihydroxyacetone phosphate (DHAP), which can enter step 4 of glycolysis (see figure 8.16). The fatty acid component goes through a process called **beta oxidation.** Fatty acids have a variable number of carbons; in beta oxidation, 2-carbon units are successively transferred to coenzyme A, creating acetyl CoA, which enters the Krebs cycle. This process can yield a large amount of energy. Oxidation of a 6-carbon fatty acid yields 50 ATPs, compared with 38 for a 6-carbon sugar.

Proteins are chains of amino acids. Enzymes called **proteases** break proteins down to their amino acid components (see figure 8.7), after which the amino groups are removed by a reaction called **deamination (figure 8.21).** This leaves a carbon compound, which is easily converted to one of several Krebs cycle intermediates.

CHECKPOINT

- Fermentation is a low energy-yielding oxidation process that takes place in the absence of oxygen.
- In fermentation, the terminal electron acceptors are organic compounds.
- Alcohol and acid are common fermentation products.
- Compounds other than glucose are also catabolized to produce energy and precursor metabolites.

8.4 Biosynthesis and the Crossing Pathways of Metabolism

Our discussion now turns from catabolism and energy extraction to anabolic functions and biosynthesis. In this section, we present aspects of intermediary metabolism, including amphibolic pathways, the synthesis of simple molecules, and the synthesis of macromolecules.

The Frugality of the Cell—Waste Not, Want Not

It must be obvious by now that cells have mechanisms for careful management of carbon compounds. Rather than being dead ends, most catabolic pathways contain strategic molecular intermediates (metabolites) that can be diverted into anabolic pathways. In this way, a given molecule can serve multiple purposes, and the maximum benefit can be derived from all nutrients and metabolites of the cell pool. The property of a system to integrate catabolic and anabolic pathways to improve cell efficiency is termed **amphibolism** (am-fee-bol'-izm).

At this point in the chapter, you can appreciate a more complex view of metabolism than that presented at the beginning, in figure 8.1. **Figure 8.22** demonstrates the amphibolic nature of intermediary metabolism. The pathways of glucose catabolism are an especially rich "metabolic marketplace." The principal sites of amphibolic interaction occur during glycolysis (glyceraldehyde-3-phosphate and pyruvic acid) and the Krebs cycle (acetyl coenzyme A and various organic acids).

Amphibolic Sources of Cellular Building Blocks

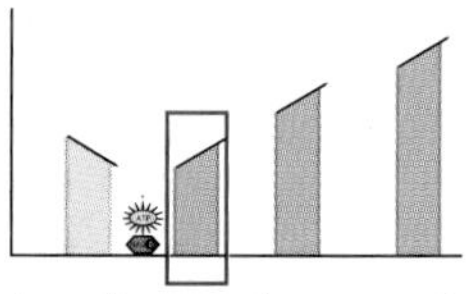

Glyceraldehyde-3-phosphate can be diverted away from glycolysis and converted into precursors for amino acid, carbohydrate, and triglyceride (fat) synthesis. (A precursor molecule is a compound that is the source of another compound.) Earlier we noted the numerous directions that pyruvic acid catabolism can take. In terms of synthesis, pyruvate also plays a pivotal role in providing intermediates for amino acids. In the event of an inadequate glucose supply, pyruvate serves as the starting point in glucose synthesis from various metabolic intermediates, a process called **gluconeogenesis** (gloo′-koh-nee′-oh-gen′-uh-sis).

The acetyl group that starts the Krebs cycle is another extremely versatile metabolite that can be fed into a number of synthetic pathways. This 2-carbon fragment can be converted as a single unit into one of several amino acids, or a number of these fragments can be condensed into hydrocarbon chains that are important building blocks for fatty acid and lipid synthesis. Note that the reverse is also true—fats can be degraded to acetyl through a process called **beta oxidation,** and thereby enter the Krebs cycle at acetyl coenzyme A.

Two metabolites of carbohydrate catabolism that the Krebs cycle produces, oxaloacetic acid and α-ketoglutaric acid, are essential intermediates in the synthesis of certain amino acids. This occurs through **amination,** the addition of an amino group to a carbon skeleton **(figure 8.23*a*).** A certain core group of amino acids can then be used to synthesize others. Amino acids and carbohydrates can be interchanged through **transamination (figure 8.23*b*).**

Pathways that synthesize the nitrogen bases (purines, pyrimidines), which are components of DNA and RNA, originate in amino acids and so can be dependent on intermediates from the Krebs

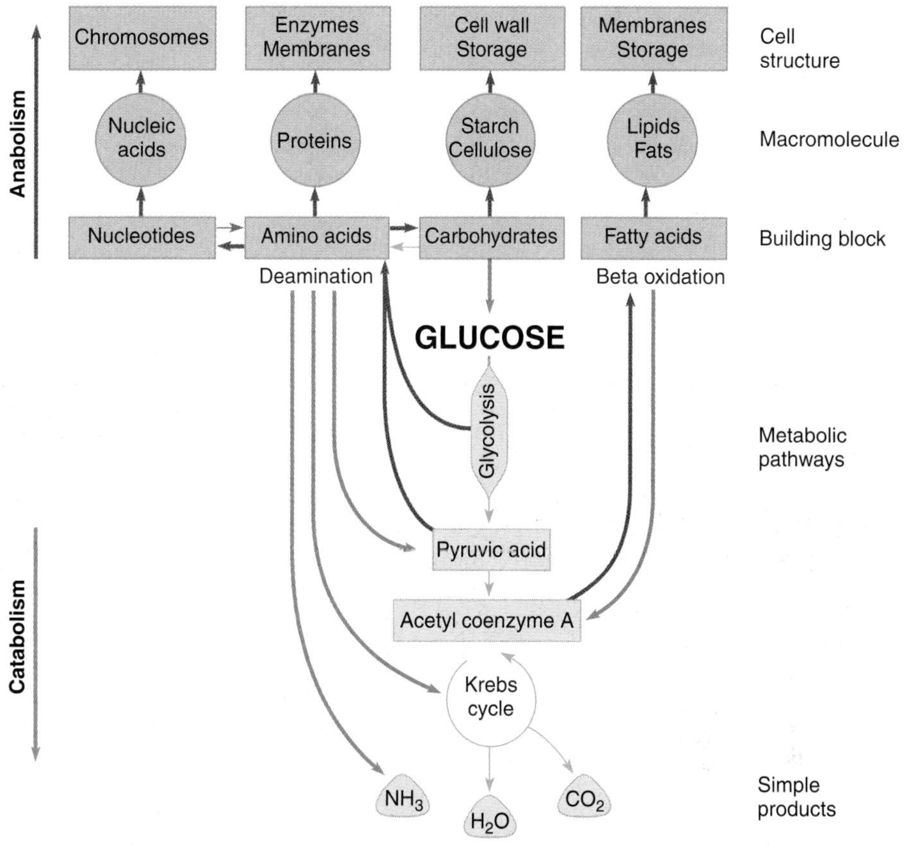

Figure 8.22 An amphibolic view of metabolism.
Intermediate compounds such as pyruvic acid and acetyl coenzyme A serve an amphibolic function. With comparatively small modifications, these compounds can be converted into other compounds and enter a different pathway. Note that catabolism of glucose (center) furnishes numerous intermediates for anabolic pathways that synthesize amino acids, fats, nucleic acids, and carbohydrates. These building blocks can serve in further synthesis of larger molecules to construct various cell components.

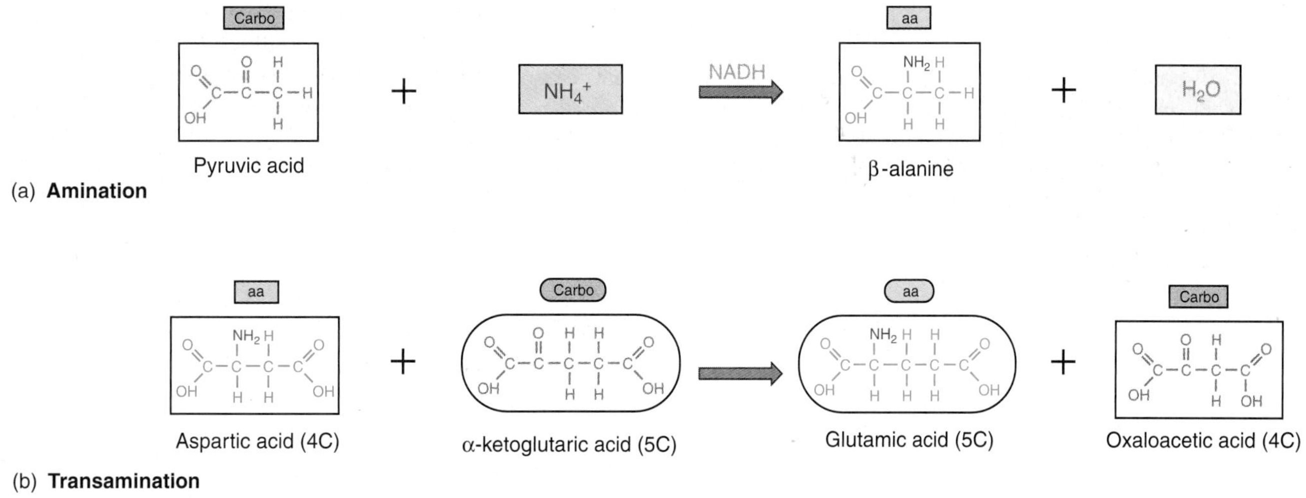

Figure 8.23 Reactions that produce and convert amino acids.
All of them require energy as ATP or NAD and specialized enzymes. **(a)** Through amination (the addition of an ammonium molecule [amino group]), a carbohydrate can be converted to an amino acid. **(b)** Through transamination (transfer of an amino group from an amino acid to a carbohydrate fragment), metabolic intermediates can be converted to amino acids that are in low supply. Contrast these to deamination in figure 8.22.

cycle as well. Because the coenzymes NAD, NADP, FAD, and others contain purines and pyrimidines similar to the nucleic acids, their synthetic pathways are also dependent on amino acids. During times of carbohydrate deprivation, organisms can likewise convert amino acids to intermediates of the Krebs cycle by **deamination** and thereby derive energy from proteins (see figure 8.21).

Anabolism: Formation of Macromolecules

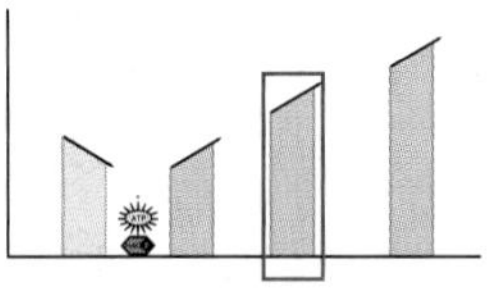

Monosaccharides, amino acids, fatty acids, nitrogen bases, and vitamins—the building blocks that make up the various macromolecules and organelles of the cell—come from two possible sources. They can enter the cell from the outside as nutrients, or they can be synthesized through various cellular pathways. The degree to which an organism can synthesize its own building blocks (simple molecules) is determined by its genetic makeup, a factor that varies tremendously from group to group. In chapter 7, you learned that autotrophs require only CO_2 as a carbon source, a few minerals to synthesize all cell substances, and no organic nutrients. Some heterotrophic organisms (*E. coli,* yeasts) are also very efficient in that they can synthesize all cellular substances from minerals and one organic carbon source such as glucose. Compare this with a strict parasite that has few synthetic abilities of its own and drains most precursor molecules from the host.

Whatever their source, once these building blocks are added to the metabolic pool, they are available for synthesis of polymers by the cell. The details of synthesis vary among the types of macromolecules, but all of them involve the formation of bonds by specialized enzymes and the expenditure of ATP.

Carbohydrate Biosynthesis

The role of glucose in bioenergetics is so crucial that its biosynthesis is ensured by several alternative pathways. Certain structures in the cell depend on an adequate supply of glucose as well. It is the major component of the cellulose cell walls of some eukaryotes and of certain storage granules (starch, glycogen). One of the intermediaries in glycolysis, glucose-6-P, is used to form glycogen. Monosaccharides other than glucose are important in the synthesis of bacterial cell walls. Peptidoglycan contains a linked polymer of muramic acid and glucosamine. Fructose-6-P from glycolysis is used to form these two sugars. Carbohydrates (deoxyribose, ribose) are also essential building blocks in nucleic acids. Polysaccharides are the predominant components of cell surface structures such as capsules and the glycocalyx, and they are commonly found in slime layers (dextran). Remember that most polymerization reactions occur via loss of a water molecule (figure 2.15) and the input of energy (figure 8.7).

Amino Acids, Protein Synthesis, and Nucleic Acid Synthesis

Proteins account for a large proportion of a cell's constituents. They are essential components of enzymes, the cell membrane, the cell wall, and cell appendages. As a general rule, 20 amino acids are needed to make these proteins. Although some organisms (*E. coli,* for example) have pathways that will synthesize all 20 amino acids, others, especially animals, lack some or all of the pathways for amino acid synthesis and must acquire the essential ones from their diets. Protein synthesis itself is a complex process that requires a genetic blueprint and the operation of intricate cellular machinery, as you will see in chapter 9.

DNA and RNA are responsible for the hereditary continuity of cells and the overall direction of protein synthesis. Because nucleic acid synthesis is a major topic of genetics and is closely allied to protein synthesis, it will likewise be covered in chapter 9.

Assembly of the Cell

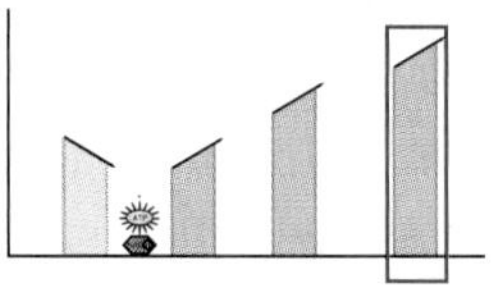

The component parts of a bacteria cell are being synthesized on a continuous basis, and catabolism is also taking place, as long as nutrients are present and the cell is in a nondormant state. When anabolism produces enough macromolecules to serve two cells, and when DNA replication produces duplicate copies of the cell's genetic material, the cell undergoes binary fission, which results in two cells from one parent cell. The two cells will need twice as many ribosomes, twice as many enzymes, and so on. The cell has created these during the initial anabolic phases we have described. Before cell division, the membrane(s) and the cell wall will have increased in size to create a cell that is almost twice as big as a "newborn" cell. Once synthesized, the phospholipid bilayer components of the membranes assemble themselves spontaneously with no energy input. But proteins and other components must be added to the membranes. Growth of the cell wall, accomplished by the addition and coupling of sugars and peptides, requires energy input. The catabolic processes provide all the energy for these complex building reactions.

CHECKPOINT

- Amphibolic compounds are the "crossroads compounds" of metabolism. They not only participate in catabolic pathways but also are precursor molecules to biosynthetic pathways.
- Biosynthetic pathways utilize building-block molecules from two sources: the environment and the cell's own catabolic pathways. Microorganisms construct macromolecules from these monomers using ATP and specialized enzymes.
- Carbohydrates are crucial as energy sources, cell wall constituents, and components of nucleotides.

- Proteins are essential macromolecules in all cells because they function as structural constituents, enzymes, and cell appendages.
- Final assembly of a new cell occurs via binary fission and requires an input of energy.

8.5 It All Starts with the Sun

As mentioned earlier, the ultimate source of all the chemical energy in cells comes from the sun. Most organisms depend either directly or indirectly on the sunlight's energy, which is converted into chemical energy through photosynthesis. (Some chemoautotrophs derive their energy and nutrients solely from inorganic substrates.) The other major products of photosynthesis are organic carbon compounds, which are produced from carbon dioxide through a process called carbon fixation.

Photosynthesis

With few exceptions, the energy that drives all life processes comes from the sun, but this source is directly available only to the cells of photosynthesizers. In the terrestrial biosphere, green plants are the primary photosynthesizers, and in aquatic ecosystems, where 80% to 90% of all photosynthesis occurs, algae, green and purple bacteria, and cyanobacteria fill this role. Other photosynthetic prokaryotes are green sulfur, purple sulfur, and purple nonsulfur bacteria.

The anatomy of photosynthetic cells is adapted to trapping sunlight, and their physiology effectively uses this solar energy to produce high-energy glucose from low- energy CO_2 and water. Photosynthetic organisms achieve this remarkable feat through a series of reactions involving light, pigment, CO_2, and water, which is used as a source for electrons.

Photosynthesis proceeds in two phases: the **light-dependent reactions,** which proceed only in the presence of sunlight, and the **light-independent reactions,** which proceed regardless of the lighting conditions (light or dark).

Solar energy is delivered in discrete energy packets called **photons** (also called quanta) that travel as waves. The wavelengths of light operating in photosynthesis occur in the visible spectrum between 400 (violet) and 700 nanometers (red). As this light strikes photosynthetic pigments, some wavelengths are absorbed, some pass through, and some are reflected. The activity that has greatest impact on photosynthesis is the absorbance of light by photosynthetic pigments. These include the **chlorophylls,** which are green; **carotenoids,** which are yellow, orange, or red; and **phycobilins,** which are red or blue-green.[4] By far the most important of these pigments are the bacterial chlorophylls, which contain a *photocenter* that consists of a magnesium atom held in the center of a complex ringed molecule called a *porphyrin.* As we will see, the chlorophyll molecule harvests the energy of photons and converts it to electron (chemical) energy. Accessory photosynthetic pigments such as carotenes trap light energy and shuttle it to chlorophyll, thereby functioning like antennae. These light-dependent reactions are catabolic (energy-producing) reactions, which pave the way for the next set of reactions, the light-independent reactions, which use the produced energy for synthesis (anabolism). During this phase, carbon atoms from CO_2 are fixed to the carbon backbones of organic molecules.

The detailed biochemistry of photosynthesis is beyond the scope of this text, but we will provide an overview of the general process as it occurs in green plants, algae, and cyanobacteria **(figure 8.24).** Many of the basic activities (electron transport and phosphorylation) are biochemically similar to certain pathways of respiration.

Light-Dependent Reactions The same systems that carry the photosynthetic pigments are also the sites for the light reactions. They occur in the **thylakoid** membranes of compartments called grana (singular-granum) in chloroplasts **(figure 8.25*a*)** and in specialized parts of the cell membranes in prokaryotes (see chapter 4). These systems exist as two separate complexes called *photosystem I* (P700) and *photosystem II*

4. The color of the pigment corresponds to the wavelength of light it reflects.

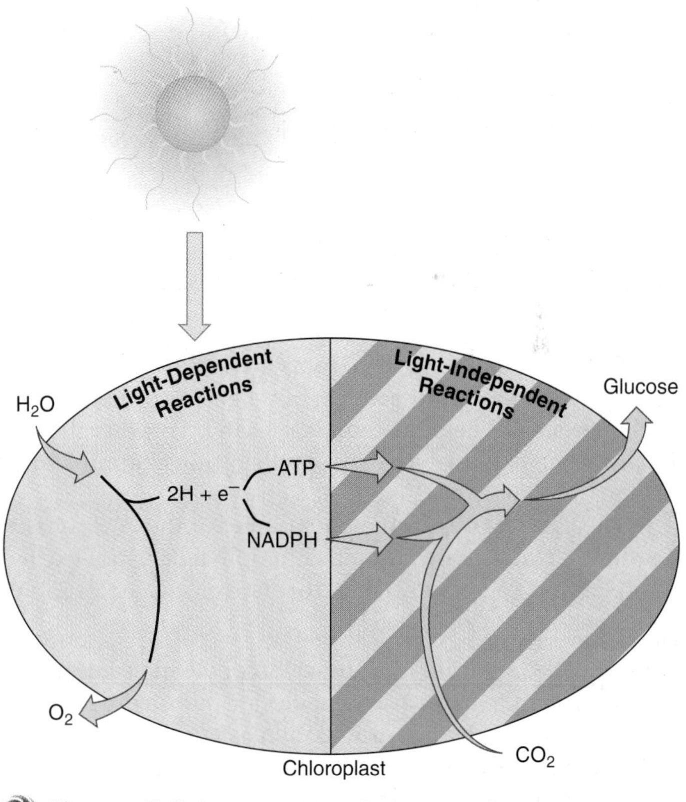

Figure 8.24 Overview of photosynthesis.
The general reactions of photosynthesis, divided into two phases called light-dependent reactions and light-independent reactions. The dependent reactions require light to activate chlorophyll pigment and use the energy given off during activation to split an H_2O molecule into oxygen and hydrogen, producing ATP and NADPH. The independent reactions, which occur either with or without light, utilize ATP and NADPH produced during the light reactions to fix CO_2 into organic compounds such as glucose.

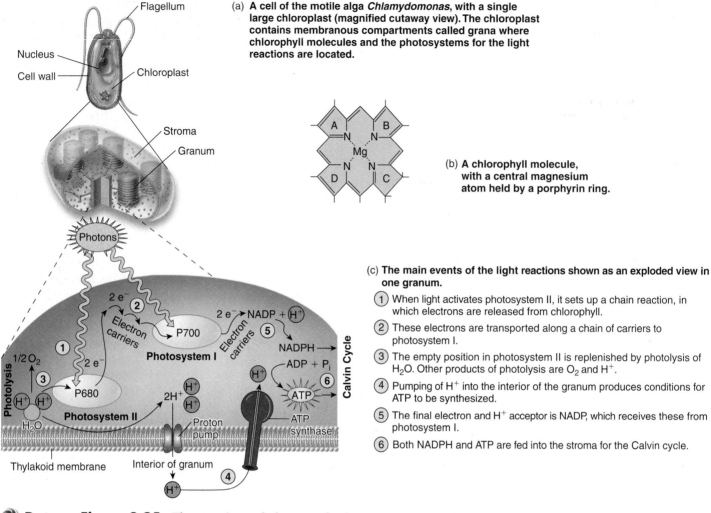

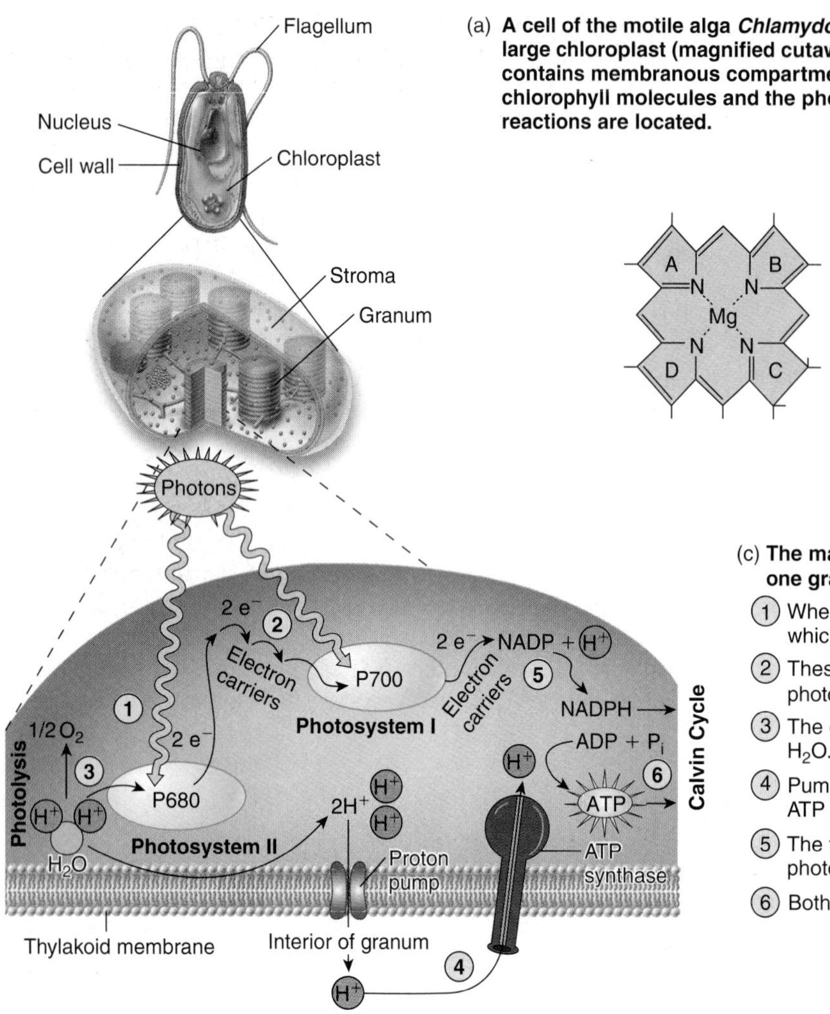

(a) **A cell of the motile alga *Chlamydomonas*, with a single large chloroplast (magnified cutaway view). The chloroplast contains membranous compartments called grana where chlorophyll molecules and the photosystems for the light reactions are located.**

(b) **A chlorophyll molecule, with a central magnesium atom held by a porphyrin ring.**

(c) **The main events of the light reactions shown as an exploded view in one granum.**

1. When light activates photosystem II, it sets up a chain reaction, in which electrons are released from chlorophyll.
2. These electrons are transported along a chain of carriers to photosystem I.
3. The empty position in photosystem II is replenished by photolysis of H_2O. Other products of photolysis are O_2 and H^+.
4. Pumping of H^+ into the interior of the granum produces conditions for ATP to be synthesized.
5. The final electron and H^+ acceptor is NADP, which receives these from photosystem I.
6. Both NADPH and ATP are fed into the stroma for the Calvin cycle.

Process Figure 8.25 The reactions of photosynthesis.

(P680)[5] **(figure 8.25c)**. Both systems contain chlorophyll and they are simultaneously activated by light, but the reactions in photosystem II help drive photosystem I. Together the systems are activated by light, transport electrons, pump hydrogen ions, and form ATP and NADPH.

When photons enter the photocenter of the P680 system (PS II), the magnesium atom in chlorophyll becomes excited and releases 2 electrons. The loss of electrons from the photocenter has two major effects:

1. It creates a vacancy in the chlorophyll molecule forceful enough to split an H_2O molecule into hydrogen (H^+) (electrons and hydrogen ions) and oxygen (O_2). This splitting of water, termed **photolysis,** is the ultimate source of the O_2 gas that is an important product of photosynthesis. The electrons released from the lysed water regenerate photosystem II for its next reaction with light.
2. Electrons generated by the first photoevent are immediately boosted through a series of carriers (cytochromes) to the P700 system. At this same time, hydrogen ions accumulate in the internal space of the thylakoid complex, thereby producing an electrochemical gradient.

The P700 system (PS I) has been activated by light so that it is ready to accept electrons generated by the PS II. The electrons it receives are passed along a second transport chain to a complex that uses electrons and hydrogen ions to reduce NADP to NADPH. (Recall that reduction in this sense entails the addition of electrons and hydrogens to a substrate.)

A second energy reaction involves synthesis of ATP by a chemiosmotic mechanism similar to that shown in figure 8.19. Channels in the thylakoids of the granum actively pump H^+ into the inner chamber, producing a charge gradient. ATP synthase located in this same thylakoid uses the energy from H^+ transport to phosphorylate ADP to ATP. Because it occurs in light, this process is termed **photophosphorylation.** Both NADPH and ATP are released into the stroma of the chloroplast, where they drive the reactions of **the Calvin cycle.**

5. The numbers refer to the wavelength of light to which each system is most sensitive.

Light-Independent Reactions The subsequent photosynthetic reactions that do not require light occur in the chloroplast stroma or the cytoplasm of cyanobacteria. These reactions use energy produced by the light phase to synthesize glucose by means of the Calvin cycle (figure 8.26).

The cycle begins at the point where CO_2 is combind with a doubly phosphorylated 5-carbon acceptor molecule called ribulose-1.5-bisphosphate **(RuBP).** This process, called **carbon fixation,** generates a 6C intermediate compound that immediately splits into two. 3-carbon molecules of 3-phosphoglyceric acid (PGA). The subsequent steps use the ATP and NADPH generated by the photosystems to form high-energy intermediates. First, ATP adds a second phosphate to 3-PGA and produces 1.3-bisphosphoglyceric acid (BPG). Then, during the same step. NADPH contributes it hydrogen to BPG, and one high-energy phosphate is removed. These events give rise to glyceraldehyde-3-phosphate (PGAL). This molecule and its isomer dihydroxyacetone phosphate (DHAP) are key molecules in hexose synthesis leading to fructose and glucose. You may notice that this pathway is very similar to glycolysis, except that it runs in reverse-(see figure 8.17). Bringing the cycle back to regenerate RuBP requires PGAL and several step not depicted in **figure 8.26.**

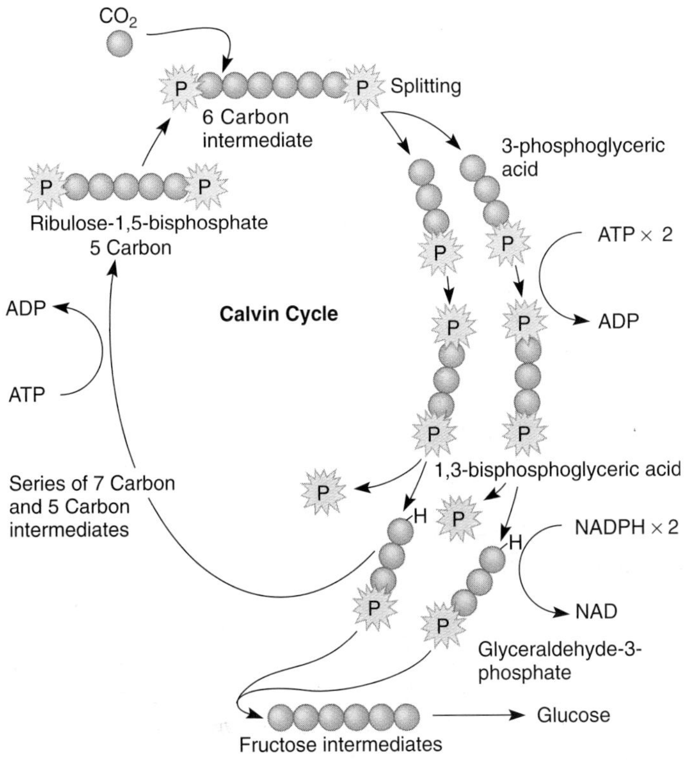

Figure 8.26 The Calvin cycle.
The main events of the reactions in photosynthesis that do not require light. It is during this cycle that carbon is fixed into organic form using the energy (ATP and NADPH) released by the light reactions. The end product glucose, can be stored as complex carbohydrates, or it can be used in various amphibolic pathways to produce other carbohydrate intermediates or amino acids.

Other Mechanisms of Photosynthesis The **oxygenic,** or oxygen-releasing, photosynthesis that occurs in plants, algae, and cyanobacteria is the dominant type on the earth. Other photosynthesizers such as green and purple bacteria possess bacteriochlorophyll, which is more versatile in capturing light. They have only a cyclic photosystem I, which routes the electrons from the photocenter to the electron carriers and back to the photosystem again. This pathway generates a relatively small amount of ATP, and it may not produce NADPH. As photolithotrophs, these bacteria use H_2, H_2S, or elemental sulfur rather than H_2O as a source of electrons and reducing power. As a consequence, they are **anoxygenic** (non-oxygen-producing), and many are strict anaerobes.

CHECKPOINT

- The sun provides most of the energy used by living cells; it is transformed to chemical energy by photosynthesis.
- The CO_2 produced in photosynthesis is converted to organic carbon compounds through carbon fixation.

CASE FILE 8 *WRAP-UP*

The process developed by O'Connor and his colleagues involves heating polystyrene to over 500°C in the absence of oxygen. This process, called pyrolysis, converts the polystyrene into styrene oil and some other compounds. *Pseudomonas putida* is able to convert the styrene oil into polyhydroxyalkanoate (PHA). Unlike polystyrene, which O'Connor calls a "dead-end product," PHA can be utilized to make other plastics and is biodegradable. Bioremediation, the utilization of microbes to clean up pollution or reduce waste, is based on the unique metabolic activity of microbes. *P. putida* and other pseudomonads are free-living organisms found in soil and aquatic areas. These organisms are considered harmless (unlike the opportunistic pathogen *Pseudomonas aeruginosa*) and have very diverse metabolic abilities. They have been widely studied as potential solutions for other hydrocarbon problems such as gasoline and oil contamination of soil. As we move toward better resource- and energy-management practices, bioremediation will continue to be an area of intense interest among researchers and ecologists.

See: O'Connor, K. E., et al. 2006. A two step chemo-biotechnological conversion of polystyrene to a biodegradable thermoplastic. Environmental Science & Technology *40:2433–2437.*

Chapter Summary with Key Terms

8.1 The Metabolism of Microbes

A. *Metabolism*

Metabolism is the sum of cellular chemical and physical activities. It is a complementary process consisting of **anabolism,** synthetic reactions that convert small molecules into large molecules, and **catabolism,** in which large molecules are degraded and energy is produced. Together, they generate thousands of intermediate molecular states, called metabolites, which are regulated at many levels.

B. *Enzymes: Metabolic Catalysts*

1. Metabolism is made possible by organic **catalysts,** or **enzymes,** that speed up reactions by lowering the **energy of activation.** Enzymes are not consumed and can be reused. Each enzyme acts specifically upon its matching molecule or **substrate.**
2. *Enzyme Specificity:* Substrate attachment occurs in the special pocket called the **active,** or **catalytic, site.** In order to fit, a substrate must conform to the active site of the enzyme. This three-dimensional state is determined by the amino acid content, sequence, and folding of the apoenzyme.
3. *Microbial Enzymes and Disease:* Many pathogens secrete enzymes or toxins, which are referred to as virulence factors, that enable them to avoid host defenses.
4. *Enzyme Sensitivity:* Enzymes are **labile** (unstable) and function only within narrow operating ranges of temperature, osmotic pressure, and pH, and they are especially vulnerable to **denaturation.** Enzymes are frequently the targets for physical and chemical agents used in control of microbes.

C. *Regulation of Enzymatic Activity*

Regulatory controls can act on enzymes directly or on the process that gives rise to the enzymes.

8.2 The Pursuit and Utilization of Energy

A. *Energy in Cells*

Energy is the capacity of a system to perform work. It is consumed in **endergonic reactions** and is released in **exergonic reactions.**

B. *A Closer Look at Biological Oxidation and Reduction*

Extracting energy requires a series of electron carriers arrayed in a redox chain between electron donors and electron acceptors.

8.3 The Pathways

A. *Catabolism*

Carbohydrates, such as glucose, are energy-rich because they can yield a large number of electrons per molecule. Glucose is dismantled in stages. **Glycolysis** is a pathway that degrades glucose to pyruvic acid without requiring oxygen. It yields pyruvic acid, which enters the second phase of reactions.

B. *Pyruvic Acid—A Central Metabolite*

1. Pyruvic acid is processed in aerobic respiration via the **Krebs cycle** and its associated electron transport chain.
2. Acetyl coenzyme A is the product of pyruvic acid processing that undergoes further oxidation and decarboxylation in the Krebs cycle, which generates ATP, CO_2, and H_2O.
3. The respiratory chain completes energy extraction.
4. The final electron acceptor in aerobic respiration is oxygen. In **anaerobic respiration,** compounds such as sulfate, nitrate, or nitrite serve this function. Bacteria serve as important agents in the nitrogen cycle **(denitrification).**

C. *Fermentation*

Fermentation is anaerobic respiration in which both the electron donor and final electron acceptors are organic compounds.

1. Fermentation enables anaerobic and facultative microbes to survive in environments devoid of oxygen. Production of alcohol, vinegar, and certain industrial solvents relies upon fermentation.

D. *Versatility of Glycolysis and the Krebs Cycle*

Glycolysis and the Krebs cycle are central pathways that link catabolic and anabolic pathways, allowing cells to break down different classes of molecules in order to synthesize compounds required by the cell.

1. Metabolites of these pathways double as building blocks and sources of energy. Intermediates such as pyruvic acid are convertible into amino acids through **amination.** Amino acids can be **deaminated** and used as precursors to glucose and other carbohydrates **(gluconeogenesis).**
2. Two-carbon acetyl molecules from pyruvate can be used in fatty acid synthesis.

8.4 Biosynthesis and the Crossing Pathways of Metabolism

A. *Amphibolism*

The ability of a cell or system to integrate catabolic and anabolic pathways to improve efficiency is called **amphibolism.**

B. *Anabolism: Formation of Macromolecules*

Macromolecules, such as proteins, carbohydrates, and nucleic acids, are made of building blocks from two possible sources: from outside the cell (preformed) or via synthesis in one of the anabolic pathways.

8.5 If All Starts with the Sun

Photosynthesis converts the sun's energy into chemical energy and organic carbon compounds, which are produced from carbon dioxide.

Multiple-Choice and True-False Questions

Multiple-Choice Questions. Select the correct answer from the answers provided.

1. Catabolism is a form of metabolism in which ____ molecules are converted into ____ molecules.
   a. large, small
   b. small, large
   c. amino acid, protein
   d. food, storage
2. An enzyme
   a. becomes part of the final products
   b. is nonspecific for substrate
   c. is consumed by the reaction
   d. is heat and pH labile
3. An apoenzyme is where the ____ is located.
   a. cofactor
   b. coenzyme
   c. redox reaction
   d. active site
4. Many coenzymes are
   a. metals
   b. vitamins
   c. proteins
   d. substrates
5. To digest cellulose in its environment, a fungus produces a/an
   a. endoenzyme
   b. exoenzyme
   c. catalase
   d. polymerase
6. Energy is carried from catabolic to anabolic reactions in the form of
   a. ADP
   b. high-energy ATP bonds
   c. coenzymes
   d. inorganic phosphate
7. A product or products of glycolysis is/are
   a. ATP
   b. H_2O
   c. CO_2
   d. both a and b
8. Fermentation of a glucose molecule has the potential to produce a net number of ____ ATPs.
   a. 4
   b. 2
   c. 40
   d. 0
9. Complete oxidation of glucose in aerobic respiration can yield a net output of ____ ATPs.
   a. 40
   b. 6
   c. 38
   d. 2
10. ATP synthase complexes can generate ____ ATPs for each NADH that enters electron transport.
   a. 1
   b. 2
   c. 3
   d. 4

True-False Questions. If statement is true, leave as is. If it is false, correct it by rewriting the sentence.

11. Metabolism is another term for biosynthesis.
12. An enzyme lowers the activation energy required for a chemical reaction.
13. Exergonic reactions occur primarily outside the cell.
14. Energy in biological systems is primarily chemical.
15. Exoenzymes are produced outside the cell.

Writing to Learn

These questions are suggested as a *writing-to-learn* experience. For each question, compose a one- or two-paragraph answer that includes the factual information needed to completely address the question.

1. a. Describe the chemistry of enzymes and explain how the apoenzyme forms.
   b. Show diagrammatically the interaction of holoenzyme and its substrate and general products that can be formed from a reaction.
2. Differentiate among the chemical composition and functions of various cofactors. Provide examples of each type.
3. Explain how oxidation of a substrate proceeds without oxygen.
4. In the following redox pairs, which compound is reduced and which is oxidized?
   a. NAD and NADH
   b. $FADH_2$ and FAD
   c. lactic acid and pyruvic acid
   d. NO_3^- and NO_2^-
   e. ethanol and acetaldehyde
5. a. Describe the roles played by ATP and NAD in metabolism.
   b. What particular features of their structure lend them to these functions?
6. a. What is meant by the concept of the "final electron acceptor"?
   b. What are the final electron acceptors in aerobic, anaerobic, and fermentative metabolism?
7. Name the major ways that substrate-level phosphorylation is different from oxidative phosphorylation.
8. Compare and contrast the location of glycolysis, Krebs cycle, and electron transport in prokaryotic and eukaryotic cells.
9. a. Outline the basic steps in glycolysis, indicating where ATP is used and given off.
   b. Where does NADH originate, and what is its fate in an aerobe?
   c. What is the fate of NADH in a fermentative organism?
10. Speculate on how organisms that live in permanently dark habitats, such as the human gut, benefit from photosynthesis.

Concept Mapping

Appendix D provides guidance for working with concept maps.

1. Supply your own linking words or phrases in this concept map, and provide the missing concepts in the empty box.

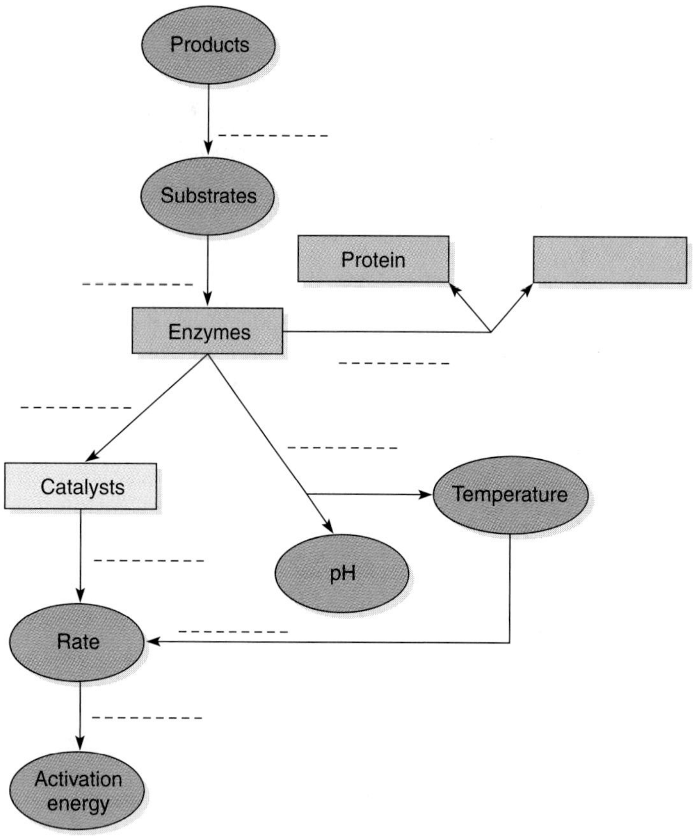

2. Construct your own concept map using the following words as the *concepts*. Supply the linking words between each pair of concepts.

anabolism
catabolism
precursor molecules
bacterial cell
nucleotides
DNA
ATP

Critical Thinking Questions

Critical thinking is the ability to reason and solve problems using facts and concepts. These questions can be approached from a number of angles, and in most cases, they do not have a single correct answer.

1. Using the simplified chart that follows, fill in a summary of the major starting compounds required and the products given off by each phase of metabolism. Use arrows to pinpoint approximately where the reactions take place.

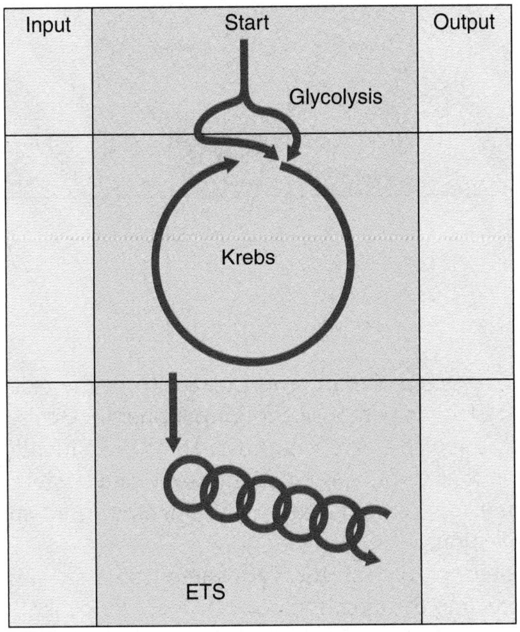

2. Using the concept of fermentation, describe the microbial (biochemical) mechanisms that cause milk to sour.
3. Explain how it is possible for certain microbes to survive and grow in the presence of cyanide, which would kill many other organisms.
4. What adaptive advantages does a fermentative metabolism confer on a microbe?
5. Describe some of the special adaptations of the enzymes found in extremophiles.
6. How many ATPs would be formed as a result of aerobic respiration if cytochrome oxidase were missing from the respiratory chain (as is the case with many bacteria)?

Visual Understanding

1. **From chapter 2, figure 2.21*d*.** Imagine this protein is an enzyme. Is it possible that amino acid #42 and amino acid #300 in the primary sequence actually end up next to each other in an active site? Explain your answer.

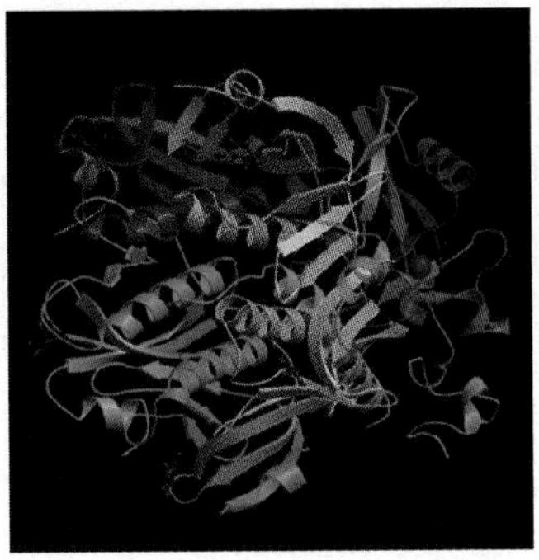

Internet Search Topics

1. Look up fermentation on a search engine, and outline some of the products made by this process.
2. Find websites that feature three-dimensional views of enzymes. Make simple models of three different enzymes indicating enzyme structure, active site, and substrate.
3. Go to www.aris.mhhe.com, and click on "microbiology" and then this textbook's author/title. Go to chapter 8, access the URLs listed under Internet Search Topics, and research the following:

   Log on to the websites listed to view animations and tutorials of the biochemical pathways and enzymes.
4. Search online for a biography of sir Hans Krebs. What other metabolic pathway is he credited with discovering?

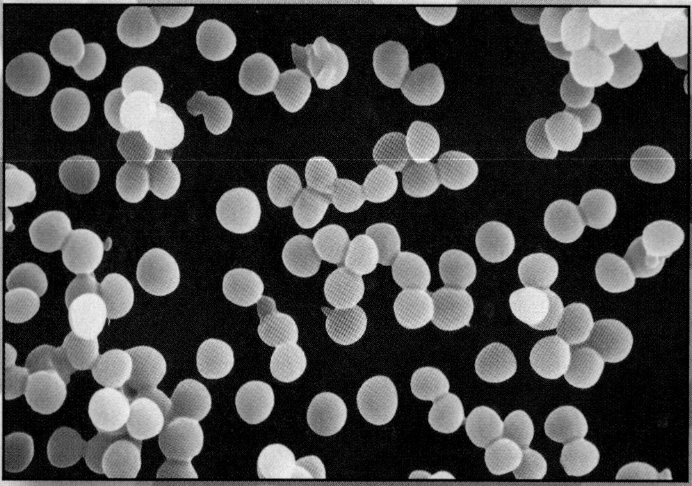

CHAPTER 9

Microbial Genetics

CASE FILE 9

Vancomycin-resistant *Staphylococcus aureus* (VRSA) was isolated from the exit site of a dialysis catheter in a 40-year-old diabetic with a history of peripheral vascular disease, chronic renal failure, and chronic foot ulcers. A few months earlier, the patient's gangrenous toe had been amputated. Following that surgery, the patient developed bacteremia with methicillin-resistant *S. aureus* from an infected hemodialysis graft. Vancomycin, rifampin, and graft removal successfully treated the infection.

A few months later, when the catheter exit site infection appeared, the area was cultured and the catheter removed, successfully treating the infection. A week later, the patient's chronic foot ulcer again appeared infected. Vancomycin-resistant *Enterococcus faecalis* (VRE) and *Klebsiella oxytoca* were cultured from the ulcer. The patient recovered after wound care and systemic treatment with trimethoprim/sulfamethoxazole.

Analysis of the VRSA isolate revealed that it contained the *van*A gene for vancomycin resistance and the *mec*A gene for oxacillin resistance.

- *How do you think the* Staphylococcus aureus *strain ended up with the gene for vancomycin resistance?*
- *What is one possible mechanism for genetic transfer of antibiotic resistance from one organism to another?*
- *Why would this particular patient be at increased risk for infection with VRSA?*

Case File 9 Wrap-Up appears on page 249.

CHAPTER OVERVIEW

- Genetics is the study of the transfer of information between biological entities. The molecules most important to this endeavor are DNA, RNA (both of which carry information), and proteins, which carry out most cellular functions and are built using the information in DNA and RNA.
- DNA is a very long molecule composed of small subunits called nucleotides. The sequence of the nucleotides contains the information needed to eventually direct the synthesis of all proteins in the cell.
- Viruses contain various forms of DNA and RNA that are translated by the genetic machinery of their host cells to form functioning viral particles.
- The genetic activities of cells are highly regulated by operons, groups of genes that interact as a unit to control the use or synthesis of metabolic substances, as well as by RNA regulatory molecules.

- The DNA molecule must be replicated for the distribution of genetic material to offspring. When replication is not faithful, permanent changes in the sequence of the DNA, called mutations, can occur. Because mutations may alter the function or expression of genes, they serve as a force in the evolution of organisms.
- Bacteria undergo genetic recombination through the transfer of small pieces of DNA between bacteria as well as the uptake of DNA from the environment.

9.1 Introduction to Genetics and Genes: Unlocking the Secrets of Heredity

Genetics is the study of the inheritance, or **heredity,** of living things. It is a wide-ranging science that explores

1. the transmission of biological properties (traits) from parent to offspring;
2. the expression and variation of those traits;
3. the structure and function of the genetic material; and
4. how this material changes.

The study of genetics takes place on several levels **(figure 9.1).** Organismal genetics observes the heredity of the whole organism or cell; chromosomal genetics examines the characteristics and actions of chromosomes; and molecular genetics deals with the biochemistry of the genes. All of these levels are useful areas of exploration, but in order to understand the expressions of microbial structure, physiology, mutations, and pathogenicity, we need to examine the operation of genes at the cellular and molecular levels. The study of microbial genetics provides a greater understanding of human genetics and an increased appreciation for the astounding advances in genetic engineering we are currently witnessing (see chapter 10).

The Nature of the Genetic Material

For a species to survive, it must have the capa replication. In single-celled microorganisms, re involves the division of the cell by means of bin or budding, but these forms of reproduction more significant activity than just simple cleav cell mass. Because the genetic material is respo inheritance, it must be accurately duplicated a rated into each daughter cell to ensure normal This genetic material itself is a long molecule that can be studied on several levels. Before we how DNA is copied, let us explore the organiz this genetic material, proceeding from the genera specific.

The Levels of Structure and Function of the Genome

The **genome** is the sum total of genetic material of a cell. Although most of the genome exists in the form of chromosomes, genetic material can appear in nonchromosomal sites as well **(figure 9.2).** For example, bacteria and some fungi contain tiny extra pieces of DNA (plasmids), and certain organelles of

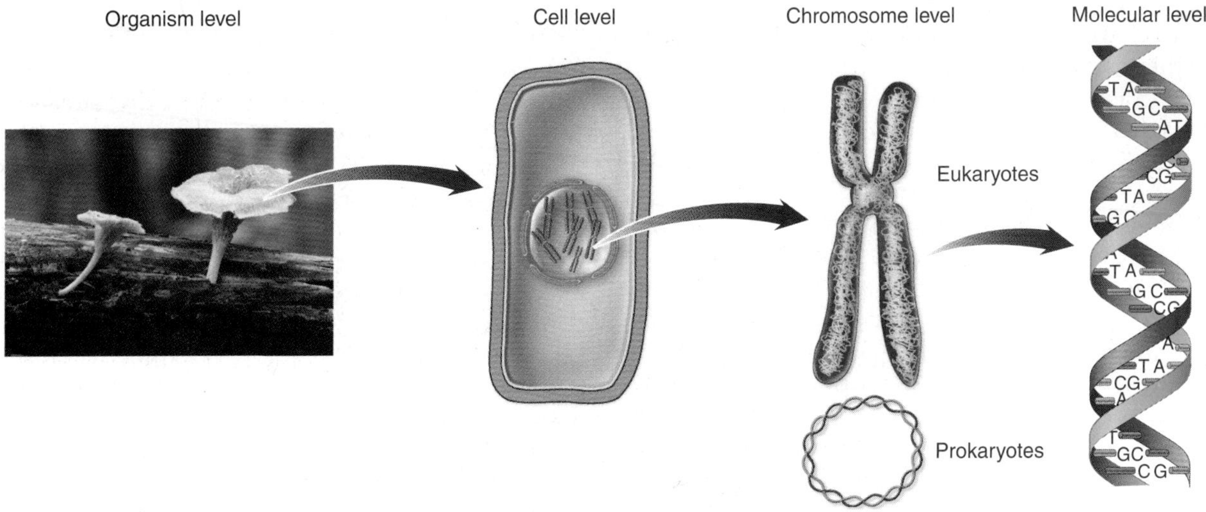

Figure 9.1 Levels of genetic study.
The operations of genetics can be observed at the levels of organism, cell, chromosome, and DNA sequence (molecular level).

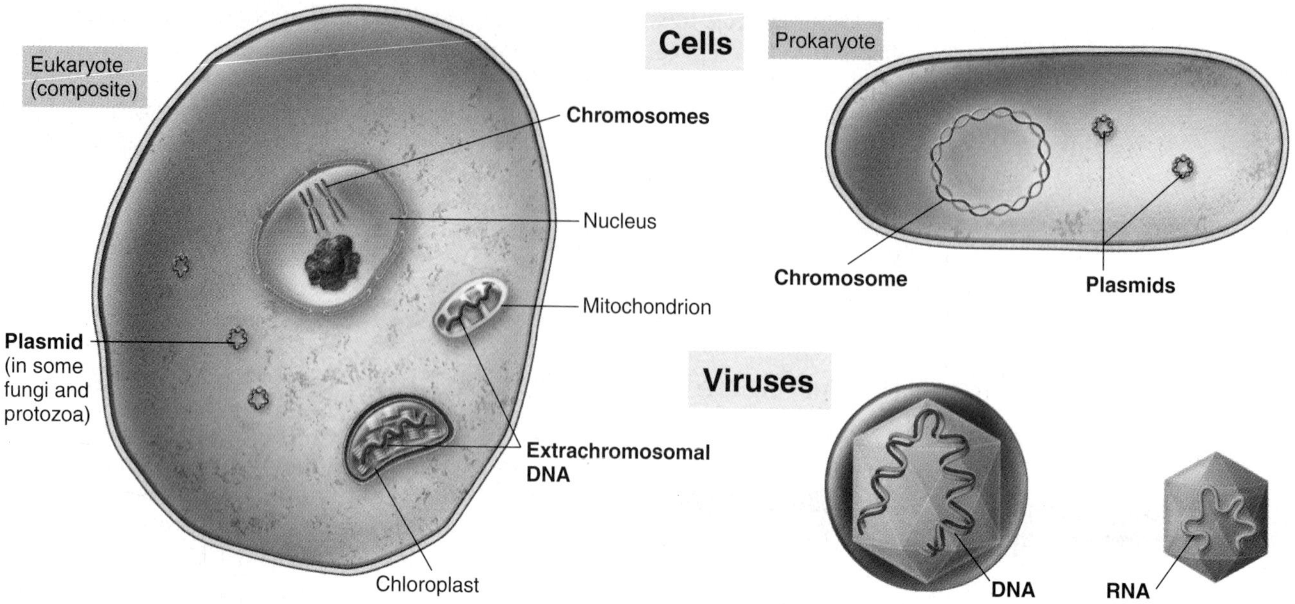

Figure 9.2 The general location and forms of the genome in selected cell types and viruses (not to scale).

eukaryotes (the mitochondria and chloroplasts) are equipped with their own genetic programs. Genomes of cells are composed exclusively of DNA, but viruses contain either DNA or RNA as the principal genetic material. Although the specific genome of an individual organism is unique, the general pattern of nucleic acid structure and function is similar among all organisms.

In general, a **chromosome** is a discrete cellular structure composed of a neatly packaged DNA molecule. The chromosomes of eukaryotes and bacterial cells differ in several respects. The structure of eukaryotic chromosomes consists of a DNA molecule tightly wound around histone proteins, whereas a bacterial chromosome is condensed and secured into a packet by means of histonelike proteins. Eukaryotic chromosomes are located in the nucleus; they vary in number from a few to hundreds; they can occur in pairs (diploid) or singles (haploid); and they appear linear. In contrast, most bacteria have a single, circular (double-stranded) chromosome, although many bacteria have multiple circular chromosomes and some have linear chromosomes.

The chromosomes of all cells are subdivided into basic informational packets called genes. A **gene** can be defined from more than one perspective. In classical genetics, the term refers to the fundamental unit of heredity responsible for a given trait in an organism. In the molecular and biochemical sense, it is a site on the chromosome that provides information for a certain cell function. More specifically still, it is a certain segment of DNA that contains the necessary code to make a **protein** or RNA molecule. This last definition of a gene will be emphasized in this chapter.

Genes fall into three basic categories: *structural genes* that code for proteins, genes that code for RNA, and *regulatory genes* that control gene expression. The sum of all of these types of genes constitutes an organism's distinctive genetic makeup, or **genotype** (jee′-noh-tīp). The expression of the genotype creates traits (certain structures or functions) referred to as the **phenotype** (fee′-noh-tīp). Just as a person inherits a combination of genes (genotype) that gives a certain eye color or height (phenotype), a bacterium inherits genes that direct the formation of a flagellum, and a virus contains genes for its capsid structure. All organisms contain more genes in their genotypes than are manifested as a phenotype at any given time. In other words, the phenotype can change depending on which genes are "turned on" (expressed).

The Size and Packaging of Genomes

Genomes vary greatly in size. The smallest viruses have four or five genes; the bacterium *Escherichia coli* has a single chromosome containing 4,288 genes, and a human cell has about 20,000 to 25,000 genes on 46 chromosomes. The chromosome of *E. coli* would measure about 1 mm if unwound and stretched out linearly, and yet this fits within a cell that measures just over 1 micron across, making the stretched-out DNA 1,000 times longer than the cell **(figure 9.3).** Still, the bacterial chromosome takes up only about one-third to one-half of the cell's volume. Likewise, if the sum of all DNA contained in the 46 human chromosomes were unraveled and laid end to end, it would measure about 6 feet. How can such elongated genomes fit into the minuscule volume of a cell, and in the case of eukaryotes, into an even smaller compartment, the nucleus? The answer lies in the regular coiling of the DNA chain (see Insight 9.1).

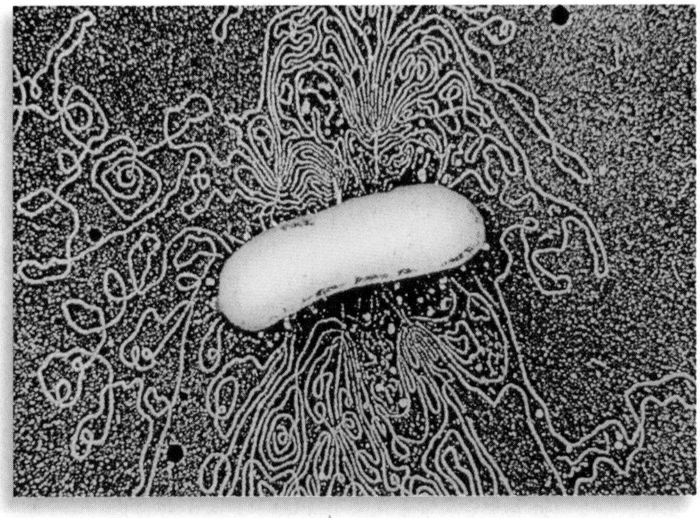

Figure 9.3 An *Escherichia coli* cell disrupted to release its DNA molecule.
The cell has spewed out its single, uncoiled DNA strand into the surrounding medium.

CASE FILE 9 *WRAP-UP*

The chapter opener described the first known case of infection with VRSA (in 2002). Analysis of this VRSA strain isolated from the infected exit site of the patient's catheter indicated the possibility that this resistant *S. aureus* strain had acquired the gene for vancomycin resistance from the resistant strain of *E. faecalis* (VRE), isolated from the patient's chronic foot ulcer. The fact that the patient was treated with vancomycin before the VRSA infection allowed for selection of vancomycin-resistant organisms. Although prior studies showed that the transfer of the *van*A gene by conjugation of *Enterococcus* species with *S. aureus* was possible *in vitro* (under laboratory conditions), this case provided good evidence that this gene transfer could occur in a living patient.

See: Staphylococcus aureus *resistant to vancomycin—United States, 2002.* MMWR *51:565–566.*
Noble, W. C., Virani, Z., and Cree, R. G. 1992. Co-transfer of vancomycin and other resistance genes from Enterococcus faecalis *NCTC 12201 to* Staphylococcus aureus. FEMS Microbiol. Lett. *93:195–198.*

The DNA Code: A Simple Yet Profound Message

Examining the function of DNA at the molecular level requires an even closer look at its structure. To do this we will imagine being able to magnify a small piece of a gene about 5 million times. What such fine scrutiny will disclose is one of the great marvels of biology. James Watson and Francis Crick put the pieces of the puzzle together in 1953 **(Insight 9.1)** to discover that DNA is a gigantic molecule, a type of nucleic acid, with two strands combined into a double helix. The general structure of DNA is universal, except in some viruses that contain single-stranded DNA. The basic unit of DNA structure is a **nucleotide,** and a chromosome in a typical bacterium consists of several million nucleotides linked end to end. Each nucleotide is composed of **phosphate, deoxyribose sugar,** and a **nitrogenous base.** The nucleotides covalently bond to form a sugar-phosphate linkage that becomes the backbone of each strand. Each sugar attaches in a repetitive pattern to two phosphates. One of the bonds is to the number 5′ (read "five prime") carbon on deoxyribose, and the other is to the 3′ carbon, which confers a certain order and direction on each strand **(figure 9.4).**

The nitrogenous bases, **purines** and **pyrimidines,** attach by covalent bonds at the 1′ position of the sugar **(figure 9.4*a*).** They span the center of the molecule and pair with appropriate complementary bases from the other strand. The paired bases are so aligned as to be joined by hydrogen bonds. Such weak bonds are easily broken, allowing the molecule to be "unzipped" into its complementary strands. This feature is of great importance in gaining access to the information encoded in the nitrogenous base sequence. Pairing of purines and pyrimidines is not random; it is dictated by the formation of hydrogen bonds between certain bases. Thus, in DNA, the purine **adenine** (A) always pairs with the pyrimidine **thymine** (T), and the purine **guanine** (G) always pairs with the pyrimidine **cytosine** (C). New research also indicates that the bases are attracted to each other in this pattern because each has a complementary three-dimensional shape that matches its pair. Although the base-pairing partners generally do not vary, the sequence of base pairs along the DNA molecule can assume any order, resulting in an infinite number of possible nucleotide sequences.

Other important considerations of DNA structure concern the nature of the double helix itself. The halves are not oriented in the same direction. One side of the helix runs in the opposite direction of the other, in what is called an *antiparallel arrangement* **(figure 9.4*b*).** The order of the bond between the carbon on deoxyribose and the phosphates is used to keep track of the direction of the two sides of the helix. Thus, one helix runs from the 5′ to 3′ direction, and the other runs from the 3′ to 5′ direction. This characteristic is a significant factor in DNA synthesis and translation.

The Significance of DNA Structure

The arrangement of nitrogenous bases in DNA has two essential effects.

1. **Maintenance of the code during reproduction.** The constancy of base-pairing guarantees that the code will be retained during cell growth and division. When the two strands are separated, each one provides a template (pattern or model) for the replication (exact copying) of a

INSIGHT 9.1 Historical

Deciphering the Structure of DNA

The search for the primary molecules of heredity was a serious focus throughout the first half of the 20th century. At first, many biologists thought that protein was the genetic material. An important milestone occurred in 1944 when Oswald Avery, Colin MacLeod, and Maclyn McCarty purified DNA and demonstrated at last that it was indeed the blueprint for life. This was followed by an avalanche of research, which continues today.

One area of extreme interest concerned the molecular structure of DNA. In 1951, American biologist James Watson and English physicist Francis Crick collaborated on solving the DNA puzzle. Although they did little of the original research, they were intrigued by several findings from other scientists. It had been determined by Erwin Chargaff that any model of DNA structure would have to contain deoxyribose, phosphate, purines, and pyrimidines arranged in a way that would provide variation and a simple way of copying itself. Watson and Crick spent long hours constructing models with cardboard cutouts and kept alert for any and every bit of information that might give them an edge.

Two English biophysicists, Maurice Wilkins and Rosalind Franklin, had been painstakingly collecting data on X-ray crystallographs of DNA for several years. With this technique, molecules of DNA bombarded by X rays produce a photographic image that can predict the three-dimensional structure of the molecule. After being allowed to view certain X-ray data, Watson and Crick noticed an unmistakable pattern: The molecule appeared to be a double helix. Gradually, the pieces of the puzzle fell into place, and a final model was assembled—a model that explained all of the qualities of DNA, including how it is copied. Although Watson and Crick were rightly hailed for the clarity of their solution, it must be emphasized that their success was due to the considerable efforts of a number of English and American scientists. This historic discovery showed that the tools of physics and chemistry have useful applications in biological systems, and it also spawned ingenious research in all areas of molecular genetics.

Since the discovery of the double helix in 1953, an extensive body of biochemical, microscopic, and crystallographic analysis has left little doubt that the model first proposed by Watson and Crick is correct. Newer techniques using scanning tunneling microscopy produce three-dimensional images of DNA magnified 2 million times. These images verify the helical shape and twists of DNA represented by models.

The men who cracked the code of life. Dr. James Watson (left) and Dr. Francis Crick (right) stand next to their model that finally explained the structure of DNA in 1953.

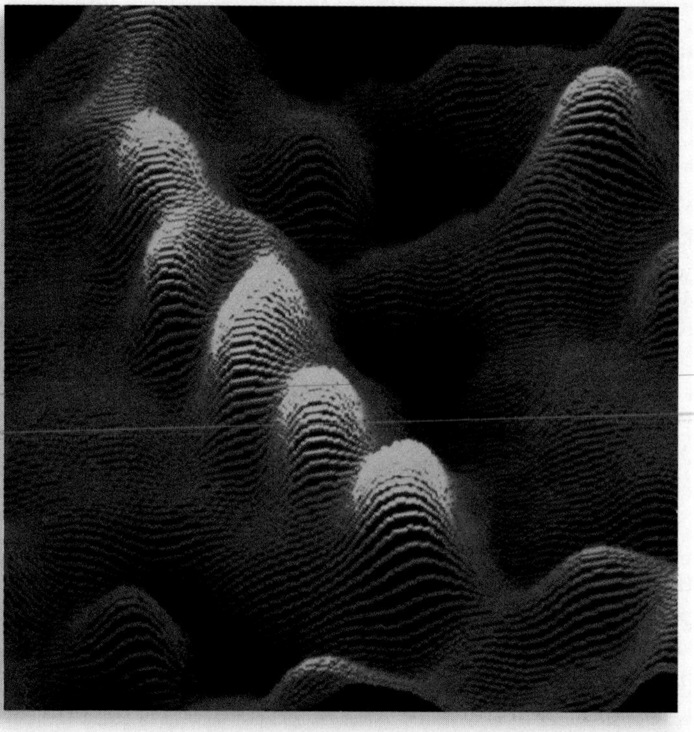

The first direct glimpse at DNA's structure. This false-color scanning tunneling micrograph of calf thymus gland DNA (2,000,000×) brings out the well-defined folds in the helix.

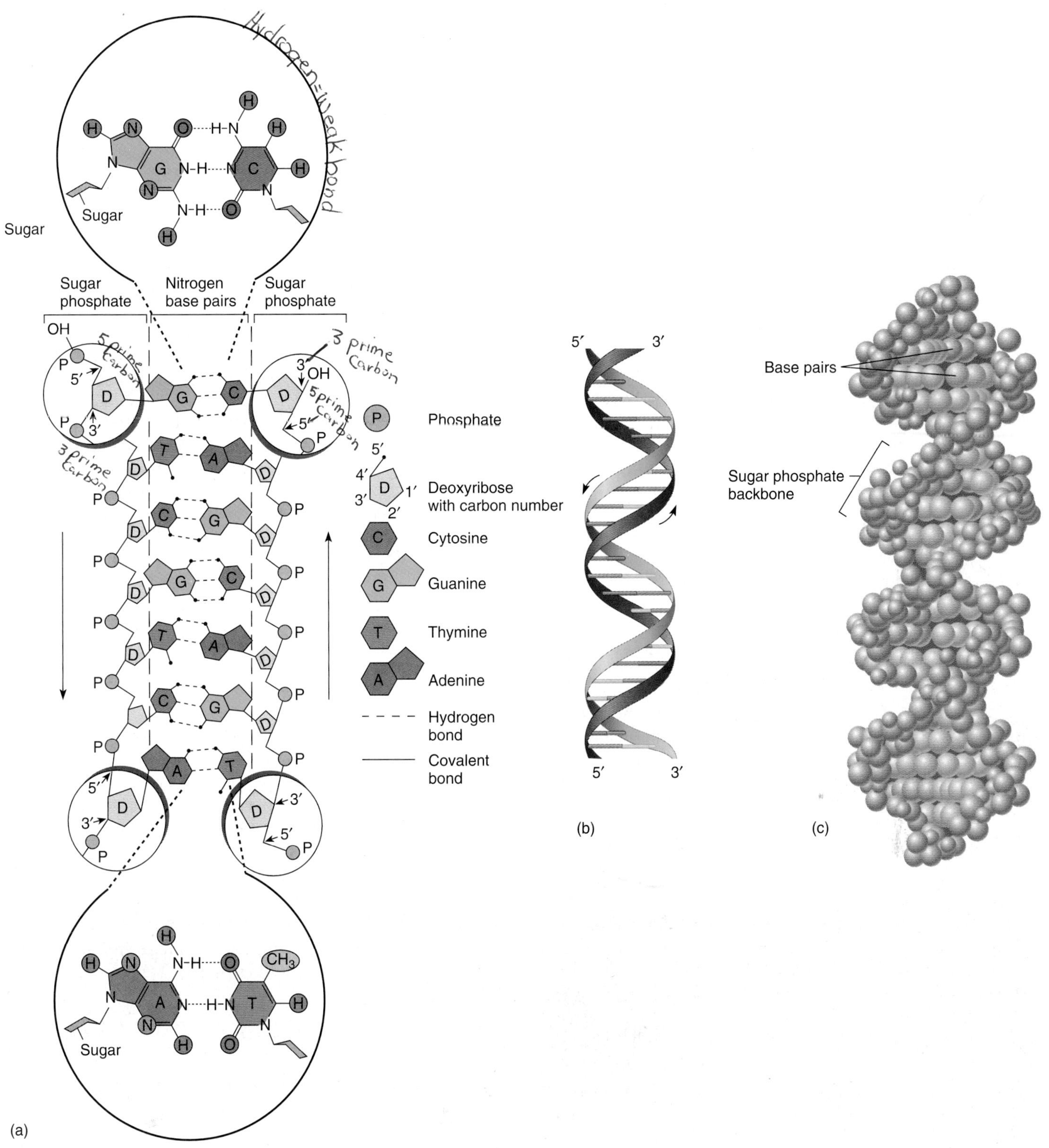

Figure 9.4 Three views of DNA structure.
(a) A schematic nonhelical model, to show the arrangement of the molecules it is made of. Note that the order of phosphate and sugar bonds differs between the two strands, going from the #5 carbon to the #3 carbon on one strand, and from the #3 carbon to the #5 carbon on the other strand. Insets show details of the nitrogen base pairs. **(b)** Simplified model that highlights the antiparallel arrangement and the major and minor grooves. **(c)** Space-filling model that more accurately depicts the three-dimensional structure of DNA.

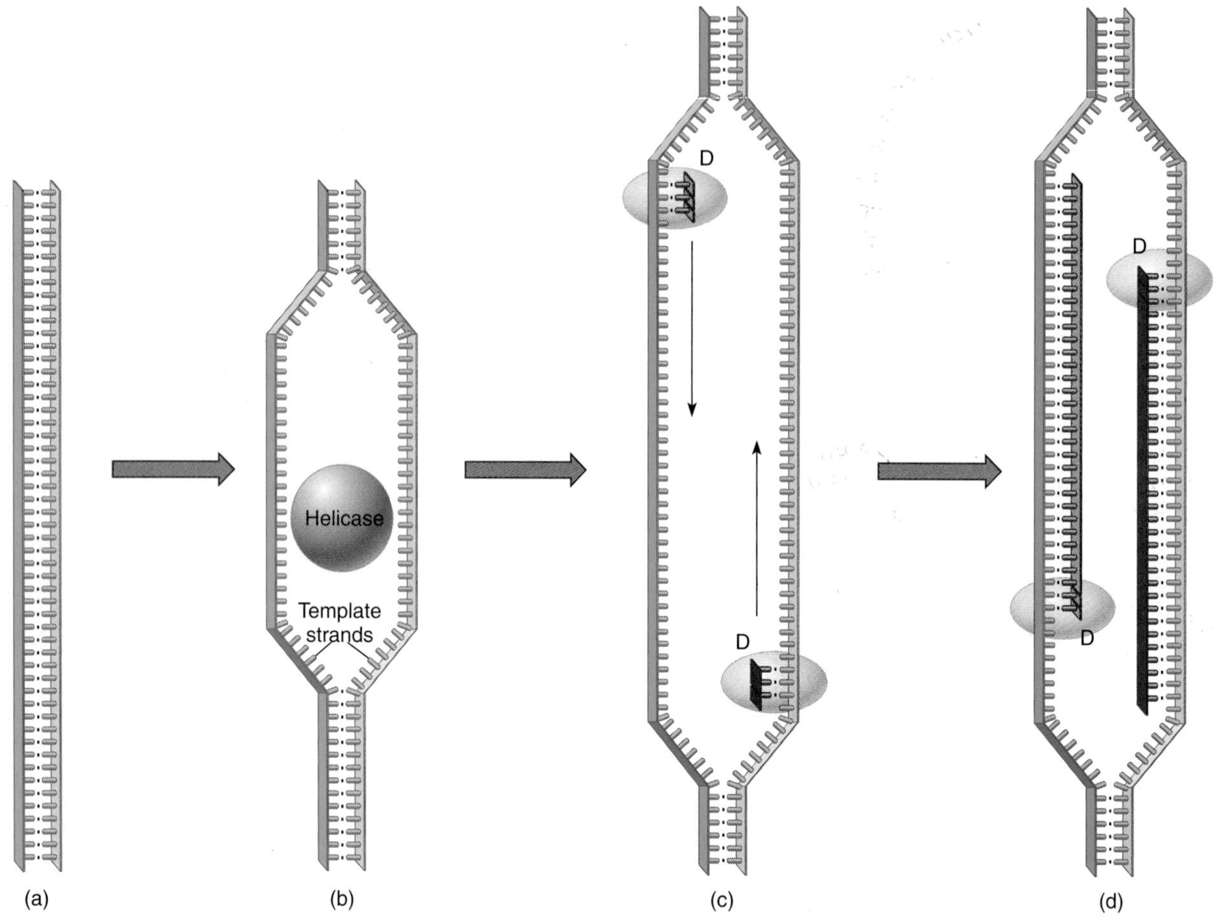

Figure 9.5 Simplified steps to show the semiconservative replication of DNA.
(a, b) The two strands of the double helix are unwound and separated by a helicase, which disrupts the hydrogen bonds and exposes the nitrogen base codes of DNA. Each single strand formed will serve as a template to synthesize a new strand of DNA. **(c)** A DNA polymerase (D) proceeds along the DNA molecule, attaching the correct nucleotides according to the pattern of the template. An A on the template will pair with a T on the new molecule, and a C will pair with a G. **(d)** The resultant new DNA molecules contain one strand of the newly synthesized DNA and the original template strand. The integrity of the code is kept intact because the linear arrangement of the bases is maintained during this process. Note that the actual details of the process are presented in figure 9.6.

new molecule **(figure 9.5).** Because the sequence of one strand automatically gives the sequence of its partner, the code can be duplicated with fidelity.

2. **Providing variety.** The order of bases along the length of the DNA strand constitutes the genetic program, or the language, of the DNA code. The message present in a gene is a precise sequence of these bases, and the genome is the collection of all DNA bases that, in an ordered combination, are responsible for the unique qualities of each organism.

It is tempting to ask how such a seemingly simple code can account for the extreme differences among forms as diverse as a virus, *E. coli,* and a human. The English language, based on 26 letters, can create an infinite variety of words, but how can an apparently complex genetic language such as DNA be based on just four nitrogen base "letters"? A mathematical example can explain the possibilities. For a segment of DNA that is 1,000 nucleotides long, there are $4^{1,000}$ different sequences possible. Carried out, this number would approximate 1.5×10^{602}, a number so huge that it provides nearly endless degrees of variation.

DNA Replication: Preserving the Code and Passing It On

The sequence of bases along the length of a gene constitutes the language of DNA. For this language to be preserved for hundreds of generations, it will be necessary for the genetic program to be duplicated and passed on to each offspring. This process of duplication is called DNA replication. In the following example, we will show replication in bacteria, but with some exceptions, it also applies to the process as it works in eukaryotes and some viruses. Early in binary fission, the metabolic machinery of a bacterium initiates the duplication of the chromosome. This DNA replication must be completed during a single generation time (around 20 minutes in *E. coli*).

The Overall Replication Process

What features allow the DNA molecule to be exactly duplicated, and how is its integrity retained? DNA replication requires a careful orchestration of the actions of 30 different enzymes (partial list in **table 9.1**), which separate the strands of the existing DNA molecule, copy its template, and produce two complete daughter molecules. A simplified version of replication is shown in figure 9.5 and includes the following:

1. uncoiling the parent DNA molecule;
2. unzipping the hydrogen bonds between the base pairs, thus separating the two strands and exposing the nucleotide sequence of each strand (which is normally buried in the center of the helix) to serve as templates; and
3. synthesizing two new strands by attachment of the correct complementary nucleotides to each single-stranded template.

A critical feature of DNA replication is that each daughter molecule will be identical to the parent in composition, but neither one is completely new; the strand that serves as a template is an original parental DNA strand. The preservation of the parent molecule in this way, termed **semiconservative replication,** helps explain the reliability and fidelity of replication.

Refinements and Details of Replication

The origin of replication is a short sequence rich in adenine and thymine that, you will recall, are held together by only two hydrogen bonds rather than three. Because the origin of replication is AT-rich, less energy is required to separate the two strands than would be required if the origin were rich in guanine and cytosine. Prior to the start of replication, enzymes called *helicases* (unzipping enzymes) bind to the DNA at the origin. These enzymes untwist the helix and break the hydrogen bonds holding the two strands together, resulting in two separate strands, each of which will be used as a template for the synthesis of a new strand.

The process of synthesizing a new daughter strand of DNA using the parental strand as a template is carried out by the enzyme DNA polymerase III. The entire process of replication does, however, depend on several enzymes and can be most easily understood by keeping in mind a few points concerning both the structure of the DNA molecule and the limitations of DNA polymerase III:

1. The nucleotides that need to be read by DNA polymerase III are buried deep within the double helix. Accessing these nucleotides requires both that the DNA molecule be unwound and that the two strands of the helix be separated from one another.
2. DNA polymerase III is unable to *begin* synthesizing a chain of nucleotides but can only continue to add nucleotides to an already existing chain.
3. DNA polymerase III can only add nucleotides in one direction, so a new strand is always synthesized 5′ to 3′.

With these constraints in mind, the details of replication can be more easily comprehended.

Replication begins when an RNA *primer* is synthesized and enters at the origin of replication **(figure 9.6, step 1).** DNA polymerase III cannot begin synthesis unless it has this short strand of RNA to serve as a starting point for adding nucleotides. Because the bacterial DNA molecule is circular, opening of the circle forms two **replication forks,** each containing its own set of replication enzymes. The DNA polymerase III is a huge enzyme complex that encircles the replication fork and adds nucleotides in accordance with the template pattern. As synthesis proceeds, the forks continually open up to expose the template for replication **(figure 9.6, steps 2, 3).**

Because DNA polymerase is correctly oriented for synthesis *only* in the 5′ to 3′ direction of the new molecule (red) strand, only one strand, called the **leading strand,** can be synthesized as a continuous, complete strand. The strand with the opposite orientation (3′ to 5′) is termed the **lagging strand (figure 9.6, steps 4, 5).** Because it cannot be synthesized continuously, the polymerase adds nucleotides a few at a time in the direction away from the fork (5′ to 3′). As the fork opens up a bit, the next segment is synthesized backward to the point of the previous segment, a process repeated at both forks until synthesis is complete. In this way, the DNA polymerase is able to synthesize the two new strands simultaneously. This manner of synthesis produces one strand containing short fragments of DNA (100 to 1,000 bases long) called **Okazaki fragments.** These fragments are attached to the growing end of the lagging strand by another enzyme called DNA ligase.

TABLE 9.1 Some Enzymes Involved in DNA Replication and Their Functions

Enzyme	Function
Helicase	Unzipping the DNA helix
Primase	Synthesizing an RNA primer
DNA polymerase III	Adding bases to the new DNA chain; proofreading the chain for mistakes
DNA polymerase I	Removing primer, closing gaps, repairing mismatches
Ligase	Final binding of nicks in DNA during synthesis and repair
Gyrase	Supercoiling

Elongation and Termination of the Daughter Molecules

The addition of nucleotides proceeds at an astonishing pace, estimated in some bacteria to be 750 bases per second at each fork! As replication proceeds, the newly produced double strand loops down **(figure 9.7*a*).** The DNA polymerase I removes the RNA primers used to initiate DNA synthesis and replaces them with DNA. When the forks come full circle and meet, ligases move along the lagging strand to begin the initial linking of the fragments and to complete synthesis and separation of the two circular daughter molecules **(figure 9.7*b*).**

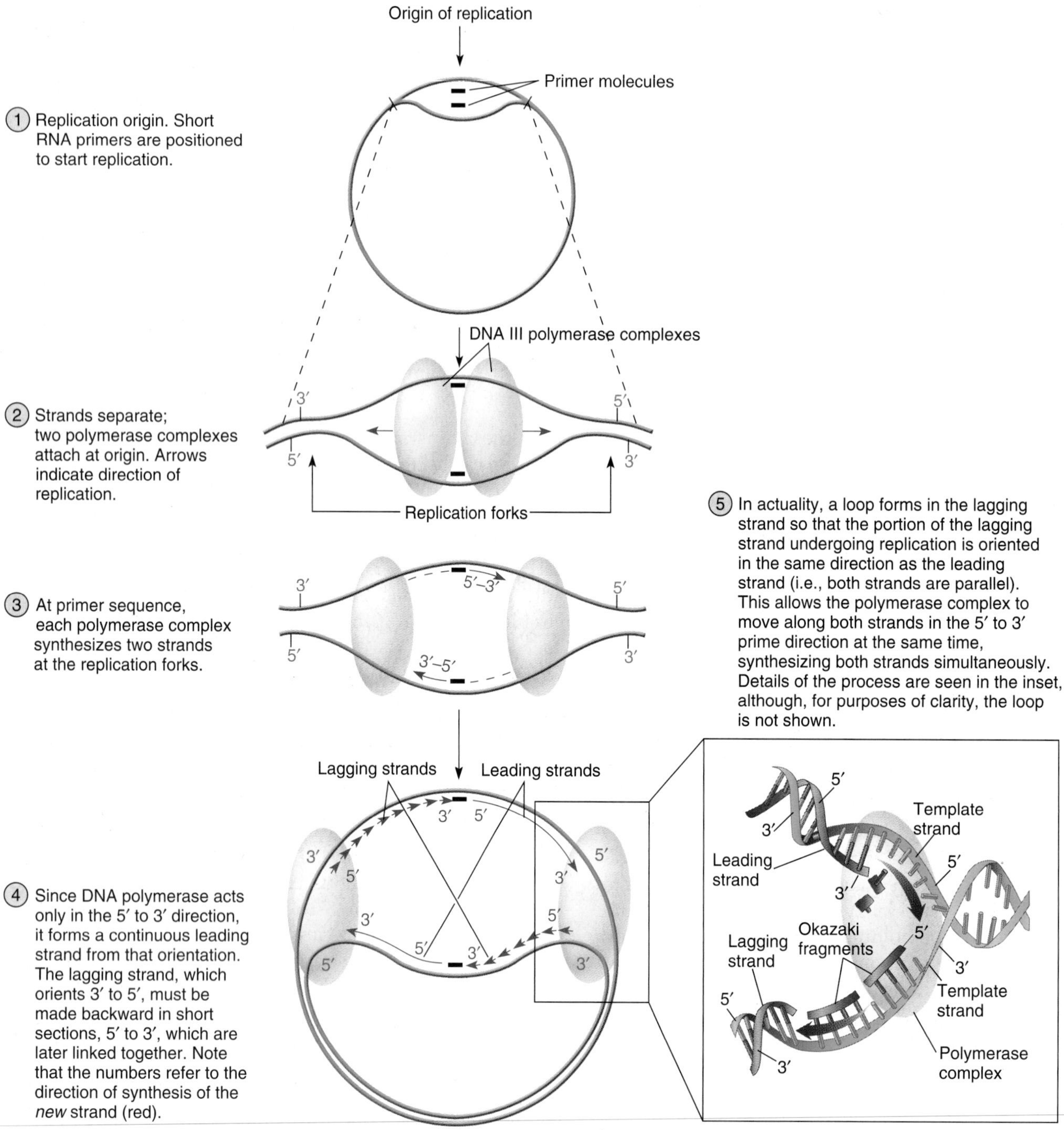

Origin of replication
Primer molecules
(1) Replication origin. Short RNA primers are positioned to start replication.
DNA III polymerase complexes
3′
5′
5′
3′
(2) Strands separate; two polymerase complexes attach at origin. Arrows indicate direction of replication.
Replication forks
3′
5′
5′–3′
3′–5′
5′
3′
(3) At primer sequence, each polymerase complex synthesizes two strands at the replication forks.
(5) In actuality, a loop forms in the lagging strand so that the portion of the lagging strand undergoing replication is oriented in the same direction as the leading strand (i.e., both strands are parallel). This allows the polymerase complex to move along both strands in the 5′ to 3′ prime direction at the same time, synthesizing both strands simultaneously. Details of the process are seen in the inset, although, for purposes of clarity, the loop is not shown.
Lagging strands
Leading strands
3′
5′
(4) Since DNA polymerase acts only in the 5′ to 3′ direction, it forms a continuous leading strand from that orientation. The lagging strand, which orients 3′ to 5′, must be made backward in short sections, 5′ to 3′, which are later linked together. Note that the numbers refer to the direction of synthesis of the *new* strand (red).
5′
3′
Template strand
Leading strand
Okazaki fragments
Lagging strand
Template strand
Polymerase complex

Process Figure 9.6 The bacterial replicon: a model for DNA synthesis.
(1) Circular DNA has a special origin site where replication originates. (2) When strands are separated, two replication forks form, and a DNA polymerase III complex enters at each fork. (3) Starting at the primer sequence, each polymerase moves along the template strands (blue), synthesizing the new strands (red) at each fork. (4) DNA polymerase works only in the 5′ to 3′ direction, necessitating a different pattern of replication at each fork. Because the leading strand orients in the 5′ to 3′ direction, it will be synthesized continuously. The lagging strand, which orients in the opposite direction, can only be synthesized in short sections, 5′ to 3′, which are later linked together. (5) Inset presents details of process at one replication fork and shows the Okazaki fragments and the relationship of the template, leading, and lagging strands.

Like any language, DNA is occasionally "misspelled" when an incorrect base is added to the growing chain. Studies have shown that such mistakes are made once in approximately 10^8 to 10^9 bases, but most of these are corrected. If not corrected, they are referred to as mutations and can lead to serious cell dysfunction and even death. Because continued cellular integrity is very dependent on accurate replication, cells have evolved their own proofreading function for DNA. DNA polymerase III, the enzyme that elongates the molecule, can also detect incorrect, unmatching

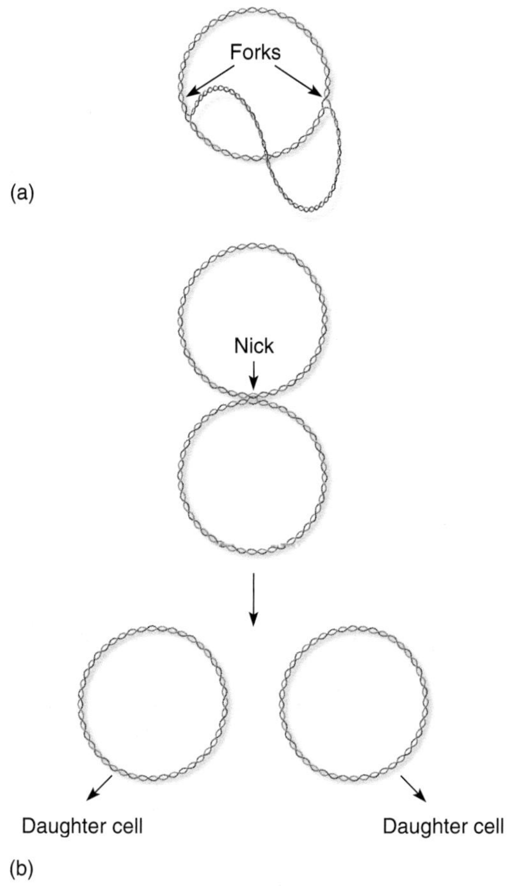

Figure 9.7 Completion of chromosome replication in bacteria.
(a) As replication proceeds, one double strand loops down. **(b)** Final separation is achieved through repair and the release of two completed molecules. The daughter cells receive these during binary fission.

bases; excise them; and replace them with the correct base. DNA polymerase I can also proofread the molecule and repair damaged DNA.

Replication in Other Biological Systems The replication pattern of eukaryotes is similar to that of prokaryotes. It also uses a variety of DNA polymerases, and replication proceeds in both directions from the point of origin.

CHECKPOINT

- Nucleic acids are molecules that contain the blueprints of life in the form of genes. DNA is the blueprint molecule for all cellular organisms. The blueprints of viruses, however, can be either DNA or RNA.
- The total amount of DNA in an organism is termed its genome (also genotype). The genome of each species contains a unique arrangement of genes that define its appearance (phenotype), metabolic activities, and pattern of reproduction.
- The genome of prokaryotes is quite small compared with the genome of eukaryotes. Bacterial DNA consists of a few thousand genes in one circular chromosome. Eukaryotic genomes range from *thousands* to *tens of thousands* of genes. Their DNA is packaged in tightly wound spirals arranged in discrete chromosomes.
- DNA copies itself just before cellular division by the process of semiconservative replication. Semiconservative replication means that each "old" DNA strand is the template upon which each "new" strand is synthesized.
- The circular bacterial chromosome is replicated at two forks as directed by DNA polymerase III. At each fork, two new strands are synthesized—one continuously and one in short fragments—and mistakes are proofread and removed.

9.2 Applications of the DNA Code: Transcription and Translation

We have explored how the genetic message in the DNA molecule is conserved through replication. Now we must consider the precise role of DNA in the cell. Given that the sequence of bases in DNA is a genetic code, just what is the nature of this code and how is it utilized by the cell? Although the genome is full of critical information, the molecule itself does not perform cell processes directly. Its stored information is conveyed to RNA molecules, which carry out instructions. The concept that genetic information flows from DNA to RNA to protein is a central theme of molecular biology **(figure 9.8*a*).** More precisely, it states that the master code of DNA is first used to synthesize an RNA molecule via a process called **transcription,** and the information contained in the RNA is then used to produce proteins in a process known as **translation.** The principal exceptions to this pattern are found in RNA viruses, which convert RNA to other RNA, and in retroviruses, which convert RNA to DNA.

This "central dogma," which outlined the primary understanding of genetics during the first half century of the genetic revolution (since the 1950s), has very recently been shown to be incomplete. While it is true that proteins are made in accordance with this central dogma, there is more to the story (**figure 9.8*b***). In addition to the RNA that is used to produce proteins, a wide variety of RNAs are used to regulate gene function. Many of the genetic malfunctions that cause human disease are in fact found in these regulatory RNA segments—and not in genes for proteins as was once thought. The DNA that codes for these very crucial RNA molecules was called "junk" DNA until very recently. We say more about this in Insight 9.2.

The Gene-Protein Connection

The Triplet Code and the Relationship to Proteins

Several questions invariably arise concerning the relationship between genes and cell function. For instance, how does gene structure lead to the expression of traits in the individual, and what features of gene expression cause one organism to be so distinctly different from another? For answers, we must turn to the correlation between gene and protein structure. We know

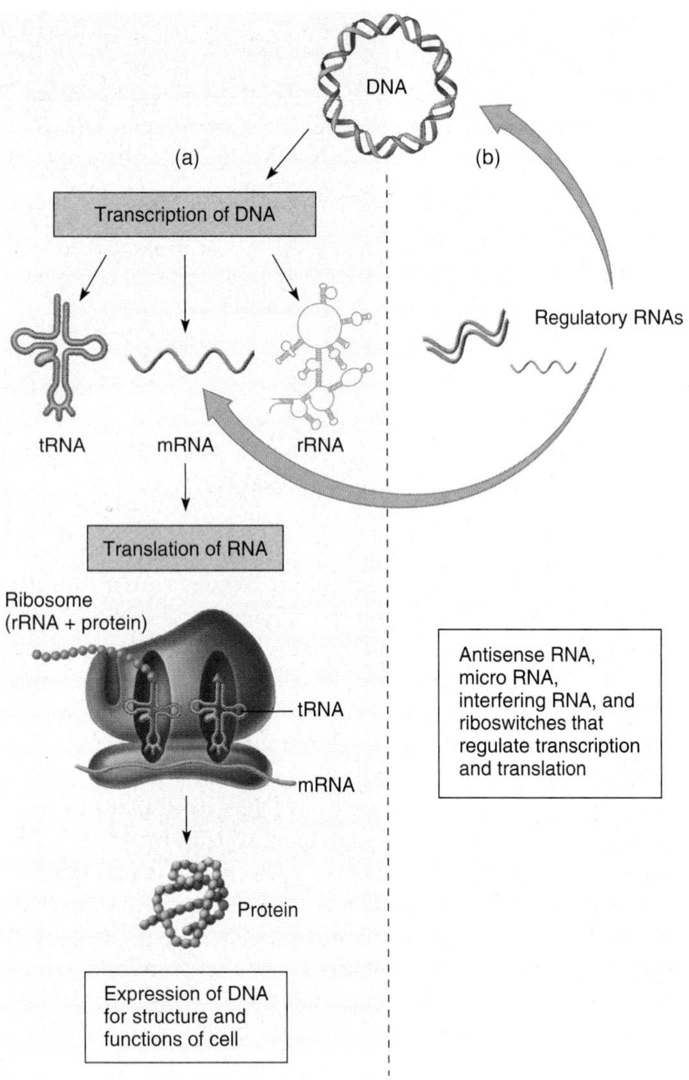

Figure 9.8 Summary of the flow of genetic information in cells.
DNA is the ultimate storehouse and distributor of genetic information. **(a)** DNA must be deciphered into a usable cell language. It does this by transcribing its code into RNA helper molecules that translate that code into protein. **(b)** Other sections of the DNA produce very important RNA molecules that regulate genes and their products.

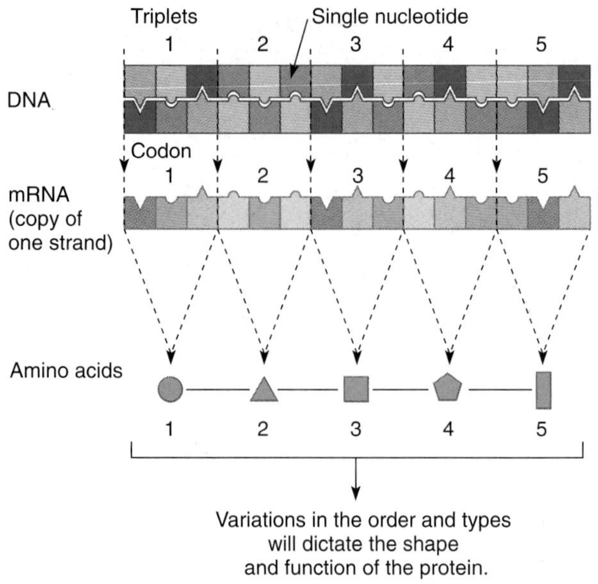

Figure 9.9 Simplified view of the DNA-protein relationship.
The DNA molecule is a continuous chain of base pairs, but the sequence must be interpreted in groups of three base pairs (a triplet). Each triplet as copied into mRNA codons will translate into one amino acid; consequently, the ratio of base pairs to amino acids is 3:1.

that each structural gene is a linear sequence of nucleotides that codes for a protein. Because each protein is different, each gene must also differ somehow in its composition. In fact, the language of DNA exists in the order of groups of three consecutive bases called triplets on one DNA strand **(figure 9.9).** Thus, one gene differs from another in its composition of triplets. An equally important part of this concept is that each triplet represents a code for a particular amino acid. When the triplet code is transcribed and translated, it dictates the type and order of amino acids in a polypeptide (protein) chain.

The final key points that connect DNA and an organism's traits are:

1. A protein's primary structure—the order and type of amino acids in the chain—determines its characteristic shape and function.
2. Proteins ultimately determine phenotype, the expression of all aspects of cell function and structure. Put more simply, living things are what their proteins make them. Regulatory RNAs help determine which proteins are made.
3. DNA is mainly a blueprint that tells the cell which kinds of proteins and RNAs to make and how to make them.

The Major Participants in Transcription and Translation

Transcription, the formation of RNA using DNA as a template, and translation, the synthesis of proteins using RNA as a template, are highly complex. A number of components participate: most prominently, messenger RNA, transfer RNA, regulatory RNAs, ribosomes, several types of enzymes, and a storehouse of raw materials. After first examining each of these components, we shall see how they come together in the assembly line of the cell.

RNAs: Tools in the Cell's Assembly Line

Ribonucleic acid is an encoded molecule like DNA, but its general structure is different in several ways:

1. It is a single-stranded molecule that exists in helical form. This single strand can assume secondary and tertiary levels of complexity due to bonds within the molecule, leading to specialized forms of RNA (tRNA and rRNA—see figure 9.8*a*).
2. RNA contains **uracil,** instead of thymine, as the complementary base-pairing mate for adenine. This does not

INSIGHT 9.2 *Discovery*

Small RNAs: An Old Dog Shows Off Some New(?) Tricks

Since the earliest days of molecular biology, RNA has been an overlooked worker of the cell, quietly ferrying the information in DNA to ribosomes to direct the formation of proteins. Current research however is showing a new, dynamic role for RNA in the cell that may forever change the reputation of this humble molecule.

Short lengths of RNA seem to have the ability to control the expression of certain genes. Some of these are called micro RNAs and some are called small interfering RNAs. They do this by folding back on themselves after being transcribed, and by doing so they activate a system inside cells that degrades dsRNA. Cells do this in order to rid themselves of invading viruses (which are organisms that might have dsRNA). The micro RNAs also bind with mRNA of certain genes, thereby causing them to be degraded as well. The repressing nature of dsRNA was discovered quite accidentally when researchers were trying to induce expression of genes by providing them in dsRNA form; instead, genes matching those RNA sequences were shut down entirely through this clever regulatory system. In 2006, the Nobel Prize for Medicine or Physiology was awarded to the two American scientists, Andrew Fire and Craig Mello, who discovered this phenomenon.

A second type of regulation seems to occur when small RNAs alter the structure of chromosomes. As DNA and proteins coil together to form chromatin, small RNAs direct how tightly or loosely the chromatin is constructed. Just as a closed book cannot be read, DNA sequences contained within tightly coiled chromatin are generally inaccessible to the cell, silencing the expression of those genes. Antisense RNA is produced from the opposite strand of the DNA that produces mRNA. This antisense molecule has the ability to pair with the "sense," or messenger, RNA and thus keep it from being transcribed. Riboswitches, RNAs that attach to a chemical with one end and only then become available for translation on the other end, were isolated for the first time in 2002. One riboswitch has been found to regulate the expression of 26 important genes in the bacterium *Bacillus subtilis*. Riboswitches have probably been around since the early days of life on the planet. So, although they are new to us, they have been used to regulate gene expression for billions of years.

These newly discovered RNA molecules have answered some vexing questions that came out of the genome sequencing studies (led by the Human Genome Project). Most of the DNA in organisms was found *not* to code for functional proteins. In humans, the "junk" percentage was 98%! Yet, in bacteria as well as humans, the junk DNA was preserved in the same form for the last millions of years of evolution, suggesting it had a very important function. We now know that much of this "junk" DNA codes for these important RNA regulatory molecules.

The RNA regulatory molecules are being heavily exploited to accomplish research tasks that were never before possible. More important, molecules such as antisense RNA are being explored for their therapeutic uses in cases where defective genes need to be shut down in order to restore a patient to health.

Our knowledge of the full role of small RNAs in the cell is just beginning. In the meantime, scientists will keep studying small RNAs while being mindful of the old adage, "Good things come in small packages."

change the inherent DNA code in any way because the uracil still follows the pairing rules.

3. Although RNA, like DNA, contains a backbone that consists of alternating sugar and phosphate molecules, the sugar in RNA is **ribose** rather than deoxyribose.

The many functional types of RNA range from small regulatory pieces to large structural ones **(table 9.2** and **Insight 9.2).** All types of RNA are formed through transcription of a DNA gene, but only mRNA is further translated into another type of molecule (protein).

TABLE 9.2 Types of Ribonucleic Acid

RNA Type	Contains Codes For	Function in Cell	Translated
Messenger (mRNA)	Sequence of amino acids in protein	Carries the DNA master code to the ribosome	Yes
Transfer (tRNA)	A cloverleaf tRNA to carry amino acids	Brings amino acids to ribosome during translation	No
Ribosomal (rRNA)	Several large structural rRNA molecules	Forms the major part of a ribosome and participates in protein synthesis	No
Micro (miRNA) antisense, riboswitch, and small interfering (siRNA)	Regulatory RNAs	Regulation of gene expression and coiling of chromatin	No
Primer	An RNA that can begin DNA replication	Primes DNA	No
Ribozymes	RNA enzymes, parts of splicer enzymes	Remove introns from other RNAs in eukaryotes	No

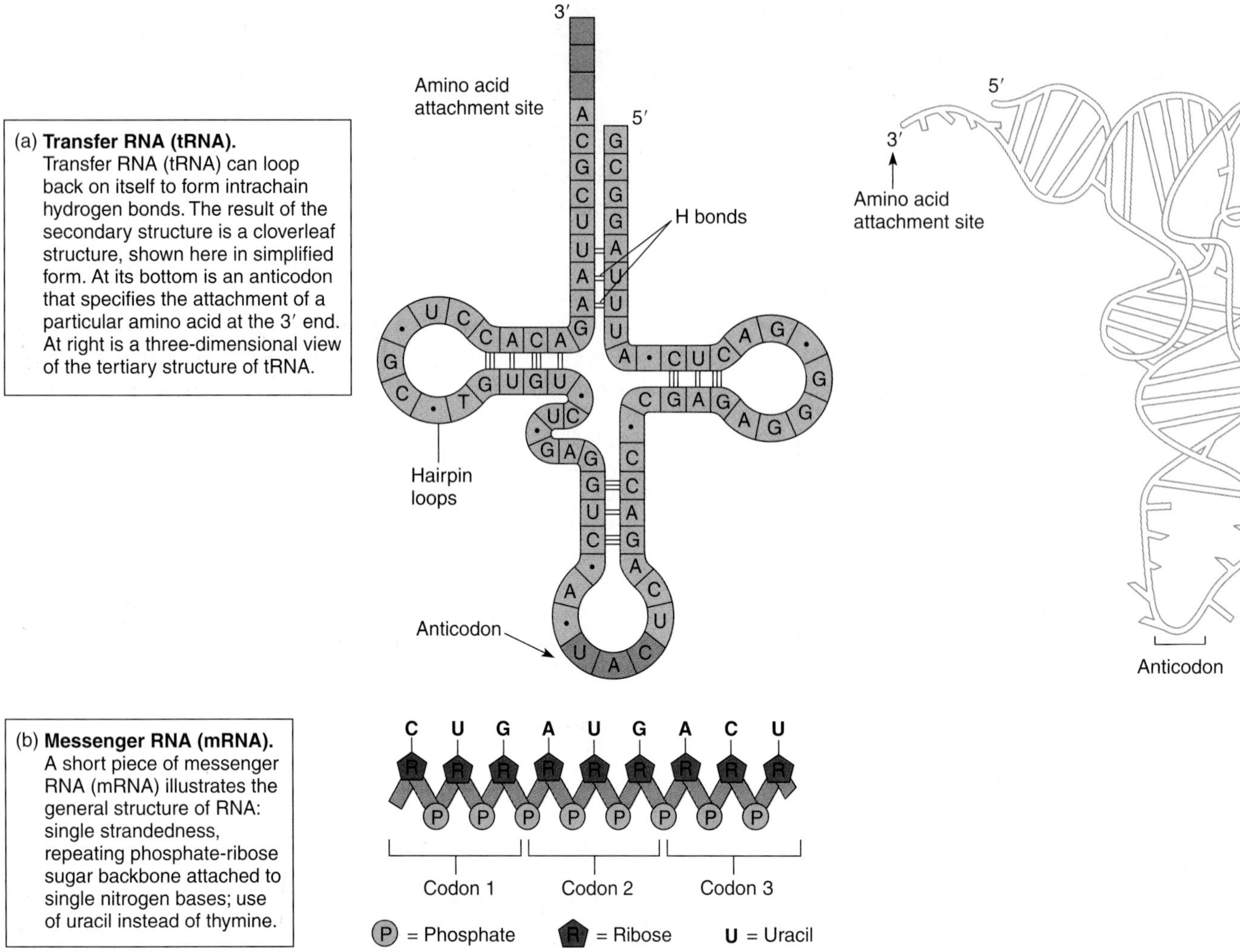

(a) Transfer RNA (tRNA).
Transfer RNA (tRNA) can loop back on itself to form intrachain hydrogen bonds. The result of the secondary structure is a cloverleaf structure, shown here in simplified form. At its bottom is an anticodon that specifies the attachment of a particular amino acid at the 3′ end. At right is a three-dimensional view of the tertiary structure of tRNA.
3′
5′
Amino acid attachment site
H bonds
Hairpin loops
Anticodon
(b) Messenger RNA (mRNA).
A short piece of messenger RNA (mRNA) illustrates the general structure of RNA: single strandedness, repeating phosphate-ribose sugar backbone attached to single nitrogen bases; use of uracil instead of thymine.
C U G A U G A C U
Codon 1
Codon 2
Codon 3
P = Phosphate
R = Ribose
U = Uracil

Figure 9.10 **Characteristics of transfer and messenger RNA.**

Messenger RNA: Carrying DNA's Message

Messenger RNA (mRNA) is a transcript (copy) of a structural gene or genes in the DNA. It is synthesized by a process similar to synthesis of the leading strand during DNA replication, and the complementary base-pairing rules ensure that the code will be faithfully copied in the mRNA transcript. The message of this transcribed strand is later read as a series of triplets called **codons (figure 9.10),** and the length of the mRNA molecule varies from about 100 nucleotides to several thousand. The details of transcription and the function of mRNA in translation will be covered shortly.

Transfer RNA: The Key to Translation

Transfer RNA (tRNA) is also a copy of a specific region of DNA; however, it differs from mRNA. It is uniform in length, being 75 to 95 nucleotides long, and it contains sequences of bases that form hydrogen bonds with complementary sections of the same tRNA strand. At these points, the molecule bends back upon itself into several *hairpin loops*, giving the molecule a secondary *cloverleaf* structure that folds even further into a complex, three-dimensional helix (figure 9.10). This compact molecule is an adaptor that converts RNA language into protein language. The bottom loop of the cloverleaf exposes a triplet, the **anticodon,** that both designates the specificity of the tRNA and complements mRNA's codons. At the opposite end of the molecule is a binding site for the amino acid that is specific for that tRNA's anticodon. For each of the 20 amino acids, there is at least one specialized type of tRNA to carry it. Binding of an amino acid to its specific tRNA, a process known as "charging" the tRNA, takes place in two enzyme-driven steps: First an ATP activates the amino acid, and then this group binds to the acceptor end of the tRNA. Because tRNA is the molecule that will convert the master code on mRNA into a protein, the accuracy of this step is crucial.

The Ribosome: A Mobile Molecular Factory for Translation

The prokaryotic (70S) ribosome is a particle composed of tightly packaged **ribosomal RNA (rRNA)** and protein. The rRNA component of the ribosome is also a long polynucleotide molecule. It forms complex three-dimensional figures that

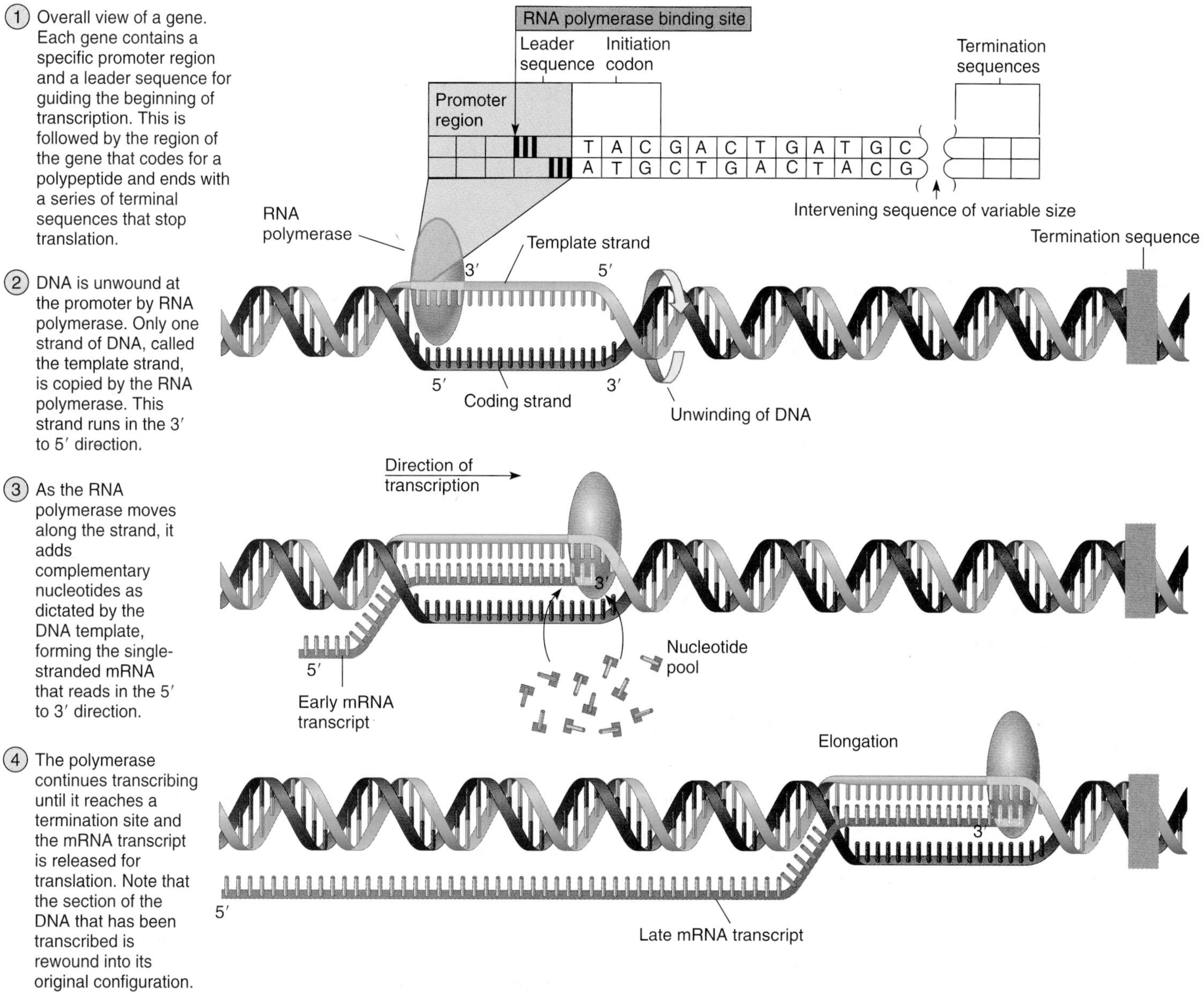
1 Overall view of a gene. Each gene contains a specific promoter region and a leader sequence for guiding the beginning of transcription. This is followed by the region of the gene that codes for a polypeptide and ends with a series of terminal sequences that stop translation.
2 DNA is unwound at the promoter by RNA polymerase. Only one strand of DNA, called the template strand, is copied by the RNA polymerase. This strand runs in the 3′ to 5′ direction.
3 As the RNA polymerase moves along the strand, it adds complementary nucleotides as dictated by the DNA template, forming the single-stranded mRNA that reads in the 5′ to 3′ direction.
4 The polymerase continues transcribing until it reaches a termination site and the mRNA transcript is released for translation. Note that the section of the DNA that has been transcribed is rewound into its original configuration.
RNA polymerase binding site
Leader sequence
Initiation codon
Termination sequences
Promoter region
T A C G A C T G A T G C
A T G C T G A C T A C G
Intervening sequence of variable size
RNA polymerase
Template strand
Termination sequence
3′
5′
5′
3′
Coding strand
Unwinding of DNA
Direction of transcription
3′
Nucleotide pool
5′
Early mRNA transcript
Elongation
3′
5′
Late mRNA transcript

Process Figure 9.11 **The major events in mRNA synthesis (transcription).**

contribute to the structure and function of ribosomes. The interactions of proteins and rRNA create the two subunits of the ribosome that engage in final translation of the genetic code (see figure 9.12). A metabolically active bacterial cell can accommodate up to 20,000 of these minuscule factories—all actively engaged in reading the genetic program, taking in raw materials, and producing proteins at an impressive rate.

Transcription: The First Stage of Gene Expression

During transcription, the DNA code is converted to RNA through several stages, directed by a huge and very complex enzyme system, **RNA polymerase (figure 9.11).** Only one strand of the DNA—the *template strand*—contains meaningful instructions for synthesis of a functioning polypeptide. The nontranscribed strand is called the *coding strand*. The strand of DNA that serves as a template varies from one gene to another.

Transcription is initiated when RNA polymerase recognizes a segment of the DNA called the *promoter region*. This region consists of two sequences of DNA just prior to the beginning of the gene to be transcribed. The first sequence, which occurs approximately 35 bases prior to the start of transcription, is tightly bound by RNA polymerase. Transcription is allowed to begin when the DNA helix begins to unwind at the second sequence, which is located about 10 bases prior to the start of transcription. As the DNA helix unwinds, the polymerase advances and begins synthesizing an RNA molecule complementary to the template strand of DNA. The nucleotide sequence of promoters differs only slightly from gene to gene, with all promoters being rich in adenine and thymine.

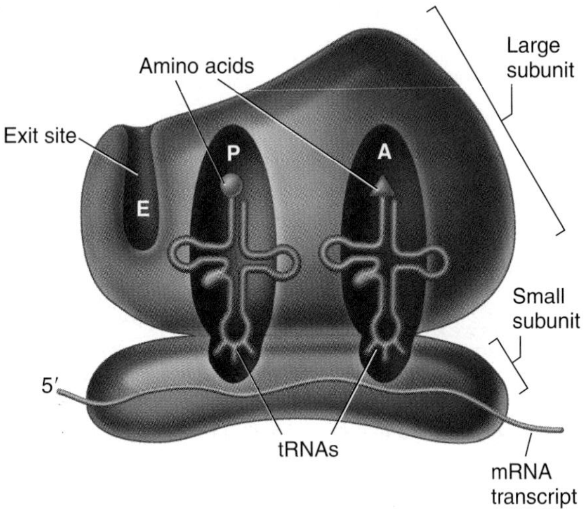

Figure 9.12 The "players" in translation.
A ribosome serves as the stage for protein synthesis. Assembly of the small and large subunits results in specific sites for holding the mRNA and two tRNAs with their amino acids.

During elongation, which proceeds in the 5′ to 3′ direction (with regard to the growing RNA molecule), the mRNA is assembled by the addition of nucleotides that are complementary to the DNA template. Remember that uracil (U) is placed as adenine's complement. As elongation continues, the part of DNA already transcribed is rewound into its original helical form. At termination, the polymerases recognize another code that signals the separation and release of the mRNA strand, also called the **transcript.** How long is the mRNA? The very smallest mRNA might consist of 100 bases; an average-size mRNA might consist of 1,200 bases; and a large one might consist of several thousand.

Translation: The Second Stage of Gene Expression

In translation, all of the elements needed to synthesize a protein, from the mRNA to the amino acids, are brought together on the ribosomes **(figure 9.12).** The process occurs in five stages: initiation, elongation, termination, protein folding, and protein processing.

Initiation of Translation

The mRNA molecule leaves the DNA transcription site and is transported to ribosomes in the cytoplasm. Ribosomal subunits are specifically adapted to assembling and forming sites to hold the mRNA and tRNAs. The ribosomes of prokaryotes and eukaryotes are different sizes. Prokaryotic ribosomes, as well as the ribosomes in chloroplasts and mitochondria of eukaryotes, are of a 70s size, made up of a 50s (large) subunit and a 30s (small) subunit. The "s" is a measurement of sedimentation rates, which is how ribosomes are characterized. It is a nonlinear measure; therefore, 30s and 50s add up to 70s. Eukaryotic ribosomes are 80s (a large subunit of 60s and a 40s small subunit). The ribosome thus recognizes these molecules and stabilizes reactions between them. The small subunit binds to the 5′ end of the mRNA, and the large subunit supplies enzymes for making peptide bonds on the protein. The ribosome begins to scan the mRNA by moving in the 5′ to 3′ direction along the mRNA. The first codon it encounters is the START codon, which is almost always AUG (and, rarely, GUG).

With the mRNA message in place on the assembled ribosome, the next step in translation involves entrance of tRNAs with their amino acids. The pool of cytoplasm contains a complete array of tRNAs, previously charged by having the correct amino acid attached. The step in which the complementary tRNA meets with the mRNA code is guided by the two sites on the large subunit of the ribosome called the P site (left) and the A site (right).[1] Think of these sites as shallow depressions in the larger subunit of the ribosome, each of which accommodates a tRNA. The ribosome also has an exit or E site where used tRNAs are released.

The Master Genetic Code: The Message in Messenger RNA

By convention, the master genetic code is represented by the mRNA codons and the amino acids they specify **(figure 9.13).** Except in a very few cases, this code is universal, whether for prokaryotes, eukaryotes, or viruses. It is worth noting that once the triplet code on mRNA is known, the original DNA sequence, the complementary tRNA code, and the types of amino acids in the protein are automatically known **(figure 9.14).** However, one cannot predict (backward) from protein structure what the exact mRNA codons are because of a factor called **redundancy,**[2] meaning that a particular amino acid can be coded for by more than a single codon.

In figure 9.13, the mRNA codons and their corresponding amino acid specificities are given. Because there are 64 different triplet codes[3] and only 20 different amino acids, it is not surprising that some amino acids are represented by several codons. For example, leucine and serine can each be represented by any of six different triplets, and only tryptophan and methionine are represented by a single codon. In such codons as leucine, only the first two nucleotides are required to encode the correct amino acid, and the third nucleotide does not change its sense. This property, called *wobble,* is thought to permit some variation or mutation without altering the message.

The Beginning of Protein Synthesis

With mRNA serving as the guide, the stage is finally set for actual protein assembly. The correct tRNA (labeled 1 on

1. P stands for peptide site; A stands for aminoacyl (amino acid) site; E stands for exit site.
2. This property is also called "degeneracy" by some books.
3. $64 = 4^3$ (the four different codons in all possible combinations of three).

First Base Position	Second Base Position: U		C		A		G		Third Base Position
U	UUU	Phenylalanine	UCU	Serine	UAU	Tyrosine	UGU	Cysteine	U
	UUC		UCC		UAC		UGC		C
	UUA	Leucine	UCA		UAA	STOP**	UGA	STOP**	A
	UUG		UCG		UAG		UGG	Tryptophan	G
C	CUU	Leucine	CCU	Proline	CAU	Histidine	CGU	Arginine	U
	CUC		CCC		CAC		CGC		C
	CUA		CCA		CAA	Glutamine	CGA		A
	CUG		CCG		CAG		CGG		G
A	AUU	Isoleucine	ACU	Threonine	AAU	Asparagine	AGU	Serine	U
	AUC		ACC		AAC		AGC		C
	AUA		ACA		AAA	Lysine	AGA	Arginine	A
	AUG	START fMethionine*	ACG		AAG		AGG		G
G	GUU	Valine	GCU	Alanine	GAU	Aspartic acid	GGU	Glycine	U
	GUC		GCC		GAC		GGC		C
	GUA		GCA		GAA	Glutamic acid	GGA		A
	GUG		GCG		GAG		GGG		G

* This codon initiates translation.
**For these codons, which give the orders to stop translation, there are no corresponding tRNAs and no amino acids.

Figure 9.13 The genetic code: codons of mRNA that specify a given amino acid. The master code for translation is found in the mRNA codons.

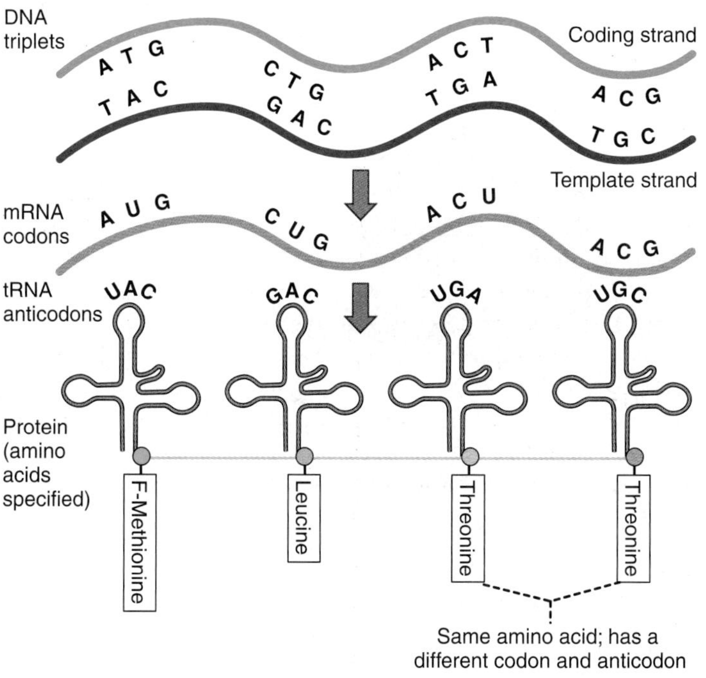

Figure 9.14 Interpreting the DNA code. If the DNA sequence is known, the mRNA codon can be surmised. If a codon is known, the anticodon and, finally, the amino acid sequence can be determined. The reverse is not possible (determining the exact codon or anticodon from amino acid sequence) due to the redundancy of the code.

figure 9.15) enters the P site and binds to the **start codon (AUG)** presented by the mRNA. Rules of pairing dictate that the anticodon of this tRNA must be complementary to the mRNA codon AUG; thus, the tRNA with anticodon UAC will first occupy site P. It happens that the amino acid carried by the initiator tRNA in bacteria is *formyl methionine* (fMet; see figure 9.13), though, in many cases, it may not remain a permanent part of the finished protein.

Continuation and Completion of Protein Synthesis: Elongation and Termination

While reviewing the dynamic process of protein assembly, you will want to remain aware that the ribosome shifts its "reading frame" to the right along the mRNA from one codon to the next. This brings the next codon into place on the ribosome and makes a space for the next tRNA to enter the A position. A peptide bond is formed between the amino acids on the adjacent tRNAs, and the polypeptide grows in length.

Elongation begins with the filling of the A site by a second tRNA (2 on **figure 9.15**). The identity of this tRNA and its amino acid is dictated by the second mRNA codon.

The entry of tRNA 2 into the A site brings the two adjacent tRNAs in favorable proximity for a peptide bond to form between the amino acids (aa) they carry. The fMet is transferred from the first tRNA to aa 2, resulting in two coupled amino acids called a dipeptide **(figure 9.15, step 2).**

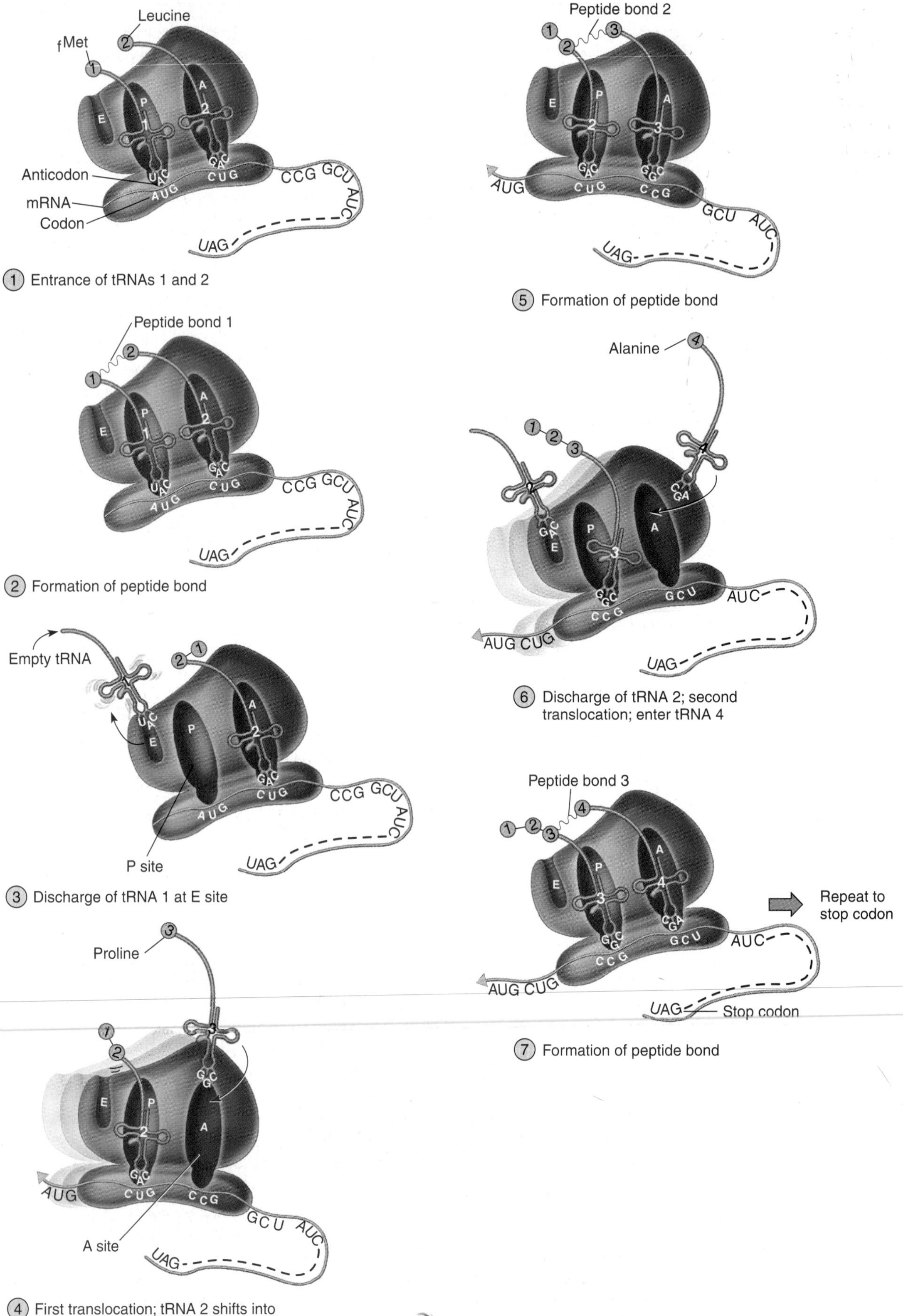

Process Figure 9.15 The events in protein synthesis.

For the next step to proceed, some room must be made on the ribosome, and the next codon in sequence must be brought into position for reading. This process is accomplished by *translocation,* the enzyme-directed shifting of the ribosome to the right along the mRNA strand, which causes the blank tRNA (1) to be discharged from the ribosome **(figure 9.15, step 3)** at the E site. This also shifts the tRNA holding the dipeptide into P position. Site A is temporarily left empty. The tRNA that has been released is now free to drift off into the cytoplasm and become recharged with an amino acid for later additions to this or another protein.

The stage is now set for the insertion of tRNA 3 at site A as directed by the third mRNA codon **(figure 9.15, step 4).** This insertion is followed once again by peptide bond formation between the dipeptide and aa 3 (making a tripeptide), splitting of the peptide from tRNA 2, and translocation. This releases tRNA 2, shifts mRNA to the next position, moves tRNA 3 to position P, and opens position A for the next tRNA (which will be called tRNA 4). From this point on, peptide elongation proceeds repetitively by this same series of actions out to the end of the mRNA.

The termination of protein synthesis is not simply a matter of reaching the last codon on mRNA. It is brought about by the presence of at least one special codon occurring just after the codon for the last amino acid. Termination codons—UAA, UAG, and UGA—are codons for which there is no corresponding tRNA. Although they are often called **nonsense codons,** they carry a necessary and useful message: *Stop here.* When this codon is reached, a special enzyme breaks the bond between the final tRNA and the finished polypeptide chain, releasing it from the ribosome.

Before newly made proteins can carry out their structural or enzymatic roles, they often require finishing touches. Even before the peptide chain is released from the ribosome, it begins folding upon itself to achieve its biologically active tertiary conformation. Other alterations, called *posttranslational* modifications, may be necessary. Some proteins must have the starting amino acid (formyl methionine) clipped off; proteins destined to become complex enzymes have cofactors added; and some join with other completed proteins to form quaternary levels of structure.

The operation of transcription and translation is machinelike in its precision. Protein synthesis in bacteria is both efficient and rapid. At 37°C, 12 to 17 amino acids per second are added to a growing peptide chain. An average protein consisting of about 400 amino acids requires less than half a minute for complete synthesis. Further efficiency is gained when the translation of mRNA starts while transcription is still occurring **(figure 9.16).** A single mRNA is long enough

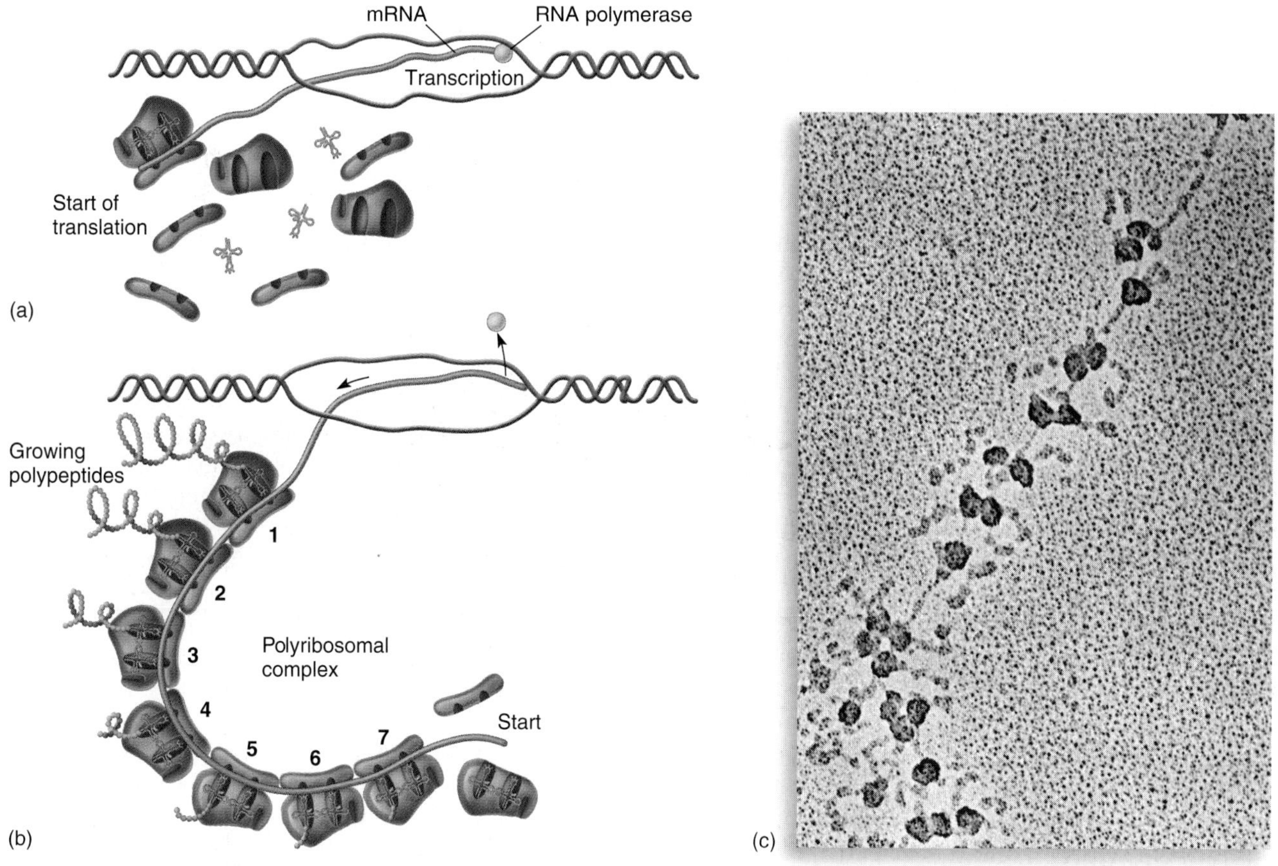

Figure 9.16 Speeding up the protein assembly line in bacteria.
(a) The mRNA transcript encounters ribosomal parts immediately as it leaves the DNA. **(b)** The ribosomal factories assemble along the mRNA in a chain, each ribosome reading the message and translating it into protein. Many products will thus be well along the synthetic pathway before transcription has even terminated. **(c)** Photomicrograph of a polyribosomal complex in action. Note that the protein "tails" vary in length depending on the stage of translation.

to be fed through more than one ribosome simultaneously. This permits the synthesis of hundreds of protein molecules from the same mRNA transcript arrayed along a chain of ribosomes. This **polyribosomal complex** is indeed an assembly line for mass production of proteins. Protein synthesis consumes an enormous amount of energy. Nearly 1,200 ATPs are required just for synthesis of an average-size protein.

Eukaryotic Transcription and Translation: Similar Yet Different

Eukaryotes and prokaryotes share many similarities in protein synthesis. The start codon in eukaryotes is also AUG, but it codes for a different form of methionine. Another difference is that eukaryotic mRNAs code for just one protein, unlike bacterial mRNAs, which often contain information from several genes in series.

There are a few differences between prokaryotic and eukaryotic gene expression. The presence of the DNA in a separate compartment (the nucleus) means that eukaryotic transcription and translation cannot be simultaneous. The mRNA transcript must pass through pores in the nuclear membrane and be carried to the ribosomes in the cytoplasm for translation.

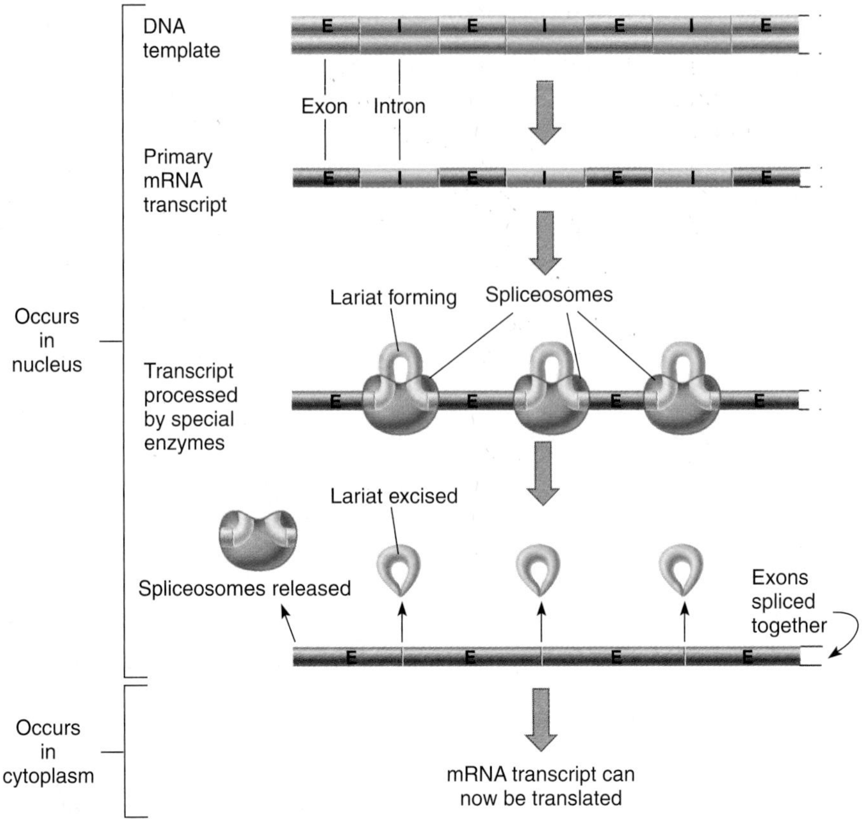

Figure 9.17 The split gene of eukaryotes. Eukaryotic genes have an additional complicating factor in their translation. Their coding sequences, or exons (E), are interrupted at intervals by segments called introns (I) that are not part of that protein's code. Introns are transcribed but not translated, which necessitates their removal by RNA splicing enzymes before translation.

We have given the simplified definition of a gene that works well for prokaryotes, but most eukaryotic genes (and, surprisingly, archaeal genes) do *not* exist as an uninterrupted series of triplets coding for a protein. A eukaryotic gene contains the code for a protein, but located along the gene are one to several intervening sequences of bases, called **introns,** that do not code for protein. Introns are interspersed between coding regions, called **exons,** that will be translated into protein **(figure 9.17).** We can use words as examples. A short section of colinear prokaryotic gene might read TOM SAW OUR DOG DIG OUT; a eukaryotic gene that codes for the same portion would read TOM SAW XZKP FPL OUR DOG QZWVP DIG OUT. The recognizable words are the exons, and the nonsense letters represent the introns.

This unusual genetic architecture, sometimes called a split gene, requires further processing before translation. Transcription of the entire gene with both exons and introns occurs first, producing a pre-mRNA. A series of adenosines is added to the mRNA molecule. This protects the molecule and eventually directs it out of the nucleus for translation. Next, a type of RNA and protein called a *spliceosome* recognizes the exon-intron junctions and enzymatically cuts through them. The action of this splicer enzyme loops the introns into lariat-shaped pieces, excises them, and joins the exons end to end. By this means, a strand of mRNA with no intron material is produced. This completed mRNA strand can then proceed to the cytoplasm to be translated.

Several different types of introns have been discovered, some of which do code for cell substances. As detailed in Insight 9.2, a great deal of non-protein-coding DNA is proving to be vital for cell function. In humans, this intron material represents 98% of the DNA and the discovery of its extreme importance has revolutionized the "genetic revolution." Another surprising finding from 2006 was the discovery of proteins that had sections that were in reverse order from the DNA sequence that encoded them. Previously it was considered a hard and fast rule that DNA sequence determined the mRNA and then the amino acid sequence once the introns were accounted for. This new data revealed that cellular machinery can flip stretches of amino acids around, essentially creating new proteins from the same gene. Many introns have been found to code for an enzyme called reverse transcriptase, which can convert RNA into DNA. Other introns are translated into endonucleases, enzymes that can snip DNA and allow insertions and deletions into the sequence.

The Genetics of Animal Viruses

The genetic nature of viruses was described in chapter 6. Viruses essentially consist of one or more pieces of DNA or RNA enclosed in a protective coating. Above all, they are genetic parasites that require access to their host cell's genetic and metabolic machinery to be replicated, transcribed, and translated; and they also have the potential for genetically changing the cells. Because they contain only those genes needed for the production of new viruses, the genomes of viruses tend to be very compact and economical. In fact, this very simplicity makes them excellent subjects for the study of gene function.

The genetics of viruses is quite diverse. In many viruses, the nucleic acid is linear in form; in others, it is circular. The genome of most viruses exists in a single molecule, though in a few, it is segmented into several smaller molecules. Most viruses contain normal double-stranded (ds) DNA or single-stranded (ss) RNA, but other patterns exist. There are ssDNA viruses, dsRNA viruses, and retroviruses, which work backward by making dsDNA from ssRNA. In some instances, viral genes overlap one another, and in a few DNA viruses, both strands contain a translatable message.

A few generalities can be stated about viral genetics. In all cases, the viral nucleic acid penetrates the cell and is introduced into the host's gene-processing machinery at some point. In successful infection, an invading virus instructs the host's machinery to synthesize large numbers of new virus particles by a mechanism specific to a particular group. With few exceptions, replication of the DNA molecule of DNA animal viruses occurs in the nucleus, where the cell's DNA replication machinery lies and the genome of RNA viruses is replicated in the cytoplasm. In all viruses, viral mRNA is translated into viral proteins on host cell ribosomes using host tRNA.

CHECKPOINT

- Information in DNA is converted to proteins by the process of transcription and translation. These proteins may be structural or functional in nature. Structural proteins contribute to the architecture of the cell while functional proteins (enzymes) control an organism's metabolic activities.
- The DNA code occurs in groups of three bases; this code is copied onto RNA as codons; the message determines the types of amino acids in a protein. This code is universal in all cells and viruses.
- DNA also contains a great number of non-protein-coding sequences. These sequences are often transcribed into RNA that serves to regulate cell function.
- The processes of transcription and translation are similar but not identical for prokaryotes and eukaryotes. Eukaryotes transcribe DNA in the nucleus, remove its introns, and translate it in the cytoplasm. Bacteria transcribe and translate simultaneously because the DNA is not sequestered in a nucleus and the bacterial DNA is free of introns.

9.3 Genetic Regulation of Protein Synthesis and Metabolism

In chapter 8, we surveyed the metabolic reactions in cells and the enzymes involved in those reactions. At that time, we mentioned that some enzymes are regulated and that one form of regulation occurs at the genetic level. Control mechanisms ensure that genes are active only when their products are required. In this way, enzymes will be produced as they are needed and prevent the waste of energy and materials in dead-end synthesis. Antisense RNAs, micro RNAs, and riboswitches (see Insight 9.2) provide regulation in both prokaryotes and eukaryotes. Prokaryotes have an additional strategy: they organize collections of genes into **operons.** Operons consist of a coordinated set of genes, all of which are regulated as a single unit. Operons are described as either inducible or repressible. The category each operon falls into is determined by how transcription is affected by the environment surrounding the cell. Many catabolic operons are inducible, meaning that the operon is turned on (induced) by the substrate of the enzyme for which the structural genes code. In this way, the enzymes needed to metabolize a nutrient (lactose, for example) are only produced when that nutrient is present in the environment. Repressible operons often contain genes coding for anabolic enzymes, such as those used to synthesize amino acids. In the case of these operons, several genes in series are turned off (repressed) by the product synthesized by the enzyme.

The Lactose Operon: A Model for Inducible Gene Regulation in Bacteria

The best understood cell system for explaining control through genetic induction is the **lactose (*lac*) operon.** This system, first described in 1961 by François Jacob and Jacques Monod, accounts for the regulation of lactose metabolism in *Escherichia coli.* Many other operons with similar modes of action have since been identified, and together they furnish convincing evidence that the environment of a cell can have great impact on gene expression.

The lactose operon has three important features **(figure 9.18):**

1. the **regulator,** composed of the gene that codes for a protein capable of repressing the operon (a **repressor**);
2. the *control locus,* composed of two areas, the **promoter** (recognized by RNA polymerase) and the **operator,** a sequence that acts as an on/off switch for transcription; and
3. the *structural locus,* made up of three genes, each coding for a different enzyme needed to catabolize lactose.

One of the enzymes, β-galactosidase, hydrolyzes the lactose into its monosaccharides; another, permease, brings lactose across the cell membrane.

The operon provides an efficient strategy that permits genes for a particular metabolic pathway to be induced or repressed in unison by a single regulatory element. The

(a) **Operon Off.** In the absence of lactose, a repressor protein (the product of a regulatory gene located elsewhere on the bacterial chromosome) attaches to the operator of the operon. This effectively locks the operator and prevents any transcription of structural genes downstream (to its right). Suppression of transcription (and consequently, of translation) prevents the unnecessary synthesis of enzymes for processing lactose.

(b) **Operon On.** Upon entering the cell, the substrate (lactose) becomes a genetic inducer by attaching to the repressor, which loses its grip and falls away. The RNA polymerase is now free to bind to the promoter and initiate transcription, and the enzymes produced by translation of the mRNA perform the necessary reactions on their lactose substrate.

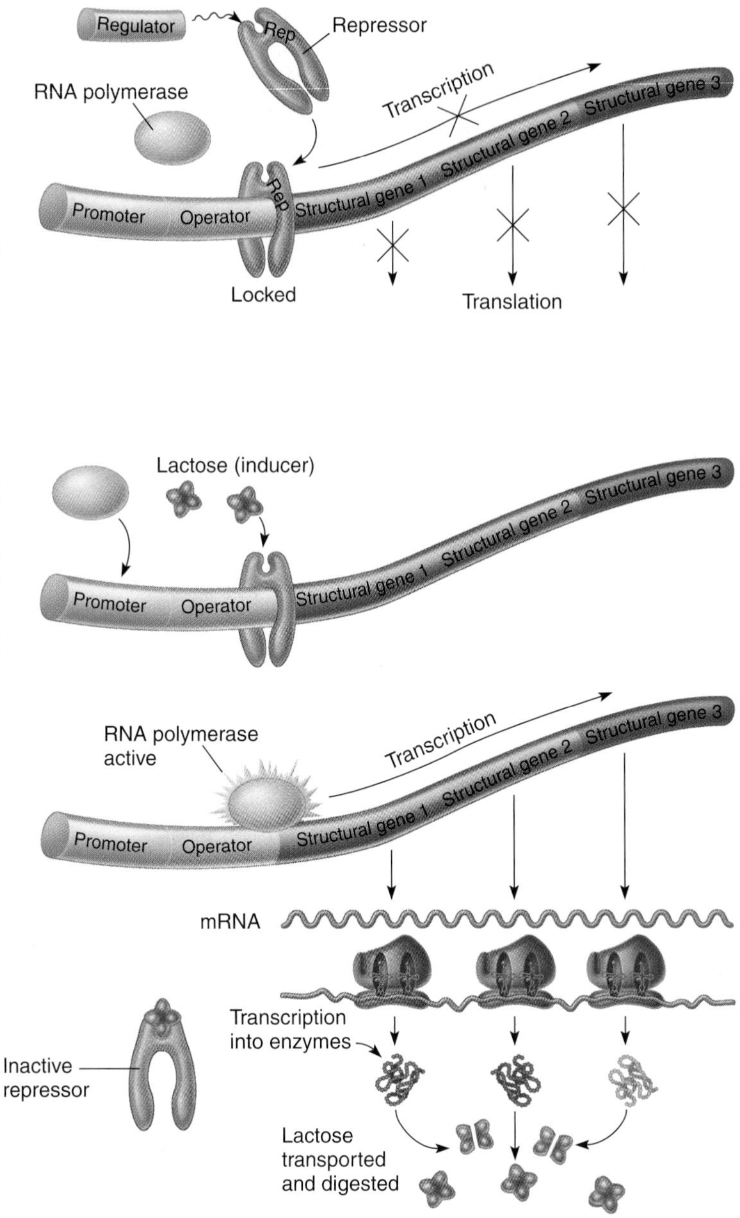

Figure 9.18 The lactose operon in bacteria: how inducible genes are controlled by substrate.

enzymes of the *lac* operon are of the inducible sort mentioned in chapter 8. The promoter, operator, and structural components lie adjacent to one another, but the regulator can be at a distant site.

In inducible systems like the *lac* operon, the operon is normally in an *off mode* and does not initiate enzyme synthesis when the appropriate substrate is absent **(figure 9.18*a*).** How is the operon maintained in this mode? The key is in the repressor protein that is coded by the regulatory gene. This relatively large molecule is **allosteric,** meaning it has two binding sites, one for the operator and another for lactose. In the absence of lactose, this repressor binds with the operator locus, thereby blocking the transcription of the structural genes lying downstream. Think of the repressor as a lock on the operator, and if the operator is locked, the structural genes cannot be transcribed. Importantly, the regulator gene lies upstream (to the left) of the operator region and is transcribed constitutively because it is not controlled in tandem with the operon.

If lactose is added to the cell's environment, it triggers several events that turn the operon *on.* The binding of lactose to the repressor protein causes a conformational change in the repressor that dislodges it from the operator segment **(figure 9.18*b*).** With the operator opened up, RNA polymerase can now bind to the promoter. The structural genes are transcribed in a single unbroken transcript coding for all three enzymes. (During translation, however, each protein is synthesized separately.) Because lactose is ultimately responsible for stimulating protein synthesis, it is called an *inducer.*

As lactose is depleted, further enzyme synthesis is not necessary, so the order of events reverses. At this point, there is no longer sufficient lactose to inhibit the repressor; hence the repressor is again free to attach to the operator.

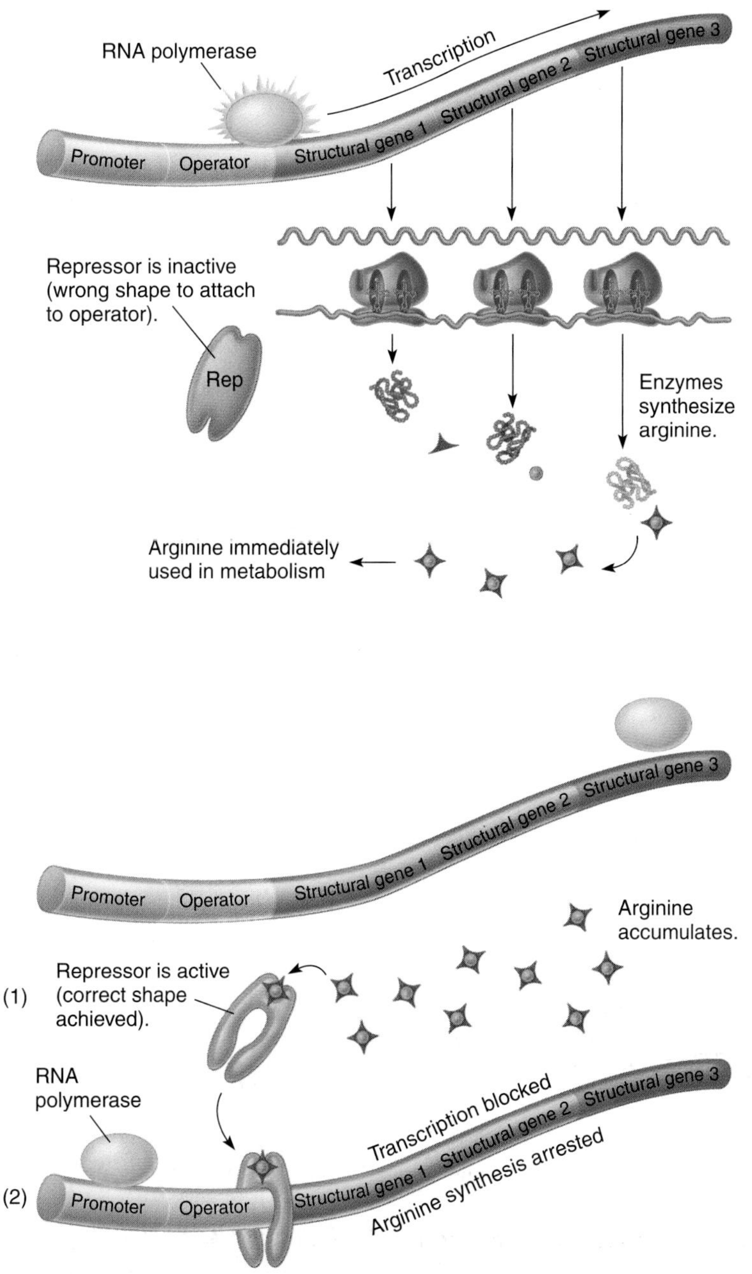

(a) **Operon On.** A repressible operon remains on when its nutrient products (here, arginine) are in great demand by the cell because the repressor is unable to bind to the operator at low nutrient levels.

(b) **Operon Off.** The operon is repressed when (1) arginine builds up and, serving as a corepressor, activates the repressor. (2) The repressor complex affixes to the operator and blocks the RNA polymerase and further transcription of genes for arginine synthesis.

Figure 9.19 Repressible operon: control of a gene through excess nutrient.

The operator is locked, and transcription of the structural genes and enzyme synthesis related to lactose both stop.

A fine but important point about the *lac* operon is that it functions only in the absence of glucose or if the cell's energy needs are not being met by the available glucose. Glucose is the preferred carbon source because it can be used immediately in growth and does not require induction of an operon. When glucose is present, a second regulatory system ensures that the *lac* operon is inactive, regardless of lactose levels in the environment.

A Repressible Operon

Bacterial systems for synthesis of amino acids, purines and pyrimidines, and many other processes work on a slightly different principle—that of repression. Similar factors such as repressor proteins, operators, and a series of structural genes exist for this operon but with some important differences. Unlike the *lac* operon, this operon is normally in the *on* mode and will be turned *off* only when this nutrient is no longer required. The excess nutrient serves as a **corepressor** needed to block the action of the operon.

A growing cell that needs the amino acid arginine (arg) effectively illustrates the operation of a repressible operon. Under these conditions, the *arg* operon is set to *on* and arginine is being actively synthesized through the action of the operon's enzymatic products **(figure 9.19*a*).** In an active cell, the arginine will be used immediately, and the repressor will remain inactive (unable to bind the operator) because there is too little free arginine to activate it. As the cell's metabolism begins to slow down, however, the synthesized arginine will no longer be used up and will accumulate. The free arginine is then available to act as a corepressor by attaching to the repressor. This reaction changes the shape of the repressor, making it capable of binding to the operator. Transcription stops; arginine is no longer synthesized **(figure 9.19*b*).**

In eukaryotic cells, gene function can be altered by intrinsic regulatory segments similar to operons. Some molecules, called transcription factors, insert on the grooves of the DNA molecule and enhance transcription of specific genes. Examples include zinc "fingers" and leucine "zippers." These transcription factors can regulate gene expression in response to environmental stimuli such as nutrients, toxin levels, or even temperature. Eukaryotic genes are also regulated during growth and development, leading to the hundreds of different tissue types found in higher multicellular organisms.

Antibiotics That Affect Transcription and Translation

Naturally occurring cell nutrients are not the only agents capable of modifying gene expression. Some infection therapy is based on the concept that certain drugs react with DNA, RNA, or ribosomes and thereby alter genetic expression (see chapter 12). Treatment with such drugs is based on an important premise: that growth of the infectious agent will be inhibited by blocking its protein-synthesizing machinery selectively, without disrupting the cell synthesis of the patient receiving the therapy.

Drugs that inhibit protein synthesis exert their influence on transcription or translation. For example, the rifamycins used in therapy for tuberculosis bind to RNA polymerase, blocking the initiation step of transcription, and are selectively more active against bacterial RNA polymerase than the corresponding eukaryotic enzyme. Actinomycin D binds to bacterial DNA and halts mRNA chain elongation, but it also binds to human DNA. For this reason, it is very toxic and never used to treat bacterial infections, though it can be applied in tumor treatment.

The ribosome is a frequent target of antibiotics that inhibit ribosomal function and ultimately protein synthesis. The value and safety of these antibiotics again depend upon the differential susceptibility of prokaryotic and eukaryotic ribosomes. One problem with drugs that selectively disrupt prokaryotic ribosomes is that the mitochondria of humans contain a prokaryotic type of ribosome, and these drugs may inhibit the function of the host's mitochondria. One group of antibiotics (including erythromycin and spectinomycin) prevents translation by interfering with the attachment of mRNA to ribosomes. Chloramphenicol, lincomycin, and tetracycline bind to the ribosome in a way that blocks the elongation of the polypeptide, and aminoglycosides (such as streptomycin) inhibit peptide initiation and elongation. It is interesting to note that these drugs have served as important tools to explore genetic events because they can arrest specific stages in these processes.

CHECKPOINT

- Gene expression must be orchestrated to coordinate the organism's needs with nutritional resources. Genes can be turned "on" and "off" by specific molecules, which expose or hide their nucleotide codes for transcribing proteins. Most, but not all, of these proteins are enzymes.
- Operons are collections of genes in bacteria that code for products with a coordinated function. They include genes for operational and structural components of the cell. Nutrients can combine with regulator gene products to turn a set of structural genes on (inducible genes) or off (repressible genes). The *lac* (lactose) operon is an example of an inducible operon. The *arg* (arginine) operon is an example of a repressible operon.
- The rifamycins, tetracyclines, and aminoglycosides are classes of antibiotics that are effective because they interfere with transcription and translation processes in microorganisms.

9.4 Mutations: Changes in the Genetic Code

As precise and predictable as the rules of genetic expression seem, permanent changes do occur in the genetic code. Indeed, genetic change is the driving force of evolution. In microorganisms, such changes may become evident in altered gene expression, such as the appearance or disappearance of anatomical or physiological traits. For example, a pigmented bacterium can lose its ability to form pigment, or a strain of the malarial parasite can develop resistance to a drug. When phenotypic changes are due to changes in the genotype, it is called a **mutation.** On a strictly molecular level, a mutation is an alteration in the nitrogen base sequence of DNA. It can involve the loss of base pairs, the addition of base pairs, or a rearrangement in the order of base pairs. Do not confuse this with genetic recombination, in which microbes transfer whole segments of genetic information between themselves.

A microorganism that exhibits a natural, nonmutated characteristic is known as a **wild type,** or wild strain. If a microorganism bears a mutation, it is called a **mutant strain.** Mutant strains can show variance in morphology, nutritional characteristics, genetic control mechanisms, resistance to chemicals, temperature preference, and nearly any type of enzymatic function. Mutant strains are very useful for tracking genetic events, unraveling genetic organization, and pinpointing genetic markers. A classic method of detecting mutant strains involves addition of various nutrients to a culture to screen for its use of that nutrient. For example, in a culture of a wild-type bacterium that is lactose-positive (meaning it has the necessary enzymes for fermenting this sugar), a small number of mutant cells have become lactose-negative, having lost the capacity to ferment this sugar. If the culture is plated on a medium containing indicators for fermentation, each colony can be observed for its fermentation reaction and the negative strain isolated. Another standard

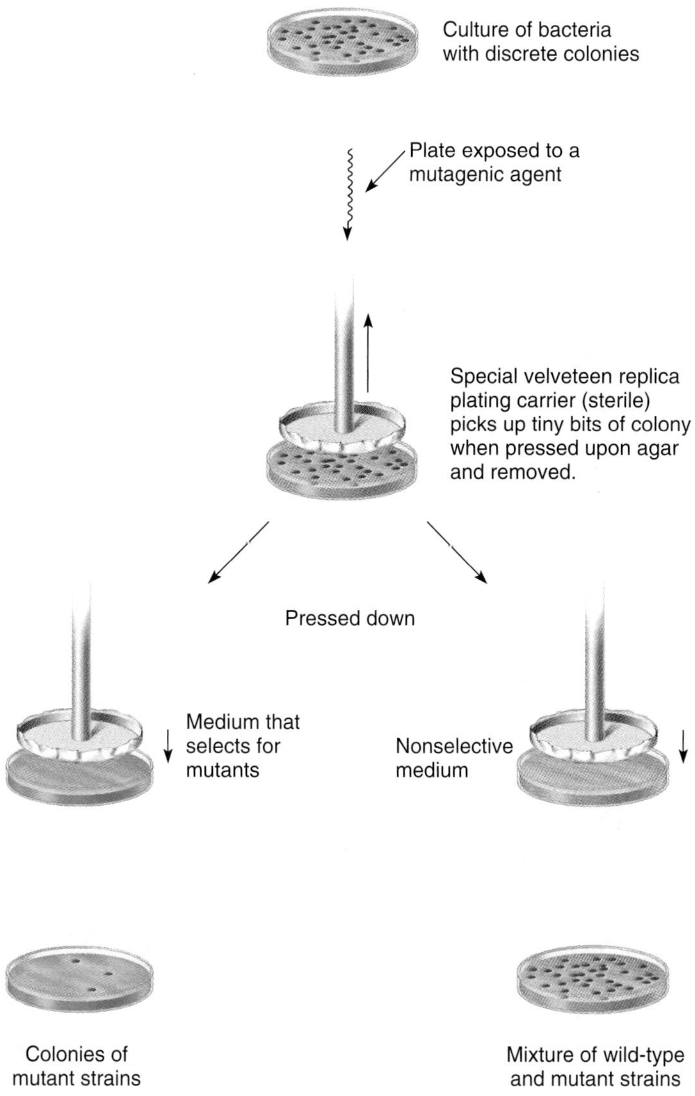

Figure 9.20 The general basis of replica plating. This method was developed by Joshua Lederberg for detecting and isolating mutant strains of microorganisms.

method of detecting and isolating microbial mutants is by replica plating **(figure 9.20).**

Causes of Mutations

A mutation is described as spontaneous or induced, depending upon its origin. A **spontaneous mutation** is a random change in the DNA arising from errors in replication that occur randomly. The frequency of spontaneous mutations has been measured for a number of organisms. Mutation rates vary tremendously, from one mutation in 10^5 replications (a high rate) to one mutation in 10^{10} replications (a low rate). The rapid rate of bacterial reproduction allows these mutations to be observed more readily in bacteria than in most eukaryotes.

Induced mutations result from exposure to known **mutagens,** which are primarily physical or chemical agents that interact with DNA in a disruptive manner **(table 9.3).** The carefully controlled use of mutagens has proved a useful way to induce mutant strains of microorganisms for study.

TABLE 9.3 Selected Mutagenic Agents and Their Effects

Agent	Effect
Chemical	
Nitrous acid, bisulfite	Removes an amino group from some bases
Ethidium bromide	Inserts between the paired bases
Acridine dyes	Cause frameshifts due to insertion between base pairs
Nitrogen base analogs	Compete with natural bases for sites on replicating DNA
Radiation	
Ionizing (gamma rays, X rays)	Form free radicals that cause single or double breaks in DNA
Ultraviolet	Causes cross-links between adjacent pyrimidines

Chemical mutagenic agents act in a variety of ways to change the DNA. Agents such as acridine dyes insert completely across the DNA helices between adjacent bases to produce mutations that distort the helix. Analogs[4] of the nitrogen bases (5-bromodeoxyuridine and 2-aminopurine, for example) are chemical mimics of natural bases that are incorporated into DNA during replication. Addition of these abnormal bases leads to mistakes in base-pairing. Many chemical mutagens are also carcinogens, or cancer-causing agents (see the discussion of the Ames test in a later section of this chapter).

Physical agents that alter DNA are primarily types of radiation. High-energy gamma rays and X rays introduce major physical changes into DNA, and it accumulates breaks that may not be repairable. Ultraviolet (UV) radiation induces abnormal bonds between adjacent pyrimidines that prevent normal replication. Exposure to large doses of radiation can be fatal, which is why radiation is so effective in microbial control; it can also be carcinogenic in animals. (The use of UV to control microorganisms is described further in chapter 11.)

Categories of Mutations

Mutations range from large mutations, in which large genetic sequences are gained or lost, to small ones that affect only a single base on a gene. These latter mutations, which involve addition, deletion, or substitution of single bases, are called **point mutations.**

4. An analog is a chemical structured very similarly to another chemical except for minor differences in functional groups.

To understand how a change in DNA influences the cell, remember that the DNA code appears in a particular order of triplets (three bases) that is transcribed into mRNA codons, each of which specifies an amino acid. A permanent alteration in the DNA that is copied faithfully into mRNA and translated can change the structure of the protein. A change in a protein can likewise change the morphology and physiology of a cell. Some mutations have a harmful effect on the cell, leading to cell dysfunction or death; these are called lethal mutations. Neutral mutations produce neither adverse nor helpful changes. A small number of mutations are beneficial in that they provide the cell with a useful change in structure or physiology.

Any change in the code that leads to placement of a different amino acid is called a **missense mutation.** A missense mutation can do one of the following:

1. create a faulty, nonfunctional (or less functional) protein,
2. produce a protein that functions in a different manner, or
3. cause no significant alteration in protein function.

A **nonsense mutation,** on the other hand, changes a normal codon into a stop codon that does not code for an amino acid and stops the production of the protein wherever it occurs. A nonsense mutation almost always results in a nonfunctional protein. A **silent mutation** alters a base but does not change the amino acid and thus has no effect. For example, because of the redundancy of the code, ACU, ACC, ACG, and ACA all code for threonine, so a mutation that changes only the last base will not alter the sense of the message in any way. A **back-mutation** occurs when a gene that has undergone mutation reverses (mutates back) to its original base composition.

Mutations also occur when one or more bases are inserted into or deleted from a newly synthesized DNA strand. This type of mutation, known as a **frameshift,** is so named because the reading frame of the mRNA has been changed. Frameshift mutations nearly always result in a nonfunctional protein because every amino acid after the mutation is different from what was coded for in the original DNA. Also note that insertion or deletion of bases in multiples of three (3, 6, 9, etc.) results in the addition or deletion of amino acids but does not disturb the reading frame. The effects of all of these types of mutations can be seen in **table 9.4.**

TABLE 9.4 Classification of Major Types of Mutations

	Example
(a) Wild type (original, nonmutated sequence)	THE BIG BAD DOG ATE THE FAT RED CAT
Substitution mutations	
(b) Missense	THE BIG BAD DOG ATE THE FIT RED CAT
(c) Nonsense	THE BIG BAD (stop)
Frameshift mutations	
(d) Insertion	THE BIG BAB DDO GAT ETH EFA TRE DCA T
(e) Deletion	THE BIG BDD OGA TET HEF ATR EDC AT

Categories of mutations based on type of DNA alteration

(a) The wild-type sequence of a gene is the DNA sequence found in most organisms and is generally considered the "normal" sequence. (b) A missense mutation causes a different amino acid to be incorporated into a protein. Effects range from unnoticeable to severe, based on how different the two amino acids are. (c) A nonsense mutation converts a codon to a stop codon, resulting in premature termination of protein synthesis. Effects of this type of mutation are almost always severe. (d, e) Insertion and deletion mutations cause a change in the reading frame of the mRNA, resulting in a protein in which every amino acid after the mutation is affected. Because of this, frameshift mutations almost always result in a nonfunctional protein.

Repair of Mutations

Earlier we indicated that DNA has a proofreading mechanism to repair mistakes in replication that might otherwise become permanent (see page 254). Because mutations are potentially life-threatening, the cell has additional systems for finding and repairing DNA that has been damaged by various mutagenic agents and processes. Most ordinary DNA damage is resolved by enzymatic systems specialized for finding and fixing such defects.

DNA that has been damaged by ultraviolet radiation can be restored by photoactivation or light repair. This repair mechanism requires visible light and a light-sensitive enzyme, DNA photolyase, which can detect and attach to the damaged areas (sites of abnormal pyrimidine binding). Ultraviolet repair mechanisms are successful only for a relatively small number of UV mutations. Cells cannot repair severe, widespread damage and will die. In humans, the genetic disease *xeroderma pigmentosa* is due to nonfunctioning genes for the enzyme photolyase. Persons suffering from this rare disorder develop severe skin cancers; this relation provides strong evidence for a link between cancer and mutations.

Mutations can be excised by a series of enzymes that remove the incorrect bases and add the correct ones. This process is known as *excision repair*. First, enzymes break the bonds between the bases and the sugar-phosphate strand at the site of the error. A different enzyme subsequently removes the defective bases one at a time, leaving a gap that will be filled in by DNA polymerase I and ligase **(figure 9.21).** A repair system can also locate mismatched bases that were missed during proofreading: for example, C mistakenly paired with A, or G with T. The base must be replaced soon after the mismatch is made, or it will not be recognized by the repair enzymes.

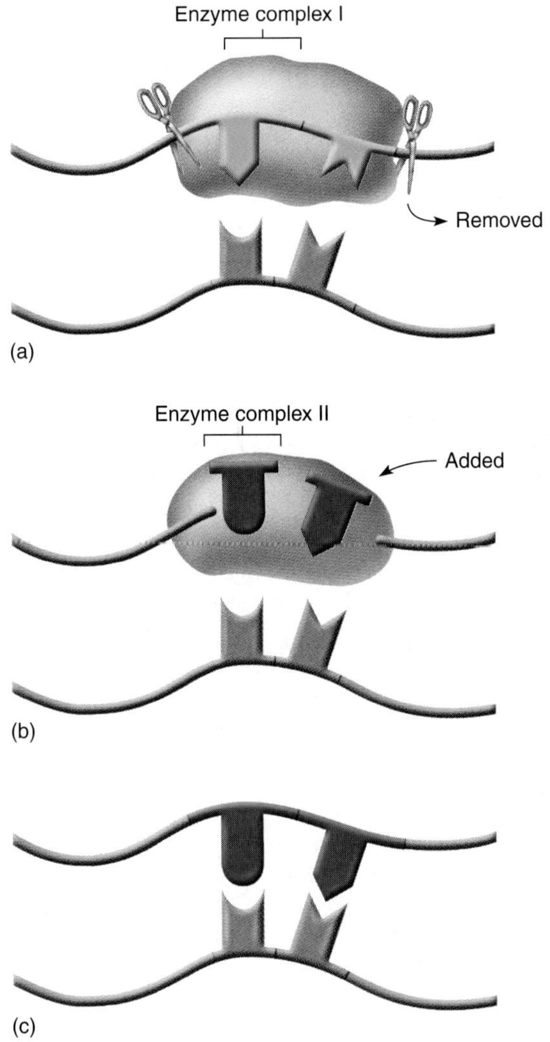

Figure 9.21 Excision repair of mutation by enzymes. **(a)** The first enzyme complex recognizes one or several incorrect bases and removes them. **(b)** The second complex (DNA polymerase I and ligase) places correct bases and seals the gaps. **(c)** Repaired DNA.

The Ames Test

New agricultural, industrial, and medicinal chemicals are constantly being added to the environment, and exposure to them is widespread. The discovery that many such compounds are mutagenic and that up to 83% of these mutagens are linked to cancer is significant. Although animal testing has been a standard method of detecting chemicals with carcinogenic potential, a more rapid screening system called the **Ames test**[5] is also commonly used. In this ingenious test, the experimental subjects are bacteria whose gene expression and mutation rate can be readily observed and monitored. The premise is that any chemical capable of mutating bacterial DNA can similarly mutate mammalian (and thus human) DNA and is therefore potentially hazardous.

One indicator organism in the Ames test is a mutant strain of ***Salmonella typhimurium***[6] that has lost the ability to synthesize the amino acid histidine, a defect highly susceptible to back-mutation because the strain also lacks DNA repair mechanisms. Mutations that cause reversion to the wild strain, which is capable of synthesizing histidine, occur spontaneously at a low rate. A test agent is considered a mutagen if it enhances the rate of back-mutation beyond levels that would occur spontaneously. One variation on this testing procedure is outlined in **figure 9.22.** The Ames test has proved invaluable for screening an assortment of environmental and dietary chemicals for mutagenicity and carcinogenicity without resorting to animal studies.

Positive and Negative Effects of Mutations

Many mutations are not repaired. How the cell copes with them depends on the nature of the mutation and the strategies available to that organism. Mutations are permanent and heritable and will be passed on to the offspring of organisms and new viruses and become a long-term part of the gene pool. Most mutations are harmful to organisms; others provide adaptive advantages.

If a mutation leading to a nonfunctional protein occurs in a gene for which there is only a single copy, as in haploid or simple organisms, the cell will probably die. This happens when certain mutant strains of *E. coli* acquire mutations in the genes needed to repair damage by UV radiation. Mutations of the human genome affecting the action of a single protein (mostly enzymes) are responsible for more than 3,500 diseases (see chapter 10).

Although most spontaneous mutations are not beneficial, a small number contribute to the success of the individual and the population by creating variant strains with alternate ways of expressing a trait. Microbes are not "aware" of this advantage and do not direct these changes; they simply respond to the environment they encounter. Those organisms with beneficial mutations can more readily adapt, survive, and reproduce. In the long-range view, mutations and the variations they produce are the raw materials for change in the population and, thus, for evolution.

Mutations that create variants occur frequently enough that any population contains mutant strains for a number of characteristics, but as long as the environment is stable, these mutants will never comprise more than a tiny percentage of the population. When the environment changes, however, it can become hostile for the survival of certain individuals, and only those microbes bearing protective mutations will be equipped to survive in the new environment. In this way, the environment naturally selects certain mutant strains that will reproduce, give rise to subsequent generations, and in time, be the dominant strain in the population. Through these means, any change that confers an advantage during selection

5. Named for its creator, Bruce Ames.

6. *S. typhimurium* inhabits the intestine of poultry and causes food poisoning in humans. It is used extensively in genetic studies of bacteria.

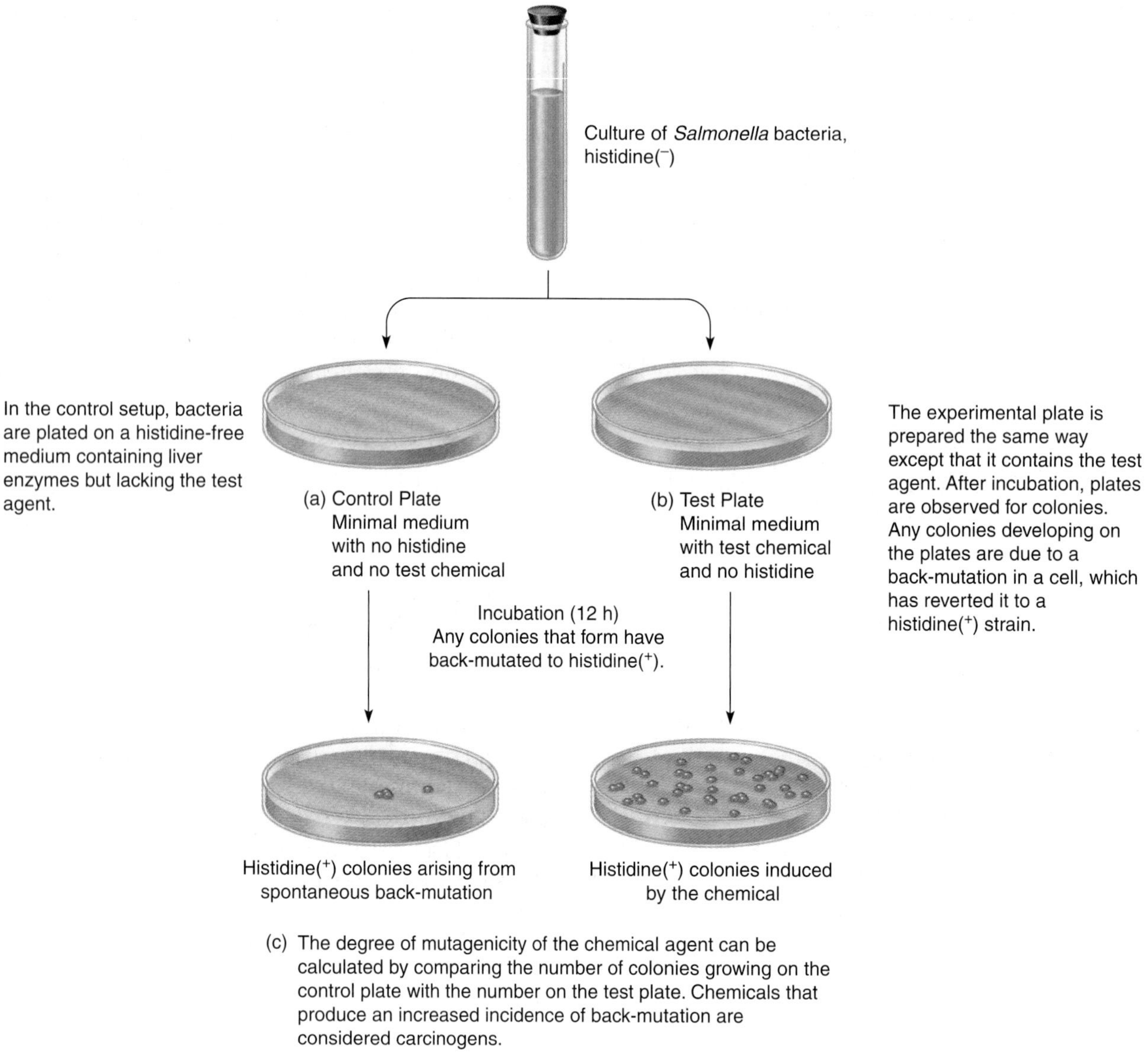
Culture of *Salmonella* bacteria, histidine(−)
In the control setup, bacteria are plated on a histidine-free medium containing liver enzymes but lacking the test agent.
(a) Control Plate Minimal medium with no histidine and no test chemical
(b) Test Plate Minimal medium with test chemical and no histidine
The experimental plate is prepared the same way except that it contains the test agent. After incubation, plates are observed for colonies. Any colonies developing on the plates are due to a back-mutation in a cell, which has reverted it to a histidine(+) strain.
Incubation (12 h) Any colonies that form have back-mutated to histidine(+).
Histidine(+) colonies arising from spontaneous back-mutation
Histidine(+) colonies induced by the chemical
(c) The degree of mutagenicity of the chemical agent can be calculated by comparing the number of colonies growing on the control plate with the number on the test plate. Chemicals that produce an increased incidence of back-mutation are considered carcinogens.

Figure 9.22 **The Ames test.**
This test is based on a strain of *Salmonella typhimurium* that cannot synthesize histidine [his(−)]. It lacks the enzymes to repair DNA so that mutations show up readily, and it has leaky cell walls that permit the ready entrance of chemicals. Many potential carcinogens (benzanthracene and aflatoxin, for example) are mutagenic agents only after being acted on by mammalian liver enzymes, so an extract of these enzymes is added to the test medium.

pressure will be retained by the population. One of the clearest models for this sort of selection and adaptation is acquired drug resistance in bacteria (see chapter 12). Bacteria have also developed a mechanism for increasing their adaptive capacity through genetic exchange, called genetic recombination.

9.5 DNA Recombination Events

Genetic recombination through sexual reproduction is an important means of genetic variation in eukaryotes. Although bacteria have no exact equivalent to sexual reproduction, they exhibit a primitive means for sharing or recombining parts of their genome. An event in which one bacterium donates DNA to another bacterium is a type of genetic transfer termed **recombination,** the end result of which is a new strain different from both the donor and the original recipient strain. Recombination in bacteria depends in part on the fact that bacteria contain extrachromosomal DNA—that is, plasmids—and are adept at interchanging genes. Genetic exchanges have tremendous effects on the genetic diversity of bacteria. They provide additional genes for resistance to drugs and metabolic poisons, new nutritional and metabolic capabilities, and increased virulence and adaptation to the environment.

In general, any organism that contains (and expresses) genes that originated in another organism is called a **recombinant.**

Transmission of Genetic Material in Bacteria

DNA transfer between bacterial cells typically involves small pieces of DNA in the form of plasmids or chromosomal fragments. Plasmids are small, circular pieces of DNA that contain their own origin of replication and therefore can replicate independently of the bacterial chromosome. Plasmids are found in many bacteria (as well as some fungi) and typically contain, at most, only a few dozen genes. Although plasmids are not necessary for bacterial survival, they often carry useful traits, such as antibiotic resistance. Chromosomal fragments that have escaped from a lysed bacterial cell are also commonly involved in the transfer of genetic information between cells. An important difference between plasmids and fragments is that while a plasmid has its own origin of replication and is stably replicated and inherited, chromosomal fragments must integrate themselves into the bacterial chromosome in order to be replicated and eventually passed to progeny cells. The process of genetic recombination is rare in nature, but its frequency can be increased in the laboratory, where the ability to shuffle genes between organisms is highly prized.

Depending upon the mode of transmission, the means of genetic recombination in bacteria is called conjugation, transformation, or transduction. **Conjugation** requires the attachment of two related species and the formation of a bridge that can transport DNA. **Transformation** entails the transfer of naked DNA and requires no special vehicle. **Transduction** is DNA transfer mediated through the action of a bacterial virus **(table 9.5).**

Conjugation: Bacterial "Sex"

Conjugation is a mode of genetic exchange in which a plasmid or other genetic material is transferred by a donor to a recipient cell via a direct connection **(figure 9.23).** Both gram-negative and gram-positive cells can conjugate. In gram-negative cells, the donor has a plasmid **(fertility,** or **F′, factor)** that allows the synthesis of a conjugative **pilus.** The recipient cell has a recognition site on its surface. A cell's role in conjugation is denoted by F^+ for the cell that has the F plasmid and by F^- for the cell that lacks it. Contact is made when a pilus grows out from the F^+ cell, attaches to the surface of the F^- cell, contracts, and draws the two cells together (**figure 9.23*a*;** see also figure 4.8). In both gram-positive and gram-negative cells, an opening is created between the connected cells, and the replicated DNA passes across from one cell to the other **(figure 9.23*b*).** Conjugation is a conservative process, in that the donor bacterium generally retains a copy of the genetic material being transferred.

There are hundreds of conjugative plasmids with some variations in their properties. One of the best understood plasmids is the F factor in *E. coli,* which exhibits these patterns of transfer:

1. The donor (F^+) cell makes a copy of its F factor and transmits this to a recipient (F^-) cell. The F^- cell is thereby changed into an F^+ cell capable of producing a pilus and conjugating with other cells **(figure 9.23*c*).** No additional donor genes are transferred at this time.
2. In high-frequency recombination (Hfr) donors, the fertility factor has been integrated into the F^+ donor chromosome.

The term *high-frequency recombination* was adopted to denote that a cell with an integrated F factor transmits its chromosomal genes at a higher frequency than other cells.

The F factor can direct a more comprehensive transfer of part of the donor chromosome to a recipient cell. This transfer occurs through duplication of the DNA, after which one strand of DNA is retained by the donor, and the other strand is transported across to the recipient cell **(figure 9.23*d*).** The F factor may not be transferred during this process. The transfer of an entire chromosome takes about 100 minutes, but the pilus bridge between cells is ordinarily broken before this time, and rarely is the entire genome of the donor cell transferred.

Conjugation has great biomedical importance. Special **resistance (R) plasmids,** or **factors,** that bear genes for resisting antibiotics and other drugs are commonly shared among bacteria through conjugation. Transfer of R factors can confer multiple resistance to antibiotics such as tetracycline,

TABLE 9.5 Types of Intermicrobial Exchange

Mode	Factors Involved	Direct or Indirect*	Examples of Genes Transferred
Conjugation	Donor cell with pilus Fertility plasmid in donor Both donor and recipient alive Bridge forms between cells to transfer DNA.	Direct	Drug resistance; resistance to metals; toxin production; enzymes; adherence molecules; degradation of toxic substances; uptake of iron
Transformation	Free donor DNA (fragment) Live, competent recipient cell	Indirect	Polysaccharide capsule; unlimited with cloning techniques
Transduction	Donor is lysed bacterial cell. Defective bacteriophage is carrier of donor DNA. Live recipient cell of same species as donor	Indirect	Toxins; enzymes for sugar fermentation; drug resistance

*Direct means the donor and recipient are in contact during exchange; indirect means they are not.

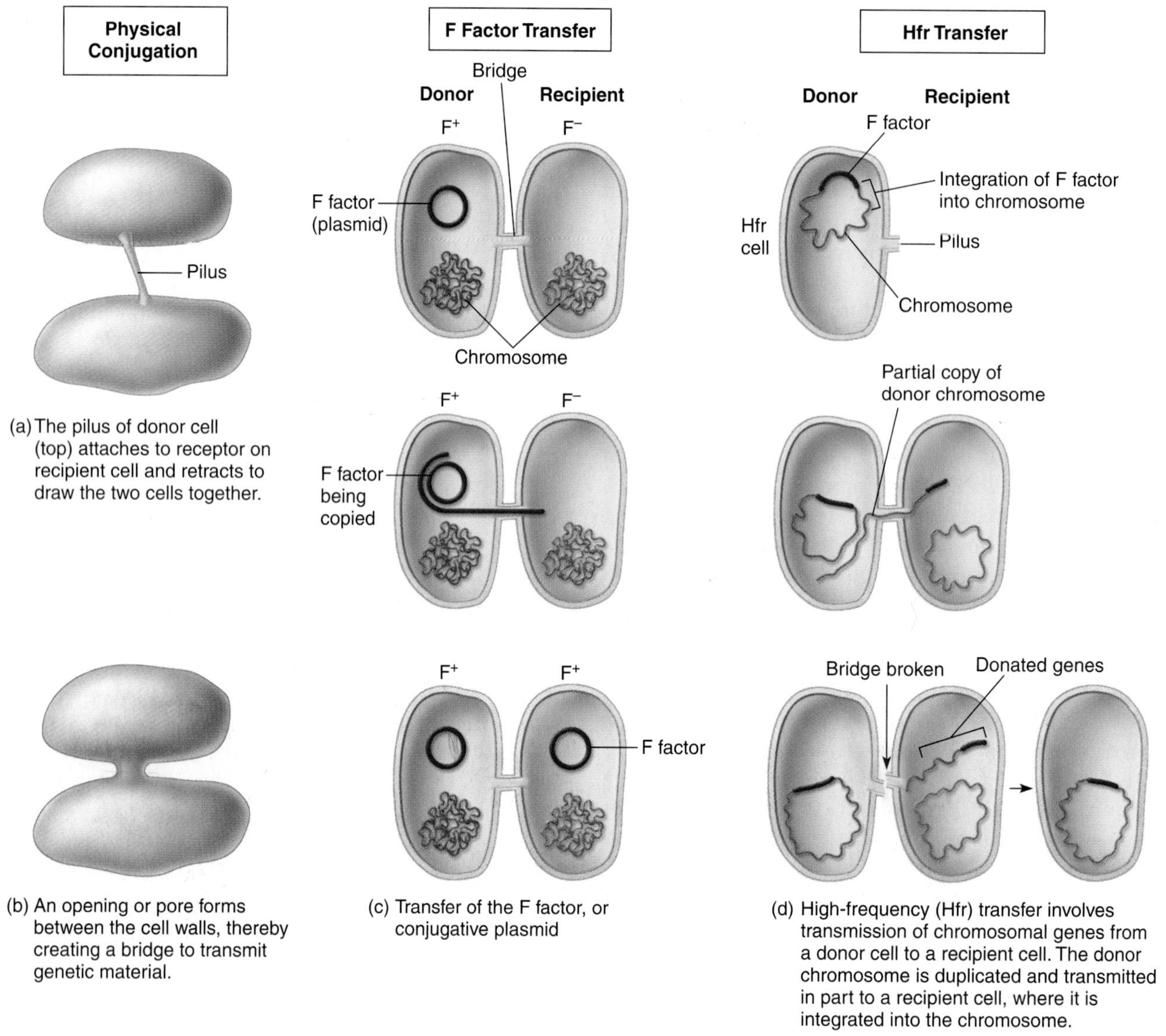

Physical Conjugation
F Factor Transfer
Hfr Transfer
Pilus
(a) The pilus of donor cell (top) attaches to receptor on recipient cell and retracts to draw the two cells together.
(b) An opening or pore forms between the cell walls, thereby creating a bridge to transmit genetic material.
Bridge
Donor
Recipient
F⁺
F⁻
F factor (plasmid)
Chromosome
F factor being copied
F factor
(c) Transfer of the F factor, or conjugative plasmid
Donor
Recipient
F factor
Integration of F factor into chromosome
Hfr cell
Pilus
Chromosome
Partial copy of donor chromosome
Bridge broken
Donated genes
(d) High-frequency (Hfr) transfer involves transmission of chromosomal genes from a donor cell to a recipient cell. The donor chromosome is duplicated and transmitted in part to a recipient cell, where it is integrated into the chromosome.

Figure 9.23 Conjugation: genetic transmission through direct contact between two cells.

chloramphenicol, streptomycin, sulfonamides, and penicillin. This phenomenon is discussed further in chapter 12. Other types of R factors carry genetic codes for resistance to heavy metals (nickel and mercury) or for synthesizing virulence factors (toxins, enzymes, and adhesion molecules) that increase the pathogenicity of the bacterial strain. Conjugation studies have also provided an excellent way to map the bacterial chromosome.

Transformation: Capturing DNA from Solution

One of the cornerstone discoveries in microbial genetics was made in the late 1920s by the English biochemist Frederick Griffith working with *Streptococcus pneumoniae* and laboratory mice. The pneumococcus exists in two major strains based on the presence of the capsule, colonial morphology, and pathogenicity. Encapsulated strains have a smooth (S) colonial appearance and are virulent; strains lacking a capsule have a rough (R) appearance and are nonvirulent. (Recall from chapter 4 that the capsule protects a bacterium from the phagocytic host defenses.) To set the groundwork, Griffith showed that when mice were injected with a live, virulent (S) strain, they soon died **(figure 9.24*a*).** Mice injected with a live, nonvirulent (R) strain remained alive and healthy **(figure 9.24*b*).** Next he tried a variation on this theme. First, he heat-killed an S strain and injected it into mice, which remained healthy **(figure 9.24*c*).** Then came the ultimate test: Griffith injected both dead S cells and live R cells into mice, with the result that the mice died from pneumococcal blood infection **(figure 9.24*d*).** If killed bacterial cells do not come back to life and the nonvirulent live strain was harmless, why did the mice die? Although he did not know it at the time, Griffith had demonstrated that dead S cells, while passing through the body of the mouse, broke open and released some of their DNA (by chance, that part containing the genes for making a capsule). A few of the live R cells subsequently

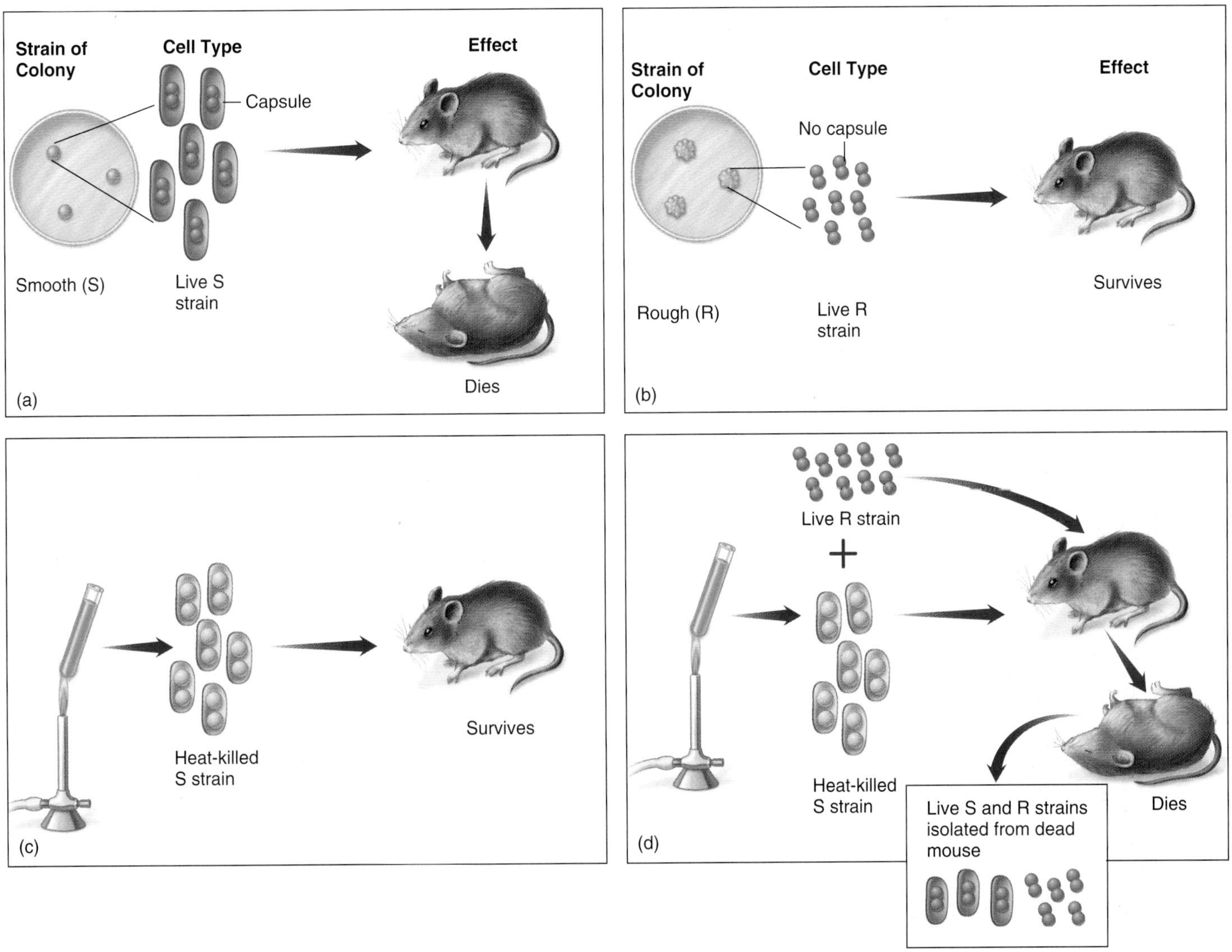

Figure 9.24 Griffith's classic experiment in transformation.
In essence, this experiment proved that DNA released from a killed cell can be acquired by a live cell. The cell receiving this new DNA is genetically transformed—in this case, from a nonvirulent strain to a virulent one.

picked up this loose DNA and were transformed by it into virulent, capsule-forming strains.

Later studies supported the concept that a chromosome released by a lysed cell breaks into fragments small enough to be accepted by a recipient cell and that DNA, even from a dead cell, retains its genetic code. This nonspecific acceptance by a bacterial cell of small fragments of soluble DNA from the surrounding environment is termed *transformation*. Transformation is apparently facilitated by special DNA-binding proteins on the cell wall that capture DNA from the surrounding medium. Cells that are capable of accepting genetic material through this means are termed *competent*. The new DNA is processed by the cell membrane and transported into the cytoplasm, where some of it is inserted into the bacterial chromosome. Transformation is a natural event found in several groups of gram-positive and gram-negative bacterial species. In addition to genes coding for the capsule, bacteria also exchange genes for antibiotic resistance and bacteriocin synthesis in this way.

Because transformation requires no special appendages and the donor and recipient cells do not have to be in direct contact, the process is useful for certain types of recombinant DNA technology. With this technique, foreign genes from a completely unrelated organism are inserted into a plasmid, which is then introduced into a competent bacterial cell through transformation. These recombinations can be carried out in a test tube, and human genes can be experimented upon and even expressed outside the human body by placing them in a microbial cell. This same phenomenon in eukaryotic cells, termed *transfection*, is an essential aspect of genetically engineered yeasts, plants, and mice, and it has been proposed as a future technique for curing genetic diseases in humans. These topics are covered in more detail in chapter 10.

Transduction: The Case of the Piggyback DNA

Bacteriophages (bacterial viruses) have been previously described as destructive bacterial parasites. Viruses can in fact serve as genetic vectors (an entity that can bring foreign DNA into a cell). The process by which a bacteriophage serves as the carrier of DNA from a donor cell to a recipient cell is transduction. Although it occurs naturally in a broad spectrum of bacteria, the participating bacteria in a single transduction event must be the same species because of the specificity of viruses for host cells.

There are two versions of transduction. In *generalized transduction* **(figure 9.25),** random fragments of disintegrating host DNA are taken up by the phage during assembly. Virtually any gene from the bacterium can be transmitted through this means. In *specialized transduction* **(figure 9.26),** a highly specific part of the host genome is regularly incorporated into the virus. This specificity is explained by the prior existence of a temperate prophage inserted in a fixed site on the bacterial chromosome. When activated, the prophage DNA separates from the bacterial chromosome, carrying a small segment of host genes with it. During a lytic cycle, these specific viral-host gene combinations are incorporated into the viral particles and carried to another bacterial cell.

Several cases of specialized transduction have biomedical importance. The virulent strains of bacteria such as *Corynebacterium diphtheriae, Clostridium* spp., and *Streptococcus pyogenes* all produce toxins with profound physiological effects, whereas nonvirulent strains do not produce toxins. It turns out that toxicity arises from the expression of bacteriophage genes that have been introduced by transduction. Only those bacteria infected with a temperate phage are toxin formers. (Details of toxin action are discussed in the organ-system-specific disease chapters.) Another instance of transduction is seen in staphylococcal transfer of drug resistance.

Transposons: "This Gene Is Jumpin'" One type of genetic transferral of great interest involves transposable elements, or **transposons.** Transposons have the distinction of shifting from one part of the genome to another and so are termed "jumping genes." When the idea of their existence in corn plants was first postulated by geneticist Barbara McClintock in 1951, it was greeted with nearly universal skepticism because it had long been believed that the location of a given gene was set and that genes did not or could not move around. Now it is evident that jumping genes are widespread among prokaryotic and eukaryotic cells and viruses.

All transposons share the general characteristic of traveling from one location to another on the genome—from one chromosomal site to another, from a chromosome to a plasmid, or from a plasmid to a chromosome **(figure 9.27).** Because transposons can occur in plasmids, they can also be transmitted from one cell to another in bacteria and a few eukaryotes. Some transposons replicate themselves before jumping to the next location, and others simply move without replicating first.

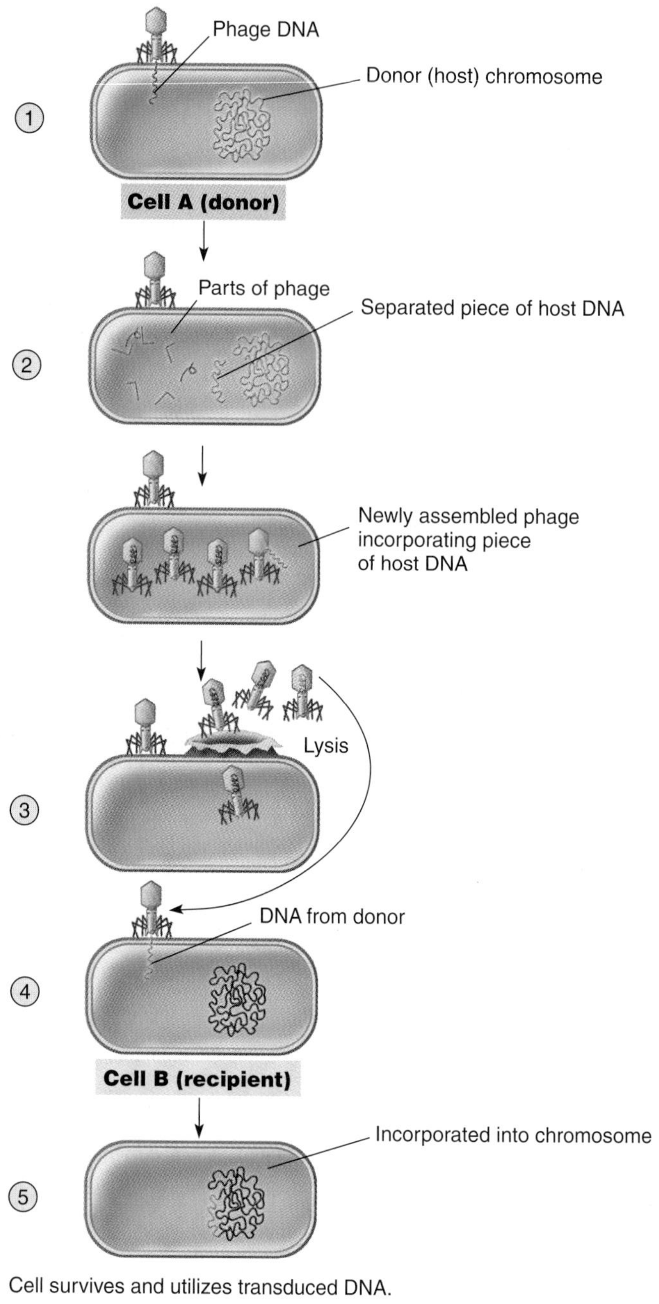

Process Figure 9.25 Generalized transduction: genetic transfer by means of a virus carrier.
① A phage infects cell A (the donor cell) by normal means. ② During replication and assembly, a phage particle incorporates a segment of bacterial DNA by mistake. ③ Cell A then lyses and releases the mature phages, including the genetically altered one. ④ The altered phage adsorbs to and penetrates another host cell (cell B), injecting the DNA from cell A rather than viral nucleic acid. ⑤ Cell B receives this donated DNA, which recombines with its own DNA. Because the virus is defective (biologically inactive as a virus), it is unable to complete a lytic cycle. The transduced cell survives and can use this new genetic material.

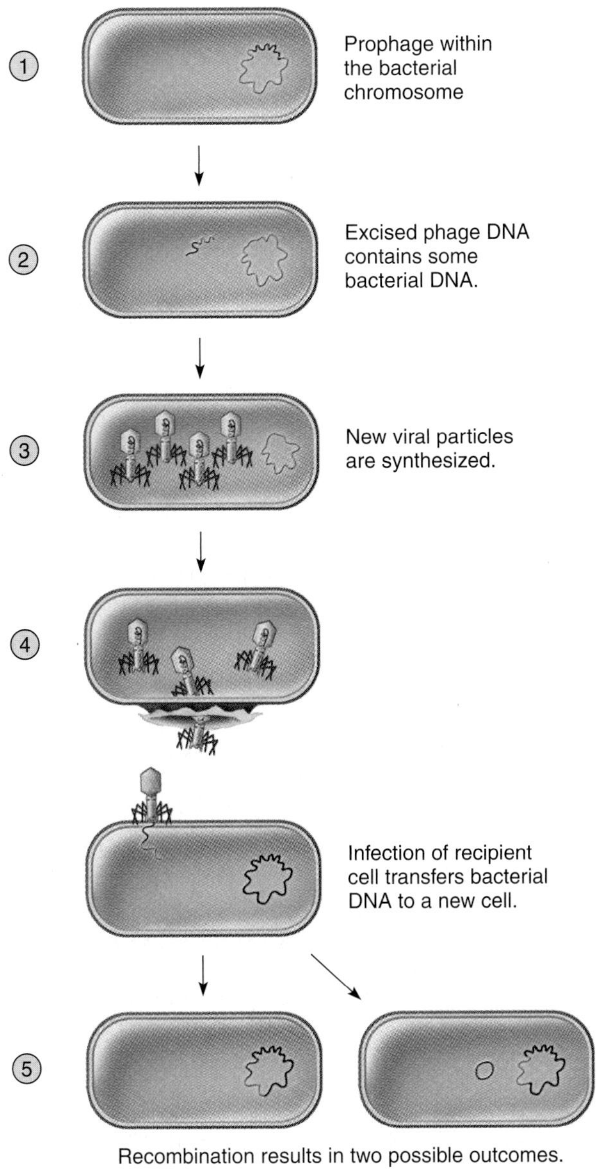

1
Prophage within the bacterial chromosome
2
Excised phage DNA contains some bacterial DNA.
3
New viral particles are synthesized.
4
Infection of recipient cell transfers bacterial DNA to a new cell.
5
Recombination results in two possible outcomes.

Process Figure 9.26 Specialized transduction: transfer of specific genetic material by means of a virus carrier.

① Specialized transduction begins with a cell that contains a prophage (a viral genome integrated into the host cell chromosome). ② Rarely, the virus enters a lytic cycle and, as it excises itself from its host cell, inadvertently includes some bacterial DNA. ③ Replication and assembly result in production of a chimeric virus, containing some bacterial DNA. ④ Release of the recombinant virus and subsequent infection of a new host result in transfer of bacterial DNA between cells. ⑤ Recombination can occur between the bacterial chromosome and the virus DNA, resulting in either bacterial DNA or a combination of viral and bacterial DNA being incorporated into the bacterial chromosome.

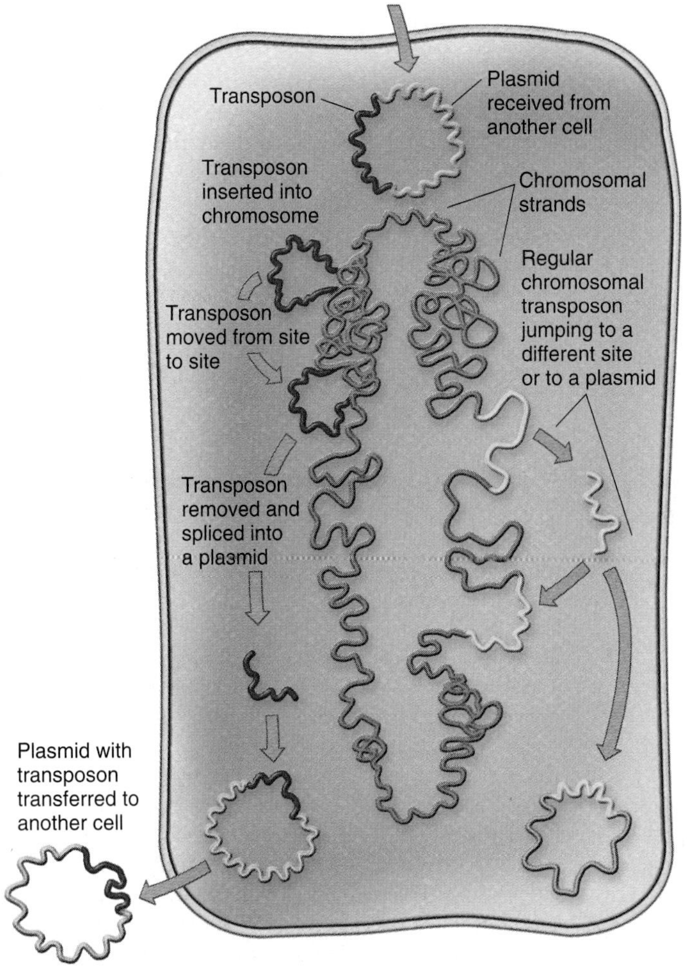

Transposon
Plasmid received from another cell
Transposon inserted into chromosome
Chromosomal strands
Regular chromosomal transposon jumping to a different site or to a plasmid
Transposon moved from site to site
Transposon removed and spliced into a plasmid
Plasmid with transposon transferred to another cell

Figure 9.27 Transposons: shifting segments of the genome.

Potential mechanisms in the movement of transposons in bacterial cells. Some transposons are plasmids that are shifted from site to site; others are regular parts of chromosomes that are moved out of one site and spliced into another.

Transposons contain DNA that codes for the enzymes needed to remove and reintegrate the transposon at another site in the genome. Flanking the coding region of the DNA are sequences called tandem repeats, which mark the point at which the transposon is removed or reinserted into the genome. The smallest transposons consist of only these two genetic sequences and are often referred to as *insertion elements.* A type of transposon called *retrotransposon* can transcribe DNA into RNA and then back into DNA for insertion in a new location. Other transposons contain additional genes that provide traits such as antibiotic resistance or toxin production.

The overall effect of transposons—to scramble the genetic language—can be beneficial or adverse, depending upon such variables as where insertion occurs in a chromosome, what kinds of genes are relocated, and the type of cell involved. In bacteria, transposons are known to be involved in

1. changes in traits such as colony morphology, pigmentation, and antigenic characteristics;
2. replacement of damaged DNA; and
3. the intermicrobial transfer of drug resistance (in bacteria).

CHECKPOINT

- Changes in the genetic code can occur by two means: mutation and recombination. Mutation means a change in the nucleotide sequence of the organism's genome.
- Recombination means the addition of genes from an outside source, such as a virus or another cell.
- Mutations can be either spontaneous or induced by exposure to some external mutagenic agent.
- All cells have enzymes that repair damaged DNA. When the degree of damage exceeds the ability of the enzymes to make repairs, mutations occur.
- Mutation-induced changes in DNA nucleotide sequencing range from a single nucleotide to addition or deletion of large sections of genetic material.
- Genetic recombination occurs in eukaryotes through sexual reproduction. In bacteria, recombination occurs through the processes of transformation, conjugation, and transduction.
- Transposons are genes that can relocate from one part of the genome to another, causing rearrangement of genetic material. Such rearrangements have either beneficial or harmful consequences for the organism involved.

Chapter Summary with Key Terms

9.1 Introduction to Genetics and Genes: Unlocking the Secrets of Heredity

A. **Genetics** is the study of **heredity** and can be studied at the level of the organism, **genome, chromosome, gene,** and DNA. Genes provide the information needed to construct **proteins,** which have structural or catalytic functions in the cell.

B. DNA is a long molecule in the form of a double helix. Each strand of the helix consists of a string of **nucleotides** that form hydrogen bonds with their counterparts on the other strand. **Adenine** base-pairs with **thymine** while **guanine** base-pairs with **cytosine.** The order of the nucleotides specifies which amino acids will be used to construct proteins during the process of translation.

C. DNA **replication** is **semiconservative** and requires the participation of several enzymes.

9.2 Applications of the DNA Code: Transcription and Translation

A. DNA is used to produce RNA **(transcription)** and RNA is then used to produce protein **(translation).** Other RNAs act in regulation.

B. RNA: Unlike DNA, RNA is single stranded, contains **uracil** instead of thymine and **ribose** instead of **deoxyribose.**
   1. Major forms of RNA found in the cell include **mRNA, tRNA,** and **rRNA.**
   2. Additional regulatory RNAs include antisense RNA, micro RNA, and ribosomids.
   3. The genetic information contained in DNA is copied to produce an RNA molecule. **Codons** in the mRNA pair with **anticodons** in the tRNA to specify what amino acids to assemble on the ribosome during translation.

C. *Transcription and Translation:* Transcription occurs when **RNA polymerase** copies the template strand of a segment of DNA. RNA is always made in the 5′ to 3′ direction. Translation occurs when the mRNA is used to direct the synthesis of proteins on the ribosome. Codons in the mRNA pair with anticodons in the tRNA to assemble a string of amino acids. This occurs until a stop codon is reached.

D. *Eukaryotic Gene Expression:* Eukaryotic genes are composed of **exons** (expressed sequences) and **introns** (intervening sequences). The introns must be removed and the exons spliced together to create the final mRNA.

E. The genetics of viruses is quite diverse.
   1. Genomes are found in many physical forms not all of which are seen in cells, including dsDNA, ssDNA, and dsRNA.
   2. DNA viruses tend to replicate in the nucleus while RNA viruses replicate in the cytoplasm. Retroviruses synthesize dsDNA from ssRNA.

9.3 Genetic Regulation of Protein Synthesis and Metabolism

Protein synthesis in prokaryotes can be regulated through gene induction or repression, as controlled by an **operon.** Operons consist of several structural genes controlled by a common regulatory element.

A. *Inducible operons* such as the lactose operon are normally off but can be turned on by a lactose inducer.

B. *Repressible operons* are usually on but can be turned off when their end product is no longer needed.

C. Many antibiotics prevent bacterial growth by interfering with transcription or translation.

9.4 Mutations: Changes in the Genetic Code

A. Permanent changes in the genome of a microorganism are known as **mutations.** Mutations may be **spontaneous** or **induced.**

B. **Point mutations** entail a change in one or a few bases and are categorized as **missense, nonsense,** or **silent mutations** or **back-mutations,** based on the effect of the change in nucleotide(s).

C. Many mutations, particularly those involving mismatched bases or damage from ultraviolet light, can be corrected using enzymes found in the cell.

D. The **Ames test** measures the mutagenicity of chemicals by determining the ability of a chemical to induce mutations in bacteria.

9.5 DNA Recombination Events

A. Intermicrobial transfer and genetic recombination permit gene sharing between bacteria. Major types of recombination include **conjugation, transformation,** and **transduction.**

B. **Transposons** are DNA sequences that regularly move to different places within the genome of a cell, as a consequence generating mutations and variations in chromosome structure.

Multiple-Choice and True-False Questions

Multiple-Choice Questions. Select the correct answer from the answers provided.

1. What is the smallest unit of heredity?
   a. chromosome
   b. gene
   c. codon
   d. nucleotide
2. The nitrogen bases in DNA are bonded to the
   a. phosphate
   b. deoxyribose
   c. ribose
   d. hydrogen
3. DNA replication is semiconservative because the ____ strand will become half of the ____ molecule.
   a. RNA, DNA
   b. template, finished
   c. sense, mRNA
   d. codon, anticodon
4. In DNA, adenine is the complementary base for ____, and cytosine is the complement for ____.
   a. guanine, thymine
   b. uracil, guanine
   c. thymine, guanine
   d. thymine, uracil
5. Transfer RNA is the molecule that
   a. contributes to the structure of ribosomes
   b. adapts the genetic code to protein structure
   c. transfers the DNA code to mRNA
   d. provides the master code for amino acids
6. As a general rule, the template strand on DNA will always begin with
   a. TAC
   b. AUG
   c. ATG
   d. UAC
7. The *lac* operon is usually in the ____ position and is activated by a/an ____ molecule.
   a. on, repressor
   b. off, inducer
   c. on, inducer
   d. off, repressor
8. Which genes can be transferred by all three methods of intermicrobial transfer?
   a. capsule production
   b. toxin production
   c. F factor
   d. drug resistance
9. Which of the following would occur through specialized transduction?
   a. acquisition of Hfr plasmid
   b. transfer of genes for toxin production
   c. transfer of genes for capsule formation
   d. transfer of a plasmid with genes for degrading pesticides

True-False Questions: If statement is true, leave as is. If it is false, correct it by rewriting the sentence.

10. The DNA pairs are held together primarily by covalent bonds.
11. Mutation usually has a negative outcome.
12. The lagging strand of DNA is replicated in short pieces because DNA polymerase can synthesize in only one direction.
13. Messenger RNA is formed by translation of a gene on the DNA template strand.
14. A nucleotide is composed of a 5-carbon sugar, a phosphatic group, and a nitrogenous base.

Writing to Learn

These questions are suggested as a *writing-to-learn* experience. For each question, compose a one- or two-paragraph answer that includes the factual information needed to completely address the question.

1. Describe what is meant by the antiparallel arrangement of DNA.
2. On paper, replicate the following segment of DNA:

   5′ A T C G G C T A C G T T C A C 3′
   3′ T A G C C G A T G C A A G T G 5′

   a. Show the direction of replication of the new strands and explain what the lagging and leading strands are.
   b. Explain how this is semiconservative replication. Are the new strands identical to the original segment of DNA?
3. Name several characteristics of DNA structure that enable it to be replicated with such great fidelity generation after generation.
4. Explain the following relationship: DNA formats RNA, which makes protein.
5. Compare the structure and functions of DNA and RNA.
6. a. Where does transcription begin?
   b. What are the template and coding strands of DNA?
   c. Why is only one strand transcribed, and is the same strand of DNA always transcribed?
7. The following sequence represents triplets on DNA:

   TAC CAG ATA CAC TCC CCT GCG ACT

   a. Give the mRNA codons and tRNA anticodons that correspond with this sequence, and then give the sequence of amino acids in the polypeptide.
   b. Provide another mRNA strand that can be used to synthesize this same protein.
   c. Looking at figure 9.13, give the type and order of the amino acids in the peptide.

8. a. Summarize how bacterial and eukaryotic cells differ in gene structure, transcription, and translation.
   b. Discuss the roles of exons and introns.
9. a. What is an operon? Describe the functions of regulators, promoters, and operators.
   b. Compare and contrast the *lac* operon with a repressible operon system.
10. a. Compare conjugation, transformation, and transduction on the basis of general method, nature of donor, and nature of recipient.
    b. Explain the differences between general and specialized transduction, using drawings.

Concept Mapping

Appendix D provides guidance for working with concept maps.

1. Supply your own linking words or phrases in this concept map, and provide the missing concepts in the empty boxes.

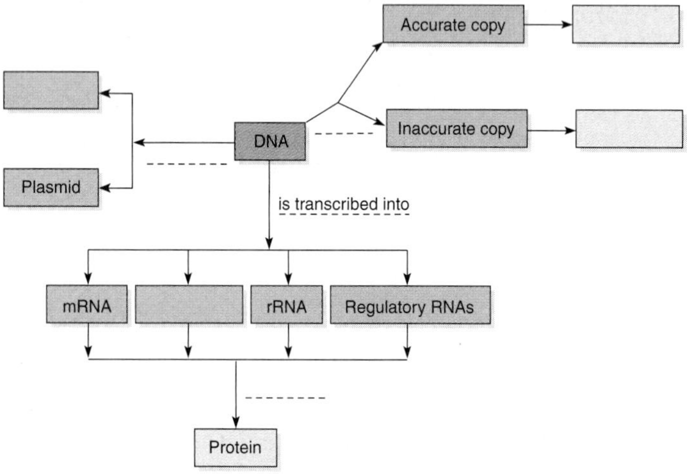

2. Construct your own concept map using the following words as the *concepts*. Supply the linking words between each pair of concepts.

ribozyme
primer
riboswitch
mRNA
tRNA
rRNA
transcription
translation
DNA

Critical Thinking Questions

Critical thinking is the ability to reason and solve problems using facts and concepts. These questions can be approached from a number of angles, and in most cases, they do not have a single correct answer.

1. Knowing that retroviruses operate on the principle of reversing the direction of transcription from RNA to DNA, propose a drug that might possibly interfere with their replication.
2. Using the piece of DNA in Writing to Learn question 7, show a deletion, an insertion, a substitution, and nonsense mutations. Which ones are frameshift mutations? Are any of your mutations nonsense? Missense? (Use the universal code to determine this.)
3. Using figure 9.13 and table 9.4, go through the steps in mutation of a codon followed by its transcription and translation that will give the end result in silent, missense, and nonsense mutations. Why is a change in the RNA code alone not really a mutation?
4. Explain the principle of "wobble" and find four amino acids that are encoded by wobble bases (figure 9.13). Suggest some benefits of this phenomenon to microorganisms.
5. The enzymes required to carry out transcription and translation are themselves produced through these same processes. Speculate which may have come first in evolution—proteins or nucleic acids—and explain your choice.
6. Explain what is meant by the expression: phenotype = genotype + environment.

Try this

A simple test you can do to demonstrate the coiling of DNA in bacteria is to open a large elastic band, stretch it taut, and twist it. First it will form a loose helix, then a tighter helix, and finally, to relieve stress, it will twist back upon itself. Further twisting will result in a series of knotlike bodies; this is how bacterial DNA is condensed.

Visual Understanding.

1. **Figure 9.15, step 3.** Label each of the parts of the illustration.

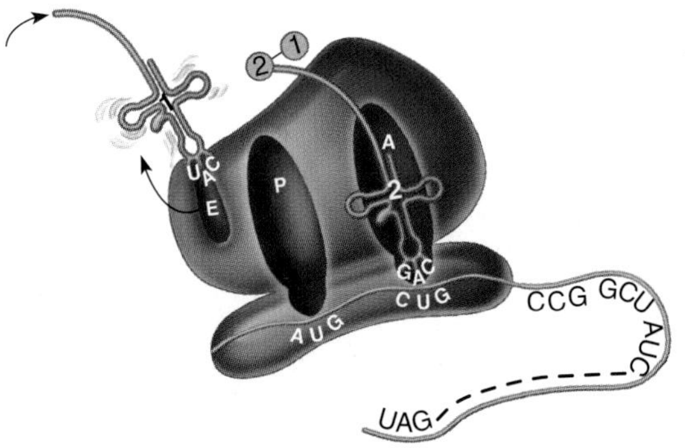

2. **From chapter 4, figure 4.7.** Speculate on why these cells contain two chromosomes (shown in blue).

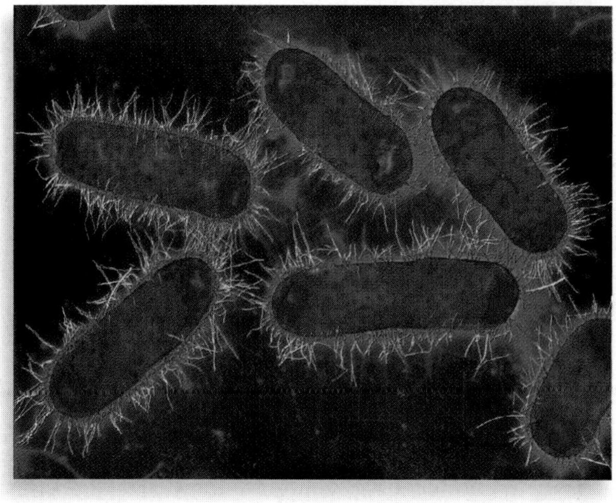

Internet Search Topics

1. Do an Internet search under the heading "DNA music." Explore several websites, discovering how this music is made and listening to some examples.
2. Find information on introns. Explain at least five current theories as to their possible functions.
3. Go to: www.aris.mhhe.com, and click on "microbiology" and then this textbook's author/title. Go to chapter 9, access the URLs listed under Internet Search Topics, and research the following:

   Log on to one or more of the listed URLs and locate animations, three-dimensional graphics, and interactive tutorials that help you to visualize replication, transcription, and translation.
4. Do an Internet search for information on Walter Benson Goad. Where did he work? What did he help create in the 1980s, the early days of molecular biology? Follow a link or do another search to find information on his NIH-associated creation.The Packaging of DNA: Winding, Twisting, and Coiling

CHAPTER 10

Genetic Engineering

A Revolution in Molecular Biology

CASE FILE

10

When the twin towers of the World Trade Center fell on the morning of September 11, 2001, it was the beginning of many epic stories and many journeys toward answers. For each family affected by that attack, the most pressing question was: Did my loved one die there? The thousands of handbills that were posted all over the city were testament to the desperation: Could my loved one have survived?

After weeks went by and a family's loved one did not return home, the answer should have been obvious; but it is hard not to hold out hope that he or she would walk in the door when there was no definitive evidence that the person had perished—no body, no memorial service, no closure. Psychologists tell us that even family members who have buried their loved one go through a phase of denial and yearn for or expect that person to someday return.

As late as 2006, fewer than half of the 2,749 victims of the World Trade Center were positively identified. In 2002, the medical examiner's office in New York had relied on tried-and-true methods of genetic identification. These methods depended on relatively long pieces of undamaged human DNA to be recovered from the site of the World Trade Center, which had collapsed catastrophically and burned like an inferno for weeks. Few remains yielded this useful DNA. Frustrated with the pace of identification, officials decided to try two experimental methods—relatively unproven but perhaps better able to provide answers to the families of the missing.

The New York State Health Department refused to certify the two new experimental methods, fearing that incorrect identification would be worse than no identification. However, the city medical examiner pushed ahead. After all, officials had a dozen refrigerated semitrailers filled with human remains outside their offices and at least 1,700 families hoping for closure.

After 2 years of work with the new techniques and millions of dollars invested, officials got relatively few additional positive identifications. They had many cases that were suggestive of identification, but only 111 rose to the level of unequivocal identification.

The researchers admitted defeat. They cataloged their data, preserved the tissue samples, and are now planning to wait until technology improves enough to identify the rest of the World Trade Center victims. As they said, "We hit the limits of science."

- ▶ *What characteristics would a new test have to possess to cause the medical examiner to feel optimistic that it would be successful?*

Case File 10 Wrap-Up appears on page 288.

CHAPTER OVERVIEW

- **Genetic engineering** deals with the ability of scientists to manipulate DNA through the use of an expanding repertoire of recombinant DNA techniques. These techniques allow DNA to be cut, separated by size, and even sequenced to determine the actual composition and order of nucleotides.
- Biotechnology, the use of an organism's biochemical processes to create a product, has been greatly advanced by the introduction of recombinant DNA technologies because organisms can now be genetically modified to accomplish goals that were previously impossible, such as bacteria that have been engineered to produce human insulin.
- Differences in DNA between organisms are being exploited so that they can be used to identify people, animals, and microbes, revolutionizing fields such as police work and epidemiology.
- Genomes can be analyzed to predict the likelihood of a particular genetic disease long before the disease strikes (and becomes less treatable). In a growing number of cases, missing or mutated genes are being replaced to correct inherited defects.
- Recombinant DNA technology is used to create genetically modified food that may have increased nutritional properties, be easier to grow, or may even vaccinate the person eating it against a host of diseases.
- Recombinant DNA techniques have given humans greater power over our own and other species than we've ever had before. Such power is not without risks, and **bioethics** is an important part of any discussion of bioengineering.

10.1 Basic Elements and Applications of Genetic Engineering

In chapter 9, we looked at the ways in which microorganisms duplicate, exchange, and use their genetic information. In scientific parlance, this is called *basic science* because no product or application is directly derived from it. Human beings being what they are, however, it is never long before basic knowledge is used to derive *applied science* or useful products and applications that owe their invention to the basic research that preceded them. As an example, basic science into the workings of the electron has led to the development of television, computers, and cell phones. None of these staples of modern life were envisioned when early physicists were deciphering the nature of subatomic particles, but without the knowledge of how electrons worked, our ability to harness them for our own uses never would have materialized.

The same scenario can be seen with regard to genetics. The knowledge of how DNA was manipulated within the cell to carry out the goals of a microbe allowed scientists to utilize these processes to accomplish goals more to the liking of human beings. Contrary to being new ideas, the methods of genetic manipulation we will review are simply more efficient ways of accomplishing goals that humans have had for thousands of years.

Examples of human goals that have been more efficiently attained through the use of modern and not-so-modern genetic technologies can be seen in each of these scenarios:

1. A farmer mates his two largest pigs in the hopes of producing larger offspring. Unfortunately, he quite often ends up with small or unhealthy animals due to other genes that are transferred during mating. Genetic manipulation allows for the transfer of specific genes so that only advantageous traits are selected.
2. Courts have, for thousands of years, relied on a description of a person's phenotype (eye color, hair color, etc.) as a means of identification. By remembering that a phenotype is the product of a particular sequence of DNA, you can quickly see how looking at someone's DNA (perhaps from a drop of blood) gives a clue as to his or her identification.
3. We have understood for a long time that many diseases are the result of a missing or dysfunctional protein, and we have generally treated the diseases by replacing the protein as best we can, usually resulting in only temporary relief and limited success. Examples include insulin-dependent diabetes, adenosine deaminase deficiency, and blood-clotting disorders. Genetic engineering offers the promise that, someday soon, fixing the underlying mutation responsible for the lack of a particular protein can treat these diseases far more successfully than we've been able to do in the past.
4. New results from whole organism sequencing show us that RNA regulatory molecules might be even more useful in permanently "fixing" many diseases.

Information on genetic engineering and its biotechnological applications is growing at such an expanding rate that some new discovery or product is disclosed almost on a daily basis. To keep this subject somewhat manageable, we present essential concepts and applications, organized under the following six topics:

- Tools and Techniques of Genetic Engineering
- Methods in Recombinant DNA Technology

- Biochemical Products of Recombinant DNA Technology
- Genetically Modified Organisms
- Genetic Treatments
- Genome Analysis

10.2 Tools and Techniques of Genetic Engineering

DNA: The Raw Material

All of the intrinsic properties of DNA hold true whether the DNA is in a bacterium or a test tube. For example, the enzyme helicase is able to unwind the two strands of the double helix just as easily in the lab as it does in a bacterial cell. But in the laboratory we can take advantage of our knowledge of DNA chemistry to make helicase unnecessary. It turns out that when DNA is heated to just below boiling (90°C to 95°C), the two strands separate, revealing the information contained in their bases. With the nucleotides exposed, DNA can be more easily identified, replicated, or transcribed. If heat-denatured DNA is then slowly cooled, complementary nucleotides will hydrogen bond with one another and the strands will renature, or regain their familiar double-stranded form **(figure 10.1*a*).** As we shall see, this process is a necessary feature of the polymerase chain reaction and in the application of nucleic acid probes described later.

Enzymes for Dicing, Splicing, and Reversing Nucleic Acids

The polynucleotide strands of DNA can also be clipped crosswise at selected positions by means of enzymes called **restriction endonucleases.**[1] These enzymes originate in bacterial cells. They recognize foreign DNA and are capable of breaking the phosphodiester bonds between adjacent nucleotides on both strands of DNA, leading to a break in the DNA strand. In the bacterial cell, this ability protects against the incompatible DNA of bacteriophages or plasmids. In the biotechnologist's lab, the enzymes can be used to cleave DNA at desired sites and are necessary for the techniques of recombinant DNA technology.

So far, hundreds of restriction endonucleases have been discovered in bacteria. They function to protect bacteria from foreign DNA such as that from phages. Each type has a known sequence of 4 to 10 base pairs as its target, so sites of cutting can be finely controlled. These enzymes have the unique property of recognizing and clipping at base sequences called **palindromes (figure 10.1*b*).** Palindromes are sequences of DNA that are identical when read from the 5′ to 3′ direction on one strand and the 5′ to 3′ direction on the other strand.

1. The meaning of restriction is that the enzymes do not act upon the bacterium's own DNA; an *endo*nuclease nicks DNA internally, not at the ends.

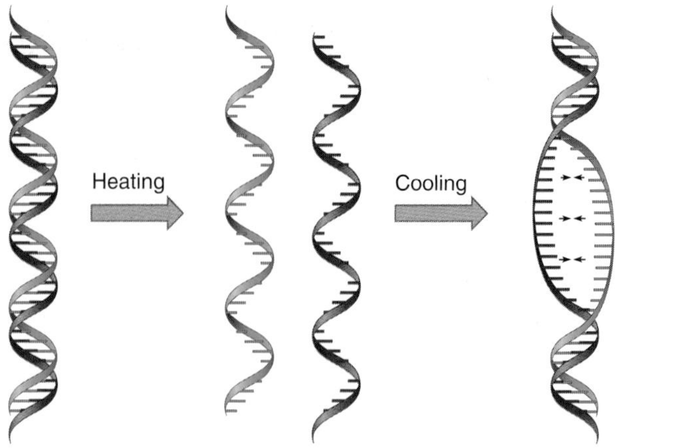

(a) **DNA heating and cooling.** DNA responds to heat by denaturing—losing its hydrogen bonding and thereby separating into its two strands. When cooled, the two strands rejoin at complementary regions. The two strands need not be from the same organisms as long as they have matching sites.

Endonuclease	*Eco*RI	*Hin*dIII	*Hae*III
Cutting pattern	G ¦ A A T T C C T T A A ¦ G	A ¦ A G C T T T T C G A ¦ A	G G ¦ C C C C ¦ G G

(b) **Examples of palindromes and cutting patterns.**

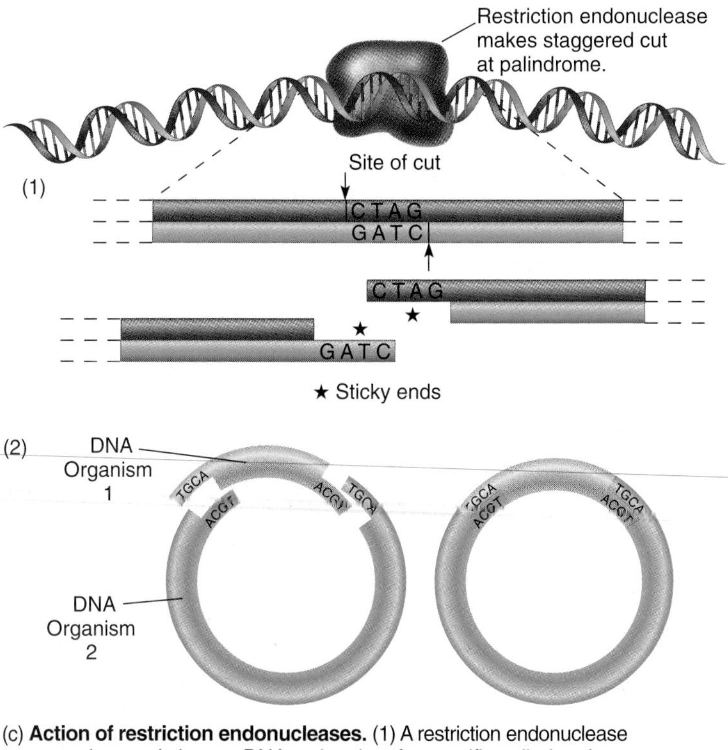

(c) **Action of restriction endonucleases.** (1) A restriction endonuclease recognizes and cleaves DNA at the site of a specific palindromic sequence. Cleavage can produce staggered tails called sticky ends that accept complementary tails for gene splicing. (2) The sticky ends can be used to join DNA from different organisms by cutting it with the same restriction enzyme, ensuring that all fragments have complementary ends.

Figure 10.1 Some useful properties of DNA.

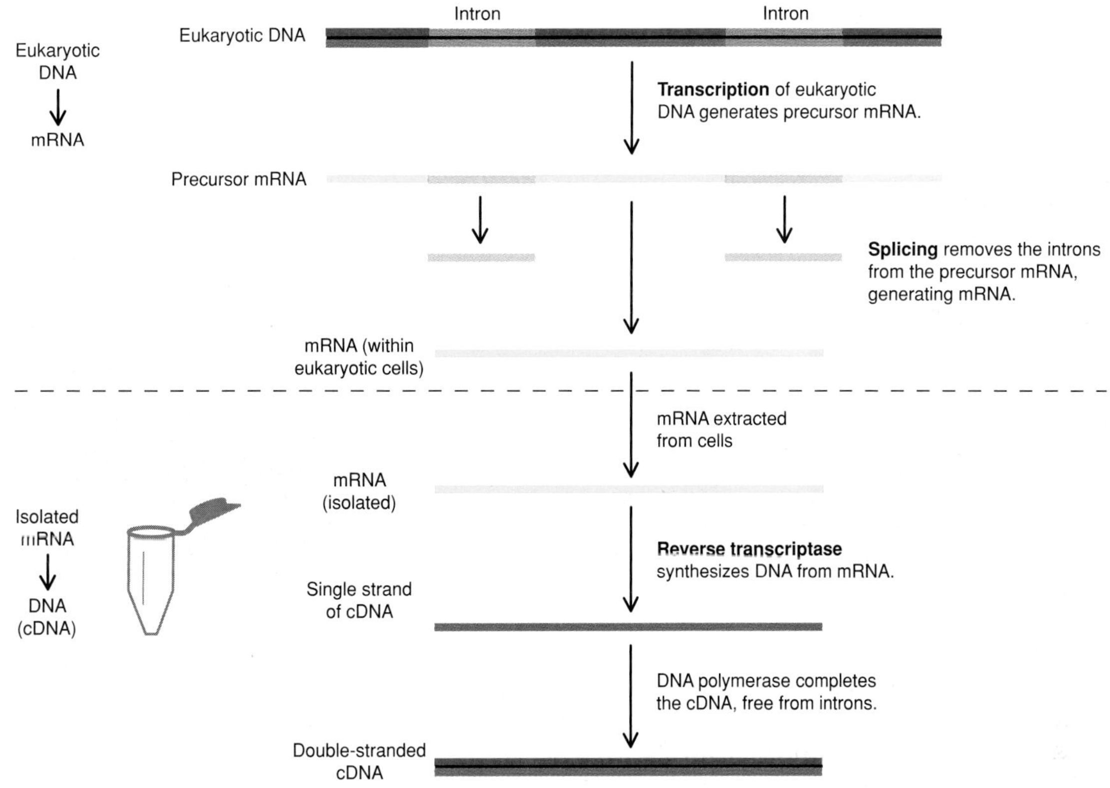

Figure 10.2 Making cDNA from eukaryotic mRNA.
In order for eukaryotic genes to be expressed by a prokaryotic cell, a copy of DNA without introns must be cloned. The cDNA encodes the same protein as the original DNA but lacks introns.

Endonucleases are named by combining the first letter of the bacterial genus, the first two letters of the species, and the endonuclease number. Thus, *Eco*RI is the first endonuclease found in *Escherichia coli* (in the R strain), and *Hin*dIII is the third endonuclease discovered in *Haemophilus influenzae* Type d (see figure 10.1*b*).

Endonucleases are used in the laboratory to cut DNA into smaller pieces for further study as well as to remove and insert sequences during recombinant DNA techniques, described in a subsequent section. Endonucleases such as *Hae*III make straight, blunt cuts on DNA. But more often, the enzymes make staggered symmetrical cuts that leave short tails called "sticky ends." Such adhesive tails will base-pair with complementary tails on other DNA fragments or plasmids. This effect makes it possible to splice genes into specific sites.

The pieces of DNA produced by restriction endonucleases are termed *restriction fragments.* Because DNA sequences vary, even among members of the same species, differences in the cutting pattern of specific restriction endonucleases give rise to restriction fragments of differing lengths, known as *restriction fragment length polymorphisms* (RFLPs). RFLPs allow the direct comparison of the DNA of two different organisms at a specific site, which, as we will see, has many uses.

Another enzyme, called a **ligase,** is necessary to seal the sticky ends together by rejoining the phosphate-sugar bonds cut by endonucleases. Its main application is in final splicing of genes into plasmids and chromosomes.

An enzyme called **reverse transcriptase** is best known for its role in the replication of the AIDS virus and other retroviruses. It also provides geneticists with a valuable tool for converting RNA into DNA. Copies called **complementary DNA,** or **cDNA,** can be made from messenger, transfer, ribosomal, and other forms of RNA. The technique provides a valuable means of synthesizing eukaryotic genes from mRNA transcripts **(figure 10.2).** The advantage is that the synthesized gene will be free of the intervening sequences (introns) that can complicate the management of eukaryotic and archaeal genes in genetic engineering. Complementary DNA can also be used to analyze the nucleotide sequence of RNAs, such as those found in ribosomes and transfer RNAs.

Analysis of DNA

One way to produce a readable pattern of DNA fragments is through **gel electrophoresis.** In this technique, samples are placed in compartments (wells) in a soft agar gel and subjected to an electrical current. The phosphate groups in DNA give the entire molecule an overall negative charge, which causes the DNA to move toward the positive pole in the gel. The rate of movement is based primarily on the size

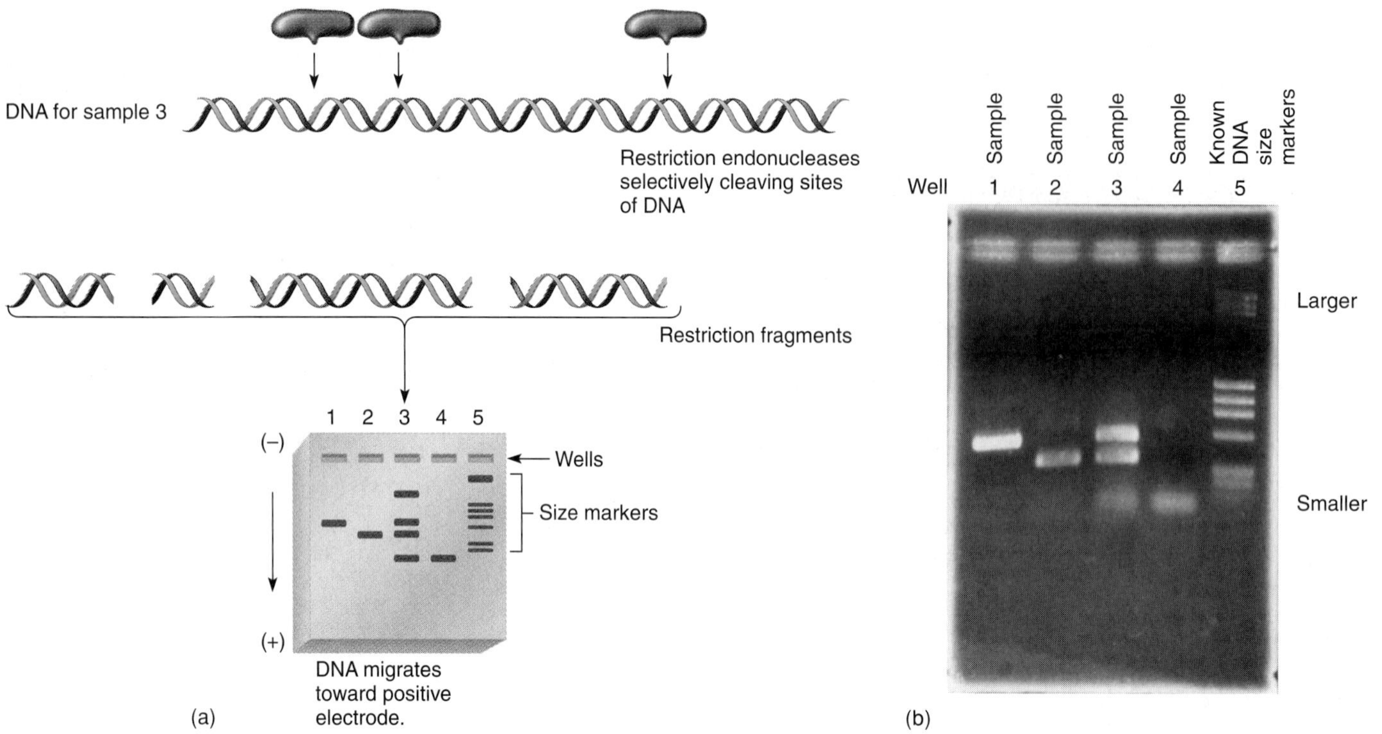

Figure 10.3 Revealing the patterns of DNA with electrophoresis.
(a) After cleavage into fragments, DNA is loaded into wells on one end of an agarose gel. When an electrical current is passed through the gel (from the negative pole to the positive pole), the DNA, being negatively charged, migrates toward the positive pole. The larger fragments, measured in numbers of base pairs, migrate more slowly and remain nearer the wells than the smaller (shorter) fragments. **(b)** An actual stained gel reveals a separation pattern of the fragments of DNA. The size of a given DNA band can be determined by comparing the distance it traveled to the distance traveled by a set of DNA fragments of known size (lane 5).

of the fragments. The larger fragments move more slowly and remain nearer the top of the gel, whereas the smaller fragments migrate faster and are positioned farther from the wells. The positions of DNA fragments are determined by staining the DNA fragments in the gel **(figure 10.3).** Electrophoresis patterns can be quite distinctive and are very useful in characterizing DNA fragments and comparing the degree of genetic similarities among samples as in a genetic fingerprint (discussed later).

Nucleic Acid Hybridization and Probes

Two different nucleic acids can **hybridize** by uniting at their complementary regions. All different combinations are possible: Single-stranded DNA can unite with other single-stranded DNA or RNA, and RNA can hybridize with other RNA. This property has allowed for the development of specially formulated oligonucleotide tracers called **gene probes.** These probes consist of a short stretch of DNA of a known sequence that will base-pair with a stretch of DNA with a complementary sequence, if one exists in the test sample. Hybridization probes have practical value because they can detect specific nucleotide sequences in unknown samples. So that areas of hybridization can be visualized, the probes carry reporter molecules such as radioactive labels, which are isotopes that emit radiation, or luminescent labels, which give off visible light. Reactions can be revealed by placing photographic film in contact with the test reaction. Fluorescent probes contain dyes that can be visualized with ultraviolet radiation, and enzyme-linked probes react with substrate to release colored dyes.

When probes hybridize with an unknown sample of DNA or RNA, they tag the precise area and degree of hybridization and help determine the nature of nucleic acid present in a sample. In a method called the **Southern blot,**[2] DNA fragments are first separated by electrophoresis and then denatured and transferred to a special filter. A DNA probe is then incubated with the sample, and wherever this probe encounters the segment for which it is complementary, it will attach and form a hybrid. Development of the hybridization pattern will show up as one or more bands **(figure 10.4).** This method is a sensitive and specific way to isolate fragments from a complex mixture and to find specific gene sequences on DNA. The Southern blot is also one of the important first steps for preparing isolated genes.

2. Named for its developer, E. M. Southern. The "northern" blot is a similar method used to analyze RNA, while the western blot detects proteins.

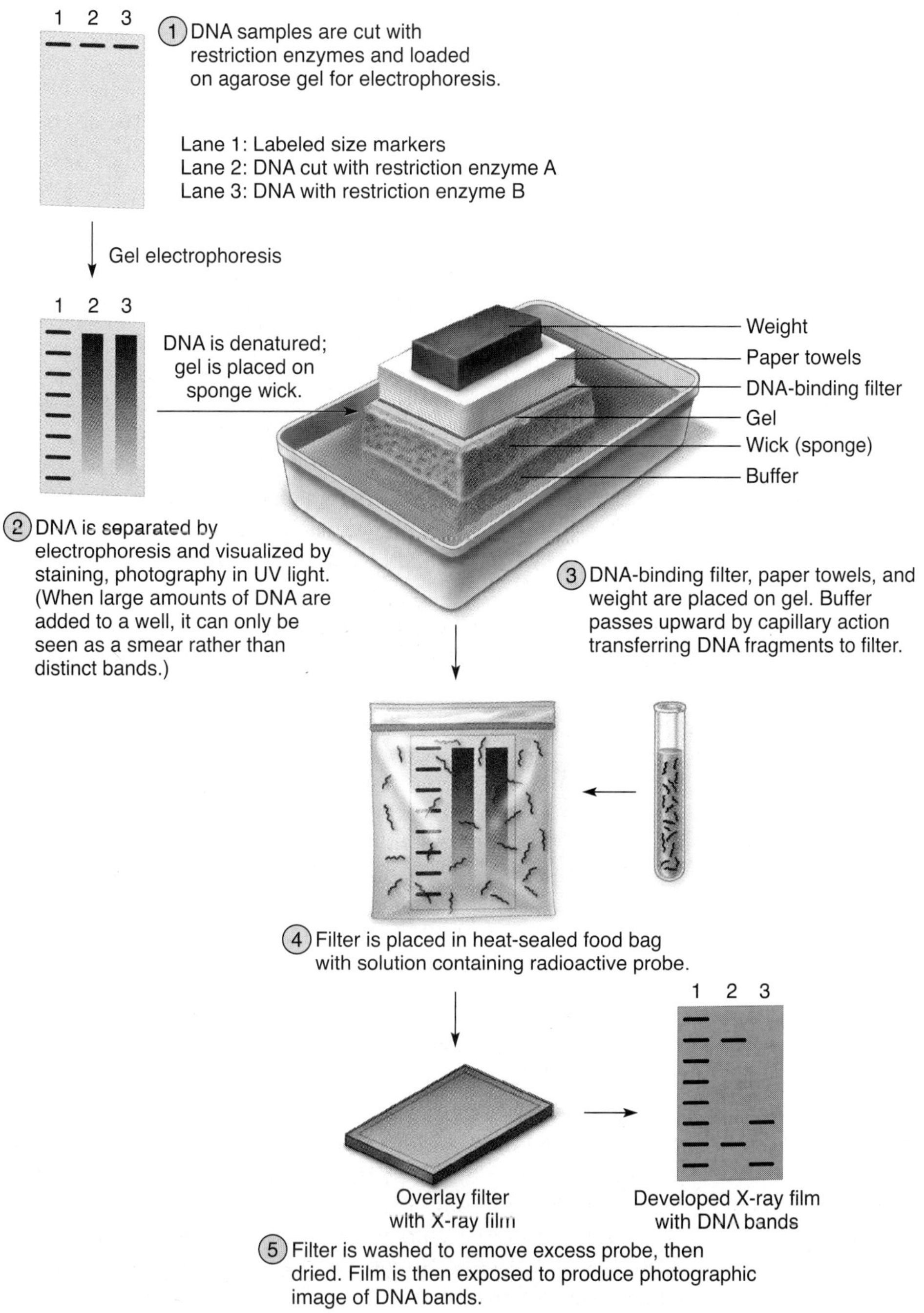

1 2 3
1 DNA samples are cut with restriction enzymes and loaded on agarose gel for electrophoresis.
Lane 1: Labeled size markers
Lane 2: DNA cut with restriction enzyme A
Lane 3: DNA with restriction enzyme B
Gel electrophoresis
1 2 3
DNA is denatured; gel is placed on sponge wick.
Weight
Paper towels
DNA-binding filter
Gel
Wick (sponge)
Buffer
2 DNA is separated by electrophoresis and visualized by staining, photography in UV light. (When large amounts of DNA are added to a well, it can only be seen as a smear rather than distinct bands.)
3 DNA-binding filter, paper towels, and weight are placed on gel. Buffer passes upward by capillary action transferring DNA fragments to filter.
4 Filter is placed in heat-sealed food bag with solution containing radioactive probe.
1 2 3
Overlay filter with X-ray film
Developed X-ray film with DNA bands
5 Filter is washed to remove excess probe, then dried. Film is then exposed to produce photographic image of DNA bands.

Process Figure 10.4 Conducting a Southern blot hybridization test. The five-step process reveals the hybridized fragments on the X-ray film.

Probes are commonly used for diagnosing the cause of an infection from a patient's specimen and identifying a culture of an unknown bacterium or virus. The method for doing this was first outlined in chapter 4. A simple and rapid method called a hybridization test does not require electrophoresis. DNA from a test sample is isolated, denatured, placed on an absorbent filter, and combined with a microbe-specific probe **(figure 10.5).** The blot is then developed and observed for areas of hybridization. Commercially available diagnostic kits are now on the market for identifying intestinal pathogens such as *Salmonella, Campylobacter, Shigella, Clostridium difficile,* rotaviruses, and adenoviruses. Other bacterial probes exist for *Mycobacterium, Legionella, Mycoplasma,* and *Chlamydia;* viral probes are available for herpes simplex and zoster, papilloma (genital warts), hepatitis A and B, and AIDS. DNA probes have also been developed for human genetic markers and some types of cancer.

With another method, called *fluorescent in situ hybridization* (FISH), probes are applied to intact cells and observed microscopically for the presence and location of specific genetic marker sequences on genes. It is a very effective way to locate genes on chromosomes. In situ techniques can also be used to identify unknown bacteria living in natural habitats without having to culture them, and they can be used to detect RNA in cells and tissues.

Methods Used to Size, Synthesize, and Sequence DNA

The relative sizes of nucleic acids are usually denoted by the number of base pairs (bp) or nucleotides they contain. For example, the palindromic sequences recognized by endonucleases are usually 4 to 10 bp in length; an average gene in *E. coli* is approximately 1,300 bp, or 1.3 kilobases (kb); and its entire genome is approximately 4,700,000 bp, 4,700 kb, or 4.7 megabases (Mb). The DNA of the human mitochondrion contains 16 kb, and the Epstein-Barr virus (a cause of infectious mononucleosis) has 172 kb. Humans have approximately 3.1 billion base pairs (bp) arrayed along 46 chromosomes. The recently completed Human Genome Project had as its goal the elucidation of the entire human genome **(Insight 10.1).**

DNA Sequencing: Determining the Exact Genetic Code

Analysis of DNA by its size, restriction patterns, and hybridization characteristics is instructive, but the most detailed information comes from determining the actual order and types of bases in DNA. This process, called **DNA sequencing,** provides the identity and order of nucleotides for all types of DNA, including genomic, cDNA, artificial chromosomes, plasmids, and cloned genes. The most common sequencing technique was developed by Frederick Sanger and is based on the synthesis and analysis of a complementary

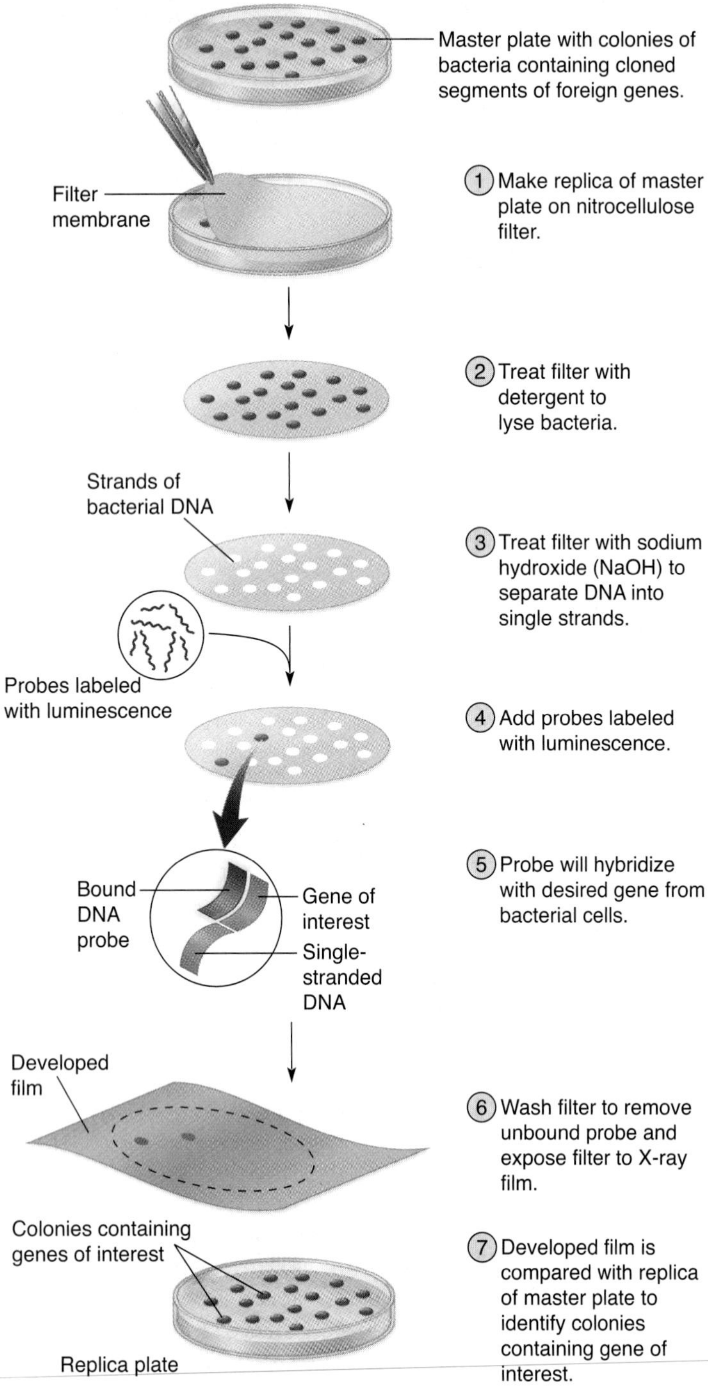
Master plate with colonies of bacteria containing cloned segments of foreign genes.
Filter membrane
1 Make replica of master plate on nitrocellulose filter.
2 Treat filter with detergent to lyse bacteria.
Strands of bacterial DNA
3 Treat filter with sodium hydroxide (NaOH) to separate DNA into single strands.
Probes labeled with luminescence
4 Add probes labeled with luminescence.
Bound DNA probe
Gene of interest
Single-stranded DNA
5 Probe will hybridize with desired gene from bacterial cells.
Developed film
6 Wash filter to remove unbound probe and expose filter to X-ray film.
Colonies containing genes of interest
Replica plate
7 Developed film is compared with replica of master plate to identify colonies containing gene of interest.

Process Figure 10.5 A hybridization test relies on the action of microbe-specific probes to identify an unknown bacterium or virus.

strand of DNA in a test tube **(figure 10.6)**. Today, the Sanger method, which is three decades old is nearing the end of its useful life. Researchers are testing different, quicker and cheaper methodology (the Sanger technique cost about $3 billion to sequence the human genome). Until then, the Sanger method is still being used.

Because most DNA being sequenced is very long, it is made more manageable by cutting it into a large number of shorter fragments and separating them. The test strands, typically several hundred nucleotides long, are then denatured to expose single strands that will serve as templates to synthesize complementary strands. The fragments are divided into four separate tubes that contain primers to set the start point for the synthesis to begin on one of the strands. The primer is labeled with a fluorescent or radioactive tag, which allows it to be detected. The nucleotides will attach at the 3′ end of the primer, using the template strands as a guide, essentially the same way that DNA synthesis occurs in a cell.

The tubes are incubated with the necessary DNA polymerase and all four of the regular nucleotides needed to carry out the process of elongating the complementary strand. Each tube also contains a single type (A, T, G, or C) of dideoxynucleotide (dd), which is critical to the sequencing process. Dideoxynucleotides have no oxygen bound to the 3′ carbon of deoxyribose. This oxygen atom is needed for the chemical attachment of the next nucleotide in the growing DNA strand. As the reaction proceeds, strands elongate by adding normal nucleotides, but a small percentage of

CASE FILE 10 *WRAP-UP*

The medical examiner in New York City had heard about two methods that were unproven but promising for his task of identifying the World Trade Center victims. The key factor was that the tests could be performed on DNA fragments that were much shorter than those required for the more traditional sequencing tests.

The two methods were called mini short tandem (DNA) repeats (mini-STR) and single nucleotide polymorphism (SNP) analysis. Both of these methods require very small pieces of DNA, a good circumstance with a catastrophic and fiery event like the World Trade Center attacks.

Mini-STR tests capitalize on and improve on more traditional tests of short tandem DNA repeats that required long stretches of DNA to provide conclusive matches between two subjects. When the 9/11 attacks occurred, the scientists who had been researching the possibility of making these tests work with much shorter pieces of DNA were contacted by the New York City Medical Examiners and they agreed to accelerate their efforts to make the tests viable.

SNP analysis relies on the fact that every person has between 3 million and 10 million variants in their 3 billion bases of DNA. The chances of finding a variation in a small piece of DNA are relatively good.

As it happened, though, the two techniques yielded few additional identifications. Families were left with uncertainty. Later, in 2006, more human remains were found in manholes that were previously unexcavated. Families were even more outraged that a thorough excavation of the site had not been completed.

INSIGHT 10.1 Discovery

OK, the Genome's Sequenced—What's Next?

In early 2001, a press conference was held to announce that the human genome had been sequenced. The 3.1 billion base pairs that make up the DNA found in (nearly) every human cell had been identified and put in the proper order. Champagne corks popped, balloons fell, bands played, reporters reported on the significance of the occasion, and . . . nothing else seemed to change. What, you may ask, has the Human Genome Project done for me lately? Here for your perusal are a few FAQs.

Q: Who sequenced the genome?
A: Francis Collins was the head of the publicly funded Human Genome Project (HGP), while Craig Venter was the head of Celera Genomics, a private company that developed a new, more powerful method of DNA sequencing and competed with the HGP. A compromise was finally reached whereby both groups took credit for sequencing the genome.
Q: How big was this project?
A: The Human Genome Project was first discussed in the mid-1980s and got under way in 1990. Although the project was to have taken at least 15 years, advances in technology led to its being completed in just over 10. The 3.1 billion base pairs of DNA code for only about 20,000 to 25,000 genes, not the 100,000 or so that the genome was thought to contain only a few years ago.
Q: Will I ever see any benefits from this project?
A: Absolutely. Sequencing the genome was only the first step. Knowing what proteins are produced in the body, what they do, and how they interact with one another is essential to understanding the workings of the human body, in both health and disease. By knowing the genetic signatures of different diseases, we will be able to design extraordinarily sensitive diagnostic tests that detect not only a disease but particular subtypes of each malady. With a precise genetic identification of, for example, a tumor, treatment can be tailored to be as effective as possible, while dramatically reducing side effects.
Q: What's next?
A: The regulatory regions are the current area of excitement. The thinking is that areas within genes that are similar in mice and humans (known as regions of homology) are most important for the function of the gene and are the most likely targets for potential drugs or genetic treatments.
Q: Were there any surprises?
A: After the human genome was sequenced, a variety of other genomes were also sequenced. The information from all of these organisms revealed that the most important sequences may well be the ones in between the protein-coding genes. These regions regulate the genes.
Q: Where could I go to check out a really cool website on the human genome?
A: http://www.ornl.gov/hgmis/

J. Craig Venter, left, and Francis Collins celebrate mapping of the human genome.

fragments will randomly incorporate the complementary dideoxynucleotide and be terminated. Eventually, all possible positions in the sequence will incorporate a terminal dideoxynucleotide, thus producing a series of strands that reflect the correct sequence.

The reaction products are placed into four wells (G, C, A, T) of a polyacrylamide gel, which is sensitive enough to separate strands that differ by only a single nucleotide in length. Electrophoresis separates the fragments in order according to both size and lane, and only the fragments carrying the labeled nucleotides will be readable on the gel. The gel indicates the comparative orientation of the bases, which graduate in size from the smallest fragments that terminate early and migrate farthest (bottom of gel) to successively longer fragments (moving stepwise to the top). Reading the order of first appearance of a given gel fragment in the G, C, A, or T lanes provides the correct sequence of bases of the complementary strand, and it allows one to infer the sequence of the template strand as well. This method of sequencing is remarkably accurate, with only about one mistake in every 1,000 bases. All of these steps have been automated so that most researchers need only insert a sample into a machine and wait for the printout of the results.

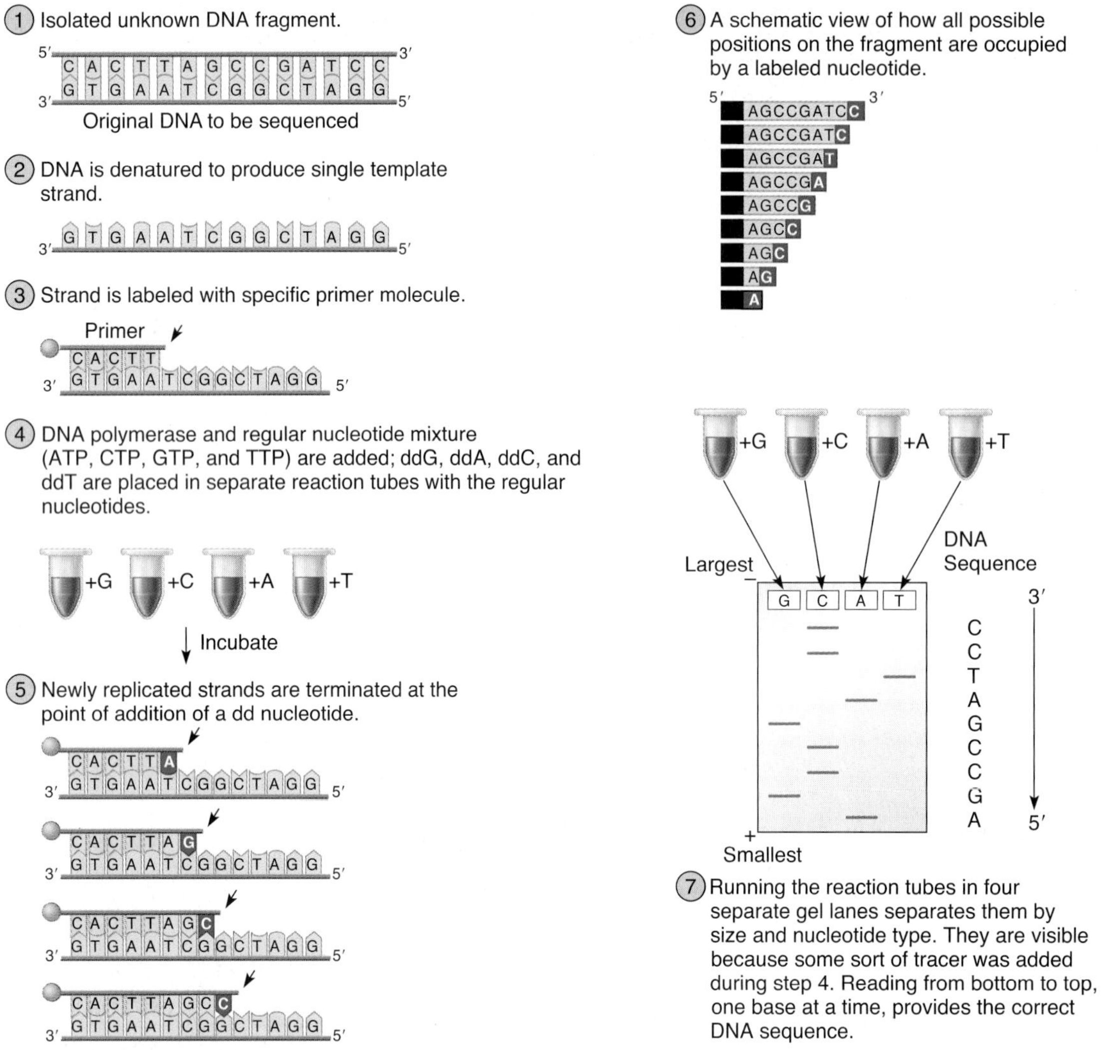

1 Isolated unknown DNA fragment.
5′ C A C T T A G C C G A T C C 3′
3′ G T G A A T C G G C T A G G 5′
Original DNA to be sequenced
2 DNA is denatured to produce single template strand.
3′ G T G A A T C G G C T A G G 5′
3 Strand is labeled with specific primer molecule.
Primer
C A C T T
3′ G T G A A T C G G C T A G G 5′
4 DNA polymerase and regular nucleotide mixture (ATP, CTP, GTP, and TTP) are added; ddG, ddA, ddC, and ddT are placed in separate reaction tubes with the regular nucleotides.
+G
+C
+A
+T
Incubate
5 Newly replicated strands are terminated at the point of addition of a dd nucleotide.
C A C T T A
C A C T T A G
C A C T T A G C
C A C T T A G C C
6 A schematic view of how all possible positions on the fragment are occupied by a labeled nucleotide.
AGCCGATCC
AGCCGATC
AGCCGAT
AGCCGA
AGCCG
AGCC
AGC
AG
A
Largest
Smallest
DNA Sequence
G C A T
C C T A G C C G A
3′
5′
7 Running the reaction tubes in four separate gel lanes separates them by size and nucleotide type. They are visible because some sort of tracer was added during step 4. Reading from bottom to top, one base at a time, provides the correct DNA sequence.

Process Figure 10.6 Steps in a Sanger DNA sequence technique.

Automation of this process was absolutely necessary for sequencing the entire genomes of humans, mice, and other organisms (see Insight 10.1).

Polymerase Chain Reaction: A Molecular Xerox Machine for DNA

Some of the techniques used to analyze DNA and RNA are limited by the small amounts of test nucleic acid available. This problem was largely solved by the invention of a simple, versatile way to amplify DNA called the **polymerase chain reaction (PCR).** This technique rapidly increases the amount of DNA in a sample without the need for making cultures or carrying out complex purification techniques. It is so sensitive that it holds the potential to detect cancer from a single cell or to diagnose an infection from a single gene copy. It is comparable to being able to pluck a single DNA "needle" out of a "haystack" of other molecules and make unlimited copies of the DNA. The rapid rate of PCR makes it possible to replicate a target DNA from a few copies to billions of copies in a few hours.

To understand the idea behind PCR, it will be instructive to review figure 9.6, which describes synthesis of DNA as it occurs naturally in cells. The PCR method uses essentially the same events, with the opening up of the double strand, using the exposed strands as templates, the addition of primers, and the action of a DNA polymerase.

Initiating the reaction requires a few specialized ingredients **(figure 10.7).** The **primers** are synthetic oligonucleotides (short DNA strands) of a known sequence of 15 to 30 bases that serve as landmarks to indicate where DNA amplification will begin. These take the place of RNA primers that would normally be synthesized by primase (in the cell). Depending upon the purposes and what is known about the DNA being replicated, the primers can be random, attaching to any sequence they may fit, or they may be highly specific and chosen to amplify a known gene. To keep the DNA strands separated, processing must be carried out at a relatively high temperature. This necessitates the use of special **DNA polymerases** isolated from thermophilic bacteria. Examples of these unique enzymes are Taq polymerase obtained from *Thermus aquaticus* and Vent polymerase from

(a) In cycle 1, the DNA to be amplified is denatured, primed, and replicated by a polymerase that can function at high temperature. The two resulting strands then serve as templates for a second cycle of denaturation, priming, and synthesis.*

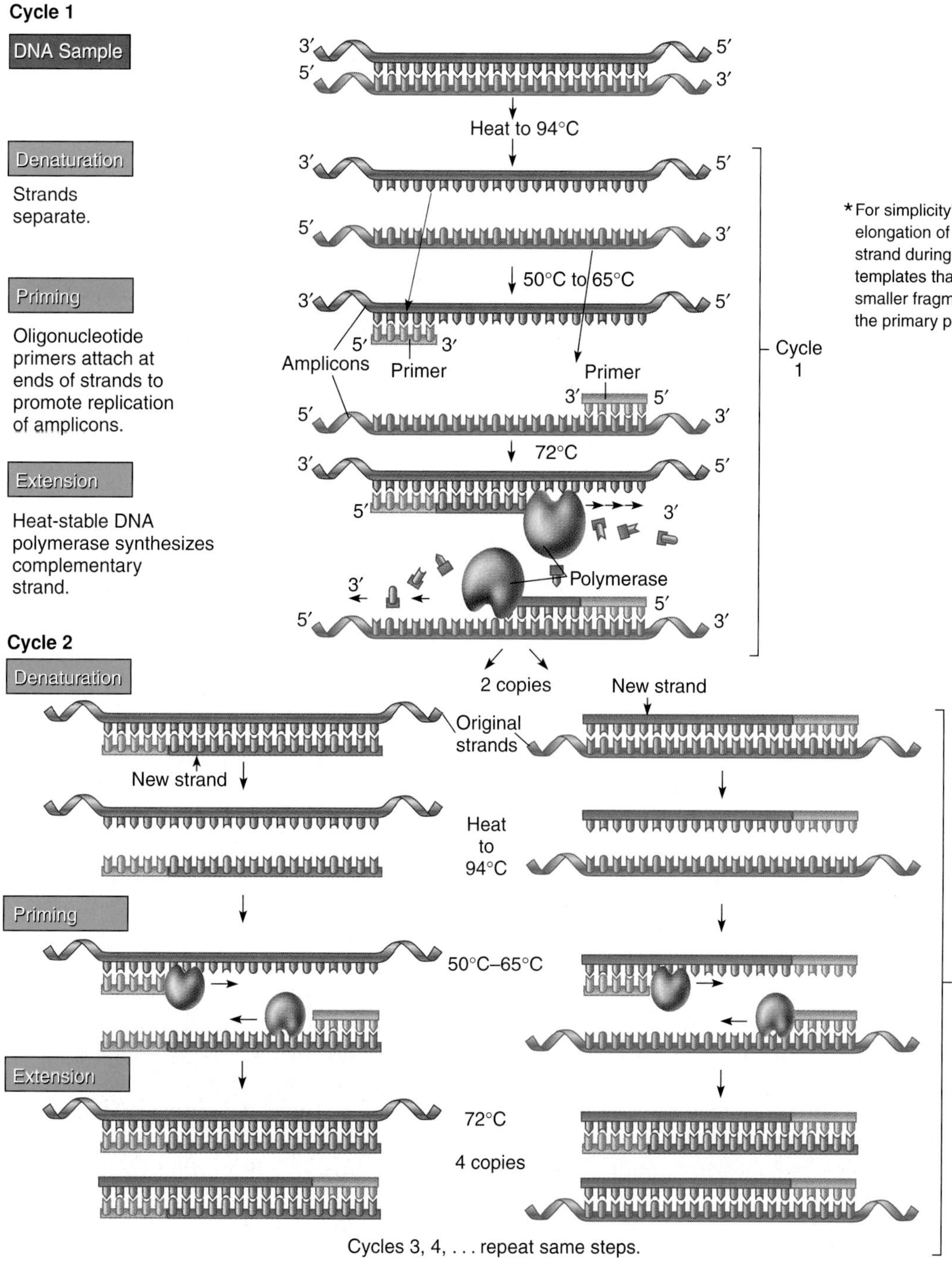

*For simplicity's sake, we have omitted the elongation of the complete original parent strand during the first cycles. Ultimately, templates that correspond only to the smaller fragments dominate and become the primary population of replicated DNA.

(b) A view of the process after 6 cycles, with 64 copies of amplified DNA. Continuing this process for 20 to 40 cycles can produce millions of identical DNA molecules.

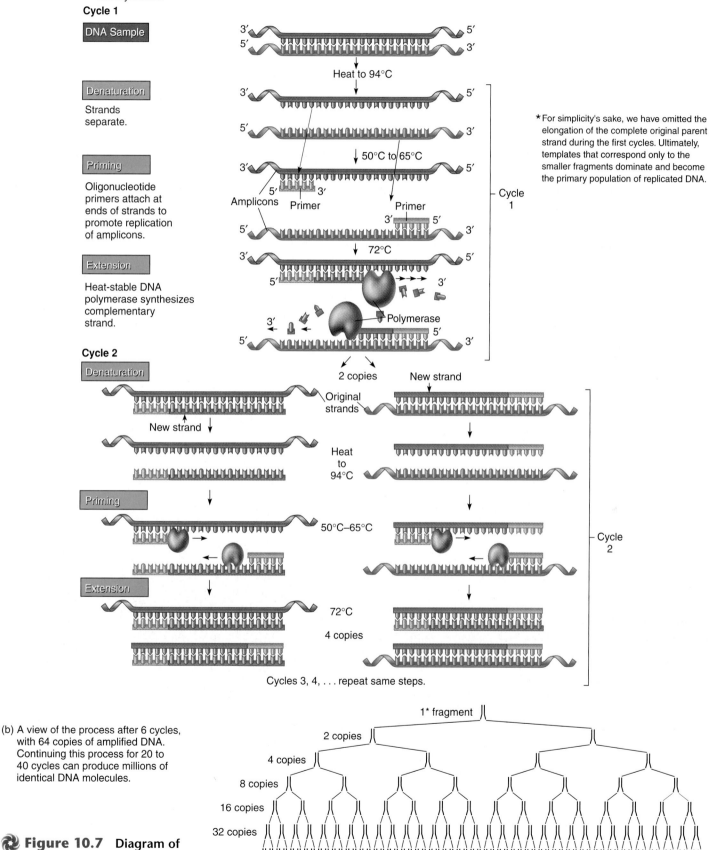

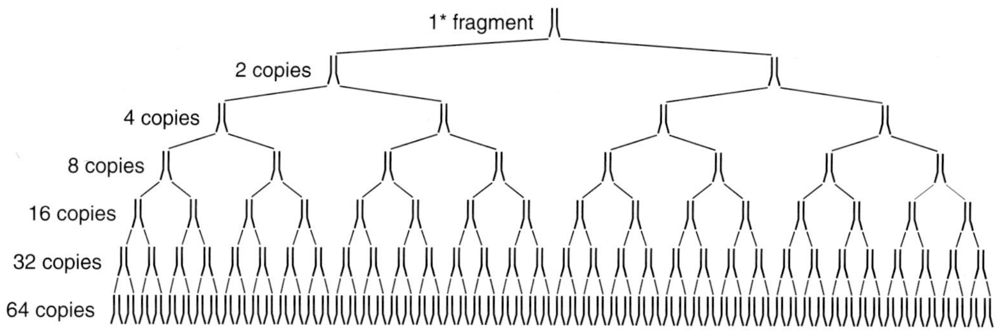

Figure 10.7 Diagram of the polymerase chain reaction.

Thermococcus litoralis. Enzymes isolated from these thermophilic organisms remain active at the elevated temperatures used in PCR. Another useful component of PCR is a machine called a thermal cycler that automatically performs the cyclic temperature changes.

The PCR technique operates by repetitive cycling of three basic steps: denaturation, priming, and extension.

1. **Denaturation.** The first step involves heating target DNA to 94°C to separate it into two strands. Next, the system is cooled to between 50°C and 65°C, depending on the exact nucleotide sequence of the primer.
2. **Priming.** Primers are added in a concentration that favors binding to the complementary strand of test DNA. This reaction prepares the two DNA strands, now called **amplicons,** for synthesis.
3. **Extension.** In the third phase, which proceeds at 72°C, DNA polymerase and raw materials in the form of nucleotides are added. Beginning at the free end of the primers on both strands, the polymerases extend the molecule by adding appropriate nucleotides and produce two complete strands of DNA.

It is through cyclic repetition of these steps that DNA becomes amplified. When the DNAs formed in the first cycle are denatured, they become amplicons to be primed and extended in the second cycle. Each subsequent cycle converts the new DNAs to amplicons and doubles the number of copies. The number of cycles required to produce a million molecules is 20, but the process is usually carried out to 30 or 40 cycles. One significant advantage of this technique has been its natural adaptability to automation. A PCR machine can perform 20 cycles on nearly 100 samples in 2 or 3 hours.

Once the PCR is complete, the amplified DNA can be analyzed by any of the techniques discussed earlier. PCR can be adapted to analyze RNA by initially converting an RNA sample to DNA with reverse transcriptase. This cDNA can then be amplified by PCR in the usual manner. It is by such means that ribosomal RNA and messenger RNA are readied for sequencing. The polymerase chain reaction has found prominence as a powerful workhorse of molecular biology, medicine, and biotechnology. It often plays an essential role in gene mapping, the study of genetic defects and cancer, forensics, taxonomy, and evolutionary studies.

For all of its advantages, PCR has some problems. A serious concern is the introduction and amplification of nontarget DNA from the surrounding environment, such as a skin cell from the technician carrying out the PCR reaction rather than material from the sample DNA that was supposed to be amplified. Such contamination can be minimized by using equipment and rooms dedicated for DNA analysis and maintained with the utmost degree of cleanliness. Problems with contaminants can also be reduced by using gene-specific primers and treating samples with special enzymes that can degrade the contaminating DNA before it is amplified.

CHECKPOINT

- The genetic revolution has produced a wide variety of industrial technologies that translate and radically alter the blueprints of life. The potential of biotechnology promises not only improved quality of life and enhanced economic opportunity but also serious ethical dilemmas. Increased public understanding is essential for developing appropriate guidelines for responsible use of these revolutionary techniques.
- Genetic engineering utilizes a wide range of methods that physically manipulate DNA for purposes of visualization, sequencing, hybridizing, and identifying specific sequences. The tools of genetic engineering include specialized enzymes, gel electrophoresis, DNA sequencing machines, and gene probes. The polymerase chain reaction (PCR) technique amplifies small amounts of DNA into larger quantities for further analysis.

10.3 Methods in Recombinant DNA Technology: How to Imitate Nature—or to "One-Up" It

The primary intent of **recombinant DNA technology** is to deliberately remove genetic material from one organism and combine it with that of a different organism. Its origins can be traced to 1970, when microbiologists first began to duplicate the clever tricks bacteria do naturally with bits of extra DNA such as plasmids, transposons, and proviruses. As mentioned earlier, humans have been trying to artificially influence genetic transmission of traits for centuries. The discovery that bacteria can readily accept, replicate, and express foreign DNA made them powerful agents for studying the genes of other organisms in isolation. The practical applications of this work were soon realized by biotechnologists. Bacteria could be genetically engineered to mass produce substances such as hormones, enzymes, and vaccines that were difficult to synthesize by the usual industrial methods.

Figure 10.8 provides an overview of the recombinant DNA procedure. An important objective of this technique is to form genetic **clones.** Cloning involves the removal of a selected gene from an animal, plant, or microorganism (the genetic donor) followed by its propagation in a different host organism. Cloning requires that the desired donor gene first be selected, excised by restriction endonucleases, and isolated. The gene is next inserted into a **vector** (usually a plasmid or a virus) that will insert the DNA into a **cloning host.** The cloning host is usually a bacterium or a yeast that can replicate the gene and translate it into the protein product for which it codes. In the next section, we examine the elements of gene isolation, vectors, and cloning hosts and show how they participate in a complete recombinant DNA procedure.

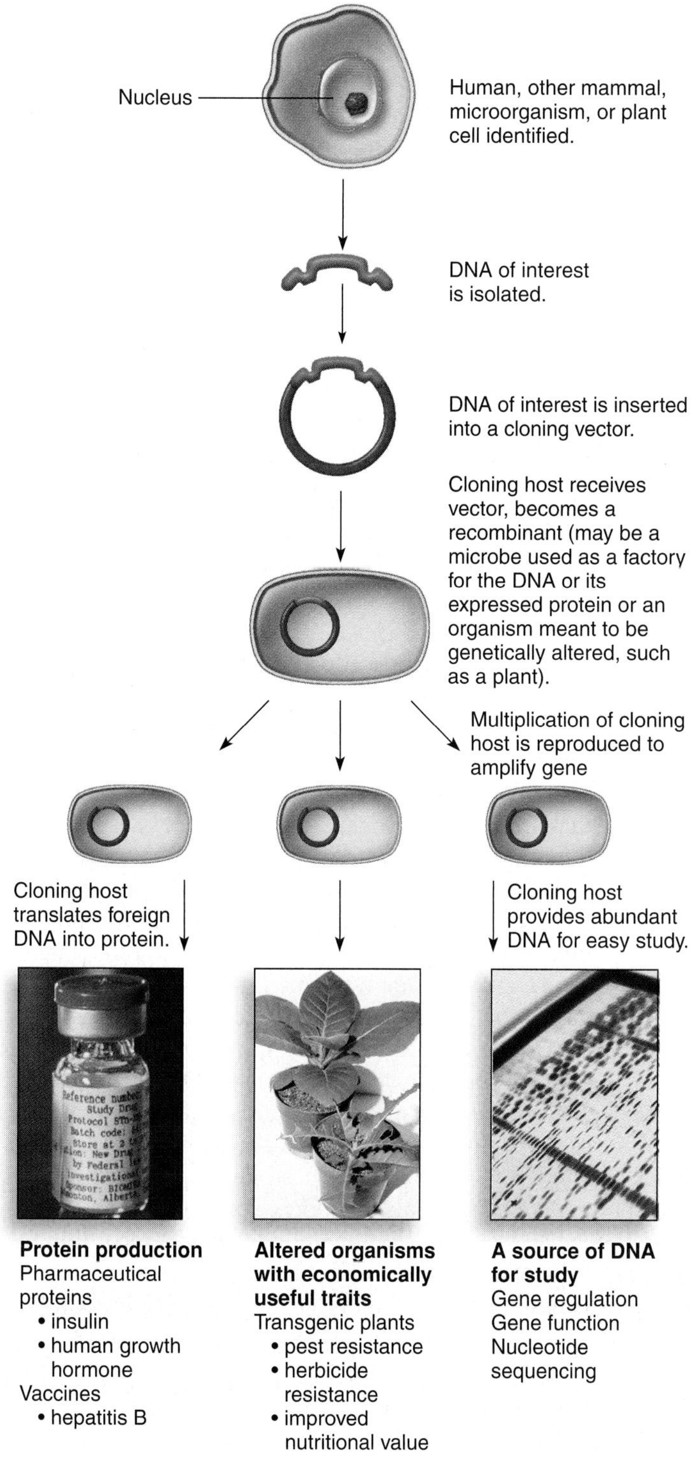
Nucleus
Human, other mammal, microorganism, or plant cell identified.
DNA of interest is isolated.
DNA of interest is inserted into a cloning vector.
Cloning host receives vector, becomes a recombinant (may be a microbe used as a factory for the DNA or its expressed protein or an organism meant to be genetically altered, such as a plant).
Multiplication of cloning host is reproduced to amplify gene
Cloning host translates foreign DNA into protein.
Cloning host provides abundant DNA for easy study.
Protein production
Pharmaceutical proteins
• insulin
• human growth hormone
Vaccines
• hepatitis B
Altered organisms with economically useful traits
Transgenic plants
• pest resistance
• herbicide resistance
• improved nutritional value
A source of DNA for study
Gene regulation
Gene function
Nucleotide sequencing

Figure 10.8 Methods and applications of genetic technology.
Practical applications of genetic engineering include the development of pharmaceuticals, genetically modified organisms, and forensic techniques.

Technical Aspects of Recombinant DNA and Gene Cloning

The first hurdles in cloning a target gene are to locate its exact site on the genetic donor's chromosome and to isolate it. Among the most common strategies for obtaining genes in an isolated state are:

1. The DNA is removed from cells and separated into fragments by endonucleases. Each fragment is then inserted into a vector and cloned. The cloned fragments undergo Southern blotting and are probed to identify desired sequences. This is a long and tedious process, because each fragment of DNA must be examined for the cloned gene.
2. A gene can be synthesized from isolated mRNA transcripts using reverse transcriptase (cDNA).
3. A gene can be amplified using PCR in many cases.

Although gene cloning and isolation can be very laborious, a fortunate outcome is that, once isolated, genes can be maintained in a cloning host and vector just like a microbial pure culture. *Genomic libraries* are collections of DNA clones that represent the entire genome of numerous organisms.

Characteristics of Cloning Vectors

A good recombinant vector has three indispensable qualities: It must be capable of carrying a significant piece of the donor DNA, it must be readily accepted by the cloning host, and it must have a promoter in front of the cloned gene. Plasmids are excellent vectors because they are small, well characterized, easy to manipulate, and can be transferred into appropriate host cells through transformation. Bacteriophages also serve well because they have the natural ability to inject DNA into bacterial hosts through transduction. A common vector in early work was an *E. coli* plasmid that carries genetic markers for resistance to antibiotics, although it is restricted by the relatively small amount of foreign DNA it can accept. A modified phage vector, the *Charon*[3] phage, is missing large sections of its genome, so it can carry a fairly large segment of foreign DNA. The simple plasmids and bacteriophages that were a staple of early recombinant DNA methodologies evolved into newer, more advanced vectors. Today, thousands of unique cloning vectors are available commercially. Although every vector has characteristics that make it ideal for a specific project, all vectors can be thought of as having three important attributes to consider **(figure 10.9):**

1. An origin of replication (ORI) is needed somewhere on the vector so that it will be replicated by the DNA polymerase of the cloning host.
2. The vector must accept DNA of the desired size. Early plasmids were limited to an insert size of less than 10 kb

3. Named for the mythical boatman in Hades who carried souls across the River Styx.

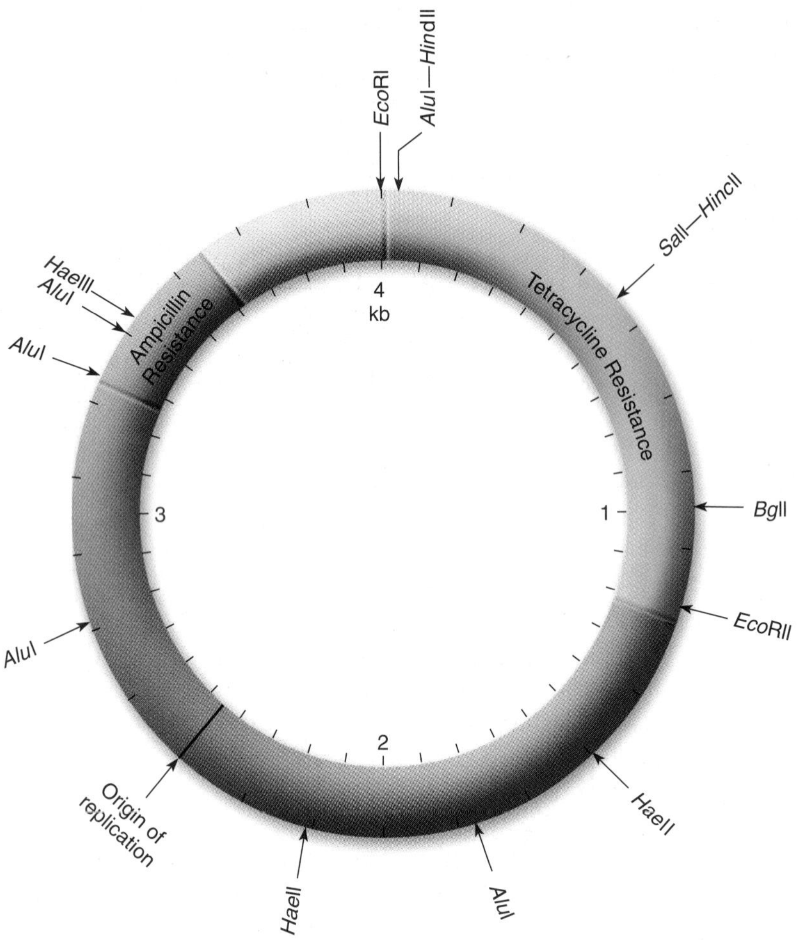

Figure 10.9 Partial map of the pBR322 plasmid of *Escherichia coli*. Arrows delineate sites cleaved by various restriction endonucleases. Numbers indicate the size of regions in kilobases (kb). Other important components are genes for ampicillin and tetracycline resistance and the origin of replication.

of DNA, far too small for most eukaryotic genes, with their sizable introns. Vectors called cosmids can hold 45 kb while complex *bacterial artificial chromosomes* (BACs) and *yeast artificial chromosomes* (YACs) can hold as much as 300 kb and 1,000 kb, respectively.

3. Vectors typically contain a gene that confers drug resistance to their cloning host. In this way, cells can be grown on drug-containing media, and only those cells that harbor a plasmid will be selected for growth.

Characteristics of Cloning Hosts

The best cloning hosts possess several key characteristics **(table 10.1).** The traditional cloning host and the one still used in the majority of experiments is *Escherichia coli.* Because this bacterium was the original recombinant host, the protocols using it are well established, relatively easy, and reliable. Hundreds of specialized cloning vectors have been developed for it. The main disadvantage with this species is that the splicing of mRNA as well as the modification of proteins that would normally occur in the eukaryotic endoplasmic reticulum and Golgi apparatus are unavailable in this prokaryotic cloning host. One alternative host for certain industrial processes and research is the yeast *Saccharomyces cerevisiae,* which, being eukaryotic, already possesses mechanisms for processing and modifying eukaryotic gene products. Certain techniques may also employ different bacteria (*Bacillus subtilis*), animal cell cultures, and even live animals and plants to serve as cloning hosts. In our coverage, we present the recombinant process as it is performed in bacteria and yeasts.

Construction of a Recombinant, Insertion into a Cloning Host, and Genetic Expression

This section illustrates one example of recombinant DNA technology, in this case, to produce a drug called alpha-2a interferon (Roferon-A). This form of interferon is used to treat cancers such as hairy-cell leukemia and Kaposi's sarcoma in AIDS patients. (Both diseases are described in chapter 20.) The human alpha interferon gene is a DNA molecule of approximately 500 bp that codes for a polypeptide of 166 amino acids. It was originally isolated and identified from human blood cells and prepared from processed mRNA transcripts that are free of introns. This step is necessary because the bacterial cloning host has none of the machinery needed to excise this nontranslated part of a gene.

The first step in cloning is to prepare the isolated interferon gene for splicing into an *E. coli* plasmid **(figure 10.10).** One way this is accomplished is to digest both the gene and the plasmid with the same restriction enzyme, resulting in complementary sticky ends on both the vector and insert DNA. In the presence of the endonuclease, the plasmid's circular molecule is nicked open and its sticky ends are exposed. When the gene and plasmid are placed together, their free ends base-pair, and a ligase makes the final covalent bonds. The resultant gene and plasmid combination is called a **recombinant.**

Following this procedure, the recombinant is introduced by transformation into the cloning host, a special laboratory

TABLE 10.1 Desirable Features in a Microbial Cloning Host
Rapid turnover, fast growth rate
Can be grown in large quantities using ordinary culture methods
Nonpathogenic
Genome that is well delineated (mapped)
Capable of accepting plasmid or bacteriophage vectors
Maintains foreign genes through multiple generations
Will secrete a high yield of proteins from expressed foreign genes

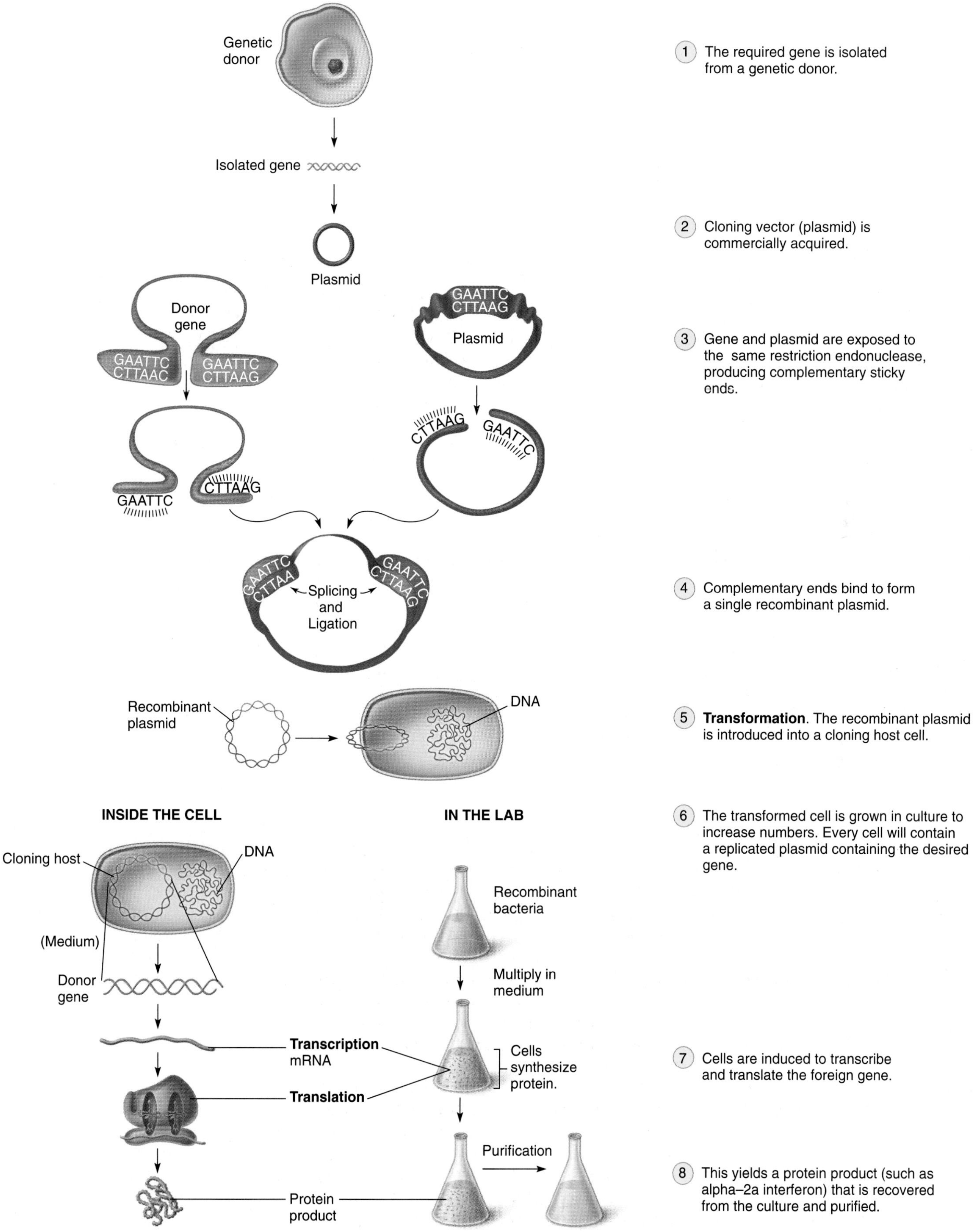
Genetic donor
Isolated gene
Plasmid
Donor gene
GAATTC CTTAAC
GAATTC CTTAAG
Plasmid
GAATTC CTTAAG
GAATTC
CTTAAG
CTTAAG
GAATTC
GAATTC CTTAA
GAATTC CTTAAG
Splicing and Ligation
Recombinant plasmid
DNA
INSIDE THE CELL
IN THE LAB
Cloning host
DNA
(Medium)
Donor gene
Recombinant bacteria
Multiply in medium
Transcription mRNA
Translation
Cells synthesize protein.
Purification
Protein product
1 The required gene is isolated from a genetic donor.
2 Cloning vector (plasmid) is commercially acquired.
3 Gene and plasmid are exposed to the same restriction endonuclease, producing complementary sticky ends.
4 Complementary ends bind to form a single recombinant plasmid.
5 Transformation. The recombinant plasmid is introduced into a cloning host cell.
6 The transformed cell is grown in culture to increase numbers. Every cell will contain a replicated plasmid containing the desired gene.
7 Cells are induced to transcribe and translate the foreign gene.
8 This yields a protein product (such as alpha–2a interferon) that is recovered from the culture and purified.

Process Figure 10.10 Using recombinant DNA for gene cloning.

strain of *E. coli* that lacks any extra plasmids that could complicate the expression of the gene. Because the recombinant plasmid enters only some of the cloning host cells, it is necessary to locate these recombinant clones. Cultures are plated out on medium containing ampicillin, and only those clones that carry the plasmid with ampicillin resistance can form colonies **(figure 10.11).** These recombinant colonies are selected from the plates and cultured. As the cells multiply, the plasmid is replicated along with the cell's chromosome. In a few hours of growth, there can be billions of cells, each containing the interferon gene. Once the gene has been successfully cloned and tested, this step does not have to be repeated—the recombinant strain can be maintained in culture for production purposes.

The bacteria's ability to express the eukaryotic gene is ensured, because the plasmid has been modified with the necessary transcription and translation recognition sequences. As the *E. coli* culture grows, it transcribes and translates the interferon gene, synthesizes the peptide, and secretes it into the growth medium. At the end of the process, the cloning cells and other chemical and microbial impurities are removed from the medium. Final processing to excise a terminal amino acid from the peptide yields the interferon product in a relatively pure form (see figure 10.10). The scale of this procedure can range from test tube size to gigantic industrial vats that can manufacture thousands of gallons of product (see chapter 25).

Although the process we have presented here produces interferon, some variation of it can be used to mass produce a variety of hormones, enzymes, and agricultural products such as pesticides. Recent advances even allow scientists to produce functions that weren't originally present in the biological world. For instance, scientists are attempting to create microbes that produce hydrogen as fuel. Engineering new genetic capabilities is called **synthetic biology.**

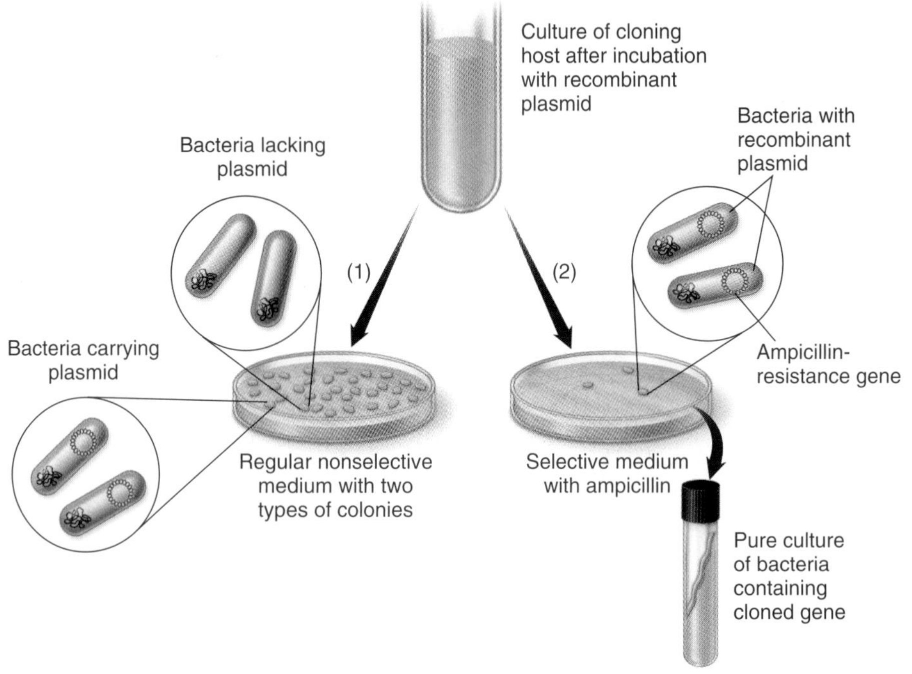

Figure 10.11 One method for screening clones of bacteria that have been transformed with the donor gene.
(1) Plating the culture on nonselective medium will not separate the transformed cells from normal cells, which lack the plasmid. (2) Plating on selective medium containing ampicillin will permit only cells containing the plasmid to multiply. Colonies growing on this medium that carry the cloned gene can be used to make a culture for gene libraries, industrial production, and other processes. (Plasmids are shown disproportionate to size of cell.)

CHECKPOINT

- Recombinant DNA techniques combine DNA from different sources to produce microorganism "factories" that produce hormones, enzymes, and vaccines on an industrial scale. Cloning is the process by which genes are removed from the original host and duplicated for transfer into a cloning host by means of cloning vectors.
- Plasmids, bacteriophages, and cosmids are types of cloning vectors used to transfer recombinant DNA into a cloning host. Cloning hosts are simply organisms that readily accept recombinant DNA, grow easily, and synthesize large quantities of specific gene product.

10.4 Biochemical Products of Recombinant DNA Technology

Recombinant DNA technology is used by pharmaceutical companies to manufacture medications that cannot be manufactured by any other means. Diseases such as diabetes and dwarfism, caused by the lack of an essential hormone, are treated by replacing the missing hormone. Insulin of animal origin was once the only form available to treat diabetes, even though such animal products can cause allergic reactions in sensitive individuals. In contrast, dwarfism cannot be treated with animal growth hormones, so originally the only source of human growth hormone (HGH) was the pituitaries of cadavers. At one time, not enough HGH was available to treat the thousands of children in need. Another serious problem with using natural human products is the potential for infection. For example, infectious agents such as the prion responsible for Creutzfeld-Jakob disease, described in chapter 19, can be transmitted in this manner. Similarly, clotting factor VIII, a protein needed for blood to clot properly, is missing in persons suffering from hemophilia A. Persons lacking factor VIII have historically received periodic injections of the missing protein, which had been collected from blood plasma. While the donated protein alleviated the symptoms of factor VIII deficiency, a tragic side effect was seen in the early 1980s, when a large percentage of the

TABLE 10.2 Examples of Current Protein Products from Recombinant DNA Technology

Immune Treatments
Interferons—peptides used to treat some types of cancer, multiple sclerosis, and viral infections such as hepatitis and genital warts
Interleukins—types of cytokines that regulate the immune function of white blood cells; used in cancer treatment
Orthoclone—an immune suppressant in transplant patients
Granulocyte-macrophage-colony-stimulating factor (GM-CSF)—used to stimulate bone marrow activity after bone marrow grafts
Tumor necrosis factor (TNF)—used to treat cancer
Granulocyte-colony-stimulating factor (Neupogen)—developed for treating cancer patients suffering from low neutrophil counts
Hormones
Erythropoietin (EPO)—a peptide that stimulates bone marrow used to treat some forms of anemia
Tissue plasminogen activating factor (tPA)—can dissolve potentially dangerous blood clots
Hemoglobin A—form of artificial blood to be used in place of real blood for transfusions
Factor VIII—needed as replacement blood-clotting factor in type A hemophilia
Relaxin—an aid to childbirth
Human growth hormone (HGH)—stimulates growth in children with dwarfism; prevents wasting syndrome
Enzymes
rH DNase (pulmozyme)—a treatment that can break down the thick lung secretions of cystic fibrosis
Antitrypsin—replacement therapy to benefit emphysema patients
PEG-SOD—a form of superoxide dismutase that minimizes damage to brain tissue after severe trauma
Vaccines
Vaccines for hepatitis B and *Haemophilus influenzae* Type b meningitis
Experimental malaria and AIDS vaccines based on recombinant surface antigens
Miscellaneous
Bovine growth hormone or bovine somatotropin (BST)—given to cows to increase milk production
Apolipoprotein—to deter the development of fatty deposits in the arteries and to prevent strokes and heart attacks
Spider silk—a light, tough fabric for parachutes and bulletproof vests

patients receiving plasma-derived factor VIII contracted HIV infections as a result of being exposed to the virus through the donated plasma.

Recombinant DNA technology changed the outcome of these and many other conditions by enabling large-scale manufacture of lifesaving hormones and enzymes of human origin. Recombinant human insulin can now be prescribed for diabetics, and recombinant HGH can now be administered to children with dwarfism and Turner syndrome. HGH is also used to prevent the wasting syndrome that occurs in AIDS and cancer patients. In all of these applications, recombinant DNA technology has led to both a safer product and one that can be manufactured in quantities previously unfathomable. Other protein-based hormones, enzymes, and vaccines produced through recombinant DNA technology are summarized in **table 10.2.**

Nucleic acid products also have a number of medical applications. A new development in vaccine formulation involves using microbial DNA as a stimulus for the immune system. So far, animal tests using DNA vaccines for AIDS and influenza indicate that this may be a breakthrough in vaccine design. Recombinant DNA could also be used to produce DNA-based drugs for the types of gene and antisense therapy discussed in the genetic treatments section.

10.5 Genetically Modified Organisms

Recombinant organisms produced through the introduction of foreign genes are called *transgenic* or genetically modified organisms (GMOs). Foreign genes have been inserted into a variety of microbes, plants, and animals through recombinant DNA techniques developed especially for them. Transgenic "designer" organisms are available for a variety of biotechnological applications. Because they are unique life forms that would never have otherwise occurred, they can be patented.

Recombinant Microbes: Modified Bacteria and Viruses

One of the first practical applications of recombinant DNA in agriculture was to create a genetically altered strain of the bacterium *Pseudomonas syringae.* The wild strain ordinarily contains a gene that promotes ice or frost formation

INSIGHT 10.2 ***Microbiology***

A Moment to Think

> We are embarking on a very potentially troublesome journey, where we begin to reduce all other animals on this planet to genetically engineered products. . . . We will increasingly think of ourselves as just gene codes and blueprints and programs that can be tinkered with.
>
> Jeremy Rifkin, Foundation on Economic Trends

> Never postpone experiments that have clearly defined future benefits for fear of dangers that can't be quantified [because] we can react rationally only to real (as opposed to hypothetical) risks.
>
> James Watson, Nobel Laureate and first head of the Human Genome Project

> I've never been less well equipped intellectually to vote on an issue than I am on this.
>
> Unnamed U.S. Senator, prior to a vote on stem cell research and human cloning

Of these three statements, the third is without a doubt the most frightening. There are always several sides to every issue and the best we can hope for is that the people regulating genetic technology are as well informed as possible. Those who will determine the limits of genetic technology include not only scientists and politicians but voters and consumers as well. History is rife with both knee-jerk rejections to new technologies that have later proven to be invaluable (vaccinations) and complacency while dangerous products were made available to an ignorant and unsuspecting public (Fen-Phen, thalidomide). Ethical choices can be properly made only from a standpoint of intellectual awareness, and in this era of advertising, polling, and focus groups, people with a stake in genetic technology know that the most effective way to drum up support (both for or against) is not by carefully educating the public as to the uses and limits of our newfound powers but rather by publicizing exaggerated claims of frightening scenarios. Enlightenment is often a casualty of these advertising campaigns, and you as a student, voter, and potential consumer of bioengineered products need to realize that the truth lies somewhere between the photograph seen here and the blissful utopia often portrayed by some proponents of biotechnology.

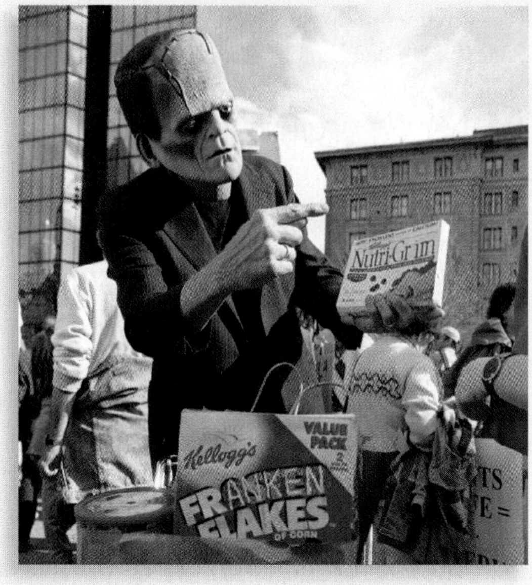

A demonstrator at a biotechnology conference protests the development and sale of genetically modified foods.

In mid-2007 the stakes were raised considerably when Craig Venter (co-sequencer of the human genome) announced he had successfully turned one bacterium into another—by completely replacing the original DNA of the original bacterium with the DNA from a donor bacterium. While the scale of this experiment is small—a single-celled organism—the implications are large. Research advances march on, and with them, the need to decide what we as a society will condone.

on moist plant surfaces. Genetic alteration of the frost gene using recombinant plasmids created a different strain that could prevent ice crystals from forming. A commercial product called Frostban has been successfully applied to stop frost damage in strawberry and potato crops. A strain of *Pseudomonas fluorescens* has been engineered with the gene from a bacterium (*Bacillus thuringiensis*) that codes for an insecticide. These recombinant bacteria are released to colonize plant roots and help destroy invading insects. All releases of recombinant microbes must be approved by the Environmental Protection Agency (EPA) and are closely monitored.

Although a number of recombinant proteins are produced by transformed bacterial hosts, many of the enzymes, hormones, and antibodies being used in drug therapy are currently being manufactured using mammalian cell cultures as the cloning and expression hosts. One of the primary advantages to this alternative procedure is that these cell cultures can modify the proteins (adding carbohydrates, for example) so that they are biologically more active. Some forms of reproductive hormones, human growth hormone, and interferon are products of cell culture.

Another very significant bioengineering interest has been to create microbes to bioremediate disturbed environments. Biotechnologists have already developed and tested several types of bacteria that clean up oil spills and degrade pesticides and toxic substances (see chapter 25). The growing power of biotechnology has caused some to wonder about possible sinister uses of this new ability **(Insight 10.2 and 10.3).**

TABLE 10.3 Examples of Engineered (Transgenic) Plants

Plant	Trait	Results	
Nicotiana tabacum (tobacco)	Herbicide resistance		Tobacco plants in the upper row have been transformed with a gene that provides protection against Buctril, a systemic herbicide. Plants in the lower row are normal and not transformed. Both groups were sprayed with Buctril and allowed to sit for 6 days. (The control plants at the beginning of each row were sprayed with a control mixture lacking Buctril.)
Pisum sativum (garden pea)	Pest protection		Pea plants were engineered with a gene that prevents digestion of the seed starch (see seeds on the left in the photo). This gene keeps tiny insects called weevils from feeding on the seeds. Seeds on the right are from plants that were not engineered and are suffering from weevil damage (note holes).
Oryza sativa (rice)	Added nutritional value		The golden rice grains seen in the photo have been genetically engineered to produce beta-carotene, a precursor to vitamin A. Lack of vitamin A leads to over 1 million deaths and 300,000 cases of blindness a year.

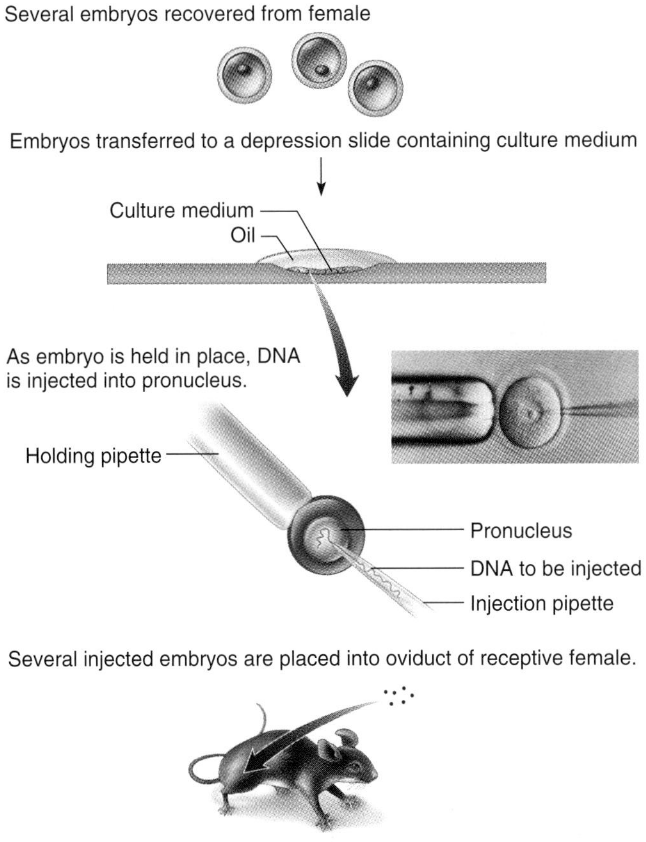

Figure 10.13 How transgenic mice are created.

fertilized mouse embryos were injected with human genes for growth hormone, producing supermice twice the size of normal mice. Laboratories have developed thousands of genetically modified mice.

The so-called knockout mouse has become a standard way of producing animals with tailor-made genetic defects. The technique involves transfecting a mouse embryo with a defective gene and cross-breeding the progeny through several generations. Eventually, mice will be born with two defective genes and will express the disease. Mouse models exist for cystic fibrosis, hardening of the arteries, Gaucher's disease (a lysosomal storage disease), Alzheimer's disease, and sickle-cell anemia. See **table 10.4** for a survey of applications of transgenic animals used in animal husbandry and the drug industry.

CHECKPOINT

- Bioengineered hormones, enzymes, and vaccines are safer and more effective than similar substances isolated directly from animals. Recombinant DNA drugs and vaccines are useful alternatives to traditional treatments for disease.
- Transgenic microorganisms are genetically designed for medical diagnosis, crop improvement, pest reduction, and bioremediation. Transgenic animals are genetically designed to model genetic therapies, improve meat yield, or synthesize specific biological products.

TABLE 10.4 Pharmaceutical Production by Some Transgenic Animals

Treatment For:	Selected Animal	Protein Expressed in Milk	Application
Hemophilia	Pig	Factor VIII and IX	Blood-clotting factor
Emphysema	Sheep	Alpha antitrypsin	Supplemental protein—lack of this protein leads to emphysema
Cancer	Rabbit	Human interleukin-2	Stimulates the production of T lymphocytes to fight selected cancers
Septicemia	Cow	Lactoferrin	Iron-binding protein inhibits the growth of bacteria and viral infection
Surgery, trauma, burns	Cow	Human albumin	Return blood volume to its normal level
GHD (growth hormone deficiency)	Goat	HGH (human) growth hormone	Supplemental protein improves bone metabolism, density, and strength
Heart attack, stroke	Goat	Tissue plasminogen activator (tPA)	Dissolves blood clots to reduce heart damage (if administered within 3 hours)

These two mice are genetically the same strain, except that the one on the right has been transfected with the human growth hormone gene (shown in its circular plasma vector).

Little piglets have been genetically engineered to synthesize human factor VIII, a clotting agent used by type A hemophiliacs. The females will produce it in their milk when mature. It is estimated that only a few hundred pigs would be required to meet the world's demand for this drug.

10.6 Genetic Treatments: Introducing DNA into the Body

Gene Therapy

We have known for decades that for certain diseases, the disease phenotype is due to the lack of a single specific protein. For example, type I diabetes is caused by a lack of insulin, leaving those with the disease unable to properly regulate their blood sugar. The initial treatment for this disease was simple: provide diabetics with insulin isolated from a different source, in most cases the pancreas of pigs or cows. While this treatment was adequate for most diabetics (especially in the short term), our increasingly sophisticated genetic engineering abilities made loitering around the slaughterhouse seem a decidedly low-tech solution to a high-tech problem. We've already discussed the first way in which genetic engineering has been used in the treatment of disease, namely producing recombinant proteins in bacteria or yeast rather than isolating the protein from animals or humans. In fact, recombinant human insulin was the first genetically engineered drug to be approved for use in humans. The next logical step is to see if we can correct or repair a faulty gene in humans suffering from a fatal or debilitating disease, a process known as **gene therapy.**

The inherent benefit of this therapy is to permanently cure the physiological dysfunction by repairing the genetic defect. There are two strategies for this therapy. In *ex vivo* therapy, the normal gene is cloned in vectors such as retroviruses (mouse leukemia virus) or adenoviruses that are infectious but relatively harmless. Tissues removed from the patient are incubated with these genetically modified viruses to transfect them with the normal gene. The transfected cells are then reintroduced into the patient's body by transfusion **(figure 10.14).** In contrast, the *in vivo* type of therapy skips the intermediate step of incubating excised patient tissue. Instead, the naked DNA or a virus vector is directly introduced into the patient's tissues.

Experimentation with various types of gene therapy, or clinical testing, is performed on human volunteers with the particular genetic condition. Over 600 of these trials have been or are being carried out in the United States and other countries. Most trials target cancer, single-gene defects, and infections, and most gene deliveries are carried out by virus vectors. So far, the therapeutic trials have been hampered by several difficulties relating to effectiveness and safety.

The first gene therapy experiment in humans was initiated in 1990 by researchers at the National Institutes of Health. The subject was a 4-year-old girl suffering from a severe immunodeficiency disease caused by the lack of

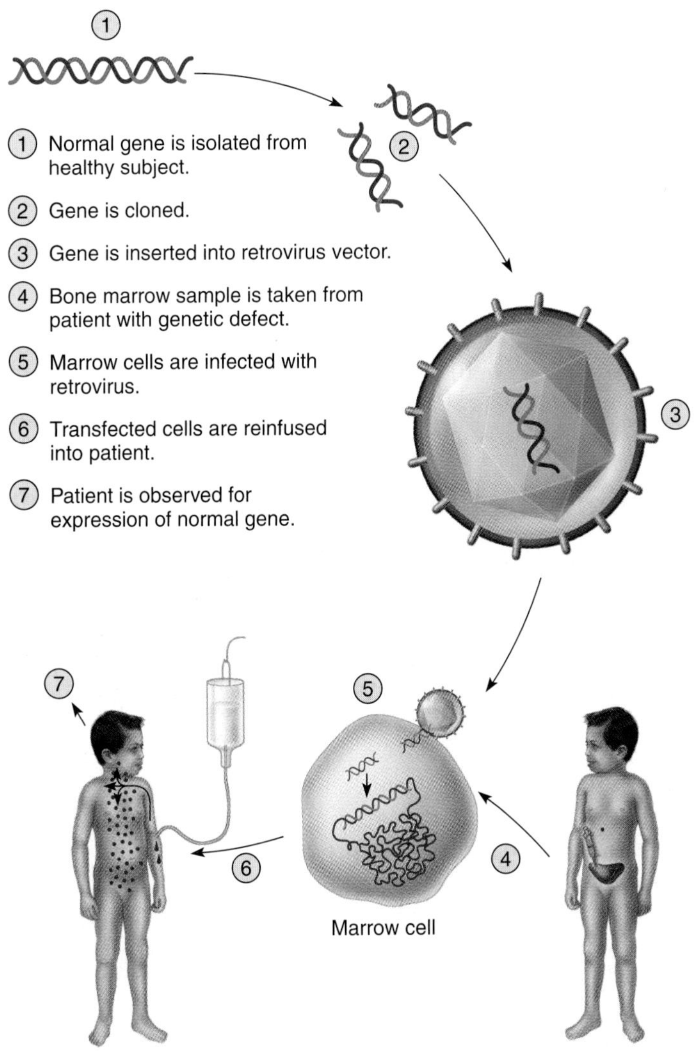

1 Normal gene is isolated from healthy subject.
2 Gene is cloned.
3 Gene is inserted into retrovirus vector.
4 Bone marrow sample is taken from patient with genetic defect.
5 Marrow cells are infected with retrovirus.
6 Transfected cells are reinfused into patient.
7 Patient is observed for expression of normal gene.
Marrow cell

Process Figure 10.14 **Protocol for the *ex vivo* type of gene therapy in humans.**

the enzyme adenosine deaminase (ADA). She was transfused with her own blood cells that had been engineered to contain a functional ADA gene. Later, other children were given the same type of therapy. So far, the children have shown remarkable improvement and continue to be healthy.

Several patients have been treated successfully for another type of severe combined immunodeficiency syndrome called X-1-linked SCID that is due to a missing enzyme required for mature immune cells. Full function has been restored to several hemophilic children, using a similar technique. In 2005, scientists successfully treated lung cancer in mice by infecting them with greatly altered HIV that contained a gene that attacked the cancer cells. The HIV was first made completely benign by removing 80% of its genes.

The ultimate sort of gene therapy is germline therapy, in which genes are inserted into an egg, sperm, or early embryo. In this type of therapy, the new gene will be present in all cells of the individual. The therapeutic gene is also heritable (that is, can be passed on to subsequent generations). Because of this last fact, germline gene therapy is not yet being pursued in humans.

DNA Technology as Genetic Medicine

Up to now, we have considered the use of genetic technology to replace a missing or faulty protein that is needed for normal cell function. A different problem arises when a disease results from the inappropriate expression of a protein. For example, Alzheimer's disease, most viral diseases, and many cancers occur when an unwanted gene is expressed rather than when a desirable gene is missing. The solution in cases such as this is to prevent transcription or translation of a gene, and scientists, as they often do, look for clues in bacteria. Bacteria have sophisticated ways of silencing genes.

Antisense DNA and RNA: Targeting Messenger RNA

Some genes in bacteria are repressed when an antisense RNA binds to them, preventing translation at the ribosome. This **antisense RNA** has bases that are complementary to the sense strand of mRNA in the area surrounding the initiation site. When the antisense RNA binds to its particular mRNA, the now double-stranded RNA is inaccessible to the ribosome, resulting in a loss of translation of that mRNA.

Note that the term *antisense* as used in molecular genetics does *not* mean "no sense." It is used to describe any nucleic acid strand with a base sequence that is complementary to the "sense" or translatable strand. For example, DNA contains a template strand that is transcribed and a matching strand that is not usually transcribed. We can also apply this terminology to RNA. Messenger RNA is considered the translatable strand, and a strand with its complementary sequence of nucleotides would be the antisense strand. To illustrate: If an mRNA sequence read AUGCGAGAC, then an antisense RNA strand for it would read UACGCUCUG.

As this process was tested in the laboratory, two significant differences between prokaryotic and eukaryotic systems emerged. The first was that single-stranded DNA was usually used as the antisense agent, as it was easier to manufacture than RNA. The second was that, for some genes, once the antisense strand bound to the mRNA, not only was the hybrid RNA not translated, it was unable even to leave the nucleus.

When an adequate dose of **antisense DNA** is delivered across the cell membrane into the cytoplasm and nucleus, it binds to specific sites on any mRNAs that are the targets of therapy. As a result, the reading of that mRNA transcript on ribosomes will be blocked and the gene product will not be synthesized **(figure 10.15).**

Early clinical trials of antisense therapeutics have been promising, with most antisense therapies dramatically

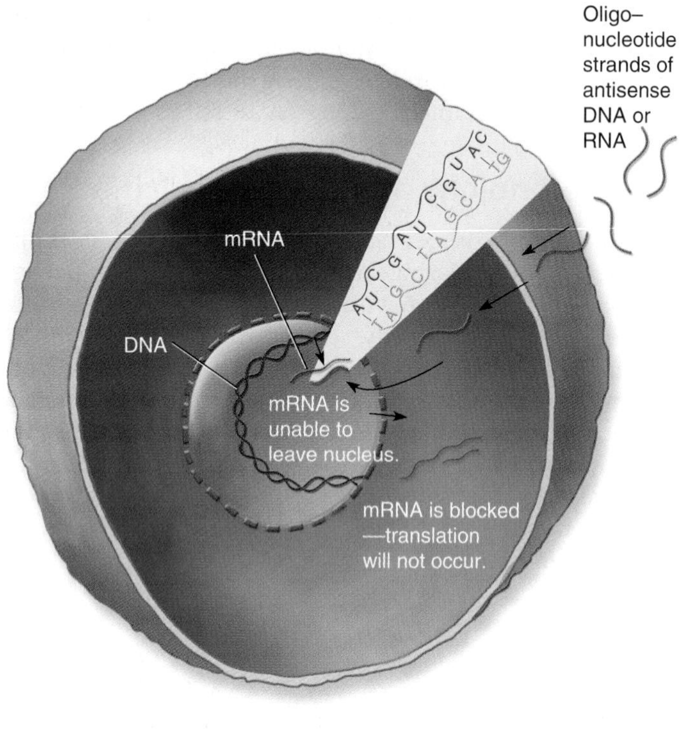

Figure 10.15 Antisense DNA.
When antisense strands enter the nucleus, they bind to complementary areas on an existing mRNA molecule. Such bound RNA will be unavailable for translation on the cell's ribosomes, and no protein will be made.

lowering the rate of synthesis of the targeted protein. Because the antisense agent can be targeted to a specific sequence of DNA, side effects are rare. Most concern has been centered on the difficulty of getting high enough levels of the drug into the nucleus of cells and the fact that these drugs must be taken for life. In 1998, Vitravene, an antisense DNA used to treat cytomegalovirus retinitis (blindness brought on by infection with cytomegalovirus in immunocompromised persons), became the first antisense therapy approved for use in humans.

In chapter 9, you read about other clever strategies bacteria (as well as some eukaryotes) can use to regulate their own genes. Small interfering RNAs and riboswitches are potentially useful in treating humans with diseases caused by the inappropriate or excessive expression of a gene.

10.7 Genome Analysis: Maps, Fingerprints, and Family Trees

As was mentioned earlier, DNA technology has allowed us to accomplish many age-old goals by new and improved means. By remembering that phenotypes, whether human, bacterial, or even viral, are the result of specific sequences of DNA, we can easily see how DNA can be used to differentiate among organisms in the same way that observing any of these phenotypes can. Additionally, DNA can be used to "see" the phenotype of an organism no longer present, as when a criminal is identified by DNA extracted from a strand of hair left behind. Finally, possession of a particular sequence of DNA may indicate an increased risk of a genetic disease. Detection of this piece of DNA (known as a marker) can identify a person as being at increased risk for cancer or Alzheimer's disease long before symptoms arise. The ability to detect diseases before symptoms arise is especially important for diseases such as cancer, for which early treatment is sometimes the difference between life and death. With examples like this in mind, let's look at several ways in which new DNA technology is allowing us to accomplish goals in ways that were only dreamed of a few years ago.

Genome Mapping and Screening: An Atlas of the Genome

We have seen a variety of methods for accessing the genomes of organisms, but the most useful information comes from knowing the sequential makeup of the genetic material. Genetic engineers find it very informative to know the **locus,** or exact position, of a particular gene on a chromosome. They also seek information on the types and numbers of **alleles,** which are sites that vary from one individual to another. The process of determining location of loci and other qualities of genomic DNA is called **mapping.** Maps vary in resolution and applications. *Linkage maps* show the relative proximity and order of genes on a chromosome and are relatively low resolution because only a few exact locations are mapped. *Physical maps* are more detailed arrays that not only depict the relative positions of distinct sections of DNA but also give the numerical size in base pairs. This technology uses restriction fragments and fluorescent hybridization probes to visualize the position of a particular selected site along a segment of DNA.

By far the most detailed maps are *sequence maps,* which are produced by the sequencers we discussed earlier. They give an exact order of bases in a plasmid, chromosome, or entire genome. Because this level of resolution is most promising for understanding the nature of the genes, what they code for, and their functions, this form of map dominates research programs.

Genome sequencing projects have been highly successful. As of 2006, the genome sequences of more than 2,000 viruses, 300 prokaryotes, and dozens of eukaryotes have been published. The eukaryotic genomes include human, mouse, *C. elegans* (a nematode worm), yeast, fruit fly, *Arabidopsis* (a small flowering plant), and rice. One of the remarkable discoveries in this huge enterprise has been how similar the genomes of relatively unrelated organisms are. Humans share around 80% of their DNA codes with mice, about 60% with rice, and even 30% with the worm *C. elegans.*

The new ease with which researchers can sequence genomes was illustrated in the spring of 2003, when a previously unknown virus began causing severe acute respiratory syndrome (SARS) in Southeast Asia. Just 2 weeks after the first virus was isolated from patients, the genome was sequenced and made available to scientists rushing to create a diagnostic test, understand its virulence, and design a vaccine.

Although sequencing provides the ultimate genetic map, it does not automatically identify the exact genes and alleles. Analyzing and storing this massive amount of new data require specialized computers. Because the data consist of information of both a biological and mathematical content, two whole new disciplines have grown up around managing these data: *genomics* (see Note) and *bioinformatics*. The job of genomics and bioinformatics is to analyze and classify genes, determine protein sequences, and ultimately determine the function of the genes. This fact is likely to bring another golden era in biology and microbiology, that of unraveling the complexity of relationships of genes and gene regulation. In time, it will provide a complete understanding of such phenomena as normal cell function, disease, development, aging, and many other issues. In addition, it will allow us to characterize the exact genetic mechanisms behind pathogens and allow new treatments to be developed against them.

A NOTE ABOUT THE NEW "-OMICS"

The ability to obtain the entire sequences of organisms has spawned new vocabulary that refers to the "total picture" of some aspect of a cell or organism.

genomics The systematic study of an organism's genes and their functions.

proteomics The study of an organism's complement of proteins (its "proteome") and functions mediated by the proteins.

metagenomics (also called "community genomics") The study of all the genomes in a particular ecological niche, as opposed to individual genomes from single species.

metabolomics The study of the complete complement of small chemicals present in a cell at any given time. Provides a snapshot of the physiological state of the cell and the end products of its metabolism.

DNA Fingerprinting: A Unique Picture of a Genome

Although DNA is based on a structure of nucleotides, the exact way these nucleotides are combined is unique for each organism. It is now possible to apply DNA technology in a manner that emphasizes these differences and arrays the entire genome in a pattern for comparison. **DNA fingerprinting** (also called DNA typing or profiling) is best known as a tool of forensic science first devised in the mid-1980s by Alex Jeffreys of Great Britain **(figure 10.16).**

Several of the methods discussed previously in this chapter are involved in the creation of a DNA fingerprint. Techniques such as the use of restriction endonucleases for cutting DNA precisely, PCR amplification for increasing the number of copies of a certain genome, electrophoresis to separate fragments, hybridization probes to locate specific loci and alleles, and the Southern blot technique for producing a visible record are all employed. Several different methods of DNA fingerprinting are available, but all depend on distinguishing one sequence of DNA from another by comparing the sequence of the strands at specific loci.

One type of analysis depends on the ability of a restriction enzyme to cut DNA at a specific recognition site. If a given strand of DNA possesses the recognition site for a particular restriction enzyme, the DNA strand is cut, resulting in two smaller pieces of DNA. If the same strand from a different person does not contain the recognition site (perhaps due to a mutation many generations ago), it is not cut by the restriction enzyme and remains as a single large piece of DNA. When each of these DNA samples is digested with restriction enzymes and separated on an electrophoretic gel, the first displays two small bands while the second displays one larger band. This is an example of a restriction fragment length polymorphism, which was discussed earlier. All methods of DNA fingerprinting depend on some variation of this strategy to ferret out differences in DNA sequence at the same location in the genome. The type of polymorphism seen previously is also sometimes referred to as *single nucleotide polymorphism* or an *SNP*, because only a single nucleotide is altered.

Tens of thousands of these differences at a single locus are known to exist throughout the genome. Other genetic **markers**, or observable variations in DNA structure, depend on small repeated sequences of nucleotides that vary in the number of repeated segments from person to person (and are hence polymorphic). Two of the most common markers of this type are *variable number of tandem repeats*, or *VNTRs* (which tend to be hundreds of nucleotides in length), and *microsatellite polymorphisms* (which tend toward dozens of nucleotides in length). In either type of marker, the number of nucleotides, rather than the identity of a single nucleotide, varies. This variation can be easily determined by cutting DNA with a restriction enzyme and then looking for changes in the size of the fragments produced on a gel.

The most powerful uses of DNA fingerprinting are in forensics, detecting genetic diseases, determining parentage, analyzing the family trees of humans and animals, identifying microorganisms, and tracing the lineage of ancient organisms.

One of the first uses of the technique of DNA fingerprinting was in forensic medicine. Besides old-fashioned fingerprints, criminals often leave other evidence at the site of a crime—a hair, a piece of skin or fingernail, semen, blood, or saliva. Because DNA fingerprinting can be combined with amplifying techniques such as PCR, it provides a way to test even specimens that have only minute amounts of DNA, or even old specimens. Regardless of the method used, the primary function of the fingerprint is to provide a snapshot of the genome, displaying it as a unique picture of an individual organism. Markers such as those just described are often used in combination to increase the odds of a correct identification.

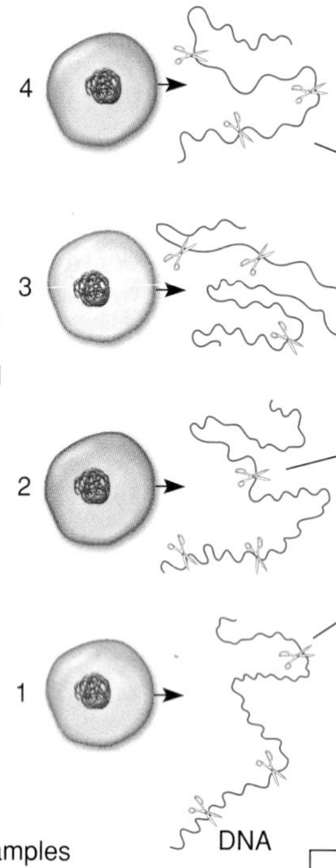

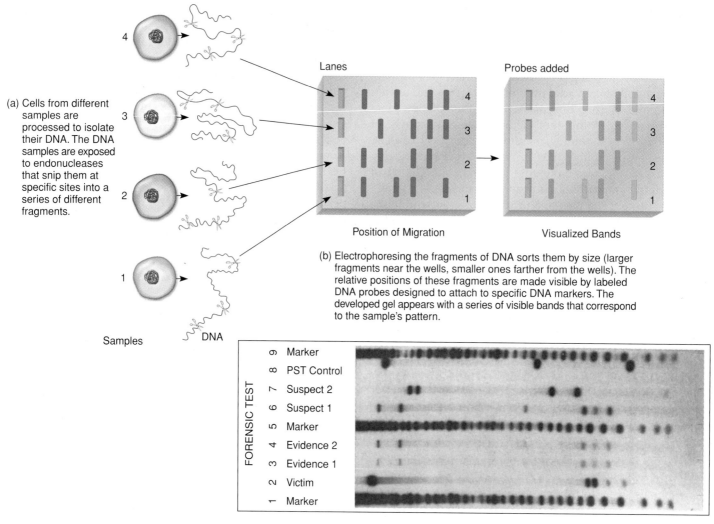

(a) Cells from different samples are processed to isolate their DNA. The DNA samples are exposed to endonucleases that snip them at specific sites into a series of different fragments.

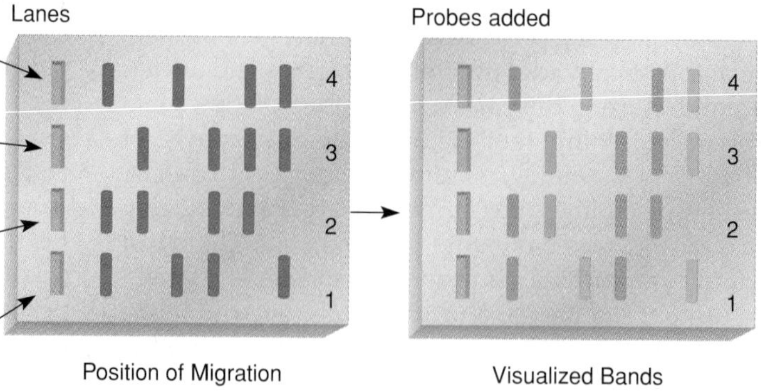

(b) Electrophoresing the fragments of DNA sorts them by size (larger fragments near the wells, smaller ones farther from the wells). The relative positions of these fragments are made visible by labeled DNA probes designed to attach to specific DNA markers. The developed gel appears with a series of visible bands that correspond to the sample's pattern.

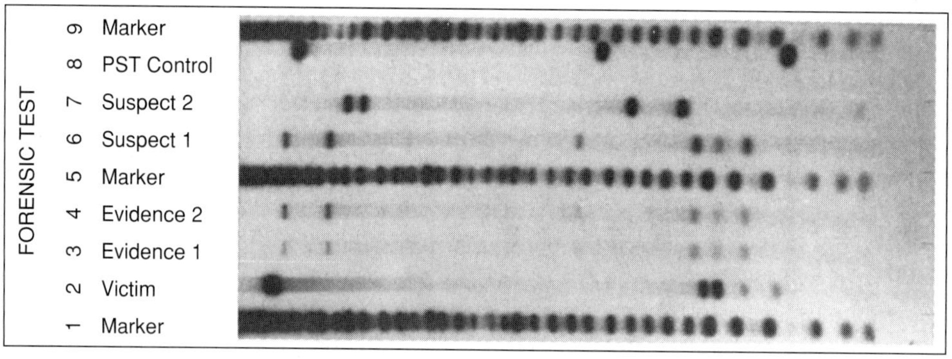

(c) An actual DNA fingerprint used in a rape trial. Control lanes with known markers are in lanes 1, 5, 8, and 9. The second lane contains a sample of DNA from the victim's blood. Evidence samples 1 and 2 (lanes 3 and 4) contain semen samples taken from the victim. Suspects 1 and 2 (lanes 6 and 7) were tested. Can you tell by comparing evidence and suspect lanes which individual committed the rape?

Figure 10.16 DNA fingerprints: the bar codes of life.

Some insight into the power of DNA typing comes from its use to identify victims of the World Trade Center attack in 2001. This same technology has also been effective in identifying remains of unknown soldiers during the Gulf War as well as the shuttle *Columbia* astronauts.

Infectious disease laboratories are developing numerous test procedures to profile bacteria and viruses. Profiling is used to identify *Neisseria gonorrhoeae*, *Chlamydia*, the syphilis spirochete, and *M. tuberculosis*. It was instrumental in identifying the anthrax strain used in the mail attacks of 2001. It is also an essential tool for determining genetic relationships between microbes, such as those involved in food-borne disease outbreaks (see Insight 22.3).

Knowing the complete nucleotide sequence of the human genome is really only half the battle. With very few exceptions, all cells in an organism contain the same DNA, so knowing the sequence of that DNA, while certainly helpful, is of very little use when comparing two cells or tissues from the same organism. Recall that genes are expressed in response to both internal needs and external stimuli, and while the DNA content of a cell is static, the mRNA (and hence protein) content at any given time provides scientists with a profile of genes currently being expressed in the cell. What truly distinguishes a liver cell from a kidney cell or a healthy cell from a diseased cell are the genes expressed in each.

Twin advances in biology and electronics have allowed biologists to view the expression of genes in any given cell using a technique called DNA *microarray analysis.* Prior to the advent of this technology, scientists were able to track the expression of, at most, a few genes at a time. Microarrays are able to track the expression of thousands of genes at once and are able to do so in a single efficient experiment. Microarrays consist of a glass "chip"[4] (occasionally a silicon chip or nylon membrane is used) onto which have been bound sequences from tens of thousands of different genes. A solution

4. It's actually just a glass slide, but all microarray setups have come to be called chips.

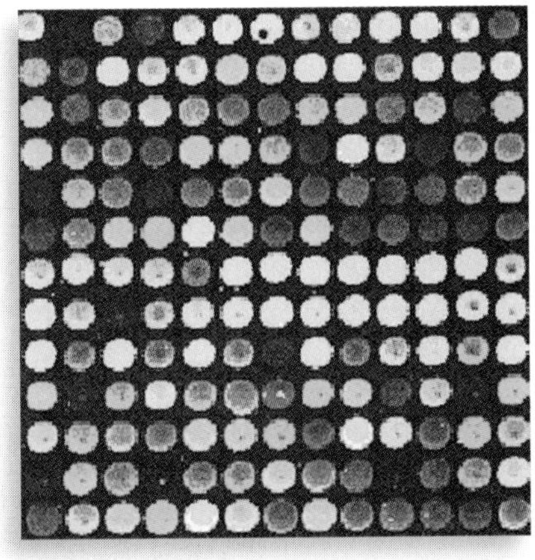

Figure 10.17 Gene expression analysis using microarrays.
The microarray seen here consists of a wide range of oligonucleotides bound to a glass slide. The gene expression of two cells (for example, one healthy and one diseased) can be compared by labeling the cDNA from each cell with either a red or green fluorescent label and allowing the cDNAs to bind to the microarray. A laser is used to excite the bound cDNAs, while a detector records those spots that fluoresce. The color of each spot reveals whether the DNA on the microarray bound to cDNAs is present in the healthy sample, the diseased sample, both samples, or neither sample.

containing fluorescently labeled cDNA, representing all of the mRNA molecules in a cell at a given time, is added to the chip. The labeled cDNA is allowed to bind to any complementary DNA bound to the chip. Bound cDNA is then detected by exciting the fluorescent tag on the cDNA with a laser and recording the fluorescence with a detector linked to a computer. The computer can then interpret this data to determine what mRNAs are present in the cell under a variety of conditions **(figure 10.17).**

Possible uses of microarrays include developing extraordinarily sensitive diagnostic tests that search for a specific pattern of gene expression. As an example, being able to identify a patient's cancer as one of many subtypes (rather than just, for instance, breast cancer) will allow pharmacologists and doctors to treat each cancer with the drug that will be most effective. Again, we see that genetic technology can be a very effective way to reach long-held goals.

CHECKPOINT

- Gene therapy is the replacement of faulty host genes with functional genes by use of cloning vectors such as viruses. This type of transfection can be used to treat genetic disorders and acquired disease. Antisense DNA and interfering RNAs are used to block expression of undesirable host genes in plants and animals as well as those of intracellular parasites.
- DNA technology has advanced understanding of basic genetic principles that have significant applications in a wide range of disciplines, particularly medicine, evolution, forensics, and anthropology.
- The Human Genome Project and other genome sequencing projects have revolutionized our understanding of organisms and led to two new biological disciplines, genomics and bioinformatics.
- DNA fingerprinting is a technique by which organisms are identified for purposes of medical diagnosis, genetic ancestry, and forensics.
- Microarray analysis can determine what genes are transcribed in a given tissue. It is used to identify and devise treatments for diseases based on the genetic profile of the disease.

Chapter Summary with Key Terms

10.1 Basic Elements and Applications of Genetic Engineering
Genetic engineering refers to the manipulation of an organism's genome and is often used in conjunction with biotechnology, the use of an organism's biochemical and metabolic pathways for industrial production of proteins.

10.2 Tools and Techniques of Genetic Engineering

A. *DNA: The Raw Material*
1. **Restriction endonucleases** are used to cut DNA at specific recognition sites, resulting in small segments of DNA that can be more easily managed in the laboratory.
2. The enzyme **ligase** is used to covalently join DNA molecules, whereas **reverse transcriptase** is used to obtain a DNA copy of a particular mRNA.
3. Short strands of DNA called oligonucleotides are used either as probes for specific DNA sequences or as primers for DNA polymerase.

B. *Nucleic Acid Hybridization and Probes*
1. The two strands of DNA can be separated (denatured) by heating. These single-stranded nucleic acids will bind, or **hybridize,** to their complementary sequence, allowing them to be used as **gene,** or **hybridization, probes.**
   a. In the Southern blot method, DNA isolated from an organism is probed to identify specific DNA sequences.
   b. A technique originally developed by Frederick Sanger allows for the determination of the exact sequence of a stretch of DNA. The process is called **DNA sequencing.**
2. The **polymerase chain reaction** is a technique used to amplify a specific segment of DNA several million times in just a few hours.

10.3 Methods in Recombinant DNA Technology: How to Imitate Nature—or to "One-Up" It
Technical Aspects of Recombinant DNA Technology

A. Recombinant DNA methods have as their goal the transfer of DNA from one organism to another. This commonly involves inserting the DNA into a **vector** (usually a plasmid or virus) and transferring the recombinant vector to a **cloning host** (usually a bacterium or yeast) for expression.

1. Examples of cloning vectors include plasmids, bacteriophages, and cosmids, as well as the larger bacterial and yeast artificial chromosomes (BACs and YACs).
2. Cloning hosts are fast-growing, nonpathogenic organisms that are genetically well understood. Examples include *E. coli* and *Saccharomyces cerevisiae*.

B. Recombinant organisms are created by cutting both the donor DNA and the plasmid DNA with restriction endonucleases and then joining them together with ligase. The vector is then inserted into a cloning host by transformation, and those cells containing a vector are selected by their resistance to antibiotics.

C. When the appropriate recombinant cell has been identified, it is grown on an industrial scale. The cloning host expresses the gene it contains, and the resulting protein is excreted from the cell where it can be recovered from the media and purified.

10.4 Biochemical Products of Recombinant DNA Technology

In many instances, the final product of recombinant DNA technology is a protein.

Medical conditions such as diabetes, dwarfism, and blood-clotting disorders can be treated with recombinant insulin, human growth factor, and clotting factors, respectively.

10.5 Genetically Modified Organisms

Recombinant organisms (transgenics) include microbes, plants, and animals that have been engineered to express genes they did not formally possess.

10.6 Genetic Treatments: Introducing DNA into the Body

Gene therapy describes a collection of techniques in which defective genes are replaced, or, more commonly, have their expression increased or decreased. Therapy is said to be *in vivo* if cells are altered within the body and *ex vivo* if the alteration is done on cells taken from the body.

10.7 Genome Analysis: Maps, Fingerprints, and Family Trees

A. Gene **mapping** is a method of identifying landmarks throughout the genome. Several types of maps are available, each with a different level of resolution. The ultimate map is a DNA sequence map, which provides the exact order of all of the nucleotides that make up the genome.

B. **DNA Fingerprinting**

Because the DNA sequence of an individual is unique, so too is the collection of DNA fragments produced by cutting that DNA with a restriction endonuclease. The collection of specific fragments produced by digestion of the DNA with a restriction endonuclease is commonly referred to as the DNA fingerprint of an organism and serves as a unique identifier of any organism.

C. *Microarray analysis* is a method of determining which genes are actively transcribed in a cell under a variety of conditions, such as health versus disease or growth versus differentiation.

Multiple-Choice and True-False Questions

Multiple-Choice Questions. Select the correct answer from the answers provided.

1. Which of the following is *not* essential to carry out the polymerase chain reaction?
   a. primers
   b. DNA polymerase
   c. gel electrophoresis
   d. high temperature
2. Which of the following is *not* a part of the Sanger method to sequence DNA?
   a. dideoxynucleotides
   b. DNA polymerase
   c. electrophoresis
   d. reverse transcriptase
3. The function of ligase is to
   a. rejoin segments of DNA
   b. make longitudinal cuts in DNA
   c. synthesize cDNA
   d. break down ligaments
4. The pathogen of plant roots that is used as a cloning host is
   a. *Pseudomonas*
   b. *Agrobacterium*
   c. *Escherichia coli*
   d. *Saccharomyces cerevisiae*
5. Which of the following sequences, when combined with its complement, could be clipped by an endonuclease?
   a. ATCGATCGTAGCTAGC
   b. AAGCTTTTCGAA
   c. GAATTC
   d. ACCATTGGTA
6. The antisense DNA strand that complements mRNA AUGCGCGAC is
   a. UACGCUCUG
   b. GTCTCGCAT
   c. TACGCTCTG
   d. DNA cannot complement mRNA
7. Which of the following is a primary participant in cloning an isolated gene?
   a. restriction endonuclease
   b. vector
   c. host organism
   d. all of these
8. **Single Matching.** Match the term with its description:
   _____ nucleic acid probe
   _____ antisense strand
   _____ template strand

_____ reverse transcriptase
_____ Taq polymerase
_____ triplex DNA
_____ primer
_____ restriction endonuclease

a. enzyme that transcribes RNA into DNA
b. DNA molecule with an extra strand inserted
c. the nontranslated strand of DNA or RNA
d. enzyme that snips DNA at palindromes
e. oligonucleotide that initiates the PCR
f. strand of nucleic acid that is transcribed or translated
g. thermostable enzyme for synthesizing DNA
h. oligonucleotide used in hybridization

True-False Questions. If statement is true, leave as is. If it is false, correct it by rewriting the sentence.

9. The synthetic unit of the polymerase chain reaction is the replica.
10. A nucleic acid probe can be used to identify unknown bacteria or viruses in clinical samples.
11. A DNA fragment with 450 bp will be closer to the top (negative pole) of an electrophoresis gel than one with 2,500 bp.
12. In order to detect recombinant cells, plasmids contain antibiotic resistance genes.
13. Plasmids are the only vector currently available for use in recombinant procedures.

Writing to Learn

These questions are suggested as a *writing-to-learn* experience. For each question, compose a one- or two-paragraph answer that includes the factual information needed to completely address the question.

1. Define genetic engineering and biotechnology, and summarize the important purposes of these fields. Review the use of the terms *genome, chromosome, gene, DNA,* and *RNA* from chapter 9.
2. a. Briefly describe the functions of DNA synthesizers and sequencers.
   b. How would you make a copy of DNA from an mRNA transcript?
   c. Show how this process would look, using base notation.
   d. What is this DNA called?
   e. Why would it be an advantage to synthesize eukaryotic genes this way?
3. a. What characteristics of plasmids and bacteriophages make them good cloning vectors?
   b. Name several types of vectors, and explain what benefits they have.
   c. List the types of genes that they can contain.
4. a. Describe the principles behind recombinant DNA technology.
   b. Outline the main steps in cloning a gene.
   c. Once cloned, how can this gene be used?
   d. Characterize several ways that recombinant DNA technology can be used.
5. a. What characteristics of bacteria make them good cloning hosts?
   b. What is one way to determine whether a bacterial culture has received a recombinant plasmid?
6. a. Briefly outline the purposes and significant steps in a gene therapy procedure.
   b. What is the main difference between *ex vivo* and *in vivo* gene therapy?
   c. Does the virus vector used in gene therapy replicate itself in the host cell? Would this be desirable or not?
   d. What are some of the main problems with gene therapy?
7. a. Describe the molecular mechanisms by which a DNA antisense molecule could work as a genetic medicine.
   b. Do the same for triplex DNA.
   c. Are the therapies permanent?
   d. Why or why not?
8. a. What is a gene map?
   b. Show by a diagram how chromosomal, physical, and sequence maps are different.
   c. Which organisms are being mapped, and what uses will these maps perform?
   d. Why did the human genome map require such a long time to complete?
   e. What are some possible effects of knowing the genetic map of humans?
9. a. Describe what a DNA fingerprint is and why and how restriction fragments can be used to form a unique DNA pattern.
   b. Discuss briefly how DNA fingerprinting is being used routinely by medicine, the law, the military, and human biology.

Concept Mapping

Appendix D provides guidance for working with concept maps.

1. Construct your own concept map using the following words as the *concepts*. Supply the linking words between each pair of concepts.

DNA
restriction endonuclease
palindrome
ligase
plasmid
vector
origin of replication
recombinant

Critical Thinking Questions

Critical thinking is the ability to reason and solve problems using facts and concepts. These questions can be approached from a number of angles, and in most cases, they do not have a single correct answer.

1. a. Give an example of a benefit of genetic engineering to society and a possible adverse outcome. Discuss.
   b. Give an example of an ecological benefit and a possible adverse side effect. Discuss.
2. a. In reference to Insight 10.2, what is your opinion of the dangers associated with genetic engineering?
   b. Most of us would agree to growth hormone therapy for a child with dwarfism, but how do we deal with parents who want to give growth hormones to their 8-year-old son so that he will be "better at sports"?
3. a. If gene probes, fingerprinting, and mapping could make it possible for you to know of future genetic diseases in you or one of your children, would you wish to use this technology to find out?
   b. What if it were used as a screen for employment or insurance?
4. a. Can you think of a reason that bacteria make restriction endonucleases?
   b. What is it about the endonucleases that prevents bacteria from destroying their own DNA?
   c. Look at figure 10.6, and determine the correct DNA sequence for the fragment's template strand. (Be careful of orientation.)
   d. Which suspect was more likely the rapist, according to the fingerprint in figure 10.16*c*?
5. You have obtained a blood sample in which only red blood cells are left to analyze.
   a. Can you conduct a DNA analysis of this blood?
   b. If no, explain why.
   c. If yes, explain what you would use to analyze it and how to do it.
6. The way that PCR amplifies DNA is similar to the doubling in a population of growing bacteria; a single DNA strand is used to synthesize 2 DNA strands, which become 4, then 8, then 16, and so on. If a complete cycle takes 3 minutes,
   a. how many strands of DNA would theoretically be present after 10 minutes?
   b. after 30 minutes?
   c. after 1 hour?
7. a. Describe any moral, ethical, or biological problems associated with eating tomatoes from an engineered plant or pork from a transgenic pig.
   b. What are the moral considerations of using transgenic animals to manufacture various human products?
8. You are on a jury to decide whether a person committed a homicide and you have to weigh DNA fingerprinting evidence. Two different sets of fingerprints were done: one that tested 5 markers and one that tested 10. Both sets match the defendant's profile. Which one is more reliable and why?
9. Who actually owns the human genome?
   a. Make cogent arguments on various sides of the question.
   b. Explain the steps required in producing a structural map (base sequence) of DNA.

Visual Understanding

1. **From chapter 6, figure 6.20.** What has happened to the bacterial DNA in this illustration? What effect can this have on a bacterium? Is this temporary or permanent?

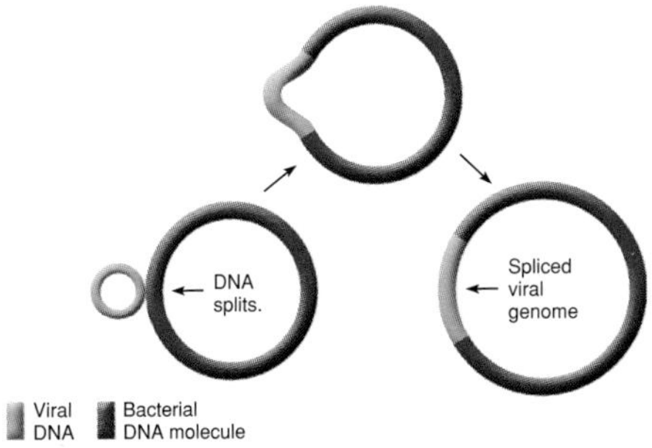

2. **From chapter 9, figure 9.25.** Study the series of events in this illustration. What do cell A (step 1) and cell B (step 5) now have in common?

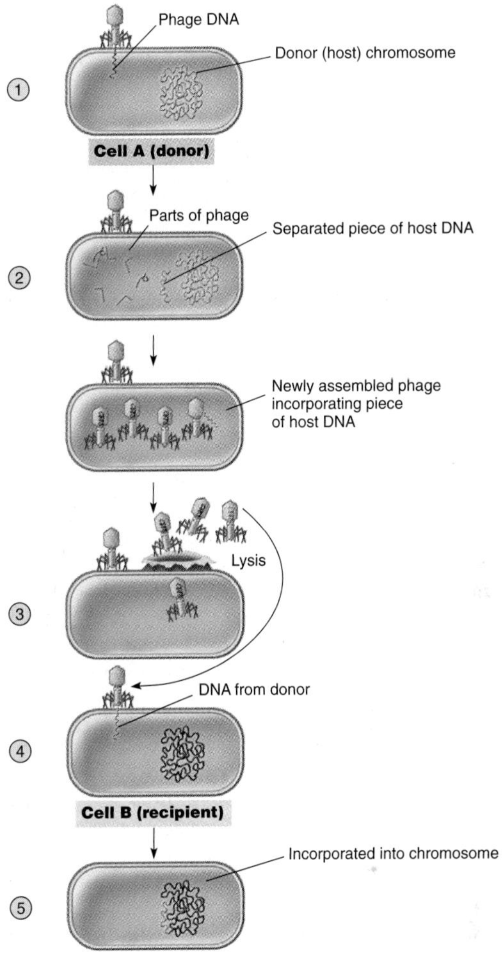

Internet Search Topics

1. If you are interested in discovering more about the human genome projects, there are dozens of websites on the Internet; for example, for human genome centers, go to http://www.ornl.gov/hgmis/, or type Genbank into a search engine such as Google.
2. For information on all aspects of biotechnology, go to the National Center for Biotechnology Information at http://www.ncbi.nlm.nih.gov. The Science Primer is a great place to start for an easy-to-follow lesson on things biotech.
3. Go to: www.aris.mhhe.com, and click on "microbiology" and then this textbook's author/title. Go to chapter 10, access the URLs listed under Internet Search Topics, and research the following:

   Use the words "identify victims World Trade Center disaster" to discover the remarkable process of identifying people who were killed in the 9/11/2001 attack.
4. Search for information on "transgenic mice." How are they produced? What kinds of alternatives are readily available? Is it possible to obtain a "custom order" mouse?
5. Enter "gene therapy" into your favorite search tool. What diseases are the target of cutting-edge research?

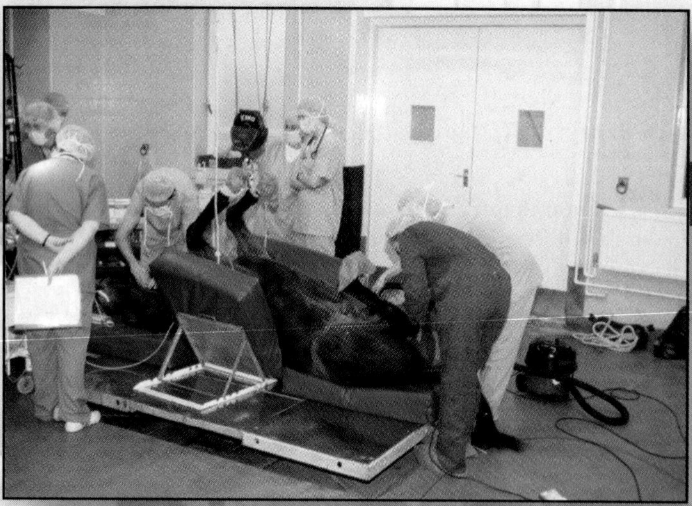

CHAPTER 11

Physical and Chemical Control of Microbes

CASE FILE 11

An outbreak of salmonellosis occurred in a large university veterinary teaching hospital. During the first 7 weeks of the outbreak, *Salmonella infantis* was isolated from 35 animals, including 28 horses, 4 cows, 1 camel, 1 goat, and 1 dog. Bacterial cultures of fecal samples collected at the time of admission were all negative for *S. infantis*. During the course of the outbreak, several infected horses developed fever and diarrhea, and some veterinary students felt that they might have been infected.

A total of 148 environmental samples were collected for bacterial culture during weeks 1 through 7 of the outbreak, and isolates of *S. infantis* were obtained from rectal thermometers, the rubber mat flooring of a horse stall, and from the hands of one hospital worker.

The large-animal portion of the veterinary teaching hospital was closed. The facility was then cleaned by high-pressure power washing and disinfected with a quaternary ammonium product. Surgical recovery stall mats were cleaned a second time with a hydrogen peroxide product. Individual stall-side thermometers were stored in 0.5% chlorhexidine solution. *Salmonella* was not isolated upon resampling of the cleaned facility, and the hospital was reopened.

The first two animals admitted after reopening, a horse and a cow, were found to be positive for *S. infantis* in their feces after only a few days. Environmental samples were again positive. A second outbreak ensued. The second outbreak was worse—over 80% of the animals with *Salmonella* in their feces also had fever or diarrhea. Two foals failed to respond to treatment and were humanely euthanized.

▶ *Why do you think the sanitizing and disinfection failed to control the* Salmonella?

Case File Wrap-Up appears on page 318.

CHAPTER OVERVIEW

- The control of microbes in the environment is a constant concern of health care and industry because microbes are the cause of infection and food spoilage, among other undesirable events.
- Microbial control is accomplished using both physical techniques and chemical agents to destroy, remove, or reduce microbes in a given area.
- Antimicrobial agents damage microbes by disrupting the structure of the cell wall or cell membrane, preventing synthesis of nucleic acids (DNA and RNA), or altering the function of cellular proteins.
- Microbicidal agents kill microbes by inflicting irreversible damage to the cell. Microbistatic agents temporarily inhibit the reproduction of microbes but do not inflict irreversible damage. Mechanical antimicrobial agents physically remove microbes from materials but do not necessarily kill or inhibit them.

- Heat is the most important physical agent in microbial control and can be delivered in both moist (steam sterilization, pasteurization) and dry (incinerators, Bunsen burners) forms.
- Radiation exposes materials to high-energy waves that can damage microbes. Examples are ionizing and ultraviolet radiation.
- Chemical antimicrobials are available for every level of microbial treatment, from low-level disinfectants to high-level sterilants. Antimicrobial chemicals include halogens, alcohols, phenolics, peroxides, heavy metals, detergents, and aldehydes.

11.1 Controlling Microorganisms

Much of the time in our daily existence, we take for granted tap water that is drinkable, food that is not spoiled, shelves full of products to eradicate "germs," and drugs to treat infections. Controlling our degree of exposure to potentially harmful microbes is a monumental concern in our lives, and it has a long and eventful history **(Insight 11.1).**

General Considerations in Microbial Control

The methods of microbial control used outside of the body will result in four possible outcomes: sterilization, disinfection, antisepsis, or decontamination. **Sterilization** is the destruction of all microbial life. **Disinfection** destroys *most* microbial life, reducing contamination on inanimate surfaces. **Antisepsis** is the same as disinfection except a living surface is involved. **Decontamination** is the mechanical removal of most microbes from an animate or inanimate surface. A flowchart **(figure 11.1)** summarizes the major applications and aims in microbial control.

Relative Resistance of Microbial Forms

The primary targets of microbial control are microorganisms capable of causing infection or spoilage that are constantly present in the external environment and on the human body. This targeted population is rarely simple or uniform; in fact, it often contains mixtures of microbes with extreme differences in resistance and harmfulness. Contaminants that can have far-reaching effects if not adequately controlled include bacterial vegetative cells and endospores, fungal hyphae and

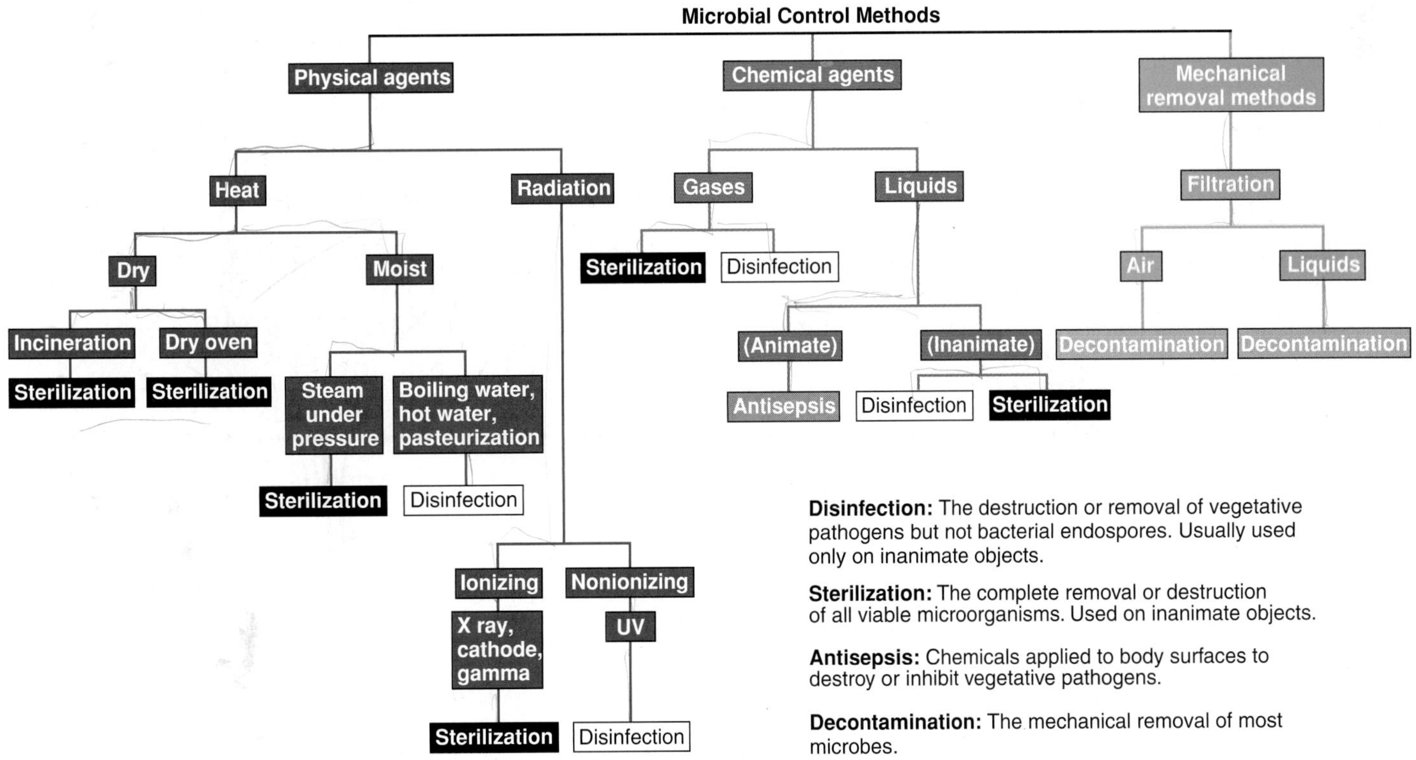

Disinfection: The destruction or removal of vegetative pathogens but not bacterial endospores. Usually used only on inanimate objects.

Sterilization: The complete removal or destruction of all viable microorganisms. Used on inanimate objects.

Antisepsis: Chemicals applied to body surfaces to destroy or inhibit vegetative pathogens.

Decontamination: The mechanical removal of most microbes.

Figure 11.1 Microbial control methods.

INSIGHT 11.1 *Historical*

Microbial Control in Ancient Times

No one knows for sure when humans first applied methods that could control microorganisms, but perhaps the starting point was the discovery and use of fire in prehistoric times. We do know that records describing simple measures to control decay and disease appear from civilizations that existed several thousand years ago. We know, too, that these ancient people had no concept that germs caused disease, but they did have a mixture of religious beliefs, skills in observing natural phenomena, and possibly, a bit of luck. This combination led them to carry out simple and sometimes rather hazardous measures that contributed to the control of microorganisms.

Salting, smoking, pickling, and drying foods and exposing food, clothing, and bedding to sunlight were prevalent practices among early civilizations. The Egyptians showed surprising sophistication and understanding of decomposition by embalming the bodies of their dead with strong salts and pungent oils. They introduced filtration of wine and water as well. The Greeks and Romans burned clothing and corpses during epidemics, and they stored water in copper and silver containers. The armies of Alexander the Great reportedly boiled their drinking water and buried their wastes. Burning sulfur to fumigate houses and applying sulfur as a skin ointment also date approximately from this era.

Illustration of protective clothing used by doctors in the 1700s to avoid exposure to plague victims. The beaklike portion of the hood contained volatile perfumes to protect against foul odors and possibly inhaling "bad air."

During the great plague pandemic of the Middle Ages, it was commonplace to bury corpses in mass graves, burn the clothing of plague victims, and ignite aromatic woods in the houses of the sick in the belief that fumes would combat the disease. In a desperate search for some sort of protection, survivors wore peculiar garments and anointed their bodies with herbs, strong perfume, and vinegar. These attempts may sound foolish and antiquated, but it now appears that they may have had some benefits. Burning wood releases formaldehyde, which could have acted as a disinfectant; herbs, perfume, and vinegar contain mild antimicrobial substances. Each of these early methods, although somewhat crude, laid the foundations for microbial control methods that are still in use today.

spores, yeasts, protozoan trophozoites and cysts, worms, viruses, and prions. This schema compares the general resistance these forms have to physical and chemical methods of control:

Highest resistance
  Prions; bacterial endospores

Moderate resistance
  Protozoan cysts; some fungal sexual spores (zygospores); some viruses. In general, naked viruses are more resistant than enveloped forms. Among the most resistant viruses are the hepatitis B virus and the poliovirus. Bacteria with more resistant vegetative cells are *Mycobacterium tuberculosis*, *Staphylococcus aureus*, and *Pseudomonas* species.

Least resistance
  Most bacterial vegetative cells; fungal spores (other than zygospores) and hyphae; enveloped viruses; yeasts; and protozoan trophozoites

Actual comparative figures on the requirements for destroying various groups of microorganisms are shown in **table 11.1.** Bacterial endospores have traditionally been considered the most resistant microbial entities, being as much as 18 times harder to destroy than their counterpart vegetative cells. Because of their resistance to microbial control, their destruction is the goal of *sterilization* because any process that kills endospores will invariably kill all less resistant microbial forms. Other methods of control (disinfection, antisepsis) act primarily upon microbes that are less hardy than endospores.

TABLE 11.1 Relative Resistance of Bacterial Endospores and Vegetative Cells to Control Agents

Method	Endospores*	Vegetative Forms*	Relative Resistance**
Heat (moist)	120°C	80°C	1.5×
Radiation (X-ray) dosage	4,000 Grays	1,000 Grays	4×
Sterilizing gas (ethylene oxide)	1,200 mg/1	700 mg/1	1.7×
Sporicidal liquid (2% glutaraldehyde)	3 h	10 min	18×

*Values are based on methods (concentration, exposure time, intensity) that are required to destroy the most resistant pathogens in each group.

**The greater resistance of spores versus vegetative cells given as an average figure.

A NOTE ABOUT PRIONS

Scientists are just beginning to understand that prions are in a class of their own when it comes to "sterilization" procedures. This chapter defines "sterile" as the absence of all viable microbial life—but none of the procedures described in this chapter are necessarily sufficient to destroy prions. Prions are extraordinarily resistant to heat and chemicals. If instruments or other objects become contaminated with these unique agents, they must either be discarded as biohazards or, if this is not possible, a combination of chemicals and heat must be applied in accordance with CDC guidelines. The guidelines themselves are constantly evolving as new information becomes available. One U.S. company has marketed a detergent that it says is effective in destroying prions when used in dishwashing devices, but its efficacy has not been extensively tested in "the field." In the meantime, this chapter discusses sterilization using bacterial endospores as the toughest form of microbial life. When tissues, fluids, or instruments are suspected of containing prions, consultation with infection control experts and/or the CDC is recommended when determining effective sterilization conditions. Chapter 19 describes prions in detail.

Terminology and Methods of Microbial Control

Through the years, a growing terminology has emerged for describing and defining measures that control microbes. To complicate matters, the everyday use of some of these terms can at times be vague and inexact. For example, occasionally one may be directed to "sterilize" or "disinfect" a patient's skin, even though this usage does not fit the technical definition of either term. To lay the groundwork for the concepts in microbial control to follow, we present here a series of concepts, definitions, and usages in antimicrobial control.

Sterilization

Sterilization is a process that destroys or removes all viable microorganisms, including viruses. Any material that has been subjected to this process is said to be **sterile.** These terms should be used only in the strictest sense for methods that have been proved to sterilize. An object cannot be slightly sterile or almost sterile—it is either sterile or not sterile. Control methods that sterilize are generally reserved for inanimate objects, because sterilizing parts of the human body would call for such harsh treatment that it would be highly dangerous and impractical.

Sterilized products—surgical instruments, syringes, and commercially packaged foods, just to name a few—are essential to human well-being. Although most sterilization is performed with a physical agent such as heat, a few chemicals called *sterilants* can be classified as sterilizing agents because of their ability to destroy spores.

At times, sterilization is neither practicable nor necessary, and only certain groups of microbes need to be controlled. Some antimicrobial agents eliminate only the susceptible vegetative states of microorganisms but do not destroy the more resistant endospore and cyst stages. Keep in mind that the destruction of spores is not always a necessity, because most of the infectious diseases of humans and animals are caused by non-spore-forming microbes.

Disinfection refers to the use of a physical process or a chemical agent (a disinfectant) to destroy vegetative pathogens but not bacterial endospores. It is important to note that disinfectants are normally used only on inanimate objects because, in the concentrations required to be effective, they can be toxic to human and other animal tissue. Disinfection processes also remove the harmful products of microorganisms (toxins) from materials. Examples of disinfection include applying a solution of 5% bleach to an examining table, boiling food utensils used by a sick person, and immersing thermometers in an iodine solution between uses.

In modern usage, **sepsis** is defined as the growth of microorganisms in the blood and other tissues. The term **asepsis** refers to any practice that prevents the entry of infectious agents into sterile tissues and thus prevents infection. Aseptic techniques commonly practiced in health care range from sterile methods that exclude all microbes to *antisepsis.* In antisepsis, chemical agents called **antiseptics** are applied directly to exposed body surfaces (skin and mucous membranes), wounds, and surgical incisions to destroy or inhibit vegetative pathogens. Examples of antisepsis include preparing the skin before surgical incisions with iodine compounds, swabbing an open root canal with hydrogen peroxide, and ordinary hand washing with a germicidal soap.

The Agents Versus the Processes

The terms *sterilization, disinfection,* and so on refer to processes. You will encounter other terms that describe the agents used in the process. Two examples of these are the terms *bactericidal* and *bacteristatic.* The root *-cide,* meaning to kill, can be combined with other terms to define an antimicrobial agent aimed at destroying a certain group of microorganisms. For example, a **bactericide** is a chemical that destroys bacteria except for those in the endospore stage. It may or may not be effective on other microbial groups. A *fungicide* is a chemical that can kill fungal spores, hyphae, and yeasts. A *virucide* is any chemical known to inactivate viruses, especially on living tissue. A *sporicide* is an agent capable of destroying bacterial endospores. A sporicidal agent can also be a sterilant because it can destroy the most resistant of all microbes. **Germicide** and **microbicide** are additional terms for chemical agents that kill microorganisms.

The Greek words *stasis* and *static* mean to stand still. They can be used in combination with various prefixes to denote a condition in which microbes are temporarily prevented from multiplying but are not killed outright. Although killing or permanently inactivating microorganisms is the usual goal of microbial control, microbistasis does have meaningful applications. **Bacteristatic** agents prevent the growth of bacteria on tissues or on objects in the environment, and *fungistatic* chemicals inhibit fungal growth. Materials used to control microorganisms in the body (antiseptics and drugs) often have **microbistatic** effects because many microbicidal compounds can be highly toxic to human cells. Note that a *-cidal* agent doesn't necessarily result in sterilization.

Decontamination

Several applications in commerce and medicine do not require actual sterilization, disinfection, or antisepsis but are based on reducing the levels of microorganisms (the microbial load) so that the possibility of infection or spoilage is greatly decreased. Restaurants, dairies, breweries, and other food industries consistently handle large numbers of soiled utensils that could readily become sources of infection and spoilage. These industries must keep microbial levels to a minimum during preparation and processing. **Sanitization** is any cleansing technique that mechanically removes microorganisms as well as other debris to reduce contamination to safe levels. A sanitizer is a compound such as soap or detergent used to perform this task.

Cooking utensils, dishes, bottles, cans, and used clothing that have been washed and dried may not be completely free of microbes, but they are considered safe for normal use (sanitary). Air sanitization with ultraviolet lamps reduces airborne microbes in hospital rooms, veterinary clinics, and laboratory installations. Note that some sanitizing processes (such as dishwashing machines) may be rigorous enough to sterilize objects, but this is not true of all sanitization methods. Also note that sanitization is often preferable to sterilization. In a restaurant, for example, you could be given a sterile fork with someone else's old food on it and a sterile glass with lipstick on the rim. On top of this, realize that the costs associated with sterilization would lead to the advent of the $50 fast-food meal. In a situation such as this, the advantage of being sanitary as opposed to sterile can be clearly seen.

It is often necessary to reduce the numbers of microbes on the human skin through **degermation.** This process usually involves scrubbing the skin or immersing it in chemicals, or both. It also emulsifies oils that lie on the outer cutaneous layer and mechanically removes potential pathogens on the outer layers of the skin. Examples of degerming procedures are the surgical handscrub, the application of alcohol wipes to the skin, and the cleansing of a wound with germicidal soap and water. The concepts of antisepsis and degermation clearly overlap, because a degerming procedure can simultaneously be antiseptic and vice versa.

Practical Concerns in Microbial Control

Numerous considerations govern the selection of a workable method of microbial control. These are among the most pressing concerns:

1. Does the application require sterilization, or is disinfection adequate? In other words, must spores be destroyed or is it necessary to destroy only vegetative pathogens?
2. Is the item to be reused or permanently discarded? If it will be discarded, then the quickest and least expensive method should be chosen.
3. If it will be reused, can the item withstand heat, pressure, radiation, or chemicals?
4. Is the control method suitable for a given application? (For example, ultraviolet radiation is a good sporicidal agent, but it will not penetrate solid materials.) Or, in the case of a chemical, will it leave an undesirable residue?
5. Will the agent penetrate to the necessary extent?
6. Is the method cost- and labor-efficient, and is it safe?

A remarkable variety of substances can require sterilization. They range from durable solids such as rubber to sensitive liquids such as serum, and even to entire office buildings, as seen in 2001 when the Hart Senate Office Building was contaminated with *Bacillus anthracis* endospores. Hundreds of situations requiring sterilization confront the network of persons involved in health care, whether technician, nurse, doctor, or manufacturer, and no universal method works well in every case.

Considerations such as cost, effectiveness, and method of disposal are all important. For example, the disposable plastic items such as catheters and syringes that are used in invasive medical procedures have the potential for infecting the tissues. These must be sterilized during manufacture by a nonheating method (gas or radiation), because heat can damage plastics. After these items have been used, it is often necessary to destroy or decontaminate them before they are discarded because of the potential risk to the handler (from needlesticks). Steam sterilization, which is quick and sure, is a sensible choice at this point, because it does not matter if the plastic is destroyed. Health care workers are held to very high standards of infection prevention.

What Is Microbial Death?

Death is a phenomenon that involves the permanent termination of an organism's vital processes. Signs of life in complex organisms such as animals are self-evident, and death is made clear by loss of nervous function, respiration, or heartbeat. In contrast, death in microscopic organisms that are composed of just one or a few cells is often hard to detect, because they reveal no conspicuous vital signs to begin with. Lethal agents (such as radiation and chemicals) do not necessarily alter the overt appearance of microbial cells. Even the loss of movement in a motile microbe cannot be used to indicate death. This fact has made it necessary to develop special qualifications that define and delineate microbial death.

The destructive effects of chemical or physical agents occur at the level of a single cell. As the cell is continuously exposed to an agent such as intense heat or toxic chemicals, various cell structures become dysfunctional, and the entire cell can sustain irreversible damage. At present, the most practical way to detect this damage is to determine if a microbial cell can still reproduce when exposed to a suitable environment. If the microbe has sustained metabolic or structural damage to such an extent that it can no longer reproduce, even under ideal environmental conditions, then it is no longer viable. The permanent loss of reproductive capability, even under optimum growth conditions, has become the accepted microbiological definition of death.

Factors That Affect Death Rate

The cells of a culture show marked variation in susceptibility to a given microbicidal agent. Death of the whole population is not instantaneous but begins when a certain threshold of microbicidal agent (some combination of time and concentration) is met. Death continues in a logarithmic manner as the time or concentration of the agent is increased **(figure 11.2).** Because many microbicidal agents target the cell's metabolic processes, active cells (younger, rapidly dividing) tend to die more quickly than those that are less metabolically active (older, inactive). Eventually, a point is reached at which survival of any cells is highly unlikely; this point is equivalent to sterilization.

The effectiveness of a particular agent is governed by several factors besides time. These additional factors influence the action of antimicrobial agents:

1. The number of microorganisms **(figure 11.2*b*).** A higher load of contaminants requires more time to destroy.
2. The nature of the microorganisms in the population **(figure 11.2*c*).** In most actual circumstances of disinfection and sterilization, the target population is not a single species of microbe but a mixture of bacteria, fungi, spores, and viruses, presenting a broad spectrum of microbial resistance.
3. The temperature and pH of the environment.
4. The concentration (dosage, intensity) of the agent. For example, UV radiation is most effective at 260 nm and most disinfectants are more active at higher concentrations.
5. The mode of action of the agent **(figure 11.2*d*).** How does it kill or inhibit the microorganism?
6. The presence of solvents, interfering organic matter, and inhibitors. Saliva, blood, and feces can inhibit the actions of disinfectants and even of heat.

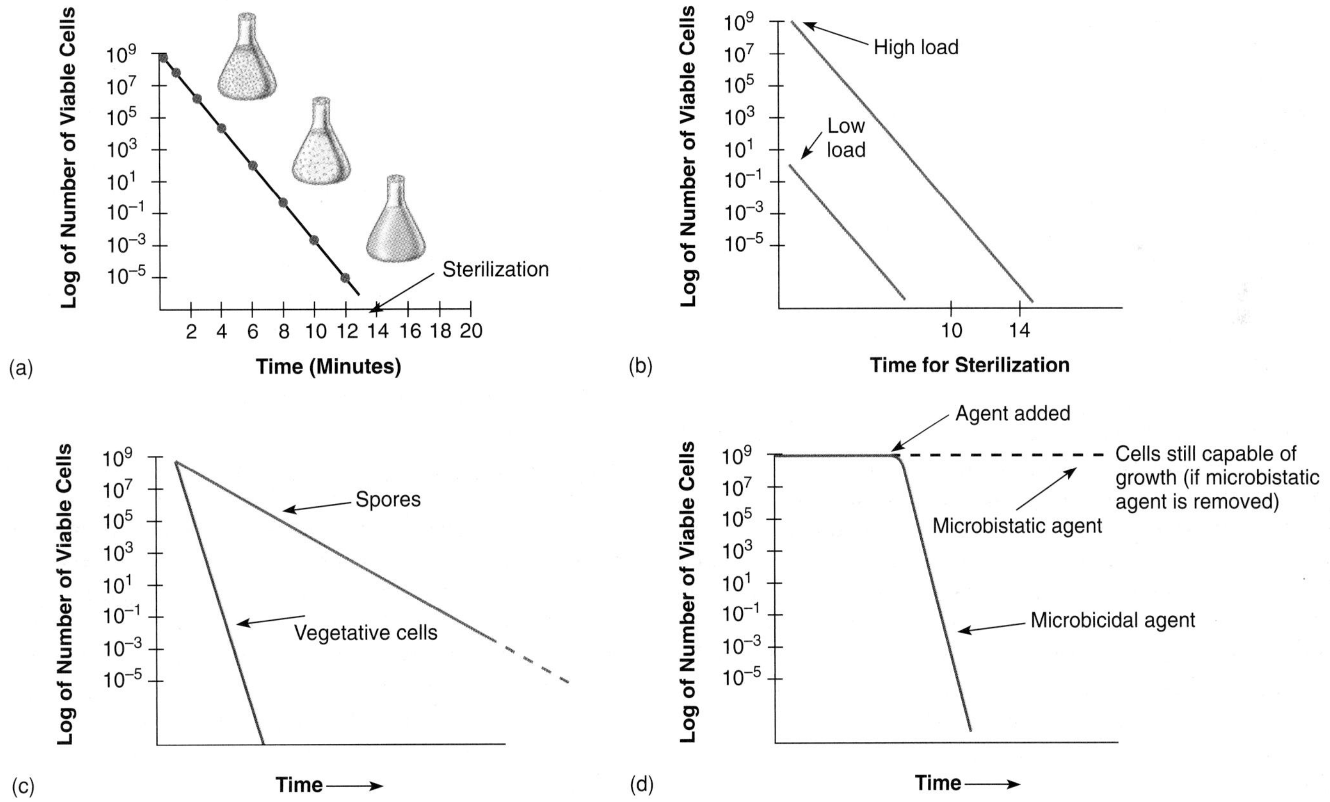

Figure 11.2 Factors that influence the rate at which microbes are killed by antimicrobial agents.
(a) Length of exposure to the agent. During exposure to a chemical or physical agent, all cells of a microbial population, even a pure culture, do not die simultaneously. Over time, the number of viable organisms remaining in the population decreases logarithmically, giving a straight-line relationship on a graph. The point at which the number of survivors is infinitesimally small is considered sterilization. **(b)** Effect of the microbial load. **(c)** Relative resistance of spores versus vegetative forms. **(d)** Action of the agent, whether microbicidal or microbistatic.

The influence of these factors is discussed in greater detail in subsequent sections.

How Antimicrobial Agents Work: Their Modes of Action

An antimicrobial agent's adverse effect on cells is known as its *mode* (or *mechanism*) *of action*. Agents affect one or more cellular targets, inflicting damage progressively until the cell is no longer able to survive. Antimicrobials have a range of cellular targets, with the agents that are least selective in their targeting tending to be effective against the widest range of microbes (examples include heat and radiation). More selective agents (drugs, for example) tend to target only a single cellular component and are much more restricted as to the microbes they are effective against.

The cellular targets of physical and chemical agents fall into four general categories:

1. the cell wall,
2. the cell membrane,
3. cellular synthetic processes (DNA, RNA), and
4. proteins.

The Effects of Agents on the Cell Wall

The cell wall maintains the structural integrity of bacterial and fungal cells. Several types of chemical agents damage the cell wall by blocking its synthesis, digesting it, or breaking down its surface. A cell deprived of a functioning cell wall becomes fragile and is lysed very easily. Detergents and alcohol can also disrupt cell walls, especially in gram-negative bacteria.

How Agents Affect the Cell Membrane

All microorganisms have a cell membrane composed of lipids and proteins, and even some viruses have an outer membranous envelope. As we learned in previous chapters, a cell's membrane provides a two-way system of transport. If this membrane is disrupted, a cell loses its selective permeability and can neither prevent the loss of vital molecules nor bar the entry of damaging chemicals. Loss of those abilities leads to cell death. Detergents called **surfactants** (sir-fak′-tunt) work as microbicidal agents because they lower the surface tension of cell membranes. Surfactants are polar molecules with hydrophilic and hydrophobic regions that can physically bind to the lipid layer and penetrate the internal hydrophobic region of membranes. In effect, this process "opens up" the once tight interface, leaving leaky spots that allow injurious chemicals to seep into the cell and important ions to seep out **(figure 11.3).**

Agents That Affect Protein and Nucleic Acid Synthesis

Microbial life depends upon an orderly and continuous supply of proteins to function as enzymes and structural molecules. As we saw in chapter 9, these proteins are synthesized via the ribosomes through a complex process called translation. For example, the antibiotic chloramphenicol binds to the ribosomes of bacteria in a way that stops peptide bonds from forming. In its presence, many bacterial cells

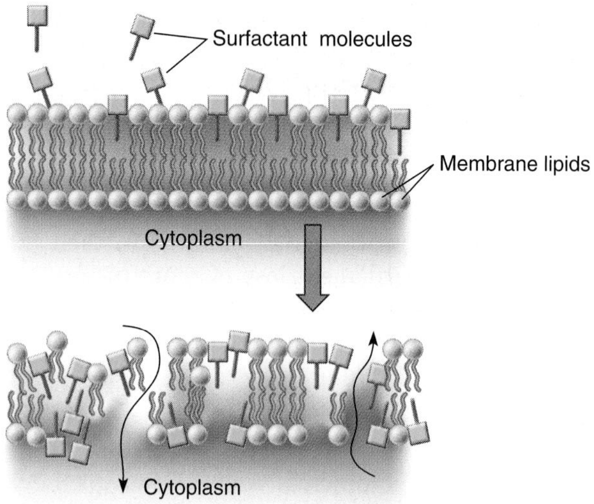

Figure 11.3 Mode of action of surfactants on the cell membrane.
Surfactants inserting in the lipid bilayer disrupt it and create abnormal channels that alter permeability and cause leakage both into and out of the cell.

CASE FILE 11 ***WRAP-UP***

The disinfection of the hospital likely failed because of the difficulty with disinfecting rough surfaces, such as concrete block walls, porous mats, and brushed concrete floors.

The problem with *Salmonella* was brought under control by several measures. First, the concrete block walls of the stalls were painted with a special epoxy product, making them smooth and easy to clean. Second, the more porous mats were discarded and replaced with smooth, solid rubber mats. The walls and mats were then cleaned with detergent containing a quaternary ammonium product and then effectively disinfected with sodium hypochlorite (bleach). Surfaces must be completely cleaned of organic material before the disinfection process is begun.

Hand scrubbing of the surfaces—including the walls, mats, brushed concrete floors, and drains—was found to be more effective than power washing. Hand cleaning and scrubbing likely increased the amount of time that the *Salmonella* was exposed to the detergent and quaternary ammonium disinfectant.

Additional hand-washing stations were installed, and faculty, staff, and students were provided education on infection control measures.

See: Tillotson, K., et al. 1997. Outbreak of Salmonella infantis *infection in a large animal veterinary teaching hospital.* J. Am. Vet. Med. Assoc. *(12):1554–1557.*

are inhibited from forming proteins required in growth and metabolism and are thus inhibited from multiplying.

The nucleic acids are likewise necessary for the continued functioning of microbes. DNA must be regularly replicated and transcribed in growing cells, and any agent that either impedes these processes or changes the genetic code is potentially antimicrobial. Some agents bind irreversibly to DNA, preventing both transcription and translation; others are mutagenic agents. Gamma, ultraviolet, or X radiation causes mutations that result in permanent inactivation of DNA. Chemicals such as formaldehyde and ethylene oxide also interfere with DNA and RNA function.

Agents That Alter Protein Function

A microbial cell contains large quantities of proteins that function properly only if they remain in a normal three-dimensional configuration called the *native state*. The antimicrobial properties of some agents arise from their capacity to disrupt, or **denature,** proteins. In general, denaturation occurs when the bonds that maintain the secondary and tertiary structure of the protein are broken. Breaking these bonds will cause the protein to unfold or create random, irregular loops and coils **(figure 11.4).** One way that proteins can be denatured is through coagulation by moist heat (the same reaction seen in the irreversible solidification of the white of an egg when boiled). Chemicals such as strong organic solvents (alcohols, acids) and phenolics also coagulate proteins. Other antimicrobial agents, such as metallic ions, attach to the active site of the protein and prevent it from interacting with its correct substrate. Regardless of the exact mechanism, such losses in normal protein function can promptly arrest metabolism. Most antimicrobials of this type are nonselective as to the microbes they affect.

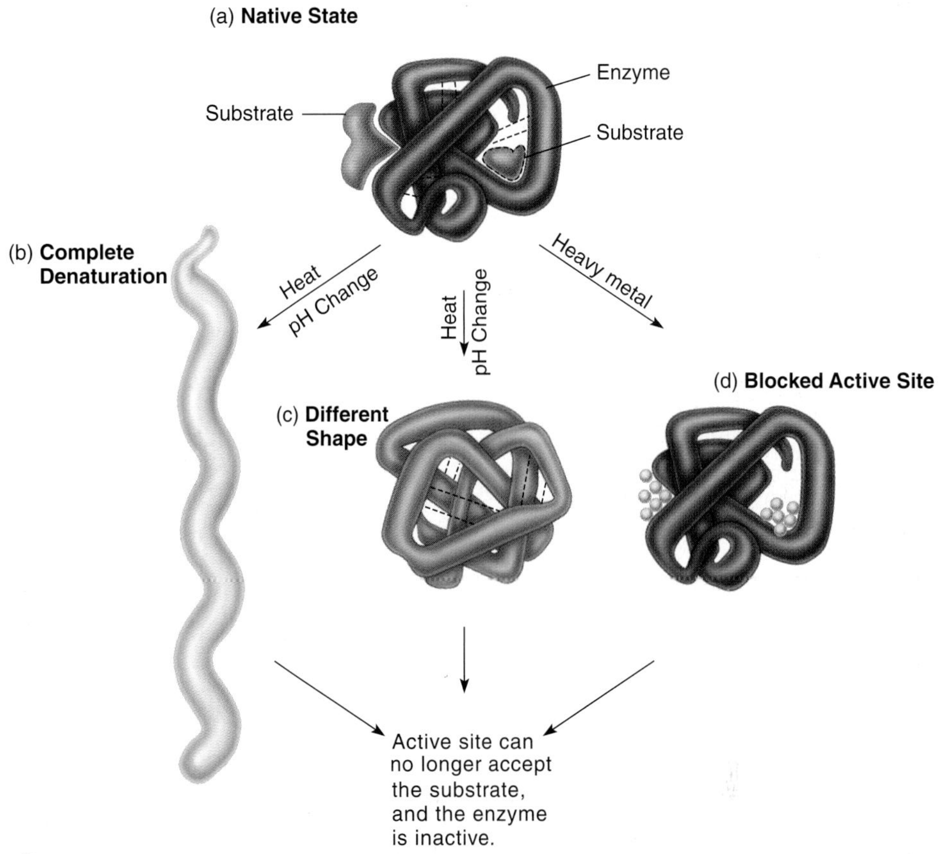

Figure 11.4 Modes of action affecting protein function. **(a)** The native (functional) state is maintained by bonds that create active sites to fit the substrate. Some agents denature the protein by breaking all or some secondary and tertiary bonds. Results are **(b)** complete unfolding or **(c)** random bonding and incorrect folding. **(d)** Some agents react with functional groups on the active site and interfere with bonding.

CHECKPOINT

- Microbial control methods involve the use of physical and chemical agents to eliminate or reduce the numbers of microorganisms from a specific environment to prevent the spread of infectious agents, retard spoilage, and keep commercial products safe.
- The population of microbes that cause spoilage or infection varies widely in species composition, resistance, and harmfulness, so microbial control methods must be adjusted to fit individual situations.
- The type of microbial control is indicated by the terminology used. Sterilization agents destroy all viable organisms, including viruses. Antisepsis, disinfection, and decontamination reduce the numbers of viable microbes to a specified level.
- Antimicrobial agents are described according to their ability to destroy or inhibit microbial growth. Microbicidal agents cause microbial death. They are described by what they are *-cidal* for: sporocides, bactericides, fungicides, viricides.
- An antiseptic agent is applied to living tissue to destroy or inhibit microbial growth.
- A disinfectant agent is used on inanimate objects to destroy vegetative pathogens but not bacterial endospores.
- Sanitization reduces microbial numbers on inanimate objects to safe levels by physical or chemical means.
- Degermation refers to the process of mechanically removing microbes from the skin.
- Microbial death is defined as the permanent loss of reproductive capability in microorganisms.
- Antimicrobial agents attack specific cell sites to cause microbial death or damage. Any given antimicrobial agent attacks one of four major cell targets: the cell wall, the cell membrane, biosynthesis pathways for DNA or RNA, or protein (enzyme) function.

11.2 Methods of Physical Control

Microorganisms have adapted to the tremendous diversity of habitats the earth provides, even severe conditions of temperature, moisture, pressure, and light. For microbes that normally withstand such extreme physical conditions, our attempts at control would probably have little effect. Fortunately for us, we are most interested in controlling microbes that flourish in the same environment in which humans live. The vast majority of these microbes are readily controlled by abrupt changes in their environment. Most prominent among antimicrobial physical agents is heat. Other less widely used agents include radiation, filtration, ultrasonic waves, and even cold. The following sections examine some of these methods and explore their practical applications in medicine, commerce, and the home.

Heat as an Agent of Microbial Control

A sudden departure from a microbe's temperature of adaptation is likely to have a detrimental effect on it. As a rule, elevated temperatures (exceeding the maximum growth temperature) are microbicidal, whereas lower temperatures (below the minimum growth temperature) are microbistatic. The two physical states of heat used in microbial control are moist and dry. *Moist heat* occurs in the form of hot water, boiling water, or steam (vaporized water). In practice, the temperature of moist heat usually ranges from 60°C to 135°C. As we shall see, the temperature of steam can be regulated by adjusting its pressure in a closed container. The expression *dry heat* denotes air with a low moisture content that has been heated by a flame or electric heating coil. In practice, the temperature of dry heat ranges from 160°C to several thousand degrees Celsius.

Mode of Action and Relative Effectiveness of Heat

Moist heat and dry heat differ in their modes of action as well as in their efficiency. Moist heat operates at lower temperatures and shorter exposure times to achieve the same effectiveness as dry heat **(table 11.2).** Although many cellular structures are damaged by moist heat, its most microbicidal effect is the coagulation and denaturation of proteins, which quickly and permanently halts cellular metabolism.

Dry heat dehydrates the cell, removing the water necessary for metabolic reactions, and it also denatures proteins. However, the lack of water actually increases the stability of some protein conformations, necessitating the use of higher temperatures when dry heat is employed as a method of microbial control. At very high temperatures, dry heat oxidizes cells, burning them to ashes. This method is the one used in the laboratory when a loop is flamed or in industry when medical waste is incinerated.

Heat Resistance and Thermal Death of Spores and Vegetative Cells

Bacterial endospores exhibit the greatest resistance, and vegetative states of bacteria and fungi are the least resistant to both moist and dry heat. Destruction of spores usually requires temperatures above boiling, although resistance varies widely.

Vegetative cells also vary in their sensitivity to heat, though not to the same extent as spores **(table 11.3).** Among bacteria, the death times with moist heat range from 50°C for 3 minutes (*Neisseria gonorrhoeae*) to 60°C for 60 minutes (*Staphylococcus aureus*). It is worth noting that vegetative cells of sporeformers are just as susceptible as vegetative cells of non-sporeformers and that pathogens are neither more nor less susceptible than nonpathogens. Other microbes, including fungi, protozoa, and worms, are rather similar in their sensitivity to heat. Viruses are surprisingly resistant to heat, with a tolerance range extending from 55°C for 2 to 5 minutes (adenoviruses) to 60°C for 600 minutes (hepatitis A virus). For practical purposes, all non-heat-resistant forms

TABLE 11.2 Comparison of Times and Temperatures to Achieve Sterilization with Moist and Dry Heat

	Temperature (°C)	Time to Sterilize (Min)
Moist heat	121	15
	125	10
	134	3
Dry heat	121	600
	140	180
	160	120
	170	60

TABLE 11.3 Average Thermal Death Times of Vegetative Stages of Microorganisms

Microbial Type	Temperature (°C)	Time (Min)
Non-spore-forming bacteria	58	28
Non-spore-forming bacteria	61	18
Vegetative stage of spore-forming bacteria	58	19
Fungal spores	76	22
Yeasts	59	19
Viruses		
Nonenveloped	57	29
Enveloped	54	22
Protozoan trophozoites	46	16
Protozoan cysts	60	6
Worm eggs	54	3
Worm larvae	60	10

of bacteria, yeasts, molds, protozoa, worms, and viruses are destroyed by exposure to 80°C for 20 minutes.

Practical Concerns in the Use of Heat: Thermal Death Measurements

Adequate sterilization requires that both temperature and length of exposure be considered. As a general rule, higher temperatures allow shorter exposure times, and lower temperatures require longer exposure times. A combination of these two variables constitutes the **thermal death time,** or TDT, defined as the shortest length of time required to kill all test microbes at a specified temperature. The TDT has been experimentally determined for the microbial species that are common or important contaminants in various heat-treated materials. Another way to compare the susceptibility of microbes to heat is the **thermal death point** (TDP), defined as the lowest temperature required to kill all microbes in a sample in 10 minutes.

Many perishable substances are processed with moist heat. Some of these products are intended to remain on the shelf at room temperature for several months or even years. The chosen heat treatment must render the product free of agents of spoilage or disease. At the same time, the quality of the product and the speed and cost of processing must be considered. For example, in the commercial preparation of canned green beans, one of the cannery's greatest concerns is to prevent growth of the agent of botulism. From several possible TDTs (that is, combinations of time and temperature) for *Clostridium botulinum* spores, the cannery must choose one that kills all spores but does not turn the beans to mush. Out of these many considerations emerges an optimal TDT for a given processing method. Commercial canneries heat low-acid foods at 121°C for 30 minutes, a treatment that sterilizes these foods. Because of such strict controls in canneries, cases of botulism due to commercially canned foods are rare.

Common Methods of Moist Heat Control

The four ways that moist heat is employed to control microbes are

1. steam under pressure,
2. nonpressurized steam,
3. pasteurization, and
4. boiling water.

Steam Under Pressure At sea level, normal atmospheric pressure is 15 pounds per square inch (psi), or 1 atmosphere. At this pressure, water will boil (change from a liquid to a gas) at 100°C, and the resultant steam will remain at exactly that temperature, which is unfortunately too low to reliably kill all microbes. In order to raise the temperature of steam, the pressure at which it is generated must be increased. As the pressure is increased, the temperature at which water boils and the temperature of the steam produced both rise. For example, at a pressure of 20 psi (5 psi above normal), the temperature of steam is 109°C. As the pressure is increased to 10 psi above normal, the steam's temperature rises to 115°C, and at 15 psi above normal (a total of 2 atmospheres), it will be 121°C. It is not the pressure by itself that is killing microbes but the increased temperature it produces.

Such pressure-temperature combinations can be achieved only with a special device that can subject pure steam to pressures greater than 1 atmosphere. Health and commercial industries use an **autoclave** for this purpose, and a comparable home appliance is the pressure cooker. An autoclave has a fundamentally similar plan: a cylindrical metal chamber with an airtight door on one end and racks to hold materials **(figure 11.5).** Its construction includes a complex network of valves, pressure and temperature gauges, and ducts for regulating and measuring pressure and conducting the steam into the chamber. Sterilization is achieved when the steam condenses against the objects in the chamber and gradually raises their temperature.

Experience has shown that the most efficient pressure-temperature combination for achieving sterilization is 15 psi, which yields 121°C. It is possible to use higher pressure to reach higher temperatures (for instance, increasing the pressure to 30 psi raises the temperature to 132°C), but doing so will not significantly reduce the exposure time and can harm the items being sterilized. It is important to avoid overpacking or haphazardly loading the chamber, which prevents steam from circulating freely around the contents and impedes the full contact that is necessary. The duration of the process is adjusted according to the bulkiness of the items in the load (thick bundles of material or large flasks of liquid) and how full the chamber is. The range of holding times varies from 10 minutes for light loads to 40 minutes for heavy or bulky ones; the average time is 20 minutes.

The autoclave is a superior choice to sterilize heat-resistant materials such as glassware, cloth (surgical dressings), rubber (gloves), metallic instruments, liquids, paper, some media, and some heat-resistant plastics. If the items are heat-sensitive (plastic Petri dishes) but will be discarded, the autoclave is still a good choice. However, the autoclave is ineffective for sterilizing substances that repel moisture (oils, waxes, powders).

Nonpressurized Steam Selected substances that cannot withstand the high temperature of the autoclave can be subjected to *intermittent sterilization,* also called **tyndallization.**[1] This technique requires a chamber to hold the materials and a reservoir for boiling water. Items in the chamber are exposed to free-flowing steam for 30 to 60 minutes. This temperature is not sufficient to reliably kill spores, so a single exposure will not suffice. On the assumption that surviving spores will germinate into less resistant vegetative cells, the items are incubated at appropriate temperatures for 23 to 24 hours, and then again subjected to steam treatment. This cycle is repeated for 3 days in a row. Because the temperature never gets above 100°C, highly resistant spores that do not germinate may survive even after 3 days of this treatment.

1. Named for the British physicist John Tyndall, who did early experiments with sterilizing procedures.

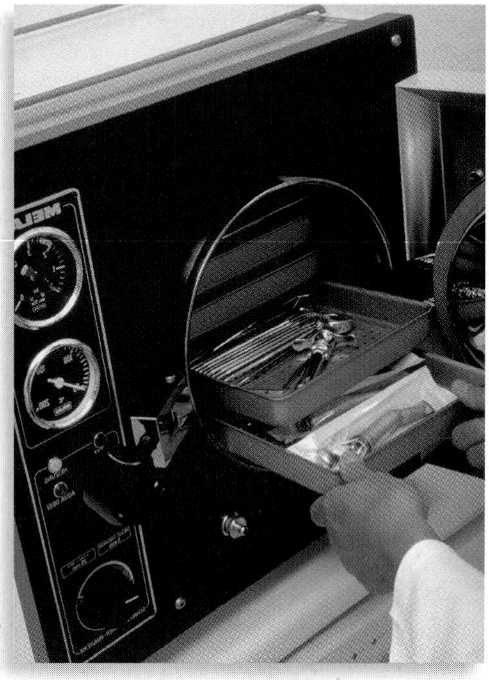
(a)

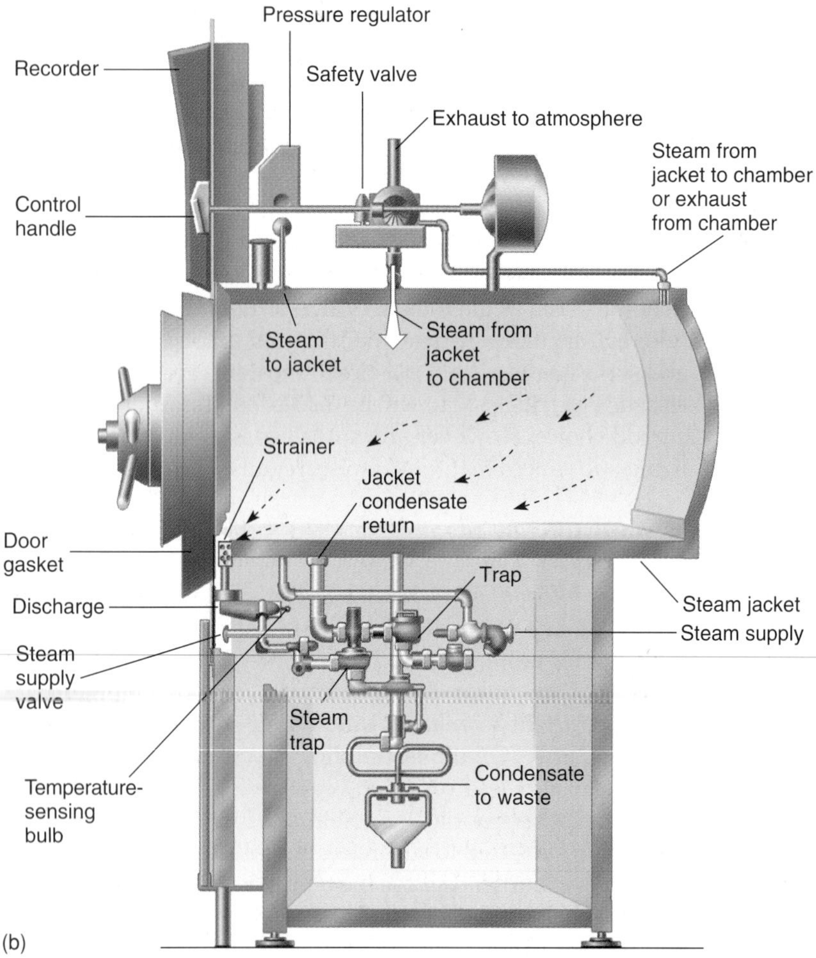

(b)

Figure 11.5 Steam sterilization with the autoclave.
(a) A table top autoclave. **(b)** Cutaway section, showing autoclave components.
Source: (b) From John J. Perkins, *Principles and Methods of Sterilization in Health Science,* 2nd ed., 1969. Courtesy of Charles C Thomas, Publisher, Springfield, Illinois.

Intermittent sterilization is used most often to process heat-sensitive culture media, such as those containing sera, egg, or carbohydrates (which can break down at higher temperatures) and some canned foods. It is probably not effective in sterilizing items such as instruments and dressings that provide no environment for spore germination, but it certainly can disinfect them.

Pasteurization: Disinfection of Beverages Fresh beverages such as milk, fruit juices, beer, and wine are easily contaminated during collection and processing. Because microbes have the potential for spoiling these foods or causing illness, heat is frequently used to reduce the microbial load and destroy pathogens. **Pasteurization** is a technique in which heat is applied to liquids to kill potential agents of infection and spoilage, while at the same time retaining the liquid's flavor and food value.

Ordinary pasteurization techniques require special heat exchangers that expose the liquid to 71.6°C for 15 seconds (flash method) or to 63°C to 66°C for 30 minutes (batch method). The first method is preferable because it is less likely to change flavor and nutrient content, and it is more effective against certain resistant pathogens such as *Coxiella* and *Mycobacterium.* Although these treatments inactivate most viruses and destroy the vegetative stages of 97% to 99% of bacteria and fungi, they do not kill endospores or **thermoduric** microbes (mostly nonpathogenic lactobacilli, micrococci, and yeasts). Milk is not sterile after regular pasteurization. In fact, it can contain 20,000 microbes per milliliter or more, which explains why even an unopened carton of milk will eventually spoil. Newer techniques can also produce *sterile milk* that has a storage life of 3 months. This milk is processed with ultrahigh temperature (UHT)—134°C—for 1 to 2 seconds.

One important aim in pasteurization is to prevent the transmission of milk-borne diseases from infected cows or milk handlers. The primary targets of pasteurization are non-spore-forming pathogens: *Salmonella* species (a common cause of food infection), *Campylobacter jejuni* (acute intestinal infection), *Listeria monocytogenes* (listeriosis), *Brucella* species (undulant fever), *Coxiella burnetii* (Q fever), *Mycobacterium bovis, M. tuberculosis,* and several enteric viruses.

Pasteurization also has the advantage of extending milk storage time, and it can also be used by some wineries and breweries to stop fermentation and destroy contaminants.

Boiling Water: Disinfection A simple boiling water bath or chamber can quickly decontaminate items in the clinic and home. Because a single processing at 100°C will not kill all resistant cells, this method can be relied on only for disinfection and not for sterilization. Exposing materials to boiling water for 30 minutes will kill most non-spore-forming pathogens, including resistant species such as the tubercle bacillus and staphylococci. Probably the greatest disadvantage with this method is that the items can be easily recontaminated when removed from the water. Boiling is also a recommended method of disinfecting unsafe drinking water. In the home, boiling water is a fairly reliable way to sanitize and disinfect materials for babies, food preparation, and utensils, bedding, and clothing from the sickroom.

Dry Heat: Hot Air and Incineration

Dry heat is not as versatile or as widely used as moist heat, but it has several important sterilization applications. The temperatures and times employed in dry heat vary according to the particular method, but in general, they are greater than with moist heat. **Incineration** in a flame or electric heating coil is perhaps the most rigorous of all heat treatments. The flame of a Bunsen burner reaches 1,870°C at its hottest point, and furnaces/incinerators operate at temperatures of 800°C to 6,500°C. Direct exposure to such intense heat ignites and reduces microbes and other substances to ashes and gas.

Incineration of microbial samples on inoculating loops and needles using a Bunsen burner is a very common practice in the microbiology laboratory. This method is fast and effective, but it is also limited to metals and heat-resistant glass materials. This method also presents hazards to the operator (an open flame) and to the environment (contaminants on needle or loop often spatter when placed in flame). Tabletop infrared incinerators **(figure 11.6)** have replaced Bunsen burners in many labs for these reasons. Large incinerators are regularly employed in hospitals and research labs for complete destruction and disposal of infectious materials such as syringes, needles, cultural materials, dressings, bandages, bedding, animal carcasses, and pathology samples.

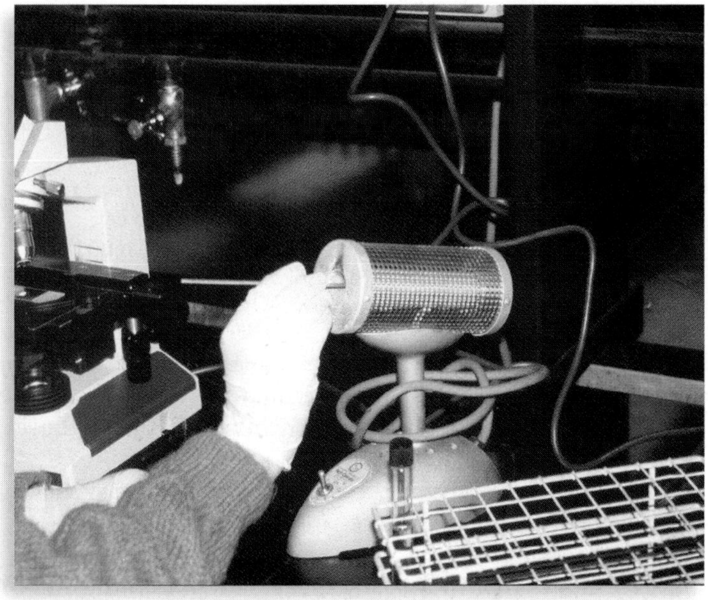

Figure 11.6 Dry heat incineration.
Infrared incinerator with shield to prevent spattering of microbial samples during flaming.

The hot-air oven provides another means of dry-heat sterilization. The so-called *dry oven* is usually electric (occasionally gas) and has coils that radiate heat within an enclosed compartment. Heated, circulated air transfers its heat to the materials in the oven. Sterilization requires exposure to 150°C to 180°C for 2 to 4 hours, which ensures thorough heating of the objects and destruction of spores.

The dry oven is used in laboratories and clinics for heat-resistant items that do not sterilize well with moist heat. Substances appropriate for dry ovens are glassware, metallic instruments, powders, and oils that steam does not penetrate well. This method is not suitable for plastics, cotton, and paper, which may burn at the high temperatures, or for liquids, which will evaporate. Another limitation is the time required for it to work.

The Effects of Cold and Desiccation

The principal benefit of cold treatment is to slow growth of cultures and microbes in food during processing and storage. *It must be emphasized that cold merely retards the activities of most microbes.* Although it is true that some microbes are killed by cold temperatures, most are not adversely affected by gradual cooling, long-term refrigeration, or deep-freezing. In fact, freezing temperatures, ranging from −70°C to −135°C, provide an environment that can preserve cultures of bacteria, viruses, and fungi for long periods. Some psychrophiles grow very slowly even at freezing temperatures and can continue to secrete toxic products. Ignorance of these facts is probably responsible for numerous cases of food poisoning from frozen foods that have been defrosted at room temperature and then inadequately cooked. Pathogens able to survive several months in the refrigerator are *Staphylococcus aureus, Clostridium* species (sporeformers), *Streptococcus* species, and several types of yeasts, molds, and viruses. Outbreaks of *Salmonella* food infection traced backed to refrigerated foods such as ice cream, eggs, and Tiramisu are testimony to the inability of freezing temperatures to reliably kill pathogens.

Vegetative cells directly exposed to normal room air gradually become dehydrated, or **desiccated.** Delicate pathogens such as *Streptococcus pneumoniae,* the spirochete of syphilis, and *Neisseria gonorrhoeae* can die after a few hours of air drying, but many others are not killed and some are even preserved. Endospores of *Bacillus* and *Clostridium* are viable for millions of years under extremely arid conditions. Staphylococci and streptococci in dried secretions and the tubercle bacillus surrounded by sputum can remain viable in air and dust for lengthy periods. Many viruses (especially nonenveloped) and fungal spores can also withstand long periods of desiccation. Desiccation can be a valuable way to preserve foods because it greatly reduces the amount of water available to support microbial growth.

It is interesting to note that a combination of freezing and drying—**lyophilization** (ly-off″-il-ih-za′-shun)—is a common method of preserving microorganisms and other cells in a viable state for many years. Pure cultures are frozen instantaneously and exposed to a vacuum that rapidly removes the water (it goes right from the frozen state into the vapor state). This method avoids the formation of ice crystals that would damage the cells. Although not all cells survive this process, enough of them do to permit future reconstitution of that culture.

As a general rule, chilling, freezing, and desiccation should not be construed as methods of disinfection or sterilization because their antimicrobial effects are erratic and uncertain, and one cannot be sure that pathogens subjected to them have been killed.

Radiation as a Microbial Control Agent

Another way in which energy can serve as an antimicrobial agent is through the use of radiation. **Radiation** is defined as energy emitted from atomic activities and dispersed at high velocity through matter or space. Although radiation exists in many states and can be described and characterized in various ways, we consider only those types suitable for microbial control: gamma rays, X rays, and ultraviolet radiation.

Modes of Action of Ionizing Versus Nonionizing Radiation

The actual physical effects of radiation on microbes can be understood by visualizing the process of **irradiation,** or bombardment with radiation, at the cellular level **(figure 11.7).** When a cell is bombarded by certain waves or particles, its molecules absorb some of the available energy, leading to one of two consequences: (1) If the radiation ejects orbital electrons from an atom, it causes ions to form; this type of radiation is termed **ionizing radiation.** It was previously believed that the most sensitive target for ionizing radiation is DNA, which undergoes mutations on a broad scale, but studies conducted in 2007 suggest that protein damage is the culprit. If proteins are not destroyed, apparently they can always repair the DNA. Secondary lethal effects appear to be chemical changes in organelles and the production of toxic substances. Gamma rays, X rays, and high-speed electrons are all ionizing in their effects. (2) **Nonionizing radiation,** best exemplified by ultraviolet (UV), excites atoms by raising them to a higher energy state, but it does not ionize them. This atomic excitation, in turn, leads to the formation of abnormal bonds within molecules such as DNA and is thus a source of mutations.

Ionizing Radiation: Gamma Rays, X Rays, and Cathode Rays

Over the past several years, ionizing radiation has become safer and more economical to use, and its applications have mushroomed. It is a highly effective alternative for sterilizing materials that are sensitive to heat or chemicals. Because it sterilizes in the absence of heat, irradiation is a type of **cold** (or low-temperature) **sterilization.**[2] Devices that emit ionizing rays include gamma-ray machines containing radioactive cobalt, X-ray machines similar to those used in medical diagnosis, and cathode-ray machines that operate like the vacuum tube in a television set. Items are placed in these machines and irradiated for a short time with a carefully chosen dosage. The dosage of radiation is measured in *Grays* (which has replaced the older term, *rads*). Depending on the application, exposure ranges from 5 to 50 kiloGrays (kGray; a kiloGray is equal to 1,000 Grays). Although all ionizing radiations can penetrate

2. This is a possibly confusing use of the word "cold." In this context, it only means the absence of heat. Beer manufacturers have sometimes used this terminology as well. When they say that their product is "cold-filtered," they mean that it has been freed of contaminants via filtration, i.e., in the absence of heat.

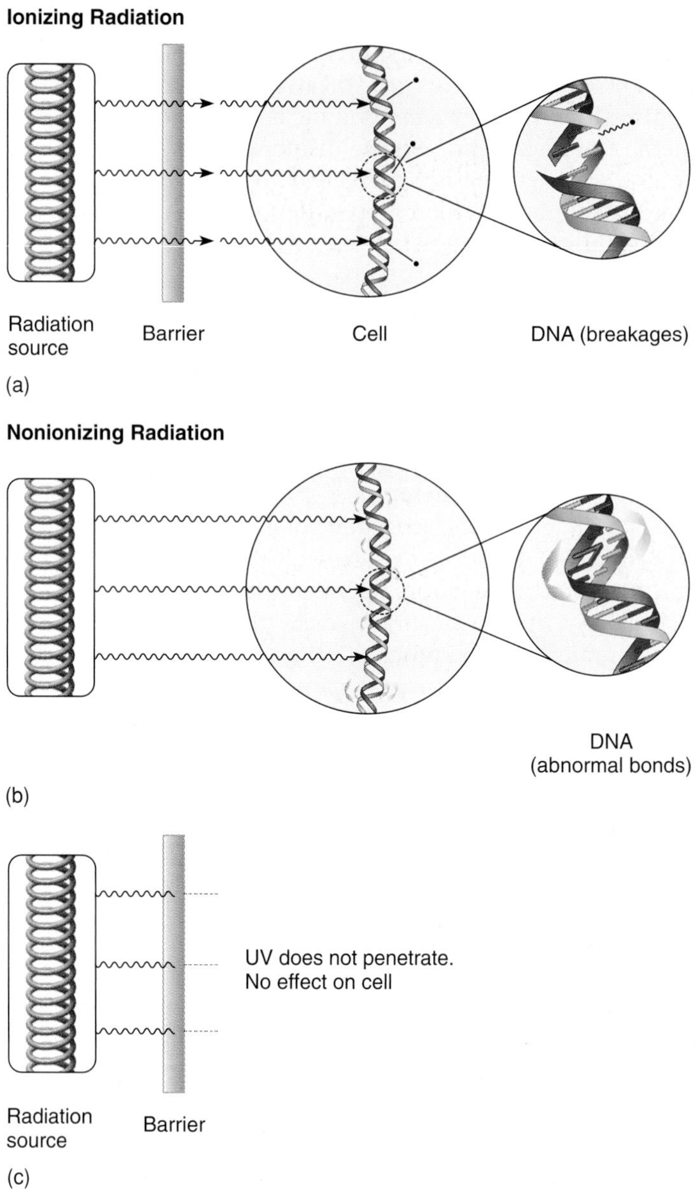

Figure 11.7 Cellular effects of irradiation.
(a) Ionizing radiation can penetrate a solid barrier, bombard a cell, enter it, and dislodge electrons from molecules. Breakage of DNA creates massive mutations and damage to proteins prevents them from repairing it. **(b)** Nonionizing radiation enters a cell, strikes molecules, and excites them. The effect on DNA is mutation by formation of abnormal bonds. **(c)** A solid barrier cannot be penetrated by nonionizing radiation.

liquids and most solid materials, gamma rays are most penetrating, X rays are intermediate, and cathode rays are least penetrating.

Applications of Ionizing Radiation

Foods have been subject to irradiation in limited circumstances for more than 50 years. From flour to pork and ground beef, to fruits and vegetables, radiation is used to kill not only bacterial pathogens but also insects and worms and even to inhibit the sprouting of white potatoes. As soon as radiation is mentioned, however, consumer concern arises that food may be made less nutritious, unpalatable, or even unsafe by its having been subjected to ionizing radiation. But irradiated food has been extensively studied, and each of these concerns has been addressed.

Irradiation may lead to a small decrease in the amount of thiamine (vitamin B1) in food, but this change is small enough to be inconsequential. The irradiation process does produce short-lived free radical oxidants, which disappear almost immediately (this same type of chemical intermediate is produced through cooking as well). Certain foods do not irradiate well and are not good candidates for this type of antimicrobial control. The white of eggs becomes milky and liquid, grapefruit gets mushy, and alfalfa seeds do not germinate properly. Lastly, it is important to remember that food is not made radioactive by the irradiation process, and many studies, in both animals and humans, have concluded that there are no ill effects from eating irradiated food. In fact, NASA relies on irradiated meat for its astronauts.

Sterilizing medical products with ionizing radiation is a rapidly expanding field **(figure 11.8).** Drugs, vaccines, medical instruments (especially plastics), syringes, surgical gloves, tissues such as bone and skin, and heart valves for grafting all lend themselves to this mode of sterilization. Since the anthrax attacks of 2001, mail delivered to certain Washington, D.C. ZIP codes has been irradiated with ionizing radiation. Its main advantages include speed, high penetrating power (it can sterilize materials through outer packages and wrappings), and the absence of heat. Its main disadvantages are potential dangers to radiation machine operators from exposure to radiation and possible damage to some materials.

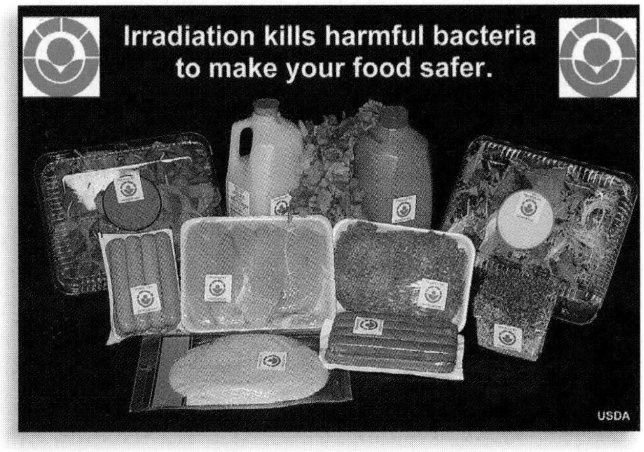

Figure 11.8 Foods commonly irradiated.
Regulations dictate that the universal symbol for irradiation must be affixed to all irradiated materials.

Nonionizing Radiation: Ultraviolet Rays

Ultraviolet (UV) radiation ranges in wavelength from approximately 100 nm to 400 nm. It is most lethal from 240 nm to 280 nm (with a peak at 260 nm). In everyday practice, the source of UV radiation is the germicidal lamp, which generates radiation at 254 nm. Owing to its lower energy state, UV radiation is not as penetrating as ionizing radiation. Because UV radiation passes readily through air, slightly through liquids, and only poorly through solids, the object to be disinfected must be directly exposed to it for full effect.

As UV radiation passes through a cell, it is initially absorbed by DNA. Specific molecular damage occurs on the pyrimidine bases (thymine and cytosine), which form abnormal linkages with each other called **pyrimidine dimers (figure 11.9).** These bonds occur between adjacent bases on the same DNA strand and interfere with normal DNA replication and transcription. The results are inhibition of growth and cellular death. In addition to altering DNA directly, UV radiation also disrupts cells by generating toxic photochemical products called free radicals. These highly reactive molecules interfere with essential cell processes by binding to DNA, RNA, and proteins. Ultraviolet rays are a powerful tool for destroying fungal cells and spores, bacterial vegetative cells, protozoa, and viruses. Bacterial spores are about 10 times more resistant to radiation than are vegetative cells, but they can be killed by increasing the time of exposure.

Applications of Ultraviolet Radiation Ultraviolet radiation is usually directed at disinfection rather than sterilization. Germicidal lamps can cut down on the concentration of airborne microbes as much as 99%. They are used in hospital rooms, operating rooms, schools, food preparation areas, and dental offices. Ultraviolet disinfection of air has proved effective in reducing postoperative infections, preventing the transmission of infections by respiratory droplets, and curtailing the growth of microbes in food-processing plants and slaughterhouses.

Ultraviolet irradiation of liquids requires special equipment to spread the liquid into a thin, flowing film that is exposed directly to a lamp. This method can be used to treat drinking water **(figure 11.10)** and to purify other liquids (milk and fruit juices) as an alternative to heat. Ultraviolet treatment has proved effective in freeing vaccines and plasma from contaminants. The surfaces of solid, nonporous materials such as walls and floors, as well as meat, nuts, tissues for grafting, and drugs, have been successfully disinfected with UV.

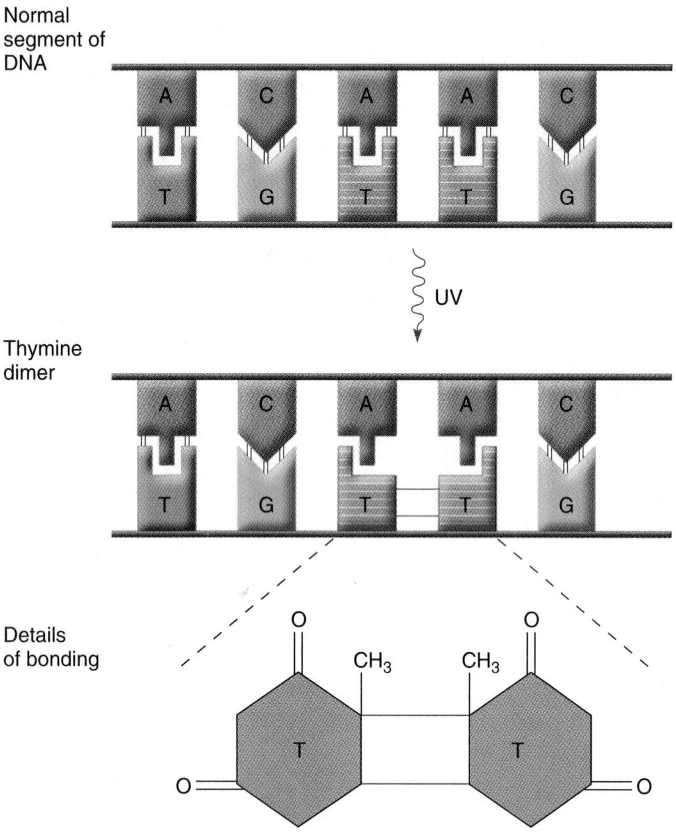

Figure 11.9 Formation of pyrimidine dimers by the action of ultraviolet (UV) radiation.
This shows what occurs when two adjacent thymine bases on one strand of DNA are induced by UV rays to bond laterally with each other. The result is a thymine dimer (shown in greater detail). Dimers can also occur between adjacent cytosines and thymine and cytosine bases. If they are not repaired, dimers can prevent that segment of DNA from being correctly replicated or transcribed. Massive dimerization is lethal to cells.

Figure 11.10 An ultraviolet (UV) treatment system for disinfection of water.
Water flows through tunnels at a water treatment plant, past racks of UV lamps. This system has a capacity of several million gallons per day and can be used as an alternative to chlorination. Home systems that fit under the sink are also available.

One major disadvantage of UV is its poor powers of penetration through solid materials such as glass, metal, cloth, plastic, and even paper. Another drawback to UV is the damaging effect of overexposure on human tissues, including sunburn, retinal damage, cancer, and skin wrinkling.

Decontamination by Filtration: Techniques for Removing Microbes

Filtration is an effective method to remove microbes from air and liquids. In practice, a fluid is strained through a filter with openings large enough for the fluid to pass through but too small for microorganisms to pass through **(figure 11.11)**.

Most modern microbiological filters are thin membranes of cellulose acetate, polycarbonate, and a variety of plastic materials (Teflon, nylon) whose pore size can be carefully controlled and standardized. Ordinary substances such as charcoal, diatomaceous earth, or unglazed porcelain are also used in some applications. Viewed microscopically, most filters are perforated by very precise, uniform pores **(figure 11.11*b*)**. The pore diameters vary from coarse (8 microns) to ultrafine (0.02 micron), permitting selection of the minimum particle size to be trapped. Those with even smaller pore diameters permit true sterilization by removing viruses, and some will even remove large proteins. A sterile liquid filtrate is typically produced by suctioning the liquid through a sterile filter into a presterilized container. These filters are also used to separate mixtures of microorganisms and to enumerate bacteria in water analysis (see chapter 24).

Applications of Filtration

Filtration is used to prepare liquids that cannot withstand heat, including serum and other blood products, vaccines, drugs, IV fluids, enzymes, and media. Filtration has been employed as an alternative method for decontaminating milk and beer without altering their flavor. It is also an important step in water purification. Its use extends to filtering out particulate impurities (crystals, fibers, and so on) that can cause severe reactions in the body. It has the disadvantage of not removing soluble molecules (toxins) that can cause disease.

Filtration is also an efficient means of removing airborne contaminants that are a common source of infection and spoilage. High-efficiency particulate air (HEPA) filters are widely used to provide a flow of decontaminated air to hospital rooms and sterile rooms. A vacuum with a HEPA filter was even used to remove anthrax spores from the Senate offices most heavily contaminated after the terrorist attack in late 2001 (see Insight 11.4).

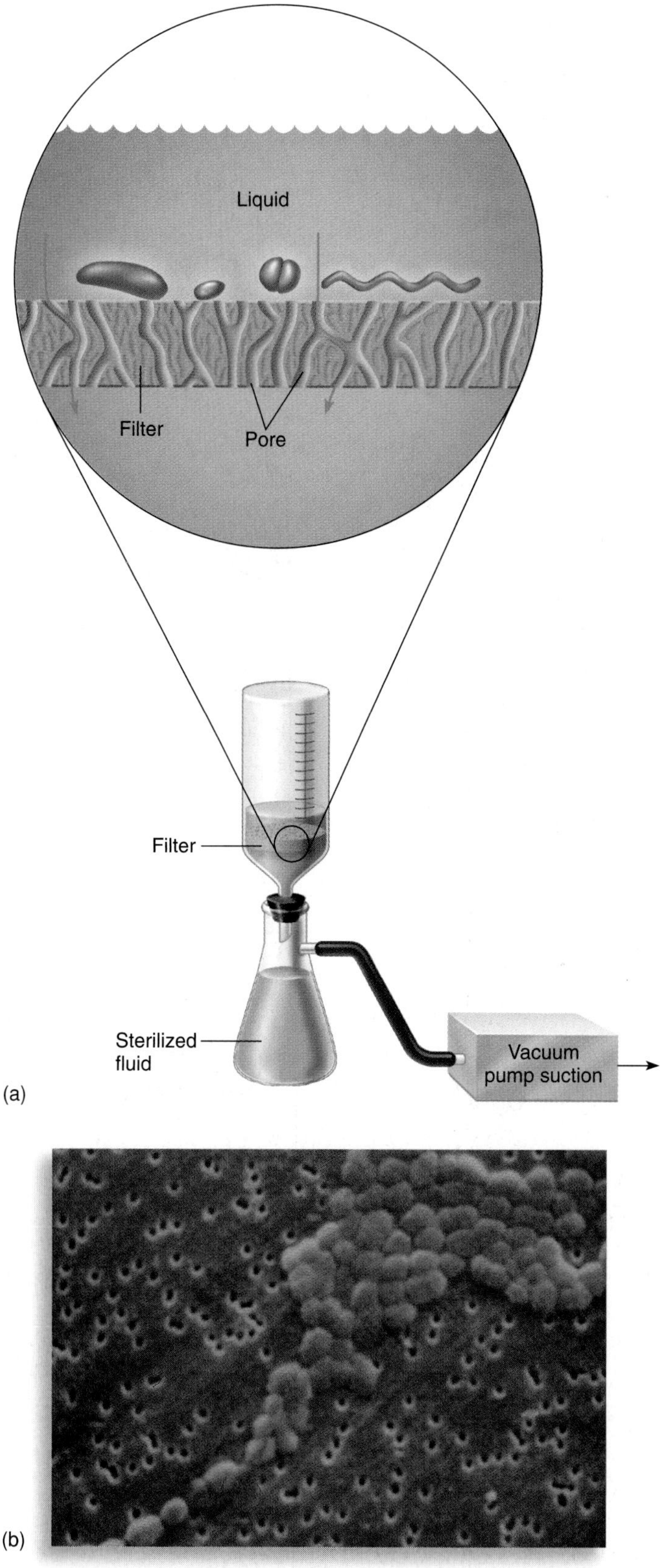

Figure 11.11 Membrane filtration.
(a) Vacuum assembly for achieving filtration of liquids through suction. Inset shows filter as seen in cross section, with tiny passageways (pores) too small for the microbial cells to enter but large enough for liquid to pass through. **(b)** Scanning electron micrograph of filter, showing relative size of pores and bacteria trapped on its surface (5,900×).

CHECKPOINT

- Physical methods of microbial control include heat, cold, radiation, drying, and filtration.
- Heat is the most widely used method of microbial control. It is used in combination with water (moist heat) or as dry heat (oven, flames).
- The thermal death time (TDT) is the shortest length of time required to kill all microbes at a specific temperature. The TDT is longest for spore-forming bacteria and certain viruses.
- The thermal death point (TDP) is the lowest temperature at which all microbes are killed in a specified length of time (10 minutes).
- Autoclaving, or steam sterilization, is the process by which steam is heated under pressure to sterilize a wide range of materials in a comparatively short time (minutes to hours). It is effective for most materials except water-resistant substances such as oils, waxes, and powders.
- Boiling water and pasteurization of beverages disinfect but do not sterilize materials.
- Dry heat is microbicidal under specified times and temperatures. Flame heat, or incineration, is microbicidal. It is used when total destruction of microbes and materials is desired.
- Chilling, freezing, and desiccation are microbistatic but not microbicidal. They are not considered true methods of disinfection because they are not consistent in their effectiveness.
- Ionizing radiation (cold sterilization) by gamma rays and X rays is used to sterilize medical products, meats, and spices. It damages DNA and cell organelles by producing disruptive ions.
- Ultraviolet light, or nonionizing radiation, has limited penetrating ability. It is therefore restricted to disinfecting air and certain liquids.
- Decontamination by filtration removes microbes from heat-sensitive liquids and circulating air. The pore size of the filter determines what kinds of microbes are removed.

11.3 Chemical Agents in Microbial Control

Chemical control of microbes probably emerged as a serious science in the early 1800s, when physicians used chloride of lime and iodine solutions to treat wounds and to wash their hands before surgery. At the present time, approximately 10,000 different antimicrobial chemical agents are manufactured; probably 1,000 of them are used routinely in the health care arena and the home. A genuine need exists to avoid infection and spoilage, but the abundance of products available to "kill germs," "disinfect," "antisepticize," "clean and sanitize," "deodorize," "fight plaque," and "purify the air" indicates a preoccupation with eliminating microbes from the environment that, at times, seems excessive **(Insight 11.2)**.

Antimicrobial chemicals occur in the liquid, gaseous, or even solid state, and they range from disinfectants and antiseptics to sterilants and preservatives (chemicals that inhibit the deterioration of substances). For the sake of convenience (and sometimes safety), many solid or gaseous antimicrobial chemicals are dissolved in water, alcohol, or a mixture of the

INSIGHT 11.2 *Microbiology*

Pathogen Paranoia: "The Only Good Microbe Is a Dead Microbe"

The sensational publicity over outbreaks of infections such as influenza, anthrax, and microbial food poisoning has monumentally influenced the public view of microorganisms. Thousands of articles have sprinkled the news services over the past 5 years. On the positive side, this glut of information has improved people's awareness of the importance of microorganisms. And, certainly, such knowledge can be seen as beneficial when it leads to well-reasoned and sensible choices, such as using greater care in hand washing, food handling, and personal hygiene. But sometimes a little knowledge can be dangerous. The trend also seems to have escalated into an obsessive fear of "germs" lurking around every corner and a fixation on eliminating microbes from the environment and the human body.

As might be expected, commercial industries have found a way to capitalize on those fears. Every year, the number of products that incorporate antibacterial or germicidal "protection" increases dramatically. A widespread array of cleansers and commonplace materials have already had antimicrobial chemicals added. First it was hand soaps and dishwashing detergents, and eventually the list grew to include shampoos, laundry aids, hand lotions, foot pads for shoes, deodorants, sponges and scrub pads, kitty litter, acne medication, cutting boards, garbage bags, toys, and toothpaste.

One chemical agent routinely added to these products is a phenolic called *triclosan* (Irgasan). This substance is fairly mild and nontoxic and does indeed kill most pathogenic bacteria. However, it does not reliably destroy viruses or fungi and has been linked to cases of skin rashes due to hypersensitivity.

One unfortunate result of the negative news on microbes is how the news fosters the feeling that all microbes are harmful. We must not forget that most human beings manage to remain healthy despite the fact that they live in continual intimate contact with microorganisms. We really do not have to be preoccupied with microbes every minute or feel overly concerned that the things we touch, drink, or eat are sterile, as long as they are somewhat clean and free of pathogens. For most of us, resistance to infection is well maintained by our numerous host defenses.

Medical experts are concerned that the widespread overuse of these antibacterial chemicals could favor the survival and growth of resistant strains of bacteria. A study reported in 2000 that many pathogens such as *Mycobacterium tuberculosis* and *Pseudomonas* are naturally resistant to triclosan and that *E. coli* and *Staphylococcus aureus* have already demonstrated decreased sensitivity to it. The widespread use of this chemical may actually select for "super microbes" that survive ordinary disinfection. Another outcome of overuse of environmental germicides is to reduce the natural contact with microbes that is required to maintain the normal resident biota and stimulate immunities. Constant use of these agents could shift the balance in the normal biota of the body by killing off harmless or beneficial microbes. And there's one more thing. More and more studies are showing that when some bacteria become resistant to antibacterial agents, including triclosan, they simultaneously become resistant to antibiotics such as tetracycline and erythromycin.

Infectious disease specialists urge a happy medium approach. Instead of filling the home with questionable germicidal products, they encourage cleaning with traditional soaps and detergents, reserving more potent products to reduce the spread of infection among household members.

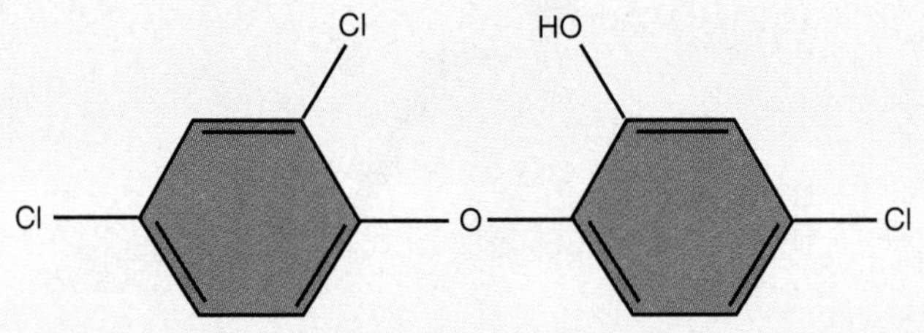

The molecular structure of triclosan, also known as Irgasan and Ster-Zac, a phenol-based chemical that destroys bacteria by disrupting cell walls and membranes.

two to produce a liquid solution. Solutions containing pure water as the solvent are termed *aqueous*, whereas those dissolved in pure alcohol or water-alcohol mixtures are termed **tinctures.**

Choosing a Microbicidal Chemical

The choice and appropriate use of antimicrobial chemical agents are of constant concern in medicine and dentistry. Although actual clinical practices of chemical decontamination vary widely, some desirable qualities in a germicide have been identified, including:

1. rapid action even in low concentrations,
2. solubility in water or alcohol and long-term stability,
3. broad-spectrum microbicidal action without being toxic to human and animal tissues,
4. penetration of inanimate surfaces to sustain a cumulative or persistent action,
5. resistance to becoming inactivated by organic matter,
6. noncorrosive or nonstaining properties,
7. sanitizing and deodorizing properties, and
8. affordability and ready availability.

As yet, no chemical can completely fulfill all of those requirements, but glutaraldehyde and hydrogen peroxide approach this ideal. At the same time, we should question the rather overinflated claims made about certain commercial agents such as mouthwashes and disinfectant air sprays.

Germicides are evaluated in terms of their effectiveness in destroying microbes in medical and dental settings. The three levels of chemical decontamination procedures are *high, intermediate,* and *low.* High-level germicides kill endospores and, if properly used, are sterilants. Materials that necessitate high-level control are medical devices—for example, catheters, heart-lung equipment, and implants—that are not heat-sterilizable and are intended to enter body tissues during medical procedures. Intermediate-level germicides kill fungal (but not bacterial) spores, resistant pathogens such as the tubercle bacillus, and viruses. They are used to disinfect items (respiratory equipment, thermometers) that come into intimate contact with the mucous membranes but are noninvasive. Low levels of disinfection eliminate only vegetative bacteria, vegetative fungal cells, and some viruses. They are used to clean materials such as electrodes, straps, and pieces of furniture that touch the skin surfaces but not the mucous membranes.

Factors That Affect the Germicidal Activity of Chemicals

Factors that control the effect of a germicide include the nature of the microorganisms being treated, the nature of the material being treated, the degree of contamination, the time of exposure, and the strength and chemical action of the germicide **(table 11.4).** The modes of action of most germicides are to attack the cellular targets discussed earlier: proteins, nucleic acids, the cell wall, and the cell membrane.

TABLE 11.4 Required Concentrations and Times for Chemical Destruction of Selected Microbes

Organism	Concentration	Time
Agent: Chlorine		
Mycobacterium tuberculosis	50 ppm	50 sec
Entamoeba cysts (protozoa)	0.1 ppm	150 min
Hepatitis A virus	3 ppm	30 min
Agent: Ethyl Alcohol		
Staphylococcus aureus	70%	10 min
Escherichia coli	70%	2 min
Poliovirus	70%	10 min
Agent: Hydrogen Peroxide		
Staphylococcus aureus	3%	12.5 sec
Neisseria gonorrhoeae	3%	0.3 sec
Herpes simplex virus	3%	12.8 sec
Agent: Quaternary Ammonium Compound		
Staphylococcus aureus	450 ppm	10 min
Salmonella typhi	300 ppm	10 min
Agent: Ethylene Oxide Gas		
Streptococcus faecalis	500 mg/l	2–4 min
Influenza virus	10,000 mg/l	25 h

A chemical's strength or concentration is expressed in various ways, depending upon convention and the method of preparation. The content of many chemical agents can be expressed by more than one notation. In dilutions, a small volume of the liquid chemical (solute) is diluted in a larger volume of solvent to achieve a certain ratio. For example, a common laboratory phenolic disinfectant such as Lysol is usually diluted 1:200; that is, one part of chemical has been added to 200 parts of water by volume. Solutions such as chlorine that are effective in very diluted concentrations are expressed in parts per million (ppm). In percentage solutions, the solute is added to water by weight or volume to achieve a certain percentage in the solution. Alcohol, for instance, is used in percentages ranging from 50% to 95%. In general, solutions of low dilution or high percentage have more of the active chemical (are more concentrated) and tend to be more germicidal, but expense and potential toxicity can necessitate using the minimum strength that is effective.

Another factor that contributes to germicidal effectiveness is the length of exposure. Most compounds require adequate contact time to allow the chemical to penetrate and to act on the microbes present. The composition of the material being treated must also be considered. Smooth, solid objects are more reliably disinfected than are those with pores or pockets that can trap soil. An item contaminated with common biological matter such as serum, blood, saliva, pus, fecal material, or urine presents a problem in disinfection. Large amounts of organic material can hinder the penetration of a disinfectant and, in some cases, can form bonds that reduce its activity. Adequate cleaning of instruments and other reusable materials ensures that the germicide or sterilant will better accomplish the job for which it was chosen.

Germicidal Categories According to Chemical Group

Several general groups of chemical compounds are widely used for antimicrobial purposes in medicine and commerce. Prominent agents include halogens, heavy metals, alcohols, phenolic compounds, oxidizers, aldehydes, detergents, and gases. These groups are surveyed in the following section from the standpoint of each agent's specific forms, modes of action, indications for use, and limitations.

The Halogen Antimicrobial Chemicals

The **halogens** are fluorine, bromine, chlorine, and iodine, a group of nonmetallic elements, all of which are found in group VII of the periodic table. These elements are highly effective components of disinfectants and antiseptics because they are microbicidal and not just microbistatic, and they are sporicidal with longer exposure. For these reasons, halogens are the active ingredients in nearly one-third of all antimicrobial chemicals currently marketed.

Chlorine and Its Compounds Chlorine has been used for disinfection and antisepsis for approximately 200 years. The major forms used in microbial control are liquid and gaseous chlorine (Cl_2), hypochlorites (OCl), and chloramines (NH_2Cl). In solution, these compounds combine with water and release hypochlorous acid (HOCl), which oxidizes the sulfhydryl (S—H) group on the amino acid cysteine and interferes with disulfide (S—S) bridges on numerous enzymes. The resulting denaturation of the enzymes is permanent and suspends metabolic reactions. Chlorine kills not only bacteria and endospores but also fungi and viruses. Chlorine compounds are less effective if exposed to light, alkaline pH, and excess organic matter.

Chlorine Compounds in Disinfection and Antisepsis Gaseous and liquid chlorine are used almost exclusively for large-scale disinfection of drinking water, sewage, and wastewater from such sources as agriculture and industry. Chlorination to a concentration of 0.6 to 1.0 parts of chlorine per million parts of water will usually ensure that water is safe to drink. This treatment rids the water of most pathogenic vegetative microorganisms without unduly affecting its taste (some persons may debate this). In chapter 22, however, you will learn about pathogenic organisms that can survive water chlorination.

Hypochlorites are perhaps the most extensively used of all chlorine compounds. The scope of applications is broad, including sanitization and disinfection of food equipment in dairies, restaurants, and canneries and treatment of swimming pools, spas, drinking water, and even fresh foods. Hypochlorites are used in the allied health areas to treat wounds and to disinfect equipment, bedding, and instruments. Common household bleach is a weak solution (5%) of sodium hypochlorite that serves as an all-around disinfectant, deodorizer, and stain remover.

Chloramines (dichloramine, halazone) are being employed more frequently as an alternative to pure chlorine in treating water supplies. Because standard chlorination of water is now believed to produce unsafe levels of cancer-causing substances such as trihalomethanes, some water districts have been directed by federal agencies to adopt chloramine treatment of water supplies. Chloramines also serve as sanitizers and disinfectants, and for treating wounds and skin surfaces.

Iodine and Its Compounds Iodine is a pungent chemical that forms brown-colored solutions when dissolved in water or alcohol. The two primary iodine preparations are *free iodine* in solution (I_2) and *iodophors*. Iodine rapidly penetrates the cells of microorganisms, where it apparently disturbs a variety of metabolic functions by interfering with the hydrogen and disulfide bonding of proteins (a mode of action similar to chlorine). All classes of microorganisms are killed by iodine if proper concentrations and exposure times are used. Iodine activity is not as adversely affected by organic matter and pH as chlorine is.

Applications of Iodine Solutions Aqueous iodine contains 2% iodine and 2.4% sodium iodide; it is used as a topical antiseptic before surgery and occasionally as a treatment for burned and infected skin. A stronger iodine solution (5% iodine and 10% potassium iodide) is used primarily as a disinfectant for plastic items, rubber instruments, cutting blades, thermometers, and other inanimate items. Iodine tincture is a 2% solution of iodine and sodium iodide in 70% alcohol that can be used in skin antisepsis. Because iodine can be extremely irritating to the skin and toxic when absorbed, strong aqueous solutions and tinctures (5% to 7%) are no longer considered safe for routine antisepsis. Iodine tablets are available for disinfecting water during emergencies or destroying pathogens in impure water supplies.

Iodophors are complexes of iodine and alcohol. This formulation allows the slow release of free iodine and increases its degree of penetration. These compounds have largely replaced free iodine solutions in medical antisepsis because they are less prone to staining or irritating tissues. Common iodophor products marketed as Betadine, Povidone (PVP), and Isodine contain 2% to 10% of available iodine. They are used to prepare skin and mucous membranes for surgery and injections, in surgical handscrubs, to treat burns, and to disinfect equipment and surfaces. Although pure iodine is toxic to the eye, a recent study showed that Betadine solution is an effective means of preventing eye infections in newborn infants, and it may replace antibiotics and silver nitrate as the method of choice.

Phenol and Its Derivatives

Phenol (carbolic acid) is an acrid, poisonous compound derived from the distillation of coal tar. First adopted by Joseph Lister in 1867 as a surgical germicide, phenol was the major antimicrobial chemical until other phenolics with fewer toxic and irritating effects were developed. Solutions of phenol are now used only in certain limited cases, but phenol remains one standard against which other phenolic disinfectants are rated. The *phenol coefficient* quantitatively compares a chemical's antimicrobic properties to those of phenol. Substances chemically related to phenol are often referred to as phenolics. Hundreds of these chemicals are now available.

Phenolics consist of one or more aromatic carbon rings with added functional groups **(figure 11.12).** Among the most important are alkylated phenols (cresols), chlorinated phenols, and bisphenols. In high concentrations, they are cellular poisons, rapidly disrupting cell walls and membranes and precipitating proteins; in lower concentrations, they inactivate certain critical enzyme systems. The phenolics are strongly microbicidal and will destroy vegetative bacteria (including the tuberculosis bacterium), fungi, and most viruses (not hepatitis B), but they are not reliably sporicidal. Their continued activity in the presence of organic matter and their detergent actions contribute to their usefulness. Unfortunately, the toxicity of many of the phenolics makes them too dangerous to use as antiseptics.

Applications of Phenolics

Phenol itself is still used for general disinfection of drains, cesspools, and animal quarters, but it is seldom applied as a medical germicide. The cresols are simple phenolic derivatives

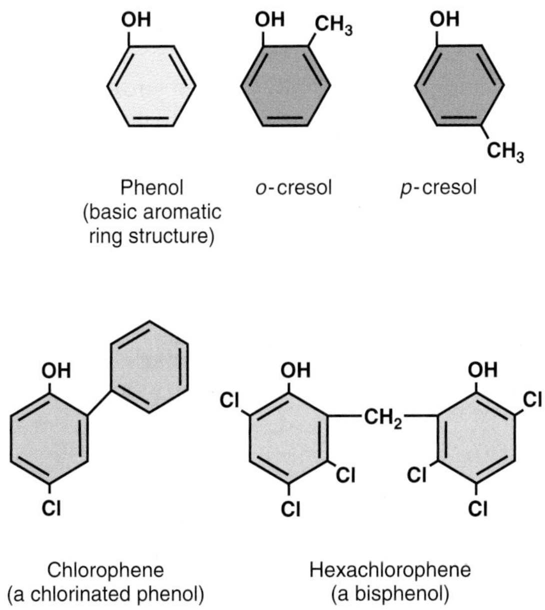

Figure 11.12 Some phenolics.
All contain a basic aromatic ring, but they differ in the types of additional compounds such as Cl and CH_3.

that are combined with soap for intermediate or low levels of disinfection in the hospital. Lysol and creolin, in a 1% to 3% emulsion, are common household versions of this type.

The bisphenols are also widely employed in commerce, clinics, and the home. One type, orthophenyl phenol, is the major ingredient in disinfectant aerosol sprays. This same phenolic is also found in some proprietary compounds (Lysol) often used in hospital and laboratory disinfection. One particular bisphenol, hexachlorophene, was once a common additive of cleansing soaps (pHisoHex) used in the hospital and home. When hexachlorophene was found to be absorbed through the skin and a cause of neurological damage, it was no longer available without a prescription. It is occasionally used to control outbreaks of skin infections.

Perhaps the most widely used phenolic is *triclosan*, chemically known as dichlorophenoxyphenol (see Insight 11.2). It is the antibacterial compound added to dozens of products, from soaps to kitty litter. It acts as both disinfectant and antiseptic and is broad-spectrum in its effects.

Chlorhexidine

The compound chlorhexidine (Hibiclens, Hibitane) is a complex organic base containing chlorine and two phenolic rings. Its mode of action targets both cell membranes (lowering surface tension until selective permeability is lost) and protein structure (causing denaturation). At moderate to high concentrations, it is bactericidal for both gram-positive and gram-negative bacteria but inactive against spores. Its effects on viruses and fungi vary. It possesses distinct advantages over many other antiseptics because of its mildness, low toxicity, and rapid action, and it is not absorbed into deeper tissues to any extent. Alcoholic or aqueous solutions of chlorhexidine are now commonly used for hand scrubbing, preparing skin sites for surgical incisions and injections, and whole-body washing. Chlorhexidine solution also serves as an obstetric antiseptic, a neonatal wash, a wound degermer, a mucous membrane irrigant, and a preservative for eye solutions.

Alcohols as Antimicrobial Agents

Alcohols are colorless hydrocarbons with one or more —OH functional groups. Of several alcohols available, only ethyl and isopropyl are suitable for microbial control. Methyl alcohol is not particularly microbicidal, and more complex alcohols are either poorly soluble in water or too expensive for routine use. Alcohols are employed alone in aqueous solutions or as solvents for tinctures (iodine, for example).

Alcohol's mechanism of action depends in part upon its concentration. Concentrations of 50% and higher dissolve membrane lipids, disrupt cell surface tension, and compromise membrane integrity. Alcohol that has entered the protoplasm denatures proteins through coagulation but only in alcohol-water solutions of 50% to 95%. Alcohol is the exception to the rule that higher concentrations of an antimicrobial chemical have greater microbicidal activity. Because water is needed for proteins to coagulate, alcohol shows a greater microbicidal activity at 70% concentration (that is, 30% water) than at 100% (0% water). Absolute alcohol (100%) dehydrates cells and inhibits their growth but is generally not a protein coagulant.

Although useful in intermediate- to low-level germicidal applications, alcohol does not destroy bacterial spores at room temperature. Alcohol can, however, destroy resistant vegetative forms, including tuberculosis bacteria and fungal spores, provided the time of exposure is adequate. Alcohol is generally more effective in inactivating enveloped viruses than nonenveloped viruses such as poliovirus and hepatitis A virus.

Applications of Alcohols Ethyl alcohol, also called ethanol or grain alcohol, is known for being germicidal, nonirritating, and inexpensive. Solutions of 70% to 95% are routinely used as skin degerming agents because the surfactant action removes skin oil, soil, and some microbes sheltered in deeper skin layers. One limitation to its effectiveness is the rate at which it evaporates. Ethyl alcohol is occasionally used to disinfect electrodes, face masks, and thermometers, which are first cleaned and then soaked in alcohol for 15 to 20 minutes. Isopropyl alcohol, sold as rubbing alcohol, is even more microbicidal and less expensive than ethanol, but these benefits must be weighed against its toxicity. It must be used with caution in disinfection or skin cleansing, because inhalation of its vapors can adversely affect the nervous system.

Hydrogen Peroxide and Related Germicides

Hydrogen peroxide (H_2O_2) is a colorless, caustic liquid that decomposes in the presence of light, metals, or catalase into water and oxygen gas. The germicidal effects of hydrogen peroxide are due to the direct and indirect actions of oxygen. Oxygen forms hydroxyl free radicals (—OH), which, like the superoxide radical (see chapter 7), are highly toxic and reactive to cells. Although most microbial cells produce catalase to inactivate the metabolic hydrogen peroxide, it cannot neutralize the amount of hydrogen peroxide entering the cell during disinfection and antisepsis. Hydrogen peroxide is bactericidal, virucidal, and fungicidal and, in higher concentrations, sporicidal.

Applications of Hydrogen Peroxide As an antiseptic, 3% hydrogen peroxide serves a variety of needs, including skin and wound cleansing, bedsore care, and mouthwashing. It is especially useful in treating infections by anaerobic bacteria because of the lethal effects of oxygen on these forms. Hydrogen peroxide is also a versatile disinfectant for soft contact lenses, surgical implants, plastic equipment, utensils, bedding, and room interiors.

A number of clinical procedures involve delicate reusable instruments such as endoscopes and dental handpieces. Because these devices can become heavily contaminated by tissues and fluids, they need to undergo sterilization, not just disinfection, between patients to prevent transmission of infections such as hepatitis, tuberculosis, and genital warts. These very effective and costly diagnostic tools (a colonoscope may cost up to $30,000) have created another dilemma. They may trap infectious agents where they cannot be easily removed, and they are delicate, complex, and difficult to clean. Traditional methods are either too harsh (heat) to protect the instruments from damage or too slow (ethylene oxide) to sterilize them in a timely fashion between patients. The need for effective rapid sterilization has led to the development of low-temperature sterilizing cabinets that contain liquid chemical sterilants **(figure 11.13).** The major types of chemical sterilants used in these machines are powerful oxidizing agents such as hydrogen peroxide (35%) and peracetic acid (35%) that penetrate into delicate machinery, kill the most resistant microbes, and do not corrode or damage the working parts.

Vaporized hydrogen peroxide is currently being used as a sterilant in enclosed areas. Hydrogen peroxide plasma sterilizers exist for those applications involving small industrial or medical items. For larger enclosed spaces, such as isolators and passthrough rooms, peroxide generators can be used to fill a room with hydrogen peroxide vapors at concentrations high enough to be sporicidal.

Another compound with effects similar to those of hydrogen peroxide is ozone (O_3), used to disinfect air, water, and industrial air conditioners and cooling towers.

Chemicals with Surface Action: Detergents

Detergents are polar molecules that act as **surfactants.** Most anionic detergents have limited microbicidal power. This includes most soaps. Much more effective are positively charged (cationic) detergents, particularly the quaternary ammonium compounds (usually shortened to *quats*).

The activity of cationic detergents arises from the amphipathic (two-headed) nature of the molecule. The positively charged end binds well with the predominantly negatively charged bacterial surface proteins while the long, uncharged hydrocarbon chain allows the detergent to disrupt the cell membrane **(figure 11.14).** Eventually the cell membrane loses selective permeability, leading to the death of the cell. Several other effects are seen but the loss of integrity of the cell membrane is most important.

The effects of detergents are varied. When used at high enough concentrations, quaternary ammonium compounds are effective against some gram-positive bacteria, viruses, fungi, and algae. In low concentrations, they exhibit only microbistatic effects. Drawbacks to the quats include their ineffectiveness against the tuberculosis bacterium, hepatitis virus, *Pseudomonas,* and spores at any concentration. Furthermore, their activity is greatly reduced in the presence of organic matter and they function best in alkaline solutions. As a result of these limitations, quats are rated only for low-level disinfection in the clinical setting.

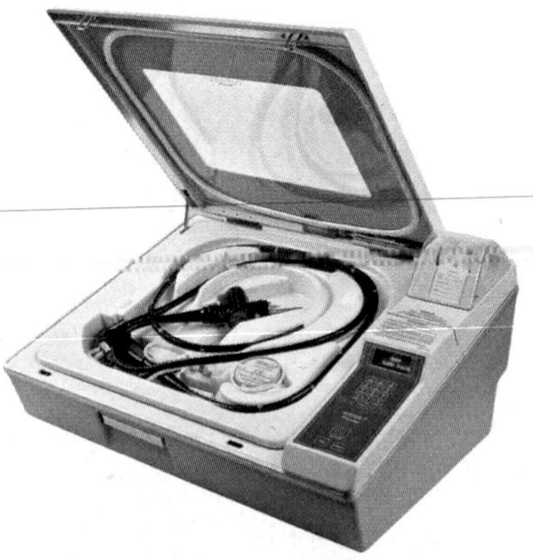

A cabinet for rapid (within 30 minutes) sterile processing of endoscopes and other microsurgical instruments

Figure 11.13 Sterile processing of invasive equipment protects patients.

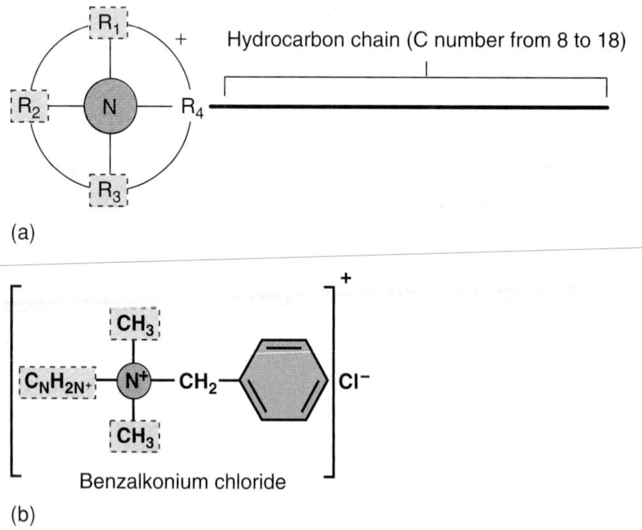

Figure 11.14 The structure of detergents.
(a) In general, detergents are polar molecules with a positively charged head and at least one long, uncharged hydrocarbon chain. The head contains a central nitrogen nucleus with various alkyl (R) groups attached. **(b)** A common quaternary ammonium detergent, benzalkonium chloride.

Applications of Detergents and Soaps Quaternary ammonium compounds **(quats)** include benzalkonium chloride, Zephiran, and cetylpyridinium chloride (Ceepryn). In dilutions ranging from 1:100 to 1:1,000, quats are mixed with cleaning agents to simultaneously disinfect and sanitize floors, furniture, equipment surfaces, and restrooms. They are used to clean restaurant eating utensils, food-processing equipment, dairy equipment, and clothing. They are common preservatives for ophthalmic solutions and cosmetics. Their level of disinfection is far too low for disinfecting medical instruments.

Soaps are alkaline compounds made by combining the fatty acids in oils with sodium or potassium salts. In usual practice, soaps are only weak microbicides, and they destroy only highly sensitive forms such as the agents of gonorrhea, meningitis, and syphilis. The common hospital pathogen *Pseudomonas* is so resistant to soap that various species grow abundantly in soap dishes.

Soaps function primarily as cleansing agents and sanitizers in industry and the home. The superior sudsing and wetting properties of soaps help to mechanically remove large amounts of surface soil, greases, and other debris that contains microorganisms. Soaps gain greater germicidal value when mixed with agents such as chlorhexidine or iodine. They can be used for cleaning instruments before heat sterilization, degerming patients' skin, routine hand washing by medical and dental personnel, and preoperative hand scrubbing. Vigorously brushing the hands with germicidal soap over a 15-second period is an effective way to remove dirt, oil, and surface contaminants as well as some resident microbes, but it will never sterilize the skin **(Insight 11.3** and **figure 11.15).**

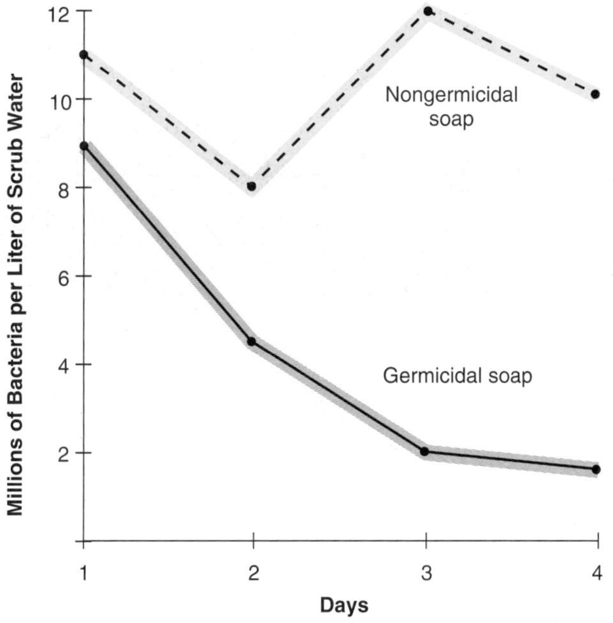

Figure 11.15 Graph showing effects of hand scrubbing. Comparison of scrubbing over several days with a nongermicidal soap versus a germicidal soap. Germicidal soap has persistent effects on skin over time, keeping the microbial count low. Without germicide, soap does not show this sustained effect.

Heavy Metal Compounds

Various forms of the metallic elements mercury, silver, gold, copper, arsenic, and zinc have been applied in microbial control over several centuries. These are often referred to as heavy metals because of their relatively high atomic weight. However, from this list, only preparations containing mercury and silver still have any significance as germicides. Although some metals (zinc, iron) are actually needed in small concentrations as cofactors on enzymes, the higher molecular weight metals (mercury, silver, gold) can be very toxic, even in minute quantities (parts per million). This property of having antimicrobial effects in exceedingly small amounts is called an **oligodynamic** (ol′-ih-goh-dy-nam′-ik) **action (figure 11.16).** Heavy metal germicides contain either an inorganic or an organic metallic salt, and they come in the form of aqueous solutions, tinctures, ointments, or soaps.

Mercury, silver, and most other metals exert microbicidal effects by binding onto functional groups of proteins and inactivating them, rapidly bringing metabolism to a standstill (see figure 11.4*c*). This mode of action can destroy many types of microbes, including vegetative bacteria, fungal cells and spores, algae, protozoa, and viruses (but not endospores).

Unfortunately, there are several drawbacks to using metals in microbial control:

1. metals can be very toxic to humans if ingested, inhaled, or absorbed through the skin, even in small quantities, for the same reasons that they are toxic to microbial cells;

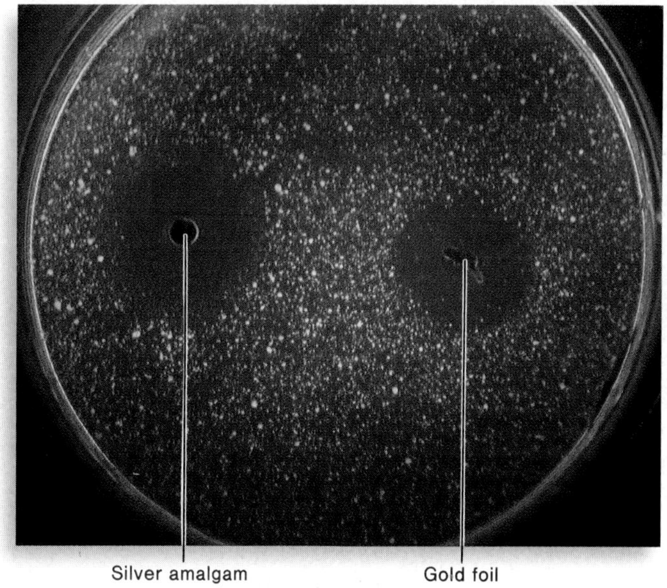

Figure 11.16 Demonstration of the oligodynamic action of heavy metals.
A pour plate inoculated with saliva has small fragments of heavy metals pressed lightly into it. During incubation, clear zones indicating growth inhibition developed around both fragments. The slightly larger zone surrounding the amalgam (used in tooth fillings) probably reflects the synergistic effect of the silver and mercury it contains.

INSIGHT 11.3 *Medical*

The Quest for Sterile Skin

More than a hundred years ago, before sterile gloves were a routine part of medical procedures, the hands remained bare during surgery. Realizing the danger from microbes, medical practitioners attempted to sterilize the hands of surgeons and their assistants to prevent surgical infections. Several stringent (and probably very painful) techniques involving strong chemical germicides and vigorous scrubbing were practiced. Here are a few examples.

In Schatz's method, the hands and forearms were first cleansed by brisk scrubbing with liquid soap for 3 to 5 minutes, then soaked in a saturated solution of permanganate at a temperature of 110°F until they turned a deep mahogany brown. Next, the limbs were immersed in saturated oxalic acid until the skin became decolorized. Then, as if this were not enough, the hands and arms were rinsed with sterile limewater and washed in warm bichloride of mercury for 1 minute.

Or, there was Park's method (more like a torture). First, the surfaces of the hands and arms were rubbed completely with a mixture of cornmeal and green soap to remove loose dirt and superficial skin. Next, a paste of water and mustard flour was applied to the skin until it began to sting. This potion was rinsed off in sterile water, and the hands and arms were then soaked in hot bichloride of mercury for a few minutes, during which the solution was rubbed into the skin.

Another method once earnestly suggested for getting rid of microorganisms was to expose the hands to a hot-air cabinet to "sweat the germs" out of skin glands. Pasteur himself advocated a quick flaming of the hands to maintain asepsis.

The old dream of sterilizing the skin was finally reduced to some basic realities: The microbes entrenched in the epidermis and skin glands cannot be completely eradicated even with the most intense efforts, and the skin cannot be sterilized without also seriously damaging it. Because this is true for both medical personnel and their patients, the chance always exists that infectious agents can be introduced during invasive medical procedures. Safe surgery had to wait until 1890, when rubber gloves were first made available for placing a sterile barrier around the hands. Of course, this did not mean that skin cleansing and antiseptic procedures were abandoned or downplayed. A thorough scrubbing of the skin, followed by application of an antiseptic, is still needed to remove the most dangerous source of infections—the superficial contaminants constantly picked up from whatever we touch.

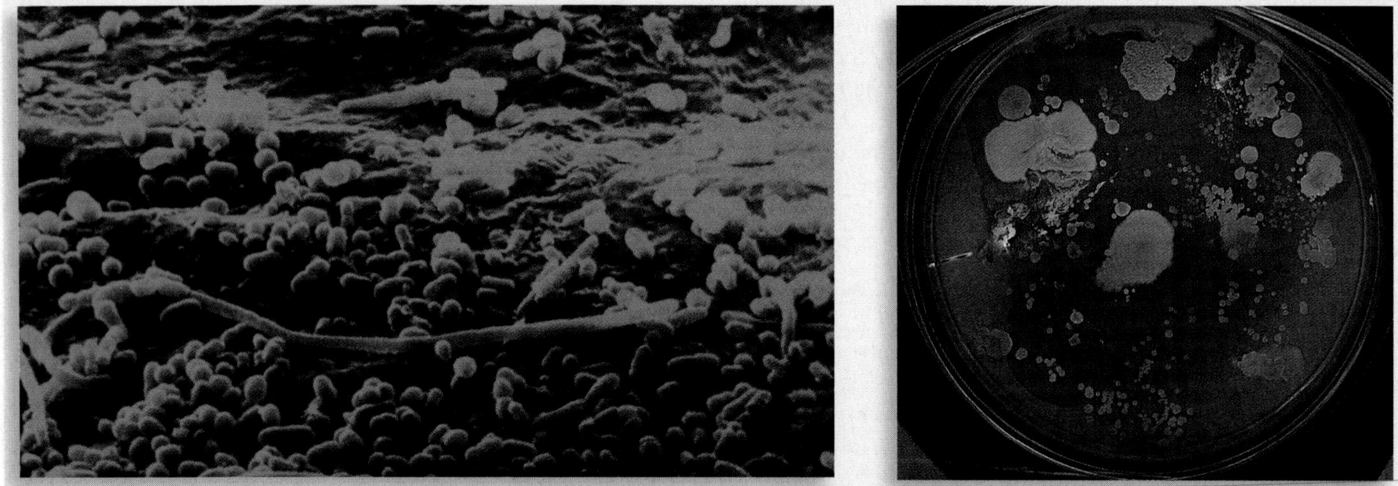

(a) (b)

Microbes on normal unwashed hands. **(a)** A scanning electron micrograph of a piece of skin from a fingertip shows clusters of bacteria perched atop a fingerprint ridge (47,000×). **(b)** Heavy growth of microbial colonies on a plate of blood agar. This culture was prepared by passing an open sterile plate around a classroom of 30 students and having each one touch its surface. After incubation, a mixed population of bacteria and fungi appeared.

2. they often cause allergic reactions;
3. large quantities of biological fluids and wastes neutralize their actions; and
4. microbes can develop resistance to metals.

Health and environmental considerations have dramatically reduced the use of metallic antimicrobial compounds in medicine, dentistry, commerce, and agriculture.

Applications of Heavy Metals Weak (0.001% to 0.2%) organic mercury tinctures such as thimerosal (Merthiolate) and nitromersol (Metaphen) are fairly effective antiseptics and infection preventives, but they should never be used on broken skin because they are harmful and can delay healing. The organic mercurials also serve as preservatives in cosmetics and ophthalmic solutions. Mercurochrome, that old staple of the medicine cabinet, is now considered among the poorest of antiseptics.

A silver compound with several applications is silver nitrate ($AgNO_3$) solution. German professor of obstetrics Carl Siegmund Franz Credé introduced it in the late 19th century for preventing gonococcal infections in the eyes of newborn infants who had been exposed to an infected birth canal (described in chapter 23). This preparation is not used as often now because many pathogens are resistant to it. It has been replaced by antibiotics in most instances. Solutions of silver nitrate (1% to 2%) can also be used as topical germicides on mouth ulcers and occasionally root canals. Silver sulfadiazine ointment, when added to dressings, effectively prevents infection in second- and third-degree burn patients, and pure silver is now incorporated into catheters to prevent urinary tract infections in the hospital. Colloidal silver preparations are mild germicidal ointments or rinses for the mouth, nose, eyes, and vagina. Silver ions are increasingly incorporated into many hard surfaces, such as plastics and steel, as a way to control microbial growth on items such as toilet seats, stethoscopes, and even refrigerator doors.

Aldehydes as Germicides

Organic substances bearing a —CHO functional group (a strong reducing group) on the terminal carbon are called aldehydes. Several common substances such as sugars and some fats are technically aldehydes. The two aldehydes used most often in microbial control are *glutaraldehyde* and *formaldehyde.*

Glutaraldehyde is a yellow acidic liquid with a mild odor. The mechanism of activity involves cross-linking protein molecules on the cell surface. In this process, amino acids are alkylated, meaning that a hydrogen atom on an amino acid is replaced by the glutaraldehyde molecule itself **(figure 11.17).** It can also irreversibly disrupt the activity of enzymes within the cell. Glutaraldehyde is rapid and broad-spectrum and is one of the few chemicals officially accepted as a sterilant and high-level disinfectant. It kills spores in 3 hours and fungi and vegetative bacteria (even *Mycobacterium* and *Pseudomonas*) in a few minutes. Viruses, including the most resistant forms, appear to be inactivated after relatively short exposure times. Glutaraldehyde retains its potency even in the presence of organic matter, is noncorrosive, does not damage plastics, and is less toxic or irritating than formaldehyde. Its principal disadvantage is that it is somewhat unstable, especially with increased pH and temperature.

Formaldehyde is a sharp, irritating gas that readily dissolves in water to form an aqueous solution called **formalin.** Pure formalin is a 37% solution of formaldehyde gas dissolved in water. The chemical is microbicidal through its attachment to nucleic acids and functional groups of amino acids. Formalin is an intermediate- to high-level disinfectant, although it acts more slowly than glutaraldehyde. Formaldehyde's extreme toxicity (it is classified as a carcinogen) and irritating effects on the skin and mucous membranes greatly limit its clinical usefulness.

A third aldehyde, ortho-phthalaldehyde (OPA) has recently been registered by the EPA as a high-level disinfectant. OPA is a pale blue liquid with a barely detectable odor and can be most directly compared to glutaraldehyde. It has a mechanism of action similar to glutaraldehyde, is stable, is nonirritating to the eyes and nasal passages, and, for most uses, is much faster acting than glutaraldehyde. It is effective against vegetative bacteria, including *Mycobacterium* and *Pseudomonas,* fungi, and viruses. Chief among its disadvantages are an inability to reliably destroy spores and, on a more practical note, its tendency to stain proteins, including those in human skin.

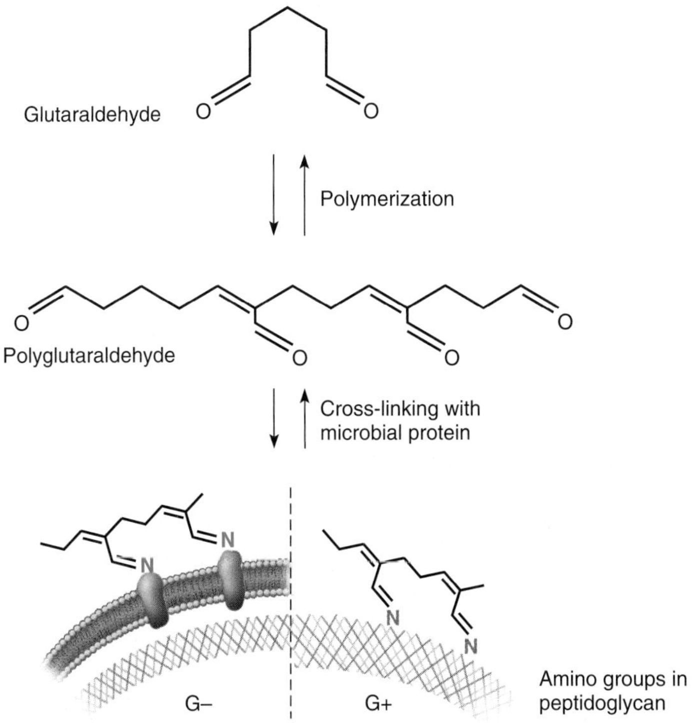

Figure 11.17 Actions of glutaraldehyde. The molecule polymerizes easily. When these alkylating polymers react with amino acids, they cross-link and inactivate proteins.

Applications of the Aldehydes Glutaraldehyde is a milder chemical for sterilizing materials that are damaged by heat. Commercial products (Cidex, Sporicidin) diluted to 2% are used to sterilize respiratory therapy equipment, hemostats, fiberoptic endoscopes (laparoscopes, arthroscopes), and kidney dialysis equipment. Glutaraldehyde is employed on dental instruments (usually in combination with autoclaving) to inactivate hepatitis B and other blood-borne viruses. It also serves to preserve vaccines, sanitize poultry carcasses, and degerm cows' teats.

Formalin tincture (8%) has limited use as a disinfectant for surgical instruments, and formalin solutions have applications in aquaculture to kill fish parasites and control growth of algae and fungi. Any object that is intended to come into intimate contact with the body must be thoroughly rinsed to neutralize the formalin residue. It is, after all, one of the active ingredients in embalming fluid.

Gaseous Sterilants and Disinfectants

Processing inanimate substances with chemical vapors, gases, and aerosols provides a versatile alternative to heat or liquid chemicals. Currently, those vapors and aerosols having the broadest applications are ethylene oxide (ETO), propylene oxide, and chlorine dioxide.

Ethylene oxide is a colorless substance that exists as a gas at room temperature. It is very explosive in air, a feature that can be eliminated by combining it with a high percentage of carbon dioxide or fluorocarbon. Like the aldehydes, ETO is a very strong alkylating agent, and it reacts vigorously with functional groups of DNA and proteins. Through these actions, it blocks both DNA replication and enzymatic actions. Ethylene oxide is one of a very few gases generally accepted for chemical sterilization because, when employed according to strict procedures, it is a sporicide. A specially designed ETO sterilizer called a *chemiclave,* a variation on the autoclave, is equipped with a chamber; gas ports; and temperature, pressure, and humidity controls. Ethylene oxide is rather penetrating but relatively slow-acting, requiring from 90 minutes to 3 hours. Some items absorb ETO residues and must be aerated with sterile air for several hours after exposure to ensure dissipation of as much residual gas as possible. For all of its effectiveness, ETO has some unfortunate features. Its explosiveness makes it dangerous to handle; it can damage the lungs, eyes, and mucous membranes if contacted directly; and it is rated as a carcinogen by the government.

Chlorine dioxide is another gas that has of late been used as a sterilant. Despite the name, chlorine dioxide works in a completely different way from the chlorine compounds discussed earlier in the chapter. It is a strong alkylating agent, which disrupts proteins and is effective against vegetative bacteria, fungi, viruses, and endospores. Although chlorine dioxide is used for the treatment of drinking water, wastewater, food processing equipment, and medical waste, its most well-known use was in the decontamination of the Senate offices after the anthrax attack of 2001 **(Insight 11.4).**

Applications of Gases and Aerosols Ethylene oxide (carboxide, cryoxide) is an effective way to sterilize and disinfect plastic materials and delicate instruments in hospitals and industries. It can safely sterilize prepackaged medical devices, surgical supplies, syringes, and disposable Petri dishes. Ethylene oxide has been used extensively to disinfect sugar, spices, dried foods, and drugs.

Propylene oxide is a close relative of ETO, with similar physical properties and mode of action, although it is less toxic. Because it breaks down into a relatively harmless substance, it is safer than ETO for sterilization of foods (nuts, powders, starches, spices).

Dyes as Antimicrobial Agents

Dyes are important in staining techniques and as selective and differential agents in media; they are also a primary source of certain drugs used in chemotherapy. Because aniline dyes such as crystal violet and malachite green are very active against gram-positive species of bacteria and various fungi, they are incorporated into solutions and ointments to treat skin infections (ringworm, for example). The yellow acridine dyes, acriflavine and proflavine, are sometimes utilized for antisepsis and wound treatment in medical and veterinary clinics. For the most part, dyes will continue to have limited applications because they stain and have a narrow spectrum of activity.

Acids and Alkalis

Conditions of very low or high pH can destroy or inhibit microbial cells; but they are limited in applications due to their corrosive, caustic, and hazardous nature. Aqueous solutions of ammonium hydroxide remain a common component of detergents, cleansers, and deodorizers. Organic acids are widely used in food preservation because they prevent spore germination and bacterial and fungal growth and because they are generally regarded as safe to eat. Acetic acid (in the form of vinegar) is a pickling agent that inhibits bacterial growth; propionic acid is commonly incorporated into breads and cakes to retard molds; lactic acid is added to sauerkraut and olives to prevent growth of anaerobic bacteria (especially the clostridia); and benzoic and sorbic acids are added to beverages, syrups, and margarine to inhibit yeasts.

For a look at the antimicrobial chemicals found in some common household products, see **table 11.5.**

CHECKPOINT

- Chemical agents of microbial control are classified by their physical state and chemical nature.
- Chemical agents can be either microbicidal or microbistatic. They are also classified as high-, medium-, or low-level germicides.
- Factors that determine the effectiveness of a chemical agent include the type and numbers of microbes involved, the material involved, the strength of the agent, and the exposure time.
- Halogens are effective chemical agents at both microbicidal and microbistatic levels. Chlorine compounds disinfect water, food, and industrial equipment. Iodine is used as either free iodine or iodophor to disinfect water and equipment. Iodophors are also used as antiseptic agents.
- Phenols are strongly microbicidal agents used in general disinfection. Milder phenol compounds, the bisphenols, are also used as antiseptics.
- Alcohols dissolve membrane lipids and destroy cell proteins. Their action depends upon their concentration, but they are generally only microbistatic.
- Hydrogen peroxide is a versatile microbicide that can be used as an antiseptic for wounds and a disinfectant for utensils. A high concentration is an effective sporicide.
- Surfactants are of two types: detergents and soaps. They reduce cell membrane surface tension, causing membrane rupture. Cationic detergents, or quats, are low-level germicides limited by the amount of organic matter present and the microbial load.
- Aldehydes are potent sterilizing agents and high-level disinfectants that irreversibly disrupt microbial enzymes.
- Ethylene oxide and chlorine dioxide are gaseous sterilants that work by alkylating protein and DNA.

INSIGHT 11.4 Microbiology

Decontaminating Congress

Choosing a microbial control technique is usually a straightforward process: cultures are autoclaved, milk is pasteurized, and medical supplies may be irradiated. When letters containing spores of *Bacillus anthracis* were opened in the Hart Office Building of the U.S. Senate, the process got just a bit trickier. Among the many concerns of the Environmental Protection Agency, which was charged with the building's remediation, were these:

- Anthrax is a lethal disease.
- *Bacillus anthracis* is a spore-forming bacterium, making eradication difficult.
- The Hart Office Building is populated by thousands of people, who could quickly and easily spread endospores from office to office.
- The area to be decontaminated included heating and air-conditioning vents, carpeting, furniture, office equipment, sensitive papers, artwork, and the various belongings of quickly evacuated workers.

The goal of the project could be simply stated: Detect and remove all traces of a lethal, highly infectious, spore-forming bacterium from an enormous space filled with all manner of easily damaged material. Easier said than done.

With this goal in mind, the EPA first set to work to determine the extent of contamination. Samples were taken from 25 buildings on Capitol Hill. Nonporous surfaces were swabbed; furniture and carpets were vacuumed into a HEPA filter; air was pumped through a filter to remove any airborne spores. Samples were placed in sterile vials and double bagged before being transferred to the laboratory. Analysis of the samples revealed the presence of spores in many areas of the Hart Office Building, including the mail-processing areas of 11 senators. Because spores were not found in the entrances to these offices, it was theorized that the spores spread primarily through the mail. The heaviest contamination was found in the office of Senator Tom Daschle, to whom the original anthrax-containing letter was addressed. Additional contamination was found in a conference room, stairwell, elevator, and restroom.

Workers in protective garments prepare the Hart Office Building for decontamination.

With the scope of the problem identified, the EPA could set out to devise a strategy for remediation, keeping in mind the difficulty of killing spore-forming organisms and the myriad contents of the building, much of it delicate, expensive, and in many cases irreplaceable. It was decided that most areas would be cleaned by a combination of HEPA vacuuming followed by either treatment with liquid chlorine dioxide or Sandia decontamination foam, an antibacterial foam that combines surfactants with oxidizing agents. For the most heavily contaminated areas (Senator Daschle's office and parts of the heating and air-conditioning system), gaseous chlorine dioxide would be used. Chlorine dioxide has been accepted as a sterilant since 1988 and is used for, among other things, treatment of medical waste. Its use in this project was based primarily on the facts that it would both work and be unlikely to harm the contents of the building. (It was even tested to ensure that it would not damage the ink making up the signature on a document.)

Before fumigation could begin, however, a method of evaluating the success of the remediation effort needed to be devised. Borrowing from a common laboratory technique, 3,000 small slips of paper covered with spores from the organism *Bacillus stearothermophilus* (which is generally considered harder to kill than *B. anthracis*) were dispersed throughout the building. If after the treatment was complete these spores were unable to germinate, then the fumigation could be considered a success.

On December 1, 2001, technicians prepared the 3,000-square-foot office for fumigation. This included constructing barriers to seal off the portion of building being fumigated and raising the humidity in the offices to approximately 75% to enable the gas to adhere to any lingering spores. Office machines (computers, copiers, for example) were turned on so that the fans inside the equipment would aid in spreading the gas. Finally, at 3:15 a.m., fumigation began. It ended 20 hours later; and, after the gas was neutralized and ventilated from the building, technicians entered to collect the test strips. Analyzing the results, it was clear that trace amounts of spores were still present but only in those areas that originally had the worst contamination (these areas were again cleaned with liquid chlorine dioxide). Senator Daschle's office got a makeover with new carpeting, paint, and furniture, and shortly thereafter the building was reoccupied. The senator also had his office equipment replaced as it was deemed too contaminated to be used. Just as antibiotics are useless against viruses, computer antivirus software doesn't work against bacteria.

TABLE 11.5 Active Ingredients of Various Commercial Antimicrobial Products

Product	Specific Chemical Agent	Antimicrobial Category
Lysol Sanitizing Wipes	Dimethyl benzyl ammonium chloride	Detergent (quat)
Clorox Disinfecting Wipes	Dimethyl benzyl ammonium chloride	Detergent (quat)
Tilex Mildew Remover	Sodium hypochlorites	Halogen
Lysol Mildew Remover	Sodium hypochlorites	Halogen
Ajax Antibacterial Hand Soap	Triclosan	Phenolic
Dawn Antibacterial Hand Soap	Triclosan	Phenolic
Dial Antibacterial Hand Soap	Triclosan	Phenolic
Lysol Disinfecting Spray	Alkyl dimethyl benzyl ammonium saccharinate/ethanol	Detergent (quats)/alcohol
ReNu Contact Lens Solution	Polyaminopropyl biguanide	Chlorhexidine
Wet Ones Antibacterial Moist Towelettes	Benzethonium chloride	Detergents (quat)
Noxzema Triple Clean	Triclosan	Phenolic
Scope Mouthwash	Ethanol	Alcohol
Purell Instant Hand Sanitizer	Ethanol	Alcohol
Pine-Sol	Phenolics and surfactant	Mixed
Allergan Eye Drops	Sodium chlorite	Halogen

Chapter Summary with Key Terms

11.1 Controlling Microorganisms

A. Microbes present at a given place and time that are undesirable or unwanted must be controlled, either by physical or chemical methods.

B. **Sterilization** is a process that destroys or removes all microbes, including viruses.

C. **Disinfection** and **antisepsis** refer to a physical or chemical process that destroys vegetative pathogens but not bacterial endospores.
   1. A **bacteri*cide*** is a chemical that destroys bacteria.
   2. **Bacteri*static*** agents inhibit or prevent the growth of bacteria on tissues or on other objects in the environment.

D. **Decontamination** is any chemical technique that removes microorganisms to "safe levels or standards."

E. Several factors influence the rate at which antimicrobial agents work. These factors are:
   1. Exposure time to the agent.
   2. Numbers of microbes present.
   3. Relative resistance of microbes (for example, endospores versus vegetative forms).
   4. Activity of the agent (microbicidal versus microbistatic).

F. How antimicrobial agents work: Agents affect cell wall synthesis, membrane permeability, and protein and nucleic acid synthesis and function.

11.2 Methods of Physical Control

A. Moist heat denatures proteins and DNA while destroying membranes.
   1. **Autoclaves** utilize steam under pressure to sterilize heat-resistant materials.
   2. Pasteurization subjects liquids to temperatures below 100°C and is used to lower the microbial load in liquids.
   3. Boiling water can be used to destroy vegetative pathogens in the home.

B. Dry heat, using higher temperatures than moist heat, can also be used to sterilize.
   1. **Incineration** can be carried out using a Bunsen burner or incinerator. Temperatures range between 600°C and 1,800°C.
   2. *Dry ovens* coagulate proteins at temperatures of 15°C to 180°C.

C. Cold temperatures are microbistatic, with refrigeration (0°C to 15°C) and freezing (below 0°C) commonly used to preserve food, media, and cultures.

D. Drying and desiccation lead to (often temporary) metabolic inhibition by reducing water in the cell.

E. *Radiation:* Energy in the form of radiation is a method of *cold sterilization*, which works by introducing mutations into the DNA of target cells.
   1. **Ionizing radiation,** such as gamma rays and X rays, has deep penetrating power and works by causing breaks in the DNA of target organisms.

2. **Nonionizing radiation** uses **ultraviolet** waves with very little penetrating power and works by creating dimers between adjacent pyrimidines, which interferes with replication.

F. Filtration involves the physical removal of microbes by passing a gas or liquid through a fine filter and can be used to disinfect or sterilize air as well as heat-sensitive liquids.

11.3 Chemical Agents in Microbial Control

A. Antimicrobial chemicals are found as solids, gases, and liquids. Liquids can be either aqueous (water based) or tinctures (alcohol based).

B. Halogens are chemicals based on elements from group VII of the periodic table.

1. Chlorine is used as chlorine gas, hypochlorites, and chloramines. All work by disrupting disulfide bonds and, given adequate time, are sporicidal.
2. Iodine is found both as free iodine (I_2) and iodophors. Iodine has a mode of action similar to chlorine and is also sporicidal, given enough time.

C. Phenolics are chemicals based on phenol, which work by disrupting cell membranes and precipitating proteins. They are bactericidal, fungicidal, and viricidal, but not sporicidal.

D. Chlorhexidine is a surfactant and protein denaturant with broad microbicidal properties, although it is not sporicidal.

E. Ethyl and isopropyl alcohol, in concentrations of 50% to 90%, are useful for microbial control. Alcohols act as **surfactants,** dissolving membrane lipids and coagulating proteins of vegetative bacterial cells and fungi. They are not sporicidal.

F. Hydrogen peroxide produces highly reactive hydroxyl-free radicals that damage protein and DNA while also decomposing to O_2 gas, which is toxic to anaerobes. Strong solutions of H_2O_2 are sporicidal.

G. Detergents and soaps

1. Cationic detergents known as quaternary ammonium compounds **(quats)** act as surfactants that alter the membrane permeability of some bacteria and fungi. They are not sporicidal.
2. Soaps have little microbicidal activity but rather function by removing grease and soil that contain microbes.

H. Heavy metals: Solutions of silver and mercury kill vegetative cells (but not spores) in exceedingly low concentrations **(oligodynamic action)** by inactivating proteins.

I. Aldehydes such as glutaraldehyde and formaldehyde kill microbes by alkylating protein and DNA molecules.

J. Gases and aerosols such as **ethylene oxide (ETO),** propylene oxide, and chlorine dioxide are strong alkylating agents, all of which are sporicidal.

K. Dyes, acids, and alkalis can also inhibit or destroy microbes.

Multiple-Choice and True-False Questions

Multiple-Choice Questions. Select the correct answer from the answers provided.

1. Microbial control methods that kill __________ are able to sterilize.
   a. viruses
   b. the tubercle bacillus
   c. endospores
   d. cysts
2. Sanitization is a process by which
   a. the microbial load on objects is reduced
   b. objects are made sterile with chemicals
   c. utensils are scrubbed
   d. skin is debrided
3. An example of an agent that lowers the surface tension of cells is
   a. phenol
   b. chlorine
   c. alcohol
   d. formalin
4. High temperatures _______ and low temperatures _______.
   a. sterilize, disinfect
   b. kill cells, inhibit cell growth
   c. denature proteins, burst cells
   d. speed up metabolism, slow down metabolism
5. Microbe(s) that is/are the target(s) of pasteurization include:
   a. *Clostridium botulinum*
   b. *Mycobacterium* species
   c. *Salmonella* species
   d. both b and c
6. The primary mode of action of nonionizing radiation is to
   a. produce superoxide ions
   b. make pyrimidine dimers
   c. denature proteins
   d. break disulfide bonds
7. The most versatile method of sterilizing heat-sensitive liquids is
   a. UV radiation
   b. exposure to ozone
   c. beta propiolactone
   d. filtration
8. _______ is the iodine antiseptic of choice for wound treatment.
   a. Eight percent tincture
   b. Five percent aqueous
   c. Iodophor
   d. Potassium iodide solution

9. A chemical with sporicidal properties is
   a. phenol
   b. alcohol
   c. quaternary ammonium compound
   d. glutaraldehyde
10. Silver nitrate is used
    a. in antisepsis of burns
    b. as a mouthwash
    c. to treat genital gonorrhea
    d. to disinfect water
11. Detergents are
    a. high-level germicides
    b. low-level germicides
    c. excellent antiseptics
    d. used in disinfecting surgical instruments
12. Which of the following is an approved sterilant?
    a. chlorhexidine
    b. betadyne
    c. ethylene oxide
    d. ethyl alcohol

True-False Questions. If statement is true, leave as is. If it is false, correct it by rewriting the sentence.

13. The process of destroying non-spore-forming organisms on inanimate objects fits within the definition of disinfection.
14. The acceptable temperature-pressure combination for an autoclave is 131°C and 9 psi.
15. Ionizing radiation dislodges protons from atoms.
16. A microbicide is an agent that destroys microorganisms.
17. Prions are easily denatured by heat.

Writing to Learn

These questions are suggested as a *writing-to-learn* experience. For each question, compose a one- or two-paragraph answer that includes the factual information needed to completely address the question.

1. a. Compare sterilization with disinfection and sanitization.
   b. Describe the relationship of the concepts of sepsis, asepsis, and antisepsis.
2. a. Briefly explain how the type of microorganisms present will influence the effectiveness of exposure to antimicrobial agents.
   b. Explain how the numbers of contaminants can influence the measures used to control them.
3. a. Precisely what is microbial death?
   b. Why does a population of microbes not die instantaneously when exposed to an antimicrobial agent?
4. Why are antimicrobial processes inhibited in the presence of extraneous organic matter?
5. Describe four modes of action of antimicrobial agents, and give a specific example of how each works.
6. a. Summarize the nature, mode of action, and effectiveness of moist and dry heat.
   b. Compare the effects of moist and dry heat on vegetative cells and spores.
   c. Explain the concepts of TDT and TDP, using examples. What are the minimum TDTs for vegetative cells and endospores?
7. How can the temperature of steam be raised above 100°C? Explain the relationship involved.
8. Explain why desiccation and cold are not reliable methods of disinfection.
9. a. What are some advantages of ionizing radiation as a method of control?
   b. What are some disadvantages?
10. a. Name one chemical for which the general rule that a higher concentration is more effective is *not* true.
    b. What is a sterilant?
    c. Name the principal sporicidal chemical agents.

Concept Mapping

Appendix D provides guidance for working with concept maps.

1. Construct your own concept map using the following words as the *concepts*. Supply the linking words between each pair of concepts.

halogens	sporicidal
oligodynamic	chemical
surfactants	physical
alcohol	silver
phenolics	

Critical Thinking Questions

Critical thinking is the ability to reason and solve problems using facts and concepts. These questions can be approached from a number of angles, and in most cases, they do not have a single correct answer.

1. What is wrong with this statement: "Prior to vaccination, the patient's skin was sterilized with alcohol"? What would be the more correct wording?
2. For each item on the following list, give a reasonable method of sterilization. You cannot use the same method more than three times; the method must sterilize, not just disinfect; and the method must not destroy the item or render it useless unless there is no other choice. After considering a workable method, think of a method that would not work. Note: Where an object containing something is given, you must sterilize everything (for example, both the jar and the Vaseline in it). Some examples of methods are autoclave, ethylene oxide gas, dry oven, and ionizing radiation.

   room air
   blood in a syringe
   serum
   a pot of soil
   plastic Petri dishes
   heat-sensitive drugs
   cloth dressings
   leather shoes from a thrift shop
   a cheese sandwich
   carcasses of cows with "mad cow" disease
   inside of a refrigerator
   wine
   a jar of Vaseline
   fruit in plastic bags
   talcum powder
   milk
   orchid seeds
   human hair (for wigs)
   a flask of nutrient agar
   an entire room (walls, floor, etc.)
   rubber gloves
   disposable syringes
   metal instruments
   mail contaminated with anthrax spores

3. a. Graph the data in table 11.3, plotting the time on the Y axis and the temperature on the X axis for three different organisms.
   b. Using pasteurization techniques as a model, compare the TDTs and explain the relationships between temperature and length of exposure.
   c. Is there any difference between the graph for a sporeformer and the graph for a non-sporeformer? Explain.
4. Can you think of situations in which the same microbe would be considered a serious contaminant in one case and completely harmless in another?
5. A supermarket/drugstore assignment: Look at the labels of 10 different products used to control microbes, and make a list of their active ingredients, their suggested uses, and information on toxicity and precautions.
6. Devise an experiment that will differentiate between bactericidal and bacteristatic effects.

Visual Understanding

1. **From chapter 2, figure 2.19.** Study this illustration of a cell membrane. In what ways could alcohol (Hint; a solvent) damage this membrane? How would that harm the cell?

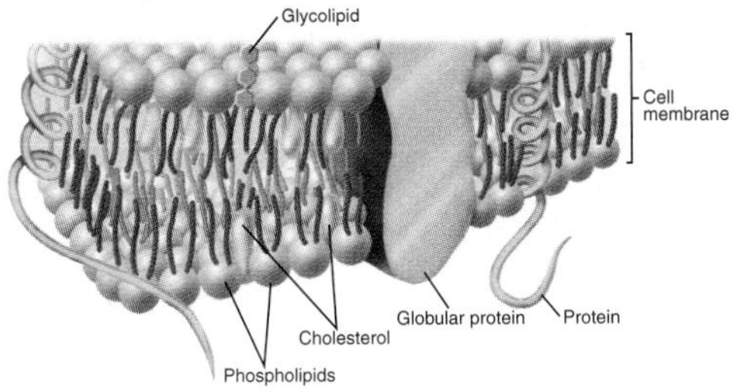

2. **From chapter 4, figure 4.21.** Why would many chemical control agents be ineffective in controlling this organism?

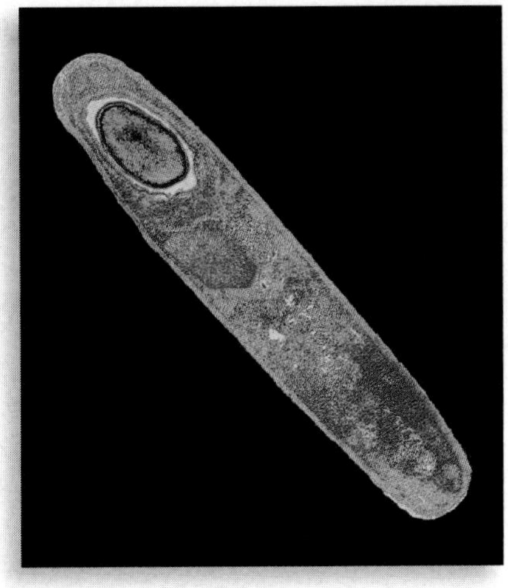

Internet Search Topics

1. Look for information on the Internet concerning triclosan-resistant bacteria. Based on what you find, do you think the widespread use of this antimicrobial is a good idea? Why or why not?
2. Find websites that discuss problems with sterilizing reusable medical instruments such as endoscopes and the types of diseases that can be transmitted with them.
3. Research "prion decontamination." What would you do if you were caring for a patient in the hospital with Creutzfeldt-Jakob disease (a prion disease) and there was a blood splatter on the floor?
4. Go to: www.aris.mhhe.com, and click on "microbiology" and then this textbook's author/title. Go to chapter 11, access the URLs listed under Internet Search Topics, and research the following:

   Hand washing is one of our main protections to ensure good health and hygiene. Log on to one of the websites listed to research information, statistics, and other aspects of hand washing.
5. Using the Internet, find a recipe for making "jerky." What steps in the process will help control microbes?

CHAPTER 12

Drugs, Microbes, Host—The Elements of Chemotherapy

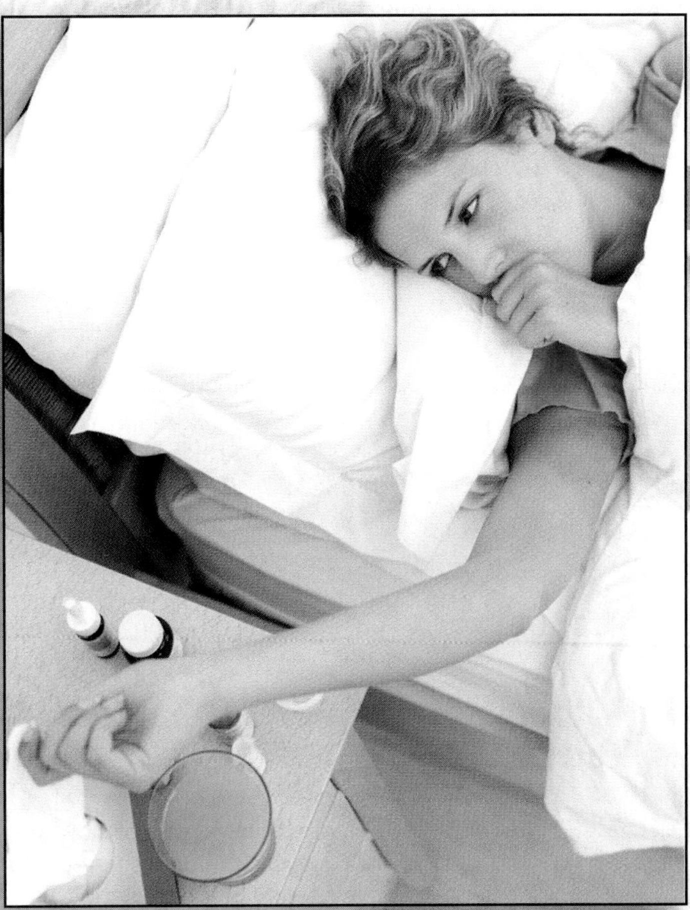

CASE FILE 12

Dana, a 21-year-old college student, was suffering from a persistent cough, fever, and constant fatigue. She was asked whether she had these symptoms in the past and if she had been treated for them at that time. She reported that she had gotten sick with similar symptoms before but never received any treatment. Since she had mild fever, and her chest X rays were normal, her doctor prescribed streptomycin and cough medicine. Her symptoms improved drastically within a week and, consequently, she decided to stop taking her medication.

After 2 months, she felt ill again. Her symptoms included coughing, loss of appetite, high fever, and night sweats. She said the cough, which had lasted for the past 3 weeks, was accompanied by a blood-stained sputum.

A preliminary diagnosis was made based on the symptoms. Chest X rays, Mantoux PPD test, and an acid-fast stain from a sputum sample were later performed. She also was tested for HIV. Based on the results of the tests and her symptoms, a cocktail of antibiotics, including isoniazid and rifampin, was prescribed. She was instructed to take the antibiotics for 9 months.

- *What disease is described here?*
- *Why might Dana's symptoms have resurfaced despite the original streptomycin treatment?*
- *Why was it necessary to place Dana on a multidrug treatment?*

Case File 12 Wrap-Up appears on page 354.

CHAPTER OVERVIEW

- Antimicrobial chemotherapy involves the use of chemicals to prevent and treat infectious diseases. These chemicals include antibiotics, which are derived from the natural metabolic processes of bacteria and fungi, as well as synthetic drugs.
- To be effective, antimicrobial therapy must disrupt a necessary component of a microbe's structure or metabolism to the extent that microbes are killed or their growth is inhibited. At the same time, the drug must be safe for humans, while not being overly toxic, causing allergies or disrupting normal biota.
- Drugs are available to treat infections by all manner of organisms, including viruses. Drug choice depends on the microbe's sensitivity, the drug's toxicity, and the health of the patient.
- Adverse side effects of drugs include damage to organs, allergies, and disruption of the host's normal biota, which can lead to other infections.
- A major problem in modern chemotherapy is the development of drug resistance. Many microbes have undergone genetic changes that allow them to circumvent the effects of a drug. The misapplication of drugs and antibiotics has worsened this problem.

12.1 Principles of Antimicrobial Therapy

A hundred years ago in the United States, one out of three children was expected to die of an infectious disease before the age of 5. Early death or severe lifelong debilitation from scarlet fever, diphtheria, tuberculosis, meningitis, and many other bacterial diseases was a fearsome yet undeniable fact of life to most of the world's population. The introduction of modern drugs to control infections in the 1930s was a medical revolution that has added significantly to the life span and health of humans. It is no wonder that for many years, antibiotics in particular were regarded as miracle drugs. In later discussions, we evaluate this perception in light of the shortcomings of chemotherapy. Although antimicrobial drugs have greatly reduced the incidence of certain infections, they have defintely not eradicated infectious disease and probably never will. In fact, in some parts of the world, mortality rates from infectious diseases are as high as before the arrival of antimicrobial drugs. Nevertheless, humans have been taking medicines to try to control diseases for thousands of years **(Insight 12.1).**

The goal of antimicrobial chemotherapy is deceptively simple: administer a drug to an infected person, which destroys the infective agent without harming the host's cells. In actuality, this goal is rather difficult to achieve, because many (often contradictory) factors must be taken into account. The ideal drug should be easily administered yet be able to reach the infectious agent anywhere in the body, be absolutely toxic to the infectious agent while simultaneously being nontoxic to the host, and remain active in the body as long as needed yet be safely and easily broken down and excreted. In short, the perfect drug does not exist; but by balancing drug characteristics against one another, a satisfactory compromise can be achieved **(table 12.1).**

Chemotherapeutic agents are described with regard to their origin, range of effectiveness, and whether they are naturally produced or chemically synthesized. A few of the more important terms you will encounter are found in **table 12.2.**

In this chapter, we describe different types of antibiotic drugs, their mechanism of action, and the types of microbes on which they are effective. The organ system chapters 18 through 23 list specific disease agents and the drugs used to treat them.

The Origins of Antimicrobial Drugs

Nature is a prolific producer of antimicrobial drugs. Antibiotics, after all, are common metabolic products of aerobic bacteria and fungi. By inhibiting the growth of other microorganisms in the same habitat (antagonism), antibiotic producers presumably enjoy less competition for nutrients and space. The greatest numbers of antibiotics are derived from bacteria in the genera *Streptomyces* and *Bacillus* and from molds in the genera *Penicillium* and *Cephalosporium.*

TABLE 12.1 Characteristics of the Ideal Antimicrobial Drug

- Selectively toxic to the microbe but nontoxic to host cells
- Microbicidal rather than microbistatic
- Relatively soluble; functions even when highly diluted in body fluids
- Remains potent long enough to act and is not broken down or excreted prematurely
- Doesn't lead to the development of antimicrobial resistance
- Complements or assists the activities of the host's defenses
- Remains active in tissues and body fluids
- Readily delivered to the site of infection
- Reasonably priced
- Does not disrupt the host's health by causing allergies or predisposing the host to other infections

INSIGHT 12.1 Historical

From Witchcraft to Wonder Drugs

Early human cultures relied on various types of primitive medications such as potions, poultices, and mudplasters. Many were concocted from plant, animal, and mineral products that had been found—usually through trial and error or accident—to have some curative effect upon ailments and complaints. In one ancient Chinese folk remedy, a fermented soybean curd was applied to skin infections. The Greeks used wine and plant resins (myrrh and frankincense), rotting wood, and various mineral salts to treat diseases. Many folk medicines were effective, but some of them either had no effect or were even harmful. It is interesting that the Greek word *pharmakeutikos* originally meant the practice of witchcraft. These ancient remedies were handed down from generation to generation, but it was not until the Middle Ages that a specific disease was first treated with a specific chemical. Syphilitic patients were dosed with toxic arsenic and mercury compounds, a practice that continued into the 20th century and may have proved the ancient axiom: *Graviora quaedum sunt remedia periculus* ("Some remedies are worse than the disease").

An enormous breakthrough in the science of drug therapy came with the germ theory of infection by Robert Koch (see chapter 1). This allowed disease treatment to focus on a particular microbe, which in turn opened the way for Paul Ehrlich to formulate the first theoretical concepts in chemotherapy in the late 1800s. Ehrlich had observed that certain dyes affixed themselves to specific microorganisms and not to animal tissues. This observation led to the profound idea that if a drug was properly selective in its actions, it would zero in on and destroy a microbial target and leave human cells unaffected. His first discovery was an arsenic-based drug that was very toxic to the spirochete of syphilis but, unfortunately, to humans as well. Ehrlich systematically altered this parent molecule, creating numerous derivatives. Finally, on the 606th try, he arrived at a compound he called *salvarsan*. This drug had some therapeutic merit and was used for a few years, but it eventually had to be discontinued because it was still not selective enough in its toxicity. Ehrlich's work had laid important foundations for many of the developments to come.

Another pathfinder in early drug research was Gerhard Domagk, whose discoveries in the 1930s launched a breakthrough in therapy that marked the true beginning of broad-scale usage of antimicrobial drugs. Domagk showed that the red dye prontosil was chemically changed by the body into a compound with specific activity against bacteria. This substance was sulfonamide—the first *sulfa* drug. In a short time, the structure of this drug was determined and it became possible to synthesize it on a wide scale and to develop scores of other sulfonamide drugs. Although these drugs had immediate applications in therapy (and still do), still another fortunate discovery was needed before the golden age of antibiotics could really blossom.

The discovery of antibiotics dramatically demonstrates how developments in science and medicine often occur through a combination of accident, persistence, collaboration, and vision. In 1928 in the London laboratory of Sir Alexander Fleming, a plate of *Staphylococcus aureus* became contaminated with the mold *Penicillium notatum*. Observing these plates, Fleming noted that the colonies of *Staphylococcus* were evidently being destroyed by some activity of the nearby mold colonies. Struck by this curious phenomenon, he extracted from the fungus a compound he called penicillin and showed that it was responsible for the inhibitory effects.

Although Fleming understood the potential for penicillin, he was unable to develop it. A decade after his discovery, English chemists Howard Florey and Ernst Chain worked out methods for industrial production of penicillin to help in the war effort. Clinical trials conducted in 1941 ultimately proved its effectiveness, and cultures of the mold were brought to the United States for an even larger-scale effort. When penicillin was made available to the world's population, it was a godsend. By the 1950s, the pharmaceutical industry had entered an era of drug research and development that soon made penicillin only one of a large assortment of antimicrobial drugs. But in time, because of extreme overuse and misunderstanding of its capabilities, it also became the model for one of the most serious drug problems—namely, drug resistance.

The father of modern antibiotics. A Scottish physician, Sir Alexander Fleming, accidently discovered penicillin when his keen eye noticed that colonies of bacteria were being lysed by a fungal contaminant. He studied the active ingredient and set the scene for the development of the drug 10 years later.

TABLE 12.2 Terminology of Chemotherapy

Chemotherapeutic Drug	Any chemical used in the treatment, relief, or prophylaxis of a disease
Prophylaxis	Use of a drug to prevent imminent infection of a person at risk
Antimicrobial Chemotherapy	The use of chemotherapeutic drugs to control infection
Antimicrobials	All-inclusive term for any antimicrobial drug, regardless of its origin
Antibiotics	Substances produced by the natural metabolic processes of some microorganisms that can inhibit or destroy other microorganisms
Semisynthetic Drugs	Drugs that are chemically modified in the laboratory after being isolated from natural sources
Synthetic Drugs	The use of chemical reactions to synthesize antimicrobial compounds in the laboratory
Narrow Spectrum (Limited Spectrum)	Antimicrobials effective against a limited array of microbial types—for example, a drug effective mainly on gram-positive bacteria
Broad Spectrum (Extended Spectrum)	Antimicrobials effective against a wide variety of microbial types—for example, a drug effective against both gram-positive and gram-negative bacteria

Not only have chemists created new drugs by altering the structure of naturally occurring antibiotics, they are actively searching for metabolic compounds with antimicrobial effects in species other than bacteria and fungi (discussed later).

CHECKPOINT

- Antimicrobial chemotherapy involves the use of drugs to control infection on or in the body.
- Antimicrobial drugs are produced either synthetically or from natural sources. They inhibit or destroy microbial growth in the infected host. Antibiotics are the subset of antimicrobials produced by the natural metabolic processes of microorganisms.
- Antimicrobial drugs are classified by their range of effectiveness. Broad-spectrum antimicrobials are effective against many types of microbes. Narrow-spectrum antimicrobials are effective against a limited group of microbes.
- Bacteria and fungi are the primary sources of most antibiotics. The molecular structures of these compounds can be chemically altered to form additional semisynthetic antimicrobials.

12.2 Interactions Between Drug and Microbe

The goal of antimicrobial drugs is either to disrupt the cell processes or structures of bacteria, fungi, and protozoa or to inhibit virus replication. Most of the drugs used in chemotherapy interfere with the function of enzymes required to synthesize or assemble macromolecules, or they destroy structures already formed in the cell. Above all, drugs should be **selectively toxic,** which means they should kill or inhibit microbial cells without simultaneously damaging host tissues. This concept of selective toxicity is central to chemotherapy, and the best drugs are those that block the actions or synthesis of molecules in microorganisms but not in vertebrate cells. Examples of drugs with excellent selective toxicity are those that block the synthesis of the cell wall in bacteria (penicillins). They have low toxicity and few direct effects on human cells because human cells lack the chemical peptidoglycan and are thus unaffected by this action of the antibiotic. Among the most toxic to human cells are drugs that act upon a structure common to both the infective agent and the host cell, such as the cell membrane (for example, amphotericin B used to treat fungal infections). As the characteristics of the infectious agent become more and more similar to those of the host cell, selective toxicity becomes more difficult to achieve, and undesirable side effects are more likely to occur. The previous example briefly illustrates this concept. We examine the subject in more detail in a later section.

A NOTE ON CHEMOTHERAPY

The word "chemotherapy" is most commonly associated with the treatment of cancer. As you see in table 12.2, its official meaning is broader than that and can also be applied to antimicrobial treatment.

Mechanisms of Drug Action

If the goal of chemotherapy is to disrupt the structure or function of an organism to the point where it can no longer survive, then the first step toward this goal is to identify the structural and metabolic needs of a living cell. Once the requirements of a living cell have been determined, methods of removing, disrupting, or interfering with these requirements can be employed as potential chemotherapeutic strategies. The metabolism of an actively dividing cell is marked by the production of new cell wall components (in most cells), DNA, RNA, proteins, and cell membrane. Consequently, antimicrobial drugs are divided into categories based on which of these metabolic targets they affect. These categories are outlined in **figure 12.1** and include:

1. inhibition of cell wall synthesis,
2. inhibition of nucleic acid (RNA and DNA) structure and function,
3. inhibition of protein synthesis,

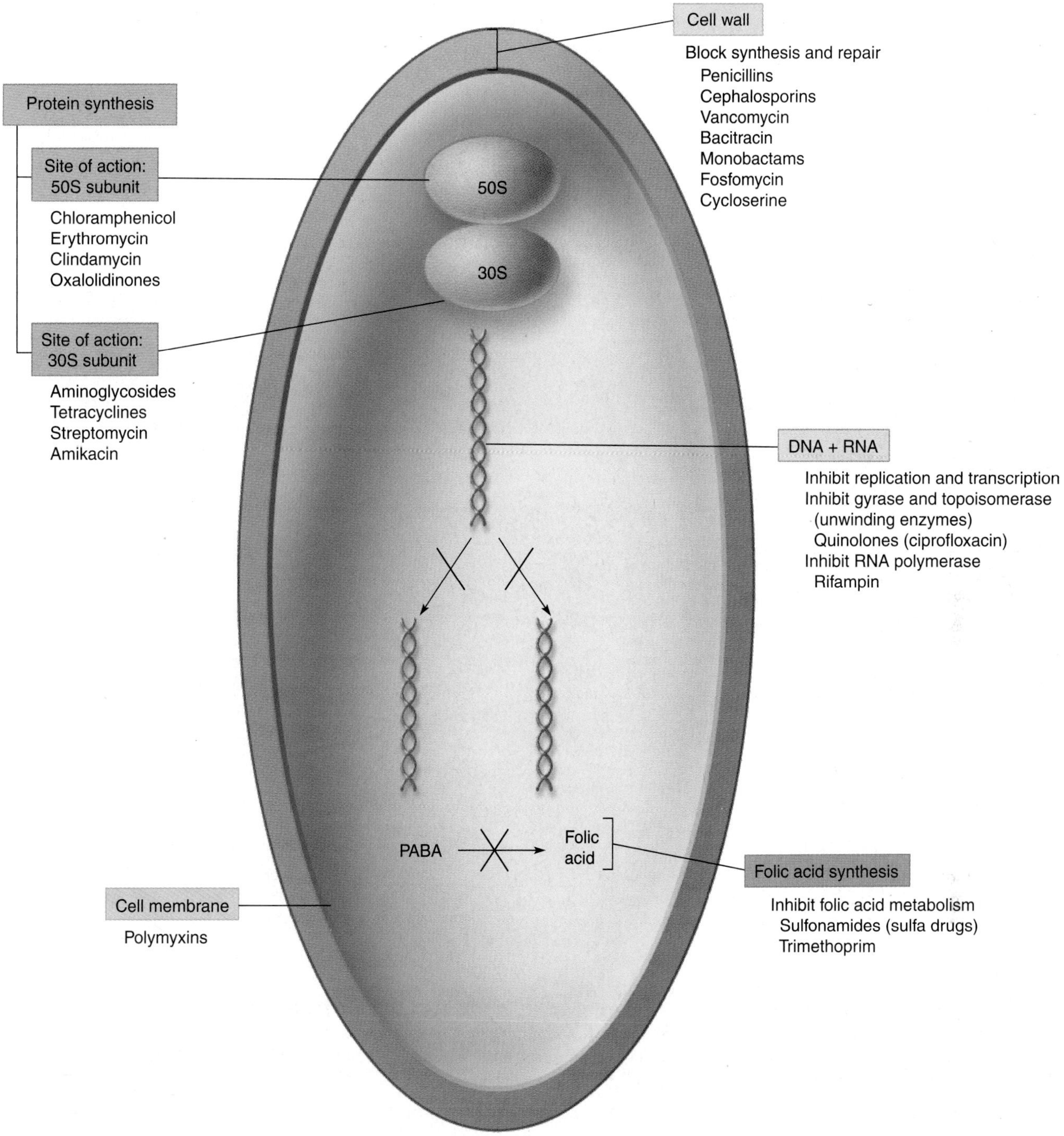

Figure 12.1 Primary sites of action of antimicrobial drugs on bacterial cells.

4. interference with cell membrane structure or function, and
5. inhibition of folic acid synthesis.

As you will see, these categories are not completely discrete, and some effects can overlap.

Antimicrobial Drugs That Affect the Bacterial Cell Wall

The cell walls of most bacteria contain a rigid girdle of peptidoglycan, which protects the cell against rupture from hypotonic environments. Active cells must constantly synthesize new peptidoglycan and transport it to its proper place in the cell envelope. Drugs such as penicillins and cephalosporins react with one or more of the enzymes required to complete this process, causing the cell to develop weak points at growth sites and to become osmotically fragile **(figure 12.2).** Antibiotics that produce this effect are considered bactericidal, because the weakened cell is subject to lysis. It is essential to note that most of these antibiotics are active only in young, growing cells, because old, inactive, or dormant cells do not synthesize

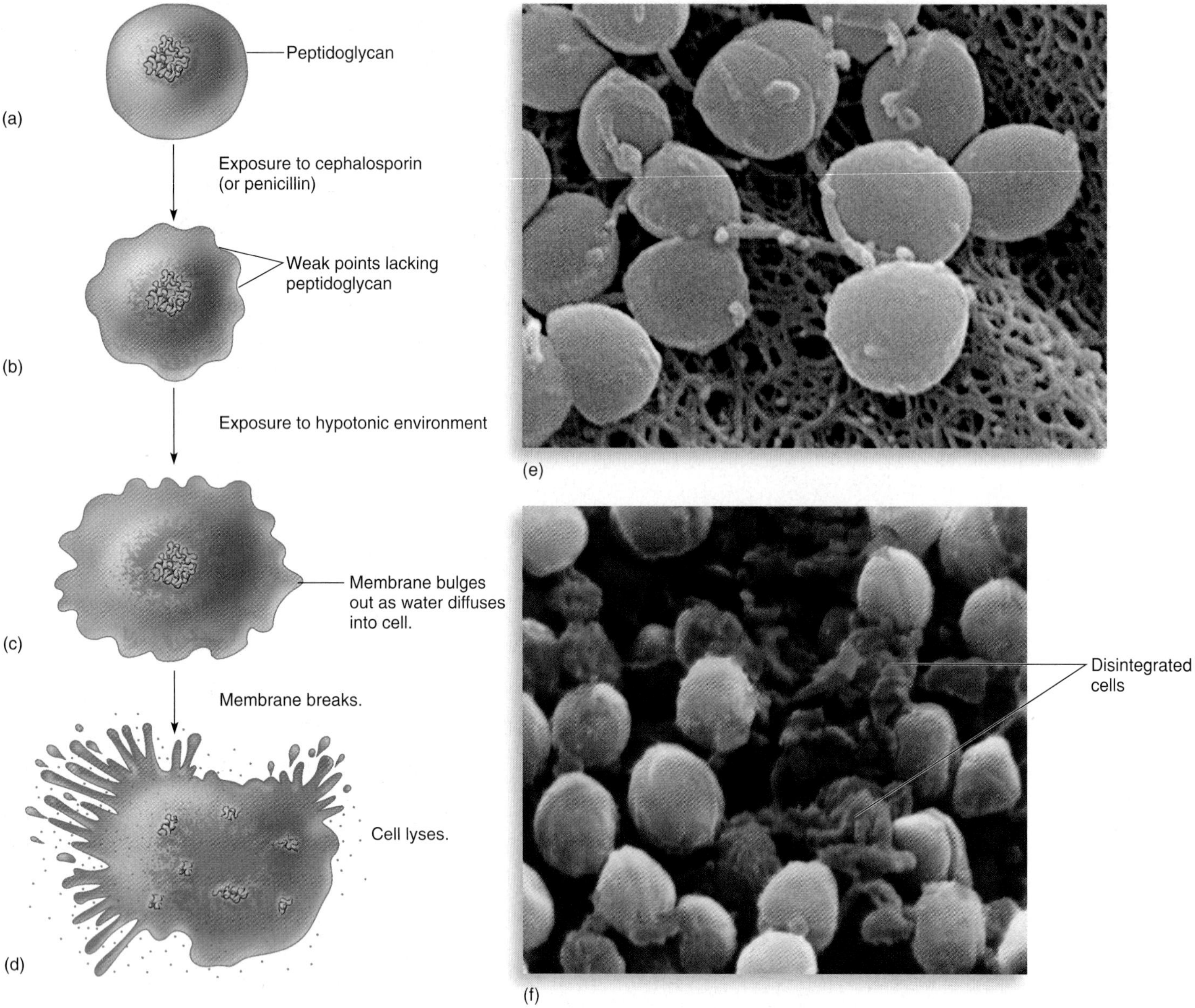

Figure 12.2 The consequences of exposing a growing cell to antibiotics that prevent cell wall synthesis.
(a)–(d) Diagram of effect of cephalosporin on a *Staphylococcus*. **(e)** Scanning electron micrograph of bacterial cells in their normal state (10,000×). **(f)** Scanning electron micrograph of the same cells in the drug-affected state, showing deformation and disintegration (10,000×). Note: These photos have been artificially colored; the different colors are not relevant.

peptidoglycan. (One exception is a new class of antibiotics called the "-penems.")

Cycloserine inhibits the formation of the basic peptidoglycan subunits, and vancomycin hinders the elongation of the peptidoglycan. Penicillins and cephalosporins bind and block peptidases that cross-link the glycan molecules, thereby interrupting the completion of the cell wall **(figure 12.3).** Penicillins that do not penetrate the outer membrane are less effective against gram-negative bacteria, but broad-spectrum penicillins and cephalosporins, such as carbenicillin or ceftriaxone, can access the cell walls of gram-negative species.

Antimicrobial Drugs That Affect Nucleic Acid Synthesis

As you learned in chapter 9, the metabolic pathway that generates DNA and RNA is a long, enzyme-catalyzed series of reactions. Like any complicated process, it is subject to breakdown at many different points along the way, and inhibition at any point in the sequence can block subsequent events. Antimicrobial drugs interfere with nucleic acid synthesis by blocking synthesis of nucleotides, inhibiting replication, or stopping transcription. Because functioning DNA and RNA are required for proper translation as well, the effects on protein metabolism can be far-reaching.

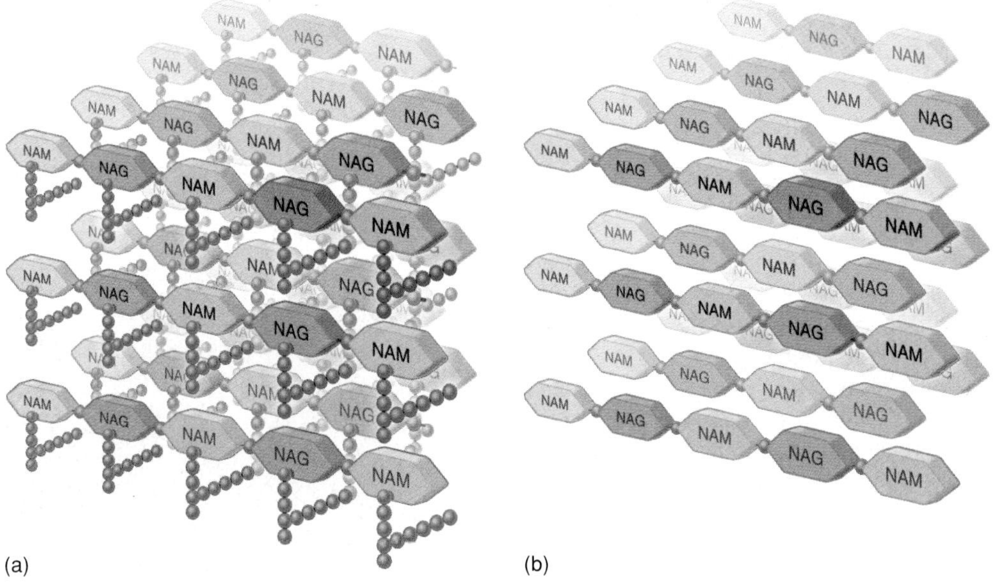

Figure 12.3 The mode of action of penicillins and cephalosporins on the bacterial cell wall.
(a) Intact peptidoglycan has chains of NAM (*N*-acetyl muramic acid) and NAG (*N*-acetyl glucosamine) glycans cross-linked by peptide bridges. **(b)** These two drugs block the peptidases that link the cross-bridges between NAMs, thereby greatly weakening the cell wall meshwork.

Other antimicrobials inhibit DNA synthesis. Chloroquine (an antimalarial drug) binds and cross-links the double helix. The newer broad-spectrum quinolones inhibit DNA unwinding enzymes or helicases, thereby stopping DNA transcription. Antiviral drugs that are analogs of purines and pyrimidines, including azidothymidine (AZT) and acyclovir, insert in the viral nucleic acid and block further replication.

Antimicrobial Drugs That Block Protein Synthesis

Most inhibitors of translation, or protein synthesis, react with the ribosome-mRNA complex. Although human cells also have ribosomes, the ribosomes of eukaryotes are different in size and structure from those of prokaryotes, so these antimicrobials usually have a selective action against bacteria. One potential therapeutic consequence of drugs that bind to the prokaryotic ribosome is the damage they can do to eukaryotic mitochondria, which contain a prokaryotic type of ribosome. Two possible targets of ribosomal inhibition are the 30S subunit and the 50S subunit **(figure 12.4).** Aminoglycosides (streptomycin, gentamicin, for example) insert on sites on the 30S subunit and cause the misreading of the mRNA, leading to abnormal proteins. Tetracyclines block the attachment of tRNA on the A acceptor site and effectively stop further protein synthesis. Other antibiotics attach to sites on the 50S subunit in a way that prevents the formation of peptide bonds (chloramphenicol) or inhibits translocation of the subunit during translation (erythromycin).

Antimicrobial Drugs That Disrupt Cell Membrane Function

A cell with a damaged membrane invariably dies from disruption in metabolism or lysis and does not even have to be actively dividing to be destroyed. The antibiotic classes that damage cell membranes have specificity for particular microbial groups, based on differences in the types of lipids in their cell membranes.

Polymyxins interact with membrane phospholipids, distort the cell surface, and cause leakage of proteins and nitrogen bases, particularly in gram-negative bacteria. The polyene antifungal antibiotics (amphotericin B and nystatin) form complexes with the sterols on fungal membranes; these complexes cause abnormal openings and seepage of small ions. Unfortunately, this selectivity is not exact, and the universal presence of membranes in microbial and animal cells alike means that most of these antibiotics can be quite toxic to humans.

Antimicrobial Drugs That Inhibit Folic Acid Synthesis

Sulfonamides and trimethoprim are drugs that act by mimicking the normal substrate of an enzyme in a process called **competitive inhibition.** They are supplied to the cell in high concentrations to ensure that a needed enzyme is constantly occupied with the **metabolic analog** rather than the true substrate of the enzyme. As the enzyme is no longer able to produce a needed product, cellular metabolism slows or stops. Sulfonamides and trimethoprim interfere with folate

INSIGHT 12.2 **Discovery**

A Quest for Designer Drugs

Once the first significant drug was developed, the world immediately witnessed a scientific scramble to find more antibiotics. This search was advanced on several fronts. Hundreds of investigators began the laborious task of screening samples from soil, dust, muddy lake sediments, rivers, estuaries, oceans, plant surfaces, compost heaps, sewage, skin, and even the hair and skin of animals for antibiotic-producing bacteria and fungi. This intense effort has paid off over the past 50 years, because more than 10,000 antibiotics have been discovered (although surprisingly, only a relatively small number have actually been used clinically). Finding a new antimicrobial substance is only a first step. The complete pathway of drug development from discovery to therapy takes between 12 and 24 years at a cost of billions of dollars.

Antibiotics are products of fermentation pathways that occur in many bacteria and fungi. The role of antibiotics in the lives of these microbes must be important because the genes for antibiotic production are preserved in evolution. Some experts theorize that antibiotic-releasing microorganisms can inhibit or destroy nearby competitors or predators; others propose that antibiotics play a part in spore formation. Whatever benefit the microbes derive, these compounds have been extremely profitable for humans. Every year, the pharmaceutical industry farms vast quantities of microorganisms and harvests their products to treat diseases caused by other microorganisms. Researchers have facilitated the work of nature by selecting mutant species that yield more abundant or useful products, by varying the growth medium, or by altering the procedures for large-scale industrial production (see chapter 25).

Another approach in the drug quest is to chemically manipulate molecules by adding or removing functional groups. Drugs produced in this way are designed to have advantages over other, related drugs. Using this **semisynthetic** method, a natural product of the microorganism is joined with various preselected functional groups. The antibiotic is reduced to its basic molecular framework (called the nucleus), and to this nucleus specially selected side chains (R groups) are added. A case in point is the metamorphosis of the semisynthetic penicillins. The nucleus is an inactive penicillin derivative called aminopenicillanic acid, which has an opening on the number 6 carbon for addition of R groups. A particular carboxylic acid (R group) added to this nucleus can "fine-tune" the penicillin, giving it special characteristics. For instance, some R groups will make the product resistant to penicillinase (methicillin), some confer a broader activity spectrum (ampicillin), and others make the product acid-resistant (penicillin V). Cephalosporins and tetracyclines also exist in several semisynthetic versions. The potential for using bioengineering techniques to design drugs seems almost limitless, and, indeed, several drugs have already been produced by manipulating the genes of antibiotic producers.

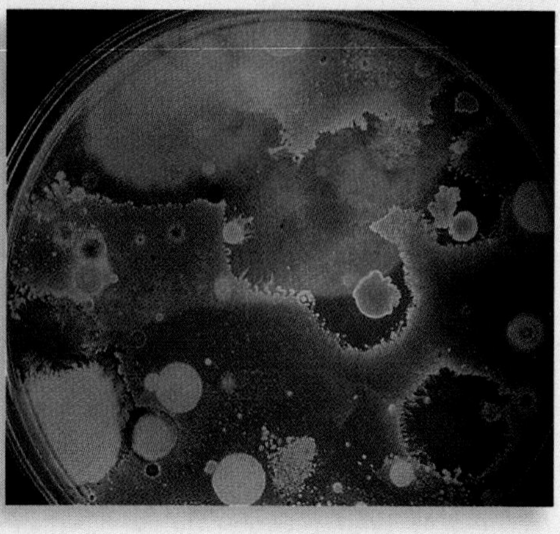

A plate with several discrete colonies of soil bacteria was sprayed with a culture of *Escherichia coli* and incubated. Zones of inhibition (clear areas with no growth) surrounding several colonies indicate species that produce antibiotics.

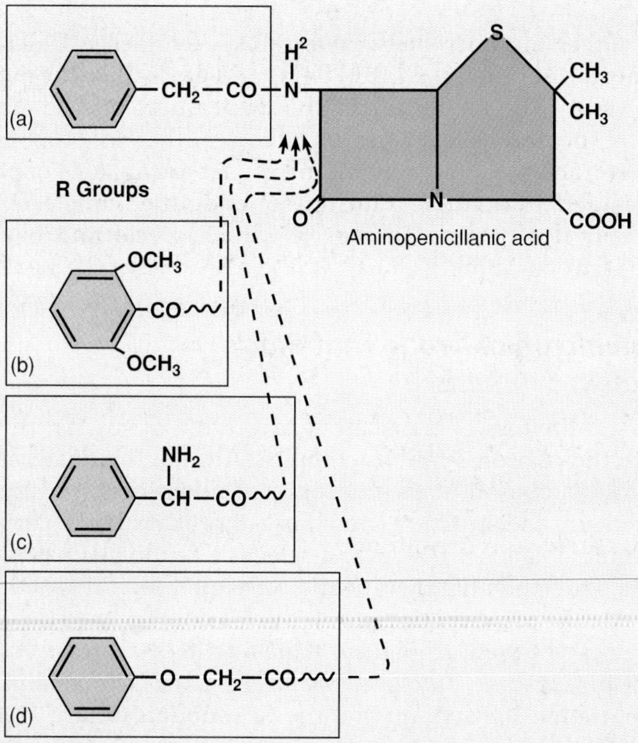

Synthesizing penicillins. **(a)** The original penicillin G molecule is a fermentation product of *Penicillium chrysogenum* that appears somewhat like a house with a removable patio on the left. This house without the patio is the basic nucleus called aminopenicillanic acid. **(b)–(d)** Various new fixtures (R groups) can be added to change the properties of the drug. These R groups will produce different penicillins: **(b)** methicillin, **(c)** ampicillin, and **(d)** penicillin V.

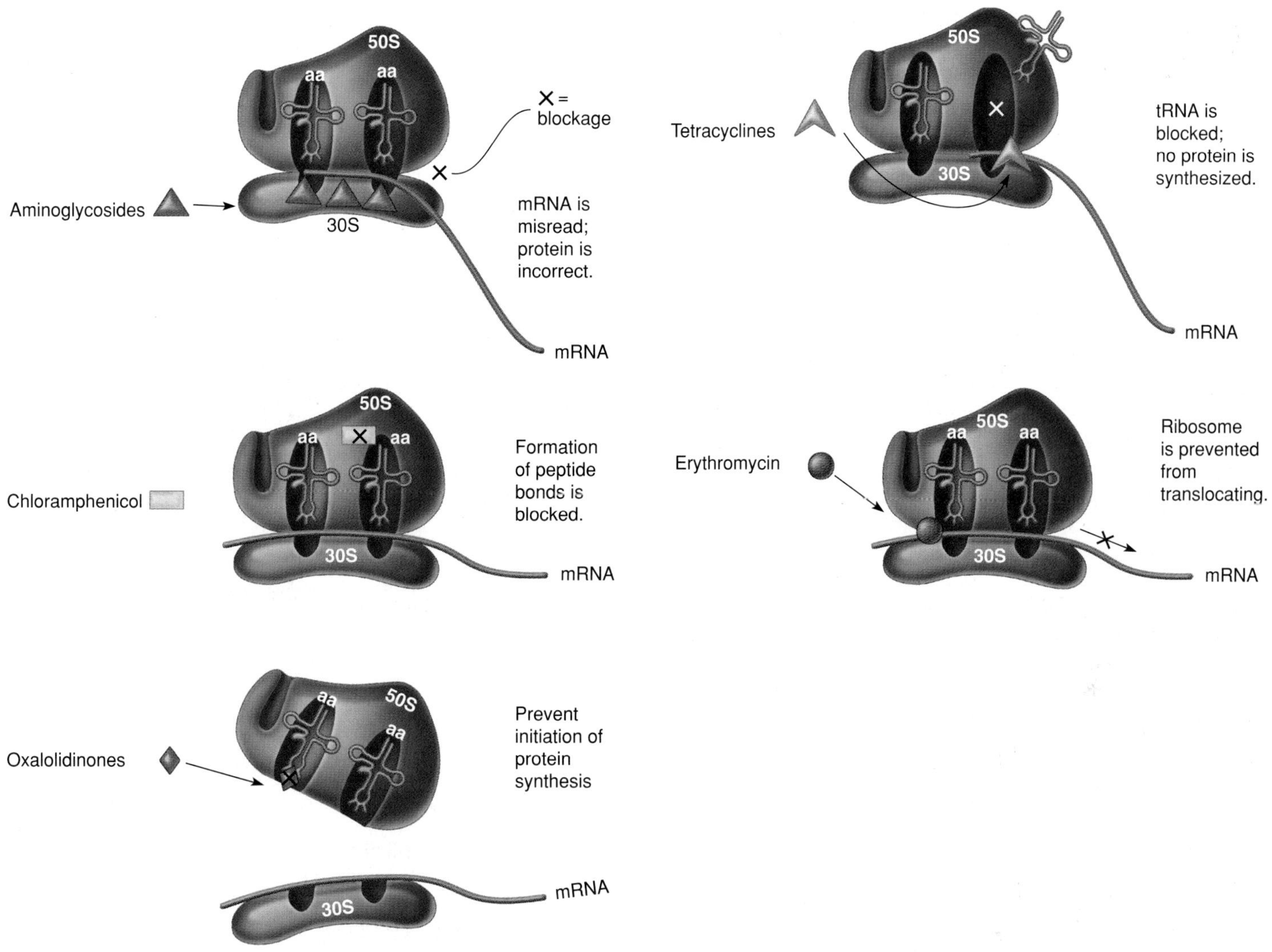

Figure 12.4 Sites of inhibition on the prokaryotic ribosome and major antibiotics that act on these sites. All have the general effect of blocking protein synthesis. Blockage actions are indicated by ×.

metabolism by blocking enzymes required for the synthesis of tetrahydrofolate, which is needed by bacterial cells for the synthesis of folic acid and the eventual production of DNA and RNA and amino acids. **Figure 12.5** illustrates sulfonamides' competition with PABA for the active site of the enzyme pteridine synthetase. Trimethoprim and sulfonamides are often given simultaneously to achieve a *synergistic effect,* which, in pharmacological terms, refers to an effect that is more than additive achieved by multiple drugs working together, thus requiring a lower dose of each.

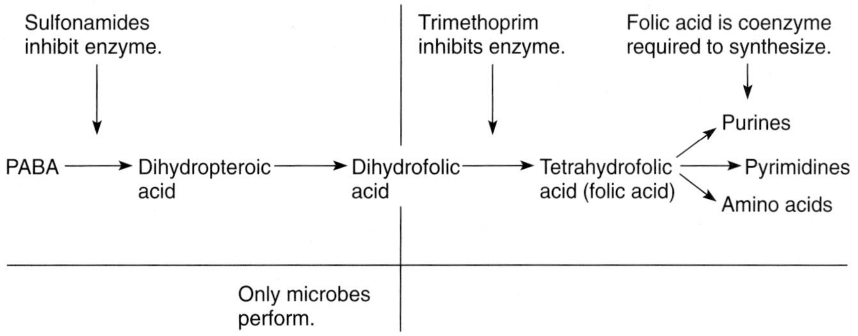

Figure 12.5 The action of sulfa drugs.
The metabolic pathway needed to synthesize tetrahydrofolic acid (THFA) contains two enzymes that are chemotherapeutic targets.

The selective toxicity of these compounds is explained by the fact that mammals derive folic acid from their diet and so do not possess this enzyme system. Therefore, inhibition of bacterial and protozoan parasites, which must synthesize folic acid, is easily accomplished while leaving the human host unaffected.

12.3 Survey of Major Antimicrobial Drug Groups

Scores of antimicrobial drugs are marketed in the United States. Although the medical and pharmaceutical literature contains a wide array of names for antimicrobials, most of them are variants of a small number of drug families. About 260 different antimicrobial drugs are currently classified in 20 drug families. Drug reference books may give the impression that there are 10 times that many because various drug companies assign different trade names to the very same generic drug. Ampicillin, for instance, is available under 50 different names. The largest number of antimicrobials is used to control bacterial infections, although we shall also consider a number of antifungal, antiviral, and antiprotozoan drugs. **Table 12.3** give examples of some major infectious agents, the diseases they cause, and the drugs of choice to treat them.

Antibacterial Drugs Targeting the Cell Wall

Penicillin and Its Relatives

The **penicillin** group of antibiotics, named for the parent compound, is a large, diverse group of compounds, most of which end in the suffix *-cillin.* Although penicillins could be completely synthesized in the laboratory from simple raw materials, it is more practical and economical to obtain natural penicillin through microbial fermentation. The natural product can then be used either in unmodified form or to make semisynthetic derivatives. *Penicillium chrysogenum* is the major source of the drug. All penicillins consist of three parts: a thiazolidine ring, a *beta-lactam* (bey′-tuh-lak′-tam) ring, and a variable side chain that dictates its microbicidal activity **(figure 12.6).**

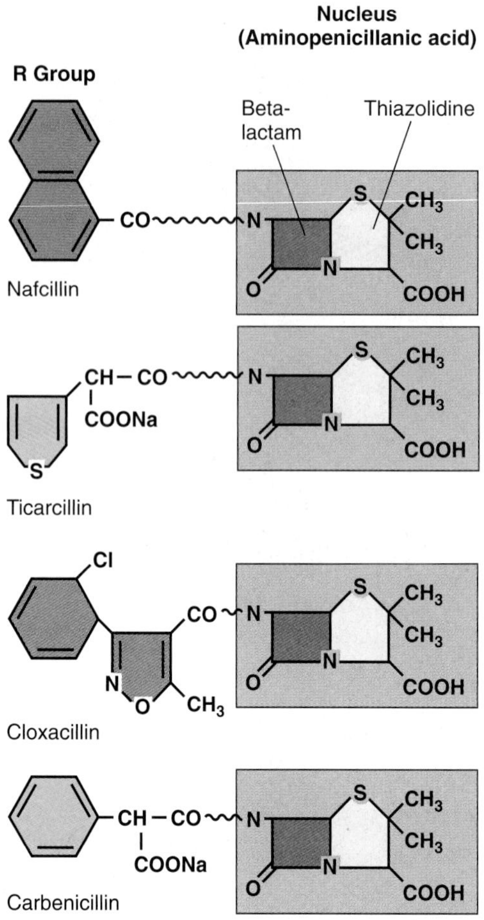

Figure 12.6 Chemical structure of penicillins. All penicillins contain a thiazolidine ring (yellow) and a beta-lactam ring (red), but each differs in the nature of the side chain (R group), which is also responsible for differences in biological activity.

Subgroups and Uses of Penicillins The characteristics of certain penicillin drugs are shown in **table 12.4.** Penicillins G and V are the most important natural forms. Penicillin is considered the drug of choice for infections by known sensitive, gram-positive cocci (most streptococci) and some gram-negative bacteria (meningococci and the syphilis spirochete).

Certain semisynthetic penicillins such as ampicillin, carbenicillin, and amoxicillin have broader spectra and thus can be used to treat infections by gram-negative enteric rods. Many bacteria produce enzymes that are capable of destroying the beta-lactam ring of penicillin. The enzymes are referred to as **penicillinases** or *beta-lactamases,* and they make the bacteria that possess them resistant to many penicillins. Penicillinase-resistant penicillins such as methicillin, nafcillin, and cloxacillin are useful in treating infections caused by some penicillinase-producing bacteria. Mezlocillin and azlocillin have such an extended spectrum that they can be substituted for combinations of antibiotics. All of the "-cillin" drugs are relatively mild and well tolerated because of their specific mode of action on cell walls (which humans lack). The primary problems in therapy include allergy, which is altogether different than toxicity, and resistant strains of pathogens. Clavulanic acid is a chemical that inhibits beta-lactamase enzymes, thereby increasing the longevity of beta-lactam antibiotics in the presence of penicillinase-producing bacteria. For this reason, clavulanic acid is often added to semisynthetic penicillins to augment their effectiveness. For example, clavamox is a combination of amoxicillin and clavulanate and is marketed under the trade name Augmentin. Zosyn is a similar combination of tazobactum, a beta-lactamase inhibitor, and piperacillin that is used for a wide variety of systemic infections.

TABLE 12.3 Selected Survey of Chemotherapeutic Agents in Infectious Diseases

Infectious Agent	Typical Infection	Drugs of Choice*
Bacteria		
Gram-positive cocci		
Staphylococcus aureus	Abscess, skin infections, toxic shock	Penicillins, vancomycin, cephalosporin
Streptococcus pyogenes	Strep throat, erysipelas, rheumatic fever	Penicillin, cephalosporin, erythromycin
Gram-positive rods		
Bacillus	Anthrax	Ciprofloxacin, doxycycline
Acid-fast rods		
Mycobacterium tuberculosis	Tuberculosis	(Isoniazid, rifampin, pyrazinamide),* ethambutol streptomycin
Gram-negative cocci		
Neisseria gonorrhoeae	Gonorrhea	Ceftriaxone, ciprofloxacin
Neisseria meningitidis	Meningitis	Penicillin G, cefotaxime
Gram-negative rods		
Escherichia coli	Sepsis, diarrhea, urinary tract infection	Cephalosporin
Haemophilus influenzae	Meningitis	Cefotaxime, cephtriaxone
Pseudomonas	Opportunistic lung and burn infections	Ticarcillin, aminoglycoside
Vibrio cholerae	Cholera	Tetracyclines, sulfamethoxazole-trimethoprim
Spirochetes		
Borrelia	Lyme disease	Doxycycline, amoxicillin
Treponema pallidum	Syphilis	Penicillin, tetracyclines
Rickettsia	Rocky Mountain spotted fever	Doxycycline
Chlamydia	Urethritis, vaginitis	Azithromycin, doxycycline
Fungi		
Systemic mycoses		
Aspergillus	Aspergillosis	Amphotericin B, azoles, flucytosine
Candida albicans	Candidiasis	Itraconazole, fluconazole
Cryptococcus neoformans	Cryptococcosis	Amphotericin B, fluconazole
Pneumocystis (carinii) jiroveci	Pneumonia (PCP)	Sulfamethoxazole-trimethoprim
Protozoa		
Giardia lamblia	Giardiasis	Quinacrine, metronidazole
Plasmodium	Malaria	Chloroquine, mefloquine
Toxoplasma gondii	Toxoplasmosis	Pyrimethamine, sulfadiazine
Trichomonas vaginalis	Trichomoniasis	Metronidazole
Helminths		
Cestodes	Tapeworm	Niclosamide, praziquantel
Various roundworm infections		Alebendazole
Viruses		
Herpesvirus	Genital herpes, oral herpes, shingles	Acyclovir, valacyclovir
HIV	AIDS	(AZT, protease inhibitors), ddI, ddC, d4T
Orthomyxovirus	Type A influenza	Amantadine, rimantidine

*More or less in order of preference.

**( ) Usually given in combination.

TABLE 12.4 Characteristics of Selected Penicillin Drugs

Name	Spectrum of Action	Uses, Advantages	Disadvantages
Penicillin G	Narrow	Best drug of choice when bacteria are sensitive; low cost; low toxicity	Can be hydrolyzed by penicillinase; allergies occur; requires injection
Penicillin V	Narrow	Good absorption from intestine; otherwise, similar to penicillin G	Hydrolysis by penicillinase; allergies
Oxacillin, dicloxacillin	Narrow	Not susceptible to penicillinase; good absorption	Allergies; expensive
Methicillin, nafcillin	Narrow	Not usually susceptible to penicillinase	Poor absorption; allergies; growing resistance
Ampicillin	Broad	Works on gram-negative bacilli	Can be hydrolyzed by penicillinase; allergies; only fair absorption
Amoxicillin	Broad	Gram-negative infections; good absorption	Hydrolysis by penicillinase; allergies
Carbenicillin	Broad	Same as ampicillin	Poor absorption; used only parenterally
Azlocillin, mezlocillin ticarcillin	Very broad	Effective against *Pseudomonas* species; low toxicity compared with aminoglycosides	Allergies, susceptible to many beta-lactamases

The Cephalosporin Group of Drugs

The **cephalosporins** are a newer group of antibiotics that currently account for a majority of all antibiotics administered. The first compounds in this group were isolated in the late 1940s from the mold *Cephalosporium acremonium*. Cephalosporins are similar to penicillins; they have a beta-lactam structure that can be synthetically altered **(figure 12.7)** and have a similar mode of action. The generic names of these compounds are often recognized by the presence of the root *cef, ceph,* or *kef* in their names.

Subgroups and Uses of Cephalosporins The cephalosporins are versatile. They are relatively broad-spectrum, resistant to most penicillinases, and cause fewer allergic reactions than penicillins. Although some cephalosporins are given orally, many are poorly absorbed from the intestine and must be administered **parenterally** (par-ehn′-tur-ah-lee), by injection into a muscle or a vein.

Four generations of cephalosporins exist, based upon their antibacterial activity. First-generation cephalosporins such as cephalothin and cefazolin are most effective against gram-positive cocci and a few gram-negative bacteria. Second-generation forms include cefaclor and cefonacid, which are more effective than the first-generation forms in treating infections by gram-negative bacteria such as *Enterobacter, Proteus,* and *Haemophilus.* Third-generation cephalosporins, such as cephalexin (Keflex) and cefotaxime, are broad-spectrum with especially well-developed activity against enteric bacteria that produce beta-lactamases. Ceftriaxone (rocephin) is a newer semisynthetic broad-spectrum drug for treating a wide variety of respiratory, skin, urinary, and nervous system infections. The fourth-generation cephalosporins include cefpirone and cefepime.

CASE FILE 12 *WRAP-UP*

Dana, the college student described on the first page of chapter 12, was sick with tuberculosis (TB). The nature of the symptoms plus positive chest X rays and Mantoux PPD (skin test) provided strong pieces of evidence that the tuberculosis microbe, *Mycobacterium tuberculosis,* was the causative agent.

Tuberculosis is a contagious disease, with a high mortality rate worldwide. It can remain dormant in a person for many years and become reactivated if the person's immune system becomes weakened. Dana's illness, with its seeming "recurrence" 2 months after the initial episode, was not a case of reactivation tuberculosis; the original case had never actually been cured. Because the physician did not recognize the infection as TB, inadequate antimicrobial therapy was prescribed. To make matters worse, Dana stopped taking her antibiotics as soon as she felt better.

Mycobacterium tuberculosis is particularly prone to becoming antibiotic resistant. Prescribing a single antibiotic to treat it often leads to situations like Dana's, because mutant bacteria that aren't affected by the drug eventually dominate the population in the lungs. TB infections are always treated with multiple drugs simultaneously for this reason. It is highly improbable that a single bacterial cell would mutate in the multiple genetic locations needed to render it resistant to more than one drug.

Other Beta-Lactam Antibiotics

Related antibiotics include imipenem, a broad-spectrum drug for infections with aerobic and anaerobic pathogens. It is

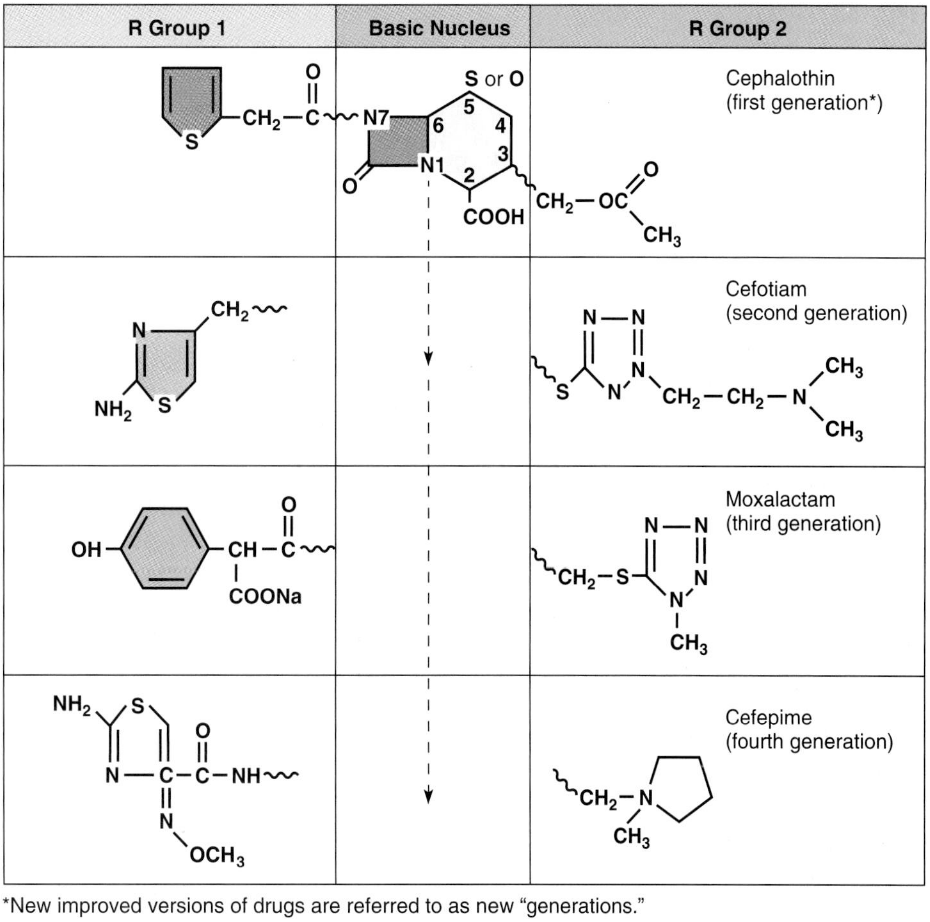

Figure 12.7 The structure of cephalosporins.
Like penicillin, they have a beta-lactam ring (red), but they have a different main ring (yellow). However, unlike penicillins, they have two sites for placement of R groups (at positions 3 and 7). This makes possible several generations of molecules with greater versatility in function and complexity in structure.

active in very low concentrations and can be taken by mouth with few side effects. Aztreonam, isolated from the bacterium *Chromobacterium violaceum*, is a newer narrow-spectrum drug for treating pneumonia, septicemia, and urinary tract infections by gram-negative aerobic bacilli. Aztreonam is especially useful when treating persons who are allergic to penicillin. Because of similarities in their chemical structure, allergies to penicillin often are accompanied by allergies to cephalosporins and carboxypenems (of which imipenem is a member). The structure of aztreonam is chemically distinct so that persons with allergies to penicillin are not usually adversely affected by treatment with aztreonam.

Other Drugs Targeting the Cell Wall

Bacitracin is a narrow-spectrum antibiotic produced by a strain of the bacterium *Bacillus subtilis*. Since it was first isolated, its greatest claim to fame has been as a major ingredient in a common drugstore antibiotic ointment (Neosporin) for combating superficial skin infections by streptococci and staphylococci. For this purpose, it is usually combined with neomycin (an aminoglycoside) and polymyxin.

Isoniazid (INH) is bactericidal to *Mycobacterium tuberculosis* but only against growing cells. Although still in use, it has been largely supplanted by rifampicin. Oral doses are indicated for both active tuberculosis and prophylaxis in cases of a positive TB test.

Vancomycin is a narrow-spectrum antibiotic most effective in treating staphylococcal infections in cases of penicillin and methicillin resistance or in patients with an allergy to penicillins. It has also been chosen to treat *Clostridium* infections in children and endocarditis (infection of the lining of the heart) caused by *Enterococcus faecalis*. Because it is very toxic and hard to administer, vancomycin is usually restricted to the most serious, life-threatening conditions.

Fosfomycin trimethamine is a phosphoric acid agent effective as alternate treatment for urinary tract infections caused by enteric bacteria. It works by inhibiting an enzyme necessary for cell wall synthesis.

Antibacterial Drugs Targeting Protein Synthesis

The Aminoglycoside Drugs

Antibiotics composed of one or more amino sugars and an aminocyclitol (6-carbon) ring are referred to as **aminoglycosides (figure 12.8).** These complex compounds are exclusively the products of various species of soil **actinomycetes** in the genera *Streptomyces* **(figure 12.9)** and *Micromonospora.*

Subgroups and Uses of Aminoglycosides The aminoglycosides have a relatively broad antimicrobial spectrum because they inhibit protein synthesis. They are especially useful in treating infections caused by aerobic gram-negative rods and certain gram-positive bacteria. Streptomycin is among the oldest of the drugs and has gradually been replaced by newer forms with less mammalian toxicity. It is still the antibiotic of choice for treating bubonic plague and tularemia and is considered a good antituberculosis agent. Gentamicin is less toxic and is widely administered for infections caused by gram-negative rods (*Escherichia, Pseudomonas, Salmonella,* and *Shigella*). Two relatively new aminoglycosides, tobramycin and amikacin, are also used for gram-negative infections and have largely replaced kanamycin.

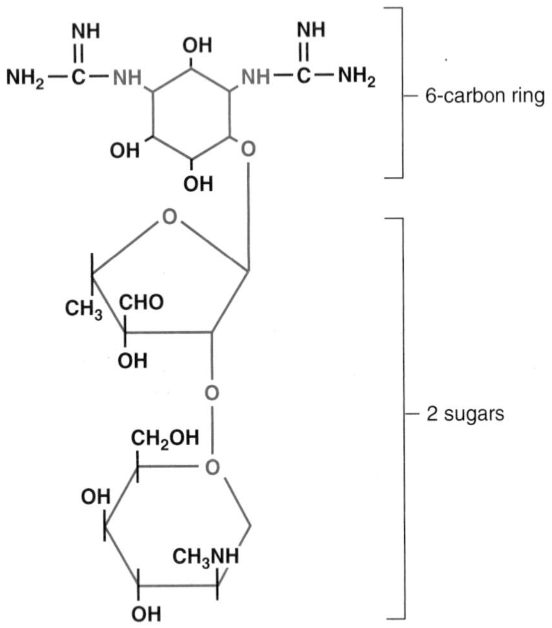

Figure 12.8 The structure of streptomycin.
Colored portions of the molecule show the general arrangement of an aminoglycoside.

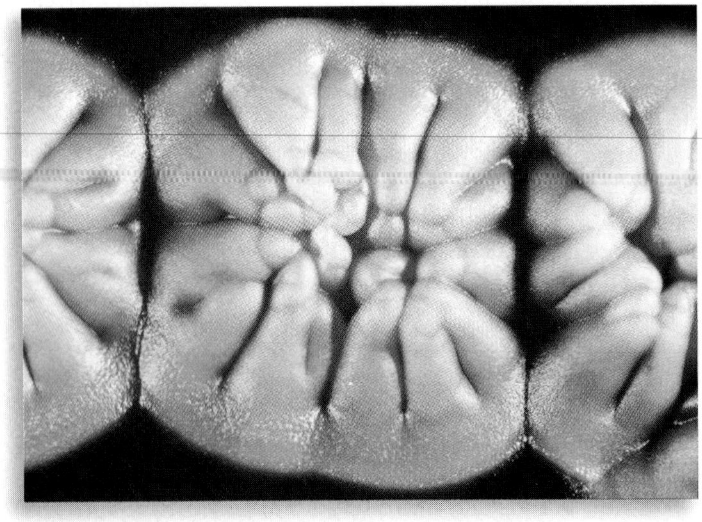

Figure 12.9 A colony of *Streptomyces*, one of nature's most prolific antibiotic producers.

Tetracycline Antibiotics

In 1948, a colony of *Streptomyces* isolated from a soil sample gave off a substance, aureomycin, with strong antimicrobial properties. This antibiotic was used to synthesize its relatives terramycin and tetracycline. These natural parent compounds and semisynthetic derivatives are known as the **tetracyclines (figure 12.10*a*).** Their action of binding to ribosomes and blocking protein synthesis accounts for the broad-spectrum effects in the group.

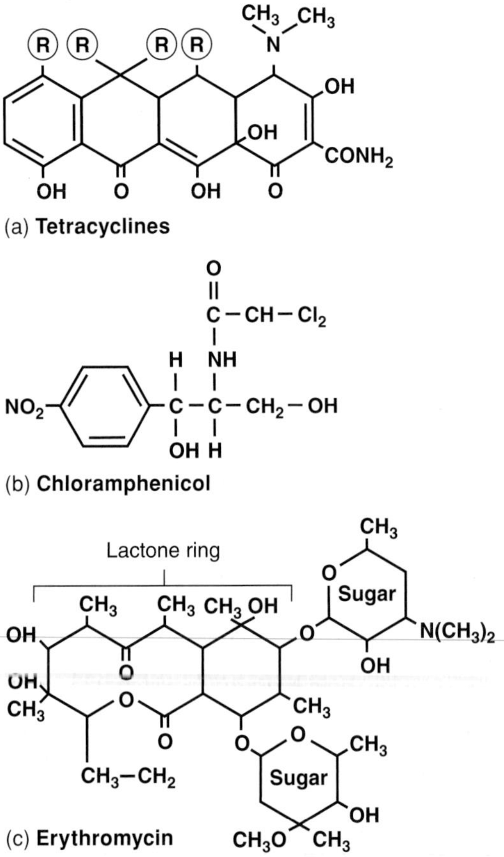

Figure 12.10 Structures of three broad-spectrum antibiotics.
(a) Tetracyclines. These are named for their regular group of four rings. The several types vary in structure and activity by substitution at the four R groups. **(b)** Chloramphenicol. **(c)** Erythromycin, an example of a macrolide drug. Its central feature is a large lactone ring to which two hexose sugars are attached.

Subgroups and Uses of Tetracyclines The scope of microorganisms inhibited by tetracyclines includes gram-positive and gram-negative rods and cocci, aerobic and anaerobic bacteria, mycoplasmas, rickettsias, and spirochetes. Tetracycline compounds such as doxycycline and minocycline are administered orally to treat several sexually transmitted diseases, Rocky Mountain spotted fever, Lyme disease, typhus, *Mycoplasma* pneumonia, cholera, leptospirosis, acne, and even some protozoan infections. Although generic tetracycline is low in cost and easy to administer, its side effects—namely, gastrointestinal disruption due to changes in the normal biota of the gastrointestinal tract and deposition in hard tissues—can limit its use (see table 12.6).

Chloramphenicol

Originally isolated in the late 1940s from *Streptomyces venezuelae*, **chloramphenicol** is a potent broad-spectrum antibiotic with a unique nitrobenzene structure **(figure 12.10*b*)**. Its primary effect on cells is to block peptide bond formation and protein synthesis. It is one type of antibiotic that is no longer derived from the natural source but is entirely synthesized through chemical processes. Although this drug is fully as broad-spectrum as the tetracyclines, it is so toxic to human cells that its uses are restricted. A small number of people undergoing long-term therapy with this drug incur irreversible damage to the bone marrow that usually results in a fatal form of aplastic anemia.[1] Its administration is now limited to typhoid fever and brain abscesses, for which an alternative therapy is not available. Chloramphenicol should never be given in large doses repeatedly over a long time period.

Erythromycin and Clindamycin

Erythromycin is a macrolide[2] antibiotic first isolated in 1952 from a strain of *Streptomyces*. Its structure consists of a large lactone ring with sugars attached **(figure 12.10*c*)**. This drug is relatively broad-spectrum and of fairly low toxicity. Its mode of action is to block protein synthesis by attaching to the ribosome. It is administered orally for *Mycoplasma* pneumonia, legionellosis, *Chlamydia* infections, pertussis, and diphtheria and as a prophylactic drug prior to intestinal surgery. It also offers a useful substitute for dealing with penicillin-resistant streptococci and gonococci and for treating syphilis and acne. Newer semisynthetic macrolides include *clarithromycin* and *azithromycin*. Both drugs are useful for middle ear, respiratory, and skin infections and have also been approved for *Mycobacterium* (MAC) infections in AIDS patients. Clarithromycin has additional applications in controlling infectious stomach ulcers (see chapter 22).

1. A failure of the blood-producing tissue that results in very low levels of blood cells.
2. Macrolide antibiotics get their name from the type of chemical ring structure (macrolide ring) they all possess.

Clindamycin is a broad-spectrum antibiotic derived from the less effective lincomycin. The tendency of clindamycin to cause adverse reactions in the gastrointestinal tract limits its applications to

1. serious infections in the large intestine and abdomen due to anaerobic bacteria (*Bacteroides* and *Clostridium*) that are unresponsive to other antibiotics,
2. infections with penicillin-resistant staphylococci, and
3. acne medications applied to the skin.

In 2004, the FDA approved the first representative of a new class of drugs called ketolides, which are similar to macrolides like erythromycin but have a different ring structure. The new drug, called Ketek, is to be used for respiratory tract infections that are suspected to be caused by antibiotic-resistant bacteria such as *Streptococcus pneumoniae*.

Synercid and Oxazolidones

Synercid is a combined antibiotic from the streptogramin group of drugs. It is effective against *Staphylococcus* and *Enterococcus* species that cause endocarditis and surgical infections, and against resistant strains of *Streptococcus*. It is one of the main choices when other drugs are ineffective due to resistance. Synercid works by binding to sites on the 50S ribosome, inhibiting translation.

A new class of synthetic antibacterial drugs, oxazolidinones, was recently developed, and the first member of that class, linezolid, was approved for use in 2000 by the FDA. These drugs work by a completely novel mechanism, inhibiting the initiation of protein synthesis. Because this class of drug is not found in nature, it is hoped that resistance among bacteria will be slow to develop. Linezolid (under the name Zyvox) is used to treat infections caused by two of the most difficult clinical pathogens, methicillin-resistant *Staphylococcus aureus* (MRSA) and vancomycin-resistant *Enterococcus* (VRE).

Antibacterial Drugs Targeting Folic Acid Synthesis

The Sulfonamides, Trimethoprim, and Sulfones

The very first modern antimicrobial drugs were the **sulfonamides**, or sulfa drugs, named for sulfanilamide, an early form of the drug **(figure 12.11)**. They are synthetic and do not originate from bacteria or fungi. Although thousands of sulfonamides have been formulated, only a few have gained any importance in chemotherapy. Because of its solubility, sulfisoxazole is the best agent for treating shigellosis, acute urinary tract infections, and certain protozoan infections. Silver sulfadiazine ointment and solution are prescribed for treatment of burns and eye infections. Another drug, trimethoprim (Septra, Bactrim), inhibits the enzymatic step immediately following the step inhibited by sulfonamides in the synthesis of folic acid. Because of this, trimethoprim is often given in combination with sulfamethoxazole to take advantage of the synergistic effect of the two drugs. This combination is one of the primary treatments for *Pneumocystis (carinii) jiroveci* pneumonia (PCP) in AIDS patients.

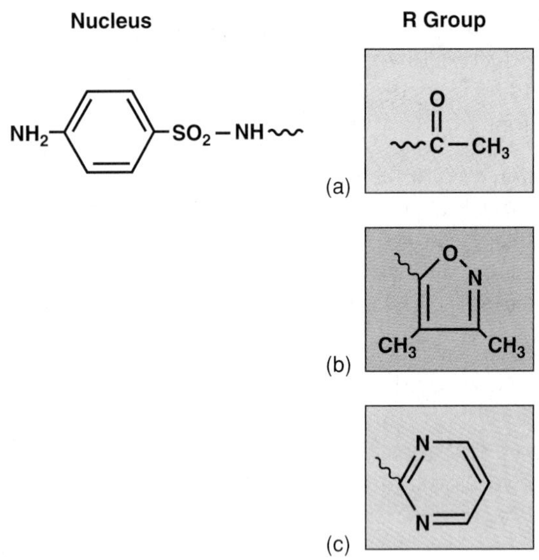

Figure 12.11 The structures of some sulfonamides. **(a)** Sulfacetamide, **(b)** sulfadiazine, and **(c)** sulfisoxazole.

Sulfones are compounds chemically related to the sulfonamides but lacking their broad-spectrum effects. This lack does not diminish their importance as key drugs in treating Hansen's disease (leprosy). The most active form is dapsone, usually given in combination with rifampin and clofazamine (an antibacterial dye) over long periods.

Antibacterial Drugs Targeting DNA or RNA

Even though nucleic acids in prokaryotes and humans are chemically similar, DNA and RNA have proved to be useful targets for antimicrobials. Much excitement was generated by a new class of synthetic drugs chemically related to quinine called **fluoroquinolones.** These drugs exhibit several ideal traits, including high potency and broad spectrum. Even in minimal concentrations, quinolones inhibit a wide variety of gram-positive and gram-negative bacterial species. In addition, they are readily absorbed from the intestine. The principal quinolones, norfloxacin and ciprofloxacin, have been successful in therapy for urinary tract infections, sexually transmitted diseases, gastrointestinal infections, osteomyelitis, respiratory infections, and soft tissue infections. Newer drugs in this category are sparfloxacin and levofloxacin. These agents are especially recommended for pneumonia, bronchitis, and sinusitis. Ciprofloxacin received a great deal of publicity recently as the drug of choice for treating anthrax. Shortly after the first attacks, however, the Centers for Disease Control and Prevention changed their recommendation from ciprofloxacin to doxycycline (a protein synthesis inhibitor). The change in preferred drug was made not because ciprofloxacin was ineffective but rather to prevent the emergence of ciprofloxacin-resistant bacteria (see Insight 12.4). Side effects that limit the use of quinolones include seizures and other brain disturbances.

Another product of the genus *Streptomyces* is rifamycin, which is altered chemically into rifampin. It is somewhat limited in spectrum because the molecule cannot pass through the cell envelope of many gram-negative bacilli. It is mainly used to treat infections by several gram-positive rods and cocci and a few gram-negative bacteria. Rifampin figures most prominently in treating mycobacterial infections, especially tuberculosis and leprosy, but it is usually given in combination with other drugs to prevent development of resistance. Rifampin is also recommended for prophylaxis in *Neisseria meningitidis* carriers and their contacts, and it is occasionally used to treat *Legionella, Brucella,* and *Staphylococcus* infections.

Antibacterial Drugs Targeting Cell Membranes

Every cell has a membrane. Some drugs target membranes, but they are not usually first-choice antimicrobials except in a few circumstances. *Bacillus polymyxa* is the source of the **polymyxins,** narrow-spectrum peptide antibiotics with a unique fatty acid component that contributes to their detergent activity. Only two polymyxins—B and E (also known as colistin)—have any routine applications, and even these are limited by their toxicity to the kidney. Either drug can be indicated to treat drug-resistant *Pseudomonas aeruginosa* and severe urinary tract infections caused by other gram-negative rods.

Daptomycin is a lipopeptide made by *Streptomyces.* It is most active against gram-positive bacteria, acting to disrupt multiple aspects of membrane function. Many experts are urging physicians to use these medications only when no other drugs are available to slow the development of drug resistance.

Agents to Treat Fungal Infections

Because the cells of fungi are eukaryotic, they present special problems in chemotherapy. For one, the great majority of chemotherapeutic drugs are designed to act on bacteria and are generally ineffective in combating fungal infections. For another, the similarities between fungal and human cells often mean that drugs toxic to fungal cells are also capable of harming human tissues. A few agents with special antifungal properties have been developed for treating systemic and superficial fungal infections. Four main drug groups currently in use are the macrolide polyene antibiotics, griseofulvin, synthetic azoles, and flucytosine **(figure 12.12).**

Polyenes bind to fungal membranes and cause loss of selective permeability. They are specific for fungal membranes because fungal membranes contain a particular sterol component called ergosterol, while human membranes do not. The toxicity of polyenes is not completely selective, however, because mammalian cell membranes contain compounds similar to ergosterol that bind polyenes to a small extent.

Macrolide polyenes, represented by amphotericin B (named for its acidic and basic—amphoteric—properties)

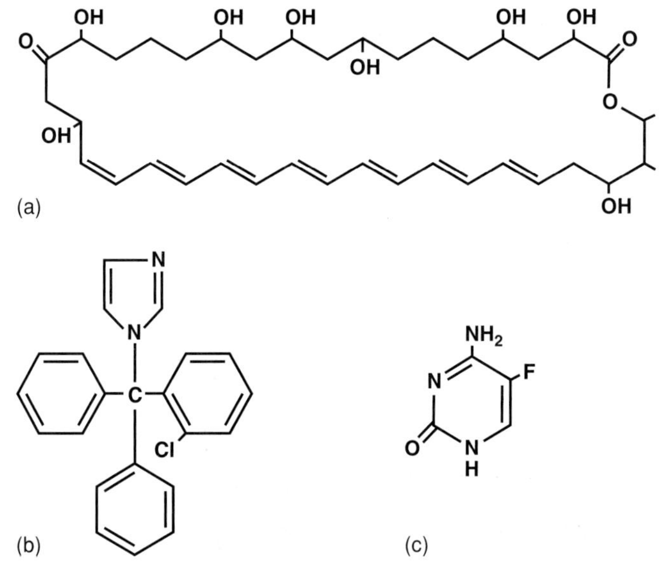

Figure 12.12 Some antifungal drug structures.
(a) Polyenes. The example shown is amphotericin B, a complex steroidal antibiotic that inserts into fungal cell membranes. **(b)** Clotrimazole, one of the azoles. **(c)** Flucytosine, a structural analog of cystosine that contains fluoride.

and nystatin (for New York State, where it was discovered), have a structure that mimics the lipids in some cell membranes. Amphotericin B is by far the most versatile and effective of all antifungals. Not only does it work on most fungal infections, including skin and mucous membrane lesions caused by *Candida albicans,* but it is one of the few drugs that can be injected to treat systemic fungal infections such as histoplasmosis and cryptococcus meningitis. Nystatin is used only topically or orally to treat candidiasis of the skin and mucous membranes, but it is not useful for subcutaneous or systemic fungal infections or for ringworm.

Griseofulvin is an antifungal product especially active in certain dermatophyte infections such as athlete's foot. The drug is deposited in the epidermis, nails, and hair, where it inhibits fungal growth. Because complete eradication requires several months and griseofulvin is relatively nephrotoxic, this therapy is given only for the most stubborn cases.

The **azoles** are broad-spectrum antifungal agents with a complex ringed structure. The most effective drugs are ketoconazole, fluconazole, clotrimazole, and miconazole. Ketoconazole is used orally and topically for cutaneous mycoses, vaginal and oral candidiasis, and some systemic mycoses. Fluconazole can be used in selected patients for AIDS-related mycoses such as aspergillosis and cryptococcus meningitis. Clotrimazole and miconazole are used mainly as topical ointments for infections in the skin, mouth, and vagina.

Flucytosine is an analog of the nucelotide cytosine that has antifungal properties. It is rapidly absorbed after oral therapy, and it is readily dissolved in the blood and cerebrospinal fluid. Alone, it can be used to treat certain cutaneous mycoses. Many fungi are resistant to flucytosine, so it is usually combined with amphotericin B to effectively treat systemic mycoses.

Antiparasitic Chemotherapy

The enormous diversity among protozoan and helminth parasites and their corresponding therapies reaches far beyond the scope of this textbook; however, a few of the more common drugs are surveyed here and described again for particular diseases in the organ systems chapters. Presently, a small number of approved and experimental drugs are used to treat malaria, leishmaniasis, trypanosomiasis, amoebic dysentery, and helminth infections, but the need for new and better drugs has spurred considerable research in this area.

Antimalarial Drugs: Quinine and Its Relatives

Quinine, extracted from the bark of the cinchona tree, was the principal treatment for malaria for hundreds of years, but it has been replaced by the synthesized quinolines, mainly chloroquine and primaquine, which have less toxicity to humans. Because there are several species of *Plasmodium* (the malaria parasite) and many stages in its life cycle, no single drug is universally effective for every species and stage, and each drug is restricted in application. For instance, primaquine eliminates the liver phase of infection, and chloroquine suppresses acute attacks associated with infection of red blood cells. Chloroquine is taken alone for prophylaxis and suppression of acute forms of malaria. Primiquine is administered to patients with relapsing cases of malaria. Mefloquine is a semisynthetic analog of quinine used to treat infections caused by chloroquine-resistant strains of *Plasmodium.*

Chemotherapy for Other Protozoan Infections

A widely used amoebicide, metronidazole (Flagyl), is effective in treating mild and severe intestinal infections and hepatic disease caused by *Entamoeba histolytica.* Given orally, it also has applications for infections by *Giardia lamblia* and *Trichomonas vaginalis* (described in chapters 22 and 23, respectively). Other drugs with antiprotozoan activities are quinicrine (a quinine-based drug), sulfonamides, and tetracyclines.

Antihelminthic Drug Therapy

Treating helminthic infections has been one of the most difficult and challenging of all chemotherapeutic tasks. Flukes, tapeworms, and roundworms are much larger parasites than other microorganisms and, being animals, have greater similarities to human physiology. Also, the usual strategy of using drugs to block their reproduction is usually not successful in eradicating the adult worms. The most effective drugs immobilize, disintegrate, or inhibit the metabolism of all stages of the life cycle.

Mebendazole and thiabendazole are broad-spectrum antiparasitic drugs used in several roundworm intestinal infestations. These drugs work locally in the intestine to inhibit the function of the microtubules of worms, eggs, and larvae, which interferes with their glucose utilization and disables them. The compounds pyrantel and piperazine paralyze the muscles of intestinal roundworms. Consequently, the worms are unable to maintain their grip on the intestinal wall and are expelled along with the feces by the normal peristaltic action of the bowel. Two newer antihelminthic drugs are praziquantel, a treatment for various tapeworm and fluke infections, and ivermectin, a veterinary drug now used for strongyloidiasis and oncocercosis in humans. Helminthic diseases are described in chapter 22 because these organisms spend at least some part of their life cycles in the digestive tract.

Antiviral Chemotherapeutic Agents

The chemotherapeutic treatment of viral infections presents unique problems. With viruses, we are dealing with an infectious agent that relies upon the host cell for the vast majority of its metabolic functions. Disrupting viral metabolism requires that we disrupt the metabolism of the host cell to a much greater extent than is desirable. Put another way, selective toxicity with regard to viral infection is almost impossible to achieve because a single metabolic system is responsible for the well-being of both virus and host. Although viral diseases such as measles, mumps, and hepatitis are routinely prevented by the use of effective vaccinations, epidemics of AIDS, influenza, and even the common cold attest to the need for more effective medications for the treatment of viral pathogens.

Over the last few years, several antiviral drugs have been developed that target specific points in the infectious cycle of viruses. Although antiviral drugs are certainly a welcome discovery, the chemotherapeutic treatment of viruses is still in its infancy and most antiviral compounds are fairly limited in their usefulness.

Most compounds have their effects on some stage of the virus cycle. Three major modes of action are

1. barring penetration of the virus into the host cell,
2. blocking the transcription and translation of viral molecules, and
3. preventing the maturation of viral particles.

Table 12.5 presents a comprehensive overview of the most widely used antiviral drugs. Hundreds of new drugs are in development. The following paragraphs provide some additional detail about the principles in table 12.5. Although antiviral drugs protect uninfected cells by keeping viruses from being synthesized and released, most are unable to destroy extracellular viruses or those in a latent state.

Fuzeon is one of the newer anti-HIV drugs; it keeps the virus from attaching to its cellular receptor and thereby prevents the initial fusion of HIV to the host cell. ***Amantadine*** and its relative, rimantidine, are restricted almost exclusively to treating infections by influenza A virus. Relenza and Tamiflu medications are effective treatments for influenza A and B and useful prophylactics as well. Because one action of these drugs is to inhibit the fusion and uncoating of the virus, they must be given rather early in an infection.

Several antiviral agents mimic the structure of nucleotides and compete for sites on replicating DNA. The incorporation of these synthetic nucleotides inhibits further DNA synthesis. **Acyclovir** (Zovirax) and its relatives are synthetic purine compounds that block DNA synthesis in a small group of viruses, particularly the herpesviruses (see chapters 18 and 23). In the topical form, they are most effective in controlling the primary attack of facial or genital herpes. Intravenous or oral acyclovir therapy can reduce the severity of primary and recurrent genital herpes episodes. Some newer relatives (valacyclovir) are more effective and require fewer doses. Famciclovir is used to treat shingles and chickenpox caused by the herpes zoster virus, and gancyclovir is approved to treat cytomegalovirus infections of the eye. An interesting aspect of some of these antiviral agents (specifically valacyclovir and famciclovir) is that they are activated by an enzyme encoded by the virus itself, activating the drug only in virally infected cells. The enzyme thymidine kinase is used by the virus to process nucleosides before incorporating them into viral RNA or DNA. When the inactive drug enters a virally infected cell, it is activated by the virus' thymidine kinase to produce a working antiviral agent. In cells without viruses, the drug is never activated and DNA replication is allowed to continue unabated.

HIV is classified as a retrovirus, meaning it carries its genetic information in the form of RNA rather than DNA (HIV and AIDS are discussed in chapter 20). Upon infection, the RNA genome is used as a template by the enzyme **reverse transcriptase** to produce a DNA copy of the virus' genetic information. Because this particular reaction is not seen outside of the retroviruses, it offers two ideal targets for chemotherapy. The first is interfering with the synthesis of the new DNA strand, which is accomplished using *nucleoside reverse transcriptase inhibitors* (nucleotide analogs), while the second involves interfering with the action of the enzyme responsible for the synthesis, which is accomplished using *nonnucleoside reverse transcriptase inhibitors.*

Azidothymidine (AZT or zidovudine) is a thymine analog that exerts its effect by incorporating itself into the growing DNA chain of HIV and terminating synthesis, in a manner analogous to that seen with acyclovir. AZT is used at all stages of HIV infection, including prophylactically with people accidentally exposed to blood or other body fluids.

Nonnucleoside reverse transcriptase inhibitors (such as nevirapine) accomplish the same goal (preventing reverse transcription of the HIV genome) by binding to the reverse transcriptase enzyme itself, inhibiting its ability to synthesize DNA.

Assembly and release of mature viral particles are also targeted in HIV through the use of protease inhibitors. These drugs (indinavir, saquinavir, nelfinavir, crixivan), usually used in combination with nucleotide analogs and reverse transcriptase inhibitors, have been shown to reduce the HIV load to undetectable levels by preventing the maturation of virus particles in the cell. Refer to table 12.5 for a summary of HIV drug mechanisms and see chapter 20 for further coverage of this topic.

A sensible alternative to artificial drugs has been a human-based substance, **interferon (IFN).** Interferon is a

TABLE 12.5 Actions of Antiviral Drugs

Mode of Action	Examples	Effects of Drug
Inhibition of Virus Entry:		
Receptor/Fusion/ Uncoating Inhibitors	Enfuvirtide (Fuzeon)	Blocks **HIV** infection by preventing the binding of viral GP-41 receptors to cell receptor ①, thereby preventing fusion of virus with cell
	Amantidine and its relatives, zanamivir (Relenza), oseltamivir (Tamiflu)	Block entry of **influenza virus** by interfering with fusion of virus with cell membrane (also release); stop the action of influenza neuraminidase, required for entry of virus into cell (also assembly) ② ③
Inhibition of Nucleic Acid Synthesis		
	Acyclovir (Zovirax), other "cyclovirs," vidarabine	Purine analogs that terminate DNA replication in **herpesviruses** ④
	Ribavirin	Purine analog, used for **respiratory syncytial virus (RSV)** and some **hemorrhagic fever viruses**
	Zidovudine (AZT), abacavir, lamivudine (3T3), didanosine (ddI), zalcitabine (ddC), and stavudine (d4T)	Nucleotide analog reverse transcriptase (RT) inhibitors; stop the action of RT in **HIV,** blocking viral DNA production ⑤
	Nevirapine, efavirenz, delaviridine	Nonnucleotide analog reverse transcriptase inhibitors; attach to **HIV** RT binding site, stopping its action ⑥
Inhibition of Viral Assembly/Release		
	Indinavir, saquinavir, nelfinavir, crixivan	Protease inhibitors; insert into **HIV** protease, stopping its action and resulting in inactive noninfectious viruses ⑦

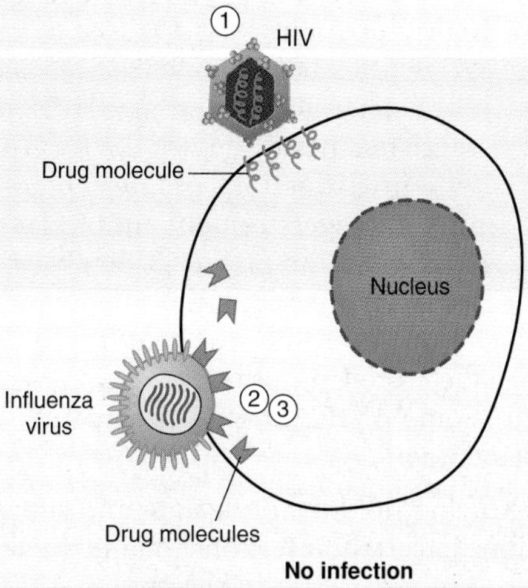

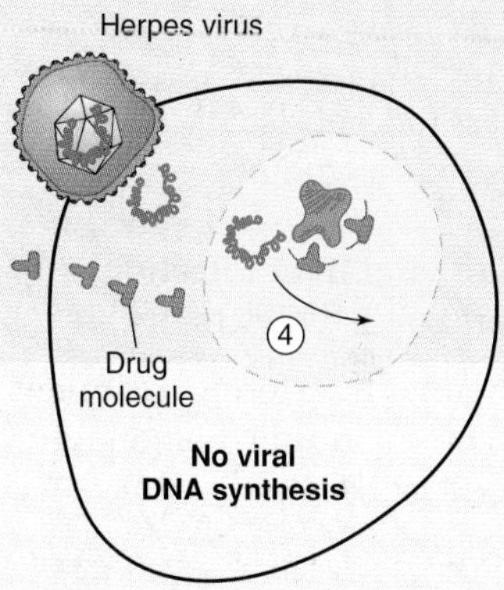

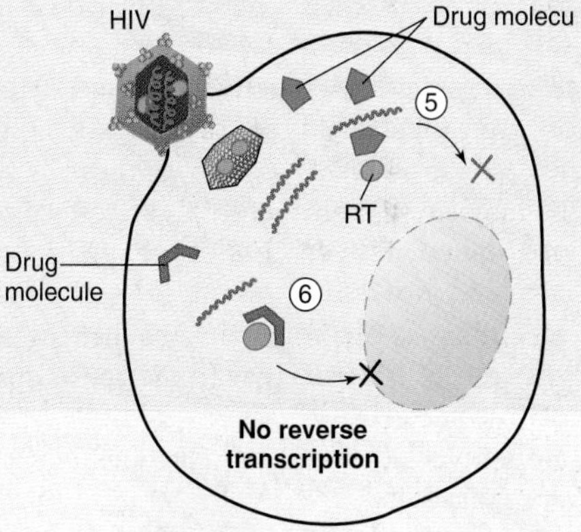

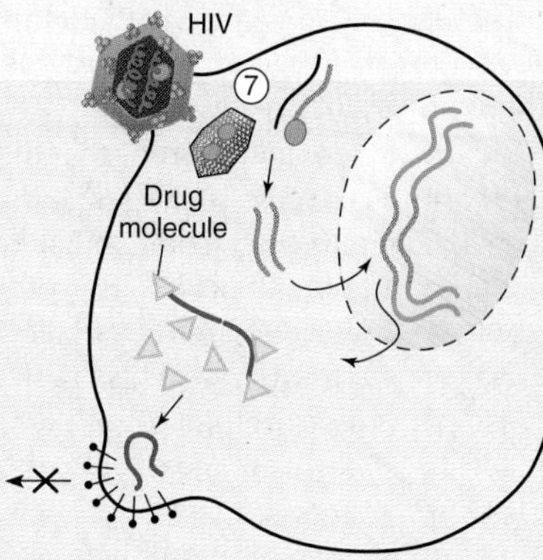

glycoprotein produced primarily by fibroblasts and leukocytes in response to various immune stimuli. It has numerous biological activities, including antiviral and anticancer properties. Studies have shown that it is a versatile part of animal host defenses, having a major role in natural immunities. (Its mechanism is discussed in chapter 14.)

The first investigations of interferon's antiviral activity were limited by the extremely minute quantities that could be extracted from human blood. Several types of interferon are currently produced by the recombinant DNA technology techniques outlined in chapter 10. Extensive clinical trials have tested its effectiveness in viral infections and cancer. Some of the known therapeutic benefits of interferon include:

1. reducing the time of healing and some of the complications in certain infections (mainly of herpesviruses);
2. preventing or reducing some symptoms of cold and papillomaviruses (warts);
3. slowing the progress of certain cancers, including bone cancer and cervical cancer, and certain leukemias and lymphomas; and
4. treating a rare cancer called hairy-cell leukemia, hepatitis C (a viral liver infection), genital warts, and Kaposi's sarcoma in AIDS patients.

Unfortunately, interferon treatment often results in serious side effects, including personality changes and decreased white blood cell counts.

CHECKPOINT

- Antimicrobials are classified into +/−20 major drug families, based on their chemical composition, source or origin, and their site of action.
- The majority of antimicrobials are effective against bacteria, but a limited number are effective against protozoa, helminths, fungi, and viruses.
- Penicillins, cephalosporins, bacitracin, vancomycin, and cycloserines block cell wall synthesis, primarily in gram-positive bacteria.
- Aminoglycosides and tetracyclines block protein synthesis in prokaryotes.
- Sulfonamides, trimethoprim, isoniazid, nitrofurantoin, and the fluoroquinolones are synthetic antimicrobials effective against a broad range of microorganisms. They block steps in the synthesis of nucleic acids.
- Polymyxins are the major drugs that disrupt cell membranes.
- Fungal antimicrobials, such as macrolide polyenes, griseofulvin, azoles, and flucytosine, must be monitored carefully because of the potential toxicity to the infected host. They promote lysis of cell membranes.
- There are fewer antiparasitic drugs than antibacterial drugs because parasites are eukaryotes like their human hosts and they have several life stages, some of which can be resistant to the drug.
- Antihelminthic drugs immobilize or disintegrate infesting helminths or inhibit their metabolism in some manner.
- Antiviral drugs interfere with viral replication by blocking viral entry into cells, blocking the replication process, or preventing the assembly of viral subunits into complete virions.
- Many antiviral agents are analogs of nucleotides. They inactivate the replication process when incorporated into viral nucleic acids. HIV antivirals interfere with reverse transcriptase or proteases to prevent the maturation of viral particles.
- Although interferon is effective *in vivo* against certain viral infections, commercial interferon is not currently effective as a broad-spectrum antiviral agent.

Interactions Between Microbes and Drugs: The Acquisition of Drug Resistance

One unfortunate outcome of the use of antimicrobials is the development of microbial **drug resistance,** an adaptive response in which microorganisms begin to tolerate an amount of drug that would ordinarily be inhibitory. The development of mechanisms for circumventing or inactivating antimicrobial drugs is due largely to the genetic versatility and adaptability of microbial populations. The property of drug resistance can be intrinsic as well as acquired. Intrinsic drug resistance can best be exemplified by the fact that bacteria must, of course, be resistant to any antibiotic that they produce. This type of resistance is limited, however, to a small group of organisms and is generally not a problem with regard to antimicrobial chemotherapy. Of much greater importance is the acquisition of resistance to a drug by a microbe that was previously sensitive to the drug. In our context, the term *drug resistance* will refer to this last type of acquired resistance.

How Does Drug Resistance Develop?

Contrary to popular belief, antibiotic resistance is not a recent phenomenon. Resistance to penicillin developed in some bacteria as early as 1940, three years before the drug was even approved for public use. The scope of the problem became apparent in the 1980s and 1990s, when scientists and physicians observed treatment failures on a large scale.

Microbes become newly resistant to a drug after one of the following events occurs: (1) spontaneous mutations in critical chromosomal genes, or (2) acquisition of entire new genes or sets of genes via transfer from another species. Chromosomal drug resistance usually results from spontaneous random mutations in bacterial populations. The chance that such a mutation will be advantageous is minimal, and the chance that it will confer resistance to a specific drug is lower still. Nevertheless, given the huge numbers of microorganisms in any population and the constant rate of mutation, such mutations do occur. The end result varies from slight changes in microbial sensitivity, which can be overcome by larger doses of the drug, to complete loss of sensitivity.

Resistance occurring through intermicrobial transfer originates from plasmids called **resistance (R) factors** that are transferred through conjugation, transformation, or transduction. Studies have shown that plasmids encoded

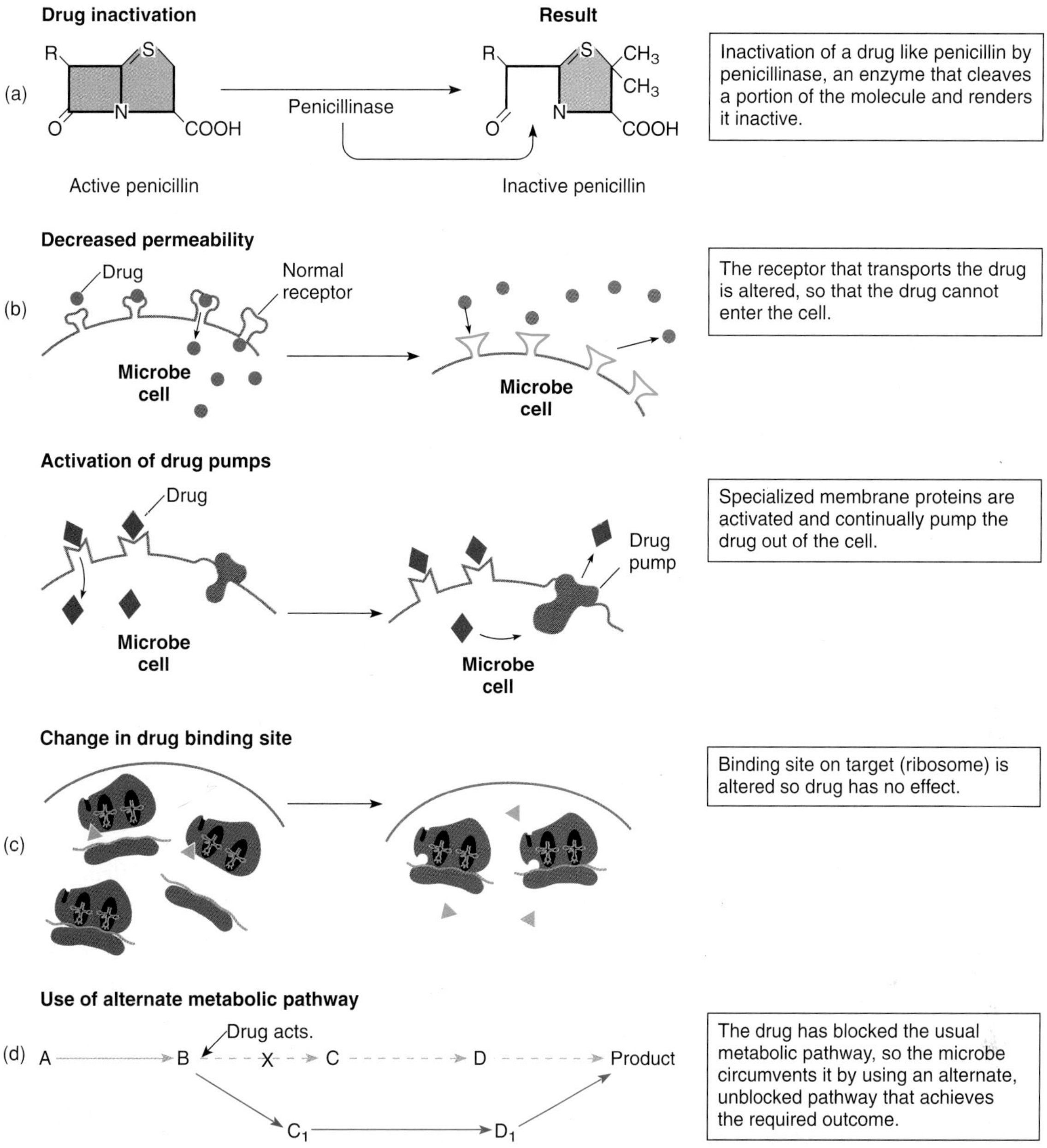

Figure 12.13 Examples of mechanisms of acquired drug resistance.

with drug resistance are naturally present in microorganisms before they have been exposed to the drug. Such traits are "lying in wait" for an opportunity to be expressed and to confer adaptability on the species. Many bacteria also maintain transposable drug resistance sequences (transposons) that are duplicated and inserted from one plasmid to another or from a plasmid to the chromosome. Chromosomal genes and plasmids containing codes for drug resistance are faithfully replicated and inherited by all subsequent progeny. This sharing of resistance genes accounts for the rapid proliferation of drug-resistant species. A growing body of evidence points to the ease and frequency of gene transfers in nature, from totally unrelated bacteria living in the body's normal biota and the environment.

Specific Mechanisms of Drug Resistance

Inside a bacterial cell, the net effect of the two events described earlier is one of the following, which actually causes the bacterium to be resistant (illustrated in **figure 12.13**):

1. New enzymes are synthesized; these inactivate the drug (only occurs when new genes are acquired) **(figure 12.13*a*).**
2. Permeability or uptake of drug into bacterium is decreased or eliminated (can occur via mutation or acquisition of new genes) **(figure 12.13*b*).**
3. Binding sites for drug are decreased in number or affinity (can occur via mutation or acquisition of new genes) **(figure 12.13*c*).**

4. An affected metabolic pathway is shut down or an alternate pathway is used (occurs due to mutation of original enzyme(s)) **(figure 12.13).**

Some bacteria can become resistant indirectly by lapsing into dormancy or, in the case of penicillin, by converting to a cell-wall-deficient form (L form) that penicillin cannot affect.

Drug Inactivation Mechanisms Microbes inactivate drugs by producing enzymes that permanently alter drug structure. One example, bacterial exoenzymes called **beta-lactamases,** hydrolyze the beta-lactam ring structure of some penicillins and cephalosporins rendering the drugs inactive. Two beta-lactamases—penicillinase and cephalosporinase—disrupt the structure of certain penicillin or cephalosporin molecules so their activity is lost. So many strains of *Staphylococcus aureus* produce penicillinase that regular penicillin is rarely a possible therapeutic choice. Now that some strains of *Neisseria gonorrhoeae,* called PPNG,[3] have also acquired penicillinase, alternative drugs are required to treat gonorrhea (see chapter 23). A large number of other gram-negative species are inherently resistant to some of the penicillins and cephalosporins because of naturally occurring beta-lactamases.

Decreased Drug Permeability or Increased Drug Elimination The resistance of some bacteria can be due to a mechanism that prevents the drug from entering the cell and acting on its target. For example, the outer membrane of the cell wall of certain gram-negative bacteria is a natural blockade for some of the penicillin drugs. Resistance to the tetracyclines can arise from plasmid-encoded proteins that pump the drug out of the cell. Aminoglycoside resistance can arise in multiple ways; one of these is change in drug permeability caused by point mutations in the transport system or LPS machinery.

Many bacteria possess multidrug-resistant (MDR) pumps that actively transport drugs and other chemicals out of cells. These pumps are proteins encoded by plasmids or chromosomes. They are stationed in the cell membrane and expel molecules by a proton-motive force similar to ATP synthesis (see figure 12.13). They confer drug resistance on many gram-positive pathogens (*Staphylococcus, Streptococcus*) and gram-negative pathogens (*Pseudomonas, E. coli*). Because they lack selectivity, one type of pump can expel a broad array of antimicrobial drugs, detergents, and other toxic substances.

Change of Drug Receptors Because most drugs act on a specific target such as protein, RNA, DNA, or membrane structure, microbes can circumvent drugs by altering the nature of this target. Bacteria can become resistant to aminoglycosides when point mutations in ribosomal proteins arise (see figure 12.13). Erythromycin and clindamycin resistance is associated with an alteration on the 50S ribosomal binding site. Penicillin resistance in *Streptococcus pneumoniae* and methicillin resistance in *Staphylococcus aureus* are related to an alteration in the binding proteins in the cell wall. Several species of enterococci have acquired resistance to vancomycin through a similar alteration of cell wall proteins. Fungi can become resistant by decreasing their synthesis of ergosterol, the principal receptor for certain antifungal drugs.

Changes in Metabolic Patterns The action of antimetabolites can be circumvented if a microbe develops an alternative metabolic pathway or enzyme. Sulfonamide and trimethoprim resistance develops when microbes deviate from the usual patterns of folic acid synthesis. Fungi can acquire resistance to flucytosine by completely shutting off certain metabolic activities.

Natural Selection and Drug Resistance

So far, we have been considering drug resistance at the cellular and molecular levels, but its full impact is felt only if this resistance occurs throughout the cell population. Let us examine how this might happen and its long-term therapeutic consequences.

Any large population of microbes is likely to contain a few individual cells that are already drug resistant because of prior mutations or transfer of plasmids **(figure 12.14*a*).** As long as the drug is not present in the habitat, the numbers of these resistant forms will remain low because they have no particular growth advantage (and often are disadvantaged relative to their nonmutated counterparts). But if the population is subsequently exposed to this drug **(figure 12.14*b*),** sensitive individuals are inhibited or destroyed, and resistant forms survive and proliferate. During subsequent population growth, offspring of these resistant microbes will inherit this drug resistance. In time, the replacement population will have a preponderance of the drug-resistant forms and can eventually become completely resistant **(figure 12.14*c*).** In ecological terms, the environmental factor (in this case, the drug) has put selection pressure on the population, allowing the more "fit" microbe (the drug-resistant one) to survive, and the population has evolved to a condition of drug resistance.

Natural selection for drug-resistant forms is apparently a common phenomenon. It takes place most frequently in various natural habitats, laboratories, and medical environments, and it also can occur within the bodies of humans and animals during drug therapy **(Insight 12.3).**

New Approaches to Antimicrobial Therapy

Researchers are constantly "pushing the envelope" of antimicrobial research. Often, the quest focuses on finding new targets in the bacterial cell and custom-designing drugs that aim for them. There are many interesting new strategies that have not yet resulted in a marketable drug—for example, (1) targeting iron-scavenging capabilities of bacteria and (2) targeting a genetic control mechanism in bacteria referred to as riboswitches.

3. Penicillinase-producing *Neisseria gonorrhoeae.*

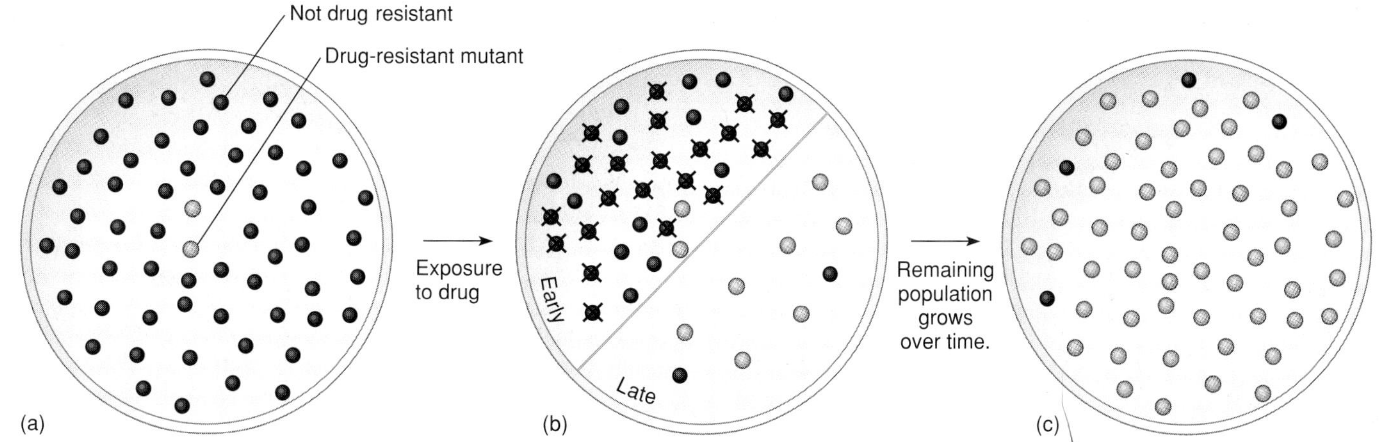

Figure 12.14 The events in natural selection for drug resistance.
(a) Populations of microbes can harbor some members with a prior mutation that confers drug resistance. **(b)** Environmental pressure (here, the presence of the drug) selects for survival of these mutants. **(c)** They eventually become the dominant members of the population.

The best example of a strategy aimed at iron-scavenging capabilities is recent work with *Staphylococcus aureus.* Scientists have found that this bacterium has a special pathway involving several proteins that punch holes in red blood cells and then "reach in" to bind the heme, strip it of its iron, and use it for their own purposes. Researchers are currently investigating inhibitory substances that block the iron-collecting pathway, which would result in inevitable bacterial death. This approach may prove effective in wiping out antibiotic-resistant strains of *S. aureus.*

Riboswitches are areas of bacterial RNA that are used to control translation of mRNA. They act by binding substances in the bacterial cell, prohibiting translation processes. Because riboswitches appear to be ubiquitous in bacteria, finding a drug that blocks their action could be a useful antimicrobial.

Other novel approaches to controlling infections include the use of **probiotics** and **prebiotics.** Probiotics are preparations of live microorganisms that are fed to animals and humans to improve the intestinal biota. This can serve to replace microbes lost during antimicrobial therapy or simply to augment the biota that is already there. This is a slightly more sophisticated application of methods that have long been used in an empiric fashion, for instance, by people who consume yogurt because of the beneficial microbes it contains. A probiotic approach is also sometimes used in female patients who have recurring urinary tract infections. These patients receive vaginal inserts or perineal swabs of *Lactobacillus* in an attempt to restore a healthy acidic environment to the region and to displace disease-causing microorganisms. Probiotics are thought to be useful for the management of food allergies; their role in the stimulation of mucosal immunity is also being investigated.

Prebiotics are nutrients that encourage the growth of beneficial microbes in the intestine. For instance, certain sugars such as fructans are thought to encourage the growth of *Bifidobacterium* in the large intestine and to discourage the growth of potential pathogens. You can be sure that you will hear more about prebiotics and probiotics as the concepts become increasingly well studied by scientists. Clearly, the use of these agents is a different type of antimicrobial strategy than we are used to, but it may have its place in a future in which traditional antibiotics are more problematic.

Another category of antimicrobial is the **lantibiotics.** Lantibiotics are short peptides produced by bacteria that inhibit the growth of other bacteria. They are distinguished by the presence of unusual amino acids that are not seen elsewhere in the cell. They exert their antimicrobial activity either by puncturing cell membranes or by inhibiting bacterial enzymes. The most well-known lantibiotic is *nisin.* Lantibiotics have long been used in food preservation, in veterinary medicine, and more recently in personal care products such as deodorants; they may soon find use in human chemotherapy. Several lantibiotics have been found to be effective against human pathogens such as *Propionibacterium acne,* methicillin-resistant *S. aureus,* and *Helicobacter pylori.*

So far, virtually all of the antimicrobials used in human infections have been derived from other microorganisms or artificially synthesized in the laboratory. But many scientists are investigating antimicrobial substances derived from plants and from animals (called "natural products"). Plant-derived antimicrobials are by no means an original idea; indigenous cultures have been using plants as medicines for centuries, as pointed out in Insight 12.1. In the fifth century BC, Hippocrates wrote about 300 to 400 medicinal plants. But phyto- (plant) chemicals have been overlooked by modern infectious disease scientists until recently. (Many widely used chemotherapeutic agents, such as Coumadin, a drug that prevents blood clots, are derived from plants, but virtually none of them are anti-infectives.) Now modern techniques have proven the antimicrobial efficacy of a wide variety of extracts from botanical sources, and the search for new ones has intensified. In addition, lantibioticlike substances have been found in amphibians and other animals.

INSIGHT 12.3 Medical

The Rise of Drug Resistance

Many people unrealistically assume that science will come to the rescue and solve the problem of drug resistance. If drug companies just keep making more and better antimicrobials, soon infectious diseases will be vanquished. This unfortunate attitude has vastly underestimated the extreme versatility and adaptability of microorganisms and the complexity of the task. It is a fact of nature that if a large number of microbes are exposed to a variety of drugs, there will always be some genetically favored individuals that survive and thrive. The AIDS virus (HIV) is so prone to drug resistance that it can become resistant during the first few weeks of therapy in a single individual. Because HIV mutates so rapidly, in most cases, it will eventually become resistant to all drugs that have been developed so far.

Ironically, thousands of patients die every year in the United States from infections that lack effective drugs, and 60% of hospital infections are caused by drug-resistant microbes. For many years, concerned observers reported the gradual development of drug resistance in staphylococci, *Salmonella*, and gonococci. But during the past decade, the scope of the problem has escalated. It is now a common event to discover microbes that have become resistant to relatively new drugs in a very short time. In fact, many strains of pathogens have multiple drug resistance, and a few are resistant to all drugs. **Table 12.A** lists some notable examples.

TABLE 12.A

Organism	Drug	Year and Prevalence of Resistance		
		1989	**2000**	**2002**
Pneumococci	Penicillin	Rare	30%	40%
	Erythromycin	0	15%	40%
	Cephalosporin	0	14%	27%
		1990	**2000**	**2002**
Gonococci	Penicillin	10%	70%	74%
	Fluoroquinolone	1.4%	10%	12%
Enterococci	Vancomycin	Rare	50%	52%
		1995	**2000**	**2002**
Campylobacter	Fluoroquinolones	Rare	10%	21%

The Hospital Factor

The clinical setting is a prolific source of drug-resistant strains of bacteria. This environment continually exposes pathogens to a variety of drugs. The hospital also maintains patients with weakened defenses, making them highly susceptible to pathogens. A classic example occurred with *Staphylococcus aureus* and penicillin. In the 1950s, hospital strains began to show resistance to this drug, and because of indiscriminate use, these strains became nearly 100% resistant in 30 years. In a short time, MRSA (methicillin-resistant *S. aureus*) strains appeared, which can tolerate nearly all antibiotics. Up until recently, MRSA has been sensitive to the drug of last resort, vancomycin. In 2002, the first cases of complete resistance to this drug were reported (VRSA). To complicate this problem, strains of MRSA are now being spread into the community.

Drugs in Animal Feeds

Another practice that has contributed significantly to growing drug resistance is the addition of antimicrobials to livestock feed, with the idea of decreasing infections and thereby improving animal health and size. This practice has had serious impact in both the United States and Europe. Enteric bacteria such as *Salmonella*,

Escherichia coli, and enterococci that live as normal intestinal biota of these animals readily share resistance plasmids and are constantly selected and amplified by exposure to drugs. These pathogens subsequently "jump" to humans and cause drug-resistant infections, oftentimes at epidemic proportions. In a deadly outbreak of *Salmonella* infection in Denmark, the pathogen was found to be resistant to seven different antimicrobials. In the United States, a strain of fluoroquinolone-resistant *Campylobacter* from chickens caused over 5,000 cases of food infection in the late 1990s. The opportunistic pathogen called VRE (vancomycin-resistant enterococcus) has been traced to the use of a vancomycinlike drug in cattle feed. It is now one of the most tenacious of hospital-acquired infections for which there are few drug choices. To attempt to curb this source of resistance, Europe and the United States have begun to ban the use of human drugs in animal feeds.

The move seems to be working. Denmark banned all agricultural antibiotic use for growth promotion in 1998. Resistance to the drugs declined dramatically among bacteria isolated from the farm animals without significant reductions in animal size or health. The European Union followed suit on January 1, 2006. There is legislation proposed in the U.S. Congress, called the Preservation of Antibiotics for Medical Treatment Act, but it has not yet been adopted.

Worldwide Drug Resistance

The drug dilemma has become a widespread problem, affecting all countries and socioeconomic groups. In general, the majority of infectious diseases, whether bacterial, fungal, protozoan, or viral, are showing increases in drug resistance. In parts of India, the main drugs used to treat cholera (furazolidone, ampicillin) have gone from being highly effective to essentially useless in 10 years. In Southeast Asia, 98% of gonococcus infections are multidrug resistant. Malaria, tuberculosis, and typhoid fever pathogens are gaining in resistance, with few alternative drugs to control them. To add to the problem, global travel and globalization of food products mean that drug resistance can be rapidly exported.

In countries with adequate money to pay for antimicrobials, most infections will be treated but at some expense. In the United States alone, the extra cost for treating the drug-resistant variety is around $10 billion per year. In many developing countries, drugs are mishandled by overuse and underuse, either of which can contribute to drug resistance. Many countries that do not regulate the sale of prescription drugs make them readily available to purchase over the counter. For example, the antituberculosis drug INH (isoniazid) is sometimes used as a "lung vitamin" to improve health, and antibiotics are taken in the wrong dose and at the wrong time for undiagnosed conditions. These countries serve as breeding grounds for drug resistance that can eventually be carried to other countries.

It is clear that we are in a race with microbes and we are falling behind. If the trend is not contained, the world may return to a time when there are few effective drugs left. We simply cannot develop them as rapidly as microbes can develop resistance. In this light, it is essential to fight the battle on more than one front. **Table 12.B** summarizes the several critical strategies to give us an edge in controlling drug resistance.

TABLE 12.B Strategies to Limit Drug Resistance by Microorganisms

Drug Usage

- Physicians have the responsibility for making an accurate diagnosis and prescribing the correct drug therapy.
- Patients must comply with and carefully follow the physician's guidelines. It is important for the patient to take the correct dosage, by the best route, for the appropriate period. This diminishes the selection for mutants that can resist low drug levels and ensures elimination of the pathogen.
- For some combinations of drugs, administration of two or more drugs together increases the chances that at least one of them will be effective and that a resistant strain of either drug will not be able to persist. The basis for this combined therapy method lies in the unlikelihood of simultaneous resistance to several drugs.

Drug Research

- Research focuses on developing shorter-term, higher-dose antimicrobials that are more effective, less expensive, and have fewer side effects.
- Pharmaceutical companies continue to seek new antimicrobial drugs with structures that are not readily inactivated by microbial enzymes or drugs with modes of action that are not readily circumvented.

Long-Term Strategies

- Proposals to reduce the abuse of antibiotics range from educational programs for health workers to requiring written justification from the physician on all antibiotics prescribed.
- Especially valuable antimicrobials may be restricted in their use to only one or two types of infections.
- The addition of antimicrobials to animal feeds must be curtailed worldwide.
- Government programs that make effective therapy available to low-income populations should be increased.
- Vaccines should be used whenever possible to provide alternative protection.

CHECKPOINT

- Microorganisms are termed drug resistant when they are no longer inhibited by an antimicrobial to which they were previously sensitive.
- Drug resistance is genetic; microbes acquire genes that code for methods of inactivating or escaping the antimicrobial, or acquire mutations that affect the drug's impact. Resistance is selected for in environments where antimicrobials are present in high concentrations, such as in hospitals.
- Microbial drug resistance develops through the selection of preexisting random mutations and through acquisition of resistance genes from other microorganisms.
- Varieties of microbial drug resistance include drug inactivation, decreased drug uptake, decreased drug receptor sites, and modification of metabolic pathways formerly attacked by the drug.
- Widespread indiscriminate use of antimicrobials has resulted in an explosion of microorganisms resistant to all common drugs.

12.4 Interaction Between Drug and Host

Until now, this chapter has focused on the interaction between antimicrobials and the microorganisms they target. During an infection, the microbe is living in or on a host; therefore, the drug is administered to the host though its target is the microbe. Therefore, the effect of the drug on the host must always be considered.

Although selective antimicrobial toxicity is the ideal constantly being sought, chemotherapy by its very nature involves contact with foreign chemicals that can harm human tissues. In fact, estimates indicate that at least 5% of all persons taking an antimicrobial drug experience some type of serious adverse reaction to it. The major side effects of drugs fall into one of three categories: direct damage to tissues through toxicity, allergic reactions, and disruption in the balance of normal microbial biota. The damage incurred by antimicrobial drugs can be short term and reversible or permanent, and it ranges in severity from cosmetic to lethal. **Table 12.6** summarizes drug groups and their major side effects.

Toxicity to Organs

Drugs can adversely affect the following organs: the liver (hepatotoxic), kidneys (nephrotoxic), gastrointestinal tract, cardiovascular system and blood-forming tissue (hemotoxic), nervous system (neurotoxic), respiratory tract, skin, bones, and teeth.

Because the liver is responsible for metabolizing and detoxifying foreign chemicals in the blood, it can be damaged by a drug or its metabolic products. Injury to liver cells can result in enzymatic abnormalities, fatty liver deposits, hepatitis, and liver failure. The kidney is involved in excreting drugs and their metabolites. Some drugs irritate the nephron tubules, creating changes that interfere with their filtration

TABLE 12.6 Major Adverse Toxic Reactions to Common Drug Groups

Antimicrobial Drug	Primary Damage or Abnormality Produced
Antibacterials	
Penicillin G	Skin
Carbenicillin	Abnormal bleeding
Ampicillin	Diarrhea and enterocolitis
Cephalosporins	Inhibition of platelet function
	Decreased circulation of white blood cells
	Nephritis
Tetracyclines	Diarrhea and enterocolitis
	Discoloration of tooth enamel
	Reactions to sunlight (photosensitization)
Chloramphenicol	Injury to red and white blood cell precursors
Aminoglycosides (streptomycin, gentamicin, amikacin)	Diarrhea and enterocolitis; malabsorption; loss of hearing, dizziness, kidney damage
Isoniazid	Hepatitis
	Seizures
	Dermatitis
Sulfonamides	Formation of crystals in kidney; blockage of urine flow
	Hemolysis
	Reduction in number of red blood cells
Polymyxin	Kidney damage
	Weakened muscular responses
Quinolones (ciprofloxacin, norfloxacin)	Headache, dizziness, tremors, GI distress
Rifampin	Damage to hepatic cells
	Dermatitis
Antifungals	
Amphotericin B	Disruption of kidney function
Flucytosine	Decreased number of white blood cells
Antiprotozoan drugs	
Metronidazole	Nausea, vomiting
Chloroquine	Vomiting
	Headache
	Itching
Antihelminthics	
Niclosamide	Nausea, abdominal pain
Pyrantel	Irritation
	Headache, dizziness
Antivirals	
Acyclovir	Seizures, confusion
	Rash
Amantadine	Nervousness, light-headedness
	Nausea
AZT	Immunosuppression, anemia

abilities. Drugs such as sulfonamides can crystallize in the kidney and form stones that can obstruct the flow of urine.

The most common complaint associated with oral antimicrobial therapy is diarrhea, which can progress to severe intestinal irritation or colitis. Although some drugs directly irritate the intestinal lining, the usual gastrointestinal complaints are caused by disruption of the intestinal microbiota (discussed in a subsequent section).

Many drugs given for parasitic infections are toxic to the heart, causing irregular heartbeats and even cardiac arrest in extreme cases. Chloramphenicol can severely depress blood-forming cells in the bone marrow, resulting in either a reversible or a permanent (fatal) anemia. Some drugs hemolyze the red blood cells, others reduce white blood cell counts, and still others damage platelets or interfere with their formation, thereby inhibiting blood clotting.

Certain antimicrobials act directly on the brain and can cause seizures. Others, such as aminoglycosides, damage nerves (very commonly, the eighth cranial nerve), leading to dizziness, deafness, or motor and sensory disturbances. When drugs block the transmission of impulses to the diaphragm, respiratory failure can result.

The skin is a frequent target of drug-induced side effects. The skin response can be a symptom of drug allergy or a direct toxic effect. Some drugs interact with sunlight to cause photodermatitis, a skin inflammation. Tetracyclines are contraindicated (not advisable) for children from birth to 8 years of age because they bind to the enamel of the teeth, creating a permanent gray to brown discoloration **(figure 12.15).** Pregnant women should avoid tetracyclines because they can cause liver damage. They also cross the placenta and can be deposited in the developing fetal bones and teeth.

Allergic Responses to Drugs

One of the most frequent drug reactions is heightened sensitivity, or **allergy.** This reaction occurs because the drug acts as an antigen (a foreign material capable of stimulating the immune system) and stimulates an allergic response. This response can be provoked by the intact drug molecule or by substances that develop from the body's metabolic alteration of the drug. In the case of penicillin, for instance, it is not the penicillin molecule itself that causes the allergic response but a product, *benzlpenicilloyl.* Allergic reactions have been reported for every major type of antimicrobial drug, but the penicillins account for the greatest number of antimicrobial allergies, followed by the sulfonamides.

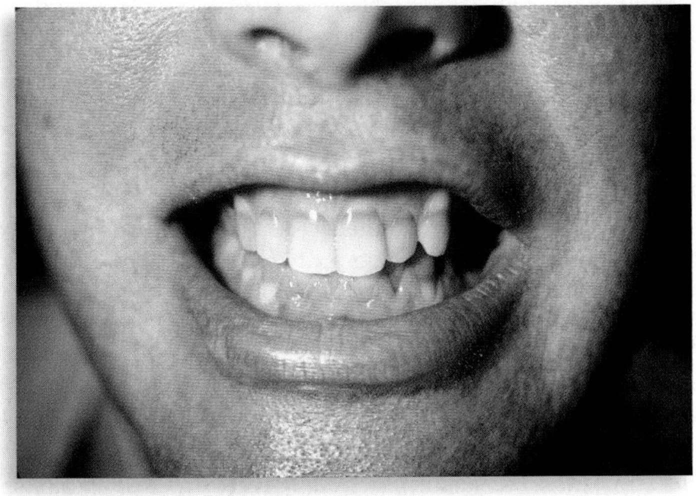

Figure 12.15 Drug-induced side effect.
An adverse effect of tetracycline given to young children is the permanent discoloration of tooth enamel.

People who are allergic to a drug become sensitized to it during the first contact, usually without symptoms. Once the immune system is sensitized, a second exposure to the drug can lead to a reaction such as a skin rash (hives), respiratory inflammation, and, rarely, anaphylaxis, an acute, overwhelming allergic response that develops rapidly and can be fatal. (This topic is discussed in greater detail in chapter 16.)

Suppression and Alteration of the Microbiota by Antimicrobials

Most normal, healthy body surfaces, such as the skin, large intestine, outer openings of the urogenital tract, and oral cavity, provide numerous habitats for a virtual "garden" of microorganisms. These normal colonists or residents, called the **biota** or microbiota, consist mostly of harmless or beneficial bacteria, but a small number can potentially be pathogens. Although we defer a more detailed discussion of this topic to chapter 13 and later chapters, here we focus on the general effects of drugs on this population.

If a broad-spectrum antimicrobial is introduced into a host to treat infection, it will destroy microbes regardless of their roles in the balance, affecting not only the targeted infectious agent but also many others in sites far removed from the original infection **(figure 12.16).** When this therapy destroys beneficial resident species, the microbes that were once in small numbers begin to overgrow and cause disease. This complication is called a **superinfection.**

Some common examples demonstrate how a disturbance in microbial biota leads to replacement biota and superinfection. A broad-spectrum cephalosporin used to treat a urinary tract infection by *Escherichia coli* will cure the infection, but it will also destroy the lactobacilli in the vagina that normally maintain a protective acidic environment there. The drug has no effect, however, on *Candida albicans,* a yeast that also resides in normal vaginas. Released from the inhibitory environment provided by lactobacilli, the yeasts proliferate and cause symptoms. *Candida* can cause similar superinfections of the oropharynx (thrush) and the large intestine.

Oral therapy with tetracyclines, clindamycin, and broad-spectrum penicillins and cephalosporins is associated with a serious and potentially fatal condition known as *antibiotic-associated colitis* (pseudomembranous colitis). This condition is due to the overgrowth in the bowel of *Clostridium difficile,* an endospore-forming bacterium that is resistant to the antibiotic. It invades the intestinal lining and releases toxins that induce diarrhea, fever, and abdominal pain. (You'll learn more about infectious diseases of the gastrointestinal tract, including *C. difficile,* in chapter 22.)

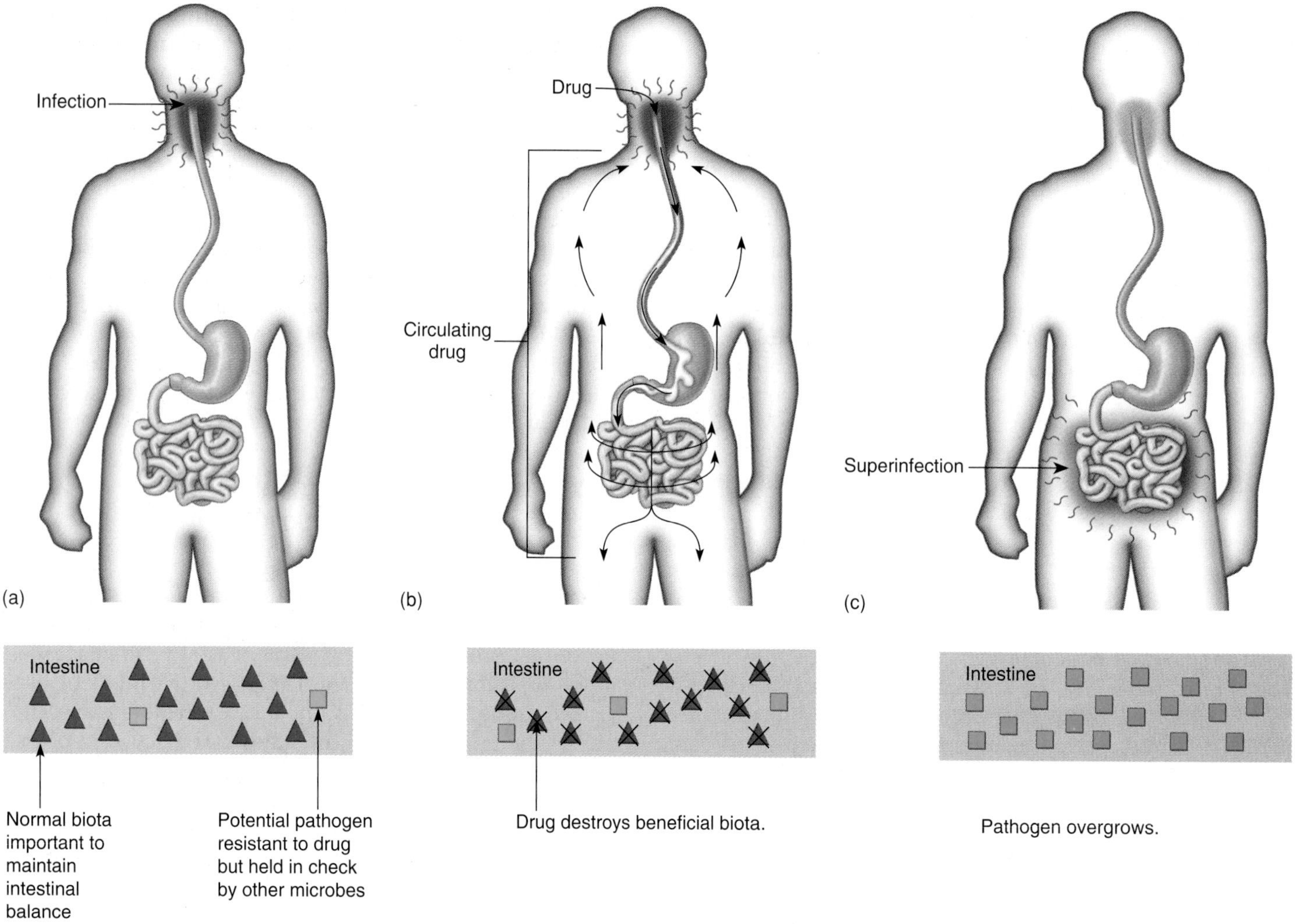

Figure 12.16 The role of antimicrobials in disrupting microbial biota and causing superinfections. **(a)** A primary infection in the throat is treated with an oral antibiotic. **(b)** The drug is carried to the intestine and is absorbed into the circulation. **(c)** The primary infection is cured, but drug-resistant pathogens have survived and create an intestinal superinfection.

CHECKPOINT

- The three major side effects of antimicrobials are toxicity to organs, allergic reactions, and problems resulting from suppression or alteration of normal biota.
- Antimicrobials that destroy most but not all normal biota allow the unaffected normal biota to overgrow, causing a superinfection.

12.5 Considerations in Selecting an Antimicrobial Drug

Before actual antimicrobial therapy can begin, it is important that at least three factors be known:

1. the nature of the microorganism causing the infection,
2. the degree of the microorganism's susceptibility (also called sensitivity) to various drugs, and
3. the overall medical condition of the patient.

Identifying the Agent

Identification of infectious agents from body specimens should be attempted as soon as possible. It is especially important that such specimens be taken before any antimicrobial drug is given, just in case the drug eliminates the infectious agent. Direct examination of body fluids, sputum, or stool is a rapid initial method for detecting and perhaps even identifying bacteria or fungi. A doctor often begins the therapy on the basis of such immediate findings. The choice of drug will be based on experience with drugs that are known to be effective against the microbe; this is called the "informed best guess." For instance, if a sore throat appears to be caused by *Streptococcus pyogenes*, the physician might prescribe penicillin, because this species seems to be almost universally sensitive to it so far. If the infectious agent is not or cannot be isolated, epidemiologic statistics may be required to predict the most likely agent in a given infection. For example, *Streptococcus pneumoniae* accounts for the majority of cases of meningitis in children, followed by *Neisseria meningitidis* (discussed in detail in chapter 19).

Testing for the Drug Susceptibility of Microorganisms

Testing is essential in those groups of bacteria commonly showing resistance, primarily *Staphylococcus species, Neisseria gonorrhoeae, Streptococcus pneumoniae,* and *Enterococcus faecalis,* and the aerobic gram-negative enteric bacilli. However, not all infectious agents require antimicrobial sensitivity testing. Drug testing in fungal or protozoan infections is difficult and is often unnecessary. When certain groups, such as group A streptococci and all anaerobes (except *Bacteroides*), are known to be uniformly susceptible to penicillin G, testing may not be necessary unless the patient is allergic to penicillin.

Selection of a proper antimicrobial agent begins by demonstrating the *in vitro* activity of several drugs against the infectious agent by means of standardized methods. In general, these tests involve exposing a pure culture of the bacterium to several different drugs and observing the effects of the drugs on growth.

The *Kirby-Bauer* technique is an agar diffusion test that provides useful data on antimicrobial susceptibility. In this test, the surface of a plate of special medium is spread with the test bacterium, and small discs containing a premeasured amount of antimicrobial are dispensed onto the bacterial lawn. After incubation, the zone of inhibition surrounding the discs is measured and compared with a standard for each drug **(table 12.7** and **figure 12.17).** The

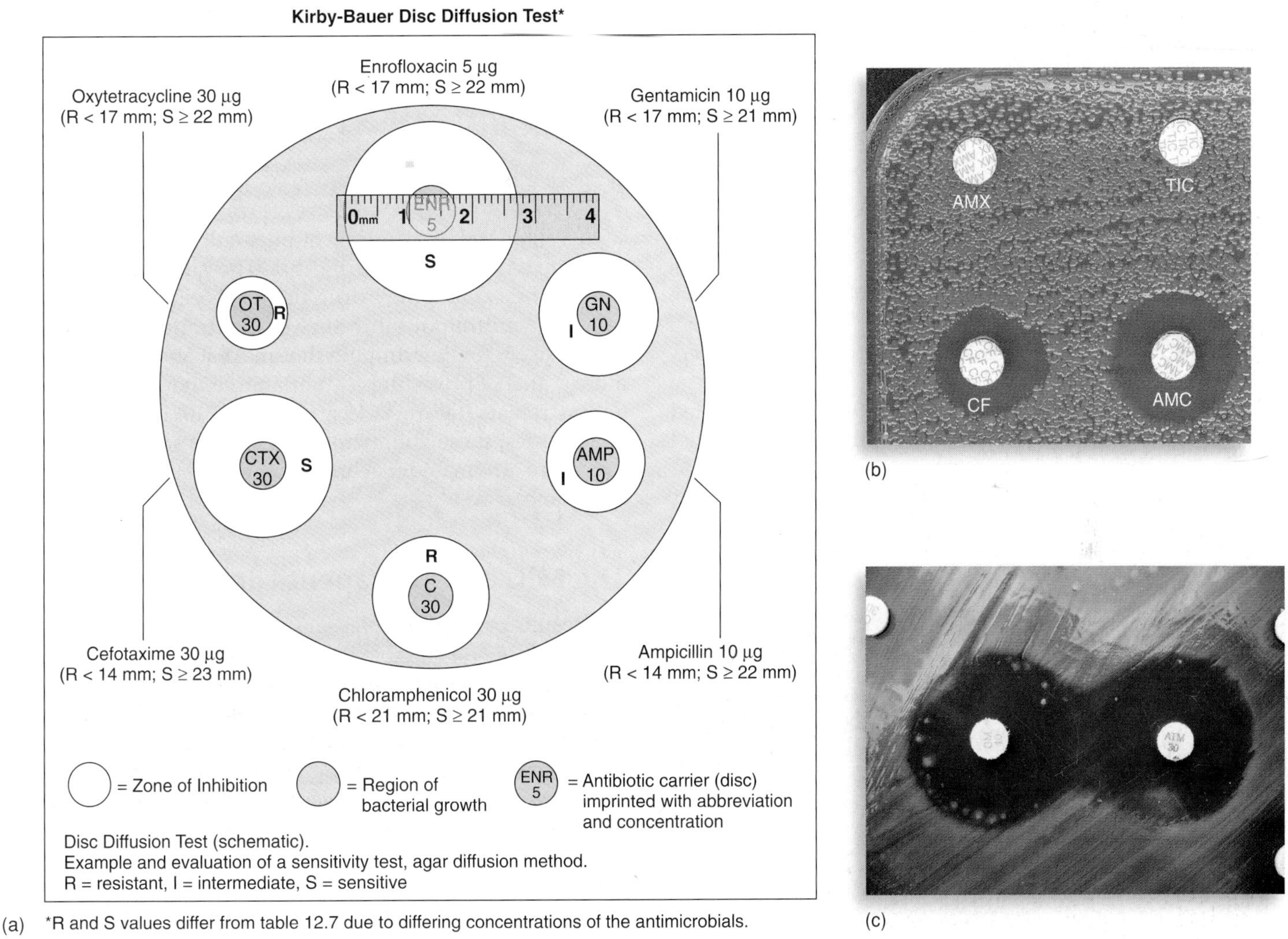

Figure 12.17 Technique for preparation and interpretation of disc diffusion tests.
(a) Standardized methods are used to seed a lawn of bacteria over the medium. A dispenser delivers several drugs onto a plate, followed by incubation. Interpretation of results: During incubation, antimicrobials become increasingly diluted as they diffuse out of the disc into the medium. If the test bacterium is sensitive to a drug, a zone of inhibition develops around its disc. Roughly speaking, the larger the size of this zone, the greater is the bacterium's sensitivity to the drug. The diameter of each zone is measured in millimeters and evaluated for susceptibility or resistance by means of a comparative standard (see table 12.7). **(b)** *Escherichia coli* showing resistance to amoxicillin (AMX) and ticarcillin (TIC); when clavulanic acid is combined with amoxicillin (AMC), the result is sensitivity (lower right) due to the combined action of the two drugs. **(c)** Results of test with *Escherichia hermannii* indicate a synergistic effect between ticarcillin (TIC) and AMC (note the expanded zone between these two drugs).

TABLE 12.7 Results of a Sample Kirby-Bauer Test

Drug	Zone Sites (mm) Required For: Susceptibility (S)	Zone Sites (mm) Required For: Resistance (R)	Actual Result (mm) for *Staphylococcus aureus*	Evaluation
Bacitracin	>13	<8	15	S
Chloramphenicol	>18	<12	20	S
Erythromycin	>18	<13	15	I
Gentamicin	>13	<12	16	S
Kanamycin	>18	<13	20	S
Neomycin	>17	<12	12	R
Penicillin G	>29	<20	10	R
Polymyxin B	>12	<8	10	R
Streptomycin	>15	<11	11	R
Vancomycin	>12	<9	15	S
Tetracycline	>19	<14	25	S

R = resistant, I = intermediate, S = sensitive

profile of antimicrobial sensitivity, or *antibiogram,* provides data for drug selection. The Kirby-Bauer procedure is less effective for bacteria that are anaerobic, highly fastidious, or slow-growing (*Mycobacterium*). An alternative diffusion system that provides additional information on drug effectiveness is the E-test **(figure 12.18).**

More sensitive and quantitative results can be obtained with tube dilution tests. First the antimicrobial is diluted serially in tubes of broth, and then each tube is inoculated with a small uniform sample of pure culture, incubated, and examined for growth (turbidity). The smallest concentration (highest dilution) of drug that visibly inhibits growth is called the **minimum inhibitory concentration,** or **MIC.** The MIC is useful in determining the smallest effective dosage of a drug and in providing a comparative index against other antimicrobials **(figure 12.19** and **table 12.8).** In many clinical laboratories, these antimicrobial testing procedures are performed in automated machines that can test dozens of drugs simultaneously.

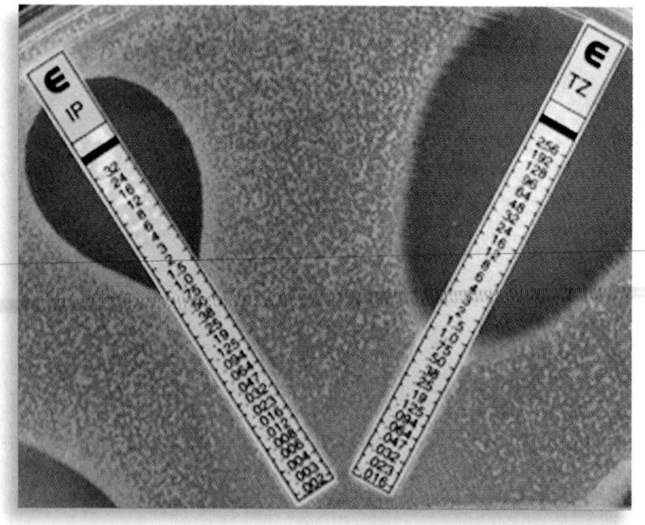

Figure 12.18 Alternative to the Kirby-Bauer procedure. Another diffusion test is the E-test, which uses a strip to produce the zone of inhibition. The advantage of the E-test is that the strip contains a gradient of drug calibrated in micrograms. This way, the MIC can be measured by observing the mark on the strip that corresponds to the edge of the zone of inhibition. (IP = imipenem and TZ = tazobactam)

The MIC and Therapeutic Index

The results of antimicrobial sensitivity tests guide the physician's choice of a suitable drug. If therapy has already commenced, it is imperative to determine if the tests bear out the use of that particular drug. Once therapy has begun, it is important to observe the patient's clinical response, because the *in vitro* activity of the drug is not always correlated with its *in vivo* effect. When antimicrobial treatment fails, the failure is due to

1. the inability of the drug to diffuse into that body compartment (the brain, joints, skin);
2. a few resistant cells in the culture that did not appear in the sensitivity test; or
3. an infection caused by more than one pathogen (mixed), some of which are resistant to the drug.

If therapy does fail, a different drug, combined therapy, or a different method of administration must be considered.

Many factors influence the choice of an antimicrobial drug besides microbial sensitivity to it. The nature and spectrum of the drug, its potential adverse effects, and the condition of the

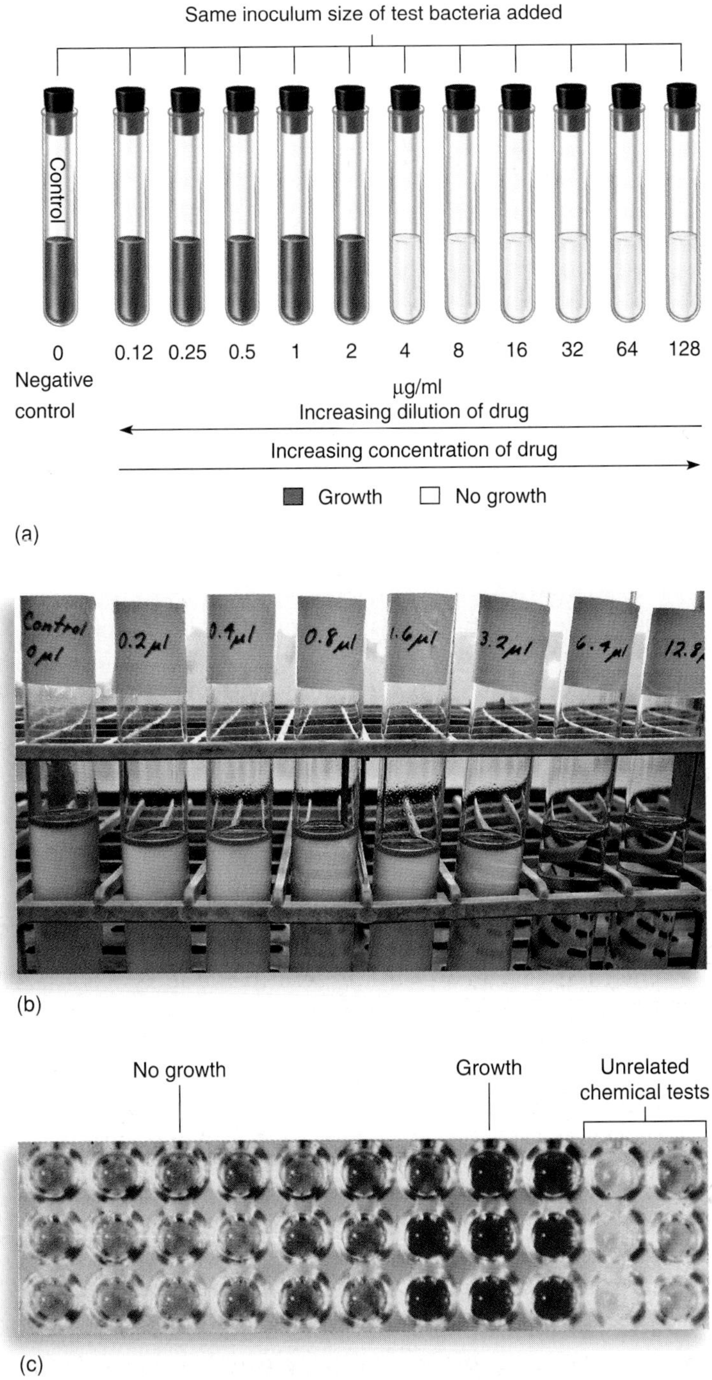

Figure 12.19 Tube dilution test for determining the minimum inhibitory concentration (MIC).
(a) The antibiotic is diluted serially through tubes of liquid nutrient from right to left. All tubes are inoculated with an identical amount of a test bacterium and then incubated. The first tube on the left is a control that lacks the drug and shows maximum growth. The dilution of the first tube in the series that shows no growth (no turbidity) is the MIC. **(b)** Photograph of MIC tube test using *Escherichia coli* and tetracycline. **(c)** Multiwell plate with an array of three tests. This system can be automated to read the MICs of several drugs simultaneously.

patient can be critically important. When several antimicrobial drugs are available for treating an infection, final drug selection advances to a new series of considerations. In general, it is better to choose the narrowest-spectrum drug of those that are effective if the causative agent is known. This decreases the potential for superinfections and other adverse reactions.

Because drug toxicity is of concern, it is best to choose the one with high selective toxicity for the infectious agent and low human toxicity. The **therapeutic index (TI)** is defined as the ratio of the dose of the drug that is toxic to humans as compared to its minimum effective (therapeutic) dose. The closer these two figures are (the smaller the ratio), the greater is the potential for toxic drug reactions. For example, a drug that has a therapeutic index of:

$$\frac{10\ \mu g/ml\text{: toxic dose}}{9\ \mu g/\mu l\ \text{(MIC)}}\quad \boxed{TI=1.1}$$

is a riskier choice than one with a therapeutic index of:

$$\frac{10\ \mu g/ml}{1\ \mu g/ml}\quad \boxed{TI=10}$$

Drug companies recommend dosages that will inhibit the microbes but not adversely affect patient cells. When a series of drugs being considered for therapy have similar MICs, the drug with the highest therapeutic index usually has the widest margin of safety.

The physician must also take a careful history of the patient to discover any preexisting medical conditions that will influence the activity of the drug or the response of the patient. A history of allergy to a certain class of drugs should preclude the administration of that drug and any drugs related to it. Underlying liver or kidney disease will ordinarily necessitate the modification of drug therapy, because these organs play such an important part in metabolizing or excreting the drug. Infants, the elderly, and pregnant women require special precautions. For example, age can diminish gastrointestinal absorption and organ function, and most antimicrobial drugs cross the placenta and could affect fetal development.

The intake of other drugs must be carefully scrutinized, because incompatibilities can result in increased toxicity or failure of one or more of the drugs. For example, the combination of aminoglycosides and cephalosporins increases nephrotoxic effects; antacids reduce the absorption of isoniazid; and the interaction of tetracycline or rifampin with oral contraceptives can abolish the contraceptive's effect. Some drugs (penicillin with certain aminoglycosides, or amphotericin B with flucytosine) act synergistically, so that reduced doses of each can be used in combined therapy. Other concerns in choosing drugs include any genetic or metabolic abnormalities in the patient, the site of infection, the route of administration, and the cost of the drug.

The Art and Science of Choosing an Antimicrobial Drug

Even when all the information is in, the final choice of a drug is not always easy or straightforward. Consider the case

TABLE 12.8 Comparitive MICs (μg/ml) for Common Drugs and Pathogens

Bacterium	Penicillin G	Ampicillin	Sulfamethoxazole	Tetracycline	Cefaclor
Staphylococcus aureus	4.0	0.05	3.0	0.3	4.0
Enterococcus faecalis	3.6	1.6	100.0	0.3	60.0
Neisseria gonorrhoeae	0.5	0.5	5.0	0.8	2.0
Escherichia coli	100.0	12.0	3.0	6–50.0	3.0
Pseudomonas aeruginosa	>500.0	>200.0	NA	>100.0	NA
Salmonella species	12.0	6.0	10.0	1.0	0.8
Clostridium	0.16	NA	NA	3.0	12.0

NA = not available

of an elderly alcoholic patient with pneumonia caused by *Klebsiella* and complicated by diminished liver and kidney function. All drugs must be given parenterally because of prior damage to the gastrointestinal lining and poor absorption. Drug tests show that the infectious agent is sensitive to third-generation cephalosporins, gentamicin, imipenem, and azlocillin. The patient's history shows previous allergy to the penicillins, so these would be ruled out. Drug interactions occur between alcohol and the cephalosporins, which are also associated with serious bleeding in elderly patients, so this may not be a good choice. Aminoglycosides such as gentamicin are nephrotoxic and poorly cleared by damaged kidneys. Imipenem causes intestinal discomfort, but it has less toxicity and would be a viable choice.

In the case of a cancer patient with severe systemic *Candida* infection, there will be fewer criteria to weigh. Intravenous amphotericin B or fluconazole are the only possible choices, despite drug toxicity and other possible adverse side effects. In a life-threatening situation in which a dangerous chemotherapy is perhaps the only chance for survival, the choices are reduced and the priorities are different.

An Antimicrobial Drug Dilemma

We began this chapter with a view of the exciting strides made in chemotherapy during the past few years, but we must end it on a note of qualification and caution. There is now a worldwide problem in the management of antimicrobial drugs. The remarkable progress in treating many infectious diseases has spawned a view of antimicrobials as a "cure-all" for infections as diverse as the common cold and acne. And, although it is true that nothing is as dramatic as curing an infectious disease with the correct antimicrobial drug, in many instances, drugs have no effect or can be harmful. The depth of this problem can perhaps be appreciated better with a few statistics:

1. Roughly 200 million prescriptions for antimicrobials are written in the United States every year. A recent study disclosed that 75% of antimicrobial prescriptions are for pharyngeal, sinus, lung, and upper respiratory infections. A fairly high percentage of these are viral in origin and will have little or no benefit from antibacterial drugs.

   Many drugs are also misprescribed as to type, dosage, or length of therapy. Such overuse of antimicrobials is known to increase the development of antimicrobial resistance, harm the patient, and waste billions of dollars.
2. Drugs are often prescribed without benefit of culture or susceptibility testing, even when such testing is clearly warranted.
3. Some physicians tend to use a "shotgun" antimicrobial therapy for minor infections, which involves administering a broad-spectrum drug instead of a more specific narrow-spectrum one. This practice can lead to superinfections and other adverse reactions. Tetracyclines and chloramphenicol are still prescribed routinely for infections that would be treated more effectively with narrower-spectrum, less toxic drugs.
4. More expensive newer drugs are chosen when a less costly older one would be just as effective. Among the most expensive drugs are cephalosporins and the longer-acting tetracyclines, yet these are among the most commonly prescribed antibiotics.
5. Tons of excess antimicrobial drugs produced in this country are exported to other countries, where controls are not as strict. Nearly 200 different antibiotics are sold over the counter in Latin America and Asian countries. It is common for people in these countries to self-medicate without understanding the correct medical indication. Drugs used in this way are largely ineffectual but, worse yet, they are known to be responsible for emergence of drug-resistant bacteria that subsequently cause epidemics.

The medical community recognizes that most physicians are motivated by important and prudent practical concerns, such as the need for immediate therapy to protect a sick patient and for defensive medicine to provide the very best care possible. But in the final analysis, every allied health professional should be critically aware not only of the admirable and utilitarian nature of antimicrobials but also of their limitations.

CHECKPOINT

- The three major considerations necessary to choose an effective antimicrobial are the nature of the infecting microbe, the microbe's sensitivity to available drugs, and the overall medical status of the infected host.
- The Kirby-Bauer test identifies antimicrobials that are effective against a specific infectious bacterial isolate.
- The MIC (minimum inhibitory concentration) identifies the smallest effective dose of an antimicrobial toxic to the infecting microbe.
- The therapeutic index is a ratio of the amount of drug toxic to the infected host and the MIC. The smaller the ratio, the greater the potential for toxic host-drug reactions.
- The effectiveness of antimicrobial drugs is being compromised by several alarming trends: inappropriate prescription, use of broad-spectrum instead of narrow-spectrum drugs, use of higher-cost drugs, sale of over-the-counter antimicrobials in other countries, and lack of sufficient testing before prescription.

Chapter Summary with Key Terms

12.1 Principles of Antimicrobial Therapy
Chemotherapeutic drugs

A. Used to control microorganisms in the body. Depending on their source, these drugs are described as antibiotics, semisynthetic or synthetic.

B. Based on their mode and spectrum of action, they are described as broad spectrum or narrow spectrum and microbistatic or microbicidal.

12.2 Interactions Between Drug and Microbe

A. The ideal antimicrobial is selectively toxic, highly potent, stable, and soluble in the body's tissues and fluids; does not disrupt the immune system or microbiota of the host; and is exempt from drug resistance.

B. Strategic approaches to the use of chemotherapeutics include
  1. Prophylaxis, where drugs are administered to *prevent* infection in susceptible people.
  2. Combined therapy, where two or more drugs are given simultaneously, either to prevent the emergence of resistant species or to achieve synergism.

C. The inappropriate use of drugs on a worldwide basis has led to numerous medical and economic problems.

12.3 Survey of Major Antimicrobial Drug Groups

A. *Antibacterial Drugs Targeting the Cell Wall*
  1. The penicillins are a large group of antibiotics originally isolated from the mold *Penicillium chrysogenum*. The natural form is penicillin G, although various semisynthetic forms, such as ampicillin and methicillin, exist that vary in their spectrum and applications.
  2. Other drugs targeting the bacterial cell wall are cephalosporins, bacitracin, isoniazid, vancomycin, and fosfomycin trimethamine.

B. *Antibacterial Drugs Targeting Protein Synthesis*
  1. Aminoglycosides include several narrow-spectrum drugs isolated from bacteria in the genus *Streptomyces*. Examples include streptomycin, gentamicin, tobramycin, and amikacin.
  2. Tetracyclines and **chloramphenicol** are very broad-spectrum drugs isolated from *Streptomyces*. Both interfere with translation but their use is limited by adverse effects.
  3. Other drugs that interfere with protein synthesis are erythromycin, clindamycin, ketolides, Synercid, and oxazolidinones.

C. *Antibacterial Drugs Targeting Folic Acid Synthesis*
The sulfonamides are broad-spectrum drugs that act as metabolic analogs, competitively inhibiting enzymes needed for folic acid synthesis. Trimethoprim is often used in combination with sulfa drugs. Sulfones (such as dapsone) are used in treating Hansen's disease (leprosy).

D. *Antibacterial Drugs Targeting DNA or RNA*
Fluoroquinolones such as ciprofloxacin and norfloxacin are a newer class of broad-spectrum synthetic drug. They exhibit high potency and are used for pneumonia, bronchitis, and other respiratory infections. Rifamycin (rifampin) interferes with RNA polymerase and is primarily used for tuberculosis and Hansen's disease.

E. *Antibacterial Drugs Targeting Cell Membranes*
  1. Polymyxin is a narrow-spectrum drug isolated from the bacterium *Bacillus*. It is also found in skin ointments and can be used to treat *Pseudomonas* infections.
  2. **Daptomycin** is a relatively new antibiotic used commonly when bacteria are resistant to older drugs. It is a lipopeptide that disrupts multiple aspects of membrane function.

F. *Drugs for Fungal Infection*
  1. Amphoteracin B and nystatin disrupt fungal membranes by detergent action.
  2. Azoles are synthetic drugs that interfere with membrane synthesis. They include ketoconazole, fluconazole, and miconazole.
  3. Flucytosine inhibits DNA synthesis. Because many fungi are now resistant, flucytosine must usually be used in conjunction with amphotericin.

G. *Drugs for Protozoan Infections*
  1. **Quinine** or the related compounds chloraquine, primaquine, or mefloquine are used to treat infections by the malarial parasite *Plasmodium*.

2. Other drugs including metronidazole, suramin, melarsopral, and nitrifurimox are used for other protozoan infections.

H. *Drugs for Helminth Infections* include mebendazole, thiabendazole, praziquantel, pyrantel, piperizine, and niclosamide, which are all used for antihelminthic chemotherapy.

I. *Drugs for Viral Infection* usually act by inhibiting viral penetration, multiplication, or assembly. Because viral and host metabolism are so closely related, toxicity is a potential adverse reaction with all antiviral drugs.
   1. Fuzeon prevents the fusion of HIV to its host receptor, thereby blocking virus entry.
   2. **Amantadine** acts early in the cycle of influenza A to prevent viral uncoating.
   3. Acyclovir, valacyclovir, famciclovir, and ribavirin act as nucleoside analogs, inhibiting viral DNA replication, especially in herpesviruses.
   4. AZT and protease inhibitors such as indinavir are used as anti-HIV drugs.
   5. Interferon is a naturally occurring protein with both antiviral and anticancer properties.

J. *The Acquisition of Drug Resistance*
   1. Microbes can lose their sensitivity to a drug through the acquisition of resistance factors. Drug resistance takes the form of:
      a. Drug inactivation.
      b. Decreased permeability to drug or increased elimination of drug from the cell.
      c. Change in drug receptors.
      d. Change of metabolic patterns.
   2. New approaches to antimicrobial therapy include the use of probiotics and prebiotics.

12.4 Interaction Between Drug and Host
Side effects of chemotherapy include organ toxicity, allergic responses, and alteration of microbiota.

12.5 Considerations in Selecting an Antimicrobial Drug
A. Rapid identification of the infectious agent is important.
B. The pathogenic microbe should be tested for its susceptibility to different antimicrobial agents. This involves using standardized methods such as the *Kirby-Bauer* and minimum inhibitory concentration (MIC) techniques.

Multiple-Choice and True-False Questions

Multiple-Choice Questions. Select the correct answer from the answers provided.

1. A compound synthesized by bacteria or fungi that destroys or inhibits the growth of other microbes is a/an
   a. synthetic drug
   b. antibiotic
   c. antimicrobial drug
   d. competitive inhibitor
2. Which statement is *not* an aim in the use of drugs in antimicrobial chemotherapy? The drug should
   a. have selective toxicity
   b. be active even in high dilutions
   c. be broken down and excreted rapidly
   d. be microbicidal
3. Drugs that prevent the formation of the bacterial cell wall are
   a. quinolones
   b. beta-lactams
   c. tetracyclines
   d. aminoglycosides
4. Microbial resistance to drugs is acquired through
   a. conjugation
   b. transformation
   c. transduction
   d. all of these
5. R factors are ____ that contain a code for ____.
   a. genes, replication
   b. plasmids, drug resistance
   c. transposons, interferon
   d. plasmids, conjugation
6. Most antihelminthic drugs function by
   a. weakening the worms so they can be flushed out by the intestine
   b. inhibiting worm metabolism
   c. blocking the absorption of nutrients
   d. inhibiting egg production
7. Which of the following modes of action would be most selectively toxic?
   a. interrupting ribosomal function
   b. dissolving the cell membrane
   c. preventing cell wall synthesis
   d. inhibiting DNA replication
8. The MIC is the ______ of a drug that is required to inhibit growth of a microbe.
   a. largest concentration
   b. standard dose
   c. smallest concentration
   d. lowest dilution
9. An antimicrobial drug with a ____ therapeutic index is a better choice than one with a _____ therapeutic index.
   a. low, high
   b. high, low

True-False Questions. If statement is true, leave as is. If it is false, correct it by rewriting the sentence.

10. Most antiviral agents work by destroying active viruses.
11. Sulfonamide drugs work by disrupting protein synthesis.
12. An antibiotic that disrupts the host's normal biota can cause superinfection.
13. Drug resistance can occur when a patient's immune system becomes reactive to a drug.

Writing to Learn

These questions are suggested as a *writing-to-learn* experience. For each question, compose a one- or two-paragraph answer that includes the factual information needed to completely address the question.

1. a. What is the major source of antibiotics?
   b. What appears to be the natural function of antibiotics?
2. a. Using the diagram below as a guide, briefly explain how the three factors in drug therapy interact.
   b. What drug characteristics will make treatment most effective?
   c. Why is it better for a drug to be microbicidal than microbistatic?

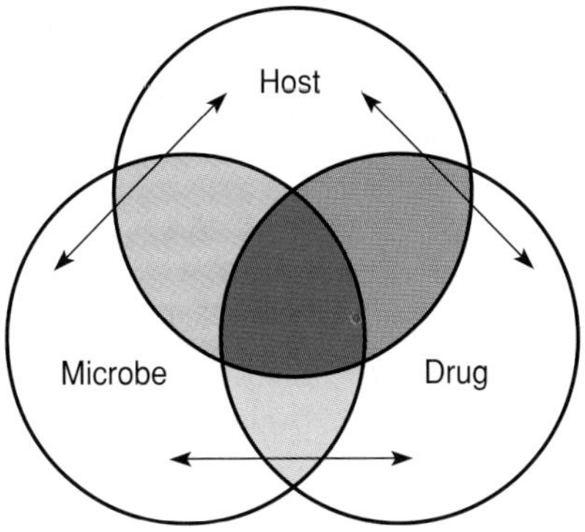

3. a. Explain the major modes of action of antimicrobial drugs, and give an example of each.
   b. What is competitive inhibition?
   c. What is the basic reason that a metabolic analog molecule can inhibit metabolism?
   d. Why does the penicillin group of drugs have milder toxicity than other antibiotics?
   e. What are the long-term effects of drugs that block transcription?
   f. Why would a drug that blocks translation on the ribosomes of bacteria also affect human cells?
   g. Why do drugs that act on bacterial and fungal membranes generally have high toxicity?
4. Construct a chart that summarizes the modes of action and applications of the major groups of antibacterial drugs (antibiotics and synthetics), antifungal drugs, antiparasitic drugs, and antiviral drugs.
5. a. Explain why there are fewer antifungal, antiparasitic, and antiviral drugs than antibacterial drugs.
   b. What effect do nitrogen-base analogs have upon viruses?
   c. Summarize the origins and biological actions of interferon.
6. Explain the phenomenon of drug resistance from the standpoint of microbial genetics (include a description of R factors).

Concept Mapping

Appendix D provides guidance for working with concept maps.

1. Supply your own linking lines *and* phrases in this concept map.

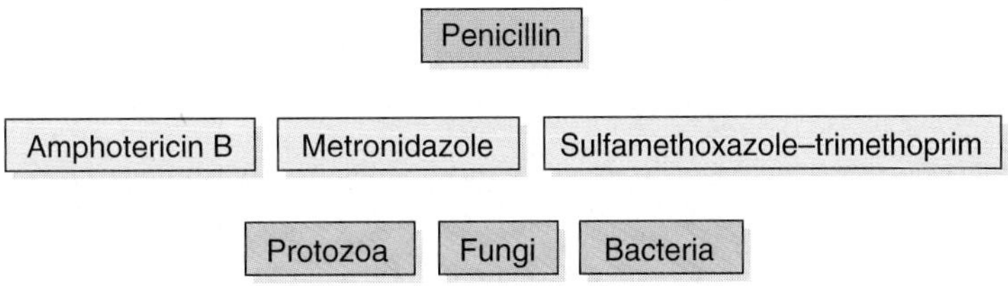

2. Use 6 to 10 bolded words of your choice from the Chapter Summary to create a concept map. Finish it by providing linking words.

Critical Thinking Questions

Critical thinking is the ability to reason and solve problems using facts and concepts. These questions can be approached from a number of angles, and in most cases, they do not have a single correct answer.

1. Occasionally, one will hear the expression that a microbe has become "immune" to a drug.
   a. What is a better way to explain what is happening?
   b. Explain a simple test one could do to determine if drug resistance was developing in a culture.
2. a. Can you think of additional ways that drug resistance can be prevented?
   b. What can health care workers do?
   c. What can one do on a personal level?

3. Drugs are often given to surgical patients, to dental patients with heart disease, or to healthy family members exposed to contagious infections.
   a. What word would you use to describe this use of drugs?
   b. What is the purpose of this form of treatment?
   c. Explain some potential undesired effects of this form of therapy.
4. a. Your pregnant neighbor has been prescribed a daily dose of oral tetracycline for acne. Do you think this therapy is advisable for her? Why or why not?
   b. A woman has been prescribed a broad-spectrum oral cephalosporin for a strep throat. What are some possible consequences in addition to cure of the infected throat?
   c. A man has a severe case of sinusitis that is negative for bacterial pathogens. A physician prescribes an oral antibacterial drug in treatment. What are your opinions of this therapy?
5. You have been directed to take a sample from a growth-free portion of the zone of inhibition in the Kirby-Bauer test and inoculate it onto a plate of nonselective medium.
   a. What does it mean if growth occurs on the new plate?
   b. What if there is no growth?
6. In cases in which it is not possible to culture or drug test an infectious agent (such as middle ear infection), how would the appropriate drug be chosen?
7. a. Refer to figure 12.18*a* and interpret the results.
   b. Give the MICs for the tests in figure 12.20*a*.
8. a. Explain the basis for combined therapy.
   b. Give reasons why it could be helpful to use combined therapy in treating HIV infection.

Visual Understanding

1. **Figure 12.6.** Where could penicillinase affect each of these antibiotics?

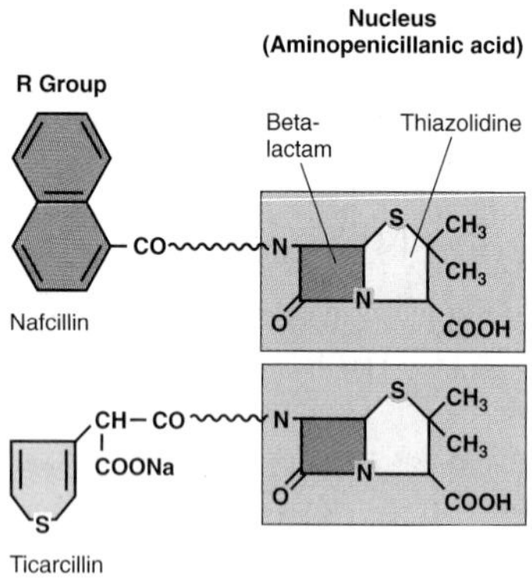

2. **From chapter 6, figure 6.12.** How could an antiviral drug interfere with the activity illustrated in the figure. How is that effective in controlling the viral cycle?

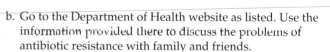

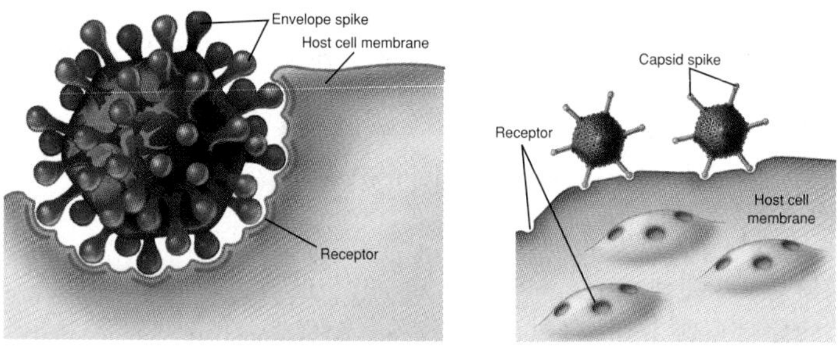

Internet Search Topics

1. Locate information on new types of antibacterial drugs called linezolid, daptomycin, and Zyvox. Determine their mode of action and indications for use.
2. Go to: www.aris.mhhe.com, and click on "microbiology" and then this textbook's author/title. Go to chapter 12, access the URLs listed under Internet Search Topics, and research the following:
   a. Go to the AIDS website listed. Look up examples of several types of anti-HIV drugs, comparing information on costs, side effects, and problems in therapy.
   b. Go to the Department of Health website as listed. Use the information provided there to discuss the problems of antibiotic resistance with family and friends.
3. Look online for information on "natural antimicrobials." Could some of these compounds be used to supplement or replace traditional antimicrobial therapies?

CHAPTER 13

Microbe-Human Interactions

Infection and Disease

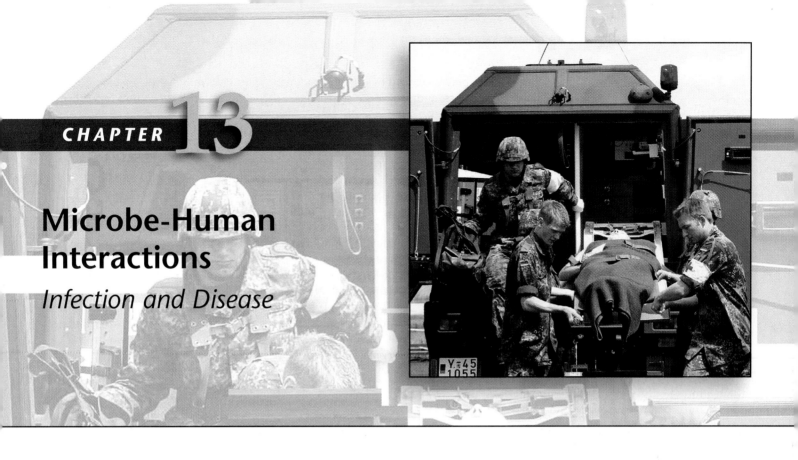

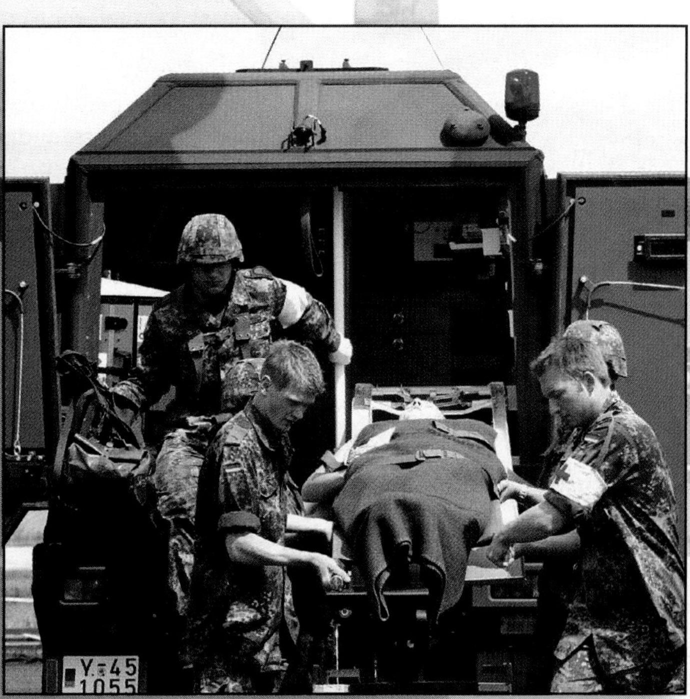

CASE FILE 13

Acinetobacter baumannii is a gram-negative bacterium commonly found in soil and water. It has occasionally been linked with health-care-associated infections. Among hospitalized patients, infections caused by *A. baumannii* have become increasingly difficult to treat due to an increasing number of isolates showing multiple drug resistance. Military medical facilities have seen an increasing number of patients with bloodstream infections caused by *A. baumannii.* Many of these patients had received traumatic injuries in Iraq, Kuwait, or Afghanistan. Antimicrobial testing of isolates from patients being treated at Landstuhl Regional Medical Center in Germany and Walter Reed Army Medical Center in the District of Columbia showed widespread resistance to antimicrobials commonly used to treat this organism.

- *What could be the source of the infections seen in the military facilities?*
- *What do we call infections that are acquired in a medical setting?*
- *What is the significance in these isolates being resistant to multiple antibiotics?*

Case File 13 Wrap-Up appears on page 402.

CHAPTER OVERVIEW

- The normal biota of humans includes bacteria, fungi, and protozoa that live on the body without causing disease. These microbes can be found in areas exposed to the outside environment, such as the gastrointestinal tract, skin, and respiratory tract, and are generally beneficial to humans.
- Pathogens are those microbes that infect the body and cause disease. Disease results when an adequate number of pathogenic cells enter the body through a specific route, grow, and disrupt tissues.
- Pathogens produce virulence factors such as toxins and enzymes that help them invade and damage the cells of their host. The effects of infection and disease are seen in the host as signs and symptoms, which may include both short- and long-term damage.

- Pathogens may be spread by direct or indirect means involving overtly infected people, carriers, vectors, and vehicles. A significant source of infection is exposure to the hospital environment.
- Pathogens may be found residing in humans, animals, food, soil, and water.
- The field of epidemiology is concerned with the patterns of disease occurrence in a population.

13.1 The Human Host

The human body exists in a state of dynamic equilibrium with microorganisms. In the healthy individual, this balance is maintained as a peaceful coexistence and lack of disease. But on occasion, the balance tips in favor of the microorganism, and an infection or disease results. In this chapter, we explore each component of the host-parasite relationship, beginning with the nature and function of normal biota, moving to the stages of infection and disease, and closing with a study of epidemiology and the patterns of disease in populations. These topics will set the scene for chapters 14 and 15, which deal with the ways the host defends itself against assault by microorganisms, and also for chapters 18 through 23, which examine diseases affecting different organ systems.

Contact, Infection, Disease—A Continuum

The body surfaces are constantly exposed to microbes. Some microbes become implanted there as colonists (normal biota), some are rapidly lost (transients), and others invade the tissues. Such intimate contact with microbes inevitably leads to **infection,** a condition in which pathogenic microorganisms penetrate the host defenses, enter the tissues, and multiply. When the cumulative effects of the infection damage or disrupt tissues and organs, the **pathologic** state that results is known as a disease. A disease is defined as any deviation from health. There are hundreds of different diseases caused by such factors as infections, diet, genetics, and aging. In this chapter, however, we discuss only **infectious disease**—the disruption of a tissue or organ caused by microbes or their products.

The pattern of the host-parasite relationship can be viewed as a series of stages that begins with contact, progresses to infection, and ends in disease. Because of numerous factors relating to host resistance and degree of pathogenicity, not all contacts lead to colonization, not all colonizations lead to infection, and not all infections lead to disease. In fact, contamination without colonization and colonization without disease are the rule.

Resident Biota: The Human as a Habitat

With its constant source of nourishment and moisture, relatively stable pH and temperature, and extensive surfaces upon which to settle, the human body provides a favorable habitat for an abundance of microorganisms. In fact, it is so favorable that, cell for cell, microbes on the human body outnumber human cells at least ten to one! The large and mixed collection of microbes adapted to the body has been variously called the **normal (resident) biota,** or indigenous biota, though some microbiologists prefer to use the terms *normal flora* and *commensals.* The normal residents include an array of bacteria, fungi, protozoa, and, to a certain extent, viruses and arthropods. These organisms have a profound effect on human biology.

Since it has become known that there are more unknown than known species that populate the human body, an effort is now underway to utilize **metagenomics** to identify the microbial profile inside and on humans. **The Human Microbiome Project** is funded by the National Institutes of Health and is being conducted at laboratories all over the country. The aim is to collect genetic sequences in the gut, respiratory tract, skin, etc., to determine which microbes are there, even when they can't be grown in the laboratory. A secondary aim is to determine what role these normal biota play in health and disease. When the project is completed, in several years, this chapter will look completely different. Keep in mind that the microbes we discuss here are those we can cultivate in the laboratory, although we know that many of them are indeed important. Keep in mind that the science of "normal biota" is in its infancy.

Acquiring Resident Biota

The human body offers a seemingly endless variety of environmental niches, with wide variations in temperature, pH, nutrients, and oxygen tension occurring from one area to another. Because the body provides such a range of habitats, it should not be surprising that the body supports a wide range of microbes. As shown in **tables 13.1** and **13.2,** most areas of the body in contact with the outside environment harbor resident microorganisms, while internal organs and tissue, along with the fluids they contain, are generally microbe-free.

TABLE 13.1 Sites That Harbor a Known Normal Biota

- Skin and its contiguous mucous membranes
- Upper respiratory tract
- Gastrointestinal tract (various parts)
- Outer opening of urethra
- External genitalia
- Vagina
- External ear canal
- External eye (lids, conjunctiva)

TABLE 13.2 Sterile (Microbe-Free) Anatomical Sites and Fluids

All internal tissues and organs
Heart and circulatory system
Liver
Kidneys and bladder
Lungs
Brain and spinal cord
Muscles
Bones
Ovaries/testes
Glands (pancreas, salivary, thyroid)
Sinuses
Middle and inner ear
Internal eye
Fluids within an organ or tissue
Blood
Urine in kidneys, ureters, bladder
Cerebrospinal fluid
Saliva prior to entering the oral cavity
Semen prior to entering the urethra
Amniotic fluid surrounding the embryo and fetus

Scientific work from the last five years has caused us to amend both of the statements in the previous sentence. The normal biota organisms listed in **table 13.3** will soon be augmented by dozens, if not hundreds, of species discovered in a large project designed to identify normal biota microbes through their genetic sequences. The microbes listed in table 13.3 were mostly identified by culturing them, a technique which we now know misses many, if not most, microbes in a habitat.

Secondly, it should be mentioned that several groups of scientists have recently reported detecting bacteria in blood from healthy humans and animals. They are bacteria that cannot be cultivated on laboratory media and were only detected using microscopy and molecular identification techniques. While these results are still somewhat controversial, they raise the possibility that microbes do exist in what we have long thought of as sterile body compartments. For the purposes of this book, we consider the blood and other sites listed in table 13.2 to be sterile. Discoveries such as these remind us that science is a continuously changing body of knowledge, and no scientific truth is ever "final."

The vast majority of microbes that come in contact with the body are removed or destroyed by the host's defenses long before they are able to colonize a particular area. Of those microbes able to establish an ongoing presence, an even smaller number are able to remain without attracting the unwanted attention of the body's defenses. This last group of organisms has evolved, along with its human hosts, to produce a complex relationship in which the effects of normal biota are generally not deleterious to the host.

A NOTE ON VIRUSES AS NORMAL BIOTA

In this and most other texts, the role of viruses as "normal biota" is not addressed. It seems microbiologists have never known how to characterize the non-disease-causing viruses and viral sequences we know are present in mammalian organisms. In 2006, researchers at Texas A & M showed that viruses called endogenous retroviruses (ERVs) are present in all mammals and are vital to healthy development of placentas in the sheep they studied.

It appears that ERVs account for about 8% to 10% of all DNA in mammals, including in humans. They seem to have originated from infections of mammals from thousands of years ago and have remained in mammals because the proteins they produce provide some real benefit to their hosts. In this case, the ERVs studied in sheep seem to be vital to healthy development of placentas and embryos. When researchers blocked the action of the ERVs, sheep miscarried at a high rate. Apparently, over time, the sheep and the viruses have coevolved to their mutual benefit.

Additionally, ERVs seem to play another beneficial role in mammals: They have been shown to be important for fighting off pathogenic viruses.

The role of "normal biota"—or coevolved—viruses in mammalian health will be an exciting area of research in coming years. Of course, viral sequences will be detected in the metagenomic search being conducted as part of the Human Microbiome Project.

Although generally stable, the biota can fluctuate to a limited extent with general health, age, variations in diet, hygiene, hormones, and drug therapy. In many cases, bacterial biota actually benefits the human host by preventing the overgrowth of harmful microorganisms. A common example is the fermentation of glycogen by lactobacilli, which keep the pH in the vagina quite acidic and prevent the overgrowth of the yeast *Candida albicans.*

The generally antagonistic effect "good" microbes have against intruder microorganisms is called **microbial antagonism.** Normal biota exist in a steady established relationship with the host and are unlikely to be displaced by incoming microbes. This antagonistic protection may simply be a result of a limited number of attachment sites in the host site, all of which are stably occupied by normal biota. Antagonism may also result from the chemical or physiological environment created by the resident biota, which is hostile to other microbes.

An increasing body of evidence suggests that the makeup of your intestinal biota can influence your tendency to be overweight. For example, scientists have found that obese humans have more bacteria in the group Firmicutes (see chapter 1) and fewer in the group Bacteroidetes. The ratios are reversed in people of normal weight. There is evidence that people harboring higher levels of Firmicutes in their intestines get more calories from their food than the other group. This area of investigation will certainly continue.

Characterizing the normal biota as beneficial or, at worst, commensal to the host presupposes that the host is in good

TABLE 13.3 Life on Humans: Sites Containing Well-Established Biota and Representative Examples

Anatomic Sites	Common Genera	Remarks
Skin	**Bacteria:** *Staphylococcus, Micrococcus, Corynebacterium, Propionibacterium, Streptococcus*	Microbes live only in upper dead layers of epidermis, glands, and follicles; dermis and layers below are sterile.
	Fungi: *Candida, Pityrosporum*	Dependent on skin lipids for growth
	Arthropods: *Demodix* mite	Present in sebaceous glands and hair follicles
Gastrointestinal Tract		
Oral cavity	**Bacteria:** *Streptococcus, Neisseria, Veillonella, Fusobacterium, Lactobacillus, Bacteroides, Actinomyces, Eikenella, Treponema, Haemophilus*	Colonize the epidermal layer of cheeks, gingiva, pharynx; surface of teeth; found in saliva in huge numbers
	Fungi: *Candida* species	Can cause thrush
	Protozoa: *Entamoeba gingivalis*	Inhabit the gingiva of persons with poor oral hygiene
Large intestine and rectum	**Bacteria:** *Bacteroides, Fusobacterium, Bifidobacterium, Clostridium,* fecal streptococci, *Lactobacillus,* coliforms (*Escherichia, Enterobacter*)	Areas of lower gastrointestinal tract other than large intestine and rectum have sparse or nonexistent biota. Biota consists predominantly of strict anaerobes; other microbes are aerotolerant or facultative.
	Fungi: *Candida*	Intestinal thrush
	Protozoa: *Entamoeba coli, Trichomonas hominis*	Feed on waste materials in the large intestine
Upper Respiratory Tract	Microbial population exists in the nasal passages, throat, and pharynx; owing to proximity, biota is similar to that of oral cavity.	Trachea and bronchi have a sparse population; bronchioles and alveoli have no normal biota and are essentially sterile.
Genital Tract	**Bacteria:** *Lactobacillus, Streptococcus, Corynebacterium, Escherichia*	In females, biota occupies the external genitalia and vaginal and cervical surfaces; internal reproductive structures normally remain sterile. Biota responds to hormonal changes during life.
	Fungi: *Candida*	Cause of yeast infections
Urinary Tract	**Bacteria:** *Staphylococcus, Streptococcus, Corynebacterium, Lactobacillus*	In females, biota exists only in the first portion of the urethral mucosa; the remainder of the tract is sterile. In males, the entire reproductive and urinary tract is sterile except for a short portion of the anterior urethra.

health, with a fully functioning immune system, and that the biota is present only in its natural microhabitat within the body. Hosts with compromised immune systems could very easily be infected by their own biota (see table 13.4). This outcome is seen when AIDS patients become sick with pneumococcal pneumonia, the causative agent of which (*Streptococcus pneumoniae*) is often carried as normal biota in the nasopharynx. **Endogenous** infections (those caused by biota that are already present in the body) can also occur when normal biota is introduced to a site that was previously sterile, as when *E. coli* enters the bladder, resulting in a urinary tract infection.

Initial Colonization of the Newborn

The uterus and its contents are normally sterile during embryonic and fetal development and remain essentially germ-free until just before birth. The event that first exposes the infant to microbes is the breaking of the fetal membranes, at which time microbes from the mother's vagina can enter the womb. Comprehensive exposure occurs during the birth process itself, when the baby unavoidably comes into intimate contact with the birth canal **(figure 13.1).** Within 8 to 12 hours after delivery, the newborn typically has been colonized by bacteria such as streptococci, staphylococci, and lactobacilli, acquired primarily from its mother. The nature of the biota initially colonizing the large intestine depends upon whether the baby is bottle- or breast-fed. Bottle-fed infants (receiving milk or a milk-based formula) tend to acquire a mixed population of coliforms, lactobacilli, enteric streptococci, and staphylococci. In contrast, the intestinal biota of breast-fed infants consists primarily of *Bifidobacterium* species whose growth is favored by a growth factor from the milk. This bacterium metabolizes sugars into acids that protect the infant from infection by certain intestinal pathogens. The skin, gastrointestinal tract, and portions of the respiratory and genitourinary tract all continue to be colonized as contact continues with family members, health care personnel, the environment, and food.

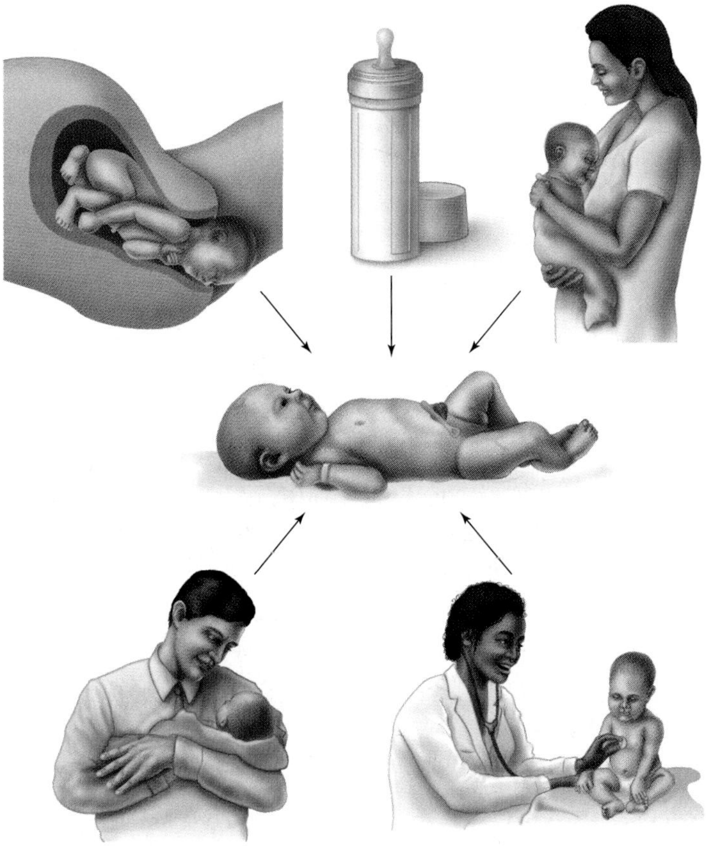

Figure 13.1 The origins of microbiota in newborns. A vaginal birth exposes babies to the biota of the mother's reproductive tract. From the moment of birth, the infant will begin to acquire microbes from its environment.

Indigenous Biota of Specific Regions

Although we tend to speak of the biota as a single unit, it is a complex mixture of hundreds of species, differing somewhat in quality and quantity from one individual to another. Studies of the biota have shown that most people harbor certain specially adapted bacteria, fungi, and protozoa. The normal, indigenous biota present in specific body sites is presented in detail in chapters 18 through 23. Table 13.3 provides an overview.

For a look into the laboratory study of resident biota, see **Insight 13.1.**

CHECKPOINT

- Humans are contaminated with microorganisms from the moment of birth onward. An infection is a condition in which contaminating microorganisms overcome host defenses, multiply, and cause disease, damaging tissues and organs.
- Normal biota reside on the skin and in the upper respiratory tract, the gastrointestinal tract, the outer parts of the urethra, the vagina, the eye, and the external ear canal.

13.2 The Progress of an Infection

A microbe whose relationship with its host is parasitic and results in infection and disease is termed a **pathogen.** The type and severity of infection depend both on the pathogenicity of the organism and the condition of the host **(figure 13.2). Pathogenicity,** you will recall, is a broad concept that describes an organism's potential to cause infection or disease, and is used to divide pathogenic microbes into one of two groups. **True pathogens** (primary pathogens) are capable of causing disease in healthy persons with normal immune defenses. They are generally associated with a specific, recognizable disease, which may vary in severity from mild (colds) to severe (malarial) to fatal (rabies). Examples of true pathogens include influenza virus, plague bacillus, and malarial protozoan.

Opportunistic pathogens cause disease when the host's defenses are compromised[1] or when they become established in a part of the body that is not natural to them. Opportunists are not considered pathogenic to a normal healthy person and, unlike primary pathogens, do not generally possess well-developed virulence properties. Examples of opportunistic pathogens include *Pseudomonas* species and *Candida albicans.* Factors that greatly predispose a person to infections, both primary and opportunistic, are shown in **table 13.4.**

The relative severity of the disease caused by a particular microorganism depends on the **virulence** of the microbe. Although the terms *pathogenicity* and *virulence* are often used interchangeably, virulence is the accurate term for describing the degree of pathogenicity. The virulence of a microbe is determined by its ability to

1. establish itself in the host, and
2. cause damage.

There is much involved in both of these steps. To establish themselves in a host, microbes must enter the host, attach firmly to host tissues, and survive the host defenses. To cause damage, microbes produce toxins or induce a host response that is actually injurious to the host. Any characteristic or structure of the microbe that contributes to the preceding activities is called a **virulence factor.** Virulence can be due to single or multiple factors. In some microbes, the causes of virulence are clearly established, but in others they are not. In the following section, we examine the effects of virulence factors while simultaneously outlining the stages in the progress of an infection.

Note that different healthy individuals have widely varying responses to the same microorganism. This is determined in part by genetic variation in the specific components of their defense systems. That is why the same infectious agent can cause severe disease in one individual and mild or no disease in another.

Why is there variation? In chapter 7, we described coevolution as changes in genetic composition by one species in response to change in another. Infectious agents evolve in

1. People with weakened immunity are often termed *immunocompromised.*

INSIGHT 13.1 Discovery

Life Without Microbiota

For years, questions lingered about how essential the microbiota is to normal life and what functions various members of the biota might serve. The need for animal models to further investigate these questions led eventually to development of laboratory strains of *germ-free*, or **axenic**, mammals and birds. The techniques and facilities required for producing and maintaining a germ-free colony are exceptionally rigorous. After the young mammals are taken from the mother aseptically by cesarean section, they are immediately transferred to a sterile isolator or incubator. The newborns must be fed by hand through gloved ports in the isolator until they can eat on their own, and all materials entering their chamber must be sterile. Rats, mice, rabbits, guinea pigs, monkeys, dogs, hamsters, and cats are some of the mammals raised in the germ-free state.

Experiments with germ-free animals are of two basic varieties: (1) general studies on how the lack of normal microbiota influences the nutrition, metabolism, and anatomy of the animal, and (2) **gnotobiotic** (noh″-toh-by-ah′-tik) studies, in which the germ-free subject is inoculated either with a single type of microbe to determine its individual effect or with several known microbes to determine interrelationships. Results are validated by comparing the germ-free group with a conventional, normal control group. **Table 13.A** summarizes some major conclusions arising from studies with germ-free animals.

A dramatic characteristic of germ-free animals is that they live longer and have fewer diseases than normal controls, as long as they remain in a sterile environment. From this standpoint, it is clear that the biota is not needed for survival and may even be the source of infectious agents. At the same time, it is also clear that axenic life is highly impractical. Additional studies have revealed important facts about the effect of the biota on various organs and systems. For example, the biota contributes significantly to the development of the immune system. When germ-free animals are placed in contact with normal control animals, they gradually develop a biota similar to that of the controls. However, germ-free subjects are less tolerant of microorganisms and can die from infections by relatively harmless species. This susceptibility is due to the immature character of the immune system of germ-free animals. These animals have a reduced number of certain types of white blood cells and slower antibody response.

Gnotobiotic experiments have clarified the dynamics of several infectious diseases. Perhaps the most striking discoveries were made in the case of oral diseases. For years, the precise involvement of microbes in dental caries had been ambiguous. Studies with germ-free rats, hamsters, and beagles confirmed that caries development is influenced by heredity, a diet high in sugars, and poor oral hygiene. Even when all these predisposing factors are present, however, germ-free animals still remain free of caries unless they have been inoculated with specific bacteria. Further discussion on dental diseases is found in chapter 22.

The ability of known pathogens to cause infection can also be influenced by normal biota, sometimes in opposing ways. Studies have indicated that germ-free animals are highly susceptible to experimental infection by the enteric pathogens *Shigella* and *Vibrio*, whereas normal animals are less susceptible, presumably because of their protective biota. In marked contrast, *Entamoeba histolytica* (the agent of amoebic dysentery) is more pathogenic in the normal animal than in the germ-free animal. One explanation for this phenomenon is that *E. histolytica* must feed on intestinal bacteria to complete its life cycle.

TABLE 13.A Effects of the Germ-Free State

Germ-Free Animals Display	Significance
Enlargement of the cecum; other degenerative diseases of the intestinal tract of rats, rabbits, chickens	Microbes are needed for normal intestinal development.
Vitamin deficiency in rats	Microbes are a significant nutritional source of vitamins.
Underdevelopment of immune system in most animals	Microbes are needed to stimulate development of certain host defenses.
Absence of dental caries and periodontal disease in dogs, rats, hamsters	Microbes are essential in caries formation and gum disease.
Heightened sensitivity to enteric pathogens (*Shigella, Salmonella, Vibrio cholerae*) and to fungal infections	Normal biota are antagonistic against pathogens.
Lessened susceptibility to amoebic dysentery	Normal biota facilitate the completion of the life cycle of the amoeba in the gut.

Sterile enclosure for rearing and handling germ-free laboratory animals.

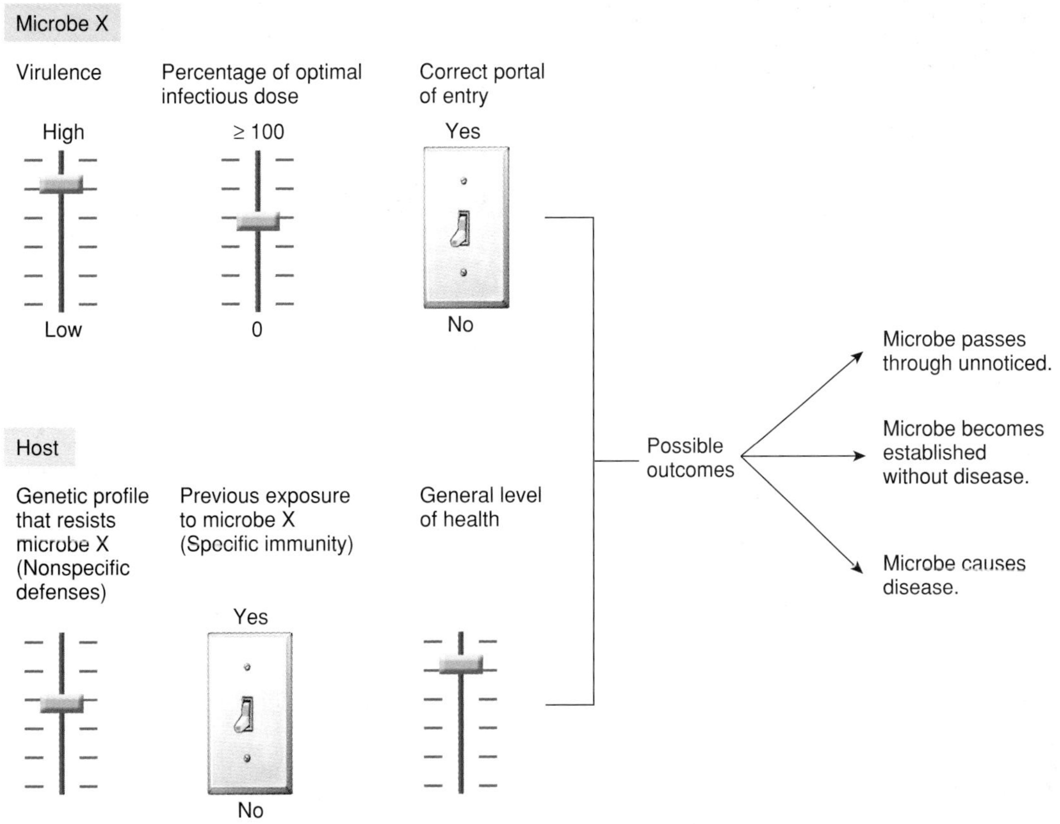

Figure 13.2 Factors determining if disease will result from an encounter between a (human) host and a microorganism. In most cases, all of the slider bars must be in the correct ranges and the microbe's toggle switch must be in the "yes" position while the host's toggle switch must be in the "no" position in order for disease to occur.

TABLE 13.4 Factors That Weaken Host Defenses and Increase Susceptibility to Infection*

- Old age and extreme youth (infancy, prematurity)
- Genetic defects in immunity and acquired defects in immunity (AIDS)
- Surgery and organ transplants
- Organic disease: cancer, liver malfunction, diabetes
- Chemotherapy/immunosuppressive drugs
- Physical and mental stress
- Other infections

*These conditions compromise defense barriers or immune responses.

response to their interaction with a host (as in the case of antibiotic resistance). Hosts evolve, too. And although their pace of change is much slower than that of a microbe, eventually changes show up in human populations due to their past experiences with pathogens. One striking example is sickle cell disease. Persons who are carriers of a mutation in their hemoglobin gene (i.e., who inherited one mutated hemoglobin gene and one normal) have few or no sickle cell disease symptoms but are more resistant to malaria than people who have no mutations in their hemoglobin genes. When a person inherits two alleles for the mutation (from both parents), they enjoy some protection from malaria but they will suffer from sickle cell disease.

People of West African descent are much more likely to have one or two sickle cell alleles. Malaria is endemic in West Africa. It seems the hemoglobin mutation is an adaptation of the human host to its long-standing relationship with the malaria protozoan.

Recognizing that classifying pathogens into only two categories may be unduly restrictive, the Centers for Disease Control and Prevention have adopted a system of biosafety categories for pathogens based on their degree of pathogenicity and the relative danger in handling them. This system assigns microbes to one of four levels or classes. Microbes not known to cause disease in humans are assigned to level 1, and highly contagious viruses that pose an extreme risk to humans are classified as level 4. Microbes of intermediate virulence are assigned to levels 2 or 3. This system is explained in more detail in **Insight 13.2.**

Becoming Established: Step One—Portals of Entry

To initiate an infection, a microbe enters the tissues of the body by a characteristic route, the **portal of entry,** usually a cutaneous or membranous boundary. The source of the infectious agent can be **exogenous,** originating from a source

INSIGHT 13.2 *Medical*

Laboratory Biosafety Levels and Classes of Pathogens

Personnel handling infectious agents in the laboratory must be protected from possible infection through special risk management or containment procedures. These involve:

1. carefully observing standard laboratory aseptic and sterile procedures while handling cultures and infectious samples;
2. using large-scale sterilization and disinfection procedures;
3. refraining from eating, drinking, and smoking; and
4. wearing personal protective items such as gloves, masks, safety glasses, laboratory coats, boots, and headgear.

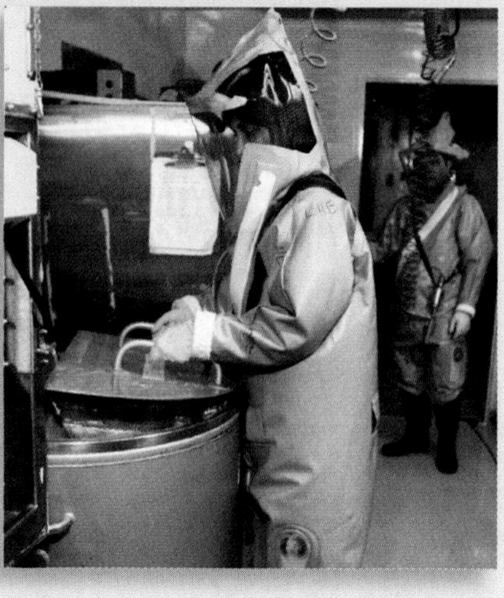

Some circumstances also require additional protective equipment such as biological safety cabinets for inoculations and specially engineered facilities to control materials entering and leaving the laboratory in the air and on personnel. **Table 13.B** summarizes the primary biosafety levels and agents of disease as characterized by the Centers for Disease Control and Prevention.

TABLE 13.B Primary Biosafety Levels and Agents of Disease

Biosafety Level	Facilities and Practices	Risk of Infection and Class of Pathogens
1	Standard, open bench, no special facilities needed; typical of most microbiology teaching labs; access may be restricted.	Low infection hazard; microbes not generally considered pathogens and will not colonize the bodies of healthy persons; *Micrococcus luteus, Bacillus megaterium, Lactobacillus, Saccharomyces.*
2	At least level 1 facilities and practices; plus personnel must be trained in handling pathogens; lab coats and gloves required; safety cabinets may be needed; biohazard signs posted; access restricted.	Agents with moderate potential to infect; class 2 pathogens can cause disease in healthy people but can be contained with proper facilities; most pathogens belong to class 2; includes *Staphylococcus aureus, Escherichia coli, Salmonella* spp., *Corynebacterium diphtheriae;* pathogenic helminths; hepatitis A, B, and rabies viruses; *Cryptococcus* and *Blastomyces.*
3	Minimum of level 2 facilities and practices; plus all manipulation performed in safety cabinets; lab designed with special containment features; only personnel with special clothing can enter; no unsterilized materials can leave the lab; personnel warned, monitored, and vaccinated against infection dangers.	Agents can cause severe or lethal disease especially when inhaled; class 3 microbes include *Mycobacterium tuberculosis, Francisella tularensis, Yersinia pestis, Brucella* spp., *Coxiella burnetii, Coccidioides immitis,* and yellow fever, WEE, and AIDS viruses.
4	Minimum of level 3 facilities and practices; plus facilities must be isolated with very controlled access; clothing changes and showers required for all people entering and leaving; materials must be autoclaved or fumigated prior to entering and leaving lab.	Agents are highly virulent microbes that pose extreme risk for morbidity and mortality when inhaled in droplet or aerosol form; most are exotic flaviviruses; arenaviruses, including Lassa fever virus; or filoviruses, including Ebola and Marburg viruses.

outside the body (the environment or another person or animal), or endogenous, already existing on or in the body (normal biota or latent infection).

For the most part, the portals of entry are the same anatomical regions that also support normal biota: the skin, gastrointestinal tract, respiratory tract, and urogenital tract. The majority of pathogens have adapted to a specific portal of entry, one that provides a habitat for further growth and spread. This adaptation can be so restrictive that if certain pathogens enter the "wrong" portal, they will not be infectious. For instance, inoculation of the nasal mucosa with the influenza virus invariably gives rise to the flu, but if this virus contacts only the skin, no infection will result. Likewise, contact with athlete's foot fungi in small cracks in the toe webs can induce an infection, but inhaling the fungus spores will not infect a healthy individual. Occasionally, an infective agent can enter by more than one portal. For instance, *Mycobacterium tuberculosis* enters through both the respiratory and gastrointestinal tracts, and pathogens in the genera *Streptococcus* and *Staphylococcus* have adapted to invasion through several portals of entry such as the skin, urogenital tract, and respiratory tract.

Infectious Agents That Enter the Skin

The skin is a very common portal of entry. The actual sites of entry are usually nicks, abrasions, and punctures (many of which are tiny and inapparent) rather than unbroken skin. Intact skin is a very tough barrier that few microbes can penetrate. *Staphylococcus aureus* (the cause of boils), *Streptococcus pyogenes* (an agent of impetigo), the fungal dermatophytes, and agents of gangrene and tetanus gain access through damaged skin. The viral agent of cold sores (herpes simplex, type 1) enters through the mucous membranes near the lips.

Some infectious agents create their own passageways into the skin using digestive enzymes. For example, certain helminth worms burrow through the skin directly to gain access to the tissues. Other infectious agents enter through bites. The bites of insects, ticks, and other animals offer an avenue to a variety of viruses, rickettsias, and protozoa. An artificial means for breaching the skin barrier is contaminated hypodermic needles by intravenous drug abusers. Users who inject drugs are predisposed to a disturbing list of well-known diseases: hepatitis, AIDS, tetanus, tuberculosis, osteomyelitis, and malaria. Contaminated needles often contain bacteria from the skin or environment that induce heart disease (endocarditis), lung abscesses, and chronic infections at the injection site.

Although the conjunctiva, the outer protective covering of the eye, is ordinarily a relatively good barrier to infection, bacteria such as *Haemophilus aegyptius* (pinkeye), *Chlamydia trachomatis* (trachoma), and *Neisseria gonorrhoeae* have a special affinity for this membrane.

The Gastrointestinal Tract as Portal

The gastrointestinal tract is the portal of entry for pathogens contained in food, drink, and other ingested substances. They are adapted to survive digestive enzymes and abrupt pH changes. The best-known enteric agents of disease are gram-negative rods in the genera *Salmonella, Shigella, Vibrio,* and certain strains of *Escherichia coli.* Viruses that enter through the gut are poliovirus, hepatitis A virus, echovirus, and rotavirus. Important enteric protozoans are *Entamoeba histolytica* (amoebiasis) and *Giardia lamblia* (giardiasis). Recent research has also shown that the intestines contain a wide variety of plant bacteria (which enter on food). It is not known whether these organisms cause disease, but scientists speculate they may be responsible for complaints that doctors can't diagnose. Although the anus is not a typical portal of entry, it becomes one in people who practice anal sex. See chapter 22 for details of these diseases.

The Respiratory Portal of Entry

The oral and nasal cavities are also the gateways to the respiratory tract, the portal of entry for the greatest number of pathogens. Because there is a continuous mucous membrane surface covering the upper respiratory tract, the sinuses, and the auditory tubes, microbes are often transferred from one site to another. The extent to which an agent is carried into the respiratory tree is based primarily on its size. In general, small cells and particles are inhaled more deeply than larger ones. Infectious agents with this portal of entry include the bacteria of streptococcal sore throat, meningitis, diphtheria, and whooping cough and the viruses of influenza, measles, mumps, rubella, chickenpox, and the common cold. Pathogens that are inhaled into the lower regions of the respiratory tract (bronchioles and lungs) can cause **pneumonia,** an inflammatory condition of the lung. Bacteria (*Streptococcus pneumoniae, Klebsiella, Mycoplasma*) and fungi (*Cryptococcus* and *Pneumocystis*) are a few of the agents involved in pneumonias. All types of pneumonia are on the increase owing to the greater susceptibility of AIDS patients and other immunocompromised hosts to them. Other agents causing unique recognizable lung diseases are *Mycobacterium tuberculosis* and fungal pathogens such as *Histoplasma.* Chapter 21 describes infections of the respiratory system.

Urogenital Portals of Entry

The urogenital tract is the portal of entry for pathogens that are contracted by sexual means (intercourse or intimate direct contact). **Sexually transmitted diseases (STDs)** account for an estimated 4% of infections worldwide, with approximately 13 million new cases occurring in the United States each year. The most recent available statistics (2006) for the estimated incidence of common STDs are provided in **table 13.5.**

The microbes of STDs enter the skin or mucosa of the penis, external genitalia, vagina, cervix, and urethra. Some can penetrate an unbroken surface; others require a cut or abrasion. The once predominant sexual diseases syphilis and gonorrhea have been supplanted by a large and growing list of STDs led by genital warts, chlamydia, and herpes. Evolving sexual practices have increased the incidence of STDs that

TABLE 13.5 Incidence of Common Sexually Transmitted Diseases

STD	Estimated Number of New Cases per Year in United States
Human papillomavirus	6,000,000
Trichomoniasis	5,000,000
Chlamydiosis	3,000,000
Herpes simplex	1,600,000
Gonorrhea	361,000
Hepatitis B	77,000
AIDS	41,002
Syphilis	32,200

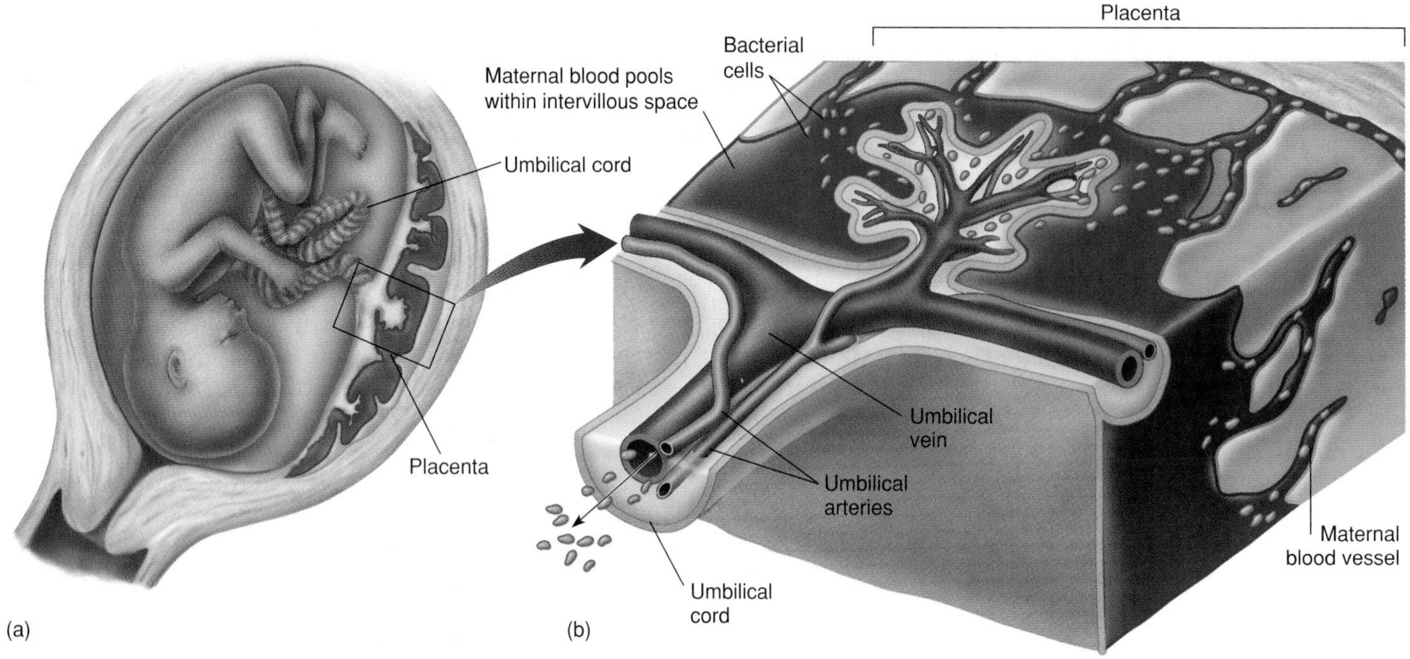

Figure 13.3 Transplacental infection of the fetus.
(a) Fetus in the womb. **(b)** In a closer view, microbes are shown penetrating the maternal blood vessels and entering the blood pool of the placenta. They then invade the fetal circulation by way of the umbilical vein.

were once uncommon, and diseases that were not originally considered STDs are now so classified.[2] Other common sexually transmitted agents are HIV (AIDS virus), *Trichomonas* (a protozoan), *Candida albicans* (a yeast), and hepatitis B virus. STDs are described in detail in chapter 23, with the exception of HIV (see chapter 20) and hepatitis B (see chapter 22).

Not all urogenital infections are STDs. Some of these infections are caused by displaced organisms (as when normal biota from the gastrointestinal tract cause urinary tract infections) or by opportunistic overgrowth of normal biota ("yeast infections").

Pathogens That Infect During Pregnancy and Birth

The placenta is an exchange organ—formed by maternal and fetal tissues—that separates the blood of the developing fetus from that of the mother yet permits diffusion of dissolved nutrients and gases to the fetus. The placenta is ordinarily an effective barrier against microorganisms in the maternal circulation. However, a few microbes such as the syphilis spirochete can cross the placenta, enter the umbilical vein, and spread by the fetal circulation into the fetal tissues **(figure 13.3).**

Other infections, such as herpes simplex, can occur perinatally when the child is contaminated by the birth canal. The common infections of fetus and neonate are grouped together in a unified cluster, known by the acronym **TORCH,** that medical personnel must monitor. TORCH stands for **t**oxoplasmosis, **o**ther diseases (hepatitis B, AIDS, and chlamydia), **r**ubella, **c**ytomegalovirus, and **h**erpes simplex virus. The most serious complications of TORCH infections are spontaneous abortion, congenital abnormalities, brain damage, prematurity, and stillbirths.

The Size of the Inoculum

Another factor crucial to the course of an infection is the quantity of microbes in the inoculating dose. For most agents, infection will proceed only if a minimum number, called the *infectious dose* (ID), is present. This number has been determined experimentally for many microbes. In general, microorganisms with smaller infectious doses have greater virulence. On the low end of the scale, the ID for rickettsia, the causative agent of Q fever, is only a single cell, and it is only about 10 infectious cells in tuberculosis, giardiasis, and coccidioidomycosis. The ID is 1,000 bacteria for gonorrhea and 10,000 bacteria for typhoid fever, in contrast to 1,000,000,000 bacteria in cholera. Numbers below an infectious dose will generally not result in an infection. But if the quantity is far in excess of the ID, the onset of disease can be extremely rapid.

Becoming Established: Step Two—Attaching to the Host

How Pathogens Attach

Adhesion is a process by which microbes gain a more stable foothold at the portal of entry. Because adhesion is

2. Amoebic dysentery, scabies, salmonellosis, and *Strongyloides* worms are examples.

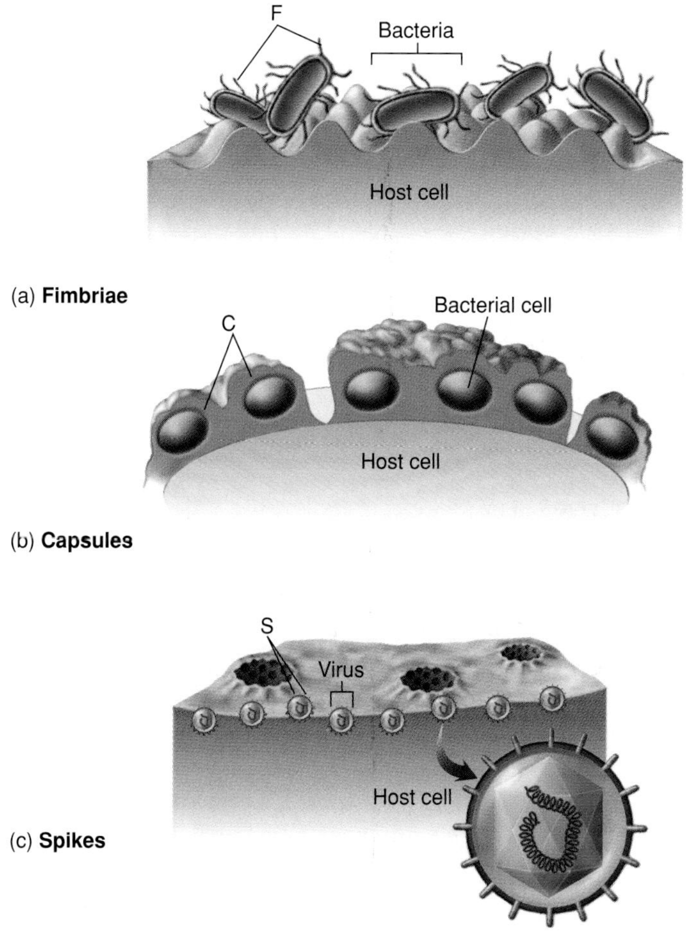

Figure 13.4 Mechanisms of adhesion by pathogens. **(a)** Fimbriae (F), minute bristlelike appendages. **(b)** Adherent extracellular capsules (C) made of slime or other sticky substances. **(c)** Viral envelope spikes (S). See table 13.6 for specific examples.

TABLE 13.6 Adhesive Properties of Microbes

Microbe	Disease	Adhesion Mechanism
Neisseria gonorrhoeae	Gonorrhea	Fimbriae attach to genital epithelium.
Escherichia coli	Diarrhea	Fimbrial adhesin
Shigella	Dysentery	Fimbriae attach to intestinal epithelium.
Mycoplasma	Pneumonia	Specialized tip at ends of bacteria fuse tightly to lung epithelium.
Pseudomonas aeruginosa	Burn, lung infections	Fimbriae and slime layer
Streptococcus pyogenes	Pharyngitis, impetigo	Lipotechoic acid and capsule anchor cocci to epithelium.
Streptococcus mutans, S. sobrinus	Dental caries	Dextran slime layer glues cocci to tooth surface after initial attachment.
Influenza virus	Influenza	Viral spikes attach to receptor on cell surface.
Poliovirus	Polio	Capsid proteins attach to receptors on susceptible cells.
HIV	AIDS	Viral spikes adhere to white blood cell receptors.
Giardia lamblia (protozoan)	Giardiasis	Small suction disc on underside attaches to intestinal surface.

dependent on binding between specific molecules on both the host and pathogen, a particular pathogen is limited to only those cells (and organisms) to which it can bind. Once attached, the pathogen is poised advantageously to invade the body compartments. Bacterial, fungal, and protozoal pathogens attach most often by mechanisms such as fimbriae (pili), surface proteins, and adhesive slimes or capsules; viruses attach by means of specialized receptors **(figure 13.4).** In addition, parasitic worms are mechanically fastened to the portal of entry by suckers, hooks, and barbs. Adhesion methods of various microbes and the diseases they lead to are shown in **table 13.6.** Firm attachment to host tissues is almost always a prerequisite for causing disease since the body has so many mechanisms for flushing microbes and foreign materials from its tissues.

Becoming Established: Step Three—Surviving Host Defenses

Microbes that are not established in a normal biota relationship in a particular body site in a host are likely to encounter resistance from host defenses when first entering, especially from certain white blood cells called **phagocytes.** These cells ordinarily engulf and destroy pathogens by means of enzymes and antimicrobial chemicals (see chapter 14).

Antiphagocytic factors are a type of virulence factor used by some pathogens to avoid phagocytes. The antiphagocytic factors of resistant microorganisms help them to circumvent some part of the phagocytic process (see figure 13.5c). The most aggressive strategy involves bacteria that kill phagocytes outright. Species of both *Streptococcus* and *Staphylococcus* produce **leukocidins,** substances that are toxic to white blood cells. Some microorganisms secrete an extracellular surface layer (slime or capsule) that makes it physically difficult for the phagocyte to engulf them. *Streptococcus pneumoniae, Salmonella typhi, Neisseria meningitidis,* and *Cryptococcus neoformans* are notable examples. Some bacteria are well adapted to survival inside phagocytes after ingestion. For instance, pathogenic species of *Legionella, Mycobacterium,* and many rickettsias are readily engulfed but are capable of avoiding further destruction. The ability to survive intracellularly in phagocytes has special significance because it provides a place for the microbes to hide, grow, and be spread throughout the body.

CHECKPOINT

- Microbial infections result when a microorganism penetrates host defenses, multiplies, and damages host tissue. The **pathogenicity** of a microbe refers to its ability to cause infection or disease. The *virulence* of a pathogen refers to the degree of damage it inflicts on the host tissues.
- *True pathogens* cause infectious disease in healthy hosts, whereas *opportunistic pathogens* become infectious only when the host immune system is compromised in some way.
- The site at which a microorganism first contacts host tissue is called the *portal of entry*. Most pathogens have one preferred portal of entry, although some have more than one.
- The respiratory system is the portal of entry for the greatest number of pathogens.
- The *infectious dose,* or ID, refers to the minimum number of microbial cells required to initiate infection in the host. The ID varies widely among microbial species.
- Fimbriae and adhesive capsules are types of adherence factors by which pathogens physically attach to host tissues.
- Antiphagocytic factors produced by microorganisms include leukocidins, capsules, and factors that resist digestion by white blood cells.

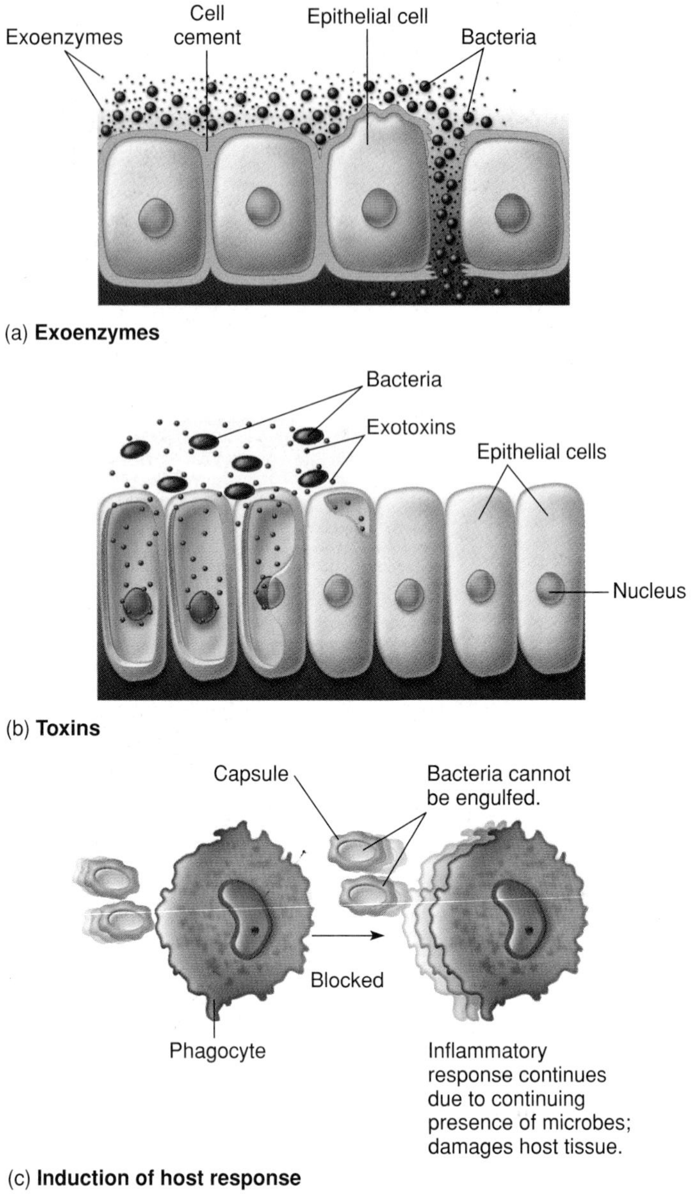

Figure 13.5 Three ways microbes damage the host. **(a)** Exoenzymes. Bacteria produce extracellular enzymes that dissolve intracellular connections and penetrate through or between cells to underlying tissues. **(b)** Toxins (primarily exotoxins) secreted by bacteria diffuse to target cells, which are poisoned and disrupted. **(c)** Bacterium has a property that enables it to escape phagocytosis and remain as an "irritant" to host defenses, which are deployed excessively.

Causing Disease

How Virulence Factors Contribute to Tissue Damage

Virulence factors from a microbe's perspective are simply adaptations it uses to invade and establish itself in the host. These same factors determine the degree of tissue damage that occurs. The effects of a pathogen's virulence factors on tissues vary greatly. Cold viruses, for example, invade and multiply but cause relatively little damage to their host. At the other end of the spectrum, pathogens such as *Clostridium tetani* or HIV severely damage or kill their host. Microorganisms either inflict direct damage on hosts through the use of exoenzymes or toxins **(figure 13.5*a,b*)**, or they cause damage indirectly when their presence causes an excessive or inappropriate host response **(figure 13.5*c*)**. For convenience, we divide the "directly damaging" virulence factors into exoenzymes and toxins. Although this distinction is useful, there is often a very fine line between enzymes and toxins because many substances called toxins actually function as enzymes.

Microbial virulence factors are often responsible for inducing the host to cause damage, as well. The capsule of *Streptococcus pneumoniae* is a good example. Its presence prevents the bacterium from being cleared from the lungs by phagocytic cells, leading to a continuous influx of fluids into the lung spaces, and the condition we know as pneumonia.

Extracellular Enzymes Many pathogenic bacteria, fungi, protozoa, and worms secrete **exoenzymes** that break down and inflict damage on tissues. Other enzymes dissolve the host's defense barriers and promote the spread of microbes to deeper tissues.

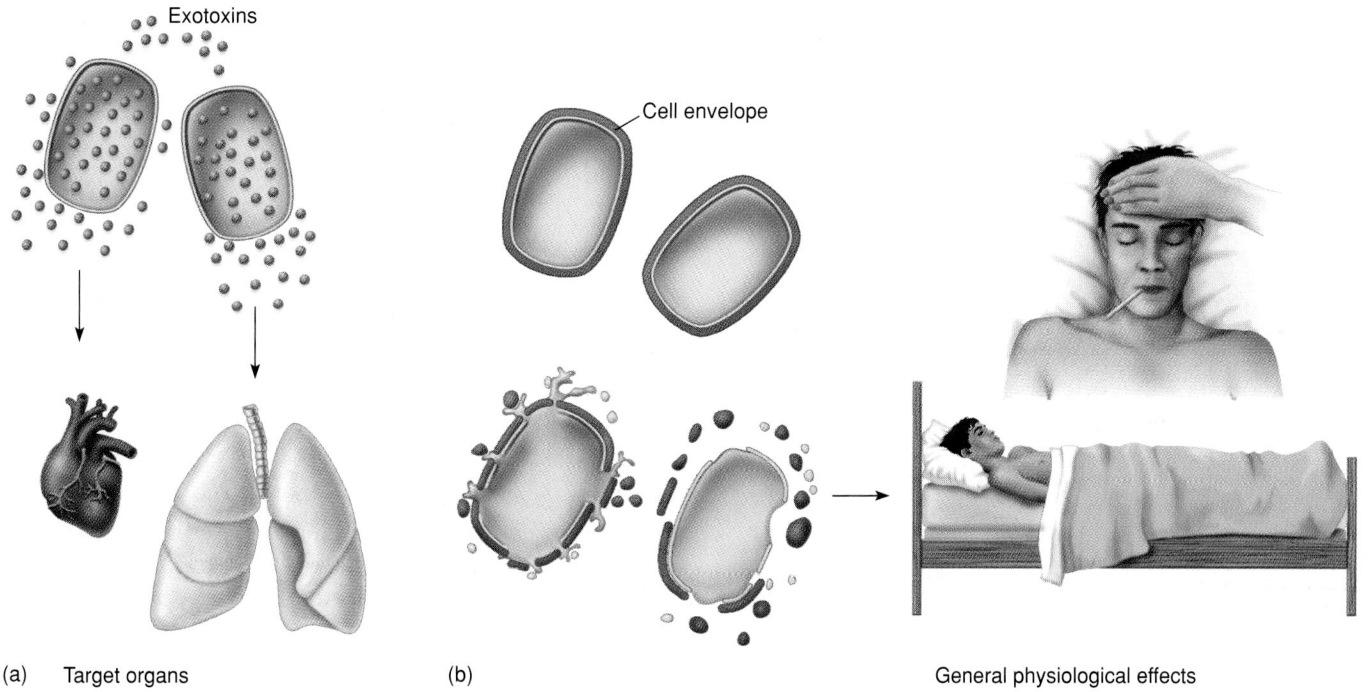

Figure 13.6 The origins and effects of circulating exotoxins and endotoxin.
(a) Exotoxins, given off by live cells, have highly specific targets and physiological effects. **(b)** Endotoxin, given off when the cell wall of gram-negative bacteria disintegrates, has more generalized physiological effects.

Examples of enzymes are:

1. mucinase, which digests the protective coating on mucous membranes and is a factor in amoebic dysentery;
2. keratinase, which digests the principal component of skin and hair, and is secreted by fungi that cause ringworm;
3. collagenase, which digests the principal fiber of connective tissue and is an invasive factor of *Clostridium* species and certain worms; and
4. hyaluronidase, which digests hyaluronic acid, the ground substance that cements animal cells together. This enzyme is an important virulence factor in staphylococci, clostridia, streptococci, and pneumococci.

Some enzymes react with components of the blood. Coagulase, an enzyme produced by pathogenic staphylococci, causes clotting of blood or plasma. By contrast, the bacterial kinases (streptokinase, staphylokinase) do just the opposite, dissolving fibrin clots and expediting the invasion of damaged tissues. In fact, one form of streptokinase (streptase) is marketed as a therapy to dissolve blood clots in patients with problems with thrombi and emboli.[3]

Bacterial Toxins: A Potent Source of Cellular Damage A **toxin** is a specific chemical product of microbes, plants, and some animals that is poisonous to other organisms. **Toxigenicity,** the power to produce toxins, is a genetically controlled characteristic of many species and is responsible for the adverse effects of a variety of diseases generally called **toxinoses.** Toxinoses in which the toxin is spread by the blood from the site of infection are called **toxemias** (tetanus and diphtheria, for example), whereas those caused by ingestion of toxins are **intoxications** (botulism). A toxin is named according to its specific target of action: Neurotoxins act on the nervous system; enterotoxins act on the intestine; hemotoxins lyse red blood cells; and nephrotoxins damage the kidneys.

A more traditional scheme classifies toxins according to their origins **(figure 13.6).** A toxin molecule secreted by a living bacterial cell into the infected tissues is an **exotoxin.** A toxin that is not actively secreted but is shed from the outer membrane is an **endotoxin.** Other important differences between the two groups are summarized in **table 13.7.**

Exotoxins are proteins with a strong specificity for a target cell and extremely powerful, sometimes deadly, effects. They generally affect cells by damaging the cell membrane and initiating lysis or by disrupting intracellular function. **Hemolysins** (hee-mahl′-uh-sinz) are a class of bacterial exotoxin that disrupts the cell membrane of red blood cells (and some other cells, too). This damage causes the red blood cells to **hemolyze**—to burst and release hemoglobin pigment. Hemolysins that increase pathogenicity include the streptolysins of *Streptococcus pyogenes* and the alpha (α) and beta (β) toxins of *Staphylococcus aureus.* When colonies of bacteria growing on blood agar produce hemolysin, distinct zones appear around the colony. The pattern of hemolysis is often used to identify bacteria and determine their degree of pathogenicity.

The exotoxins of diphtheria, tetanus, and botulism, among others, attach to a particular target cell, become internalized,

3. These conditions are intravascular blood clots that can cause circulatory obstructions.

TABLE 13.7 Differential Characteristics of Bacterial Exotoxins and Endotoxin

Characteristic	Exotoxins	Endotoxin
Toxicity	Toxic in minute amounts	Toxic in high doses
Effects on the Body	Specific to a cell type (blood, liver, nerve)	Systemic: fever, inflammation
Chemical Composition	Small proteins	Lipopolysaccharide of cell wall
Heat Denaturation at 60°C	Unstable	Stable
Toxoid Formation	Can be converted to toxoid*	Cannot be converted to toxoid
Immune Response	Stimulate antitoxins**	Does not stimulate antitoxins
Fever Stimulation	Usually not	Yes
Manner of Release	Secreted from live cell	Released by cell via shedding or during lysis
Typical Sources	A few gram-positive and gram-negative	All gram-negative bacteria

*A toxoid is an inactivated toxin used in vaccines.

**An antitoxin is an antibody that reacts specifically with a toxin.

and interrupt an essential cell pathway. The consequences of cell disruption depend upon the target. One toxin of *Clostridium tetani* blocks the action of certain spinal neurons; the toxin of *Clostridium botulinum* prevents the transmission of nerve-muscle stimuli; pertussis toxin inactivates the respiratory cilia; and cholera toxin provokes profuse salt and water loss from intestinal cells. More details of the pathology of exotoxins are found in later chapters on specific diseases.

In contrast to the category *exotoxin,* which contains many specific examples, the word *endotoxin* refers to a single substance. Endotoxin is actually a chemical called lipopolysaccharide (LPS), which is part of the outer membrane of gram-negative cell walls. Gram-negative bacteria release these LPS molecules into tissues or into the circulation. Endotoxin differs from exotoxins in having a variety of systemic effects on tissues and organs. Depending upon the amounts present, endotoxin can cause fever, inflammation, hemorrhage, and diarrhea. Blood infection by gram-negative bacteria such as *Salmonella, Shigella, Neisseria meningitidis,* and *Escherichia coli* are particularly dangerous, in that it can lead to fatal endotoxic shock.

Inducing an Injurious Host Response Despite the extensive discussion on direct virulence factors, such as enzymes and toxins, it is probably the case that more microbial diseases are the result of indirect damage, or the host's excessive or inappropriate response to a microorganism. This is an extremely important point because it means that pathogenicity is not a trait inherent in microorganisms, but is really a consequence of the interplay between microbe and host.

A NOTE ABOUT TERMINOLOGY

Words in medicine have great power and economy. A single technical term can often replace a whole phrase or sentence, thereby saving time and space in patient charting. The beginning student may feel overwhelmed by what seems like a mountain of new words. However, having a grasp of a few root words and a fair amount of anatomy can help you learn many of these words and even deduce the meaning of unfamiliar ones. Some examples of medical shorthand follow.

The suffix *-itis* means an inflammation and, when affixed to the end of an anatomical term, indicates an inflammatory condition in that location. Thus, meningitis is an inflammation of the meninges surrounding the brain; encephalitis is an inflammation of the brain itself; hepatitis involves the liver; vaginitis, the vagina; gastroenteritis, the intestine; and otitis media, the middle ear. Although not all inflammatory conditions are caused by infections, many infectious diseases inflame their target organs.

The suffix *-emia* is derived from the Greek word *haeima,* meaning blood. When added to a word, it means "associated with the blood." Thus, septicemia means sepsis (infection) of the blood; bacteremia, bacteria in the blood; viremia, viruses in the blood; and fungemia, fungi in the blood. It is also applicable to specific conditions such as toxemia, gonococcemia, and spirochetemia.

The suffix *-osis* means "a disease or morbid process." It is frequently added to the names of pathogens to indicate the disease they cause: for example, listeriosis, histoplasmosis, toxoplasmosis, shigellosis, salmonellosis, and borreliosis. A variation of this suffix is *-iasis,* as in trichomoniasis and candidiasis.

The suffix *-oma* comes from the Greek word *onkomas* (swelling) and means tumor. Although it is often used to describe cancers (sarcoma, melanoma), it is also applied in some infectious diseases that cause masses or swellings (tuberculoma, leproma).

Of course, it is easier to study and characterize the microbes that cause direct damage through toxins or enzymes. For this reason, these true pathogens were the first to be fully understood as the science of microbiology progressed. But in the last 15 to 20 years, microbiologists have come to appreciate exactly how important the relationship between microbe and host is, and this has greatly expanded our understanding of infectious diseases.

The various ways in which a microbe/host interaction causes damage are described in the chapters describing specific diseases (see chapters 18–23).

INSIGHT 13.3 — Medical

The Classic Stages of Clinical Infections

As the body of the host responds to the invasive and toxigenic activities of a parasite, it passes through four distinct phases of infection and disease: the incubation period, the prodrome, the period of invasion, and the convalescent period.

The **incubation period** is the time from initial contact with the infectious agent (at the portal of entry) to the appearance of the first symptoms. During the incubation period, the agent is multiplying at the portal of entry but has not yet caused enough damage to elicit symptoms. Although this period is relatively well defined and predictable for each microorganism, it does vary according to host resistance, degree of virulence, and distance between the target organ and the portal of entry (the farther apart, the longer the incubation period). Overall, an incubation period can range from several hours in pneumonic plague to several years in leprosy. The majority of infections, however, have incubation periods ranging between 2 and 30 days.

The earliest notable symptoms of infection appear as a vague feeling of discomfort, such as head and muscle aches, fatigue, upset stomach, and general malaise. This short period (1–2 days) is known as the **prodromal stage.** The infectious agent next enters a **period of invasion,** during which it multiplies at high levels, exhibits its greatest toxicity, and becomes well established in its target tissue. This period is often marked by fever and other prominent and more specific signs and symptoms, which can include cough, rashes, diarrhea, loss of muscle control, swelling, jaundice, discharge of exudates, or severe pain, depending on the particular infection. The length of this period is extremely variable.

As the patient begins to respond to the infection, the symptoms decline—sometimes dramatically, other times slowly. During the recovery that follows, called the **convalescent period,** the patient's strength and health gradually return owing to the healing nature of the immune response. An infection that results in death is called terminal.

The transmissibility of the microbe during these four stages must be considered on an individual basis. A few agents are released mostly during incubation (measles, for example); many are released during the invasive period (*Shigella*); and others can be transmitted during all of these periods (hepatitis B).

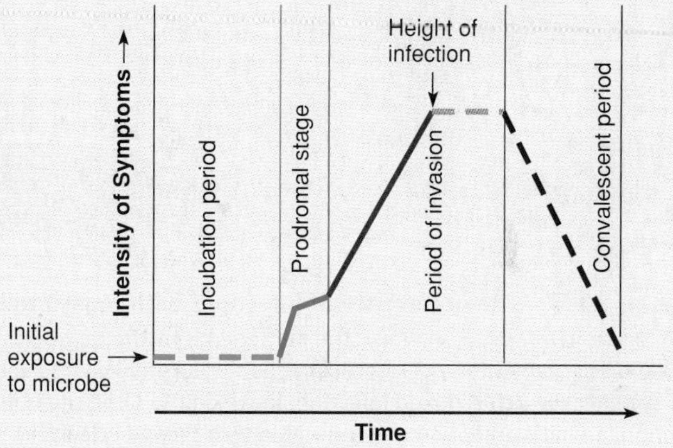

Stages in the course of infection and disease. Dashed lines represent periods with a variable length.

The Process of Infection and Disease

Establishment, Spread, and Pathologic Effects

Aided by virulence factors, microbes eventually settle in a particular target organ and continue to cause damage at the site. The type and scope of injuries inflicted during this process account for the typical stages of an infection **(Insight 13.3),** the patterns of the infectious disease, and its manifestations in the body.

In addition to the adverse effects of enzymes, toxins, and other factors, multiplication by a pathogen frequently weakens host tissues. Pathogens can obstruct tubular structures such as blood vessels, lymphatic channels, fallopian tubes, and bile ducts. Accumulated damage can lead to cell and tissue death, a condition called **necrosis.** Although viruses do not produce toxins or destructive enzymes, they destroy cells by multiplying in and lysing them. Many of the cytopathic effects of viral infection arise from the impaired metabolism and death of cells (see chapter 6).

Patterns of Infection Patterns of infection are many and varied. In the simplest situation, a **localized infection,** the microbe enters the body and remains confined to a specific tissue **(figure 13.7*a*).** Examples of localized infections are boils, fungal skin infections, and warts.

Many infectious agents do not remain localized but spread from the initial site of entry to other tissues. In fact, spreading is necessary for pathogens such as rabies and hepatitis A virus, whose target tissue is some distance from the site of entry. The rabies virus travels from a bite wound along nerve tracts to its target in the brain, and the hepatitis A virus moves from the intestine to the liver via the circulatory system. When an infection spreads to several sites and tissue fluids, usually in the bloodstream, it is called a **systemic infection (figure 13.7*b*).** Examples of systemic infections are viral diseases (measles, rubella, chickenpox, and AIDS); bacterial diseases (brucellosis, anthrax, typhoid fever, and syphilis); and fungal diseases (histoplasmosis and cryptococcosis). Infectious agents can also travel to their targets by means of nerves (as in rabies) or cerebrospinal fluid (as in meningitis).

A **focal infection** is said to exist when the infectious agent breaks loose from a local infection and is carried into other tissues **(figure 13.7*c*).** This pattern is exhibited by tuberculosis or

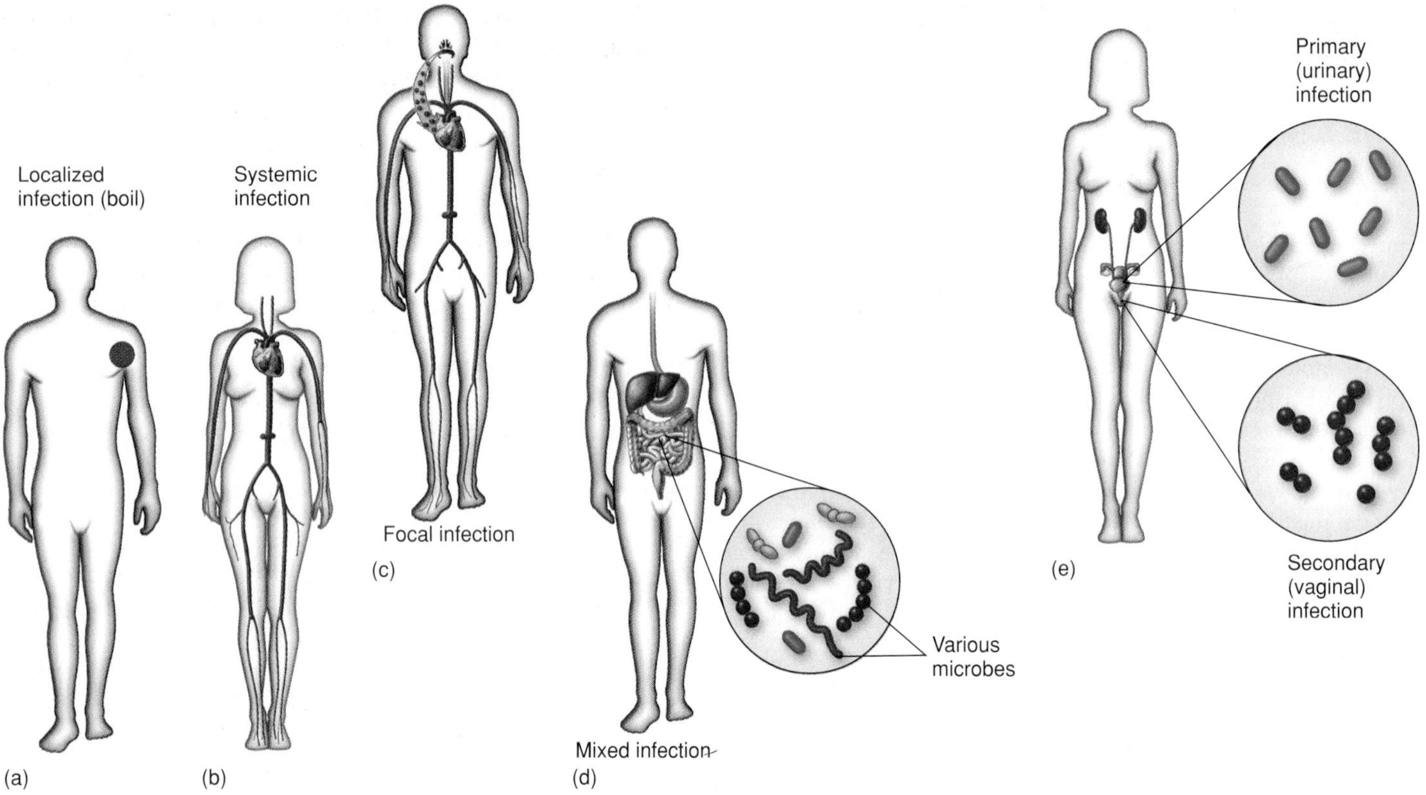

Figure 13.7 The occurrence of infections with regard to location, type of microbe, and order of infection. **(a)** A localized infection, in which the pathogen is restricted to one specific site. **(b)** Systemic infection, in which the pathogen spreads through circulation to many sites. **(c)** A focal infection occurs initially as a local infection, but circumstances cause the microbe to be carried to other sites systemically. **(d)** A mixed infection, in which the same site is infected with several microbes at the same time. **(e)** In a primary-secondary infection, an initial infection is complicated by a second one in the same or a different location and caused by a different microbe.

by streptococcal pharyngitis, which gives rise to scarlet fever. In the condition called toxemia,[4] the infection itself remains localized at the portal of entry, but the toxins produced by the pathogens are carried by the blood to the actual target tissue. In this way, the target of the bacterial cells can be different from the target of their toxin.

An infection is not always caused by a single microbe. In a **mixed infection,** several agents establish themselves simultaneously at the infection site **(figure 13.7*d*).** In some mixed or synergistic infections, the microbes cooperate in breaking down a tissue. In other mixed infections, one microbe creates an environment that enables another microbe to invade. Gas gangrene, wound infections, dental caries, and human bite infections tend to be mixed. These are sometimes called **polymicrobial** diseases.

Some diseases are described according to a sequence of infection. When an initial, or **primary, infection** is complicated by another infection caused by a different microbe, the second infection is termed a **secondary infection (figure 13.7*e*).** This pattern often occurs in a child with chickenpox (primary infection) who may scratch his pox and infect them with *Staphylococcus aureus* (secondary infection). The secondary infection need not be in the same site as the primary infection, and it usually indicates altered host defenses.

Infections that come on rapidly, with severe but short-lived effects, are called **acute infections.** Infections that progress and persist over a long period of time are **chronic infections.**

Signs and Symptoms: Warning Signals of Disease

When an infection causes pathologic changes leading to disease, it is often accompanied by a variety of signs and symptoms. A **sign** is any objective evidence of disease as noted by an observer; a **symptom** is the subjective evidence of disease as sensed by the patient. In general, signs are more precise than symptoms, though both can have the same underlying cause. For example, an infection of the brain might present with the sign of bacteria in the spinal fluid and symptom of headache. Or a streptococcal infection might produce a sore throat (symptom) and inflamed pharynx (sign). Disease indicators that can be sensed and observed can qualify as either a sign or a symptom. When a disease can be identified or defined by a certain complex of signs and symptoms, it is

4. Not to be confused with toxemia of pregnancy, which is a metabolic disturbance and not an infection.

TABLE 13.8 Common Signs and Symptoms of Infectious Diseases

Signs	Symptoms
Fever	Chills
Septicemia	Pain, ache, soreness, irritation
Microbes in tissue fluids	Malaise
Chest sounds	Fatigue
Skin eruptions	Chest tightness
Leukocytosis	Itching
Leukopenia	Headache
Swollen lymph nodes	Nausea
Abscesses	Abdominal cramps
Tachycardia (increased heart rate)	Anorexia (lack of appetite)
Antibodies in serum	Sore throat

termed a **syndrome.** Signs and symptoms with considerable importance in diagnosing infectious diseases are shown in **table 13.8.** Specific signs and symptoms for particular infectious diseases are covered in chapters 18 through 23.

Signs and Symptoms of Inflammation

The earliest symptoms of disease result from the activation of the body defense process called **inflammation.** The inflammatory response includes cells and chemicals that respond nonspecifically to disruptions in the tissue. This subject is discussed in greater detail in chapter 14, but as noted earlier, many signs and symptoms of infection are caused by the mobilization of this system. Some common symptoms of inflammation include fever, pain, soreness, and swelling. Signs of inflammation include **edema,** the accumulation of fluid in an afflicted tissue; **granulomas** and **abscesses,** walled-off collections of inflammatory cells and microbes in the tissues; and **lymphadenitis,** swollen lymph nodes.

Rashes and other skin eruptions are common symptoms and signs in many diseases, and because they tend to mimic each other, it can be difficult to differentiate among diseases on this basis alone. The general term for the site of infection or disease is **lesion.** Skin lesions can be restricted to the epidermis and its glands and follicles, or they can extend into the dermis and subcutaneous regions. The lesions of some infections undergo characteristic changes in appearance during the course of disease and thus fit more than one category (see Insight 18.3).

Signs of Infection in the Blood

Changes in the number of circulating white blood cells, as determined by special counts, are considered to be signs of possible infection. **Leukocytosis** (loo″-koh′-sy-toh′-sis) is an increase in the level of white blood cells, whereas **leukopenia** (loo″-koh-pee′-nee-uh) is a decrease. Other signs of infection revolve around the occurrence of a microbe or its products in the blood. The clinical term for blood infection, **septicemia,** refers to a general state in which microorganisms are multiplying in the blood and are present in large numbers. When small numbers of bacteria or viruses are found in the blood, the correct terminology is **bacteremia** or **viremia,** which means that these microbes are present in the blood but are not necessarily multiplying.

During infection, a normal host will invariably show signs of an immune response in the form of antibodies in the serum or some type of sensitivity to the microbe. This fact is the basis for several serological tests used in diagnosing infectious diseases such as AIDS or syphilis. Such specific immune reactions indicate the body's attempt to develop specific immunities against pathogens. We concentrate on this role of the host defenses in chapters 14 and 15.

Infections That Go Unnoticed

It is rather common for an infection to produce no noticeable symptoms, even though the microbe is active in the host tissue. In other words, although infected, the host does not manifest the disease. Infections of this nature are known as **asymptomatic, subclinical,** or *inapparent* because the patient experiences no symptoms or disease and does not seek medical attention. However, it is important to note that most infections are attended by some sort of sign. In the section on epidemiology, we further address the significance of subclinical infections in the transmission of infectious agents.

The Portal of Exit: Vacating the Host

Earlier, we introduced the idea that a parasite is considered *unsuccessful* if it does not have a provision for leaving its host and moving to other susceptible hosts. With few exceptions, pathogens depart by a specific avenue called the **portal of exit (figure 13.8).** In most cases, the pathogen is shed or released from the body through secretion, excretion, discharge, or sloughed tissue. The usually very high number of infectious agents in these materials increases the likelihood that the pathogen will reach other hosts. In many cases, the portal of exit is the same as the portal of entry, but a few pathogens use a different route. As we see in the next section, the portal of exit concerns epidemiologists because it greatly influences the dissemination of infection in a population.

Respiratory and Salivary Portals

Mucus, sputum, nasal drainage, and other moist secretions are the media of escape for the pathogens that infect the lower or upper respiratory tract. The most effective means of releasing these secretions are coughing and sneezing (see figure 13.12), although they can also be released during talking and laughing. Tiny particles of liquid released into the air form aerosols or droplets that can spread the infectious agent to other people. The agents of tuberculosis, influenza, measles, and chickenpox most often leave the host through airborne droplets. Droplets of saliva are the exit route for several viruses, including those of mumps, rabies, and infectious mononucleosis.

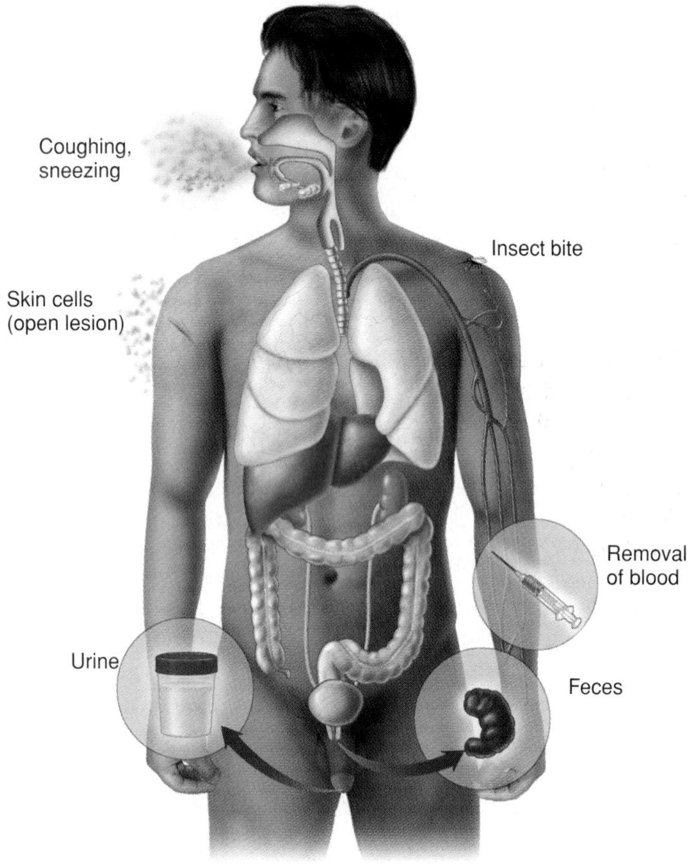

Figure 13.8 Major portals of exit of infectious diseases.

Skin Scales

The outer layer of the skin and scalp is constantly being shed into the environment. A large proportion of household dust is actually composed of skin cells. A single person can shed several billion skin cells a day, and some persons, called shedders, disseminate massive numbers of bacteria into their immediate surroundings. Skin lesions and their exudates can serve as portals of exit in warts, fungal infections, boils, herpes simplex, smallpox, and syphilis.

Fecal Exit

Feces are a very common portal of exit. Some intestinal pathogens grow in the intestinal mucosa and create an inflammation that increases the motility of the bowel. This increased motility speeds up peristalsis, resulting in diarrhea, and the more fluid stool provides a rapid exit for the pathogen. A number of helminth worms release cysts and eggs through the feces (see chapter 22). Feces containing pathogens are a public health problem when allowed to contaminate drinking water or when used to fertilize crops.

Urogenital Tract

A number of agents involved in sexually transmitted infections leave the host in vaginal discharge or semen. This is also the source of neonatal infections such as herpes simplex, *Chlamydia*, and *Candida albicans*, which infect the infant as it passes through the birth canal. Less commonly, certain pathogens that infect the kidney are discharged in the urine: for instance, the agents of leptospirosis, typhoid fever, tuberculosis, and schistosomiasis.

Removal of Blood or Bleeding

Although the blood does not have a direct route to the outside, it can serve as a portal of exit when it is removed or released through a vascular puncture made by natural or artificial means. Blood-feeding animals such as ticks and fleas are common transmitters of pathogens (see Insight 20.2). The AIDS and hepatitis viruses are transmitted by shared needles or through small gashes in a mucous membrane caused by sexual intercourse. Blood donation is also a means for certain microbes to leave the host, though this means of exit is now unusual because of close monitoring of the donor population and blood used for transfusions.

The Persistence of Microbes and Pathologic Conditions

The apparent recovery of the host does not always mean that the microbe has been completely removed or destroyed by the host defenses. After the initial symptoms in certain chronic infectious diseases, the infectious agent retreats into a dormant state called **latency.** Throughout this latent state, the microbe can periodically become active and produce a recurrent disease. The viral agents of herpes simplex, herpes zoster, hepatitis B, AIDS, and Epstein-Barr can persist in the host for long periods. The agents of syphilis, typhoid fever, tuberculosis, and malaria also enter into latent stages. The person harboring a persistent infectious agent may or may not shed it during the latent stage. If it is shed, such persons are chronic carriers who serve as sources of infection for the rest of the population.

Some diseases leave **sequelae** in the form of long-term or permanent damage to tissues or organs. For example, meningitis can result in deafness, a strep throat can lead to rheumatic heart disease, Lyme disease can cause arthritis, and polio can produce paralysis.

CHECKPOINT

- Exoenzymes, toxins, and the ability to induce injurious host responses are the three main types of *virulence factors* pathogens utilize to combat host defenses and damage host tissue.
- Exotoxins and endotoxins differ in their chemical composition and tissue specificity.
- Characteristics or structures of microbes that induce extreme host responses are a major factor in most infectious diseases.
- Patterns of infection vary with the pathogen or pathogens involved. They range from local and focal to systemic.
- A mixed infection is caused by two or more microorganisms simultaneously.

- Infections can be characterized by their sequence as primary or secondary and by their duration as either acute or chronic.
- An infectious disease is characterized by both objective signs and subjective symptoms.
- Infectious diseases that are asymptomatic or subclinical nevertheless often produce clinical signs.
- The portal of exit by which a pathogen leaves its host is often but not always the same as the portal of entry.
- The portals of exit and entry determine how pathogens spread in a population.
- Some pathogens persist in the body in a latent state.

Reservoirs: Where Pathogens Persist

In order for an infectious agent to continue to exist and be spread, it must have a permanent place to reside. The **reservoir** is the primary habitat in the natural world from which a pathogen originates. Often it is a human or animal carrier, although soil, water, and plants are also reservoirs. The reservoir can be distinguished from the infection **source,** which is the individual or object from which an infection is actually acquired. In diseases such as syphilis, the reservoir and the source are the same (the human body). In the case of hepatitis A, the reservoir (a human carrier) is usually different from the source of infection (contaminated food).

Living Reservoirs

Persons or animals with frank symptomatic infection are obvious sources of infection, but a **carrier** is, by definition, an individual who *inconspicuously* shelters a pathogen and spreads it to others without any notice. Although human carriers are occasionally detected through routine screening (blood tests, cultures) and other epidemiological devices, they are unfortunately very difficult to discover and control. As long as a pathogenic reservoir is maintained by the carrier state, the disease will continue to exist in that population, and the potential for epidemics will be a constant threat. The duration of the carrier state can be short or long term, and the carrier may or may not have experienced disease due to the microbe.

Several situations can produce the carrier state. **Asymptomatic** (apparently healthy) **carriers** are infected, but as previously indicated, they show no symptoms **(figure 13.9*a*).** A few asymptomatic infections (gonorrhea and genital warts, for instance) can carry out their entire course without overt manifestations. **Figure 13.9*b*** demonstrates three types of carriers who have had or will have the disease but do not at the time they transmit the organism. *Incubation carriers* spread the infectious agent during the incubation period. For example, AIDS patients can harbor and spread the virus for months and years before their first symptoms appear. Recuperating patients without symptoms are considered *convalescent carriers* when they continue to shed viable microbes and convey the infection to others. Diphtheria patients, for example, spread the microbe for up to 30 days after the disease has subsided.

An individual who shelters the infectious agent for a long period after recovery because of the latency of the infectious agent is a *chronic carrier.* Patients who have recovered from tuberculosis, hepatitis, and herpes infections frequently carry the agent chronically. About one in 20 victims of typhoid fever continues to harbor *Salmonella typhi* in the gallbladder for several years, and sometimes for life. The most infamous of these was "Typhoid Mary,"a cook who spread the infection to hundreds of victims in the early 1900s. (*Salmonella* infection is described in chapter 22.)

The **passive carrier** state is of great concern during patient care (see a later section on nosocomial infections). Medical and dental personnel who must constantly handle materials that are heavily contaminated with patient secretions and blood risk picking up pathogens mechanically and accidently transferring them to other patients **(figure 13.9*c*).** Proper handwashing, handling of contaminated materials, and aseptic techniques greatly reduce this likelihood.

Animals as Reservoirs and Sources Up to now, we have lumped animals with humans in discussing living reservoirs or carriers, but animals deserve special consideration as vectors of infections. The word **vector** is used by epidemiologists to indicate a live animal that transmits an infectious agent from one host to another. (The term is sometimes misused to include any object that spreads disease.) The majority of vectors are arthropods such as fleas, mosquitoes, flies, and ticks, although larger animals can also spread infection—for example, mammals (rabies), birds (psittacosis), or lizards (salmonellosis).

By tradition, vectors are placed into one of two categories, depending upon the animal's relationship with the microbe **(figure 13.10).** A **biological vector** actively participates in a pathogen's life cycle, serving as a site in which it can multiply or complete its life cycle. A biological vector communicates the infectious agent to the human host by biting, aerosol formation, or touch. In the case of biting vectors, the animal can

1. inject infected saliva into the blood (the mosquito) **(figure 13.10*a*),**
2. defecate around the bite wound (the flea), or
3. regurgitate blood into the wound (the tsetse fly).

A detailed discussion of arthropod vectors is found in Insight 20.2.

Mechanical vectors are not necessary to the life cycle of an infectious agent and merely transport it without being infected. The external body parts of these animals become contaminated when they come into physical contact with a source of pathogens. The agent is subsequently transferred to humans indirectly by an intermediate such as food or, occasionally, by direct contact (as in certain eye infections). Houseflies **(figure 13.10*b*)** are noxious mechanical vectors. They feed on decaying garbage and feces, and while they

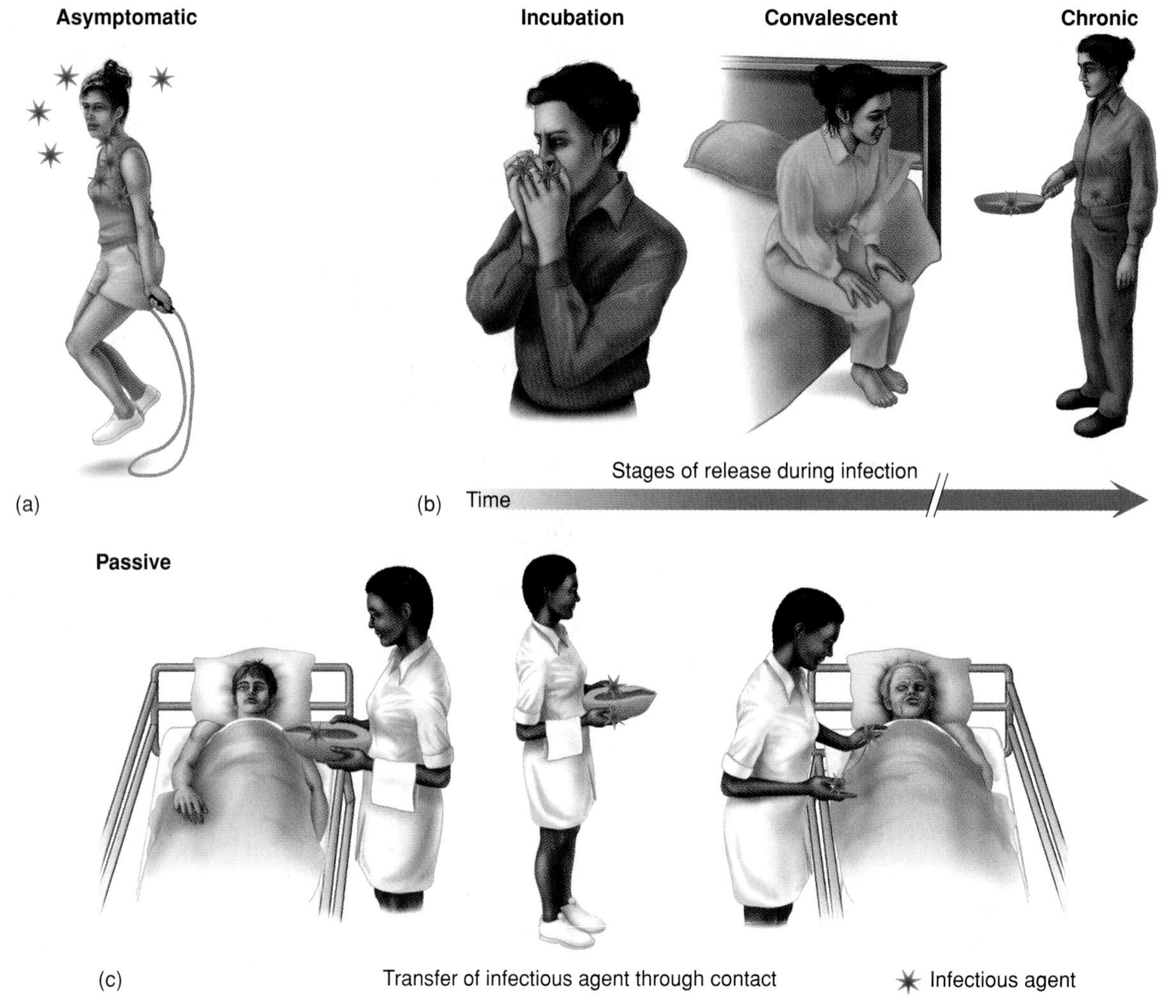

Figure 13.9
(a) An asymptomatic carrier is infected without symptoms. **(b)** Incubation, convalescent, and chronic carriers can transmit the infection either before or after the period of symptoms. **(c)** A passive carrier is contaminated but not infected.

are feeding, their feet and mouthparts easily become contaminated. They also regurgitate juices onto food to soften and digest it. Flies spread more than 20 bacterial, viral, protozoan, and worm infections. Other nonbiting flies transmit tropical ulcers, yaws, and trachoma. Cockroaches, which have similar unsavory habits, play a role in the mechanical transmission of fecal pathogens as well as contributing to allergy attacks in asthmatic children.

Many vectors and animal reservoirs spread their own infections to humans. An infection indigenous to animals

(a) Biological vectors are infected. Example: The *Anopheles* mosquito carries the malaria protozoan in its gut and salivary glands and transmits it to humans when it bites.

(b) Mechanical vectors are not infected. Example: Flies can transmit cholera by landing on feces then landing on food or a drinking glass.

Figure 13.10 Two types of vectors.
(a) Biological vectors serve as hosts during pathogen development. One example is the mosquito, a carrier of malaria. **(b)** Mechanical vectors such as the housefly transport pathogens on their feet and mouthparts.

TABLE 13.9 Common Zoonotic Infections

Disease	Primary Animal Reservoirs
Viruses	
Rabies	All mammals
Yellow fever	Wild birds, mammals, mosquitoes
Viral fevers	Wild mammals
Hantavirus	Rodents
Influenza	Chickens, birds, swine
West Nile virus	Wild birds, mosquitoes
Bacteria	
Rocky Mountain spotted fever	Dogs, ticks
Psittacosis	Birds
Leptospirosis	Domestic animals
Anthrax	Domestic animals
Brucellosis	Cattle, sheep, pigs
Plague	Rodents, fleas
Salmonellosis	Variety of mammals, birds, and rodents
Tularemia	Rodents, birds, arthropods
Miscellaneous	
Ringworm	Domestic mammals
Toxoplasmosis	Cats, rodents, birds
Trypanosomiasis	Domestic and wild mammals
Trichinosis	Swine, bears
Tapeworm	Cattle, swine, fish

but naturally transmissible to humans is a **zoonosis** (zoh″-uh-noh′-sis). In these types of infections, the human is essentially a dead-end host and does not contribute to the natural persistence of the microbe. Some zoonotic infections (rabies, for instance) can have multihost involvement, and others can have very complex cycles in the wild (see plague in chapter 20). Zoonotic spread of disease is promoted by close associations of humans with animals, and people in animal-oriented or outdoor professions are at greatest risk. At least 150 zoonoses exist worldwide; the most common ones are listed in **table 13.9.** Zoonoses make up a full 70% of all new emerging diseases worldwide. It is worth noting that zoonotic infections are impossible to completely eradicate without also eradicating the animal reservoirs. Attempts have been made to eradicate mosquitoes and certain rodents, and in 2004 China slaughtered tens of thousands of civet cats who were thought to be a source of SARS.

A 2005 United Nations study warned that one of the most troublesome trends is the increase in infectious diseases due to environmental destruction. Deforestation and urban sprawl cause animals to find new habitats, often leading to new patterns of disease transmission. For example, the fatal Nipahvirus seems to have begun to infect humans although it previously only infected Asian fruit bats. The bats were pushed out of their forest habitats by the creation of palm plantations. They encountered domesticated pigs, passing the virus to them, who in turn transmitted it to their human handlers.

Nonliving Reservoirs

Clearly, microorganisms have adapted to nearly every habitat in the biosphere. They thrive in soil and water and often find their way into the air. Although most of these microbes are saprobic and cause little harm and considerable benefit to humans, some are opportunists and a few are regular pathogens. Because human hosts are in regular contact with these environmental sources, acquisition of pathogens from natural habitats is of diagnostic and epidemiological importance.

Soil harbors the vegetative forms of bacteria, protozoa, helminths, and fungi, as well as their resistant or developmental stages such as spores, cysts, ova, and larvae. Bacterial pathogens include the anthrax bacillus and species of *Clostridium* that are responsible for gas gangrene, botulism, and tetanus. Pathogenic fungi in the genera *Coccidioides* and *Blastomyces* are spread by spores in the soil and dust. The invasive stages of the hookworm *Necator* occur in the soil. Natural bodies of water carry fewer nutrients than soil does but still support pathogenic species such as *Legionella, Cryptosporidium,* and *Giardia.*

The Acquisition and Transmission of Infectious Agents

Infectious diseases can be categorized on the basis of how they are acquired. A disease is **communicable** when an infected host can transmit the infectious agent to another host and establish infection in that host. (Although this terminology is standard, one must realize that it is not the disease that is communicated but the microbe. Also be aware that the word *infectious* is sometimes used interchangeably with the word *communicable,* but this is not precise usage.) The transmission of the agent can be direct or indirect, and the ease with which the disease is transmitted varies considerably from one agent to another. If the agent is highly communicable, especially through direct contact, the disease is **contagious.** Influenza and measles move readily from host to host and thus are contagious, whereas leprosy is only weakly communicable. Because they can be spread through the population, communicable diseases are our main focus in the following sections.

In contrast, a **noncommunicable** infectious disease does *not* arise through transmission of the infectious agent from host to host. The infection and disease are acquired through some other, special circumstance. Noncommunicable infections occur primarily when a compromised person is invaded by his or her own microbiota (as with certain pneumonias, for example) or when an individual has accidental contact with a microbe that exists in a nonliving reservoir such as soil. Some examples are certain mycoses, acquired through inhalation of fungal spores, and tetanus, in which *Clostridium tetani* spores from a soiled object enter a cut or wound. Persons thus infected do not become a source of disease to others.

Patterns of Transmission in Communicable Diseases

The routes or patterns of disease transmission are many and varied. The spread of diseases is by direct or indirect contact with animate or inanimate objects and can be horizontal or vertical. The term *horizontal* means the disease is spread through a population from one infected individual to another; *vertical* signifies transmission from parent to offspring via the ovum, sperm, placenta, or milk. The extreme complexity of transmission patterns among microorganisms makes it very difficult to generalize. However, for easier organization, we will divide microorganisms into two major groups, as shown in **figure 13.11:** transmission by some form of direct contact or transmission by indirect routes, in which some vehicle is involved.

Modes of Contact Transmission In order for microbes to be directly transferred, some type of contact must occur between the skin or mucous membranes of the infected person and that of the infectee. It may help to think of this route as the portal of exit meeting the portal of entry without the involvement of an intermediate object, substance, or space. Most sexually transmitted diseases are spread directly. In addition, infections that result from kissing or bites by biological vectors are direct. Most obligate parasites are far too sensitive to survive for long outside the host and can be transmitted only through direct contact. Diseases transmitted vertically from mother to baby fit in this contact category also. The trickiest type of "contact" transmission is droplet contact, in which fine droplets are sprayed directly upon a person during sneezing or coughing (as distinguished from droplet nuclei that are transmitted some distance by air). While there is some space between the infecter and the infectee, it is still considered a form of contact because the two people have to be in each other's presence, as opposed to indirect forms of contact.

Routes of Indirect Transmission For microbes to be indirectly transmitted, the infectious agent must pass from an

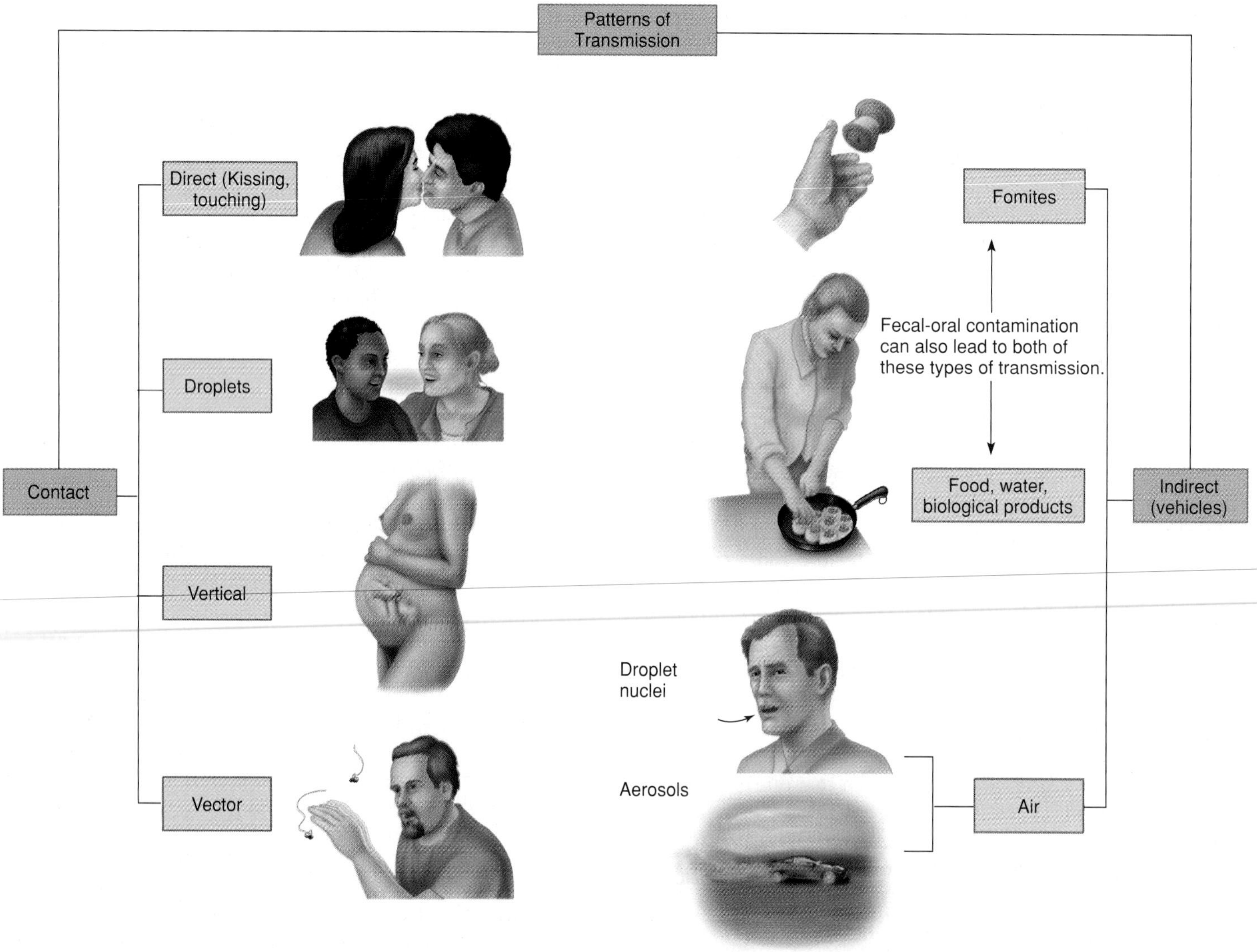

Figure 13.11 Summary of how communicable infectious diseases are acquired.

infected host to an intermediate conveyor and from there to another host. This form of communication is especially pronounced when the infected individuals contaminate inanimate objects, food, or air through their activities. The transmitter of the infectious agent can be either openly infected or a carrier.

Indirect Spread by Vehicles: Contaminated Materials The term **vehicle** specifies any inanimate material commonly used by humans that can transmit infectious agents. A *common vehicle* is a single material that serves as the source of infection for many individuals. Some specific types of vehicles are food, water, various biological products (such as blood, serum, and tissue), and fomites. A **fomite** is an inanimate object that harbors and transmits pathogens. The list of possible fomites is as long as your imagination allows. Probably highest on the list would be objects commonly in contact with the public such as doorknobs, telephones, push buttons, and faucet handles that are readily contaminated by touching. Shared bed linens, handkerchiefs, toilet seats, toys, eating utensils, clothing, personal articles, and syringes are other examples. Although paper money is impregnated with a disinfectant to inhibit microbes, pathogens are still isolated from bills as well as coins.

Outbreaks of food poisoning often result from the role of food as a common vehicle. The source of the agent can be soil, the handler, or a mechanical vector. Because milk provides a rich growth medium for microbes, it is a significant means of transmitting pathogens from diseased animals, infected milk handlers, and environmental sources of contamination. The agents of brucellosis, tuberculosis, Q fever, salmonellosis, and listeriosis are transmitted by contaminated milk. Water that has been contaminated by feces or urine can carry *Salmonella*, *Vibrio* (cholera) viruses (hepatitis A, polio), and pathogenic protozoans (*Giardia*, *Cryptosporidium*).

In the type of transmission termed the *oral-fecal route*, a fecal carrier with inadequate personal hygiene contaminates food during handling, and an unsuspecting person ingests it. Hepatitis A, amoebic dysentery, shigellosis, and typhoid fever are often transmitted this way. Oral-fecal transmission can also involve contaminated materials such as toys and diapers. It is really a special category of indirect transmission, which specifies that the way in which the vehicle became contaminated was through contact with fecal material and that it found its way to someone's mouth.

Indirect Spread by Vehicles: Air as a Vehicle Unlike soil and water, outdoor air cannot provide nutritional support for microbial growth and seldom transmits airborne pathogens. On the other hand, indoor air (especially in a closed space) can serve as an important medium for the suspension and dispersal of certain respiratory pathogens via droplet nuclei and aerosols. **Droplet nuclei** are dried microscopic residues created when microscopic pellets of mucus and saliva are ejected from the mouth and nose.

Figure 13.12 The explosiveness of a sneeze. Special photography dramatically captures droplet formation in an unstifled sneeze. Even the merest attempt to cover a sneeze with one's hand will reduce this effect considerably. When such droplets dry and remain suspended in air, they are droplet nuclei.

They are generated forcefully in an unstifled sneeze or cough **(figure 13.12)** or mildly during other vocalizations. The larger beads of moisture settle rapidly. If these settle in or on another person, it is considered droplet contact, as described earlier; but the smaller particles evaporate and remain suspended for longer periods. They can be encountered by a new host who is geographically or chronologically distant; thus, they are considered indirect contact. Droplet nuclei are implicated in the spread of hardier pathogens such as the tubercle bacillus and the influenza virus. **Aerosols** are suspensions of fine dust or moisture particles in the air that contain live pathogens. Q fever is spread by dust from animal quarters, and psittacosis is spread by aerosols from infected birds. An unusual outbreak of coccidioidomycosis (a lung infection) occurred during the 1994 southern California earthquake. Epidemiologists speculate that disturbed hillsides and soil gave off clouds of dust containing the spores of *Coccidioides*.

In the disease chapters of this book (see chapters 18–23), the modes of transmission appearing in the pink boxes in figure 13.11 will be used to describe the diseases.

Nosocomial Infections: The Hospital as a Source of Disease

Infectious diseases that are acquired or develop during a hospital stay are known as **nosocomial** (nohz″-oh-koh′-mee-al) **infections.** This concept seems incongruous at first thought, because a hospital is regarded as a place to get treatment for a disease, not a place to acquire a disease. Yet it is not uncommon for a surgical patient's incision to become infected or a burn patient to develop a case of pneumonia in the clinical setting. The rate of nosocomial infections can be as low as 0.1% or as high as 20% of all admitted patients depending on the clinical setting, with an average of about 5%. In light of the number of admissions, this adds up to 2 to 4 million cases a year, which result in nearly 90,000 deaths. Nosocomial infections cost time and money as well as suffering. By one estimate, they amount to 8 million additional days of hospitalization a year and an increased cost of $5 to $10 billion.

So many factors unique to the hospital environment are tied to nosocomial infections that a certain number of infections are virtually unavoidable. After all, the hospital both attracts and creates compromised patients, and it serves as a collection point for pathogens. Some patients become infected when surgical procedures or lowered defenses permit resident biota to invade their bodies. Other patients acquire infections directly or indirectly from fomites, medical equipment, other patients, medical personnel, visitors, air, and water.

The health care process itself increases the likelihood that infectious agents will be transferred from one patient to another. Treatments using reusable instruments such as respirators and thermometers constitute a possible source of infectious agents. Indwelling devices such as catheters, prosthetic heart valves, grafts, drainage tubes, and tracheostomy tubes form ready portals of entry and habitats for infectious agents. Because such a high proportion of the hospital population receives antimicrobial drugs during their stay, drug-resistant microbes are selected for at a much greater rate than is the case outside the hospital.

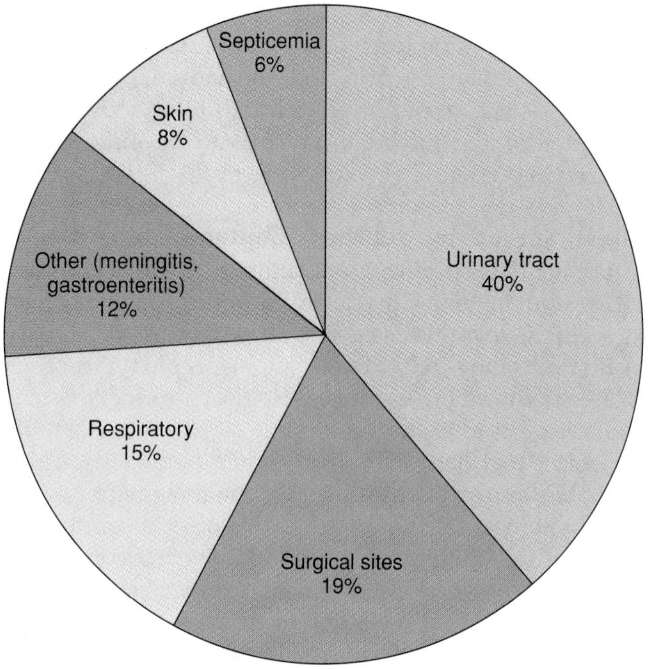

Figure 13.13 Most common nosocomial infections. Relative frequency by target area.

The most common nosocomial infections involve the urinary tract, the respiratory tract, and surgical incisions **(figure 13.13).** Gram-negative intestinal biota (*Escherichia coli, Klebsiella, Pseudomonas*) are cultured in more than half of patients with nosocomial infections. Gram-positive bacteria (staphylococci and streptococci) and yeasts make up most of the remainder. True pathogens such as *Mycobacterium tuberculosis, Salmonella,* hepatitis B, and influenza virus can be transmitted in the clinical setting as well.

The news is not all bad. The potential seriousness and impact of nosocomial infections have spurred hospitals to action. In the summer of 2005, a group called The Institute for Healthcare Improvement in Boston, Massachusetts, launched a campaign to improve six areas of hospital care, including the rate of nosocomial infections. In the first 6 months of the program, 80 hospitals in Michigan and New Jersey virtually eliminated ventilator-associated pneumonias and blood infections from neck and groin catheters. Other hospitals are showing similar results. By the summer of 2006, 3,000 of the nation's 5,500 hospitals were part of the program.

CASE FILE 13 *WRAP-UP*

Several things need to be considered in attempting to determine the source of infection in these patients. The organism has been associated with nosocomial (hospital-acquired) infection in nonmilitary patients. However, according to a recent national survey, *A. baumannii* accounted for only 1.3% of bloodstream infections in nonmilitary patients. Patients being treated in military facilities had received traumatic wounds that may have been environmentally contaminated. Further evidence of an environmental source comes from a different era and another war. *A. baumannii* was found to be the most commonly isolated gram-negative bacillus from traumatic wounds in the Vietnam War.

A microbial survey has been initiated to determine the prominence of *A. baumannii* in soil samples and medical facilities. Molecular techniques are being used to characterize environmental and patient isolates.

Care facilities have increased surveillance for patients colonized with *A. baumannii.* Infection control measures have been implemented or revised. Increased use of alcohol-based hand sanitizers has been implemented.

The multidrug resistance of *A. baumannii* is cause for concern in both military and civilian facilities. It further emphasizes the need for increased caution in caring for colonized patients. It challenges clinicians to carefully consider how they treat infections and to consider new combinations of antimicrobials. It further challenges researchers to consider new combinations of antimicrobials. It further challenges researchers and the pharmaceutical industry to continue searching for new and innovative ways to treat infectious diseases.

SEE: CDC. 2004. Acinetobacter baumannii *infections among patients at military medical facilities treating injured U.S. service members, 2002–2004.* MMWR *53:1063–1066.*

Medical asepsis includes practices that lower the microbial load in patients, caregivers, and the hospital environment. These practices include proper hand washing, disinfection, and sanitization, as well as patient isolation. The goal of these procedures is to limit the spread of infectious agents from person to person. An even higher level of stringency is seen with *surgical asepsis,* which involves all of the strategies listed previously plus ensuring that all surgical procedures are conducted under sterile conditions. This includes sterilization of surgical instruments, dressings, sponges, and the like, as well as clothing personnel in sterile garments and scrupulously disinfecting the room surfaces and air.

Hospitals generally employ an *infection control officer* who not only implements proper practices and procedures throughout the hospital but is also charged with tracking potential outbreaks, identifying breaches in asepsis, and training other health care workers in aseptic technique. Among those most in need of this training are nurses and other caregivers whose work, by its very nature, exposes them to needlesticks, infectious secretions, blood, and physical contact with the patient. The same practices that interrupt the routes of infection in the patient can also protect the health care worker. It is for this reason that most hospitals have adopted universal precautions that recognize that all secretions from all persons in the clinical setting are potentially infectious and that transmission can occur in either direction.

Universal Blood and Body Fluid Precautions

Medical and dental settings require stringent measures to prevent the spread of nosocomial infections from patient to patient, from patient to worker, and from worker to patient. But even with precautions, the rate of such infections is rather high. Recent evidence indicates that more than one-third of nosocomial infections could be prevented by consistent and rigorous infection control methods.

Previously, control guidelines were disease-specific, and clearly identified infections were managed with particular restrictions and techniques. With this arrangement, personnel tended to handle materials labeled *infectious* with much greater care than those that were not so labeled. The AIDS epidemic spurred a reexamination of that policy. Because of the potential for increased numbers of undiagnosed HIV-infected patients, the Centers for Disease Control and Prevention laid down more stringent guidelines for handling patients and body substances. These guidelines have been termed **universal precautions (UPs),** because they are based on the assumption that all patient specimens could harbor infectious agents and so must be treated with the same degree of care. They also include body substance isolation (BSI) techniques to be used in known cases of infection.

It is worth mentioning that these precautions are designed to protect all individuals in the clinical setting—patients, workers, and the public alike. In general, they include techniques designed to prevent contact with pathogens and contamination and, if prevention is not possible, to take purposeful measures to decontaminate potentially infectious materials.

The universal precautions recommended for all health care settings are:

1. Barrier precautions, including masks and gloves, should be taken to prevent contact of skin and mucous membranes with patients' blood or other body fluids. Because gloves can develop small invisible tears, double gloving decreases the risk further. For protection during surgery, venipuncture, or emergency procedures, gowns, aprons, and other body coverings should be worn. Dental workers should wear eyewear and face shields to protect against splattered blood and saliva.
2. More than 10% of health care personnel are pierced each year by sharp (and usually contaminated) instruments. These accidents carry risks not only for AIDS but also for hepatitis B, hepatitis C, and other diseases. Preventing inoculation infection requires vigilant observance of proper techniques. All disposable needles, scalpels, or sharp devices from invasive procedures must immediately be placed in puncture-proof containers for sterilization and final discard. Under no circumstances should a worker attempt to recap a syringe, remove a needle from a syringe, or leave unprotected used syringes where they pose a risk to others. Reusable needles or other sharp devices must be heat-sterilized in a puncture-proof holder before they are handled. If a needlestick or other injury occurs, immediate attention to the wound, such as thorough degermation and application of strong antiseptics, can prevent infection.
3. Dental handpieces should be sterilized between patients, but if this is not possible, they should be thoroughly disinfected with a high-level disinfectant (peroxide, hypochlorite). Blood and saliva should be removed completely from all contaminated dental instruments and intraoral devices prior to sterilization.
4. Hands and other skin surfaces that have been accidently contaminated with blood or other fluids should be scrubbed immediately with a germicidal soap. Hands should likewise be washed after removing rubber gloves, masks, or other barrier devices.
5. Because saliva can be a source of some types of infections, barriers should be used in all mouth-to-mouth resuscitations.
6. Health care workers with active, draining skin or mucous membrane lesions must refrain from handling patients or equipment that will come into contact with other patients. Pregnant health care workers risk infecting their fetuses and must pay special attention to these

guidelines. Personnel should be protected by vaccination whenever possible.

Isolation procedures for known or suspected infections should still be instituted on a case-by-case basis.

Which Agent Is the Cause? Using Koch's Postulates to Determine Etiology

An essential aim in the study of infection and disease is determining the precise **etiologic,** or causative, **agent.** In our modern technological age, we take for granted that a certain infection is caused by a certain microbe, but such has not always been the case. More than a century ago, Robert Koch realized that in order to prove the germ theory of disease he would have to develop a standard for determining causation that would stand the test of scientific scrutiny. Out of his experimental observations on the transmission of anthrax in cows came a series of proofs, called **Koch's postulates,** that established the principal criteria for etiologic studies **(figure 13.14).** These postulates direct an investigator to

1. find evidence of a particular microbe in every case of a disease,
2. isolate that microbe from an infected subject and cultivate it in pure culture in the laboratory,
3. inoculate a susceptible healthy subject with the laboratory isolate and observe the same resultant disease, and
4. reisolate the agent from this subject.

Valid application of Koch's postulates requires attention to several critical details. Each isolated culture must be pure, observed microscopically, and identified by means of characteristic tests; the first and second isolates must be identical; and the pathologic effects, signs, and symptoms of the disease in the first and second subjects must be the same. Once established, these postulates were rapidly put to the test, and within a short time, they had helped determine the causative agents of tuberculosis, diphtheria, and plague. Today, most infectious diseases have been directly linked to a known infectious agent.

Koch's postulates continue to play an essential role in modern epidemiology. Every decade, new diseases challenge the scientific community and require application of the postulates. Prominent examples are toxic shock syndrome, AIDS, Lyme disease, and Legionnaires disease (named for the American Legion members who first contracted a mysterious lung infection in Philadelphia).

Koch's postulates are reliable for many infectious diseases, but they cannot be completely fulfilled in certain situations. For example, some infectious agents are not readily isolated or grown in the laboratory. If one cannot elicit a similar infection by inoculating it into an animal, it is very difficult to prove the etiology. In the past, scientists have attempted to circumvent this problem by using human subjects **(Insight 13.4).**

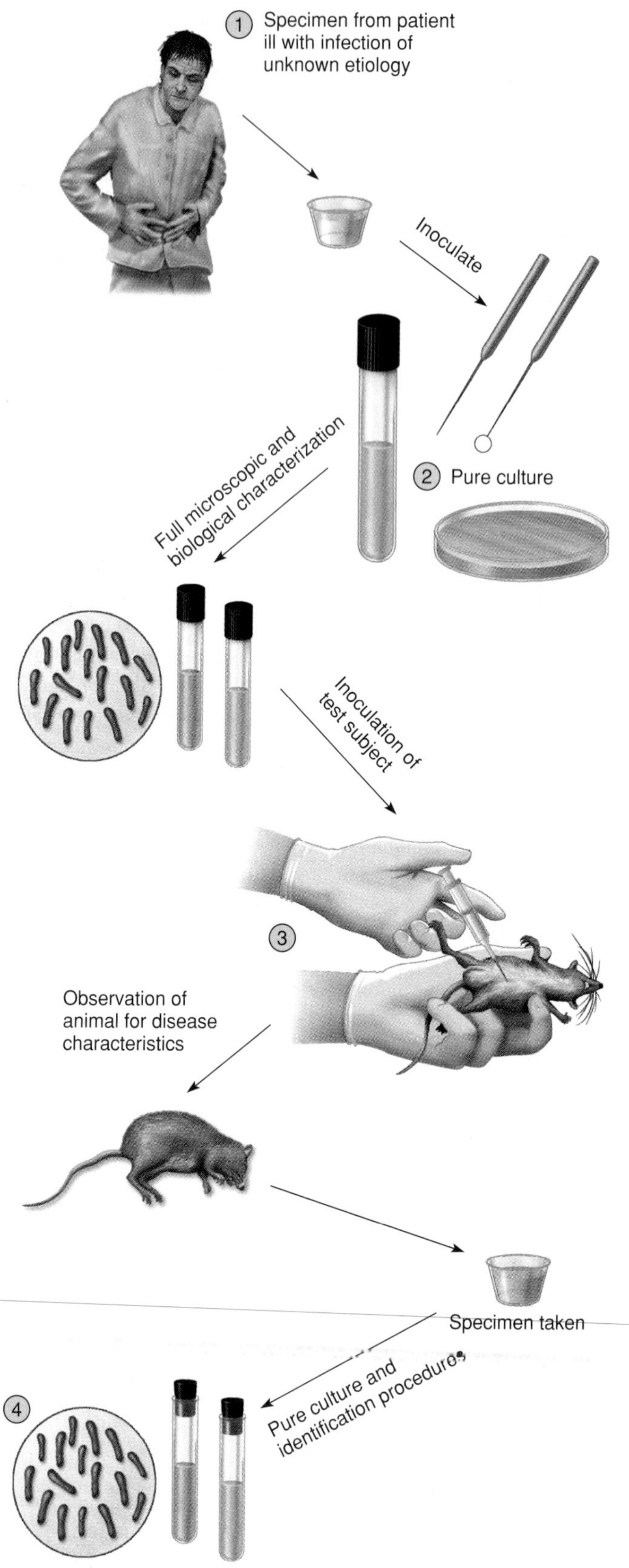

Process Figure 13.14 Koch's postulates: Is this the etiologic agent?
The microbe in the initial and second isolations and the disease in the patient and experimental animal must be identical for the postulates to be satisfied.

INSIGHT 13.4 Historical

The History of Human Guinea Pigs

These days, human beings are not used as subjects for determining the cause of infectious disease, but in earlier times they were. A long tradition of human experimentation dates well back into the 18th century, with the subject frequently the experimenter himself. In some studies, mycologists inoculated their own skin and even that of family members with scrapings from fungal lesions to demonstrate that the disease was transmissible. In the early days of parasitology, it was not uncommon for a brave researcher to swallow worm eggs in order to study the course of his infection and the life cycle of the worm.

In a sort of reverse test, a German colleague of Koch's named Max von Petenkofer believed so strongly that cholera was *not* caused by a bacterium that he and his assistant swallowed cultures of *Vibrio cholerae.* Fortunately for them, they acquired only a mild form of the disease. Many self-experimenters have not been so fortunate.

One of the most famous cases is that of Jesse Lazear, a Cuban physician who worked with Walter Reed on the etiology of yellow fever in 1900. Dr. Lazear was convinced that mosquitoes were directly involved in the spread of yellow fever, and by way of proof, he allowed himself and two volunteers to be bitten by mosquitoes infected with the blood of yellow fever patients. Although all three became ill as a result of this exposure, Dr. Lazear's sacrifice was the ultimate one—he died of yellow fever. Years later, paid volunteers were used to completely fulfill the postulates.

Dr. Lazear was not the first martyr in this type of cause. Fifteen years previously, a young Peruvian medical student, Daniel Carrion, attempted to prove that a severe blood infection, Oroya fever, had the same etiology as *verucca peruana,* an ancient disfiguring skin disease. After inoculating himself with fluid from a skin lesion, he developed the severe form and died, becoming a national hero. In his honor, the disease now identified as bartonellosis is also sometimes called Carrion's disease. Eventually, the microbe (*Bartonella bacilliformis*) was isolated, a monkey model was developed, and the sand fly was shown to be the vector for this disease.

The incentive for self-experimentation has continued, but present-day researchers are more likely to test experimental vaccines than dangerous microbes. One exception to this was Dr. J. Robert Warren and Barry Marshall who tested the effects of *Helicobacter pylori* by swallowing a culture. Although they didn't get ulcers, they helped to establish the pathogenicity of the microbe. Marshall and Warren were awarded the Nobel Prize in Medicine in 2005 for their discovery.

It is difficult to satisfy Koch's postulates for viral diseases because viruses usually have a very narrow host range. Human viruses may only cause disease in humans, or perhaps in primates, though the disease symptoms in apes will often be different. To address this, T. M. Rivers proposed modified postulates for viral infections. These were used in 2003 to definitively determine the corona virus cause of SARS **(Insight 13.5).**

It is also usually not possible to use Koch's postulates to determine causation in polymicrobial diseases. Diseases such as periodontitis and soft tissue abscesses are caused by complex mixtures of microbes. While theoretically possible to isolate each member and to re-create the exact proportions of individual cultures for step 3, it is not yet routine.

13.3 Epidemiology: The Study of Disease in Populations

So far, our discussion has revolved primarily around the impact of an infectious disease in a single individual. Let us now turn our attention to the effects of diseases on the community—the realm of **epidemiology.** By definition, this term involves the study of the frequency and distribution of disease and other health-related factors in defined human populations. It involves many disciplines—not only microbiology but also anatomy, physiology, immunology, medicine, psychology, sociology, ecology, and statistics—and it considers all forms of disease, including heart disease, cancer, drug addiction, and mental illness. The techniques of epidemiology are also used to track behaviors, such as exercise or smoking. The epidemiologist is a medical sleuth who collects clues on the causative agent, pathology, sources and modes of transmission and tracks the numbers and distribution of cases of disease in the community. In fulfilling these demands, the epidemiologist asks who, when, where, how, why, and what about diseases. The outcome of these studies helps public health departments develop prevention and treatment programs and establish a basis for predictions.

Who, When, and Where? Tracking Disease in the Population

Epidemiologists are concerned with all of the factors covered earlier in this chapter: virulence, portals of entry and exit, and the course of disease. But they are also interested in surveillance—that is, collecting, analyzing, and reporting data on the rates of occurrence, mortality, morbidity, and transmission of infections. Surveillance involves keeping data for a large number of diseases seen by the

INSIGHT 13.5 — Historical

Koch's Postulates Still Critical in SARS Era

SARS (severe acute respiratory syndrome) hit the news in the winter of 2002, and though it was deadly, ultimately killing hundreds of people of the 8,000 or so it infected, it was contained in a period of 7 months, even though it was new to the medical community. The epidemic was brought to a halt quickly because the response by the scientific and medical personnel was lightning fast. By April of 2003, scientists had sequenced the entire genome of the suspected agent, a coronavirus. In May of 2003, Dutch scientists published a paper in the journal *Nature* with the title "Aetiology: Koch's Postulates Fulfilled for SARS Virus."

The set of Koch's postulates used in this study was that modified by Rivers in 1937 for viral diseases. There are six postulates in the modified version, not four as in the original Koch's postulates. In their paper, the scientists noted that the first three postulates had been met by other researchers. The final three were examined in the work described in the article. The scientists inoculated two macaque monkeys with the SARS virus that had been isolated from a fatal human case and cultivated in cell culture. The two macaques became lethargic. One of them suffered respiratory distress, and both of them excreted virus from their noses and throats. At autopsy, the macaques were found to have histological signs of pneumonia that were indistinguishable from human cases (postulate 4). The virus that was recovered from the monkeys was shown by PCR and electron microscopy to be identical to the one used for inoculation (postulate 5). Finally, 2 weeks after infection, the macaques' blood tested positive for antibody to the SARS virus. This fulfilled the last postulate and gave scientists the proof they needed to rapidly design interventions targeted at this particular coronavirus.

Koch's Postulates as Modified by Rivers

1. Virus must be isolated from each diseased host.
2. Virus must be cultivated in cell culture.
3. Virus must be filterable, that is, must pass through pores small enough to impede bacteria and other microorganisms.
4. Virus must produce comparable disease when inoculated into the original host species or a related one.
5. The same virus must be reisolated from the new host.
6. There must be a specific immune response to the original virus in the new host.

medical community and reported to public health authorities. By law, certain **reportable,** or notifiable, **diseases** must be reported to authorities; others are reported on a voluntary basis.

A well-developed network of individuals and agencies at the local, district, state, national, and international levels keeps track of infectious diseases. Physicians and hospitals report all notifiable diseases that are brought to their attention. These reports are either made about individuals or in the aggregate, depending on the disease.

Local public health agencies first receive the case data and determine how they will be handled. In most cases, health officers investigate the history and movements of patients to trace their prior contacts and to control the further spread of the infection as soon as possible through drug therapy, immunization, and education. In sexually transmitted diseases, patients are asked to name their partners so that these persons can be notified, examined, and treated. It is very important to maintain the confidentiality of the persons in these reports. The principal government agency responsible for keeping track of infectious diseases nationwide is the Centers for Disease Control and Prevention (CDC) in Atlanta, Georgia; the CDC is a part of the U.S. Public Health Service. The CDC publishes a weekly notice of diseases (the *Morbidity and Mortality Report*) that provides weekly and cumulative summaries of the case rates and deaths for about 50 notifiable diseases, highlights important and unusual diseases, and presents data concerning disease occurrence in the major regions of the United States. It is available to anyone at http://www.cdc.gov/mmwr/. Ultimately, the CDC shares its statistics on disease with the World Health Organization (WHO) for worldwide tabulation and control.

Epidemiological Statistics: Frequency of Cases

The **prevalence** of a disease is the total number of existing cases with respect to the entire population. It is often thought of as a snapshot and is usually reported as the percentage of the population having a particular disease at any given time. Disease **incidence** measures the number of new cases over a certain time period. This statistic, also called the case, or morbidity, rate, indicates both the rate and the risk of infection. The equations used to figure these rates are:

$$\text{Prevalence} = \frac{\text{Total number of cases in population}}{\text{Total number of persons in population}} \times 100 = \%$$

$$\text{Incidence} = \frac{\text{Number of new cases}}{\text{Total number of susceptible persons}} \quad \text{(Usually reported per 100,000 persons)}$$

The changes in incidence and prevalence are usually followed over a seasonal, yearly, and long-term basis and are helpful in predicting trends **(figure 13.15).** Statistics of concern to the epidemiologist are the rates of disease with regard to sex, race, or geographic region. Also of importance is the **mortality rate,** which measures the total number of deaths in a population due to a certain disease. Over the past century, the overall death rate from infectious diseases has dropped, although the number of persons afflicted with infectious diseases (the **morbidity rate**) has remained relatively high.

Monitoring statistics also makes it possible to define the frequency of a disease in the population. An infectious disease that exhibits a relatively steady frequency over a long time period in a particular geographic locale is **endemic (figure 13.16*a*).** For example, Lyme disease is endemic to certain areas of the United States where the tick vector is found. A certain number of new cases are expected in these areas every year. When a disease is **sporadic,** occasional cases are reported at irregular intervals in random locales **(figure 13.16*b*).** Tetanus and diphtheria are reported sporadically in the United States (fewer than 50 cases a year).

When statistics indicate that the prevalence of an endemic or sporadic disease is increasing beyond what is expected for that population, the pattern is described as an **epidemic (figure 13.16*c*).** The time period is not defined—it can range from hours in food poisoning to years in syphilis—nor is an exact percentage of increase needed before an outbreak can qualify as an epidemic. Several epidemics occur every year in the United States, most recently among STDs such as chlamydia and gonorrhea. The spread of an epidemic across continents is a **pandemic,** as exemplified by AIDS and influenza **(figure 13.16*d*).**

One important epidemiological truism might be called the "iceberg effect," which refers to the fact that only a small portion of an iceberg is visible above the surface of the ocean, with a much more massive part lingering unseen below the surface. Regardless of case reporting and public health screening, a large number of cases of infection in the community go undiagnosed and unreported. (For a list of reportable diseases in the United States, see **table 13.10.**) In the instance of salmonellosis, approximately 40,000 cases are reported each year. Epidemiologists estimate that the actual number is more likely somewhere between 400,000 and 4,000,000. The iceberg effect can be even more lopsided for sexually transmitted diseases or for infections that are not brought to the attention of reporting agencies.

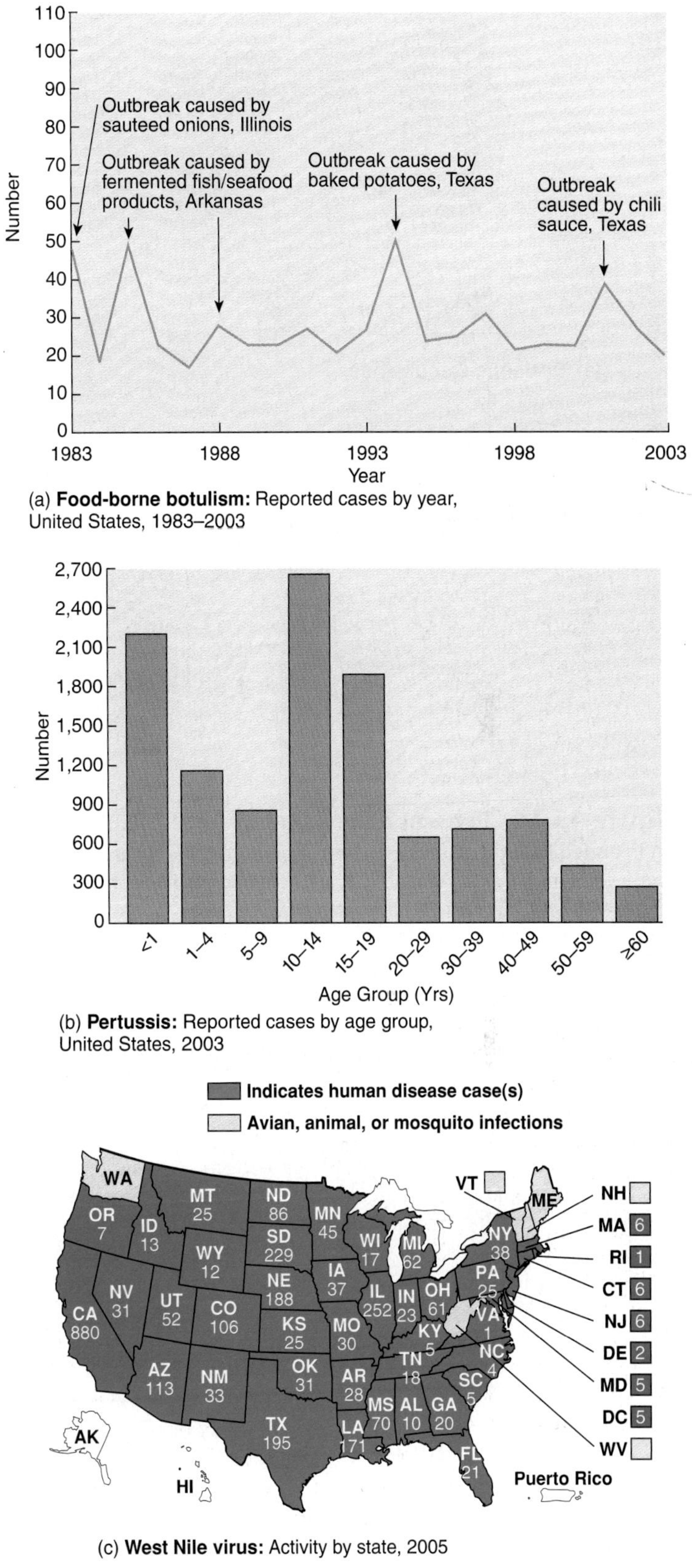

(a) **Food-borne botulism:** Reported cases by year, United States, 1983–2003

(b) **Pertussis:** Reported cases by age group, United States, 2003

(c) **West Nile virus:** Activity by state, 2005

Figure 13.15 Graphical representation of epidemiological data.

The Centers for Disease Control and Prevention collect epidemiological data that are analyzed with regard to **(a)** time frame, **(b)** age, and **(c)** geographic region.

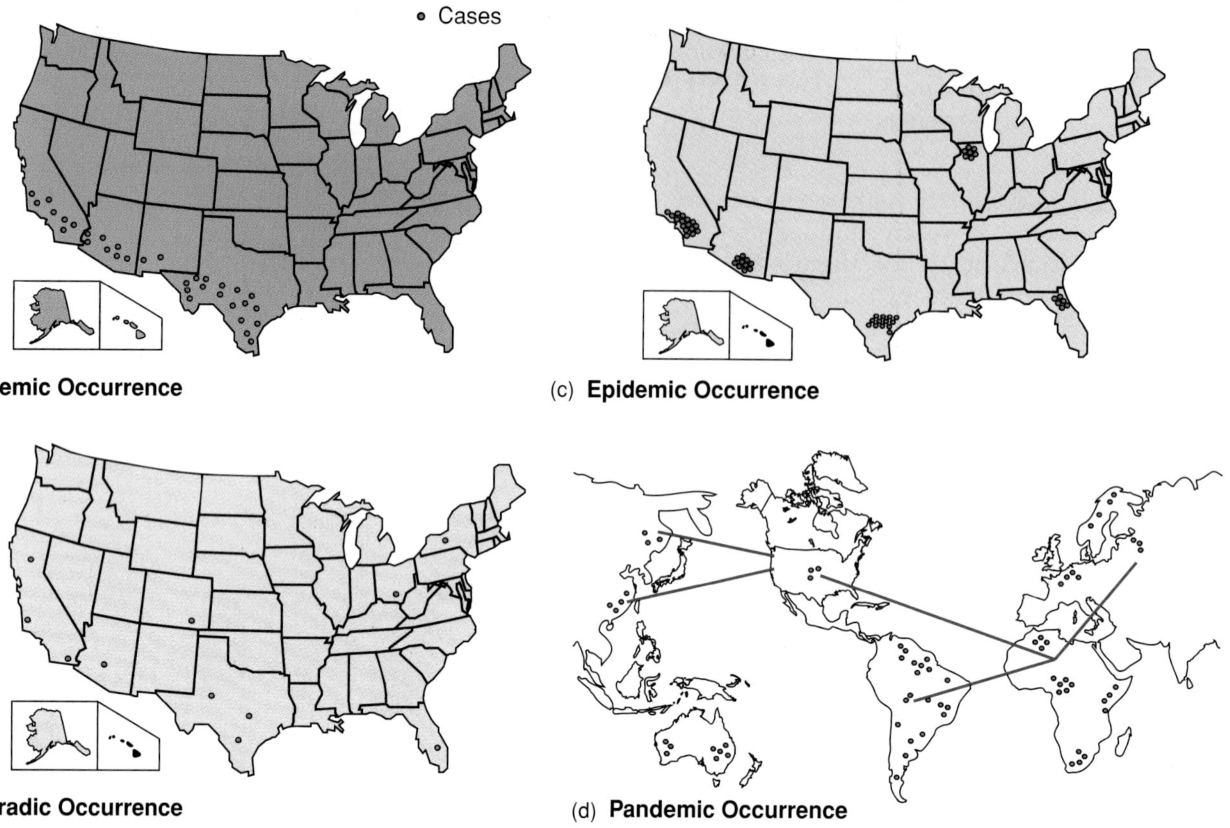

Figure 13.16 Patterns of infectious disease occurrence.
(a) In endemic occurrence, cases are concentrated in one area at a relatively stable rate. **(b)** In sporadic occurrence, a few cases occur randomly over a wide area. **(c)** An epidemic is an increased number of cases that often appear in geographic clusters. The clusters may be local, as in the case of a restaurant-related food-borne epidemic, or nationwide, as is the case with *Chlamydia.* **(d)** Pandemic occurrence means that an epidemic ranges over more than one continent.

CHECKPOINT

- The primary habitat of a pathogen is called its reservoir. A human reservoir is also called a carrier.
- Animals can be either reservoirs or vectors of pathogens. An infected animal is a biological vector. Uninfected animals, especially insects, that transmit pathogens mechanically are called mechanical vectors.
- Soil and water are nonliving reservoirs for pathogenic bacteria, protozoa, fungi, and worms.
- A communicable disease can be transmitted from an infected host to others, but not all infectious diseases are communicable.
- The spread of infectious disease from person to person is called horizontal transmission. The spread of infectious disease from parent to offspring is called vertical transmission.
- Infectious diseases are spread by either contact or indirect routes of transmission. Vehicles of indirect transmission include soil, water, food, air, and fomites (inanimate objects).
- Nosocomial infections are acquired in a hospital from surgical procedures, equipment, personnel, and exposure to drug-resistant microorganisms.
- Causative agents of infectious disease must be isolated and identified according to Koch's postulates.
- Epidemiology is the study of the determinants and distribution of all diseases in populations. The study of infectious disease in populations is just one aspect of this field.
- Data on specific, reportable diseases is collected by local, national, and worldwide agencies.
- The *prevalence* of a disease is the percentage of existing cases in a given population. The disease *incidence,* or *morbidity rate,* is the number of newly infected members in a population during a specified time period.
- Disease frequency is described as sporadic, epidemic, pandemic, or endemic.

TABLE 13.10 Reportable Diseases in the United States*

- Acquired immunodeficiency syndrome (AIDS)
- Anthrax
- Botulism
- Brucellosis
- Chancroid
- *Chlamydia trachomatis* genital infections
- Cholera
- Coccidioidomycosis
- Cryptosporidiosis
- Cyclosporiasis
- Diphtheria
- Ehrlichiosis
- Encephalitis/meningitis, arboviral
    - Encephalitis/meningitis, California serogroup viral
    - Encephalitis/meningitis, eastern equine
    - Encephalitis/meningitis, Powassan
    - Encephalitis/meningitis, St. Louis
    - Encephalitis/meningitis, western equine
    - Encephalitis/meningitis, West Nile
- Giardiasis
- Gonorrhea
- *Haemophilus influenzae* invasive disease
- Hansen's disease (leprosy)
- Hantavirus pulmonary syndrome
- Hemolytic uremic syndrome
- Hepatitis, viral, acute
    - Hepatitis A, acute
    - Hepatitis B, acute
    - Hepatitis B virus, perinatal infection
    - Hepatitis C, acute
- Hepatitis, viral, chronic
    - Chronic hepatitis B
    - Hepatitis C virus infection (past or present)
- HIV infection
    - HIV infection, adult (>13 years)
    - HIV infection, pediatric (<13 years)
- Influenza-associated pediatric mortality
- Legionellosis
- Listeriosis
- Lyme disease
- Malaria
- Measles
- Meningococcal disease
- Mumps
- Novel influenza A infections
- Pertussis
- Plague
- Poliomyelitis, paralytic
- Poliovirus infection
- Psittacosis
- Q fever
- Rabies
    - Rabies, animal
    - Rabies, human
- Rocky Mountain spotted fever
- Rubella
- Rubella, congenital syndrome
- Salmonellosis
- Severe acute respiratory syndrome–associated coronavirus (SARS-CoV) disease
- Shigellosis
- Smallpox
- Streptococcal disease, invasive, group A
- Streptococcal toxic shock syndrome
- *Streptococcus pneumoniae,* drug-resistant, invasive disease
- *Streptococcus pneumoniae,* invasive in children <5 years
- Syphilis
- Syphilis, congenital
- Tetanus
- Toxic shock syndrome
- Trichinellosis
- Tuberculosis
- Tularemia
- Typhoid fever
- Vancomycin-intermediate *Staphylococcus aureus* (VISA)
- Vancomycin-resistant *Staphylococcus aureus*
- Varicella
- Vibriosis
- Yellow fever

*Reportable to the CDC; other diseases may be reportable to State Departments of Health.
**New as of mid 2007.
Source: Centers for Disease Control and Prevention, 2007.

Chapter Summary with Key Terms

13.1 The Human Host

A. The human body is in constant contact with microbes, some of which invade the body and multiply. Microbes are classified as either part of the normal biota of the body or as **pathogens,** depending on whether or not their colonization of the body results in infection.

B. *Resident Biota: The Human as Habitat*

1. **Resident biota** or microbiota consists of a huge and varied population of microbes that reside permanently on all surfaces of the body exposed to the environment.
2. Colonization begins just prior to birth and continues throughout life. Variations in biota occur in response to changes in an individual's age, diet, hygiene, and health.
3. Biota often provides some benefit to the host but is also capable of causing disease, especially when the immune system of the host is compromised or when biota invades a normally sterile area of the body.
4. An important benefit provided by normal biota is **microbial antagonism.**

C. *Indigenous Biota of Specific Regions*
Resident normal biota includes bacteria, fungi, and protozoa. This normal biota occupies the skin, mouth, gastrointestinal tract, large intestine, respiratory tract, and genitourinary tract.

13.2 The Progress of an Infection

A. **Pathogenicity** is the ability of microorganisms to cause infection and disease; **virulence** refers to the degree to which a microbe can invade and damage host tissues.
   1. **True pathogens** are able to cause disease in a normal healthy host with intact immune defenses.
   2. **Opportunistic pathogens** can cause disease only in persons whose host defenses are compromised by *predisposing conditions.*

B. *Becoming Established: Step One—Portals of Entry*
   1. Microbes typically enter the body through a specific **portal of entry.** Portals of entry are generally the same as those areas of the body that harbor microbiota.
   2. The minimum number of microbes needed to produce an infection is termed the infectious dose, and these pathogens may be classified as **exogenous** or **endogenous,** based on their source.

C. *Becoming Established: Step Two—Attaching to the Host*
   Microbes attach to the host cell, a process known as **adhesion,** by means of fimbriae, capsules, or receptors.

D. *Becoming Established: Step Three—Surviving Host Defenses*
   Microbes that persist in the human host have developed mechanisms of disabling early defensive strategies, especially phagocytes.

E. *Causing Disease*
   Virulence factors are structures or properties of a microbe that lead to pathologic effects on the host. While some virulence factors are those that enable entry or adhesion, many virulence factors lead directly to damage. These fall in three categories: enzymes, toxins, and the induction of a damaging host response.
   1. **Exoenzymes** digest epithelial tissues and permit invasion of pathogens.
   2. **Toxigenicity** refers to a microbe's capacity to produce **toxins** at the site of multiplication. Toxins are divided into **endotoxins** and **exotoxins,** and diseases caused by toxins are classified as **intoxications, toxemias,** or **toxinoses.**
   3. Microbes have a variety of ways of causing the host to damage itself.

F. *The Process of Infection and Disease*
   1. Infectious diseases follow a typical pattern consisting of an *incubation period, prodromal stage, period of invasion,* and *convalescent period.* Each stage is marked by symptoms of a specific intensity.
   2. Infections are classified by:
      a. whether they remain localized at the site of inoculation **(localized infection, systemic infection, focal infection).**
      b. number of microbes involved in an infection and the order in which they infect the body **(mixed infection, primary infection, secondary infection).**
      c. persistence of the infection (**acute infection, chronic infection,** subacute infection).
   3. **Signs** of infection refer to *objective* evidence of infection, and **symptoms** refer to *subjective* evidence. A **syndrome** is a disease that manifests as a predictable complex of symptoms.
   4. After the initial infection pathogens may remain in the body in a **latent** state and may later cause recurrent infections. Long-term damage to host cells is referred to as **sequelae.**

G. **The Portal of Exit**
   Microbes are released from the body through a specific **portal of exit** that allows them access to another host.

H. *Reservoirs Where Pathogens Persist*
   Microbes reside in both living and nonliving reservoirs.

I. *The Acquisition and Transmission of Infectious Agents*
   1. Agents of disease are either **communicable** or **noncommunicable,** with highly communicable diseases referred to as **contagious.**
   2. Direct transmission of a disease occurs when a portal of exit meets a portal of entrance.
   3. Indirect transmission occurs when an intermediary such as a **vehicle, fomite,** or **droplet nuclei** connect portals of entrance and exit.

J. **Nosocomial** infections are infectious diseases acquired in a hospital setting. Special vigilance is required to reduce their occurrence.

K. **Koch's postulates** are a series of proofs used to determine the **etiologic** agent of a specific disease.

13.3 Epidemiology: The Study of Disease in Populations

A. Epidemiologists are involved in the surveillance of reportable diseases in populations. Diseases are tracked with regard to their **prevalence** and **incidence,** while populations are tracked with regard to **morbidity** and **mortality.**

B. Diseases are described as being **endemic, sporadic, epidemic,** or **pandemic** based on the frequency of their occurrence in a population.

Multiple-Choice and True-False Questions

Multiple-Choice Questions. Select the correct answer from the answers provided.

1. The best descriptive term for the resident biota is
   a. commensals
   b. parasites
   c. pathogens
   d. mutualists

2. Resident biota is absent from the
   a. pharynx
   b. lungs
   c. intestine
   d. hair follicles

3. Virulence factors include
   a. toxins
   b. enzymes
   c. capsules
   d. all of these
4. The specific action of hemolysins is to
   a. damage white blood cells
   b. cause fever
   c. damage red blood cells
   d. cause leukocytosis
5. The ____ is the time that lapses between encounter with a pathogen and the first symptoms.
   a. prodrome
   b. period of invasion
   c. period of convalescence
   d. period of incubation
6. A short period early in a disease that manifests with general malaise and achiness is the
   a. period of incubation
   b. prodrome
   c. sequela
   d. period of invasion
7. A/an ____ is a passive animal transporter of pathogens.
   a. zoonosis
   b. biological vector
   c. mechanical vector
   d. asymptomatic carrier
8. An example of a noncommunicable infection is
   a. measles
   b. leprosy
   c. tuberculosis
   d. tetanus
9. A positive antibody test for HIV would be a _____ of infection.
   a. sign
   b. symptom
   c. syndrome
   d. sequela

True-False Questions. If statement is true, leave as is. If it is false, correct it by rewriting the sentence.

10. The presence of a few bacteria in the blood is called septicemia.
11. Resident microbiota is commonly found in the urethra.
12. A subclinical infection is one that is acquired in a hospital or medical facility.
13. The general term that describes an increase in the number of white blood cells is leukopenia.

Writing to Learn

These questions are suggested as a *writing-to-learn* experience. For each question (except #9), compose a one- or two-paragraph answer that includes the factual information needed to completely address the question.

1. Differentiate between contamination, infection, and disease. What are the possible outcomes in each?
2. How are infectious diseases different from other diseases?
3. Explain several ways that true pathogens differ from opportunistic pathogens.
4. a. Distinguish between pathogenicity and virulence.
   b. Define virulence factors, and give examples of them in gram-positive and gram-negative bacteria, viruses, and parasites.
5. Describe the course of infection from contact with the pathogen to its exit from the host.
6. Differentiate between exogenous and endogenous infections.
7. a. What factors possibly affect the size of the infectious dose?
   b. Name four factors involved in microbial adhesion.
8. Compare and contrast: systemic versus local infections; primary versus secondary infections; infection versus intoxication.
9. a. Outline the science of epidemiology and the work of an epidemiologist.
   b. Using the following statistics, based on number of reported cases, determine which show endemic, sporadic, or epidemic patterns. Explain how you can determine each type.

United States Region	Cases	
	2000	**2001**
Chlamydiosis		
Northeast	93,116	115,467
Midwest	168,332	184,111
South	286,136	307,405
West	161,868	176,259
Total cases	709,452	783,242
Lyme disease		
Northeast	14,932	14,435
Midwest	1,343	1,260
South	1,319	1,198
West	136	136
Total cases	17,730	17,029
Rubella		
Northeast	27	9
Midwest	3	5
South	123	5
West	15	4
Total cases	168	23

   c. Explain what would have to occur for these diseases to have a pandemic distribution.
10. a. List the main features of Koch's postulates.
    b. Why is it so difficult to prove them for some diseases?

Concept Mapping

Appendix D provides guidance for working with concept maps.

1. Supply your own linking words or phrases in this concept map, and provide the missing concepts in the empty boxes.

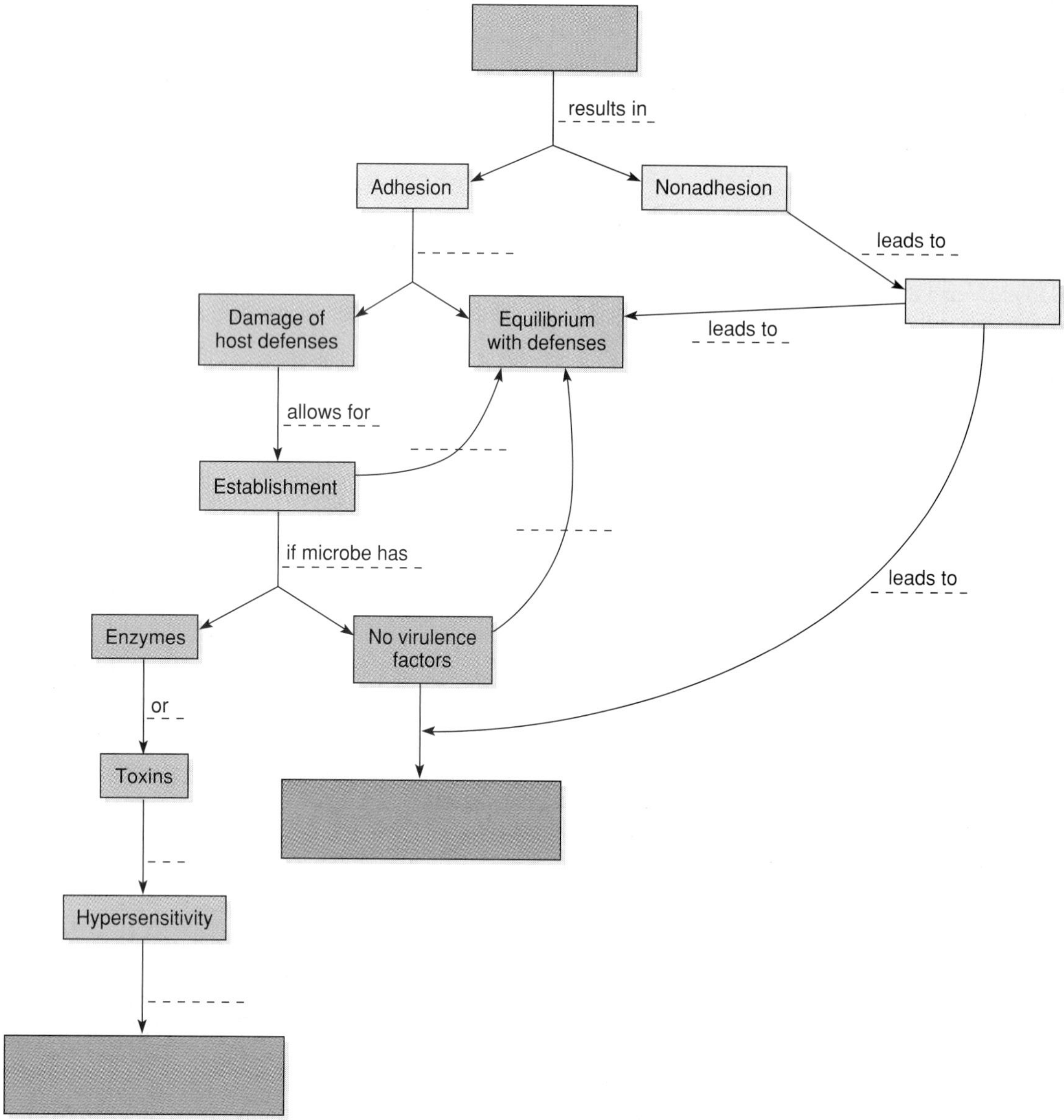

2. Use 6 to 10 bolded words of your choice from the Chapter Summary to create a concept map. Finish it by providing linking words.

Critical Thinking Questions

Critical thinking is the ability to reason and solve problems using facts and concepts. These questions can be approached from a number of angles, and in most cases, they do not have a single correct answer.

1. a. Discuss the relationship between the vaginal residents and the colonization of the newborn.
   b. Can you think of some serious medical consequences of this relationship?
   c. Why would normal biota cause some infections to be more severe and other infections to be less severe?
2. If the following patient specimens produced positive cultures when inoculated and grown on appropriate media, indicate

whether this result indicates a disease state and why or why not:

Urine	Throat
Lung biopsy	Feces
Saliva	Blood
Cerebrospinal fluid	Urine from bladder
Liver biopsy	Semen

What are the important clinical implications of positive blood or cerebrospinal fluid?

3. Explain how endotoxin gets into the bloodstream of a patient with endotoxic shock.
4. Describe each of the following infections using correct technical terminology. (Descriptions may fit more than one category.) Use terms such as *primary, secondary, nosocomial, STD, mixed, latent, toxemia, chronic, zoonotic, asymptomatic, local, systemic, -itis, -emia.*

   Caused by needlestick in dental office
   Pneumocystis pneumonia in AIDS patient
   Bubonic plague from rat flea bite
   Diphtheria
   Undiagnosed chlamydiosis
   Acute necrotizing gingivitis
   Syphilis of long duration
   Large numbers of gram-negative rods in the blood
   A boil on the back of the neck
   An inflammation of the meninges
   Scarlet fever
5. Name 10 fomites that you came into contact with today.
6. Describe what parts of Koch's postulates were unfulfilled by Dr. Lazear's experiment described in Insight 13.4.
7. a. Suggest several reasons why urinary tract, respiratory tract, and surgical infections are the most common nosocomial infections.
   b. Name several measures that health care providers must exercise at all times to prevent or reduce nosocomial infections.

Visual Understanding

1. **From chapter 3, figure 3.7*a*.** What chemical is the organism in this illustration producing? How does this add to an organism's pathogenicity?

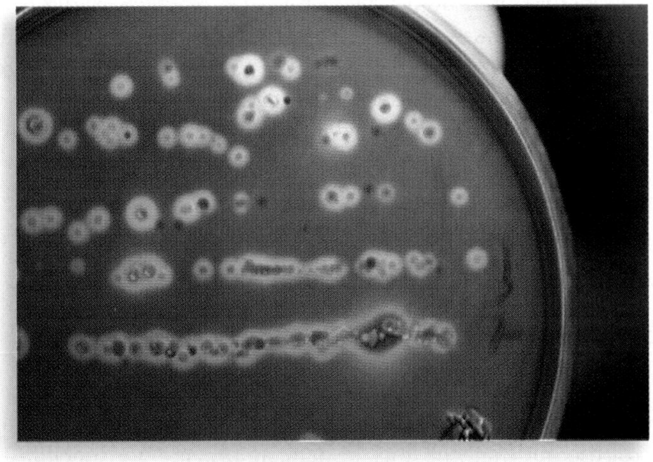

2. **From chapter 4, figure 4.11.** What type of infection is this illustration likely associated with? What could be the source of the microorganism?

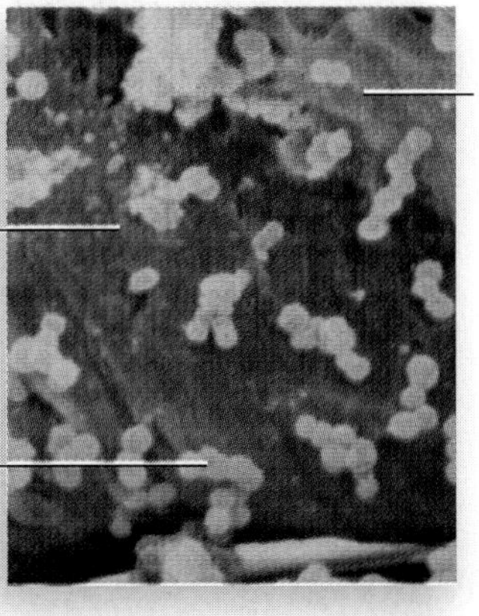

Internet Search Topics

1. Use an Internet search engine to look up the following topics. Write a short summary of what you discover on your searches.

   Sentinel chickens
   Tuskegee syphilis experiment
   Koch's postulates verified for AIDS and HIV
   Newcastle virus in chicken
2. Go to: www.aris.mhhe.com, and click on "microbiology" and then this textbook's author/title. Go to chapter 13, access the URLs listed under Internet Search Topics, and research the following:

   To gain additional insight into the science of epidemiology, log on to the websites listed for an overview of infection, disease, and their monitoring plus solving epidemiological mysteries.
3. Go to http://nobelprize.org and find the presentation lecture and banquet lecture associated with the 2005 Prize in Physiology or Medicine.

CHAPTER 14

Host Defenses I

Overview and Nonspecific Defenses

CASE FILE

14

An infant boy, 6 days of age, developed severe diaper rash. Nine days later, in spite of topical treatment and oral antibiotics, he was hospitalized for evaluation of sepsis because of fever and pustules. His history was remarkable only in that there was a papule at the base of the fifth finger, which was present at birth.

The infant's chest and abdominal CT scans revealed lesions in the liver and a lung nodule. Upon biopsy, neutrophils were noted in the liver abscesses and *Serratia marcescens,* a gram-negative bacillus, grew after culture of the liver specimens. Biopsy of the lung nodule revealed evidence of acute pneumonitis, and *Aspergillus,* a fungus, was identified in the lung specimen. Fortunately, after appropriate antibiotic therapy, the lesions healed and the child recovered.

Further testing revealed that the infant's neutrophils had a negative reaction in two tests designed to assess phagocyte activity (the nitroblue tetrazolium test (NBT) and in the neutrophil cytochrome b_{558} assay). These tests monitor the ability of phagocytes to produce toxic oxygen products during a process called the *respiratory burst.* Based on these data, as well as the boy's history, he was diagnosed with X-linked, cytochrome-negative, chronic granulomatous disease (CGD) of childhood.

- *What are some of the signs and symptoms of CGD that this infant exhibited?*
- *What is typically the problem with phagocytic function in someone diagnosed with CGD?*
- *Individuals with this condition are more susceptible to what type of microbial agents?*

Case File 14 Wrap-Up appears on page 433.

CHAPTER OVERVIEW

- The human body possesses a complex series of overlapping defenses that protect it against invasion. The first two of the body's three lines of defense provide nonspecific protection against anything (living or not) regarded as foreign to the body. The first line consists of barriers to foreign matter, and the second is charged with protecting the body once foreign matter has entered.
- White blood cells formed in the bone marrow are responsible for most of the reactions of the immune system, including antibody production, phagocytosis, and many aspects of inflammation. Lymphoid organs such as the spleen, lymph nodes, and thymus are also intimately involved in these defense mechanisms.
- Cells of the immune system are able to travel freely among different areas of the body because of the interrelationship between the blood, the lymphatic system, and the reticuloendothelial system.
- Communication between cells of the immune system is facilitated by the use of chemical messengers such as cytokines. These chemicals are released by cells of the immune system to increase blood flow, stimulate the migration of white blood cells, initiate fever, or induce the death of virally infected cells.

- The inflammatory response is a complex reaction to infection that works to fight foreign agents and limit further damage to the body. As part of the inflammatory response, white blood cells known as phagocytes help to clear foreign organisms from the body. The complement system acts to lyse cells that have been identified as foreign.
- Fever and complement are further components of the second line of defense.
- The induction of B and T lymphocytes (the third line of defense) occurs if the body's innate immunities are incapable of resolving an infection. B and T cells provide acquired, long-term protection against specific microbes.

14.1 Defense Mechanisms of the Host in Perspective

The survival of the host depends upon an elaborate network of defenses that keep harmful microbes and other foreign materials from penetrating the body. Should they penetrate, additional host defenses are summoned to prevent them from becoming established in tissues. Defenses involve barriers, cells, and chemicals, and they range from nonspecific to specific and from inborn or innate to acquired. This chapter introduces the main lines of defense intrinsic to all humans. Topics included in this survey are the anatomical and physiological systems that detect, recognize, and destroy foreign substances and the general adaptive responses that account for an individual's long-term immunity or resistance to infection and disease.

In chapter 13, we explored the host-parasite relationship, with emphasis on the role of microorganisms in disease. In this chapter, we examine the other side of the relationship—that of the host defending itself against microorganisms. As previously stated, whether an encounter between a human and a microbe results in disease is dependent on many factors (see figure 13.2). The encounters occur constantly. In the battle against all sorts of invaders, microbial and otherwise, the body erects a series of barriers, sends in an army of cells, and emits a flood of chemicals to protect tissues from harm.

The host defenses are a multilevel network of innate, nonspecific protections and specific **immunities** referred to as the *first, second, and third lines of defense* **(figure 14.1).** The interaction and cooperation of these three levels of defense normally provide complete protection against infection. The *first line of defense* includes any barrier that blocks invasion at the portal of entry. This mostly nonspecific line of defense limits access to the internal tissues of the body. However, it is not considered a true immune response because it does not involve recognition of a specific foreign substance but is very general in action. The *second line of defense* is a more internalized system of protective cells and fluids that includes inflammation and phagocytosis. It acts rapidly at both the local and systemic levels once the first line of defense has been circumvented. The highly specific *third line of defense* is acquired on an individual basis as each foreign substance is encountered by white blood cells called lymphocytes. The reaction with each different microbe produces unique protective substances and cells that can come into play if that microbe is encountered again. The third line of defense provides long-term immunity. It is discussed in chapter 15. This chapter focuses on the first and second lines of defense.

The human systems are armed with various levels of defense that do not operate in a completely separate fashion; most defenses overlap and are even redundant in some of their effects. This literally bombards microbial invaders with an entire assault force, making their survival unlikely. Because of the interwoven nature of host defenses, we will introduce basic concepts of structure and function that will prepare you for later information on specific reactions of the immune system (see chapter 15).

Barriers at the Portal of Entry: A First Line of Defense

A number of defenses are a normal part of the body's anatomy and physiology. These inborn, nonspecific defenses can be divided into physical, chemical, and genetic barriers that impede the entry of not only microbes but any foreign agent, whether living or not **(figure 14.2).**

Physical or Anatomical Barriers at the Body's Surface

The skin and mucous membranes of the respiratory and digestive tracts have several built-in defenses. The outermost layer (stratum corneum) of the skin is composed of epithelial cells that have become compacted, cemented together, and impregnated with an insoluble protein, keratin. The result is a thick, tough layer that is highly impervious and waterproof. Few pathogens can penetrate this unbroken barrier, especially in regions such as the soles of the feet or the palms of the hands, where the stratum corneum is much thicker than on other parts of the body. It is so obvious as to be overlooked: the skin separates our inner bodies from the microbial assaults of the environment. It is a surprisingly tough and sophisticated barrier. The top layer of cells is packed with keratin, a protective and waterproofing protein. In addition, outer layers of skin are constantly sloughing off, taking associated microbes with them. Other cutaneous barriers include hair follicles and skin glands. The hair shaft is periodically extruded, and the follicle cells are **desquamated** (des'-kwuh-mayt-ud). The flushing effect of sweat glands also helps remove microbes.

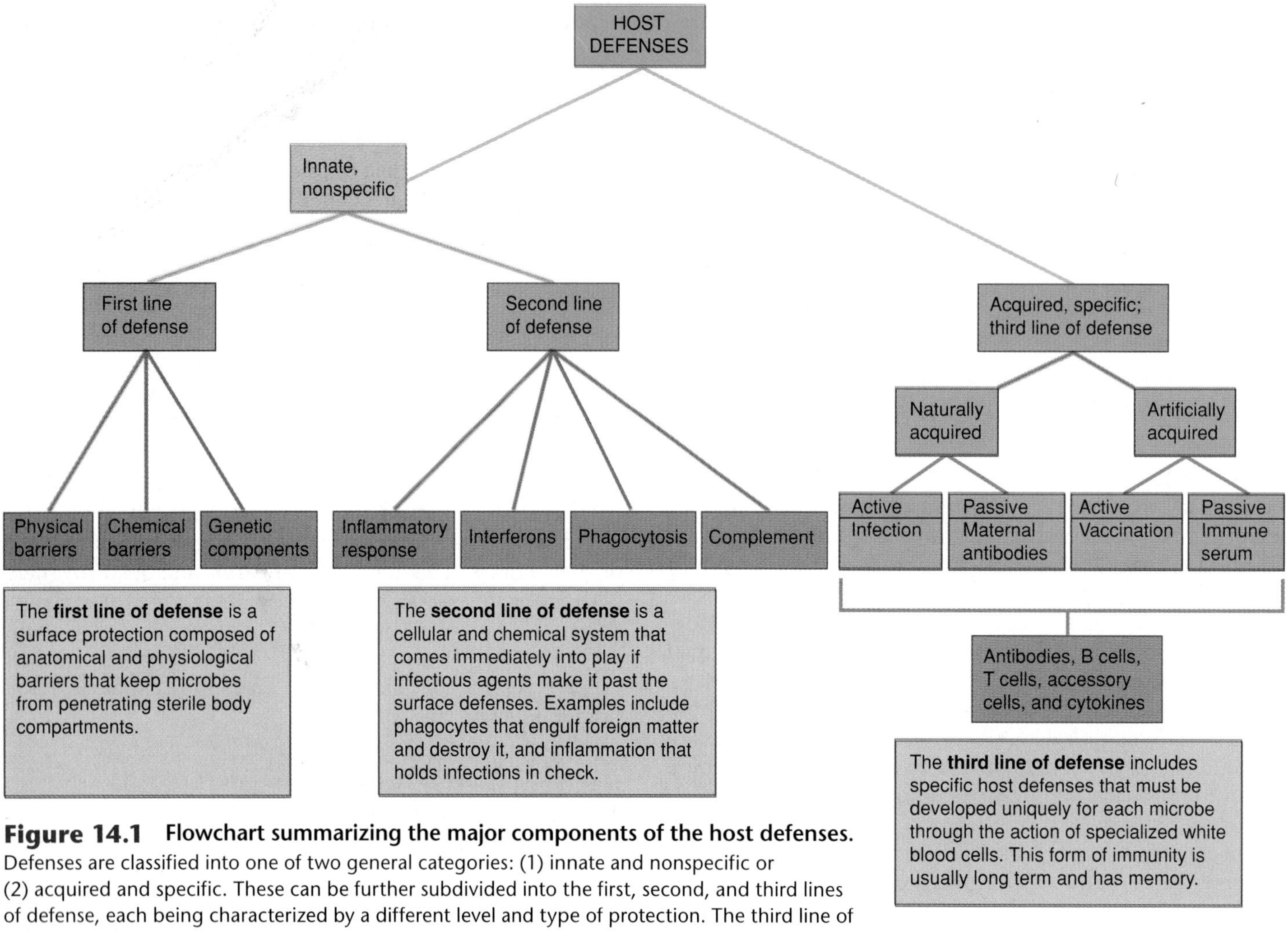

Figure 14.1 Flowchart summarizing the major components of the host defenses. Defenses are classified into one of two general categories: (1) innate and nonspecific or (2) acquired and specific. These can be further subdivided into the first, second, and third lines of defense, each being characterized by a different level and type of protection. The third line of defense is the most complex and is responsible for specific immunity.

The mucous membranes of the digestive, urinary, and respiratory tracts and of the eye are moist and permeable. They do provide barrier protection but without a keratinized layer. The mucous coat on the free surface of some membranes impedes the entry and attachment of bacteria. Blinking and tear production (lacrimation) flush the eye's surface with tears and rid it of irritants. The constant flow of saliva helps carry microbes into the harsh conditions of the stomach. Vomiting and defecation also evacuate noxious substances or microorganisms from the body.

The respiratory tract is constantly guarded from infection by elaborate and highly effective adaptations. Nasal hair traps larger particles. The copious flow of mucus and fluids that occurs in allergy and colds exerts a flushing action. In the respiratory tree (primarily the trachea and bronchi), a ciliated epithelium (called the ciliary escalator) conveys foreign particles entrapped in mucus toward the pharynx to be removed **(figure 14.3).** Irritation of the nasal passage reflexively initiates a sneeze, which expels a large volume of air at high velocity. Similarly, the acute sensitivity of the bronchi, trachea, and larynx to foreign matter triggers coughing, which ejects irritants.

The genitourinary tract derives partial protection from the continuous trickle of urine through the ureters and from periodic bladder emptying that flushes the urethra.

The composition of resident microbiota and its protective effect were discussed in chapter 13. Even though the resident biota does not constitute an anatomical barrier, its presence can block the access of pathogens to epithelial surfaces and can create an unfavorable environment for pathogens by competing for limited nutrients or by altering the local pH.

A great deal of research in recent years has highlighted the importance of the gut microbiota on the development of nonspecific defenses (described in this chapter) as well as specific immunity. It appears that the presence of a robust commensal biota "trains" host defenses in such a way that commensals are kept in check and pathogens are eliminated. Evidence suggests that interruptions in this process, which may include frequent antibiotic treatments that affect the gut, can lead to immunologic disturbances in the gut. Some scientists believe that inflammatory bowel disease, which has been increasing in Western countries especially, may well be a result of our overzealous attempts to free our environment of microbes and to overtreat ourselves with antibiotics. The result, they say, is an "ill-trained" gut defense system that inappropriately responds to commensal biota.

Nonspecific Chemical Defenses

The skin and mucous membranes offer a variety of chemical defenses. Sebaceous secretions exert an antimicrobial effect, and specialized glands such as the meibomian glands of the eyelids lubricate the conjunctiva with an antimicrobial secretion. An additional defense in tears and saliva is **lysozyme,** an enzyme that hydrolyzes the peptidoglycan in the cell wall of bacteria. The high lactic acid and electrolyte concentrations of sweat and the skin's acidic pH and fatty acid content are also inhibitory to many microbes. Likewise, the hydrochloric acid in the stomach renders protection against many pathogens that are swallowed, and the intestine's digestive juices and bile are potentially destructive to microbes. Even semen contains an antimicrobial chemical that inhibits bacteria, and the vagina has a protective acidic pH maintained by normal biota.

Genetic Differences in Susceptibility

Some hosts are genetically immune to the diseases of other hosts. One explanation for this phenomenon is that some pathogens have such great specificity for one host species that they are incapable of infecting other species. One way of putting it is: "Humans can't acquire distemper from cats, and cats can't get mumps from humans." This specificity is particularly true of viruses, which can invade only by attaching to a specific host receptor. But it does not hold true for zoonotic infectious agents that attack a broad spectrum of animals. Genetic differences in susceptibility can also exist within members of one species, as described in chapter 13. Often these differences arise from mutations in

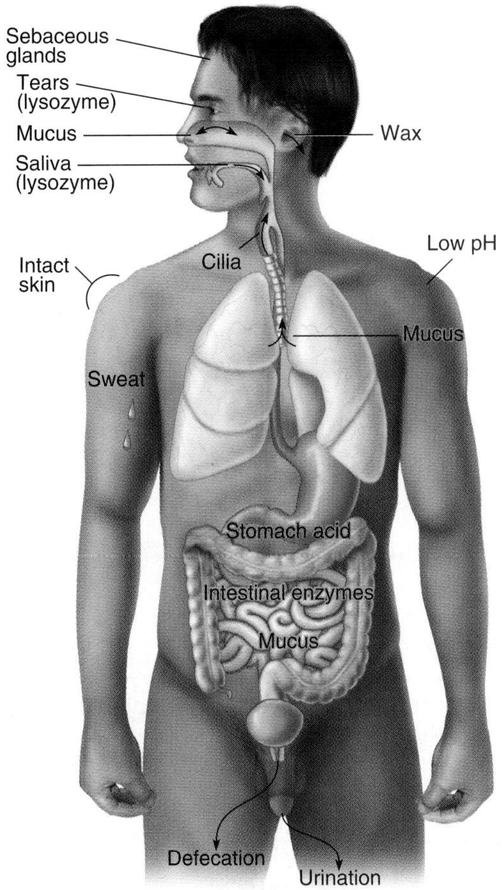

Figure 14.2 The primary physical and chemical defense barriers.

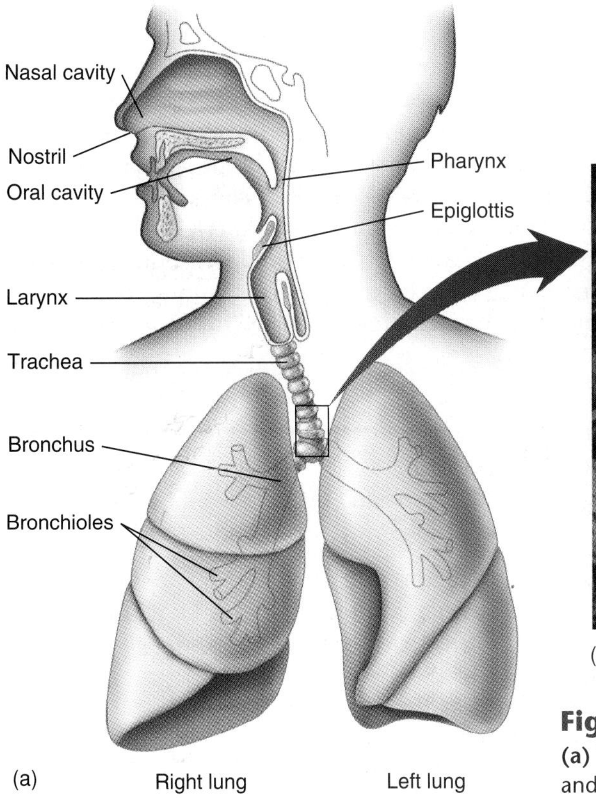

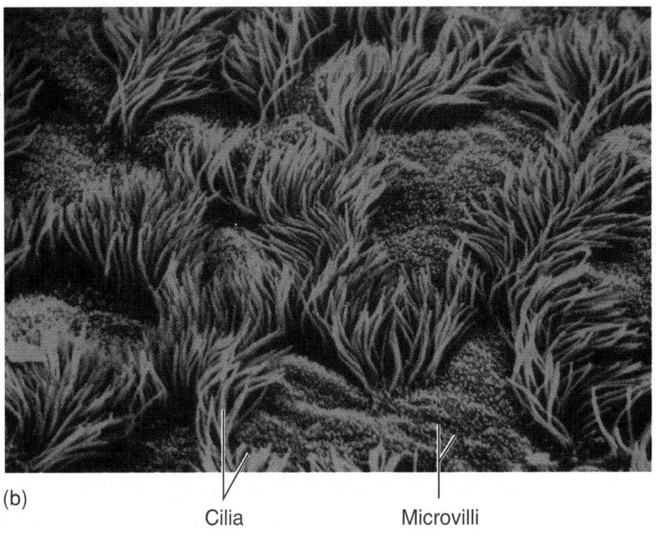

Figure 14.3 The ciliary defense of the respiratory tree. **(a)** The epithelial lining of the airways contains a brush border of cilia to entrap and propel particles upward toward the pharynx. **(b)** Tracheal mucosa (5,000×).

the genes that code for components described in this chapter and the next, such as complement proteins, cytokines, and T-cell receptors.

The vital contribution of barriers is clearly demonstrated in people who have lost them or never had them. Patients with severe skin damage due to burns are extremely susceptible to infections; those with blockages in the salivary glands, tear ducts, intestine, and urinary tract are also at greater risk for infection. But as important as it is, the first line of defense alone is not sufficient to protect against infection. Because many pathogens find a way to circumvent the barriers by using their virulence factors (discussed in chapter 13), a whole new set of defenses—inflammation, phagocytosis, specific immune responses—are brought into play.

CHECKPOINT

- The multilevel, interconnecting network of host protection against microbial invasion is organized into three lines of defense.
- The first line consists of physical and chemical barricades provided by the skin and mucous membranes.
- The second line encompasses all the nonspecific cells and chemicals found in the tissues and blood.
- The third line, the specific immune response, is customized to react to specific antigens of a microbial invader. This response is designed to immobilize and destroy the invader every time it appears in the host.

14.2 The Second and Third Lines of Defense: An Overview

Immunology encompasses the study of all features of the body's second and third lines of defense. Although this chapter is concerned, not surprisingly, with infectious microbial agents, be aware that immunology is central to the study of fields as diverse as cancer (at least partly the result of an underactive immune system) and allergy (an overactive immune system). In chapter 17, you will see that many of the most powerful aspects of immunology have been developed for use in health care, laboratory, or commercial settings.

In the body, the mandate of the immune system can be easily stated. A healthy functioning immune system is responsible for

1. surveillance of the body,
2. recognition of foreign material, and
3. destruction of entities deemed to be foreign **(figure 14.4).**

Because infectious agents could potentially enter through any number of portals, the cells of the immune system constantly move about the body, searching for potential pathogens. This process is carried out primarily by white blood cells, which have been trained to recognize body cells (so-called **self**) and differentiate them from any foreign material in the body, such as an invading bacterial cell **(nonself).** The ability to evaluate cells as either self or nonself is central to the functioning of the immune system. While foreign cells must be recognized as a potential threat and dealt with appropriately, self cells must not come under attack by the immune defenses.

The immune system evaluates cells by examining certain molecules on their surfaces called **markers.**[1] These markers, which generally consist of proteins and/or sugars, can be thought of as the cellular equivalent of facial characteristics in humans and allow the cells of the immune system to identify whether or not a newly discovered cell poses a threat. While cells deemed to be self are left alone, cells and other objects designated as foreign are marked for destruction by a number of methods, the most common of which is phagocytosis.

CHECKPOINT

- The immune system operates first as a surveillance system that discriminates between the host's self identity markers and the nonself identity markers of foreign cells. As far as the immune system is concerned, if an antigen is not self, it is foreign, does not belong, and must be destroyed.

14.3 Systems Involved in Immune Defenses

Unlike many systems, the immune system does not exist in a single, well-defined site; rather, it encompasses a large, complex, and diffuse network of cells and fluids that permeate every organ and tissue. It is this very arrangement that promotes the surveillance and recognition processes that help screen the body for harmful substances.

The body is partitioned into several fluid-filled spaces called the intracellular, extracellular, lymphatic, cerebrospinal, and circulatory compartments. Although these compartments are physically separated, they have numerous connections. Their structure and position permit extensive interchange and communication. Among the body compartments that participate in immune function are

1. the *reticuloendothelial* (reh-tik″-yoo-loh-en″-doh-thee′-lee-al) *system (RES),*
2. the spaces surrounding tissue cells that contain *extracellular fluid (ECF),*
3. the *bloodstream,* and
4. the *lymphatic system.*

In the following section, we consider the anatomy of these main compartments and how they interact in the second and third lines of defense.

1. The term *marker* is also employed in genetics in a different sense—that is, to denote a detectable characteristic of a particular genetic mutant. A genetic marker may or may not be a surface marker.

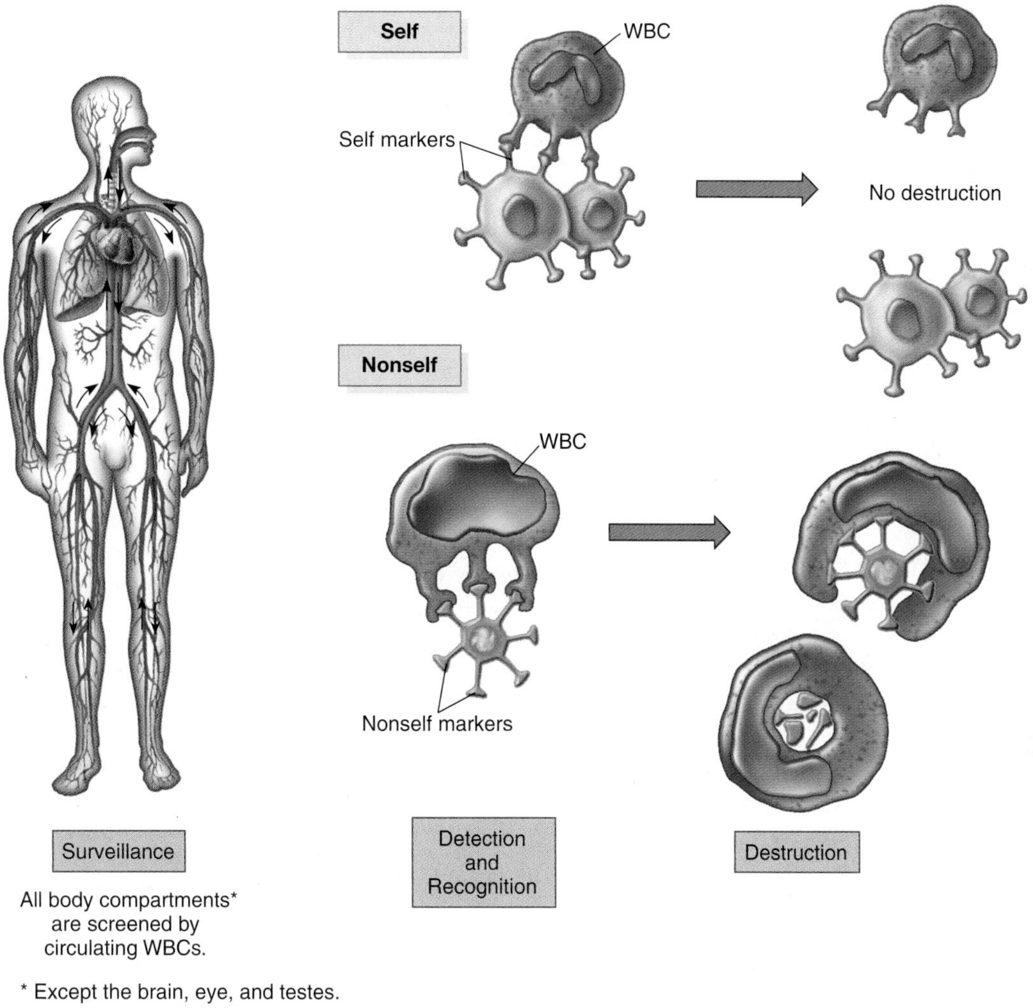

Figure 14.4 Search, recognize, and destroy is the mandate of the immune system.
White blood cells are equipped with a very sensitive sense of "touch." As they sort through the tissues, they feel surface markers that help them determine what is self and what is not. When self markers are recognized, no response occurs. However, when nonself is detected, a reaction to destroy it is mounted.

The Communicating Body Compartments

For effective immune responsiveness, the activities in one fluid compartment must be conveyed to other compartments. Let us see how this occurs by viewing tissue at the microscopic level **(figure 14.5).** At this level, clusters of tissue cells are in direct contact with the reticuloendothelial system (RES), which is described shortly, and the extracellular fluid (ECF). Other compartments (vessels) that penetrate at this level are blood and lymphatic capillaries. This close association allows cells and chemicals that originate in the RES and ECF to diffuse or migrate into the blood and lymphatics; any products of a lymphatic reaction can be transmitted directly into the blood through the connection between these two systems; and certain cells and chemicals originating in the blood can move through the vessel walls into the extracellular spaces and migrate into the lymphatic system.

The flow of events among these systems depends on where an infectious agent or foreign substance first intrudes. A typical progression might begin in the extracellular spaces and RES, move to the lymphatic circulation, and ultimately end up in the bloodstream. Regardless of which compartment is first exposed, an immune reaction in any one of them will eventually be communicated to the others at the microscopic level. An obvious benefit of such an integrated system is that no cell of the body is far removed from competent protection, no matter how isolated. Let us take a closer look at each of these compartments.

Immune Functions of the Reticuloendothelial System

The tissues of the body are permeated by a support network of connective tissue fibers, or a *reticulum,* that originates in the cellular basal lamina, interconnects nearby cells, and meshes with the massive connective tissue network surrounding all organs. This network, called the **reticuloendothelial system (figure 14.6)** is intrinsic to the immune function because it provides a passageway within and between tissues and organs. It also

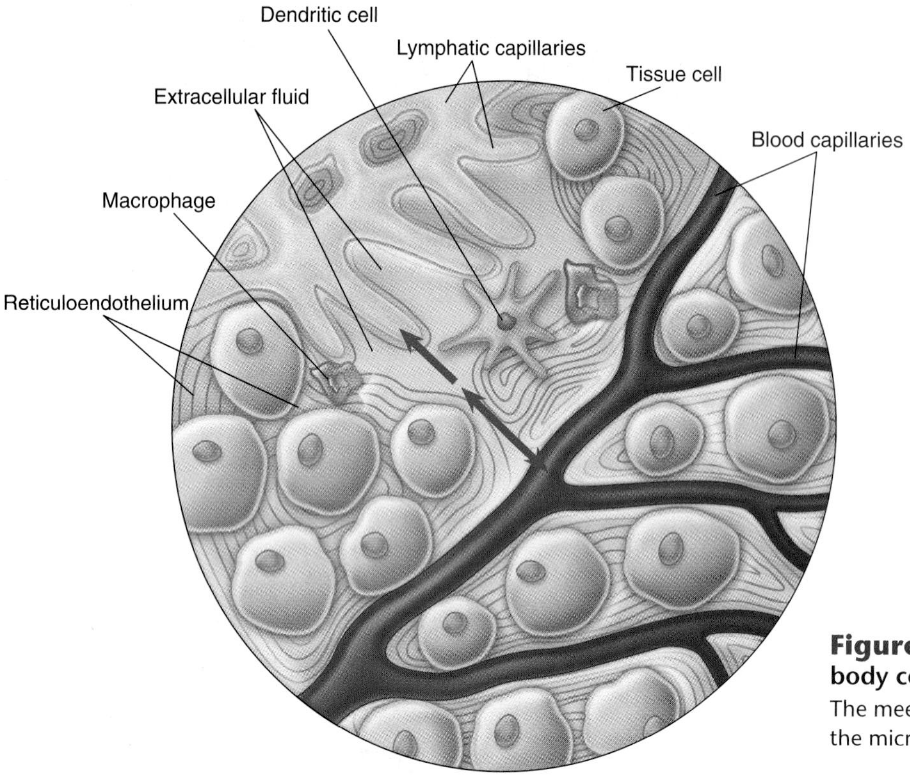

Figure 14.5 Connections between the body compartments.
The meeting of the major fluid compartments at the microscopic level.

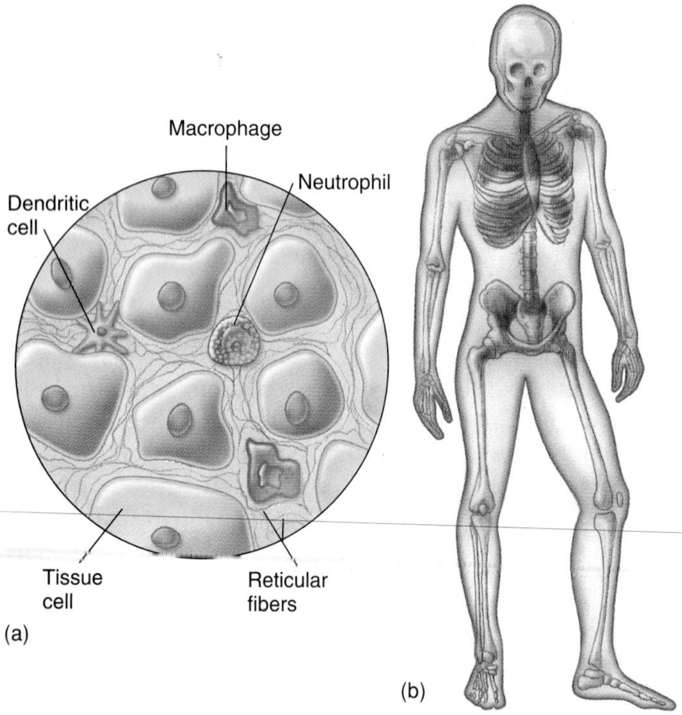

Figure 14.6 The reticuloendothelial system occurs as a pervasive, continuous connective tissue framework throughout the body.
(a) This system begins at the microscopic level with a fibrous support network (reticular fibers) enmeshing each cell. This web connects one cell to another within a tissue or organ and provides a niche for phagocytic white blood cells, which can crawl within and between tissues. **(b)** The degrees of shading in the body indicate variations in phagocyte concentration (darker = greater).

coexists with and helps form a niche for a collection of phagocytic cells termed the **mononuclear phagocyte system.** The RES is heavily endowed with white blood cells called macrophages waiting to attack passing foreign intruders as they arrive in the skin, lungs, liver, lymph nodes, spleen, and bone marrow.

Origin, Composition, and Functions of the Blood

The circulatory system consists of the circulatory system proper, which includes the heart, arteries, veins, and capillaries that circulate the blood, and the lymphatic system, which includes lymphatic vessels and lymphatic organs (lymph nodes) that circulate lymph. As you will see, these two circulations parallel, interconnect with, and complement one another.

The substance that courses through the arteries, veins, and capillaries is **whole blood,** a liquid connective tissue consisting of **blood cells** (formed elements) suspended in **plasma.** One can visualize these two components with the naked eye when a tube of *unclotted* blood is allowed to sit or is spun in a centrifuge. The cells' density causes them to settle into an opaque layer at the bottom of the tube, leaving the plasma, a clear, yellowish fluid, on top **(figure 14.7).** In chapter 15, we introduce the concept of **serum.** This substance is essentially the same as plasma, except it is the clear fluid from clotted blood. Serum is often used in immune testing and therapy.

Fundamental Characteristics of Plasma Plasma contains hundreds of different chemicals produced by the liver, white blood cells, endocrine glands, and nervous system and absorbed from the digestive tract. The main component of this fluid is water (92%), and the remainder consists of proteins such as albumin and globulins (including antibodies);

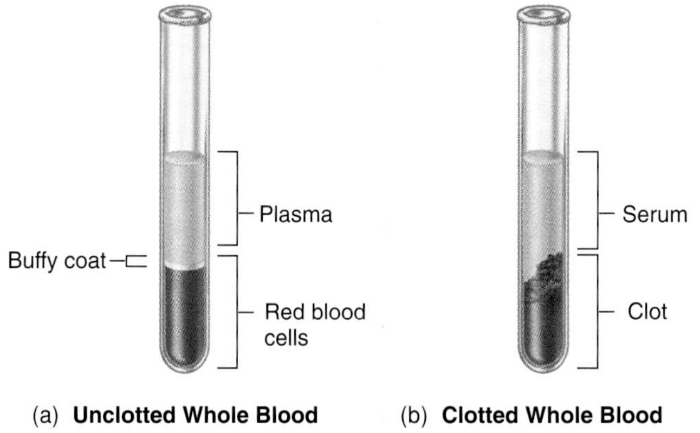

Figure 14.7 The macroscopic composition of whole blood.
(a) When blood containing anticoagulants is allowed to sit for a period, it stratifies into a clear layer of plasma; a thin layer of off-white material called the buffy coat (which contains the white blood cells); and a layer of red blood cells in the bottom, thicker layer. **(b)** Serum is the clear fluid that separates from clotted blood.

other immunochemicals; fibrinogen and other clotting factors; hormones; nutrients (glucose, amino acids, fatty acids); ions (sodium, potassium, calcium, magnesium, chloride, phosphate, bicarbonate); dissolved gases (O_2 and CO_2); and waste products (urea). These substances support the normal physiological functions of nutrition, development, protection, homeostasis, and immunity. We return to the subject of plasma and its function in immune interactions later in this chapter and in chapter 15.

A Survey of Blood Cells The production of blood cells, or **hematopoiesis** (hee″-mat-o-poy-ee′-sis), begins early in embryonic development in the yolk sac (an embryonic membrane). Later, it is taken over by the liver and lymphatic organs, and is finally assumed entirely and permanently by the red bone marrow **(figure 14.8).** Although much of a newborn's red marrow is devoted to hematopoietic function, the active marrow sites gradually recede, and by the age of 4 years, only the ribs, sternum, pelvic girdle, flat bones of the skull and spinal column, and proximal portions of the humerus and femur are devoted to blood cell production.

The relatively short life of blood cells demands a rapid turnover that is continuous throughout a human life span. The primary precursor of new blood cells is a pool of undifferentiated cells called pluripotential **stem cells**[2] maintained in the marrow. During development, these stem cells proliferate and *differentiate*—meaning that immature or unspecialized cells develop the specialized form and function of mature cells. The primary lines of cells that arise from this process produce red blood cells (RBCs, or erythrocytes), white blood cells (WBCs, or leukocytes), and platelets (thrombocytes). The white blood cell lines are programmed to develop into several secondary lines of cells during the final process of differentiation **(figure 14.9).** These committed lines of WBCs are largely responsible for immune function.

2. Pluripotential stem cells can develop into several different types of blood cells; unipotential cells have already committed to a specific line of development.

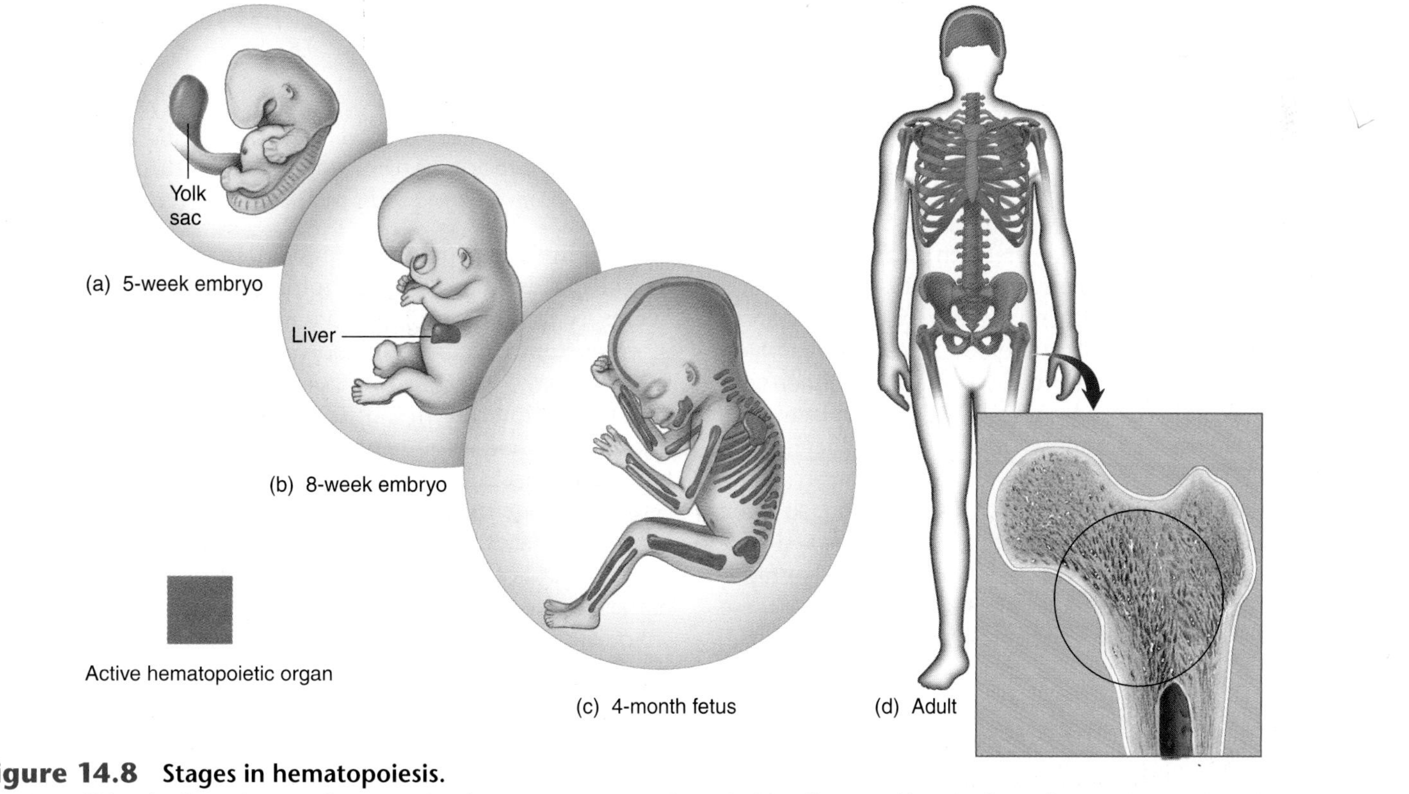

Figure 14.8 Stages in hematopoiesis.
The sites of blood cell production change as development progresses from **(a, b)** yolk sac and liver in the embryo to **(c)** extensive bone marrow sites in the fetus and **(d)** selected bone marrow sites in the child and adult. **(Inset)** Red marrow occupies the spongy bone (circle) in these areas.

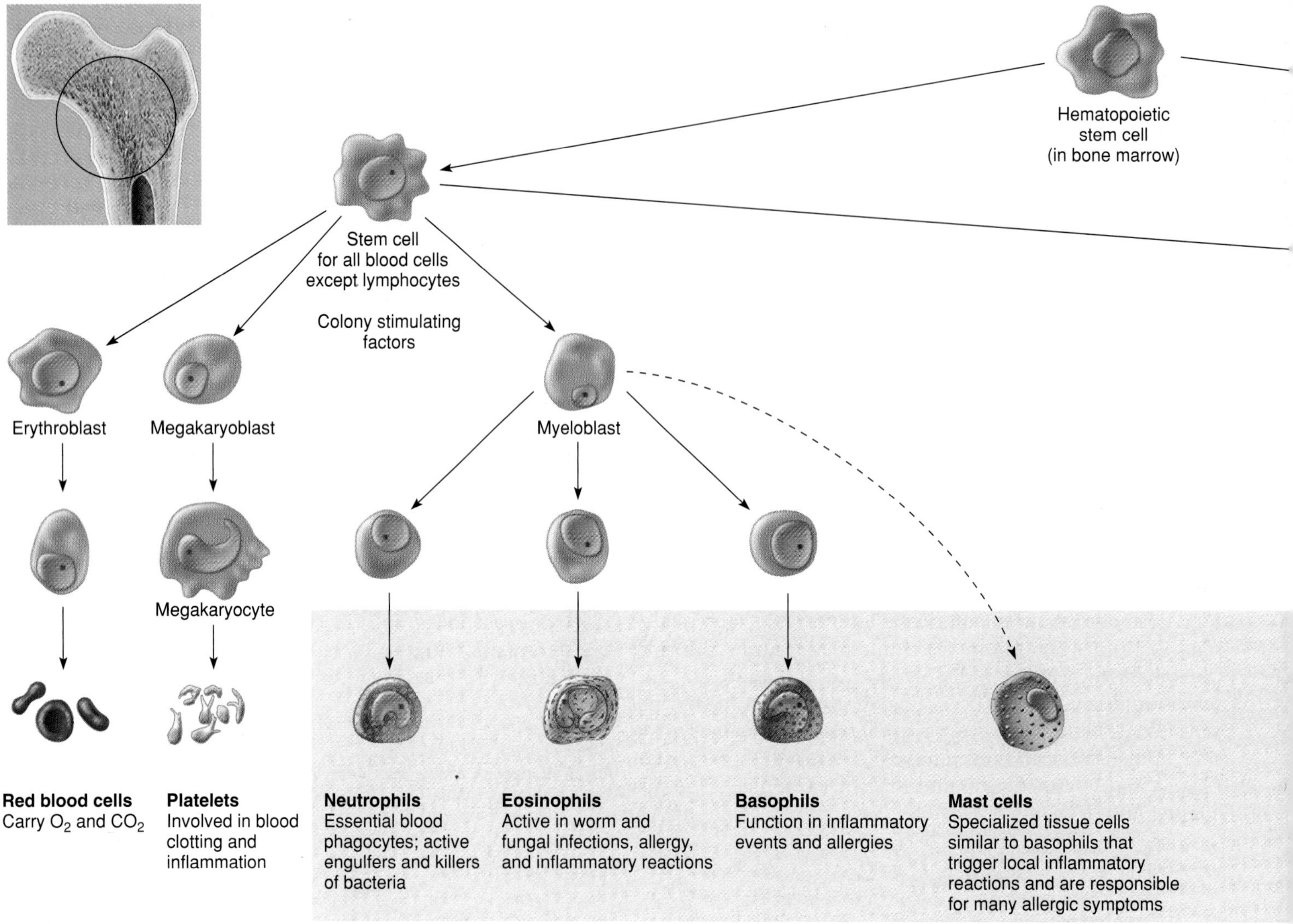

Figure 14.9 The development of blood cells and platelets.
Undifferentiated stem cells in the red marrow differentiate to give rise to several different cell lines that become increasingly specialized until mature cells (bottom row) are released into circulation. Some cells migrate into the tissues to achieve fully functional status. The shaded areas indicate leukocytes.

The *white blood cells,* or **leukocytes,** are traditionally evaluated by their reactions with hematologic stains that contain a mixture of dyes and can differentiate cells by color and morphology. When this stain used on blood smears is evaluated using the light microscope, the leukocytes appear either with or without noticeable colored granules in the cytoplasm and, on that basis, are divided into two groups: **granulocytes** and **agranulocytes.** Greater magnification reveals that even the agranulocytes have tiny granules in their cytoplasm, so some hematologists also use the appearance of the nucleus to distinguish them. Granulocytes have a lobed nucleus, and agranulocytes have an unlobed, rounded nucleus. Note both of these characteristics in circulating leukocytes shown in figure 14.9.

Granulocytes The types of granular leukocytes present in the bloodstream are neutrophils, eosinophils, and basophils. All three are known for prominent cytoplasmic granules that stain with some combination of acidic dye (eosin) or basic dye (methylene blue). Although these granules are useful diagnostically, they also function in numerous physiological events. Refer to figure 14.9 to view the cell types described.

Neutrophils, also called polymorphonuclear neutrophils (PMNs), make up 55% to 90% of the circulating leukocytes about 25 billion cells in the circulation at any given moment. The main work of the neutrophils is in phagocytosis. Their high numbers in both the blood and tissues suggest a constant challenge from resident microbiota and environmental sources. Most of the cytoplasmic granules carry digestive enzymes and other chemicals that degrade the phagocytosed materials (see the discussion of phagocytosis later in this chapter). The average neutrophil lives only about 8 days, spending much of this time in the tissues and only about 6 to 12 hours in circulation.

The role of the **eosinophil** (ee″-oh-sin′-oh-fil) in the immune system is not fully defined, although several functions have been suggested. Their granules contain peroxidase, lysozyme,

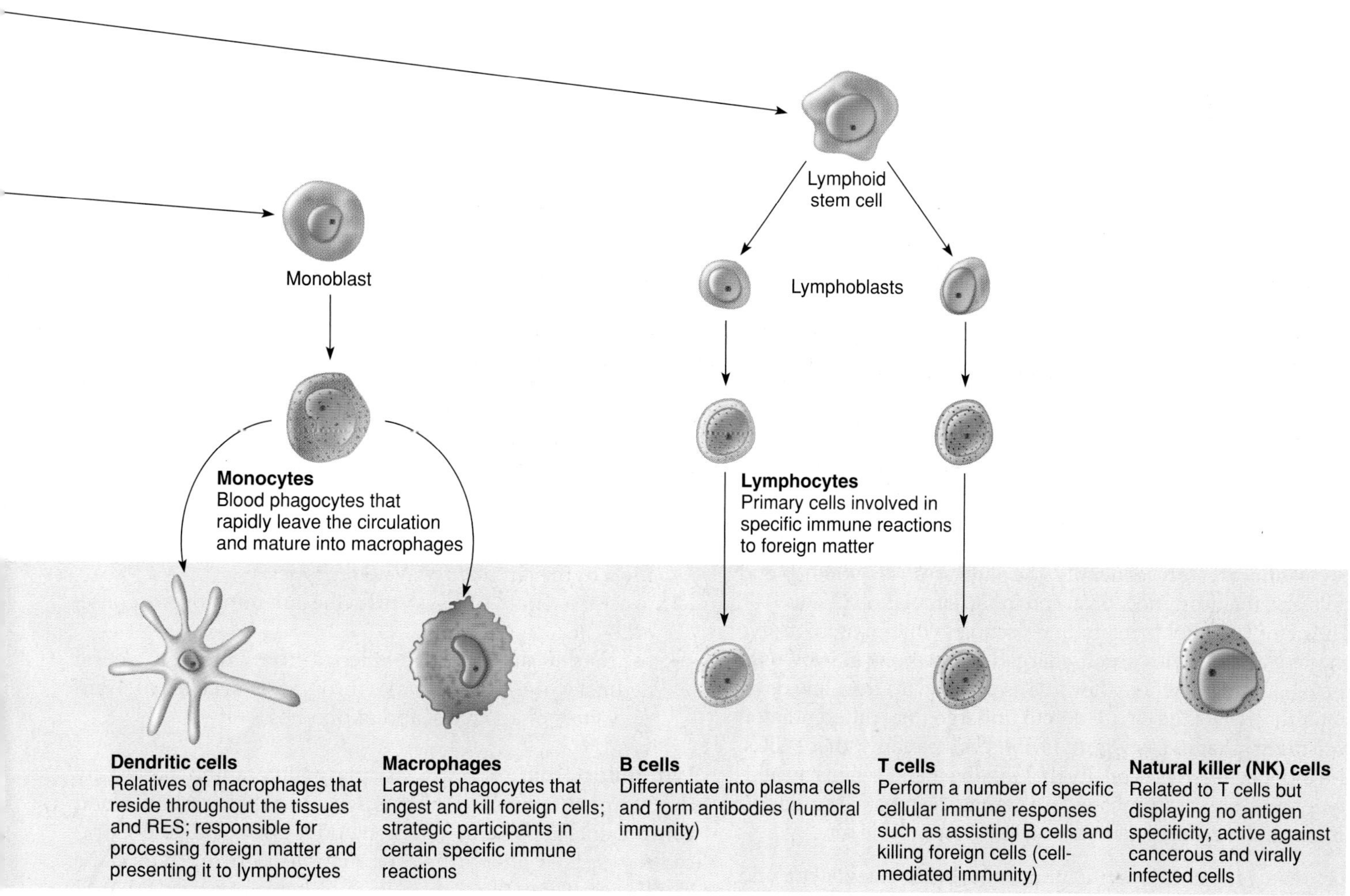
Lymphoid
stem cell
Monoblast
Lymphoblasts
Monocytes
Blood phagocytes that rapidly leave the circulation and mature into macrophages
Lymphocytes
Primary cells involved in specific immune reactions to foreign matter
Dendritic cells
Relatives of macrophages that reside throughout the tissues and RES; responsible for processing foreign matter and presenting it to lymphocytes
Macrophages
Largest phagocytes that ingest and kill foreign cells; strategic participants in certain specific immune reactions
B cells
Differentiate into plasma cells and form antibodies (humoral immunity)
T cells
Perform a number of specific cellular immune responses such as assisting B cells and killing foreign cells (cell-mediated immunity)
Natural killer (NK) cells
Related to T cells but displaying no antigen specificity, active against cancerous and virally infected cells

and other digestive enzymes, as well as toxic proteins and inflammatory chemicals. The protective action of eosinophils is to attack and destroy large eukaryotic pathogens, but they are also involved in inflammation and allergic reactions. Among their most important targets are helminth worms and fungi. Eosinophils are among the earliest cells to accumulate near sites of inflammation and allergic reactions, where they attract other leukocytes and release chemical mediators.

Basophils are the scarcest type of leukocyte, making up less than 0.5% of the total circulating WBCs in a normal individual. They share some morphological and functional similarities with widely distributed tissue cells called **mast cells.** Although these two cell types were once regarded as identical, mast cells are nonmotile elements bound to connective tissue around blood vessels, nerves, and epithelia, and basophils are motile elements derived from bone marrow.

Basophils parallel eosinophils in many of their actions, because they also contain granules with potent chemical mediators. Mast cells are first line defenders against the local invasion of pathogens, they recruit other inflammatory cells, and they are directly responsible for the release of histamine and other allergic stimulants during immediate allergies (see chapter 16).

Agranulocytes Agranular leukocytes have globular, nonlobed nuclei and lack prominent cytoplasmic granules when viewed with the light microscope. The two general types are monocytes and lymphocytes.

Although lymphocytes are the cornerstone of the third line of defense, which is the subject of chapter 15, their origin and morphology are described here so their relationship to the other blood components is clear. **Lymphocytes** are the second most common WBC in the blood, comprising 20% to 35% of the total circulating leukocytes. The fact that their overall number throughout the body is among the highest of all cells indicates how important they are to immunity. One estimate suggests that about one-tenth of all adult body cells are lymphocytes, exceeded only by erythrocytes and fibroblasts. Lymphocytes exist as two functional types—the bursal-equivalent, or **B lymphocytes** (**B cells,** for short), and the thymus-derived, or **T lymphocytes** (**T cells,** for short). B cells were first demonstrated in and named for a special lymphatic gland of chickens called the *bursa of Fabricius,* the site for their maturation in birds. In humans, B cells mature in special bone marrow sites; humans do not have a bursa of fabricius. T cells mature in the thymus gland in all birds and mammals. Both populations of cells are transported by the bloodstream and lymph and move about freely between lymphoid organs and connective tissue.

Lymphocytes are the key cells of the third line of defense and the specific immune response. When stimulated by foreign substances (antigens), lymphocytes are transformed into activated cells that neutralize and destroy that foreign substance. The contribution of B cells is mainly in **humoral immunity,** defined as protective molecules carried in the fluids of the body. When activated B cells divide, they form specialized **plasma cells,** which produce **antibodies,** large protein molecules that interlock with an antigen and participate in its destruction. Activated T cells engage in a spectrum of immune functions characterized as **cell-mediated immunity** in which T cells modulate immune functions and kill foreign cells. The action of both classes of lymphocytes accounts for the recognition and memory typical of immunity. Lymphocytes are so important to the defense of the body that most of chapter 15 is devoted to their reactions.

Monocytes are generally the largest of all white blood cells and the third most common in the circulation (3% to 7%). The cytoplasm holds many fine vacuoles containing digestive enzymes. Monocytes are discharged by the bone marrow into the bloodstream, where they live as phagocytes for a few days. Later they leave the circulation to undergo final differentiation into **macrophages** (see figure 14.15). Unlike many other WBCs, the monocyte series is relatively long-lived and retains an ability to multiply. Macrophages are among the most versatile and important of cells. In general, they are responsible for

1. many types of specific and nonspecific phagocytic and killing functions (they assume the job of cellular housekeepers, "mopping up the messes" created by infection and inflammation);
2. processing foreign molecules and presenting them to lymphocytes; and
3. secreting biologically active compounds that assist, mediate, attract, and inhibit immune cells and reactions.

We touch upon these functions in several ensuing sections.

Another product of the monocyte cell line is **dendritic cells,** named for their long, thin cell processes. Immature dendritic cells move from the blood to the RES and lymphatic tissues, where they trap pathogens. Ingestion of bacteria and viruses stimulates dendritic cells to migrate to lymph nodes and the spleen. Here, they mature into highly effective processors and presenters of foreign proteins (see chapter 15).

Erythrocyte and Platelet Lines These elements stay in the circulatory system proper. Their development is also shown in figure 14.9.

Erythrocytes develop from stem cells in the bone marrow and lose their nucleus just prior to entering the circulation. The resultant red blood cells are simple, biconcave sacs of hemoglobin that transport oxygen and carbon dioxide to and from the tissues. These are the most numerous of circulating blood cells, appearing in stains as small pink circles. Red blood cells do not ordinarily have immune functions, though they can be the target of immune reactions (see chapter 17).

Platelets are formed elements in circulating blood that are *not* whole cells. In stains, platelets are blue-gray with fine red granules and are readily distinguished from cells by their small size. Platelets function primarily in hemostasis (plugging broken blood vessels to stop bleeding) and in releasing chemicals that act in blood clotting and inflammation.

Components and Functions of the Lymphatic System

The **lymphatic system** is a compartmentalized network of vessels, cells, and specialized accessory organs **(figure 14.10).** It begins in the farthest reaches of the tissues as tiny capillaries that transport a special fluid (lymph) through an increasingly larger tributary system of vessels and filters (lymph nodes), and it leads to major vessels that drain back into the regular circulatory system. Some major functions of the lymphatic system are

1. to provide an auxiliary route for the return of extracellular fluid to the circulatory system proper;
2. to act as a "drain-off" system for the inflammatory response; and
3. to render surveillance, recognition, and protection against foreign materials through a system of lymphocytes, phagocytes, and antibodies.

Lymphatic Fluid Lymph is a plasmalike liquid carried by the lymphatic circulation. It is formed when certain blood components move out of the blood vessels into the extracellular spaces and diffuse or migrate into the lymphatic capillaries. Thus, the composition of lymph parallels that of plasma in many ways. It is made up of water, dissolved salts, and 2% to 5% protein (especially antibodies and albumin). Like blood, it also transports numerous white blood cells (especially lymphocytes) and miscellaneous materials such as fats, cellular debris, and infectious agents that have gained access to the tissue spaces. Unlike blood, red blood cells are not normally found in lymph.

Lymphatic Vessels The system of vessels that transports lymph is constructed along the lines of blood vessels. As the lymph is never subjected to high pressure, the lymphatic vessels appear most similar to thin-walled veins rather than thicker-walled arteries. The tiniest vessels, lymphatic capillaries, accompany the blood capillaries and permeate all parts of the body except the central nervous system and certain organs such as bone, placenta, and thymus. Their thin walls are easily permeated by extracellular fluid that has escaped from the circulatory system. Lymphatic vessels are found in particularly high numbers in the hands, feet, and around the areola of the breast.

Two overriding differences between the bloodstream and the lymphatic system should be mentioned. First, because one of the main functions of the lymphatic system is returning lymph to the circulation, the flow of lymph is in one direction

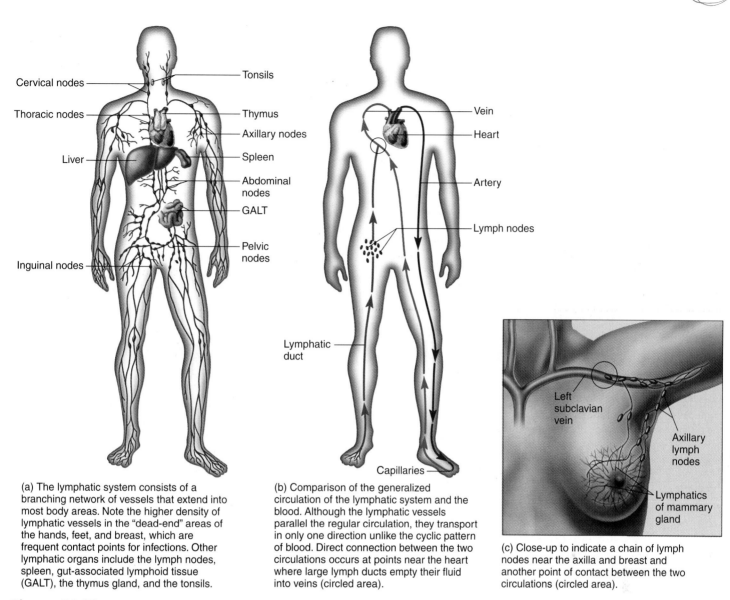

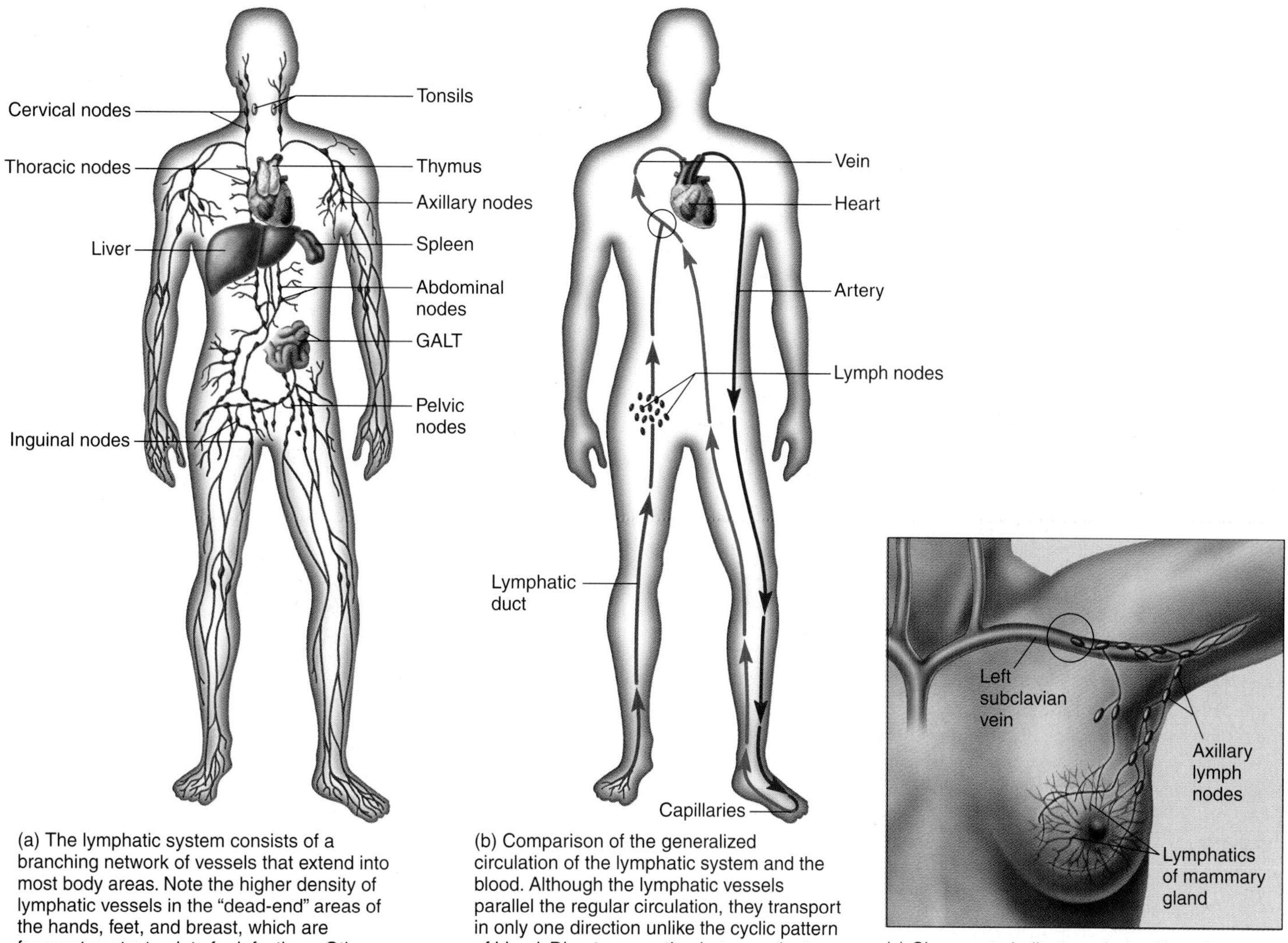

(a) The lymphatic system consists of a branching network of vessels that extend into most body areas. Note the higher density of lymphatic vessels in the "dead-end" areas of the hands, feet, and breast, which are frequent contact points for infections. Other lymphatic organs include the lymph nodes, spleen, gut-associated lymphoid tissue (GALT), the thymus gland, and the tonsils.

(b) Comparison of the generalized circulation of the lymphatic system and the blood. Although the lymphatic vessels parallel the regular circulation, they transport in only one direction unlike the cyclic pattern of blood. Direct connection between the two circulations occurs at points near the heart where large lymph ducts empty their fluid into veins (circled area).

(c) Close-up to indicate a chain of lymph nodes near the axilla and breast and another point of contact between the two circulations (circled area).

Figure 14.10 General components of the lymphatic system.

only with lymph moving from the extremities toward the heart. Eventually, lymph will be returned to the bloodstream through the thoracic duct or the right lymphatic duct to the subclavian vein near the heart. The second difference concerns how lymph travels through the vessels of the lymphatic system. While blood is transported through the body by means of a dedicated pump (the heart), lymph is moved only through the contraction of the skeletal muscles through which the lymphatic ducts wend their way. This dependence on muscle movement helps to explain the swelling of the hands and feet that sometimes occurs during the night (when muscles are inactive) yet dissipates soon after waking.

Lymphoid Organs and Tissues Other organs and tissues that perform lymphoid functions are the lymph nodes (glands), thymus, spleen, and clusters of tissues in the gastrointestinal tract (gut-associated lymphoid tissue; GALT) and the pharynx (the tonsils, for example). A trait common to these organs is a loose connective tissue framework that houses aggregations of lymphocytes, the important class of white blood cells mentioned previously.

Lymph Nodes Lymph nodes are small, encapsulated, bean-shaped organs stationed, usually in clusters, along lymphatic channels and large blood vessels of the thoracic and abdominal cavities (see figure 14.10). Major aggregations of nodes occur in the loose connective tissue of the armpit (axillary nodes), groin (inguinal nodes), and neck (cervical nodes). Both the location and architecture of these nodes clearly specialize them for filtering out materials that have entered the lymph and providing appropriate cells and niches for immune reactions.

Spleen The spleen is a lymphoid organ in the upper left portion of the abdominal cavity. It is somewhat similar to a lymph node except that it serves as a filter for blood instead of lymph. While the spleen's primary function is to remove

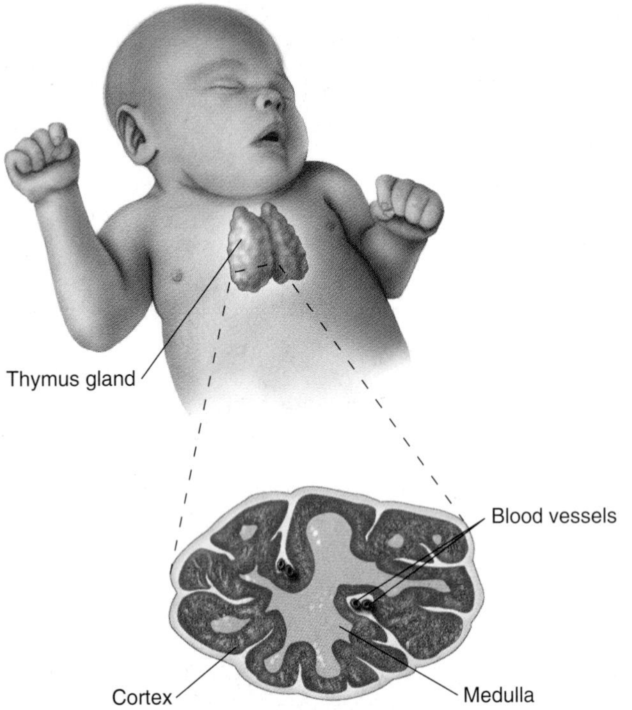

Figure 14.11 The thymus gland.
Immediately after birth, the thymus is a large organ that nearly fills the region over the midline of the upper thoracic region. In the adult, however, it is proportionately smaller (to compare, see figure 14.10*a*). Section shows the main anatomical regions of the thymus. Immature T cells enter through the cortex and migrate into the medulla as they mature.

worn-out red blood cells from circulation, its most important immunologic function centers on the filtering of pathogens from the blood and their subsequent phagocytosis by resident macrophages. Although adults whose spleens have been surgically removed can live a relatively normal life, asplenic children are severely immunocompromised.

The Thymus: Site of T-Cell Maturation The **thymus** originates in the embryo as two lobes in the pharyngeal region that fuse into a triangular structure. The size of the thymus is greatest proportionately at birth **(figure 14.11),** and it continues to exhibit high rates of activity and growth until puberty, after which it begins to shrink gradually through adulthood. Under the influence of thymic hormones, thymocytes develop specificity and are released into the circulation as mature T cells. The T cells subsequently migrate to and settle in other lymphoid organs (for example, the lymph nodes and spleen), where they occupy the specific sites described previously.

Children born without a thymus (DiGeorge syndrome, see chapter 16) or who have had their thymus surgically removed are severely immunodeficient and fail to thrive. Adults have developed enough mature T cells that removal of the thymus or reduction in its function has milder effects. Do not confuse the thymus with the thyroid gland, which is located nearby but has an entirely different function.

Miscellaneous Lymphoid Tissue At many sites on or just beneath the mucosa of the gastrointestinal and respiratory tracts lie discrete bundles of lymphocytes. The positioning of this diffuse system provides an effective first-strike potential against the constant influx of microbes and other foreign materials in food and air. In the pharynx, a ring of tissues called the tonsils provides an active source of lymphocytes. The breasts of pregnant and lactating women also become temporary sites of antibody-producing lymphoid tissues (see colostrum, Insight 15.2). The intestinal tract houses the best-developed collection of lymphoid tissue, called **gut-associated lymphoid tissue,** or **GALT.** Examples of GALT include the appendix, the lacteals (special lymphatic vessels stationed in each intestinal villus), and **Peyer's patches,** compact aggregations of lymphocytes in the ileum of the small intestine. GALT provides immune functions against intestinal pathogens and is a significant source of some types of antibodies. Other, less well-organized collections of secondary lymphoid tissue include the *mucosal-associated lymphoid tissue (MALT), skin-associated lymphoid tissue (SALT), and bronchial-associated lymphoid tissue (BALT).*

CHECKPOINT

- The immune system is a complex collection of fluids and cells that penetrate every organ, tissue space, fluid compartment, and vascular network of the body. The four major subdivisions of this system are the RES, the ECF, the blood vascular system, and the lymphatic system.
- The RES, or reticuloendothelial system, is a network of connective tissue fibers inhabited by macrophages ready to attack and ingest microbes that have managed to bypass the first line of defense.
- The ECF, or extracellular fluid, compartment surrounds all tissue cells and is penetrated by both blood and lymph vessels, which bring all components of the second and third line of defense to attack infectious microbes.
- The blood contains both specific and nonspecific defenses. Nonspecific cellular defenses include the granulocytes, macrophages, and dendritic cells. The two components of the specific immune response are the T lymphocytes, which provide specific cell-mediated immunity, and the B lymphocytes, which produce specific antibody or humoral immunity.
- The lymphatic system has three functions: (1) It returns tissue fluid to general circulation; (2) it carries away excess fluid in inflamed tissues; (3) it concentrates and processes foreign invaders and initiates the specific immune response. Important sites of lymphoid tissues are lymph nodes, spleen, thymus, tonsils, and GALT.

14.4 The Second Line of Defense

Now that we have introduced the principal anatomical and physiological framework of the immune system, we address some mechanisms that play important roles in host defenses: (1) inflammation, (2) phagocytosis, (3) interferon, and (4) complement. Because of the generalized nature of these defenses, they are primarily nonspecific in their effects, but they also support and interact with the specific immune responses described in chapter 15.

INSIGHT 14.1 *Medical*

When Inflammation Gets Out of Hand

Not every aspect of inflammation is protective or results in the proficient resolution of tissue damage. As one looks over a list of diseases, it is rather striking how many of them are due in part or even completely to an overreactive or dysfunctional inflammatory response.

Some "itis" reactions mentioned in chapter 13 are a case in point. Inflammatory exudates that build up in the brain in African trypanosomiasis, cryptococcosis, and other brain infections can be so injurious to the nervous system that impairment is permanent. Frequently, an inflammatory reaction that walls off the pathogen leads to an abscess, a swollen mass of neutrophils and dead, liquefied tissue that can harbor live pathogens in the center. Abscesses are a prominent feature of staphylococcal, amoebic, and enteric infections.

Other pathologic manifestations of chronic diseases—for example, the tubercles of tuberculosis, the lesions of late syphilis, the disfiguring nodules of leprosy, and the cutaneous ulcers of leishmaniasis—are due to an aberrant tissue response called *granuloma formation*. Granulomas develop not only in response to microbes but also in response to inanimate foreign bodies (sutures and mineral grains that are difficult to break down). This condition is initiated when neutrophils ineffectively and incompletely phagocytose the pathogens or materials involved in an inflammatory reaction. The macrophages then enter to clean up and attempt to phagocytose the dead neutrophils and foreign substances, but they fail to completely manage them. They respond by storing these ingested materials in vacuoles and becoming inactive. Over a given time period, large numbers of adjacent macrophages fuse into giant, inactive multinucleate cells called foreign body giant cells. These sites are further infiltrated with lymphocytes. The resultant collections make the tissue appear granular—hence, the name. A granuloma can exist in the tissue for months, years, or even a lifetime.

Medical science is rapidly searching for new applications for the massive amount of new information on inflammatory mediators. One highly promising area appears to be the use of chemokine inhibitors that could reduce chemotaxis and the massive, destructive influx of leukocytes. Such therapy could ultimately be used for certain cancers, hardening of arteries, and Alzheimer's disease.

The Inflammatory Response: A Complex Concert of Reactions to Injury

At its most general level, the inflammatory response is a reaction to any traumatic event in the tissues. It is so commonplace that most of us manifest inflammation in some way every day. It appears in the nasty flare of a cat scratch, the blistering of a burn, the painful lesion of an infection, and the symptoms of allergy. It is readily identifiable by a classic series of signs and symptoms characterized succinctly by four Latin terms: *rubor, calor, tumor,* and *dolor*. Rubor (redness) is caused by increased circulation and vasodilation in the injured tissues; calor (warmth) is the heat given off by the increased flow of blood; tumor (swelling) is caused by increased fluid escaping into the tissues; and dolor (pain) is caused by the stimulation of nerve endings **(figure 14.12).** A fifth symptom, loss of function, has been added to give a complete picture of the effects of inflammation. Although these manifestations can be unpleasant, they serve an important warning that injury has taken place and set in motion responses that save the body from further injury.

Factors that can elicit inflammation include trauma from infection (the primary emphasis here), tissue injury or necrosis due to physical or chemical agents, and specific immune reactions. Although the details of inflammation are very complex, its chief functions can be summarized as follows:

1. to mobilize and attract immune components to the site of the injury,
2. to set in motion mechanisms to repair tissue damage and localize and clear away harmful substances, and
3. to destroy microbes and block their further invasion **(figure 14.13).**

The inflammatory response is a powerful defensive reaction, a means for the body to maintain stability and restore itself after an injury. But when it is chronic, it has the potential to actually *cause* tissue injury, destruction, and disease **(Insight 14.1).**

The Stages of Inflammation

The process leading to inflammation is a dynamic, predictable sequence of events that can be acute, lasting from a few minutes or hours, to chronic, lasting for days, weeks, or years. Once the initial injury has occurred, a chain reaction takes place at the site of damaged tissue, summoning beneficial

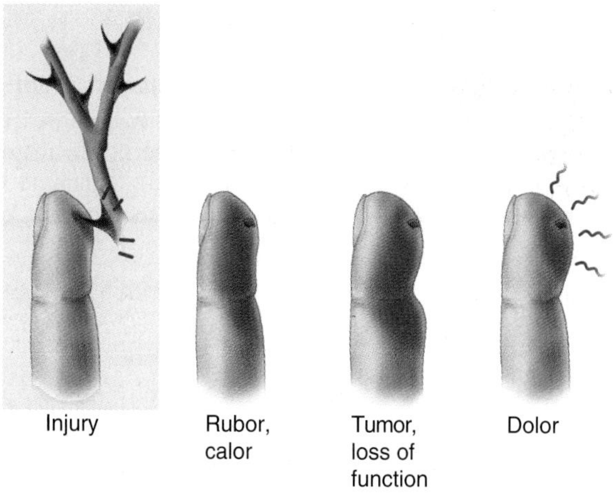

Figure 14.12 The response to injury.
This classic checklist encapsulates the reactions of the tissues to an assault. Each of the events is an indicator of one of the mechanisms of inflammation described in this chapter.

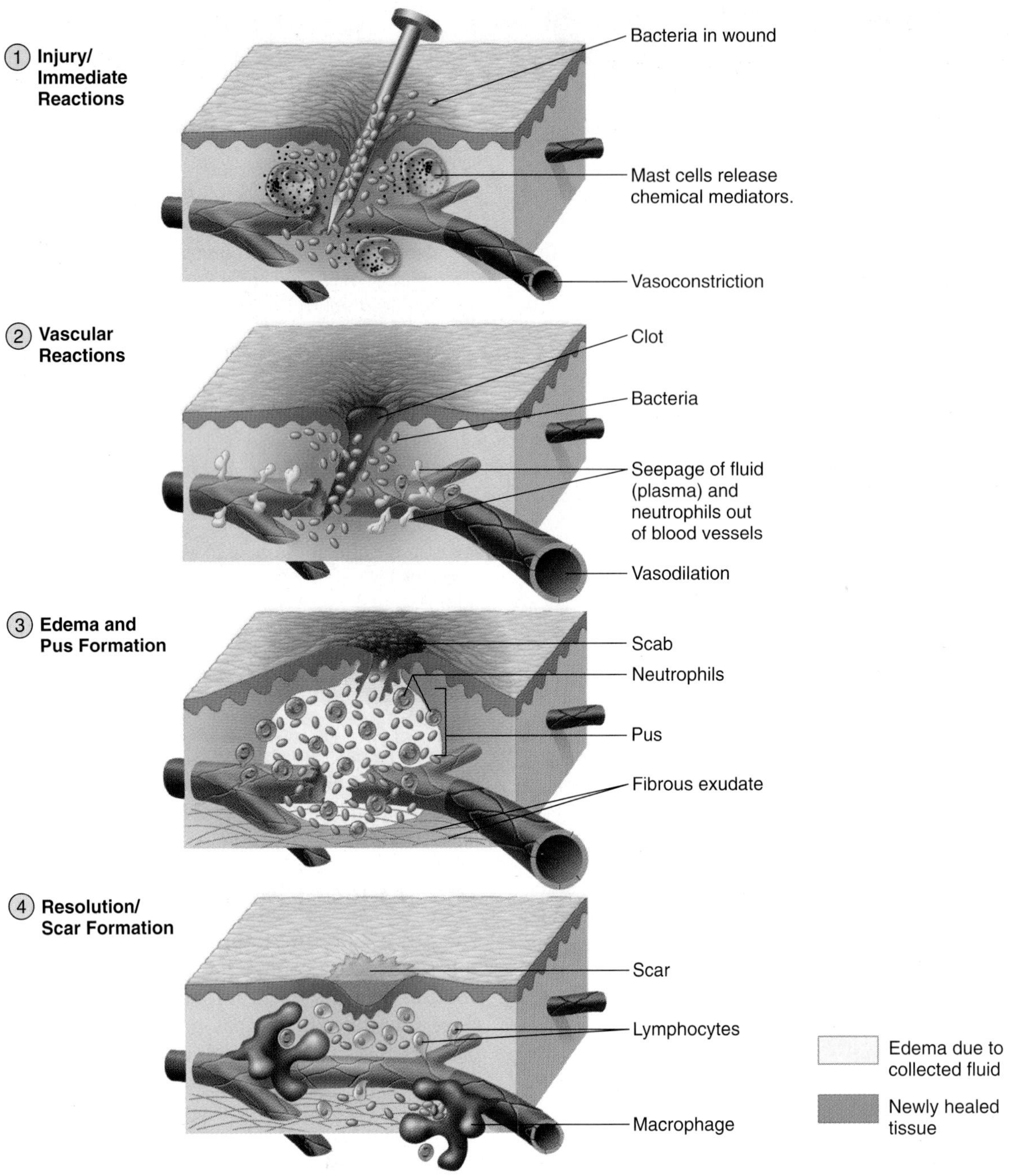

Process Figure 14.13 The major events in inflammation.
(1) Injury → Reflex narrowing of the blood vessels (vasoconstriction) lasting for a short time → Release of chemical mediators into area. (2) Increased diameter of blood vessels (vasodilation) → Increased blood flow → Increased vascular permeability → Leakage of fluid (plasma) from blood vessels into tissues (exudate formation). (3) Edema → Infiltration of site by neutrophils and accumulation of pus. (4) Macrophages and lymphocytes → Repair, either by complete resolution and return of tissue to normal state or by formation of scar tissue.

cells and fluids into the injured area. As an example, we will look at an injury at the microscopic level and observe the flow of major events (see figure 14.13).

Vascular Changes: Early Inflammatory Events

Following an injury, some of the earliest changes occur in the vasculature (arterioles, capillaries, venules) in the vicinity of the damaged tissue. These changes are controlled by nervous stimulation, **chemical mediators,** and **cytokines** released by blood cells, tissue cells, and platelets in the injured area. Some mediators are *vasoactive*—that is, they affect the endothelial cells and smooth muscle cells of blood vessels, and others are **chemotactic factors,** also called **chemokines,** that affect white blood cells. Inflammatory mediators cause fever, stimulate lymphocytes, prevent virus spread, and cause allergic symptoms **(Insight 14.2).** Although the constriction of arterioles is stimulated first, it lasts for only a few seconds or minutes

INSIGHT 14.2 Medical

The Dynamics of Inflammatory Mediators

Just as the nervous system is coordinated by a complex communications network, so too is the immune system. Hundreds of small, active molecules are constantly being secreted to regulate, stimulate, suppress, and otherwise control the many aspects of cell development, inflammation, and immunity. These substances are the products of several types of cells, including monocytes, macrophages, lymphocytes, fibroblasts, mast cells, platelets, and endothelial cells of blood vessels. Their effects may be local or systemic, short term or long-lasting, nonspecific or specific, protective or pathologic.

In recent times, the field of cytokines has become so increasingly complex that we can include here only an overview of the major groups of important cytokines and other mediators. The major functional types can be categorized into

1. cytokines that mediate nonspecific immune reactions such as inflammation and phagocytosis,
2. cytokines that regulate the growth and activation of lymphocytes,
3. cytokines that activate immune reactions during inflammation,
4. hematopoiesis factors for white blood cells,
5. vasoactive mediators, and
6. miscellaneous inflammatory mediators.

Nonspecific Mediators of Inflammation and Immunity

- *Tumor necrosis factor (TNF),* a substance from macrophages, lymphocytes, and other cells that increases chemotaxis and phagocytosis and stimulates other cells to secrete inflammatory cytokines. It also serves as an endogenous pyrogen that induces fever, increases blood coagulation, suppresses bone marrow, and causes wasting of the body called cachexia.
- *Interferon (IFN), alpha and beta,* produced by leukocytes fibroblasts, and other cells, inhibits virus replication and cell division and increases the action of certain lymphocytes that kill other cells.
- *Interleukin* (IL) 1,* a product of macrophages and epithelial cells that has many of the same biological activities as TNF, such as inducing fever and activation of certain white blood cells.
- *Interleukin-6,* secreted by macrophages, lymphocytes, and fibroblasts. Its primary effects are to stimulate the growth of B cells and to increase the synthesis of liver proteins.
- *Various chemokines.* By definition, chemokines are cytokines that stimulate the movement and migration of white blood cells (chemotactic factors). Included among these are complement C5A, interleukin 8, and platelet-activating factor.

**Interleukin* is a term that refers to a group of small peptides originally isolated from leukocytes. There are currently more than 30 named interleukins. We now know that other cells besides leukocytes can synthesize them and that they have a variety of biological activities. Functions of some selected examples are presented in chapter 15.

Cytokines That Regulate Lymphocyte Growth and Activation

- *Interleukin-2,* the primary growth factor from T cells. Interestingly, it acts on the same cells that secrete it. It stimulates mitosis and secretion of other cytokines. In B cells, it is a growth factor and stimulus for antibody synthesis.
- *Interleukin-4,* a stimulus for the production of allergy antibodies; inhibits macrophage actions; favors development of T cells.
- *Granulocyte colony-stimulating factor (G-CSF),* produced by T cells, macrophages, and neutrophils. It stimulates the activation and differentiation of neutrophils.
- *Macrophage colony-stimulating factor (M-CSF),* produced by a variety of cells. M-CSF promotes the growth and development of macrophages from undifferentiated precursor cells.

Cytokines That Activate Specific Immune Reactions

- *Interferon gamma,* a T-cell-derived mediator whose primary function is to activate macrophages. It also promotes the differentiation of T and B cells, activates neutrophils, and stimulates diapedesis.
- *Interleukin-5* activates eosinophils and B cells; *interleukin-10* inhibits macrophages and B cells; and *interleukin-12* activates T cells and killer cells.

Vasoactive Mediators

- **Histamine,** a vasoactive mediator produced by mast cells and basophils that causes vasodilation, increased vascular permeability, and mucus production. It functions primarily in inflammation and allergy.
- **Serotonin,** a mediator produced by platelets and intestinal cells that causes smooth muscle contraction, inhibits gastric secretion, and acts as a neurotransmitter.
- **Bradykinin,** a vasoactive amine from the blood or tissues that stimulates smooth muscle contraction and increases vascular permeability, mucus production, and pain. It is particularly active in allergic reactions.

Miscellaneous Inflammatory Mediators

- **Prostaglandins,** produced by most body cells; complex chemical mediators that can have opposing effects (for example, dilation or constriction of blood vessels) and are powerful stimulants of inflammation and pain.
- **Leukotrienes** stimulate the contraction of smooth muscle and enhance vascular permeability. They are implicated in the more severe manifestations of immediate allergies (constriction of airways).
- **Platelet-activating factor,** a substance released from basophils, causes the aggregation of platelets and the release of other chemical mediators during immediate allergic reactions.

and is followed in quick succession by the opposite reaction, vasodilation. The overall effect of vasodilation is to increase the flow of blood into the area, which facilitates the influx of immune components and also causes redness and warmth.

Edema: Leakage of Vascular Fluid into Tissues

Some vasoactive substances cause the endothelial cells in the walls of postcapillary venules to contract and form gaps through which blood-borne components exude into the extracellular spaces. The fluid part that escapes is called the *exudate.* Accumulation of this fluid in the tissues gives rise to local swelling and hardness called **edema.** The edematous exudate contains varying amounts of plasma proteins, such as globulins, albumin, the clotting protein fibrinogen, blood cells, and cellular debris. Depending upon its content, the exudate may be clear (called *serous*), or it may contain red blood cells or pus. Pus is composed mainly of white blood cells and the debris generated by phagocytosis. In some types of edema, the fibrinogen is converted to fibrin threads that enmesh the injury site. Within an hour, multitudes of neutrophils responding chemotactically to special signaling molecules converge on the injured site (see figure 14.13, step 3).

Unique Dynamic Characteristics of White Blood Cells In order for WBCs to leave the blood vessels and enter the tissues, they adhere to the inner walls of the smaller blood vessels. From this position, they are poised to migrate out of the blood into the tissue spaces by a process called **diapedesis** (dye″-ah-puh-dee′-sis).

Diapedesis, also known as transmigration, is aided by several related characteristics of WBCs. For example, they are actively motile and readily change shape. This phenomenon is also assisted by the nature of the endothelial cells lining the venules. They contain complex adhesive receptors that capture the WBCs and participate in their transport from the venules into the extracellular spaces **(figure 14.14).**

Another factor in the migratory habits of these WBCs is **chemotaxis,** defined as the tendency of cells to migrate in response to a specific chemical stimulus given off at a site of injury or infection (see inflammation and phagocytosis later in this chapter). Through this means, cells swarm from many compartments to the site of infection and remain there to perform general and specific immune functions. These basic properties are absolutely essential for the sort of intercommunication and deployment of cells required for most immune reactions (see figure 14.14).

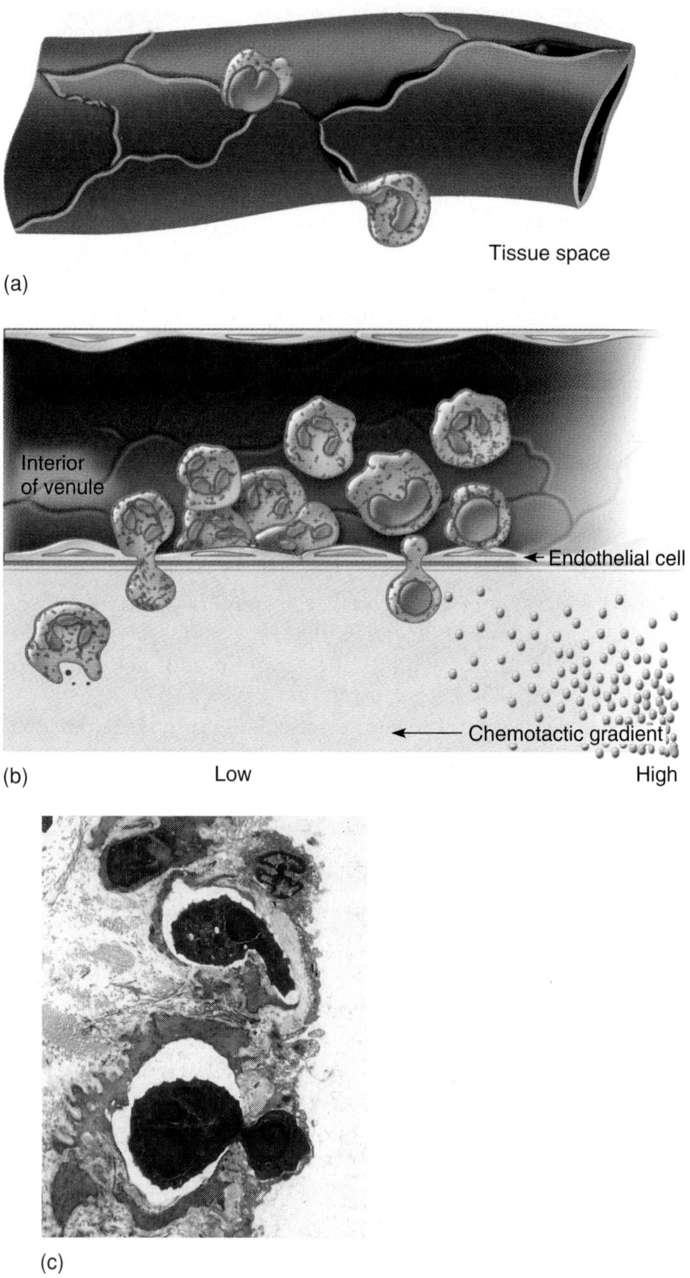

Figure 14.14 Diapedesis and chemotaxis of leukocytes. **(a)** View of a venule depicts white blood cells squeezing themselves between spaces in the blood vessel wall through diapedesis. **(b)** This process, shown in cross section, indicates how the pool of leukocytes adheres to the endothelial wall. From this site, they are poised to migrate out of the vessel into the tissue space. **(c)** This photograph captures neutrophils in the process of diapedesis.

The Benefits of Edema and Chemotaxis Both the formation of edematous exudate and the infiltration of neutrophils are physiologically beneficial activities. The influx of fluid dilutes toxic substances, and the fibrin clot can effectively trap microbes and prevent their further spread. The neutrophils that aggregate in the inflamed site are immediately involved in phagocytosing and destroying bacteria, dead tissues, and particulate matter (by mechanisms discussed in a later section on phagocytosis). In some types of inflammation, accumulated phagocytes contribute to **pus,** a whitish mass of cells, liquefied cellular debris, and bacteria. Certain bacteria (streptococci, staphylococci, gonococci, and meningococci) are especially powerful attractants for neutrophils and are thus termed **pyogenic,** or pus-forming, bacteria.

Late Reactions of Inflammation Sometimes a mild inflammation can be resolved by edema and phagocytosis.

INSIGHT 14.3 Medical

Some Facts About Fever

Fever is such a prevalent reaction that it is a prominent symptom of hundreds of diseases. For thousands of years, people believed fever was part of an innate protective response. Hippocrates offered the idea that it was the body's attempt to burn off a noxious agent. Sir Thomas Sydenham wrote in the 17th century: "Why, fever itself is Nature's instrument!" So widely held was the view that fever could be therapeutic that pyretotherapy (treating disease by inducing an intermittent fever) was once used to treat syphilis, gonorrhea, leishmaniasis (a protozoan infection), and cancer. This attitude fell out of favor when drugs for relieving fever (aspirin) first came into use in the early 1900s, and an adverse view of fever began to dominate.

Changing Views of Fever

In recent times, the medical community has returned to the original concept of fever as more healthful than harmful. Experiments with vertebrates indicate that fever is a universal reaction, even in cold-blooded animals such as lizards and fish. A study with febrile (feverish) mice and frogs indicated that fever increases the rate of antibody synthesis. Work with tissue cultures showed that increased temperatures stimulate the activities of T cells and increase the effectiveness of interferon. Artificially infected rabbits and pigs allowed to remain febrile survive at a higher rate than those given suppressant drugs. Fever appears to enhance phagocytosis of staphylococci by neutrophils in guinea pigs and humans. In recent years, malaria-induced fevers have been experimentally studied as a way to treat HIV infection.

Hot and Cold: Why Do Chills Accompany Fever?

Fever almost never occurs as a single response; it is usually accompanied by chills. What causes this oddity—that a person flushed with fever periodically feels cold and trembles uncontrollably? The explanation lies in the natural physiological interaction between the thermostat in the hypothalamus and the temperature of the blood. For example, if the thermostat has been set (by pyrogen) at 102°F but the blood temperature is 99°F, the muscles are stimulated to contract involuntarily (shivering) as a means of producing heat. In addition, the vessels in the skin constrict, creating a sensation of cold, and the piloerector muscles in the skin cause "goose bumps" to form.

Inflammatory reactions that are more long-lived attract a collection of monocytes, lymphocytes, and macrophages to the reaction site. Clearance of pus, cellular debris, dead neutrophils, and damaged tissue is performed by macrophages, the only cells that can engulf and dispose of such large masses. At the same time, B lymphocytes react with foreign molecules and cells by producing specific antimicrobial proteins (antibodies), and T lymphocytes kill intruders directly. Late in the process, the tissue is completely repaired, if possible, or replaced by connective tissue in the form of a scar (see figure 14.13, step 4). If the inflammation cannot be relieved or resolved in this way, it can become chronic and create a long-term pathologic condition.

Fever: An Adjunct to Inflammation

An important systemic component of inflammation is fever, defined as an abnormally elevated body temperature. Although fever is a nearly universal symptom of infection, it is also associated with certain allergies, cancers, and other organic illnesses. Fevers whose causes are unknown are called fevers of unknown origin, or FUO.

The body temperature is normally maintained by a control center in the hypothalamus region of the brain. This thermostat regulates the body's heat production and heat loss and sets the core temperature at around 37°C (98.6°F) with slight fluctuations (1°F) during a daily cycle. Fever is initiated when a circulating substance called pyrogen (py′-roh-jen) resets the hypothalamic thermostat to a higher setting. This change signals the musculature to increase heat production and peripheral arterioles to decrease heat loss through vasoconstriction **(Insight 14.3).** Fevers range in severity from low-grade (37.7°C to 38.3°C, or 100°F to 101°F) to moderate (38.8°C to 39.4°C, or 102°F to 103°F) to high (40.0°C to 41.1°C, or 104°F to 106°F). Pyrogens are described as *exogenous* (coming from outside the body) or *endogenous* (originating internally). Exogenous pyrogens are products of infectious agents such as viruses, bacteria, protozoans, and fungi. One well-characterized exogenous pyrogen is endotoxin, the lipopolysaccharide found in the cell walls of gram-negative bacteria. Blood, blood products, vaccines, or injectable solutions can also contain exogenous pyrogens. Endogenous pyrogens are liberated by monocytes, neutrophils, and macrophages during the process of phagocytosis and appear to be a natural part of the immune response. Two potent pyrogens released by macrophages are interleukin-1 (IL-1) and tumor necrosis factor (TNF).

Benefits of Fever The association of fever with infection strongly suggests that it serves a beneficial role, a view still being debated but gaining acceptance. Aside from its practical and medical importance as a sign of a physiological disruption, increased body temperature has additional benefits:

- Fever inhibits multiplication of temperature-sensitive microorganisms such as the poliovirus, cold viruses, herpes zoster virus, systemic and subcutaneous fungal pathogens, *Mycobacterium* species, and the syphilis spirochete.
- Fever impedes the nutrition of bacteria by reducing the availability of iron. It has been demonstrated that during

fever, the macrophages stop releasing their iron stores, which could retard several enzymatic reactions needed for bacterial growth.
- Fever increases metabolism and stimulates immune reactions and naturally protective physiological processes. It speeds up hematopoiesis, phagocytosis, and specific immune reactions.

Treatment of Fever With this revised perspective on fever, whether to suppress it or not can be a difficult decision. Some advocates feel that a slight to moderate fever in an otherwise healthy person should be allowed to run its course, in light of its potential benefits and minimal side effects. All medical experts do agree that high and prolonged fevers or fevers in patients with cardiovascular disease, seizures, and respiratory ailments are risky and must be treated immediately with fever-reducing drugs. The classic therapy for fever is an antipyretic drug such as aspirin or acetaminophen (Tylenol) that lowers the setting of the hypothalamic center and restores normal temperature. Any physical technique that increases heat loss (tepid baths, for example) can also help reduce the core temperature.

Phagocytosis: Cornerstone of Inflammation and Specific Immunity

By any standard, a phagocyte represents an impressive piece of living machinery, meandering through the tissues to seek, capture, and destroy a target. The general activities of phagocytes are

1. to survey the tissue compartments and discover microbes, particulate matter (dust, carbon particles, antigen-antibody complexes), and injured or dead cells;
2. to ingest and eliminate these materials; and
3. to extract immunogenic information (antigens) from foreign matter.

It is generally accepted that all cells have some capacity to engulf materials, but *professional phagocytes* do it for a living. The three main types of phagocytes are neutrophils, monocytes, and macrophages.

Neutrophils and Eosinophils

As previously stated, neutrophils are general-purpose phagocytes that react early in the inflammatory response to bacteria and other foreign materials and to damaged tissue. A common sign of bacterial infection is a high neutrophil count in the blood (neutrophilia), and neutrophils are also a primary component of pus. Eosinophils are attracted to sites of parasitic infections and antigen-antibody reactions, though they play only a minor phagocytic role.

Macrophage: King of the Phagocytes

After emigrating out of the bloodstream into the tissues, monocytes are transformed by various inflammatory mediators into macrophages. This process is marked by an increase in size and by enhanced development of lysosomes and other organelles **(figure 14.15).** At one time, macrophages were classified as either fixed (adherent to tissue) or wandering, but this terminology can be misleading. All macrophages retain the capacity to move about. Whether they reside in a specific organ or wander depends upon their stage of development and the immune stimuli they receive. Specialized macrophages called **histiocytes** migrate to a certain tissue and remain there during their life span. Examples are alveolar (lung) macrophages; the Kupffer cells in the liver; Langerhans cells in the skin **(figure 14.16);** and macrophages in the spleen, lymph nodes, bone marrow, kidney, bone, and brain. Other macrophages do not reside permanently in a particular tissue and drift nomadically throughout the RES. Not only are macrophages dynamic scavengers, but they also process foreign substances and prepare them for reactions with B and T lymphocytes (see chapter 15).

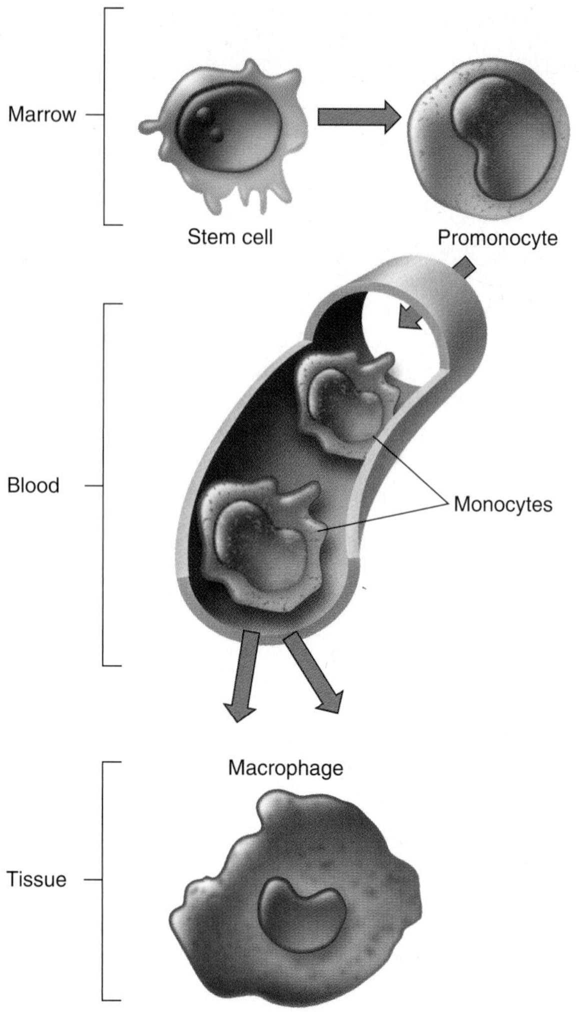

Figure 14.15 The developmental stages of monocytes and macrophages.
The cells progress through maturational stages in the bone marrow and peripheral blood. Once in the tissues, a macrophage can remain nomadic or take up residence in a specific organ.

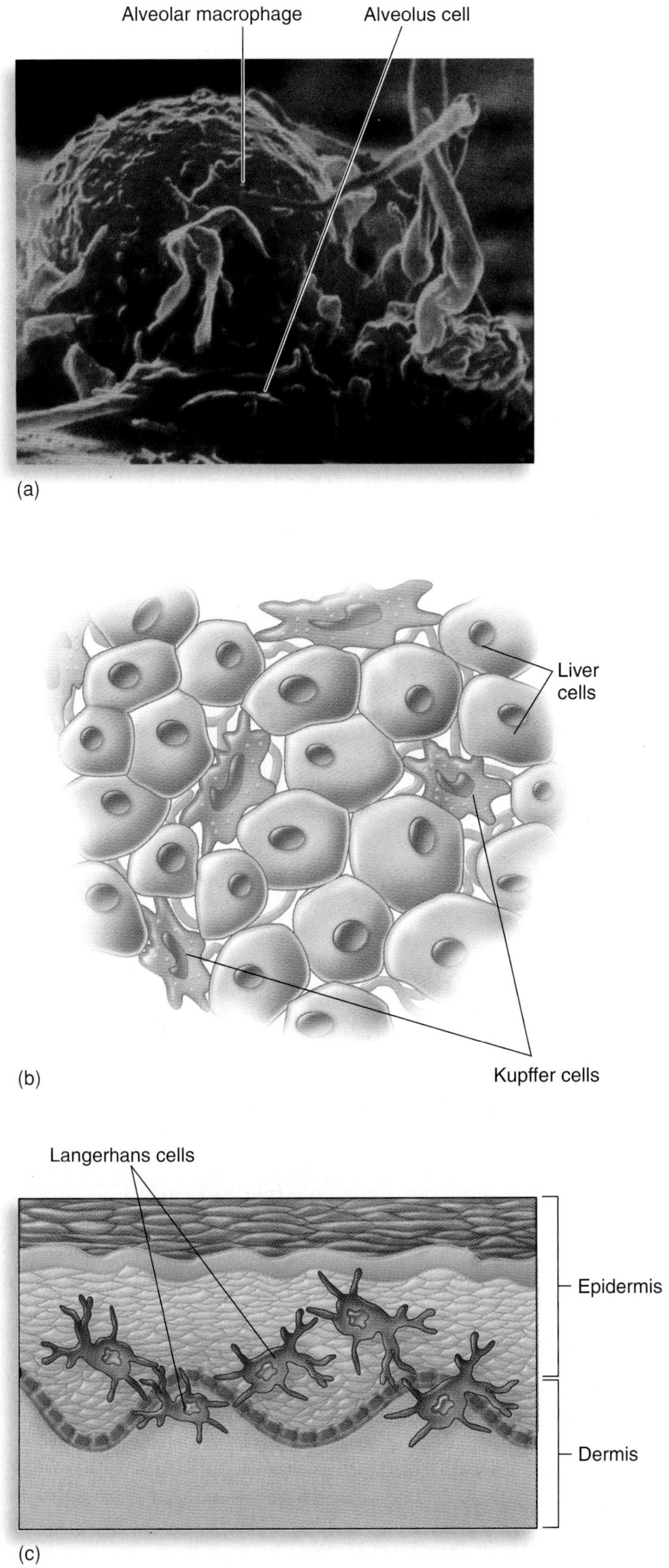

Figure 14.16 Sites containing macrophages.
(a) Scanning electron micrograph view of a lung with an alveolar macrophage. **(b)** Liver tissue with Kupffer cells. **(c)** Langerhans cells deep in the epidermis.

Mechanisms of Phagocytic Recognition, Engulfment, and Killing

The term **phagocytosis** literally means "eating cell." But phagocytosis is more than just the physical process of engulfment, because phagocytes also actively attack and dismantle foreign cells with a wide array of antimicrobial substances. Phagocytosis can occur as an isolated event performed by a lone phagocytic cell responding to a minor irritant in its area or as part of the orchestrated events of inflammation described in the previous section. The events in phagocytosis include chemotaxis, ingestion, phagolysosome formation, destruction, and excretion **(figure 14.17).**

Chemotaxis and Ingestion Phagocytes migrate into a region of inflammation with a deliberate sense of direction, attracted by a gradient of stimulant products from the parasite and host tissue at the site of injury. Phagocytes are now known to be able to recognize some microorganisms as foreign because of signal molecules that the microbes have on their surfaces. These are called pathogen-associated molecular patterns, or PAMPs. They are usually molecules shared by many microorganisms, but not present in mammals, and therefore serve as "red flags" for phagocytes and other cells of innate immunity.

Bacterial PAMPs include peptidoglycan and lipopolysaccharide. Double-stranded RNA, which is found only in

CASE FILE 14 *WRAP-UP*

At the beginning of the chapter, an infant was described who was diagnosed with chronic granulomatous disease (CGD). He exhibited typical signs and symptoms of CGD in that the majority of these patients experience serious or recurrent infections with catalase-positive pathogens such as *Serratia* and *Aspergillus.* This finding is usually coupled with the fact that their neutrophils are cytochrome b_{558} negative and nitroblue tetrazolium test (NBT) negative. These tests indicate that the phagocytic cells lack the respiratory burst required to kill certain organisms that these phagocytes have ingested. In fact, these phagocytes cannot kill catalase-positive organisms because these organisms lack toxic oxygen metabolites such as superoxide and hydrogen peroxide. Interestingly, the phagocytes of those with CGD can kill ingested catalase-negative organisms (such as *Streptococcus pneumoniae*) because these organisms generate their own hydrogen peroxide. Individuals with CGD typically are diagnosed before the age of 2 as they present with serious or recurrent infections. Life expectancy in these patients has been increased due to long-term antibiotic therapy and use of interferon gamma.

See: Herman, T. E., and Siegel, M. J. 2002. Chronic granulomatous disease of childhood: Neonatal Serratia, *hepatic abscesses and pulmonary aspergillosis.* J. Perinatol. *22:255–256.*

Goldblatt, D., and Thrasher, A. J. 2000. Chronic granulomatous disease. Clin. Exper. Immunol. *122:1–9.*

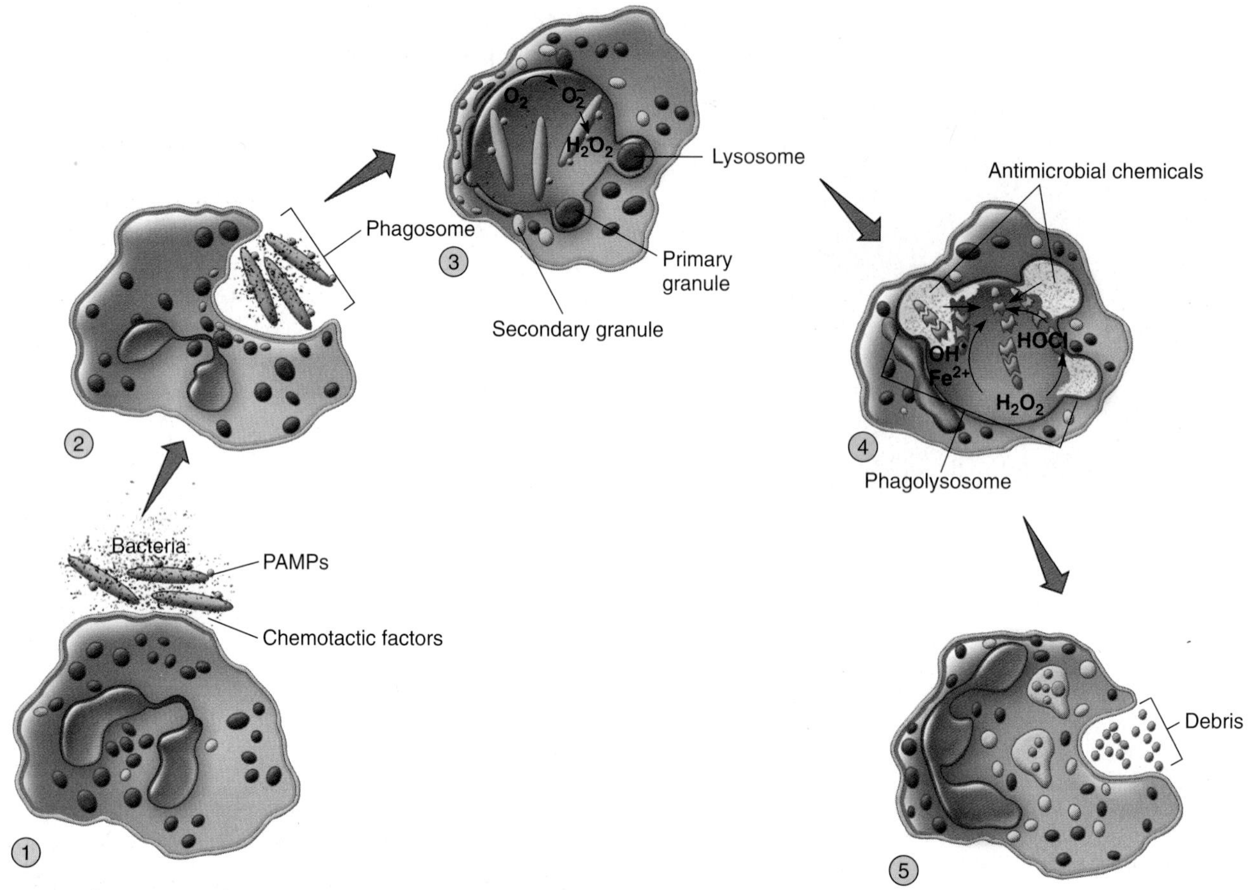

Process Figure 14.17 The phases in phagocytosis.
(1) Chemotaxis. (2) Contact and ingestion (forming a phagosome). (3) Formation of phagolysosome (granules fuse with phagosome). (4) Killing, digestion of the microbe. (5) Release of debris (exocytosis).

some viruses, is also thought to serve as a PAMP. Phagocytes possess molecules on their surfaces that recognize and bind PAMPs. Unlike the cells of specific immunity, phagocytes possess these PAMP receptors all the time, whether they have encountered PAMPs before or not.

On the scene of an inflammatory reaction, phagocytes often trap cells or debris against the fibrous network of connective tissue or the wall of blood and lymphatic vessels. Once the phagocyte has made contact with its prey, it extends pseudopods that enclose the cells or particles in a pocket and internalize them in a vacuole called a *phagosome.* It also secretes more cytokines to further amplify the innate response.

Phagolysosome Formation and Killing In a short time, **lysosomes** migrate to the scene of the phagosome and fuse with it to form a **phagolysosome.** Other granules containing antimicrobial chemicals are released into the phagolysosome, forming a potent brew designed to poison and then dismantle the ingested material. The destructiveness of phagocytosis is evident by the death of bacteria within 30 minutes after contacting this battery of antimicrobial substances.

Destruction and Elimination Systems Two separate systems of destructive chemicals await the microbes in the phagolysosome. The oxygen-dependent system (known as the respiratory burst, or oxidative burst) elaborates several substances that were described in chapters 7 and 11. Myeloperoxidase, an enzyme found in granulocytes, forms halogen ions (OCl^-) that are strong oxidizing agents. Other products of oxygen metabolism such as hydrogen peroxide, the superoxide anion ($O_2^{\bullet-}$), activated or so-called singlet oxygen ($^1O^{\bullet-}$), and the hydroxyl free radical ($HO^\bullet$) separately and together have formidable killing power. Other mechanisms that come into play are the liberation of lactic acid, lysozyme, and *nitric oxide* (NO), a powerful mediator that kills bacteria and inhibits viral replication. Cationic proteins that injure bacterial cell membranes and a number of proteolytic and other hydrolytic enzymes complete the job. The small bits of undigestible debris are released from the macrophage by exocytosis.

Interferon: Antiviral Cytokines and Immune Stimulants

Interferon (IFN) was described in chapter 12 as a small protein produced naturally by certain white blood and tissue cells that is used in therapy against certain viral infections and cancer. Although the interferon system was originally

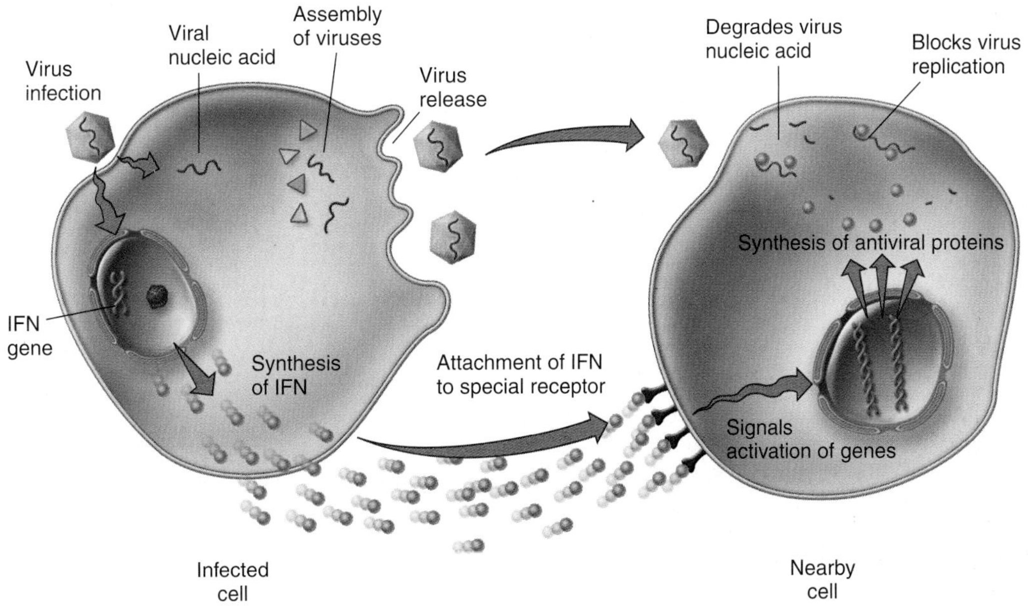

Figure 14.18 The antiviral activity of interferon.
When a cell is infected by a virus, its nucleus is triggered to transcribe and translate the interferon (IFN) gene. Interferon diffuses out of the cell and binds to IFN receptors on nearby uninfected cells, where it induces production of proteins that eliminate viral genes and block viral replication. Note that the original cell is not protected by IFN and that IFN does not prevent viruses from invading the protected cells.

thought to be directed exclusively against viruses, it is now known to be involved also in defenses against other microbes and in immune regulation and intercommunication. Three major types are *interferon alpha* and *beta*, products of many cells, including lymphocytes, fibroblasts, and macrophages; and *interferon gamma*, a product of T cells.

All three classes of interferon are produced in response to viruses, RNA, immune products, and various antigens. Their biological activities are extensive. In all cases, they bind to cell surfaces and induce changes in genetic expression, but the exact results vary. In addition to antiviral effects discussed in the next section, all three IFNs can inhibit the expression of cancer genes and have tumor suppressor effects. IFN alpha and beta stimulate phagocytes, and IFN gamma is an immune regulator of macrophages and T and B cells.

Characteristics of Antiviral Interferon

When a virus binds to the receptors on a host cell, a signal is sent to the nucleus that directs the cell to synthesize interferon. After transcribing and translating the interferon gene, newly synthesized interferon molecules are rapidly secreted by the cell into the extracellular space, where they bind to other host cells. The binding of interferon to a second cell induces the production of proteins in that cell that inhibit viral multiplication either by degrading the viral RNA or by preventing the translation of viral proteins **(figure 14.18).** Interferon is not virus-specific, so its synthesis in response to one type of virus will also protect against other types. Because this protein is an inhibitor of viruses, it has been a valuable treatment for a number of virus infections.

Other Roles of Interferon

Interferons are also important immune regulatory cytokines that activate or instruct the development of white blood cells. For example, interferon alpha produced by T lymphocytes activates a subset of cells called natural killer (NK) cells. In addition, one type of interferon beta plays a role in the maturation of B and T lymphocytes and in inflammation. Interferon gamma inhibits cancer cells, stimulates B lymphocytes, activates macrophages, and enhances the effectiveness of phagocytosis.

Complement: A Versatile Backup System

Among its many overlapping functions, the immune system has another complex and multiple-duty system called **complement** that, like inflammation and phagocytosis, is brought into play at several levels. The complement system, named for its property of "complementing" immune reactions, consists of at least 26 blood proteins that work in concert to destroy bacteria and certain viruses. Some knowledge of this important system will help in your understanding of topics in chapter 15.

The concept of a cascade reaction is helpful in understanding how complement functions. A *cascade reaction* is a sequential physiological response like that of blood clotting, in which the first substance in a chemical series activates the next substance, which activates the next, and so on, until a desired end product is reached. There are three different complement pathways, distinguished by how they become activated. The final stages of the three pathways are the same and yield a similar end result. The *classical pathway*

TABLE 14.1 Complement Pathways

Pathway	Activators	Host Components That Initially Bind	Complement Proteins Involved	
Classical (Rapid, efficient)	Complement-fixing antibodies (IgG, IgM) (sometimes microbe surface components)	C1 complex	C1 complex, C4, C2, C3 →	C5, C6, C7, C8, C9 Membrane attack complex
Lectin (Enters classical pathway)	Mannans	Mannose-binding lectin →	C2, C4 →	
Alternative (Slower, less efficient)	Bacterial or fungal cell wall, Viruses, Parasite surfaces	C3	C3, Factor B, Factor D, Properdin →	

is activated mainly by the presence of antibody bound to microorganisms, although in some cases it can occur in the absence of antibody. The *lectin pathway* is activated when a host serum protein binds a sugar called mannan present in the walls of fungi and other microbes. (Proteins that bind carbohydrates are called lectins.) The *alternative pathway* begins when complement proteins bind to normal cell wall and surface components of microbes. For our discussion, we will focus on one pathway and point out how the others differ in **table 14.1.** Note that because the complement numbers (C1–C9) are based on the order of their discovery, factors C1–C4 do not appear in numerical order during activation (see table 14.1).

Overall Stages in the Complement Cascade

In general, the complement cascade includes the four stages of *initiation, amplification and cascade, polymerization,* and *membrane attack.* At the outset, an initiator (such as microbes, or antibodies, see table 14.1) reacts with the first complement chemical, which propels the reaction on its course. There is a recognition site on the surface of the target cell where the initial C components will bind. Through a stepwise series, each component reacts with another on or near the recognition site. In the C2–C5 series, enzymatic cleavage produces several inflammatory mediators. Other details of the pathways differ, but whether classical, lectin, or alternative, the functioning end product is a large ring-shaped protein termed the *membrane attack complex.* This complex can digest holes in the cell membranes of bacteria, cells, and enveloped viruses, thereby destroying them **(figure 14.19, steps 4 and 5).**

The **classical pathway** is a part of the specific immune response covered in chapter 15. It is initiated either by the foreign cell membrane of a parasite or a surface antibody. The first chemical, C1, is a large complex of three molecules, C1q, C1r, and C1s. When the C1q subunit has recognized and bound to surface receptors on the membrane, the C1r subunit cleaves the C1s proenzyme, and an activated enzyme emerges. During amplification, the C1s enzyme has as its primary targets proenzymes C4 and C2. Through the enzyme's action, C4 is converted into C4a and C4b, and C2 is converted into C2a and C2b. C4b and C2a fragments remain attached as an enzyme, C3 convertase, whose substrate is factor C3. The cleaving of C3 yields subunits C3a and C3b. C3b has the property of binding strongly with the cell membrane in close association with the component C5, and it also forms an enzyme complex with C4b–C2b that converts C5 into two fragments, C5a and C5b. C5b will form the nucleus for the membrane attack complex. This is the point at which the two pathways merge. From this point on, C5b reacts with C6 and C7 to form a stable complex inserted in the membrane. Addition of C8 to the complex causes the polymerization of several C9 molecules into a giant ring-shaped membrane attack complex that bores ring-shaped holes in the membrane. If the target is a cell, this reaction causes it to disintegrate (see figure 14.19, steps 4 and 5). If the target is an enveloped virus, the envelope is perforated and the virus inactivated. The end result of complement action is multifaceted. Gram-negative bacteria and infected host cells may be lysed. Phagocytes will be attracted to the site in greater numbers. Overall inflammation will be amplified by the action of complement. In recent years, the excessive actions of complement have been implicated as aggravators of several autoimmune diseases, such as lupus, rheumatoid arthritis, and myasthenia gravis.

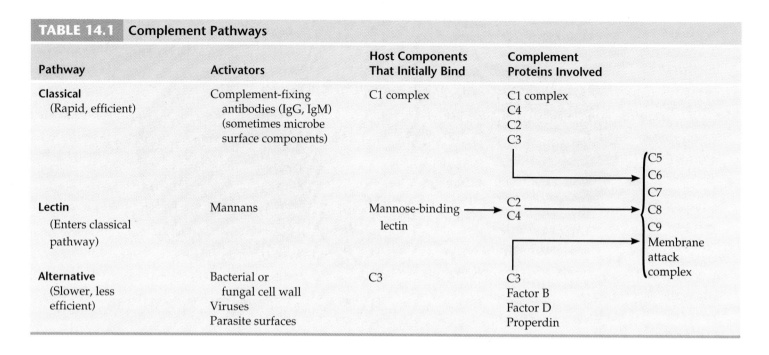

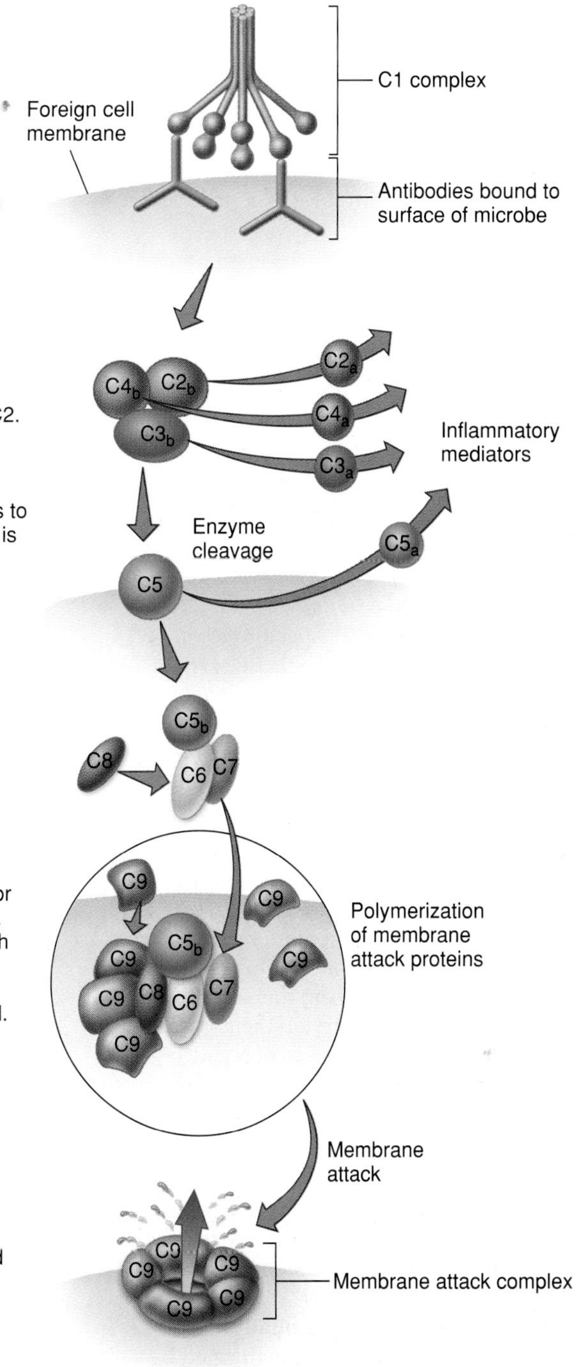

(1) **Initiation.** The classical pathway begins when C1 components bind to antibodies bound to a foreign cell.

(2) **Amplification and cascade.** The C1 complex is an enzyme that activates a second series of components, C4 and C2. When these have been enzymatically cleaved into separate molecules, they become a second enzyme complex that activates C3. At this same site, C3 binds to C5 and cleaves it to form a product that is tightly bound to the membrane.

(3) **Polymerization.** $C5_b$ is a reactive site for the final assembly of an attack complex. In series, C6, C7, and C8 aggregate with C5 and become integrated into the membrane. They form a substrate upon which the final component, C9, can bind. Up to 15 of these C9 units ring the central core of the complex.

(4) **Membrane attack.** The final product of these reactions is a large, donut-shaped enzyme that punctures small pores through the membrane, leading to cell lysis.

(5) An electron micrograph (187,000×) of a cell reveals multiple puncture sites over its surface. The lighter, ringlike structures are the actual enzyme complex.

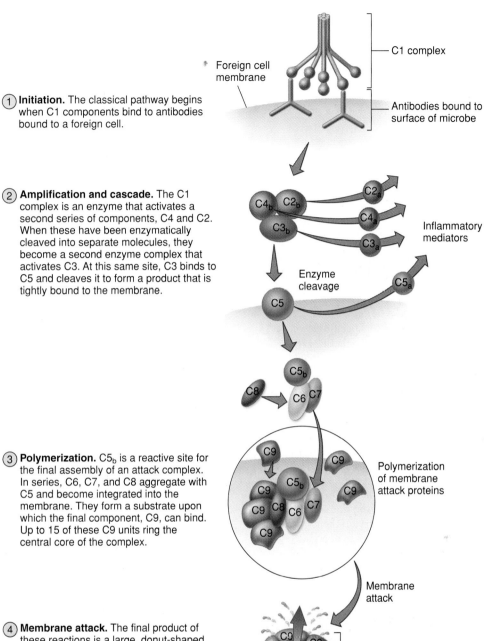

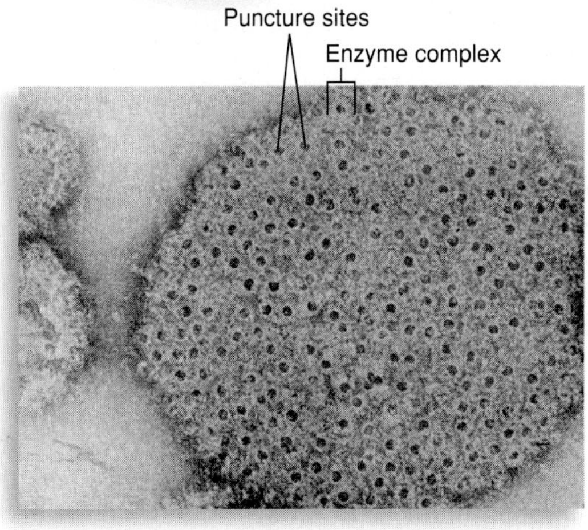

Process Figure 14.19 Steps in the classical complement pathway at a single site. All complement pathways function in a similar way, but the details differ. The classical pathway is illustrated here.

CHECKPOINT

- Nonspecific immune reactions are generalized responses to invasion, regardless of the type. These include inflammation, phagocytosis, interferon, and complement.
- The four symptoms of inflammation are rubor (redness), calor (heat), tumor (edema), and dolor (pain). Loss of function often accompanies these.
- Fever is another component of nonspecific immunity. It is caused by both endogenous and exogenous pyrogens. Fever increases the rapidity of the host immune responses and reduces the viability of many microbial invaders.
- Macrophages are activated monocytes. Along with neutrophils (PMNs), they are the key phagocytic agents of nonspecific response to disease.
- The plasma contains complement, a nonspecific group of chemicals that works on its own or with the third line of defense to attack foreign cells.

Chapter Summary with Key Terms

14.1 Defense Mechanisms of the Host in Perspective
The *first line of defense* is an inborn, nonspecific system composed of anatomical, chemical, and genetic barriers that block microbes at the portal of entry. The *second line of defense* is also inborn and nonspecific and includes protective cells and fluids in tissues, and the *third line of defense* is acquired and specific and is dependent on the function of T and B cells.

14.2 The Second and Third Lines of Defense: An Overview
Immunology is the study of **immunity,** which in turn refers to the development of resistance to infectious agents by the body. Immunity is marked by white blood cells (WBCs), which conduct surveillance of the body, recognizing and differentiating **self** from **nonself** (foreign) cells by virtue of the **markers** they carry on their surface. Under normal conditions, self cells are left alone but nonself cells are destroyed.

14.3 Systems Involved in Immune Defenses

A. For purposes of immunologic study, the body is divided into three compartments: the blood, lymphatics, and **reticuloendothelial system.** A fourth component, the extracellular fluid, surrounds the first three and allows constant communication between all areas of the body.

B. **Circulatory System: Blood and Lymphatics**
   1. **Whole blood** consists of **blood cells** (formed by **hematopoiesis** in the bone marrow) dispersed in **plasma. Stem cells** in the bone marrow differentiate to produce white blood cells **(leukocytes),** red blood cells **(erythrocytes),** and megakaryocytes, which give rise to **platelets.**
   2. Leukocytes are the primary mediators of immune function and can be divided into **granulocytes (neutrophils, eosinophils,** and **basophils)** and **agranulocytes** (**monocytes** and **lymphocytes**) based on their appearance. Monocytes later differentiate to become **macrophages.** Lymphocytes are divided into **B cells,** which produce **antibodies** as part of **humoral immunity,** and **T cells,** which participate in **cell-mediated immunity.** Leukocytes display both **chemotaxis** and **diapedesis** in response to chemical mediators of the immune system.

C. *Lymphatic System*
   1. The **lymphatic system** parallels the circulatory system and transports lymph while also playing host to cells of the immune system. Lymphoid organs and tissues include lymph nodes, the spleen, **thymus,** as well as areas of less-well-organized immune tissues such as **GALT,** SALT, MALT, and BALT.

14.4 The Second Line of Defense

A. The inflammatory response is a complex reaction to tissue injury marked by *redness, heat, swelling,* and *pain.*
   1. Blood vessels narrow and then dilate in response to **chemical mediators** and **cytokines.**
   2. **Edema** swells tissues, helping prevent the spread of infection.
   3. WBCs, microbes, debris, and fluid collect to form **pus.**
   4. Pyrogens may induce fever.

B. Macrophages and neutrophils engage in **phagocytosis,** engulfing microbes in a phagosome. Uniting the phagosome with a **lysosome** results in destruction of the phagosome contents.

C. **Interferon (IFN)** is a family of proteins produced by leukocytes and fibroblasts that inhibit the reproduction of viruses by degrading viral RNA or blocking the synthesis of viral proteins.

D. **Complement** is a complex defense system that results, by way of a cascade mechanism, in the formation of a membrane attack complex that kills cells by creating holes in their membranes.

Multiple-Choice and True-False Questions

Multiple-Choice Questions. Select the correct answer from the answers provided.

1. An example of a nonspecific chemical barrier to infection is
   a. unbroken skin
   b. lysozyme in saliva
   c. cilia in respiratory tract
   d. all of these
2. Which nonspecific host defense is associated with the trachea?
   a. lacrimation
   b. ciliary lining
   c. desquamation
   d. lactic acid
3. Which of the following blood cells function primarily as phagocytes?
   a. eosinophils
   b. basophils
   c. lymphocytes
   d. neutrophils
4. Which of the following is not a lymphoid tissue?
   a. spleen
   b. thyroid gland
   c. lymph nodes
   d. GALT
5. What is included in GALT?
   a. thymus
   b. Peyer's patches
   c. tonsils
   d. breast lymph nodes
6. Monocytes are ____ leukocytes that develop into ____.
   a. granular, phagocytes
   b. agranular, mast cells
   c. agranular, macrophages
   d. granular, T cells
7. An example of an exogenous pyrogen is
   a. interleukin-1
   b. complement
   c. interferon
   d. endotoxin
8. ____ interferon is secreted by ____ and is involved in destroying viruses.
   a. Gamma, fibroblasts
   b. Beta, lymphocytes
   c. Alpha, natural killer cells
   d. Beta, fibroblasts
9. Which of the following substances is *not* produced by phagocytes to destroy engulfed microorganisms?
   a. hydroxyl radicals
   b. superoxide anion
   c. hydrogen peroxide
   d. bradykinin
10. Which of the following is the end product of the complement system?
    a. properdin
    b. cascade reaction
    c. membrane attack complex
    d. complement factor C9

True-False Questions. If statement is true, leave as is. If it is false, correct it by rewriting the sentence.

11. The liquid component of clotted blood is called plasma.
12. Pyogenic bacteria are commonly associated with fever.
13. Communication between cells of the immune system is accomplished using chemical signals.
14. Lysozyme is an enzyme found in tears and saliva that hydrolyzes peptidoglycan in bacterial cell walls.
15. The immune system uses DNA content to distinguish self from nonself.

Writing to Learn

These questions are suggested as a *writing-to-learn* experience. For each question, compose a one- or two-paragraph answer that includes the factual information needed to completely address the question.

1. a. Explain the functions of the three lines of defense.
   b. Which is the most essential to survival?
   c. What is the difference between nonspecific host defenses and immune responses?
2. a. Use the pointers in the figure to describe the major components of the first line of defense.
   b. What effects do these defenses have on microbes?

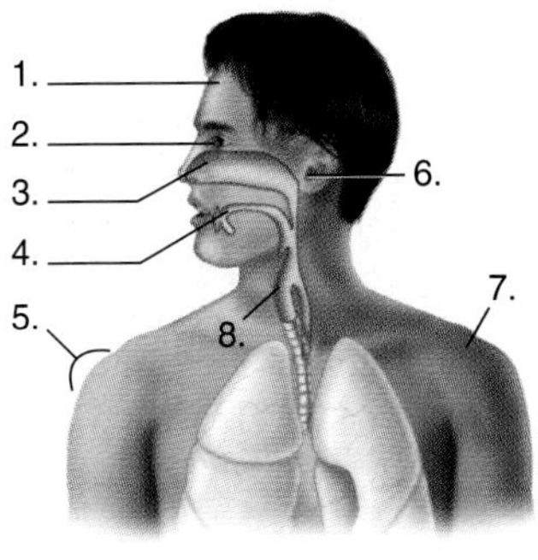

3. a. Describe the main elements of the process through which the immune system distinguishes self from nonself.
   b. How is surveillance of the tissues carried out?
   c. What is responsible for it?
   d. What does the term *foreign* mean in reference to the immune system?
4. a. Differentiate between granulocytes and agranulocytes.
   b. Describe the main cell types in each group, their functions, and their incidence in the circulation.
5. a. What is the principal function of lymphocytes?
   b. Differentiate between the two lymphocyte types and between humoral and cell-mediated immunity.
6. a. What is lymph, and how is it formed?
   b. Why are white cells but not red cells normally found in it?
   c. What are the functions of the lymphatic system?
   d. Explain the filtering action of a lymph node.
   e. What is GALT, and what are its functions?
7. a. Outline the major phases of phagocytosis.
   b. In what ways is a phagocyte a tiny container of disinfectants?
8. a. Briefly describe the three major types of interferon, their sources, and their biological effects.
   b. Describe the mechanism by which interferon acts as an antiviral compound.
9. a. Describe the general complement reaction in terms of a cascade.
   b. What are some functions of complement components?

Concept Mapping

Appendix D provides guidance for working with concept maps.

1. Construct your own concept map using the following words as the *concepts*. Supply the linking words between each pair of concepts.

defenses	inflammation
leukocytes	antibodies
lymphocytes	neutrophils
monocytes	fever
macrophages	

Critical Thinking Questions

Critical thinking is the ability to reason and solve problems using facts and concepts. These questions can be approached from a number of angles, and in most cases, they do not have a single correct answer.

1. Suggest some reasons that there is so much redundancy of action and there are so many interacting aspects of immune responses.
2. a. What are some possible elements missing in children born without a functioning lymphocyte system?
   b. What is the most important component extracted in bone marrow transplants?
3. a. What is the likelihood that plants have some sort of immune protection?
   b. Explain your reasoning.
4. a. What actions of the inflammatory and immune defenses account for swollen lymph nodes and leukocytosis?
   b. What is pus, and what does it indicate?
   c. In what ways can edema be beneficial?
   d. In what ways is it harmful?
5. An obsolete treatment for syphilis involved inducing fever by deliberately infecting patients with the agent of relapsing fever. A recent experimental AIDS treatment involved infecting patients with malaria to induce high fevers. Can you provide some possible explanations behind these peculiar forms of treatment?
6. Patients with a history of tuberculosis often show scars in the lungs and experience recurrent infection. Account for these effects on the basis of the inflammatory response.
7. Macrophages perform the final job of removing tissue debris and other products of infection. Indicate some of the possible effects when these scavengers cannot successfully complete the work of phagocytosis.
8. a. Knowing that fever is potentially both harmful and beneficial, what are some possible guidelines for deciding whether to suppress it or not?
   b. What is the specific target organ of a fever-suppressing drug?

Visual Understanding

1. **From chapter 10, figure 10.8; and from chapter 14, figure 14.18.** Study the two illustrations. Is there any way you could apply the technique illustrated on the top to achieve the same outcome seen in figure 14.18 on the bottom? Explain your answer.

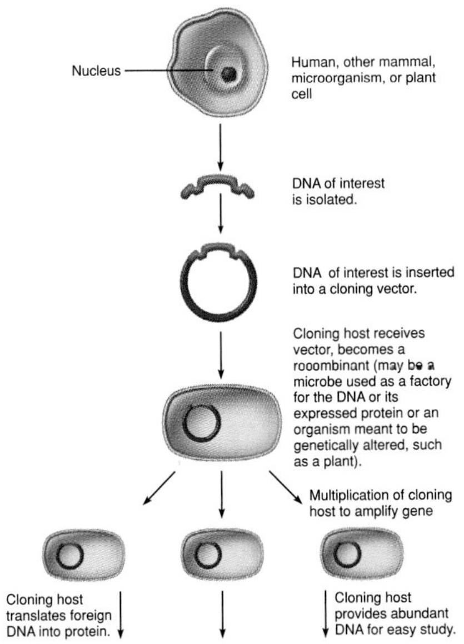

Nucleus
Human, other mammal, microorganism, or plant cell
DNA of interest is isolated.
DNA of interest is inserted into a cloning vector.
Cloning host receives vector, becomes a recombinant (may be a microbe used as a factory for the DNA or its expressed protein or an organism meant to be genetically altered, such as a plant).
Multiplication of cloning host to amplify gene
Cloning host translates foreign DNA into protein.
Cloning host provides abundant DNA for easy study.

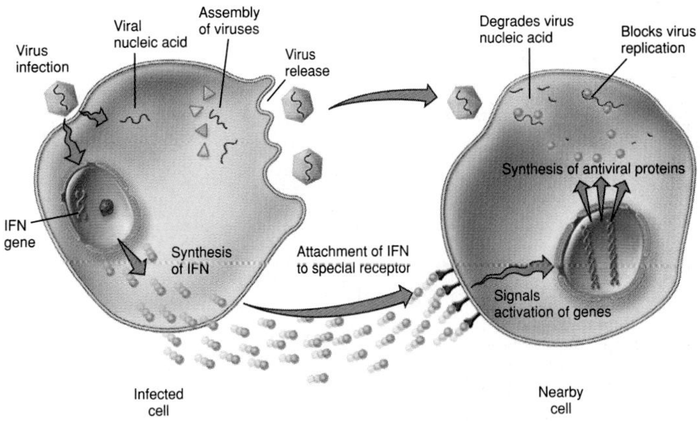

Virus infection
Viral nucleic acid
Assembly of viruses
Virus release
Degrades virus nucleic acid
Blocks virus replication
Synthesis of antiviral proteins
IFN gene
Synthesis of IFN
Attachment of IFN to special receptor
Signals activation of genes
Infected cell
Nearby cell

Internet Search Topics

Go to: www.aris.mhhe.com, and click on "microbiology" and then this textbook's author/title. Go to chapter 14, access the URLs listed under Internet Search Topics, and research the following:

1. Find a website that deals with the uses of interferon in therapy. Name viral infections and cancers it is used to treat. Are there any adverse side effects?
2. Review a good overview of basic immune function.
3. Use a search engine to locate information on dendritic cells. Outline their origins, distribution, and functions. How are they involved in HIV infection?
4. Observe movies showing chemotaxis and phagocytosis; or as an alternate, type the terms "chemotaxis" and "phagocytosis" plus "animation" or "movie" into a search engine.

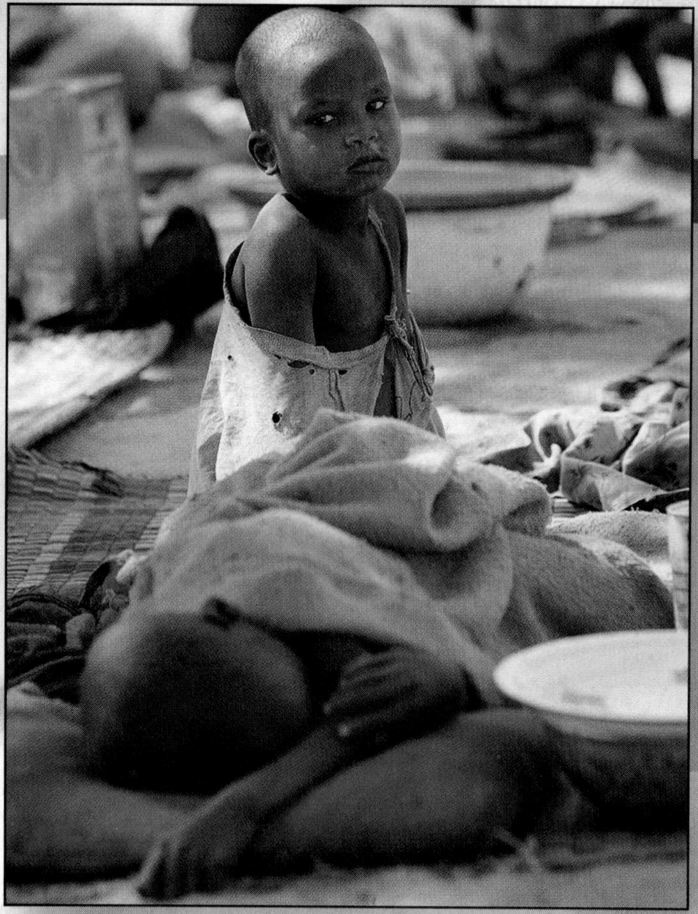

CHAPTER 15

Host Defenses II

Specific Immunity and Immunization

CASE FILE 15

The Darfur region of Sudan has been the focus of much international attention recently. Civil war has ravaged much of the region, creating severe security and health concerns for the more than 6 million residents. An estimated 1 million people have been forced to leave their homes to live in camps. Nearly 170,000 others who fled the country moved to refugee camps in neighboring Chad. The violent conflict and widespread displacement of citizens have disrupted World Health Organization (WHO) and United Nations Children's Fund (UNICEF) immunization efforts in the region.

In March and April 2004, the Federal Ministry of Health (FMOH) in Sudan began receiving reports of measles outbreaks among the residents of many of the camps. Fearing that a widespread epidemic of measles would make a desperate situation worse, the FMOH, WHO, and UNICEF determined that it was feasible to reach approximately 83% of the children in the region and initiated efforts to resume measles immunization.

- *What causes measles?*
- *Why does a once common childhood disease pose a major threat to children in Darfur?*

Case File 15 Wrap-Up appears on page 467.

CHAPTER OVERVIEW

- The most specific host defenses are derived from a dual system of lymphocytes that are genetically programmed to react with foreign substances (antigens) found in microbes and other organisms.
- These cells carry glycoprotein receptors that dictate their specificity and reactivity.
- Both B lymphocytes and T lymphocytes have antigen receptors on their surfaces.
- B cells and T cells arise in the bone marrow, B cells develop in a compartment of the bone marrow, and T cells develop in the thymus.

- Differentiation of lymphocytes creates billions of genetically different cells that each have a unique specificity for antigen.
- Both types of lymphocytes home (migrate) to separate sites in lymphoid tissue where they serve as a constant source of immune cells primed to respond to their correct antigen.
- Antigens are foreign cells, viruses, and molecules that meet a required size and complexity. They are capable of triggering immune reactions by lymphocytes.
- The B and T cells react with antigens through a complex series of cooperative events that involve the presentation of antigens by macrophages and the assistance of helper T cells and cytokine stimulants.
- B cells activated by antigen or by T-cell presentation of antigen enter the cell cycle and mitosis, giving rise to plasma cells that secrete antibodies (humoral immunity) and long-lived memory cells.
- Antibodies have binding sites that affix tightly to an antigen and hold it in place for agglutination, opsonization, complement fixation, and neutralization.
- The amount of antibodies increases during the initial contact with antigen and rises rapidly during subsequent exposures due to memory cells ready for immediate reactions.
- T cells have various receptors that signal their ability to respond as helper cells and cytotoxic cells that kill pathogens, infected cells, and cancer cells.
- Acquired immunities fall into the categories of natural or artificial and active or passive.

15.1 Specific Immunity: The Third and Final Line of Defense

In chapter 14, we described the capacity of the immune system to survey, recognize, and react to foreign cells and molecules and we overviewed the characteristics of nonspecific host defenses, blood cells, phagocytosis, inflammation, and complement. In addition, we introduced the concepts of acquired immunity and specificity. In this chapter, we take a closer look at those topics.

When host barriers and nonspecific defenses fail to control an infectious agent, a person with a normally functioning immune system has a mechanism to resist the pathogen—the third, specific line of immunity. Immunity is the resistance developed after contracting childhood ailments such as chickenpox or measles that provides long-term protection against future attacks. This sort of immunity is not innate but adaptive; it is acquired only after an immunizing event such as an infection. The absolute need for acquired or adaptive immunity is impressively documented in children who have genetic defects in this system or in AIDS patients who have lost it. Even with heroic measures to isolate the patient, combat infection, or restore lymphoid tissue, the victim is constantly vulnerable to life-threatening infections.

Acquired specific immunity is the product of a dual system that we have previously mentioned—the B and T lymphocytes. During development, these lymphocytes undergo a selective process that specializes them for reacting only to one specific antigen or immunogen. During this time, **immunocompetence,** the ability of the body to react with myriad foreign substances, develops. An infant is born with the theoretical potential to react to millions of different immunogens.

Antigens or immunogens figure very prominently in specific immunity. They are defined as molecules that stimulate a response by T and B cells. They are usually protein or polysaccharide molecules on or inside all cells and viruses, including our own. (Environmental chemicals can also be antigens. These are mentioned in chapter 16.) In fact, any exposed or released protein or polysaccharide is potentially an antigen, even those on our own cells. For reasons we discuss later, our own antigens do not usually evoke a response from our own immune systems. So it is acceptable to think of antigens as *foreign* molecules that stimulate an immune response.

In chapter 14, we discussed pathogen-associated molecular patterns (PAMPs) that stimulate responses by phagocytic cells during an innate defense response. PAMPs are molecules shared by many types of microbes that stimulate a nonspecific response; antigens are highly individual and stimulate specific immunity. The two types of molecules do share two characteristics: (1) they are "parts" of foreign cells (microbes), and (2) they provoke a defensive reaction from the host.

Two features that most characterize this third line of defense are specificity and **memory.** Unlike mechanisms such as anatomical barriers or phagocytosis, acquired immunity is highly selective. For example, the antibodies produced during an infection against the chickenpox virus will function against that virus and not against the measles virus. The property of memory pertains to the rapid mobilization of lymphocytes that have been programmed to "recall" their first engagement with the invader and rush to the attack once again.

The elegance and complexity of immune function are largely due to lymphocytes working closely together with macrophages. To simplify and clarify the network of immunologic development and interaction, we present it here as a series of five stages, with each stage covered in a separate section **(figure 15.1).** The principal stages include these:

I. lymphocyte development and differentiation;
II. the presentation of antigens;

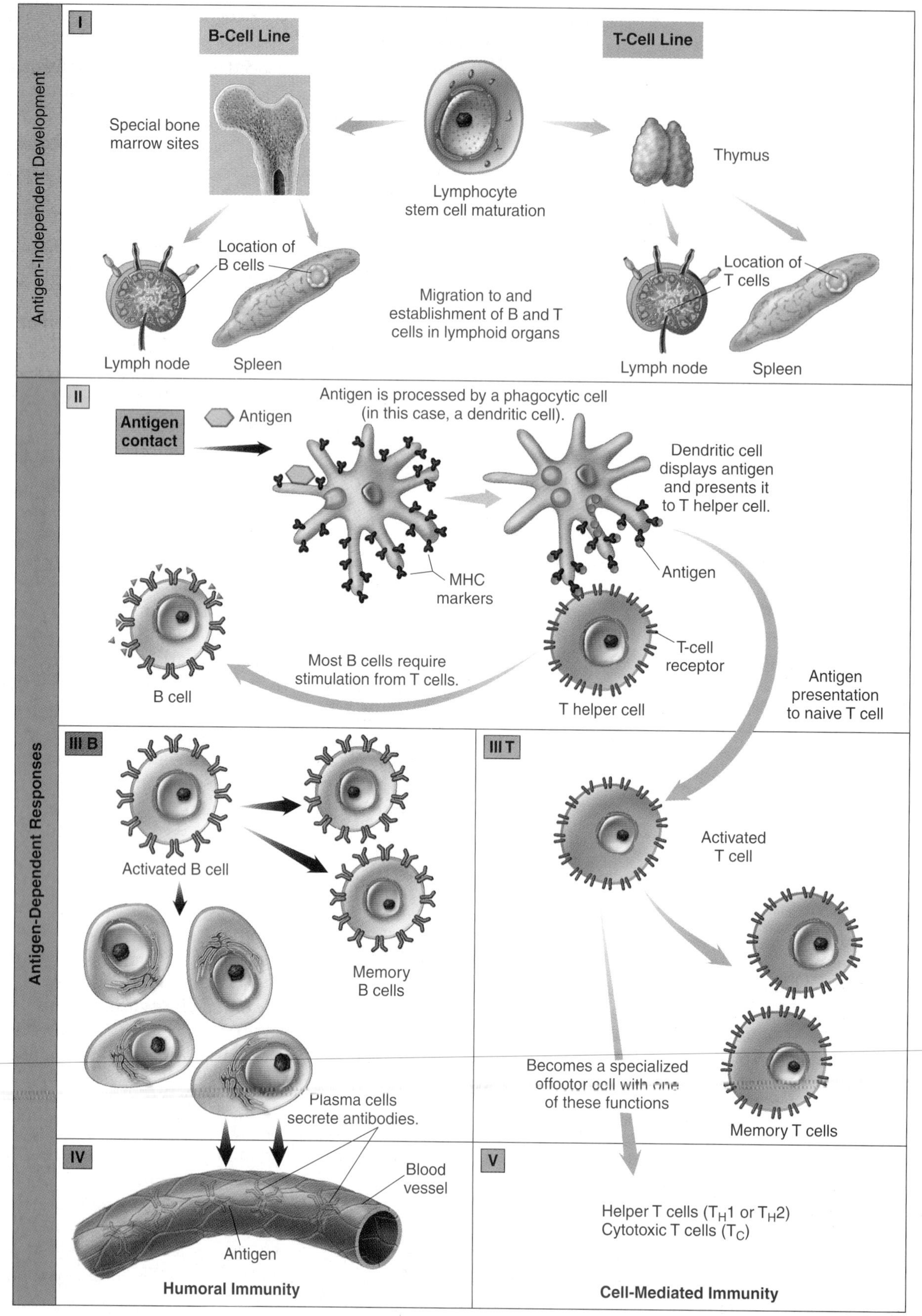

Figure 15.1 Overview of the stages of lymphocyte development and function.
I. Development of B- and T-lymphocyte specificity and migration to lymphoid organs. **II.** Antigen processing by dendritic cell and presentation to lymphocytes; assistance to B cells by T cells. **III.** Lymphocyte activation, clonal expansion, and formation of memory B and T cells. **IV.** Humoral immunity, B-cell line produces antibodies to react with the original antigen. **V.** Cell-mediated immunity. Activated T cells perform various functions, depending on the signal and type of antigen. Details of these processes are covered in each corresponding section heading.

III. the challenge of B and T lymphocytes by antigens;
IV. B lymphocytes and the production and activities of antibodies; and
V. T-lymphocyte responses.

As each stage is covered, we simultaneously follow the sequence of an immune response.

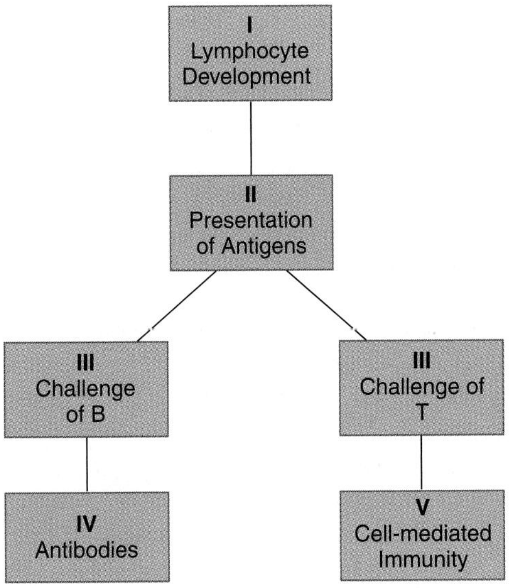

15.2 An Overview of Specific Immune Responses

Development of the Dual Lymphocyte System

Lymphocytes are central to immune responsiveness. They undergo development that begins in the embryonic yolk sac and shifts to the liver and bone marrow. Although all lymphocytes arise from the same basic stem cell type, at some point in development they diverge into two distinct types. Final maturation of B cells occurs in specialized bone marrow sites, and that of T cells occurs in the thymus. This process commits each individual B cell or T cell to one specificity. Both cell types subsequently migrate to separate areas in the lymphoid organs (for instance, nodes and spleen). B and T cells constantly recirculate through the circulatory system and lymphatics, migrating into and out of the lymphoid organs.

Entrance and Presentation of Antigens and Clonal Selection

When foreign cells, like pathogens (antigens), cross the first line of defense and enter the tissue, resident phagocytes migrate to the site. Tissue macrophages ingest the pathogen and induce an inflammatory response in the tissue if appropriate. Tissue dendritic cells ingest the antigen and migrate to the nearest lymphoid organ (often the draining lymph nodes). Here they process and present antigen to T lymphocytes. Pieces of the pathogen also drain into these lymph nodes. These antigens activate B cells. In most cases, the response of B cells also requires the additional assistance of special classes of T cells called T helper cells.

Activation of Lymphocytes and Clonal Expansion

When challenged by immunogen,[1] both B cells and T cells further proliferate and differentiate. The multiplication of a particular lymphocyte creates a clone, or group of genetically identical cells, some of which are memory cells that will ensure future reactiveness against that antigen. Because the B-cell and T-cell responses depart notably from this point in the sequence, they are summarized separately.

Products of B Lymphocytes: Antibody Structure and Functions

The progeny of a dividing B-cell clone are called plasma cells. These cells are programmed to synthesize and secrete antibodies into the tissue fluid and blood. When these antibodies attach to the antigen for which they are specific, the antigen is marked for destruction or neutralization. Because secreted antibody molecules circulate freely in the tissue fluids, lymph, and blood, the immunity they provide is humoral.

How T Cells Respond to Antigen: Cell-Mediated Immunity (CMI)

T-cell types and responses are extremely varied. When activated (sensitized) by antigen, a T cell gives rise to one of three different types of progeny, each involved in a cell-mediated immune function. The three main functional types of T cells are:

1. T_H1 cells that activate macrophages and help activate T_C cells,
2. T_H2 cells that assist B-cell processes, and
3. T_C cells that lead to the destruction of infected host cells and other "foreign" cells.

Although T cells secrete cytokines that help destroy pathogens or regulate immune responses, they do not produce antibodies.

CHECKPOINT

- Acquired specific immunity is an elegant but complex matrix of interrelationships between lymphocytes and macrophages consisting of several stages.
- Stage I. Lymphocytes originate in hematopoietic tissue but go on to diverge into two distinct types: B cells, which produce antibody, and T cells, which produce cytokines that mediate and coordinate the entire immune response.

1. "Antigen" and "immunogen" are used interchangeably; for simplicity's sake, we use "antigen" for the remainder of the chapter.

- Stage II. Antigen-presenting cells detect invading pathogens and present these antigens to lymphocytes, which recognize the antigen and initiate the specific immune response.
- Stage III. Lymphocytes proliferate, producing clones of progeny that include groups of responder cells and memory cells.
- Stage IV. Activated B lymphocytes become plasma cells that produce and secrete large quantities of antibodies.
- Stage V. Activated T lymphocytes are one of three subtypes, which regulate and participate directly in the specific immune responses.

Essential Preliminary Concepts for Understanding Immune Reactions

Before we examine lymphocyte development and function in greater detail, we must initially review concepts such as the unique structure of molecules (especially proteins), the characteristics of cell surfaces (membranes and envelopes), the ways that genes are expressed, and immune recognition and identification of self and nonself. Ultimately, the shape and function of protein receptors and markers protruding from the surfaces of cells are the result of genetic expression, and these molecules are responsible for specific immune recognition and, thus, immune reactions.

Receptors on Cell Surfaces Involved in Recognition of Self and Nonself

Chapter 14 touched on the fundamental idea that cell markers or receptors confer specificity and identity. A given cell can express several different receptors, each type playing a distinct and significant role in detection, recognition, and cell communication. Major functions of immune system receptors are:

1. attachment to nonself or foreign antigens;
2. binding to cell surface receptors that indicate self, such as MHC molecules;
3. receiving and transmitting chemical messages to coordinate the response;
4. aiding in cellular development.

Because of their importance in the immune response, we concentrate here on the major receptors of lymphocytes and macrophages.

Major Histocompatibility Complex

One set of genes that codes for human cell receptors is the major histocompatibility complex (MHC). This gene complex gives rise to a series of glycoproteins (called MHC molecules) found on all cells except red blood cells. The MHC is also known as the human leukocyte antigen (HLA) system. This receptor complex plays a vital role in recognition of self by the immune system and in rejection of foreign tissue.

Three classes of MHC genes have been identified. Class I genes code for markers that display unique characteristics of self and allow for the recognition of self molecules and the regulation of immune reactions. The system is rather complicated in its details, but in general, each human being inherits a particular combination of class I MHC (HLA) genes in a relatively predictable fashion. Although millions of different combinations and variations of these genes are possible among humans, the closer the relationship, the greater the probability for similarity in MHC profile. Individual differences in the exact inheritance of MHC genes, however, make it fairly unlikely that even closely related persons will express an identical MHC profile. This fact necessitates testing for MHC and other antigens when blood is transfused and organs are transplanted (see graft rejection, chapter 16).

Class II MHC genes also code for immune regulatory receptors. These receptors are found on macrophages, dendritic cells, and B cells and are involved in presenting antigens to T cells during cooperative immune reactions. Class III MHC genes encode proteins involved with the complement system among others. We'll focus on class I and II in this chapter. See **figure 15.2** for depictions of the first two MHC classes.

Lymphocyte Receptors and Specificity to Antigen

The part lymphocytes play in immune surveillance and recognition emphasizes the essential role of their receptors.

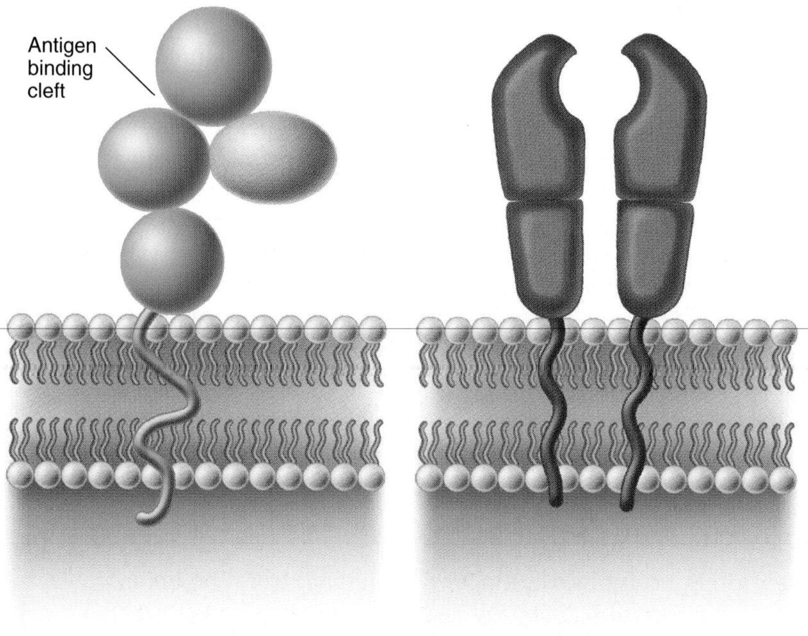

Figure 15.2 Class I and II of molecules the human major histocompatibility complex.

B cells have receptors that bind antigens, and T cells have receptors that bind processed antigens plus MHC molecules on the cells that present antigens to them. Antigen molecules exist in great diversity; there are potentially millions and even billions of unique types. The many sources of antigens include microorganisms as well as an awesome array of chemical compounds in the environment. One of the most fascinating questions in immunology is: How can the lymphocyte receptors be varied to react with such a large number of different antigens? After all, it is generally accepted that there will have to be a different lymphocyte receptor for each unique antigen. Some questions that naturally follow are: How can a cell accommodate enough genetic information to respond to millions or even billions of antigens? When, where, and how does the capacity to distinguish self from foreign tissue arise? To answer these questions, we must first introduce a central theory of immunity.

The Origin of Diversity and Specificity in the Immune Response

The Clonal Selection Theory and Lymphocyte Development

Research findings have shown that lymphocytes use slightly more than 500 gene segments to produce the tremendous repertoire of specific receptors they must display for antigens. The most widely accepted explanation for how this diversity is generated is called the **clonal selection theory.** According to this theory, early undifferentiated lymphocytes in the embryo, fetus, and adult bone marrow undergo a continuous series of divisions and genetic changes that generate hundreds of millions of different types of B and T cells, each carrying a particular receptor specificity **(figure 15.3).**

The mechanism, generally true for both B and T cells, can be summarized as follows: In the bone marrow, stem cells

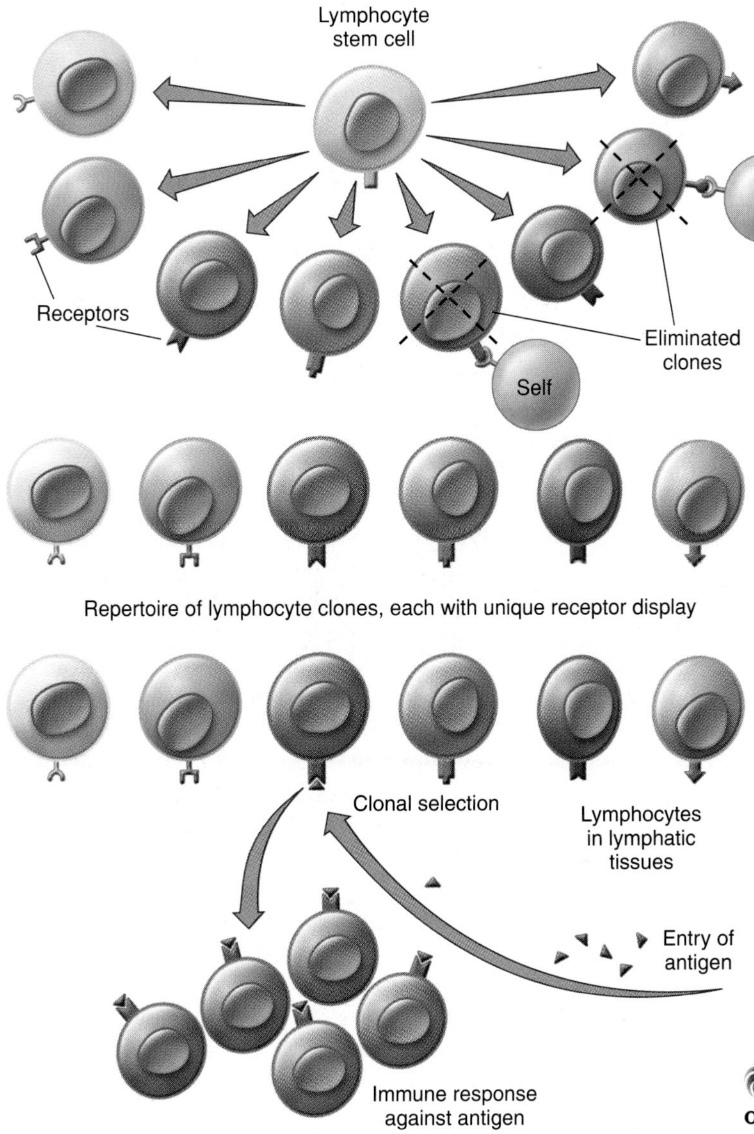

(a) **Antigen-Independent Period**

1. During development of early lymphocytes from stem cells, a given stem cell undergoes rapid cell division to form numerous progeny.

During this period of cell differentiation, random rearrangements of the genes that code for cell surface protein receptors occur. The result is a large array of genetically distinct cells, each bearing a different receptor that is specific to react with only a single type of foreign molecule or antigen.

2. At this same time, any lymphocytes that develop a specificity for self molecules and could be harmful are eliminated from the pool of diversity. This is called immune tolerance.

3. The specificity for a single antigen molecule is programmed into the lymphocyte and is set for the life of a given cell. The end result is an enormous pool of mature but naive lymphocytes that are ready to further differentiate under the influence of certain organs and immune stimuli.

(b) **Antigen-Dependent Period**

4. Mature lymphocytes come to populate the lymphatic organs, where they will finally encounter antigens. These antigens will become the stimulus for the lymphocytes' final activation and immune function. Entry of a specific antigen selects only the lymphocyte that carries matching surface receptors. This will trigger proliferation, which results in large numbers of cells bearing identical antigen-specific receptors.

Figure 15.3 Overview of the clonal selection theory of lymphocyte development and diversity.

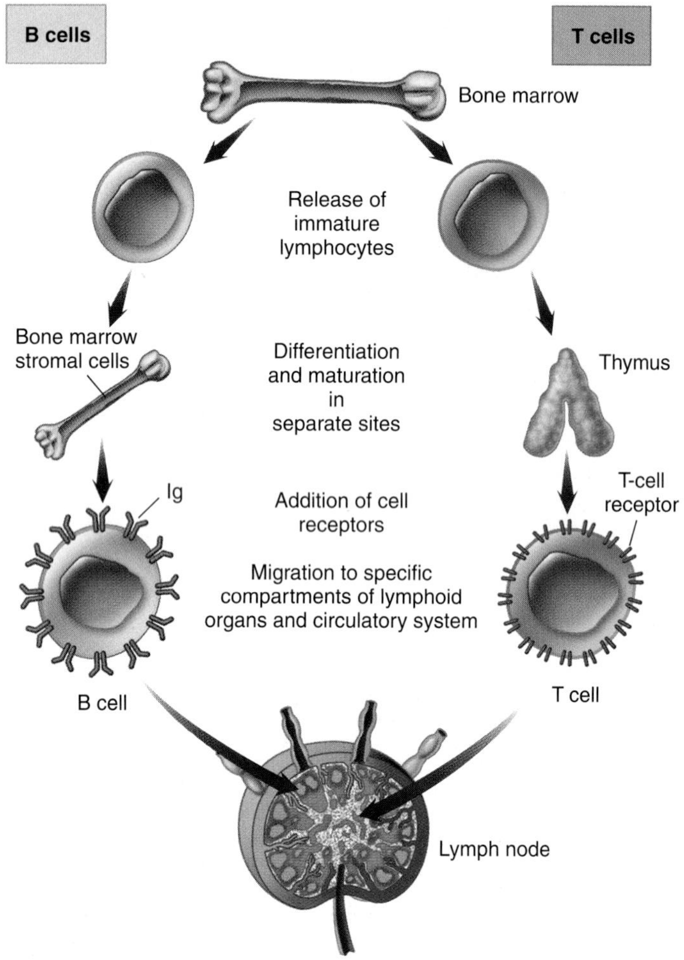

Figure 15.4 Major stages in the development of B and T cells.

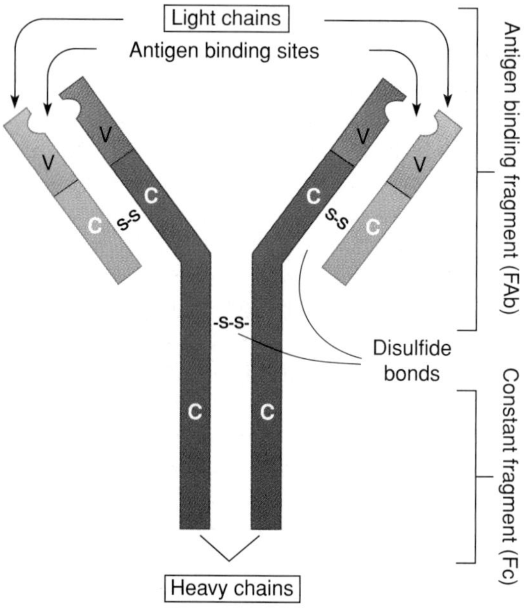

Figure 15.5 Simplified structure of an immunoglobulin molecule on the surface of B cells. The main components are four polypeptide chains—two identical light chains and two identical heavy chains bound by disulfide bonds as shown. Each chain consists of a variable region (V) and a constant region (C). The variable regions of light and heavy chains form a binding site for antigen.

can become granulocytes, monocytes, or lymphocytes. The lymphocytic cells then become either T cells or B cells. Cells destined to become B cells stay in the bone marrow; T cells migrate to the thymus. Here they build their unique antigen receptor. Both B and T cells then migrate to secondary lymphoid tissues **(figure 15.4).** The secondary lymphoid tissues are resupplied with B and T cells because some self-destruct if they are not used and others become activated and leave.

By the time T and B cells reach the lymphoid tissues, each one is already equipped to respond to a single unique antigen. This amazing diversity is generated by extensive rearrangements of the gene segments that code for the proteinaceous antigen receptors on the T and B cells **(figure 15.5).** In time, every possible recombination occurs, leading to a huge assortment of lymphocytes.[2] Each genetically unique line of lymphocytes arising from these recombinations is termed a **clone.** Keep in mind that the rearranged genetic code is expressed as a protein receptor of unique configuration on the surface of the lymphocyte, something like a "sign post" announcing its specificity and reactivity for an antigen.

2. Estimates of the theoretical number of possible variations that may be created vary from 10^{14} to 10^{18} different specificities.

This *proliferative* stage of lymphocyte development does not require the actual presence of foreign antigens.

The second stage of development—*clonal selection and expansion*—does require stimulation by an antigen such as a microbe. When this antigen enters the immune surveillance system, it encounters specific lymphocytes ready to recognize it. Such contact stimulates that clone to undergo mitotic divisions and expands it into a larger population of lymphocytes all bearing the same specificity. This increases the capacity of the immune response to that antigen. Two important generalities one can derive from the clonal selection theory are (1) lymphocyte specificity is preprogrammed, existing in the genetic makeup before an antigen has ever entered the tissues, and (2) each genetically distinct lymphocyte expresses only a single specificity and can react to only one type of antigen. Other important features of the lymphocyte response system are expanded in later sections.

One potentially problematic outcome of random genetic assortment is the development of clones of lymphocytes able to react to *self.* This outcome could lead to severe damage when the immune system actually perceives self molecules as foreign and mounts a harmful response against the host's tissues. According to a corollary of the clonal selection theory, any such clones are destroyed during development through clonal *deletion.* The removal of such potentially harmful clones is the basis of immune tolerance or tolerance to self. But because humans are exposed to many new antigenic substances during their lifetimes, T cells and B cells in the periphery of the body also have mechanisms for *not* reacting to innocuous antigens. Some diseases (autoimmunity) are thought to be caused by the loss of immune tolerance through the survival of certain "forbidden clones" or failure of these other systems (see chapter 16).

The Specific B-Cell Receptor: An Immunoglobulin Molecule

In the case of B lymphocytes, the receptor genes that undergo the recombination described are those governing **immunoglobulin** (im″-yoo-noh-glahb′-yoo-lin) **(Ig)** synthesis. Immunoglobulins are large glycoprotein molecules that serve as the antigen receptors of B cells and, when secreted, as antibodies. The basic immunoglobulin molecule is a composite of four polypeptide chains: a pair of identical heavy (H) chains and a pair of identical light (L) chains (see figure 15.5). One light chain is bonded to one heavy chain, and the two heavy chains are bonded to one another with disulfide bonds, creating a symmetrical, Y-shaped arrangement.

The ends of the forks formed by the light and heavy chains contain pockets, called the **antigen binding sites.** These sites can be highly variable in shape to fit a wide range of antigens. This extreme versatility is due to **variable regions (V)** in antigen binding sites, where amino acid composition is highly varied from one clone of B lymphocytes to another. The remainder of the light chains and heavy chains consist of constant regions (C) whose amino acid content does not vary greatly from one antibody to another.

T-Cell Receptors

The T-cell receptor for antigen belongs to the same protein family as the B-cell receptor. It is similar to B cells in being formed by genetic modification, having variable and constant regions, being inserted into the membrane, and having an antigen binding site formed from two parallel polypeptide chains **(figure 15.6).** Unlike the immunoglobulins, the T-cell receptor is relatively small and is never secreted. Various other receptors that are not antigen-specific are described in a later section.

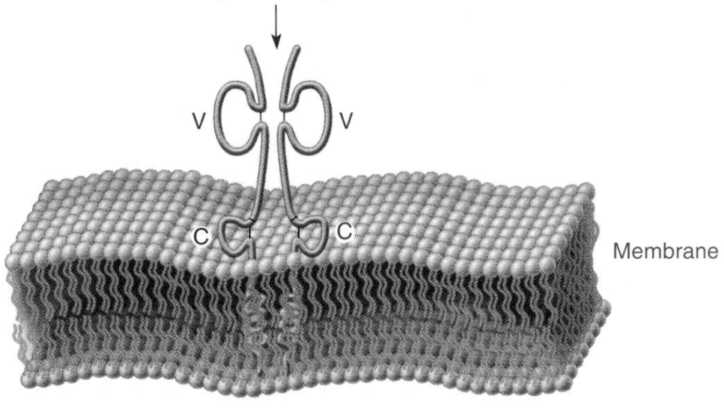

Figure 15.6 Proposed structure of the T-cell receptor for antigen.
The structure of this polypeptide is similar to that of an immunoglobulin. V stands for variable region and C for constant region.

CHECKPOINT

- The surfaces of all cell membranes contain protein receptors. These receptors function in identification, communication, and cell development. Some are also self identity markers.
- Human identity markers are genetically determined by MHCs. MHC-I codes for identity markers on all nucleated host cells. MHC-II codes for identity receptors on dendritic cells, macrophages, and B cells.
- The clonal selection theory explains that during development, both B and T cells develop millions of genetically different clones. Together these clones possess enough genetic variability to respond to many millions of different antigens. Each clone, however, can respond to only one specific antigen.

15.3 The Lymphocyte Response System in Depth

Now that you have a working knowledge of some factors in the development of immune specificity, let us look at each stage of an immune response as originally outlined in figure 15.1.

Specific Events in B-Cell Maturation

The site of B-cell maturation was first discovered in birds, which have an organ in the intestine called the bursa. For some time, the human bursal equivalent was not established. Now it is known to be certain bone marrow sites that harbor *stromal cells.* These huge cells nurture the lymphocyte stem cells and provide chemical signals that initiate B-cell development. As a result of gene modification and selection, hundreds of millions of distinct B cells develop. These naive lymphocytes circulate through the blood, "homing" to specific sites in the lymph nodes, spleen, and gut-associated lymphoid tissue (GALT), where they adhere to specific binding molecules. Here they will come into contact with antigens throughout life. B cells have immunoglobulins as surface receptors **(table 15.1).**

Specific Events in T-Cell Maturation

The maturation of T cells and the development of their specific receptors are directed by the thymus gland and its hormones. In addition to the antigen-specific T-cell receptor, mature T lymphocytes express either CD4 or CD8 coreceptors. CD4 binds to MHC class II and is expressed on T helper cells. CD8 is found on cytotoxic T cells, and it binds MHC class I molecules. Like B cells, T cells also constantly circulate between the lymphatic and general circulatory system, migrating to specific T-cell areas of the lymph nodes and spleen. It has been estimated that 25×10^9 T cells pass between the lymphatic and general circulation per day.

TABLE 15.1 Contrasting Properties of B Cells and T Cells

	B Cells	T Cells
Site of Maturation	Bone marrow	Thymus
Specific Surface Markers	Immunoglobulin Several CD molecules	T-cell receptor Several CD molecules
Circulation in Blood	Low numbers	High numbers
Receptors for Antigen	B-cell receptor (immunoglobulin)	T-cell receptor
Distribution in Lymphatic Organs	Cortex (in follicles)	Paracortical sites (interior to the follicles)
Require Antigen Presented with MHC	No	Yes
Product of Antigenic Stimulation	Plasma cells and memory cells	Several types of sensitized T cells and memory cells
General Functions	Production of antibodies to inactivate, neutralize, target antigens	Cells function in helping other immune cells, suppressing, killing abnormal cells; hypersensitivity; synthesize cytokines

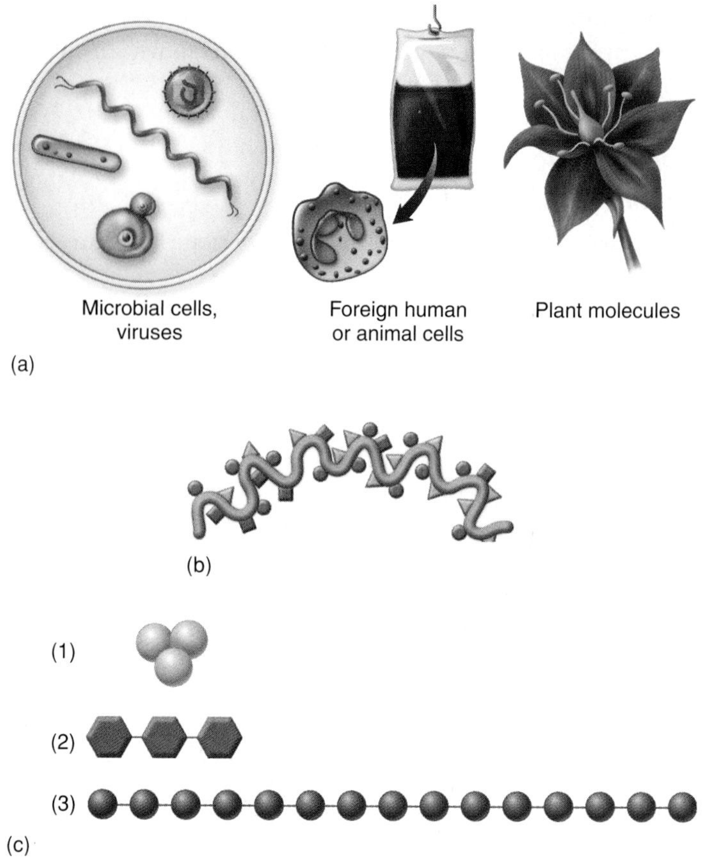

Figure 15.7 Characteristics of antigens.
(a) Whole cells and viruses make good immunogens. **(b)** Complex molecules with several epitopes make good immunogens. **(c)** Poor immunogens include small molecules not attached to a carrier molecule (*1*), simple molecules (*2*), and large but repetitive molecules (*3*).

Entrance and Processing of Antigens and Clonal Selection

Having reviewed the characteristics of lymphocytes, let us more deeply examine the properties of antigens, the substances that cause them to react. As discussed earlier, an **antigen (Ag)** is a substance that provokes an immune response in specific lymphocytes. The property of behaving as an antigen is called **antigenicity.** The term **immunogen** is another term of reference for a substance that can elicit an immune response. To be perceived as an antigen or immunogen, a substance must meet certain requirements in foreignness, shape, size, and accessibility.

Characteristics of Antigens

One important characteristic of an antigen is that it be perceived as foreign, meaning that it is not a normal constituent of the body. Whole microbes or their parts, cells, or substances that arise from other humans, animals, plants, and various molecules all possess this quality of foreignness and thus are potentially antigenic to the immune system of an individual **(figure 15.7).** Molecules of complex composition such as proteins and protein-containing compounds prove to be more immunogenic than repetitious polymers composed of a single type of unit. Most materials that serve as antigens fall into these chemical categories:

- Proteins and polypeptides (enzymes, cell surface structures, hormones, exotoxins)
- Lipoproteins (cell membranes)
- Glycoproteins (blood cell markers)
- Nucleoproteins (DNA complexed to proteins but not pure DNA)
- Polysaccharides (certain bacterial capsules) and lipopolysaccharides

Effects of Molecular Shape and Size To initiate an immune response, a substance must also be large enough to "catch the attention" of the surveillance cells. Molecules with a molecular weight (MW) of less than 1,000 are seldom complete antigens, and those between 1,000 MW and 10,000 MW are weakly so. Complex macromolecules approaching 100,000 MW are the most immunogenic, a category also dominated by large proteins. Note that large size alone is not sufficient for antigenicity; glycogen, a polymer of glucose with a highly repetitious structure, has a molecular weight over 100,000 and is not normally antigenic, whereas insulin, a protein with a molecular weight of 6,000, can be antigenic.

A lymphocyte's capacity to discriminate differences in molecular shape is so fine that it recognizes and responds to only a portion of the antigen molecule. This molecular fragment, called the **epitope,** is the primary signal that the molecule is foreign **(figure 15.7*b*).** The particular tertiary structure and shape of this determinant must conform like a key to the receptor "lock" of the lymphocyte, which then responds to it. Certain amino acids accessible at the surface of proteins or protruding carbohydrate side chains are typical examples. Many foreign cells and molecules are very complex antigenically, with numerous component parts, each of which will elicit a separate and different lymphocyte response. Examples of these multiple, or *mosaic,* antigens include bacterial cells containing cell wall, membrane, flagellar, capsular, and toxin antigens; and viruses (T-cell antigen receptors recognize small pieces of antigens—epitopes in combination with MHC molecules).

Small foreign molecules that consist only of a determinant group and are too small by themselves to elicit an immune response are termed **haptens.** However, if such an incomplete antigen is linked to a larger carrier molecule, the combination develops immunogenicity **(figure 15.8).** The carrier group contributes to the size of the complex and enhances the proper spatial orientation of the determinative group, while the hapten serves as the epitope. Haptens include such molecules as drugs; metals; and ordinarily innocuous household, industrial, and environmental chemicals. Many haptens develop antigenicity in the body by combining with large carrier molecules such as serum proteins (see allergy in chapter 16).

Because each human being is genetically and biochemically unique (except for identical twins), the proteins and other molecules of one person can be antigenic to another. **Alloantigens** are cell surface markers and molecules that occur in some members of the same species but not in others. Alloantigens are the basis for an individual's blood group and major histocompatibility profile, and they are responsible for incompatibilities that can occur in blood transfusion or organ grafting.

Some bacterial toxins, which belong to a group of immunogens called **superantigens,** are potent stimuli for T cells. Their presence in an infection activates T cells at a rate 100 times greater than ordinary antigens. The result can be an overwhelming release of cytokines and cell death. Such diseases as toxic shock syndrome and certain autoimmune diseases are associated with this class of antigens.

Antigens that evoke allergic reactions, called **allergens,** are characterized in detail in chapter 16.

15.4 Cooperation in Immune Reactions to Antigens

The basis for most immune responses is the encounter between antigens and white blood cells. Microbes and other foreign substances enter most often through the respiratory or gastrointestinal mucosa and less frequently through other mucous membranes, the skin, or across the placenta. Antigens introduced intravenously become localized in the liver, spleen, bone marrow, kidney, and lung. If introduced by some other route, antigens are carried in lymphatic fluid and concentrated by the lymph nodes. The lymph nodes and spleen are important in concentrating the antigens and circulating them thoroughly through all areas populated by lymphocytes so that they come into contact with the proper clone.

The Role of Antigen Processing and Presentation

In most immune reactions, the antigen must be further acted upon and formally presented to lymphocytes by cells called **antigen-presenting cells (APCs).** Three different cells can serve as APCs: macrophages, B cells, and **dendritic** (den'-drih-tik) **cells.** Dendritic cells engulf the antigen and modify it so that it will be more immunogenic and recognizable to

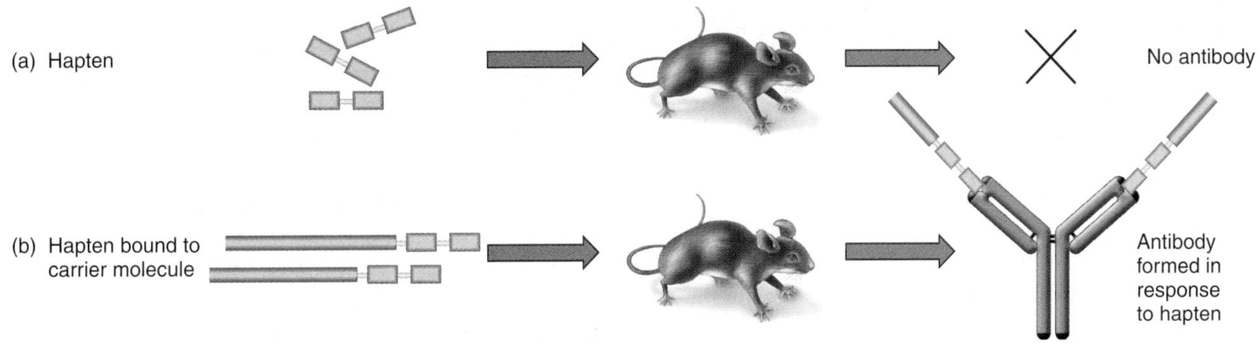

Figure 15.8 The hapten-carrier phenomenon.
(a) Haptens are too small to be discovered by an animal's immune system; no response. **(b)** A hapten bound to a large molecule will serve as an epitope and stimulate a response and an antibody that is specific for it.

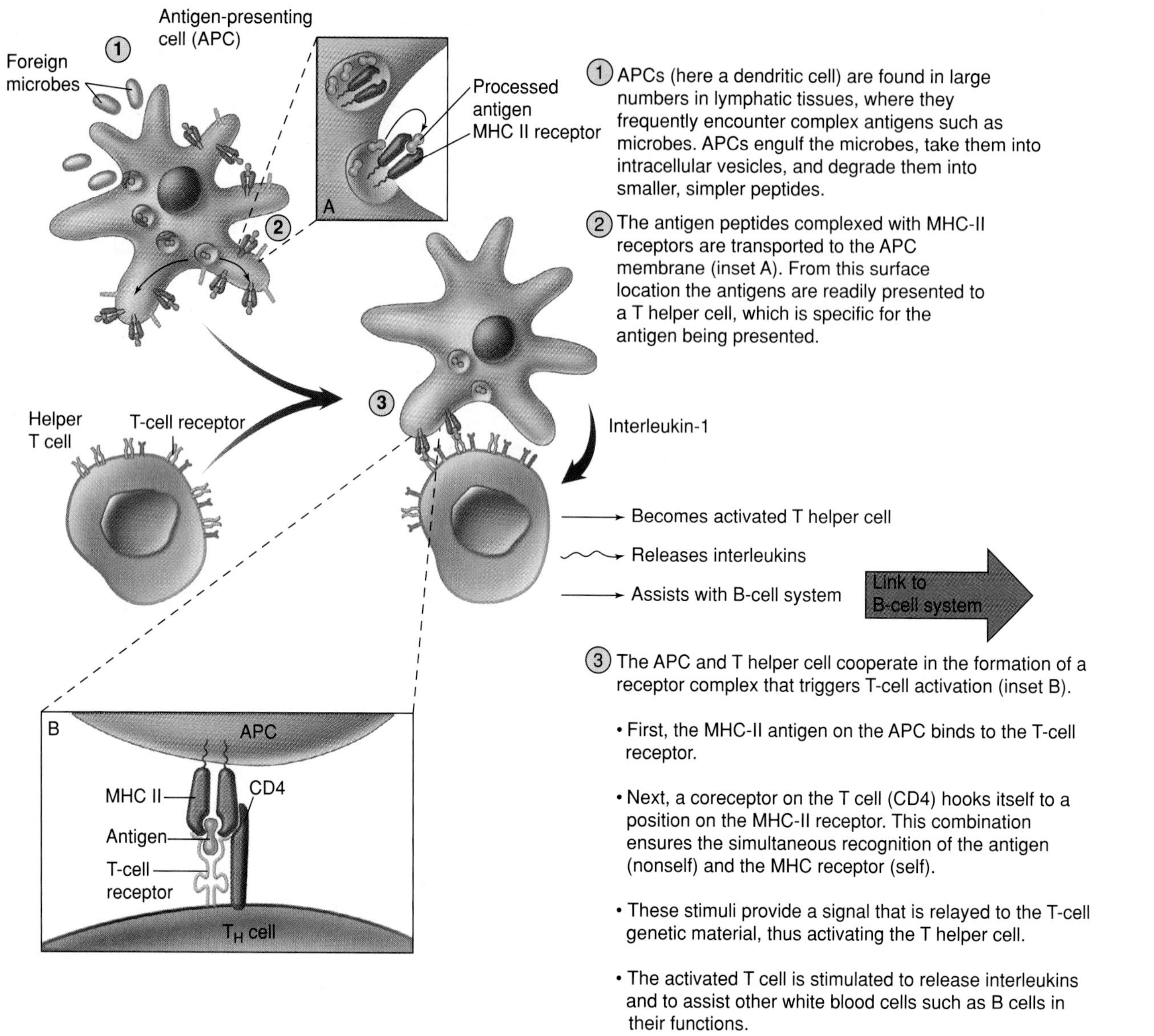

Antigen-presenting cell (APC)
Foreign microbes
Processed antigen
MHC II receptor
A
1 APCs (here a dendritic cell) are found in large numbers in lymphatic tissues, where they frequently encounter complex antigens such as microbes. APCs engulf the microbes, take them into intracellular vesicles, and degrade them into smaller, simpler peptides.
2 The antigen peptides complexed with MHC-II receptors are transported to the APC membrane (inset A). From this surface location the antigens are readily presented to a T helper cell, which is specific for the antigen being presented.
Helper T cell
T-cell receptor
Interleukin-1
Becomes activated T helper cell
Releases interleukins
Assists with B-cell system
Link to B-cell system
3 The APC and T helper cell cooperate in the formation of a receptor complex that triggers T-cell activation (inset B).
• First, the MHC-II antigen on the APC binds to the T-cell receptor.
• Next, a coreceptor on the T cell (CD4) hooks itself to a position on the MHC-II receptor. This combination ensures the simultaneous recognition of the antigen (nonself) and the MHC receptor (self).
• These stimuli provide a signal that is relayed to the T-cell genetic material, thus activating the T helper cell.
• The activated T cell is stimulated to release interleukins and to assist other white blood cells such as B cells in their functions.
B
APC
MHC II
CD4
Antigen
T-cell receptor
TH cell

Process Figure 15.9 Interactions between antigen-presenting cells (APCs) and T helper (CD4) cells required for T-cell activation.
For T cells to recognize foreign antigens, they must have the antigen processed and presented by a professional APC such as a dendritic cell.

T lymphocytes. After processing is complete, the antigen is bound to the MHC receptor and moved to the surface of the APC so that it will be readily accessible to T lymphocytes during presentation **(figure 15.9).**

Presentation of Antigen to the Lymphocytes and Its Early Consequences

For lymphocytes to respond to the APC-bound antigen, certain conditions must be met. T-cell-dependent antigens, usually protein-based, require recognition steps between the APC, antigen, and lymphocytes. APCs (often dendritic cells that have engulfed antigen) activate CD4 T helper cells in the lymph nodes. This class of T cell bears an antigen-specific T-cell receptor that binds to MHC class II plus a piece of the antigen (epitope) and the CD4 molecule, which also binds to MHC class II (see figure 15.9). Once identification has occurred, a molecule on the APC activates this T helper cell. The T_H cell, in turn, produces a cytokine, **interleukin-2 (IL-2),** which is a growth factor for the T helper cells and cytotoxic T cells. These T helper cells can then help activate B cells. The manner in which B and T cells subsequently become activated by the APC–T helper cell complex and their individual responses to antigen are addressed in later sections.

A few antigens can trigger a response from B lymphocytes without the cooperation of APCs or T helper cells. These T-cell-independent antigens are usually simple molecules such as carbohydrates with many repeating and invariable determinant groups. Examples include lipopolysaccharide from the cell wall of *Escherichia coli,* polysaccharide from the capsule of *Streptococcus pneumoniae,* and molecules from rabies and Epstein-Barr virus. Because so few antigens are of this type, most B-cell reactions require T helper cells.

CHECKPOINT

- Immature lymphocytes released from hematopoietic tissue migrate (home) to one of two sites for further development. B cells mature in the stromal cells of the bone marrow. T cells mature in the thymus.
- Antigens or immunogens are proteins or other complex molecules of high molecular weight that trigger the immune response in the host.
- Lymphocytes respond to a specific portion of an antigen called the epitope. A given microorganism has many such epitopes, all of which stimulate individual specific immune responses.
- Haptens are molecules that are too small to trigger an immune response alone but can be immunogenic when they attach to a larger substance, such as host serum protein.
- Antigen-presenting cells (APCs) such as dendritic cells and macrophages engulf and process foreign antigen and bind the epitope to MHC class II molecules on their cell surface for presentation to CD4 T lymphocytes. Physical contact between the APC, T cells, and B cells activates these lymphocytes to proceed with their respective immune responses.

15.5 B-Cell Response

Activation of B Lymphocytes: Clonal Expansion and Antibody Production

The immunologic activation of most B cells requires a series of events **(figure 15.10).**

1. **Clonal selection and binding of antigen.** In this case, a precommitted B cell of a particular clonal specificity binds the antigen on its Ig receptors.
2. **Antigen processing and presentation.** The antigen is endocytosed by the B cell and degraded into smaller peptide determinants. The antigen is then bound to the MHC II receptors on the surface of the B cell.
3. **B-cell/T-cell recognition and cooperation.** The MHC/Ag receptor is recognized and bound by a T_H cell. The B cell receives chemical signals from macrophages and T cells (interleukins).
4. **B-cell activation.** The combination of these stimuli on the membrane receptors causes a signal to be transmitted internally to the B-cell nucleus. These events trigger B-cell activation. An activated B cell undergoes an increase in DNA synthesis, organelle bulk, and overall size in preparation for entering the cell cycle and dividing.

5–6. **Clonal expansion.** A stimulated B cell multiplies through successive mitotic divisions and produces a large population of genetically identical daughter cells. Some cells that stop short of becoming fully differentiated are **memory cells,** which remain for long periods to react with that same antigen at a later time. This reaction also expands the clone size, so that subsequent exposure to that antigen provides more cells with that specificity. This expansion of the clone size accounts for the increased memory response. The most numerous progeny are large, specialized, terminally differentiated B cells called **plasma cells.**

7. **Antibody production and secretion.** The primary action of plasma cells is to secrete into the surrounding tissues copious amounts of antibodies with the same specificity as the original receptor (figure 15.10). Although an individual plasma cell can produce around 2,000 antibodies per second, production does not continue indefinitely. The plasma cells do not survive for long and deteriorate after they have synthesized antibodies.

Products of B Lymphocytes: Antibody Structure and Functions

The Structure of Immunoglobulins

Earlier we saw that a basic immunoglobulin (Ig) molecule contains four polypeptide chains connected by disulfide bonds. Let us view this structure once again, using an IgG molecule as a model. Two functionally distinct segments called *fragments* can be differentiated. The two "arms" that bind antigen are termed antigen binding fragments (FAbs), and the rest of the molecule is the crystallizable fragment (Fc), so called because it was the first to be crystallized in pure form. The amino-terminal end of each FAb fragment (consisting of the variable regions of the heavy and light chains) folds into a groove that will accommodate one epitope. The presence of a special region at the site of attachment between the FAb and Fc fragments allows swiveling of the FAb fragments. In this way, they can change their angle to accommodate nearby antigen sites that vary slightly in distance and position. The Fc fragment is involved in binding to various cells and molecules of the immune system itself. **Figure 15.11** shows three views of antibody structure.

Antibody-Antigen Interactions and the Function of the FAb

The site on the antibody where the epitope binds is composed of a *hypervariable region* whose amino acid content can be extremely varied. Antibodies differ somewhat in the exactness of this groove for antigen, but a certain complementary fit is necessary for the antigen to be held effectively **(figure 15.12).** The specificity of antigen binding sites for antigens is very similar to enzymes and substrates (in fact, some antibodies are used as enzymes, as you learned in Insight 8.2). So specific are some immunoglobulins for antigen that they can distinguish between a single functional group of a few atoms. Because the specificity of the two FAb sites is identical, an Ig molecule can bind epitope on the same cell or on two separate cells and thereby link them.

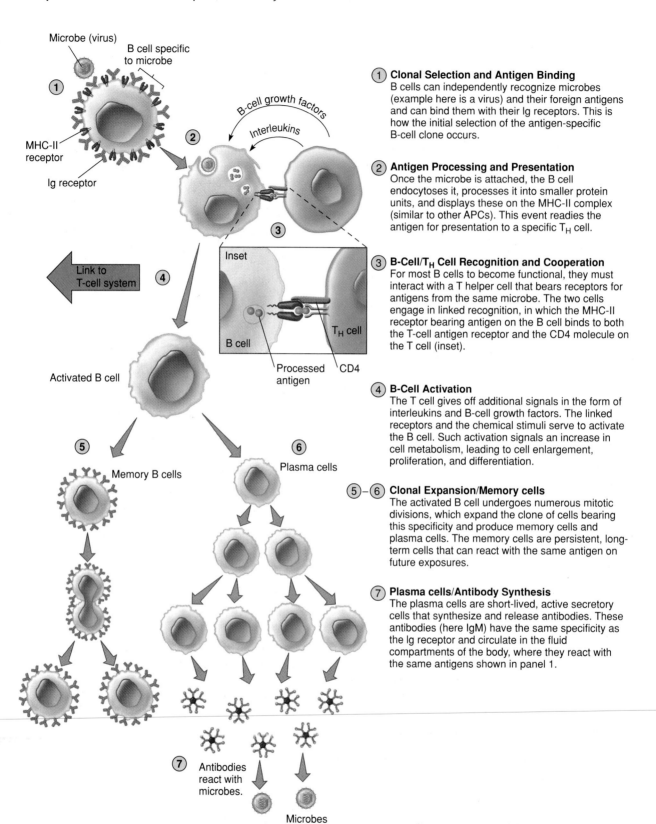

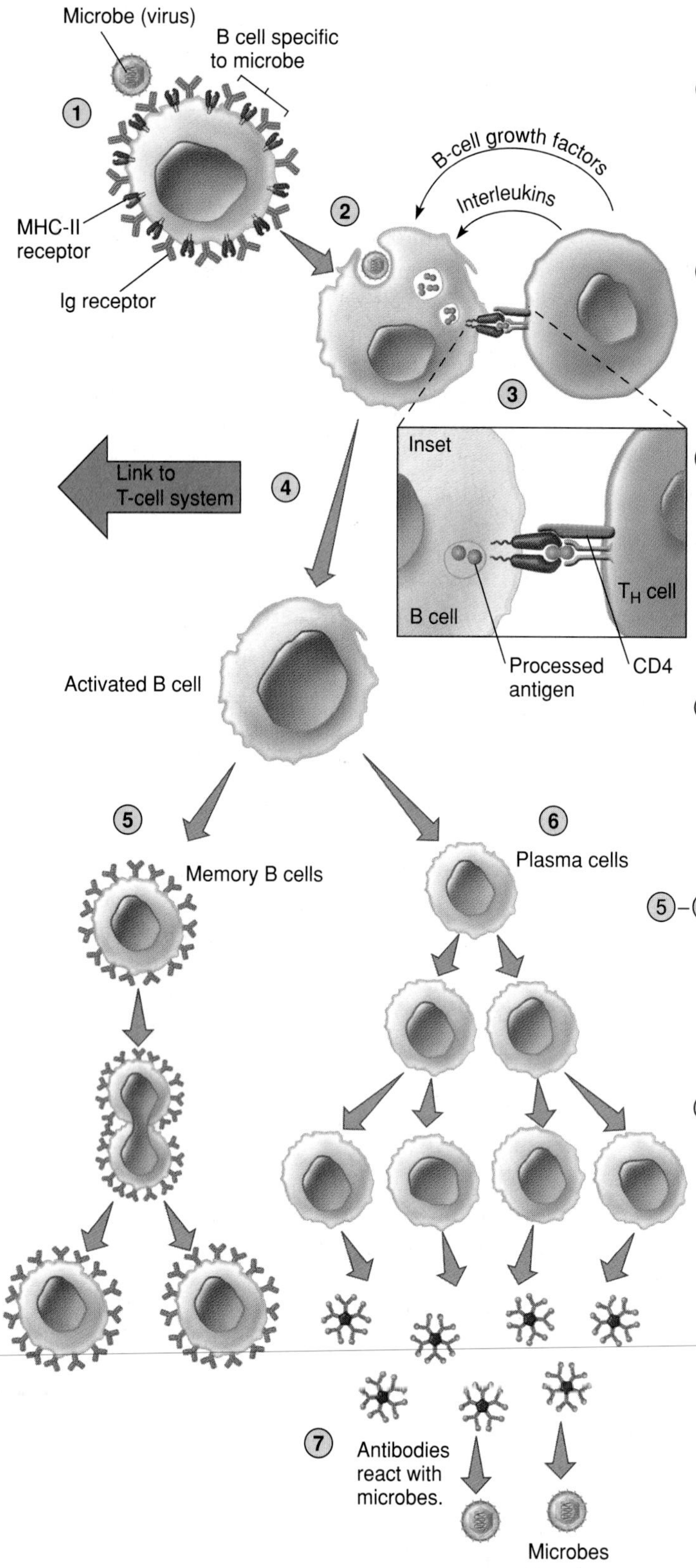

(1) **Clonal Selection and Antigen Binding**
B cells can independently recognize microbes (example here is a virus) and their foreign antigens and can bind them with their Ig receptors. This is how the initial selection of the antigen-specific B-cell clone occurs.

(2) **Antigen Processing and Presentation**
Once the microbe is attached, the B cell endocytoses it, processes it into smaller protein units, and displays these on the MHC-II complex (similar to other APCs). This event readies the antigen for presentation to a specific T_H cell.

(3) **B-Cell/T_H Cell Recognition and Cooperation**
For most B cells to become functional, they must interact with a T helper cell that bears receptors for antigens from the same microbe. The two cells engage in linked recognition, in which the MHC-II receptor bearing antigen on the B cell binds to both the T-cell antigen receptor and the CD4 molecule on the T cell (inset).

(4) **B-Cell Activation**
The T cell gives off additional signals in the form of interleukins and B-cell growth factors. The linked receptors and the chemical stimuli serve to activate the B cell. Such activation signals an increase in cell metabolism, leading to cell enlargement, proliferation, and differentiation.

(5)–(6) **Clonal Expansion/Memory cells**
The activated B cell undergoes numerous mitotic divisions, which expand the clone of cells bearing this specificity and produce memory cells and plasma cells. The memory cells are persistent, long-term cells that can react with the same antigen on future exposures.

(7) **Plasma cells/Antibody Synthesis**
The plasma cells are short-lived, active secretory cells that synthesize and release antibodies. These antibodies (here IgM) have the same specificity as the Ig receptor and circulate in the fluid compartments of the body, where they react with the same antigens shown in panel 1.

Process Figure 15.10 **Events in B-cell activation and antibody synthesis.**

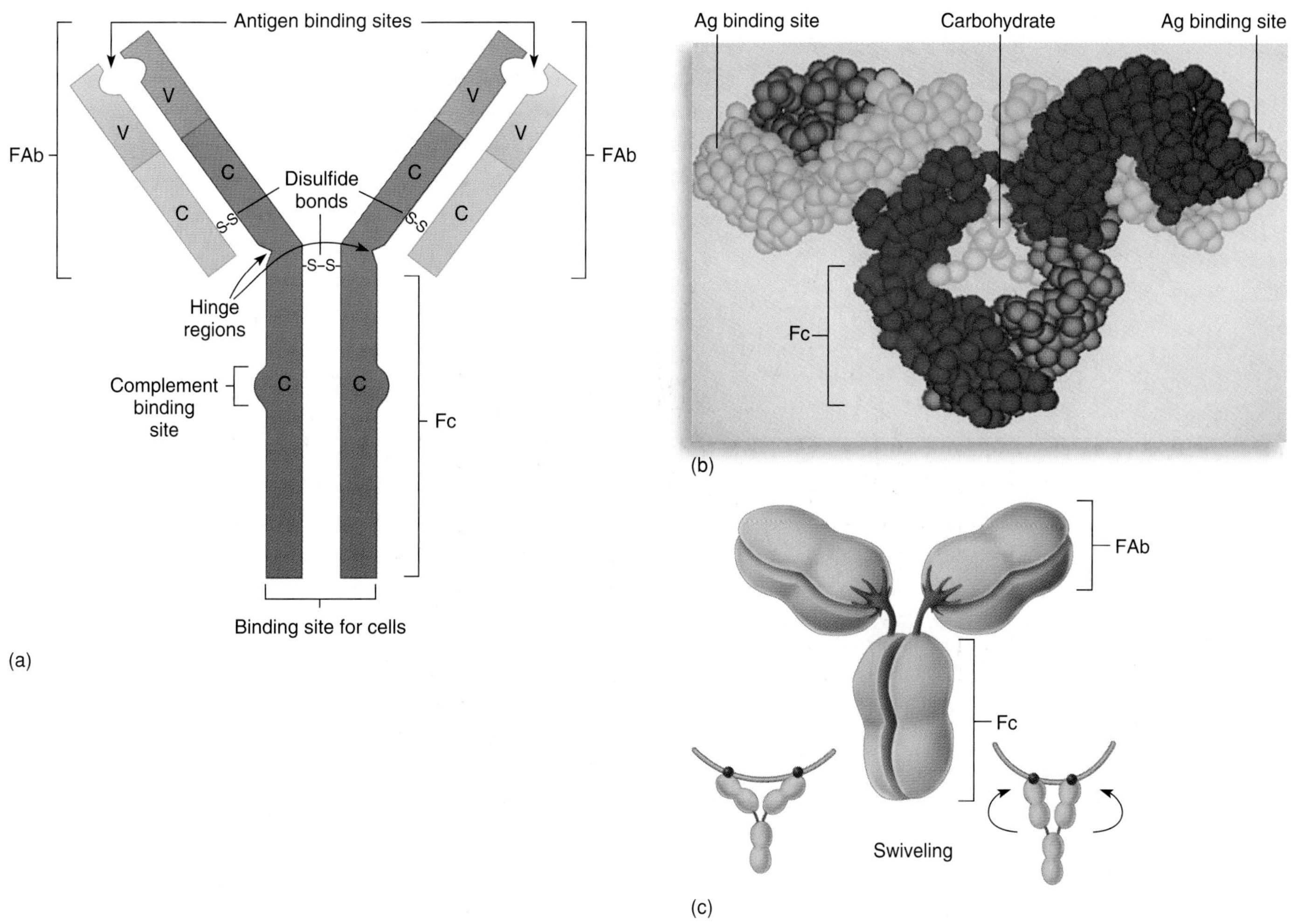

Figure 15.11 Working models of antibody structure.
(a) Diagrammatic view of IgG depicts the principal functional areas (FAbs and Fc) of the molecule. **(b)** Realistic model of immunoglobulin shows the tertiary and quaternary structure achieved by additional intrachain and interchain bonds and the position of the carbohydrate component. **(c)** The "peanut" model of IgG helps illustrate swiveling of Fabs relative to one another and to Fc.

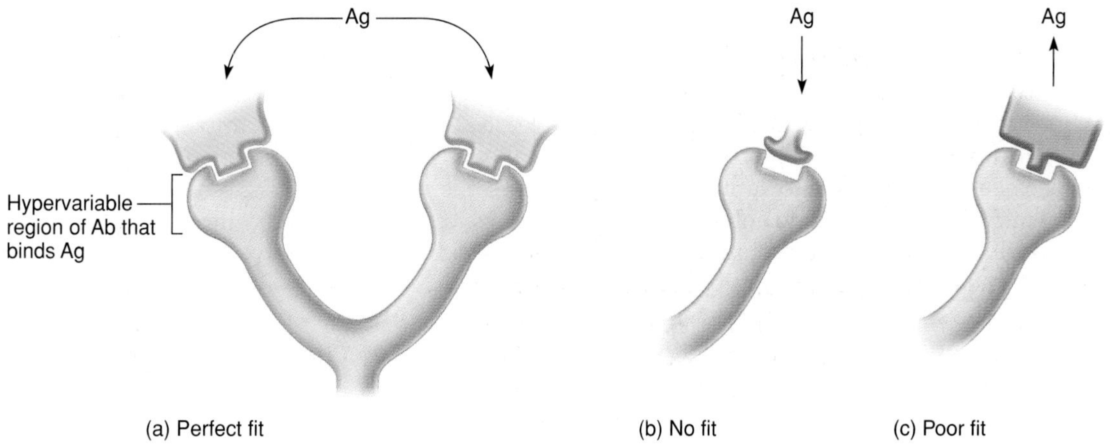

Figure 15.12 Antigen-antibody binding.
The union of antibody (Ab) and antigen (Ag) is characterized by a certain degree of fit and is supported by a multitude of weak linkages, especially hydrogen bonds and electrostatic attraction. The better the fit (i.e., antigen in **(a)** versus antigen in **(c)**), the stronger the stimulation of the lymphocyte during the activation stage.

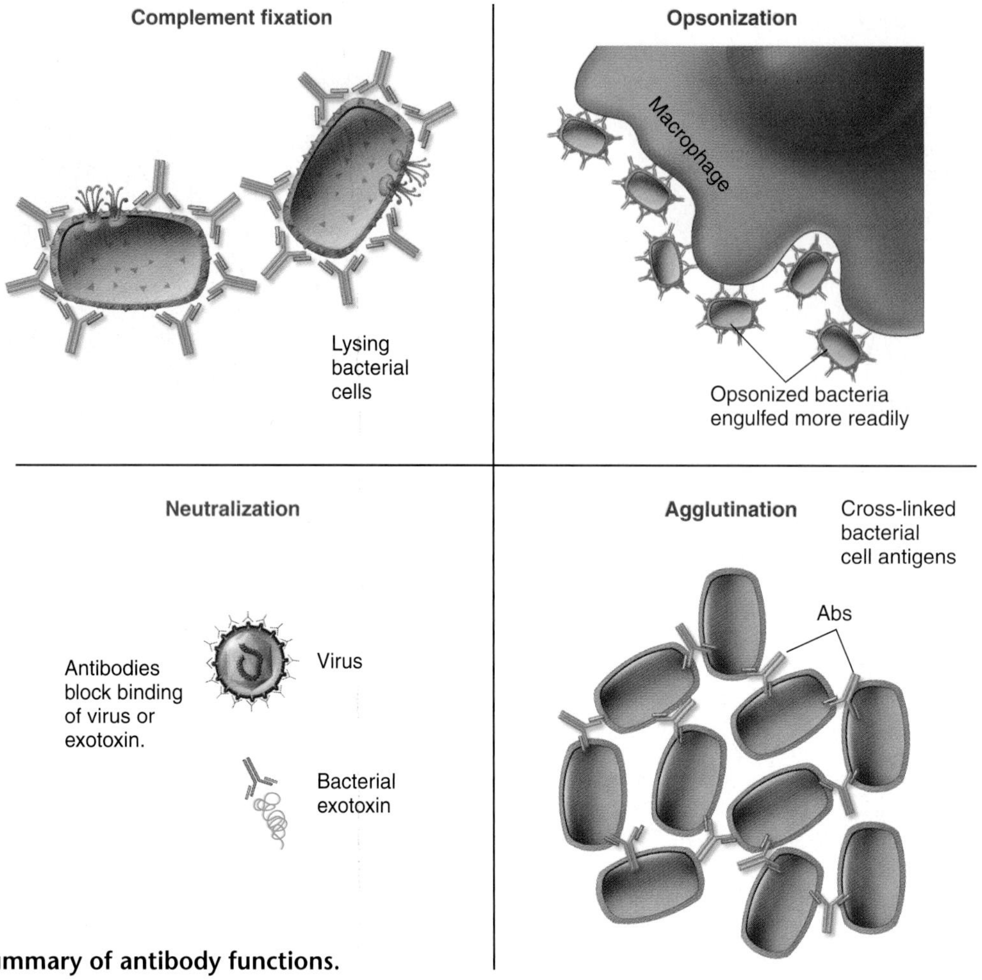

Figure 15.13 Summary of antibody functions.

The principal activity of an antibody is to unite with, immobilize, call attention to, or neutralize the antigen for which it was formed **(figure 15.13).** Antibodies called opsonins stimulate **opsonization** (ahp″-son-uh-zaz′-shun), a process in which microorganisms or other particles are coated with specific antibodies so that they will be more readily recognized by phagocytes, which dispose of them. Opsonization has been likened to putting handles on a slippery object to provide phagocytes a better grip. The capacity for antibodies to aggregate, or **agglutinate,** antigens is the consequence of their cross-linking cells or particles into large clumps. Agglutination renders microbes immobile and enhances their phagocytosis. This is a principle behind certain immune tests discussed in chapter 17. The interaction of an antibody with complement can result in the specific rupturing of cells and some viruses. In **neutralization** reactions, antibodies fill the surface receptors on a virus or the active site on a microbial enzyme to prevent it from attaching normally. An **antitoxin** is a special type of antibody that neutralizes bacterial exotoxins. Note that not all antibodies are protective; some neither benefit nor harm, and a few actually cause diseases.

Functions of the Fc Fragment

Although the FAb fragments bind antigen, the Fc fragment has a different binding function. In most classes of immunoglobulin, the Fc end contains an effector portion that can bind to receptors on the membranes of cells, such as macrophages, neutrophils, eosinophils, mast cells, basophils, and lymphocytes. The effect of an antibody's Fc fragment binding to a cell depends upon that cell's role. In the case of opsonization, the attachment of antibody to foreign cells and viruses exposes the Fc fragments to phagocytes. Certain antibodies have regions on the Fc portion for fixing complement; and in some immune reactions, the binding of Fc causes the release of cytokines. For example, the Fc end of the antibody of allergy (IgE) binds to basophils and mast cells, which causes the release of allergic mediators such as histamine. The size and amino acid composition of Fc also determine an antibody's permeability, its distribution in the body, and its class.

Accessory Molecules on Immunoglobulins

All antibodies contain molecules in addition to the basic polypeptides. Varying amounts of carbohydrates are affixed to the constant regions in most instances **(table 15.2).** Two additional accessory molecules are the *J chain,* which joins the monomers[3] of IgA and IgM, and the *secretory component,* which helps move IgA across mucous membranes.

3. "Monomer" means "one unit" or "one part." Accordingly, "dimer" means "two units," pentamer means "five units," and polymer means "many units."

TABLE 15.2 Characteristics of the Immunoglobulin (Ig) Classes

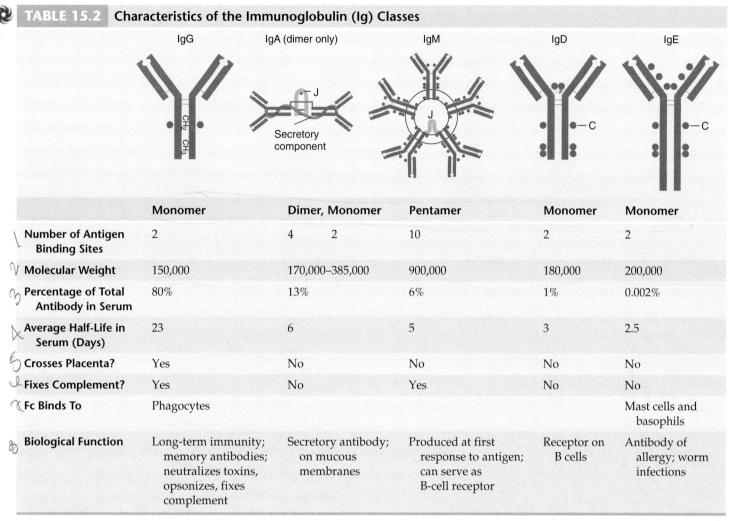

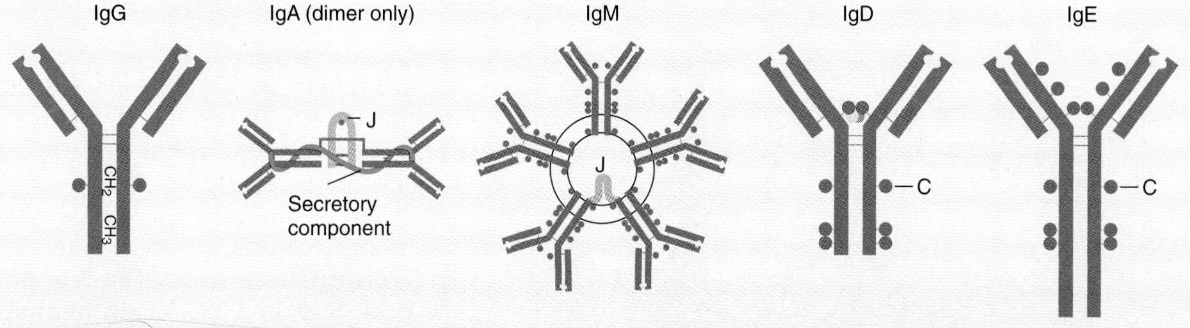

	IgG	IgA (dimer only)	IgM	IgD	IgE
	Monomer	**Dimer, Monomer**	**Pentamer**	**Monomer**	**Monomer**
Number of Antigen Binding Sites	2	4 2	10	2	2
Molecular Weight	150,000	170,000–385,000	900,000	180,000	200,000
Percentage of Total Antibody in Serum	80%	13%	6%	1%	0.002%
Average Half-Life in Serum (Days)	23	6	5	3	2.5
Crosses Placenta?	Yes	No	No	No	No
Fixes Complement?	Yes	No	Yes	No	No
Fc Binds To	Phagocytes				Mast cells and basophils
Biological Function	Long-term immunity; memory antibodies; neutralizes toxins, opsonizes, fixes complement	Secretory antibody; on mucous membranes	Produced at first response to antigen; can serve as B-cell receptor	Receptor on B cells	Antibody of allergy; worm infections

C = carbohydrate. J = J chain.

The Classes of Immunoglobulins

Immunoglobulins exist as structural and functional classes called *isotypes* (compared and contrasted in table 15.2). The differences in these classes are due primarily to variations in the Fc fragment. The classes are differentiated with shorthand names (Ig, followed by a letter: IgG, IgA, IgM, IgD, IgE).

The structure of IgG has already been presented. It is a monomer produced by plasma cells in a primary response and by memory cells responding the second time to a given antigenic stimulus. It is by far the most prevalent antibody circulating throughout the tissue fluids and blood. It has numerous functions: It neutralizes toxins, opsonizes, and fixes complement; and it is the only antibody capable of crossing the placenta.

The two forms of IgA are (1) a monomer that circulates in small amounts in the blood and (2) a dimer that is a significant component of the mucous and serous secretions of the salivary glands, intestine, nasal membrane, breast, lung, and genitourinary tract. The dimer, called secretory IgA, is formed in a plasma cell by two monomers held together by a J chain. To facilitate the transport of IgA across membranes, a secretory piece is later added. IgA coats the surface of these membranes and appears free in saliva, tears, colostrum, and mucus. It confers the most important specific local immunity to enteric, respiratory, and genitourinary pathogens. Its contribution in protecting newborns who derive it passively from nursing is mentioned in Insight 15.2.

IgM is a huge molecule composed of five monomers (making it a pentamer) attached by the Fc portions to a central J chain. With its 10 binding sites, this molecule has tremendous capacity for binding antigen. IgM is the first class synthesized following the host's first encounter with antigen. Its complement-fixing qualities make it an important antibody in many immune reactions. It circulates mainly in the blood and does not cross the placental barrier.

IgD is a monomer found in minuscule amounts in the serum, and it does not fix complement, opsonize, or cross the placenta. Its main function is to serve as a receptor for antigen on B cells, usually along with IgM. It seems to be the triggering molecule for B-cell activation.

IgE is also an uncommon blood component unless one is allergic or has a parasitic worm infection. Its Fc region interacts with receptors on mast cells and basophils. Its biological significance is to stimulate an inflammatory response through the release of potent physiological substances by the basophils and mast cells. Because inflammation would enlist blood cells such as eosinophils and lymphocytes to the site of

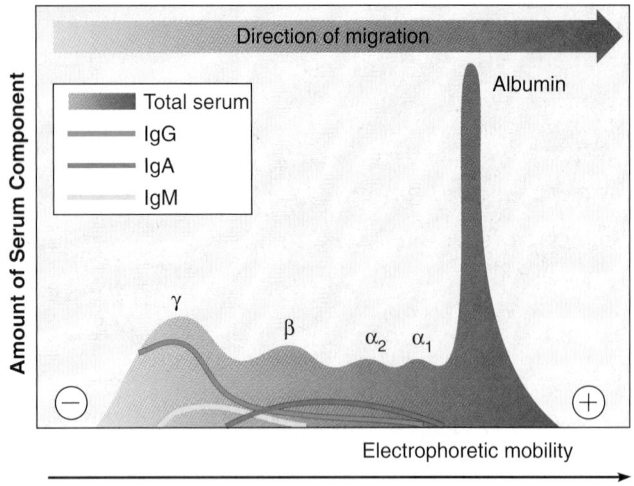

Figure 15.14 Pattern of human serum after electrophoresis.
When antiserum is subjected to electrical current, the various proteinaceous components are separated into bands. This is a means of separating the different antibodies and serum proteins as well as quantifying them.

infection, it would certainly be one defense against parasites. Unfortunately, IgE has another, more insidious effect—that of mediating anaphylaxis, asthma, and certain other allergies (explained in chapter 16).

Evidence of Antibodies in Serum

Regardless of the site where antibodies are first secreted, a large quantity eventually ends up in the blood by way of the body's communicating networks. If one subjects a sample of **antiserum** (serum containing specific antibodies) to electrophoresis, the major groups of proteins migrate in a pattern consistent with their mobility and size **(figure 15.14).** The albumins show up in one band, and the globulins in four bands called alpha-1 (α_1), alpha-2 (α_2), beta (β), and gamma (γ) globulins. Most of the globulins represent antibodies, which explains how the term *immunoglobulin* was derived. **Gamma globulin** is composed primarily of IgG, whereas beta and alpha-2 globulins are a mixture of IgG, IgA, and IgM.

Monitoring Antibody Production over Time: Primary and Secondary Responses to Antigens

We can learn a great deal about how the immune system reacts to an antigen by studying the levels of antibodies in serum over time **(figure 15.15).** This level is expressed quantitatively as the **titer** (ty′-tur), or concentration of antibodies. Upon the first exposure to an antigen, the system undergoes a **primary response.** The earliest part of this response, the *latent period,* is marked by a lack of antibodies for that antigen, but much activity is occurring. During this time, the antigen is being concentrated in lymphoid tissue and is being processed by the correct clones of B lymphocytes. As plasma cells synthesize antibodies, the serum titer increases to a certain plateau and then tapers off to a low level over a few weeks or months. When the class of antibodies produced during this response is tested, an important characteristic of the response is uncovered. It turns out that, early in the primary response, most of the antibodies are the IgM type, which is the first class to be secreted by plasma cells. Later, the class of the antibodies (but not their specificity) is switched to IgG or some other class (IgA or IgE).

When the immune system is exposed again to the same immunogen within weeks, months, or even years, a **secondary response** occurs. The rate of antibody synthesis, the peak titer, and the length of antibody persistence are greatly increased over the primary response. The rapidity and amplification seen in this response are attributable to the memory B cells that were formed during the primary response. Because of its association with recall, the secondary response is also called the **anamnestic response.** The advantage of this response is evident: It provides a quick and potent strike against subsequent exposures to infectious agents. This memory effect is the fundamental basis for vaccination, which we discuss later.

CHECKPOINT

- B cells produce five classes of antibody: IgM, IgG, IgA, IgD, and IgE. IgM and IgG predominate in plasma. IgA predominates in body secretions. IgD is expressed on B cells as an antigen receptor. IgE binds to most cells and basophils in tissues, promoting inflammation.
- Antibodies bind physically to the specific antigen that stimulates their production, thereby immobilizing the antigen and enabling it to be destroyed by other components of the immune system.
- The memory response means that the second exposure to antigen calls forth a much faster and more vigorous response than the first.

15.6 T-Cell Response

Cell-Mediated Immunity (CMI)

During the time that B cells have been actively responding to antigens, the T cell limb of the system has been similarly engaged. The responses of T cells, however, are **cell-mediated** immunities, which require the direct involvement of T lymphocytes throughout the course of the reaction. These reactions are among the most complex and diverse in the immune system and involve several subsets of T cells whose particular actions are dictated by the APCs that activate them. T cells are restricted; that is, they require some type of MHC (self) recognition before they can be activated, and all produce cytokines with a spectrum of biological effects.

T cells have notable differences in function from B cells. Rather than making antibodies to control foreign antigens, T cells stimulate other T cells, B cells, and phagocytes.

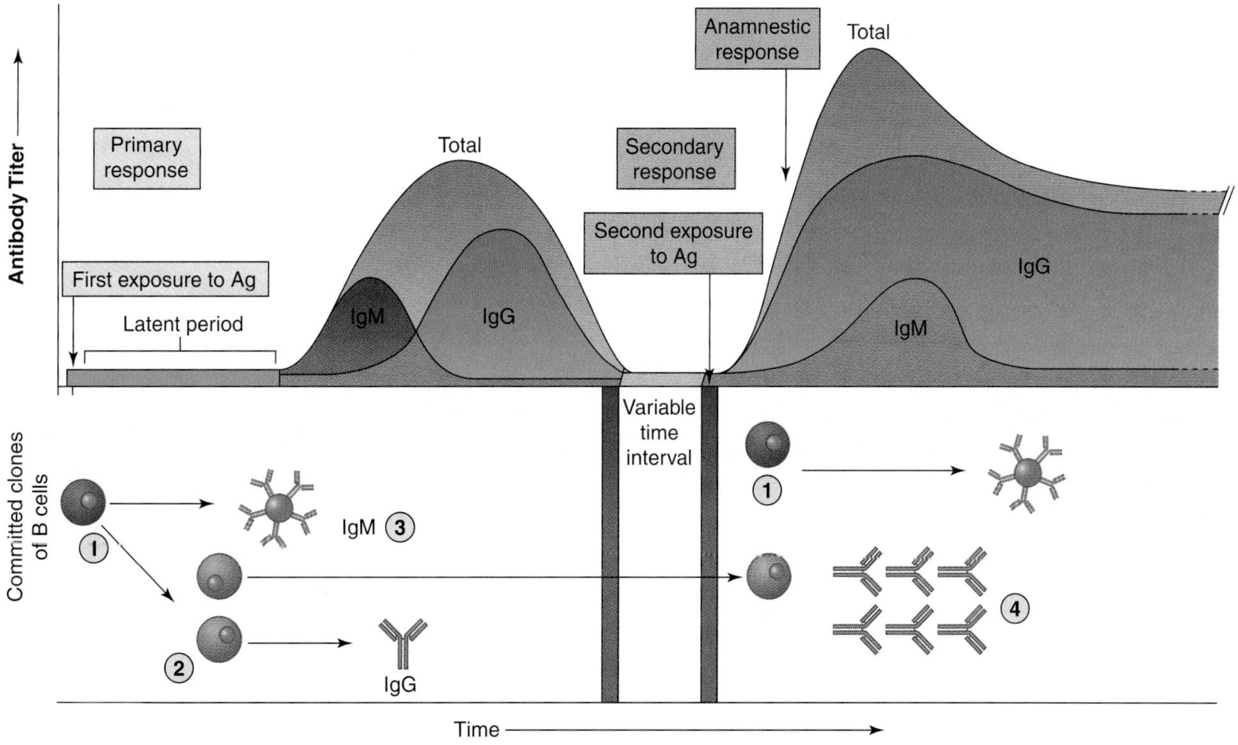

Process Figure 15.15 **Primary and secondary responses to antigens.**
(Top) The pattern of antibody titer and subclasses as monitored during initial and subsequent exposure to the same antigen. *(Bottom)* A view of the B-cell responses that account for the pattern. Depicted are (1) clonal selection, (2) production of memory cells, and the predominant antibody class occurring at (3) first and (4) second contact with antigen. Note that residual memory cells remaining from the primary response are ready to act immediately, which produces the rapid rise of antibody levels early in the secondary period.

The Activation of T Cells and Their Differentiation into Subsets

The mature T cells in lymphoid organs are primed to react with antigens that have been processed and presented to them by dendritic cells and macrophages. They recognize an antigen only when it is presented in association with an MHC carrier (see figure 15.9). T cells with CD4 receptors recognize endocytosed peptides presented on MHC-II, and T cells with CD8 receptors recognize peptides presented on MHC-I.

A T cell is initially sensitized when an antigen/MHC complex is bound to its receptors. As with B cells, activated T cells transform in preparation for mitotic divisions, and they differentiate into one of the subsets of effector cells and memory cells that can respond quickly to MHC plus antigen on APCs upon subsequent contact **(table 15.3).** Memory T cells are some of the longest-lived blood cells known (70 years in one well-documented case).

T Helper (T_H) Cells Helper cells play a central role in regulating immune reactions to antigens, including those of B cells and other T cells. They are also involved in activating macrophages. They do this directly by receptor contact and indirectly by releasing cytokines like interferon gamma (IFNγ). T helper cells secrete interleukin-2, which stimulates the primary growth and activation of many types of T cells,

TABLE 15.3 Characteristics of Subsets of T Cells

Types	Primary Receptors on T Cell	Functions/Important Features
T helper cell 1 (T_H1)	CD4	Activates the cell-mediated immunity pathway, secretes tumor necrosis factor and interferon gamma, also responsible for delayed hypersensitivity (allergy occurring several hours or days after contact)
T helper cell 2 (T_H2)	CD4	Drives B-cell proliferation, secretes IL-4, IL-5, IL-6, IL-10; can dampen T_H1 activity
T cytotoxic cell (T_C)	CD8	Destroys a target foreign cell by lysis; important in destruction of complex microbes, cancer cells, virus-infected cells; graft rejection; requires MHC-I for function

INSIGHT 15.1 Medical

Monoclonal Antibodies: Variety Without Limit

The value of antibodies as tools for locating or identifying antigens is well established. For many years, antiserum extracted from human or animal blood was the main source of antibodies for tests and therapy, but most antiserum has a basic problem. It contains **polyclonal antibodies,** meaning that it is a mixture of different antibodies because it reflects dozens of immune reactions from a wide variety of B-cell clones. This characteristic is to be expected, because several immune reactions may be occurring simultaneously, and even a single species of microbe can stimulate several different types of antibodies. Certain applications in immunology require a pure preparation of **monoclonal antibodies (MAbs)** that originate from a single clone and have a single specificity for antigen.

The technology for producing monoclonal antibodies is possible by hybridizing cancer cells and activated B cells in the lab. This technique began with the discovery that tumors isolated from multiple myelomas in mice consist of identical plasma cells. These monoclonal plasma cells secrete a strikingly pure form of antibodies with a single specificity and continue to divide indefinitely. Immunologists recognized the potential in these plasma cells and devised a **hybridoma** approach to creating MAbs. The basic idea behind this approach is to hybridize or fuse a myeloma cell with a normal plasma cell from a mouse spleen to create an immortal cell that secretes a supply of functional antibodies with a single specificity.

The introduction of this technology has the potential for numerous biomedical applications. Monoclonal antibodies have provided immunologists with excellent standardized tools for studying the immune system and for expanding disease diagnosis and treatment. Most of the successful applications thus far use MAbs in diagnostic testing and research. Although injecting monoclonal antibodies to treat human disease is an exciting prospect, so far this therapy has been stymied because most MAbs are of mouse origin and many humans will develop hypersensitivity to them. The development of human MAbs and other novel approaches using genetic engineering is currently under way.

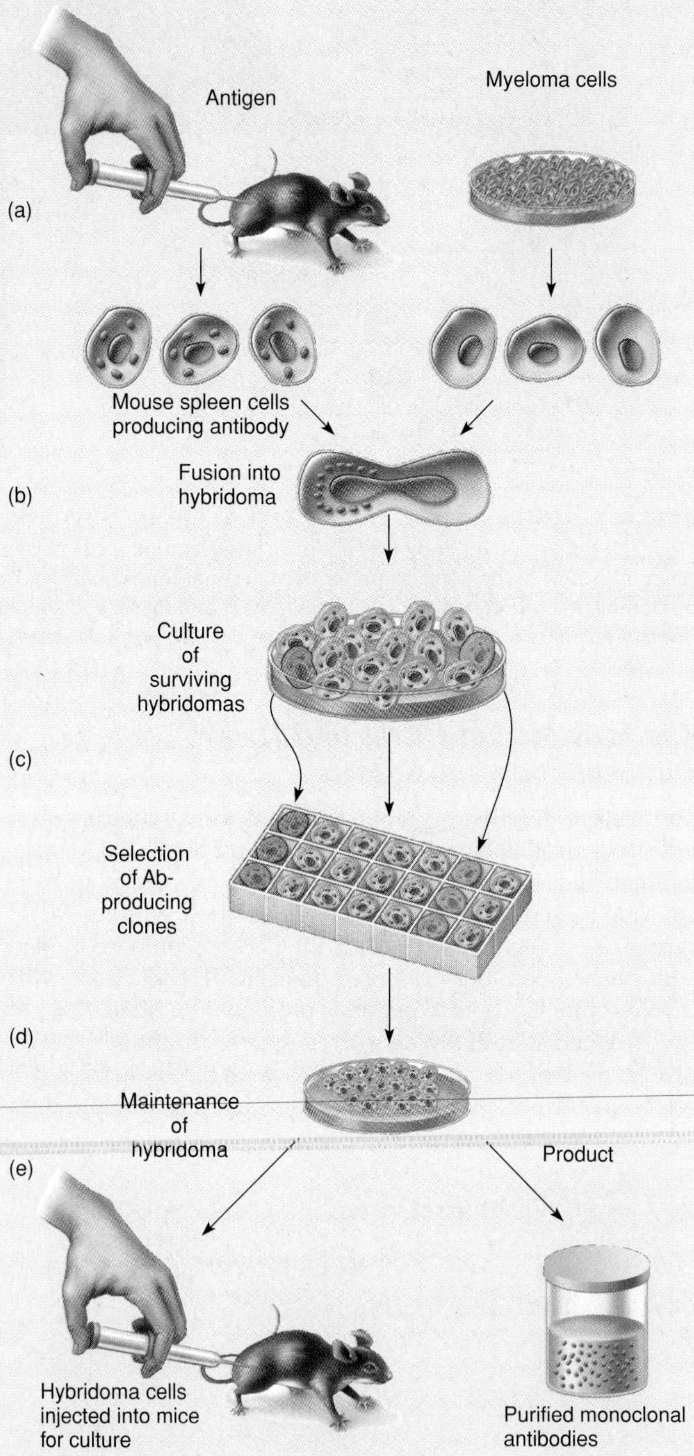

Summary of the technique for producing monoclonal antibodies by hybridizing myeloma tumor cells with normal plasma cells. **(a)** A normal mouse is inoculated with an antigen having the desired specificity, and activated cells are isolated from its spleen. A special strain of mouse provides the myeloma cells. **(b)** The two cell populations are mixed with polyethylene glycol, which causes some cells in the mixture to fuse and form hybridomas. **(c)** Surviving cells are cultured and separated into individual wells. **(d)** Tests are performed on each hybridoma to determine the specificity of the antibody (Ab) it secretes. **(e)** A hybridoma with the desired specificity is grown in tissue culture; antibody product is then isolated and purified. The hybridoma is maintained in a susceptible mouse for future use.

including cytotoxic T cells. Some T helper cells secrete interleukins-4, -5, and -6, which stimulate various activities of B cells. T helper cells are the most prevalent type of T cell in the blood and lymphoid organs, making up about 65% of this population. The severe depression of this class of T cells (with CD4 receptors) by HIV is what largely accounts for the immunopathology of AIDS.

When T helper (CD4) cells are stimulated by antigen/ MHC complex, they differentiate into either T helper 1 (T_H1) cells, or T helper 2 (T_H2) cells, probably depending on what type of cytokines the antigen-presenting cells secrete. It is thought that if the dendritic (APC) cell secretes IL-12 and/or interferon gamma, the T cell will become a T_H1 cell. A T_H1 cell will activate phagocytic cells to be better at inducing inflammation, resulting in a delayed hypersensitivity reaction. (Inappropriate delayed hyporsensitivity is a type of response to allergens, distinct from immediate allergies such as hay fever and anaphylaxis. Both of these reactions will be discussed in chapter 16.)

If the APC secretes another set of cytokines (some think IL-4 is involved), the T cell will differentiate into a T_H2 cell. These cells have the functions of secreting substances that influence B-cell differentiation and enhancing the antibody response. They may also be able to dampen the response of T_H1 cells when necessary.

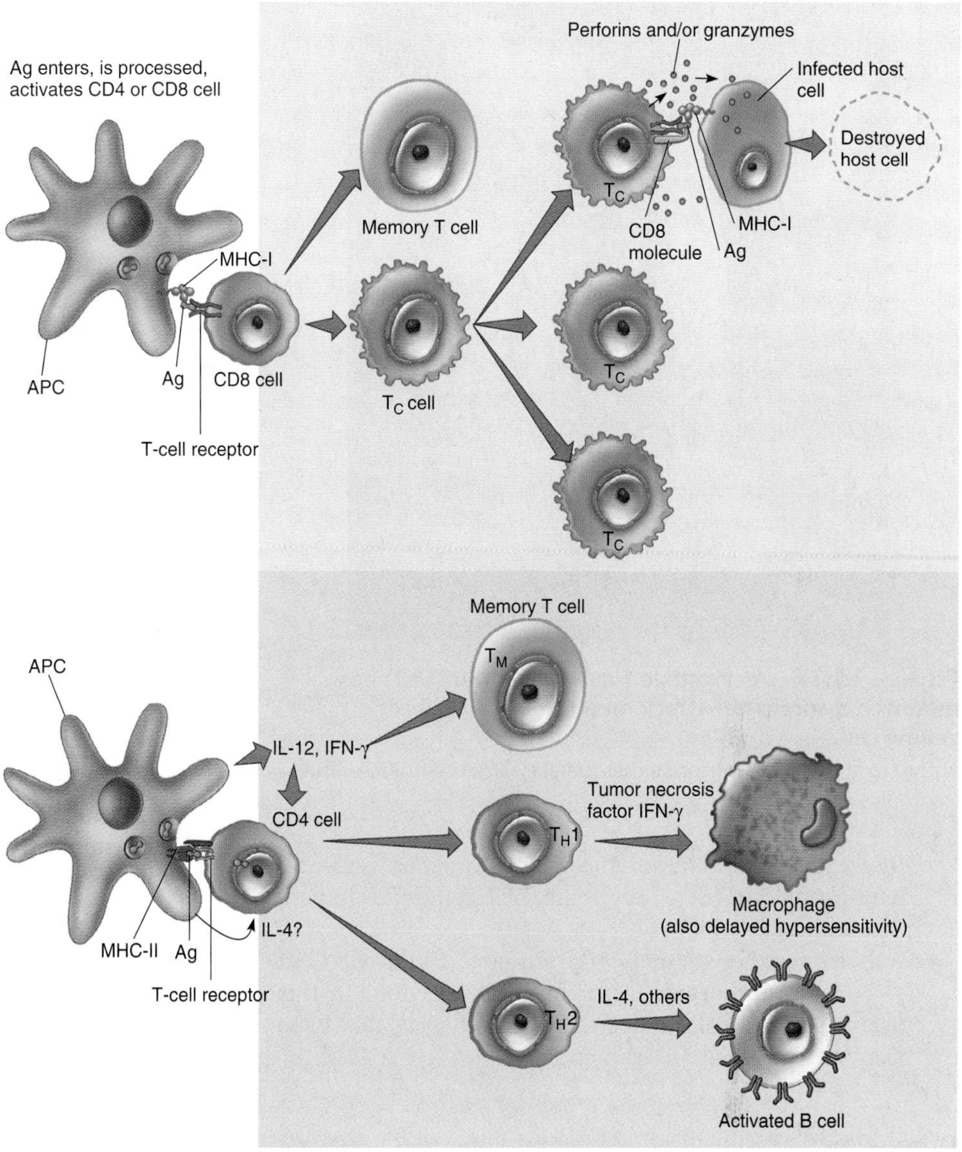

Figure 15.16 Overall scheme of T-cell activation and differentiation into different types of T cells.
Antigen-presenting cells present antigenic peptides to T cells bearing either CD4 or CD8 markers. Upon binding antigen/MHC-I complex, CD8 cells lead to the apoptosis of those cells. CD4 cells bind antigen/MHC-II complexes on APCs and, depending on the type of cytokine released by the APC, become either T_H1 or T_H2 cells. T_H1 cells influence macrophages to destroy ingested microbes or to become more active. T_H2 cells secrete cytokines that enhance B-cell activation.

Cytotoxic T (T_C) Cells: Cells That Kill Other Cells **Cytotoxicity** is the capacity of certain T cells to kill a specific target cell. It is a fascinating and powerful property that accounts for much of our immunity to foreign cells and cancer, and yet, under some circumstances, it can lead to disease. For a CD8 **killer T cell** to become activated, it must recognize a foreign peptide complexed with self MHC-I presented to it and mount a direct attack upon the target cell. After activation, the T_C cell severely injures the target cell **(figure 15.16).** This process involves the secretion of **perforins**[4] and **granzymes.** Perforins are proteins that can punch holes in the membranes of target cells. Granzymes are enzymes that attack proteins of target cells. The action of the perforins causes ions to leak out of target cells and creates a passageway for granzymes to enter. These events are usually followed by targeted cell death through a process called *apoptosis.*

Target cells that T_C cells can destroy include the following:

- Virally infected cells (figure 15.16). Cytotoxic cells recognize these because of telltale virus peptides expressed on their surface. Cytotoxic defenses are an essential protection against viruses.
- Cancer cells. T cells constantly survey the tissues and immediately attack any abnormal cells they encounter

4. **perforin** From the term *perforate* or to penetrate with holes.

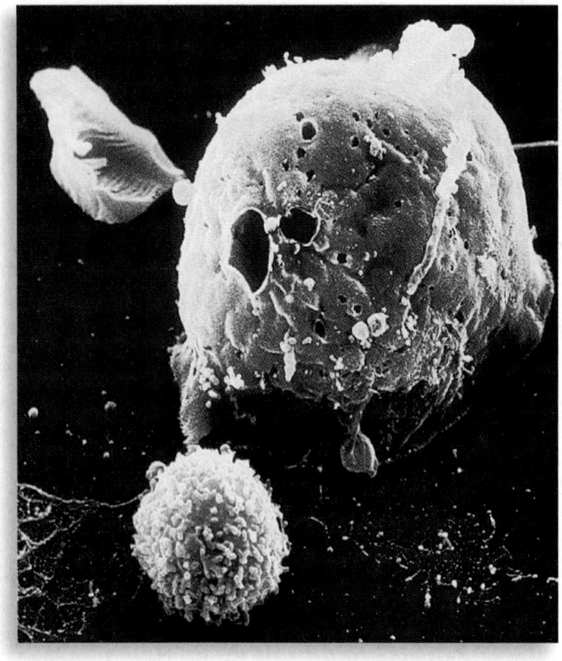

Figure 15.17 A cytotoxic T cell (lower blue cell) has mounted a successful attack on a tumor cell (larger yellow cell).
These small killer cells perforate their cellular targets with holes that lead to lysis and death.

(figure 15.17). The importance of this function is clearly demonstrated in the susceptibility of T-cell-deficient people to cancer (chapter 16).

- Cells from other animals and humans. Cytotoxic CMI is the most important factor in graft rejection. In this instance, the T_C cells attack the foreign tissues that have been implanted into a recipient's body.

Other Types of Killer Cells Natural killer (NK) cells are a type of lymphocyte related to T cells that lack specificity for antigens. They circulate through the spleen, blood, and lungs and are probably the first killer cells to attack cancer cells and virus-infected cells. They destroy such cells by similar mechanisms as T cells. They are not considered part of specific cell-mediated immunity although their activities are acutely sensitive to cytokines such as interleukin-12 and interferon.

As you can see, the T-cell system is very complex. In summary, T cells differentiate into three different types of cells, each of which contributes to the orchestrated immune response, under the influence of a multitude of cytokines.

15.7 A Practical Scheme for Classifying Specific Immunities

The means by which humans acquire immunities can be conveniently encapsulated within four interrelated categories: active, passive, natural, and artificial.

Active immunity occurs when an individual receives an immune stimulus (antigen) that activates the B and T cells, causing the body to produce immune substances such as antibodies. Active immunity is marked by several characteristics: (1) It is an essential attribute of an immunocompetent individual; (2) it creates a memory that renders the person ready for quick action upon reexposure to that same antigen; (3) it requires several days to develop; and (4) it lasts for a relatively long time, sometimes for life. Active immunity can be stimulated by natural or artificial means.

Passive immunity occurs when an individual receives immune substances (antibodies) that were produced actively in the body of another human or animal donor. The recipient is protected for a time even though he or she has not had prior exposure to the antigen. It is characterized by (1) lack of memory for the original antigen; (2) lack of production of new antibodies against that disease; (3) immediate onset of protection; and (4) short-term effectiveness, because antibodies have a limited period of function and, ultimately, the recipient's body disposes of them. Passive immunity can also be natural or artificial in origin.

Natural immunity encompasses any immunity that is acquired during the normal biological experiences of an individual rather than through medical intervention.

Artificial immunity is protection from infection obtained through medical procedures. This type of immunity is induced by immunization with vaccines and immune serum.

Figure 15.18 illustrates the various possible combinations of acquired immunities.

Natural Active Immunity: Getting the Infection

After recovering from infectious disease, a person may be actively resistant to reinfection for a period that varies according to the disease. In the case of childhood viral infections such as measles, mumps, and rubella, this natural active stimulus provides nearly lifelong immunity. Other diseases result in a less extended immunity of a few months to years (such as pneumococcal pneumonia and shigellosis), and reinfection is possible. Even a subclinical infection can stimulate natural active immunity. This probably accounts for the fact that some people are immune to an infectious agent without ever having been noticeably infected with or vaccinated for it.

Natural Passive Immunity: Mother to Child

Natural, passively acquired immunity occurs only as a result of the prenatal and postnatal, mother-child relationship. During fetal life, IgG antibodies circulating in the maternal bloodstream are small enough to pass or be actively transported across the placenta. Antibodies against tetanus,

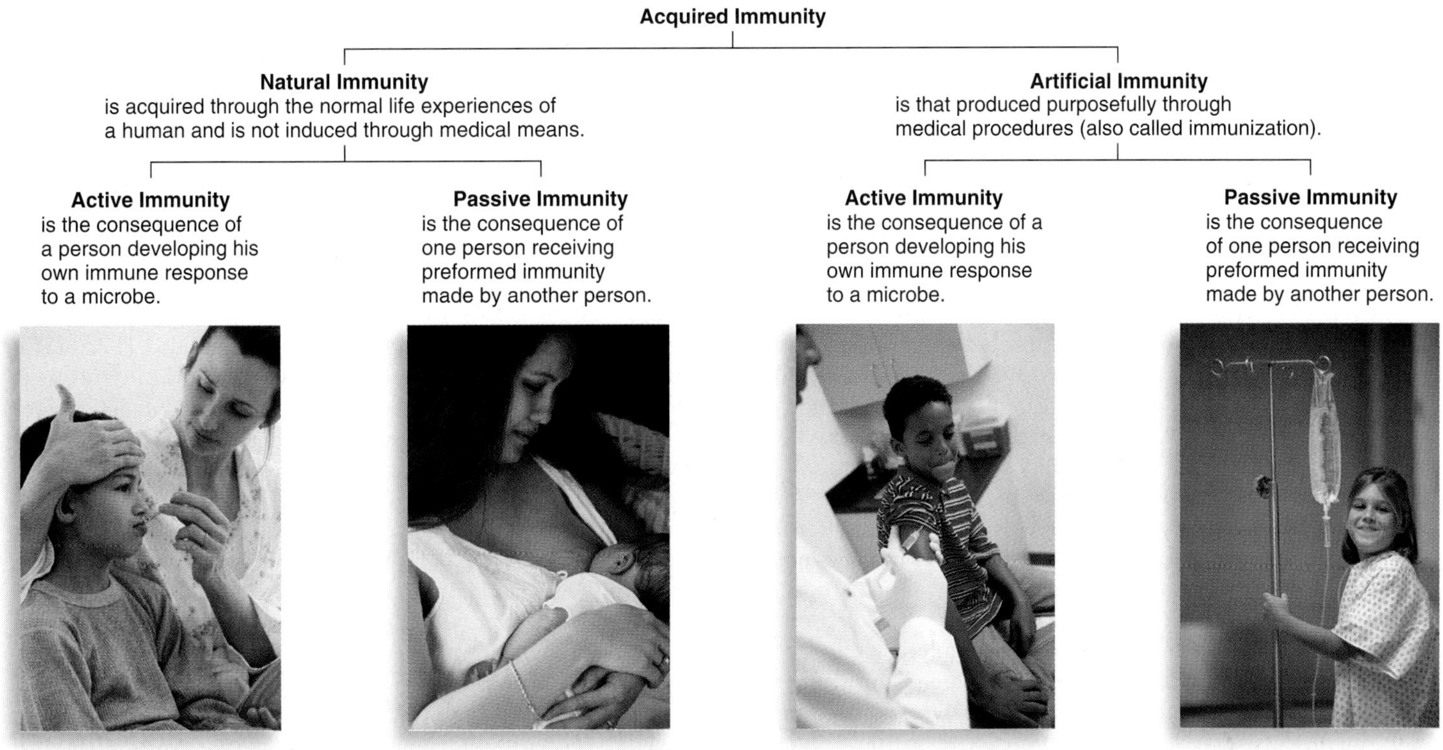

Acquired Immunity
Natural Immunity
is acquired through the normal life experiences of a human and is not induced through medical means.
Artificial Immunity
is that produced purposefully through medical procedures (also called immunization).
Active Immunity
is the consequence of a person developing his own immune response to a microbe.
Passive Immunity
is the consequence of one person receiving preformed immunity made by another person.
Active Immunity
is the consequence of a person developing his own immune response to a microbe.
Passive Immunity
is the consequence of one person receiving preformed immunity made by another person.

Figure 15.18 Categories of acquired immunities.
Natural immunities, which occur during the normal course of life, are either active (acquired from an infection and then recovering) or passive (antibodies donated by the mother to her child). Artificial immunities are acquired through medical practices and can be active (vaccinations with antigen to stimulate an immune response) or passive (immune therapy with a serum containing antibodies).

diphtheria, pertussis, and several viruses regularly cross the placenta. This natural mechanism provides an infant with a mixture of many maternal antibodies that can protect it for the first few critical months outside the womb, while its own immune system is gradually developing active immunity. Depending upon the microbe, passive protection lasts anywhere from a few months to a year. But eventually, the infant's body clears the antibody. Most childhood vaccinations are timed so that there is no lapse in protection against common childhood infections.

Another source of natural passive immunity comes to the baby by way of mother's milk **(Insight 15.2).** Although the human infant acquires 99% of natural passive immunity in utero and only about 1% through nursing, the milk-borne antibodies provide a special type of intestinal protection that is not available from transplacental antibodies.

Artificial Immunity: Immunization

Immunization is any clinical process that produces immunity in a subject. Because it is often used to give advance protection against infection, it is also called *immunoprophylaxis.* The use of these terms is sometimes imprecise, thus it should be stressed that active immunization, in which a person is administered antigen, is synonymous with vaccination, and that passive immunization, in which a person is given antibodies, is a type of immune therapy.

Vaccination: Artificial Active Immunization

The term *vaccination* originated from the Latin word *vacca* (cow), because the cowpox virus was used in the first preparation for active immunization against smallpox (see Insight 15.3). Vaccination exposes a person to a specially prepared microbial (antigenic) stimulus, which then triggers the immune system to produce antibodies and lymphocytes to protect the person upon future exposure to that microbe. As with natural active immunity, the degree and length of protection vary. Commercial vaccines are currently available for many diseases. Also, vaccines are being developed for cancer. Some of these vaccines are not targeted to microbes at all. Also, some are designed to be used as treatment rather than prevention (see A Note About New Vaccines on page 470).

Immunotherapy: Artificial Passive Immunization

In immunotherapy, a patient at risk for acquiring a particular infection is administered a preparation that contains specific antibodies against that infectious agent. In the past, these therapeutic substances were obtained by vaccinating animals (horses, in particular), then taking blood and extracting the serum. However, horse serum is now used only in limited situations because of the potential for hypersensitivity

INSIGHT 15.2 *Historical*

Breast Feeding: The Gift of Antibodies

An advertising slogan from the past claims that cow's milk is "nature's most nearly perfect food." One could go a step further and assert that human milk is nature's *perfect* food for young humans. Clearly, it is loaded with essential nutrients, not to mention being available on demand from a readily portable, hygienic container that does not require refrigeration or warming. But there is another and perhaps even greater benefit. During lactation, the breast becomes a site for the proliferation of lymphocytes that produce IgA, a special class of antibody that protects the mucosal surfaces from local invasion by microbes. The very earliest secretion of the breast, a thin, yellow milk called **colostrum,** is very high in IgA. These antibodies form a protective coating in the gastrointestinal tract of a nursing infant that guards against infection by a number of enteric pathogens (*Escherichia coli, Salmonella,* poliovirus, rotavirus). Protection at this level is especially critical because an infant's own IgA and natural intestinal barriers are not yet developed. As with immunity in utero, the necessary antibodies will be donated only if the mother herself has active immunity to the microbe through a prior infection or vaccination.

In recent times, the ready availability of artificial formulas and the changing lifestyles of women have reduced the incidence of breast feeding. Where adequate hygiene and medical care prevail, bottle-fed infants get through the critical period with few problems, because the foods given them are relatively sterile and they have received protection against some childhood infections in utero. Mothers in developing countries with untreated water supplies or poor medical services are strongly discouraged from using prepared formulas, because they can actually inoculate the baby's intestine with pathogens from the formula. Millions of neonates suffer from severe and life-threatening diarrhea that could have been prevented by the hygienic practice of nursing.

In the mid-1900s, baby formula manufacturers tried to introduce the widespread use of prepared formula to developing countries. They provided free samples of formula, and once babies were weaned from breast milk, mothers were forced to buy formula. The health effects were so damaging that in the 1980s the World Health Organization issued an "International Code on the Marketing of Breast Milk Substitutes," discouraging formula use in the developing world.

to it. Pooled human serum from donor blood (gamma globulin) and immune serum globulins containing high quantities of antibodies are more frequently used. Immune serum globulins are used to protect people who have been exposed to hepatitis, measles, and rubella. More specific immune serum, obtained from patients recovering from a recent infection, is useful in preventing and treating hepatitis B, rabies, pertussis, and tetanus.

An outline summarizing the system of host defenses covered in chapters 14 and 15 was presented in figure 14.1. You may want to use this resource to review major aspects of immunity and to guide you in answering certain questions (see multiple-choice question 10).

CHECKPOINT

- T cells do not produce antibodies. Instead, they produce different cytokines that play diverse roles in the immune response. Each subset of T cell produces a distinct set of cytokines that stimulate lymphocytes or destroy foreign cells.
- Active immunity means that your body produces antibodies to a disease agent. If you contract the disease, you can develop natural active immunity. If you are vaccinated, your body will produce artificial active immunity.
- In passive immunity, you receive antibodies from another person. Natural passive immunity comes from the mother. Artificial passive immunity is administered medically.

15.8 Immunization: Methods of Manipulating Immunity for Therapeutic Purposes

Methods that actively or passively immunize people are widely used in disease prevention and treatment. In the case of passive immunization, a patient is given preformed antibodies, which is actually a form of **immunotherapy.** In the case of active immunization, a patient is vaccinated with a microbe or its antigens, providing a form of advance protection.

Passive Immunization

As mentioned earlier, the first attempts at passive immunization involved the transfusion of horse serum containing antitoxins to prevent tetanus and to treat patients exposed to diphtheria. Since then, antisera from animals have been replaced with products of human origin that function with various degrees of specificity. Immune serum globulin (ISG), sometimes called *gamma globulin,* contains immunoglobulin extracted from the pooled blood of at least 1,000 human donors. The method of processing ISG concentrates the antibodies to increase potency and eliminates potential pathogens (such as the hepatitis B and HIV viruses). It is a treatment of choice in preventing measles and hepatitis A and in replacing antibodies in immunodeficient patients. Most forms of ISG are injected intramuscularly to minimize adverse reactions, and the protection it provides lasts 2 to 3 months.

INSIGHT 15.3 Historical

The Lively History of Active Immunization

The basic notion of immunization has existed for thousands of years. It probably stemmed from the observation that persons who had recovered from certain communicable diseases rarely if ever got a second case. Undoubtedly, the earliest crude attempts involved bringing a susceptible person into contact with a diseased person or animal. The first recorded attempt at immunization occurred in sixth-century China. It consisted of drying and grinding up smallpox scabs and blowing them with a straw into the nostrils of vulnerable family members. By the 10th century, this practice had changed to the deliberate inoculation of dried pus from the smallpox pustules of one patient into the arm of a healthy person, a technique later called **variolation** (variola is the smallpox virus). This method was used in parts of the Far East for centuries before Lady Mary Montagu brought it to England in 1721. Although the principles of the technique had some merit, unfortunately many recipients and their contacts died of smallpox. This outcome vividly demonstrates a cardinal rule for a workable vaccine: It must contain an antigen that will provide protection but not cause the disease. Variolation was so controversial that any English practitioner caught doing it was charged with a felony.

Eventually, this human experimentation paved the way for the first really effective vaccine, developed by the English physician Edward Jenner in 1796. Jenner conducted the first scientifically controlled study, one that had a tremendous impact on the advance of medicine. His work gave rise to the words **vaccine** and *vaccination* (from L., *vacca*, cow), which now apply to any immunity obtained by inoculation with selected antigens. Jenner was inspired by the case of a dairymaid who had been infected by a pustular infection called cowpox. This related virus afflicts cattle but causes a milder condition in humans. She explained that she and other milkmaids had remained free of smallpox. Other residents of the region expressed a similar confidence in the cross-protection of cowpox. To test the effectiveness of this new vaccine, Jenner prepared material from human cowpox lesions and inoculated a young boy. When challenged 2 months later with an injection of crusts from a smallpox patient, the boy proved immune.

Detail from "The Cowpock," an 1808 etching that caricatured the worst fears of the English public concerning Edward Jenner's smallpox vaccine.

Jenner's discovery—that a less pathogenic agent could confer protection against a more pathogenic one—is especially remarkable in view of the fact that microscopy was still in its infancy and the nature of viruses was unknown. At first, the use of the vaccine was regarded with some fear and skepticism (see illustration). When Jenner's method proved successful and word of its significance spread, it was eventually adopted in many other countries. Eventually, the original virus mutated into a unique strain *vaccinia* (virus) that became the basis of the current vaccine. In 1973, the World Health Organization declared that smallpox had been eradicated. As a result, smallpox vaccination had been discontinued until recently, due to the threat of bioterrorism.

Other historical developments in vaccination included using heat-killed bacteria in vaccines for typhoid fever, cholera, and plague and techniques for using neutralized toxins for diphtheria and tetanus. Throughout the history of vaccination, there have been vocal opponents and minimizers, but numbers do not lie. Whenever a vaccine has been introduced, the prevalence of that disease has declined dramatically.

A preparation called specific immune globulin (SIG) is derived from a more defined group of donors. Companies that prepare SIG obtain serum from patients who are convalescing and in a hyperimmune state after such infections as pertussis, tetanus, chickenpox, and hepatitis B. These globulins are preferable to ISG because they contain higher titers of specific antibodies obtained from a smaller pool of patients. Although useful for prophylaxis in persons who have been exposed or may be exposed to infectious agents, these sera are often limited in availability.

When a human immune globulin is not available, antisera and antitoxins of animal origin can be used. Sera produced in horses are available for diphtheria, botulism, and spider and snake bites. Unfortunately, the presence of horse antigens can stimulate allergies such as serum sickness or anaphylaxis (see chapter 16). Although donated immunities only last a relatively short time, they act immediately and can protect patients for whom no other useful medication or vaccine exists.

Artificial Active Immunity: Vaccination

Active immunity can be conferred artificially by **vaccination**—exposing a person to material that is antigenic but not pathogenic. The discovery of vaccination was one of the farthest reaching and most important developments in medical science **(Insight 15.3).** The basic principle behind vaccination is to stimulate a primary and secondary anamnestic response that primes the immune system for future exposure to a virulent pathogen. If this pathogen enters the body, the immune response will be immediate, powerful, and sustained.

Vaccines have profoundly reduced the prevalence and impact of many infectious diseases that were once common and often deadly. In this section, we survey the principles of

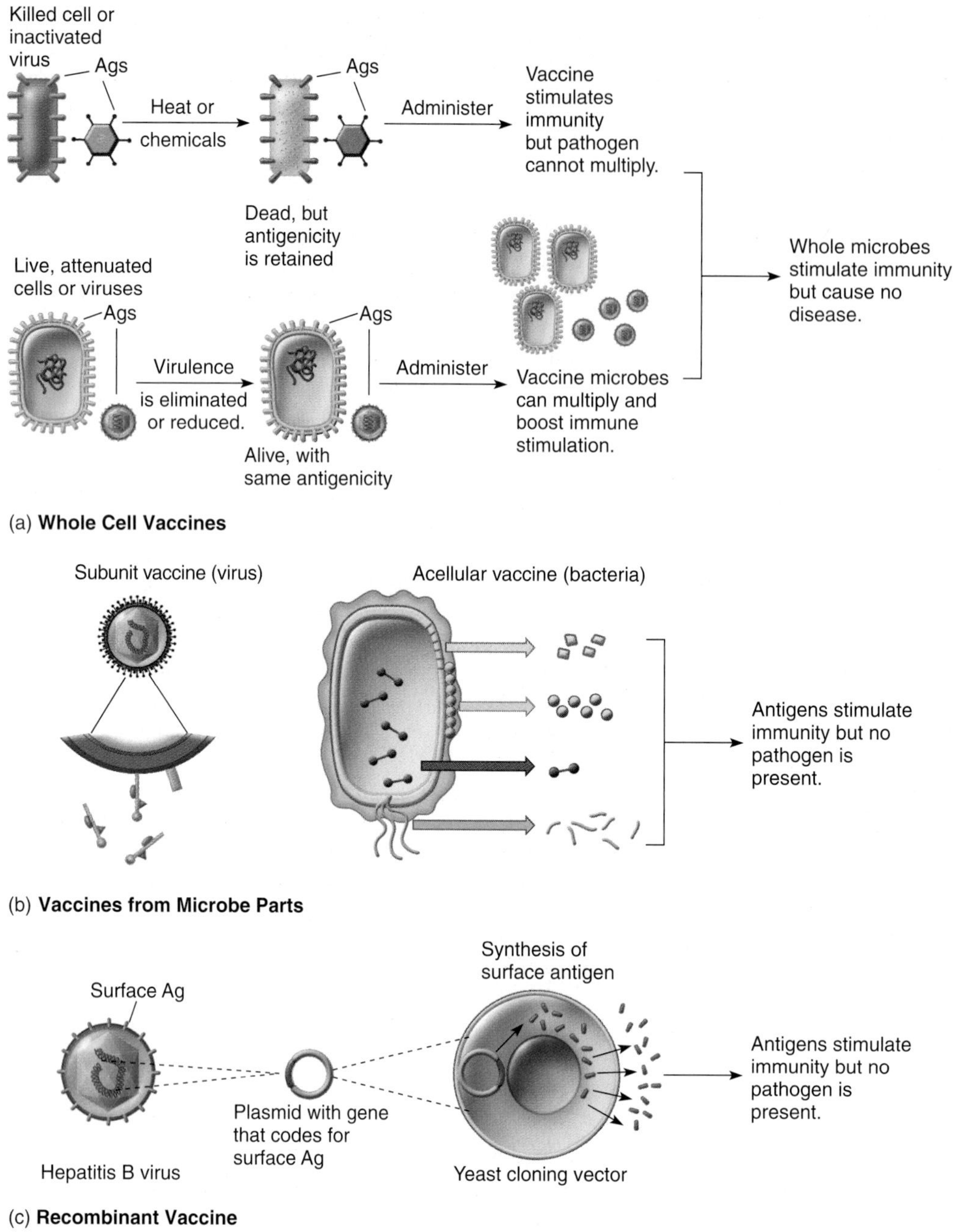

Figure 15.19 Strategies in vaccine design.
(a) Whole cells or viruses, killed or attenuated. **(b)** Acellular or subunit vaccines are made by disrupting the microbe to release various molecules or cell parts that can be isolated and purified. **(c)** Recombinant vaccines are made by isolating a gene for the antigen from the pathogen (here a hepatitis virus) and splicing it into a plasmid. Insertion of the recombinant plasmid into a cloning host (yeast) results in the production of large amounts of viral surface antigen to use in vaccine preparation.

vaccine preparation and important considerations surrounding vaccine indication and safety. (Vaccines are also given specific consideration in later chapters on infectious diseases and organ systems.)

Principles of Vaccine Preparation

A vaccine must be considered from the standpoints of antigen selection, effectiveness, ease in administration, safety, and cost. In natural immunity, an infectious agent stimulates a relatively long-term protective response. In artificial active immunity, the objective is to obtain this same response with a modified version of the microbe or its components. Qualities of an effective vaccine are listed in **table 15.4.** Most vaccine preparations are based on one of the following antigen preparations **(figure 15.19):**

1. killed whole cells or inactivated viruses;
2. live, attenuated cells or viruses;
3. antigenic molecules derived from bacterial cells or viruses; or
4. genetically engineered microbes or microbial antigens.

TABLE 15.4 Checklist of Requirements for an Effective Vaccine

- It should have a low level of adverse side effects or toxicity and not cause serious harm.
- It should protect against exposure to natural, wild forms of pathogen.
- It should stimulate both antibody (B-cell) response and cell-mediated (T-cell) response.
- It should have long-term, lasting effects (produce memory).
- It should not require numerous doses or boosters.
- It should be inexpensive, have a relatively long shelf life, and be easy to administer.

A survey of the major licensed vaccines and their indications is presented in **table 15.5.**

Large, complex antigens such as whole cells or viruses are very effective immunogens. Depending on the vaccine, these are either killed or attenuated. **Killed or inactivated vaccines** are prepared by cultivating the desired strain or strains of a bacterium or virus and treating them with formalin, radiation, heat, or some other agent that does not destroy antigenicity. One type of vaccine for the bacterial disease cholera is of this type (see chapter 22). The Salk polio vaccine and one form of influenza vaccine contain inactivated viruses. Because the microbe does not multiply, killed vaccines often require a larger dose and more boosters to be effective.

A number of vaccines are prepared from live, **attenuated** microbes. Attenuation is any process that substantially lessens or negates the virulence of viruses or bacteria. It is usually achieved by modifying the growth conditions or manipulating microbial genes in a way that eliminates virulence factors. Attenuation methods include long-term cultivation, selection of mutant strains that grow at colder temperatures (cold mutants), passage of the microbe through unnatural hosts or tissue culture, and removal of virulence genes. The vaccine for tuberculosis (BCG) was obtained after 13 years of subculturing the agent of bovine tuberculosis. Vaccines for measles, mumps, polio (Sabin), and rubella contain live, nonvirulent viruses. The advantages of live preparations are:

1. Viable microorganisms can multiply and produce infection (but not disease) like the natural organism.
2. They confer long-lasting protection.
3. They usually require fewer doses and boosters than other types of vaccines.
4. They are particularly effective at inducing cell-mediated immunity.

Disadvantages of using live microbes in vaccines are that they require special storage facilities, can be transmitted to other people, and can conceivably mutate back to a virulent strain.

If the exact epitopes that stimulate immunity are known, it is possible to produce a vaccine based on a selected component of a microorganism. These vaccines for bacteria are called **acellular** or **subcellular vaccines.** For viruses, they are called **subunit vaccines.** The antigen used in these vaccines may be taken from cultures of the microbes, produced by genetic engineering or synthesized chemically.

Examples of component antigens currently in use are the capsules of the pneumococcus and meningococcus, the protein surface antigen of anthrax, and the surface proteins of hepatitis B virus. A special type of vaccine is the **toxoid,** which consists of a purified bacterial exotoxin that has been chemically denatured. By eliciting the production of antitoxins that can neutralize the natural toxin, toxoid vaccines provide protection against toxinoses such as diphtheria and tetanus.

CASE FILE 15 *WRAP-UP*

Measles is caused by a paramyxovirus, a member of the genus *Morbillivirus.* It is a highly infectious disease and is transmitted by respiratory droplets. In otherwise healthy individuals, measles produces a skin rash and mild to moderate upper respiratory symptoms (see chapter 18). Serious disease is rare but can produce subacute sclerosing panencephalitis, which often results in death. Crowding, malnutrition, and lack of medical care in the camps in Darfur and Chad created an ideal environment for a serious epidemic. From March to June 2004, there were 725 reported cases of measles in the region resulting in over 100 deaths. Health experts feared the worst was yet to come.

In June 2004, a massive immunization effort was initiated. More than 6,000 workers staffed over 1,500 immunization centers. Mobile centers were dispatched to remote locations. Their efforts were successful in immunizing about 77% of the susceptible population. Increased stability and security in the region will help the FMOH in its plan to continue the measles immunization program. This continued effort has the potential to make measles a disease of the past in the region.

See CDC. 2004. Emergency measles control activities—Darfur, Sudan, October 2004. MMWR *53(38):897–899.*

Development of New Vaccines

Despite considerable successes, dozens of bacterial, viral, protozoan, and fungal diseases still remain without a functional vaccine. At the present time, no reliable vaccines are available for malaria, HIV/AIDS, various diarrheal diseases, respiratory diseases, and worm infections that affect over 200 million people per year worldwide. Worse than that, even those vaccines that are available are out of reach for much of the world's population **(Insight 15.4).**

One group of infections that are difficult to design a vaccine for is the latent or persistent viral infections, such as some herpesviruses and cytomegaloviruses. In these cases, the host's natural immunity is not capable of clearing the infection so artificial immunity must actually outperform the host's response to a natural infection. This has proved to be very difficult. Of all of the challenges facing vaccine specialists, probably the most difficult has been choosing vaccine antigens that are safe and that properly stimulate immunity.

TABLE 15.5 Currently Approved Vaccines

Disease/Vaccine Preparation	Route of Administration	Recommended Usage/Comments
Contain killed whole bacteria		
Cholera	Subcutaneous (SQ) injection	For travelers; effect not long term
Plague	SQ	For exposed individuals and animal workers; variable protection
Contain live, attenuated bacteria		
Tuberculosis (BCG)	Intradermal (ID) injection	For high-risk occupations only; protection variable
Typhoid	Oral	For travelers only; low rate of effectiveness
Acellular vaccines (capsular polysaccharides or proteins)		
Anthrax	SQ	For protection in military recruits, occupationally exposed
Meningitis (meningococcal)	SQ	For protection in high-risk infants, military recruits; short duration
Meningitis (*Haemophilus influenzae*)	IM	For infants and children; may be administered with DTaP
Pneumococcal pneumonia	IM or SQ	Important for people at high risk: the young, elderly, and immunocompromised; moderate protection
Pertussis (aP)	IM	For newborns and children; contains recombinant protein antigens
Toxoids (formaldehyde-inactivated bacterial exotoxins)		
Diphtheria	IM	A routine childhood vaccination; highly effective in systemic protection
Tetanus	IM	A routine childhood vaccination; highly effective
Pertussis	IM	A routine childhood vaccination; highly effective
Botulism	IM	Only for exposed individuals such as laboratory personnel
Contain inactivated whole viruses		
Poliomyelitis (Salk)	IM	Routine childhood vaccine; now used as first choice
Rabies	IM	For victims of animal bites or otherwise exposed; effective
Influenza	IM	For high-risk populations; requires constant updating for new strains; immunity not durable
Japanese encephalitis	SQ	For those residing in endemic areas, lab workers
Hepatitis A	IM	Protection for travelers, institutionalized people
Contain live, attenuated viruses		
Adenovirus infection	Oral	For immunizing military recruits
Measles (rubeola)	SQ	Routine childhood vaccine; very effective
Mumps (parotitis)	SQ	Routine childhood vaccine; very effective
Poliomyelitis (Sabin)	Oral	Routine childhood vaccine; very effective but can cause polio
Rubella	SQ	Routine childhood vaccine; very effective
Chickenpox (varicella)	SQ	Routine childhood vaccine; immunity can diminish over time; same vaccine under name Zostavax® now approved to prevent shingles in older adults
Smallpox (live vaccinia virus, not attenuated variola)	MP*	Since 2003, offered on voluntary basis for health care workers, some military
Yellow fever	SQ	Travelers, military personnel in endemic areas
Influenza	Inhaled	Same as for inactivated
Rotavirus	Oral	2006, replaces withdrawn vaccine
Subunit viral vaccines		
Hepatitis B	IM	Recommended for all children, starting at birth; also for health workers and others at risk
Influenza	IM	See influenza, inactivated, above
Recombinant vaccines		
Hepatitis B	IM	Used more often than subunit, but for same groups
Pertussis	IM	See acellular above
Human papilloma virus	IM	Approved in 2006

*MP = multiple puncture method.

INSIGHT 15.4 Discovery

They Said It Couldn't Be Done

Two major factors make it difficult for the developing world to enjoy the life-saving benefits of vaccines, and they both involve *scarcity*. The first is a scarcity of refrigeration in many areas of the world; the second is a scarcity of money in these developing countries.

In 2003, the Bill and Melinda Gates Foundation contributed hundreds of millions of dollars to research that explored ways to make vaccines easier to administer and more heat-resistant. One scientist, Dr. Jeffery Griffiths at Tufts University, is working on a measles vaccine that requires no refrigeration and no needles—also a bonus because clean needles can be in short supply in the developing world. As it is now, half a million children in the world die of measles every year, even though the developed world has been largely spared this disease since the vaccine became available in the 1960s. At only 10 cents a dose, cost is not the problem with this vaccine. But it can't last for more than a week without refrigeration. Dr. Griffiths hopes to change this.

Another vaccine that is badly needed in the developing world is the hepatitis B vaccine. This one can cost up to $25 a dose and, therefore, is prohibitively expensive in many countries. One man decided to tackle this problem head-on. Krishna Ella is a native of India but had been educated in the United States and was working here as a molecular biologist. He designed a new method for purifying the hepatitis B surface protein that resulted in much less waste and much greater efficiency than the one used by the current manufacturer. He tried to secure money to set up a production facility in India, promising that he could produce the vaccine for $1 a dose. When he proposed his idea to investors, they laughed him out of their offices.

Undeterred, Dr. Ella and his wife sold everything, borrowed money from friends and colleagues, and moved back to India to realize their dream. When they tried to secure support in India, banks and investors questioned why an Indian expatriate with such a successful career in the United States would return to India. They wondered if he had run into trouble with the law, for instance. Eventually, the couple overcame these obstacles and started a company that did, in fact, produce a low-cost, effective hepatitis B vaccine. It costs pennies. The Indian government reports that without Dr. Ella's vaccine, no one in India would be getting vaccinated against hepatitis B.

Now the Ellas' company has received funding from the Bill and Melinda Gates Foundation to develop a malaria vaccine and a cheap vaccine for rotavirus infection. These two diseases kill millions of people every year in poor countries.

Currently, much attention is being focused on newer strategies for vaccine preparation that employs antigen synthesis, recombinant DNA, and gene cloning technology.

When the exact composition of an epitope is known, it is sometimes possible to artificially synthesize it. This ability permits preservation of antigenicity while greatly increasing antigen purity and concentration. The malaria vaccine currently being used in areas of South America and Africa is composed of three synthetic peptides from the parasite. In recent years, biotechnology companies tried to use plants to mass produce vaccine antigens. Tomatoes, potatoes, and bananas were genetically engineered to synthesize proteins from cholera, hepatitis, papillomavirus, and *E. coli* pathogens. This strategy aimed to deliver vaccines to populations that otherwise would not have access to them, by making them part of the food supply. However, by 2006, most of these strategies were abandoned because of fears of these edible vaccine plants accidentally finding their way into the general food supply.

Genetically Engineered Vaccines

Some of the genetic engineering concepts introduced in chapter 10 offer novel approaches to vaccine development. These methods are particularly effective in designing vaccines for obligate parasites that are difficult or expensive to culture, such as the syphilis spirochete or the malaria parasite. This technology provides a means of isolating the genes that encode various microbial antigens, inserting them into plasmid vectors, and cloning them in appropriate hosts. The outcome of recombination can be varied as desired. For instance, the cloning host can be stimulated to synthesize and secrete a protein product (antigen), which is then harvested and purified **(figure 15.19*c*).** Certain vaccines for hepatitis are currently being prepared in this way. Antigens from the agents of syphilis, *Schistosoma*, and influenza have been similarly isolated and cloned and are currently being considered as potential vaccine material.

Another ingenious technique using genetic recombination has been nicknamed the *Trojan horse* vaccine. The term derives from an ancient legend in which the Greeks sneaked soldiers into the fortress of their Trojan enemies by hiding them inside a large, mobile wooden horse. In the microbial equivalent, genetic material from a selected infectious agent is inserted into a live carrier microbe that is nonpathogenic. In theory, the recombinant microbe will multiply and express the foreign genes, and the vaccine recipient will be immunized against the microbial antigens. Vaccinia, the virus originally used to vaccinate for smallpox, and adenoviruses have proved practical agents for this technique. Vaccinia is used as the carrier in one of the experimental vaccines for AIDS, herpes simplex 2, leprosy, and tuberculosis.

DNA vaccines are being hailed as the most promising of all of the newer approaches to immunization. The technique in these formulations is very similar to gene therapy as described in chapter 10, except in this case, microbial (not human) DNA is inserted into a plasmid vector and inoculated into a recipient **(figure 15.20).** The expectation is that the human cells will take up some of the plasmids and

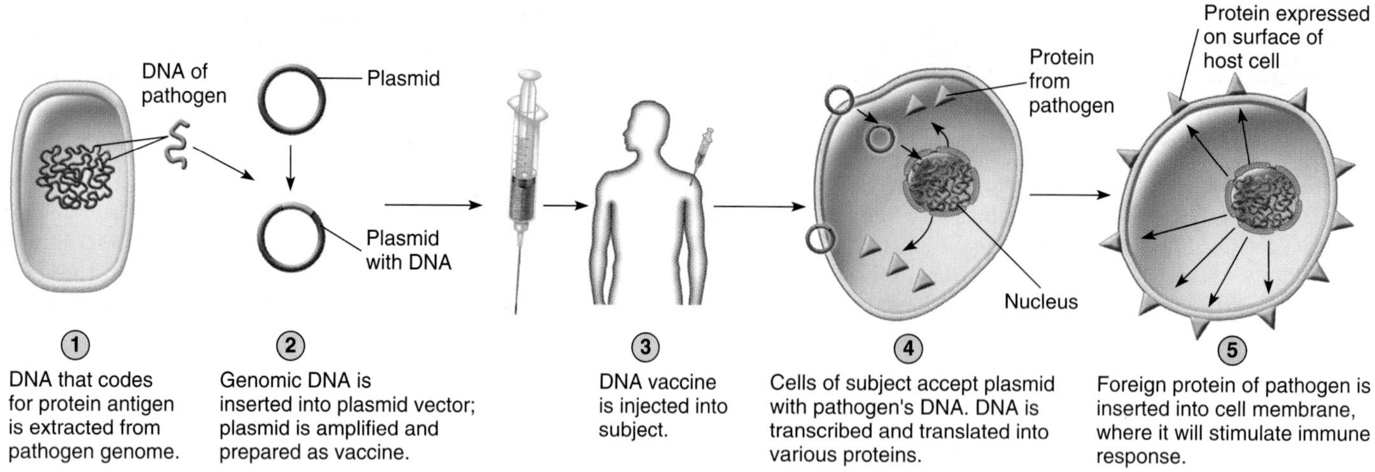
DNA of pathogen
Plasmid
Plasmid with DNA
Protein from pathogen
Protein expressed on surface of host cell
Nucleus
1
DNA that codes for protein antigen is extracted from pathogen genome.
2
Genomic DNA is inserted into plasmid vector; plasmid is amplified and prepared as vaccine.
3
DNA vaccine is injected into subject.
4
Cells of subject accept plasmid with pathogen's DNA. DNA is transcribed and translated into various proteins.
5
Foreign protein of pathogen is inserted into cell membrane, where it will stimulate immune response.

Process Figure 15.20 DNA vaccine preparation.
DNA vaccines contain all or part of the pathogen's DNA, which is used to "infect" a recipient's cells. Processing of the DNA leads to production of an antigen protein that can stimulate a specific response against that pathogen.

express the microbial DNA in the form of proteins. Because these proteins are foreign, they will be recognized during immune surveillance and cause B and T cells to be sensitized and form memory cells.

Experiments with animals have shown that these vaccines are very safe and that only a small amount of the foreign antigen need be expressed to produce effective immunity. Another advantage to this method is that any number of potential microbial proteins can be expressed, making the antigenic stimulus more complex and improving the likelihood that it will stimulate both antibody and cell-mediated immunity. At the present time, over 30 DNA-based vaccines are being tested in animals. Vaccines for HIV, Lyme disease, hepatitis C, herpes simplex, influenza, tuberculosis, papillomavirus, malaria, and SARS are undergoing clinical trials, most with encouraging results.

Concerns about potential terrorist release of Class A bioterrorism agents have created pressure to reintroduce or improve vaccines for smallpox, anthrax, botulism, plague, and tularemia. Smallpox vaccination is now being given on a voluntary basis to first responders (health care providers, the military). Earlier vaccines already exist for anthrax, tularemia, botulism, and plague, but they have been difficult to give or not effective. New vaccines are being developed and tested to immunize military and other personnel in the event of a disaster.

Route of Administration and Side Effects of Vaccines

Most vaccines are injected by subcutaneous, intramuscular, or intradermal routes. Oral vaccines are available for only a few diseases (see table 15.5), but they have some distinct advantages. An oral dose of a vaccine can stimulate protection (IgA) on the mucous membrane of the portal of entry. Oral vaccines are also easier to give, more readily accepted, and well tolerated.

Some vaccines require the addition of a special binding substance, or **adjuvant** (ad'-joo-vunt). An adjuvant is any compound that enhances immunogenicity and prolongs antigen retention at the injection site. The adjuvant precipitates the antigen and holds it in the tissues so that it will be released gradually. Its gradual release presumably facilitates contact with antigen-presenting cells and lymphocytes. Common adjuvants are alum (aluminum hydroxide salts), Freund's adjuvant (emulsion of mineral oil, water, and extracts of mycobacteria), and beeswax.

A NOTE ABOUT NEW VACCINES

Some new vaccines in development seem to violate two important principles of immunization: (1) they are not targeted at microbes at all, and (2) they are not designed to prevent a disease. The new "vaccines" are cancer vaccines. Note that two approved cancer vaccines do have the traits of traditional vaccines. In 2006, a vaccine for human papilloma virus (HPV) was approved. This vaccine can prevent cervical cancer, because most cervical cancers are caused by HPV. The vaccine against hepatitis B can also prevent liver cancer. In that sense, they bear the designation *cancer vaccines*. They prevent the cancer by preventing infection with cancer-causing viruses. But other vaccines are in development that are aimed at cancers that are not associated with microbial infection. A vaccine called Provenge was close to being approved in 2007, but at the last minute the FDA asked for more tests. This vaccine is therapeutic rather than preventive. It is intended for patients who already have prostate cancer. It stimulates the immune system to more effectively fight malignant cells. Studies found that it extended the lives of patients by several months.

It is clear that the definition of "vaccine" is changing due to medical advances.

Vaccines must go through many years of trials in experimental animals and human volunteers before they are licensed for general use. Even after they have been approved, like all therapeutic products, they are not without complications. The most common of these are local reactions at the injection site, fever, allergies, and other adverse reactions. Relatively rare reactions (about 1 case out of 220,000 vaccinations) are panencephalitis (from measles vaccine), back-mutation to a virulent strain (from polio vaccine), disease due to contamination with dangerous viruses or chemicals, and neurological effects of unknown cause (from pertussis and swine flu vaccines). Some patients experience allergic reactions to the medium (eggs or tissue culture) rather than to vaccine antigens. Some recent studies have attempted to link childhood vaccinations to later development of diabetes, asthma, and autism. After thorough examination of records, epidemiologists have found no convincing evidence for a vaccine connection to these diseases.

When known or suspected adverse effects have been detected, vaccines are altered or withdrawn. Recently, the whole-cell pertussis vaccine was replaced by the acellular capsule (aP) form when it was associated with adverse neurological effects. The first oral rotavirus vaccine had to be withdrawn when children experienced intestinal blockage. An improved version was licensed in 2006. Polio vaccine was switched from live oral to inactivated when too many cases of paralytic disease occurred from back-mutated vaccine stocks. Vaccine companies have also phased out certain preservatives, such as thimerosal, that are thought to cause allergies and other potential side effects.

Professionals involved in giving vaccinations must understand their inherent risks but also realize that the risks from the infectious disease almost always outweigh the chance of an adverse vaccine reaction. The greatest caution must be exercised in giving live vaccines to immunocompromised or pregnant patients, the latter because of possible risk to the fetus.

To Vaccinate: Why, Whom, and When?

Vaccination confers long-lasting, sometimes lifetime, protection in the individual, but an equally important effect is to protect the public health. Vaccination is an effective method of establishing **herd immunity** in the population. According to this concept, individuals immune to a communicable infectious disease will not harbor it, thus reducing the occurrence of that pathogen. With a larger number of immune individuals in a population (herd), it will be less likely that an unimmunized member of the population will encounter the agent. In effect, collective immunity through mass immunization confers indirect protection on the nonimmune members (such as children). Herd immunity maintained through immunization is an important force in preventing epidemics.

Until recently, vaccination was recommended for all typical childhood diseases for which a vaccine is available and for adults only in certain special circumstances (health workers, travelers, military personnel). It has become apparent to public health officials that vaccination of adults is often needed in order to boost an older immunization, protect against "adult" infections (such as pneumonia in elderly people), or provide special protection in people with certain medical conditions.

In **table 15.6,** the current recommended schedule for childhood and adolescent immunizations is provided, with complete footnotes from the Centers for Disease Control and Prevention. **Table 15.7** contains the recommended adult immunization schedule. Some vaccines are mixtures of antigens from several pathogens, notably Pediatrix (DTaP, IPV, and HB). It is also common for several vaccines to be given simultaneously, as occurs in military recruits who receive as many as 15 injections within a few minutes and children who receive boosters for DTaP and polio at the same time they receive the MMR vaccine. Experts doubt that immune interference (inhibition of one immune response by another) is a significant problem in these instances, and the mixed vaccines are carefully balanced to prevent this eventuality. The main problem with simultaneous administration is that side effects can be amplified.

In July of 2004, the Centers for Disease Control and Prevention reported that 79% of children in the United States were being vaccinated on time. While this represents a record high, it means that at least 1 million children in this country have not received adequate immunization.

CHECKPOINT

- Knowledge of the specific immune response has a practical application: commercial production of antisera and vaccines.
- Artificial passive immunity usually involves administration of antiserum and, occasionally, B and T cells. Antibodies collected from donors (human or otherwise) are injected into people who need protection immediately. Examples include ISG (immune serum globulin) and SIG (specific immune globulin).
- Artificial active agents are vaccines that provoke a protective immune response in the recipient but do not cause the actual disease. Vaccination is the process of challenging the immune system with a specially selected antigen. Examples are (1) killed or inactivated microbes; (2) live, attenuated microbes; (3) subunits of microbes; and (4) genetically engineered microbes or microbial parts.
- Vaccination programs seek to protect the individual directly through raising the antibody titer and indirectly through the development of herd immunity.

TABLE 15.6 Recommended Immunization Schedule United States · 2007

Persons Aged 0–6 Years

Vaccine ▼ / Age ▶	Birth	1 month	2 months	4 months	6 months	12 months	15 months	18 months	19–23 years	2–3 months	4–6 years
Hepatitis B	HepB	HepB				HepB				HepB Series	
Rotavirus			Rota	Rota	Rota						
Diphtheria, Tetanus, Pertussis			DTaP	DTaP	DTaP		DTaP				DTaP
Haemophilus influenzae type b			Hib	Hib	Hib	Hib			Hib		
Pneumococcal			PCV	PCV	PCV	PCV				PCV PPV	
Inactivated Poliovirus			IPV	IPV		IPV					IPV
Influenza								Influenza (Yearly)			
Measles, Mumps, Rubella						MMR					MMR
Varicella						Varicella					Varicella
Hepatitis A							HepA (2 doses)			HepA Series	
Meningococcal										MPSV4	

Persons Aged 7–18 Years

Vaccine ▼ / Age ▶	7–10 years	11–12 years	13–14 years	15 years	16–18 years
Diphtheria, Tetanus, Pertussis		Tdap		Tdap	
Human Papillomavirus		HPV (3 doses)		HPV Series	
Meningococcal	MPSV4	MCV4		MCV4 MCV4	
Pneumococcal		PPV			
Influenza		Influenza (Yearly)			
Hepatitis A		HepA Series			
Hepatitis B		HepB Series			
Inactivated Poliovirus		IPV Series			
Measles, Mumps, Rubella		MMR Series			
Varicella		Varicella Series			

Range of recommended ages

Catch-up immunization

Certain high-risk groups

TABLE 15.7 Recommended Immunization Schedule United States 2007

Adults

Vaccine ▼ / Age ▶	19–49 years	50–64 years	≥ 65 years
Diphtheria, Tetanus, Pertussis (Td/Tdap)	1–dose Td booster every 10 yrs Substitute 1 dose of Tdap for Td		
Human Papillomavirus (HPV)	3 doses (females)		
Measles, Mumps, Rubella (MMR)	1 or 2 doses	1 dose	
Varicella	2 doses (0, 4–8 wks)	2 doses (0, 4–8 wks)	
Influenza	1 dose annually	1 dose annually	
Pneumococcal (polysaccharide)	1–2 doses		1 dose
Hepatitis A	2 doses (0, 6–12 mos, or 0, 6–18 mos)		
Hepatitis B	3 doses (0, 1–2, 4–6 mos)		
Meningococcal	1 or more doses		

Chapter Summary with Key Terms

15.1 Specific Immunity: The Third and Final Line of Defense Development of Lymphocyte Specificity/Receptors

Acquired immunity involves the reactions of B and T lymphocytes to foreign molecules, or **antigens.** Before they can react, each lymphocyte must undergo differentiation into its final functional type by developing protein receptors for antigen, the specificity of which is randomly generated and unique for each lymphocyte.

15.2 An Overview of Specific Immune Responses

A. Genetic recombination and mutation during embryonic and fetal development produce billions of different lymphocytes, each bearing a different receptor.
B. *Tolerance to self* occurs during this time.
C. The antigen receptors on B cells are **immunoglobulin (Ig)** molecules, and receptors on T cells are unrelated glycoprotein molecules.
D. T-cell receptors bind epitopes on MHC molecules, which in humans are sometimes called **human leukocyte antigens (HLA).**

15.3 The Lymphocyte Response System in Depth

An antigen (Ag), also known as immunogen, is any substance that stimulates an immune response.

A. Requirements for **antigenicity** include foreignness (recognition as nonself), large size, and complexity of cell or molecule.
B. Foreign molecules less than 1,000 MW **(haptens)** are not antigenic unless attached to a larger carrier molecule.
C. The epitope is the small molecular group of the foreign substance that is recognized by lymphocytes. Cells, viruses, and large molecules can have numerous epitopes.

15.4 Cooperation in Immune Reactions to Antigens

A. T-cell-dependent antigens must be processed by large phagocytes such as dendritic cells or macrophages called **antigen-presenting cells (APCs).**
B. The presentation of a single antigen usually involves a direct collaboration between an APC, a T helper (T_H), and an antigen-specific B or T cell.

15.5 B-Cell Response

A. Once B cells process the antigen, interact with T_H cells, and are stimulated by B-cell growth and differentiation factors, they enter the cell cycle in preparation for mitosis and clonal expansion.
B. Divisions give rise to **plasma cells** that secrete antibodies and **memory cells** that can react to that same antigen later.
   1. *Nature of Antibodies* (Immunoglobulins): A single immunoglobulin molecule (monomer) is a large Y-shaped protein molecule consisting of four polypeptide chains. It contains two identical regions **(FAb)** with ends that form the active site that binds with a unique specificity to an antigen. The **Fc** region determines the location and the function of the antibody molecule.

2. *Antigen-Antibody (Ag-Ab) Reactions* include **opsonization, neutralization, agglutination,** and complement fixation.
3. The five **antibody classes,** which differ in size and function, are **IgG, IgA** (secretory Ab), **IgM, IgD,** and **IgE.**
4. *Antibodies in Serum* **(Antiserum):**
   a. The first introduction of an Ag to the immune system produces a **primary response,** with a gradual increase in Ab titer.
   b. The second contact with the same Ag produces a **secondary,** or **memory response,** due to memory cells produced during initial response.

15.6 T-Cell Response

A. T_H cells (CD4) secrete cytokines that activate macrophages to kill phagocytosed antigens and activate cytotoxic T (T_C) cells to kill infected cells and to become memory T_C cells. These cytokines suppress humoral immunity (antibody production).
B. T helper 2 (T_H2) CD4 cells activate B cells to make IgG, IgM, and IgE antibodies and become memory B cells. T_H2 cells suppress cellular immunity.
C. Cytotoxic T (T_C) CD8 cells recognize and kill infected cells, tumor cells, and foreign transplant cells.

15.7 A Practical Scheme for Classifying Specific Immunities

Immunities acquired through B and T lymphocytes can be classified by a simple system.

A. Natural immunity is acquired as part of normal life experiences.
B. Artificial immunity is acquired through medical procedures such as immunization.
C. **Active immunity** results when a person is challenged with antigen that stimulates production of antibodies. It creates memory, takes time, and is lasting.
D. In **passive immunity,** preformed antibodies are donated to an individual. It does not create memory, acts immediately, and is short term.

15.8 Immunization: Methods of Manipulating Immunity for Therapeutic Purposes

A. Passive immunotherapy includes administering immune serum globulin and specific immune globulins pooled from donated serum to prevent infection and disease in those at risk; antisera and antitoxins from animals are occasionally used.
B. Active immunization is synonymous with **vaccination;** provides an antigenic stimulus that does not cause disease but can produce long-lasting, protective immunity. **Vaccines** are made with
   1. **Killed** whole cells or **inactivated viruses** that do not reproduce but are antigenic.
   2. Live, **attenuated** cells or viruses that are able to reproduce but have lost virulence.
   3. **Acellular** or **subunit** components of microbes such as surface antigen or neutralized toxins **(toxoids).**
   4. Genetic engineering techniques, including cloning of antigens, recombinant attenuated microbes, and **DNA**-based **vaccines.**
C. Boosters (additional doses) are often required.
D. Vaccination increases herd immunity, protection provided by mass immunity in a population.

Multiple-Choice and True-False Questions

Multiple-Choice Questions. Select the correct answer from the answers provided.

1. The primary B-cell receptor is
   a. IgD c. IgE
   b. IgA d. IgG
2. In humans, B cells mature in the ____ and T cells mature in the ____.
   a. GALT, liver c. bone marrow, thymus
   b. bursa, thymus d. lymph nodes, spleen
3. Small, simple molecules are ____ antigens.
   a. poor c. good
   b. never d. heterophilic
4. The cross-linkage of antigens by antibodies is known as
   a. opsonization c. agglutination
   b. a cross-reaction d. complement fixation
5. T____ cells assist in the functions of certain B cells and other T cells.
   a. sensitized c. helper
   b. cytotoxic d. natural killer
6. T_C cells are important in controlling
   a. virus infections c. autoimmunity
   b. allergy d. all of these
7. Which of the following can serve as antigen-presenting cells (APCs)?
   a. T cells d. dendritic cells
   b. B cells e. b, c, and d
   c. macrophages
8. A vaccine that contains parts of viruses is called
   a. acellular c. subunit
   b. recombinant d. attenuated
9. Widespread immunity that protects the population from the spread of disease is called
   a. seropositivity c. epidemic prophylaxis
   b. cross-reactivity d. herd immunity
10. **Multiple Matching.** (Summarizes information from chapters 14 and 15 [see figure 14.1].) In the blanks on the left, place the letters of all of the host defenses and immune responses in the right column that can fit the description.

____ vaccination for tetanus	a. active
____ lysozyme in tears	b. passive
____ immunization with horse serum	c. natural
____ in utero transfer of antibodies	d. artificial

____ booster injection for diphtheria
____ recovery from a case of mumps
____ colostrum
____ interferon
____ action of neutrophils
____ injection of gamma globulin
____ recovery from a case of mumps
____ edema
____ humans having protection from canine distemper virus
____ stomach acid
____ cilia in trachea
____ asymptomatic chickenpox
____ complement

e. acquired
f. innate, inborn
g. chemical barrier
h. mechanical barrier
i. genetic barrier
j. specific
k. nonspecific
l. inflammatory response
m. second line of defense
n. none of these

True-False Questions. If statement is true, leave as is. If it is false, correct it by rewriting the sentence.

11. Antibodies are secreted by monocytes.
12. Vaccination could be described as artificial passive immunity.
13. IgE antibodies are found in body secretions.
14. The process of reducing the virulence of microbes so that they can be used in vaccines is called denaturation.
15. An adjuvant is used to increase the response to an immunogen.

Writing to Learn

These questions are suggested as a *writing-to-learn* experience. For each question, compose a one- or two-paragraph answer that includes the factual information needed to completely address the question.

1. a. What function do receptors play in specific immune responses?
   b. How can receptors be made to vary so widely?
2. Describe the major histocompatibility complex, and explain how it participates in immune reactions.
3. a. Explain the clonal selection theory of antibody specificity and diversity.
   b. Why must the body develop tolerance to self?
4. Describe three ways that B cells and T cells are similar and at least five major ways in which they are different.
5. a. Describe the actions of an antigen-presenting cell.
   b. What is the difference between a T-cell-dependent and a T-cell-independent response?
6. a. Trace the immune response system, beginning with the entry of a T-cell-dependent antigen, antigen processing, presentation, the cooperative response among the macrophage and lymphocytes, and the reactions of activated B and T cells.
   b. What are the actions of interleukins-1 and -2?
7. a. On what basis is a particular B-cell clone selected?
   b. How are B cells activated, and what events are involved in this process?
   c. What happens after B cells are activated?
   d. What are the functions of plasma cells, clonal expansion, and memory cells?
8. a. Describe the structure of immunoglobulin.
   b. What are the functions of the Fab and Fc portions?
   c. Describe four or five ways that antibodies function in immunity.
   d. Describe the attachment of Abs to Ags. (What eventually happens to the Ags?)
9. a. Contrast the primary and secondary response to Ag.
   b. Explain the type, order of appearance, and amount of immunoglobulin in each response and the reasons for them.
   c. What causes the latent period? The anamnestic response?
10. a. Describe the concept of herd immunity.
    b. How does vaccination contribute to its development in a community?
    c. Give some possible explanations for recent epidemics of diphtheria and whooping cough.

Concept Mapping

Appendix D provides guidance for working with concept maps.

1. Supply your own linking lines, as well as the linking words or phrases, in this concept map, and provide the missing concepts in the empty boxes.

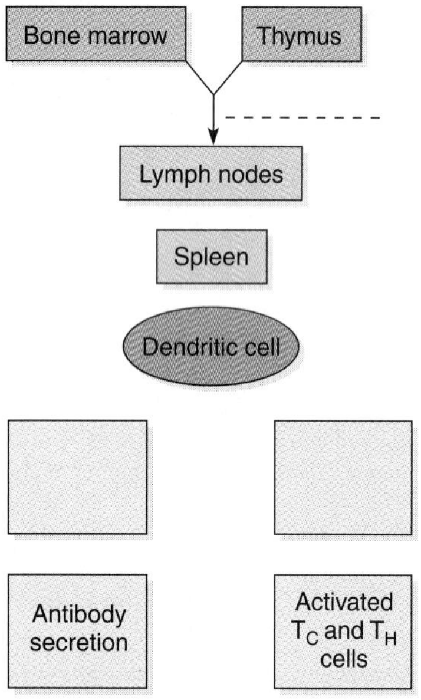

2. Construct your own concept map using the following words as the *concepts*. Supply the links and the linking words between each pair of concepts.

active immunity
passive immunity
natural immunity
artificial immunity
innate immunity
vaccines
interferon
inflammation
memory

Critical Thinking Questions

Critical thinking is the ability to reason and solve problems using facts and concepts. These questions can be approached from a number of angles, and in most cases, they do not have a single correct answer.

1. What is the advantage of having lymphatic organs screen the body fluids, directly and indirectly?
2. Cells contain built-in suicide genes to self-destruct by apoptosis under certain conditions. Can you explain why development of the immune system might depend in part on this sort of adaptation?
3. a. Give some possible explanations for the need to have immune tolerance.
   b. Why would it be necessary for the T cells to bind both antigen and self (MHC) receptors?
4. Explain how it is possible for people to give a false-positive reaction in blood tests for syphilis, AIDS, and infectious mononucleosis.
5. Explain why most immune reactions result in a polyclonal collection of antibodies.
6. Describe the relationship between an antitoxin, a toxin, and a toxoid.
7. It is often said that a vaccine does not prevent infection; rather, it primes the immune system to undergo an immediate response to prevent an infection from spreading. Explain what is meant by this statement, and outline what is happening at the cellular/molecular level from the time of vaccination until subsequent contact with the infectious agent actually occurs.
8. a. Determine the vaccines you have been given and those for which you will require periodic boosters.
   b. Suggest vaccines you may need in the future.
9. When traders and missionaries first went to the Hawaiian Islands, the natives there experienced severe disease and high mortality rates from smallpox, measles, and certain STDs.
   a. Explain what factors are involved in the sudden outbreaks of disease in previously unexposed populations.
   b. Explain the ways in which vaccination has been responsible for the worldwide eradication of diseases such as smallpox and polio.

Visual Understanding

1. **Figure 15.14.** What would be the benefit in giving pooled human gamma globulin to a health care worker following an accidental needlestick? Explain your answer.

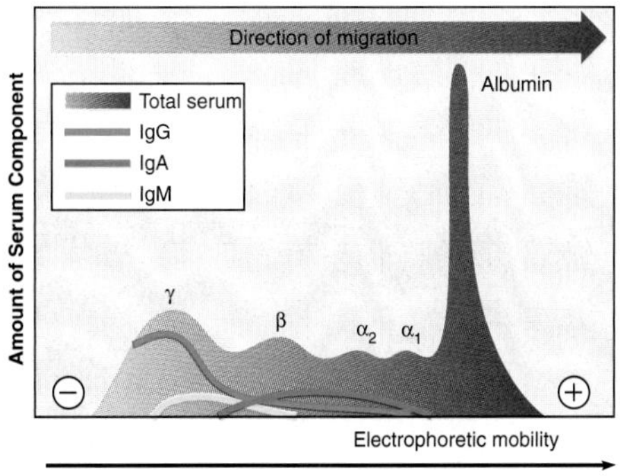

Internet Search Topics

1. Find information on new monoclonal antibodies being developed and used for medical purposes. Write a short description summarizing your research into this topic.
2. Go to: www.aris.mhhe.com, and click on "microbiology" and then this textbook's author/title. Go to chapter 15, access the URLs listed under Internet Search Topics, and research the following:

   Access the microbiology website listed. Review the diagrams, animations, and interactive images.
3. Search the Internet for information about multiple myeloma. What component of the immune system does this disease affect? What impact does the disease have on an individual's overall immunity?

CHAPTER 16

Disorders in Immunity

CASE FILE 16

In July of 2002, a 55-year-old woman was brought into a Lane County, Oregon, hospital emergency room. She was unconscious, with low blood pressure and an irregular heart rhythm. Emergency room personnel observed that she had hives all over her body, that she had swelling or edema in her arms and legs, and that her tongue was swollen.

The woman's daughter reported that she and her mother had been doing yard work outside, trimming a hedge beside the mother's house. Her mother gave a sudden shout, took three steps, and collapsed. The daughter immediately called 911, and paramedics arrived within 5 minutes.

Soon after doctors started attending to the woman, her heart stopped. After 30 minutes of treatment, her heart rhythm returned, but the time without blood flow caused her to suffer a massive stroke and she never regained consciousness. Her family elected to disconnect her from life support 4 days later.

▶ *What happened to the woman that led to her death? Why would this event cause death?*

Case File 16 Wrap-Up appears on page 487.

CHAPTER OVERVIEW

- The immune system is subject to several types of dysfunctions termed immunopathologies.
- Some dysfunctions are due to abnormally heightened or incorrectly targeted responses (allergies, hypersensitivities, and autoimmunities).
- Some dysfunctions are due to the reduction or loss in protective immune reactions (immunodeficiencies).
- Some immune damage is caused by normal actions that are directed at foreign tissues placed in the body for therapy (transfusions and transplants).
- Hypersensitivities are divided into immediate, antibody-mediated, immune complex, and delayed allergies.
- Allergens are the foreign molecules that cause a hypersensitive or allergic response.
- Most hypersensitivities require an initial sensitizing event, followed by a later contact that causes symptoms.
- The immediate type of allergy is mediated by special types of B cells, IgE, and mast cells that release allergic chemicals such as histamine that stimulate symptoms.
- Examples of immediate allergies are atopy, asthma, food allergies, and anaphylaxis.
- Hypersensitivity can arise from the action of other antibodies (IgG and IgM) that fix complement and lyse foreign cells.
- Immune complex reactions are caused by large amounts of circulating antibodies accumulating in tissues and organs.
- T-cell responses to certain allergens and foreign molecules cause the diseases known as delayed-type hypersensitivities and graft rejection.

- Autoimmune diseases, such as rheumatoid arthritis and multiple sclerosis, are due to B and T cells that are abnormally sensitized to react with the body's natural molecules and thus can damage cells and tissues.
- Immunodeficiencies occur when B and T cells and other immune cells are missing or destroyed. They may be inborn and genetic or acquired.
- The primary outcome of immunodeficiencies is manifested in recurrent infections and lack of immune competence.

16.1 The Immune Response: A Two-Sided Coin

Humans possess a powerful and intricate system of defense, which by its very nature also carries the potential to cause injury and disease. In most instances, a defect in immune function is expressed in commonplace but miserable symptoms such as those of hay fever and dermatitis. But abnormal or undesirable immune functions are also actively involved in debilitating or life-threatening diseases such as asthma, anaphylaxis, rheumatoid arthritis, and graft rejection.

With few exceptions, our previous discussions of the immune response have centered around its numerous beneficial effects. The precisely coordinated system that seeks out, recognizes, and destroys an unending array of foreign materials is clearly protective, but it also presents another side—a side that promotes rather than prevents disease. In this chapter, we survey **immunopathology,** the study of disease states associated with overreactivity or underreactivity of the immune response **(figure 16.1).** In the cases of allergies and *autoimmunity,* the

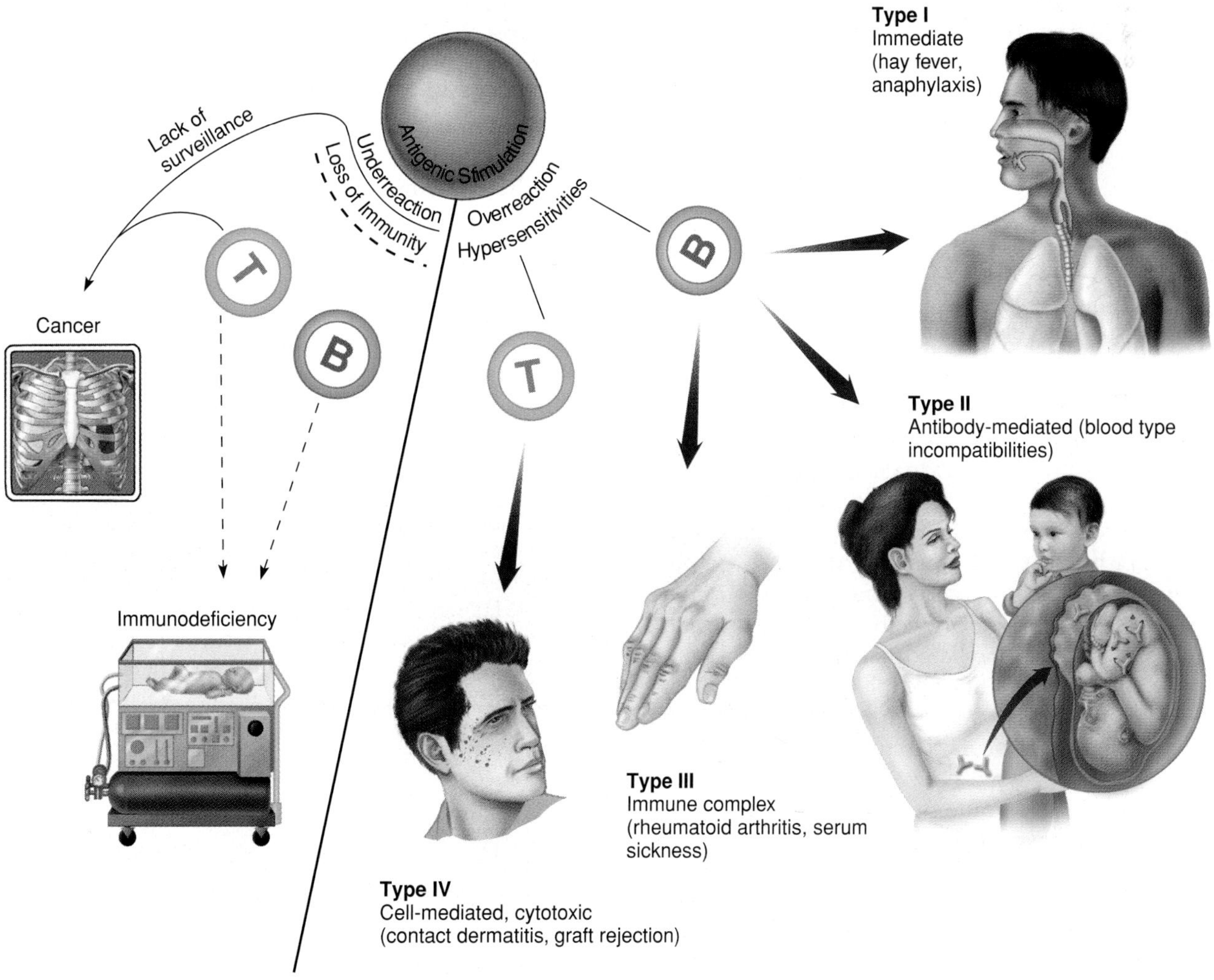

Figure 16.1 Overview of diseases of the immune system.
Just as the system of T cells and B cells provides necessary protection against infection and disease, the same system can cause serious and debilitating conditions by overreacting or underreacting to immune stimuli.

TABLE 16.1 Hypersensitivity States

Type		Systems and Mechanisms Involved	Examples
I	Immediate hypersensitivity	IgE-mediated; involves mast cells, basophils, and allergic mediators	Anaphylaxis, allergies such as hay fever, asthma
II	Antibody-mediated	IgG, IgM antibodies act upon cells with complement and cause cell lysis; includes some autoimmune diseases	Blood group incompatibility, pernicious anemia; myasthenia gravis
III	Immune complex-mediated	Antibody-mediated inflammation; circulating IgG complexes deposited in basement membranes of target organs; includes some autoimmune diseases	Systemic lupus erythematosus; rheumatoid arthritis; serum sickness; rheumatic fever
IV	T-cell-mediated	Delayed hypersensitivity and cytotoxic reactions in tissues	Infection reactions; contact dermatitis; graft rejection; some types of autoimmunity

tissues are innocent bystanders attacked by immunologic functions that can't distinguish one's own tissues from those expressing foreign material. In **grafts** and **transfusions,** a recipient reacts to the foreign tissues and cells of another individual. In **immunodeficiency** diseases, immune function is incompletely developed, suppressed, or destroyed. Cancer falls into a special category, because it is both a cause and an effect of immune dysfunction. As we shall see, one fascinating by-product of studies of immune disorders has been our increased understanding of the basic workings of the immune system.

Overreactions to Antigens: Allergy/Hypersensitivity

The term **allergy** means a condition of altered reactivity or exaggerated immune response that is manifested by inflammation. Although it is sometimes used interchangeably with hypersensitivity, some experts refer to immediate reactions such as hay fever as allergies and to delayed reactions as hypersensitivities. Allergic individuals are acutely sensitive to repeated contact with antigens, called **allergens,** that do not noticeably affect nonallergic individuals. Although the general effects of hypersensitivity are detrimental, we must be aware that it involves the very same types of immune reactions as those at work in protective immunities. These include humoral and cell-mediated actions, the inflammatory response, phagocytosis, and complement. Such an association means that all humans have the potential to develop hypersensitivity under particular circumstances.

Originally, allergies were defined as either immediate or delayed, depending upon the time lapse between contact with the allergen and onset of symptoms. Subsequently, they were differentiated as humoral versus cell-mediated. But as information on the nature of the allergic immune response accumulated, it became evident that, although useful, these schemes oversimplified what is really a very complex spectrum of reactions. The most widely accepted classification, first introduced by immunologists P. Gell and R. Coombs, includes four major categories: type I ("common" allergy and anaphylaxis), type II (IgG- and IgM-mediated cell damage), type III (immune complex), and type IV (delayed hypersensitivity) **(table 16.1).** In general, types I, II, and III involve a B-cell–immunoglobulin response, and type IV involves a T-cell response (see figure 16.1). The antigens that elicit these reactions can be exogenous, originating from outside the body (microbes, pollen grains, and foreign cells and proteins), or endogenous, arising from self tissue (autoimmunities).

One of the reasons allergies are easily mistaken for infections is that both involve damage to the tissues and thus trigger the inflammatory response, as described in chapter 14. Many symptoms and signs of inflammation (redness, heat, skin eruptions, edema, and granuloma) are prominent features of allergies.

CHECKPOINT

- Immunopathology is the study of diseases associated with excesses and deficiencies of the immune response. Such diseases include allergies, autoimmunity, grafts, transfusions, immunodeficiency disease, and cancer.
- An allergy or hypersensitivity is an exaggerated immune response that injures or inflames tissues.
- There are four categories of hypersensitivity reactions: type I (allergy and anaphylaxis), type II (IgG and IgM tissue destruction), type III (immune complex reactions), and type IV (delayed hypersensitivity reactions).
- Antigens that trigger hypersensitivity reactions are allergens. They can be either exogenous (originate outside the host) or endogenous (involve the host's own tissue).

16.2 Type I Allergic Reactions: Atopy and Anaphylaxis

All type I allergies share a similar physiological mechanism, are immediate in onset, and are associated with exposure to specific antigens. However, it is convenient to recognize two levels of severity: **Atopy** is any chronic local allergy such as hay fever or asthma; **anaphylaxis** (an″-uh-fih-lax′-us) is a systemic, sometimes fatal reaction that involves airway obstruction and circulatory collapse. In the following sections, we consider the epidemiology of type I allergies, allergens and routes of inoculation, mechanisms of disease, and specific syndromes.

Epidemiology and Modes of Contact with Allergens

Allergies exert profound medical and economic impact. Allergists (physicians who specialize in treating allergies) estimate that about 10% to 30% of the population is prone to atopic allergy. It is generally acknowledged that self-treatment with over-the-counter medicines accounts for significant underreporting of cases. The 35 million people afflicted by hay fever (15% to 20% of the population) spend about half a billion dollars annually for medical treatment. The monetary loss due to employee debilitation and absenteeism is immeasurable. The majority of type I allergies are relatively mild, but certain forms such as asthma and anaphylaxis may require hospitalization and can cause death. Millions of people in the United States suffer from asthma.

The predisposition for type I allergies has a strong familial association. Be aware that what is hereditary is a generalized *susceptibility*, not the allergy to a specific substance. For example, a parent who is allergic to ragweed pollen can have a child who is allergic to cat hair. The prospect of a child's developing atopic allergy is at least 25% if one parent is atopic, increasing up to 50% if grandparents or siblings are also afflicted. The actual basis for atopy appears to be a genetic program that favors allergic antibody (IgE) production, increased reactivity of mast cells, and increased susceptibility of target tissue to allergic mediators. Allergic persons often exhibit a combination of syndromes, such as hay fever, eczema, and asthma.

Other factors that affect the presence of allergy are age, infection, and geographic locale. New allergies tend to crop up throughout an allergic person's life, especially as new exposures occur after moving or changing lifestyle. In some persons, atopic allergies last for a lifetime; others "outgrow" them, and still others suddenly develop them later in life. Some features of allergy are not yet completely explained.

The Nature of Allergens and Their Portals of Entry

As with other antigens, allergens have certain immunogenic characteristics. Not unexpectedly, proteins are more allergenic than carbohydrates, fats, or nucleic acids. Some allergens are haptens, nonproteinaceous substances with a molecular weight of less than 1,000 that can form complexes with carrier molecules in the body (shown in figure 15.8). Organic and inorganic chemicals found in industrial and household products, cosmetics, food, and drugs are commonly of this type. **Table 16.2** lists a number of common allergenic substances.

Allergens typically enter through epithelial portals in the respiratory tract, gastrointestinal tract, and skin. The mucosal surfaces of the gut and respiratory system present a thin, moist surface that is normally quite penetrable. The dry, tough keratin coating of skin is less permeable, but access still occurs through tiny breaks, glands, and hair follicles. It is worth noting that the organ of allergic expression may or may not be the same as the portal of entry.

Airborne environmental allergens such as pollen, house dust, dander (shed skin scales), or fungal spores are termed *inhalants*. Each geographic region harbors a particular combination of airborne substances that varies with the season and humidity **(figure 16.2*a*)**. Pollen, the most common offender, is given off seasonally by the reproductive structures of pines and flowering plants (weeds, trees, and grasses). Unlike pollen, mold spores are released throughout the year and are especially profuse in moist areas of the home and garden. Airborne animal hair and dander (skin flakes), feathers, and the saliva of dogs and cats are common sources of allergens. The component of house dust that appears to account for most dust allergies is not soil or other debris but the decomposed bodies and feces of tiny mites that commonly live in this dust **(figure 16.2*b*)**. Some people are allergic to their work, in the sense that they are exposed to allergens on the job. Examples include florists, woodworkers, farmers, drug processors, welders, and plastics manufacturers whose work can aggravate inhalant and contact allergies.

Allergens that enter by mouth, called *ingestants*, often cause food allergies: *Injectant* allergies are an important adverse side effect of drugs or other substances used in diagnosing, treating, or preventing disease. A natural source of injectants is venom from stings by hymenopterans, a family of insects that includes honeybees and wasps. *Contactants* are allergens that enter through the skin. Many contact allergies are of the type IV, delayed variety discussed later in this chapter. It is also possible to be exposed to certain allergens, penicillin among them, during sexual intercourse due to the presence of allergens in the semen.

Mechanisms of Type I Allergy: Sensitization and Provocation

What causes some people to sneeze and wheeze every time they step out into the spring air, while others suffer no ill

TABLE 16.2 Common Allergens, Classified by Portal of Entry

Inhalants	Ingestants	Injectants	Contactants
Pollen	Food	Hymenopteran	Drugs
Dust	(milk, peanuts,	venom (bee,	Cosmetics
Mold spores	wheat, shellfish,	wasp)	Heavy metals
Dander	soybeans,	Drugs	Detergents
Animal hair	nuts, eggs,	Vaccines	Formalin
Insect parts	fruits)	Serum	Rubber
Formalin	Food additives	Enzymes	Glue
Drugs	Drugs (aspirin,	Hormones	Solvents
Enzymes	penicillin)		Dyes

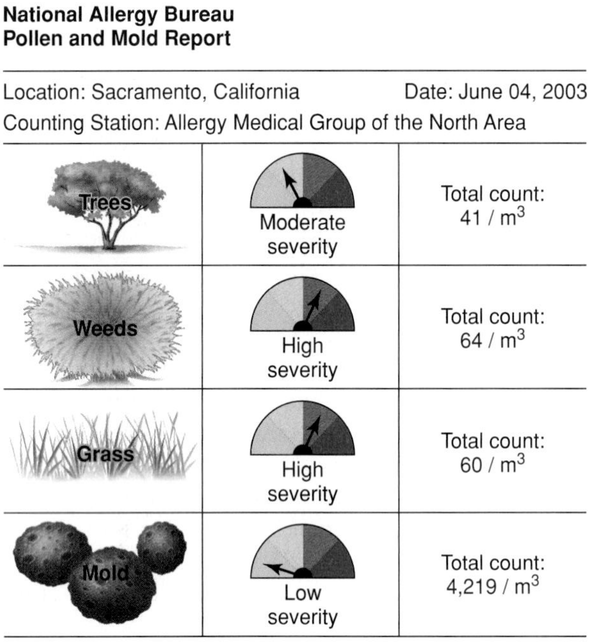
National Allergy Bureau
Pollen and Mold Report
Location: Sacramento, California
Date: June 04, 2003
Counting Station: Allergy Medical Group of the North Area
Trees
Moderate severity
Total count: 41 / m3
Weeds
High severity
Total count: 64 / m3
Grass
High severity
Total count: 60 / m3
Mold
Low severity
Total count: 4,219 / m3
(a)

(b)

(c)

Figure 16.2 **Monitoring airborne allergens.**
(a) The air in heavily vegetated places with a mild climate is especially laden with allergens such as pollen and mold spores. These counts vary seasonally. **(b)** Because the dust mite *Dermatophagoides* feeds primarily on human skin cells in house dust, these mites are found in abundance in bedding and carpets. Airborne mite feces and particles from their bodies are an important source of allergies. **(c)** Scanning electron micrograph of a single pollen grain from a rose (6,000x). Millions of these are released from a single flower.

effects? In order to answer this question, we must examine what occurs in the tissues of the allergic individual that does not occur in the normal person. In general, type I allergies develop in stages **(figure 16.3).** The initial encounter with an allergen provides a **sensitizing dose** that primes the immune system for a subsequent encounter with that allergen but generally elicits no signs or symptoms. The memory cells and immunoglobulin are then ready to react with a subsequent provocative dose of the same allergen. It is this dose that precipitates the signs and symptoms of allergy. Despite numerous anecdotal reports of people showing an allergy upon first contact with an allergen, it is generally believed that these individuals unknowingly had contact at some previous time. Fetal exposure to allergens from the mother's bloodstream is one possibility, and foods can be a prime source of "hidden" allergens such as penicillin.

The Physiology of IgE-Mediated Allergies

During primary contact and sensitization, the allergen penetrates the portal of entry **(figure 16.3*a*).** When large particles such as pollen grains, hair, and spores encounter a moist membrane, they release molecules of allergen that pass into the tissue fluids and lymphatics. The lymphatics then carry the allergen to the lymph nodes, where specific clones of B cells recognize it, are activated, and proliferate into plasma cells. These plasma cells produce immunoglobulin E (IgE), the antibody of allergy. IgE is different from other immunoglobulins in having an Fc region with great affinity for mast cells and basophils. The binding of IgE to these cells in the tissues sets the scene for the reactions that occur upon repeated exposure to the same allergen **(figure 16.3*b*).**

The Role of Mast Cells and Basophils

The most important characteristics of mast cells and basophils relating to their roles in allergy are:

1. Their ubiquitous location in tissues. Mast cells are located in the connective tissue of virtually all organs, but particularly high concentrations exist in the lungs, skin, gastrointestinal tract, and genitourinary tract. Basophils circulate in the blood but migrate readily into tissues.
2. Their capacity to bind IgE during sensitization (see figure 16.3). Each cell carries 30,000 to 100,000 cell receptors that bind 10,000 to 40,000 IgE antibodies.
3. Their cytoplasmic granules (secretory vesicles), which contain physiologically active cytokines (histamine, serotonin—introduced in chapter 14).
4. Their tendency to **degranulate** (see figures 16.3*b* and 16.4), or release the contents of the granules into the tissues when triggered by a specific allergen through the IgE bound to them.

Let us now see what occurs when sensitized cells are challenged with allergen a second time.

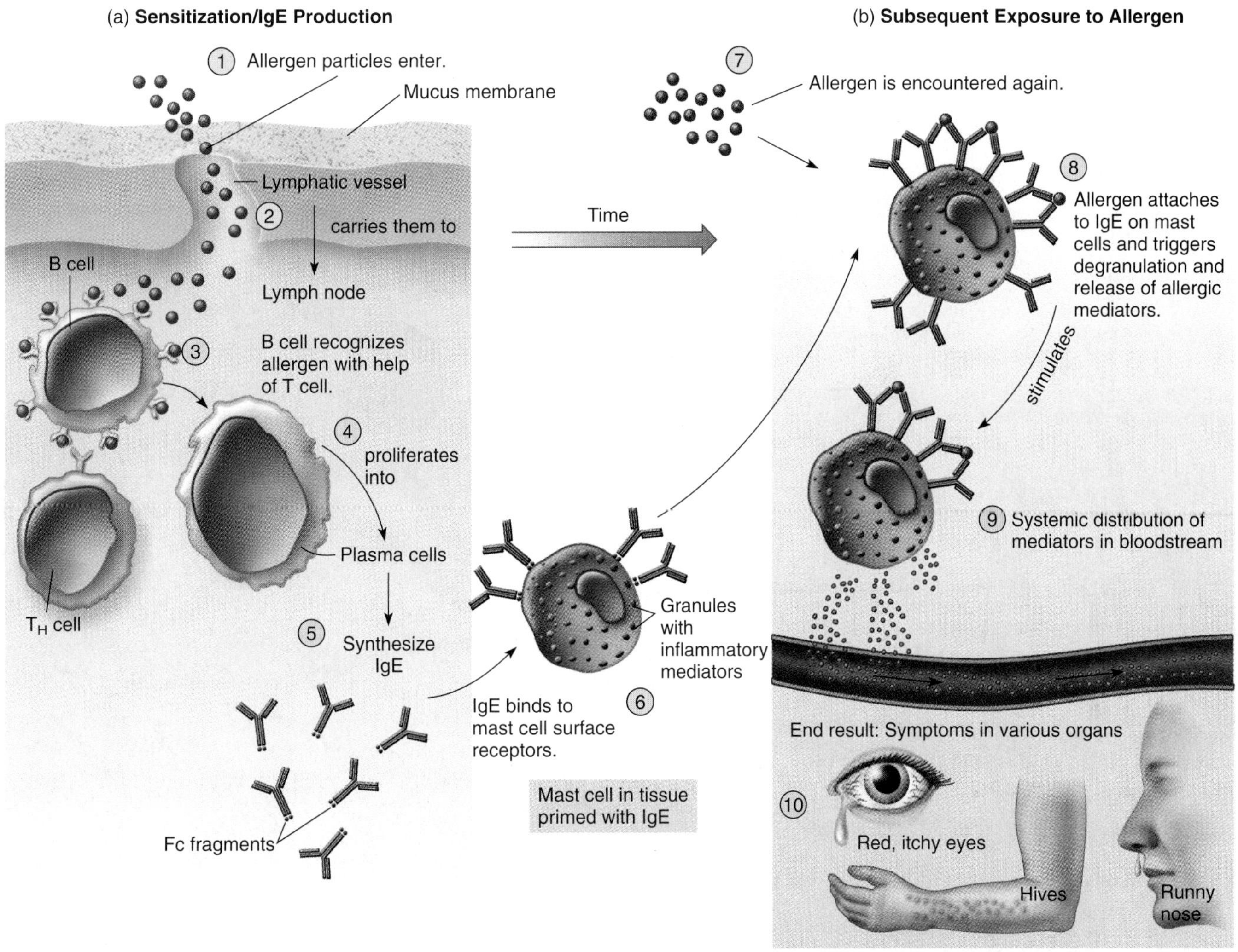
(a) Sensitization/IgE Production
(1) Allergen particles enter.
Mucus membrane
Lymphatic vessel
(2) carries them to
Lymph node
B cell
(3) B cell recognizes allergen with help of T cell.
(4) proliferates into
Plasma cells
T_H cell
(5) Synthesize IgE
Fc fragments
IgE binds to mast cell surface receptors. (6)
Granules with inflammatory mediators
Mast cell in tissue primed with IgE
Time
(b) Subsequent Exposure to Allergen
(7) Allergen is encountered again.
(8) Allergen attaches to IgE on mast cells and triggers degranulation and release of allergic mediators.
stimulates
(9) Systemic distribution of mediators in bloodstream
End result: Symptoms in various organs
(10)
Red, itchy eyes
Hives
Runny nose

Process Figure 16.3 A schematic view of cellular reactions during the type I allergic response.
(a) Sensitization (initial contact with sensitizing dose), (1)–(6). **(b)** Provocation (later contacts with provocative dose), (7)–(10).

The Second Contact with Allergen

After sensitization, the IgE-primed mast cells can remain in the tissues for years. Even after long periods without contact, a person can retain the capacity to react immediately upon reexposure. The next time allergen molecules contact these sensitized cells, they bind across adjacent receptors and stimulate degranulation. As chemical mediators are released, they diffuse into the tissues and bloodstream. Cytokines give rise to numerous local and systemic reactions, many of which appear quite rapidly (see figure 16.3*b*). The symptoms of allergy are not caused by the direct action of allergen on tissues but by the physiological effects of mast cell mediators on target organs.

Cytokines, Target Organs, and Allergic Symptoms

Numerous substances involved in mediating allergy (and inflammation) have been identified. The principal chemical mediators produced by mast cells and basophils are histamine, serotonin, leukotriene, platelet-activating factor, prostaglandins, and bradykinin **(figure 16.4).** These chemicals, acting alone or in combination, account for the tremendous scope of allergic symptoms. For some theories pertaining to this function of the allergic response, see **Insight 16.1.** Targets of these mediators include the skin, upper respiratory tract, gastrointestinal tract, and conjunctiva. The general responses of these organs include rashes, itching, redness, rhinitis, sneezing, diarrhea, and shedding of tears. Systemic targets include smooth muscle, mucus glands, and nervous tissue. Because smooth muscle is responsible for regulating the size of blood vessels and respiratory passageways, changes in its activity can profoundly alter blood flow, blood pressure, and respiration. Pain, anxiety, agitation, and lethargy are also attributable to the effects of mediators on the nervous system.

Histamine is the most profuse and fastest-acting allergic mediator. It is a potent stimulator of smooth muscle, glands, and eosinophils. Histamine's actions on smooth muscle vary with location. It *constricts* the smooth muscle layers of the

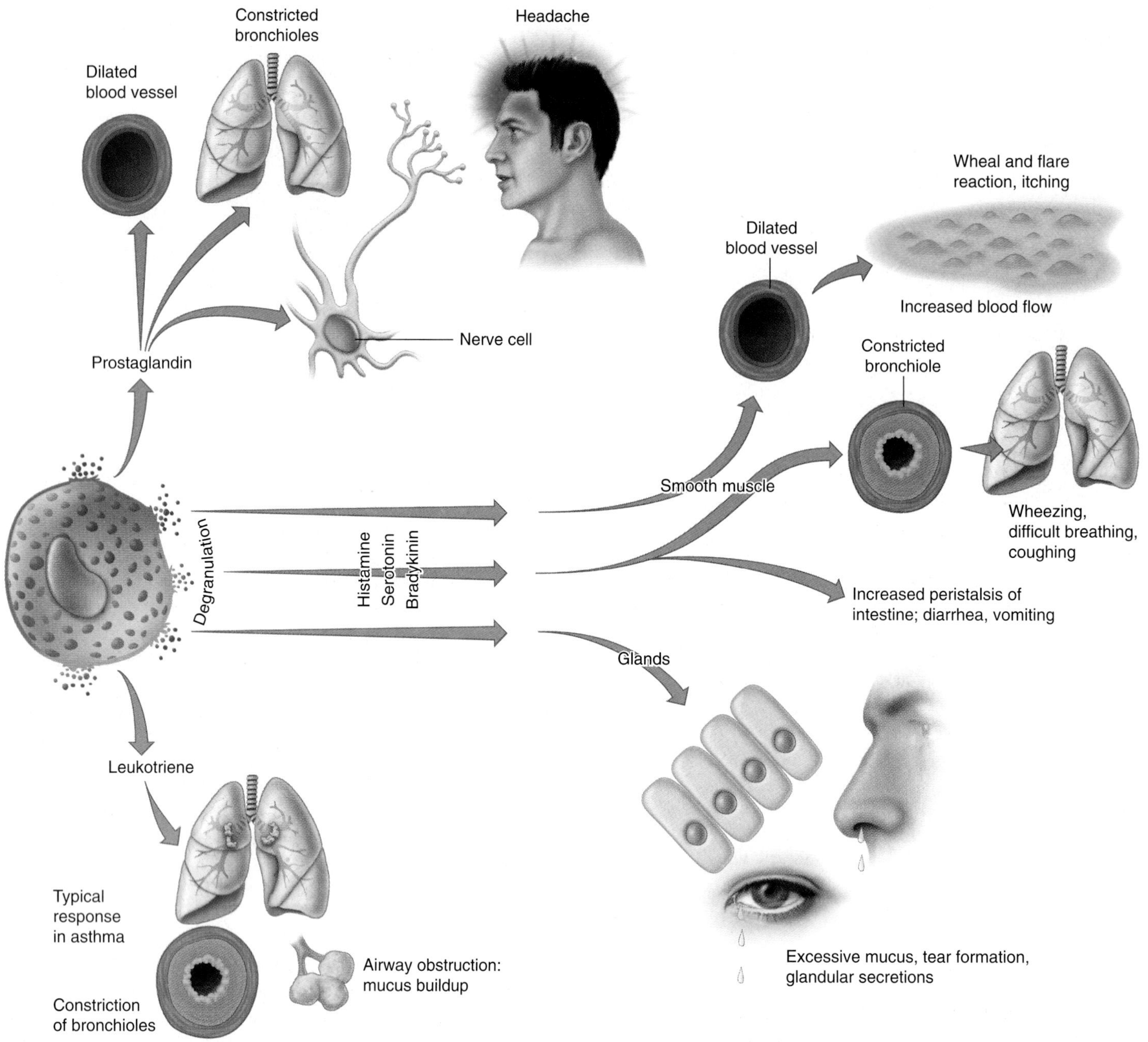

Figure 16.4 The spectrum of reactions to inflammatory cytokines released by mast cells and the common symptoms they elicit in target tissues and organs.
Note the extensive overlapping effects.

small bronchi and intestine, thereby causing labored breathing and increased intestinal motility. In contrast, histamine *relaxes* vascular smooth muscle and dilates arterioles and venules. It is responsible for the *wheal and flare* reaction in the skin (see figure 16.6), pruritis (itching), and headache. More severe reactions such as anaphylaxis can be accompanied by edema and vascular dilation, which lead to hypotension, tachycardia, circulatory failure, and, frequently, shock. Salivary, lacrimal, mucus, and gastric glands are also histamine targets.

Although the role of **serotonin** in human allergy is uncertain, its effects appear to complement those of histamine. In experimental animals, serotonin increases vascular permeability, capillary dilation, smooth muscle contraction, intestinal peristalsis, and respiratory rate, but it diminishes central nervous system activity.

Before the specific types were identified, **leukotriene** (loo″-koh-try′-een) was known as the "slow-reacting substance of anaphylaxis" for its property of inducing gradual contraction of smooth muscle. This type of leukotriene is responsible for the prolonged bronchospasm, vascular permeability, and mucus secretion of the asthmatic individual. Other leukotrienes stimulate the activities of polymorphonuclear leukocytes.

INSIGHT 16.1 *Medical*

Of What Value Is Allergy?

Why would humans and other mammals evolve an allergic response that is capable of doing so much harm and even causing death? It is unlikely that this limb of immunity exists merely to make people miserable; it must have a role in protection and survival. What are the underlying biological functions of IgE, mast cells, and the array of potent cytokines? Analysis has revealed that, although allergic persons have high levels of IgE, trace quantities are present even in the sera of nonallergic individuals, just as mast cells and inflammatory chemicals are also part of normal human physiology. It is generally believed that one important function of this system is to defend against helminth worms and other multicellular organisms that are ubiquitous human parasites. In chapter 14, you learned that inflammatory mediators serve valuable functions, such as increasing blood flow and vascular permeability to summon essential immune components to an injured site. They are also responsible for increased mucus secretion, gastric motility, sneezing, and coughing, which help expel noxious agents. The difference is that, in allergic persons, the quantity and quality of these reactions are excessive and uncontrolled.

Platelet-activating factor is a lipid released by basophils, neutrophils, monocytes, and macrophages. The physiological response to stimulation by this factor is similar to that of histamine, including increased vascular permeability, pulmonary smooth muscle contraction, pulmonary edema, hypotension, and a wheal and flare response in the skin.

Prostaglandins are a group of powerful inflammatory agents. Normally, these substances regulate smooth muscle contraction (for example, they stimulate uterine contractions during delivery). In allergic reactions, they are responsible for vasodilation, increased vascular permeability, increased sensitivity to pain, and bronchoconstriction. Certain anti-inflammatory drugs work by preventing the actions of prostaglandins.

Bradykinin is related to a group of plasma and tissue peptides known as kinins that participate in blood clotting and chemotaxis. In allergy, it causes prolonged smooth muscle contraction of the bronchioles, dilatation of peripheral arterioles, increased capillary permeability, and increased mucus secretion.

Specific Diseases Associated with IgE- and Mast-Cell-Mediated Allergy

The mechanisms just described are basic to hay fever, allergic asthma, food allergy, drug allergy, eczema, and anaphylaxis. In this section, we cover the main characteristics of these conditions, followed by methods of detection and treatment.

Atopic Diseases

Hay fever is a generic term for **allergic rhinitis,** a seasonal reaction to inhaled plant pollen or molds, or a chronic, year-round reaction to a wide spectrum of airborne allergens or inhalants (see table 16.2). The targets are typically respiratory membranes, and the symptoms include nasal congestion; sneezing; coughing; profuse mucus secretion; itchy, red, and teary eyes; and mild bronchoconstriction.

Asthma is a respiratory disease characterized by episodes of impaired breathing due to severe bronchoconstriction. The airways of asthmatic people are exquisitely responsive to minute amounts of inhalant allergens, food, or other stimuli, such as infectious agents. The symptoms of asthma range from occasional, annoying bouts of difficult breathing to fatal suffocation. Labored breathing, shortness of breath, wheezing, cough, and ventilatory **rales** are present to one degree or another. The respiratory tract of an asthmatic person is chronically inflamed and severely overreactive to allergy chemicals, especially leukotrienes and serotonin from pulmonary mast cells. Other pathologic components are thick mucus plugs in the air sacs and lung damage that can result in long-term respiratory compromise. An imbalance in the nervous control of the respiratory smooth muscles is apparently involved in asthma, and the episodes are influenced by the psychological state of the person, which strongly supports a neurological connection.

The number of asthma sufferers in the United States is estimated at more than 10 million, with nearly one-third of them children. For reasons that are not completely understood, asthma is on the increase, and deaths from it have doubled since 1982, even though effective agents to control it are more available now than they have ever been before. It has been suggested that more highly insulated buildings, mandated by energy efficiency regulations, have created indoor air conditions that harbor higher concentrations of contaminants, including insect remains and ozone.

Atopic dermatitis is an intensely itchy inflammatory condition of the skin, sometimes also called **eczema.** Sensitization occurs through ingestion, inhalation, and, occasionally, skin contact with allergens. It usually begins in infancy with reddened, vesicular, weeping, encrusted skin lesions **(figure 16.5).** It then progresses in childhood and adulthood to a dry, scaly, thickened skin condition. Lesions can occur on the face, scalp, neck, and inner surfaces of the limbs and trunk. The itchy, painful lesions cause considerable discomfort, and they are often predisposed to secondary bacterial infections. An anonymous writer once aptly described eczema as "the itch that rashes" or "one scratch is too many but one thousand is not enough."

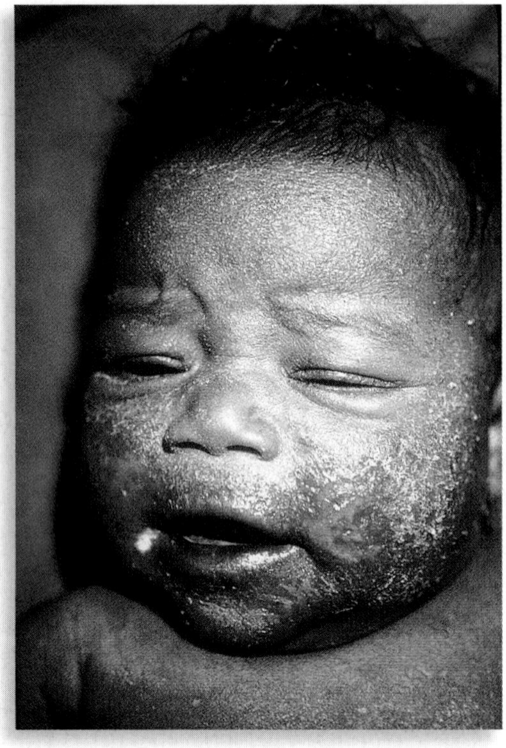

Figure 16.5 Atopic dermatitis, or eczema. Vesicular, encrusted lesions are typical in afflicted infants. This condition is prevalent enough to account for 1% of pediatric care.

Food Allergy

The ordinary diet contains a vast variety of compounds that are potentially allergenic. Although the mode of entry is intestinal, food allergies can also affect the skin and respiratory tract. Gastrointestinal symptoms include vomiting, diarrhea, and abdominal pain. In severe cases, nutrients are poorly absorbed, leading to growth retardation and failure to thrive in young children. Other manifestations of food allergies include eczema, hives, rhinitis, asthma, and occasionally, anaphylaxis. Classic food hypersensitivity involves IgE and degranulation of mast cells, but not all reactions involve this mechanism. The most common food allergens come from peanuts, fish, cow's milk, eggs, shellfish, and soybeans.[1]

Drug Allergy

Modern chemotherapy has been responsible for many medical advances. Unfortunately, it has also been hampered by the fact that drugs are foreign compounds capable of stimulating allergic reactions. In fact, allergy to drugs is one of the most common side effects of treatment (present in 5% to 10% of hospitalized patients). Depending upon the allergen, route of entry, and individual sensitivities, virtually any tissue of the body can be affected, and reactions range from mild atopy to fatal anaphylaxis. Compounds implicated most often are antibiotics (penicillin is number one in prevalence), synthetic antimicrobials (sulfa drugs), aspirin, opiates, and contrast dye used in X rays. The actual allergen is not the intact drug itself but a hapten given off when the liver processes the drug. Some forms of penicillin sensitivity are due to the presence of small amounts of the drug in meat, milk, and other foods and to exposure to *Penicillium* mold in the environment.

1. Do not confuse food allergy with food intolerance. Many people are lactose intolerant, for example, due to a deficiency in the enzyme that degrades the milk sugar.

Anaphylaxis: An Overpowering Systemic Reaction

The term **anaphylaxis,** or anaphylactic shock, was first used to denote a reaction of animals injected with a foreign protein. Although the animals showed no response during the first contact, upon reinoculation with the same protein at a later time, they exhibited acute symptoms—itching, sneezing, difficult breathing, prostration, and convulsions—and many died in a few minutes. Two clinical types of anaphylaxis are seen in humans. *Cutaneous anaphylaxis* is the wheal and flare inflammatory reaction to the local injection of allergen. *Systemic anaphylaxis,* on the other hand, is characterized by sudden respiratory and circulatory disruption that can be fatal in a few minutes. In humans, the allergen and route of entry are variable, though bee stings and injections of antibiotics or serum are implicated most often. Bee venom is a complex material containing several allergens and enzymes that can create a sensitivity that can last for decades after exposure.

The underlying physiological events in systemic anaphylaxis parallel those of atopy, but the concentration of chemical mediators and the strength of the response are greatly amplified. The immune system of a sensitized person exposed to a provocative dose of allergen responds with a sudden, massive release of chemicals into the tissues and blood, which act rapidly on the target organs. Anaphylactic persons have been known to die in 15 minutes from complete airway blockage.

Diagnosis of Allergy

Because allergy mimics infection and other conditions, it is important to determine if a person is actually allergic. If possible or necessary, it is also helpful to identify the specific allergen or allergens. Allergy diagnosis involves several levels of tests, including nonspecific, specific, *in vitro*, and *in vivo* methods.

A new test that can distinguish whether a patient has experienced an allergic attack measures elevated blood levels of tryptase, an enzyme released by mast cells that increases during an allergic response. Several types of specific *in vitro* tests can determine the allergic potential of a patient's blood sample. A differential blood cell count can indicate the levels of basophils and eosinophils—a higher level of these indicates allergy. The leukocyte histamine-release test measures the amount of histamine released from the patient's basophils when exposed to a specific allergen. Serological tests that use radioimmune assays (see chapter 17) to reveal the quantity and type of IgE are also clinically helpful.

CASE FILE 16 *WRAP-UP*

Further investigation by health care personnel revealed that the woman was allergic to wasp stings; a sting mark was discovered on her fingertip, and a nest of yellow jackets was found in the ground below the hedge she had been trimming.

Her death was caused by a severe, systemic allergic reaction to protein in the wasp venom, which resulted in a condition called anaphylaxis or anaphylactic shock. Anaphylactic shock initiates a host of conditions simultaneously throughout the body, causing airway blockage and circulatory collapse within a matter of minutes.

Approximately 1% to 3% of the population has some type of allergy to insect venom, usually to the venom produced by bees and wasps. Reactions can range from minor allergic reactions resulting in hives, headache, and stomach upset to severe allergic reactions that cause anaphylactic shock. People with known allergies to insect venom should wear a medical ID bracelet stating their allergy, as well as talk to their doctors about getting an insect sting allergy kit to carry with them at all times.

See: Li, J. T., and Yunginger, J. W. 1992. Management of insect sting hypersensitivity. Mayo Clin. Proc. *67:188–194.*

Muellman, R. L., Lindzon, R. D., and Silvers, N. S. 1998. Allergy, hypersensitivity and anaphylaxis. In P. Rosen, editor. Emergency medicine, concepts and clinical practice. *4th ed. St. Louis: Mosby Year Book.*

Pumphrey, R. S., and Roberts, I. S. 2000. Postmortem findings after fatal anaphylactic reactions. J. Clin. Pathol. *53:273–276.*

Skin Testing

A useful *in vivo* method to detect precise atopic or anaphylactic sensitivities is skin testing. With this technique, a patient's skin is injected, scratched, or pricked with a small amount of a pure allergen extract. There are hundreds of these allergen extracts containing common airborne allergens (plant and mold pollen) and more unusual allergens (mule dander, theater dust, bird feathers). Unfortunately, skin tests for food allergies using food extracts are unreliable in most cases. In patients with numerous allergies, the allergist maps the skin on the inner aspect of the forearms or back and injects the allergens intradermally according to this predetermined pattern **(figure 16.6*a*)**. Approximately 20 minutes after antigenic challenge, each site is appraised for a wheal response indicative of histamine release. The diameter of the wheal is measured and rated on a scale of 0 (no reaction) to 4+ (greater than 15 mm). **Figure 16.6*b*** shows skin test results for a person with extreme inhalant allergies.

Treatment and Prevention of Allergy

In general, the methods of treating and preventing type I allergy involve:

1. avoiding the allergen, although this may be very difficult in many instances;
2. taking drugs that block the action of lymphocytes, mast cells, or chemical mediators; and
3. undergoing desensitization therapy.

It is not possible to completely prevent initial sensitization, because there is no way to tell in advance if a person will develop

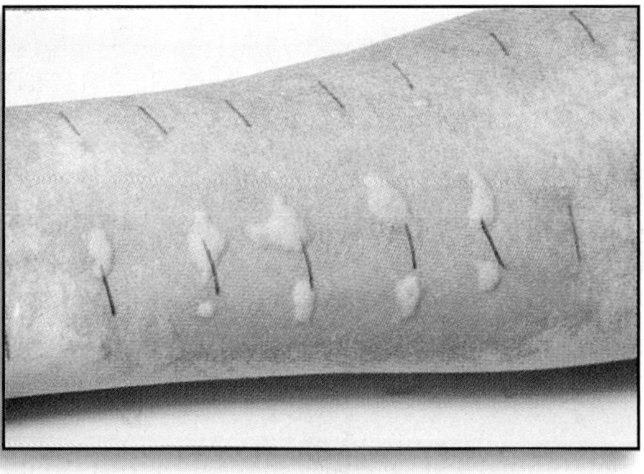

(a)

Environmental Allergens

No. 1 Standard Series

	ID 8/85	
	+++	1. Acacia gum
	+++	2. Cat dander
	++++	3. Chicken feathers
	++++	4. Cotton lint
	++	5. Dog dander
	+	6. Duck feathers
	+	7. Glue, animal
	++	8. Horse dander
	×	9. Horse serum
	+++	10. House dust #1
	+	11. Kapok
	+	12. Mohair (goat)
	+	13. Paper
	++++	14. Pyrethrum
	+++	15. Rug pad, ozite
	+	16. Silk dust
	+	17. Tobacco dust
	+	18. Tragacanth gum
	+++++	19. Upholstery dust
	+++	20. Wool

No. 2 Airborne Particles

	ID 8/85	
	+++	1. Ant
	+++++	2. Aphid
	++++	3. Bee
	++++	4. Housefly
	×	5. House mite
	+++	6. Mosquito
	++++	7. Moth
	+++	8. Roach
	++	9. Wasp
	0	10. Yellow jacket
		Airborne mold spores
	++	11. *Alternaria*
	+++	12. *Aspergillus*
	++	13. *Cladosporium*
	+++	14. *Hormodendrum*
	0	15. *Penicillium*
	+	16. *Phoma*
	+++	17. *Rhizopus*
		18. ....

× - not done
0 - no reaction
+ - slight reaction
++ - mild reaction
+++ - moderate reaction
++++ - severe reaction

(b)

Figure 16.6 A method for conducting an allergy skin test.
The forearm (or back) is mapped and then injected with a selection of allergen extracts. The allergist must be very aware of potential anaphylaxis attacks triggered by these injections. **(a)** Close-up of skin wheals showing a number of positive reactions (dark lines are measurer's marks). **(b)** An actual skin test record for some common environmental allergens [not related to **(a)**].

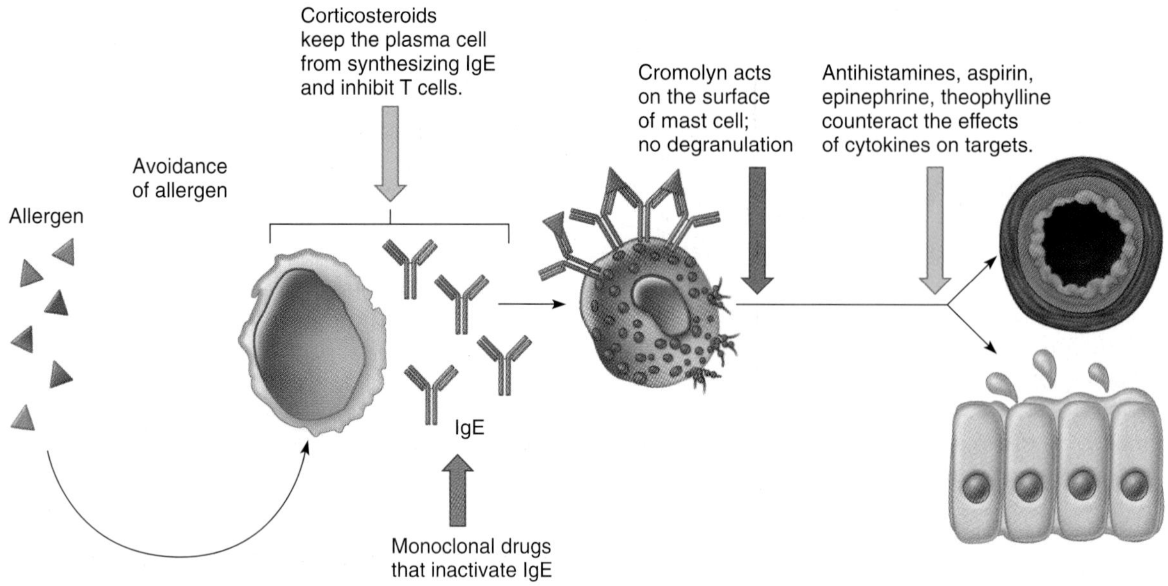

Figure 16.7 Strategies for circumventing allergy attacks.

an allergy to a particular substance. The practice of delaying the introduction of solid foods apparently has some merit in preventing food allergies in children, although even breast milk can contain allergens ingested by the mother. Rigorous cleaning and air conditioning can reduce contact with airborne allergens, but it is not feasible to isolate a person from all allergens, which is the reason drugs are so important in control.

Therapy to Counteract Allergies

The aim of antiallergy medication is to block the progress of the allergic response somewhere along the route between IgE production and the appearance of symptoms **(figure 16.7).** Oral anti-inflammatory drugs such as corticosteroids inhibit the activity of lymphocytes and thereby reduce the production of IgE, but they also have dangerous side effects and should not be taken for prolonged periods. Some drugs block the degranulation of mast cells and reduce the levels of inflammatory cytokines. The most effective of these are diethylcarbamazine and cromolyn. Asthma and rhinitis sufferers can find relief with a drug that blocks synthesis of leukotriene and a monoclonal antibody that inactivates IgE (Xolair).

Widely used medications for preventing symptoms of atopic allergy are **antihistamines,** the active ingredients in most over-the-counter allergy-control drugs. Antihistamines interfere with histamine activity by binding to histamine receptors on target organs. Most of them have major side effects, however, such as drowsiness. Newer antihistamines lack this side effect because they do not cross the blood-brain barrier. Other drugs that relieve inflammatory symptoms are aspirin and acetaminophen, which reduce pain by interfering with prostaglandin, and theophylline, a bronchodilator that reverses spasms in the respiratory smooth muscles. Persons who suffer from anaphylactic attacks are urged to carry at all times injectable epinephrine (adrenaline) and an identification tag indicating their sensitivity. An aerosol inhaler containing epinephrine can also provide rapid relief. Epinephrine reverses constriction of the airways and slows the release of allergic mediators.

Approximately 70% of allergic patients benefit from controlled injections of specific allergens as determined by skin tests. This technique, called **desensitization** or **hyposensitization,** is a therapeutic way to prevent reactions between allergen, IgE, and mast cells. The allergen preparations contain pure, preserved suspensions of plant antigens, venoms, dust mites, dander, and molds (but so far, hyposensitization for foods has not proved very effective). The immunologic basis of this treatment is open to differences in interpretation. One theory suggests that injected allergens stimulate the formation of high levels of allergen-specific IgG **(figure 16.8)** instead of IgE. It has been proposed that these IgG **blocking antibodies** remove allergen from the system before it can bind to IgE, thus preventing the degranulation of mast cells. It is also possible that allergen delivered in this fashion combines with the IgE itself and takes it from circulation before it can react with the mast cells.

CHECKPOINT

- Type I hypersensitivity reactions result from excessive IgE production in response to an exogenous antigen.
- The two kinds of type I hypersensitivities are atopy, a chronic, local allergy, and anaphylaxis, a systemic, potentially fatal allergic response.
- The predisposition to type I hypersensitivities is inherited, but age, geographic locale, and infection also influence allergic response.
- Type I allergens include inhalants, ingestants, injectants, and contactants.

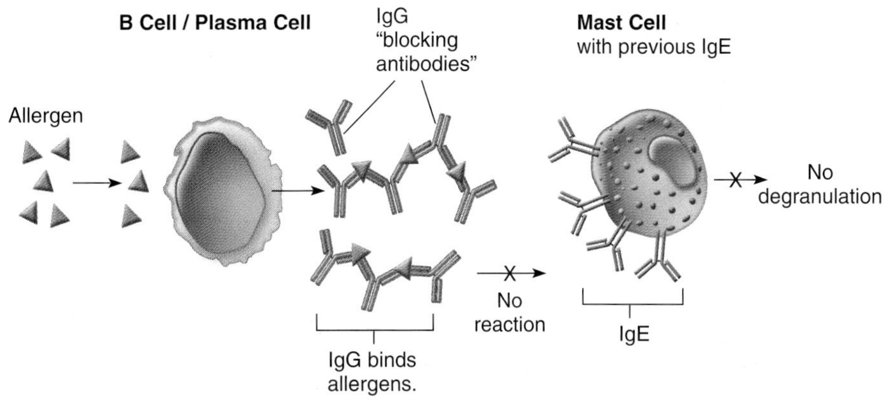

Figure 16.8 The blocking antibody theory for allergic desensitization.
An injection of allergen causes IgG antibodies to be formed instead of IgE; these blocking antibodies cross-link and effectively remove the allergen before it can react with the IgE in the mast cell.

- The portals of entry for type I antigens are the skin, respiratory tract, gastrointestinal tract, and genitourinary tract.
- Type I hypersensitivities are set up by a sensitizing dose of allergen and expressed when a second provocative dose triggers the allergic response. The time interval between the two can be many years.
- The primary participants in type I hypersensitivities are IgE, basophils, mast cells, and agents of the inflammatory response.
- Allergies are diagnosed by a variety of *in vitro* and *in vivo* tests that assay specific cells, IgE, and local reactions.
- Allergies are treated by medications that interrupt the allergic response at certain points. Allergic reactions can often be prevented by desensitization therapy.

16.3 Type II Hypersensitivities: Reactions That Lyse Foreign Cells

The diseases termed type II hypersensitivities are a complex group of syndromes that involve complement-assisted destruction (lysis) of cells by antibodies (IgG and IgM) directed against those cells' surface antigens. This category includes transfusion reactions and some types of autoimmunities (discussed in a later section). The cells targeted for destruction are often red blood cells, but other cells can be involved.

Chapters 14 and 15 described the functions of unique surface markers on cell membranes. Ordinarily, these molecules play essential roles in transport, recognition, and development, but they become medically important when the tissues of one person are placed into the body of another person. Blood transfusions and organ donations introduce alloantigens (molecules that differ in the same species) on donor cells that are recognized by the lymphocytes of the recipient. These reactions are not really immune dysfunctions as allergy and autoimmunity are. The immune system is in fact working normally, but it is not equipped to distinguish between the desirable foreign cells of a transplanted tissue and the undesirable ones of a microbe.

The Basis of Human ABO Antigens and Blood Types

The existence of human blood types was first demonstrated by an Austrian pathologist, Karl Landsteiner, in 1904. While studying incompatibilities in blood transfusions, he found that the serum of one person could clump the red blood cells of another. Landsteiner identified four distinct types, subsequently called the **ABO blood groups.**

Like the MHC antigens on white blood cells, the ABO antigen markers on red blood cells are genetically determined and composed of glycoproteins. These ABO antigens are inherited as two (one from each parent) of three alternative **alleles:** A, B, or O. A and B alleles are dominant over O and codominant with one another. As **table 16.3** indicates, this mode of inheritance gives rise to four blood types (phenotypes), depending on the particular combination of genes. Thus, a person with an *AA* or *AO* genotype has type A blood; genotype *BB* or *BO* gives type B; genotype *AB* produces type AB; and genotype *OO* produces type O. Some important points about the blood types are:

1. they are named for the dominant antigen(s),
2. the RBCs of type O persons have antigens but not A and B antigens, and
3. tissues other than RBCs carry A and B antigens.

A diagram of the AB antigens and blood types is shown in **figure 16.9.** The A and B genes each code for an enzyme that adds a terminal carbohydrate to RBC surface molecules during maturation. RBCs of type A contain an enzyme that adds *N*-acetylgalactosamine to the molecule; RBCs of type B have an enzyme that adds D-galactose; RBCs of type AB contain both enzymes that add both carbohydrates; and RBCs of type O lack the genes and enzymes to add a terminal molecule.

Antibodies Against A and B Antigens

Although an individual does not normally produce antibodies in response to his or her own RBC antigens, the serum can

TABLE 16.3 Characteristics of ABO Blood Groups

Genotype	Blood Type	Antigen Present on Erythrocyte Membranes	Antibody in Plasma	Incidence of Type in United States: Among Whites (%)	Among Asians (%)	Among Those of African and Caribbean Descent (%)
AA, AO	A	A	Anti-B	41	28	27
BB, BO	B	B	Anti-A	10	27	20
AB	AB	A and B	Neither anti-A nor anti-B	4	5	7
OO	O	Neither A nor B	Anti-A and anti-B	45	40	46

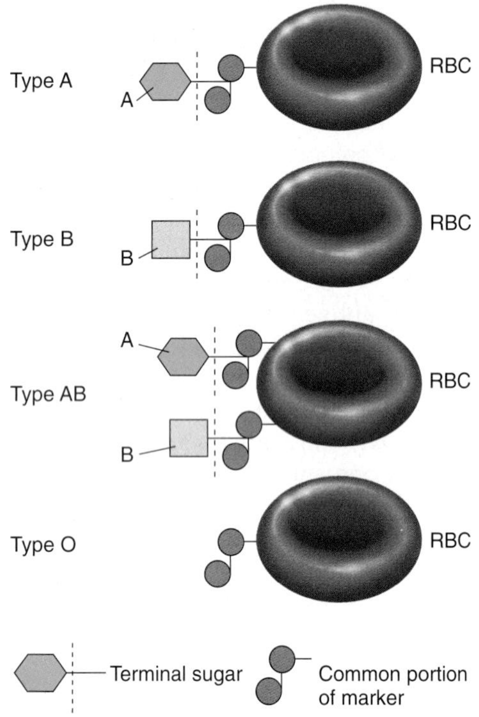

Figure 16.9 The genetic/molecular basis for the A and B antigens (receptors) on red blood cells.
In general, persons with blood types A, B, and AB inherit a gene for the enzyme that adds a certain terminal sugar to the basic RBC receptor. Type O persons do not have such an enzyme and lack the terminal sugar.

contain antibodies that react with blood of another antigenic type even though contact with this other blood type has *never* occurred. These preformed antibodies account for the immediate and intense quality of transfusion reactions. As a rule, type A blood contains antibodies (anti-B) that react against the B antigens on type B and AB red blood cells. Type B blood contains antibodies (anti-A) that react with A antigen on type A and AB red blood cells. Type O blood contains antibodies against both A and B antigens. Type AB blood does not contain antibodies against either A or B antigens[2] (see table 16.3). What is the source of these anti-A and anti-B antibodies? It appears that they develop in early infancy because of exposure to certain antigens that are widely distributed in nature. These antigens are surface molecules on bacteria and plant cells that mimic the structure of A and B antigens. Exposure to these sources stimulates the production of corresponding antibodies.

Clinical Concerns in Transfusions

The presence of ABO antigens and A, B antibodies underlie several clinical concerns in giving blood transfusions. First, the individual blood types of donor and recipient must be determined. By use of a standard technique, drops of blood are mixed with antisera that contain antibodies against the A and B antigens and are then observed for the evidence of agglutination **(figure 16.10).**

Knowing the blood types involved makes it possible to determine which transfusions are safe to do. The general rule of compatibility is that the RBC antigens of the donor must not be agglutinated by antibodies in the recipient's blood **(figure 16.11).** The ideal practice is to transfuse blood that is a perfect match (A to A, B to B). But even in this event, blood samples must be cross-matched before the transfusion because other blood group incompatibilities can exist. This test involves mixing the blood of the donor with the serum of the recipient to check for agglutination.

Under certain circumstances (emergencies, the battlefield), the concept of universal transfusions can be used. To appreciate how this works, we must apply the rule stated in the previous paragraph. Type O blood lacks A and B antigens and will not be agglutinated by other blood types, so it could theoretically be used in any transfusion. Hence, a person with this blood type is called a **universal donor.** Because type AB blood lacks agglutinating antibodies, an individual with this blood could conceivably receive any type of blood. Type AB persons are consequently called *universal recipients.*

2. Why would this be true? The answer lies in the first sentence of the paragraph.

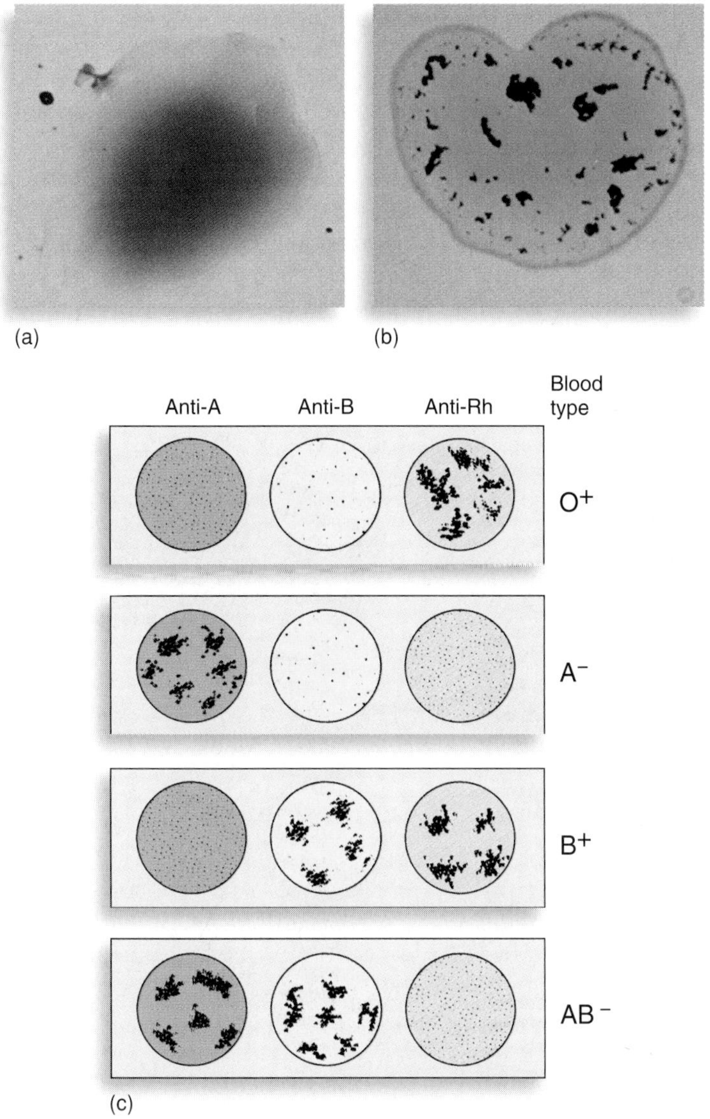

Figure 16.10 Interpretation of blood typing.
In this test, a drop of blood is mixed with a specially prepared antiserum known to contain antibodies against the A, B, or Rh antigens. **(a)** If that particular antigen is not present, the red blood cells in that droplet do not agglutinate and form an even suspension. **(b)** If that antigen is present, agglutination occurs and the RBCs form visible clumps. **(c)** Several patterns and their interpretations. Anti-A, anti-B, and anti-Rh are shorthand for the antiserum applied to the drops. (In general, O^+ is the most common blood type, and AB^- is the rarest.)

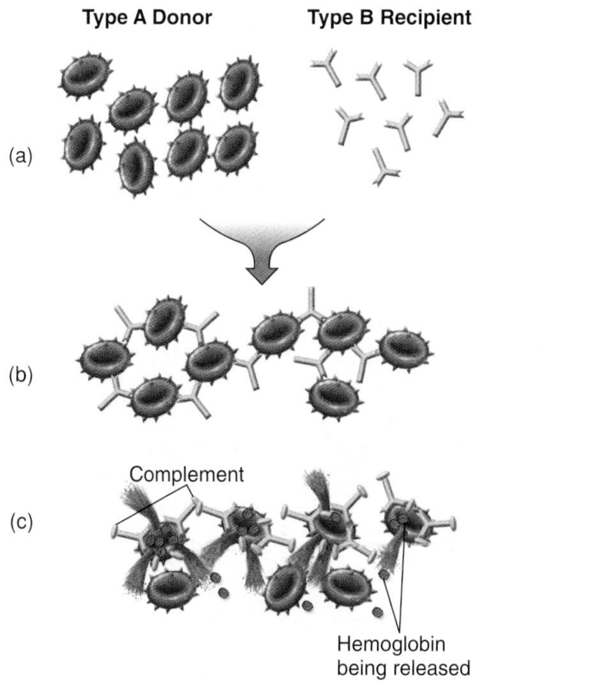

Figure 16.11 Microscopic view of a transfusion reaction.
(a) Incompatible blood. The red blood cells of the type A donor contain antigen A, while the serum of the type B recipient contains anti-A antibodies that can agglutinate donor cells. **(b)** Agglutination complexes can block the circulation in vital organs. **(c)** Activation of the complement by antibody on the RBCs can cause hemolysis and anemia. This sort of incorrect transfusion is very rare because of the great care taken by blood banks to ensure a correct match.

Although both types of transfusions involve antigen-antibody incompatibilities, these are of less concern because of the dilution of the donor's blood in the body of the recipient. Additional RBC markers that can be significant in transfusions are the Rh, MN, and Kell antigens (see next sections).

Transfusion of the wrong blood type causes differing degrees of adverse reaction. The severest reaction is massive hemolysis when the donated red blood cells react with recipient antibody and trigger the complement cascade (see figure 16.11). The resultant destruction of red cells leads to systemic shock and kidney failure brought on by the blockage of glomeruli (blood-filtering apparatus) by cell debris. Death is a common outcome. Other reactions caused by RBC destruction are fever, anemia, and jaundice. A transfusion reaction is managed by immediately halting the transfusion, administering drugs to remove hemoglobin from the blood, and beginning another transfusion with red blood cells of the correct type.

The Rh Factor and Its Clinical Importance

Another RBC antigen of major clinical concern is the **Rh factor** (or D antigen). This factor was first discovered in experiments exploring the genetic relationships among animals. Rabbits inoculated with the RBCs of rhesus monkeys produced an antibody that also reacted with human RBCs. Further tests showed that this monkey antigen (termed Rh for rhesus) was present in about 85% of humans and absent in the other 15%. The details of Rh inheritance are more complicated than those of ABO, but in simplest terms, a person's Rh type results from a combination of two possible alleles—a dominant one that codes for the factor and a recessive one that does not. A person inheriting at least one Rh gene will be Rh^+; only those persons inheriting two recessive genes are Rh^-. The "+" or "−" appearing after a blood type refers to the Rh status of the person, as in O^+ or AB^- (see figure 16.10*c*). However, unlike the ABO antigens, exposure to environmental antigens does not sensitize Rh^- persons to the Rh factor. The only ways one can develop antibodies against this factor are through placental sensitization or transfusion.

Hemolytic Disease of the Newborn and Rh Incompatibility

The potential for placental sensitization occurs when a mother is Rh^- and her unborn child is Rh^+. The obvious intimacy between mother and fetus makes it possible for fetal RBCs to leak into the mother's circulation during childbirth, when the detachment of the placenta creates avenues for fetal blood to enter the maternal circulation. The mother's immune system detects the foreign Rh factors on the fetal RBCs and is sensitized to them by producing antibodies and memory B cells. The first Rh^+ child is usually not affected because the process begins so late in pregnancy that the child is born before maternal sensitization is completed. However, the mother's immune system has been strongly primed for a second contact with this factor in a subsequent pregnancy **(figure 16.12*a*).**

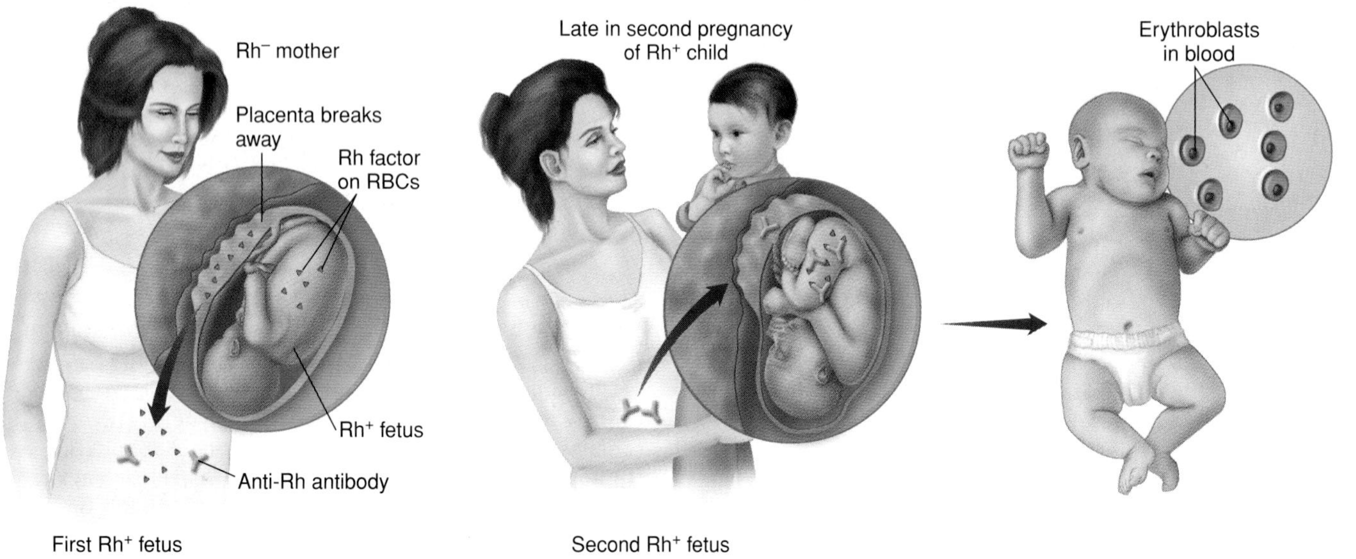

(a) **The development and aftermath of Rh sensitization**
Initial sensitization of the maternal immune system to fetal Rh^+ factor occurs when fetal cells leak into the Rh^- mother's circulation late in pregnancy or during delivery when the placenta tears away. The child will escape hemolytic disease in most instances, but the mother, now sensitized, will be capable of an immediate reaction to a second Rh^+ fetus and its Rh-factor antigen. At that time, the mother's anti-Rh antibodies pass into the fetal circulation and elicit severe hemolysis in the fetus and neonate.

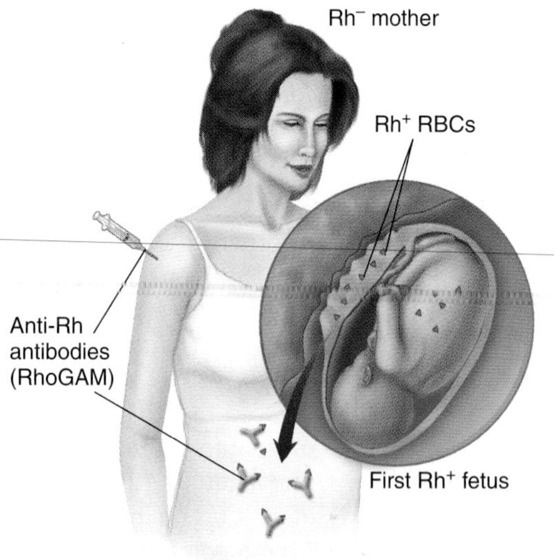

(b) **Prevention of erythroblastosis fetalis with anti-Rh immune globulin (RhoGAM)**
Injecting a mother who is at risk with RhoGAM during her first Rh^+ pregnancy helps to inactivate and remove the fetal Rh^+ cells before her immune system can react and develop sensitivity.

Figure 16.12 Development and control of Rh incompatibility.

INSIGHT 16.2 *Medical*

Why Doesn't a Mother Reject Her Fetus?

Think of it: Even though mother and child are genetically related, the father's genetic contribution guarantees that the fetus will contain molecules that are antigenic to the mother. In fact, with the recent practice of implanting one woman with the fertilized egg of another woman, the surrogate mother is carrying a fetus that has no genetic relationship to her. Yet, even with this essentially foreign body inside the mother, dangerous immunologic reactions such as Rh incompatibility are rather rare. In what ways do fetuses avoid the surveillance of the mother's immune system? The answer appears to lie in the placenta and embryonic tissues. The fetal components that contribute to these tissues are not strongly antigenic, and they form a barrier that keeps the fetus isolated in its own antigen-free environment. The placenta is surrounded by a dense, many-layered envelope that prevents the passage of maternal cells, and it actively absorbs, removes, and inactivates circulating antigens.

In the next pregnancy with an Rh^+ fetus, fetal blood cells escape into the maternal circulation late in pregnancy and elicit a memory response. The fetus is at risk when the maternal anti-Rh antibodies cross the placenta into the fetal circulation, where they affix to fetal RBCs and cause complement-mediated lysis. The outcome is a potentially fatal **hemolytic disease of the newborn (HDN)** called *erythroblastosis fetalis* (eh-rith″-roh-blas-toh′-sis fee-tal′-is). This term is derived from the presence of immature nucleated RBCs called erythroblasts in the blood. They are released into the infant's circulation to compensate for the massive destruction of RBCs stimulated by maternal antibodies. Additional symptoms are severe anemia, jaundice, and enlarged spleen and liver.

Maternal-fetal incompatibilities are also possible in the ABO blood group, but adverse reactions occur less frequently than with Rh sensitization because the antibodies to these blood group antigens are IgM rather than IgG and are unable to cross the placenta in large numbers. In fact, the maternal-fetal relationship is a fascinating instance of foreign tissue not being rejected, despite the extensive potential for contact **(Insight 16.2).**

Preventing Hemolytic Disease of the Newborn

Once sensitization of the mother to Rh factor has occurred, all other Rh^+ fetuses will be at risk for hemolytic disease of the newborn. Prevention requires a careful family history of an Rh^- pregnant woman. It can predict the likelihood that she is already sensitized or is carrying an Rh^+ fetus. It must take into account other children she has had, their Rh types, and the Rh status of the father. If the father is also Rh^- the child will be Rh^- and free of risk; but if the father is Rh^+, the probability that the child will be Rh^+ is 50% or 100%, depending on the exact genetic makeup of the father. If there is any possibility that the fetus is Rh^+, the mother must be passively immunized with antiserum containing antibodies against the Rh factor (*Rh_0 [D] immune globulin,* or RhoGAM[3]). This antiserum, injected at 28 to 32 weeks and again immediately after delivery, reacts with any fetal RBCs that have escaped into the maternal circulation, thereby preventing the sensitization of the mother's immune system to Rh factor **(figure 16.12*b*).** Anti-Rh antibody must be given with each pregnancy that involves an Rh^+ fetus. It is ineffective if the mother has already been sensitized by a prior Rh^+ fetus or an incorrect blood transfusion, which can be determined by a serological test.

As in ABO blood types, the Rh factor should be matched for a transfusion, although it is acceptable to transfuse Rh^- blood if the Rh type is not known.

Other RBC Antigens

Although the ABO and Rh systems are of greatest medical significance, about 20 other red blood cell antigen groups have been discovered. Examples are the *MN, Ss, Kell,* and *P* blood groups. Because of incompatibilities that these blood groups present, transfused blood is screened to prevent possible cross-reactions. The study of these blood antigens (as well as ABO and Rh) has given rise to other useful applications. For example, they can be useful in forensic medicine (crime detection), studying ethnic ancestry, and tracing prehistoric migrations in anthropology. Many blood cell antigens are remarkably hardy and can be detected in dried blood stains, semen, and saliva. Even the 3,300-year-old mummy of King Tutankhamen has been typed A_2MN!

CHECKPOINT

- Type II hypersensitivity reactions occur when preformed antibodies react with foreign cell-bound antigens. The most common type II reactions occur when transfused blood is mismatched to the recipient's ABO type. IgG or IgM antibodies attach to the foreign cells, resulting in complement fixation. The resultant formation of membrane attack complexes lyses the donor cells.
- Complement, IgG, and IgM antibodies are the primary mediators of type II hypersensitivities.

3. RhoGAM: Immunoglobulin fraction of human anti-Rh serum, prepared from pooled human sera.

- The concepts of universal donor (type O) and universal recipient (type AB) apply only under emergency circumstances. Cross-matching donor and recipient blood is necessary to determine which transfusions are safe to perform.
- Type II hypersensitivities can also occur when Rh^- mothers are sensitized to Rh^+ RBCs of their unborn babies and the mother's anti-Rh antibodies cross the placenta, causing hemolysis of the newborn's RBCs. This is called hemolytic disease of the newborn, or erythroblastosis fetalis.

16.4 Type III Hypersensitivities: Immune Complex Reactions

Type III hypersensitivity involves the reaction of soluble antigen with antibody and the deposition of the resulting complexes in basement membranes of epithelial tissue. It is similar to type II, because it involves the production of IgG and IgM antibodies after repeated exposure to antigens and the activation of complement. Type III differs from type II because its antigens are not attached to the surface of a cell. The interaction of these antigens with antibodies produces free-floating complexes that can be deposited in the tissues, causing an **immune complex reaction** or disease. This category includes therapy-related disorders (serum sickness and the Arthus reaction) and a number of autoimmune diseases (such as glomerulonephritis and lupus erythematosus).

Mechanisms of Immune Complex Disease

After initial exposure to a profuse amount of antigen, the immune system produces large quantities of antibodies that circulate in the fluid compartments. When this antigen enters the system a second time, it reacts with the antibodies to form antigen-antibody complexes **(figure 16.13).** These complexes summon various inflammatory components such as complement and neutrophils, which would ordinarily eliminate Ag-Ab complexes as part of the normal immune response. In an immune complex disease, however, these complexes are so abundant that they deposit in the **basement membranes**[4]

4. **Basement membranes** are basal partitions of epithelia that normally filter out circulating antigen-antibody complexes.

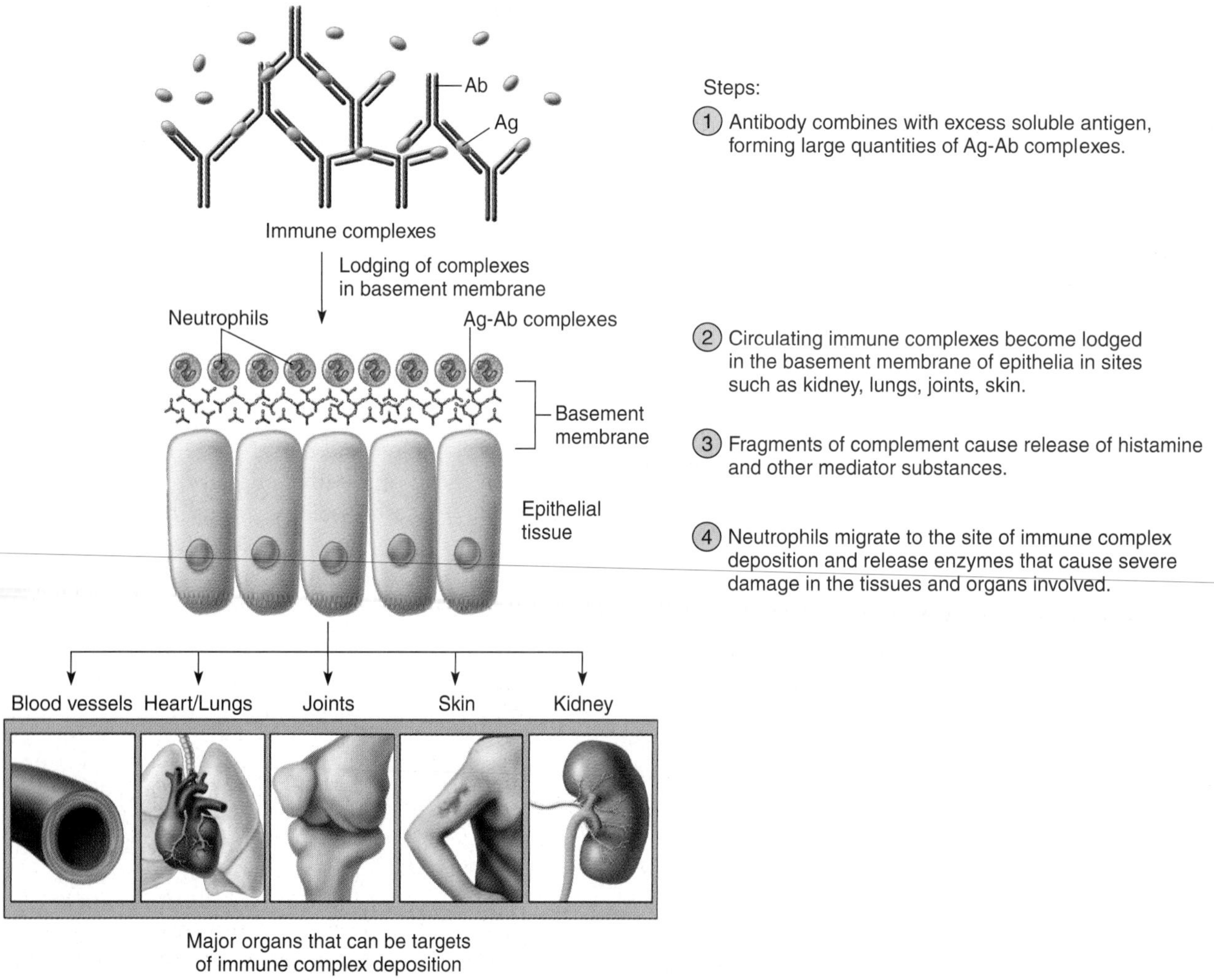
Ab
Ag
Immune complexes
Lodging of complexes in basement membrane
Neutrophils
Ag-Ab complexes
Basement membrane
Epithelial tissue
Blood vessels
Heart/Lungs
Joints
Skin
Kidney
Major organs that can be targets of immune complex deposition
Steps:
1 Antibody combines with excess soluble antigen, forming large quantities of Ag-Ab complexes.
2 Circulating immune complexes become lodged in the basement membrane of epithelia in sites such as kidney, lungs, joints, skin.
3 Fragments of complement cause release of histamine and other mediator substances.
4 Neutrophils migrate to the site of immune complex deposition and release enzymes that cause severe damage in the tissues and organs involved.

Process Figure 16.13 Pathogenesis of immune complex disease.

of epithelial tissues and become inaccessible. In response to these events, neutrophils release lysosomal granules that digest tissues and cause a destructive inflammatory condition. The symptoms of type III hypersensitivities are due in great measure to this pathologic state.

Types of Immune Complex Disease

During the early tests of immunotherapy using animals, hypersensitivity reactions to serum and vaccines were common. In addition to anaphylaxis, two syndromes, the **Arthus reaction**[5] and **serum sickness,** were identified. These syndromes are associated with certain types of passive immunization (especially with animal serum).

Serum sickness and the Arthus reaction are like anaphylaxis in requiring sensitization and preformed antibodies. Characteristics that set them apart from anaphylaxis are:

1. they depend upon IgG, IgM, or IgA (precipitating antibodies) rather than IgE;
2. they require large doses of antigen (not a minuscule dose as in anaphylaxis); and
3. their symptoms are delayed (a few hours to days).

The Arthus reaction and serum sickness differ from each other in some important ways. The Arthus reaction is a *localized* dermal injury due to inflamed blood vessels in the vicinity of any injected antigen. Serum sickness is a *systemic* injury initiated by antigen-antibody complexes that circulate in the blood and settle into membranes at various sites.

The Arthus Reaction

The Arthus reaction is usually an acute response to a second injection of vaccines (boosters) or drugs at the same site as the first injection. In a few hours, the area becomes red, hot to the touch, swollen, and very painful. These symptoms are mainly due to the destruction of tissues in and around the blood vessels and the release of histamine from mast cells and basophils. Although the reaction is usually self-limiting and rapidly cleared, intravascular blood clotting can occasionally cause necrosis and loss of tissue.

Serum Sickness

Serum sickness was named for a condition that appeared in soldiers after repeated injections of horse serum to treat tetanus. It can also be caused by injections of animal hormones and drugs. The immune complexes enter the circulation; are carried throughout the body; and are eventually deposited in blood vessels of the kidney, heart, skin, and joints (see figure 16.13). The condition can become chronic, causing symptoms such as enlarged lymph nodes, rashes, painful joints, swelling, fever, and renal dysfunction.

5. Named after Maurice Arthus, the physiologist who first identified this localized inflammatory response.

CHECKPOINT

- Type III hypersensitivities are induced when a profuse amount of antigen enters the system and results in large quantities of antibody formation.
- Type III hypersensitivity reactions occur when large quantities of antigen react with host antibody to form small, soluble immune complexes that settle in tissue cell membranes, causing chronic destructive inflammation. The reactions appear hours or days after the antigen challenge.
- The mediators of type III hypersensitivity reactions include soluble IgA, IgG, or IgM, and agents of the inflammatory response.
- Two kinds of type III hypersensitivities are localized (Arthus) reactions and systemic (serum sickness). Arthus reactions occur at the site of injected drugs or booster immunizations. Systemic reactions occur when repeated antigen challenges cause systemic distribution of the immune complexes and subsequent inflammation of joints, lymph nodes, and kidney tubules.

16.5 Type IV Hypersensitivities: Cell-Mediated (Delayed) Reactions

The adverse immune responses we have covered so far are explained primarily by B-cell involvement and antibodies. A notable difference exists in type IV hypersensitivity, which involves primarily the T-cell branch of the immune system. Type IV immune dysfunction has traditionally been known as delayed hypersensitivity because the symptoms arise one to several days following the second contact with an antigen. In general, type IV diseases result when T cells respond to antigens displayed on self tissues or transplanted foreign cells. Examples of type IV hypersensitivity include delayed allergic reactions to infectious agents, contact dermatitis, and graft rejection.

Delayed-Type Hypersensitivity

Infectious Allergy

A classic example of a delayed-type hypersensitivity occurs when a person sensitized by tuberculosis infection is injected with an extract (tuberculin) of the bacterium *Mycobacterium tuberculosis.* The so-called tuberculin reaction is an acute skin inflammation at the injection site appearing within 24 to 48 hours. So useful and diagnostic is this technique for detecting present or prior tuberculosis that it is the chosen screening device (see chapter 21). Other infections that use similar skin testing are leprosy, syphilis, histoplasmosis, toxoplasmosis, and candidiasis. This form of hypersensitivity arises from time-consuming cellular events involving a specific class of T cells (T_H1) that receive the processed allergens from dendritic cells. Activated T_H cells release cytokines that recruit various inflammatory cells such as macrophages, neutrophils,

INSIGHT 16.3 Medical

Pretty, Pesky, Poisonous Plants

As a cause of allergic contact dermatitis (affecting about 10 million people a year), nothing can compare with a single family of plants belonging to the genus *Toxicodendron.* At least one of these plants—either poison ivy, poison oak, or poison sumac—flourishes in the forests, woodlands, or along the trails of most regions of America. The allergen in these plants, an oil called urushiol, has such extreme potency that a pinhead-size amount could spur symptoms in 500 people, and it is so long lasting that botanists must be careful when handling 100-year-old plant specimens. Although degrees of sensitivity vary among individuals, it is estimated that 85% of all Americans are potentially hypersensitive to this compound. Some people are so acutely sensitive that even the most minuscule contact, such as handling pets or clothes that have touched the plant or breathing vaporized urushiol, can trigger an attack.

Humans first become sensitized by contact during childhood. Individuals at great risk (firefighters, hikers) are advised to determine their degree of sensitivity using a skin test, so that they can be adequately cautious and prepared. Some odd remedies include skin potions containing bleach, buttermilk, ammonia, hair spray, and meat tenderizer. Commercial products are available for blocking or washing away the urushiol. Allergy researchers are currently testing oral vaccines containing a form of urushiol, which seem to desensitize experimental animals. An effective method using poison ivy desensitization injection is currently available to people with extreme sensitivity.

Poison oak

Poison sumac

Poison ivy

Learning to identify these common plants can prevent exposure and sensitivity. One old saying that might help warns, "Leaves of three, let it be; berries white, run with fright."

and eosinophils. The buildup of fluid and cells at the site gives rise to a red papule (for example, see **figure 16.14**). In a chronic infection (tertiary syphilis, for example), extensive damage to organs can occur through granuloma formation.

Figure 16.14 Positive tuberculin test.
Intradermal injection of tuberculin extract in a person sensitized to tuberculosis yields a slightly raised red bump greater than 10 mm in diameter.

Contact Dermatitis

The most common delayed allergic reaction, contact dermatitis, is caused by exposure to resins in poison ivy or poison oak **(Insight 16.3),** to simple haptens in household and personal articles (jewelry, cosmetics, elasticized undergarments), and to certain drugs. Like immediate atopic dermatitis, the reaction to these allergens requires a sensitizing and a provocative dose. The allergen first penetrates the outer skin layers, is processed by Langerhans cells (skin dendritic cells), and is presented to T cells. When subsequent exposures attract lymphocytes and macrophages to this area, these cells give off enzymes and in-

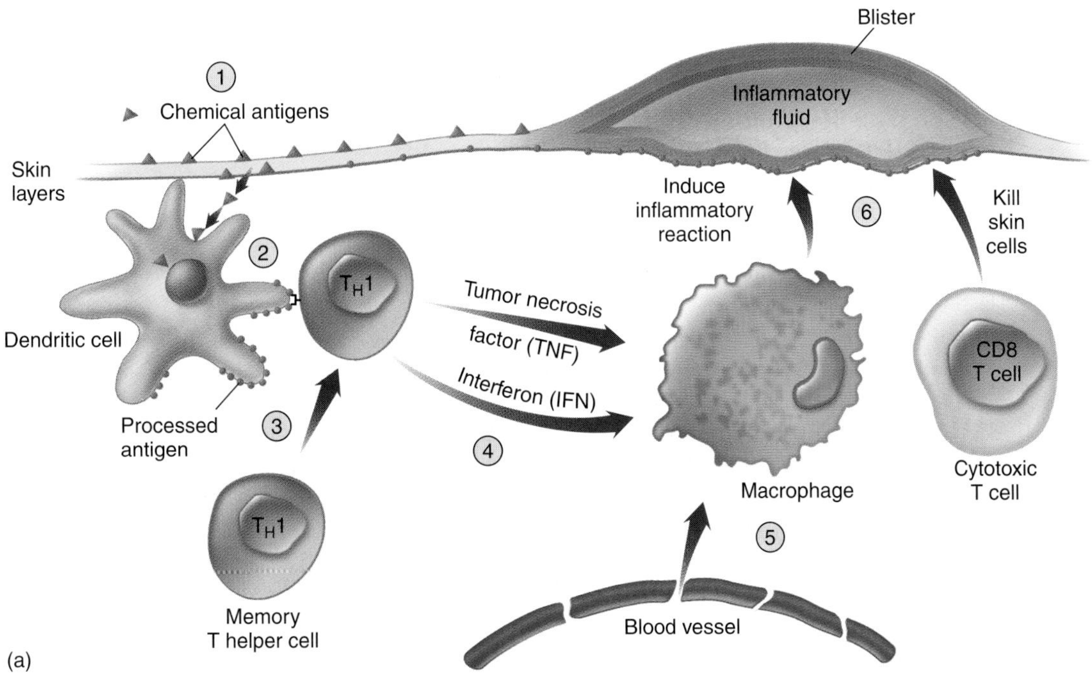

(a)

(1) Lipid-soluble chemicals are absorbed by the skin.

(2) Dendritic cells close to the epithelium pick up the allergen, process it, and display it on MHC receptors.

(3) Previously sensitized T_H1 (CD4) cells recognize the presented allergen.

(4) Sensitized T_H1 cells are activated to secrete cytokines (IFN, TNF) that

(5) attract macrophages and cytotoxic T cells to the site.

(6) Macrophage releases mediators that stimulate a strong, local inflammatory reaction. Cytotoxic T cells directly kill cells and damage the skin. Fluid-filled blisters result.

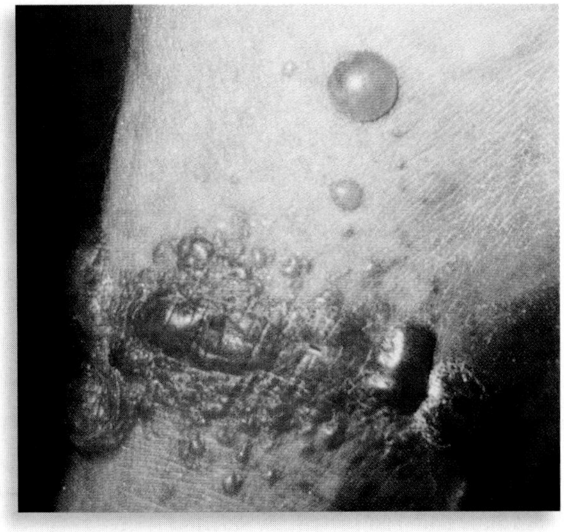

(b)

Process Figure 16.15 Contact dermatitis.
(a) Genesis of contact dermatitis. **(b)** Contact dermatitis from poison oak, showing various stages of involvement: blister, scales, and thickened patches.

flammatory cytokines that severely damage the epidermis in the immediate vicinity **(figure 16.15*a*).** This response accounts for the intensely itchy papules and blisters that are the early symptoms **(figure 16.15*b*).** As healing progresses, the epidermis is replaced by a thick, horny layer. Depending upon the dose and the sensitivity of the individual, the time from initial contact to healing can be a week to 10 days.

T Cells and Their Role in Organ Transplantation

Transplantation or grafting of organs and tissues is a common medical procedure. Although it is life-giving, this technique is plagued by the natural tendency of lymphocytes to seek out foreign antigens and mount a campaign to destroy them. The bulk of the damage that occurs in graft rejections can be attributed to expression of cytotoxic T cells and other killer cells. This section covers the mechanisms involved in graft rejection, tests for transplant compatibility, reactions against grafts, prevention of graft rejection, and types of grafts.

The Genetic and Biochemical Basis for Graft Rejection

In chapter 15, we discussed the role of major histocompatibility (MHC or HLA) genes and surface markers in immune function. In general, the genes and markers in MHC classes I and II are extremely important in recognizing self and in regulating the immune response. These molecules also set the events of graft rejection in motion. The MHC genes of humans are inherited from among a large pool of genes, so the cells of each person can exhibit variability in the pattern of cell surface molecules. The pattern is identical in different cells of the same person and can be similar in related siblings and parents, but the more distant the relationship, the less likely that the MHC genes and markers will be similar. When donor tissue (a graft) displays surface molecules of a different MHC class, the T cells of the recipient (called the host) will recognize its foreignness and react against it.

T-Cell-Mediated Recognition of Foreign MHC Receptors

Host Rejection of Graft When the cytotoxic T cells of a host recognize foreign class I MHC markers on the surface of grafted cells, they release interleukin-2 as part of a general immune mobilization. Receipt of this stimulus amplifies helper and cytotoxic T cells specific to the foreign antigens on the donated cells. The cytotoxic cells bind to the grafted tissue and secrete lymphokines that begin the rejection process within 2 weeks of transplantation. Late in this process, antibodies formed against the graft tissue contribute to immune damage. A final blow is the destruction of the vascular supply, promoting death of the grafted tissue.

Graft Rejection of Host In certain severe immunodeficiencies, the host cannot or does not reject a graft. But this failure may not protect the host from serious damage because graft incompatibility is a two-way phenomenon. Some grafted tissues (especially bone marrow) contain an indigenous population called passenger lymphocytes. This makes it quite possible for the graft to reject the host, causing **graft versus host disease (GVHD).** Because any host tissue bearing MHC markers foreign to the graft can be attacked, the effects of GVHD are widely systemic and toxic. A papular, peeling skin rash is the most common symptom. Other organs affected are the liver, intestine, muscles, and mucus membranes. Previously, GVHD occurred in approximately 30% of bone marrow transplants within 100 to 300 days of the graft. This percentage is declining as better screening and selection of tissues are developed.

Classes of Grafts Grafts are generally classified according to the genetic relationship between the donor and the recipient. Tissue transplanted from one site on an individual's body to another site on his or her body is known as an **autograft.** Typical examples are skin replacement in burn repair and the use of a vein to fashion a coronary artery bypass. In an **isograft,** tissue from an identical twin is used. Because isografts do not contain foreign antigens, they are not rejected, but this type of grafting has obvious limitations. **Allografts,** the most common type of grafts, are exchanges between genetically different individuals belonging to the same species (two humans). A close genetic correlation is sought for most allograft transplants (see next section). A **xenograft** is a tissue exchange between individuals of different species. Until rejection can be better controlled, most xenografts are experimental or for temporary therapy only.

Avoiding and Controlling Graft Incompatibility

Graft rejection can be averted or lessened by directly comparing the tissue of the recipient with that of potential donors. Several tissue matching procedures are used. In the *mixed lymphocyte reaction* (*MLR*), lymphocytes of the two individuals are mixed and incubated. If an incompatibility exists, some of the cells will become activated and proliferate. *Tissue typing* is similar to blood typing, except that specific antisera are used to disclose the HLA antigens on the surface of lymphocytes. In most grafts (one exception is bone marrow transplants), the ABO blood type must also be matched. Although a small amount of incompatibility is tolerable in certain grafts (liver, heart, kidney), a closer match is more likely to be successful, so the closest match possible is sought.

Types of Transplants

Today, transplantation is a recognized medical procedure whose benefit is reflected in the fact that more than 25,000 transplants take place each year in the United States. It has been performed on every major organ, including parts of the brain. The most frequent transplant operations involve skin, liver, heart, kidney, coronary artery, cornea, and bone marrow. The sources of organs and tissues are live donors (kidney, skin, bone marrow, liver), cadavers (heart, kidney, cornea), and fetal tissues. In the past decade, we have witnessed some unusual types of grafts. For instance, the fetal pancreas has been implanted as a potential treatment for diabetes, and fetal brain tissues have been implanted for Parkinson disease. Part of a liver has been transplanted from a live parent to a child, and parents have donated a lobe from their lungs to help restore function in their children with severe cystic fibrosis.

Recent advances in stem cell technology have made it possible to isolate stem cells directly from the blood of donors without bone marrow sampling. Another potential source is the umbilical cord blood from a newborn infant. These have expanded the possibilities for treatment and survival.

Bone marrow transplantation is a rapidly growing medical procedure for patients with immune deficiencies, aplastic anemia, leukemia and other cancers, and radiation damage. This procedure is extremely expensive, costing up to $400,000 per patient. Before bone marrow from a closely matched donor can be infused **(Insight 16.4),** the patient is pretreated with chemotherapy and whole-body irradiation, a procedure designed to destroy the person's own blood stem cells and thus prevent rejection of the new marrow cells. Within 2 weeks to a month after infusion, the grafted cells are established in the host. Because donor lymphoid cells can still cause GVHD, antirejection drugs may be necessary. An amazing consequence of bone marrow transplantation is that a recipient's blood type may change to the blood type of the donor.

INSIGHT 16.4 Medical

The Mechanics of Bone Marrow Transplantation

In some ways, bone marrow is the most exceptional form of transplantation. It does not involve invasive surgery in either the donor or recipient, and it permits the removal of tissue from a living donor that is fully replaceable. While the donor is sedated, a bone marrow/blood sample is aspirated by inserting a special needle into an accessible marrow cavity. The most favorable sites are the crest and spine of the ilium (major bone of the pelvis). During this procedure, which lasts 1 to 2 hours, 3% to 5% of the donor's marrow is withdrawn in 20 to 30 separate extractions. Between 500 and 800 milliliters of marrow are removed. The donor may experience some pain and soreness, but there are rarely any serious complications. In a few weeks, the depleted marrow will naturally replace itself. Implanting the harvested bone marrow is rather convenient, because it is not necessary to place it directly into the marrow cavities of the recipient. Instead, it is dripped intravenously into the circulation, and the new marrow cells automatically settle in the appropriate bone marrow regions. The survival and permanent establishment of the marrow cells are increased by administering various growth factors and stem cell stimulants to the patient.

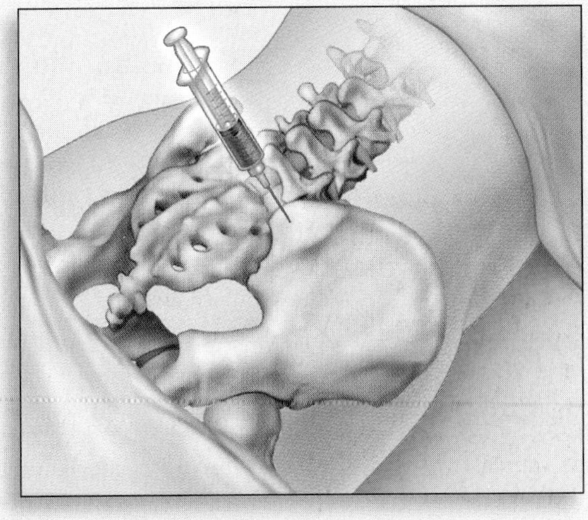

Removal of a bone marrow sample for transplantation. Samples are removed by inserting a needle into the spine or crest of the ilium. (The ilium is a prolific source of bone marrow.)

CHECKPOINT

- Type IV hypersensitivity reactions occur when cytotoxic T cells attack either self tissue or transplanted foreign cells. Type IV reactions are also termed delayed hypersensitivity reactions because they occur hours to days after the antigenic challenge.
- Type IV hypersensitivity reactions are mediated by T lymphocytes and are carried out against foreign cells that show both a foreign MHC and a nonself receptor site.
- Examples of type IV reactions include the tuberculin reaction, contact dermatitis, and mismatched organ transplants (host rejection and GVHD reactions).
- The four classes of transplants or grafts are determined by the degree of MHC similarity between graft and host. From most to least similar, these are: autografts, isografts, allografts, and xenografts.
- Graft rejection can be minimized by tissue matching procedures, immunosuppressive drugs, and use of tissues that do not provoke a type IV response.

16.6 An Inappropriate Response Against Self, or Autoimmunity

The immune diseases we have covered so far are all caused by foreign antigens. In the case of autoimmunity, an individual actually develops hypersensitivity to him or herself. This pathologic process accounts for **autoimmune diseases,** in which **autoantibodies,** T cells, and, in some cases, both mount an abnormal attack against self antigens. The scope of autoimmune diseases is extremely varied. In general, they are either *systemic,* involving several major organs, or *organ-specific,* involving only one organ or tissue. They usually fall into the categories of type II or type III hypersensitivity, depending upon how the autoantibodies bring about injury. There are more than 80 recognized autoimmune diseases. Some major diseases, their targets, and basic pathology are presented in **table 16.4.** (For a reminder of hypersensitivity types, refer to table 16.1.)

Genetic and Gender Correlation in Autoimmune Disease

In most cases, the precipitating cause of autoimmune disease remains obscure, but we do know that susceptibility is determined by genetics and influenced by gender. Cases cluster in families, and even unaffected members tend to develop the autoantibodies for that disease. More direct evidence comes from studies of the major histocompatibility gene complex. Particular genes in the class I and II major histocompatibility complex coincide with certain autoimmune diseases. For example, autoimmune joint diseases such as rheumatoid arthritis and ankylosing spondylitis are more common in persons with the B-27 HLA type; systemic lupus erythematosus, Graves disease, and myasthenia gravis are associated with the B-8 HLA antigen. Why autoimmune diseases (except ankylosing spondylitis) afflict more females than males also remains a mystery. Females are more susceptible during childbearing years than before puberty or after menopause, suggesting a possible hormonal relationship.

TABLE 16.4 Selected Autoimmune Diseases

Disease	Target	Type of Hypersensitivity	Characteristics
Systemic lupus erythematosus (SLE)	Systemic	III	Inflammation of many organs; antibodies against red and white blood cells, platelets, clotting factors, nucleus DNA
Rheumatoid arthritis and ankylosing spondylitis	Systemic	III and IV	Vasculitis; frequent target is joint lining; antibodies against other antibodies (rheumatoid factor)
Scleroderma	Systemic	II	Excess collagen deposition in organs; antibodies formed against many intracellular organelles
Hashimoto's thyroiditis	Thyroid	II	Destruction of the thyroid follicles
Graves disease	Thyroid	II	Antibodies against thyroid-stimulating hormone receptors
Pernicious anemia	Stomach lining	II	Antibodies against receptors prevent transport of vitamin B_{12}.
Myasthenia gravis	Muscle	II	Antibodies against the acetylcholine receptors on the nerve-muscle junction alter function.
Type I diabetes	Pancreas	IV	T cells attack insulin-producing cells.
Multiple sclerosis	Myelin	II and IV	T cells and antibodies sensitized to myelin sheath destroy neurons.
Goodpasture syndrome (glomerulonephritis)	Kidney	II	Antibodies to basement membrane of the glomerulus damage kidneys.
Rheumatic fever	Heart	II	Antibodies to group A *Streptococcus* cross-react with heart tissue.

The Origins of Autoimmune Disease

Because otherwise healthy individuals show (very low levels of) autoantibodies, it is suspected that there is a function for them. A moderate, regulated amount of autoimmunity is probably required to dispose of old cells and cellular debris. Disease apparently arises when this regulatory or recognition apparatus goes awry. Attempts to explain the origin of autoimmunity include the following theories.

The *sequestered antigen theory* explains that during embryonic growth, some tissues are immunologically privileged; that is, they are sequestered behind anatomical barriers and cannot be scanned by the immune system. Examples of these sites are regions of the central nervous system, which are shielded by the meninges and blood-brain barrier; the lens of the eye, which is enclosed by a thick sheath; and antigens in the thyroid and testes, which are sequestered behind an epithelial barrier. Eventually the antigen becomes exposed by means of infection, trauma, or deterioration and is perceived by the immune system as a foreign substance.

According to the **clonal selection theory,** the immune system of a fetus develops tolerance by eradicating all self-reacting lymphocyte clones, called *forbidden clones,* while retaining only those clones that react to foreign antigens. Some of these forbidden clones may survive; and because they have not been subjected to this tolerance process, they can attack tissues with self antigens.

The *theory of immune deficiency* proposes that mutations in the receptor genes of some lymphocytes render them reactive to self or that a general breakdown in the normal suppression of the immune response sets the scene for inappropriate immune responses.

Inappropriate expression of MHC II markers on cells that don't normally express them has been found to cause abnormal immune reactions to self. In a related phenomenon, T-cell activation may incorrectly "turn on" B cells that can react with self antigens. This phenomenon is called the *bystander effect.*

Some autoimmune diseases appear to be caused by *molecular mimicry,* in which microbial antigens bear molecular determinants similar to normal human cells. An infection could cause formation of antibodies that can cross-react with tissues. This is one purported explanation for the pathology of rheumatic fever. Another probable example of mimicry leading to autoimmune disease is the skin condition psoriasis. Although the etiology of this condition is complex and involves the inheritance of certain types of MHC alleles, infection with group A streptococci also plays a role. Scientists report that T cells primed to react with streptococcal surface proteins also react with keratin cells in the skin, causing them to proliferate. For this reason, psoriasis patients often report flare-ups after a strep throat infection.

Autoimmune disorders such as type I diabetes and multiple sclerosis are possibly triggered by *viral infection.* Viruses can noticeably alter cell receptors, thereby causing immune cells to attack the tissues bearing viral receptors.

The most recent theory of autoimmunity involves a protein called the **autoimmune regulator,** known as AIRE. In healthy subjects, this protein directs the transcription of many self antigens in the thymus. Their expression there instructs the immune system not to respond to them. Many patients with autoimmune diseases display defects in this protein. In those cases, the immune response is not instructed to ignore those self antigens.

Examples of Autoimmune Disease

Systemic Autoimmunities

One of the most severe chronic autoimmune diseases is systemic lupus erythematosus (SLE, or lupus). This name originated from the characteristic butterfly-shaped rash that drapes across the nose and cheeks **(figure 16.16*a*).** Apparently, ancient physicians thought the rash resembled a wolf bite (*lupus* in Latin for wolf). Although the manifestations of the disease vary considerably, all patients produce autoantibodies against a great variety of organs and tissues. The organs most involved are the kidneys, bone marrow, skin, nervous system, joints, muscles, heart, and GI tract. Antibodies to intracellular materials such as the nucleoprotein of the nucleus and mitochondria are also common.

In SLE, autoantibody-autoantigen complexes appear to be deposited in the basement membranes of various organs. Kidney failure, blood abnormalities, lung inflammation, myocarditis, and skin lesions are the predominant symptoms. One form of chronic lupus (called discoid) is influenced by exposure to the sun and primarily afflicts the skin. The etiology of lupus is still a puzzle. It is not known how such a generalized loss of self-tolerance arises, though viral infection or loss of normal immune response suppression are suspected. The fact that women of childbearing years account for 90% of cases indicates that hormones may be involved. The diagnosis of SLE can usually be made with blood tests. Antibodies against the nucleus and various tissues (detected by indirect fluorescent antibody or radioimmune assay techniques) are common, and a positive test for the lupus factor (an antinuclear factor) is also very indicative of the disease.

Rheumatoid arthritis, another systemic autoimmune disease, incurs progressive, debilitating damage to the joints. In some patients, the lung, eye, skin, and nervous system are also involved. In the joint form of the disease, autoantibodies form immune complexes that bind to the synovial membrane of the joints and activate phagocytes and stimulate release of cytokines. Chronic inflammation leads to scar tissue and joint destruction. The joints in the hands and feet are affected first, followed by the knee and hip joints **(figure 16.16*b*).** The precipitating cause in rheumatoid arthritis is not known, though infectious agents such as Epstein-Barr virus have been suspected. The most common feature of the disease is the presence of an IgM antibody, called rheumatoid factor (RF), directed against other antibodies. This does not cause the disease but is used mainly in diagnosis. The symptoms are complicated by a type IV delayed hypersensitivity response.

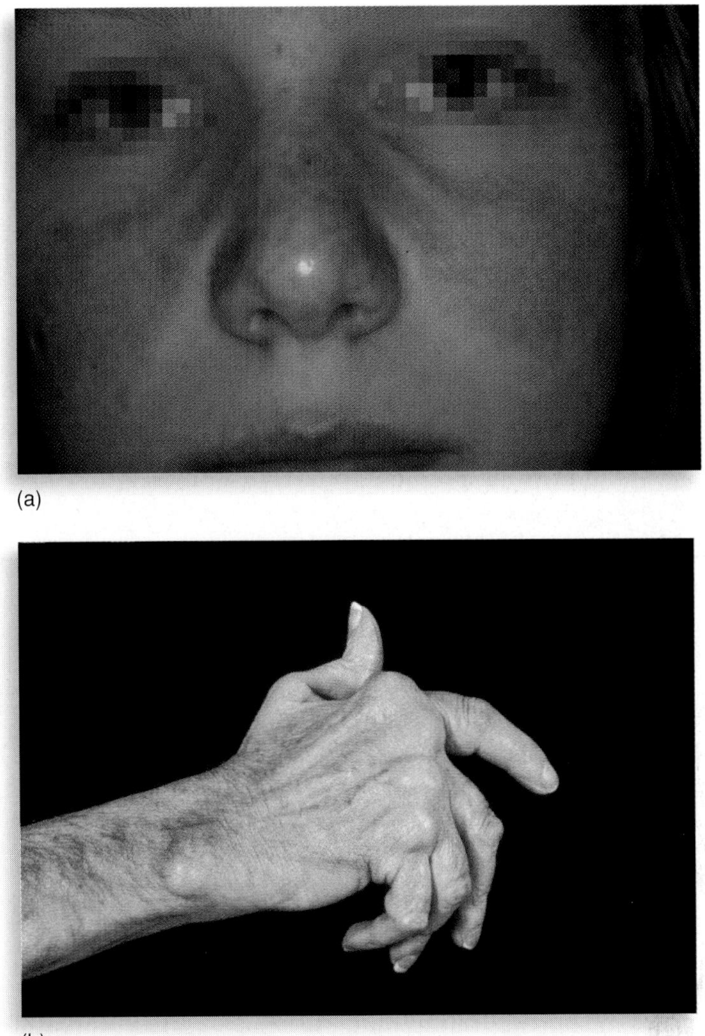

(a)

(b)

Figure 16.16 Common autoimmune diseases.
(a) Systemic lupus erythematosus. One symptom is a prominent rash across the bridge of the nose and on the cheeks. These papules and blotches can also occur on the chest and limbs. **(b)** Rheumatoid arthritis commonly targets the synovial membrane of joints. Over time, chronic inflammation causes thickening of this membrane, erosion of the articular cartilage, and fusion of the joint. These effects severely limit motion and can eventually swell and distort the joints.

Autoimmunities of the Endocrine Glands

On occasion, the thyroid gland is the target of autoimmunity. The underlying cause of **Graves' disease** is the attachment of autoantibodies to receptors on the follicle cells that secrete the hormone thyroxin. The abnormal stimulation of these cells causes the overproduction of this hormone and the symptoms of hyperthyroidism. In **Hashimoto's thyroiditis,** both autoantibodies and T cells are reactive to the thyroid gland, but in this instance, they reduce the levels of thyroxin by destroying follicle cells and by inactivating the hormone. As a result of these reactions, the patient suffers from hypothyroidism.

The pancreas and its hormone, insulin, are other autoimmune targets. Insulin, secreted by the beta cells in the pancreas, regulates and is essential to the utilization of glucose by cells. **Diabetes mellitus** is caused by a dysfunction in insulin production or utilization. Type I diabetes is associated with sensitized T cells that damage the beta cells. A complex inflammatory reaction leading to lysis of these cells greatly reduces the amount of insulin secreted.

A recent experimental study actually "cured" people of their type 1 diabetes by wiping out their immune systems with powerful drugs, after some of their bone marrow stem cells had been removed. When scientists then re-infused these patients with their own stem cells, a functional immune system was rebuilt that did not attack pancreatic cells. We can expect to see more examples of these types of treatments as our understanding of autoimmunity increases.

Neuromuscular Autoimmunities

Myasthenia gravis is named for the pronounced muscle weakness that is its principal symptom. Although the disease afflicts all skeletal muscle, the first effects are usually felt in the muscles of the eyes and throat. Eventually, it can progress to complete loss of muscle function and death. The classic syndrome is caused by autoantibodies binding to the receptors for acetylcholine, a chemical required to transmit a nerve impulse across the synaptic junction to a muscle **(figure 16.17).** The immune attack so severely damages the muscle cell membrane that transmission is blocked and paralysis ensues. Current treatment usually includes immunosuppressive drugs and therapy to remove the autoantibodies from the circulation. Experimental therapy using immunotoxins to destroy lymphocytes that produce autoantibodies shows some promise.

Multiple sclerosis (MS) is a paralyzing neuromuscular disease associated with lesions in the insulating myelin sheath that surrounds neurons in the white matter of the central nervous system. The underlying pathology involves damage to the sheath by both T cells and autoantibodies that severely compromises the capacity of neurons to send impulses. The principal motor and sensory symptoms are muscular weakness and tremors, difficulties in speech and vision, and some degree of paralysis. Most MS patients first experience symptoms as young adults, and they tend to experience remissions (periods of relief) alternating with recurrences of disease throughout their lives. Convincing evidence from studies of the brain tissue of MS patients points to a strong connection between the disease and infection with human herpesvirus 6. The disease can be treated passively with monoclonal antibodies that target T cells, and a vaccine containing the myelin protein has shown beneficial effects. Immunosuppressants such as cortisone and interferon beta may also alleviate symptoms.

CHECKPOINT

- Autoimmune hypersensitivity reactions occur when autoantibodies or host T cells mount an abnormal attack against self antigens. Autoimmune antibody responses can be either local or systemic type II or type III hypersensitivity reactions. Autoimmune T-cell responses are type IV hypersensitivity reactions.
- Susceptibility to autoimmune disease appears to be influenced by gender and by genes in the MHC complex.
- Autoimmune disease may be an excessive response of a normal immune function, the appearance of sequestered antigens, "forbidden" clones of lymphocytes that react to self antigens, or the result of alterations in the immune response caused by infectious agents, particularly viruses.
- Examples of autoimmune diseases include systemic lupus erythematosus, rheumatoid arthritis, diabetes mellitus, myasthenia gravis, and multiple sclerosis.

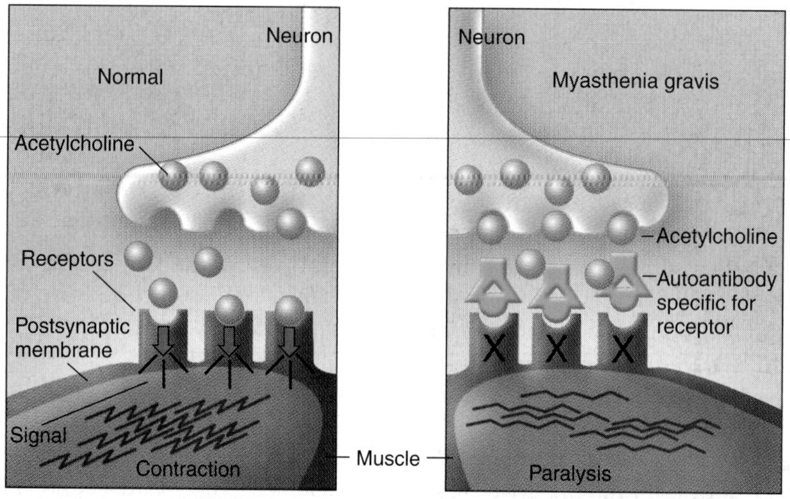

Figure 16.17 Mechanism for involvement of autoantibodies in myasthenia gravis.
Antibodies developed against receptors on the postsynaptic membrane block them so that acetylcholine cannot bind and muscle contraction is inhibited.

16.7 Immunodeficiency Diseases: Hyposensitivity of the Immune System

It is a marvel that development and function of the immune system proceed as normally as they do. On occasion, however, an error occurs and a person is born with or develops weakened immune responses. In many cases, these "experiments" of nature have provided penetrating insights into the exact functions of certain cells, tissues, and organs because of the specific signs and symptoms shown by the immunodeficient individuals. The predominant consequences of immunodeficiencies are recurrent, overwhelming infections, often with opportunistic microbes. Immunodeficiencies fall into two general categories: *primary diseases,* present at birth (congenital) and usually stemming from genetic errors, and *secondary diseases,* acquired after birth and caused by natural or artificial agents **(table 16.5).**

Primary Immunodeficiency Diseases

Deficiencies affect both specific immunities such as antibody production and less-specific ones such as phagocytosis. Consult **figure 16.18** to survey the places in the normal sequential development of lymphocytes where defects can occur and the possible consequences. In many cases, the deficiency is due to an inherited abnormality, though the exact nature of the abnormality is not known for a number of diseases. Because the development of B cells and T cells departs at some point, an individual can lack one or both cell lines. It must be emphasized, however, that some deficiencies affect other cell functions. For example, a T-cell deficiency can affect B-cell function because of the role of T helper cells. In some deficiencies, the lymphocyte in question is completely absent or is present at very low levels, whereas in others, lymphocytes are present but do not function normally.

Clinical Deficiencies in B-Cell Development or Expression

Genetic deficiencies in B cells usually appear as an abnormality in immunoglobulin expression. In some instances, only certain immunoglobulin classes are absent; in others, the levels of all types of immunoglobulins (Ig) are reduced. A significant number of B-cell deficiencies are X-linked (also called sex-linked) recessive traits, meaning that the gene occurs on the X chromosome and the disease appears primarily in male children.

The term **agammaglobulinemia** literally means the absence of gamma globulin, the fraction of serum that contains immunoglobulins. Because it is very rare for Ig to be completely absent, some physicians prefer the term **hypogammaglobulinemia.** T-cell function in these patients is usually normal. The symptoms of recurrent, serious bacterial infections usually appear about 6 months after birth. The bacteria most often implicated are pyogenic cocci, *Pseudomonas,* and *Haemophilus influenzae;* and the most common infection sites are the lungs, sinuses, meninges, and blood. Many Ig-deficient patients can have recurrent infections with viruses and protozoa, as well. Patients often manifest a wasting syndrome and have a reduced life span, but modern therapy has improved their prognosis. The current treatment for this condition is passive immunotherapy with immune serum globulin and continuous antibiotic therapy.

The lack of a particular class of immunoglobulin is a relatively common condition. Although genetically controlled, its underlying mechanisms are not yet clear. IgA deficiency is the most prevalent, occurring in about one person in 600. Such persons have normal quantities of B cells and other

TABLE 16.5 General Categories of Immunodeficiency Diseases with Selected Examples

PRIMARY IMMUNE DEFICIENCIES (GENETIC)
B-cell defects (low levels of B cells and antibodies)
Agammaglobulinemia (X-linked, non-sex-linked)
Hypogammaglobulinemia
Selective immunoglobulin deficiencies
T-cell defects (lack of all classes of T cells)
Thymic aplasia (DiGeorge syndrome)
Chronic mucocutaneous candidiasis
Combined B-cell and T-cell defects (usually caused by lack or abnormality of lymphoid stem cell)
Severe combined immunodeficiency disease (SCID)
X-SCIDI due to an interleukin defect
Adenosine deaminase (ADA) deficiency
Wiskott-Aldrich syndrome
Ataxia-telangiectasia
Phagocyte defects
Chédiak-Higashi syndrome
Chronic granulomatous disease of children (see Case File 14, chapter 14)
Lack of surface adhesion molecules
Complement defects
Lacking one of C components
Hereditary angioedema
Associated with rheumatoid diseases
SECONDARY IMMUNE DEFICIENCIES (ACQUIRED)
From natural causes
Infections (AIDS) or cancers
Nutrition deficiencies
Stress
Pregnancy
Aging
From immunosuppressive agents
Irradiation
Severe burns
Steroids (cortisones)
Drugs to treat graft rejection and cancer
Removal of spleen

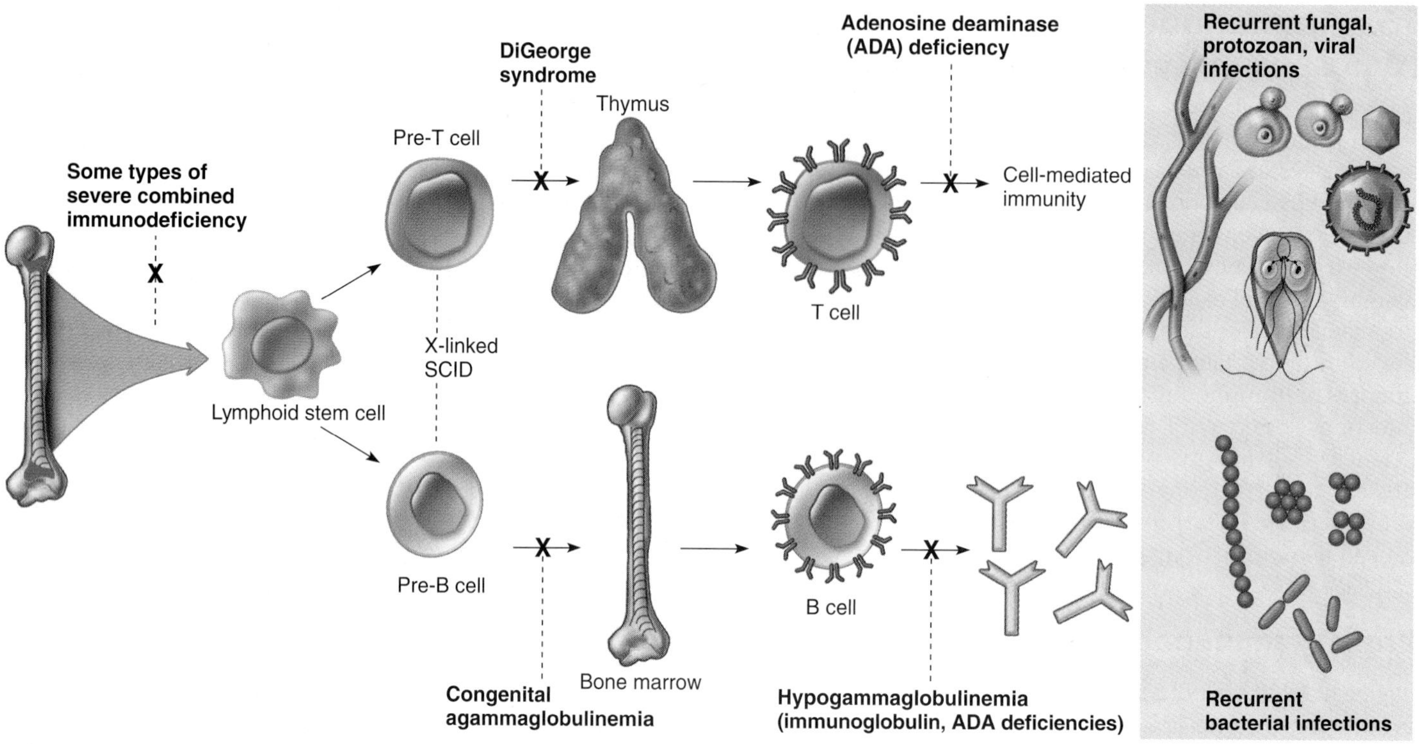

Figure 16.18 The stages of development and the functions of B cells and T cells, whose failure causes immunodeficiencies. Dotted lines represent the phases in development where breakdown can occur.

immunoglobulins, but they are unable to synthesize IgA. Consequently, they lack protection against local microbial invasion of the mucous membranes and suffer recurrent respiratory and gastrointestinal infections. The usual treatment using Ig replacement does not work, because conventional preparations are high in IgG, not IgA.

Clinical Deficiencies in T-Cell Development or Expression

Due to their critical role in immune defenses, a genetic defect in T cells results in a broad spectrum of disease, including severe opportunistic infections, wasting, and cancer. In fact, a dysfunctional T-cell line is usually more devastating than a defective B-cell line because T helper cells are required to assist in most specific immune reactions. The deficiency can occur anywhere along the developmental spectrum, from thymus to mature, circulating T cells.

Abnormal Development of the Thymus The most severe of the T-cell deficiencies involve the congenital absence or immaturity of the thymus gland. Thymic aplasia, or **DiGeorge syndrome,** results when the embryonic third and fourth pharyngeal pouches fail to develop. Some cases are associated with a deletion in chromosome 22. The accompanying lack of cell-mediated immunity makes children highly susceptible to persistent infections by fungi, protozoa, and viruses. Common, usually benign childhood infections such as chickenpox, measles, or mumps can be overwhelming and fatal in these children. Even vaccinations using attenuated microbes pose a danger. Other symptoms of thymic failure are reduced growth, wasting of the body, unusual facial characteristics **(figure 16.19),** and an increased incidence of lymphatic cancer. These children can have reduced antibody levels, and they are unable to reject transplants. The major therapy for them is a transplant of thymus tissue.

Severe Combined Immunodeficiencies: Dysfunction in B and T Cells

Severe combined immunodeficiencies (SCIDs) are the most dire and potentially lethal of the immunodeficiency diseases because they involve dysfunction in both lymphocyte systems. Some SCIDs are due to the complete absence of the lymphocyte stem cell in the marrow; others are attributable to the dysfunction of B cells and T cells later in development. Infants with SCID usually manifest the T-cell deficiencies within days after birth by developing candidiasis, sepsis, pneumonia, or systemic viral infections. This debilitating condition appears to have several forms. In the two most common forms, Swiss-type agammaglobulinemia and thymic alymphoplasia, the numbers of all types of lymphocytes are extremely low, the blood antibody content is greatly diminished, and the thymus and cell-mediated immunity are poorly developed. Both diseases are due to a genetic defect in the development of the lymphoid cell line.

A rarer form of SCID is **adenosine deaminase (ADA) deficiency,** which is caused by an autosomal recessive defect

INSIGHT 16.5 **Discovery**

An Answer to the Bubble Boy Mystery

David Vetter, the most famous SCID child, lived all but the last 2 weeks of his life in a sterile environment to isolate him from the microorganisms that could have quickly ended his life. When medical tests performed before birth had indicated that David might inherit this disease, he was delivered by cesarean section and immediately placed in a sterile isolette. From that time, he lived in various plastic chambers—ranging from room-size to a special suit that allowed him to walk outside. Remarkably, he developed into a well-adjusted child, even though his only physical contact with others was through special rubber gloves. When he was 12, his doctors decided to attempt a bone marrow transplant that might allow him to live free of his bubble prison. David was transplanted with his sister's bone marrow, but the marrow harbored a common herpesvirus called Epstein-Barr virus. Because he lacked any form of protective immunities against this oncogenic virus, a cancer spread rapidly through his body. Despite the finest medical care available, David died a short time later from the metastatic cancer.

In 1993, after several years of study, researchers discovered the basis for David's immunodeficiency. He had inherited a form that arises from a defective genetic code in the receptors for interleukin-2, interleukin-4, and interleukin-7. The defect prevents the receptors on T cells and B cells from receiving the appropriate interleukin signals for growth, development, and reactivity. The end result is that both cytotoxic immunities and antibody-producing systems are shut down, leaving the body defenseless against infections and cancer.

David Vetter, the boy in the plastic bubble.

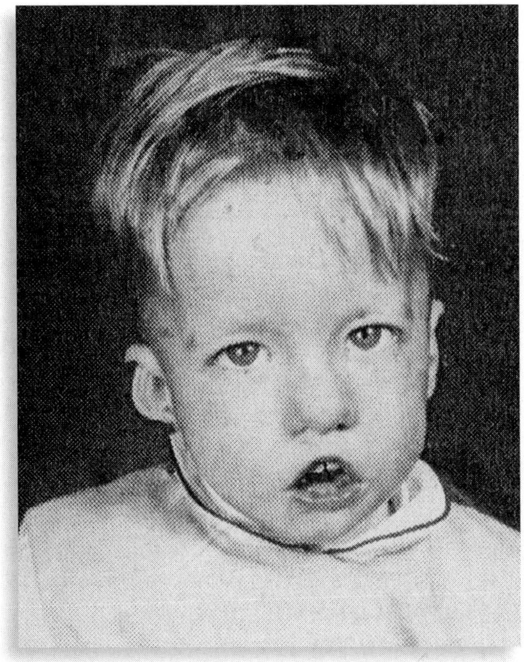

Figure 16.19 Facial characteristics of a child with DiGeorge syndrome.

Typical defects include low-set, deformed earlobes; wide-set, slanted eyes; a small, bowlike mouth; and the absence of a philtrum (the vertical furrow between the nose and upper lip).

in the metabolism of adenosine. In this case, lymphocytes develop but a metabolic product builds up abnormally and selectively destroys them. Infants with ADA deficiency are subject to recurrent infections and severe wasting typical of severe deficiencies. A small number of SCID cases are due to a developmental defect in receptors for B and T cells. An X-linked deficiency in interleukin receptors was responsible for the disease of David, the child in the "plastic bubble" **(Insight 16.5).** Another newly identified condition, *bare lymphocyte syndrome,* is caused by the lack of genes that code for class II MHC receptors.

Because of their profound lack of specific adaptive immunities, SCID children require the most rigorous kinds of aseptic techniques to protect them from opportunistic infections. Aside from life in a sterile plastic bubble, the only serious option for their longtime survival is total replacement or correction of dysfunctional lymphoid cells. Some infants can benefit from fetal liver or stem cell grafts. Although transplanting compatible bone marrow has been about 50% successful in curing the disease, it is complicated by graft versus host disease. The condition of some ADA-deficient patients has been partly corrected by periodic transfusions of blood containing large amounts of the normal enzyme. A more lasting treatment for both X-linked and ADA types of SCID would be gene therapy—insertion of normal genes to replace the defective genes (see chapter 10). Although several children have benefited by transfecting their bone marrow stem cells with the normal gene for ADA, this technique is still being worked out.

Secondary Immunodeficiency Diseases

Secondary acquired deficiencies in B cells and T cells are caused by one of four general agents:

1. infection,
2. organic disease,
3. chemotherapy, or
4. radiation.

The most recognized infection-induced immunodeficiency is **AIDS.** This syndrome is caused when several types of immune cells, including T helper cells, monocytes, macrophages, and antigen presenting cells, are infected by the human immunodeficiency virus (HIV). It is generally thought that the depletion of T helper cells and functional impairment of immune responses ultimately account for the cancers and opportunistic protozoan, fungal, and viral infections associated with this disease. See chapter 20 for an extensive discussion of AIDS. Other infections that can deplete immunities are measles, leprosy, and malaria.

Cancers that target the bone marrow or lymphoid organs can be responsible for extreme malfunction of both humoral and cellular immunity. In leukemia, the massive number of cancer cells compete for space and literally displace the normal cells of the bone marrow and blood. Plasma cell tumors produce large amounts of nonfunctional antibodies, and thymus gland tumors cause severe T-cell deficiencies.

An ironic outcome of lifesaving medical procedures is the possible suppression of a patient's immune system. For instance, some immunosuppressive drugs that prevent graft rejection by T cells can likewise suppress beneficial immune responses. Although radiation and anticancer drugs are the first line of therapy for many types of cancer, both agents are extremely damaging to the bone marrow and other body cells.

CHECKPOINT

- Immunodeficiency diseases occur when the immune response is reduced or absent.
- Primary immune diseases are genetically induced deficiencies of B cells, T cells, the thymus gland, or combinations of these.
- Secondary immune diseases are caused by infection, organic disease, chemotherapy, or radiation.
- The best-known infection-induced immunodeficiency is AIDS.

Chapter Summary with Key Terms

16.1 The Immune Response: A Two-Sided Coin
The study of disease states involving the malfunction of the immune system is called **immunopathology.**

A. **Allergy,** or hypersensitivity, is an exaggerated, misdirected expression of certain immune responses.
B. Autoimmunity involves abnormal responses to self antigens. A deficiency or loss in immune function is called **immunodeficiency.**

16.2 Type I Allergic Reactions: Atopy and Anaphylaxis

A. Immediate-onset allergies involve contact with **allergens,** antigens that affect certain people; susceptibility is inherited.
B. On first contact with allergen, specific B cells react with allergen and form a special antibody class called IgE, which affixes by its Fc receptor to mast cells and basophils.
   1. This **sensitizing dose** primes the allergic response system.
   2. Upon subsequent exposure with a provocative dose, the same allergen binds to the IgE–mast cell complex.
   3. This causes **degranulation,** release of intracellular granules containing mediators **(histamine, serotonin, leukotriene, prostaglandin),** with physiological effects such as vasodilation and bronchoconstriction.
   4. Symptoms are rash, itching, redness, increased mucus discharge, pain, swelling, and difficulty in breathing.
C. Diagnosis of allergy can be made by a histamine release test on basophils; serological assays for IgE; and skin testing, which injects allergen into the skin and mirrors the degree of reaction.
D. Control of allergy involves drugs to interfere with the action of histamine, inflammation, and release of cytokines from mast cells. **Desensitization** therapy involves the administration of purified allergens.

16.3 Type II Hypersensitivities: Reactions That Lyse Foreign Cells

A. Type II reactions involve the interaction of antibodies, foreign cells, and complement, leading to lysis of the foreign cells. In transfusion reactions, humans may become sensitized to special antigens on the surface of the red blood cells of other humans.
B. The **ABO blood groups** are genetically controlled: Type A blood has A antigens on the RBCs; type B has B antigens; type AB has both A and B antigens; and type O has neither antigen. People produce antibodies against A or B antigens if they lack these antigens. Antibodies can react with antigens if the wrong blood type is transfused.
C. **Rh factor** is another RBC antigen that becomes a problem if an Rh^- mother is sensitized by an Rh^+ fetus. A second fetus can receive antibodies she has made against the factor and develop **hemolytic disease of the newborn.** Prevention involves therapy with Rh immune globulin.

16.4 **Type III Hypersensitivities: Immune Complex Reactions**
Exposure to a large quantity of soluble foreign antigens (serum, drugs) stimulates antibodies that produce small, soluble Ag-Ab complexes. These **immune complexes** are trapped in various organs and tissues, which incites a damaging inflammatory response.

16.5 **Type IV Hypersensitivities: Cell-Mediated (Delayed) Reactions**
A delayed response to antigen involving the activation of and damage by T cells.

A. *Delayed allergic response:* Skin response to allergens, including infectious agents. Example is tuberculin reaction; contact dermatitis is caused by exposure to plants (ivy, oak) and simple environmental molecules (metals, cosmetics); cytotoxic T cells acting on allergen elicit a skin reaction.
B. *Graft rejection:* Reaction of cytotoxic T cells directed against foreign cells of a grafted tissue; involves recognition of foreign HLA by T cells and rejection of tissue.
   1. Host may reject graft; graft may reject host.
   2. Types of grafts include: **autograft,** from one part of body to another; **isograft,** grafting between identical twins; **allograft,** between two members of same species; **xenograft,** between two different species.
   3. All major organs may be successfully transplanted.
   4. Allografts require tissue match (HLA antigens must correspond); rejection is controlled with drugs.

16.6 **An Inappropriate Response Against Self, or Autoimmunity**

A. In certain type II and III hypersensitivities, the immune system has lost tolerance to self molecules (autoantigens) and forms **autoantibodies** and sensitized T cells against them. Disruption of function can be systemic or organ specific.
B. **Autoimmune diseases** are genetically influenced and more common in females.

16.7 **Immunodeficiency Diseases: Hyposensitivity of the Immune System**
Components of the immune response system are absent. Deficiencies involve B and T cells, phagocytes, and complement.

A. Primary immunodeficiency is genetically based, congenital; defect in inheritance leads to lack of B-cell activity, T-cell activity, or both.
B. B-cell defect is called **agammaglobulinemia;** patient lacks antibodies; serious recurrent bacterial infections result. In Ig deficiency, one of the classes of antibodies is missing or deficient.
C. In T-cell defects, the thymus is missing or abnormal. In **DiGeorge syndrome,** the thymus fails to develop; afflicted children experience recurrent infections with eukaryotic pathogens and viruses; immune response is generally underdeveloped.
D. In **severe combined immunodeficiency (SCID),** both limbs of the lymphocyte system are missing or defective; no adaptive immune response exists; fatal without replacement of bone marrow or other therapies.
E. Secondary (acquired) immunodeficiency is due to damage after birth (infections, drugs, radiation). **AIDS** is the most common of these; T helper cells are main target; deficiency manifests in numerous opportunistic infections and cancers.

Multiple-Choice and True-False Questions

Multiple-Choice Questions. Select the correct answer from the answers provided.

1. Pollen is which type of allergen?
   a. contactant c. injectant
   b. ingestant d. inhalant
2. B cells are responsible for which allergies?
   a. asthma c. tuberculin reactions
   b. anaphylaxis d. both a and b
3. The contact with allergen that results in symptoms is called the
   a. sensitizing dose c. provocative dose
   b. degranulation dose d. desensitizing dose
4. The direct, immediate cause of allergic symptoms is the action of
   a. the allergen directly on smooth muscle
   b. the allergen on B lymphocytes
   c. allergic mediators released from mast cells and basophils
   d. IgE on smooth muscle
5. Theoretically, type _____ blood can be donated to all persons because it lacks _____.
   a. AB, antibodies
   b. O, antigens
   c. AB, antigens
   d. O, antibodies
6. An example of a type III immune complex disease is
   a. serum sickness c. graft rejection
   b. contact dermatitis d. atopy
7. Type II hypersensitivities are due to
   a. IgE reacting with mast cells
   b. activation of cytotoxic T cells
   c. IgG-allergen complexes that clog epithelial tissues
   d. complement-induced lysis of cells in the presence of antibodies
8. Production of autoantibodies may be due to
   a. emergence of forbidden clones of B cells
   b. production of antibodies against sequestered tissues
   c. infection-induced change in receptors
   d. all of these are possible
9. Rheumatoid arthritis is an ____ that affects the ____.
   a. immunodeficiency disease, muscles
   b. autoimmune disease, nerves

c. allergy, cartilage
d. autoimmune disease, joints

10. Which disease would be most similar to AIDS in its pathology?
a. X-linked agammaglobulinemia
b. SCID
c. ADA deficiency
d. DiGeorge syndrome

True-False Questions. If statement is true, leave as is. If it is false, correct it by rewriting the sentence.

11. T cells are associated with type IV allergies.
12. A positive tuberculin skin test is an example of antibody-mediated inflammation.
13. Contact dermatitis can be caused by proteins found in foods.
14. Antibody-mediated degranulation of most cells is involved in anaphylaxis.

Writing to Learn

These questions are suggested as a *writing-to-learn* experience. For each question, compose a one- or two-paragraph answer that includes the factual information needed to completely address the question.

1. a. Define allergy and hypersensitivity.
b. What accounts for the reactions that occur in these conditions?
c. What does it mean when a reaction is immediate or delayed?
d. Give examples of each type.
2. Describe several factors that influence types and severity of allergic responses.
3. a. How are atopic allergies similar to anaphylaxis?
b. How are they different?
4. a. How do allergens gain access to the body?
b. What are some examples of allergens that enter by these portals?
5. a. Trace the course of a pollen grain through sensitization and provocation in type I allergies.
b. Include in the discussion the role of mast cells, basophils, IgE, and allergic mediators.
c. Outline the target organs and symptoms of the principal atopic diseases and their diagnosis and treatment.
6. a. Describe the allergic response that leads to anaphylaxis. Include its usual causes, how it is diagnosed and treated, and two effective physiological targets for treatment.
b. Explain how hyposensitization is achieved and suggest two mechanisms by which it might work.
7. a. What is the mechanism of type II hypersensitivity?
b. Why are the tissues of some people antigenic to others?
c. Would we be concerned about this problem if it were not for transfusions?
d. What is the actual basis of the four ABO and Rh blood groups?
e. Where do we derive our natural hypersensivities to the A or B antigens that we do not possess?
f. How does a person become sensitized to Rh factor? List consequences.
8. Explain the rules of transfusion. Illustrate what will happen if type A blood is accidently transfused into a type B person.
9. a. Contrast type II and type III hypersensitivities with respect to type of antigen, antibody, and manifestations of disease.
b. What is immune complex disease?
10. a. In general, what causes primary immunodeficiencies?
b. Acquired immunodeficiencies?
c. Why can T-cell deficiencies have greater impact than B-cell deficiencies?
d. What kinds of symptoms accompany a B-cell defect?
e. A T-cell defect?
f. Combined defects?
g. Give examples of specific diseases that involve each type of defect.

Concept Mapping

Appendix D provides guidance for working with concept maps.

1. Construct your own concept map using the following terms as the *concepts*. Supply the linking words between each pair of concepts.

lysed cells
degranulation release of mediators
immune complexes
damage by T cells
allergens
cell-bound antibody
processed antigen
soluble antigen

Critical Thinking Questions

Critical thinking is the ability to reason and solve problems using facts and concepts. These questions can be approached from a number of angles, and in most cases, they do not have a single correct answer.

1. a. Discuss the reasons that the immune system is sometimes called a double-edged sword.
   b. Suggest a possible function of allergy.
2. A 3-week-old neonate develops severe eczema after being given penicillin therapy for the first time. Can you explain what has happened?
3. Can you explain why a person would be allergic to strawberries when he eats them but shows a negative skin test to them?
4. a. Where in the course of type I allergies do antihistamine drugs work? Exactly what do they do?
   b. Cortisone?
   c. Desensitization?
5. Why would it be necessary for an Rh^- woman who has had an abortion, miscarriage, or an ectopic pregnancy to be immunized against the Rh factor?
6. a. Describe three circumstances that might cause antibodies to develop against self tissues.
   b. Can you explain how people with autoimmunity could develop antibodies against intracellular components (nucleus, mitochondria, and DNA)?
7. Would a person show allergy to poison oak upon first contact? Explain why or why not.
8. Why are primary immunodeficiencies considered experiments of nature?
9. a. Explain why babies with agammaglobulinemia do not develop opportunistic infections until about 6 months after birth.
   b. Explain why people with B-cell deficiencies can benefit from artificial passive immunotherapy. Explain whether vaccination would work for them.

Visual Understanding

1. **From chapter 15, figure 15.16.** How would a person's immunity be affected if he or she had a deficiency in CD8 cells? Would a deficiency in CD4 cells have a greater or lesser effect? Explain your answer.

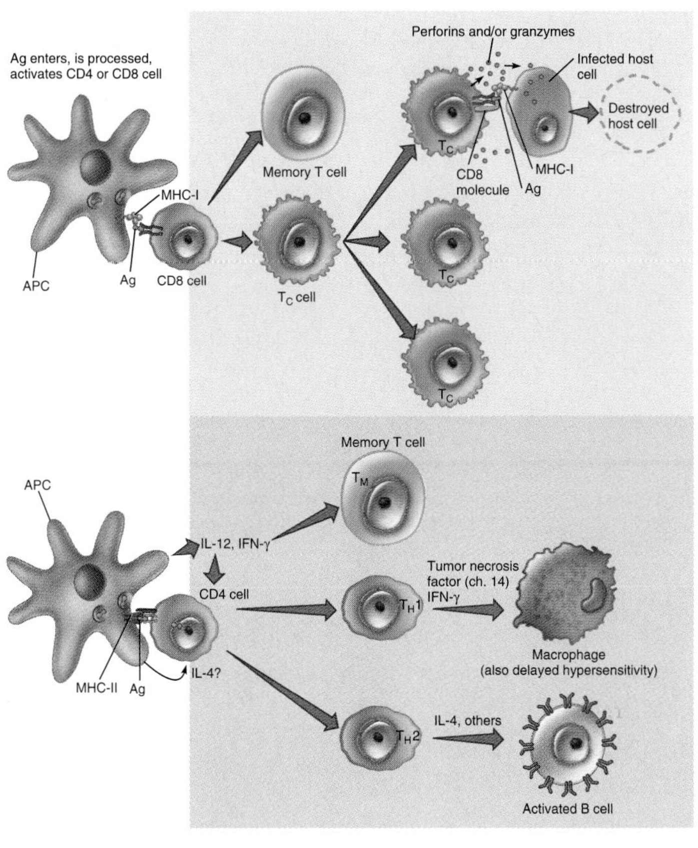

2. **Figure 16.10*c*.** Draw the agglutination patterns for the other four common blood types.

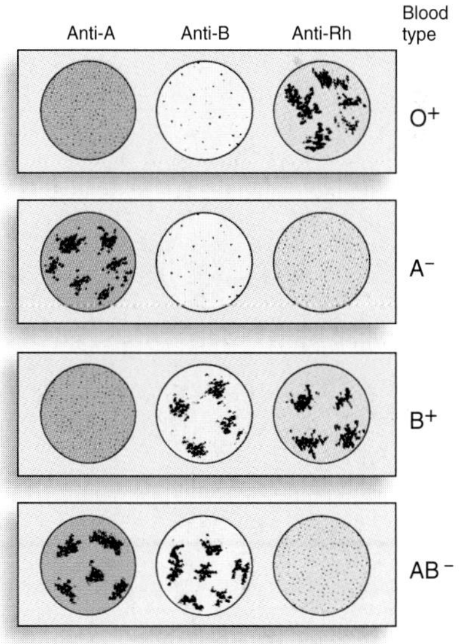

Internet Search Topics

Go to: www.aris.mhhe.com, and click on "microbiology" and then this textbook's author/title. Go to chapter 16, access the URLs listed under Internet Search Topics, and research the following:

1. Bone marrow transplants. Locate information on registering to donate marrow and the process surrounding testing and transplantation.
2. Go to the website listed and access the pollen count section. Look up the data for the allergen content at the station closest to you. Determine how the counts change between fall, winter, spring, and summer.
3. Go to the website listed for food allergies. Determine which foods are most implicated in allergies. What measures must be taken by people with food allergies?
4. Research allergies, autoimmunity, and immunodeficiencies on the websites listed.

CHAPTER 17

Diagnosing Infections

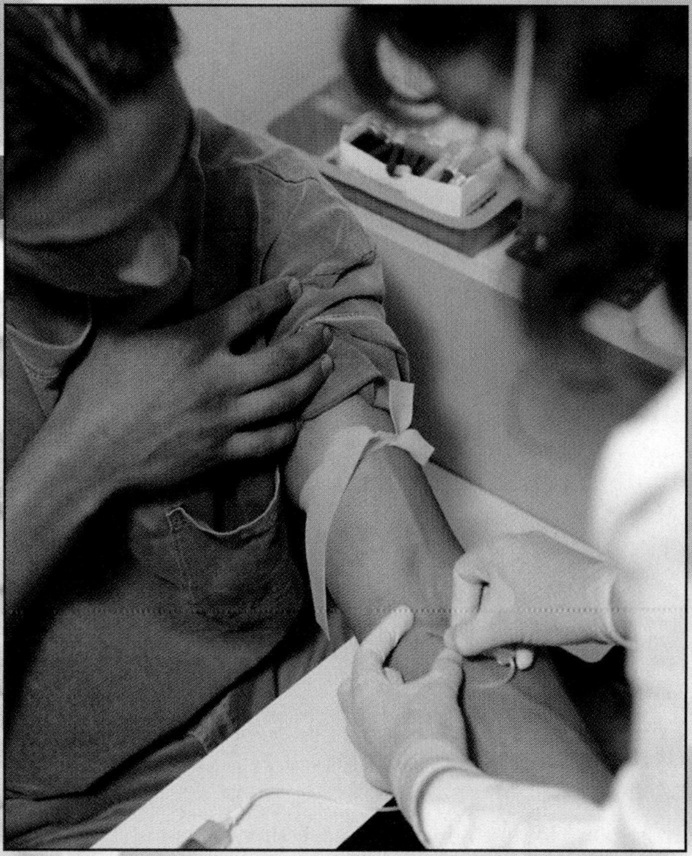

CASE FILE 17

While attempting to start an intravenous line on an AIDS patient, a registered nurse punctured her finger with a visibly bloody needle. She was observing proper universal precautions, including wearing two layers of gloves and having proper needle disposal equipment nearby. Despite these precautions, the needlestick injury occurred. Fearing that she would receive a poor performance evaluation for making such a mistake, she did not report the incident to her supervisor, rather electing to simply scrub the wound with soap. Had she reported the incident, she would have been given postexposure prophylaxis consisting of two reverse transcriptase inhibitors and a protease inhibitor for approximately 1 month to prevent an HIV infection. Three months after the incident, fear of becoming HIV positive led her to consult with her family physician. The family physician drew a blood sample from the nurse and sent it to a qualified testing facility.

The testing facility employed a test called an enzyme-linked immunosorbent assay (ELISA) (see p. 527) commonly used to screen blood for exposure to HIV by detecting antibodies to the virus. This screening test yielded a positive result—therefore, the test was repeated with the same result. As a confirmatory step, Western blot analysis of the nurse's blood demonstrated an indeterminate result. Based on the Western blot results, a second confirmatory test, called the p24 antigen capture assay, was used to test the blood. This test utilizes an antibody sandwich ELISA method to detect HIV p24 protein in the blood specimen. The p24 antigen capture assay yielded negative results, confirming that the nurse was not HIV infected.

- *Why might the nurse test positive by the ELISA screening test but negative by the p24 antigen capture assay confirmatory test? What is another term for this ELISA test result?*
- *How common are "sharps" injuries among health care workers?*
- *How likely is it that a person would become HIV positive after being punctured by an HIV-contaminated needle?*

Case File 17 Wrap-Up appears on page 527.

CHAPTER OVERVIEW

- The ability to identify microbes that are responsible for a patient's symptoms is central to infectious disease microbiology. Diagnosis might be considered an art and a science, involving multiple health care providers and clinical personnel.
- Accurate specimen collection is the cornerstone of accurate diagnosis.
- Phenotypic identification methods assess a microbe's appearance, growth characteristics, and/or arsenal of chemicals and enzymes. Direct examination of specimens, cultivation of specimens, and biochemical testing are tools used for phenotypic methods of diagnosis.
- Genotypic methods examine the genetic content of a microbe by analyzing its G + C content, its DNA, or its rRNA sequence. Polymerase chain reaction is used to increase the amount of DNA in a sample so that it can be analyzed.
- Immunologic methods of identification can be used to probe the antigenic makeup of a microbe or to identify the presence of antibodies to a microbe in a patient's blood. There are many variations on immunologic methods, including introducing antigens into a patient to detect the presence of an immune response. These methods are often referred to as *serological* tests.
- Accurate diagnosis depends on the ability to differentiate the causative organism from normal biota or contaminating microbes.

17.1 Preparation for the Survey of Microbial Diseases

In chapters 18 through 23, the most clinically significant bacterial, fungal, parasitic, and viral diseases are covered. The chapters survey the most prevalent infectious conditions and the organisms that cause them. This chapter gets us started with an introduction to the how-to of diagnosing the infections.

For many students (and professionals), the most pressing topic in microbiology is *how to identify unknown bacteria* in patient specimens or in samples from nature. Methods microbiologists use to identify bacteria to the level of genus and species fall into three main categories: *phenotypic,* which includes a consideration of morphology (microscopic and macroscopic) as well as bacterial physiology or biochemistry; *immunologic,* which entails serological analysis; and *genotypic* (or genetic) techniques. Data from a cross section of such tests can produce a unique profile of each bacterium. Increasingly, genetic means of identification are being used as a sole resource for identifying bacteria. As universally used databases become more complete because of submissions from scientists and medical personnel worldwide, genetic analyses provide a more accurate and speedy way of identifying microbes than was possible even a decade ago. There are still many organisms, however, that must be identified in the "old-fashioned" way—via biochemical, serological, and morphological means. Serology is so reliable for some diseases that it may never be replaced. All of these methods—phenotypic, genotypic, and serological—are described in this chapter.

Phenotypic Methods

Microscopic Morphology

Traits that can be valuable aids to identification are combinations of cell shape and size; Gram stain reaction; acid-fast reaction; and special structures, including endospores, granules, and capsules. Electron microscope studies can pinpoint additional structural features (such as the cell wall, flagella, pili, and fimbriae).

Macroscopic Morphology

Traits that can be assessed with the naked eye are also useful in diagnosis. These include the appearance of colonies, including texture, size, shape, pigment, speed of growth, and patterns of growth in broth and gelatin media.

Physiological/Biochemical Characteristics

These have been the traditional mainstay of bacterial identification. Enzymes and other biochemical properties of bacteria are fairly reliable and stable expressions of the chemical identity of each species. Dozens of diagnostic tests exist for determining the presence of specific enzymes and to assess nutritional and metabolic activities. Examples include tests for fermentation of sugars; capacity to digest or metabolize complex polymers such as proteins and polysaccharides; production of gas; presence of enzymes such as catalase, oxidase, and decarboxylases; and sensitivity to antimicrobic drugs. Special rapid identification test systems that record the major biochemical reactions of a culture have streamlined data collection.

Chemical Analysis

This involves analyzing the types of specific structural substances that the microorganism contains, such as the chemical composition of peptides in the cell wall and lipids in membranes.

Genotypic Methods

Examining the genetic material itself has revolutionized the identification and classification of bacteria. There are many advantages of genotypic methods over phenotypic methods,

INSIGHT 17.1 **Discovery**

The Uncultured

By the 1990s, it was clear to microbiologists that culture-based (phenotypic) methods for identifying bacteria were becoming inadequate. This was first confirmed by environmental researchers, who came to believe that at most 1% (and in some environments it was 0.001%) of microbes present in lakes, soil, and saltwater environments could be grown in laboratories and, therefore, were unknown and unstudied. These microbes are termed **viable nonculturable,** or **VNC.**

Although it took microbiologists many years to come to this realization, once they did, it made sense. Scientists had spent several decades (since microbes could first routinely be grown) concocting recipes for media and having great success in growing all kinds of bacteria from all kinds of environments. They had plenty to do, just in identifying and studying those. By the 1990s, the advent of non-culture-dependent tools, such as gene probing and PCR, revealed vast numbers of species that had never before turned up on a culture dish. That this vast zoo of microbes was revealed in environmental samples was not surprising due to the huge array of microenvironments that would have had to have been reproduced in media for them to be grown in the lab.

But medical microbiologists felt fairly confident that they could culture microbes from a human, since the "environment" of human tissues is well understood. Although some human-inhabiting microbes cannot be grown in culture, we never suspected that we were missing a large proportion of them. In 1999, three Stanford University scientists applied PCR techniques to a collection of subgingival plaque harvested from one of their own mouths. They used a wide library of DNA fragments as probes, essentially "fishing" for new isolates. Oral biologists had previously recovered about 500 bacterial strains from this site; the Stanford scientists found 30 species that had never before been cultured or described. This discovery shook the medical world and led to increased investigation of "normal" human biota, using non-culture-based methods to find VNCs in the human body.

The new realization that our bodies are hosts to a wide variety of microbes about which we know nothing has several implications. As evolutionary microbiologist Paul Ewald has said, "What are all those microbes doing in there?" He points out that many oral microbes previously assumed to be innocuous are now associated with cancer and heart disease. Many of the diseases that we currently think of as noninfectious will likely be found to have an infectious cause once we learn to look for VNCs. Another question is: *How did the microbes get there?* One organism found in the oral cavity was a metal-oxidizing soil bacterium. The researchers speculate that it may have been obtained from drinking water. Ewald suggests that many of our oral residents (and gastrointestinal biota), including those that may turn out to be stealthily pathogenic, have been obtained from a very common activity—kissing. This is an activity that may not be as innocuous as previously thought.

when they are available. The primary advantage is that actually culturing the microorganisms is not always necessary. In recent decades, scientists have come to realize that there are many more microorganisms that we can't grow in the lab compared with those that we can **(Insight 17.1).** Another advantage is that genotypic methods are increasingly automated, and results are obtained very quickly, often with more precision than with phenotypic methods.

Immunologic Methods

Bacteria and other microbes have surface and other molecules called antigens that are recognized by the immune system. One immune response to antigens is the production of molecules called antibodies that are designed to bind tightly to the antigens. The nature of the antibody response is also exploited for diagnosis when a patient's blood (or other tissue) is tested for the presence of specific antibodies to a suspected pathogen. This is often easier than testing for the microbe itself, especially in the case of viral infections. Most HIV testing entails examination of a person's blood for presence of antibody to the virus. Laboratory kits based on this technique are available for immediate identification of a number of pathogens.

17.2 On the Track of the Infectious Agent: Specimen Collection

Regardless of the method of diagnosis, specimen collection is the common point that guides the health care decisions of every member of a clinical team. Indeed, the success of identification and treatment depends on how specimens are collected, handled, and stored. Specimens can be taken by a clinical laboratory scientist or medical technologist, nurse, physician, or even by the patient. However, it is imperative that general aseptic procedures be used, including sterile sample containers and other tools to prevent contamination

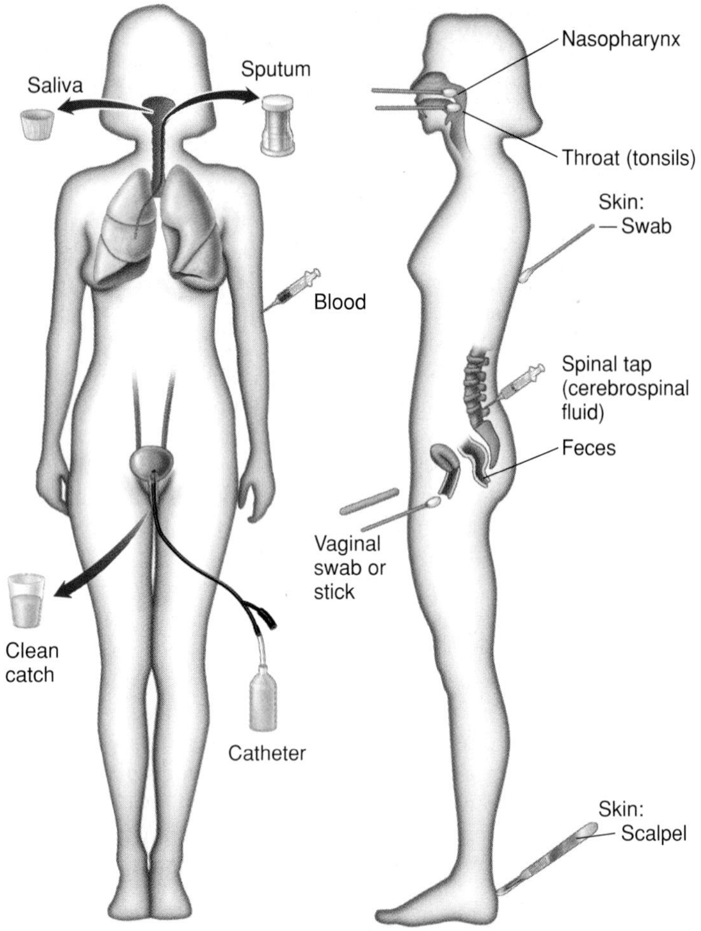

Figure 17.1 Sampling sites and methods of collection for clinical laboratories.

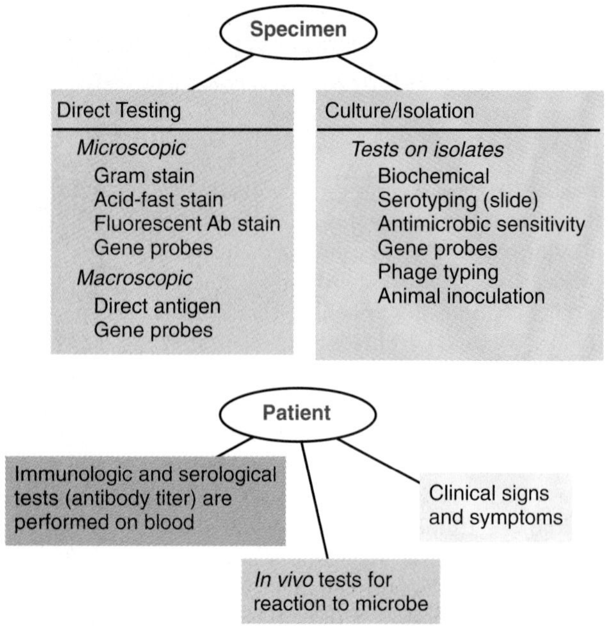

Figure 17.2 A scheme of specimen isolation and identification.

from the environment or the patient. **Figure 17.1** delineates the most common sampling sites and procedures.

In sites that normally contain resident microbiota, care should be taken to sample only the infected site and not surrounding areas. For example, throat and nasopharyngeal swabs should not touch the tongue, cheeks, or saliva. Saliva is an especially undesirable contaminant because it contains millions of bacteria per milliliter, most of which are normal biota. Saliva samples are occasionally taken for dental diagnosis by having the patient expectorate into a container. Depending on the nature of the lesion, skin can be swabbed or scraped with a scalpel to expose deeper layers. The mucus lining of the vagina, cervix, or urethra can be sampled with a swab or applicator stick.

Urine is taken aseptically from the bladder with a thin tube called a catheter. Another method, called a "clean catch," is taken by washing the external urethra and collecting the urine midstream. The latter method inevitably incorporates a few normal biota into the sample, but these can usually be differentiated from pathogens in an actual infection. Sometimes diagnostic techniques require first-voided "dirty catch" urine. Sputum, the mucus secretion that coats the lower respiratory surfaces, especially the lungs, is discharged by coughing or taken by catheterization to avoid contamination with saliva. Sterile materials such as blood, cerebrospinal fluid, and tissue fluids must be taken by sterile needle aspiration. Antisepsis of the puncture site is extremely important in these cases. Additional sources of specimens are the eye, ear canal, nasal cavity (all by swab), and diseased tissue that has been surgically removed (biopsied).

After proper collection, the specimen is promptly transported to a lab and stored appropriately (usually refrigerated) if it must be held for a time. Nonsterile samples in particular, such as urine, feces, and sputum, are especially prone to deterioration at room temperature. Special swab and transport systems are designed to collect the specimen and maintain it in stable condition for several hours. These devices contain nonnutritive maintenance media (so the microbes survive but do not grow), a buffering system, and an anaerobic environment to prevent possible destruction of oxygen-sensitive bacteria.

Overview of Laboratory Techniques

The routes taken in specimen analysis are the following: (1) direct tests using microscopic, immunologic, or genetic methods that provide immediate clues as to the identity of the microbe or microbes in the sample; and (2) cultivation, isolation, and identification of pathogens using a wide variety of general and specific tests **(figure 17.2).** Most test results fall into two categories: presumptive data, which place the isolated microbe (isolate) in a preliminary category such as a genus, and more specific, confirmatory data, which provide more definitive evidence of a species. Some tests are more important for some groups of bacteria than for others. The total time required for analysis ranges from a few minutes in a streptococcal sore throat to several weeks in tuberculosis.

MICROBIOLOGY UNIT

DATE, TIME & PERSON COLLECTING	SPECIMEN NUMBER	ANTIBIOTIC THERAPY	TENTATIVE DIAGNOSIS

SOURCE OF SPECIMEN

- ☐ THROAT
- ☐ SPUTUM
- ☐ STOOL
- ☐ CERVIX
- ☐ AEROSOL INDUCED SPUTUM
- ☐ BLOOD
- ☐ URINE - CLEAN CATCH
- ☐ URINE - CATH
- ☐ BRONCHIAL WASHING
- ☐ WOUND - SPECIFY SITE ____
- ☐ OTHER - SPECIFY ____

TEST REQUEST

- ☐ GRAM STAIN
- ☐ ROUTINE CULTURE
- ☐ SENSITIVITY
- ☐ MIC
- ☐ ANAEROBIC CULTURE
- ☐ R/O GROUP A STREP
- ☐ WRIGHT STAIN (WBC)
- ☐ ACID FAST SMEAR
- ☐ ACID FAST CULTURE
- ☐ FUNGUS WET MOUNT
- ☐ FUNGUS CULTURE
- ☐ PARASITE STUDIES
- ☐ OCCULT BLOOD
- ☐ PCR ANALYSIS

DO NOT WRITE BELOW THIS LINE -# FOR LAB USE ONLY

GRAM STAIN (4+ NUMEROUS; 3+ MANY; 2+ MODERATE; 1+FEW; 0 NONE SEEN)

COCCI: GRAM POS.____ GRAM NEG.____ W B C____

BACILLI: GRAM POS.____ GRAM NEG.____ EPITHELIAL CELLS____

INTRACELLULAR & EXTRACELLULAR GRAM-NEGATIVE DIPLOCOCCI____

YEAST____ ☐ No organisms seen.

FUNGUS: WET MOUNT ☐ No mycotic elements or budding structures seen.

☐ ____

CULTURE ☐ ____

AFB: SMEAR ☐ No acid fast bacilli seen.

☐ ____

CULTURE ☐ ____

PARASITE STUDIES: DIRECT:____

CONCENTRATE:____

PERMANENT:____

OCCULT BLOOD:

APPEARANCE OF STOOL:____

OCCULT BLOOD:____

COLONY COUNT: Urine organisms/ml. ☐ ____ ☐ > 100,000

MISCELLANEOUS RESULTS:

- ☐ NO GROWTH IN: ☐ 2 DAYS ☐ 3 DAYS ☐ 5 DAYS ☐ 7 DAYS
- ☐ NORMAL FLORA ISOLATED
- ☐ NO ENTEROPATHOGENS ISOLATED
- ☐ SPUTUM UNACCEPTABLE FOR CULTURE — REPRESENTS SALIVA — NEW SPECIMEN REQUESTED
- ☐ URINE > 2 COLONY TYPES PRESENT REPRESENT CONTAMINATION — NEW SPECIMEN REQUESTED

CULTURE RESULTS

1+ FEW 3+ MANY

2+ MODERATE 4+ NUMEROUS

ANAEROBES	☐ BACTEROIDES ☐ CLOSTRIDIUM ☐ PEPTOSTREPTOCOCCUS ☐ ☐
ENTERICS	☐ ESCHERICHIA COLI ☐ ENTEROBACTER ☐ KLEBSIELLA ☐ PROTEUS ☐
STAPHYLOCOCCUS	☐ AUREUS ☐ EPIDERMIDIS ☐ SAPROPHYTICUS
STREPTOCOCCUS	☐ GROUP A ☐ GROUP B ☐ GROUP D ENTEROCOCCI ☐ GROUP D NON ENTEROCOCCI ☐ PNEUMONIAE ☐ VIRIDANS ☐
YEAST	☐ CANDIDA ☐
OTHER ISOLATES	☐ PSEUDOMONAS ☐ HAEMOPHILUS ☐ GARDNERELLA VAGINALIS ☐ NEISSERIA ☐ CL. DIFFICILE ☐

SENSITIVITY TESTS

NOTE: Bacteria with intermediate susceptibility may not respond satisfactorily to therapy.	AMIKACIN	AMPICILLIN	BETA LACTAMASE PRODUCTION	CARBENICILLIN	CEFAZOLIN	CEFOTAXIME	CEFOXITIN	CEFUROXIME	CHLORAMPHENICOL	CLINDAMYCIN	ERYTHROMYCIN	GENTAMICIN	METHICILLIN	METRONIDAZOLE	NITROFURANTOIN	PENICILLIN	IMMUNOLOGY	TETRACYCLINE	TOBRAMYCIN	TRIMETHO PRIM SULFAME THOXAZOLE	VANCOMYCIN	CIPROFLOX
A																						
B																						
C																						

☐ COMMENTS: ____

DATE ____ TECHNOLOGIST ____

706-30A

Figure 17.3 Example of a clinical form used to report data on a patient's specimens.

Results of specimen analysis are entered in a summary patient chart **(figure 17.3)** that can be used in assessment and treatment regimens. The type of antimicrobial drugs chosen for testing varies with the type of microorganism isolated.

Some diseases are diagnosed without the need to identify microbes from specimens. Serological tests on a patient's serum can detect signs of an antibody response. One method that clarifies whether a positive test indicates current or prior infection is to take two samples several days apart to see if the antibody titer is rising. Skin testing can pinpoint a delayed allergic reaction to a microorganism. These tests are also important in screening the general population for exposure to an infectious agent such as rubella or tuberculosis.

Because diagnosis is both a science and an art, the ability of the practitioner to interpret signs and symptoms of disease can be very important. AIDS, for example, is usually diagnosed by serological tests and a complex of signs and symptoms without ever isolating the virus. Some diseases (athlete's foot, for example) are diagnosed purely by the typical presenting symptoms and may require no lab tests at all.

17.3 Phenotypic Methods

Immediate Direct Examination of Specimen

Direct microscopic observation of a fresh or stained specimen is one of the most rapid methods of determining presumptive and sometimes confirmatory characteristics. Stains most often employed for bacteria are the Gram stain (see Insight 4.2) and the acid-fast stain (see figure 21.16). For many species these ordinary stains are useful, but they do not work with certain organisms. Direct fluorescence antibody (DFA) tests can highlight the presence of the microbe in patient specimens by means of labeled antibodies **(figure 17.4).** DFA tests are particularly useful for bacteria, such as the syphilis spirochete, that are not readily cultivated in the laboratory or if rapid diagnosis is essential for the survival of the patient.

Another way that specimens can be analyzed is through *direct antigen testing,* a technique similar to direct fluorescence in that known antibodies are used to identify antigens on the surface of bacterial isolates. But in direct antigen testing, the reactions can be seen with the naked eye. Quick test kits that greatly speed clinical diagnosis are available for *Staphylococcus aureus, Streptococcus pyogenes, Neisseria gonorrhoeae, Haemophilus influenzae,* and *Neisseria meningitidis.* However, when the microbe is very sparse in the specimen, direct testing is like looking for a needle in a haystack, and more sensitive methods are necessary.

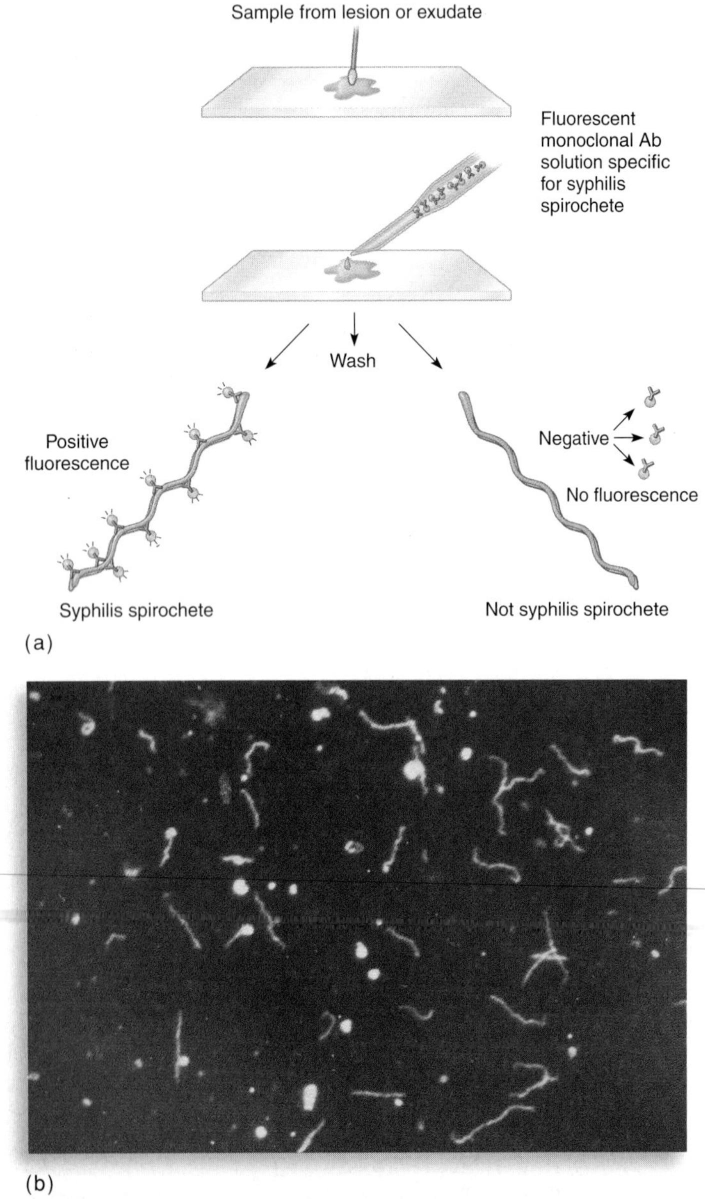

Figure 17.4 Direct fluorescence antigen test.
(a) Results for *Treponema pallidum*, the syphilis spirochete, and an unrelated spirochete. **(b)** Photomicrograph of this technique used on a blood sample from a syphilitic patient.

Cultivation of Specimen

Isolation Media

Such a wide variety of media exist for microbial isolation that a certain amount of preselection must occur, based on the nature of the specimen. In cases in which the suspected pathogen is present in small numbers or is easily overgrown, the specimen can be initially enriched with specialized media. In specimens such as urine and feces that have high bacterial counts and a diversity of species, selective media are used. In most cases, specimens are also inoculated into differential media that define such characteristics as reactions in blood (blood agar) and fermentation patterns (mannitol salt and MacConkey agar). A patient's blood is usually cultured in a special bottle of broth that can be periodically sampled for growth. Numerous other examples of isolation, differential, and biochemical media were presented in chapter 3. So that subsequent steps in identification will be as accurate as possible, all work must be done from isolated colonies or pure cultures, because working with a mixed or contaminated culture gives misleading and inaccurate results. From such isolates, clinical microbiologists obtain information about a pathogen's microscopic morphology and staining reactions, cultural appearance, motility, oxygen requirements, and biochemical characteristics.

Biochemical Testing

The physiological reactions of bacteria to nutrients and other substrates provide excellent indirect evidence of the types of enzyme systems present in a particular species. Many of these tests are based on an enzymatic reaction (a step in the bacterium's metabolic pathway) that is visualized by a color change.

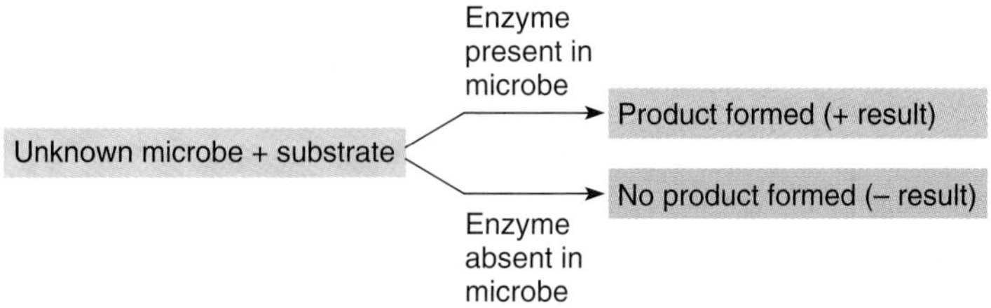

The microbe is cultured in a medium with a special substrate and then tested for a particular end product. The presence of the end product (made visible by a color dye) indicates that the enzyme is expressed in that species; its absence means it lacks the enzyme for utilizing the substrate in that particular way. These types of reactions are particularly meaningful in bacteria, which are haploid and generally express their genes for utilizing a given nutrient.

Among the prominent biochemical tests are carbohydrate fermentation (acid and/or gas); hydrolysis of gelatin, starch, and other polymers; enzyme actions such as catalase, oxidase, and coagulase; and various by-products of metabolism. Many are presently performed with rapid, miniaturized systems that can simultaneously determine up to 23 characteristics in small individual cups or spaces **(figure 17.5).** An important plus, given the complexity of biochemical profiles, is that such systems are readily adapted to computerized analysis.

Common schemes exist for identifying bacteria. These are based on easily recognizable characteristics such as motility, oxygen requirements, Gram stain reactions, shape, spore formation, and various biochemical reactions. Schemes can be set up as flowcharts **(figure 17.6)** that trace a route of identification by offering pairs of opposing characteristics (positive versus negative, for example) from which to select. Flowcharts that offer two choices at each level are called **dichotomous keys.** Eventually, an endpoint is reached, and the name of a genus or species that fits that particular combination of characteristics appears. Diagnostic tables that provide more complete information are preferred by many laboratories because variations from the general characteristics used on the flowchart can be misleading.

Miscellaneous Tests

When morphological and biochemical tests are insufficient to complete identification, other tests come into play.

Bacteria host viruses called bacteriophages that are very species- and strain-specific. Such selection by a virus for its host is useful in typing some bacteria, primarily *Staphylococcus* and *Salmonella.* The technique of phage typing involves inoculating a lawn of cells onto a Petri dish, mapping it off into blocks, and applying a different phage to each block. Cleared areas corresponding to lysed cells indicate sensitivity to that phage. Phage typing is chiefly used for tracing strains of bacteria in epidemics.

Animals must be inoculated to cultivate bacteria such as *Mycobacterium leprae* and *Treponema pallidum,* whereas avian embryos and cell cultures are used to grow rickettsias, chlamydias, and viruses. Animal inoculation is also occasionally used to test bacterial or fungal virulence.

Antimicrobial sensitivity tests are not only important in determining the drugs to be used in treatment (see figure 12.17), but the patterns of sensitivity can also be used in presumptive identification of some species of *Streptococcus, Pseudomonas,* and *Clostridium.* Antimicrobials are also used as selective agents in many media.

Determining Clinical Significance of Cultures

Questions that can be difficult but necessary to answer in this era of debilitated patients and opportunists are, Is an isolate clinically important? and How do you decide whether it is a contaminant or just part of the normal biota? The number of microbes in a sample is one useful criterion. For example, a few colonies of *Escherichia coli* in a urine sample can simply indicate normal biota, whereas several hundred can mean active infection. In contrast, the presence of a single colony of a true pathogen such as *Mycobacterium tuberculosis* in a sputum culture or an opportunist in sterile sites such as cerebrospinal fluid or blood is highly suggestive of its role in disease. Furthermore, the repeated isolation of a relatively pure culture of any microorganism can mean it is an agent of disease, though care must be taken in this diagnosis.

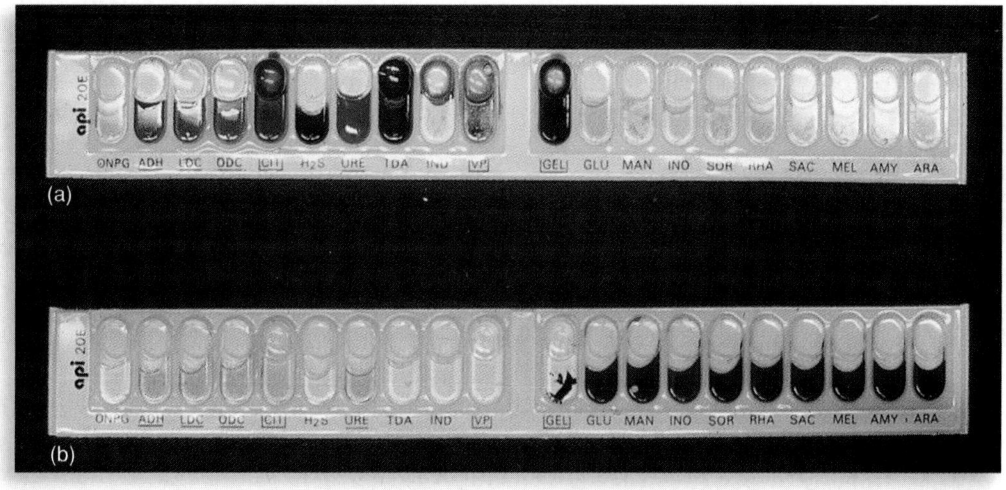

Figure 17.5 Rapid tests.
The API 20E manual biochemical system for microbial identification. Samples of a single bacterial culture are placed in the 20 different cups, which already contain chemicals designed to test for a particular enzyme. Different bacterial cultures were used in strip a and strip b. The culture in **(a)** was positive for every tested enzyme; the culture in **(b)** was negative for each one.

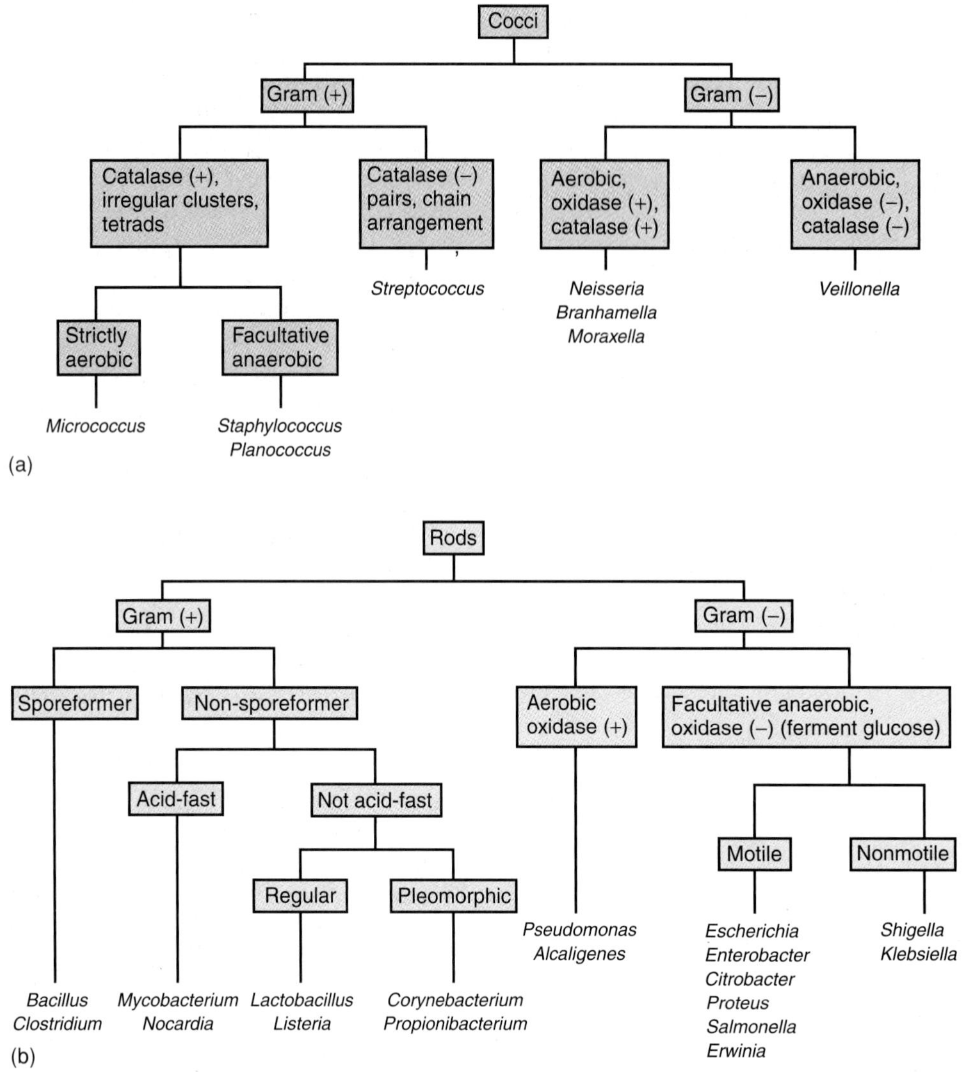

Figure 17.6 Flowchart to separate primary genera of gram-positive and gram-negative bacteria. **(a)** Cocci and **(b)** rods commonly involved in human diseases.

Another problem facing clinical laboratory personnel is that of differentiating a pathogen from species in the normal biota that are similar in morphology to their more virulent relatives.

17.4 Genotypic Methods

DNA Analysis Using Genetic Probes

The exact order and arrangement of the DNA code is unique to each organism. With a technique called *hybridization*, it is possible to identify a bacterial species by analyzing segments of its DNA. This requires small fragments of single-stranded DNA (or RNA) called **probes** that are known to be complementary to the specific sequences of DNA from a particular microbe. The test is conducted by extracting unknown test DNA from cells in specimens or cultures and binding it to special blotter paper. After several different probes have been added to the blotter, it is observed for visible signs that the probes have become fixed (hybridized) to the test DNA. The binding of probes onto several areas of the test DNA indicates close correspondence and makes positive identification possible **(figure 17.7)**.

Nucleic Acid Sequencing and rRNA Analysis

One of the most viable indicators of evolutionary relatedness and affiliation is comparison of the sequence of nitrogen bases in ribosomal RNA, a major component of ribosomes. Ribosomes have the same function (protein synthesis) in all cells, and they tend to remain more or less stable in their nucleic acid content over long periods. Thus, any major difference in the sequence, or "signature," of the rRNA is likely to indicate some distance in ancestry. This technique

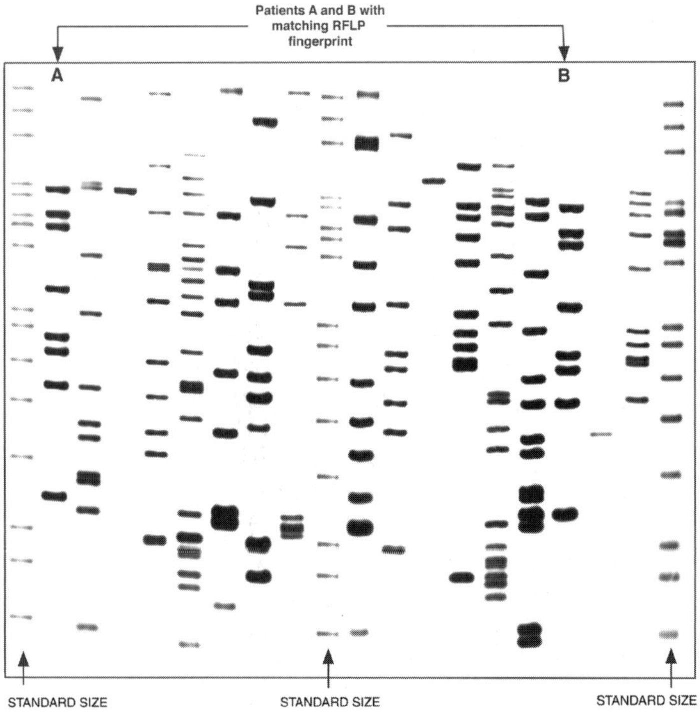

Figure 17.7 DNA typing of restriction fragment length polymorphisms (RFLPs) for *Mycobacterium tuberculosis*.
The pattern shows results for various strains isolated from 17 patients. Bands were developed by DNA hybridization using probes specific to genes from several *M. tuberculosis* strains. Lanes 1, 10, and 20 provide size markers for reference. Patients A and B are infected with the same common strain of the pathogen.

is powerful at two levels: It is effective for differentiating general group differences (it was used to separate the three superkingdoms of life discussed in chapter 1), and it can be fine-tuned to identify at the species level (for example, in *Mycobacterium* and *Legionella*).

Polymerase Chain Reaction

Many nucleic acid assays use the polymerase chain reaction **(PCR).** This method can amplify DNA present in samples even in tiny amounts, which greatly improves the sensitivity of the test (see figure 10.7). PCR tests are being used or developed for a wide variety of bacteria, viruses, protozoa, and fungi.

Since 9/11, very rapid identification of pathogens has been a research priority in the United States. This has led to a type of microbial identification system called a **biosensor.** These apparatuses often employ PCR techniques. Samples can either be manually loaded into a machine so that rapid PCR is conducted, or the machines can "sample" the environment by retrieving air or fluid samples on a regular basis and subjecting them to genomic techniques so personnel can continuously monitor the pathogen census in an environment.

CHECKPOINT

- Microbiologists use three categories of techniques to diagnose infections: *phenotypic*, *genotypic*, and *immunologic*.
- The first step in clinical diagnosis (after observing the patient) is obtaining a sample. If this step is not performed correctly, the test will not be accurate no matter how "sensitive."
- The main phenotypic methods include the direct examination of specimens, observing the growth of specimen cultures on special media, and biochemical testing of specimen cultures.
- The use of genotypic methods has been increasing rapidly. These include genetic probing, nucleic acid sequencing, rRNA analysis and PCR-based methods.

17.5 Immunologic Methods

The antibodies formed during an immune reaction are important in combating infection, but they hold additional practical value. Characteristics of antibodies (such as their quantity or specificity) can reveal the history of a patient's contact with microorganisms or other antigens. This is the underlying basis of serological testing. **Serology** is the branch of immunology that traditionally deals with *in vitro* diagnostic testing of serum. Serological testing is based on the familiar concept that antibodies have extreme specificity for antigens, so when a particular antigen is exposed to its specific antibody, it will fit like a hand in a glove. The ability to visualize this interaction by some means provides a powerful tool for detecting, identifying, and quantifying antibodies—or for that matter, antigens. The scheme works both ways, depending on the situation. One can detect or identify an unknown antibody using a known antigen, or one can use an antibody of known specificity to help detect or identify an unknown antigen **(figure 17.8).** Modern serological testing has grown into a field that tests more than just serum. Urine, cerebrospinal fluid, whole tissues, and saliva can also be used to determine the immunologic status of patients. These and other immune tests are helpful in confirming a suspected diagnosis or in screening a certain population for disease.

General Features of Immune Testing

The strategies of immunologic tests are diverse, and they underline some of the brilliant and imaginative ways that antibodies and antigens can be used as tools. We summarize them under the headings of agglutination, precipitation, immunodiffusion, complement fixation, fluorescent antibody tests, and immunoassay tests. First we overview the general characteristics of immune testing, and we then look at each type separately.

The most effective serological tests have a high degree of specificity and sensitivity **(figure 17.9).** *Specificity* is the property of a test to focus upon only a certain antibody or antigen and not to react with unrelated or distantly related

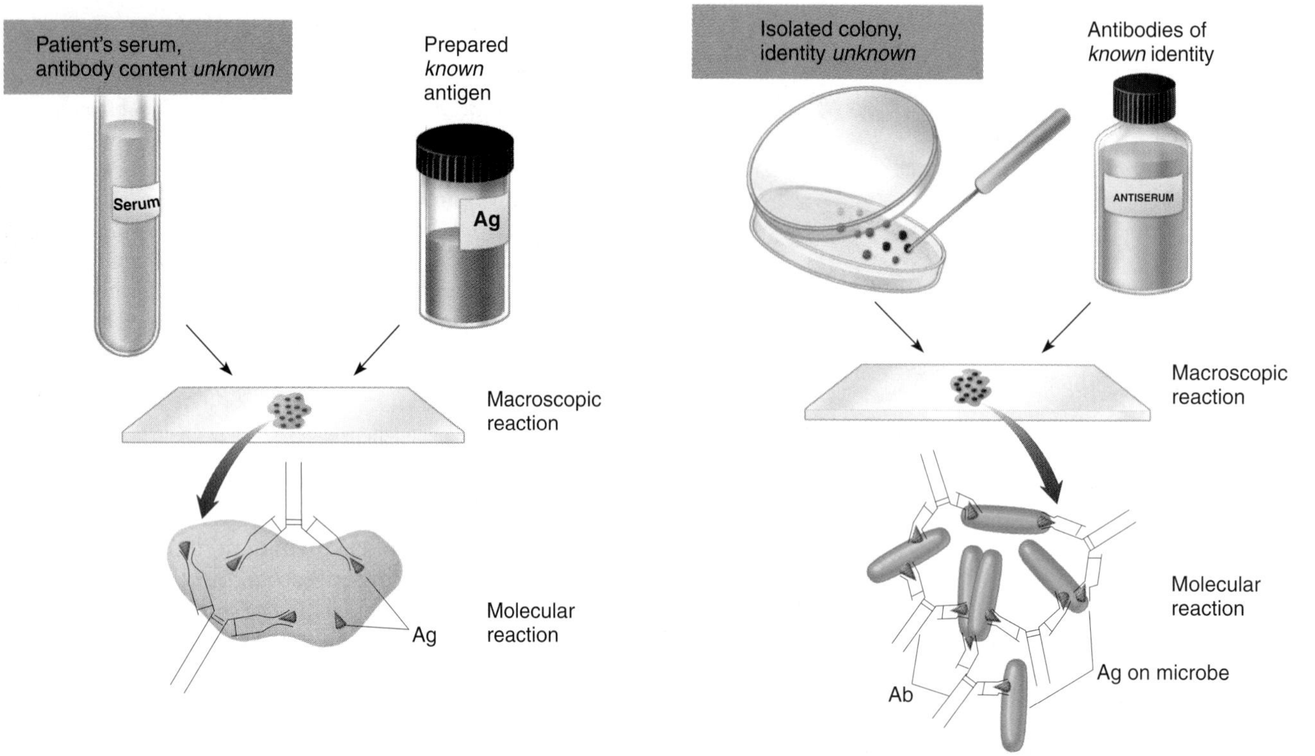

(a) In serological diagnosis of disease, a blood sample is scanned for the presence of antibody using an antigen of known specificity. A positive reaction is usually evident as some visible sign, such as color change or clumping, that indicates a specific interaction between antibody and antigen. (The reaction at the molecular level is rarely observed.)

(b) An unknown microbe is mixed with serum containing antibodies of known specificity, a procedure known as serotyping. Microscopically or macroscopically observable reactions indicate a correct match between antibody and antigen and permit identification of the microbe.

Figure 17.8 Basic principles of serological testing using antibodies and antigens.

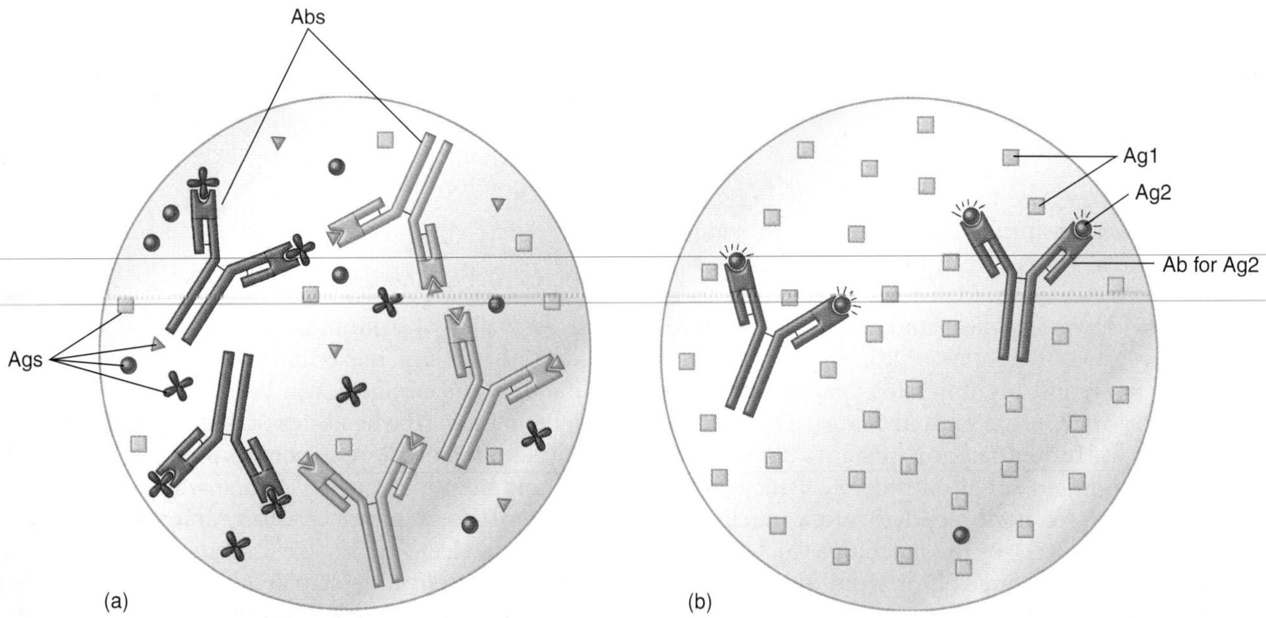

Figure 17.9 Specificity and sensitivity in immune testing.
(a) This test shows specificity in which an antibody (Ab) attaches with great exactness with only one type of antigen (Ag). **(b)** Sensitivity is demonstrated by the fact that Ab can pick up antigens even when the antigen is greatly diluted.

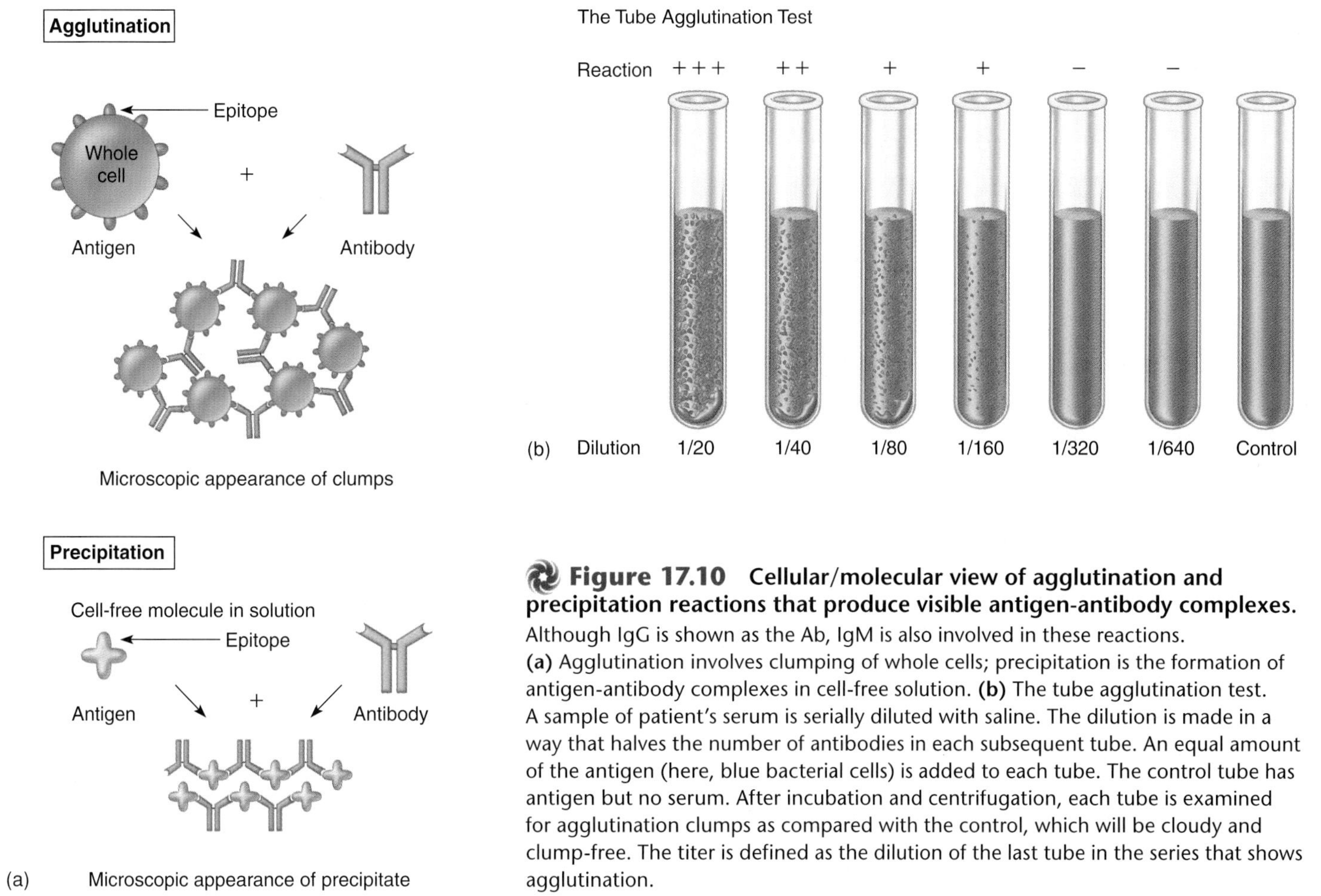

Figure 17.10 Cellular/molecular view of agglutination and precipitation reactions that produce visible antigen-antibody complexes. Although IgG is shown as the Ab, IgM is also involved in these reactions. **(a)** Agglutination involves clumping of whole cells; precipitation is the formation of antigen-antibody complexes in cell-free solution. **(b)** The tube agglutination test. A sample of patient's serum is serially diluted with saline. The dilution is made in a way that halves the number of antibodies in each subsequent tube. An equal amount of the antigen (here, blue bacterial cells) is added to each tube. The control tube has antigen but no serum. After incubation and centrifugation, each tube is examined for agglutination clumps as compared with the control, which will be cloudy and clump-free. The titer is defined as the dilution of the last tube in the series that shows agglutination.

ones. *Sensitivity* means that the test can detect even very small amounts of antibodies or antigens that are the targets of the test. New systems using monoclonal antibodies have greatly improved specificity; and those using radioactivity, enzymes, and electronics have improved sensitivity.

Visualizing Antigen-Antibody Interactions

The primary basis of most tests is the binding of an antibody (Ab) to a specific molecular site on an antigen (Ag). Because this reaction cannot be readily seen without an electron microscope, tests involve some type of endpoint reaction visible to the naked eye or with regular magnification that tells whether the result is positive or negative. In the case of large antigens such as cells, Ab binds to Ag and creates large clumps or aggregates that are visible macroscopically or microscopically **(figure 17.10*a*).** Smaller Ag-Ab complexes that do not result in readily observable changes will require special indicators in order to be visualized. Endpoints are often revealed by dyes or fluorescent reagents that can tag molecules of interest. Similarly, radioactive isotopes incorporated into antigens or antibodies constitute sensitive tracers that are detectable with photographic film.

An antigen-antibody reaction can be used to read a **titer,** or the quantity of antibodies in the serum. Titer is determined by serially diluting a sample in tubes or in a multiple-welled microtiter plate and mixing it with antigen **(figure 17.10*b*).** It is expressed as the highest dilution of serum that produces a visible reaction with an antigen. The more a sample can be diluted and yet still react with antigen, the greater is the concentration of antibodies in that sample and the higher is its titer. Interpretation of testing results is discussed in **Insight 17.2.**

Agglutination and Precipitation Reactions

The essential differences between agglutination and precipitation are in size, solubility, and location of the antigen. In agglutination, the antigens are whole cells such as red blood cells or bacteria with determinant groups on the surface. In precipitation, the antigen is a soluble molecule. In both instances, when Ag and Ab are optimally combined so that neither is in excess, one antigen is interlinked by several antibodies to form an insoluble, three-dimensional aggregate so large that it cannot remain suspended and it settles out.

INSIGHT 17.2 Medical

When Positive Is Negative: How to Interpret Serological Test Results

What if a patient's serum gives a positive reaction—is **seropositive**—in a serological test? In most situations, it means that antibodies specific for a particular microbe have been detected in the sample. But one must be cautious in proceeding to the next level of interpretation. The mere presence of antibodies does not necessarily indicate that the patient has a disease, but only that he or she has possibly had contact with a microbe or its antigens through infection or vaccination. In screening tests for determining a patient's history (rubella, for instance), knowing that a certain titer of antibodies is present can be significant because it shows that the person has some protection.

When the test is being used to diagnose current disease, however, a series of tests to show a rising titer of antibodies is necessary. The accompanying figure indicates how such a test can be used to diagnose patients who have nonspecific symptoms that could fit several diseases. Lyme disease, for instance, can be mistaken for arthritis or viral infections. In the first group, note that the antibody titer against *Borrelia burgdorferi*, the causative agent, increased steadily over a 6-week period. A control group that shared similar symptoms did not exhibit a rise in titer for antibodies to this microbe. Clinicians call samples collected early and late in an infection *acute* and *convalescent* sera.

Another important consideration in testing is the occasional appearance of biological false positives. These are results in which a patient's serum shows a positive reaction, even though, in reality, he or she is not or has not been infected by the microbe. False positives, such as those in syphilis and AIDS testing, arise when antibodies or other substances present in the serum cross-react with the test reagents, producing a positive result. Such false results may require retesting by a method that greatly minimizes cross-reactions.

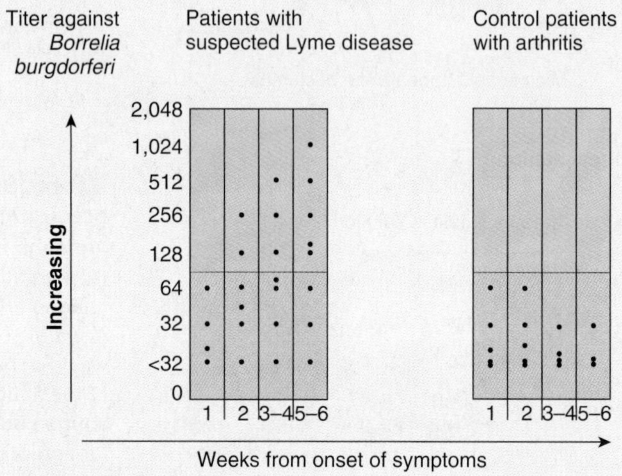

Agglutination Testing

Agglutination is discernible because the antibodies cross-link the antigens to form visible clumps. **Agglutination** tests are performed routinely by blood banks to determine ABO and Rh (Rhesus) blood types in preparation for transfusions. In this type of test, antisera containing antibodies against the blood group antigens on red blood cells are mixed with a small sample of blood and read for the presence or absence of clumping. The **Widal test** is an example of a tube agglutination test for diagnosing salmonelloses and undulant fever. In addition to detecting specific antibody, it also gives the serum titer.

Numerous variations of agglutination testing exist. The rapid plasma reagin (RPR) test is one of several tests commonly used to test for antibodies to syphilis. The cold agglutinin test, named for antibodies that react only at lower temperatures (4°C to 20°C), was developed to diagnose *Mycoplasma* pneumonia. The *Weil-Felix reaction* is an agglutination test sometimes used in diagnosing rickettsial infections.

In some tests, special agglutinogens have been prepared by affixing antigen to the surface of an inert particle. In *latex agglutination* tests, the inert particles are tiny latex beads. Kits using latex beads are available for assaying pregnancy hormone in the urine, identifying *Candida* yeasts and bacteria (staphylococci, streptococci, and gonococci), and diagnosing rheumatoid arthritis.

Precipitation Tests

In precipitation reactions, the soluble antigen is precipitated (made insoluble) by an antibody. This reaction is observable in a test tube in which antiserum has been carefully laid over an antigen solution. At the point of contact, a cloudy or opaque zone forms.

One example of this technique is the VDRL (Veneral Disease Research Lab) test that also detects antibodies to syphilis. Although it is a good screening test, it contains a heterophilic antigen (cardiolipin) that may give rise to false positive results. Although precipitation is a useful detection tool, the precipitates are so easily disrupted in liquid media that most precipitation reactions are carried out in agar gels. These substrates are sufficiently soft to allow the reactants (Ab and Ag) to freely diffuse, yet firm enough to hold the Ag-Ab precipitate in place. One technique with applications in microbial identification and diagnosis of disease is the double diffusion (Ouchterlony) method **(figure 17.11).** It is called double diffusion because it involves diffusion of both antigens and antibodies. The test is performed by punching a pattern of small wells into an agar medium and filling them with test antigens and antibodies. A band forming between two wells indicates that antibodies from one well have met and reacted with antigens from the other well. Variations on this technique provide a means of identifying unknown antibodies or antigens.

(a)

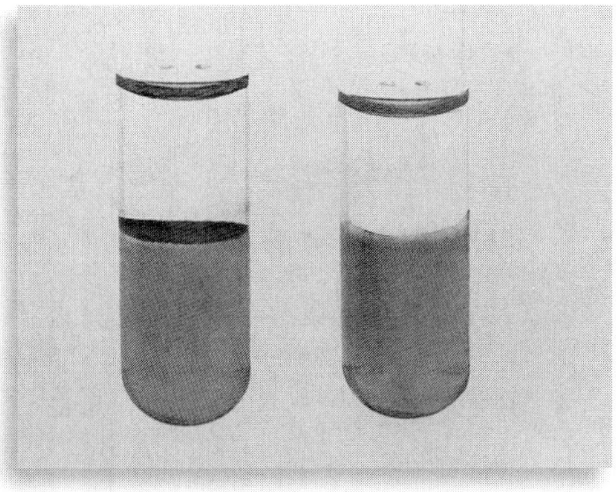

I.

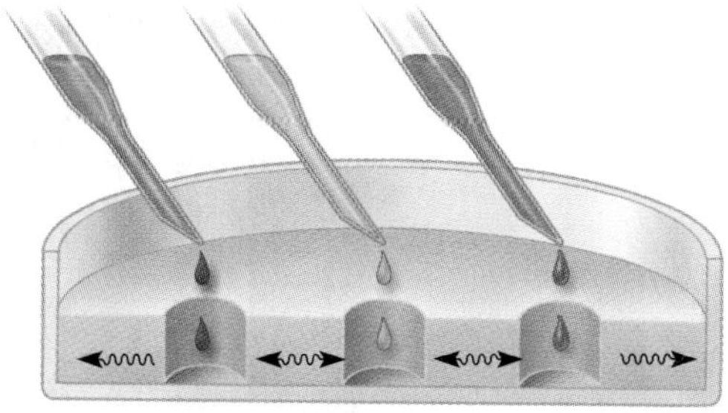

I. In one method of setting up a double-diffusion test, wells are punctured in soft agar, and antibodies (Ab) and antigens (Ag) are added in a pattern. As the contents of the wells diffuse toward each other, a number of reactions can result, depending on whether antibodies meet and precipitate antigens.

II.

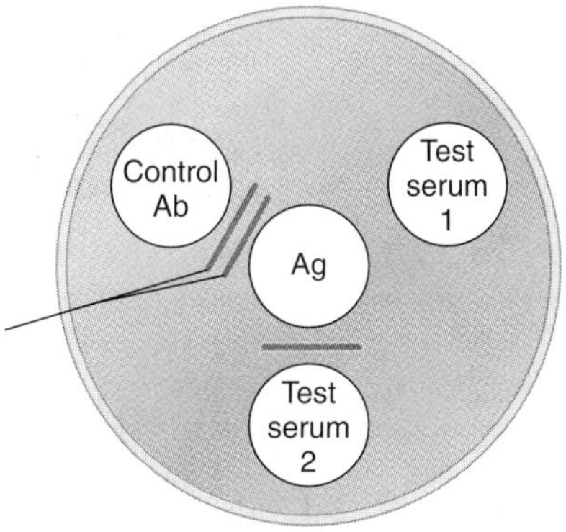

II. Example of test pattern and results. Antigen (Ag) is placed in the center well and antibody (Ab) samples are placed in outer wells. The control contains known Abs to the test Ag. Note bands that form where Ab and Ag meet. The other wells (1, 2) contai unknown test sera. One is positive and the other is negative. Double bands indicate more than one antigen and antibody that can react.

III.

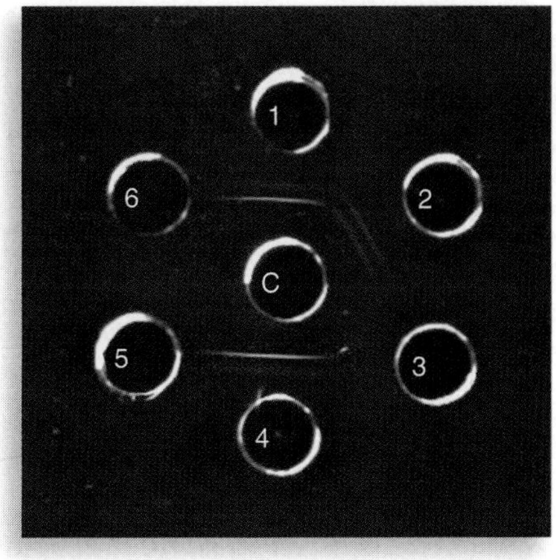

III. Actual test results for detecting infection with the fungal pathogen *Histoplasma*. Numbers 1 and 4 are controls and 2, 3, 5, and 6 are patient test sera. Can you determine which patients have the infection and which do not?

(b)

Figure 17.11 Precipitation reactions.
(a) A tube precipitation test for streptococcal group antigens. Specific antiserum has been placed in the bottom of tubes and antigen solution carefully overlaid to form a zone of contact. The left-hand tube has developed a band of precipitate; the right-hand tube is negative.
(b) Double-diffusion (Ouchterlony) tests in a semisolid matrix.

Step 1   Serum sample separated next to trough

(+)   Direction of electrophoresis ◄———————— (−)

Serum sample

Step 2   Antiserum added to trough

Albumin   Alpha globulin   IgM   Gamma globulin
           (α globulin)   (γ globulin)(IgG)

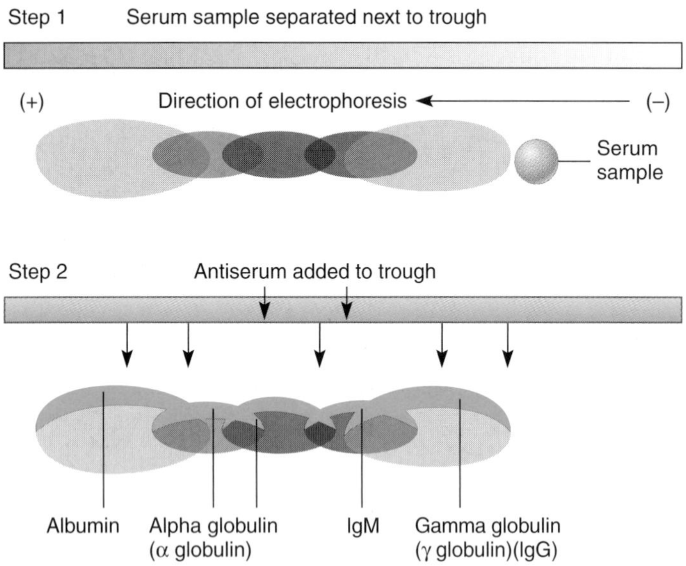

Figure 17.12   **Immunoelectrophoresis of normal human serum.**

Step 1. Proteins are separated by electrophoresis on a gel.
Step 2. To identify the bands and increase visibility, antiserum containing multiple antibodies specific for serum proteins is placed in a trough and allowed to diffuse toward the bands. This diffusion produces a pattern of numerous arcs representing major serum components.

Immunoelectrophoresis constitutes yet another refinement of diffusion and precipitation in agar. With this method, a serum sample is first electrophoresed to separate the serum proteins as previously described in chapter 15. Antibodies that react with specific serum proteins are placed in a trough parallel to the direction of migration, forming reaction arcs specific for each protein **(figure 17.12).** This test is widely used to detect disorders in the production of antibodies.

The Western Blot for Detecting Proteins

The **Western blot** test is somewhat similar to the previous tests because it involves the electrophoretic separation of proteins, followed by an immunoassay to detect these proteins. This test is a counterpart of the Southern blot test for identifying DNA, described in chapter 10. It is a highly specific and sensitive way to identify or verify a particular protein (antibody or antigen) in a sample **(figure 17.13).** First, the test material is electrophoresed in a gel to separate out particular bands. The gel is then transferred to a special blotter that binds the reactants in place. The blot is developed by incubating it with a solution of antigen or antibody that has been labeled with radioactive, fluorescent, or luminescent labels. Sites of specific binding will appear as a pattern of bands that can be compared with known positive and negative samples. This is currently the second (verification) test for people who are antibody-positive for

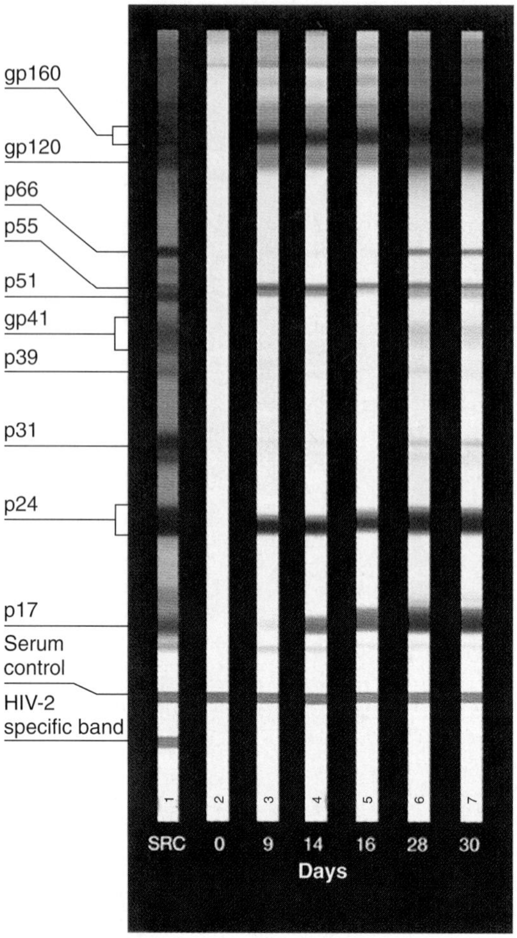

Figure 17.13   **The Western blot procedure.**
The example shown here tests one patient's blood for antibodies to specific HIV antigens. Samples were taken at six different time periods after suspected exposure. The test strips are prepared by electrophoresing several of the major HIV surface and core antigens and then blotting them onto special filters. The test strips are incubated with a patient's serum and developed with a radioactive or colorimetric label. Sites where HIV antigens have bound antibodies show up as bands. Patients' sera are then compared with a positive control strip (SRC) containing antibodies for all HIV antigens. Certain criteria must be met to consider the result positive.

Interpretation of Bands

Labels correspond with glycoproteins (GP) or proteins (P) that are part of HIV-1 antigen structure.
- The test is considered positive if bands occur at two locations: gp160 or gp120 and p31 or p24.
- The test is considered negative if no bands are present for any HIV antigen.
- The test is considered indeterminate if bands are present but not at the criteria locations. This result may require retesting at a later date.

HIV in the ELISA test (described in a later section), because it tests more types of antibodies and is less subject to misinterpretation than are other antibody tests. The technique has significant applications for detecting microbes and their antigens in specimens.

✓ CHECKPOINT

- Serological tests can test for either antigens or antibodies. Most are *in vitro* assessments of antigen-antibody reactivity from a variety of body fluids. The basis of these tests is an antigen-antibody reaction made visible through the processes of agglutination, precipitation, immunodiffusion, complement fixation, fluorescent antibody, and immunoassay techniques.
- One measurement is the *titer,* described as the concentration of antibody in serum. It is the highest dilution of serum that gives a visible antigen-antibody reaction. The higher the titer, the greater the level of antibody present.
- Agglutination reactions occur between antibody and antigens bound to cells. This results in visible clumps caused by large antibody-antigen complexes. In viral hemagglutination testing, the antibody reacts with the antigen and inhibits it from agglutinating red blood cells.
- In precipitation reactions, soluble antigen and antibody react to form insoluble, visible precipitates. Precipitation reactions can also be visualized by adding radioactive or enzyme markers to the antigen-antibody complex.
- In immunoelectrophoresis techniques such as the Western blot, proteins that have been separated by electrical current are identified by labeled antibodies. HIV infections are verified with this method.

Complement Fixation

An antibody that requires complement to complete the lysis of its antigenic target cell is termed a **lysin** or cytolysin (see chapter 15 for a discussion of complement). When lysins act in conjunction with the intrinsic complement system on red blood cells, the cells hemolyze (lyse and release their hemoglobin). This lysin-mediated hemolysis is the basis of a group of tests called complement fixation, or CF **(figure 17.14).**

Complement fixation testing uses four components—antibody, antigen, complement, and sensitized sheep red blood cells—and it is conducted in two stages. In the first stage, the test antigen is allowed to react with the test antibody (at least one must be of known identity) in the absence of complement. If the Ab and Ag are specific for each other, they form complexes. To this mixture, purified complement proteins from guinea pig blood are added. If antibody and antigen have complexed during the previous step, they attach, or fix, the complement to them, thus preventing it from participating in further reactions. The extent of this complement fixation is determined in the second stage by means of sheep RBCs with surface lysin molecules. The sheep RBCs serve as an indicator complex that can also fix complement. Contents of the stage 1 tube are mixed with the stage 2 tube and observed for hemolysis, which can be observed with the naked eye as a clearing

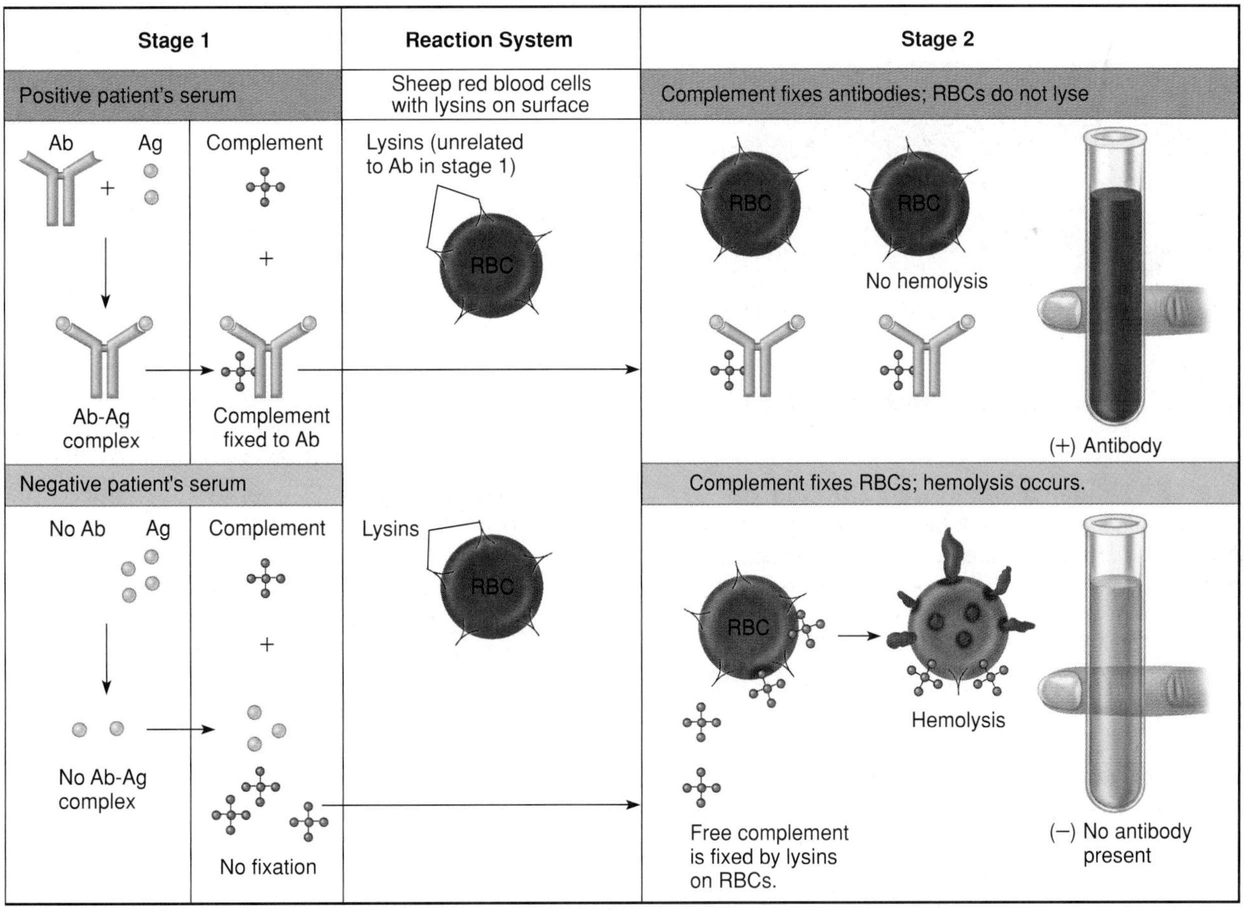

Figure 17.14 Complement fixation test.
In this example, two serum samples are being tested for antibodies to a certain infectious agent. In reading this test, one observes the cloudiness of the tube. If it is cloudy, the RBCs are not hemolyzed and the test is positive. If it is clear and pink, the RBCs are hemolyzed and the test is negative.

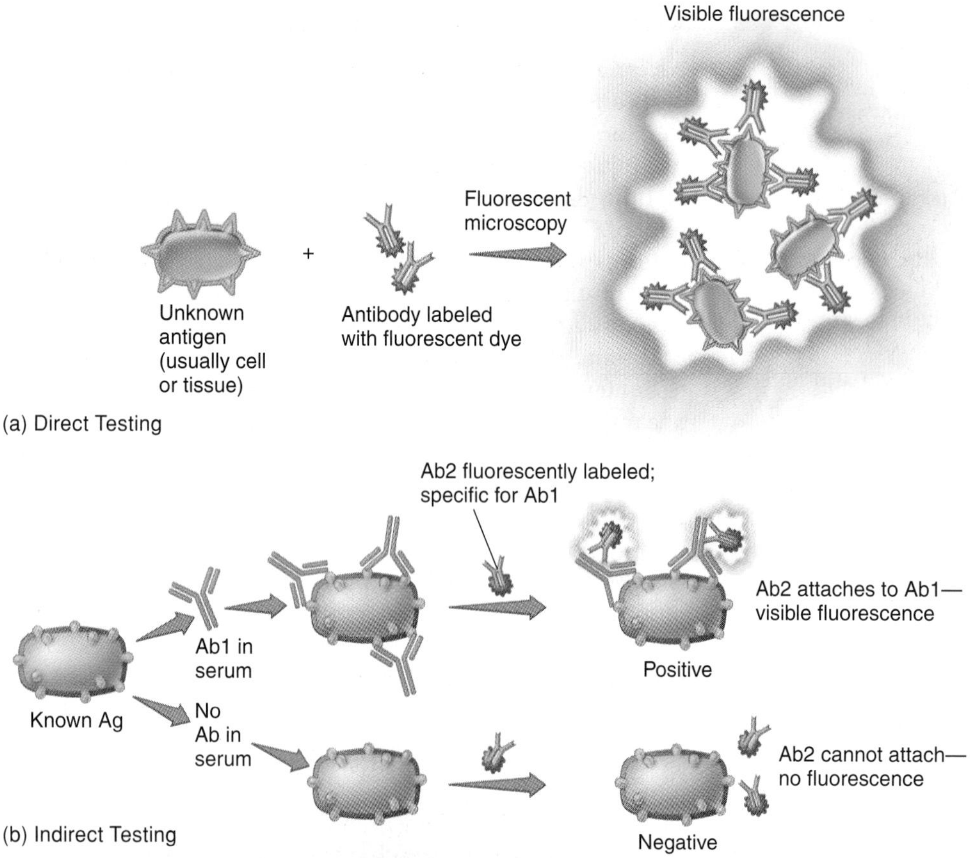

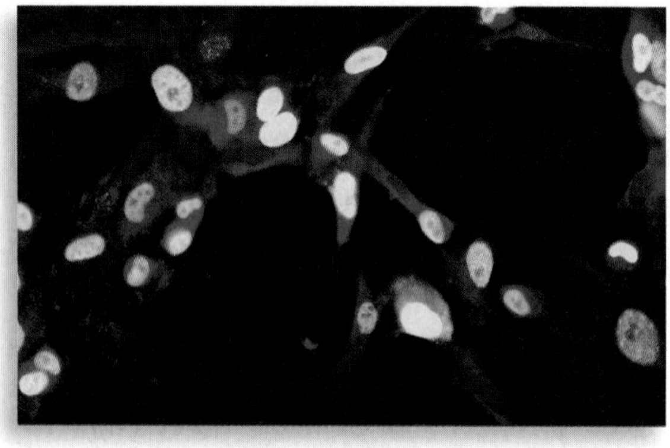

(c) Indirect Immunofluorescence Testing

Figure 17.15 Immunofluorescence testing.
(a) Direct: Unidentified antigen (Ag) is directly tagged with fluorescent Ab. **(b)** Indirect: Ag of known identity is used to assay unknown Ab; a positive reaction occurs when the second Ab (with fluorescent dye) affixes to the first Ab. **(c)** An indirect immunofluorescent stain of cells infected with two different viruses. Cells fluorescing green contain cytomegalovirus; cells fluorescing yellow contain adenovirus.

of the solution. If hemolysis *does not* occur, it means that the complement was used up by the first stage Ag-Ab complex and that the unknown antigen or antibody was indeed present. This result is considered positive. If hemolysis *does* occur, it means that unfixed complement from tube 1 reacted with the RBC complex instead, thereby causing lysis of the sheep RBCs. This result is negative for the antigen or antibody that was the target of the test. Complement fixation tests are invaluable in diagnosing influenza, polio, and various fungi.

The antistreptolysin O (ASO) titer test measures the levels of antibody against the streptolysin toxin, an important hemolysin of group A streptococci. It employs a technique related to complement fixation. A serum sample is exposed to known suspensions of streptolysin and then allowed to incubate with RBCs. Lack of hemolysis indicates antistreptolysin antibodies in the patient's serum that have neutralized the streptolysin and prevented hemolysis. This is an important verification procedure for scarlet fever, rheumatic fever, and other related streptococcal syndromes (see chapter 21).

Miscellaneous Serological Tests

A test that relies on changes in cellular activity as seen microscopically is the *Treponema pallidum immobilization* (TPI) test for syphilis. The impairment or loss of motility of the *Treponema* spirochete in the presence of test serum and complement indicates that the serum contains anti-*Treponema pallidum* antibodies. In *toxin neutralization* tests, a test serum is incubated with the microbe that produces the toxin. If the serum inhibits the growth of the microbe, one can conclude that antitoxins are present.

Serotyping is an antigen-antibody technique for identifying, classifying, and subgrouping certain bacteria into categories called serotypes, using antisera for cell antigens such as the capsule, flagellum, and cell wall. It is widely used in typing *Salmonella* species and strains and is the basis for identifying the numerous serotypes of streptococci. The Quellung test, which identifies serotypes of the pneumococcus, involves a precipitation reaction in which antibodies react with the capsular polysaccharide. Although the reaction makes the capsule seem to swell, it is actually creating a zone of Ag-Ab complex on the cell's surface.

Fluorescent Antibodies and Immunofluorescence Testing

The property of dyes such as fluorescein and rhodamine to emit visible light in response to ultraviolet radiation was discussed in chapter 3. This property of fluorescence has found numerous applications in diagnostic immunology. The fundamental tool in immunofluorescence testing is a fluorescent antibody—a monoclonal antibody labeled by a fluorescent dye (fluorochrome).

The two ways that fluorescent antibodies (FAbs) can be used for diagnosis are shown in **figure 17.15.** In *direct testing*, an unknown test specimen or antigen is fixed to a slide and exposed to a fluorescent antibody solution of known composition. If the antibodies are complementary to antigens in the material, they will bind to it. After the slide is rinsed to remove unattached antibodies, it is observed with the fluorescent microscope. Fluorescing cells or particles indicate the presence of Ab-Ag complexes and a positive result. These tests are valuable for identifying and locating antigens on the surfaces of cells or in tissues and in identifying the disease agents of syphilis, gonorrhea, chlamydiosis, whooping cough, Legionnaires' disease, plague, trichomoniasis, meningitis, and listeriosis.

In *indirect testing* methods, the fluorescent antibodies are antibodies made to react with the Fc region of another antibody (remember that antibodies can be antigenic). In this scheme, an antigen of known character (a bacterial cell, for example) is combined with a test serum of unknown antibody content. The fluorescent antibody solution that can react with the unknown antibody is applied and rinsed off to visualize whether the serum contains antibodies that have affixed to the antigen. A positive test shows fluorescing aggregates or cells, indicating that the fluorescent antibodies have combined with the unlabeled antibodies. In a negative test, no fluorescent complexes will appear. This technique is frequently used to diagnose syphilis (FTA-ABS) and various viral infections.

Immunoassays

The elegant tools of the microbiologist and immunologist are being used increasingly in athletics, criminology, government, and business to test for trace amounts of substances such as hormones, metabolites, and drugs. But traditional techniques in serology are not refined enough to detect a few molecules of these chemicals. Extremely sensitive alternative methods that permit rapid and accurate measurement of trace antigen or antibody are called **immunoassays.** Examples of the technology for detecting an antigen or antibody in minute quantities include radioactive isotope labels, enzyme labels, and sensitive electronic sensors. Many of these tests are based on specifically formulated monoclonal antibodies.

Radioimmunoassay (RIA)

Antibodies or antigens labeled with a radioactive isotope can be used to pinpoint minute amounts of a corresponding antigen or antibody. Although very complex in practice, these assays compare the amount of radioactivity present in a sample before and after incubation with a known, labeled antigen or antibody. The labeled substance competes with its natural, nonlabeled partner for a reaction site. Large amounts of a bound radioactive component indicate that the unknown test substance was not present. The amount of radioactivity is measured with an isotope counter or a photographic emulsion (autoradiograph). Radioimmunoassay has been employed to measure the levels of insulin and other hormones and to diagnose allergies, chiefly by the radioimmunosorbent test (RIST) for measurement of IgE in allergic patients and the radioallergosorbent test (RAST) to standardize allergenic extracts.

Enzyme-Linked Immunosorbent Assay (ELISA)

The **ELISA test,** also known as enzyme immunoassay (EIA), contains an enzyme-antibody complex that can be used as a color tracer for antigen-antibody reactions. The enzymes used most often are horseradish peroxidase and alkaline phosphatase, both of which release a dye (chromogen) when exposed to their substrate. This technique also relies on a solid support such as a plastic microtiter plate that can *adsorb* (attract on its surface) the reactants **(figure 17.16).**

The *indirect ELISA* test can detect antibodies in a serum sample. As with other indirect tests, the final positive reaction is achieved by means of an antibody-antibody reaction. The indicator antibody is complexed to an enzyme that produces a color change with positive serum samples **(figure 17.16*a,b*).** The starting reactant is a known antigen that is adsorbed to the surface of a well. To this, an unknown serum is added. After rinsing, an enzyme-Ab reagent that can react with the unknown test antibody is placed in the well. The substrate for the enzyme is then added, and the wells are scanned for color changes. Color development indicates that all the components reacted and

CASE FILE 17 *WRAP-UP*

The nurse described at the beginning of the chapter was tested for HIV infection. Initially, she tested positive by the ELISA screening test; however, by the confirmatory test, she was shown to be HIV negative. As a result, the ELISA test is referred to as a false positive. Although the ELISA test is very sensitive, its specificity is much lower than many of the confirmatory tests. Almost 9 out of 10 low-risk individuals (such as blood donor volunteers) that test positive for HIV by the ELISA method are shown to be HIV negative by the confirmatory test methods. Patients that have delivered multiple babies, or those with certain autoimmune diseases, can test positive by the ELISA method in the absence of HIV infection.

Up to 800,000 U.S. health care workers annually experience needlestick or other "sharps" injuries. Based on data analysis, only 0.3% of those exposed to HIV-contaminated blood, after the skin had been pierced through a sharps injury, became productively infected with HIV.

Interestingly, most ELISA tests will detect antibodies to both HIV-1 and HIV-2, whereas the Western blot method is initially performed with HIV-1, because the virus is more common in the United States. An HIV-2 Western blot can be performed if the HIV-1 Western blot yields negative results.

CDC. 1989. *Interpretation and use of the Western blot assay for serodiagnosis of human immunodeficiency virus type 1 infections.* MMWR 38:S4–S6.

CDC. 2001. *Revised guidelines for HIV counseling, testing, and referral.* MMWR 50(No. RR-19).

CDC. 2001. *Updated U.S. Public Health Service guidelines for the management of occupational exposures to HBV, HCV, and HIV and recommendations for post-exposure prophylaxis.* MMWR 50(No. RR-11).

(a) **Indirect ELISA,** comparing a positive versus negative reaction. This is the basis for HIV screening tests.

Well A   Well B

Known antigen is adsorbed to well.

Sample A   Sample B

Serum samples with unknown antibodies A B

Well is rinsed to remove unbound (nonreactive) antibodies.

Indicator antibody linked to enzyme attaches to any bound antibody.

Wells are rinsed to remove unbound indicator antibody. A colorless substrate for enzyme is added.

Enzymes linked to indicator Ab hydrolyze the substrate, which releases a dye. Wells that develop color are positive for the antibody; colorless wells are negative.

(+)   (−)

(b) **Microtiter ELISA Plate with 96 Tests for HIV Antibodies.** Colored wells indicate a positive reaction.

(c) **Capture or Antibody Sandwich ELISA Method.** Note that an antigen is trapped between two antibodies. This test is used to detect hantavirus and measles virus.

Antibody is adsorbed to well.

Test antigen is added; if complementary, antigen binds to antibody.

Enzyme

Enzyme-linked antibody specific for test antigen then binds to another antigen, forming a sandwich.

Enzyme's substrate (□) is added, and reaction produces a visible color change (●).

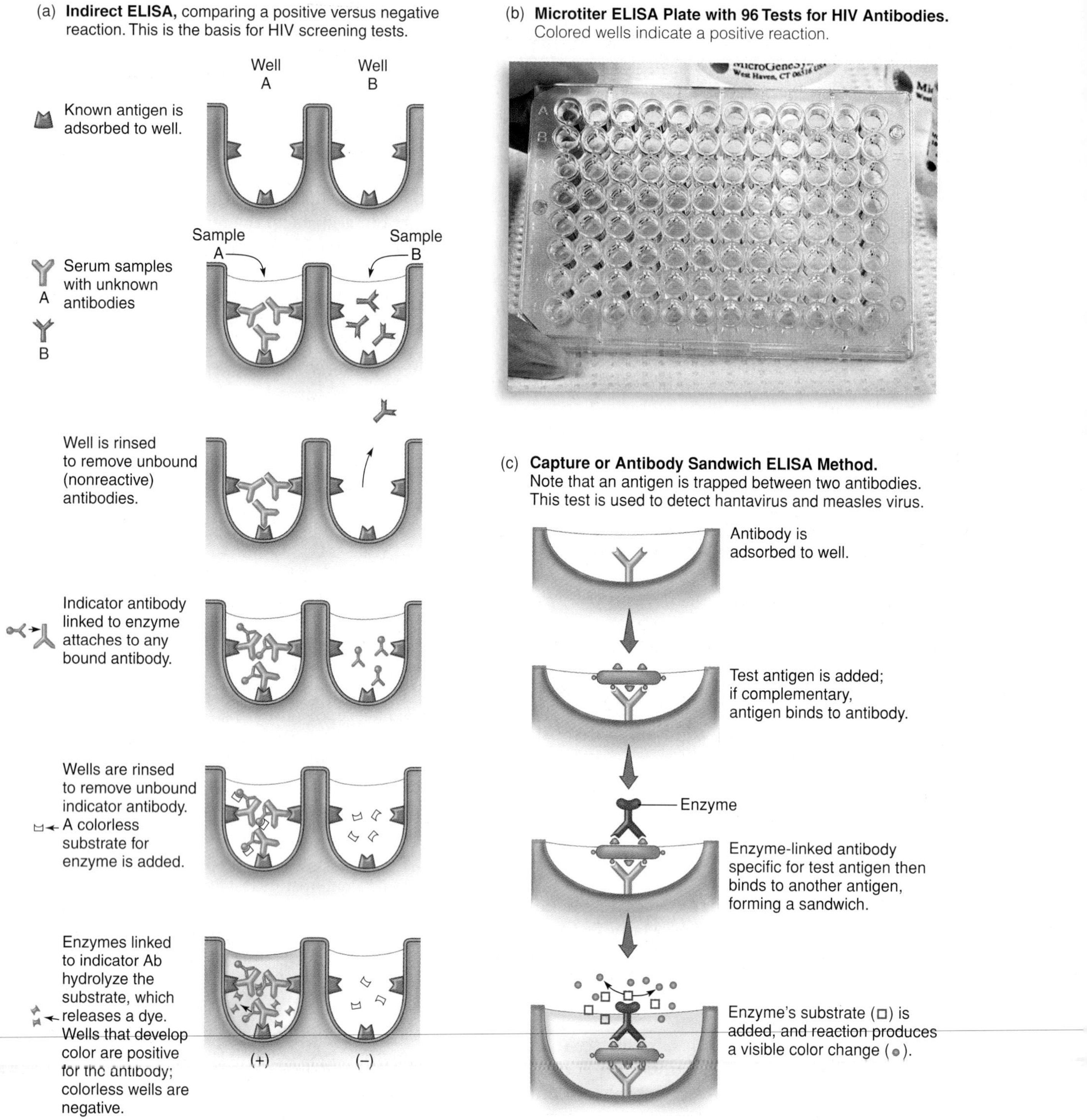

Figure 17.16   Methods of ELISA testing.

that the antibody was present in the patient's serum. This is the common screening test for the antibodies to HIV, various rickettsial species, hepatitis A and C, the cholera vibrio, and *Helicobacter,* a cause of gastric ulcers. Because false positives can occur, a verification test may be necessary (such as Western blot for HIV).

In *direct ELISA* (or sandwich) tests, a known antibody is adsorbed to the bottom of a well and incubated with a

solution containing unknown antigen **(figure 17.16c).** After excess unbound components have been rinsed off, an enzyme-antibody indicator that can react with the antigen is added. If antigen is present, it will attract the indicator-antibody and hold it in place. Next, the substrate for the enzyme is placed in the wells and incubated. Enzymes affixed to the antigen will hydrolyze the substrate and release a colored dye. Thus, any color developing in the wells is a positive result. Lack of color means that the antigen was not present and that the subsequent rinsing removed the enzyme-antibody complex. The direct technique is used to detect antibodies to hantavirus, rubella virus, and *Toxoplasma*.

A newer technology uses electronic monitors that directly read out antigen-antibody reactions. Without belaboring the technical aspects, these systems contain computer chips that sense the minute changes in electrical current given off when an antibody binds to antigen. The potential for sensitivity is extreme; it is thought that amounts as small as 12 molecules of a substance can be detected in a sample. In another procedure, antibody substrate molecules are incubated with sample and then exposed to the enzyme alkaline phosphatase. If the antibody is bound, the enzyme reacts with the substrate and causes visible light to be emitted. The light can be detected by machines or photographic films.

Tests That Differentiate T Cells and B Cells

So far, we have concentrated on tests that identify antigens and antibodies in samples, but techniques also exist that differentiate between B cells and T cells and can quantify subsets of each. Information on the types and numbers of lymphocytes in blood and other samples is a common way to evaluate immune dysfunctions such as those in AIDS, immunodeficiencies, and cancer. A simple method for identifying T cells is to mix them with untreated sheep red blood cells. Receptors on the T cells bind the RBCs into a flowerlike cluster called a **rosette formation (figure 17.17*a*).** Rosetting can also occur in B cells if one uses Ig-coated bovine RBCs or mouse erythrocytes.

For routine identification of lymphocytes, rosette formation has been replaced by fluorescent techniques. These techniques can differentiate between T cells and B cells as well as subgroup them **(figure 17.17*b*).** These subgroup tests utilize monoclonal antibodies produced in response to specific cell markers. B-cell tests categorize different stages in B-cell development and are very useful in characterizing B-cell cancers. Tests that can help differentiate the CD4, CD8, and other T-cell subsets are important in monitoring AIDS and other immunodeficiency diseases.

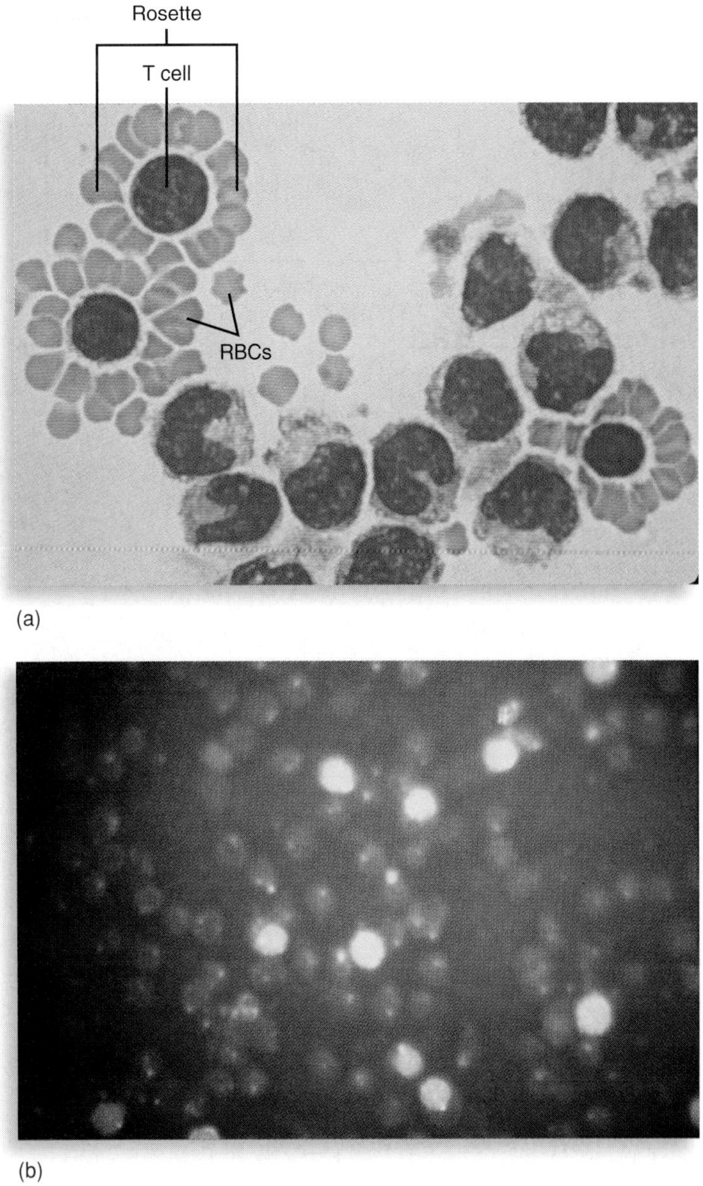

Figure 17.17 Tests for characterizing T cells and B cells.
(a) Photomicrograph of rosette formation that identifies T cells.
(b) Plasma cells highlighted by fluorescent antibodies.

In Vivo Testing

Probably the first immunologic tests were performed not in a test tube but on the body itself. A classic example of one such technique is the **tuberculin test,** which uses a small amount of purified protein derivative (PPD) from *Mycobacterium tuberculosis* injected into the skin. The appearance of a red, raised, thickened lesion in 48 to 72 hours can indicate previous exposure to tuberculosis (shown in figure 16.14). In practice, *in vivo* tests employ principles similar to serological tests, except in this case an antigen or an antibody is introduced into a patient to elicit some sort of visible reaction. Like the tuberculin test, some of these diagnostic skin tests are useful for evaluating infections due to fungi (coccidioidin and histoplasmin tests, for example) or allergens. Allergic reactions and other immune system disorders are the topics of chapter 16.

A Viral Example

All of the methods discussed so far—phenotypic, genotypic, and immunologic—are applicable to the different types of microorganisms. Viruses sometimes present special difficulties because they are not cells and they are more labor intensive to culture in the laboratory. **Figure 17.18** presents an overview of various techniques that might be used to diagnose viral infections. It provides one example of the variety of methods that can be employed for many infections regardless of their cause.

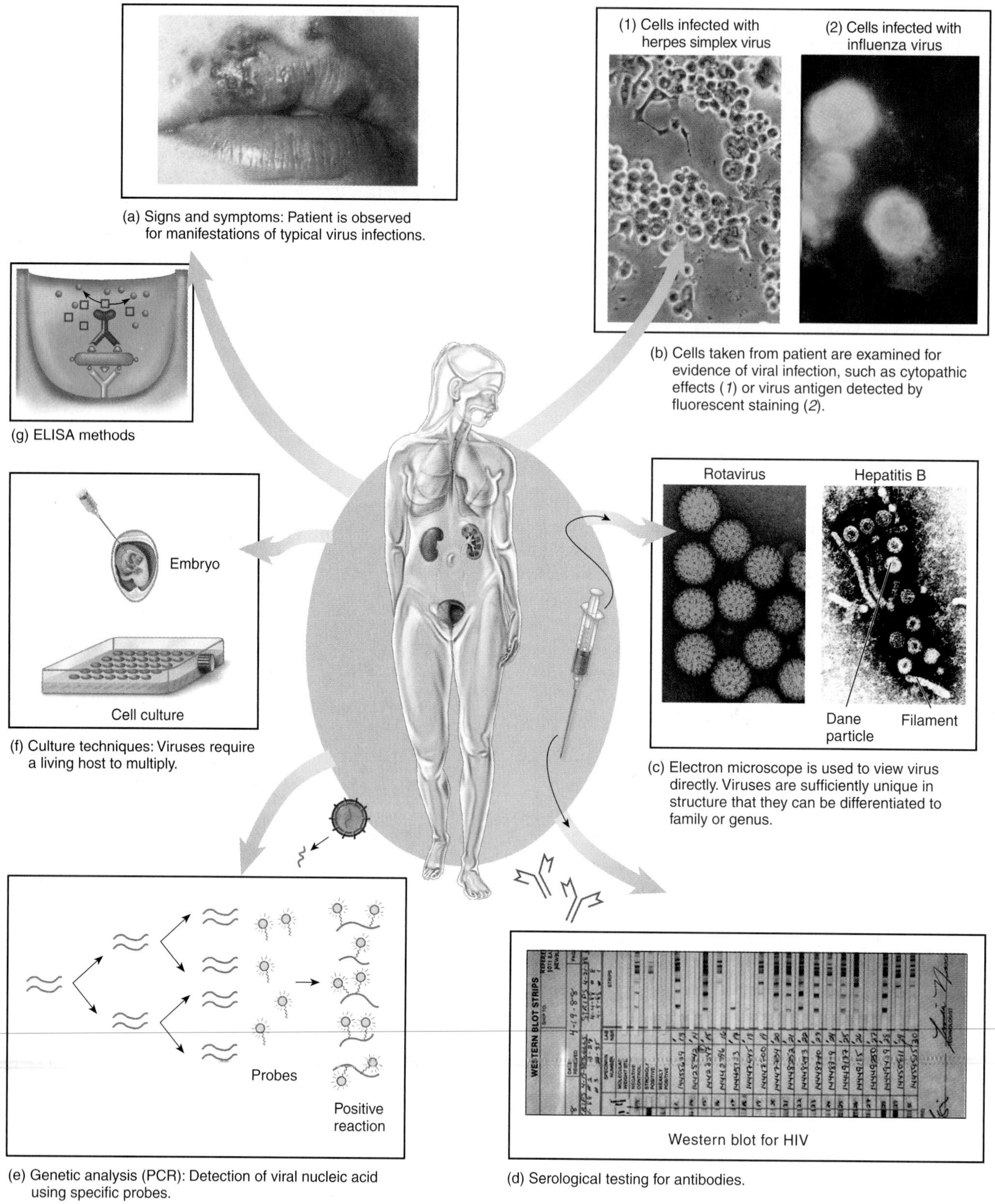
(a) Signs and symptoms: Patient is observed for manifestations of typical virus infections.
(1) Cells infected with herpes simplex virus
(2) Cells infected with influenza virus
(b) Cells taken from patient are examined for evidence of viral infection, such as cytopathic effects (*1*) or virus antigen detected by fluorescent staining (*2*).
(g) ELISA methods
Embryo
Cell culture
(f) Culture techniques: Viruses require a living host to multiply.
Rotavirus
Hepatitis B
Dane particle
Filament
(c) Electron microscope is used to view virus directly. Viruses are sufficiently unique in structure that they can be differentiated to family or genus.
Probes
Positive reaction
(e) Genetic analysis (PCR): Detection of viral nucleic acid using specific probes.
WESTERN BLOT STRIPS
Western blot for HIV
(d) Serological testing for antibodies.

Figure 17.18 **Summary of methods used to diagnose viral infections.**

CHECKPOINT

- Complement fixation involves a two-part procedure in which complement fixes to a specific antibody if present or to red blood cell antigens if antibody is absent. Lack of RBC hemolysis is indicative of a positive test.
- Serological tests can measure the degree to which host antibody binds directly to antigens, such as disease agents or toxins. This is the principle behind tests for syphilis and rheumatic fever.
- Direct fluorescent antibody tests indicate presence of an antigen and are useful in identifying infectious agents. Indirect fluorescent tests indicate the presence of a particular antibody and can diagnose infection.
- Immunoassays can detect very small quantities of antigen, antibody, or other substances. Radioimmunoassay uses radioisotopes to detect trace amounts of biological substances.
- The ELISA test uses enzymes and dyes to detect antigen-antibody complexes. It is widely used to detect viruses, bacteria, and antibodies in HIV infection.
- Technicians use precise assays to differentiate between B and T cells and to identify subgroups of these cells for disease diagnosis.
- *In vivo* serological testing, such as the tuberculin test, involves subcutaneous injection of antigen to elicit a visible immune response in the host.

Chapter Summary with Key Terms

17.1 Preparation for the Survey of Microbial Diseases

A. Phenotypic methods are those that assess microscopic morphology, macroscopic morphology, physiological and biochemical characteristics, and chemical composition.
B. Genotypic methods examine the genetic composition of a microorganism.
C. Immunologic methods exploit the host's antibody reaction to microbial antigens for purposes of diagnosis.

17.2 On the Track of the Infectious Agent: Specimen Collection

A. Diagnosis begins with accurate specimen collection, which means the sampling of body sites or fluids that are suspected to contain the infective agent.
B. Laboratory methods yield either presumptive data or confirmatory data. Sometimes pathogens are diagnosed based solely on signs and symptoms in the patient.

17.3 Phenotypic Methods

A. The most obvious phenotypic characteristic of a microbe is what it looks like. Many infections can be diagnosed by microscopic examination of fresh or stained microorganisms from specimens. Direct antigen or antibody testing can also be performed on fresh specimens.
B. Cultivation of specimens allows the examination of colony morphology and/or growth characteristics, which are useful for identifying microbes.
C. Biochemical tests determine whether microbes possess particular enzymes or biochemical pathways that can serve to identify them.
D. A variety of other tests are available to test specific phenotypic characteristics of particular suspected microorganisms.
E. It is of utmost importance to consider whether an identified microbe is actually causing the disease or is simply a bystander, or member of the normal biota.

17.4 Genotypic Methods

A. Genetic methods of identification are being used increasingly in diagnostic microbiology. Tests can determine DNA or rRNA sequences.
B. An increasingly useful procedure is the use of **polymerase chain reaction** to amplify DNA present in a sample in order to detect small quantities of microbial sequences.

17.5 Immunologic Methods

A. **Serology** is a science that attempts to detect signs of infection in a patient's serum such as antibodies specific for a microbe.
B. The basis of serological tests is that Abs specifically bind to Ag *in vitro*. An Ag of known identity will react with antibodies in an unknown serum sample. The reverse is also true; known antibodies can be used to detect and type antigens.
C. These Ag-Ab reactions are visible in the form of obvious clumps and precipitates, color changes, or the release of radioactivity. Test results are read as positive or negative.
D. Desirable properties of tests are high specificity and sensitivity.
E. Types of Tests
   1. In **agglutination** tests, antibody cross-links whole-cell antigens, forming complexes that settle out and form visible clumps in the test chamber; examples are tests for blood type, some bacterial diseases, and viral diseases.
   2. Double diffusion precipitation tests involve the diffusion of Ags and Abs in a soft agar gel, forming zones of precipitation where they meet.
   3. In immunoelectrophoresis, migration of serum proteins in gel is combined with precipitation by antibodies.

4. The **Western blot** test separates antigen (protein) into bands. After the gel is affixed to a blotter, it is reacted with a test specimen and developed by radioactivity or with dyes.
5. Complement fixation tests detect *lysins*—antibodies that fix complement and can lyse target cells. It involves first mixing test Ag and Ab with complement and then with sensitized sheep RBCs. If the complement is fixed by the Ag-Ab, the RBCs remain intact and the test is positive. If RBCs are hemolyzed, specific antibodies are lacking.
6. Immunofluorescence testing uses fluorescent antibodies (FABs tagged with fluorescent dye) either directly or indirectly to visualize cells or cell aggregates that have reacted with the FABs.
   a. In direct assays, known marked Ab is used to detect unknown Ag (microbe).
   b. In indirect testing, known Ag reacts with unknown Ab, and the reaction is made visible by a second Ab that can affix to and identify the unknown Ab.
7. **Immunoassays** are highly sensitive tests for Ag and Ab.
   a. In *radioimmunoassay*, Ags or Abs are labeled with radioactive isotopes and traced.
   b. The **enzyme-linked immunosorbent assay (ELISA)** can detect unknown Ag or Ab by direct or indirect means. A positive result is visualized when a colored product is released by an enzyme-substrate reaction.
   c. Tests are also available to differentiate B cells from T cells and their subtypes.
8. With *in vivo* testing, Ags are introduced into the body directly to determine the patient's immunologic history.

Multiple-Choice and True-False Questions

Multiple-Choice Questions. Select the correct answer from the answers provided.

1. **Multiple Matching.** Evaluate and match each of the following culture results to the most likely interpretation.
   a. probable infection
   b. normal biota
   c. contamination

   ___ 1. Isolation of two colonies of *E. coli* on a plate streaked from the urine sample
   ___ 2. Isolation of 50 colonies of *Streptococcus pneumoniae* on a plate streaked with sputum
   ___ 3. A mixture of 80 colonies of various streptococci on a culture from a throat swab
   ___ 4. Colonies of black bread mold on selective media used to isolate bacteria from stool
   ___ 5. Blood culture bottle with heavy growth

2. Which of the following methods can identify different strains of a microbe?
   a. microscopic examination
   b. radioimmunoassay
   c. DNA typing
   d. agglutination test

3. In agglutination reactions, the antigen is a ___; in precipitation reactions, it is a ___.
   a. soluble molecule, whole cell
   b. whole cell, soluble molecule
   c. bacterium, virus
   d. protein, carbohydrate

4. Which reaction requires complement?
   a. hemagglutination
   b. precipitation
   c. hemolysis
   d. toxin neutralization

5. A patient with a ___ titer of antibodies to an infectious agent generally has greater protection than a patient with a ___ titer.
   a. high, low
   b. low, high
   c. negative, positive
   d. old, new

6. Direct immunofluorescence tests use a labeled antibody to identify ___.
   a. an unknown microbe
   b. an unknown antibody
   c. fixed complement
   d. agglutinated antigens

7. The Western blot test can be used to identify
   a. unknown antibodies
   b. unknown antigens
   c. specific DNA
   d. both a and b

True-False Questions. If statement is true, leave as is. If it is false, correct it by rewriting the sentence.

8. Complement fixation is an example of an *in vivo* serological test.
9. The antistreptolysin O titer (Asotiter) is a test that measures a person's immunity to streptococcus O.
10. DNA probes are used to search for complementary segments of DNA.
11. Biochemical identification methods are based on a microbe's utilization of nutrients.
12. All microorganisms that grow from a clinical sample should be considered significant.

Writing to Learn

These questions are suggested as a *writing-to-learn* experience. For each question, compose a one- or two-paragraph answer that includes the factual information needed to completely address the question.

1. Why do specimens need to be taken aseptically even when nonsterile sites are being sampled and selective media are to be used?
2. Explain the general principles in specimen collection.
3. a. What is involved in direct specimen testing?
   b. In presumptive and confirmatory tests?
   c. In cultivating and isolating the pathogen?
   d. In biochemical testing?
   e. In gene probes?
4. Differentiate between the serological tests used to identify isolated cultures of pathogens and those used to diagnose disease from patients' serum.
5. Why is it important to prevent microbes from growing in specimens?
6. Why is speed so important in the clinical laboratory?
7. Summarize the important points in determining if a clinical isolate is involved in infection.
8. a. What is the basis of serology and serological testing?
   b. Differentiate between specificity and sensitivity.
   c. Describe several general ways that Ag-Ab reactions are detected.
9. a. What does seropositivity mean?
   b. What is a false positive test result and what are some possible causes?
   c. What is meant by a false negative result and what might account for it?
   d. What does the titer of serum tell us about the immune status?
10. a. Explain the differences between direct and indirect procedures in serological or immunoassay tests.
    b. How is fluorescence detected?
    c. How is the reaction in a radioimmunoassay detected?
    d. How does a positive reaction in an ELISA test appear? How many wells are positive in figure 17.16*b*?

Concept Mapping

Appendix D provides guidance for working with concept maps.

1. Supply your own linking words or phrases in this concept map, and provide the missing concepts in the empty boxes.

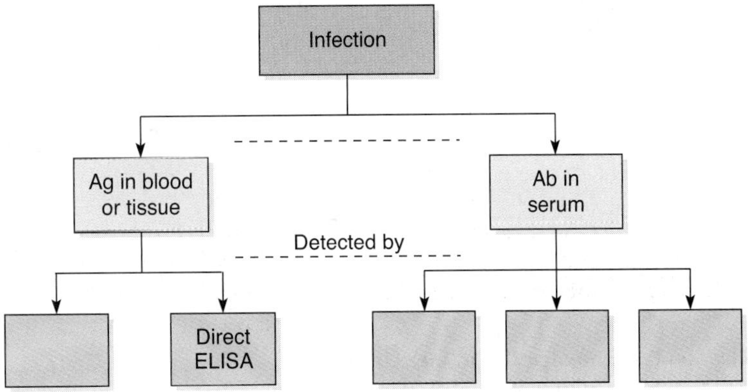

Critical Thinking Questions

Critical thinking is the ability to reason and solve problems using facts and concepts. These questions can be approached from a number of angles, and in most cases, they do not have a single correct answer.

1. See Insight 17.1. How would you explain to a junior high biology class that in the next decade some diseases currently thought to be noninfectious will probably be found to be caused by microbes?
2. In what way could the extreme sensitivity of the PCR method be a problem when working with clinical specimens?
3. Why do some tests for antibody in serum (such as for HIV and syphilis) require backup verification with additional tests at a later date?
4. Why do we interpret positive hemolysis in the complement fixation test to mean negative for the test substance?
5. Observe figure 17.16 and make note of the several steps in the indirect ELISA test. What four essential events are necessary to develop a positive reaction (besides having antibody A)? Hint: What would happen without rinses?
6. Using the criteria for band interpretation in figure 17.13, does this patient test positive for HIV? Why or why not?
7. Explain how an immunoassay method could use monoclonal antibodies to differentiate between B and T cells and between different subsets of T cells.

Visual Understanding Questions

1. **From chapter 3, figure 3.8*b*.** What biochemical characteristic does this figure illustrate? How could this characteristic be used to begin the identification of these two organisms? Explain your answer.

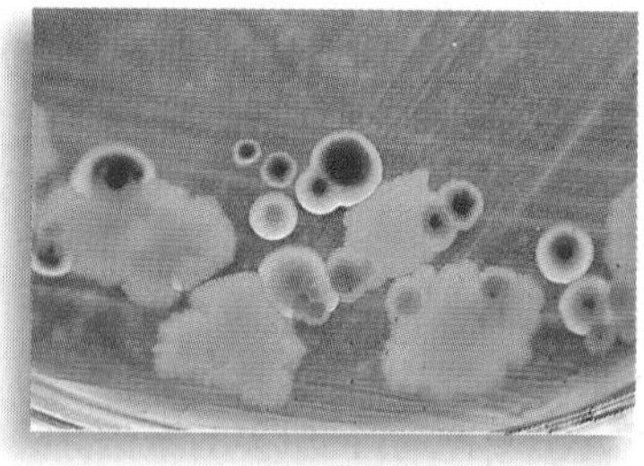

2. **From chapter 16, figure 16.15*b*.** Imagine that this patient is being seen by his or her physician for this unknown rash. What rapid phenotypic test could suggest that this condition is caused by a bacterium? Could a rapid immunoassay or fluorescent procedure be used to identify a specific viral cause? Explain your answer.

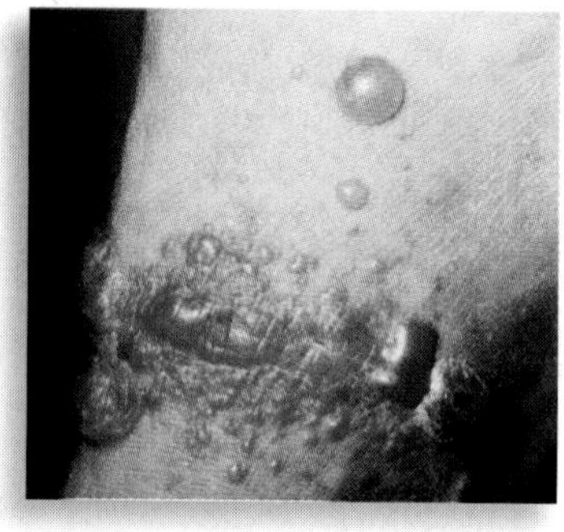

Internet Search Topics

Go to: www.aris.mhhe.com, and click on "microbiology" and then this textbook's author/title. Go to chapter 17, access the URLs listed under Internet Search Topics, and research the following:

1. Access the websites listed for excellent graphics and explanations of immune tests.
2. Search the Internet for information on urine cultures. What criteria are used to evaluate a urine sample for the presence/absence of a UTI? Are these criteria absolute? Explain your answer.
3. Use a search engine to look for information on rapid streptococcus tests. Select a site that provides information about one or more streptococcal test products. What is the principle of the test? What methodology does it use? What is being defected? What is reported specificity of the test?

CHAPTER 18

Infectious Diseases Affecting the Skin and Eyes

CASE FILE 18

In February of 2001, a 10-month-old baby in Texas became ill with fever, conjunctivitis, and a maculopapular rash. The baby's parents reported that the fever had begun during their recent flight from China, where they had adopted the child. The hospital where the child was taken conducted a thorough physical exam, making note of low-grade fever, cough, runny nose, and conjunctivitis. Raised bumps were apparent on the mucosal lining inside the cheeks.

A presumptive diagnosis was made based on the symptoms; blood was drawn, and the diagnosis was confirmed when high levels of a specific IgM were found. The diagnosis set off a flurry of activity because the disease is relatively serious and highly contagious. Furthermore, the baby could have potentially exposed hundreds of people during the incubation period preceding the flight home from China. First, of course, there were the other children and staff at the orphanage where the child lived. Also potentially exposed were 63 families that had traveled together to and from China on this trip and dozens of staff members at the Chinese hospital where medical exams were performed, as well as staff at the U.S. Consulate and passengers and crew on the two flights (from China to Los Angeles and from Los Angeles to Houston).

The Centers for Disease Control and Prevention worked together with the Central China Adoption Agency to identify contacts, isolate additional cases, and supply vaccinations where needed. Eventually, 14 cases of this disease were found in the United States, most of them babies from a single orphanage in China, or their U.S. contacts. Officials at the orphanage acknowledged that recent arrivals at the orphanage had not been adequately immunized against the disease. A vaccine against this disease has been widely used in the United States since the 1960s, but less developed countries may have low vaccination rates.

- *What is the disease described here?*
- *If a vaccine has been widely used in the United States for 40 years, why did American contacts of the child also become infected?*

Case File 18 Wrap-Up appears on page 555.

CHAPTER OVERVIEW

- The skin is organized in layers, from the deepest layer, called the stratum basale, to the uppermost layer, which is the epidermis. The epidermis is composed of cells packed with the protein keratin, which protects the skin and "waterproofs" it. Keratin also provides protection against microbial invasion. Other defenses include low pH, high salt, and antimicrobial peptides. Both the skin and the eye are protected by lysozyme.

- The skin has a diverse array of microbes as its normal biota. Gram-positive bacteria are most common. Inhabitants of the skin must be tolerant of high-salt, low-moisture conditions. The eye is also home to an array of (mostly gram-positive) bacteria but in lower numbers than on the skin.
- Infectious diseases with their most visible manifestations on the surface of the body (or on the skin) range from acne and athlete's foot to leprosy and gangrene.
- The surfaces of the eye exposed to microbial infection are the conjunctiva and the cornea. The flushing action of the tears, which contain lysozyme and lactoferrin, is the major protective feature of the eye.
- Infectious diseases with their major manifestations in the eye include neonatal eye infections as well as conditions afflicting all ages, such as keratitis and trachoma.

18.1 The Skin and Eyes

The skin makes contact directly with the environment—not only with solid objects but also with water and other fluids and with the atmosphere. What's more, many infectious diseases include skin eruptions or lesions as part of the course of illness and often as a major symptom, even if the infective agent does not enter via the skin. Prior to more sophisticated diagnostic methods, the appearance of a skin rash was often the best clue to the type of disease being experienced by a patient. This is still true in many instances.

The eye surface, like the skin, is also exposed constantly to the environment. For this reason, we include diseases of both organ systems in this chapter.

18.2 The Skin and Its Defenses

The organ under consideration in this chapter is the boundary between the organism and the environment. The skin, together with the hair, nails, and sweat and oil glands, forms the **integument.** The skin has a total surface area of 1.5 to 2 square meters. Its thickness varies from 1.5 millimeters at places such as the eyelids to 4 millimeters on the soles of the feet. Several distinct layers can be found in this thickness, and we summarize them here. Follow **figure 18.1** as you read.

The outermost portion of the skin is the epidermis, which is further subdivided into four or five distinct layers. On top is a thick layer of epithelial cells called the stratum corneum, about 25 cells thick. The cells in this layer are

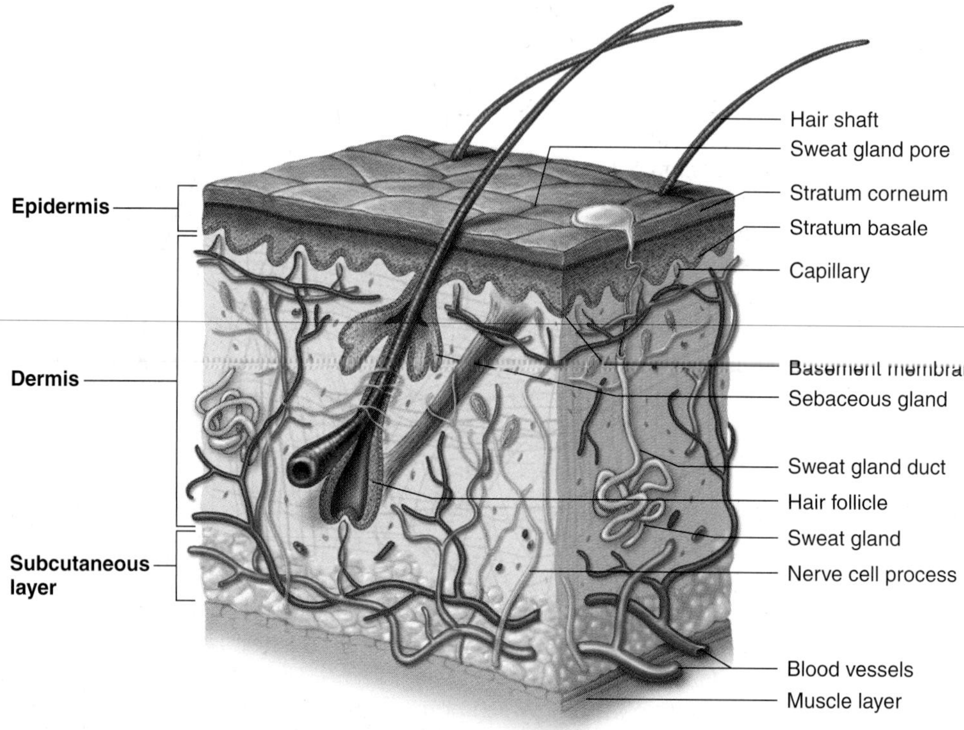

Figure 18.1 A cross section of skin.

dead and have migrated from the deeper layers during the normal course of cell division. They are packed with a protein called keratin, which the cells have been producing ever since they arose from the deepest level of the epidermis. Because this process is continuous, the entire epidermis is replaced every 25 to 45 days. Keratin gives the cells their ability to withstand damage, abrasion, and water penetration; the surface of the skin is termed *keratinized* for this reason. Below the stratum corneum are three or four more layers of epithelial cells. The lowest layer, the stratum basale, or basal layer, is attached to the underlying dermis and is the source for all of the cells that make up the epidermis.

The dermis, underneath the epidermis, is composed of connective tissue instead of epithelium. This means that it is a rich matrix of fibroblast cells and fibers such as collagen, and it contains macrophages and mast cells. The dermis also harbors a dense network of nerves, blood vessels, and lymphatic vessels. Damage to the epidermis generally does not result in bleeding, whereas damage deep enough to penetrate the dermis results in broken blood vessels. Blister formation, the result of friction trauma or burns, causes a separation between the dermis and epidermis.

The "roots" of hairs, called follicles, are in the dermis. **Sebaceous** (oil) **glands** and scent glands are associated with the hair follicle. Separate sweat glands are also found in this tissue. All of these glands have openings on the surface of the skin, so they pass through the epidermis as well.

It could be said that the skin is its own defense—in other words, the very nature of its keratinized surface prevents most microorganisms from penetrating into sensitive deeper tissues. Millions of cells from the stratum corneum slough off every day, and attached microorganisms slough off with them. The skin is also brimming with antimicrobial substances. The sebaceous glands' secretion, called **sebum,** has a low pH, which makes the skin inhospitable to most microorganisms. Sebum is oily due to its high concentration of lipids. The lipids can serve as nutrients for normal microbiota, but breakdown of the fatty acids contained in lipids leads to toxic by-products that inhibit the growth of microorganisms not adapted to the skin environment. This mechanism helps control the growth of potentially pathogenic bacteria. Sweat is also inhibitory to microorganisms, because of both its low pH and its high salt concentration. **Lysozyme** is an enzyme found in sweat (and tears and saliva) that specifically breaks down peptidoglycan, which you learned in chapter 4 is a unique component of eubacterial cell walls.

Perhaps the most effective skin defense against infection is the one most recently discovered. In the past 15 years, small molecules called **antimicrobial peptides** have been identified in epithelial cells. These are positively charged chemicals that act by disrupting (negatively charged) membranes of bacteria. There are many different types of these peptides, and they seem to be chiefly responsible for keeping the microbial count on skin relatively low.

18.3 Normal Biota of the Skin

Microbes that do live on the skin surface as normal biota must be capable of living in the dry, salty conditions they find there. Microbes are rather sparsely distributed over dry, flat areas of the body such as on the back, but they can grow into dense populations in moist areas and skin folds, such as the underarm and groin areas. The normal microbiota also live in the protected environment of the hair follicles and glandular ducts.

As discussed in chapter 13, we don't know how many species call the skin "home" because the majority of them are probably not cultivable. But we are aware of some of them. Three main categories of microorganisms reside on the skin: the diphtheroids, the micrococci (including staphylococci and streptococci), and yeasts. The diphtheroids are club-shaped bacteria that resemble *Corynebacterium diphtheriae*. They are gram-positive and can be aerobic, aerotolerant, or anaerobic. Unlike *C. diphtheriae,* they are not usually virulent. One very prominent member of this group is *Propionibacterium acnes*, which is aerotolerant or even anaerobic. It lives on healthy skin, but its metabolic activities can contribute to the development of acne (discussed later in the chapter).

The micrococcus group includes the genus *Staphylococcus* as well as the genus *Micrococcus*. It has been said that *S. epidermidis* is the bacterial species most well adapted to life on the human body. It is present on the skin of every human. *S. aureus*, by contrast, could be called the bacterium best adapted to damage its host. It is found living on the skin of at least 20% of the human population, but it can hardly be called "normal biota" because it is such a potentially dangerous pathogen. Sometimes it is referred to as "abnormal biota," indicating that although it can inhabit skin without causing disease, it can also cause dangerous infections **(Insight 18.1).**

S. epidermidis is sometimes referred to as "CNS" or coagulase-negative staphylococcus, highlighting its lack of the enzyme **coagulase,** which is indeed found in *S. aureus*. The staphylococci are particularly well adapted to life on the skin because they can tolerate high salt concentrations. In fact, a common culturing method to select for staphylococci employs a high-salt agar, which inhibits the growth of most nonstaphylococci.

Alpha-hemolytic and nonhemolytic streptococci are also found on the skin.

Defenses and Normal Biota of the Skin

	Defenses	Normal Biota
Skin	Keratinized surface, sloughing, low pH, high salt, lysozyme	*Corynebacterium, Propionibacterium, Staphylococcus epidermidis* and *S. aureus, Micrococcus,* alpha-hemolytic and nonhemolytic streptococci, *Candida,* and *Malassezia*

INSIGHT 18.1 *Medical*

The Skin Predators: *Staphlyococcus* and *Streptococcus*

The relatively hostile environment of the skin makes it difficult for many microorganisms to set up shop there. But two genera of gram-positive bacteria, *Staphylococcus* and *Streptococcus,* are uniquely suited to living and sometimes causing disease there. Many of these diseases are described in this chapter. Here we discuss some very common skin conditions caused by the two bacteria.

Staphylococcus aureus: Folliculitis, Furuncles, and Carbuncles

Currently, 31 species have been placed in the genus *Staphylococcus.* Of these, the most important human pathogen is probably *S. aureus.* Elsewhere in this chapter, we have discussed some of the other staphylococcal skin conditions (impetigo, cellulitis, and scalded skin syndrome), but other common skin diseases have *S. aureus* as a cause. **Folliculitis** is a mild, superficial inflammation of hair follicles or glands. Although these lesions are usually resolved with no complications, they can lead to infections of subcutaneous tissues. An *abscess* is a more serious localized staphylococcal skin infection, which appears as an inflamed, fibrous lesion enclosing a core of pus. There are two types: furuncles and carbuncles. A **furuncle** results when the inflammation of a single hair follicle or sebaceous gland progresses into a large, red, and extremely tender abscess or pustule. Furuncles often occur in clusters on parts of the body such as the buttocks, axillae, and back of the neck, where skin rubs against other skin or clothing. They are also commonly called *boils.* A **carbuncle** is a larger and deeper lesion, sometimes as big as a baseball, created by aggregation and interconnection of a cluster of furuncles. It is usually found in areas of thick, tough skin such as on the back of the neck. Carbuncles are extremely painful and can even be fatal in elderly patients when they give rise to systemic disease.

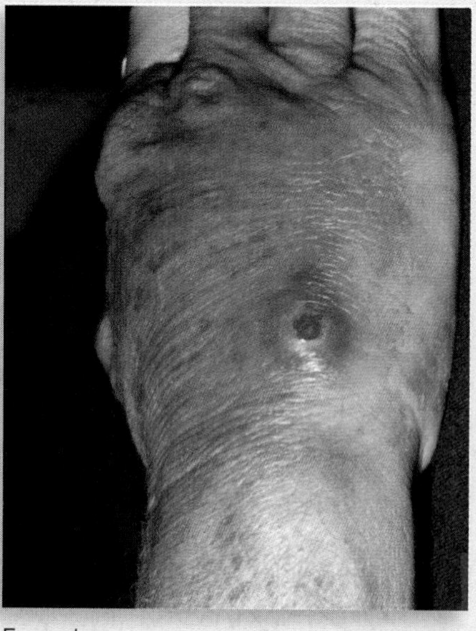

Furuncle

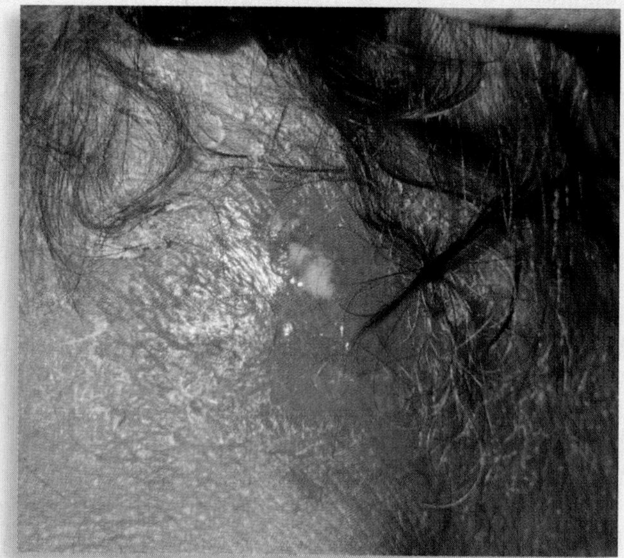

Carbuncle

Streptococcus pyogenes: Erysipelas

In addition to impetigo and cellulitis, at least two other important *S. pyogenes* diseases begin on the skin. One fairly invasive manifestation is **erysipelas.** The pathogen usually enters through a small wound or incision on the face or extremities and eventually spreads

Various yeast groups grow in low numbers on the skin; they can cause opportunistic disease. These groups include familiar species, such as *Candida albicans,* and less familiar types, such as *Malassezia.*

18.4 Skin Diseases Caused by Microorganisms

Acne

The term *acne* encompasses all follicle-associated lesions, from the isolated pimple to severe widespread acne. Normally, the sebaceous glands associated with hair follicles (see figure 18.1) are a self-contained system for protecting, softening, and lubricating the skin. As hair and skin grow, dead epidermal cells and sebum work their way upward and are discharged from the pore to the skin surface.

Skin prone to pimples and acne has a structure that traps the mass of sebum and dead cells, clogging the pores. An exaggerated process of keratinization occurs in skin cells in and around the follicle, which also helps to block the pore. An added factor is overproduction of sebum when the sebaceous gland is stimulated by hormones (especially male). *Propionibacterium acnes* present in the follicle releases lipases to digest this surplus of oil. The combination of digestive products (fatty acids) and bacterial antigens stimulates an intense local inflammation that to the dermis and subcutaneous tissues. Early symptoms are edema and redness of the skin near the portal of entry, and fever and chills. The lesion begins to spread outward, producing a slightly elevated edge that is noticeably red and hot. Depending on the depth of the lesion and how the infection progresses, cutaneous lesions can remain superficial or can produce long-term systemic complications. Severe cases involving large areas of skin are occasionally fatal.

Both *Streptococcus pyogenes* and *Staphylococcus aureus:* Necrotizing Fasciitis

As you will read in this chapter, both *S. pyogenes* and *S. aureus* can lead to impetigo and cellulitis. Often, both of the pathogens are isolated from lesions. The same is true for a very invasive infection called necrotizing fasciitis. The disease has been known for hundreds of years, but in recent years small outbreaks of the disease have received heavy publicity as the "flesh-eating disease." Cases of this disease are rather rare, but its potential for harm is high. It can begin with an innocuous cut in the skin and spread rapidly into nearby tissue, causing severe disfigurement and even death.

There is really no mystery to the pathogenesis of necrotizing fasciitis. It begins very much like impetigo and other skin infections: Streptococci and/or staphylococci on the skin are readily introduced into small abrasions or cuts, where they begin to grow rapidly. The particular strains of bacteria that cause this condition have great toxigenicity and invasiveness because of special enzymes and toxins. The enzymes digest the connective tissue in skin, and the toxins poison the epidermal and dermal tissue. As the flesh is killed, it separates and sloughs off, forming a pathway for the bacteria to spread into deeper tissues such as muscle. More dangerous cases involve polymicrobial infections that can include anaerobic bacteria and the systemic spread of the toxin to other organs. Some patients have lost parts of their limbs and faces, and others have suffered amputation, but early diagnosis and treatment can prevent these complications.

Erysipelas

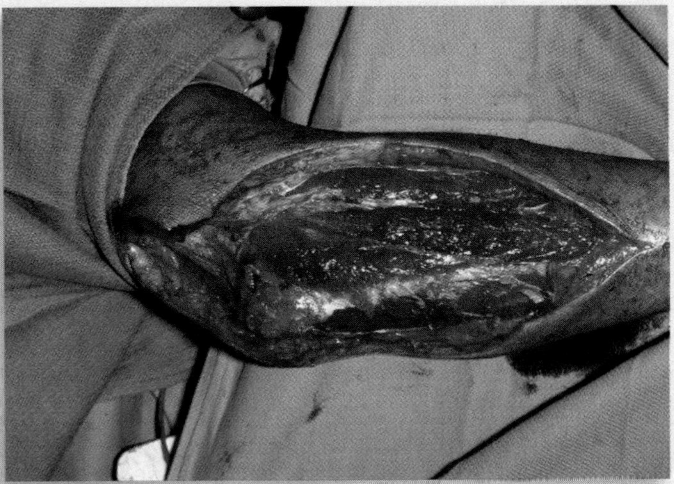

Necrotizing fasciitis

eventually can burst the follicle. In time, the lesion can erupt on the surface.

Different types of lesions are associated with this process. When the skin initially swells over the pore leading out of a hair follicle, it is called a *comedo.* If the pore is closed, this comedo is commonly called a whitehead. If the pore remains open to the surface but is blocked with a dark plug of sebum, it appears as a blackhead. When the lesion erupts on the surface, it is called a pustule or papule. At this point, the lesion contains sebum and pus, a collection of bacteria, dead skin cells, and white blood cells from the inflammatory reaction. Pustules that come to involve deeper layers of skin are called cysts, and they can be quite painful. Widespread lesions of this type are called cystic acne.

Causative Agent

Propionibacterium acnes is the bacterium associated with acne, but the "cause" of acne is multifactorial, requiring other conditions to be just right before the presence of this otherwise benign bacterium results in acne.

The bacterium is an anaerobic or aerotolerant gram-positive rod arranged in short chains or clumps. It releases a variety of enzymes that contribute to its virulence. The most important of these appears to be lipase, although it also releases proteases, neuraminidase, and a hyaluronidase. In addition, it secretes a low molecular weight protein that is a strong attractant for white blood cells (contributing to inflammation).

The complete genome sequence of the bacterium was published in 2004. This will allow researchers to identify additional virulence factors and to design more precise therapies for it.

Transmission and Epidemiology

As already noted, *P. acnes* is normal biota on human skin, so it is not a transmissible infection. The epidemiology of a condition refers to its distribution in populations and usually takes into consideration the mode of transmission of the microorganism, the degree of susceptibility of different hosts, environmental parameters such as climate and geography, and even the behavior of current and potential hosts. When speaking of the epidemiology of acne we are really considering what groups have the combination of factors that can result in acne rather than the distribution of *P. acnes*. Almost 100% of adolescents and young adults experience acne of some degree at some time in their lives. More severe forms of adolescent acne are more common in males than females, probably because male hormones, or androgens, aggravate the condition. Females produce male hormones as well, but during adolescence males have a higher incidence of moderate to severe acne. Evidence exists that acne extending into adulthood, or beginning in adulthood, is more common in women.

Prevention and Treatment

There is no effective prevention of acne; it is not the result of poor hygiene or even of eating the wrong foods. For many years, the only treatment options were (1) topical agents that enhanced the sloughing of skin cells, which could help to prevent comedo formation or to keep comedos from becoming pustules or papules; and (2) either topical or oral antibiotics, such as erythromycin or tetracycline. It has become apparent that such long-term use of antibiotics causes a high rate of antibiotic resistance in skin bacteria (in the case of topical application) or in whole-body normal biota. This result should have been predicted because oral antibiotics are typically given for long periods of time in low, sublethal doses—the perfect set of conditions for creating antibiotic resistance in bacteria. It has even been shown that live-in family members of people taking antibiotics for their acne also eventually harbor skin bacteria resistant to the same antibiotic the acne patient is taking. In this way, resistant bacteria can spread beyond the original host.

Recently, females have been prescribed oral contraceptive pills (containing estrogen) to treat their acne. This treatment is controversial because of the dangers of estrogen use. For patients with severe acne and for whom other treatment options have failed, isotretinoin (Accutane) may be prescribed. But this drug can have severe side effects, including psychological depression, and patients must be closely monitored. The drug's mechanism of action is not clearly understood, although there is a significant decrease in sebaceous gland activity and *P. acnes* growth.

Impetigo

Impetigo is a superficial bacterial infection that causes the skin to flake or peel off (**figure 18.2;** see also Insight 18.1). It is not a serious disease but is highly contagious, and children are the primary victims. Impetigo can be caused by either *Staphylococcus aureus* or *Streptococcus pyogenes*, and some cases are probably caused by a mixture of the two. It has been suggested that *S. pyogenes* begins all cases of the disease, and in some cases *S. aureus* later takes over and becomes the predominant bacterium cultured from lesions. Because *S. aureus* produces a bacteriocin (toxin) that can destroy *S. pyogenes*, it is possible that *S. pyogenes* is often missed in culture-based diagnosis.

CHECKPOINT 18.1	**Acne**
Causative Organism(s)	*Propionibacterium acnes*
Most Common Mode(s) of Transmission	Endogenous
Virulence Factors	Lipase, inflammatory mediator, other enzymes
Culture/Diagnosis	Based on clinical picture
Prevention	None
Treatment	Antibiotics (topical or oral), isotretinoin

Signs and Symptoms

The "lesion" of impetigo looks variously like peeling skin, crusty and flaky scabs, or honey-colored crusts. Lesions are most often found around the mouth, face, and extremities, though they can occur anywhere on the skin. It is very superficial and it itches. The symptomology does not indicate whether the infection is caused by *Staphylococcus* or *Streptococcus*.

Impetigo Caused by Staphylococcus aureus

Staphylococcus aureus This bacterium is one of the most exquisitely tuned microorganisms for causing disease in humans. It is responsible for a long list of different diseases in

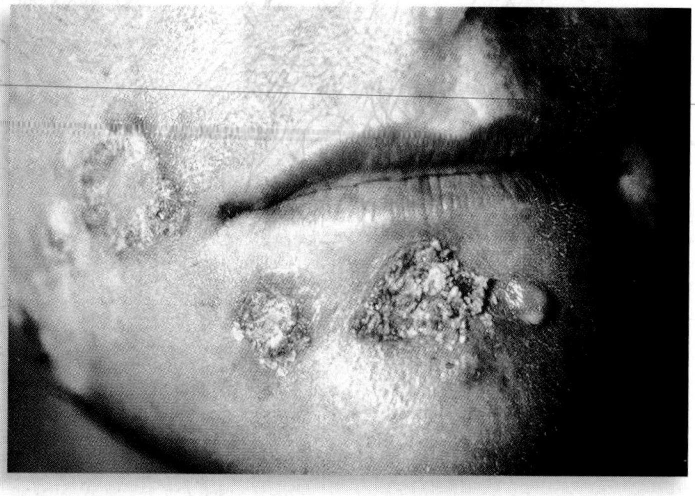

Figure 18.2 Impetigo lesions on the face.

addition to the ones highlighted in this chapter. *S. aureus* can cause pneumonias, food poisoning, serious bloodstream infections, bone infections, toxic shock syndrome, and meningitis. It is a gram-positive coccus that grows in clusters, like a bunch of grapes **(figure 18.3*a*).** It is nonmotile.

S. aureus in culture produces large, round, opaque colonies **(figure 18.3*b*)** at an optimum of 37°C, although it can grow at any temperature between 10°C and 46°C. The species is a facultative anaerobe whose growth is enhanced in the presence of O_2 and CO_2. Its nutrient requirements can be satisfied by routine laboratory media, and most strains are metabolically versatile—that is, they can digest proteins and lipids and ferment a variety of sugars. This species is considered the sturdiest of all non-spore-forming pathogens, with well-developed capacities to withstand high salt (7.5% to 10%), extremes in pH, and high temperatures (up to 60°C for 60 minutes). *S. aureus* also remains viable after months of air drying and resists the effect of many disinfectants and antibiotics. These properties contribute to the reputation of *S. aureus* as a troublesome hospital pathogen.

Pathogenesis and Virulence Factors The most important virulence factors relevant to *S. aureus* impetigo are exotoxins called exfoliative toxins A and B, which are coded for by a phage that infects some *S. aureus* strains. At least one of the toxins attacks a protein that is very important for epithelial cell-to-cell binding in the outermost layer of the skin. Breaking up this protein leads to the characteristic blistering seen in the condition. The breakdown of skin architecture also facilitates the spread of the bacterium. All pathogenic *S. aureus* strains typically produce **coagulase,** an enzyme that coagulates plasma and blood. The precise importance of coagulase to the disease process remains uncertain. It may be that coagulase causes fibrin to be deposited around staphylococcal cells. Fibrin can stop the action of host defenses, such as phagocytosis, or it may promote staphylococcal adherence to tissues. Because 97% of all human isolates of *S. aureus* produce this enzyme, its presence is considered the most diagnostic species characteristic.

Other enzymes expressed by *S. aureus* include hyaluronidase, which digests the intercellular "glue" (hyaluronic acid) that binds connective tissue in host tissues; staphylokinase, which digests blood clots; a nuclease that digests DNA (DNase); and lipases that help the bacteria colonize oily skin surfaces. Most hospital strains also have penicillinase, which makes them resistant to that group of antibiotics.

Culture and/or Diagnosis Doctors usually diagnose impetigo by looking at it, and they treat it with antibiotics that target both probable causes of it. But when the etiologic agent requires identification (for instance, if initial treatment fails), well-established methods exist for looking for *S. aureus*. Primary isolation of *S. aureus* is achieved by inoculation on sheep or rabbit blood agar. For heavily contaminated specimens, selective media such as mannitol salt agar are used. In addition to culturing, the specimen can be Gram stained, and irregular clusters of gram-positive cocci can be observed. Because differentiating among gram-positive cocci, including *S. epidermidis*, is not possible using colonial and morphological characteristics alone, other tests are required. The production of catalase, an enzyme that breaks down hydrogen peroxide accumulated during oxidative metabolism, can be used to differentiate the staphylococci, which produce it, from the streptococci, which do not. The ability of *Staphylococcus* to grow anaerobically and to ferment sugars separates it from *Micrococcus*, a nonpathogenic genus that is a common specimen contaminant.

One key technique for separating *S. aureus* from other species of *Staphylococcus* is the coagulase test **(figure 18.4).** By definition, any isolate that coagulates plasma is *S. aureus*; all others are coagulase negative. Rapid multitest systems are used routinely to collect other physiological information **(figure 18.5).** An important confirming identification of *S. aureus* can be made with a latex bead agglutination test. It is based on *S. aureus* surface protein A that can bind to IgG antibodies. PCR tests can also detect *S. aureus*, as well as methicillin-resistant strains of *S. aureus* (discussed later).

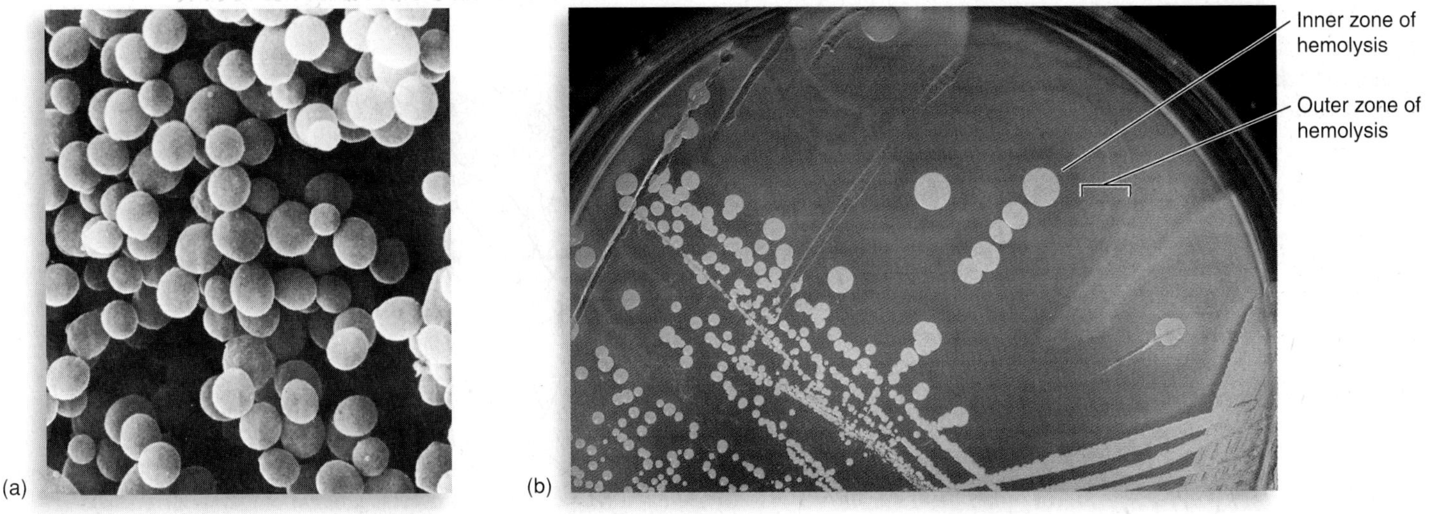

Figure 18.3 ***Staphylococcus aureus.***
(a) Scanning electron micrograph of *S. aureus*. **(b)** Blood agar plate growing *S. aureus*. Some strains show two zones of hemolysis, caused by two different hemolysins. The inner zone is clear, whereas the outer zone is fuzzy and appears only if the plate has been refrigerated after growth.

A NOTE ABOUT THE CHAPTER ORGANIZATION

Beginning in this chapter, we discuss all the conditions caused by microbial infection. The chapter organization mirrors the clinical experience. Patients present themselves to health care practitioners with a set of symptoms, and the health care team makes an "anatomical" diagnosis—such as a *generalized vesicular rash.* The anatomical diagnosis allows practitioners to narrow down the list of possible causes to microorganisms that are known to be capable of creating such a condition. Then the proper tests can be performed to arrive at an etiologic diagnosis (that is, determining the exact microbial cause). So the order of events is (1) anatomical diagnosis based on signs and symptoms; (2) consideration of a number of agents that are known to cause disease in that anatomical location (often called the differential diagnosis); followed by (3) the etiologic diagnosis. In practice, this process may be shortened. For instance, if a patient has a disease such as Hansen's disease (leprosy), the distinctive signs and symptoms of that disease may allow the practitioner to make the anatomical and the etiologic diagnosis at the same time, followed by confirmation of the etiology through laboratory methods, if necessary. In other cases, such as the common cold, the physician may consider only the anatomical diagnosis and never advance to the etiologic diagnosis because a cold is a mild self-limiting disease.

The chapters are organized by anatomical diagnosis (for example, Microbial Diseases of the Skin and Eyes). Specific diseases and the microorganisms that cause them then are detailed. Some diseases are the result of infection by a single type of microorganism. Hansen's disease is an example of this type of disease. In other cases, single diseases or conditions can be caused by many different microorganisms, including bacteria, viruses, and so on. The classic examples are pneumonia in the respiratory tract, meningitis in the central nervous system, and diarrhea in the gastrointestinal system. In this chapter, for instance, a maculopapular rash may be caused by the measles virus, the rubella virus, or a parvovirus. A table at the end of each disease/condition makes it clear whether a single microorganism or multiple agents are to be considered in the diagnosis.

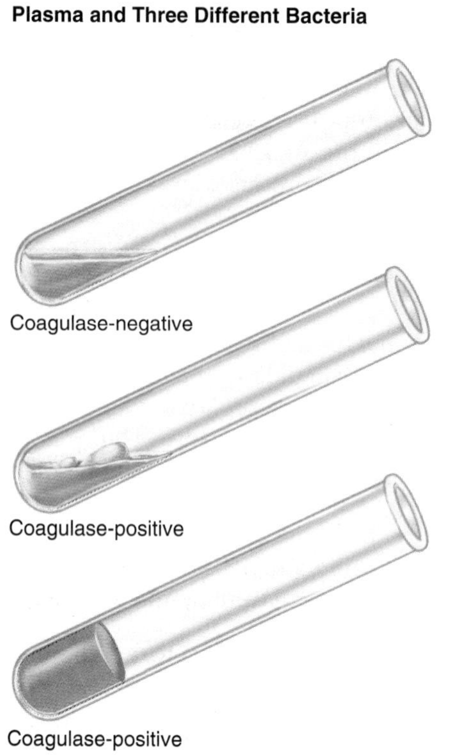

Figure 18.4 The coagulase test.
Staphylococcal coagulase is an enzyme that reacts with factors in plasma to initiate clot formation. In the coagulase test, a tube of plasma is inoculated with the bacterium. If it remains liquid, the test is negative. If the plasma develops a lump or becomes completely clotted, the test is positive.

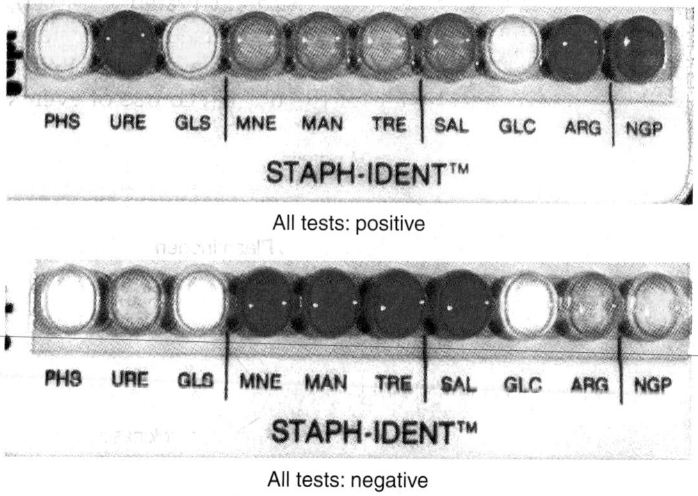

Figure 18.5 Miniaturized test system used in further identification of *Staphylococcus* isolates.
A single isolate is used to inoculate all the cupules on a strip. The cupules contain substrates that detect phosphatase production (PHS), urea hydrolysis (URE), glucosidase production (GLS), mannose fermentation (MNE), mannitol fermentation (MAN), trehalose fermentation (TRE), salicin fermentation (SAL), glucuronidase production (GLC), arginine hydrolysis (ARG), and galactosidase production (NGP).

Impetigo Caused by Streptococcus pyogenes

Streptococcus pyogenes This bacterium is thoroughly described in chapter 21. The important features are briefly summarized here and the features pertinent to impetigo are listed in **Checkpoint 18.2.**

S. pyogenes is a gram-positive coccus in Lancefield group A and is beta-hemolytic on blood agar. In addition to impetigo, it causes streptococcal pharyngitis (strep throat), scarlet fever, pneumonia, puerperal fever, necrotizing fasciitis, serious bloodstream infections, and poststreptococcal conditions such as rheumatic fever.

CHECKPOINT 18.2 Impetigo

Causative Organism(s)	*Staphylococcus aureus*	*Streptococcus pyogenes*
Most Common Modes of Transmission	Direct contact, indirect contact	Direct contact, indirect contact
Virulence Factors	Exfoliative toxin A, coagulase, other enzymes	Streptokinase, plasminogen-binding ability, hyaluronidase, M protein
Culture/Diagnosis	Routinely based on clinical signs, when necessary, culture and Gram stain, coagulase and catalase tests, multitest systems, PCR	Routinely based on clinical signs, when necessary, culture and Gram stain, coagulase and catalase tests, multitest systems, PCR
Prevention	Hygiene practices	Hygiene practices
Treatment	Topical mupirocin, oral cephalexin	Topical mupirocin, oral cephalexin
Distinguishing Features	Seen more often in older children, adults	Seen more often in newborns; may have some involvement in all impetigo (preceding *S. aureus* in staphylococcal impetigo)

If the precise etiologic agent must be identified, there are well-established methods for identifying group A streptococci. Refer to chapter 21.

Pathogenesis and Virulence Factors The symptoms of *S. pyogenes* impetigo are indistinguishable from that caused by *S. aureus.* Like *S. aureus,* this bacterium possesses a huge arsenal of enzymes and toxins. As mentioned earlier, it anchors itself to surfaces (including skin) using a variety of adhesive elements on its surface (LTA, M protein and other proteins, and a hyaluronic acid capsule). M protein also protects it from phagocytosis. Like *S. aureus,* it possesses hyaluronidase.

S. pyogenes has a clever system to exploit host factors to increase its ability to spread in tissues. The bacterium's M protein has a high-affinity binding site for plasminogen—a host plasma protein that, when activated (cleaved), becomes plasmin **(figure 18.6).** Plasmin itself is a protein-splitting enzyme that digests fibrin and other tissue proteins. But *S. pyogenes* doesn't wait for the normal course of events in which another host protein activates the plasminogen. Instead, it secretes an enzyme called streptokinase, which is a plasminogen activator. So the bacterium coats itself with host plasminogen, then uses its own enzyme (streptokinase) to activate it. It turns itself into a tissue degrader.

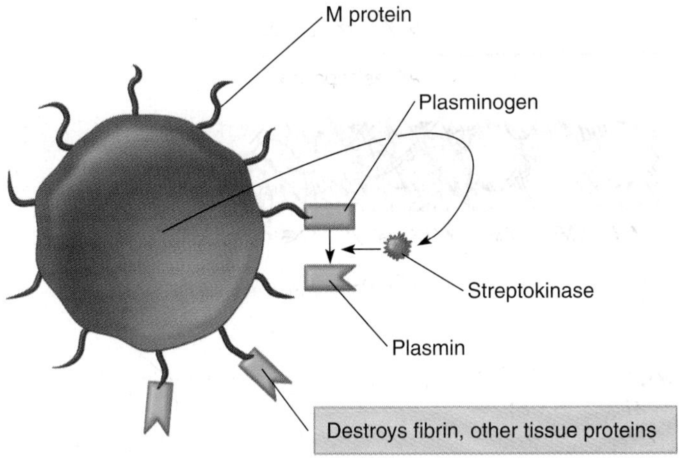

Figure 18.6 Plasmin activation by *S. pyogenes*.
The bacterium binds host plasminogen, then secretes an enzyme (streptokinase) that cleaves it, creating plasmin, which has tissue-degrading power. In the figure, bacterial components are red and host components are green.

Rarely, impetigo caused by *S. pyogenes* can be followed by acute poststreptococcal glomerulonephritis (see chapter 21). The strains that cause impetigo never cause rheumatic fever, however.

Transmission and Epidemiology of Impetigo

Impetigo, whether it is caused by *S. pyogenes, S. aureus,* or both, is highly contagious and transmitted through direct contact but also via fomites and mechanical vector transmission. It affects mostly preschool children, but all ages can acquire the disease. The peak incidence is in the summer and fall. *S. pyogenes* is more often the cause of impetigo in newborns, and *S. aureus* is more often the cause of impetigo in older children, but both can cause infection in either age group.

Prevention

The only current prevention for impetigo is good hygiene. Vaccines are in development for both of the etiologic agents, but none are currently available.

Treatment

Impetigo is usually treated with a drug that will kill either bacterium, *S. pyogenes* or *S. aureus,* eliminating the need to determine the exact etiologic agent. The drug of choice is topical mupirocin (brand name Bactroban), a protein synthesis inhibitor. In cases of widespread skin involvement, oral antibiotics such as cephalexin may be used.

The treatment of nonimpetigo *S. pyogenes* infections is usually straightforward because this organism is sensitive to penicillin. *S. aureus* infections are another story. Although *S. aureus* impetigo is usually easily treated, other *S. aureus* diseases can be very difficult to treat effectively. We discuss these difficulties later in this chapter.

A NOTE ABOUT MRSA

Everyone who has any recent work experience in health care knows the term *MRSA*, which stands for methicillin-resistant *Staphylococcus aureus.* MRSAs are *S. aureus* strains that are resistant to penicillin derivatives. (Methicillin is a penicillin derivative that is used only in the laboratory for testing purposes.) The first strain of MRSA appeared in hospitals in the 1960s, but the strains became common in the 1990s. Now it is very common for nosocomial infections to be caused by MRSA. These MRSA strains must be treated very aggressively with vancomycin in most cases, because they are resistant to multiple first-line and second-line antibiotics.

During the period between 2004 and 2006, a disturbing *S. aureus* trend that had been bubbling under the surface emerged into the public eye. While the health care establishment had grown accustomed to (though not complacent about) hospital-acquired MRSA (HA-MRSA), people who had no recent history in hospitals or health care facilities started turning up with MRSA infections. These infections are classified as *community-acquired* and given the acronym CA-MRSA. They turned up in tattoo parlors in Ohio and Kentucky, in the locker rooms of professional football teams, and in prisons. Over the past 2 years, the Washington Redskins football team has experienced four MRSA infections. The humid and sweaty conditions of locker rooms favor survival of the organism, and many sports teams are undergoing new "hygiene" training to try to stem the incidence of the disease.

The infection can begin as a boil or may invade a tiny cut. Many people report having had a "spider bite" because of the red swollen appearance of an early lesion. The good news is that CA-MRSA is susceptible to a wider array of antibiotics than the HA-MRSA. But how long will that last?

Cellulitis

Cellulitis is a condition caused by a fast-spreading infection in the dermis and in the subcutaneous tissues below. It causes pain, tenderness, swelling, and warmth. Fever and swelling of the lymph nodes draining the area may also occur. Frequently, red lines leading away from the area are visible (a phenomenon called *lymphangitis*); this symptom is the result of microbes and inflammatory products being carried by the lymphatic system. Bacteremia could develop with this disease, but uncomplicated cellulitis has a good prognosis.

Cellulitis generally follows introduction of bacteria or fungi into the dermis, either through trauma or by subtle means (with no obvious break in the skin). Symptoms take several days to develop. The most common causes of the condition in healthy people are *Staphylococcus aureus* and *Streptococcus pyogenes*, although almost any bacterium and some fungi can cause this condition in an immunocompromised patient. In infants, group B streptococci are a frequent cause (see chapter 23).

People who are immunocompromised or who have cardiac insufficiency are at higher risk for this condition than are healthy persons. They also risk complications, such as spread to the bloodstream, rapid spreading through adjacent tissues, and, especially in children, meningitis. Occasionally, cellulitis is a complication of varicella (chickenpox) infections.

Mild cellulitis responds well to oral antibiotics chosen to be effective against both *S. aureus* and *S. pyogenes.* More involved infections and infections in immunocompromised people require intravenous antibiotics. If there are extensive areas of tissue damage, surgical debridement (duh-breed'-munt) is warranted **(Checkpoint 18.3).**

Staphylococcal Scalded Skin Syndrome (SSSS)

This syndrome is another **dermolytic** condition caused by *Staphylococcus aureus.* It affects mostly newborns and babies, although children and adults can experience the infection. Newborns are susceptible when sharing a nursery with another newborn who is colonized with *S. aureus.* Transmission may occur when caregivers carry the bacterium from one baby to another. Adults in the nursery can also directly transfer *S. aureus* because approximately 30% of adults are asymptomatic carriers. Carriers can harbor the bacteria in the nasopharynx,

CHECKPOINT 18.3 Cellulitis

Causative Organism(s)	*Staphylococcus aureus*	*Streptococcus pyogenes*	Other bacteria or fungi
Most Common Modes of Transmission	Parenteral implantation	Parenteral implantation	Parenteral implantation
Virulence Factors	Exfoliative toxin A, coagulase, other enzymes	Streptokinase, plasminogen-binding ability, hyaluronidase, M protein	–
Culture/Diagnosis	Based on clinical signs	Based on clinical signs	Based on clinical signs
Prevention	–	–	–
Treatment	Oral or IV antibiotic (cephalexin); surgery sometimes necessary	Oral or IV antibiotic penicillin; surgery sometimes necessary	Aggressive treatment with oral or IV antibiotic (cephalexin or penicillin); surgery sometimes necessary
Distinguishing Features	–	–	More common in immunocompromised

axilla, perineum, and even the vagina. (Fortunately, only about 5% of *S. aureus* strains are lysogenized by the type of phage that codes for the toxins responsible for this disease.)

This condition can be thought of as a systemic form of impetigo. Like impetigo, it is an exotoxin-mediated disease. The phage-encoded exfoliative toxins A and B are responsible for the damage. Unlike impetigo, the toxins enter the bloodstream from some focus of infection (the throat, the eye, or sometimes an impetigo infection) and then travel to the skin throughout the body. These toxins cause **bullous** lesions, which often appear first around the umbilical cord (in neonates) or in the diaper or axilla area. The lesions begin as red areas, take on the appearance of wrinkled tissue paper, and then form very large blisters. Fever may precede the skin manifestations. Eventually, the top layers of epidermis peel off completely. The split occurs in the epidermal tissue layers just above the stratum basale (see figure 18.1). Widespread **desquamation** of the skin follows, leading to the burned appearance referred to in the name **(figure 18.7).**

At this point, the protective keratinized layer is gone, and the patient is vulnerable to secondary infections, cellulitis, and bacteremia. In the absence of these complications, young patients nearly always recover if treated promptly. Adult patients have a higher mortality rate—as high as 50%. Once a tentative diagnosis of SSSS is made, immediate antibiotic therapy should be instituted, using cloxacillin or cephalexin.

It is important, however, to differentiate this disease from a similar skin condition called *toxic epidermal necrolysis* (*TEN*), which is caused by a reaction to antibiotics, barbiturates, or other drugs. TEN has a significant mortality rate. The treatments for the two diseases are very different, so it is important to distinguish between them before instituting therapy. In TEN, the split in skin tissue occurs *between* the dermis and the epidermis, not within the epidermis as is the case with SSSS. Histological examination of tissue from a lesion is usually a better way to diagnose the disease than reliance on culture. Because SSSS is caused by the dissemination of exotoxin, *S. aureus* may not be found in lesions. Nevertheless, culture should be attempted so that antibiotic sensitivities can be established.

CHECKPOINT 18.4	Scalded Skin Syndrome
Causative Organism(s)	*Staphylococcus aureus*
Most Common Modes of Transmission	Direct contact, droplet contact
Virulence Factors	Exfoliative toxins A and B
Culture/Diagnosis	Histological sections; culture performed but false negatives common because toxins alone are sufficient for disease
Prevention	Eliminate carriers in contact with neonates
Treatment	Immediate systemic antibiotics (cloxacillin or cephalexin)
Distinguishing Features	Split in skin occurs *within* epidermis

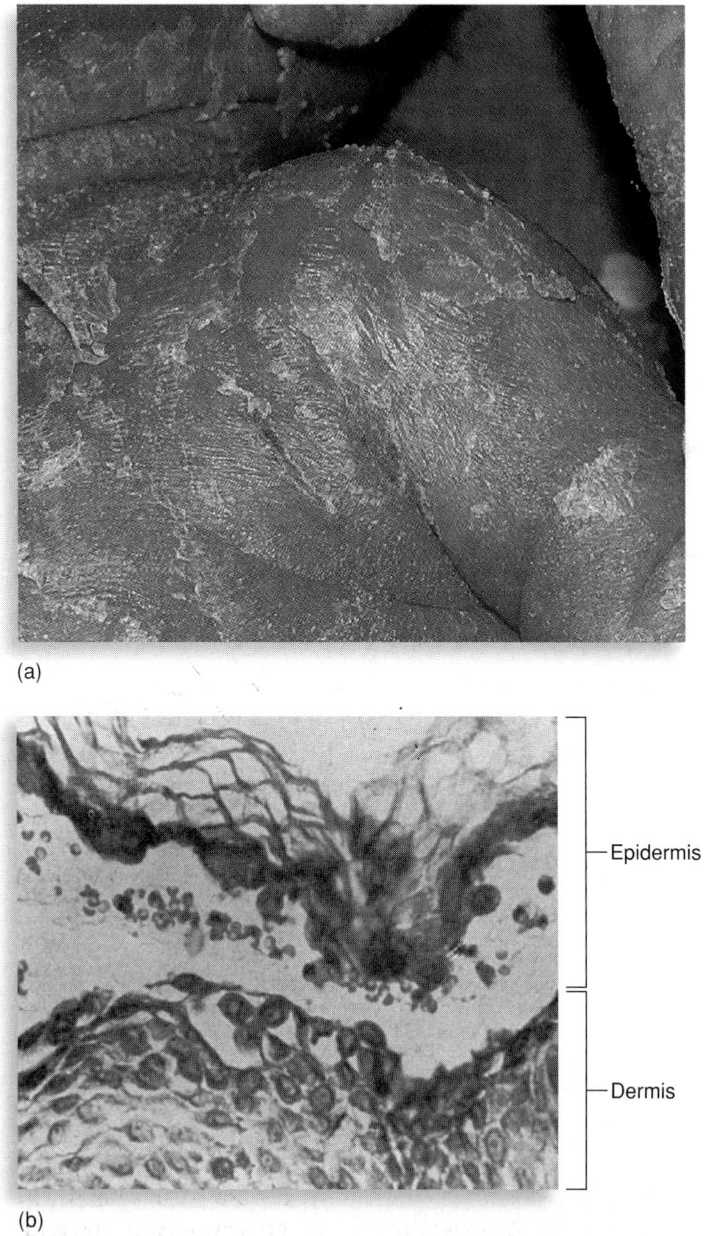

Figure 18.7 Staphylococcal scalded skin syndrome (SSSS) in a newborn child.
(a) Exfoliative toxin produced in local infections causes blistering and peeling away of the outer layer of skin. **(b)** Photomicrograph of a segment of skin affected with SSSS. The point of epidermal shedding, or desquamation, is in the epidermis. The lesions will heal well because the level of separation is so superficial.

Gas Gangrene

Clostridium perfringens, a gram-positive endospore-forming bacterium, as well as some related species, cause a serious condition called **gas gangrene,** or clostridial **myonecrosis** (my″-oh-neh-kro′-sis). The spores of these species can be found in soil, on human skin, and in the human intestine and vagina. The bacteria are anaerobic, and they require anaerobic conditions to manufacture and release the exotoxins that mediate the damage in the disease.

Signs and Symptoms

Two forms of gas gangrene have been indentified. In anaerobic cellulitis, the bacteria spread within damaged necrotic muscle tissue, producing toxin and gas, but the infection remains localized and does not spread into healthy tissue. The pathology of true myonecrosis is more destructive. Toxins produced in large muscles, such as the thigh, shoulder, and buttocks, diffuse into nearby healthy tissue and cause local necrosis there. This damaged tissue then serves as a focus for continued clostridial growth, toxin formation, and gas production. The disease can quickly progress through an entire limb or body area, destroying tissues as it goes **(figure 18.8).** Initial symptoms of pain, edema, and a bloody exudate in the lesion are followed by fever, tachycardia, and blackened necrotic tissue filled with bubbles of gas. Gangrenous infections of the uterus, due to septic abortions, and clostridal septicemia are particularly serious complications. If treatment is not begun early, the disease is invariably fatal.

Pathogenesis and Virulence Factors

Because clostridia are not highly invasive, infection requires damaged or dead tissue that supplies growth factors, and an anaerobic environment. The low-oxygen environment results from an interrupted blood supply and the presence of aerobic bacteria that deplete oxygen. Such conditions stimulate spore germination, rapid vegetative growth in the dead tissue, and release of exotoxins. *C. perfringens* produces several physiologically active exotoxins; the most potent one, *alpha toxin,* causes red blood cell rupture, edema, and tissue destruction **(figure 18.9).** Additional virulence factors that enhance tissue destruction are collagenase, hyaluronidase, and DNase. The gas formed in tissues, resulting from fermentation of muscle carbohydrates, can also destroy muscle structure.

Transmission and Epidemiology

The conditions that may predispose a person to gangrene are surgical incisions, compound fractures, diabetic ulcers, septic abortions, puncture and gunshot wounds, and crushing injuries contaminated by spores from the body or the environment.

Prevention and Treatment

One of the most effective ways to prevent clostridial wound infections is immediate and rigorous cleansing and surgical repair of deep wounds, decubitus ulcers (bedsores), compound fractures, and infected incisions. Debridement of diseased tissue eliminates the conditions that promote the spread of gangrenous infection. This procedure is most difficult in the intestine or body cavity, where only limited amounts of tissue can be removed. Surgery is supplemented by large doses of clindamycin or penicillin to control infection. Hyperbaric oxygen therapy, in which the affected part is exposed to an increased oxygen mix in a pressurized chamber, can also lessen the severity of infection.

Extensive myonecrosis of a limb may call for amputation. Because there are so many different antigenic subtypes in this bacterial group, active immunization is not possible **(Checkpoint 18.5).**

Vesicular or Pustular Rash Diseases

There are two diseases that present as generalized "rashes" over the body in which the individual lesions contain fluid. The lesions are often called *pox*, and the two diseases are chickenpox and smallpox. Chickenpox is very common and mostly benign, but even a single case of smallpox constitutes a public health emergency. Both are viral diseases.

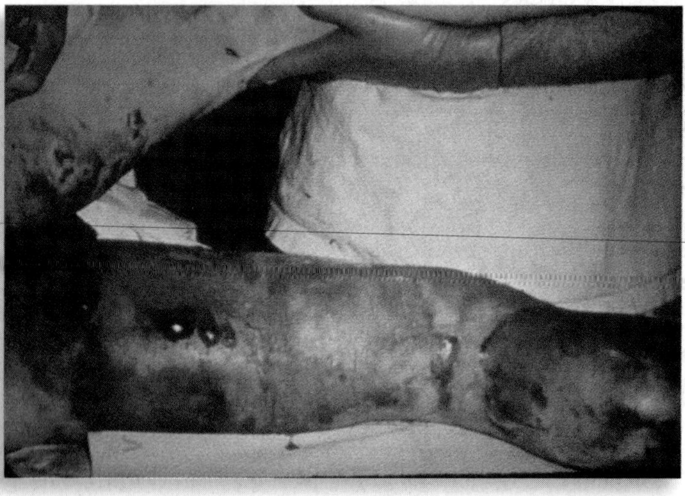

Figure 18.8 The clinical appearance of myonecrosis in a compound fracture of the leg.

Necrosis has traveled from the main site of the break to other areas of the leg. Note the severe degree of involvement, with blackening, general tissue destruction, and bubbles on the skin caused by gas formation in underlying tissue.

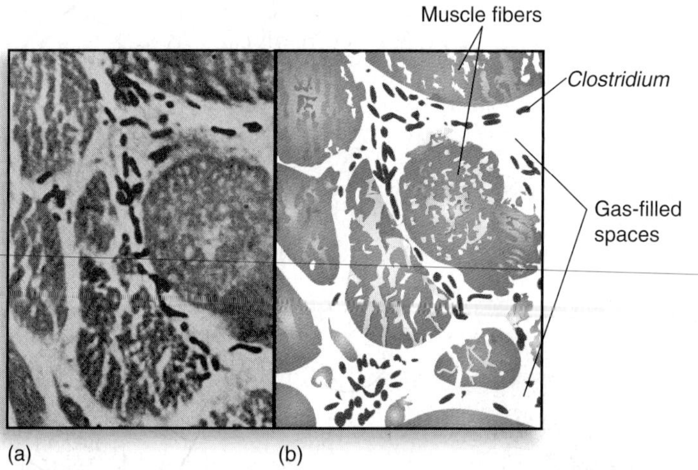

Figure 18.9 Growth of *Clostridium perfringens* (plump rods), causing gas formation and separation of the fibers.

(a) A microscopic analysis of clostridial myonecrosis, showing a histological section of gangrenous skeletal muscle. **(b)** A schematic drawing of the same section.

(a) From N. A. Boyd et al., *Journal of Medical Microbiology,* 5:459, 1972. Reprinted by permission of Longman Group, Ltd.

CHECKPOINT 18.5	Gas Gangrene
Causative Organism(s)	*Clostridium perfringens*, other species
Most Common Modes of Transmission	Vehicle (soil), endogenous transfer from skin, GI tract, reproductive tract
Virulence Factors	Alpha toxin, other exotoxins, enzymes, gas formation
Culture/Diagnosis	Gram stain, CT scans (abdominal infections), X ray, clinical picture
Prevention	Clean wounds, debride dead tissue
Treatment	Penicillin & Clindamycin, surgical removal, oxygen therapy

Chickenpox

Most people think of chickenpox as a mild disease, and in most people it is. In immunocompromised people it can be life-threatening, however. Before the introduction of the vaccine in 1995, it was not unheard of for some families to hold "chickenpox parties." When one child in a group of acquaintances had chickenpox, other children would be brought together to play with them so that all the children could get chickenpox and be done with it. Parents wanted to ensure that their children got the disease while they were young because they knew that getting the disease at an older age could lead to more serious disease.

Signs and Symptoms After an incubation period of 10 to 20 days, the first symptoms to appear are fever and an abundant rash that begins on the scalp, face, and trunk and radiates in sparse crops to the extremities. Skin lesions progress quickly from macules and papules to itchy vesicles filled with a clear fluid. In several days, they encrust and drop off, usually healing completely but sometimes leaving a tiny pit or scar. Lesions number from a few to hundreds and are more abundant in adolescents and adults than in young children. **Figure 18.10*a*** contains images of the chickenpox lesions in a child and in an adult. The lesion distribution is *centripetal*, meaning that there are more in the center of the body and fewer on the extremities, in contrast to the distribution

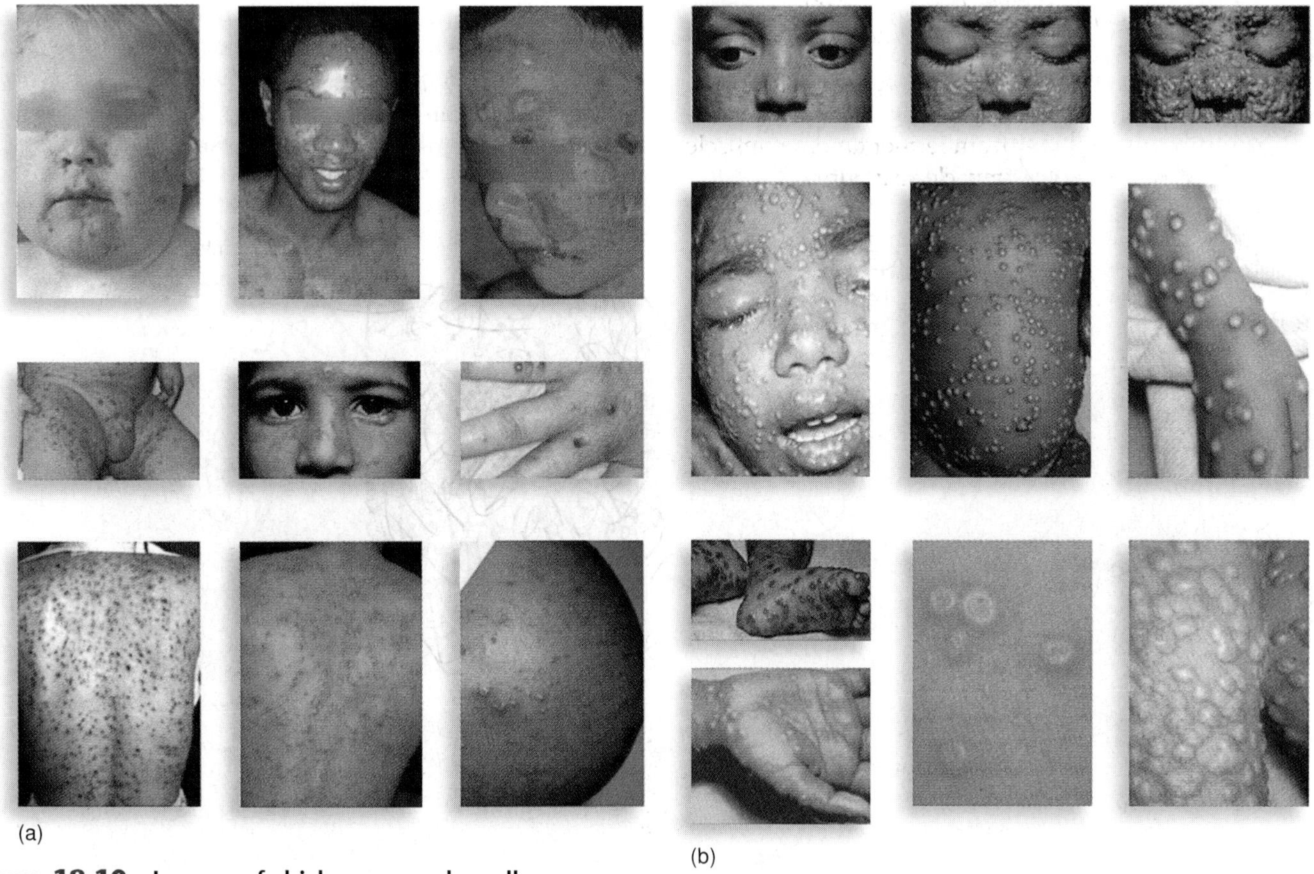

Figure 18.10 Images of chickenpox and smallpox.
(a) Chickenpox. **(b)** Smallpox.

seen with smallpox. The illness usually lasts 4 to 7 days; new lesions stop appearing after about 5 days. Patients are considered contagious until all of the lesions have crusted over.

Most cases resolve without event within 2 to 3 weeks of onset. Some patients may experience secondary infections of the lesions with group A streptococci or staphylococci, and these require antibiotic therapy. Immunocompromised patients, as well as some adults and adolescents, may experience pneumonia as a result of chickenpox. The immunocompromised may also experience infection of the heart, liver, and kidney, resulting in a 20% mortality rate for this population.

Approximately 0.1% of chickenpox cases are followed by encephalopathy, or inflammation of the brain caused by the virus. It can be fatal, but in most cases recovery is complete.

Women who become infected with chickenpox during the early months of pregnancy are at risk for infecting the fetus. These babies may be born with serious birth defects such as missing limbs and cataracts. Also, women who develop chickenpox just before or after giving birth may have passed the infection to the baby just before birth, resulting in serious infection in the newborn infant.

Shingles In many individuals, recuperation from chickenpox is associated with the entry of the virus into the sensory endings that innervate dermatomes, regions of the skin supplied by the cutaneous branches of nerves, especially the thoracic **(figure 18.11*a*)** and trigeminal nerves. From here it becomes latent in the ganglia and may reemerge as **shingles** (also known as **herpes zoster**) with its characteristic asymmetrical distribution on the skin of the trunk or head **(figure 18.11*b*).**

Shingles develops abruptly after reactivation by such stimuli as psychological stress, X-ray treatments, immunosuppressive and other drug therapy, surgery, or a developing malignancy. The virus is believed to migrate down the ganglion to the skin, where multiplication resumes and produces crops of tender, persistent vesicles. Inflammation of the ganglia and the pathways of nerves can cause pain and tenderness that can last for several months. Involvement of cranial nerves can lead to eye inflammation and ocular and facial paralysis.

Causative Agent Human herpesvirus 3 (HHV-3, also called **varicella** (var″-ih′sel′-ah)) causes chickenpox, as well as the condition called herpes zoster or shingles. The virus is sometimes referred to as the varicella-zoster virus (VZV). Like other herpesviruses, it is an enveloped DNA virus.

Pathogenesis and Virulence Factors HHV-3 enters the respiratory tract, attaches to respiratory mucosa, and then invades and enters the bloodstream. The viremia disseminates the virus to the skin, where the virus causes adjacent cells to fuse and eventually lyse, resulting in the characteristic lesions. The virus enters the sensory nerves at this site, traveling to the ganglia.

The ability of HHV-3 to remain latent in ganglia is an important virulence factor, because resting in this site protects it from attack by the immune system and provides a reservoir of virus for the reactivation condition of shingles.

Transmission and Epidemiology Humans are the only natural hosts for HHV-3. The virus is harbored in the respiratory tract but is communicable from both respiratory droplets and the fluid of active skin lesions. People can acquire a chickenpox infection by being exposed to the fluid of shingles lesions. (It is not possible to "get" shingles from someone with shingles. If you are not immune to HHV-3, you can acquire HHV-3, which will manifest as chickenpox or, very occasionally, as an asymptomatic infection. Once you have the virus, whether you experience shingles or not is dependent on your own host factors.)

Infected persons are most infectious a day or two prior to the development of the rash. Only in rare instances will a person acquire chickenpox more than once. Chickenpox is so contagious that if you are exposed to it you almost certainly will get it. Some people have a subclinical case of it, meaning that their lesions never erupt. But they will have lifelong immunity (and will likely harbor the virus in their ganglia

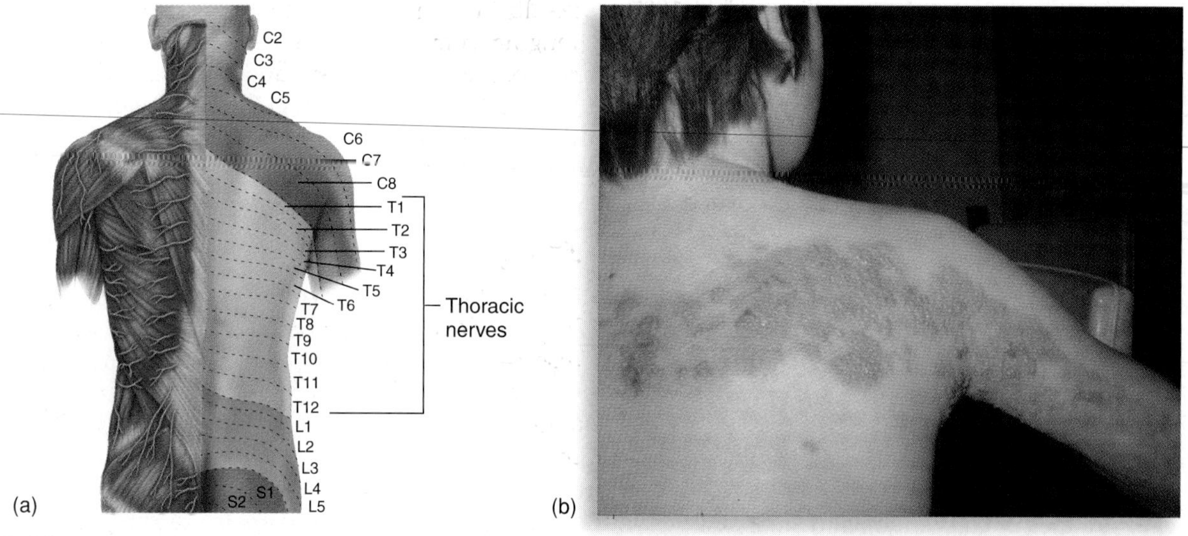

Figure 18.11 Varicella-zoster virus reemergence as shingles.
(a) Dermatomes served by the thoracic nerves. **(b)** Clinical appearance of shingles lesions.

INSIGHT 18.2 Historical

Smallpox: An Ancient Scourge Revisited

In earlier editions of this book, you could have read that smallpox had been eliminated from the earth and that this feat was one of the greatest triumphs of modern medicine. Today, soldiers, health care workers, and even the president of the United States are being vaccinated against this "vanquished" disease. What happened to cause this shift in thought, and how concerned should each of us be?

Historians note that smallpox epidemics have occurred for thousands of years. The 20th century saw one of humankind's greatest achievements when, through a massive worldwide health campaign that focused on immunization and isolation, the last case of smallpox was seen in Somalia in 1977. In 1980, with the war against smallpox "won," a committee of World Health Organization (WHO) experts recommended that laboratories worldwide destroy their stocks of variola virus or transfer them to one of two laboratories, the Institute of Virus Preparation in Moscow or the Centers for Disease Control and Prevention in the United States. At the time, all countries reported full compliance with the WHO request. Although the WHO committee recommended that even the two remaining stockpiles of virus be destroyed in 1999, scientific and governmental organizations balked at destroying a potential useful research subject. Besides, many had reservations about carrying out the first *intentional* extinction of another species.

In the late 1990s, a more immediate concern arose. Ken Alibek, a former director of the Soviet Union's civilian bioweapons, alleged that since 1980 the Soviet government had had in place a successful program to produce many tons of smallpox virus annually. Furthermore, strains of the virus had been selected to be more virulent, and contagious, and able to survive delivery in bombs and intercontinental ballistic missiles. As Russia's economic situation worsened in the late 1990s, support for research fell considerably and the concern arose that a country or individual with enough money could acquire Russian knowledge, equipment, or even viral strains rather easily. Other concerns have since been expressed about the existence of smallpox stocks in other regions of the world. In fact, Iraqi prisoners captured in the 1991 Gulf War are reported to have had high levels of antibody to smallpox, which suggested they had been immunized, perhaps as protection against Iraqi biological weapons. And since the anthrax bioterror incident, which followed closely on the heels of the events of September 11, 2001, the U.S. government has taken the possibility of smallpox bioterrorism very seriously.

If smallpox is being contemplated as a weapon, it would not be the first time. During the French and Indian Wars (1754–1767), British soldiers were instructed to distribute blankets that had been used by smallpox patients to Native Americans in an attempt to infect them. Whether they actually did or not, bioterror and biowarfare are not new ideas, after all.

---

and be subject to shingles). When people think they have never had chickenpox, yet they don't seem to get it when exposed to infected persons, it is likely that they have had a subclinical case at some time in their lives.

Epidemics of the disease used to occur in winter and early spring. The introduction of the vaccine in 1995 reduced the occurrence of the disease, so it now occurs only sporadically.

Prevention Live attenuated vaccine was licensed in 1995. It consists of a weakened form of the Oka strain of the virus, which was isolated from a Japanese boy named Oka. It is recommended as a single dose between the ages of 12 to 18 months.

In 2006, the FDA approved a unique vaccine called Zostavax®. It is intended for adults ages 60 and over and is for the prevention of shingles.

Treatment Uncomplicated varicella is self-limiting and requires no therapy aside from alleviation of discomfort. Secondary bacterial infection, as just noted, is treated with topical or systemic antibiotics. Oral acyclovir should be administered to people considered to be at risk for serious complications within 24 hours of onset of the rash. The acyclovir may diminish viral load and prevent complications.

Special Note About Reye's Syndrome In the 1980s, researchers made a connection between aspirin and a serious condition that occurred in children, usually following a febrile (feb′-ruhl) viral infection, especially influenza or chickenpox. It seems that the condition is the result of some interaction of three factors: recent viral infection, age less than 15 years, and the use of salicylates (common aspirin).

The syndrome can be mild or severe. It usually begins with vomiting and nausea and is followed by a sudden change in mental status caused by encephalopathy. This condition results in a variety of central nervous system symptoms, also ranging from mild to severe, such as amnesia, disorientation, seizures, coma, and respiratory arrest. The incidence has decreased tenfold (from about 300 cases per year to 20 or 30 in the United States) since the public became aware that they should not administer salicylates to children with fever.

Smallpox

Largely through the World Health Organization's comprehensive global efforts, naturally occurring smallpox is now a disease of the past. However, after the terrorist attacks on the United States on September 11, 2001, and the anthrax bioterrorism shortly thereafter, the U.S. government began taking the threat of smallpox bioterrorism very seriously. Vaccination, which had been discontinued, was once again offered to certain U.S. populations. After languishing in obscurity since its elimination from humans in the 1970s, smallpox was back in the news **(Insight 18.2)**.

Signs and Symptoms Infection begins with fever and malaise, and later a rash begins in the pharynx, spreads to the face, and progresses to the extremities. Initially the rash is *macular*, evolving in turn to *papular, vesicular*, and *pustular* before eventually crusting over, leaving nonpigmented sites pitted with scar tissue (see Insight 18.3 for a description of these terms). There are two principal forms of smallpox, variola minor and variola major. Variola major is a highly virulent form that causes toxemia, shock, and intravascular coagulation. People who have survived any form of smallpox nearly always develop lifelong immunity.

It is vitally important for health care workers to be able to recognize the early signs of smallpox (figure 18.10*b*). The diagnosis of even a single suspected case must be treated as a health and law enforcement emergency. The symptoms of variola major progress as follows: After the prodrome period of high fever and malaise, a rash emerges, first in the mouth. Severe abdominal and back pain sometimes accompany this phase of the disease. As lesions develop, they break open and spread virus into the mouth and throat, making the patient highly contagious. A rash appears on the skin and spreads throughout the body within 24 hours. A distribution of the rash on the body is shown in figure 18.10*b*.

By the third or fourth day of the rash, the bumps become larger and fill with a thick opaque fluid. A major distinguishing feature of this disease is that the pustules are indented in the middle (see Checkpoint 18.6). Also, patients report that the lesions feel as if they contain a BB pellet. Within a few days, these pustules begin to scab over. After 2 weeks, most of the lesions will have crusted over; the patient remains contagious until the last scabs fall off because the crusts contain the virus. During the entire rash phase, the patient is very ill.

A patient with variola minor has a rash that is less dense and is generally less ill than someone with variola major.

Causative Agent The causative agent of smallpox, the variola virus, is an orthopoxvirus, an enveloped DNA virus. Variola is shaped like a brick and is 200 nanometers in diameter. Other members of this group are the monkeypox virus and the vaccinia virus from which smallpox vaccine is made. Variola is a hardy virus, surviving outside the host longer than most viruses.

Pathogenesis and Virulence Factors The infection begins by implantation of the virus in the nasopharynx. The virus invades the mucosa and multiplies in the regional lymph nodes, leading to viremia. Variola multiplies within white blood cells and then travels to the small blood vessels in the dermis. The lesions occur at the dermal level, which is the reason that scars remain after the lesions are healed.

Much of the research on smallpox was suspended after its eradication from the human population. However, the virus genome has been sequenced, and scientists are once again studying how it causes damage to the host. It is turning out to be more difficult than expected to determine why it is so virulent. Scientists discovered the puzzling fact that the much tamer vaccinia virus has *more* genes that code for immune-evasion proteins than does the virulent variola virus.

Transmission and Epidemiology Before the eradication of smallpox, almost everyone contracted the disease over the course of their lifetime, either surviving with lifelong immunity or dying. It is spread primarily through droplets, although fomites such as contaminated bedding and clothing can also spread it. Traditionally the incidence of smallpox was highest in the winter and early spring.

In the early 1970s, smallpox was endemic in 31 countries. Every year, 10 to 15 million people contracted the disease, and approximately 2 million people died from it. By 1977, after 11 years of intensive effort by the world health community, the last natural case occurred in Somalia.

Prevention In the 18th century, English physician Edward Jenner noticed that milkmaids who contracted a limited disease called cowpox, or vaccinia (from vacca, the Latin word for cow), seemed to be unaffected when smallpox swept through a locale. He no doubt also knew of the ages-old practice of *variolation*, the purposeful introduction of actual smallpox pus or scabs into healthy people, either through injection or inhalation, to protect them from natural infections. Variolation had been practiced for centuries in places such as Africa, India, Turkey, and China, because it was evident that those who recovered from the natural disease were protected from reinfection. This crude precursor of vaccination was dangerous, however; up to 10% of people who were variolated came down with severe smallpox, and some died from it. Still, with smallpox a constant threat, many people thought it worth the risk.

Jenner must have been aware of this practice; he combined his knowledge with his observation that the milkmaids' more limited lesions resembled those of smallpox and that these women then seemed to escape smallpox infection. He tested his theory by inoculating a young boy with material from cowpox lesions. That boy proved to be immune to both cowpox and smallpox—and the name vaccination was given to the practice of immunization.

To this day, the vaccination for smallpox is based on the vaccinia virus. Immunizations were stopped in the United States in 1972, and from then until 2002 no new vaccine was manufactured. After the security threats of 2001, the U.S. government tested stored vaccine and even found that it could be diluted and still be effective, ensuring that enough vaccine would be available to immunize every man, woman, and child if necessary. In the meantime, new vaccine based on the vaccinia virus was developed.

Beginning in late 2002, the vaccine was offered on a voluntary basis to health care workers who might be part of a "smallpox response team" in the event of a bioterror incident. Also, President George W. Bush ordered that certain members of the military be vaccinated. He also took the vaccination himself. Any American who wants the vaccine may take it, though no one (besides the military) is required to do so.

CHECKPOINT 18.6 Vesicular/Pustular Rash Diseases

Disease	Chickenpox	Smallpox
Causative Organism(s)	Human herpesvirus 3 (varicella-zoster virus)	Variola virus
Most Common Modes of Transmission	Droplet contact, inhalation of aerosolized lesion fluid	Droplet contact, indirect contact
Virulence Factors	Ability to fuse cells, ability to remain latent in ganglia	Ability to dampen, avoid immune response
Culture/Diagnosis	Based largely on clinical appearance	Based largely on clinical appearance
Prevention	Live attenuated vaccine; there is also vaccine to prevent reactivation of latent virus (shingles)	Live virus vaccine (vaccinia virus)
Treatment	None in uncomplicated cases; acyclovir for high risk	–
Distinguishing Features	No fever prodrome; lesions are superficial; in centripetal distribution (more in center of body)	Fever precedes rash, lesions are deep and in centrifugal distribution (more on extremities)
Appearance of Lesion	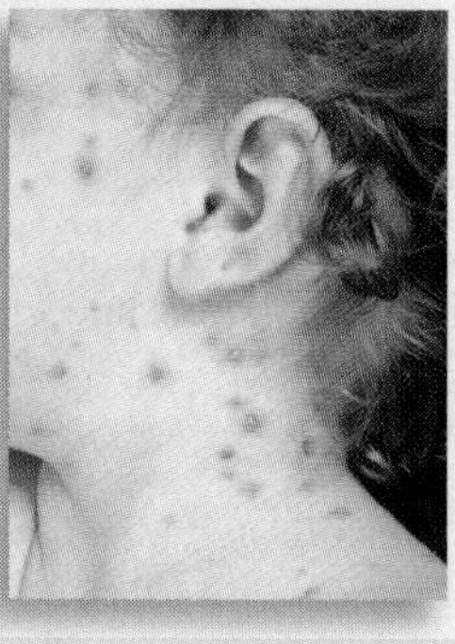	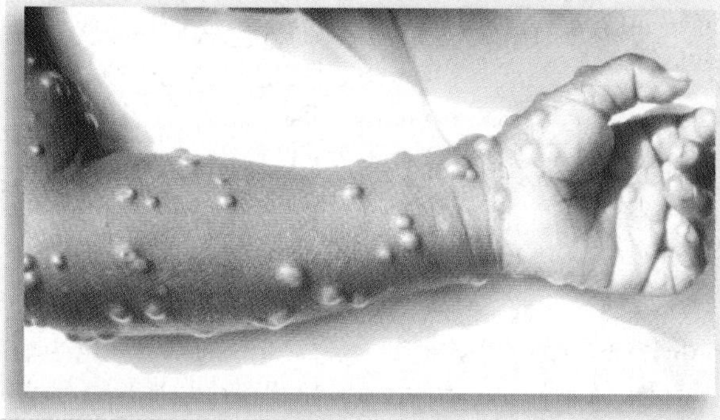

The vaccine for smallpox is considered the most problematic of all modern vaccines because it has a relatively high rate of side effects. It is estimated that for every million people vaccinated, one or two may die as a result. One thousand may have non-life-threatening but serious reactions, such as severe reactions at the site of vaccination or spread of the virus to the bloodstream, causing generalized illness. Once vaccination began in December 2002, several people developed cardiac symptoms, so people at risk for heart problems were advised not to take the vaccine. Since vaccinations began, there have been a handful of cases of inadvertent transmission of the vaccinia strain, resulting in localized symptoms in persons contacting those who had been immunized. In 2007 a new vaccine was approved by the Food and Drug Administration, it is called ACAM 2000.

Vaccination is also useful for postexposure prophylaxis, meaning that it can prevent or lessen the effects of the disease after you have already been infected with it.

Another chapter was added to the smallpox story in 2003, when dozens of people came down with a disease called monkeypox, caused by the monkey variant of the smallpox virus. They had apparently caught the disease from their pet prairie dogs, which had seemingly caught it from an exotic species of African rat. Both of these animals were imported to the United States as part of the exotic pet trade. The U.S. government recommended that people exposed to infected animals be vaccinated with the vaccinia vaccine.

Treatment There is no treatment for smallpox. Antiviral treatment is not effective, although an experimental drug, designed for cancer treatment, saved the lives of mice treated with it in experiments conducted in 2005. If lesions become infected secondarily with bacteria, antibiotics can be used for that complication **(Checkpoint 18.6).**

Maculopapular Rash Diseases

Insight 18.3 contains a description of the different infectious conditions that can result in a rash of some sort on the skin. The infectious conditions described in this section are those with their major manifestations on the skin. (Meningococcal meningitis, for instance, can result in a diffuse rash on the skin, but its major manifestations are in the central nervous system, so it is discussed in chapter 19.) In this section, we examine measles, rubella, "fifth disease," and roseola. They all cause skin eruptions classified as maculopapular.

Measles

Most of us living in the United States don't think twice about measles. It is just another vaccination we get when we are children. But every year hundreds of thousands of children

INSIGHT 18.3 *Medical*

Naming Skin Lesions

There seems to be no end to the types of lesions or irregularities that can occur on the skin. Dermatology, the study of the skin, is a branch of medicine that relies heavily on visual characteristics for initial diagnoses. The many types of skin lesions or irregularities have been given specific descriptive names for this purpose. None of these names points to an exact etiologic cause, but because certain infectious agents generally cause distinctive types of lesions, the list of possible causes can be narrowed considerably once the "style" of irregularity is identified and named. **Table 18.A** contains a list of the more common descriptors of skin bumps, lesions, and irregularities.

TABLE 18.A Skin Terms

Descriptive Name	Appearance	Examples
Macule	Flat, well-demarcated lesion characterized mainly by color change	Freckle, tinea versicolor (fungus infection)
Papule	Small elevated, solid bump	Warts, cutaneous leishmaniasis
Maculopapular Rash	Flat to slightly raised colored bump	Measles, rubella, fifth disease, roseola
Plaque	Elevated flat-topped lesion larger than 1 cm (i.e., a wider papule)	Psoriasis
Vesicle	Elevated lesion filled with clear fluid	Chickenpox
Bulla	Large (wide) vesicle	Blister, gas blisters in gangrene
Pustule	Small elevated lesion filled with purulent fluid (pus)	Acne, smallpox, mucocutaneous leishmanisasis, cutaneous anthrax
Cyst	Raised, encapsulated lesion, usually solid or semisolid when palpated	Severe acne
Purpura	Reddish-purple discoloration due to blood in small areas of tissue; does not blanch when pressed	Meningococcal bloodstream infection (see chapter 19)
Petechiae	Small purpura	Meningococcal bloodstream infection
Scale	Flaky portions of skin separated from deeper portions	Ringworm of body and scalp, athlete's foot

in the developing world die from this disease, even though an extremely effective vaccine has been available since 1964. Health campaigns all over the world seek to make measles vaccine available to all, but there is much work to be done. Many scientists and public health advocates hope that once polio is eradicated (see chapter 19), measles will be the next disease targeted globally for eradication.

Measles is also known as **rubeola.** Be very careful not to confuse it with the next maculopapular rash disease, rubella.

Signs and Symptoms The initial symptoms of measles are sore throat, dry cough, headache, conjunctivitis, lymphadenitis, and fever. In a short time, unusual oral lesions called *Koplik's spots* appear as a prelude to the characteristic red maculopapular **exanthem** (eg-zan′-thum) that erupts on the head and then progresses to the trunk and extremities until most of the body is covered **(figure 18.12).** The rash gradually coalesces into red patches that fade to brown.

In a small number of cases, children develop laryngitis, bronchopneumonia, and bacterial secondary infections such as ear and sinus infections. Children afflicted with leukemia or thymic deficiency are especially predisposed to pneumonia because of their lack of the natural T-cell defense. Undernourished children may experience severe diarrhea and abdominal discomfort that adds to their debilitation.

In a small percentage of cases, the virus can cause pneumonia. Affected patients are very ill and often have a characteristic dusky skin color from lack of oxygen. Occasionally (1 in 100 cases), measles progresses to encephalitis, resulting

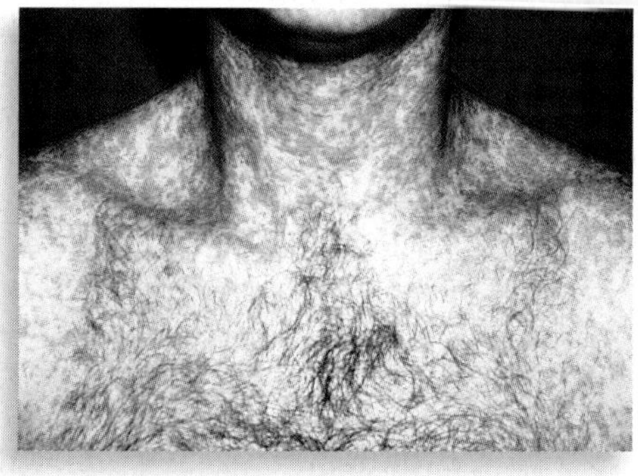

Figure 18.12 **The rash of measles.**

in various CNS changes ranging from disorientation to coma. Permanent brain damage or epilepsy can result.

A large number of measles patients experience secondary bacterial infections with *Haemophilus influenzae, Streptococcus pneumoniae*, or other streptococci or staphylococci. These can also lead to pneumonia or upper respiratory tract complications.

The most serious complication is **subacute sclerosing panencephalitis (SSPE),** a progressive neurological degeneration of the cerebral cortex, white matter, and brain stem. Its incidence is approximately one case in a million measles infections, and it afflicts primarily male children and adolescents. The pathogenesis of SSPE appears to involve a defective virus, one that has lost its ability to form a capsid and be released from an infected cell. Instead, it spreads unchecked through the brain by cell fusion, gradually destroying neurons and accessory cells and breaking down myelin. The disease is known for profound intellectual and neurological impairment. The course of the disease invariably leads to coma and death in a matter of months or years.

Measles during pregnancy has been associated with spontaneous miscarriage and low-birthweight babies, but severe birth defects have not been reported.

Causative Agent The measles virus is a member of the *Morbillivirus* genus. It is a single-stranded enveloped RNA virus in the Paramyxovirus family.

Pathogenesis and Virulence Factors The virus implants in the respiratory mucosa and infects the tracheal and bronchial cells. From there it travels to the lymphatic system, where it multiplies and then enters the bloodstream. Viremia carries the virus to the skin and to various organs.

The measles virus induces the cell membranes of adjacent host cells to fuse into large **syncytia** (sin-sish′-uh), giant cells with many nuclei. These cells no longer perform their proper function. The virus seems proficient at disabling many aspects of the host immune response, especially cell-mediated immunity and delayed-type hypersensitivity. The host may be left vulnerable for many weeks after infection; this immune response disruption is one of the reasons that secondary bacterial infections are so common.

Transmission and Epidemiology Measles is one of the most contagious infectious diseases, transmitted principally by respiratory droplets. Epidemic spread is favored by crowding, low levels of herd immunity, malnutrition, and inadequate medical care. There is no reservoir other than humans, and a person is infectious during the periods of incubation, prodrome phase, and the skin rash but usually not during convalescence. Only relatively large, dense populations of susceptible individuals can sustain the continuous chain necessary for transmission. Japan and India have experienced prolonged outbreaks in recent years. In the United States, the incidence of measles is sporadic, usually less than 100 cases per year. In the 1980s and 1990s, a significant proportion of U.S. measles cases occurred among college students, perhaps because of communal living conditions and perhaps due to a waning of their childhood immunity. Now at least 32 states have laws requiring that students present proof of two measles immunizations before they can enroll in college.

Culture and Diagnosis The disease can be diagnosed on clinical presentation alone; but if further identification is required, an ELISA test is available that tests for patient IgM to measles antigen, indicating a current infection. For best results, blood should be drawn on the third day of onset or later, because before that time titers of IgM may not be high enough to be detected by the test. Also, the method of comparing acute and convalescent sera may be used to confirm a measles infection after the fact. As you may recall from chapter 17, much higher IgG titers 14 days after onset when compared to titers at day 1 or 2 are a clear indication of current or recent infection. This knowledge allows health care providers to be on the lookout for complications and to be ahead of the game if a person who has had contact with the patient presents with similar symptoms.

Prevention The MMR vaccine (for measles, mumps, and rubella) contains live attenuated measles virus, which confers protection for about 20 years. Measles immunization is recommended for all healthy children at the age of 12 to 15 months, with a booster before the child enters school. Failing that, the preadolescent health check serves as a good time to get the second dose of measles vaccine.

Treatment Treatment relies on reducing fever, suppressing cough, and replacing lost fluid. Complications require additional remedies to relieve neurological and respiratory symptoms and to sustain nutrient, electrolyte, and fluid levels. Therapy includes antibiotics for bacterial complications and doses of immune globulin. Vitamin A supplements are recommended by some physicians; they have been found effective in reducing the symptoms and decreasing the rate of complications.

Rubella

This disease is also known as German measles. Rubella is derived from the Latin for "little red," and that's a good way to remember it because it causes a relatively minor rash disease with few complications. Sometimes it is called the 3-day measles. The only exception to this mild course of events is when a fetus is exposed to the virus while in its mother's womb (in utero). Serious damage can occur, and for that reason women of childbearing years must be sure to have been vaccinated well before they plan to conceive.

Signs and Symptoms The two clinical forms of rubella are referred to as postnatal infection, which develops in children or adults, and **congenital** (prenatal) infection of the fetus, expressed in the newborn as various types of birth defects.

Postnatal Rubella During an incubation period of 2 to 3 weeks, the rubella virus multiplies in the respiratory epithelium, infiltrates local lymphoid tissue, and enters the bloodstream. Early symptoms include malaise, mild fever, sore throat, and lymphadenopathy. The rash of pink macules and papules first appears on the face and progresses down the trunk and toward the extremities, advancing and resolving in about 3 days. The rash is milder looking than the measles rash (see Checkpoint 18.7). Adult rubella is often accompanied by joint inflammation and pain rather than a rash. Very occasionally, complications such as arthralgia/arthritis, or even encephalitis, can occur but more often in adults than in children.

Congenital Rubella Rubella is a strongly **teratogenic** (ter-at'-oh-jen"-ik) virus. Transmission of the rubella virus to a fetus in utero can result in a serious complication called **congenital rubella (figure 18.13).** The mother is able to transmit the virus even if she is asymptomatic. Fetal injury varies according to the time of infection. It is generally accepted that infection in the first trimester is most likely to induce miscarriage or multiple permanent defects in the newborn. The most common of these is deafness and may be the only defect seen in some babies. Other babies may experience cardiac abnormalities, ocular lesions, deafness and mental and physical retardation in varying combinations. Less drastic sequelae that usually resolve in time are anemia, hepatitis, pneumonia, carditis, and bone infection.

Causative Agent The rubella virus is a *Rubivirus,* in the family Togavirus. It is a nonsegmented single-stranded RNA virus with a loose lipid envelope. There is only one known serotype of the virus, and humans are the only natural host. Its envelope contains two different viral proteins.

Pathogenesis and Virulence Factors The course of disease in postnatal rubella is mostly unremarkable. But when exposed to a fetus, the virus creates havoc. It has the ability to stop mitosis, which is an important process in a rapidly developing embryo and fetus. It also induces apoptosis of normal tissue cells. This inappropriate cell death can do irreversible harm to organs it affects. And last, the virus damages vascular endothelium, leading to poor development of many organs. Studies have shown that the earlier in gestation that the infection process begins, the more devastating its effects.

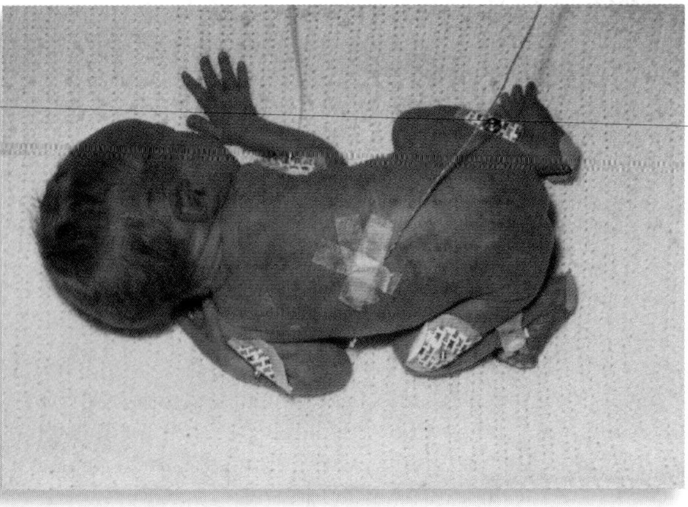

Figure 18.13 **An infant born with congenital rubella can manifest a papular pink or purple rash.**

Transmission and Epidemiology Rubella is an endemic disease with worldwide distribution. Infection is initiated through contact with respiratory secretions and occasionally urine. The virus is shed during the prodromal phase and up to a week after the rash appears. Congenitally infected infants are contagious for a much longer period of time. Because the virus is only moderately communicable, close living conditions are required for its spread. This disease is well-controlled in the United States, with fewer than 10 cases reported in each of the last several years. Most cases are reported among adolescents and young adults in military training camps, colleges, and summer camps. The greatest concern is that nonimmune women of childbearing age might be caught up in this cycle, raising the prospect of congenital rubella.

Culture and Diagnosis Diagnosing rubella relies on the same twin techniques discussed earlier for measles. Because it mimics other diseases, rubella should not be diagnosed on clinical grounds alone. IgM antibody to rubella virus can be detected early using an ELISA technique or a latex-agglutination card. Other conditions and infections can lead to false positives, however, and the IgM test should be augmented by an acute and convalescent measurement of IgG antibody. It is important to know whether the infection is indeed rubella, especially in women, because if so, they will be immune to reinfection.

Prevention The attenuated rubella virus vaccine is usually given to children in the combined form (MMR vaccination) at 12 to 15 months and a booster at 4 or 6 years of age. The vaccine for rubella can be administered on its own, without the measles and mumps components.

Many health care providers recommend screening adult women of childbearing age for antibodies to rubella, which would indicate either that they had had the infection or that they had been immunized. The current recommendation for nonpregnant, antibody-negative women is immediate immunization. Because the vaccine contains live virus, and because a teratogenic effect is theoretically possible, the vaccine is administered on the condition that the patient not become pregnant for 3 months afterward. The vaccine is not given to pregnant women.

Treatment Postnatal rubella is generally benign and requires only symptomatic treatment. No specific treatment is available for the congenital manifestations.

Fifth Disease

This disease, more precisely called *erythema infectiosum,* is so named because about 100 years ago it was the fifth of the diseases recognized by doctors to cause rashes in children. The first four were scarlet fever (see chapter 21), measles, rubella, and another rash that was thought to be distinct but was probably not. Fifth disease is a very mild disease that often results in a characteristic "slapped-cheek" appearance because of a confluent reddish rash that begins on the face. Within 2 days, the rash spreads on the body but is most prominent on the arms, legs, and trunk. The rash is maculopapular and the blotches tend to run together rather than to appear as distinct bumps. The illness is rather mild, featuring low-grade fever and malaise and lasting 5 to 10 days. The rash may persist for days to weeks, and it tends to recur under stress or with exposure to sunlight. As with almost any infectious agent, it can cause more serious disease in people with underlying immune disease.

The causative agent is parvovirus B19. You may have heard of "parvo" as a disease of dogs, but strains of this virus group infect humans as well. Fifth disease is usually diagnosed by the clinical presentation, but sometimes it is helpful to rule out rubella by testing for IgM against rubella. Specific serological tests for fifth disease are available if they are considered necessary.

This infection is very contagious. It is transmitted through respiratory droplets or even direct contact. It can be transmitted through the placenta, with a range of possible effects, from no symptoms to stillbirth. There is no vaccine and no treatment for this usually mild disease.

Roseola

This disease is common in young children and babies. It sometimes results in a maculopapular rash, but a high percentage (up to 70%) of cases proceed without the rash stage. Children sick with this disease exhibit a high fever (up to 41°C, or 105°F) that comes on quickly and lasts for up to 3 days. Seizures may occur during this period, but other than that patients remain alert and do not act terribly ill. On the fourth day, the fever disappears, and it is at this point that a rash can appear, first on the chest and trunk and less prominently on the face and limbs. By the time the rash appears, the disease is almost over.

Roseola is caused by a human herpesvirus called HHV-6, and sometimes by HHV-7. Like all herpesviruses, it can remain latent in its host indefinitely after the disease has cleared. Very occasionally, the virus reactivates in childhood or adulthood, leading to mononucleosislike or hepatitislike symptoms. Immunocompetent hosts generally do not experience reactivation. It is thought that 100% of the U.S. population is infected with this virus by adulthood. Some people experienced the disease roseola when they became infected, and some of them did not. The suggestion has been made that this virus causes other disease conditions later in life, such as multiple sclerosis or chronic fatigue syndrome, but so far no convincing connection has been demonstrated. No vaccine and no treatment exist for roseola.

The two HHV viruses can cause severe disseminated disease in AIDS patients and other people with compromised immunity.

Scarlet Fever

To complete our survey of infections that can cause maculopapular rashes, we include a disease that has primary symptoms elsewhere but can produce a distinctive red rash on the skin as well. **Scarlet fever** is most often the result of a respiratory infection with *Streptococcus pyogenes* (most often, pharyngitis). Occasionally, scarlet fever will follow a streptococcal skin infection, such as impetigo or cellulitis. If the *S. pyogenes* strain contains a bacteriophage carrying a gene for an exotoxin called erythrogenic toxin, scarlet fever can result. More details on scarlet fever are given in chapter 21; it is included here mainly for purposes of differentiating the rash from the others in this group **(Checkpoint 18.7).**

Wartlike Eruptions

All types of warts are caused by viruses. Most common warts you have seen on yourself and others are probably caused by one of more than 80 human papillomaviruses, or HPVs. HPVs are also the cause of genital warts, described in chapter 23. Another virus in the poxvirus family causes a condition called **molluscum contagiosum,** which causes bumps that may look like warts.

CASE FILE 18 *WRAP-UP*

The baby from China described at the beginning of the chapter was sick with measles. Hospital staff looked inside the baby's mouth for Koplik's spots. Finding these spots provides some assurance that, among all the maculopapular diseases, measles is the likely diagnosis.

The U.S. incidence of measles is low (usually less than 100 per year), but sporadic outbreaks do occur, even among people who were fully vaccinated as children, probably due to waning of their artificially acquired active immunity. This case highlights the need for constant vigilance and continued immunization, especially as the world "shrinks" and we come in contact with people from other parts of the world where different levels of immunization are achieved.

This story provides an example of some unintended consequences of an otherwise joyous event: international adoptions. In 1992, about 6,000 babies from foreign countries were brought to the United States and adopted. In 2001, nearly 19,000 infants were adopted by happy U.S. families, and the trend is likely to continue.

See: CDC. 2003. Measles outbreak among internationally adopted children arriving in the United States, February–March 2001. MMWR *51:1115–1116.*

CHECKPOINT 18.7 Maculopapular Rash Diseases

Disease	Measles (Rubeola)	Rubella	Fifth Disease	Roseola	Scarlet Fever
Causative Organism(s)	Measles virus	Rubella virus	Parvovirus B19	Human herpesvirus 6 or 7	*Streptococcus pyogenes* (lysogenized)
Most Common Modes of Transmission	Droplet contact	Droplet contact	Droplet contact, direct contact	?	Droplet or direct contact
Virulence Factors	Syncytium formation, ability to suppress CMI	In fetuses: inhibition of mitosis, induction of apoptosis, and damage to vascular endothelium	–	Ability to remain latent	Erythrogenic toxin
Culture/ Diagnosis	ELISA for IgM, acute/ convalescent IgG	Acute IgM, acute/ convalescent IgG	Usually diagnosed clinically	Usually diagnosed clinically	Examination of skin lesions, throat culture (beta-hemolytic on blood agar, sensitive to bacitracin, rapid antigen tests)
Prevention	Live attenuated vaccine (MMR)	Live attenuated vaccine (MMR)	–	–	Hygiene practices
Treatment	No antivirals; vitamin A, antibiotics for secondary bacterial infections	–	–	–	Penicillin, cephalexin in penicillin-allergic
Distinguishing Features of the Rashes	Starts on head, spreads to whole body, lasts over a week	Milder red rash, lasts approximately 3 days	"Slapped-face" rash first, spreads to limbs and trunk, tends to be confluent rather than distinct bumps	High fever precedes rash stage—rash not always present	Sandpaper feel to affected skin; severe sore throat
Appearance of Lesions	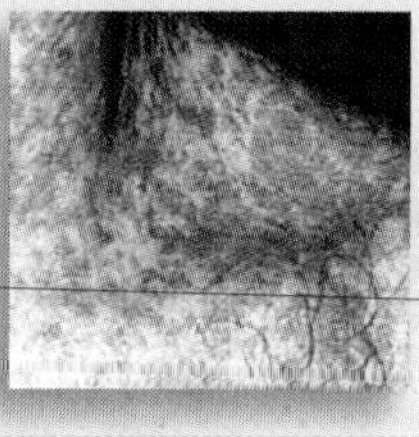	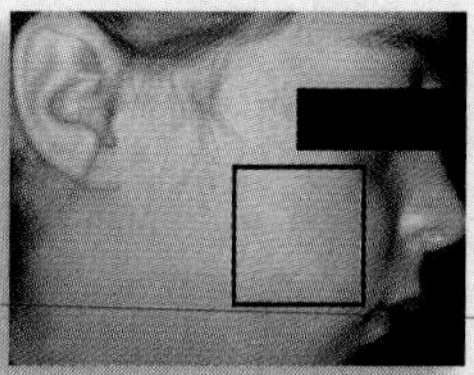	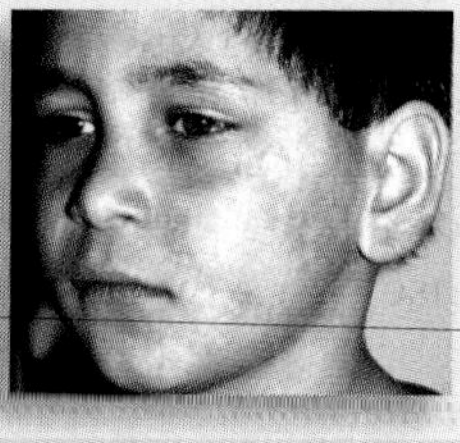	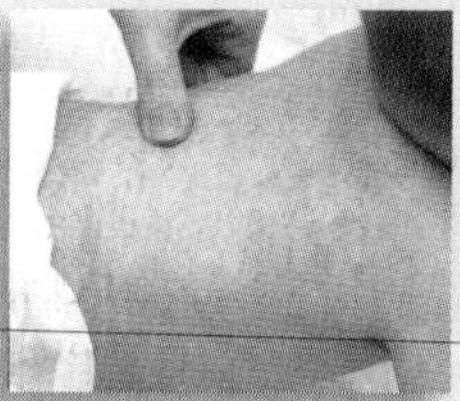	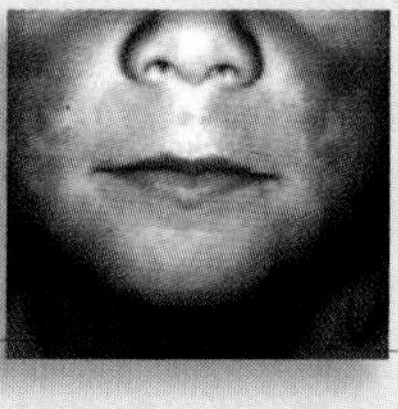

Warts

Warts, also known as **papillomas,** afflict nearly everyone. Children seem to get them more frequently than adults, and there is speculation that people gradually build up immunity to the various HPVs that they encounter over time, as is the case with the viruses that cause the common cold.

The warts are benign, squamous epithelial growths. Some HPVs can infect mucus membranes; others invade skin. The appearance and seriousness of the infection vary somewhat from one anatomical region to another. Painless, elevated, rough growths on the fingers and occasionally on other body parts are called common, or seed, warts (see Checkpoint 18.8). These growths commonly occur in children and young adults. Just as certain types of HPVs are associated with particular outcomes in the genital area, common warts are most often caused by HPV 2, 4, 27, and 29. **Plantar warts** are often caused by HPV 1. They are deep, painful papillomas on the soles of the feet. Flat warts (HPV types 3, 10, 28, and 49) are smooth, skin-colored lesions that develop on the face, trunk, elbows, and knees.

CHECKPOINT 18.8 Wart and Wartlike Eruptions

Disease	Warts	Molluscum contagiosum
Causative Organism(s)	Human papillomaviruses	Molluscum contagiosum viruses
Most Common Modes of Transmission	Direct contact, autoinoculation, indirect contact	Direct contact, including sexual contact, autoinoculation
Virulence Factors	–	–
Culture/Diagnosis	Clinical diagnosis, also histology, microscopy, PCR	Clinical diagnosis, also histology, microscopy, PCR
Prevention	Avoid contact	Avoid contact
Treatment	Home treatments, cryosurgery (virus not eliminated)	Usually none, although mechanical removal can be performed (virus not eliminated)
Appearance of Lesions	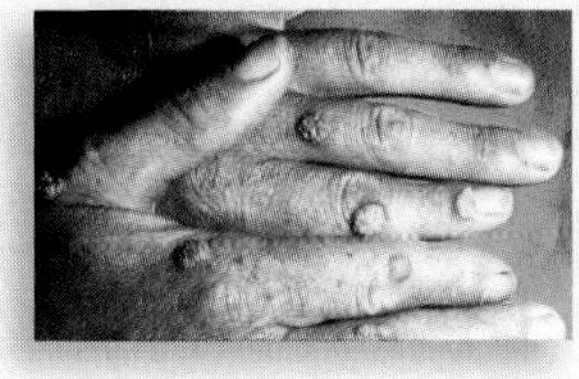	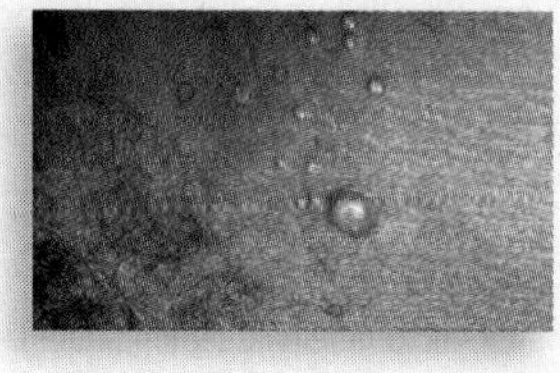

The warts contain variable amounts of virus. Transmission occurs through direct contact, and often warts are transmitted from one part of the body to another by autoinoculation. Because the viruses are fairly stable in the environment, they can also be transmitted indirectly from towels or from a shower stall, where they persist inside the protective covering of sloughed-off keratinized skin cells. The incubation period can be from 1 to 8 months. Almost all nongenital warts are harmless, and they tend to resolve themselves over time. Rarely, a wart can become malignant when caused by a particular type of HPV.

The warts caused by papillomaviruses are usually distinctive enough to permit reliable clinical diagnosis without much difficulty. However, a biopsy and histological examination can help clarify ambiguous cases. Warts disappear on their own 60% to 70% of the time, usually over the course of 2 to 3 years. Physicians do approve of home remedies for resolving warts. These include nonprescription salicylic acid preparations, as well as the use of adhesive tape. Yes, you read that right: well-controlled medical studies have shown that adhesive tape (even duct tape!) can cause warts to disappear, presumably because the tape creates an airtight atmosphere that stops virus reproduction. But a psychological component, similar to a placebo effect, cannot be ruled out. (Neither of these treatments should be used for genital warts; see chapter 23.) Physicians have other techniques for removing warts, including a number of drugs and/or cryosurgery. No treatment guarantees that the viruses are eliminated; therefore, warts can always grow back.

Molluscum contagiosum

This disease is distributed throughout the world, with highest incidence occurring on certain Pacific islands, although its incidence in North America has been increasing since the 1980s. Skin lesions take the form of smooth, waxy nodules on the face, trunk, and limbs. The firm nodules may be indented in the middle (see Checkpoint 18.8), and they contain a milky fluid containing epidermal cells filled with viruses in intracytoplasmic inclusion bodies. This condition is common in children, where it most often causes nodules on the face, arms, legs, and trunk. In adults, it appears mostly in the genital areas. In immunocompromised patients, the lesions can be more disfiguring and more widespread on the body. It is particularly common in AIDS patients and often presents as facial lesions.

The molluscum contagiosum virus is a poxvirus, containing double-stranded DNA and possessing an envelope. It is spread via direct contact and also through fomites. Adults who acquire this infection usually acquire it through sexual contact. Autoinoculation can spread the virus from existing lesions to new places on the body, resulting in new nodules.

The condition may be diagnosed on clinical appearance alone, or a skin biopsy may be performed and histological analysis undertaken. A clinician can perform a more simple "squash procedure," in which fluid from the lesion is extracted onto a microscope slide, squashed by another microscope slide, stained, and examined for the presence of the characteristic inclusion bodies in the epithelial cells. PCR can also be used to detect the virus in skin lesions. In most cases, no treatment is indicated, although a physician may remove the lesions or treat them with a topical chemical. Treatment of lesions does not ensure elimination of the virus **(Checkpoint 18.8).**

Large Pustular Skin Lesions

Leishmaniasis

Two infections that result in large lesions (greater than a few millimeters across) deserve mention in this chapter on skin infections. The first is leishmaniasis, a zoonosis transmitted among various mammalian hosts by female sand flies.

This infection can express itself in several different forms, depending on which species of the protozoan *Leishmania* is involved. Cutaneous leishmaniasis is a localized infection of the capillaries of the skin caused by *L. tropica*, found in Mediterranean, African, and Indian regions. A form of mucocutaneous leishmaniasis called espundia is caused by *L. brasiliensis*, endemic to parts of Central and South America. It affects both the skin and mucus membranes. Another form of this infection is systemic leishmaniasis.

Leishmania is transmitted to the mammalian host by the sand fly when it ingests the host's blood. The disease is endemic to equatorial regions that provide favorable conditions for the sand fly. Numerous wild and domesticated animals, especially dogs, serve as reservoirs for the protozoan. Although humans are usually accidental hosts, the flies freely feed on them. At particular risk are travelers or immigrants who have never had contact with the protozoan and lack specific immunity.

Leishmania infection begins when an infected fly injects the motile forms of the protozoan into the host while feeding. After being engulfed by macrophages, the parasite converts to a nonmotile reproductive form and multiplies in the macrophage. The manifestations of the disease vary with the fate of the macrophages. If they remain fixed, the infection stays localized in the skin or mucus membranes, but if the infected macrophages migrate, systemic disease occurs.

In cutaneous leishmaniasis, a small red papule occurs at the site of the bite and spreads laterally into a large ulcer (see Checkpoint 18.9). The edges of the ulcer are raised and the base is moist. It can be filled with a serous/purulent exudate or covered with a crust. Satellite lesions may occur. Mucocutaneous leishmaniasis usually begins with a skin lesion on the head or face and then progresses to single or multiple lesions, usually in the mouth and nose. Lesions can be quite extensive, eventually involving and disfiguring the hard palate, the nasal septum, and the lips.

There is no vaccine; avoiding the sand fly is the only prevention. The disease can be treated with chemicals such as sodium stibogluconate. Other antimicrobials may be indicated for secondary infections of the lesions.

Cutaneous Anthrax

This form of anthrax is the most common and least dangerous version of infection with *Bacillus anthracis*. (The spectrum of anthrax disease is discussed fully in chapter 20.) It is caused by endospores entering the skin through small cuts or abrasions. Germination and growth of the pathogen in the skin are marked by the production of a papule that becomes increasingly necrotic and later ruptures to form a painless, black **eschar** (ess′-kar) (see Checkpoint 18.9). In the fall of 2001, 11 cases of cutaneous anthrax occurred in the United States as a result of bioterrorism (along with 11 cases of inhalational anthrax). Mail workers and others contracted the infection when endospores were sent through the mail. The infection can be naturally transmitted by contact with hides of infected animals (especially goats).

Left untreated, even the cutaneous form of anthrax is fatal approximately 20% of the time. A vaccine exists but is recommended only for high-risk persons and the military. Upon suspicion of cutaneous anthrax, ciprofloxacin, levofloxacin and/or doxycycline should be used initially. If the isolate is found to be sensitive to penicillin, patients can be switched to that drug **(Checkpoint 18.9).**

CHECKPOINT 18.9 Large Pustular Skin Lesions

Disease	Leishmaniasis	Cutaneous Anthrax
Causative Organism(s)	*Leishmania* spp.	*Bacillus anthracis*
Most Common Modes of Transmission	Biological vector	Direct contact with endospores
Virulence Factors	Multiplication within macrophages	Endospore formation; capsule, lethal factor, edema factor (see chapter 20)
Culture/Diagnosis	Culture of protozoa, microscopic visualization	Culture on blood agar; serology, PCR performed by CDC
Prevention	Avoiding sand fly	Avoid contact; vaccine available but not widely used
Treatment	Sodium stibogluconate	Ciprofloxacin, doxycycline, levofloxacin
Distinguishing Features	Mucocutaneous and systemic forms	Can be fatal
Appearance of Lesions	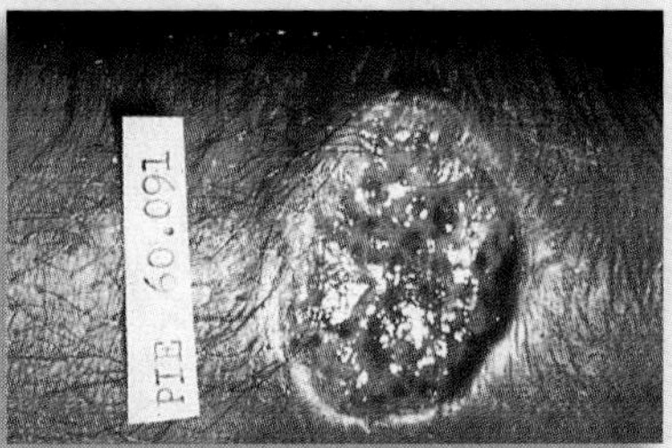	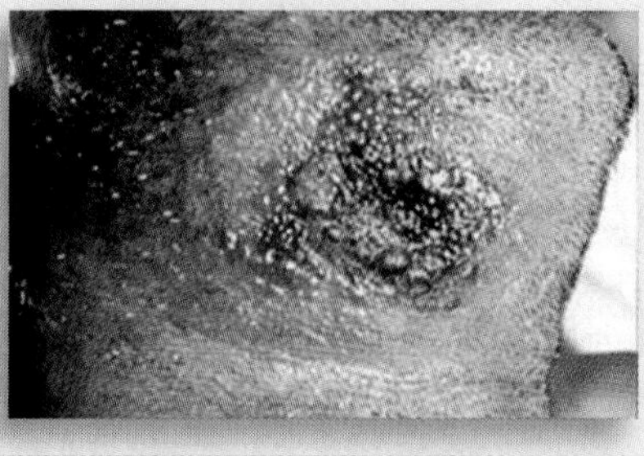

Ringworm (Cutaneous Mycoses)

A group of fungi that is collectively termed **dermatophytes** causes a constellation of integument conditions. These mycoses are strictly confined to the nonliving epidermal tissues (stratum corneum) and their derivatives (hair and nails). All these conditions have different names that begin with the word **tinea** (tin′-ee-ah), which derives from the erroneous belief that they were caused by worms. That misconception is also the reason these diseases are often called *ringworm*—ringworm of the scalp (tinea capitis), beard (tinea barbae), body (tinea corporis), groin (tinea cruris), foot (tinea pedis), and hand (tinea manuum). (Don't confuse these "tinea" terms with genus and species names. It is simply an old practice for naming conditions.) Most of these conditions are caused by one of three different dermatophytes, which are discussed here.

One fungal infection is even more superficial than the others; it infects only the most superficial layers of the stratum corneum and causes a condition called **tinea versicolor.** It is not a ringworm but is nevertheless included at the end of this section.

Signs and Symptoms of the Cutaneous Mycoses

Ringworm of the Scalp (Tinea Capitis) This mycosis results from the fungal invasion of the scalp and the hair of the head, eyebrows, and eyelashes **(figure 18.14).** Very common in children, tinea capitis is acquired from other children and adults or from domestic animals. Manifestations range from small scaly patches to a severe inflammatory reaction to destruction of the hair follicle and temporary or permanent hair loss.

Ringworm of the Beard (Tinea Barbae) This tinea, also called *barber's itch,* affects the chin and beard of adult males. Although once a common aftereffect of unhygienic barbering, it is now contracted mainly from animals.

Ringworm of the Body (Tinea Corporis) This extremely prevalent infection of humans can appear nearly anywhere on the body's glabrous (smooth and bare) skin. The principal sources are other humans, animals, and soil, and it is transmitted primarily by direct contact and fomites (clothing, bedding). The infection usually appears as one or more scaly reddish rings on the trunk, hip, arm, neck, or face **(figure 18.15).** The ringed pattern is formed when the infection radiates from the original site of invasion into the surrounding skin. Depending on the causal species and the health and hygiene of the patient, lesions vary from mild and diffuse to florid and pustular.

Ringworm of the Groin (Tinea Cruris) Sometimes known as *jock itch,* crural ringworm occurs mainly in males on the groin, perianal skin, scrotum, and, occasionally, the penis. The fungus thrives under conditions of moisture and humidity created by sweating. It is transmitted primarily from human to human and is pervasive among athletes and persons living in close quarters (ships, military installations).

Ringworm of the Foot (Tinea Pedis) Tinea pedis has more colorful names as well, including athlete's foot and jungle rot. The disease is clearly connected to wearing shoes because it is uncommon in cultures where people customarily go barefoot. Conditions that encase the feet in a closed, warm, moist environment increase the possibility of infection. Tinea pedis is a known hazard in shared facilities such as shower stalls, public floors, and locker rooms. Infections begin with blisters between the toes that burst, crust over, and can spread to the rest of the foot and nails **(figure 18.16*a*).**

Ringworm of the Hand (Tinea Manuum) Infection of the hand by dermatophytes is nearly always associated with concurrent infection of the foot. Lesions usually occur on the fingers and palms of one hand, and they vary from white and patchy to deep and fissured.

Ringworm of the Nail (Tinea Unguium) Fingernails and toenails, being masses of keratin, are often sites for persistent fungus colonization. The first symptoms are usually superficial white patches in the nail bed. A more invasive form causes thickening, distortion, and darkening of the nail **(figure 18.16*b*).** Nail problems caused by dermatophytes are on the rise as

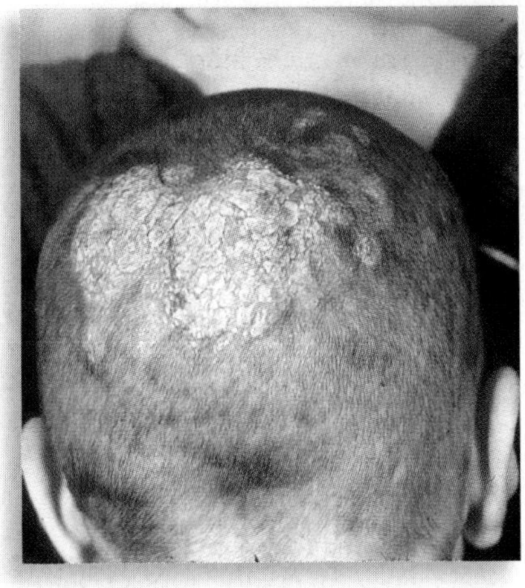

Figure 18.14 Ringworm of the scalp.
Hair loss can accompany these lesions.

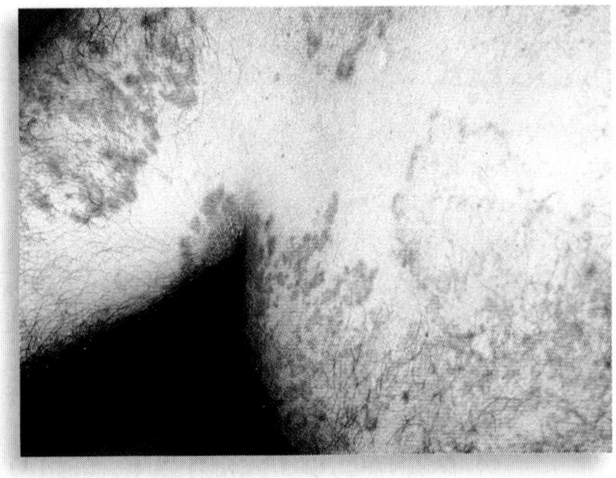

Figure 18.15 Ringworm of the body.

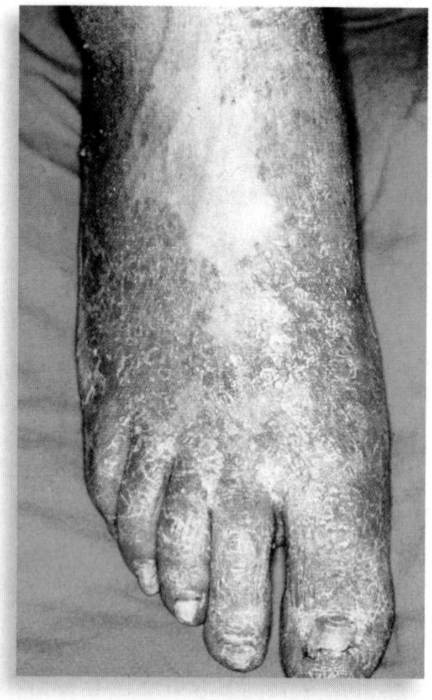

(a)

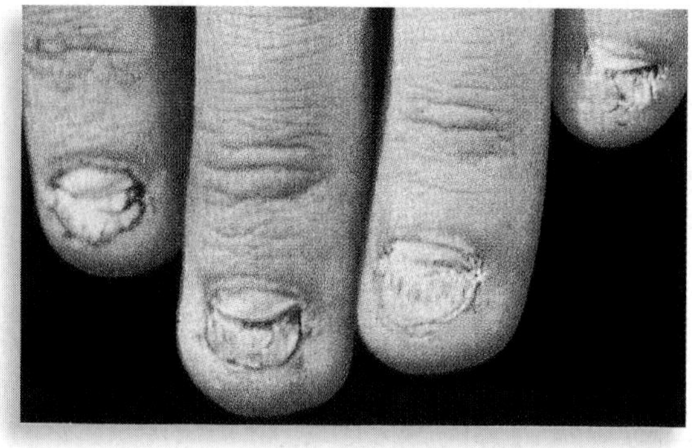

(b)

Figure 18.16 Ringworm of the extremities.
(a) *Trichophyton* infection spreading over the foot in a "moccasin" pattern. The chronicity of tinea pedis is attributed to the lack of fatty-acid-forming glands in the feet. **(b)** Ringworm of the nails. Invasion of the nail bed causes some degree of thickening, accumulation of debris, cracking, and discoloration; nails can be separated from underlying structures as shown here.

more women wear artificial fingernails, which can provide a portal of entry into the nail bed.

Causative Agents

There are about 39 species in the genera *Trichophyton, Microsporum,* and *Epidermophyton* that can cause the preceding conditions. The causative agent of a given type of ringworm varies from one geographic location to another and is not restricted to a particular genus and species. These fungi are so closely related and morphologically similar that they can be difficult to differentiate. Various species exhibit unique macroconidia, microconidia, and unusual types of hyphae. In general, *Trichophyton* produces thin-walled, smooth macroconidia and numerous microconidia **(figure 18.17*a*)**; *Microsporidium* produces thick-walled, rough macroconidia and sparser microconidia **(figure 18.17*b*)**; and *Epidermophyton* has

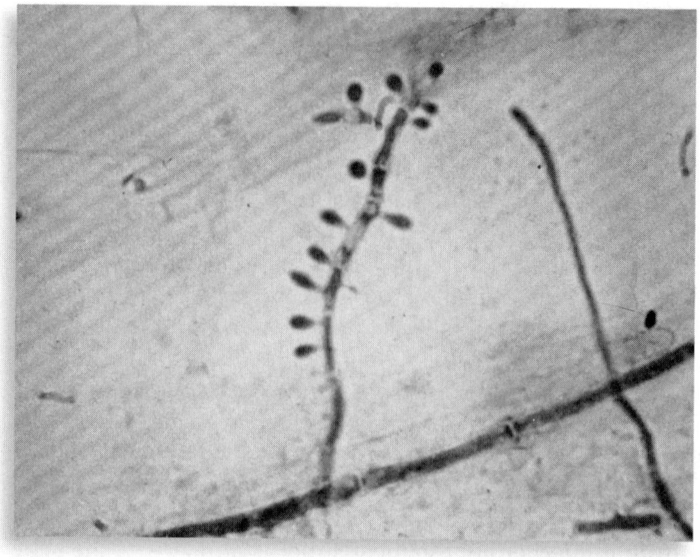

(a)

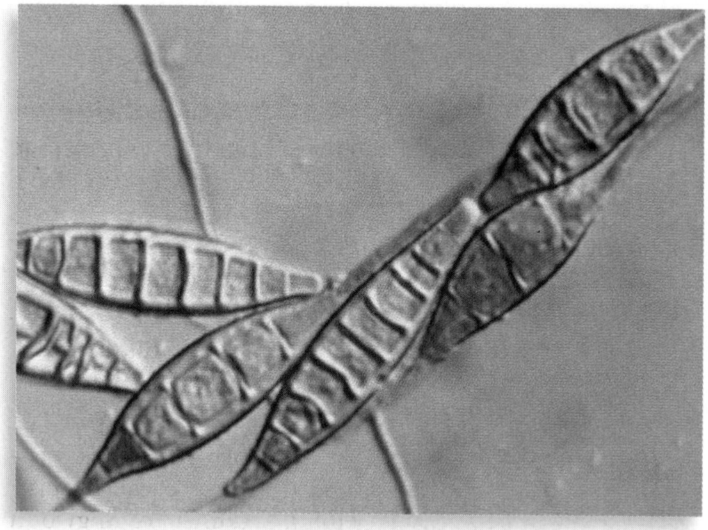

(b)

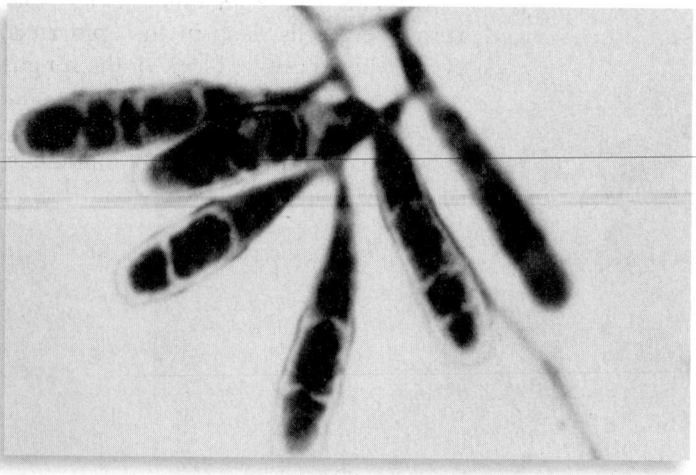

(c)

Figure 18.17 Examples of dermatophyte spores.
(a) Regular, numerous microconidia of *Trichophyton*.
(b) Macroconidia of *Microsporum canis*, a cause of ringworm in cats, dogs, and humans. **(c)** Smooth-surfaced macroconidia in clusters characteristic of *Epidermophyton*.

ovoid, smooth, clustered macroconidia and no microconidia **(figure 18.17*c*).**

The presenting symptoms of a cutaneous mycosis occasionally are so dramatic and suggestive of these genera that no further testing is necessary. In most cases, however, direct microscopic examination and culturing are required. Diagnosis of tinea of the scalp caused by some species is aided by use of a long-wave ultraviolet lamp that causes infected hairs to fluoresce. Samples of hair, skin scrapings, and nail debris treated with heated potassium hydroxide (KOH) show a thin, branching fungal mycelium if infection is present.

Pathogenesis and Virulence Factors

The dermatophytes have the ability to invade and digest keratin, which is naturally abundant in the cells of the stratum corneum. The fungi do not invade deeper epidermal layers. Important factors that promote infection are the hardiness of the dermatophyte spores (they can last for years on fomites); presence of abraded skin, and intimate contact. Most infections exhibit a long incubation period (months), followed by localized inflammation and allergic reactions to fungal proteins. As a general rule, infections acquired from animals and soil cause more severe reactions than do infections acquired from other humans, and infections eliciting stronger immune reactions are resolved faster.

Transmission and Epidemiology

Transmission of the fungi that cause these diseases is direct and indirect contact with other humans or with infected animals. Some of these fungi can be acquired from the soil.

Prevention and Treatment

The only way to prevent these infections is to avoid contact with the dermatophytes, which is impractical. Keeping susceptible skin areas dry is helpful. Treatment of ringworm is based on the knowledge that the dermatophyte is feeding on dead epidermal tissues. These regions undergo constant replacement from living cells deep in the epidermis, so if multiplication of the fungus can be blocked, the fungus will eventually be sloughed off along with the skin or nail. Unfortunately, this takes time. By far the most satisfactory choice for therapy is a topical antifungal agent. Ointments containing tolnaftate, miconazole, itraconazole, terbinafine, or thiabendazine are applied regularly for several weeks. Some drugs work by speeding up loss of the outer skin layer. Intractable infections can be treated with griseofulvin, but placing a patient on this drug, which is toxic to the liver and the kidneys, for the long periods needed is probably too risky in most cases. Gentle debridement of skin can have some benefit.

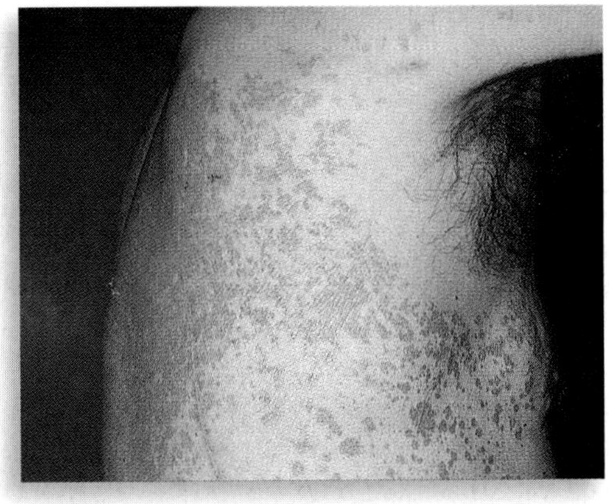

Figure 18.18 **Tinea versicolor.**
Mottled, discolored skin pigmentation is characteristic of superficial skin infection by *Malassezia furfur.*

Superficial Mycoses

Agents of **superficial mycoses** involve the outer epidermal surface and are ordinarily innocuous infections with cosmetic rather than inflammatory effects. Tinea versicolor is caused by the yeast *Malassezia furfur,* a normal inhabitant of human skin that feeds on the high oil content of the skin glands. Even though this yeast is very common (carried by nearly 100% of humans tested), in some people its growth elicits mild, chronic scaling and interferes with production of pigment by melanocytes. The trunk, face, and limbs may take on a mottled appearance **(figure 18.18).** The disease is most pronounced in young people who are frequently exposed to the sun, because the area affected doesn't tan well. Other skin conditions in which *M. furfur* is implicated are folliculitis, psoriasis, and seborrheic dermatitis. It is also occasionally associated with systemic infections and catheter-associated sepsis in compromised patients **(Checkpoint 18.10).**

CHECKPOINT 18.10 Cutaneous and Superficial Mycoses

Disease	Cutaneous Infections	Superficial Infections (Tinea Versicolor)
Causative Organism(s)	*Trichophyton, Microsporum, Epidermophyton*	*Malassezia furfur*
Most Common Modes of Transmission	Direct and indirect contact, vehicle (soil)	Endogenous "normal biota"
Virulence Factors	Ability to degrade keratin, invoke hypersensitivity	–
Culture/Diagnosis	Microscopic examination, KOH staining, culture	Usually clinical, KOH can be used
Prevention	Avoid contact	None
Treatment	Topical tolnaftate, itraconazole, terbinafine, miconazole, thiabendazine	Topical antifungals

18.5 The Surface of the Eye and Its Defenses

The eye is a complex organ with many different tissue types, but for the purposes of this chapter we consider only its exposed surfaces, the *conjunctiva* and the *cornea* **(figure 18.19).** The **conjunctiva** is a very thin membranelike tissue that covers the eye (except for the cornea) and lines the eyelids. It secretes an oil- and mucus-containing fluid that lubricates and protects the eye surface. The **cornea** is the dome-shaped central portion of the eye lying over the iris (the colored part of the eye). It has five to six layers of epithelial cells that can regenerate quickly if they are superficially damaged. It has been called "the windshield of the eye."

The eye's best defense is the film of tears, which consists of an aqueous fluid, oil, and mucus. The tears are formed in the lacrimal gland at the outer and upper corner of each eye **(figure 18.20),** and they drain into the lacrimal duct at the inner corner. The aqueous portion of tears contains sugars, lysozyme, and lactoferrin. These last two substances have antimicrobial properties. The mucus layer contains proteins and sugars and plays a protective role. And, of course, the flow of the tear film prevents the attachment of microorganisms to the eye surface.

Because the eye's primary function is vision, anything that hinders vision would be counterproductive. For that reason, inflammation does not occur in the eye as readily as it does elsewhere in the body. Flooding the eye with fluid containing a large number of light-diffracting objects such as lymphocytes and phagocytes in response to every irritant would mean almost constantly blurred vision. So even though the eyes are relatively vulnerable to infection (not being covered by keratinized epithelium), the evolution of the vertebrate eye has of necessity favored reduced innate immunity. This characteristic is sometimes known as *immune privilege*.

The specific immune response, involving B and T cells, is also somewhat restricted in the eye. The anterior chamber (see figure 18.19) is largely cut off from the blood supply. Lymphocytes that do gain access to this area are generally less active than lymphocytes elsewhere in the body.

18.6 Normal Biota of the Eye

The normal biota of the eye is generally sparse. When people are tested, up to 20% have no recoverable bacteria in their eyes. The few bacteria that are found resemble the normal biota of the skin—namely, diphtheroids, coagulase-negative staphylococci, *Micrococcus*, nonhemolytic streptococci, and some yeast. *Neisseria* species can also live on the surface of the eye.

Defenses and Normal Biota of the Eyes

	Defenses	Normal Biota
Eyes	Mucus in conjunctiva and in tears, lysozyme and lactoferrin in tears	Sparsely populated with *Staphylococcus aureus, Staphylococcus epidermidis*, and *Corynebacterium* species.

18.7 Eye Diseases Caused by Microorganisms

In this section, we cover the infectious agents that cause diseases of the surface structures of the eye—namely, the cornea and conjunctiva.

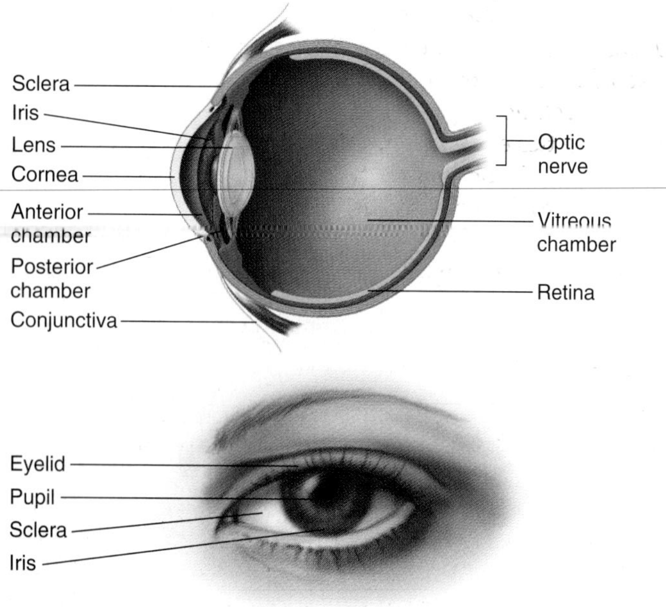

Figure 18.19 The anatomy of the eye.

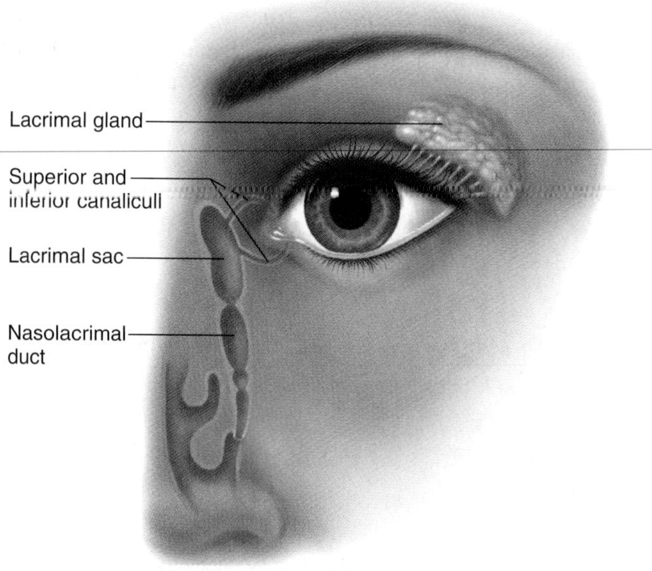

Figure 18.20 The lacrimal apparatus of the eye.

Conjunctivitis

Infection of the conjunctiva is relatively common. It can be caused by specific microorganisms that have a predilection for eye tissues, by contaminants that proliferate due to the presence of a contact lens or an eye injury, or by accidental inoculation of the eye by a traumatic event.

Signs and Symptoms

Just as there are many different causes of conjunctivitis, there are many different clinical presentations. Inflammation of this tissue almost always causes a discharge of some sort. Most bacterial infections produce a milky discharge, whereas viral infections tend to produce a clear exudate. It is typical for a patient to wake up in the morning with an eye "glued" shut by secretions that have accumulated and solidified through the night. Some conjunctivitis cases are caused by an allergic response, and these often produce copious amounts of clear fluid as well. The pain generally is mild, although often patients report a gritty sensation in their eye(s). Redness and eyelid swelling are common, and in some cases patients report photophobia (sensitivity to light). The informal name for common conjunctivitis is pinkeye.

Causative Agents and Their Transmission

Cases of neonatal eye infection with *Neisseria gonorrhoeae* or *Chlamydia trachomatis* are usually transmitted vertically from a genital tract infection in the mother (discussed in chapter 23). Either one of these eye infections can lead to serious eye damage if not treated promptly **(figure 18.21).** Note that herpes simplex can also cause neonatal conjunctivitis, but it is usually accompanied by generalized herpes infection (covered in chapter 23).

Bacterial conjunctivitis in other age groups is most commonly caused by *Staphylococcus epidermidis, Streptococcus pyogenes,* or *Streptococcus pneumoniae,* although *Haemophilus infuenzae* and *Moraxella* species are also frequent causes. *N. gonorrhoeae* and *C. trachomatis* can also cause conjunctivitis in adults. These infections may result from autoinoculation from a genital infection or from sexual activity, although *N. gonorrhoeae* can be part of the normal biota in the respiratory tract. A wide variety of bacteria, fungi, and protozoa can contaminate contact lenses and lens cases and then be transferred to the eye, resulting in disease that may be very serious. This means of infection is considered vehicle transmission, with the lens or the solution being the vehicle.

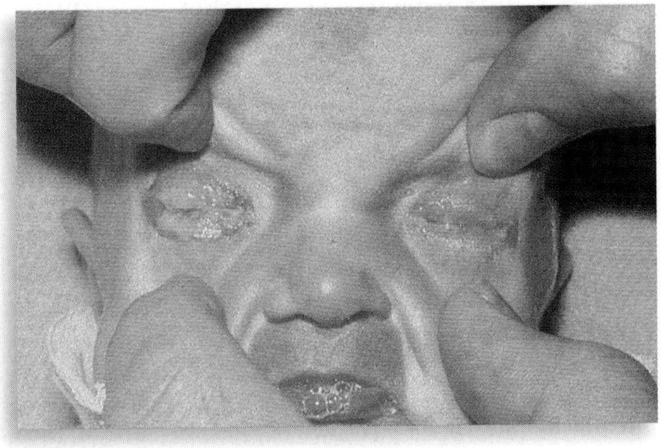

Figure 18.21 Neonatal conjunctivitis.

Viral conjunctivitis is commonly caused by adenoviruses, although other viruses may be responsible. (Herpesvirus infection of the eye is discussed later on.) Both bacterial and viral conjunctivitis are transmissible by direct and even indirect contact and are usually highly contagious.

A NOTE ABOUT *STIES*

A sty in the eye is the most common condition seen by eye doctors. What most people call a sty may be one of two different conditions. A *hordeolum* is an infection of an oil gland at the edge of the eyelid (much like a pimple). A *chalazion* is a noninfectious inflammation of an oil gland. Both of these conditions are distinct from conjunctivitis.

Prevention and Treatment

Good hygiene is the only way to prevent conjunctivitis in adults and children other than neonates. Newborn children in the United States are administered antimicrobials in their eyes after delivery to prevent neonatal conjunctivitis from either *N. gonorrhoeae* or *C. trachomatis.* Treatment of those infections, if they are suspected, is started before lab results are available and usually is accomplished with erythromycin, both topical and oral. If *N. gonorrhoeae* is confirmed, oral therapy is usually switched to ceftriaxone. If antibacterial therapy is prescribed for other conjunctivitis cases, it should cover all possible bacterial pathogens. Ciprofloxacin eyedrops are a common choice. Erythromycin or gentamicin are also often used. Because conjunctivitis is usually diagnosed based on clinical signs, a physician may prescribe prophylactic antibiotics even if a viral cause is suspected. If symptoms don't begin improving within 48 hours, more extensive diagnosis may be performed. **Checkpoint 18.11** lists the most common causes of conjunctivitis; keep in mind that other microorganisms can also cause conjunctival infections.

Trachoma

Ocular trachoma is a chronic *Chlamydia trachomatis* infection of the epithelial cells of the eye. It is an ancient disease and a major cause of blindness in certain parts of the world. Although a few cases occur annually in the United States, several million cases occur endemically in parts of Africa and Asia. Transmission is favored by contaminated fingers, fomites, fleas, and a hot, dry climate. It is caused by a different *C. trachomatis* strain than the one that causes simple conjunctivitis. Ongoing infection or

CHECKPOINT 18.11 Conjunctivitis

Disease	Neonatal Conjunctivitis	Bacterial Conjunctivitis	Viral Conjunctivitis
Causative Organism(s)	*Chlamydia trachomatis* or *Neisseria gonorrhoeae*	*Streptococcus pyogenes, Streptococcus pneumoniae, Staphylococcus aureus, Haemophilus influenzae, Moraxella,* and also *Neisseria gonorrhoeae, Chlamydia trachomatis*	Adenoviruses and others
Most Common Modes of Transmission	Vertical	Direct, indirect contact	Direct, indirect contact
Virulence Factors	–	–	–
Culture/Diagnosis	Gram stain and culture	Clinical diagnosis	Clinical diagnosis
Prevention	Screen mothers, apply antibiotic or silver nitrate to newborn eyes	Hygiene	Hygiene
Treatment	Topical and oral antibiotics	Broad-spectrum topical antibiotic, often ciprofloxacin	None, although antibiotics often given because type of infection not distinguished
Distinguishing Features	In babies <28 days old	Mucopurulent discharge	Serous (clear) discharge

many recurrent infections with this strain eventually lead to chronic inflammatory damage and scarring.

The first signs of infection are a mild conjunctival discharge and slight inflammation of the conjunctiva. These symptoms are followed by marked infiltration of lymphocytes and macrophages into the infected area. As these cells build up, they impart a pebbled (rough) appearance to the inner aspect of the upper eyelid **(figure 18.22).** In time, a vascular pseudomembrane of exudates and inflammatory leukocytes forms over the cornea, a condition called *pannus*, which lasts a few weeks. Chronic and secondary infections can lead to corneal damage and impaired vision. Early treatment of this disease with azithromycin is highly effective and prevents all of the complications. It is a tragedy that in this day of sophisticated preventive medicine, millions of children worldwide will develop blindness for lack of a few dollars' worth of antibiotics.

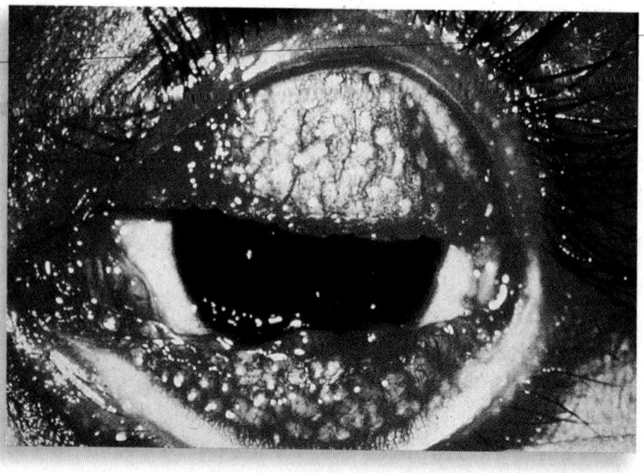

Figure 18.22 Ocular trachoma caused by *C. trachomatis.*

CHECKPOINT 18.12 Trachoma

Causative Organism(s)	*C. trachomatis* serovars A–C
Most Common Modes of Transmission	Indirect contact, mechanical vector
Virulence Factors	Intracellular growth
Culture/Diagnosis	Detection of inclusion bodies in stained preparations
Prevention	Hygiene, vector control, prompt treatment of initial infection
Treatment	Azithromycin or topical erythromycin

Keratitis

Keratitis is a more serious eye infection than conjunctivitis. Invasion of deeper eye tissues occurs and can lead to complete corneal destruction. Any microorganism can cause this condition, especially after trauma to the eye, but this section focuses on one of the more common causes: herpes simplex virus. It can cause keratitis in the absence of predisposing trauma.

The usual cause of herpetic keratitis is a "misdirected" reactivation of (oral) herpes simplex virus type 1 (HSV-1). The virus, upon reactivation, travels into the ophthalmic rather than the mandibular branch of the trigeminal nerve. Infections with HSV-2 can also occur as a result of a sexual encounter with the virus or transfer of the virus from the genital to eye area or if an individual has a recurrent oral infection with HSV-2. Preliminary symptoms are a gritty

☑ CHECKPOINT 18.13	Keratitis	
Causative Organism(s)	Herpes simplex virus	Miscellaneous microorganisms
Most Common Modes of Transmission	Reactivation of latent virus, although primary infections can occur in the eye	Often traumatic introduction (parenteral)
Virulence Factors	Latency	Various
Culture/Diagnosis	Usually clinical diagnosis; viral culture or PCR if needed	Various
Prevention	–	–
Treatment	Topical trifluridine and/or oral acyclovir	Specific antimicrobials

feeling in the eye, conjunctivitis, sharp pain, and sensitivity to light. Some patients develop characteristic branched or opaque corneal lesions as well. In 25% to 50% of cases, this keratitis is recurrent and chronic and can interfere with vision. Blindness due to herpes is the leading infectious cause of blindness in the United States.

The viral condition is treated with trifluridine or acyclovir or both. Keratitis resulting from trauma and subsequent bacterial infection is treated with appropriate antibiotics. Most physicians will prescribe antibiotics for prophylactic reasons when there is damage to the eye, even if the original cause is viral (Checkpoint 18.13).

River Blindness

River blindness is a chronic parasitic (helminthic) infection. It is endemic in dozens of countries in Latin America, Africa, Asia, and the Middle East. At any given time, tens of millions of people are infected with the worm called *Onchocerca volvulus* (ong"-koh'ser'-kah'volv'-yoo'lus). This organism is a filarial (threadlike) helminthic worm transmitted by small biting vectors called *black flies*. These voracious flies often attack in large numbers, and it is not uncommon in endemic areas to be bitten several hundred times a day. The disease gets its name from the habitat where these flies are most often found, rural settlements along rivers bordered with overhanging vegetation.

The *Onchocerca* larvae are deposited into a bite wound and develop into adults in the immediate subcutaneous tissues, where disfiguring nodules form within 1 to 2 years after initial contact. Microfilariae given off by the adult female migrate via the bloodstream to many locations but especially to the eyes. While the worms are in the blood, they can be transmitted to other feeding black flies.

Some cases of onchocerciasis result in a severe itchy rash that can last for years. It was previously thought that the condition was caused by degeneration of the worms and the inflammation and granulomatous lesion formation that result from the release of their antigens. It is in fact the case that the worms eventually invade the entire eye, producing much inflammation and permanent damage to the retina and optic nerve. In 1999, researchers first discovered large colonies of bacteria called *Wolbachia* living

inside the *Onchocerca* worms. By 2002, scientists felt very strongly that the damage caused to human tissues was induced by the bacteria rather than by the worms. Of course, the worms serve as the delivery system to the human as it does not appear that the bacteria can infect humans on their own. These bacteria enjoy a mutualistic relationship with their hosts; they are essential for normal *Onchocerca* development.

In regions of high prevalence, it is not unusual for an ophthalmologist to see microfilariae wiggling in the anterior chamber during a routine eye checkup. Microfilariae die in several months, but adults can exist for up to 15 years in skin nodules.

River blindness has been a serious problem in many areas of Africa. In some villages, nearly half of the residents are affected by the disease. A campaign to eradicate onchocerciasis by 2007 is currently underway, supported by the Carter Center, an organization run by former U.S. President Jimmy Carter. The approach is to treat people with *ivermectin*, a potent antifilarial drug and to use insecticides to control the black flies. This approach need not be changed because of the new information about *Wolbachia*; eliminating the protozoan will still eliminate the disease. The drug company that manufactures ivermectin has promised to provide the drug for free for as long as the need for it exists.

☑ CHECKPOINT 18.14	River Blindness
Causative Organism(s)	*Wolbachia* plus *Onchocerca volvulus*
Most Common Modes of Transmission	Biological vector
Virulence Factors	Induction of inflammatory response
Culture/Diagnosis	"Skin snips": small piece of skin in NaCl solution examined under microscope and microfilariae counted
Prevention	Avoiding black fly
Treatment	Ivermectin
Distinguishing Features	Worms often visible in eye

SUMMING UP

Taxonomic Organization Microorganisms Causing Diseases of the Skin and Eyes

Microorganism	Disease	Chapter Location
Gram-positive bacteria		
Propionibacterium acnes	Acne	Acne, p. 538
Staphylococcus aureus	Impetigo, cellulitis, scalded skin syndrome, folliculitis, abscesses (furuncles and carbuncles), necrotizing fasciitis	Impetigo, p. 541 Cellulitis, p. 544 Scalded Skin Syndrome, p. 544, Insight 18.1, p. 538, Note on p. 544
Streptococcus pyogenes	Impetigo, cellulitis, erysipelas, necrotizing fasciitis, scarlet fever	Impetigo, p. 542 Cellulitis, p. 544, Insight 18.1, p. 538
Clostridium perfringens	Gas gangrene	Gas gangrene, p. 545
Bacillus anthracis	Cutaneous anthrax	Large pustular skin lesions, p. 558
Gram-negative bacteria		
Neisseria gonorrhoeae	Neonatal conjunctivitis	Conjunctivitis, p. 563
Chlamydia trachomatis	Neonatal conjunctivitis, trachoma	Conjunctivitis, p. 563 Trachoma, p. 563
Wolbachia (in combination with *Onchocerca*)	River blindness	River blindness, p. 565
DNA viruses		
Human herpesvirus 3 (varicella) virus	Chickenpox	Vesicular or pustular rash diseases, p. 547
Variola virus	Smallpox	Vesicular or pustular rash diseases, p. 549
Parvovirus B 19	Fifth disease	Maculopapular rash diseases, p. 555
Human herpesvirus 6 and 7	Roseola	Maculopapular rash diseases, p. 555
Human papillomavirus	Warts	Warts and wartlike eruptions, p. 555
Molluscum contagiosum virus	Molluscum contagiosum	Warts and wartlike eruptions, p. 555
Herpes simplex virus	Keratitis	Keratitis, p. 564
RNA viruses		
Measles virus	Measles	Maculopapular rash diseases, p. 551
Rubella virus	Rubella	Maculopapular rash diseases, p. 553
Fungi		
Trichophyton	Ringworm	Ringworm, p. 559
Microsporum	Ringworm	Ringworm, p. 559
Epidermophyton	Ringworm	Ringworm, p. 559
Malassezia furfur	Superfical mycosis	Superficial mycoses, p. 561
Protozoa		
Leishmania spp.	Leishmaniasis	Large pustular skin lesions, p. 557
Helminths		
Onchocerca volvulus (in combination with *Wolbachia*)	River blindness	River blindness, p. 565

*There is some debate about the gram status of the genus *Mycobacterium;* it is generally not considered gram-positive or gram-negative.

INFECTIOUS DISEASES AFFECTING
The Skin and Eyes

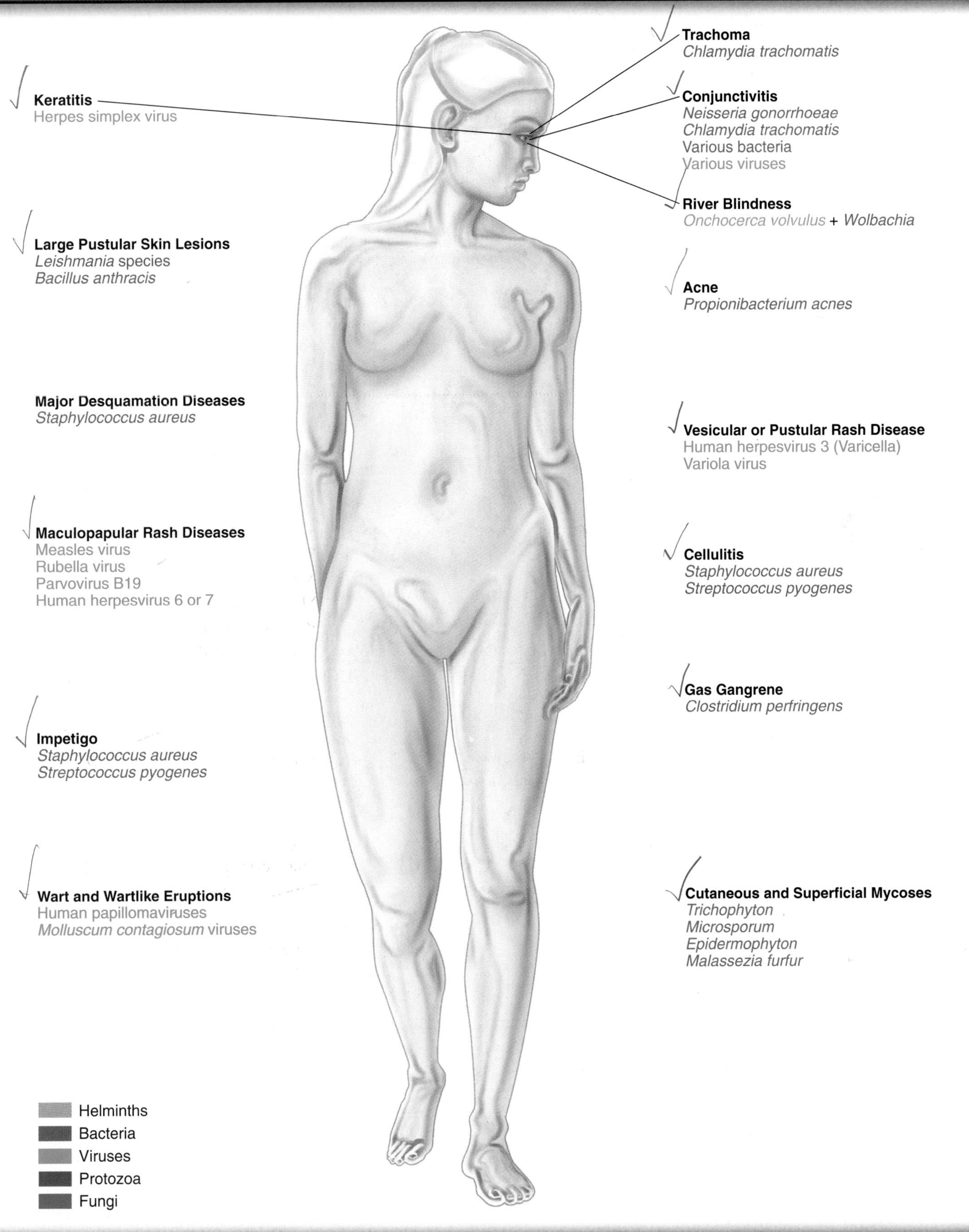

Trachoma
Chlamydia trachomatis
Keratitis
Herpes simplex virus
Conjunctivitis
Neisseria gonorrhoeae
Chlamydia trachomatis
Various bacteria
Various viruses
River Blindness
Onchocerca volvulus + Wolbachia
Large Pustular Skin Lesions
Leishmania species
Bacillus anthracis
Acne
Propionibacterium acnes
Major Desquamation Diseases
Staphylococcus aureus
Vesicular or Pustular Rash Disease
Human herpesvirus 3 (Varicella)
Variola virus
Maculopapular Rash Diseases
Measles virus
Rubella virus
Parvovirus B19
Human herpesvirus 6 or 7
Cellulitis
Staphylococcus aureus
Streptococcus pyogenes
Gas Gangrene
Clostridium perfringens
Impetigo
Staphylococcus aureus
Streptococcus pyogenes
Wart and Wartlike Eruptions
Human papillomaviruses
Molluscum contagiosum viruses
Cutaneous and Superficial Mycoses
Trichophyton
Microsporum
Epidermophyton
Malassezia furfur
Helminths
Bacteria
Viruses
Protozoa
Fungi

System Summary Figure 18.23

Chapter Summary with Key Terms

18.1 The Skin and Eyes

The skin is organized in layers, from the deepest layer called the stratum basale, to the uppermost epidermis.

18.2 The Skin and Its Defenses

A. The epidermal cells contain the protein keratin, which "waterproofs" the skin and protects it from microbial invasion.

B. Other defenses include low pH sebum, high salt and lysozyme in sweat, and antimicrobial peptides.

18.3 Normal Biota of the Skin

The skin has a diverse array of microbes as its normal biota, especially the diphtheroids, micrococci, other gram-positives, and yeasts.

18.4 Skin Diseases Caused by Microorganisms

A. **Acne:** These are follicle-associated lesions caused by microbial digestion of excess sebum trapped in the pores of the skin. *Propionibacterium acnes* is the main causative agent.

B. **Impetigo:** A highly contagious superficial bacterial infection that can cause the skin to peel or flake off that is transmitted both by direct contact and via fomites and mechanical vectors. Causative organisms can be either *Staphylococcus aureus* or *Streptococcus pyogenes* or both.

C. **Cellulitis:** This condition results from a fast-spreading infection of the dermis and subcutaneous tissue below. Most commonly, it is caused by the introduction of *S. aureus* or *S. pyogenes* into the dermis.

D. **Staphylococcal Scalded Skin Syndrome (SSSS):** This dermolytic condition is caused by *S. aureus*. It affects mostly newborns and babies and is similar to a systemic form of impetigo. *Toxic epidermal necrolysis (TEN)* is a similar manifestation caused by a reaction to antibiotics, barbiturates, or other drugs.

E. **Gas Gangrene:** Also called clostridial myonecrosis, this disease can be manifested in two forms: anaerobic cellulitis or myonecrosis. The spore-forming anaerobe, *Clostridium perfringens,* is the most common causative organism.

F. **Vesicular or Pustular Rash Diseases**

1. **Chickenpox:** Skin lesions progress quickly from macules and papules to itchy vesicles filled with a clear fluid. Patients are considered contagious until all of the lesions have crusted over.
2. **Shingles:** Recuperation from chickenpox is associated with the virus becoming latent in the ganglia and may reemerge as **shingles.** Human herpesvirus 3, an enveloped DNA virus, causes chickenpox, as well as herpes zoster or shingles. The virus is sometimes referred to as the varicella-zoster virus (VZV).
3. **Smallpox:** Naturally occurring smallpox has been eradicated from the world. This infection manifests as a rash in the pharynx, spreads to the face, and progresses to the extremities. Variola major is a highly virulent form of smallpox that causes toxemia, shock, and intravascular-coagulation. A patient with variola minor has a milder form of the disease. The causative agent of smallpox, the variola virus, is an orthopoxvirus, an enveloped DNA virus.

G. **Maculopapular Rash Diseases**

1. **Measles:** Measles or **rubeola** results in oral lesions called *Koplik's spots* and characteristic red maculopapular **exanthem** that erupts on the head and then progresses to the trunk and extremities until most of the body is covered. The most serious complication is **subacute sclerosing panencephalitis (SSPE),** a progressive neurological degeneration of the cerebral cortex, white matter, and brain stem. The measles virus is a member of the *Morbillivirus* genus. It is a single-stranded enveloped RNA virus in the Paramyxovirus family. The MMR vaccine (measles, mumps, and rubella) contains live attenuated measles virus.
2. **Rubella:** This disease is also known as German measles and can appear in two forms: postnatal and **congenital** (prenatal) infection of the fetus. The rubella virus is a nonsegmented single-stranded RNA virus with a loose lipid envelope called *Rubivirus,* in the family Togavirus. The MMR vaccination contains protection from rubella.
3. **Fifth Disease:** Also called *erythema infectiosum,* fifth disease is a very mild but highly contagious disease that often results in a characteristic "slapped-cheek" appearance because of a confluent reddish rash that begins on the face. The causative agent is parvovirus B19.
4. **Roseola:** This disease can result in a maculopapular rash and is caused by a human herpesvirus called HHV-6 and sometimes by HHV-7.
5. **Scarlet fever:** May accompany infection of throat or skin with *Streptococcus pyogenes.*

H. **Wartlike Eruptions:** Viruses cause virtually all warts. Most common warts are caused by human papillomavirus or a poxvirus, **molluscum contagiosum,** which causes bumps that may look like warts. Warts, or **papillomas,** are benign, squamous epithelial growths. Virus can infect mucus membranes or invade skin. Rarely, a wart can become malignant when caused by a particular type of HPV.

I. **Larger Pustular Skin Lesions**

1. **Leishmaniasis:** This is a zoonosis transmitted by the female sand fly when it ingests the host's blood. A protozoan causes this equatorial disease, and the infection can either be localized in the skin or mucus membranes, or systemic.
2. **Cutaneous Anthrax:** This form of anthrax is the most common and least dangerous version of infection with *Bacillus anthracis.* The skin shows formation of a papule that becomes necrotic and later ruptures to form a painless, black **eschar.**

J. **Ringworm (Cutaneous Mycoses):** A group of fungi that are collectively termed **dermatophytes** cause mycoses that are confined to the nonliving epidermal tissues, hair, and nails. These diseases are often called "ringworm"—ringworm of the scalp (tinea capitis), beard (tinea barbae), body (tinea corporis), groin (tinea cruris), foot (tinea pedis), and hand (tinea manuum). Species in the genera *Trichophyton, Microsporum,* and *Epidermophyton* cause all of the cutaneous mycoses.

K. **Superficial Mycosis:** Agents of **superficial mycoses** involve the outer epidermis. Tinea versicolor is caused by the yeast *Malassezia furfur,* a normal inhabitant of human skin that feeds on the high oil content of the skin glands.

18.5 The Surface of the Eye and Its Defenses

The flushing action of the tears, which contain lysozyme and lactoferrin, is the major protective feature of the eye.

18.6 Normal Biota of the Eye

The eye has similar microbes as the skin but in lower numbers.

18.7 Eye Diseases Caused by Microorganisms

A. **Conjunctivitis:** Infection of the conjunctiva (commonly called pinkeye) has many different clinical presentations. Neonatal eye infection is usually associated with *Neisseria gonorrhoeae* or *Chlamydia trachomatis*; they are transmitted vertically via a genital tract infection in the mother. Bacterial conjunctivitis in other age groups is most commonly caused by *Staphylococcus epidermidis* or by *Streptococcus pyogenes, Streptococcus pneumoniae, Haemophilus influenzae,* or *Moraxella* species. Viral conjunctivitis is commonly caused by adenoviruses. Both bacterial and viral conjunctivitis are highly contagious.

B. **Trachoma:** Ocular trachoma is a chronic *Chlamydia trachomatis* infection of the epithelial cells of the eye and a major cause of blindness in certain parts of the world. Trachoma and simple conjunctivitis are caused by different strains of *C. trachomatis.*

C. **Keratitis:** Keratitis is a more serious eye infection than conjunctivitis. Herpes simplex viruses (HSV-1 and HSV-2) have been implicated in this condition.

D. **River Blindness:** River blindness is a chronic parasitic helminth infection that is endemic in dozens of countries in Latin America, Africa, Asia, and the Middle East. The condition is caused by a symbiotic pair, the bacterium *Wolbachia* living inside the helminth *Onchocerca.* The worm is transmitted to humans by small biting black flies.

Multiple-Choice and True-False Questions

Multiple-Choice Questions. Select the correct answer from the answers provided.

1. An effective treatment for a cutaneous mycosis like tinea pedis would be
   a. penicillin
   b. miconazole
   c. griseofulvin
   d. doxycycline
2. What is the antimicrobial enzyme found in sweat, tears, and saliva that can specifically break down peptidoglycan?
   a. lysozyme
   b. beta-lactamase
   c. catalase
   d. coagulase
3. Which of the following have been used as treatments for acne?
   a. erythromycin
   b. tetracycline
   c. oral contraceptives
   d. Accutane
   e. all of the above
4. Name the organism(s) most commonly associated with cellulitis.
   a. *Staphylococcus aureus*
   b. *Propionibacterium acnes*
   c. *Streptococcus pyogenes*
   d. both a and b
   e. both a and c
5. Due to a highly successful vaccination program, the WHO has managed the worldwide eradication of the naturally occurring disease:
   a. chickenpox
   b. anthrax
   c. smallpox
   d. German measles
6. Warts are caused by
   a. human herpesvirus 3
   b. papillomavirus
   c. herpes simplex virus
   d. morbillivirus
7. Herpesviruses can cause all of the following diseases, except
   a. chickenpox
   b. shingles
   c. keratitis
   d. smallpox
   e. roseola
8. Which disease is incorrectly matched with the causative agent?
   a. viral conjunctivitis—adenovirus
   b. river blindness—*Onchocerca volvulus*
   c. smallpox—variola virus
   d. gas gangrene—*Staphylococcus aureus*
9. Dermatophytes are fungi that infect the epidermal tissue by invading and attacking
   a. collagen
   b. keratin
   c. fibroblasts
   d. sebaceous glands

True-False Questions. If statement is true, leave as is. If it is false, correct it by rewriting the sentence.

10. The enzyme catalase is associated with pathogenic strains of *Staphylococcus aureus.*
11. Fifth disease can be treated with acyclovir and prevented by immunization.
12. Measles can potentially be eradicated because humans are the only reservoir.
13. The blistering and peeling of the skin in scalded skin syndrome are due to the ability of *Staphylococcus aureus* to produce catalase.
14. *Streptococcus pyogenes* is also known as Lancefield group A strep.

Writing to Learn

These questions are suggested as a *writing-to-learn* experience. For each question, compose a one- or two-paragraph answer that includes the factual information needed to completely address the question.

1. Name and describe the three main categories of normal microbiota found on the skin surface.
2. The cause of acne appears to be multifactorial. Describe the factors necessary that may result in an outbreak of acne.
3. a. What are some of the most common conditions that can lead to the onset of gas gangrene?
   b. Describe the key physiological characteristics of the causative organism that mediate the damage observed in this disease.
4. a. What is the causative agent of chickenpox?
   b. How is the occurrence of shingles related to chickenpox?
5. a. Name the three genera of fungi associated with tinea.
   b. Discuss the treatment options available for treating tinea.
6. Why would antibiotics such as penicillin be ineffective in treating fungal infections?
7. a. How are warts contracted? List the different forms of transmission.
   b. Describe the causative agent(s).
   c. What are some of the methods used to treat warts?
8. a. Name the person credited with developing the first vaccine.
   b. What microorganism was used to create this vaccine?
   c. What was the disease that this vaccine was made against?
   d. What is "variolation"?

Concept Mapping

Appendix D provides guidance for working with concept maps.

1. Supply your own linking words or phrases in this concept map, and provide the missing concepts in the empty boxes.
2. Use 6 to 10 bolded words of your choice from the Chapter Summary to create a concept map. Finish it by providing linking words.

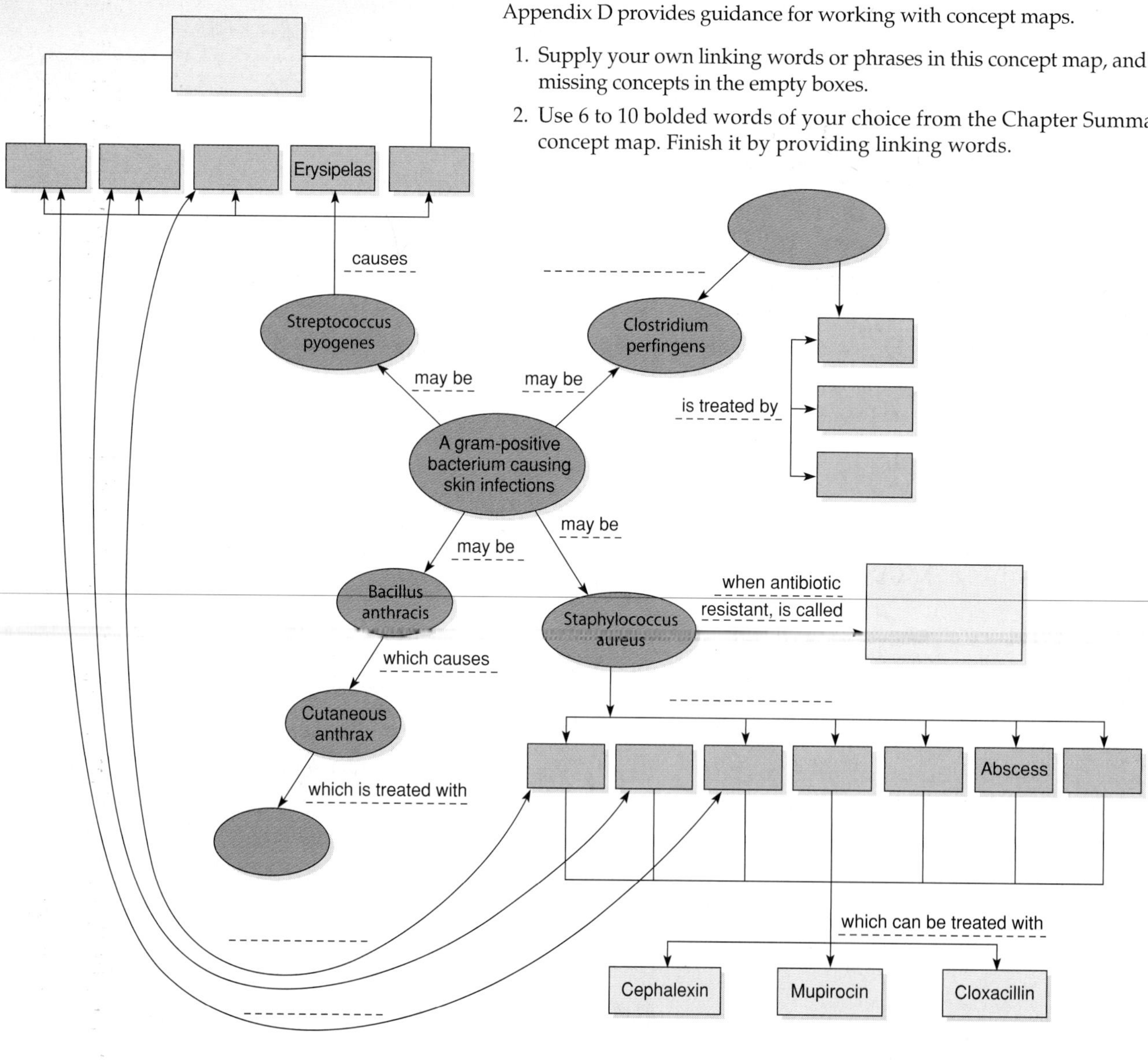

Critical Thinking Questions

Critical thinking is the ability to reason and solve problems using facts and concepts. These questions can be approached from a number of angles, and in most cases, they do not have a single correct answer.

1. Once a person gets chickenpox, he or she usually develops immunity to the disease. Why then does the person get shingles later on?
2. Why is it that people under the age of 20 rarely get shingles—even if they have had chickenpox? Discuss how age factors may influence the latency of the virus.
3. Smallpox has been widely reported as a possible bioterror weapon. Given what you know about the etiology of the disease and the current state of the world's immunity to smallpox, discuss how effective (or ineffective) a smallpox weapon might be. What kind of defense could be mounted against such an attack?
4. Describe a strategy for eradicating river blindness worldwide. Would a vaccination program like that used for smallpox be suitable? Explain.
5. Despite the availability of the measles vaccine, outbreaks of measles still occur. Discuss some of the reasons for these occurrences.

Visual Understanding Questions

1. **From chapter 13, figure 13.5*a*.** How does this figure help explain impetigo caused by *Staphylococcus aureus* or *Streptococcus pyogenes*?

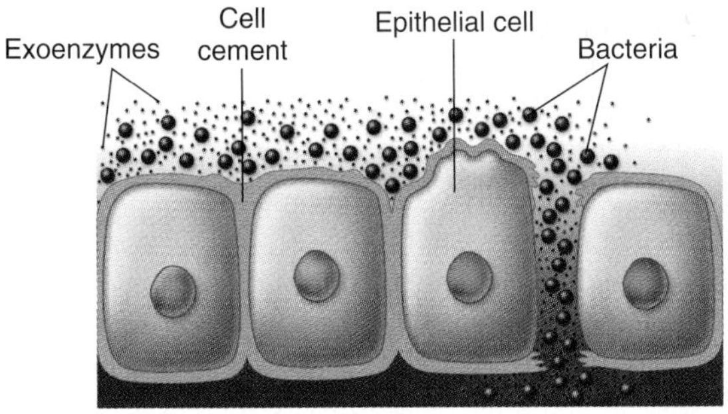

2. **From chapter 1, and chapter 15.** Remember Tyler's disease? Describe a scenario in which streptococcal *superantigens* might have been involved in Tyler's outcomes.

Internet Search Topics

1. You be the detective. Access several Web sources on the subject of monkeypox. Determine the proposed events in the transmission of the virus to the United States. What are the differences between African and American outbreaks? How is monkeypox related to smallpox?
2. Go to: www.aris.mhhe.com, and click on "microbiology" and then this textbook's author/title. Go to chapter 18, access the URLs listed under Internet Search Topics, and research the following:

   Go to the website for the Centers for Disease Control and Prevention. Find all of the conditions that would make a person a poor candidate for the smallpox vaccination. What are all the potential side effects of the vaccine?
3. Search the Internet for information on the rubella titer test. What information does this test provide? How is this information used in patient care? What is the principle of the test?

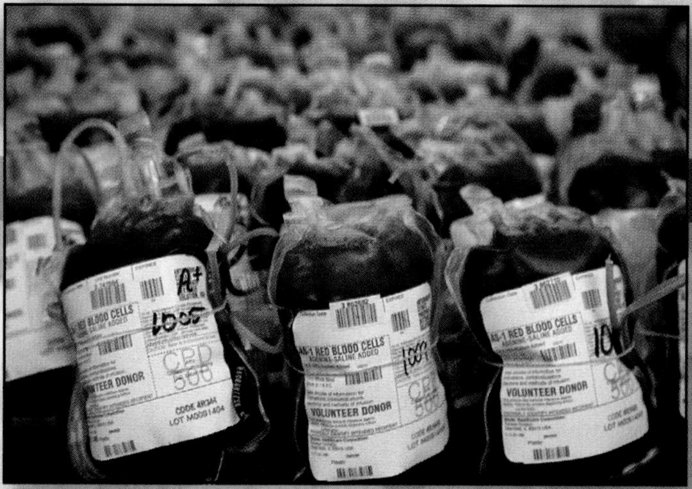

CHAPTER 19

Infectious Diseases Affecting the Nervous System

CASE FILE 19

In the 1980s, cows in Great Britain started dying of a strange disease that turned their brains into mush. The disease, called *bovine spongiform encephalopathy* (the "spongiform" refers to the spongelike nature of the diseased brain) or BSE, is more commonly known as mad cow disease. Since that time, at least 180,000 cows have been diagnosed with the condition, and billions more have been slaughtered as a preventive measure. In the 1990s, some humans in Britain contracted the human form of the disease (called variant Creutzfeldt-Jakob disease, or vCJD) from eating beef from diseased cows, although this number turned out to be much lower than originally predicted.

While transmission of vCJD from one human to another has been documented, until recently these cases were the result of implantation of contaminated corneal grafts; the use of contaminated human growth hormone; or, notably, the use of surgical instruments contaminated from previous neurosurgical procedures on vCJD-infected patients. The situation changed dramatically in 2003, when Britain reported the first case of vCJD transmitted by a blood transfusion. As of early 2007, a total of 66 people in the United Kingdom were known to have received blood from donors who later developed vCJD. Three of these people now have been confirmed to have vCJD.

Variant Creutzfeldt-Jakob disease is invariably fatal; there are no effective treatments for it. Because of this, it is considered controversial to test people for the disease since no hope of recovery can be offered to them. For that reason, of the 66 people known to have received blood from infected donors, only eight have been tested for vCJD (34 have already died from other causes). Of these, three have tested positive.

- *What does it say about the disease agent if "sterilized" surgical instruments have transmitted it?*
- *The infection did not appear to be highly transmissible via contaminated beef products. Can we say the same about transmission via contaminated human blood?*

Case File 19 Wrap-Up appears on page 590.

CHAPTER OVERVIEW

- The nervous system has two parts: the central nervous system (CNS), made up of the brain and spinal cord, and the peripheral nervous system, made of the spinal and cranial nerves. The nervous system has a somewhat muted immune response but is protected by the blood-brain barrier, which limits the passage of substances from the bloodstream to the brain.
- Infections in the nervous system generally affect the brain, the meninges, and/or the peripheral nerves. Brain infections are called encephalitis; meningeal infections are called meningitis. Some infections affect multiple sites, as in meningoencephalitis.
- Infections in the central nervous system are often very serious, because the tissues affected have lowered defenses and can be permanently damaged by inflammation.

19.1 The Nervous System and Its Defenses

The nervous system can be thought of as having two component parts: the central nervous system (CNS), consisting of the brain and spinal cord, and the peripheral nervous system (PNS), which contains the nerves that emanate from the brain and spinal cord to sense organs and to the periphery of the body **(figure 19.1).** The nervous system performs three important functions—sensory, integrative, and motor. The sensory function is fulfilled by sensory receptors at the ends of peripheral nerves. They generate nerve impulses that are transmitted to the central nervous system. There, the impulses are translated, or integrated, into sensation or thought, which in turn drives the motor function. The motor function necessarily involves structures outside of the nervous system, such as muscles and glands.

The brain and the spinal cord are dense structures made up of cells called *neurons*. They are both surrounded by bone. The brain is situated inside the skull, and the spinal cord lies within the spinal column **(figure 19.2),** which is composed of a stack of interconnected bones called vertebrae. The soft tissue of the brain and spinal cord is encased within a tough casing of three membranes called the **meninges.** The layers of membranes, from outer to inner, are the dura mater, the arachnoid mater, and the pia mater. Between the arachnoid mater and pia mater is the subarachnoid space (that is, the space under the arachnoid mater). The subarachnoid space is filled with a clear serumlike fluid called cerebrospinal fluid (CSF). The CSF provides nutrition to the CNS, while also providing a liquid cushion for the sensitive brain and spinal cord. The meninges are a common site of infection, and microorganisms can often be found in the CSF when meningeal infection **(meningitis)** occurs.

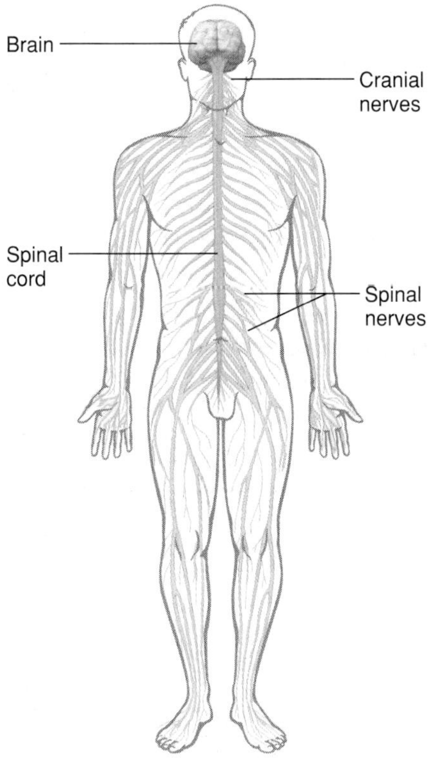

Figure 19.1 Nervous system.
The central nervous system and the peripheral nerves.

The PNS consists of cranial and spinal nerves (see figure 19.1). Nerves, or neurons, are bundles of cellular fibers in the form of axons and dendrites that receive and transmit nerve signals. The axons and dendrites of adjacent neurons communicate with each other over a very small space, called a synapse. Chemicals called neurotransmitters are released from one cell and act on the next cell in the synapse.

The defenses of the nervous system are mainly structural. The bony casings of the brain and spinal cord protect them from traumatic injury. The cushion of surrounding CSF also serves a protective function. The entire nervous system is served by the vascular system, but the interface between the blood vessels serving the brain and the brain itself is different from that of other areas of the body and provides a third structural protection. The cells that make up the walls of the blood vessels allow very few molecules to pass through. In other parts of the body, there is freer passage of ions, sugars, and other metabolites through the walls of blood vessels. The restricted permeability of blood vessels in the brain is called the **blood-brain barrier,** and it prohibits most microorganisms from passing into the central nervous system. The drawback of this phenomenon is that drugs and antibiotics are difficult to introduce into the CNS when needed.

The CNS is considered an "immunologically privileged" site. These sites are able to mount only a partial, or at least a different, immune response when exposed to immunologic challenge. The functions of the CNS are so vital for the life of an organism that even temporary damage that could potentially result from "normal" immune responses would be very detrimental. The uterus and parts of the eye are other immunologically privileged sites. Cells in the CNS express lower levels of MHC antigens. Complement proteins are also in much lower quantities in the CNS. However, specialized cells in the central nervous system perform defensive functions. Microglia are a type of cell having phagocytic capabilities, and brain macrophages also exist in the CNS, although the activity of both of these types of cells is thought to be reduced when compared with phagocytic cells elsewhere in the body.

19.2 Normal Biota of the Nervous System

There is no normal biota in either the CNS or PNS. Finding microorganisms of any type in these tissues represents a deviation from the healthy state. Viruses such as herpes

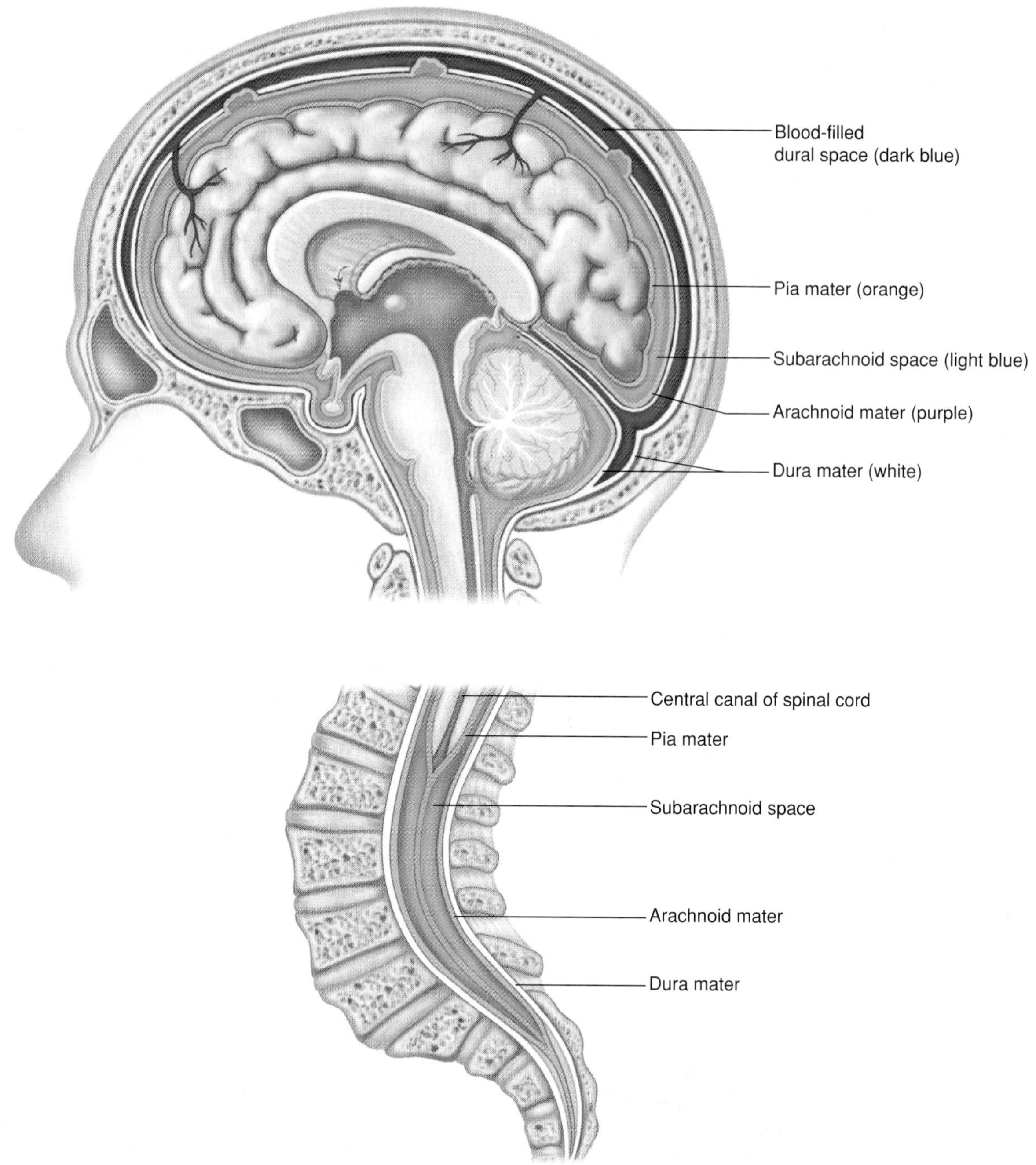

Figure 19.2 Detailed anatomy of the brain and spinal cord.

simplex live in a dormant state in the nervous system between episodes of acute disease, but they are not considered normal biota.

Nervous System Defenses and Normal Biota

	Defenses	Normal Biota
Nervous System	Bony structures, blood-brain barrier, microglial cells, and macrophages	None

19.3 Nervous System Diseases Caused by Microorganisms

Meningitis

Meningitis, an inflammation of the meninges, is an excellent example of an anatomical syndrome. Many different microorganisms can cause an infection of the meninges, and they produce a similar constellation of symptoms. Noninfectious causes of meningitis exist as well, but they are much less common than the infections listed here.

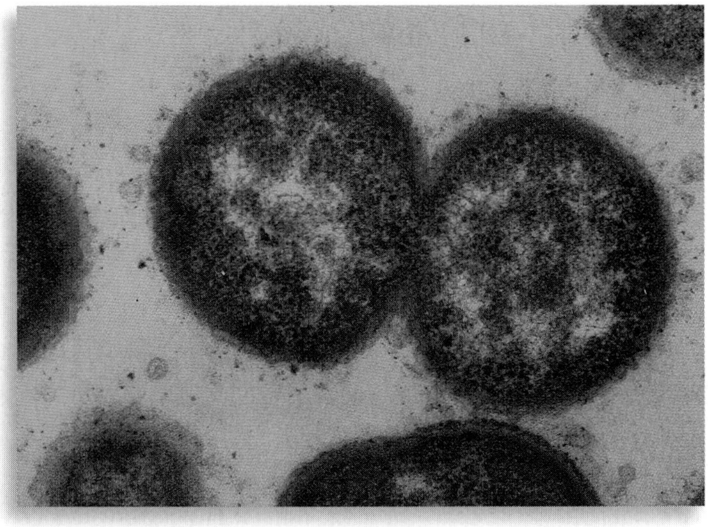

Figure 19.3 Transmission electron micrograph of *Neisseria* (52,000×).
This cross section makes the bacteria appear more spherical than usual.

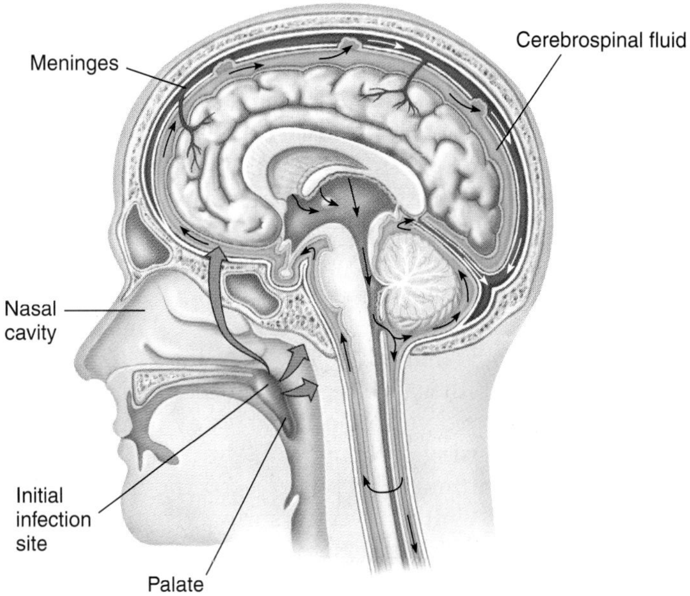

Figure 19.4 Dissemination of the meningococcus from a nasopharyngeal infection.
Bacteria spread to the roof of the nasal cavity, which borders a highly vascular area at the base of the brain. From this location, they can enter the blood and escape into the cerebrospinal fluid. Infection of the meninges leads to meningitis and an inflammatory purulent exudate over the brain surface.

The more serious forms of acute meningitis are caused by bacteria, but it is thought that their entrance to the CNS is often facilitated by coinfection or previous infection with respiratory viruses. Meningitis in neonates is most often caused by different microorganisms, and therefore it is described separately in the following section.

Whenever meningitis is suspected, lumbar puncture (spinal tap) is performed to obtain cerebrospinal fluid (CSF), which is then examined by Gram stain and/or culture. Most physicians will begin treatment with a broad-spectrum antibiotic immediately and shift treatment if necessary after a diagnosis has been confirmed.

Signs and Symptoms

No matter the cause, meningitis results in these typical symptoms: headache, painful or stiff neck, fever, and usually an increased number of white blood cells in the CSF. Specific microorganisms may cause additional, and sometimes characteristic, symptoms, which are described in the individual sections that follow.

Like many other infectious diseases, meningitis can manifest as acute or chronic disease. Some microorganisms are more likely to cause acute meningitis, and others are more likely to cause chronic disease.

In a normal healthy patient, it is very difficult for microorganisms to gain access to the nervous system. Those that are successful usually have specific virulence factors.

Neisseria meningitidis

Neisseria meningitidis appears as gram-negative diplococci lined up side by side **(figure 19.3)** and is commonly known as the meningococcus. It is often associated with epidemic forms of meningitis. This organism causes the most serious form of acute meningitis, and it is responsible for about 25% of all meningitis cases, most of them in children younger than age 2. Although 12 different strains of capsular antigens exist, serotypes A, B, and C are responsible for most cases of infection.

Pathogenesis and Virulence Factors Bacteria entering the blood vessels rapidly permeate the meninges and produce symptoms of meningitis. Meningitis is marked by fever, sore throat, headache, stiff neck, convulsions, and vomiting. The most serious complications of meningococcal infection are due to meningococcemia **(figure 19.4),** which can accompany meningitis but can also occur on its own. The pathogen releases endotoxin into the generalized circulation, which is a potent stimulus for certain white blood cells. Damage to the blood vessels caused by cytokines released by the white blood cells leads to vascular collapse, hemorrhage, and crops of lesions called **petechiae** (pee-tee'-kee-ay) on the trunk and appendages.

In a small number of cases, meningococcemia becomes a fulminant disease with a high mortality rate. Recent evidence suggests that persons who experience meningitis, rather than mild infection, have a genetic predisposition to it. These patients have changes in the genes that encode toll-like receptors (see chapter 14). The changes make it less likely that the host will initiate an early defensive response to the bacterium.

The disease has a sudden onset, marked by fever higher than 40°C, chills, delirium, severe widespread ecchymosis (ek″-ih'moh'seez) (areas of bleeding under the skin), shock, and coma. Generalized intravascular clotting, cardiac failure, damage to the adrenal glands, and death can occur within a few hours. The bacterium has an IgA protease and a capsule, both of which counter the body's defenses.

Transmission and Epidemiology Because meningococci do not survive long in the environment, these bacteria are usually acquired through close contact with secretions or droplets. Upon reaching its portal of entry in the nasopharynx, the meningococci attach there using pili. In many people, this can result in simple asymptomatic colonization. In the more vulnerable individual, however, the meningococci are engulfed by epithelial cells of the mucosa and penetrate into the nearby blood vessels, along the way damaging the epithelium and causing pharyngitis.

Meningococcal meningitis has a sporadic or epidemic incidence in late winter or early spring. The continuing reservoir of infection is humans who harbor the pathogen in the nasopharynx. The carriage state, which can last from a few days to several months, exists in 3% to 30% of the adult population and can exceed 50% in institutional settings. The scene is set for transmission when carriers live in close quarters with nonimmune individuals, as might be expected in families, day care facilities, college dormitories, and military barracks. The highest risk groups are young children (6 to 36 months old) and older children and young adults (10 to 20 years old).

In 2007, a meningococcal meningitis outbreak killed almost 2,000 people and infected almost 20,000 in Sub-Saharan Africa.

Culture and Diagnosis Suspicion of bacterial meningitis constitutes a medical emergency, and differential diagnosis must be done with great haste and accuracy. It is most important to confirm (or rule out) meningococcal meningitis, because it can be rapidly fatal. Treatment (described in the following section) is usually begun with this bacterium in mind until it can be ruled out. Cerebrospinal fluid, blood, or nasopharyngeal samples are stained and observed directly for the typical gram-negative diplococci. Cultivation may be necessary to differentiate the bacterium from other species. Specific rapid tests are also available for detecting the capsular polysaccharide or the cells directly from specimens without culturing.

It is usually necessary to differentiate this species from normal *Neisseria* that also live in the human body and can be present in infectious fluids. Immediately after collection, specimens are streaked on Modified Thayer-Martin medium (MTM) or chocolate agar and incubated in a high CO_2 atmosphere. Presumptive identification of the genus is obtained by a Gram stain and oxidase testing on isolated colonies **(figure 19.5)**. Further testing may be necessary to differentiate *N. meningitidis* and *N. gonorrhoeae* from one another, from other oxidase-positive species, and from normal biota of the oropharynx that can be confused with the pathogens. Several rapid-method identification kits have been developed for this purpose.

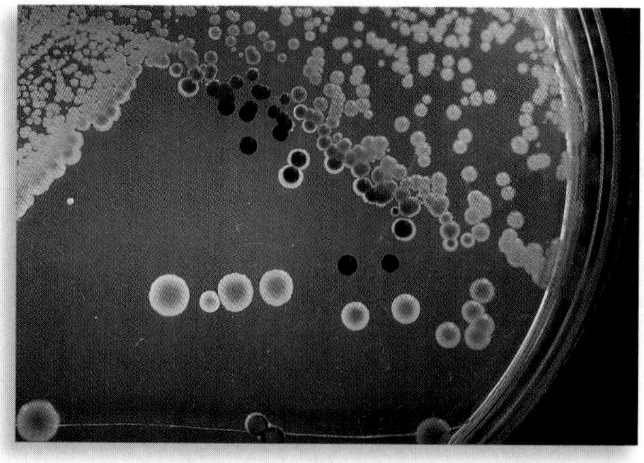

Figure 19.5 The oxidase test.
A drop of oxidase reagent is placed on a suspected *Neisseria* or *Branhamella* colony. If the colony reacts with the chemical to produce a purple to black color, it is oxidase-positive; those that remain white to tan are oxidase-negative. Because several species of gram-negative rods are also oxidase-positive, this test is presumptive for these two genera only if a Gram stain has verified the presence of gram-negative cocci.

Prevention and Treatment The infection rate in most populations is about 1%, so well-developed natural immunity to the meningococcus appears to be the rule. A sort of natural immunization occurs during the early years of life as one is exposed to the meningococcus and its close relatives. Resistance is due to opsonizing antibodies that develop against the capsular polysaccharides in groups A and C and against membrane antigens to group B. Because even treated meningococcemial disease has a mortality rate of up to 15%, it is vital that chemotherapy begin as soon as possible with one or more drugs. Penicillin G is the most potent of the drugs available for meningococcal infections; it is generally given in high doses intravenously. Patients may also require treatment for shock and intravascular clotting.

When family members, medical personnel, or children in day care or school have come in close contact with infected people, preventive therapy with rifampin or tetracycline may be warranted. In the past, meningococcal vaccines that contain specific purified capsular antigens were available to protect high-risk groups, especially during epidemics. A new vaccine was licensed in 2005 and is recommended for children during their preadolescent visit. The vaccine is effective against groups A, C, Y, and W-135 and is conjugated to diphtheria toxoid as an adjuvant. It is thought to provide protection for 10 years. Until this practice has been established for 6 or 7 years, students entering college are urged to have the vaccine as well.

In European countries, serogroup B is more common than A, C, Y, and W-135. There is no vaccine for this serogroup.

Streptococcus pneumoniae

Because *Streptococcus pneumoniae* causes the majority of bacterial pneumonias (see chapter 21), it is also referred to as the **pneumococcus.** Pneumococcal meningitis is also caused by this bacterium; indeed, it is the most frequent cause of community-acquired meningitis and is also very severe. It does not cause the petechiae associated with meningococcal meningitis, and that difference is useful diagnostically. As many as 25% of pneumococcal meningitis patients will also have pneumococcal pneumonia. Pneumococcal meningitis is most likely to occur in patients with underlying susceptibility, such as alcoholic patients and patients with sickle-cell disease or those with absent or defective spleen function.

This bacterium is covered thoroughly in chapter 21, because it is a common cause of ear infections and pneumonia. It obviously has the potential to be highly pathogenic, while at the same time appearing as normal biota in many people. It can penetrate the respiratory mucosa; gain access to the bloodstream; and then, under certain conditions, enter the meninges.

Like the meningococcus, this bacterium has a polysaccharide capsule that protects it against phagocytosis. It also produces an alpha-hemolysin and hydrogen peroxide, both of which have been shown to induce damage in the CNS. It also appears capable of inducing brain cell apoptosis (for a discussion of apoptosis, see chapter 14).

The bacterium is a small gram-positive flattened coccus that appears in end-to-end pairs. It has a distinctive appearance in a Gram stain of cerebrospinal fluid. Staining or culturing the nasopharynx is not useful because it is often normal biota there. It is also alpha-hemolytic on blood agar. Treatment requires a drug to which the bacterium is not resistant; penicillin is therefore not a good choice. Cefotaxime is often used, but drug susceptibilities must always be tested. It is recommended that a steroid be administered 20 minutes prior to antibiotic administration. This will dampen the inflammatory response to cell wall components that are released by antibiotic treatment of the gram-positive bacterium.

As mentioned in chapter 21, two vaccines are available for *S. pneumoniae:* a seven-valent conjugated vaccine (Prevnar), which is now recommended as part of the childhood immunization schedule, and a 23-valent polysaccharide vaccine (Pneumovax) is available for adults.

Haemophilus influenzae

Haemophilus influenzae was originally named when it was isolated from patients with "flu" about 100 years ago. For over 40 years, it was erroneously proclaimed the causative agent until the real agent, the influenza virus, was discovered. This species eventually was shown to be an agent of acute bacterial meningitis in humans. *Haemophilus* cells are tiny (0.5×0.8 mm) gram-negative pleomorphic rods sometimes confused with the genus *Neisseria* in clinical samples. The members of this group tend to be fastidious and are sensitive to drying, temperature extremes, and disinfectants. Even though their name means blood-loving, none of these organisms can be grown on blood agar alone without special techniques. They require certain factors from blood, namely Factor X (hemin), a necessary component of cytochromes, catalase, and peroxidase, and Factor V, nicotinamide adenine dinucleotide (NAD or NADP), an important coenzyme. Media such as chocolate agar (a form of cooked blood agar) and Fildes medium provide these factors.

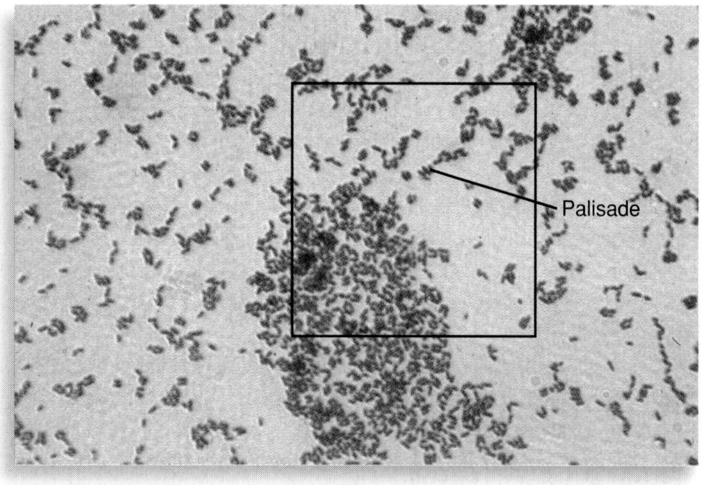

Figure 19.6 *Listeria monocytogenes.*
The bacterium is generally rod shaped. In Gram stains, individual cells tend to stack up in structures called palisades.

The meningitis caused by this bacterium is severe. The disease is caused primarily by the B serotype and was once most common in children between 3 months and 5 years of age. The case rates have declined because of an increased emphasis on vaccination. In contrast to meningococcal meningitis, *Haemophilus* meningitis is not associated with epidemics in the general population but tends to occur as sporadic outbreaks in day care and family settings. It is transmitted by close contact and nose and throat discharges. Healthy adult carriers are the usual reservoirs of the bacterium.

Haemophilus meningitis is very similar to meningococcal meningitis, with symptoms of fever, vomiting, stiff neck, and neurological impairment. Untreated cases have a fatality rate of nearly 90%, and even with prompt diagnosis and aggressive treatment, about 33% of children sustain residual damage.

Haemophilus infections are usually treated with Cefotaxime. Outbreaks of disease in families and day care centers may necessitate rifampin prophylaxis for all contacts. Routine vaccination with a subunit vaccine (Hib containing capsular polysaccharide conjugated to a protein is recommended for all children, beginning at age 2 months, with three follow-up boosters).

Listeria monocytogenes

Listeria monocytogenes is a gram-positive bacterium that ranges in morphology from coccobacilli to long filaments in palisades formation **(figure 19.6).** Cells do not produce capsules or spores and have from one to four flagella. *Listeria* is not fastidious and is resistant to cold, heat, salt, pH extremes, and bile. It grows inside host cells and can moved directly from an infected host cell to an adjacent healthy cell.

Listeriosis in normal adults is often a mild or subclinical infection with nonspecific symptoms of fever, diarrhea, and sore throat. However, listeriosis in elderly or imunocompromised patients, fetuses, and neonates (described later) usually affects the brain and meninges and results in septicemia. The death rate is around 20%. Pregnant women are especially susceptible to infection, which can be transmitted to the infant prenatally when the microbe crosses the placenta or postnatally through the birth canal. Intrauterine infections are widely systemic and usually result in premature abortion and fetal death.

The distribution of *L. monocytogenes* is so broad that its reservoir has been difficult to determine. It has been isolated all over the world from water, soil, plant materials, and the intestines of healthy mammals (including humans), birds, fish, and invertebrates. Apparently, the primary reservoir is soil and water, and animals, plants, and food are secondary sources of infection. Most cases of listeriosis are associated with ingesting contaminated dairy products, poultry, and meat. Recent epidemics have spurred an in-depth investigation into the prevalence of *L. monocytogenes* in these sources. A 2003 U.S. government report concluded that consumers are exposed to low to moderate levels of *L. monocytogenes* on a regular basis. The pathogen has been isolated in 10% to 15% of ground beef and in 25% to 30% of chicken and turkey carcasses and is also present in 5% to 10% of luncheon meats, hot dogs, and cheeses. Aged cheeses made from raw milk are of special concern because *Listeria* readily survives long storage and can grow during refrigeration.

In late 2002, *Listeria* contamination of a poultry processing plant in Pennsylvania led to the recall of 27.4 million pounds of processed chicken and turkey, the largest meat recall in U.S. history.

Except in cases of pregnancy, human-to-human transmission is probably not a significant factor. A predisposing factor in listeriosis seems to be the weakened condition of host defenses in the intestinal mucosa, because studies have shown that immunocompetent individuals are somewhat resistant to infection.

Diagnosing listeriosis is hampered by the difficulty in isolating it. The chances of isolation, however, can be improved by using a procedure called *cold enrichment*, in which the specimen is held at 4°C and periodically plated onto media, but this procedure can take 4 weeks. Rapid diagnostic kits using ELISA, immunofluorescence, and gene probe technology are now available for direct testing of dairy products and cultures. Antibiotic therapy should be started as soon as listeriosis is suspected. Ampicillin and trimethoprimsulfamethoxazole are the first choices, followed by erythromycin. Prevention can be improved by adequate pasteurization temperatures and by proper washing, refrigeration, and cooking of foods that are suspected of being contaminated with animal manure or sewage. Pregnant women are cautioned by the U.S. Food and Drug Administration not to eat soft, unpasteurized cheeses.

Cryptococcus neoformans

The fungus *Cryptococcus neoformans* causes a more chronic form of meningitis with a more gradual onset of symptoms, although in AIDS patients the onset may be fast and the course of the disease more acute. It is sometimes classified as a meningoencephalitis. Headache is the most common symptom, but nausea and neck stiffness are very common. This fungus is a widespread resident of human habitats. It has a spherical to ovoid shape, with small, constricted buds and a large capsule that is important in its pathogenesis **(figure 19.7).**

Transmission and Epidemiology The primary ecological niche of *C. neoformans* is the bird population. It is prevalent in urban areas where pigeons congregate, and it proliferates in the high-nitrogen environment of droppings that accumulate on pigeon roosts. Masses of dried yeast cells are readily scattered into the air and dust. Its role as an opportunist is supported by evidence that healthy humans have strong resistance to it and that frank infection occurs primarily in debilitated patients. Most cryptococcal infections cause symptoms in the respiratory and central nervous systems.

By far the highest rates of cryptococcal meningitis occur among patients with AIDS. This meningitis is frequently fatal. Other conditions that predispose individuals to infection are

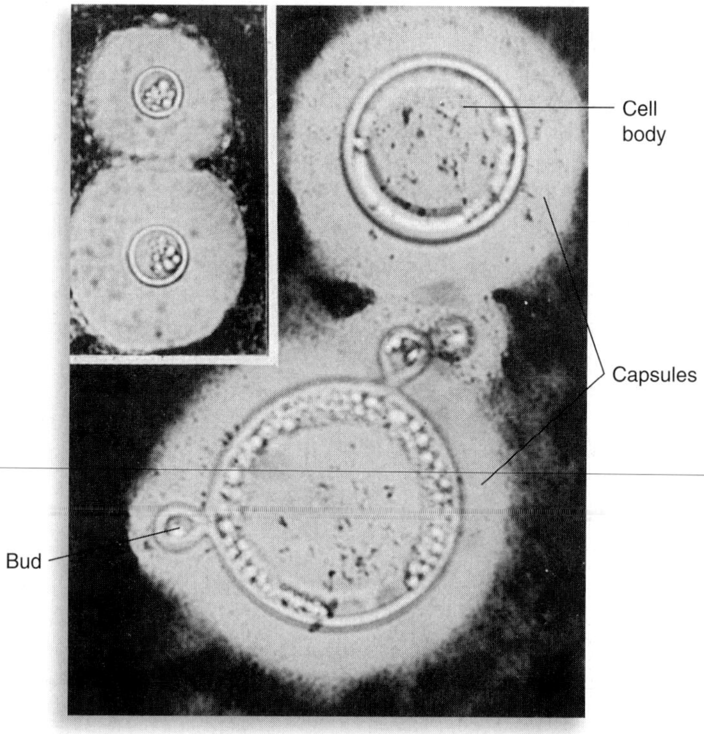

Figure 19.7 ***Cryptococcus neoformans*** **from infected spinal fluid stained negatively with India ink.**
Halos around the large spherical yeast cells are thick capsules. Also note the buds forming on one cell. Encapsulation is a useful diagnostic sign for cryptococcosis, although the capsule is fragile and may not show up in some preparations (150×).

steroid treatment, diabetes, and cancer. It is not considered communicable among humans.

The primary portal of entry for *C. neoformans* is the respiratory tract, but most lung infections are subclinical and rapidly resolved.

Pathogenesis and Virulence Factors The escape of the yeasts into the blood is intensified by weakened host defenses and results in severe complication. *Cryptococcus* shows an extreme affinity for the meninges and brain. The tumorlike masses formed in these locations can cause headache, mental changes, coma, paralysis, eye disturbances, and seizures. In some cases, the infection disseminates into the skin, bones, and viscera **(figure 19.8).**

Culture and/or Diagnosis The first step in diagnosis of cryptococcosis is negative staining of specimens to detect encapsulated budding yeast cells that do not occur as pseudohyphae. Isolated colonies can be used to perform screening tests that presumptively differentiate *C. neoformans* from other cryptococcal species. Confirmatory results include a negative nitrate assimilation, pigmentation on birdseed agar, and fluorescent antibody tests. Cryptococcal antigen can be detected in a specimen by means of serological tests, and DNA probes can make a positive genetic identification.

Prevention and Treatment Systemic cryptococcosis requires immediate treatment with amphotericin B and fluconazole over a period of weeks or months. There is no prevention.

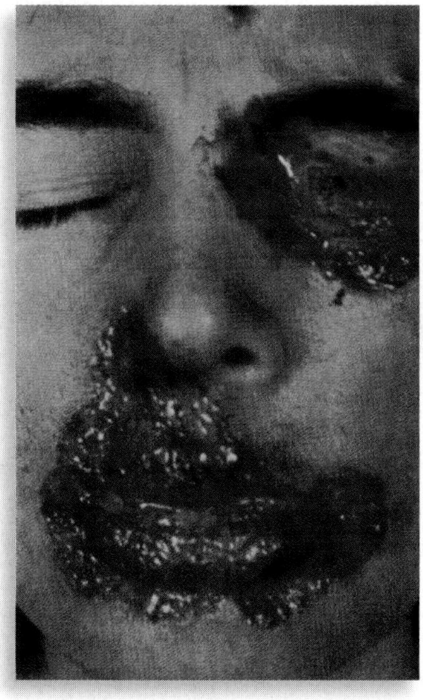

Figure 19.8 Cryptococcosis.
A late disseminated case of cutaneous cryptococcosis in which fungal growth produces a gelatinous exudate. The texture is due to the capsules surrounding the yeast cells.

Coccidioides immitis

Although the fungus *Coccidioides immitis* has probably lived in soil for millions of years, human encounters with it are relatively recent and coincide with increased exposure to soil and encroachment of humans into its habitat, through, for example, expanded agricultural practices. The morphology of *Coccidioides immitis* is very distinctive. At 25°C, it forms a moist white to brown colony with abundant, branching, septate hyphae. These hyphae fragment into thick-walled, blocklike **arthroconidia** (arthrospores) at maturity **(figure 19.9).** On special media incubated at 37°C to 40°C, an arthrospore germinates into the parasitic phase, a small, spherical cell called a spherule, which can be found in infected tissues as well. This structure swells into a giant sporangium that cleaves internally to form numerous endospores that look like bacterial endospores but lack their resistance traits.

Pathogenesis and Virulence Factors This is a true systemic fungal infection of high virulence, as opposed to an opportunistic infection. It usually begins with pulmonary infection but can disseminate quickly throughout the body. Coccidioidomycosis of the meninges is the most serious manifestation. All persons inhaling the arthrospores probably develop some degree of infection, but certain groups have a genetic susceptibility that gives rise to more serious disease.

Transmission and Epidemiology *C. immitis* occurs endemically in various natural reservoirs and casually in areas where it has been carried by wind and animals. Conditions favoring its settlement include high carbon and salt content and a semiarid, relatively hot climate. The fungus has been isolated from soils, plants, and a large number of vertebrates. The natural history of *C. immitis* follows a cyclic pattern—a period of dormancy in winter and spring, followed by growth in summer and fall. Growth and spread are greatly increased by cycles of drought and heavy rains.

Skin testing has disclosed that the highest incidence of coccidioidomycosis, estimated at 100,000 cases per year, occurs in the southwestern United States **(figure 19.10),** although it also occurs in Mexico and parts of Central and South America. Especially concentrated reservoirs exist in the San Joaquin Valley of California and in southern Arizona. Outbreaks are usually associated with farming activity, archeological digs, construction, and mining. A highly unusual outbreak of coccidioidomycosis was traced to the Northridge, California, earthquake. Clouds of dust bearing loosened spores were given off by landslides, and local winds then carried the dust into the outlying residential areas.

Culture and Diagnosis Diagnosis of coccidioidomycosis is straightforward when the highly distinctive spherules are found in sputum, spinal fluid, and biopsies. This finding is further supported by isolation of typical mycelia

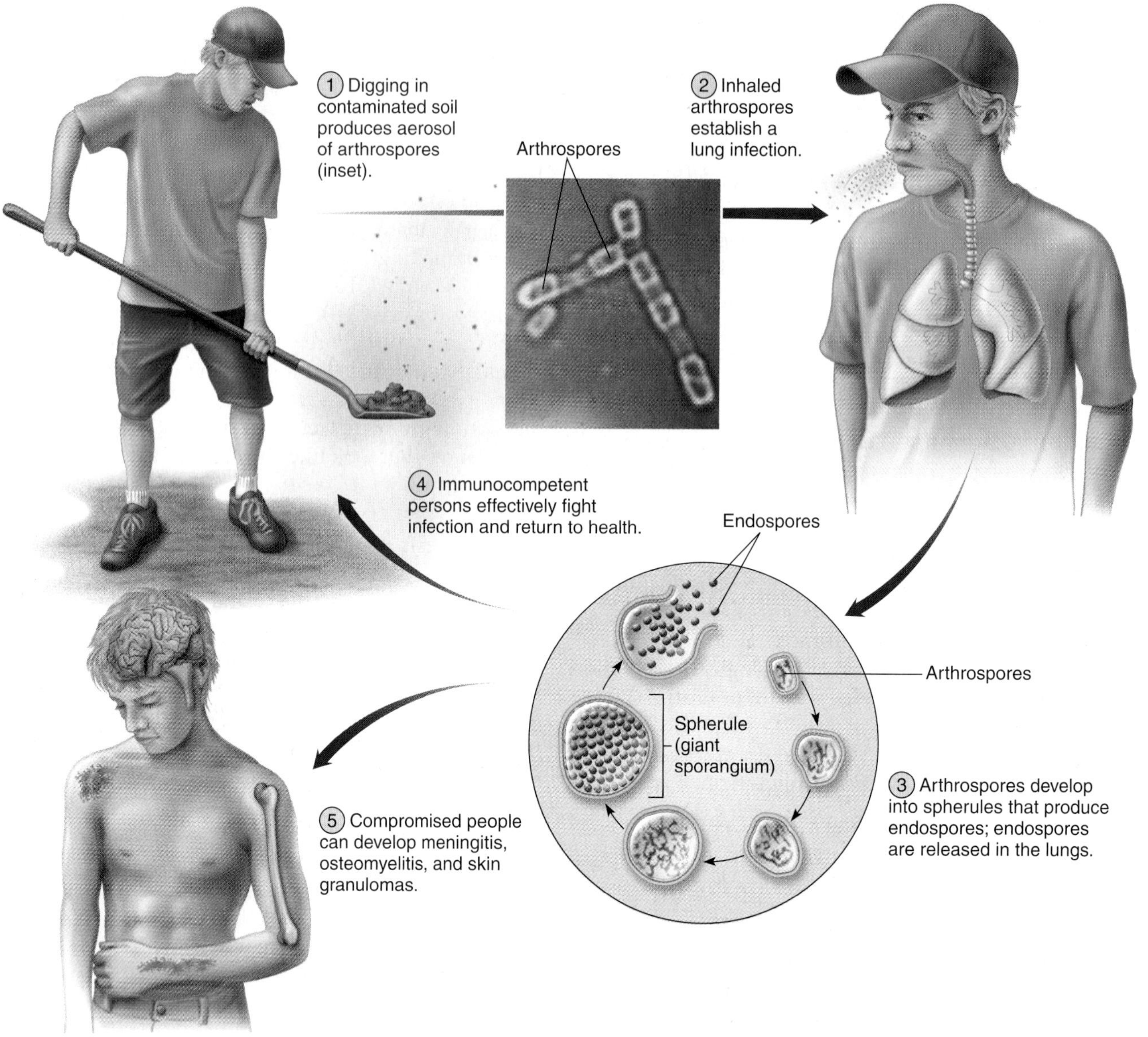

1 Digging in contaminated soil produces aerosol of arthrospores (inset).
Arthrospores
2 Inhaled arthrospores establish a lung infection.
4 Immunocompetent persons effectively fight infection and return to health.
Endospores
Arthrospores
Spherule (giant sporangium)
3 Arthrospores develop into spherules that produce endospores; endospores are released in the lungs.
5 Compromised people can develop meningitis, osteomyelitis, and skin granulomas.

Process Figure 19.9 Events in *Coccidioides* infection and coccidioidomycosis.

and arthrospores on Sabouraud's agar. Newer specific antigen tests have been effective tools to identify and differentiate *Coccidioides* from other fungi. All cultures must be grown in closed tubes or bottles and opened in a biological containment hood to prevent laboratory infections. Immunodiffusion and latex agglutination tests on serum samples are excellent screens for detecting early infection. Skin tests using an extract of the fungi are of primary importance in epidemiological studies.

Prevention and Treatment The majority of patients do not require treatment. In people with disseminated disease, however, amphotericin B is administered intravenously; alternatively, oral or IV itraconazole is used. Minimizing contact with the fungus in its natural habitat has been of some value. For example, oiling dirt roads and planting vegetation help reduce spore aerosols, and using dust masks while excavating soil prevents workers from inhaling spores.

Viruses

A wide variety of viruses can cause meningitis. Because no bacteria or fungi are found in the CSF in viral meningitis, the condition is often called *aseptic meningitis*, although the microorganisms previously described can occasionally cause aseptic meningitis as well when no microbes can be found in the CSF. This may happen when bacterial or fungal meningitis is incompletely treated with antimicrobials; aseptic meningitis may also have noninfectious causes.

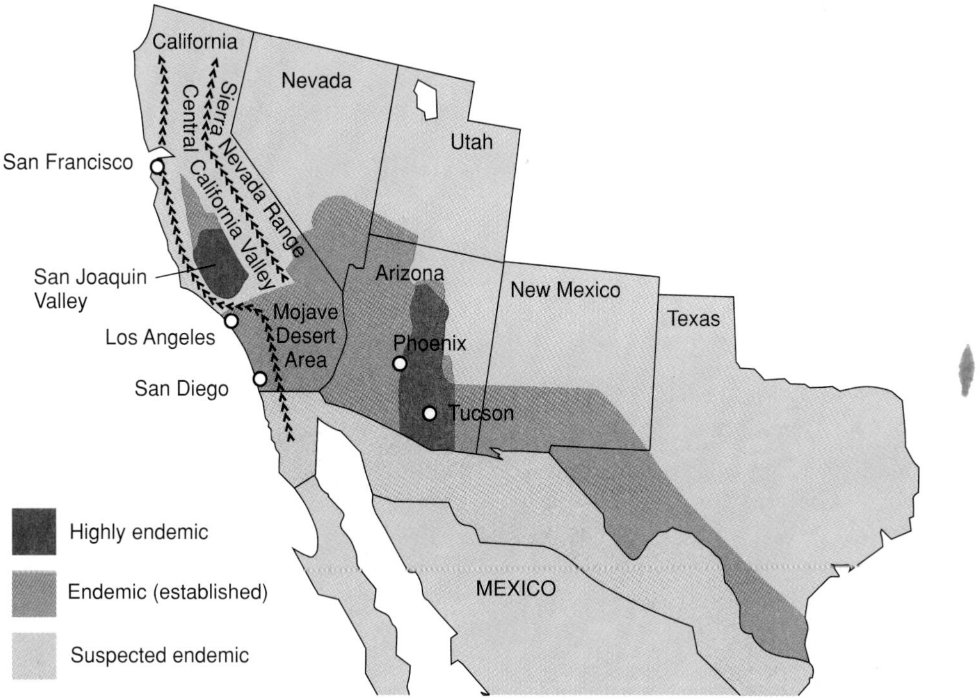

Figure 19.10 **Areas in the United States endemic for *Coccidioides immitis*.**

By far the majority of cases of viral meningitis occur in children, and 90% are caused by enteroviruses. But many other viruses also gain access to the central nervous system on occasion. An initial infection with herpes simplex type 2 is sometimes known to cause meningitis. Also, other herpesviruses such as HHV-6 and HHV-7 and also HHV-3 (the chickenpox virus) and cytomegalovirus (CMV) can infect the meninges, resulting in symptoms. Arboviruses, arenaviruses, and adenoviruses have also been found in meningitis cases. Finally, HIV infection may manifest as meningitis.

Viral meningitis is generally milder than bacterial or fungal meningitis, and it is usually resolved within 2 weeks. The mortality rate is less than 1%. Diagnosis begins with the failure to find bacteria, fungi, or protozoa in CSF and can be confirmed, depending on the virus, by viral culture or specific antigen tests. In most cases, no treatment is indicated. Acyclovir can be used when the causative agent is a herpesvirus; and, of course, if HIV is the cause the entire HIV antiviral regimen is called for (HIV is discussed in chapter 20). **Checkpoint 19.1** summarizes the agents causing meningitis.

Neonatal Meningitis

Meningitis in newborns is almost always a result of infection transmitted by the mother, either in utero or (more frequently) during passage through the birth canal (although **Insight 19.1** describes a troubling exception to this trend). As more premature babies survive, the rates of neonatal meningitis increase, because the condition is favored in patients with immature immune systems. The two most common causes are *Streptococcus agalactiae* and *Escherichia coli*. *Listeria monocytogenes* is also found frequently in neonates. It has already been covered here but is included in Checkpoint 19.2 as a reminder that it can cause neonatal cases as well.

Streptococcus agalactiae

This species of *Streptococcus* belongs to group B of the streptococci. It colonizes 10% to 30% of female genital tracts and is the most frequent cause of neonatal meningitis (for details about this condition in women, see chapter 23). The treatment for neonatal disease is penicillin G, sometimes supplemented with an aminoglycoside.

Escherichia coli

The K1 strain of *Escherichia coli* is the second most common cause of neonatal meningitis. Most babies who suffer from this infection are premature, and their prognosis is poor. Twenty percent of them die, even with aggressive antibiotic treatment, and those who survive often have brain damage.

The bacterium is usually transmitted from the mother's birth canal. It causes no disease in the mothers but can infect the vulnerable tissues of a neonate. It seems to have a predilection for the tissues of the central nervous system. Cefotaxime is usually administered intravenously, in combination with aminoglycosides **(Checkpoint 19.2).**

CHECKPOINT 19.1 Meningitis

Causative Organism(s)	*Neisseria meningitidis*	*Streptococcus pneumoniae*	*Haemophilus influenzae*	*Listeria monocytogenes*	*Cryptococcus neoformans*	*Coccidioides immitis*	Viruses
Most Common Modes of Transmission	Droplet contact	Droplet contact	Droplet contact	Vehicle (food)	Vehicle (air, dust)	Vehicle (air, dust, soil)	Droplet contact
Virulence Factors	Capsule, endotoxin, IgA protease	Capsule, induction of apoptosis, hemolysin and hydrogen peroxide production	Capsule	Intracellular growth	Capsule, melanin production	Granuloma (spherule) formation	Lytic infection of host cells
Culture/ Diagnosis	Gram stain/ culture of CSF, blood, rapid antigenic tests	Gram stain/ culture of CSF	Culture on chocolate agar	Cold enrichment, rapid methods	Negative staining, biochemical tests, DNA probes	Identification of spherules, cultivation on Sabouraud's agar	Initially, absence of bacteria/fungi/ protozoa, followed by viral culture or antigen tests
Prevention	Conjugated vaccine; rifampin or tetracycline used to protect contacts	Two vaccines: Prevnar (children), and Pneumovax (adults)	Hib vaccine	Cooking food, avoiding unpasteurized dairy products	–	Avoiding airborne spores	–
Treatment	Penicillin G or Cefotaxime	Cefotaxime check for resistance (add vancomycin in that case)	Cefotaxime	Ampicillin, trimethoprim-sulfamethox-azole	Amphotericin B and fluconazole	Amphotericin B or oral or IV itraconazole	Usually none unless specific virus identified and specific antiviral exists)
Distinctive Features	Petechiae, meningo-coccemia	Serious, acute, most common meningitis in adults	Serious, acute, less common since vaccine became available	Asymptomatic in healthy adults, meningitis in neonates, elderly and immuno-compromised	Acute or chronic, most common in AIDS patients	Almost exlusively in endemic regions	Generally milder than bacterial or fungal

CHECKPOINT 19.2 Neonatal Meningitis

Causative Organism(s)	*Streptococcus agalactiae*	*Escherichia coli*, strain K1	*Listeria monocytogenes*
Most Common Modes of Transmission	Vertical (during birth)	Vertical (during birth)	Vertical
Virulence Factors	Capsule	–	Intracellular growth
Culture/Diagnosis	Culture mother's genital tract on blood agar; CSF culture of neonate	CSF Gram stain/culture	Cold enrichment, rapid methods
Prevention	Culture and treatment of mother	–	Cooking food, avoiding unpasteurized dairy products
Treatment	Penicillin G plus aminoglycosides	Cefotaxime plus aminoglycoside	Ampicillin, trimethoprim-sulfamethoxazole
Distinctive Features	Most common; positive culture of mother confirms diagnosis	Suspected if infant is premature	Suspected if infant is premature

INSIGHT 19.1 — *Discovery*

Baby Food and Meningitis

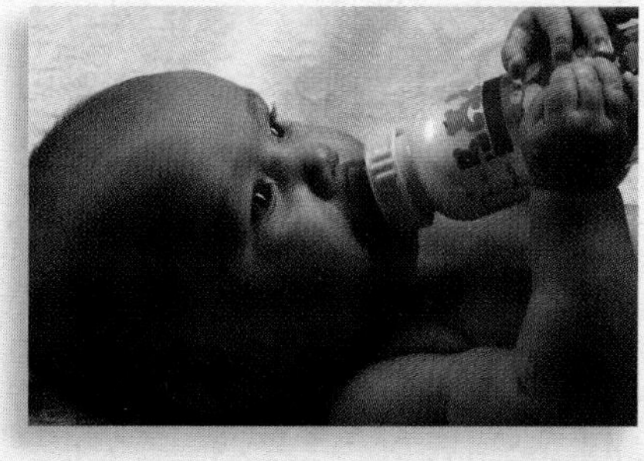

It will come as no surprise to you that there are potentially dangerous bacteria in a wide variety of foods we consume. Cases of *E. coli* O157:H7 disease associated with hamburgers and hepatitis contracted by customers of a Mexican restaurant get a lot of media attention. It may surprise you to learn that pathogenic bacteria are found in dried infant formula and dried baby food, as well.

In 2001, an outbreak of meningitis in a neonatal intensive care unit in Tennessee was traced to a batch of powdered formula. The manufacturer recalled the product after the Centers for Disease Control and Prevention issued a warning. In 2004, scientists in England investigated 110 different types of baby foods. Ten percent of the powdered formula samples and 25% of the dried infant foods contained intestinal bacteria. In many of the samples, they found a bacterium called *Enterobacter sakazakii,* an intestinal bacterium that has been linked to several fatal outbreaks of meningitis in children's hospitals.

The disease is rare but almost always associated with infant foods and has a 33% fatality rate, with up to 80% of infants suffering permanent neurological damage. So what is to be done? In this case, it is "consumer beware." Manufacturers have never claimed that their formulas and foods are sterile. The scientists who conducted the infant food study also investigated ideal conditions for preparing and storing the products. They noted that the bacterial doubling time in prepared formula is 10 hours when refrigerated, and 30 minutes at room temperature. This means that leaving prepared formula in your diaper bag or on the kitchen counter for even a few hours could lead to high levels of bacteria in the bottle.

Powdered formula is made by manufacturing the nutritious liquid and then freeze-drying it. It is sterile as a liquid but bacteria can be introduced during the freeze-drying and packaging phases. Since the outbreaks, the FDA has recommended that the powder be reconstituted with boiling water. The CDC has not supported this recommendation because of many problems with it including the risk of destroying important nutrients and the lack of data that boiling it would be sufficient to kill *E. sakazakii.* Hospitals are advised to use ready-to-feed or concentrated liquid formulas.

Meningoencephalitis

Up to this point, we have described microorganisms causing meningitis (inflammation of the meninges). Next we discuss microorganisms that cause **encephalitis,** inflammation of the brain. Because the brain and the spinal cord (and the meninges) are so closely connected, infections of one of these structures may also involve the other.

But two microorganisms cause a distinct disease called *meningoencephalitis,* and they are both amoebas. *Naegleria fowleri* and *Acanthamoeba* are accidental parasites that invade the body only under unusual circumstances.

Naegleria fowleri

The trophozoite of *Naegleria* is a small, flask-shaped amoeba that moves by means of a single, broad pseudopod **(figure 19.11).** It forms a rounded, thick-walled, uninucleate cyst that is resistant to temperature extremes and mild chlorination.

Most cases of *Naegleria* infection reported worldwide occur in people who have been swimming in warm, natural bodies of fresh water. One epidemic in Australia was due to contamination of a public water supply; and in Belgium, an outbreak was reported among people who had bathed in polluted canal water. Infection can begin when amoebas are forced into human nasal passages as a result of swimming, diving, or other aquatic activities. Once the amoeba is inoculated into the favorable habitat of the nasal mucosa, it burrows in, multiplies, and subsequently migrates into the brain and surrounding structures. The result is primary amoebic meningoencephalitis (PAM), a rapid, massive destruction of brain and spinal tissue that causes hemorrhage and coma and invariably ends in death within a week or so. We should note that this organism is very common—children often

Figure 19.11 Scanning electron micrograph of *Naegleria fowleri.*
The "eyes" and "mouth" of its facelike appearance are its attachment and feeding structures.

CHECKPOINT 19.3 Meningoencephalitis

	Primary Amoebic Meningoencephalitis	Granulomatous Amoebic Meningoencephalitis
Causative Organism(s)	*Naegleria fowleri*	*Acanthamoeba*
Most Common Modes of Transmission	Vehicle (exposure while swimming in water)	Direct contact
Virulence Factors	Invasiveness	Invasiveness
Culture/Diagnosis	Examination of CSF; brain imaging, biopsy	Examination of CSF; brain imaging, biopsy
Prevention	Avoid warm fresh water	–
Treatment	Amphotericin B; mostly ineffective	Surgical excision of granulomas; Ketoconazole may help

carry the amoeba as harmless biota, especially during the summer months, and the series of events leading to disease is exceedingly rare.

Unfortunately, *Naegleria* meningoencephalitis advances so rapidly that treatment usually proves futile. Studies have indicated that early therapy with amphotericin B, sulfadiazine, or tetracycline in some combination can be of some benefit. Because of the wide distribution of the amoeba and its hardiness, no general means of control exists. Public swimming pools and baths must be adequately chlorinated and checked periodically for the amoeba.

Acanthamoeba

This protozoan has a large, amoeboid trophozoite with spiny pseudopods and a double-walled cyst. It differs from *Naegleria* in its portal of entry; it invades broken skin, the conjunctiva, and occasionally the lungs and urogenital epithelia. Although it causes a meningoencephalitis somewhat similar to that of *Naegleria*, the course of infection is lengthier. The disease is called granulomatous amoebic meningoencephalitis (GAM). At special risk for infection are people with traumatic eye injuries, contact lens wearers, and AIDS patients exposed to contaminated water. Ocular infections can be avoided by carefully tending to injured eyes and by using sterile solutions to store and clean contact lenses. Cutaneous and CNS infections with this organism are occasional complications in AIDS **(Checkpoint 19.3).**

Acute Encephalitis

Encephalitis can present as acute or **subacute.** It is always a serious condition, as the tissues of the brain are extremely sensitive to damage by inflammatory processes. Acute encephalitis is almost always caused by viral infection. One category of viral encephalitis is caused by viruses borne by insects (arboviruses), including West Nile virus. Alternatively, other viruses, such as members of the herpes family, are causative agents. Bacteria such as those covered under meningitis can also cause encephalitis, but the symptoms are almost always more pronounced in the meninges than in the brain.

The signs and symptoms of encephalitis vary, but they may include behavior changes or confusion because of inflammation. Decreased consciousness and seizures frequently occur. Symptoms of meningitis are often also present. Few of these agents have specific treatments, but because swift initiation of acyclovir therapy can save the life of a patient suffering from herpesvirus encephalitis, most physicians will begin empiric therapy with acyclovir in all seriously ill neonates and most other patients showing evidence of encephalitis. Treatment will, in any case, do no harm in patients who are infected with other agents.

Arboviruses

Wherever there are arthropods, there are also arboviruses so, collectively, their distribution is worldwide. The vectors and viruses tend to be clustered in the tropics and subtropics, but many temperate zones report periodic epidemics. A given arbovirus type may have very restricted distribution, even to a single isolated region, but some types range over several continents, and others can spread along with their vectors **(figure 19.12).**

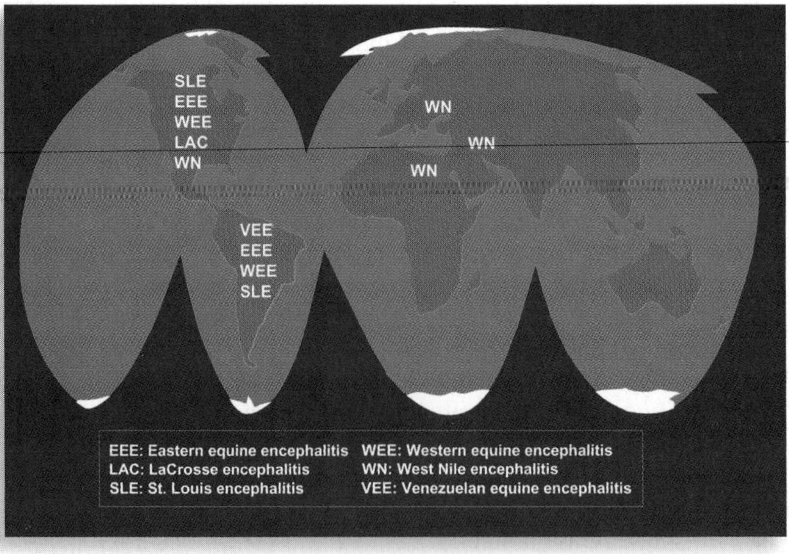

Figure 19.12 Worldwide distribution of major arboviral encephalitides.

INSIGHT 19.2 — Medical

A Long Way from Egypt: West Nile Virus in the United States

In 1999, the first cases of West Nile encephalitis were seen in several northeastern states. Over the next few years, West Nile virus spread westward across the country, resulting in at least 1 million infections and over 664 deaths by mid-2005. In the summer of 2003, the Centers for Disease Control and Prevention reported that they had detected the virus in 600 blood donors in the United States. Government officials were made aware of the need to test for the virus after 23 patients acquired West Nile from blood transfusions in 2002.

The arrival of a deadly new disease, especially at a time of public nervousness concerning potential biological warfare, led to a great deal of media attention concerning West Nile virus, much of it sensational, even if not entirely accurate.

West Nile virus is an arbovirus commonly found in Africa, the Middle East, and parts of Asia, but until mid-1999 had not been detected in the Americas. The virus is known to infect a host of mammals (including humans), as well as birds and mosquitoes. Mosquitoes generally become infected when they feed on birds infected with the virus, and they can then bite and transmit the virus to humans. If infection results, the illness is generally characterized by flulike symptoms that last just a few days and have no long-term consequences. Less than 1% of infected persons will suffer the potentially lethal inflammation of the brain known as West Nile encephalitis.

Because transmission of the virus depends on the presence of mosquitoes to act as vectors, viral control is synonymous with vector control. Insect repellant, long-sleeved shirts and long pants, and control of mosquito breeding grounds (primarily stagnant water) all decrease the spread of the virus. Because the virus is blood-borne, person-to-person transmission is unlikely. There has been one confirmed case, however, in which the virus was transmitted from mother to fetus, as well as another case in which an infected organ donor passed on the virus to four organ recipients.

Epidemiologists are also closely monitoring the magnitude of the disease in animals. So far, the virus has been isolated in nearly 200 bird species as well as cats, dogs, rodents, horses, and even alligators.

Most arthropods that serve as infectious disease vectors feed on the blood of hosts, a process that infects them for varying time periods. Infections show a peak incidence when the arthropod is actively feeding and reproducing, usually from late spring through early fall. Warm-blooded vertebrates also maintain the virus during the cold and dry seasons. Humans can serve as dead-end, accidental hosts, as in equine encephalitis, or they can be a maintenance reservoir, as in yellow fever (discussed in chapter 20).

Arboviral diseases have a great impact on humans. Although exact statistics are unavailable, it is believed that millions of people acquire infections each year and thousands of them die. One common outcome of arboviral infection is an acute fever, often accompanied by rash. Viruses that primarily cause these symptoms are covered in chapter 20.

The arboviruses discussed in this chapter can cause encephalitis, and we consider them as a group because the symptoms and management are similar. The transmission and epidemiology of individual viruses are different, however, and are discussed for each virus. **Insight 19.2** discusses West Nile virus, an arbovirus that has spread across North America in recent years.

Pathogenesis and Virulence Factors Arboviral encephalitis begins with an arthropod bite, the release of the virus into tissues, and its replication in nearby lymphatic tissues. Prolonged viremia establishes the virus in the brain, where inflammation can cause swelling and damage to the brain, nerves, and meninges. Symptoms are extremely variable and can include coma, convulsions, paralysis, tremor, loss of coordination, memory deficits, changes in speech and personality, and heart disorders. In some cases, survivors experience some degree of permanent brain damage. Young children and the elderly are most sensitive to injury by arboviral encephalitis.

The virulence of these viruses is not well understood, but much research has focused on proteins that the virus uses to attach to host tissues or to induce fusion with host cell membranes. Both of these functions facilitate invasion of the virus.

Culture and Diagnosis Except during epidemics, detecting arboviral infections can be difficult. The patient's history of travel to endemic areas or contact with vectors, along with serum analysis, is highly supportive of a diagnosis. Rapid serological tests are available for some of the viruses.

Prevention and Treatment No satisfactory treatment exists for any of the arboviral encephalitides (plural of *encephalitis*). As mentioned earlier, empiric acyclovir treatment may be begun in case the infection is actually caused by either herpes simplex virus or varicella zoster. Treatment of the other infections relies entirely on support measures to control fever, convulsions, dehydration, shock, and edema.

Most of the control safeguards for arbovirus disease are aimed at the arthropod vectors. Mosquito abatement by eliminating breeding sites and by broadcasting insecticides has been highly effective in restricted urban settings. Birds play a role as reservoirs of the virus, but direct transmission between birds and humans does not occur.

Live attenuated vaccines are available for Western equine encephalitis and Eastern equine encephalitis and are administered to laboratory workers, veterinarians, ranchers, and horses.

Western Equine Encephalitis (WEE) This disease occurs sporadically in the western United States and Canada, appearing first in horses and later in humans. The mosquito that carries the virus emerges in the early summer when irrigation begins in rural areas and breeding sites are abundant. The disease is extremely dangerous to infants and small children, with a case fatality rate of 3% to 7%.

Eastern Equine Encephalitis (EEE) EEE is endemic to an area along the eastern coast of North America and Canada. The usual pattern is sporadic, but occasional epidemics can occur in humans and horses. High periods of rainfall in the late summer increase the chance of an outbreak, and disease usually appears first in horses and caged birds. The case fatality rate can be very high (70%).

California Encephalitis This condition may be caused by two different viral strains. The California strain occurs occasionally in the western United States and has little impact on humans. The LaCrosse strain is widely distributed in the eastern United States and Canada and is a prevalent cause of viral encephalitis in North America. Children living in rural areas are the primary target group, and most of them exhibit mild, transient symptoms. Fatalities are rare.

St. Louis Encephalitis (SLE) St. Louis encephalitis may be the most common of all American viral encephalitides. Cases appear throughout North and South America, but epidemics in the United States occur most often in the Midwest and South. Inapparent infection is very common, and the total number of cases is probably thousands of times greater than the 50 to 100 reported each year. The seasons of peak activity are spring and summer, depending on the region and species of mosquito. In the eastern United States, mosquitoes breed in stagnant or polluted water in urban and suburban areas during the summer. In the West, mosquitoes frequent spring floodwaters in rural areas.

West Nile Encephalitis The West Nile virus is a close relative of the SLE virus. It emerged in the United States in 1999, and by mid-2005 more than 1 million people had been infected. See Insight 19.2 for details.

Herpes Simplex *Virus*

Herpes simplex type 1 and 2 viruses can cause encephalitis in newborns born to HSV-positive mothers. In this case, the virus is disseminated and the prognosis is poor. Older children and young adults (ages 5 to 30), as well as older adults (over 50 years old), are also susceptible to herpes simplex encephalitis caused most commonly by HSV-1. In these cases, the HSV encephalitis represents a reactivation of dormant HSV from the trigeminal ganglion.

It should be noted the varicella-zoster virus (see chapter 18) can also reactivate from the dormant state, and it is responsible for rare cases of encephalitis.

JC *Virus*

The **JC virus (JCV)** gets its name from the initials of the patient in whom it was first diagnosed as the cause of illness. Serological studies indicate that infection with this polyoma virus is commonplace. In patients with immune dysfunction, especially in those with AIDS, it can cause a condition called **progressive multifocal leukoencephalopathy** (loo″-koh-en-sef″uh-lop′-uh-thee) **(PML).** This uncommon but generally fatal infection is a result of JC virus attack of accessory brain cells. The infection demyelinizes certain parts of the cerebrum. This virus should be considered when encephalitis symptoms are observed in AIDS patients. Recently a few deaths from this condition have been prevented with high doses of zidovudine.

Other Virus-Associated Encephalitides

Infection with measles and other childhood rash diseases can result 1 to 2 weeks later in an inappropriate immune response with consequences in the CNS. The condition is called postinfection encephalitis (PIE), and it is thought to be a result of immune system action and not of direct viral invasion of neural tissue. Very rarely PIE can occur after immunization with live attenuated vaccines against viral infections. Note that PIE is distinct from another possible sequela of measles virus infection called SSPE (discussed later in this chapter) **(Checkpoint 19.4).**

Subacute Encephalitis

When encephalitis symptoms take longer to show up and when the symptoms are less striking, the condition is termed **subacute encephalitis.** The most common cause of subacute encephalitis is the protozoan *Toxoplasma.* Another form of subacute encephalitis can be caused by persistent measles virus as many as 7 to 15 years after the initial infection. Finally, a class of infectious agents known as prions can cause a condition called spongiform encephalopathy.

Toxoplasma gondii

Toxoplasma gondii is a flagellated parasite with such extensive cosmopolitan distribution that some experts estimate it affects the majority of the world's population at some time in their lives. In most of these cases, toxoplasmosis goes unnoticed, but disease in the fetus and in immunodeficient people, especially those with AIDS, is severe and often fatal. *T. gondii* is a very successful parasite with so little host specificity that it can attack at least 200 species of birds and mammals. However, its primary reservoir and hosts are members of the feline family, both domestic and wild.

Signs and Symptoms As just mentioned, most cases of toxoplasmosis are asymptomatic or marked by mild symptoms such as sore throat, lymph node enlargement, and low-grade fever. In patients whose immunity is suppressed by infection, cancer, or drugs, the outlook may be grim. The infection

CHECKPOINT 19.4	Encephalitis			
Causative Organism(s)	Arboviruses (viruses causing WEE, EEE, California encephalitis, SLE, West Nile encephalitis)	Herpes simplex 1 or 2	JC virus	Immunologic reaction to other viral infections
Most Common Modes of Transmission	Vector (arthropod bites)	Vertical or reactivation of latent infection	? Ubiquitous	Sequelae of measles, other viral infections and occasionally, vaccination
Virulence Factors	Attachment, fusion, invasion capabilities	–	–	–
Culture/Diagnosis	History, rapid serological tests	Clinical presentation, PCR, Ab tests, growth of virus in cell culture	PCR of cerebrospinal fluid	History of viral infection or vaccination
Prevention	Insect control, vaccines for WEE and EEE available	Maternal screening for HSV	None	–
Treatment	None	Acyclovir	Zidovudine or other antivirals	Steroids, anti-inflammatory agents
Distinctive Features	History of exposure to insect important	In infants, disseminated disease present; rare between 30 and 50 years	In severely immunocompromised, especially AIDS	History of virus/vaccine exposure critical

causes a more chronic or subacute form of encephalitis than do most viruses, often producing extensive brain lesions and fatal disruptions of the heart and lungs. A pregnant woman with toxoplasmosis has a 33% chance of transmitting the infection to her fetus. Congenital infection occurring in the first or second trimester is associated with stillbirth and severe abnormalities such as liver and spleen enlargement, liver failure, hydrocephalus, convulsions, and damage to the retina that can result in blindness.

Pathogenesis and Virulence Factors *Toxoplasma* is an obligate intracellular parasite, making its ability to invade host cells an important factor for virulence.

Transmission and Epidemiology To follow the transmission of toxoplasmosis, we must first look at the general stages of the *Toxoplasma* life cycle in the cat **(figure 19.13*a*).** The parasite undergoes a sexual phase in the intestine and is then released in feces, where it becomes an infective *oocyst* that survives in moist soil for several months. Ingested oocysts release an invasive asexual tissue phase called a *tachyzoite* that infects many different tissues and often causes disease in the cat. These forms eventually enter an asexual cyst state in tissues, called a pseudocyst. Most of the time, the parasite does not cycle in cats alone and is spread by oocysts to intermediate hosts, usually rodents and birds. The cycle returns to cats when they eat these infected prey animals.

In 2007, scientists at Stanford University found that the protozoan crowds into a part of the rat brain that usually directs the rat to avoid the smell of cat urine (a natural defense against a domestic rat's major predator). When *Toxoplasma* infects rat brains, the rats lose their fear of cats. Infected rats are then easily eaten by cats, ensuring the continuing *Toxoplasma* life cycle. All other neurological functions in the rat are left intact.

Other vertebrates become a part of this transmission cycle **(figure 19.13*b*).** Herbivorous animals such as cattle and sheep ingest oocysts that persist in the soil of grazing areas and then develop pseudocysts in their muscles and other organs. Carnivores such as canines are infected by eating pseudocysts in the tissues of carrier animals.

Humans appear to be constantly exposed to the pathogen. The rate of prior infections, as detected through serological tests, can be as high as 90% in some populations. Many cases are caused by ingesting pseudocysts in contaminated meats. A common source is raw or undercooked meat. The grooming habits of cats spread fecal oocysts on their body surfaces, and unhygienic handling of them presents an opportunity to ingest oocysts. Infection can also occur when oocysts are inhaled in air or dust contaminated with cat droppings and when tachyzoites cross the placenta to the fetus.

Culture and Diagnosis This infection can be differentiated from viral encephalitides by means of serological tests that detect antitoxoplasma antibodies, especially those for IgM, which appears early in infection. Disease can also be diagnosed by culture and histological analysis.

Prevention and Treatment The most effective drugs are pyrimethamine and sulfadiazine alone or in combination.

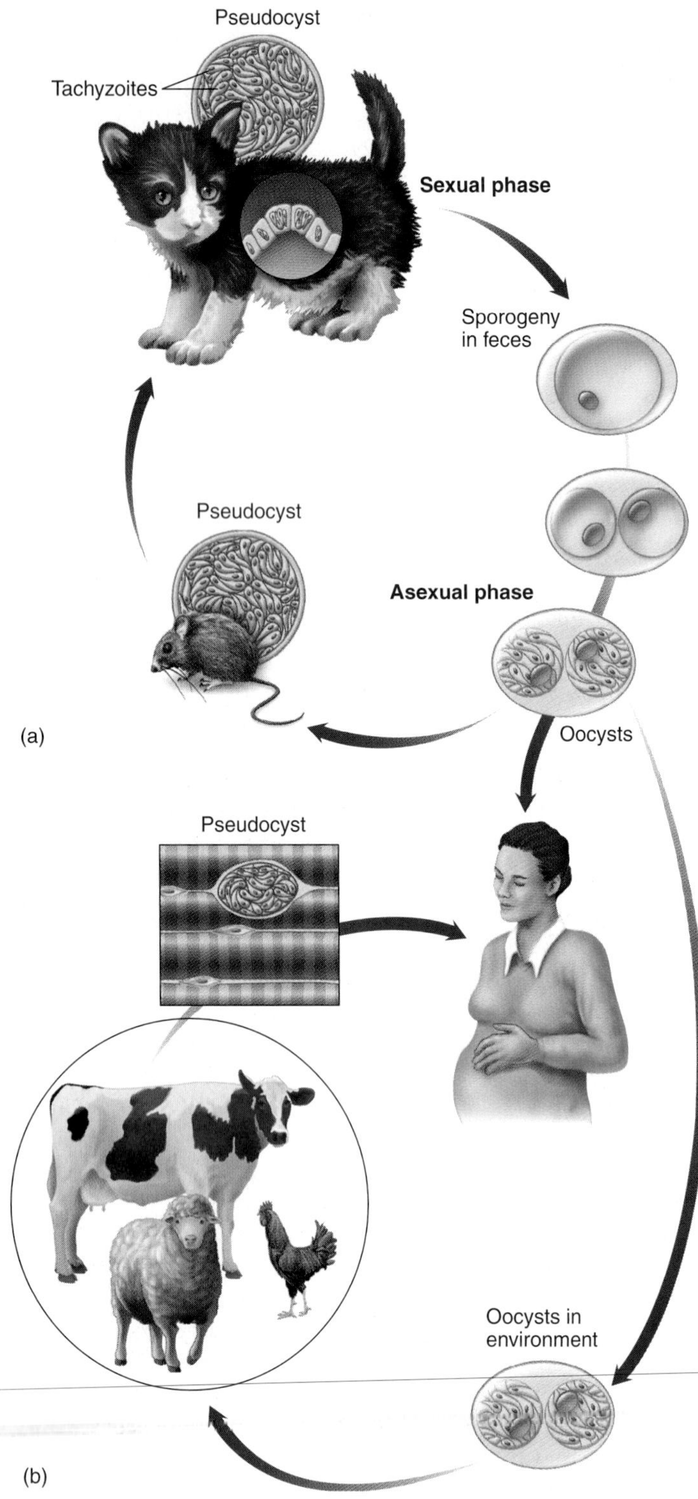

Figure 19.13 The life cycle and morphological forms of *Toxoplasma gondii.*
(a) The cycle in cats and their prey. **(b)** The cycle in other animal hosts. The zoonosis has a large animal reservoir (domestic and wild) that becomes infected through contact with oocysts in the soil. Humans can be infected through contact with cats or ingestion of pseudocysts in animal flesh. Infection in pregnant women is a serious complication because of the potential damage to the fetus.

Because these drugs do not destroy the cyst stage, they must be given for long periods to prevent recurrent infection.

In view of the fact that the oocysts are so widespread and resistant, hygiene is of paramount importance in controlling toxoplasmosis. There is no such thing as a safe form of raw meat, even salted or spiced. Adequate cooking or freezing below −20°C destroys both oocysts and tissue cysts. Oocysts can also be avoided by washing the hands after handling cats or soil possibly contaminated with cat feces, especially sandboxes and litter boxes. Pregnant women should be especially attentive to these rules and should never clean the cat's litter box.

Measles Virus: *Subacute Sclerosing Panencephalitis*

Subacute sclerosing panencephalitis (SSPE) is sometimes called a "slow virus infection." It occurs years after an initial measles episode and is different from immune-mediated postinfectious encephalitis, described earlier. SSPE seems to be caused by direct viral invasion of neural tissue. It is not clear what factors lead to persistence of the virus in some people. See chapter 18 for more details. SSPE's important features are listed in **Checkpoint 19.5.**

Prions

As you read in chapter 6, prions are *proteinaceous infectious particles* containing, apparently, no genetic material. They are known to cause diseases called **transmissible spongiform encephalopathies (TSEs),** neurodegenerative diseases with long incubation periods but rapid progressions once they begin. The human TSEs are **Creutzfeldt-Jakob disease (CJD),** Gerstmann-Strussler-Scheinker disease, and fatal familial insomnia. TSEs are also found in animals and include a disease called **scrapie** in sheep and goats, transmissible mink encephalopathy, and bovine spongiform encephalopathy (BSE). This last disease is commonly known as mad cow disease and has been in the headlines in recent years due to its apparent link to a variant form of CJD human disease in Great Britain.

Signs and Symptoms of CJD Symptoms of CJD include altered behavior, dementia, memory loss, impaired senses, delirium, and premature senility. Uncontrollable muscle contractions continue until death, which usually occurs within 1 year of diagnosis.

Causative Agent of CJD The transmissible agent in CJD is a prion. In some forms of the disease, a normal host protein (called PrP) found in mammalian brains has undergone a mutation that alters its structure. Once this happens, the abnormal PrP itself becomes catalytic and able to spontaneously convert other normal human PrP proteins into the abnormal form. This becomes a self-propagating chain reaction that creates a massive accumulation of altered PrP, leading to plaques, spongiform damage (that is, holes in the brain), and severe loss of brain function.

Using the term *transmissible agent* may be a bit misleading, however, as scientists believe that some cases of CJD

CHECKPOINT 19.5 Subacute Encephalitis			
Causative Organism(s)	*Toxoplasma gondii*	Subacute sclerosing panencephalitis	Prions
Most Common Modes of Transmission	Vehicle (meat) or fecal-oral	Persistence of measles virus	CJD = direct/parenteral contact with infected tissue; or inherited vCJD = vehicle (meat, parenteral)
Virulence Factors	Intracellular growth	Cell fusion, evasion of immune system	Avoidance of host immune response
Culture/Diagnosis	Serological detection of IgM, culture, histology	EEGs, MRI, serology (Ab versus measles virus)	Biopsy, image of brain
Prevention	Personal hygiene, food hygiene	None	Avoiding tissue
Treatment	Pyrimethamine and/or sulfadiazine	None	None
Distinctive Features	Subacute, slower development of disease	History of measles	Long incubation period; fast progression once it begins

arise through genetic mutation of the PrP gene, which can be a heritable trait. So it seems that although one can acquire a defective PrP protein via transmission, one can also have an altered PrP gene passed on through heredity. It is thought that 10% to 15% of CJD cases are inherited in this way. They are termed familial CJD. Another form of CJD is termed sporadic CJD, which is the most mysterious and the most common. Up to 85% of CJD cases are of this type.

Prions are incredibly hardy "pathogens." They are highly resistant to chemicals, radiation, and heat. They can withstand prolonged autoclaving.

Pathogenesis and Virulence Factors Autopsies of the brain of CJD patients reveal spongiform lesions as well as tangled protein fibers (neurofibrillary tangles) and enlarged astroglial cells **(figure 19.14)**. These changes affect the gray matter of the CNS and seem to be caused by the massive accumulation of altered PrP, which may be toxic to neurons. The altered PrPs apparently stimulate no host immune response.

Transmission and Epidemiology CJD is not a communicable disease, in that ordinary contact with infected people will not allow transmission of the disease. Direct or indirect contact with infected brain tissue or cerebrospinal fluid has been thought to be necessary for prion transmission. The tonsils also harbor large amounts of the altered PrP, a fact that has been used in diagnosis. It also raises the possibility that oral fluids can transmit the disease, but this has not been proven. Familial CJD and sporadic CJD are most common in elderly people.

In the late 1990s, it became apparent that humans were contracting a variant form of CJD (vCJD) after ingesting

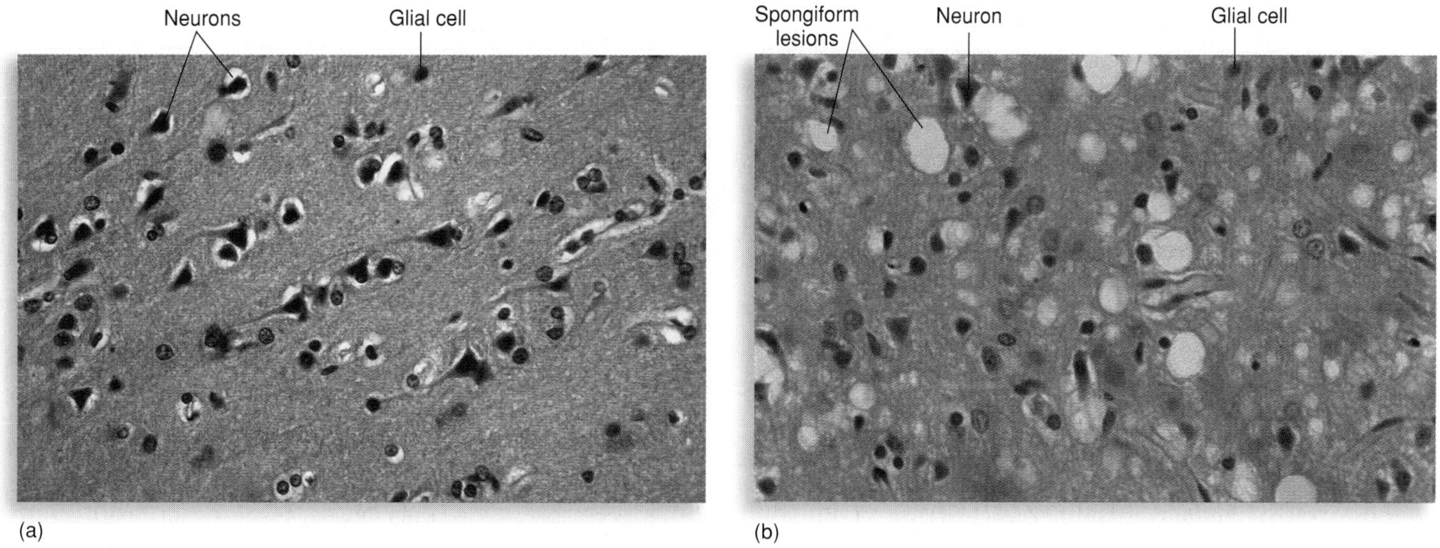

Figure 19.14 The microscopic effects of spongiform encephalopathy.
(a) Normal cerebral cortex section, showing neurons and glial cells. **(b)** Section c cortex in CJD patient shows numerous round holes, producing a "spongy" appearance. This destroys brain architecture and causes massive loss of neurons and glial cells.

meat from cattle that had been afflicted by bovine spongiform encephalopathy. Presumably meat products had been contaminated with fluid or tissues infected with the prion, although the exact food responsible for the transmission has not been pinpointed. When experimenters purposely infect cattle with prions, they subsequently detect the agent in the retina, dorsal root ganglia, parts of the digestive tract, and the bone marrow of the animals. Even so, the risk of contracting vCJD from the ingestion of meat is extremely small, even in countries such as Great Britain, where a significant number of livestock have been found to have BSE. There the risk of infection is estimated to be one case per 10 billion meat servings. In May 2003, the first North American case of mad cow disease appeared in Canada; in December of that year, a single cow in the United States was found to have BSE. As of spring of 2002, a total of 125 cases of vCJD had occurred worldwide. The median age at death of patients with vCJD is 28 years. In contrast, the median age at death of patients with other forms of CJD is 68 years.

Health care professionals should be aware of the possibility of CJD in patients, especially when surgical procedures are performed, as cases have been reported of transmission of CJD via contaminated surgical instruments. Due to the heat and chemical resistance of prions, normal disinfection and sterilization procedures are usually not sufficient to eliminate them from instruments and surfaces. The latest CDC guidelines for handling of CJD patients in a health care environment should be consulted. CJD has also been transmitted through corneal grafts and administration of contaminated human growth hormone. In 2003, a British patient died of CJD after receiving a blood transfusion in 1996 from a donor who had CJD. Experiments suggest that vCJD seems to be more transmissible through blood than classic CJD. For that reason, blood donation programs screen for possible exposure to BSE by asking about travel and residence history.

Culture and Diagnosis It is very difficult to diagnose CJD. Definitive diagnosis requires examination of biopsied brain or nervous tissue, and this procedure is usually considered too risky because of both the trauma induced in the patient and the undesirability of contaminating surgical instruments and operating rooms. Electroencephalograms and magnetic resonance imaging can provide important clues. A new test to distinguish abnormally folded proteins from correctly folded proteins may yield improved diagnostics for cattle very soon and within a few years for humans. The tests can only be performed after death, however.

Prevention and/or Treatment Prevention of this disease relies on avoiding infected tissues. Avoiding vCJD entails not ingesting tainted meats. No known treatment exists for either form of CJD, and patients inevitably die. Medical intervention focuses on easing symptoms and making the patient as comfortable as possible (see Checkpoint 19.5).

CASE FILE 19 *WRAP-UP*

BSE and vCJD are caused by prions, which have proven to be highly resistant to normal sterilization and disinfection procedures. Ideally vCJD-contaminated instruments should be discarded after use, but this is not always practical. The Centers for Disease Control and Prevention currently recommend some variation of autoclaving the instruments while immersed in 1N sodium hydroxide or a bleach solution. Again, these solutions can be damaging to surgical instruments. Problems arise, of course, if the patient being operated on is not suspected of having vCJD but is in fact infected.

Scientists in England feared the worst when BSE was discovered in cattle there. They estimate that nearly every person living in Britain has ingested meat from BSE-infected cows. (The cattle population in Britain now is virtually free of BSE.) Knowing this, it was feared that thousands if not tens of thousands of Britons would acquire vCJD. In reality, fewer than 200 cases occurred, leading scientists to believe that transmission via ingestion of meat from infected cattle had a very low probability. However, the fact that three of the eight people who were tested after exposure to tainted blood were infected suggests a much grimmer scenario for transfusion transmissions. British officials are taking aggressive steps to ensure the safety of their blood supply, including banning people who have ever received transfusions from donating blood.

Rabies

Rabies is a slow, progressive zoonotic disease characterized by a fatal encephalitis. It is so distinctive in its pathogenesis and its symptoms that we discuss it separately from the other encephalitides. It is distributed nearly worldwide, except for 34 countries that have remained rabies-free by practicing rigorous animal control.

Signs and Symptoms

The average incubation period of rabies is 1 to 2 months or more, depending upon the wound site, its severity, and the inoculation dose. The incubation period is shorter in facial, scalp, or neck wounds because of closer proximity to the brain. The prodromal phase begins with fever, nausea, vomiting, headache, fatigue, and other nonspecific symptoms.

In the form of rabies termed *furious,* the first acute signs of neurological involvement are periods of agitation, disorientation, seizures, and twitching. Spasms in the neck and pharyngeal muscles lead to severe pain upon swallowing, leading to a symptom known as *hydrophobia* (fear of water). Throughout this phase, the patient is fully coherent and alert. With the *dumb* form of rabies, a patient is not hyperactive but is paralyzed, disoriented, and stuporous. Ultimately, both forms progress to the coma phase, resulting in death from cardiac or respiratory arrest. Until recently, humans were never known to survive rabies. But a handful of patients have recovered in recent years after receiving intensive, long-term treatment.

INSIGHT 19.3 Medical

Cheating Death

For decades, it has been assumed that people contracting rabies who were not administered the vaccine before symptoms appeared were destined for certain death. But the case of a 15-year-old girl from Wisconsin in 2004 changed that. Jeanna Giese was bitten by a rabid bat during church but didn't seek treatment. If she had reported a bat bite, she would have certainly been given the rabies vaccination regimen. A month later, she began to show symptoms and was hospitalized 2 days later. Doctors told her parents to prepare for her death. But then they decided to try something that had never been tried before: they decided to put her into a coma and to administer a mix of antiviral drugs. By doing this, they hoped to let her live long enough so that her immune system could conquer the infection. No one was more surprised than these doctors when it worked. In late December, Jeanna was released from the hospital and she returned to school in March. In the fall of 2005, she had some residual neurological effects but was able to obtain her driver's license and was considered a miracle patient. Doctors were initially optimistic that use of the relatively common drugs they used in this case could revolutionize rabies treatment, especially in the developing world where most human rabies cases occur, but so far several attempts to repeat Jeanna's miracle cure have been unsuccessful.

Causative Agent

The rabies virus is in the family Rhabdoviridiae, genus *Lyssavirus*. The particles of this virus have a distinctive bulletlike appearance, round on one end and flat on the other. Additional features are a helical nucleocapsid and spikes that protrude through the envelope **(figure 19.15).** The family contains about 60 different viruses, but only the rabies *Lyssavirus* infects humans.

Pathogenesis and Virulence Factors

Infection with rabies virus typically begins when an infected animal's saliva enters a puncture site. The virus occasionally is inhaled or inoculated through the membranes of the eye. The rabies virus remains up to a week at the trauma site, where it multiplies. The virus then gradually enters nerve endings and advances toward the ganglia, spinal cord, and brain. Viral multiplication throughout the brain is eventually followed by migration to such diverse sites as the eye, heart, skin, and oral cavity. The infection cycle is completed when the virus replicates in the salivary gland and is shed into the saliva. Clinical rabies proceeds through several distinct stages that almost inevitably end in death, unless vaccination is performed before symptoms begin **(Insight 19.3).**

Scientists have discovered that virulence is associated with an envelope glycoprotein that seems to give the virus

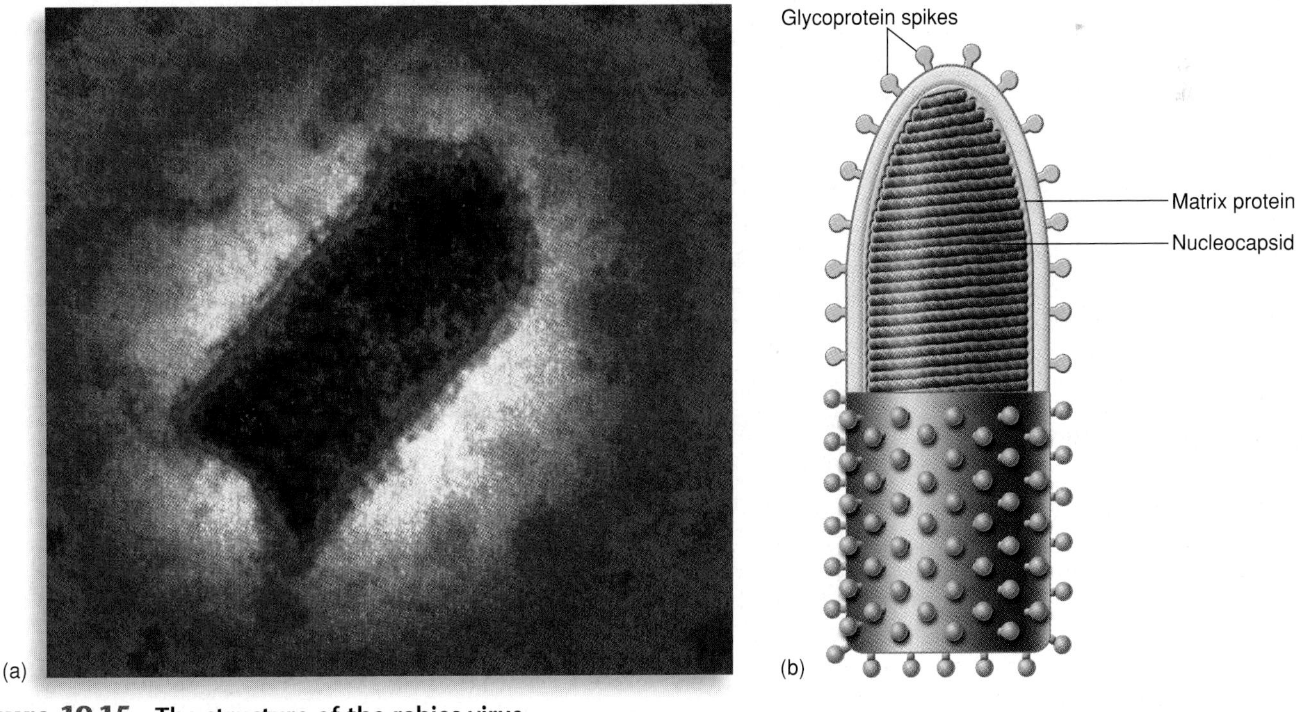

Figure 19.15 The structure of the rabies virus.
(a) Color-enhanced virion shows internal serrations, which represent the tightly coiled nucleocapsid. **(b)** A schematic model of the virus, showing its major features.

its ability to spread in the CNS and to invade certain types of neural cells.

Transmission and Epidemiology

The primary reservoirs of the virus are wild mammals such as canines, skunks, raccoons, badgers, cats, and bats that can spread the infection to domestic dogs and cats. Both wild and domestic mammals can spread the disease to humans through bites, scratches, and inhalation of droplets. The annual worldwide total for human rabies is estimated at about 30,000 cases, but only a tiny number of these cases occur in the United States. Most U.S. cases of rabies occur in wild animals (about 6,000 to 7,000 cases per year), while dog rabies has declined.

The epidemiology of animal rabies in the United States varies. The most common wild animal reservoir host has changed from foxes to skunks to raccoons. Regional differences in the dominant reservoir also occur. Rats, skunks, and bobcats are the most common carriers of rabies in California, raccoons are the predominant carriers in the East, and coyotes dominate in Texas.

In 2004, the first cases of rabies in recipients of donated organs occurred. The lungs, kidneys, and liver of a man were donated to four patients; three of them died of rabies (the fourth died of surgical complications). The virus has previously been transmitted through cornea transplants.

Culture and Diagnosis

When symptoms appear after an attack by a rabid animal, the disease is readily diagnosed. But the diagnosis can be obscured when contact with an infected animal is not clearly defined or when symptoms are absent or delayed. Anxiety, agitation, and depression can pose as a psychoneurosis; muscle spasms resemble tetanus; and encephalitis with convulsions and paralysis mimics a number of other viral infections. Often the disease is diagnosed only at autopsy. The direct fluorescent antibody test is the standard for postmortem identification **(figure 19.16).**

Diagnosis before death requires multiple tests. Reverse transcription PCR is used with saliva samples but must be accompanied by detection of antibodies to the virus in serum or spinal fluid. Skin biopsies are also used.

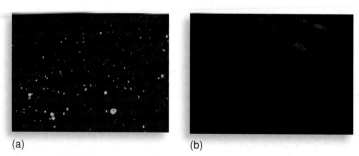

(a) (b)

Figure 19.16 Direct fluorescent antibody test for rabies. Brain tissue is smeared on a slide and fluorescent antibodies to the rabies virus are incubated with it. **(a)** If the rabies antigen is present, fluorescent antibody will be detected after washing. **(b)** Brain tissue negative for rabies.

Prevention and Treatment

A bite from a wild or stray animal demands assessment of the animal, meticulous care of the wound, and a specific treatment regimen. A wild mammal, especially a skunk, raccoon, fox, or coyote that bites without provocation, is presumed to be rabid, and therapy is immediately begun. If the animal is captured, brain samples and other tissue are examined for verification of rabies. Healthy domestic animals are observed closely for signs of disease and sometimes quarantined. Preventive therapy is initiated if any signs of rabies appear.

After an animal bite, the wound should be scrupulously washed with soap or detergent and water, followed by debridement and application of an antiseptic such as alcohol or peroxide. Rabies is one of the few infectious diseases for which a combination of passive and active postexposure immunization is indicated (and successful). Initially the wound is infused with human rabies immune globulin (HRIG) to impede the spread of the virus, and globulin is also injected intramuscularly to provide immediate systemic protection. A full course of vaccination is started simultaneously. The current vaccine of choice is the **human diploid cell vaccine (HDCV).** This potent inactivated vaccine is cultured in human embryonic fibroblasts. The routine postexposure vaccination entails intramuscular or intradermal injection on the 1st, 3rd, 7th, 14th, 28th, and 60th days, with two additional boosters. High-risk groups such as veterinarians, animal handlers, laboratory personnel, and travelers should receive three doses to protect against possible exposure. A DNA vaccine for rabies is in development.

Control measures such as vaccination of domestic animals, elimination of strays, and strict quarantine practices have helped reduce the virus reservoir. In recent years, the United States and other countries have utilized a live oral vaccine made with a vaccinia virus that carries the gene for the rabies virus surface antigen. The vaccine has been incorporated into bait (sometimes peanut butter sandwiches!) placed in the habitats of wild reservoir species such as skunks and raccoons.

CHECKPOINT 19.6	Rabies
Causative Organism(s)	Rabies virus
Most Common Modes of Transmission	Parenteral (bite trauma), droplet contact
Virulence Factors	Envelope glycoprotein
Culture/Diagnosis	RT-PCR of saliva; Ab detection of serum or CSF; skin biopsy
Prevention	HDCV—inactivated vaccine
Treatment	Postexposure passive and active immunization

Poliomyelitis

Poliomyelitis (poh″-lee-oh′my″-eh′ly′tis) (polio) is an acute enteroviral infection of the spinal cord that can cause neuromuscular paralysis. Because it often affects small children, in the past it was called infantile paralysis **(Insight 19.4).** No

INSIGHT 19.4 **Historical**

Polio

Canadian member of Parliament Ellen Fairclough led this "Mothers' March " in 1951.

Polio is a disease that in some ways defined the 20th century. Large epidemics of paralytic poliomyelitis started appearing around 1916 in the United States. Waves of epidemics continued throughout the first half of the 1900s. The disease seemed to strike in summer and early fall, and during those times children were cautioned not to drink from public water fountains and were not allowed to have slumber parties, because their parents feared polio so much. It was many years before scientists understood how the virus was transmitted, how it traveled in the body, and why it started causing a devastating form of paralysis in the early 1900s, because it had clearly been around for hundreds, if not thousands, of years.

Scientists eventually discovered that the rise in paralytic cases—dubbed *infantile paralysis* because it affected mainly small children—was probably due to *increased* public sanitation and hygiene. The poliovirus is spread through contaminated vehicles and through the fecal-oral route. During earlier times of poor sanitation, nearly everyone was exposed to the virus at a very early age—before the age of 6 months. Because babies still enjoy passive protection from maternal antibodies during this period, the poliovirus was kept in check. Most cases were confined to the gastrointestinal tract or limited to mild viremias. When these children were later exposed, they were protected by their naturally acquired active immunity.

Once water sources became more pure and the importance of personal hygiene and hand washing was understood, babies often escaped exposure for months or years. When they did encounter the virus, they had to battle it on their own; maternal antibodies were long gone. Many of these infections progressed into the paralytic form of the disease memorialized by so many photographs of children in iron lungs and with leg braces and crutches.

In 1921, a young adult became ill with a disease that was assumed to be polio—and his condition would change the history of the disease forever. The young man was Franklin Delano Roosevelt, and he would go on to become president of the United States in 1932. In 1938, he founded an organization called the National Foundation for Infantile Paralysis, putting his former law partner Basil O'Connor in charge. Every year, the Foundation held lavish fund-raising events on FDR's birthday, called "The Birthday Balls." Another fund-raising event was called the March of Dimes, which began with a nationwide call to send dimes to the president. Later it evolved into an event organized by mothers. On a single night, mothers would "march" through their neighborhoods, going door to door collecting donations from homes who signaled their willingness to contribute by leaving their porch lights on. Hundreds of thousands of dollars were raised this way. The money went toward treatment and support of victims of polio, as well as toward research into a vaccine.

In 1954, massive field trials of an experimental killed virus vaccine developed by Jonas Salk were conducted. When it was deemed to be effective, nationwide vaccination was begun. By 1962, the Salk vaccine had been replaced by Albert Sabin's live attenuated vaccine, which was administered orally. And by 1964 the annual case rate had fallen from a high of 58,000 in 1952 to only 121. The last case of polio in the United States occurred in 1979; the World Health Organization is still trying to eradicate polio from the rest of the world.

Ironically, some modern scientists have suggested that FDR did not in fact have polio but was a victim of Guillain-Barré syndrome (GBS), a neuromuscular syndrome that can be brought on by viral or bacterial infection or by vaccination. Diagnosing either disease was very difficult in the 1920s (indeed, the existence of GBS as a distinct syndrome was still being debated in the 1920s), and FDR's paralysis was attributed to the poliovirus. But for thousands of polio victims who benefited from the president's commitment to their cause—and for a world that benefited from the vaccines—it hardly matters whether it was truly polio or not.

civilization or culture has escaped the devastation of polio. The efforts of a WHO campaign, called National Vaccination Days, have significantly reduced the global incidence of polio. It was the campaign's goal to eradicate all of the remaining wild polioviruses by 2005. In 2003, most cases were confined to a few pockets in Africa, India, and parts of the Middle East. But that year religious leaders in Nigeria urged a boycott of polio vaccinations, claiming that the vaccines were a U.S. plot to infect the world's Muslims with HIV or to make them infertile. Polio cases began to spread from Nigeria to Saudi Arabia and Yemen and Indonesia. Sixteen previously polio-free countries eventually had new cases again.

Signs and Symptoms

Most infections are contained as short-term, mild viremia. Some persons develop mild nonspecific symptoms of fever, headache, nausea, sore throat, and myalgia. If the viremia persists, viruses can be carried to the central nervous system through its blood supply. The virus then spreads along specific pathways in the spinal cord and brain. Being **neurotropic,** the virus infiltrates the motor neurons of the anterior horn of the spinal cord, although it can also attack spinal ganglia, cranial nerves, and motor nuclei. Nonparalytic disease involves the invasion but not the destruction of nervous tissue. It gives rise to muscle pain and spasm, meningeal inflammation, and vague hypersensitivity.

In paralytic disease, invasion of motor neurons causes various degrees of flaccid paralysis over a period of a few hours to several days. Depending on the level of damage to motor neurons, paralysis of the muscles of the legs, abdomen, back, intercostals, diaphragm, pectoral girdle, and bladder can result. In rare cases of **bulbar poliomyelitis,** the brain stem, medulla, or even cranial nerves are affected. This situation leads to loss of control of cardiorespiratory regulatory centers, requiring mechanical respirators. In time, the unused muscles begin to atrophy, growth is slowed, and severe deformities of the trunk and limbs develop. Common sites of deformities are the spine, shoulder, hips, knees, and feet. Because motor function but not sensation is compromised, the crippled limbs are often very painful.

In recent times, a condition called post-polio syndrome (PPS) has been diagnosed in long-term survivors of childhood infection. PPS manifests as a progressive muscle deterioration that develops in about 25% to 50% of patients several decades after their original polio attack.

Causative Agent

The poliovirus is in the family Picornaviridae, genus *Enterovirus*—named for its small (*pico*) size and its RNA core **(figure 19.17).** It is nonenveloped and nonsegmented. The naked capsid of the virus confers chemical stability and resistance to acid, bile, and detergents. By this means, the virus survives the gastric environment and other harsh conditions, which contributes to its ease of transmission.

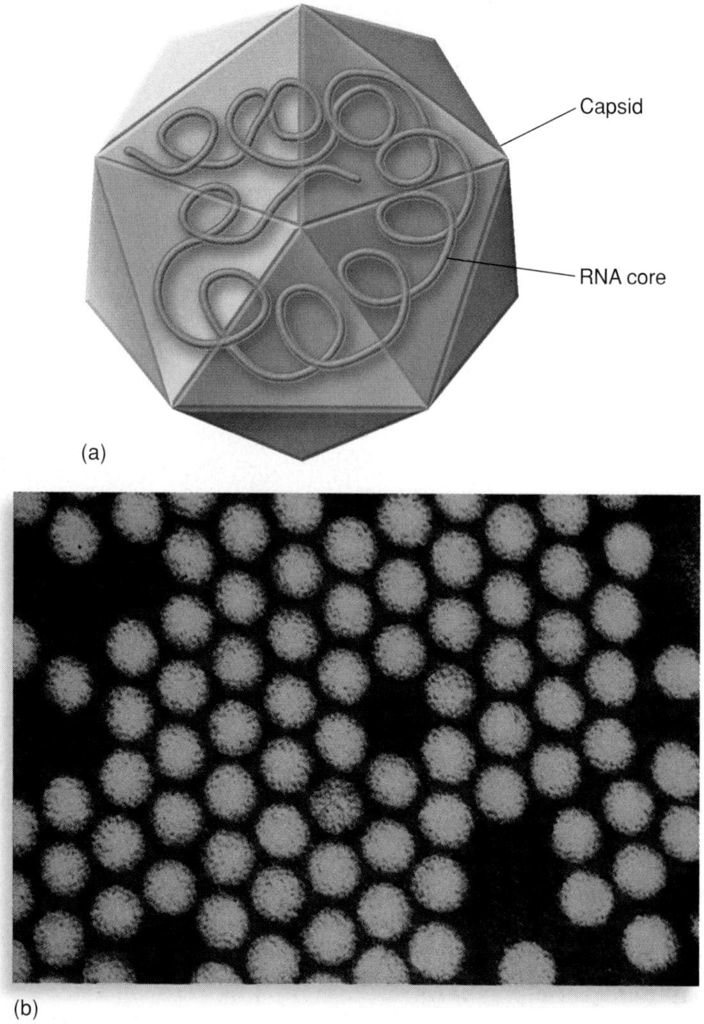

Figure 19.17 Typical structure of a picornavirus. **(a)** A poliovirus, a type of picornavirus that is one of the simplest and smallest viruses (30 nm). It consists of an icosahedral capsid shell around a molecule of RNA. **(b)** A crystalline mass of stacked poliovirus particles in an infected host cell.

Pathogenesis and Virulence Factors

After being ingested, polioviruses adsorb to receptors of mucosal cells in the oropharynx and intestine **(figure 19.18).** Here, they multiply in the mucosal epithelia and lymphoid tissue. Multiplication results in large numbers of viruses being shed into the throat and feces, and some of them leak into the blood. The sequence of events just described then ensues. Depending on the number of viruses in the blood and their duration of stay there, an individual may exhibit no symptoms, mild nonspecific symptoms such as fever or short-term muscle pain, or devastating paralysis **(figure 19.19).** Scientists studying poliovirus virulence focus on components of the virus that allow attachment and penetration of host cells.

Transmission and Epidemiology

Sporadic cases of polio can break out at any time of the year, but its incidence is more pronounced during the summer

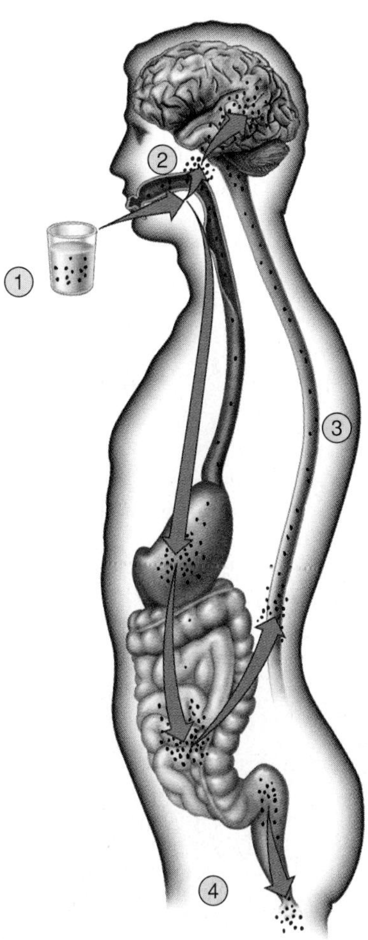

Process Figure 19.18 The stages of infection and pathogenesis of poliomyelitis.
(1) First, the virus is ingested and carried to the throat and intestinal mucosa. (2) The virus then multiplies in the tonsils. Small numbers of viruses escape to the regional lymph nodes and blood. (3) The viruses are further amplified and cross into certain nerve cells of the spinal column and central nervous system. (4) Last, the intestine actively sheds viruses.

and fall. The virus is passed within the population through food, water, hands, objects contaminated with feces, and mechanical vectors. Although the 20th century saw a very large rise in paralytic polio cases, it was also the century during which effective vaccines were developed. The infection was eliminated from the Western Hemisphere in the late 20th century. Progress was reversed in 2003 and 2004

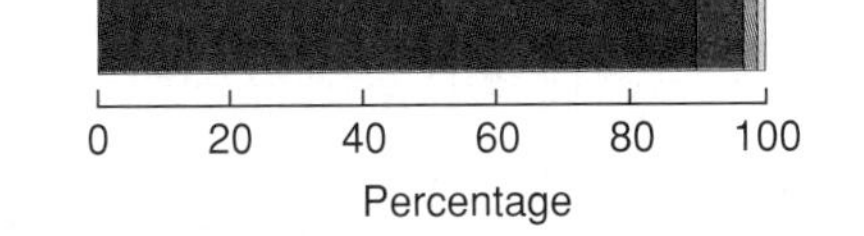

Figure 19.19 Diagrammatic representation of possible outcomes of poliovirus infection.

when 16 countries in Africa previously declared polio-free became reinfected. Worldwide, there were 784 cases in 2003. In 2005, there were 1,940 cases. In most countries, cases are declining again thanks to renewed vaccination programs in these countries. India, which had just 3% of the cases in the world in 2005, had 26% of the cases in 2006. They launched an emergency vaccination campaign in late 2006.

Culture and Diagnosis

Polio is mainly suspected when epidemics of neuromuscular disease occur in the summer in temperate climates. Poliovirus can usually be isolated by inoculating cell cultures with stool or throat washings in the early part of the disease. Viruses are sometimes then subjected to RNA fingerprinting to determine if they are wild strains or vaccine strains. The stage of the patient's infection can also be demonstrated by testing serum samples for the type and amount of antibody.

Prevention and Treatment

Treatment of polio rests largely on alleviating pain and suffering. During the acute phase, muscle spasm, headache, and associated discomfort can be alleviated by pain-relieving drugs. Respiratory failure may require artificial ventilation maintenance. Prompt physical therapy to diminish crippling deformities and to retrain muscles is recommended after the acute febrile phase subsides.

The mainstay of polio prevention is vaccination as early in life as possible, usually in four doses starting at about 2 months of age. Adult candidates for immunization are travelers and members of the armed forces. The two forms of vaccine currently in use are inactivated poliovirus vaccine (IPV), developed by Jonas Salk in 1954, and oral poliovirus vaccine (OPV), developed by Albert Sabin in the 1960s. Both are prepared from animal cell cultures and are trivalent (combinations of the three serotypes of the poliovirus). Both vaccines are effective, but one may be favored over the other under certain circumstances.

For many years, the oral vaccine was used in the United States because it is easily administered by mouth, but it is not free of medical complications. It contains an attenuated virus that can multiply in vaccinated people and be spread to others. In very rare instances, the attenuated virus reverts to a neurovirulent strain that causes disease rather than protects against it. Numerous instances of paralytic polio have occurred among children with hypogammaglobulinemia who have been mistakenly vaccinated. There is a tiny risk (about one case in 4 million) that an unvaccinated family member will acquire infection and disease from a vaccinated child. For these reasons, public health officials have revised their recommendations and now require the use of IPV for all childhood vaccinations. OPV is no longer generally available in the United States, although it is still used in the eradication campaigns in developing countries. When the WHO goal of polio eradication is achieved, it may be possible to discontinue polio vaccination, as has been done for smallpox.

CHECKPOINT 19.7	Poliomyelitis
Causative Organism(s)	Poliovirus
Most Common Modes of Transmission	Fecal-oral, vehicle
Virulence Factors	Attachment mechanisms
Culture/Diagnosis	Viral culture, serology
Prevention	Live attenuated (developing world) or inactivated vaccine (developed world)
Treatment	None, palliative, supportive

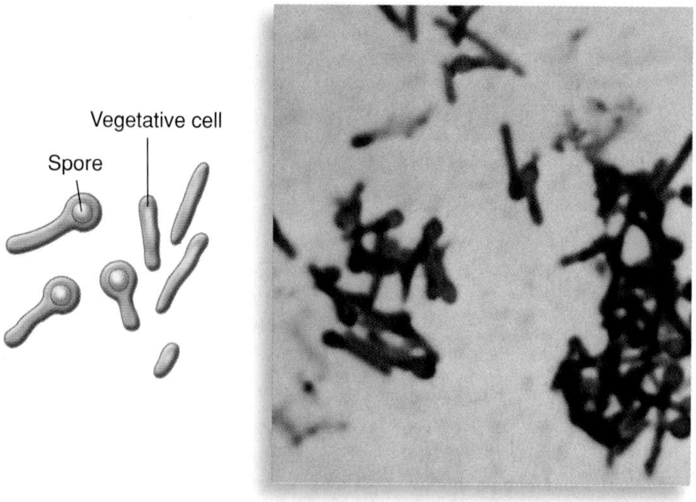

Figure 19.20 C. tetani.
Its typical tennis racket morphology is created by terminal spores that swell the end of the cell (170×).

Tetanus

Tetanus is a neuromuscular disease whose alternate name, lockjaw, refers to an early effect of the disease on the jaw muscle. The etiologic agent, *Clostridium tetani,* is a common resident of cultivated soil and the gastrointestinal tracts of animals. It is a gram-positive, spore-forming rod. The endospores it produces often swell the vegetative cell **(figure 19.20).** Spores are produced only under anaerobic conditions.

Signs and Symptoms

C. tetani releases a powerful neurotoxin, **tetanospasmin,** that binds to target sites on peripheral motor neurons, spinal cord and brain, and in the sympathetic nervous system. The toxin acts by blocking the inhibition of muscle contraction. Without inhibition of contraction, the muscles contract uncontrollably, resulting in spastic paralysis. The first symptoms are clenching of the jaw, followed in succession by extreme arching of the back, flexion of the arms, and extension of the legs **(figure 19.21).** Lockjaw confers the bizarre appearance of *risus sardonicus* (sardonic grin), which looks eerily as though the person is smiling **(figure 19.22).** Death most often occurs due to paralysis of the respiratory muscles and respiratory arrest.

Pathogenesis and Virulence Factors

The mere presence of spores in a wound is not sufficient to initiate infection because the bacterium is unable to invade damaged tissues readily. It is also a strict anaerobe, and the spores cannot become established unless tissues at the site of the wound are necrotic and poorly supplied with blood, conditions that favor germination.

As the vegetative cells grow, various metabolic products are released into the infection site, including the tetanospasmin toxin. The toxin spreads to nearby motor nerve endings in the injured tissue, binds to them, and travels via axons to the ventral horns of the spinal cord (see figure 19.22). The toxin blocks the release of neurotransmitter, and only a small amount is required to initiate the symptoms. The incubation period varies from 4 to 10 days, and shorter incubation periods signify a more serious condition.

Figure 19.21   Neonatal tetanus.
Baby with neonatal tetanus, showing spastic paralysis of the paravertebral muscles, which locks the back into a rigid, arched position. Also note the abnormal flexion of the arms and legs.

The muscle contractions are intermittent and extremely painful, and they may be forceful enough to break bones, especially the vertebrae. The fatality rate, ranging from 10% to 70%, is highest in cases involving delayed medical attention, a short incubation time, or head wounds. Full recovery requires a few weeks, and no permanent damage to the muscles usually remains.

Transmission and Epidemiology

Spores usually enter the body through accidental puncture wounds, burns, umbilical stumps, frostbite, and crushed body parts. The incidence of tetanus is low in North America. Most cases occur among geriatric patients and intravenous drug abusers. The incidence of neonatal tetanus—predominantly the result of an infected umbilical stump or circumcision—is higher in cultures that apply dung, ashes, or mud to these sites to arrest bleeding or as a customary ritual. The disease accounts for several hundred thousand infant deaths a year worldwide.

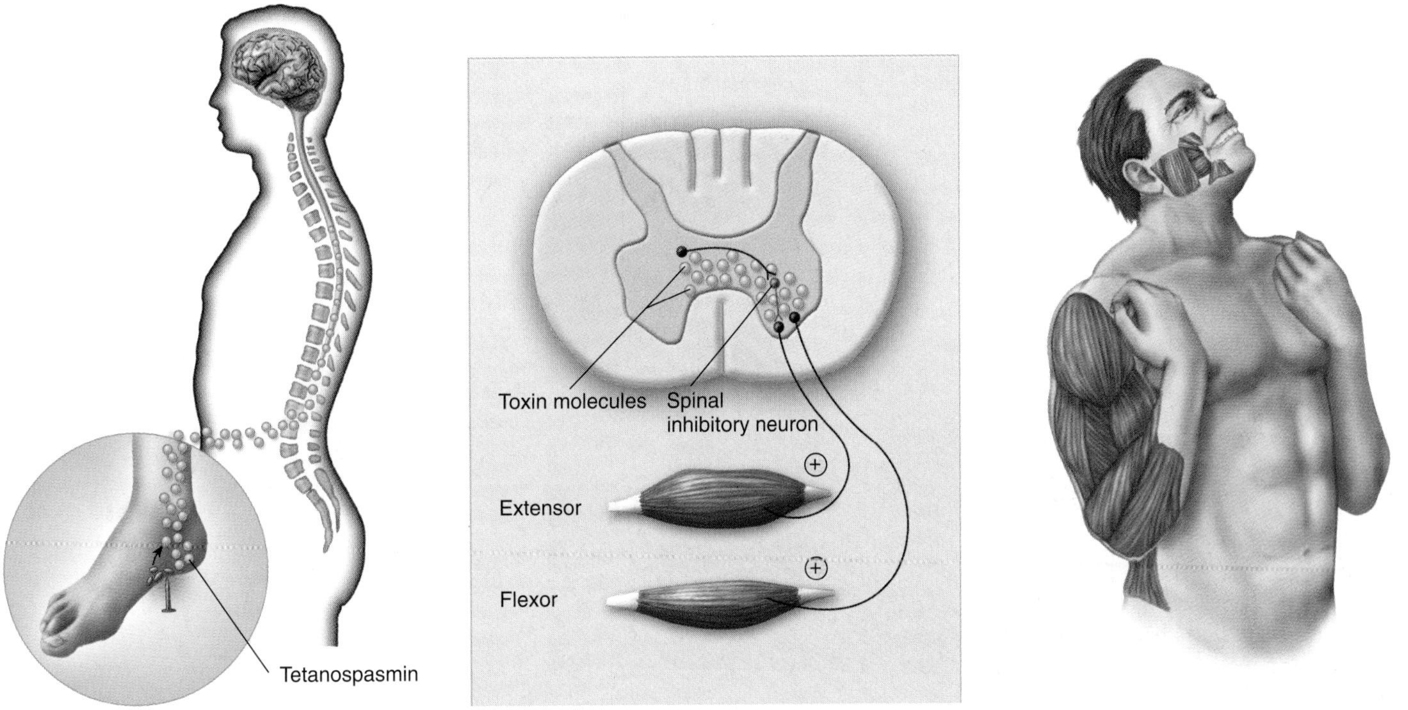

Figure 19.22 The events in tetanus.
(a) After traumatic injury, bacteria infecting the local tissues secrete tetanospasmin, which is absorbed by the peripheral axons and is carried to the target neurons in the spinal column. **(b)** In the spinal cord, the toxin attaches to the junctions of regulatory neurons that inhibit inappropriate contraction. Released from inhibition, the muscles, even opposing members of a muscle group, receive constant stimuli and contract uncontrollably. **(c)** Muscles contract spasmodically, without regard to regulatory mechanisms or conscious control. Note the clenched jaw typical of *risus sardonicus.*

Prevention and Treatment

Tetanus treatment is aimed at deterring the degree of toxemia and infection and maintaining patient homeostasis. A patient with a clinical appearance suggestive of tetanus should immediately receive antitoxin therapy with human tetanus immune globulin (TIG). Although the antitoxin inactivates circulating toxin, it will not counteract the effect of toxin already bound to neurons. Other methods include thoroughly cleansing and removing the afflicted tissue, controlling infection with penicillin or tetracycline, and administering muscle relaxants. The patient may require the assistance of a respirator, and a *tracheostomy*[1] is sometimes performed to prevent respiratory complications such as aspiration pneumonia or lung collapse.

Tetanus is one of the world's most preventable diseases, chiefly because of an effective vaccine containing tetanus toxoid. The recommended vaccination series for 1- to 3-month-old babies consists of three injections given 2 months apart, followed by booster doses about 1 and 4 years later. Children thus immunized probably have protection for 10 years. Additional protection against neonatal tetanus may be achieved by vaccinating pregnant women, whose antibodies will be passed to the fetus. Toxoid should also be given to injured persons who have never been immunized, have not completed the series, or whose last booster was received more than 10 years previously. The vaccine can be given simultaneously with passive TIG immunization to achieve immediate as well as long-term protection.

1. The surgical formation of an air passage by perforation of the trachea.

CHECKPOINT 19.8	Tetanus
Causative Organism(s)	*Clostridium tetani*
Most Common Modes of Transmission	Parenteral, direct contact
Virulence Factors	Tetanospasm exotoxin
Culture/Diagnosis	Symptomatic
Prevention	Tetanus toxoid immunization
Treatment	Combination of passive antitoxin and tetanus toxoid active immunization, supportive

Botulism

Botulism is an *intoxication* (that is, caused by an exotoxin) associated with eating poorly preserved foods, although it can also occur as a true infection. Until recent times, it was relatively common and frequently fatal, but modern techniques of food preservation and medical treatment have reduced both its incidence and its fatality rate. However, botulism is

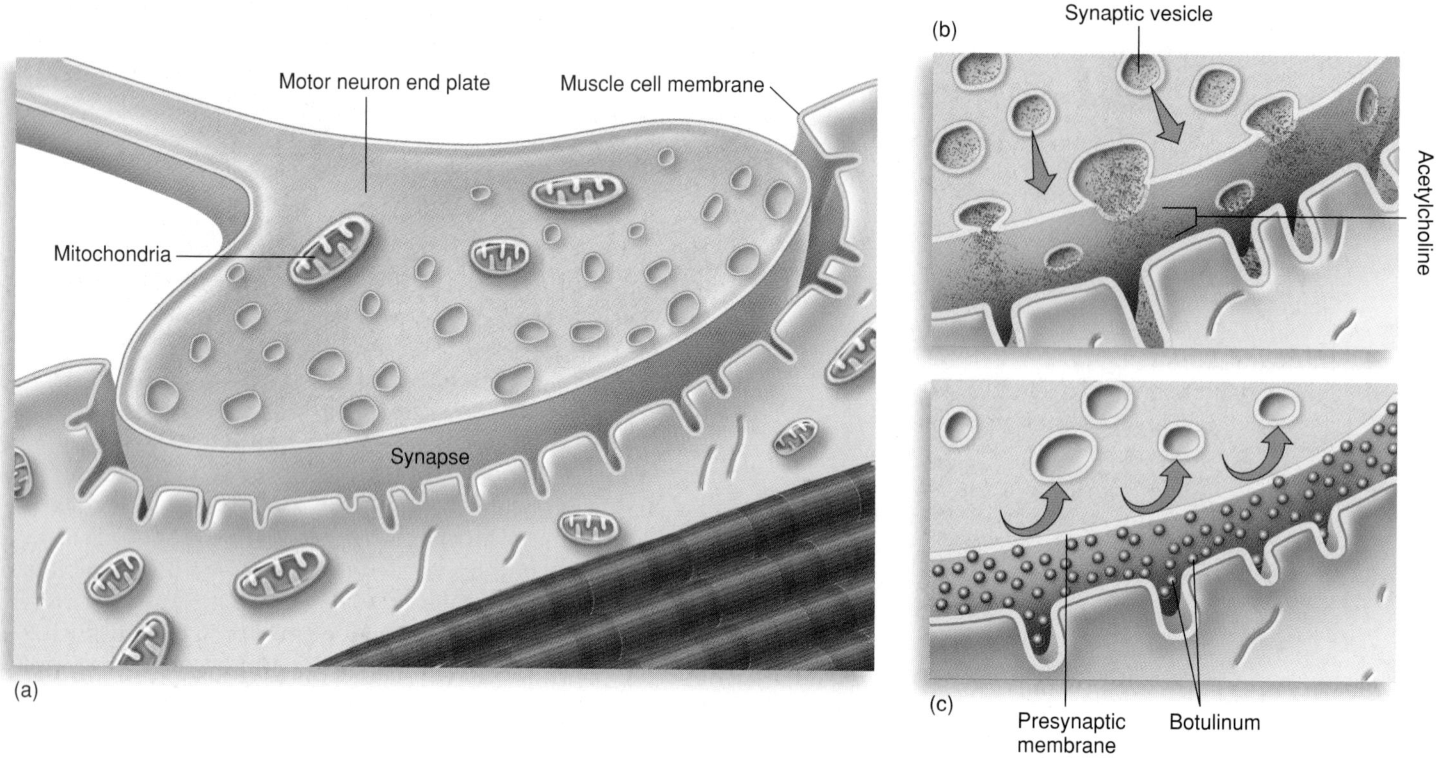

Figure 19.23 The physiological effects of botulism toxin (botulinum).
(a) The relationship between the motor neuron and the muscle at the neuromuscular junction. **(b)** In the normal state, acetylcholine released at the synapse crosses to the muscle and creates an impulse that stimulates muscle contraction. **(c)** In botulism, the toxin enters the motor end plate and attaches to the presynaptic membrane, where it blocks release of the chemical. This prevents impulse transmission and keeps the muscle from contracting.

a common cause of death in livestock that have grazed on contaminated food and in aquatic birds that have eaten decayed vegetation. In the United States, there are on average 110 cases of human botulism a year.

Signs and Symptoms

There are three major forms of botulism, distinguished by their means of transmission and the population they affect. These are **food-borne botulism** (in children and adults), **infant botulism,** and **wound botulism.** Food-borne botulism in children and adults is an intoxication resulting from the ingestion of preformed toxin; the other two types of botulism are infections that are followed by the entrance of an exotoxin called **botulinum** toxin into the bloodstream (that is, toxemia). The symptoms are largely the same in all three forms, however. From the circulatory system, the toxin travels to its principal site of action, the neuromuscular junctions of skeletal muscles **(figure 19.23).** The effect of botulinum is to prevent the release of the neurotransmitter substance, acetylcholine, that initiates the signal for muscle contraction. The usual time before onset of symptoms is 12 to 72 hours, depending on the size of the dose. Neuromuscular symptoms first affect the muscles of the head and include double vision, difficulty in swallowing, and dizziness, but there is no sensory or mental lapse. Although nausea and vomiting can occur at an early stage, they are not common. Later symptoms are descending muscular paralysis and respiratory compromise. In the past, death resulted from respiratory arrest, but mechanical respirators have reduced the fatality rate to about 10%.

Causative Agent

Clostridium botulinum, like *Clostridium tetani,* is a spore-forming anaerobe that does its damage through the release of an exotoxin. *C. botulinum* commonly inhabits soil and water and occasionally the intestinal tract of animals. It is distributed worldwide but occurs most often in the Northern Hemisphere. The species has eight distinctly different types (designated A, B, C_α, C_β, D, E, F, and G), which vary in distribution among animals, regions of the world, and types of exotoxin. Human disease is usually associated with types A, B, E, and F, and animal disease with types A, B, C, D, and E.

Both *C. tetani* and *C. botulinum* produce neurotoxins; but tetanospasmin, the toxin made by *C. tetani,* results in spastic paralysis (uncontrolled muscle contraction). In contrast, botulinum, the *C. botulinum* neurotoxin, results in flaccid paralysis, a loss of ability to contract the muscles.

Pathogenesis and Virulence Factors

As just described, the symptoms are caused entirely by the exotoxin botulinum.

Transmission and Epidemiology of Food-Borne Botulism in Children and Adults

There is a high correlation between cultural dietary preferences and food-borne botulism. In the United States, the disease is often associated with low-acid vegetables (green beans, corn); fruits; and occasionally meats, fish, and dairy products. Most botulism outbreaks occur in home-processed foods, including canned vegetables, smoked meats, and cheese spreads. The demand for prepackaged convenience foods, such as vacuum-packed cooked vegetables and meats, has created a new source of risk, but most commercially canned foods are held to very high standards of preservation and are only rarely a source of botulism.

Several factors in food processing can lead to botulism. Spores are present on the vegetables or meat at the time of gathering and are difficult to remove completely. When contaminated food is put in jars and steamed in a pressure cooker that does not reach reliable pressure and temperature, some spores survive (botulinum spores are highly heat-resistant). At the same time, the pressure is sufficient to evacuate the air and create anaerobic conditions. Storage of the jars at room temperature favors spore germination and vegetative growth, and one of the products of the cell's metabolism is botulinum, the most potent microbial toxin known.

Bacterial growth may not be evident in the appearance of the jar or can or in the food's taste or texture, and only minute amounts of toxin may be present. Botulism is never transmitted person-to-person. Of the more than 100 cases of botulism every year in the United States, about one-quarter are food-borne.

Transmission and Epidemiology of Infant Botulism

Infant botulism was first described in the late 1970s in children between the ages of 2 weeks and 6 months who had ingested spores. It is currently the most common type of botulism in the United States, with approximately 80 cases reported annually. The exact food source is not always known, although raw honey has been implicated in some cases, and the spores are common in dust and soil. Apparently, the immature state of the neonatal intestine and microbial biota allows the spores to gain a foothold, germinate, and give off neurotoxin. As in adults, babies exhibit flaccid paralysis, usually manifested as a weak sucking response, generalized loss of tone (the "floppy baby syndrome"), and respiratory complications. Although adults can also ingest botulinum spores in contaminated vegetables and other foods, the adult intestinal tract normally inhibits this sort of infection.

Transmission and Epidemiology of Wound Botulism

Perhaps three or four cases of wound botulism occur each year in the United States. In this form of the disease, spores enter a wound or puncture, much as in tetanus, but the symptoms are similar to those of food-borne botulism. Increased cases of this form of botulism are being reported in intravenous drug users as a result of needle puncture.

Culture and Diagnosis

Diagnostic standards are slightly different for the three different presentations of botulism. In food-borne botulism, some laboratories attempt to identify the toxin in the offending food. Alternatively, if multiple patients present with the same symptoms after ingesting the same food, a presumptive diagnosis can be made. The cultivation of *C. botulinum* in feces is considered confirmation of the diagnosis since the carrier rate is very low.

In infant botulism, finding the toxin or the organism in the feces confirms the diagnosis. In wound botulism, the toxin should be demonstrated in the serum, or the organism should be grown from the wound. Because minute amounts of the toxin are highly dangerous, laboratory testing should only be performed by experienced personnel. A suspected case of botulism should trigger a phone call to the state health department or the CDC before proceeding with diagnosis or treatment.

Prevention and Treatment

The CDC maintains a supply of type A, B, and E trivalent horse antitoxin, which, when administered early, can prevent the worst outcomes of the disease. Patients are also managed with respiratory and cardiac support systems. Antitoxin therapy is generally not administered to infants with botulism; supportive care is primary. In all cases, hospitalization is required and recovery takes weeks. There is an overall 5% mortality rate.

CHECKPOINT 19.9	Botulism
Causative Organism(s)	*Clostridium botulinum*
Most Common Modes of Transmission	Vehicle (food-borne toxin, airborne organism); direct contact (wound); parenteral (injection)
Virulence Factors	Botulinum exotoxin
Culture/Diagnosis	Culture of organism; demonstration of toxin
Prevention	Food hygiene; toxoid immunization available for laboratory professionals
Treatment	Antitoxin, supportive care

African Sleeping Sickness

This condition is caused by *Trypanosoma brucei,* a member of the protozoan group known as hemoflagellates because of their propensity to live in the blood and tissues of the human host. The disease, also called **trypanosomiasis,** has greatly affected the living conditions of Africans since ancient times.

INSIGHT 19.5 *Discovery*

Botox: No Wrinkles. No Headaches. No Worries?

In 2002, nearly 2 million people paid good money (and lots of it) to have one of the most potent toxins on earth injected into their faces. The toxin, of course, is Botox, short for botulinum toxin, and the story of how these injections came to be the most popular cosmetic procedure in the United States is a fascinating one.

Scientists have long known that death from *Clostridium botulinum* infection results from paralysis of the respiratory muscles. In fact, researchers had even determined that botulinum toxin causes death by interfering with the release of acetylcholine, a neurotransmitter that causes the contraction of skeletal muscles. The trick was finding a practical application for this knowledge.

In 1989 Botox was first approved to treat cross-eyes and uncontrollable blinking, two conditions resulting from the inappropriate contracting of muscles around the eye. Success in this first arena led to Botox treatment for a variety of neurological disorders that cause painful contraction of neck and shoulder muscles. A much wider use of Botox occurred in so-called "off label" uses, as doctors found that injecting facial muscles with the toxin inhibited contraction of these muscles and consequent wrinkling of the overlying skin. The "lunch-hour facelift" went over exactly as most people would imagine, becoming the most popular cosmetic procedure even before winning official FDA approval (which it did in 2002). In a surprise twist, patients undergoing Botox treatment for wrinkles reported fewer headaches, especially migraines. Clinical trials have shown this result to be widespread and reproducible, but the exact mechanism by which Botox works to prevent headaches is unknown. Most recently, a doctor claims to have cured clinical depression in patients, presumably by removing their ability to frown and thereby relieving their depression.

So then, Botox is perfect and we need never have, as George Orwell once said, ". . . the face he (or she) deserves"? Not so fast. In the rush to embrace the admittedly dramatic results of Botox injections, most patients have paid scant attention to potential problems that can arise from the use of a potent paralytic agent. The most common problem arising from Botox treatment is excessive paralysis of facial muscles, resulting from poorly targeted injections. Depending on the site of the injection, results such as drooping eyelids, facial paralysis, slurred speech, and drooling are possible. Even if the treatment works perfectly, the wrinkle-free visage is a result of muscle paralysis, meaning that patients are often unable to move their eyebrows or in some cases to frown or squint.

Finally, Botox is not a permanent solution; as the effects of the toxin wear off, the wrinkles (or headaches, as the case may be) return. Every 4 to 6 months, the treatment must be repeated. This last fact has been a boon to doctors for two reasons. The first is obvious—namely, that a permanent solution to wrinkles would rule out any repeat clientele. The second advantage has to do with malpractice lawsuits. By the time a patient has consulted a lawyer and made it through the legal system, any consequences of a botched treatment may have already worn off.

Today at least 50 million people are at risk, and 30,000 to 40,000 new cases occur each year. It imposes an additional hardship when it attacks domestic and wild mammals.

Signs and Symptoms

Trypanosomiasis affects the lymphatics and areas surrounding blood vessels. Usually a long asymptomatic period precedes onset of symptoms. Symptoms include intermittent fever, enlarged spleen, swollen lymph nodes, and joint pain. There are two variants of the disease, caused by two different subspecies of the protozoan. In both forms, the central nervous system is affected, the initial signs being personality and behavioral changes that progress to lassitude and sleep disturbances. The disease is commonly called *sleeping sickness,* but in fact, uncontrollable sleepiness occurs primarily in the day and is followed by sleeplessness at night. Signs of advancing neurological deterioration are muscular tremors, shuffling gait, slurred speech, seizures, and local paralysis. Death results from coma, secondary infections, or heart damage.

Causative Agent

Trypanosoma brucei is a flagellated protozoan, an obligate parasite that is spread by a blood-sucking insect called the tsetse fly, which serves as its intermediate host. It shares a complicated life cycle with other hemoflagellates. In chapter 5, we first described the trypanosome life cycle using the example of *T. cruzi,* the agent that causes Chagas disease.

Transmission and Epidemiology

The cycle begins when a tsetse fly becomes infected after feeding on an infected reservoir host, such as a wild animal (antelope, pig, lion, hyena), domestic animal (cow, goat), or human **(figure 19.24).** In the fly's gut, the trypanosome multiplies, migrates to the salivary glands, and develops into the infectious stage. When the fly bites a new host, it releases the large, fully formed stage of the parasite into the wound. At this site, the trypanosome multiplies and produces a sore called the *primary chancre.* From there, the pathogen moves into the lymphatics and the blood (see figure 19.24). The trypanosome can also cross the placenta and damage a developing fetus.

Two variants of sleeping sickness are the Gambian (West African) strain, caused by the subspecies *Trypanosoma brucei gambiense,* and the Rhodesian (East African) strain, caused by *T. b. rhodesiense* (see figure 19.24). These geographically isolated types are associated with different ecological niches of the principal tsetse fly vectors. In the West African form, the fly inhabits the dense vegetation along rivers and forests typical of that region, whereas the East African form is adapted to savanna woodlands and lakefront thickets.

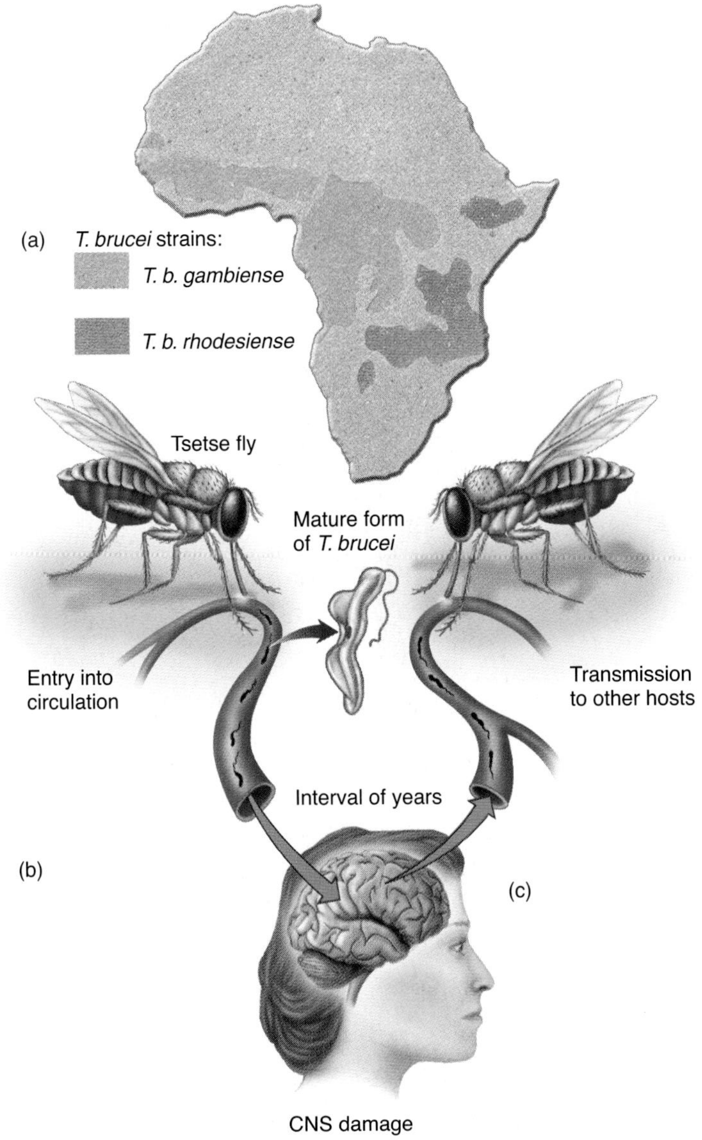

Figure 19.24 The generalized cycle between humans and the tsetse fly vector.
(a) The distribution of African trypanosomiasis. **(b)** The saliva of a fly infected with *T. brucei* inoculates the human bloodstream. The parasite matures and invades various organs. In time, its cumulative effects cause central nervous system (CNS) damage. **(c)** The trypanosome is spread to other hosts through another fly in whose alimentary tract the parasite completes a series of developmental stages.

African sleeping sickness occurs only in Sub-Saharan Africa. Tsetse flies exist elsewhere, and it is not known why they do not support *Trypanosoma* in other regions. In some parts of equatorial Africa, *T. b. gambiense* has recently undergone a resurgence. Epidemics are reported in areas of the Sudan and the Democratic Republic of the Congo, where a significant segment of the population is infected. These outbreaks were partly due to a civil war and an accompanying disruption in medical services.

Pathogenesis and Virulence Factors

The protozoan manages to flourish in the blood even though it stimulates a strong immune response. The immune response is counteracted by an unusual adaptation of the trypanosome. As soon as the host begins manufacturing IgM antibodies to the trypanosome, surviving organisms change the structure of their surface glycoprotein antigens. This change in specificity (sometimes referred to as an *antigenic shift*) renders the existing IgM ineffective, so that the parasite eludes control and multiplies in the blood. The host responds by producing IgM of a new specificity, but the protozoan changes its antigens again. The host eventually becomes exhausted and overwhelmed by repeated efforts to catch up with this trypanosome masquerade. This cycle has tremendous impact on the pathology and control of the disease. The presence of the trypanosome in the blood and the severity of symptoms follow a wavelike pattern.

Culture and Diagnosis

Sleeping sickness may be suspected if the patient has been bitten by a distinctive fly while living or traveling in an endemic area. Trypanosomes are readily demonstrated in blood smears, as well as in spinal fluid or lymph nodes.

Prevention and Treatment

Control of trypanosomiasis in western Africa, where humans are the main reservoir hosts, involves eliminating tsetse flies by applying insecticides, trapping flies, or destroying the shelter and breeding sites. In eastern regions, where cattle herds and large wildlife populations are reservoir hosts, control is less practical because large mammals are the hosts, and flies are less concentrated in specific sites. The antigenic shifting practiced by the trypanosome makes the development of a vaccine very difficult.

Chemotherapy is most successful if administered prior to nervous system involvement. Two different drugs are available for the early stages of the disease. Suramin works against *T. b. rhodesiense,* and pentamidine is used for *T. b. gambiense.* Brain infection must be treated with drugs that can cross the blood-brain barrier. One of these is a highly toxic arsenic-based drug called melarsoprol. It causes nervous system symptoms itself, sometimes permanent, and is itself fatal in a small percentage of cases. An alternative was developed in 1990, but the company stopped making it and gave the license to the World Health Organization. The WHO is seeking a new company to make the drug.

CHECKPOINT 19.10	African Sleeping Sickness
Causative Organism(s)	*Trypanosoma brucei* subspecies *gambiense* or *rhodesiense*
Most Common Modes of Transmission	Vector, vertical
Virulence Factors	Immune evasion by antigen shifting
Culture/Diagnosis	Microscopic examination of blood, CSF
Prevention	Vector control
Treatment	Suramin or pentamidine (early), melarsoprol (late)

SUMMING UP

Taxonomic Organization **Microorganisms Causing Disease in the Nervous System**

Microorganism	Disease	Chapter Location
Gram-positive endospore-forming bacteria		
Clostridium botulinum	Botulism	Botulism, p. 597
Clostridium tetani	Tetanus	Tetanus, p. 596
Gram-positive bacteria		
Streptococcus agalactiae	Neonatal meningitis	Neonatal meningitis, p. 581
Streptococcus pneumoniae	Meningitis	Meningitis, p. 576
Listeria monocytogenes	Meningitis, neonatal meningitis	Meningitis, p. 577 Neonatal meningitis, p. 581
Gram-negative bacteria		
Escherichia coli	Neonatal meningitis	Neonatal meningitis, p. 581
Haemophilus influenzae	Meningitis	Meningitis, p. 577
Neisseria meningitidis	Meningococcal meningitis	Meningitis, p. 575
DNA viruses		
Herpes simplex virus 1 and 2	Encephalitis	Encephalitis, p. 586
JC virus	Progressive multifocal leukoencephalopathy	Encephalitis, p. 586
RNA viruses		
Arboviruses		
Western equine encephalitis virus, Eastern equine encephalitis virus, California encephalitis virus (California and LaCrosse strains), St. Louis encephalitis virus, West Nile virus	Encephalitis	Encephalitis, p. 584
Measles virus	Subacute sclerosing panencephalitis	Subacute encephalitis, p. 588
Poliovirus	Poliomyelitis	Poliomyelitis, p. 592
Rabies virus	Rabies	Rabies, p. 590
Fungi		
Cryptococcus neoformans	Meningitis	Meningitis, p. 578
Coccidioides immitis	Meningitis	Meningitis, p. 579
Prions		
Creutzfeldt-Jakob prion	Creutzfeldt-Jakob disease	Subacute encephalitis, p. 588
Protozoa		
Acanthamoeba	Meningoencephalitis	Meningoencephalitis, p. 584
Naegleria fowleri	Meningoencephalitis	Meningoencephalitis, p. 583
Toxoplasma gondii	Subacute encephalitis	Subacute encephalitis, p. 586
Trypanosoma brucei subspecies *gambiense* and *rhodesiense*	African sleeping sickness	African sleeping sickness, p. 599

INFECTIOUS DISEASES AFFECTING
The Nervous System

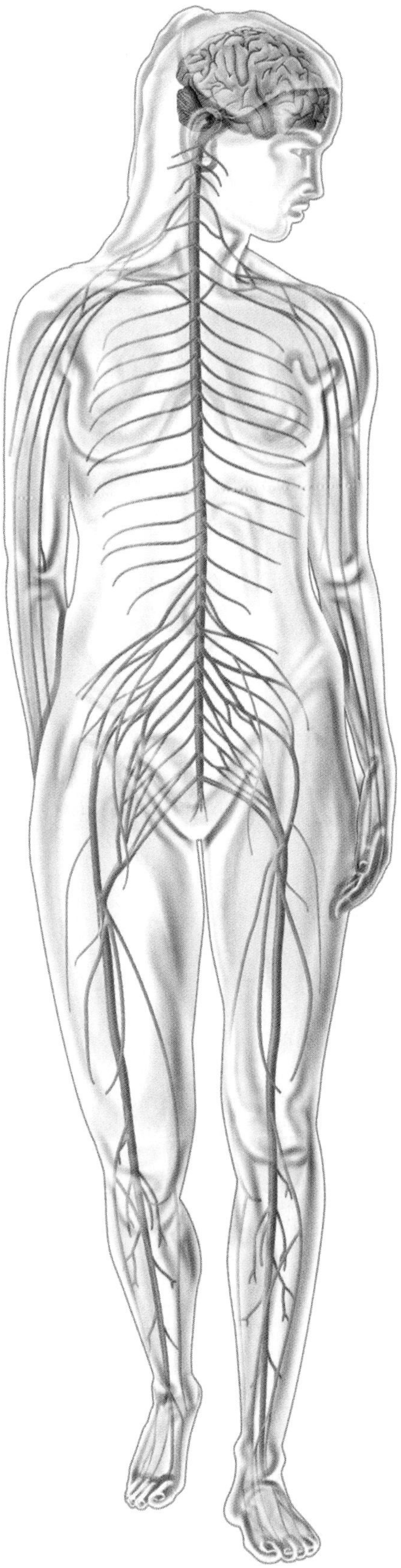

Encephalitis
Arboviruses
Herpes simplex virus 1 or 2
JC virus

Subacute Encephalitis
Toxoplasma gondii
Measles virus
Prions

Rabies
Rabies virus

Tetanus
Clostridium tetani

African Sleeping Sickness
Trypanosoma brucei

Creutzfeldt-Jakob Disease
Prion

Meningoencephalitis
Naegleria fowleri
Acanthamoeba

Meningitis
Neisseria meningitidis
Streptococcus pneumoniae
Haemophilus influenzae
Listeria monocytogenes
Cryptococcus neoformans
Coccidioides immitis
Various viruses

Neonatal Meningitis
Streptococcus agalactiae
Escherichia coli
Listeria monocytogenes

Polio
Poliovirus

Botulism
Clostridium botulinum

- Bacteria
- Viruses
- Protozoa
- Fungi
- Prions

System Summary Figure 19.25

Chapter Summary with Key Terms

19.1 The Nervous System and Its Defenses

A. The nervous system has two parts: the central nervous system (the brain and spinal cord), and the peripheral nervous system (spinal and cranial nerves). The soft tissue of the brain and spinal cord is encased within a tough casing of three membranes called the **meninges.** The subarachnoid space (under the arachnoid mater) is filled with a clear serumlike fluid called cerebrospinal fluid (CSF). The meninges are a common site of infection, and microorganisms can often be found in the CSF when meningeal infection **(meningitis)** occurs.

B. The nervous system is protected by the **blood-brain barrier,** which limits the passage of substances from the bloodstream to the brain.

19.2 Normal Biota of the Nervous System

There is no normal biota in either the CNS or PNS.

19.3 Nervous System Diseases Caused by Microorganisms

A. **Meningitis** is an inflammation of the meninges. Symptoms include headache, painful or stiff neck, fever, and usually an increased number of white blood cells in the CSF. The more serious forms of acute meningitis are caused by bacteria, often facilitated by coinfection or previous infection with respiratory viruses.

1. *Neisseria meningitidis:* This gram-negative diplococcus is commonly known as the meningococcus and causes the most serious form of acute meningitis. The most serious complications of meningococcal infection are due to meningococcemia.
2. *Streptococcus pneumoniae:* Also known as the **pneumococcus** this gram-positive coccus is the most frequent cause of community-acquired pneumococcal meningitis.
3. *Haemophilus influenzae:* This bacterium is a gram-negative pleomorphic rod and an agent of acute bacterial meningitis in humans. The disease is caused primarily by the B serotype and was once most common in children between 3 months and 5 years of age, although these rates have declined because of vaccination.
4. *Listeria monocytogenes: L. monocytogenes* is a gram-positive bacterium that ranges in morphology from coccobacilli to long filaments in palisades formation. Pregnant women are especially susceptible to infection, which can result in premature abortion and fetal death. Most cases of listeriosis are associated with ingesting contaminated dairy products, poultry, and meat.
5. *Cryptococcus neoformans:* The fungus *C. neoformans* causes a more chronic form of meningitis with a more gradual onset of symptoms. The primary ecological niche of *C. neoformans* is the bird population. Most cryptococcal infections cause symptoms in the respiratory and central nervous systems, with the highest rates of cryptococcal meningitis occurring among patients with AIDS; frequently, it is fatal.
6. *Coccidioides immitis:* This is a true systemic fungal infection that begins with pulmonary infection but can disseminate quickly throughout the body and can lead to coccidioidomycosis of the meninges. The highest incidence of coccidioidomycosis occurs in the southwestern United States, Mexico, and parts of Central and South America.
7. *Viruses:* A wide variety of viruses can cause meningitis, particularly in children, and 90% are caused by enteroviruses.

B. **Neonatal Meningitis:** Meningitis in newborns is usually transmitted by the mother, either in utero or during passage through the birth canal.

1. *Streptococcus agalactiae:* This species of *Streptococcus* belongs to the group B streptococci and is the most frequent cause of neonatal meningitis.
2. *Escherichia coli:* The K1 strain of *E. coli* is the second most common cause of neonatal meningitis. Most babies who suffer from this infection are premature, and their prognosis is poor.

C. **Meningoencephalitis:** Because the brain and the spinal cord (and the meninges) are so closely connected, infections of one of these structures may also involve the other. Two amoebas, *Naegleria fowleri* and *Acanthamoeba,* are parasites that cause meningoencephalitis.

D. **Acute encephalitis** is almost always caused by viral infection.

1. Arboviruses: Arthropods serve as infectious disease vectors for many arboviruses. Arboviral encephalitis begins with an arthropod bite, the release of the virus into tissues, and its replication in nearby lymphatic tissues.
   a. **Western equine encephalitis (WEE)** occurs sporadically in the western United States and Canada and is carried by a mosquito.
   b. **Eastern equine encephalitis (EEE)** is endemic to an area along the eastern coast of North America and Canada.
   c. **California encephalitis** may be caused by two different viral strains. The California strain occurs occasionally in the western United States and has little impact on humans. The LaCrosse strain is widely distributed in the eastern United States and Canada.
   d. **St. Louis encephalitis (SLE)** may be the most common of all American viral encephalitides. Cases appear throughout North and South America, but epidemics occur most often in the Midwest and South.
   e. **West Nile encephalitis:** The West Nile virus is a close relative of the SLE virus. It emerged in the United States in 1999.
2. Herpes simplex virus: Herpes simplex type 1 and 2 viruses can cause encephalitis in newborns born to HSV-positive mothers, as well as older children and young adults (ages 5 to 30), and even older adults (over 50 years old).

3. JC virus: The JC virus (JCV) can cause a condition called **progressive multifocal leukoencephalopathy (PML),** particularly in immunocompromised individuals. This is a fatal infection.

E. **Subacute Encephalitis:** When encephalitis symptoms take longer to show up, it is termed subacute encephalitis.
   1. *Toxoplasma gondii* is a protozoan that causes toxoplasmosis, the most common form of subacute encephalitis. Although relatively asymptomatic in the general population, the disease in pregnant women and in immunodeficient people is severe and often fatal. *T. gondii* has primary reservoir and hosts in members of the feline family, both domestic and wild.
   2. Measles virus can produce **subacute sclerosing panencephalitis (SSPE),** which occurs years after an initial measles infection.
   3. Prions are proteinaceous infectious particles containing no genetic material. They cause diseases called transmissible spongiform encephalopathies (TSEs), neurodegenerative diseases with long incubation periods but rapid progressions once they begin. The human TSEs are **Creutzfeldt-Jakob disease (CJD),** Gerstmann-Strussler-Scheinker disease, and fatal familial insomnia. Prions are very resistant to chemicals, radiation heat, and autoclaving.

F. **Rabies** is a slow, progressive zoonotic disease characterized by fatal encephalitis. The rabies virus is in the family Rhabdoviridiae, genus *Lyssavirus.* The particles of this virus have a distinctive bulletlike appearance, round on one end and flat on the other. The primary reservoirs of the virus are wild mammals. An effective vaccine regimen is available.

G. **Poliomyelitis:** Polio is an acute enterovirus infection of the spinal cord that can cause neuromuscular paralysis. Because it often affects small children, in the past it was called infantile paralysis. Prevention is by vaccination using one of two forms of vaccine currently in use—inactivated Salk poliovirus vaccine (IPV), and attenuated oral Sabin poliovirus vaccine (OPV).

H. **Tetanus** is a neuromuscular disease, also called lockjaw, and is caused by *Clostridium tetani,* a gram-positive, spore-forming rod. *C. tetani* releases a powerful neurotoxin, **tetanospasmin,** which binds to target sites on spinal neurons and blocks the inhibition of muscle contraction. Without inhibition of contraction, the muscles contract uncontrollably, which can be fatal.

I. **Botulism** is often an *intoxication* (that is, caused by an exotoxin) associated with eating poorly preserved foods, although it can also occur as a true infection. There are three major forms of botulism: food-borne botulism (in children and adults), infant botulism, and wound botulism. The causative agent is *Clostridium botulinum,* a spore-forming anaerobe that does its damage through the release of an exotoxin. Symptoms include double vision and flaccid paralysis.

J. **African sleeping sickness** is caused by a protozoan, *Trypanosoma brucei.* Trypanosomiasis affects the central nervous system, leading to neurological deterioration: muscular tremors, shuffling gait, slurred speech, seizures, and local paralysis. Death results from coma, secondary infections, or heart damage.

Multiple-Choice and True-False Questions

Multiple-Choice Questions. Select the correct answer from the answers provided.

1. Which of the following organisms does *not* cause meningitis?
   a. *Haemophilus influenzae*
   b. *Streptococcus pneumoniae*
   c. *Neisseria meningitidis*
   d. *Clostridium tetani*
2. The first choice antibiotic for bacterial meningitis is the broad-spectrum
   a. cephalosporin
   b. penicillin
   c. ampicillin
   d. vancomycin
3. Meningococcal meningitis is caused by
   a. *Haemophilus influenzae*
   b. *Streptococcus pneumoniae*
   c. *Neisseria meningitidis*
   d. *Listeria monocytogenes*
4. Which of the following neurological diseases is not caused by a prion?
   a. Creutzfeldt-Jakob disease
   b. scrapie
   c. mad cow disease
   d. St. Louis encephalitis
5. *Cryptococcus neoformans* is primarily transmitted by
   a. direct contact
   b. bird droppings
   c. fomites
   d. sexual activity
6. Which of the following is *not* caused by an arbovirus?
   a. St. Louis encephalitis
   b. Eastern equine encephalitis
   c. West Nile encephalitis
   d. PAM
7. CJD is caused by a(n)
   a. arbovirus
   b. prion
   c. protozoan
   d. bacterium
8. What food should you avoid feeding a child under 1 year old because of potential botulism?
   a. honey
   b. milk
   c. apple juice
   d. applesauce
9. *Naegleria fowleri* meningoencephalitis is commonly acquired by
   a. bird droppings
   b. swimming in ponds and streams
   c. mosquito bites
   d. chickens

10. Which organism is responsible for progressive multifocal leukoencephalopathy?
    a. JC virus
    b. herpesvirus
    c. *E. coli*
    d. *Haemophilus influenzae*

True-False Questions. If statement is true, leave as is. If it is false, correct it by rewriting the sentence.

11. *Toxoplasma gondii* is a bacterium.
12. Penicillin G is the first line of treatment for coccidiomycosis.
13. A diagnosis of bacterial meningitis can be made by analyzing cerebral spinal fluid (CSF).
14. In the United States, dogs are a common reservoir for rabies.

Writing to Learn

These questions are suggested as a *writing-to-learn* experience. For each question, compose a one- or two-paragraph answer that includes the factual information needed to completely address the question.

1. Describe the components of the human nervous system.
2. a. What is meningitis?
   b. Describe the symptoms of this condition.
   c. Define petechiae.
   d. What is the most frequent cause of community-acquired meningitis?
   e. Name the clinical symptom that can distinguish between this organism and the agent of meningococcal meningitis.
3. What is the common mode(s) of transmission in neonatal meningitis?
4. What is the difference between meningitis and encephalitis?
5. What sterilization methods are most effective against prions?
6. Discuss the transmission of CJD.
7. Name the infectious viral disease for which postexposure passive and active immunization is indicated. Describe the procedure.
8. a. Name and describe the three major forms of botulism.
   b. What is the causative agent of this disease?

Concept Mapping

Appendix D provides guidance for working with concept maps.

1. Construct your own concept map using the following words as the *concepts*. Supply the linking words between each pair of concepts. Add other concepts if needed.

   bacteria
   viruses
   fungi
   droplets
   vehicles
   vaccines
   meningitis
   colonization
   transmission
   vaccination

2. Use 6 to 10 bolded words of your choice from the Chapter Summary to create a concept map. Finish it by providing linking words.

Critical Thinking Questions

Critical thinking is the ability to reason and solve problems using facts and concepts. These questions can be approached from a number of angles, and in most cases, they do not have a single correct answer.

1. Why is there no normal biota associated with the nervous system?
2. What organisms cause aseptic meningitis? Why is this condition referred to as aseptic meningitis?
3. Discuss the roles of arthropods and birds in arbovirus infections.
4. How did West Nile encephalitis spread to the United States?
5. Why should pregnant women be careful around cats and their litter boxes?
6. Why is the Sabin oral polio vaccine no longer recommended for childhood vaccinations?
7. Why is botulism associated with preserved canned foods?
8. How should trypanosomiasis (African sleeping sickness) be controlled or eradicated?
9. Why are young children so susceptible to infant botulism?

Visual Understanding

1. **From chapter 3, figure 3.25.** Study the negative stain of *Cryptococcus neoformans* shown in the figure. Write a simple procedure that could be used to make a rapid (10–15 min), presumptive diagnosis of cryptococcal meningitis. How could the presumptive diagnosis be confirmed?

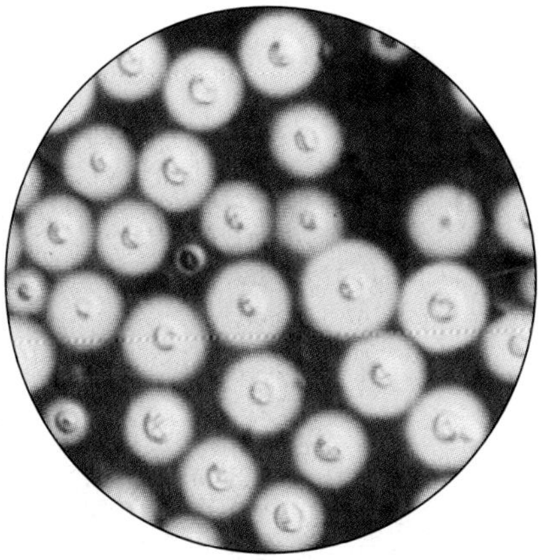

2. **From chapter 15, figure 15.19.** A vaccine used to immunize individuals against meningococcal meningitis is described as containing "meningococcal capsular polysaccharide antigens." Which of the vaccine production strategies shown in this illustration could be used to produce this vaccine? Explain your answer. (Flip back to page 466 to see the figure in more detail.)

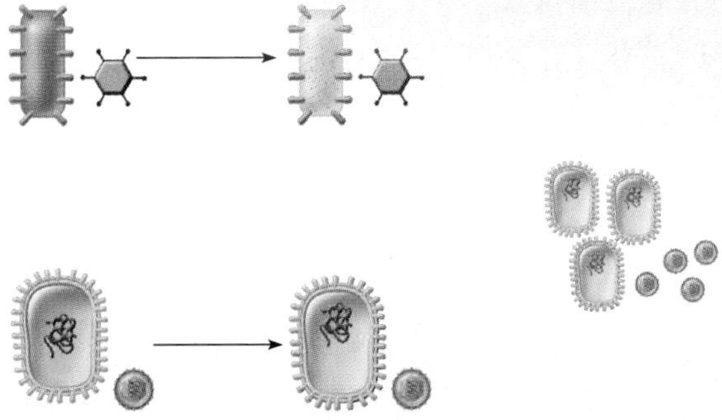

(a) **Whole Cell Vaccines**

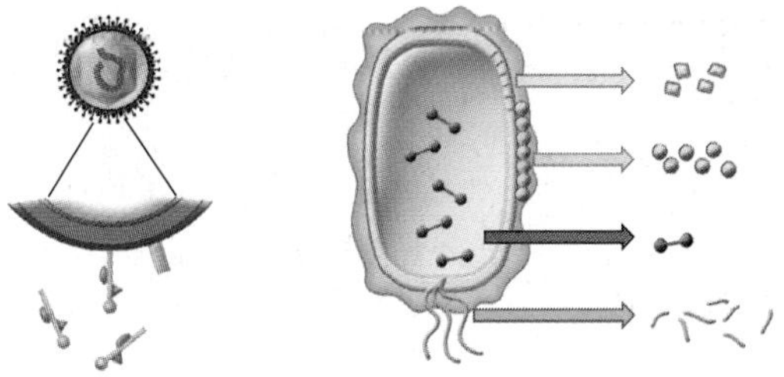

(b) **Vaccines from Microbe Parts**

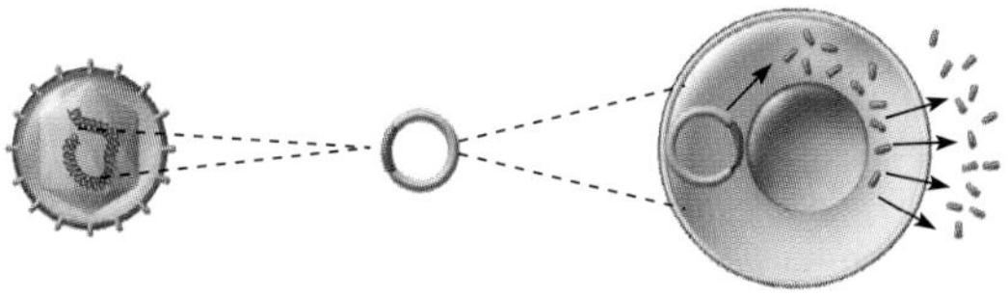

(c) **Recombinant Vaccine**

Internet Search Topics

1. Be the detective and use search engines to investigate the link between increased risk for meningococcal meningitis and college students or military personnel. Write a summary of your findings.
2. Find information about the study in which polioviruses were synthesized in a test tube. Answer questions about the why, how, and potential dangers of this technology.
3. Go to: www.aris.mhhe.com, and click on "microbiology" and then this textbook's author/title. Go to chapter 19, access the URLs listed under Internet Search Topics, and research the following:
   a. Compare the toxicity, molecular effects, physiological effects, and other features of botulin and tetanospasmin.
   b. Determine the reservoir, source, case rate, and other features of the 2002 outbreak of listeriosis. You may also research with a search engine.
4. Search for information on post-polio syndrome. Whom does this condition affect? What are the symptoms? What is thought to be the cause?

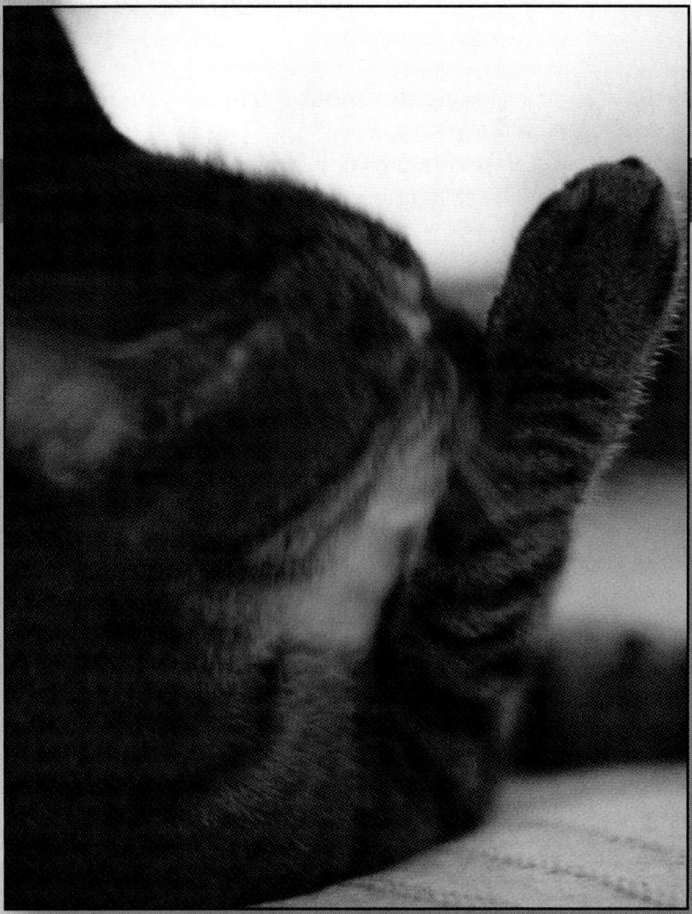

CHAPTER 20

Infectious Diseases Affecting the Cardiovascular and Lymphatic Systems

CASE FILE 20

In August of 1996, a teenage girl living in Colorado was seen in the local hospital emergency department. Her presenting symptoms were numbness in her left arm accompanied by pain in the axilla region. Her temperature, pulse rate, blood pressure, and chest film were all within normal ranges. She reported recently having fallen from a trampoline. She was released from the emergency department with a diagnosis of possible nerve injury to her left arm and a prescription for painkillers.

Four days later, she was back in the emergency room in very bad shape. She was semiconscious, had a fever of 102.5°F, and pulse of 170 beats per minute. A new chest X ray revealed bilateral pulmonary edema. She experienced respiratory failure while in the emergency room and was intubated. Blood and cerebrospinal fluid (CSF) samples were taken for culture, cell count, glucose and protein analysis, and for Gram staining. Gram-positive diplococci were detected in the blood. At this point, she was diagnosed with septicemia, disseminated intravascular coagulation, adult respiratory distress syndrome, and possible meningitis. A Gram stain of sputum revealed no bacteria. Treatment for gram-positive sepsis was initiated.

The patient died later that day. Two days later, her blood and CSF cultures revealed an unidentified gram-negative rod and *Streptococcus pneumoniae.* As part of the epidemiological investigation after her death, health officials visited her home. The only possibly relevant information they found was a family cat with a healing abscess in its jaw and evidence of an extensive prairie dog die-off in the vicinity.

- *What do you suppose was the cause of death in this case?*
- *How is the pain in the left arm and underarm related to her infection?*

Case File 20 Wrap-Up appears on page 616.

CHAPTER OVERVIEW

- The cardiovascular system consists of the heart and the blood vessels. Its function is to deliver nutrients and oxygen to all parts of the body.
- The lymphatic system is a network of vessels that returns fluid from the tissues to the bloodstream. It houses major portions of the immune system and serves to filter deleterious agents from the body.
- There is a wide variety of immune cells and substances in both the cardiovascular and lymphatic system.
- There is no normal biota in the cardiovascular or lymphatic systems.
- Infections in the bloodstream have access to the whole body and they often result in systemic effects.
- Various bacteria can colonize parts of the heart, especially the heart valves.
- Some bloodstream infections lead to leakage of blood from the vessels and are called *hemorrhagic* fever diseases.
- Other bloodstream infections lead to prolonged or repeated fevers, with no hemorrhagic effects.
- One protozoan bloodstream infection, malaria, leads to at least 2 million deaths in the world per year.
- HIV infection cripples vital components of the blood and lymphatic system.

20.1 The Cardiovascular and Lymphatic Systems and Their Defenses

The Cardiovascular System

The cardiovascular system is the pipeline of the body. It is composed of the blood vessels, which carry blood to and from all regions of the body, and the heart, which pumps the blood. This system moves the blood in a closed circuit, and it is therefore known as the *circulatory system*. The cardiovascular system provides tissues with oxygen and nutrients and carries away carbon dioxide and waste products, delivering them to the appropriate organs for removal. A closely related but largely separate system, the **lymphatic system** is a major source of immune cells and fluids, and it serves as a one-way passage, returning fluid from the tissues to the cardiovascular system.

The heart is a fist-size muscular organ that pumps blood through the body. It is divided into two halves, each of which is divided into an upper and lower chamber **(figure 20.1).** The upper chambers are called atria (singular, atrium), and the lower are ventricles. The entire organ is encased in a fibrous covering, the pericardium, which is an occasional site of infection. The actual wall of the heart has three layers: from outer to inner, they are the epicardium, the myocardium, and the endocardium. The endocardium also covers the valves of the heart, and it is a relatively common target of microbial infection.

The atria receive blood coming from the body. This blood, which is low in oxygen and high in carbon dioxide, enters the right atrium and passes through to the right ventricle. From there it is pumped to the pulmonary arteries in the lung, where it becomes oxygenated and reenters the heart through the left atrium. Finally, the blood moves into the left ventricle and is pumped into the aorta and the rest of the body. The movement of blood into and out of the chambers of the heart is controlled by valves.

The blood vessels consist of *arteries, veins,* and *capillaries.* Arteries carry oxygenated blood away from the heart under relatively high pressure. They branch into smaller vessels called arterioles. Veins actually begin as smaller venules in the periphery of the body and coalesce into veins. The smallest blood vessels, the capillaries, connect arterioles to venules. Both arteries and veins have walls made of three layers of tissue. The innermost layer is composed of a smooth epithelium called endothelium. Its smooth surface encourages the smooth flow of cells and platelets through the system. The next layer is composed of connective tissue and muscle fibers. The outside layer is a thin layer of connective tissue. Capillaries, the smallest vessels, have walls made of only one layer of endothelium. **Figure 20.2** illustrates the complete cardiovascular system.

The Lymphatic System

Chapter 14 provided a detailed description of the lymphatic system; you may wish to review page 424 and figure 14.10 before continuing. In short, the lymphatic system consists mainly of the lymph vessels, which roughly parallel the blood vessels; lymph nodes, which cluster at body sites such as the groin, neck, armpit, and intestines; and the spleen. It serves to collect fluid that has left the blood vessels and entered tissues, filter it of impurities and infectious agents, and return it to the blood.

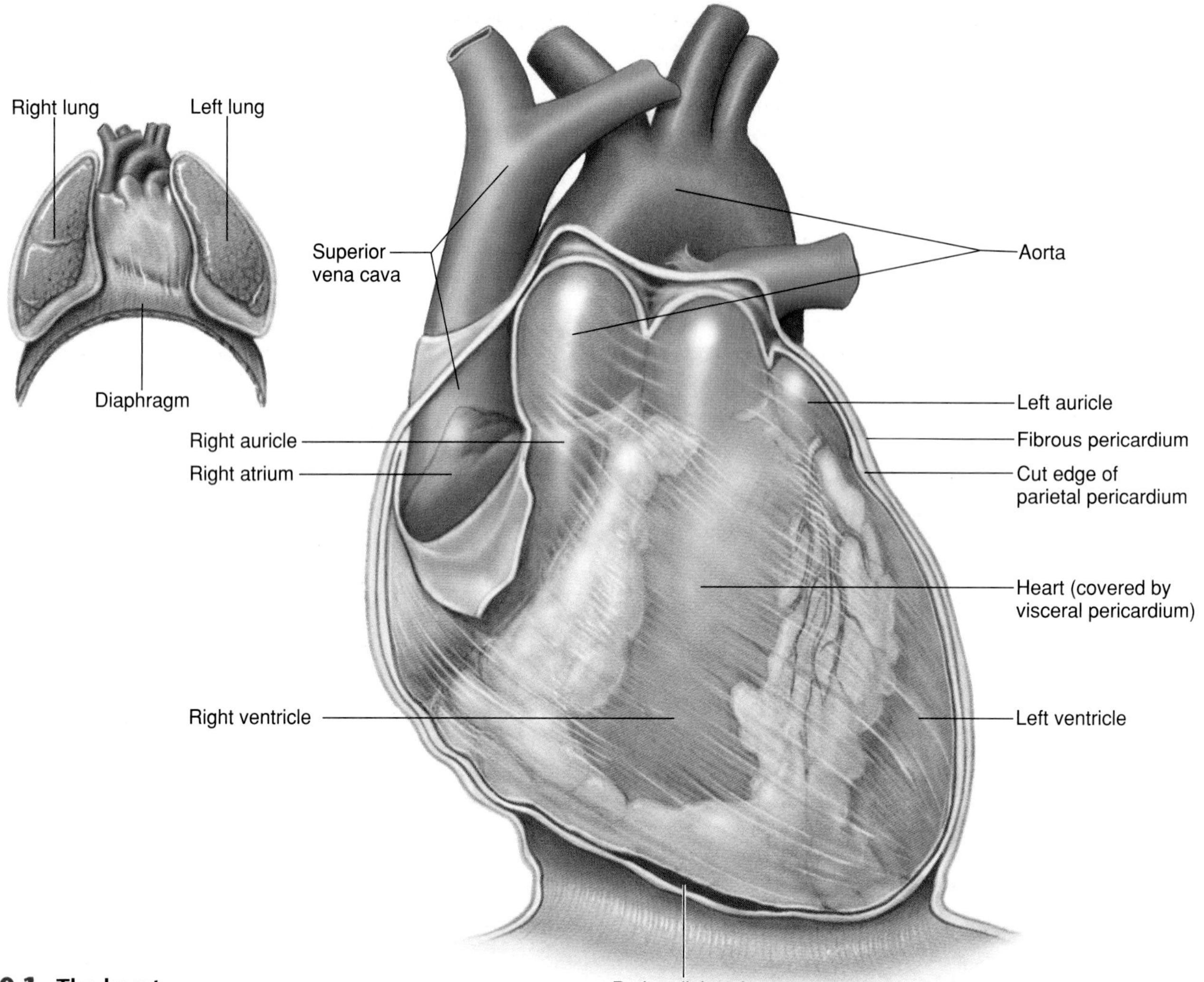

Figure 20.1 **The heart.**

Defenses of the Cardiovascular and Lymphatic Systems

The cardiovascular system is highly protected from microbial infection. Microbes that successfully invade the system, however, gain access to every part of the body, and every system may potentially be affected. For this reason, bloodstream infections are called **systemic infections.**

Multiple defenses against infection reside in the bloodstream. The blood is full of leukocytes, with approximately 5,000 to 10,000 white blood cells per milliliter of blood. The various types of white blood cells include the lymphocytes, responsible for specific immunity, and the phagocytes, which are so critical to nonspecific as well as specific immune responses. Very few microbes can survive in the blood with so many defensive elements. That said, a handful of infectious agents have nonetheless evolved exquisite mechanisms for avoiding blood-borne defenses.

Medical conditions involving the blood often have the suffix *-emia*. For instance, viruses that cause meningitis can travel to the nervous system via the bloodstream. Their presence in the blood is called **viremia.** When fungi are in the blood, the condition is termed **fungemia,** and bacterial presence is called **bacteremia,** a general term denoting only their *presence.* Although the blood contains no normal biota (see next section), bacteria frequently are introduced into the bloodstream during the course of daily living. Brushing your teeth or tearing a hangnail can introduce bacteria from the mouth or skin into the bloodstream; this situation is usually temporary. But when bacteria flourish and grow in the bloodstream, the condition is termed **septicemia.** Septicemia can very quickly lead to cascading immune responses, resulting in decreased systemic blood pressure that can lead to **septic shock,** a life-threatening condition.

20.2 Normal Biota of the Cardiovascular and Lymphatic Systems

Like the nervous system, the cardiovascular and lymphatic systems are "closed" systems with no normal access to the external environment. Therefore they possess no normal biota. In the absence of disease, microorganisms may be transiently present in either system as just described. The

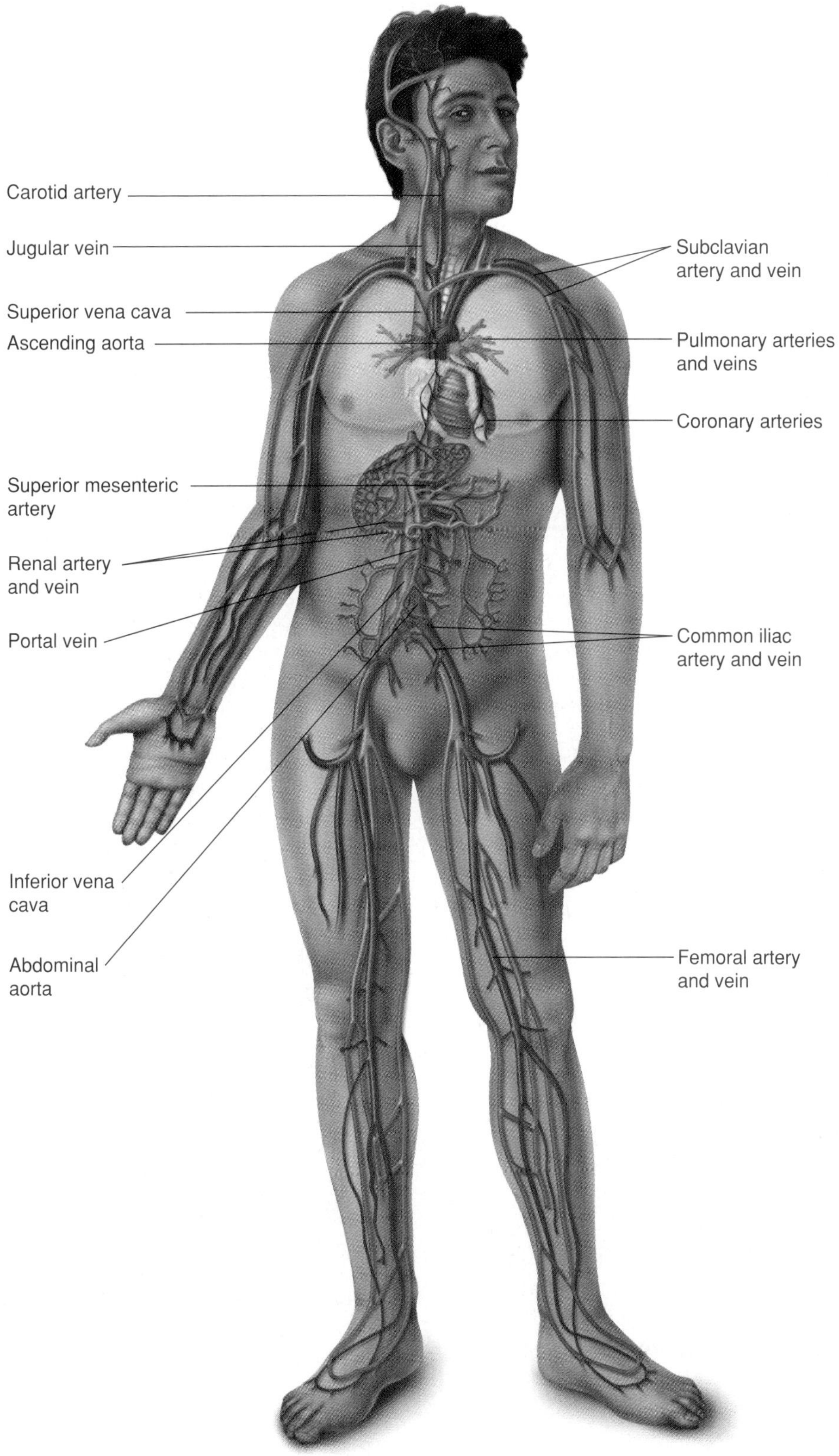

Figure 20.2 **The anatomy of the cardiovascular system.**

lymphatic system serves to filter microbes and their products out of tissues. Thus, in the healthy state, no microorganisms *colonize* either the lymphatic or cardiovascular systems. Of course, this is biology, and it is never quite that simple. Recent studies have suggested that the bloodstream is not completely sterile, even during periods of apparent health. It is tempting to speculate that these low-level microbial "infections" may contribute to diseases for which no etiology has previously been found or for conditions currently thought to be noninfectious **(Insight 20.1).**

INSIGHT 20.1 **Medical**

Atherosclerosis

Atherosclerosis is a condition you may not associate with infection. In atherosclerosis, plaques form on the inner endothelium of the arteries, decreasing the flexibility of the arterial walls and possibly leading to obstruction. The well-established cause of plaque formation seems to be an elevation in low-density lipoproteins (LDLs) in the plasma (originating in the diet) accompanied by chronic endothelial injury events. Injury to the endothelium results in adhesion of platelets to the surface, accumulation of other blood components, and release of growth factors that cause the proliferation of smooth muscle cells. This begs the question: What causes the chronic endothelial injury? The answer is complex and can include such things as nicotine in the bloodstream from cigarette smoke or high blood levels of insulin as seen in diabetics. But recent findings suggest that another of the causes may be chronic bloodstream infection with the bacterium *Chlamydophila pneumoniae* or even various viruses.

Recent studies have not lent support to the idea that bacterial infection influences heart-related conditions. One group administered antibiotics once a week for 1 year to thousands of people across the United States. The patients were then followed for as long as 4 years. No difference was found in the cardiac outcomes of the patients who had taken the antibiotics compared to those who had not. The roles of *Chlamydia* and *Chlamydophila* continue to be investigated. Other researchers have found that reproductive tract infections with *Chlamydia trachomatis* may predispose people to heart disease. These studies found that the immune response to proteins on the surface of *C. trachomatis* can cross-react with myosin, a protein in heart muscle. This phenomenon is called molecular mimicry, referring to the similarity between *Chlamydia* proteins and heart proteins. The misdirected immune response may injure the heart muscle.

Obviously, much remains to be discovered about heart disease, atherosclerosis, and the role infection may play.

Cardiovascular and Lymphatic Systems Defenses and Normal Biota

	Defenses	Normal Biota
Cardiovascular System	Blood-borne components of nonspecific and specific immunity—including phagocytosis, specific immunity	None
Lymphatic System	Numerous immune defenses reside here.	None

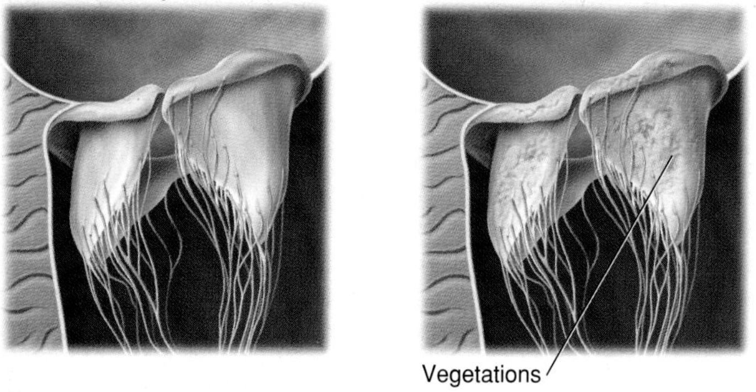

Figure 20.3 Endocarditis.
Infected valves don't work as well as healthy ones.

20.3 Cardiovascular and Lymphatic System Diseases Caused by Microorganisms

Categorizing cardiovascular and lymphatic infections according to clinical presentation is somewhat difficult because most of these conditions are systemic, with effects on multiple organ systems. We begin with infections involving the heart and the blood in general and then discuss conditions with more specific causes.

Endocarditis

Endocarditis is an inflammation of the endocardium, or inner lining of the heart. Most of the time, endocarditis refers to an infection of the valves of the heart, often the mitral or aortic valve **(figure 20.3).** Two variations of infectious endocarditis have been described: acute and subacute. Each has distinct groups of possible causative agents. Rarely, endocarditis can also be caused by vascular trauma or by circulating immune complexes in the absence of infectious agents.

The surgical innovation of prosthetic valves presents a new hazard for development of endocarditis. Patients with prosthetic valves can acquire acute endocarditis if bacteria are introduced during the surgical procedure; alternatively, the prosthetic valves can serve as infection sites for the subacute form of endocarditis long after the surgical procedure. Because the symptoms and the diagnostic procedures are similar for both forms of endocarditis, they are discussed first; then the specific aspects of acute and subacute endocarditis are addressed.

Signs and Symptoms

The signs and symptoms are similar for both types of endocarditis, except that in the subacute condition they develop more slowly and are less pronounced than with the acute disease. Symptoms include fever, anemia, abnormal heartbeat, and sometimes symptoms similar to myocardial infarction (heart attack). Abdominal or side pain is sometimes reported. The patient may look very ill and may have petechiae (small red-to-purple discolorations) over the upper half of the body and under the fingernails. In subacute cases, an enlarged spleen may have developed over time; cases of extremely long duration can lead to clubbed fingers and toes.

Culture and Diagnosis

The diagnostic procedures for the two forms of endocarditis are essentially the same. One of the most important diagnostic tools is a high index of suspicion. A history of risk factors, or behaviors, such as abnormal valves, intravenous drug use, recent surgery, or bloodstream infections, should lead one to consider endocarditis when the symptoms just described are observed. Blood cultures, if positive, are the gold standard for diagnosis, but negative blood cultures do not rule out endocarditis. If it is possible to obtain the agent, it is very important to determine its antimicrobial susceptibilities.

In acute endocarditis, the symptoms may be magnified. The patient may also display central nervous system symptoms suggestive of meningitis, such as stiff neck or headache.

Acute Endocarditis

Acute endocarditis is most often the result of an overwhelming bloodstream challenge with bacteria. Certain of these bacteria seem to have the ability to colonize normal heart valves. Accumulations of bacteria on the valves (vegetations) hamper their function and can lead directly to cardiac malfunction and death. Alternatively, pieces of the bacterial vegetation can break off and create emboli (blockages) in vital organs. The bacterial colonies can also provide a constant source of bloodborne bacteria, with the accompanying systemic inflammatory response and shock. Bacteria that are attached to surfaces bathed by blood (such as heart valves) quickly become covered with a mesh of fibrin and platelets that protects them from the immune components in the blood.

Causative Agents The acute form of endocarditis is most often caused by *Staphylococcus aureus*. Other agents that cause it are *Streptococcus pyogenes*, *Streptococcus pneumoniae*, and *Neisseria gonorrhoeae*, as well as a host of other bacteria. Each of these bacteria is described elsewhere in this book; all are pathogenic.

Transmission and Epidemiology The most common route of transmission for acute endocarditis is parenteral—that is, via direct entry into the body. Intravenous or subcutaneous drug users have been a growing risk group for the condition. Traumatic injuries and surgical procedures can also introduce the large number of bacteria required for the acute form of endocarditis.

Prevention and Treatment Prevention is based on avoiding the introduction of bacteria into the bloodstream during surgical procedures or injections. Untreated, this condition is invariably fatal. Treatment depends on the identity and the antimicrobial susceptibility of the causative agent. In the case of gram-positive cocci, the drug of choice is penicillin for susceptible strains. Otherwise, vancomycin plus an aminoglycoside may be used. High, continuous blood levels of antibiotics are required to resolve the infection because the bacteria exist in biofilm vegetations. In addition to the decreased access of antibiotics to bacteria deep in the biofilm, these bacteria often express a phenotype of lower susceptibility to antibiotics. Surgical debridement of the valves, accompanied by antibiotic therapy, is sometimes required.

Subacute Endocarditis

Subacute forms of this condition are almost always preceded by some form of damage to the heart valves or by congenital malformation. Irregularities in the valves encourage the attachment of bacteria, which then form biofilms and impede normal function, as well as provide an ongoing source of bacteria to the bloodstream. People who have suffered rheumatic fever and the accompanying damage to heart valves are particularly susceptible to this condition (see chapter 21 for a complete discussion of rheumatic fever).

Causative Agents Most commonly, subacute endocarditis is caused by bacteria of low pathogenicity, often originating in the oral cavity. Alpha-hemolytic streptococci, such as *Streptococcus sanguis*, *S. oralis*, and *S. mutans*, are most often responsible, although normal biota from the skin and other bacteria can also colonize abnormal valves and lead to this condition.

Transmission and Epidemiology Minor disruptions in the skin or mucous membranes, such as those induced by vigorous toothbrushing, dental procedures, or relatively minor cuts and lacerations, can introduce bacteria into the bloodstream and lead to valve colonization. The bacteria are not, therefore, transmitted from other people or from the environment. The average age of onset for subacute endocarditis has increased in recent decades from the mid-20s to the mid-50s. Males are slightly more likely to experience it than females.

Prevention and Treatment The practice of prophylactic antibiotic therapy in advance of surgical and dental procedures on patients with underlying valve irregularities has decreased the incidence of this infection. When it occurs, treatment is similar to treatment for the acute form of the disease, described earlier **(Checkpoint 20.1).**

Septicemias

Septicemia occurs when organisms are actively multiplying in the blood. Many different bacteria (and a few fungi) can cause this condition. Patients suffering from these infections are sometimes described as "septic." One infection that should be considered in cases of aggressive septicemia, especially if respiratory symptoms are also present, is anthrax.

☑ CHECKPOINT 20.1	Endocarditis	
Disease	**Acute Endocarditis**	**Subacute Endocarditis**
Causative Organism(s)	*Staphylococcus aureus, Streptococcus pyogenes, S. pneumoniae, Neisseria gonorrhoeae*, others	Alpha-hemolytic streptococci, others
Most Common Modes of Transmission	Parenteral	Endogenous transfer of normal biota to bloodstream
Virulence Factors	Attachment	Attachment
Culture/Diagnosis	Blood culture	Blood culture
Prevention	Aseptic surgery, injections	Prophylactic antibiotics before invasive procedures
Treatment	Penicillin, or vancomycin plus aminoglycoside; surgery may be necessary	Penicillin, or vancomycin plus aminoglycoside; surgery may be necessary
Distinctive Features	Acute onset, high fatality rate	Slower onset

Signs and Symptoms

Fever is a prominent feature of septicemia. The patient appears very ill and may have an altered mental state, shaking chills, and gastrointestinal symptoms. Often an increased breathing rate is exhibited, accompanied by respiratory alkalosis (increased tissue pH due to breathing disorder). Low blood pressure is a hallmark of this condition and is caused by the inflammatory response to infectious agents in the bloodstream, which leads to a loss of fluid from the vasculature. This condition is the most dangerous feature of the disease, often culminating in death.

Causative Agents

The vast majority of septicemias are caused by bacteria, and they are approximately evenly divided between gram-positives and gram-negatives. Perhaps 10% are caused by fungal infections. Polymicrobic bloodstream infections increasingly are being identified in which more than one microorganism is causing the infection.

Pathogenesis and Virulence Factors

Gram-negative bacteria multiplying in the blood release large amounts of endotoxin into the bloodstream, stimulating a massive inflammatory response mediated by a host of cytokines. This response invariably leads to a drastic drop in blood pressure, a condition called **endotoxic shock**. Gram-positive bacteria can instigate a similar cascade of events when fragments of their cell walls are released into the blood.

Transmission and Epidemiology

In many cases, septicemias can be traced to parenteral introduction of the microorganisms via intravenous lines or surgical procedures. Other infections may arise from serious urinary tract infections or from renal, prostatic, pancreatic, or gallbladder abscesses. Patients with underlying spleen malfunction may be predisposed to multiplication of microbes in the bloodstream. Meningeal infections or pneumonia occasionally can lead to sepsis. Approximately half a million cases occur each year in the United States, resulting in more than 100,000 deaths.

Culture and Diagnosis

Because the infection is in the bloodstream, a blood culture is the obvious route to diagnosis. A full regimen of media should be inoculated to ensure isolation of the causative microorganism. Antibiotic susceptibilities should be assessed. Empiric therapy should be started immediately before culture and susceptibility results are available. The choice of antimicrobial agent should be informed by knowledge of any suspected source of the infection, such as an intravenous catheter (in which case, skin biota should be considered), urinary tract infections (in which case, gram-negatives and *Streptococci* should be considered), and so forth.

Prevention and Treatment

Empiric therapy, which is begun immediately after blood cultures are taken, often begins with a broad-spectrum antibiotic. Once the organism is identified and its antibiotic susceptibility is known, treatment can be adjusted accordingly.

☑ CHECKPOINT 20.2	Septicemia
Causative Organism(s)	Bacteria or fungi
Most Common Modes of Transmission	Parenteral, endogenous transfer
Virulence Factors	Cell wall or membrane components
Culture/Diagnosis	Blood culture
Prevention	–
Treatment	Broad-spectrum antibiotic until identification and susceptibilities tested

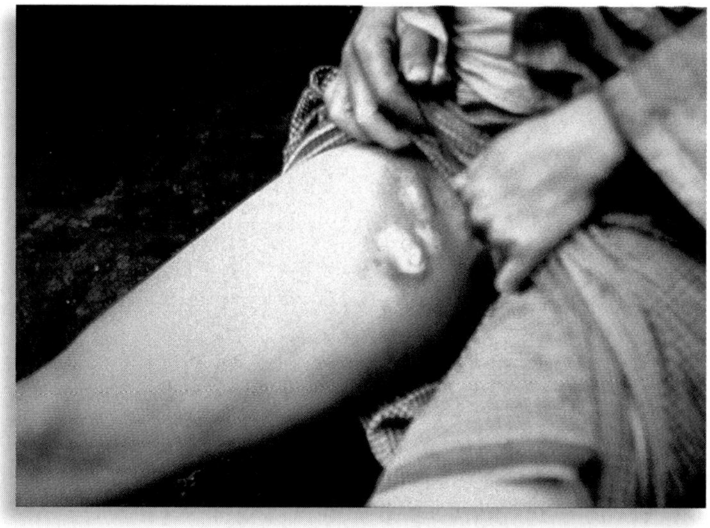

Figure 20.4 A classic inguinal bubo of bubonic plague.
This hard nodule is very painful and can rupture onto the surface.

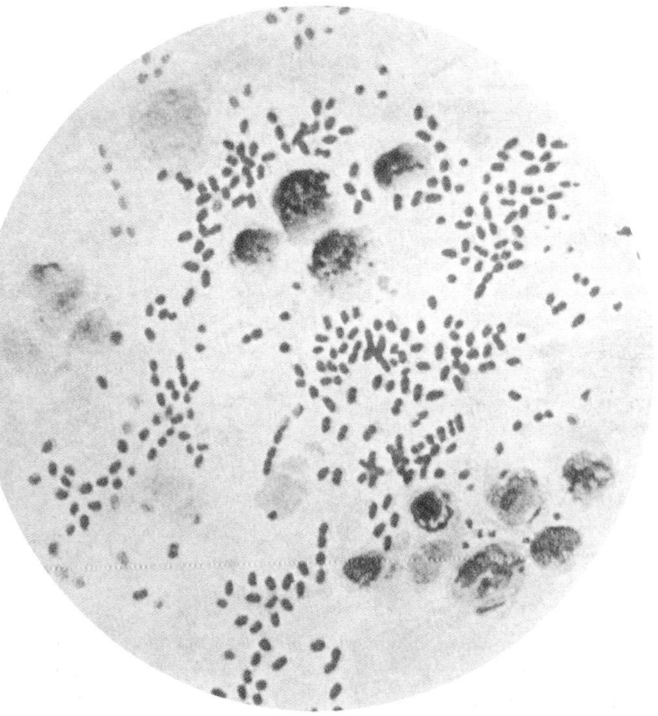

Figure 20.5 ***Yersinia pestis.***
Note the more darkly stained poles of the bacterium, lending it a "safety pin" appearance.

Plague

The word **plague**[1] conjures up visions of death and morbidity unlike any other infectious disease. Although pandemics of plague have probably occurred since antiquity, the first one that was reliably chronicled killed an estimated 100 million people in the sixth century AD. The last great pandemic occurred in the late 1800s and was transmitted around the world, primarily by rat-infested ships. The disease was brought to the United States through the port of San Francisco around 1906. Infected rats eventually mingled with native populations of rodents and gradually spread the disease throughout the West and Midwest.

Signs and Symptoms

Three possible manifestations of infection occur with the bacterium causing plague. **Pneumonic plague** is a respiratory disease, described in chapter 21. In **bubonic plague,** the bacterium, which is injected by the bite of a flea, enters the lymph and is filtered by a local lymph node. Infection causes inflammation and necrosis of the node, resulting in a swollen lesion called a **bubo,** usually in the groin or axilla **(figure 20.4).** The incubation period lasts 2 to 8 days, ending abruptly with the onset of fever, chills, headache, nausea, weakness, and tenderness of the bubo. Mortality rates, even with treatment, are greater than 15%.

These cases often progress to massive bacterial growth in the blood termed septicemic plague. The presence of the bacteria in the blood results in disseminated intravascular coagulation, subcutaneous hemorrhage, and purpura that may degenerate into necrosis and gangrene. Mortality rates, once the disease has progressed to this point, are 30% to 50% with treatment and 100% without treatment. Because of the visible darkening of the skin, the plague has often been called the "black death."

1. From the Latin *plaga,* meaning to strike, infect, or afflict with disease, calamity, or some other evil.

Causative Agent

The cause of this dreadful disease is a tiny, harmless-looking gram-negative rod, ***Yersinia pestis,*** a member of the Family Enterobacteriaceae. Other species members are *Y. enterocolitica* and *Y. pseudotuberculosis.* These species cause gastrointestinal tract diseases in humans. *Y. pestis* displays unusual bipolar staining that makes it look like a safety pin **(figure 20.5).**

Pathogenesis and Virulence Factors

The number of bacteria required to initiate a plague infection is small—perhaps only 3 to 50 cells. Much research has been conducted on the differences between the two *Yersinia* species that cause GI tract disease and this *Yersinia,* because it has such different effects on the host. Scientists have discovered that *Y. pestis* carries three plasmids: One is common to all three *Yersinia* species; two are unique to *Y. pestis.* All three plasmids carry genes important for pathogenesis. The plasmid all *Yersinia* carry contains genes for a system called the Yop system, a series of proteins the bacteria use to attach to host cells and inject proteins into them that short circuit the immune response. The two plasmids unique to *Y. pestis* carry genes that help it to cause disease in mice and to survive in the flea vector. Examples of these genes include a gene for capsule formation and a gene for plasminogen activation (similar to the streptokinase expressed by *S. pyogenes*) (see chapter 18).

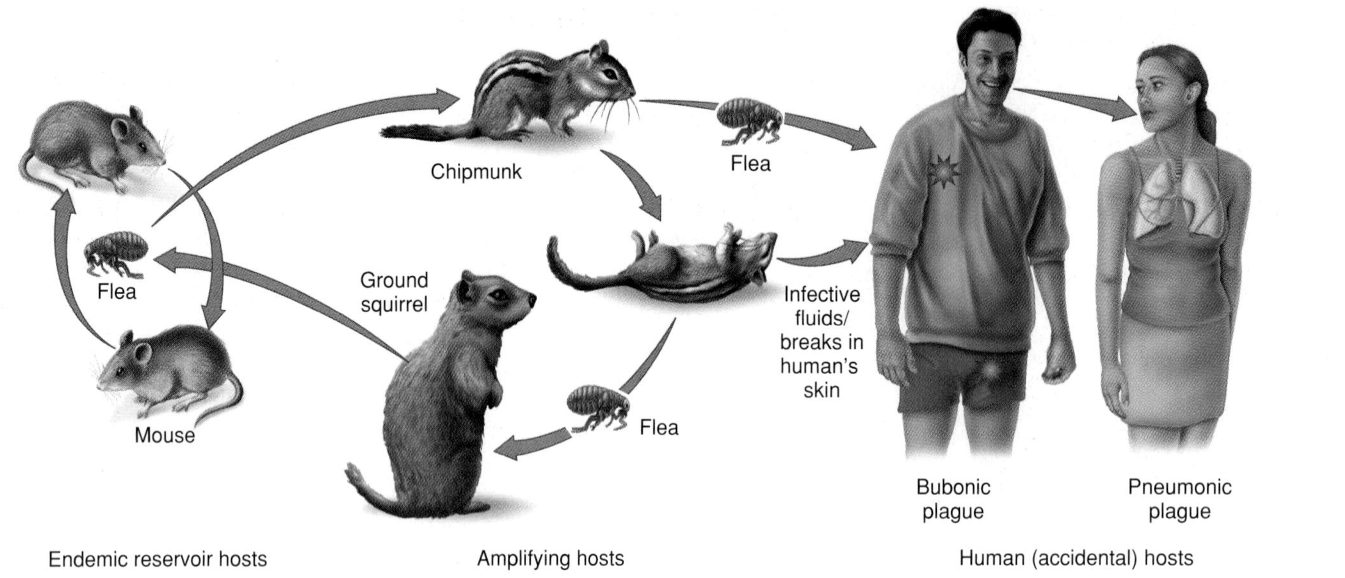

Figure 20.6 The infection cycle of *Yersinia pestis*.

Plasminogen activation leads to clotting, which helps the microbe resist phagocytosis.

Transmission and Epidemiology

The principal agents in the transmission of the plague bacterium are fleas. These tiny, bloodsucking insects have a special relationship with the bacterium. After a flea ingests a blood meal from an infected animal, the bacteria multiply in its gut. In fleas that effectively transmit the bacterium, the esophagus becomes blocked due to coagulation factors produced by the pathogen. Being unable to feed properly, the ravenous flea jumps from animal to animal in a futile attempt to get nourishment. During this process, regurgitated infectious material is inoculated into the bite wound.

The plague bacterium exists naturally in many animal hosts, and its distribution is extensive. Although the incidence of disease has been reduced in the developed world, it has actually been increasing in Africa and other parts of the world. Plague still exists endemically in large areas of Africa, South America, the Mideast, Asia, and the former Soviet Union, and it sometimes erupts into epidemics such as the outbreak in India in the 1990s that infected hundreds of residents. This new surge of cases was attributed to increased populations of rats following the monsoon floods. In the United States, sporadic cases (usually less than 10 per year) occur as a result of contact with wild and domestic animals. This disease is considered endemic in U.S. western and southwestern states. Persons most at risk for developing plague are veterinarians and people living and working near woodlands and forests. Dogs and cats can be infected with the plague, often from contact with infected wild animals such as prairie dogs. Human cases have been traced to a chain of events involving a flea from a prairie dog moving to a domestic cat, and then a flea from the cat moving to a human.

The epidemiology of plague is among the most complex of all diseases. It involves several different types of vertebrate hosts and flea vectors, and its exact cycle varies from one region to another. A general scheme of the cycle is presented in **figure 20.6.** Humans can develop plague through contact with the fleas of wild or domestic or semidomestic animals.

CASE FILE 20 *WRAP-UP*

The girl in the chapter opener was suffering from septicemic plague. "But wait!" you say. The original Gram stain revealed gram-*positive* diplococci, later identified as *Streptococcus pneumoniae,* and as a result she was treated for gram-positive septicemia.

This story is an excellent example of the importance of using multiple indicators for presumptive diagnosis. It is not clear why *S. pneumoniae* was present in large numbers in the patient's blood, but apparently its presence indicated an infection secondary to plague. Any patient with clinical signs of sepsis and a history that suggests possible plague exposure should immediately be treated empirically with antibiotics known to be highly effective against *Y. pestis,* such as streptomycin or gentamicin.

Testing revealed that four of the five family dogs and the cat with the abscess were all positive for *Y. pestis.* They had presumably been infected with fleas shared with the local prairie dog colony. The victim had cared for the sick family cat in recent days. In this case, patient history (place of residence, recent exposure to a sick animal) was a better indicator than laboratory tests.

The presenting symptom of pain in the arm, and especially in the axilla, was probably an early sign of *Y. pestis* infection and multiplication in the axillary lymph nodes. The earlier trampoline injury only served to deflect attention away from the true cause of the axillary pain.

See: CDC.1997. Fatal human plague—Arizona and Colorado, 1996. MMWR 46:617–620.

Contact with infected body fluids can also spread the disease. If a person has breaks in the skin on his or her hands, handling infected animals or animal skins is a possible means of transmission. (Persons with the pneumonic form of the disease can spread *Y. pestis* through respiratory droplets.)

The Animal Reservoirs The plague bacillus occurs in 200 different species of mammals. The primary long-term *endemic reservoirs* are various rodents, such as mice and voles, that harbor the organism but do not develop the disease. These hosts spread the disease to other mammals called *amplifying hosts* that become infected with the bacterium and experience massive die-offs during epidemics. These hosts, including rats, ground squirrels, chipmunks, and rabbits, are the usual sources of human plague. The particular mammal that is most important in this process depends on the area of the world. Other mammals (camels, sheep, coyotes, deer, dogs, and cats) can also be involved in the transmission cycle.

Culture and Diagnosis

Because death can ensue as quickly as 2 to 4 days after the appearance of symptoms, prompt diagnosis and treatment of plague are imperative. The patient's history, including recent travel to endemic regions, can help establish a diagnosis. Culture of the organism is the definitive method of diagnosis, although a Gram stain of aspirate from buboes often reveals the presence of the safety-pin-shaped bacteria.

Prevention and Treatment

Plague is one of a handful of internationally quarantinable diseases (others are cholera and yellow fever). In addition to quarantine during epidemics, plague is controlled by trapping rodents and by poisoning their burrows with insecticide to kill fleas. These methods, however, cannot begin to control the reservoir hosts, so the potential for plague will always be present in endemic areas, especially as humans encroach into rodent habitats. A killed or attenuated vaccine that protects against the disease for a few months is given to military personnel, veterinarians, and laboratory workers.

Streptomycin or gentamicin are the drugs of choice.

CHECKPOINT 20.3	Plague
Causative Organism(s)	*Yersinia pestis*
Most Common Modes of Transmission	Vector, biological; also droplet contact (pneumonic) and direct contact with body fluids
Virulence Factors	Capsule, Yop system, plasminogen activator
Culture/Diagnosis	Culture or Gram stain of blood or bubo aspirate
Prevention	Flea and or animal control; vaccine available for high-risk individuals
Treatment	Streptomycin or gentamicin

Tularemia

The causative agent of tularemia is a facultative intracellular gram-negative bacterium called *Francisella tularensis.* It has several characteristics in common with *Yersinia pestis,* and the two species were previously often included in a single genus called *Pasteurella.* It is a zoonotic disease of assorted mammals endemic to the Northern Hemisphere. Because it has been associated with outbreaks of disease in wild rabbits, it is sometimes called rabbit fever. It is currently listed as a pathogen of concern on the lists of bioterrorism agents (see Insight 21.3 for details).

Tularemia is abundantly distributed through numerous animal reservoirs and vectors in northern Europe, Asia, and North America but not in the tropics. This disease is noteworthy for its complex epidemiology and spectrum of symptoms. Although rabbits and rodents (muskrats and ground squirrels) are the chief reservoirs, other wild animals (skunks, beavers, foxes, opossums) and some domestic animals are implicated as well. The chief route of transmission in the past had been through the activity of skinning rabbits, but with the decline of rabbit hunting, transmission via tick bites is more common. Ticks are the most frequent arthropod vector, followed by biting flies, mites, and mosquitoes.

Tularemia is strikingly varied in its portals of entry and disease manifestations. Although bites by a vector are the most common source of infection, in many cases infection results when the skin or eye is inoculated through contact with infected animals, animal products, contaminated water, and dust. Pulmonary forms of the infection can result from aerosolized soils or animal fluids and also from spread of the bacterium in the bloodstream. The disease is not communicated from human to human. With an estimated infective dose of between 10 and 50 organisms, *F. tularensis* is often considered one of the most infectious of all bacteria. The term "lawnmower" tularemia refers to tularemia acquired while performing grass-mowing or brush-cutting chores. Cases of tularemia have appeared in people who have accidentally run over dead rabbits while lawn mowing, presumably from inhaling aerosolized bacteria. In 2004, three lab workers in Boston became infected. They thought they were working with an attenuated strain of the bacterium and were not using appropriate biosafety precautions. All three survived.

After an incubation period ranging from a few days to 3 weeks, acute symptoms of headache, backache, fever, chills, malaise, and weakness appear. Further clinical manifestations are tied to the portal of entry. They include ulcerative skin lesions, swollen lymph glands, conjunctival inflammation, sore throat, intestinal disruption, and pulmonary involvement. The death rate in the most serious forms of disease is 30%, but proper treatment with gentamicin or streptomycin reduces mortality to almost zero. Because the intracellular persistence of *F. tularensis* can lead to relapses, antimicrobial therapy must not be discontinued prematurely. Protection is available in the form of a live attenuated vaccine. Laboratory workers and other occupationally exposed personnel must wear gloves, masks, and eyewear.

CHECKPOINT 20.4	Tularemia
Causative Organism(s)	*Francisella tularensis*
Most Common Modes of Transmission	Vector, biological; also direct contact with body fluids from infected animal; airborne
Virulence Factors	Intracellular growth
Culture/Diagnosis	Culture dangerous to lab workers and not reliable; serology most often used
Prevention	Live attenuated vaccine for high-risk individuals
Treatment	Gentamicin or streptomycin

Lyme Disease

In the 1970s, an enigmatic cluster of arthritis cases appeared in the town of Old Lyme, Connecticut. The phenomenon caught the attention of nonprofessionals and professionals alike, whose persistence and detective work ultimately disclosed the unusual nature and epidemiology of Lyme disease. The process of discovery began in the home of Polly Murray, who, along with her family, was beset for years by recurrent bouts of stiff neck, swollen joints, malaise, and fatigue that seemed vaguely to follow a rash from tick bites. When Mrs. Murray's son was diagnosed as having juvenile rheumatoid arthritis, she became skeptical. Conducting her own literature research, she began to discover inconsistencies. Rheumatoid arthritis was described as a rare, noninfectious disease, yet over an 8-year period, she found that 30 of her neighbors had experienced similar illnesses. Ultimately, this cluster of cases and others were reported to state health authorities. Eventually Lyme disease was shown to be caused by *Borrelia burgdorferi*. It is now recognized that Lyme disease has been around for centuries.

Signs and Symptoms

Lyme disease is nonfatal, but it often evolves into a slowly progressive syndrome that mimics neuromuscular and rheumatoid conditions. An early symptom in 70% of cases is a rash at the site of a tick bite. The lesion, called *erythema migrans,* looks something like a bull's-eye, with a raised erythematous (reddish) ring that gradually spreads outward and a pale central region **(figure 20.7).** Other early symptoms are fever, headache, stiff neck, and dizziness. If not treated or if treated too late, the disease can advance to the second stage, during which cardiac and neurological symptoms, such as facial palsy, can develop. After several weeks or months, a crippling polyarthritis can attack joints. Some people acquire chronic neurological complications that are severely disabling.

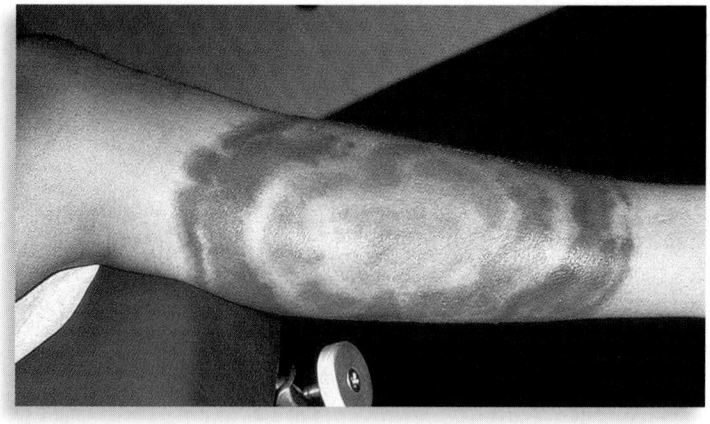

Figure 20.7 Lesions of Lyme disease on the lower leg. Note the flat, reddened rings in the form of a bull's-eye.

Causative Agent

Borrelia burgdorferi was discovered in 1981 by Dr. Willy Burgdorfer, although he did not realize at that time its connection with disease. Borrelia are spirochetes, but they are morphologically distinct from other pathogenic spirochetes. They are comparatively larger, ranging from 0.2 to 0.5 micrometer in width and from 10 to 20 micrometer in length, and they contain 3 to 10 irregularly spaced and loose coils **(figure 20.8).** The nutritional requirements of *Borrelia* are so complex that the bacterium can be grown in artificial media only with difficulty.

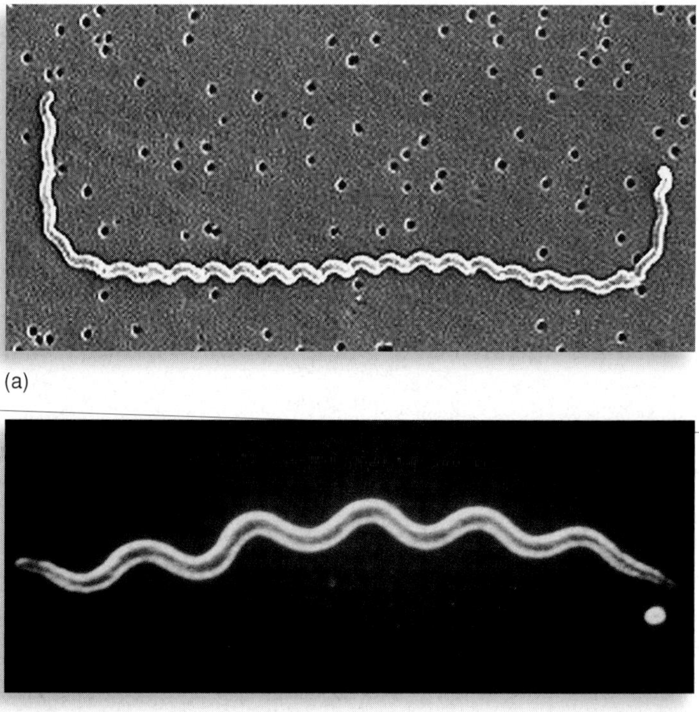

(a)

(b)

Figure 20.8 Spirochetes.
(a) *Leptospira* has numerous fine, regular coils and one or both ends curved. **(b)** *Borrelia* has 3 to 10 loose, irregular coils.

INSIGHT 20.2 *Medical*

The Arthropod Vectors of Infectious Disease

Many bacterial pathogens have evolved with and made complex adaptations to the bodies of arthropods, particularly insects and arachnids. In their role as biological vectors, they are an important source of zoonotic infections in humans. In this chapter alone, you learn about the flea transmitting plague; lice transmitting trench fever; and the very busy tick transmitting tularemia, Lyme disease, and ehrlichioses to humans, while playing a part in keeping Q fever cycling among animal species. Here we describe some main groups implicated in disease.

Arthropod vectors.
(a) Hard (ixodid) tick. **(b)** An engorged soft tick. **(c)** The body louse. **(d)** Cat flea.

Ticks

There are over 810 species of ticks throughout the world. About 100 of them are vectors of infectious disease. Ticks are arachnids, as compared with fleas and lice, which are insects.

Hard (ixodid) ticks have adapted to a wide-ranging lifestyle, hitchhiking along as their hosts wander through forest, savanna, or desert regions. Depending upon the species, ticks feed during larval, nymph, and adult metamorphic stages. The longevity of ticks is formidable; metamorphosis can extend for 2 years, and adults can survive for 4 years away from a host without feeding. The tiny unengorged ticks humans pick up from vegetation crawl on the body, embed their mouthparts in the skin, and fill with blood, expanding to hundreds of times in size. Ixodid ticks are implicated in Rocky Mountain spotted fever and Q fever, as well as the ehrlichioses.

Fleas

Fleas are laterally flattened, wingless insects with well-developed jumping legs and a prominent proboscis for piercing the skin of warm-blooded animals. They are known for their extreme longevity and resistance, and many are notorious in their nonspecificity, passing with ease from wild or domesticated mammals to humans. In response to mechanical stimulation and warmth, fleas jump onto their targets and crawl about, feeding as they go. A well-known example is the oriental rat flea that transmits *Rickettsia typhi,* the cause of murine typhus. The flea harbors the pathogen in its gut and periodically contaminates the environment with virulent bacteria by defecating. This same flea is involved in the transmission of plague.

Lice

Lice (singular, louse) are small, flat insects equipped with biting or sucking mouthparts. The lice of humans usually occupy head and body hair or pubic, chest, and axillary hair. They feed by gently piercing the skin and sucking blood and tissue fluid. Infection develops when the louse (or its feces) is inadvertently squashed and rubbed into wounds, skin, eyes, or mucous membranes. See the discussion on trench fever to read about a disease transmitted by lice.

Pathogenesis and Virulence Factors

The bacterium is a master of immune evasion. It changes its surface antigens while it is in the tick and again after it has been transmitted to a mammalian host. It provokes a strong humoral and cellular immune response, but this response is mainly ineffective, perhaps because of the bacterium's ability to switch its antigens. Indeed, it is possible that the immune response contributes to the pathology of the infection.

B. burgdorferi also has multiple proteins for attachment to host cells; these are considered virulence factors as well.

Transmission and Epidemiology

B. burgdorferi is transmitted primarily by hard ticks of the genus *Ixodes.* (See **Insight 20.2** for a discussion of ticks and other arthropod vectors of diseases.) In the northeastern part of the United States, *Ixodes scapularis* (the black-legged deer tick **figure 20.9**) passes through a complex 2-year cycle that involves two principal hosts **(figure 20.10).** As a larva or nymph, it feeds on either the white-footed mouse or birds or racoons, where it picks up the infectious agent. The nymph is relatively nonspecific and will try to feed on nearly any type of vertebrate—thus, it is the form most likely to bite humans. The adult tick reproductive phase of the cycle is completed

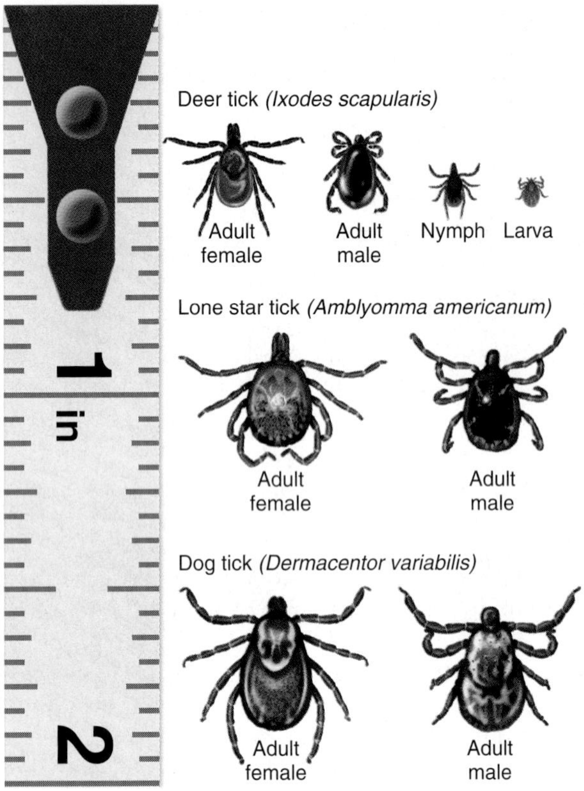

(b)

Figure 20.9
(a) Relative sizes of deer ticks (*Ixodes scapularis*), which are known to transmit Lyme disease, and the other common ticks that are not known to transmit the disease. **(b)** The nymph stage, which is infective to humans shown actual size.

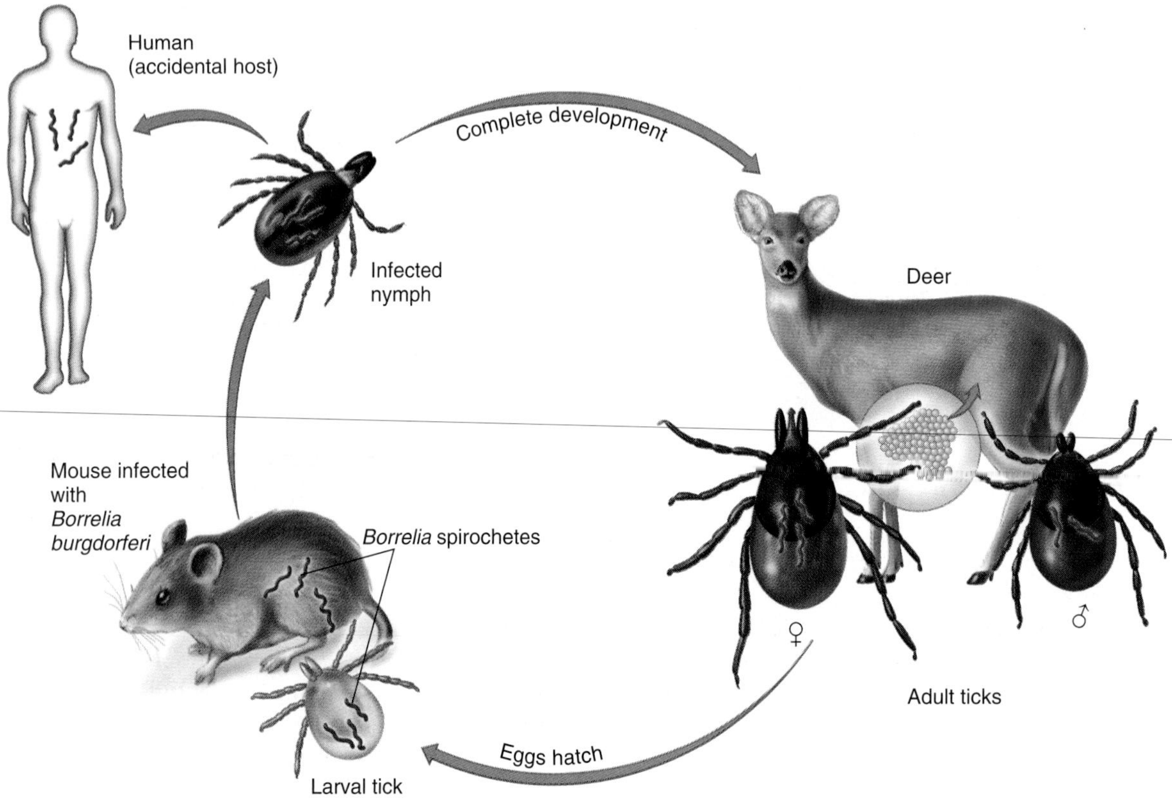

Figure 20.10 The cycle of Lyme disease in the northeastern United States. The exact reservoir hosts vary from region to region in the United States and worldwide.

on deer. In California, the transmission cycle involves *Ixodes pacificus,* another black-legged tick, and the dusky-footed woodrat as reservoir.

The incidence of Lyme disease showed a gradual upward trend from about 10,000 cases per year in 1991 to 18,000 in 2002. This increase may be partly due to improved diagnosis, but it also reflects changes in the numbers of hosts and vectors. The greatest concentrations of Lyme disease are found in areas having high deer populations. Most of the cases have occurred in New York, Pennsylvania, Connecticut, New Jersey, Rhode Island, and Maryland, but the numbers in the Midwest and West are growing. Highest risk groups include hikers, backpackers, and people living in newly developed communities near woodlands and forests.

Culture and Diagnosis

Diagnosis of Lyme disease can be difficult because of the range of symptoms it presents. Most suggestive are the ring-shaped lesions, isolation of spirochetes from the patient, and serological testing with an ELISA method that tracks a rising antibody titer. Tests for spirochetal DNA in specimens is especially helpful for late-stage diagnosis.

Prevention and Treatment

A vaccine for Lyme disease was available for a brief period of time, but it was withdrawn from the market in early 2002 because of controversy over its possible side effects. Other vaccines are in development. Because dogs can also acquire the disease, there is a vaccine for them. Anyone involved in outdoor activities should wear protective clothing, boots, leggings, and insect repellant containing DEET.[2] Individuals exposed to heavy infestation should routinely inspect their bodies for ticks and remove ticks gently without crushing, preferably with forceps or fingers protected with gloves, because it is possible to become infected by tick feces or body fluids.

Early, prolonged (3 to 4 weeks) treatment with doxycycline and amoxicillin is effective, and other antibiotics such as ceftriaxone and penicillin are used in late Lyme disease therapy.

CHECKPOINT 20.5	Lyme Disease
Causative Organism(s)	*Borrelia burgdorferi*
Most Common Modes of Transmission	Vector, biological
Virulence Factors	Antigenic shifting, adhesins
Culture/Diagnosis	ELISA for Ab, PCR
Prevention	Tick avoidance
Treatment	Doxycycline and/or amoxicillin (3–4 weeks), also cephalosporins and penicillin

2. N,N-Diethyl-M-toluamide—the active ingredient in OFF! and Cutter brand insect repellants.

Infectious Mononucleosis

This lymphatic system disease, which is often simply called "mono" or the "kissing disease," can be caused by a number of bacteria or viruses, but the vast majority of cases are caused by the **Epstein-Barr virus (EBV)** and most of the remainder are caused by cytomegalovirus (CMV). Both of these viruses are in the herpes family.

Signs and Symptoms

The symptoms of mononucleosis are sore throat, high fever, and cervical lymphadenopathy, which develop after a long incubation period (30 to 50 days). Many patients also have a gray-white exudate in the throat, a skin rash, and enlarged spleen and liver. A notable sign of mononucleosis is sudden leukocytosis, consisting initially of infected B cells and later T cells. Fatigue is a hallmark of the disease. Patients remain fatigued for a period of weeks. During that time, they are advised not to engage in strenuous activity due to the possibility of injuring their enlarged spleen (or liver).

Eventually, the strong, cell-mediated immune response is decisive in controlling the infection and preventing complications. But after recovery, people usually remain chronically infected with EBV and CMV.

Epstein-Barr Virus

Although "mono" was first described more than a century ago, its most frequent cause was finally discovered through a series of accidental events starting in 1958, when Michael Burkitt discovered an unusual malignant tumor in African children (Burkitt's lymphoma) that appeared to be infectious. Later, Michael Epstein and Yvonne Barr cultured a virus from tumors that showed typical herpesvirus morphology. Evidence that the two diseases had a common cause was provided when a laboratory technician accidentally acquired mononucleosis while working with the Burkitt's lymphoma virus. The Epstein-Barr virus shares morphological and antigenic features with other herpesviruses; and in addition, it contains a circular form of DNA that is readily spliced into the host cell DNA.

Scientists have long suspected a link between chronic EBV infection and illnesses such as chronic fatigue syndrome, but the connection is still controversial. In 2003, a report in the *New England Journal of Medicine* presented strong evidence that chronic EBV infection was necessary, although probably not sufficient, to cause certain forms of Hodgkin's lymphoma.

Pathogenesis and Virulence Factors The latency of the virus and its ability to splice its DNA into host cell DNA make it an extremely versatile virus that can avoid the host's immune response.

Transmission and Epidemiology More than 90% of the world's population is infected with EBV. In general, the virus causes no noticeable symptoms, but the time of life when the virus is first encountered seems to matter. In the case of EBV, infection during the teen years seems to result in disease, whereas infection before or after this period is usually asymptomatic. You will soon see that infection with CMV during the fetal period can lead to severe disease.

Direct oral contact and contamination with saliva are the principal modes of transmission, although transfer through blood transfusions, sexual contact, and organ transplants is possible.

Culture and Diagnosis A differential blood count that shows excess lymphocytes, reduced neutrophils, and large, atypical lymphocytes with lobulated nuclei and vacuolated cytoplasm is suggestive of EBV infection **(figure 20.11).** A test called the "Monospot test" detects *heterophile antibodies*—which are antibodies that are not directed against EBV but are seen when a person has an EBV infection. This test is not reliable in children younger than age 4, in which case a specific EBV antigen/antibody test is conducted.

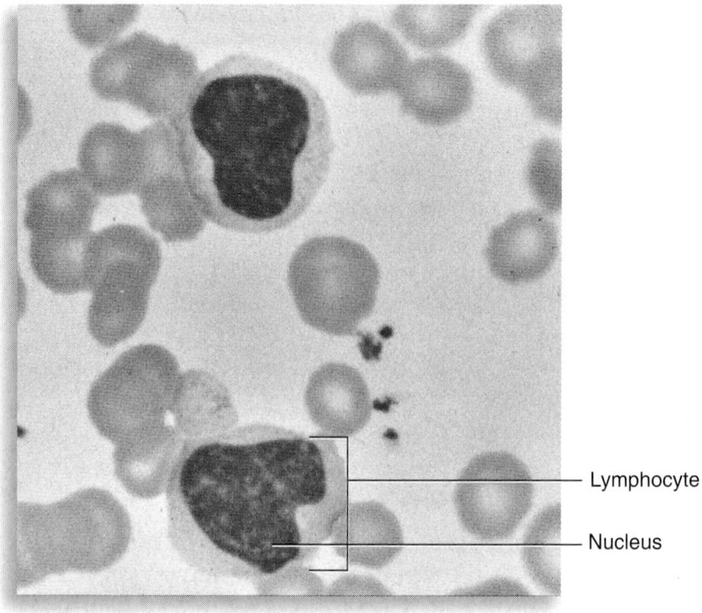

Figure 20.11 Evidence of Epstein-Barr infection in the blood smear of a patient with infectious mononucleosis. Note the abnormally large lymphocytes containing indented nuclei with light discolorations.

Prevention and Treatment The usual treatments for infectious mononucleosis are directed at symptomatic relief of fever and sore throat. Hospitalization is rarely needed. Occasionally, rupture of the spleen necessitates immediate surgery to remove it.

Cytomegalovirus

Cytomegalovirus (CMV) is also a herpesvirus. It is generally distinguished by its ability to produce giant (megalo) cells (cyto) with nuclear and cytoplasmic inclusion bodies. Like other herpesviruses, both EBV and CMV have a tendency to become latent in host cells. Infections are likely to be permanent. The viruses do not reemerge the way herpes simplex viruses do, unless a patient becomes severely immunocompromised. (CMV ocular symptoms are a common complication of AIDS—affecting up to 40% of all AIDS patients.)

Pathogenesis and Virulence Factors The ability of the virus to fuse cells and its latency both contribute to its virulence.

Transmission and Epidemiology Like EBV, CMV is ubiquitous in humans. Unlike EBV, CMV generally causes disease only in fetuses, newborns, and immunodeficient adults. Although not covered here, CMV infection of fetuses affects up to 5,000 babies a year and can cause long-term neurological and sensory disturbances.

CMV is transmitted in saliva, respiratory mucus, milk, urine, semen, cervical secretions, and feces. Transmission usually involves intimate contact such as sex, vaginal birth, transplacental infection, blood transfusion, and organ transplantation.

Culture and Diagnosis During CMV mononucleosis, the virus can be isolated from virtually all organs as well as from epithelial tissue. Cell enlargement and prominent inclusions in the cytoplasm and nucleus are suggestive of CMV. The virus can be cultured and tested with monoclonal antibody against a CMV protein called *early nuclear antigen.* Direct ELISA tests and DNA probe analysis are also useful in diagnosis. Testing serum for antibodies may fail to diagnose infection in neonates and in the immunocompromised, so it is less reliable.

Prevention and Treatment Drug therapy is generally reserved for serious disease in immunosuppressed patients, and not for CMV mononucleosis. The three main drugs are ganciclovir, valacyclovir, and foscarnet, which have toxic side effects and cannot be administered for long periods. The development of a vaccine is hampered by the lack of an animal than can be infected with human cytomegalovirus. One crucial concern is whether vaccine-stimulated antibodies would be protective, because patients already seropositive can become naturally reinfected. Despite these odds, clinical trials began in late 2003 for an experimental CMV vaccine **(Checkpoint 20.6).**

Hemorrhagic Fever Diseases

A number of agents that infect the blood and lymphatics cause extreme fevers, some of which are accompanied by internal hemorrhaging. The diseases are grouped into the category of "hemorrhagic fevers" and are covered in this section. The following section deals with diseases in which the main symptom is fever—without the hemorrhagic part.

CHECKPOINT 20.6	Infectious Mononucleosis	
Causative Organism(s)	Epstein-Barr virus (EBV)	Cytomegalovirus (CMV)
Most Common Modes of Transmission	Direct, indirect contact, parenteral	Direct, indirect contact, parenteral, vertical
Virulence Factors	Latency, ability to incorporate into host DNA	Latency, ability to fuse cells
Culture/Diagnosis	Differential blood count, Monospot test for heterophile antibody, specific ELISA	Virus isolation and growth, ELISA or PCR tests
Prevention	–	Vaccine in trials
Treatment	Supportive	Only for immunosuppressed patients, not usually for mononucleosis
Distinctive Features	Most common in teens	More common in adults, dangerous to fetus

All hemorrhagic fever diseases described here are caused by viruses in one of three families: Arenaviridae, Filoviridae, and Flaviviridae. Bunyaviridae is a fourth family with members that cause hemorrhagic fevers, but we do not discuss examples of these here. All of these viruses are RNA enveloped viruses, the distribution of which is restricted to their natural host's distribution.

Yellow Fever

This disease is caused by an arbovirus, a single-stranded RNA flavivirus that is generally called the yellow fever virus. It currently occurs only in parts of Africa and South America. Two patterns of transmission are seen in nature. One is an urban cycle between humans and the mosquito *Aedes aegypti*, which reproduces in standing water in cities. The other is a sylvan (forest) cycle, maintained between forest monkeys and mosquitoes.

The presence of the virus in the bloodstream causes capillary fragility and disrupts the blood-clotting system, which can lead to localized bleeding and shock. Infection begins acutely with fever, headache, and muscle pain. In some patients, the disease progresses to oral hemorrhage, nosebleed, vomiting, jaundice, and liver and kidney damage with significant mortality rates. Most cases occur during the rainy season.

Dengue Fever

Dengue fever is caused by a single-stranded RNA flavivirus and is also carried by *Aedes* mosquitoes. Although mild infection is the usual pattern, a form called dengue hemorrhagic shock syndrome can be lethal. Dengue fever is also called "breakbone fever" because of the severe pain it induces in muscles and joints (it does not actually cause fractures). The illness is endemic to Southeast Asia and India, and several epidemics have occurred in South America and Central America, the Caribbean, and Mexico. The Pan American Health Organization has reported an ongoing epidemic of dengue fever in the Americas that increased from 390,000 cases in 1984 to more than 1 million cases in 2002. In Mexico, cases have increased 600% since 2001.

Researchers in Thailand, where dengue fever is one of the leading causes of child mortality, have developed a live attenuated vaccine, which is being tested in clinical trials. A low-tech approach has led to big successes in Vietnam. There, health officials urged local citizens to round up tiny crustaceans that are common in natural water sources and to put them in water tanks and wells. The crustaceans, which are not harmful to humans, eat the mosquitoes that carry dengue. Officials reported a complete elimination of the disease in communities where the strategy was used.

Ebola and Marburg

Unlike the two viruses causing yellow fever and dengue fever, the Ebola and Marburg viruses are filoviruses (Family Filoviridae). The two viruses are related and cause similar symptoms, although Ebola has received the greatest share of media attention. Its gruesome symptoms are extreme manifestations of the same kind of hemorrhagic events described for yellow fever and dengue fever. The virus in the bloodstream leads to extensive capillary fragility and disruption of clotting. Patients bleed from their orifices, even from their mucous membranes, and experience massive internal and external hemorrhage. Very often they manifest a rash on their trunk in early stages of the disease. The mortality rate is between 25% and 100%, and there is no effective treatment.

It is not known how humans acquire these viruses. They are both indigenous to Africa. Until 2007 their natural reservoir was unknown. In August of 2007 researchers found the virus in a cave-dwelling fruit bat. Direct contact with an infected person or with their body fluids will transmit the virus. Hospital workers caring for Ebola patients are at high risk of becoming infected.

In 2002 and 2003, Ebola caused a catastrophic epidemic among lowland gorillas in Central Africa. During that time dozens of humans also contracted the disease, probably from handling dead gorilla carcasses or from using primate meat for food. The disease was then transmitted from human to human.

Outbreaks with Marburg virus are also rare, but individuals have been infected sporadically since it was first recognized in 1967. In 2005, the largest Marburg outbreak

CHECKPOINT 20.7 Hemorrhagic Fevers

Disease	Yellow Fever	Dengue Fever	Ebola and/or Marburg	Lassa Fever
Causative Organism(s)	Yellow fever virus	Dengue fever virus	Ebola virus, Marburg virus	Lassa fever virus
Most Common Modes of Transmission	Biological vector	Biological vector	Direct contact, body fluids	Droplet contact (aerosolized rodent excretions), direct contact with infected fluids
Virulence Factors	Disruption of clotting factors	Disruption of clotting factors	Disruption of clotting factors	Disruption of clotting factors
Culture/Diagnosis	ELISA, PCR	Rise in IgM titers	PCR, viral culture (conducted at CDC)	ELISA
Prevention	Live attenuated vaccine available	Live attenuated vaccine being tested	–	Avoiding rats, safe food storage
Treatment	Supportive	Supportive	Supportive	Ribavirin
Distinctive Features	Accompanied by jaundice	"Breakbone fever"—so named due to severe pain	Massive hemorrhage; rash sometimes present	Chest pain, deafness as long-term sequelae

in history occurred in and around a hospital in Angola. Sixty-three people died during the 5-month outbreak. Symptoms are similar to Ebola virus infection.

There is no treatment and no vaccine for Ebola or Marburg, though some promising research is being conducted. An antisense molecule (see chapter 10) has been shown to be an effective treatment in guinea pigs and monkeys.

Lassa Fever

The Lassa fever virus is an arenavirus. Several related arenaviruses cause the diseases Argentine hemorrhagic fever, Bolivian hemorrhagic fever, and lymphocytic choriomeningitis (an infection of the brain and meninges). Lassa fever virus is found in West Africa. In most cases infection with this virus is asymptomatic, but in 20% of the cases a severe hemorrhagic syndrome develops. The syndrome includes chest pain, hemorrhaging, sore throat, back pain, vomiting, diarrhea, and sometimes encephalitis. Patients who recover suffer from deafness at a significant rate.

The reservoir of the virus is a rodent found in Africa called the multimammate rat. It is spread to humans through aerosolization of rat droppings, urine, hair, and so forth. Eating food contaminated by rat excretions also transmits the virus. Infected persons can spread it to other people through their own secretions. Vertical transmission also occurs, and the disease leads to spontaneous abortions in 95% of infected pregnant women.

This hemorrhagic fever has been shown to respond to the antiviral agent ribavirin, especially if administered in the early stages of infection. There is no vaccine **(Checkpoint 20.7).**

Nonhemorrhagic Fever Diseases

In this section, we examine some infectious diseases that result in a syndrome characterized by high fever but without the capillary fragility that leads to hemorrhagic symptoms. All of the diseases in this section are caused by bacteria.

Brucellosis

This disease goes by several different names (besides brucellosis): Malta fever, undulant fever, and Bang's disease.[3] It is on the CDC list of possible bioterror agents, though it is not designated as being "of highest concern."

Signs and Symptoms The *Brucella* bacteria responsible for this disease live in phagocytic cells. These cells carry the bacteria into the bloodstream, creating focal lesions in the liver, spleen, bone marrow, and kidney. The cardinal manifestation of human brucellosis is a fluctuating pattern of fever, which is the origin of the common name *undulant fever* **(figure 20.12).** It is also accompanied by chills, profuse sweating, headache, muscle pain and weakness, and weight loss. Fatalities are not common, although the syndrome can last for a few weeks to a year, even with treatment.

Causative Agent The bacterial genus *Brucella* contains tiny, aerobic gram-negative coccobacilli. Two species can cause this disease in humans: *B. abortus* (common in cattle) and *B. suis* (from pigs). Humans can become infected with either of these bacteria and experience severe disease. Even though a principal manifestation of the disease in animals is an infection of the placenta and fetus, human placentas do not become infected.

Pathogenesis and Virulence Factors *Brucella* enters through damaged skin or via mucous membranes of the digestive tract, conjunctiva, and respiratory tract. From there it is taken up by phagocytic cells. Because it is able to avoid destruction in the

3. After B. L. Bang, a Danish physician.

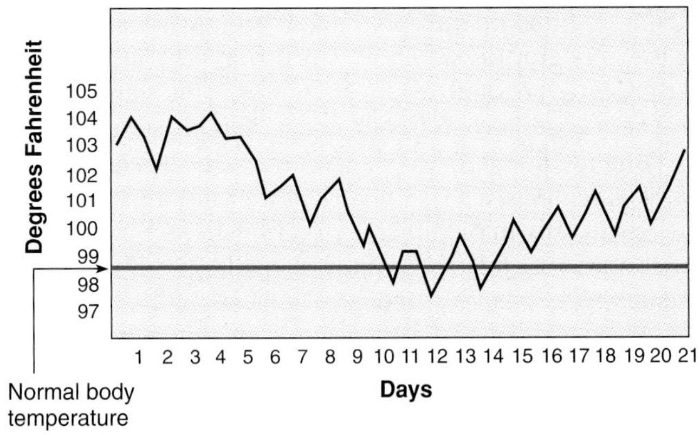

Figure 20.12 The temperature cycle in classic brucellosis. Body temperature undulates between day and night and between fever, normal, and subnormal.
Source: A. Smith, *Principles of Microbiology,* 10th ed., 1985.

phagocytes, the bacterium is transported easily through the bloodstream and to various organs, such as the liver, kidney, breast tissue, or joints. Scientists suspect that the up-and-down nature of the fever is related to unusual properties of the bacterial lipopolysaccharide.

Transmission and Epidemiology Brucellosis occurs worldwide, with concentrations in Europe, Africa, India, and Latin America. It is associated predominantly with occupational contact in slaughterhouses, livestock handling, and the veterinary profession. Infection takes place through contact with blood, urine, placentas, and through consumption of raw milk and cheese. Human-to-human transmission is rare. Needlesticks are one of the more common modes of transmission in the United States. In 2007, a researcher in a university lab that studied possible bioweapons agents contracted brucellosis while cleaning a chamber used to infect mice.

Brucellosis is also a common disease of wild herds of bison and elk. Cattle that share grazing land with these wild herds often suffer severe outbreaks of the placental infections (called Bang's disease).

Culture and Diagnosis The patient's history can be very helpful in diagnosis, as are serological tests of the patient's blood and blood culture of the pathogen. In areas where *Brucella* is endemic, serology is of limited use because significant proportions of the population already display antibodies to the bacterium. Blood culture is positive in less than 40% of cases; Gram staining of biopsy material from lymph nodes or bone marrow (from the sternum) is considered more reliable.

Prevention and Treatment Prevention is effectively achieved by testing and elimination of infected animals, quarantine of imported animals, and pasteurization of milk. Although several types of animal vaccines are available, those developed so far for humans are ineffective or unsafe. The status of this pathogen as a potential germ warfare agent makes a reliable vaccine even more urgent.

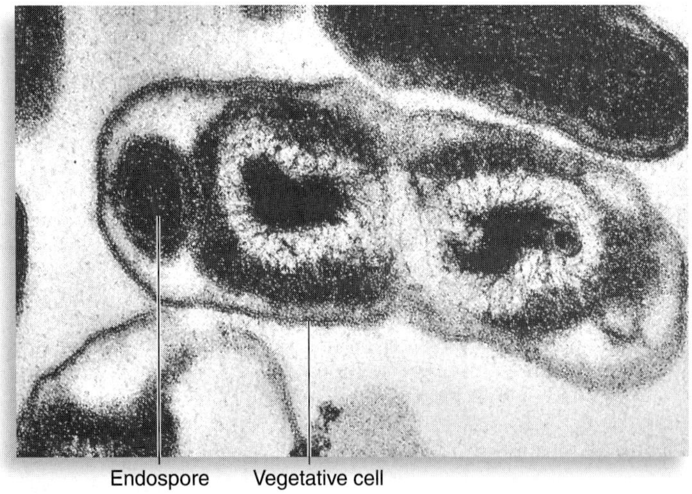

Figure 20.13 The agent of Q fever. The vegetative cells of *Coxiella burnetii* produce unique endospore-like structures that are released when the cell disintegrates. Free spores survive outside the host and are important in transmission.

A combination of doxycycline and gentamicin or streptomycin taken for 3 to 6 weeks is usually effective in controlling infection.

Q Fever

The name of this disease arose from the frustration created by not being able to identify its cause. The Q stands for "query." Its cause, a bacterium called *Coxiella burnetii,* was finally identified in the mid-1900s. The clinical manifestations of acute Q fever are abrupt onset of fever, chills, head and muscle ache, and, occasionally, a rash. The disease is sometimes complicated by pneumonitis (30% of cases), hepatitis, and endocarditis. About a quarter of the cases are chronic rather than acute and result in vascular damage and endocarditis-like symptoms.

C. burnetii is a very small pleomorphic gram-negative bacterium, and for a time it was considered a rickettsia. It is an intracellular parasite, but it is much more resistant to environmental pressures because it produces an unusual type of endosporelike structure **(figure 20.13).** *C. burnetii* is apparently harbored by a wide assortment of vertebrates and arthropods, especially ticks, which play an essential role in transmission between wild and domestic animals. Ticks do not transmit the disease to humans, however. Humans acquire infection largely by means of environmental contamination and airborne spread. Birth products, such as placentas, of infected domestic animals contain large numbers of bacteria. Other sources of infectious material include urine, feces, milk, and airborne particles from infected animals. The primary portals of entry are the lungs, skin, conjunctiva, and gastrointestinal tract.

C. burnetii has been isolated from most regions of the world. California and Texas have the highest case rates in the United States, although most cases probably go undetected. People at highest risk are farm workers, meat cutters, veterinarians, laboratory technicians, and consumers of raw milk products.

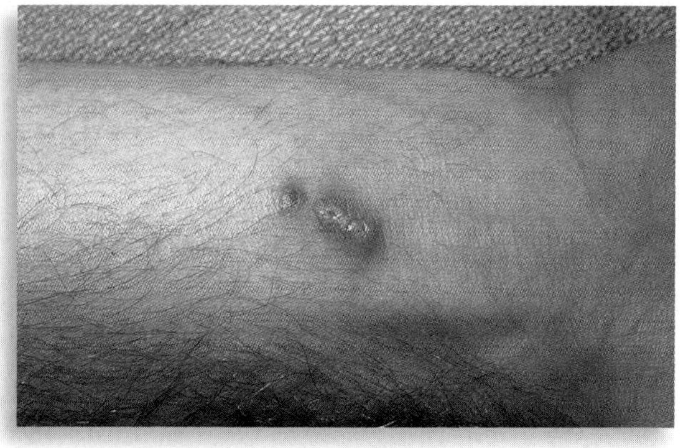

Figure 20.14 Cat-scratch disease.
A primary nodule appears at the site of the scratch in about 21 days. In time, large quantities of pus collect and the regional lymph nodes swell.

Mild or subclinical cases resolve spontaneously, and more severe cases respond to doxycycline therapy. A vaccine is available in many parts of the world and is used for U.S. military and laboratory workers. Q fever is of potential concern as a bioterror agent because it is very resistant to heat and drying, it can be inhaled, and even a single bacterium is enough to cause disease. It is an organism that the U.S. military worked with during the period when potential biowarfare agents were being developed in this country (the 1950s and 1960s).

Cat-Scratch Disease

This disease is one of a pair of diseases caused by different species of the small gram-negative rod *Bartonella*. *Bartonella* species are considered to be emerging pathogens. They are fastidious but not obligate intracellular parasites, so they will grow on blood agar. In addition to cat-scratch disease and trench fever, discussed next, *Bartonella* species cause a particulary nasty cutaneous and systemic infection in AIDS patients called bacillary angiomatosis.

Bartonella henselae is the agent of cat-scratch disease (CSD), an infection connected with being clawed or bitten by a cat. The pathogen is present in over 40% of cats, especially kittens. There are approximately 25,000 cases per year in the United States, 80% of them in children 2 to 14 years old. The symptoms start after 1 to 2 weeks, with a cluster of small papules at the site of inoculation **(figure 20.14)**. In a few weeks, the lymph nodes along the lymphatic drainage swell and can become pus-filled. Only about one-third of patients experience high fever. Most infections remain localized and resolve in a few weeks, but drugs such as azithromycin, erythromycin, and rifampin can be effective therapies. The disease can be prevented by thorough antiseptic cleansing of a cat bite or scratch.

Trench Fever

This disease has a long history. Trench fever was once a common condition of soldiers in battle. The causative agent, *Bartonella quintana*, is carried by lice. Most cases occur in endemic regions of Europe, Africa, and Asia, although the disease is beginning to show up in poverty-stricken areas of large cities in the developed world. This version of the disease is called "urban trench fever." Highly variable symptoms can include a 5- to 6-day fever (the species epithet, *Quintana*, refers to a 5-day fever). Symptoms also include leg pains, especially in the tibial region (the disease is sometimes called "shinbone fever"), headache, chills, and muscle aches. A macular rash can also occur. (See Insight 18.3 for definitions of skin lesions.) Endocarditis can develop, especially in the urban version of the disease. The microbe can persist in the blood long after convalescence and is responsible for later relapses.

Trench fever may be treated with doxycycline or erythromycin.

HGA and HME

There are two similar tick-borne, fever-producing diseases caused by members of the genus *Ehrlichia* and *Anaplasma*. The causative organisms for the two diseases were thought to be in the single species *Ehrlichia* until 2005, when a reclassification identified the two different genera.

Members of the two genera are small intracellular bacteria, and they share many characteristics with rickettsia, including a strict parasitic existence and association with ticks. *Ehrlichia chaffeensis* causes human monocytic ehrlichiosis (HME). *Anaplasma phagocytophilum* causes human granulocytic anaplasmosis (HGA). Another species, *Ehrlichia ewingii*, can cause either syndrome. The diseases are sometimes referred to as "spotless" Rocky Mountain spotted fever.

Both bacteria spend part of their life cycle in ticks in the genus *Ixodes*. The species of tick varies with the various regions of the United States and Europe. Serology samples taken from residents of endemic areas suggest that between 15 to 36% of them have been infected with *A. phagocytophilum*, mostly without symptoms. Both HME and HGA are showing increased incidence, probably due to improved diagnosis.

The signs and symptoms of HGA and HME are similar: an acute febrile state manifesting headache, muscle pain, and rigors. Most patients recover rapidly with no lasting effects, but around 5% of older chronically ill patients die from disseminated infection. Rapid diagnosis is enabled by PCR tests and indirect fluorescent antibody tests. It can be critical to differentiate or detect coinfection with Lyme disease *Borrelia*, which is carried by the same tick. Doxycycline will clear up most infections within 7 to 10 days.

Rocky Mountain Spotted Fever (RMSF)

This disease is named for the region in which it was first detected in the United States—the Rocky Mountains of Montana and Idaho. In spite of its name, the disease occurs infrequently in the western United States. The majority of cases are concentrated in the Southeast and eastern seaboard regions **(figure 20.15)**. It also occurs in Canada and Central and South America. Infections occur most frequently in the spring and

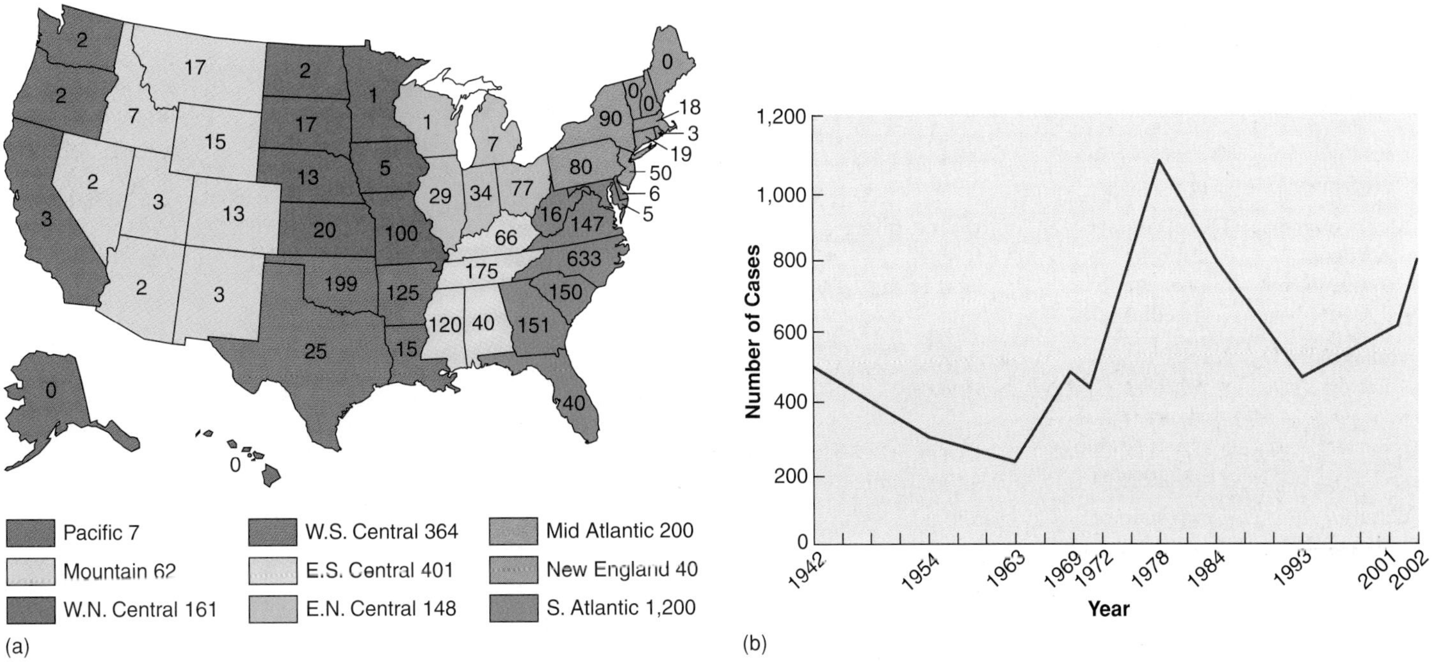

Figure 20.15 Trends in infection for Rocky Mountain spotted fever.
(a) A summary map showing the distribution of Rocky Mountain spotted fever cases reported over a 4-year period. **(b)** Reported cases of Rocky Mountain spotted fever in the United States, 1942–2002. The spike in cases in the late 1970s may be due to increased reporting.

summer, when the tick vector is most active. The yearly rate of RMSF is 20 to 40 cases per 10,000 population, with fluctuations coinciding with weather and tick infestations.

RMSF is caused by a bacterium called *Rickettsia rickettsii* transmitted by hard ticks such as the wood tick (*Dermacentor andersoni*), the American dog tick (*D. variabilis*, among others), and the Lone Star tick (*Ambylomma americanum*). The dog tick is probably most responsible for transmission to humans because it is the major vector in the southeastern United States.

After 2 to 4 days of incubation, the first symptoms are sustained fever, chills, headache, and muscular pain. A distinctive spotted rash usually comes on within 2 to 4 days after the prodrome **(figure 20.16).** Early lesions are slightly mottled like measles, but later ones are macular, maculopapular, and even petechial. In the most severe untreated cases, the enlarged lesions merge and can become necrotic, predisposing to gangrene of the toes or fingertips.

Although the spots are the most obvious symptom of the disease, the most grave manifestations are cardiovascular disruption, including hypotension, thrombosis, and hemorrhage. Conditions of restlessness, delirium, convulsions, tremor, and coma are signs of the often overwhelming effects on the central nervous system. Fatalities occur in an average of 20% of untreated cases and 5% to 10% of treated cases.

Suspected cases of RMSF require immediate treatment even before laboratory confirmation. A recent aid to early diagnosis is a method for staining rickettsias directly in a tissue biopsy using fluorescent antibodies. Isolating rickettsias from the patient's blood or tissues is desirable, but it is expensive and requires specially qualified lab personnel and lab facilities. Specimens taken from the rash lesions are suitable for PCR assay, which is very specific and sensitive and can circumvent the need for culture.

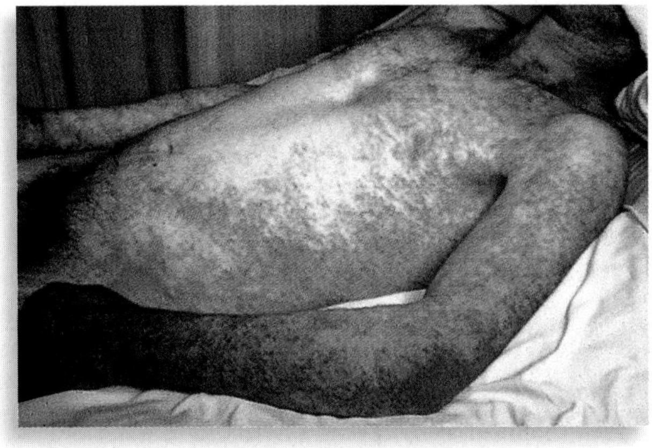

Figure 20.16 Late generalized rash of Rocky Mountain spotted fever.
In some cases, lesions become hemorrhagic and may become gangrenous in the extremities.

The drug of choice for suspected and known cases is doxycycline administered for 1 week. Other preventive measures parallel those for Lyme disease: wearing protective clothing, using insect sprays, and fastidiously removing ticks **(Checkpoint 20.8).**

Malaria

Throughout human history, including prehistoric times, malaria has been one of the greatest afflictions, in the same rank as bubonic plague, influenza, and tuberculosis. Even now,

CHECKPOINT 20.8 Nonhemorrhagic Fever Diseases

Disease	Brucellosis	Q fever	Cat-Scratch Disease	Trench Fever	Ehrlichioses	Rocky Mountain Spotted Fever
Causative Organism(s)	*Brucella abortus* or *B. suis*	*Coxiella burnetii*	*Bartonella henselae*	*Bartonella quintana*	*Ehrlichia* species	*Rickettsia rickettsii*
Most Common Modes of Transmission	Direct contact, airborne, parenteral (needlesticks)	Airborne, direct contact, food-borne	Parenteral (cat scratch or bite)	Biological vector (lice)	Biological vector (tick)	Biological vector (tick)
Virulence Factors	Intracellular growth; avoidance of destruction by phagocytes	Endosporelike structure	Endotoxin	Endotoxin	–	Induces apoptosis in cells lining blood vessels
Culture/Diagnosis	Gram stain of biopsy material	Serological tests for antibody	Biopsy of lymph nodes plus Gram staining; ELISA (performed by CDC)	ELISA (performed by CDC)	PCR, indirect antibody test	Fluorescent antibody, PCR
Prevention	Animal control, pasteurization of milk	Vaccine for high-risk population	Clean wound sites	Avoid lice	Avoid ticks	Avoid ticks
Treatment	Doxycycline plus (gentamicin or streptomycin)	Doxycycline	Azithromycin	Doxycycline or erythromycin	Doxycycline	Doxycycline
Distinguishing Characteristics	Undulating fever, muscle aches	Airborne route of transmission, variable disease presentation	History of cat bite or scratch; fever not always present	Endocarditis common, 5-day fever	Seasonal occurrence (April–Oct.)	Most common in east and southeast United States

as the dominant protozoan disease, it threatens 40% of the world's population every year. The origin of the name is from the Italian words *mal*, bad, and *aria*, air. The superstitions of the Middle Ages alleged that evil spirits or mists and vapors arising from swamps caused malaria, because many victims came down with the disease after this sort of exposure. We now know that a swamp was mainly involved as a habitat for the mosquito vector.

Signs and Symptoms

After a 10- to 16-day incubation period, the first symptoms are malaise, fatigue, vague aches, and nausea with or without diarrhea, followed by bouts of chills, fever, and sweating. These symptoms occur at 48- or 72-hour intervals, as a result of the synchronous rupturing of red blood cells. The interval, length, and regularity of symptoms reflect the type of malaria (described next). Patients with falciparum malaria, the most virulent type, often manifest persistent fever, cough, and weakness for weeks without relief. Complications of malaria are hemolytic anemia from lysed blood cells and organ enlargement and rupture due to cellular debris that accumulates in the spleen, liver, and kidneys. One of the most serious complications of falciparum malaria is termed *cerebral malaria*. In this condition, small blood vessels in the brain become obstructed due to their increased ability to adhere to vessel walls (a condition called *cytoadherence* induced by the infecting protozoan). The resulting decrease in oxygen in brain tissue can result in coma and death. In general, malaria has the highest death rate in the acute phase, especially in children. Certain kinds of malaria are subject to relapses because some infected liver cells harbor dormant protozoans for up to 5 years.

Causative Agent

Plasmodium species are protozoans in the sporozoan group. They are **apicomplexans,** which live in animal hosts and lack locomotor organelles in the mature state (chapter 5 describes protozoan classification). Apicomplexans alternate between sexual and asexual phases, often in different animal hosts. The genus *Plasmodium* contains four species: *P. malariae*, *P. vivax*, *P. falciparum*, and *P. ovale*. Humans are the primary vertebrate hosts for most of the species. The four species show variations in the pattern and severity of disease.

Development of the malarial parasite is divided into two distinct phases: the asexual phase, carried out in the human, and the sexual phase, carried out in the mosquito

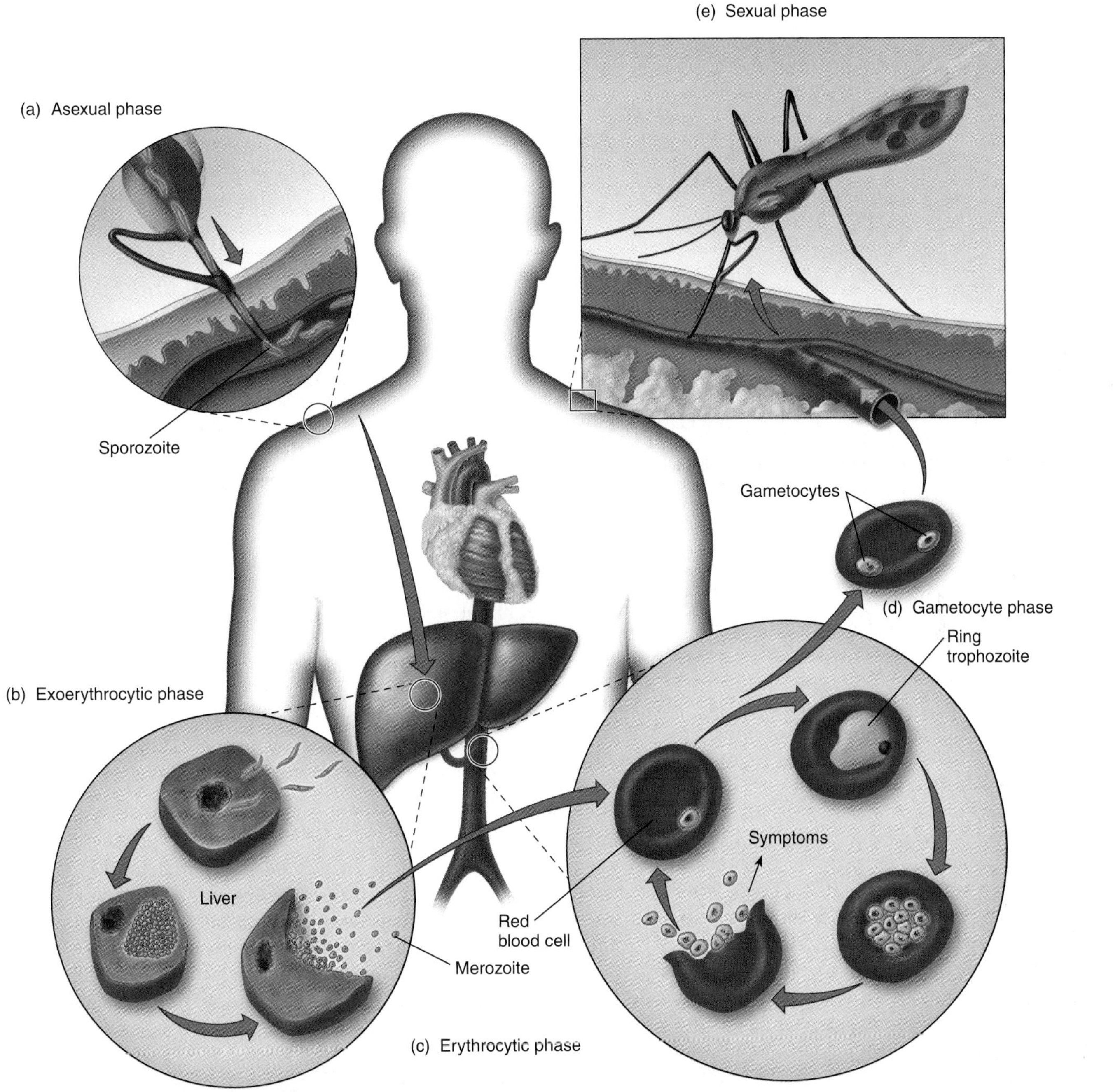

Figure 20.17 The life and transmission cycle of *Plasmodium*, the cause of malaria.
(a) In the asexual phase in humans, sporozoites enter a capillary through the saliva of a feeding mosquito. **(b)** Exoerythrocytic (liver) phase. Sporozoites invade the liver cells and develop into large numbers of merozoites. **(c)** Erythrocytic phase. Merozoites released into the circulation enter red blood cells. Initial infection is marked by a ring trophozoite; schizogony of the ringed form produces additional merozoites that burst out and infect other red blood cells. **(d)** Gametocytes that develop in certain infected red blood cells are ingested by another mosquito. **(e)** The sexual phase of fertilization and sporozoite formation occurs in the mosquito.

(figure 20.17). The *asexual phase* (and infection) begins when an infected female *Anopheles* mosquito injects saliva containing anticoagulant into a capillary in preparation for taking a blood meal. In the process, she inoculates the blood with motile, spindle-shaped asexual cells called **sporozoites** (Gr. *sporo*, seed, and *zoon*, animal). The sporozoites circulate through the body and migrate to the liver in a short time. Within liver cells, the sporozoites undergo asexual division called *schizogony* (Gr. *schizo*, to divide, and *gone*, seed), which generates numerous daughter parasites, or *merozoites*. This phase of *pre-erythrocytic development* lasts from 5 to 16 days, depending upon the species of *Plasmodium*. Its end is marked by eruption of the liver cell, which releases from 2,000 to 40,000 mature merozoites into the circulation.

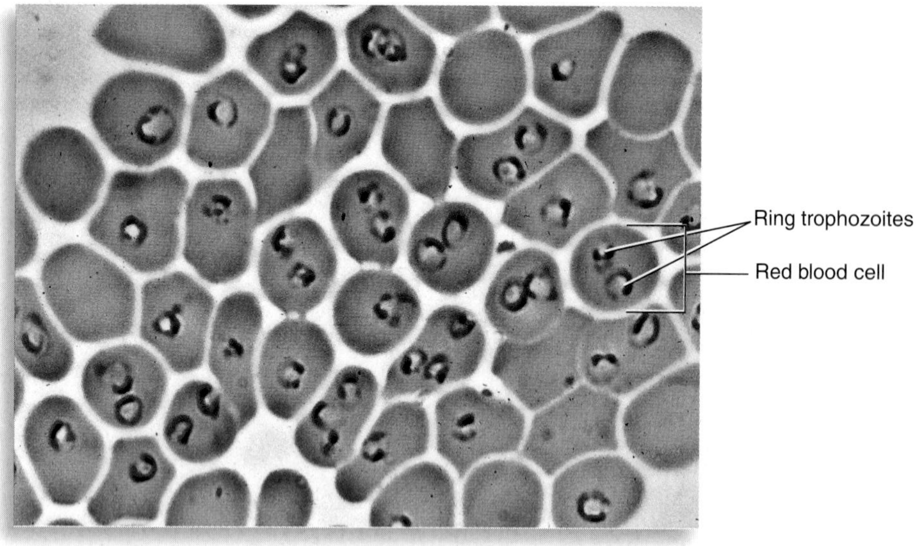

Figure 20.18 The ring trophozoite stage in a *Plasmodium falciparum* infection.
A smear of peripheral blood shows ring forms in red blood cells. Some RBCs have multiple trophozoites.

P. vivax and *P. ovale* have a propensity to persist in the liver; and without sufficient treatment, they can reemerge over the course of several years to cause recurrent bouts of malarial symptoms.

The fact that the protozoan has several different life stages within a host helps it escape immune responses mounted against any single life stage.

During the *erythrocytic phase,* merozoites attach to special receptors on red blood cells (RBCs) and invade them, converting in a short time to circular (ring-shaped) trophozoites **(figure 20.18).** This stage feeds upon hemoglobin, grows, and undergoes multiple divisions to produce a cell called a *schizont,* which is filled with more merozoites. Bursting RBCs liberate merozoites to infect more red cells. Eventually, certain merozoites differentiate into two types of specialized gametes called *macrogametocytes* (female) and *microgametoctyes* (male). Because the human does not provide a suitable environment for the next phase of development, this is the end of the cycle in humans.

The *sexual phase* (sporogony) occurs when a mosquito draws infected red blood cells into her stomach. In the stomach, the microgametocyte releases spermlike gametes that fertilize the larger macrogametocytes. The resultant diploid cell (ookinete) implants into the stomach wall of the mosquito, becoming an oocyst, which undergoes multiple mitotic divisions, ultimately releasing sporozoites that migrate to the salivary glands and lodge there. This event completes the sexual cycle and makes the sporozoites available for infecting the next victim.

Pathogenesis and Virulence Factors

The invasion of the merozoites into RBCs leads to the release of fever-inducing chemicals into the bloodstream. Chills and fevers often occur in a cyclic pattern. *Plasmodium* also metabolizes glucose at a very high rate, leading to hypoglycemia in the human host. The damage to RBCs results in anemia. The accumulation of malarial products in the liver and the immune stimulation in the spleen can lead to enlargement of these organs. The inducement of RBC adhesion to blood vessels in the brain (cytoadherence) adds to its virulence. A surface protein called GPI (glycosyl-phosphatidyl inositol) is thought to be responsible for the fever seen in malaria.

Transmission and Epidemiology

All forms of malaria are spread primarily by the female *Anopheles* mosquito and occasionally by shared hypodermic needles and blood transfusions. Although malaria was once distributed throughout most of the world, the control of mosquitoes in temperate areas has successfully restricted it mostly to a belt extending around the equator. Despite this achievement, approximately 300 million to 500 million new cases are still reported each year, about 90% of them in Africa. The most frequent victims are children and young adults, of whom at least 2 million die annually. A particular form of the malarial protozoan causes damage to the placenta in pregnant women, leading to excess mortality among fetuses and newborns. The total case rate in the United States is about 1,000 to 2,000 new cases a year, most of which occur in immigrants.

Culture and Diagnosis

Malaria can be diagnosed definitively by the discovery of a typical stage of *Plasmodium* in stained blood smears (see figure 20.18). Newer serological procedures have made diagnosis more accurate while requiring less skill to perform. Other indications are knowledge of the patient's residence or travel in endemic areas and symptoms such as recurring chills, fever, and sweating.

Prevention

As recently as 15 to 20 years ago, eradicating malaria seemed possible, but since then, morbidity and mortality have increased in several regions. The two main reasons involve the sheer adaptive and survival capacity of both the parasite and the vector. Standard chemotherapy with antimalarial drugs has selected for drug-resistant strains of the malarial parasite, and mosquitoes have developed resistance to some common insecticides used to control them.

Malaria prevention is attempted through long-term mosquito abatement and human chemoprophylaxis. Abatement includes elimination of standing water that could serve as a breeding site and spraying of insecticides to reduce populations of adult mosquitoes, especially in and near human dwellings. Scientists have also tried introducing sterile male mosquitoes into endemic areas in an attempt to decrease

INSIGHT 20.3 — Historical

Computer Geek + Investment Guru = World Health Revolution

Bill Gates

In 1968, a 13-year-old student at the Lakeside School in Seattle, Washington, discovered a love for programming computers. That event, improbably, eventually led to the most significant chance for the improvement of global health that the world has ever known. That student was Bill Gates; and by the time he was 20, he and a friend, Paul Allen, had begun a company called Microsoft. By the end of 2005, Microsoft had revenues of $40 billion dollars and employed 61,000 people in 102 countries.

Microsoft's huge success put Bill Gates in a position in 2000 to start a charitable foundation with his wife Melinda French Gates. The Gates Foundation has the following two core values: (1) that all lives have equal value; and (2) to whom much has been given, much is expected. It organizes its programs in three catgories: local, national, and international. In the local Seattle area, it focuses on improving the lives of low-income families. In the United States, it seeks to give all people access to a great education and to technology in public libraries. Its worldwide focus is to improve health and to reduce extreme poverty, as well as to increase access to technology.

As of late 2005, the Gates Foundation had committed more than $3.6 billion to world health initiatives. Then, in the summer of 2006, the second wealthiest man in the world (behind Gates), investor Warren Buffett announced he would give away his $42 billion fortune, with most of it going to the Gates Foundation. Buffett's gift constitutes the largest charitable donation in U.S. history.

The Gates Foundation has a list of priority diseases that they are battling **(table 20.A)**, diseases that have only minor impacts on developed countries but that take a devastating toll in the developing world. Remember the first principle of the Gates Foundation: All lives—no matter where they are being led—have equal value.

As one example, the Gates Foundation is providing $258 million for malaria research alone: $108 million of that is targeted to a Malaria Vaccine Initiative, which focuses on testing the vaccine described in the section on malaria; $100 million is devoted to therapeutic treatments for malaria; and $50 million goes to new approaches to mosquito control.

TABLE 20.A Gates Foundation Priority Diseases

Acute diarrheal diseases
Acute lower respiratory diseases
HIV/AIDS
Child health
Malaria
Poor nutrition
Reproductive and maternal health
Tuberculosis
Vaccine-preventable diseases
Other infectious diseases

mosquito populations. Humans can reduce their risk of infection considerably by using netting, screens, and repellants; by remaining indoors at night; and by taking weekly doses of prophylactic drugs. (Western travelers to endemic areas are often prescribed antimalarials for the duration of their trips.) People with a recent history of malaria must be excluded from blood donations. The WHO and other international organizations focus on efforts to distribute bed nets and to teach people how to dip the nets twice a year into an insecticide **(figure 20.19)**. The use of bed nets has been estimated to reduce childhood mortality from malaria by 20%. Even with massive efforts undertaken by the WHO, the prevalence of malaria in endemic areas is still high.

The best protection would come from a malaria vaccine, and scientists have struggled for decades to develop one. A successful malaria vaccine must be capable of striking a diverse and rapidly changing target. Not only are there four different species, each having different sporozoite, merozoite, and gametocyte stages, but each species can also have different antigenic types of sporozoites and merozoites. A sporozoite vaccine would prevent liver infection, whereas a merozoite vaccine would prevent infection of red blood cells and diminish the pathology that is most responsible for morbidity and mortality. In December of 2006, the World Health Organization issued a challenge to researchers: Have a vaccine licensed by 2015.

The most promising malaria vaccine candidate is in large-scale testing now (supported by funding through the Gates Foundation, **Insight 20.3**). It contains a *P. falciparum* protein

Figure 20.19 A public demonstration of impregnating bed nets with insecticide.
Part of the events on Africa Malaria Day 2002 in Benin.

with a hepatitis B surface antigen to improve immunogenicity. It targets the stage of the protozoa that is injected by the mosquito.

Another potentially powerful strategy is the use of interfering RNAs in the mosquitoes to render them resistant to *Plasmodium* infection.

Treatment

Quinine has long been a mainstay of malaria treatment. It is a compound originally found in the bark of the cinchona tree, which grows in the Andes, but is now synthesized in various forms. Quinine is an ingredient in tonic water; for that reason tonic water was popularized among 19th-century British colonists in India who believed that the tonic water could protect them from malaria. (In reality, the concentration of quinine is too low to be of medicinal use.) Chloroquine, the least toxic type, is used in nonresistant forms of the disease. In areas of the world where resistant strains of *P. falciparum* and *P. vivax* predominate, a course of mefloquine, artemisinin, or Fansidar (pyrimethamine plus sulfadoxine) is indicated.

Eliminating the parasite from the liver and preventing relapses can be managed with long-term therapy with primaquine or proguanil. Scientists recently sequenced the entire genome of *P. falciparum*, which will enable them to develop more effective drug treatments and to more effectively circumvent drug resistance in the protozoan.

CHECKPOINT 20.9	Malaria
Causative Organism(s)	*Plasmodium falciparum, P. vivax, P. ovale, P. malariae*
Most Common Modes of Transmission	Biological vector (mosquito), vertical
Virulence Factors	Multiple life stages; multiple antigenic types, ability to scavenge glucose, GPI, cytoadherence
Culture/Diagnosis	Blood smear; serological methods
Prevention	Mosquito control; use of bed nets; no vaccine yet available; prophylactic antiprotozoal agents
Treatment	Chloroquine, mefloquine, artemisinin, Fansidar, quinine, or proguanil

Anthrax

Anthrax is discussed in other chapters as well as this one (for example, Insight 21.3 examines pulmonary anthrax). And because one of the possible sites of anthrax infection is the skin, cutaneous anthrax is described in chapter 18. We discuss anthrax in this chapter because it multiplies in large numbers in the blood and because septicemic anthrax is a possible outcome of all forms of anthrax.

For centuries, anthrax has been known as a zoonotic disease of herbivorous livestock (sheep, cattle, and goats). It has an important place in the history of medical microbiology because Robert Koch used anthrax as a model for developing his postulates in 1877 and, later, Louis Pasteur used the disease to prove the usefulness of vaccination.

Signs and Symptoms

As just noted, anthrax infection can exhibit its primary symptoms in various locations of the body: on the skin (cutaneous anthrax), in the lungs (pulmonary anthrax), in the gastrointestinal tract (acquired through ingestion of contaminated foods), and in the central nervous system (anthrax meningitis). The cutaneous and pulmonary forms of the disease are the most common. In all of these forms, the anthrax bacterium gains access to the bloodstream, and death, if it occurs, is usually a result of an overwhelming septicemia. Pulmonary anthrax—and the accompanying pulmonary edema and hemorrhagic lung symptoms—can sometimes be the primary cause of death, although it is difficult to separate the effects of septicemia from the effects of pulmonary infection.

In addition to symptoms specific to the site of infection, septicemic anthrax results in headache, fever, and malaise. Bleeding in the intestine and from mucous membranes and orifices may occur in late stages of septicemia.

Causative Agent

Bacillus anthracis is a gram-positive endospore-forming rod that is among the largest of all bacterial pathogens. It is composed of block-shaped, angular rods 3 to 5 micrometer long and 1 to 1.2 micrometer wide. Central spores develop under all growth conditions except in the living body of the host **(figure 20.20)**. The genus *Bacillus* is aerobic and catalase-positive, and none of the species are fastidious. *Bacillus* as

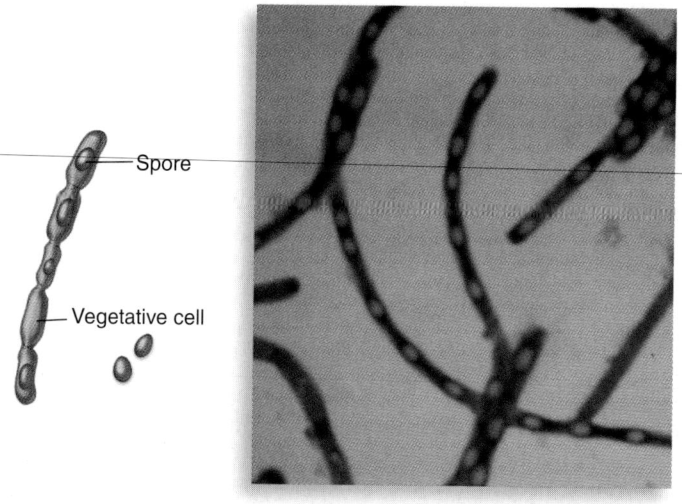

Figure 20.20 ***Bacillus anthracis.***
Note the centrally placed endospores and streptobacillus arrangement (600×).

a group is noted for its versatility in degrading complex macromolecules, and it is also a common source of antibiotics. Because the primary habitat of many species, including *B. anthracis*, is the soil, spores are continuously dispersed by means of dust into water and onto the bodies of plants and animals.

Pathogenesis and Virulence Factors

The main virulence factors of *B. anthracis* are its polypeptide capsule and what is referred to as a "tripartite" toxin—an exotoxin "complex" composed of three separate proteins. One of the proteins is called *edema factor*, an enzyme that acts as an adenylyl cyclase, interfering with cellular metabolism by causing the production of high levels of cyclic AMP. Excess cyclic AMP leads to excess cellular secretion and other pathologic effects. Another part of the toxin is *protective antigen*, so named because it is a good target for vaccination, not because it protects the bacterium or the host directly. It helps the edema factor get to its target site. The third exotoxin is called *lethal factor*. It also uses enzymatic action to inhibit important cellular processes. The end result of lethal factor action is massive inflammation and initiation of shock.

The *B. anthracis* exotoxin complex is like other bacterial "A-B toxins," which are described in detail in chapter 21. Most A-B toxins have two components: a "B" component that binds to host cells and an "A," or active, component that enters the cell and exerts some toxic effect. *B. anthracis* is a bit different; its protective antigen is the B component, and both lethal factor and edema factor are A components. The bacteria that cause cholera, shigellosis, pertussis, and diphtheria all use A-B exotoxins.

Additional virulence factors for *B. anthracis* include hemolysins and other enzymes that damage host membranes.

Transmission and Epidemiology

The anthrax bacillus is a facultative parasite that undergoes its cycle of vegetative growth and sporulation in the soil. Animals become infected while grazing on grass contaminated with spores. When the pathogen is returned to the soil in animal excrement or carcasses, it can sporulate and become a long-term reservoir of infection for the animal population. The majority of natural anthrax cases are reported in livestock from Africa, Asia, and the Middle East. Most recent (natural) cases in the United States have occurred in textile workers handling imported animal hair or hide or products made from them. Because of effective control procedures, the number of cases in the United States is extremely low (fewer than 10 per year).

As a result of the terrorist attacks of 2001, anthrax has dominated the public consciousness as never before. The anthrax attack aimed at two senators and several media outlets focused a great deal of attention on the threat of bioterrorism. During that attack, 22 people acquired anthrax and 5 people died. The attacks led to permanent and drastic changes to our public health response system.

Culture and Diagnosis

Diagnosis requires a high index of suspicion. This means that anthrax must be present as a possibility in the clinician's mind or it is likely not to be diagnosed, because it is such a rare disease in the developed world and because, in all of its manifestations, it can mimic other infections that are not so rare. (A very astute public health clinician in Florida first suspected anthrax in the attacks of 2001 and called for the proper tests.) First-level (presumptive) diagnosis begins with culturing the bacterium on blood agar and performing a Gram stain. Further tests can be performed to provide evidence regarding presence of *B. anthracis* as opposed to other *Bacillus* species. These tests include motility (*B. anthracis* is nonmotile) and a lack of hemolysis on blood agar. Ultimately, samples should be handled by the Centers for Disease Control and Prevention, which will perform confirmatory tests, usually involving direct fluorescent antibody testing and phage lysis tests.

Prevention and Treatment

A vaccine containing live spores and a toxoid prepared from a special strain of *B. anthracis* are used to protect livestock in areas of the world where anthrax is endemic. Humans should be vaccinated with the purified toxoid if they have occupational contact with livestock or products such as hides and bone or if they are members of the military. Effective vaccination requires six inoculations given over 1.5 years, with yearly boosters. The cumbersome nature of vaccination has spurred research and development of more manageable vaccines. Persons who are suspected of being exposed to the bacterium are given prophylactic antibiotics, which seem to be effective at preventing disease even after exposure.

Carcasses of animals that have died from anthrax must be burned or chemically decontaminated before burial to prevent establishing the microbe in the soil. Imported items containing animal hides, hair, and bone should be gas sterilized.

The recommended treatment for anthrax is penicillin, doxycycline, or ciprofloxacin. During the attacks in 2001, initial treatment of exposed and sick persons was with ciprofloxacin because of fear that the *B. anthracis* strains used in the attacks could have been penicillin-resistant, either through intentional genetic engineering or due to the natural presence of beta-lactamase genes in the bacterium. Ciprofloxacin treatment continued for the course of the 2001 incident. The CDC is now recommending the use of doxycycline instead of ciprofloxacin, because ciprofloxacin is often used for empirical treatment of all types of infections of unknown etiology. More frequent use of ciprofloxacin could lead to higher levels of antibiotic resistance in bacteria in the U.S. population, which would render ciprofloxacin less effective as an empirical agent.

CHECKPOINT 20.10	Anthrax
Causative Organism(s)	*Bacillus anthracis*
Most Common Modes of Transmission	Vehicle (air, soil) indirect contact (animal hides), vehicle (food)
Virulence Factors	Triple exotoxin, capsule
Culture/Diagnosis	Culture, direct fluorescent antibody tests
Prevention	Vaccine for high-risk population, postexposure antibiotic prophylaxis
Treatment	Doxycycline, ciprofloxacin, penicillin

HIV Infection and AIDS

The sudden emergence of AIDS in the early 1980s focused an enormous amount of public attention, research studies, and financial resources on the virus and its disease.

The first cases of AIDS were seen by physicians in Los Angeles, San Francisco, and New York City. They observed clusters of young male patients with one or more of a complex of symptoms: severe pneumonia caused by *Pneumocystis (carinii) jiroveci* (ordinarily a harmless fungus), a rare vascular cancer called Kaposi's sarcoma, sudden weight loss, swollen lymph nodes, and general loss of immune function. Another common feature was that all of these young men were homosexuals. Early hypotheses attempted to explain the disease as a consequence of a "homosexual lifestyle" or as a result of immune suppression by chronic drug abuse or infections. Soon, however, cases were reported in nonhomosexual patients who had been transfused with blood or blood products. Eventually, virologists at the Pasteur Institute in France isolated a novel retrovirus, later named the **human immunodeficiency virus (HIV).** This cluster of symptoms was therefore clearly a communicable infectious disease, and the medical community termed it **acquired immunodeficiency syndrome,** or **AIDS.**

One important question about HIV seems to have been answered: *Where did it come from?* Researchers have been comparing the genetics of HIV with the various African monkey viruses, called simian immunodeficiency viruses, or SIVs. The genetic sequences in these various viruses led them to conclude that HIV is a hybrid virus, with genetic sequences from two separate monkey SIVs. One of the SIVs has as its natural host the greater spot-nosed monkey, and the other infects red-capped mangabeys. Apparently, one or more chimpanzees became coinfected with the two viruses after making a meal of both of the smaller monkeys. Within the chimpanzee, a third type of virus emerged that contained genetic sequences from both SIVs. This new type of SIV was probably transmitted to humans when they captured chimps, butchered them, and used them for food. So humans originally acquired HIV from eating or skinning chimps; chimps got SIV from eating monkeys. The crossover into humans probably occurred in the early part of the 1900s; the earliest record we have of human infection is a blood sample preserved from an African man who died in 1959.

HIV probably remained in small isolated villages causing sporadic cases and mutating into more virulent strains that were readily transmitted from human to human. When this pattern was combined with changing social and sexual practices and increased immigration and travel, a pathway was opened up for rapid spread of the virus to the rest of the world.

Signs and Symptoms

A spectrum of clinical disease is associated with HIV infection. To understand the progression, follow **figure 20.21** closely. Symptoms in HIV infection are directly tied to two things: the level of virus in the blood and the level of T cells in the blood. (The figure shows two different lines that correspond to virus and T cells.) Note also that the figure depicts the course of HIV infection in the absence of medical intervention or chemotherapy.

Initial infection is often attended by vague, mononucleosis-like symptoms that soon disappear. This phase corresponds to the initial high levels of virus (the green line in the figure), and the subsequent drop in virus load. Within days of infection, about 50% of the T helper cells with memory for the virus

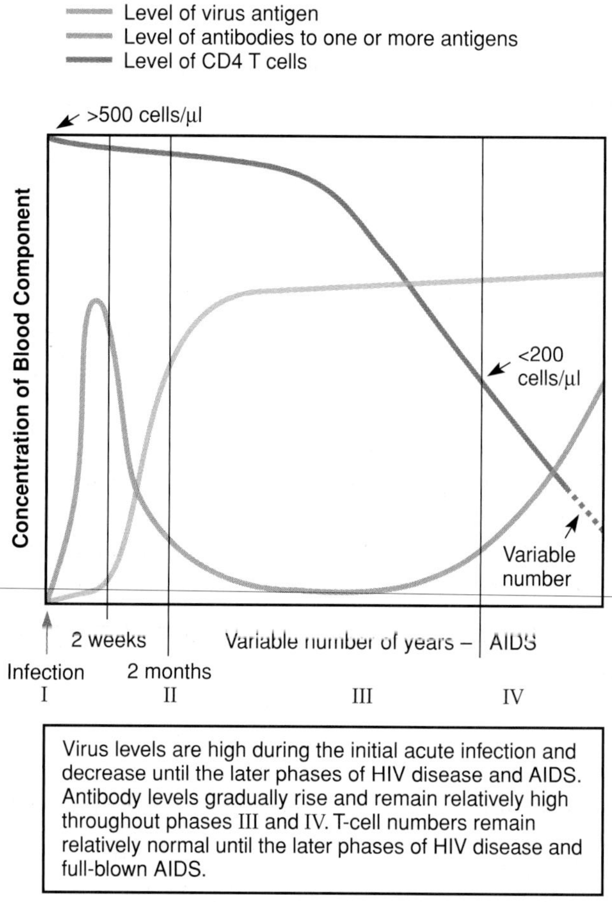
Level of virus antigen
Level of antibodies to one or more antigens
Level of CD4 T cells
>500 cells/μl
Concentration of Blood Component
<200 cells/μl
Variable number
2 weeks
Variable number of years –
AIDS
Infection
I
2 months
II
III
IV
Virus levels are high during the initial acute infection and decrease until the later phases of HIV disease and AIDS. Antibody levels gradually rise and remain relatively high throughout phases III and IV. T-cell numbers remain relatively normal until the later phases of HIV disease and full-blown AIDS.

Figure 20.21 Dynamics of virus antigen, antibody, and T cells in circulation.
The figure depicts a generalized curve. Specifics vary in actual infections.

are destroyed. Note that antibody levels (the orange line) rise at the same time that virus load is dropping; the immune response is responsible for the decreasing numbers of virus in the blood. This initial state is followed by a period of (mostly) asymptomatic infection that varies in length from 2 to 15 years (the average is 10). The lymphadenopathy that attended the initial infection usually persists throughout the entire infection. Note that during the mid- to late-asymptomatic periods, the number of T cells in the blood is steadily decreasing (purple line). Once the T cells reach low enough levels, symptoms of AIDS ensue.

Initial symptoms may be fatigue, diarrhea, weight loss, and neurological changes, but most patients first notice this phase of infection because of one or more opportunistic infections or neoplasms. These are detailed in **Insight 20.4.** Other disease-related symptoms appear to accompany severe immune deregulation, hormone imbalances, and metabolic disturbances. Pronounced wasting of body mass is a consequence of weight loss, diarrhea, and poor nutrient absorption. Protracted fever, fatigue, sore throat, and night sweats are significant and debilitating. Both a rash and generalized lymphadenopathy in several chains of lymph nodes are presenting symptoms in many AIDS patients.

Some of the most virulent complications are neurological. Lesions occur in the brain, meninges, spinal column, and peripheral nerves. Patients with nervous system involvement show some degree of withdrawal, persistent memory loss, spasticity, sensory loss, and progressive AIDS dementia.

Causative Agent

HIV is a retrovirus, in the genus lentivirus. Most retroviruses have the potential to cause cancer and produce dire, often fatal diseases and are capable of altering the host's DNA in profound ways. They are named "retroviruses" because they reverse the usual order of transcription. They contain an unusual enzyme called **reverse transcriptase (RT)** that catalyzes the replication of double-stranded DNA from single-stranded RNA. The association of retroviruses with their hosts can be so intimate that viral genes are permanently integrated into the host genome. In fact, as the technology of DNA probes for detecting retroviral genes is employed, it becomes increasingly evident that retroviral sequences are integral parts of host chromosomes. Not only can this retroviral DNA be incorporated into the host genome as a provirus that can be passed on to progeny cells, but some retroviruses also transform cells and regulate certain host genes.

The most prominent human retroviruses are the T-cell lymphotropic viruses I and II (HTLV-I and HTLV-II) and HTLV-III. Types I and II are associated with leukemia (discussed in a later section) and lymphoma; type III is now called HIV. There are two types of HIV, namely HIV-1, which is the dominant form in most of the world, and HIV-2.

HIV and other retroviruses display structural features typical of enveloped RNA viruses **(figure 20.22*a*).** The outermost component is a lipid envelope with transmembrane glycoprotein spikes and knobs that mediate viral adsorption to the host cell. HIV can only infect host cells that present the required receptors, which is a combination receptor consisting of the CD4 marker plus a coreceptor. The virus uses these receptors to gain entrance to several types of leukocytes and tissue cells **(figure 20.22*b*).**

Pathogenesis and Virulence Factors

As summarized in **figure 20.23,** HIV enters a mucus membrane or the skin and travels to dendritic cells, a type of phagocyte living beneath the epithelium. In the dendritic cell, the virus

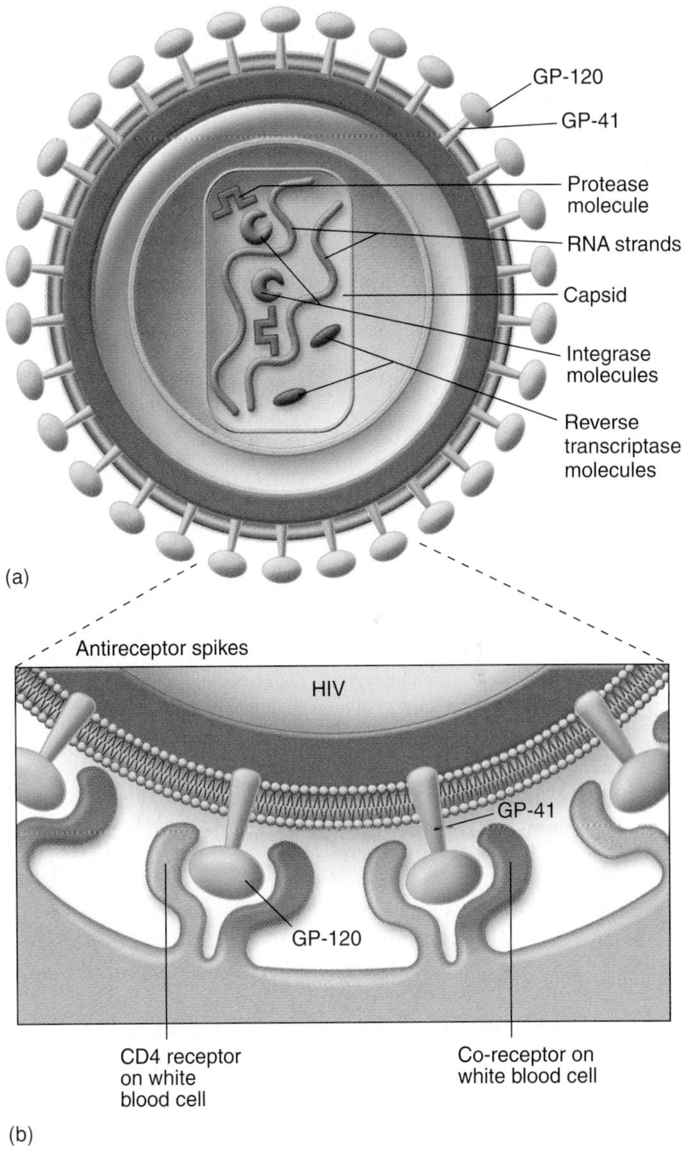

Figure 20.22 The general structure of HIV.
(a) The envelope contains two types of glycoprotein (GP) spikes, two identical RNA strands, and several molecules of reverse transcriptase, protease, and integrase encased in a protein capsid. **(b)** The snug attachment of HIV glycoprotein molecules (GP-41 and GP-120) to their specific receptors on a human cell membrane. These receptors are CD4 and a co-receptor called CCR-5 (fusin) that permit docking with the host cell and fusion with the cell membrane.

INSIGHT 20.4 *Medical*

AIDS-Defining Illnesses (ADIs)

In AIDS patients who do not receive or do not comply with antiretroviral therapy (and even in some who do), the slow destruction of the immune system results in a wide variety of infectious and noninfectious conditions called AIDS-associated illnesses or AIDS-defining illnesses (ADIs). It is almost always one or more of these conditions that causes death in AIDS patients.

Because the virus eventually collapses the immune system like a house of cards, it is not surprising that the body is beset by normally harmless microorganisms, many of which have been living in or on the host for decades without causing disease. The spectrum of AIDS-associated illnesses also provides insight into how vital the immune system is in controlling or mitigating cancerous changes in our cells. AIDS patients are at increased risk for Burkitt's lymphoma, Kaposi's sarcoma (KS), and invasive cervical carcinomas, all of which are associated with viral infections.

Since the beginning of the AIDS epidemic in the early 1980s, the CDC has maintained a list of conditions that are part of the case definition. The list has been modified periodically over the two decades of its existence. One of the ways that people currently meet the case definition for AIDS is if they are positive for the virus *and* experience one or more of these ADIs. The ADIs are listed in **table 20.B.** The diseases are listed according to the organ system where the presenting symptoms might be found. (Some of the conditions may be listed in more than one column.) You can see that most of them—or at least, the way they occur in AIDS patients—are very rare in the otherwise healthy population.

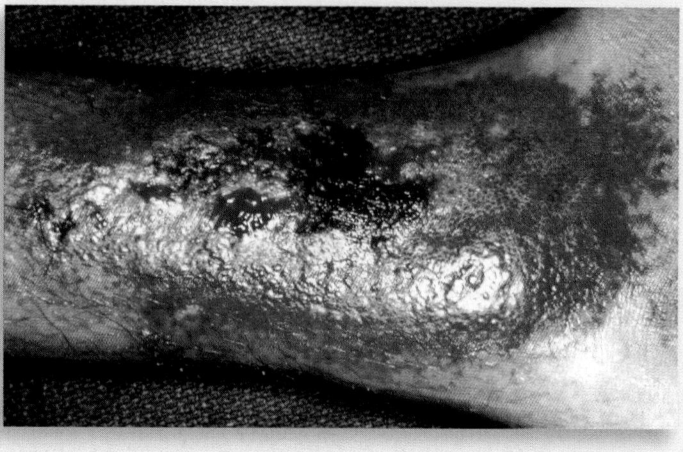

Kaposi's sarcoma lesion on the arm.
The flat, purple tumors occur in almost any tissue and are frequently multiple.

TABLE 20.B AIDS-Defining Illnesses

Skin and/or Mucous Membranes (includes eyes)	Nervous System	Cardiovascular and Lymphatic System or Multiple Organ Systems	Respiratory Tract	Gastrointestinal Tract	Genitourinary and/or Reproductive Tract
Cytomegalovirus retinitis (with loss of vision) Herpes simplex chronic ulcers (>1 month duration) Kaposi's sarcoma	Cryptococcosis, extrapulmonary HIV encephalopathy Lymphoma primarily in brain Progressive multifocal leukoencephalopathy Toxoplasmosis of the brain	Coccidiomycosis, disseminated or extrapulmonary Cytomegalovirus (other than liver, spleen, nodes) Histoplasmosis, disseminated or extrapulmonary Burkitt's lymphoma Immunoblastic lymphoma *Mycobacterium kansasii*, disseminated or extrapulmonary *Mycobacterium tuberculosis*, disseminated or extrapulmonary *Salmonella* septicemia, recurrent Wasting syndrome	Candidiasis of trachea, bronchi, or lungs Herpes simplex bronchitis or pneumonitis *Mycobacterium avium* complex Tuberculosis (*Mycobacterium tuberculosis*) *Pneumocystis (carinii) jiroveci* pneumonia Pneumonia, recurrent in 12-month period	Candidiasis of esophagus, GI tract Herpes simplex chronic ulcers (>1 month duration) or esophagitis Isosporiasis, (diarrhea caused by *Isospora*) chronic intestinal (>1 month duration) Cryptosporidiosis, chronic intestinal (>1 month duration)	Invasive cervical carcinoma (HPV) Herpes simplex chronic ulcers (>1 month duration)

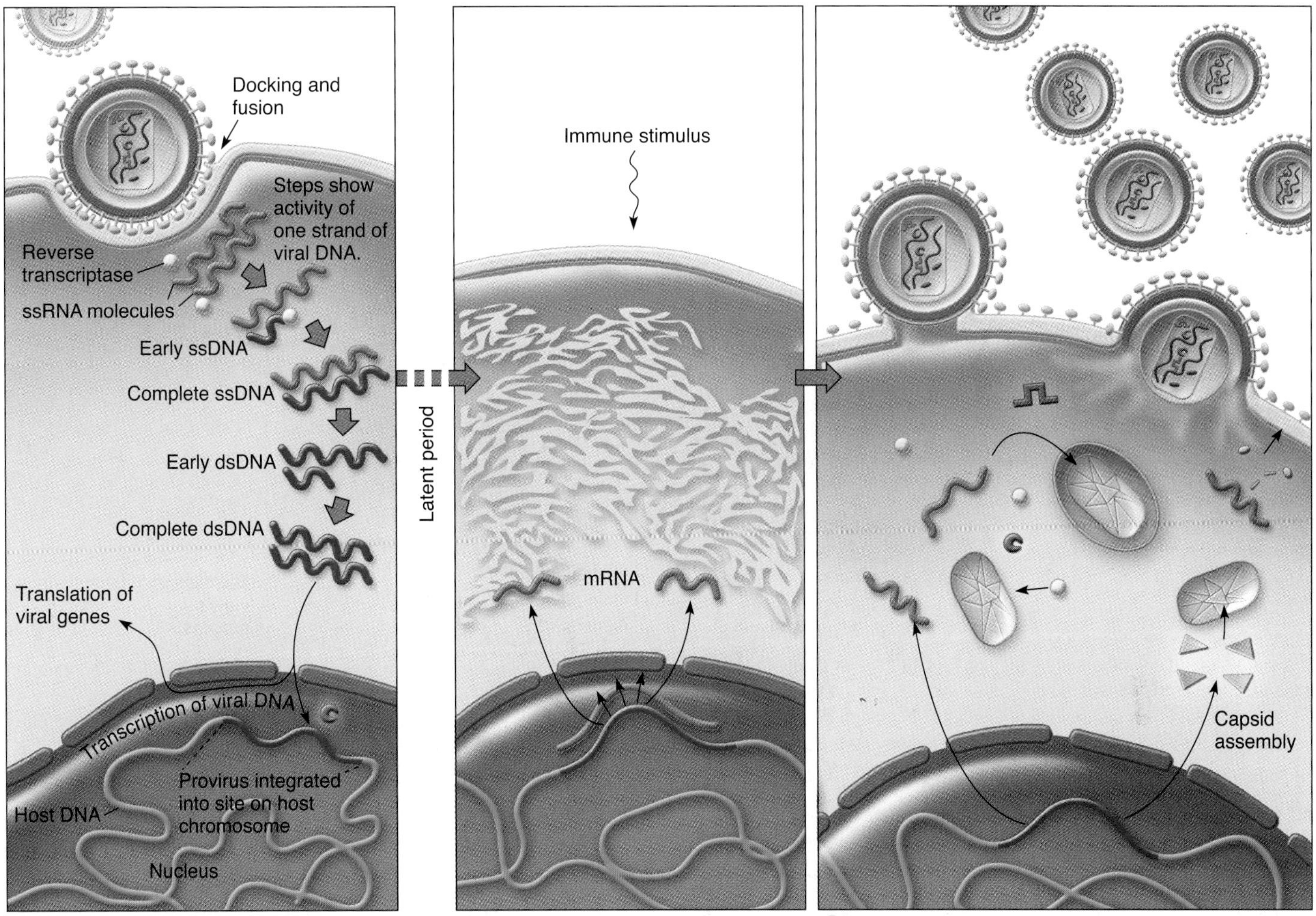

① The virus is adsorbed and endocytosed, and the twin RNAs are uncoated. Reverse transcriptase catalyzes the synthesis of a single complementary strand of DNA (ssDNA). This single strand serves as a template for synthesis of a double strand (ds) of DNA. In latency, dsDNA is inserted into the host chromosome as a provirus.

② After a latent period, various immune activators stimulate the infected cell, causing reactivation of the provirus genes and production of viral mRNA.

③ HIV mRNA is translated by the cell's synthetic machinery into virus components (capsid, reverse transcriptase, spikes), and the viruses are assembled. Budding of mature viruses lyses the infected cell.

Process Figure 20.23 The general multiplication cycle of HIV.

grows and is shed from the cell without killing it. The virus is amplified by macrophages in the skin, lymph organs, bone marrow, and blood. One of the great ironies of HIV is that it infects and destroys many of the very cells needed to combat it, including the helper (T4 or CD4) class of lymphocytes, monocytes, macrophages, and even B lymphocytes. The virus is adapted to docking onto its host cell's surface receptors (see figure 20.22). It then induces viral fusion with the cell membrane and creates syncytia.

Once the virus is inside the cell, its reverse transcriptase makes its RNA into DNA. Although initially it can produce a lytic infection, in many cells it enters a latent period in the nucleus of the host cell and integrates its DNA into host DNA (see figure 20.23). This latency accounts for the lengthy course of the disease. Despite being described as a "latent" stage, research suggests that new viruses are constantly being produced and new T cells are constantly being manufactured, in an ongoing race that ultimately the host cells lose (in the absence of treatment).

The primary effects of HIV infection—those directly due to viral action—are harm to T cells and the central nervous system. The death of T cells and other white blood cells results in extreme **leukopenia** and loss of essential T4 memory clones and stem cells. The viruses also cause formation of giant T cells and other syncytia, which allow the spread of viruses directly from cell to cell, followed by mass destruction of the syncytia. The destruction of T4 lymphocytes paves the way for invasion by opportunistic agents and malignant cells. The central nervous system is affected when infected macrophages cross the blood-brain barrier and release viruses, which then invade nervous tissue. Studies have indicated that some of the viral envelope proteins can have a direct toxic effect on the

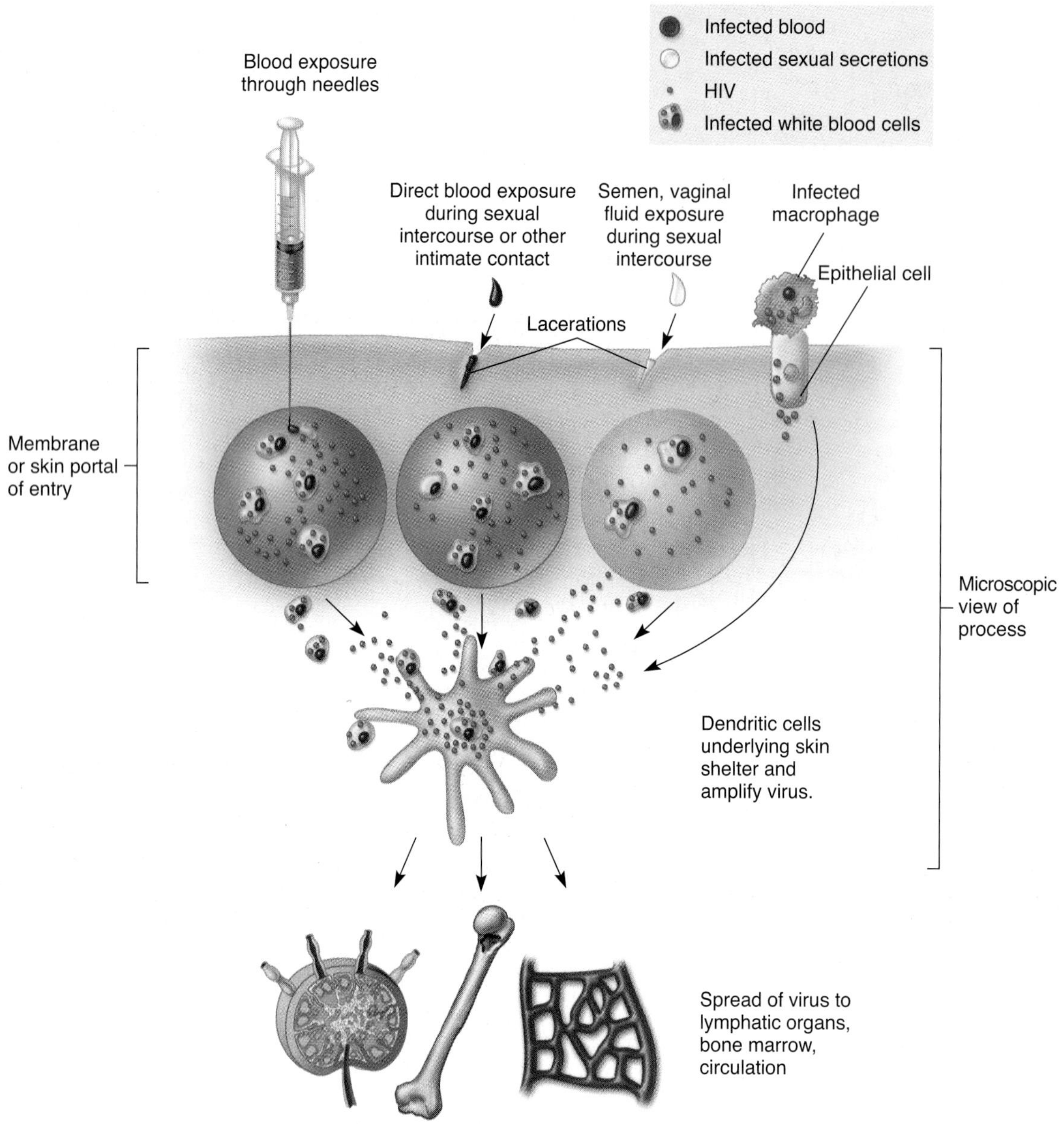

Figure 20.24 Primary sources and suggested routes of infection by HIV.

brain's glial cells and other cells. Other research has shown that some peripheral nerves become demyelinated and the brain becomes inflamed.

The secondary effects of HIV infection are the opportunistic infections and malignancies that occur as the immune system becomes progressively crippled by viral attack. These are summarized in Insight 20.4.

Transmission

HIV transmission occurs mainly through two forms of contact: sexual intercourse and transfer of blood or blood products **(figure 20.24).** Babies can also be infected before or during birth, as well as through breast feeding. The mode of transmission is similar to that of hepatitis B virus, except that the AIDS virus does not survive for as long outside the host and it is far more sensitive to heat and disinfectants. And HIV is not transmitted through saliva, as hepatitis B can be.

In general, HIV is spread only by direct and rather specific routes. Because the blood of HIV-infected people harbors high levels of free virus in both very early and very late stages of infection and high levels of infected leukocytes throughout infection, any form of intimate contact involving transfer of blood (trauma, needle sharing) can be a potential source of infection. Semen and vaginal secretions also harbor free virus and infected white blood cells, and thus they are significant factors in sexual transmission. The virus can be isolated from urine, tears, sweat, and saliva but in such small numbers that these fluids are not considered sources of infection. Because breast milk contains significant numbers of leukocytes, neonates who have escaped infection prior to and during birth can still become infected through nursing.

Epidemiology

HIV infection and AIDS have been reported in every country. Parts of Africa and Asia are especially devastated by it. The best estimate of the number of individuals currently infected with HIV (in 2005) is 39 million worldwide, with approximately 1 million (the range is 800,000 to 2 million) in the United States. A large number of these people have not yet begun to show symptoms.

AIDS first became a notifiable disease at the national level in 1984, and it has continued in an epidemic pattern, although the number of AIDS cases in the United States has decreased since 1994. This decrease is occurring at the same time that new HIV infections are being reported at the rate of 65,000 a year. This situation is due to the advent of effective therapies that prevent the progression to AIDS. But in developing countries, which are hardest hit by the HIV epidemic, access to these lifesaving drugs is still very limited. And even in the United States, despite treatment advances, HIV infection/AIDS is the sixth most common cause of death among people ages 25 to 44, although it has fallen out of the top 10 list for causes of death overall.

Table 20.1 spells out some shifts in the behaviors that result in HIV infection in the United States. Throughout the course of the epidemic, close to half (47%) of all cases can be traced to male-to-male sexual contact. For those cases of AIDS identified in 2004, that proportion had fallen to 41%. Significantly more infections were acquired through heterosexual sex (31% versus 17%).

In large metropolitan areas especially, as many as 60% of intravenous drug users (IDUs) can be HIV carriers. Infection from contaminated needles is growing more rapidly than any other mode of transmission, and it is another significant factor in the spread of HIV to the heterosexual population.

TABLE 20.1 AIDS Cases in the United States by Exposure Category**

Exposure Category	Percentage of New AIDS Cases in 2004	Percentage of All AIDS Cases Through 2004
Male-to-Male Sexual Contact	41	47
Injection Drug Use	22	27
Male-to-Male Sexual Contact and Injection Drug Use	5	7
Heterosexual Contact	31	17
Other*	1	2

*Includes hemophilia, blood transfusion, perinatal, and risk not reported or identified.

**Data from the Centers for Disease Control and Prevention.

In most parts of the world, heterosexual intercourse is the primary mode of transmission. In the industrialized world, the overall rate of heterosexual infection has increased dramatically in the past several years, especially in adolescent and young adult women. In the United States, about 31% of HIV infections arise from unprotected sexual intercourse with an infected partner of the opposite sex.

Now that donated blood is routinely tested for antibodies to the AIDS virus, transfusions are no longer considered a serious risk. Because there can be a lag period of a few weeks to several months before antibodies appear in an infected person, it is remotely possible to be infected through donated blood. Rarely, organ transplants can carry HIV, so they too should be tested. Other blood products (serum, coagulation factors) were once implicated in AIDS. Thousands of hemophiliacs died from the disease in the 1980s and 1990s. It is now standard practice to heat-treat any therapeutic blood products to destroy all viruses.

A small percentage of AIDS cases occur in people without apparent risk factors. This does not mean that some other unknown route of spread exists. Factors such as patient denial, unavailability of history, death, or uncooperativeness make it impossible to explain every case.

We should note that not everyone who becomes infected or is antibody-positive develops AIDS. About 5% of people who are antibody-positive remain free of disease, indicating that functioning immunity to the virus can develop. Any person who remains healthy despite HIV infection is termed a *nonprogressor*. These people are the object of intense scientific study. Some have been found to lack the cytokine receptors that HIV requires. Others are infected by a weakened virus mutant.

Treatment of HIV-infected mothers with AZT has dramatically decreased the rate of maternal-to-infant transmission of HIV during pregnancy. Current treatment regimens result in a transmission rate of approximately 11%, with some studies of multidrug regimens claiming rates as low as 5%. Evidence suggests that giving mothers protease inhibitors can reduce the transmission rate to around 1%. (Untreated mothers pass the virus to their babies at the rate of 33%.) The cost of perinatal prevention strategies (approximately $1,000 per pregnancy) and the scarcity of medical counseling in underserved areas has led to an increase in maternal transmission of HIV in developing parts of the world, at the same time that the developed world has seen a marked decrease.

Medical and dental personnel are not considered a high-risk group, although several hundred medical and dental workers are known to have acquired HIV or become antibody-positive as a result of clinical accidents. A health care worker involved in an accident in which gross inoculation with contaminated blood occurs (as in the case of a needlestick) has a less than 1 in 1,000 chance of becoming infected. We should emphasize that transmission of HIV will not occur through casual contact or routine patient care procedures and that universal precautions for infection control (see chapter 13) were designed to give full protection for both worker and patient.

Culture and Diagnosis

First, let's define some terms. A person is diagnosed as having HIV infection if he or she has tested positive for the human immunodeficiency virus. This diagnosis is not the same as having AIDS.

In late 2006, the CDC issued new recommendations that HIV testing become much more routine. The guidelines call for testing all patients accessing health care facilities and for HIV testing to be included in the routine panel of prenatal screening for pregnant women. In both cases, patients can opt out of the test, although no separate consent will be solicited besides the general consent for medical care.

Most viral testing is based on detection of antibodies specific to the virus in serum or other fluids, which allows for the rapid, inexpensive screening of large numbers of samples. Testing usually proceeds at two levels. The initial screening tests include the older ELISA and newer latex agglutination and rapid antibody tests. The favored initial screening test is an "oral swab test" that provides results within minutes. It is licensed only for use by health care professionals. Even though saliva can be tested with this kit, blood is more often used in the test because of increased viral load in blood relative to saliva. The advantage of this test is that results are available within minutes instead of the days to weeks previously required for the return of results from ELISA testing. One kit has been licensed by the Food and Drug Administration for home use; in this test, a blood spot is applied to a card and then sent to a testing laboratory to perform the actual procedure. The client can call an automated phone service, enter his or her anonymous testing code, and receive results over the phone.

Although the approved tests just described are largely accurate, around 1% of results are false positives, and they always require follow-up with a more specific test called *Western blot* analysis (see p. 524). This test detects several different anti-HIV antibodies and can usually rule out false positive results.

Another inaccuracy can be false negative results that occur when testing is performed before the onset of detectable antibody production. To rule out this possibility, persons who test negative but feel they may have been exposed should be tested a second time 3 to 6 months later.

Blood and blood products are sometimes tested for HIV antigens (rather than for HIV antibodies) to close the window of time between infection and detectable levels of antibodies during which contamination could be missed by antibody tests. The American Red Cross is currently participating in a program to test its blood supply with a DNA probe for HIV.

In the United States, people are diagnosed with AIDS if they meet the following criteria: (1) they are positive for the virus, *and* (2) they fulfill one of these additional criteria:

- They have a CD4 (helper T cell) count of fewer than 200 cells per microliter of blood.
- Their CD4 cells account for fewer than 14% of all lymphocytes.
- They experience one or more of a CDC-provided list of AIDS-defining illnesses (ADIs).

The list of ADIs is long and includes opportunistic infections such as *Pneumocystis (carinii) jiroveci* pneumonia and *Cryptosporidium* diarrhea; neoplasms such as Kaposi's sarcoma and invasive cervical cancer; and other conditions such as wasting syndrome and neuropathy (see Insight 20.4).

Prevention

Avoidance of sexual contact with infected persons is a cornerstone of HIV prevention. Abstaining from sex is an obvious prevention method, although those who are sexually active can also take steps to decrease their risk. Epidemiologists cannot overemphasize the need to screen prospective sex partners and to follow a monogamous sexual lifestyle. And monogamous or not, a sexually active person should consider every partner to be infected unless proven otherwise. This may sound harsh, but it is the only sure way to avoid infection during sexual encounters. Barrier protection (condoms) should be used when having sex with anyone whose HIV status is not known with certainty to be negative. Although avoiding intravenous drugs is an obvious deterrent, many drug addicts do not, or cannot, choose this option. In such cases, risk can be decreased by not sharing syringes or needles or by cleaning needles with bleach and then rinsing before another use.

From the very first years of the AIDS epidemic, the potential for creating a vaccine has been regarded as slim, because the virus presents many seemingly insurmountable problems. Among them, HIV becomes latent in cells; its cell surface antigens mutate rapidly; and although it does elicit immune responses, it is apparently not completely controlled by them. In view of the great need for a vaccine, however, none of those facts has stopped the medical community from moving ahead.

Currently, dozens of potential HIV vaccines are in clinical trials. To protect against as many strains as possible, most vaccine designers have concentrated their efforts on proteins that seem to be highly conserved (show very few changes) between strains. Currently two vaccines have completed Phase III trials—the last step before possible FDA approval. While earlier trials investigate vaccine safety and whether the vaccine causes antibody production, Phase III trials examine the ability of the vaccine to protect humans from becoming infected. Neither of the first two candidate vaccines were able to prevent infection. Two more vaccines are currently in Phase III trials.

Treatment

It must be clearly stated: There is no cure for HIV. None of the therapies do more than prolong life or diminish symptoms.

Clear-cut guidelines exist for treating people who test HIV-positive. These guidelines are updated regularly, and they differ depending on whether a person is completely asymptomatic or experiencing some of the manifestations

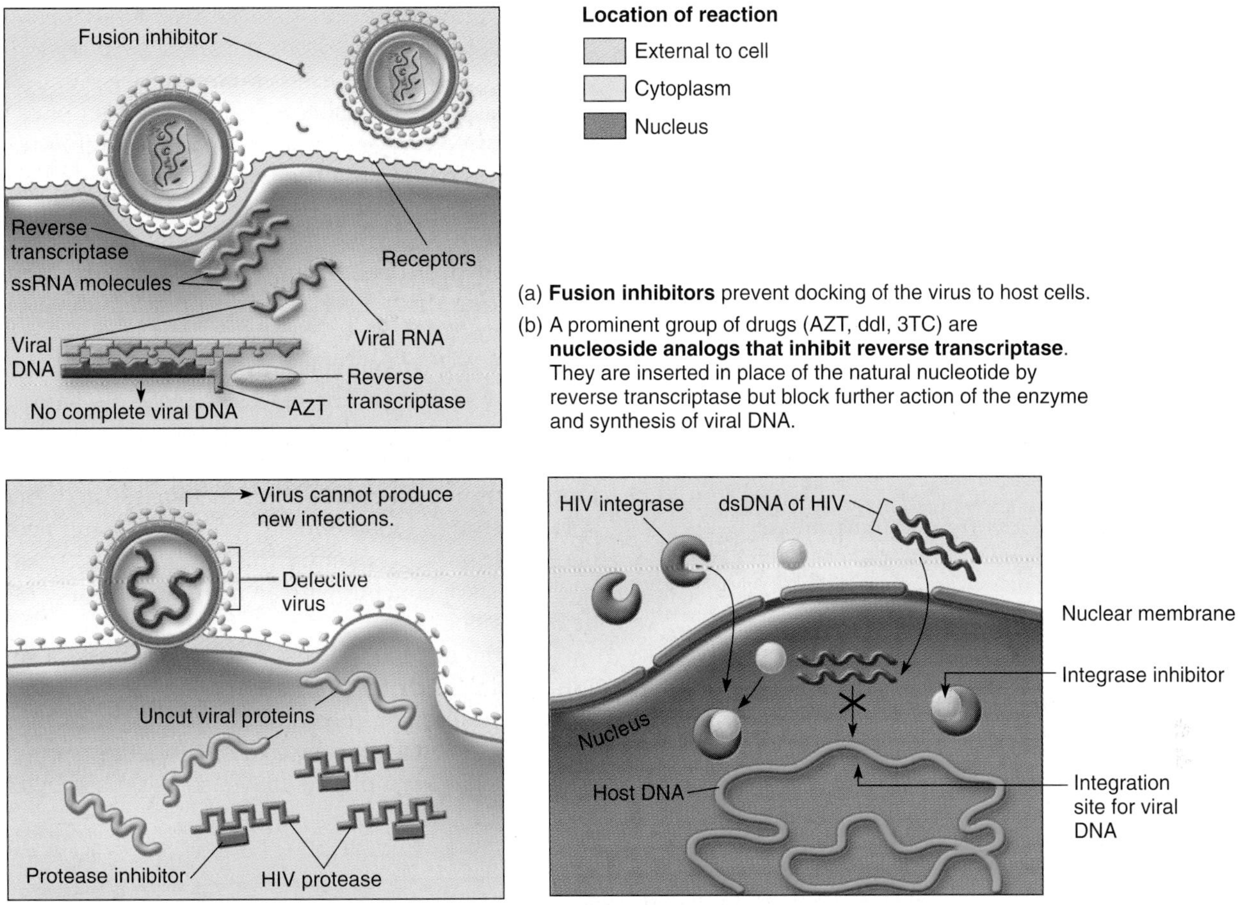

(a) **Fusion inhibitors** prevent docking of the virus to host cells.

(b) A prominent group of drugs (AZT, ddI, 3TC) are **nucleoside analogs that inhibit reverse transcriptase**. They are inserted in place of the natural nucleotide by reverse transcriptase but block further action of the enzyme and synthesis of viral DNA.

(c) **Protease inhibitors** plug into the active sites on HIV protease.This enzyme is necessary to cut elongate HIV protein strands and produce functioning smaller protein units. Because the enzyme is blocked, the proteins remain uncut, and abnormal defective viruses are formed.

(d) **Integrase inhibitors** are a new class of experimental drugs that attach to the enzyme required to splice the dsDNA from HIV into the host genome. This will prevent formation of the provirus and block future virus multiplication in that cell.

Figure 20.25 Mechanisms of action of anti-HIV drugs.

of HIV infection and also whether a person has previously been treated with antiretroviral agents. A person diagnosed with AIDS receives treatment for the HIV infection and also a wide array of drugs to prevent or treat a variety of opportunistic infections and other ADIs such as wasting disease. These treatment regimens vary according to each patient's profile and needs.

The first effective drugs developed were the synthetic nucleoside analogs (reverse transcriptase inhibitors) azidothymidine (AZT), didanosine (ddI), Epivir (3TC), and stavudine (d4T). They interrupt the HIV multiplication cycle by mimicking the structure of actual nucleosides and being added to viral DNA by reverse transcriptase. Because these drugs lack all of the correct binding sites for further DNA synthesis, viral replication and the viral cycle are terminated **(figure 20.25*a*).** Other reverse transcriptase inhibitors that are not nucleosides are nevirapine and sustiva, both of which bind to the enzyme and restructure it. Another important class of drugs is the protease inhibitors **(figure 20.25*c*)** that block the action of the HIV enzyme (protease) involved in the final assembly and maturation of the virus. Examples of these drugs include Crixivan, Norvir, and Agenerase. Integrase inhibitors are currently under investigation as a means to stop virus multiplication **(figure 20.25*d*).**

The latest addition to the arsenal is Fuzeon, a drug classified as a fusion inhibitor. It prevents the virus from fusing with the membrane of target cells, thereby stopping infection altogether **(figure 20.25*b*).**

A regimen that has proved to be extremely effective in controlling AIDS and inevitable drug resistance is **HAART,** short for *highly active antiretroviral therapy*. By combining two reverse transcriptase inhibitors and one protease inhibitor in a "cocktail," the virus is interrupted in two different phases of its cycle. This therapy has been successful in reducing viral load to undetectable levels and facilitating the improvement of immune function. It has also reduced the incidence of viral drug resistance, because the virus would have to undergo three separate mutations simultaneously, at nearly impossible odds. Patients who are HIV-positive but asymptomatic can remain healthy with this therapy as well. The primary drawbacks are high cost, toxic side effects, drug failure due to patient noncompliance, and an inability to completely eradicate the virus.

CHECKPOINT 20.11	HIV Infection and AIDS
Causative Organism(s)	Human immunodeficiency virus 1 or 2
Most Common Modes of Transmission	Direct contact (sexual), parenteral (blood-borne), vertical (perinatal and via breast milk)
Virulence Factors	Attachment, syncytia formation, reverse transcriptase, high mutation rate
Culture/Diagnosis	Initial screening for antibody followed by Western blot confirmation of antibody
Prevention	Avoidance of contact with infected sex partner, contaminated blood, breast milk
Treatment	HAART (reverse transcriptase inhibitors plus protease inhibitors), Fuzeon, nonnucleoside RT inhibitors

Adult T-Cell Leukemia and Hairy-Cell Leukemia

Leukemia is the general name for at least four different malignant diseases of the white blood cell forming elements originating in the bone marrow. Some forms of leukemia are acute and others are chronic. Leukemias have many causes, only two of which are thought to be viral. The retrovirus HTLV-I is associated with a form of leukemia called adult T-cell leukemia. HTLV-II is thought by some to cause hairy-cell leukemia.

The signs and symptoms of all leukemias are similar and include easy bruising or bleeding, paleness, fatigue, and recurring minor infections. These symptoms are associated with the underlying pathologies of anemia, platelet deficiency, and immune dysfunction brought about by the disturbed lymphocyte ratio and function. In some cases of adult T-cell leukemia, cutaneous T-cell lymphoma is the prime clinical manifestation, accompanied by dermatitis, with thickened, scaly, ulcerative, or tumorous skin lesions. Other complications are lymphadenopathy and dissemination of the tumors to the lung, spleen, and liver.

The possible mechanisms by which retroviruses stimulate cancer are not entirely clear. One hypothesis is that the virus carries an oncogene that, when spliced into a host's chromosome and triggered by various carcinogens, can immortalize the cell and deregulate the cell division cycle. One of HTLV's genetic targets seems to be the gene and receptor for interleukin-2, a potent stimulator of T cells.

Adult T-cell leukemia was first described by physicians working with a cluster of patients in southern Japan. Later, a similar clinical disease was described in Caribbean immigrants. In time, it was shown that these two diseases were the same. Although more common in Japan, Europe, and the Caribbean, a small number of cases occur in the United States. The disease is not highly transmissible; studies among families show that repeated close or intimate contact is required. Because the virus is thought to be transferred in infected blood cells, blood transfusions and blood products are potential agents of transmission. Intravenous drug users could spread it through needle sharing.

Hairy-cell leukemia is a rare form of cancer that may be caused by HTLV-II. The disease derives its name from the appearance of the afflicted lymphocytes, which have fine cytoplasmic projections that resemble hairs. The disease is more common in males than in females and is probably spread through blood and shared syringes and needles. Unlike adult T-cell leukemia this form of leukemia exhibits overall leukopenia but with an increased number of neoplastic B lymphocytes.

Treatment of these diseases may include a number of antineoplastic drugs, radiation therapy, and transplants. Alpha-interferon has been used with some effectiveness.

There is some evidence to suggest that some forms of non-Hodgkin's lymphoma may also be caused by HTLV-I.

CHECKPOINT 20.12	Adult T-Cell Leukemia and Hairy-Cell Leukemia	
Disease	**Adult T-Cell Leukemia**	**Hairy-Cell Leukemia**
Causative Organism(s)	HTLV-I	(Possibly) HTLV-II
Most Common Modes of Transmission	Unclear—blood-borne transmission implicated	Unclear—blood-borne transmission implicated
Virulence Factors	Induction of malignant state	Induction of malignant state
Culture/Diagnosis	Differential blood count followed by histological examination of excised lymph node tissue	Differential blood count followed by histological examination of excised lymph node tissue
Prevention	–	–
Treatment	Antineoplastic drugs, interferon alpha	Antineoplastic drugs, interferon alpha

SUMMING UP

Taxonomic Organization **Summing Up Microorganisms Causing Disease in the Cardiovascular and Lymphatic System**

Microorganism	Disease	Chapter Location
Gram-positive endospore-forming bacteria		
Bacillus anthracis	Anthrax	Anthrax, p. 632
Gram-positive bacteria		
Staphylococcus aureus	Acute endocarditis	Endocarditis, p. 612
Streptococcus pyogenes	Acute endocarditis	Endocarditis, p. 612
Streptococcus pneumoniae	Acute endocarditis	Endocarditis, p. 612
Gram-negative bacteria		
Yersinia pestis	Plague	Plague, p. 615
Francisella tularensis	Tularemia	Tularemia, p. 617
Borellia burgdorferi	Lyme disease	Lyme disease, p. 618
Brucella abortus, B. suis	Brucellosis	Nonhemorrhagic fever diseases, p. 624
Coxiella burnetii	Q fever	Nonhemorrhagic fever diseases, p. 625
Bartonella henselae	Cat-scratch disease	Nonhemorrhagic fever diseases, p. 626
Bartonella quintana	Trench fever	Nonhemorrhagic fever diseases, p. 626
Ehrlichia chaffeensis, E. phagocytophila, E. ewingii	Ehrlichiosis	Nonhemorrhagic fever diseases, p. 626
Neisseria gonorrhoeae	Acute endocarditis	Endocarditis, p. 612
Rickettsia rickettsii	Rocky Mountain spotted fever	Nonhemorrhagic fever diseases, p. 626
DNA viruses		
Epstein-Barr virus	Infectious mononucleosis	Infectious mononucleosis, p. 621
Cytomegalovirus	Infectious mononucleosis	Infectious mononucleosis, p. 622
RNA viruses		
Yellow fever virus	Yellow fever	Hemorrhagic fevers, p. 623
Dengue fever virus	Dengue fever	Hemorrhagic fevers, p. 623
Ebola and Marburg viruses	Ebola and Marburg hemorrhagic fevers	Hemorrhagic fevers, p. 623
Lassa fever virus	Lassa fever	Hemorrhagic fevers, p. 624
Retroviruses		
Human immunodeficiency virus 1 and 2	HIV infection and AIDS	HIV infection and AIDS, p. 634
Human T-cell lymphotropic virus I	Adult T-cell leukemia	Leukemias, p. 642
Human T-cell lymphotropic virus II	Hairy-cell leukemia (?)	Leukemias, p. 642
Protozoa		
Plasmodium falciparum, P. vivax, P. ovale, P. malariae	Malaria	Malaria, p. 627

INFECTIOUS DISEASES AFFECTING
The Cardiovascular and Lymphatic Systems

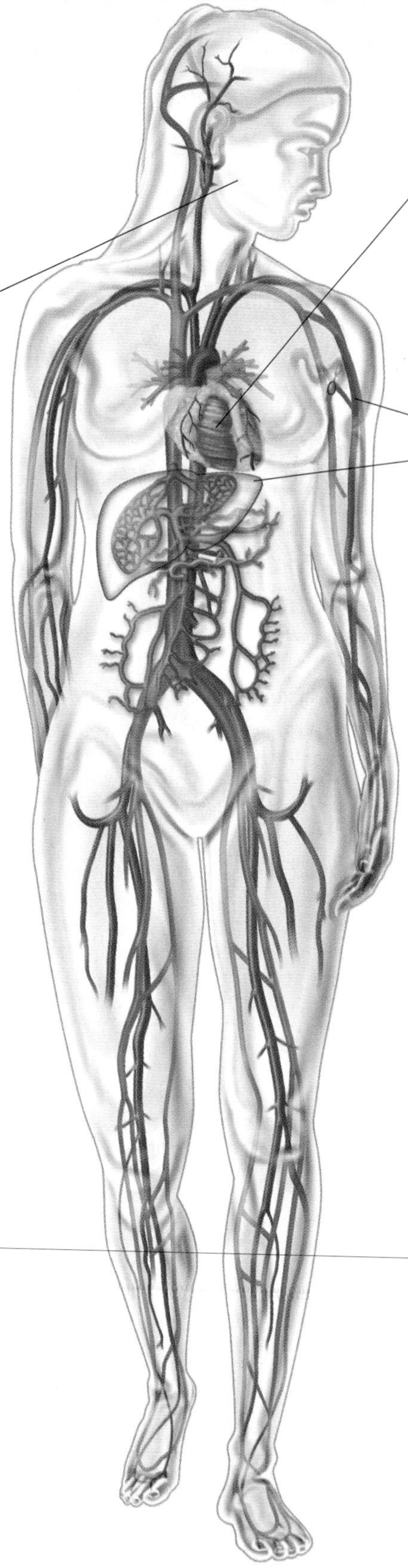

System Summary Figure 20.26

Chapter Summary with Key Terms

20.1 The Cardiovascular and Lymphatic Systems and Their Defenses

A. *The cardiovascular system* is composed of the blood vessels, which carry blood to and from all regions of the body, and the heart, which pumps the blood. The cardiovascular system provides tissues with oxygen and nutrients and carries away carbon dioxide and waste products.

B. *The Lymphatic System:* A major source of immune cells and fluids, the **lymphatic system** serves as a one-way passage, returning fluid from the tissues to the cardiovascular system. It collects fluid that has left the blood vessels and entered tissues, filters it of impurities and infectious agents, and returns it to the blood.

C. *Defenses of the Cardiovascular and Lymphatic Systems:* The cardiovascular system is highly protected from microbial infection.

1. Microbes that successfully invade the system give rise to **systemic infections.** The various types of white blood cells include the lymphocytes, responsible for specific immunity, and the phagocytes, essential in defending the body from microbial invasion. The same defenses are present in the lymphatic system, whose very existence is centered on host immunity.
2. Viruses present in the blood lead to **viremia.** When fungi are in the blood, the condition is termed fungemia, and bacterial presence is called **bacteremia,** but when bacteria flourish and grow in the bloodstream, the condition is termed **septicemia.**

20.2 Normal Biota of the Cardiovascular and Lymphatic Systems

The cardiovascular and lymphatic systems are "closed" systems with no normal biota.

20.3 Cardiovascular and Lymphatic System Diseases Caused by Microorganisms

A. **Endocarditis** is an inflammation of the endocardium, usually due to an infection of the valves of the heart. Two variations of infectious endocarditis have been described: acute and subacute.

1. **Acute endocarditis** is most often caused by *Staphylococcus aureus*, as well as group A streptococci, *Streptococcus pneumoniae*, and *Neisseria gonorrhoeae*.
2. **Subacute forms of endocarditis** are almost always preceded by some form of damage to the heart valves or by congenital malformation. Alpha-hemolytic streptococci, such as *Streptococcus sanguis*, *S. oralis*, and *S. mutans*, are most often responsible, although normal biota from the skin and other bacteria can also colonize abnormal valves and lead to this condition.

B. **Septicemias** occur when organisms are actively multiplying in the blood. Most are caused by bacteria, and to a lesser extent fungi. Gram-negative bacteria multiplying in the blood release large amounts of endotoxin into the bloodstream, resulting in **endotoxic shock.** Gram-positive bacteria can instigate a similar cascade of events when fragments of their cell walls are released into the blood.

C. **Plague** can manifest in three different ways: **Pneumonic plague** is a respiratory disease; **bubonic plague** causes inflammation and necrosis of the lymph nodes, resulting in a swollen lesion called a **bubo;** septicemic plague is the result of multiplication of bacteria in the blood. *Yersinia pestis* is the causative organism—a tiny, gram-negative rod. The principal agents in the transmission of the plague bacterium are fleas.

D. **Tularemia's** causative agent is a facultative intracellular gram-negative bacterium called *Francisella tularensis*. This disease is often called rabbit fever, because it is associated with outbreaks of disease in wild rabbits.

E. **Lyme disease** is caused by *Borrelia burgdorferi*. This nonfatal syndrome mimics neuromuscular and rheumatoid conditions. An early symptom is a bull's-eye rash at the site of a tick bite. *B. burgdorferi* is a unique spirochete that is transmitted primarily by *Ixodes* ticks.

F. **Infectious mononucleosis:** The vast majority of cases are caused by the herpesviruses **Epstein-Barr virus (EBV)** and cytomegalovirus (CMV). A strong, cell-mediated immune response can control the infection, but people usually remain chronically infected with EBV and CMV.

G. **Hemorrhagic fever diseases** are extreme fevers that are often accompanied by internal hemorrhaging. All hemorrhagic fever diseases described here are caused by RNA enveloped viruses in one of three families: Arenaviridae, Filoviridae, and Flaviviridae.

1. **Yellow fever** is caused by an arbovirus, a single-stranded RNA flavivirus that is transmitted by the mosquito, *Aedes aegypti*. Infection begins acutely with fever, headache, and muscle pain.
2. **Dengue fever** is caused by a single-stranded RNA flavivirus and is also carried by *Aedes* mosquitoes. Although mild infection is the usual pattern, a form called dengue hemorrhagic shock syndrome can be lethal.
3. *Ebola and Marburg* viruses are filoviruses (Family Filoviridae) endemic to Central Africa. The virus in the bloodstream leads to extensive capillary fragility and disruption of clotting. Marburg outbreaks are less common than Ebola fever.
4. The *Lassa fever* virus is an arenavirus found in West Africa. The reservoir of the virus is a rodent found in Africa called the multimammate rat.

H. **Nonhemorrhagic Fever Diseases:** These infectious diseases are characterized by high fever but without the capillary fragility that leads to hemorrhagic symptoms.

1. **Brucellosis** is also called Malta fever, undulant fever, and Bang's disease. The bacterial genus *Brucella* contains tiny, aerobic gram-negative coccobacilli. Two species can cause this disease in humans: *B. abortus* (common in cattle) and *B. suis* (from pigs).
2. **Q fever** is caused by *Coxiella burnetii*, is a very small pleomorphic gram-negative bacterium, and an intracellular parasite. *C. burnetii* is apparently

harbored by a wide assortment of vertebrates and arthropods, especially ticks. Ticks, however, do not transmit the disease to humans, who acquire infection mainly by environmental contamination and airborne transmission.

3. **Cat-Scratch Disease:** *Bartonella henselae* is the agent of **cat-scratch disease (CSD),** an infection connected with being clawed or bitten by a cat. The pathogen is present in over 40% of cats, especially kittens.
4. **Trench fever's** causative agent, *Bartonella quintana,* is carried by lice. Highly variable symptoms can include a 5- to 6-day fever, leg pains, headache, chills, and muscle aches.
5. **Ehrlichioses:** There are four tick-borne, fever-producing diseases caused by members of the genus *Ehrlichia*. Members of the genus *Ehrlichia* are small intracellular bacteria and are strict parasites.
6. **Rocky Mountain Spotted Fever:** Another tick-borne disease that causes a distinctive rash. It is caused by *Rickettsia riskettsii.*

I. **Malaria:** This protozoan disease exhibits symptoms of malaise, fatigue, vague aches, and nausea, followed by bouts of chills, fever, and sweating. These symptoms occur at 48- or 72-hour intervals, as a result of the synchronous rupturing of red blood cells. The causative organisms are *Plasmodium* species: *P. malariae, P. vivax, P. falciparum,* and *P. ovale.* The *sexual phase* of the protozoan is carried out in the *Anopheles* mosquito.

J. **Anthrax** infection can exhibit its primary symptoms in various locations of the body: on the skin (cutaneous anthrax), in the lungs (pulmonary anthrax), in the gastrointestinal tract, and in the central nervous system (anthrax meningitis). In all of these forms, the anthrax bacterium gains access to the bloodstream, and death, if it occurs, is usually a result of an overwhelming septicemia. *Bacillus anthracis* is a gram-positive endospore-forming rod that is found in the soil.

K. **HIV Infection and AIDS: Human immunodeficiency virus (HIV)** is the organism responsible for the cluster of symptoms known as **acquired immunodeficiency syndrome,** or **AIDS.** Symptoms in HIV infection are directly tied to two things: the level of virus in the blood and the level of T cells in the blood.

1. HIV is a retrovirus, in the genus lentivirus. It contains **reverse transcriptase** that catalyzes the replication of double-stranded DNA from single-stranded RNA. Retroviral DNA is incorporated into the host genome as a provirus that can be passed on to progeny cells in a latent state. There are two types of HIV: HIV-1, which is the dominant form in most of the world, and HIV-2.
2. The primary effects of HIV infection—those directly due to viral action—are harm to T cells and harm to the central nervous system. The death of T cells and other white blood cells results in extreme **leukopenia** and loss of essential T4 memory clones and stem cells. The destruction of T4 lymphocytes paves the way for invasion by opportunistic agents and malignant cells.
3. HIV transmission occurs mainly through sexual intercourse and transfer of blood or blood products.
4. A regimen that has proved to be extremely effective in controlling AIDS and inevitable drug resistance is **HAART,** short for *highly active antiretroviral therapy.*

L. **Adult T-Cell Leukemia and Hairy-Cell Leukemia:** Leukemia is the general name for at least four different malignant diseases of the white blood cell forming elements originating in the bone marrow. The retrovirus HTLV-I is associated with a form of leukemia called adult T-cell leukemia. HTLV-II is associated with hairy-cell leukemia.

Multiple-Choice and True-False Questions

Multiple-Choice Questions. Select the correct answer from the answers provided.

1. When bacteria flourish and grow in the bloodstream, this is referred to as
   a. viremia c. septicemia
   b. bacteremia d. fungemia
2. Which of the following is *not* a symptom of septicemia?
   a. fever c. shaking chills
   b. respiratory alkalosis d. high blood pressure
3. The plague bacterium, *Yersinia pestis,* is transmitted mainly by
   a. mosquitoes c. dogs
   b. fleas d. birds
4. Rabbit fever is caused by
   a. *Yersinia pestis* c. *Borrelia burgdorferi*
   b. *Francisella tularensis* d. *Chlamydia bunnyensis*
5. A distinctive bull's-eye rash results from a tick bite transmitting
   a. Lyme disease c. Q fever
   b. tularemia d. Rocky Mountain spotted fever
6. Cat-scratch disease is caused by
   a. *Coxiella burnetii* c. *Bartonella quintana*
   b. *Bartonella henselae* d. *Brucella abortus*
7. The bite of the Lone Star tick, *Ixodes scapularis,* can cause
   a. ehrlichioses d. both a and b
   b. Lyme disease e. both b and c
   c. trench fever
8. Cat-scratch disease is effectively treated with
   a. rifampin c. amoxicillin
   b. penicillin d. acyclovir
9. Wool-sorter's disease is caused by
   a. *Brucella abortus* c. *Coxiella burnetii*
   b. *Bacillus anthracis* d. *Rabies virus*
10. Which of the following is *not* a hemorrhagic fever?
    a. Lassa fever c. Ebola fever
    b. Marburg fever d. Trench fever

True-False Questions. If statement is true, leave as is. If it is false, correct it by rewriting the sentence.

11. Brucellosis can be transmitted to humans by drinking contaminated milk.
12. Respiratory tract infection with *Bartonella henselae* is considered an AIDS defining condition.
13. Lyme disease is caused by *Rickettsia rickettsii*.
14. Yellow fever is caused by a protozoan transmitted by fleas.
15. A distinctive bull's-eye rash is associated with Q fever.

Writing to Learn

These questions are suggested as a *writing-to-learn* experience. For each question, compose a one- or two-paragraph answer that includes the factual information needed to completely address the question.

1. What is endotoxic shock?
2. a. Name the agent(s) responsible for infectious mononucleosis.
   b. What other diseases are associated with these agents?
3. Describe the infectious cycle of HIV.
4. a. What is the causative organism of Q fever?
   b. What characteristic(s) of this organism labels it as an "atypical" bacterium?
5. Describe the life cycle of the malarial parasite, including the significant events of sexual and asexual reproduction.
6. What criteria are used in the United States to diagnose a person with AIDS?
7. a. What are retroviruses? Where does the name come from?
   b. Name some retroviruses implicated in human diseases.
8. a. What are the different locations in the human body that anthrax infection can be exhibited?
   b. Which of these are the most common forms of the disease?
   c. What organism(s) cause this disease?

Concept Mapping

Appendix D provides guidance for working with concept maps.

1. Provide the missing concepts in this map. Supply as many links as you can among all the concepts, and provide linking words or phrases.

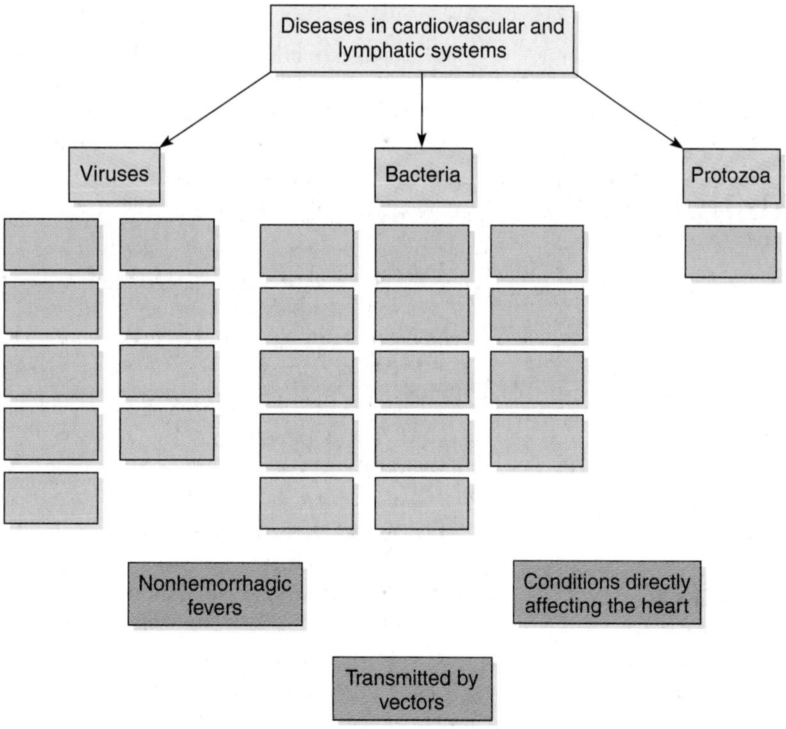

Critical Thinking Questions

Critical thinking is the ability to reason and solve problems using facts and concepts. These questions can be approached from a number of angles, and in most cases, they do not have a single correct answer.

1. What is the significance of biofilms in subacute bacterial endocarditis?
2. Why is the bubonic plague called the "black death"?
3. Explain how *Yersinia pestis* infection of the flea vector enhances the spread of the plague disease.
4. Why do you think there are people who are HIV-positive but remain healthy and never develop AIDS—so-called nonprogressors?
5. What characteristics make tularemia a potential bioweapon?
6. What are some of the challenges facing current strategies to eliminate malaria?
7. How are retroviruses thought to cause cancers?

Visual Understanding

1. a. **From chapter 14, figure 14.14.** Imagine that the WBCs shown in this illustration are unable to control the microorganisms. Could the change that has occurred in the vessel wall help the organism spread to other locations? If so, how?

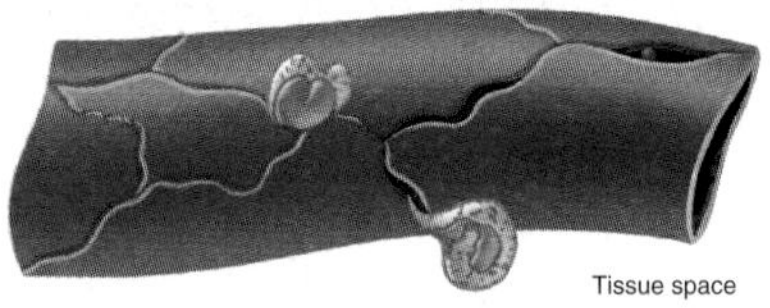

(b)

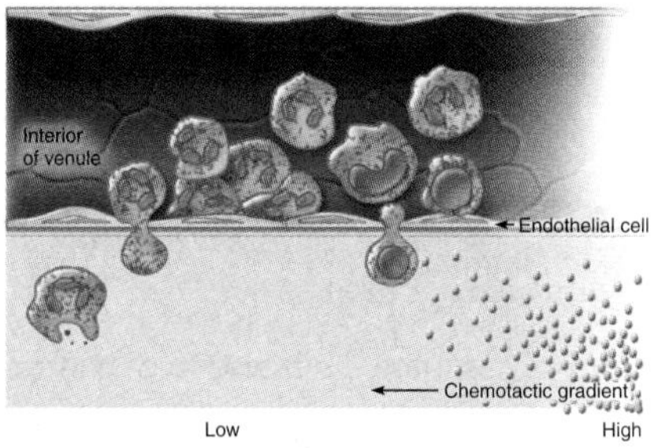

(c)

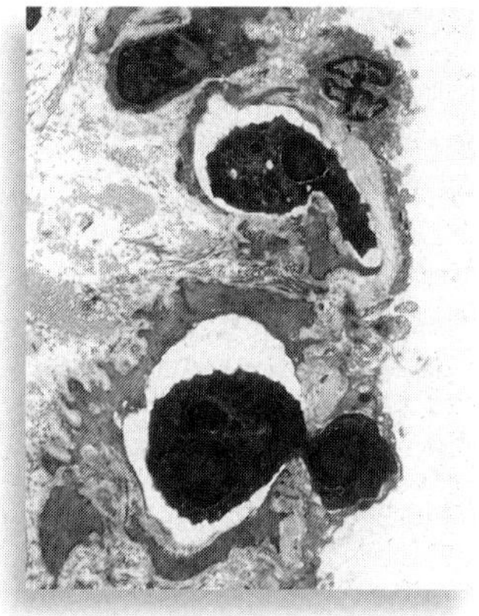

b. If the organisms are able to survive the process shown in (c), how could that impact the progress of this disease? Explain your answer.

Internet Search Topics

1. Find websites that give information on common infections of AIDS patients and the effects they have on health and survival.
2. a. Go to a search engine to access your home state's Department of Public Health website. Locate the list of reportable diseases and determine the incidence of Lyme disease, ehrlichiosis, and plague.
   b. At the same site, try to find information about what measures are being taken in your state to monitor potential bioterrorist actions.
3. Go to: www.aris.mhhe.com, and click on "microbiology" and then this textbook's author/title. Go to chapter 20, access the URLs listed under Internet Search Topics, and research the following:
   a. What is the initiative called Roll Back Malaria and what are its goals?
   b. Access the site for Rapid Diagnostic Tests. Why are malaria Rapid Diagnostic Tests needed? Draw out the steps of the RDT, using terminology you learned in this class (in chapter 17 and elsewhere).
   http://www.rbm.who.int/
   http://www.wpro.who.int/rdt/
4. Search the Internet for information about the David John Differential Test. What information does this test provide? How is this information more useful than a "monospot" test result?

CHAPTER 21

Infectious Diseases Affecting the Respiratory System

CASE FILE 21

In recent years vaccine-preventable diseases have caused a number of outbreaks in the United States. Such was the case in an Amish community in Kent County in southern Delaware. Between September 2004 and February 2005, a total of 345 cases of pertussis (also known as whooping cough) were identified in this close-knit community of 1,711 people. Nurses from the Delaware Department of Public Health detected the first cases among patients that were being seen in the Southern Health Services clinic that serves the area. In order to control the spread of the disease, special pertussis clinics were set up in the Amish schools within the community. The clinics provided a means to educate community members about prevention, control, and treatment measures. In addition, local doctors and health care providers were notified and provided with diagnostic kits. To determine the full extent of the outbreak, surveillance strategies were developed. These included self-administered questionnaires and household interviews.

- *What causes pertussis and how is it diagnosed?*
- *What measures can be taken to prevent outbreaks such as this one?*

Case File 21 Wrap-Up appears on page 661.

CHAPTER OVERVIEW

- The respiratory tract is divided into upper and lower portions. The lungs, in the lower portion, are the sites for oxygen exchange. Anatomical features provide protection, including nasal hair, the ciliary escalator, and mucus. There are macrophages in the alveoli of the lungs and secretory IgA on the epithelial surfaces.
- The upper respiratory tract has a diverse normal biota dominated by gram-positive bacteria. The lower respiratory tract has no normal biota.
- Upper respiratory tract (URT) infections are extremely common and generally (but not always) milder than lower respiratory tract infections.
- Sore throats, ear infections, and sinusitis are common URT infections that can be caused by multiple microorganisms.
- Diphtheria is a serious URT that has a single causative organism; it is controlled by the DTaP vaccination.
- Whooping cough, respiratory syncytial virus infection, and influenza affect the lower URT and the lower respiratory tract (LRT).
- Tuberculosis and pneumonia are two major LRT infections.
- Many different organisms can cause pneumonia, including the newly discovered SARS virus.

21.1 The Respiratory Tract and Its Defenses

The respiratory tract is the most common place for infectious agents to gain access to the body. We breathe 24 hours a day, and anything in the air we breathe passes at least temporarily into this organ system.

The structure of the system is illustrated in **figure 21.1*a*.** Most clinicians divide the system into two parts, the *upper* and *lower respiratory tracts.* The upper respiratory tract includes the mouth, the nose, nasal cavity and sinuses above it, the throat or pharynx, and the epiglottis and larynx. The lower respiratory tract begins with the trachea, which feeds into the bronchi and bronchioles in the lungs. Attached to the bronchioles are small balloonlike structures called alveoli, which inflate and deflate with inhalation and exhalation. These are the site of oxygen exchange in the lungs.

Several anatomical features of the respiratory system protect it from infection. As described in chapter 14, nasal hair serves to trap particles. Cilia **(figure 21.1*b*)** on the epithelium of the trachea, and bronchi (the ciliary escalator) propel particles upward and out of the respiratory tract. Mucus on the surface of the mucous membranes lining the respiratory tract is a natural trap for invading microorganisms. Once the microorganisms are trapped, involuntary responses such as coughing, sneezing, and swallowing can move them out of sensitive areas. These are first-line defenses.

The second and third lines of defense also help protect the respiratory tract. Macrophages inhabit the alveoli of the lungs and the clusters of lymphoid tissue (tonsils) in the throat. Secretory IgA against specific pathogens can be found in the mucus secretions as well.

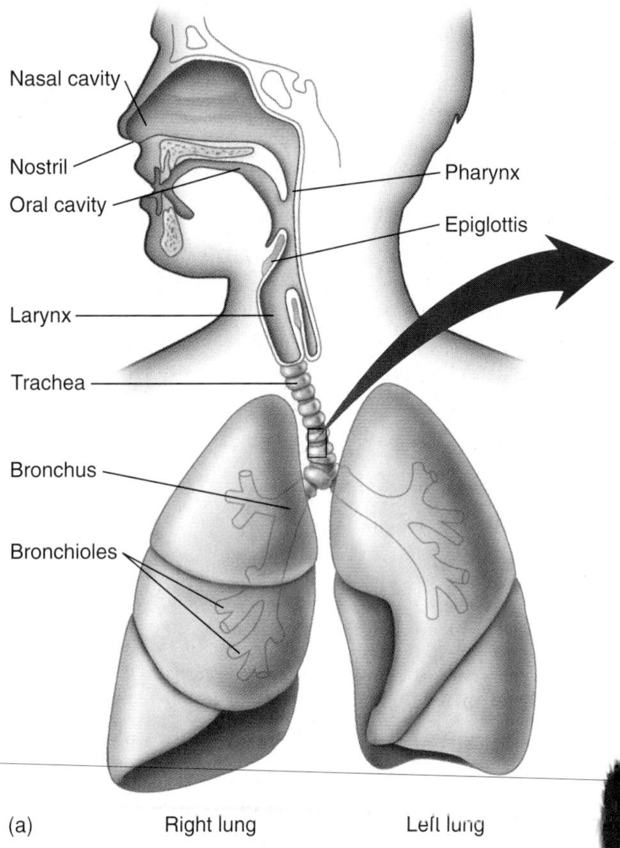

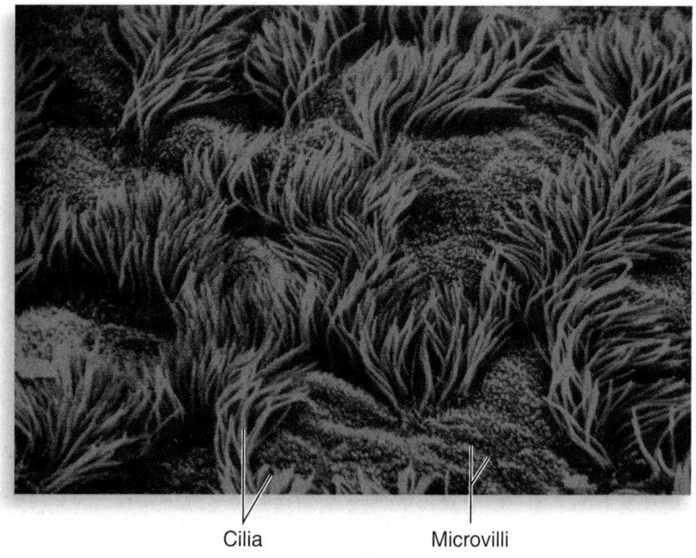

(b) Ciliary defense of the tracheal mucosa (5,000x)

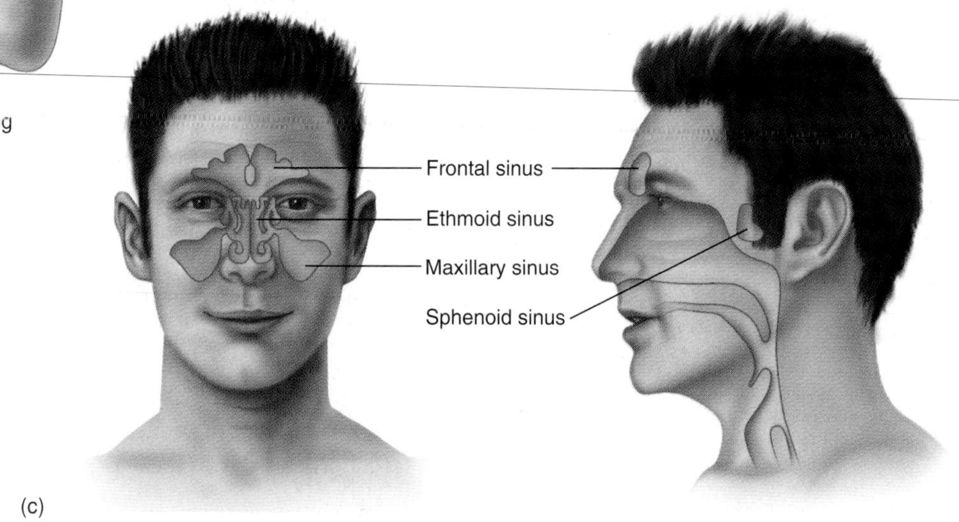

Figure 21.1 The respiratory tract. **(a)** Important structures in the upper and lower respiratory tract. **(b)** Ciliary defense of the respiratory tract. **(c)** The four pairs of sinuses in the face and skull.

21.2 Normal Biota of the Respiratory Tract

Because of its constant contact with the external environment, the respiratory system harbors a large number of commensal microorganisms. The normal biota is generally limited to the upper respiratory tract, and gram-positive bacteria such as streptococci and staphylococci are very common. Note that some bacteria that can cause serious disease are frequently present in the upper respiratory tract as "normal" biota; these include *Streptococcus pyogenes, Haemophilus influenzae, Streptococcus pneumoniae, Neisseria meningitidis,* and *Staphylococcus aureus.* These bacteria can potentially cause disease if their host becomes immunocompromised for some reason, and they can cause disease in other hosts when they are innocently transferred to them. Other normal biota bacteria include nonhemolytic and alpha-hemolytic streptococci, *Moraxella* species, and *Corynebacterium* species (often called diphtheroids). Yeasts, especially *Candida albicans,* also colonize the mucosal surfaces of the mouth.

In the respiratory system, as in some other organ systems, the normal biota performs the important function of microbial antagonism (see chapter 13). This reduces the chances of pathogens establishing themselves in the same area by competing with them for resources and space. As is the case with the other body sites harboring normal biota, the microbes reported here are those we have been able to culture in the laboratory. More microbes will come to light as scientists catalog the genetic sequences in the Human Microbiome project.

21.3 Upper Respiratory Tract Diseases Caused by Microorganisms

Rhinitis, or the Common Cold

In the course of a year, people in the United States suffer from about 1 billion colds, called rhinitis because *rhin-* means nose and *-itis* means inflammation. Many people have several episodes a year. Economists estimate that this fairly innocuous infection costs the United States $40 billion a year in trips to the doctors, medications, and lost work time.

Signs and Symptoms

Everyone is familiar with the symptoms of rhinitis: sneezing, scratchy throat, and runny nose (rhinorrhea), which usually begin 2 or 3 days after infection. An uncomplicated cold generally is not accompanied by fever, although children can experience low fevers (less than 102°F). The incubation period is usually 2 to 5 days.

Causative Agents

The common cold is caused by one of over 200 different kinds of viruses. The particular virus is almost never identified, and the symptoms and handling of the infection are the same no matter which of the viruses is responsible.

The most common type of virus leading to rhinitis is the group called rhinoviruses. Coronaviruses probably are in second place. Most viruses causing the common cold never lead to any serious consequences, but some of them can be serious for some patients. For instance, the respiratory syncytial virus (RSV) causes colds in most people, but in some, especially children, they can lead to more serious respiratory tract symptoms. (RSV is discussed later in the chapter.) In this section, we consider all cold-causing viruses together as a group because they are treated similarly.

Viral infection of the upper respiratory tract can predispose a patient to secondary infections by other microorganisms, such as bacteria. Secondary infections may explain why some people report that their colds improved when they were given antibiotics. The cold was caused by viruses; bacterial infection may have followed.

Pathogenesis and Virulence Factors

Viruses that induce rhinitis do not have many virulence mechanisms. They must penetrate the mucus that coats the respiratory tract and then find firm attachment points. Once they are attached, they use host cells to produce more copies of themselves (see chapter 6). The symptoms we experience as the common cold are mainly the result of our body fighting back against the viral invaders. Virus-infected cells in the upper respiratory tract release chemicals that attract certain types of white blood cells to the site, and these cells release cytokines and other inflammatory mediators,

Respiratory Tract Defenses and Normal Biota

	Defenses	Normal Biota
Upper Respiratory Tract	Nasal hair, ciliary escalator, mucus, involuntary responses such as coughing and sneezing, secretory IgA	*Moraxella,* nonhemolytic and alpha-hemolytic streptococci, *Corynebacterium* and other diphtheroids, *Candida albicans* Note: *Streptococcus pyogenes, Streptococcus pneumoniae, Haemophilus influenzae, Neisseria meningitidis,* and *Staphylococcus aureus* often present as "normal" biota.
Lower Respiratory Tract	Mucus, alveolar macrophages, secretory IgA	None

inflammatory mediators, as described earlier in chapters 14 and 16. These mediators generate a localized inflammatory reaction, characterized by swelling and inflammation of the nasal mucosa, leakage of fluid from capillaries and lymph vessels, and the increased production of mucus. The similarity of these symptoms to those of inhalant allergies illustrates that the same immune reactions are involved in both conditions.

Transmission and Epidemiology

Cold viruses are transmitted by droplet contact, but indirect transmission may be more common, such as when a healthy person touches a fomite and then touches one of his or her own vulnerable surfaces, such as the mouth, nose, or an eye. In some cases, the viruses can remain airborne in droplet nuclei and aerosols and can be transmitted in that way.

The epidemiology of the common cold is fairly simple: Practically everybody gets them, and fairly frequently. Children have more frequent infections than adults, probably because nearly every virus they encounter is a new one and they have no secondary immunity to it. People can acquire some degree of immunity to a cold virus that they have encountered before, but because there are more than 200 viruses, this immunity doesn't provide much overall protection.

Prevention

There is no vaccine for rhinitis. A traditional vaccine would need to contain antigens from about 200 viruses to provide complete protection. Researchers are studying novel types of immunization strategies, however. Because most of the viruses causing rhinitis use only a few different chemicals on host epithelium for their attachment site, some scientists have proposed developing a vaccine that would stimulate antibody to the docking site on the host. Other approaches include inducing antibody to the sites of action for the inflammatory mediators. But for now, the best prevention is to stop the transmission between hosts. The best way to prevent transmission is frequent hand washing, followed closely by stopping droplets from traveling away from the mouth and nose by covering them when sneezing or coughing. It is better to do this by covering the face with the crook of the arm rather than the hand, because subsequent contact with surfaces is less likely.

Treatment

No chemotherapeutic agents cure the common cold. A wide variety of over-the-counter agents, such as antihistamines and decongestants, improve symptoms by blocking inflammatory mediators and their action. The use of these agents may also cut down on transmission to new hosts, because fewer virus-loaded secretions are produced.

CHECKPOINT 21.1	Rhinitis
Causative Organism(s)	Approximately 200 viruses
Most Common Modes of Transmission	Indirect contact, droplet contact
Virulence Factors	Attachment proteins; most symptoms induced by host response
Culture/Diagnosis	Not necessary
Prevention	Hygiene practices
Treatment	For symptoms only

Sinusitis

Commonly called a *sinus infection*, this inflammatory condition of any of the four pairs of sinuses in the skull **(figure 21.1*c*)** can actually be caused by allergy (most common), infections, or simply by structural problems such as narrow passageways or a deviated nasal septum. The infectious agents that may be responsible for the condition commonly include a variety of viruses or bacteria and, less commonly, fungi. Infections of the sinuses generally follow a bout with the common cold. But the inflammatory symptoms produce a large amount of fluid and mucus and when trapped in the sinuses, these secretions provide an excellent growth medium for bacteria or fungi. So viral rhinitis is frequently followed by sinusitis caused by bacteria or fungi.

Signs and Symptoms

A person suffering from any form of sinusitis experiences nasal congestion, pressure above the nose or in the forehead, and sometimes the feeling of a headache or a toothache. Facial swelling and tenderness are common. Discharge from the nose and mouth appears opaque and has a green or yellow color in the case of bacterial infections. Discharge caused by an allergy is usually clear, and the symptoms may be accompanied by itchy, watery eyes.

Causative Agents

Bacteria Any number of bacteria that are normal biota in the upper respiratory tract may cause sinus infections. Many cases are caused by *Streptococcus pneumoniae*, *Streptococcus pyogenes*, *Staphylococcus aureus*, and *Haemophilus influenzae*. The causative organism is usually not identified, but treatment is begun empirically, based on the symptoms.

The bacteria that cause these infections are most often normal biota in the host and don't have an arsenal of virulence factors that lead to their ability to cause disease. The pathogenesis of this condition is brought about by the confluence of several factors: predisposition to infection because of underlying (often viral) infection; buildup of fluids, providing a rich environment for bacterial multiplication; and sometimes the anatomy of the sinuses, which can contribute to entrapment of mucus and bacterial growth.

Bacterial sinusitis is not a communicable disease. Of course, the virus originally causing rhinitis is transmissible, but the host takes it from there by creating the conditions favorable for respiratory tract microorganisms to multiply in the sinus spaces, which normally do not harbor microorganisms to any significant extent.

Sinusitis is extremely common, resulting in approximately 11.5 million office visits a year in the United States. A large proportion of these cases are allergic sinusitis episodes, but approximately 30% of them are caused by bacterial overgrowth in the sinuses. Women and residents of the southern United States have slightly higher rates. As with many upper respiratory tract infections, smokers have higher rates of infection than nonsmokers. Children who are exposed to large amounts of secondhand smoke are also more susceptible.

Broad-spectrum antibiotics may be prescribed when the physician feels that the sinusitis is bacterial in origin (that is, when allergic sinusitis and fungal sinusitis are ruled out).

Fungi Fungal sinusitis is rare, but it is often recognized when antibacterial drugs fail to alleviate symptoms. Simple fungal infections may normally be found in the maxillary sinuses and are noninvasive in nature. These colonies are generally not treated with antifungal agents but instead simply mechanically removed by a physician. *Aspergillus fumigatus* is a common fungus involved in this type of infection. The growth of fungi in this type of sinusitis may be encouraged by trauma to the area.

More serious invasive fungal infections of the sinuses may be found in severely immunocompromised patients. Fungi such as *Aspergillus* and *Mucor* species may invade the bony structures in the sinuses and even travel to the brain or eye. These infections are treated aggressively with a combination of surgical removal of the fungus and intravenous antifungal therapy **(Checkpoint 21.2).**

Acute Otitis Media (Ear Infection)

This condition is another common sequela of rhinitis, or the common cold, and for reasons similar to the ones described for sinusitis. Viral infections of the upper respiratory tract lead to inflammation of the eustachian tubes and the buildup of fluid in the middle ear, which can lead to bacterial multiplication in those fluids. Although the middle ear normally has no biota, bacteria can migrate along the eustachian tube from the upper respiratory tract **(figure 21.2).** When bacteria encounter mucus and fluid buildup in the middle ear, they multiply rapidly. Their presence increases the inflammatory response, leading to pus production and continued fluid secretion. This fluid is referred to as *effusion.*

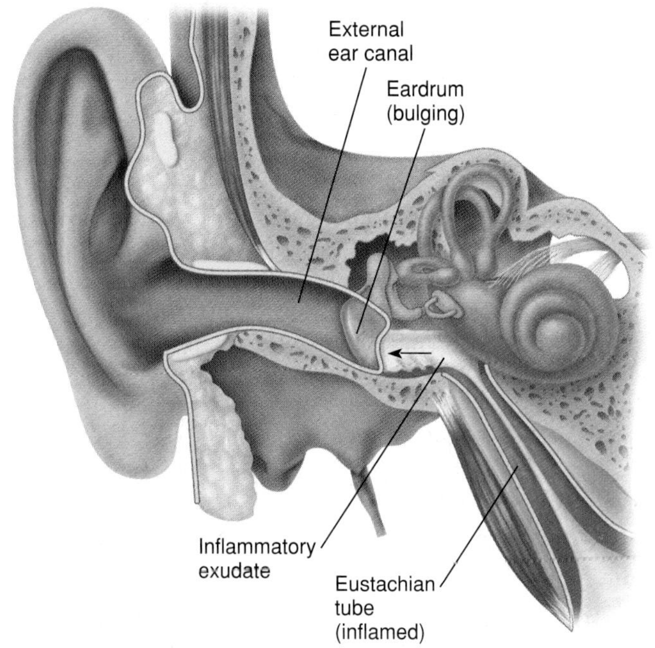

Figure 21.2 An infected middle ear.

Another condition, known as chronic otitis media, occurs when fluid remains in the middle ear for indefinite periods of time. Until recently, physicians considered it to be the result of a noninfectious immune reaction because they could not culture bacteria from the site and because antibiotics were not effective. New data suggest that this form of otitis media is caused by a mixed biofilm of bacteria that are attached to

CHECKPOINT 21.2 Sinusitis

Causative Organism(s)	Various bacteria, often mixed infection	Various fungi
Most Common Modes of Transmission	Endogenous (opportunism)	Introduction by trauma *or* opportunistic overgrowth
Virulence Factors	–	–
Culture/Diagnosis	Culture not usually performed; diagnosis based on clinical presentation, occasionally X rays or other imaging technique used	Same
Prevention	–	–
Treatment	Broad-spectrum antibiotics	Physical removal of fungus; in severe cases antifungals used
Distinctive Features	Much more common than fungal	Suspect in immunocompromised patients

the membrane of the inner ear. Biofilm bacteria generally are less susceptible to antibiotics (as discussed in chapter 4), and their presence in biofilm form would explain the inability to culture them from ear fluids.

Signs and Symptoms

Otitis media may be accompanied by a sensation of fullness or pain in the ear and loss of hearing. Younger children may exhibit irritability, fussiness, and difficulty in sleeping, eating, or hearing. Severe or untreated infections can lead to rupture of the eardrum because of pressure of pus buildup, or to internal breakthrough of these infected fluids, which can lead to more serious conditions such as mastoiditis, meningitis, or intracranial abscess.

Causative Agents

Many different viruses and bacteria can cause acute otitis media, but the most common cause is *Streptococcus pneumoniae* (also discussed in the section on pneumonia later in this chapter). *Haemophilus influenzae* is another common cause of this condition; however, the incidence of all types of infections with this bacterium was significantly reduced with the introduction of a childhood vaccine against it in the 1980s.

Streptococcus pneumoniae appears as pairs of elongated, gram-positive cocci joined end to end. It is often called by the familiar name *pneumococcus,* and diseases caused by it are termed *pneumococcal.*

Transmission and Epidemiology

Otitis media is a sequela of upper respiratory tract infection and is not communicable, although the upper respiratory infection preceding it is. Children are particularly susceptible, and boys have a slightly higher incidence than do girls.

Prevention

A vaccine against *S. pneumoniae* has been a part of the recommended childhood vaccination schedule since 2000. The vaccine (Prevnar) is a seven-valent conjugated vaccine (see chapter 15). It contains polysaccharide capsular material from seven different strains of the bacterium complexed with a chemical that makes it more antigenic. It is distinct from another vaccine for the same bacterium (Pneumovax), which is primarily targeted to the older population to prevent pneumococcal pneumonia.

Treatment

Until the late 1990s, broad-spectrum antibiotics were routinely prescribed for otitis media. When it became clear that frequently treating children with these drugs was producing a bacterial biota with high rates of antibiotic resistance, the treatment regimen was reexamined.

The current recommendation for uncomplicated acute otitis media with a fever below 104°F is "watchful waiting" for 72 hours to allow the body to clear the infection, avoiding the use of antibiotics.

Children who experience frequent recurrences of ear infections sometimes have small tubes placed through the tympanic membranes into their middle ears to provide a means of keeping fluid out of the site when inflammation occurs **(Checkpoint 21.3).**

Pharyngitis

Signs and Symptoms

The name says it all—this is an inflammation of the throat, which the host experiences as pain and swelling. The severity of pain can range from moderate to severe, depending on the causative agent. Viral sore throats are generally mild and sometimes lead to hoarseness. Sore throats caused by group A streptococci are generally more painful than those caused by viruses, and they are more likely to be accompanied by fever, headache, and nausea.

Clinical signs of a sore throat are reddened mucosa, swollen tonsils, and sometimes white packets of inflamma-

CHECKPOINT 21.3 Otitis Media

Causative Organism(s)	*Streptococcus pneumoniae*	*Haemophilus influenzae*	Other bacteria
Most Common Modes of Transmission	Endogenous (may follow upper respiratory tract infection by *S. pneumoniae* or other microorganisms)	Endogenous (follows upper respiratory tract infection)	Endogenous
Virulence Factors	Capsule, hemolysin	Capsule, fimbriae	–
Culture/Diagnosis	Usually relies on clinical symptoms and failure to resolve within 72 hours	Same	Same
Prevention	Pneumococcal conjugate vaccine (heptavalent)	Hib vaccine	None
Treatment	Wait for resolution; if needed, amoxicillin (are high rates of resistance) or amoxicillin + clavalanate or cefuroxine	Same as for *S. pneumoniae*	Wait for resolution; if needed, a broad-spectrum antibiotic (azithromycin) might be used in absence of etiologic diagnosis
Distinctive Features	–	–	Suspect if fully vaccinated against other two

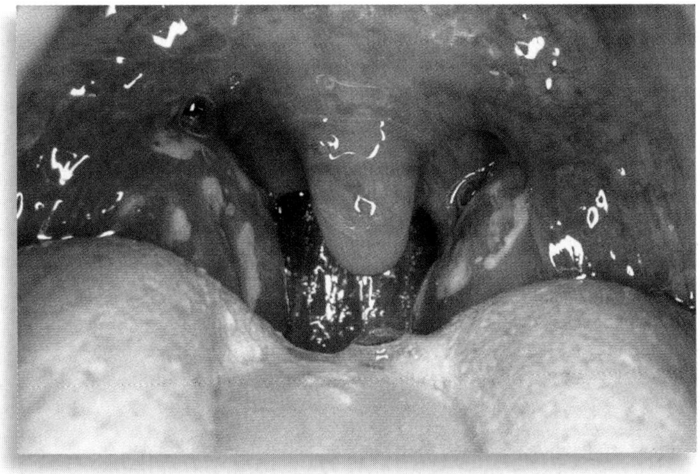

Figure 21.3 The appearance of the throat in pharyngitis and tonsillitis.
The pharynx and tonsils become bright red and suppurative. Whitish pus nodules may also appear on the tonsils.

tory products visible on the walls of the throat, especially in streptococcal disease **(figure 21.3).** The mucous membranes may be swollen, affecting speech and swallowing. Often pharyngitis results in foul-smelling breath. The incubation period for most sore throats is generally 2 to 5 days.

Causative Agents

A sore throat is most commonly caused by the same viruses causing the common cold. It can also accompany other diseases, such as infectious mononucleosis (described in chapter 20). Pharyngitis may simply be the result of mechanical irritation from prolonged shouting or from drainage of an infected sinus cavity. The most serious cause of pharyngitis is *Streptococcus pyogenes.* The possibility of this bacterium being involved is the reason that even moderately severe sore throats should be monitored and diagnosed.

S. pyogenes is a gram-positive coccus that grows in chains. It does not form spores, is nonmotile, and forms capsules and slime layers. *S. pyogenes* is a facultative anaerobe that ferments a variety of sugars. It does not form catalase, but it does have a peroxidase system for inactivating hydrogen peroxide, which allows its survival in the presence of oxygen.

Pathogenesis

Untreated streptococcal throat infections occasionally can result in serious complications, either right away or days to weeks after the throat symptoms subside. These complications include scarlet fever, rheumatic fever, and glomerulonephritis. More rarely, invasive and deadly conditions such as necrotizing fasciitis can result from infection by *S. pyogenes.* These invasive conditions are described in chapter 18.

Scarlet Fever Scarlet fever is the result of infection with an *S. pyogenes* strain that is itself infected with a bacteriophage. This lysogenic virus confers on the streptococcus the ability to produce erythrogenic toxin, described in the section on

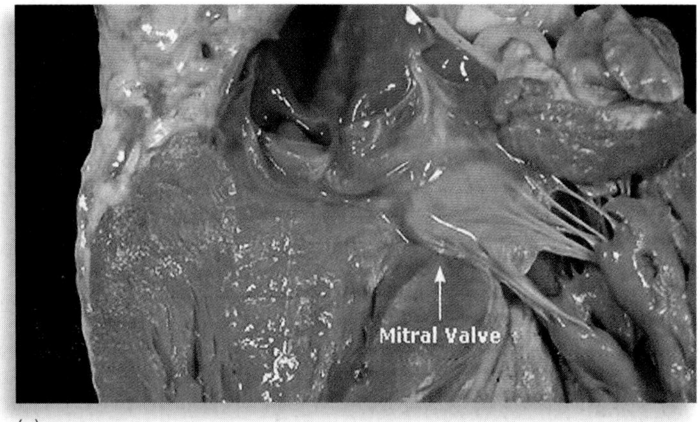

(a)

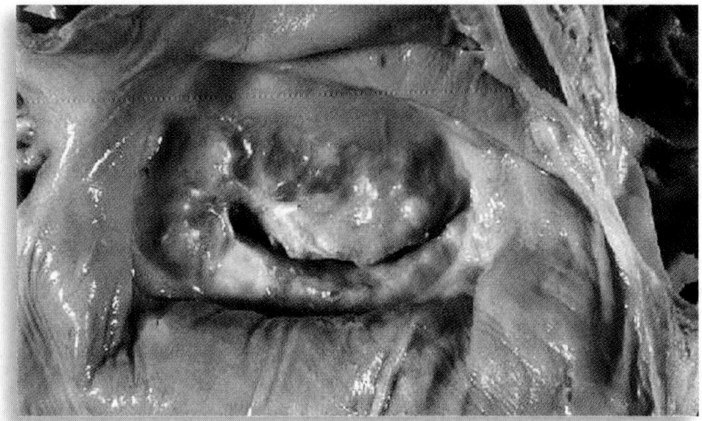

(b)

Figure 21.4 The cardiac complications of rheumatic fever.
Pathologic processes of group A streptococcal infection can extend to the heart. In this example, it is believed that cross-reactions between streptococcal-induced antibodies and heart proteins have a gradual destructive effect on the atrioventricular valves (especially the mitral valve) or semilunar valves. Scarring and deformation change the capacity of the valves to close and shunt the blood properly. **(a)** A normal valve, viewed from above. **(b)** Higher magnification view of a scarred mitral valve.

virulence. Scarlet fever is characterized by a sandpaperlike rash, most often on the neck, chest, elbows, and inner surfaces of the thighs. High fever accompanies the rash. It most often affects school-age children, and was a source of great suffering in the United States in the early part of the 20th century. In epidemic form, the disease can have a fatality rate of up to 95%. Most cases seen today are mild. They are easily recognizable and amenable to antibiotic therapy. Because of the fear elicited by the name "scarlet fever," the disease is often called scarlatina in North America.

Rheumatic Fever Rheumatic fever is thought to be due to an immunologic cross-reaction between the streptococcal M protein and heart muscle. It tends to occur approximately 3 weeks after pharyngitis has subsided. It can result in permanent damage to heart valves **(figure 21.4).** Other symptoms include arthritis in multiple joints and the appearance of nodules over bony surfaces just under the skin. Rheumatic fever is completely preventable if the original streptococcal infection is treated with antibiotics. Nevertheless, it is still a serious problem today in many parts of the world.

Glomerulonephritis Glomerulonephritis is thought to be the result of streptococcal proteins participating in the formation of antigen-antibody complexes, which then are deposited in the basement membrane of the glomeruli of the kidney. It is characterized by **nephritis** (appearing as swelling in the hands and feet and low urine output), blood in the urine, increased blood pressure, and occasionally heart failure. It can result in permanent kidney damage. The incidence of poststreptococcal glomerulonephritis has been declining in the United States, but it is still common in Africa, the Caribbean, and South America.

Toxic shock syndrome and necrotizing fasciitis are other, less frequent, consequences of streptococcal infections, and are discussed in chapter 18.

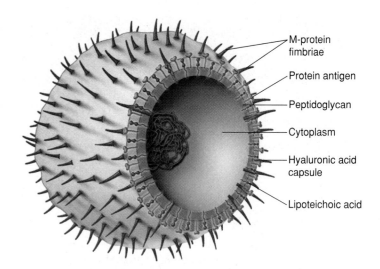

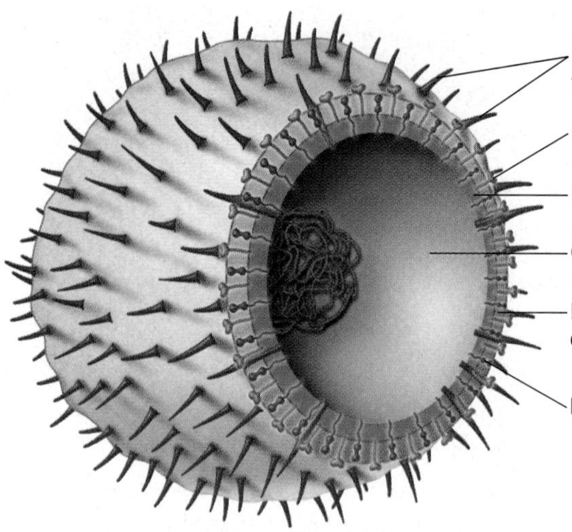

Figure 21.5 Cutaway view of group A streptococcus.

Virulence Factors

The virulence of *S. pyogenes* is partly due to the substantial array of surface antigens, toxins, and enzymes it can generate.

Streptococci display numerous surface antigens **(figure 21.5).** Specialized polysaccharides on the surface of the cell wall help to protect the bacterium from being dissolved by the lysozyme of the host. Lipoteichoic acid (LTA) contributes to the adherence of *S. pyogenes* to epithelial cells in the pharynx. A spiky surface projection called *M protein* contributes to virulence by resisting phagocytosis and possibly by contributing to adherence. A capsule made of hyaluronic acid (HA) is formed by most *S. pyogenes* strains. It probably contributes to the bacterium's adhesiveness. Because this HA is chemically indistinguishable from HA found in human tissues, it does not provoke an immune response from the host.

Extracellular Toxins Group A streptococci owe some of their virulence to the effects of hemolysins called **streptolysins.** The two types are streptolysin O (SLO) and streptolysin S (SLS).[1] Both types cause beta-hemolysis of sheep blood agar (see "Culture and Diagnosis"). Both hemolysins rapidly injure many cells and tissues, including leukocytes and liver and heart muscle (in other forms of streptococcal disease).

A key toxin in the development of scarlet fever is **erythrogenic** (eh-rith″-roh-jen′-ik) **toxin.** This toxin is responsible for the bright red rash typical of this disease, and it also induces fever by acting upon the temperature regulatory center in the brain. Only lysogenic strains of *S. pyogenes* that contain genes from a temperate bacteriophage can synthesize this toxin. (For a review of the concept of lysogeny, see chapter 6.)

Some of the streptococcal toxins (erythrogenic toxin and streptolysin O) contribute to increased tissue injury by acting as *superantigens*. These toxins elicit excessively strong reactions from monocytes and T lymphocytes. When activated, these cells proliferate and produce *tumor necrosis factor* (*TNF*), which leads to a cascade of immune responses resulting in vascular injury. This is the likely mechanism for the severe pathology of toxic shock syndrome and necrotizing fasciitis.

Transmission and Epidemiology

Physicians estimate that 30% of sore throats may be caused by *S. pyogenes*, adding up to several million cases each year. Most transmission of *S. pyogenes* is via respiratory droplets or direct contact with mucus secretions. This bacterium is carried as "normal" biota by 15% of the population, but transmission from this reservoir is less likely than from a person who is experiencing active disease from the infection because of the higher number of bacteria present in the disease condition. It is less common but possible to transmit this infection via fomites. Humans are the only significant reservoir of *S. pyogenes*.

More than 80 serotypes of *S. pyogenes* exist, and thus people can experience multiple infections throughout their lives because immunity is serotype-specific. Even so, only a minority of encounters with the bacterium result in disease. An immunocompromised host is more likely to suffer from strep pharyngitis as well as serious sequelae of the throat infection.

Although most sore throats caused by *S. pyogenes* can resolve on their own, they should be treated with antibiotics because serious sequelae are a possibility.

Culture and Diagnosis

S. pyogenes is classified as a group A streptococcus. The group designation of streptococci was developed by Rebecca Lancefield **(figure 21.6)** in the 1930s. She discovered that the cell wall carbohydrates (antigens) of various cultures stimulated formation of antibodies with differing specificities. She characterized 14 different groups using an alphabetic system (A, B, C).

1. In SLO, O stands for oxygen because the substance is inactivated by oxygen. In SLS, S stands for serum because the substance has an affinity for serum proteins. SLS is oxygen-stable.

Figure 21.6 **Rebecca C. Lancefield, M.D.**
She became known worldwide for the Lancefield classification of streptococci, which she developed in the first half of the 20th century.

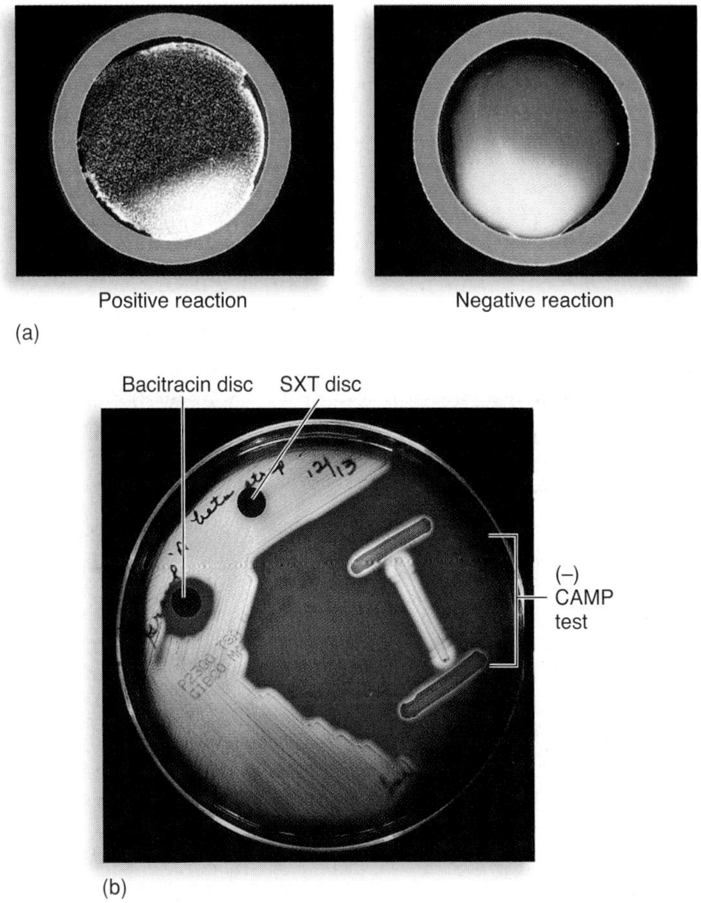

Figure 21.7 **Streptococcal tests.**
(a) A rapid, direct test kit for diagnosis of group A infections. With this method, a patient's throat swab is introduced into a system composed of latex beads and monoclonal antibodies. *(Left)* In a positive reaction, the C-carbohydrate on group A streptococci produces visible clumps. *(Right)* A smooth, milky reaction is negative. **(b)** Bacitracin disc test. With very few exceptions, only *Streptococcus pyogenes* is sensitive to a minute concentration (0.02 μg) of bacitracin. Any zone of inhibition around the B disc is interpreted as a presumptive indication of this species. (Note: Group A streptococci are negative for sulfameth oxazole-trimethoprim (SXT) sensitivity and the CAMP test.)

The failure to recognize group A streptococcal infections can have devastating effects. Rapid cultivation and diagnostic techniques to ensure proper treatment and prevention measures are essential. Several different rapid diagnostic test kits are used in clinics and doctors' offices to detect group A streptococci from pharyngeal swab samples. These tests are based on antibodies that react with the outer carbohydrates of group A streptococci **(figure 21.7*a*).**

Complete identification is accomplished by cultivating a specimen on sheep blood agar plates. *S. pyogenes* displays a beta-hemolytic pattern due to its streptolysins (and hemolysins) **(figure 21.7*b*).** If the pharyngitis is caused by a virus, the blood agar dish will show a variety of colony types, representing the normal bacterial biota. Active infection with *S. pyogenes* will yield a plate with a majority of beta-hemolytic colonies. Group A streptococci are by far the most common beta-hemolytic isolates in human diseases, but lately an increased number of infections by group B streptococci (also beta-hemolytic), as well as the existence of beta-hemolytic enterococci, have made it important to use differentiation tests. A positive bacitracin disc test **(figure 21.7*b*)** provides additional evidence for group A.

Prevention

No vaccine exists for group A streptococci, although many researchers are working on the problem. A vaccine against this bacterium would also be a vaccine against rheumatic fever, and thus it is in great demand. In the meantime, infection can be prevented by good hand washing, especially after coughing and sneezing and before preparing foods or eating.

Treatment

The antibiotic of choice for *S. pyogenes* is penicillin; many group A streptococci have become resistant to erythromycin, a macrolide antibiotic. In patients with penicillin allergies, a first-generation cephalosporin, such as cephalexin, is prescribed **(Checkpoint 21.4).**

Diphtheria

For hundreds of years, diphtheria was a significant cause of morbidity and mortality, but in the last 50 years, both the number of cases and the fatality rate have steadily declined throughout the world. In the United States in recent years, only one or two cases have been reported each year. But

CHECKPOINT 21.4 Pharyngitis

Causative Organism(s)	*Streptococcus pyogenes*	Viruses
Most Common Modes of Transmission	Droplet or direct contact	All forms of contact
Virulence Factors	LTA, M protein, hyaluronic acid capsule, SLS and SLO, superantigens	–
Culture/Diagnosis	Beta-hemolytic on blood agar, sensitive to bacitracin, rapid antigen tests	Goal is to rule out *S. pyogenes*, further diagnosis usually not performed
Prevention	Hygiene practices	Hygiene practices
Treatment	Penicillin, cephalexin in penicillin-allergic	Symptom relief only
Distinctive Features	Generally more severe than viral pharyngitis	Hoarseness frequently accompanies viral pharyngitis

when healthy people are screened for the presence of the bacterium, it is found in a significant percentage of them, indicating that the lack of cases is due to the protection afforded by immunization with the diphtheria toxoid, which is part of the childhood immunization series.

Indeed, during the 1990s, a diphtheria epidemic occurred in the former Soviet Union in which 157,000 people became ill with diphtheria and 5,000 people died. This upsurge of cases was attributed to a breakdown in immunization practices and production of vaccine, which followed the breakup of the Soviet Union. These examples emphasize the importance of maintaining vaccination, even for diseases that have long been kept under control.

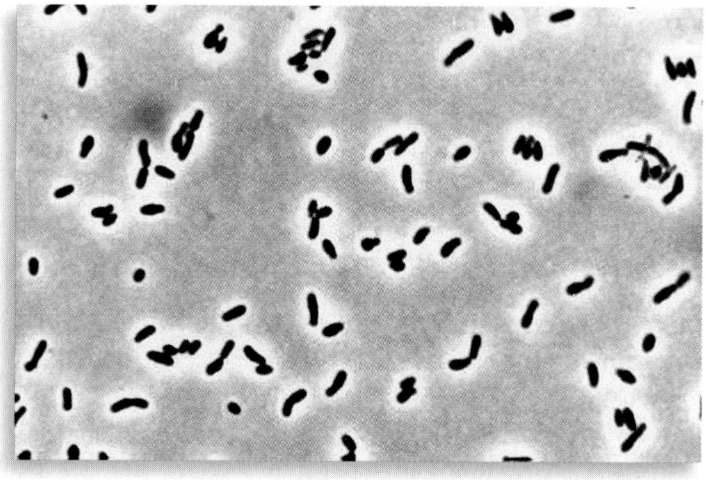

Figure 21.8 *Corynebacterium diphtheriae.*

Signs, Symptoms, and Causative Organism

The disease is caused by *Corynebacterium diphtheriae,* a non-spore-forming, gram-positive club-shaped bacterium **(figure 21.8).** The symptoms of diphtheria are experienced initially in the upper respiratory tract. At first the patient experiences a sore throat, lack of appetite, and low-grade fever. A characteristic membrane, usually referred to as a pseudomembrane, forms on the tonsils or pharynx **(figure 21.9).** The membrane is formed by the bacteria and consists of bacterial cells, fibrin, lymphocytes, and dead tissue cells; and it may be quite extensive. It adheres to tissues and cannot easily be removed. It may eventually completely block respiration. The patient may recover after this crisis. Alternatively, exotoxin manufactured by the bacterium may penetrate the bloodstream and travel throughout the body.

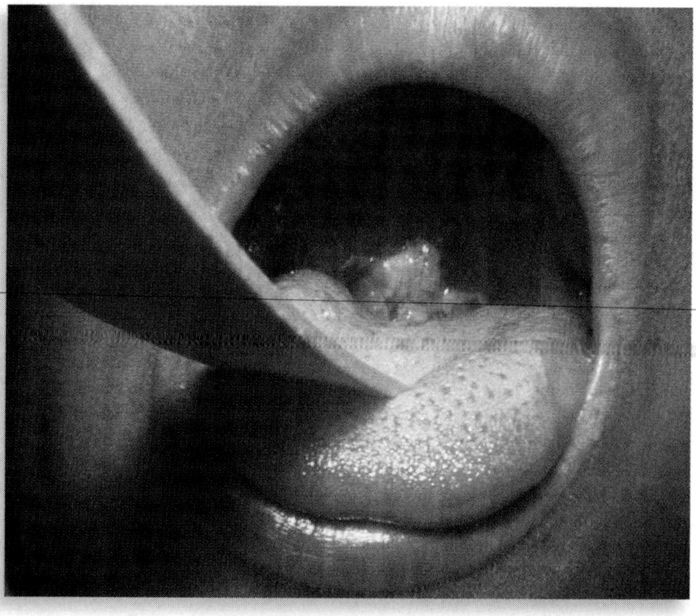

Figure 21.9 Diagnosing diphtheria.
The clinical appearance in diphtheria infection includes gross inflammation of the pharynx and tonsils marked by grayish patches (a pseudomembrane) and swelling over the entire area.

Pathogenesis and Virulence Factors

The exotoxin is encoded by a bacteriophage of *C. diphtheriae.* Strains of the bacterium that are not lysogenized by this phage do not cause serious disease. The exotoxin is of a type called **A-B toxin.** It is illustrated in **figure 21.10** and explained briefly here. A-B toxins are so named because they consist of two parts, an A (active) component and a B (binding) component. The B component binds to a receptor molecule on the surface of the host cell. The next step is for the A component to be moved across the host cell membrane. The A components of most A-B toxins then catalyze a reaction by which they remove a sugar derivative called the ADP-ribosyl group from the coenzyme NAD and attach it to one host cell protein or another. This process is called *ADP-ribosylation.* This process disrupts the normal function of that host protein, resulting in some type of symptom for the patient.

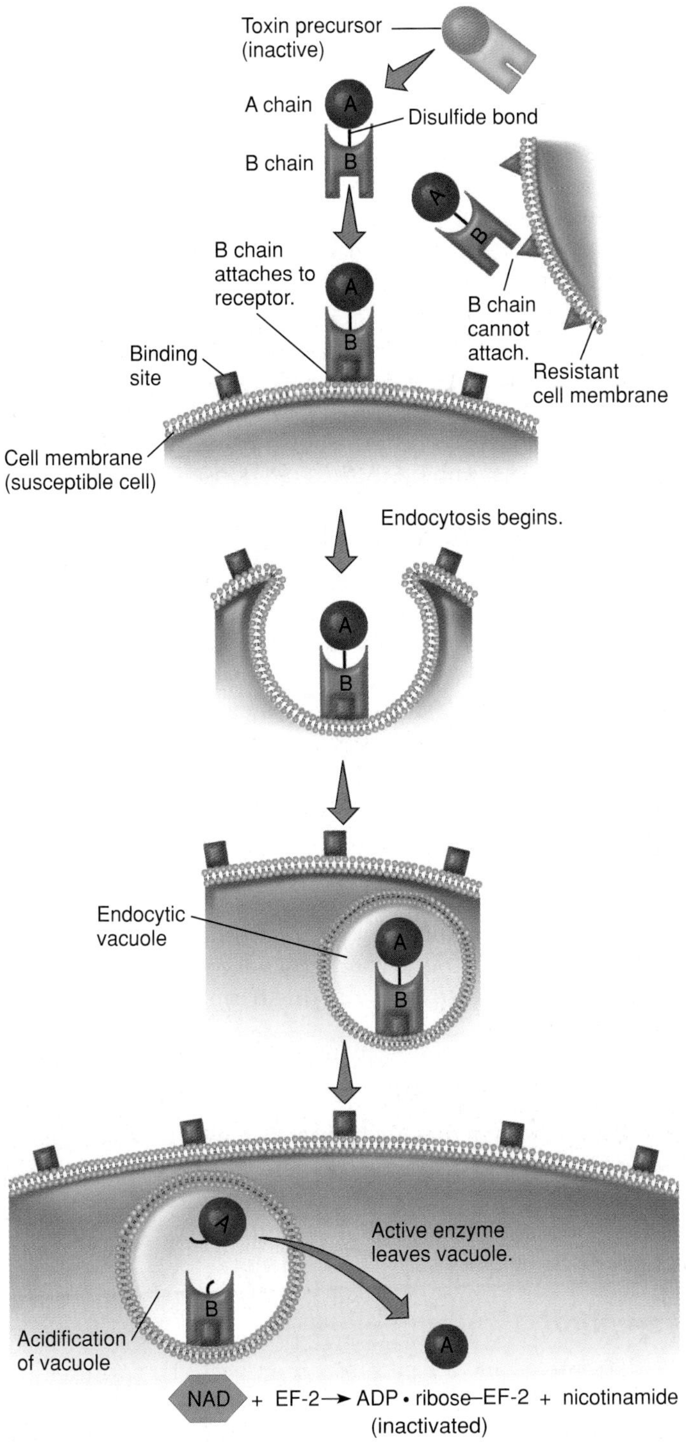

Figure 21.10 A-B toxin of *Corynebacterium diphtheriae.*
The B chain attaches to host cell membrane, then the toxin enters the cell. The two chains separate and the A chain enters the cytoplasm as an active enzyme that ADP-ribosylates a protein (EF-2) needed for protein synthesis. Cell death follows.

The release of diphtheria toxin in the blood leads to complications in distant organs, especially myocarditis and neuritis. Myocarditis can cause abnormal cardiac rhythms and in the worst cases can lead to heart failure. Neuritis affects motor nerves and may result in temporary paralysis of limbs, the soft palate, and even the diaphragm, a condition that can predispose a patient to other lower respiratory tract infections.

Prevention and Treatment

Diphtheria can easily be prevented by a series of vaccinations with toxoid, usually given as part of a mixed vaccine against tetanus and pertussis as well, called the *DTaP* (for diphtheria, tetanus, and acellular pertussis). If a patient has diphtheria, and it has progressed to the bloodstream, the adverse effects of toxemia are treated with diphtheria antitoxin derived from horses. Prior to injection, the patient must be tested for allergy to horse serum and be desensitized if necessary. The infection itself may be treated with antibiotics from the penicillin or erythromycin family. Bed rest, heart medication, and tracheostomy or bronchoscopy to remove the membrane (sometimes called a pseudomembrane) may be indicated.

Adults and adolescents should receive a DTaP booster.

CHECKPOINT 21.5	Diphtheria
Causative Organism(s)	*Corynebacterium diphtheriae*
Most Common Modes of Transmission	Droplet contact, direct contact or indirect contact with contaminated fomites
Virulence Factors	Exotoxin: diphtheria toxin
Culture/Diagnosis	Tellurite medium—gray/black colonies, club-shaped morphology on Gram stain; *treatment begun before definitive identification*
Prevention	Diphtheria toxoid vaccine (part of DTaP)
Treatment	Antitoxin plus penicillin or erythromycin

21.4 Diseases Caused by Microorganisms Affecting the Upper and Lower Respiratory Tract

A number of infectious agents affect both the upper and lower respiratory tract regions. We discuss the more well-known diseases in this section; specifically, they are whooping cough, respiratory syncytial virus (RSV), and influenza.

Whooping Cough

Whooping cough is also known as *pertussis* (the suffix *-tussis* is Latin for cough). A vaccine for this potentially serious infection has been available since 1926. The disease is still troubling to the public health community because its incidence is increasing in the United States, despite improvements in the vaccine. In addition, in the recent past there has been concern over the vaccine among the general public. For these reasons, it is an important disease for health care professionals to understand.

Signs and Symptoms

The disease has two distinct symptom phases called the catarrhal and paroxysmal stages, which are followed by a long recovery (or convalescent) phase, during which a patient is particularly susceptible to other respiratory infections. After an incubation period of from 3 to 21 days, the **catarrhal** stage begins when bacteria present in the respiratory tract cause what appear to be cold symptoms, most notably a runny nose. This stage lasts 1 to 2 weeks. The disease worsens in the second **(paroxysmal)** stage, which is characterized by severe and uncontrollable coughing (a *paroxysm* can be thought of as a convulsive attack). The common name for the disease comes from the whooping sound a patient makes as he or she tries to grab a breath between uncontrollable bouts of coughing. The violent coughing spasms can result in burst blood vessels in the eyes or even vomiting. In the worst cases, seizures result from small hemorrhages in the brain.

As in any disease, the **convalescent phase** is the time when numbers of bacteria are decreasing and no longer cause ongoing symptoms. But the active stages of the disease damage the cilia on respiratory tract epithelial cells, and complete recovery of these surfaces requires weeks or even months. During this time, other microorganisms can more easily colonize and cause secondary infection.

Causative Agent

Bordetella pertussis is a very small gram-negative rod. Sometimes it looks like a coccobacillus. It is strictly aerobic and fastidious, having specific nutritional requirements for successful culture.

Pathogenesis and Virulence Factors

The progress of this disease can be clearly traced to the virulence mechanisms of the bacterium. It is absolutely essential for the bacterium to attach firmly to the epithelial cells of the mouth and throat, and it does so using specific adhesive molecular structures on its surface. One of these structures is called *filamentous hemagglutinin* (*FHA*). It is a fibrous structure that surrounds the bacterium like a capsule and is also secreted in soluble form. In that form, it can act as a bridge between the bacterium and the epithelial cell.

Once the bacteria are attached in large numbers, production of mucus increases and localized inflammation ensues, resulting in the early stages of the disease. Then the real damage begins: The bacteria release multiple exotoxins that damage ciliated respiratory epithelial cells and cripple other components of the host defense, including phagocytic cells.

The two most important exotoxins are *pertussis toxin* and *tracheal cytotoxin*. Pertussis toxin is a classic A-B toxin, like the diphtheria toxin illustrated in figure 21.10. In the case of pertussis toxin, the host protein affected by the process of ADP-ribosylation is one that normally limits the production of cyclic AMP. Cyclic AMP is a critical molecule that regulates numerous functions inside host cells. The excessive amounts of cyclic AMP result in copious production of mucus and a variety of other effects in the respiratory tract and the immune system.

Tracheal cytotoxin results in more direct destruction of ciliated cells. The cells are no longer capable of clearing mucus and secretions, leading to the extraordinary coughing required to get relief. Another important contributor to the pathology of the disease is *B. pertussis* endotoxin. As always with endotoxins, its release leads to the production of a host of cytokines that have direct and indirect effects on physiological processes and on the host response.

Transmission and Epidemiology

B. pertussis is transmitted via respiratory droplets. It is highly contagious during both the catarrhal and paroxysmal stages. The disease manifestations are most serious in infants. Twenty-five percent of infections occur in older children and adults, who generally have milder symptoms. The disease results in 300,000 to 500,000 deaths annually worldwide.

Pertussis outbreaks continue to occur in the United States and elsewhere. Even though it is estimated that approximately 85% of U.S. children are vaccinated against pertussis, it continues to be spread, perhaps by adults whose own immunity has dwindled. These adults may experience mild, unrecognized disease and unwittingly pass it to others. It has also been found that fully vaccinated children can experience the disease, possibly due to antigenic changes in the bacterium.

Culture and Diagnosis

This disease is often diagnosed based solely on its symptoms because they are so distinctive. When culture confirmation is desired, nasopharyngeal swabs can be inoculated on specific media—Bordet-Gengou (B-G) medium, charcoal agar, or potato-glycerol agar.

Prevention

The current vaccine for pertussis is an acellular formulation of important *B. pertussis* antigens. It results in far fewer side effects than the previous whole-cell vaccine, which was used until the mid-1990s. It is generally given in the form of the DTaP vaccine.

A second prevention strategy is the administration of antibiotics to contacts of people who have been diagnosed with the disease. Erythromycin or trimethoprim-sulfamethoxazole is given for 14 days to prevent disease in those who may have been infected.

Treatment

Treating someone who is already ill with pertussis is focused on supportive care; antibiotics may or may not shorten the course of the disease, which is often the case when major symptoms of a condition are the result of exotoxin secretion. Antibiotics (erythromycin) are sometimes administered because they do decrease the contagiousness of the patient.

CHECKPOINT 21.6	Pertussis (Whooping Cough)
Causative Organism(s)	*Bordetella pertussis*
Most Common Modes of Transmission	Droplet contact
Virulence Factors	FHA (adhesion), pertussis toxin and tracheal cytotoxin, endotoxin
Culture/Diagnosis	Grown on B-G, charcoal, or potato-glycerol agar; diagnosis can be made on symptoms
Prevention	Acellular vaccine (DTaP), erythromycin or trimethoprim-sulfamethoxazole for contacts
Treatment	Mainly supportive; erythromycin to decrease communicability

CASE FILE 21 *WRAP-UP*

Pertussis, commonly known as whooping cough, is caused by the tiny, gram-negative bacillus *Bordetella pertussis*. (See text for a complete description of the disease.) It can be cultured from nasopharyngeal secretions on special culture media. Newer molecular-based techniques allow for the detection of *B. pertussis* DNA. In this outbreak, 30 of 49 samples from symptomatic patients yielded positive DNA results. Some samples that were sent to the CDC had positive culture results indicating that *B. pertussis* was indeed present in the community.

Surveillance data found that 19% of individuals living in households that submitted questionnaires reported having symptoms consistent with pertussis. Interviews revealed that 72% of the 123 children between 6 months and 5 years of age living in the surveyed households had no immunization records. In 96 households that reported cases of pertussis, 45% indicated they had not vaccinated any of their children.

Pertussis can be treated with appropriate antibiotics such as erythromycin. Among the 96 interviewed households, 49% of the individuals had not taken any antibiotics. In some households, antibiotics had been used but not for the recommended 14 days.

Outbreaks such as this one indicate a need for increased efforts in the prevention and treatment of diseases such as pertussis. Immunization in the United States has reduced the death rate from pertussis from 5,000 to 10,000 per year to less than 30. The Amish community has no restrictions regarding immunization or the use of antibiotics.

See: CDC. 2006. Pertussis outbreak in an Amish community—Kent County, Delaware, September 2004–February 2005. MMWR *55(30):817–821.*

Respiratory Syncytial Virus Infection

As its name indicates, respiratory syncytial virus (RSV) infects the respiratory tract and produces giant multinucleated cells (syncytia). It is a member of the paramyxovirus family and contains single-stranded negative-sense RNA. It is an enveloped virus. Outbreaks of droplet-spread RSV disease occur regularly throughout the world, with peak incidence in the winter and early spring. Children 6 months of age or younger, as well as premature babies, are especially susceptible to serious disease caused by this virus. RSV is the most prevalent cause of respiratory infection in the newborn age group, and nearly all children have experienced it by age 2. An estimated 100,000 children are hospitalized with RSV infection each year in the United States. The mortality rate is highest for children with complications such as prematurity, congenital disease, and immunodeficiency. Infection in older children and adults usually manifests as a cold.

The first symptoms are fever that lasts for approximately 3 days, rhinitis, pharyngitis, and otitis. More serious infections progress to the bronchial tree and lung parenchyma, giving rise to symptoms of croup that include acute bouts of coughing, wheezing, difficulty in breathing (called **dyspnea**), and abnormal breathing sounds (called rales). (Note: This condition is often called croup and also bronchiolitis; be aware that both of these terms are clinical descriptions of diseases caused by a variety of viruses [in addition to RSV] and sometimes by bacteria.)

The virus is highly contagious and is transmitted through droplet contact but also through fomite contamination. Diagnosis of RSV infection is more critical in babies than in older children or adults. The afflicted child is conspicuously ill, with signs typical of pneumonia and bronchitis. The best diagnostic procedures are those that demonstrate the viral antigen directly from specimens (direct and indirect fluorescent staining, ELISA, and DNA probes).

There is no RSV vaccine available yet, but an effective passive antibody preparation is used as prevention in high-risk children and babies born prematurely. Ribavirin, an antiviral drug, can be administered as an inhaled aerosol to very sick children, although the clinical benefit is uncertain.

CHECKPOINT 21.7	RSV Disease
Causative Organism(s)	Respiratory syncytial virus (RSV)
Most Common Modes of Transmission	Droplet and indirect contact
Virulence Factors	Syncytia formation
Culture/Diagnosis	Direct antigen testing
Prevention	Passive antibody in high-risk children
Treatment	Ribavirin in severe cases

Influenza

The "flu" is a very important disease to study for several reasons. First of all, everyone is familiar with the cyclical increase of influenza infections occurring during the winter months in the United States. Second, many conditions are erroneously termed the "flu," while in fact only diseases caused by influenza viruses are actually the flu. Third, the way that influenza viruses behave provides an excellent illustration of the way other viruses can, and do, change to cause more serious diseases than they did previously.

Signs and Symptoms

Influenza begins in the upper respiratory tract but in serious cases may also affect the lower respiratory tract. There is a 1- to 4-day incubation period, after which symptoms begin very quickly. These include headache, chills, dry cough, body aches, fever, stuffy nose, and sore throat. Even the sum of all these symptoms can't describe how a person actually feels: lousy. The flu is known to "knock you off your feet." Extreme fatigue can last for a few days or even a few weeks. An infection with influenza can leave patients vulnerable to secondary infections, often bacterial. Influenza infection alone occasionally leads to a pneumonia that can cause rapid death, even in young healthy adults.

Patients with emphysema or cardiopulmonary disease, along with very young, elderly, or pregnant patients, are more susceptible to serious complications.

Causative Agent

Influenza is caused by one of three influenza viruses: A, B, or C. They belong to the family Orthomyxoviridae. They are spherical particles with an average diameter of 80 to 120 nanometers. Each virion is covered with a lipoprotein envelope that is studded with glycoprotein spikes acquired during viral maturation **(figure 21.11).** The two glycoproteins that make up the spikes of the envelope and contribute to virulence are called hemagglutinin (H) and neuraminidase (N). The name hemagglutinin is derived from this glycoprotein's agglutinating action on red blood cells, which is the basis for viral assays used to identify the viruses. Hemagglutinin contributes to infectivity by binding to host cell receptors of the respiratory mucosa, a process that facilitates viral penetration. Neuraminidase breaks down the protective mucous coating of the respiratory tract, assists in viral budding and release, keeps viruses from sticking together, and participates in host cell fusion.

The ssRNA genome of the influenza virus is known for its extreme variability. It is subject to constant genetic changes that alter the structure of its envelope glycoproteins. Research has shown that genetic changes are very frequent in the area of the glycoproteins recognized by the host immune response but very rare in the areas of the glycoproteins used for attachment to the host cell **(figure 21.12).** In this way, the virus can continue to attach to host cells while managing to decrease the effectiveness of the host response to its presence. This constant mutation of the glycoproteins is called **antigenic drift**—the antigens gradually change their amino acid composition, resulting in decreased ability of host memory cells to recognize them.

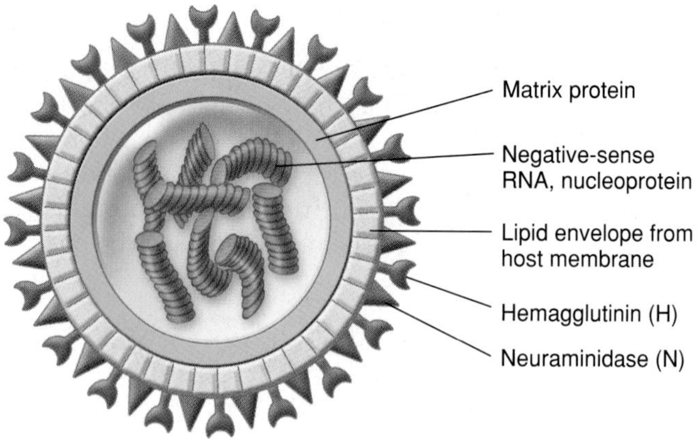

Figure 21.11 Schematic drawing of influenza virus.

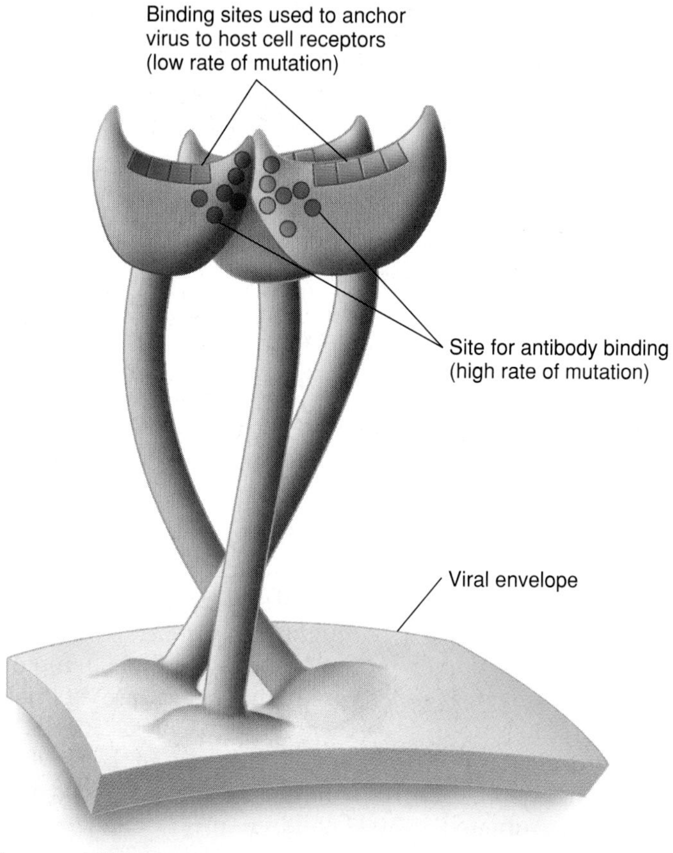

Figure 21.12 Schematic drawing of hemagglutinin (HA) of influenza virus.
Blue boxes depict site used to attach virus to host cells; green circles depict sites for anti-influenza antibody binding.

An even more serious phenomenon is known as **antigenic shift.** The genome of the virus consists of just 10 genes, encoded on 8 separate RNA strands. Antigenic shift is the swapping out of one of those genes or strands with a gene or strand from a different influenza virus. Some explanation is

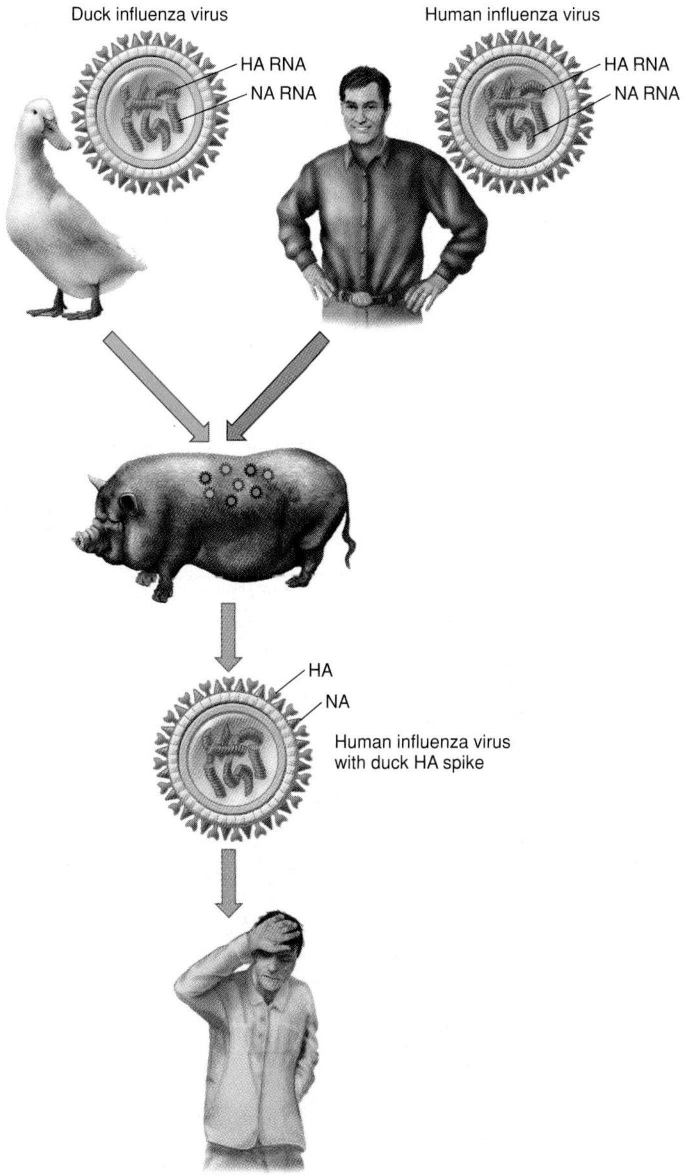

Figure 21.13 Antigenic shift event.
Where ducks and swine and humans live close together, the swine can serve as a melting pot for creating "hybrid" influenza viruses that are not recognized by the human immune system.

in order. First, we know that certain influenza viruses infect both humans and swine. Other influenza viruses infect birds (or ducks) and swine. All of these viruses have 10 genes coding for the same important influenza proteins (including H and N)—but the actual sequence of the genes is different in the different types of viruses. Second, when the two viruses just described infect a single swine host, with both virus types infecting the same host cell, the viral packaging step can accidentally produce a human influenza virus that contains seven human influenza virus RNA strands plus a single duck influenza virus RNA strand **(figure 21.13).** When that virus infects a human, no immunologic recognition of the protein that came from the duck virus occurs. Experts have traced the flu pandemics of 1918, 1957, 1968, and 1977 to strains of a virus that came from pigs (swine flu). Influenza A viruses are named according to the different types of H and N spikes they display on their surfaces. For instance, in 2004 the most common circulating subtypes of influenza A viruses were H1N1 and H3N2. Influenza B viruses are not divided into subtypes because they are thought only to undergo antigenic drift and not antigenic shift. Influenza C viruses are thought to cause only minor respiratory disease and are probably not involved in epidemics.

Scientists have also recently found that antigenic drift and shift are not even required to make an influenza virus deadly. It appears that a minor genetic alteration in another influenza virus gene, one that seems to produce an enzyme used to manufacture new viruses in the host cell, can make the difference between a somewhat pathogenic influenza virus and a lethal one. It is still not clear exactly how many of these minor changes can lead to pandemic levels of infection and a catastrophe for the public health.

Starting in 1997, a form of influenza that infects birds began spreading around Asia. Early on, it infected humans who had contact with poultry. By the beginning of 2005, the strain, termed H5N1, had spread through nine countries, wiping out entire flocks of chickens and even killing humans who contracted it from birds. By mid-2007, 189 humans had acquired the disease and up to half of those had died. The high death rate indicates that the virus is highly virulent. The disease has spread to Europe, appearing in birds in Sweden, Germany, Denmark, the Czech Republic, and the United Kingdom.

The good news at this point is that the virus is not easily transmitted human-to-human. The human cases so far were mainly in people who had close association with chickens. But the name of the game for influenza viruses of all types is *change.* Antigenic shift in the virus could result in a new form that is spread person-to-person. If that occurs, the predictions are dire. Optimistic experts estimate that 7 million people worldwide would die from such a pandemic; more pessimistic (some would say "realistic") scientists suggest the death toll could be as high as 1.5 billion **(Insight 21.1).**

Certainly, the evidence has convinced many governmental agencies to take aggressive steps to prepare for bird flu. In November of 2005, U.S. President George Bush announced a $7.1 billion plan to prepare for the pandemic. It also includes $3 billion to speed up the development of new technologies for vaccine manufacture. One billion dollars will go to stockpiling antiviral drugs such as Tamiflu and Relenza. The H5N1 virus has been found to be resistant to two other influenza drugs, amantadine and rimantadine.

In the spring of 2007, the United States approved a vaccine specifically designed for H5N1. It improves on the former vaccine but is still only expected to be effective in about 45% of its recipients.

Pathogenesis and Virulence Factors

The influenza virus binds primarily to ciliated cells of the respiratory mucosa. Infection causes the rapid shedding of

INSIGHT 21.1 ***Medical***

The World Health Organization Responds to the Threat of Bird Flu

In the fall of 2005, the threat of avian influenza, or bird flu, had grown so worrisome that the World Health Organization created a list of facts to help educate the world's citizens about the possible pandemic. They are reprinted here.

Ten Things You Need to Know About Pandemic Influenza

1. **Pandemic influenza is different from avian influenza.** Avian influenza refers to a large group of different influenza viruses that primarily affect birds. On rare occasions, these bird viruses can infect other species, including pigs and humans. The vast majority of avian influenza viruses do not infect humans. An influenza pandemic happens when a new subtype emerges that has not previously circulated in humans.

   For this reason, avian H5N1 is a strain with pandemic potential, since it might ultimately adapt into a strain that is contagious among humans. Once this adaptation occurs, it will no longer be a bird virus—it will be a human influenza virus. Influenza pandemics are caused by new influenza viruses that have adapted to humans.
2. **Influenza pandemics are recurring events.** An influenza pandemic is a rare but recurrent event. Three pandemics occurred in the previous century: "Spanish influenza" in 1918, "Asian influenza" in 1957, and "Hong Kong influenza" in 1968. The 1918 pandemic killed an estimated 40–50 million people worldwide. That pandemic, which was exceptional, is considered one of the deadliest disease events in human history. Subsequent pandemics were much milder, with an estimated 2 million deaths in 1957 and 1 million deaths in 1968.

   A pandemic occurs when a new influenza virus emerges and starts spreading as easily as normal influenza—by coughing and sneezing. Because the virus is new, the human immune system will have no preexisting immunity. This makes it likely that people who contract pandemic influenza will experience more serious disease than that caused by normal influenza.
3. **The world may be on the brink of another pandemic.** Should H5N1 evolve to a form as contagious as normal influenza, a pandemic could begin.
4. **All countries will be affected.** Once a fully contagious virus emerges, its global spread is considered inevitable. Countries might, through measures such as border closures and travel restrictions, delay arrival of the virus but cannot stop it. The pandemics of the previous century encircled the globe in 6 to 9 months, even when most international travel was by ship. Given the speed and volume of international air travel today, the virus could spread more rapidly, possibly reaching all continents in less than 3 months.
5. **Widespread illness will occur.** Because most people will have no immunity to the pandemic virus, infection and illness rates are expected to be higher than during seasonal epidemics of normal influenza. Current projections for the next pandemic estimate that a substantial percentage of the world's population will require some form of medical care. Few countries have the staff, facilities, equipment, and hospital beds needed to cope with large numbers of people who suddenly fall ill.
6. **Medical supplies will be inadequate.** Supplies of vaccines and antiviral drugs—the two most important medical interventions for reducing illness and deaths during a pandemic—will be inadequate in all countries at the start of a pandemic and for many months thereafter. Inadequate supplies of vaccines are of particular concern, as vaccines are considered the first line of defense for protecting populations. On present trends, many developing countries will have no access to vaccines throughout the duration of a pandemic.

these cells along with a load of viruses. Stripping the respiratory epithelium to the basal layer eliminates protective ciliary clearance and leads to severe inflammation and irritation. The illness is further aggravated by fever, headache, and the other symptoms just described. The viruses tend to remain in the respiratory tract rather than spread to the bloodstream. As the normal ciliated epithelium is restored in a week or two, the symptoms subside.

As just noted, the glycoproteins and their structure are important virulence determinants. First of all, they mediate the adhesion of the virus to host cells. Second, they change gradually and sometimes suddenly, evading immune recognition.

Transmission and Epidemiology

Inhalation of virus-laden aerosols and droplets constitutes the major route of influenza infection, although fomites can play a secondary role. Transmission is greatly facilitated by crowding and poor ventilation in classrooms, barracks, nursing homes, dormitories, and military installations in the late fall and winter. It is highly contagious and affects people of all ages. Annually, there are approximately 36,000 U.S. deaths from influenza and its complications, mainly among the very young and the very old.

Culture and Diagnosis

Very often, physicians will diagnose influenza based on symptoms alone. But there is a wide variety of culture-based and nonculture-based methods to diagnose the infection. Rapid influenza tests (such as PCR, ELISA-type assays, or immunofluorescence) provide results within 24 hours; viral culture provides results in 3 to 10 days. Cultures are not typically performed at the point of care; they

A worker sprays disinfectant at ducks in Thailand.
Picture: Reuters

7. **Large numbers of deaths will occur.**
Historically, the number of deaths during a pandemic has varied greatly. Death rates are largely determined by four factors: the number of people who become infected, the virulence of the virus, the underlying characteristics and vulnerability of affected populations, and the effectiveness of preventive measures. Accurate predictions of mortality cannot be made before the pandemic virus emerges and begins to spread. All estimates of the number of deaths are purely speculative.
WHO has used a relatively conservative estimate—from 2 million to 7.4 million deaths—because it provides a useful and plausible planning target. This estimate is based on the comparatively mild 1957 pandemic. Estimates based on a more virulent virus, closer to the one seen in 1918, have been made and are much higher. However, the 1918 pandemic was considered exceptional.
8. **Economic and social disruption will be great.**
High rates of illness and worker absenteeism are expected, and these will contribute to social and economic disruption. Past pandemics have spread globally in two and sometimes three waves. Not all parts of the world or of a single country are expected to be severely affected at the same time. Social and economic disruptions could be temporary but may be amplified in today's closely interrelated and interdependent systems of trade and commerce. Social disruption may be greatest when rates of absenteeism impair essential services, such as power, transportation, and communications.
9. **Every country must be prepared.**
WHO has issued a series of recommended strategic actions for responding to the influenza pandemic threat. The actions are designed to provide different layers of defense that reflect the complexity of the evolving situation. Recommended actions are different for the present phase of pandemic alert, the emergence of a pandemic virus, and the declaration of a pandemic and its subsequent international spread.
10. **WHO will alert the world when the pandemic threat increases.**
WHO works closely with ministries of health and various public health organizations to support countries' surveillance of circulating influenza strains. A sensitive surveillance system that can detect emerging influenza strains is essential for the rapid detection of a pandemic virus.
Six distinct phases have been defined to facilitate pandemic preparedness planning, with roles defined for governments, industry, and WHO. The present situation is categorized as phase 3: a virus new to humans is causing infections but does not spread easily from one person to another.

Reference: http://www.who.int/csr/disease/influenza/pandemic10things/en/index.html, 2006.

must be sent to diagnostic laboratories, and they require up to 10 days for results. Despite these disadvantages, culture can be useful to identify which subtype of influenza is causing infections, which is important for public health authorities to know. In 2006, the FDA approved a new PCR-based test that identifies H5 strains of influenza from patients that yields results in 4 hours. Positive samples are then further tested to determine if they are H5N1.

Prevention

Preventing influenza infections and epidemics is one of the top priorities for public health officials. The standard vaccine contains inactivated dead viruses that had been grown in embryonated eggs. It has an overall effectiveness of 70% to 90%. The vaccine consists of three different influenza viruses (usually two influenza A and one influenza B) that have been judged to most resemble the virus variants likely to cause infections in the coming flu season. Because of the changing nature of the antigens on the viral surface, annual vaccination is considered the best way to avoid infection. Anyone over the age of 6 months can take the vaccine, and it is recommended for anyone in a high-risk group or for people who have a high degree of contact with the public.

A new vaccine called FluMist is a nasal mist vaccine consisting of the three strains of influenza virus in live attenuated form. It is designed to stimulate secretory immunity in the upper respiratory tract. Its safety and efficacy have so far been demonstrated only for persons between the ages of 5 and 49. It is not advised for immunocompromised individuals, and it is significantly more expensive than the injected vaccine.

Companies are now actively engaged in new vaccine development spurred on by the possibility of a pandemic.

Several of the anti-influenza drugs listed in the following section can be used for preventive purposes, especially in epidemics.

Treatment

Influenza is one of the first viral diseases for which effective antiviral drugs became available. The drugs must be taken early in the infection, preferably by the second day. This requirement is an inherent difficulty because most people do not realize until later that they may have the flu. Amantadine and rimantadine can be used to treat and prevent some influenza type A infections, but they do not work against influenza type B viruses and do not seem to work against the new H5N1 strain.

Zanamivir (Relenza) is an inhaled drug that works against influenza A and B. Oseltamivir (Tamiflu) is available in capsules or as a powdered mix to be made into a drink. It can also be used for prevention of influenza A and B.

CHECKPOINT 21.8	Influenza
Causative Organism(s)	Influenza A, B, and C viruses
Most Common Modes of Transmission	Droplet contact, direct contact, some indirect contact
Virulence Factors	Glycoprotein spikes, overall ability to change genetically
Culture/Diagnosis	Viral culture (3–10 days) or rapid antigen-based or PCR tests
Prevention	Killed injected vaccine or inhaled live attenuated vaccine—taken annually
Treatment	Amantadine, rimantadine, zanamivir, or oseltamivir

21.5 Lower Respiratory Tract Diseases Caused by Microorganisms

In this section, we consider microbial diseases that affect the lower respiratory tract primarily—namely, the bronchi, bronchioles, and lungs, with minimal involvement of the upper respiratory tract. Our discussion focuses on tuberculosis and pneumonia.

Tuberculosis

Mummies from the Stone Age, ancient Egypt, and Peru provide unmistakable evidence that tuberculosis (TB) is an ancient human disease. In fact, historically it has been such a prevalent cause of death that it was called "Captain of the Men of Death" and "White Plague." After the discovery of streptomycin in 1943, the rates of tuberculosis in the developed world declined rapidly. But since the mid-1980s, it has reemerged as a serious threat. In many regions of the world, the rates of TB are so high that the World Health Organization has requested emergency aid. The cause of tuberculosis is primarily the bacterial species *Mycobacterium tuberculosis*, informally called the tubercle bacillus.

Signs and Symptoms

A clear-cut distinction can be made between infection with the tubercle TB bacterium and the disease it causes. In general, humans are rather easily infected with the bacterium but are resistant to the disease. Estimates project that only about 5% of infected people actually develop a clinical case of tuberculosis. Untreated tuberculosis progresses slowly, and people with the disease may have a normal life span, with periods of health alternating with episodes of morbidity. The majority (85%) of TB cases are contained in the lungs, even though disseminated TB bacteria can give rise to tuberculosis in any organ of the body. Clinical tuberculosis is divided into primary tuberculosis, secondary (reactivation or reinfection) tuberculosis, and disseminated tuberculosis.

Primary Tuberculosis The minimum infectious dose for lung infection is around 10 cells. Alveolar macrophages phagocytose these cells, but they are not killed and continue to multiply inside the macrophages. This period of hidden infection is asymptomatic or is accompanied by mild fever. Some bacteria escape from the lungs into the blood and lymphatics. After 3 to 4 weeks, the immune system mounts a complex, cell-mediated assault against the bacteria. The large influx of mononuclear cells into the lungs plays a part in the formation of specific infection sites called **tubercles.** Tubercles are granulomas that consist of a central core containing TB bacteria in enlarged macrophages and an outer wall made of fibroblasts, lymphocytes, and neutrophils **(figure 21.14).** Although this response further checks spread of infection and helps prevent the disease, it also carries a potential for damage. Frequently, the centers of tubercles break down into necrotic **caseous** (kay'-see-us) **lesions** that gradually heal by *calcification*—normal lung tissue is replaced by calcium deposits. The response of T cells to *M. tuberculosis* proteins also causes a cell-mediated immune response evident in the **tuberculin reaction,** a valuable diagnostic and epidemiological tool **(figure 21.15).**

Secondary (Reactivation) Tuberculosis Although the majority of TB patients recover more or less completely from the primary episode of infection, live bacteria can remain dormant and become reactivated weeks, months, or years later, especially in people with weakened immunity. In chronic tuberculosis, tubercles filled with masses of bacteria expand and drain into the bronchial tubes and upper respiratory tract. The patient gradually experiences more severe symptoms, including violent coughing, greenish or bloody sputum, low-grade fever, anorexia, weight loss, extreme fatigue, night sweats, and chest pain. It is the gradual

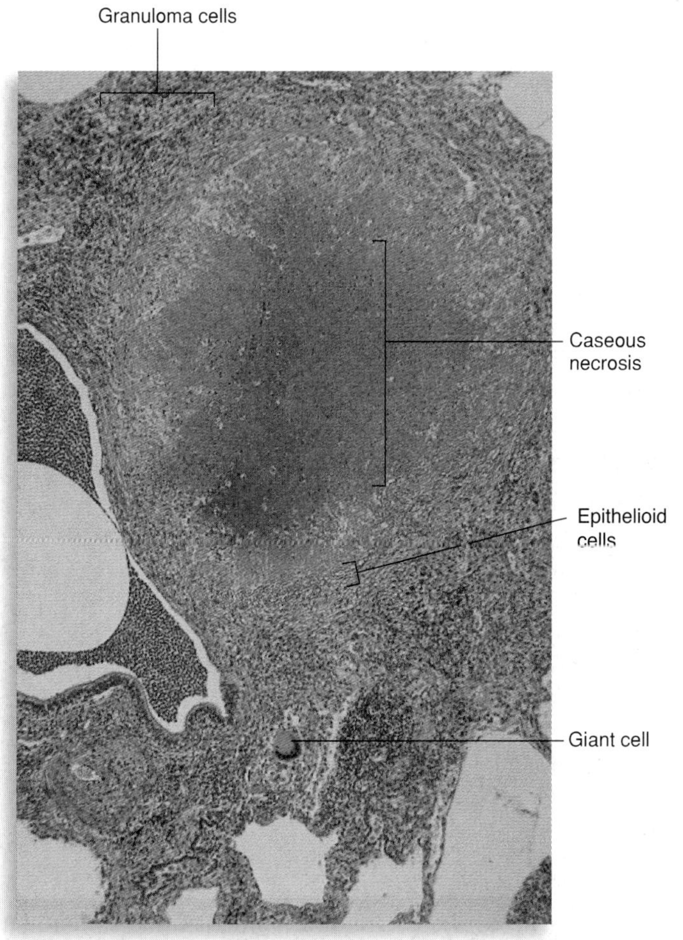

Figure 21.14 Tubercle formation.
Photomicrograph of a tubercle (16×). The massive granuloma infiltrate has obliterated the alveoli and set up a dense collar of fibroblasts, lymphocytes (granuloma cells), and epithelioid cells. The core of this tubercle is a caseous (cheesy) material containing the bacilli.

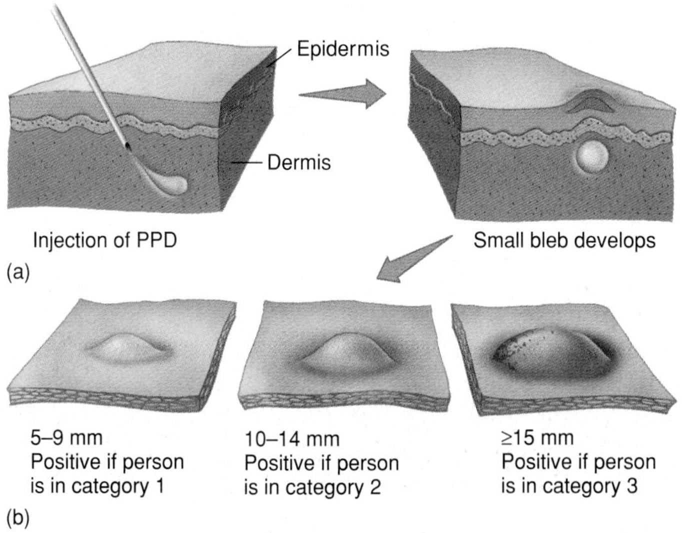

Figure 21.15 Skin testing for tuberculosis.
(a, b) The Mantoux test. Tuberculin is injected into the dermis. A small bleb from the injected fluid develops but will be absorbed in a short time. After 48 to 72 hours, the skin reaction is rated by the degree (or size) of the raised area. The surrounding red area is *not* counted in the measurement. A reaction of less than 5 mm is negative in all cases. See also figure 16.14.

wasting of the body that accounts for an older name for tuberculosis—*consumption*. Untreated secondary disease has nearly a 60% mortality rate.

Extrapulmonary Tuberculosis TB infection outside of the lungs is more common in immunosuppressed patients and young children. Organs most commonly involved in **extrapulmonary TB** are the regional lymph nodes, kidneys, long bones, genital tract, brain, and meninges. Because of the debilitation of the patient and the high load of TB bacteria, these complications are usually grave. Renal tuberculosis results in necrosis and scarring of the kidney and the pelvis, ureters, and bladder. This damage is accompanied by painful urination, fever, and the presence of blood and the TB bacterium in urine. Genital tuberculosis in males damages the prostate gland, epididymis, seminal vesicle, and testes; and in females, the fallopian tubes, ovaries, and uterus. Tuberculosis of the bones and joints is a common complication. The spine is a frequent site of infection, although the hip, knee, wrist, and elbow can also be involved. Advanced infiltration of the vertebral column produces degenerative changes that collapse the vertebrae, resulting in abnormal curvature of the thoracic region (humpback) or of the lumbar region (swayback). Neurological damage stemming from compression on nerves can cause extensive paralysis and sensory loss.

Tubercular meningitis is the result of an active brain lesion seeding bacteria into the meninges. Over a period of several weeks, the infection of the cranial compartments can create mental deterioration, permanent retardation, blindness, and deafness. Untreated tubercular meningitis is invariably fatal, and even treated cases can have a 30% to 50% mortality rate.

Causative Agents

M. tuberculosis is the cause of tuberculosis in most patients. It is an acid-fast rod, long and thin. It is a strict aerobe, and technically speaking, there is still debate about whether it is a gram-positive or a gram-negative organism. It is rarely called gram anything, however, because its acid-fast nature is much more relevant in a clinical setting **(figure 21.16).** It grows very slowly. With a generation time of 15 to 20 hours, a period of up to 6 weeks is required for colonies to appear in culture. (Note: The prefix *Myco-* might make you think of fungi, but this is a bacterium. The prefix in the name came from the mistaken impression that colonies growing on agar **(figure 21.17)** resembled fungal colonies. And be sure to differentiate this bacterium from *Mycoplasma*—they are unrelated.)

Robert Koch identified that *M. tuberculosis* often forms serpentine cords while growing, and he called the unknown substance causing this style of growth *cord factor*. Cord factor appears to be associated with virulent strains, and it is a lipid component of the mycobacterial cell wall. All mycobacterial

Figure 21.16 A fluorescent acid-fast stain of *Mycobacterium tuberculosis* from sputum.
Smears are evaluated in terms of the number of AFB (acid-fast bacteria) seen per field. This quantity is then applied to a scale ranging from 0 to 4+, 0 being no AFB observed and 4+ being more than 9 AFB per field.

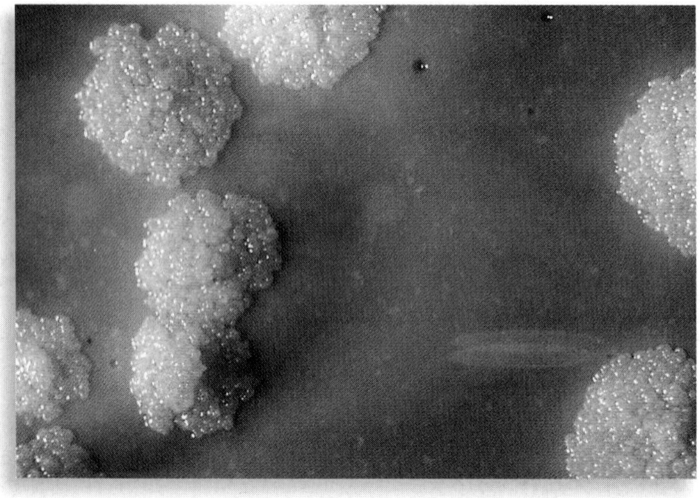

Figure 21.17 Cultural appearance of *Mycobacterium tuberculosis*.
Colonies with a typical granular, waxy pattern of growth.

species have walls that have a very high content of complex lipids, including mycolic acid and waxes. This chemical characteristic makes them relatively impermeable to stains and difficult to decolorize (acid-fast) once they are stained. The lipid wall of the bacterium also influences its virulence and makes it resistant to drying and disinfectants.

In recent decades, tuberculosis-like conditions caused by *Mycobacterium avium* and related mycobacterial species (sometimes referred to as the *M. avium* complex, or MAC) have been found in AIDS patients and other immunocompromised people. In this section, we consider only *M. tuberculosis*, although *M. avium* is discussed briefly near the conclusion.

Before routine pasteurization of milk, humans acquired bovine TB, caused by a species called *Mycobacterium bovis*, from the milk they drank. It is very rare today, but in 2004, six people in a nightclub acquired bovine TB from a fellow reveler. One person died from her infection.

Pathogenesis and Virulence Factors

The course of the infection—and all of its possible variations—was previously described under "Signs and Symptoms." Important characteristics of the bacterium that contribute to its virulence are its waxy surface (contributing both to its survival in the environment and its survival within macrophages) and its ability to stimulate a strong cell-mediated immune response that contributes to the pathology of the disease.

Transmission and Epidemiology

The agent of tuberculosis is transmitted almost exclusively by fine droplets of respiratory mucus suspended in the air. The TB bacterium is highly resistant and can survive for 8 months in fine aerosol particles. Although larger particles become trapped in mucus and are expelled, tinier ones can be inhaled into the bronchioles and alveoli. This effect is especially pronounced among people sharing small closed rooms with limited access to sunlight and fresh air.

The epidemiological patterns of *M. tuberculosis* infection vary with the living conditions in a community or an area of the world. Factors that significantly affect people's susceptibility to tuberculosis are inadequate nutrition, debilitation of the immune system, poor access to medical care, lung damage, and their own genetics. People in developing countries are often infected as infants and harbor the microbe for many years until the disease is reactivated in young adulthood. Estimates indicate that possibly one-third of the world's population and 15 million people in the United States carry the TB bacterium.

Cases in the United States show a strong correlation with the age, sex, and recent immigration history of a patient. The highest case rates occur in non-White males over 30 years of age and non-White females over 60. High rates also occur in new immigrants from certain areas of Indochina, South America, and Africa and in AIDS patients.

Culture and Diagnosis

You are probably familiar with several methods of detecting tuberculosis in humans. Clinical diagnosis of tuberculosis relies on four techniques: (1) tuberculin testing, (2) chest X rays, (3) direct identification of acid-fast bacilli (AFB) in sputum or other specimens, and (4) cultural isolation and antimicrobial susceptibility testing.

Tuberculin Sensitivity and Testing Because infection with the TB bacillus can lead to delayed hypersensitivity to tuberculoproteins, testing for hypersensitivity has been an important way to screen populations for tuberculosis infection and disease. Although there are newer methods available, the most widely used test is still the tuberculin test, called the **Mantoux test.** It involves local injection of purified protein derivative (PPD), a standardized solution taken from culture

INSIGHT 21.2 Medical

Fungal Lung Diseases

Increasingly, the microorganisms that cause pulmonary infections are fungi. Although still much rarer than bacterial lung infections, fungal pneumonias have shown a remarkable rise in incidence. One hospital in the Midwest reported an overall 20-fold increase in fungal infections (of all types) in the 10 years between the late 1970s and the late 1980s. And a great many of those infections occur in the lungs. As you read in chapter 5, two broad categories of fungi cause human infections: those considered to be *primary pathogens,* which readily cause disease even in healthy hosts, and *opportunists,* which cause disease primarily in hosts that are weakened due to underlying illness, advanced age, immune deficiency, or chemotherapy of some sort.

The primary pathogens usually have restricted geographic distributions. **Table 21.A** describes major characteristics of these fungi. As you can imagine, when primary pathogens invade people with weakened immune systems, the results can be disastrous.

In contrast to the primary pathogens, the opportunists are more likely to be ubiquitous and can affect weakened patients indiscriminately. **Table 21.B** lists some of the most common opportunistic fungal infections of the lungs. These opportunistic fungal infections are the ones increasing at a steady rate in the modern era, for several reasons:

- Fungi and their spores are everywhere. They constantly enter our respiratory tracts. They live in our GI tracts and on our skin.
- Antibiotic use decreases the bacterial count in our bodies, leaving fungi unhindered and able to flourish.
- More invasive procedures are being employed in hospitals and for outpatient procedures, opening pathways for fungi to access "sterile" areas of the body.
- The number of patients who are immunosuppressed (or otherwise "weakened") is constantly increasing.

For these reasons, health care professionals should be particularly vigilant for symptoms of fungal diseases in patients who are hospitalized, are HIV-positive, or have other underlying health problems. Invasive fungal infections are extremely difficult to treat effectively; there is a significant mortality rate for patients suffering from opportunistic fungal infections in the lungs.

TABLE 21.A Primary Fungal Pathogens of the Lungs

Pathogen	Geographic Distribution	Disease and Symptoms
Histoplasma capsulatum	All continents except Australia; highest rates in U.S. Ohio Valley	Histoplasmosis (see p. 677); aches, pains, and coughing; more severe symptoms include fever, night sweats, and weight loss
Blastomyces dermatitidis	Forest soils, areas of decaying wood and organic matter; worldwide distribution, in United States most common on East Coast and in Midwest	Blastomycosis—cough, chest pain, hoarseness, fever; severe cases involve skin and other organs; lung abscesses resemble malignant tumors; skin nodules, bone infections, involvement of central nervous system possible
Coccidioides immitis	Semiarid, hot climates; Mexico, Central and South America; southwest U.S., especially California and southern Arizona	Coccidioidomycosis—fever, chest pain, headaches, malaise, chronic infection can lead to pulmonary nodular growths and cavity formation in lungs
Paracoccidioides brasiliensis	Tropical and semitropical regions of South and Central America	Paracoccidioidomycosis—infections of lung and skin; in severe cases, fungus can invade lungs, skin, and lymphatic organs

TABLE 21.B Opportunistic Fungi in the Lungs

Fungus	Disease
Pneumocystis (carinii) jiroveci	"PCP" pneumonia (see p. 678); cough, fever, shallow respiration, and cyanosis
Aspergillus spp.	Aspergillosis; fungus balls form in the lungs and other tissues, necrotic pneumonia, dissemination to the brain, heart, skin
Geotrichum candidum	Geotrichosis; secondary infections in tuberculosis or very ill patients
*Cryptococcus neoformans**	Cryptococcosis; lung infections followed often by brain and meninges involvement
Candida albicans	Candidal lung infections; in HIV-positive and lung transplant patients

**Cryptococcus* could fit in either category—primary or opportunistic pathogen—but its array of virulence factors are (individually) less potent than most of those expressed by primary pathogens. However, it often causes disease in otherwise healthy patients.

fluids of *M. tuberculosis*. The injection is done intradermally into the forearm to produce an immediate small bleb. After 48 and 72 hours, the site is observed for a red wheal called an **induration,** which is measured and interpreted as positive or negative according to size (see figure 21.15).

The accepted practices for tuberculin testing are currently limited to selected groups known to have higher risk for tuberculosis infection. It is no longer used as a routine screening method among populations of children or adults who are not within the target groups. The reasoning behind this change is to allow more focused screening and to reduce expensive and unnecessary follow-up tests and treatments. Guidelines for test groups and methods of interpreting tests are listed in the following summary.[2]

Category 1. Induration (skin reaction) that is equal to or greater than *5 millimeters* is classified as positive in persons:

- Who have had contact with actively infected TB patients
- Who are HIV positive or have risk factors for HIV infection
- With past history of tuberculosis as determined through chest X rays

Category 2. Induration that is equal to or greater than *10 millimeters* is classified as positive in persons who are not in category 1 but who fit the following high-risk groups:

- HIV-negative intravenous drug users
- Persons with medical conditions that put them at risk for progressing from latent TB infection to active TB
- Persons who live or work in high-risk residences such as nursing homes, jails, or homeless shelters
- New immigrants from countries with high rates of TB
- Low-income populations lacking access to adequate medical care
- High-risk adults from ethnic minority populations as determined by local public health departments
- Children who have contact with members of high-risk adult populations

Category 3. Induration that is equal to or greater than *15 millimeters* is classified as positive in persons who do not meet criteria in categories 1 or 2.

A positive reaction in a person from one of the risk groups is fairly reliable evidence of recent infection or reactivation of a prior latent infection. Because the test is not 100% specific, false positive reactions will occasionally occur in patients who have recently been vaccinated with the BCG vaccine. Because BCG vaccination can also stimulate delayed hypersensitivity, clinicians must weigh a patient's vaccine history, especially among individuals who have immigrated from countries where the vaccine is routinely given. Another cause of a false positive reaction is the presence of an infection with a closely related species of *Mycobacterium*.

2. See the entire guidelines at http://www.thoracic.org/adobe/statements/tbadult1-20.pdf; http://medicine.iupui.edu/pulmonary/ppdguide.htm

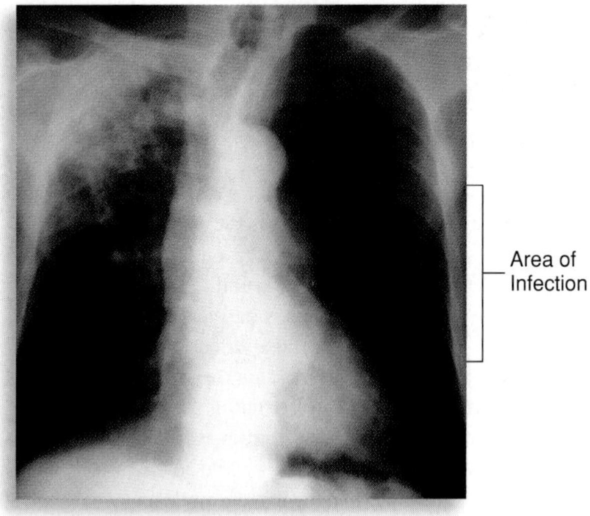

Figure 21.18 **Primary tuberculosis.**

A negative skin test usually indicates that ongoing TB infection is not present. In some cases, it may be a false negative, meaning that the person is infected but is not yet reactive. One cause of a false negative test may be that it is administered too early in the infection, requiring retesting at a later time. Subgroups with severely compromised immune systems, such as those with AIDS, advanced age, and chronic disease, may be unable to mount a reaction even though they are infected. Skin testing may not be a reliable diagnostic indicator in these populations.

X Rays Chest X rays can help verify TB when other tests have given indeterminate results, and they are generally used after a positive test for further verification. X-ray films reveal abnormal radiopaque patches, the appearance and location of which can be very indicative. Primary tubercular infection presents the appearance of fine areas of infiltration and enlarged lymph nodes in the lower and central areas of the lungs **(figure 21.18).** Secondary tuberculosis films show more extensive infiltration in the upper lungs and bronchi and marked tubercles. Scars from older infections often show up on X rays and can furnish a basis for comparison when trying to identify newly active disease.

Acid-Fast Staining The diagnosis of tuberculosis in people with positive skin tests or X rays can be backed up by acid-fast staining of sputum or other specimens. Several variations on the acid-fast stain are currently in use. The Ziehl-Neelsen stain produces bright red acid-fast bacilli (AFB) against a blue background **(figure 21.19).** Fluorescence staining shows luminescent yellow-green bacteria against a dark background (see figure 21.16).

Diagnosis that differentiates between *M. tuberculosis* and other mycobacteria must be accomplished as rapidly as possible so that appropriate treatment and isolation precautions can be instituted. The newer fast-identification techniques such as fluorescent staining (see figure 21.16), high-performance liquid chromatography (HPLC) analysis of mycolic acids, and

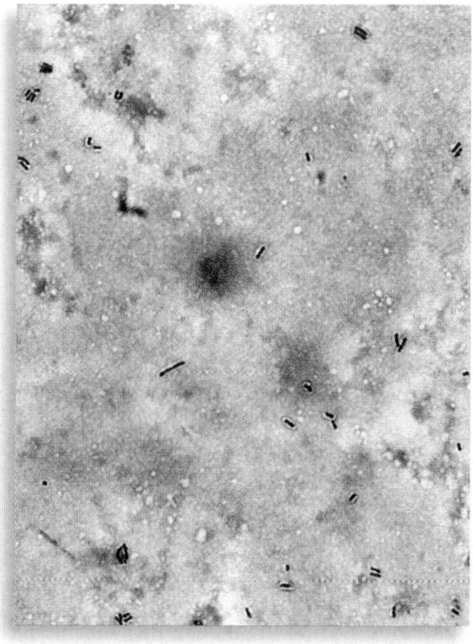

Figure 21.19 **Ziehl-Neelsen staining of *Mycobacterium tuberculosis.***
Red rods are *M. tuberculosis.*

PCR diagnosis can and should be used to identify isolates as *Mycobacterium.* Even though newer cultivation schemes exist that shorten the incubation period from 6 weeks to several days, this delay is unacceptable for beginning treatment or isolation precautions. But culture still must be performed because growing colonies are required to determine antibiotic sensitivities.

Because the specimens are often contaminated with rapid-growing bacteria that will interfere with the isolation of *M. tuberculosis,* they are pretreated with chemicals to remove contaminants and are plated onto selective medium (such as Lowenstein-Jensen medium). *M. tuberculosis* colonies are depicted in figure 21.17.

Prevention

Preventing TB in the United States is accomplished by limiting exposure to infectious airborne particles. Extensive precautions, such as isolation in negative-pressure rooms, are used in health care settings when a person with active TB is identified. Vaccine is generally not used in the United States, although an attenuated vaccine, called BCG, is used in many countries. Remember that persons vaccinated with BCG may respond positively to a tuberculin skin test.

In the past, prevention in the context of tuberculosis referred to preventing a person with latent TB from experiencing reactivation. This strategy is more accurately referred to as treatment of latent infection and is considered in the next section.

Treatment

Treatment of latent TB infection is effective in preventing full-blown disease in persons who have positive tuberculin skin tests and who are at risk for reactivated TB. Treatment with isoniazid for 9 months or with a combination of rifampin plus an additional antibiotic called pyrazinamide for 2 months is recommended.

Treatment of active TB infection when the microorganism has been found to have no antibiotic resistance consists of 9 months of treatment with isoniazid plus rifampin, with pyrazinamide also taken for the first 2 months. If there is evidence of extrapulmonary tubercular disease, the treatment should be extended to 12 months.

When the bacterium is resistant to one or more of the preceding agents, at least three additional antibiotics must be added to the treatment regimen and the duration of treatment should be extended.

One of the biggest problems with TB therapy is noncompliance on the part of the patient. It is very difficult, even under the best of circumstances, to keep to a regimen of multiple antibiotics daily for months. And most TB patients are not living under the best of circumstances. But failure to adhere to the antibiotic regimen leads to antibiotic resistance in the slow-growing microorganism, and in fact many *M. tuberculosis* isolates are now found to be **MDRTB,** or multi-drug-resistant TB. For this reason, it has been recommended that all patients with TB be treated by directly observed therapy (DOT), in which ingestion of medications is observed by a responsible person. The rate of drug-resistant TB and the rate of tuberculosis relapse have been shown to be decreased in communities where DOT is used.

In 2006, a new strain of *M. tuberculosis* was identified in Africa. It is particularly lethal for HIV-infected people and has been named **XDRTB** (extensively drug-resistant TB). It made the news in 2007 when an American man traveled to Europe and back after doctors advised him not to so that he would not expose fellow travelers to his XDRTB. The CDC sought to test those sharing his flights, even though the risk of transmission in that setting is very low.

Mycobacterium avium Complex (MAC)

Before the introduction of effective HIV treatments, described in chapter 20, disseminated tuberculosis infection with MAC was one of the biggest killers of AIDS patients. It mainly affects patients with CD4 counts below 50 cells per milliliter of blood. Antibiotics to prevent this condition should be given to all patients with AIDS **(Checkpoint 21.9).**

Pneumonia

Pneumonia is a classic example of an *anatomical diagnosis.* It is defined as an inflammatory condition of the lung in which fluid fills the alveoli. The set of symptoms that we call pneumonia can be caused by a wide variety of different microorganisms. In a sense, the microorganisms need only to have appropriate characteristics to allow them to circumvent the host's defenses and to penetrate and survive in the lower respiratory tract. In particular, the microorganisms must avoid being phagocytosed by alveolar macrophages, or at least avoid being killed once inside the macrophage. Bacteria and a wide variety of viruses

CHECKPOINT 21.9 Tuberculosis		
Causative Organism(s)	*Mycobacterium tuberculosis*	*Mycobacterium avium* complex
Most Common Modes of Transmission	Vehicle (airborne)	Vehicle (airborne)
Virulence Factors	Lipids in wall, ability to stimulate strong cell-mediated immunity (CMI)	–
Culture/Diagnosis	Rapid methods plus culture; initial tests are skin testing and chest X ray	Positive blood culture
Prevention	Avoiding airborne *M. tuberculosis*, BCG vaccine in other countries	Rifabutin or azithromycin given to AIDS patients at risk
Treatment	Isoniazid, rifampin, and pyrazinamide + ethambutol or streptomycin for varying lengths of time (always lengthy); if resistant, two other drugs added to regimen	Azithromycin or clarithromycin plus one additional antibiotic
Distinctive Features	Responsible for nearly all TB except for HIV-positive patients	Suspect this in HIV-positive patients

can cause pneumonias. Viral pneumonias are usually, but not always, milder than those caused by bacteria. At the same time, some bacterial pneumonias are very serious and others are not. In addition, fungi such as *Histoplasma* can also cause pneumonia. Overall, U.S. residents experience 2 to 3 million cases of pneumonia and more than 45,000 deaths due to this condition every year. It is much more common in the winter.

Physicians distinguish between community-acquired pneumonias and nosocomial pneumonias, because different bacteria are more likely to be causing the two types. Community-acquired pneumonias are those experienced by persons in the general population. Nosocomial pneumonias are those acquired by patients in hospitals and other health care residential facilities. All pneumonias have similar symptoms, which we describe next, followed by separate sections for each type of pneumonia.

Signs and Symptoms

Pneumonias of all types usually begin with upper respiratory tract symptoms, including runny nose and congestion. Headache is common. Fever is often present, and the onset of lung symptoms follows. These symptoms are chest pain, fever, cough, and the production of discolored sputum. Because of the pain and difficulty of breathing, the patient appears pale and presents an overall sickly appearance.

The severity and speed of onset of the symptoms varies according to the etiologic agent.

Causative Agents of Community-Acquired Pneumonia

Streptococcus pneumoniae (often called pneumococcus) accounts for about two-thirds of community-acquired bacterial pneumonia cases. It causes more lethal pneumonia cases than any other microorganism. *Legionella* is a less common but serious cause of the disease. *Haemophilus influenzae* had been a major cause of community-acquired pneumonia, but the introduction of the Hib vaccine in 1988 has reduced its incidence. A number of bacteria cause a milder form of pneumonia that is often referred to as "walking pneumonia." Two of these are *Mycoplasma pneumoniae* and *Chlamydophila pneumoniae* (formerly known as *Chlamydia pneumoniae*).[3] *Histoplasma capsulatum* is a fungus that infects many people but causes a pneumonialike disease in relatively few. Two viruses cause pneumonias that can be very serious: hantavirus and the new variant coronavirus, which emerged in the spring of 2003. Pneumonia may be a secondary effect of influenza disease. Some physicians treat pneumonia empirically, meaning they do not determine the etiologic agent.

The rest of this section covers pneumonias caused by *S. pneumoniae*, *Legionella*, *Mycoplasma*, the hantavirus, the SARS-associated coronavirus, and the fungi *Histoplasma* and *Pneumocystis* in more detail.

Streptococcus pneumoniae This bacterium, which is often simply called the pneumococcus, is a small gram-positive flattened coccus that often appears in pairs, lined up end to end **(figure 21.20*a*)**. It is alpha-hemolytic on blood agar **(figure 21.20*b*)**. *S. pneumoniae* is normal biota in the upper respiratory tract of from 5% to 50% of healthy people. Infection can occur when the bacterium is inhaled into deep areas of the lung or by transfer of the bacterium between two people via respiratory droplets. *S. pneumoniae* is very delicate and does not survive long out of its habitat. Factors that favor the ability of the pneumococcus to cause disease are old age, the season (rate of infection is highest in the winter), underlying viral respiratory disease, diabetes, and chronic abuse of alcohol or narcotics. Healthy people commonly inhale this

3. The genus formerly known as *Chlamydia* contains two important human pathogens *Chlamydia pneumoniae* and *Chlamydia trachomatis*. The latter remains "*Chlamydia*," but the respiratory pathogen is now *Chlamydophila pneumoniae*.

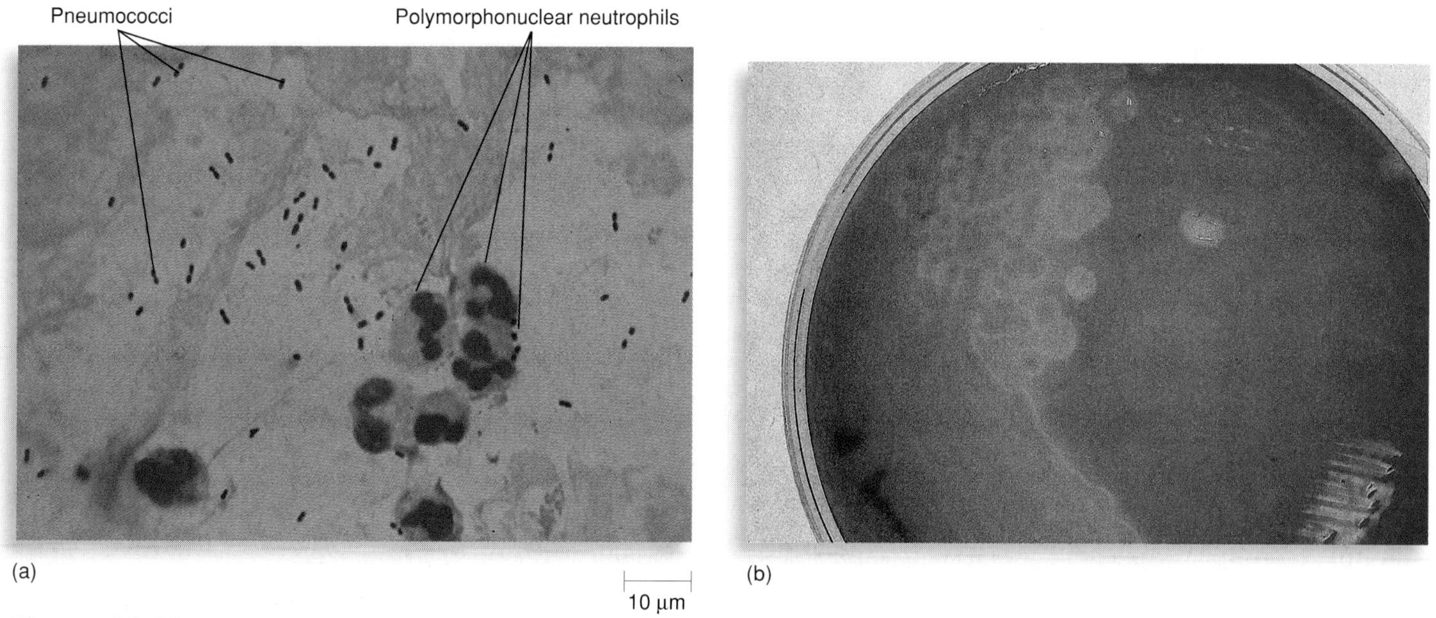

Figure 21.20 ***Streptococcus pneumoniae.***
(a) Gram stain of sputum. **(b)** Alpha-hemolysis of *S. pneumoniae* on blood agar.

and other microorganisms into the respiratory tract without serious consequences because of the host defenses present there.

Pneumonia is likely to occur when mucus containing a load of bacterial cells passes into the bronchi and alveoli. The pneumococci multiply and induce an overwhelming inflammatory response. The polysaccharide capsule of the bacterium prevents efficient phagocytosis, with the result that edematous fluids are continuously released into the lungs. In one form of pneumococcal pneumonia, termed **lobar pneumonia,** in which the infection is focused in and eventually totally fills an entire lobe of the lung, this fluid accumulates in the alveoli along with red and white blood cells. As the infection and inflammation spread rapidly through the lung, the patient can actually "drown" in his or her own secretions. If this mixture of exudates, cells, and bacteria solidifies in the air spaces, a condition known as *consolidation* **(figure 21.21)** occurs.

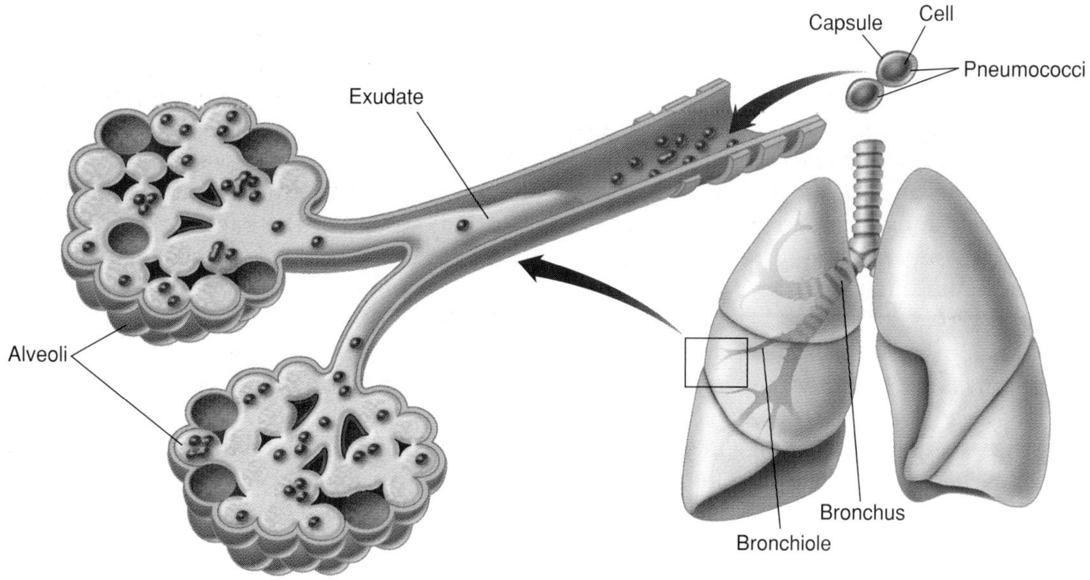

Figure 21.21 The course of bacterial pneumonia.
As the pneumococcus traces a pathway down the respiratory tree, it provokes intense inflammation and exudate formation. The blocking of the bronchioles and alveoli by consolidation of inflammatory cells and products is evident.

In infants and the elderly, the areas of infection are usually spottier and centered more in the bronchi than in the alveoli (bronchial pneumonia). Systemic complications of pneumonia are pleuritis and endocarditis, but pneumococcal bacteremia and meningitis are the greatest danger to the patient.

Because the pneumococcus is such a frequent cause of pneumonia in older adults, this population is encouraged to seek immunization with the older pneumococcal polysaccharide vaccine, which stimulates immunity to the capsular polysaccharides of 23 different strains of the bacterium. Active disease is treated with antibiotics, but the choice of antibiotic is often difficult. Many isolates of *S. pneumoniae* are resistant to penicillin and its derivatives, as well as to the macrolides, so often cephalosporins are now prescribed. A new drug called ketek, in a new class of drugs called ketolides, has recently been developed for drug-resistant *S. pneumoniae*. This bacterium is clearly capable of rapid development of resistance, and effective treatment requires that the practitioner be familiar with local resistance trends.

Legionella pneumophila *Legionella* is a weakly gram-negative bacterium that has a range of shapes, from coccus to filaments. Several species or subtypes have been characterized, but *L. pneumophila* (lung-loving) is the one most frequently isolated from infections.

Although the organisms were originally described in the late 1940s, they were not clearly associated with human disease until 1976. The incident that brought them to the attention of medical microbiologists was a sudden and mysterious epidemic of pneumonia that afflicted 200 American Legion members attending a convention in Philadelphia and killed 29 of them. After 6 months of painstaking analysis, epidemiologists isolated the pathogen and traced its source to contaminated air-conditioning vents in the Legionnaires' hotel.

Legionella's ability to survive and persist in natural habitats has been something of a mystery, yet it appears to be widely distributed in aqueous habitats as diverse as tap water, cooling towers, spas, ponds, and other fresh waters. It is resistant to chlorine. The bacteria can live in close association with free-living amoebas **(figure 21.22).** It is released during aerosol formation and can be carried for long distances. Cases have been traced to supermarket vegetable sprayers, hotel fountains, and even the fallout from the Mount St. Helens volcano eruption in 1980.

Although this bacterium can cause another disease called Pontiac fever, pneumonia is the more serious disease, with a fatality rate of 3% to 30%. *Legionella* pneumonia is thought of as an opportunistic disease, usually affecting elderly people and rarely being seen in children and healthy adults. It is difficult to diagnose, even with specific antibody tests. It is not transmitted person to person. Curiously, urine is often used for antigen testing with this microorganism.

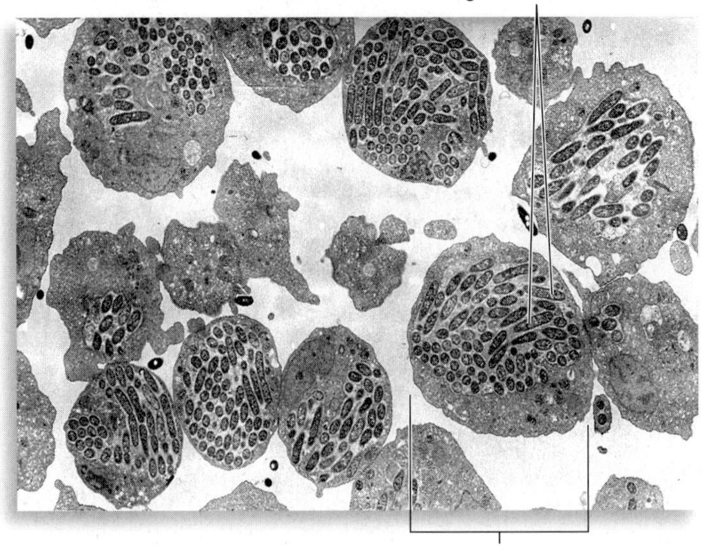

Figure 21.22 ***Legionella*** **living intracellularly in the amoeba *Hartmanella*.**
Amoebas inhabiting natural waters appear to be the reservoir for this pathogen and a means for it to survive in rather hostile environments. The pathogenesis of *Legionella* in humans is likewise dependent on its uptake by and survival in phagocytes.

Mycoplasma pneumoniae Mycoplasmas, as you learned in chapter 4, are among the smallest known self-replicating microorganisms. They naturally lack a cell wall and are therefore irregularly shaped. They may resemble cocci, filaments, doughnuts, clubs, or helices. They are free-living but fastidious, requiring complex medium to grow in the lab. (This genus should not be confused with *Mycobacterium*.)

Pneumonias caused by *Mycoplasma* (as well as those caused by *Chlamydia* and some other microorganisms) are often called atypical pneumonia—atypical in the sense that the symptoms do not resemble those of pneumococcal or other severe pneumonias. *Mycoplasma* pneumonia is transmitted by aerosol droplets among people confined in close living quarters, especially families, students, and the military.

The bacterium binds very tightly to specific receptors of the respiratory epithelium and inhibits ciliary action. Gradual spread of the bacteria over the next 2 to 3 weeks disrupts the cilia and damages the respiratory epithelium. The first symptoms—fever, malaise, sore throat, and headache—are not suggestive of pneumonia. A cough is not a prominent early symptom, and when it does appear it is mostly unproductive. As the disease progresses, nasal symptoms, chest pain, and earache can develop. The lack of acute illness in most patients has given rise to the name "walking pneumonia." For some reason, there is an increase in *Mycoplasma* pneumonias every 3 to 6 years in the United States.

Diagnosis of *Mycoplasma* may begin with ruling out other bacteria or viral agents. Serological or PCR tests confirm the diagnosis. These bacteria do not stain with Gram's stain and are not visible in direct smears of sputum.

Hantavirus In 1993, hantavirus suddenly burst into the American consciousness. A cluster of unusual cases of severe lung edema among healthy young adults arose in the Four Corners area of New Mexico. Most of the patients died within a few days. They were later found to have been infected with hantavirus, an agent that had previously only been known to cause severe kidney disease and hemorrhagic fevers in other parts of the world. The new condition was named hantavirus pulmonary syndrome (HPS). Since 1993, the disease has occurred sporadically, but it has a mortality rate of at least 33%. It is considered an emerging disease.

Symptoms, Pathogenesis, and Virulence Factors Common features of the prodromal phase of this infection include fever, chills, myalgias (muscle aches), headache, nausea, vomiting, and diarrhea or a combination of these symptoms. A cough is common but is not a prominent early feature. Initial symptoms resemble those of other common viral infections. Soon a severe pulmonary edema occurs and causes acute respiratory distress (ARDS, or acute respiratory distress syndrome, has many microbial and nonmicrobial causes; this is but one of them).

The acute lung symptoms appear to be due to the presence of large amounts of hantavirus antigen, which becomes disseminated throughout the bloodstream (including the capillaries surrounding the alveoli of the lung). Massive amounts of fluid leave the blood vessels and flood the alveolar spaces in response to the inflammatory stimulus, causing severe breathing difficulties and a drop in blood pressure. The propensity to cause a massive inflammatory response could be considered a virulence factor for this organism.

Transmission and Epidemiology Very soon after the initial cases in 1993, it became clear that the virus was associated with the presence of mice in close proximity to the victims. Investigators eventually determined that the virus, an enveloped virus of the bunyavirus family, is transmitted via airborne dust contaminated with the urine, feces, or saliva of infected rodents. Deer mice and other rodents can carry the virus with few apparent symptoms. Small outbreaks of the disease are usually correlated with increases in the local rodent population. Epidemiologists suspect that rodents have been infected with this pathogen for centuries. It has no doubt been the cause of sporadic cases of unexplained pneumonia in humans for decades, but the incidence seems to be increasing, especially in areas of the United States west of the Mississippi River **(figure 21.23).**

Treatment and Prevention The diagnosis is established by detection of IgM to hantavirus in the patient's blood or by using PCR techniques to find hantavirus genetic material in clinical specimens. Treatment consists mainly of supportive care. Mechanical ventilation is often required.

An older, inactivated vaccine for hantavirus has been used in Asia to prevent the renal hemorrhagic form of infection with this virus, but it has never been used in the United States. Several research groups are working on a newer hantavirus vaccine that could be used for people in high-risk areas to prevent HPS.

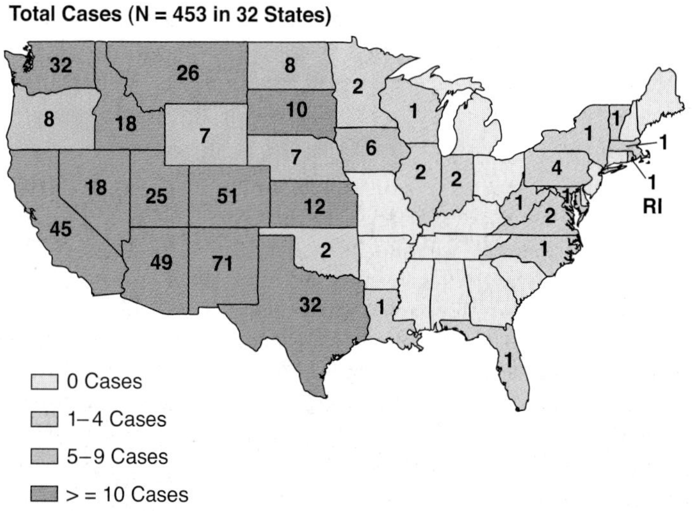

Figure 21.23 Hantavirus pulmonary syndrome cases. Cumulative data through May 2006.

Severe Acute Respiratory Syndrome–Associated Coronavirus In the winter of 2002, reports of an acute respiratory illness, originally termed an *atypical pneumonia,* started to filter in from Asia. In March of 2003, the World Health Organization issued a global health alert about the new illness. By mid-April, scientists had sequenced the entire genome of the causative virus, making the creation of diagnostic tests possible and paving the way for intensive research on the virus. The epidemic was contained by the end of July 2003, but in less than a year it had sickened more than 8,000 people **(figure 21.24).** About 9% of those died. The disease was given the name **SARS,** for **severe acute respiratory syndrome** (see chapter 1). It was concentrated in China and Southeast Asia, although several dozen countries, from Australia and Canada to the United States, have reported cases. Most of the cases seem to have originated in people who had traveled to Asia or who had close contact with people from that region. Close contact (direct or droplet) seems to be required for its transmission. In 2004, scientists found the virus in the tears of infected people and suggested that it might also be transferred through contact with that fluid. The virus is a previously unknown strain of coronavirus (family Coronaviridae).

Symptoms begin with a fever of above 38°C (100.4°F) and progress to body aches and an overall feeling of malaise. Early in the infection, there seems to be little virus in the patient and a low probability of transmission. Within a week, viral numbers surge and transmissibility is very high. After 3 weeks, if the patient survives, viral levels decrease significantly and symptoms subside. Patients may or may not experience classic respiratory symptoms. They may develop breathing problems. Severe cases of the illness can result in respiratory distress and death.

INSIGHT 21.3 **Discovery**

Bioterror in the Lungs

After the terrorist attacks of September 11, 2001, and the anthrax attacks via the U.S. Postal Service that occurred later that fall, the U.S. government renewed its interest in preparing for bioterror or biowarfare attacks of all kinds. The U.S. Public Health Service designated six infectious diseases as "Category A," meaning that they have the highest priority in research and funding. Category A agents have the following characteristics:

1. They can be easily disseminated or transmitted from person to person.
2. They result in high mortality rates and have the potential for major public health impact.
3. They have the ability to cause public panic and social disruption.
4. They require special action for public health preparedness.

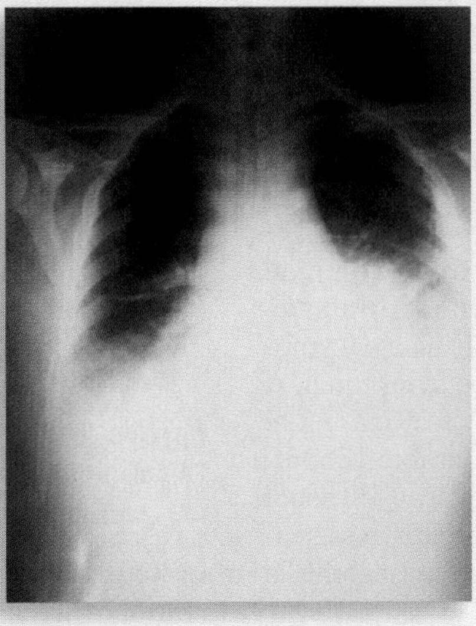

X ray showing the widened mediastinum in inhalation anthrax.

Of the six diseases, three of them can have their primary effects in the respiratory tract: pulmonary anthrax, pneumonic plague, and tularemia. The other three diseases on the A list are botulism, smallpox, and viral hemorrhagic fevers.

One of the most important components of a successful bioterror prevention strategy is early detection of infected persons. Because most of the conditions on the A list are rarely seen in the United States, clinicians' index of suspicion may be low. Here are the symptoms of the three agents that cause overt respiratory symptoms.

Pulmonary Anthrax (or Inhalation Anthrax)

This disease is the result of lung infection with *Bacillus anthracis* (see chapter 20). It should be considered when there is lung congestion accompanied by fever, malaise, and headache. Chest X rays are very useful because a widened mediastinum (the interpleural space that appears as the dark divider in the center of most chest X rays) is **pathognomic** (path-oh-nōm-ik) for this disease. Typical bronchopneumonia does not occur. In about half of patients, a hemorrhagic meningitis accompanies the pneumonitis. It is not transmitted from person to person, but because the bacterium forms endospores, these are easily disseminated through a variety of methods.

The most useful test for this disease is blood culture, because the organism is abundant in blood. Treatment is with penicillin, doxycycline, or ciprofloxacin. People presumed to have been exposed to the agent are also treated with one of these antibiotics for 30 to 60 days, because the endospores may persist in the respiratory tract for several weeks before germinating and becoming susceptible to antibiotics.

A vaccine for anthrax is currently administered only to military personnel and to some with occupational exposure to livestock.

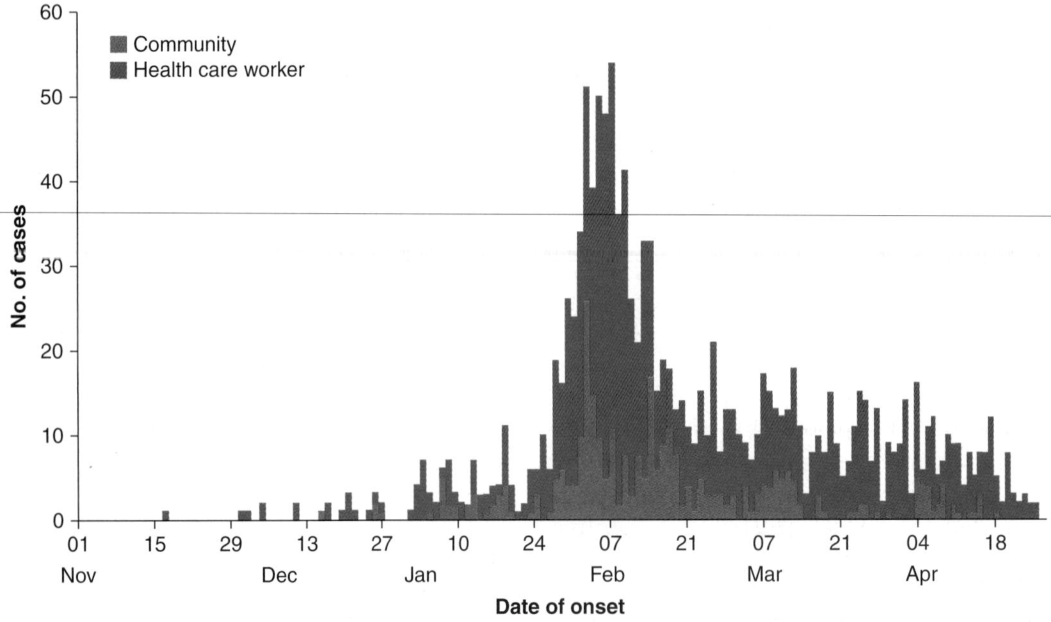

Figure 21.24 Epidemic curve of the SARS outbreak in 2002–2003.

Pneumonic Plague

This pneumonia illness is caused by *Yersinia pestis*, the same agent responsible for bubonic plague (chapter 20). The first signs of the pneumonic form are fever, headache, weakness, and rapidly developing pneumonia. Sometimes sputum is bloody or watery. Within 2 to 4 days, respiratory failure and shock can ensue. The incidence of plague in the United States is low and generally of the bubonic type, which is transmitted by fleas from a small mammal host. *Y. pestis* used as a bioterror agent would likely be disseminated as an aerosol, leading to large numbers of pneumonic cases. Gram staining of sputum, blood, or lymph node aspirates would reveal gram-negative rods, and additional staining with Wright or Giemsa stain would result in rods with characteristic bipolar staining.

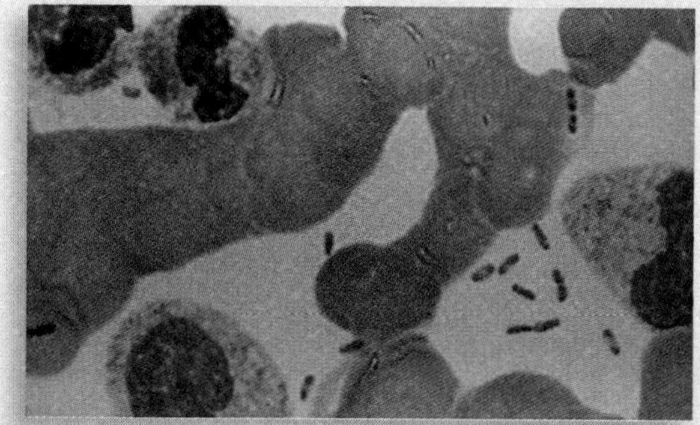

Wright-Giemsa stain of *Yersinia pestis* from peripheral blood.

Without treatment, patients die within 2 to 6 days; but swift antibiotic therapy with streptomycin, gentamicin, tetracyclines, or sulfonamides can save lives. A vaccine exists, but it does not protect against the pneumonic form of the disease and is no longer available in the United States.

Tularemia

This infection, caused by *Francisella tularensis*, is not widely known in the United States (see chapter 20). It can cause skin and bloodstream infections, lung disease, and severe ocular infections. The infectious dose is extremely low; as few as 10 bacteria can initiate serious disease. As a bioterror weapon, it would most probably be disseminated via the aerosol route and most of the infections would no doubt be of the respiratory variety. The abrupt appearance of large numbers of people with acute pneumonitis that progresses rapidly to sepsis would be the first sign that a tularemia bioterror incident has occurred. Because *F. tularensis* does not seem to be transmitted person to person, it would be unusual to find large numbers of infected people over a short period of time, which would raise the possibility that there was an intentional release.

Tularemia is difficult to diagnose, and the first steps in a suspected bioterror incident would be to rule out plague or anthrax pneumonic disease. The bacterium is extremely dangerous to laboratory workers, so caution must be used if *Francisella* is suspected. Antibiotics such as tetracycline and gentamicin can prevent death in most cases. An investigational vaccine has been developed, but its use is not approved.

As you can see, one of the greatest difficulties associated with managing a bioterror incident is that initial symptoms in patients are nonspecific. The time it takes for public health officials to begin to suspect one of these unusual etiologic agents (as opposed to common community-acquired respiratory infections) may make the difference between life and death for large numbers of people. We already have one advantage, however. Since the fall of 2001, U.S. health practitioners are much more alert to the possibility of intentional dissemination of infectious agents.

Diagnosis of the disease relies first on exclusion of other likely agents, using a Gram stain and attempted identification of influenza and RSV viruses. Acute and convalescent sera should be collected so that rise in antibody against the coronavirus can be documented. Specimens can be sent to reference labs where PCR will be performed to confirm the diagnosis. There is no specific treatment other than supportive care.

Histoplasma capsulatum Pulmonary infections with this dimorphic fungus have probably afflicted humans since antiquity, but it was not described until 1905 by Dr. Samuel Darling. Through the years, it has been known by various names: Darling's disease, Ohio Valley fever, and spelunker's disease. Certain aspects of its current distribution and epidemiology suggest that it has been an important disease for as long as humans have practiced agriculture. (See Insight 21.2 for other important fungal lung pathogens.)

Pathogenesis and Virulence Factors Histoplasmosis presents a formidable array of manifestations. It can be benign or severe, acute or chronic, and it can show pulmonary, systemic, or cutaneous lesions. Inhaling a small dose of microconidia into the deep recesses of the lung establishes a primary pulmonary infection that is usually asymptomatic. Its primary location of growth is in the cytoplasm of phagocytes such as macrophages. It flourishes within these cells and is carried to other sites. Some people experience mild symptoms such as aches, pains, and coughing; but a few develop more severe symptoms, including fever, night sweats, and weight loss.

The most serious systemic forms of histoplasmosis occur in patients with defective cell-mediated immunity such as AIDS patients. In these cases, the infection can lead to lesions in the brain, intestines, heart, liver, spleen, bone marrow, and skin. Persistent colonization of patients with emphysema and bronchitis causes *chronic pulmonary histoplasmosis*, a complication that has signs and symptoms similar to those of tuberculosis.

Transmission and Epidemiology The organism is endemically distributed on all continents except Australia. Its highest rates of incidence occur in the eastern and central regions of the United States, especially in the Ohio Valley. This fungus

CHECKPOINT 21.10 Pneumonia

Causative Organism(s)	*Streptococcus pneumoniae*	*Legionella* species	*Mycoplasma pneumoniae*
Most Common Modes of Transmission	Droplet contact or endogenous transfer	Vehicle (water droplets)	Droplet contact
Virulence Factors	Capsule	–	Adhesins
Culture/Diagnosis	Gram stain often diagnostic, alpha-hemolytic on blood agar	Requires selective charcoal yeast extract agar; serology unreliable	Rule out other etiologic agents
Prevention	Pneumococcal polysaccharide vaccine (23-valent)	–	No vaccine, no permanent immunity
Treatment	Cefotaxime, ceftriaxone, ketek; much resistance	Fluoroquinolone, azithromycin, clarithromycin	Recommended not to treat in most cases, doxycycline or macrolides may be used if necessary
Distinctive Features	Patient usually severely ill	Mild pneumonias in healthy people; can be severe in elderly or immunocompromised	Usually mild; "walking pneumonia"

appears to grow most abundantly in moist soils high in nitrogen content, especially those supplemented by bird and bat droppings **(figure 21.25).**

A useful tool for determining the distribution of *H. capsulatum* is to inject a fungal extract into the skin and monitor for allergic reactions (much like the TB skin test). Application of this test has verified the extremely widespread distribution of the fungus. In high-prevalence areas such as southern Ohio, Illinois, Missouri, Kentucky, Tennessee, Michigan, Georgia, and Arkansas, 80% to 90% of the population show signs of prior infection. Histoplasmosis prevalence in the United States is estimated at about 500,000 cases per year, with several thousand of them requiring hospitalization and a small number resulting in death.

Figure 21.25 Sign in wooded area in Kentucky.
The sign is covered in bird droppings. Up to 90% of the population in the Ohio Valley show evidence of past infection with *Histoplasma.*

People of both sexes and all ages incur infection, but adult males experience the majority of symptomatic cases. The oldest and youngest members of a population are most likely to develop serious disease.

Culture and Diagnosis Discovering *Histoplasma* in clinical specimens is a substantial diagnostic indicator. Usually it appears as spherical, "fish-eye" yeasts intracellularly in macrophages and occasionally as free yeasts in samples of sputum and cerebrospinal fluid. Complement fixation and immunodiffusion serological tests can support a diagnosis by showing a rising antibody titer. (Because a positive histoplasmin (skin) test does not indicate a new infection, this test is not useful in diagnosis.) Fluorescent antibody to the fungus is also a useful diagnostic tool.

Prevention and Treatment Avoiding the fungus is the only way to prevent this infection, and in many parts of the country this is impossible. Luckily, undetected or mild cases of histoplasmosis resolve without medical management. Chronic or disseminated disease calls for systemic antifungal chemotherapy. Amphotericin B and itraconazole are considered the drugs of choice and are usually administered in daily intravenous doses for up to several weeks. Surgery to remove affected masses in the lungs or other organs is sometimes also useful.

Pneumocystis (carinii) jiroveci Although *Pneumocystis jiroveci* (formerly called *P. carinii*) was discovered in 1909, it remained relatively obscure until it was suddenly propelled into clinical prominence as the agent of *Pneumocystis* pneumonia (called PCP because of the old name of the

Hantavirus	SARS-associated coronavirus	*Histoplasma capsulatum*	*Pneumocystis jiroveci*
Vehicle—airborne virus emitted from rodents	Droplet, direct contact	Vehicle—inhalation of contaminated soil	Droplet contact
Ability to induce inflammatory response	?	Survival in phagocytes	–
Serology (IgM), PCR identification of antigen in tissue	Rule out other agents, serology, PCR	Usually serological (rising Ab titers)	Immunofluorescence
Avoid mouse habitats and droppings	–	Avoid contaminated soil/ bat, bird droppings	Antibiotics given to AIDS patients to prevent this
Supportive	Supportive	Amphotericin B and/or itraconazole	Trimethoprim-sulfamethoxazole
Rapid onset; high mortality rate	Rapid onset	Many infections asymptomatic	Vast majority occur in AIDS patients

fungus). PCP is the most frequent opportunistic infection in AIDS patients, most of whom will develop one or more episodes during their lifetimes.

Symptoms, Pathogenesis, and Virulence Factors In people with intact immune defenses, *P. jiroveci* is usually held in check by lung phagocytes and lymphocytes; but in those with deficient immune systems, it multiplies intracellularly and extracellularly. The massive numbers of fungi adhere tenaciously to the lung pneumocytes and cause an inflammatory condition. The lung epithelial cells slough off, and a foamy exudate builds up. Symptoms are nonspecific and include cough, fever, shallow respiration, and **cyanosis** (sī-əh-nō-sis).

Transmission and Epidemiology Unlike most of the human fungal pathogens, little is known about the life cycle or epidemiology of *Pneumocystis.* It is probably spread in droplet form between humans. Contact with the agent is so widespread that in some populations a majority of people show serological evidence of infection by the age of 3 or 4. Until the AIDS epidemic, symptomatic infections by this organism were very rare, occurring only among elderly people, premature infants, or patients that were severely debilitated or malnourished.

Culture and Diagnosis Although conventional microscopy performed on sputum or lavage fluids is often used, immunofluorescence using monoclonal antibodies against the organism has a higher sensitivity.

Prevention and Treatment Traditional antifungal drugs are ineffective against *Pneumocystis* pneumonia because the chemical makeup of the organism's cell wall differs from that of most fungi. The primary treatment is trimethoprim-sulfamethoxazole. This combination should be administered even if disease appears mild or is only suspected. It is sometimes given to patients with low T-cell counts to prevent the disease. The airways of patients in the active stage of infection often must be suctioned to reduce the symptoms **(Checkpoint 21.10).**

Causative Agents of Nosocomial Pneumonia

About 1% of hospitalized or institutionalized people experience the complication of pneumonia. It is the second most common nosocomial infection, behind urinary tract infections. The mortality rate is quite high, between 30% and 50%. Although *Streptococcus pneumoniae* is frequently responsible, in addition it is very common to find a gram-negative bacterium called *Klebsiella pneumoniae* as well as anaerobic bacteria or even coliform bacteria in nosocomial pneumonia. Futher complicating matters, many nosocomial pneumonias appear to be polymicrobial in origin—meaning that there are multiple microorganisms multiplying in the alveolar spaces.

In nosocomial infections, bacteria gain access to the lower respiratory tract through abnormal breathing and aspiration of the normal upper respiratory tract biota (and occasionally the stomach) into the lungs. Stroke victims have high rates of nosocomial pneumonia. Mechanical ventilation is another route of entry for microbes. Once there, the organisms take advantage of the usual lowered immune response in a hospitalized patient and cause pneumonia symptoms.

Diagnosis and Culture Culture of sputum or of tracheal swabs is not very useful in diagnosing nosocomial pneumonia, because the condition is usually caused by normal biota. Obtaining cultures of fluids obtained through endotracheal tubes or from bronchoalveolar lavage provide better information but are fairly intrusive. It is also important to remember that if the patient has already received antibiotics, culture results will be affected.

Prevention and Treatment Because most nosocomial pneumonias are caused by microorganisms aspirated from the upper respiratory tract, measures that discourage the transfer of microbes into the lungs are very useful for preventing the condition. Elevating patients' heads to a 45-degree angle helps reduce aspiration of secretions. Good preoperative education of patients about the importance of deep breathing and frequent coughing can reduce postoperative infection rates. Proper care of mechanical ventilation and respiratory therapy equipment is essential as well.

Studies have shown that delaying antibiotic treatment of suspected nosocomial pneumonia leads to a greater likelihood of death. Even in this era of conservative antibiotic use, empiric therapy should be started as soon as nosocomial pneumonia is suspected, using multiple antibiotics that cover both gram-negative and gram-positive organisms.

CHECKPOINT 21.11	Nosocomial Pneumonia
Causative Organism(s)	Gram-negative and gram-positive bacteria from upper respiratory tract or stomach
Most Common Modes of Transmission	Endogenous (aspiration)
Virulence Factors	–
Culture/Diagnosis	Culture of lung fluids
Prevention	Elevating patient's head, preoperative education, care of respiratory equipment
Treatment	Broad-spectrum antibiotics

SUMMING UP

Taxonomic Organization Microorganisms Causing Disease in the Respiratory Tract

Microorganism	Disease	Chapter Location
Gram-positive bacteria		
Streptococcus pneumoniae	Otitis media, pneumonia	Otitis media, p. 653 Pneumonia, p. 672
S. pyogenes	Pharyngitis	Pharyngitis, p. 655
Corynebacterium diphtheriae	Diphtheria	Diphtheria, p. 657
Gram-negative bacteria		
Haemophilus influenzae	Otitis media	Otitis media, p. 653
Bordetella pertussis	Whooping cough	Whooping cough, p. 660
Mycobacterium tuberculosis, M. avium* complex	Tuberculosis	Tuberculosis, p. 666
Legionella spp.	Pneumonia	Pneumonia, p. 674
Other bacteria		
Mycoplasma pneumoniae	Pneumonia	Pneumonia, p. 674
RNA viruses		
Respiratory syncytial virus	RSV disease	RSV disease, p. 661
Influenza virus A, B, and C	Influenza	Influenza, p. 662
Hantavirus	Hantavirus pulmonary syndrome	Pneumonia, p. 675
SARS-associated coronavirus	SARS	Pneumonia, p. 675
Fungi		
Pneumocystis jiroveci	*Pneumocystis* pneumonia	Pneumonia, p. 678
Histoplasma capsulatum	Histoplasmosis	Pneumonia, p. 677

*There is some debate about the gram status of the genus *Mycobacterium;* it is generally not considered gram-positive or gram-negative.

INFECTIOUS DISEASES AFFECTING
The Respiratory System

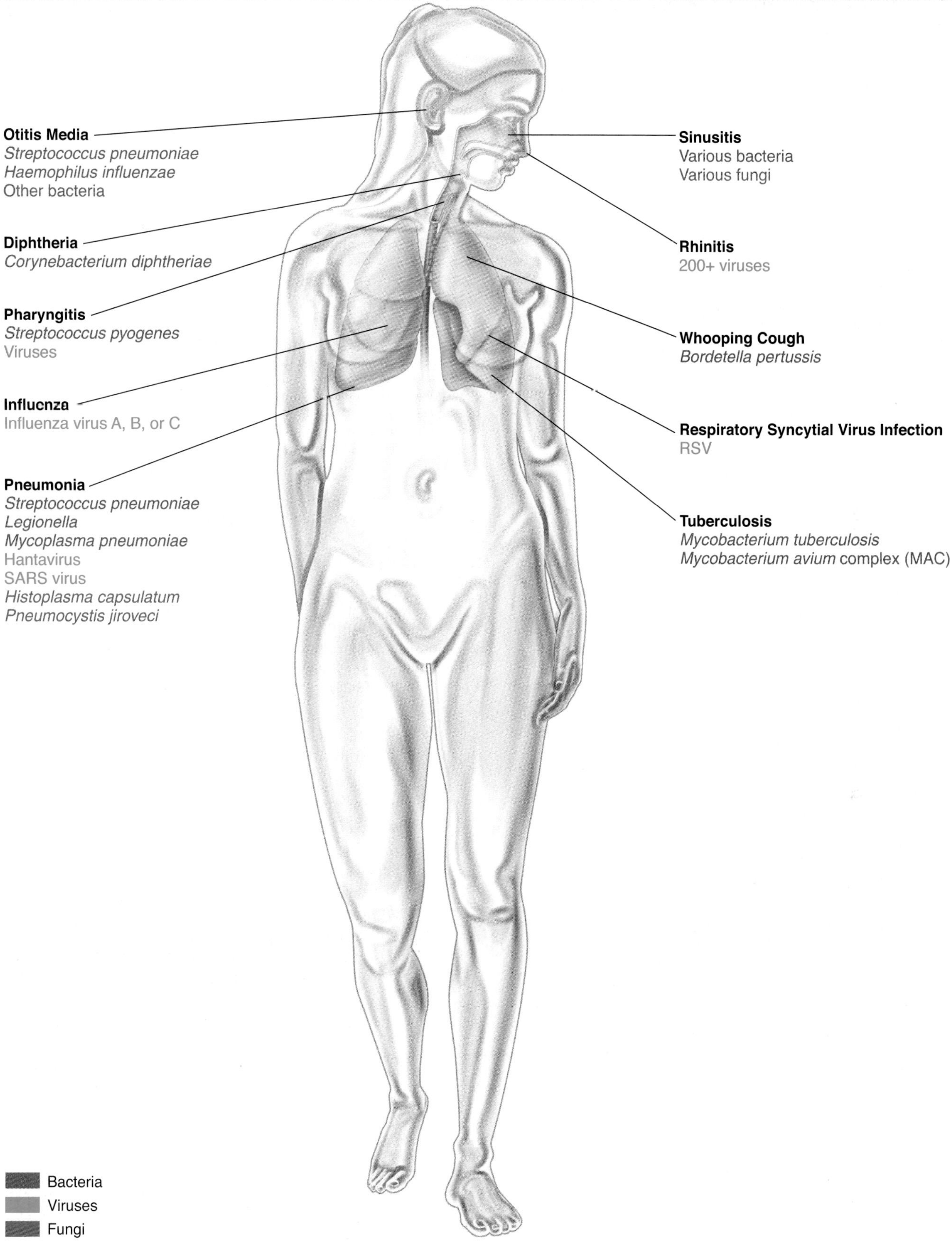

System Summary Figure 21.26

Chapter Summary with Key Terms

21.1 The Respiratory Tract and Its Defenses

A. The respiratory tract is the most common place for infectious agents to gain access to the body. The upper respiratory tract includes the mouth, the nose, nasal cavity and sinuses above it, the throat or pharynx, and the epiglottis and larynx. The lower respiratory tract begins with the trachea, which feeds into the bronchi and bronchioles in the lungs. Attached to the bronchioles are small balloonlike structures called alveoli, the site of oxygen exchange in the lungs.

B. The ciliary escalator propels particles upward and out of the respiratory tract. Mucus on the surface of the mucous membranes traps microorganisms, and involuntary responses such as coughing, sneezing, and swallowing can move them out of sensitive areas. Macrophages inhabit the alveoli of the lungs and the clusters of lymphoid tissue (tonsils) in the throat. Secretory IgA against specific pathogens can be found in the mucus secretions as well.

21.2 Normal Biota of the Respiratory Tract

A. These include *Streptococccus pyogenes, Haemophilus influenzae, Streptococcus pneumoniae, Neisseria meningitidis,* and *Staphylococcus aureus.* These bacteria can potentially cause disease if their host becomes immunocompromised, and they can cause disease in other hosts when they are innocently transferred to them.

B. Other normal biota include *Moraxella* and *Corynebacterium* species and *Candida albicans.*

21.3 Upper Respiratory Tract Diseases Caused by Microorganisms

A. *Rhinitis, or the Common Cold:* The common cold is caused by one of over 200 different kinds of viruses, most commonly the rhinoviruses, followed by the coronaviruses. Respiratory syncytial virus (RSV) causes colds in many people, but in some, especially children, they can lead to more serious respiratory tract symptoms.
   1. Viral infection of the upper respiratory tract can lead to secondary infections by bacteria.
   2. Cold viruses are transmitted by droplet contact and airborne transmission, but indirect transmission may be more common.

B. *Sinusitis:* This inflammatory condition of the sinuses in the skull is most commonly caused by allergy (most common), infections, or simply by structural problems. The infectious agents that may be responsible for the condition commonly include a variety of viruses or bacteria and, less commonly, fungi.

C. *Acute Otitis Media (Ear Infection):* Viral infections of the upper respiratory tract lead to inflammation of the eustachian tubes and the buildup of fluid in the middle ear, which can lead to bacterial multiplication in those fluids. The most common causes are *Streptococcus pneumoniae* and *Haemophilus influenzae.* Vaccines exist for both microorganisms.

D. *Pharyngitis:* The same viruses causing the common cold commonly cause an inflammation of the throat. However, the most serious cause of pharyngitis is *Streptococcus pyogenes,* a gram-positive coccus that grows in chains. *Streptococcus pyogenes* is classified as a group A streptococcus that produces hemolysins called **streptolysins.** Untreated streptococcal throat infections can result in complications including scarlet fever, rheumatic fever, glomerulonephritis, and necrotizing fasciitis.
   1. Scarlet fever, characterized by a sandpaperlike rash, is the result of infection with an *S. pyogenes* strain that is itself infected with a bacteriophage.
   2. Rheumatic fever is thought to be due to an immunologic cross-reaction between the streptococcal M-protein and heart muscle and can result in permanent damage to heart valves.
   3. Glomerulonephritis is the result of streptococcal proteins participating in the formation of antigen-antibody complexes, which then are deposited in the basement membrane of the glomerulus of the kidney.
   4. Toxic shock syndrome and necrotizing fasciitis are other, less frequent, sequelae of streptococcal infections.

E. *Diphtheria:* This disease is caused by *Corynebacterium diphtheriae,* a non-spore-forming, gram-positive club-shaped bacterium. The exotoxin is encoded by a bacteriophage of *C. diptheriae.* Diphtheria can easily be prevented by a series of vaccinations with toxoid, usually given as part of a mixed vaccine against tetanus and pertussis, as well, called the *DTaP.*

21.4 Diseases Caused by Microorganisms Affecting the Upper and Lower Respiratory Tract

A. *Whooping Cough:* This disease has two distinct symptom phases called the **catarrhal** and **paroxysmal stages,** which are followed by a long recovery (or **convalescent**) **phase** during which a patient is particularly susceptible to other respiratory infections. The causative agent, *Bordetella pertussis,* is a very small gram-negative rod. The bacterium releases multiple exotoxins—*pertussis toxin* and *tracheal cytotoxin*—that damage ciliated respiratory epithelial cells and cripple other components of the host defense, including phagocytic cells. The current vaccine for pertussis is an acellular formulation of important *B. pertussis* antigens and is usually given in the form of the DTaP vaccine.

B. Respiratory syncytial virus (RSV) infects the respiratory tract and produces giant multinucleated cells (syncytia). RSV is the most prevalent cause of respiratory infection in the newborn age group. The virus is highly contagious and is transmitted through droplet contact but also through fomite contamination.

C. *Influenza:* This disease begins in the upper respiratory tract but may also affect the lower respiratory tract. Patients with emphysema or cardiopulmonary disease, along with very young, elderly, or pregnant patients, are more susceptible to serious complications.

Influenza is caused by one of three influenza viruses: A, B, or C. Each virion is covered with a lipoprotein envelope that is studded with glycoprotein spikes called hemagglutinin (HA) and neuraminidase (NA) that contribute to virulence. The ssRNA genome of the influenza virus is subject to constant genetic changes that alter the structure of its envelope glycoprotein. This constant mutation of the glycoprotein is called **antigenic drift**—resulting in decreased ability of host memory cells to recognize them. **Antigenic shift,** where the eight separate RNA strands are involved in the swapping out of one of those genes or strands with a gene or strand from a different influenza virus, is even more serious. Inhalation of virus-laden aerosols and droplets is the main means of influenza infection, although fomites can play a secondary role. The influenza vaccine consists of three different influenza viruses that have been judged to most resemble the virus variants likely to cause infections in the coming flu season. Because of the changing nature of the antigens on the viral surface, annual vaccination is considered the best way to avoid infection.

21.5 Lower Respiratory Tract Diseases Caused by Microorganisms

A. *Tuberculosis:* The cause of tuberculosis is primarily the bacterial species *Mycobacterium tuberculosis.* Clinical tuberculosis is divided into primary tuberculosis, secondary tuberculosis, and disseminated tuberculosis.

Clinical diagnosis of tuberculosis relies on four techniques: (1) tuberculin testing, (2) chest X rays, (3) direct identification of acid-fast bacilli (AFB) in sputum or other specimens, and (4) cultural isolation and antimicrobial susceptibility testing.

Vaccine is generally not used in the United States, although an attenuated vaccine, called BCG, is used in many countries.

Mycobacterium avium Complex: Before the introduction of effective HIV treatments, disseminated tuberculosis infection with MAC was one of the biggest killers of AIDS patients.

B. **Pneumonia** is an inflammatory condition of the lung in which fluid fills the alveoli, caused by a wide variety of different microorganisms.

Community-acquired pneumonias are those experienced by persons in the general population. Nosocomial pneumonias are those acquired by patients in hospitals and other health care residential facilities. Pneumonias of all types usually begin with upper respiratory tract symptoms, including runny nose and congestion.

Streptococcus pneumoniae is the main agent for community-acquired bacterial pneumonia cases. *Legionella* is a less common but serious cause of the disease. *Haemophilus influenzae* used to be a major cause of community-acquired pneumonia, but use of the Hib vaccine has reduced its incidence. Other bacteria that cause pneumonia are *Mycoplasma pneumoniae* and *Chlamydophila pneumoniae. Histoplasma capsulatum* is a fungus that can cause a pneumonialike disease. Two viruses cause pneumonias that can be very serious: hantavirus, which causes a condition named hantavirus pulmonary syndrome (HPS), and the new variant coronavirus, which emerged in the spring of 2003 and has been responsible for a **severe acute respiratory syndrome (SARS).** Pneumonia may be a secondary effect of influenza disease. Some physicians treat pneumonia empirically, meaning they do not determine the etiologic agent.

Causative Agents of Nosocomial Pneumonia: Streptococcus pneumoniae and a gram-negative bacterium called *Klebsiella pneumoniae* are commonly responsible. Furthermore, many nosocomial pneumonias appear to be polymicrobial in origin.

Multiple-Choice and True-False Questions

Multiple-Choice Questions. Select the correct answer from the answers provided.

1. The two most common groups of virus associated with the common cold are
   a. rhinoviruses
   b. coronaviruses
   c. influenza viruses
   d. both a and b
   e. both a and c

2. Which of the following conditions are associated with *Streptococcus pyogenes*?
   a. pharyngitis
   b. scarlet fever
   c. rheumatic fever
   d. all of the above

3. Which is not a characteristic of *Streptococcus pyogenes*?
   a. group A streptococcus
   b. alpha-hemolytic
   c. sensitive to bacitracin
   d. gram-positive
4. The common stain used to identify *Mycobacterium* species is
   a. Gram stain
   b. acid-fast stain
   c. negative stain
   d. spore stain
5. Which of the following techniques are used to diagnose tuberculosis?
   a. tuberculin testing
   b. chest X rays
   c. cultural isolation and antimicrobial testing
   d. all of the above
6. The DTaP vaccine provides protection against the following diseases, *except*
   a. diphtheria
   b. pertussis
   c. pneumonia
   d. tetanus
7. Amphotericin B and itraconazole would be effective drug treatments for pneumonia caused by
   a. hantavirus
   b. *Legionella*
   c. *Pneumocystis jiroveci*
   d. *Mycoplasma pneumoniae*
8. The vast majority of pneumonias caused by this organism occur in AIDS patients.
   a. hantavirus
   b. *Histoplasma capsulatum*
   c. *Pneumocystis jiroveci*
   d. *Mycoplasma pneumoniae*
9. The beta-hemolysis of blood agar observed with *Streptococcus pyogenes* is due to the presence of
   a. streptolysin
   b. M-protein
   c. hyaluronic acid
   d. catalase

True-False Questions. If statement is true, leave as is. If it is false, correct it by rewriting the sentence.

10. *Bordetella pertussis* is the causative agent for whooping cough.
11. *Mycoplasma pneumoniae* causes "atypical" pneumonia and is diagnosed by sputum culture.
12. BCG vaccine is used in other countries to prevent Legionnaires' disease.
13. Respiratory syncytial virus (RSV) is a respiratory infection associated with elderly people.

Writing to Learn

These questions are suggested as a *writing-to-learn* experience. For each question, compose a one- or two-paragraph answer that includes the factual information needed to completely address the question.

1. Discuss the anatomical features of the respiratory system that form part of the body's defense against infection.
2. What organism(s) is (are) responsible for the common cold?
3. Discuss the most common causative agents of otitis media.
4. What two vaccines are available for treating *Streptococcus pneumoniae,* and what are their target populations?
5. List some anti-influenza remedies and preventions.
6. What parts of the body are affected by extrapulmonary tuberculosis?
7. a. What type of vaccine is used against *Corynebacterium diphtheria?*
   b. What is the characteristic toxin produced by this microorganism?
   c. What treatment is suggested for a diphtheria infection?
8. a. Name the organisms responsible for the flu.
   b. To what family do these viruses belong?
   c. Describe the genome of this virus.

Concept Mapping

Appendix D provides guidance for working with concept maps.

1. Construct your own concept map using the following words as the *concepts.* Supply the linking words between each pair of concepts.

FHA
coughing
multiplication
pertussis toxin
tracheal cytotoxin
endotoxin
cilia
mucus
Bordetella pertussis

Critical Thinking Questions

Critical thinking is the ability to reason and solve problems using facts and concepts. These questions can be approached from a number of angles, and in most cases, they do not have a single correct answer.

1. What are some of the likely explanations if you are not responding to antibiotic treatment for sinusitis?
2. Can there be a vaccine against the common cold?
3. A 5-year-old boy is diagnosed with otitis media. He has severe pain in his left ear and a fever of 101°F. Inspection of the eardrum reveals that both membranes are red but intact. His history reveals that he seldom has ear infections. How would you treat this patient?
4. Why do you think that despite full vaccinations, some immunized people may be susceptible to whooping cough?
5. What are antigenic shift and antigenic drift?
6. A graduate student from Namibia tests positive in the tuberculin skin test. Upon reading the patient history, the doctor determines that the test is a false positive and does not pursue further treatment. What is the possible explanation for the false positive skin test?
7. Why is noncompliance during TB therapy such a big concern?
8. Why do we need to take the flu vaccine every year? Why does it not confer long-term immunity to the flu like other vaccines?

Visual Understanding Questions

1. **Figure 21.2.** Some doctors suggest that gently forcing one's ears to "pop" is an effective way to treat or even prevent ear infection. Use the following illustration to explain how this could work.

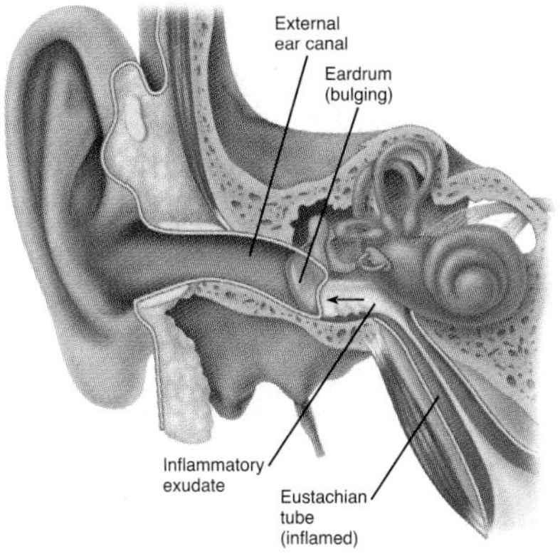

2. **From chapter 3, figure 3.25.** Although there are many different organisms present in the respiratory tract, an acid-fast stain of sputum like the one shown here along with patient symptoms can establish a presumptive diagnosis of tuberculosis. Explain why.

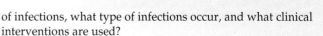

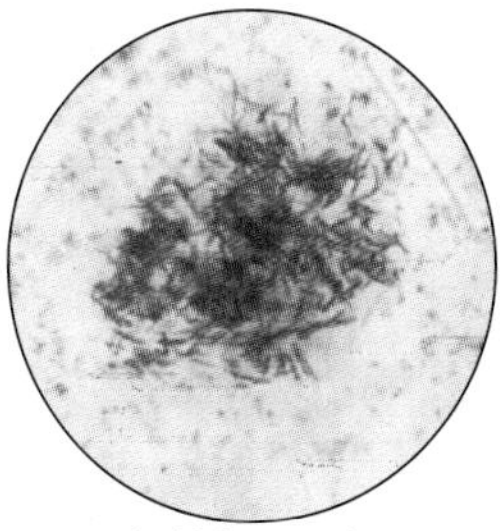

Acid-fast stain
Red cells are acid-fast.
Blue cells are non-acid-fast.

Internet Search Topics

Go to: www.aris.mhhe.com, and click on "microbiology" and then this textbook's author/title. Go to chapter 21, access the URLs listed under Internet Search Topics, and research the following:

1. You be the detective. Surf the Web to find some possible answers to these mysteries:
   a. What was the origin of the SARS coronavirus?
   b. What factors gave rise to the severity of the influenza pandemic of 1918?
2. Explore the Internet to locate information on outbreaks of fungal infections in hospitals. What are some common sources of infections, what type of infections occur, and what clinical interventions are used?
3. The famous dogsled race in Alaska called the Iditarod has its origins in a 1925 infectious disease event. Investigate this on the Web and consider the differences in medical treatments available in 1925 and today.
4. Search the Internet for the details surrounding the pneumonia outbreak at the American Legion Convention in 1976. Suggest a simple addition to routine maintenance procedures for commercial air-conditioning systems that could prevent this from happening again.

CHAPTER 22

Infectious Diseases Affecting the Gastrointestinal Tract

CASE FILE 22

Following Hurricane Katrina, many evacuees from the region were provided food and shelter in a variety of locations. Of the estimated 240,000 evacuees, approximately 24,000 were provided temporary shelter in the Reliant Park Sports and Convention Center in Houston, Texas. The complex, which includes the Astrodome, was renamed Reliant City and given its own ZIP Code by the U.S. Postal Service. Nearly 60,000 staff members and volunteers from around the country were involved in the operation of the center. Evacuees were provided with cots, necessary bedding, food, water, toilet, and shower facilities. A temporary medical clinic was set up to serve the needs of the residents. It was staffed by personnel from the Harris County Hospital District, Baylor College of Medicine, and Texas Children's Hospital. Over the next several weeks, 1,169 individuals (18% of the total clinic visitors) presented with symptoms of acute gastroenteritis. The symptoms included diarrhea, vomiting, or both. In addition, some of the volunteers, police officers, and medical personnel developed similar symptoms.

- *What organisms are associated with acute gastroenteritis?*
- *How did this outbreak likely begin? How did it probably spread?*

Case File 22 Wrap-Up appears on page 706.

CHAPTER OVERVIEW

- The GI tract consists of a tube extending from mouth to anus. Associated organs are the salivary glands, liver, gallbladder, and pancreas. It has a wide variety of gut-associated lymphoid tissue (GALT), IgA, and other secretions that protect it from pathogenic invaders.
- Diseases of the oral cavity (teeth, gums, and mucous membranes) are extremely common; and although they are often considered minor, they can have a huge impact on quality of life. They may influence disease in other parts of the body, such as the cardiovascular system.
- Numerous infections result in diarrhea; they range from the bacterial (*Salmonella, Vibrio cholerae, E. coli*) to viral (Rotavirus) to protozoal (*Giardia, Entamoeba*) to helminthic (many different worms).
- Some diarrheal diseases are accompanied by vomiting and other neurological symptoms.
- Hepatitis and mumps are diseases of accessory organs of the gastrointestinal tract.
- Helminthic GI tract diseases have distinct patterns of symptoms; they are extremely common in most of the world.

22.1 The Gastrointestinal Tract and Its Defenses

The gastrointestinal (GI) tract can be thought of as a long tube, extending from mouth to anus. It is a very sophisticated delivery system for nutrients, composed of *eight* main sections and augmented by *four* accessory organs. The eight sections are the mouth, pharynx, esophagus, stomach, small intestine, large intestine, rectum, and anus. Along the way, the salivary glands, liver, gallbladder, and pancreas add digestive fluids and enzymes to assist in digesting and processing the food we take in **(figure 22.1).** The GI tract is often called the *digestive tract* or the *alimentary tract.*

Anything inside the GI tract is in some ways not "inside" the body; it is passing through an internal tube, called a **lumen,** and only those chemicals that are absorbed through the walls of the GI tract actually gain entrance to the internal portions of the body. Food begins to be broken down into absorbable subunits as soon as it enters the mouth, where the teeth begin to mechanically break down solid particles and where enzymes in saliva break the food down chemically. The swallowed food

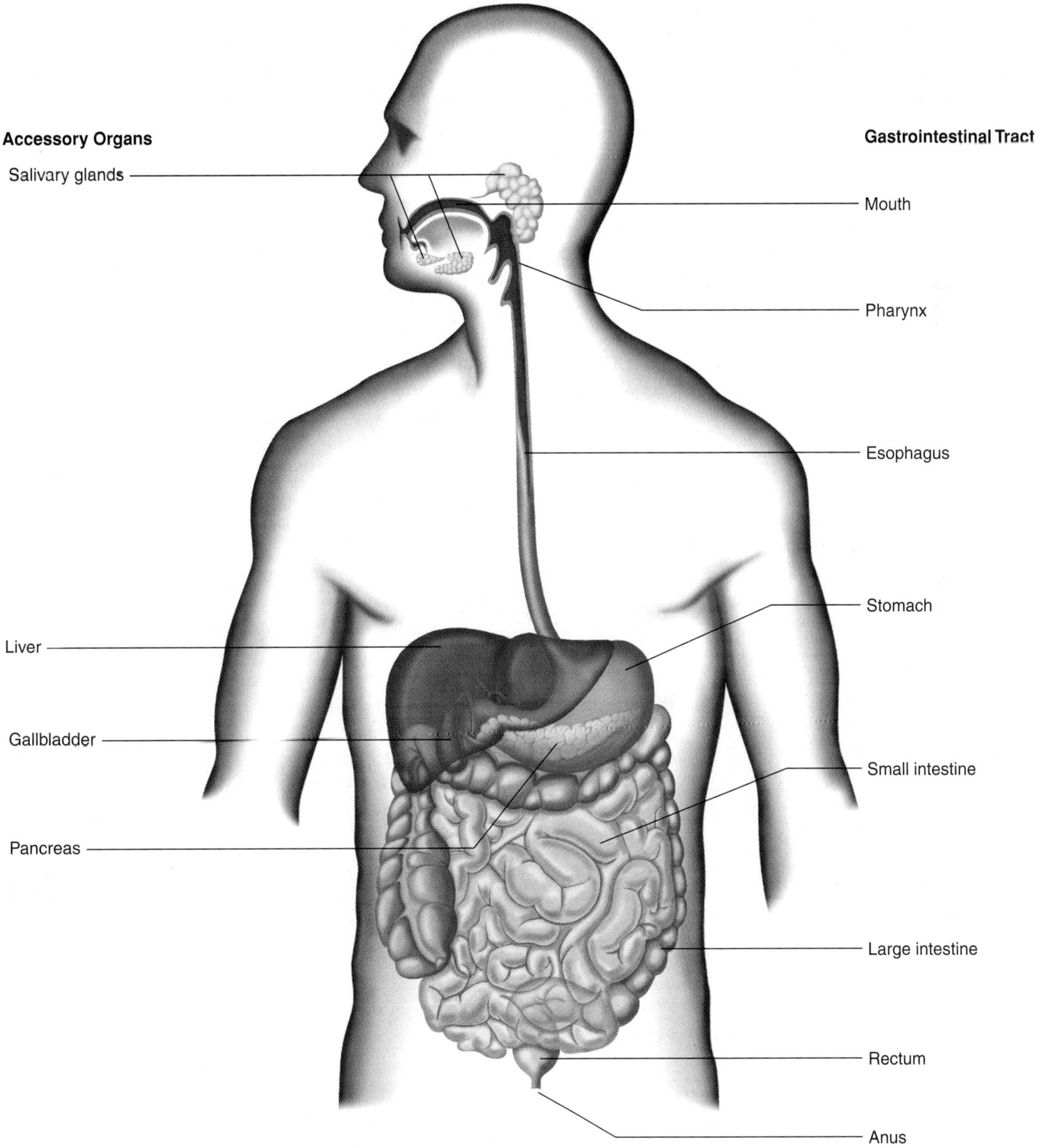

Figure 22.1 Major organs of the digestive system.

travels through the pharynx and into the esophagus, emptying into the stomach. Here the food is mixed with gastric juice, which has a very low pH and contains the important gastric enzyme pepsin, which breaks down proteins (peptides). From here the food travels to the small intestine, a long, tightly coiled portion of the lumen where most nutrient absorption takes place. The small intestine is divided into the duodenum (leading directly out of the stomach), the jejunum (most of the coiled part), and the ileum (connecting the coils to the large intestine). The pancreas secretes a variety of digestive enzymes into the small intestine, and the liver and the gallbladder work together to add bile.

Once food leaves the small intestine, it enters the large intestine, which is divided into the cecum, the colon, the rectum, and the anus. In the large intestine, water and electrolytes are absorbed from any undigested food. What is left combines with mucus and bacteria from the large intestine, becoming fecal material. Forty to sixty percent of the mass of fecal material is composed of bacteria.

The GI tract has a very heavy load of microorganisms, and it encounters millions of new ones every day. Because of this, defenses against infection are extremely important. All intestinal surfaces are coated with a layer of mucus, which confers mechanical protection. Secretory IgA can also be found on most intestinal surfaces. The muscular walls of the GI tract keep food (and microorganisms) moving through the system through the action of peristalsis. Various fluids in the GI tract have antimicrobial properties. Saliva contains the antimicrobial proteins lysozyme and lactoferrin. The stomach fluid is antimicrobial by virtue of its extremely high acidity. Bile is also antimicrobial.

The entire system is outfitted with cells of the immune system, collectively called gut-associated lymphoid tissue (GALT). The tonsils and adenoids in the oral cavity and pharynx, small areas of lymphoid tissue in the esophagus, Peyer's patches in the small intestine, and the appendix are all packets of lymphoid tissue consisting of T and B cells as well as cells of nonspecific immunity. One of their jobs is to produce IgA, but they perform a variety of other immune functions.

A huge population of commensal organisms lives in this system, especially in the large intestine. They provide the protection of microbial antagonism and avoid immune destruction through various mechanisms, including cloaking themselves with host sugars they find on the intestinal walls.

22.2 Normal Biota of the Gastrointestinal Tract

As just mentioned, the GI tract is home to a large variety of normal biota. The oral cavity alone is populated by more than 550 known species of microorganisms, including *Streptococcus, Neisseria, Veillonella, Staphylococcus, Fusobacterium, Lactobacillus, Corynebacterium, Actinomyces,* and *Treponema* species. Fungi such as *Candida albicans* are also numerous. A few protozoa (*Trichomonas tenax, Entamoeba gingivalis*) also call the mouth "home." Bacteria live on the teeth as well as the soft structures in the mouth. Numerous species of normal biota bacteria live on the teeth in large accretions called dental plaque, which is a kind of biofilm (see chapter 4). Bacteria are held in the biofilm by specific recognition molecules. Alpha-hemolytic streptococci are generally the first colonizers of the tooth surface after it has been cleaned. The streptococci attach specifically to proteins in the **pellicle,** a mucinous glycoprotein covering on the tooth. Then other species attach specifically to proteins or sugars on the surface of the streptococci, and so on.

The pharynx contains a variety of microorganisms, which were described in chapter 21. The esophagus and stomach are much more sparsely populated. Although the stomach was previously thought to be sterile, researchers have found that a very small number of bacteria have mechanisms for overcoming the extreme acidity of the stomach fluid and can survive there. On average, approximately 1,000 (10^3) bacteria can be found per gram of stomach contents. The small intestine is also sparsely populated with "normal" biota, due to the large variety of antimicrobial substances found in the stomach and the short residence time of food in the small intestine. The large intestine, in contrast, is a haven for billions of microorganisms (10^{11} per gram of contents), including the bacteria *Bacteroides, Fusobacterium, Bifidobacterium, Clostridium, Streptococcus, Peptostreptococcus, Lactobacillus, Escherichia,*

Gastrointestinal Tract Defenses and Normal Biota

	Defenses	Normal Biota
Oral cavity	IgA, lysozyme, lactoferrin, saliva, lymphoid tissue in tonsils, adenoids	*Streptococcus, Veillonella, Moraxella, Bacteroides, Actinomyces, Treponema, Candida, Entamoeba, Eikenella, Haemophilus*
Upper GI (esophagus, stomach)	IgA, low pH of stomach fluid, GALT (Peyer's patches)	Sparsely populated
Lower GI (small and large intestine)	IgA, GALT, bile, large commensal population	*Bacteroides, Fusobacterium, Bifidobacterium, Streptococcus, Clostridium, Lactobacillus, Escherichia, Enterobacter, Candida, Entamoeba, Trichomonas hominis*

and *Enterobacter;* the fungus *Candida;* and several protozoa as well. You may be surprised to learn that anaerobic bacteria outnumber the aerobic bacteria in the large intestine by several orders of magnitude.

The normal biota in the gut provide a protective function, but they also perform other jobs as well. Some of them help with digestion. Some provide nutrients that we can't produce ourselves. *E. coli,* for instance, synthesizes vitamin K. Their mere presence in the large intestine seems to be important for the proper formation of epithelial cell structure. And the normal biota in the gut plays an important role in "teaching" our immune system to react to microbial antigens. Some scientists believe that the mix of microbiota in the healthy gut can influence a host's chances for obesity or autoimmune diseases.

The accessory organs (salivary glands, gallbladder, liver, and pancreas) are free of microorganisms, just as all internal organs are.

22.3 Gastrointestinal Tract Diseases Caused by Microorganisms

Tooth and Gum Infections

It is difficult to pinpoint exactly when the "normal biota biofilm" just described becomes a "pathogenic biofilm." If left undisturbed, the biofilm structure eventually contains anaerobic bacteria that can damage the soft tissues and bones (referred to as the periodontium) surrounding the teeth. Also, the introduction of carbohydrates to the oral cavity can result in breakdown of hard tooth structure (the dentition) due to the production of acid by certain oral streptococci in the biofilm. These two separate circumstances are discussed here.

Dental Caries (Tooth Decay)

Dental caries is the most common infectious disease of human beings. The process involves the dissolution of solid tooth surface due to the metabolic action of bacteria. (**Figure 22.2** depicts the structure of a tooth.) The symptoms are often not noticeable but range from minor disruption in the outer (enamel) surface of the tooth to complete destruction of the enamel and then destruction of deeper layers **(figure 22.3).** Deeper lesions can result in infection to the soft tissue inside the tooth, called the pulp, which contains blood vessels and nerves. These deeper infections lead to pain, referred to as a "toothache."

Causative Agent

Two representatives of oral alpha-hemolytic streptococci, *Streptococcus mutans* and *Streptococcus sobrinus,* seem to be the main causes of dental caries, although a mixed species consortium, consisting of other *Streptococcus* species and some lactobacilli, is probably the best route to caries. Note that in the absence of dietary carbohydrates bacteria do not cause decay.

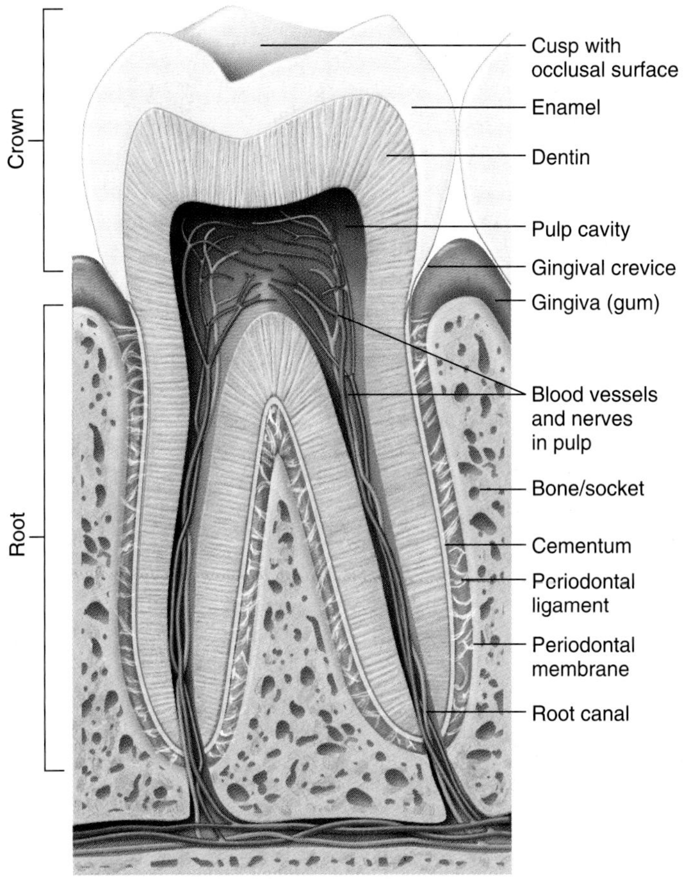

Figure 22.2 The anatomy of a tooth.

Pathogenesis and Virulence Factors

In the presence of sucrose and, to a lesser extent, other carbohydrates, *S. sobrinus* and *S. mutans* produce sticky polymers of glucose called fructans and glucans. These adhesives help bind them to the smooth enamel surfaces and contribute to the sticky bulk of the plaque biofilm **(figure 22.4).** If mature plaque is not removed from sites that readily trap food, it can result in a carious lesion. This is due to the action of the streptococci and other bacteria that produce acid as they ferment the carbohydrates. If the acid is immediately flushed from the plaque and diluted in the mouth, it has little effect. However, in the denser regions of plaque, the acid can accumulate in direct contact with the enamel surface and lower the pH to below 5, which is acidic enough to begin to dissolve (decalcify) the calcium phosphate of the enamel in that spot. This initial lesion can remain localized in the enamel and can be repaired with various inert materials (fillings). Once the deterioration has reached the level of the dentin, tooth destruction speeds up and the tooth can be rapidly destroyed. Exposure of the pulp leads to severe tenderness and toothache, and the chance of saving the tooth is diminished.

Teeth become vulnerable to caries as soon as they appear in the mouth at around 6 months of age. Early childhood caries, defined as caries in a child between birth and 6 years of age, can extensively damage a child's primary teeth and affect the proper eruption of the permanent teeth. The practice

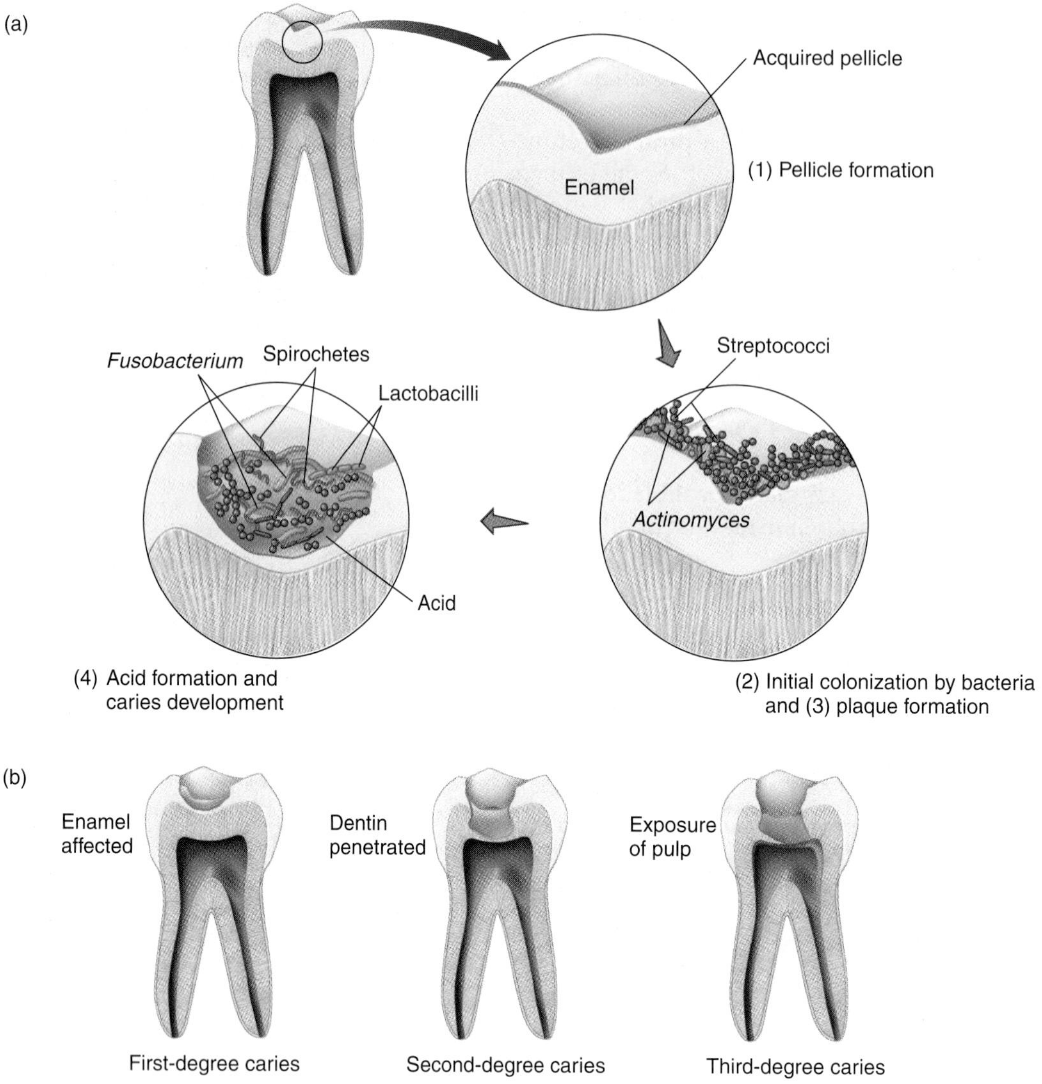

Figure 22.3 Stages in plaque development and cariogenesis.
(a) A microscopic view of pellicle and plaque formation, acidification, and destruction of tooth enamel. **(b)** Progress and degrees of cariogenesis.

Figure 22.4 The macroscopic and microscopic appearance of plaque.
(a) Disclosing tablets containing vegetable dye stain heavy plaque accumulations at the junction of the tooth and gingiva. **(b)** Scanning electron micrograph of plaque with long filamentous forms and "corn cobs" that are mixed bacterial aggregates.

of putting a baby down to nap with a bottle of fruit juice or formula can lead to rampant dental caries in the vulnerable primary dentition. This condition is called *nursing bottle caries.*

Transmission and Epidemiology

The bacteria that cause dental caries are transmitted to babies and children by their close contacts, especially the mother or closest caregiver. There is evidence for transfer of oral bacteria between children in day care centers, as well. Although it was previously believed that humans don't acquire *S. mutans* or *S. sobrinus* until the eruption of teeth in the mouth, it now seems likely that both of these species may survive in the infant's oral cavity prior to appearance of the first teeth.

Dental caries has a worldwide distribution. Its incidence varies according to many factors, including amount of carbohydrate consumption, hygiene practices, and host genetic factors. Susceptibility to caries generally decreases with age, possibly due to the fact that grooves and fissures—common sites of dental caries—tend to become more shallow as teeth are worn down. As the population ages, and natural teeth are retained for longer periods, the caries rate may well increase in the elderly, because receding gums expose the more susceptible root surfaces.

In the Western world, the 20th century saw huge increases in the overall caries incidence, probably due to increased refined sugar consumption. Since the 1970s, there has been an overall decrease in the population's caries rate, for reasons that aren't entirely clear. Improved awareness and hygiene no doubt play a role. Fluoride (in water and supplements) has also been important. Unfortunately, as the oral health of many groups is increasing, the disparity in oral health between higher and lower socioeconomic populations is growing.

Culture and Diagnosis

Dental professionals diagnose caries based on the tooth condition. Culture of the lesion is not routinely performed.

Prevention and Treatment

The best way to prevent dental caries is through dietary restriction of sucrose and other refined carbohydrates. Regular brushing and flossing to remove plaque are also important. Most municipal communities in the United States add trace amounts of fluoride to their drinking water, because fluoride, when incorporated into the tooth structure, can increase tooth (as well as bone) hardness. Fluoride can also encourage the remineralization of teeth that have begun the demineralization process. These and other proposed actions of fluoride could make teeth less susceptible to decay. Fluoride is also added to toothpastes and mouth rinses and can be applied in gel form. Many European countries do not fluoridate their water due to concerns over additives in drinking water. Also, evidence suggests that chewing sugarless gums, especially those sweetened with xylitol, can actually reduce the risk of caries. One experimental prevention strategy involves the use of probiotics. Subjects rinse their mouths twice a day with a solution containing a mix of bacteria found in healthy mouths. Early results suggest that the levels of *S. mutans*, as well as bacteria associated with periodontitis, were significantly reduced.

There are several vaccines being tested to prevent dental caries. Some utilize the proteins that bacteria use for initial attachment; others consist of the enzyme streptococci use to produce glucans. One of the more promising experimental approaches is the oral application of IgA antibody directed to bacterial attachment proteins (that is, passive immunization).

Treatment of a carious lesion involves removal of the affected part of the tooth (or the whole tooth in the case of advanced caries), followed by restoration of the tooth structure with an artificial material.

CHECKPOINT 22.1	Dental Caries
Causative Organism(s)	*Streptococcus mutans, Streptococcus sobrinus*, others
Most Common Modes of Transmission	Direct contact
Virulence Factors	Adhesion, acid production
Culture/Diagnosis	–
Prevention	Oral hygiene, fluoride supplementation
Treatment	Removal of diseased tooth material

Periodontal Diseases

Periodontal disease is so common that 97% to 100% of the population has some manifestation of it by age 45. Most kinds are due to bacterial colonization and varying degrees of inflammation that occur in response to gingival damage.

Periodontitis

Signs and Symptoms

The initial stage of periodontal disease is **gingivitis,** the signs of which are swelling, loss of normal contour, patches of redness, and increased bleeding of the gingiva. Spaces or pockets of varying depth also develop between the tooth and the gingiva. If this condition persists, a more serious disease called periodontitis results. This is the natural extension of the disease into the periodontal membrane and cementum. The deeper involvement increases the size of the pockets and can cause bone resorption severe enough to loosen the tooth in its socket. If the condition is allowed to progress, the tooth can be lost **(figure 22.5).**

Causative Agent

Dental scientists stop short of stating that particular bacteria cause periodontal disease, because not all of the criteria for establishing causation have been satisfied. In fact, dental diseases (in particular, periodontal disease) provide an excellent model of disease mediated by communities of

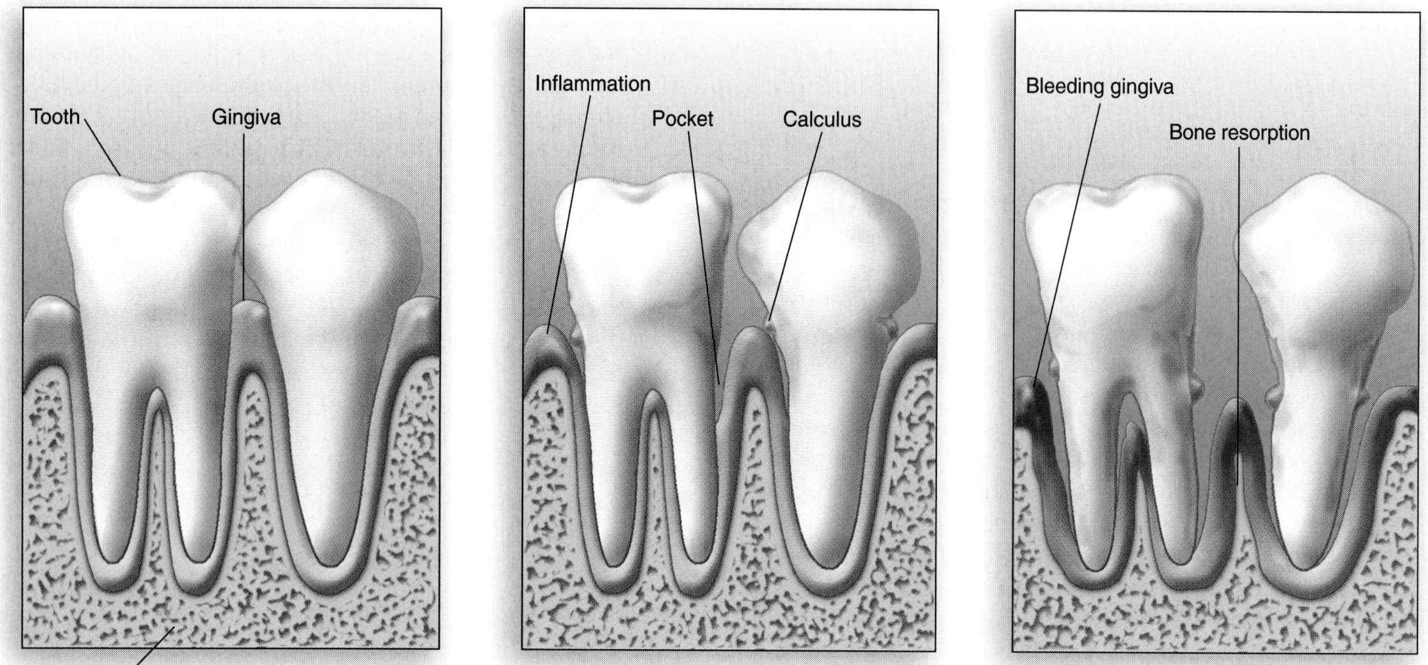

(a) Normal, nondiseased state of tooth, gingiva, and bone

(b) Calculus buildup and early gingivitis

(c) Late-stage periodontitis, with tissue destruction, deep pocket formation, loosening of teeth, and bone loss

Figure 22.5 Stages in soft-tissue infection, gingivitis, and periodontitis.

microorganisms rather than single organisms. When the polymicrobial biofilms consist of the right combination of bacteria, such as the anaerobes *Tannerella forsythus* (formerly *Bacteroides forsythus*), *Actinobacillus actinomycetemcomitans*, *Porphyromonas gingivalis*, and perhaps *Fusobacterium* and spirochete species, the periodontal destruction process begins. Evidence indicates that the presence of archaeal species in the gingival crevice is an important contributor to disease. If this is true, it will be the first link found between archaea and human disease. Scientists even suspect that aggressive versus chronic forms of periodontitis are mediated by communities that have different members or even different orders of succession. (Succession refers to the order in which microbes become part of the biofilm.) Other factors are also important in the development of periodontal disease, such as behavioral and genetic influences, as well as tooth position. The most common predisposing condition occurs when the plaque becomes mineralized (calcified) with calcium and phosphate crystals. This process produces a hard, porous substance called **calcium** above and below the gingival margin (edge) that can induce varying degrees of periodontal damage **(figure 22.6).** The presence of calculus leads to a series of inflammatory events that probably allow the bacteria to cause disease.

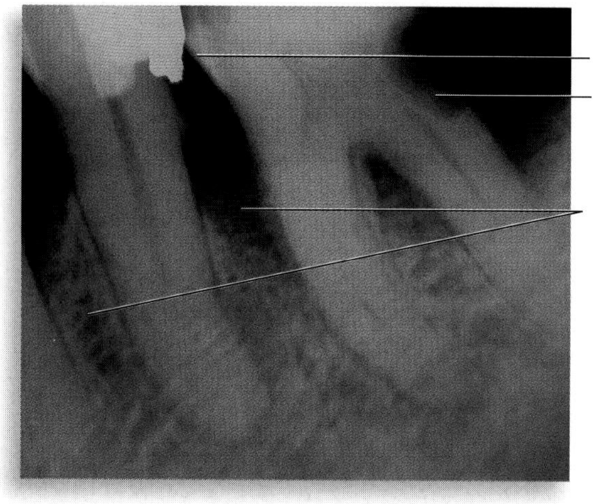

Figure 22.6 The nature of calculus.
Radiograph of mandibular premolar and molar, showing calculus on the top and a caries lesion on the right. Bony defects caused by periodontitis affect both teeth.

Pathogenesis and Virulence Factors

Calculus and plaque accumulating in the gingival sulcus cause abrasions in the delicate gingival membrane, and the chronic trauma causes a pronounced inflammatory reaction. The damaged tissues become a portal of entry for a variety of bacterial residents. The bacteria have an arsenal of enzymes, such as proteases, that destroy soft oral tissues. In response to the mixed infection, the damaged area becomes infiltrated by neutrophils and macrophages and, later, by lymphocytes, which cause further inflammation and tissue damage. There is some evidence to suggest that people with high numbers of the bacteria associated with periodontitis also have thicker carotid arteries, a risk factor for stroke and heart disease.

Transmission and Epidemiology

As with caries, the resident oral bacteria, acquired from close oral contact, are responsible for periodontal disease. Dentists refer to a wide range of risk factors associated with periodontal disease, especially deficient oral hygiene. But because it is so common in the population, it is evident that most of us could use some improvement in our oral hygiene.

Culture and Diagnosis

Like caries, periodontitis is generally diagnosed by the appearance of the oral tissues.

Prevention and Treatment

Regular brushing and flossing to remove plaque automatically reduce both caries and calculus production. Mouthwashes are relatively ineffective in controlling plaque formation because of the high bacterial content of saliva and the relatively short-acting time of the mouthwash. Once calculus has formed on teeth, it cannot be removed by brushing but can be dislodged only by special mechanical procedures (scaling) in the dental office.

Most periodontal disease is treated by removal of calculus and plaque and maintenance of good oral hygiene. Often, surgery to reduce the depth of periodontal pockets is required. Antibiotic therapy, either systemic or applied in periodontal packings, may also be utilized. There is some evidence that exposing the periodontium to blue light (similar to that used to whiten teeth) can selectively kill disease-causing anaerobes while leaving normal biota intact.

Necrotizing Ulcerative Gingivitis and Periodontitis

The most destructive periodontal diseases are necrotizing ulcerative gingivitis (NUG) and necrotizing ulcerative periodontitis (NUP). The two diseases were formerly lumped under one name, acute necrotizing ulcerative gingivitis, or ANUG. These diseases are synergistic infections involving *Treponema vincentii, Prevotella intermedia,* and *Fusobacterium* species. These pathogens together produce several invasive factors that cause rapid advancement into the periodontal tissues. The condition is associated with severe pain, bleeding, pseudomembrane formation, and necrosis. Scientists believe that NUP may be an extension of NUG, but the conditions can be distinguished by the advanced bone destruction that results from NUP. Both diseases seem to result from poor oral hygiene, altered host defenses, or prior gum disease rather than being communicable. The diseases are common in AIDS patients and other immunocompromised populations. Diabetes and cigarette smoking can predispose people to these conditions. NUG and NUP usually respond well to broad-spectrum antibiotics, after debridement of damaged periodontal tissue **(Checkpoint 22.2).**

Mumps

The word *mumps* is Old English for lump or bump. The symptoms of this viral disease are so distinctive that Hippocrates clearly characterized it in the fifth century BC as a self-limited, mildly epidemic illness associated with painful swelling at the angle of the jaw **(figure 22.7).**

Signs and Symptoms

After an average incubation period of 2 to 3 weeks, symptoms of fever, nasal discharge, muscle pain, and malaise develop. These may be followed by inflammation of the salivary glands (especially the parotids), producing the classic gopherlike swelling of the cheeks on one or both sides (see figure 22.7). Swelling of the gland is called parotitis, and it can cause considerable discomfort. Viral multiplication in salivary glands is followed by invasion of other organs, especially the testes, ovaries, thyroid gland, pancreas, meninges, heart, and kidney. Despite the invasion of multiple organs, the prognosis of most infections is complete, uncomplicated recovery with permanent immunity.

CHECKPOINT 22.2 Periodontal Diseases

Disease	Periodontitis	Necrotizing Ulcerative Gingivitis and Periodontitis
Causative Organism(s)	Polymicrobial community including some or all of: *Tannerella forsythus, Actinobacillus actinomycetemcomitans, Porphyromonas gingivalis,* others?	Polymicrobial community (*Treponema vincentii, Prevotella intermedia, Fusobacterium* species)
Most Common Modes of Transmission	–	–
Virulence Factors	Induction of inflammation, enzymatic destruction of tissues	Inflammation, invasiveness
Culture/Diagnosis	–	–
Prevention	Oral hygiene	Oral hygiene
Treatment	Removal of plaque and calculus, gum reconstruction, tetracycline	Debridement of damaged tissue, metronidazole, clindamycin

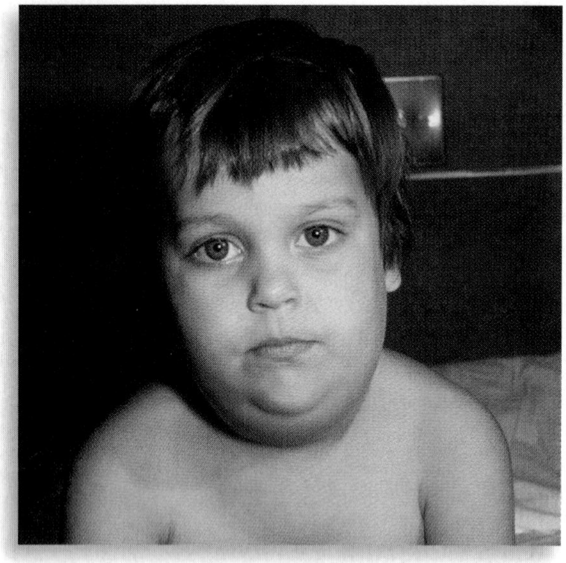

Figure 22.7 The external appearance of swollen parotid glands in mumps (parotitis).

Complications in Mumps In 20% to 30% of young adult males, mumps infection localizes in the epididymis and testis, usually on one side only. The resultant syndrome of orchitis and epididymitis may be rather painful, but no permanent damage usually occurs. The popular belief that mumps readily causes sterilization of adult males is still held, despite medical evidence to the contrary. Perhaps this notion has been reinforced by the tenderness that continues long after infection and by the partial atrophy of one testis that occurs in about half the cases. Permanent sterility due to mumps is very rare.

In mumps pancreatitis, the virus replicates in beta cells and pancreatic epithelial cells. Viral meningitis, characterized by fever, headache, and stiff neck, appears 2 to 10 days after the onset of parotitis, lasts for 3 to 5 days, and then dissipates, leaving few or no adverse side effects. Another rare event is infection of the inner ear that can lead to deafness.

Causative Agent

Mumps is caused by an enveloped single-stranded RNA virus (mumps virus) from the genus *Paramyxovirus,* which is part of the family Paramyxoviridae. Other members of this family that infect humans are *Morbillivirus* (measles virus) and the respiratory syncytial virus. The envelopes of paramyxoviruses possess spikes that have specific functions.

Pathogenesis and Virulence Factors

A virus-infected cell is modified by the insertion of the HN spikes into its cell membrane. The HN spikes immediately bind an uninfected neighboring cell, and in the presence of F spikes, the two cells permanently fuse. A chain reaction of multiple cell fusions then produces a *syncytium* (sin-sish′-yum) with cytoplasmic inclusion bodies, which is a diagnostically useful cytopathic effect **(figure 22.8).** The ability to induce the formation of syncytia is characteristic of the family Paramyxoviridae.

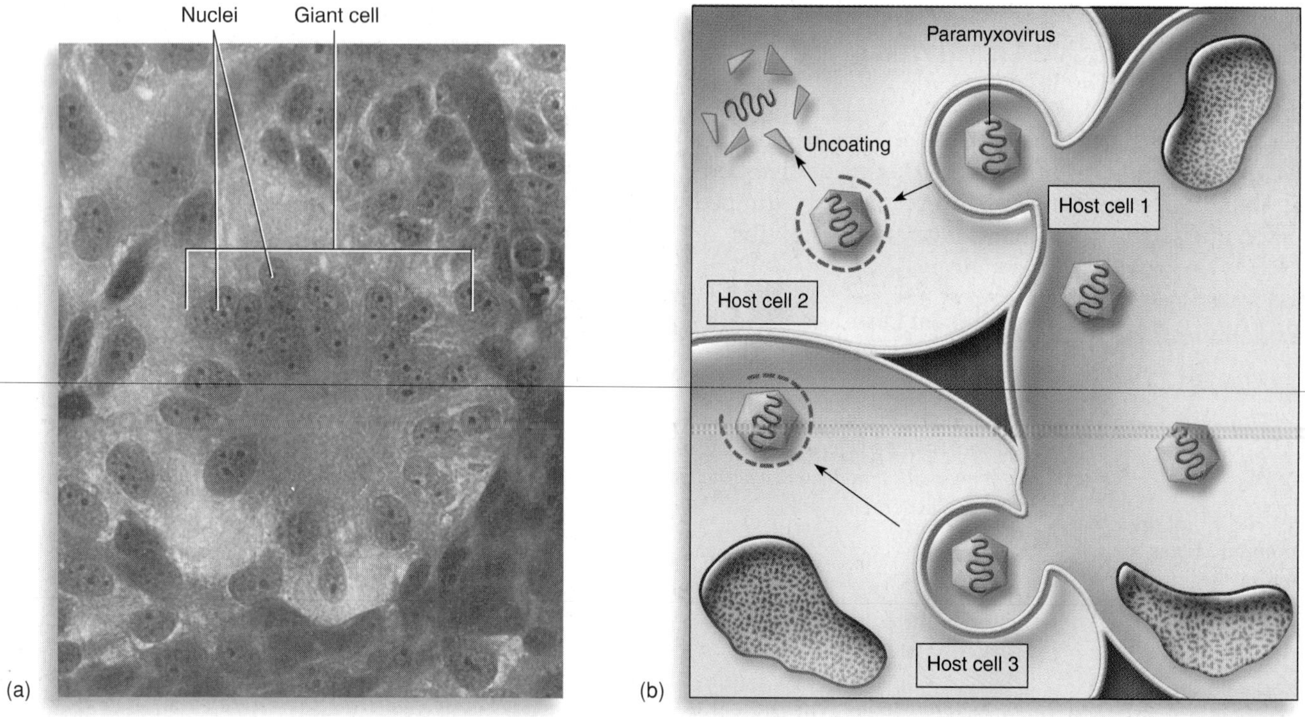

Figure 22.8 The effects of paramyxoviruses.
(a) When they infect a host cell, paramyxoviruses induce the cell membranes of adjacent cells to fuse into large multinucleate giant cells, or syncytia. **(b)** This fusion allows direct passage of viruses from an infected cell to uninfected cells by communicating membranes. Through this means, the virus evades antibodies.

Transmission and Epidemiology of Mumps Virus

Humans are the exclusive natural hosts for the mumps virus. It is communicated primarily through salivary and respiratory secretions. Infection occurs worldwide, with increases in the late winter and early spring in temperate climates. High rates of infection arise among crowded populations or communities with poor herd immunity. Most cases occur in children under the age of 15, and as many as 40% are subclinical. Because lasting immunity follows any form of mumps infection, no long-term carrier reservoir exists in the population. The incidence of mumps had been reduced in the United States to around 300 cases per year. In the first four months of 2006, the United States saw 2,600 cases of mumps, resulting in the recommendation that all children and health care workers receive two doses of the vaccine (see "Prevention and Treatment").

Culture and Diagnosis

Diagnosis is usually based on the clinical sign of swollen parotid glands and known exposure 2 or 3 weeks previously. Because parotitis is not always present, and the incubation period can range from 7 to 23 days, a practical diagnostic alternative is to perform a direct fluorescent test for viral antigen or an ELISA test on a patient's serum.

Prevention and Treatment

The general pathology of mumps is mild enough that symptomatic treatment to relieve fever, dehydration, and pain is usually adequate. The new vaccine recommendations call for a dose of MMR at 12 to 15 months and a second dose at 4 to 6 years. Health care workers and college students who haven't already had both doses are advised to do so.

CHECKPOINT 22.3	Mumps
Causative Organism(s)	Mumps virus (genus *Paramyxovirus*)
Most Common Modes of Transmission	Droplet contact
Virulence Factors	Spike-induced syncytium formation
Culture/Diagnosis	Clinical, fluorescent Ag tests, ELISA for Ab
Prevention	MMR live attenuated vaccine
Treatment	Supportive

Gastritis and Gastric Ulcers

Although the human stomach has been regarded as a hostile habitat for microorganisms, an unusual vibrio, *Helicobacter pylori*, has found its own special niche there. Not only does it thrive in the acidic environment, but evidence has also clearly linked it to a variety of gastrointestinal ailments.

The curved cells of *Helicobacter* were first detected by J. Robin Warren in 1979 in stomach biopsies from ulcer patients. He and an assistant, Barry J. Marshall, isolated the microbe in culture and even served as guinea pigs by swallowing a good-size inoculum to test its effects. Both developed transient gastritis.

Signs and Symptoms

Gastritis is experienced as sharp or burning pain emanating from the abdomen. Gastric ulcers are actual lesions in the mucosa of the stomach (gastric ulcers) or in the uppermost portion of the small intestine (duodenal ulcer). Both of these conditions are also called *peptic ulcers.* Severe ulcers can be accompanied by bloody stools, vomiting, or both. The symptoms are often worse at night, after eating, or under conditions of psychological stress.

The second most common cancer in the world is stomach cancer (although it has been declining in the United States), and ample evidence suggests that long-term infection with *H. pylori* is a major contributing factor.

Causative Agent

Helicobacter pylori is a curved gram-negative rod, closely related to *Campylobacter,* which we study later in this chapter.

Pathogenesis and Virulence Factors

Once the bacterium passes into the gastrointestinal tract, it bores through the outermost mucous layer that lines the stomach epithelial tissue. Then it attaches to specific binding sites on the cells and entrenches itself. One receptor specific for *Helicobacter* is the same molecule on human cells that confers the O blood type. This finding accounts for the higher rate of ulcers in people with this blood type. Another protective adaptation of the bacterium is the formation of urease, an enzyme that converts urea into ammonium and bicarbonate, both alkaline compounds that can neutralize stomach acid. As the immune system recognizes and attacks the pathogen, infiltrating white blood cells damage the epithelium to some degree, leading to chronic active gastritis. In some people, these lesions lead to deeper erosions and ulcers that can lay the groundwork for cancer to develop.

Before the bacterium was discovered, spicy foods, high-sugar diets (which increase acid levels in the stomach), and psychological stress were considered to be the cause of gastritis and ulcers. Now it appears that these factors merely aggravate the underlying infection.

Transmission and Epidemiology

The mode of transmission of this bacterium remains a mystery. Studies have revealed that the pathogen is present in a large proportion of the human population. It occurs in the stomachs of 25% of healthy middle-age adults and in more than 60% of adults over 60 years of age. *H. pylori* is probably transmitted from person to person by the oral-oral or fecal-oral route. It seems to be acquired early in life and carried asymptomatically until its activities begin to damage the digestive mucosa. Because other animals are also susceptible to *H. pylori* and even develop chronic gastritis, it has been proposed that the disease is a zoonosis transmitted from an animal reservoir. The bacterium has also been found in water sources.

Approximately two-thirds of the world's population are infected with *H. pylori*. It is not known what causes some people to experience symptoms, although it is most likely that those with the right combination of aggravating factors are those who experience disease.

Culture and Diagnosis

Diagnosis has typically been accomplished with endoscopy, a procedure in which a long flexible tube **(figure 22.9)** is inserted through the throat into the stomach to visualize any lesions there. The urea breath test is sometimes used. In this test, patients ingest urea that has a radioactive tag on its carbon molecule. If *Helicobacter* is present in a patient's stomach, the bacterium's urease breaks down the urea and the patient exhales radioactively labeled carbon dioxide. In the absence of urease, the intact urea molecule passes through the digestive system. Patients whose breath is positive for the radioactive carbon are considered positive for *Helicobacter*.

A blood test is also available that uses ELISA technology to find antibodies to *H. pylori*. Because many people are asymptomatically infected with *H. pylori*, it is sometimes useful to perform the endoscopic procedure to look for pathology in the stomach.

Prevention and Treatment

The only preventive approaches available currently are those that diminish some of the aggravating factors just mentioned. Limiting spicy foods and decreasing the sugar content of the diet might reduce the risk for overt disease. Many over-the-counter remedies offer symptom relief; most of them act to neutralize stomach acid. The best treatment is a course of antibiotics augmented by acid suppressors. The antibiotics most prescribed are clarithromycin or metronidazole. Bismuth subsalicylate (Pepto-Bismol) or the prescription medication omeprazole is the most frequently administered acid suppressor.

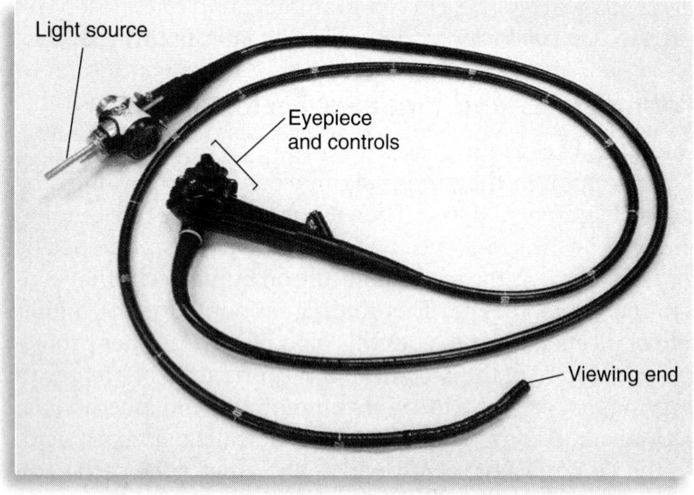

(a)

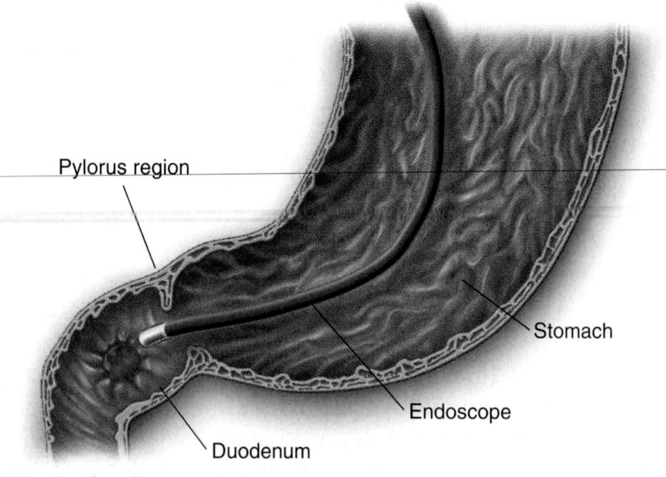

(b)

Figure 22.9 Endoscopy.
(a) A flexible tube is inserted through the mouth into the stomach, **(b)** acting as a camera to visualize the stomach surface.

CHECKPOINT 22.4	Gastritis and Gastric Ulcers
Causative Organism(s)	*Helicobacter pylori*
Most Common Modes of Transmission	?
Virulence Factors	Adhesions, urease
Culture/Diagnosis	ELISA, endoscopy
Prevention	None
Treatment	Antibiotics plus acid suppressors (clarithromycin or metronidazole plus omeprazole or bismuth subsalicylate)

Acute Diarrhea

Diarrhea needs little explanation. In recent years, on average, citizens of the United States experienced 1.2 to 1.9 cases of diarrhea per person per year, and among children that number is twice as high. The incidence of diarrhea is even higher among children attending day care centers. In tropical countries, children may experience more than 10 episodes of diarrhea a year. In fact, more than 3 million children a year, mostly in developing countries, die from a diarrheal disease (see Insight 22.2). In developing countries, the high mortality rate is not the only issue. Children who survive dozens of bouts with diarrhea during their developmental years are likely to have permanent physical and cognitive effects. The effect on the overall well-being of these children is hard to estimate, but it is very significant.

In the United States, up to a third of all acute diarrhea is transmitted by contaminated food (a case of diarrhea is usually defined as three or more loose stools in a 24-hour period). In recent years, consumers have become much more aware of the possibility of *E. coli*–contaminated hamburgers or *Salmonella*-contaminated ice cream. New food safety measures are being implemented all the time, but it is still necessary for the consumer to be aware and to practice good food handling. As just mentioned, the increased use of day care centers has also led to increased transmission of diarrheal agents.

For a disease that exacts such a high price, there is relatively little consensus on how to manage a patient with acute diarrhea. Although most diarrhea episodes are self-limiting and therefore do not require treatment, others (such as *E. coli* O157:H7) can have devastating effects. In most diarrheal illnesses, antimicrobial treatment is contraindicated (inadvisable), but some, such as shigellosis, call for quick treatment with antibiotics. For public health reasons, it is important to know which agents are causing diarrhea in the community, but in most cases identification of the agent is not performed.

In this section, we describe acute diarrhea having infectious agents as the cause. In the sections following this one, we discuss acute diarrhea and vomiting caused by toxins, commonly known as food poisoning, and chronic diarrhea and its causes.

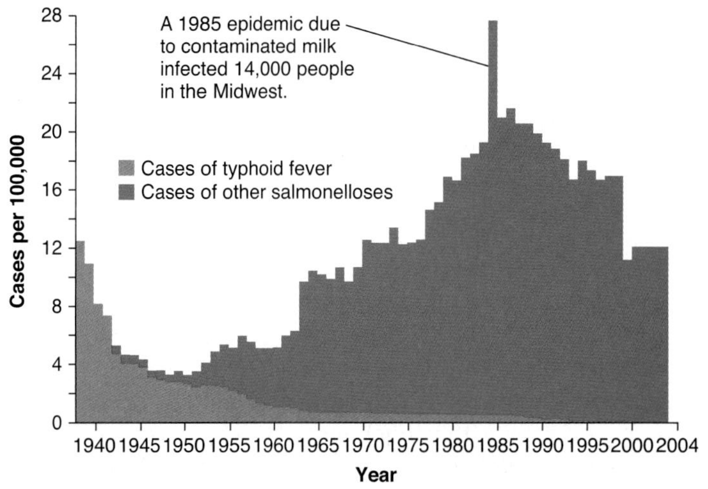

Figure 22.10 Data on the prevalence of typhoid fever and other salmonelloses from 1940 to 2004. Nontyphoidal salmonelloses did occur before 1940, but the statistics are not available.

Salmonella

A decade ago, one of every three chickens destined for human consumption was contaminated with *Salmonella,* but the rate is now about 10%. Other poultry, such as ducks and turkeys, is also affected. Eggs are infected as well because the bacteria may actually enter the egg while the shell is being formed in the chicken. In 2007, peanut butter was found to be the source of a *Salmonella* outbreak in the United States. *Salmonella* is a very large genus of bacteria, but only one species is of interest to us: *S. enterica* is divided into many variant, based on variation in the major surface antigens.

As mentioned in chapter 4, serotype or variant analysis aids in bacterial identification. Many gram-negative enteric bacteria are named and designated according to the following antigens: **H,** the flagellar antigen; **K,** the capsular antigen; and **O,** the cell wall antigen. Not all enteric bacteria carry the H and K antigens, but all have O, the polysaccharide portion of the lipopolysaccharide implicated in endotoxic shock (see chapter 20). Most species of gram-negative enterics exhibit a variety of subspecies, variant, or serotypes caused by slight variations in the chemical structure of the HKO antigens. Some bacteria in this chapter (for example, *E. coli* O157:H7) are named according to their surface antigens; however, we will use Latin variant names for *Salmonella.*

Salmonellae are motile; they ferment glucose with acid and sometimes gas; and most of them produce hydrogen sulfide (H_2S) but not urease. They grow readily on most laboratory media and can survive outside the host in inhospitable environments such as fresh water and freezing temperatures. These pathogens are resistant to chemicals such as bile and dyes, which are the basis for isolation on selective media.

Signs and Symptoms The genus *Salmonella* causes a variety of illnesses in the GI tract and beyond. Until fairly recently, its most severe manifestation was typhoid fever, which is discussed shortly. Since the mid-1900s, a milder disease usually called *salmonellosis* has been much more common **(figure 22.10).** Sometimes the condition is also called enteric fever or gastroenteritis. Whereas typhoid fever is caused by the *typhi* variant, gastroenteritises are generally caused by the variant known as *paratyphi, hirschfeldii,* and *typhimurium.* Another variant, which is sometimes called *Arizona hinshawii* (even though it is still a *Salmonella*) is a pathogen found in the intestines of reptiles. Most of these strains come from animals, unlike the typhi strain, which infects humans exclusively. *Salmonella* bacteria are normal intestinal biota in cattle, poultry, rodents, and reptiles.

Salmonellosis can be relatively severe, with an elevated body temperature and septicemia as more prominent features than GI tract disturbance. But it can also be fairly mild, with gastroenteritis—vomiting, diarrhea, and mucosal irritation—as its major feature. Blood can appear in the stool. In otherwise healthy adults, symptoms spontaneously subside after 2 to 5 days; death is infrequent except in debilitated persons.

Typhoid fever is so named because it bears a superficial resemblance to typhus, a rickettsial disease, even though the two diseases are otherwise very different. In the United States, the incidence of typhoid fever has remained at a steady rate for the last 30 years, appearing sporadically (see figure 22.10). Of the 50 to 100 cases reported annually, roughly half are imported from endemic regions. In other parts of the world, typhoid fever is still a serious health problem, responsible for 25,000 deaths each year and probably millions of cases.

Typhoid fever, caused by the typhi variant of *S. enterica,* is characterized by a progressive, invasive infection that leads eventually to septicemia. Symptoms are fever, diarrhea, and abdominal pain. The bacterium infiltrates the mesenteric lymph nodes and the phagocytes of the liver and spleen. In some people, the small intestine develops areas of ulceration that are vulnerable to hemorrhage, perforation, and peritonitis. Its presence in the circulatory system may lead to nodules or abscesses in the liver or urinary tract.

Because it is so rare compared with the less severe salmonellosis, the rest of this section refers mainly to salmonellosis and not to typhoid fever.

Pathogenesis and Virulence Factors The ability of *Salmonella* to cause disease seems to be highly dependent on its ability to adhere effectively to the gut mucosa. Recent research has uncovered an "island" of genes in *Salmonella* that seems to confer virulence on the bacterium. There are other pathogenicity islands, but this one is directly related to attachment. This island was discovered when those genes were inactivated, and the bacterium was no longer capable of causing disease in an experimental model. Researchers weren't sure what the functions of those genes were, but when they *injected* the inactivated strain or the wild-type strain into experimental animals rather than transmitting the bacteria orally, both were still capable of causing disease. This result indicated to the researchers that the "virulence genes" were most important for entry, adhesion, or invasion into the host. It is also believed that endotoxin is an important virulence factor for *Salmonella.*

Transmission and Epidemiology Animal products such as meat and milk can be readily contaminated with *Salmonella* during slaughter, collection, and processing. Inherent risks are involved in eating poorly cooked beef or unpasteurized fresh or dried milk, ice cream, and cheese. A 2001 U.S. outbreak was traced to green grapes. A particular concern is the contamination of foods by rodent feces. Several outbreaks of infection have been traced to unclean food storage or to food-processing plants infested with rats and mice.

Most cases are traceable to a common food source such as milk or eggs. Some cases may be due to poor sanitation. In one outbreak, about 60 people became infected after visiting the Komodo dragon exhibit at the Denver zoo. They picked up the infection by handling the rails and fence of the dragon's cage. In 2002, two people apparently acquired salmonellosis from a blood transfusion, and one of them died. The blood donor, who had an asymptomatic infection with *Salmonella,* had contracted the infection from his pet snake.

In recent years, many cancer patients and HIV-positive people have become deathly ill from ingesting a folk remedy called "rattlesnake pill," sometimes known as Pulvo de Vibora. It is particularly common in California, the Southwest, and in Mexico. The CDC has investigated these pills and found that they can contain the *Arizona hinshawii* variant of *Salmonella* (found in the intestines of reptiles) as well as many other pathogens. Extreme care should be taken by immunocompromised persons when considering such unlicensed alternative therapies, and health care professionals should be alert to the possibility of such exposures when they see unusual infections in patients.

Prevention and Treatment The only prevention for salmonellosis is avoiding contact with the bacterium. In 1998, a vaccine was approved for use in poultry, making it the first "food safety" vaccine. A vaccine for humans is undergoing testing as well.

Uncomplicated cases of salmonellosis are treated with fluid and electrolyte replacement; if the patient has underlying immunocompromise or if the disease is severe, trimethoprim-sulfamethoxazole is recommended.

Typhoid fever, by contrast, is always treated with antibiotics, in part to clear the patient of the typhi strain, which has a tendency to be shed for weeks after recovery. A small number of people chronically carry the bacterium for longer periods in the gallbladder; from this site, the bacteria are constantly released into the intestine and feces. In some people, gallbladder removal is necessary to stop the shedding. Two vaccines are available for the typhi strain and are recommended for people traveling to endemic areas.

Shigella

The *Shigella* bacteria are gram-negative straight rods, nonmotile and non-spore-forming. They do not produce urease or hydrogen sulfide, traits that help in their identification. They are primarily human parasites, though they can infect apes. All produce a similar disease that can vary in intensity. These bacteria resemble some types of pathogenic *E. coli* very closely. Diagnosis is complicated by the fact that several alternative candidates can cause bloody diarrhea, such as *E. coli* and others. Isolation and identification follow the usual protocols for enterics. Stool culture is still the gold standard for identification in the case of *Shigella* infections **(Insight 22.1).**

Although *Shigella dysenteriae* causes the most severe form of dysentery, it is uncommon in the United States and occurs primarily in the Eastern Hemisphere. In the past decade, the prevalent agents in the United States have been *Shigella sonnei* and *Shigella flexneri,* which cause approximately 20,000 to 25,000 cases each year, half of them in children.

Signs and Symptoms The symptoms of shigellosis include frequent, watery stools, as well as fever, and often intense abdominal pain. Nausea and vomiting are common. Stools often contain obvious blood and even more often are found to have occult (not visible to the naked eye) blood. Diarrhea containing blood is also called **dysentery.** Mucus from the GI tract will also be present in the stools.

Pathogenesis and Virulence Factors Shigellosis is different from many GI tract infections in that *Shigella* invades the villus cells of the large intestine rather than the small intestine. In addition, it is not as invasive as *Salmonella* and does not perforate the intestine or invade the blood. It enters the intestinal mucosa by means of lymphoid cells in Peyer's patches. Once in the mucosa, *Shigella* instigates an inflammatory response that causes extensive tissue destruction. The release of endotoxin causes fever. **Enterotoxin,** an exotoxin that affects the enteric (or GI) tract, damages the mucosa and villi. Local areas of erosion give rise to bleeding and heavy secretion of mucus **(figure 22.11).** *Shigella dysenteriae* (and

INSIGHT 22.1 Medical

Stools: To Culture or Not to Culture?

The practice of diagnosing GI tract infections is really at a crossroads in the early 21st century. For decades, clinical microbiologists have relied on stool cultures complemented with a battery of biochemical tests to try to tease out the single pathogenic bacterium among the multitude of normal strains that reside in the intestinal tract. Now many physicians feel that stool cultures are not necessary except in certain circumstances. Indeed, some studies show that as few as 2% of routinely ordered stool cultures come back positive for anything. When we consider that some of these cultures can cost as much as $1,000, it is easy to see their point.

It seems that the best guideline to use is this: *Will the results of the culture change the therapy?* Physicians generally agree that when fever is present, when there is blood in the stools or pain suggesting appendicitis, or if the patient gives a history that suggests possible exposure to *E. coli* O157:H7, stool cultures should be ordered. In other cases, physicians should use a variety of other indicators, both clinical and epidemiological, to diagnose diarrhea and other gastrointestinal disorders. Newer technologies that can test for specific pathogens without culturing, such as ELISA and PCR tests, may eventually make costly and slow culture techniques obsolete.

perhaps some of the other species) produces a heat-labile exotoxin called **shiga toxin,** which seems to be responsible for the more serious damage to the intestine as well as any systemic effects, including injury to nerve cells. It is an A-B toxin (see figure 21.10). To review, the B portion of the toxin attaches to host cells, and the whole toxin is internalized. Once inside, the A portion of the toxin exerts its effect. In the case of the shiga toxin, the A portion of the toxin binds to ribosomes, interrupting protein synthesis and leading to the damage just described. You'll encounter shiga toxin again when we discuss *E. coli* O157:H7.

Transmission and Epidemiology In addition to the usual oral route, shigellosis is also acquired through direct person-to-person contact, largely because of the small infectious dose required (from 10 to 200 bacteria). The disease is mostly associated with lax sanitation, malnutrition, and crowding; and it is spread epidemically in day care centers, prisons, mental institutions, nursing homes, and military camps. *Shigella* was responsible for some cruise ship outbreaks in the mid-1990s (later cruise ship outbreaks were caused by viruses). As in other enteric infections, *Shigella* can establish a chronic carrier condition in some people that lasts several months.

Prevention and Treatment The only prevention of this and most other diarrheal diseases is good hygiene and avoiding contact with infected persons. Although some experts say that bloody diarrhea in this country should not be treated with antibiotics (which is generally accepted for *E. coli* O157:H7 infections), most physicians recommend prompt treatment of shigellosis with trimethoprim-sulfamethoxazole (TMP-SMZ).

E. coli O157:H7 *(EHEC)*

In January of 1993, this awkwardly named bacterium burst into the public's consciousness when three children died after eating undercooked hamburgers at a fast-food restaurant in Washington State. The cause of their illness was determined to be this particular strain of *E. coli,* which had actually been recognized since the 1980s. Since then, it has led to approximately 73,000 illnesses and about 50 deaths each year in the United States. It is considered an emerging pathogen.

Dozens of different strains of *E. coli* exist, many of which cause no disease at all. A handful of them cause various degrees of intestinal symptoms as described in this and the following section. Some of them cause urinary tract infections (see chapter 23). *E. coli* O157:H7 and its close relatives are the most virulent of them all. The group of *E. coli* of which this strain is the most famous representative is generally referred to as **enterohemorrhagic *E. coli*,** or **EHEC.**

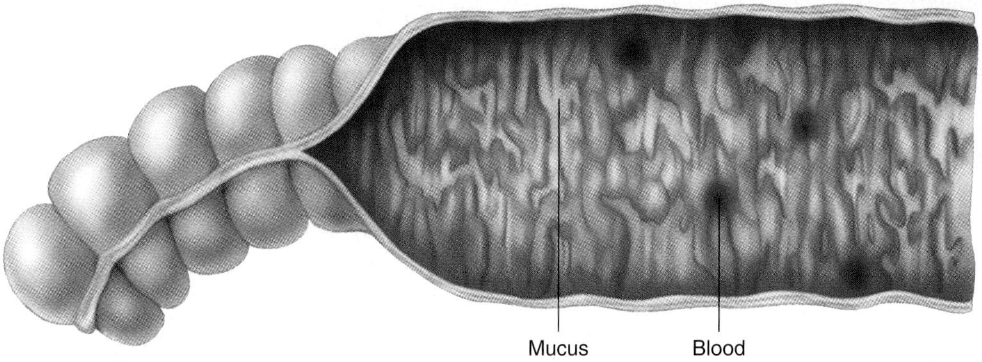

Figure 22.11 The appearance of the large intestinal mucosa in *Shigella* dysentery. Note the patches of blood and mucus, the erosion of the lining, and the absence of perforation.

Signs and Symptoms *E. coli* O157:H7 is the agent of a spectrum of conditions, ranging from mild gastroenteritis with fever to bloody diarrhea. About 10% of patients develop **hemolytic uremic syndrome (HUS),** a severe hemolytic anemia that can cause kidney damage and failure. Neurological symptoms such as blindness, seizure, and stroke (and long-term debilitation) are also possible. These serious manifestations are most likely to occur in children younger than 5 and in elderly people.

Pathogenesis and Virulence Factors This bacterium owes much of its virulence to shiga toxins (so named because they are identical to the shiga exotoxin secreted by virulent *Shigella* species). Sometimes this *E. coli* is referred to as STEC (shiga-toxin-producing *E. coli*). For simplicity, EHEC is used here. The shiga toxin genes are present on bacteriophage in *E. coli* but are on the chromosome of *Shigella dysenteriae*, suggesting that the *E. coli* acquired the virulence factor through phage-mediated transfer. As described earlier for *Shigella*, the shiga toxin interrupts protein synthesis in its target cells. It seems to be responsible especially for the systemic effects of this infection.

Another important virulence determinant for EHEC is the ability to efface (rub out or destroy) enterocytes, which are gut epithelial cells. This is accomplished with a set of bacterial proteins, one of which is called *intimin*—used for "intimate" attachment to host cells. Another set of proteins enables the bacterium to construct a complex bridging system between *E. coli* and host cell membranes, which allows *E. coli* to insert its products into the host cell. This system, displayed by many types of bacteria, is called the *Type III secretion system* **(figure 22.12).**

The bacterium also produces a set of proteins that are actually passed through the apparatus in figure 22.12, including the protein that does the damage to host cells. The most startling discovery, however, has been that one of the products sent through the Type III "pipeline" was a protein that the bacterium inserts into the host cell membrane so that it will become a receptor for the bacterial intimin protein. Essentially, the bacterium is sending over the lock into which it can insert its key—ensuring a very tight bond indeed.

The net effect of the action of these products is a lesion in the gut (effacement), usually in the large intestine. The microvilli are lost from the gut epithelium, and the lesions produce bloody diarrhea.

Transmission and Epidemiology The most common mode of transmission for EHEC is the ingestion of contaminated and undercooked beef, although other foods and beverages can be contaminated as well. Ground beef is more dangerous than steaks or other cuts of meat, for several reasons. Consider the way that the beef becomes contaminated in the first place. The bacterium is a natural inhabitant of the GI tracts of cattle. Contamination occurs when intestinal contents contact the animal carcass, so bacteria are confined to the surface of meats. Because high heat destroys this bacterium, even a brief trip under the broiler is usually sufficient to kill *E. coli* on the surface of steaks or roasts. But in ground beef, the "surface" of meat is mixed and ground up throughout a batch, meaning any bacteria are mixed in also. This mixing explains why hamburgers should be cooked all the way through. Hamburger is also a common vehicle because meat processing plants tend to grind meats from several cattle sources together, thereby contaminating large amounts of hamburger with meat from one animal carrier.

Other farm products may also become contaminated by cattle feces. Products that are eaten raw, such as lettuce, vegetables, and apples used in unpasteurized cider, are particularly problematic. The disease can also be spread via the fecal-oral route of transmission, especially among young children in group situations. Even touching surfaces contaminated with cattle feces can cause disease, since ingesting as few as 10 organisms has been found to be sufficient to initiate this disease.

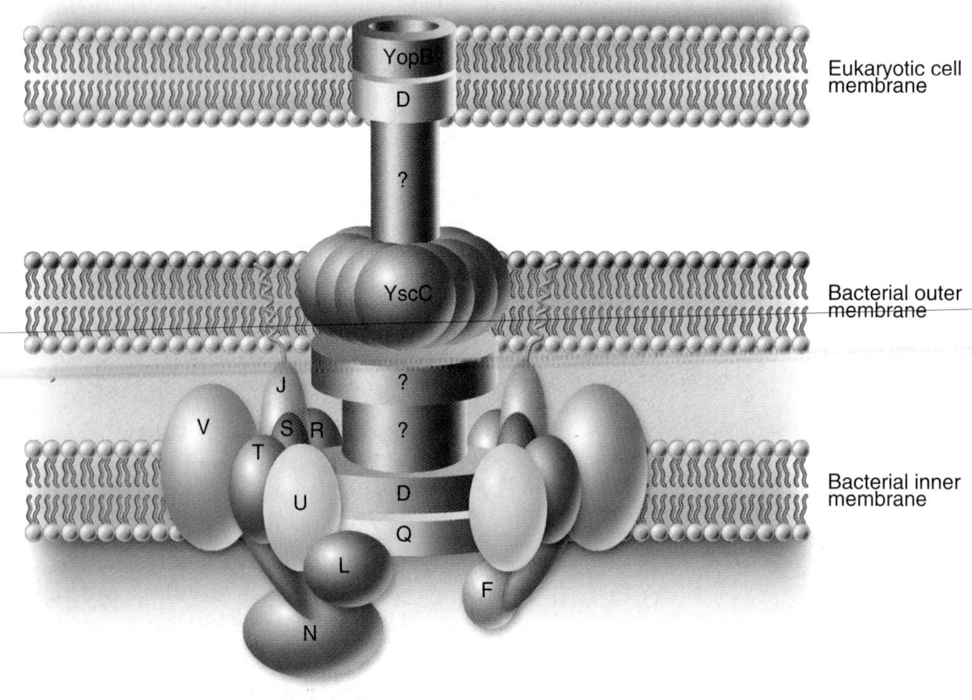

Figure 22.12 Type III secretion system.
This multiprotein "pipeline" is situated in the cytoplasmic and outer membranes of *E. coli*. Each of the differently shaped objects is a different protein.

Culture and Diagnosis Infection with this type of *E. coli* should be confirmed with stool culture or with newer techniques such as ELISA or PCR.

Prevention and Treatment The best prevention for this disease is never to eat raw or even rare hamburger. The shiga toxin is heat-labile and the *E. coli* is killed by heat as well. If you are thinking "I used to be able to eat rare hamburgers," you are correct, but things have changed. The emergence of this pathogen in the early 1980s, probably resulting from a regular *E. coli* picking up the shiga toxin from *Shigella*, has changed the rules.

No vaccine exists for *E. coli* O157:H7. A great deal of research is directed at vaccinating livestock to break the chain of transmission to humans.

Antibiotics are contraindicated for this infection. Even with severe disease manifestations, antibiotics have been found to be of no help, and they may increase the pathology. It is also recommended that antimotility drugs (to limit the diarrhea) not be used. Supportive therapy is the only option.

Other *E. coli*

At least four other categories of *E. coli* can cause diarrheal diseases. Scientists call these **enterotoxigenic** *E. coli,* **enteroinvasive** *E. coli,* **enteropathogenic** *E. coli,* and **enteroaggregative** *E. coli.* In clinical practice, most physicians are interested in differentiating shiga-toxin-producing *E. coli* (EHEC) from all the others. Each of these is considered separately and briefly here; in Checkpoint 22.5, the non-shiga-toxin-producing *E. coli* are grouped together in one column.

Enterotoxigenic *E. coli* (ETEC) The presentation varies depending on which type of *E. coli* is causing the disease. **Traveler's diarrhea,** characterized by watery diarrhea, low-grade fever, nausea, and vomiting, is usually caused by enterotoxigenic *E. coli* (ETEC). These strains also cause a great deal of illness in infants in developing countries.

The bacterium is transmitted through the fecal-oral route or via contaminated vehicles or even fomites (such as a dirty glass). Travelers are susceptible to these strains because they are likely to be new to their immune systems. People living in endemic areas probably encounter the bacteria as infants. As the name suggests, the virulence of the bacterium derives from its ability to secrete two types of exotoxins that act on the enteric tract (enterotoxin). One toxin is a heat-labile A-B toxin, and it acts like the cholera toxin, described later. Another toxin, actually a group of toxins, is heat-stable. These toxins are very small proteins that alter host cell function in order to cause large amounts of fluid secretion into the intestinal tract. The bacterium mainly affects the small intestine.

Most infections with ETEC are self-limiting, however miserable they make you feel. They are treated only with fluid replacement. In infants, ETEC can be life-threatening, and fluid replacement is vital to survival.

Enteroinvasive *E. coli* (EIEC) These strains cause a disease that is very similar to *Shigella* dysentery. The bacteria invade gut mucosa and cause widespread destruction. Blood and pus will be found in the stool. Significant fever is often present. EIEC does not produce the heat-labile or heat-stable exotoxins just described and does not have a shiga toxin, despite the clinical similarity to *Shigella* disease. EIEC does seem to have a protein that is expressed inside host cells, which leads to its destruction.

Disease caused by this bacterium is more common in developing countries. It is transmitted primarily through contaminated food and water. Treatment is supportive (including rehydration).

Enteropathogenic *E. coli* (EPEC) These strains result in a profuse, watery diarrhea. Fever and vomiting are also common. The EPEC bacteria are very similar to the EHEC *E. coli* described earlier—they produce effacement of gut surfaces. The important difference between EPEC and EHEC is that EPEC does not produce a shiga toxin and, therefore, does not produce the systemic symptoms characteristic of those bacteria.

EPEC has been known to cause outbreaks in hospital nurseries in this country but is more notorious for causing diarrhea in infants in developing countries.

Most disease is self-limiting. As with any other diarrhea, however, it can be life-threatening in young babies. Rehydration is the main treatment.

Enteroaggregative *E. coli* (EAEC) These bacteria are most notable for their ability to cause chronic diarrhea in young children and in AIDS patients. EAEC is considered in the section on chronic diarrhea.

Campylobacter

Although you may never have heard of *Campylobacter,* it is considered to be the most common bacterial cause of diarrhea in the United States. It probably causes more diarrhea than *Salmonella* and *Shigella* combined, with 2 million cases of diarrhea credited to it per year.

The symptoms of campylobacteriosis are frequent watery stools, fever, vomiting, headaches, and severe abdominal pain. The symptoms may last longer than most acute diarrheal episodes, sometimes extending beyond 2 weeks. They may subside and then recur over a period of weeks.

Campylobacter jejuni is the most common cause, although there are other *Campylobacter* species. Campylobacters are slender, curved or spiral gram-negative bacteria propelled by polar flagella at one or both poles, often appearing in S-shaped or gull-winged pairs **(figure 22.13).** These bacteria tend to be microaerophilic inhabitants of the intestinal tract,

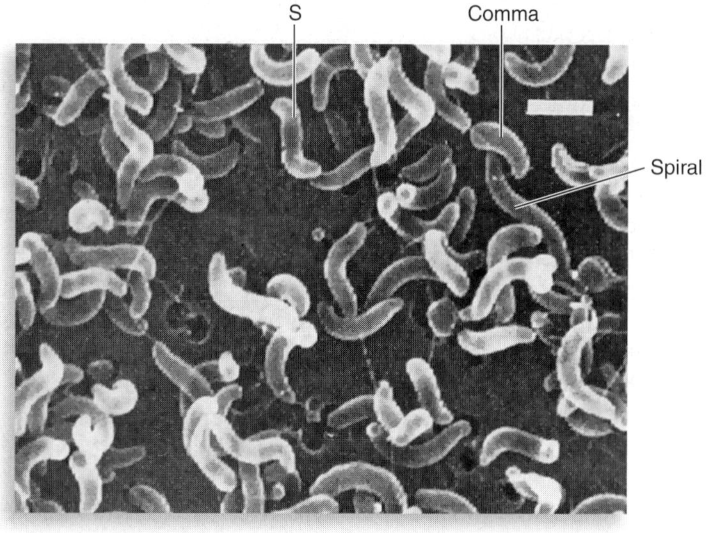

Figure 22.13 Scanning micrograph of *Campylobacter jejuni,* showing comma, S, and spiral forms.

genitourinary tract, and oral cavity of humans and animals. A close relative, *Helicobacter pylori,* is the causative agent of most stomach ulcers (described earlier). Transmission of this pathogen takes place via the ingestion of contaminated beverages and food, especially water, milk, meat, and chicken.

Once ingested, *C. jejuni* cells reach the mucosa at the last segment of the small intestine (ileum) near its junction with the colon; they adhere, burrow through the mucus, and multiply. Symptoms commence after an incubation period of 1 to 7 days. The mechanisms of pathology appear to involve a heat-labile enterotoxin that stimulates a secretory diarrhea like that of cholera. In a small number of cases, infection with this bacterium can lead to a serious neuromuscular paralysis called Guillain-Barré syndrome.

Guillain-Barré syndrome (GBS) is the leading cause of acute paralysis in the United States since the eradication of polio here. The good news is that many patients recover completely from this paralysis. The condition is still mysterious in many ways, but it seems to be an autoimmune reaction that can be brought on by infection with viruses and bacteria, by vaccination in rare cases, and even by surgery. The single most common precipitating event for the onset of GBS is *Campylobacter* infection. Twenty to forty percent of GBS cases are preceded by infection with *Campylobacter.* The reasons for this are not clear. (Note that even though 20% to 40% of GBS cases are preceded by *Campylobacter* infection, only about 1 in 1,000 cases of *Campylobacter* infection results in GBS.)

Diagnosis of *C. jejuni* enteritis requires isolation of the bacterium from stool samples and occasionally from blood samples. More rapid presumptive diagnosis can be obtained from direct examination of feces with a dark-field microscope, which accentuates the characteristic curved rods and darting motility. This procedure is difficult to perform and not often used except in specialized labs. Resolution of infection occurs in most instances with simple, nonspecific rehydration and electrolyte balance therapy. In more severely affected patients, it may be necessary to administer erythromycin. Antibiotic resistance is growing in these bacteria. Because vaccines are yet to be developed, prevention depends on rigid sanitary control of water and milk supplies and care in food preparation.

Yersinia Species

Yersinia is a genus of gram-negative bacteria that includes the infamous plague bacterium, *Yersinia pestis* (discussed in chapter 20). There are two species that cause GI tract disease: *Y. enterocolitica* and *Y. pseudotuberculosis.* The infections are most notable for the high degree of abdominal pain they cause. This symptom is accompanied by fever. Often the symptoms are mistaken for appendicitis.

The disease is uncommon in the United States, but outbreaks do occasionally occur. Food and beverages can become contaminated with these bacteria, which inhabit the intestines of farm animals, pets, and wild animals. Transmission also occurs when people handle raw food and then touch fomites such as toys or baby bottles without washing their hands.

The bacteria invade the small intestinal mucosa, and some enter the lymphatics and are harbored intracellularly in phagocytes. Inflammation of the ileum and mesenteric lymph nodes gives rise to severe abdominal pain. The infection occasionally spreads to the bloodstream, but systemic effects are rare. Two to three percent of patients experience joint pain a month following the diarrhea episode. This symptom resolves spontaneously within a few months. Infections with *Y. pseudotuberculosis* tend to be milder than those with *Y. enterocolitica* and center on lymph node inflammation rather than mucosal involvement.

Simple rules of food hygiene are usually sufficient to prevent the spread of this infection. Antibiotics are not usually prescribed for this disease, unless bacteremia is documented. In that case, doxycycline or TMP-SMZ is used.

Clostridium difficile

Clostridium difficile is a gram-positive endospore-forming rod found as normal biota in the intestine. It was once considered relatively harmless but now is known to cause a condition called pseudomembranous colitis. It is also sometimes called antibiotic-associated colitis. In most cases, this infection seems to be precipitated by therapy with broad-spectrum antibiotics such as ampicillin, clindamycin, or cephalosporins. It is a major cause of diarrhea in hospitals. Also, new studies suggest that the use of gastric acid inhibitors for the treatment of heartburn can predispose patients to this infection. Although *C. difficile* is relatively noninvasive, it is able to superinfect the large intestine when drugs have disrupted the normal biota. It produces two enterotoxins, toxins A and B, that cause areas of necrosis in the wall of the intestine. The predominant symptom is diarrhea commencing late in therapy or even after therapy has stopped. More severe cases exhibit abdominal cramps, fever, and leukocytosis. The colon is inflamed and gradually sloughs off loose, membranelike patches called pseudomembranes consisting of fibrin and cells **(figure 22.14).** If the condition is not stopped, perforation of the cecum and death can result.

Mild, uncomplicated cases respond to withdrawal of antibiotics and replacement therapy for lost fluids and electrolytes. More severe infections are treated with oral vancomycin or metronidazole for several weeks until the intestinal biota returns to normal. Because infected persons often shed large numbers of spores in their stools, increased precautions are necessary to prevent spread of the agent to other patients who may be on antimicrobial therapy. Some new techniques on the horizon are vaccination with *C. difficile* toxoid and restoration of normal biota by ingestion of a mixed culture of lactobacilli and yeasts.

Vibrio cholerae

Cholera has been a devastating disease for centuries. It is not an exaggeration to say that the disease has shaped a good deal of human history in Asia and Latin America, where it has been endemic. These days we have come to expect outbreaks of cholera to occur after natural disasters, war, or large refugee movements, especially in underdeveloped parts of the world.

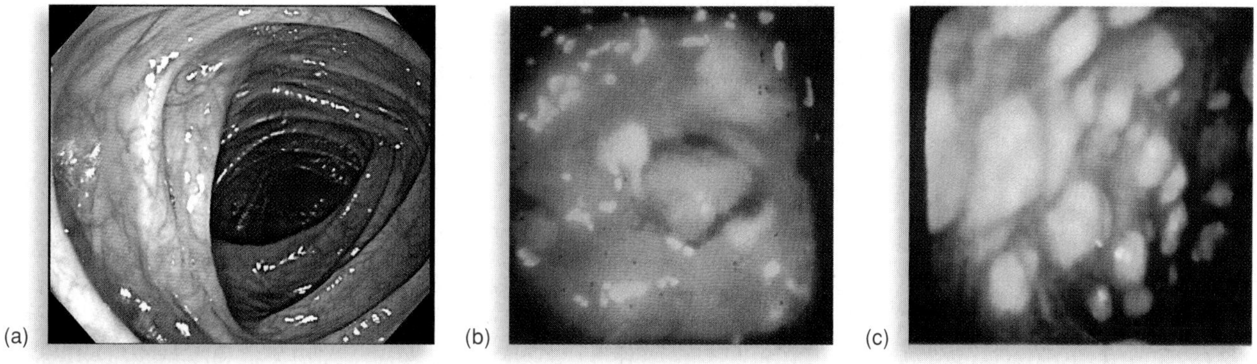

Figure 22.14 Antibiotic-associated colitis.
(a) Normal colon. **(b)** A mild form with diffuse, inflammatory patches. **(c)** Heavy yellow plaques, or pseudomembranes, typical of more severe cases. Photographs were made by a sigmoidoscope, an instrument capable of photographing the interior of the colon.

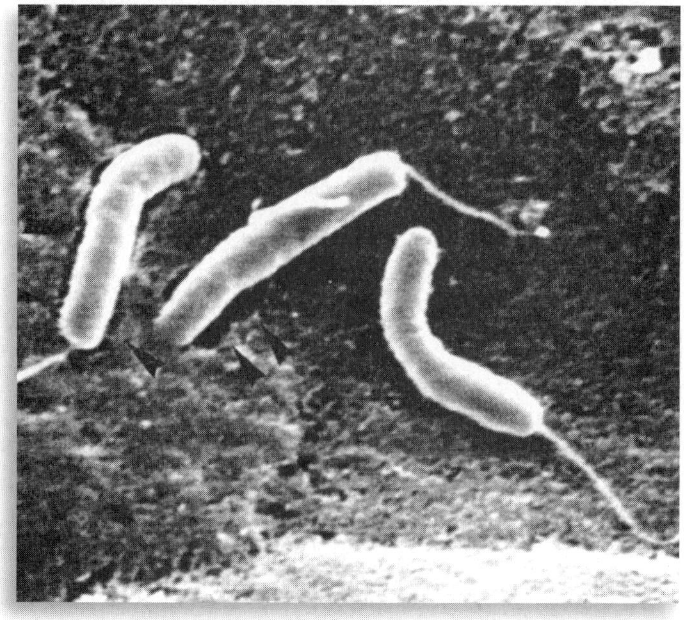

Figure 22.15 ***Vibrio cholerae.***
Note the characteristic curved shape and single polar flagellum.

Vibrios are comma-shaped rods with a single polar flagellum. They belong to the family Vibrionaceae. A freshly isolated specimen of *Vibrio cholerae* reveals quick, darting cells that slightly resemble a cooked hot dog or a comma **(figure 22.15).** *Vibrio* shares many cultural and physiological characteristics with members of the Enterobacteriaceae, a closely related family. Vibrios are fermentative and grow on ordinary or selective media containing bile at 37°C. They possess unique O and H antigens and membrane receptor antigens that provide some basis for classifying members of the family. There are two major biotypes, called classic and *El Tor.*

Signs and Symptoms After an incubation period of a few hours to a few days, symptoms begin abruptly with vomiting, followed by copious watery feces called secretory diarrhea. The intestinal contents are lost very quickly, leaving only secreted fluids. This voided fluid contains flecks of mucus, hence the description "rice-water stool." Fluid losses of nearly 1 liter per hour have been reported in severe cases, and an untreated patient can lose up to 50% of body weight during the course of this disease. The diarrhea causes loss of blood volume, acidosis from bicarbonate loss, and potassium depletion, which manifest in muscle cramps, severe thirst, flaccid skin, sunken eyes, and in young children, coma and convulsions. Secondary circulatory consequences can include hypotension, tachycardia, cyanosis, and collapse from shock within 18 to 24 hours. If cholera is left untreated, death can occur in less than 48 hours, and the mortality rate approaches 55%.

Pathogenesis and Virulence Factors After being ingested with food or water, *V. cholerae* encounters the potentially destructive acidity of the stomach. This hostile environment influences the size of the infectious dose (10^8 cells), although certain types of food shelter the pathogen more readily than others. At the junction of the duodenum and jejunum, the vibrios penetrate the mucus barrier using their flagella, adhere to the microvilli of the epithelial cells, and multiply there. The bacteria never enter the host cells or invade the mucosa. The virulence of *V. cholerae* is due entirely to an enterotoxin called cholera toxin (CT), which disrupts the normal physiology of intestinal cells. It is a typical A-B type toxin as previously described for *Shigella.* When this toxin binds to specific intestinal receptors, a secondary signaling system is activated. Under the influence of this system, the cells shed large amounts of electrolytes into the intestine, an event accompanied by profuse water loss. Most cases of cholera are mild or self-limited, but in children and weakened individuals, the disease can be deadly.

Transmission and Epidemiology Although the human intestinal tract was once thought to be the primary reservoir, it is now known that the parasite lives in certain endemic regions. The pattern of cholera transmission and the onset of epidemics are greatly influenced by the season of the year and the climate. Cold, acidic, dry environments inhibit the migration and survival of *Vibrio,* whereas warm, monsoon, alkaline, and saline conditions favor them. The bacteria survive in water sources for long periods of time. Recent outbreaks in several

INSIGHT 22.2 **Discovery**

A Little Water, Some Sugar, and Salt Save Millions of Lives

Volunteers in front of an Oral Rehydration Clinic in the Philippines. ORT clinics are commonplace in developing countries.

In 1970, a clinical trial was conducted on a very low-tech solution to the devastating problem of death from diarrhea, especially among children in the developing world. Until that time, the treatment, if a child could get it, was rehydration through an IV drip. This treatment usually required traveling to the nearest clinic, often miles or days away. Most children received no treatment at all, and 3 million of them died every year. Then scientists tested a simple sugar-salt solution that patients could drink. They tested it first in India, where cholera was rampant, and found that mortality rates were greatly decreased. After more testing in Bangladesh, Turkey, the Philippines, and the United States, oral-rehydration therapy (ORT) became the treatment of choice for diarrhea from all causes. The WHO and UNICEF began providing packages of the sugar and salt mixture and instructions for mixing it with boiled water to dozens of countries. They also oversaw training of individuals who could in turn teach townspeople and villagers about ORT.

The relatively simple solution, developed by the WHO, consists of a mixture of the electrolytes sodium chloride, sodium bicarbonate, potassium chloride, and glucose or sucrose dissolved in water. When administered early in amounts ranging from 100 to 400 milliliters per hour, the solution can restore patients in 4 hours, often bringing them literally back from the brink of death. Infants and small children who once would have died now survive so often that the mortality rate for treated cases of cholera is near zero. This therapy has several advantages, especially for countries with few resources. It does not require medical facilities, high-technology equipment, or complex medication protocols. It also eliminates the need for clean needles, which is a pressing issue in many parts of the world.

In 1978, the British Medical journal *The Lancet* called ORT "potentially the most important medical advance this century." With estimates of at least a million lives saved every year since its introduction, this statement seems to have been proven correct.

parts of the world have been traced to giant cargo ships that pick up ballast water in one port and empty it in another elsewhere in the world. Cholera ranks among the top seven causes of morbidity and mortality, affecting several million people in endemic regions of Asia and Africa.

In nonendemic areas such as the United States, the microbe is spread by water and food contaminated by asymptomatic carriers, but it is relatively uncommon. Sporadic outbreaks occur along the Gulf of Mexico, and *V. cholerae* is sometimes isolated from shellfish in that region.

Culture and Diagnosis During epidemics of this disease, clinical evidence is usually sufficient to diagnose cholera. But confirmation of the disease is often required for epidemiological studies and detection of sporadic cases. *V. cholerae* can be readily isolated and identified in the laboratory from stool samples. Direct dark-field microscopic observation reveals characteristic curved cells with brisk, darting motility as confirmatory evidence. Immobilization or fluorescent staining of feces with group-specific antisera is supportive as well. Difficult cases can be traced by detecting a rising antitoxin titer in the serum.

Prevention and Treatment Effective prevention is contingent upon proper sewage treatment and water purification. Detecting and treating carriers with mild or asymptomatic cholera are serious goals, but they are difficult to accomplish because of inadequate medical provisions in those countries where cholera is endemic. Vaccines are available for travelers and people living in endemic regions. One vaccine contains killed *V. cholerae* but protects for only 6 months or less. An oral vaccine containing live, attenuated bacteria was developed to be a more effective alternative, but evidence suggests it also confers only short-term immunity. It is not available in the United States.

The key to cholera therapy is prompt replacement of water and electrolytes, because their loss accounts for the severe morbidity and mortality. This therapy can be accomplished by various rehydration techniques that replace the lost fluid and electrolytes. One of these, oral rehydration therapy (ORT), is described in **Insight 22.2.**

Cases in which the patient is unconscious or has complications from severe dehydration require intravenous replenishment as well. Oral antibiotics such as tetracycline

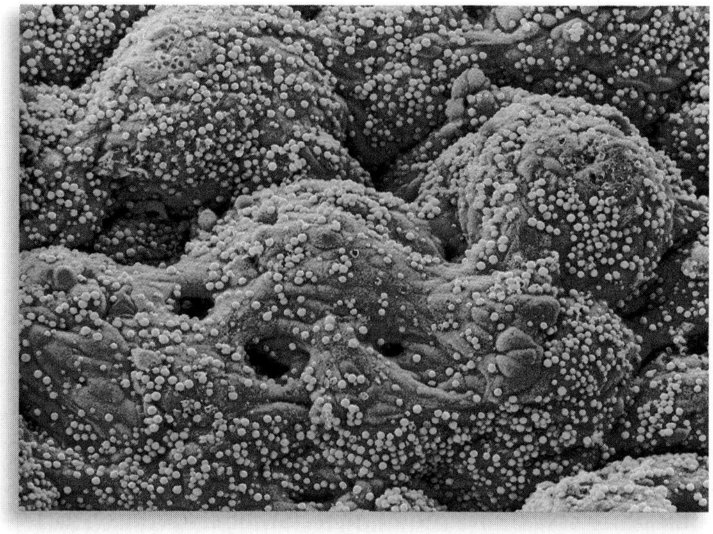

Figure 22.16 Scanning electron micrograph of *Cryptosporidium* attached to the intestinal epithelium.

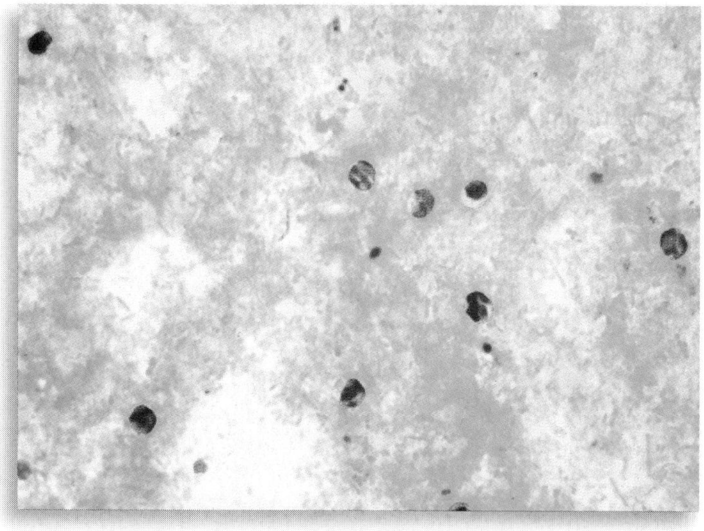

Figure 22.17 Acid-fast stain of *Cryptosporidium*. Oocysts of *Cryptosporidium* stain bright red or purple.

and drugs such as trimethoprim-sulfamethoxazole can terminate the diarrhea in 48 hours. They also diminish the period of vibrio excretion.

Cryptosporidium

Cryptosporidium is an intestinal protozoan of the apicomplexan type (see chapter 5) that infects a variety of mammals, birds, and reptiles. For many years, cryptosporidiosis was considered an intestinal ailment exclusive to calves, pigs, chickens, and other poultry, but it is clearly a zoonosis as well. The organism's life cycle includes a hardy intestinal oocyst as well as a tissue phase. Humans accidentally ingest the oocysts with water or food that has been contaminated by feces from infected animals. The oocyst "excysts" once it reaches the intestines and releases sporozoites that attach to the epithelium of the small intestine **(figure 22.16).** The organism penetrates the intestinal cells and lives intracellularly in them. It undergoes asexual and sexual reproduction in the gut and produces more oocysts, which are excreted from the host and after a short time become infective again. The oocysts are highly infectious and extremely resistant to treatment with chlorine and other disinfectants.

The prominent symptoms mimic other types of gastroenteritis, with headache, sweating, vomiting, severe abdominal cramps, and diarrhea. AIDS patients may experience chronic persistent cryptosporidial diarrhea that can be used as a criterion to help diagnose AIDS. The agent can be detected in fecal samples with indirect immunofluorescence and by acid-fast staining of biopsy tissues **(figure 22.17).** Stool cultures should be performed to rule out other (bacterial) causes of infection.

Cryptosporidiosis has a cosmopolitan distribution. Its highest prevalence is in areas with unreliable water and food sanitation. The carrier state occurs in 3% to 30% of the population in developing countries. The susceptibility of the general public to this pathogen has been amply demonstrated by several large-scale epidemics. In 1993, 370,000 people developed *Cryptosporidium* gastroenteritis from the municipal water supply in Milwaukee, Wisconsin. Other mass outbreaks of this sort have been traced to contamination of the local water reservoir by livestock wastes. Half of the outbreaks of diarrhea associated with swimming pools are caused by *Cryptosporidium*. Because chlorination is not entirely successful in eradicating the cysts, most treatment plants use filtration to remove them, but even this method can fail.

Treatment is not usually required for otherwise healthy patients. Antidiarrheal agents (antimotility drugs) may be used. Although no curative antimicrobial agent exists for *Cryptosporidium*, physicians will often try paromomycin, an aminoglycoside that can be effective against protozoa.

Rotavirus

Rotavirus is a member of the *Reovirus* group, which consists of an unusual double-stranded RNA genome with both an inner and an outer capsid. Globally, rotavirus is the primary viral cause of morbidity and mortality resulting from diarrhea, accounting for nearly 50% of all cases. It is estimated that there are 1 million cases of rotavirus infection in the United States every year, leading to 70,000 hospitalizations. Peak occurrences of this infection are seasonal; in the U.S. Southwest the peak is often in the late fall, and in the Northeast the peak comes in the spring.

Diagnosis of rotavirus infections is usually not performed, as it is treated symptomatically. Nevertheless, studies are often conducted so that public health officials can maintain surveillance of how prevalent the infection is. Stool samples from infected persons contain large amounts of virus, which is readily visible using an electron microscope **(figure 22.18).** The virus gets its name from its physical appearance, which is said to resemble a "spoked wheel." An ELISA test is also available.

The virus is transmitted by the fecal-oral route, including through contaminated food, water, and fomites. For this reason, disease is most prevalent in areas of the world with

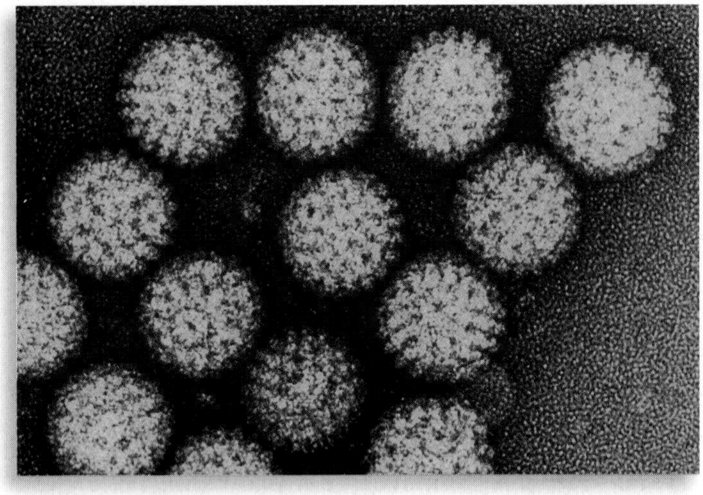

Figure 22.18 ***Rotavirus*** **visible in a sample of feces from a child with gastroenteritis.**
Note the unique "spoked-wheel" morphology of the virus.

poor sanitation. In the United States, rotavirus infection is relatively common but its course is generally mild.

The effects of infection vary with the age, nutritional state, general health, and living conditions of the patient. Babies from 6 to 24 months of age lacking maternal antibodies have the greatest risk for fatal disease. These children present symptoms of watery diarrhea, fever, vomiting, dehydration, and shock. The intestinal mucosa can be damaged in a way that chronically compromises nutrition, and long-term or repeated infections can retard growth. Newborns seem to be protected by maternal antibodies. Adults can also acquire this infection, but it is generally mild and self-limiting.

Children are treated with oral replacement fluid and electrolytes. A vaccine was introduced in 1998 but was withdrawn 9 months later because of a side effect called intussusception, a form of intestinal blockage that seemed to be associated with immunization. A new oral live virus vaccine, RotaTeq™ has been available since 2006.

Other Viruses

A bewildering array of viruses can cause gastroenteritis, including adenoviruses, noroviruses (sometimes known as Norwalk viruses), and astroviruses. They are extremely common in the United States and around the world. They are usually "diagnosed" when no other agent (such as those just described) is identified.

Transmission is fecal-oral or via contamination of food and water. Viruses generally cause a profuse, watery diarrhea of 3 to 5 days duration. Vomiting may accompany the disease, especially in the early phases. Mild fever is often seen.

In 2002, a series of gastroenteritis outbreaks occurred on cruise ships, most of which were ascribed to viruses other than rotavirus.

Treatment of these infections always focuses on rehydration **(Checkpoint 22.5).**

CASE FILE 22 *WRAP-UP*

Stool samples or rectal swabs from 44 of the symptomatic patients were tested, and norovirus was identified in 22 of these samples. No other enteric pathogen was isolated. Noroviruses are often identified as the causative agent in outbreaks of gastroenteritis in the United States. Such outbreaks are often associated with contaminated food or water. In addition, outbreaks can be associated with persons living in crowded living conditions such as those present at the Reliant Park Complex.

Norovirus is highly contagious (I.D. < 100 organisms) and is easily spread person to person and by contact with contaminated materials. The typical incubation period is 24 to 48 hours, and the resulting symptoms persist for 12 to 60 hours. In this situation, it is likely that one or more individuals were infected with the virus when they arrived at the shelter. Although the source of the initial infection was unknown, contact with contaminated floodwaters was certainly a possibility. The infection spread quickly due to the crowded living conditions and shared facilities. Implementation of infection-control measures including isolation of symptomatic individuals, distribution of gel hand sanitizer, and education of staff and evacuees quickly brought the outbreak under control.

See: CDC. 2005. Norovirus outbreak among evacuees from Hurricane Katrina—Houston, Texas, September 2005. MMWR *53(40):1016–1018.*

Acute Diarrhea with Vomiting (Food Poisoning)

If a patient presents with severe nausea, frequent vomiting accompanied by diarrhea, and reports that companions with whom he or she shared a recent meal (within the last 1 to 6 hours) are suffering the same fate, food poisoning should be suspected. **Food poisoning** refers to symptoms in the gut that are caused by a preformed toxin of some sort. In many cases, the toxin comes from *Staphylococcus aureus*. In others, the source of the toxin is *Bacillus cereus* or *Clostridium perfringens*. The toxin occasionally comes from nonmicrobial sources such as fish, shellfish, or mushrooms. In any case, if the symptoms are violent and the incubation period is very short, *intoxication* (the effects of a toxin) rather than *infection* should be considered. (**Insight 22.3** has information about outbreak investigations in general.)

Staphylococcus aureus Exotoxin

This illness is associated with eating foods such as custards, sauces, cream pastries, processed meats, chicken salad, or ham that have been contaminated by handling and then left unrefrigerated for a few hours. Because of the high salt

INSIGHT 22.3 Medical

Microbes Have Fingerprints, Too

Until recently, epidemiological investigations of outbreaks of disease relied primarily on careful examination of oral case histories and reports from the patients themselves, which might provide clues about the source of exposure. If organisms could be isolated and identified in the laboratory, they could provide evidence to support or negate a hypothetical exposure, but usually they could not provide definitive proof.

When more sophisticated molecular methods for identifying microbial strains became available, the situation changed. A wide variety of techniques, including PCR, Southern blot analysis, and ribotyping, allowed the identification of bacteria below the species level, allowing the movement of a particular microbe to be traced through various hosts and environments. The most useful of these techniques for public health purposes seems to be the process called pulsed-field gel electrophoresis, or PFGE.

PFGE is a technique for macrorestriction analysis. Pathogens are isolated from a patient, and their DNA is harvested. The DNA is then cut up with restriction enzymes specifically chosen so that they find only a few places to cut into the organism's genome. The result is just a few very large pieces of DNA rather than the many small ones obtained with older methods of restriction analysis. The DNA fragments are then separated using the pulsed-field method of gel electrophoresis. This method involves constantly changing the direction of (pulsing) the electrical field during electrophoresis. You can think of it as teasing out the DNA pieces from one another in the gel matrix. This method allows effective separation of the large pieces. Once the electrophoresis is finished, the fragments of different lengths can be seen as dark bands after the gel is immersed in a special stain. The lengths of the fragments and, thus, the pattern revealed by each microbe will be different—even for different strains of the same microbial species—because the enzymes are cut in different places on the genome where small DNA changes exist, corresponding to different strain types. This pattern is also called a DNA fingerprint, much like that used in forensic studies.

In 1993, the CDC used PFGE for the first time to trace an outbreak of food-borne illness in the United States. They determined that the strain of *E. coli* O157:H7 found in patients had the same PFGE pattern as the strain found in the suspected hamburger patties that had been served at a fast-food restaurant. The use of the technique led to the creation of a national database called PulseNet, which contains the PFGE patterns of common food-borne pathogens that have been implicated in outbreaks. Participating PulseNet laboratories all around the country can compare PFGE patterns they obtain from patients or suspected foods to patterns in the centralized database. In this way, outbreaks that are geographically dispersed (for instance, those caused by contaminated meat that may have been distributed nationally) can be identified quickly. When new patterns come in, they are also archived so that other laboratories submitting the same patterns will quickly realize that the cases are related.

PulseNet currently tracks outbreaks of the following food-borne bacteria: *E. coli* O157:H7, *Campylobacter, Listeria, Salmonella, Shigella, Vibrio,* and *Yersinia enterocolitica*. Patterns are also available representing the food-borne protozoa *Cyclospora* and *Cryptosporidium*. The fingerprints of many more microbial culprits soon will be available to help public health officials solve food outbreak mysteries in record time.

A pulsed-field gel electrophoresis "fingerprint."
The identity of the microbe is revealed in this pattern.

tolerance of *S. aureus*, even foods containing salt as a preservative are not exempt. The toxins produced by the multiplying bacteria do not noticeably alter the food's taste or smell. The exotoxin (which is an enterotoxin) is heat-stable; inactivation requires 100°C for at least 30 minutes. Thus, heating the food after toxin production may not prevent disease. The ingested toxin acts upon the gastrointestinal epithelium and stimulates nerves with acute symptoms of cramping, nausea, vomiting, and diarrhea. Recovery is also rapid, usually within 24 hours. The disease is not transmissible person to person. Often, a single source will contaminate several people, leading to a mini-outbreak.

The illness is caused by the toxin and does not require *S. aureus* to be present or alive in the contaminated food. If the bacterium is allowed to multiply in the food, it produces its exotoxin. Even if the bacteria are subsequently destroyed by heating, the preformed toxin will act quickly once it is ingested.

As you learned earlier, many diarrheal diseases have symptoms caused by bacterial exotoxins. In most cases, the bacteria take up temporary residence in the gut and then start producing exotoxin, so the incubation period is longer than the 1 to 6 hours seen with *S. aureus* food poisoning. Because this toxin is heat-stable, mishandling of food, such as allowing bacteria to multiply and then heating or reheating, can provide the perfect conditions for food poisoning to occur.

This condition is almost always self-limiting, and antibiotics are definitely not warranted.

CHECKPOINT 22.5 Acute Diarrhea

	Bacterial Causes				
Causative Organism(s)	*Salmonella*	*Shigella*	Shiga-toxin-producing *E. coli* O157:H7 (EHEC)	Other *E. coli* (non-shiga-toxin-producing)	*Campylobacter*
Most Common Modes of Transmission	Vehicle (food, beverage), fecal-oral	Fecal-oral, direct contact	Vehicle (food, beverage), fecal-oral	Vehicle, fecal-oral	Vehicle (food, water), fecal-oral
Virulence Factors	Adhesins, endotoxin	Endotoxin, enterotoxin, shiga toxins in some strains	Shiga toxins; proteins for attachment, secretion, effacement	Various: proteins for attachment, secretion, effacement; heat-labile and/or heat-stable exotoxins; invasiveness	Adhesins, exotoxin, induction of autoimmunity
Culture/ Diagnosis	Stool culture, not usually necessary	Stool culture; antigen testing for shiga toxin	Stool culture, antigen testing for shiga toxin	Stool culture not usually necessary in absence of blood, fever	Stool culture not usually necessary; dark-field microscopy
Prevention	Food hygiene and personal hygiene	Food hygiene and personal hygiene	Avoid live *E. coli* (cook meat and clean vegetables)	Food and personal hygiene	Food and personal hygiene
Treatment	Rehydration; no antibiotic for uncomplicated disease	TMP-SMZ, rehydration	Antibiotics contraindicated, supportive measures	Rehydration	Rehydration, erythromycin in severe cases (antibiotic resistance rising)
Fever Present	Usually	Often	Often	Sometimes	Usually
Blood in Stool	Sometimes	Often	Usually	Sometimes	No
Distinctive Features	Often associated with chickens, reptiles	Very low ID_{50}	Hemolytic uremic syndrome	EIEC, ETEC, EPEC	Guillain-Barré syndrome

Bacillus cereus Exotoxin

Bacillus cereus is a sporulating gram-positive bacterium that is naturally present in soil. As a result, it is a common resident on vegetables and other products in close contact with soil. It produces two exotoxins, one of which causes a diarrheal-type disease, the other of which causes an **emetic** (ee-met'-ik) or vomiting disease. The type of disease that takes place is influenced by the type of food that is contaminated by the bacterium. The emetic form is most frequently linked to fried rice, especially when it has been cooked and kept warm for long periods of time. These conditions are apparently ideal for the expression of the low-molecular-weight, heat-stable exotoxin having an emetic effect. Outbreaks are often associated with Chinese restaurants, although a notable outbreak occurred at two day care centers in 1993.

The diarrheal form of the disease is usually associated with cooked meats or vegetables that are held at a warm temperature for long periods of time. These conditions apparently favor the production of the high-molecular-weight, heat-labile exotoxin. The symptom in these cases is a watery, profuse diarrhea that lasts only for about 24 hours.

Diagnosis of the emetic form of the disease is accomplished by finding the bacterium in the implicated food

			Nonbacterial Causes		
Yersinia	*Clostridium difficile*	*Vibrio cholerae*	*Cryptosporidium*	Rotavirus	Other viruses
Vehicle (food, water), fecal-oral, indirect contact	Endogenous (normal biota)	Vehicle (water and some foods), fecal-oral	Vehicle (water, food), fecal-oral	Fecal-oral, vehicle, fomite	Fecal-oral, vehicle
Intracellular growth	Enterotoxins A and B	Cholera toxin (CT)	Intracellular growth	–	–
Cold-enrichment stool culture	Stool culture, PCR, ELISA demonstration of toxins in stool	Clinical diagnosis, microscopic techniques, serological detection of antitoxin	Acid-fast staining, ruling out bacteria	Usually not performed	Usually not performed
Food and personal hygiene	–	Water hygiene	Water treatment, proper food handling	Oral live virus vaccine	Hygiene
None in most cases, doxycycline or TMP-SMZ for bacteremia	Withdrawal of antibiotic, in severe cases metronidazole or vancomycin	Rehydration, in severe cases tetracycline, TMP-SMZ	None, paromomycin used sometimes	Rehydration	Rehydration
Usually	Sometimes	No	Often	Often	Sometimes
Occasionally	Not usually; mucus prominent	No	Not usually	No	No
Severe abdominal pain	Antibiotic-associated diarrhea	Rice-water stools	Resistant to chlorine disinfection	Severe in babies	–

source. Microscopic examination of stool samples is used to diagnose the diarrheal form of the disease. Of course, in everyday practice, diagnosis as well as treatment is not performed because of the short duration of the disease.

In both cases, the only prevention is the proper handling of food.

Clostridium perfringens Exotoxin

Another sporulating gram-positive bacterium that causes intestinal symptoms is *Clostridium perfringens*. You first read about this bacterium as the causative agent of gas gangrene in chapter 18. Endospores from *C. perfringens* can also contaminate many kinds of foods. Those most frequently implicated in disease are animal flesh (meat, fish) and vegetables such as beans that have not been cooked thoroughly enough to destroy endospores. When these foods are cooled, spores germinate, and the germinated cells multiply, especially if the food is left unrefrigerated. If the food is eaten without adequate reheating, live *C. perfringens* cells enter the small intestine and release exotoxin. The toxin, acting upon epithelial cells, initiates acute abdominal pain, diarrhea, and nausea in 8 to 16 hours. Recovery is rapid, and deaths are extremely rare.

CHECKPOINT 22.6	Acute Diarrhea with Vomiting (Food Poisoning)		
Causative Organism(s)	*Staphylococcus aureus* exotoxin	*Bacillus cereus*	*Clostridium perfringens*
Most Common Modes of Transmission	Vehicle (food)	Vehicle (food)	Vehicle (food)
Virulence Factors	Heat-stable exotoxin	Heat-stable toxin, heat-labile toxin	Heat-labile toxin
Culture/Diagnosis	Usually based on epidemiological evidence	Microscopic analysis of food or stool	Detection of toxin in stool
Prevention	Proper food handling	Proper food handling	Proper food handling
Treatment	None	None	None
Fever Present	Not usually	Not usually	Not usually
Blood in Stool	No	No	No
Distinctive Features	Suspect in foods with high salt or sugar content	Two forms: emetic and diarrheal	Acute abdominal pain

C. perfringens also causes an enterocolitis infection similar to that caused by *C. difficile*. This infectious type of diarrhea is acquired from contaminated food, or it may be transmissible by inanimate objects **(Checkpoint 22.6)**.

Chronic Diarrhea

Chronic diarrhea is defined as lasting longer than 14 days. It can have infectious causes or can reflect noninfectious conditions. Most of us are familiar with diseases that present a constellation of bowel syndromes, such as irritable bowel syndrome, Crohn's disease, and ulcerative colitis, none of which are directly caused by a microorganism as far as we know. They may indeed represent an overreaction to the presence of an infectious agent or another irritant, but the host response seems to be responsible for the pathology. When the presence of an infectious agent is ruled out by a negative stool culture or other tests, these conditions are suspected.

People suffering from AIDS almost universally suffer from chronic diarrhea. Most of the patients who are not taking antiretroviral drugs have diarrhea caused by a variety of opportunistic microorganisms, including *Cryptosporidium*, *Mycobacterium avium*, and so forth. Recently, investigators have found that patients who are aggressively treating their HIV infection with the cocktail of drugs known as HAART (see chapter 20) still suffer from chronic diarrhea at a high rate. The causes for this diarrhea are not completely understood. A patient's HIV status should be considered if he or she presents with chronic diarrhea.

Next we examine a few of the microbes that can be responsible for chronic diarrhea in otherwise healthy people. Keep in mind that practically any disease of the intestinal tract has a sexual mode of transmission in addition to the ones that are commonly stated. For example, any kind of oral-anal sexual contact efficiently transfers pathogens to the "oral" partner. This mode is more commonly seen in cases of chronic illness than it is in patients experiencing acute diarrhea, for obvious reasons.

Enteroaggregative E. coli (EAEC)

In the section on acute diarrhea, you read about the various categories of *E. coli* that can cause disease in the gut. One type, the enteroaggregative *E. coli* (EAEC), is particularly associated with chronic disease, especially in children. This bacterium was first recognized in 1987. It secretes neither the heat-stable nor heat-labile exotoxins previously described for enterotoxigenic *E. coli* (ETEC). It is distinguished by its ability to adhere to human cells in aggregates rather than as single cells **(figure 22.19)**. Its presence appears to stimulate secretion of large amounts of mucus in the gut, which may be part of its role in causing chronic diarrhea. The bacterium also seems capable of exerting toxic effects on the gut epithelium, although the mechanisms are not well understood.

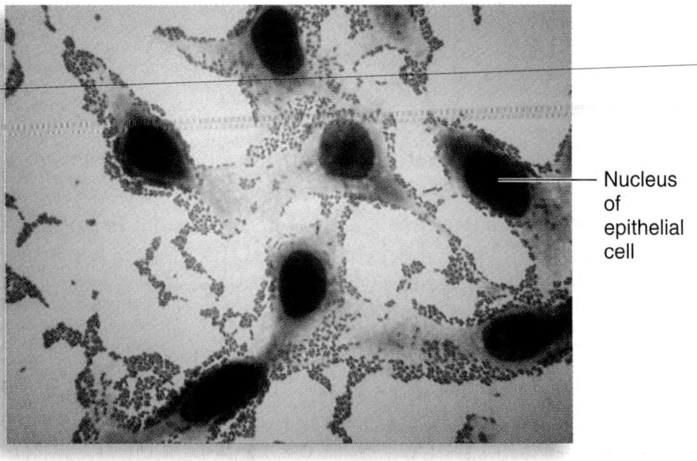

Figure 22.19 Enteroaggregative *E. coli* adhering to epithelial cells.

Transmission of the bacterium is through contaminated food and water. It is difficult to diagnose in a clinical lab because EAEC is not easy to distinguish from other *E. coli,* including normal biota. And the designation EAEC is not actually a serotype but is functionally defined as an *E. coli* that adheres in an aggregative pattern.

This bacterium seems to be associated with chronic diarrhea in people who are malnourished. It is not exactly clear whether the malnutrition predisposes patients to this infection or whether this infection contributes to malnutrition. Probably both possibilities are operating in patients, who are usually children in developing countries. More recently, the bacterium has been associated with acute diarrhea in industrialized countries, perhaps providing a clue to this question. It may be that in well-nourished hosts the bacterium produces acute, self-limiting disease.

Cyclospora

Cyclospora cayetanensis is an emerging protozoan pathogen. Since the first occurrence in 1979, hundreds of outbreaks have been reported in the United States and Canada. Its mode of transmission is fecal-oral, and most cases have been associated with consumption of fresh produce and water presumably contaminated with feces. This disease occurs worldwide, and although primarily of human origin, it is not spread directly from person to person. Outbreaks have been traced to imported raspberries, salad made with fresh greens, and drinking water. The parasite has also been identified as a significant cause of diarrhea in travelers.

The organism is 8 to 10 micrometers in diameter and stains variably in an acid-fast stain. Diagnosis can be complicated by the lack of recognizable oocysts in the feces. Techniques that improve identification of the parasite are examination of fresh preparations under a fluorescent microscope and an acid-fast stain of a processed stool specimen **(figure 22.20)**. A PCR-based test can also be used to identify *Cyclospora* and differentiate it from other parasites. This form of analysis is more sensitive and can detect protozoan genetic material even in the absence of actual cysts.

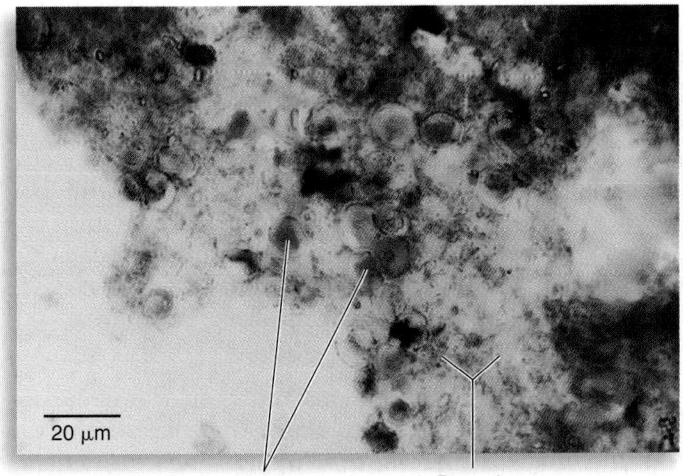

Figure 22.20 An acid-fast stain of *Cyclospora* in a human fecal sample.
The large (8–10 µm) cysts stain pink to red and have a wrinkled outer wall. Bacteria stain blue.

The disease begins when oocysts enter the small intestine and release invasive sporozoites that invade the mucosa. After an incubation period of about 1 week, symptoms of watery diarrhea, stomach cramps, bloating, fever, and muscle aches appear. Patients with prolonged diarrheal illness experience anorexia and weight loss.

Most cases of infection have been effectively controlled with trimethoprim-sulfamethoxazole lasting 1 week. Traditional antiprotozoan drugs are not effective. Some cases of disease may be prevented by cooking or freezing food to kill the oocysts.

Giardia

Giardia lamblia is a pathogenic flagellated protozoan first observed by Antonie van Leeuwenhoek in his own feces. For 200 years, it was considered a harmless or weak intestinal pathogen; and only since the 1950s has its prominence as a cause of diarrhea been recognized. In fact, it is the most common flagellate isolated in clinical specimens. Observed straight on, the trophozoite has a unique symmetrical heart shape with organelles positioned in such a way that it resembles a face **(figure 22.21)**. Four pairs of flagella emerge from the ventral surface, which is concave and acts like a suction cup for attachment to a substrate. *Giardia* cysts are small, compact, and contain four nuclei.

Signs and Symptoms Typical symptoms include diarrhea of long duration, abdominal pain, and flatulence. Stools have a greasy, malodorous quality to them. Fever is usually not present.

Pathogenesis and Virulence Factors Ingested *Giardia* cysts enter the duodenum, germinate, and travel to the jejunum to feed and multiply. Some trophozoites remain on the surface, while others invade the deeper crypts to varying degrees. Superficial invasion by trophozoites causes damage

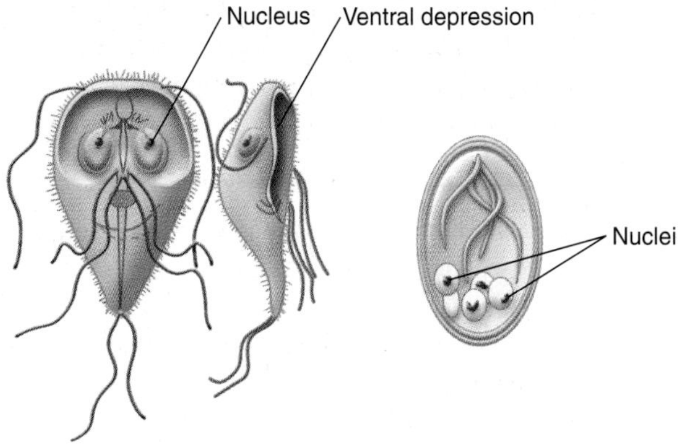

Figure 22.21 The "face" of a *Giardia lamblia* trophozoite.
The cyst form is also shown.

to the epithelial cells, edema, and infiltration by white blood cells, but these effects are reversible. The presence of the protozoan leads to maladsorption (especially of fat) in the digestive tract and can cause significant weight loss.

Transmission and Epidemiology of Giardiasis Giardiasis has a complex epidemiological pattern. The protozoan has been isolated from the intestines of beavers, cattle, coyotes, cats, and human carriers, but the precise reservoir is unclear at this time. Although both trophozoites and cysts escape in the stool, the cysts play a greater role in transmission. Unlike other pathogenic flagellates, *Giardia* cysts can survive for 2 months in the environment. Cysts are usually ingested with water and food or swallowed after close contact with infected people or contaminated objects. Infection can occur with a dose of only 10 to 100 cysts.

Outbreaks of giardiasis point to a spectrum of possible modes of transmission. Community water supplies in areas throughout the United States have been implicated as common vehicles of infection. *Giardia* epidemics have been traced to water from fresh mountain streams as well as chlorinated municipal water supplies in several states. Infections are not uncommon in hikers and campers who used what they thought was clean water from ponds, lakes, and streams in remote mountain areas. Because wild mammals such as muskrats and beavers are intestinal carriers, they could account for cases associated with drinking water from these sources. Checking water for purity by its appearance obviously is unreliable, because the cysts are too small to be detected.

Cases of fecal-oral transmission have been documented in day care centers; food contaminated by infected persons has also transmitted the disease. Anal-oral sex also has been shown to transmit the disease.

Culture and Diagnosis Diagnosis of **giardiasis** can be difficult because the organism is shed in feces only intermittently. Sometimes ELISA tests are used to screen fecal samples for *Giardia* antigens, and PCR tests are available, although they are mainly used for detection of the protozoan in environmental samples.

Prevention and Treatment There is a vaccine against *Giardia* that can be given to animals, including dogs. No human vaccine is available. Avoiding drinking from freshwater sources is the major preventive measure that can be taken. Even municipal water is at some risk; water agencies have had to rethink their policies on water maintenance and testing. The agent is killed by boiling, ozone, and iodine; but unfortunately, the amount of chlorine used in municipal water supplies does not destroy the cysts.

Treatment is with quinacrine or metronidazole.

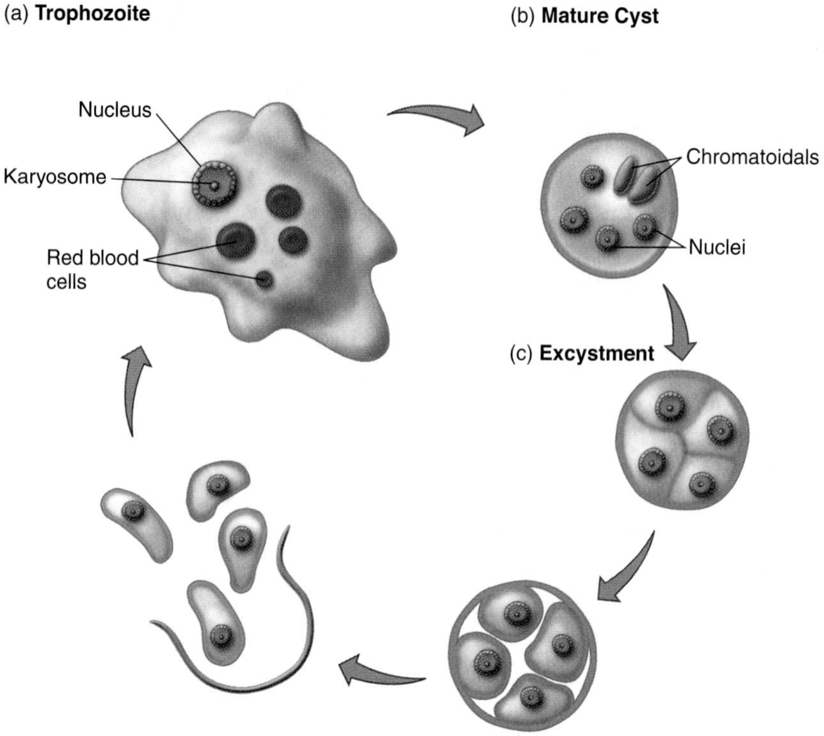

Figure 22.22 **Cellular forms of *Entamoeba histolytica*.** **(a)** A trophozoite containing a single nucleus, a karyosome, and red blood cells. **(b)** A mature cyst with four nuclei and two blocky chromatoidals. **(c)** Stages in excystment. Divisions in the cyst create four separate cells, or metacysts, that differentiate into trophozoites and are released.

Entamoeba

Amoebas are widely distributed in aqueous habitats and are frequent parasites of animals, but only a small number of them have the necessary virulence to invade tissues and cause serious pathology. One of the most significant pathogenic amoebas is *Entamoeba histolytica* (en″-tah-mee′-bah his″-toh-lit′-ihkuh). The relatively simple life cycle of this parasite alternates between a large trophozoite that is motile by means of pseudopods and a smaller, compact, nonmotile cyst **(figure 22.22).** The trophozoite lacks most of the organelles of other eukaryotes, and it has a large single nucleus that contains a prominent nucleolus called a *karyosome*. Amoebas from fresh specimens are often packed with food vacuoles containing host cells and bacteria. The mature cyst is encased in a thin yet tough wall and contains four nuclei as well as distinctive cigar-shaped bodies called *chromatoidal bodies*, which are actually dense clusters of ribosomes.

Signs and Symptoms As hinted by its species name, tissue damage is one of the formidable characteristics of untreated *E. histolytica* infection. Clinical amoebiasis exists in intestinal and extraintestinal forms. The initial targets of intestinal amoebiasis are the cecum, appendix, colon, and rectum. The amoeba secretes enzymes that dissolve tissues, and it actively penetrates deeper layers of the mucosa, leaving erosive ulcerations **(figure 22.23).** This phase is marked

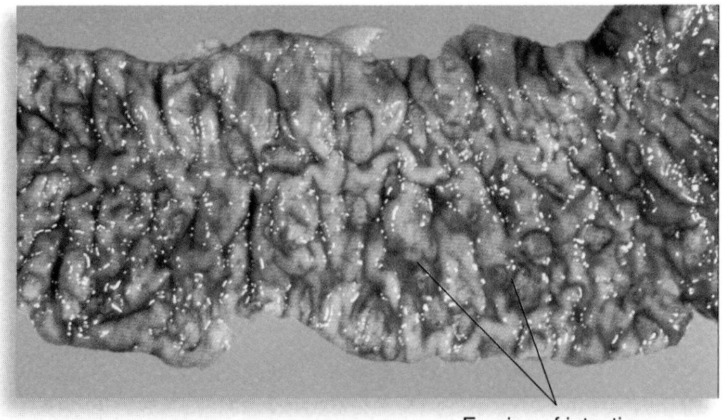

Figure 22.23 Intestinal amoebiasis and dysentery of the cecum.
Red patches are sites of amoebic damage to the intestinal mucosa.

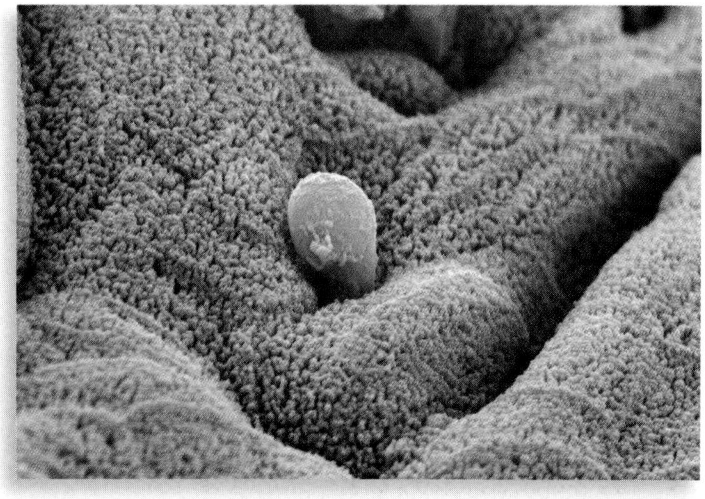

Figure 22.24 Trophozoite of *Entamoeba histolytica.*
Note the fringe of very fine pseudopods it uses to invade and feed on tissue.

by dysentery (bloody, mucus-filled stools), abdominal pain, fever, diarrhea, and weight loss. The most life-threatening manifestations of intestinal infection are hemorrhage, perforation, appendicitis, and tumorlike growths called amoebomas. Lesions in the mucosa of the colon have a characteristic flask-like shape.

Extraintestinal infection occurs when amoebas invade the viscera of the peritoneal cavity. The most common site of invasion is the liver. Here, abscesses containing necrotic tissue and trophozoites develop and cause amoebic hepatitis. Another rarer complication is pulmonary amoebiasis. Other infrequent targets of infection are the spleen, adrenals, kidney, skin, and brain. Severe forms of the disease result in about a 10% fatality rate.

Pathogenesis and Virulence Factors Amoebiasis begins when viable cysts are swallowed and arrive in the small intestine, where the alkaline pH and digestive juices of this environment stimulate excystment. Each cyst releases four trophozoites, which are swept into the cecum and large intestine. There, the trophozoites attach by fine pseudopods **(figure 22.24),** multiply, actively move about, and feed. In about 90% of patients, infection is asymptomatic or very mild, and the trophozoites do not invade beyond the most superficial layer. The severity of the infection can vary with the strain of the parasite, inoculum size, diet, and host resistance.

The secretion of lytic enzymes by the amoeba seems to induce apoptosis of host cells. This means that the host is contributing to the process by destroying its own tissues on cue from the protozoan. The invasiveness of the amoeba is also a clear contributor to its pathogenicity.

Transmission and Epidemiology of Amoebiasis *Entamoeba* is harbored by chronic carriers whose intestines favor the encystment stage of the life cycle. Cyst formation cannot occur in active dysentery because the feces are so rapidly flushed from the body; but after recuperation, cysts are continuously shed in feces.

Humans are the primary hosts of *E. histolytica.* Infection is usually acquired by ingesting food or drink contaminated with cysts released by an asymptomatic carrier. The amoeba is thought to be carried in the intestines of one-tenth of the world's population, and it kills up to 100,000 people a year. Its geographic distribution is partly due to local sewage disposal and fertilization practices. Occurrence is highest in tropical regions (Africa, Asia, and Latin America), where night soil (human excrement) or untreated sewage is used to fertilize crops and sanitation of water and food can be substandard. Although the prevalence of the disease is lower in the United States, as many as 10 million people could harbor the agent.

Epidemics of amoebiasis are infrequent but have been documented in prisons, hospitals, juvenile care institutions, and communities where water supplies are polluted. Amoebic infections can also be transmitted by anal-oral sexual contact.

Culture and Diagnosis Diagnosis of this protozoal infection relies on a combination of tests, including microscopic examination of stool for the characteristic cysts or trophozoites, ELISA tests of stool for *E. histolytica* antigens, and serological testing for the presence of antibodies to the pathogen. PCR testing is currently being refined. It is important to differentiate *E. histolytica* from the similar *Entamoeba coli* and *Entamoeba dispar,* which occur as normal biota.

Prevention and Treatment No vaccine yet exists for *E. histolytica,* although several are in development. Prevention of the disease therefore relies on purification of water. Because regular chlorination of water supplies does not kill cysts, more rigorous methods such as boiling or iodine are required.

Effective treatment usually involves the use of drugs such as iodoquinol, which acts in the feces, and metronidazole or

CHECKPOINT 22.7 Chronic Diarrhea

Causative Organism(s)	Enteroaggregative *E. coli* (EAEC)	*Cyclospora cayetanensis*	*Giardia lamblia*	*Entamoeba histolytica*
Most Common Modes of Transmission	Vehicle (food, water), fecal-oral	Fecal-oral, vehicle	Vehicle, fecal-oral, direct and indirect contact	Vehicle, fecal-oral
Virulence Factors	?	Invasiveness	Attachment to intestines alters mucosa	Lytic enzymes, induction of apoptosis, invasiveness
Culture/Diagnosis	Difficult to distinguish from other *E. coli*	Stool examination, PCR	Stool examination, ELISA	Stool examination, ELISA, serology
Prevention	?	Washing, cooking food, personal hygiene	Water hygiene, personal hygiene	Water hygiene, personal hygiene
Treatment	None, or ciprofloxacin	TMP-SMZ	Quinacrine, metronidazole	Iodoquinol plus metronidazole or chloroquine, Flagyl
Fever Present	No	Usually	Not usually	Yes
Blood in Stool	Sometimes, mucus also	No	No, mucus present (greasy and malodorous)	Yes
Distinctive Features	Chronic in the malnourished	–	Frequently occurs in backpackers, campers	–

chloroquine, which work in the tissues. Flagyl is used as well. Dehydroemetine is used to control symptoms, but it will not cure the disease. Other drugs are given to relieve diarrhea and cramps, while lost fluid and electrolytes are replaced by oral or intravenous therapy. Infection with *E. histolytica* provokes antibody formation against several antigens, but permanent immunity is unlikely and reinfection can occur **(Checkpoint 22.7).**

Hepatitis

When certain viruses infect the liver, they cause **hepatitis,** an inflammatory disease marked by necrosis of hepatocytes and a mononuclear response that swells and disrupts the liver architecture. This pathologic change interferes with the liver's excretion of bile pigments such as bilirubin into the intestine. When bilirubin, a greenish-yellow pigment, accumulates in the blood and tissues, it causes **jaundice,** a yellow tinge in the skin and eyes. The condition can be caused by a variety of different viruses. They are all named hepatitis viruses but only because they all can cause this inflammatory condition in the liver.

Note that noninfectious conditions can also cause inflammation and disease in the liver, including some autoimmune conditions, drugs, and alcohol overuse.

Hepatitis A Virus

Hepatitis A virus (HAV) is a nonenveloped, single-stranded RNA enterovirus. It belongs to the family Picornaviridae. In general, HAV disease is far milder and shorter term than the other forms.

Signs and Symptoms Most infections by this virus are either subclinical or accompanied by vague, flulike symptoms. In more overt cases, the presenting symptoms may include jaundice and swollen liver. Darkened urine is often seen in this and other hepatitises. Jaundice is present in only about 10% of the cases. Hepatitis A occasionally occurs as a fulminating disease and causes liver damage, but this manifestation is quite rare. The virus is not **oncogenic** (cancer causing), and complete uncomplicated recovery results.

Pathogenesis and Virulence Factors The hepatitis A virus is generally of low virulence. Most of the pathogenic effects are thought to be the result of host response to the presence of virus in the liver.

Transmission and Epidemiology There is an important distinction between this virus and hepatitis B and C viruses: Hepatitis A virus is spread through the fecal-oral route (and is sometimes known as infectious hepatitis). In general, the disease is associated with deficient personal hygiene and lack of public health measures. In countries with inadequate sewage control, most outbreaks are associated with fecally contaminated water and food. The United States has a yearly reported incidence of 15,000 to 20,000 cases. Most of these result from close institutional contact, unhygienic food handling, eating shellfish, sexual transmission, or travel to other countries. In 2003, the largest single hepatitis A outbreak to date in the United States was traced to contaminated green onions used in salsa dips at a Mexican restaurant. At least 600 people who had eaten at the restaurant fell ill with hepatitis A.

Hepatitis A occasionally can be spread by blood or blood products, but this is the exception rather than the rule. In developing countries, children are the most common victims, because exposure to the virus tends to occur early in life, whereas in North America and Europe, more cases appear in adults. Because the virus is not carried chronically, the principal reservoirs are asymptomatic, short-term carriers (often children) or people with clinical disease.

Culture and Diagnosis Diagnosis of the disease is aided by detection of anti-HAV IgM antibodies produced early in the infection and by tests to identify HA antigen or virus directly in stool samples.

Prevention and Treatment Prevention of hepatitis A is based primarily on immunization. An inactivated viral vaccine (Havrix) is currently approved, and an oral vaccine based on an attenuated strain of virus is in development. Short-term protection can be conferred by passive immune globulin. This treatment is useful for people who have come in contact with HAV-infected individuals, or who have eaten at a restaurant that was the source of a recent outbreak. In the 2003 green onion outbreak, 9,000 patrons of the Mexican restaurant received passive immunization as a precaution. A combined hepatitis A/hepatitis B vaccine, called Twinrix, is recommended for people who may be at risk for both diseases, such as people with chronic liver disfunction, intravenous drug users, and men who have sex with men. Travelers to areas with high rates of both diseases should obtain vaccine coverage as well.

No specific medicine is available for hepatitis A once the symptoms begin. Drinking lots of fluids and avoiding liver irritants such as aspirin or alcohol will speed recovery. Patients who receive immune globulin early in the disease usually experience milder symptoms than patients who do not receive it.

A NOTE ABOUT HEPATITIS E

Another RNA virus, called hepatitis E, causes a type of hepatitis very similar to that caused by hepatitis A. It is transmitted by the fecal-oral route, although it does not seem to be transmitted person to person. It is usually self-limiting, except in the case of pregnant women for whom the fatality rate is 15% to 25%. It is more common in developing countries, and almost all of the cases reported in the United States occur in people who have traveled to these regions. There is currently no vaccine.

Hepatitis B Virus

Hepatitis B virus (HBV) is an enveloped DNA virus in the family Hepadnaviridae. Intact viruses are often called Dane particles. An antigen of clinical and immunologic significance is the surface (or S) antigen. The genome is partly double-stranded and partly single-stranded.

Signs and Symptoms In addition to the direct damage to liver cells just outlined, the spectrum of hepatitis disease may include fever, chills, malaise, anorexia, abdominal discomfort, diarrhea, and nausea. Rashes may appear and arthritis may occur. Hepatitis B infection can be very serious, even life-threatening. A small number of patients develop glomerulonephritis and arterial inflammation. Complete liver regeneration and restored function occur in most patients; however, a small number of patients develop chronic liver disease in the form of necrosis or **cirrhosis** (permanent liver scarring and loss of tissue). In some cases, chronic HBV infection can lead to a malignant condition.

Patients who become infected as children have significantly higher risks of long-term infection and disease. In fact, 90% of neonates infected at birth develop chronic infection, as do 30% of children infected between the ages of 1 and 5, but only 6% of persons infected after the age of 5. This finding is one of the major justifications for the routine vaccination of children. The mortality rate is 15% to 25% for people with chronic infection.

The association of HBV with **hepatocellular carcinoma** is based on these observations:

1. Certain hepatitis B antigens are found in malignant cells and are often detected as integrated components of the host genome.
2. Persistent carriers of the virus are more likely to develop this cancer.
3. People from areas of the world with a high incidence of hepatitis B (Africa and the Far East) are more frequently affected by liver cancer.

In addition, investigators have found that mass vaccination against HBV in Taiwan, begun 18 years ago, has resulted in a significant decrease in liver cancer in that country. (Taiwan previously had one of the highest rates of this cancer.) It is speculated that cancer is probably a result of infection early in life and the long-term carrier state. In general, people with chronic hepatitis are 200 times more likely to develop liver cancer, though the exact role of the virus is still the object of molecular analysis.

Some patients infected with hepatitis B are coinfected with a particle called the delta agent, sometimes also called a hepatitis D virus. This agent seems to be a defective RNA virus that cannot produce infection unless a cell is also infected with HBV. Hepatitis D virus invades host cells by "borrowing" the outer receptors of HBV. When HBV infection is accompanied by the delta agent, the disease becomes more severe and is more likely to progress to permanent liver damage.

Pathogenesis and Virulence Factors The hepatitis B virus enters the body through a break in the skin or mucous membrane or by injection into the bloodstream. Eventually, it reaches the liver cells (hepatocytes) where it multiplies and releases viruses into the blood during an incubation period of 4 to 24 weeks (7 weeks average). Surprisingly, the majority of those infected exhibit few overt symptoms and eventually develop an immunity to HBV, but some people experience the symptoms described earlier. The precise mechanisms of virulence are not clear. The ability of HBV to remain latent in some patients contributes to its pathogenesis.

Transmission and Epidemiology An important factor in the transmission pattern of hepatitis B virus is that it multiplies exclusively in the liver, which continuously seeds the blood with viruses. Electron microscopic studies have revealed up to 10^7 virions per milliliter of infected blood. Even a minute amount of blood (a *millionth* of a milliliter) can transmit infection. The abundance of circulating virions is so high and the minimal dose so low that such simple practices as sharing a toothbrush or a razor can transmit the infection. Over the past 10 years, HBV has also been detected in semen and vaginal secretions, and it can be transmitted by these fluids. Spread of the virus by means of close contact in families or institutions is also well documented. Vertical transmission is possible, and it predisposes the child to development of the carrier state and increased risk of liver cancer. It is sometimes known as *serum hepatitis.*

Hepatitis B is an ancient disease that has been found in all populations, although the incidence and risk are highest among people living under crowded conditions, drug addicts, the sexually promiscuous, and those in certain occupations, including people who conduct medical procedures involving blood or blood products.

This virus is one of the major infectious concerns for health care workers. Needle sticks can easily transmit the virus, and therefore most workers are required to have the full series of HBV vaccinations. Unlike the more notorious HIV, HBV remains infective for days in dried blood, for months when stored in serum at room temperature, and for decades if frozen. Although it is not inactivated after 4 hours of exposure to 60°C, boiling for the same period can destroy it. Disinfectants containing chlorine, iodine, and glutaraldehyde show potent anti–hepatitis B activity.

Cosmetic manipulation such as tattooing and ear or body piercing can expose a person to infection if the instruments are not properly sterilized. The only reliable method for destroying HBV on reusable instruments is autoclaving.

Culture and Diagnosis Serological tests can detect either virus antigen or antibodies. Radioimmunoassay and ELISA testing permit detection of the important surface antigen of HBV very early in infection. These same tests are essential for screening blood destined for transfusions, semen in sperm banks, and organs intended for transplant. Antibody tests are most valuable in patients who are negative for the antigen.

Prevention and Treatment Since 1981, the primary prevention for HBV infection is vaccination. The most widely used vaccines are recombinant, containing the pure surface antigen cloned in yeast cells. Vaccines are given in three doses over 18 months, with occasional boosters. Vaccination is a must for medical and dental workers and students, patients receiving multiple transfusions, immunodeficient persons, and cancer patients. The vaccine is also now strongly recommended for all newborns as part of a routine immunization schedule. As just mentioned, a combined vaccine for HAV/HBV may be appropriate for certain people.

Passive immunization with hepatitis B immune globulin (HBIG) gives significant immediate protection to people who have been exposed to the virus through needle puncture, broken blood containers, or skin and mucosal contact with blood. Another group for whom passive immunization is highly recommended is neonates born to infected mothers.

Mild cases of hepatitis B are managed by symptomatic treatment and supportive care. Chronic infection can be controlled with recombinant human interferon, adefovir dipivoxil, lamivudine (another nucleotide analog best known for its use in HIV patients), or a newly approved drug called entecavir (Baraclude™). All of these can help to stop virus multiplication and prevent liver damage in many but not all patients. None of the drugs are considered curative.

Hepatitis C Virus

Hepatitis C is sometimes referred to as the "silent epidemic" because more than 4 million Americans are infected with the virus, but it takes many years to cause noticeable symptoms. In the United States, at least 35,000 new infections occur every year. Liver failure from hepatitis C is one of the most common reasons for liver transplants in this country. Hepatitis C is an RNA virus in the Flaviviridae family. It used to be known as "non-A non-B" virus. It is usually diagnosed with a blood test for antibodies to the virus.

Signs and Symptoms People have widely varying experiences with this infection. It shares many characteristics of hepatitis B disease, but it is much more likely to become chronic. Of those infected, 75% to 85% will remain infected indefinitely. (In contrast, only about 6% of persons who acquire hepatitis B after the age of 5 will be chronically infected.) With HCV infection, it is possible to have severe symptoms without permanent liver damage but it is more common to have chronic liver disease even if there are no overt symptoms. Cancer may also result from chronic HCV infection. Worldwide, HBV infection is the most common cause of liver cancer but in the United States it is more likely to be caused by HCV.

Pathogenesis and Virulence Factors The virus is so adept at establishing chronic infections that researchers are studying the ways that it evades immunologic detection and destruction. The virus's core protein seems to play a role in the suppression of cell-mediated immunity as well as in the production of various cytokines.

Transmission and Epidemiology This virus is acquired in similar ways to HBV. It is more commonly transmitted through blood contact (both "sanctioned," such as in blood transfusions, and "unsanctioned," such as needle sharing by injecting drug users) than through transfer of other body fluids. Vertical transmission is also possible.

Before a test was available to test blood products for this virus, it seems to have been frequently transmitted through blood transfusions. Hemophiliacs who were treated with clotting factor prior to 1985 were infected at a high rate with

CHECKPOINT 22.8 Hepatitis			
Causative Organism(s)	Hepatitis A or E virus	Hepatitis B virus	Hepatitis C virus
Most Common Modes of Transmission	Fecal-oral, vehicle	Parenteral (blood contact), direct contact (especially sexual), vertical	Parenteral (blood contact), vertical
Virulence Factors	–	Latency	Core protein suppresses immune function?
Culture/Diagnosis	IgM serology	Serology (ELISA, radioimmunoassay)	Serology
Prevention	Hepatitis A vaccine or combined HAV/HBV vaccine	HBV recombinant vaccine	–
Treatment	Immune globulin	Interferon, nucleoside analogs	(Pegylated) interferon with or without ribavirin
Long-Term Consequences	None	Chronic infection, liver cancer, death	Chronic infection and liver disease very common; cancer, death
Incubation Period	2–7 weeks	1–6 months	2–8 weeks

HCV. Once blood began to be tested for HIV (in 1985) and screened for so-called "non-A non-B" hepatitis, the risk of contracting HCV from blood was greatly reduced. The current risk for transfusion-associated HCV is thought to be 1 in 100,000 units transfused.

Because HCV was not recognized sooner, a relatively large percentage of the population is infected. Eighty percent of the 4 million affected in this country are suspected to have no symptoms. It has a very high prevalence in parts of South America, Central Africa, and in China.

Prevention and Treatment There is currently no vaccine for hepatitis C. Various treatment regimens have been attempted; most include the use of therapeutic interferon and a more effective derivative of interferon called pegylated interferon. Some clinicians also prescribe ribavirin to try to suppress viral multiplication. The treatments are not curative, but they may prevent or lessen damage to the liver **(Checkpoint 22.8).**

Helminthic Intestinal Infections

Helminths that parasitize humans are amazingly diverse, ranging from barely visible roundworms (0.3 mm) to huge tapeworms (25 m long). In the introduction to these organisms in chapter 5, we grouped them into three categories: nematodes (roundworms), trematodes (flukes), and cestodes (tapeworms), and we discussed basic characteristics of each group. You may wish to review those sections before continuing. In this section, we examine the intestinal diseases caused by helminths. Although they can cause symptoms that might be mistaken for some of the diseases discussed elsewhere in this chapter, helminthic diseases are usually accompanied by an additional set of symptoms that arise from the host response to helminths. Worm infection usually provokes an increase in granular leukocytes called eosinophils, which have a specialized capacity to destroy worms. This increase, termed **eosinophilia,** is a hallmark of helminthic infection and is detectable in blood counts. If the following symptoms occur coupled with eosinophilia, helminthic infection should be suspected.

Helminthic infections may be acquired through the fecal-oral route or through penetration of the skin, but most of them spend part of their lives in the intestinal tract. (**Figure 22.25** depicts the four different types of life cycles of the helminths.) While the worms are in the intestines, they can produce a gamut of intestinal symptoms. Some of them also produce symptoms outside of the intestines; they are considered in separate categories.

General Clinical Considerations

This section on helminthic intestinal infections is organized a bit differently: We talk about diagnosis, pathogenesis and prevention, and treatment of the helminths as a group in the next subsections. Each type of infection is then described in the sections that follow.

Pathogenesis and Virulence Factors in General In most cases, helminths that infect humans do not have sophisticated virulence factors. They do have numerous adaptations that allow them to survive in their hosts. They have specialized mouthparts for attaching to tissues and for feeding, enzymes with which they liquefy and penetrate tissues, and a cuticle or other covering to protect them from host defenses. In addition, their organ systems are usually reduced to the essentials: getting food and processing it, moving, and reproducing. The damage they cause in the host is very often the result of the host's response to the presence of the invader.

Many helminths have more than one host during their lifetimes. If this is the case, the host in which the adult worm is found is called the **definitive host** (usually a vertebrate). Sometimes the actual definitive host is not the host usually used by the parasite but an accidental bystander. Humans often become the accidental definitive hosts for helminths whose normal definitive host is a cow, pig, or fish. Larval

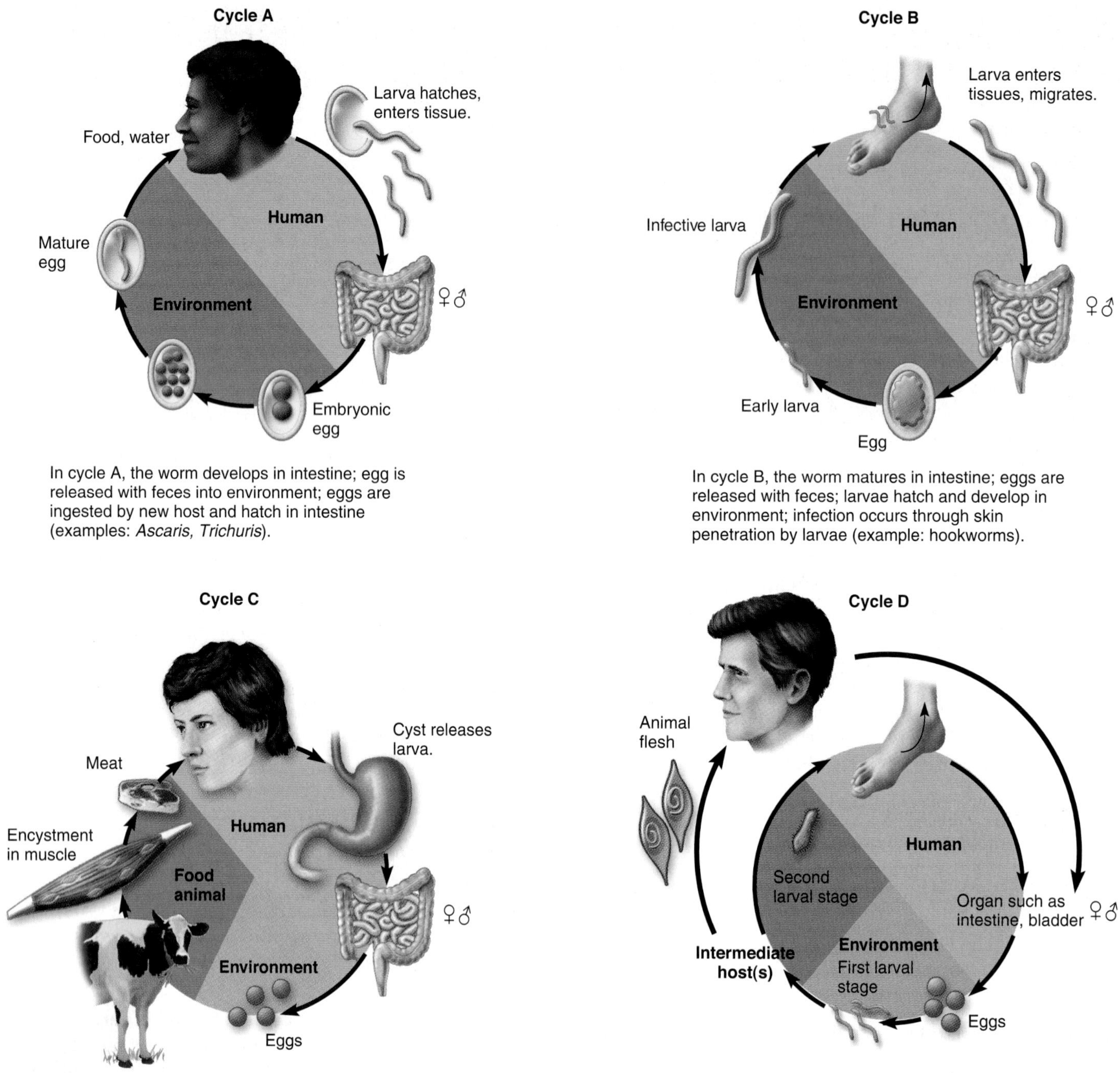

In cycle A, the worm develops in intestine; egg is released with feces into environment; eggs are ingested by new host and hatch in intestine (examples: *Ascaris, Trichuris*).

In cycle B, the worm matures in intestine; eggs are released with feces; larvae hatch and develop in environment; infection occurs through skin penetration by larvae (example: hookworms).

In cycle C, the adult matures in human intestine; eggs are released into environment; eggs are eaten by grazing animals; larval forms encyst in tissue; humans eating animal flesh are infected (example: *Taenia*).

In cycle D, eggs are released from human; humans are infected through ingestion or direct penetration by larval phase (examples: *Opisthorchis* and *Schistosoma*).

Figure 22.25 **Four basic helminth life and transmission cycles.**

stages of helminths are found in intermediate hosts. Humans can serve as intermediate hosts, too. Helminths may require no intermediate host at all or may need one or more intermediate hosts for their entire life cycle.

Diagnosis in General Diagnosis of almost all helminthic infections follows a similar series of steps. A differential blood count showing eosinophilia and serological tests indicating sensitivity to helminthic antigens all provide indirect evidence of worm infection. A history of travel to the tropics or immigration from those regions is also helpful, even if it occurred years ago, because some flukes and nematodes persist for decades. The most definitive evidence, however, is the discovery of eggs, larvae, or adult worms in stools or other tissues. The worms are sufficiently distinct in morphology that positive identification can be based on any stage, including eggs. That said, not all of these diseases result in eggs or larval stages that can easily be found in stool.

INSIGHT 22.4 *Discovery*

Treating Inflammatory Bowel Disease with Worms?

Probably every one of us knows someone who suffers from an inflammatory bowel condition such as Crohn's disease or ulcerative colitis. Currently, there are two competing theories for the cause of these conditions. Some studies suggest that *Mycobacterium* species are responsible. The work described here supports a different etiology.

Many recent epidemiological investigations have revealed that inflammatory bowel disease (IBD) is most common in Western industrialized countries and is very rare in developing countries. More specifically, the prevalence of IBD in any given country is inversely proportional to the prevalence of helminthic infections in that country. Looking at the picture in this country, the incidence of helminthic infections decreased dramatically between the 1930s and the 1950s; the incidence of IBD began its continuous rise in the 1950s. Scientists suspect a connection here: that the *absence* of exposure to helminthic infection predisposes a person to IBD.

These researchers have developed a hypothesis that the parts of the immune system that are activated during helminthic infection begin to "malfunction" when left idle, eventually resulting in damage to host tissue. Researchers wondered whether they could "treat" IBD by exposing patients to an intestinal helminthic infection. The first studies were conducted in mice, and the results looked promising. Then researchers at the University of Iowa conducted studies in human volunteers. They selected eight patients with either Crohn's disease or ulcerative colitis and administered to them Gatorade containing 2,500 eggs of the pig whipworm *Trichuris suis*. They chose this worm because it colonizes the intestines for a few weeks and then is completely eliminated without treatment. It does not invade tissues, and the eggs that are shed in the stools are not infective.

The researchers found marked improvement in the inflammatory bowel conditions in all of the patients. They determined that the effects were of short duration and asked several of the patients to continue in the study, receiving fresh doses of *T. suis* every 3 weeks. All of these patients experienced significant and long-lasting remission of their IBD symptoms. What's more, they indicated that they would be willing to continue the treatments indefinitely. There is also evidence that this approach could work for other autoimmune disorders, such as multiple sclerosis. And scientists in England are now studying whether controlled infections with hookworms can ease the respiratory allergy symptoms of allergy sufferers. It seems that occasional contact with helminths keeps the complicated network of immunoregulatory mechanisms in good working order. This story serves to remind us about the intimate association between humans and their parasites.

Prevention and Treatment in General Preventive measures are aimed at minimizing human contact with the parasite or interrupting its life cycle. In areas where the worm is transmitted by fecally contaminated soil and water, disease rates are significantly reduced through proper sewage disposal, using sanitary latrines, avoiding human feces as fertilizer, and disinfection of the water supply. In cases where the larvae invade through the skin, people should avoid direct contact with infested water and soil. Food-borne disease can be avoided by thoroughly washing and cooking vegetables and meats. Also, because adult worms, larvae, and eggs are sensitive to cold, freezing foods is a highly satisfactory preventive measure. These methods work best if humans are the sole host of the parasite; if they are not, control of reservoirs or vector populations may be necessary.

Although several useful antihelminthic medications exist, the cellular physiology of the eukaryotic parasites resembles that of humans, and drugs toxic to them can also be toxic to us. Some antihelminthic drugs suppress a metabolic process that is more important to the worm than to the human. Others inhibit the worm's movement and prevent it from maintaining its position in a certain organ. Therapy is also based on a drug's greater toxicity to the more vulnerable helminths or on the local effects of oral drugs in the intestine. Antihelminthic drugs of choice and their effects are given in **table 22.1.** Note that some helminths have developed resistance to the drugs used to treat them. In some cases, surgery may be necessary to remove worms or larvae, although this procedure can be difficult if the parasite load is high or is not confined to one area.

Intestinal Distress as the Primary Symptom

Both tapeworms and roundworms can infect the intestinal tract in such a way as to cause primary symptoms there. The pork tapeworm (*Taenia solium*) and the fish tapeworm (*Diphyllobothrium latum*) are highlighted, as well as two nematodes (roundworms): the whipworm *Trichuris trichiura* and the pinworm *Enterobius vermicularis*. Both of the roundworms are deposited in the small intestine and migrate to the large intestine. We start with these.

TABLE 22.1 Antihelminthic Therapeutic Agents and Their Effects

Drug	Effect
Piperazine	Paralyzes worm so it can be expelled in feces
Pyrantel	Paralyzes worm so it can be expelled in feces
Mebendazole	Blocks key step in worm metabolism
Thiabendazole	Blocks key step in worm metabolism
Praziquantel	Interferes with worm metabolism
Niclosamide	Inhibits ATP formation in worm; destroys proglottids but not eggs

Trichuris trichiura The common name for this nematode—whipworm—refers to its likeness to a miniature buggy whip. Its life cycle and transmission is of the cycle A type (see figure 22.25). Humans are the sole host. Trichuriasis has its highest incidence in areas of the tropics and subtropics that have poor sanitation. Embryonic eggs deposited in the soil are not immediately infective and continue development for 3 to 6 weeks in this habitat. Ingested eggs hatch in the small intestine, where the larvae attach, penetrate the outer wall, and go through several molts. The mature adults move to the large intestine and gain a hold with their long, thin heads, while the thicker tail dangles free in the intestinal lumen. Following sexual maturation and fertilization, the females eventually lay 3,000 to 5,000 eggs daily into the bowel. The entire cycle requires about 90 days, and untreated infection can last up to 2 years.

Symptoms of this infection may include localized hemorrhage of the bowel caused by worms burrowing and piercing intestinal mucosa. This can also provide a portal of entry for secondary bacterial infection. Heavier infections can cause dysentery, loss of muscle tone, and rectal prolapse, which can prove fatal in children.

Enterobius vermicularis This nematode is often called the pinworm, or seatworm. It is the most common worm disease of children in temperate zones. Some estimates put the prevalence of this infection in the United States at 5% to 15%, although most experts feel that this has declined in recent years. The transmission of this roundworm is of the cycle A type. Freshly deposited eggs have a sticky coating that causes them to lodge beneath the fingernails and to adhere to fomites. Upon drying, the eggs become airborne and settle in house dust. Worms are ingested from contaminated food or drink and from self-inoculation from one's own fingers. Eggs hatch in the small intestine and release larvae that migrate to the large intestine. There the larvae mature into adult worms and mate.

The symptoms of this condition are pronounced anal itching when the mature female emerges from the anus and lays eggs. Although infection is not fatal and most cases are asymptomatic, the afflicted child can suffer from disrupted sleep and sometimes nausea, abdominal discomfort, and diarrhea. A simple rapid test can be performed by pressing a piece of transparent adhesive tape against the anal skin and then applying it to a slide for microscopic examination. When one member of the family is diagnosed, the entire family should be tested and/or treated because it is likely that multiple members are infected.

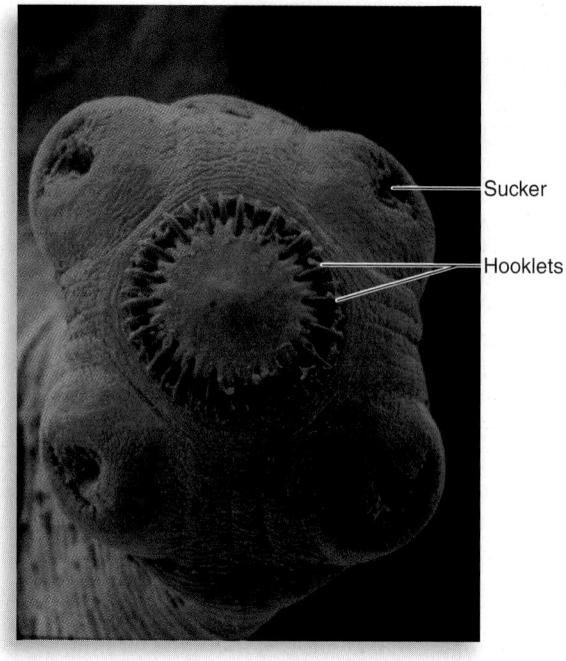

(a)

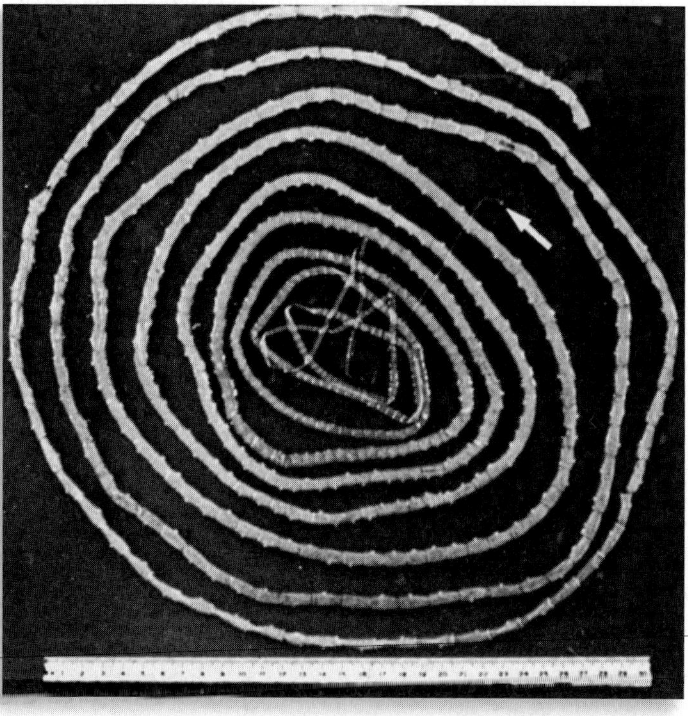

(b)

Figure 22.26 Tapeworm characteristics.
(a) Tapeworm scolex showing sucker and hooklets. **(b)** Adult *Taenia saginata.* The arrow points to the scolex; the remainder of the tape, called the strobila, has a total length of 5 meters.

Taenia solium In contrast to the last two helminths, this one is a tapeworm. Adult worms are usually around 5 meters long and have a scolex with hooklets and suckers to attach to the intestine **(figure 22.26).** Taeniasis caused by the *T. solium* (the pig tapeworm) is distributed worldwide but is mainly concentrated in areas where humans live in close proximity with pigs or eat undercooked pork. In pigs, the eggs hatch in the small intestine and the released larvae migrate throughout the organs. Ultimately, they encyst in the muscles, becoming *cysticerci,* young tapeworms that are the infective stage for humans. When humans ingest a live cysticercus in pork, the coat is digested and the organism is flushed into the intestine, where it firmly attaches by the scolex and develops into an adult tapeworm. Infection with *T. solium* can take

CHECKPOINT 22.9 Intestinal Distress

Causative Organism(s)	*Trichuris trichiura* (whipworm)	*Enterobius vermicularis* (pinworm)	*Taenia solium* (pork tapeworm)	*Diphyllobothrium latum* (fish tapeworm)	*Hymenolepis nana* and *H. diminuta*
Most Common Modes of Transmission	Cycle A: vehicle (soil)/fecal-oral	Cycle A: vehicle (food, water), fomites, self-inoculation	Cycle C: vehicle (pork)—also fecal-oral	Cycle C: vehicle (seafood)	Cycle C: vehicle (ingesting insects)—also fecal-oral
Virulence Factors	Burrowing and invasiveness	–	–	Vitamin B12 usage	–
Culture/ Diagnosis	Blood count, serology, egg or worm detection	Adhesive tape method	Blood count, , serology, egg or worm detection	Blood count, serology, egg or worm detection	Blood count, serology, egg or worm detection
Prevention	Hygiene, sanitation	Hygiene	Cook meat, avoid pig feces	Cook meat	Hygienic environment
Treatment	Mebendazole	Piperazine, pyrantel	Praziquantel, Niclosamide	Praziquantel, Niclosamide	Praziquantel
Distinctive Features	Humans sole host	Common in United States	Tapeworm; intermediate host is pigs	Large tapeworm; anemia	Most common tapeworm infection

another form when humans ingest the tapeworm eggs rather than cysticerci. Although humans are not the usual intermediate hosts, the eggs can still hatch in the intestine, releasing tapeworm larvae that migrate to all tissues. They form bladderlike sacs throughout the body that can cause serious damage. This transmission and life cycle are shown in cycle C in figure 22.25. The pork tapeworm is not the same as the more commonly known pork helminthic infection, *trichinosis*. It is discussed in a later section.

For such a large organism, it is remarkable how few symptoms a tapeworm causes. Occasionally, a patient discovers proglottids in his or her stool, and some patients complain of vague abdominal pain and nausea.

Other tapeworms of the genus *Taenia* infect humans. One of them is the beef tapeworm, *Taenia saginata*. It usually causes similar general symptoms of helminthic infection. But humans are not known to acquire *T. saginata* infection by ingesting the eggs.

Diphyllobothrium latum This tapeworm has an intermediate host in fish. It is common in the Great Lakes, Alaska, and Canada. Humans are its definitive host. It develops in the intestine and can cause long-term symptoms. It can be transmitted in raw food such as sushi and sashimi made from salmon. (Reputable sushi restaurants employ authentic sushi chefs who are trained to carefully examine fish for larvae and other signs of infection.)

As is the case with most tapeworms, symptoms are minor and usually vague and include possible abdominal discomfort or nausea. The tapeworm seems to have the ability to absorb and use the vitamin B12, making it unavailable to its human host. Anemia is therefore sometimes reported with this infection.

***Hymenolepis* species** These relatively small tapeworms are the most common tapeworm infections in the world. There are two species: *H. nana*, known as the dwarf tapeworm because it is only 15 to 40 mm in length, and *H. diminuta*, the rat tapeworm, which is usually 20 to 60 cm in length as an adult.

The life cycle of these tapeworms often involves insects as well as the definitive host, which may be a rodent or a human. When eggs are passed in the feces of a rodent or human, they can be ingested by various insects, which are in turn accidentally ingested by humans (in cereals or other foods). Alternatively, eggs in the environment can be directly ingested by humans. Tapeworms become established in the small intestine and eggs can be released after proglottids break off from the attached worms.

Symptoms are mild, and the treatment of choice is praziquantel **(Checkpoint 22.9).**

Intestinal Distress Accompanied by Migratory Symptoms

A diverse group of helminths enter the body as larvae or eggs, mature to the worm stage in the intestine, and then migrate into the circulatory and lymphatic systems, after which they travel to the heart and lungs, migrate up the respiratory tree to the throat, and are swallowed. This journey returns the mature worms to the intestinal tract where they then take up residence. All of these conditions, in addition to causing symptoms in the digestive tract, may induce inflammatory reactions along their migratory routes, resulting in eosinophilia and, during their lung stage, pneumonia. Three different examples of this type of infection follow.

Ascaris lumbricoides *Ascaris lumbricoides* is a giant intestinal roundworm (up to 300 mm long) that probably accounts for the greatest number of worm infections (estimated at 1 billion cases worldwide). Most reported cases in the United States occur in the southeastern states. *Ascaris* spends its larval and adult stages in humans and releases embryonic eggs in feces, which are then spread to other humans through food, drink, or contaminated objects placed in the mouth. The eggs thrive in warm, moist soils and resist cold and chemical disinfectants, but they are sensitive to sunlight, high temperatures, and drying. After ingested eggs hatch in the human intestine, the larvae embark upon an odyssey in the tissues. First, they penetrate the intestinal wall and enter the lymphatic and circulatory systems. They are swept into the heart and eventually arrive at the capillaries of the lungs. From this point, the larvae migrate up the respiratory tree to the glottis. Worms entering the throat are swallowed and returned to the small intestine, where they reach adulthood and reproduce, producing up to 200,000 fertilized eggs a day.

Even as adults, male and female worms are not attached to the intestine and retain some of their exploratory ways. They are known to invade the biliary channels of the liver and gallbladder, and on occasion the worms emerge from the nose and mouth. Severe inflammatory reactions mark the migratory route; and allergic reactions such as bronchospasm, asthma, or skin rash can occur. Heavy worm loads can retard the physical and mental development of children. One possibility with intestinal worm infections is self-reinoculation due to poor personal hygiene.

Necator americanus* and *Ancylostoma duodenale These two different nematodes are called by the common name hookworm. *Necator americanus* (nee-kay'-tor ah-mer"-ih-cah'-nus) is endemic to the New World, and *Ancylostoma duodenale* (an'-kih-los'-toh-mah doo-oh-den-ah'-lee) is endemic to the Old World, although the two species overlap in parts of Latin America. Otherwise, with respect to transmission, life cycle, and pathology, they are usually lumped together. The *hook* refers to the adult's oral cutting plates, by which it anchors to the intestinal villi, and its curved anterior end **(figure 22.27).**

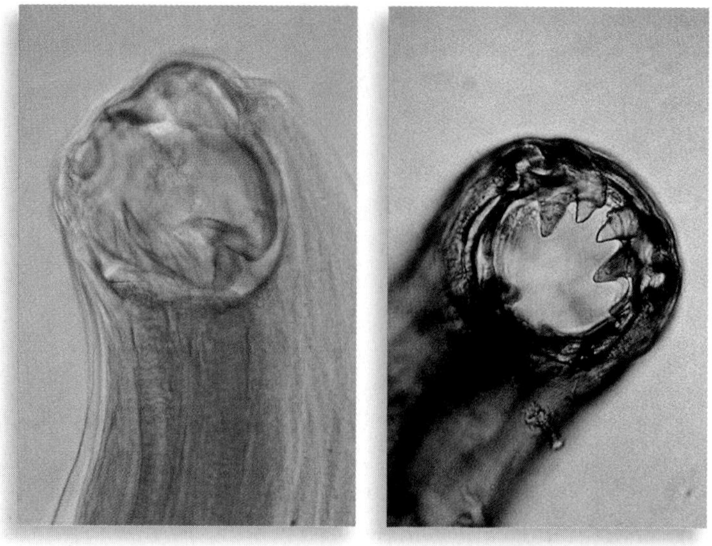

Figure 22.27 **Cutting teeth on the mouths of (left) *Necator americanus* and (right) *Ancylostoma duodenale.***

Unlike other intestinal worms, hookworm larvae hatch outside the body and infect by penetrating the skin. Hookworm transmission is described by cycle B (see figure 22.25). Ordinarily, the parasite is present in soil contaminated with human feces. It enters sites on bare feet such as hair follicles, abrasions, or the soft skin between the toes, but cases have occurred via mud that was splattered on the ankles of

CHECKPOINT 22.10 Intestinal Distress plus Migratory Symptoms

Causative Organism(s)	*Ascaris lumbricoides* (intestinal roundworm)	*Necator americanus* and *Ancylostoma duodenale* (hookworms)	*Strongyloides stercoralis* (threadworm)
Most Common Modes of Transmission	Cycle A: vehicle (soil/fecal-oral), fomites, self-inoculation	Cycle B: vehicle (soil), fomite	Cycle B: vehicle (soil), fomite
Virulence Factors	Induction of hypersensitivity, adult worm migration, and abdominal obstruction	Induction of hypersensitivity, adult worm migration, and abdominal obstruction	Induction of hypersensitivity, adult worm migration, and abdominal obstruction
Culture/Diagnosis	Blood count, serology, egg or worm detection	Blood count, serology, egg or worm detection	Blood count, serology, egg or worm detection
Prevention	Hygiene	Sanitation	Sanitation
Treatment	Alebendazole	Alebendazole	Invermectin or thiabendazole
Distinctive Features	Roundworm; 1 billion persons infected	Penetrates skin, serious intestinal symptoms	Penetrates skin, severe for immunocompromised

people wearing shoes. Infection has even been reported in people handling soiled laundry.

On contact, the hookworm larvae actively burrow into the skin. After several hours, they reach the lymphatic or blood circulation and are immediately carried into the heart and lungs. The larvae proceed up the bronchi and trachea to the throat. Most of the larvae are swallowed with sputum and arrive in the small intestine, where they anchor, feed on blood, and mature. Eggs first appear in the stool about 6 weeks after the time of entry, and the untreated infection can last about 5 years.

Symptoms from these infections follow the progress of the worm in the body. A localized dermatitis called *ground itch* may be caused by the initial penetration of larvae. The transit of the larvae to the lungs is ordinarily brief, but it can cause symptoms of pneumonia and eosinophilia. The potential for injury is greatest during the intestinal phase, when heavy worm burdens can cause nausea, vomiting, cramps, and bloody diarrhea. Because blood loss is significant, iron-deficient anemia develops, and infants are especially susceptible to hemorrhagic shock. Chronic fatigue, listlessness, apathy, and anemia worsen with chronic and repeated infections.

Hookworm infections are treated with antihelminthic drugs, but frequent reinfection is a problem. U.S. and Brazilian researchers are testing a vaccine against *Necator americanus*. In 2000, the Bill and Melinda Gates Foundation, recognizing the impact of worldwide hookworm infections, contributed $18 million to the development of a hookworm vaccine.

Strongyloides stercoralis The agent of strongyloidiasis, or threadworm infection, is *Strongyloides stercoralis* (stron′-jih-loy-deez ster″-kor-ah′-lis). This nematode is exceptional because of its minute size and its capacity to complete its life cycle either within the human body or outside in moist soil. It shares a similar distribution and life cycle to hookworms and afflicts an estimated 100 to 200 million people worldwide. Infection occurs when soil larvae penetrate the skin (cycle B in figure 22.25). The worm then enters the circulation, is carried to the respiratory tract and swallowed, and then enters the small intestine to complete development. Although adult *S. stercoralis* lays eggs in the gut just as hookworms do, the eggs hatch into larvae in the colon and can remain entirely in the host's body to complete the cycle. The larval form of the organism can likewise exit with feces and go through an environmental cycle. These numerous alternative life cycles greatly increase the chance of transmission and the likelihood for chronic infection.

The first symptom of threadworm infection is usually a red, intensely itchy skin rash at the site of entry. Mild migratory activity in an otherwise normal person can escape notice, but heavy worm loads can cause symptoms of pneumonitis and eosinophilia. The nematode activities in the intestine produce bloody diarrhea, liver enlargement, and malabsorption. In immunocompromised patients, there is a risk of disseminated infection involving numerous organs **(figure 22.28).** Hardest hit are AIDS patients, transplant patients on immunosuppressant drugs, and cancer patients receiving irradiation therapy, who can die if not treated promptly.

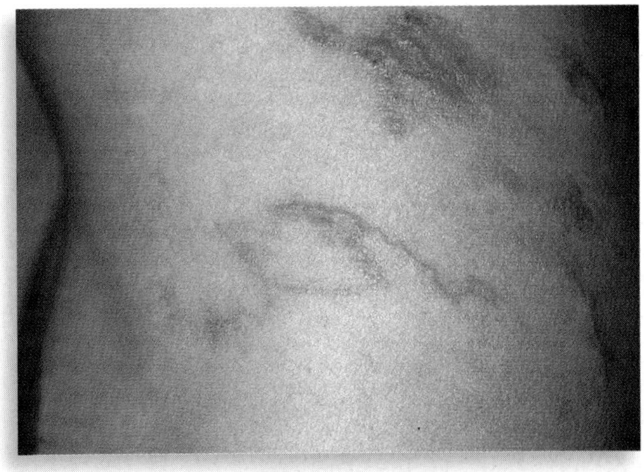

Figure 22.28 A patient with disseminated *Strongyloides* infection.

Trails under the skin indicate the migration tracks of the worms.

Liver and Intestinal Disease

One group of worms that lands in the intestines has a particular affinity for the liver. Two of these worms are trematodes (flatworms), and they are categorized as liver flukes.

Opisthorchis sinensis* and *Clonorchis sinensis *Opisthorchis sinensis* and *Clonorchis sinensis* are two worms known as Chinese liver flukes. They complete their sexual development in mammals such as humans, cats, dogs, and swine. Their intermediate development occurs in snail and fish hosts. Humans ingest cercariae in inadequately cooked or raw freshwater fish (see cycle D in figure 22.25). Larvae hatch and crawl into the bile duct, where they mature and shed eggs into the intestinal tract. Feces containing eggs are passed into standing water that harbors the intermediate snail host. The cycle is complete when infected snails release cercariae that invade fish living in the same water.

Symptoms of *Opisthorchis* and *Clonorchis* infection are slow to develop but include thickening of the lining of the bile duct and possible granuloma formation in areas of the liver if eggs enter the stroma of the liver. If the infection is heavy, the bile duct could be blocked.

Fasciola hepatica This liver fluke is a common parasite in sheep, cattle, goats, and other mammals and is occasionally transmitted to humans **(figure 22.29).** Periodic outbreaks in temperate regions of Europe and South America are associated with eating wild watercress. The life cycle is very complex, involving the mammal as the definitive host, the release of eggs in the feces, the hatching of eggs in the water into *miracidia,* invasion of freshwater snails, development

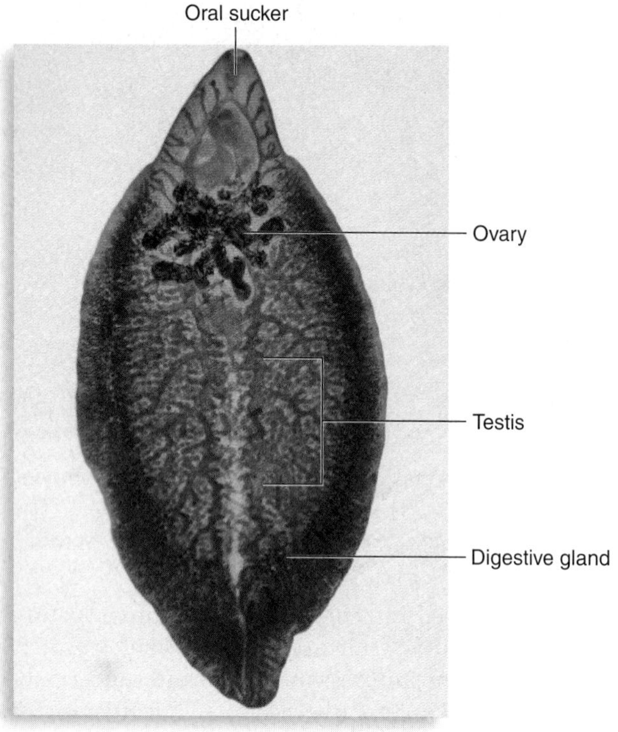

Figure 22.29 *Fasciola hepatica,* **the sheep liver fluke (2×).**

and release of cercariae, encystment of cercariae on a water plant, and ingestion of the cyst by a mammalian host eating the plant. The cysts release young flukes into the intestine that wander to the liver, lodge in the gallbladder, and develop into adults. Humans develop symptoms of vomiting, diarrhea, hepatomegaly, and bile obstruction only if they are chronically infected by a large number of flukes.

Muscle and Neurological Symptoms

Trichinosis is an infection transmitted by eating pork (and sometimes other wildlife) that have the cysts of *Trichinella* species embedded in the meat. The life cycle of this nematode is spent entirely within the body of a mammalian host such as a pig, bear, cat, dog, or rat. In nature, the parasite is maintained in an encapsulated (encysted) larval form in the muscles of these animal reservoirs and is transmitted when other animals prey upon them. The disease cannot be transmitted from one human to another except in the case of cannibalism.

Because all wild and domesticated mammals appear to be susceptible to *Trichinella* species, one might expect human trichinosis to be common worldwide. But in reality, it is more common in the United States and in Europe than in the rest of the world. This distribution appears to be related to regional or ethnic customs of eating raw or rare pork dishes or wild animal meats. Bear meat is the source of up to one-third of the cases in the United States. Home or small-scale butchering enterprises that do not carefully inspect pork can spread the parasite, although commercial pork can also be a source. Practices such as tasting raw homemade pork sausage or serving rare pork or pork-beef mixtures have been responsible for sporadic outbreaks.

The cyst envelope is digested in the stomach and small intestine, which liberates the larvae. After burrowing into the intestinal mucosa, the larvae reach adulthood and mate. The larvae that result from this union penetrate the intestine and enter the lymphatic channels and blood. All tissues are at risk for invasion, but final development occurs when the coiled larvae are encysted in the skeletal muscle. At maturity, the cyst is about 1 mm long and can be observed by careful inspection of meat. Although larvae can deteriorate over time, they have also been known to survive for years.

Symptoms may be unnoticeable or they could be life-threatening, depending on how many larvae were ingested in the tainted meat. The first symptoms, when present, mimic influenza or viral fevers, with diarrhea, nausea, abdominal pains, fever, and sweating. The second phase, brought on by the mass migration of larvae and their entrance into muscle, produces puffiness around the eyes, intense muscle and joint pain, shortness of breath, and pronounced eosinophilia. The most serious life-threatening manifestations are heart and

CHECKPOINT 22.11 Liver and Intestinal Disease

Causative Organism(s)	*Opisthorchis sinensis, Clonorchis sinensis*	*Fasciola hepatica*
Most Common Modes of Transmission	Cycle D: vehicle (fish or crustaceans)	Cycle D: vehicle (water and water plants)
Virulence Factors	–	–
Culture/Diagnosis	Blood count, serology, egg or worm detection	Blood count, serology, egg or worm detection
Prevention	Cook food, sanitation of water	Sanitation of water
Treatment	Praziquantel	Triclabendazole
Distinctive Features	Live in bile duct	Live in liver and gallbladder

CHECKPOINT 22.12 Muscle and Neurological Symptoms

Causative Organism(s)	*Trichinella* species
Most Common Modes of Transmission	Vehicle (food)
Virulence Factors	–
Culture/Diagnosis	Serology combined with clinical picture; muscle biopsy
Prevention	Cook meat
Treatment	Mebendazole, steroids
Distinctive Features	Brain and heart involvement can be fatal

brain involvement. Although the symptoms eventually subside, a cure is not available once the larvae have encysted in muscles.

The most effective preventive measures for trichinosis are to adequately store and cook pork and wild meats.

Liver Disease

When liver swelling or malfunction is accompanied by eosinophilia, **schistosomiasis** should be suspected. Schistosomiasis has afflicted humans for thousands of years. The disease is caused by the blood flukes *Schistosoma mansoni* or *S. japonicum,* species that are morphologically and geographically distinct but share similar life cycles, transmission methods, and general disease manifestations. It is one of the few infectious agents that can invade intact skin.

Signs and Symptoms The first symptoms of infection are itchiness in the area where the worm enters the body, followed by fever, chills, diarrhea, and cough. The most severe consequences, associated with chronic infection, are hepatomegaly and liver disease and splenomegaly. Other serious conditions caused by a different schistosome occur in the urinary tract—bladder obstruction and blood in the urine. This condition is discussed in chapter 23 (genitourinary tract diseases). Occasionally, eggs from the worms are carried into the central nervous system and heart and create a severe granulomatous response. Adult flukes can live for many years and, by eluding the immune defenses, cause a chronic affliction.

Causative Agent Schistosomes are trematodes, or flukes (see chapter 5), but they are more cylindrical than flat **(figure 22.30*b*).** They are often called *blood flukes.* Flukes have digestive, excretory, neuromuscular, and reproductive systems, but they lack circulatory and respiratory systems. Humans are the definitive hosts for the blood fluke, and snails are the intermediate host.

Pathogenesis and Virulence Factors This parasite is clever indeed. Once inside the host, it coats its outer surface with proteins from the host's bloodstream, basically "cloaking" itself from the host defense system. This coat reduces its surface antigenicity and allows it to remain in the host indefinitely.

Other virulence attributes are the organism's ability to invade intact skin and attach to vascular endothelium, to sequester iron from the bloodstream, and to induce a granulomatous response.

Transmission and Epidemiology The life cycle of the schistosome is complex (see figure 22.30). The cycle begins when infected humans release eggs into irrigated fields or ponds, either by deliberate fertilization with excreta or by defecating or urinating directly into the water. The egg hatches in the water and gives off an actively swimming ciliated larva called a **miracidium (figure 22.30*a*),** which

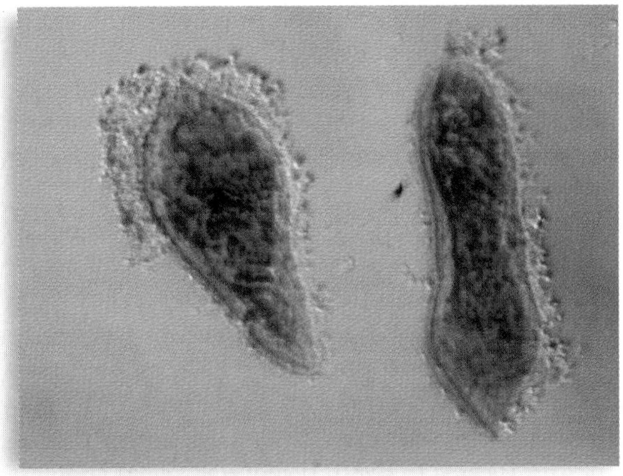

(a) The miracidium phase, which infects the snail.

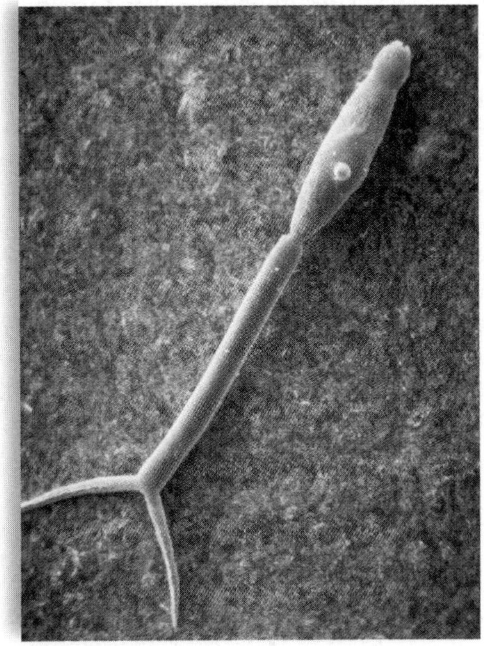

(b) The cercaria phase, which is released by snails and burrows into the human host.

(c) An electron micrograph of normal mating position of adult worms. The male worm holds the female in a groove on his ventral surface.

Figure 22.30 **Stages in the life cycle of *Schistosoma*.**

instinctively swims to a snail and burrows into a vulnerable site, shedding its ciliated covering in the process. In the body of the snail, the miracidium multiplies into a larger, fork-tailed swimming larva called a **cercaria** (see figure 22.30*b*). Cercariae are given off by the thousands into the water by infected snails.

Upon contact with a human wading or bathing in water, cercariae attach themselves to the skin by ventral suckers and penetrate into hair follicles. They pass into small blood and lymphatic vessels and are carried to the liver. Here, the schistosomes achieve sexual maturity, and the male and female worms remain permanently entwined to facilitate mating **(figure 22.30*c*).** In time, the pair migrates to and lodges in small blood vessels at specific sites. *Schistosoma mansoni* and *S. japonicum* end up in the mesenteric venules of the small intestine. While attached to these intravascular sites, the worms feed upon blood and the female lays eggs that are eventually voided in feces or urine.

The disease is endemic to 74 countries located in Africa, South America, the Middle East, and the Far East. *S. mansoni* is found throughout these regions but not in the Far East. *S. japonicum* has a much smaller geographical distribution than *S. mansoni,* only being found in the Far East. Schistosomiasis (including the urinary tract form) is the second most prominent parasitic disease after malaria, probably affecting 200 million people at any one time worldwide. Recent increases in its occurrence in Africa have been attributed to new dams on the Nile River, which have provided additional habitat for snail hosts.

Culture and Diagnosis Diagnosis depends on identifying the eggs in urine or feces. The clinical picture of hepatomegaly, splenomegaly, or both also contribute to the diagnosis.

Prevention and Treatment The cycle of infection cannot be broken as long as people are exposed to untreated sewage in their environment. It is quite common for people to be cured and then to be reinfected because their village has no sewage treatment. A vaccine would provide widespread control of the disease, but so far none is licensed. More than one vaccine is in development, however.

Praziquantel is the drug treatment of choice. It works by crippling the worms, making them more antigenic and thereby allowing the host immune response to eliminate them. Clinicians use an "egg hatching test" to determine whether an infection is current and whether treatment is actually killing the eggs. Urine or feces containing eggs are placed in room temperature water, and if miracidia emerge, the infection is still "active."

CHECKPOINT 22.13	Liver Disease
Causative Organism(s)	*Schistosoma mansoni, S. japonicum*
Most Common Modes of Transmission	Cycle D: vehicle (contaminated water)
Virulence Factors	Antigenic "cloaking"
Culture/Diagnosis	Identification of eggs in feces, scarring of intestines detected by endoscopy
Prevention	Avoiding contaminated vehicles
Treatment	Praziquantel
Distinctive Features	Penetrates skin, lodges in blood vessels of intestine, damages liver

SUMMING UP

Taxonomic Organization **Microorganisms Causing Disease in the GI Tract**

Microorganism	Disease	Chapter Location
Gram-positive endospore-forming bacteria		
Clostridium difficile	Antibiotic-associated diarrhea	Acute diarrhea, p. 702
Clostridium perfringens	Food poisoning	Acute diarrhea and/or vomiting, p. 709
Bacillus cereus	Food poisoning	Acute diarrhea and/or vomiting, p. 708
Gram-positive bacteria		
Streptococcus mutans	Dental caries	Dental caries, p. 689
Streptococcus sobrinus	Dental caries	Dental caries, p. 689
Staphylococcus aureus	Food poisoning	Acute diarrhea and/or vomiting, p. 706
Gram-negative bacteria		
Campylobacter jejuni	Acute diarrhea	Acute diarrhea, p. 701
Helicobacter pylori	Gastritis/gastric ulcers	Gastritis/gastric ulcers, p. 695
Escherichia coli O157:H7	Acute diarrhea plus hemolytic syndrome	Acute diarrhea, p. 699
Other *E. coli*	Acute or chronic diarrhea	Acute diarrhea, p. 701 Chronic diarrhea, p. 710
Salmonella	Acute diarrhea or typhoid fever	Acute diarrhea, p. 697
Shigella	Acute diarrhea and dysentery	Acute diarrhea, p. 698
Vibrio cholerae	Cholera	Acute diarrhea, p. 702
Yersinia enterocolitica and *Y. pseudotuberculosis*	Acute diarrhea	Acute diarrhea, p. 702
Tannerella forsythus, Actinobacillus actinomycetemcomitans, Porphyromonas gingivalis, Treponema vincentii, Prevotella intermedia, Fusobacterium	Periodontal disease	Periodontal disease, p. 691
DNA viruses		
Hepatitis B virus	"Serum" hepatitis	Hepatitis, p. 715
RNA viruses		
Hepatitis A virus	"Infectious" hepatitis	Hepatitis, p. 714
Hepatitis C virus	"Serum" hepatitis	Hepatitis, p. 716
Hepatitis E virus	"Infectious" hepatitis	Hepatitis, p. 715
Mumps virus	Mumps	Mumps, p. 693
Rotavirus	Acute diarrhea	Acute diarrhea, p. 705
Protozoa		
Entamoeba histolytica	Chronic diarrhea	Chronic diarrhea, p. 712
Cryptosporidium	Acute diarrhea	Acute diarrhea, p. 705
Cyclospora	Chronic diarrhea	Chronic diarrhea, p. 711
Giardia lamblia	Chronic diarrhea	Chronic diarrhea, p. 711
Helminths—nematodes		
Ascaris lumbricoides	Intestinal distress plus migratory symptoms	Intestinal distress plus migratory symptoms, p. 722
Enterobius vermicularis	Intestinal distress	Intestinal distress, p. 720
Trichuris trichiura	Intestinal distress	Intestinal distress, p. 720
Necator americanus and *Ancylostoma duodenale*	Intestinal distress plus migratory symptoms	Intestinal distress plus migratory symptoms, p. 722
Strongyloides stercoralis	Intestinal distress plus migratory symptoms	Intestinal distress plus migratory symptoms, p. 723
Trichinella spp.	Muscle and neurological symptoms	Muscle and neurological symptoms, p. 724
Helminths—cestodes		
Hymenolepis	Intestinal distress	Intestinal distress, p. 721
Taenia solium	Intestinal distress	Intestinal distress, p. 720
Diphyllobothrium latum	Intestinal distress	Intestinal distress, p. 721
Opisthorchis sinensis and *Clonorchis sinensis*	Liver and intestinal disease	Liver and intestinal disease, p. 723
Fasciola hepatica	Liver and intestinal disease	Liver and intestinal disease, p. 723
Helminths—trematodes		
Schistosoma mansoni, S. japonicum	Schistosomiasis	Helminthic liver disease, p. 725

INFECTIOUS DISEASES AFFECTING The Gastrointestinal Tract

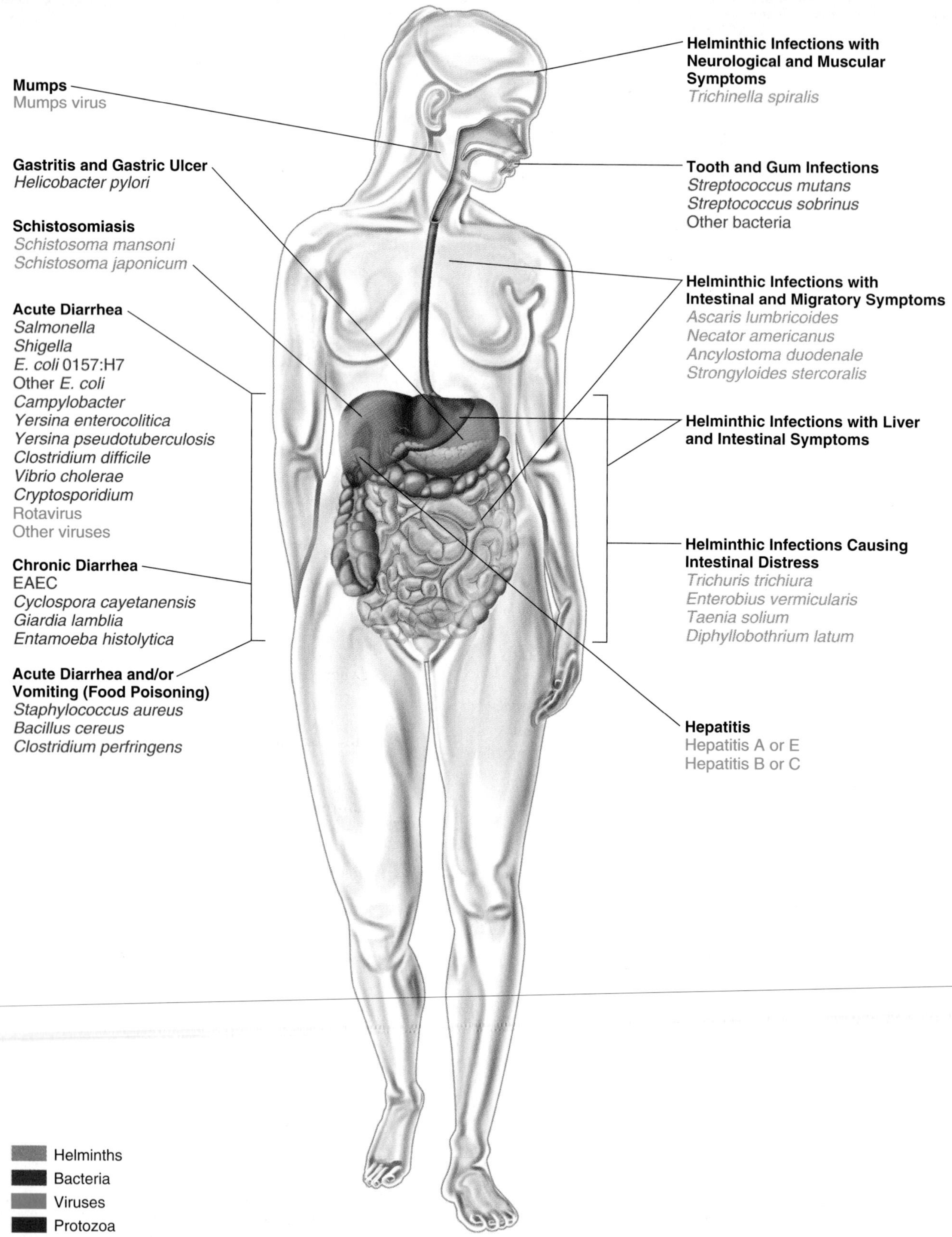

Mumps
Mumps virus
Gastritis and Gastric Ulcer
Helicobacter pylori
Schistosomiasis
Schistosoma mansoni
Schistosoma japonicum
Acute Diarrhea
Salmonella
Shigella
E. coli 0157:H7
Other E. coli
Campylobacter
Yersina enterocolitica
Yersina pseudotuberculosis
Clostridium difficile
Vibrio cholerae
Cryptosporidium
Rotavirus
Other viruses
Chronic Diarrhea
EAEC
Cyclospora cayetanensis
Giardia lamblia
Entamoeba histolytica
Acute Diarrhea and/or Vomiting (Food Poisoning)
Staphylococcus aureus
Bacillus cereus
Clostridium perfringens
Helminthic Infections with Neurological and Muscular Symptoms
Trichinella spiralis
Tooth and Gum Infections
Streptococcus mutans
Streptococcus sobrinus
Other bacteria
Helminthic Infections with Intestinal and Migratory Symptoms
Ascaris lumbricoides
Necator americanus
Ancylostoma duodenale
Strongyloides stercoralis
Helminthic Infections with Liver and Intestinal Symptoms
Helminthic Infections Causing Intestinal Distress
Trichuris trichiura
Enterobius vermicularis
Taenia solium
Diphyllobothrium latum
Hepatitis
Hepatitis A or E
Hepatitis B or C
Helminths
Bacteria
Viruses
Protozoa

System Summary Figure 22.31

Chapter Summary with Key Terms

22.1 The Gastrointestinal Tract and Its Defenses

A. The gastrointestinal (GI) tract can be thought of as a long tube, extending from mouth to anus. It is composed of *eight* main sections—the mouth, pharynx, esophagus, stomach, small intestine, large intestine, rectum, and anus, and *four* accessory organs—the salivary glands, liver, gallbladder, and pancreas.

B. The GI tract has a very heavy load of microorganisms, and it encounters millions of new ones every day. Therefore, there are significant immune defenses in the form of mechanical, chemical, and antimicrobial measures to combat microbial invasion.

22.2 Normal Biota of the Gastrointestinal Tract

Bacteria live on the teeth as well as the soft structures in the mouth. The pharynx contains a variety of microorganisms, while the esophagus and stomach have less. The small intestine is also sparsely populated with "normal" biota, but the large intestine contains billions of microorganisms—most of which are anaerobic.

22.3 Gastrointestinal Tract Diseases Caused by Microorganisms

A. **Tooth and Gum Infections:** Oral alpha-hemolytic "viridans" streptococci, *Streptococcus mutans* and *Streptococcus sobrinus* are the main causes of dental caries. Periodontal disease is mainly due to bacterial colonization and varying degrees of inflammation that occur in response to gingival damage.

B. **Periodontitis:** The anaerobic bacteria *Bacteroides forsythus, Actinobacillus actinomycetemcomitans, Porphyromonas, Fusobacterium,* and spirochete species are causative agents.

C. **Necrotizing Ulcerative Gingivitis and Periodontitis:** The most destructive periodontal diseases are necrotizing ulcerative **gingivitis** (NUG) and necrotizing ulcerative periodontitis (NUP), collectively called acute necrotizing ulcerative gingivitis, or ANUG. These diseases are synergistic infections involving *Treponema vincentii, Prevotella intermedia,* and *Fusobacterium* species.

D. **Mumps:** The classic swelling of the cheeks on one or both sides is due to swelling of the salivary gland—a condition called parotitis. Viral multiplication in salivary glands is followed by invasion of other organs, especially the testes, ovaries, thyroid gland, pancreas, meninges, heart, and kidneys. Mumps is caused by an enveloped, single-stranded RNA virus (mumps virus) from the genus *Paramyxovirus.*

E. **Gastritis and Gastric Ulcers: Gastritis** is experienced as sharp or burning pain emanating from the abdomen. Gastric ulcers are actual lesions in the mucosa of the stomach (gastric ulcers) or in the uppermost portion of the small intestine (duodenal ulcer). *Helicobacter pylori,* a curved gram-negative rod, is the causative agent of this condition.

F. **Acute Infectious Diarrhea:** In the United States, up to a third of all acute diarrhea is transmitted by contaminated food.

1. *Salmonella: Salmonella enteriditis* is divided into many serotypes, based on variation in the major surface antigens. Animal and dairy products are often contaminated with the bacterium, and undercooked or unpasteurized products provide risks for Salmonellosis. **Typhoid fever,** caused by *S. enteriditis* variant *typhi,* is characterized by a progressive, invasive infection that leads eventually to septicemia. Symptoms are fever, diarrhea, and abdominal pain.
2. *Shigella* are primarily human parasites and can give symptoms of frequent, watery, bloody stools, fever, and often intense abdominal pain. Diarrhea containing blood and mucus is also called **dysentery.** The bacterium *Shigella dysenteriae* produces a heat-labile exotoxin called shiga toxin.
3. Dozens of different strains of *E. coli* exist: *E. coli* O157:H7 and its close relatives are the most virulent. This group of *E. coli* is referred to as **enterohemorrhagic *E. coli,*** or **EHEC.** *E. coli* O157:H7 is the agent of a spectrum of conditions, ranging from mild gastroenteritis with fever to bloody diarrhea. About 10% of patients develop **hemolytic uremic syndrome (HUS),** a severe hemolytic anemia that can cause kidney damage and failure. Virulence is due to **shiga toxins** (often called STEC—shiga-toxin-producing *E. coli*). The most common means of contamination is from undercooked meats and contaminated water.
4. Other *E. coli:* At least four other categories of *E. coli* can cause diarrheal diseases. These are **enterotoxigenic** *E. coli* (traveler's diarrhea), **enteroinvasive** *E. coli,* **enteropathogenic** *E. coli,* and **enteroaggregative** *E. coli.*
5. *Campylobacter:* The symptoms of campylobacteriosis are frequent watery stools, fever, vomiting, headaches, and severe abdominal pain. *Campylobacter jejuni* is transmitted via the ingestion of contaminated beverages and food. Infrequently, infection with this bacterium can lead to a serious neuromuscular paralysis called *Guillain-Barré syndrome.*
6. *Yersinia enterocolitica* and *Y. pseudotuberculosis* are both agents of GI disease. Food and beverages can become contaminated with these bacteria.
7. *Clostridium difficile* is known to cause a condition called pseudomembranous colitis (antibiotic-associated colitis), where the infection is precipitated by therapy with broad-spectrum antibiotics such as ampicillin, clindamycin, or cephalosporin. It is a major cause of diarrhea in hospitals.
8. *Vibrio cholerae:* Cholera symptoms of secretory diarrhea and severe fluid loss can lead to death in less than 48 hours. *V. cholerae* produces an enterotoxin called cholera toxin (CT), which disrupts the normal physiology of intestinal cells.
9. *Cryptosporidium* is an intestinal waterborne protozoan that infects a variety of mammals,

birds, and reptiles. AIDS patients may experience chronic persistent cryptosporidial diarrhea that can be used as a criterion to help diagnose AIDS.

10. Rotavirus is the primary viral cause of morbidity and mortality resulting from diarrhea, accounting for nearly 50% of all cases. The virus is transmitted by the fecal-oral route, including through contaminated food, water, and fomites.

G. **Acute Diarrhea with Vomiting: Food poisoning** refers to symptoms in the gut that are caused by a preformed toxin.
   1. *Staphylococcus aureus* exotoxin: The heat-stable enterotoxin requires 100°C for at least 30 minutes to achieve inactivation. Thus, heating the food after toxin production may not prevent disease. The ingested toxin acts upon the gastrointestinal epithelium and stimulates nerves, with acute symptoms of cramping, nausea, vomiting, and diarrhea.
   2. *Bacillus cereus* exotoxin: *Bacillus cereus* is a common resident on vegetables and other products in close contact with soil. It produces two exotoxins, one of which causes a diarrheal-type disease, the other of which causes an **emetic** disease.
   3. *Clostridium perfringens* exotoxin: The toxin, acting upon epithelial cells, initiates acute abdominal pain, diarrhea, and nausea in 8 to 16 hours.

H. **Chronic Diarrhea**
   1. Enteroaggregative *E. coli* (EAEC) is particularly associated with chronic disease, especially in children. Transmission of the bacterium is through contaminated food and water.
   2. *Cyclospora cayetanensis* is an emerging protozoan pathogen that is transmitted via the fecal-oral route and has been associated with consumption of fresh produce and water.
   3. *Giardia lamblia* is a protozoan that can cause diarrhea of long duration, abdominal pain, and flatulence. Freshwater supplies are common vehicles of infection.
   4. *Entamoeba histolytica* is a freshwater parasite that causes intestinal amoebiasis, which targets the cecum, appendix, colon, and rectum, leading to dysentery, abdominal pain, fever, diarrhea, and weight loss.

I. **Hepatitis** is an inflammatory disease marked by necrosis of hepatocytes and a mononuclear response that swells and disrupts the liver architecture, causing **jaundice,** a yellow tinge in the skin and eyes. The condition can be caused by a variety of different viruses.
   1. **Hepatitis A virus (HAV)** is a nonenveloped, single-stranded RNA enterovirus of low virulence. Hepatitis A virus is spread through the fecal-oral route. An inactivated viral vaccine (Havrix) is currently approved, and an oral vaccine based on an attenuated strain of virus is in development.
   2. **Hepatitis B virus (HBV)** is an enveloped DNA virus in the family Hepadnaviridae. Hepatitis B infection can be very serious, even life-threatening; some patients develop chronic liver disease in the form of necrosis or cirrhosis. HBV is also associated with **hepatocellular carcinoma.** Some patients infected with hepatitis B are coinfected with a particle called the delta agent, sometimes also called a hepatitis D virus. HBV is transmitted by blood and other bodily fluids. Thus, this virus is one of the major infectious concerns for health care workers.
   3. *Hepatitis C virus:* Hepatitis C is an RNA virus in the Flaviviridae family. It shares many characteristics of hepatitis B disease, but it is much more likely to become chronic. It is more commonly transmitted through blood contact than through transfer of other body fluids.

J. **Helminthic Intestinal Infections:** Intestinal distress as the primary symptom. Both tapeworms and roundworms can infect the intestinal tract in such a way as to cause primary symptoms there.
   1. *Trichuris trichiura:* Symptoms of infection may include localized hemorrhage of the bowel, caused by worms burrowing and piercing intestinal mucosa.
   2. *Enterobius vermicularis:* This pinworm is the most common worm disease of children in temperate zones. Infection is not fatal and most cases are asymptomatic.
   3. *Taenia solium:* This tapeworm is transmitted to humans by the consumption of raw or undercooked pork. Other tapeworms of the genus *Taenia* infect humans. One of them is the beef tapeworm, *Taenia saginata.*
   4. *Diphyllobothrium latum:* The intermediate host for this tapeworm is fish, and it can be transmitted in raw food such as sushi and sashimi made from salmon.

K. **Helminthic Intestinal Infections:** Intestinal distress accompanied by migratory symptoms.
   1. *Ascaris lumbricoides* is an intestinal roundworm that releases eggs in feces, which are then spread to other humans through fecal-oral routes.
   2. *Necator americanus* and *Ancylostoma duodenale:* These two different nematodes are called by the common name "hookworm." Hookworm larvae hatch outside the body in soil contaminated with feces and infect by penetrating the skin.
   3. *Strongyloides stercoralis:* This nematode infection occurs when soil larvae penetrate the skin, similar to hookworm infestations. The most susceptible are AIDS patients, transplant patients on immunosuppressant drugs, and cancer patients receiving radiation therapy.

L. **Liver and Intestinal Disease:** One group of worms that appear in the intestines has a particular affinity for the liver—liver flukes.
   1. *Opisthorchis sinensis* and *Clonorchis sinensis:* Humans are infested by eating inadequately cooked or raw freshwater fish and crustaceans.
   2. *Fasciola hepatica:* This liver fluke is a common parasite in sheep, cattle, goats, and other mammals and is occasionally transmitted to humans. Humans develop symptoms only if they are chronically infected by a large number of flukes.

M. **Muscle and Neurological Symptoms**
   1. **Trichinosis** is an infection transmitted by eating undercooked pork that has the cysts of *Trichinella* species embedded in the meat.
   2. **Schistosomiasis** in the intestines is caused by the blood flukes *Schistosoma mansoni* and *S. japonicum* species. Symptoms of infection include fever, chills, diarrhea, hepatomegaly and liver disease, and splenomegaly.

Multiple-Choice and True-False Questions

Multiple-Choice Questions. Select the correct answer from the answers provided.

1. Food moves down the GI tract through the action of
   a. cilia
   b. peristalsis
   c. gravity
   d. microorganisms
2. The microorganism(s) most associated with acute necrotizing ulcerative periodontitis (ANUP) is (are)
   a. *Treponema vincentii*
   b. *Prevotella intermedia*
   c. *Fusobacterium*
   d. all of the above
3. Gastric ulcers are caused by
   a. *Treponema vincentii*
   b. *Prevotella intermedia*
   c. *Helicobacter pylori*
   d. all of the above
4. Virus family Paramyxoviridae contains viruses that cause which of the following diseases?
   a. measles
   b. mumps
   c. influenza
   d. both a and b
   e. both b and c
5. Which of these microorganisms is considered the most common cause of diarrhea in the United States?
   a. *E. coli*
   b. *Salmonella*
   c. *Campylobacter*
   d. *Shigella*
6. Which of these microorganisms is associated with Guillain-Barré syndrome?
   a. *E. coli*
   b. *Salmonella*
   c. *Campylobacter*
   d. *Shigella*
7. This microorganism is commonly associated with fried rice and produces an emetic (vomiting) toxin.
   a. *Bacillus cereus*
   b. *Clostridium perfringens*
   c. *Shigella*
   d. *Staphylococcus aureus*
8. This sporeformer contaminates meats as well as vegetables and is also the causative agent of gas gangrene.
   a. *Bacillus cereus*
   b. *Clostridium perfringens*
   c. *Shigella*
   d. *Staphylococcus aureus*
9. This hepatitis virus is an enveloped DNA virus.
   a. hepatitis A virus
   b. hepatitis B virus
   c. hepatitis C virus
   d. hepatitis E virus

True-False Questions. If statement is true, leave as is. If it is false, correct it by rewriting the sentence.

10. Mumps is a disease that affects humans and several other species.
11. *Cyclospora cayetanensis* is a water-borne, flagellated protozoan often associated with chronic diarrhea.
12. Pseudomembranous colitis or antibiotic-associated colitis is caused by *Clostridium difficile.*
13. *Enterobius vermicularis,* commonly known as the pinworm, is a common cause of anal itching in young children in the United States.

Writing to Learn

These questions are suggested as a *writing-to-learn* experience. For each question, compose a one- or two-paragraph answer that includes the factual information needed to completely address the question.

1. a. Which microorganism(s) is (are) the major culprit(s) associated with tooth decay?
   b. How do these microorganisms facilitate tooth decay?
2. a. What is food poisoning?
   b. What are some likely microbial culprits associated with food poisoning?
   c. List some nonmicrobial sources of toxins involved in food poisoning.
3. *Entamoeba histolytica* can cause three different forms of amoebiasis. Discuss them.
4. How can hepatitis A infections be prevented?
5. a. What are the most common means of transmission of the hepatitis C virus?
   b. What is the current treatment for hepatitis C?
6. Describe the definitive diagnosis of most helminthic infections.
7. Compare the methods of transmission of hepatitis A and hepatitis B.
8. Discuss five different types of *E. coli* associated with diarrheal diseases.

Concept Mapping

Appendix D provides guidance for working with concept maps.

1. Use 6 to 10 bolded words of your choice from the Chapter Summary to create a concept map. Finish it by providing linking words.

Critical Thinking Questions

Critical thinking is the ability to reason and solve problems using facts and concepts. These questions can be approached from a number of angles, and in most cases, they do not have a single correct answer.

1. There is a commonly held belief that a mumps infection in adult males can cause sterility and impotence. Discuss the validity of this belief.
2. Embryonated eggs are often used as incubators for virus culture and vaccine production because they are a sterile source of living cells. Is this always true? Can the sterility of the shelled egg be breached?
3. Why is a hamburger a greater risk for *E. coli* contamination than a steak?
4. Describe your strategy for treating a cholera patient.
5. Why is heating food contaminated with *Staphylococcus aureus* no guarantee that the associated food poisoning will be prevented?
6. What are some of the ways we can prevent or slow down the spread of helminthic diseases?
7. Which members of the population are most at risk for hepatitis B? Why?

Visual Understanding Questions

1. **From chapter 13, figure 13.6*b*.** Imagine for a minute that the organism in this illustration is *E. coli* O157:H7. What would be one reason not to treat a patient having this infection with powerful antibiotics?

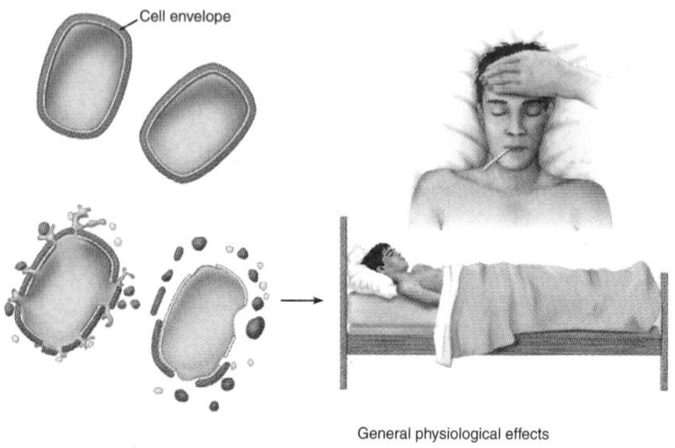

2. **From chapter 12, figure 12.15.** Assume the growth on plate "a" represents normal intestinal microbiota. How could you use these illustrations to explain the development of *C. difficile*–associated colitis?

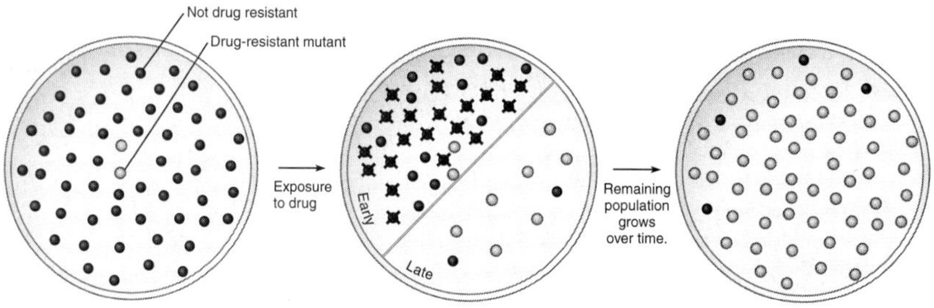

Internet Search Topics

1. Use the Internet to locate information on salmonellosis and shigellosis. Make a comparison table of the two pathogens, including basic characteristics, epidemiology, pathology, and symptoms.
2. Go to: www.aris.mhhe.com, and click on "microbiology" and then this textbook's author/title. Go to chapter 22, access the URLs listed under Internet Search Topics, and research the following:
   a. Find the case studies in enteric diseases. Try your hand at diagnosis.
   b. Look at the site for the Schistosomiasis Control Initiative. Use the information you find there to write a short (2- to 3-paragraph) news story for a magazine intended for middle-school science classes.
3. You be the detective: Use search engines to discover the causes behind the epidemic of cholera in Peru in the late 1990s. What is the current status of this disease worldwide?
4. Mount Healthy, a city in southwestern Ohio, owes its name to a regional epidemic that occurred in 1950. Do a search to find the story behind the name. Based on your knowledge of this disease, what could explain this city's good fortune?

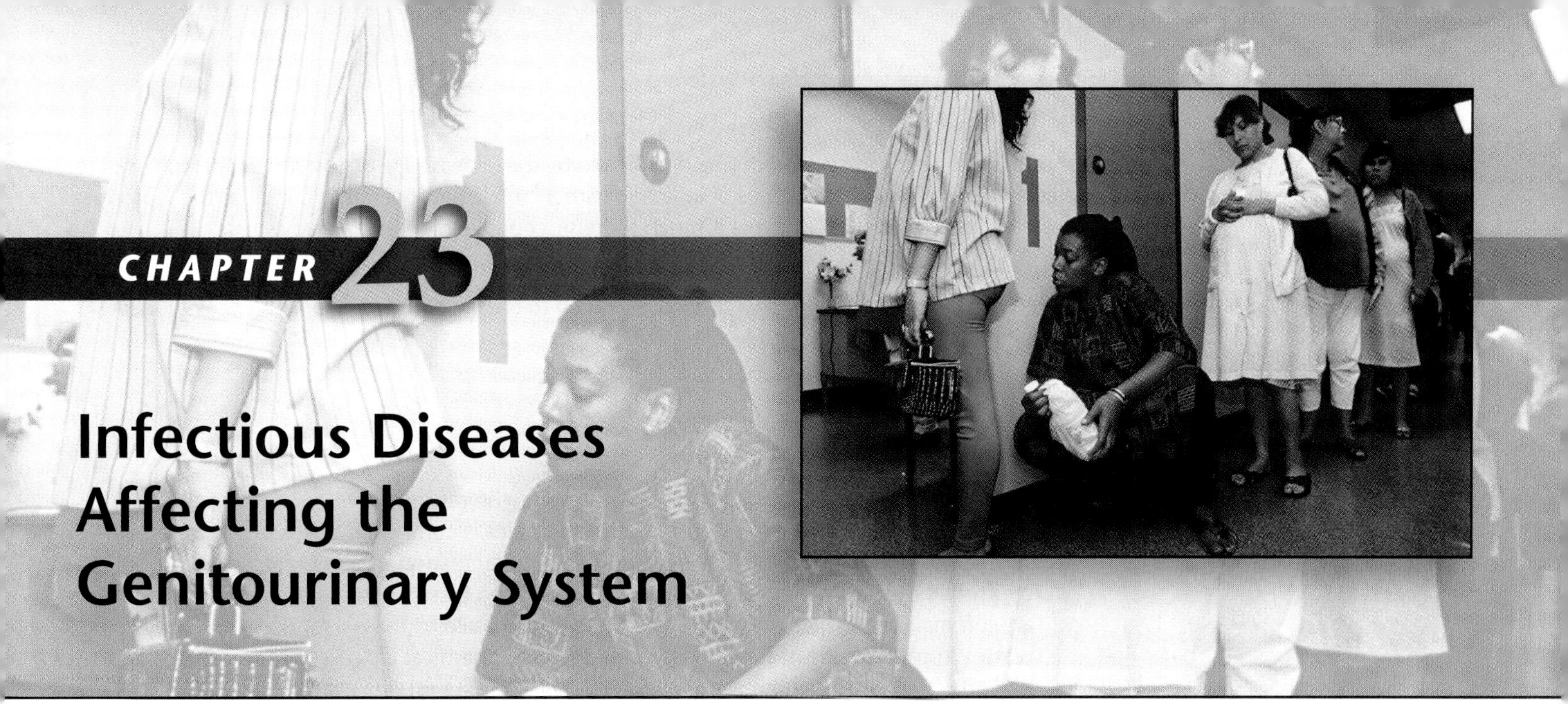

CHAPTER 23

Infectious Diseases Affecting the Genitourinary System

CASE FILE 23

A women's clinic in a downtown neighborhood of Detroit, Michigan, served a large population of inner-city residents. The clinic had recently hired a new supervising physician. When Dr. Mott began working at the clinic, she began to systematically study clinic records from the previous few years. She found it surprising that in 2002, the rate of new human immunodeficiency virus (HIV) infections was actually higher than in 1997. She found this troubling because she knew that much progress had been made in educating the public about behaviors that put them at high risk for HIV transmission. Dr. Mott knew that some of the major risk factors for women becoming infected with HIV were (1) illicit use of injected drugs; (2) numerous sexual partners; and (3) infection with other sexually transmitted diseases, which made the reproductive tract more susceptible to transmission of the virus.

Dr. Mott was determined to find out why HIV infection rates were increasing among her patients. She gathered information about risk factors in her own clinic population. First, she looked at the statistics for injecting-drug use among clinic patrons. The number of patients that had been referred for drug treatment or counseling had actually decreased steadily over the past 3 years. Admissions to the local hospital for drug overdoses were also down. She surmised that drug use was not a major factor contributing to the increased HIV infection rate.

In the mid-1990s, Detroit, like other American cities, had experienced a syphilis epidemic. A massive public health campaign had successfully brought down the rate of syphilis infection in the city, and it remained low. Dr. Mott thought this also indicated that partner exchange rates had decreased and that the second risk factor was therefore not a major contributor to the increased HIV infection rate. She turned to the third possibility, that other underlying sexually transmitted diseases (STDs) were making women more susceptible to HIV infection.

Dr. Mott found that in recent years a successful public health campaign in this neighborhood had encouraged women to visit the clinic as soon as they suspected they had a reproductive tract infection. This campaign was designed to prevent pelvic inflammatory disease (PID) and its long-term consequences. The clinic had won a citywide award for this effort, as its rate of PID decreased more than any other clinic in the city. Dr. Mott took this as an indicator that bacterially caused STDs were being treated promptly and probably were not contributing to an increased susceptibility to HIV.

Dr. Mott found this problem puzzling. By initiating an aggressive screening campaign, however, she eventually managed to tease out the answer. She discovered that the increased rate of HIV infection was probably caused by underlying infection with a microorganism that "slips under the radar" by often not causing overt symptoms, not being on the watch list of organisms that might cause PID or other long-term effects, and not being affected by the increased use of antibiotic therapy used to prevent PID. She found high rates of infection with a "mild" pathogen that damages the reproductive tract mucosa enough to make it much more susceptible to penetration by viruses.

- *Which microorganisms are probably ruled out by the facts of this case?*
- *Which microorganism do you think contributed to the increased rates of HIV infection seen among these clinic patients?*

Case File Wrap-Up appears on page 742.

CHAPTER OVERVIEW

- The "genitourinary system" is really two systems, the reproductive system and the urinary system. The reproductive tract in males and females is composed of structures and substances that allow for sexual intercourse and the creation of a new fetus; it is protected by normal mucosal defenses as well as by specialized features (such as the low pH of the adult female reproductive tract). The urinary system allows the excretion of fluid and wastes from the body. It has mechanical as well as chemical defense mechanisms.
- Both the genital and the urinary systems have normal biota only in their most distal regions. Normal biota in the male reproductive and urinary systems are found in the distal part of the urethra and resemble skin biota. The same is generally true for the female urinary system. The female reproductive tract has a normal biota that changes over the course of a woman's lifetime.
- Urinary tract infections are most often caused by normal biota from the gastrointestinal tract; *E. coli* is the most common etiologic agent.
- Leptospirosis is a bacterium that infects animals; when it is excreted in their urine and transmitted to humans it can cause a wide array of neurological and urinary tract symptoms.
- One species of *Schistosoma* deposits its eggs in the bladder. Long-term infection with the parasite can lead to severe damage to that organ.
- Not all genital tract diseases are sexually transmitted. For example, vaginal yeast infections are caused by overgrowth of the normal biota *Candida albicans.*
- Sexually transmitted infections of the genital tract can be placed in three groups: discharge diseases (such as gonorrhea), ulcer diseases (such as herpes), and wart diseases (HPV and others).
- Group B streptococci that colonize the adult female genital tract can cause a life-threatening infection in newborn babies exposed to the bacteria.

23.1 The Genitourinary Tract and Its Defenses

As suggested by the name, the structures considered in this chapter are really two distinct organ systems. The *urinary tract* has the job of removing substances from the blood, regulating certain body processes, and forming urine and transporting it out of the body. The *genital system* has reproduction as its major function. It is also called the *reproductive system.*

The urinary tract includes the kidneys, ureters, bladder, and the urethra **(figure 23.1).** The kidneys remove metabolic wastes from the blood, acting as a sophisticated filtration system. Ureters are tubular organs extending from each kidney to the bladder. The bladder is a collapsible organ that stores urine and empties it into the urethra, which is the conduit of urine to the exterior of the body. In males, the urethra is also the terminal organ of the reproductive tract, but in females the urethra is separate from the vagina, which is the outermost organ of the reproductive tract.

Several defenses are present in the urinary system that help to prevent infection when microorganisms are introduced. The most obvious defensive mechanism is the flushing action of the urine flowing out of the system. The flow of urine also encourages the **desquamation** (shedding) of the epithelial cells lining the urinary tract. For example, each time a person urinates, he or she loses hundreds of thousands of epithelial cells! Any microorganisms attached to them are also shed, of course. Probably the most common microbial threat to the urinary tract is the group of microorganisms that comprise the normal biota in the gastrointestinal tract, because the two organ systems are in close proximity. But the cells of the epithelial lining of the urinary tract have different chemicals on their surfaces than do those lining the GI tract. For that reason, most bacteria that are adapted to adhere to the chemical structures in the GI tract cannot gain a foothold in the urinary tract.

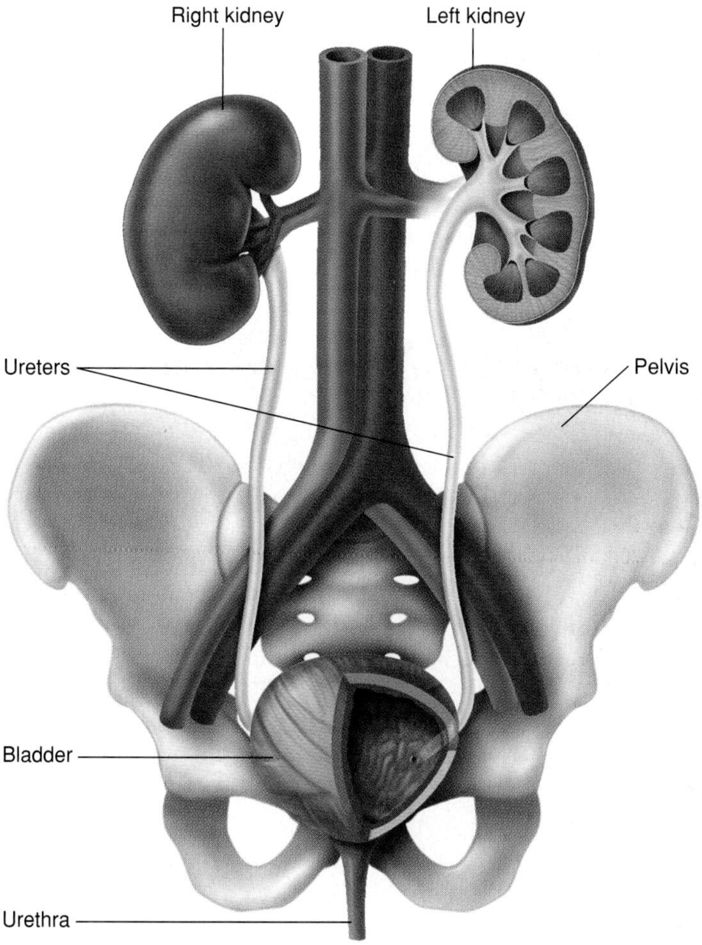

Figure 23.1 The urinary system.

Urine, in addition to being acidic, also contains two antibacterial proteins, lysozyme and lactoferrin. You may recall that lysozyme is an enzyme that breaks down peptidoglycan. Lactoferrin is an iron-binding protein that inhibits bacterial growth. Finally, secretory IgA specific for previously encountered microorganisms can be found in the urine.

The male reproductive system produces, maintains, and transports sperm cells and is the source of male sex hormones. It consists of the *testes,* which produce sperm cells and hormones, and the *epididymis,* which is a coiled tube leading out of the testes. The epididymis terminates in the *vas deferens,* which combines with the seminal vesicle and terminates in the ejaculatory duct **(figure 23.2).** The contents of the ejaculatory duct empty into the urethra during ejaculation. The *prostate gland* is a walnut-shaped structure at the base of the urethra. It also contributes to the released fluid (semen). The external organs are the scrotum, containing the testes, and the *penis,* a cylindrical organ that houses the urethra. As for its innate defenses, the male reproductive system also benefits from the flushing action of the urine, which helps move microorganisms out of the system.

The female reproductive system consists of the *uterus,* the *fallopian tubes* (also called uterine tubes), *ovaries,* and *vagina* **(figure 23.3).** During childbearing years, an egg is released from one of the ovaries approximately every 28 days. It enters the fallopian tubes, where fertilization by sperm may take place if sperm are present. The fertilized egg moves through the fallopian tubes to the uterus, where it is implanted in the uterine lining. If fertilization does not occur, the lining of the uterus degenerates and sloughs off; this is the process of menstruation. The terminal portion of the female reproductive tract is the vagina, which is a tube about 9 cm long. The vagina is the exit tube for fluids from the uterus, the channel for childbirth, and the receptive chamber for the penis during sexual intercourse. One very important tissue of the female reproductive tract is the *cervix,* which is the lower one-third of the uterus and the part that connects to the vagina. The opening of the uterus is part of the cervix. The cervix is a common site of infection in the female reproductive tract.

The natural defenses of the female reproductive tract vary over the lifetime of the woman. The vagina is lined with mucous membranes and, thus, has the protective covering of secreted mucus. During childhood and after menopause, this mucus is the major nonspecific defense of this system. Secretory IgA antibodies specific for any previously encountered infections would be present on these surfaces. During a woman's reproductive years, a major portion of the defense is provided by changes in the pH of the vagina brought about by the release of estrogen. This hormone stimulates the vaginal mucosa to secrete glycogen, which certain bacteria can ferment into acid, lowering the pH of the vagina to about 4.5. Before puberty, a girl produces little estrogen and little glycogen and has a vaginal pH of about 7. The change in pH beginning in adolescence results in a vastly different normal biota in the vagina, described later. The biota of women in their childbearing years is thought to prevent the establishment and invasion of microbes that might have the potential to harm a developing fetus.

23.2 Normal Biota of the Urinary Tract

In both genders, the outer region of the urethra harbors some normal biota. The kidney, ureters, bladder, and upper urethra are presumably kept sterile by urine flow and regular bladder emptying (urinating). The principal known residents of the urethra are the nonhemolytic streptococci, staphylococci, corynebacteria, and some lactobacilli. Because the urethra in women is so short (about 3.5 cm long) and is in such close proximity to the anus, it can act as a pipeline for bacteria from the GI tract to the bladder, resulting in urinary tract infections.

Normal Biota of the Male Genital Tract

Because the terminal "tube" of the male genital tract is the urethra, the normal biota of the male genital tract (that is, in the urethra) is comprised of the same residents as just described.

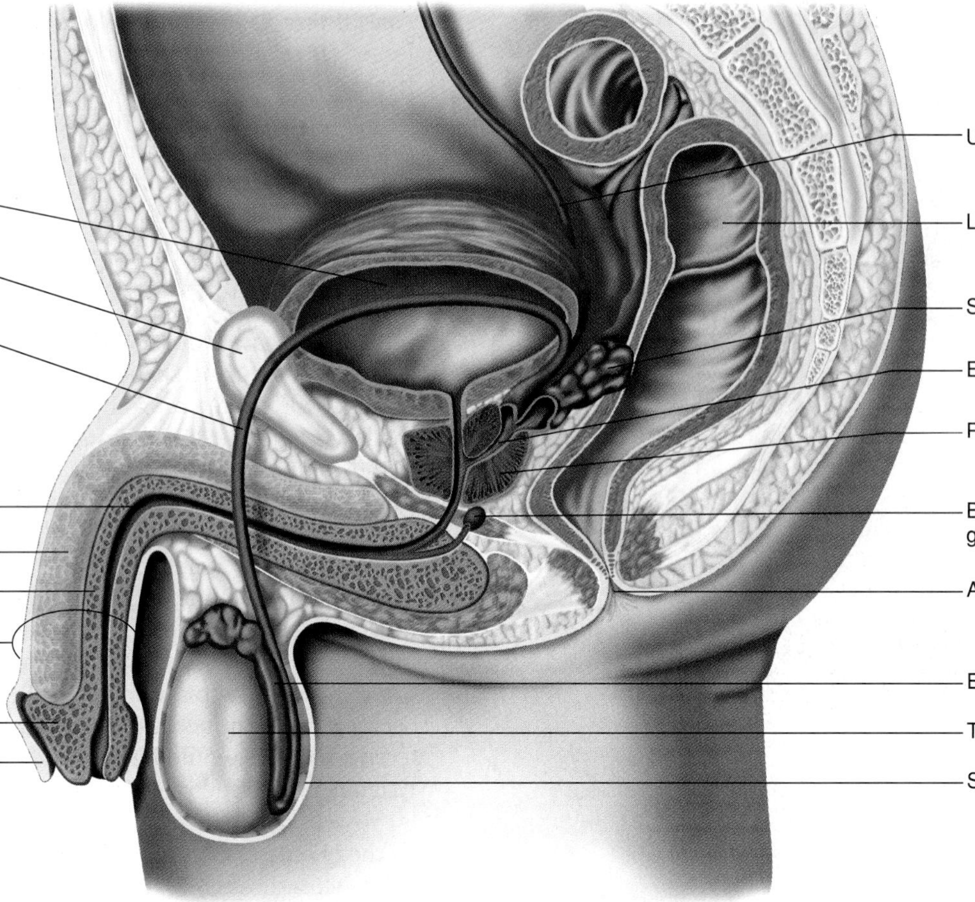

Figure 23.2 The male reproductive system.

Genitourinary Tract Defenses and Normal Biota

	Defenses	Normal Biota
Urinary Tract (both genders)	Flushing action of urine; specific attachment sites not recognized by most nonnormal biota; shedding of urinary tract epithelial cells, secretory IgA, lysozyme, and lactoferrin in urine	Nonhemolytic *Streptococcus*, *Staphylococcus*, *Corynebacterium*, *Lactobacillus*
Female Genital Tract (childhood and postmenopause)	Mucus secretions, secretory IgA	Same as for urinary tract
Femal Genital Tract (childbearing years)	Acidic pH, mucus secretions, secretory IgA	Predominantly *Lactobacillus* but also *Candida*
Male Genital Tract	Same as for urinary tract	Same as for urinary tract

Normal Biota of the Female Genital Tract

In the female genital tract, only the vagina harbors a normal population of microbes. Starting at the cervix and for all organs above it, there is no normal biota. As just mentioned, before puberty and after menopause, the pH of the vagina is close to neutral and the vagina harbors a biota that is similar to that found in the urethra. After the onset of puberty, estrogen production leads to glycogen release in the vagina, resulting in an acidic pH. *Lactobacillus* species thrive in the acidic environment and contribute to it, converting sugars to acid. Their predominance in the vagina, combined with the acidic environment discourages the growth of many microorganisms. The estrogen-glycogen effect continues, with minor disruptions, throughout the childbearing years until menopause, when the biota gradually returns to a mixed population similar to that of prepuberty. Note that the very common fungus *Candida albicans* is also present at low levels in the healthy female reproductive tract.

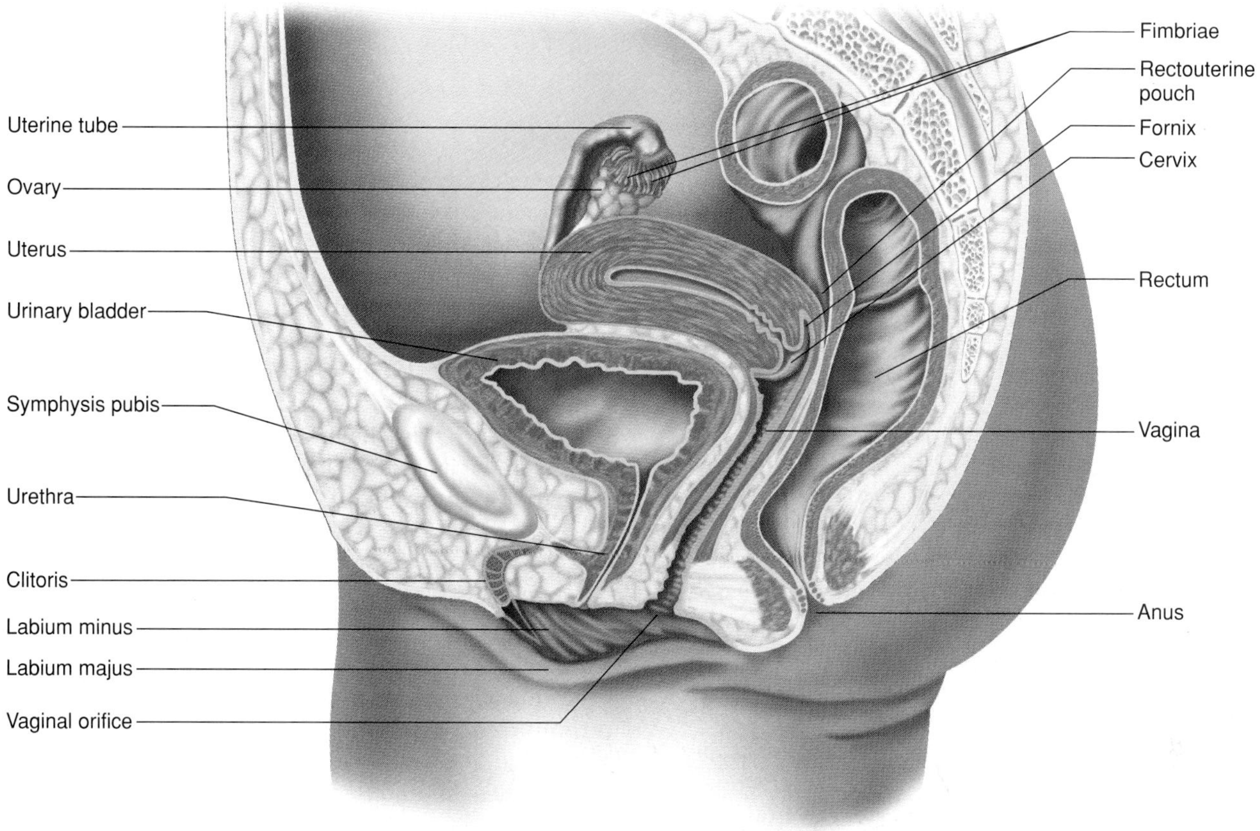

Figure 23.3 **The female reproductive system.**

23.3 Urinary Tract Diseases Caused by Microorganisms

We consider two types of diseases in this section. **Urinary tract infections (UTIs)** result from invasion of the urinary system by bacteria or other microorganisms. Leptospirosis, by contrast, is a spirochete-caused disease transmitted by contact of broken skin or mucous membranes with contaminated animal urine.

Urinary Tract Infections (UTIs)

Even though the flushing action of urine helps to keep infections to a minimum in the urinary tract, urine itself is a good growth medium for many microorganisms. When urine flow is reduced or bacteria are accidentally introduced into the bladder, an infection of that organ (known as *cystitis*) can occur. Occasionally, the infection can also affect the kidneys, in which case it is called *pyelonephritis*. If an infection is limited to the urethra, it is called *urethritis*. In practice, urethritis is not a very useful term when referring to urinary tract infections; females often don't notice urinary tract infections if they are limited to the urethra. And a male presenting with urethritis could be experiencing a sexually transmitted infection (covered later in the chapter).

Signs and Symptoms

Cystitis is a disease of sudden onset. Symptoms include pain in the pubic area, frequent urges to urinate even when the bladder is empty, and burning pain accompanying urination (called *dysuria*). The urine can be cloudy due to the presence of bacteria and white blood cells. It may have an orange tinge from the presence of red blood cells (*hematuria*). Fever and nausea are frequently present. If back pain is present, it is an indication that the kidneys may also be involved (pyelonephritis). Inadequately treated pyelonephritis may result in septicemia, especially in the immunocompromised. If only the bladder is involved, the condition is sometimes called acute uncomplicated UTI.

Causative Agents

In 95% of cystitis and pyelonephritis cases, the cause is bacteria that are normal biota in the gastrointestinal tract. *Escherichia coli* is by far the most common of these. *Staphylococcus saprophyticus* and *Proteus mirabilis* are also common culprits. These last two are only referenced in **Checkpoint 23.1** following the discussion of *E. coli*.

The *E. coli* species that cause UTIs are ones that exist as normal biota in the gastrointestinal tract. They are not the ones that cause diarrhea and other digestive tract diseases.

Pathogenesis and Virulence Factors

E. coli secure themselves in the gastrointestinal tract using specific adhesins on the ends of long fimbriae. They can also use these adhesins to attach to slightly different chemicals present on the epithelial lining of the urinary tract. Many *E. coli* that cause disease in the urinary tract also have different fimbriae with adhesins that recognize chemicals only present on cells lining the ureters and kidney. These *E. coli* exhibit a motility that allows them to travel along mucosal surfaces, so they seem to be specially adapted to ascending the urinary system. Their presence in these normally sterile areas induces an inflammatory response that we experience as symptoms and that may lead to scarring in the ureters and kidneys.

Transmission and Epidemiology

Community-acquired UTIs are nearly always "transmitted" *not* from one person to another but from one organ system to another namely from the GI tract to the urinary system. They are much more common in women than in men because of the shorter length of the female urethra and because of nearness of the female urethral opening to the anus (see figure 23.3). Many women experience what have been referred to as "recurrent urinary tract infections," although it is now known that some *E. coli* can invade the deeper tissue of the urinary tract and therefore avoid being destroyed by antibiotics. They can emerge later to cause symptoms again. It is not clear how many "recurrent" infections are actually infections that reactivate in this way.

Note that urinary tract infections are also the most common of nosocomial infections. Patients of both sexes who have urinary catheters are susceptible to infections with a variety of microorganisms, not just the three mentioned here.

Prevention

A vaccine currently is in development based on the fimbrial adhesion of *E. coli* that can cause UTIs. The vaccine is made of the bacterial adhesin, so that the immune system will make an antibody to it, thereby blocking its attachment. But for now, prevention of all UTIs relies on more basic practices, such as emptying the bladder frequently and (for females) wiping from front to back after a bowel movement. People who are predisposed to UTIs often drink cranberry juice to prevent the disease. Scientists have found that there are multiple compounds in the juice that help to discourage the attachment of *E. coli* to urinary epithelium.

Treatment

Ampicillin, amoxicillin, or sulfa drugs such as trimethoprim-sulfamethoxazole are most often used for UTIs of various etiologies. Often another nonantibiotic drug called Pyridium is administered simultaneously. This drug relieves the very uncomfortable symptoms of burning and urgency. A large percentage of *E. coli* strains is resistant to penicillin derivatives, so these should be avoided **(Checkpoint 23.1).**

Leptospirosis

This infection is a zoonosis associated with wild animals and domesticated animals. It can affect the kidneys, liver, brain, and eyes. It is considered in this section because it can have its major effects on the kidneys and because its presence in

CHECKPOINT 23.1 Urinary Tract Infections (Cystitis, Pyelonephritis)

Causative Organism(s)	*Escherichia coli*	*Staphylococcus saprophyticus*	*Proteus mirabilis*
Most Common Modes of Transmission	Endogenous transfer from GI tract (opportunism)	Opportunism	Opportunism
Virulence Factors	Adhesins, motility	–	Urease enzyme, leads to kidney stone formation
Culture/Diagnosis	Often "bacterial infection" diagnosed on basis of increased white cells in urinalysis; if culture performed, bacteria may or may not be identified to species level	Often "bacterial infection" diagnosed on basis of increased white cells in urinalysis; if culture performed, bacteria may or may not be identified to species level	Often "bacterial infection" diagnosed on basis of increased white cells in urinalysis; if culture performed, bacteria may or may not be identified to species level
Prevention	Vaccine may be available soon; hygiene practices	Hygiene practices	Hygiene practices
Treatment	Cephalosporin	Ampicillin, amoxicillin, trimethoprim-sulfamethoxazole	Ampicillin or cephalosporins
Distinctive Features	—	—	Kidney stones and severe pain may ensue

animal urinary tracts causes it to be shed into the environment through animal urine.

Signs and Symptoms

Leptospirosis has two phases. During the early, or leptospiremic, phase, the pathogen appears in the blood and cerebrospinal fluid. Symptoms are sudden high fever, chills, headache, muscle aches, conjunctivitis, and vomiting. During the second phase (called the immune phase), the blood infection is cleared by natural defenses. This period is marked by milder fever; headache due to leptospiral meningitis; and *Weil's syndrome,* a cluster of symptoms characterized by kidney invasion, hepatic disease, jaundice, anemia, and neurological disturbances. Long-term disability and even death can result from damage to the kidneys and liver, but they occur primarily with the most virulent strains and in elderly persons.

Causative Agent

Leptospires are typical spirochete bacteria marked by tight, regular, individual coils with a bend or hook at one or both ends **(figure 23.4).** *Leptospira interrogans* (lep″-toh-spy′-rah in-terr′-oh-ganz) is the species that causes leptospirosis in humans and animals. There are nearly 200 different serotypes of this species distributed among various animal groups, which accounts for extreme variations in the disease manifestations in humans.

Pathogenesis and Virulence Factors

In 2003, Chinese scientists sequenced the entire genome of this bacterium and found a series of genes that code for virulence factors such as adhesins and invasion proteins. Because it appears that the bacterium evolved from its close relatives, which are free-living and cause no disease, finding out how the bacterium acquired these genes will be useful in understanding its pathogenesis.

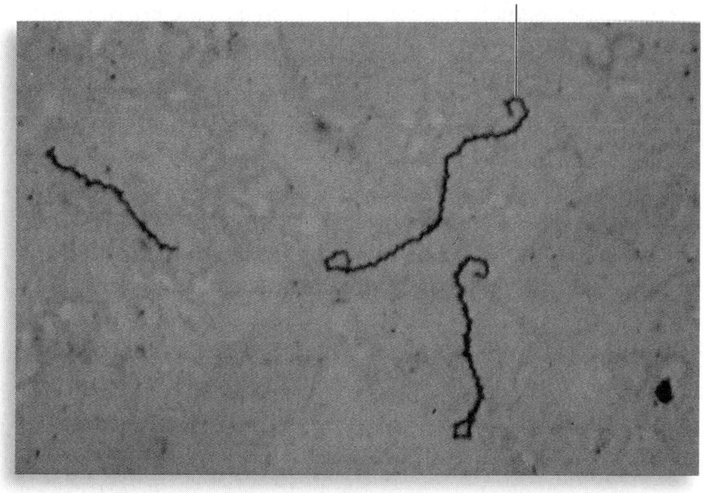

Figure 23.4 ***Leptospira interrogans,* the agent of leptospirosis.**
Note the curved hook at the ends of the spirochete.

Transmission and Epidemiology

Leptospirosis is a zoonosis, affecting wild animals such as rodents, skunks, raccoons, and foxes and some domesticated animals, particularly horses, dogs, cattle, and pigs. It is found throughout the world, although it is more common in the tropics. It is an occupational hazard of people who work with animals or in the outdoors. Leptospires shed in the urine of an infected animal can survive for several months in neutral or alkaline soil or water. Infection occurs almost entirely through contact of skin abrasions or mucous membranes with animal urine or some environmental source containing urine. In 1998, dozens of athletes competing in the swimming phase of a triathlon in Illinois contracted leptospirosis from the water. The disease is not transmissible person to person.

Prevention

Vaccines are available, each of which is targeted to a specific strain of the bacterium, so they are of limited use for widespread protection. They are mainly used for military troops training in jungle regions and animal care and livestock workers. The new DNA sequence data should reveal new targets for vaccines that will be more broadly useful. For now, the best prevention is to wear protective footwear and clothing and to avoid swimming and wading in natural water sources that are frequented by livestock.

Treatment

Early treatment with amoxicillin or doxycycline rapidly reduces symptoms and shortens the course of disease, but delayed therapy is less effective. Other spirochete diseases, such as syphilis (described later), exhibit this same pattern of being susceptible to antibiotics early in the infection but less so later.

CHECKPOINT 23.2	Leptospirosis
Causative Organism(s)	*Leptospira interrogans*
Most Common Modes of Transmission	Vehicle—contaminated soil or water
Virulence Factors	Adhesins? Invasion proteins?
Culture/Diagnosis	Slide agglutination test of patient's blood for antibodies
Prevention	Strain-specific vaccine available to limited populations; avoiding contaminated vehicles
Treatment	Doxycycline, amoxicillin

Urinary Schistosomiasis

In chapter 22, we talked about schistosomiasis, because one of its two distinct disease manifestations occurs in the liver and spleen, both parts of the digestive system. One particular species of the trematode (helminth) lodges in the blood vessels of the bladder. This may or may not result in symptoms. Alternatively, blood in the urine and, eventually, bladder obstruction can occur.

Signs and Symptoms

As with the other forms of schistosomiasis, the first symptoms of infestation are itchiness in the area where the worm enters the body, followed by fever, chills, diarrhea, and cough. Urinary tract symptoms occur at a later date. Remember that adult flukes can live for many years and, by eluding the immune defenses, cause chronic infection.

Causative Agent

The urinary manifestations occur if a host is infected with a particular species of schistosome, *Schistosoma haematobium*. It is found throughout Africa, the Caribbean, and the Middle East. (*S. mansoni* and *S. japonicum* are the species responsible for liver manifestations.) *Schistosomes* are trematodes, or flukes (illustrated in figure 22.30). Humans are the definitive hosts for schistosomes, and snails are the intermediate host.

Pathogenesis and Virulence Factors

Like the other species, *S. haematobium* is able to invade intact skin and attach to vascular endothelium. It engages in the same antigenic cloaking behavior as the other two species. The disease manifestations occur when the eggs in the bladder induce a massive granulomatous response that leads to leakage in the blood vessels and blood in the urine. Significant portions of the bladder eventually can be filled with granulomatous tissue and scar tissue. Function of the bladder is decreased or halted altogether. Chronic infection with *S. haematobium* can also lead to bladder cancer.

Transmission and Epidemiology

The life cycle of the schistosome is described completely in chapter 22. After the worms pass into small blood and lymphatic vessels, they are carried to the liver. Eventually *S. haematobium* enters the venous plexus of the bladder. While attached to these intravascular sites, the worms feed upon blood, and the female lays eggs that are eventually voided in urine.

Culture and Diagnosis

Diagnosis depends on identifying the eggs in urine.

Prevention and Treatment

The cycle of infection cannot be broken as long as people are exposed to untreated sewage in their environment. It is quite common for people to be cured and then to be reinfected because their village has no sewage treatment. A vaccine would provide widespread control of the disease, but so far none is licensed. More than one vaccine is in development, however.

Praziquantel is the drug treatment of choice and is quite effective at eliminating the worms.

CHECKPOINT 23.3	Urinary Schistosomiasis
Causative Organism(s)	*Schistosoma haematobium*
Most Common Modes of Transmission	Vehicle (contaminated water)
Virulence Factors	Antigenic "cloaking," induction of granulomatous response
Culture/Diagnosis	Identification of eggs in urine
Prevention	Avoiding contaminated vehicles
Treatment	Praziquantel

23.4 Reproductive Tract Diseases Caused by Microorganisms

We saw earlier that reproductive tract diseases in men almost always involve the urinary tract as well, and this is sometimes but not always the case with women. Note that although many of the infectious diseases of the reproductive tract are transmitted through sexual contact, not all of them are.

We begin this section with a discussion of infections that are symptomatic primarily in women: *vaginitis* and *vaginosis*. Men may also harbor these infections with or without symptoms. We next consider three broad categories of **sexually transmitted diseases (STDs):** *discharge diseases* in which increased fluid is released in male and female reproductive tracts; *ulcer diseases* in which microbes cause distinct open lesions; and the *wart diseases*. The section concludes with a neonatal disease caused by group B *Streptococcus* colonization.

Vaginitis and Vaginosis

Signs and Symptoms

Vaginitis, an inflammation of the vagina, is a condition characterized by some degree of vaginal itching, depending on the etiologic agent. Symptoms may also include burning, and sometimes a discharge, which may take different forms as well. From the name, it is obvious that vaginitis only affects women, but most of the agents can also colonize the male reproductive tract.

Causative Agents

The most common cause of vaginitis is *Candida albicans*. The vaginal condition caused by this fungus is known as a *yeast infection*. Most women experience this condition one or multiple times during their lives. Other causes can be bacterial, as in the case of *Gardnerella*, or even protozoal, as in the case of *Trichomonas*. We describe each of these agents here.

Candida albicans *C. albicans* is a dimorphic fungus that is normal biota in from 50% to 100% of humans, living in low numbers on many mucosal surfaces such as the mouth, gastrointestinal tract, vagina, and so on. The vaginal condition

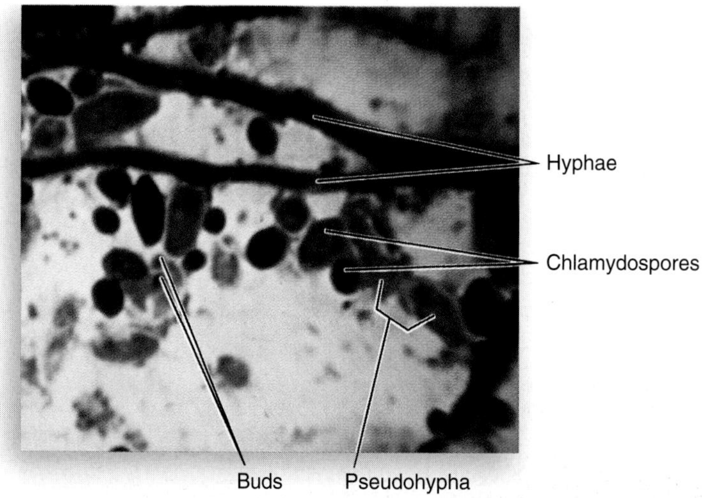

Figure 23.5 Gram stain of *Candida albicans* in a vaginal smear.

it causes is often called *vulvovaginal candidiasis.* The yeast is easily detectable on a wet prep or a Gram stain of material obtained during a pelvic exam **(figure 23.5).** The presence of pseudohyphae in the smear is a clear indication that the yeast is growing rapidly and causing a yeast infection.

Pathogenesis and Virulence Factors The fungus grows in thick curdlike colonies on the walls of the vagina. The colony debris contributes to a white vaginal discharge. In otherwise healthy people, the fungus is not invasive and limits itself to this surface infection. Please note, however, that *Candida* infections of the bloodstream do occur and they have high mortality rates. They do not normally stem from vaginal infections with the fungus, however, but are seen most frequently in hospitalized patients.

Transmission and Epidemiology Vaginal infections with this organism are nearly always opportunistic. Disruptions of the normal bacterial biota or even minor damage to the mucosal epithelium in the vagina can lead to overgrowth by this fungus. Disruptions may be mechanical, such as wearing very tight pants, or they may be chemical, as when broad-spectrum antibiotics taken for some other purpose temporarily diminish the vaginal bacterial population. Diabetics and pregnant women are also predisposed to vaginal yeast overgrowths. Some women are prone to this condition during menstruation. The term "infection" is really a misnomer—because this condition is not the result of a new infection but rather an increased rate of growth of a member of the normal biota.

It is possible to transmit this yeast through sexual contact, especially if a woman is experiencing an overgrowth of it. The recipient's immune system may well subdue the yeast so that it acts as normal biota in them. But the yeast may be passed back to the original partner during further sexual contact after treatment. By that time, the circumstances that led to it becoming dominant in the vagina may have returned to normal and its growth would be limited by the normal bacterial biota. So the sexual route of transmission is difficult to assess. Nevertheless, it is recommended that a patient's sexual partner also be treated to short-circuit the possibility of retransmission. The important thing to remember is that *Candida* is an opportunistic fungus.

Women with HIV infection experience frequently recurring yeast infections. Also, a small percentage of women with no underlying immune disease experience chronic or recurrent vaginal infection with *Candida* for reasons that are not clear.

Prevention and Treatment No vaccine is available for *C. albicans.* Topical and oral azole drugs are used to treat vaginal candidiasis, and some of them are now available over the counter. If infections recur frequently or fail to resolve, it is important to see a physician for evaluation.

***Gardnerella* Species** The bacterium *Gardnerella* is associated with a particularly common condition in women in their childbearing years. This condition is usually called vaginosis rather than vaginitis because it doesn't appear to induce inflammation in the vagina. It is also known as BV, or bacterial vaginosis. Despite the absence of an inflammatory response, a vaginal discharge is associated with the condition, which is said to have a very fishy odor, especially after sex. Itching is common. But it is also true that many women have this condition with no noticeable symptoms.

Vaginosis is most likely a result of a shift from a predominance of "good bacteria" (lactobacilli) in the vagina to a predominance of "bad bacteria," and one of those is *Gardnerella vaginalis.* This genus of bacteria is aerotolerant and gram-positive, although in a Gram stain it usually appears gram-negative. Probably a mixed infection leads to the condition, however. Anaerobic streptococci and other bacteria, particularly a genus known as *Mobiluncus,* that are normally found in low numbers in a healthy vagina can also often be found in high numbers in this condition. The often-mentioned fishy odor comes from the metabolic by-products of anaerobic metabolism by these bacteria.

Pathogenesis and Virulence Factors The mechanism of damage in this disease is not well understood. But some of the outcomes are. Besides the symptoms just mentioned, vaginosis can lead to complications such as **pelvic inflammatory disease (PID;** to be discussed later in the chapter), infertility, and more rarely, ectopic pregnancies. Babies born to some mothers with vaginosis have low birth weights.

Transmission and Epidemiology This mixed infection is not considered to be sexually transmitted, although women who have never had sex rarely develop the condition. It is very common in sexually active women. It may be that the condition is *associated* with sex but not transmitted by it. This situation could occur if the act of penetration or the presence of semen (or saliva) causes changes in the vaginal epithelium, or in the vaginal biota. We do not know exactly what causes the increased numbers of *Gardnerella* and other normally rare biota. The low pH typical of the vagina is usually higher in vaginosis. It is not clear whether this causes or is caused by the change in bacterial biota.

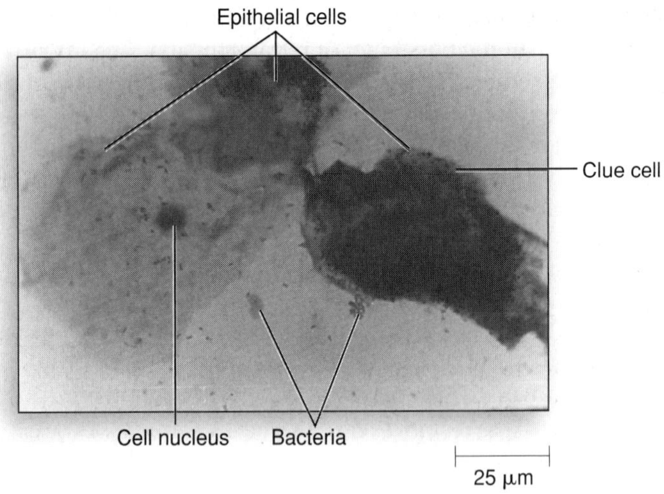

Figure 23.6 **Clue cell in bacterial vaginosis.**
These epithelial cells came from a pelvic exam. The one on the right is completely covered with bacteria.

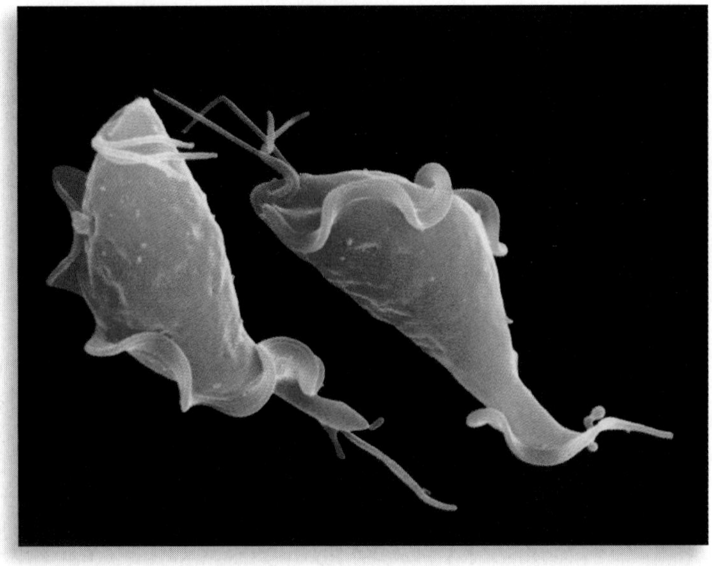

Figure 23.7 ***Trichomonas vaginalis.***

Culture and Diagnosis The condition can be diagnosed by a variety of methods. Sometimes a simple stain of vaginal secretions is used to examine sloughed vaginal epithelial cells. In vaginosis, some cells will appear to be nearly covered with adherent bacteria. In normal times, vaginal epithelial cells are sparsely covered with bacteria. These cells are called clue cells and are a helpful diagnostic indicator **(figure 23.6).** They can also be found on Pap smears.

Prevention and Treatment No known prevention exists. Asymptomatic cases are generally not treated. Women who find the condition uncomfortable or who are planning on becoming pregnant should be treated. Women who use intrauterine devices (IUDs) for contraception should also be treated because IUDs can provide a passageway for the bacteria to gain access to the upper reproductive tract. The usual treatment is oral or topical metronidazole or clindamycin.

Trichomonas vaginalis Trichomonads are small, pear-shaped protozoa with four anterior flagella and an undulating membrane **(figure 23.7).** *Trichomonas vaginalis* seems to cause asymptomatic infections in approximately 50% of females *and* males, despite its species name. Trichomonads are considered asymptomatic infectious agents rather than normal biota because of evidence that some people experience long-term negative effects. Even though *Trichomonas* is a protozoan, it has no cyst form and it does not survive long out of the host.

Pathogenesis and Virulence Factors Many cases are asymptomatic, and men seldom have symptoms. Women often have vaginitis symptoms, which can include a white to green frothy discharge. Chronic infection can make a person more susceptible to other infections, including HIV. Also, women who become infected during pregnancy are predisposed to premature labor and low-birth-weight infants. Chronic infection may also lead to infertility.

CASE FILE 23 *WRAP-UP*

The infection that predisposed many women at the Detroit clinic to being infected with HIV was *Trichomonas.* This protozoan would not be affected by antibiotics prescribed to prevent and treat PID. The most common microorganisms leading to PID are *Neisseria gonorrhoeae* and *Chlamydia trachomatis,* although other vaginitis and vaginosis infections have also been found to lead to this condition. Whether PID is treated on an inpatient or an outpatient basis, broad-spectrum antibiotics are administered **(Insight 23.1).** Both of these bacteria would answer the question "Which bacteria are ruled out by the facts of this case?" In the clinic situation, patients were aggressively treated with antibacterial agents to prevent PID.

Trichomonas has been found in some studies roughly to double the chances that a woman exposed to HIV will become infected. The organism is thought to expand the portal of entry for HIV because it can cause small mucosal lesions and also results in a large migration of CD4 lymphocytes and macrophages to the area of infection. These cells are the target cells of HIV, so bringing a lot of them to the mucosal surface increases a woman's susceptibility.

Because *Trichomonas* infection itself is relatively mild and doesn't affect fertility or birth outcomes, it has not been the subject of intensive study or even of public health control programs. Although data are relatively scarce, the prevalence of *Trichomonas* infection in women in the United States are thought to range from 3% to 58%. The vast majority of these infections are asymptomatic. For reasons that are unclear, African-American women have higher rates (1.5 to 4 times higher) than any other ethnic group studied. The emerging link between *Trichomonas* infection and HIV susceptibility makes this infection an important area of study.

See: Sorvillo, F., Smith, L., Kerndt, P., and Ash, L. 2001. Trichomonas vaginalis, *HIV, and African-Americans.* Emerg. Infect. Dis. *7:927–932.*

INSIGHT 23.1 — Medical

Pelvic Inflammatory Disease and Infertility

The National Center for Health Statistics estimates that more than 6 million women in the United States have impaired fertility. There are many different reasons for infertility, but the leading cause is pelvic inflammatory disease, or PID. PID is caused by infection of the upper reproductive structures in women, namely the uterus, fallopian tubes, and ovaries. These organs have no normal biota, and when bacteria from the vagina are transported higher in the tract, they start a chain of inflammatory events that may or may not be noticeable to the patient. The inflammation can be acute, resulting in pain, abnormal vaginal discharge, fever, and nausea, or it can be chronic, with less noticeable symptoms. In acute cases, women usually seek care; in some ways, these can be considered the lucky ones. If the inflammation is curbed at an early stage by using antibiotics to kill the bacteria, chances are better that the long-term sequelae of PID can be avoided.

The most notable long-term consequence is tubal infertility, caused by the repair step of inflammation. Inflammatory repair processes, especially in the fallopian tubes, can lead to the deposition of scar tissue that narrows the lumen in the tubes, in some cases closing them off completely. But if the lumen is only narrowed, fertilization may occur. A fertilized egg could then be unable to travel through the tube and implant in the uterine wall. In some cases, fertilized eggs implant in the tube walls or even leave the fallopian tubes and implant elsewhere in the abdominal cavity. Both of these situations are known as ectopic pregnancies. Women with a history of PID have a seven- to ten-fold greater chance of experiencing an ectopic pregnancy than other women. Ectopic pregnancy is a life-threatening situation. An embryo growing in the tube usually causes the tube to rupture in about 12 weeks, and an embryo in the abdominal cavity can cause the same complication as a tumor. Surgical intervention is usually required in either case to eliminate the embryo and save the woman's life.

Chlamydia infection is the leading cause of PID, followed closely by *N. gonorrhoeae* infection. But other bacteria, perhaps also including normal biota of the reproductive tract, can also cause PID if they are traumatically introduced into the uterus. Intercourse, tampon usage, the use of an intrauterine contraceptive device, and even douching can encourage the transmission of bacteria into the upper genital tract. (In addition to being a risk factor for PID, douching can also temporarily ease the symptoms of a reproductive tract infection, which could result in dangerous delays in seeking treatment.)

With the relatively high rates of infertility in the developed world, the message needs to be loud and clear: PID is a preventable condition. Women who suspect for any reason that they may have a reproductive tract infection should always seek diagnosis and treatment from health care professionals.

Transmission and Epidemiology Because *Trichomonas* is common biota in so many people, it is easily transmitted through sexual contact. It has been called the most common nonviral sexually transmitted infection. It does not appear to undergo opportunistic shifts within its host (that is, becomes symptomatic under certain conditions), but rather, the protozoan causes symptoms when transmitted to a noncarrier. Some debate exists over whether the protozoan can be transmitted through communal bathing, public facilities, and from mother to child, but if this type of transmission happens, it is only rarely.

Prevention and Treatment There is no vaccine for *Trichomonas*. The antiprotozoal drug metronidazole is the drug of choice, although some isolates are resistant to it **(Checkpoint 23.4).**

CHECKPOINT 23.4	Vaginitis/Vaginosis		
Causative Organism(s)	*Candida albicans*	Mixed infection, usually including *Gardnerella*	*Trichomonas vaginalis*
Most Common Modes of Transmission	Opportunism	Opportunism?	Direct contact (STD)
Virulence Factors	–	–	–
Culture/Diagnosis	Wet prep or Gram stain	Visual exam of vagina, or clue cells seen in Pap smear or other smear	Protozoa seen on Pap smear or Gram stain
Prevention	–	–	Barrier use during intercourse
Treatment	Topical or oral azole drugs, some over-the-counter drugs	Metronidazole or clindamycin	Metronidazole
Distinctive Features	White curdlike discharge	Discharge may have fishy smell	Discharge may be greenish

Prostatitis

Prostatitis is an inflammation of the prostate gland (see figure 23.2). It can be acute or chronic. Acute prostatitis is virtually always caused by bacterial infection. The bacteria are usually normal biota from the intestinal tract or may have caused a previous urinary tract infection. Chronic prostatitis is also often caused by bacteria. Researchers have found that chronic prostatitis, often unresponsive to antibiotic treatment, can be caused by mixed biofilms of bacteria in the prostate. Some forms of chronic prostatitis have no known microbial cause, though many infectious disease specialists feel that one or more bacteria are involved, but they are simply not culturable with current techniques.

The symptoms of prostatitis are pain in the pelvic area, lower back, or genital area; frequent urge to urinate; blood in the urine; and/or painful ejaculation. Acute prostatitis is accompanied by fever and chills and flulike symptoms. Patients appear to be quite ill with the acute form of the disease.

Treatment involves antibiotics when bacteria are indicated. Also, muscle relaxers or drugs called alpha blockers, which relax the neck of the bladder, may be prescribed.

CHECKPOINT 23.5	Prostatitis
Causative Organism(s)	GI tract biota
Most Common Modes of Transmission	Endogenous transfer from GI tract; otherwise unknown
Virulence Factors	Various
Culture/Diagnosis	Digital rectal exam to examine prostate; culture of urine or semen
Prevention	None
Treatment	Antibiotics, muscle relaxers, alpha blockers
Distinctive Features	Pain in genital area and/or back, difficulty urinating

Discharge Diseases with Major Manifestation in the Genitourinary Tract

Discharge diseases are those in which the infectious agent causes an increase in fluid discharge in the male and female reproductive tracts. Examples are trichomoniasis, HIV, gonorrhea, and *Chlamydia* infection. The causative agents are transferred to new hosts when the fluids in which they live contact the mucosal surfaces of the receiving partner. As noted, HIV is discussed in chapter 20. Trichomoniasis has been described in the preceding section because its disease manifestations are considered to be a vaginitis. In this section, we cover the other two major discharge diseases: gonorrhea and *Chlamydia*.

A NOTE ABOUT HIV AND HEPATITIS B AND C

This chapter is about diseases *whose major (presenting) symptoms occur in the genitourinary tract.* But some sexually transmitted diseases do not have their major symptoms in this system. HIV and hepatitis B and C can all be transmitted in several ways, one of them being through sexual contact. HIV is considered in chapter 20 because its major symptoms occur in the cardiovascular and lymphatic systems. Because the major disease manifestations of hepatitis B and C occur in the gastrointestinal tract, these diseases are discussed in chapter 22. Anyone diagnosed with any sexually transmitted disease should also be tested for HIV.

Gonorrhea

Gonorrhea has been known as a sexually transmitted disease since ancient times. Its name originated with the Greek physician Claudius Galen, who thought that it was caused by an excess flow of semen. For a fairly long period in history, gonorrhea was confused with syphilis. Later, microbiologists went on to cultivate *N. gonorrhoeae,* also known as the **gonococcus,** and to prove conclusively that it alone was the etiologic agent of gonorrhea. It has traditionally been called *the clap*.

Signs and Symptoms In the male, infection of the urethra elicits **urethritis,** painful urination and a yellowish discharge, although a relatively large number of cases are asymptomatic. In most cases, infection is limited to the distal urogenital tract, but it can occasionally spread from the urethra to the prostate gland and epididymis (refer to figure 23.2). Scar tissue formed in the spermatic ducts during healing of an invasive infection can render a man infertile. This outcome is becoming increasingly rare with improved diagnosis and treatment regimens.

In the female, it is likely that both the urinary and genital tracts will be infected during sexual intercourse. A mucopurulent (containing mucus and pus) or bloody vaginal discharge occurs in about half of the cases, along with painful urination if the urethra is affected. Major complications occur when the infection ascends from the vagina and cervix to higher reproductive structures such as the uterus and fallopian tubes **(figure 23.8).** One disease resulting from this progression is **salpingitis** (sal″-pin-jy′-tis). This inflammation of the fallopian tubes may be isolated, or it may also include inflammation of other parts of the upper reproductive tract, termed pelvic inflammatory disease (PID). It is not unusual for the microbe that initiates PID to become involved in mixed infections with anaerobic bacteria. The buildup of scar tissue from PID can block the fallopian tubes, causing sterility or ectopic pregnancies (see Insight 23.1).

Serious consequences of gonorrhea can occur outside of the reproductive tract. In a small number of cases, the gonococcus enters the bloodstream and is disseminated to the

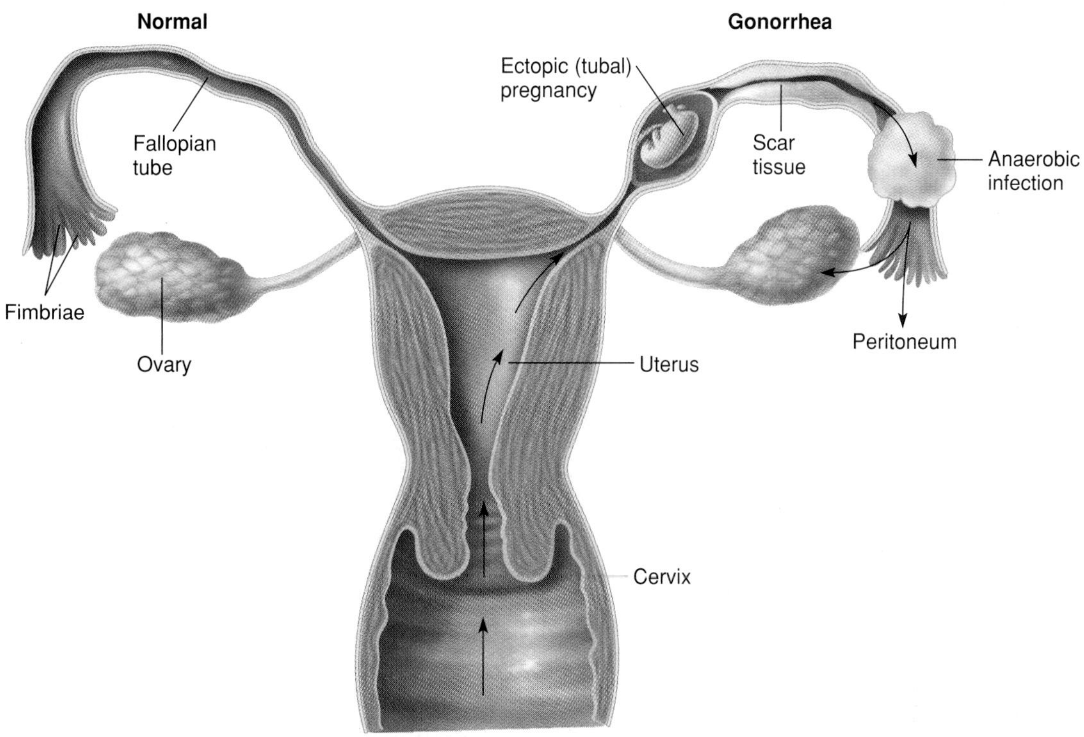

Figure 23.8 Invasive gonorrhea in women.
(*Left*) Normal state. (*Right*) In ascending gonorrhea, the gonococcus is carried from the cervical opening up through the uterus and into the fallopian tubes. Pelvic inflammatory disease (PID) is a serious complication that can lead to scarring in the fallopian tubes, ectopic pregnancies, and mixed anaerobic infections.

joints and skin. Involvement of the wrist and ankle can lead to chronic arthritis and a painful, sporadic, papular rash on the limbs. Rare complications of gonococcal bacteremia are meningitis and endocarditis.

Children born to gonococcus carriers are also in danger of being infected as they pass through the birth canal. Because of the potential harm to the fetus, physicians usually screen pregnant mothers for its presence. Gonococcal eye infections are very serious and often result in keratitis, ophthalmia neonatorum, and even blindness **(figure 23.9).** A universal precaution to prevent such complications is the use of antibiotic eyedrops for newborn babies. The pathogen may also infect the pharynx and respiratory tract of neonates. Finding gonorrhea in children other than neonates is strong evidence of sexual abuse by infected adults, and it calls for child welfare consultation along with thorough bacteriologic analysis.

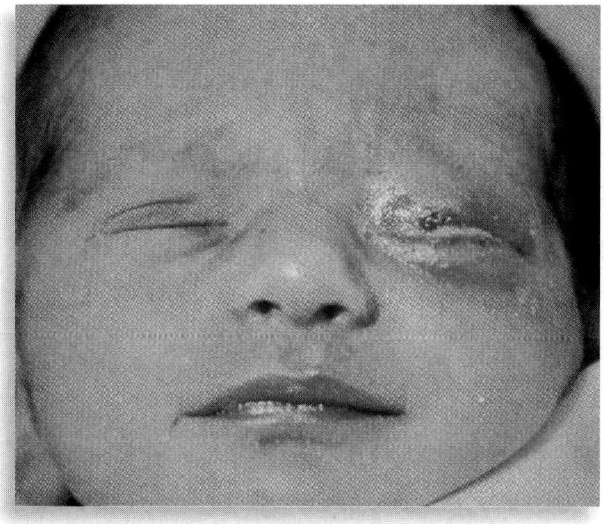

Figure 23.9 Gonococcal ophthalmia neonatorum in a week-old infant.
The infection is marked by intense inflammation and edema; if allowed to progress, it causes damage that can lead to blindness. Fortunately, this infection is completely preventable and treatable.

Causative Agent *N. gonorrhoeae* is a pyogenic gram-negative diploccoccus. It appears as pairs of kidney bean–shaped bacteria, with their flat sides touching **(figure 23.10).**

Pathogenesis and Virulence Factors Successful attachment is key to the organism's ability to cause disease. Gonococci use specific chemicals on the tips of fimbriae to anchor themselves to mucosal epithelial cells. They only attach to nonciliated cells of the urethra and the cervix, for example. Once the bacterium attaches, it invades the cells and multiplies on the basement membrane.

The fimbriae may also play a role in slowing down effective immunity. The fimbrial proteins are controlled by genes that can be turned on or off, depending on the bacterium's situation. This phenotypic change is called phase variation. In addition, the genes can rearrange themselves to put together fimbriae of different configurations. This antigenic variation confuses the body's immune system. Antibodies that previously recognized fimbrial proteins may not recognize them once they are rearranged.

The gonococcus also possesses an enzyme called IgA protease, which can cleave IgA molecules stationed on mucosal surfaces. In addition, it pinches off pieces of its outer membrane. These "blebs," containing endotoxin, probably play a role in pathogenesis because they can stimulate portions of the nonspecific defense response, resulting in localized damage.

Transmission and Epidemiology *N. gonorrhoeae* does not survive more than 1 or 2 hours on fomites and is most infectious when transferred to a suitable mucous membrane.

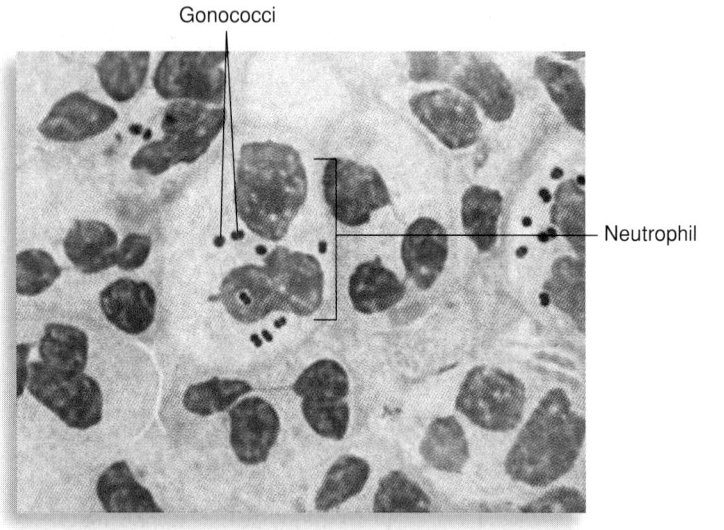

Figure 23.10 Gram stain of urethral pus from a male patient with gonorrhea (1,000×).
Note the intracellular (phagocytosed) gram-negative diplococci (arranged side-to-side) in polymorphonuclear leukocytes (neutrophils).

Except for neonatal infections, the gonococcus spreads through some form of sexual contact. The pathogen requires an appropriate portal of entry that is genital or extragenital (rectum, eye, or throat).

Gonorrhea is a strictly human infection that occurs worldwide and ranks among the most common sexually transmitted diseases. Although about 500,000 cases are reported in the United States each year, it is estimated that the actual incidence is much higher—in the millions if one counts asymptomatic infections. Figures on the prevalence of gonorrhea and syphilis over the past 60 years show a fluctuating pattern apparently corresponding to periods of social and political upheaval when number of sex partners tends to increase **(figure 23.11).** One interesting effect occurred during the "sexual revolution" of the 1960s, when oral contraceptives were introduced and began to be used more commonly than condoms to prevent pregnancy. This contraceptive strategy increased the transmission of the gonococcus (and other STDs as well).

It is important to consider the reservoir of asymptomatic males and females when discussing the transmission of the infection. Because approximately 10% of infected males and 50% of infected females experience no symptoms, it is often spread unknowingly.

Culture and Diagnosis In males, it is easy to diagnose this disease; a Gram stain of urethral discharge is diagnostic. The normal biota of the male urethra is so sparse that it is easy to see the diplococcus inside of phagocytes (see figure 23.10). In females, other methods, such as ELISA or PCR tests, are called for. Alternatively, the bacterium can be cultured on Thayer-Martin agar, a rich chocolate agar base with added antibiotics that inhibit competing bacteria.

N. gonorrhoeae grows best in an atmosphere containing increased CO_2. Because *Neisseria* is so fragile, it is best to inoculate it onto media directly from the patient rather than using a transport tube. Gonococci produce catalase, enzymes for fermenting various carbohydrates, and the enzyme

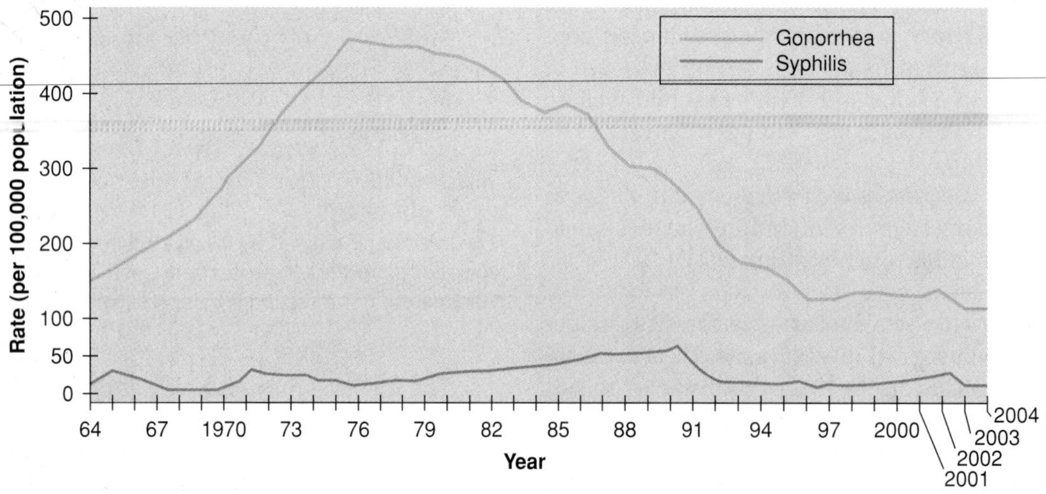

Figure 23.11 Gonorrhea and syphilis—reported rates: United States, 1964–2004.

cytochrome oxidase that can be used for identification as well. Gonorrhea is a reportable disease.

Prevention Currently, no vaccine is available for gonorrhea, although finding one is a priority for government health agencies. The development of a vaccine is hampered by the fact that no good animal model exists for the disease. Using condoms is an effective way to avoid transmission of this and other discharge diseases.

Treatment The CDC runs a program called the Gonococcal Isolate Surveillance Project (GISP) to monitor the occurrence of antibiotic resistance in *N. gonorrhoeae.* Penicillin was traditionally the drug of choice, but a large percentage of isolates now are able to produce penicillinase. They are called PPNG, or penicillinase-producing *N. gonorrhoeae.* Others are tetracycline resistant (called TRNG). As alternatives, practitioners have been using quinolones (like ciprofloxacin) or cephalosporins. In 2002, the CDC advised doctors in California and Hawaii not to use quinolones because such a high incidence of resistance to that antibiotic had developed in those regions. Instead, ceftriaxone (a cephalosporin) was recommended there. This development highlights the need for practitioners to be aware of local resistance patterns before prescribing antibiotics for gonorrhea.

Chlamydia

Genital chlamydial infection is the most common reportable infectious disease in the United States. Annually, 850,000 to 1 million cases are reported but the actual infection rate may be 5 to 7 times that number. The overall prevalence among young adults in the United States is 4%. It is at least two to three times as common as gonorrhea. The vast majority of cases are asymptomatic. When we consider the serious consequences that may follow *Chlamydia* infection, those facts are very disturbing.

Signs and Symptoms In males who experience *Chlamydia* symptoms, the bacterium causes an inflammation of the urethra (a condition formerly called *nongonococcal urethritis*). The symptoms mimic gonorrhea, namely discharge and painful urination. Untreated infections may lead to epididymitis. Females who experience symptoms have cervicitis, a discharge, and often salpingitis. Pelvic inflammatory disease is a frequent sequela of female chlamydial infection. A woman is even more likely to experience PID as a result of a *Chlamydia* infection than as a result of gonorrhea. (**Figure 23.12** depicts *Chlamydia* bacteria adhering inside a fallopian tube.) Most cases of *Chlamydia* infection are asymptomatic, which puts women at risk for developing PID because they don't seek treatment for initial infections. The PID itself may be acute and painful, or it may be relatively asymptomatic, allowing damage to the upper reproductive tract to continue unchecked.

Certain strains of *C. trachomatis* can invade the lymphatic tissues, resulting in another condition called lymphogranuloma venereum. This condition is accompanied by headache, fever, and muscle aches. The lymph nodes near the lesion begin to fill with granuloma cells and become enlarged and tender. These "nodes" can cause long-term lymphatic obstruction that leads to chronic, deforming edema of the genitalia or anus. The disease is endemic to South America, Africa, and Asia but occasionally occurs in other parts of the world. Its incidence in the United States is about 500 cases per year.

Babies born to mothers with *Chlamydia* infections can develop eye infections and also pneumonia if they become infected during passage through the birth canal. Infant conjunctivitis caused by contact with maternal *Chlamydia* infection is the most prevalent form of conjunctivitis in the United States (100,000 cases per year). Antibiotic drops or ointment applied to newborns' eyes are chosen to eliminate both *Chlamydia* and *N. gonorrhoeae.*

Causative Agent *C. trachomatis* is a very small bacterium, technically gram-negative. It lives inside host cells as an obligate intracellular parasite. All *Chlamydia* species alternate between two distinct stages: (1) a small, metabolically inactive infectious form called the elementary body, which is released by the infected host cell; and (2) a larger, noninfectious, actively dividing form called the reticulate body, which grows within the host cell vacuoles **(figure 23.13).** Elementary bodies are tiny, dense spheres shielded by a rigid, impervious envelope that ensures survival outside the eukaryotic host cell. Studies of reticulate bodies indicate that they are "energy parasites," entirely lacking enzyme systems for synthesizing ATP, although they do possess ribosomes and mechanisms for synthesizing proteins, DNA, and RNA. Reticulate bodies ultimately become elementary bodies during their life cycle.

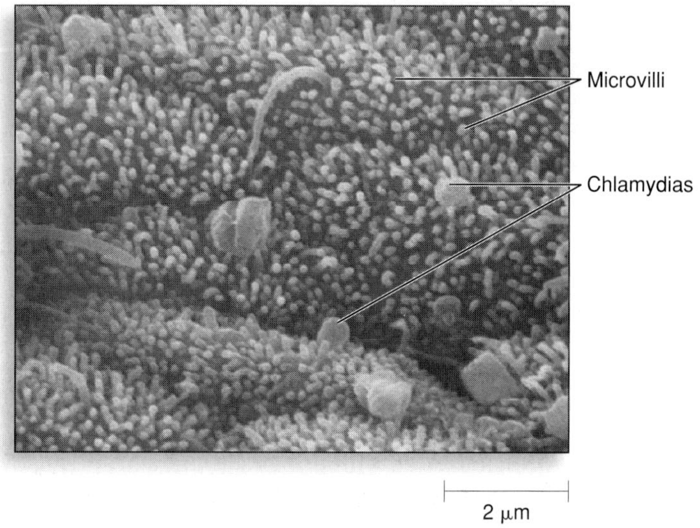

Figure 23.12 ***Chlamydia trachomatis*** **adhering to mucosa of fallopian tube.**

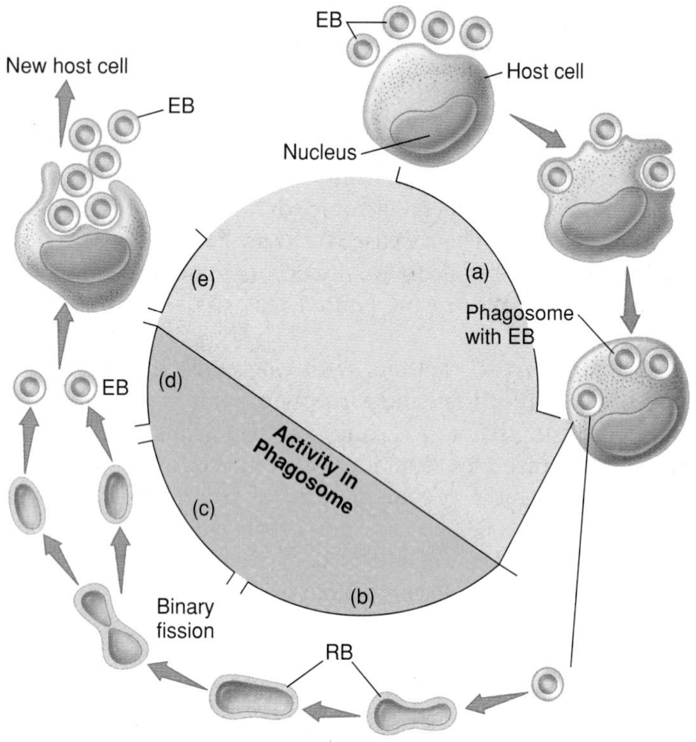

Figure 23.13 The life cycle of *Chlamydia*.
(a) The infectious stage, or elementary body (EB), is taken into phagocytic vesicles by the host cell. **(b)** In the phagosome, each elementary body develops into a reticulate body (RB). **(c)** Reticulate bodies multiply by regular binary fission. **(d)** Mature RBs become reorganized into EBs. **(e)** Completed EBs are released from the host cell.

Pathogenesis and Virulence Factors *Chlamydia*'s ability to grow intracellularly contributes to its virulence because it escapes certain aspects of the host's immune response. Also, the bacterium has a unique cell wall that apparently prevents the phagosome from fusing with the lysosome inside phagocytes. The presence of the bacteria inside cells causes the release of cytokines that provoke intense inflammation. This defensive response leads to most of the actual tissue damage in *Chlamydia* infection. Of course, the last step of inflammation is repair, which often results in scarring as described in Insight 23.1. This can have disastrous effects on a narrow tube like the fallopian tube.

Transmission and Epidemiology The reservoir of pathogenic strains of *C. trachomatis* is the human body. The microbe shows an astoundingly broad distribution within the population. Adolescent women are more likely than older women to harbor the bacterium because it prefers to infect cells that are particularly prevalent on the adolescent cervix. It is transmitted through sexual contact and also vertically. Fifty percent of babies born to infected mothers will acquire conjunctivitis (more common) or pneumonia (less common).

Culture and Diagnosis Infection with this microorganism is usually detected initially using a rapid technique such as PCR or ELISA. Direct fluorescent antibody detection is sometimes used. Serology is not always reliable. In addition, antibody to *Chlamydia* is very common in adults and often indicates past, not present, infection. Isolating the bacterium and growing it in cell culture is the best method for detecting this bacterium, but because it is time-consuming and expensive, it is performed only in cases where 100% accuracy is required—such as in rape or child abuse cases.

Prevention As yet, no vaccine exists for *Chlamydia*. Researchers have developed several types of experimental vaccines, including a DNA vaccine, but none has been approved for use to date. Avoiding contact with infected tissues and secretions through abstinence or barrier protection (condoms) is the only means of prevention.

Treatment Treatment for this infection relies on being aware of it, so part of the guidelines issued by the CDC is a recommendation for annual screening of young women for presence of the bacterium. It is also recommended that older women with some risk factor (new sexual partner, for instance) also be screened. If infection is found, treatment is usually with azithromycin, a macrolide antibiotic. Note that according to public health officials, many patients become reinfected soon after treatment; therefore, the recommendation is that patients be rechecked for *Chlamydia* infection 3 to 4 months after treatment. (Repeated infections with *Chlamydia* increase the likelihood of PID and other serious sequelae.) **(Checkpoint 23.6.)**

Genital Ulcer Diseases

Three common infectious conditions can result in lesions on a person's genitals: syphilis, chancroid, and genital herpes. In this section, we consider each of these. One very important fact to remember about the ulcer diseases is that having one of them increases the chances of infection with HIV because of the open lesions.

Syphilis

The origin of **syphilis**[1] is an obscure yet intriguing topic of speculation. The disease was first recognized at the close of the 15th century in Europe, a period coinciding with the return of Columbus from the West Indies. From this, some medical scholars have concluded that syphilis was introduced to Europe from the New World. However, a more

1. The term *syphilis* first appeared in a poem entitled "Syphilis sive Morbus Gallicus" by Fracastorius (1530), about a mythical shepherd whose name eventually became synonymous with the disease from which he suffered.

CHECKPOINT 23.6	Genital "Discharge" Diseases (in Addition to Vaginitis/Vaginosis)	
	Gonorrhea	**Chlamydia**
Causative Organism(s)	*Neisseria gonorrhoeae*	*Chlamydia trachomatis*
Most Common Modes of Transmission	Direct contact (STD), also vertical	Direct contact (STD), vertical
Virulence Factors	Fimbrial adhesions, antigenic variation, IgA protease, membrane blebs/endotoxin	Intracellular growth resulting in avoiding immune system and cytokine release, unusual cell wall preventing phagolysosome fusion
Culture/Diagnosis	Gram stain in males, rapid tests (PCR, ELISA) for females, culture on Thayer-Martin agar	PCR or ELISA, can be followed by cell culture
Prevention	Avoid contact; condom use	Avoid contact; condom use
Treatment	Many strains resistant to various antibiotics; local and current guidelines must be consulted	Azithromycin, doxycycline and follow-up to check for reinfection
Distinctive Features	Rare complications include arthritis, meningitis, endocarditis	More commonly asymptomatic than gonorrhea
Effects on Fetus	Eye infections, blindness	Eye infections, pneumonia

probable explanation contends that the spirochete that causes the disease evolved from a related subspecies, perhaps an endemic bacterium already present in the Mediterranean basin. The combination of the immunologically naive population of Europe, the European wars, and sexual promiscuity set the stage for worldwide transmission of syphilis that continues to this day.

A disturbing chapter of syphilis history in the United States is worth noting here. Beginning in 1932, the U.S. government conducted a study called the Tuskegee Study of Untreated Syphilis in the Negro Male, which eventually involved 399 indigent African-American men living in the South. Infected men were recruited into the study, which sought to document the natural progression of the disease. These men were never told that they had syphilis and were never treated for it, even after penicillin was shown to be an effective cure. The study ended in 1972 after it became public. In 1997, President Bill Clinton issued a public apology on behalf of the U.S. government for the study, and the government has paid millions of dollars in compensation to the victims and their heirs.

Signs and Symptoms Untreated syphilis is marked by distinct clinical stages designated as *primary, secondary,* and *tertiary syphilis*. The disease also has latent periods of varying duration during which it is quiescent. The spirochete appears in the lesions and blood during the primary and secondary stages and, thus, is transmissible at these times. During the early latency period between secondary and tertiary syphilis, it is also transmissible. Syphilis is largely nontransmissible during the "late latent" and tertiary stages. Symptoms of each of these stages and congenital syphilis are briefly described here.

Primary Syphilis The earliest indication of syphilis infection is the appearance of a hard **chancre** (shang-ker) at the site of entry of the pathogen (see Checkpoint 23.7 for photos of all three types of genital lesions). A chancre appears after an incubation period that varies from 9 days to 3 months. The chancre begins as a small, red, hard bump that enlarges and breaks down, leaving a shallow crater with firm margins. The base of the chancre beneath the encrusted surface swarms with spirochetes. Most chancres appear on the internal and external genitalia, but about 20% occur on the lips, oral cavity, nipples, fingers, or around the anus. Because these ulcers tend to be painless, they may escape notice, especially when they are on internal surfaces. Lymph nodes draining the affected region become enlarged and firm, but systemic symptoms are absent at this point. The chancre heals spontaneously without scarring in 3 to 6 weeks, but the healing is deceptive because the spirochete has escaped into the circulation and is entering a period of tremendous activity.

Secondary Syphilis About 3 weeks to 6 months after the chancre heals, the secondary stage appears. By then, many systems of the body have been invaded and the signs and symptoms are more profuse and intense. Initial symptoms are fever, headache, and sore throat, followed by lymphadenopathy and a peculiar red or brown rash that breaks out on all skin surfaces, including the palms of the hands and the soles of the feet **(figure 23.14).** A person's hair often falls out. Like the chancre, the lesions contain viable spirochetes and disappear spontaneously in a few weeks. The major complications of this stage, occurring in the bones, hair follicles, joints, liver, eyes, and brain, can linger for months and years.

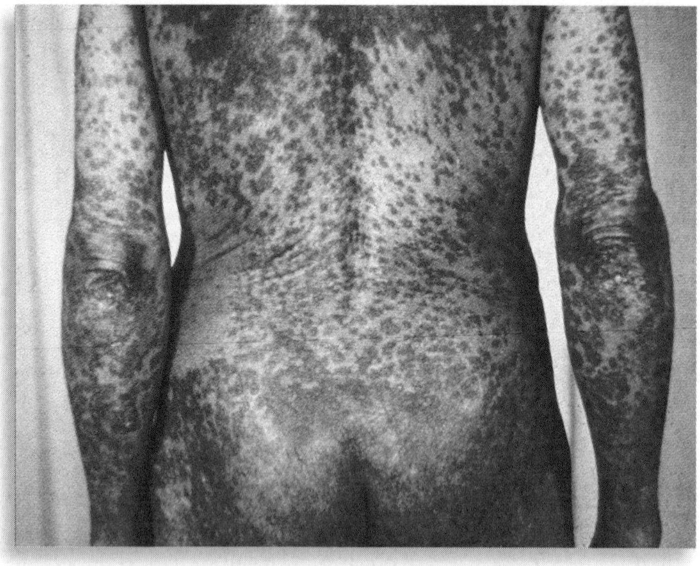

Figure 23.14 Symptom of secondary syphilis.
The skin rash in secondary syphilis can form on the trunk, arms, and even palms and soles (this latter location is particularly diagnostic). The rash does not hurt or itch and can persist for months.

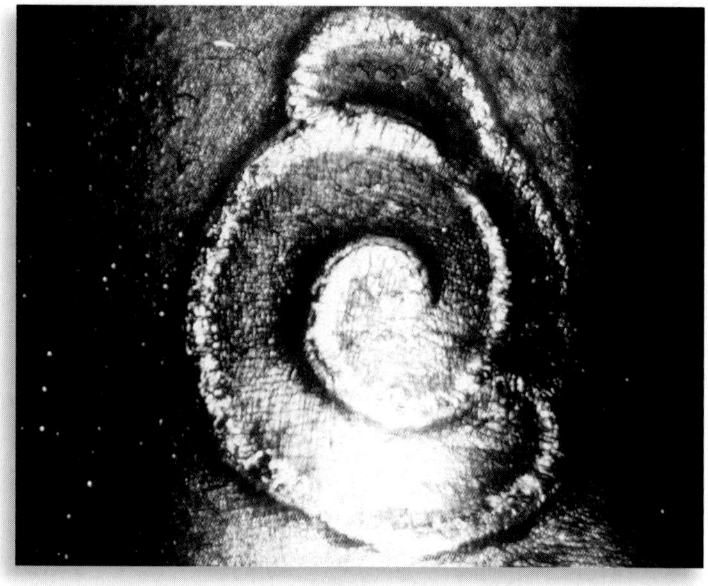

Figure 23.15 The pathology of late, or tertiary, syphilis.
A ring-shaped erosive gumma appears on the arm of this patient. Other gummas can be internal.

Latency and Tertiary Syphilis After resolution of secondary syphilis, about 30% of infections enter a highly varied latent period that can last for 20 years or longer. During latency, although antibodies to the bacterium are readily detected, the bacterium itself is not. The final stage of the disease, tertiary syphilis, is relatively rare today because of widespread use of antibiotics. But it is so damaging that it is important to recognize. By the time a patient reaches this phase, numerous pathologic complications occur in susceptible tissues and organs. Cardiovascular syphilis results from damage to the small arteries in the aortic wall. As the fibers in the wall weaken, the aorta is subject to distension and fatal rupture. The same pathologic process can damage the aortic valves, resulting in insufficiency and heart failure.

In one form of tertiary syphilis, painful swollen syphilitic tumors called **gummas** (goo-mahz′) develop in tissues such as the liver, skin, bone, and cartilage **(figure 23.15).** Gummas are usually benign and only occasionally lead to death, but they can impair function. Neurosyphilis can involve any part of the nervous system, but it shows particular affinity for the blood vessels in the brain, cranial nerves, and dorsal roots of the spinal cord. The diverse results include severe headaches, convulsions, atrophy of the optic nerve, blindness, dementia, and a sign called the Argyll-Robertson pupil—a condition caused by adhesions along the inner edge of the iris that fix the pupil's position into a small irregular circle.

Congenital Syphilis The syphilis bacterium can pass from a pregnant woman's circulation into the placenta and can be carried throughout the fetal tissues. An infection leading to **congenital syphilis** can occur in any of the three trimesters, but it is most common in the second and third. The pathogen inhibits fetal growth and disrupts critical periods of development with varied consequences, ranging from mild to the extremes of spontaneous miscarriage or stillbirth. Early congenital syphilis encompasses the period from birth to 2 years of age and is usually first detected 3 to 8 weeks after birth. Infants often demonstrate such signs as profuse nasal discharge **(figure 23.16*a*),** skin eruptions, bone deformation, and nervous system abnormalities. The late form gives rise to an unusual assortment of problems in the bones, eyes, inner ear, and joints and causes the formation of Hutchinson's teeth **(figure 23.16*b*).** The number of congenital syphilis cases is closely tied to the incidence in adults.

Causative Agent *Treponema pallidum*, a spirochete, is a thin, regularly coiled cell with a gram-negative cell wall. It is a strict parasite with complex growth requirements that necessitate cultivating it in living host cells. Most spirochete bacteria are nonpathogenic; *Treponema* and *Leptospira*, described earlier, are among the pathogens of this group.

Syphilis is a complicated disease to diagnose. Not only do the stages each mimic other diseases, but their appearance can also be so separated in time as to seem unrelated. The chancre and secondary lesions must be differentiated from bacterial, fungal, and parasitic infections; tumors; and even allergic reactions. Overlapping symptoms of sexually transmitted infections that the patient is concurrently experiencing, such as gonorrhea or chlamydiosis, can further complicate diagnosis. The disease can be diagnosed using two different strategies: either by detecting the bacterium in patient lesions or by looking for antibodies in the patient's blood.

Pathogenesis and Virulence Factors Brought into direct contact with mucous membranes or abraded skin, *T. pallidum*

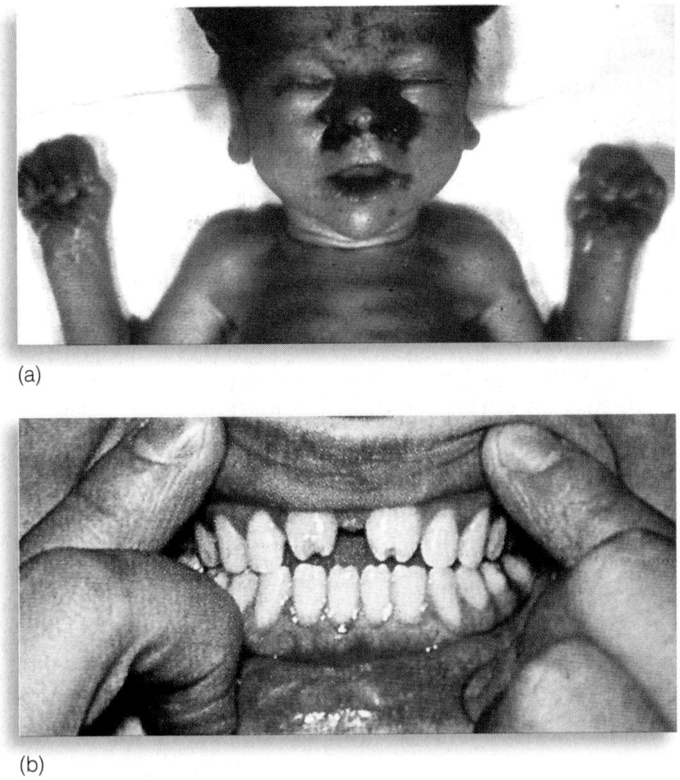

Figure 23.16 Congenital syphilis.
(a) An early sign is snuffles, a profuse nasal discharge that obstructs breathing. **(b)** A common characteristic of late congenital syphilis is notched, barrel-shaped incisors (Hutchinson's teeth).

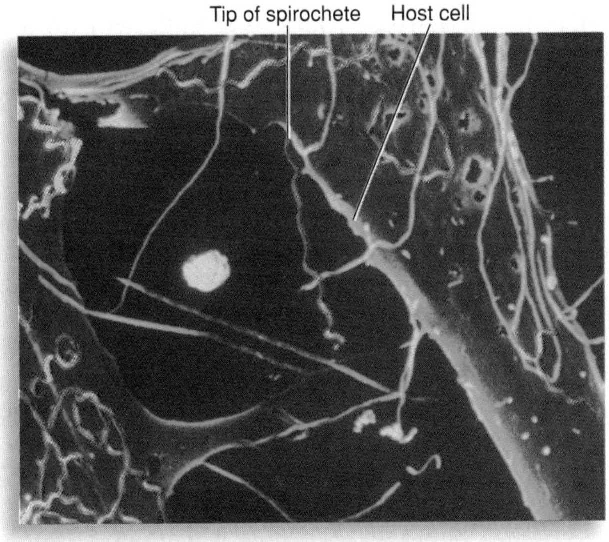

Figure 23.17 Electron micrograph of the syphilis spirochete attached to cells.

binds avidly by its hooked tip to the epithelium **(figure 23.17).** At the binding site, the spirochete multiplies and penetrates the capillaries nearby. Within a short time, it moves into the circulation, and the body is literally transformed into a large receptacle for incubating the pathogen. Virtually any tissue is a potential target.

The specific factor that accounts for the virulence of the syphilis spirochete appears to be outer membrane lipoproteins. These molecules appear to stimulate a strong inflammatory response, which is helpful in clearing the organism but can produce damage as well. *T. pallidum* produces no toxins and does not appear to kill cells directly. Studies have shown that, although phagocytes seem to act against it and several types of antitreponemal antibodies are formed, immune responses are unable to contain it. The primary lesion occurs when the spirochetes invade the spaces around arteries and stimulate an inflammatory response. Organs are damaged when granulomas form at these sites and block circulation.

Transmission and Epidemiology Humans are evidently the sole natural hosts and source of *T. pallidum*. The bacterium is extremely fastidious and sensitive and cannot survive for long outside the host, being rapidly destroyed by heat, drying, disinfectants, soap, high oxygen tension, and pH changes. It survives a few minutes to hours when protected by body secretions and about 36 hours in stored blood. Research with human subjects has demonstrated that the risk of infection from an infected sexual partner is 12% to 30% per encounter. The bacterium can also be transmitted to the fetus in utero. Syphilis infection through blood transfusion or exposure to fomites is rare.

For centuries, syphilis was a common and devastating disease in the United States, so much so that major medical centers had "Departments of Syphilology." Its effect on social life was enormous. This effect diminished quickly when antibiotics were discovered. In the 20th and 21st centuries, syphilis, like other STDs, has experienced periodic increases during times of social disruption. Most cases tend to be concentrated in larger metropolitan areas among prostitutes, their contacts, and crack cocaine users. Syphilis continues to be a serious problem worldwide, especially in Africa and Asia. As mentioned previously, persons with syphilis often suffer concurrent infections with other STDs. Coinfection with the AIDS virus can be an especially deadly combination with a rapidly fatal course.

Culture and Diagnosis Syphilis can be detected in patients most rapidly by using dark-field microscopy of a suspected lesion. The lesions are gently squeezed or scraped to extract clear fluid. A wet mount is then observed for the characteristic size, shape, and motility of *T. pallidum* **(figure 23.18).** A single negative test is not enough to exclude syphilis because the patient may have removed the organisms by washing, so follow-up tests are recommended. Another microscopic test for discerning the spirochete directly in samples is direct immunofluorescence staining with monoclonal antibodies.

Very commonly, blood tests are used for this diagnosis. These tests are based upon detection of antibody formed in response to *T. pallidum* infection. Two kinds of antibodies are formed: those that specifically react with treponemal antigens and, perhaps surprisingly, those that are formed

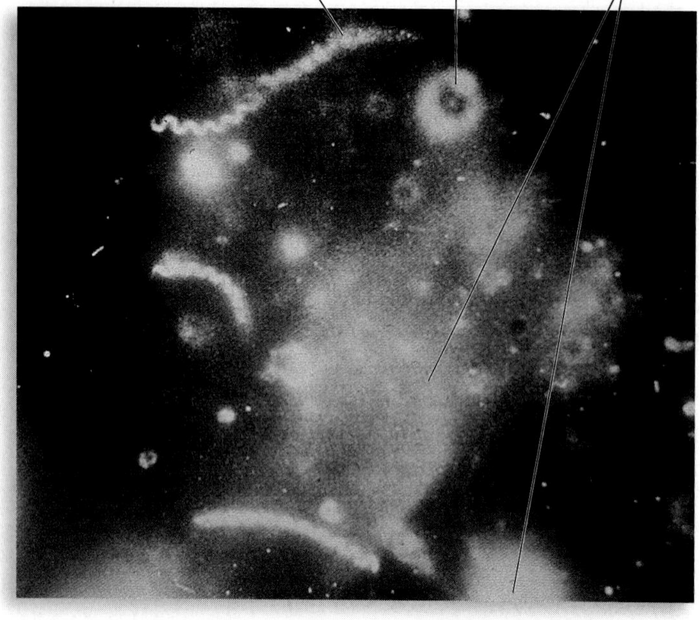

Figure 23.18 *Treponema pallidum* **from a syphilitic chancre, viewed with dark-field illumination.**
Its tight spirals are highlighted next to human cells and tissue debris.

against nontreponemal antigens. After infection with *T. pallidum,* the body abnormally produces antibodies to a natural constituent of human cells called *cardiolipin,* and the presence of these cardiolipin antibodies is also indicative of *T. pallidum* infection. Several different tests detect these antibodies, such as rapid plasma reagin (RPR), VDRL, Kolmer, and the Wasserman test.

More specific tests are available when considered necessary. One of these is the indirect immunofluorescent method called the FTA-ABS (fluorescent treponemal antibody absorbance) test. The test serum is first allowed to react with treponemal cells and then reacted with antihuman globulin antibody labeled with fluorescent dyes. If antibodies to the treponeme are present, the fluorescently labeled antibody will bind to the human antibody bound to the treponemal cells. The result is highly visible with a fluorescence microscope. A PCR test is available for syphilis, but its accuracy is dependent on the type of tissue being tested.

Prevention The core of an effective prevention program depends upon detection and treatment of the sexual contacts of syphilitic patients. Public health departments and physicians are charged with the task of questioning patients and tracing their contacts. All individuals identified as being at risk, even if they show no signs of infection, are given immediate prophylactic penicillin in a single long-acting dose.

The barrier effect of a condom provides excellent protection during the primary phase. Protective immunity apparently does arise in humans, allowing the prospect of an effective immunization program in the future, although no vaccine exists currently.

Treatment Throughout most of history, the treatment for syphilis was a dose of mercury or even a "mercurial rub" applied to external lesions. In 1906, Paul Ehrlich discovered that a derivative of arsenic called salvarsan could be very effective. The fact that toxic compounds like mercury and arsenic were used to treat syphilis gives some indication of how dreaded the disease was and to what lengths people would go to rid themselves of it. In 1918, Paul A. O'Leary formalized the practice of infecting syphilis patients with malaria as a therapeutic approach. The patients were allowed to have a dozen or so episodes of high fever and then were cured of the malaria with quinine. This procedure proved to be effective in curing syphilis. ("Malaria therapy" has also been investigated in recent years as an alternative treatment for HIV infection.)

Once penicillin became available, it replaced all other treatments, and penicillin G retains its status as a wonder drug in the treatment of all stages and forms of syphilis. It is given parenterally in large doses with benzathine or procaine. The goal is to maintain a blood level lethal to the spirochete for at least 7 days. Alternative drugs (tetracycline and erythromycin) are less effective, and they are indicated only if penicillin allergy has been documented. It is important that patients be monitored for successful clearance of the spirochete.

Chancroid

This ulcerative disease is not caused by a spirochete and has no systemwide effects. Infection usually begins as a soft papule, or bump, at the point of contact. It develops into a "soft chancre" (in contrast to the hard syphilis chancre), which is very painful in men, but may be unnoticed in women (see Checkpoint 23.7). Inguinal lymph nodes can become very swollen and tender.

Chancroid is caused by a **pleomorphic** gram-negative rod called *Haemophilus ducreyi.* Recent research indicates that a hemolysin (exotoxin) is important in the pathogenesis of chancroid disease. It is very common in the tropics and subtropics and is becoming more common in the United States. Chancroid is transmitted exclusively through direct contact, especially sexually. This disease is associated with prostitutes and poor hygiene; uncircumcised men seem to be more commonly infected than those who have been circumcised. People may carry this bacterium asymptomatically.

No vaccine exists. Prevention of chancroid is the same as for other sexually transmitted diseases: Avoid contact with infected tissues, either by abstaining from sexual contact or by proper use of barrier protection.

Antibiotics such as azithromycin and ceftriaxone are effective, but patients should be reexamined after a course of treatment to ensure that the bacterium has been eliminated.

Genital Herpes

Virtually everyone becomes infected with a herpesvirus at some time, because this large family of viruses can infect a wide range of host tissues. (We studied three herpesviruses in chapter 21 alone.) Genital herpes is caused by herpes simplex viruses (HSVs). Two types of HSV have been identified, HSV-1 and HSV-2. Other members of the herpes family are herpes zoster (causing chickenpox and shingles), cytomegalovirus (associated with congenital disease and also with HIV-associated disease), Epstein-Barr virus (causing infection of the lymphoid tissue as in infectious mononucleosis), and more recently identified viruses (herpesvirus-6, -7, and -8).

Genital herpes is much more common than most people think.

Signs and Symptoms Genital herpes infection has multiple presentations. After initial infection, a person may notice no symptoms. Alternatively, herpes could cause the appearance of single or multiple vesicles on the genitalia, perineum, thigh, and buttocks. The vesicles are small and are filled with a clear fluid (see Checkpoint 23.7). They are intensely painful to the touch. The appearance of lesions the first time you get them can be accompanied by malaise, anorexia, fever, and bilateral swelling and tenderness in the groin. Occasionally central nervous system symptoms such as meningitis or encephalitis can develop. Thus we see that initial infection can either be completely asymptomatic or be serious enough to require hospitalization.

After recovery from initial infection, a person may have recurrent episodes of lesions. They are generally less severe than the original symptoms, although the whole gamut of possible severity is seen here as well. Some people never have recurrent lesions. Others have nearly constant outbreaks with little recovery time between them. On average, the number of recurrences is four or five a year. Their frequency tends to decrease over the course of years.

In most cases, patients remain asymptomatic or experience recurrent "surface" infections indefinitely. Very rarely, complications can occur. Every year, one or two persons per million with chronic herpes infections develop encephalitis. The virus disseminates along nerve pathways to the brain (although it can also infect the spinal cord). The effects on the central nervous system begin with headache and stiff neck and can progress to mental disturbances and coma. The fatality rate in untreated cases is 70%, although treatment with acyclovir is effective. Patients with underlying immunodeficiency are more prone to severe, disseminated herpes infection than are immunocompetent patients. Of greatest concern are patients receiving organ grafts, cancer patients on immunosuppressive therapy, those with congenital immunodeficiencies, and AIDS patients.

Herpes of the Newborn Although HSV infections in healthy adults are annoying and unpleasant, only rarely are they life-threatening. However, in the neonate and the fetus **(figure 23.19),** HSV infections are very destructive and can be fatal. Most cases occur when infants are contaminated by the mother's reproductive tract immediately before or during birth, but they have also been traced to hand transmission from the mother's lesions to the baby. Because HSV-2 is more often associated with genital infections, it is more frequently involved; however, HSV-1 infection has similar complications. In infants whose disease is confined to the mouth, skin, or eyes, the mortality rate is 30%, but disease affecting the central nervous system has a 50% to 80% mortality rate.

Because of the danger of herpes to fetuses and newborns and also because of the increase in the number of cases of genital herpes, it is now standard procedure to screen pregnant women for the herpesvirus early in their prenatal care. (Don't forget that most women who are infected do not even know it.) Pregnant women with a history of recurrent infections must be constantly monitored for any signs of viral shedding, especially in the last 4 weeks of pregnancy. If no evidence of recurrence is seen, vaginal birth is indicated, but any evidence of an outbreak at the time of delivery necessitates a cesarean section.

Causative Agent Both HSV-1 and HSV-2 can cause genital herpes if the virus contacts the genital epithelium, although HSV-1 is thought of as a virus that infects the oral mucosa, resulting in "cold sores" or "fever blisters" **(figure 23.20),**

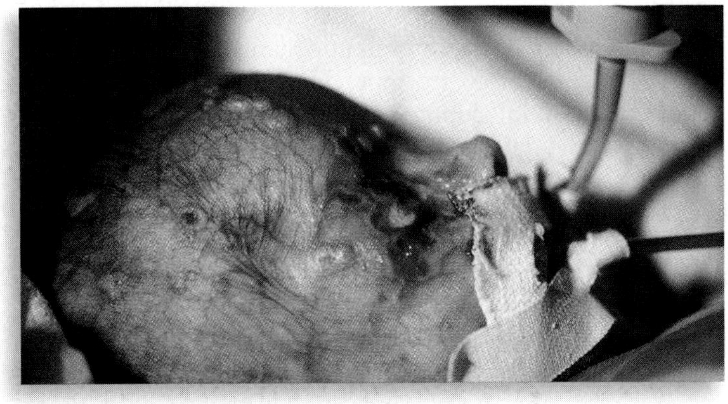

Figure 23.19 Neonatal herpes simplex.
This premature infant was born with the classic "cigarette burn" pattern of HSV infection. Babies can be born with the lesions or develop them 1 to 2 weeks after birth.

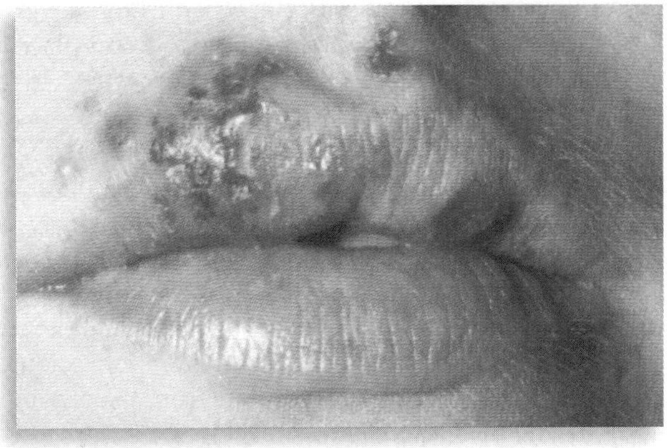

Figure 23.20 Oral herpes infection.
Tender itchy papules erupt around the mouth and progress to vesicles that burst, drain, and scab over. These sores and fluid are highly infectious and should not be touched.

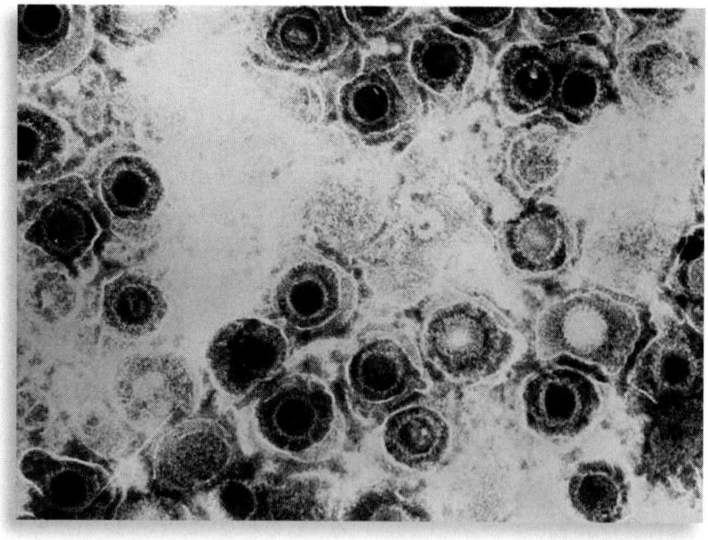

Figure 23.21 **Transmission electron micrograph of herpes simplex virus.**

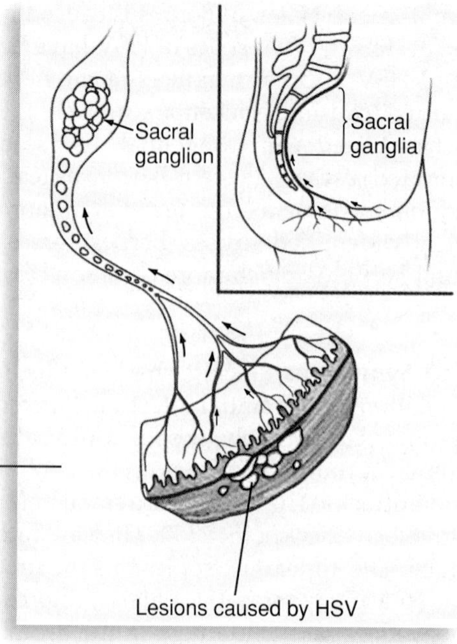

Figure 23.22 **HSV latent in lumbosacral ganglion.** The ganglion is the nerve root near the base of the spine. When the virus is reactivated, it travels down the neuron to the body's surface.

and HSV-2 is thought of as the genital virus. In reality, either virus can infect either region, depending on the type of contact. It is generally believed that HSV-1 infections in the genital region are less symptomatic than are HSV-2 and that HSV-1 oral infections are more likely to cause symptoms than HSV-2 oral infections.

HSV-1 and HSV-2 are DNA viruses with icosahedral capsids and envelopes containing glycoprotein spikes **(figure 23.21).** Like other enveloped viruses, herpesviruses are prone to deactivation by organic solvents or detergents and are unstable outside the host's body.

Pathogenesis and Virulence Factors Herpesviruses have a tendency to become latent. The molecular basis of latency is not entirely clear. It may involve the incorporation of viral nucleic acid into the host genome, but recent evidence points instead to the viral DNA simply coexisting with host DNA inside the nucleus of the host cell. During latency, some type of signal causes most of the HSV genome not to be transcribed. This allows the virus to be maintained within cells of the nervous system between episodes. It is further suggested that in some peripheral cells, viral replication takes place at a constant, slow rate, resulting in constant low-level shedding of the virus without lesion production.

HSV-2 (or HSV-1, if it has infected the genital region) usually becomes latent in the ganglion of the lumbosacral spinal nerve trunk **(figure 23.22).** Reactivation of the virus can be triggered by a variety of stimuli, including stress, UV radiation (sunlight), injury, menstruation, or another microbial infection. At that point, the virus begins manufacturing large numbers of entire virions, which cause new lesions on the surface of the body served by the neuron, usually in the same site as previous lesions.

HSV-1 (or HSV-2 if it is in the oral region) behaves in a similar way, but it becomes latent in the trigeminal nerve, which has extensive innervations in the oral region.

Transmission and Epidemiology Herpes simplex infection occurs globally in all seasons and among all age groups. Because these viruses are relatively sensitive to the environment, transmission is primarily through direct exposure to secretions containing the virus. People with active lesions are the most significant source of infection, but studies indicate that genital herpes can be transmitted even when no lesions are present (due to the constant shedding just referred to).

As with all sexually transmitted diseases, many different figures are cited as to its prevalence in society. The terminology associated with STDs can be confusing. One reason is that some of the STDs are officially "reportable" diseases (chlamydia, gonorrhea). Earlier in this chapter, you read that chlamydia is the most common *reported* infectious disease in the United States. Elsewhere you might hear that gonorrhea is one of the most common reportable *STDs* in the United States. Both statements are true. It is also true that genital herpes is much more common than either of these diseases. Herpes, however, is not an officially *reportable* disease.

It is estimated that about 20% of American adults have genital herpes. That estimate would put the number of infected people in this country at around 42 million. Scientists think that the vast majority of people who are infected don't even know it, either because they have rare symptoms that they fail to recognize or because they have no symptoms at all.

Culture and Diagnosis These two viruses are sometimes diagnosed based on the characteristic lesions alone. PCR tests are available to test for these viruses directly from lesions.

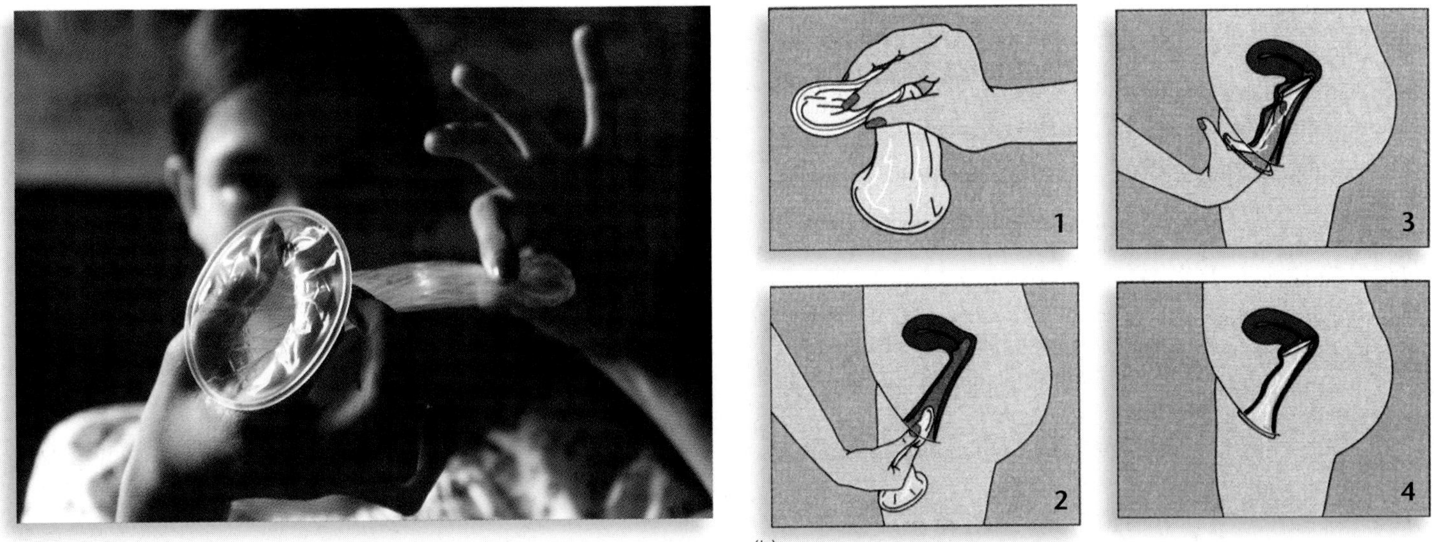

(a) (b)

Figure 23.23 The female condom.
The condom has a closed ring that fits over the cervix and an open ring that rests on the external genitalia.

Alternatively, antibody to either of the viruses can be detected from blood samples. Detecting antibody to either HSV-1 or HSV-2 in blood does not necessarily indicate whether the infection is oral or genital or whether the infection is new or preexisting.

Laboratory culture and specific tests are essential for diagnosing severe or complicated herpes infections. They are also used when screening pregnant women for the presence of virus on the vaginal mucosa. A specimen of tissue or fluid is inoculated into a primary cell culture line and is then observed for cytopathic effects, which are characteristic for specific viruses.

Prevention No vaccine is currently licensed for HSV, but more than one is being tested in clinical trials, meaning that vaccines may become available very soon. In the meantime, avoiding contact with infected body surfaces is the only way to avoid HSV. Condoms provide good protection when they actually cover the site where the lesion is, but lesions can occur outside of the area covered by a condom. Women with herpes are sometimes counseled to use the female condom **(figure 23.23)** because these cover a substantial portion of the female external genitalia. In general, people experiencing active lesions should avoid sex. Because the virus can be shed when no lesions are present, barrier protection should be practiced at all times by persons infected with HSV.

Mothers with cold sores should be careful in handling their newborns; they should never kiss their infants on the mouth. Hospital attendants with active oral herpes infection should be barred from the newborn nursery.

Some of the drugs used to "treat" genital herpes really function to prevent recurrences of lesions. In this way, they serve as prevention for potential partners of people with herpes.

Treatment Several agents are available for treatment. These agents often result in reduced viral shedding and a decrease in the frequency of lesion occurrence. They are not curative. Acyclovir and its derivatives (Zovirax, Valtrex) are very effective. Topical formulations can be applied directly to lesions, and pills are available as well. Sometimes medicines are prescribed on an ongoing basis to decrease the frequency of recurrences, and sometimes they are prescribed to be taken at the beginning of a recurrence to shorten it **(Checkpoint 23.7).**

Wart Diseases

In this section, we describe two viral STDs that cause wartlike growths. The more serious disease is caused by *human papillomavirus* (*HPV*); the other condition, called *molluscum contagiosum*, apparently has no serious effects outside of the growths themselves.

Human Papillomavirus

These viruses are the causative agents of genital warts. But an individual can be infected with these viruses without having any warts, while still risking serious consequences.

Signs and Symptoms Symptoms, if present, may manifest as warts—outgrowths of tissue on the genitals **(Checkpoint 23.8).** In females, these growths can occur on the vulva and in and around the vagina. In males, the warts can occur in or on the penis and the scrotum. In both sexes, the warts can appear in or on the anus and even on the skin around the groin, such as the area between the thigh and the pelvis. The warts themselves range from tiny, flat, inconspicuous bumps to extensively branching, cauliflowerlike masses called **condyloma acuminata.** The warts are unsightly and

CHECKPOINT 23.7 Genital Ulcer Diseases

	Syphilis	Chancroid	Herpes
Causative Organism(s)	*Treponema pallidum*	*Haemophilus ducreyi*	Herpes simplex 1 and 2
Most Common Modes of Transmission	Direct contact and vertical	Direct contact (vertical transmission *not* documented)	Direct contact, vertical
Virulence Factors	Lipoproteins	Hemolysin (exotoxin)	Latency
Culture/Diagnosis	Direct tests (immunofluorescence, dark-field microscopy), blood tests for treponemal and nontreponemal antibodies, PCR	Culture from lesion	Clinical presentation, PCR, Ab tests, growth of virus in cell culture
Prevention	Antibiotic treatment of all possible contacts, avoiding contact	Avoiding contact	Avoiding contact, antivirals can reduce recurrences
Treatment	Penicillin G	Azithromycin, ceftriaxone	Acyclovir and derivatives
Distinctive Features	Three stages of disease plus latent period, possibly fatal	No systemic effects	Ranges from asymptomatic to frequent recurrences
Effects on Fetus	Congenital syphilis	None	Blindness, disseminated herpes infection
Appearance of Lesions	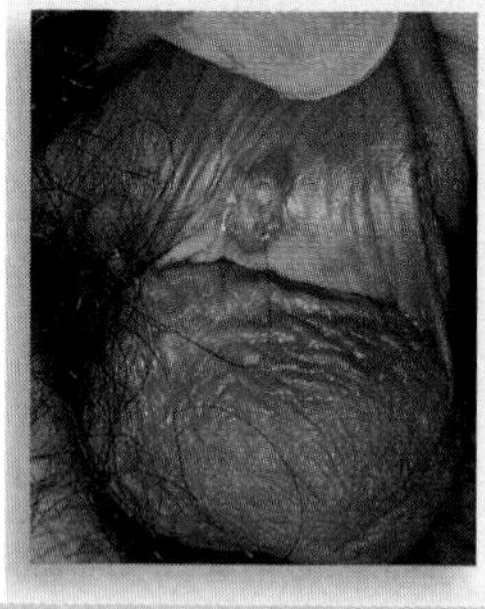	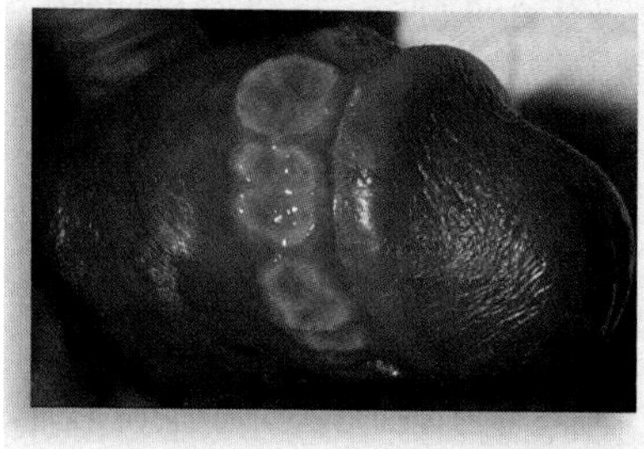	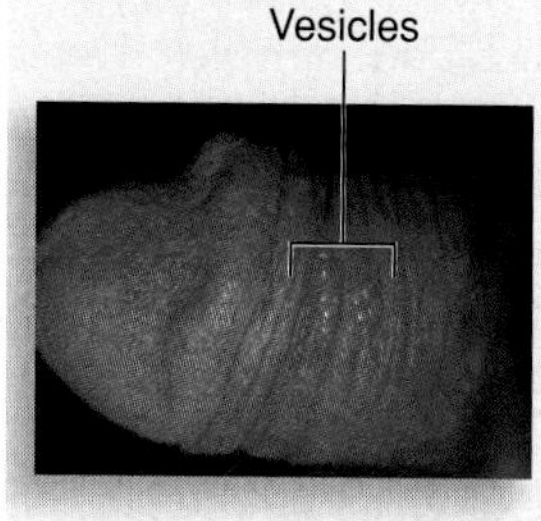

can be obstructive, but they don't generally lead to more serious symptoms.

Other types of HPV can lead to more subtle symptoms. Certain types of the virus infect cells on the female cervix. This infection may be "silent," or it may lead to abnormal cell changes in the cervix. Some of these cell changes can eventually result in malignancies of the cervix. The vast majority of cervical cancers are caused by HPV infection. (It is possible that chronic infections with other microorganisms cause a very small percentage of cervical malignancies.) Approximately 4,000 women die each year in the United States from cervical cancer. Also, data released in 2007 indicate a link between having had more than five oral sex partners and a greatly increased risk of throat cancer, presumably due to HPV.

Males can also get cancer from infection with these viruses. The sites most often affected are the penis and the anus. These cases are much less common than cervical cancer.

Causative Agent The human papillomaviruses are a group of nonenveloped DNA viruses belonging to the Papoviridae family. There are more than 90 different types of HPV. Some types are specific for the mucous membranes; others invade the skin. Some of these viruses are the cause of plantar warts, which often occur on the soles of the feet. Other HPVs cause the common or "seed" warts and flat warts. In this chapter, we are concerned only with the HPVs that colonize the genital tract.

Among the HPVs that infect the genital tract, some are more likely to cause the appearance of warts. Others that have a preference for growing on the cervix can lead to cancerous changes. Two types in particular, HPV-16 and HPV-18, appear to be very closely associated with development of cervical cancer.

Pathogenesis and Virulence Factors Scientists are working hard to understand how viruses cause the growths we know as warts and also how some of them can cause cancer. The major virulence factors for cancer-causing HPVs are **oncogenes,** which code for proteins that interfere with normal host cell function, resulting in uncontrolled growth.

Transmission and Epidemiology It is estimated that the majority of people who are sexually active are infected with one or more types of this virus. But because there are dozens

CHECKPOINT 23.8 Wart Diseases		
	HPV	**Molluscum Contagiosum**
Causative Organism(s)	Human papillomaviruses	Poxvirus, sometimes called the molluscum contagiosum virus (MCV)
Most Common Modes of Transmission	Direct contact (STD)—also autoinoculation, indirect contact	Direct contact (STD), also indirect and autoinoculation
Virulence Factors	Oncogenes (in the case of malignant types of HPV)	–
Culture/Diagnosis	PCR tests for certain HPV types, clinical diagnosis	Clinical diagnosis, also histology, PCR
Prevention	Vaccine available; avoid direct contact; prevent cancer by screening cervix	Avoid direct contact
Treatment	Warts or precancerous tissue can be removed; virus not treatable	Warts can be removed; virus not treatable
Distinguishing Features	Infection may or may not result in warts; infection may result in malignancy	Wartlike growths are only known consequence of infection
Effects on Fetus	May cause laryngeal warts	–
Appearance of Growths	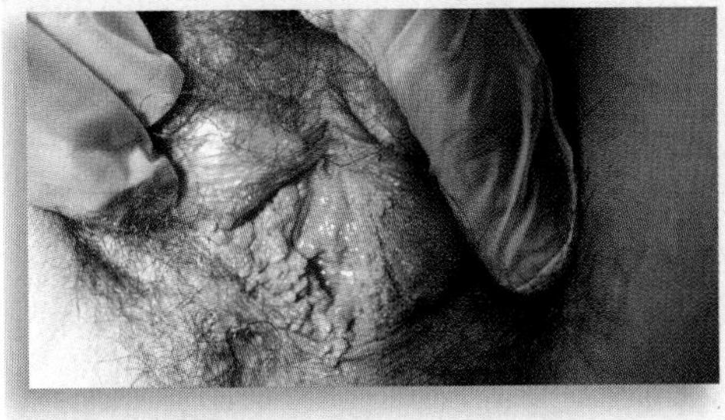	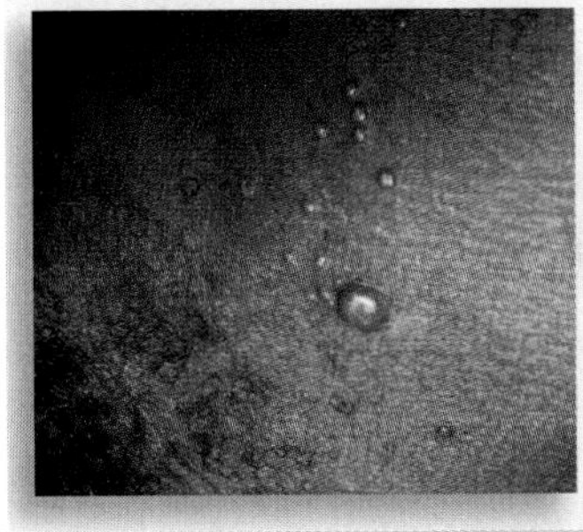

of different virus types, we do not know how many people are actually at risk of serious disease. The CDC conducted a study in 2000 that found that 18% of women and 8% of men are infected with HPV-16—one of the most dangerous types. Some experts assert that HPV is the most common STD in the United States. It is difficult to know whether genital herpes or HPV is more common, but it is probably safe to assume that any unprotected sex carries a good chance of encountering either HSV or HPV.

The mode of transmission is direct contact. Autoinoculation is also possible—meaning that the virus can be spread to other parts of the body by touching warts. Indirect transmission occurs but is more common for nongenital warts caused by HPV.

Culture and Diagnosis PCR-based screening tests can be used to test samples from a pelvic exam for the presence of dangerous HPV types. These tests are now recommended for women over the age of 30.

Prevention When discussing HPV prevention, we must consider two possibilities. One of these is infection with the viruses, which is prevented the same way other sexually transmitted infections are prevented—by avoiding direct, unprotected contact, but also by a new vaccine approved in 2006 called Gardasil. The vaccine prevents infection by four types of HPV and is recommended in girls as young as age 9.

The second issue is the prevention of cervical cancer. Even though women now have access to the vaccine, cancer can still result from HPV types not included in the vaccine. The good news is that cervical cancer is slow in developing, so that even if a woman is infected with a malignant HPV type, regular screening of the cervix can detect abnormal changes early. The standardized screen for cervical cell changes is the Pap smear **(Insight 23.2).** Precancerous changes show up very early, and the development process can be stopped by removal of the affected tissue. Women should have their first Pap smear by age 21 or within 3 years of their first sexual activity, whichever comes first. New Pap smear technologies have been developed; and depending on which one your physician uses, it is now possible that you need to be screened only once every 2 or 3 years. But you should base your screening practices on the sound advice of a physician.

Treatment Infection with any HPV is incurable. Genital warts can be removed through a variety of methods, some of which can be used at home. But the virus causing them will most likely remain with you. It is possible for the viral infection to resolve itself, but this is very unpredictable.

INSIGHT 23.2 *Medical*

The Pap Smear

In the early part of the 20th century, a Greek-born physician named George Papanicolaou, who taught at Cornell University and collaborated with hospital physicians there, became interested in the cytological changes that take place in precancerous and cancerous tissue of the female reproductive tract. He developed a technique for evaluating "vaginal smears" for precancerous changes and in 1943 published a paper that would change women's lives forever. The title was "Diagnosis of Uterine Cancer by the Vaginal Smear." The test came to be known as the Pap smear.

The Pap smear is still the single best screening procedure available for cervical cancer, a disease that claims the lives of over 4,000 women every year in the United States. This incidence has decreased 74% since 1955, almost entirely due to the increased use of the Pap smear. The procedure is simple and painless: During a pelvic exam, a sample of cells is taken from the cervix using a wooden spatula or a small cervical brush. Then the sample is "smeared" onto a glass microscope slide and preserved with a fixative. In a newer method, the brush or spatula is rinsed with preservative fluid, the fluid is saved, and later it is automatically applied in a thin layer to a microscope slide. Whether the slide was made as a "smear" or as a "thin prep," it is then viewed microscopically by a technician or, in newer methods, by a computer so that abnormal cells can be detected.

A variety of "abnormal" results can be found and reported to the patient after a Pap smear. Here are some words that may appear on the Pap report:

- **Dysplasia**—abnormal cells found, not cancer but with a slight potential for developing into very early cancer of the cervix, depending on the degree of dysplasia (mild, moderate, severe, or the most severe form called *carcinoma in situ*).
- **Squamous intraepithelial lesion (SIL)**—a term that refers to the type of cells (squamous) that form the outer surface of the cervix. The "intraepithelial" designation refers to the observation that abnormal cells are only present on the surface of the cervix and not in the deeper tissue.
- **Cervical intraepithelial neoplasia (CIN)**—another term referring to abnormal cells. "Neoplasia" means an abnormal growth of cells. There will often be a number after the CIN (that is, CIN-1 or CIN-3). The number corresponds with how far the abnormal cells extend into the cervix.
- **Atypical squamous cells**—cells appear abnormal, but the nature and degree of abnormality are unclear.

Cervical cancer is nearly always caused by infection with human papillomavirus, as detailed in the section on HPV in this chapter. A study performed in 2003 in England found that among a relatively well-educated group of 1,000 women, less than a third had even heard of HPV; even fewer knew that it was associated with cervical cancer. Because some types of HPV are shown to be more strongly associated with cervical cancer, a physician may perform a PCR test on cervical material to look for the presence of these HPV types. A negative HPV test can provide reassurance that the abnormalities detected on a Pap smear do not point to a cancerous or precancerous condition.

Other follow-up procedures that may be performed once an abnormal Pap smear is reported include colposcopy, in which a microscopelike instrument is inserted into the cervix to look more directly for cell changes; biopsy, in which a tiny piece of cervical tissue is removed to be examined histologically; and endocervical curettage, which involves scraping cells from inside the endocervical canal with a small spoon-shaped instrument. If abnormalities are confirmed, some form of treatment is warranted, such as the removal of abnormal tissue using cryotherapy (freezing), laser excision, and so forth.

In practice, most abnormal Pap smears are simply followed up with an additional Pap smear within 3 months. No screening method is 100% accurate, and the Pap smear can give both false positive and false negative results. Due to the slowly progressing nature of cervical cancer, waiting for this length of time is not risky. Multiple abnormal smears trigger further investigation and treatment as needed. Nearly all cervical cancer can be prevented if women get Pap smears on the recommended schedule.

Thanks to the relatively simple Pap smear, countless women have avoided not only early deaths from cancer but also hysterectomies, which later stages of cervical cancer require.

Treatment of cancerous cell changes is an important part of HPV therapy, and it can only be instituted if the changes are detected through Pap smears. Again, the *results* of the infection are treated (cancerous cells removed), but the viral infection is not amenable to treatment.

Molluscum Contagiosum

An unclassified virus in the pox family can cause a condition called molluscum contagiosum. This disease can take the form of skin lesions, and it can also be transmitted sexually. The wartlike growths that result from this infection can be found on the mucous membranes or the skin of the genital area (see Checkpoint 23.8). Few problems are associated with these growths beyond the warts themselves. In severely immunocompromised people, the disease can be more serious.

The virus causing these growths can also be transmitted through fomites such as clothing or towels and through autoinoculation. For a more detailed description of this condition, see chapter 18 **(Checkpoint 23.8).**

Group B *Streptococcus* "Colonization"—Neonatal Disease

Ten to forty percent of women in the United States are colonized, asymptomatically, by a beta-hemolytic *Streptococcus* in Lancefield group B. Nonpregnant women experience no ill effects from this colonization. But when these women become pregnant and give birth, about half of their infants become colonized by the bacterium during passage through the birth canal or by ascension of the bacteria through ruptured membranes; thus, this colonization is considered a reproductive tract disease.

A small percentage of infected infants experience life-threatening bloodstream infections, meningitis, or pneumonia. If they recover from these acute conditions, they may have permanent disabilities such as developmental disabilities, hearing loss, or impaired vision. In some cases, the mothers also experience disease, such as amniotic infection or subsequent stillbirths.

In 2002, the CDC recommended that all pregnant women be screened for group B *Streptococcus* colonization at 35 to 37 weeks of pregnancy. In late 2002, the FDA approved a rapid DNA-based test that enables earlier treatment, important for preventing long-term consequences. Women positive for the bacterium should be treated with penicillin or ampicillin unless the bacterium is found to be resistant to these and unless allergy to penicillin is present.

CHECKPOINT 23.9	Group B *Streptococcus* Colonization
Causative Organism(s)	Group B *Streptococcus*
Most Common Modes of Transmission	Vertical
Virulence Factors	–
Culture/Diagnosis	Culture of mother's genital tract
Prevention/Treatment	Treat mother with penicillin/ampicillin

SUMMING UP

Taxonomic Organization Microorganisms Causing Disease in the Genitourinary Tract

Microorganism	Disease	Chapter Location
Gram-positive bacteria		
Staphylococcus saprophyticus	Urinary tract infection	UTI, p. 737
Gardnerella (note: stains gram-negative)	Vaginosis	Vaginitis or vaginosis, p. 741
Group B *Streptococcus*	Neonatal disease	Group B strep neonatal disease, p. 759
Gram-negative bacteria		
Escherichia coli	Urinary tract infection	UTI, p. 737
Leptospira interrogans (spirochete)	Leptospirosis	Leptospirosis, p. 738
Proteus mirabilis	Urinary tract infection plus kidney stones	UTI, p. 737
Neisseria gonorrhoeae	Gonorrhea	Discharge diseases, p. 744
Chlamydia trachomatis	"Chlamydia"	Discharge diseases, p. 747
Treponema pallidum (spirochete)	Syphilis	Genital ulcer diseases, p. 748
Haemophilus ducreyi	Chancroid	Genital ulcer diseases, p. 752
DNA viruses		
Herpes simplex viruses 1 and 2	Genital herpes	Genital ulcer diseases, p. 752
Human papillomaviruses	Genital warts, cervical carcinoma	Wart diseases, p. 755
Poxviruses	Molluscum contagiosum	Wart diseases, p. 758
Fungi		
Candida albicans	Vaginitis	Vaginitis or vaginosis, p. 740
Protozoa		
Trichomonas vaginalis	Trichomoniasis (vaginitis)	Vaginitis or vaginosis, p. 742
Helminth—trematode		
Schistosoma haematobium	Urinary schistosomiasis	Urinary schistosomiasis, p. 739

INFECTIOUS DISEASES AFFECTING
The Genitourinary System

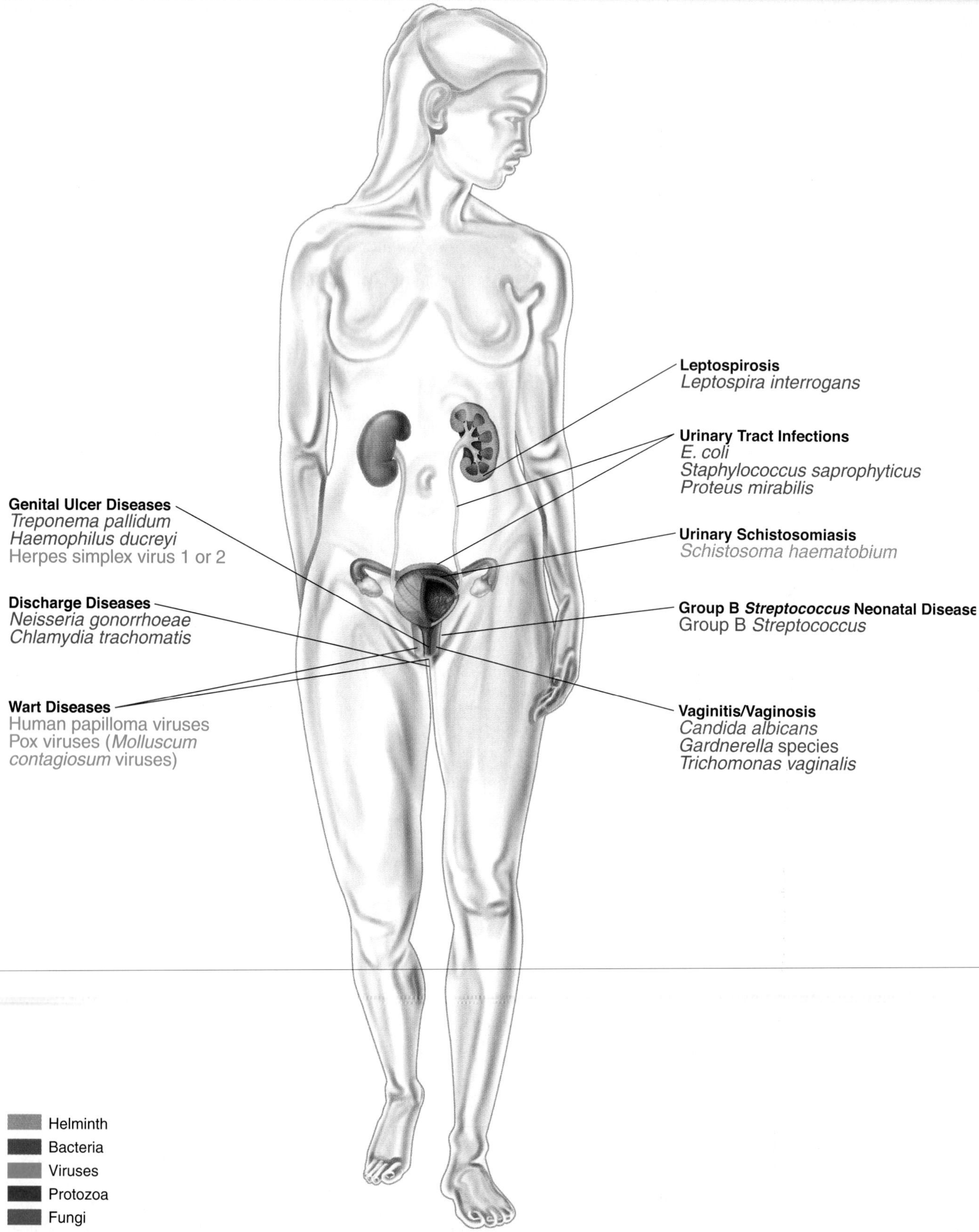

System Summary Figure 23.24

INFECTIOUS DISEASES AFFECTING
The Genitourinary System

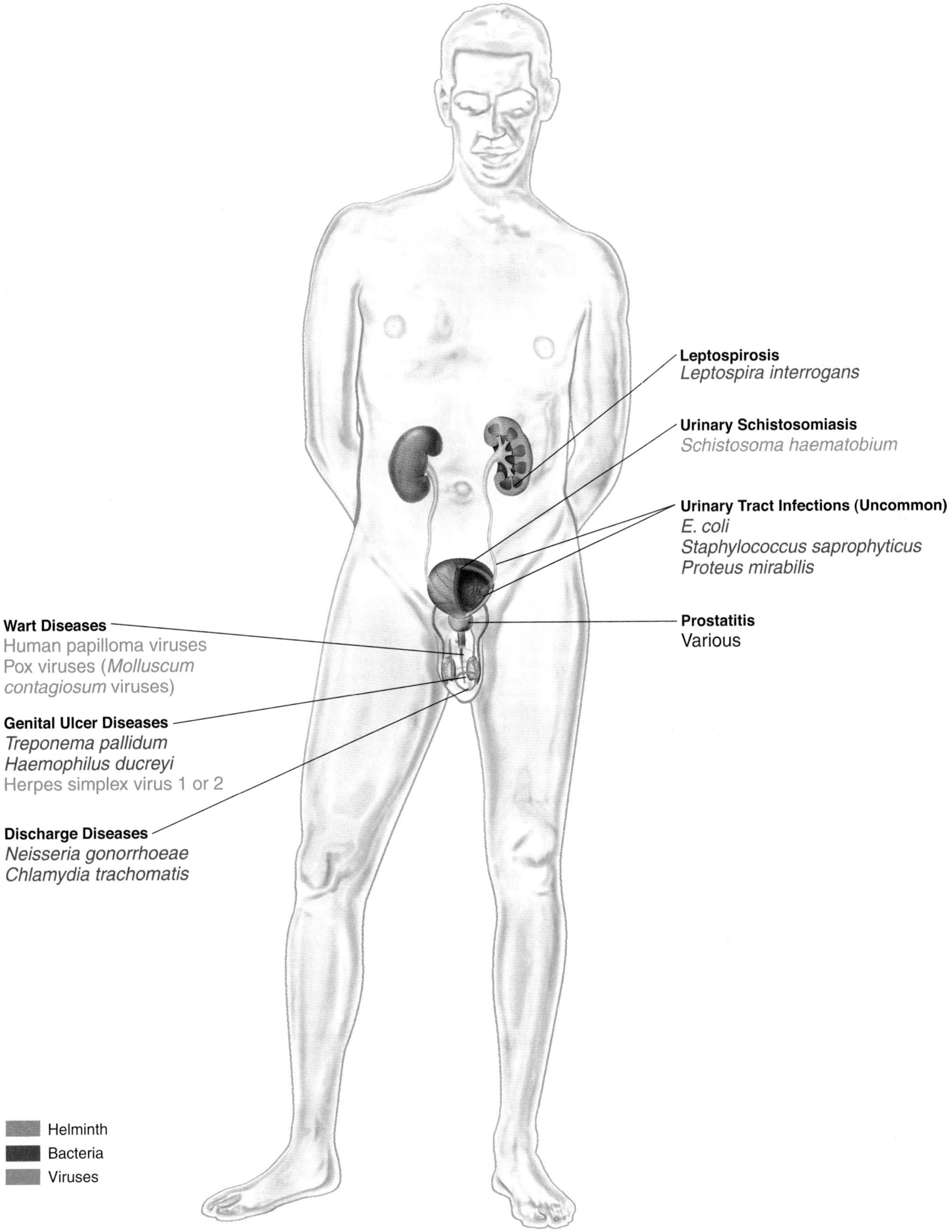

Chapter Summary with Key Terms

23.1 The Genitourinary Tract and Its Defenses
This "system" is really two systems, the reproductive system and the urinary system.

A. The reproductive tract in males and females is composed of structures and substances that allow for sexual intercourse and the creation of a new fetus; it is protected by normal mucosal defenses as well as specialized features (such as the low pH of the adult female reproductive tract).

B. The urinary system allows the excretion of fluid and wastes from the body. It has mechanical as well as chemical defense mechanisms.

23.2 Normal Biota of the Genitourinary Tract
As far as is known, the genital and the urinary systems have normal biota only in their most distal regions. Normal biota in the male reproductive and urinary systems are found in the distal part of the urethra and resemble skin biota. The same is generally true for the female urinary system. The normal biota in the female reproductive tract changes over the course of a woman's lifetime.

23.3 Urinary Tract Diseases Caused by Microorganisms

A. **Urinary Tract Infections (UTIs):** Infection can occur at a number of sites; the bladder (*cystitis*), the kidneys (*pyelonephritis*), and the urethra (*urethritis*). In most cystitis and pyelonephritis cases, the cause is bacteria that are normal biota in the gastrointestinal tract—most commonly, *Escherichia coli, Staphylococcus saprophyticus,* and *Proteus mirabilis*. Community-acquired UTIs are most often transmitted from the GI tract to the urinary system. UTIs are the most common of nosocomial infections. Patients of both sexes who have urinary catheters are susceptible to infections with a variety of microorganisms.

B. **Leptospirosis** is a zoonosis associated with wild animals that can affect the kidneys, liver, brain, and eyes. The causative agent is the *Leptospira interrogans* spirochete.

C. **Urinary Schistosomiasis:** This form of schistosomiasis is caused by *S. haematobium*. The bladder is damaged by trematode eggs and the granulomatous response they induce.

23.4 Reproductive Tract Diseases Caused by Microorganisms

A. **Vaginitis and Vaginosis**
1. Vaginitis is an inflammation of the vagina, most commonly caused by *Candida albicans*. This vulvovaginal candidiasis is nearly always an opportunistic infection.
2. The bacterium *Gardnerella* is associated with vaginosis that has a discharge but no inflammation in the vagina. Vaginosis could lead to complications such as **pelvic inflammatory disease (PID).**
3. *Trichomonas vaginalis* causes mostly asymptomatic infections in females and males. *Trichomonas* is a flagellated protozoan and is easily transmitted through sexual contact.

B. **Prostatitis:** Inflammation of the prostate, which can be acute or chronic. Not all cases have been established to have microbial cause, but most have.

C. **Discharge Diseases with Major Manifestation in the Genitourinary Tract:** Discharge diseases are those in which the infectious agent causes an increase in fluid discharge in the male and female reproductive tracts.
1. **Gonorrhea** elicits *urethritis* in males, but many cases are asymptomatic. In females, both the urinary and genital tracts will be infected during sexual intercourse. Major complications occur when the infection reaches uterus and fallopian tubes. One disease resulting from this progression is **salpingitis,** which can lead to pelvic inflammatory disease (PID). The causative agent, *Neisseria gonorrhoeae,* is a gram-negative diploccoccus.
2. **Chlamydia:** Genital chlamydial infection is the most common reportable infectious disease in the United States. In males, the bacterium causes an inflammation of the urethra (NGU). Females have cervicitis, a discharge, salpingitis, and frequently PID as a result of a *Chlamydia* infection.

   Certain strains of *Chlamydia trachomatis* can invade the lymphatic tissues, resulting in another condition called lymphogranuloma venereum.

D. **Genital Ulcer Diseases**
1. **Syphilis** is caused by *Treponema pallidum,* a spirochete; it is a thin, regularly coiled cell with a gram-negative cell wall. There are three distinct clinical stages, designated as *primary, secondary,* and *tertiary syphilis,* with a latent period of quiescence. The spirochete appears in the lesions and blood during the primary and secondary stages and thus is transmissible at these times. During the early latency period between secondary and tertiary syphilis, it is also transmissible. Syphilis is largely nontransmissible during the "late latent" and tertiary stages.

   The syphilis bacterium can pass from a pregnant woman's circulation into the placenta and can be carried throughout the fetal tissues, leading to **congenital syphilis.** The pathogen inhibits fetal growth and disrupts critical periods of development, which can lead to spontaneous miscarriage or stillbirth.
2. **Chancroid:** This ulcerative disease is caused by a **pleomorphic** gram-negative rod called *Haemophilus ducreyi*. Chancroid is transmitted exclusively through direct—mainly sexual—contact.
3. **Genital herpes** is caused by **herpes simplex viruses (HSVs).** Two types of HSV have been identified, HSV-1 and HSV-2. After infection, there may be no symptoms, or there may be fluid-filled, painful vesicles on the genitalia, perineum, thigh, and buttocks. In severe cases, meningitis or encephalitis can develop. Patients that recover remain asymptomatic or experience recurrent

"surface" infections indefinitely. HSV infections in the neonate and the fetus can be fatal.

HSV-1 and HSV-2 are DNA viruses with icosahedral capsids and envelopes containing glycoprotein spikes. Herpesviruses can become latent, most likely by the incorporation of viral nucleic acid into the host genome. The virus becomes latent in the nerve cells of the body. People with active lesions are the most contagious.

E. **Wart Diseases**
   1. *Human papillomaviruses* are the causative agents of genital warts. Certain types of the virus infect cells on the female cervix that can eventually result in malignancies of the cervix. Males can also get cancer from infection with these viruses.

      The human papillomaviruses are a group of nonenveloped DNA viruses belonging to the Papoviridae family. Two types, HPV-16 and HPV-18, appear to be very closely associated with development of cervical cancer.

      Infection with any HPV is incurable. Genital warts can be removed, but the virus will remain. Treatment of cancerous cell changes is an important part of HPV therapy, and it can only be instituted if the changes are detected through Pap smears. A vaccine for four types of HPV is now available.
   2. A pox family virus causes a condition called molluscum contagiosum. This disease can take the form of skin lesions from wartlike growths in the membranes of the genitalia, and it can also be transmitted sexually.

F. **Group B *Streptococcus* "Colonization"—Neonatal Disease:** Asymptomatic colonization of women by a beta-hemolytic *Streptococcus* in Lancefield group B is very common. But when these women become pregnant and give birth, about half of their infants become colonized by the bacterium during passage through the birth canal or by ascension of the bacteria through ruptured membranes; some infected infants experience life-threatening bloodstream infections, meningitis, or pneumonia. In 2002, the CDC recommended that all pregnant women be screened for group B *Streptococcus* colonization and treated with antibiotics prior to childbirth.

Multiple-Choice and True-False Questions

Multiple-Choice Questions. Select the correct answer from the answers provided.

1. Cystitis is an infection of the
   a. bladder
   b. urethra
   c. kidney
   d. vagina
2. Nongonococcal urethritis (NGU) is caused by
   a. *Neisseria gonorrhoeae*
   b. *Chlamydia trachomatis*
   c. *Treponema pallidum*
   d. *Trichomonas vaginalis*
3. Leptospirosis is transmitted to humans by
   a. person to person
   b. fomites
   c. mosquitoes
   d. contaminated soil or water
4. Syphilis is caused by
   a. *Treponema pallidum*
   b. *Neisseria gonorrhoeae*
   c. *Trichomonas vaginalis*
   d. *Haemophilus ducreyi*
5. Bacterial vaginosis is commonly associated with the following organism:
   a. *Candida albicans*
   b. *Gardnerella*
   c. *Trichomonas*
   d. all of the above
   e. none of the above
6. This dimorphic fungus is a common cause of vaginitis.
   a. *Candida albicans*
   b. *Gardnerella*
   c. *Trichomonas*
   d. all of the above
7. There are estimates that approximately ______ % of adult Americans have genital herpes.
   a. 2
   b. 10
   c. 20
   d. 50
8. Genital herpes transmission can be reduced or prevented by all of the following except
   a. condom
   b. abstinence
   c. contraceptive pill
   d. female condom
9. This protozoan can be treated with the drug Flagyl.
   a. *Neisseria gonorrhoeae*
   b. *Chlamydia trachomatis*
   c. *Treponema pallidum*
   d. *Trichomonas vaginalis*

True-False Questions. If the statement is true, leave as is. If it is false, correct it by rewriting the sentence.

10. Genital herpes can be treated with acyclovir.
11. Chancroid is caused by a fungus.
12. The majority of cervical cancers are caused by human papillomavirus.

Writing to Learn

These questions are suggested as a *writing-to-learn* experience. For each question, compose a one- or two-paragraph answer that includes the factual information needed to completely address the question.

1. Besides *E. coli,* name two other microorganisms associated with cystitis and pyelonephritis.
2. Describe the symptoms of Weil's syndrome.
3. Describe the common treatments for gonorrhea.
4. a. What is PID?
   b. What are the two most common microorganisms associated with this disease?
   c. Describe the long-term consequences of untreated PID.
5. Describe the life cycle of *Chlamydia.*
6. What are some of the stimuli that can trigger reactivation of a latent herpesvirus infection? Speculate on why.
7. a. Human papillomavirus is associated with what condition?
   b. Name some of the different sites on the body that can be affected by this virus.
8. What are the clinical stages of syphilis?
9. a. What is the standard screening for cervical cancer?
   b. In this screening technique, cervical cells are screened for abnormalities. What are some of the terms used to describe these abnormalities?

Concept Mapping

Appendix D provides guidance for working with concept maps.

1. Construct your own concept map using the following words as the *concepts.* Supply the linking words between each pair of concepts.

genital warts
discharge
herpes
chancroid
bacterium
molluscum contagiosum
virus
curable
ulcers
warts
syphilis
incurable
cancer

Critical Thinking Questions

Critical thinking is the ability to reason and solve problems using facts and concepts. These questions can be approached from a number of angles, and in most cases, they do not have a single correct answer.

1. What characteristics of *Neisseria gonorrhoeae* allow its effective sexual transmission?
2. What is the concern regarding sexually transmitted diseases such as genital herpes and syphilis and an increased risk of HIV infections?
3. What is a reportable disease?
4. Why is *Chlamydia* considered an "atypical" bacterium?
5. What has "malaria therapy" had to do with curing syphilis?
6. It has been stated that the actual number of people in the United States who have genital herpes may be a lot higher than official statistics depict. What are some possible reasons for this discrepancy?
7. Why is herpes of the newborn of particular concern, and what can be done to prevent this type of transmission?
8. Why are urinary tract infections such common nosocomial infections?

Visual Understanding

1. a. **From chapters 20 and 23, figures 20.16 and 23.14.** Compare these two rashes. What kind of information would help you determine the diagnosis in both cases?

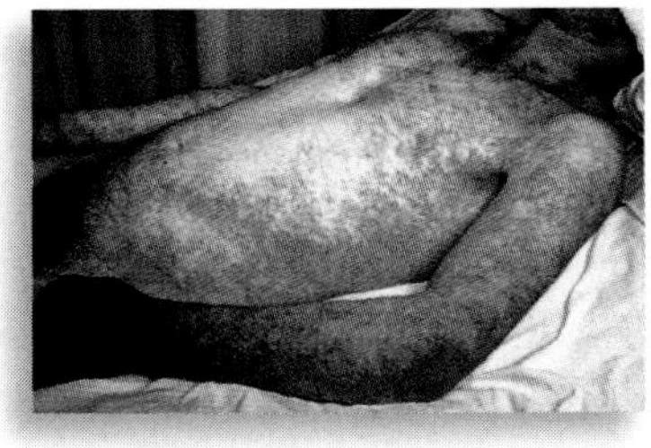

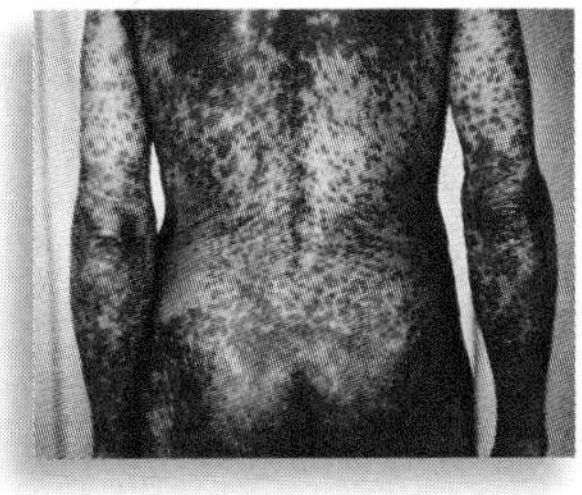

   b. Now compare both of these to the rashes summarized in **Checkpoint 18.7** (p. 556). Which of the diseases in Checkpoint 18.8 most resembles the rashes in the preceding question and how would you distinguish among the three?

Internet Search Topics

ARIS

1. Investigate the Aswan Dam in Egypt. What were the unintended infectious disease consequences of building the dam?
2. Go to: www.aris.mhhe.com, and click on "microbiology" and then this textbook's author/title. Go to chapter 23, access the URLs listed under Internet Search Topics, and research the following:
   a. Follow the epidemiology of gonorrhea and syphilis (case rates, status, current outbreaks).
   b. Observe an animation of herpes simplex infection cycle.
   c. Visit the syphilis history site and read the article there. Comment on what aspect of the skeletal remains led investigators to believe that the person had syphilis.
   d. Study the site about famous people who have had syphilis. Speculate on why, during certain periods of history, a link was suspected between genius and the infection.

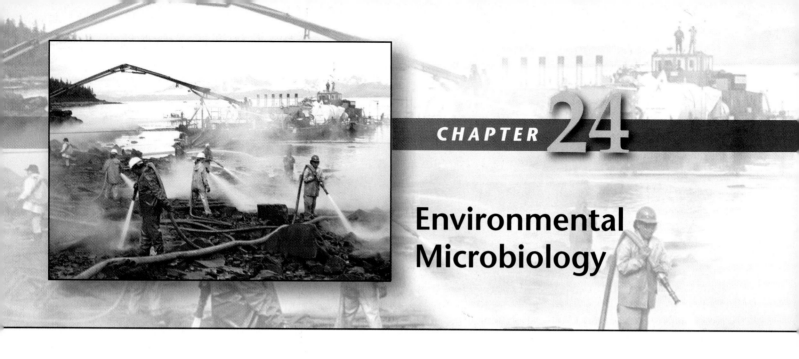

CHAPTER 24

Environmental Microbiology

CASE FILE 24

On March 23, 1989, the oil tanker *Exxon Valdez* ran aground on Bligh Reef in Prince William Sound in Alaska. Almost 41 million liters of crude oil were spilled into the beautiful, pristine wilderness of the Sound. The news media rushed to the scene to cover this tragic event, the largest oil spill in U.S. history. Americans watched in horror as about 2,000 kilometers of some of the most spectacular shores in the country were reduced to oil-covered graves for indigenous flora and fauna.

As anyone who has washed clothes knows, oil stains can be difficult to remove. Extricating spilled oil from the natural environment is a far more arduous and complex task than removing oil stains from laundry. Similar to cleaning heavily stained clothes, though, hot water was used to remove oil from the Sound. Specifically, steam was applied under high pressure to remove oil from the shores. This technique initially seemed to work. Today, the shores superficially appear as they were before the oil spill; however, closer examination reveals that oil remains. The high-pressure steam cleaning forced much of the oil deeper into the rocky shores of the Sound.

In conjunction with the steam-cleaning approach, an additional approach that relied on microorganisms was employed to remove the oil. Many microorganisms—even those inhabiting the rocks on the shores of the Sound—have the capacity to utilize oil as a source of carbon and energy, simultaneously transforming it into harmless water and carbon dioxide. This process is called bioremediation. Crude oil is composed of hydrocarbons, which are rich sources of carbon for microorganisms; however, microorganisms require nutrients in addition to carbon. In fact, without additional nutrients, bacterial metabolism and bioremediation often do not proceed very quickly. Rapid cleanup of the Sound was imperative to minimize further negative impacts of the spill on this once-pristine environment.

- *Do you think steam cleaning was beneficial to the cleanup process?*
- *Can you think of approaches environmental microbiologists might have used to speed up bioremediation in Prince William Sound?*

Case File 24 Wrap-Up appears on page 770.

CHAPTER OVERVIEW

- Earth is a microbial planet. Hence, microorganisms contribute in profound ways to its structure and function and, therefore, to the survival of all other life forms.
- Microorganisms exist in complex associations with both living and nonliving components of their environment.

- Microbes have adapted to specific habitats and niches from which they derive food, energy, shelter, and other essential components of the biosphere.
- Microbes maintain and cycle the biologically important elements, such as carbon, nitrogen, and phosphorus, that exist only in certain reservoirs.
- Microbes constitute the beginning and the end of every energy pyramid as primary producers and as decomposers, respectively.

This chapter emphasizes microbial activities that help maintain, sustain, and control the life support systems on the earth. This subject is explored from the standpoint of the natural roles of microorganisms in the environment and their contributions to the ecological balance, including soil, water, and mineral cycles **(figure 24.1).**

Figure 24.1 A sample of water from a deep cavern as imaged by scanning electron microscopy.
This view shows a bacterial biofilm that actively forms mineral deposits of zinc and sulfate (light green and yellow). This single image brings focus to several themes of this chapter: (1) microbes work together in mixed communities, (2) microbes can alter the chemistry of the nonliving environment, and (3) microbes can be used to control undesirable wastes created by humans.

24.1 Ecology: The Interconnecting Web of Life

The study of microbes in their natural habitats is known as **microbial ecology;** the study of the practical uses of microbes in food processing, industrial production, and biotechnology is known as industrial or **applied microbiology.** The two areas actually overlap to a considerable degree—largely because most natural habitats have been altered by human activities. Human intervention in natural settings has changed the earth's warming and cooling cycles, increased wastes in soil, polluted water, and altered some of the basic relationships between microbial, plant, and animal life. Now that humans are also beginning to release new, genetically recombined microbes into the environment and to alter the genes of plants, animals, and even themselves, what does the future hold? Although this question may be imponderable, we know one thing for certain: Microbes—the most vast and powerful resource of all—will be silently working in nature.

In chapter 7, we first touched upon the widespread distribution of microorganisms and their adaptations to most habitats of the world, from extreme to temperate. Regardless of their exact location or type of adaptation, microorganisms necessarily are exposed to and interact with their environment in complex and extraordinary ways. Microbial ecology studies interactions between microbes and their environment and the effects of those interactions on the earth. Unlike studies that deal with the activities of a single organism or its individual characteristics in the laboratory, ecological studies are aimed at the interactions taking place between organisms and their environment at many levels at any given moment. Therefore, ecology is a broad-based science that merges many subsciences of biology as well as geology, physics, and chemistry.

Ecological studies deal with both the biotic and the abiotic components of an organism's environment. **Biotic** factors are defined as any living or dead organisms[1] that occupy an organism's habitat. **Abiotic** factors include nonliving components such as atmospheric gases, minerals, water, temperature, and light. You may recall these from chapters 7 and 11 as

1. Biologists make a distinction between nonliving and dead. A nonliving thing has never been alive, whereas a dead thing was once alive but no longer is.

factors that affect microbial growth. A collection of organisms together with its surrounding physical and chemical factors is defined as an **ecosystem.**

The Organization of Ecosystems

The earth initially may seem like a random, chaotic place, but it is actually an incredibly organized, fine-tuned machine. Ecological relationships exist at several levels, ranging from the entire earth all the way down to a single organism **(figure 24.2).** The most all-encompassing of these levels, the **biosphere,** contains all physical locations on earth that support life, including the thin envelope of life that surrounds the earth's surface and extending several miles below. This global ecosystem comprises the **hydrosphere** (water), the **lithosphere** (a few miles into the soil), and the **atmosphere** (a few miles into the air). The biosphere maintains or creates the conditions of temperature, light, gases, moisture, and minerals required for life processes. The biosphere can be naturally subdivided into terrestrial and aquatic realms. The terrestrial realm is usually distributed into particular climatic regions called **biomes** (by'-ohmz), each of which is characterized by a dominant plant form, altitude, and latitude. Particular biomes include grassland, desert, mountain, and tropical rain forest. The aquatic biosphere is generally divisible into freshwater and marine realms. We have also very recently learned that the earth's crust also supports a vast and diverse number of life forms, estimated to be equal to or even greater than life as we know it in aquatic and terrestrial realms.

Biomes and aquatic ecosystems are generally composed of mixed assemblages of organisms that live together at the same place and time and that usually exhibit well-defined nutritional or behavioral interrelationships. These clustered associations are called **communities.** Although most communities are identified by their easily visualized dominant plants and animals, they also contain a complex assortment of bacteria, fungi, algae, protozoa, and even viruses. The basic units of community structure are **populations,** groups of organisms of the same kind. For organisms with sexual reproduction, this level is the species. In contrast, prokaryotes are classified using taxonomic units such as "strain." The organizational unit of a population is the individual organism, and each multicellular organism, in turn, has its own levels of organization (organs, tissues, cells).

Ecosystems are generally balanced, with each organism existing in its particular habitat and niche. The **habitat** is the physical location in the environment to which an organism has adapted. In the case of microorganisms, the habitat is frequently a *microenvironment,* where particular qualities of oxygen, light, or nutrient content are somewhat stable. The **niche** is the overall role that a species (or population) serves in a community. This includes such activities as nutritional intake (what it eats), position in the community structure (what is eating it), and rate of population growth. A niche can be broad (such as scavengers that feed on nearly any organic food source) or narrow (microbes that decompose cellulose in forest litter or that fix nitrogen).

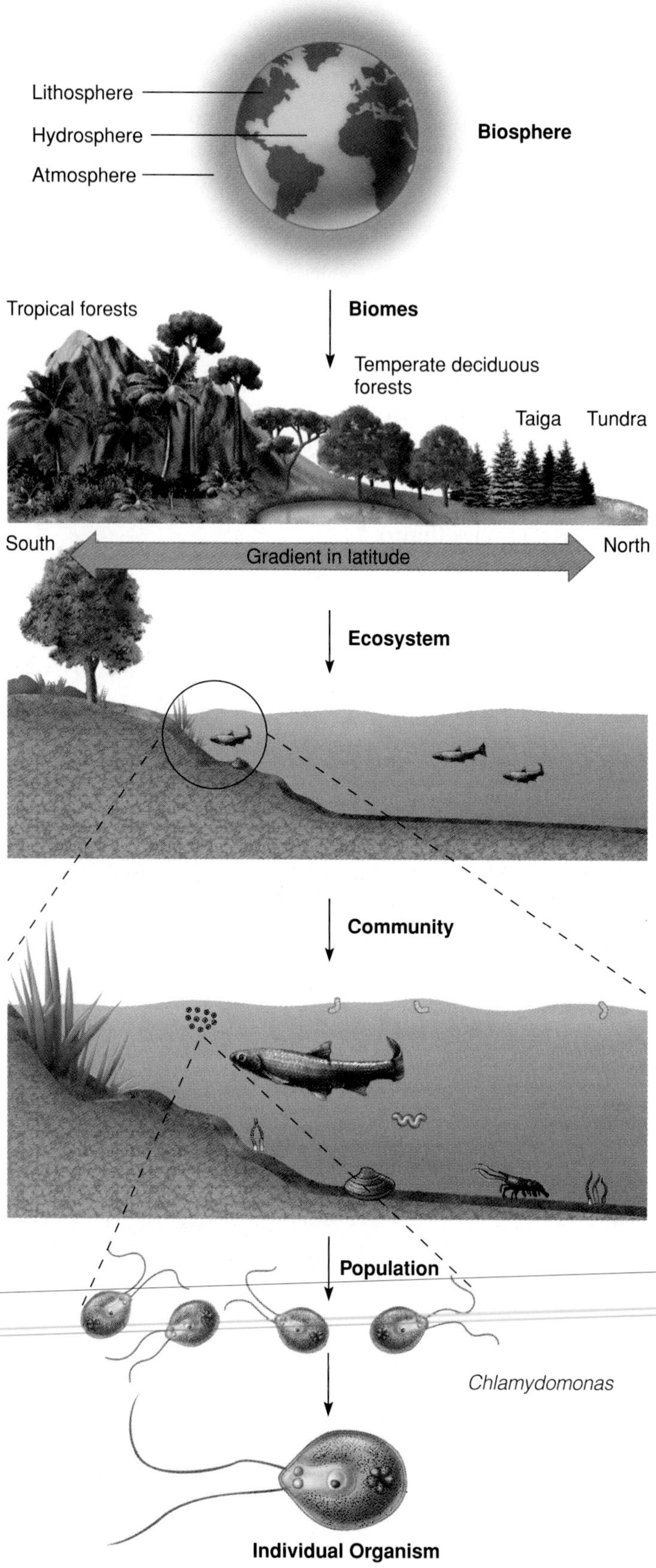

Figure 24.2 Levels of organization in an ecosystem, ranging from the biosphere to the individual organism.

Note that microbes exist as communities in and on plants and animals as well, including humans. Pure cultures are seldom found anywhere in nature.

Energy and Nutritional Flow in Ecosystems

All living things must obtain nutrients and a usable form of energy from their abiotic and biotic environments. The energy and nutritional relationships in ecosystems can be described in a number of convenient ways. A **food chain** or **energy pyramid** provides a simple summary of the general trophic (feeding) levels, designated as producers, consumers, and decomposers, and traces the flow and quantity of available energy from one level to another **(figure 24.3).** It is worth noting that microorganisms are the only living beings that exist at all three major trophic levels. The nutritional roles of microorganisms in ecosystems are summarized in **table 24.1.**

Life would not be possible without **producers,** because they provide the fundamental energy source that drives the trophic pyramid. Producers are the only organisms in an ecosystem that can produce organic carbon compounds such as glucose by assimilating (fixing) inorganic carbon (CO_2) from the atmosphere. If CO_2 is the sole source from which they can obtain carbon for growth, these organisms are called **autotrophs.** Most producers are photosynthetic organisms, such as plants and cyanobacteria, that convert the sun's energy into chemical bond energy. Photosynthesis was covered in chapter 8. A smaller but not less important amount of CO_2 assimilation is brought about by bacteria called lithotrophs. These organisms derive energy from simple inorganic compounds such as ammonia, sulfides, and hydrogen by using redox reactions. In certain ecosystems (see thermal vents, Insight 7.5), lithotrophs are the sole supporters of the energy pyramid as primary producers.

Consumers feed on other living organisms and obtain energy from bonds present in the organic substrates they contain. The category includes animals, protozoa, and a few bacteria and fungi. A pyramid usually has several levels of

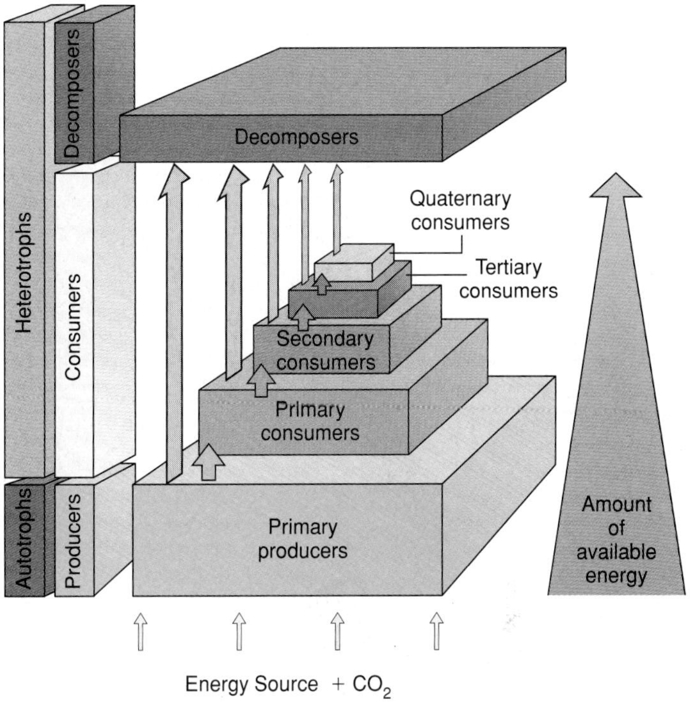

Figure 24.3 A trophic and energy pyramid.
The relative size of the blocks indicates the number of individuals that exist at a given trophic level. The orange arrow on the right indicates the amount of usable energy from producers to top consumers. Both the number of organisms and the amount of usable energy decrease with each trophic level. Decomposers are an exception to this pattern but only because they can feed from all trophic levels (gray arrows). Blocks shown on the left indicate the general nutritional types and levels that correspond with the pyramid.

TABLE 24.1 The Major Roles of Microorganisms in Ecosystems

Role	Description of Activity	Examples of Microorganisms Involved
Primary producers	Photosynthesis Chemosynthesis	Algae, bacteria, sulfur bacteria Chemolithotrophic bacteria in thermal vents
Consumers	Predation	Free-living protozoa that feed on algae and bacteria; some fungi that prey upon nematodes
Decomposers	Degradation of plant and animal matter and wastes Mineralization of organic nutrients	Soil saprobes (primarily bacteria and fungi) that degrade cellulose, lignin, and other complex macromolecules Soil bacteria that reduce organic compounds to inorganic compounds such as CO_2 and minerals
Cycling agents for biogeochemical cycles	Recycling compounds containing carbon, nitrogen, phosphorus, sulfur	Specialized bacteria that transform elements into different chemical compounds to keep them cycling from the biotic to the abiotic and back to the biotic phases of the biosphere
Parasites	Living and feeding on hosts	Viruses, bacteria, protozoa, fungi, and worms that play a role in population control

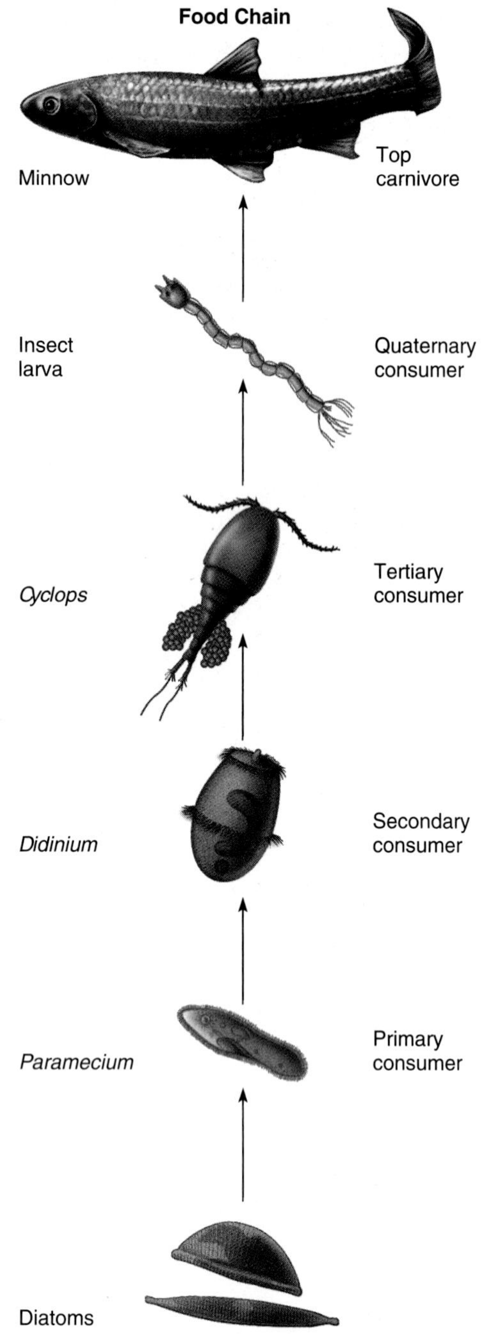

Figure 24.4 Food chain.

A food chain is the simplest way to present specific feeding relationships among organisms, but it may not reflect the total nutritional interactions in a community (figure not to scale).

CASE FILE 24 *WRAP-UP*

Bioremediation, as discussed in the beginning of this chapter and also in the next chapter, relies upon microorganisms to mineralize pollutants, such as oil spilled from an oil tanker. As with all microorganisms, those involved in bioremediation require nutrients. Oil is a hydrocarbon that is rich in carbon that microorganisms can utilize, but it lacks other essential nutrients, such as nitrogen and phosphorus. For this reason, environmental microbiologists attempted to accelerate bioremediation of the *Exxon Valdez* oil spill by applying fertilizers containing nitrogen and phosphorus. Approximately 50,000 kilograms of nitrogen and 5,000 kilograms of phosphorus were applied between 1989 and 1992. Overall, these enormous applications appeared to have the desired effect: Bacteria from fertilized beaches mineralized components of oil up to 18 times faster than bacteria from beaches that did not receive fertilizer. Bioremediation rates tended to increase with nitrogen levels.

Steam cleaning the shores was beneficial in some aspects, including aesthetics and quick removal of large quantities of oil, but the cleaning may have killed many of the bacteria that could have facilitated a more rapid cleanup of the oil.

See: Bragg, J. R., Prince, R. C., Harner, E. J., and Atlas, R. M. 1994. Effectiveness of bioremediation for the Exxon Valdez *oil spill.* Nature *368:413–418.*

consumers, ranging from *primary consumers* (grazers or herbivores), which consume producers; to *secondary consumers* (carnivores), which feed on primary consumers; to *tertiary consumers,* which feed on secondary consumers; and up to *quaternary consumers* (usually the last level), which feed on tertiary consumers. **Figures 24.4** and **24.5** show specific organisms at these levels.

Decomposers, primarily microbes inhabiting soil and water, break down and absorb the organic matter of dead organisms, including plants, animals, and other microorganisms. Because of their biological function, decomposers are active at all levels of the food pyramid. Without this important nutritional class of saprobes, the biosphere would stagnate and die. The work of decomposers is to reduce organic matter into inorganic minerals and gases that can be cycled back into the ecosystem, especially for the use of primary producers. This process, also termed **mineralization,** is so efficient that almost all biological compounds can be reduced by some type of decomposer. Numerous microorganisms decompose cellulose and lignin, polysaccharides from plant cell walls that account for the vast bulk of detritus in soil and water. Complex macromolecules from animal bodies are also broken down by an assortment of bacteria and fungi. Surprisingly, decomposers can also break down most man-made compounds that are not naturally found on earth. This process is referred to as **bioremediation.** Often, bioremediation involves more than one kind of microbe and the collection of participating microbes in this process is known as a **consortium.**

The pyramid in figure 24.3 illustrates several limitations of ecosystems with regard to energy. Unlike nutrients, which can be passed among trophic levels, recycled, and reused, energy does not cycle. Maintenance of complex interdependent trophic relationships such as those shown in figures 24.4 and 24.5 requires a constant input of energy at the producer level. As energy is transferred to the next level, a large proportion (as high as 90%) of the energy will be lost in a form

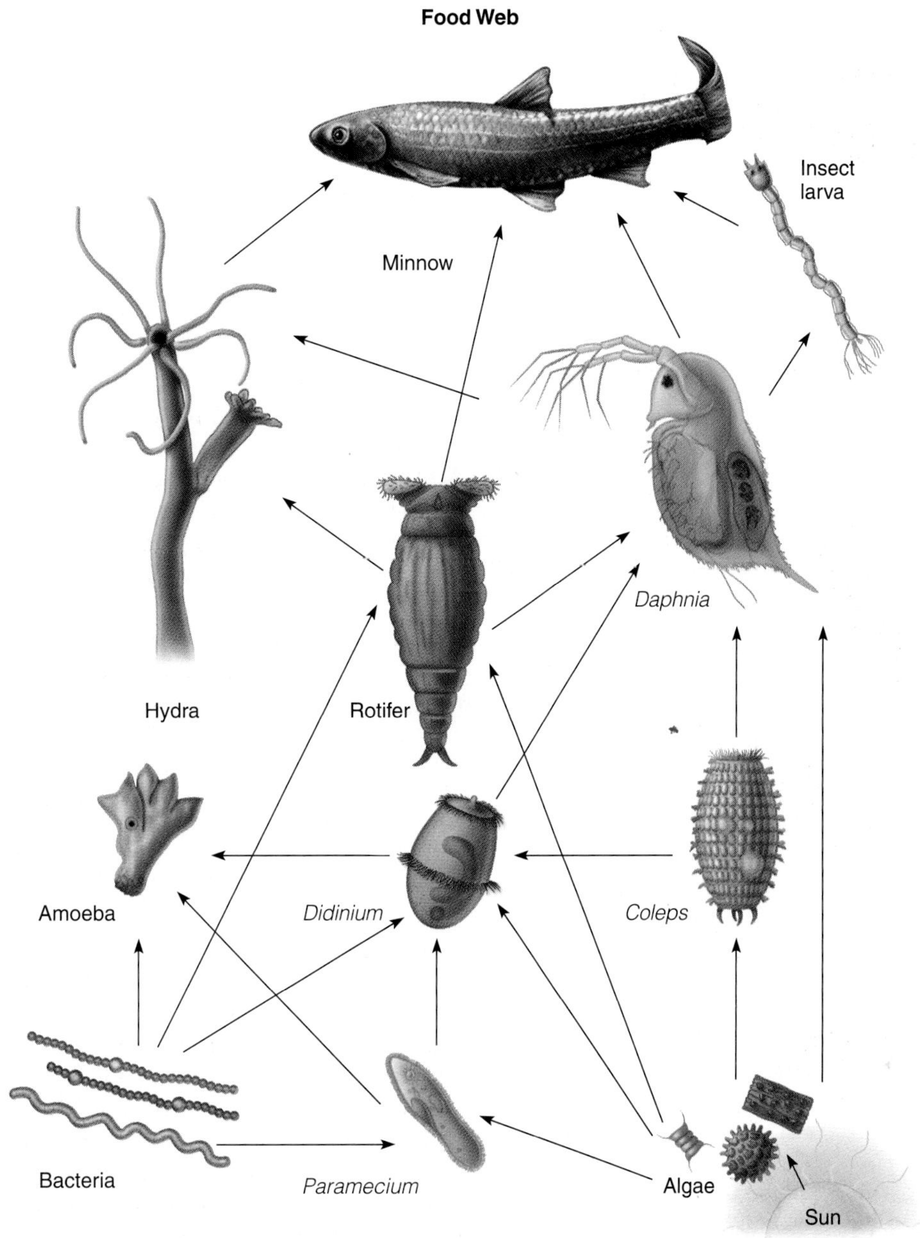

Figure 24.5 Food web.
More complex trophic patterns are accurately depicted by a food web, which traces the multiple feeding options that exist for most organisms. Note: Arrows point toward the consumers. Compare this pattern of feeding with the chain in figure 24.4 (organisms not to scale).

(primarily heat) that cannot be fed back into the system. Thus, the amount of energy available decreases at each successive trophic level. This energy loss also decreases the actual number of individuals that can be supported at each successive level.

The most basic image of a feeding pathway can be provided by a food chain. Although it is a somewhat simplistic way to describe feeding relationships, a food chain helps identify the types of organisms that are present at a given trophic level in a natural setting (see figure 24.4). Feeding relationships in communities are more accurately represented by a multichannel food chain, or a **food web** (see figure 24.5). A food web reflects the actual nutritional structure of a community. It can help to identify feeding patterns typical of herbivores (plant eaters), carnivores (flesh eaters), and omnivores (feed on both plants and flesh).

Ecological Interactions Between Organisms in a Community

Whenever complex mixtures of organisms associate, they develop various dynamic interrelationships based on nutrition and shared habitat. These relationships, some of which were described in earlier chapters, include mutualism, commensalism, parasitism, competition, synergism (cross-feeding), predation, and scavenging.

Mutually beneficial associations (mutualism), such as that of protozoans living in the termite intestine, are so well evolved that the two members require each other for survival. In contrast, **commensalism** is one-sided and independent. (These terms describing relationships between organisms echo the terms we described in chapter 13.) Although the action of one microbe favorably affects another, the first microbe receives no benefit. Many commensal unions involve *co-metabolism,* meaning that the waste products of the first microbe are useful nutrients for the second one, a process called **syntrophy.** In **synergism,** two organisms that are usually independent cooperate to break down a nutrient neither one could have metabolized alone. *Parasitism* is an intimate relationship whereby a parasite derives its nutrients and habitat from a host that is usually harmed in the process. In *competition,* one microbe gives off antagonistic substances that inhibit or kill susceptible species sharing its habitat. A *predator* is a form of consumer that actively seeks out and ingests live prey (protozoa that prey on algae and bacteria). *Scavengers* are nutritional jacks-of-all-trades; they feed on a variety of food sources, ranging from live cells to dead cells and wastes.

CHECKPOINT

- The study of ecology includes both living (biotic) and nonliving (abiotic) components of the earth. Applied microbiology studies their utilization for commercial purposes.
- Ecosystems are organizations of living populations in specific habitats. Environmental ecosystems require a continuous outside source of energy for survival and a nonliving habitat consisting of soil, water, and air.
- A living community is composed of populations that show a pattern of energy and nutritional relationships called a food web. Microorganisms are essential producers and decomposers in any ecosystem.
- The relationships between populations in a community are described according to the degree of benefit or harm they pose to one another. These relationships include mutualism, commensalism, predation, parasitism, synergism, scavenging, and competition.

24.2 The Natural Recycling of Bioelements

Environmental ecosystems are exposed to the sun, which constantly infuses them with a renewable source of energy. In contrast, the bioelements and nutrients that are essential components of cells and multicellular organisms are supplied exclusively by sources somewhere in the biosphere and are not being continually replenished from outside the earth. In fact, the lack of a required nutrient in the immediate habitat is one of the chief factors limiting organismic and population growth. Because of the finite source of life's building blocks, the long-term sustenance of the biosphere requires continuous **recycling** of elements and nutrients. Essential elements such as carbon, nitrogen, sulfur, phosphorus, oxygen, and iron are cycled through biological, geologic, and chemical mechanisms called **biogeochemical cycles.** Although these cycles vary in certain specific characteristics, they share several general qualities, as summarized in the following list:

- All elements ultimately originate from a nonliving, long-term reservoir in the atmosphere, the lithosphere, or the hydrosphere. They cycle in pure form (N_2) or as compounds (PO_4). Their cycling is facilitated by redox reactions.
- Elements make the rounds between the abiotic environment and the biotic environment.
- Recycling maintains a necessary balance of nutrients in the biosphere so that they do not build up or become unavailable.
- Cycles are complex systems that rely upon the interplay of producers, consumers, and decomposers. Often the waste products of one organism become a source of energy or building material for another.
- All organisms participate directly in recycling, but only certain categories of microorganisms have the metabolic pathways for converting inorganic compounds from one nutritional form to another.

The English biologist James Lovelock has postulated a concept called the **Gaia** (guy'-uh) **Theory,** after the mythical Greek goddess of earth. This hypothesis proposes that the biosphere contains a diversity of habitats and niches favorable to life because living things have made it that way. Not only does the earth shape the character of living things, but living things shape the character of the earth. After all, we know that the chemical compositions of the aquatic environment, the atmosphere, and even the soil would not exist as they do without the actions of living things. Organisms are also very active in evaporation and precipitation cycles, formation of mineral deposits, and rock weathering.

For billions of years, microbes have played prominent roles in the formation and maintenance of the earth's crust, the development of rocks and minerals, and the formation of fossil fuels. This revolution in understanding the biological involvement in geologic processes has given rise to a new field called *geomicrobiology.* A logical extension of this discipline is **astromicrobiology.**

In the next several sections, we examine how, jointly and over a period of time, the varied microbial activities affect and are themselves affected by the abiotic environment.

Atmospheric Cycles

The Carbon Cycle

Because carbon is the fundamental atom in all biomolecules and accounts for at least one-half of the dry weight of biomass, the **carbon cycle** is more intimately associated with the energy transfers and trophic patterns in the biosphere than are other elements. Carbon exists predominantly in the mineral

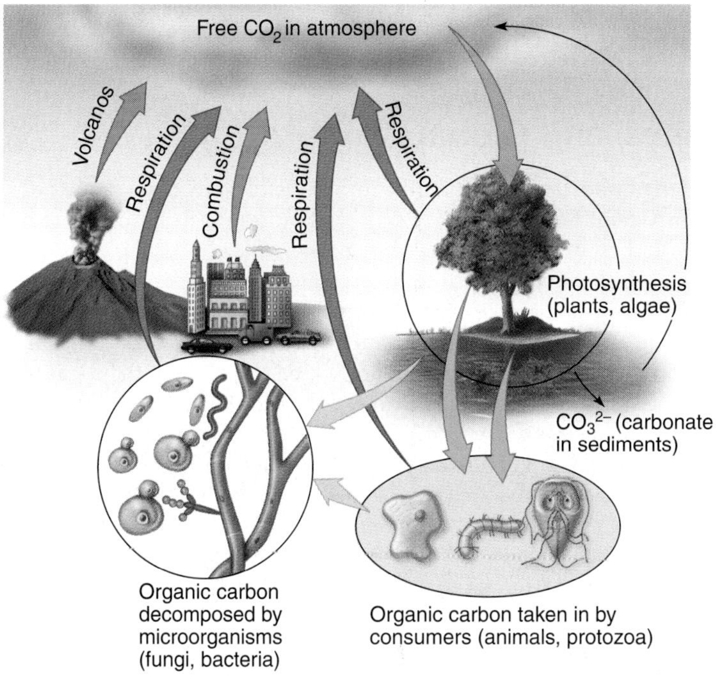

Figure 24.6 The carbon cycle.
This cycle traces carbon from the CO_2 pool in the atmosphere to the primary producers (green) where it is fixed into protoplasm. Organic carbon compounds are taken in by consumers (blue) and decomposers (yellow) that produce CO_2 through respiration and return it to the atmosphere (pink). Combustion of fossil fuels and volcanic eruptions also add to the CO_2 pool. Some of the CO_2 is carried into inorganic sediments by organisms that synthesize carbonate (CO_3) skeletons. In time, natural processes acting on exposed carbonate skeletons can liberate CO_2.

state and as an organic reservoir in the bodies of organisms. A much smaller amount of carbon also exists in the gaseous state as carbon dioxide (CO_2), carbon monoxide (CO), and methane (CH_4). In general, carbon is recycled through ecosystems via carbon fixation, respiration, or fermentation of organic molecules, limestone decomposition, and methane production. A convenient starting point from which to trace the movement of carbon is with carbon dioxide, which occupies a central position in the cycle and represents a large common pool that diffuses into all parts of the ecosystem **(figure 24.6).** As a general rule, the cycles of oxygen and hydrogen are closely allied to the carbon cycle.

The principal users of the atmospheric carbon dioxide pool are photosynthetic autotrophs (photoautotrophs) such as plants, algae, and bacteria. An estimated 165 billion tons of organic material per year are produced by terrestrial and aquatic photosynthesis. Although we don't yet know exactly how many autotrophs exist in the earth's crust, a small amount of CO_2 is used by these bacteria (chemolithoautotrophs) that derive their energy from bonds in inorganic chemicals. A review of the general equation for photosynthesis in figure 8.24 reveals that phototrophs use energy from the sun to fix CO_2 into organic compounds such as glucose that can be used in synthesis. Photosynthesis is also the primary means by which the atmospheric supply of O_2 is regenerated.

Just as photosynthesis removes CO_2 from the atmosphere, other modes of generating energy, such as respiration and fermentation, can be used to remove and return it. As you may recall from the discussion of aerobic respiration in chapter 8, in the presence of O_2, organic compounds such as glucose are degraded completely to CO_2, with the release of energy and the formation of H_2O. Carbon dioxide is also released by anaerobic respiration and by certain types of fermentation reactions.

A small but important phase of the carbon cycle involves certain limestone deposits composed primarily of calcium carbonate ($CaCO_3$). Limestone is produced when marine organisms such as mollusks, corals, protozoans, and algae form hardened shells by combining carbon dioxide and calcium ions from the surrounding water. When these organisms die, the durable skeletal components accumulate in marine deposits. As these immense deposits are gradually exposed by geologic upheavals or receding ocean levels, various decomposing agents liberate CO_2 and return it to the CO_2 pool of the water and atmosphere.

The complementary actions of photosynthesis and respiration, along with other natural CO_2-releasing processes such as limestone erosion and volcanic activity, have maintained a relatively stable atmospheric pool of carbon dioxide. Recent figures show that this balance is being disturbed as humans burn *fossil fuels* and other organic carbon sources. Fossil fuels, including coal, oil, and natural gas, were formed through millions of years of natural biological and geologic activities. Humans are so dependent upon this energy source that, within the past 25 years, the proportion of CO_2 in the atmosphere has steadily increased from 32 ppm to 36 ppm. Although this increase may seem slight and insignificant, most scientists now feel it has begun to disrupt the delicate temperature balance of the biosphere **(Insight 24.1).**

Compared with carbon dioxide, methane gas (CH_4) plays a secondary part in the carbon cycle, though it can be a significant product in anaerobic ecosystems dominated by **methanogens** (methane producers). In general, when methanogens reduce CO_2 by means of various oxidizable substrates, they give off CH_4. The practical applications of methanogens are covered in chapter 25 in a section on sewage treatment, and their contribution to the greenhouse effect is also discussed in Insight 24.1.

The Nitrogen Cycle

Nitrogen (N_2) gas is the most abundant component of the atmosphere, accounting for nearly 79% of air volume. As we will see, this extensive reservoir in the air is largely unavailable to most organisms. Only about 0.03% of the earth's nitrogen is combined (or fixed) in some other form such as nitrates (NO_3), nitrites (NO_2), ammonium ion (NH_4^+), and organic nitrogen compounds (proteins, nucleic acids).

The **nitrogen cycle** is relatively more intricate than other cycles because it involves such a diversity of specialized microbes to maintain the flow of the cycle. In many ways, it is actually more of a nitrogen "web" because of the array

INSIGHT 24.1 Discovery

Greenhouse Gases, Fossil Fuels, Cows, Termites, and Global Warming

The sun's radiant energy does more than drive photosynthesis; it also helps maintain the stability of the earth's temperature and climatic conditions. As radiation impinges on the earth's surface, much of it is absorbed, but a large amount of the infrared (heat) radiation bounces back into the upper levels of the atmosphere. For billions of years, the atmosphere has been insulated by a layer of gases (primarily CO_2; CH_4; water vapor; and nitrous oxide, N_2O) formed by natural processes such as respiration and decomposition, which are part of biogeochemical cycles. This layer traps a certain amount of the reflected heat yet also allows some of it to escape into space. As long as the amounts of heat entering and leaving are balanced, the mean temperature of the earth will not rise or fall in an erratic or life-threatening way. Although this phenomenon, called the **greenhouse effect,** is popularly viewed in a negative light, it must be emphasized that its function for eons has been primarily to foster life.

The greenhouse effect has recently been a matter of concern because *greenhouse gases* appear to be increasing at a rate that could disrupt the temperature balance. In effect, a denser insulation layer will trap more heat energy and gradually heat the earth. The amount of CO_2 released collectively by respiration, anaerobic microbial activity, fuel combustion, and volcanic activity has increased more than 30% since the beginning of the industrial era. By far the greatest increase in CO_2 production results from human activities such as combustion of fossil fuels, burning forests to clear agricultural land, and manufacturing. Deforestation has the added impact of removing large areas of photosynthesizing plants that would otherwise consume some of the CO_2.

Originally, experts on the greenhouse effect were concerned primarily about increasing CO_2 levels, but it now appears that the other greenhouse gases combined may have a greater contribution than CO_2, and they, too, are increasing. One of these gases, methane (CH_4) released from the gastrointestinal tract of ruminant animals such as cattle, goats, and sheep, has doubled over the past century. The gut of termites also harbors wood-digesting bacteria and methanogenic archaea. Even the human intestinal tract can support methanogens. Methane traps 21 times more heat than does carbon dioxide. Other greenhouse gases such as nitrous oxide and sulfur dioxide (SO_2) are also increasing through automobile and industrial pollution.

There is not yet complete agreement as to the extent and effects of global warming. It has been documented that the mean temperature of the earth has increased by ±1.0°C since 1860. If the rate of increase continues, by 2050 a rise in the average temperature of 4°C to 5°C will begin to melt the polar ice caps and raise the levels of the ocean 2 to 3 feet. Some experts predict more serious effects, including massive flooding of coastal regions, changes in rainfall patterns, expansion of deserts, and long-term climatic disruptions.

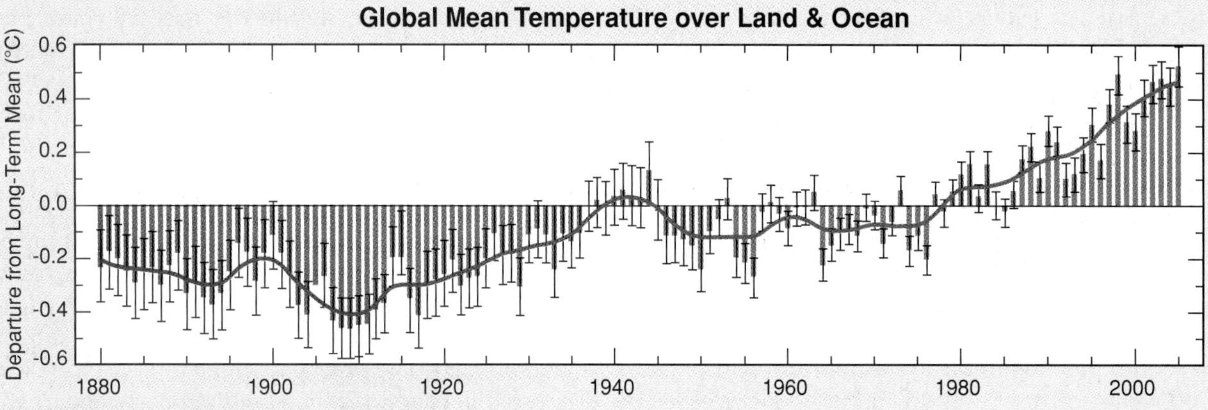

of adaptations that occur. Higher plants can utilize NO_3^- and NH_4^+; animals must receive nitrogen in organic form from plants or other animals; however, microorganisms can use all forms of nitrogen: NO_2^-, NO_3^-, NH_4^+, N_2, and organic nitrogen. The cycle includes four basic types of reactions: nitrogen fixation, ammonification, nitrification, and denitrification **(figure 24.7).**

Nitrogen Fixation The biosphere is most dependent upon the only process that can remove N_2 from the air and convert it to a form usable by living beings. This process, called **nitrogen fixation,** is the beginning step in the synthesis of virtually all nitrogenous compounds. Nitrogen fixation is brought about primarily by nitrogen-fixing bacteria in soil and water, though a small amount is formed through nonliving processes involving lightning. Nitrogen-fixing microbes have developed a unique enzyme system capable of breaking the triple bonds of the N_2 molecule and reducing the N atoms, an anaerobic process that requires the expenditure of considerable ATP. The primary product of nitrogen fixation is the ammonium ion, NH_4^+. Nitrogen-fixing bacteria live free or in a symbiotic relationship with plants. Among the common free-living nitrogen fixers are the aerobic *Azotobacter* and

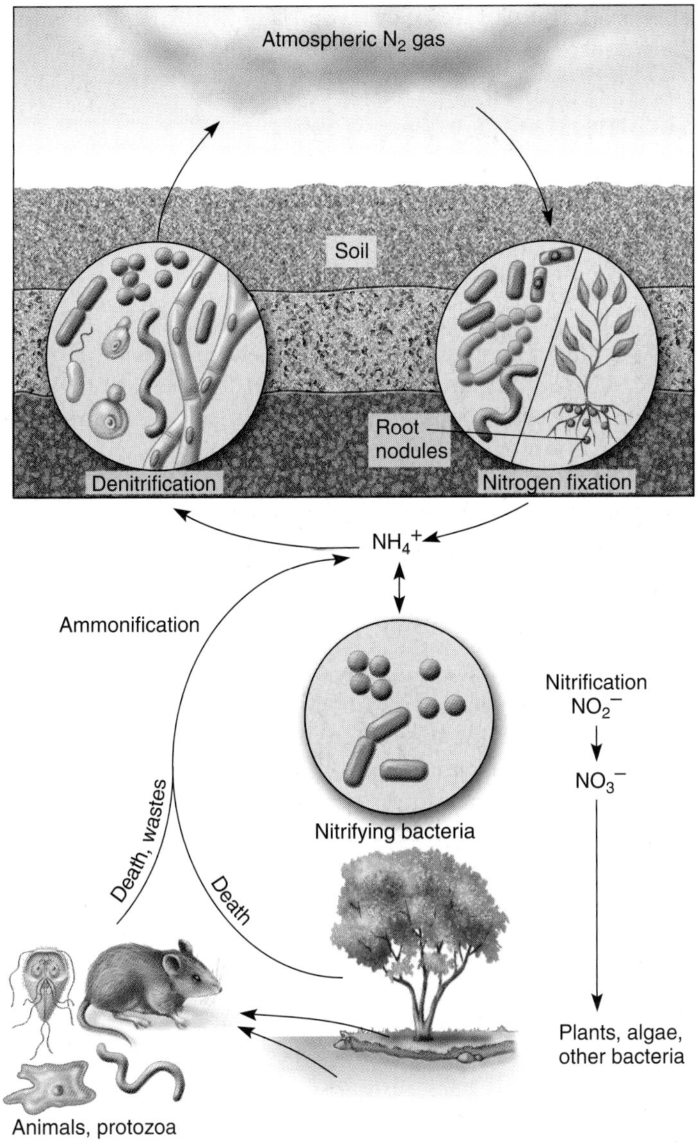

Figure 24.7 The simplified events in the nitrogen cycle. In nitrogen fixation, gaseous nitrogen (N_2) is acted on by nitrogen-fixing bacteria, which give off ammonia (NH_3). Ammonia is converted to nitrite (NO_2^-) and nitrate (NO_3^-) by nitrifying bacteria in nitrification. Plants, algae, and bacteria use nitrates to synthesize nitrogenous organic compounds (proteins, amino acids, nucleic acids). Organic nitrogen compounds are used by animals and other consumers. In ammonification, nitrogenous macromolecules from wastes and dead organisms are converted to NH_4^+ by ammonifying bacteria. NH_4^+ can be either directly recycled into nitrates or returned to the atmospheric N_2 form by denitrifying bacteria (denitrification).

Azospirillum and certain members of the anaerobic genus *Clostridium*. Other free-living nitrogen fixers are the cyanobacteria *Anabaena* and *Nostoc*.

Root Nodules: Natural Fertilizer Factories A significant symbiotic association occurs between **rhizobia** (ry-zoh'-bee-uh) (bacteria in the genera such as *Rhizobium*, *Bradyrhizobium*, and *Azorhizobium*) and **legumes** (plants

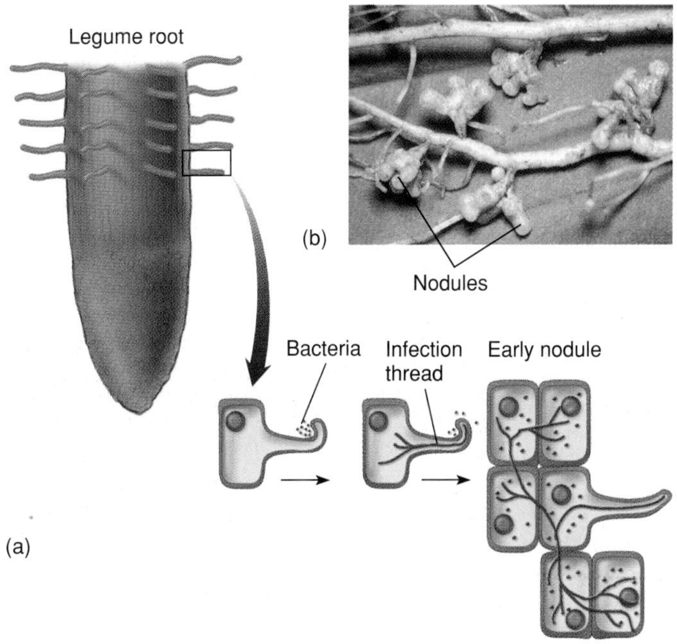

Figure 24.8 Nitrogen fixation through symbiosis. **(a)** Events leading to formation of root nodules. Cells of the bacterium *Rhizobium* attach to a legume root hair and cause it to curl. Invasion of the legume root proper by *Rhizobium* initiates the formation of an infection thread that spreads into numerous adjacent cells. The presence of bacteria in cells causes nodule formation. **(b)** Mature nodules that have developed in a sweet clover plant.

such as soybeans, peas, alfalfa, and clover that characteristically produce seeds in pods). The infection of legume roots by these gram-negative, motile, rod-shaped bacteria causes the formation of special nitrogen-fixing organs called **root nodules (figure 24.8).** Nodulation begins when rhizobia colonize specific sites on root hairs. From there, the bacteria invade deeper root cells and induce the cells to form tumorlike masses. The bacterium's enzyme system supplies a constant source of reduced nitrogen to the plant, and the plant furnishes nutrients and energy for the activities of the bacterium. The legume uses the NH_4^+ to aminate (add an amino group to) various carbohydrate intermediates and thereby synthesize amino acids and other nitrogenous compounds that are used in plant and animal synthesis.

Plant–bacteria associations have great practical importance in agriculture, because an available source of nitrogen is often a limiting factor in the growth of crops. The self-fertilizing nature of legumes makes them valuable food plants in areas with poor soils and in countries with limited resources. It has been shown that crop health and yields can be improved by inoculating legume seeds with pure cultures of rhizobia, because the soil is often deficient in the proper strain of bacteria for forming nodules **(figure 24.9).**

Ammonification, Nitrification, and Denitrification In another part of the nitrogen cycle, nitrogen-containing organic matter is decomposed by various bacteria (*Clostridium*,

(a) (b)

Figure 24.9 Inoculating legume seeds with *Rhizobium* bacteria increases the plant's access to nitrogen.
The legumes in **(a)** were inoculated and are healthy. The poor growth and yellowish color of the uninoculated legumes in **(b)** indicate a lack of nitrogen.

Proteus, for example) that live in the soil and water. Organic detritus consists of large amounts of protein and nucleic acids from dead organisms and nitrogenous animal wastes such as urea and uric acid. The decomposition of these substances splits off amino groups and produces NH_4^+. This process is thus known as **ammonification.** The ammonium released can be reused by certain plants or converted to other nitrogen compounds, as discussed next.

The oxidation of NH_4^+ to NO_2^- and NO_3^- is a process called **nitrification.** It is an essential conversion process for generating the most oxidized form of nitrogen (NO_3). This reaction occurs in two phases and involves two different kinds of lithotrophic bacteria in soil and water. In the first phase, certain gram-negative genera such as *Nitrosomonas, Nitrosospira* and *Nitrosococcus* oxidize NH_3 to NO_2^- as a means of generating energy. Nitrite is rapidly acted upon by a second group of nitrifiers, including *Nitrobacter, Nitrosospira,* and *Nitrococcus,* which perform the **final** oxidation of NO_2^- to NO_3^-. Nitrates can be assimilated through several routes by a variety of organisms (plants, fungi, and bacteria). Nitrate and nitrite are also important in anaerobic respiration where they serve as terminal electron acceptors; some bacteria use them as a source of oxygen as well.

The nitrogen cycle is complete when nitrogen compounds are returned to the reservoir in the air by a reaction series that converts NO_3^- through intermediate steps to atmospheric nitrogen. The first step, which involves the reduction of nitrate to nitrite, is so common that hundreds of different bacterial species can do it. Several genera such as *Bacillus, Pseudomonas, Spirillum,* and *Thiobacillus* can carry out this **denitrification process** to completion as follows:

$$NO_3^- \rightarrow NO_2^- \rightarrow NO \rightarrow N_2O \rightarrow N_2 \text{ (gas)}$$

This process illustrates that incomplete denitrification is the main source of the greenhouse gas nitrous oxide (N_2O).

Sedimentary Cycles

The Sulfur Cycle

The sulfur cycle resembles the carbon cycle more than the nitrogen cycle in that sulfur is mostly in solid form and originates from natural sedimentary deposits in rocks, oceans, lakes, and swamps rather than from the atmosphere. Sulfur exists in the elemental form (S) and as hydrogen sulfide gas (H_2S), sulfate (SO_4), and thiosulfate (S_2O_3). Most of the oxidations and reductions that convert one form of inorganic sulfur to another are accomplished by bacteria. Plants and many microorganisms can assimilate only SO_4, and animals must have an organic source. Organic sulfur occurs in the amino acids cystine, cysteine, and methionine, which contain sulfhydryl (—SH) groups and form disulfide (S—S) bonds that contribute to the stability and configuration of proteins.

One of the most remarkable contributors to the cycling of sulfur in the biosphere are the thiobacilli. These gram-negative, motile rods flourish in mud, sewage, bogs, mining drainage, and brackish springs that can be inhospitable to organisms that require complex organic nutrients. But the metabolism of these specialized lithotrophic bacteria is adapted to extracting energy by oxidizing elemental sulfur, sulfides, and thiosulfate. One species, *T. thiooxidans,* is so efficient at this process that it secretes large amounts of sulfuric acid into its environment, as shown by the following equation:

$$Na_2S_2O_3 + H_2O + O_2 \rightarrow Na_2SO_4 + H_2SO_4 \text{ (sulfuric acid)} + 4S$$

The marvel of this bacterium is its ability to create and survive in the most acidic habitats on the earth. It also plays an essential part in the phosphorus cycle, and its relative, *T. ferrooxidans,* participates in the cycling of iron. Other bacteria that can oxidize sulfur to sulfates are the photosynthetic sulfur bacteria mentioned in the section on photosynthesis.

The sulfates formed from oxidation of sulfurous compounds are assimilated into biomass by a wide variety of organisms. The sulfur cycle reaches completion when inorganic and organic sulfur compounds are reduced. Bacteria in the genera *Desulfovibrio* and *Desulfuromonas* anaerobically reduce sulfates to hydrogen sulfide or metal sulfide as the final step in electron transport. Sites in ocean sediments and mud where these bacteria live usually emanate a strong, rotten-egg stench from H_2S and may be blackened by the iron they contain.

The Phosphorus Cycle

Phosphorus is an integral component of DNA, RNA, and ATP, and all life depends upon a constant supply of it. It cycles between the abiotic and biotic environments almost exclusively as inorganic phosphate (PO_4) rather than its

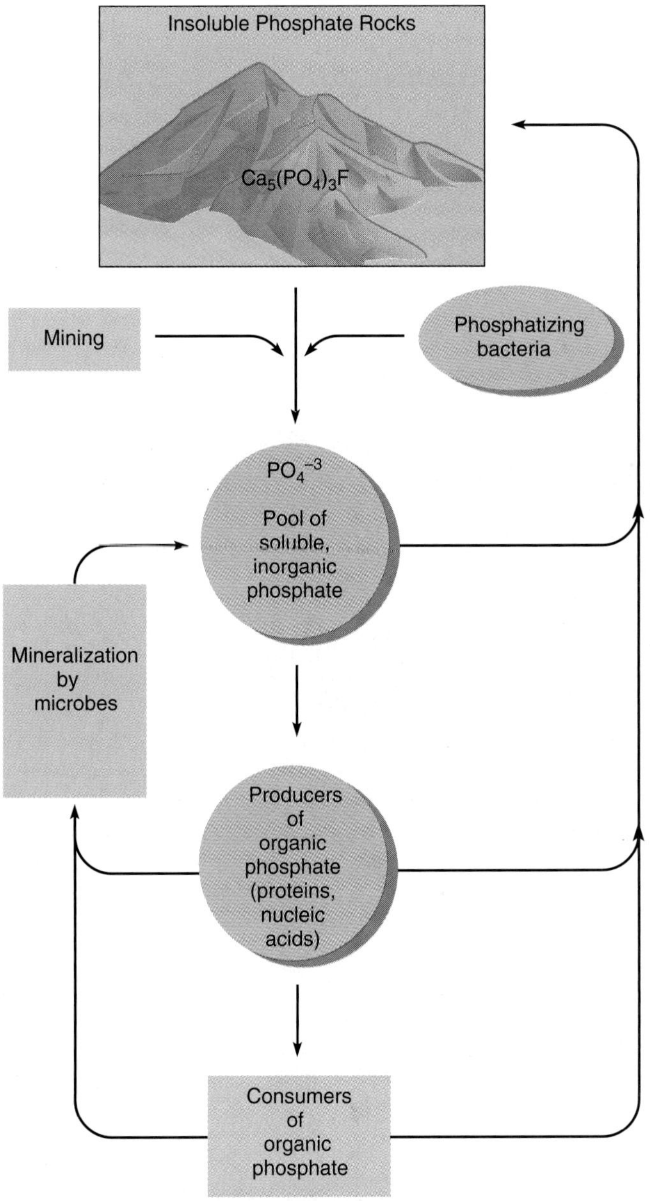

Figure 24.10 The phosphorus cycle.
The pool of phosphate existing in sedimentary rocks is released into the ecosystem either naturally by erosion and microbial action or artificially by mining and the use of phosphate fertilizers. Soluble phosphate (PO_4^{3-}) is cycled through producers, consumers, and decomposers back into the soluble pool of phosphate, or it is returned to sediment in the aquatic biosphere.

elemental form **(figure 24.10).** The chief inorganic reservoir is phosphate rock, which contains the insoluble compound fluorapatite, $Ca_5(PO_4)_3F$. Before it can enter biological systems, this mineral must be *phosphatized*—converted into more soluble PO_4^{3-} by the action of acid. Phosphate is released naturally when the sulfuric acid produced by *Thiobacillus* dissolves phosphate rock. Soluble phosphate in the soil and water is the principal source for autotrophs, which fix it onto organic molecules and pass it on to heterotrophs in this form. Organic phosphate is returned to the pool of soluble phosphate by decomposers, and it is finally cycled back to the mineral reservoir by slow geologic processes such as sedimentation. Because the low phosphate content of many soils can limit productivity, phosphate is added to soil to increase agricultural yields. The excess runoff of fertilizer into the hydrosphere is often responsible for overgrowth of aquatic pests (see eutrophication in a subsequent section on aquatic habitats).

Other Forms of Cycling

The involvement of microbes in cycling elements and compounds can be escalated by the introduction of toxic substances into the environment. Such toxic elements as arsenic, chromium, lead, and mercury as well as hundreds of thousands of synthetic chemicals introduced into the environment over the past hundred years are readily caught up in cycles by microbial actions. Some of these chemicals will be converted into less harmful substances, but others, such as PCB and heavy metals, persist and flow along with nutrients into all levels of the biosphere. If such a pollutant accumulates in living tissue and is not excreted, it can be accumulated by living things through the natural trophic flow of the ecosystem. This process is known as bioaccumulation. Microscopic producers such as bacteria and algae begin the accumulation process. With each new level of the food chain, the consumers gather an increasing amount of the chemical, until the top consumers can contain toxic levels **(Insight 24.2).**

One example of this is mercury compounds used in household antiseptics and disinfectants, agriculture, and industry. Elemental mercury precipitates proteins by attaching to functional groups and is most toxic in the ethyl or methyl mercury form. Recent studies have disclosed increased mercury content in fish taken from oceans and freshwater lakes in North America and even in canned tuna, adding to the risk in consumption of these products.

CHECKPOINT

- Nutrients and minerals necessary to communities and ecosystems must be continuously recycled. These biogeochemical cycles involve transformation of elements from inorganic to organic forms usable by many populations in the community and back again. Specific types of microorganisms are needed to convert many nutrients from one form to another.
- The sun is the primary energy source for most surface ecosystems. Photosynthesis captures this energy, which can be used for carbon fixation by producer populations.
- Elements of critical importance to all ecosystems that cycle through various forms are carbon, nitrogen, sulfur, phosphorus, and water. Carbon and nitrogen are part of the atmospheric cycle. Sulfur and phosphorus are part of the sedimentary cycling of nutrients.

INSIGHT 24.2 ***Discovery***

Cute Killer Whale—Or Swimming Waste Dump?

In the early 1990s, Keiko the killer whale stole hearts as the star of the movie *Free Willy*. Eleven years later, Keiko died of pneumonia in a fjord in Norway, never having fully adjusted to being back in the wild. Even though whales that die close to shore are usually towed out to sea, Keiko was buried on the beach where he was found, probably because of the close connection humans felt with him. That was not the end of the story, however. Environmental groups in Norway raised concerns about burying the animal onshore due to the high probability that his tissues contained high amounts of PCBs. It was nothing personal against Keiko; whales all over the world have been found to have bioaccumulated this toxic chemical.

PCBs (polychlorinated biphenyls) are very stable manufactured compounds that were heavily used in industrial settings from the 1930s to the 1970s. They found widespread use as insulating fluids in electrical applications. They are highly soluble in lipid compounds, and for that reason they bioaccumulate in the fat tissues of animals. The bioaccumulation seems to be worst near the poles of the earth, where many higher animals (including humans) use contaminated fish as a major part of their diets. Complicating this fact is the concentration of volatile PCBs in the atmosphere. Atmospheric circulation carries PCBs to the poles, and the cold temperatures cause the pollutants to condense and fall to the surface, where they further contaminate the food chain. In 1998, a group of polar bears in Norway was found to have bizarre developmental deformities. Seven bears of a group of 450 surveyed (approximately 2%) possessed both male and female reproductive organs—a bizarre mutation that was attributed to PCB accumulation in the bears' bodies.

PCB contamination of wildlife is not limited to the poles, however. In Belgium in 1994, four sperm whales were stranded and died in coastal waters. All were found to have 30 parts per million (ppm) PCBs in their kidneys and blubber. Beluga whales in the Gulf of St. Lawrence in eastern Canada have been found to have 3,200 ppm PCBs in their tissues—a level 1,600 times higher than the level of contamination that triggers EPA regulations requiring the incineration of any materials found to have that concentration of PCB. A bottlenose dolphin in Cape Cod was recently found to have 6,800 ppm PCBs. "This animal was, by definition, a swimming toxic waste dump," says Roger Payne, author of *Among Whales*.

And that brings us back to Keiko. For weeks after he was buried in a quiet ceremony, local schoolchildren came to place rocks on his grave in a Viking tradition of respect. The stark contrast between that loving act and the fact that many people feel he should have been dug up and incinerated highlights the conflicted relationship we have with nature. We love it, but are we ignoring the damage we inflict upon it?

Keiko was buried on the shore of the Taknes Bay in Norway, December 15, 2003.

24.3 Microbes on Land and in Water

Soil Microbiology: The Composition of the Lithosphere

Descriptions such as "soiled" or "dirty" may suggest to some that soil is an undesirable, possibly harmful substance; or its appearance might suggest a somewhat homogenous, inert substance. At the microscopic level, however, soil is a dynamic ecosystem that supports complex interactions between numerous geologic, chemical, and biological factors. This rich region, called the lithosphere, teems with microbes, serves a dynamic role in biogeochemical cycles, and is an important repository for organic detritus and dead terrestrial organisms.

The abiotic portion of soil is a composite of mineral particles, water, and atmospheric gas. The development of soil begins when geologic sediments are mechanically disturbed and exposed to weather and microbial action.

Rock decomposition releases various-size particles ranging from rocks, pebbles, and sand grains to microscopic morsels that lie in a loose aggregate **(figure 24.11).** The porous structure of soil creates various-size pockets or spaces that provide numerous microhabitats. Some spaces trap moisture and form a liquid phase in which mineral ions and

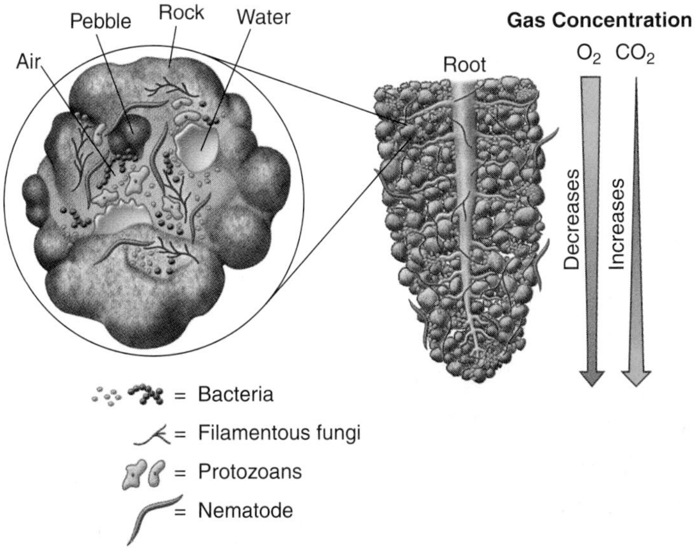

Figure 24.11 The structure of the rhizosphere and the microhabitats that develop in response to soil particles, moisture, air, and gas content.
The microbes contained in even a tiny sample form complex mixed communities in biofilms on the surfaces of rocks and pebbles.

Figure 24.12 Mycorrhizae, symbiotic associations between fungi and plant roots, favor the absorption of water and minerals from the soil.

other nutrients are dissolved. Other spaces trap air that will provide gases to soil microbes, plants, and animals. Because both water and air compete for these pockets, the water content of soil is directly related to its oxygen content. Water-saturated soils contain less oxygen, and dry soils have more. Gas tensions in soil can also vary vertically. In general, the concentration of O_2 decreases and that of CO_2 increases with the depth of soil. Aerobic and facultative organisms tend to occupy looser, drier soils, whereas anaerobes are adapted to waterlogged, poorly aerated soils.

Within the superstructure of the soil are varying amounts of humus, the slowly decaying organic litter from plant and animal tissues. This soft, crumbly mixture holds water like a sponge. It is also an important habitat for microbes that decompose the complex litter and gradually recycle nutrients. The humus content varies with climate, temperature, moisture and mineral content, and microbial action. Warm, tropical soils have a high rate of humus production and microbial decomposition. Because nutrients in these soils are swiftly released and used up, they do not accumulate. Fertilized agricultural soils in temperate climates build up humus at a high rate and are rich in nutrients. The very low content of humus and moisture in desert soils greatly reduces its microbial biota, rate of decomposition, and nutrient content. Bogs are likewise nutrient-poor due to a slow rate of decomposition of the humus caused by high acid content and lack of oxygen. Humans can artificially increase the amount of humus by mixing plant refuse and animal wastes with soil and allowing natural decomposition to occur, a process called *composting*. Composting is a very active metabolic process that generates a great deal of heat. The temperature inside a well-maintained compost can reach 80°C to 100°C.

Living Activities in Soil

The rich culture medium of the soil supports a fantastic array of microorganisms (bacteria, fungi, algae, protozoa, and viruses). A gram of moist loam soil with high humus content can have a microbe count as high as 10 billion, each competing for its own niche and microhabitat. Some of the most distinctive biological interactions occur in the **rhizosphere,** the zone of soil surrounding the roots of plants, which contains associated bacteria, fungi, and protozoa (see figure 24.11). Plants interact with soil microbes in a truly synergistic fashion. Studies have shown that a rich microbial community grows in a biofilm around the root hairs and other exposed surfaces. Their presence stimulates the plant to exude growth factors such as carbon dioxide, sugars, amino acids, and vitamins. These nutrients are released into fluid spaces, where they can be readily captured by microbes. Bacteria and fungi likewise contribute to plant survival by releasing hormone-like growth factors and protective substances. They are also important in converting minerals into forms usable by plants. We saw numerous examples in the nitrogen, sulfur, and phosphorus cycles.

We previously observed that plants can form close symbiotic associations with microbes to fix nitrogen. Other mutualistic partnerships between plant roots and microbes are **mycorrhizae** (my″-koh-ry′-zee). These associations occur when various species of basidiomycetes, ascomycetes, or zygomycetes attach themselves to the roots of vascular plants **(figure 24.12).** The plant feeds the fungus through photosynthesis, and the fungus sustains the relationship in several ways. By extending its mycelium into the rhizosphere, it helps anchor the plant and increases the surface area for capturing water from dry soils and minerals from poor soils. Plants with mycorrhizae can inhabit severe habitats more successfully than plants without them.

The topsoil, which extends a few inches to a few feet from the surface, supports a host of burrowing animals such as nematodes, termites, and earthworms. Many of these animals are decomposer-reducer organisms that break down organic nutrients through digestion and also mechanically reduce or fragment the size of particles so that they are more readily mineralized by microbes. Aerobic bacteria initiate the digestion of organic matter into carbon dioxide and water and generate minerals such as sulfate, phosphate, and nitrate, which can be further degraded by anaerobic bacteria. Fungal enzymes increase the efficiency of soil decomposition by hydrolyzing complex natural substances such as cellulose, keratin, lignin, chitin, and paraffin.

The soil is also a repository for agricultural, industrial, and domestic wastes such as insecticides, herbicides, fungicides, manufacturing wastes, and household chemicals. Applied microbiologists, using expertise from engineering, biotechnology, and ecology, work to explore the feasibility of harnessing indigenous soil microbes to break down undesirable hydrocarbons and pesticides (see chapter 25).

Aquatic Microbiology

Water occupies nearly three-fourths of the earth's surface. In the same manner as minerals, the earth's supply of water is continuously cycled between the hydrosphere, atmosphere, and lithosphere **(figure 24.13).** The **hydrologic cycle** begins when surface water (lakes, oceans, rivers) exposed to the sun and wind evaporates and enters the vapor phase of the atmosphere. Living beings contribute to this reservoir by various activities. Plants lose moisture through transpiration (evaporation through leaves), and all aerobic organisms give off water during respiration. Airborne moisture accumulates in the atmosphere, most conspicuously as clouds.

Water is returned to the earth through condensation or precipitation (rain, snow). The largest proportion of precipitation falls back into surface waters, where it circulates rapidly between running water and standing water. Only about 2% of water seeps into the earth or is bound in ice, but these are very important reservoirs. **Table 24.2** shows how water is distributed in the various surface compartments. Surface water collects in extensive subterranean pockets produced by the underlying layers of rock, gravel, and sand. This process forms a deep groundwater source called an **aquifer.** The water in aquifers circulates very slowly and is an important replenishing source for surface water. It can resurface through springs, geysers, and hot vents, and it is also tapped as the primary supply for one-fourth of all water used by humans.

Although the total amount of water in the hydrologic cycle has not changed over millions of years, its distribution and quality have been greatly altered by human activities. Two serious problems have arisen with aquifers. First, as a result of increased well drilling, land development, and persistent local droughts, the aquifers in many areas have not been replenished as rapidly as they have been depleted. As these reserves are used up, humans will have to rely on other delivery systems such as pipelines, dams, and reservoirs, which can further disrupt the cycling of water. Second, because water picks up materials when falling through air or percolating through the ground, aquifers are also important collection points for pollutants. As we will see, the proper management of water resources is one of the greatest challenges of this century.

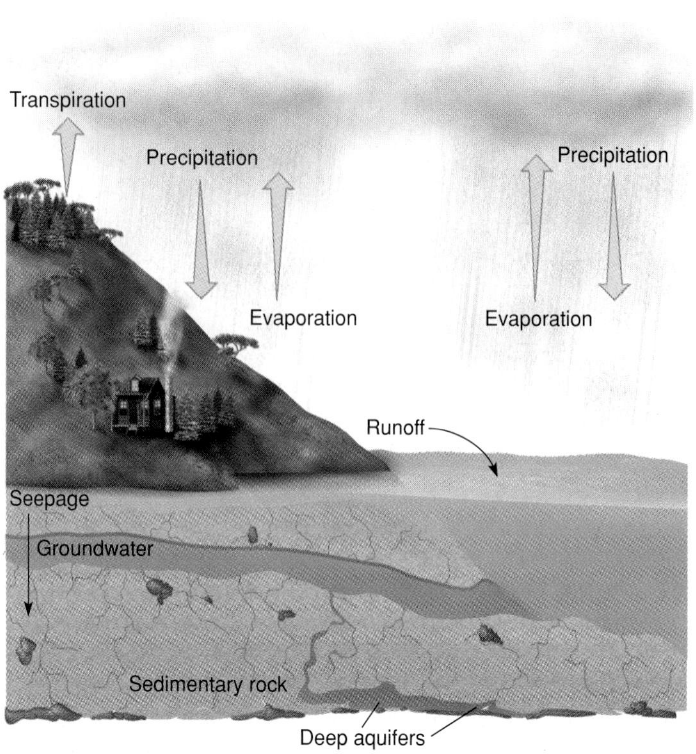

Figure 24.13 The hydrologic cycle.
The largest proportion of water cycles through evaporation, transpiration, and precipitation between the hydrosphere and the atmosphere. Other reservoirs of water exist in the groundwater or deep storage aquifers in sedimentary rocks. Plants add to this cycle by releasing water through transpiration, and heterotrophs release it through respiration.

TABLE 24.2 Distribution of Water on Earth's Surface

Water Source	Water Volume, in Cubic Miles	Percentage of Total Water
Oceans	317,000,000	97.24
Icecaps, glaciers	7,000,000	2.14
Groundwater	2,000,000	0.61
Freshwater lakes	30,000	0.009
Inland seas	25,000	0.008
Soil moisture	16,000	0.005
Atmosphere	3,100	0.001
Rivers	300	0.0001
		100

Source: U.S. Geological Survey.

Marine Environments

The ocean exhibits extreme variations in salinity, depth, temperature, hydrostatic pressure, and mixing. Even so, it supports a great abundance of bacteria and viruses, the extent of which has only been appreciated in very recent years. It contains a unique zone where the river meets the sea called an *estuary.* This region fluctuates in salinity, is very high in nutrients, and supports a specialized microbial community. It is often dominated by salt-tolerant species of *Pseudomonas* and *Vibrio.* Another important factor is the tidal and wave action that subjects the coastal habitat to alternate periods of submersion and exposure. The deep ocean, or **abyssal zone,** is poor in nutrients and lacks sunlight for photosynthesis, and its tremendous depth (up to 10,000 meters) makes it oxygen-poor and cold (average temperature 4°C). This zone supports communities with extreme adaptations, including halophilic, psychrophilic, barophilic, and anaerobic lifestyles.

Aquatic Communities

The freshwater environment is a site of tremendous microbiological activity. Microbial distribution is associated with sunlight, temperature, oxygen levels, and nutrient availability. The uppermost portion is the most productive self-sustaining region because it contains large numbers of **plankton,** a floating microbial community that drifts with wave action and currents. A major member of this assemblage is the phytoplankton, containing a variety of photosynthetic algae and cyanobacteria. The phytoplankton provide nutrition for **zooplankton,** microscopic consumers such as protozoa and invertebrates that filter, feed, prey, or scavenge. The plankton supports numerous other trophic levels such as larger invertebrates and fish. With its high nutrient content, the deeper regions also support an extensive variety and concentration of organisms, including aquatic plants, aerobic bacteria, and anaerobic bacteria actively involved in recycling organic detritus.

Larger bodies of standing water develop gradients in temperature or thermal stratification, especially during the summer **(figure 24.14).** The upper region, called the *epilimnion,* is warmest, and the deeper *hypolimnion* is cooler. Between these is a buffer zone, the **thermocline,** that ordinarily prevents the mixing of the two. Twice a year, during the warming cycle of spring and the cooling cycle of fall, temperature changes in the water column break down the thermocline and cause the water from the two strata to mix. Mixing disrupts the stratification and creates currents that bring nutrients up from the sediments. This process, called *upwelling,* is associated with increased activity by certain groups of microbes and is one explanation for the periodic emergence of *red tides* in oceans **(figure 24.15)** caused by toxin-producing dinoflagellates. A recent outbreak of fish and human disease on the eastern seaboard has been attributed to the overgrowth of certain species of these algae in polluted water. These algae produce a potent muscle toxin that can be concentrated by shellfish through filtration feeding. When humans eat clams, mussels, or oysters that contain the toxin, they develop paralytic shellfish poisoning. People living in coastal areas are cautioned not to eat shellfish during those months of the year associated with red tides (varies from one area to another).

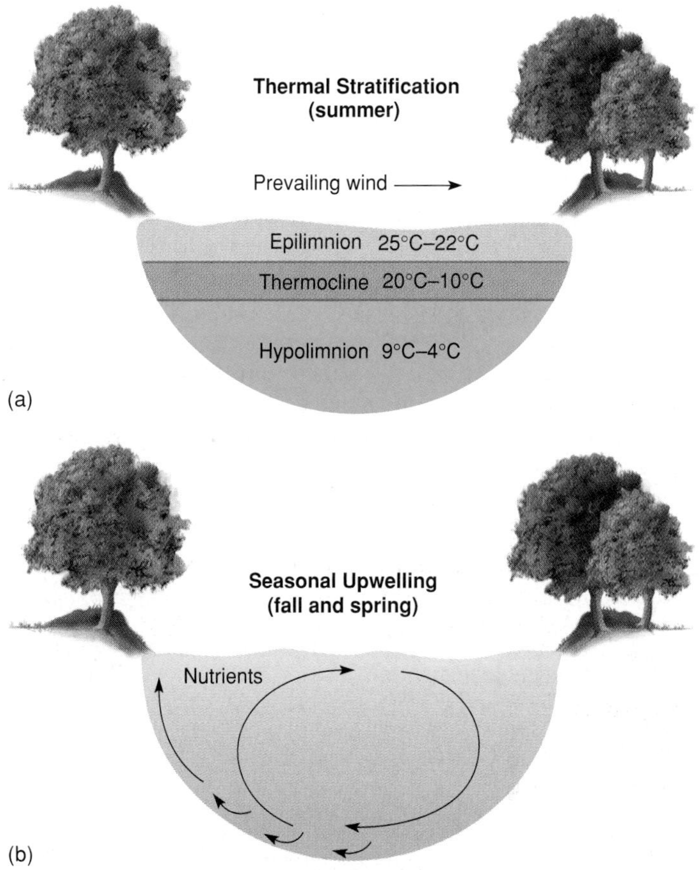

Figure 24.14 Profiles of a lake.
(a) During summer, a lake becomes stabilized into three major temperature strata. **(b)** During fall and spring, cooling or heating of the water disrupts the temperature strata and causes upwelling of nutrients from the bottom sediments.

Because oxygen is not very soluble in water and is rapidly used up by the plankton, its concentration forms a gradient from highest in the epilimnion to lowest at the bottom. In general, the amount of oxygen that can be dissolved is dependent on temperature. Warmer strata on the surface tend to carry lower levels of this gas. But of all the characteristics of water, the greatest range occurs in nutrient levels. Nutrient-deficient aquatic ecosystems are called **oligotrophic** (ahl″-ih-goh-trof′-ik). Species that can make a living on such starvation rations are *Hyphomicrobium* and *Caulobacter.* These bacteria have special stalks that capture even minuscule amounts of hydrocarbons present in oligotrophic habitats. At one time, it was thought that viruses were present only in very low levels in aquatic habitats, but researchers have now discovered that there are anywhere from 2 to 10 times as many viruses as bacteria in marine and freshwater communities. Oceans and lakes contain anywhere from 1 to 125 viruses per milliliter. Most of these viruses pose no danger to humans, but as parasites of bacteria, they appear to be a natural control mechanism for these populations.

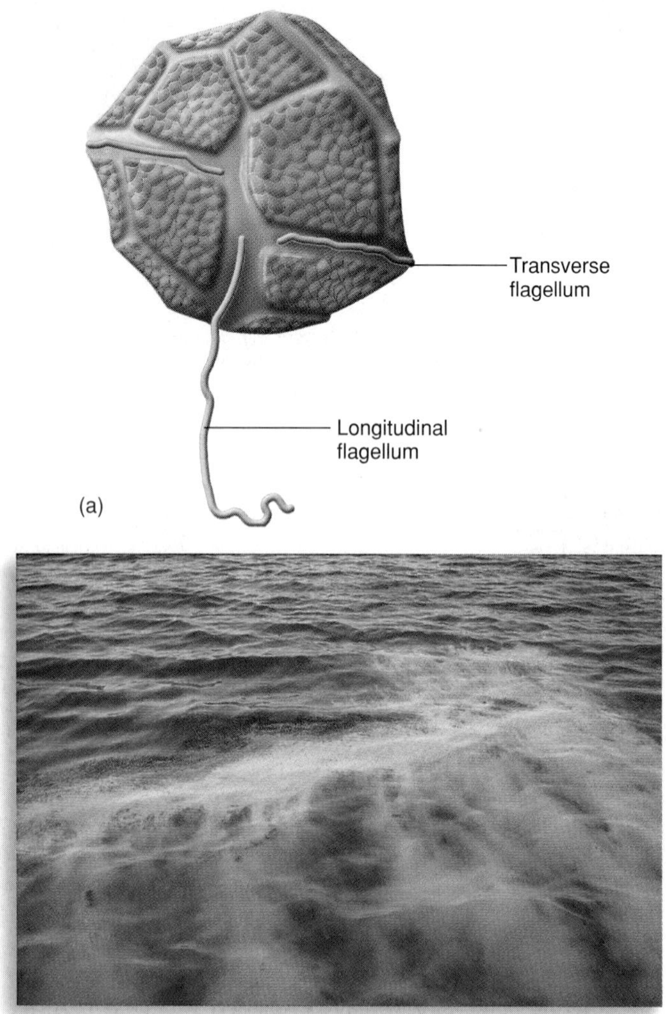

Figure 24.15 Red tides.
(a) Single-celled red algae called dinoflagellates (*Gymnodinium* shown here) bloom in high-nutrient, warm seawater and impart a noticeable red color to it, as shown in (b). **(b)** An aerial view of California coastline in the midst of a massive red tide.

Figure 24.16 Heavy surface growth of algae and cyanobacteria in a eutrophic pond.

At the other extreme are waters overburdened with organic matter and dissolved nutrients. Some nutrients are added naturally through seasonal upwelling and disasters (floods or typhoons), but the most significant alteration of natural waters comes from effluents from sewage, agriculture, and industry that contain heavy loads of organic debris or nitrate and phosphate fertilizers. The addition of excess quantities of nutrients to aquatic ecosystems, termed **eutrophication,** often wreaks havoc on the communities involved. The sudden influx of abundant nutrients along with warm temperatures encourages a heavy surface growth of cyanobacteria and algae called a bloom **(figure 24.16).** This heavy mat of biomass effectively shuts off the oxygen supply to the lake below. The oxygen content below the surface is further depleted by aerobic heterotrophs that actively decompose the organic matter. The lack of oxygen greatly disturbs the ecological balance of the community. It causes massive die-offs of strict aerobes (fish, invertebrates), and only anaerobic or facultative microbes will survive.

Water Monitoring to Prevent Disease

Microbiology of Drinking Water Supplies We do not have to look far for overwhelming reminders of the importance of safe water. Worldwide epidemics of cholera have killed thousands of people, and an outbreak of *Cryptosporidium* in Wisconsin affecting 370,000 people was traced to a contaminated municipal water supply. In a large segment of the world's population, the lack of sanitary water is responsible for billions of cases of diarrheal illness that kills 3 million children each year (see chapter 22). In the United States, nearly 1 million people develop water-borne illness every year.

Good health is dependent upon a clean, potable (drinkable) water supply. This means the water must be free of pathogens; dissolved toxins; and disagreeable turbidity, odor, color, and taste. As we shall see, water of high quality does not come easily, and we must look to microbes as part of the problem and part of the solution.

Through ordinary exposure to air, soil, and effluents, surface waters usually acquire harmless, saprobic microorganisms. But along its course, water can also pick up pathogenic contaminants. Among the most prominent water-borne pathogens of recent times are the protozoans *Giardia* and *Cryptosporidium;* the bacteria *Campylobacter, Salmonella, Shigella, Vibrio,* and *Mycobacterium;* and hepatitis A and Norwalk viruses. Some of these agents (especially encysted protozoans) can survive in natural waters for long periods without a human host, whereas others are present only transiently and are rapidly lost. The microbial content of drinking water must be continuously monitored to ensure that the water is free of infectious agents.

Attempting to survey water for specific pathogens can be very difficult and time-consuming, so most assays of water purity are more focused on detecting fecal contamination. High fecal levels can mean the water contains pathogens and is consequently unsafe to drink. Thus, wells, reservoirs,

INSIGHT 24.3 *Historical*

The Waning Days of a Classic Test?

Keeping water and the seafood we harvest from it free from fecal contamination is absolutely imperative in making it safe to ingest. In the late 1800s, it was suggested that a good way to determine if water or its products had been exposed to feces was to test for *E. coli*. Although most *E. coli* strains are not pathogenic, they almost always come from a mammal's intestinal tract so their presence in a sample is a clear indicator of fecal contamination.

Because at the time it was too difficult to differentiate *E. coli* from the closely related species of *Citrobacter*, *Klebsiella*, and *Enterobacter*, laboratories instead simply reported whether a sample contained one of these isolates. (All of these bacteria ferment lactose and are phenotypically similar.) The terminology adopted was "*coliform-*" (*E. coli*-like) positive or negative. In other words, one of these bacteria was present in the sample but it was not necessarily *E. coli*.

The use of this coliform assay has been the standard procedure since 1914, and it is still in widespread use. Pick up a newspaper in the summer, and you will likely find a report about a swimming pool or a river with a high coliform count. Coliform counts are also used to regulate food production and to trace the causes of food-borne outbreaks. Recently, microbiologists have noted serious problems with its use to indicate fecal contamination. The main issue is that the three other bacterial species already mentioned, among others, are commonly found growing in nonfecal environments such as fresh water and plants that eventually become food. In other words, if you're not looking specifically for *E. coli*, you can't be sure you're looking for feces.

In 1995, there was a minor panic when media outlets reported that iced tea from restaurants contained significant numbers of "fecal coliforms." The public was outraged. One headline read, "Iced Tea Worse Than River Water." Restaurants were named, and their reputations were damaged. When scientists did more detailed testing, they found that the predominant species found were *Klebsiella* and *Enterobacter*, both of which are normal colonizers of plants, such as tea leaves. Furthermore, despite the reports of widespread contamination with large numbers of "fecal coliforms," no one ever became sick from drinking iced tea.

Microbiologists are now advocating that *E. coli* alone—not the outdated grouping of coliforms—be used as an indicator of fecal contamination. Newer identification techniques make this as simple, if not simpler, than the standard coliform tests. But old habits die hard, and regulatory and public laboratories are proving slow to convert to the *E. coli* standard.

and other water sources can be analyzed for the presence of various **indicator bacteria.** These species are intestinal residents of birds and mammals, and they are readily identified using routine lab procedures.

Enteric bacteria most useful in the routine monitoring of microbial pollution are gram-negative rods called *coliforms* and enteric *streptococci*, which survive in natural waters but do not multiply there. Finding them in high numbers thus implicates recent or high levels of fecal contamination. Environmental Protection Agency standards for water sanitation are based primarily on the levels of coliforms, which are described as gram-negative, lactose-fermenting, gas-producing bacteria such as *Escherichia coli*, *Enterobacter*, and *Citrobacter*. Fecal contamination of marine waters that poses a risk for gastrointestinal disease is more readily correlated with gram-positive cocci, primarily in the genus *Enterococcus*. Occasionally, coliform bacteriophages and reoviruses (the Norwalk virus) are good indicators of fecal pollution, but their detection is more difficult and more technically demanding.

Water Quality Assays A rapid method for testing the total bacterial levels in water is the standard plate count. In this technique, a small sample of water is spread over the surface of a solid medium. The numbers of colonies that develop provide an estimate of the total viable population without differentiating coliforms from other species. This information is particularly helpful in evaluating the effectiveness of various water purification stages. Another general indicator of water quality is the level of dissolved oxygen it contains. It is established that water containing high levels of organic matter and bacteria will have a lower oxygen content because of consumption by aerobic respiration.

Coliform Enumeration Water quality departments employ some standard assays for routine detection and quantification of coliforms. The techniques available are

- simple tests, such as presence-absence broth, that detect coliform activity but do not quantify it;
- rapid tests that isolate coliform colonies and provide quantities of coliforms present; and
- rapid tests that identify specific coliforms and determine numbers.

In many circumstances (drinking water, for example), it is important to *differentiate* between facultative coliforms (*Enterobacter*) that are often found in other habitats (soil, water) *and* true **fecal coliforms** that live mainly in human and animal intestines. Microbiologists are calling for the discontinuation of the use of coliforms as an indicator of fecal contamination **(Insight 24.3).** But its use is still widespread so we cover its principles here.

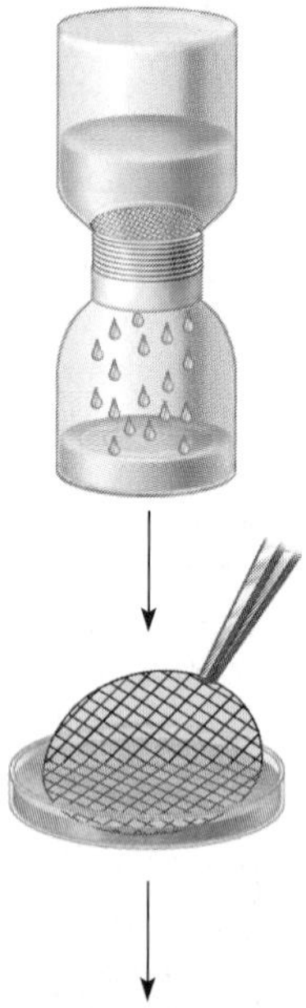

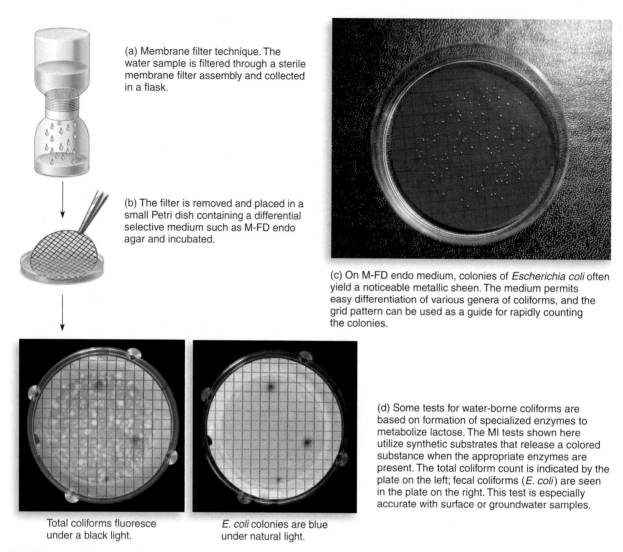

(a) Membrane filter technique. The water sample is filtered through a sterile membrane filter assembly and collected in a flask.

(b) The filter is removed and placed in a small Petri dish containing a differential selective medium such as M-FD endo agar and incubated.

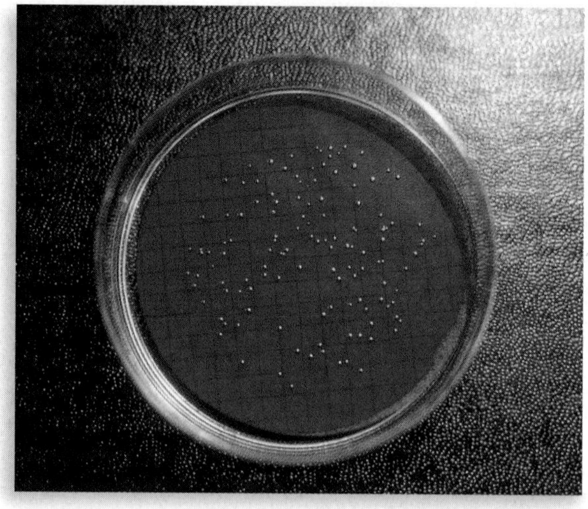

(c) On M-FD endo medium, colonies of *Escherichia coli* often yield a noticeable metallic sheen. The medium permits easy differentiation of various genera of coliforms, and the grid pattern can be used as a guide for rapidly counting the colonies.

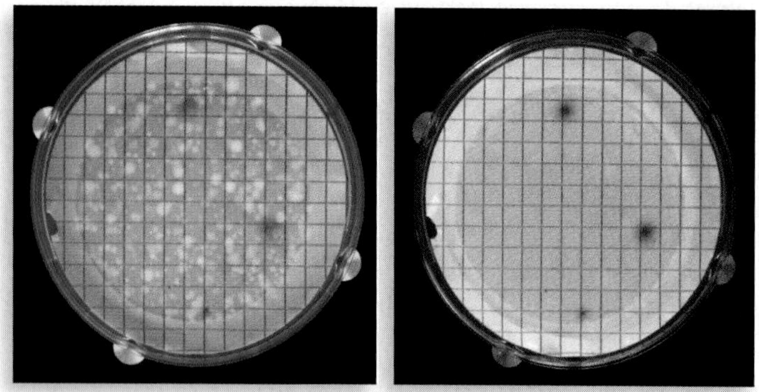

Total coliforms fluoresce under a black light.

E. coli colonies are blue under natural light.

(d) Some tests for water-borne coliforms are based on formation of specialized enzymes to metabolize lactose. The MI tests shown here utilize synthetic substrates that release a colored substance when the appropriate enzymes are present. The total coliform count is indicated by the plate on the left; fecal coliforms (*E. coli*) are seen in the plate on the right. This test is especially accurate with surface or groundwater samples.

Figure 24.17 Rapid methods of water analysis for coliform contamination.

The membrane filter method is a widely used rapid method that can be used in the field or lab to process and test larger quantities of water. This method is more suitable for dilute fluids, such as drinking water, that are relatively free of particulate matter, and it is less suitable for water containing heavy microbial growth or debris. This technique is related to the method described in chapter 11 for sterilizing fluids by filtering out microbial contaminants, except that in this system, the filter containing the trapped microbes is the desired end product. The steps in membrane filtration are diagrammed in **figure 24.17*a,b*.** After filtration, the membrane filter is placed in a Petri dish containing selective broth. After incubation, both nonfecal and fecal coliform colonies can be counted and often presumptively identified by their distinctive characteristics on these media **(figure 24.17*c,d*).**

Another more time-consuming but useful technique is the **most probable number (MPN)** procedure, which detects coliforms by a series of *presumptive, confirmatory,* and *completed* tests. The presumptive test involves three subsets of fermentation tubes, each containing different amounts of lactose or lauryl tryptose broth. The three subsets are inoculated with various-size water samples. After 24 hours of incubation, the tubes are evaluated for gas production. A positive test for gas formation is presumptive evidence of coliforms; negative for gas means no coliforms. The number of positive tubes in each subset is tallied, and this set of numbers is applied to a statistical table to estimate the most likely or probable concentration of coliforms.

It does not specifically detect fecal coliforms. When a test is negative for coliforms, the water is considered generally fit for human consumption. But even slight coliform levels are allowable under some circumstances. For example, municipal waters can have a maximum of 4 coliforms per 100 ml; private wells can have an even higher count. There is no acceptable level for fecal coliforms, enterococci, viruses, or pathogenic protozoans in drinking water. Waters that will not be consumed but are used for fishing or swimming are permitted to have counts of 70 to 200 coliforms per 100 ml. If the coliform level of recreational water reaches 1,000 coliforms per 100 ml, health departments usually bar its usage.

CHECKPOINT

- The lithosphere, or soil, is an ecosystem in which mineral-rich rocks are decomposed to organic humus, the base for the soil community. Soil ecosystems vary according to the kinds of rocks and amount of water, air, and nutrients present. The rhizosphere is the most ecologically active zone of the soil.
- The food web of the aquatic community is built on phytoplankton and zooplankton. The nature of the aquatic community varies with the temperature, depth, minerals, and amount of light present in each zone.
- Aquatic ecosystems are readily contaminated by chemical pollutants and pathogens because of industry, agriculture, and improper disposal of human wastes.
- Significant water-borne pathogens include protozoans, bacteria, and viruses. *Giardia* and *Cryptosporidium* are the most significant protozoan pathogens. *Campylobacter, Salmonella,* and *Vibrio* are the most significant bacterial pathogens. Hepatitis A and Norwalk virus are the most significant viral pathogens.
- Water quality assays assess the most probable number of microorganisms in a water sample and screen for the presence of enteric pathogens using coliforms as the indicator organisms.

Chapter Summary with Key Terms

24.1 Ecology: The Interconnecting Web of Life

A. Microbial ecology deals with the interaction between the environment and microorganisms. The environment is composed of **biotic** (living or once-living) and **abiotic** (nonliving) components. The combination of organisms and the environment make up an **ecosystem.**

B. **Ecosystem Organization**
1. Living things inhabit only that area of the earth called the **biosphere,** which is made up of the **hydrosphere** (water), the **lithosphere** (soil), and the **atmosphere** (air).
2. The biosphere consists of terrestrial ecosystems **(biomes)** and aquatic ecosystems.
3. Biomes contain **communities,** assemblages of coexisting organisms.
4. Communities consist of **populations,** groups of like organisms of the same species.
5. The space within which an organism lives is its **habitat;** its role in community dynamics is its **niche.**

C. **Energy and Nutrient Flow**
1. Organisms derive nutrients and energy from their habitat.
2. Their collective trophic status relative to one another is summarized in a **food** or **energy pyramid.**
3. At the beginning of the chain or pyramid are **producers**—organisms that synthesize large, complex organic compounds from small, simple inorganic molecules.
4. The levels above producer are occupied by **consumers,** organisms that feed upon other organisms.
5. **Decomposers** are consumers that obtain nutrition from the remains of dead organisms and help recycle and **mineralize** nutrients.
6. **Bioremediation** is the process by which microbes, or **consortia** of microbes, decompose chemicals that are harmful to the environment and its inhabitants.

24.2 The Natural Recycling of Bioelements

A. **Atmospheric Cycles**
Key compounds in the **carbon cycle** include carbon dioxide, methane, and carbonates.
1. Carbon is fixed when autotrophs (photosynthesizers) add carbon dioxide to organic carbon compounds.
2. The **nitrogen cycle** requires four processes and several types of microbes.
   a. In **nitrogen fixation,** atmospheric N_2 gas (the primary reservoir) is converted to NO_2^-, NO_3^-, or NH_4^+ salts.
   b. **Ammonification** is a stage in the degradation of nitrogenous organic compounds (proteins, nucleic acids) by bacteria to ammonium.
   c. Some bacteria **nitrify** NH_4^+ by converting it to NO_2^- and to NO_3^-.
   d. **Denitrification** is a multistep microbial conversion of various nitrogen salts back to atmospheric N_2.

B. **Sedimentary Cycles**
1. In the sulfur cycle, environmental sulfurous compounds are converted into useful substrates and returned to the inorganic reservoir through the action of microbes.
2. The chief compound in the phosphorus cycle is phosphate (PO_4) found in certain mineral rocks. Microbial action on this reservoir makes it available to be incorporated into organic phosphate forms.
3. Microorganisms often cycle and help accumulate heavy metals and other toxic pollutants that have been added to habitats by human activities.

24.3 Microbes on Land and in Water

A. **Soil Microbiology**
Soil is a dynamic, complex ecosystem that accommodates a vast array of microbes, animals, and plants coexisting among rich organic debris, water and air spaces, and minerals.

B. **Aquatic Microbiology**
   1. The surface water, atmospheric moisture, and groundwater are linked through a **hydrologic cycle** that involves evaporation and precipitation. Living things contribute to the cycle through respiration and transpiration.
   2. The diversity and distribution of water communities are related to sunlight, temperature, aeration, and dissolved nutrients. Phytoplankton and zooplankton drifting in the uppermost zone constitute a microbial community that supports the aquatic ecosystem.
   3. Water Monitoring
      a. Providing potable water is central to prevention of water-borne disease.
      b. Water is constantly surveyed for certain **indicator bacteria** (coliforms and enterococci) that signal fecal contamination.
      c. Assays for possible water contamination include the standard plate count and membrane filter tests to enumerate coliforms.

Multiple-Choice and True-False Questions

Multiple-Choice Questions. Select the correct answer from the answers provided.

1. Which of the following is *not* a major subdivision of the biosphere?
   a. hydrosphere c. stratosphere
   b. lithosphere d. atmosphere
2. A/an ____ is defined as a collection of populations sharing a given habitat.
   a. biosphere c. biome
   b. community d. ecosystem
3. The quantity of available nutrients ____ from the lower levels of the energy pyramid to the higher ones.
   a. increases c. remains stable
   b. decreases d. cycles
4. Which of the following is considered a greenhouse gas?
   a. CO_2 c. N_2O
   b. CH_4 d. all of these
5. Root nodules contain ____, which can ____.
   a. *Azotobacter*, fix N_2
   b. *Nitrosomonas*, nitrify NH_3^-
   c. rhizobia, fix N2
   d. *Bacillus*, denitrify NO_3^-
6. Which element(s) has/have an inorganic reservoir that exists primarily in sedimentary deposits?
   a. nitrogen c. sulfur
   b. phosphorus d. both b and c
7. Which of the following bacteria would be the most accurate indicator of fecal contamination?
   a. *Enterobacter* c. *Escherichia*
   b. *Thiobacillus* d. *Staphylococcus*

True-False Questions. If the statement is true, leave as is. If it is false, correct it by rewriting the sentence.

8. Pure cultures are very common in the biosphere.
9. Bioremediation usually involves more than one type of microorganism.
10. The production of all nitrogenous compounds begins with the process called nitrogen fixation.
11. The high mercury content found in some fish is the result of a process called bioaccumulation.

Writing to Learn

These questions are suggested as a *writing-to-learn* experience. For each question, compose a one- or two-paragraph answer that includes the factual information needed to completely address the question.

1. a. Present in outline form the levels of organization in the biosphere. Define the term *biome*.
   b. Compare autotrophs and heterotrophs; producers and consumers.
   c. Where in the energy and trophic schemes do decomposers enter?
   d. Compare the concepts of habitat and niche using *Chlamydomonas* (figure 24.2) as an example. What is mineralization, and which organisms are responsible for it?
2. a. Outline the general characteristics of a biogeochemical cycle.
   b. What are the major sources of carbon, nitrogen, phosphorus, and sulfur?
3. a. In what major forms is carbon found? Name three ways carbon is returned to the atmosphere.
   b. Name a way it is fixed into organic compounds.
   c. What form is the least available for the majority of living things?
4. a. Describe nitrogen fixation, ammonification, nitrification, and denitrification.
   b. What form of nitrogen is required by plants? By animals?
5. a. Describe the structure of the soil and the rhizosphere.
   b. What is humus?
   c. Compare and contrast root nodules with mycorrhizae.

6. a. Outline the modes of cycling water through the lithosphere, hydrosphere, and atmosphere.
   b. What are the roles of precipitation, condensation, respiration, transpiration, surface water, and aquifers?

7. a. What causes the formation of the epilimnion, hypolimnion, and thermocline?
   b. What is upwelling?
   c. In what ways are red tides and eutrophic algal blooms similar and different?

Concept Mapping

Appendix D provides guidance for working with concept maps.

1. Supply your own linking words or phrases in this concept map, and provide the missing concepts in the empty boxes.

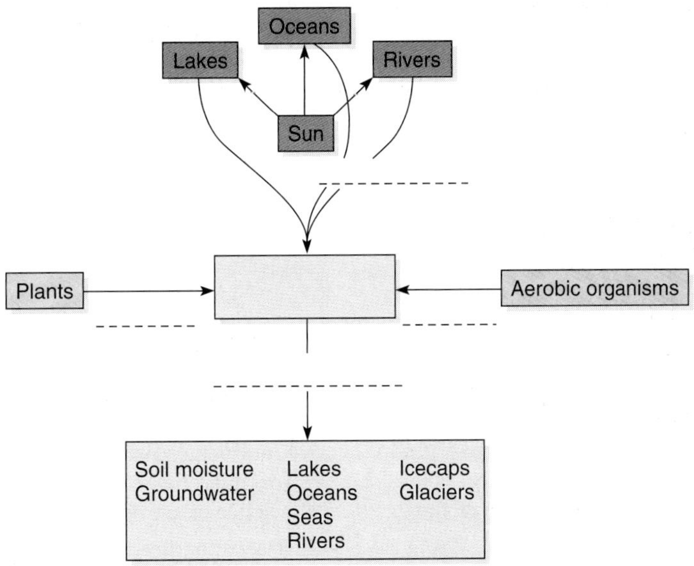

Critical Thinking Questions

Critical thinking is the ability to reason and solve problems using facts and concepts. These questions can be approached from a number of angles, and in most cases, they do not have a single correct answer.

1. a. What factors cause energy to decrease with each trophic level?
   b. How is it possible for energy to be lost and the ecosystem to still run efficiently?
   c. Are the nutrients on the earth a renewable resource? Why, or why not?
2. Give specific examples from biogeochemical cycles that support the Gaia Theory.
3. Biologists can set up an ecosystem in a small, sealed aquarium that continues to function without maintenance for years. Describe the minimum biotic and abiotic components it must contain to remain balanced and stable.
4. a. Is the greenhouse effect harmful under ordinary circumstances?
   b. What occurrence has made it dangerous to the global ecosystem?
   c. What could each person do on a daily basis to cut down on the potential for disrupting the delicate balance of the earth?
5. Why are organisms in the abyssal zone of the ocean necessarily halophilic, psychrophilic, barophilic, and anaerobic?
6. a. What eventually happens to the nutrients that run off into the ocean with sewage and other effluents?
   b. Why can high mountain communities usually dispense with water treatment?

Visual Understanding Questions

1. **From chapter 3, figure 3.8*b*.** If this MacConkey agar plate was inoculated with well water, would you report that coliforms were present in the well?

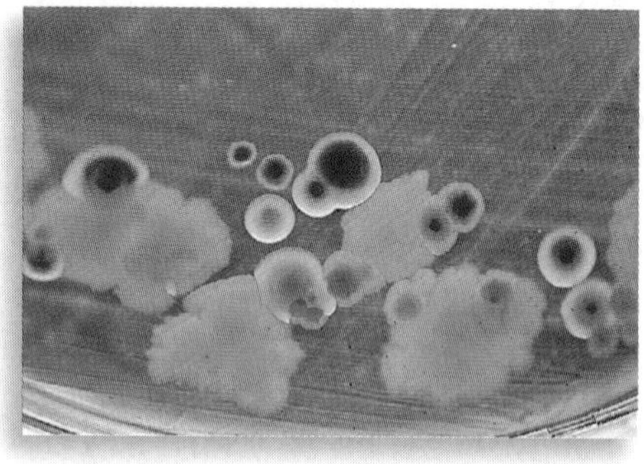

2. **From chapter 8, figure 8.24.** What process does this represent? How does it link to the biogeochemical cycles from this chapter?

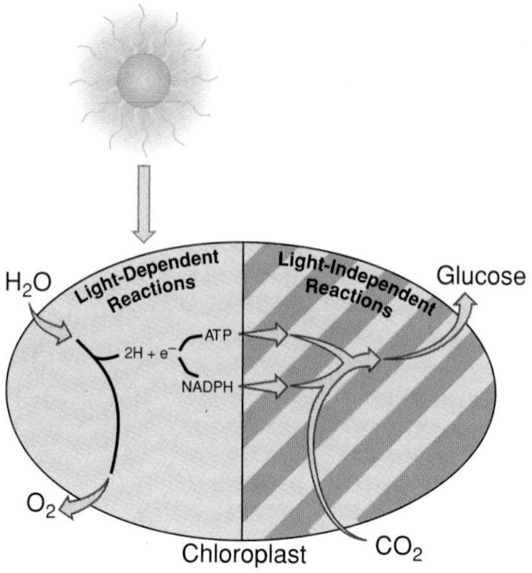

Internet Search Topics

Go to: www.aris.mhhe.com, and click on "microbiology" and then this textbook's author/title. Go to chapter 24, access the URLs listed under Internet Search Topics, and research the following:

1. Look up information on techniques for testing water. Explain how several of the tests work and their uses.
2. Find information on red tide outbreaks and illness in humans.
3. Conduct a search using the term "interplanetary transfer (of) microbes." What role does this phenomenon play in the science of astrobiology?
4. Search for "iced tea coliforms" and critically analyze what you find. Be sure to examine multiple sites.

CHAPTER 25

Applied and Industrial Microbiology

CASE FILE 25

On September 14, 2006, the U.S. Food and Drug Administration (FDA) issued an alert to consumers about an outbreak of *E. coli* O157:H7 in multiple states that may have been associated with the consumption of produce. At the time, preliminary epidemiological evidence suggested that bagged fresh spinach may have been a possible cause of this outbreak. Based on this information, the FDA advised that consumers not eat bagged fresh spinach at this time and urged individuals who believed that they may have experienced symptoms of illness after consuming bagged spinach to contact their health care provider. The FDA's Center for Food Safety and Applied Nutrition (CFSAN) immediately exchanged information with the U.S. Centers for Disease Control and Prevention (CDC) as well as state and local agencies to determine the cause and scope of the problem. *Escherichia coli* O157:H7 usually causes diarrhea, often with bloody stools. Although not lethal to most infected hosts, some people can develop a form of kidney failure called "hemolytic uremic syndrome" (HUS). HUS is most likely to occur in young children and elderly people. The condition can lead to serious kidney damage and even death. By October 6, 2006, 199 persons had been reported to the CDC from 26 states as infected with the outbreak strain of *E. coli* O157:H7.

To resolve the case, the FDA, the CDC, and other involved agencies used sophisticated modern technology based on research by microbial ecologists, molecular biologists, and bioengineers to identify a specific DNA fingerprint of the outbreak strain, which could then be compared with DNA isolated from packaged and fresh spinach obtained from a variety of sources. It was possible to isolate *E. coli* O157:H7 from all 13 packages of spinach supplied by patients living in 10 states, and the DNA fingerprints of all 13 of these *E. coli* matched that of the outbreak strain.

- *Do you think that radiated food is safer for consumption?*
- *Think about the warning labels on liquor and tobacco products. Now, what do you think about a required label informing that the lack of food sterilization may expose the consumer to infection by causal agents of disease?*

Case File 25 Wrap-Up appears on page 802.

CHAPTER OVERVIEW

- Microbes play significant roles in practical endeavors related to agriculture, food production, industrial processes, and waste treatment.
- Water quality is greatly dependent on its microbial and chemical content. Water is made safe by treatment methods that remove pathogenic microbes and toxic wastes.
- Biotechnology creates industrial, agricultural, nutritional, or medical products through microbial activities.
- Food fermentations are used to make a variety of milk products (cheeses, yogurt), alcoholic beverages (beer, wine, spirits), and pickles.
- Large-scale industrial fermentations employ microbial metabolism to manufacture drugs, hormones, enzymes, vaccines, and vitamins.

This chapter emphasizes the artificial applications of microbes in the remediation of communal wastes; the treatment of water; and the artificial applications of microbes in the food, medical, biochemical, drug, and agricultural industries. Key to the application of microbes is the understanding of their ecology (see chapters 7, 8, and 24) and the structure of their natural microenvironments. Microbes have evolved by responding to functional pressures, such as when nutrients are limited, nutrients are unevenly available, or there is competition for nutrients by other organisms. Applied and industrial microbiologists have learned from microbes' own survival mechanisms and have learned how to manipulate them in some way that will be useful to people.

25.1 Applied Microbiology and Biotechnology

Never underestimate the power of the microbe.

—Jackson W. Foster

The profound and sweeping involvement of microbes in the natural world is inescapable. Although our daily encounters with them usually go unnoticed, human and microbial life are clearly intertwined on many levels. It is no wonder that long ago humans realized the power of microbes and harnessed them for specific metabolic tasks. The practical applications of microorganisms in manufacturing products or carrying out a particular decomposition process belong to the large and diverse area of **biotechnology.** Biotechnology has an ancient history, dating back nearly 6,000 years to those first observant humans who discovered that grape juice left sitting produced wine or that bread dough properly infused with a starter would rise. Today, biotechnology has become a fertile ground for hundreds of applications in industry, medicine, agriculture, food sciences, and environmental protection, and it has even come to include the genetic alterations of microbes and other organisms.

Most biotechnological systems involve the actions of bacteria, yeasts, molds, and algae that have been selected or altered to synthesize a certain food, drug, organic acid, alcohol, or vitamin. Many such food and industrial end products are obtained through **fermentation,** a general term used here to refer to the mass, controlled culture of microbes to produce desired organic compounds. It also includes the use of microbes in sewage control, pollution control, metal mining, and bioremediation **(Insight 25.1).** A single section cannot cover this diverse area of microbiology in its entirety, but we touch on some of its more important applications in water treatment, food technology, and industrial processes.

Microorganisms in Water and Wastewater Treatment

Most drinking water comes from rivers, aquifers, and springs. Only in remote, undeveloped, or high mountain areas is this water used in its natural form. In most cities, it must be treated before it is supplied to consumers. Water supplies such as deep wells that are relatively clean and free of contaminants require less treatment than those from surface sources laden with wastes. The stepwise process in water purification as carried out by most cities is shown in **figure 25.1.** Treatment begins with the impoundment of water in a large reservoir such as a dam or catch basin that serves the dual purpose of storage and sedimentation. The access to reservoirs is controlled to avoid contamination by animals, wastes, and runoff water. In addition, overgrowth of cyanobacteria and algae that add undesirable qualities to the water is prevented by pretreatment with copper sulfate (0.3 ppm). Sedimentation to remove large particulate matter is also encouraged during this storage period.

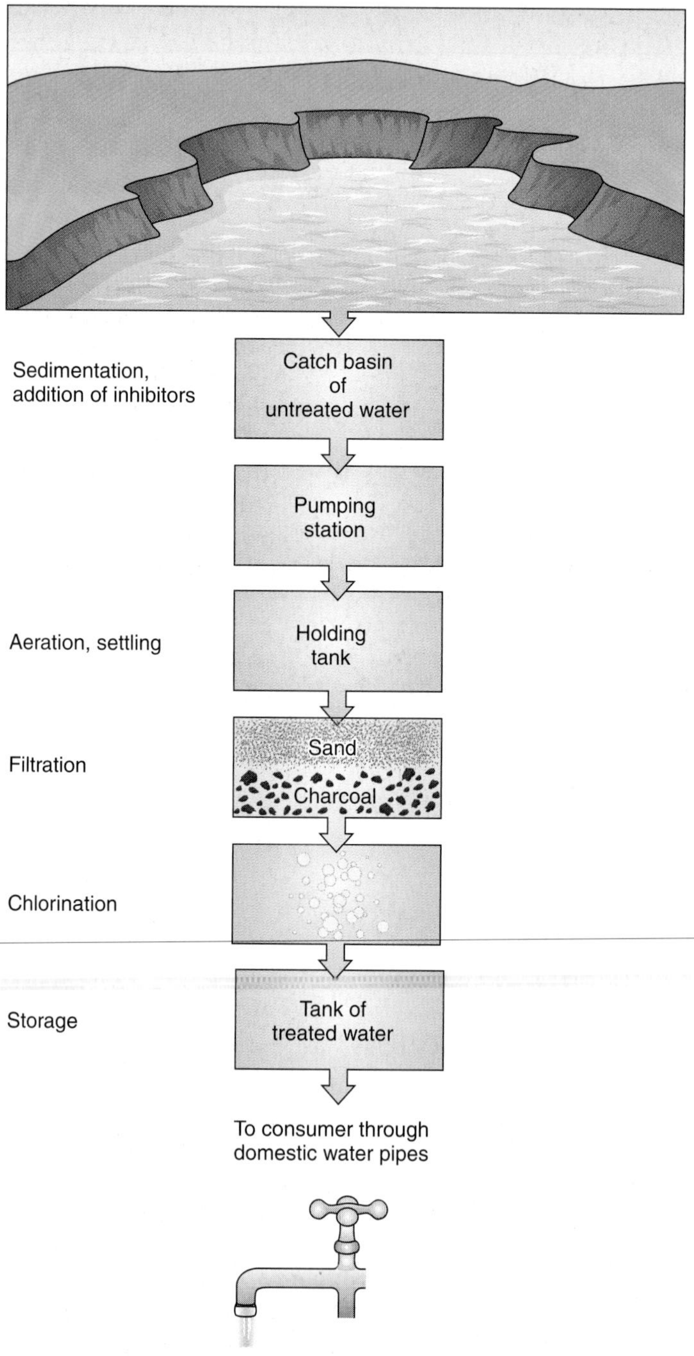

Figure 25.1 The major steps in water purification as carried out by a modern municipal treatment plant.

INSIGHT 25.1 ***Microbiology***

Bioremediation: The Pollution Solution?

The soil and water of the earth have long been considered convenient repositories for solid and liquid wastes. Humans have been burying solid wastes for thousands of years, but the process has escalated in the past 50 years. Every year, about 300 metric tons of pollutants, industrial wastes, and garbage are deposited into the natural environment. Often, this dumping is done with the mistaken idea that naturally occurring microbes will eventually biodegrade (break down) waste material.

Landfills currently serve as a final resting place for hundreds of castoffs from an affluent society, including yard wastes, paper, glass, plastics, wood, textiles, rubber, metal, paints, and solvents. This conglomeration is dumped into holes and is covered with soil. Although it is true that many substances are readily biodegradable, materials such as plastics and glass are not. Successful biodegradation also requires a compost containing specific types of microorganisms, adequate moisture, and oxygen. The environment surrounding buried trash provides none of these conditions. Large, dry, anaerobic masses of plant materials, paper, and other organic materials will not be successfully attacked by the aerobic microorganisms that dominate in biodegradation. As we continue to fill up hillsides with waste, the future of these landfills is a prime concern. One of the most serious of these concerns is that they will be a source of toxic compounds that seep into the ground and water.

Pollution of groundwater, the primary source of drinking water for 100 million people in the United States, is an increasing problem. Because of the extensive cycling of water through the hydrosphere and lithosphere, groundwater is often the final collection point for hazardous chemicals released into lakes, streams, oceans, and even garbage dumps. Many of these chemicals are pesticide residues from agriculture (dioxin, selenium, 2,4-D), industrial hydrocarbon wastes (PCBs), and hydrocarbon solvents (benzene, toluene). They are often hard to detect and, if detected, are hard to remove.

For many years, polluted soil and water were simply sealed off or dredged and dumped in a different site, with no attempt to get rid of the pollutant. But now, with greater awareness of toxic wastes, many Americans are adopting an attitude known as NIMBY (not in my backyard!), and environmentalists are troubled by the long-term effects of contaminating the earth.

In a search for solutions, waste management has turned to **bioremediation**—using microbes to break down or remove toxic wastes in water and soil. Some of these waste-eating microbes are natural soil and water residents with a surprising capacity to decompose even artificial substances. Because the natural, unaided process occurs too slowly, most cleanups are accomplished by commercial bioremediation services that treat the contaminated soil with oxygen, nutrients, and water to increase the rate of microbial action. Through these actions, levels of pesticides such as 2,4-D can be reduced to 96% of their original levels, and solvents can be reduced from 1 million parts per billion (ppb) to 10 ppb or less. Bacteria are also being used to help break up and digest oil from spills and refineries.

Among the most important bioremedial microbes are species of *Pseudomonas* and *Bacillus* and various toxin-eating fungi. Although much recent work has focused on creating "superbugs" through genetic engineering, public resistance to releasing genetically modified organisms (or GMOs) in the environment is high. Thankfully, naturally occurring biodegraders are plentiful, and efforts to optimize their performance are also very successful.

So far, about 35 recombinant microbes have been created for bioremediation. Species of *Rhodococcus* and *Burkholderia* have been engineered to decompose PCBs, and certain forms of *Pseudomonas* now contain genes for detoxifying heavy metals, carbon tetrachloride, and naphthalene. With over 3,000 toxic waste sites in the United States alone, the need for effective bioremediation is a top priority.

The genome sequence of the high-powered PCB degrader, *Burkholderia xenovorans* LB400 has just been worked out. This knowledge will provide additional tools for the cleanup of polluted environments.

Source: 2006. *Proceedings of the National Academy of Science* 103:15280–15287.

This marsh had been used to dump oil refinery waste. The level of certain pollutants was over 130,000 ppm.

After bioremediation with nutrients and microbes, the levels were reduced to less than 300 ppm in 4 months. It is bioremediated to the point that the land may be used for growing plants.

Next, the water is pumped to holding ponds or tanks, where it undergoes further settling, aeration, and filtration. The water is filtered first through sand beds or pulverized diatomaceous earth to remove residual bacteria, viruses, and protozoans and then through activated charcoal to remove undesirable organic contaminants. Pipes coming from the filtration beds collect the water in storage tanks. The final step in treatment is chemical disinfection by bubbling chlorine gas through the tank until it reaches a concentration of 1 to 2 ppm (some municipal plants use chloramines for this purpose) (see chapter 11). A few pilot plants in the United States are using ozone or peroxide for final disinfection, but these methods are expensive and cannot sustain an antimicrobial effect over long storage times. The final quality varies, but most tap water has a slight odor or taste from disinfection.

In many parts of the world, the same water that serves as a source of drinking water is also used as a dump for solid and liquid wastes **(figure 25.2).** Continued pressure on the finite water resources may require reclaiming and recycling of contaminated water such as sewage. Sewage is the used wastewater draining out of homes and industries that contains a wide variety of chemicals, debris, and microorganisms. The dangers of typhoid, cholera, and dysentery linked to the unsanitary mixing of household water and sewage have been a threat for centuries. In current practice, some sewage is treated to reduce its microbial load before release, but a large quantity is still being emptied raw (untreated) into the aquatic environment primarily because heavily contaminated waters require far more stringent and costly methods of treatment than are currently available to most cities.

Sewage contains large amounts of solid wastes, dissolved organic matter, and toxic chemicals that pose a health risk. To remove all potential health hazards, treatment typically requires three phases: The primary stage separates out large matter; the secondary stage reduces remaining matter and can remove some toxic substances; and the tertiary stage completes the purification of the water **(figure 25.3).** Microbial activity is an integral part of the overall process. The systems for sewage treatment are massive engineering marvels.

In the **primary phase** of treatment, floating bulkier materials such as paper, plastic waste, and bottles are skimmed off. The remaining smaller, suspended particulates are allowed to settle. Sedimentation in settling tanks usually takes 2 to 10 hours and leaves a mixture rich in organic matter. This aqueous residue is carried into a **secondary phase** of active microbial decomposition, or biodegradation. In this phase, a diverse community of natural bioremediators (bacteria, algae, and protozoa) aerobically decomposes the remaining particles of wood, paper, fabrics, petroleum, and organic molecules inside a large digester tank **(figure 25.4).** This forms a suspension of material called *sludge* that tends to settle out and slow the process. To hasten aerobic decomposition of the sludge, most processing plants have systems to *activate* it by injecting air, mechanically stirring it, and recirculating it. A large amount of organic matter is mineralized into sulfates, nitrates, phosphates, carbon dioxide, and water. Certain volatile gases such as hydrogen sulfide, ammonia, nitrogen, and methane may also be released. Water from this process is siphoned off and carried to the tertiary phase, which involves further filtering and chlorinating prior to discharge. Such reclaimed sewage water is usually used to water golf courses and parks rather than for drinking, or it is gradually released into large bodies of water.

In some cases, the solid waste that remains after aerobic decomposition is harvested and reused. Its rich content of nitrogen, potassium, and phosphorus makes it a useful fertilizer. But if the waste contains large amounts of nondegradable or toxic substances, it must be disposed of properly.

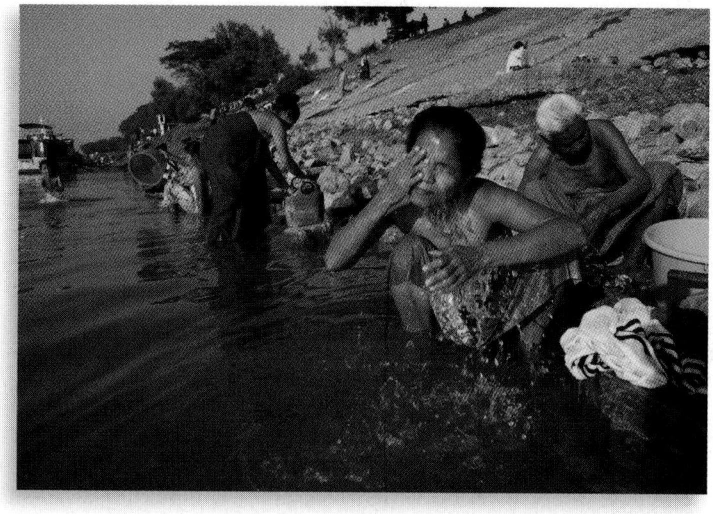

Figure 25.2 Water: one source, many uses.

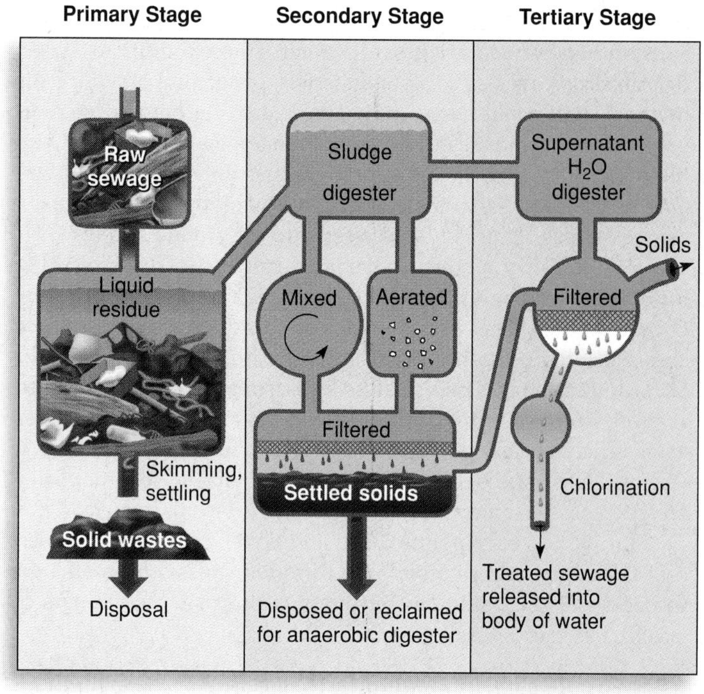

Figure 25.3 The primary, secondary, and tertiary stages in sewage treatment.

Figure 25.4 Treatment of sewage and wastewater. **(a)** Digester tanks used in the primary phase of treatment; each tank can process several million gallons of raw sewage a day. **(b)** View inside the secondary reactor shows the large stirring paddle that mixes the sludge to aerate it to encourage microbial decomposition.

In many parts of the world, the sludge, which still contains significant amounts of simple but useful organic matter is used as a secondary source of energy. Further digestion is carried out by microbes in sealed chambers called bioreactors, or **anaerobic digesters.** The digesters convert components of the sludge to swamp gas, primarily methane with small amounts of hydrogen, carbon dioxide, and other volatile compounds. Swamp gas can be burned to provide energy to run the sewage processing facility itself or to power small industrial plants.

Recently, scientists found a way to harness the bacteria found in sewage to construct a microbial fuel cell to produce usable energy. In these experiments, wastewater bacteria form biofilms on special rods inserted in the sewage that is being treated. These biofilms generate electrons that are transferred via copper wires to cathodes, producing electricity. Considering the mounting waste disposal and energy shortage problems, these technologies should gain momentum.

25.2 Microorganisms and Food

All human food—from vegetables to caviar to cheese—comes from some other organism, and rarely is it obtained in a sterile, uncontaminated state. Food is but a brief stopover in the overall scheme of biogeochemical cycling. This means that microbes and humans are in direct competition for the nutrients in food, and we must be constantly aware that microbes' fast growth rates give them the winning edge. Somewhere along the route of procurement, processing, or preparation, food becomes contaminated with microbes from the soil, the bodies of plants and animals, water, air, food handlers, or utensils. The final effects depend upon the types and numbers of microbes and whether the food is cooked or preserved. In some cases, specific microbes can even be added to food to obtain a desired effect. The effects of microorganisms on food can be classified as beneficial, detrimental, or neutral to humans, as summarized by the following outline:

Beneficial Effects
Food is fermented or otherwise chemically changed by the addition of microbes or microbial products to alter or improve flavor, taste, or texture.
Microbes can serve as food.
Detrimental Effects
Food poisoning or food-borne illness
Food spoilage
- Growth of microbes makes food unfit for consumption; adds undesirable flavors, appearance, and smell; destroys food value

Neutral Effects
The presence or growth of microbes that do not cause disease or change the nature of the food

As long as food contains no harmful substances or organisms, its suitability for consumption is largely a matter of taste. But what tastes like rich flavor to some may seem like decay to others. The test of whether certain foods are edible is guided by culture, experience, and preference. The flavors, colors, textures, and aromas of many cultural delicacies are supplied by bacteria and fungi. Poi, pickled cabbage, Norwegian fermented fish, and Limburger cheese are notable examples. If you examine the foods of most cultures, you will find some foods that derive their delicious flavor from microbes.

Microbial Fermentations in Food Products from Plants

In contrast to methods that destroy or keep out unwanted microbes, many culinary procedures deliberately add microorganisms and encourage them to grow. Common substances such as bread, cheese, beer, wine, yogurt, and pickles are the result of **food fermentations.** These reactions actively encourage biochemical activities that impart a particular taste, smell, or appearance to food. The microbe or microbes can occur naturally on the food substrate, as in sauerkraut, or they can be added as pure or mixed samples of known bacteria, molds, or yeasts called **starter cultures.** Many food fermentations are synergistic, with a series of microbes acting in concert to convert a starting substrate to the desired end product. Because large-scale production of fermented milk, cheese, bread, alcoholic brews, and vinegar depends upon inoculation with starter cultures, considerable effort is spent selecting, maintaining, and preparing these cultures and excluding contaminants that can spoil the fermentation. Most starting raw materials are of plant origin (grains, vegetables, beans) and, to a lesser extent, of animal origin (milk, meat).

Bread

Microorganisms accomplish three functions in bread making:

1. leavening the flour-based dough,
2. imparting flavor and odor, and
3. conditioning the dough to make it workable.

Leavening is achieved primarily through the release of gas to produce a porous and spongy product. Without leavening, bread dough remains dense, flat, and hard. Although various microbes and leavening agents can be used, the most common ones are various strains of the baker's yeast *Saccharomyces cerevisiae.* Other gas-forming microbes such as coliform bacteria, certain *Clostridium* species, heterofermentative lactic acid bacteria, and wild yeasts can be employed, depending on the type of bread desired.

Yeast metabolism requires a source of fermentable sugar such as maltose or glucose. Because the yeast respires aerobically in bread dough, the chief products of maltose fermentation are carbon dioxide and water rather than alcohol (the main product in beer and wine). Other contributions to bread texture come from kneading, which incorporates air into the dough, and from microbial enzymes, which break down flour proteins (gluten) and give the dough elasticity.

Besides carbon dioxide production, bread fermentation generates other volatile organic acids and alcohols that impart delicate flavors and aromas. These are especially well developed in homebaked bread, which is leavened more slowly than commercial bread. Yeasts and bacteria can also impart unique flavors, depending upon the culture mixture and baking techniques used. The pungent flavor of rye bread, for example, comes in part from starter cultures of lactic acid bacteria such as *Lactobacillus plantarum, L. brevis, L. bulgaricus, Leuconostoc mesenteroides,* and *Streptococcus thermophilus.* Sourdough bread gets its unique tang from *Lactobacillus sanfrancisco.*

Beer and Other Alcoholic Beverages

The production of alcoholic beverages takes advantage of another useful property of yeasts. By fermenting carbohydrates in fruits or grains anaerobically, they produce ethyl alcohol, as shown by this equation:

$$C_6H_{12}O_6 \rightarrow 2C_2H_5OH + 2CO_2$$
$$\text{(Yeast + Sugar = Ethanol + Carbon dioxide)}$$

Depending upon the starting materials and the processing method, alcoholic beverages vary in alcohol content and flavor. The principal types of fermented beverages are beers, wines, and spirit liquors.

The earliest evidence of beer brewing appears in ancient tablets by the Sumerians and Babylonians around 6000 BC. The starting ingredients for both ancient and present-day versions of beer, ale, stout, porter, and other variations are water, malt (barley grain), hops, and special strains of yeasts. The steps in brewing include malting, mashing, adding hops, fermenting, aging, and finishing.

For brewer's yeast to convert the carbohydrates in grain into ethyl alcohol, the barley must first be sprouted and softened to make its complex nutrients available to yeasts. This process, called **malting,** releases amylases that convert starch to dextrins and maltose, and proteases that digest proteins. Other sugar and starch supplements added in some forms of beer are corn, rice, wheat, soybeans, potatoes, and sorghum. After the sprouts have been separated, the remaining malt grain is dried and stored in preparation for mashing.

The malt grain is soaked in warm water and ground up to prepare a **mash.** Sugar and starch supplements are then introduced to the mash mixture, which is heated to a temperature of about 65°C to 70°C. During this step, the starch is hydrolyzed by amylase and simple sugars are released. Heating this mixture to 75°C stops the activity of the enzymes. Solid particles are next removed by settling and filtering. **Wort,** the clear fluid that comes off, is rich in dissolved carbohydrates. It is boiled for about 2.5 hours with **hops,** the dried scales of the female flower of *Humulus lupulus* **(figure 25.5),** to extract the bitter acids and resins that give aroma and flavor to the finished product. Boiling also caramelizes the sugar and imparts a golden or brown color, destroys any bacterial contaminants that can destroy flavor, and concentrates the mixture. The filtered and cooled supernatant is then ready for the addition of yeasts and fermentation.

Fermentation begins when wort is inoculated with a species of *Saccharomyces* that has been specially developed

Figure 25.5 Hops.
Female flowers of hops, the herb that gives beer some of its flavor and aroma.

for beer making. Top yeasts such as *Saccharomyces cerevisiae* function at the surface and are used to produce the higher alcohol content of *ales*. Bottom yeasts such as *S. uvarum* (*carlsbergensis*) function deep in the fermentation vat and are used to make other beers. In both cases, the initial inoculum of yeast starter is aerated briefly to promote rapid growth and increase the load of yeast cells. Shortly thereafter, an insulating blanket of foam and carbon dioxide develops on the surface of the vat and promotes anaerobic conditions **(figure 25.6).** During 8 to 14 days of fermentation, the wort sugar is converted chiefly to ethanol and carbon dioxide. The diversity of flavors in the finished product is partly due to the release of small amounts of glycerol, acetic acid, and esters. Fermentation is self-limited, and it essentially ceases when a concentration of 3% to 6% ethyl alcohol is reached.

Freshly fermented, or "green," beer is **lagered,** meaning it is held for several weeks to months in vats near 0°C. During this maturation period, yeast, proteins, resin, and other materials settle, leaving behind a clear, mellow fluid. Lager beer is subjected to a final filtration step to remove any residual yeasts that could spoil it. Finally, it is carbonated with carbon dioxide collected during fermentation and packaged in kegs, bottles, or cans.

Wine and Liquors

Wine is traditionally considered any alcoholic beverage arising from the fermentation of grape juice, but practically any fruit can be rendered into wine. The essential starting point is the preparation of **must,** the juice given off by crushed fruit that is used as a substrate for fermentation. In general, grape wines are either white or red. The color comes from the skins of the grapes, so white wine is prepared either from white-skinned grapes or from red-skinned grapes that have had the skin removed. Red wine comes from the red- or purple-skinned varieties. Major steps in making wine include must preparation (crushing), fermentation, storage, and aging **(figure 25.7).**

Figure 25.6 Anaerobic conditions in homemade beer production.
A layer of carbon dioxide foam keeps oxygen out.

For proper fermentation, must should contain 12% to 25% glucose or fructose, so the art of wine making begins in the vineyard. Grapes are harvested when their sugar content reaches 15% to 25%, depending on the type of wine to be made. Grapes from the field carry a mixed biofilm on their surface called the *bloom* that can serve as a source of wild yeasts. Some wine makers allow these natural yeasts to dominate, but many wineries inoculate the must with a special strain of *Saccharomyces cerevisiae,* variety *ellipsoideus.* To discourage yeast and bacterial spoilage agents, wine makers sometimes treat grapes with sulfur dioxide or potassium metabisulfite. The inoculated must is thoroughly aerated and mixed to promote rapid aerobic growth of yeasts, but when the desired level of yeast growth is achieved, anaerobic alcoholic fermentation is begun.

The temperature of the vat during fermentation must be carefully controlled to facilitate alcohol production. The length of fermentation varies from 3 to 5 days in red wines and from 7 to 14 days in white wines. The initial fermentation yields ethanol concentrations reaching 7% to 15% by volume, depending upon the type of yeast, the source of the juice, and ambient conditions. The fermented juice (raw wine) is decanted and transferred to large vats to settle and clarify. Before the final aging process, it is flash-pasteurized to kill microorganisms and filtered to remove any remaining yeasts and sediments. Wine is aged in wooden casks for varying time periods (months to years), after which it is bottled and stored for further aging. During aging, nonmicrobial changes produce aromas and flavors (the bouquet) characteristic of a particular wine.

The fermentation processes discussed thus far can only achieve a maximum alcoholic content of 17%, because concentrations above this level inhibit the metabolism of the yeast. The fermentation product must be distilled to obtain higher concentrations such as those found in liquors. During distillation, heating the liquor separates the more volatile alcohol from the less volatile aqueous phase. The alcohol is then condensed and collected. The alcohol content of distilled liquors is rated by *proof,* a measurement that is usually two times the alcohol content. Thus, 80 proof vodka contains 40% ethyl alcohol.

Distilled liquors originate through a process similar to wine making, although the starting substrates can be extremely diverse. In addition to distillation, liquors can be subjected to special treatments such as aging to provide unique flavor or color. Vodka, a colorless liquor, is usually prepared from fermented potatoes, and rum is distilled from fermented sugarcane. Assorted whiskeys are derived from fermented grain mashes; rye whiskey is produced from rye mash, and bourbon from corn mash. Brandy is distilled grape, peach, or apricot wine.

Other Fermented Plant Products

Fermentation provides an effective way of preserving vegetables, as well as enhancing flavor with lactic acid and salt. During pickling fermentations, vegetables are immersed in an anaerobic salty solution (brine) to extract sugar and nutrient-laden juices. The salt also disperses bacterial clumps, and its high osmotic pressure inhibits proteolytic bacteria and sporeformers that can spoil the product.

Sauerkraut is the fermentation product of cabbage. Cabbage is washed, wilted, shredded, salted, and packed tightly into a fermentation vat. Weights cover the cabbage mass and squeeze out its juices. The fermentation is achieved by natural cabbage microbiota or by an added culture. The initial agent of fermentation is *Leuconostoc mesenteroides*, which grows rapidly in the brine and produces lactic acid. It is followed by *Lactobacillus plantarum*, which continues to raise the acid content to as high as 2% (pH 3.5) by the end of fermentation. The high acid content restricts the growth of spoilage microbes.

Fermented cucumber pickles come chiefly in salt and dill varieties. Salt pickles are prepared by washing immature cucumbers, placing them in barrels of brine, and allowing them to ferment for 6 to 9 weeks. The brine can be inoculated with *Pediococcus cerevisiae* and *Lactobacillus plantarum* to avoid unfavorable qualities caused by natural microbiota and to achieve a more consistent product. Fermented dill pickles are prepared in a somewhat more elaborate fashion, with the addition of dill herb, spices, garlic, onion, and vinegar.

Natural vinegar is produced when the alcohol in fermented plant juice is oxidized to acetic acid, which is responsible for the pungent odor and sour taste. Although a reasonable facsimile of vinegar could be made by mixing about 4% acetic acid and a dash of sugar in water, this preparation would lack the traces of various esters, alcohol, glycerin, and volatile oils that give natural vinegar its pleasant character. Vinegar is actually produced in two stages. The first stage is similar to wine or beer making, in which

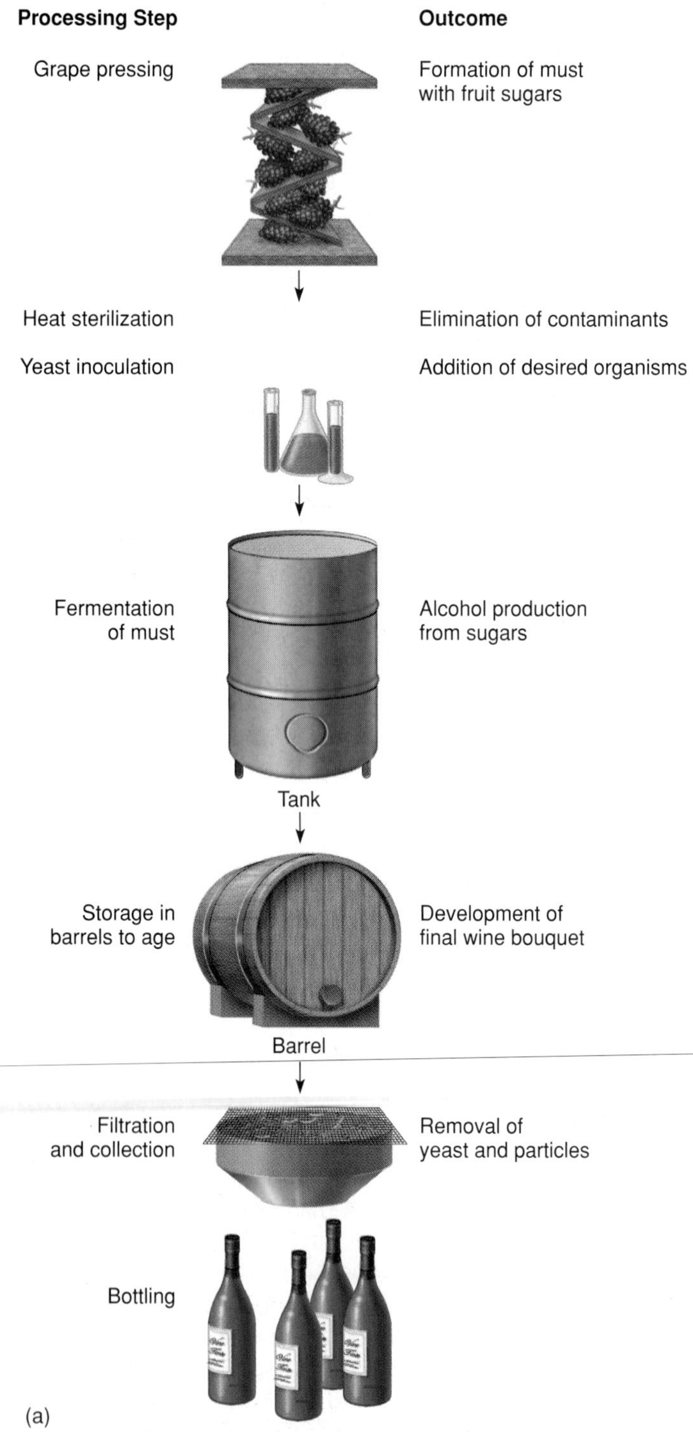

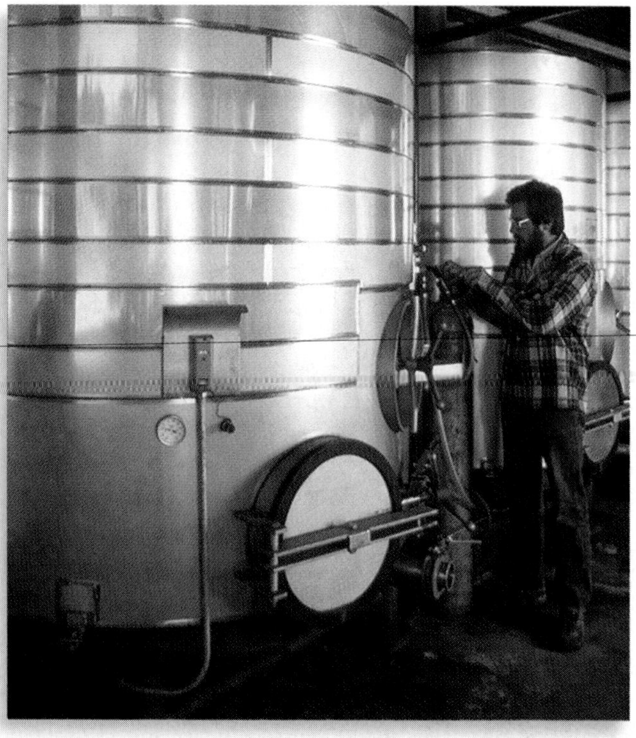

Figure 25.7 Wine making.
(a) General steps in wine making. **(b)** Wine fermentation vats in a large commercial winery.

a plant juice is fermented to alcohol by *Saccharomyces*. The second stage involves an aerobic fermentation carried out by acetic acid bacteria in the genera *Acetobacter* and *Gluconobacter*. These bacteria oxidize the ethanol in a two-step process, as shown here:

$$\underset{\text{Ethanol}}{2C_2H_5OH} + \tfrac{1}{2}\,O_2 \rightarrow \underset{\text{Acetaldehyde}}{CH_3CHO} + H_2O$$

$$\underset{\text{Acetaldehyde}}{CH_3CHO} + \tfrac{1}{2}\,O_2 \rightarrow \underset{\text{Acetic acid}}{CH_3COOH}$$

The abundance of oxygen necessary in commercial vinegar making is furnished by exposing inoculated raw material to air by arranging it in thin layers in open trays, allowing it to trickle over loosely packed beechwood twigs and shavings, or aerating it in a large vat. Different types of vinegar are derived from substrates such as apple cider (cider vinegar), malted grains (malt vinegar), and grape juice (wine vinegar).

Microbes in Milk and Dairy Products

Milk has a highly nutritious composition. It contains an abundance of water and is rich in minerals, protein (chiefly casein), butterfat, sugar (especially lactose), and vitamins. It starts its journey in the udder of a mammal as a sterile substance, but as it passes out of the teat, it is inoculated by the animal's normal biota. Other microbes can be introduced by milking utensils. Because milk is a nearly perfect culture medium, it is highly susceptible to microbial growth. When raw milk is left at room temperature, a series of bacteria ferment the lactose, produce acid, and alter the milk's content and texture **(figure 25.8*a*)**. This progression can occur naturally, or it can be induced, as in the production of cheese and yogurt.

In the initial stages of milk fermentation, lactose is rapidly attacked by *Streptococcus lactis* and *Lactobacillus* species **(figure 25.8*b*)**. The resultant lactic acid accumulation and lowered pH cause the milk proteins to coagulate into a solid mass called the **curd.** Curdling also causes the separation of a watery liquid called **whey** on the surface. Curd can be produced by microbial action or by an enzyme, **rennin** (casein coagulase), which is isolated from the stomach of unweaned calves.

Cheese

Since 5000 BC, various forms of cheese have been produced by spontaneous fermentation of cow, goat, or sheep milk. Present-day, large-scale cheese production is carefully controlled and uses only freeze-dried samples of pure cultures. These are first inoculated into a small quantity of pasteurized milk to form an active starter culture. This amplified culture is subsequently inoculated into a large vat of milk, where rapid curd development takes place. Such rapid growth is desired because it promotes the overgrowth of the desired inoculum and prevents the activities of undesirable contaminants. Rennin is usually added to increase the rate of curd formation.

After its separation from whey, the curd is rendered to produce one of the 20 major types of soft, semisoft, or hard cheese **(figure 25.9).** The composition of cheese is varied by adjusting water, fat, acid, and salt content. Cottage and cream cheese are examples of the soft, more perishable variety. After light salting and the optional addition of cream, they are ready for consumption without further processing. Other cheeses acquire their character from "ripening," a complex curing process involving bacterial, mold, and enzyme reactions that develop the final flavor, aroma, and other features characteristic of particular cheeses.

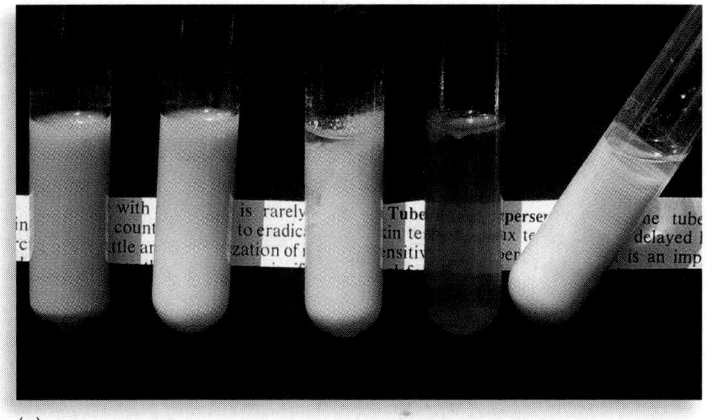

(a)

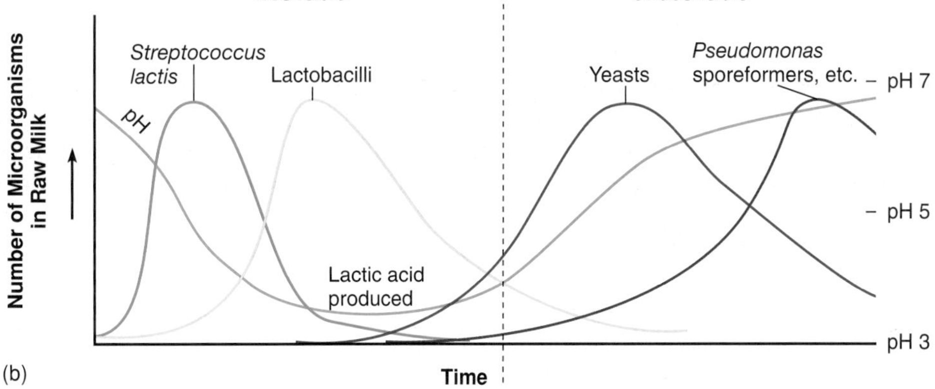

(b)

Figure 25.8 Microbes at work in milk products.
(a) Litmus milk is a medium used to indicate pH and consistency changes in milk resulting from microbial action. The first tube is an uninoculated, unchanged control. The second tube has a white, decolorized zone indicative of litmus reduction. The third tube has become acidified (pink), and its proteins have formed a loose curd. In the fourth tube, digestion of milk proteins has caused complete clarification or peptonization of the milk. The fifth tube shows a well-developed solid curd overlaid by a clear fluid, the whey. **(b)** Chart depicting spontaneous changes in the number and type of microorganism and the pH of raw milk as it incubates.
Source: Chart from Philip L. Carpenter, *Microbiology,* 3rd ed., copyright © 1972 by Holt, Rinehart and Winston, Inc. Reprinted by permission of the publisher.

Figure 25.9 Cheese making.
The curd-cutting stage in the making of cheddar cheese.

The distinctive traits of soft cheeses such as Limburger, Camembert, and Liederkranz are acquired by ripening with a reddish-brown mucoid coating of yeasts, micrococci, and molds. The microbial enzymes permeate the curd and ferment lipids, proteins, carbohydrates, and other substrates. This process leaves assorted acids and other by-products that give the finished cheese powerful aromas and delicate flavors. Semisoft varieties of cheese such as Roquefort, bleu, or Gorgonzola are infused and aged with a strain of *Penicillium roqueforti* mold. Hard cheeses such as Swiss, cheddar, and Parmesan develop a sharper flavor by aging with selected bacteria. The pockets in Swiss cheese come from entrapped carbon dioxide formed by *Propionibacterium*, which is also responsible for its bittersweet taste.

Other Fermented Milk Products

Yogurt is formed by the fermentation of milk by *Lactobacillus bulgaricus* and *Streptococcus thermophilus*. These organisms produce organic acids and other flavor components and can grow in such numbers that a gram of yogurt regularly contains 100 million bacteria. Live cultures of *Lactobacillus acidophilus* are an important additive to acidophilus milk, which is said to benefit digestion and to help maintain the normal biota of the intestine. Fermented milks such as kefir, koumiss, and buttermilk are a basic food source in many cultures.

Microorganisms as Food

At first, the thought of eating bacteria, molds, algae, and yeasts may seem odd or even unappetizing. We do eat their macroscopic relatives such as mushrooms, truffles, and seaweed, but we are used to thinking of the microscopic forms as agents of decay and disease or, at most, as food flavorings. The consumption of microorganisms is not a new concept. In Germany during World War II, it became necessary to supplement the diets of undernourished citizens by adding yeasts and molds to foods. At present, most countries are able to produce enough food for their inhabitants, but in the future, countries with exploding human populations and dwindling arable land may need to consider microbes as a significant source of protein, fat, and vitamins. Several countries already commercially mass-produce food yeasts, bacteria, and in a few cases, algae. Although eating microbes has yet to win total public acceptance, their use as feed supplements for livestock is increasing. A technology that shows some promise in increasing world food productivity is single-cell protein (SCP). This material is produced from waste materials such as molasses from sugar refining, petroleum by-products, and agricultural wastes. In England, an animal feed called pruteen is produced by mass culture of the bacterium *Methylophilus methylotrophus*. Mycoprotein, a product made from the fungus *Fusarium graminearum*, is also sold there. The filamentous texture of this product makes it a likely candidate for producing meat substitutes for human consumption.

Health food stores carry bottles of dark green pellets or powder that are a culture of a spiral-shaped cyanobacterium called *Spirulina*. This microbe is harvested from the surface of lakes and ponds, where it grows in great mats. In some parts of Africa and Mexico, *Spirulina* has become a viable alternative to green plants as a primary nutrient source. It can be eaten in its natural form or added to other foods and beverages.

Microbial Involvement in Food-Borne Diseases

The CDC estimates that several million people suffer each year from some form of food infection (see chapter 22). Until very recently, reports of food poisoning were escalating rapidly in the United States and worldwide. Outbreaks attributed to common pathogens (*Salmonella, E. coli, Vibrio,* hepatitis A, *Listeria, Campylobacter,* and various protozoa) had doubled in the past 20 years. A major factor in the escalation was the mass production and distribution of processed food such as raw vegetables, fruits, and meats. Improper handling can lead to gross contamination of these products with soil or animal wastes.

Growing concerns about food safety **(Table 25.1)** led to a new approach to regulating the food industry. The system is called Hazard Analysis and Critical Control Point, or HACCP, and it is adapted from procedures crafted for the space program in the 1970s. It involves principles that are more systematic and scientific than previous random-sampling quality

TABLE 25.1 Estimated Incidence of Food-Borne Illness in the United States

Illnesses	76,000,000 cases
Hospitalizations	325,000 cases
Deaths	5,200 cases

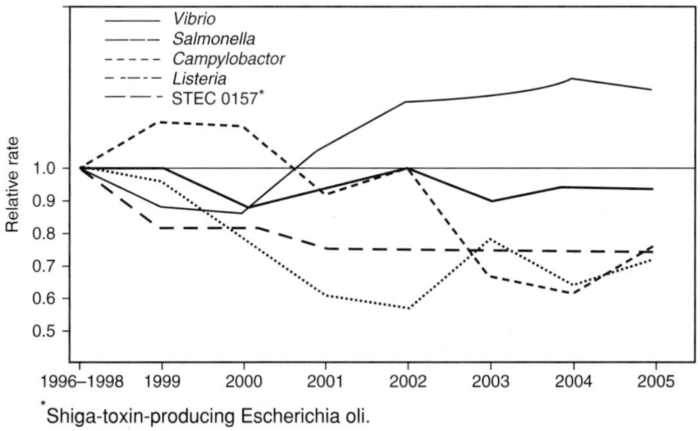

Figure 25.10 Recent trends in food-borne disease. The lines indicate the degree to which each disease has deviated from the levels in the 1996–1998 period.

procedures. The program focuses on the identification, evaluation, control, and prevention of hazards at all stages of the food production process. Since 1998 HACCP has been phased in by the U.S. Department of Agriculture for meat and poultry processing plants and by the Food and Drug Administration for seafood and juice plants. Pilot HACCP projects are taking place in facilities that process cheese, breakfast cereals, salad dressings, and bread. As the procedures become more widespread, microbiology-trained HACCP coordinators will be in high demand. Some 2005 data show that some foodborne illnesses have begun to decline, due in part to implementation of HACCP procedures **(figure 25.10).**

Many reported food poisoning outbreaks occur where contaminated food has been served to large groups of people,[1] but most cases probably occur in the home and are not reported.

Data collected by food microbiologists indicate that the most common bacterial food-borne pathogens are *Campylobacter, Salmonella, Listeria, Clostridium, Vibrio,* and *Staphylococcus aureus.* The dominant protozoa causing food infections are *Giardia, Cryptosporidium,* and *Toxoplasma.* The top viruses are Norwalk and hepatitis A viruses.

Prevention Measures for Food Poisoning and Spoilage

It will never be possible to avoid all types of food-borne illness because of the ubiquity of microbes in air, water, food, and the human body. But most types of food poisoning require the growth of microbes in the food. In the case of food infections, an infectious dose (sufficient cells to initiate infection) must be present, and in food intoxication, enough cells to produce the toxin must be present. Thus, food poisoning or spoilage can be prevented by proper food handling, preparation, and storage. The methods shown in **figure 25.11**

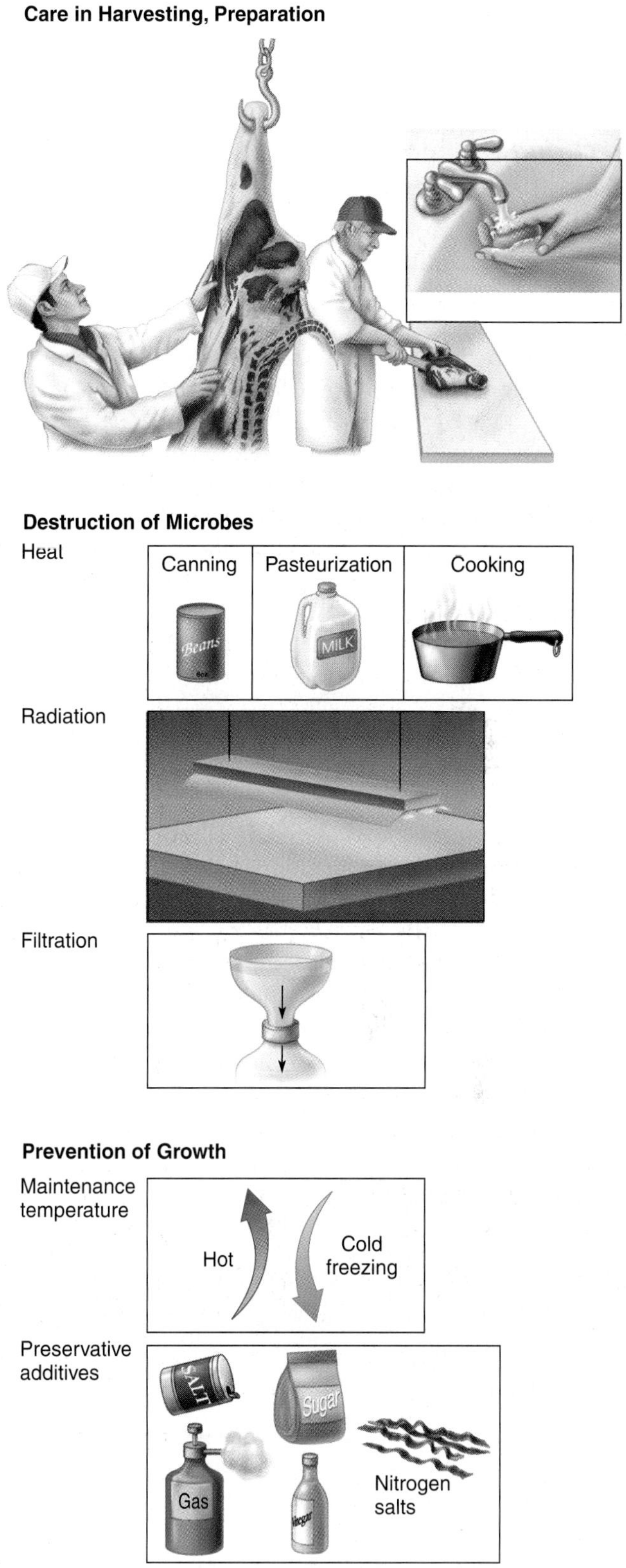

Figure 25.11 The primary methods to prevent food poisoning and food spoilage.

1. One-third of all reported cases result from eating restaurant food.

are aimed at preventing the incorporation of microbes into food, removing or destroying microbes in food, and keeping microbes from multiplying.

Preventing the Incorporation of Microbes into Food

Most agricultural products such as fruits, vegetables, grains, meats, eggs, and milk are naturally exposed to microbes. Vigorous washing reduces the levels of contaminants in fruits and vegetables, whereas meat, eggs, and milk must be taken from their animal source as aseptically as possible. Aseptic techniques are also essential in the kitchen. Contamination of foods by fingers can be easily remedied by hand washing and proper hygiene, and contamination by flies or other insects can be stopped by covering foods or eliminating pests from the kitchen. Care and common sense also apply in managing utensils. It is important to avoid cross-contaminating food by using the same cutting board for meat and vegetables without disinfecting it between uses. The subject of cutting board safety is discussed in **Insight 25.2.**

Preventing the Survival or Multiplication of Microbes in Food

Because it is not possible to eliminate all microbes from certain types of food by clean techniques alone, a more efficient approach is to preserve the food by physical or chemical methods. Hygienically preserving foods is especially important for large commercial companies that process and sell bulk foods and must ensure that products are free from harmful contaminants. Regulations and standards for food processing are administered by two federal agencies: the Food and Drug Administration (FDA) and the U.S. Department of Agriculture (USDA).

Temperature and Food Preservation

Heat is a common way to destroy microbial contaminants or to reduce the load of microorganisms. Commercial canneries preserve food in hermetically sealed containers that have been exposed to high temperatures over a specified time period. The temperature used depends upon the type of food, and it can range from 60°C to 121°C, with exposure times ranging from 20 minutes to 115 minutes. The food is usually processed at a thermal death time (TDT; see chapter 11) that will destroy the main spoilage organisms and pathogens but will not alter the nutrient value or flavor of the food. For example, tomato juice must be heated to between 121°C and 132°C for 20 minutes to ensure destruction of the spoilage agent *Bacillus coagulans.* Most canning methods are rigorous enough to sterilize the food completely, but some only render the food "commercially sterile," which means it contains live bacteria that are unable to grow under normal conditions of storage.

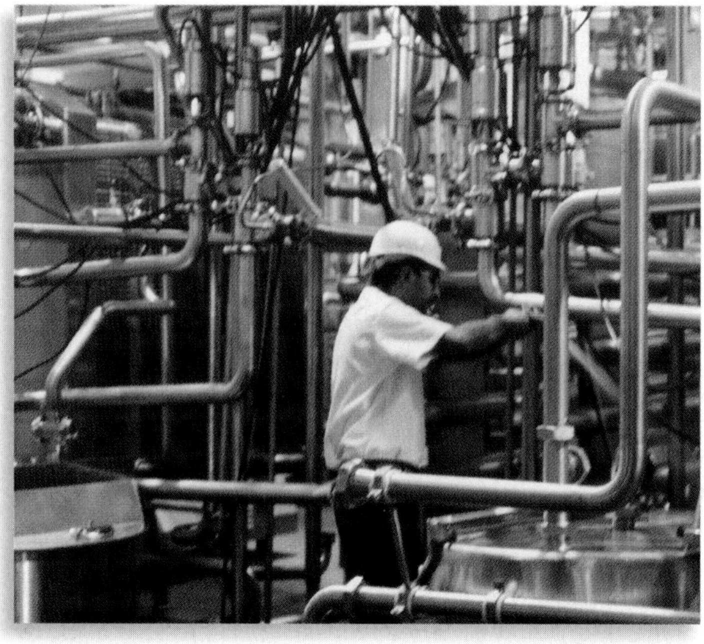

Figure 25.12 A modern flash pasteurizer, a system used in dairies for high-temperature short-time (HTST) pasteurization.
Source: Photo taken at Alta Dena Dairy, City of Industry, California.

Another use of heat is **pasteurization,** usually defined as the application of heat below 100°C to destroy nonresistant bacteria and yeasts in liquids such as milk, wine, and fruit juices. The heat is applied in the form of steam, hot water, or even electrical current. The most prevalent technology is the *high-temperature short-time* (*HTST*), or flash method, using extensive networks of tubes that expose the liquid to 72°C for 15 seconds **(figure 25.12).** An alternative method, ultra-high-temperature (UHT) pasteurization, steams the product until it reaches a temperature of 134°C for at least 1 second. Although milk processed this way is not actually sterile, it is often marketed as sterile, with a shelf life of up to 3 months. Older methods involve large bulk tanks that hold the fluid at a lower temperature for a longer time, usually 62.3°C for 30 minutes.

Cooking temperatures used to boil, roast, or fry foods can render them free or relatively free of living microbes if carried out for sufficient time to destroy any potential pathogens. A quick warming of chicken or an egg is inadequate to kill microbes such as *Salmonella.* In fact, any meat is a potential source of infectious agents and should be adequately cooked. Because most meat-associated food poisoning is caused by nonsporulating bacteria, heating the center of meat to at least 80°C and holding it there for 30 minutes is usually sufficient to kill pathogens. Roasting or frying food at temperatures of at least 200°C or boiling it will achieve a satisfactory degree of disinfection.

Any perishable raw or cooked food that could serve as a growth medium must be stored to prevent the multiplication of bacteria that have survived during processing or handling. Because most food-borne bacteria and molds

INSIGHT 25.2 — Discovery

Wood or Plastic: On the Cutting Edge of Cutting Boards

Inquiring cooks have long been curious for the final word on which type of cutting board is the better choice for food safety. When the USDA recommended plastic cutting boards, it seemed the logical, reasonable choice. After all, plastic is nonabsorbent and easy to clean, presumably making it less likely to harbor bacteria and other microorganisms on its surface than wood is. But this recommendation was never based on evidence from scientific tests. Recently, two separate research groups turned their attention to this important kitchen question. What emerged from these studies came as rather a surprise—the two groups came up with exactly opposite conclusions.

First came the study by a team of microbiologists from the University of Wisconsin. They experimented with hardwood chopping blocks and acrylic plastic boards inoculated with pathogens such as *Salmonella, Escherichia coli,* and *Listeria monocytogenes.* One of the most unexpected results was that the wooden boards actually killed 99.9% of the bacteria within a few minutes. The team concluded from the lack of viable cells that wood must contain some antibacterial substances, although they were unable to isolate them. The plastic boards did not similarly reduce the numbers of pathogens and they failed to live up to expectations in other ways. For instance, they continued to harbor bacteria if left unwashed for a given time period. If they were scored from extensive use, even after scrubbing with soap and water, they still held live bacteria. In contrast, even heavily used wooden boards did not grow microorganisms and had a far lower bacterial count. The Wisconsin researchers concluded that the grounds for advocating plastic are questionable and that wood is as safe as plastic, if not superior to it.

In the other study, researchers from the Food and Drug Administration performed an electron microscope study of wood. They found that pathogens such as *E. coli* O157:H7 and *Campylobacter* became trapped in the porous spaces of wooden boards and were able to survive for 2 hours to several days, depending on the moisture content of the wood. They continue to recommend the use of plastic because bacteria trapped in wood would be difficult to remove and could be released during use.

What is a chef to do? Although these contradictory studies seem not to provide a definitive answer, they can serve to emphasize an important point. The solution still exists in simple, commonsense guidelines that are the crux of good kitchen practices. It is apparent that both boards can be safe if properly handled and their limitations are taken into account. All boards should be scrubbed with soap and hot water and disinfected between uses, especially if meats, poultry, or fish have been cut on them. Boards should be replaced if their surface has become too roughened with use, and wooden boards must not be left moist for any period of time.

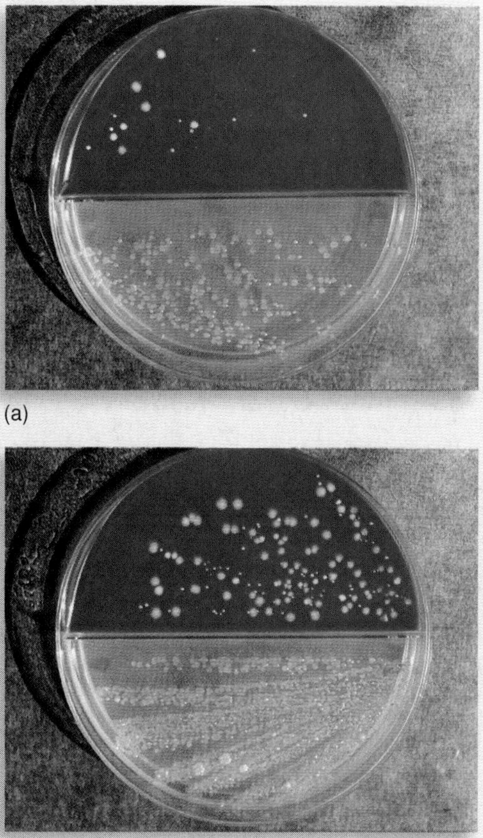

(a) (b)

Double-sided plates of blood agar (top) and MacConkey agar (bottom) after swabbing with samples from cutting boards. The boards were equally contaminated with a fresh chicken carcass, and the samples were taken 10 minutes later. Results appear in **(a)** for the wooden board and in **(b)** for the plastic board. Note that, in this case, the wooden board yielded significantly fewer colonies on both types of media.

that are agents of spoilage or infection can multiply at room temperature, manipulation of the holding temperature is a useful preservation method **(figure 25.13).** A good general directive is to store foods at temperatures below 4°C or above 60°C.

Regular refrigeration reduces the growth rate of most mesophilic bacteria by 10 times, although some psychrotrophic microbes can continue to grow at a rate that causes spoilage. This factor limits the shelf life of milk, because even at 7°C, a population could go from a few cells to a billion in 10 days. Pathogens such as *Listeria monocytogenes* and *Salmonella* can also continue to grow in refrigerated foods. Freezing is a longer-term method for cold preservation. Foods can be either slow-frozen for 3 to 72 hours at −15°C to −23°C or rapidly frozen for 30 minutes at −17°C to −34°C. Because freezing cannot be counted upon to kill microbes, rancid, spoiled,

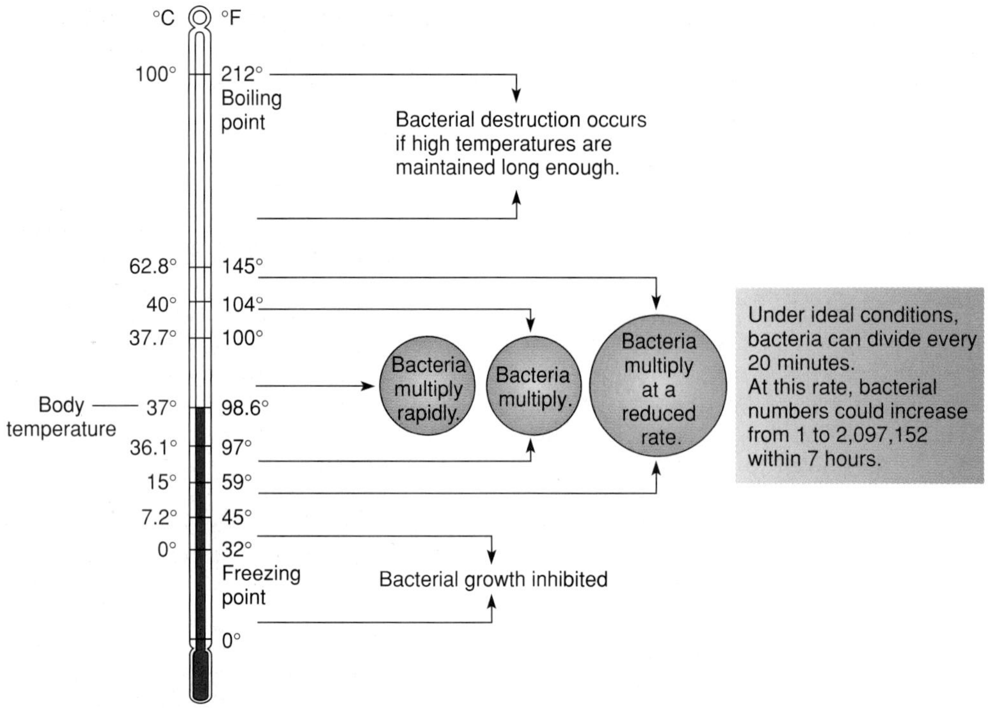

Figure 25.13 Temperatures favoring and inhibiting the growth of microbes in food.

Most microbial agents of disease or spoilage grow in the temperature range of 15°C to 40°C. Preventing unwanted growth in foods in long-term storage is best achieved by refrigeration or freezing (4°C or lower). Preventing microbial growth in foods intended to be consumed warm in a few minutes or hours requires maintaining the foods above 60°C.

Source: From Ronald Atlas, *Microbiology: Fundamentals and Applications,* 2nd ed., © 1998, p. 475. Reprinted by permission of Prentice Hall, Upper Saddle River, New Jersey.

or infectious foods will still be unfit to eat after freezing and defrosting. *Salmonella* is known to survive several months in frozen chicken and ice cream, and *Vibrio parahaemolyticus* can survive in frozen shellfish. For this reason, frozen foods should be defrosted rapidly and immediately cooked or reheated. However, even this practice will not prevent staphylococcal intoxication if the toxin is already present in the food before it is heated.

Foods such as soups, stews, gravies, meats, and vegetables that are generally eaten hot should not be maintained at warm or room temperatures, especially in settings such as cafeterias, banquets, and picnics. The use of a hot plate, chafing dish, or hot water bath will maintain foods above 60°C, well above the incubation temperature of food-poisoning agents.

As a final note about methods to prevent food poisoning, remember the simple axiom: "When in doubt, throw it out."

Radiation

Ultraviolet (nonionizing) lamps are commonly used to destroy microbes on the surfaces of foods or utensils, but they do not penetrate far enough to sterilize bulky foods or food in packages. Food preparation areas are often equipped with UV radiation devices that are used to destroy spores on the surfaces of cheese, breads, and cakes and to disinfect packaging machines and storage areas.

Food itself is usually sterilized by gamma or cathode radiation because these ionizing rays can penetrate denser materials. It must also be emphasized that this method does not cause the targets of irradiation to become radioactive.

Concerns have been raised about the possible secondary effects of radiation that could alter the safety and edibility of foods. Experiments over the past 30 years have demonstrated some side reactions that affect flavor, odor,

CASE FILE 25 *WRAP-UP*

The investigation of the spinach-associated *E. coli* outbreak determined that wild pigs had access to the spinach fields from which the bagged spinach originated. The pigs seem to have contaminated spinach with *E. coli*–contaminated excrement.

As mentioned in the chapter, radiated food is safer from a microbiological standpoint. It is also certified safe by the Food and Drug Administration and is endorsed by the American Medical Association. It is approved in this country for the treatment of meat, fruits, and vegetables. It is no substitute for good food production and handling techniques but can provide another measure of safety.

and vitamin content, but it is currently thought that irradiated foods are relatively free of toxic by-products. The government has currently approved the use of radiation in sterilizing beef, pork, poultry, fish, spices, grain, and some fruits and vegetables. Less than 10% of these products are sterilized this way, but outbreaks of food-borne illness have increased its desirability for companies and consumers. It also increases the shelf life of perishable foods, thus lowering their cost.

Other Forms of Preservation

The addition of chemical preservatives to many foods can prevent the growth of microorganisms that could cause spoilage or disease. Preservatives include natural chemicals such as salt (NaCl) or table sugar and artificial substances such as ethylene oxide. The main classes of preservatives are organic acids, nitrogen salts, sulfur compounds, oxides, salt, and sugar.

Organic acids, including lactic, benzoic, and propionic acids, are among the most widely used preservatives. They are added to baked goods, cheeses, pickles, carbonated beverages, jams, jellies, and dried fruits to reduce spoilage from molds and some bacteria. Nitrites and nitrates are used primarily to maintain the red color of cured meats (hams, bacon, and sausage). By inhibiting the germination of *Clostridium botulinum* spores, they also prevent botulism intoxication, but their effects against other microorganisms are limited. Sulfite prevents the growth of undesirable molds in dried fruits, juices, and wines and retards discoloration in various foodstuffs. Ethylene and propylene oxide gases disinfect various dried foodstuffs. Their use is restricted to fruit, cereals, spices, nuts, and cocoa.

The high osmotic pressure contributed by hypertonic levels of salt plasmolyzes bacteria and fungi and removes moisture from food, thereby inhibiting microbial growth. Salt is commonly added to brines, pickled foods, meats, and fish. However, it does not retard the growth of pathogenic halophiles such as *Staphylococcus aureus,* which grows readily even in 7.5% salt solutions. The high sugar concentrations of candies, jellies, and canned fruits also exert an osmotic preservative effect. Other chemical additives that function in preservation are alcohols and antibiotics. Alcohol is added to flavoring extracts, and antibiotics are approved for treating the carcasses of chickens, fish, and shrimp.

Food can also be preserved by **desiccation,** a process that removes moisture needed by microbes for growth by exposing the food to dry, warm air. Solar drying was traditionally used for fruits and vegetables, but modern commercial dehydration is carried out in rapid-evaporation mechanical devices. Drying is not a reliable microbicidal method, however. Numerous resistant microbes such as micrococci, coliforms, staphylococci, salmonellae, and fungi survive in dried milk and eggs, which can subsequently serve as agents of spoilage and infections.

In 2006, the Food and Drug Administration approved the spraying of bacteriophages onto ready-to-eat meat products. The bacteriophages are specific for *Listeria* and will act to kill the bacteria that would not otherwise be killed because the cold cuts and poultry are usually not cooked before consumption.

CHECKPOINT

- The use of microorganisms for practical purposes to benefit humans is called biotechnology.
- Wastewater or sewage is treated in three stages to remove organic material, microorganisms, and chemical pollutants. The primary phase removes physical objects from the wastewater. The secondary phase removes the organic matter by biodegradation. The tertiary phase disinfects the water and removes chemical pollutants.
- Microorganisms can compete with humans for the nutrients in food. Their presence in food can be beneficial, detrimental, or of neutral consequence to human consumers.
- Food fermentation processes utilize bacteria or yeast to produce desired components such as alcohols and organic acids in foods and beverages. Beer, wine, yogurt, and cheeses are examples of such processes.
- Some microorganisms are used as a source of protein. Examples are single-cell protein, mycoprotein, and *Spirulina*. Microbial protein could replace meat as a major protein source.
- Food-borne disease can be an intoxication caused by microbial toxins produced as by-products of microbial decomposition of food. Or it can be a food infection when pathogenic microorganisms in the food attack the human host after being consumed.
- Heat, radiation, chemicals, and drying are methods used to limit numbers of microorganisms in food. The type of method used depends on the nature of the food and the type of pathogens or spoilage agents it contains.

25.3 General Concepts in Industrial Microbiology

Virtually any large-scale commercial enterprise that enlists microorganisms to manufacture consumable materials is part of the realm of industrial microbiology. Here the term pertains primarily to bulk production of organic compounds such as antibiotics, hormones, vitamins, acids, solvents **(table 25.2),** and enzymes **(table 25.3).** Many of the processing steps involve fermentations similar to those described in food technology, but industrial processes usually occur on a much larger scale, produce a specific compound, and involve numerous complex stages. The aim of industrial microbiology is to produce chemicals that can be purified and packaged for sale or for use in other commercial processes. Thousands of tons of organic chemicals worth several billion dollars are produced by this industry every year. To create just one of these products, an industry must determine which microbes, starting compounds, and growth conditions work best. The research and development involved requires an investment of 10 to 15 years and billions of dollars.

TABLE 25.2 Industrial Products of Microorganisms

Chemical	Microbial Source	Substrate	Applications
Pharmaceuticals			
Cephalosporins	*Cephalosporium*	Glucose	Antibacterial antibiotic, broad spectrum
Pencillins	*Penicillium chrysogenum*	Lactose	Antibacterial antibiotics, broad and narrow spectrum
Vitamin B_{12}	*Pseudomonas*	Molasses	Dietary supplement
Steroids (hydrocortisone)	*Rhizopus, Cunninghamella*	Deoxycholic acid, stigmasterol	Treatment of inflammation, allergy; hormone replacement therapy
Food additives and amino acids			
Citric acid	*Aspergillus, Candida*	Molasses	Acidifier in soft drinks; used to set jam; candy additive; fish preservative; retards discoloration of crabmeat; delays browning of sliced peaches
Xanthan	*Xanthomonas*	Glucose medium	Food stabilizer; not digested by humans
Acetic acid	*Acetobacter*	Any ethylene source, ethanol	Food acidifer; used in industrial processes
Miscellaneous			
Ethanol	*Saccharomyces*	Beet, cane, grains, wood, wastes	Additive to gasoline (gasohol)
Acetone	*Clostridium*	Molasses, starch	Solvent for lacquers, resins, rubber, fat, oil
Glycerol	Yeast	By-product of alcohol fermentation	Explosive (nitroglycerine)
Dextran	*Klebsiella, Acetobacter, Leuconostoc*	Glucose, molasses, sucrose	Polymer of glucose used as adsorbents, blood expanders, and in burn treatment; a plasma extender; used to stabilize ice cream, sugary syrup, candies

TABLE 25.3 Industrial Enzymes and Their Uses

Enzyme	Source	Application
Amylase	*Aspergillus, Bacillus, Rhizopus*	Flour supplement, desizing textiles, mash preparation, syrup manufacture, digestive aid, precooked foods, spot remover in dry cleaning
Cellulase	*Aspergillus, Trichoderma*	Denim finishing ("stone-washing"), digestive aid, increase digestibility of animal feed, degradation of wood or wood by-products
Hyaluronidase	Various bacteria	Medical use in wound cleansing, preventing surgical adhesions
Keratinase	*Streptomyces*	Hair removal from hides in leather preparation
Pectinase	*Aspergillus, Sclerotina*	Clarifies wine, vinegar, syrups, and fruit juices by degrading pectin, a gelatinous substance; used in concentrating coffee
Proteases	*Aspergillus, Bacillus, Streptomyces*	To clear and flavor rice wines, process animal feed, remove gelatin from photographic film, recover silver, tenderize meat, unravel silkworm cocoon, remove spots
Rennet	*Mucor*	To curdle milk in cheese making
Streptokinase	*Streptococcus*	Medical use in clot digestion, as a blood thinner

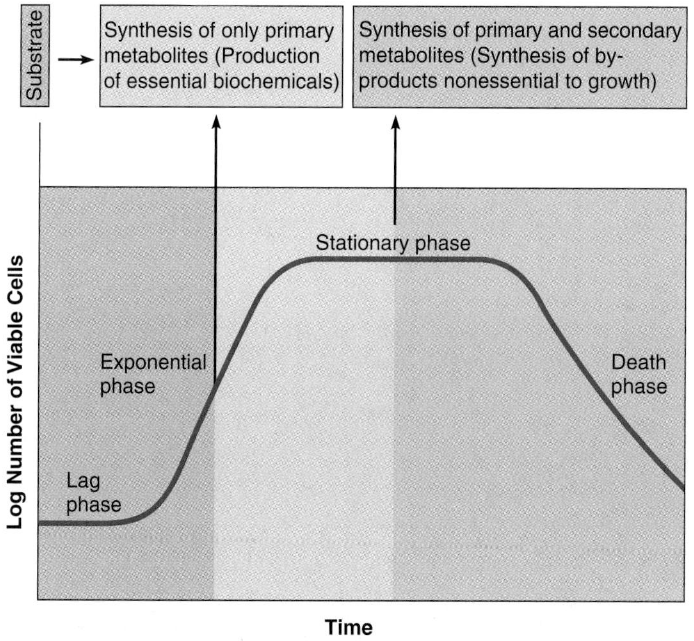

Figure 25.14 The origins of primary and secondary microbial metabolites harvested by industrial processes.

Figure 25.15 A cell culture vessel used to mass-produce pharmaceuticals.
Such elaborate systems require the highest levels of sterility and clean techniques.

The microbes used by fermentation industries are mutant strains of fungi or bacteria that selectively synthesize large amounts of various metabolic intermediates, or **metabolites.** Two basic kinds of metabolic products are harvested by industrial processes: (1) *Primary metabolites* are produced during the major metabolic pathways and are essential to the microbe's function. (2) *Secondary metabolites* are by-products of metabolism that may not be critical to the microbe's function **(figure 25.14).** In general, primary products are compounds such as amino acids and organic acids synthesized during the logarithmic phase of microbial growth, and secondary products are compounds such as vitamins, antibiotics, and steroids synthesized during the stationary phase (see chapter 7). Most strains of industrial microorganisms have been chosen for their high production of a particular primary or secondary metabolite. Certain mutated strains of yeasts and bacteria can produce 20,000 times more metabolite than a wild strain of that same microbe.

Industrial microbiologists have several tricks to increase the amount of the chosen end product. First, they can manipulate the growth environment to increase the synthesis of a metabolite. For instance, adding lactose instead of glucose as the fermentation substrate increases the production of penicillin by *Penicillium.* Another strategy is to select microbial strains that genetically lack a feedback system to regulate the formation of end products, thus encouraging mass accumulation of this product. Many syntheses occur in sequential fashion, wherein the waste products of one organism become the building blocks of the next. During these *biotransformations,* the substrate undergoes a series of slight modifications, each of which gives off a different by-product. The production of an antibiotic such as tetracycline requires several microorganisms and 72 separate metabolic steps.

From Microbial Factories to Industrial Factories

Industrial fermentations begin with microbial cells acting as living factories. When exposed to optimum conditions, they multiply in massive numbers and synthesize large volumes of a desired product. Producing appropriate levels of growth and fermentation requires cultivation of the microbes in a carefully controlled environment. This process is basically similar to culturing bacteria in a test tube of nutrient broth. It requires a sterile medium containing appropriate nutrients, protection from contamination, provisions for introduction of sterile air or total exclusion of air, and a suitable temperature and pH **(figure 25.15).**

Many commercial fermentation processes have been worked out on a small scale in a lab and then *scaled up* to a large commercial venture. An essential component for scaling up is a **fermentor,** a device in which mass cultures are grown, reactions take place, and product develops. Some fermentors are large tubes, flasks, or vats, but most industrial types are metal cylinders with built-in mechanisms for stirring, cooling, monitoring, and harvesting product **(figure 25.16).** Fermentors are made of materials that can withstand pressure and are rust-proof, nontoxic, and leakproof. They range in holding capacity from small, 5-gallon systems used in research labs to larger, 5,000- to 100,000-gallon vessels and, in some industries, to tanks of 250 million to 500 million gallons.

For optimum yield, a fermentor must duplicate the actions occurring in a tiny volume (a test tube) on a massive scale. Most microbes performing fermentations have

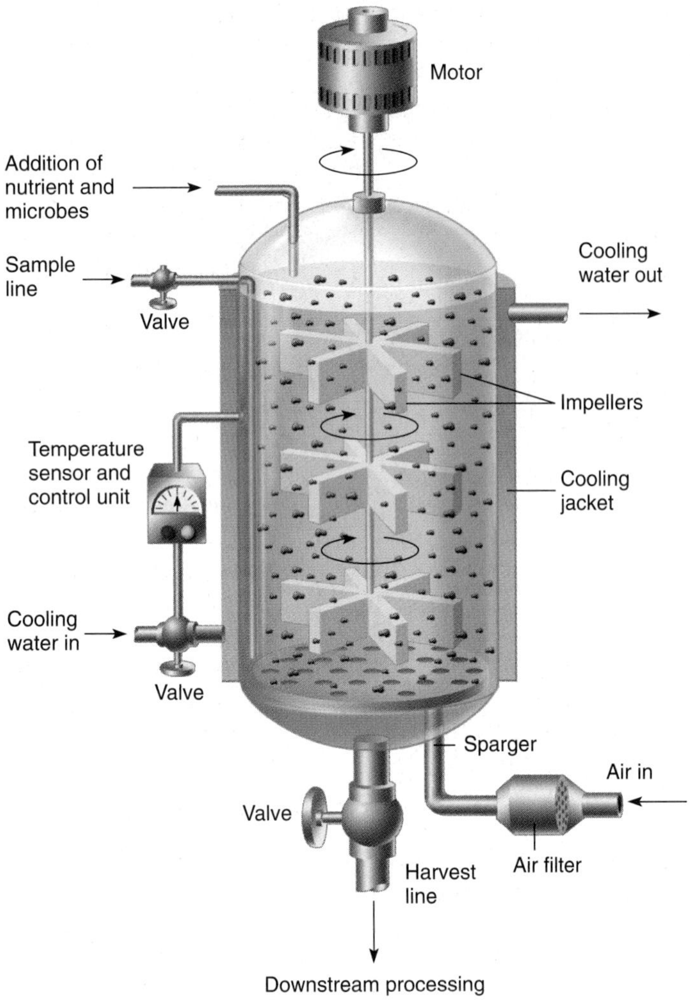

Figure 25.16 A schematic diagram of an industrial fermentor for mass culture of microorganisms. Such instruments are equipped to add nutrients and cultures; to remove product under sterile or aseptic conditions; and to aerate, stir, and cool the mixture automatically.

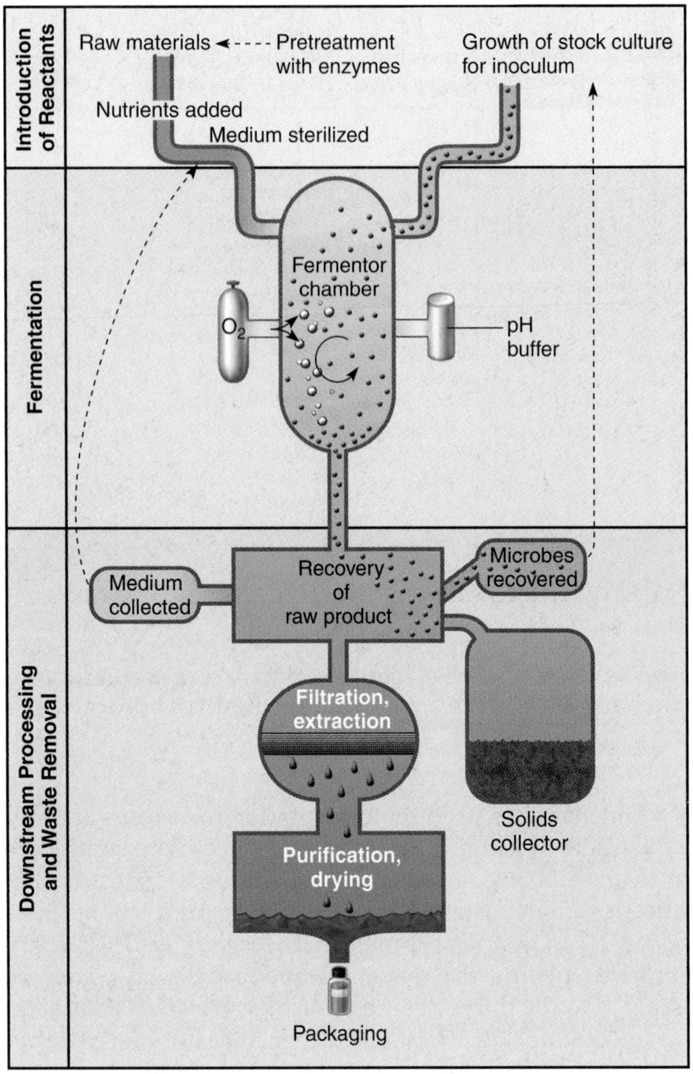

Figure 25.17 The general layout of a fermentation plant for industrial production of drugs, enzymes, fuels, vitamins, and amino acids.

an aerobic metabolism, and the large volumes make it difficult to provide adequate oxygen. Fermentors have a built-in device called a *sparger* that aerates the medium to promote aerobic growth. Paddles (*impellers*) located in the central part of the fermentor increase the contact between the microbe and the nutrients by vigorously stirring the fermentation mixture. Their action also maintains its uniformity.

Substance Production

The general steps in mass production of organic substances in a fermentor are illustrated in **figure 25.17.** These can be summarized as

1. introduction of microbes and sterile media into the reaction chamber;
2. fermentation;
3. *downstream processing* (recovery, purification, and packaging of product); and
4. removal of waste.

All phases of production must be carried out aseptically and monitored (usually by computer) for rate of flow and quality of product. The starting raw substrates include crude plant residues, molasses, sugars, fish and meat meals, and whey. Additional chemicals can be added to control pH or to increase the yield. In *batch fermentation,* the substrate is added to the system all at once and taken through a limited run until product is harvested. In *continuous feed systems,* nutrients are continuously fed into the reactor and the product is siphoned off throughout the run.

Ports in the fermentor allow the raw product and waste materials to be recovered from the reactor chamber when fermentation is complete. The raw product is recovered by settling, precipitation, centrifugation, filtration, or cell lysis. Some products come from this process ready to package, whereas others require further purification, extraction, concentration, or drying. The end product is usually in a powder, cake, granular, or liquid form that is placed in sterilized containers. The waste products can be siphoned off to be used in other processes or discarded, and the residual microbes and nutrients from the fermentation chamber can be recycled back into the system or removed for the next run.

Fermentation technology for large-scale cultivation of microbes and production of microbial products is versatile. Table 25.2 itemizes some of the major pharmaceutical substances, food additives, and solvents produced by microorganisms. Some newer technologies employ extremophiles and their enzymes to run the processes at high or low temperatures or in high-salt conditions. Hyperthermophiles have been adapted for high-temperature detergent and enzyme production. Psychrophiles are used for cold processing of reagents for molecular biology and medical tests. Halophiles are effective for processing of salted foods and dietary supplements.

Pharmaceutical Products

Health care products derived from microbial biosynthesis are antibiotics, hormones, vitamins, and vaccines. The first mass-produced antimicrobic was penicillin, which came from *Penicillium chrysogenum,* a mold first isolated from a cantaloupe in Wisconsin. The current strain of this species has gone through 40 years of selective mutation and screening to increase its yield. (The original wild *P. chrysogenum* synthesized 60 mg/ml of medium, and the latest isolate yields 85,000 mg/ml.) The semisynthetic penicillin derivatives are produced by introducing the assorted side-chain precursors to the fermentation vessel during the most appropriate phase of growth. These experiences with penicillin have provided an important model for the manufacture of other antibiotics.

Several steroid hormones used in therapy are produced industrially. Corticosteroids of the adrenal cortex, cortisone and cortisol (hydrocortisone), are invaluable for treating inflammatory and allergic disorders, and female hormones such as progesterone or estrogens are the active ingredients in birth control pills. For years, the production of these hormones was tedious and expensive because it involved purifying them from slaughterhouse animal glands or chemical syntheses. In time, it was shown that, through biotransformation, various molds could convert a precursor compound called diogenin into cortisone. By the same means, stigmasterol from soybean oil could be transformed into progesterone.

Some vaccines are also adaptable to mass production through fermentation. Vaccines for *Bordetella pertussis, Salmonella typhi, Vibrio cholerae,* and *Mycobacterium tuberculosis* are produced in large batch cultures. *Corynebacterium diphtheriae* and *Clostridium tetani* are propagated for the synthesis of their toxins, from which toxoids for the DT vaccines are prepared.

Miscellaneous Products

An exciting innovation has been the development and industrial production of natural *biopesticides* using *Bacillus thuringiensis.* During sporulation, these bacteria produce intracellular crystals that can be toxic to certain insects. When the insect ingests this endotoxin, its digestive tract breaks down and it dies, but the material is relatively nontoxic to other organisms. Commercial dusts are now on the market to suppress caterpillars, moths, and worms on various agricultural crops and trees. A strain of this bacterium is also being considered to control the mosquito vector of malaria (chapter 20) and the black fly vector of onchocerciasis (river blindness; chapter 18).

Enzymes are critical to chemical manufacturing, the agriculture and food industries, textile and paper processing, and even laundry and dry cleaning. The advantage of enzymes is that they are very specific in their activity and are readily produced and released by microbes. Microbes and their enzymes are even proving to be useful in preserving our cultural heritage **(Insight 25.3).** Mass quantities of proteases, amylases, lipases, oxidases, and cellulases are produced by fermentation technology (see table 25.3). The wave of the future appears to be custom designing enzymes to perform a specific task by altering their amino acid content. Other compounds of interest that can be mass-produced by microorganisms are amino acids, organic acids, solvents, and natural flavor compounds to be used in air fresheners and foods.

CHECKPOINT

- Industrial microbiology refers to the bulk production of any organic compound derived from microorganisms. Currently, these include antibiotics, hormones, vitamins, acids, solvents, and enzymes.

INSIGHT 25.3 Discovery

Microbes Degrade—and Repair—Ancient Works of Art

It has long been understood that microorganisms cause the deterioration of old books, statues, and paintings. The process of **biodegradation,** deterioration by living organisms (microbes or insects), is responsible for the crumbling of stone monuments and buildings all over the world, such as the Mayan temple depicted in the photo. Both bacteria and fungi are notorious for colonizing old books, paintings, stone, mortar, and concrete. This happens because microbes release chemicals that damage stone, and they can use cellulose (paper), glues (book binding), and other organic chemicals (paints, pigments) as nutrients. Watercolor and oil paintings are particularly vulnerable to microbial attack.

A newer and more encouraging finding is that microbes can actually be used to *restore* works of art. In 2004, a group of Italian scientists managed to uncover a painting that had been obscured by a technique that earlier curators had used to "preserve" the fresco when it was no longer prudent for it to hang in its building in Pisa, Italy. When the 14th-century painting *The Conversion of Saint Efisio and Battle* by Spinello Aretino was damaged in a bomb that fell during World War II, technicians used an animal-based glue to apply gauze to the front of it and lifted it off of the wall onto the gauze matrix. They then applied a canvas to the back of the painting, attached it to a supporting sheet, and stored it.

The idea behind that strategy in the 1940s was that the gauze on the front of the painting could later be removed by using solvents that would dissolve the glue. But over time, the glue formed complexes with other chemicals in the painting and became resistant to all known solvents. After these failed attempts, the scientists determined the chemical structure of the glue (see illustration a). Luckily, the glue was purely organic, and the paints used by the artist were purely inorganic. The scientists decided to apply a paste of whole bacteria that contained a mix of enzymes that they predicted would dissolve the glue but not the paint. Cultures of *Pseudomonas stutzeri* (b) were applied to the painting on saturated cotton wool. The scientists also added an extra protease, which served to "clean up" the organic residues that were left after the cleanup process. It was successful! The glue dissolved and the gauze was removed, revealing the painting underneath (c).

Finally, after centuries of only damaging precious artworks, microbes, with a little help from humans, are repairing them.

(a)

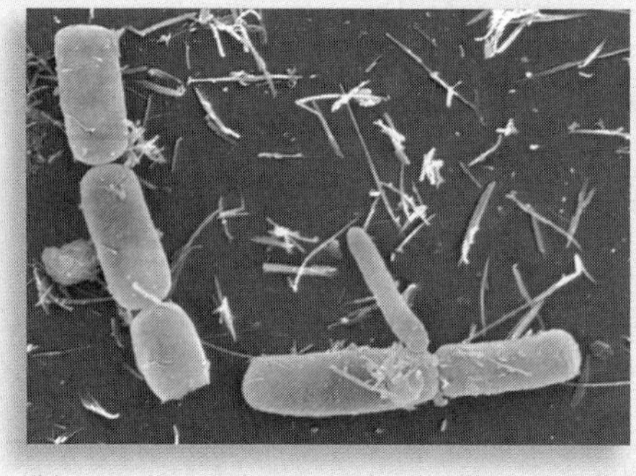

(b)

(c)

Chapter Summary with Key Terms

25.1 Applied Microbiology and Biotechnology
Biotechnology is the practical application of microbiology in the manufacture of food, industrial chemicals, drugs, and other products. Many of these processes use mass, controlled microbial **fermentations** and bioengineered microorganisms.

A. Microorganisms in Water and Wastewater Treatment
1. Drinking water is rendered safe by a purification process that involves storage, sedimentation, settling, aeration, filtration, and disinfection.
2. Sewage or used wastewater can be processed to remove solid matter, dangerous chemicals, and microorganisms.
3. Microbes biodegrade the waste material or sludge. Solid wastes are further processed in **anaerobic digesters.**

25.2 Microorganisms and Food

A. Microbes and humans compete for the rich nutrients in food. Many microbes are present on food as harmless contaminants; some are used to create flavors and nutrients; others may produce unfavorable reactions.
1. The beneficial effects of microorganisms on food are based mainly on their ability to ferment or chemically change it to alter flavor, taste, or texture. Microbes can also serve as food themselves.
2. The detrimental effects of microorganisms are their ability to cause food-borne illness and to spoil food, making it unsuitable for consumption.
3. Microbes in or on food may result in no consequences at all. In that case, their effect on food is considered neutral.

B. Fermentations in foods: Microbes can impart desirable aroma, flavor, or texture to foods. Bread, alcoholic beverages, some vegetables, and some dairy products are infused with pure microbial strains to yield the necessary fermentation products.
1. Baker's yeast, *Saccharomyces cerevisiae,* is used to **leaven** bread dough by giving off CO_2.
2. Beer making involves the following steps: Barley is sprouted **(malted)** to generate digestive enzymes, transformed to **mash,** and heated with **hops** to produce a **wort.** Wort is fermented by yeast and **lagered** in large tanks before it is carbonated and packaged.
3. Wine is started by fermentation of **must** (fruit juices). Whiskey, vodka, brandy, and other alcoholic beverages are distilled to increase their alcohol content.
4. Vegetable products, including sauerkraut, pickles, and soybean derivatives, can be pickled in the presence of salt or sugar.
5. Vinegar is produced by fermenting plant juices first to alcohol and then to acetaldehyde and acetic acid.
6. Most dairy products are produced by microbes acting on nutrients in milk. In cheese production, milk proteins are coagulated to a solid **curd** that separates from the watery **whey.**
7. Mass-cultured microbes such as yeasts, molds, and bacteria can serve as food. Single-cell protein, pruteen, and mycoprotein are currently added to animal feeds.

C. Food-borne disease: Some microbes cause spoilage and food poisoning.

D. Precautions for food:
1. Microbial growth that leads to spoilage and food poisoning can be avoided by high temperature and pressure treatment (canning) and **pasteurization** for disinfecting milk and other heat-sensitive beverages.
2. Refrigeration and freezing inhibit microbial growth.
3. Irradiation sterilizes or disinfects foods for longer-term storage.
4. Alternative preservation methods include additives (salt, nitrites), treatment with ethylene oxide gas, and drying.

25.3 General Concepts in Industrial Microbiology

A. Industrial microbiology involves the large-scale commercial production of organic compounds such as antibiotics, vitamins, amino acids, enzymes, and hormones using specific microbes in carefully controlled fermentation settings.

B. Microbes are chosen for their production of a desired **metabolite;** several different species can be used to biotransform raw materials in a stepwise series of metabolic reactions.

C. Fermentations are conducted in massive culture devices called **fermentors** that have special mechanisms for adding nutrients, stirring, oxygenating, altering pH, cooling, monitoring, and harvesting product.

Multiple-Choice and True-False Questions

Multiple-Choice Questions. Select the correct answer from the answers provided.

1. Drinking water utilities monitor their production system for the occurrence of
   a. methanogens
   b. coliform bacteria
   c. nematodes
   d. yeasts

2. Milk is usually pasteurized by
   a. the high-temperature short-time method
   b. ultrapasteurization
   c. batch method
   d. electrical currents

3. Substances given off by yeasts during fermentation are
   a. alcohol
   b. carbon dioxide
   c. organic acids
   d. all of these
4. Which of the following is added to facilitate milk curdling during cheese making?
   a. lactic acid
   b. salt
   c. *Lactobacillus*
   d. rennin
5. Secondary metabolites of microbes are formed during the ____ phase of growth.
   a. exponential
   b. stationary
   c. trophophase
   d. idiophase

True-False Questions. If the statement is true, leave as is. If it is false, correct it by rewriting the sentence.

6. Raw sewage is still being dumped into the aquatic environment in many places around the world.
7. Food products should always be kept completely free of microorganisms.
8. Alcoholic beverages are produced by the fermentation of sugar to ethanol and carbon dioxide.
9. The incidence of many food-borne illnesses has been declining for some years now.
10. Refrigerating food prevents the growth of all bacteria.

Writing to Learn

These questions are suggested as a *writing-to-learn* experience. For each question, compose a one- or two-paragraph answer that includes the factual information needed to completely address the question.

1. a. Draw a diagram of the flow of water in a water utility plant.
   b. Describe the three phases of sewage treatment.
   c. What is activated sludge?
2. a. Explain the meaning of fermentation from the standpoint of industrial microbiology.
   b. Describe five types of fermentations.
3. When are microbes on food harmless?
4. a. Which microbes are used as starter cultures in bread, beer, wine, cheeses, and sauerkraut?
   b. Outline the steps in beer making.
   c. List the steps in wine making.
   d. What are curds and whey, and what causes them?
5. a. Describe the aims of industrial microbiology.
   b. Differentiate between primary and secondary metabolites.
   c. Describe a fermentor.
   d. How is it scaled up for industrial use?
   e. What are specific examples of products produced by these processes?

Concept Mapping

Appendix D provides guidance for working with concept maps.

1. Construct your own concept map using the following words as the *concepts*. Supply the linking words between each pair of concepts. You may add additional concepts if desired.

primary metabolites
secondary metabolites
fermentation
microbes
downstream processing
substrate
pH
biotransformations

Critical Thinking Questions

Critical thinking is the ability to reason and solve problems using facts and concepts. These questions can be approached from a number of angles, and in most cases, they do not have a single correct answer.

1. Every year, supposedly safe municipal water supplies cause outbreaks of enteric illness.
   a. How in the course of water analysis and treatment might these pathogens be missed?
   b. What kinds of microbes are they most likely to be?
   c. Why is there less tolerance for a fecal coliform in drinking or recreational water than other bacteria?

2. Describe four food-preparation and food-maintenance practices in your own kitchen that could expose people to food poisoning and explain how to prevent them.
   a. What is the purpose of boiling the wort in beer preparation?
   b. What are hops used for?
   c. If fermentation of sugars to produce alcohol in wine is anaerobic, why do winemakers make sure that the early phase of yeast growth is aerobic?
3. Predict the differences in the outcome of raw milk that has been incubated for 48 hours versus pasteurized milk that has been incubated for the same length of time.
4. Explain the ways that co-metabolism and biotransformations of microorganisms are harnessed in industrial microbiology.
5. Review chapter 10 and describe several ways that recombinant DNA technology can be used in biotechnology processes.

Visual Understanding Questions

1. **From chapter 7, figure 7.4.** What method of food preservation exploits the principles that are illustrated here?

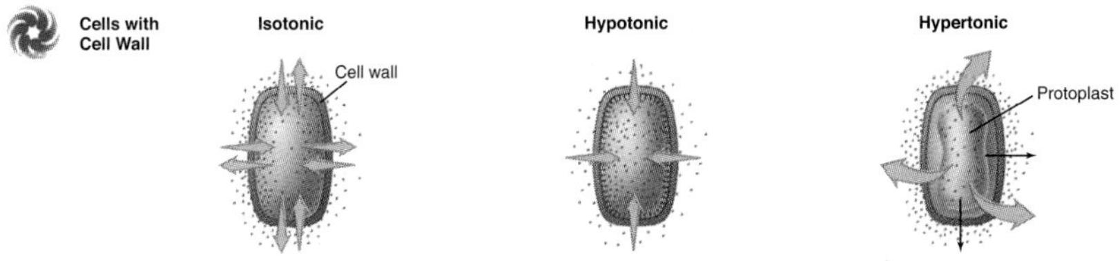

2. **From Insight 25.2, Illustration a.** This is a plate with blood agar on the top and MacConkey agar on the bottom. It has been inoculated with samples from a wooden cutting board that had been exposed to a raw chicken carcass. Knowing what you know about the properties of these two agars, say as much as you can about the bacteria growing on them.

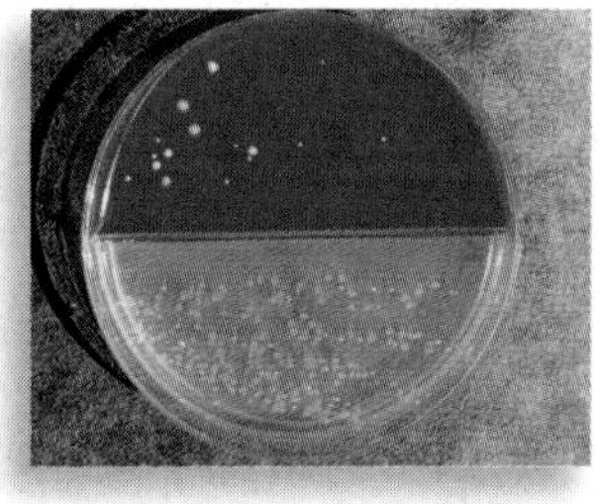

Internet Search Topics

Go to: www.aris.mhhe.com, and click on "microbiology" and then this textbook's author/title. Go to chapter 25, access the URLs listed under Internet Search Topics, and research the following:

1. Research the subject of bioremediation. What sorts of toxic substances are being cleaned up and what types of microbes are involved?
2. Go to websites to research wine making. How are different types of wines made and how do they vary in color, flavor, alcohol content, and other features?
3. Log on to the URL of your city's water company and read its mission statement and how it achieves its goals. Is your water treated? Does your water carry certain odors and tastes?

APPENDIX A

Exponents

Dealing with concepts such as microbial growth often requires working with numbers in the billions, trillions, and even greater. A mathematical shorthand for expressing such numbers is with exponents. The exponent of a number indicates how many times (designated by a superscript) that number is multiplied by itself. These exponents are also called common *logarithms*, or logs. The following chart, based on multiples of 10, summarizes this system.

Exponential Notation for Base 10

Number	Quantity	Exponential Notation*	Number Arrived at By:	One Followed By:
1	One	10^0	Numbers raised to zero power are equal to one	No zeros
10	Ten	10^{1**}	10×1	One zero
100	Hundred	10^2	10×10	Two zeros
1,000	Thousand	10^3	$10 \times 10 \times 10$	Three zeros
10,000	Ten thousand	10^4	$10 \times 10 \times 10 \times 10$	Four zeros
100,000	Hundred thousand	10^5	$10 \times 10 \times 10 \times 10 \times 10$	Five zeros
1,000,000	Million	10^6	10 times itself 6 times	Six zeros
1,000,000,000	Billion	10^9	10 times itself 9 times	Nine zeros
1,000,000,000,000	Trillion	10^{12}	10 times itself 12 times	Twelve zeros
1,000,000,000,000,000	Quadrillion	10^{15}	10 times itself 15 times	Fifteen zeros
1,000,000,000,000,000,000	Quintillion	10^{18}	10 times itself 18 times	Eighteen zeros

Other large numbers are sextillion (10^{21}), septillion (10^{24}), and octillion (10^{27}).

*The proper way to say the numbers in this column is 10 raised to the *n*th power, where *n* is the exponent. The numbers in this column can also be represented as 1×10^n, but for brevity, the $1 \times$ can be omitted.

**The exponent 1 is usually omitted.

Converting Numbers to Exponent Form

As the chart shows, using exponents to express numbers can be very economical. When simple multiples of 10 are used, the exponent is always equal to the number of zeros that follow the 1, but this rule will not work with numbers that are more varied. Other large whole numbers can be converted to exponent form by the following operation: First, move the decimal (which we assume to be at the end of the number) to the left until it sits just behind the first number in the series (example: 3568. = 3.568). Then count the number of spaces (digits) the decimal has moved; that number will be the exponent. (The decimal has moved from 8. to 3., or 3 spaces.) In final notation, the converted number is multiplied by 10 with its appropriate exponent: 3568 is now 3.568×10^3.

Rounding Off Numbers

The notation in the previous example has not actually been shortened, but it can be reduced further by rounding off the decimal fraction to the nearest thousandth (three digits), hundredth (two digits), or tenth (one digit). To round off a number, drop its last digit and either increase the one next to it or leave it as it is. If the number dropped is 5, 6, 7, 8, or 9, the subsequent digit is increased by one (rounded up); if it is 0, 1, 2, 3, or 4, the subsequent digit remains as is. Using the example of 3.528, removing the 8 rounds off the 2 to a 3 and produces 3.53 (two digits). If further rounding is desired, the same rule of thumb applies, and the number becomes 3.5 (one digit). Other examples of exponential conversions follow.

Number	Is the Same As	Rounded Off, Placed in Exponent Form
16,825.	$1.6825 \times 10 \times 10 \times 10 \times 10$	1.7×10^4
957,654.	$9.57654 \times 10 \times 10 \times 10 \times 10 \times 10$	9.58×10^5
2,855,000.	$2.855000 \times 10 \times 10 \times 10 \times 10 \times 10 \times 10$	2.86×10^6

Negative Exponents

The numbers we have been using so far are greater than 1 and are represented by positive exponents. But the correct notation for numbers less than 1 involves negative exponents (10 raised to a negative power, or 10^{-n}). A negative exponent says that the number has been divided by a certain power of 10 (10, 100, 1,000). This usage is handy when working with concepts such as pH that are based on very small numbers otherwise needing to be represented by large decimal fractions—for example, 0.003528. Converting this and other such numbers to exponential notation is basically similar to converting positive numbers, except that you work from left to right and the exponent is negative. Using the example of 0.003528, first convert the number to a whole integer followed by a decimal fraction and keep track of the number of spaces the decimal point moves (example: 0.003528 = 3.528). The decimal has moved three spaces from its original position, so the finished product is 3.528×10^{-3}. Other examples follow.

Number	Is the Same As	Rounded Off, Expressed with Exponents
0.0005923	$\frac{5.923}{10 \times 10 \times 10 \times 10}$	5.92×10^{-4}
0.00007295	$\frac{7.295}{10 \times 10 \times 10 \times 10 \times 10}$	7.3×10^{-5}

APPENDIX **B**

Significant Events in Microbiology

Date	Discovery/People Involved
1546	Italian physician Girolamo Fracastoro suggests that invisible organisms may be involved in disease.
1660	Englishman Robert Hooke explores various living and nonliving matter with a compound microscope that uses reflected light.
1668	Francesco Redi, an Italian naturalist, conducts experiments that demonstrate the fallacies in the spontaneous generation theory.
1676	Antonie van Leeuwenhoek, a Dutch linen merchant, uses a simple microscope of his own design to observe bacteria and protozoa.
1796	English surgeon Edward Jenner introduces a vaccination for smallpox.
1838	Phillipe Ricord, a French physician, inoculates 2,500 human subjects to demonstrate that syphilis and gonorrhea are two separate diseases.
1847–1850	The Hungarian physician Ignaz Semmelweis substantiates his theory that childbed fever is a contagious disease transmitted to women by their physicians during childbirth.
1853–1854	John Snow, a London physician, demonstrates the epidemic spread of cholera through a water supply contaminated with human sewage.
1857	French bacteriologist Louis Pasteur shows that fermentations are due to microorganisms and originates the process now known as pasteurization.
1861	Louis Pasteur completes the definitive experiments that finally lay to rest the theory of spontaneous generation.
1867	The English surgeon Joseph Lister publishes the first work on antiseptic surgery, beginning the trend toward modern aseptic techniques in medicine.
1876–1877	German bacteriologist Robert Koch* studies anthrax in cattle and implicates the bacterium *Bacillus anthracis* as its causative agent.
1881	Pasteur develops a vaccine for anthrax in animals.
	Koch introduces the use of pure culture techniques for handling bacteria in the laboratory.
1882	Koch identifies the causative agent of tuberculosis.
1884	Koch outlines his postulates.
	Elie Metchnikoff,* a Russian zoologist, lays groundwork for the science of immunology by discovering phagocytic cells.
	The Danish physician Hans Christian Gram devises the Gram stain technique for differentiating bacteria.
1885	Pasteur develops a special vaccine for rabies.
1892	A Russian, D. Ivanovski, is the first to isolate a virus (the tobacco mosaic virus) and show that it could be transmitted in a cell-free filtrate.
1898	R. Ross* and G. Grassi demonstrate that malaria is transmitted by the bite of female mosquitoes.
1899	Dutch microbiologist Martinus Beijerinck further elucidates the viral agent of tobacco mosaic disease and postulates that viruses have many of the properties of living cells and that they reproduce within cells.
1903	American pathologist James Wright and others demonstrate the presence of antibodies in the blood of immunized animals.
1905	Syphilis is shown to be caused by *Treponema pallidum*, through the work of German bacteriologists Fritz Schaudinn and E. Hoffman.
1908	The German Paul Ehrlich* becomes the pioneer of modern chemotherapy by developing salvarsan to treat syphilis.
1910	An American pathologist, Francis Rous,* discovers viruses that can induce cancer.
1928	Frederick Griffith lays the foundation for modern molecular genetics by his discovery of transformation in bacteria.
1929	A Scottish bacteriologist, Alexander Fleming,* discovers and describes the properties of the first antibiotic, penicillin.
1933–1938	Germans Ernst Ruska* and B. von Borries develop the first electron microscope.
1935	Gerhard Domagk,* a German physician, discovers the first sulfa drug and paves the way for the era of antimicrobic chemotherapy.
1941	Australian Howard Florey* and Englishman Ernst Chain* develop commercial methods for producing penicillin; this first antibiotic is tested and put into widespread use.
1944	Oswald Avery, Colin MacLeod, and Maclyn McCarty show that DNA is the genetic material.
	Joshua Lederberg* and E. L. Tatum* discover conjugation in bacteria.
	The Russian Selman Waksman* and his colleagues discover the antibiotic streptomycin.
1953	James Watson,* Francis Crick,* Rosalind Franklin, and Maurice Wilkins* determine the structure of DNA.
1954	Jonas Salk develops the first polio vaccine.
1959–1960	Gerald Edelman* and Rodney Porter* determine the structure of antibodies.
1972	Paul Berg* develops the first recombinant DNA in a test tube.
1973	Herb Boyer and Stanley Cohen clone the first DNA using plasmids.
1982	Development of first hepatitis B vaccine using virus isolated from human blood.
1983	Isolation and characterization of human immunodeficiency virus (HIV) by Luc Montagnier of France and Robert Gallo of the United States.
	The polymerase chain reaction is invented by Kary Mullis.*
	First release of recombinant strain of *Pseudomonas* to prevent frost formation on strawberry plants.
1989	Cancer-causing genes called oncogenes are characterized by J. Michael Bishop, Robert Huber, Hartmut Michel, and Harold Varmus.
1990	First clinical trials in gene therapy testing.
	Vaccine for *Haemophilus influenzae*, a cause of meningitis, is introduced.
1994	Human breast cancer gene isolated.
1995	First bacterial genome fully sequenced, for *Haemophilus influenzae*.
2000	A rough version of the human genome is mapped.
2001	Mailed anthrax spores cause major bioterrorism event.
2003	New roles for small nuclear RNAs discovered.
2006	New vaccine for a persistent microbe, human papillomavirus (HPV), is introduced.

*These scientists were awarded Nobel prizes for their contributions to the field.

Answers to Multiple-Choice and Selected True-False and Matching Questions

Chapter 1

1. d
2. c
3. d
4. d
5. a
6. c
7. c
8. c
9. d
10. Ist col: 3, 7, 4, 2 2nd col: 8, 5, 6, 1
11. c
12. F: Organisms in the same family are more closely related than those in the same order.
14. T
16. T

Chapter 2

1. c
2. c
3. b
4. d
5. c
6. a
7. c
8. b
9. b
10. d
11. b
12. c
13. a
14. d
15. b
16. T
18. F: A compound is called "organic" if it contains both carbon and hydrogen bonded together in various combinations.
20. F: Membranes are mainly composed of macromolecules called phospholipids.

Chapter 3

1. b
2. c
3. b
4. b
5. c
6. c
7. a
8. b
9. c
10. abf, df, abf, ef, af, bef, ac, bef
11. d
12. b
13. F: Agar is not easily decomposed by microorganisms (gelatin can be).
15. F: The factor that most limits the clarity of an image in a microscope is the *resolution*.
17. F: The best stain to use to visualize a microorganism with a large capsule is a negative stain.

Chapter 4

1. d
2. a
3. c
4. a
5. c
6. b
7. d
8. b
9. d
10. b
11. c
12. F: One major difference in the envelope structure between gram-positive bacteria and gram-negative bacteria is the presence or absence of an outer membrane.
14. T
16. T

Chapter 5

1. b
2. d
3. d
4. b
5. b
6. d
7. d
8. b
9. c
10. Matching: b, e, c, h, g, j, i, d, a, f
12. F: Ribosomes are attached to the rough endoplasmic reticulum.
14. F: Both the trophozoite and the cyst stages of protozoans can be infective.
15. F: In humans, fungi can infect skin, mucous membranes, lungs, and other areas.
16. F: Fungi generally derive nutrients by digesting organic substrates.

Chapter 6

1. c
2. d
3. d
4. a
5. a
6. b
7. b
8. d
9. a
13. F: A viral capsid is composed of subunits called capsomeres.
15. F: The nucleic acid of an animal virus enters the host cell through endocytosis (engulfment) or through fusion.

Chapter 7

1. a
2. a
3. c
4. b
5. a
6. a
7. b
8. c
9. c
10. b
11. F: Active transport of a substance across a membrane requires energy.
13. F: The time required for a cell to undergo binary fission is called the generation time.
15. F: A facultative anaerobe can grow with or without oxygen.

Chapter 8

1. a
2. d
3. d
4. b
5. b
6. b
7. d
8. b
9. c
10. c
11. F: Anabolism is another name for biosynthesis.
13. F: Exergonic reactions release potential energy.
15. F: Exoenzymes are produced inside a cell then released to the outside.

Chapter 9

1. b
2. b
3. b
4. c
5. b
6. a
7. b
8. d
9. b
10. F: The DNA base pairs are held together primarily by hydrogen bonds.
12. T
14. T

Chapter 10

1. c
2. d
3. a
4. b
5. c
6. c
7. d
8. Matching: h, c, f, a, g, b, e, d
9. F: The synthetic unit of the polymerase chain reaction is the amplicon.
11. F: A DNA fragment with 450 bp will migrate farther toward the positive pole (away from the origin) than one with 2,500 bp.
13. F: Plasmids and bacteriophages are commonly used as cloning vectors.

Chapter 11

1. c
2. a
3. c
4. b
5. d
6. b
7. d
8. c
9. d
10. a
11. b
12. c
13. T
15. F: Ionizing radiation dislodges electrons from atoms.
17. F: Prions are highly resistant to denaturation by heat.

Chapter 12

1. b
2. c
3. b
4. d
5. b
6. a
7. c
8. c
9. b
10. F: Most antiviral agents work by blocking an essential viral activity.
12. T

Chapter 13

1. a
2. b
3. d
4. c
5. d
6. b
7. c
8. d
9. a
10. F: The presence of a few bacteria in the blood is called bacteremia.
12. F: A nosocomial infection is one that is acquired in a hospital or medical facility.

Chapter 14

1. b
2. b
3. d
4. b
5. b
6. c
7. d
8. c
9. d
10. c
11. F: The liquid component of unclotted blood is called plasma.
13. T
15. F: The immune system uses markers on the surface of cells to distinguish self from nonself.

Chapter 15

1. a
2. c
3. a
4. c
5. c
6. a
7. e
8. c

9. d
11. F: Antibodies are secreted by plasma cells.
13. F: IgA antibodies are found in body secretions.
15. T

Chapter 16

1. d
2. d
3. c
4. c
5. b
6. a
7. d
8. d
9. d
10. d
11. T
13. F: Contact dermatitis can be caused by chemicals absorbed through the skin.

Chapter 17

1. c, a, b, c, a
2. c
3. b
4. c
5. a
6. a
7. d
8. F: The tuberculin skin test is an example of an *in vivo* serological test.
10. T
12. F: Micro organisms that are grown from clinical samples should be evaluated to determine their clinical significance.

Chapter 18

1. b
2. a
3. e
4. e
5. c
6. b
7. d
8. d
9. b
10. F: The enzyme coagulase is associated with pathogenic strains of *Staphylococcus aureus*.
12. T
14. T

Chapter 19

1. d
2. a
3. c
4. d
5. b
6. d
7. b
8. a
9. b
10. a
11. F: *Toxoplasma gondii* is a protozoan.
13. T

Chapter 20

1. c
2. d
3. b
4. b
5. a
6. b
7. d
8. a
9. b
10. d
11. T
13. F: Lyme disease is caused by *Borrelia burgdorferi*.
15. F: A distinctive bull's-eye rash is associated with Lyme disease.

Chapter 21

1. d
2. d
3. b
4. b
5. d
6. c
7. c
8. c
9. a
10. T
12. BC6 vaccine is used in other countries to prevent TB.

Chapter 22

1. b
2. d
3. c
4. d
5. c
6. c
7. a
8. b
9. b
10. F: Humans are the only natural host for the mumps virus.
12. T

Chapter 23

1. a
2. b
3. d
4. a
5. b
6. a
7. c
8. c
9. d
10. T
12. T

Chapter 24

1. c
2. b
3. b
4. d
5. c
6. d
7. c
8. F: Pure cultures are very rare in the biosphere.
10. T

Chapter 25

1. b
2. c
3. a
4. b
5. b
6. T
8. T
10. F: Refrigerating food prevents the growth of many bacteria, but some pathogens, such as *Listeria* and *Salmonella*, can continue to grow at low temperatures.

An Introduction to Concept Mapping

Concept maps are visual tools for presenting and organizing what you have learned. They can take the place of an outline, though for most people they contain much more meaning and can illustrate connections and interconnections in ways that ordinary outlines cannot. They are also very flexible. If you are creating a concept map, there is a nearly infinite number of ways that they can be put together and still be "correct." Concept maps are also a way to incorporate and exploit your own creative impulses, so that you are not stuck inside a rigid framework but can express your understanding of concepts and their connections in ways that make sense to you.

This is an example of a relatively large concept map:

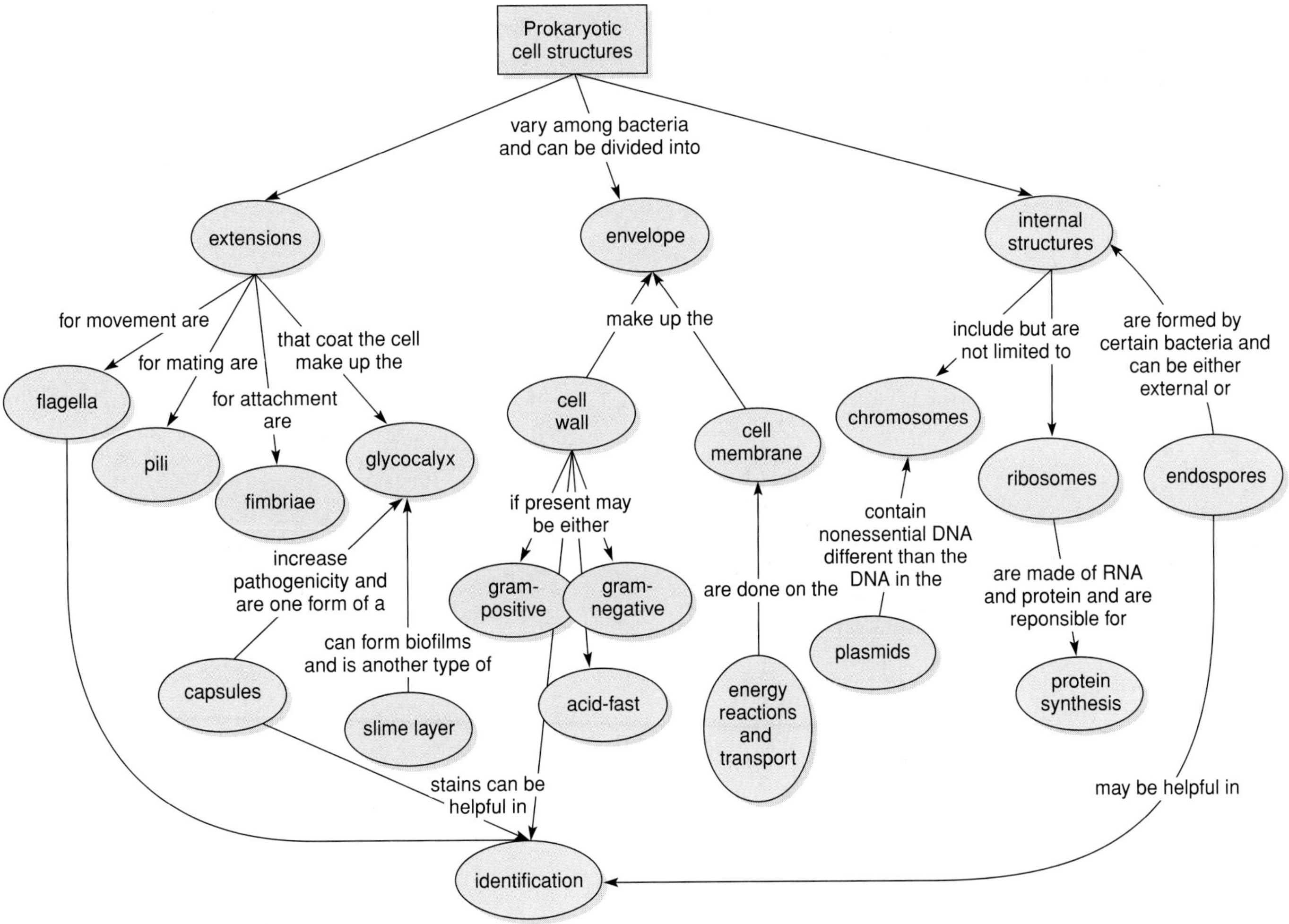

There is a wide variety of different ways to work with concept maps, such as using them as an introductory overview of material or using them as an evaluation tool. There are even software programs that enable concept mappers to create elaborate maps, complete with sound bytes and photos. Some of these will even convert an outline into a concept map for you. In the end-of-chapter materials in this book, we use only three different methods, all of them fairly simple. These three are explained and illustrated here.

All concept maps are made of two basic components:

1. Boxes or circles, each containing a single *concept*, which is most often a noun. The boxes are arranged on the page in vertical, horizontal, or diagonal rows or arrangements. They may also be arranged in a more free-form manner.
2. Connecting lines that join each concept box to at least one other box. Each connecting line has a word or a phrase associated with it—a linking word. These words/phrases are almost never nouns—but are verbs (like "requires") or adjectives or adverbs (like "underneath").

In the end, a picture is created that maps what you know about a subject. It illustrates which concepts are bigger and which are details. It illustrates that multiple concepts may be connected. Experts say that concept maps almost always lead us to conclude that all concepts in a subject can be connected in some way. This is true! And nowhere is it truer than in biology. The trick is to get used to finding the right connecting word to show how two concepts are, indeed, related. When you succeed, you will know the material in a deeper way than is possible by simply answering a single question or even a series of questions.

The first kind of concept map used in this book is the "fill-in-the-blank" version. In these concept maps, you are provided with all the boxes and most of the concepts in the boxes. Some boxes may be blank for you to fill in with the appropriate concept. You will do this by looking at the concepts close to the box and examining the connecting word. In these maps, you will also encounter blanks for linking words/phrases. Sometimes all the blanks will be filled in, but there will be no connecting lines or phrases and you will have to supply these. In a few of these maps, you may be asked to draw the linking lines themselves.

This is an exercise most like answering a simple question. In the example below, for instance, say to yourself "Enzymes are __________ by pH and temperature." You might ask yourself, "What relation do pH and temperature have with enzymes?" Either way, you would probably end up with a linking phrase like *are affected by* or *can be regulated by*. There is some variation in what is a correct answer, but not wide variation.

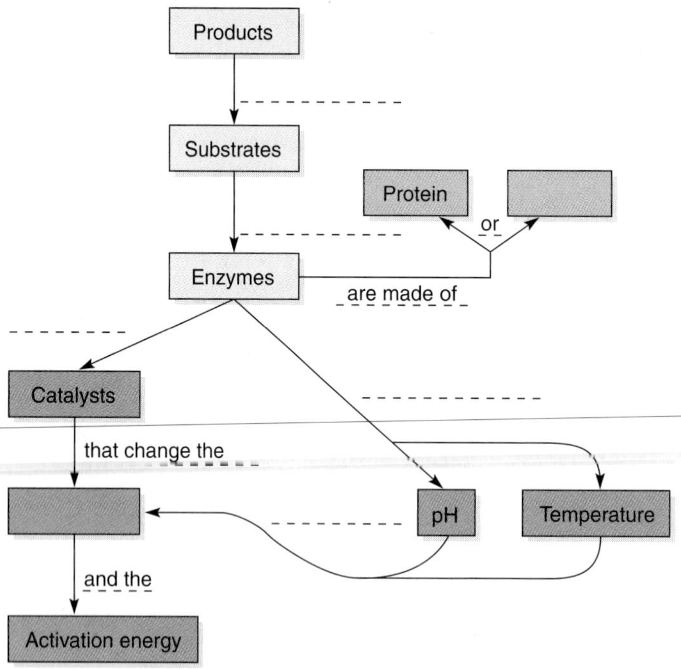

Acknowledgment: Pat Johnson, Palm Beach Community College, supplied information and a concept map for this Appendix.

The second kind of concept map you will see is one in which you will be provided a list of words to be used as concepts. You will be asked to draw the boxes and put the words in them in some way that makes sense. Here, there will be a lot of variability based on your view of how the concepts might relate to each other. After you put the concepts in your own boxes, you will need to add linking words/phrases. By the time you have drawn your boxes and added the concepts, you will have many ideas about what kind of linkers you want.

The last type of concept map in this book is the "freestyle" version. You will simply be asked to choose 6 to 10 key words from the Chapter Summary and create a map—complete with linking words.

Many students report that their first experiences with concept mapping can be frustrating. But when they have invested some time in their first few concept maps, many of them find they can never "go back" to organizing information in linear ways. Maps can make the time fly when you're studying. And creating concept maps with a partner or a group is also a great way to review material in a meaningful way. Give concept maps a try. Let your creative side show!

Glossary

A

A-B toxin A class of bacterial exotoxin consisting of two components: a binding (B) component and an active (A) or enzymatic component.
abiogenesis The belief in spontaneous generation as a source of life.
abiotic Nonliving factors such as soil, water, temperature, and light that are studied when looking at an ecosystem.
ABO blood group system Developed by Karl Landsteiner in 1904; the identification of different blood groups based on differing isoantigen markers characteristic of each blood type.
abscess An inflamed, fibrous lesion enclosing a core of pus.
abyssal zone The deepest region of the ocean; a sunless, high-pressure, cold, anaerobic habitat.
acellular vaccine A vaccine preparation that contains specific antigens such as the capsule or toxin from a pathogen and not the whole microbe. Acellular (without a cell).
acid-fast A term referring to the property of mycobacteria to retain carbol fuchsin even in the presence of acid alcohol. The staining procedure is used to diagnose tuberculosis.
acidic A solution with a pH value below 7 on the pH scale.
acidic fermentation An anaerobic degradation of pyruvic acid that results in organic acid production.
acquired immunodeficiency syndrome See *AIDS*.
actin Long filaments of protein arranged like ribbons under the cell membrane of some bacteria; contribute to cell shape.
actin cytoskeleton A scaffoldlike structure made of protein that lies under the cytoplasmic membrane of some bacteria.
actinomycetes A group of filamentous, funguslike bacteria.
active immunity Immunity acquired through direct stimulation of the immune system by antigen.
active site The specific region on an apoenzyme that binds substrate. The site for reaction catalysis.
active transport Nutrient transport method that requires carrier proteins in the membranes of the living cells and the expenditure of energy.
acute Characterized by rapid onset and short duration.
acyclovir A synthetic purine analog that blocks DNA synthesis in certain viruses, particularly the herpes simplex viruses.
adenine (A) One of the nitrogen bases found in DNA and RNA, with a purine form.
adenosine deaminase (ADA) deficiency An immunodeficiency disorder and one type of SCIDS that is caused by an inborn error in the metabolism of adenine. The accumulation of adenine destroys both B and T lymphocytes.
adenosine triphosphate (ATP) A nucleotide that is the primary source of energy to cells.
adhesion The process by which microbes gain a more stable foothold at the portal of entry; often involves a specific interaction between the molecules on the microbial surface and the receptors on the host cell.
adjuvant In immunology, a chemical vehicle that enhances antigenicity, presumably by prolonging antigen retention at the injection site.
adsorption A process of adhering one molecule onto the surface of another molecule.
aerobe A microorganism that lives and grows in the presence of free gaseous oxygen (O_2).
aerobic respiration Respiration in which the final electron acceptor in the electron transport chain is oxygen (O_2).
aerosols Suspensions of fine dust or moisture particles in the air that contain live pathogens.
aerotolerant The state of not utilizing oxygen but not being harmed by it.
aflatoxin From *Aspergillus flavus* toxin, a mycotoxin that typically poisons moldy animal feed and can cause liver cancer in humans and other animals.
agammaglobulinemia Also called hypogammaglobulinemia. The absence of or severely reduced levels of antibodies in serum.
agar A polysaccharide found in seaweed and commonly used to prepare solid culture media.
agglutination The aggregation by antibodies of suspended cells or similar-size particles (agglutinogens) into clumps that settle.
agranulocyte One form of leukocyte (white blood cells) having globular, nonlobed nuclei and lacking prominent cytoplasmic granules.
AIDS Acquired immunodeficiency syndrome. The complex of signs and symptoms characteristic of the late phase of human immunodeficiency virus (HIV) infection.
alcoholic fermentation An anaerobic degradation of pyruvic acid that results in alcohol production.
algae Photosynthetic, plantlike organisms that generally lack the complex structure of plants; they may be single-celled or multicellular and inhabit diverse habitats such as marine and freshwater environments, glaciers, and hot springs.
allele A gene that occupies the same location as other alternative (allelic) genes on paired chromosomes.
allergen A substance that provokes an allergic response.
allergy The altered, usually exaggerated, immune response to an allergen. Also called hypersensitivity.
alloantigen An antigen that is present in some but not all members of the same species.
allograft Relatively compatible tissue exchange between nonidentical members of the same species. Also called homograft.
allosteric Pertaining to the altered activity of an enzyme due to the binding of a molecule to a region other than the enzyme's active site.
amantadine Antiviral agent used to treat influenza; prevents fusion and uncoating of virus.
Ames test A method for detecting mutagenic and potentially carcinogenic agents based upon the genetic alteration of nutritionally defective bacteria.
amination The addition of an amine (—NH_2) group to a molecule.
amino acids The building blocks of protein. Amino acids exist in 20 naturally occurring forms that impart different characteristics to the various proteins they compose.
aminoglycoside A complex group of drugs derived from soil actinomycetes that impairs ribosome function and has antibiotic potential. Example: streptomycin.
ammonification Phase of the nitrogen cycle in which ammonia is released from decomposing organic material.
amphibolism Pertaining to the metabolic pathways that serve multiple functions in the breakdown, synthesis, and conversion of metabolites.
amphipathic Relating to a compound that has contrasting characteristics, such as hydrophilic-hydrophobic or acid-base.
amphitrichous Having a single flagellum or a tuft of flagella at opposite poles of a microbial cell.

amplicon DNA strand that has been primed for replication during polymerase chain reaction.
anabolism The energy-consuming process of incorporating nutrients into protoplasm through biosynthesis.
anaerobe A microorganism that grows best, or exclusively, in the absence of oxygen.
anaerobic digesters Closed chambers used in a microbial process that converts organic sludge from waste treatment plants into useful fuels such as methane and hydrogen gases. Also called bioreactors.
anaerobic respiration Respiration in which the final electron acceptor in the electron transport chain is an inorganic molecule containing sulfate, nitrate, nitrite, carbonate, and so on.
analog In chemistry, a compound that closely resembles another in structure.
anamnestic response In immunology, an augmented response or memory related to a prior stimulation of the immune system by antigen. It boosts the levels of immune substances.
anaphylaxis The unusual or exaggerated allergic reaction to antigen that leads to severe respiratory and cardiac complications.
anion A negatively charged ion.
anoxygenic Non-oxygen-producing.
antagonism Relationship in which microorganisms compete for survival in a common environment by taking actions that inhibit or destroy another organism.
antibiotic A chemical substance from one microorganism that can inhibit or kill another microbe even in minute amounts.
antibody A large protein molecule evoked in response to an antigen that interacts specifically with that antigen.
anticodon The trinucleotide sequence of transfer RNA that is complementary to the trinucleotide sequence of messenger RNA (the codon).
antigen Any cell, particle, or chemical that induces a specific immune response by B cells or T cells and can stimulate resistance to an infection or a toxin. See *immunogen.*
antigen binding site Specific region at the ends of the antibody molecule that recognize specific antigens. These sites have numerous shapes to fit a wide variety of antigens.
antigenic drift Minor antigenic changes in the influenza A virus due to mutations in the spikes' genes.
antigenic shift Major changes in the influenza A virus due to recombination of viral strains from two different host species.
antigenicity The property of a substance to stimulate a specific immune response such as antibody formation.
antigen-presenting cell (APC) A macrophage or dendritic cell that ingests and degrades an antigen and subsequently places the antigenic determinant molecules on its surface for recognition by CD4 T lymphocytes.
antihistamine A drug that counters the action of histamine and is useful in allergy treatment.
antimetabolite A substance such as a drug that competes with, substitutes for, or interferes with a normal metabolite.
antimicrobial A special class of compounds capable of destroying or inhibiting microorganisms.
antimicrobial peptides Short protein molecules found in epithelial cells; have the ability to kill bacteria.
antisense DNA A DNA oligonucleotide that binds to a specific piece of RNA, thereby inhibiting translation; used in gene therapy.
antisense RNA An RNA oligonucleotide that binds to a specific piece of RNA, thereby inhibiting translation; used in gene therapy.
antisepsis Chemical treatments to kill or inhibit the growth of all vegetative microorganisms on body surfaces.
antiseptic A growth-inhibiting agent used on tissues to prevent infection.
antiserum Antibody-rich serum derived from the blood of animals (deliberately immunized against infectious or toxic antigen) or from people who have recovered from specific infections.
antitoxin Globulin fraction of serum that neutralizes a specific toxin. Also refers to the specific antitoxin antibody itself.
apicomplexans A group of protozoans that lack locomotion in the mature state.
apoenzyme The protein part of an enzyme, as opposed to the nonprotein or inorganic cofactors.
apoptosis The genetically programmed death of cells that is both a natural process of development and the body's means of destroying abnormal or infected cells.
appendages Accessory structures that sprout from the surface of bacteria. They can be divided into two major groups: those that provide motility and those that enable adhesion.
applied microbiology The study of the practical uses of microorganisms.
aquifer A subterranean water-bearing stratum of permeable rock, sand, or gravel.
archaea Prokaryotic single-celled organisms of primitive origin that have unusual anatomy, physiology, and genetics and live in harsh habitats; when capitalized **(Archaea),** the term refers to one of the three domains of living organisms as proposed by Woese.
arthroconidia Reproductive body of *Coccidioides immitis*, also *arthrospore.*
Arthus reaction An immune complex phenomenon that develops after repeat injection. This localized inflammation results from aggregates of antigen and antibody that bind, complement, and attract neutrophils.
artificial immunity Immunity that is induced as a medical intervention, either by exposing an individual to an antigen or administering immune substances to him or her.
ascospore A spore formed within a saclike cell (ascus) of Ascomycota following nuclear fusion and meiosis.
ascus Special fungal sac in which haploid spores are created.
asepsis A condition free of viable pathogenic microorganisms.
aseptic technique Methods of handling microbial cultures, patient specimens, and other sources of microbes in a way that prevents infection of the handler and others who may be exposed.
assembly (viral) The step in viral multiplication in which capsids and genetic material are packaged into virions.
astromicrobiology A branch of microbiology that studies the potential for and the possible role of microorganisms in space and on other planets.
asymptomatic An infection that produces no noticeable symptoms even though the microbe is active in the host tissue.
asymptomatic carrier A person with an inapparent infection who shows no symptoms of being infected yet is able to pass the disease agent on to others.
atmosphere That part of the biosphere that includes the gaseous envelope up to 14 miles above the earth's surface. It contains gases such as carbon dioxide, nitrogen, and oxygen.
atom The smallest particle of an element to retain all the properties of that element.
atomic number (AN) A measurement that reflects the number of protons in an atom of a particular element.
atomic weight The average of the mass numbers of all the isotopic forms for a particular element.
atopy Allergic reaction classified as type I, with a strong familial relationship; caused by allergens such as pollen, insect venom, food, and dander; involves IgE antibody; includes symptoms of hay fever, asthma, and skin rash.
ATP synthase A unique enzyme located in the mitochondrial cristae and chloroplast grana that harnesses the flux of hydrogen ions to the synthesis of ATP.
attenuate To reduce the virulence of a pathogenic bacterium or virus by passing it through a non-native host or by long-term subculture.
AUG (start codon) The codon that signals the point at which translation of a messenger RNA molecule is to begin.
autoantibody An "anti-self" antibody having an affinity for tissue antigens of the subject in which it is formed.
autoclave A sterilization chamber that allows the use of steam under pressure to sterilize materials. The most common temperature/pressure combination for an autoclave is 121°C and 15 psi.
autograft Tissue or organ surgically transplanted to another site on the same subject.

autoimmune disease The pathologic condition arising from the production of antibodies against autoantigens. Example: rheumatoid arthritis. Also called autoimmunity.

autoimmune regulator (AIRE) A protein that regulates the transcription of self antigens in the thymus; defects in AIRE can lead to inappropriate responses to self antigens.

autotroph A microorganism that requires only inorganic nutrients and whose sole source of carbon is carbon dioxide.

axenic A sterile state such as a pure culture. An axenic animal is born and raised in a germ-free environment. See *gnotobiotic.*

axial filament A type of flagellum (called an endoflagellum) that lies in the periplasmic space of spirochetes and is responsible for locomotion. Also called periplasmic flagellum.

azole Five-membered heterocyclic compounds typical of histidine, which are used in antifungal therapy.

B

B lymphocyte (B cell) A white blood cell that gives rise to plasma cells and antibodies.

bacillus Bacterial cell shape that is cylindrical (longer than it is wide).

bacitracin Antibiotic that targets the bacterial cell wall; component of over-the-counter topical antimicrobial ointments.

back-mutation A mutation that counteracts an earlier mutation, resulting in the restoration of the original DNA sequence.

bacteremia The presence of viable bacteria in circulating blood.

bacteremic Bacteria present in the bloodstream.

Bacteria When capitalized can refer to one of the three domains of living organisms proposed by Woese, containing all nonarchaea prokaryotes.

bacteria (plural of bacterium) Category of prokaryotes with peptidoglycan in their cell walls and circular chromosome(s). This group of small cells is widely distributed in the earth's habitats.

bacterial chromosome A circular body in bacteria that contains the primary genetic material. Also called nucleoid.

bactericide An agent that kills bacteria.

bacteriocin Proteins produced by certain bacteria that are lethal against closely related bacteria and are narrow spectrum compared with antibiotics; these proteins are coded and transferred in plasmids.

bacteriophage A virus that specifically infects bacteria.

bacteristatic Any process or agent that inhibits bacterial growth.

bacterium A tiny unicellular prokaryotic organism that usually reproduces by binary fission and usually has a peptidoglycan cell wall, has various shapes, and can be found in virtually any environment.

barophile A microorganism that thrives under high (usually hydrostatic) pressure.

basement membrane A thin layer (1–6 μm) of protein and polysaccharide found at the base of epithelial tissues.

basic A solution with a pH value above 7 on the pH scale.

basidiospore A sexual spore that arises from a basidium. Found in basidiomycota fungi.

basidium A reproductive cell created when the swollen terminal cell of a hypha develops filaments (sterigmata) that form spores.

basophil A motile polymorphonuclear leukocyte that binds IgE. The basophilic cytoplasmic granules contain mediators of anaphylaxis and atopy.

beta oxidation The degradation of long-chain fatty acids. Two-carbon fragments are formed as a result of enzymatic attack directed against the second or beta carbon of the hydrocarbon chain. Aided by coenzyme A, the fragments enter the Krebs cycle and are processed for ATP synthesis.

beta-lactamase An enzyme secreted by certain bacteria that cleaves the beta-lactam ring of penicillin and cephalosporin and thus provides for resistance against the antibiotic. See *penicillinase.*

binary fission The formation of two new cells of approximately equal size as the result of parent cell division.

binomial system Scientific method of assigning names to organisms that employs two names to identify every organism—genus name plus species name.

biochemistry The study of organic compounds produced by (or components of) living things. The four main categories of biochemicals are carbohydrates, lipids, proteins, and nucleic acids.

biodegradation The breaking down of materials through the action of microbes or insects.

bioenergetics The study of the production and use of energy by cells.

bioethics The study of biological issues and how they relate to human conduct and moral judgment.

biofilm A complex association that arises from a mixture of microorganisms growing together on the surface of a habitat.

biogenesis Belief that living things can only arise from others of the same kind.

biogeochemical cycle A process by which matter is converted from organic to inorganic form and returned to various nonliving reservoirs on earth (air, rocks, and water) where it becomes available for reuse by living things. Elements such as carbon, nitrogen, and phosphorus are constantly cycled in this manner.

biological vector An animal that not only transports an infectious agent but plays a role in the life cycle of the pathogen, serving as a site in which it can multiply or complete its life cycle. It is usually an alternate host to the pathogen.

biomes Particular climate regions in a terrestrial realm.

bioremediation Decomposition of harmful chemicals by microbes or consortia of microbes.

biosensor A device used to detect microbes or trace amounts of compounds through PCR, genome techniques, or electrochemical signaling.

biosphere Habitable regions comprising the aquatic (hydrospheric), soil-rock (lithospheric), and air (atmospheric) environments.

biota Beneficial or harmless resident bacteria commonly found on and/or in the human body.

biotechnology The use of microbes or their products in the commercial or industrial realm.

biotic Living factors such as parasites, food substrates, or other living or once-living organisms that are studied when looking at an ecosystem.

blast cell An immature precursor cell of B and T lymphocytes. Also called a lymphoblast.

blocking antibody The IgG class of immunoglobulins that competes with IgE antibody for allergens, thus blocking the degranulation of basophils and mast cells.

blood cells Cellular components of the blood consisting of red blood cells, primarily responsible for the transport of oxygen and carbon dioxide, and white blood cells, primarily responsible for host defense and immune reactions.

blood-brain barrier Decreased permeability of the walls of blood vessels in the brain, restricting access to that compartment.

botulinum *Clostridium botulinum* toxin. Ingestion of this potent exotoxin leads to flaccid paralysis.

bradykinin An active polypeptide that is a potent vasodilator released from IgE-coated mast cells during anaphylaxis.

broad spectrum Denotes drugs that have an effect on a wide variety of microorganisms.

Brownian movement The passive, erratic, nondirectional motion exhibited by microscopic particles. The jostling comes from being randomly bumped by submicroscopic particles, usually water molecules, in which the visible particles are suspended.

brucellosis A zoonosis transmitted to humans from infected animals or animal products; causes a fluctuating pattern of severe fever in humans as well as muscle pain, weakness, headache, weight loss, and profuse sweating. Also called undulant fever.

bubo The swelling of one or more lymph nodes due to inflammation.

bubonic plague The form of plague in which bacterial growth is primarily restricted to the lymph and is characterized by the appearance of a swollen lymph node referred to as a bubo.

budding See *exocytosis.*

bulbar poliomyelitis Complication of polio infection in which the brain stem, medulla, or cranial nerves are affected. Leads to loss of respiratory control and paralysis of the trunk and limbs.
bullous Consisting of fluid-filled blisters.

C

calculus Dental deposit formed when plaque becomes mineralized with calcium and phosphate crystals. Also called tartar.
Calvin cycle A series of reactions in the second phase of photosynthesis that generates glucose.
cancer Any malignant neoplasm that invades surrounding tissue and can metastasize to other locations. A carcinoma is derived from epithelial tissue, and a sarcoma arises from proliferating mesodermal cells of connective tissue.
capsid The protein covering of a virus's nucleic acid core. Capsids exhibit symmetry due to the regular arrangement of subunits called capsomers. See *icosahedron*.
capsomer A subunit of the virus capsid shaped as a triangle or disc.
capsule In bacteria, the loose, gel-like covering or slime made chiefly of polysaccharides. This layer is protective and can be associated with virulence.
carbohydrate A compound containing primarily carbon, hydrogen, and oxygen in a 1:2:1 ratio.
carbon cycle That pathway taken by carbon from its abiotic source to its use by producers to form organic compounds (biotic), followed by the breakdown of biotic compounds and their release to a nonliving reservoir in the environment (mostly carbon dioxide in the atmosphere).
carbon fixation Reactions in photosynthesis that incorporate inorganic carbon dioxide into organic compounds such as sugars. This occurs during the Calvin cycle and uses energy generated by the light reactions. This process is the source of all production on earth.
carbuncle A deep staphylococcal abscess joining several neighboring hair follicles.
carotenoid Yellow, orange, or red photosynthetic pigments.
carrier A person who harbors infections and inconspicuously spreads them to others. Also, a chemical agent that can accept an atom, chemical radical, or subatomic particle from one compound and pass it on to another.
caseous lesion Necrotic area of lung tubercle superficially resembling cheese. Typical of tuberculosis.
catabolism The chemical breakdown of complex compounds into simpler units to be used in cell metabolism.
catalyst A substance that alters the rate of a reaction without being consumed or permanently changed by it. In cells, enzymes are catalysts.
catalytic site The niche in an enzyme where the substrate is converted to the product (also active site).
catarrhal A term referring to the secretion of mucus or fluids; term for the first stage of pertussis.
cation A positively charged ion.
cell An individual membrane-bound living entity; the smallest unit capable of an independent existence.
cell wall In bacteria, a rigid structure made of peptidoglycan that lies just outside the cytoplasmic membrane; eukaryotes also have a cell wall but may be composed of a variety of materials.
cell-mediated immunity The type of immune responses brought about by T cells, such as cytotoxic and helper effects.
cellulitis The spread of bacteria within necrotic tissue.
cellulose A long, fibrous polymer composed of β-glucose; one of the most common substances on earth.
cephalosporins A group of broad-spectrum antibiotics isolated from the fungus *Cephalosporium*.
cercaria The free-swimming larva of the schistosome trematode that emerges from the snail host and can penetrate human skin, causing schistosomiasis.
cestode The common name for tapeworms that parasitize humans and domestic animals.
chancre The primary sore of syphilis that forms at the site of penetration by *Treponema pallidum*. It begins as a hard, dull red, painless papule that erodes from the center.
chancroid A lesion that resembles a chancre but is soft and is caused by *Haemophilus ducreyi*.
chemical bond A link formed between molecules when two or more atoms share, donate, or accept electrons.
chemical mediators Small molecules that are released during inflammation and specific immune reactions that allow communication between the cells of the immune system and facilitate surveillance, recognition, and attack.
chemiosmosis The generation of a concentration gradient of hydrogen ions (called the proton motive force) by the pumping of hydrogen ions to the outer side of the membrane during electron transport.
chemoautotroph An organism that relies upon inorganic chemicals for its energy and carbon dioxide for its carbon. Also called a chemolithotroph.
chemoheterotroph Microorganisms that derive their nutritional needs from organic compounds.
chemokine Chemical mediators (cytokines) that stimulate the movement and migration of white blood cells.
chemostat A growth chamber with an outflow that is equal to the continuous inflow of nutrient media. This steady-state growth device is used to study such events as cell division, mutation rates, and enzyme regulation.
chemotactic factors Chemical mediators that stimulate the movement of white blood cells. See *chemokines*.
chemotaxis The tendency of organisms to move in response to a chemical gradient (toward an attractant or to avoid adverse stimuli).
chemotherapy The use of chemical substances or drugs to treat or prevent disease.
chemotroph Organism that oxidizes compounds to feed on nutrients.
chitin A polysaccharide similar to cellulose in chemical structure. This polymer makes up the horny substance of the exoskeletons of arthropods and certain fungi.
chloramphenicol Antibiotic that inhibits protein synthesis by binding to the 50s subunit of the ribosome.
chlorophyll A group of mostly green pigments that are used by photosynthetic eukaryotic organisms and cyanobacteria to trap light energy to use in making chemical bonds.
chloroplast An organelle containing chlorophyll that is found in photosynthetic eukaryotes.
cholesterol Best-known member of a group of lipids called steroids. Cholesterol is commonly found in cell membranes and animal hormones.
chromatin The genetic material of the nucleus. Chromatin is made up of nucleic acid and stains readily with certain dyes.
chromosome The tightly coiled bodies in cells that are the primary sites of genes.
chronic Any process or disease that persists over a long duration.
cilium (plural: *cilia*) Eukaryotic structure similar to flagella that propels a protozoan through the environment.
class In the levels of classification, the division of organisms that follows phylum.
classical pathway Pathway of complement activation initiated by a specific antigen-antibody interaction.
clonal selection theory A conceptual explanation for the development of lymphocyte specificity and variety during immune maturation.
clone A colony of cells (or group of organisms) derived from a single cell (or single organism) by asexual reproduction. All units share identical characteristics. Also used as a verb to refer to the process of producing a genetically identical population of cells or genes.
cloning host An organism such as a bacterium or a yeast that receives and replicates a foreign piece of DNA inserted during a genetic engineering experiment.
coagulase A plasma-clotting enzyme secreted by *Staphylococcus aureus*. It contributes to virulence and is involved in forming a fibrin wall that surrounds staphylococcal lesions.

coccobacillus An elongated coccus; a short, thick, oval-shaped bacterial rod.
coccus A spherical-shaped bacterial cell.
codon A specific sequence of three nucleotides in mRNA (or the sense strand of DNA) that constitutes the genetic code for a particular amino acid.
coenzyme A complex organic molecule, several of which are derived from vitamins (e.g., nicotinamide, riboflavin). A coenzyme operates in conjunction with an enzyme. Coenzymes serve as transient carriers of specific atoms or functional groups during metabolic reactions.
coevolution A biological process whereby a change in the genetic composition in one organism leads to a change in the genetics of another organism.
cofactor An enzyme accessory. It can be organic, such as coenzymes, or inorganic, such as Fe^{+2}, Mn^{+2}, or Zn^{+2} ions.
cold sterilization The use of nonheating methods such as radiation or filtration to sterilize materials.
coliform A collective term that includes normal enteric bacteria that are gram-negative and lactose-fermenting.
colony A macroscopic cluster of cells appearing on a solid medium, each arising from the multiplication of a single cell.
colostrum The clear yellow early product of breast milk that is very high in secretory antibodies. Provides passive intestinal protection.
commensalism An unequal relationship in which one species derives benefit without harming the other.
communicable infection Capable of being transmitted from one individual to another.
community The interacting mixture of populations in a given habitat.
competitive inhibition Control process that relies on the ability of metabolic analogs to control microbial growth by successfully competing with a necessary enzyme to halt the growth of bacterial cells.
complement In immunology, serum protein components that act in a definite sequence when set in motion either by an antigen-antibody complex or by factors of the alternative (properdin) pathway.
complementary DNA (cDNA) DNA created by using reverse transcriptase to synthesize DNA from RNA templates.
compounds Molecules that are a combination of two or more different elements.
concentration The expression of the amount of a solute dissolved in a certain amount of solvent. It may be defined by weight, volume, or percentage.
condyloma acuminata Extensive, branched masses of genital warts caused by infection with human papillomavirus.
congenital Transmission of an infection from mother to fetus.
congenital rubella Transmission of the rubella virus to a fetus in utero. Injury to the fetus is generally much more serious than it is to the mother.
congenital syphilis A syphilis infection of the fetus or newborn acquired from maternal infection in utero.
conidia Asexual fungal spores shed as free units from the tips of fertile hyphae.
conidiospore A type of asexual spore in fungi; not enclosed in a sac.
conjugation In bacteria, the contact between donor and recipient cells associated with the transfer of genetic material such as plasmids. Can involve special (sex) pili. Also a form of sexual recombination in ciliated protozoans.
conjunctiva The thin fluid-secreting tissue that covers the eye and lines the eyelid.
consortium A group of microbes that includes more than one species.
constitutive enzyme An enzyme present in bacterial cells in constant amounts, regardless of the presence of substrate. Enzymes of the central catabolic pathways are typical examples.
consumer An organism that feeds on producers or other consumers. It gets all nutrients and energy from other organisms (also called heterotroph). May exist at several levels, such as primary (feeds on producers) and secondary (feeds on primary consumers).
contagious Communicable; transmissible by direct contact with infected people and their fresh secretions or excretions.
contaminant An impurity; any undesirable material or organism.
contaminated culture A medium that once held a pure (single or mixed) culture but now contains unwanted microorganisms.
convalescence Recovery; the period between the end of a disease and the complete restoration of health in a patient.
corepressor A molecule that combines with inactive repressor to form active repressor, which attaches to the operator gene site and inhibits the activity of structural genes subordinate to the operator.
covalent A type of chemical bond that involves the sharing of electrons between two atoms.
covalent bond A chemical bond formed by the sharing of electrons between two atoms.
Creutzfeldt-Jakob disease (CJD) A spongiform encephalopathy caused by infection with a prion. The disease is marked by dementia, impaired senses, and uncontrollable muscle contractions.
crista The infolded inner membrane of a mitochondrion that is the site of the respiratory chain and oxidative phosphorylation.
cryptosporidiosis A gastrointestinal disease caused by *Cryptosporidium parvum*, a protozoan.
culture The visible accumulation of microorganisms in or on a nutrient medium. Also, the propagation of microorganisms with various media.
curd The coagulated milk protein used in cheese making.
cutaneous Second level of skin, including the stratum corneum and occasionally the upper dermis.
cyanosis Blue discoloration of the skin or mucous membranes indicative of decreased oxygen concentration in blood.
cyst The resistant, dormant but infectious form of protozoans. Can be important in spread of infectious agents such as *Entamoeba histolytica* and *Giardia lamblia*.
cystine An amino acid, $HOOC—CH(NH_2)—CH_2—S—S—CH_2—CH(NH_2)COOH$. An oxidation product of two cysteine molecules in which the OSH (sulfhydryl) groups form a disulfide union. Also called dicysteine.
cytochrome A group of heme protein compounds whose chief role is in electron and/or hydrogen transport occurring in the last phase of aerobic respiration.
cytokine A chemical substance produced by white blood cells and tissue cells that regulates development, inflammation, and immunity.
cytopathic effect The degenerative changes in cells associated with virus infection. Examples: the formation of multinucleate giant cells (Negri bodies), the prominent cytoplasmic inclusions of nerve cells infected by rabies virus.
cytoplasm Dense fluid encased by the cell membrane; the site of many of the cell's biochemical and synthetic activities.
cytoplasmic membrane Lipid bilayer that encloses the cytoplasm of bacterial cells.
cytosine (C) One of the nitrogen bases found in DNA and RNA, with a pyrimidine form.
cytotoxicity The ability to kill cells; in immunology, certain T cells are called cytotoxic T cells because they kill other cells.

D

daptomycin A lipopetide antibiotic that disrupts the cytoplasmic membrane.
deamination The removal of an amino group from an amino acid.
death phase End of the cell growth due to lack of nutrition, depletion of environment, and accumulation of wastes. Population of cells begins to die.
debridement Trimming away devitalized tissue and foreign matter from a wound.
decomposer A consumer that feeds on organic matter from the bodies of dead organisms. These microorganisms feed from all levels of the food pyramid and are responsible for recycling elements (also called saprobes).
decomposition The breakdown of dead matter and wastes into simple compounds that can be directed back into the natural cycle of living things.
decontamination The removal or neutralization of an infectious, poisonous, or injurious agent from a site.
deduction Problem-solving process in which an individual constructs a hypothesis, tests its validity by outlining particular events that are predicted by the hypothesis, and

then performs experiments to test for those events.
deductive approach Method of investigation that uses **deduction.** See *deduction.*
definitive host The organism in which a parasite develops into its adult or sexually mature stage. Also called the final host.
degerm To physically remove surface oils, debris, and soil from skin to reduce the microbial load.
degranulation The release of cytoplasmic granules, as when cytokines are secreted from mast cell granules.
dehydration synthesis During the formation of a carbohydrate bond, the step in which one carbon molecule gives up its OH group and the other loses the H from its OH group, thereby producing a water molecule. This process is common to all polymerization reactions.
denaturation The loss of normal characteristics resulting from some molecular alteration. Usually in reference to the action of heat or chemicals on proteins whose function depends upon an unaltered tertiary structure.
dendritic cell A large, antigen-processing cell characterized by long, branchlike extensions of the cell membrane.
denitrification The end of the nitrogen cycle when nitrogen compounds are returned to the reservoir in the air.
dental caries A mixed infection of the tooth surface that gradually destroys the enamel and may lead to destruction of the deeper tissue.
deoxyribonucleic acid (DNA) The nucleic acid often referred to as the "double helix." DNA carries the master plan for an organism's heredity.
deoxyribose A 5-carbon sugar that is an important component of DNA.
dermatophytes A group of fungi that cause infections of the skin and other integument components. They survive by metabolizing keratin.
dermolytic Capable of damaging the skin.
desensitization See *hyposensitization.*
desiccation To dry thoroughly. To preserve by drying.
desquamate To shed the cuticle in scales; to peel off the outer layer of a surface.
diabetes mellitus A disease involving compromise in insulin function. In one form, the pancreatic cells that produce insulin are destroyed by autoantibodies, and in another, the pancreas does not produce sufficient insulin.
diapedesis The migration of intact blood cells between endothelial cells of a blood vessel such as a venule.
dichotomous keys Flow charts that offer two choices or pathways at each level.
differential medium A single substrate that discriminates between groups of microorganisms on the basis of differences in their appearance due to different chemical reactions.
differential stain A technique that utilizes two dyes to distinguish between different microbial groups or cell parts by color reaction.
diffusion The dispersal of molecules, ions, or microscopic particles propelled down a concentration gradient by spontaneous random motion to achieve a uniform distribution.
DiGeorge syndrome A birth defect usually caused by a missing or incomplete thymus gland that results in abnormally low or absent T cells and other developmental abnormalities.
dimorphic In mycology, the tendency of some pathogens to alter their growth form from mold to yeast in response to rising temperature.
diplococci Spherical or oval-shaped bacteria, typically found in pairs.
direct or total cell count 1. Counting total numbers of individual cells being viewed with magnification. 2. Counting isolated colonies of organisms growing on a plate of media as a way to determine population size.
disaccharide A sugar containing two monosaccharides. Example: sucrose (fructose + glucose).
disease Any deviation from health, as when the effects of microbial infection damage or disrupt tissues and organs.
disinfection The destruction of pathogenic nonsporulating microbes or their toxins, usually on inanimate surfaces.
division In the levels of classification, an alternate term for phylum.
DNA See *deoxyribonucleic acid.*
DNA fingerprint A pattern of restriction enzyme fragments that is unique for an individual organism.
DNA polymerase Enzyme responsible for the replication of DNA. Several versions of the enzyme exist, each completing a unique portion of the replication process.
DNA sequencing Determining the exact order of nucleotides in a fragment of DNA. Most commonly done using the Sanger dideoxy sequencing method.
DNA vaccine A newer vaccine preparation based on inserting DNA from pathogens into host cells to encourage them to express the foreign protein and stimulate immunity.
domain In the levels of classification, the broadest general category to which an organism is assigned. Members of a domain share only one or a few general characteristics.
doubling time Time required for a complete fission cycle—from parent cell to two new daughter cells. Also called generation time.
droplet nuclei The dried residue of fine droplets produced by mucus and saliva sprayed while sneezing and coughing. Droplet nuclei are less than 5 μm in diameter (large enough to bear a single bacterium and small enough to remain airborne for a long time) and can be carried by air currents. Droplet nuclei are drawn deep into the air passages.
drug resistance An adaptive response in which microorganisms begin to tolerate an amount of drug that would ordinarily be inhibitory.
dysentery Diarrheal illness in which stools contain blood and/or mucus.
dyspnea Difficulty in breathing.

E

ecosystem A collection of organisms together with its surrounding physical and chemical factors.
ectoplasm The outer, more viscous region of the cytoplasm of a phagocytic cell such as an amoeba. It contains microtubules, but not granules or organelles.
eczema An acute or chronic allergy of the skin associated with itching and burning sensations. Typically, red, edematous, vesicular lesions erupt, leaving the skin scaly and sometimes hyperpigmented.
edema The accumulation of excess fluid in cells, tissues, or serous cavities. Also called swelling.
electrolyte Any compound that ionizes in solution and conducts current in an electrical field.
electron A negatively charged subatomic particle that is distributed around the nucleus in an atom.
electrophoresis The separation of molecules by size and charge through exposure to an electrical current.
electrostatic Relating to the attraction of opposite charges and the repulsion of like charges. Electrical charge remains stationary as opposed to electrical flow or current.
element A substance comprising only one kind of atom that cannot be degraded into two or more substances without losing its chemical characteristics.
ELISA Abbreviation for **e**nzyme-**l**inked **i**mmuno**s**orbent **a**ssay, a very sensitive serological test used to detect antibodies in diseases such as AIDS.
emerging disease Newly identified diseases that are becoming more prominent.
emetic Inducing to vomit.
encephalitis An inflammation of the brain, usually caused by infection.
endemic disease A native disease that prevails continuously in a geographic region.
endergonic reaction A chemical reaction that occurs with the absorption and storage of surrounding energy. Antonym: exergonic.
endocytosis The process whereby solid and liquid materials are taken into the cell through membrane invagination and engulfment into a vesicle.
endoenzyme An intracellular enzyme, as opposed to enzymes that are secreted.
endogenous Originating or produced within an organism or one of its parts.
endoplasmic reticulum (ER) An intracellular network of flattened sacs or tubules with or without ribosomes on their surfaces.

endospore A small, dormant, resistant derivative of a bacterial cell that germinates under favorable growth conditions into a vegetative cell. The bacterial genera *Bacillus* and *Clostridium* are typical sporeformers.

endosymbiosis Relationship in which a microorganism resides within a host cell and provides a benefit to the host cell.

endotoxic shock A massive drop in blood pressure caused by the release of endotoxin from gram-negative bacteria multiplying in the bloodstream.

endotoxin A bacterial toxin that is not ordinarily released (as is exotoxin). Endotoxin is composed of a phospholipid-polysaccharide complex that is an integral part of gram-negative bacterial cell walls. Endotoxins can cause severe shock and fever.

energy of activation The minimum energy input necessary for reactants to form products in a chemical reaction.

energy pyramid An ecological model that shows the energy flow among the organisms in a community. It is structured like the food pyramid but shows how energy is reduced from one trophic level to another.

enriched medium A nutrient medium supplemented with blood, serum, or some growth factor to promote the multiplication of fastidious microorganisms.

enteric Pertaining to the intestine.

enteroaggregative The term used to describe certain types of intestinal bacteria that tend to stick to each other in large clumps.

enterohemorrhagic *E. coli* (EHEC) A group of *E. coli* species that induce bleeding in the intestines and also in other organs; *E. coli* 0157:H7 belongs to this group.

enteroinvasive Predisposed to invade the intestinal tissues.

enteropathogenic Pathogenic to the alimentary canal.

enterotoxigenic Having the capacity to produce toxins that act on the intestinal tract.

enterotoxin A bacterial toxin that specifically targets intestinal mucous membrane cells. Enterotoxigenic strains of *Escherichia coli* and *Staphylococcus aureus* are typical sources.

enveloped virus A virus whose nucleocapsid is enclosed by a membrane derived in part from the host cell. It usually contains exposed glycoprotein spikes specific for the virus.

enzyme A protein biocatalyst that facilitates metabolic reactions.

enzyme induction One of the controls on enzyme synthesis. This occurs when enzymes appear only when suitable substrates are present.

enzyme repression The inhibition of enzyme synthesis by the end product of a catabolic pathway.

eosinophil A leukocyte whose cytoplasmic granules readily stain with red eosin dye.

eosinophilia An increase in eosinophil concentration in the bloodstream, often in response to helminth infection.

epidemic A sudden and simultaneous outbreak or increase in the number of cases of disease in a community.

epidemiology The study of the factors affecting the prevalence and spread of disease within a community.

epitope The precise molecular group of an antigen that defines its specificity and triggers the immune response.

Epstein-Barr virus (EBV) Herpesvirus linked to infectious mononucleosis, Burkitt's lymphoma, and nasopharyngeal carcinoma.

erysipelas An acute, sharply defined inflammatory disease specifically caused by hemolytic *Streptococcus*. The eruption is limited to the skin but can be complicated by serious systemic symptoms.

erythroblastosis fetalis Hemolytic anemia of the newborn. The anemia comes from hemolysis of Rh-positive fetal erythrocytes by anti-Rh maternal antibodies. Erythroblasts are immature red blood cells prematurely released from the bone marrow.

erythrocytes (red blood cells) Blood cells involved in the transport of oxygen and carbon dioxide.

erythrogenic toxin An exotoxin produced by lysogenized group A strains of β-hemolytic streptococci that is responsible for the severe fever and rash of scarlet fever in the nonimmune individual. Also called a pyrogenic toxin.

eschar A dark, sloughing scab that is the lesion of anthrax and certain rickettsioses.

essential nutrient Any ingredient such as a certain amino acid, fatty acid, vitamin, or mineral that cannot be formed by an organism and must be supplied in the diet. A growth factor.

ester bond A covalent bond formed by reacting carboxylic acid with an OH group:

$$(\mathrm{R}-\overset{\overset{\displaystyle \mathrm{O}}{\|}}{\mathrm{C}}-\mathrm{O}-\mathrm{R}')$$

Olive and corn oils, lard, and butter fat are examples of triacylglycerols—esters formed between glycerol and three fatty acids.

ethylene oxide A potent, highly water-soluble gas invaluable for gaseous sterilization of heat-sensitive objects such as plastics, surgical and diagnostic appliances, and spices.

etiologic agent The microbial cause of disease; the pathogen.

eubacteria Term used for nonarchaea prokaryotes, means "true bacteria."

Eukarya One of the three domains (sometimes called superkingdoms) of living organisms, as proposed by Woese; contains all eukaryotic organisms.

eukaryotic cell A cell that differs from a prokaryotic cell chiefly by having a nuclear membrane (a well-defined nucleus), membrane-bounded subcellular organelles, and mitotic cell division.

eutrophication The process whereby dissolved nutrients resulting from natural seasonal enrichment or industrial pollution of water cause overgrowth of algae and cyanobacteria to the detriment of fish and other large aquatic inhabitants.

evolution Scientific principle that states that living things change gradually through hundreds of millions of years, and these changes are expressed in structural and functional adaptations in each organism. Evolution presumes that those traits that favor survival are preserved and passed on to following generations, and those traits that do not favor survival are lost.

exanthem An eruption or rash of the skin.

exergonic A chemical reaction associated with the release of energy to the surroundings. Antonym: endergonic.

exfoliative toxin A poisonous substance that causes superficial cells of an epithelium to detach and be shed. Example: staphylococcal exfoliatin. Also called an epidermolytic toxin.

exocytosis The process that releases enveloped viruses from the membrane of the host's cytoplasm.

exoenzyme An extracellular enzyme chiefly for hydrolysis of nutrient macromolecules that are otherwise impervious to the cell membrane. It functions in saprobic decomposition of organic debris and can be a factor in invasiveness of pathogens.

exogenous Originating outside the body.

exon A stretch of eukaryotic DNA coding for a corresponding portion of mRNA that is translated into peptides. Intervening stretches of DNA that are not expressed are called introns. During transcription, exons are separated from introns and are spliced together into a continuous mRNA transcript.

exotoxin A toxin (usually protein) that is secreted and acts upon a specific cellular target. Examples: botulin, tetanospasmin, diphtheria toxin, and erythrogenic toxin.

exponential Pertaining to the use of exponents, numbers that are typically written as a superscript to indicate how many times a factor is to be multiplied. Exponents are used in scientific notation to render large, cumbersome numbers into small workable quantities.

exponential growth phase The period of maximum growth rate in a growth curve. Cell population increases logarithmically.

extrapulmonary tuberculosis A condition in which tuberculosis bacilli have spread to organs other than the lungs.

extremophiles Organisms capable of living in harsh environments, such as extreme heat or cold.

F

facilitated diffusion The passive movement of a substance across a plasma membrane from an area of higher concentration to an area of lower concentration utilizing specialized carrier proteins.

facultative Pertaining to the capacity of microbes to adapt or adjust to variations; not obligate. Example: the presence of oxygen is not obligatory for a facultative anaerobe to grow. See *obligate.*

family In the levels of classification, a midlevel division of organisms that groups more closely related organisms than previous levels. An order is divided into families.

fastidious Requiring special nutritional or environmental conditions for growth. Said of bacteria.

fecal coliforms Any species of gram-negative lactose-positive bacteria (primarily *Escherichia coli*) that live primarily in the intestinal tract and not the environment. Finding evidence of these bacteria in a water or food sample is substantial evidence of fecal contamination and potential for infection (see *coliform*).

feedback inhibition Temporary end to enzyme action caused by an end product molecule binding to the regulatory site and preventing the enzyme's active site from binding to its substrate.

fermentation The extraction of energy through anaerobic degradation of substrates into simpler, reduced metabolites. In large industrial processes, fermentation can mean any use of microbial metabolism to manufacture organic chemicals or other products.

fermentor A large tank used in industrial microbiology to grow mass quantities of microbes that can synthesize desired products. These devices are equipped with means to stir, monitor, and harvest products such as drugs, enzymes, and proteins in very large quantities.

fertility (F′) factor Donor plasmid that allows synthesis of a pilus in bacterial conjugation. Presence of the factor is indicated by F^+, and lack of the factor is indicated by F^-.

filament A helical structure composed of proteins that is part of bacterial flagella.

fimbria A short, numerous-surface appendage on some bacteria that provides adhesion but not locomotion.

Firmicutes Taxonomic category of bacteria that have gram-positive cell envelopes.

flagellum A structure that is used to propel the organism through a fluid environment.

fluid mosaic model A conceptualization of the molecular architecture of cellular membranes as a bilipid layer containing proteins. Membrane proteins are embedded to some degree in this bilayer, where they float freely about.

fluorescence The property possessed by certain minerals and dyes to emit visible light when excited by ultraviolet radiation. A fluorescent dye combined with specific antibody provides a sensitive test for the presence of antigen.

fluoroquinolones Synthetic antimicrobial drugs chemically related to quinine. They are broad spectrum and easily adsorbed from the intestine.

focal infection Occurs when an infectious agent breaks loose from a localized infection and is carried by the circulation to other tissues.

folliculitis An inflammatory reaction involving the formation of papules or pustules in clusters of hair follicles.

fomite Virtually any inanimate object an infected individual has contact with that can serve as a vehicle for the spread of disease.

food chain A simple straight-line feeding sequence among organisms in a community.

food fermentations Addition to and growth of known cultures of microorganisms in foods to produce desirable flavors, smells, or textures. Includes cheeses, breads, alcoholic beverages, and pickles.

food poisoning Symptoms in the intestines (which may include vomiting) induced by preformed exotoxin from bacteria.

food web A complex network that traces all feeding interactions among organisms in a community (see *food chain*). This is considered to be a more accurate picture of food relationships in a community than a food chain.

formalin A 37% aqueous solution of formaldehyde gas; a potent chemical fixative and microbicide.

fosfomycin trimethamine Antibiotic that inhibits an enzyme necessary for cell wall synthesis.

frameshift mutation An insertion or deletion mutation that changes the codon reading frame from the point of the mutation to the final codon. Almost always leads to a nonfunctional protein.

fructose One of the carbohydrates commonly referred to as sugars. Fructose is commonly fruit sugars.

functional group In chemistry, a particular molecular combination that reacts in predictable ways and confers particular properties on a compound. Examples: —COOH, —OH, —CHO.

fungemia The condition of fungi multiplying in the bloodstream.

fungi Macroscopic and microscopic heterotrophic eukaryotic organisms that can be uni- or multicellular.

fungus Heterotrophic unicellular or multicellular eukaryotic organism that may take the form of a larger macroscopic organism, as in the case of mushrooms, or a smaller microscopic organism, as in the case of yeasts and molds.

furuncle A boil; a localized pyogenic infection arising from a hair follicle.

fuzeon Anti-HIV drug that inhibits viral attachment to host cells.

G

Gaia Theory The concept that biotic and abiotic factors sustain suitable conditions for one another simply by their interactions. Named after the mythical Greek goddess of earth.

gamma globulin The fraction of plasma proteins high in immunoglobulins (antibodies). Preparations from pooled human plasma containing normal antibodies make useful passive immunizing agents against pertussis, polio, measles, and several other diseases.

gas gangrene Disease caused by a clostridial infection of soft tissue or wound. The name refers to the gas produced by the bacteria growing in the tissue. Unless treated early, it is fatal. Also called myonecrosis.

gastritis Pain and/or nausea, usually experienced after eating; result of inflammation of the lining of the stomach.

gel electrophoresis A laboratory technique for separating DNA fragments according to length by employing electricity to force the DNA through a gel-like matrix typically made of agarose. Smaller DNA fragments move more quickly through the gel, thereby moving farther than larger fragments during the same period of time.

gene A site on a chromosome that provides information for a certain cell function. A specific segment of DNA that contains the necessary code to make a protein or RNA molecule.

gene probe Short strands of single-stranded nucleic acid that hybridize specifically with complementary stretches of nucleotides on test samples and thereby serve as a tagging and identification device.

gene therapy The introduction of normal functional genes into people with genetic diseases such as sickle cell anemia and cystic fibrosis. This is usually accomplished by a virus vector.

generation time Time required for a complete fission cycle—from parent cell to two new daughter cells. Also called doubling time.

genetic engineering A field involving deliberate alterations (recombinations) of the genomes of microbes, plants, and animals through special technological processes.

genetics The science of heredity.

genital warts A prevalent STD linked to some forms of cancer of the reproductive organs. Caused by infection with human papillomavirus.

genome The complete set of chromosomes and genes in an organism.

genomics The systematic study of an organism's genes and their functions.

genotype The genetic makeup of an organism. The genotype is ultimately responsible for an organism's phenotype, or expressed characteristics.

genus In the levels of classification, the second most specific level. A family is divided into several genera.

geomicrobiology A branch of microbiology that studies the role of microorganisms in the earth's crust.

germ free See *axenic.*

germ theory of disease A theory first originating in the 1800s that proposed that microorganisms can be the cause of diseases.

The concept is actually so well established in the present time that it is considered a fact.

germicide An agent lethal to non-endospore-forming pathogens.

giardiasis Infection by the *Giardia* flagellate. Most common mode of transmission is contaminated food and water. Symptoms include diarrhea, abdominal pain, and flatulence.

gingivitis Inflammation of the gum tissue in contact with the roots of the teeth.

gluconeogenesis The formation of glucose (or glycogen) from noncarbohydrate sources such as protein or fat. Also called glyconeogenesis.

glucose One of the carbohydrates commonly referred to as sugars. Glucose is characterized by its 6-carbon structure.

glycerol A 3-carbon alcohol, with three OH groups that serve as binding sites.

glycocalyx A filamentous network of carbohydrate-rich molecules that coats cells.

glycogen A glucose polymer stored by cells.

glycolysis The energy-yielding breakdown (fermentation) of glucose to pyruvic or lactic acid. It is often called anaerobic glycolysis because no molecular oxygen is consumed in the degradation.

glycosidic bond A bond that joins monosaccharides to form disaccharides and polymers.

gnotobiotic Referring to experiments performed on germ-free animals.

Golgi apparatus An organelle of eukaryotes that participates in packaging and secretion of molecules.

gonococcus Common name for *Neisseria gonorrhoeae,* the agent of gonorrhea.

Gracilicutes Taxonomic category of bacteria that have gram-negative envelopes.

graft Live tissue taken from a donor and transplanted into a recipient to replace damaged or missing tissues such as skin, bone, blood vessels.

graft versus host disease (GVHD) A condition associated with a bone marrow transplant in which T cells in the transplanted tissue mount an immune response against the recipient's (host) normal tissues.

Gram stain A differential stain for bacteria useful in identification and taxonomy. Gram-positive organisms appear purple from crystal violet mordant retention, whereas gram-negative organisms appear red after loss of crystal violet and absorbance of the safranin counterstain.

gram-negative A category of bacterial cells that describes bacteria with an outer membrane, a cytoplasmic membrane, and a thin cell wall.

gram-positive A category of bacterial cells that describes bacteria with a thick cell wall and no outer membrane.

grana Discrete stacks of chlorophyll-containing thylakoids within chloroplasts.

granulocyte A mature leukocyte that contains noticeable granules in a Wright stain. Examples: neutrophils, eosinophils, and basophils.

granuloma A solid mass or nodule of inflammatory tissue containing modified macrophages and lymphocytes. Usually a chronic pathologic process of diseases such as tuberculosis or syphilis.

granzymes Enzymes secreted by cytotoxic T cells that damage proteins of target cells.

Graves' disease A malfunction of the thyroid gland in which autoantibodies directed at thyroid cells stimulate an overproduction of thyroid hormone (hyperthyroidism).

greenhouse effect The capacity to retain solar energy by a blanket of atmospheric gases that redirects heat waves back toward the earth.

group translocation A form of active transport in which the substance being transported is altered during transfer across a plasma membrane.

growth curve A graphical representation of the change in population size over time. This graph has four periods known as lag phase, exponential or log phase, stationary phase, and death phase.

growth factor An organic compound such as a vitamin or amino acid that must be provided in the diet to facilitate growth. An essential nutrient.

guanine (G) One of the nitrogen bases found in DNA and RNA in the purine form.

Guillain-Barré syndrome A neurological complication of infection or vaccination.

gumma A nodular, infectious granuloma characteristic of tertiary syphilis.

gut-associated lymphoid tissue (GALT) A collection of lymphoid tissue in the gastrointestinal tract that includes the appendix, the lacteals, and Peyer's patches.

gyrase The enzyme responsible for supercoiling DNA into tight bundles; a type of topoisomerase.

H

HAART Highly active antiretroviral therapy; three-antiviral treatment for HIV infection.

habitat The environment to which an organism is adapted.

halogens A group of related chemicals with antimicrobial applications. The halogens most often used in disinfectants and antiseptics are chlorine and iodine.

halophile A microbe whose growth is either stimulated by salt or requires a high concentration of salt for growth.

Hansen's disease A chronic, progressive disease of the skin and nerves caused by infection by a mycobacterium that is a slow-growing, strict parasite. Hansen's disease is the preferred name for leprosy.

hapten An incomplete or partial antigen. Although it constitutes the determinative group and can bind antigen, hapten cannot stimulate a full immune response without being carried by a larger protein molecule.

Hashimoto's thyroiditis An autoimmune disease of the thyroid gland that damages the thyroid follicle cells and results in decreased production of thyroid hormone (hypothyroidism).

hay fever A form of atopic allergy marked by seasonal acute inflammation of the conjunctiva and mucous membranes of the respiratory passages. Symptoms are irritative itching and rhinitis.

helical Having a spiral or coiled shape. Said of certain virus capsids and bacteria.

helminth A term that designates all parasitic worms.

helper T cell A class of thymus-stimulated lymphocytes that facilitate various immune activities such as assisting B cells and macrophages. Also called a T helper cell.

hemagglutinin A molecule that causes red blood cells to clump or agglutinate. Often found on the surfaces of viruses.

hematopoiesis The process by which the various types of blood cells are formed, such as in the bone marrow.

hemolysin Any biological agent that is capable of destroying red blood cells and causing the release of hemoglobin. Many bacterial pathogens produce exotoxins that act as hemolysins.

hemolytic disease Incompatible Rh factor between mother and fetus causes maternal antibodies to attack the fetus and trigger complement-mediated lysis in the fetus.

hemolytic uremic syndrome (HUS) Severe hemolytic anemia leading to kidney damage or failure; can accompany *E. coli* O157:H7 intestinal infection.

hemolyze When red blood cells burst and release hemoglobin pigment.

hepatitis Inflammation and necrosis of the liver, often the result of viral infection.

hepatitis A virus (HAV) Enterovirus spread by contaminated food responsible for short-term (infectious) hepatitis.

hepatitis B virus (HBV) DNA virus that is the causative agent of serum hepatitis.

hepatocellular carcinoma A liver cancer associated with infection with hepatitis B virus.

herd immunity The status of collective acquired immunity in a population that reduces the likelihood that nonimmune individuals will contract and spread infection. One aim of vaccination is to induce herd immunity.

heredity Genetic inheritance.

hermaphroditic Containing the sex organs for both male and female in one individual.

herpes zoster A recurrent infection caused by latent chickenpox virus. Its manifestation on the skin tends to correspond to dermatomes and to occur in patches that "girdle" the trunk. Also called shingles.

heterotroph An organism that relies upon organic compounds for its carbon and energy needs.

hexose A 6-carbon sugar such as glucose and fructose.

hierarchies Levels of power. Arrangement in order of rank.

histamine A cytokine released when mast cells and basophils release their granules. An important mediator of allergy, its effects include smooth muscle contraction, increased vascular permeability, and increased mucus secretion.

histiocyte Another term for macrophage.

histone Proteins associated with eukaryotic DNA. These simple proteins serve as winding spools to compact and condense the chromosomes.

HLA An abbreviation for **h**uman **l**eukocyte **a**ntigens. This closely linked cluster of genes programs for cell surface glycoproteins that control immune interactions between cells and is involved in rejection of allografts. Also called the major histocompatibility complex (MHC).

holoenzyme An enzyme complete with its apoenzyme and cofactors.

hops The ripe, dried fruits of the hop vine (*Humulus lupulus*) that are added to beer wort for flavoring.

host Organism in which smaller organisms or viruses live, feed, and reproduce.

host range The limitation imposed by the characteristics of the host cell on the type of virus that can successfully invade it.

human diploid cell vaccine (HDCV) A vaccine made using cell culture that is currently the vaccine of choice for preventing infection by rabies virus.

human immunodeficiency virus (HIV) A retro virus that causes acquired immunodeficiency syndrome (AIDS).

human papillomavirus (HPV) A group of DNA viruses whose members are responsible for common, plantar, and genital warts.

humoral immunity Protective molecules (mostly B lymphocytes) carried in the fluids of the body.

hybridization A process that matches complementary strands of nucleic acid (DNA-DNA, RNA-DNA, RNA-RNA). Used for locating specific sites or types of nucleic acids.

hybridoma An artificial cell line that produces monoclonal antibodies. It is formed by fusing (hybridizing) a normal antibody-producing cell with a cancer cell, and it can produce pure antibody indefinitely.

hydration The addition of water as in the coating of ions with water molecules as ions enter into aqueous solution.

hydrogen bond A weak chemical bond formed by the attraction of forces between molecules or atoms—in this case, hydrogen and either oxygen or nitrogen. In this type of bond, electrons are not shared, lost, or gained.

hydrologic cycle The continual circulation of water between hydrosphere, atmosphere, and lithosphere.

hydrolysis A process in which water is used to break bonds in molecules. Usually occurs in conjunction with an enzyme.

hydrophilic The property of attracting water. Molecules that attract water to their surface are called hydrophilic.

hydrophobic The property of repelling water. Molecules that repel water are called hydrophobic.

hydrosphere That part of the biosphere that encompasses water-containing environments such as oceans, lakes, rivers.

hypertonic Having a greater osmotic pressure than a reference solution.

hyphae The tubular threads that make up filamentous fungi (molds). This web of branched and intertwining fibers is called a mycelium.

hypogammaglobulinemia An inborn disease in which the gamma globulin (antibody) fraction of serum is greatly reduced. The condition is associated with a high susceptibility to pyogenic infections.

hyposensitization A therapeutic exposure to known allergens designed to build tolerance and eventually prevent allergic reaction.

hypothesis A tentative explanation of what has been observed or measured.

hypotonic Having a lower osmotic pressure than a reference solution.

I

icosahedron A regular geometric figure having 20 surfaces that meet to form 12 corners. Some virions have capsids that resemble icosahedral crystals.

immune complex reaction Type III hypersensitivity of the immune system. It is characterized by the reaction of soluble antigen with antibody, and the deposition of the resulting complexes in basement membranes of epithelial tissue.

immunity An acquired resistance to an infectious agent due to prior contact with that agent.

immunoassays Extremely sensitive tests that permit rapid and accurate measurement of trace antigen or antibody.

immunocompetence The ability of the body to recognize and react with multiple foreign substances.

immunodeficiency Immune function is incompletely developed, suppressed, or destroyed.

immunodeficiency disease A form of immunopathology in which white blood cells are unable to mount a complete, effective immune response, which results in recurrent infections. Examples would be AIDS and agammaglobulinemia.

immunogen Any substance that induces a state of sensitivity or resistance after processing by the immune system of the body.

immunoglobulin (Ig) The chemical class of proteins to which antibodies belong.

immunology The study of the system of body defenses that protect against infection.

immunopathology The study of disease states associated with overreactivity or underreactivity of the immune response.

immunotherapy Preventing or treating infectious diseases by administering substances that produce artificial immunity. May be active or passive.

in utero Literally means "in the uterus"; pertains to events or developments occurring before birth.

in vitro Literally means "in glass," signifying a process or reaction occurring in an artificial environment, as in a test tube or culture medium.

in vivo Literally means "in a living being," signifying a process or reaction occurring in a living thing.

incidence In epidemiology, the number of new cases of a disease occurring during a period.

incineration Destruction of microbes by subjecting them to extremes of dry heat. Microbes are reduced to ashes and gas by this process.

inclusion A relatively inert body in the cytoplasm such as storage granules, glycogen, fat, or some other aggregated metabolic product.

inclusion body One of a variety of different storage compartments in bacterial cells.

incubate To isolate a sample culture in a temperature-controlled environment to encourage growth.

incubation period The period from the initial contact with an infectious agent to the appearance of the first symptoms.

indicator bacteria In water analysis, any easily cultured bacteria that may be found in the intestine and can be used as an index of fecal contamination. The category includes coliforms and enterococci. Discovery of these bacteria in a sample means that pathogens may also be present.

induced mutation Any alteration in DNA that occurs as a consequence of exposure to chemical or physical mutagens.

inducible enzyme An enzyme that increases in amount in direct proportion to the amount of substrate present.

inducible operon An operon that under normal circumstances is not transcribed. The presence of a specific inducer molecule can cause transcription of the operon to begin.

induction The process whereby a bacteriophage in the prophage state is activated and begins replication and enters the lytic cycle.

induration Area of hardened, reddened tissue associated with the tuberculin test.

infection The entry, establishment, and multiplication of pathogenic organisms within a host.

infectious disease The state of damage or toxicity in the body caused by an infectious agent.

inflammation A natural, nonspecific response to tissue injury that protects the host from further damage. It stimulates immune reactivity and blocks the spread of an infectious agent.

inoculation The implantation of microorganisms into or upon culture media.

inorganic chemicals Molecules that lack the basic framework of the elements of carbon and hydrogen.
integument The outer surfaces of the body: skin, hair, nails, sweat glands, and oil glands.
interferon (IFN) Natural human chemical that inhibits viral replication; used therapeutically to combat viral infections and cancer.
interferon gamma A protein produced by a virally infected cell that induces production of antiviral substances in neighboring cells. This defense prevents the production and maturation of viruses and thus terminates the viral infection.
interleukins A class of chemicals released from host cells that have potent effects on immunity.
intoxication Poisoning that results from the introduction of a toxin into body tissues through ingestion or injection.
intron The segments on split genes of eukaryotes that do not code for polypeptide. They can have regulatory functions. See *exon.*
iodophor A combination of iodine and an organic carrier that is a moderate-level disinfectant and antiseptic.
ion An unattached, charged particle.
ionic bond A chemical bond in which electrons are transferred and not shared between atoms.
ionization The aqueous dissociation of an electrolyte into ions.
ionizing radiation Radiant energy consisting of short-wave electromagnetic rays (X ray) or high-speed electrons that cause dislodgment of electrons on target molecules and create ions.
irradiation The application of radiant energy for diagnosis, therapy, disinfection, or sterilization.
irritability Capacity of cells to respond to chemical, mechanical, or light stimuli. This property helps cells adapt to the environment and obtain nutrients.
isograft Transplanted tissue from one monozygotic twin to the other; transplants between highly inbred animals that are genetically identical.
isolation The separation of microbial cells by serial dilution or mechanical dispersion on solid media to create discrete colonies.
isoniazid Older drug that targets the bacterial cell wall; used against *M. tuberculosis.*
isotonic Two solutions having the same osmotic pressure such that, when separated by a semipermeable membrane, there is no net movement of solvent in either direction.
isotope A version of an element that is virtually identical in all chemical properties to another version except that their atoms have slightly different atomic masses.

J

jaundice The yellowish pigmentation of skin, mucous membranes, sclera, deeper tissues, and excretions due to abnormal deposition of bile pigments. Jaundice is associated with liver infection, as with hepatitis B virus and leptospirosis.
JC virus (JCV) Causes a form of encephalitis (progressive multifocal leukoencephalopathy), especially in AIDS patients.

K

Kaposi sarcoma A malignant or benign neoplasm that appears as multiple hemorrhagic sites on the skin, lymph nodes, and viscera and apparently involves the metastasis of abnormal blood vessel cells. It is a clinical feature of AIDS.
keratin Protein produced by outermost skin cells that provide protection from trauma and moisture.
killed or inactivated vaccine A whole cell or intact virus preparation in which the microbes are dead or preserved and cannot multiply but are still capable of conferring immunity.
killer T cells A T lymphocyte programmed to directly affix cells and kill them. See *cytotoxicity.*
kingdom In the levels of classification, the second division from more general to more specific. Each domain is divided into kingdoms.
Koch's postulates A procedure to establish the specific cause of disease. In all cases of infection: (1) The agent must be found; (2) inoculations of a pure culture must reproduce the same disease in animals; (3) the agent must again be present in the experimental animal; and (4) a pure culture must again be obtained.
Koplik's spots Tiny red blisters with central white specks on the mucosal lining of the cheeks. Symptomatic of measles.
Krebs cycle or tricarboxylic acid cycle (TCA) The second pathway of the three pathways that complete the process of primary catabolism. Also called the citric acid cycle.

L

L form L-phase variants; wall-less forms of some bacteria that are induced by drugs or chemicals. These forms can be involved in infections.
labile In chemistry, molecules, or compounds that are chemically unstable in the presence of environmental changes.
lactose One of the carbohydrates commonly referred to as sugars. Lactose is commonly found in milk.
lactose (*lac*) operon Control system that manages the regulation of lactose metabolism. It is composed of three DNA segments, including a regulator, a control locus, and a structural locus.
lag phase The early phase of population growth during which no signs of growth occur.
lager The maturation process of beer, which is allowed to take place in large vats at a reduced temperature.
lagging strand The newly forming 5′ DNA strand that is discontinuously replicated in segments (Okazaki fragments).
lantibiotics Short peptides produced by bacteria that inhibit the growth of other bacteria.
latency The state of being inactive. Example: a latent virus or latent infection.
leading strand The newly forming 3′ DNA strand that is replicated in a continuous fashion without segments.
leaven To lighten food material by entrapping gas generated within it. Example: the rising of bread from the CO_2 produced by yeast or baking powder.
Legionnaire's disease Infection by *Legionella* bacterium. Weakly gram-negative rods are able to survive in aquatic habitats. Some forms may be fatal.
legumes Plants that produce seeds in pods. Examples include soybeans and peas.
lepromas Skin nodules seen on the face of persons suffering from lepromatous leprosy. The skin folds and thickenings are caused by the overgrowth of *Mycobacterium leprae.*
lepromatous leprosy Severe, disfiguring leprosy characterized by widespread dissemination of the leprosy bacillus in deeper lesions.
leprosy See *Hansen's disease.*
lesion A wound, injury, or some other pathologic change in tissues.
leukocidin A heat-labile substance formed by some pyogenic cocci that impairs and sometimes lyses leukocytes.
leukocytes White blood cells. The primary infection-fighting blood cells.
leukocytosis An abnormally large number of leukocytes in the blood, which can be indicative of acute infection.
leukopenia A lower-than-normal leukocyte count in the blood that can be indicative of blood infection or disease.
leukotriene An unsaturated fatty acid derivative of arachidonic acid. Leukotriene functions in chemotactic activity, smooth muscle contractility, mucus secretion, and capillary permeability.
ligase An enzyme required to seal the sticky ends of DNA pieces after splicing.
light-dependent reactions The series of reactions in photosynthesis that are driven by the light energy (photons) absorbed by chlorophyll. They involve splitting of water into hydrogens and oxygen, transport of electrons by NADP, and ATP synthesis.
light-independent reactions The series of reactions in photosynthesis that can proceed with or without light. It is a cyclic system that uses ATP from the light reactions to incorporate or fix carbon dioxide into organic compounds, leading to the production of glucose and other carbohydrates (also called the Calvin cycle).
lipase A fat-splitting enzyme. Example: triacylglycerol lipase separates the fatty acid chains from the glycerol backbone of triglycerides.

lipid A term used to describe a variety of substances that are not soluble in polar solvents such as water but will dissolve in nonpolar solvents such as benzene and chloroform. Lipids include triglycerides, phospholipids, steroids, and waxes.

lipopolysaccharide A molecular complex of lipid and carbohydrate found in the bacterial cell wall. The lipopolysaccharide (LPS) of gram-negative bacteria is an endotoxin with generalized pathologic effects such as fever.

lithoautotroph Bacteria that rely on inorganic minerals to supply their nutritional needs. Sometimes referred to as chemoautotrophs.

lithosphere That part of the biosphere that encompasses the earth's crust, including rocks and minerals.

lithotroph An autotrophic microbe that derives energy from reduced inorganic compounds such as N_2S.

lobar pneumonia Infection involving whole segments (lobes) of the lungs, which may lead to consolidation and plugging of the alveoli and extreme difficulty in breathing.

localized infection Occurs when a microbe enters a specific tissue, infects it, and remains confined there.

locus A site on a chromosome occupied by a gene. Plural: loci.

log phase Maximum rate of cell division during which growth is geometric in its rate of increase. Also called exponential growth phase.

lophotrichous Describing bacteria having a tuft of flagella at one or both poles.

lumen The cavity within a tubular organ.

lymphadenitis Inflammation of one or more lymph nodes. Also called lymphadenopathy.

lymphatic system A system of vessels and organs that serve as sites for development of immune cells and immune reactions. It includes the spleen, thymus, lymph nodes, and GALT.

lymphocyte The second most common form of white blood cells.

lyophilization A method for preserving microorganisms (and other substances) by freezing and then drying them directly from the frozen state.

lyse To burst.

lysin A complement-fixing antibody that destroys specific targeted cells. Examples: hemolysin and bacteriolysin.

lysis The physical rupture or deterioration of a cell.

lysogenic conversion A bacterium acquires a new genetic trait due to the presence of genetic material from an infecting phage.

lysogeny The indefinite persistence of bacteriophage DNA in a host without bringing about the production of virions.

lysosome A cytoplasmic organelle containing lysozyme and other hydrolytic enzymes.

lysozyme An enzyme found in sweat, tears, and saliva that breaks down bacterial peptidoglycan.

M

macromolecules Large, molecular compounds assembled from smaller subunits, most notably biochemicals.

macronutrient A chemical substance required in large quantities (phosphate, for example).

macrophage A white blood cell derived from a monocyte that leaves the circulation and enters tissues. These cells are important in nonspecific phagocytosis and in regulating, stimulating, and cleaning up after immune responses.

macroscopic Visible to the naked eye.

malt The grain, usually barley, that is sprouted to obtain digestive enzymes and dried for making beer.

maltose One of the carbohydrates referred to as sugars. A fermentable sugar formed from starch.

Mantoux test An intradermal screening test for tuberculin hypersensitivity. A red, firm patch of skin at the injection site greater than 10 mm in diameter after 48 hours is a positive result that indicates current or prior exposure to the TB bacillus.

mapping Determining the location of loci and other qualities of genomic DNA.

marine microbiology A branch of microbiology that studies the role of microorganisms in the oceans.

marker Any trait or factor of a cell, virus, or molecule that makes it distinct and recognizable. Example: a genetic marker.

mash In making beer, the malt grain is steeped in warm water, ground up, and fortified with carbohydrates to form mash.

mass number (MN) Measurement that reflects the number of protons and neutrons in an atom of a particular element.

mast cell A nonmotile connective tissue cell implanted along capillaries, especially in the lungs, skin, gastrointestinal tract, and genitourinary tract. Like a basophil, its granules store mediators of allergy.

matrix The dense ground substance between the cristae of a mitochondrion that serves as a site for metabolic reactions.

matter All tangible materials that occupy space and have mass.

maximum temperature The highest temperature at which an organism will grow.

MDRTB Multidrug-resistant tuberculosis.

mechanical vector An animal that transports an infectious agent but is not infected by it, such as houseflies whose feet become contaminated with feces.

medium (plural, *media*) A nutrient used to grow organisms outside of their natural habitats.

meiosis The type of cell division necessary for producing gametes in diploid organisms. Two nuclear divisions in rapid succession produce four gametocytes, each containing a haploid number of chromosomes.

membrane In a single cell, a thin double-layered sheet composed of lipids such as phospholipids and sterols and proteins.

memory (immunologic memory) The capacity of the immune system to recognize and act against an antigen upon second and subsequent encounters.

memory cell The long-lived progeny of a sensitized lymphocyte that remains in circulation and is genetically programmed to react rapidly with its antigen.

Mendosicutes Taxonomic category of bacteria that have unusual cell walls; archaea.

meninges The tough tri-layer membrane covering the brain and spinal cord. Consists of the dura mater, arachnoid mater, and pia mater.

meningitis An inflammation of the membranes (meninges) that surround and protect the brain. It is often caused by bacteria such as *Neisseria meningitidis* (the meningococcus) and *Haemophilus influenzae.*

merozoite The motile, infective stage of an apicomplexan parasite that comes from a liver or red blood cell undergoing multiple fission.

mesophile Microorganisms that grow at intermediate temperatures.

messenger RNA (mRNA) A single-stranded transcript that is a copy of the DNA template that corresponds to a gene.

metabolic analog Enzyme that mimics the natural substrate of an enzyme and vies for its active site.

metabolism A general term for the totality of chemical and physical processes occurring in a cell.

metabolites Small organic molecules that are intermediates in the stepwise biosynthesis or breakdown of macromolecules.

metabolomics The study of the complete complement of small chemicals present in a cell at any given time.

metachromatic Exhibiting a color other than that of the dye used to stain it.

metachromatic granules A type of inclusion in storage compartments of some bacteria that stain a contrasting color when treated with colored dyes.

metagenomics The study of all the genomes in a particular ecological niche, as opposed to individual genomes from single species.

methanogens Methane producers.

MHC Major histocompatibility complex. See *HLA.*

MIC Abbreviation for **m**inimum **i**nhibitory **c**oncentration. The lowest concentration of antibiotic needed to inhibit bacterial growth in a test system.

microaerophile An aerobic bacterium that requires oxygen at a concentration less than that in the atmosphere.

microbe See *microorganism.*

microbial antagonism Relationship in which microorganisms compete for survival in a common environment by taking actions that inhibit or destroy another organism.

microbial ecology The study of microbes in their natural habitats.

microbicides Chemicals that kill microorganisms.

microbiology A specialized area of biology that deals with living things ordinarily too small to be seen without magnification, including bacteria, archaea, fungi, protozoa, and viruses.

microfilaments Cellular cytoskeletal element formed by thin protein strands that attach to cell membrane and form a network through the cytoplasm. Responsible for movement of cytoplasm.

micronutrient A chemical substance required in small quantities (trace metals, for example).

microorganism A living thing ordinarily too small to be seen without magnification; an organism of microscopic size.

microscopic Invisible to the naked eye.

microscopy Science that studies structure, magnification, lenses, and techniques related to use of a microscope.

microtubules Long hollow tubes in eukaryotic cells; maintain the shape of the cell and transport substances from one part of cell to another; involved in separating chromosomes in mitosis.

miliary tuberculosis Rapidly fatal tuberculosis due to dissemination of mycobacteria in the blood and formation of tiny granules in various organs and tissues. The term *miliary* means resembling a millet seed.

mineralization The process by which decomposers (bacteria and fungi) convert organic debris into inorganic and elemental form. It is part of the recycling process.

minimum inhibitory concentration (MIC) The smallest concentration of drug needed to visibly control microbial growth.

minimum temperature The lowest temperature at which an organism will grow.

miracidium The ciliated first-stage larva of a trematode. This form is infective for a corresponding intermediate host snail.

missense mutation A mutation in which a change in the DNA sequence results in a different amino acid being incorporated into a protein, with varying results.

mitochondrion A double-membrane organelle of eukaryotes that is the main site for aerobic respiration.

mitosis Somatic cell division that preserves the somatic chromosome number.

mixed acid fermentation An anaerobic degradation of pyruvic acid that results in more than one organic acid being produced (e.g., acetic acid, lactic acid, succinic acid).

mixed culture A container growing two or more different, known species of microbes.

mixed infection Occurs when several different pathogens interact simultaneously to produce an infection. Also called a synergistic infection.

molecule A distinct chemical substance that results from the combination of two or more atoms.

molluscum contagiosum Poxvirus-caused disease that manifests itself by the appearance of small lesions on the face, trunk, and limbs. Can be associated with sexual transmission.

monoclonal antibodies (MAbs) Antibodies that have a single specificity for a single antigen and are produced in the laboratory from a single clone of B cells.

monocyte A large mononuclear leukocyte normally found in the lymph nodes, spleen, bone marrow, and loose connective tissue. This type of cell makes up 3% to 7% of circulating leukocytes.

monomer A simple molecule that can be linked by chemical bonds to form larger molecules.

mononuclear phagocyte system A collection of monocytes and macrophages scattered throughout the extracellular spaces that function to engulf and degrade foreign molecules.

monosaccharide A simple sugar such as glucose that is a basic building block for more complex carbohydrates.

monotrichous Describing a microorganism that bears a single flagellum.

morbidity A diseased condition.

morbidity rate The number of persons afflicted with an illness under question or with illness in general, expressed as a numerator, with the denominator being some unit of population (as in $x/100{,}000$).

mordant A chemical that fixes a dye in or on cells by forming an insoluble compound and thereby promoting retention of that dye. Example: Gram's iodine in the Gram stain.

morphology The study of organismic structure.

mortality rate The number of persons who have died as the result of a particular cause or due to all causes, expressed as a numerator, with the denominator being some unit of population (as in $x/100{,}000$).

most probable number (MPN) Test used to detect the concentration of contaminants in water and other fluids.

motility Self-propulsion.

mumps Viral disease characterized by inflammation of the parotid glands.

must Juices expressed from crushed fruits that are used in fermentation for wine.

mutagen Any agent that induces genetic mutation. Examples: certain chemical substances, ultraviolet light, radioactivity.

mutant strain A subspecies of microorganism that has undergone a mutation, causing expression of a trait that differs from other members of that species.

mutation A permanent inheritable alteration in the DNA sequence or content of a cell.

mutualism Organisms living in an obligatory but mutually beneficial relationship.

mycelium The filamentous mass that makes up a mold. Composed of hyphae.

mycoplasma A genus of bacteria; contain no peptidoglycan/cell wall, but the cytoplasmic membrane is stabilized by sterols.

mycorrhizae Various species of fungi adapted in an intimate, mutualistic relationship to plant roots.

mycosis Any disease caused by a fungus.

myonecrosis Death of muscle tissue.

N

NAD/NADH Abbreviations for the oxidized/reduced forms of nicotinamide adenine dinucleotide, an electron carrier. Also known as the vitamin niacin.

nanobacteria (also *nanobes*) Bacteria that are up to 100 times smaller than average bacteria.

nanobes Cell-like particles found in sediments and other geologic deposits that some scientists speculate are the smallest bacteria. Short for nanobacteria.

narrow spectrum Denotes drugs that are selective and limited in their effects. For example, they inhibit either gram-negative or gram-positive bacteria but not both.

natural immunity Any immunity that arises naturally in an organism via previous experience with the antigen.

natural selection A process in which the environment places pressure on organisms to adapt and survive changing conditions. Only the survivors will be around to continue the life cycle and contribute their genes to future generations. This is considered a major factor in evolution of species.

necrosis A pathologic process in which cells and tissues die and disintegrate.

negative stain A staining technique that renders the background opaque or colored and leaves the object unstained so that it is outlined as a colorless area.

nematode A common name for helminths called roundworms.

nephritis Inflammation of the kidney.

neurotropic Having an affinity for the nervous system. Most likely to affect the spinal cord.

neutralization The process of combining an acid and a base until they reach a balanced proportion, with a pH value close to 7.

neutron An electrically neutral particle in the nuclei of all atoms except hydrogen.

neutrophil A mature granulocyte present in peripheral circulation, exhibiting a multilobular nucleus and numerous cytoplasmic granules that retain a neutral stain. The neutrophil is an active phagocytic cell in bacterial infection.

niche In ecology, an organism's biological role in or contribution to its community.

nitrification Phase of the nitrogen cycle in which ammonium is oxidized.

nitrogen base A ringed compound of which pyrimidines and purines are types.

nitrogen cycle The pathway followed by the element nitrogen as it circulates from inorganic sources in the nonliving environment to living things and back to the nonliving environment. The longtime reservoir is nitrogen gas in the atmosphere.

nitrogen fixation A process occurring in certain bacteria in which atmospheric N_2 gas is converted to a form (NH_4) usable by plants.

nitrogenous base A nitrogen-containing molecule found in DNA and RNA that provides the basis for the genetic code. Adenine, guanine, and cytosine are found in both DNA and RNA while thymine is found exclusively in DNA and uracil is found exclusively in RNA.

nomenclature A set system for scientifically naming organisms, enzymes, anatomical structures, and so on.

noncommunicable An infectious disease that does not arrive through transmission of an infectious agent from host to host.

noncompetitive inhibition Form of enzyme inhibition that involves binding of a regulatory molecule to a site other than the active site.

nonionizing radiation Method of microbial control, best exemplified by ultraviolet light, that causes the formation of abnormal bonds within the DNA of microbes, increasing the rate of mutation. The primary limitation of nonionizing radiation is its inability to penetrate beyond the surface of an object.

nonpolar A term used to describe an electrically neutral molecule formed by covalent bonds between atoms that have the same or similar electronegativity.

nonself Molecules recognized by the immune system as containing foreign markers, indicating a need for immune response.

nonsense codon A triplet of mRNA bases that does not specify an amino acid but signals the end of a polypeptide chain.

nonsense mutation A mutation that changes an amino-acid-producing codon into a stop codon, leading to premature termination of a protein.

normal biota The native microbial forms that an individual harbors.

nosocomial infection An infection not present upon admission to a hospital but incurred while being treated there.

nucleocapsid In viruses, the close physical combination of the nucleic acid with its protective covering.

nucleoid The basophilic nuclear region or nuclear body that contains the bacterial chromosome.

nucleolus A granular mass containing RNA that is contained within the nucleus of a eukaryotic cell.

nucleosome Structure in the packaging of DNA. Formed by the DNA strands wrapping around the histone protein to form nucleus bodies arranged like beads on a chain.

nucleotide The basic structural unit of DNA and RNA; each nucleotide consists of a phosphate, a sugar (ribose in RNA, deoxyribose in DNA), and a nitrogenous base such as adenine, guanine, cytosine, thymine (DNA only), or uracil (RNA only).

numerical aperture In microscopy, the amount of light passing from the object and into the object in order to maximize optical clarity and resolution.

nutrient Any chemical substance that must be provided to a cell for normal metabolism and growth. Macronutrients are required in large amounts, and micronutrients in small amounts.

nutrition The acquisition of chemical substances by a cell or organism for use as an energy source or as building blocks of cellular structures.

O

obligate Without alternative; restricted to a particular characteristic. Example: an obligate parasite survives and grows only in a host; an obligate aerobe must have oxygen to grow; an obligate anaerobe is destroyed by oxygen.

Okazaki fragment In replication of DNA, a segment formed on the lagging strand in which biosynthesis is conducted in a discontinuous manner dictated by the 5′ → 3′ DNA polymerase orientation.

oligodynamic action A chemical having antimicrobial activity in minuscule amounts. Example: certain heavy metals are effective in a few parts per billion.

oligonucleotides Short pieces of DNA or RNA that are easier to handle than long segments.

oligotrophic Nutrient-deficient ecosystem.

oncogene A naturally occurring type of gene that when activated can transform a normal cell into a cancer cell.

oncovirus Mammalian virus capable of causing malignant tumors.

oocyst The encysted form of a fertilized macrogamete or zygote; typical in the life cycles of apicomplexan parasites.

operator In an operon sequence, the DNA segment where transcription of structural genes is initiated.

operon A genetic operational unit that regulates metabolism by controlling mRNA production. In sequence, the unit consists of a regulatory gene, inducer or repressor control sites, and structural genes.

opportunistic In infection, ordinarily nonpathogenic or weakly pathogenic microbes that cause disease primarily in an immunologically compromised host.

opsonization The process of stimulating phagocytosis by affixing molecules (opsonins such as antibodies and complement) to the surfaces of foreign cells or particles.

optimum temperature The temperature at which a species shows the most rapid growth rate.

orbitals The pathways of electrons as they rotate around the nucleus of an atom.

order In the levels of classification, the division of organisms that follows class. Increasing similarity may be noticed among organisms assigned to the same order.

organelle A small component of eukaryotic cells that is bounded by a membrane and specialized in function.

organic chemicals Molecules that contain the basic framework of the elements carbon and hydrogen.

osmophile A microorganism that thrives in a medium having high osmotic pressure.

osmosis The diffusion of water across a selectively permeable membrane in the direction of lower water concentration.

osteomyelitis A focal infection of the internal structures of long bones, leading to pain and inflammation. Often caused by *Staphylococcus aureus*.

outer membrane An additional membrane possessed by gram-negative bacteria; a lipid bilayer containing specialized proteins and polysaccharides. It lies outside of the cell wall.

oxidation In chemical reactions, the loss of electrons by one reactant.

oxidation-reduction Redox reactions, in which paired sets of molecules participate in electron transfers.

oxidative phosphorylation The synthesis of ATP using energy given off during the electron transport phase of respiration.

oxidizing agent An atom or a compound that can receive electrons from another in a chemical reaction.

oxygenic Any reaction that gives off oxygen; usually in reference to the result of photosynthesis in eukaryotes and cyanobacteria.

P

palindrome A word, verse, number, or sentence that reads the same forward or backward. Palindromes of nitrogen bases in DNA have genetic significance as transposable elements, as regulatory protein targets, and in DNA splicing.

palisades The characteristic arrangement of *Corynebacterium* cells resembling a row of fence posts and created by snapping.

pandemic A disease afflicting an increased proportion of the population over a wide geographic area (often worldwide).

papilloma Benign, squamous epithelial growth commonly referred to as a wart.

parasite An organism that lives on or within another organism (the host), from which it obtains nutrients and enjoys protection. The parasite produces some degree of harm in the host.

parasitism A relationship between two organisms in which the host is harmed in some way while the colonizer benefits.

parenteral Administering a substance into a body compartment other than through the gastrointestinal tract, such as via intravenous, subcutaneous, intramuscular, or intramedullary injection.

paroxysmal Events characterized by sharp spasms or convulsions; sudden onset of a symptom such as fever and chills.

passive carrier Persons who mechanically transfer a pathogen without ever being infected by it. For example, a health care worker who doesn't wash his/her hands adequately between patients.

passive immunity Specific resistance that is acquired indirectly by donation of preformed immune substances (antibodies) produced in the body of another individual.

passive transport Nutrient transport method that follows basic physical laws and does not require direct energy input from the cell.

pasteurization Heat treatment of perishable fluids such as milk, fruit juices, or wine to destroy heat-sensitive vegetative cells, followed by rapid chilling to inhibit growth of survivors and germination of spores. It prevents infection and spoilage.

pathogen Any agent (usually a virus, bacterium, fungus, protozoan, or helminth) that causes disease.

pathogenicity The capacity of microbes to cause disease.

pathognomic Distinctive and particular to a single disease, suggestive of a diagnosis.

pathologic Capable of inducing physical damage on the host.

pathology The structural and physiological effects of disease on the body.

pellicle A membranous cover; a thin skin, film, or scum on a liquid surface; a thin film of salivary glycoproteins that forms over newly cleaned tooth enamel when exposed to saliva.

pelvic inflammatory disease (PID) An infection of the uterus and fallopian tubes that has ascended from the lower reproductive tract. Caused by gonococci and chlamydias.

penetration (viral) The step in viral multiplication in which virus enters the host cell.

penicillinase An enzyme that hydrolyzes penicillin; found in penicillin-resistant strains of bacteria.

penicillins A large group of naturally occurring and synthetic antibiotics produced by *Penicillium* mold and active against the cell wall of bacteria.

pentose A monosaccharide with five carbon atoms per molecule. Examples: arabinose, ribose, xylose.

peptide Molecule composed of short chains of amino acids, such as a dipeptide (two amino acids), a tripeptide (three), and a tetrapeptide (four).

peptide bond The covalent union between two amino acids that forms between the amine group of one and the carboxyl group of the other. The basic bond of proteins.

peptidoglycan A network of polysaccharide chains cross-linked by short peptides that forms the rigid part of bacterial cell walls. Gram-negative bacteria have a smaller amount of this rigid structure than do gram-positive bacteria.

perforin Proteins released by cytotoxic T cells that produce pores in target cells.

perinatal In childbirth, occurring before, during, or after delivery.

period of invasion The period during a clinical infection when the infectious agent multiplies at high levels, exhibits its greatest toxicity, and becomes well established in the target tissues.

periodontal Involving the structures that surround the tooth.

periplasmic space The region between the cell wall and cell membrane of the cell envelopes of gram-negative bacteria.

peritrichous In bacterial morphology, having flagella distributed over the entire cell.

petechiae Minute hemorrhagic spots in the skin that range from pinpoint- to pinhead-size.

Peyer's patches Oblong lymphoid aggregates of the gut located chiefly in the wall of the terminal and small intestine. Along with the tonsils and appendix, Peyer's patches make up the gut-associated lymphoid tissue that responds to local invasion by infectious agents.

pH The symbol for the negative logarithm of the H ion concentration; p (power) or $[H^+]_{10}$. A system for rating acidity and alkalinity.

phage A bacteriophage; a virus that specifically parasitizes bacteria.

phagocyte A class of white blood cells capable of engulfing other cells and particles.

phagocytosis A type of endocytosis in which the cell membrane actively engulfs large particles or cells into vesicles.

phagolysosome A body formed in a phagocyte, consisting of a union between a vesicle containing the ingested particle (the phagosome) and a vacuole of hydrolytic enzymes (the lysosome).

phenotype The observable characteristics of an organism produced by the interaction between its genetic potential (genotype) and the environment.

phosphate An acidic salt containing phosphorus and oxygen that is an essential inorganic component of DNA, RNA, and ATP.

phospholipid A class of lipids that compose a major structural component of cell membranes.

phosphorylation Process in which inorganic phosphate is added to a compound.

photoactivation (light repair) A mechanism for repairing DNA with ultraviolet-light-induced mutations using an enzyme (photolyase) that is activated by visible light.

photoautotroph An organism that utilizes light for its energy and carbon dioxide chiefly for its carbon needs.

photolysis The splitting of water into hydrogen and oxygen during photosynthesis.

photon A subatomic particle released by electromagnetic sources such as radiant energy (sunlight). Photons are the ultimate source of energy for photosynthesis.

photophosphorylation The process of electron transport during photosynthesis that results in the synthesis of ATP from ADP.

photosynthesis A process occurring in plants, algae, and some bacteria that traps the sun's energy and converts it to ATP in the cell. This energy is used to fix CO_2 into organic compounds.

phototrophs Microbes that use photosynthesis to feed.

phycobilin Red or blue-green pigments that absorb light during photosynthesis.

phylum In the levels of classification, the third level of classification from general to more specific. Each kingdom is divided into numerous phyla. Sometimes referred to as a division.

physiology The study of the function of an organism.

phytoplankton The collection of photosynthetic microorganisms (mainly algae and cyanobacteria) that float in the upper layers of aquatic habitats where sun penetrates. These microbes are the basis of aquatic food pyramids and, together with zooplankton, make up the plankton.

pili Small, stiff filamentous appendages in gram-negative bacteria that function in DNA exchange during bacterial conjugation.

pilus A hollow appendage used to bring two bacterial cells together to transfer DNA.

pinocytosis The engulfment, or endocytosis, of liquids by extensions of the cell membrane.

plague Zoonotic disease caused by infection with *Yersinia pestis.* The pathogen is spread by flea vectors and harbored by various rodents.

plankton Minute animals (zooplankton) or plants (phytoplankton) that float and drift in the limnetic zone of bodies of water.

plantar warts Deep, painful warts on the soles of the feet as a result of infection by human papillomavirus.

plaque In virus propagation methods, the clear zone of lysed cells in tissue culture or chick embryo membrane that corresponds to the area containing viruses. In dental application, the filamentous mass of microbes that adheres tenaciously to the tooth and predisposes to caries, calculus, or inflammation.

plasma The carrier fluid element of blood.

plasma cell A progeny of an activated B cell that actively produces and secretes antibodies.

plasmids Extrachromosomal genetic units characterized by several features. A plasmid is a double-stranded DNA that is smaller than and replicates independently of the cell chromosome; it bears genes that are not essential for cell growth; it can bear genes that code for adaptive traits; and it is transmissible to other bacteria.

platelet-activating factor A substance released from basophils that causes release of allergic mediators and the aggregation of platelets.

platelets Formed elements in the blood that develop when megakaryocytes disintegrate. Platelets are involved in hemostasis and blood clotting.

pleomorphism Normal variability of cell shapes in a single species.

pluripotential Stem cells having the developmental plasticity to give rise to more than one type. Example: undifferentiated blood cells in the bone marrow.

pneumococcus Common name for *Streptococcus pneumoniae,* the major cause of bacterial pneumonia.

pneumonia An inflammation of the lung leading to accumulation of fluid and respiratory compromise.

pneumonic plague The acute, frequently fatal form of pneumonia caused by *Yersinia pestis.*

point mutation A change that involves the loss, substitution, or addition of one or a few nucleotides.

polar Term to describe a molecule with an asymmetrical distribution of charges. Such a molecule has a negative pole and a positive pole.

poliomyelitis An acute enteroviral infection of the spinal cord that can cause neuromuscular paralysis.

polyclonal In reference to a collection of antibodies with mixed specificities that arose from more than one clone of B cells.

polyclonal antibodies A mixture of antibodies that were stimulated by a complex antigen with more than one antigenic determinant.

polymer A macromolecule made up of a chain of repeating units. Examples: starch, protein, DNA.

polymerase An enzyme that produces polymers through catalyzing bond formation between building blocks (polymerization).

polymerase chain reaction (PCR) A technique that amplifies segments of DNA for testing. Using denaturation, primers, and heat-resistant DNA polymerase, the number can be increased several-million-fold.

polymicrobial Involving multiple distinct microorganisms.

polymorphonuclear leukocytes (PMNLs) White blood cells with variously shaped nuclei. Although this term commonly denotes all granulocytes, it is used especially for the neutrophils.

polymyxin A mixture of antibiotic polypeptides from *Bacillus polymyxa* that are particularly effective against gram-negative bacteria.

polypeptide A relatively large chain of amino acids linked by peptide bonds.

polyribosomal complex An assembly line for mass production of proteins composed of a chain of ribosomes involved in mRNA transcription.

polysaccharide A carbohydrate that can be hydrolyzed into a number of monosaccharides. Examples: cellulose, starch, glycogen.

population A group of organisms of the same species living simultaneously in the same habitat. A group of different populations living together constitutes the community level.

porin Transmembrane proteins of the outer membrane of gram-negative cells that permit transport of small molecules into the periplasmic space but bar the penetration of larger molecules.

portal of entry Route of entry for an infectious agent; typically a cutaneous or membranous route.

portal of exit Route through which a pathogen departs from the host organism.

positive stain A method for coloring microbial specimens that involves a chemical that sticks to the specimen to give it color.

potable Describing water that is relatively clear, odor-free, and safe to drink.

PPNG Penicillinase-producing *Neisseria gonorrhoeae.*

prebiotics Nutrients used to stimulate the growth of favorable biota in the intestine.

prevalence The total number of cases of a disease in a certain area and time period.

primary infection An initial infection in a previously healthy individual that is later complicated by an additional (secondary) infection.

primary response The first response of the immune system when exposed to an antigen.

primary structure Initial protein organization described by type, number, and order of amino acids in the chain. The primary structure varies extensively from protein to protein.

primers Synthetic oligonucleotides of known sequence that serve as landmarks to indicate where DNA amplification will begin.

prion A concocted word to denote "proteinaceous infectious agent"; a cytopathic protein associated with the slow-virus spongiform encephalopathies of humans and animals.

probes Small fragments of single-stranded DNA (RNA) that are known to be complementary to the specific sequence of DNA being studied.

probiotics Preparations of live microbes used as a preventive or therapeutic measure to displace or compete with potential pathogens.

prodromal stage A short period of mild symptoms occurring at the end of the period of incubation. It indicates the onset of disease.

producer An organism that synthesizes complex organic compounds from simple inorganic molecules. Examples would be photosynthetic microbes and plants. These organisms are solely responsible for originating food pyramids and are the basis for life on earth (also called autotroph).

product(s) In a chemical reaction, the substance(s) that is(are) left after a reaction is completed.

proglottid The egg-generating segment of a tapeworm that contains both male and female organs.

progressive multifocal leukoencephalopathy (PML) An uncommon, fatal complication of infection with JC virus (polyoma virus).

prokaryotic cell Small cells, lacking special structures such as a nucleus and organelles. All prokaryotes are microorganisms.

promastigote A morphological variation of the trypanosome parasite responsible for leishmaniasis.

promoter Part of an operon sequence. The DNA segment that is recognized by RNA polymerase as the starting site for transcription.

promoter region The site composed of a short signaling DNA sequence that RNA polymerase recognizes and binds to commence transcription.

prophage A lysogenized bacteriophage; a phage that is latently incorporated into the host chromosome instead of undergoing viral replication and lysis.

prophylactic Any device, method, or substance used to prevent disease.

prostaglandin A hormonelike substance that regulates many body functions. Prostaglandin comes from a family of organic acids containing 5-carbon rings that are essential to the human diet.

protease Enzymes that act on proteins, breaking them down into component parts.

protease inhibitors Drugs that act to prevent the assembly of functioning viral particles.

protein Predominant organic molecule in cells, formed by long chains of amino acids.

proteomics The study of an organism's complement of proteins (its *proteome*) and functions mediated by the proteins.

proton An elementary particle that carries a positive charge. It is identical to the nucleus of the hydrogen atom.

protoplast A bacterial cell whose cell wall is completely lacking and that is vulnerable to osmotic lysis.

protozoa A group of single-celled, eukaryotic organisms.

pseudohypha A chain of easily separated, spherical to sausage-shaped yeast cells partitioned by constrictions rather than by septa.

pseudomembrane A tenacious, noncellular mucous exudate containing cellular debris that tightly blankets the mucosal surface in infections such as diphtheria and pseudomembranous enterocolitis.

pseudopodium A temporary extension of the protoplasm of an amoeboid cell. It serves both in amoeboid motion and for food gathering (phagocytosis).

pseudopods Protozoan appendage responsible for motility. Also called "false feet."

psychrophile A microorganism that thrives at low temperature (0°C–20°C), with a temperature optimum of 0°C–15°C.

pulmonary Occurring in the lungs. Examples include pulmonary anthrax and pulmonary nocardiosis.

pure culture A container growing a single species of microbe whose identity is known.

purine A nitrogen base that is an important encoding component of DNA and RNA. The two most common purines are adenine and guanine.

pus The viscous, opaque, usually yellowish matter formed by an inflammatory infection. It consists of serum exudate, tissue debris, leukocytes, and microorganisms.

pyogenic Pertains to pus formers, especially the pyogenic cocci: pneumococci, streptococci, staphylococci, and neisseriae.

pyrimidine Nitrogen bases that help form the genetic code on DNA and RNA. Uracil, thymine, and cytosine are the most important pyrimidines.

pyrimidine dimer The union of two adjacent pyrimidines on the same DNA strand, brought about by exposure to ultraviolet light. It is a form of mutation.

pyrogen A substance that causes a rise in body temperature. It can come from pyrogenic microorganisms or from polymorphonuclear leukocytes (endogenous pyrogens).

Q

quaternary structure Most complex protein structure characterized by the formation of large, multiunit proteins by more than one of the polypeptides. This structure is typical of antibodies and some enzymes that act in cell synthesis.

quats A word that pertains to a family of surfactants called quaternary ammonium compounds. These detergents are only weakly microbicidal and are used as sanitizers and preservatives.

quinine A substance derived from cinchona trees that was used as an antimalarial treatment; has been replaced by synthetic derivatives.

quinolone A class of synthetic antimicrobic drugs with broad-spectrum effects.

R

rabies The only rhabdovirus that infects humans. Zoonotic disease characterized by fatal meningoencephalitis.

radiation Electromagnetic waves or rays, such as those of light given off from an energy source.

radioactive isotopes Unstable isotopes whose nuclei emit particles of radiation. This emission is called radioactivity or radioactive decay. Three naturally occurring emissions are alpha, beta, and gamma radiation.

rales Sounds in the lung, ranging from clicking to rattling; indicate respiratory illness.

reactants Molecules entering or starting a chemical reaction.

real image An image formed at the focal plane of a convex lens. In the compound light microscope, it is the image created by the objective lens.

receptor Cell surface molecules involved in recognition, binding, and intracellular signaling.

recombinant An organism that contains genes that originated in another organism, whether through deliberate laboratory manipulation or natural processes.

recombinant DNA technology A technology, also known as genetic engineering, that deliberately modifies the genetic structure of an organism to create novel products, microbes, animals, plants, and viruses.

recombination A type of genetic transfer in which DNA from one organism is donated to another.

recycling A process that converts unusable organic matter from dead organisms back into their essential inorganic elements and returns them to their nonliving reservoirs to make them available again for living organisms. This is a common term that means the same as mineralization and decomposition.

redox Denoting an oxidation-reduction reaction.

reducing agent An atom or a compound that can donate electrons in a chemical reaction.

reduction In chemistry, the gain of electrons.

redundancy The property of the genetic code that allows an amino acid to be specified by several different codons.

refraction In optics, the bending of light as it passes from one medium to another with a different index of refraction.

regulated enzymes Enzymes whose extent of transcription or translation is influenced by changes in the environment.

regulator DNA segment that codes for a protein capable of repressing an operon.

regulatory site The location on an enzyme where a certain substance can bind and block the enzyme's activity.

rennin The enzyme casein coagulase, which is used to produce curd in the processing of milk and cheese.

replication In DNA synthesis, the semiconservative mechanisms that ensure precise duplication of the parent DNA strands.

replication fork The Y-shaped point on a replicating DNA molecule where the DNA polymerase is synthesizing new strands of DNA.

reportable disease Those diseases that must be reported to health authorities by law.

repressible operon An operon that under normal circumstances is transcribed. The buildup of the operon's amino acid product causes transcription of the operon to stop.

repressor The protein product of a repressor gene that combines with the operator and arrests the transcription and translation of structural genes.

reservoir In disease communication, the natural host or habitat of a pathogen.

resident biota The deeper, more stable microbiota that inhabit the skin and exposed mucous membranes, as opposed to the superficial, variable, transient population.

resistance (R) factor Plasmids, typically shared among bacteria by conjugation, that provide resistance to the effects of antibiotics.

resolving power The capacity of a microscope lens system to accurately distinguish between two separate entities that lie close to each other. Also called resolution.

respiratory chain A series of enzymes that transfer electrons from one to another, resulting in the formation of ATP. It is also known as the electron transport chain. The chain is located in the cell membrane of bacteria and in the inner mitochondrial membrane of eukaryotes.

respiratory syncytial virus (RSV) An RNA virus that infects the respiratory tract. RSV is the most prevalent cause of respiratory infection in newborns.

restriction endonuclease An enzyme present naturally in cells that cleaves specific locations on DNA. It is an important means of inactivating viral genomes, and it is also used to splice genes in genetic engineering.

reticuloendothelial system Also known as the mononuclear phagocyte system, it pertains to a network of fibers and phagocytic cells (macrophages) that permeates the tissues of all organs. Examples: Kupffer cells in liver sinusoids, alveolar phagocytes in the lung, microglia in nervous tissue.

retrovirus A group of RNA viruses (including HIV) that have the mechanisms for converting their genome into a double strand of DNA that can be inserted on a host's chromosome.

reverse transcriptase (RT) The enzyme possessed by retroviruses that carries out the reversion of RNA to DNA—a form of reverse transcription.

Reye's syndrome A sudden, usually fatal neurological condition that occurs in children after a viral infection. Autopsy shows cerebral edema and marked fatty change in the liver and renal tubules.

Rh factor An isoantigen that can trigger hemolytic disease in newborns due to incompatibility between maternal and infant blood factors.

rhizobia Bacteria that live in plant roots and supply supplemental nitrogen that boosts plant growth.

rhizosphere The zone of soil, complete with microbial inhabitants, in the immediate vicinity of plant roots.

ribonucleic acid (RNA) The nucleic acid responsible for carrying out the hereditary program transmitted by an organism's DNA.

ribose A 5-carbon monosaccharide found in RNA.

ribosomal RNA (rRNA) A single-stranded transcript that is a copy of part of the DNA template.

ribosome A bilobed macromolecular complex of ribonucleoprotein that coordinates the codons of mRNA with tRNA anticodons and, in so doing, constitutes the peptide assembly site.

ribozyme A part of an RNA-containing enzyme in eukaryotes that removes intervening sequences of RNA called introns and splices together the true coding sequences (exons) to form a mature messenger RNA.

rickettsias Medically important family of bacteria, commonly carried by ticks, lice, and fleas. Significant cause of important emerging diseases.

ringworm A superficial mycosis caused by various dermatophytic fungi. This common name is actually a misnomer.

RNA polymerase Enzyme process that translates the code of DNA to RNA.

rolling circle An intermediate stage in viral replication of circular DNA into linear DNA.

root nodules Small growths on the roots of legume plants that arise from a symbiotic association between the plant tissues and bacteria (Rhizobia). This association allows fixation of nitrogen gas from the air into a usable nitrogen source for the plant.

rosette formation A technique for distinguishing surface receptors on T cells by reacting them with sensitized indicator sheep red blood cells. The cluster of red cells around the central white blood cell resembles a little rose blossom and is indicative of the type of receptor.

rough endoplasmic reticulum (RER) Microscopic series of tunnels that originates in the outer membrane of the nuclear envelope and is used in transport and storage. Large numbers of ribosomes, partly attached to the membrane, give the rough appearance.

rubeola (red measles) Acute disease caused by infection with Morbillivirus.

S

saccharide Scientific term for sugar. Refers to a simple carbohydrate with a sweet taste.

salpingitis Inflammation of the fallopian tubes.

sanitize To clean inanimate objects using soap and degerming agents so that they are safe and free of high levels of microorganisms.

saprobe A microbe that decomposes organic remains from dead organisms. Also known as a saprophyte or saprotroph.

sarcina A cubical packet of 8, 16, or more cells; the cellular arrangement of the genus *Sarcina* in the family Micrococcaceae.

satellitism A commensal interaction between two microbes in which one can grow in the vicinity of the other due to nutrients or protective factors released by that microbe.

saturation The complete occupation of the active site of a carrier protein or enzyme by the substrate.

schistosomiasis Infection by blood fluke, often as a result of contact with contaminated water in rivers and streams. Symptoms appear in liver, spleen, or urinary system depending on species of *Schistosoma*. Infection may be chronic.

schizogony A process of multiple fission whereby first the nucleus divides several times, and subsequently the cytoplasm is subdivided for each new nucleus during cell division.

scientific method Principles and procedures for the systematic pursuit of knowledge, involving the recognition and formulation of a problem, the collection of data through observation and experimentation, and the formulation and testing of a hypothesis.

scolex The anterior end of a tapeworm characterized by hooks and/or suckers for attachment to the host.

sebaceous glands The sebum- (oily, fatty) secreting glands of the skin.

sebum Low pH, oil-based secretion of the sebaceous glands.

secondary infection An infection that compounds a preexisting one.

secondary response The rapid rise in antibody titer following a repeat exposure to an antigen that has been recognized from a previous exposure. This response is brought about by memory cells produced as a result of the primary exposure.

secondary structure Protein structure that occurs when the functional groups on the outer surface of the molecule interact by forming hydrogen bonds. These bonds cause the amino acid chain to either twist, forming a helix, or to pleat into an accordion pattern called a β-pleated sheet.

secretory antibody The immunoglobulin (IgA) that is found in secretions of mucous membranes and serves as a local immediate protection against infection.

selective media Nutrient media designed to favor the growth of certain microbes and to inhibit undesirable competitors.

selectively toxic Property of an antimicrobial agent to be highly toxic against its target microbe while being far less toxic to other cells, particularly those of the host organism.

self Natural markers of the body that are recognized by the immune system.

self-limited Applies to an infection that runs its course without disease or residual effects.

semiconservative replication In DNA replication, the synthesis of paired daughter strands, each retaining a parent strand template.

semisolid media Nutrient media with a firmness midway between that of a broth (a liquid medium) and an ordinary solid medium; motility media.

semisynthetic Drugs that, after being naturally produced by bacteria, fungi, or other living sources, are chemically modified in the laboratory.

sensitizing dose The initial effective exposure to an antigen or an allergen that stimulates an immune response. Often applies to allergies.

sepsis The state of putrefaction; the presence of pathogenic organisms or their toxins in tissue or blood.

septic shock Blood infection resulting in a pathological state of low blood pressure accompanied by a reduced amount of blood circulating to vital organs. Endotoxins of all gram-negative bacteria can cause shock, but most clinical cases are due to gram-negative enteric rods.

septicemia Systemic infection associated with microorganisms multiplying in circulating blood.

septum A partition or cellular cross wall, as in certain fungal hyphae.

sequela A morbid complication that follows a disease.

sequencing Determining the actual order and types of bases in a segment of DNA.

serology The branch of immunology that deals with *in vitro* diagnostic testing of serum.

seropositive Showing the presence of specific antibody in a serological test. Indicates ongoing infection.

serotonin A vasoconstrictor that inhibits gastric secretion and stimulates smooth muscle.

serotyping The subdivision of a species or subspecies into an immunologic type, based upon antigenic characteristics.

serum The clear fluid expressed from clotted blood that contains dissolved nutrients, antibodies, and hormones but not cells or clotting factors.

serum sickness A type of immune complex disease in which immune complexes enter circulation, are carried throughout the body, and are deposited in the blood vessels of the kidney, heart, skin, and joints. The condition may become chronic.

severe acute respiratory syndrome (SARS) A severe respiratory disease caused by infection with a newly described coronavirus.

severe combined immunodeficiencies A collection of syndromes occurring in newborns caused by a genetic defect that knocks out both B and T cell types of immunity. There are several versions of this disease, termed SCIDS for short.

sex pilus A conjugative pilus.

sexually transmitted disease (STD) Infections resulting from pathogens that enter the body via sexual intercourse or intimate, direct contact.

shiga toxin Heat-labile exotoxin released by some *Shigella* species and by *E. coli* 0157:H7; responsible for worst symptoms of these infections.

shingles Lesions produced by reactivated human herpesvirus 3 (chickenpox) infection; also known as herpes zoster.

sign Any abnormality uncovered upon physical diagnosis that indicates the presence of disease. A sign is an objective assessment of disease, as opposed to a symptom, which is the subjective assessment perceived by the patient.

silent mutation A mutation that, because of the degeneracy of the genetic code, results in a nucleotide change in both the DNA and mRNA but not the resultant amino acid and thus, not the protein.

simple stain Type of positive staining technique that uses a single dye to add color to cells so that they are easier to see. This technique tends to color all cells the same color.

smooth endoplasmic reticulum (SER) A microscopic series of tunnels lacking ribosomes that functions in the nutrient processing function of a cell.

solute A substance that is uniformly dispersed in a dissolving medium or solvent.

solution A mixture of one or more substances (solutes) that cannot be separated by filtration or ordinary settling.

solvent A dissolving medium.

somatic (O or cell wall antigen) One of the three major antigens commonly used to differentiate gram-negative enteric bacteria.

source The person or item from which an infection is directly acquired. See *reservoir*.

Southern blot A technique that separates fragments of DNA using electrophoresis and identifies them by hybridization.

species In the levels of classification, the most specific level of organization.

specificity Limited to a single, precise characteristic or action.

spheroplast A gram-negative cell whose peptidoglycan, when digested by lysozyme, remains intact but is osmotically vulnerable.

spike A receptor on the surface of certain enveloped viruses that facilitates specific attachment to the host cell.

spirillum A type of bacterial cell with a rigid spiral shape and external flagella.

spirochete A coiled, spiral-shaped bacterium that has endoflagella and flexes as it moves.

spontaneous generation Early belief that living things arose from vital forces present in nonliving, or decomposing, matter.

spontaneous mutation A mutation in DNA caused by random mistakes in replication and not known to be influenced by any mutagenic agent. These mutations give rise to an organism's natural, or background, rate of mutation.

sporadic Description of a disease that exhibits new cases at irregular intervals in unpredictable geographic locales.

sporangiospore A form of asexual spore in fungi; enclosed in a sac.

sporangium A fungal cell in which asexual spores are formed by multiple cell cleavage.

spore A differentiated, specialized cell form that can be used for dissemination, for survival in times of adverse conditions, and/or for reproduction. Spores are usually unicellular and may develop into gametes or vegetative organisms.

sporicide A chemical agent capable of destroying bacterial endospores.

sporozoite One of many minute elongated bodies generated by multiple division of the oocyst. It is the infectious form of the malarial parasite that is harbored in the salivary gland of the mosquito and inoculated into the victim during feeding.

sporulation The process of spore formation.

start codon The nucleotide triplet AUG that codes for the first amino acid in protein sequences.

starter culture The sizable inoculation of pure bacterial, mold, or yeast sample for bulk processing, as in the preparation of fermented foods, beverages, and pharmaceuticals.

stasis A state of rest or inactivity; applied to nongrowing microbial cultures. Also called microbistasis.

stationary growth phase Survival mode in which cells either stop growing or grow very slowly.

stem cells Pluripotent, undifferentiated cells.

sterile Completely free of all life forms, including spores and viruses.

sterilization Any process that completely removes or destroys all viable microorganisms, including viruses, from an object or habitat. Material so treated is sterile.

STORCH Acronym for common infections of the fetus and neonate. Storch stands for **s**yphilis, **t**oxoplasmosis, **o**ther diseases (hepatitis B, AIDS and chlamydiosis), **r**ubella, **c**ytomegalovirus, and **h**erpes simplex virus.

strain In microbiology, a set of descendants cloned from a common ancestor that retain the original characteristics. Any deviation from the original is a different strain.

streptolysin A hemolysin produced by streptococci.

strict or obligate anaerobe An organism that does not use oxygen gas in metabolism and cannot survive in oxygen's presence.

stroma The matrix of the chloroplast that is the site of the dark reactions.

structural gene A gene that codes for the amino acid sequence (peptide structure) of a protein.

subacute Indicates an intermediate status between acute and chronic disease.

subacute sclerosing panencephalitis (SSPE) A complication of measles infection in which progressive neurological degeneration of the cerebral cortex invariably leads to coma and death.

subcellular vaccine A vaccine preparation that contains specific antigens such as the capsule or toxin from a pathogen and not the whole microbe.

subclinical A period of inapparent manifestations that occurs before symptoms and signs of disease appear.

subculture To make a second-generation culture from a well-established colony of organisms.

subcutaneous The deepest level of the skin structure.

substrate The specific molecule upon which an enzyme acts.

subunit vaccine A vaccine preparation that contains only antigenic fragments such as surface receptors from the microbe. Usually in reference to virus vaccines.

sucrose One of the carbohydrates commonly referred to as sugars. Common table or cane sugar.

sulfonamide Antimicrobial drugs that interfere with the essential metabolic process of bacteria and some fungi.

superantigens Bacterial toxins that are potent stimuli for T cells and can be a factor in diseases such as toxic shock.

superficial mycosis A fungal infection located in hair, nails, and the epidermis of the skin.

superinfection An infection occurring during antimicrobial therapy that is caused by an overgrowth of drug-resistant microorganisms.

superoxide A toxic derivative of oxygen; (O_2^-).

surfactant A surface-active agent that forms a water-soluble interface. Examples: detergents, wetting agents, dispersing agents, and surface tension depressants.

sylvatic Denotes the natural presence of disease among wild animal populations. Examples: sylvatic (sylvan) plague, rabies.

symbiosis An intimate association between individuals from two species; used as a synonym for mutualism.

symptom The subjective evidence of infection and disease as perceived by the patient.

syncytium A multinucleated protoplasmic mass formed by consolidation of individual cells.

syndrome The collection of signs and symptoms that, taken together, paint a portrait of the disease.

synergism The coordinated or correlated action by two or more drugs or microbes that results in a heightened response or greater activity.

syngamy Conjugation of the gametes in fertilization.

synthesis (viral) The step in viral multiplication in which viral genetic material and proteins are made through replication and transcription/translation.

synthetic biology The use of known genes to produce new applications.

syntrophy The productive use of waste products from the metabolism of one organism by a second organism.

syphilis A sexually transmitted bacterial disease caused by the spirochete *Treponema pallidum*.

systemic Occurring throughout the body; said of infections that invade many compartments and organs via the circulation.

T

T lymphocyte (T cell) A white blood cell that is processed in the thymus gland and is involved in cell-mediated immunity.

Taq polymerase DNA polymerase from the thermophilic bacterium *Thermus aquaticus* that enables high-temperature replication of DNA required for the polymerase chain reaction.

tartar See *calculus*.

taxa Taxonomic categories.

taxonomy The formal system for organizing, classifying, and naming living things.

temperate phage A bacteriophage that enters into a less virulent state by becoming incorporated into the host genome as a prophage instead of in the vegetative or lytic form that eventually destroys the cell.

template The strand in a double-stranded DNA molecule that is used as a model to synthesize a complementary strand of DNA or RNA during replication or transcription.

Tenericutes Taxonomic category of bacteria that lack cell walls.

teratogenic Causing abnormal fetal development.

tertiary structure Protein structure that results from additional bonds forming between functional groups in a secondary structure, creating a three-dimensional mass.

tetanospasmin The neurotoxin of *Clostridium tetani*, the agent of tetanus. Its chief action is directed upon the inhibitory synapses of the anterior horn motor neurons.

tetracyclines A group of broad-spectrum antibiotics with a complex 4-ring structure.

tetrads Groups of four.

theory A collection of statements, propositions, or concepts that explains or accounts for a natural event.

therapeutic index The ratio of the toxic dose to the effective therapeutic dose that is used to assess the safety and reliability of the drug.

thermal death point The lowest temperature that achieves sterilization in a given quantity of broth culture upon a 10-minute exposure. Examples: 55°C for *Escherichia coli*, 60°C for *Mycobacterium tuberculosis*, and 120°C for spores.

thermal death time The least time required to kill all cells of a culture at a specified temperature.

thermocline A temperature buffer zone in a large body of water that separates the warmer water (the epilimnion) from the colder water (the hypolimnion).

thermoduric Resistant to the harmful effects of high temperature.

thermophile A microorganism that thrives at a temperature of 50°C or higher.

thrush *Candida albicans* infection of the oral cavity.

thylakoid Vesicles of a chloroplast formed by elaborate folding of the inner membrane to form "discs." Solar energy trapped in the thylakoids is used in photosynthesis.

thymine (T) One of the nitrogen bases found in DNA but not in RNA. Thymine is in a pyrimidine form.

thymus Butterfly-shaped organ near the tip of the sternum that is the site of T-cell maturation.

tincture A medicinal substance dissolved in an alcoholic solvent.

tinea Ringworm; a fungal infection of the hair, skin, or nails.

tinea versicolor A condition of the skin appearing as mottled and discolored skin pigmentation as a result of infection by the yeast *Malassezia furfur*.

titer In immunochemistry, a measure of antibody level in a patient, determined by agglutination methods.

tonsils A ring of lymphoid tissue in the pharynx that acts as a repository for lymphocytes.

topoisomerases Enzymes that can add or remove DNA twists and thus regulate the degree of supercoiling.

toxemia Condition in which a toxin (microbial or otherwise) is spread throughout the bloodstream.

toxigenicity The tendency for a pathogen to produce toxins. It is an important factor in bacterial virulence.

toxin A specific chemical product of microbes, plants, and some animals that is poisonous to other organisms.

toxinosis Disease whose adverse effects are primarily due to the production and release of toxins.

toxoid A toxin that has been rendered nontoxic but is still capable of eliciting the formation of protective antitoxin antibodies; used in vaccines.

trace elements Micronutrients (zinc, nickel, and manganese) that occur in small amounts and are involved in enzyme function and maintenance of protein structure.

transamination The transfer of an amino group from an amino acid to a carbohydrate fragment.

transcript A newly transcribed RNA molecule.

transcription mRNA synthesis; the process by which a strand of RNA is produced against a DNA template.

transduction The transfer of genetic material from one bacterium to another by means of a bacteriophage vector.

transfer RNA (tRNA) A transcript of DNA that specializes in converting RNA language into protein language.

transformation In microbial genetics, the transfer of genetic material contained in "naked" DNA fragments from a donor cell to a competent recipient cell.

transfusion Infusion of whole blood, red blood cells, or platelets directly into a patient's circulation.

translation Protein synthesis; the process of decoding the messenger RNA code into a polypeptide.

transposon A DNA segment with an insertion sequence at each end, enabling it to migrate to another plasmid, to the bacterial chromosome, or to a bacteriophage.

traveler's diarrhea A type of gastroenteritis typically caused by infection with enterotoxigenic strains of *E. coli* that are ingested through contaminated food and water.

trematode A category of helminth; also known as flatworm or fluke.

trichinosis Infection by the *Trichinella spiralis* parasite, usually caused by eating the meat of an infected animal. Early symptoms include fever, diarrhea, nausea, and abdominal pain that progress to intense muscle and joint pain and shortness of breath. In the final stages, heart and brain function are at risk, and death is possible.

trichomoniasis Sexually transmitted disease caused by infection by the trichomonads, a group of protozoa. Symptoms include urinary pain and frequency and foul-smelling vaginal discharge in females or recurring urethritis, with a thin milky discharge, in males.

triglyceride A type of lipid composed of a glycerol molecule bound to three fatty acids.

triplet See *codon*.

trophozoite A vegetative protozoan (feeding form) as opposed to a resting (cyst) form.

true pathogen A microbe capable of causing infection and disease in healthy persons with normal immune defenses.

trypomastigote The infective morphological stage transmitted by the tsetse fly or the reduviid bug in African trypanosomiasis and Chagas disease.

tubercle In tuberculosis, the granulomatous well-defined lung lesion that can serve as a focus for latent infection.

tuberculin A glycerinated broth culture of *Mycobacterium tuberculosis* that is evaporated and filtered. Formerly used to treat tuberculosis, tuberculin is now used chiefly for diagnostic tests.

tuberculin reaction A diagnostic test in which PPD, or purified protein derivative (of *M. tuberculosis*), is injected superficially under the skin and the area of reaction measured; also called the Mantoux test.

tuberculoid leprosy A superficial form of leprosy characterized by asymmetrical, shallow skin lesions containing few bacterial cells.

turbid Cloudy appearance of nutrient solution in a test tube due to growth of microbe population.

tyndallization Fractional (discontinuous, intermittent) sterilization designed to destroy spores indirectly. A preparation is exposed to flowing steam for an hour, and then the mineral is allowed to incubate to permit spore germination. The resultant vegetative cells are destroyed by repeated steaming and incubation.

typhoid fever Form of salmonellosis. It is highly contagious. Primary symptoms include fever, diarrhea, and abdominal pain. Typhoid fever can be fatal if untreated.

U

ubiquitous Present everywhere at the same time.

ultraviolet (UV) radiation Radiation with an effective wavelength from 240 nm to 260 nm. UV radiation induces mutations readily but has very poor penetrating power.

uncoating The process of removal of the viral coat and release of the viral genome by its newly invaded host cell.

undulant fever See *brucellosis*.

universal donor In blood grouping and transfusion, a group O individual whose erythrocytes bear neither agglutinogen A nor B.

universal precautions (UPs) Centers for Disease Control and Prevention guidelines for health care workers regarding the prevention of disease transmission when handling patients and body substances.

uracil (U) One of the nitrogen bases in RNA but not in DNA. Uracil is in a pyrimidine form.

urinary tract infection (UTI) Invasion and infection of the urethra and bladder by bacterial residents, most often *E. coli.*

V

vaccination Exposing a person to the antigenic components of a microbe without its pathogenic effects for the purpose of inducing a future protective response.

vaccine Originally used in reference to inoculation with the cowpox or vaccinia virus to protect against smallpox. In general, the term now pertains to injection of whole microbes (killed or attenuated), toxoids, or parts of microbes as a prevention or cure for disease.

vacuoles In the cell, membrane-bounded sacs containing fluids or solid particles to be digested, excreted, or stored.

valence The combining power of an atom based upon the number of electrons it can either take on or give up.

van der Waals forces Weak attractive interactions between molecules of low polarity.

vancomycin Antibiotic that targets the bacterial cell wall; used often in antibiotic-resistant infections.

variable region The antigen binding fragment of an immunoglobulin molecule, consisting of a combination of heavy and light chains whose molecular conformation is specific for the antigen.

varicella Informal name for virus responsible for chickenpox as well as shingles; also known as human herpesvirus 3 (HHV-3).

variolation A hazardous, outmoded process of deliberately introducing smallpox material scraped from a victim into the nonimmune subject in the hope of inducing resistance.

vector An animal that transmits infectious agents from one host to another, usually a biting or piercing arthropod like the tick, mosquito, or fly. Infectious agents can be conveyed mechanically by simple contact or biologically whereby the parasite develops in the vector. A genetic element such as a plasmid or a bacteriophage used to introduce genetic material into a cloning host during recombinant DNA experiments.

vegetative In describing microbial developmental stages, a metabolically active feeding and dividing form, as opposed to a dormant, seemingly inert, nondividing form. Examples: a bacterial cell versus its spore; a protozoan trophozoite versus its cyst.

vehicle An inanimate material (solid object, liquid, or air) that serves as a transmission agent for pathogens.

vesicle A blister characterized by a thin-skinned, elevated, superficial pocket filled with serum.

viable nonculturable (VNC) Describes microbes that cannot be cultivated in the laboratory but that maintain metabolic activity (i.e., are alive).

vibrio A curved, rod-shaped bacterial cell.

viremia The presence of viruses in the bloodstream.

virion An elementary virus particle in its complete morphological and thus infectious form. A virion consists of the nucleic acid core surrounded by a capsid, which can be enclosed in an envelope.

viroid An infectious agent that, unlike a virion, lacks a capsid and consists of a closed circular RNA molecule. Although known viroids are all plant pathogens, it is conceivable that animal versions exist.

virtual image In optics, an image formed by diverging light rays; in the compound light microscope, the second, magnified visual impression formed by the ocular from the real image formed by the objective.

virucide A chemical agent that inactivates viruses, especially on living tissue.

virulence In infection, the relative capacity of a pathogen to invade and harm host cells.

virulence factors A microbe's structures or capabilities that allow it to establish itself in a host and cause damage.

virus Microscopic, acellular agent composed of nucleic acid surrounded by a protein coat.

virus particle A more specific name for a virus when it is outside of its host cells.

vitamins A component of coenzymes critical to nutrition and the metabolic function of coenzyme complexes.

W

wart An epidermal tumor caused by papillomaviruses. Also called a verruca.

Western blot test A procedure for separating and identifying antigen or antibody mixtures by two-dimensional electrophoresis in polyacrylamide gel, followed by immune labeling.

wheal A welt; a marked, slightly red, usually itchy area of the skin that changes in size and shape as it extends to adjacent area. The reaction is triggered by cutaneous contact or intradermal injection of allergens in sensitive individuals.

whey The residual fluid from milk coagulation that separates from the solidified curd.

whitlow A deep inflammation of the finger or toe, especially near the tip or around the nail. Whitlow is a painful herpes simplex virus infection that can last several weeks and is most common among health care personnel who come in contact with the virus in patients.

whole blood A liquid connective tissue consisting of blood cells suspended in plasma.

Widal test An agglutination test for diagnosing typhoid.

wild type The natural, nonmutated form of a genetic trait.

wort The clear fluid derived from soaked mash that is fermented for beer.

X

XDRTB Extensively drug-resistant tuberculosis (worse than multidrug-resistant tuberculosis).

xenograft The transfer of a tissue or an organ from an animal of one species to a recipient of another species.

Z

zoonosis An infectious disease indigenous to animals that humans can acquire through direct or indirect contact with infected animals.

zooplankton The collection of nonphotosynthetic microorganisms (protozoa, tiny animals) that float in the upper regions of aquatic habitat and together with phytoplankton comprise the plankton.

zygospore A thick-walled sexual spore produced by the zygomycete fungi. It develops from the union of two hyphae, each bearing nuclei of opposite mating types.

Credits

Photographs

Chapter 1

Opener: Courtesy of Mark Wiesmann; **1.2a:** © Doug Sokeli/Tom Stack & Associates; **1.2b:** Jack Dykinga/USDA, ARS, IS Photo Unit; **1.3a:** © Corale L. Brierley/Visuals Unlimited; **1.3b:** © Science VU/SIM, NBS/Visuals Unlimited; **1.3c:** NOAA; **Insight 1.1 (left):** National Institutes of Health (NIH)/U.S. National Library of Medicine; **Insight 1.1 (right):** © Ron Edmonds/AP Photo; **1.6 (top, left):** © Janice Carr/Public Health Image Library; **1.6 (top, center):** © Tom Volk; **1.6 (top, right):** © T.E. Adams/Visuals Unlimited; **1.6 (bottom, left):** Public Health Image Library; **1.6 (bottom, center):** © Carolina Biological Supply/Phototake; **1.6 (bottom, right):** Public Health Image Library; **1.8:** © Bettmann/Corbis; **1.9a, inset:** © Kathy Park Talaro/Visuals Unlimited; **1.9b:** © Science VU/Visuals Unlimited; **1.11:** © AKG/Photo Researchers; **Insight 1.3:** David McKay/NASA;

Chapter 2

Opener: © Todd Trigsted, www.ttstrategicmedia.com; **2.6:** © Kathy Park Talaro; **2.10 (all):** © John W. Hole; **Insight 2.2a:** © Don Facett/Visuals Unlimited; **2.21d:** RCSB Protein Data Bank;

Chapter 3

Opener: © Royalty-Free/Corbis/Vol. 52; **3.3b, d, f:** © Kathy Park Talaro; **Insight 3.1:** Charles River Lab; **3.4a:** © Fundamental Photographs; **3.4b, 3.5b, 3.6a, 3.6b, 3.8a, 3.8b, 3.9a, 3.9b:** © Kathy Park Talaro; **3.10:** Harold J. Benson; **3.11 (all):** © Kathy Park Talaro; **3.12:** © Kathy Park Talaro/Visuals Unlimited; **3.13:** Leica Microsystems Inc.; **3.18a:** © Abbey/Visuals Unlimited; **3.18b:** © George J. Wilder/Visuals Unlimited; **3.18c:** © Molecular Probes, Inc.; **3.19:** © Abbey/Visuals Unlimited; **3.20:** © Molecular Probes, Inc.; **3.21a:** Courtesy of Leica Microsystems; **3.21b:** Courtesy of Dr. Jeremy Allen/University of Salford, Biosciences Research Institute; **3.21c:** Anne Fleury; **3.22a:** © Billy Curran, Department of Veterinary Science. Queen's University Belfast; **3.22b:** J.P. Dubley et al., Clinical Microbiology Reviews, © ASM, April 1998, Vol. II, #2, 281. Image courtesy of Dr. Jitender P. Dubey; **Insight 3.2a:** Courtesy of IBM Research/Almaden Research Center; **Insight 3.2b:** Courtesy of Dr. Karl-Heinz Rieder, Institut für Experimentalphysik, Berlin; **3.23 (both):** © Dennis Kunkel/CNRI/Phototake; **3.25a (top):** © Kathy Park Talaro; **3.25a (bottom):** Harold J. Benson; **3.25b (top, middle):** © Jack Bostrack/Visuals Unlimited; **3.25b (bottom):** © Manfred Kage/Peter Arnold, Inc.; **3.25c (top):** © A.M. Siegelman/Visuals Unlimited; **3.25c (bottom):** © David Frankhauser; **Visual Understanding, question 1:** © Kathy Park Talaro; **Visual Understanding, question 2 (top, left):** © Janice Carr/Public Health Image Library; **Visual Understanding, question 2 (top, right):** © Tom Volk; **Visual Understanding, question 2 (bottom, left):** Public Health Image Library; **Visual Understanding, question 2 (bottom, right):** © Carolina Biological Supply/Phototake;

Chapter 4

Opener: © Royalty-Free/Corbis; **4.3a:** Dr. Jeffrey C. Burnham; **4.3b:** From Reichelt and Baumann, *Arch. Microbiol.* 94:283–330. © Springer-Verlag, 1973; **4.3c:** From Noel R. Krieg in *Bacteriological Reviews*, March 1976, Vol. 40(1):87 fig 7; **4.3d:** From Preer et al., *Bacteriological Review*, June 1974, 38(2):121, fig 7. © ASM; **4.6c:** Stanley F. Hayes, Rocky Mountain Laboratories, NIAID, NIH; **4.7a:** © Eye of Science/Photo Researchers, Inc.; **4.7b:** from D.R. Lloyd and S. Knurron, *Infection and Immunity*, January 1987, p. 86–92. © ASM; **4.8:** © L. Caro/SPL/Photo Researchers, Inc.; **4.10a:** © John D. Cunningham/Visuals Unlimited; **4.10b:** Courtesy of Graham C. Walker; **4.11:** © Science VU-Charles W. Stratton/Visuals Unlimited; **4.12a:** © S.C. Holt/Biological Photo Service; **4.12b:** © T.J. Beveridge/Biological Photo Service; **4.15:** © David M. Phillips/Visuals Unlimited; **4.17:** © E.S. Anderson/Photo Researchers, Inc.; **4.19:** © Paul W. Johnson/Biological Photo Service; **4.20:** © Rut CARBALLIDO-LOPEZ/I.N.R.A. Jouy-en-Josas, Laboratoire de Génétique Microbienne; **4.21:** © George Chapman/Visuals Unlimited; **Table 4.1 (top):** Kit Pogliano and Marc Sharp/UCSD; **Table 4.1 (bottom):** © Lee D. Simon/Photo Researchers, Inc.; **4.23a, b:** © David M. Phillips/Visuals Unlimited; **4.23c:** From *Microbiological Reviews*, 55(1); 25, fig 2b, March 1991. Courtesy of Jorge Benach; **4.23d:** © R.G. Kessel-G. Shih/Visuals Unlimited; **4.24:** © A.M. Siegelman/Visuals Unlimited; **4.26:** Baca and Paretsky, *Microbiological Reviews*, 47(20):133, fig 16, June 1983 ©-ASM; **Insight 4.3:** © Max Planck Institute for Marine Microbiology, Germany; **4.27a, b:** © T.E. Adams/Visuals Unlimited; **4.27c:** John Waterbury, Woods Hole Oceanographic Institute; **4.28:** From *ASM News*, 53(2), Feb. 1987. © ASM, H. Kaltwasser; **4.29a:** © Wayne P. Armstrong, Palomar College; **4.29b:** Dr. Mike Dyall-Smith, University of Melbourne; **Visual Understanding, question 2:** Harold J. Benson;

Chapter 5

Opener: © Royalty-Free/Corbis/Vol. 9; **5.1a, b:** © Andrew Knoll; **Insight 5.1:** Courtesy of Lynn Margulis; **5.3a, b:** © RMF/FDF/Visuals Unlimited; **5.5a:** © John J. Cardamone, Jr./Biological Photo Service. All rights reserved,; **.5.6:** © Don Facett/Visuals Unlimited; **5.7b:** © Science VU/Visuals Unlimited; **5.11b:** Courtesy of Dr. Catherine Chia, School of Biological Sciences, University of Nebraska at Lincoln; **5.12b:** © Don Facett/Visuals Unlimited; **5.14b:** © Albert Tousson/Photo Take; **5.15a, b:** Dr. Judy A. Murphy, San Joaquin Delta College, Department of Microscopy, Stocton, CA; **5.16a:** © David M. Phillips/Visuals Unlimited; **5.17a:** © Kathy Park Talaro; **5.17b:** © Everett S. Beneke/Visuals Unlimited; **Insight 5.2 (top):** © A.M. Siegelman/Visuals Unlimited; **Insight 5.2 (bottom):** © John D. Cunningham/Visuals Unlimited; **5.20a:** © Kathy Park Talaro; **5.20b:** © William Marin, Jr./The Image Works; **5.21a:** © A.M. Siegelman/Visuals Unlimited; **5.21b:** © John D. Cunningham/Visuals Unlimited; **5.21c:** Neil Cartson, Department of Environmental Health and Safety; **5.23b:** © David M. Phillips/Visuals Unlimited; **5.24b:** © Stephen Durr; **5.25a:** © BioMEDIA ASSOCIATES; **5.25b:** © Yuuji Tsukii, Protist Information Server, http://protist.i.hosei.ac.jp/protist_menuE.html; **5.26b:** Michael Riggs et al., infection and immunity, Vol. 62, #5, May 1994, p. 1931 © ASM; **5.29b:** © Carol Geake/Animals Animals; **5.29d:** © Arthur Siegelman/Visuals Unlimited; **Visual Understanding, question 1 (top):** From ASM News, 53(2), Feb. 1987. © ASM, H. Kaltwasser; **Visual Understanding, question 1 (bottom):** © Wayne P. Armstrong, Palomar College;

Chapter 6

Opener: © AFP PHOTO/Peter PARKS/Getty; **6.2a:** © K.G. Murti/Visuals Unlimited; **6.2b:** © CDC/Phototake; **6.2c:** © A.B. Dowsette/SPL/Photo Researchers; **6.3:** © Omnikron/Photo Researchers; **Insight 6.1:** © Science VU/Wayside/Visuals Unlimited; **6.6b:** © Dennis Kunkel/CNRI/Phototake; **6.6d:** © K.G. Murti/Visuals Unlimited; **6.8a:** © Lennart Nilsson/Boehringer Ingelheim International GMBH; **6.8b:** © Kathy Park Talaro; **6.9b:** © Harold Fisher; **6.14:** © K.G. Murti/Visuals Unlimited; **6.15b:** © Chris Bjornberg/Photo Researchers, Inc.; **6.16a:** © Patricia Barber/Custom Medical Stock; **6.16b:** Massimo Battaglia, INeMM CNR, Rome, Italy; **6.18b:** © Lee D. Simon/Photo Researchers, Inc.; **6.19:** © K.G. Murti/Visuals Unlimited; **6.21a:** Ted Heald, State of Iowa Hygienic Laboratory; **6.22a:** © E.S. Chan/Visuals Unlimited; **6.22b, c:** © Jack W. Frankel; **Insight 6.3:** © Dr. Dennis Kunkel/Visuals Unlimited;

Chapter 7

Opener: © Royalty-Free/Corbis/Vol. 52; **7.1a:** © Ralph Robinson/Visuals Unlimited; **7.1b:** Jack Jones, US EPA; **Insight 7.3a:** E.E. Adams, Montana State University/National Science Foundation, Office of Polar Programs; **Insight 7.3b:** © Ralph Robinson/Visuals Unlimited; **7.9a:** © Pat Armstrong/Visuals Unlimited; **7.9b:** © Philip Sze/Visuals Unlimited; **Insight 7.4:** © Michael Milstein; **7.10a:** Sheldon Manufacturing, Inc.; **7.11:** © Terese M. Barta, Ph.D.; **Insight 7.5:** © Mike Abbey/Visuals Unlimited; **7.12:** © Science VU-Fred Marsik/Visuals Unlimited; **7.16a:** © Kathy Park Talaro/Visuals Unlimited;

Chapter 8

Opener: © InCommunicado/iStockphoto; **Insight 8.5:** © David Tietz/Editorial Image, LLC; **8.21d:** RCSB Protein Data Bank;

Chapter 9

Opener: © Jean Claude Revy-ISM/Phototake—All rights reserved.; **9.1:** PhotoDisc VO6, Nature, Wildlife and the Environment; **9.3:** © K.G. Murti/Visuals Unlimited; **Insight 9.1 (left):** © A. Barrington Brown/Photo Researchers Inc.; **Insight 9.1 (right):** © Lawrence Livermore Laboratory/SPL/Custom Medical Stock Photo; **9.16c:** Steven McKnight and Oscar L. Miller, Department of Biology, University of Virginia; **p. 281:** © Eye of Science/Photo Researchers, Inc.;

Chapter 10:

Opener: Photo by G.N. Miller/Rex USA, courtesy Everett Collection; **10.3b:** © Kathy Park Talaro; **Insight 10.1:** © STEPHEN JAFFE/AFP/Getty; **10.8 (left):** © Koester Axel/Corbis; **10.8 (middle):** © Milton P. Gordon, Department of Biochemistry, University of Washington; **10.8 (right):** © Andrew Brookes/Corbis; **Insight 10.2:** © Marilyn Humphries; **Table 10.3 (top):** David M. Stalker; **Table 10.3 (middle):** Richard Shade, Purdue University; **Table 10.3 (bottom):** Peter Beyer, UNI-Freiburg; **10.13:** Brigid Hogan, Howard Hughes Medical Institute, Vanderbilt University; **TA 10.4 (left):** R.L. Brinster, School of Veterinary Medicine, University of Pennsylvania; **TA 10.4 (right):** © Karen Kasmauski/National Geographic Image Collection; **10.16c:** Dr. Michael Baird; **10.17:** Tyson Clark, University of California, Santa Cruz;

Chapter 11

Opener: © Anthony Reynolds/Cordaiy Photo Library Ltd./CORBIS; **Insight 11.1:** © Bettmann/Corbis; **11.5a:** © Science VU/Visuals Unlimited; **11.6:** © Raymond B. Otero/Visuals Unlimited; **11.8:** Photo courtesy of Dr. Brendan A. Niemira/U.S. Department of Agriculture; **11.10:** © Tom Pantages; **11.11b:** © Fred Hossler/Visuals Unlimited; **11.13:** STERIS Corporation. System 1 ® is registered trademark of STERIS Corporation; **11.16:** © Kathy Park Talaro/Visuals Unlimited; **Insight 11.3a:** © David M. Phillips/Visuals Unlimited; **Insight 11.3b:** © Kathy Park Talaro; **Insight 11.4:** © Stephen JaffeAFP/Getty; **p. 349:** © George Chapman/Visuals Unlimited;

Chapter 12

Opener: © Fotosearch; **Insight 12.1:** © Bettmann/Corbis; **12.2e:** © David Scharf/Peter Arnold; **12.2f:** © CNRI/PHOTO RESEARCHERS, INC.; **Insight 12.2:** © Kathy Park Talaro/Visuals Unlimited; **12.9:** © Cabisco/Visuals Unlimited; **12.15:** © Kenneth E. Greer/Visuals Unlimited; **12.17b:** © Alain Philippon; **12.1c:** © Kathy Park Talaro; **12.18:** Etest® is a registered trademark belonging to AB BIODISK, Sweden, and the product and underlying technologies are patented by AB BIODISK in all major markets; **12.19b:** © Kathy Park Talaro; **12.19c:** © Alain Philippon;

Chapter 13

Opener: © Winfried Rothermel/AP Photos; **Insight 13.1:** © Eurelios/Phototake; **Insight 13.2:** © PHIL image 1965; **13.12:** Marshall W. Jennison, Massachusetts Institute of Technology, 1940; **Visual Understanding, question 2:** © Science VU-Charles W. Stratton/Visuals Unlimited;

Chapter 14

Opener: © David Grossman/Photo Researchers, Inc.; **14.3b:** © Ellen R. Dirksen/Visuals Unlimited; **14.14c:** Steve Kunkel; **14.16a:** © David M. Phillips/Visuals Unlimited; **14.19:** Reproduced from *The Journal of Experimental Medicine,* 1966, Vol. 123, p. 969–984. Copyright 1966 Rockefeller University Press;

Chapter 15

Opener: © Christophe Ena/AP Photo; **15.11b:** © R. Feldman/Rainbow; **15.17:** © Lennart Nilsson, "The Body Victorious," Bonnier Fakta; **15.18 (left), 15.18b (middle, left), 15.18c (middle, right):** © Photodisc/Getty; **15.18d (right):** © Creatas/PictureQuest; **Insight 15.3:** James Gillroy, British, 1757–1815. "The CowPock," engraving, 1802, William McCallin McKee Memorial Collection, 1928. 1407. © 1991, The Art Institute of Chicago. All rights reserved;

Chapter 16

Opener: © PhotoLink/Getty; **16.2b:** © SPL/Photo Researchers; **16.2c:** © David M. Phillips/Visuals Unlimited; **16.5:** © Kenneth E. Greer/Visuals Unlimited; **16.6a:** © STU/Custom Medical Stock; **16.10 (both):** © Stuart I. Fox; **Insight 16.3 (left):** © Renee Lynn/Photo Researchers; **Insight 16.3 (middle):** © Walter H. Hodge/Peter Arnold; **Insight 16.3 (right):** © Runk/Schoenberger/Grant Heilman Photography; **16.14:** © Kathy Park Talaro; **16.15b:** © Kenneth E. Greer/Visuals Unlimited; **16.16a:** © Diepgen TL, Yihume G et al. Dermatology Online Atlas published online at: www.dermis.net. Reprinted with permission.; **16.16b:** © SIU/Visuals Unlimited; **Insight 16.5:** Baylor College of Medicine, Public Affairs; **16.19:** Reprinted from R. Kretchmer, New England Journal Of Medicine, 279:1295, 1968 Massachusetts Medical Society. All rights reserved.;

Chapter 17

Opener: © Keith Brofsky/Getty; **Insight 17.1:** © Photodisc/Getty; **17.4b:** © Fred Marsik/Visuals Unlimited; **17.5 (both):** Analytab Products, a division of Sherwood Medical; **17.7:** Wadsworth Center, NYS Department of Public Health; **17.11a:** No credit in previous edition.; **17.11b:** Immuno-Mycologics, Inc.; **17.13:** Genelabs Diagnostics Pte Ltd.; **17.15c:** CHEMICON® International, Inc.; **17.16b:** © Hank Morgan/Science Source/Photo Researchers; **17.17a:** © PhotoTake; **17.17b:** © Custom Medical Stock Photo, Inc.; **17.18a:** © Carroll H. Weiss/Camera M.D. Studios; **17.18b1:** © A.M. Siegelman/Visuals Unlimited; **17.18b2:** © Science VU/CDC/Visuals Unlimited; **17.18c1:** © K.G. Murti/Visuals Unlimited; **Visual Understanding, question 1:** © Kathy Park Talaro; **Visual Understanding, question 2:** © Kenneth E. Greer/Visuals Unlimited;

Chapter 18

Opener: © Lynsey Addario/Corbis; **Insight 18.1, p. 552 (left), Insight 18.1, p. 552 (right):** © Carroll H. Weiss/Camera M.D. Studios; **Insight 18.1, p. 553 (left):** © ISM/Phototake; **Insight 18.1, p. 553 (right):** © Biomedical Communications/Custom Medical Stock Photo; **18.2:** © Dr. Ken Greer/Visuals Unlimited; **18.3a:** © David M. Phillips/Visuals Unlimited; **18.3b:** © Kathy Park Talaro/Visuals Unlimited; **18.5:** Analytab Products, a division of Sherwood Medical; **18.7a:** National Institute Slide Bank/The Welcome Centre for Medical Sciences; **18.7b:** © Dr. Ken Greer/Visuals Unlimited; **18.8:** © Science VU-Charles W. Stratton/Visuals Unlimited; **18.9a:** M.A. Boyd et al., *Journal of Medical Microbiology,* 5:459, 1972. Reprinted by permission of Longman Group, Ltd. © Pathological Society of Great Britain and Ireland; **18.10 (both):** © Centers for Disease Control; **18.11b:** © Kenneth E. Greer/Visuals Unlimited; **Checkpoint 18.7 (left):** © Phil Degginger; **Checkpoint 18.6 (right):** © World Health Org./Peter Arnold; **18.12:** © Kenneth E. Greer/Visuals Unlimited; **18.15:** Public Health Image Library; **Checkpoint 18.7 (left):** © Lennart Nilsson/Boehringer Ingelheim International GMBH: **Checkpoint 18.7 (middle, left):** © Centers for Disease Control; **18.7 (middle, right):** © Jack Ballard/Visuals Unlimited; **Checkpoint 18.7 (right):** © Custom Medical Stock Photo, Inc.; **Checkpoint 18.8 (left):** © Kenneth E. Greer/Visuals Unlimited; **Checkpoint 18.8 (right):** © Charles Stoer/Camera M.D. Studios; **Checkpoint 18.9 (left):** © Centers for Disease Control; **Checkpoint 18.9 (right):** © Science VU-Charles W. Stratton/Visuals Unlimited; **18.14:** © Everett S. Beneke/Visuals Unlimited; **18.15, 18.16a:** © Kenneth E. Greer/Visuals Unlimited; **18.16b:** Reprinted from J. Walter Wilson, Fungous Diseases of Man, Plate 42 (middle right), © 1965, The Regents of the University of California; **18.17a:** From Elmer W. Koneman and Roberts, Practical Laboratory Mycology, 1985, pages 133, 134 © Williams and Wilkins Co., Baltimore, MD; **18.17b:** © A. M. Siegelman/Visuals Unlimited; **18.17c:** From Elmer W. Koneman and Roberts, Practical Laboratory Mycology, 1985, pages 133, 134 © Williams and Wilkins Co., Baltimore, MD; **18.18:** © Carroll H. Weiss/Camera M.D. Studios; **18.21:** © Science VU-Bascom Palmer Institute/Visuals Unlimited; **18.22:** Armed Forces Institute of Pathology; **Visual Understanding, question 2:** Courtesy of Mark Wiesmann;

Chapter 19

Opener: © ER Productions/Corbis; **19.5:** © Kathy Park Talaro/Visuals Unlimited; **19.6:** © Louis De Vos; **19.7:** © Gordon Love, M.D. VA, North CA Healthcare System, Martinez, CA; **19.8:** Reprinted from J. Walter Wilson, Fungous Diseases of Man, Plate 21, © 1965, The Regents of the University of California; **Insight 19.1:** © Royalty-Free/CORBIS/Vol. 124; **19.11:** © Science VU-David John/Visuals Unlimited; **19.12:** © Centers for Disease Control; **19.14a:** © M. Abbey/Photo Researchers, Inc.; **19.14b:** © Pr. J.J. Hauw/ISM/Phototake; **19.15a:** © Lennart Nilsson/Boehringer Ingelheim International GMBH; **19.16 (both):** © Centers for Disease Control; **Insight 19.4:** Marching Mothers ® Photo courtesy of Ontario March of Dimes; **19.17b:** © Science VU-Charles W. Stratton/Visuals Unlimited; **19.20:** © John D. Cunningham/Visuals Unlimited; **19.21:** Dr. T.F. Sellers, Jr.;

Chapter 20

Opener: © Javier Pierini/CORBIS; **20.4, 20.5:** © Centers for Disease Control; **20.7:** © Centers for Disease Control/Peter Arnold; **20.8a:** CDC/NCID/HIP/Janice Carr; **20.8b:** © Science VU-Charles W. Stratton/Visuals Unlimited; **Insight 20.2a:** © Dwight Kuhn; **Insight 20.2b:** © Science VU/Visuals Unlimited; **Insight 20.2c:** © A.M. Siegelman/Visuals Unlimited; **Insight 20.2d:** © George D. Lepp/CORBIS; **20.9b:** Dr. Jeremy Burgess/Photo Researchers, Inc.; **20.11:** Barbara O'Connor; **20.13:** McCaul and Williams, "Development Cycle of C. Burnetii," *Journal of Bacteriology,* 147:1063, 1981. Reprinted with permission of American Society for Microbiology; **20.14:** © Kenneth E. Greer/Visuals Unlimited; **20.16:** Department of Health and Human Resources, Courtesy of Dr. W. Burgdorfer; **20.18:** Stephen B. Aley, PhD., University of Texas at El Paso; **Insight 20.3:** © AP Photo/Kristie Bull/Graylock.com; **20.19:** © Roll Back Malaria Partnership; **20.20:** © A.M. Siegelman/Visuals Unlimited; **Insight 20.4:** © Science VU/Visuals Unlimited; **Visual Understanding, question 1c:** Steve Kunkel;

Chapter 21

Opener: © Peter Finger/Corbis; **21.1b:** © Ellen R. Dirksen/Visuals Unlimited; **21.3:** © Pulse Picture Library/CMP Images/PhotoTake; **21.4a:** Multimedia Library, Congenital Heart Disease, Children's Hospital, Boston. Editor. Robert Geggel, MD. www.childrenshospital.org/mml/cvp; **21.6:** Courtesy of Wellesley College Archives; **21.7a (both):** Diagnostic Products Corporation; **21.7b:** © Dr. David Schlaes/John D. Cunningham/Visuals Unlimited; **21.8:** Federal Agriculture Research Centre; **21.9:** Centers for Disease Control; **Insight 21.1:** REUTERS/Sukree Sukplang; **21.14:** © John D. Cunningham/Visuals Unlimited; **21.16:** © Elmer Koneman/Visuals Unlimited; **21.17:** © CDC/PhotoTake; **21.18:** © SIU Bio Med/Custom Medical Stock Photo; **21.19:** © Dr. Leonid Heifets, National Jewish Medical Research Center; **21.20a:** From Nester et al., Microbiology: A Human Perspective, 4th ed. © Evans Roberts; **21.20b:** © L.M. Pope and D.R. Grote/Biological Photo Service; **Insight 21.3 (left):** Centers for Disease Control; **Insight 21.3 (right):** © JAMA; **21.25:** © Tom Volk; **Visual Understanding, question 2:** © Jack Bostrack/Visuals Unlimited;

Chapter 22

Opener: FEMA hoto/Andrea Booher; **22.4a:** © R. Gottsegen/Peter Arnold, Inc.; **22.4b:** © Stanley Flegler/Visuals Unlimited; **22.6:** © Science VU-Max A. Listgarten/Visuals Unlimited; **22.7:** CDC; **22.8a:** Exeen M. Morgan and Fred Rapp, "Measles Virus and Its Associates Disease," Bacteriological Reviews, 41(3):636–666, 1977. Reprinted by permission of American Society for Microbiology; **22.9a:** © PhotoTake; **22.13:** R.R. Colwell and D.M. Rollins, "Viable but Nonculturable Stage of *Campylobacter jejuni* and Its Role in Survival in the Natural Aquatic Environment," *Applied and Environmental Microbiology*, 52(3):531–538, 1986. Reprinted with permission of American Society for Microbiology; **22.14a:** © David Musher/Photo Researchers, Inc.; **22.14b:** Fred Pittman; **22.14c:** Farrar and Lambert: *Pocket Guide for Nurses: Infectious Diseases.* © 1984, Williams and Wilkins, Baltimore, MD; **22.15:** Centers for Disease Control; **Insight 22.2:** © Kathleen Jagger; **22.16:** © Moredun Animal Health Ltd./Photo Researchers, Inc.; **22.17:** Original image from DPDx-Identification and Diagnosis of Parasites of Public Health Concern; **22.18:** © K.G. Murti/Visuals Unlimited; **Insight 22.3:** © Tom Pantages; **22.19:** © Iruka Okeke; **22.20:** © Ynes R. Ortega; **22.23:** © Science VU-Charles W. Stratton/Visuals Unlimited; **22.24:** © Eye of Science/Photo Researchers, Inc.; **22.26a:** © Stanley Flegler/Visuals Unlimited; **22.26b:** Katz et al., "Parasitic Diseases," © Springer-Verlag; **22.27a:** © R. Calentine/Visuals Unlimited; **22.27b:** © Science VU-Fred Marsik/Visuals Unlimited; **22.28:** © Carroll H. Weiss/Camera M.D. Studios; **22.29:** © A.M. Siegelman/Visuals Unlimited; **22.30a:** © Cabisco/Visuals Unlimited; **22.30b:** Harvey Blankespoor; **22.30c:** © Science VU/Visuals Unlimited;

Chapter 23

Opener: © Annie Griffiths Belt/CORBIS; **23.4:** Science VU/Fred Marsik/Visuals Unlimited; **23.5:** © Raymond B. Otero/Visuals Unlimited; **23.6:** © Mary Stallone/Medical Images, Inc.; **23.7:** © David M. Phillips/The Population Council/Photo Researchers, Inc.; **23.9:** James Bingham, *Pocket Guide for Clinical Medicine.* © 1984 Williams and Wilkins Co., Baltimore, MD; **23.10:** © George J. Wilder/Visuals Unlimited; **23.12:** Courtesy Morris D. Cooper, Ph.D., Professor of Medical Microbiology, Southern Illinois University School of Medicine, Springfield, IL; **23.14:** © Science VU/Visuals Unlimited; **23.15:** Kenneth E. Greer/Visuals Unlimited; **23.16 (both):** © Science VU/CDC/Visuals Unlimited; **23.17:** © Custom Medical Stock Photo, Inc.; **23.18:** © Science VU/CDC/Visuals Unlimited; **23.19:** © Kenneth E. Greer/Visuals Unlimited; **23.20:** © Carroll H. Weiss/Camera M.D. Studios; **23.21:** Public Health Image Library; **23.23a:** © CHOR SOKUNTHEA/Reuters/Corbis; **23.23b:** © Tatiana Markow/Sygma/CORBIS; **Checkpoint 23.7 (left):** © Carroll H. Weiss/Camera M.D. Studios; **Checkpoint 23.7 (middle):** © Carroll H. Weiss/Camera M.D. Studios; **Checkpoint 23.7 (right):** © Carroll H. Weiss/Camera M.D. Studios; **Checkpoint 23.8 (left):** © Kenneth E. Greer/Visuals Unlimited; **Checkpoint 23.8 (right):** © Charles Stoer/Camera M.D. Studios; **Visual Understanding, question 1 (left):** © Kenneth E. Greer/Visuals Unlimited; **Visual Understanding, question 1 (right):** © Science VU/Visuals Unlimited;

Chapter 24

Opener: © Vanessa Vick/Photo Researchers; **24.1:** Reprinted cover image from December 1, 2000 *Science* with permission from Jillian Banfield, Vol. 290, 12/1/2000. © 2000 American Association for the Advancement of Science; Image courtesy of Jillian Banfield; **24.8b:** © John D. Cunningham/Visuals Unlimited; **24.9a:** © Sylvan Wittwer/Visuals Unlimited; **24.9b:** © Sylvan Wittwer/Visuals Unlimited; **Insight 24.2 (top):** © Kevin Schafer/Peter Arnold Inc.; **Insight 24.2 (bottom):** © Gorm Kallested/AP Photo; **24.12:** © John D. Cunningham/Visuals Unlimited; **24.15b:** © Carleton Ray/Photo Researchers, Inc.; **24.16:** © John D. Cunningham/Visuals Unlimited; **Insight 24.3:** © Renee Comet/National Cancer Institute; **24.17c:** Reprinted from EPA Method 1604 (EPA-821-R-02-024) courtesy of Dr. Kristen Brenner from the Microbial Exposure Research Branch, Microbiological and Chemical Exposure Assessment Research Division, National Exposure Research Laboratory, Office of Research and Development, U.S. Environmental Protection Agency; **24.17d (left):** © Kathy Park Talaro; **24.17d (right):** Reprinted from EPA Method 1604 (EPA-821-R-02-024) courtesy of Dr. Kristen Brenner from the Microbial Exposure Research Branch, Microbiological and Chemical Exposure Assessment Research Division, National Exposure Research Laboratory, Office of Research and Development, U.S. Environmental Protection Agency; **Visual Understanding, question 1:** © Kathy Park Talaro;

Chapter 25

Opener: AP Photo/Jeff Chiu; **Insight 25.1 (both):** © Carl Oppenheimer; **25.2:** © Paula Bronstein/Getty Images; **25.4 (both):** Sanitation Districts of Los Angeles County; **25.5:** © John D. Cunningham/Visuals Unlimited; **25.6:** © Ryan Hatch; **25.7b:** © Kevin Schafer/Peter Arnold Inc.; **25.8:** © Kathy Park Talaro/Visuals Unlimited; **25.9:** © Joe Munroe/Photo Researchers; **25.12, Insight 25.2 (both):** © Kathy Park Talaro; **25.15:** © J.T. MacMillan; **Insight 25.3 (top):** © R.H. Productions/Robert Harding World Imagery/Getty Images; **Insight 25.3a, b, c:** © Giancarlo Ranalli/Journal of Applied Microbiology; **Visual Understanding, question 2:** © Kathy Park Talaro;

Line Art

Chapter 6

Figure 6.9: From Westwood et al., *Journal of Microbiology*, 34:67, 1964. Reprinted by permission of The Society for General Microbiology, United Kingdom.

Chapter 11

Figure 11.05b: From John J. Perkins, *Principles and Methods of Sterilization in Health Sciences*, 2/e 1969. Courtesy of Charles C. Thomas Publisher, Ltd., Springfield, Illinois.
Figure 11.15: From Nolte, et al., *Oral Microbiology*, 4e. © 1982 Mosby.

Chapter 15

Figure 15.15: From Joseph A. Bellanti, MD, *Immunology III.* (Philadelphia, PA: W.B. Saunders, 1985). Reprinted by permission of Joseph A. Bellanti, MD.

Chapter 23

Figure 23.22: Image from www.infectiousdiseasenews.com/200007/alexander1aCREAM.gif. Reprinted by permission.

Cover

Inside back cover: All graphics courtesy of Centers for Disease Control and Prevention. Summary of notifiable diseases—United States, 2002. Published April 30, 2004, for *MMWR* 2002; 51(No. 53):[36–49].

Front Matter

Kelly Cowan author photo, page iii
Courtesy of Michael Williams, Miami University Middletown.

Index

Note: Page numbers followed by a lower-case *t* refer to tables, *f* to figures, *n* to footnotes, and *b* to boxed material. Page numbers in *italics* refer to definitions or introductory discussions.

Abacavir, 361t
Abatement programs, and malaria, 630–31
Abiogenesis, 12b, 13b
Abiotic factors, *767*
ABO blood groups, 489, 490f
Abscesses, 353t, 395, 427b, 538b, 566t
A-B toxins, 633, 658–59, 699, 701, 703
Abyssal zone, 781
Acanthamoeba, 584, 602t
Accessory molecules, 456
Accessory organs, 687
Acellular vaccines, 467
Acetic acid, 336, 804t
Acetobacter, 235b, 797
Acetone, 804t
Acetylcholine, 600b
Acetyl coenzyme A (acetyl CoA), 227, 229
Acid(s), as antimicrobial agents, 336. *See also* Acidity; Nucleic acids; Organic acids
Acid-fast bacilli (AFB), 668
Acid-fast stain, 79, 80f, 96, 670–71
Acidic fermentation, 235
Acidity, 37–38, 186b
Acidophiles, 196
Acinebacter baumannii, 379, 402b
Acne, 537, 538–40, 566t
Acquired immunodeficiency syndrome (AIDS). *See also* Human immunodeficiency virus (HIV)
  AIDS-defining illnesses, 636b
  antiviral drugs, 174, 353t
  causative agent of, 643t
  chronic diarrhea, 710
  classification of viruses, 159t
  cryptococcal meningitis, 578
  Cryptococcus neoformans infections, 187
  cryptosporidial diarrhea, 705
  cytomegalovirus infections, 622
  diagnosis of, 515
  drug resistance, 366b
  epidemiology, 639
  histoplasmosis, 677
  incidence of, 387t
  normal biota as pathogenic, 382
  pathogenesis and virulence factors, 635–38
  recombinant human growth hormone, 297
  reportable diseases, 409t
  secondary immunodeficiency diseases, 503t, 506
  signs and symptoms, 634–35
  susceptibility to infectious disease, 387
  syphilis, 751
  treatment of, 174, 353t, 640–41
  universal blood and body fluid precautions, 403
Acriflavine, 336
Actin cytoskeleton, 88f, 101
Actinobacillus actinomycetemcomitans, 692, 727b
Actinomycetes, 356
Actinomycin D, 268
Activation energy, 213
Active immunity, 462–63
Active site, 214, 215f
Active transport, 190–92
Acute diarrhea, 696–97, 727b
Acute encephalitis, 584
Acute endocarditis, 612, 613, 614t, 643t
Acute gastroenteritis, 686
Acute infections, 394, 522b
Acute necrotizing ulcerative gingivitis (ANUG), 693
Acute otitis media, 653–54
Acute poststreptococcal glomerulonephritis, 543
Acute prostatitis, 744
Acute respiratory distress (ARDS), 675
Acyclovir
  applications and effectiveness of, 353t, 360
  complications of chickenpox, 549
  genital herpes, 755
  nucleic acid synthesis, 349, 361t
  toxic reactions, 368t
  viral meningitis, 581
Adaptation, to temperature, 193–94
Adenine (A), 49, 50f, 249
Adeno-associated virus (AAV), 173
Adenosine deaminase (ADA), 303
Adenosine deaminase (ADA) deficiency disease, 503t, 504–505
Adenosine diphosphate. *See* ADP
Adenosine monophosphate (AMP), 223
Adenosine triphosphate. *See* ATP
Adenoviridae, 159t
Adenovirus infection, 159t, 166t, 468t, 706
Adhesion, of infectious agent to host, 388–89
Adjuvants, for vaccines, 470
Administration, of vaccines, 470–71
Adolescents. *See also* Age; Children
  acne, 540
  Chlamydia infections, 748
ADP (adenosine diphosphate), 51
ADP-ribosylation, 659
Adsorption, of viruses, 160, 161f, 169t
Adults
  recommended immunization schedule, 473t
  staphylococcal scalded skin syndrome, 545
Adult T-cell leukemia, 642, 643t
Adverse reactions. *See* Side effects
Aedes aegypti, 623
Aerobe, *195*
Aerobic bacteria, 106, 780
Aerobic respiration, 185, 224–27, 232–33
Aerosols, 401
Aerotolerant anaerobes, 196
Africa
  AIDS, 634
  Ebola and Marburg virus, 623
  malaria, 630
  polio, 594, 595
  river blindness, 565
  schistosomiasis, 726
African sleeping sickness, 599–601, 602t
Agammaglobulinemia, 503, 504
Agar, *42*, 62
Age. *See also* Adolescents; Adults; Children; Elderly; Infants
  allergies, 481
  dental caries, 691
  host defenses and susceptibility to infection, 385t
  secondary immune deficiencies, 503t
selection of antibiotics, 373
Agglutination and agglutination testing, 456, 521, 522
Agranulocytes, 422, 423–24
Agricultural microbiology, *2*
Agriculture, 135, 300–301, 366–67b, 775, 798. *See also* Agricultural microbiology; Cattle; Department of Agriculture; Insecticides; Plants; Poultry; Sheep; Swine
Agrobacterium rhizogenes, 299
Agrobacterium tumefaciens, 299, 300f
AIDS. *See* Acquired immunodeficiency syndrome
AIDS-defining illnesses (ADIs), 636b, 640, 641
Air, and transmission of infectious disease, 401
Airborne allergens, 481, 482f, 487f
Airborne contaminants, and filtration, 326
AIRE, 501
Alaska, and *Exxon Valdez* oil spill, 766
Alcohol. *See also* Beer; Wine
  antimicrobial chemicals, 329, 331
  fermentation, 234–35
  food preservation, 803
  molecular formula, 40t
Aldehydes, 40t, 335
Ales, 795
Alexander the Great, 314b
Algae, 122, 128f, 136–37, 782. *See also* Red tide
Alibek, Ken, 549b
Alimentary tract, 687. *See also* Gastrointestinal tract
Alkali(s), as antimicrobial agents, 336
Alkalinity, 37–38, 186b
Alkalinophiles, 196
Alkylated phenols, 330
Alkyl dimethyl benzyl ammonium saccharinate/ethanol, 338t
Alleles, 489

Allen, Paul, 631b
Allergens, 451, 480, 481
Allergic rhinitis, 485
Allergic sinusitis, 653
Allergy. *See also* Allergens
  adaptive value of, 485b
  antimicrobial drugs, 369, 374
  atopy and anaphylaxis, 480–88
  case study, 478, 487b
  fungi, 135
  immunology and study of, 3
  in vivo testing, 529
  overreactions to antigens and hypersensitivity, 480
  vaccines, 471
Alloantigens, 451
Allografts, 498
Allosteric molecule, 266
Alpha antiptrypsin, 302t
Alpha blockers, 744
Alpha-ketoglutaric acid, 229b
Alpha toxin, 546
Alpha-2a interferon (Roferon-A), 294
Alternative pathway, of complement, 436
Amantadine, 353t, 360, 361t, 368t, 663, 666
American Red Cross, 640
American Society for Microbiology, 299b
American Type Culture Collection, 69, 194b
Ames test, 271, 272f
Amikacin, 356
Amination, 237
Amino acids, 40t, 46–47, 238, 260, 261f, 804t
Aminoglycosides
  acute endocarditis, 613
  applications of, 353t, 356
  drug interactions, 373
  drug permeability, 364
  E. coli infections, 581
  protein synthesis, 349, 356
  toxic reactions, 368t
Aminotransferases, 217
Ammonification, 775–76
Ammonium hydroxide, 336
Amoebas, 139, 181b, 712–14. *See also Entamoeba* spp.; Granulomatous amoebic meningoencephalitis (GAM)
Amoebiasis (amoebic dysentery), 141–42, 712–14. *See also Entamoeba histolytica*
Amoebic hepatitis, 713
Amoeboid motion, 128
Amoxicillin, 352, 354t, 621, 738, 739
Amphibolism, 236–37
Amphipathic ions, 37
Amphitrichous flagellum, 89
Amphotericin B, 349, 353t, 358–59, 368t, 579, 580, 678
Ampicillin
  applications of, 352, 353t
  cell wall, 352
  characteristics of, 354t
  group B *Streptococcus* colonization of infants, 759
  listeriosis, 578
  minimum inhibitory concentration (MIC), 374t
  recombinant DNA technology, 296
  toxic reactions, 368t
  urinary tract infections, 738
Amplicons, 292
Amplification, of complement cascade, 436, 437f
Amplifying hosts, 617
Amylase, 216b, 804t
Anabaena, 775
Anabolic pathways, 224
Anabolism, 212, 224, 238
Anaerobes, 195
Anaerobic bacteria, 106, 780
Anaerobic digesters, 793
Anaerobic respiration, 225, 233
Analog, 269n
Anamnestic response, 458
Anaphase, of mitosis, 124f
Anaphylaxis, *369*, 480, 484, 486, 487b
Anatomical diagnosis, 542b, 671
Ancylostoma duodenale, 722–23, 727b
Anemia, 721
Aniline dyes, 336
Animal(s). *See also* Animal inoculation; Animal rights movement; Horse serum; Zoonosis; *specific animals*
  allergies to, 481
  axenic, 384b
  drug resistance and antibiotics in feeds, 366–67b
  genetic engineering, 300–301
  giardiasis, 712
  methane emissions, 774b
  monkeypox and exotic pet trade, 551
  plague, 617
  pruteen as feed, 798
  rabies, 590–92
  reservoirs for infectious diseases, 397–99
  salmonellosis, 312, 318b, 698
Animal inoculation, 62b, 170, 517
Animal rights movement, 62b
Animal viruses, 265
Anions, 34
Ankylosing spondylitis, 499, 500t
Anoxygenic photosynthesis, 4, 241
Anopheles mosquito, 629, 630–31
Antacids, 373
Antagonism, 199
Antarctica, 186b
Anthrax. *See also Bacillus anthracis*
  bioterrorism, 57, 81b, 103, 299b, 306, 316, 337b, 676b
  causative agent, 632–34
  chemotherapy for, 353t, 358
  ciprofloxacin, 299b, 358
  cutaneous form of, 558
  HEPA filtration, 326
  irradiation, 325
  reportable diseases, 409t
  signs and symptoms, 632
  vaccine, 468t
  zoonotic infections, 399t
Anthropology, and blood groups, 493
Antibiogram, 372
Antibiosis, 199
Antibiotic(s). *See also* Antibiotic resistance; Antimicrobial therapy; Cephalosporins; Penicillin; Tetracyclines; Vancomycin
  case study of tuberculosis, 343
  designer drugs, 350b
  discovery of, 345b
  food preservation, 803
  immunologic disturbances, 416–17
  terminology, 346t
  transcription and translation, 268
  viral infections, 174
Antibiotic-associated colitis, 369, 702, 703f
Antibiotic resistance
  acne, 540
  acquisition of, 362–64
  ciprofloxin-resistant anthrax, 299b, 633
  genetic transfer of, 246, 249b, 273–74
  malaria, 630
  natural selection, 364, 365f
  Neisseria gonorrhoeae, 747
  otitis media, 654
  overuse of antimicrobial chemicals, 328b
  rise in, 366–67b, 374
  specific mechanisms of, 363–64
  strategies to limit, 367t
  tuberculosis, 354b
Antibodies, *49*. *See also* Blocking antibody theory
  A and B antigens, 489–91
  antigen interactions, 453–56, 521
  B lymphocytes, 445, 453
  immune system, 424
  molecular weight, 39
  monoclonal, 460b
Antibody-mediated hypersensitivity state, 480t
Anticodon, 258
Antifungal drugs, 358–59, 368t
Antigen(s), 445, 446–47, 450–56, 480, 489–91, 521. *See also* Antigen binding sites; Antigenic drift; Antigenic shift
Antigen binding fragments (Fabs), 453, 455f, 456
Antigen binding sites, 449
Antigenic drift, 662, 663
Antigenicity, 450
Antigenic shift, 601, 662–63
Antigen-presenting cells (APCs), 451–52
Antihelminthic drugs, 359–60, 368t, 719
Antihistamines, 488
Antimalarial drugs, 359
Antimicrobial chemicals, in consumer products, 328b
Antimicrobial peptides, 537
Antimicrobial sensitivity tests, 517
Antimicrobial therapy. *See also* Antibiotic(s)
  allergies, 486
  biofilms, 92b
  characteristics of ideal, 344t
  culture media, 65
  dilemma in use of, 374
  drug/host interaction, 368–69
  fungal infections, 358–59
  goals of, 344
  host defenses and susceptibility to infection, 385t
  infectious diseases treated by, 353t
  mechanisms of action, 95, 346–52
  new approaches to, 364–65
  normal microbiota, 369
  origins of, 344–46
  parasitic infections, 359–60
  selection of drug, 370–74
  survey of specific drugs, 352, 354–58
  terminology for, 346t
Antiparallel arrangement, 249, 251f
Antiphagocytic factors, 389–90
Antiprotozoan drugs, 368t
Antisense DNA, 303–304
Antisense RNA, 257b, 303–304
Antiseptics, *315*. *See also* Asepsis and aseptic techniques
Antiserum, 458
Antistreptolysin O (ASO) titer test, 526
Antitoxin, 456, 597, 599, 659
Antitrypsin, 297t
Antiviral drugs, 174, 349, 360–62, 368t, 666
Antiviral interferon, 435
Apicomplexa, 140, 628
API 203 manual biochemical system, 517f
Aplastic anemia, 357
Apoenzymes, 214, 215f
Apolipoprotein, 297t
Apoptosis, 461

Appendages, of cell, 88–91, 121
Applied microbiology, *767*. *See also* Biotechnology
  bioremediation of pollution, 780, 791b
  microorganisms and food, 793–803
  water and wastewater treatment, 790, 792–93
Applied science, 283
Aquaspirillum, 89f
Aquatic microbiology, 2, 780–85
Aquifer, 780
Arachnoid mater, 573
Arboviruses, 584–86, 587t, 602t
Archaea, 18b, 21–22, 96–97, 111–12, 185, 186b, 692
Arenaviridae, 159t, 623
Aretino, Spinello, 808b
Argentine hemorrhagic fever, 624
Arginine (arg), 267–68
Argyll-Robertson pupil, 750
Arizona hinshawii, 697, 698
Arrangements, of bacterial cells, 103–105
Art, and biodegradation, 808b
Artemisinin, 632
Arteries, 609
Arterioles, 609
Arthroconidia, 579
Arthropods, 108n, 619b. *See also* Insects
Arthus, Maurice, 495n
Arthus reaction, 495
Artificial immunity, 462, 463–64, 465–67
Artificial viruses, 172b
Ascariosis, 144t
Ascaris spp., 144. *See also* Ascariosis; Roundworms
  A. lumbricoides, 144t, 722, 727b
Asepsis and aseptic techniques, 16, *58n*, *315*, 800. *See also* Antiseptics
Aseptic meningitis, 580
Asexual spore formation, 133, 134f
Asian influenza (1957), 664b
A site, of mRNA, 260
Aspergillosis, 353t, 669b
Aspergillus spp., 353t, 669b. *See also* Aspergillosis
  A. flavus, 135
  A. fumigatus, 653
Aspirin, 549
Assay media, 65–66
Assembly, of viruses, 161f, 162, 169t
Asthma, 485, 488
Astromicrobiology, 3, 18b, 772
Astroviruses, 706
Asymptomatic carriers, 397, 398f
Asymptomatic infections, 395
Ataxia-telangiectasia, 503t
Atherosclerosis, 612b
Athlete's foot, 559
Atmosphere, 4, 768
Atmospheric cycles, 772–77
Atom, 28–30
Atomic force microscope (AFM), 73t, 77b
Atopic diseases, 485
Atopy, and allergic reactions, 480–88
ATP (adenosine triphosphate), 51, 212, 221, 222–24, 231f, 232
ATP synthase, 232
Atria, 609
Attachment, of infectious agent to host, 388–89
Attenuated microbes, 467
Atypical pneumonia, 674, 675
Atypical squamous cells, 758
Australia, 299b, 583
Autoantibodies, 499, 502f
Autoclave, 321, 322f
Autograft, 498
Autoimmune diseases, 499–502
Autoimmune regulator, 501
Autoimmunity, 3, 479, 499–502
Autotroph, *182*, 183t, 184–85, 769
Avery, Oswald, 250b, B2
Avian influenza, 172, 664–65b
Axenic animals, 384b
Axenic culture, 68
Axial filaments, 90
Azidothymide (AZT), 174, 349, 360, 368t, 639, 641
Azithromycin, 353t, 357, 564, 748
Azlocillin, 352, 354t
Azobacter, 774
Azoles, 353t, 359
Azorhizobium, 775
Azospirillum, 775
Aztreonam, 355

Baby food, and meningitis, 583b
Baby formula, 464b, 583b
Babylonians, and beer, 794
Bacillus, and bacterial shapes, *103*, 104f, 105
Bacillus spp.
  antimicrobial drugs, 353t
  bioremediation, 791b
  characteristics of, 632–33
  fossil endospores, 102
  nitrogen cycle, 776
  B. anthracis, 566t, 632–34, 676b. *See also* Anthrax
  B. cereus, 708–709, 710b, 727b
  B. coagulans, 800
  B. polymyxa, 358
  B. stearothermophilus, 337b
  B. subtilis, 257b, 355
  B. thuringiensis, 101f, 298, 807
Bacitracin, 355, 372t
Bacitracin disc test, 657f
Back-mutation, 270, 271
Bacteremia, 395, 610, 674
Bacteria. *See also* Normal microbiota; Prokaryotes
  Archaea and Eukarya compared to, 111t
  in blood, 381
  cell envelope, 94–99
  cell structure, 87, 88f
  classification, 22
  DNA recombination events, 273–77
  external structures, 88–93
  free-living nonpathogenic, 110–11
  genetically modified organisms, 297–98
  internal structure, 99–103
  lactose operon, 265–67
  plants and symbiotic associations, 775
  repressible operon, 267–68
  shapes and arrangements, 103–105
  sizes, 104–105, 109b, 151f
  thermal death times, 320t
  viruses infecting, 167–69
  zoonotic infections, 399t
Bacterial artificial chromosomes (BACs), 294
Bacterial chromosome, 99–100
Bacterial conjunctivitis, 564t
Bacterial vaginosis (BV), 741
Bactericide, 315
Bacteriophages, 153b, 167–69, 276, 803
Bacteriostatic agents, 316
Bacteroidetes, 381
Balantidiosis, 141t
Bang, B. L., 624n
Barber's itch, 559
Bare lymphocyte syndrome, 505
Barophiles, 197
Barr, Yvonne, 621
Barrier(s), and portals of entry, 415–18
Barrier precautions, 403
Bartonella spp., 626. *See also* Bartonellosis
  B. henselae, 626, 643t
  B. quintana, 626, 643t
Bartonellosis, 405b. *See also Bartonella* spp.
Basal layer, 537
Basement membranes, 494–95
Basic science, 283
Basic solution, 37
Basophils, 422f, 423, 482
Batch culture method, 203b
Batch fermentation, 806
B cells. *See* B lymphocytes
BCG vaccine, 670, 671
Bdellovibrio, 89f
Bears, and trichinosis, 724
Beavers, and giardiasis, 712
Beer, 5, 234–35, 322, 324n, 794–95
Beijerinck, Martinus, 150, B1
Belgium
  Naegleria infection and water pollution, 583
  PCB contamination of wildlife, 778b
Beluga whales, 778b
Benzathine, 752
Benzene, 35f
Benzethonium chloride, 338t
Benzlpenicilloyl, 369
Benzoic acid, 336
Berg, Paul, B2
Bergey's Manual of Determinative Bacteriology and *Bergey's Manual of Systematic Bacteriology*, 106
Berkeley Pit Lake (Montana), 27
Berkelic acid, 27, 33b
Beta-carotene, 301b
Betadine, 330
Beta-lactam antibiotics, 352, 354–55
Beta-lactamases, 352, 364
Beta oxidation, 236, 237
Bifidobacterium, 365, 382
Bile salts, 65
Binary fission, and population growth, 200
Binomial system, 20
Bioaccumulation, 777, 778b
Biochemical tests, for identification, 69, 512, 516–17
Biochemistry, 40, 512
Biodegradation, 791b, 792, 808b. *See also* Bioremediation
Bioelements, 772–77
Bioengineering, and designer drugs, 350b
Biofilms
  acute otitis media, 653
  aquatic environment, 767f
  dental plaque, 93, 688, 689
  environmental role of, 92b
  fuel cells, 793
  medical implications, 92b
  plant roots, 779
  Staphylococcus aureus, 94f
  subacute endocarditis, 613
Biogenesis, 12–13b
Biogeochemical cycles, 769t, 772
Bioinformatics, 305
Biological vector, 397
Biomedicine. *See* Medical microbiology
Biomes, *768*
Biopesticides, 807
Biopsy, 758b
Bioremediation, *6*, *770*. *See also* Biodegradation
  genetic engineering, 298
  oil spills, 5f, 766, 770b
  pseudomonads, 241b

soil and water pollution, 791b
water pollution from mining wastes, 33b
Biosafety, and classes of pathogens, 386b
Biosensor systems, 299b, 519
Biosphere, 768
Biosynthesis, 129t, 236–38
Biotechnology, 2, 469, 790. *See also* Applied microbiology; Genetic engineering
Bioterrorism. *See also* Terrorism
anthrax, 57, 81b, 103, 299b, 470, 558, 633, 676b
biotechnology and genetic engineering, 299b
brucellosis, 624
Q fever, 626
respiratory infections, 676–77b
smallpox, 465b, 470, 549
tularemia, 617
vaccine development, 470
Biotic factors, *767*
Biotransformations, 805, 807
Bird(s). *See also* Avian influenza; Poultry
Cryptococcus neoformans, 578
embryos and cultivation of viruses, 170–71
influenza viruses, 663
Birdseed agar, 67t
Bishop, J. Michael, B2
Bismuth subsalicylate, 696
Bisphenols, 330, 331
1,3-Bisphosphoglyceric acid (BPG), 241
"Black death," 615
Black flies, and river blindness, 565
Bladder, 734
Blastomyces dermatitidis, 135t, 669b
Blastomycosis, 135t, 669b
Blocking antibody theory, 488, 489f
Blood. *See also* ABO blood groups; Blood transfusions; Red blood cells; White blood cells
bacteria in, 381
Candida infections, 741
immune system, 420–24
portals of exit for infection, 396
signs of infection, 395
Blood agar, 65f, 67t
Blood-brain barrier, 573
Blood flukes, 144t, 725
Blood transfusions, 396, 490–91, 585b, 590, 639, 698, 716–17
Blood types, 489, 490f
Bloom, on grapes, 795
Blue-green bacteria, 110–11
B lymphocytes (B cells), 423–24, 445, 449, 450t, 453–58, 503, 529
Body substance isolation (BSI) techniques, 403
Bogs, 779
Boiling water, and disinfection, 322–23
Bolivian hemorrhagic fever, 624
Bone marrow transplantation, 498, 499b, 505
Bordet, Jules, B1
Bordetella pertussis, 660, 680b, 807. *See also* Pertussis
Borna agent, 7
Borrelia spp., 353t
B. burgdorferi, 104f, 522b, 618–21, 643t. *See also* Lyme disease
Botox, 600b
Bottlenose dolphin, 778b
Botulinum, 598, 600b
Botulism. *See also Clostridium botulinum*
bioterrorism, 676b
canning of foods, 321
epidemiology of, 407f
lysogenic conversion, 169
medical significance of bacterial spores, 103
reportable diseases, 409t
vaccine, 468t
Bourbon, 795
Bovine growth hormone, 297t
Bovine somatotropin (BST), 297t
Bovine spongiform encephalopathy (BSE), 173, 572, 590
Bovine tuberculosis, 668
Boyer, Herb, B2
Bradykinin, 429b, 485
Bradyrhizobium, 775
Brain
anatomy of, 574
antimicrobial drugs, 369
infection, 141t
nervous system, 573
Brain heart infusion broth, 64t
Brandy, 795
Bread, 794
Breakbone fever, 623
Breast-feeding, 382, 463, 464b
Brewer's yeast, 794
Bright-field microscopy, 72–73
Brilliant green dye, 65
Broad spectrum drugs, 346t
Brock, Thomas, 194b
Bronchial-associated lymphoid tissue (BALT), 426
Bronchiolitis, 661
Bronchitis, 159t
Brownian movement, 189
Brucella spp., 322, 624. *See also* Brucellosis
B. abortus, 624, 643t
B. suis, 624, 643t
Brucellosis, 399t, 409t, 624–25, 628t, 643t. *See also Brucella* spp.
Bubble boy mystery, 505b
Bubo, 615
Bubonic plague, 615. *See also* Plague
Buctril, 301b
Budding, of viruses, 162
Buffett, Warren, 631b
Bulbar poliomyelitis, 594
Bulk transport, 192t
Bulla, 552t
Bullous lesions, 545
Bunsen burner, 323
Bunyavirus, 159t, 623, 675
Burgdorfer, Willy, 618
Burkholderia spp., 791b
B. xenovorans, 791b
Burkitt, Michael, 621
Burkitt's lymphoma, 621, 636b
Burns and burn infections, 302t, 353t, 503t
Bursa of Fabricus, 424
Bush, George W., 550, 663
Bystander effect, 500

Calcification, of tubercular lesions, 666
Calcium, 29t, 180t, 183
Calcium carbonate, 773
Calciviridae, 159t
Calculus, 692, 693
California. *See* Earthquakes; Owens Lake
California encephalitis, 159t, 409t, 586, 602t
Calor, 427
Calvin cycle, 240
Campylobacter spp., 367b, 701–702, 708b, 782. *See also* Campylobacteriosis
C. jejuni, 20, 322, 701–702, 727b
Campylobacteriosis, 701–702, 708b. *See also Campylobacter* spp.
Canada
bioaccumulation in marine mammals, 778b
bovine spongiform encephalopathy, 590
Cancer. *See also* Carcinogens; Cervical cancer; Oncogenic viruses; Prostate cancer; Stomach cancer; Throat cancer
antimicrobial drugs, 374
genetic engineering and treatment of, 297, 302t
genetic mutations, 270, 271
hepatitis B virus, 715
hepatitis C virus, 716
human growth hormone, 297
human papillomavirus, 756, 757
interferon, 362
microarray analysis, 307
retroviruses, 642
salmonellosis, 698
secondary immune deficiencies, 503t, 506
T cells, 461–62
vaccines, 470b
viruses, 7, 153b, 167
Candida albicans, 135t. *See also* Candidiasis; Yeast infections
chemotherapy, 353t
lung diseases, 669b
normal microbiota, 369, 736
vaginitis, 740–41, 743b, 759b
Candidiasis, 135t, 353t, 669b. *See also Candida albicans*
Canning, of foods, 103, 321, 599, 800
Capnophiles, 196
Capsid, 152, 154–55, 156
Capsomers, *154*
Capsule, of prokaryotic cell, *93*
Capsule staining, 80
Carbenicillin, 352, 354t, 368t
Carbohydrase, 216b
Carbohydrate(s), 40–44, 238
Carbohydrate fermentation media, 66
Carbon. *See also* Carbon cycle
atomic structure, 28f
chemistry of, 38–39
elements of life, 29t, 30
fixation, 241
sources of, 180t, 182
Carbon cycle, 772–73
Carbon dioxide, 36f, 196, 773, 774b, 779
Carbon monoxide, 773
Carbonyl, 40t
Carboxyl, 40t
Carbuncles, 538b, 566t
Carcinogens, 269
Cardinal temperatures, 193
Cardiolipin, 752
Cardiovascular syphilis, 750
Cardiovascular system. *See also* Heart and heart attack
anatomy, 609, 611f
defense mechanisms, 610
infectious diseases, 612–44
Carotenoids, 239
Carotid arteries, 692
Carrier(s), of infectious disease, 397, 699, 713
Carrier-mediated active transport, 191f, 192t
Carrion, Daniel, 405b
Carter, Jimmy, 565
Carter Center, 565
Cascade reaction, 435
Caseous lesions, 666
Cat, and toxoplasmosis, 587, 588f. *See also* Cat-scratch disease
Catabolism, 212, 224–25, 236

Catalase, 214t, 233
Catalysts, 212
Catalytic site, 214
Catarrhal stage, of pertussis, 660
Catheter, 514
Cathode rays, 324–25
Cation(s), 34
Cationic detergents, 332
Cat-scratch disease (CSD), 626, 628t, 643t
Cattle. *See also* Brucellosis
  bovine spongiform encephalopathy, 572, 590b
  bovine tuberculosis, 668
  tapeworms, 721
  toxoplasmosis, 587, 588t
Caudovirales, 158t
Caulobacter, 781
Cefaclor, 354, 374t
Cefazolin, 354
Cefonacid, 354
Cefotaxime, 353t, 577, 581
Ceftriaxone, 353t, 354, 563, 747
Celera Genomics, 289b
Cell. *See also* Bacteria; Cell envelope; Cell membrane; Cell wall; Eukaryotes; Prokaryotes
  catabolism and assembly of, 238
  chemistry of, 51
  energy, 221
  fundamental characteristics of, 52
  organization of, 7–8
  structure of eukaryotic, 120
  structure of prokaryotic, 87, 88f
Cell culture, 171
Cell envelope, of bacteria, 94–99
Cell-mediated (delayed) reactions, 495–98
Cell-mediated immunity (CMI), 424, 445, 458–59, 461–62
Cell membrane
  antimicrobial drugs, 349, 358
  eukaryotic cell, 122
  microbial control, 318
  prokaryotic cell, 88f
  structure and functions of, 46b
Cellulase, 804t
Cellulitis, 544, 566t
Cellulose, 42, 43f
Cell wall
  antimicrobial drugs, 347–48, 352, 354–55
  bacterial, 94–97, 107t
  eukaryotes, 122
  extracellular digestion, 187f
  microbial control, 318
  mycobacterial, 667–68
  prokaryotes, 87, 88f
Cell-wall-deficient bacteria, 97–98
Center for National Security and Arms Control, 299b
Centers for Disease Control and Prevention (CDC)
  AIDS epidemic, 636b
  anthrax, 358, 633
  artificial viruses, 172b
  baby food and meningitis, 583b
  biosafety categories for pathogens, 385
  botulism, 599
  China, 535
  Chlamydia infections, 748
  Creutzfeldt-Jakob disease, 590
  E. coli O157:H7 infections, 789
  food-borne diseases, 789, 798
  gonorrhea, 747
  group *B Streptococcus*, 759
  guidelines for sterilization, 315b
  HIV testing, 640
  human papillomavirus, 757
  infectious diseases and human condition, 6
  pneumococcal pneumonia, 98b
  recommended vaccine schedules, 471
  reportable diseases, 406
  role in public health, 3, 406
  Salmonella contamination of folk remedies, 698
  smallpox, 549b, 551
  tuberculosis, 671
  universal precautions (UPs), 403
  West Nile Virus, 585b
Center for Food Safety and Applied Nutrition (CFSAN), 789
"Central dogma," of genetics, 255
Central nervous system (CNS), 573. *See also* Nervous system
Centripetal lesions, 547
Centromere, 124f
Cephalosporin(s)
  applications of, 353t, 354
  bioengineering, 350b, 804b
  cell wall, 348
  drug interactions, 373
  drug resistance, 366t
  normal microbiota, 369
  Streptococcus pyogenes, 657
  structure of, 355f
  toxic reactions, 368t
Cephalosporinase, 364
Cephalosporium acremonium, 354
Cephalothin, 354
Cercaria, 724, 725f, 726
Cerebral malaria, 628
Cerebrospinal fluid (CSF), 573
Cervical cancer, 756, 757, 758b
Cervical intraepithelial neoplasia (CIN), 758b
Cervix, 735. *See also* Cervical cancers
Cestodes, 143, 144t, 353t
Chad, and measles epidemic, 467b
Chagas disease, 141, 142f
Chain, Ernst, 345b, B2
Chalazion, 563b
Chancre, 749
Chancroid, 409t, 752, 756b, 759b
Chargaff, Erwin, 250b
Charon phage, 293
Chédiak-Higashi syndrome, 503t
Cheeses, 5, 235, 797–98
Chemical agents, in microbial control, 327–36
Chemical analysis, and identification, 512
Chemical bonds, 30–39
Chemical composition, of media, 60t
Chemical mediators, 428
Chemical mutagenic agents, 269
Chemiclave, 336
Chemiosmosis, 231f, 232
Chemistry. *See also* Chemical analysis
  atoms, bonds, and molecules, 28–39
  cells, 51–52
  composition of *Escherichia coli* cell, 181t
  composition of microbial cytoplasm, 181–82
  culture media, 63–64
  macromolecules, 40–51
Chemoautotrophs, 184
Chemoheterotroph, 183t, 185
Chemokine(s), 428, 429b
Chemokine inhibitors, 427b
Chemostat, 204
Chemotactic factors, in inflammation, 428
Chemotaxis, 90, 430
Chemotherapy, 346. *See also* Antimicrobial therapy
Chemotrophs, 183
Chest X-rays, 670
Chicago disease, 135t
Chickenpox, 159t, 468t, 544, 546, 547–49, 551t, 566t
Children. *See also* Infants
  cat-scratch disease, 626
  dental caries, 689, 691
  diarrheal diseases, 696
  gonorrhea, 745
  immunizations, 471, 472t
  meningitis, 576
  mumps virus, 695
  recurring ear infections, 654
  respiratory syncytial virus infection, 661
  rotavirus, 706
Chills, and fever, 431b
Chimpanzee, 634
China
  history of antimicrobial therapy, 345b
  history of immunization, 465b
  measles, 535, 555b
  severe acute respiratory syndrome (SARS), 149, 166b, 399
Chinese liver flukes, 723
Chitin, 42
Chlamydial infections, 747–48, 749b, 759b. *See also* Chlamydiosis
Chlamydia trachomatis. *See also* Chlamydiosis; Trachoma
  genitourinary tract diseases, 749b, 759b
  heart disease, 612b
  lymphogranuloma venereum, 747
  medical impact of, 110
  pelvic inflammatory disease, 743b
  reportable diseases, 409t
  transmission of, 748
Chlamydiosis, 110, 387t, 409t. *See also Chlamydia trachomatis*
Chlamydomonas nivalis, 193f
Chlamydophila pneumoniase, 110, 612b, 672
Chloramines, 330
Chloramphenicol, 268, 349, 353t, 357, 368t, 369, 372t
Chlorhexidine, 331
Chloride, 180t
Chlorinated phenols, 330
Chlorination, of water, 330
Chlorine
  antimicrobial chemicals, 329, 330, 336
  elements of life, 29t
  ionic bonds, 32–33, 34
Chlorine dioxide, 336, 337b
Chlorophylls, 239, 240
Chloroplasts, 119b, 126–28
Chloroquine, 349, 353t, 359, 368t, 632, 714
Chocolate agar, 65f, 576
Cholera, *702–705*. *See also Vibrio cholerae*
  chemotherapy for, 353t
  drug resistance, 367b
  human experimentation, 405b
  lysogenic conversion, 169
  reportable diseases, 409t
  vaccine, 468t
Cholera toxin (CT), 703
Cholesterol, 45
Chordata, 19
CHROMagar Orientation, 67f
Chromatin, 123
Chromatoidal bodies, 712
Chromobacterium violaceum, 355
Chromosomal drug resistance, 362
Chromosome, 88f, 123, *248*. *See also* Bacterial chromosome
Chronic carriers, 397, 398f
Chronic diarrhea, 710–26, 727b

Chronic fatigue syndrome, 621
Chronic granulomatous disease (CGD) of childhood, 414, 433b, 503t
Chronic latent state, of infection, 166–67
Chronic mucocutaneous candidiasis, 503t
Chronic otitis media, 653
Chronic prostatitis, 744
Chronic pulmonary histoplasmosis, 677
Chroococcus, 110f
Cigarette smoking, and respiratory infections, 653
Ciguatera, 136
Cilia, 121, 138, 416, 417f, 650
Ciliophora, 140
Ciprofloxacin, 299b, 358, 558, 563, 633
Circulatory system, 609
Circumcision, and chancroid, 752
Cirrhosis, 715
Citric acid, 229b, 804t
Citrobacter, 783b
Clarithromycin, 357, 696
Class, and taxonomy, 18–20
Classical pathway, of complement, 435–36, 437f
Classic biotype, of *Vibrio cholerae*, 703
Classification, 17. *See also* Nomenclature; Taxonomy; Terminology
  helminths, 145
  levels of taxonomic, 17–20
  prokaryotes, 106–108
  specific immunities, 462–64
  viruses, 157–58
Clavamox, 352
Clavulanic acid, 352
"Clean catch," for urine samples, 514
Clindamycin, 357, 364, 742
Clinical infections, stages of, 393b
Clinton, Bill, 749
Clonal deletion, of lymphocytes, 448
Clonal expansion, of lymphocytes, 445, 448, 453, 454f
Clonal selection and clonal selection theory, 445, 447–49, 450, 453, 454f, 500
Clones and cloning. *See also* Clonal expansion; Clonal selection and clonal selection theory
  lymphocytes, 448
  recombinant DNA technology, 292, 293–96
  viruses, 172b
Cloning host, 292, 294–96
Clonorchis sinensis, 723, 724b, 727b
Clostridium spp., 103, 775
  C. botulinum, 103, 169, 197b, 321, 598, 602t, 800. *See also* Botulism
  C. difficile, 369, 702, 709b, 727b
  C. perfringens, 103, 218, 545–46, 566t, 709–710, 727b. *See also* Gas gangrene
  C. tetani, 103, 598, 602t, 807. *See also* Tetanus
Clotrimazole, 359
Clotting factor VIII, 296–97
Cloverleaf structure, of tRNA, 258
Cloxacillin, 352
CMV infections, 159t
Coagulase, 391, 537, 541
Coagulase-negative staphylococcus, 537
Coagulase test, 541
Cobalt, 29t
Coccidioides immitis, 135t, 579–80, 581f, 582t, 602t, 669b. *See also* Coccidiomycosis
Coccidioidomycosis, 135t, 409t, 579, 669b. *See also Coccidioides immitis*
Coccobacillus, *103*, 104f
Coccus, *103*, 104f, 105
Cockroaches, 398
Coding strand, of DNA, 259
Codons, 258, 263
Coenzymes, 214, 215
Coevolution, 199b, 383, 385
Cofactors, 214, 215
Cohen, Stanley, B2
Cohn, Ferdinand, 16
Cold. *See also* Freezing; Refrigeration; Temperature
  environments and extremophiles, 186b
  microbial control, 323–24
Cold enrichment, 578
Cold sores, 753, 755
Cold sterilization, *324*
Coliforms and coliform enumeration, 783–84
Colistin, 358
Collagenase, 391
Collins, Francis, 289b
Colonization, and group B *Streptococcus*, 759
Colony, 58
Colony-forming unit (CFU), 203b
Colorado tick fever, 159t
Colostrum, 464b
Colposcopy, 758b
Columbia (space shuttle), 306
Combined therapy, with antimicrobial drugs, 373
Comedo, 539
Co-metabolism, 772
Commensals and commensalism, 197–98, 380, 688, 772
Commercial antimicrobial products, 328b, 338t
Common cold, 159t, 651–52
Common names, 20
Common vehicle, 401
Communicable disease, 399, 400–401
Communities, and ecology, *768*, 771–72, 781–82
Community-acquired MRSA (CA-MRSA), 544b
Community-acquired UTIs, 738
Competent DNA, 275
Competition, and ecosystems, 772
Competitive inhibition, 219, 220, 349
Complement, 435–37, 503t
Complementary DNA (cDNA), 285, 307
Complement fixation test, 525–26
Completed tests, for fecal coliforms, 784
Complexity, and evolution, 21
Complex nonsynthetic media, *64*
Composting, 779
Compounds, *30*
Computers. *See* Web sites
Concentration, of solution, 37, 329
Condensing vesicles, 125–26
Concept mapping, D1–D3
Condoms, and prevention of sexually transmitted diseases, 752, 755
Condyloma acuminata, 755
Confirmatory data, 514, 784
Confocal microscopy, 73t, 75f
Congenital (prenatal) infection, 553
Congenital rubella, 554
Congenital syphilis, 750
Conidia, 133, 134f
Conidiospores, 133, 134f
Conjugated enzymes, 213–14
Conjugation, 91, 139, 273–74
Conjunctiva, of eye, 387, 562
Conjunctivitis, 563, 564t, 566t, 747
Consolidation, and pneumonia, 673
Consortium, 770
Constitutive enzymes, 216
Consumers, in ecosystems, 769–70
Consumption, and tuberculosis, 667
Contactants, and allergens, 481
Contact dermatitis, 496–97
Contact transmission, 400
Contagious disease, 399
Contaminants, of cultures, 69
Contaminated culture, 68–69
Contaminated materials, and disease transmission, 401
Continuous culture system, 204
Continuous feed systems, for fermentation, 806
Contraceptive pills, 807. *See also* Oral contraceptives
Control locus, of lactose operon, 265
Convalescent carriers, 397, 398f
Convalescent period, of infection, 393b, 522b
Convalescent stage, of pertussis, 660
Conversion of Saint Efisio and Battle (painting), 808b
Coombs, R., 480
Copper, 29t
Copper sulfate, 790
Cord factor, 96, 667
Corepressor, 267
Cornea, 562
Coronary artery disease, 7
Coronaviridae, 159t, 675
Coronavirus, 159t, 651, 675. *See also* Severe acute respiratory syndrome (SARS)
Corticosteroids, 488, 807
Corynebacterium diphtheriae, 99, 103, 169, 658–59, 680b, 807. *See also* Diphtheria
Cosmetic treatments
  botox, 600b
  hepatitis B virus, 716
Cottage cheese, 797
Coughing, and disease transmission, 395, 400, 401, 416
Coulter counter, 205
Coumadin, 365
Counterstain, 79
Covalent bonds, 30–32
Cowpox, 159t, 463, 465b, 550. *See also* Vaccinia virus
Coxiella burnetii, 322, 625, 643t
Coxsackie virus, 7, 159t
Cranberry juice, 738
Credé, Carl Siegmund Franz, 335
Creolin, 331
Cresols, 330–31
Creutzfeldt-Jakob disease, 173, 296, 588–90, 602t
Crick, Francis, 249, 250b, B2
Crime, and forensic medicine, 305
Crimean-Congo hemorrhagic fever (CCHF), 159t
Cristae, 126
Crixivan, 361t
Crohn's disease, 710, 719b
Cromolyn, 488
Croup, 661
Crown gall disease, 299
Cryotherapy, 758b
Cryptococcosis, 135f, 353t, 427b, 579, 669b
Cryptococcus neoformans, 187, 578–79, 582t, 602t. *See also* Cryptococcosis
Cryptosporidiosis, 137b, 141t, 409t, 705. *See also Cryptosporidium* spp.
Cryptosporidium spp., 705, 709b, 727b, 782. *See also* Cryptosporidiosis
  C. parvum, 137b
Crystallizable fragment (Fc), 453, 455f
Crystal violet, 95b, 336
Culture(s). *See also* Media
  of anaerobes, 195f
  Clostridium spp., 197b
  Coccidioides immitis, 580

Culture(s) *(continued)*
of fungi, 134
intermittent sterilization, 322
methods of, 58–69
of protozoa, 141
specimen analysis, 516–18
Staphylococcus aureus, 541
of stool samples, 699b
of viruses, 170–72, 530f
"Culture bias," and bacterial reproduction, 201b
Curd, and milk fermentation, 797
Cutaneous anaphylaxis, 486
Cutaneous anthrax, 558, 566t, 632
Cutaneous mycoses, 559–61
Cutaneous T-cell lymphoma, 642
Cutting boards, 801b
Cyanide, 233
Cyanobacteria, 110–11, 782
Cyanosis, 679
Cyclic AMP, 633, 660
Cyclohexane, 35f
Cycloserine, 348
Cyclospora cayetanensis, 711, 727b
Cyclosporiasis, 141t, 409t
Cyst, *138,* 552t
Cysteine, 40t, 49
Cystic acne, 539
Cysticerci, 720
Cystitis, 737
Cytoadherence, 628
Cytochromes, 229, 233
Cytokines, 428, 429b, 435, 483–85
Cytolysin, 525
Cytomegalovirus (CMV), 581, 622, 623t, 643t
Cytomegalovirus retinitis, 304
Cytopathic effects (CPEs), 163, 166
Cytoplasm, 88f, 99–101, 181–82
Cytoplasmic membrane, 98–99, 122
Cytopathic effects (CPEs), 172
Cytosine (C), 49, 50f, 249
Cytoskeleton, 101, 128–29
Cytotoxicity, 461
Cytotoxic T (TC) cells, 461–62

Dairy microbiology, 2, 797–98. *See also* Cheeses; Milk
Dapsone, 358
Daptomycin, 358
Darfur (Sudan), 442, 467b
Dark-field microscopy, 73
Darling, Samuel, 677
Darwin, Charles, 21, 199b
Daschle, Tom, 337b
Daughter molecules, 253–55
Dead Sea, 189
Deamination, 236, 238
Death. *See also* Microbial death; Mortality rates; Thermal death point; Thermal death time
leading causes of, 6t, 7f
nonliving as distinct from, 767n
phases of population growth, 204
rate of and microbial control, 317–18
Decarboxylases, 217
Decomposers, 769t, 770
Decomposition, 4
Decomposition reactions, 36
Decontamination, 313, 316
Deductive approach, 14
Deer, and Lyme disease, 620
Defense mechanisms. *See also* Immune system
barriers at portal of entry, 415–18
cardiovascular system, 610
eye, 562
gastrointestinal tract, 688
inflammation, 427–32
lymphatic system, 610
lymphocyte response system, 449–51
major components of, 416f
nervous system, 573
respiratory tract, 650
skin, 536–37
specific immunity, 443–49
system involved in, 418–26
Definitive host, 144, 717
Deforestation, 399, 774b
Degermation, 316
Degranulation, and allergies, 482
Dehydration reactions, 216–17, 218f
Dehydration synthesis, 41
Dehydroemetine, 714
Dehydrogenases, 216b
Delaviridine, 361t
Delaware (state) Department of Public Health, 649
Delaware City County Health Department (Ohio), 117
Delayed-type hypersensitivity, 495–96
Deletion, and genetic mutations, 270t
Delta agent, 173–74, 715
Denaturation, 218, 292
Denatured proteins, 49, 319
Dendritic cells, 423f, 424, 451–52
Dengue fever, 159t, 623, 624t, 643t
Denitrification, 233, 775–76
Denmark, 367b
Dental care, and HIV transmission, 639
Dental caries, 195, 384b, 689–91, 727b
Deoxyribonuclease, 216b
Deoxyribose, 50, 249
Department of Agriculture (USDA), 300, 799, 800, 801b
Department of Defense, 172b
Dermatology, 552b
Dermatophytes, 559
Dermatophytosis, 135t
Dermis, 537
Dermolytic conditions, 544
Desensitization, 488
Deserts, 186b, 779
Desiccation
food preservation, 803
microbial control, 323–24
Designer drugs, 350b
Desquamated cells, 415
Desquamation, 545, 734
Desulfovibrio, 776
Desulfuromonas, 776
Detergents, and antimicrobial chemicals, 328b, 332 33
Detroit (Michigan), 733
Developing countries. *See* Africa; Emerging diseases; Globalization; India; Third World
Dextran, 42, 804t
D'Herelle, Felix, 167, B2
Dhurnadha, Nikhil, 173b
Diabetes, 7, 296, 297, 302, 500t, 502
Diagnosis. *See also* Culture; Identification; Infectious diseases; *specific diseases*
AIDS, 640
allergy, 486
genotypic methods, 512–13, 518–19
immunologic methods, 513, 519–30
phenotypic methods, 512, 513b, 516–18
specimen collection, 513–15
Diagnostic scheme, for classification of bacteria, 106, 107t
Diagnostic tables, 517
Diapedesis, 430
Diarrhea, 353t, 369, 696–97, 727b
Diatoms, 136f
Dichotomous keys, 517
Dicloxacillin, 354t
Didanosine (ddI), 361t, 641
Dideoxynucleotides, 288–89
Diet. *See also* Food; Nutrition
cultural preferences and food-borne botulism, 599
prevention of dental caries, 691
Diethylcarbamazine, 488
Differential interference contrast (DIC) microscope, 73t, 74
Differential media, 65–66, 67t
Differential permeability, 187
Differential stains, 79–80
Diffusion, 187–88, 189–92
Digestive tract, 687. *See also* Gastrointestinal tract
DiGeorge syndrome, 426, 503t, 504, 505f
Dihydroxyacetone phosphate (DHAP), 241
Dimethyl benzyl ammonium chloride, 338t
Dimorphic fungi, 130
Dinoflagellates, 782f
Diogenin, 807
Diphyllobothrium latum, 144t, 719, 721, 727b
Dipicolinc acid, 102
Diplococci, 104f
Diplodia maydis, 130f
Diphtheria, 169, 409t, 468t, 472t, 473t, 657–59, 680b. *See also Corynebacterium diphtheriae*
Direct antigen testing, 516
Direct cell count, 205
Direct ELISA test, 528–29
Direct examination, and identification of infectious agent, 370
Direct fluorescence antibody (DFA) tests, 516, 592
Directly observed therapy (DOT), 671
Direct testing, of fluorescent antibodies, 527
Disaccharide, 40, 41
Disc diffusion tests, 371f
Discharge diseases, and sexually transmitted diseases, 740, 744
Disease. *See* Infectious diseases
Disinfectants, 335–36
Disinfection, 313, 315, 318b, 322–23, 792. *See also* Antiseptics; Asepsis and aseptic techniques
Disposal, of cultures, 69
Disulfide, 776
Diversity, in immune response, 447–49
Division, taxonomic, 18–20
DNA (deoxyribonucleic acid). *See also* Genetics; Nucleic acids; Recombinant DNA technology
analysis of, 285–86
antimicrobial drugs, 349, 358
chemical structure, 49–50
double helix, 50–51
eukaryotic transcription and translation, 264
genetic code, 249
genetic medicine, 303–304
genetic significance of structure, 249–52
hybridization and probes, 286–87
macromolecules, 41t
nucleus, 123
polymerase chain reaction, 290–92
prokaryotes, 87
recombination events, 272–77

repair of mutations, 270
replication, 252–55
restriction endonucleases, 284–85
sequencing, 287–90
triplet code, 255–56
viruses, 157
DNA analysis, 518
DNA fingerprinting, 305–307, 707b
DNA polymerase, 214t, 216t, 253, 290–92
DNAse, 216b
DNA sequencing, 287–90
DNA vaccines, 469–70, 592
DNA viruses, 159t, 566t
Dobell, C., 11f
Dogs, and Lyme vaccine, 621
Dolor, 427
Domagk, Gerhard, 345b, B2
Domain, 18–20, 22
Double diffusion (Ouchterlony) method, 522, 523f
Double helix, of DNA, 50–51
Double-stranded RNA (dsRNA), 157, 164b, 265
Doubling time, 200–201
Downstream processing, and fermentation, 806
Doxycycline
anthrax, 358, 558, 633
applications of, 353t, 357
brucellosis, 625
leptospirosis, 739
Lyme disease, 621
Q fever, 626
Rock Mountain spotted fever, 627
Dracunculus medinensis, 144t
Droplet contact, 400, 401
Drug abuse, and AIDS, 639, 640
Drug allergy, 486
Drug inactivation mechanisms, 364
Drug interactions, 373, 374
Drug resistance, 362–64, 366–67b, 374, 379, 402. *See also* Antibiotic resistance
Drug susceptibility, testing for, 371–72
Dry habitats, 186b, 197. *See also* Deserts; Desiccation
Dry heat, and microbial control, 320, 323
Dry oven, 323
DTaP vaccine, 659, 660
Dura mater, 573
Dust mites, 481, 482f
Dutch elm disease, 132b
Dwarfism, 296, 297
Dyes. *See also* Stained smears
antimicrobial agents, 336
culture media, 65, 66
Gram staining, 95b
Dysentery, 698, 712–14. *See also* Amoebiasis; *Shigella dysenteriae*
Dysplasia, 758b
Dyspnea, 661
Dysuria, 737

Ear infection, 653–54
Early nuclear antigen, 622
Earthquakes, and transmission of infectious disease, 401, 579
Eastern equine encephalitis (EEE), 159t, 409t, 585, 586, 602t
Ebola virus, 152f, 159t, 623–24, 643t
E. coli. See Escherichia coli
Ecological associations, among microorganisms, 197–99
Ecology, 132b, 767–72. *See also* Environment
Ecosystems, 4n, 153b, 768–71
Ectopic pregnancies, 743b
Ectoplasm, 137
Eczema, 485, 486f
Edelman, Gerald, B2
Edema, 395, 430–31
Edema factor, 633
Education, and ethical decisions on genetic engineering, 298b
Efavirenz, 361t
Effusion, and ear infection, 653
Egg hatching test, for schistosomiasis, 726
Egypt, ancient, 5, 314b
Ehrlich, Paul, 345b, 752, B2
Ehrlichia spp., 626. *See also* Ehrlichiosis
E. chaffeensis, 626, 643t
E. ewingii, 626, 643t
E. phagocytophila, 643t
E. sennetsu, 626
Ehrlichiosis, 409t, 619b, 626, 628t, 643t
Elderly. *See also* Age
antimicrobial drugs, 373, 374
pneumonia, 86, 98b, 674
Electrolytes, 34, 704b
Electromagnetic radiation, 196–97
Electron(s), *28,* 30, 31f. *See also* Electron transfer; Electron transport system
Electron carriers, 222
Electronegativity, 31n
Electron microscopy, 76, 152
Electron transfer, 222
Electron transport system (ETS), 229, 230f, 232
Elements, *28,* 29t, 30, 180t
ELISA tests. *See* Enzyme-linked immunosorbent assay
Ella, Krishna, 469b
Elongation, and protein synthesis, 261, 263–64
El Tor biotype, of *Vibrio cholerae,* 703
Emergency, meningitis as medical, 576
Emerging diseases, 7, 8b, 399
Emetic disease, *708*
-Emia (suffix), 392b, 610
Emphysema, 302t
Encapsulated bacteria, 93
Encephalitis, 409t, 552–53, 583–92, 602t, 753
Encephalopathy, 548, 549
Encystment, 138
Endemic disease, 407, 408f
Endemic reservoirs, 617
Endergonic reaction, 221
Endocarditis, 612–13, 614t, 643t, 673
Endocardium, 609
Endocervical curettage, 758b
Endocrine glands, 501–502
Endocytosis, 160, 191f, 192
Endoenzymes, 216
Endogenous infections, 382
Endogenous pyrogens, 431
Endogenous retroviruses (ERVs), 381
Endonucleases, 285
Endoplasm, 137
Endoplasmic reticulum (ER), 123, 125, 126
Endoscopy, 696
Endospore(s), 101–103, 314, 320, 709
Endospore stain, 79–80
Endosymbiosis, 119b
Endothelium, 612b
Endotoxic shock, 614
Endotoxins, 98, 391, 392, 660. *See also* Toxins and toxicity
Energy. *See also* Energy cascade; Energy flow; Energy pyramid
metabolism, 221–24
mitochondria, 126
nutritional flow in ecosystems, 769–71
photosynthesis, 239–41
role of ATP in storage and release of, 51
Energy of activation, 213
Energy cascade, 229, 232
Energy flow, 4–5
Energy pyramid, 769
Enfuvirtide, 361t
England, and bovine spongiform encephalopathy, 572, 590
Enriched medium, 64, 65f
Entamoeba spp., 141–42, 712–14. *See also* Amoebas; Amoebiasis
E. coli, 713
E. dispar, 713
E. histolytica, 138, 141–42, 384b, 712–14, 727b. *See also* Amoebiasis
Entecavir, 716
Enteric bacteria, 783
Enteric fever, 697
Enteroaggregative *Escherichia coli* (EAEC), 701, 710–11
Enterobacter spp., 783
E. sakazakii, 583b
Enterobiasis, 145
Enterobius vermicularis, 144t, 145, 719, 720, 721b, 727b
Enterococcus spp., 783
E. faecalis, 193
Enterococcus faecalis broth, 66t
Enterocolitis infection, 710
Enterohemorrhagic *Escherichia coli* (EHEC), 409t, 699, 708b
Enteroinvasive *Escherichia coli* (EIEC), 701
Enteropathogenic *Escherichia coli* (EPEC), 701
Enterotoxigenic *Escherichia coli* (ETEC), 701, 710
Enterotoxin, 698–99
Enterovirus, 594
Enumeration, of bacteria, 205
Enumeration media, 67
Enveloped virus, 153f, 155–56, 161f
Environment. *See also* Ecology; Environmentalism; Environmental microbiology
biofilms and development of terrestrial and aquatic, 92b
emerging diseases, 399
enzymes, 218
extreme, 27, 33b, 112
factors influencing microbes, 192–99
fungal control agents, 132b
osmotic variations, 188–89
Environmentalism, and long-term effects of pollution, 791b
Environmental microbiology. *See also* Environment
principles of ecology, 767–72
recycling of bioelements, 772–77
soil and composition of lithosphere, 778–80
Environmental Protection Agency (EPA), 211, 298, 337b, 783
Enzyme(s), *49, 212*
apoenzymes, 214
characteristics of, 212t
classification of functions, 215, 216b
cofactors, 215
DNA replication, 253t
environmental sensitivity, 218
genetic engineering, 284–85
industrial products, 804t, 807
infectious diseases, 390–91
location and regularity of action, 215–18
mechanisms of action, 212–13
names of, 216b

Enzyme(s) *(continued)*
  recombinant DNA technology, 297t
  regulation of metabolic pathways and, 218–21
  structure of, 213–14
  substrate interactions, 214–15
  unconventional forms of, 214b
  viruses, 157
Enzyme-linked immunosorbent assay (ELISA), 511, 527b, 527–29, 530f, 553, 554, 696
Eosinophils, 422–23, 432, 717
Epicardium, 609
Epidemics, 407, 408f, 593b, 616, 675, 705, 713
Epidemiology, *3*, 404, 405–409, 491. *See also specific diseases*
Epidermis, 536–37
Epidermophyton, 135t, 560–61, 566t
Epididymis, 735
Epilimnion, 781
Epinephrine, 488
Epitope, 451
Epivir (3TC), 641
Epstein-Barr virus (EBV), 505, 621–22, 623t, 643t
Epulopiscium fishelsoni, 109b
Equations, 35–36
Ergosterol, 364
Erysipelas, 353t, 538–39b, 566t
Erythema infectiosum, 159t, 555
Erythema migrans, 618
Erythroblastosis fetalis, 492f, 493
Erythrocytes, 421, 424
Erythrocytic phase, of malaria, 630
Erythrogenic toxin, 656
Erythromycin
  applications of, 353t, 357
  conjunctivitis, 563
  drug receptors, 364
  drug resistance, 364
  Kirby-Bauer test, 372t
  listeriosis, 578
  protein synthesis, 349, 357
  ribosomes and translation, 268
  trench fever, 626
  whooping cough, 660
Erythropoietin (EPO), 297t
Eschar, 558
Escherichia coli. *See also* Enteroaggregative *Escherichia coli* (EAEC); Enterohemorrhagic *Escherichia coli* (EHEC); Enteroinvasive *Escherichia coli* (EIEC); Enteropathogenic *Escherichia coli* (EPEC); Enterotoxigenic *Escherichia coli* (ETEC); *Escherichia coli* 0157:H7
  adhesion, 389t
  anaerobic respiration, 233
  antimicrobial drugs, 353t
  bacteriophages, 167
  chemical composition of cell, 181t
  cloning hosts, 294
  commensalism, 199
  diplobacillus arrangement, 104f
  enzyme induction, 221
  fimbriae, 90
  food-borne illnesses, 696, 701, 708b
  genetic mutations, 271
  genome size, 157, 248
  meningitis, 581, 582t, 602t
  recombinant DNA technology, 296, 299b
  urinary tract infections, 737–38, 759b
  vitamin K synthesis, 689
  water contamination, 783
Escherichia coli 0157:H7, 8b, 583b, 699–701, 707b, 727b, 789
Essential nutrient, *180*, 182–87
Ester, 40t
Estrogen, 735, 736
E-test, for drug susceptibility, 372f
Ethambutol, 353t
Ethanol, 804t
Ethics, and genetic engineering, 298b
Ethyl alcohol (ethanol), 329t, 331
Ethylene oxide (ETO), 329t, 336, 803
Etiologic diagnosis, 542b
Etiology, of infectious disease, 404–405. *See also specific diseases*
Euglena mutabilis, 196
Eukarya, 22
Eukaryotes, *3*
  cellular organization, 7–8, 9f
  characteristics of cells, 52
  evolutionary history of, 118–20
  external structures, 120–22
  internal structures, 123–29
  prokaryotes compared to, 111t, 129t
  viruses compared to, 129t, 151f
European Union, 367b
Eustachian tube, 653
Eutrophication, 782
Evolution, *21*. *See also* Adaptation; Coevolution; Natural selection
  eukaryotes, 118–20
  genetic mutations, 271–72
  origin of microorganisms and, 20–21
  prokaryotes, 87, 106
  time line, 3f
  tree diagrams, 21–23
  vertebrate eye, 562
  viruses, 153b
Ewald, Paul, 513b
Exanthem, 552
Exchange reactions, 36
Excision repair, 270
Exergonic reaction, 221
Exocytosis, 162
Exoenzymes, 216, 390
Exogenous infections, 385–86
Exogenous pyrogens, 431
Exons, 264
Exotoxins, 391–92, 545, 633, 658, 701, 706–710
Experiments, and scientific method, 14, 15f
Exponent(s), A1–A2
Exponential growth, 201, 202
Extension, and polymerase chain reaction, 292
Extensively drug-resistant TB (XDRTB), 671
Extracellular enzymes, 390–91
Extracellular fluid (ECF), 418, 419
Extracellular source, of carbon, 182
Extrapulmonary tuberculosis, 667
Extreme habitats, 27, 33b, 111, 186b
Extremophiles, 111, 186b, 807
Extremozymes, 194b, 214b
Ex vivo therapy, and genetic treatments, 302, 303f
Exxon Valdez (ship), 766, 770b
Eye and eye infections. *See also* Conjunctivitis
  Chlamydia infections, 747
  defense mechanisms, 562
  gonorrhea, 745
  infectious diseases, 387, 416, 562–67
  normal biota, 562

Facilitated diffusion, 190, 192t
Factor VIII, 297t, 302t
Factor IX, 302t
Facultative, definition of, *185b*
Facultative aerobe, 195
Facultative halophiles, 196
Facultative parasite, 185
Facultative psychrophiles, 193
Fairclough, Edith, 593b
Falciparum malaria, 628
Fallopian tubes, 735
False negatives, in tuberculin tests, 670
False positives, in serological tests, 522b, 640
Famciclovir, 360
Family, and taxonomy, 18–20
Fasciola hepatica, 723, 727b
Fastidious bacteria, 64
Fatty acids, 40t
Fc. *See* Crystallizable fragment
Fecal coliforms, 783–84
Feces, as portal of exit for infection, 396
Female condom, 755
Female reproductive system, 735, 737f
Fermentation, 234–36, *790*, 793–97, 805–807
Fermentor, 805–807
Fertility (F factor), 273
Fertilizers, 770b
Fetus. *See* Infants; Pregnancy
Fever, 431–32. *See also* Hemorrhagic fever viruses
Fever blisters, 753
Fevers of unknown origin (FUO), 431
Field, of optical microscope, 72
Fifth disease, 555, 556t, 566t
Filament, 88
Filamentous hemagglutinin (FHA), 660
Filamentous rods, 104f
"Fill-in-the-blank" version, of concept mapping, D2
Filoviridae, 159t, 623
Filterable virus, 150
Filtration, and microbial control, 326, 327f
Fimbriae, 88f, 90, 91f, 745–46
Fingerprinting. *See* DNA fingerprinting
Fire, Andrew, 257b
Firmibacteria, 106
Firmicutes, 106, 381
Fish
  Diphyllobothrium latum, 721
  mercury and bioaccumulation, 777
  Pfiesteria infection, 137
Five-kingdom system, 21
Fixed, stained smears, 78–80
Flagella (flagellum), 88–90, 121, 138
Flagellar staining, 80
Flagyl, 714
Flash method, of pasteurization, 800
Flatworms, 143, 144t
Flavin adenine dinucleotide (FAD), 217
Flaviviridae, 159t, 623, 716
Fleas, 616, 619b
"Flesh-eating disease," 539b
Fleming, Alexander, 345b, B2
Floppy baby syndrome, 197b, 599
Florey, Howard, 345b, B2
Flow charts, 517, 518f
Flow cytometer, 205
Fluconazole, 353t, 359, 579
Flucytosine, 353t, 359, 368t
Fluid mosaic model, 46b
Flukes. *See* Blood flukes; Liver flukes
FluMist, 665
Fluorescence microscopy, 73t, 74–75
Fluorescent antibodies, 75, 526–27
Fluorescent in situ hybridization (FISH), 287
Fluoride, 691
Fluoroquinolones, 358, 366t
Focal infection, 393–94

Folic acid synthesis, and antimicrobial drugs, 349, 351–52, 357–58
Folk medicines, 345b, 365. *See also* Home remedies
Folliculitis, 538b, 566t
Fomite, 401
Fonseca pedrosoi, 131f
Food(s). *See also* Dairy microbiology; Diet; Food chain; Food microbiology; Food poisoning; Food web; Nutrition
  additives, 804t
  allergies to, 486
  beneficial and detrimental effects of microorganisms, 793
  campylobacteriosis, 702
  Creutzfeldt-Jakob disease, 590
  cryptosporidiosis, 705
  Cyclospora cayetanensis, 711
  E. coli infections, 700, 789
  fermentations, 793–94
  helminthic infections, 719
  hemolytic uremic syndrome, 789
  hepatitis, 714
  human uses of microorganisms, 5
  irradiation, 325
  listeriosis, 578
  meningitis, 583b
  safety and disease prevention, 798–99
  salmonellosis, 696–98
  toxoplasmosis, 587, 588
  trichinosis, 724
  Yersina spp., 702
Food-borne botulism, 598, 599
Food chain, 769, 770f, 771
Food and Drug Administration (FDA), 470b, 789, 799, 800, 801b, 802b, 803
Food microbiology, 2. *See also* Dairy microbiology
Food poisoning, 323, 401, 706–10, 727b, 799–803. *See also* Botulism; Salmonellosis
Food web, 771
Forbidden clones, 500
Forensic medicine, 305. *See also* DNA fingerprinting; Medical microbiology
Formaldehyde, 335
Formalin, 335
Formulas, 35–36
Formyl methionine, 261
Foscarnet, 622
Fosfomycin trimethamine, 355
Fossil(s), 102, 118
Fossil fuels, 773, 774b
Fox, George, 21–22
Fracastoro, Girolamo, B1
Frameshift mutations, 270
Francisella tularensis, 617, 643t, 677b. *See also* Tularemia
Franklin, Rosalind, 250b, B2
Free energy, 182
Free iodine, 330
Free-living nonpathogenic bacteria, 110–11
"Freestyle" version, of concept mapping, D3
Freezing, and food preservation, 801–802
French and Indian Wars (1754–1767), 549b
Fresh, living preparations, 78
Frosch, Paul, 150, B1
Frostban, 298
Fructans, 689
Fructose, 35, 41, 43f
FTA-ABS test, 752
Fumaric acid, 229b
Fumigation, 337b
Functional groups, of organic compounds, 39, 40t
Functional type, of media, 60t
Fungemia, 610
Fungicide, 315
Fungistatic chemicals, 316
Fungus (fungi)
  antimicrobial drugs, 358–59, 368t
  cell walls, 122
  characteristics of, 130
  drug resistance, 364
  ecological importance of, 132b
  immunologic testing, 529
  lung diseases, 669b
  nutrition, 130–31
  organization of, 131, 133
  roles in nature and industry, 134–35
  sinusitis, 653
  skin and eye diseases, 566t
  soil decomposition, 780
  thermal death times, 320t
Furuncles, 538b, 566t
Fusarium spp., 132b
  F. graminearum, 798
Fusion inhibitor, 641
Fusobacterium, 693, 727b
Fuzeon, 360, 641

Gaia theory, *772*
Gajdusek, D. Carleton, B2
Galactose, 35
Galen, Claudius, 744
Gallbladder, and typhoid fever, 698
Gallo, Robert, B2
Gamma globulin, 458, 464
Gamma rays, 269, 324–25
Gancyclovir, 353t, 360, 622
Gardasil, 757
Gardnerella spp., 740, 741–42, 743b, 759b
  G. vaginalis, 741
Gas(es), and microbial growth, 194–96, 355–36. *See also* Carbon dioxide; Carbon monoxide; Hydrogen; Methane; Nitrous oxide; Oxygen; Swamp gas
Gas gangrene, 103, 545–46, 566t. *See also* *Clostridium perfringens*
Gastric ulcers, 7, 695–96, 727b
Gastritis, 695–96, 727b
Gastroenteritis, 697
Gastrointestinal distress, and recreational water activities, 117, 137b
Gastrointestinal (GI) tract
  anatomy, 687–88
  defense mechanisms, 688
  infectious diseases, 689–726
  normal biota, 382t, 688–89, 734
  portals of entry for infection, 387
Gates Foundation, 469b, 631b, 723
Gelatin media, 62
Gel electrophoresis, 285–86
Gell, P., 480
Gender, and autoimmune disease, 499
Gene(s), *248*. *See also* Gene therapy; Genetic(s)
Gene probes, 286–87
Generalized transduction, 276
Generalized vesicular rash, 542b
Generation time, 200–201
Gene therapy, 153b, 302–303. *See also* Genetic diseases
Genetic(s), *247*. *See also* DNA; Gene(s); Gene therapy; Genetic engineering; Genome; Nucleic acid(s); RNA
  allergies, 481
  animal viruses, 265
  autoimmune disease, 499
  DNA code, 249
  DNA recombination events, 272–77
  DNA replication, 252–55
  DNA structure, 249–52
  eukaryotes and prokaryotes compared, 129t
  eukaryotic transcription and translation, 264
  evolution, 21
  familial Creutzfeldt-Jakob disease, 589
  gene-protein connection, 255–56
  host defenses and susceptibility to infection, 385t, 417–18
  identification of crime victims, 282, 288b
  identification of microorganisms, 69
  influenza virus, 662–63
  meningitis, 575
  mutations, 268–72
  nature of material, 247–48
  regulation of protein synthesis and metabolism, 265–68
  Rh factor, 491–93
  transcription and gene expression, 259–60
  translation and gene expression, 260–64
Genetically modified organisms (GMOs), 5, 297–301, 791b
Genetic analysis, of viruses, 530f
Genetic code, 260, 261f
Genetic diseases, viruses and treatment of, 153b. *See also* Gene therapy; Xeroderma pigmentosa
Genetic engineering, *2–3*. *See also* Recombinant DNA technology
  basic elements and applications, 283–84
  gene therapy and genetic medicine, 302–304
  genetically modified organisms, 297–301
  genome analysis, 304–307
  human use of microorganisms, 5
  recombinant DNA technology, 292–97
  tools and techniques of, 284–92
  transfection, 275
  vaccines, 469–70
Genetic markers, 305, 418n
Genetic probes, 518
Genetic revolution, 264
Genital herpes, 159t, 353t, 752–55, 759b
Genital tract. *See also* Reproduction and reproductive system; Urinary tract
  defense mechanisms, 734–35
  normal biota, 382t
  portals of entry for infection, 387, 416
  portals of exit for infection, 396
Genital tuberculosis, 667
Genital ulcer diseases, 748
Genome, 156, 247–48, 304–307. *See also* Genomic libraries
Genome analysis, 304–307
Genome mapping, 304–305
Genomic(s), 305
Genomic libraries, 293
Genotype, 248
Genotypic methods, of diagnosis, 512–13, 518–19
Gentamicin, 356, 372t, 563, 617, 625
Genus, 18–20
Geomicrobiology, 3, 772
Geotrichosis, 669b
Geotrichum candidum, 669b
Germicidal lamps, 325
Germicides, 315, 329, 335
Germination, of endospores, 103
Germline therapy, 303
Germ theory of disease, 15, 16–17
Giardia lamblia, 20, 138, 353t, 389t, 711–12, 727b, 782. *See also* Giardiasis

Giardiasis, 141t, 353t, 409t, 711–12. *See also* *Giardia lamblia*
Giese, Jeanna, 591b
Gingivitis, 691, 692f
Globalization, and drug resistance, 367b, 374
Global warming, 112, 774b. *See also* Greenhouse gases and greenhouse effect
Glomerulonephritis, 500t, 656
Glucans, 689
Gluconeogenesis, 237
Gluconobacter, 235b, 797
Glucose
  aerobic respiration, 185, 225, 233t
  energy production, 222f
  glycosidic bond, 43f
  hexoses, 41
  lac operon, 267
  molecular formula, 35
  three-dimensional model, 36f
Glutaraldehyde, 328, 335
Glycan chains, 95
Glyceraldehyde-3-phosphate (PGAL), 241
Glycerol, 44, 804t
Glycocalyx, 42, 88f, 93, 121–22
Glycogen, 44, 451
Glycolysis, 224, 225–27
Glycoproteins, 662
Glycosidic bonds, 41, 43f
Gnotobiotic studies, 384b
Golgi, C., 125n
Golgi apparatus, 125–26
Gonococcal Isolate Surveillance Project (GISP), 747
Gonorrhea, 353t, 387t, 409t, 744–47, 749b, 759b. *See also Neisseria gonorrhoeae*
Goodpasture syndrome, 500t
Gorillas, and Ebola virus, 623
Government. *See* Centers for Disease Control and Prevention; Department of Agriculture; Department of Defense; Environmental Protection Agency; Food and Drug Administration; U.S. Public Health Service
GPI (glycosyl-phosphatidyl inositol), 630
Gracilicutes, 106
Grady Memorial General Hospital (Ohio), 117
Graft rejection, 462, 497–98
Graft versus host disease (GVHD), 498
Gram, Hans Christian, 79, 94, 95b, B1
Gram-negative bacteria, 79, 94, 95b, 96, 107t, 566t
Gram-positive bacteria, 79, 94, 95b, 96, 107t, 566t
Gram's iodine, 95b
Gram staining, 79, 80f, 94, 95b
Grana, 127
Granules, 88f, 100–101. *See also* Inclusion bodies
Granulocyte(s), 422
Granulocyte-colony-stimulating factor (G-CSF), 297t, 429b
Granulocyte-macrophage-colony-stimulating factor (GM-CSF), 297t
Granulomas, 395, 427b
Granulomatous amoebic meningoencephalitis (GAM), 584
Granzymes, 461
Grapes, and wine, 795
Grassi, G., B1
Graves' disease, 499, 500t, 501
Grays, of radiation, 324
Great Salt Lake (Utah), 189
Greece, ancient, 314b, 345b
Greenhouse gases and greenhouse effect, 4, 112, 773, 774b, 776. *See also* Global warming
Green sulfur bacteria, 111
Griffith, Frederick, 274, B2
Griffiths, Jeffery, 469b
Griseofulvin, 359, 561
Ground itch, 723
Groundwater, 791b
Group A streptococcal infections, 656, 657
Group B streptococcal infections, 657, 759
Group translocation, 191f, 192
Growth, microbial, 200–205
Growth curve, 202, 204
Growth factors, 64, 183
Growth hormone deficiency (GHD), 302t
Guanine (G), 49, 50f, 249
Guillain-Barré syndrome (GBS), 593b, 702
Guinea worm, 144t
Gulf War, 306, 549b
Gummas, 750
Gut-associated lymphoid tissue (GALT), 425, 426, 688
Gymnodinium, 782f
Gyrase, 253t

HAART (highly active antiretroviral therapy), 641, 710
Habitat(s), 137, 768. *See also* Extreme habitats; Osmotic pressure; pH; Temperature
Haeckel, Ernst, 21
Haemophilus ducreyi, 752, 756b, 759b
Haemophilus influenzae. *See also* Influenza
  chemotherapy of, 353t
  growth factors, 183
  immunization, 472t
  otitis media, 654, 680b
  meningitis, 577, 582t, 602t
  pneumonia, 672
  reportable diseases, 409t
  satellitism, 199f
Hair follicles, 537
Hairpin loops, of tRNA, 258
Hairy-cell leukemia, 642, 643t
Halobacteria, 189, 196
Halococcus, 196
Halogens, 329
Halophiles, 112, 196, 807
Hand-foot-mouth disease, 159t
Handscrubbing. *See* Surgical handscrub
Hansen's disease (leprosy), 358, 409t, 427b, 542b
Hantavirus, 399, 409t, 675, 679b, 680b
Hantavirus pulmonary syndrome (HPS), 675
Haptens, 451
Hashimoto's thyroiditis, 500t, 501
Hay fever, 481, 485
Hazard Analysis and Critical Control Point (HACCP), 798–99
Headaches, and botox, 600b
Health care personnel. *See also* Health care products; Hospitals; Medical microbiology; Public health
  Creutzfeldt-Jakob disease, 590
  HIV transmission, 527b, 639
  signs of smallpox, 550
  universal blood and body fluid precautions, 403, 511
Health care products, and industrial microbiology, 807
Heart and heart attack, 302t, 369, 609. *See also* Acute endocarditis; Cardiovascular system
Heat, and microbial control, 320–23. *See also* Temperature
Heavy metals, 333–35, 777. *See also* Mercury
Hektoen enteric (HE) agar, 67t
Helical capsids, 154
Helicases, 253
Heliobacter spp. 7
  H. pylori, 695–96, 702, 727b
Helium, 30
Helminths, 143–45, 353t, 359–60, 368t, 566t, 717–26, 727b
Hemagglutinin, 662
Hematopoiesis, 421, 422f
Hematuria, 737
Hemoglobin, 297t, 385
Hemolysins, 391, 656
Hemolytic disease of the newborn, 492–93
Hemolytic uremic syndrome (HUS), 409t, 699–700, 789
Hemolyze, 391
Hemophilia, 296–97, 302t, 716–17
Hemorrhagic fever viruses, 361t, 622–24, 676b
Hepadnaviridae, 159t, 715
Hepatitis, 714–17
Hepatitis A virus (HAV)
  classification of viruses, 159t
  immunization, 472t, 473t
  "infectious" hepatitis, 714–15, 717b, 727t
  reportable diseases, 409t
  water contamination, 782
Hepatitis B immune globulin (HBIG), 716
Hepatitis B virus (HBV)
  classification of viruses, 159t
  HIV, 744
  incidence of, 387t
  reportable diseases, 409t
  "serum" hepatitis, 714–16, 717b, 727b
  vaccine, 469b, 470b, 472t, 473t
Hepatitis C virus, 714, 716–17, 727b, 744
Hepatitis D virus, 715
Hepatitis E virus, 717b, 727b
Hepatocellular carcinoma, 715
Herbicide, 301b
Herd immunity, 471
Hereditary angioedema, 503t
Heredity, *247*
Hermaphroditism, 144
Herpes simplex virus, 159t, 166t, 387t, 566t, 754. *See also* Herpesviruses
Herpes simplex virus type 1 (HSV-1)
  classification of viruses, 159t
  encephalitis, 586, 587t, 602t
  genital herpes, 753, 754, 756b, 759b
  herpetic keratitis, 564
Herpes simplex virus type 2 (HSV-2)
  classification of viruses, 159t
  encephalitis, 586, 587t, 602t
  genital herpes, 753, 754, 756b, 759b
  herpetic keratitis, 564
  meningitis, 581
Herpesviruses. *See also* Herpes simplex virus; Herpes simplex virus type 1; Herpes simplex virus type 2; Human herpesvirus 6; Human herpesvirus 7
  antiviral drugs and nucleic acid synthesis, 361t
  chemotherapy for, 353t
  classification of viruses, 159t
  encephalitis, 586, 587t, 602t
  genital herpes, 753, 754, 756b, 759b
  meningitis, 581
  roseola, 555
Herpes zoster, 548
Hesse, Walther & Fanny, B1
Heterolactic fermentation, 235
Heterophile antibodies, 622
Heterotroph, *182*, 183t, 185, 187

Heterotrophic fungi, 130
Hexachlorophene, 331
Hexokinase, 214t
Hexoses, 41
Hierarchies, taxonomic, 20
High-efficiency particulate air (HEPA) filters, 326
High-frequency recombination, 273
High-level germicides, 329
High-temperature method, of pasteurization, 800
Hippocrates, 365, 693
Histamine, 429b, 483–84
Histidine, 272f
Histiocytes, 432
Histones, 87, 123
Histoplasma capsulatum, 135t, 669b, 672, 677–78, 679b, 680b. *See also* Histoplasmosis
Histoplasmin test, 678
Histoplasmosis, 135t, 669b, 677, 680b. *See also* *Histoplasma capsulatum*
History
  antimicrobial therapy, 345b
  beer brewing, 794
  foundations of microbiology, 11–17
  immunization, 465b
  malaria, 628
  microbial control, 314b
  microorganisms as food, 798
  plague, 314b, 615
  significant events in microbiology, B1–B2
  smallpox, 549b
  syphilis, 748–49
  tuberculosis, 666
HIV. *See* Human immunodeficiency virus
Hives, 478
HKO antigens, 697
Hoffman, E., B2
Holmes, Oliver Wendell, 16
Holoenzyme, 214
Home remedies, 557, 738
Hominoidea, 20
Homolactic fermentation, 235
Homo sapiens, 20
Honey, and infant botulism, 599
Hong Kong influenza, 664b
Hooke, Robert, 11, B1
Hookworm, 722
Hops, 794
Hordeolum, 563b
Horizontal infection transmission, 400
Hormones, and recombinant DNA technology, 297t
Horse serum, 312, 464, 465, 495, 659
Hospital(s). *See also* Health care personnel
  drug resistance, 366t
  nosocomial infections, 401–403, 679–80
Hospital-acquired MRSA (HA-MRSA), 544b
Host, *11*. *See also* Cloning host
  graft rejection, 498
  infectious disease and human, 380–83
Host defenses, 389–90. *See also* Defense mechanisms
Host range, 160
Host response, 392
Hot air, 323
Hot environments, 186b
Hot springs, 194b
Houseflies, 397–98
Huber, Robert, B2
Human(s)
  experimentation, 405b, 465b, 749
  infectious diseases and human condition, 6–7
  interrelationships with microbes, 199
  size of genome, 248
  taxonomy, 19–20
  use of microorganisms, 5–6
Human albumin, 302t
Human diploid cell vaccine (HDCV), 592
Human Genome Project, 257b, 289b
Human granulocytic ehrlichiosis (HGE), 626
Human growth hormone (HGH), 296, 297, 302t
Human herpesvirus 3 (HHV-3), 548, 566t, 581
Human herpesvirus 6 (HHV-6), 581
Human herpesvirus 7 (HHV-7), 581
Human immunodeficiency virus (HIV), 643t. *See also* Acquired immunodeficiency disease (AIDS)
  adhesion, 389t
  attachment mode, 160
  cancer and cancer therapy, 167, 303
  Candida infections, 741
  causative agent, 635, 643t
  chemotherapy, 353t, 360, 361t
  chronic diarrhea, 710
  culture and diagnosis, 640
  discovery of, 634
  drug resistance, 366b
  health care workers and accidental exposure, 511, 527b
  hepatitis B and C, 744
  HIV types 1 and 2 (HIV-1 and HIV-2), 635, 643t
  meningitis, 581
  pelvic inflammatory disease (PID), 733–34
  prevention, 640
  protease inhibitors, 174
  recombinant DNA technology, 297
  reportable diseases, 409t
  salmonellosis, 698
  secondary immunodeficiency diseases, 506
  structure of, 635f
  transmission, 638–39
  Trichomonas infections, 742b
Human interleukin-2, 302t
Human leukocyte antigen (HLA) system, 446
Human monocytic ehrlichiosis (HME), 626
Human papillomavirus (HPV). *See also* Sexually transmitted diseases
  classification of viruses, 159t
  genital warts and cervical carcinoma, 566t, 755–58, 759b
  incidence of, 387t
  vaccine, 468t, 470b, 472t, 473t
Human rabies immune globulin (HRIG), 592
Human rhinovirus, 159t
Human T-cell lymphotropic viruses I and II (HTLV-I and HTLV-II), 635, 642, 643t
Humoral immunity, 424
Humulus lupulus, 794
Humus, 779
Hurricane Katrina, 686
Hutchinson's teeth, 750, 751f
Hyaluronic acid (HA), 656
Hyaluronidase, 391, 804t
Hybridization probes, 286–87, 288f, 518
Hybridoma, 460b
Hydrated ions, 36
Hydrocortisone, 804t, 807
Hydrogen
  atomic structure, 28f
  covalent bonds, 32f
  elements of life, 29t
  sources of, 180t, 182
Hydrogen bond, 32f, 34, 35f
Hydrogen peroxide, 195, 328, 329t, 331–32
Hydrogen sulfide, 776
Hydrolases, 216b
Hydrologic cycle, 780
Hydrolysis reaction, 44, 216–17, 218f
Hydrophilic ions, 36
Hydrophobia, 590
Hydrophobic ions, 36
Hydrosphere, 768
Hydrostatic pressure, 197
Hydrothermal vents, 186b, 198b
Hydroxyl, 40t
Hymenolepsis spp., 721, 727b
  H. diminuta, 721
  H. nana, 721
Hypersensitivity, 3, 480, 489–93
Hyperthermophiles, 112, 807
Hypertonic conditions, 188
Hypervariable region, 453
Hyphae, 130, 131, 133f
Hyphomicrobium, 781
Hypochlorites, 330
Hypogammaglobulinemia, 503
Hypolimnion, 781
Hyposensitization, of immune system, 488, 503–506
Hypothesis, 13–14, 15f
Hypotonic conditions, 188

"Iceberg effect," in epidemiology, 407
Icosahedral viruses, 155f
Icosahedron, 154
Identification. *See also* Diagnosis
  antimicrobial drug selection, 370
  fungi, 134
  Gram stain, 95b
  helminths, 145
  laboratory methods, 58, 59f, 68–69
  protozoa, 141
  taxonomy, 17
  viruses, 170–72
IgE-mediated allergies, 482, 485–86
IgG blocking antibodies, 488
Illinois, and leptospirosis, 739
Imipenem, 354–55
Immediate hypersensitivity, 480t
Immune complex-mediated hypersensitivity, 480t
Immune complex reactions, 494–95
Immune deficiency, theory of, 500
Immune disorders. *See* Allergy; Autoimmune diseases; Autoimmunity; Immunodeficiency diseases
Immune interference, and mixed vaccines, 471
Immune privilege, 562, 573
Immune serum globulin (ISG), 464–65
Immune system. *See also* Defense mechanisms; Immunology; Immunotherapy; Specific immunity
  blood and bloodstream, 418, 420–24
  communicating body compartments, 419–26
  extracellular fluid, 418, 419
  germ-free animals, 384b
  lymphatic system, 418, 424–26
  responsibilities of, 418
  reticuloendothelial system, 418, 419–20
  systems involved in, 418
Immune testing, 519, 521
Immune treatments, and recombinant DNA technology, 297t. *See also* Immunotherapy
Immunities, 415
Immunization, 442, 467b, 463, 464–73. *See also* Vaccination; Vaccines
Immunoassays, 527
Immunocompetence, 443

Immunodeficiency diseases, 302–303, 503–506
Immunoelectrophoresis, 524
Immunofluorescence testing, 526–27
Immunogen, 450
Immunoglobulin, 449, 453, 457–58, 482, 488
Immunologic methods, of diagnosis, 513, 519–30
Immunologic testing, and identification, 69
Immunology, 3, 418
Immunopathology, *479*
Immunoprophylaxis, 463
Immunosuppressive agents, 503t, 506
Immunotherapy, 463–64. *See also* Immune treatments
Impellers, 806
Impetigo, 540–42, 545, 566t
Inactivated poliovirus vaccine (IPV), 595
Inappropriate expression, of MHC II markers, 500
Incidence, of disease, 406–407
Incineration, and microbial control, 323
Inclusion bodies, 88f, 100–101, 166. *See also* Granules
Incubation, and laboratory methods, 58, 59f, 68–69
Incubation carriers, 397, 398f
Incubation period, of infection, 393
India
  cholera, 704b
  measles, 553
  plague, 616
  polio, 595
  vaccination programs, 469b
India ink, 79
Indicator bacteria, 783
Indinavir, 361t
Indirect ELISA test, 528
Indirect testing, of fluorescent antibodies, 527
Indirect transmission, 400–401
Indonesia, 594
Induced mutations, 269
Inducer, of lactose operon, 266
Inducible genes, 265–67
Induction
  of enzymes, 221
  of lysogeny, 167
Induration, 670
Industry and industrial microbiology, *803*. *See also* Applied microbiology; Pharmaceutical industry
  commercial antimicrobial products, 328b
  enzymes, 804t
  examples of industrial products, 804t
  fermentation, 235–36, 805–807
  fungi, 135
Infant(s). *See also* Baby food; Baby formula; Breast-feeding; Children; Pregnancy
  antimicrobial drugs, 373
  birth process and infections, 388
  botulism, 179, 197b, 598, 599
  chronic granulomatous disease (CGD), 414, 433b
  congenital syphilis, 750
  dental caries, 690–91
  group B *Streptococcus*, 759
  hemolytic disease of the newborn, 492–93
  herpes infections, 753
  HIV transmission, 638–39
  initial colonization with microbes, 382, 383f
  natural passive immunity, 463
  neonatal eye infections, 563
  neonatal meningitis, 581–83
  pneumonia, 674
  rotavirus, 706
  tetanus, 596
Infant botulism, 598
Infantile paralysis, 593b
Infection, *380*
Infection control officer, 403
Infectious allergy, 495–96
Infectious diseases, *380*. *See also* Diagnosis; Epidemics; Medical microbiology; Sexually transmitted diseases; *specific diseases*
  acquisition and transmission of, 399–401
  anaerobes, 195
  antimicrobial drugs, 353t
  attachment to host, 388–89
  cardiovascular system, 612–44
  encephalitis, 583–92
  epidemiology, 405–409
  establishment, spread, and pathologic effects, 393–94
  eye, 562–67
  gastrointestinal tract, 689–726
  genetic treatments, 302–304
  hospitals and nosocomial infections, 401–403
  host defenses, 389–90
  human condition, 6–7
  human host, 380–83
  inflammation, 427b
  Koch's postulates and etiology of, 404–405
  lower respiratory tract, 666–80
  lower and upper respiratory tracts, 659–66
  lymphatic system, 612–44
  meningitis, 574–83
  microbial enzymes, 218
  persistence of, 396
  portals of entry, 385–88
  portals of exit, 395–96
  progress of, 383, 385–405
  reproductive system, 740–59
  reservoirs, 397–99
  signs and symptoms, 394–95
  skin, 538–61
  terminology, 392b
  universal blood and body fluid precautions, 403–404
  upper respiratory tract, 651–59
  urinary tract, 737–40
  virulence factors, 390–92
  water supplies and monitoring, 782–83
Infectious dose (ID), 388
Infectious hepatitis, 714, 727b
Infertility, and pelvic inflammatory disease, 743b
Inflammation, 395, 427–32
Inflammatory bowel disease, 416–17, 719b
Inflammatory mediators, 429b
Inflammatory response, 427
Influenza, 662–66. *See also Haemophilus influenzae*
  artificial viruses, 172b
  attachment mode, 160
  causative agent, 159t, 680b
  chemotherapy for, 353t, 361t
  cytopathic changes in cells, 166t
  reportable diseases, 409t
  vaccine, 468t, 472t, 473t
  zoonotic infections, 399
Influenza-associated pediatric mortality, 409t
"Informed best guess," and identification of infectious agent, 370
Ingestants, and allergens, 481
Inhalants, and allergens, 481
Inhalation anthrax, 676b
Initiation, of complement cascade, 436, 437f
Inoculating loop, 58, 60
Inoculation, 58, 59f
Inorganic chemicals, 38–39, 181t
Insect(s). *See also* Arthropods; Black flies; Fleas; Houseflies; Mosquito; Sand fly; Ticks
  allergic reactions to venom, 478, 487b
  as disease vectors, 397–98
Insecticides, 298, 585, 601, 807. *See also* Pesticides
Insertion, and genetic mutations, 270t
Insertion elements, 277
Inspection, and laboratory methods, 58, 59f, 68–69
Institute for Healthcare Improvement, 402
Insulin, 296, 297, 302, 501–502
Integrase inhibitors, 641
Integrative function, of nervous system, 573
Integument, 536
Interferons
  antiviral cytokines and immune stimulants, 434–35
  applications and effectiveness of, 360, 362
  hepatitis B, 716
  hepatitis C, 717
  inflammatory mediators, 429b
  prevention of viral infections, 174
  recombinant DNA technology, 297t
Interleukins, 297t, 429b, 452
Intermediate hosts, 144, 718
Intermediate-level germicides, 329
Intermittent sterilization, 321–22
International Committee on the Taxonomy of Viruses, 157–58
Internet. *See* Web sites
Interphase, of mitosis, 124f
Intimin, 700
Intoxication, *391*, 597
Intracellular function, of carbon, 182
Intrachain bonding, 49
Intrauterine devices (IUDs), 742
Intravenous drug users (IDUs), and AIDS, 639, 640
Introns, 264
Intussusception, 706
Invasive cervical carcinomas, 636b
In vitro activity, of antimicrobial drugs, 372
In vitro allergy tests, 486
In vitro culture methods, *170*
In vivo activity, of antimicrobial drugs, 372
In vivo allergy tests, 486
In vivo culture methods, *170*
In vivo genetic treatments, 302
In vivo immunologic testing, 529
Iodine, 29t, 330
Iodophors, 330
Iodoquionol, 713
Ion(s), 34
Ionic bonds, 32–36
Ionization, 33–34, 35f
Ionizing radiation, 197, 324–25
Iron, 29t, 180t
Iron-scavenging capabilities, of *Staphylococcus aureus*, 365
Irradiation, 324, 325, 503t
Irritable bowel syndrome, 710
Isaacs, Alick, B2
Isocitric acid, 229b
Isodine, 330
Isograft, 498
Isolation, and culture methods, 58, 59f, 60, 61f, 516
Isomerases, 216b
Isoniazid (INH), 343, 353t, 355, 367b, 368t, 373, 671
Isopropyl alcohol, 331
Isotonic conditions, 188

Isotopes, 30
Isotretinoin (Accutane), 540
Isotypes, 457
-Itis (suffix), 392b
Itraconazole, 561, 580, 678
Ivanovski, D., 150, B1
Ivermectin, 360, 565
Ixodes pacificus, 621
Ixodes scapularis, 619, 620f

Jablot, Louis, 12b
Jacob, François, 265
Janssen, Zaccharias, B1
Japan
  ehrlichioses, 626
  measles outbreaks, 553
Japanese encephalitis, 468t
Jaundice, 714
J chain, 456
JC virus (JCV), 586, 602t
Jeffreys, Alex, 305
Jenner, Edward, 465b, 550, B1
Jerne, Niels Kai, B2
Jock itch, 559
"Jumping genes," 276
Junk DNA, 255, 257b

Kajander, Olavi, 109b
Kanamycin, 356, 372t
Kaposi's sarcoma, 634, 636b
Karyosome, 712
Kazakhstan, 299b
Keiko (killer whale), 778b
Keratin, 537
Keratinase, 391, 804t
Keratitis, 564–65, 566t
Ketoconazole, 359
Ketolides, 357, 674
Ketones, 40
Kidneys
  antimicrobial drugs, 368–69
  urinary tract, 734
Killed or inactivated vaccine, 467
Killer T cell, 461
Killer whale, 778b
Kingdom, and classification, 18–20
Kirby-Bauer technique, 371–72
Kitasato, Shibasaburo, B1
Klebsiella oxytoca, 246
Klebsiella pneumoniae, 679
Knockout mouse, 301
Koch, Robert, 16, 17, 78, 345b, 632, 667, B1
Koch's postulates, 16–17, 62b, 404–405
Kohler, Georges, B2
Komodo dragon, 698
Koplik's spots, 552
Krebs, Hans, 224n
Krebs cycle, 227–29, 237
Kupffer cells, 433f

Lability, of enzymes, 218
Laboratory. *See also* Cultures; Incubation; Media; Microscopes and microscopy
  biosafety levels, 386b
  specimen analysis techniques, 514–15
LaCrosse strain, of California encephalitis, 586, 602t
Lactase, 216t
Lactate dehydrogenase, 216t
Lactic acid, 235, 336
Lactobacillus spp., 199, 736, 797
  L. brevis, 794
  L. bulgaricus, 794
  L. plantarum, 794, 796
  L. sanfrancisco, 20, 794
Lactoferrin, 302t, 735
Lactose, 41
Lactose intolerance, 486n
Lactose operon, 265–67
Lagered beer, 795
Lagging strand, 253
Lag phase, of growth curve, 202
Lakes, and microbes, 781
Lamivudine, 361t, 716
Lamtibiotics, 365
Lancefield, Rebecca, 656, 657f
Lancefield group B, 759
Lancet, The (journal), 704b
Landfills, 791b
Landsteiner, Karl, 489, B2
Landstuhl Regional Medical Center (Germany), 379
Langerhans cells, 433f
Large intestine, 688–89
Laser excision, of abnormal tissue, 758b
Lassa fever, 159t, 624, 643t
Latency, of infection, 396, 749, 750, 754
Latent period, of antibody production, 458
Latex agglutination tests, 522, 541, 554
"Lawnmower" tularemia, 617
Lazear, Jesse, 405b
Leading strand, 253
Leavening, 794
Lectin pathway, 436
Lederberg, Joshua, B2
Legionella spp., 672, 678b
  L. pneumophila, 674
Legionellosis, 409t, 674
Legumes, 775
Leishmania brasiliensis, 558, 566t
Leishmania tropica, 558, 566t
Leishmaniasis, 141t, 427b, 557–58, 566t
Leprosy. *See* Hansen's disease
Leptospira interrogans, 739, 759b
Leptospirosis, 399t, 738–39, 759b
Lesion, *395*
Lethal factor, of anthrax, 633
Leucine, 260, 268
Leuconostoc mesenteroides, 794, 796
Leukemia, 506, 642, 643t
Leukocidins, *389*
Leukocytes, 421, 422
Leukocytosis, 395
Leukopenia, 395, 637
Leukotrienes, 429b, 484
Levofloxacin, 358, 558
L forms, 97–98
Lice, 619b, 626
Life, elements of, 29t, 30
Life cycle
  Chlamydia, 748f
  helminths, 144, 718f
  liver fluke, 723–24
  protozoa, 138
  tapeworms, 721
  Toxoplasma gondii, 588f
Lifestyles, of microorganisms, 11
Ligases, 216b, 253t, 285
Light-dependent reactions, 239–40
Light-driven organic synthesis, 184b
Light-independent reactions, 239, 241
Light microscopy, 70–72
Limestone, 773
Lincomycin, 268
Lindenmann, Jean, B2
Linezolid, 357
Linkage maps, 304
Linnaeus, 17
Lipases, 216b, 236, 539
Lipids, 40t, 41t, 44–46
Lipmann, F. A., 224n
Lipopolysaccharide (LPS), 42, 98, 392
Lipoteichoic acid (LTA), 96, 656
Liquid(s), and ultraviolet irradiation, 325
Liquid media, 61–62
Liquors, and distilling, 795
Lister, Joseph, 16, 330, B1
Listeria monocytogenes, 193, 322, 577–78, 582t, 602t, 801
Listeriosis, 322, 409t, 578
Lithoautotrophs, 184
Lithosphere, *768*, 778–80
Lithotrophs, 769
Liver
  antimicrobial drugs, 368
  gastrointestinal infections, 713, 714–17, 727b
Liver flukes, 143f, 723–24
Live samples, for microscopy, 78
Living media, 62b
Lobar pneumonia, 673
Localized infection, 393
Lockjaw, 596
Locomotion, of protozoa, 138
Loeffler, Friedrich, 150, B1
Logarithm(s), A1
Logarithmic (log) phase, 202
Loop dilution method, 61f
Lophotrichous flagellum, 88, 89f
Louisiana State University, 173b
Lovelock, James, 772
Low-density lipoproteins (LDLs), 612b
Lowenstein-Jensen medium, 66t
Lower respiratory tract, 650, 659–80
Low-level germicides, 329
Low-temperature sterilizing cabinets, 332
Lumbosacral spinal nerve trunk, 754
Lumen, 687
Lung diseases, fungal, 669b
Lyases, 216b
Lyme disease, 353t, 409t, 618–21, 643t. *See also* *Borrelia burgdorferi*
Lymph and lymphatic system
  anatomy and functions, 418, 424–26, 609
  defense mechanisms, 610
  infectious diseases, 612–44
Lymphadenitis, 395
Lymphangitis, 544
Lymph nodes, 609
Lymphocytes, 423, 424f, 444f, 445, 446–53. *See also* B lymphocytes; T lymphocytes
Lymphocytic choriomeningitis, 624
Lymphogranuloma venereum, 747
Lyophilization, 323–24
Lysin, 525
Lysis, 95, 167
Lysogenic conversion, 169
Lysogeny, 167–69
Lysol, 331
Lysosome, 126, 127f, 434
Lysozyme, 95, 417, 537, 735
Lyssavirus, 591. *See also* Rabies

MacConkey agar, 66, 67t
MacLeod, Colin, 250b, B2
Macrogametocytes, 630
Macrolide polyenes, 358–59
Macromolecules, 40–51, 238

Macronutrients, 180
Macrophage(s), 423f, 424, 432, 433f, 650
Macrophage colony-stimulating factor (M-CSF), 429b
Macroscopic fungi, 130
Macroscopic morphology, 512
Macular rash, 550
Macule, 552t
Maculopapular rash diseases, 551–56
"Mad cow disease," 173, 590
Magnesium, 29t, 30, 180t
Magnetotactic bacteria, 101
Magnification, and microscope design, 69–70
Maintenance, of cultures, 69
Major histocompatibility complex (MHC), 446, 497–98
Malachite green dye, 336
Malaria. *See also Plasmodium* spp.
  asexual reproduction, 138
  chemotherapy for, 353t, 359
  developing countries and prevention efforts, 6
  drug resistance, 367b
  protozoan pathogens, 141t
  reportable diseases, 409t
  sickle cell disease, 385
  treatment of syphilis, 752
  vaccine, 469, 631
Malaria Vaccine Institute, 631b
Malassezia furfur, 135t, 561, 566t
Male reproductive system, 735, 736f
Malic acid, 229b
Malnutrition, and parasitic worms, 145. *See also* Nutrition
Malpractice lawsuits, 600b
Maltase, 216b
Malting, of beer, 794
Maltose, 41, 43f
Mammalia, 19
Manganese, 29t, 180t
Mannitol salt agar (MSA), 65, 66f, 67t
Mantoux test, 668, 670
Mapping, of genomes, 304–305
Marburg virus, 623–24, 643t
March of Dimes, 593b
Margulis, Lynn, 19b
Marine environments, 781
Marine microbiology, 3
Markers, and immune system, 418
Mars, 18b
Marshall, Barry J., 405b, 695
Mash, and beer, 794
Mast cell(s), 422f, 423, 482
Mast-cell-mediated allergy, 485–86
Mastigophora, 139
Mathematics, and exponents, A1–A2
Matrix, 126
Matter, *28*
Maturation
  of B lymphocytes and T lymphocytes, 449
  of virus, 163f
Maximum temperature, 193
Maya, and art, 808b
McCarty, Maclyn, 250b, B2
McClintock, Barbara, 276
Measles. *See also Morbillivirus;* Subacute sclerosing panencephalitis
  causative agent, 159t, 566t, 602t
  China, 555b
  cytopathic changes in cells, 166t
  encephalitis, 588, 589t
  immunization, 442, 467b, 468t, 469b, 472t, 473t
  reportable diseases, 409t
Meat, and food safety, 800
Mebendazole, 353t, 359–60, 719t
Mechanical vectors, 397–98
Mechanical ventilation, and nosocomial pneumonia, 679, 680
Media, for culturing of microorganisms, 60–67, 322
Medical asepsis, 403
Medical ecology, 3
Medical microbiology, *16–17. See also* Infectious diseases
  algae, 136
  bacterial spores, 103
  biofilms, 92b
  classification of bacteria, 106, 107t
  fungi, 132b
  importance of viruses, 173
  protozoa, 139
  RNA viruses, 158t
  unusual forms of bacteria, 108, 110
  virus families and genera, 159t
Medicine. *See* Antimicrobial therapy; Gene therapy; Health care personnel; Home remedies; Hospitals; Medical microbiology; Prevention; Public health
Mefloquine, 353t, 359, 632
Meiosis, 123
Melarsoprol, 601
Mello, Craig, 257b
Membrane. *See* Cell membrane
Membrane attack, and complement cascade, 436, 437f
Membrane filtration, 327f, 784
Membrane lipids, 45
Memory, and specific immunity, 443
Memory cells, 453, 454f
Mendosicutes, 106
Meninges, 573
Meningitis, 574–83. *See also Coccidioides immitis; Cryptococcus neoformans; Haemophilus influenza; Listeria monocytogenes;* Meningococcal meningitis; *Neisseria meningitidis;* Neonatal meningitis; *Streptococcus pneumoniae*
  causative agents, 602t
  cerebrospinal fluid (CSF), 573
  chemotherapy for, 353t
  reportable diseases, 409t
  Streptococcus pneumoniae, 674
  tuberculosis, 667
  vaccines, 468t
Meningococcal meningitis, 575–76, 602t
Meningococcal vaccine, 472t, 473t
Meningococcemia, 575
Meningoencephalitis, 583, 584t, 602t. *See also Naegleria fowleri*
Menopause, 736
Menstruation, 735
Mercurochrome, 334
Mercury
  bioaccumulation, 777
  microbial control, 333, 334, 752
Merozoites, 629, 630
Mesophiles, 194
Messenger RNA (mRNA), 51, 257t, 258, 260, 303–304
Metabolic analog, 349
Metabolism, *212, 224*
  aerobic respiration, 225–27
  anaerobic respiration, 233
  antibiotic resistance, 364
  ATP, 222–24
  biosynthesis, 236–38
  catabolism, 224–25, 236
  energy, 221–24
  enzymes, 212–21
  fermentation, 234–36
  genetic regulation of, 265–68
  Krebs cycle, 227–29
  nutrients important in, 183
  organic acids, 38
  pyruvic acid, 227
  respiratory chain, 229–32
Metabolites, *805*
Metabolomics, 305b
Metachromatic granules, 101
Metagenomics, 305b
Metals. *See* Copper; Elements; Heavy metals; Iron
Metaphase, of mitosis, 124f
Metchnikoff, Elie, B1
Methane, 32f, 39, 112, 185, 773, 774b
Methanococcus jannaschii, 184f
Methanogens, 112, 184–85, 773
Methicillin, 352, 354t, 364
Methicillin-resistant *Staphylococcus aureus* (MRSA), 357, 366b, 544b
Methyl alcohol, 331
Methylophilus methylotrophus, 798
Methyltransferases, 217
Metronidazole
  amoebic dysentery, 713
  applications of, 353t
  Clostridium difficile, 702
  gastritis and gastric ulcers, 696
  giardiasis, 712
  protozoan infections, 359
  toxic reactions, 368t
  Trichomonas infections, 743b
  vaginosis, 742, 743b
Mexico, and dengue fever, 623
Mezlocillin, 352, 354t
MIC. *See* Minimum inhibitory concentration
Mice, and genetic engineering, 300–301. *See also* Rodents
Michel, Hartmut, B2
Michigan, and HIV infection rates, 733. *See also* Detroit
Miconazole, 359, 561
Microaerophile, 195
Microarray analysis, 306–307
Microbes, 2. *See also* Microorganisms
  dimensions of, 10–11
  environmental factors influencing, 192–99
  impact of on earth, 3–5
  interrelationships with humans, 199
Microbial antagonism, 381, 651
Microbial control
  chemical agents, 327–36
  cold and desiccation, 323–24
  commercial products, 328b, 338t
  filtration, 326, 327f
  general considerations in, 313
  heat as agent of, 320–23
  history of, 314b
  microbial death, 316–18
  modes of action, 318–19
  radiation, 324–26
  relative resistance of microbial forms, 313–14
  terminology and methods of, 315–16
Microbial death, 316–18
Microbicide, 315

Microbiology, 2. *See also* Agricultural microbiology; Applied microbiology; Aquatic microbiology; Astromicrobiology; Environmental microbiology; Geomicrobiology; Industry and industrial microbiology; Medical microbiology
  historical foundations of, 11–17
  scope of, 2–3
Microenvironment, 768
Microfilaments, 128
Microfilariae, 565
Microflora. *See* Normal microbiota
Microgametocytes, 630
Micronutrients, 180
Microorganisms, 2. *See also* Bacteria; Fungus; Microbes; Viruses
  general characteristics of, 7–11
  human use of, 5–6
  soil, 779–80
  taxonomy, 17–23
Micro RNAs, 257b
Microsatellite polymorphisms, 305
Microscopes and microscopy, 11–13, 14f, 69–80
Microscopic fungi, 130
Microscopic morphology, 512
Microsoft Corp., 631b
Microsporum, 135t, 560, 561t, 566t
Microtubules, 121, 128–29
Middle Ages, and history of microbiology, 314b, 345b, 628
Miescher, Johann, B1
Migraines, and botox, 600b
Mildew, 132b
Military. *See also* Bioterrorism
  medical facilities and *A baumannii* infections, 379, 402b
  smallpox vaccine, 550
Milk, 322, 625, 668, 797–98. *See also* Dairy microbiology
Milstein, Cesar, B2
Milwaukee (Wisconsin), and cryptosporidiosis, 137b, 705. *See also* Wisconsin
Mineralization, *770*
Minimum inhibitory concentration (MIC), 372–74
Minimum temperature, 193
Mini short tandem (DNA) repeats (mini-STR), 288b
Minocycline, 357
Miracidium, 725–26
Missense mutation, 270
Mitochondria, 126, 127f
Mitosis, 123, 124f
Mixed acid fermentation, 235
Mixed culture, 68
Mixed infection, 394
Mixed lymphocyte reaction (MLR), 498
MMR vaccine, 553, 695
Mobiluncus, 741
Modified Thayer-Martin medium (MTM), 576
Moist heat, 320, 321–23
Molarity, 37
Mole, 37
Molecular biology, 21
Molecular formulas, 35–36
Molecular weight (MW), 450
Molecules
  chemical bonds, 30–39
  immune response, 450–51
  structure of ATP, 223–24
Mollicutes, 106
Molluscum contagiosum, 555, 557, 566t, 757b, 758, 759b
Monkey(s), and HIV, 634
Monkeypox, 550, 551
Monoclonal antibodies (MAbs), 460b
Monocytes, 423f, 424
Monod, Jacques, 265
Monolayer, 171
Monomers, 40
Mononegavirales, 158t
Mononuclear phagocyte system, 420
Mononucleosis, 621–22, 623t, 643t, 655
Monosaccharide, 40, 41, 238
Monospot test, 622
Monotrichous flagellum, 88, 89f
Montagnier, Luc, B2
Montagu, Mary, 465b
Morbidity and Mortality Weekly Report (CDC), 3, 406
Morbidity rate, 407
Morbillivirus, 467b, 553. *See also* Measles
Mordant, 95b
Morphology, and evolution, 21
Mortality rates, *407*. *See also* Death
  cholera and oral rehydration therapy, 704b
  influenza pandemics, 665b
Mosaic antigen, 451
Mosquito, 398f, 585, 623, 629, 630–31
Most probable number (MPN), 784
Motility
  eukaryotes, 121, 129t
  prokaryotes, 88, 89, 129t
  viruses, 129t
Motility test medium, 62
Motor function, of nervous system, 573
Mount St. Helens volcano eruption (1980), 674
Mount Sinai School of Medicine, 172b
Mousepox virus, 299b
Mucinase, 391
Mucor spp., 653
Mucosal-associated lymphoid tissue (MALT), 426
Mucous membranes, 416, 735
Mucus, 650, 735
Mueller tellurite, 66t
Mullis, Kary, 194b, B2
Multidrug-resistant (MDR) pumps, 364
Multidrug-resistant TB (MDRTB), 671
Multiple sclerosis, 7, 500t, 502
Multiplication, viral, 160–69
Mumps, 159t, 409t, 468t, 472t, 473t, 693–95, 727b
Mupirocin, 543
Murray, Polly, 618
Mushrooms, 135
Must, and wine, 795
Mutagens, 269
Mutant strain, 268
Mutations, genetic, 254–55, 268–72
Mutualism, 197, 198b, 772
Myasthenia gravis, 499, 500t, 502
Mycelium, 131, 133
Myceteae, 130
Mycobacterium spp., 782
  M. avium, 668. *See also Mycobacterium avium* complex (MAC) infections
  M. bovis, 322, 668
  M. leprae, 200–201. *See also* Hansen's disease
  M. tuberculosis. *See also* Tuberculosis
    acid-fast stain, 79, 670–71
    antimicrobial drugs, 343, 353t, 354b
    causative agent of tuberculosis, 666, 667–68, 672b, 680b
    delayed-type hypersensitivity, 495
    in vivo testing, 529
    pasteurization, 322
    vaccine, 807
Mycobacterium avium complex (MAC) infections, 357, 668, 671, 680b
Mycolic acid, 96
Mycoplasma(s), 97–98
Mycoplasma spp., 99, 389t
  M. pneumoniae, 97, 672, 674, 678b, 680b. *See also* Pneumonia
Mycoprotein, 798
Mycorrhizae, 779
Mycoses, 130–31, 135
Myeloperoxidase, 434
Myocarditis, 659
Myocardium, 609
Myonecrosis, 545, 546

NADH, 222, 227, 229, 232
Naegleria fowleri, 583–84, 602t. *See also* Meningoencephalitis
Nafcillin, 352, 354t
Naked viruses, 152, 153f, 156f
Nanobacteria, 109b
Nanobes, 109b
Nanotechnology, 77b
Narrow spectrum drugs, 346t
National Aeronautics and Space Administration (NASA), 18b, 299b, 325
National Cancer Institute, 27
National Center for Health Statistics, 743b
National Foundation for Infantile Paralysis, 593b
National Institutes of Health, 302–303
National Oceanic and Atmospheric (NOAA), 5f
National Park Service, 194b
National Vaccination Days, 594
Native state, of protein, 49, 319
Natural immunity, 462–63
Natural killer (NK) cells, 424f, 462
Natural selection, 21, 364, 365f
Necator americanus, 727b
Necrosis, 393
Necrotizing fasciitis, 566t, 655, 656
Necrotizing ulcerative gingivitis (NUG), 693
Necrotizing ulcerative periodontitis (NUP), 693
Negative exponents, A2
Negative-sense RNA, 157
Negative stain, 79
Neisseria gonorrhoeae. *See also* Gonorrhea
  acute endocarditis, 613, 643t
  adhesive properties, 389t
  chemotherapy, 353t
  drug inactivation mechanisms, 364
  heat resistance, 320
  neonatal eye infection, 563
  pelvic inflammatory disease, 743b
Neisseria meningitidis, 353t, 575–76, 582t. *See also* Meningitis
Nelfinavir, 361t
Nematodes, 143, 144t, 722
Neomycin, 355, 372t
Neonatal conjunctivitis, 564t, 566t
Neonatal eye infections, 563
Neonatal meningitis, 581–83, 602t
Neoplasia, 758b
Nephritis, 656
Nerve damage, and antimicrobial drugs, 368
Nervous system. *See also* Central nervous system; Peripheral nervous system
  African sleeping sickness, 599–601
  botulism, 597–99
  defense mechanisms, 573

Nervous system *(continued)*
  encephalitis, 583–92
  meningitis, 574–83
  poliomyelitis, 592–96
  tetanus, 596–97
Neuritis, 659
Neuromuscular autoimmunities, 502
Neurons, 573
Neurosyphilis, 750
Neurotropic virus, 594
Neutralization reactions, 38, 456
Neutral red, 66
Neutrons, *28*
Neutrophil(s), 422, 432
Neutrophil cytochrome b_{558} assay, 414
Nevirapine, 361t
New England Journal of Medicine, 621
New York State Health Department, 282, 288b
Niche, and ecosystem, *768*
Nicholson, C. K., 46b
Niclosamide, 353t, 368t, 719t
Nicotiana tabacum, 301b
Nicotinamide adenine dinucleotide (NAD), 217, 222, 223f
Nidovirales, 158t
Nigeria, and polio vaccinations, 594
Nigrosin, 79
NIMBY (not in my backyard), 791b
Nipah virus, 399
Nisin, 365
Nitrate(s), 776, 803
Nitrate reductase, 214t
Nitric oxide (NO), 434
Nitrification, 775–76
Nitrites, 776, 803
Nitrobacter, 776
Nitroblue tetrazolium test (NBT), 414
Nitrogen. *See also* Nitrogen cycle
  elements of life, 29t
  fixation, 774–75
  sources of, 180t, 182
Nitrogen base, 49
Nitrogen cycle, 773–76
Nitrogenous base, 249
Nitromersol, 334
Nitrosococcus, 776
Nitrosomonas, 776
Nitrosospira, 776
Nitrous oxide, 774b, 776
Nomenclature, *17, 20*. *See also* Terminology
  enzymes, 216b
  viruses, 157–58
Noncellular infectious agents, 173–74
Noncommunicable disease, 399
Noncompetitive inhibition, 219, 220f
Nongonococcal urethritis, 747
Nonhemorrhagic fever diseases, 624–27
Nonionizing radiation, 324, 325–26
Nonliquefiable solid media, 63
Nonliving reservoirs, 399
Nonnucleoside reverse transcriptase inhibitors, 360
Nonpressurized steam, and microbial control, 321–22
Nonprogressor, and HIV infection, 639
Nonself, 418, 446
Nonsense codons, 263
Nonsense mutation, 270
Nonseptate hyphae, 131
Nonspecific chemical defenses, 417
Norfloxacin, 358
Normal microbiota
  acquisition of, 380–82
  axenic animals, 384b
  defense mechanisms, 416–17
  eye, 562
  gastrointestinal tract, 688–89, 734
  pathogens, 384b
  respiratory tract, 651
  skin, 537–38
  Staphylococcus pneumoniae, 672
  Streptococcus pyogenes, 656
  superinfections, 370f
  suppression and alternation of by microbials, 369
  urinary tract, 382t, 735–36
Noroviruses, 706
Norwalk viruses, 159t, 706, 782, 783
Norway, and bioaccumulation, 778b
Nosocomial infections, 401–403, 738
Nosocomial pneumonia, 679–80
Nostoc, 775
Notifiable diseases. *See* Reportable diseases
Nucleic acid(s). *See also* DNA; RNA
  anabolism and synthesis of, 238
  antimicrobial drugs and synthesis of, 348–49
  microbial control, 318–19
  molecular structure of, 40t, 41t, 49–50
  viruses, 156–57
Nucleic acid hybridization, 286–87
Nucleic acid sequencing, 518–19
Nucleocapsid, 152, 153f
Nucleoid, 100
Nucleolus, 123
Nucleoside reverse transcriptase inhibitors, 360
Nucleotide(s), 49, 249
Nucleotide analog reverse transcriptase (RT) inhibitors, 361t
Nucleus, of eukaryotic cell, 123, 126
Numerical aperture, 72
Nursing bottle caries, 691
Nutrient(s), *180*, 224. *See also* Nutrition
Nutrient agar, 62
Nutrient broth, 61
Nutrient flow, 4–5, 769–71
Nutrition, *180*. *See also* Diet; Food; Malnutrition; Vitamins
  chemical analysis of cytoplasm, 181–82
  deficiencies, 503t
  definitions of terms, 180
  diffusion and transport, 189–92
  ecological associations, 197–99
  endocytosis, 192
  environmental factors, 196–97
  fungi, 130–31
  gas requirements, 194–96
  osmosis, 187–89
  osmotic pressure, 196
  pH scale 196
  protozoa, 137
  sources of essential nutrients, 182–83
  transport mechanisms, 187
  types, 183–84
Nystatin, 349, 359

Obesity, 173b, 381
Obligate, *185b*
Obligate aerobe, 195
Obligate anaerobe, 195
Obligate halophiles, 196
Obligate intracellular parasites, 108, 110, 150
Obligate parasites, 187
Obligate saprobes, 185
Obsessive compulsive disorder, 7
Occupations, and allergies, 481
Oceans. *See also* Aquatic microbiology
  bioaccumulation, 777, 778b
  hydrostatic pressure, 197
  hydrothermal vents, 186b, 198b
  marine environments, 781
O'Connor, Basil, 593b
O'Connor, Kevin, 211, 241b
Ohio Valley fever, 135t, 677
Oil immersion lenses, 72b
Oil spills, and bioremediation, 5f, 6, 766, 770b, 770b
Okazaki fragments, 253
O'Leary, Paul A., 752
Oligodynamic action, *333*
Oligotrophic ecosystems, 781
-Oma (suffix), 392b
Omeprazole, 696
Onchocerca volvulus, 144t, 565, 566t. *See also* River blindness
Oncogenic viruses, 167, 714
Oncoviruses, 167
Oocysts, 587, 588
Operator, of lactose operon, 265
Opisthorchis sinensis, 723, 724b, 727b
Opportunistic pathogens, 185, 187, 383, 669b
Opsonization, 456
Optical microscope, 72
Optimum temperature, 193
Oral cavity, and normal biota, 688
Oral contraceptives, 373, 540, 746
Oral-fecal route, for disease transmission, 401
Oral herpes, 353t
Oral hygiene, 693
Oral poliovirus vaccine (OPV), 595
Oral rehydration therapy (ORT), 704
Oral swab test, for HIV, 640
Oral vaccines, 470, 592, 595
Orbitals, electron, 30
Order, and taxonomy, 18–20
Orfvirus, 152f
Organ(s), *118*. *See also* Organ transplantation
Organ donation, and rabies, 592
Organelles, 7–8, 52, 118
Organic acids, 803
Organic chemicals, 39
Organic compounds, 38–39, 40t, 181t
Organ-specific autoimmune diseases, 499
Organ transplantation, 497–98, 639. *See also* Bone marrow transplantation
Origin of replication (ORI), 293
Oroya fever, 405b
Orthoclone, 297t
Orthomyxoviridae, 159t, 662
Ortho-phthaladehyde (OPA), 335
Orthopoxvirus, 550
Orwell, George, 600b
Oryza sativa, 301b
Oscillatoria, 110f
Oseltamivir, 361t, 666
-Osis (suffix), 392b
Osmosis, 187–89
Osmotic pressure, 196
Otitis media, 653–54, 680b
Outer membrane (OM), 98
Ovaries, 735
Owens Lake (California), 112f
Oxacillin, 354t
Oxaloacetic acid, 229b
Oxazolidones, 357
Oxidase, 214t, 216t
Oxidase detection test, 233
Oxidation-reduction reactions, 34b, 217, 221–22

Oxidative phosphorylation, 224, 232
Oxidized compound, 217
Oxidizing agent, 34b
Oxidoreductases, 216b
Oxygen
  aerobic respiration, 233
  elements of life, 29t
  microbial processing of, 195–96
  shell of electrons, 30
  soil composition, 779
  sources of, 180t, 182
Oxygenic photosynthesis, 241
Ozone, 332

Paleontology, 118
Palindromes, 284
Palisades arrangement, 104f, 105
Pan American Health Organization, 623
Pancreatitis, 694
Pandemic disease, 407, 408f, 615, 663, 664–65b
Pannus, 564
Papanicolaou, George, 758b
Papillomas, 556
Papillomaviruses, 557. *See also* Human papillomavirus (HPV)
Papovaviridae, 159t
Pap smear, 742, 757, 758b
Papular rash, 550
Papule, 552t
Paracoccidioides brasiliensis, 135t, 669b
Paracoccidioidomycosis, 135t, 669b
Parainfluenza, 159t
Paralytic shellfish poisoning, 136
Paramecium, 89f
Paramyxovirus, 159t, 467b, 694. *See also* Measles; Mumps
Parasites, 11. *See also* Helminths; Malaria; Parasitism; Parasitology
  fungi, 130
  nutritional types, 183t, 185, 187, 769t
  protozoa, 141
  worms, 145
Parasitism, 198, 772
Parasitology, 141, 405b
Parenteral administration, of antibiotics, 354
Park's method, of handscrubbing, 334b
Paromomycin, 705
Parotitis, 694
Paroxysmal stage, of pertussis, 660
Parvoviruses, 157, 159t, 555, 566t
Passive carriers, 397, 398f
Passive immunity, 462, 463, 464–65
Passive transport, 192t
Pasteur, Louis, 7, 13b, 16, 150, 234, 235b, 334b, 632, B1
Pasteur Institute, 634
Pasteurization, 235b, 322, 625, 668, 800
Pathogen(s), *6, 383. See also* Infectious diseases; Pathogenicity; specific diseases
  biosafety levels and classes of, 386b
  discovery of, 16–17
  fungi, 135
  parasites, 187
  protozoa, 141–42
Pathogen-associated molecular patterns (PAMPs), 433–34, 443
Pathogenicity, *383*
Pathognomic disease, 676b
Pathologic state, *380*
Patient chart, 515
Payne, Roger, 778b
PCBs (polychlorinated biphenyls), 777, 778b
Pectinase, 804t
Pediococcus cerevisae, 796
PEG-SOD, 297t
Pellicle, 688
Pelvic inflammatory disease (PID), 733, 741, 742b, 743b, 744, 747
Penetration
  of bacterial cell by bacteriophage, 168f
  of viruses, 160, 161f, 162, 169t
Penicillin
  cell wall, 348, 352
  characteristics of selected forms, 354t
  discovery of, 345b
  drug resistance, 362, 364, 366b
  drug susceptibility tests, 372t
  group B *Streptococcus,* 759
  industrial production of, 804t, 807
  infectious diseases, 353t
  meningococcal meningitis, 576
  minimum inhibitory concentration, 374t
  semisynthetic method, 350b
  Streptococcus pyogenes, 657
  syphilis, 752
  toxic reactions, 368t
Penicillinase, 216t, 218, 352, 364
Penicillinase-producing *Neisseria gonorrhoeae* (PPNG), 364, 747
Penicillium chrysogenum, 352, 807
Penicillium roqueforti, 798
Penis, 735
Pentoses, 41, 49, 50f
Peptic ulcers, 695
Peptidase, 216b
Peptide, 47
Peptide bond, 47
Peptidoglycan (PG), 42, 87, 95, 238
Peracetic acid, 332
Perforins, 461
Period of invasion, 393b
Periodontal diseases, 691, 727b
Periodontitis, 691–93
Peripheral nervous system (PNS), 573. *See also* Nervous system
Periplasmic flagella, 90
Periplasmic space, 96
Peritrichous flagellum, 89
Pernicious anemia, 500t
Peroxide generators, 332
Persistent infections, 166
Pertussis, 409t, 468t, 471, 472t, 473t, 649, 660–61. *See also Bordetella pertussis*
Pertussis toxin, 660
Pesticides, 791b, 807. *See also* Insecticides
Petechiae, 552b, *575*
Petri, Julius, B1
Peyer's patches, 426
Pfiesteria piscicida, 136f, 137
pH. *See also* Acidity; Alkalinity
  hydrogen as essential nutrient, 182
  microbial control, 317
  microbial growth and survival, 196
  negative exponents, A2
  normal biota of vagina, 735, 736
  scale, 37–38
Phage typing, 517
Phagocytes, 93, *389,* 503t
Phagocytosis, 93, 127f, 192, 432–34
Phagolysosome, 434
Phagosome, 434
Pharmaceutical industry, 350b, 804t, 807. *See also* Designer drugs; Industry and industrial microbiology
Pharyngitis, 654–57, 680b. *See also Streptococcus pyogenes*
Pharynx, 688
Phase-contrast microscope, 73t, 74
Phenol, 330–31
Phenol coefficient, 330
Phenolics, 338t
Phenotype, 248, 283
Phenotypic methods, of diagnosis, 512, 513b, 516–18
Phenotypic traits, of bacteria, 106
Phenylethanol agar, 66t
Phosphate, 40t, 49, 249, 776–77
Phosphatized minerals, 777
3-Phosphoglyceric acid (PGA), 241
Phospholipids, 41t, 45
Phosphorus, 29t, 180t, 182–83
Phosphorus cycle, 776–77
Phosphorylate, 221
Phosphotransferases, 217
Photoautotrophs, 184, 773
Photocenter, 239
Photodermatitis, 369
Photoheterotroph, 183t
Photolysis, 240
Photons, 239
Photophosphorylation, 224, 240
Photophobia, 563
Photosynthesis, *4*
  carbon cycle, 773
  chloroplasts, 126–28
  energy production, 239–41
  equations, 184b
  eukaryotic and prokaryotic cells compared, 129t
  nutritional flow in ecosystems, 769
Photosynthetic bacteria, 110
Photosystems I and II, 239–40, 241
Phototaxis, 90
Phototrophs, 183
Phycobilins, 239
Phylogeny, 21
Phylum, 18–20
Physical barriers, to infection, 415–17
Physical maps, of genomes, 304
Physical state, of media, 60t, 61–63
Physiology
  bacterial identification, 512
  evolution, 21
Phytoplankton, 781
Pia mater, 573
Pickles, 796
Picornaviridae, 159t, 594
Pigs. *See* Swine
Pilin, 90
Pilus, 88f, 90, 273
Pinkeye, 563
Pinocytosis, 192
Pinworm, 144t, 145, 720
Piperacillin, 352
Piperazine, 353t, 360, 719t
Pisum sativum, 301b
Placenta, 388, 493b
Plague
  bioterrorism, 677b
  causative agent of, 643t
  culture and diagnosis, 616
  history of, 314b, 615
  pathogenesis and virulence factors, 615–16
  prevention and treatment, 617
  reportable diseases, 409t
  signs and symptoms, 615
  transmission and epidemiology, 616

Plague *(continued)*
  vaccine, 468t
  zoonotic infections, 399t
Plankton, 136, 781
Plantar warts, 556
Plants. *See also* Photosynthesis
  antimicrobial drugs, 365
  development of vaccines, 469
  fungal control agents, 132b
  poisonous, 496b
  soil microbes, 779
  symbiotic associations with bacteria, 775
  transgenic, 301b
  viroids, 174
Plaques
  cultivation of viruses, 172
  skin lesions, 552t
  teeth, 92b, 93, 689, 690f, 692, 693
Plasma cell tumors, 506
Plasma and plasma cells, 420–21, 424, 453, 454f
Plasmids, 88f, 99–100, 362–63, 615
Plasmodium spp., 140, 353t, 628. *See also* Malaria
  P. falciparum, 628, 632, 643t
  P. malariae, 628, 643t
  P. ovale, 628, 630, 643t
  P. vivax, 628, 630, 632, 643t
Plate count, 203b
Platelet(s), 424
Platelet-activating factor, 429b, 485
Pleomorphic gram-negative rod, 752
Pleomorphism, *103,* 105f
Pleuritis, 674
Pneumococcal meningitis, 576–77
Pneumococcal pneumonia, 468t, 472t, 473t
Pneumococcal polysaccharide vaccine (PPV), 86, 98b
Pneumocystis (carinii) jiroveci pneumonia (PCP), 353t, 357, 634, 669b, 678–79, 680b
Pneumonia, *387. See also* Pneumococcal pneumonia; *Pneumocystis (carinii) jiroveci* pneumonia (PCP)
  causative agents, 672–75, 677–79, 680b
  chemotherapy, 353t
  measles, 552
  nosocomial forms of, 679–80
  signs and symptoms, 672
  types of, 671–72
Pneumonic plague, 615, 676b, 677b
Point mutations, 269
Poisonous plants, 496b
Polar bears, 778b
Polarity, 32, 33f
Poliomyelitis, 159t, 409t, 467, 468t, 472t, 592–96, 602t
Poliovirus, 166t, 389t, 471, 602t
Pollen, 481, 482f
Pollution. *See also* Airborne contaminants; Solid waste disposal; Water and water supplies
  bioremediation, 5f, 6, 33b, 791b
  environmental recycling, 777
  extreme environments, 27, 33b
Polyaminopropyl biguanide, 338t
Polyclonal antibodies, 460b
Polyenes, 358
Polyhydroxy aldehydes, 40
Polyhydroxyalkanoate (PHA), 211, 241b
Polymer(s), 40
Polymerase(s), 157, 216b
Polymerase chain reaction (PCR), 194b, 205, 290–92, 519, 530f, 707b, 752
Polymerization, and complement cascade, 436, 437f
Polymicrobial diseases, 394
Polymyxins, 349, 355, 368t, 372t
Polypeptide, 47
Polyphosphate granules, 100–101
Polyribosomal complex, 264
Polysaccharides, 40–44, 238
Polystyrene foam, 211, 241b
Pontiac fever, 674
Population, and ecology, *768*
Population growth, 200–205
Porin proteins, 98
Porphyrin, 239
Porphyromonas gingivalis, 692, 727b
Portals of entry
  for allergens, 481
  for infection, 385–88, 415–18
Portals of exit, for infection, 395–96
Porter, J. R., 11f
Porter, Rodney, B2
Positive-sense RNA viruses, 157, 165b
Positive stain, 79
Postinfection encephalitis (PIE), 586
Postnatal rubella, 553–54
Post-polio syndrome (PPS), 594
Posttranslational modifications, 263
Potassium, 29t, 180t, 183
Poultry, and salmonellosis, 697, 698. *See also* Birds
Povidine (PVP), 330
Poxviruses, 159t, 759b
Prairie dogs, 551, 616
Praziquantel, 353t, 360, 719t, 726, 740
Prebiotics, 365
Precipitation, and hydrologic cycle, 780
Precipitation reactions, 521, 522–24
Precursor molecules, 224
Predator, 772
Pre-erythrocytic development, of malaria, 629
Pregnancy. *See also* Breast-feeding; Infants
  antimicrobial drugs, 373
  chickenpox, 548
  HIV transmission, 639
  listeriosis, 578
  measles, 553
  pathogenic infections, 388
  rubella, 553, 554
  secondary immune deficiencies, 503t
  toxoplasmosis, 587
Presence-absence broths, and coliform enumeration, 783
Preservation of Antibiotics for Medical Treatment Act, 367b
Preservatives, for food, 803
Pressure, and steam sterilization, 321
Presumptive data, 514, 784
Prevalence, of disease, 406–407
Prevention. *See also* Centers for Disease Control and Prevention; Public Health
  of allergy, 487–88
  food safety and food-borne illnesses, 798–803
  of hemolytic disease of the newborn, 493
  of HIV transmission, 640
Prevotella intermedia, 693, 727b
Primaquine, 359
Primary amoebic meningoencephalitis (PAM), 583, 584t
Primary cell cultures, 171–72
Primary consumers, 770
Primary dye, 79
Primary immunodeficiency diseases, 503–506
Primary infection, 394
Primary metabolites, 805
Primary pathogens, 669b
Primary producers, 769t
Primary response, and antibody production, 458
Primary stage, of sewage treatment, 792
Primary structure, of proteins, 47–49
Primary syphilis, 749
Primary tuberculosis, 666, 670f
Primase, 253t
Primates, 19
Primers, for DNA and RNA, 257t, 290, 292
Prince William Sound (Alaska), 766
Prions, *173,* 315b, 588–90, 602t
Probiotics, 365, 691
Procaine, 752
Prodomal stage, of infection, 393b
Producers, ecological, 769
Products, 36
Proflavine, 336
Progressive multifocal leukoencephalopathy (PML), 159t, 586, 602t
Prokaryotes, *3. See also* Bacteria
  cell structure, 87, 88f
  cellular organization, 7–8, 9f
  characteristics of, 52, 87
  classification of, 106–108
  eukaryotes compared to, 111t, 129t
  ribosomes, 260
  unusual characteristics, 108–12
  viruses compared to, 129t
Proliferative stage, of lymphocyte development, 448
Promoter, of lactose operon, 265
Promoter region, 259
Proof, of liquor, 795
Prophage stage, of lysogeny, 167
Prophase, of mitosis, 124f
Prophylaxis, 346t
Propionibacterium spp. 235, 798
P. acnes, 537, 538–40, 566t
Propylene oxide, 336, 803
Prostaglandins, 45–46, 429b, 485
Prostate cancer, 470b
Prostate gland, 735
Prostatitis, 744
Protease(s), 216b, 236, 804t
Protease inhibitors, 174, 361t, 641
Protective antigen, 633
Protein(s)
  anabolism and synthesis, 238
  antimicrobial drugs and synthesis, 349, 356–57
  enzyme structure, 213–14
  gene-protein connection, 255–56
  genetic code, 248, 260–64
  genetic regulation of synthesis, 265–68
  macromolecules, 41t
  microbial control methods, 318–19
  molecular structure of, 46–49
  organic compounds, 40t
  ribosomes and synthesis, 100, 128
  RNA and synthesis, 51
Proteinase, 216b
Proteomics, 305b
Proteus spp., 196
  P. mirabilis, 737, 738b, 759b
Prosthetic valves, and endocarditis, 612
Protista, 19, 136–42. *See also* Algae; Protozoa
Proton(s), *28*
Proton motive force (PMF), 232
Protoplast, 98
Prototheca, 136
Protozoa
  antimicrobial drugs, 368t
  characteristics of, 137–42

cilia, 121
gastrointestinal infections, 727b
taxonomy, 19–20
thermal death times, 320t
Provocation, and allergy, 481, 483f
Pruteen, 798
Pseudomembrane, 658, 659
Pseudomembranous colitis, 702
Pseudomonas spp.
antimicrobial drugs, 353t
bioremediation, 791b
marine environments, 781
microbial control, 333
nitrogen cycle, 776
P. aeruginosa, 187, 218, 389t
P. fluorescens, 298
P. putida, 211, 241b
P. stutzeri, 808b
P. syringae, 297–98
Pseudophypha, 130
Pseudopods, 138
P site, 260
Psittacosis, 399t, 409t
Psychrophile, 112, *193*, 807
Psychrotrophs, 193
P24 antigen capture assay, 511
Public health. *See also* Centers for Disease Control and Prevention; Prevention; U.S. Public Health Service
bioterrorism, 677b
Chlamydia infections, 748
diarrheal diseases, 697
microbiology, *3*
pertussis, 660
side effects of vaccines, 471
syphilis, 752
Trichomonas infections, 742b
Pulmonary anthrax, 632, 676b
Pulmonary infections, and fungi, 669b
Pulsed-field gel electrophoresis (PFGE), 707b
PulseNet, 707b
Pulvo de Vibora, 698
Pure culture, 58n, 68
Purified protein derivative (PPD), 668, 670
Purines, 49, 50f, 249
Purple sulfur bacteria, 111
Purpura, 552t
Pus, 430
Pustular rash diseases, 546–51
Pustular skin lesions, 557–58
Pustules, 539, 552t
Pyelonephritis, 737
Pyogenic bacteria, 430
Pyrantel, 360, 368t, 719t
Pyrazinamide, 353t, 671
Pyretotherapy, 431b
Pyridium, 738
Pyrimethamine, 353t, 587–88, 632
Pyrimidine(s), 49, 50f, 249
Pyrimidine dimers, 325
Pyrolysis, 241b
Pyruvate, 236–37
Pyruvic acid, 227

Q fever, 322, 409t, 619b, 628t, 643t
Quarantines, 617
Quaternary ammonium compounds (quats), 329t, 332, 333
Quaternary consumers, 770
Quaternary structure, of proteins, 48f, 49
Quellung test, 526
Quick test kits, 516
Quinacrine, 353t, 359, 712
Quinine, 353t, 359, 632
Quinolones, 349, 353t, 358, 368t, 747

Rabbits, and tularemia, 617
Rabies. *See also Lyssavirus*
attachment mode, 160
causative agent, 150, 602t
cytopathic changes in cell, 166t
reportable diseases, 409t
vaccine, 468t
zoonotic infections, 399
Radiation
food preservation, 802–803
genetic mutations, 269
microbial control, 324–26
Radioactive isotopes, 30
Radioimmunoassay (RIA), 527
Radioimmunosorbent test (RIST), 527
Rales, 485
Rapid influenza tests, 664
Rapid-method identification kits, 576
Rapid PCR, 519
Rapid plasma reagin (RPR) test, 522
Rapid Syndrome Validation Project (RSVP), 299b
Rapid tests, for coliform enumeration, 783. *See also* Rapid influenza tests
Rashes, 546
Rats, 587, 616, 624. *See also* Rodents
"Rattlesnake pill," 698
Reactants, 36
Real image, 70
Real-time PCR, 205
Receptors
drug resistance, 364
specific immunity, 446–47
Recombinant, definition of, *294*
Recombinant DNA technology, *2–3*. *See also* Genetic engineering
biochemical products of, 296–97
human use of microorganisms, 5
methods in, 292–96
recombination events, 272–77
Recombinant organisms, *272*
Recurrent urinary tract infections, 738
Recycling, in environment, 772–77
Red blood cells (RBCs), 421, 630
Redi, Francesco, 12b, B1
Redox pair, 221
Redox reactions, 34b, 221–22
Red pigment, 112
Red Sea, 112
Red snow, 193f
Red tide, 136, 781, 782f
Reduced compound, 217
Reducing agent, 34b
Reducing medium, 66
Redundancy, in genetic code, 260
Reduviid bug, 141
Reed, Walter, B2
Refraction, 69
Refrigeration, and microbial control, 323, 801
Regulated enzymes, 216
Regulation. *See also* Regulatory genes
enzymatic activity and metabolic pathway, 218–21
genetic of protein synthesis and metabolism, 265–68
Regulator, of lactose operon, 265
Regulatory genes, 248
Relaxin, 297t
Release, of virus, 161f, 162–63, 169t
Relenza, 360, 663, 666
Renal tuberculosis, 667
Renin, 797
Rennet, 804t
Reoviruses, 157, 159t, 166t, 783
Replica plating, 269
Replication
of DNA, 252–55
of viruses, 164–65b
Replication forks, 253
Reportable diseases, 406, 409t, 754
Repressible operon, 267–68
Repression, of enzymes, 219–21
Repressor, of lactose operon, 265
Reproduction and reproductive system. *See also* Female reproductive system; Genital tract; Male reproductive system; Sexually transmitted diseases
"culture bias" and bacterial, 201b
DNA code, 249, 252
eukaryotic and prokaryotic cells compared, 129t
fungi and spore formation, 133–34
helminths, 144
infectious diseases, 740–59
protozoa, 138–39
Reservoirs, for infectious disease, 397–99, 617
Resident biota. *See* Normal microbiota
Resistance. *See also* Antibiotic resistance
endospore formation, 102
microbial control, 313–14, 320–21
Resistance (R) factors, 273–74, 362–63
Resolution, of microscope, 71
Resolving power, 71
Respiration, of eukaryotic and prokaryotic cells compared, 129t. *See also* Respiratory system
Respiratory burst, 414, 434
Respiratory chain, 229–32
Respiratory distress syndrome, 159t
Respiratory synctial virus (RSV), 361t, 661, 680b
Respiratory system
defense mechanisms, 650
infectious diseases of lower respiratory tract, 666–80
infectious diseases affecting lower and upper tracts, 659–66
infectious diseases of upper respiratory tract, 651–59
normal microbiota, 382t, 651
portals of entry for infection, 387, 416
portals of exit for infection, 395
Respiratory syncytial virus (RSV), 651
Restriction endonucleases, 284
Restriction enzyme, 305
Restriction fragment(s), 285
Restriction fragment length polymorphisms (RFLPs), 285
Reticuloendothelial system (RES), 418, 419–20
Retrotransposon, 277
Retroviruses, 157, 159t, 165b, 265, 635, 642
Reverse transcriptase (RT), 157, 165b, 285, 360, 635
Reverse transcriptase inhibitors, 641
Reverse transcription PCR, 592
Reversible reactions, 36
Reversible solid media, 62
Reye's syndrome, 549
R factors, 273–74
Rhabdoviridae, 159t, 591
rH Dnase (pulmozyme), 297t
Rheumatoid arthritis, 499, 500t, 618
Rh factor, 491–93

Rheumatic fever, 353t, 500t, 501, 613, 655–56
Rhinitis, 651–52
Rhinoviruses, 651. *See also* Human rhinovirus
Rhizobia, 775
Rhizobium, 775
Rhizoids, 133f
Rhizopus, 131f
Rhizosphere, 779
Rhodococcus spp., 791b
RhoGAM, 492f, 493
Ribavirin, 361t, 624, 661, 717
Ribose, 50
Ribosomal RNA (rRNA), 51, 100, 106, 257t, 258–59, 518–19
Ribosomes
- antibiotics, 268, 349, 351f
- eukaryotic, 260
- prokaryotic, 88f, 106, 260
- protein synthesis, 100, 128
- translation, 258–259

Riboswitches, 257b, 365
Ribotyping, 707b
Ribozymes, 214b, 257t
Ribulose-1,5-bisphosphate (RuBP), 241
Rice, 301b
Ricketts, Howard, 108n, B2
Rickettsia rickettsii, 110, 627, 643t. *See also* Rickettsias; Rocky Mountain spotted fever
Rickettsia typhi, 110, 619b. *See also* Typhus
Rickettsias, 108, 110. *See also Rickettsia rickettsii*
Ricord, Phillipe, B1
Rifampin, 343, 353t, 358, 368t, 373, 577, 671
Rifamycins, 268, 358
Riftia, 198
Rift Valley fever, 159t
Rimantidine, 353t, 360, 663, 666
Ringworm, 135t, 399t, 559–61, 566t
Ripening, of cheese, 797
Risus sardonicus, 596
River blindness, 144t, 565. *See also Onchocerca volvulus*
RNA. *See also* Genetic(s); Nucleic acids
- antimicrobial drugs, 358
- chemical structure, 49–50
- macromolecules, 41t
- protein synthesis, 51
- transcription and translation, 256–59
- types of, 257t
- viruses, 157

RNA polymerase, 259
RNA primer, 253
RNA viruses, 158t, 159t, 162, 164–65b, 566t
Rock decomposition, 778
Rocky Mountain spotted fever, *110, 626–27, 628t. See also Rickettsia rickettsii*
- chemotherapy for, 353t
- reportable diseases, 409t
- ticks as vector, 619b
- zoonotic infections, 399t

Rodents, and hantavirus, 675. *See also* Mice; Rats
Romans, ancient, 314b
Roosevelt, Franklin Delano, 593b
Root nodules, 775
Roseola, 555, 556t, 566t
Rosette formation, 529
Ross, R., B1
Rotavirus, 159t, 468t, 469b, 471, 705–706, 709b, 727b
Rough endoplasmic reticulum (RER), 123, 125f
Rounding off numbers, A1–A2
Roundworms, 143, 144t, 353t, 722
Rous, Francis, B2
rRNA analysis, 518–19
Rubella, *553–54*
- causative agent, 566t
- classification of viruses, 159t
- reportable diseases, 409t
- vaccine and immunization, 468t, 472t, 473t

Rubeola, 552
Rubor, 427
Rum, 795
Run, 90
Ruska, Ernst, B2
Russia, 299b, 549b, 658

Sabin, Albert, 593b, 595
Sabouraud's agar, 66t
Saccharide, 40
Saccharomyces spp., 794–95, 797
- *S. cerevisiae*, 131f, 135, 794, 795
- *S. cerevisiae (ellipsoideus)*, 795
- *S. uvarum (carlsbergensis)*, 795

Safranin, 95b
Safety, and universal blood and body fluid precautions, 403–404
St. Louis encephalitis (SLE), 159t, 409t, 586, 602t
Salinity, of ocean, 781
Saliva and salivary glands, 395, 416, 514
Salk, Jonas, 593b, 595, B2
Salmonella spp. *See also* Salmonellosis
- antimicrobial drugs, 353t
- drug resistance, 367b
- food infection, 323, 696, 697–98, 708b, 727b
- pasteurization, 322
- refrigerated and frozen foods, 801, 802
- water contaminants, 782
- *S. enteritidis*, 201, 697. *See also* Typhoid fever
- *S. infantis*, 312, 318b
- *S. typhimurium*, 271, 272f, 807

Salmonella/Shigella (SS) agar, 66t
Salmonellosis, 312, 318b, 353t, 399t, 409t
Salpingitis, 744
Salts. *See also* Sodium
- extreme habitats, 186b
- food preservation, 803
- osmotic pressure, 196

Salvarsan, 345b, 752
Sand fly, 558
Sanger, Frederick, 287
Sanger method, of DNA sequencing, 287–88, 290f
Sanitization, 316
San Joaquin Valley fever, 135t. *See also* Coccidioidomycosis
Saprobe, 130, 183t, 185, 187
Saquinavir, 361t
Sarcinae, 104f, 105
Sarcodina, 139
Satcher, David, 8b
Satellitism, 198, 199f
Saturation, and facilitated diffusion, 190
Saudi Arabia, 594
Sauerkraut, 796
Scanning confocal microscope, 75
Scanning electron microscope (SEM), 73t, 76, 78f
Scanning probe microscopes, 77b
Scanning tunneling microscope (STM), 73t, 77b
Scarcity, of vaccines, 469b
Scarlet fever, 555, 556t, 566t, 655, 656
Scavengers, 772
Schatz's method, of handscrubbing, 334b
Schaudinn, Fritz, B2
Schistosoma haematobium, 740, 759b
Schistosoma japonicum, 144t, 725, 726, 727b
Schistosoma mansoni, 725, 726, 727b. *See also* Schistosomiasis
Schistosomes, 740
Schistosomiasis, 725–26, 727b, 739–40. *See also Schistosoma mansoni*
Schizogony, 629
Schizont, 630
Schizophrenia, 7
Schleiden, Matthias, B1
Schultz, Heide, 109b
Schwann, Theodor, 13b, B1
Scientific Applications International Corporation, 299b
Scientific method, 13–16
Scleroderma, 500t
Scotobacteria, 106
Sebaceous glands, 537
Sebum, 537
Secondary consumers, 770
Secondary immune deficiencies, 503t
Secondary immunodeficiency diseases, 506
Secondary infections, 394, 651, 662
Secondary metabolites, 805
Secondary response, and antibody production, 458
Secondary stage, of sewage treatment, 792
Secondary structure, of proteins, 47–49
Secondary syphilis, 749
Secondary (reactivation) tuberculosis, 666–67
Secondhand smoke, and respiratory infections, 653
Secretion, and cell membrane, 99
Secretory component, 456
Sedimentary cycles, 776–77
Sedimentation, and water treatment, 790
Segmented RNA, 157
Selective permeability, of cell membrane, 99, 187, 188f
Selective media, 64–65, 66t
Selective toxicity, of antimicrobial drugs, 346, 352
Selenite, 65
Self, and immune system, 418, 446, 448
Self-experimentation, 405b
Semiconservative replication, 252f, 253
Semisolid media, 62, 63f
Semisynthetic drugs, 346t, 350b
Semmelweis, Ignaz, 16, B1
Senate Office Building, and anthrax contamination, 316, 326, 337b
Sensitization, and allergy, 481–82, 483f, 520f
Sensitizing dose, of allergen, 482
Sensory function, of nervous system, 573
Sepsis, 315, 353t
Septa, 131
September 11, 2001 (terrorist attacks), 282, 288b, 306, 325, 633, 676b
Septicemia, 302t, 395, 610, 613–14
Septicemic anthrax, 632
Septic shock, 14b, 610
Sequelae, of infectious disease, 396
Sequence maps, 304
Sequestered antigen theory, 500
Serology, 519, 520f, 522b, 530f
Seropositive serological tests, 522b
Serotonin, 429b, 484
Serotype and serotyping, 108, 526
Serratia marcescens, 414
Serum, of blood, 420, 458
Serum hepatitis, 159t, 716, 727b
Serum sickness, 495
Severe acute respiratory syndrome (SARS), *675*
- classification of viruses, 159t
- epidemic curve, 676f
- genome sequencing, 304–305

Koch's postulates, 406b
pneumonia, 679b, 680b
reportable diseases, 409t
transmission of, 149, 166b
zoonotic infections, 399
Severe combined immunodeficiencies (SCIDs), 504–505
Sewage treatment, 713, 726, 740, 792
Sex pilus, 90
Sexual behavior, and transmission of HIV, 639, 640
"Sexual revolution," of 1960s, 746
Sexually transmitted diseases (STDs), 110, 138, 387–88, 733, 740. *See also* Gonorrhea; Syphilis
Sexual spore formation, 133–34
Shadowcasting techniques, 152
Shapes, of bacteria, 103–105, 129t
Sheep, and toxoplasmosis, 587
Shells, electron, 30, 31f
Shiga toxin, 699, 700, 708b
Shigella spp., 389t, 727b, 782. *See also* Shigellosis
S. dysenteriae, 698, 700. *See also* Dysentery
S. flexneri, 698
S. sonnei, 698
Shigellosis, 409t, 697, 698–99, 708b. *See also* *Shigella* spp.
Shingles, 159t, 353t, 548
Shultze, Franz, 13b
"Sick building syndrome," 132b
Sickle cell disease, 385
Side effects
of antimicrobial drugs, 368
of vaccines, 470–71, 551
Signs, of infection, 394–95
Silent mutation, 270
Silver compounds, and microbial control, 333, 335
Silver nitrate, 335
Silver sulfadiazine ointment, 335, 357
Simian immunodeficiency viruses (SIVs), 634
Simple enzymes, 213
Simple stains, 79, 80f
Singer, S. J., 46b
Single-cell protein (SCP), 798
Single nucleotide polymorphism (SNP) analysis, 288b, 305
Single-stranded DNA (ssDNA), 157, 265
Singlet oxygen, 195
Sinusitis, 652–53
Siroketals, 27
Size
bacteria, 104–105, 109b
eukaryotic and prokarytoic cells and viruses compared, 129t, 151f
genome, 248
microbial dimensions, 10–11
molecules, 450–51
viruses, 129t, 151–52
Skin. *See also* Rashes; Skin lesions
allergy tests, 487, 495
antimicrobial drugs for infection, 353t
contact dermatitis, 496–97
defense mechanisms, 536–37
infectious diseases of, 538–61
microbial control, 316, 333, 334b
normal microbiota, 369, 382t, 537–38
portals of entry for infection, 387, 415
portals of exit for infection, 396
side effects of antimicrobial drugs, 369
specimen collection, 514
Skin-associated lymphoid tissue (SALT), 426
Skin lesions, 552b
Sleeping sickness, 141, 600
Sludge, 792–93
Small interfering RNA (siRNA), 257t
Small intestine, 688
Smallpox, *549–51. See also* Variola
bioterrorism, 299b, 676b
causative agent, 159t, 566t
cytopathic changes in cell, 166t
history of, 549b
as public health emergency, 546
reportable diseases, 409t
vaccine and immunization, 465b, 468t
Smooth endoplasmic reticulum (SER), 123
Sneezing, and disease transmission, 395, 400, 401, 416
Snow, John, B1
Soaps, and antimicrobial chemicals, 328b, 332, 333
Sodium. *See also* Salts
elements of life, 29t
inorganic reservoirs of, 180t
ionic bonds, 32–33, 34
metabolism, 183
Sodium azide, 65
Sodium chloride, 32–33
Sodium chlorite, 338t
Sodium hypochlorites, 338t
Software programs, for concept mapping, D2
Soil
Bacillus anthracis, 633
botulism, 599
composition and microbiology of, 778–80
helminths, 719
phosphate content, 777
as reservoir for infectious disease, 399
Solar energy, 239–41
Solid media, 62–63
Solid waste disposal
biodegradable plastics, 211, 241b
bioremediation, 6, 791b
sewage treatment, 792–93
soil microbes, 780
Solutes, 36
Solutions, 36–37, 329
Solvent, 36, 317
Somalia, 549b, 550
Sorbic acid, 336
Southern, E. M., 286n
Southern blot analysis, 286, 287f, 707b
Spallanzani, Lazzaro, B1
Spanish influenza (1918), 664b
Sparfloxacin, 358
Sparger, 806
Specialized colonies, 118
Specialized transduction, 276
Species. *See also* Subspecies
of bacteria, 108
taxonomy, 18–20
of viruses, 158
Specific immune globulin (SIG), 465
Specific immunity. *See also* Immune system
B-cell response, 453–58
classification, 462–64
cooperation in immune reactions to antigens, 451–53
immunization, 464–73
lymphocyte response system, 449–51
T cell response, 459–62
as third and final line of defense, 443–49
Specificity
antigens, 446–47
apoenzyme, 215f
facilitated diffusion, 190
immune response, 447–49
immune testing, 519, 520f, 521
Specimen collection and preparation, 59f, 78–80, 513–15
Spectinomycin, 268, 353t
Speed, of enzymes, 213
Spelunker's disease, 677
Sperm whales, 778b
Spheroplast, 98
Spherule, 579
Spider silk, 297t
Spikes, 155–56
Spinal cord, and nervous system, 573, 574f
Spirillum, *103,* 104f, 105, 776
Spirochetes, 90, 104f, 105
Spirogyra, 136f
Spirulina, 798
Spleen, 425–26, 503t
Spliceosome, 264
Split gene, 264
Spoilage, of food, 799–803
Spongiform encephalopathy, 173, 589
Spontaneous generation, 12b
Spontaneous mutations, 269, 271, 362
Sporadic disease, 407, 408f
Sporangiospores, 133, 134f
Sporangium, 102, 133, 134f
Spore(s). *See also* Endospores
discovery of, 16
fungi, 133–34
terminology, 102b
Spore-forming rods, 104f
Spore stain, 80f
Sporozoa, 140
Sporozoites, 140, 629
Sporulation, 102
Spread plate technique, 60, 61f
Sputum, and sample collection, 514
Squamous intraepithelial lesion (SIL), 758b
ssu rRNA, 21–22, 662
Stained smears, 78–80
Stanford University, 513b
Stanley, Wendell, B2
Staphylococcal scalded skin syndrome (SSSS), 544–45, 566t
Staphylococcus aureus
antibiotic resistance, 246, 249b, 357, 364, 365, 366b
antimicrobial drugs, 353t
cellulitis, 544b
culture media, 64t
doubling time, 201
endocarditis, 613, 643t
follicles, furuncles, and carbuncles, 538b
food poisoning, 706–707, 727b, 803
impetigo, 540–41
microbial control, 320
necrotizing fasciitis, 539b
normal biota of skin, 537
origin of species name, 20
osmotic pressure, 196
salt solutions, 803
satellitism, 199f
skin diseases, 566t
staphylococcal scalded skin syndrome, 544
temperature range, 193
Staphylococcus epidermis, 537
Staphylococcus saprophyticus, 737, 738b, 759b
Starch, 43f
Start codon, 261
Starter cultures, 794
State University of New York at Stony Brook, 172b
Stationary growth phase, 202

Stavudine (d4T), 361t, 641
Steam-cleaning approach, to oil pollution, 766, 770
Steam sterilization, 316
Steam under pressure, and microbial control, 321
STEC (shiga-toxin-producing *E. coli*), 700
Stem cells, 421, 422f, 498
Sterilants, 315
Sterile milk, 322
Sterilization, *313*. *See also* Asepsis and aseptic techniques
  discovery of, 16
  endospores, 314
  heat, 320t
  methods, 315, 320t
  skin, 334b
Steroids, 41t, 45, 503t, 804t, 807
Sterols, 122
Stierle, Andrea, 27, 33b
Sties, 563b
Stock cultures, 69
Stomach, and normal biota, 688. *See also* Gastrointestinal tract
Stomach cancer, 695
Stool cultures, 699b
Stop, of dark-field microscope, 73
Strains, of bacteria, 108
Stratum corneum, 537
Streak plate method, 61f
"Strep" infections, 1, 14b, 353t
Streptococcal disease, 409t
Streptococcus spp., 233, 783
  S. agalactiae, 581, 582t, 602t
  S. lactis, 797
  S. mutans, 389t, 613, 689, 691, 727b
  S. oralis, 613
  S. pneumoniae. *See also* Pneumonia
    acute endocarditis, 613, 643t
    meningitis, 576–77, 582t, 602t
    otitis media, 654, 680b
    pneumococcal polysaccharide vaccine (PPV), 86, 98b
    reportable diseases, 409t
    transformation, 274
  S. pyogenes. *See also* Pharyngitis; Scarlet fever
    adhesive properties, 389t
    cell envelope structure, 99
    cellulitis, 544
    chemotherapy, 353t
    endocarditis, 613, 643t
    enzymes and pathogenicity, 218
    impetigo, 540, 542–43
    skin diseases, 538–39b, 566t
    "strep" infections, 14b
  S. sanguis, 613
  S. sobrinus, 389t, 689, 691, 727b
  S. thermophilus, 794
Streptokinase, 391, 543, 804t
Streptolysins, 218, 656
Streptomyces venezuelae, 357
Streptomycin, 268, 353t, 356, 372t, 617, 625
Stress
  secondary immune disorders, 503t
  susceptibility to infection, 385t
Strict anaerobe, 195
Stroke, 302t, 679
Stroma, 127
Stromal cells, 449
Strongyloides stercoralis, 722b, 723, 727b
Structural formulas, 35–36
Structural genes, 248
Structural locus, of lactose operon, 265
Styrofoam, 211
Subacute encephalitis, 584, 586, 602t
Subacute endocarditis, 612, 613, 614t
Subacute sclerosing panencephalitis (SSPE), 467b, 553, 588, 589t, 602t
Subarachnoid space, 573
Subcellular vaccines, 467
Subclinical infections, 395
Subclass, 19
Subculture, 68
Subspecies, of bacteria, 108
Substitution mutations, 270t
Substrate(s), 130, 213, 214–15
Substrate-level phosphorylation, 224
Subunit vaccines, 467
Succinic acid, 229b
Succinyl CoA, 229b
Sucrose, 41, 43f
Sudan, and crisis in Darfur, 442, 467b
Suffixes, and terminology, 392b, 610
Sugars
  chemical composition of, 40–44
  food preservation, 803
Sulfadiazine, 353t, 587–88
Sulfadoxine, 632
Sulfa drugs, 345b, 351f, 357, 738
Sulfamethoxazole-trimethoprim (TMP-SMZ), 353t, 374t, 660, 679, 698, 699, 711
Sulfate, 776
Sulfhydryl, 40t, 776
Sulfite, 803
Sulfite polymyxin sulfadiazine (SPS) medium, 197b
Sulfolobus, 186b
Sulfonamides, 349, 351, 357, 359, 364, 368t, 369
Sulfones, 358
Sulfur, 29t, 180t, 183. *See also* Sulfur cycle; Sulfur granules
Sulfur cycle, 776
Sulfur dioxide, 36, 774b
Sulfur granules, 100
Sulfur indole motility medium (SIM), 62, 67t
Sumerians, ancient, 794
Sun, and energy, 239–41
S (Svedberg) units, 100
Superantigens, 451, 656
Superficial mycoses, 561, 566t
Superinfection, 369, 370f
Superoxide dismutase, 195, 233
Superphylum, 19
Suramin, 601
Surface waters, 780
Surfactants, 318
Surgeon general, 8b
Surgery, and genetic engineering, 302t
Surgical asepsis, 403
Surgical handscrub, 316, 333, 334b
Susceptibility, to allergy, 481
Svedberg, T., 100n
Swamp gas, 793
Sweat glands, 537
Swimming pools, and outbreaks of infectious disease, 117, 137b, 705
Swine
  influenza viruses, 663
  taeniasis, 720–21
  trichinosis, 724
Swiss cheese, 798
Sydenham, Thomas, 431b
Symbionts, 197
Symbiosis, 118, 197, 198b
Symptoms. *See also specific diseases*
  of allergy, 483–85
  of infection, 394–95
  of meningitis, 575
Syncytium, 166, 553, *694*
Syndrome, 395
Synercid, 357
Synergism, 198, 351, 772
Synthesis
  of amino acids, protein, and nucleic acid, 238
  of enzymes, 216–17, 219–21
  of virus, 161f, 162, 169t
Synthesis reaction, 36
Synthetase, 216b
Synthetic biology, 296
Synthetic drugs, 346t
Synthetic media, 63, 64t
Syntrophy, 772
Syphilis, *748–52*. *See also Treponema pallidum*
  chemotherapy, 353t
  epidemiology, 387t, 733, 746
  genital ulcer diseases, 756b
  granuloma formation, 427b
  history of medicine, 345b
  reportable diseases, 409t
  Treponema pallidum immobilization (TPI) test, 526
Systemic anaphylaxis, 486
Systemic autoimmune diseases, 499, 501
Systemic infections, 393, 610
Systemic lupus erythematosus (SLE), 499, 500t, 501

Tachyzoite, 587
Taenia saginata, 721
Taenia solium, 719, 720–21, 727b
Taeniasis, 720–21
Taiwan, and hepatitis B virus, 715
Tamiflu, 360, 663, 666
Tannerella forsythus, 692, 727b
Tapeworms, 143, 144t, 353t, 399t, 720–21. *See also Taenia solium*
Taq polymerase, 194b, 290, 292
Target organs, of allergy, 483–85
Target population, and microbial control, 317
Tatum, E. L., B2
Taxa, 17
Taxonomy, of microorganisms, 17–23, 106. *See also* Classification; Nomenclature
Tazobactum, 352
T cell(s). *See* T lymphocytes
T-cell independent antigens, 452–53
T-cell leukemia, 159t, 642
T-cell-mediated hypersensitivity, 480t
Tears, and eye, 562
Teeth, and plaque, 92b, 93
Teichoic acid, 96
Telophase, of mitosis, 124f
Temperate phages, 167
Temperature. *See also* Cold; Heat
  extreme habitats, 186b
  food preservation, 800–802
  microbial adaptations, 193–94
  microbial control, 317
Template strand, 259
Tenericutes, 106
Teratogenicity, *554*
Terbinafine, 561
Terminal step, of aerobic respiration, 233
Termination, and protein synthesis, 261, 263–64
Termination codons, 263
Terminology. *See also* Classification; Nomenclature
  antimicrobial therapy, 346t
  infectious disease, 392b, 610

microbial adaptations, 185b
skin lesions, 552t
spores, 102b
Termites, 774b
Terrorism, 325, 326, 337b. *See also* Bioterrorism; September 11, 2001
Terry, Luther, 8f
Tertiary consumers, 770
Tertiary stage, of sewage treatment, 792
Tertiary structure, of proteins, 48f, 49
Tertiary syphilis, 749, 750
Testes, 735
Tetanospasmin, 596
Tetanus, 409t, 464, 468t, 472t, 473t, 596–97, 602t. *See also Clostridium tetani*
Tetanus immune globulin (TIG), 597
Tetracyline
applications of, 353t, 356
bioengineering, 350b
children and, 369
drug interactions, 373
drug permeability, 364
Kirby-Bauer test, 372t
minimum inhibitory concentration, 374t
protozoan infections, 359
ribosomes and protein synthesis, 268, 349
toxic reactions, 368t, 369
Tetrads, 104f, 105
Tetrahydrofolate, 351
T-even bacteriophages, 167
Texas A & M University, 381
Thailand, and dengue fever, 623
Thallobacteria, 106
T helper (TH) cells, 459, 461
Theophylline, 488
Theory, 14–16
Therapeutic index, 372–74
Thermal death, 320–21
Thermal death point (TDP), *321*
Thermal death time (TDT), *321*, 800
Thermal vent symbionts, 198b
Thermocline, *781*
Thermococcus litoralis, 292
Thermoduric microbes, 194, 322
Thermophile, 194
Thermoplasma, 112, 196
Thermus aquaticus, 194b, 290
Thiabendazole, 359–60, 561, 719t
Thiamine, 325
Thimerosal, 334, 471
Thiobacillus spp., 776, 777
T. ferrooxidans, 776
T. thiooxidans, 776
Thioglycollate broth, 196f
Thiomargarita namibia, 109b
Thiosulfate, 776
Third World, and marketing of baby formulas, 464b. *See also* Africa; India
Threadworm, 722b, 723
Three-dimensional models, 36f
Three-domain system, of classification, 22–23
Throat cancers, 756
Thykaloids, 110, 127, 239
Thymic alymphoplasia, 504
Thymic aplasia, 503t
Thymidine kinase, 360
Thymine (T), 49, 50f, 249
Thymus, 426, 504
Ticarcillin, 353t, 354t
Ticks, 617, 619–20, 627
Tinea, 559
Tinea versicolor, 135t, 559, 561
Tissue(s), *118*
Tissue culture, 171
Tissue plasminogen activating factor (tPA), 297t, 302t
Tissue typing, 498
Titer, of antibodies, 458, 521
T lymphocytes (T cells)
B cells compared to, 450t
cell-mediated immunity, 445, 458–59, 461–62
HIV infection, 634–35
immunodeficiency diseases, 503t, 504
organ transplantation, 497–98
receptors for antigen, 449
tests differentiating B cells from, 529
thymus and maturation, 426, 449
Tobacco, 301b. *See also* Cigarette smoking
Tobacco mosaic virus, 154
Tobramycin, 356
Togaviridae, 159t
Tolnaftate, 561
Tomato juice agar, 66t
Tonegawa, Susumu, B2
Tonsils, 425, 650
Tooth decay, 689–91
Topsoil, 780
Total cell count, 205
Toxemia, 391, 394
Toxic epidermal necrolysis (TEN), 545
Toxicodendron, 496b
Toxic shock syndrome, 353t, 409t, 656
Toxic waste sites, 791b
Toxigenicity, *391*
Toxin neutralization tests, 526
Toxins and toxicity. *See also* Endotoxins; Exotoxins
antimicrobial drugs, 368–69
bacteriophages, 169
cellular damage, 391–92
environmental recycling, 777
heavy metal compounds, 333
staphylococcal scalded skin syndrome, 545
streptococcal, 656
Toxinoses, 391
Toxoid vaccine, 467, 597
Toxoplasma gondii, 140, 353t, 586–88, 589t, 602t
Toxoplasmosis, 138, 140, 141t, 353t, 399t, 586–88, 589t
Tracheal cytotoxin, 660
Trace elements, 180
Tracheostomy, 597
Trachoma, 110, 563–64, 566t. *See also Chlamydia trachomatis*
Transamination, 237
Transcript, of mRNA, 260
Transcription
antibiotics, 268
DNA code, 255
eukaryotic translation, 264
gene expression, 259–60
RNA and cell assembly, 256–59
RNA viruses, 164–65b
Transduction, 273, 276–77
Transfection, 275
Transferases, 216b
Transfer reactions, by enzymes, 217
Transfer RNA (tRNA), 51, 257t, 258, 260
Transformation, 167, 273, 274–75
Transgenic animals, 300–301
Transgenic plants, 301b
Transitional vesicles, 125
Translation
antibiotics, 268
DNA code, 255
eukaryotic transcription, 264
gene expression, 260–64
RNA and cell assembly, 256–59
RNA viruses, 164–65b
Translocation, 262f, 263
Transmissible agent, 588
Transmissible spongiform encephalopathies (TSEs), 588
Transmission, of infectious disease, 399–401
Transmission electron microscope (TEM), 73t, 76
Transport
cell membrane, 99
diffusion, 189–92
nutrient absorption, 187
Transport media, 66
Transposons, 276–77, 363
Transverse fission, 200
Trauma, genetic engineering and treatment of, 302t
Traveler's diarrhea, 701
Tree diagrams, 21–23
Trematodes, 143, 144t
Trench fever, 619b, 626, 628t, 643t
Treponema pallidum, 353t, 750–51, 756b, 759b. *See also* Syphilis
Treponema pallidum immobilization (TPI) test, 526
Treponema vincentii, 693, 727b
Trichinella spp., 724–25, 727b. *See also* Trichinellosis
T. solium, 144t
T. spiralis, 144t. *See also* Trichinosis
Trichinellosis, 399t, 409t
Trichinosis, 721, 724–25. *See also Trichinella spiralis*
Trichomonas vaginalis, 138, 139f, 353t, 742–43, 759b
Trichomoniasis, 141t, 353t, 387t, 759b
Trichophyton, 135t, 560, 561t, 566t
Trichuriasis, 720. *See also Trichuris trichiura*
Trichuris suis, 719b
Trichuris trichiura, 719, 720, 721b, 727b. *See also* Whipworm
Triclosan (Irgasan), 328b, 331, 338t
Triglycerides, 41t, 44
Trimethoprim
Cyclospora infections, 711
folic acid synthesis, 349, 351, 357, 364
listeriosis, 578
metabolic patterns, 364
Pneumocystis pneumonia, 679
salmonellosis, 698
shigellosis, 699
whooping cough, 660
Triple-sugar iron agar (TSIA), 67t
Triplet code, 255–56, 260
Trojan horse vaccine, 469
Trophozoite, 138, 140, 713
Tropisms, 160
True pathogens, *385*
Trypanosoma brucei, 141, 599, 600. *See also* African sleeping sickness
Trypanosoma brucei gambiense, 600, 601f, 602t
Trypanosoma brucei rhodesiense, 600, 601f, 602t
Trypanosoma cruzi, 141
Trypanosomes, 141
Trypanosomiasis, 399t, 427b, 599
Trypticase soy agar (TSA), 64
Tsetse fly, 600, 601f
Tubercles, 666
Tubercular meningitis, 667
Tuberculin test, 495, 496f, 529, 666, 668, 670

Tuberculosis, *666–68*. *See also Mycobacterium tuberculosis*
acid-fast staining and diagnosis, 670–71
antibiotic resistance, 354b, 367b
case study of, 343, 354b
chemotherapy for, 343, 353t
delayed-type hypersensitivity, 495
granuloma formation, 427b
reportable diseases, 409t
vaccine, 467, 468t
Tularemia, 399t, 409t, 617–18, 619b, 643t, 676b, 677b. *See also Francisella tularensis*
Tulips, and viruses, 153b
Tumbles, 90
Tumor, and inflammation, 427
Tumor necrosis factor (TNF), 297t, 429b, 656
Turbidity, and microbial growth, 204
Turgidity, and osmosis, 188
Turner syndrome, 297
Tuskegee Study (1930s), 749
Twinrix, 715
Twort, Frederick, 167, B2
Tyndall, John, 16, 321n
Tyndallization, *321*
Type I allergic reactions, 480–88
Type II hypersensitivities, 489–93
Type III hypersensitivities, 494–95
Type III secretion system, 700
Type IV hypersensitivities, 495–98
Types, of bacteria, 108
Typhoid fever, 353t, 367b, 397, 409t, 468t, 697–98. *Salmonella enteritidis*
"Typhoid Mary," 397
Typhus, 110, 619b. *See also Rickettsia typhi*

Ubiquity, of microbes, 4
Ulcerative colitis, 710, 719t
Ulcer diseases, and sexually transmitted diseases, 740, 748
Ultrahigh temperature (UHT), 322
Ultraviolet (UV) radiation, 197, 269, 270, 317, 325–26
Uncoated virus, 160
Uncoating, of animal viruses, 160, 162
Undulant fever, 322, 624
Undulating membrane, 138
United Nations
cholera and oral rehydration therapy, 704b
emerging diseases, 399
immunization programs, 442
U.S. Public Health Service (USPHS), 3, 676b
Universal blood and body fluid precautions (UPs), 403–404, 511, 639
Universal blood donor, 490
University of Iowa, 719b
University of Wisconsin, 801b
Upper respiratory tract, 650, 651–66
Upwelling, in oceans, 781
Uracil (U), 49, 50f, 256–57
Urea, 196
Urea breath test, 696
Urea broth, 67t
Urease, 214t, 695
Ureters, 734
Urethra, 734
Urethritis, 353t, 737, 744
Urinary schistosomiasis, 739–40, 759b
Urinary tract. *See also* Genital tract
antibiotics for infections of, 353t
defense mechanisms, 734–35
infectious diseases, 737–40
normal microbiota, 382t, 735–36
portals of entry for infection, 387, 416
portals of exit for infection, 396
specimen collection, 514
Urinary tract infections (UTIs), 737–38, 759b
Urushiol, 496b
USDA. *See* Department of Agriculture
Uterus, 735

Vaccination, 463, 465–67. *See also* Immunization
Vaccine(s). *See also* Immunization; Vaccination; *specific diseases*
administration and side effects of, 470–71
animal inoculation, 62b
cancer, 470b
currently approved, 468t
dental caries, 691
development of new, 467, 469
history of, 465b
industrial production of, 807
recombinant DNA technology, 297
recommended schedules for, 472–73t
requirements for effective, 467t
viruses, 153b, 174
Vaccinia virus, 152f, 156f, 159t, 550. *See also* Cowpox
Vacuoles, 126
Vagina, 735, 736
Vaginitis, 353t, 740–43, 759b
Vaginosis, 740–43, 759b
Valacyclovir, 360, 622
Valence, *30*
Vampirovibrio chlorellavorus, 20
Vancomycin
acute endocarditis, 613
antibiotic resistance, 246, 249b, 357, 366b
applications of, 353t
cell wall, 348, 355, 364
Clostridium difficile, 702
drug receptors, 364
Kirby-Bauer test, 372t
Vancomycin-resistant *Enterococcus faecalis* (VRE), 246, 357, 367b
Vancomycin-resistant *Staphylococcus aureus* (VRSA), 246, 249b, 366b, 409t
Van der Waals forces, 34
Van Leeuwenhoek, Antonie, 11–13, 69, 70, 73, 711, B1
Vaporized hydrogen peroxide, 332
Variable number of tandem repeats (VNTRs), 305
Variable regions (V), 449
Variant Creutzfeldt-Jakob disease (VCJD), 572, 589–90
Varicella zoster virus (VZV), 159t, 409t, 472t, 473t, 548, 586
Variola, 159t, 550, 566t. *See also* Smallpox
Variolation, 465b, 550
Varmus, Harold, B2
Vascular changes, and inflammation, 428, 430
Vas deferens, 735
Vasoactive mediators, 428, 429b
VDRL (Venereal Disease Research Lab) test, 522
Vectors
arthropods, 619b
cloning, 292, 293–94
infectious disease, 397–99
Vegetative cells, 80, 101–102, 320
Vegetative hyphae, 131, 133
Vehicles, for transmission of disease, 401
Veins, 609
Venter, Craig, 289b, 298b
Vent polymerase, 290, 292
Vertical infection transmission, 400
Vesicle, 552b
Vesicular rash diseases, 546–51
Veterinary medicine, and salmonellosis, 312, 318b
Vetter, David, 505b
Vibrio spp., *103*, 104f, 781, 782
V. cholerae, 169, 353t, 702–705, 709b, 727b, 807. *See also* Cholera
V. fischeri, 89f
V. parahaemolyticus, 802
Vibrionaceae, 703
Vidarabine, 361t
Vietnam, and dengue fever, 623
Viking Explorer, 18b
Vinegar, 235b, 796–97
Viral conjunctivitis, 563, 564t
Viral envelope, 155–56
Viral hemorrhagic fevers, 676b
Viral persistence, 169t
Virchow, Rudolf, B1
Viremia, 395, 610
Virion, 10, 152
Viroids, 174
Virology, 150
Virtual image, 70
Virucide, 315
Virulence, *383*
Virulence factors, *383*, 390–92
Viruses, 2. *See also specific diseases*
aquatic environments, 781
autoimmune disorders, 500
characteristics of, 9–10
chemotherapeutic agents, 353t, 360–62, 368t
classification, 157–58
common cold, 651
culture and identification of, 170–72
diagnostic methods, 530
discovery of, 150
encephalitis, 584, 587t
general structure of, 151–57
genetically modified organisms, 297–98
genetics of, 265
heat resistance, 320
infectious diseases, 7, 173–74
interferon, 435
meningitis, 580–81, 582t, 694
multiplication, 160–69
normal microbiota, 381b
position of in biological spectrum, 150
positive view of, 153b
properties of, 151t
zoonotic infections, 399t
Vision, and eye, 562
Vitamins
coenzymes, 215
E. coli and synthesis of, 689
industrial products, 804t
transgenic plants, 301b
Vitravene, 304
Vodka, 795
Von Behring, Emil, B1
Von Borries, B., B2
Von Linné, Carl, 17
Von Petenkofer, Max, 405b
Vulvovaginal candidiasis, 741

Waksman, Selman, B2
Walking pneumonia, 672, 674
Walter Reed Army Medical Center (Washington, DC), 379
Warren, J. Robert, 405b, 695
Warts, 555, 556–57, 566t, 740, 755, 759b

Wasserman, August, B2
Waste disposal. *See* Sewage treatment; Solid waste disposal; Toxic waste sites
Water quality assays, 783
Water and water supplies. *See also* Aquatic microbiology; Oceans; Pollution
  applied microbiology and treatment of, 790, 792–93
  amoebic dysentery, 713
  chlorination, 330
  cholera, 703–704
  contamination of and outbreaks of infectious disease, 117, 137b
  cryptosporidiosis, 705
  environmental requirements, 197
  filtration and purification of, 326
  giardiasis, 712
  helminths, 719
  hepatitis, 714
  hydrogen bonding, 35f
  infectious diseases, 399, 401, 782–83
  leptospirosis, 739
  monitoring of, 782–83
  Naegleria infection, 583
  pollution of groundwater, 791b
  soil composition, 779
  solvents, 36
  synthesis reactions, 36
  three-dimensional model, 36f
  ultraviolet radiation treatment, 326f
Watson, James, 249, 250b, B2
Wavelength, of light, 71
Waxes, 41t, 46
Web sites
  Centers for Disease Control and Prevention, 3, 406, 551
  Human Genome Project, 289b
Weil-Felix reaction, 522
Weil's syndrome, 739
Western blot analysis, 511, 524, 640
Western equine encephalitis (WEE), 159t, 409t, 585, 586, 602t
West Nile virus, 159t, 399t, 407f, 409t , 585b, 586, 602t
Whales, 778b
Wheal and flare reaction, 484
Whey, 797
Whipworm, 720
Whiskeys, 795
White blood cells (WBCs), 421, 422, 430
Whittaker, Robert, 21, 22f
Whole blood, 420
Whooping cough, 660–61, 680b
Widal test, 522
Wild type, of mutation, 268, 270t
Wilkins, Maurice, 250b, B2
Wine, 5, 234–35, 322, 324n, 795, 796f
Wisconsin, and *Cryptosporidium* outbreak, 782. *See also* Milwaukee
Wiscott-Aldrich syndrome, 503t
Witchcraft, 345b
Wobble, in genetic code, 260
Woese, Carl, 21–22
Wolbachia, 565, 566t
Word lists, and concept mapping, D3
World Health Organization (WHO)
  avian influenza, 664–65b
  baby formula use in Third World, 464b
  Darfur crisis in Sudan, 442
  immunization programs, 7, 594
  infectious diseases and human condition, 6
  malaria, 631
  oral rehydration therapy (ORT), 704b
  polio, 593b, 594
  role in public health, 3, 406
  severe acute respiratory syndrome (SARS), 675
  smallpox, 465b, 549
  trypanosomiasis, 601
  tuberculosis, 666
Wort, 794
Wound botulism, 598, 599
Wright, James, B2

Xanthan, 804t
Xenograft, 498
Xeroderma pigmentosa, 270
XLD agar, 67t
X-1-linked SCID, 303, 503t
X rays, 269, 324–25, 670

Yeast(s). *See also Saccharomyces;* Yeast infections
  bread and leavening, 794
  human uses of, 5
  microscopic morphology of, 131f
  mitosis, 124f
  thermal death times, 320t
Yeast artificial chromosomes (YACs), 294
Yeast infections, 740. *See also Candida albicans*
Yellow fever, *623*
  classification of viruses, 159t
  hemorrhagic fevers, 624t
  human experimentation, 405b
  reportable diseases, 409t
  vaccine, 468t
  zoonotic infections, 399
Yellowstone National Park, 194b
Yemen, 594
Yersina spp., 702, 709b
  Y. enterocolitica, 615, 702, 727b
  Y. pestis, 299b, 615, 643t, 677b, 702
  Y. pseudotuberculosis, 615, 702, 727b

Zalcitabine, 361t
Zanamivir, 361t, 666
Zidovudine. *See* Azidothymide
Ziehl-Neelsen stain, 670
Zinc, 29t, 183, 333
Zinc "fingers," 268
Zoonosis, *399*, 739
Zooplankton, 781

Figures with Animation Quizzes

Image in the text	Animation *(view on the website or download to your portable device)*
Fig 1.2b	Food Spoilage
Fig 1.5	Entry of Virus into Host Cell
	Lamda Phage Replication Cycle
	Life Cycle of T2 Phage
	Steps in the Replication of T4 Phage in E. coli
Fig 1.6	Prokaryotic Cell Shapes
Fig 2.22	DNA Structure
Fig 3.25	Gram Stain
Fig 4.3	Bacterial Locomotion
	Appendaged Bacteria
	Bdellorlbria
Fig 4.5	Chemotaxis in E. coli
Fig 4.8	Bacterial Conjugation
	Bacterial Conjugation—Transfer of a Plasmid
Fig 4.11	Biofilm Formation
	Quorum Sensing
Fig 4.12a	Gram Stain
Fig 4.13	Peptidoglycan Synthesis
Table 4.1	Bacterial Endospore Formation
Fig 4.22	Prokaryotic Cell Shapes
Fig 5.7	How the Cell Cycle Works
	Control of the Cell Cycle
	Mitosis and Cytokinesis
Fig 5.11	Lysosomes
Fig 6.11	Mechanism for Releasing Enveloped Viruses
Fig 6.13	Entry of Virus into Host Cell
Fig 6.15	Mechanism for Releasing Enveloped Viruses
Fig 6.17	Lamda Phage Replication Cycle
	Replication of a Positive (+) Sense Strand of Lytic RNA Phage
	Lifecycles of T2 Phage
	Steps in the Replication of T4 Phage in E. coli
Fig 7.3	How Osmosis Works
Fig 7.5	How Diffusion Works
Fig 7.7	How the Sodium-Potassium Pump Works
	Sodium-Potassium Exchange Pump (Quiz 2, Quiz 3)
Fig 7.8	Food Spoilage
Fig 7.13	Binary Fission
	Bacterial Cell Cycle
Fig 7.14	Food Pathogens and Temperature
Fig 8.4	How Enzymes Work
Fig 8.8	A Biochemical Pathway
Fig 8.12	How NAD+ Works
Fig 8.16	How Glycolysis Works
Insight 8.3	How the Krebs Cycle Works
Fig 8.18	Electron Transport System and ATP Synthesis
	Electron Transport System and Formation of ATP
Fig 8.19	Proton Pump
Fig 8.23	The Calvin Cycle
Fig 8.24	Photosynthetic Electron Transport and ATP Synthesis
Fig 8.25	Cyclic and Noncyclic Photophosphorylation

Image in the text	Animation *(view on the website or download to your portable device)*
Fig 9.4	DNA Structure
Fig 9.5	Structural Basis of DNA Replication
Fig 9.6	Bidirectional DNA Replication (Quiz 1, Quiz 2)
	How Nucleotides are added in DNA Replication
	DNA Replication Fork (Quiz 1-3)
	DNA Replication (Quiz 1-3)
Fig 9.7	Bidirectional Replication (Quiz 1, 2)
Fig 9.8	How Translation Works
Fig 9.9	Simple Gene Expression
Fig 9.11	Stages of Transcription
	mRNA Synthesis (Transcription) (Quiz 1, 2)
Fig 9.15	Protein Synthesis (Quiz 1-4)
	Translation Elongation
	How Translation Works
	Translation Termination
Fig 9.16	Processing of Gene Information: Procaryotes v. Eucaryotes
Fig 9.17	Processing of Gene Information: Procaryotes v. Eucaryotes
Fig 9.18	The lac Operon
	Regulatory Proteins: Regulation by Repression (Quiz 1, 2)
Fig 9.19	The trp Operon
	Regulatory Proteins: Regulation by Repression
Fig 9.21	Proofreading Function of DNA Polymerase
	Direct Repair (Quiz 1, 2)
Fig 9.23	Bacterial Conjunction: Transfer of a Plasmid
	Conjunction: Transfer of Chromosomal DNA (Quiz 1,2)
	Conjunction: Transfer of the F Plasmid
	Bacterial Conjunction
	Rolling Circle Mechanisms of Replication
Fig 9.24	Bacterial Transformation
Fig 9.25	Transduction (Generalized) (Quiz 1, 2)
Fig 9.26	Specialized Transduction
Fig 9.27	Transposons: Shifting Segments of the Genome (Quiz 1, 2)
Fig 10.1	Construction of a Plasmid Vector
	Restriction Endonucleases
Fig 10.2	cDNA
Fig 10.4	Southern Blot
Fig 10.5	DNA Probe
Fig 10.6	Sanger Sequencing
Fig 10.7	Polymerase Chain Reaction (Quiz 1-3)
	PCR Reactions
Fig 10.10	Integration and Excision of a Plasmid
	Steps in Cloning a Gene (Quiz 1-4)
	Early Genetic Engineering Experiment
Fig 10.16	DNA Fingerprinting
	Restriction Fragment Length Polymorphisms
Fig 10.17	Microarrays
Visual Understanding Chapter 10	Transduction (Generalized) (Quiz 1, 2)
Fig 11.4	Protein Denaturation

Image in the text	Animation *(view on the website or download to your portable device)*
Fig 12.1	Antibiotic Inhibition of Protein Synthesis
	Cell Wall Antibiotics
Table 12.5	Antiviral Agents
Fig 14.13	The Inflammatory Response
Fig 14.17	Phagocytosis
Fig 14.19	Complement Function
	Activation of Complement
	Complement Activation
Visual Understanding Chapter 14	Steps in Cloning a Gene (Quiz 1-4)
Fig 15.1	The Immune Response
Fig 15.3	Clonal Selection
Fig 15.5	Antibody Diversity
Fig 15.7	Antigenic Determinants
Fig 15.9	Antigen Processing
Fig 15.10	Clonal Selection
Table 15.2	Diversity of Antibodies
Insight 15.1	Monoclonal Antibody Production (Quiz 1, 2)
Fig 15.16	Cytotoxic T-Cell Activity against Target Cells (Quiz 1, 2)
	T-Cell Dependent Antigen (Quiz 1, 2)
Fig 15.19	Constructing Vaccines (Quiz 1, 2)
Fig 16.1	IgE Mediated (Type I) Hypersensitivity (Quiz 1, 2)
	Cytotoxic (Type II) Hypersensitivity
	Delayed (Type IV) Hypersensitivity
	Immune Complex (Type III) Hypersensitivity
Fig 16.3	IgE Mediated (Type I) Hypersensitivity (Quiz 1, 2)
Fig 16.11	Cytotoxic (Type II) Hypersensitivity
Fig 16.13	Immune Complex (Type III) Hypersensitivity
Fig 16.15	Delayed (Type IV) Hypersensitivity
Visual Understanding Chapter 16	Cytotoxic T-Cell Activity against Target Cells (Quiz 1, 2)
	T-Cell Dependent Antigens
Fig 17.10	Agglutination and Precipitation
Fig 17.11	Immuno-Diffusion and –Electrophoresis
Fig 17.12	Immuno-Diffusion and –Electrophoresis
Fig 17.14	Complement Fixation Test
Fig 17.15	Immunoflourescence and RIA
Fig 17.16	ELISA Enzyme-Linked Immunosorbent Assay
Fig 19.25	Prion Diseases
	How Prions Arise
Visual Understanding Chapter 19	Construction Vaccines (Quiz 1, 2)
Fig 20.17	Malaria: Life Cycle of Plasmodium
Fig 20.23	HIV Replication (Quiz 1, 2)
Fig 20.25	Treatment of HIV Infection
Fig 21.10	A-B Exotoxins (Diphtheria Exotoxin)
Fig 22.1	Organs of Digestion
Fig 24.1	Biofilm Formation
Fig 24.8	Root Nodule Formation
Visual Understanding Chapter 25	How Osmosis Works

Displaying Disease Statistics

Infectious disease specialists use a number of different methods to visually represent the numbers of disease cases or deaths.

ACQUIRED IMMUNODEFICIENCY SYNDROME (AIDS). Incidence* — United States† and U.S. territories, 2005

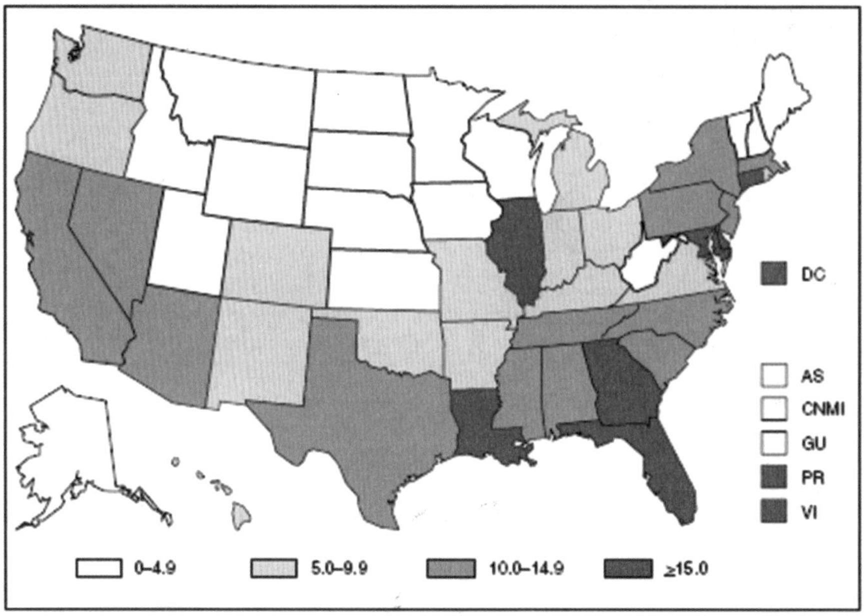

* Per 100,000 population.
† Includes 209 persons with unknown state of residence.

LYME DISEASE. Number of reported cases, by county — United States, 2005

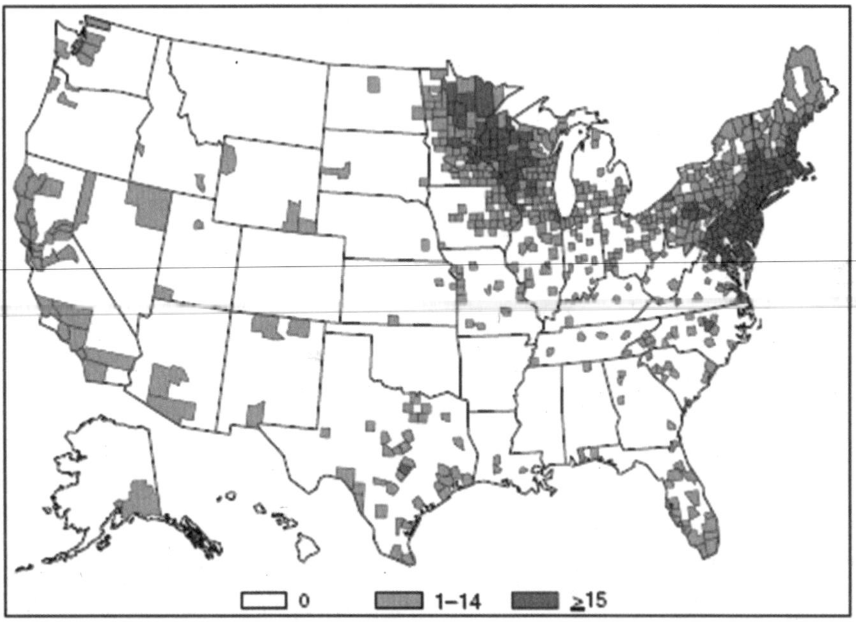

Some methods emphasize the geographical distribution of disease.